Mass	$1\text{ g} = 10^{-3}\text{ kg}$
	$1\text{ kg} = 10^3\text{ g}$
	$1\text{ u} = 1.66 \times 10^{-24}\text{ g} = 1.66 \times 10^{-27}\text{ kg}$
	$1\text{ metric ton} = 1000\text{ kg}$

Length
$1\text{ nm} = 10^{-9}\text{ m}$
$1\text{ cm} = 10^{-2}\text{ m} = 0.394\text{ in.}$
$1\text{ m} = 10^{-3}\text{ km} = 3.28\text{ ft} = 39.4\text{ in.}$
$1\text{ km} = 10^3\text{ m} = 0.621\text{ mi}$
$1\text{ in.} = 2.54\text{ cm} = 2.54 \times 10^{-2}\text{ m}$
$1\text{ ft} = 0.305\text{ m} = 30.5\text{ cm}$
$1\text{ mi} = 5280\text{ ft} = 1609\text{ m} = 1.609\text{ km}$

Area
$1\text{ cm}^2 = 10^{-4}\text{ m}^2 = 0.155\,0\text{ in}^2$
$\quad = 1.08 \times 10^{-3}\text{ ft}^2$
$1\text{ m}^2 = 10^4\text{ cm}^2 = 10.76\text{ ft}^2 = 1550\text{ in}^2$
$1\text{ in}^2 = 6.94 \times 10^{-3}\text{ ft}^2 = 6.45\text{ cm}^2$
$\quad = 6.45 \times 10^{-4}\text{ m}^2$
$1\text{ ft}^2 = 144\text{ in}^2 = 9.29 \times 10^{-2}\text{ m}^2 = 929\text{ cm}^2$

Volume
$1\text{ cm}^3 = 10^{-6}\text{ m}^3 = 3.53 \times 10^{-5}\text{ ft}^3$
$\quad = 6.10 \times 10^{-2}\text{ in}^3$
$1\text{ m}^3 = 10^6\text{ cm}^3 = 10^3\text{ L} = 35.3\text{ ft}^3$
$\quad = 6.10 \times 10^4\text{ in}^3 = 264\text{ gal}$
$1\text{ liter} = 10^3\text{ cm}^3 = 10^{-3}\text{ m}^3 = 1.056\text{ qt}$
$\quad = 0.264\text{ gal} = 0.035\,3\text{ ft}^3$
$1\text{ in}^3 = 5.79 \times 10^{-4}\text{ ft}^3 = 16.4\text{ cm}^3$
$\quad = 1.64 \times 10^{-5}\text{ m}^3$
$1\text{ ft}^3 = 1728\text{ in}^3 = 7.48\text{ gal} = 0.028\,3\text{ m}^3$
$\quad = 28.3\text{ L}$
$1\text{ qt} = 2\text{ pt} = 946\text{ cm}^3 = 0.946\text{ L}$
$1\text{ gal} = 4\text{ qt} = 231\text{ in}^3 = 0.134\text{ ft}^3 = 3.785\text{ L}$

Time
$1\text{ h} = 60\text{ min} = 3600\text{ s}$
$1\text{ day} = 24\text{ h} = 1440\text{ min} = 8.64 \times 10^4\text{ s}$
$1\text{ y} = 365\text{ days} = 8.76 \times 10^3\text{ h}$
$\quad = 5.26 \times 10^5\text{ min} = 3.16 \times 10^7\text{ s}$

Angle
$1\text{ rad} = 57.3°$

$1° = 0.0175\text{ rad}$	$60° = \pi/3\text{ rad}$
$15° = \pi/12\text{ rad}$	$90° = \pi/2\text{ rad}$
$30° = \pi/6\text{ rad}$	$180° = \pi\text{ rad}$
$45° = \pi/4\text{ rad}$	$360° = 2\pi\text{ rad}$

$1\text{ rev/min} = (\pi/30)\text{ rad/s} = 0.104\,7\text{ rad/s}$

Speed
$1\text{ m/s} = 3.60\text{ km/h} = 3.28\text{ ft/s}$
$\quad = 2.24\text{ mi/h}$
$1\text{ km/h} = 0.278\text{ m/s} = 0.621\text{ mi/h}$
$\quad = 0.911\text{ ft/s}$
$1\text{ ft/s} = 0.682\text{ mi/h} = 0.305\text{ m/s}$
$\quad = 1.10\text{ km/h}$
$1\text{ mi/h} = 1.467\text{ ft/s} = 1.609\text{ km/h}$
$\quad = 0.447\text{ m/s}$
$60\text{ mi/h} = 88\text{ ft/s}$

Force
$1\text{ N} = 0.225\text{ lb}$
$1\text{ lb} = 4.45\text{ N}$
Equivalent weight of a mass of 1 kg
$\quad$ on Earth's surface $= 2.2\text{ lb} = 9.8\text{ N}$

Pressure
$1\text{ Pa (N/m}^2) = 1.45 \times 10^{-4}\text{ lb/in}^2$
$\quad = 7.5 \times 10^{-3}\text{ torr (mm Hg)}$
$1\text{ torr (mm Hg)} = 133\text{ Pa (N/m}^2)$
$\quad = 0.02\text{ lb/in}^2$
$1\text{ atm} = 14.7\text{ lb/in}^2 = 1.013 \times 10^5\text{ N/m}^2$
$\quad = 30\text{ in. Hg} = 76\text{ cm Hg}$
$1\text{ lb/in}^2 = 6.90 \times 10^3\text{ Pa (N/m}^2)$
$1\text{ bar} = 10^5\text{ Pa}$
$1\text{ millibar} = 10^2\text{ Pa}$

Energy
$1\text{ J} = 0.738\text{ ft·lb} = 0.239\text{ cal}$
$\quad = 9.48 \times 10^{-4}\text{ Btu} = 6.24 \times 10^{18}\text{ eV}$
$1\text{ kcal} = 4186\text{ J} = 3.968\text{ Btu}$
$1\text{ Btu} = 1055\text{ J} = 778\text{ ft·lb} = 0.252\text{ kcal}$
$1\text{ cal} = 4.186\text{ J} = 3.97 \times 10^{-3}\text{ Btu}$
$\quad = 3.09\text{ ft·lb}$
$1\text{ ft·lb} = 1.36\text{ J} = 1.29 \times 10^{-3}\text{ Btu}$
$1\text{ eV} = 1.60 \times 10^{-19}\text{ J}$
$1\text{ kWh} = 3.6 \times 10^6\text{ J}$

Power
$1\text{ W} = 0.738\text{ ft·lb/s} = 1.34 \times 10^{-3}\text{ hp}$
$\quad = 3.41\text{ Btu/h}$
$1\text{ ft·lb/s} = 1.36\text{ W} = 1.82 \times 10^{-3}\text{ hp}$
$1\text{ hp} = 550\text{ ft·lb/s} = 745.7\text{ W}$
$\quad = 2545\text{ Btu/h}$

Mass–Energy Equivalents
$1\text{ u} = 1.66 \times 10^{-27}\text{ kg} \leftrightarrow 931.5\text{ MeV}$
$1\text{ electron mass} = 9.11 \times 10^{-31}\text{ kg}$
$\quad = 5.49 \times 10^{-4}\text{ u} \leftrightarrow 0.511\text{ MeV}$
$1\text{ proton mass} = 1.672\,62 \times 10^{-27}\text{ kg}$
$\quad = 1.007\,276\text{ u} \leftrightarrow 938.27\text{ MeV}$
$1\text{ neutron mass} = 1.674\,93 \times 10^{-27}\text{ kg}$
$\quad = 1.008\,665\text{ u} \leftrightarrow 939.57\text{ MeV}$

Temperature
$T_F = \frac{9}{5}T_C + 32$
$T_C = \frac{5}{9}(T_F - 32)$
$T_K = T_C + 273$

cgs Force
$1\text{ dyne} = 10^{-5}\text{ N} = 2.25 \times 10^{-6}\text{ lb}$

cgs Energy
$1\text{ erg} = 10^{-7}\text{ J} = 7.38 \times 10^{-6}\text{ ft·lb}$

YOUR ACCESS TO SUCCESS

Welcome to the Companion Website for College Physics.

College Physics
Sixth Edition

by Jerry Wilson, Anthony Buffa, and Bo Lou

Welcome to the Companion Website with GradeTracker for *College Physics, Sixth Edition*. Built to complement the text as part of an integrated course package, this site is a useful study tool for your physics course.

To enter the site, select a chapter from the chapter menu above and click Go.

This website hosts contributions from leaders in physics education research, providing students with a variety of interactive explorations of each chapter's topics, and easily accommodating differences in learning styles. Resources include:

- **Warm Ups, Puzzles, and "What Is Physics Good For?" Applications** from the **Just in Time Teaching** method.
- **Multiple-choice Practice Questions**, with hints and feedback.
- *Physlet® Physics* **Problems**, including links to Java applets that illustrate physics concepts through animation.
- *Ranking Task Exercises*, which are gradable conceptual exercises that require students to rank a number of situations or variations of a situation.
- **MCAT Study Guide** questions by Kaplan Test Prep and Admissions.

This site incorporates Java elements. Check your computer for these plug-ins by clicking on the "Browser Tuneup" below.

(Browser Tuneup)

Students with new copies of Wilson/Buffa/Lou *College Physics, 6th Edition* have full access to the book's Companion Website with GradeTracker—a 24/7 study tool with quizzes, self study tools, and other features designed to help you make the most of your limited study time.

Just follow the easy website registration steps listed below...

Registration Instructions for the Wilson/Buffa/Lou, *College Physics, Sixth Edition Companion Website with GradeTracker*

1. Go to *www.prenhall.com/wilson*.
2. Click the cover for Wilson/Buffa/Lou, *College Physics, Sixth Edition*.
3. Click on the link to the *Companion Website with GradeTracker*.
4. Click "*Register*".
5. Using a coin, scratch off the metallic coating below to reveal your Access Code.
6. Complete the online registration form, choosing your own personal Login Name and Password.
7. Enter your pre-assigned Access Code exactly as it appears below.
8. Complete the online registration form by entering your School Location information.
9. After your personal Login Name and Password are confirmed by e-mail, repeat steps 1-3 above, and enter your new Login Name and Password to log in.

Your Access Code is:

If there is no metallic coating covering the access code above, the code may no longer be valid and you will need to purchase online access using a major credit card to use the website. To do so, go to www.prenhall.com/wilson, click the cover for Wilson/Buffa/Lou, *College Physics, Sixth Edition*, and follow the instructions to purchase access to the Companion Website with Grade Tracker.

Important: Please read the Subscription and End-User License Agreement located on the "Log In" screen before using the Wilson/Buffa/Lou, *College Physics, Sixth Edition Companion Website with Grade Tracker*. By using the website, you indicate that you have read, understood and accepted the terms of the agreement.

Minimum system requirements

PC Operating Systems:
Windows 2000/XP

Pentium II 233 MHz processor. 64 MB RAM In addition to the minimum memory required by your OS

Internet Explorer (TM) 5.5 or 6, Netscape(TM) 7, Firefox 1.0

Macintosh Operating Systems:
Macintosh Power PC with OS X (10.2 and 10.3)

In addition to the RAM required by your OS, this application requires 64 MB RAM, with 40 MB Free RAM, with Virtual Memory enabled

Netscape(TM) 7, Safari 1.3, Firefox 1.0

Acrobat Reader 6.0.1

4x CD-ROM drive

800 x 600 pixel screen resolution

Technical Support Call 1-800-677-6337. Phone support is available Monday-Friday, 8am to 8pm and Sunday 5pm to 12am, Eastern time. Visit our support site at *http://247.pearsoned.com* E-mail support is available 24/7.

SIXTH EDITION

Annotated Instructor's Edition

COLLEGE PHYSICS

Jerry D. Wilson
Lander University
Greenwood, SC

Anthony J. Buffa
California Polytechnic State University
San Luis Obispo, CA

Bo Lou
Ferris State University
Big Rapids, MI

PEARSON
Prentice
Hall

Upper Saddle River, NJ 07458

Senior Editor: Erik Fahlgren
Associate Editor: Christian Botting
Editor in Chief, Science: Dan Kaveney
Executive Managing Editor: Kathleen Schiaparelli
Assistant Managing Editor: Beth Sweeten
Manufacturing Buyer: Alan Fischer
Manufacturing Manager: Alexis Heydt-Long
Director of Creative Services: Paul Belfanti
Creative Director: Juan López
Art Director: Heather Scott
Director of Marketing, Science: Patrick Lynch
Media Editor: Michael J. Richards
Senior Managing Editor, Art Production and Management: Patricia Burns
Manager, Production Technologies: Matthew Haas
Managing Editor, Art Management: Abigail Bass
Art Editor: Eric Day
Art Studio: ArtWorks
Image Coordinator: Cathy Mazzucca
Mgr Rights & Permissions: Zina Arabia
Photo Researchers: Alexandra Truitt & Jerry Marshall
Research Manager: Beth Brenzel
Interior and Cover Design: Tamara Newnam
Cover Image: Greg Epperson/Index Stock Imagery
Managing Editor, Science Media: Nicole Jackson
Media Production Editors: William Wells, Dana Dunn
Editorial Assistant: Jessica Berta
Production Assistant: Nancy Bauer
Production Supervision/Composition: Prepare, Inc.

© 2007, 2003, 2000, 1997, 1994, 1990 by Pearson Education, Inc.
Pearson Prentice Hall
Pearson Education, Inc.
Upper Saddle River, New Jersey 07458

Printed in the United States of America
10 9 8 7 6 5 4 3 2 1

ISBN 0-13-149706-5

Pearson Education LTD., *London*
Pearson Education Australia PTY, Limited, *Sydney*
Pearson Education Singapore, Pte. Ltd
Pearson Education North Asia Ltd. *Hong Kong*
Pearson Education Canada, Ltd., *Toronto*
Pearson Educación de Mexico, S.A. de C.V.
Pearson Education — Japan, *Tokyo*
Pearson Education Malaysia, Pte. Ltd

ABOUT THE AUTHORS

Jerry D. Wilson, a native of Ohio, is Emeritus Professor of Physics and former chair of the Division of Biological and Physical Sciences at Lander University in Greenwood, South Carolina. He received a B.S. degree from Ohio University, an M.S. degree from Union College, and, in 1970, a Ph.D. from Ohio University. He earned his M.S. degree while employed as a Materials Behavior physicist.

As a doctoral graduate student, Professor Wilson held the faculty rank of Instructor and began teaching physical science courses. During this time, he coauthored a physical science text that is now in its eleventh edition. In conjunction with his teaching career, Professor Wilson continued his writing and has authored or coauthored six titles. Having retired from full-time teaching, he continues to write, producing, among other works, *The Curiosity Corner*, a weekly column for local newspapers that can also be found on the Internet.

Anthony J. Buffa received his B.S. degree in physics from Rensselaer Polytechnic Institute in Troy, New York, and M.S. and Ph.D. degrees in physics from the University of Illinois, Urbana–Champaign. In 1970, Professor Buffa joined the faculty at California Polytechnic State University, San Luis Obispo. Recently retired, he now teaches halftime at Cal Poly as an Emeritus Professor of Physics. During his career he was involved in nuclear physics research at several accelerators, including LAMPF at Los Alamos National Laboratory. On campus, he was a research associate with the physics department's radioanalytical facility for sixteen years.

Professor Buffa's main interest remains teaching. During his tenure at Cal Poly, he taught courses ranging from introductory physical science to quantum mechanics, developed and revised many laboratory experiments, and taught physics to local K-12 teachers at an NSF-sponsored workshop. Combining physics with his interests in art and architecture, Dr. Buffa develops his own artwork and sketches, which he uses to increase his effectiveness in teaching physics. In addition to continued teaching, during his (partial) retirement, he and his wife intend to travel more and hopefully enjoy their future grandkids for a long time to come.

Bo Lou is currently Professor of Physics at Ferris State University in Michigan. His primary teaching responsibilities are undergraduate introductory physics lectures and laboratories. Professor Lou emphasizes the importance of conceptual understanding of the basic laws and principles of physics and their practical applications to the real world. He is also an enthusiastic advocate of using technology in teaching and learning.

Professor Lou received B.S. and M.S. degrees in optical engineering from Zhejiang University (China) in 1982 and 1985, respectively, and a Ph.D. in condensed matter physics from Emory University in 1989.

Dr. Lou, his wife Lingfei, and their daughter Alina reside in Big Rapids, Michigan. The family enjoys travel, nature, and tennis.

BRIEF CONTENTS

CONTENTS

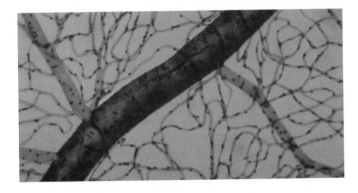

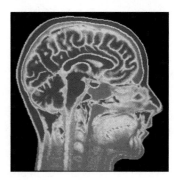

LEARN BY DRAWING

Applications (Insights appear in **boldface**, and "**(bio)**" indicates a biomedical application)

PREFACE

We believe there are two basic goals in any introductory physics course: (1) to impart an understanding of the basic concepts of physics and (2) to enable students to use these concepts to solve a variety of problems.

These goals are linked. We want students to apply their conceptual understanding as they solve problems. Unfortunately, students often begin the problem-solving process by searching for an equation. There is the temptation to try to plug numbers into equations before visualizing the situation or considering the physical concepts that could be used to solve the problem. In addition, students often do not check their numerical answer to see if it matches their understanding of the relevant physical concept.

We feel, and users agree, that the strengths of this textbook are as follows:

Conceptual Basis. Giving students a secure grasp of physical principles will almost invariably enhance their problem-solving abilities. We have organized discussions and incorporated pedagogical tools to ensure that conceptual insight drives the development of practical skills.

Concise Coverage. To maintain a sharp focus on the essentials, we have avoided topics of marginal interest. We do not derive relationships when they shed no additional light on the principle involved. It is usually more important for students in this course to understand what a relationship means and how it can be used than to understand the mathematical or analytical techniques employed to derive it.

Applications. *College Physics* is known for the strong mix of applications related to medicine, science, technology, and everyday life in its text narrative and Insight boxes. While the Sixth Edition continues to have a wide range of applications, we have also increased the number of biological and biomedical applications, in recognition of the high percentage of premed and allied health majors who take this course. A complete list of applications, with page references, is found on pages X-XIII.

The Sixth Edition

While we worked to reduce the total number of pages in this edition, we have added material to further student understanding and make physics more relevant, interesting, and memorable for students.

Physics Facts. Each chapter begins with four to six interesting facts about discoveries or everyday phenomena applicable to the chapter.

Visual Summary. Each end-of-chapter Summary includes visual representations of the key concepts from the chapter to serve as a reminder for students as they review.

---

130 CHAPTER 4 Force and Motion

Chapter Review

- A **force** is something that is capable of changing an object's state of motion. To produce a change in motion, there must be a nonzero net, or unbalanced, force:

$$\vec{F}_{net} = \Sigma \vec{F}_i$$

- **Newton's first law of motion** is also called the *law of inertia*, where inertia is the natural tendency of an object to maintain its state of motion. It states that in the absence of a net applied force, a body at rest remains at rest, and a body in motion remains in motion with constant velocity.

- **Newton's second law** relates the net force acting on an object or system to the (total) mass and the resulting acceleration. It defines the cause-and-effect relationship between force and acceleration:

$$\Sigma \vec{F}_i = \vec{F}_{net} = m\vec{a} \qquad (4.1)$$

A nonzero net force accelerates the crate: $a = F/m$

The equation for **weight** in terms of mass is a form of Newton's second law:

$$w = mg \qquad (4.2)$$

- **Newton's third law** states that for every force, there is an equal and opposite reaction force. The opposing forces of a third-law force pair always act on different objects.

- An object is said to be in **translational equilibrium** when it either is at rest or moves with a constant velocity. When remaining at rest, an object is said to be in *static translational equilibrium*. The condition for translational equilibrium is represented as

$$\Sigma \vec{F}_i = 0 \qquad (4.4)$$

or

$$\Sigma F_x = 0 \quad \text{and} \quad \Sigma F_y = 0 \qquad (4.5)$$

- **Friction** is the resistance to motion that occurs between contacting surfaces. (In general, friction occurs for all types of media—solids, liquids, and gases.)

- The frictional force between surfaces is characterized by coefficients of friction (μ), one for the static case and one for the kinetic (moving) case. In many cases, $f = \mu N$, where N is the normal force—the force perpendicular to the surface (that is, the force exerted *by* the surface on th...

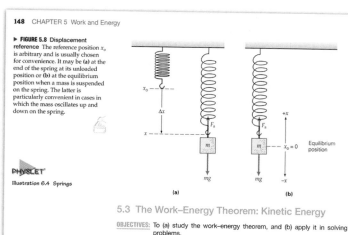

▶ **FIGURE 5.8 Displacement reference** The reference position x_o is arbitrary and is usually chosen for convenience. It may be (a) at the end of the spring at its unloaded position or (b) at the equilibrium position when a mass is suspended on the spring. The latter is particularly convenient in cases in which the mass oscillates up and down the spring.

PHYSLET

Illustration 6.4 Springs

5.3 The Work–Energy Theorem: Kinetic Energy

OBJECTIVES: To (a) study the work–energy theorem, and (b) apply it in solving problems.

Now that we have an operational definition of work, let's take a look at how work is related to energy. Energy is one of the most important concepts in science. We describe it as a quantity that objects or systems possess. Basically, work is something that is *done on* objects, whereas energy is something that objects *have*—the ability to do work.

One form of energy that is closely associated with work is *kinetic energy*. (Another form of energy, *potential energy*, will be described in Section 5.4.) Consider an object at rest on a frictionless surface. Let a horizontal force act on the object and set it in motion. Work is done *on* the object, but where does the work "go," so to speak? It goes into setting the object into motion, or changing its *kinetic* conditions. Because of its motion, we say the object has gained energy—kinetic energy, which gives it the capability to do work.

PHYSLET

Exploration 6.1 An Operational Definition of Work

PHYSLET

Illustration 4.2 Free-Body Diagrams

LEARN BY DRAWING

Forces on an Object on an Inclined Plane and Free-body Diagrams

4.5 More on Newton's Laws: Free-Body Diagrams and Translational Equilibrium

OBJECTIVES: To (a) apply Newton's laws in analyzing various situations using free-body diagrams, and (b) understand the concept of translational equilibrium.

Now that you have been introduced to Newton's laws and some applications in analyzing motion, the importance of these laws should be evident. They are so simply stated, yet so far-reaching. The second law is probably the most often applied, because of its mathematical relationship. However, the first and third laws are often used in qualitative analysis, as our continuing study of the different areas of physics will reveal.

In general, we will be concerned with applications that involve constant forces. Constant forces result in constant accelerations and allow the use of the kinematic equations from Chapter 2 in analyzing the motion. When there is a variable force, Newton's second law holds for the *instantaneous* force and acceleration, but the acceleration will vary with time, requiring advanced mathematics to analyze. We will generally limit ourselves to constant accelerations and forces. Several examples of applications of Newton's second law are presented in this section so that you can become familiar with its use. This small but powerful equation will be used again and again throughout this text.

There is still one more item in the problem-solving arsenal that is a great help with force applications—free-body diagrams. These are explained in the following Problem-Solving Strategy.

Problem-Solving Strategy: Free-Body Diagrams

In illustrations of physical situations, sometimes called *space diagrams*, force vectors may be drawn at different locations to indicate their points of application. However, because we are presently concerned only with linear motions, vectors in *free-body diagrams* (FBD) may be shown as emanating from a common point, which is usually chosen as the origin of the *x*–*y* axes. One of the axes is generally chosen along the direction of the net force acting on a body, since that is the direction in which the body will accelerate. Also, it is often important to resolve force vectors into components, and properly chosen *x*–*y* axes simplify this task.

In a free-body diagram, the vector arrows do not have to be drawn exactly to scale. However, the diagram should clearly show whether there is a net force and whether forces balance each other in a particular direction. When the forces aren't balanced, we know from Newton's second law that there must be an acceleration.

In summary, the general steps in constructing and using free-body diagrams are as follows (refer to the accompanying Learn by Drawing as you read):

1. Make a sketch, or space diagram, of the situation (if one is not already available) and identify the forces acting on each body of the system. A space diagram is an illustration of the physical situation that identifies the force vectors.

2. Isolate the body for which the free-body diagram is to be constructed. Draw a set of Cartesian axes, with the origin at a point through which the forces act and with one of the axes along the direction of the body's acceleration. (The acceleration will be in the direction of the net force, if there is one.)

3. Draw properly oriented force vectors (including angles) on the diagram, emanating from the origin of the axes. If there is an unbalanced force, assume a direction of acceleration and indicate it with an acceleration vector. Be sure to include only those forces that act on the isolated body of interest.

4. Resolve any forces that are not directed along the *x*- or *y*-axis into *x*- or *y*-components (use plus and minus signs to indicate direction). Use the free-body diagram and force components to analyze the situation in terms of Newton's second law of motion. (*Note:* If you assume that the acceleration is in one direction, and in the solution it comes out with the opposite sign, then the acceleration is actually in the opposite direction from that assumed. For example, if you assume that $\vec{a}$ is in the +*x*-direction, but you get a negative answer, then $\vec{a}$ is in the −*x*-direction.)

Integration of Physlet® Physics. Physlets are Java-based applets that illustrate physics concepts through animation. *Physlet Physics* is a best-selling book and CD-ROM containing more than 800 Physlets in three different formats: Physlet Illustrations, Physlet Explorations, and Physlet Problems. In the Sixth Edition of *College Physics*, the Physlets from *Physlet Physics* are denoted by an icon so students know when an alternate explanation and animation are available to further their understanding. The *Physlet Physics* CD-ROM is included with the purchase of a new textbook.

Biological Applications. We have greatly increased the number and scope of biological and biomedical applications. Examples of new biological applications include "People Power: Using Body Energy," "Osteoporosis and Bone Mineral Density (BMD)," and the Magnetic Force in Future Medicine.

We have enhanced the following pedagogical features in the Sixth Edition:

Learn by Drawing Boxes. Visualization is one of the most important steps in problem solving. In many cases, if students can make a sketch of a problem, they can solve it. "Learn by Drawing" features offer students specific help on making certain types of sketches and graphs that will provide key insights into a variety of physical situations.

Suggested Problem-Solving Procedure. Section 1.7 provides a framework for thinking about problem solving. This section includes the following:

- An overview of problem-solving strategies

- A six-step procedure that is general enough to apply to most problems in physics, but is easily used in specific situations

- Examples that illustrate the detailed problem-solving process, showing how the general procedure is applied in practice

Problem-Solving Strategies and Hints. The initial treatment of problem solving is followed throughout with an abundance of suggestions, tips, cautions, shortcuts, and useful techniques for solving specific kinds of problems. These strategies and hints help students apply general principles to specific contexts as well as avoid common pitfalls and misunderstandings.

Conceptual Examples. These Examples ask students to think about a physical situation and conceptually solve a question or choose the correct prediction out of a set of possible outcomes, on the basis of an understanding of relevant principles. The discussion that follows ("Reasoning and Answer") explains clearly how the correct answer can be identified, as well as why the other answers are wrong.

Worked Examples. We have tried to make in-text Examples as clear and detailed as possible. The aim is not merely to show students which equations to use, but to explain the strategy being employed and the role of each step in the overall plan. Students are encouraged to learn the "why" of each step along with the "how." Our goal is to provide a model for students to use as they solve problems. Each worked Example includes the following:

- *Thinking It Through* focuses students on the critical thinking and analysis they should undertake before beginning to use equations.
- *Given* and *Find* are provided as the first part of every *Solution* to remind students the importance of identifying what is known and what needs to be solved.
- *Follow-Up Exercises* at the end of each Conceptual Example and each worked Example further reinforce the importance of conceptual understanding and offer additional practice. (Answers to Follow-Up Exercises are given at the back of the text.)

Integrated Examples. In order to further emphasize the connection between conceptual understanding and quantitative problem solving, we have developed Integrated Examples for each chapter. These Examples work through a physical situation both qualitatively and quantitatively. The qualitative portion is solved by conceptually choosing the correct answer from a set of possible answers. The quantitative portion involves a mathematical solution related to the conceptual part, demonstrating how conceptual understanding and numerical calculations go hand in hand.

End-of-Chapter Exercises. Each section of the end-of-chapter material begins with multiple-choice questions (**MC**) to allow students a quick self-test for that section. These are followed by short-answer conceptual questions (CQ) that test students' conceptual understanding and ask students to reason from principles. Quantitative problems round out the Exercises in each section. *College Physics* provides short answers to all odd-numbered Exercises (quantitative *and* conceptual) at the back of the text, so students can check their understanding.

Paired Exercises. To encourage students to work problems on their own, most sections include at least one set of Paired Exercises that deal with similar situations. The first problem in a pair is solved in the *Student Study Guide and Solutions Manuals*; the second problem, which explores a similar situation to that presented in the first problem, has only an answer at the back of the book.

Integrated Exercises. Like the Integrated Examples in the chapter, Integrated Exercises (IE) ask students to solve a problem quantitatively as well as answer a conceptual question dealing with the exercise. By answering both parts, students can see if their numerical answer matches their conceptual understanding.

Comprehensive Exercises. To ensure that students can synthesize concepts, each chapter concludes with a section of comprehensive exercises drawn from all sections of the chapter and perhaps basic principles from previous chapters.

Absolutely Zero Tolerance for Errors Club (AZTEC)

We have continued to ensure accuracy through the Absolutely Zero Tolerance for Errors Club (AZTEC). Bo Lou of Ferris State University, new co-author of the text and author of our *Instructor's Solutions Manual* and *Student Study Guide and Solutions Manuals*, headed the AZTEC team and was supported by the text's co-authors as well as Billy Younger, College of the Albemarle; Mike LoPresto, Henry Ford Community College; David Curott, University of North Alabama; and Daniel Lottis, Panconsult Ciências Físicas e Naturais, Curitiba, Brazil. Every end-of-chapter Exercise was worked by three of the five team members individually and independently. The results were collected and discrepancies were resolved by a team discussion. While there has never been a text that was absolutely free of errors, that was our goal; we worked very hard to make the book error-free.

The Sixth Edition is supplemented by a media and print ancillary package developed to address the needs of both students and instructors.

For the Instructor

Annotated Instructor's Edition **(0-13-149706-5).** The margins of the *Annotated Instructor's Edition* (*AIE*) contain an abundance of suggestions for classroom demonstrations and activities, along with teaching tips (points to emphasize, discussion suggestions, and common misunderstandings to avoid). In addition, the *AIE* contains the following:

- Icons that identify each illustration reproduced as a transparency in the *Transparency Pack*
- Answers to end-of-chapter exercises (following each exercise)

Instructor's Resource Manual and Instructor Notes on ConcepTest Questions **(0-13-149711-1).** This manual has two parts. The first part, prepared by Katherine Whatley and Judith Beck (both of the University of The North Carolina–Asheville), contains sample syllabi, lecture outlines, notes, demonstration suggestions, readings, and additional references and resources. The second part, prepared by Cornelius Bennhold and Gerald Feldman (both of The George Washington University) contains an overview of the development and implementation of ConcepTests, as well as instructor notes for each ConcepTest found on the *Instructor Resource Center on CD-ROM*.

Instructor's Solutions Manual **(0-13-149710-3).** Prepared by Bo Lou of Ferris State University, the *Instructor's Solutions Manual* supplies answers with complete, worked-out solutions to all end-of-chapter exercises. This manual is also available electronically on the *Instructor Resource Center on CD-ROM*.

Test Item File **(0-13-149713-8).** Fully revised by Delena Gatch (University of North Alabama), the *Test Item File* offers more than 2800 multiple-choice, true/false, and short-answer/essay questions, approximately 50% conceptual. The questions are organized and referenced by chapter section and by question type. This manual is also available both within TestGenerator and as chapter-by-chapter Microsoft Word files on the *Instructor Resource Center on CD-ROM* (see below).

Transparency Pack **(0-13-149708-1).** The *Transparency Pack* contains more than 400 full-color acetates of text illustrations.

"Physics You Can See" Video Demonstrations **(0-205-12393-7).** Each segment, two to five minutes long, demonstrates a classical physics experiment. Eleven segments are included, such as "Coin & Feather" (acceleration due to gravity), "Monkey & Gun" (rate of vertical free fall), "Swivel Hips" (force pairs), and "Collapse a Can" (atmospheric pressure). Digital versions of the videos are available on the *Instructor Resource Center on CD-ROM*.

Instructor Resource Center on CD-ROM **(0-13-149712-X).** This multiple CD-ROM set, new to the Sixth Edition, provides virtually every electronic asset you'll need in and out of the classroom. Though you can navigate freely through the CDs to find the resources you want, the software allows you to browse and search through the catalog of assets. The CD-ROMs are organized by chapter and include all text illustrations and tables from the Sixth Edition in JPEG and PowerPoint formats. The IRC/CDs also contain *TestGenerator*, a powerful dual-platform, fully networkable software program for creating tests ranging from short quizzes to long exams. Questions from the Sixth Edition *Test Item File*, including randomized versions, are supplied, and professors can use the Question Editor to modify existing questions or create new questions. The IRC/CDs also contain the instructor's version of *Physlet Physics*, chapter-by-chapter lecture outlines in PowerPoint, *ConcepTest*

"Clicker" Questions in PowerPoint, chapter-by-chapter Microsoft Word files of all numbered equations, the eleven *Physics You Can See* demonstration videos, and Microsoft Word and PDF versions of the *Test Item File*, the *Instructor's Solutions Manual*, the *Instructor's Resource Manual*, and the end-of-chapter exercises from *College Physics*, Sixth Edition.

Peer Instruction **(0-13-656441-6).** Authored by Eric Mazur (Harvard University), this manual explains an interactive teaching style that actively involves students in the learning process by focusing their attention on underlying concepts through interactive "ConcepTests," reading quizzes, and conceptual exam questions.

Just-in-Time Teaching: Blending Active Learning with Web Technology **(0-13-085034-9).** Gregor Novak (Indiana University–Purdue University, Indianapolis), Evelyn Patterson (United States Air Force Academy), Andrew Gavrin (Indiana University–Purdue University, Indianapolis), and Wolfgang Christian (Davidson College) authored this resource book for educators. Just-in-time teaching (JITT) is a teaching and learning methodology designed to engage students. Using feedback from preclass Web assignments, instructors can adjust classroom lessons so that students receive rapid response to the specific questions and problems they are having.

Physlets: Teaching Physics with Interactive Curricular Material **(0-13-029341-5).** Authored by Wolfgang Christian and Mario Belloni (both of Davidson College), this text is a teacher's resource book with an accompanying CD for instructors who are interested in incorporating Physlets into their physics courses. The book and CD discuss the pedagogy behind the use of Physlets.

For the Student

Student Study Guide and Selected Solutions Manual **(Volume 1: 0-13-149716-2, Volume 2: 0-13-174405-4).** Written by Bo Lou of Ferris State University, the *Student Study Guide and Selected Solutions Manual* presents chapter-by-chapter reviews, chapter summaries, additional worked examples, practice quizzes, and solutions to paired and selected exercises.

Student Pocket Guide **(0-13-149718-9).** Written by Biman Das (State University of New York–Potsdam), this easy-to-carry 5" x 7" paperback contains a summary of the entire text, including all key concepts and equations, as well as tips and hints.

E&M TIPERs: Electricity & Magnetism Tasks Inspired by Physics Education Research **(0-13-185499-2).** Curtis J. Hieggelke (Joliet Junior College), David P. Maloney (Indiana University Purdue University Fort Wayne), Stephen E. Kanim (New Mexico State University), and Thomas L. O'Kuma (Lee College) created this comprehensive set of conceptual exercises for electricity and magnetism based on the results of education research into how students learn physics. This workbook contains more than 300 tasks in eleven different task formats.

Tutorials in Introductory Physics **(0-13-097069-7).** Written by Lillian C. McDermott, Peter S. Schaffer, and the Physics Education Group at the University of Washington, this landmark book presents a series of physics tutorials designed by a leading physics education research group. Emphasizing the development of concepts and scientific reasoning skills, the tutorials focus on the specific conceptual and reasoning difficulties that students tend to encounter. The tutorials cover a range of topics in Mechanics, E & M, and Waves and Optics.

Ranking Task Exercises in Physics: Student Edition **(0-13-144851-X).** Developed by Thomas L. O'Kuma (Lee College), David P. Maloney (Indiana University–Purdue University, Fort Wayne), and Curtis J. Hieggelke (Joliet Junior College), ranking tasks are an innovative type of conceptual exercises that ask students to make comparative judgments about a set of variations on a particular physical situation. This text is a unique resource for physics instructors who are looking for tools to incorporate more conceptual analysis in their courses. This supplement contains 218 ranking task exercises that cover all classical physics topics.

Interactive Physics Player Workbook, Second Edition (0-13-067108-8). Created by Cindy Schwarz (Vassar College), John P. Ertel (U.S. Naval Academy), and MSC Software, this interactive workbook, tutorial-oriented worksheets, and CD-ROM package are designed to help students visualize and work with specific physics problems through simulations created with Interactive Physics files. Forty problems of varying degrees of difficulty require students to make predictions, change variables, run, and visualize motion on the computer. The accompanying workbook/study guide provides instructions, physics review, hints, and questions. The CD-ROM contains everything students need to run the simulations.

Mathematics for College Physics (0-13-141427-5). Written by Biman Das (SUNY–Potsdam), this text, for students who need help with the necessary mathematical tools, shows how mathematics is directly applied to physics and discusses how to overcome math anxiety.

MCAT Physics Study Guide (0–13–627951–1). Because most MCAT questions require more thought and reasoning than simply plugging numbers into an equation, this study guide, by Joseph Boone (California Polytechnic State University–San Luis Obispo), is designed to refresh students' memory about the topics they've covered in class. Additional review, practice problems, and review questions are included.

Companion Website with GradeTracker (http://www.prenhall.com/wilson)

This text-specific Web site provides students and instructors a wealth of innovative online materials for use with *College Physics*, Sixth Edition. The Companion Website with GradeTracker includes the following:

- Integration of Just-in-Time Teaching (JiTT) Warm-Ups, Puzzles, & Applications by Gregor Novak and Andrew Gavrin (Indiana University–Purdue University, Indianapolis): Warm-Up questions are short-answer questions based on important concepts in the text chapters. Puzzles are more complex questions and often require the integration of more than one concept. Thus professors can assign Warm-Up questions as a reading quiz before the class lecture on that topic, and Puzzle questions as follow-up assignments submitted after class. The Applications modules answer the question, "What is physics good for?" by connecting physics concepts to real-world phenomena and new developments in science and technology. Each Application module contains short-answer/essay questions.

- Practice Questions: A module of twenty to thirty gradable multiple-choice practice questions are available for review with each chapter.

- Ranking Task Exercises, edited by Thomas O'Kuma (Lee College), David Maloney (Indiana University–Purdue University, Fort Wayne), and Curtis Hieggelke (Joliet Junior College): Ranking tasks are gradable conceptual exercises that require students to rank a number of situations or variations of a situation.

- *Physlet Physics* Problems: Physlets are Java-based applets that illustrate physics concepts through animation. *Physlet Physics* is a best-selling book and CD-ROM containing more than 800 Physlets. Physlet problems are interactive versions of the kind of exercises typically assigned for homework. Gradable problems from *Physlet Physics* are available for students to self-test. To access the Physlets, students use their copy of the *Physlet Physics* CD-ROM, included with the purchase of a new copy of this text.

- *MCAT Study Guide*, by Kaplan Test Prep and Admissions: The MCAT Study Guide module provides students with ten quizzes on topics and concepts covered on the MCAT exam.

Blackboard. Blackboard is a comprehensive and flexible software platform that delivers a course management system, customizable institution-wide portals, online communities, and an advanced architecture that allows for Web-based integration of multiple administrative systems. Its features include the following:

- Progress tracking, class and student management, grade book, communication, assignments, and reporting tools.
- Testing programs that enable instructors to create online quizzes and tests from the *College Physics*, Sixth Edition course content and automatically grade and track results. All tests can be entered into the grade book for easy course management.
- Communications tools such as the Virtual Classroom (chat rooms, whiteboard, slides), document sharing, and bulletin boards.

CourseCompass, Powered by Blackboard. With the highest level of service, support, and training available today, CourseCompass combines tested, quality online course resources for *College Physics*, Sixth Edition with easy-to-use online course management tools. CourseCompass is designed to address the individual needs of instructors, who will be able to create an online course without any special technical skills or training. Its features include the following:

- Higher flexibility—Professors can adapt Prentice Hall content to match their own teaching goals, with little or no outside assistance needed.
- Assessment, customization, class administration, and communication tools.
- Point-and-click access—*College Physics*, Sixth Edition course resources are available to instructors at a click of the mouse.
- A nationally hosted and fully supported system that relieves individuals and institutions of the burdens of troubleshooting and maintenance.

WebCT. WebCT offers a powerful set of tools that enables instructors to create practical Web-based educational programs—ideal resources to enhance a campus course or to construct one entirely online. The WebCT shell and tools, integrated with the *College Physics*, Sixth Edition content, results in a versatile, course-enhancing teaching and learning system. Its features include the following:

- Page tracking, progress tracking, class and student management, grade book, communication, calendar, and reporting tools.
- Communication tools including chat rooms, bulletin boards, private e-mail, and whiteboard.
- Testing tools that help create and administer timed online quizzes and tests and automatically grade and track all results.

WebAssign (http://www.webassign.com). WebAssign's homework delivery service gives you the freedom to get back to create assignments from a database of exercises from *College Physics*, Sixth Edition, or write and customize your own exercises. You have complete control over the homework your students receive, including due date, content, feedback, and question formats. Its features include the following:

- Create, post, and review assignments twenty-four hours a day, seven days a week.
- Deliver, collect, grade, and record assignments instantly.
- Offer more practice exercises, quizzes, homework, labs, and tests.
- Randomize numerical values or phrases to create unique questions.
- Assess student performance to keep abreast of individual progress.
- Grade algebraic formulas with enhanced math-type display.
- Capture the attention of your online and distance-learning students.

Acknowledgments

The members of AZTEC—Billy Younger, Michael LoPresto, David Curott, and Daniel Lottis—as well as accuracy reviewers Michael Ottinger, and Mark Sprague deserve more than a special thanks for their tireless, timely, and extremely thorough review of this book.

Dozens of other colleagues, listed in the upcoming section, helped us identify ways to make the sixth edition a better learning tool for students. We are indebted to them, as their thoughtful and constructive suggestions benefited the book greatly.

We owe many thanks to the editorial and production team at Prentice Hall including Erik Fahlgren, Senior Acquisitions Editor, Heather Scott, Art Director, Christian Botting, Associate Editor, and Jessica Berta, Editorial Assistant. In particular, the authors wish to acknowledge the outstanding performance of Simone Lukashov, Production Editor. His courteous, coscientious, and cheerful manner made for an efficient and enjoyable production process. Also, we thank Karen Karlin, Prentice Hall Development Editor, for her helpful assistance and editing.

In addition, I (Tony Buffa) once again extend many thanks to my co-authors, Jerry Wilson and Bo Lou, for their cheerful helpfulness and professional approach to the work on this edition. As always, several colleagues of mine at Cal Poly gave of their time for fruitful discussions. Among them are Professors Joseph Boone, Ronald Brown, and Theodore Foster. My family—my wife, Connie, and daughters, Jeanne and Julie—was, as always, a continuous and welcomed source of support. I also acknowledge the support from my father, Anthony Buffa, Sr., and my aunt, Dorothy Abbott. Last, I thank the students in my classes who contributed excellent ideas over the past few years.

Finally, we would like to urge anyone using the book—student or instructor—to pass on to us any suggestions that you have for its improvement. We look forward to hearing from you.

—Jerry D. Wilson
jwilson@greenwood.net
—Anthony J. Buffa
abuffa@calpoly.edu
—Bo Lou
loub@ferris.edu

Reviewers of the Sixth Edition:

David Aaron
South Dakota State University

E. Daniel Akpanumoh
Houston Community College, Southwest

Ifran Azeem
Embry-Riddle Aeronautical University

Raymond D. Benge
Tarrant County College

Frederick Bingham
University of North Carolina, Wilmington

Timothy C. Black
University of North Carolina, Wilmington

Mary Boleware
Jones County Junior College

Art Braundmeier
Southern Illinois University, Edwardsville

Michael L. Broyles
Collin County Community College

Debra L. Burris
Oklahoma City Community College

Jason Donav
University of Puget Sound

Robert M. Drosd
Portland Community College

Bruce Emerson
Central Oregon Community College

Milton W. Ferguson
Norfolk State University

Phillip Gilmour
Tri-County Technical College

Allen Grommet
East Arkansas Community College

Brian Hinderliter
North Dakota State University

Ben Yu-Kuang Hu
University of Akron

Porter Johnson
Illinois Institute of Technology

Andrew W. Kerr
University of Findlay

Jim Ketter
Linn-Benton Community College

Terrence Maher
Alamance Community College

Kevin McKone
Copiah Lincoln Community College

Kenneth L. Menningen
University of Wisconsin, Stevens Point

Michael Mikhaiel
Passaic County Community College

Ramesh C. Misra
Minnesota State University, Mankato

Sandra Moffet
Linn Benton Community College

Michael Ottinger
Missouri Western State College

James Palmer
University of Toledo

Kent J. Price
Morehead State University

Salvatore J. Rodano
Harford Community College

John B. Ross
Indiana University-Purdue University, Indianapolis

Terry Scott
University of Northern Colorado

Rahim Setoodeh
Milwaukee Area Technical College

Martin Shingler
Lakeland Community College

Mark Sprague
East Carolina State University

Steven M. Stinnett
McNeese State University

John Underwood
Austin Community College

Tristan T. Utschig
Lewis-Clark State College

Steven P. Wells
Louisiana Technical University

Christopher White
Illinois Institute of Technology

Anthony Zable
Portland Community College

John Zelinsky
Community College of Baltimore County, Essex

Reviewers of Previous Editions

William Achor
Western Maryland College

Alice Hawthorne Allen
Virginia Tech

Arthur Alt
College of Great Falls

Zaven Altounian
McGill University

Frederick Anderson
University of Vermont

Charles Bacon
Ferris State College

Ali Badakhshan
University of Northern Iowa

Anand Batra
Howard University

Michael Berger
Indiana University

William Berres
Wayne State University

James Borgardt
Juniata College

Hugo Borja
Macomb Community College

Bennet Brabson
Indiana University

Jeffrey Braun
University of Evansville

Michael Browne
University of Idaho

David Bushnell
Northern Illinois University

Lyle Campbell
Oklahoma Christian University

James Carroll
Eastern Michigan State University

Aaron Chesir
Lucent Technologies

Lowell Christensen
American River College

Philip A. Chute
University of Wisconsin–Eau Claire

Robert Coakley
University of Southern Maine

Lawrence Coleman
University of California–Davis

Lattie F. Collins
East Tennessee State University

Sergio Conetti
University of Virginia, Charlottesville

James Cook
Middle Tennessee State University

David M. Cordes
Belleville Area Community College

James R. Crawford
Southwest Texas State University

William Dabby
Edison Community College

Purna Das
Purdue University

J. P. Davidson
University of Kansas

Donald Day
Montgomery College

Richard Delaney
College of Aeronautics

James Ellingson
College of DuPage

Donald Elliott
Carroll College

Arnold Feldman
University of Hawaii

John Flaherty
Yuba College

Rober J. Foley
University of Wisconsin–Stout

Lewis Ford
Texas A&M University

Donald Foster
Wichita State University

Donald R. Franceschetti
Memphis State University

Frank Gaev
ITT Technical Institute–Ft. Lauderdale

Rex Gandy
Auburn University

Simon George
California State–Long Beach

Barry Gilbert
Rhode Island College

Richard Grahm
Ricks College

Tom J. Gray
University of Nebraska

Douglas Al Harrington
Northeastern State University

Gary Hastings
Georgia State University

Xiaochun He
Georgia State University

J. Erik Hendrickson
University of Wisconsin–Eau Claire

Al Hilgendorf
University of Wisconsin–Stout

Joseph M. Hoffman
Frostburg State University

Andy Hollerman
University of Louisiana, Layfayette

Jacob W. Huang
Towson University

Randall Jones
Loyola University

Omar Ahmad Karim
University of North Carolina–Wilmington

S. D. Kaviani
El Camino College

Victor Keh
ITT Technical Institute–Norwalk, California

John Kenny
Bradley University

James Kettler
Ohio University, Eastern Campus

Dana Klinck
Hillsborough Community College

Chantana Lane
University of Tennessee–Chattanooga

Phillip Laroe
Carroll College

Rubin Laudan
Oregon State University

Bruce A. Layton
Mississippi Gulf Coast Community College

R. Gary Layton
Northern Arizona University

Kevin Lee
University of Nebraska

Paul Lee
California State University, Northridge

Federic Liebrand
Walla Walla College

Mark Lindsay
University of Louisville

Bryan Long
Columbia State Community College

Michael LoPresto
Henry Ford Community College

Dan MacIsaac
Northern Arizona University

Robert March
University of Wisconsin

Trecia Markes
University of Nebraska–Kearney

Aaron McAlexander
Central Piedmont Community College

William McCorkle
West Liberty State University

John D. McCullen
University of Arizona

Michael McGie
California State University–Chico

Paul Morris
Abilene Christian University

Gary Motta
Lassen College

J. Ronald Mowrey
Harrisburg Area Community College

Gerhard Muller
University of Rhode Island

K. W. Nicholson
Central Alabama Community College

Erin O'Connor
Allan Hancock College

Anthony Pitucco
Glendale Community College

William Pollard
Valdosta State University

R. Daryl Pedigo
Austin Community College

T. A. K. Pillai
University of Wisconsin–La Crosse

Darden Powers
Baylor University

Donald S. Presel
University of Massachusetts–Dartmouth

E. W. Prohofsky
Purdue University

Dan R. Quisenberry
Mercer University

W. Steve Quon
Ventura College

David Rafaelle
Glendale Community College

George Rainey
California State Polytechnic University

Michael Ram
SUNY–Buffalo

William Riley
Ohio State University

William Rolnick
Wayne State University

Robert Ross
University of Detroit–Mercy

Craig Rottman
North Dakota State University

Gerald Royce
Mary Washington College

Roy Rubins
University of Texas, Arlington

Sid Rudolph
University of Utah

Om Rustgi
Buffalo State College

Anne Schmiedekamp
Pennsylvania State University–Ogontz

Cindy Schwarz
Vassar College

Ray Sears
University of North Texas

Mark Semon
Bates College

Bartlett Sheinberg
Houston Community College

Jerry Shi
Pasadena City College

Peter Shull
Oklahoma State University

Thomas Sills
Wilbur Wright College

Larry Silva
Appalachian State University

Michael Simon
Housatonic Community Technical College

Christopher Sirola
Tri-County Technical College

Gene Skluzacek
St. Petersburg College

Soren P. Sorensen
University of Tennessee–Knoxville

Ross Spencer
Brigham Young University

Dennis W. Suchecki
San Diego Mesa College

Frederick J. Thomas
Sinclair Community College

Jacqueline Thornton
St. Petersburg Junior College

Anthony Trippe
ITT Technical Institute–San Diego

Gabriel Umerah
Florida Community College–Jacksonville

Lorin Vant-Hull
University of Houston

Pieter B. Visscher
University of Alabama

Karl Vogler
Northern Kentucky University

John Walkup
California Polytechnic State University

Arthur J. Ward
Nashville State Technical Institute

Larry Weinstein
Old Dominion University

John C. Wells
Tennessee Technical University

Arthur Wiggins
Oakland Community College

Kevin Williams
ITT Technical Institute–Earth City

Linda Winkler
Appalachian State University

Jeffery Wragg
College of Charleston

Rob Wylie
Carl Albert State University

John Zelinsky
Southern Illinois University

Dean Zollman
Kansas State University

MEASUREMENT AND PROBLEM SOLVING

PHYSICS FACTS

- Tradition holds that in the twelfth century King Henry I of England decreed that the yard should be the distance from the tip of his royal nose to the thumb of his out-stretched arm. (Had King Henry's arm been 3.37 inches longer, the yard and the meter would be equal in length.)

- The abbreviation for the pound, *lb*, comes from the Latin word *libra*, which was a Roman unit of weight approximately equal to a pound. The word *pound* comes from the Latin *pondero*, "to weigh". Libra is also a sign of the zodiac and is symbolized by a set of scales (used for weight measurement).

- Thomas Jefferson suggested that the length of a pendulum with a period of one second be used as a length standard.

- Is the old saying, "A pint's a pound the world around," true? It depends on what you are talking about. The saying is a good approximation for water and other similar liquids. Water weighs 8.3 pounds per gallon, so one-eighth of that, or a pint, weighs 1.04 lb.

- Pi (π), the ratio of the circumference of a circle to its diameter is always the same number, no matter which circle. Pi is an irrational number; that is, it cannot be written as the ratio of two whole numbers and is an infinite, nonrepeating decimal. Computers have calculated π to billions of digits. According to the *Guinness Book of Records* (2004), π has been calculated to 1 241 100 000 000 decimal places.

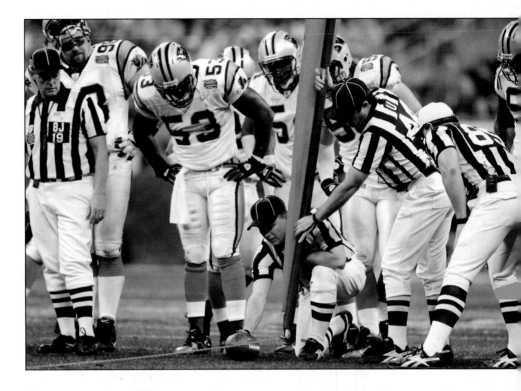

$\mathbf{I}$s it first and ten? A measurement is needed, as with many other things in our lives. Length measurements tell us how far it is between cities, how tall you are, and as in the photo, if it's first and ten (yards to go). Time measurements tell you how long it is until the class ends, when the semester or quarter begins, and how old you are. Drugs taken because of illnesses are given in measured doses. Lives depend on various measurements made by doctors, medical technologists, and pharmacists in the diagnosis and treatment of disease.

Measurements enable us to compute quantities and solve problems. Units of measurement are also important in measurements and problem solving. For example, in finding the volume of a rectangular box, if you measure its dimensions in inches, the volume would have units of in^3 (cubic inches); if measured in centimeters, then it would be cm^3 (cubic centimeters). Measurement and problem solving are part of our lives. They play a particularly central role in our attempts to describe and understand the physical world, as we shall see in this chapter. But first, let's look at why one should study physics (Insight 1.1).

INSIGHT 1.1 WHY STUDY PHYSICS?

The question, "Why study physics?" occurs to many students sometime during their college careers. The truth is that there are probably as many answers as there are students, much as with any other subject. However, the answers can usually be arranged into several general groups as follows.

You are probably not a *physics major*, but for them, the answer is obvious. Introductory physics provides the foundation of their career. The fundamental goal of physics is to understand where the universe came from, how it has evolved and is still evolving, and the rules ("laws") that govern the phenomena we observe. These students will use their knowledge of physics continually during their careers. As an example of physics research, consider the discovery of the transistor in a special area of research known as solid-state physics in the late 1940s.

You are also probably not an *engineering "applied physics" major*. For them physics provides the basis of the engineering principles used to solve technological (applied and practical) problems. Some of these students may not use physics directly in their careers, but a good understanding of physics is crucial to the problem-solving needed in technological advances. For example, after the discovery of the transistor by physicists, engineers then developed uses for it. Decades later it evolved into the modern computer chip, which is an electrical computing network containing millions of tiny transistor elements.

More than likely you are a *life science* or *technology major* (premedicine, pre-physical therapy, preveterinary, industrial

technology, and so on). In this case, physics can provide a background understanding of the principles involved in your work. Although the applications of the laws of physics may not be immediately obvious, understanding them can be a valuable tool in your career. If you are a medical professional, for example, it may be necessary to evaluate MRI (magnetic resonance imaging) results, a procedure that is now commonplace. Would you be surprised to know that MRI scans are based on a physical phenomenon called *nuclear magnetic resonance*, first discovered by physicists and still used for measuring nuclear and solid-state properties?

If you are a student in a *nontechnical major*, the physics requirement is intended to provide a well-rounded education; that is, the ability to evaluate technology in the context of societal needs. For example, you may be called on to vote on tax benefits for an energy production source, and you might want to evaluate the pros and cons of the process. Or you might be tempted to vote for an official who has strong views on nuclear waste disposal. Are these views scientifically correct? To evaluate them, a knowledge of physics is necessary.

So as you can see, there is no one answer to the question, "Why study physics?" There is however, one overriding theme: Knowledge of the laws of physics can provide an excellent background for your career or an understanding of the world around you, or it can simply help make you a better and more well-rounded citizen.

1.1 Why and How We Measure

OBJECTIVES: To distinguish standard units and systems of units.

Imagine that someone is giving you directions to her house. Would you find it helpful to be told, "Drive along Elm Street for a little while, and turn right at one of the lights. Then keep going for quite a long way"? Or would you want to deal with a bank that sent you a statement at the end of the month saying, "You still have some money left in your account. Not a great deal, though."

Measurement is important to all of us. It is one of the concrete ways in which we deal with our world. This concept is particularly true in physics. *Physics is concerned with the description and understanding of nature* and measurement is one of its most important tools.

There are ways of describing the physical world that do not involve measurement. For instance, we might talk about the color of a flower or a dress. But the perception of color is subjective; it may vary from one person to another. Indeed, many people are color-blind and cannot tell certain colors apart. Light received by our eyes can be described in terms of wavelengths and frequencies. Different wavelengths are associated with different colors because of the physiological response of our eyes to light. But unlike the sensations or perceptions of color, the wavelengths can be measured. They are the same for everyone. In other words, measurements are objective. *Physics attempts to describe nature in an objective way through measurement.*

Standard Units

Measurements are expressed in terms of unit values, or units. As you are probably aware, a large variety of units are used to express measured values. Some of the earliest units of measurement, such as the foot, were originally referenced to parts of the human body. (Even today, the hand is still used as a unit to measure the height of horses. One hand is equal to 4 in.) If a unit becomes officially accepted, it is called a **standard unit**. Traditionally, a government or international body establishes standard units.

A group of standard units and their combinations is called a **system of units**. Two major systems of units are in use today—the metric system and the British system. The latter is still widely used in the United States, but has virtually disappeared in the rest of the world, having been replaced by the metric system.

Different units in the same system or units of different systems can be used to describe the same thing. For example, your height can be expressed in inches, feet, centimeters, meters—or even miles, for that matter (although this unit would not be very convenient). It is always possible to convert from one unit to another, and such conversions are sometimes necessary. However, it is best, and certainly most practical, to work consistently within the same system of units, as you will see.

Teaching tip: Ask students to bring to class a ruler calibrated in centimeters. A plastic ruler displaying both centimeters and inches is best.

1.2 SI Units of Length, Mass, and Time

OBJECTIVES: To (a) describe the SI, and (b) specify the references for the three main base quantities of this system.

Length, mass, and time are fundamental physical quantities that describe a great many quantities and phenomena. In fact, the topics of mechanics (the study of motion and force) covered in the first part of this book require *only* these physical quantities. The system of units used by scientists to represent these and other quantities is based on the metric system.

Historically, the metric system was the outgrowth of proposals for a more uniform system of weights and measures in France during the seventeenth and eighteenth centuries. The modern version of the metric system is called the **International System of Units**, officially abbreviated as **SI** (from the French *Système International des Unités*).

The SI includes *base quantities* and *derived quantities*, which are described by base units and derived units, respectively. **Base units**, such as the meter and the kilogram, are defined by standards. Other quantities that are expressed in terms of combinations of base units are called **derived units**. (Think of how we commonly measure the length of a trip in miles and the amount of time the trip takes in hours. To express how fast, or the rate we travel, the derived unit of miles per hour is used, which represents distance traveled per unit of time, or length per time.)

One of the refinements of the SI was the adoption of new standard references for some base units, including those of length and of time.

Length

Length is the base quantity used to measure distances or dimensions in space. We commonly say that length is the distance between two points. But the distance between any two points depends on how the space between them is traversed, which may be in a straight or a curved path.

The SI unit of length is the **meter (m)**. The meter was originally defined as 1/10 000 000 of the distance from the North Pole to the equator along a meridian

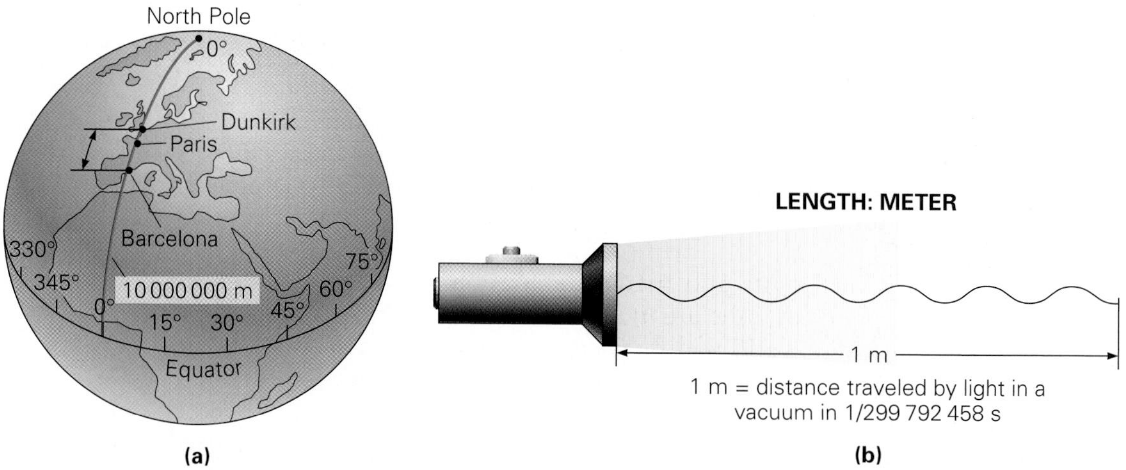

North Pole

Dunkirk
Paris

Barcelona

330°
345°
0°
10 000 000 m
15° 30° 45°
60°
75°
0°

Equator

(a)

LENGTH: METER

1 m

1 m = distance traveled by light in a
vacuum in 1/299 792 458 s

(b)

▲ **FIGURE 1.1 The SI length standard: the meter** **(a)** The meter was originally defined as 1/10 000 000 of the distance from the North Pole to the equator along a meridian running through Paris, of which a portion was measured between Dunkirk and Barcelona. A metal bar (called the Meter of the Archives) was constructed as a standard. **(b)** The meter is currently defined in terms of the speed of light.

running through Paris (▲Fig. 1.1a).* A portion of this meridian between Dunkirk, France, and Barcelona, Spain, was surveyed to establish the standard length, which was assigned the name *metre*, from the Greek word *metron*, meaning "a measure." (The American spelling is *meter*.) A meter is 39.37 in.—slightly longer than a yard.

The length of the meter was initially preserved in the form of a material standard: the distance between two marks on a metal bar (made of a platinum–iridium alloy) that was stored under controlled conditions and called the Meter of the Archives. However, it is not desirable to have a reference standard that changes with external conditions, such as temperature. In 1983, the meter was redefined in terms of a more accurate standard, an unvarying property of light: the length of the path traveled by light in a vacuum during an interval of 1/299 792 458 of a second (Fig. 1.1b). In other words, light travels 299 792 458 m in a second, and the speed of light in a vacuum is defined to be $c = 299\,792\,458$ m/s. (c is the common symbol for the speed of light.) Note that the length standard is referenced to time, which can be measured with great accuracy.

Mass

Mass is the base quantity used to describe amounts of matter. The more massive an object, the more matter it contains. (We will encounter more discussions of mass in Chapters 4 and 7.)

The SI unit of mass is the **kilogram** (**kg**). The kilogram was originally defined in terms of a specific volume of water, but is now referenced to a specific material standard: the mass of a prototype platinum–iridium cylinder kept at the International Bureau of Weights and Measures in Sèvres, France (▶Fig. 1.2). The United States has a duplicate of the prototype cylinder. The duplicate serves as a reference for secondary standards that are used in everyday life and commerce. It is hoped that the kilogram may eventually be referenced to something other than a material standard.

---

*Note that this book and most physicists have adopted the practice of writing large numbers with a thin space for three-digit groups—for example, 10 000 000 (not 10,000,000). This is done to avoid confusion with the European practice of using a comma as a decimal point. For instance, 3.141 in the United States would be written as 3,141 in Europe. Large decimal numbers, such as 0.537 84, may also be separated, for consistency. Spaces are generally used for numbers with more than four digits on either side of the decimal point.

You may have noticed that the phrase *weights and measures* is generally used instead of *masses and measures*. In the SI, mass is a base quantity, but in the British system, weight is used instead to describe amounts of mass—for example, weight in pounds instead of mass in kilograms. The weight of an object is the gravitational attraction that the Earth exerts on the object. For example, when you weigh yourself on a scale, your weight is a measure of the downward gravitational force exerted on you by the Earth. We can use weight as a measure of mass in this way because near the Earth's surface, because weight and mass are directly proportional to each other.

But treating weight as a base quantity creates some problems. A base quantity should have the same value everywhere. This is the case with mass—an object has the same mass, or amount of matter, regardless of its location. *But this is not true of weight.* For example, the weight of an object on the Moon is less than its weight on the Earth. This is because the Moon is less massive than the Earth, and the gravitational attraction exerted on an object by the Moon (the object's weight) is less than that exerted by the Earth. That is, an object with a given amount of mass has a particular weight on the Earth, but on the Moon, the same amount of mass will weigh only about one sixth as much. Similarly, the weight of an object would vary for different planets.

For now, keep in mind that in a given location, such as on the Earth's surface, *weight is related to mass, but they are not the same.* Since the weight of an object of a certain mass can vary with location, it is much more useful to take mass as the base quantity, as the SI does. Base quantities should remain the same regardless of where they are measured, under normal or standard conditions. The distinction between mass and weight will be more fully explained in a later chapter. Our discussion until then will be chiefly concerned with mass.

Time

Time is a difficult concept to define. A common definition is that time is the continuous, forward flow of events. This statement is not so much a definition but an observation that time has never been known to run backward, as it might appear to do when you view a film run backward in a projector. Time is sometimes said to be a fourth dimension, accompanying the three dimensions of space (x, y, z, t). That is, if something exists in space, it also exists in time. In any case, events can be used to mark time measurements. The events are analogous to the marks on a meterstick used for measurements of length. (See Insight 1.2 on What Is Time?)

The SI unit of time is the **second (s)**. The solar "clock" was originally used to define the second. A solar day is the interval of time that elapses between two successive crossings of the same longitude line (meridian) by the Sun. A second was fixed as 1/86 400 of this apparent solar day (1 day = 24 h = 1440 min = 86 400 s). However, the elliptical path of the Earth's motion around the Sun causes apparent solar days to vary in length.

As a more precise standard, an average, or mean, solar day was computed from the lengths of the apparent solar days during a solar year. In 1956, the second was referenced to this mean solar day. But the mean solar day is not exactly the same for each yearly period, because of minor variations in the Earth's motions and a steady slowing of its rate of rotation due to tidal friction. So scientists kept looking for something better.

In 1967, an atomic standard was adopted as a better reference. The second was defined by the radiation frequency of the cesium-133 atom. This "atomic clock" used a beam of cesium atoms to maintain our time standard, with a variation of about one second in 300 years. In 1999, another cesium-133 atomic clock was adopted, the atomic fountain clock, which, as the name implies, is based on the radiation frequency of a fountain of cesium atoms rather than a beam (▼Fig. 1.3). The variation of this timepiece is less than one second per 20 million years!*

---

*An even more precise clock, the all-optical atomic clock, is under development. It is so named because it uses laser technology and measures the shortest time interval ever recorded—0.000 01 s. This new clock does not use cesium atoms, but rather a single cooled ion of liquid mercury linked to a laser oscillator. The frequency of the mercury ion is 100 000 times the frequency of cesium atoms, hence the shorter, more precise time interval.

MASS: KILOGRAM

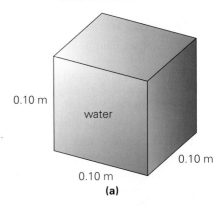

0.10 m

water

0.10 m

0.10 m

(a)

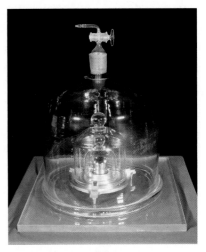

(b)

▲ **FIGURE 1.2 The SI mass standard: the kilogram (a)** The kilogram was originally defined in terms of a specific volume of water, that of a cube 0.10 m on a side, thereby associating the mass standard with the length standard. **(b)** The standard kilogram is now defined by a metal cylinder. The international prototype of the kilogram is kept at the French Bureau of Weights and Measures. It was manufactured in the 1880s of an alloy of 90% platinum and 10% iridium. Copies have been made for use as 1-kg national prototypes, one of which is the mass standard for the United States. It is kept at the National Institute of Standards and Technology (NIST) in Gaithersburg, MD.

Teaching tip: Many students now have digital watches that can be used as stopwatches. This use of the watches is a good way to get students to participate in simple demonstrations involving measurements of time.

INSIGHT   1.2 WHAT IS TIME?

For centuries the question, "What is time?" has been debated and attempted answers made, often philosophical in nature. But the definition of time still remains elusive to some degree. If you were asked to define or explain time, what would you say? General definitions may seem a bit vague. We commonly say:

> Time is the continuous, forward flow of events.

Other thoughts on time include the following.
Plato, the great Greek philosopher observed:

> The sun, moon, and . . . . the planets were made for defining and preserving the numbers of time.

St. Augustine pondered time too:

> What then is time? If no one asks, I know; if I want to explain it to a questioner, I do not know.

Marcus Aurelius, the Roman emperor and philosopher, wrote:

> Time is sort of a river of passing events, and strong is its current.

The Mad Hatter of Lewis Carroll's *Alice's Adventures in Wonderland* thought he knew time:

> If you know Time as well as I do, you wouldn't talk about wasting it. It's him.... Now, if you only kept on good

terms with him, he'd do almost anything you liked with the clock. For instance, suppose it were nine o'clock in the morning, just time to begin lessons; you'd only have to whisper a hint to Time, and around goes the clock in a twinkling; Half-past one, time for dinner.

The "forward" flow of time implies a direction, and this is sometimes described as *the arrow of time*. Events do not happen as they appear when a movie projector is run backward. You put cold milk into hot, black coffee and get a drinkable brown mixture—but you cannot get cold milk and hot, black coffee back out of the same brown mixture. Such is the irreversible arrow of a physical process (and time)—never the reverse of the coffee unmixing into cold and hot parts. This arrow of time will be described in Chapter 12 in terms of entropy, which tells how a thermodynamic process will "flow."

The question "What is time?" gives some insight as to what is meant by a *fundamental* physical quantity, such as mass, length, or time. Basically, these are the simplest properties of which we can think to describe nature. So, the safe answer is:

> Time is a fundamental physical quantity.

This sort of masks our ignorance, and physics goes on from there, using time to describe and explain what we observe.

SI Base Units

The SI has seven *base units* for seven base quantities, which are assumed to be mutually independent. In addition to the meter, kilogram, and second for (1) length, (2) mass, and (3) time, SI units include (4) electric current (charge/second) in amperes (A), (5) temperature in kelvins (K), (6) amount of substance in moles (mol), and (7) luminous intensity in candelas (cd). See Table 1.1.

The foregoing quantities are thought to compose the smallest number of base quantities needed for a full description of everything observed or measured in nature.

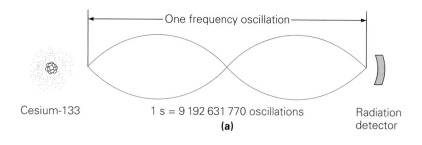

Cesium-133   1 s = 9 192 631 770 oscillations   Radiation detector

One frequency oscillation

(a)

(b)

▲ **FIGURE 1.3 The SI time standard: the second** The second was once defined in terms of the average solar day. **(a)** It is now defined by the frequency of the radiation associated with an atomic transition. **(b)** The atomic fountain "clock" shown here, at NIST, is the time standard for the United States. The variation of this "timepiece" is less than one second per 20 million years.

TABLE 1.1	The Seven Base Units of the SI
Name of Unit (abbreviation)	*Property Measured*
meter (m)	length
kilogram (kg)	mass
second (s)	time
ampere (A)	electric current
kelvin (K)	temperature
mole (mol)	amount of substance
candela (cd)	luminous intensity

1.3 More about the Metric System

OBJECTIVES: To learn to use (a) metric prefixes, and (b) nonstandard metric units.

The metric system involving the standard units of length, mass, and time, now incorporated into the SI, was once called the **mks system** (for *meter–kilogram–second*). Another metric system that has been used in dealing with relatively small quantities is the **cgs system** (for *centimeter–gram–second*). In the United States, the system still generally in use is the British (or English) engineering system, in which the standard units of length, mass, and time are foot, slug, and second, respectively. You may not have heard of the slug, because, as we mentioned earlier, gravitational force (weight) is commonly used instead of mass—pounds instead of slugs—to describe quantities of matter. As a result, the British system is sometimes called the **fps system** (for *foot–pound–second*).

The metric system is predominant throughout the world and is coming into increasing use in the United States. Because it is simpler mathematically, the SI is the preferred system of units for science and technology. SI units are used throughout most of this book. All quantities can be expressed in SI units. However, some units from other systems are accepted for limited use as a matter of practicality—for example, the time unit of hour and the temperature unit of degree Celsius. British units will sometimes be used in the early chapters for comparison purposes, since these units are still employed in everyday activities and for many practical applications.

The increasing worldwide use of the metric system means that you should be familiar with it. One of the greatest advantages of the metric system is that it is a decimal, or base-10, system. This means that larger or smaller units are obtained by multiplying or dividing, respectively, a base unit by powers of 10. A list of some multiples and corresponding prefixes for metric units is given in Table 1.2.

Teaching tip: Students often have the misconception that the basic unit of mass in the SI is the gram. Hold up a nickel coin and explain that its mass is about 5 g, or 1/200 kg.

Teaching tip: In general, SI units will be used throughout the text. British units will be used in a few exercises. Make sure that students understand the advantages of the SI.

| TABLE 1.2 | Some Multiples and Prefixes for Metric Units* |

Multiple†	Prefix (and Abbreviation)	Pronunciation	Multiple†	Prefix (and Abbreviation)	Pronunciation
10^{12}	tera- (T)	ter'a (as in *terrace*)	10^{-2}	centi- (c)	sen'ti (as in *senti*mental)
10^{9}	giga- (G)	jig'a (*jig* as in *jiggle*, *a* as in *a*bout)	10^{-3}	milli- (m)	mil'li (as in *milli*tary)
10^{6}	mega- (M)	meg'a (as in *mega*phone)	10^{-6}	micro- (μ)	mi'kro (as in *micro*phone)
10^{3}	kilo- (k)	kil'o (as in *kilo*watt)	10^{-9}	nano- (n)	nan'o (*an* as in *an*nual)
10^{2}	hecto- (h)	hek'to (*heck-toe*)	10^{-12}	pico- (p)	pe'ko (*peek-oh*)
10	deka- (da)	dek'a (*deck* plus *a* as in *a*bout)	10^{-15}	femto- (f)	fem'to (*fem* as in *fem*inine)
10^{-1}	deci- (d)	des'i (as in *deci*mal)	10^{-18}	atto- (a)	at'toe (as in *an*atomy)

*For example, 1 gram (g) multiplied by 1000, or 10^3, is 1 kilogram (kg); 1 gram multiplied by 1/1000, or 10^{-3}, is 1 milligram (mg).

†The most commonly used prefixes are printed in blue. Note that the abbreviations for the multiples 10^6 and greater are capitalized, whereas the abbreviations for the smaller multiples are lowercased.

Illustration 1.2 Animations, Units, and Measurement

Demonstration/activity: A ruler that has both centimeters and inches will show the difference between a base-10 system and the British system of quarters, eighths, and so on. Display a clear plastic ruler on the overhead projector.

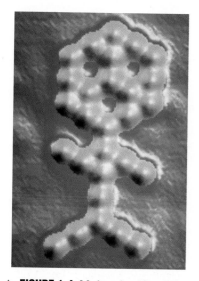

▲ **FIGURE 1.4 Molecular Man** This figure was crafted by moving 28 molecules, one at a time. Each of the gold-colored peaks is the image of a carbon monoxide molecule. The molecules rest on a single crystal platinum surface. "Molecular Man" measures 5 nm tall and 2.5 nm wide (hand to hand). More than 20 000 such figures, linked hand to hand, would be needed to span a single human hair. The molecules in the figure were positioned using a special microscope at very low temperatures.

Note: *Liter* is sometimes abbreviated as a lowercase "ell" (l), but a capital "ell" (L) is preferred in the United States so that the abbreviation is less likely to be confused with the numeral one. (Isn't 1 L clearer than 1 l ?)

In decimal measurements, the prefixes *micro-, milli-, centi-, kilo-,* and *mega-* are the ones most commonly used—for example, microsecond (μs), millimeter (mm), centimeter (cm), kilogram (kg), and megabyte (MB), as for computer disk or CD storage sizes). The decimal characteristics of the metric system make it convenient to change measurements from one size of metric unit to another. In the British system, different conversion factors must be used, such as 16 for converting pounds to ounces and 12 for converting feet to inches. The British system developed historically and not very scientifically.

You are already familiar with one base-10 system—U.S. currency. Just as a meter can be divided into 10 decimeters, 100 centimeters, or 1000 millimeters, the "base unit" of the dollar can be broken down into 10 "decidollars" (dimes), 100 "centidollars" (cents), or 1000 "millidollars" (tenths of a cent, or mills, used in figuring property taxes and bond levies). Since all the metric prefixes are powers of 10, there are no metric analogues for quarters or nickels.

The official metric prefixes help eliminate confusion. For example, in the United States, a billion is a thousand million (10^9); in Great Britain, a billion is a million million (10^{12}). The use of metric prefixes eliminates any confusion, since *giga-* indicates 10^9 and *tera-* stands for 10^{12}. You will probably be hearing more about *nano-,* the prefix that indicates 10^{-9}, with respect to nanotechnology (*nanotech* for short).

In general, nanotechnology is any technology done on the nanometer scale. A nanometer is one billionth (10^{-9}) of a meter, about the width of three to four atoms. Basically, nanotechnology involves the manufacture or building of things one atom or molecule at a time, so the nanometer is the appropriate scale. One atom or molecule at a time? That may sound a bit farfetched, but it's not (see ◄Fig. 1.4).

The chemical properties of atoms and molecules are well understood. For example, rearranging the atoms in coal can produce a diamond. (We can already do this task without nanotechnology, using heat and pressure.) Nanotechnology presents the possibility of constructing novel molecular devices or "machines" with extraordinary properties and abilities; for example, in medicine. Nanostructures might be injected into the body to go to a particular site, such as a cancerous growth, and deliver a drug directly. Other organs of the body would then be spared any effects of the drug. (This process might be considered nanochemotherapy.)

It is difficult for us to grasp or visualize the new concept of nanotechnology. Even so, keep in mind that a nanometer is one billionth of a meter. The diameter of a human hair is about 200 000 nanometers—huge compared with the new nanoapplications. The future should be an exciting nanotime.

Volume

In the SI, the standard unit of volume is the cubic meter (m^3)—the three-dimensional derived unit of the meter base unit. Because this unit is rather large, it is often more convenient to use the nonstandard unit of volume (or capacity) of a cube 10 cm (centimeters) on a side. This volume was given the name *litre*, which is spelled **liter (L)** in the United States. The volume of a liter is 1000 cm^3 (10 cm × 10 cm × 10 cm). Since 1 L = 1000 mL (milliliters), it follows that 1 mL = 1 cm^3. See ►Fig. 1.5a. (The cubic centimeter is sometimes abbreviated as cc, particularly in chemistry and biology. Also, the milliliter is sometimes abbreviated as ml, but the capital L is preferred (mL) so as not to be confused with the numeral one, 1.)

Recall from Fig. 1.2 that the standard unit of mass, the kilogram, was originally defined to be the mass of a cubic volume of water 10 cm, or 0.10 m, on a side, or the mass of one liter of water*. That is, *1 L of water has a mass of 1 kg* (Fig. 1.5b).

---

*This is specified at 4°C. A volume of water changes slightly with temperature (thermal expansion, Chapter 10). For our purposes here, we will consider a volume of water to remain constant under normal temperature conditions.

Also, since 1 kg = 1000 g and 1 L = 1000 cm³, then *1 cm³ (or 1 mL) of water has a mass of 1 g.*

Example 1.1 ■ The Metric Ton (or Tonne): Another Unit of Mass

As we have seen, the metric unit of mass was originally related to the length standard, with a liter (1000 cm³) of water having a mass of 1 kg. The standard metric unit of volume is the cubic meter (m³) and this volume of water was used to define a larger unit of mass called the *metric ton* (or *tonne*, as it is sometimes spelled). A metric ton is equivalent to how many kilograms?

Thinking It Through. A cubic meter is a relatively large volume and holds a large amount of water (more than a cubic yard; why?). The key is to find how many cubic volumes measuring 10 cm on a side (liters) are in a cubic meter. We expect, therefore, a large number.

Solution. Each liter of water has a mass of 1 kg, so we must find out how many liters are in 1 m³. Since there are 100 cm in a meter, a cubic meter is simply a cube with sides 100 cm in length. Therefore, a cubic meter (1 m³) has a volume of $10^2 \, \text{cm} \times 10^2 \, \text{cm} \times 10^2 \, \text{cm} = 10^6 \, \text{cm}^3$. Since $1 \, \text{L} = 10^3 \, \text{cm}^3$, there must be $(10^6 \, \text{cm}^3)/(10^3 \, \text{cm}^3/\text{L}) = 1000 \, \text{L}$ in 1 m³. Thus, 1 metric ton is equivalent to 1000 kg.

Note that this line of reasoning can be expressed very concisely in a single ratio:

$$\frac{1 \, \text{m}^3}{1 \, \text{L}} = \frac{100 \, \text{cm} \times 100 \, \text{cm} \times 100 \, \text{cm}}{10 \, \text{cm} \times 10 \, \text{cm} \times 10 \, \text{cm}} = 1000 \quad \text{or} \quad 1 \, \text{m}^3 = 1000 \, \text{L}$$

Follow-Up Exercise. What would be the length of the sides of a cube that contained a metric kiloton of water *(Answers to all Follow-Up Exercises are at the back of the text.)**

You are probably more familiar with the liter than you think. The use of the liter is becoming quite common in the United States, as ▾Fig. 1.6 indicates.

Because the metric system is coming into increasing use in the United States, you may find it helpful to have an idea of how metric and British units compare. The relative sizes of some units are illustrated in ▸Fig. 1.7. The mathematical conversion from one unit to another will be discussed shortly.

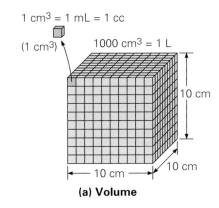

1 cm³ = 1 mL = 1 cc

(1 cm³)

1000 cm³ = 1 L

10 cm
10 cm
10 cm

(a) Volume

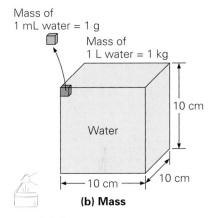

Mass of 1 mL water = 1 g

Mass of 1 L water = 1 kg

Water

10 cm
10 cm
10 cm

(b) Mass

▲ **FIGURE 1.5 The liter and the kilogram** Other metric units are derived from the meter. **(a)** A unit of volume (capacity) was taken to be the volume of a cube 10 cm, or 0.10 m, on a side and was given the name *liter* (L). **(b)** The mass of a liter of water was defined to be 1 kg. Note that the decimeter cube contains 1000 cm³, or 1000 mL. Thus, 1 cm³, or 1 mL, of water has a mass of 1 g.

◂ **FIGURE 1.6 Two, three, one, and one-half liters** The liter is now a common volume unit for soft drinks.

*The Answers to Follow-Up Exercises section after the appendices contains the answers—and, for Conceptual Exercises, the reasoning—for all Follow-Up Exercises in this book.

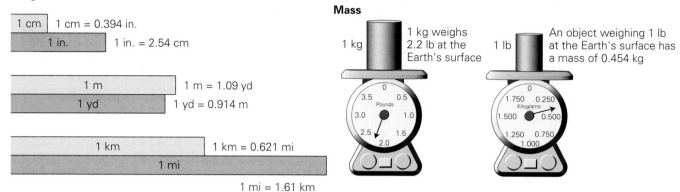

Volume

1 L = 1.06 qt
1 qt = 0.947 L

Length

1 cm = 0.394 in.
1 in. = 2.54 cm

1 m = 1.09 yd
1 yd = 0.914 m

1 km = 0.621 mi

1 mi = 1.61 km

Mass

1 kg weighs 2.2 lb at the Earth's surface

An object weighing 1 lb at the Earth's surface has a mass of 0.454 kg

▲ **FIGURE 1.7 Comparison of some SI and British units** The bars illustrate the relative magnitudes of each pair of units. (*Note*: The comparison scales are different in each case.)

1.4 Unit Analysis

OBJECTIVES: To explain the advantages of, and apply, unit analysis.

The fundamental, or base, quantities used in physical descriptions are called *dimensions*. For example, length, mass, and time are dimensions. You could measure the distance between two points and express it in units of meters, centimeters, or feet, but the quantity would still have the dimension of length.

Dimensions provide a procedure by which the consistency of equations may be checked. In practice, it is convenient to use specific units, such as m, s, and kg. (See Table 1.3.) Such units can be treated as algebraic quantities and can be canceled. Using units to check equations is called **unit analysis**, which shows the consistency of units and whether an equation is dimensionally correct.

You have used equations and know that an equation is a mathematical equality. Since physical quantities used in equations have units, *the two sides of an equation must be equal not only in numerical value, but also in units (dimensions).* For example, suppose you had the length quantities $a = 3.0$ m and $b = 4.0$ m. Inserting these values into the equation $a \times b = c$ gives 3.0 m $\times 4.0$ m $= 12$ m^2.

Teaching tip: Make sure that students understand why dimensional analysis is such a helpful tool in problem solving.

TABLE 1.3	Some Units of Common Quantities
Quantity	*Unit*
mass	kg
time	s
length	m
area	m^2
volume	m^3
velocity (v)	$\dfrac{m}{s}$
acceleration (a or g)	$\dfrac{m}{s^2}$

Both sides of the equation are numerically equal ($3 \times 4 = 12$), and both sides have the same units, $m \times m = m^2 = (length)^2$. If an equation is correct by unit analysis, it must be dimensionally correct. Example 1.2 demonstrates the further use of unit analysis.

Example 1.2 ■ Checking Dimensions: Unit Analysis

A professor puts two equations on the board: (a) $v = v_o + at$ and (b) $x = v/2a$, where x is a distance in meters (m); v and v_o are velocities in meters/second (m/s); a is acceleration in (meters/second)/second, or meters/second^2 (m/s^2); and t is time in seconds (s). Are the equations dimensionally correct? Use unit analysis to find out.

Thinking It Through. Simply insert the units for the quantities in each equation, cancel, and check the units on both sides.

Solution.

(a) The equation is

$$v = v_o + at$$

Inserting units for the physical quantities gives (Table 1.3)

$$\frac{m}{s} = \frac{m}{s} + \left(\frac{m}{s^2} \times s\right) \quad \text{or} \quad \frac{m}{s} = \frac{m}{s} + \left(\frac{m}{s \times s} \times s\right)$$

Notice that units cancel like numbers in a fraction. Then, we have

$$\frac{m}{s} = \frac{m}{s} + \frac{m}{s} \quad \text{(dimensionally correct)}$$

The equation is dimensionally correct, since the units on each side are meters per second. (The equation is also a correct relationship, as we shall see in Chapter 2.)

(b) Using unit analysis, the equation

$$x = \frac{v}{2a}$$

is

$$m = \frac{\left(\dfrac{m}{s}\right)}{\left(\dfrac{m}{s^2}\right)} = \frac{m}{s} \times \frac{s^2}{m} \quad \text{or} \quad m = s \quad \text{(not dimensionally correct)}$$

The meter (m) is not the same unit as the second (s), so in this case, the equation is not dimensionally correct (length $\neq$ time), and therefore is also not physically correct.

Follow-Up Exercise. Is the equation $ax = v^2$ dimensionally correct? (*Answers to all Follow-Up Exercises are at the back of the text.*)

Unit analysis will tell you if an equation is dimensionally correct, but a dimensionally consistent equation may not correctly express the real relationship of quantities. For example, in terms of units,

$$x = at^2$$

is

$$m = (m/s^2)(s^2) = m$$

This equation is dimensionally correct (length = length). But, as you will see in Chapter 2, it is not physically correct. The correct form of the equation—both dimensionally and physically—is $x = \frac{1}{2}at^2$. (The fraction $\frac{1}{2}$ has no dimensions; it is a dimensionless number.) Unit analysis cannot tell you if an equation is correct, only whether or not it is dimensionally consistent.

Mixed Units

Unit analysis also allows you to check for mixed units. In general, when working problems, you should always use the same system of units and the same unit for a given dimension throughout an exercise.

Suppose you wanted to buy a rug to fit a rectangular area, and you measure the sides to be 4.0 yd × 3.0 m. The area of the rug would then be $A = l \times w = 4.0 \text{ yd} \times 3.0 \text{ m} = 12 \text{ yd} \cdot \text{m}$, which might cause a problem at the carpet store. Note that this equation is dimensionally correct, $(\text{length})^2 = (\text{length})^2$, but the units are inconsistent or mixed. So, unit analysis will point out *mixed units*. Note that it is possible for an equation to be dimensionally correct, even if the units are mixed.

Let's look at mixed units in an equation. Suppose that you used centimeters as the unit for x in the equation

$$v^2 = v_0^2 + 2ax$$

and the units for the other quantities as in Example 1.2. In terms of units, this equation would give

$$\left(\frac{m}{s}\right)^2 = \left(\frac{m}{s}\right)^2 + \left(\frac{m \times cm}{s^2}\right)$$

or

$$\frac{m^2}{s^2} = \frac{m^2}{s^2} + \frac{m \times cm}{s^2}$$

which is dimensionally correct, $(\text{length})^2/(\text{time})^2$—on both sides of the equation. But the units are mixed (m and cm). The terms on the right-hand side should not be added together without centimeters first being converted to meters.

Determining the Units of Quantities

Another aspect of unit analysis that is very important in physics is the determination of the units of quantities from defining equations. For example, the **density (ρ)** of an object (represented by the Greek letter rho, ρ) is defined by the equation

$$\rho = \frac{m}{V} \quad (density) \tag{1.1}$$

where m is its mass and V its volume. (Density is the mass per unit volume and is a measure of the compactness of the mass of an object or substance.) What are the units of density? In SI units, mass is measured in kilograms and volume in cubic meters. Hence, the defining equation

$$\rho = m/V \quad (kg/m^3)$$

gives the derived SI unit for density as kilograms per cubic meter.

What are the units of π? The relationship between the circumference (c) and the diameter (d) of a circle is given by the equation $c = \pi d$, so $\pi = c/d$. If lengths are measured in meters, then unitwise we have

$$\pi = \frac{c}{d}\left(\frac{\cancel{m}}{\cancel{m}}\right)$$

Thus, π has no units. It is unitless, or a dimensionless, constant—but one with a lot of digits, as pointed out in the Physics Facts at the beginning of the chapter.

1.5 Unit Conversions

OBJECTIVES: To (a) explain conversion-factor relationships, and (b) apply them in converting units within a system or from one system of units to another.

Because units in different systems, or even different units in the same system, can be used to express the same quantity, it is sometimes necessary to convert the units of a quantity from one unit to another. For example, we may need to convert feet to yards or convert inches to centimeters. You already know how to do many unit conversions. If a room is 12 ft long, what is its length in yards? Your immediate answer is 4 yd.

How did you do this conversion? Well, you must have known a relationship between the units of foot and yard. That is, you know that 3 ft = 1 yd. This is what we call an *equivalence statement*. As was seen in Section 1.4, the numerical values and units on both sides of an equation must be the same. In equivalence statements, we commonly use an equal sign to indicate that 1 yd and 3 ft stand for the *same*, or *equivalent, length*. The numbers are different because they stand for different *units* of length.

Mathematically, to change units we use **conversion factors**, which are simply equivalence statements expressed in the form of ratios—for example, 1 yd/3 ft or 3 ft/1 yd. (The "1" is often omitted in the denominators of such ratios for convenience—for example, 3 ft/yd.) To understand why such ratios are useful, note the expression 1 yd = 3 ft in ratio form:

$$\frac{1 \text{ yd}}{3 \text{ ft}} = \frac{3 \text{ ft}}{3 \text{ ft}} = 1 \quad \text{or} \quad \frac{3 \text{ ft}}{1 \text{ yd}} = \frac{1 \text{ yd}}{1 \text{ yd}} = 1$$

As you can see from this example, a conversion factor has an actual value of unity or one—and you can multiply any quantity by one without changing its value or size. Thus, *a conversion factor simply lets you express a quantity in terms of other units without changing its physical value or size.*

The manner in which 12 ft is converted to yards may be expressed mathematically as follows:

$$12 \text{ ft} \times \frac{1 \text{ yd}}{3 \text{ ft}} = 4 \text{ yd} \quad \textit{(units cancel)}$$

Using the appropriate conversion-factor form, the units cancel, as shown by the slash marks, giving the correct unit analysis, yd = yd.

Suppose you are asked to convert 12.0 in. to centimeters. You may not know the conversion factor in this case, but you can get it from a table (such as the one that appears inside the front cover of this book) that gives the needed relationships: 1 in. = 2.54 cm or 1 cm = 0.394 in. It makes no difference which of these equivalence statements you use. The question, once you have expressed the equivalence statement as a conversion factor, is whether to multiply or divide by that factor to make the conversion. *In doing unit conversions, take advantage of unit analysis*—that is, let the units determine the appropriate form of conversion factor.

Note that the equivalence statement 1 in. = 2.54 cm can give rise to two forms of the conversion factor: 1 in./2.54 cm or 2.54 cm/in. When changing inches to centimeters, the appropriate form for multiplying is 2.54 cm/in. When changing centimeters to inches, use the form 1 in./2.54 cm. (The inverse forms could be used in each case, but the quantities would have to be *divided* by the conversion factors for proper unit cancellation.) In general, the multiplication form of conversion factors will be used throughout this book.

A few commonly used equivalence statements are not dimensionally or physically correct; for example, consider 1 kg = 2.2 lb, which is used for quickly determining the weight of an object near the Earth's surface, given its mass. The kilogram is a unit of mass, and the pound is a unit of weight. This means that 1 kg is *equivalent* to 2.2 lb; that is, a 1-kg *mass* has a *weight* of 2.2 lb. Since mass and weight are directly proportional, we can use the dimensionally incorrect conversion factor 1 kg/2.2 lb (but *only* near the Earth's surface).

Teaching tip: Students often ask, "Do we have to show all of the work?" Insist that they show the work, and encourage them to analyze the unit cancellations. Also insist that students perform the unit conversions using a horizontal division line as shown in the displayed conversion equations—not a slanted division line, which obscures the numerator and denominator positions.

Note: 1 kg of mass has an equivalent weight of 2.2 lb near the surface of the Earth.

ELEVATION
1000 ft
305 m

(a)

Thinkmetric
MAXIMUM **30** km/h WAS MAXIMUM **20** MPH

(b)

▲ **FIGURE 1.8** Unit conversion
Signs sometimes list both the British and metric units, as shown here for elevation and speed.

Example 1.3 ■ Converting Units: Use of Conversion Factors

(a) A basketball player is 6.5 ft tall. What is the player's height in meters? (b) How many seconds are in a 30-day month? (c) What is 50 mi/h in meters per second? (See the table of conversion factors inside the front cover of this book.)

Thinking It Through. If we use the correct conversion factors, the rest is arithmetic.

Solution.

(a) From the conversion table, we have 1 ft = 0.305 m, so

$$6.5 \text{ ft} \times \frac{0.305 \text{ m}}{1 \text{ ft}} = 2.0 \text{ m}$$

Another foot–meter conversion is shown in ◄Fig. 1.8. Is it correct?

(b) The conversion factor for days and seconds is available from the table (1 day = 86 400 s), but you may not always have a table handy. You can always use several better-known conversion factors to get the result:

$$30 \frac{\text{days}}{\text{month}} \times \frac{24 \text{ h}}{\text{day}} \times \frac{60 \text{ min}}{\text{h}} \times \frac{60 \text{ s}}{\text{min}} = \frac{2.6 \times 10^6 \text{ s}}{\text{month}}$$

Note how unit analysis checks the conversion factors for you. The rest is simple arithmetic.

(c) In this case, from the conversion table, 1 mi = 1609 m and 1 h = 3600 s. (The latter is easily computed.) These ratios are used to cancel the units that are to be changed, leaving behind the ones that are wanted:

$$\frac{50 \text{ mi}}{1 \text{ h}} \times \frac{1609 \text{ m}}{1 \text{ mi}} \times \frac{1 \text{ h}}{3600 \text{ s}} = 22 \text{ m/s}$$

Follow-Up Exercise. (a) Convert 50 mi/h directly to meters per second by using a single conversion factor, and (b) show that this single conversion factor can be derived from those in part (c) of this Example. *(Answers to all Follow-Up Exercises are at the back of the text.)*

Example 1.4 ■ More Conversions: A Really Long Capillary System

Capillaries, the smallest blood vessels of the body, connect the arterial system with the venous system and supply our tissues with oxygen and nutrients (▼Fig. 1.9). It is estimated that if all of the capillaries of an average adult were unwound and spread out end to end, they would extend to a length of about 64 000 km. (a) How many miles is this length? (b) Compare this length with the circumference of the Earth.

Thinking It Through. (a) This conversion is straightforward—just use the appropriate conversion factor. (b) How do we calculate the circumference of a circle or sphere?

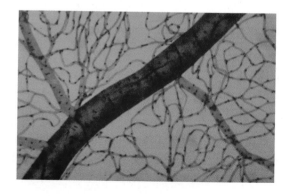

► **FIGURE 1.9** Capillary system
Capillaries connect the arterial and venous systems in our bodies. They are the smallest blood vessels, but their total length is impressive.

There is an equation to do so, but you must know the radius or diameter of the Earth. (If you do not remember one of these values, see the solar system data table inside the back cover of this book.)

Solution.

(a) We see in the conversion table that 1 km = 0.621 mi, so

$$64\,000 \ \cancel{\text{km}} \times \frac{0.621 \ \text{mi}}{1 \ \cancel{\text{km}}} = 40\,000 \ \text{mi} \quad (\textit{rounded off})$$

(b) A length of 40 000 mi is substantial. To see how this length compares with the circumference (c) of the Earth, recall that the radius of the Earth is approximately 4000 mi, so the diameter (d) is 8000 mi. The circumference of a circle is given by $c = \pi d$, and

$$c = \pi d \approx 3 \times 8000 \ \text{mi} \approx 24\,000 \ \text{mi} \quad (\textit{rounded off})$$

[To make a general comparison, $\pi \ (= 3.14\ldots)$ is rounded off to 3. The $\approx$ symbol means "approximately equal to."]
  So,

$$\frac{\text{capillary length}}{\text{Earth's circumference}} = \frac{40\,000 \ \text{mi}}{24\,000 \ \text{mi}} = 1.7$$

The capillaries of your body have a total length that would extend 1.7 times around the world. Wow!

Follow-Up Exercise. Taking the average distance between the East Coast and West Coast of the continental United States to be 4800 km, how many times would the total length of your body's capillaries cross the country? (*Answers to all Follow-Up Exercises are at the back of the text.*)

Example 1.5 ■ Converting Units of Area: Choosing the Correct Conversion Factor

A hall bulletin board has an area of 2.5 m². What is this area in square centimeters (cm²)?

Thinking It Through. This problem is a conversion of area units, and we know that 1 m = 100 cm. So, some squaring must be done to get square meters related to square centimeters.

Solution. A common error in such conversions is the use of incorrect conversion factors. Because 1 m = 100 cm, it is sometimes assumed that 1 m² = 100 cm², which is *wrong*. The correct area conversion factor may be obtained directly from the correct linear conversion factor, 100 cm/1 m, or 10^2 cm/1 m, by *squaring* the linear conversion factor:

$$\left(\frac{10^2 \ \text{cm}}{1 \ \text{m}}\right)^2 = \frac{10^4 \ \text{cm}^2}{1 \ \text{m}^2}$$

Hence, 1 m² = 10^4 cm² (= 10 000 cm²). We can therefore write the following:

$$2.5 \ \text{m}^2 \times \left(\frac{10^2 \ \text{cm}}{1 \ \text{m}}\right)^2 = 2.5 \ \cancel{\text{m}^2} \times \frac{10^4 \ \text{cm}^2}{1 \ \cancel{\text{m}^2}} = 2.5 \times 10^4 \ \text{cm}^2$$

Follow-Up Exercise. How many cubic centimeters are in one cubic meter? (*Answers to all Follow-Up Exercises are at the back of the text.*)

Throughout this textbook, you will be presented with various Conceptual Examples. These examples show the reasoning used in applying particular concepts, often with little or no mathematics.

Conceptual Example 1.6 ■ Comparing Speeds Using Unit Conversions

Two students disagree on which speed is faster, (a) 1 km/h or (b) 1 m/s. Which would you choose? *Clearly establish the reasoning used in determining your answer before checking it next. That is,* **why** *did you select your answer?*

Reasoning and Answer. To answer this, the quantities should be compared in the same units, so unit conversion is involved, and one looks for the easiest conversions. Observing the prefix *kilo-*, we know that 1 km is 1000 m. Also, 1 h is quickly expressed as 3600 s. Then the numerical ratio of km/h is less than 1, and 1 km/h < 1 m/s, so the answer is (b). [1 km/h = 1000 m/3600 s = 0.3 m/s]

Follow-Up Exercise. An American and a European are comparing the gas mileage they get with their RVs. The American calculates that he gets 10 mi/gal, and the European gives his as 10 km/L. Who is getting the better gas mileage? *(Answers to all Follow-Up Exercises are at the back of the text.)*

Some examples of the importance of unit conversion are given in the accompanying Insight 1.3.

INSIGHT   1.3 IS UNIT CONVERSION IMPORTANT?

The answer to this question is, you bet! Here are a couple of cases in point. In 1999, the $125 million Mars Climate Orbiter was making a trip to the Red Planet to investigate its atmosphere (Fig. 1). The spacecraft approached the planet in September, but suddenly contact between the Orbiter and personnel on Earth was lost, and the Orbiter was never heard from again. Investigations showed that the orbiter had approached Mars at a far lower altitude than planned. Instead of passing 147 km (87 mi) above the Martian surface, tracking data showed that the Orbiter was on a trajectory that would have taken it as close as 57 km (35 mi) from the surface. As a result, the spacecraft either burned up in the Martian atmosphere or crashed into the surface.

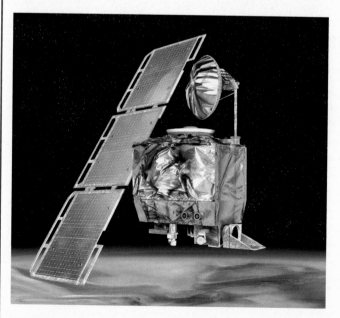

FIGURE 1 Mars Climate Orbiter An artist's conception of the orbiter near the surface of Mars. The actual orbiter either burned up in the Martian atmosphere or crashed into the surface. The cause was attributed to a mix-up in units, and a $125 million spacecraft was lost.

How could this have happened? Investigations showed that the failure of the Orbiter was primarily a problem of unit conversions. At Lockheed Martin Astronautics, which built the spacecraft, the engineers calculated the navigational information in British units. When scientists at NASA's Jet Propulsion Laboratory received the data, they assumed that the information was in metric units, as called for in the mission specifications. The unit conversions weren't made, and a $125 million spacecraft was lost on the Red Planet—causing more than a few red faces.

Closer to Earth, in 1983 Air Canada Flight 143 was on course from Montréal to Edmonton, Canada, with sixty-one passengers in a new Boeing 767, at the time the most advanced jetliner in the world. Almost halfway into the flight, a warning light came on for a fuel pump, then for another, and finally for all four pumps. The engines quit, and this advanced plane was now a glider, about 100 mi from the nearest major airport, at Winnipeg. Without engines, Flight 143's descent would bring it down 10 mi short of the airport, so it was diverted to an old Royal Canadian Air Force landing field at Gimli. The pilot maneuvered the powerless plane to a landing, stopping just short of a barrier. Did the plane, which was dubbed "The Gimli Glider," have bad fuel pumps? No, it had run out of fuel!

This near-disaster was caused by another conversion problem. The fuel computers weren't working properly so the mechanics had used the old procedure of measuring the fuel in the tanks with a dipstick. The length of the stick that is wet is used to determine the volume of fuel by means of conversion values in tables. Air Canada had for years computed the amount of fuel in pounds, whereas the new 767's fuel consumption was expressed in kilograms. Even worse, the dipstick procedure gave the amount of fuel onboard in liters instead of pounds or kilograms. The result was that the aircraft was loaded with 22 300 lb of fuel instead of the required 22 300 kg. Since 1 lb has a mass of 0.45 kg, the plane had less than half the required fuel.

These incidents underscore the importance of using appropriate units, making correct unit conversions, and working consistently in the same system of units. Several exercises at the end of the chapter will challenge you to develop your skills in accurate unit conversions.

1.6 Significant Figures

OBJECTIVES: To (a) determine the number of significant figures in a numerical value, and (b) report the proper number of significant figures after performing simple calculations.

Most of the time, you will be given numerical data when asked to solve a problem. In general, such data are either exact numbers or measured numbers (quantities). **Exact numbers** are numbers without any uncertainty or error. This category includes numbers such as the 100 used to calculate a percentage and the 2 in the equation $r = d/2$ relating the radius and diameter of a circle. **Measured numbers** are numbers obtained from measurement processes and thus generally have some degree of uncertainty or error.

When calculations are done with measured numbers, the error of measurement is *propagated*, or carried along, by the mathematical operations. The question of how to report the error for a result arises. For example, suppose that you are asked to find time (t) from the equation $x = vt$ and are given that $x = 5.3$ m and $v = 1.67$ m/s. Then

$$t = \frac{x}{v} = \frac{5.3 \text{ m}}{1.67 \text{ m/s}} = ?$$

Doing the division operation on a calculator yields a result such as 3.173 652 695 seconds (▶Fig. 1.10). How many figures, or digits, should you report in the answer?

The error or uncertainty of the result of a mathematical operation may be computed by statistical methods. A simpler and more widely used procedure for estimating this uncertainty involves the use of **significant figures** (**sf**), sometimes called *significant digits*. The degree of accuracy of a measured quantity depends on how finely divided the measuring scale of the instrument is. For example, you might measure the length of an object as 2.5 cm with one instrument and 2.54 cm with another. The second instrument provides more significant figures and thus a greater degree of accuracy.

Basically, *the significant figures in any measurement are the digits that are known with certainty, plus one digit that is uncertain.* This set of digits is usually defined as all of the digits that can be read directly from the instrument used to make the measurement, plus one uncertain digit that is obtained by estimating the fraction of the smallest division of the instrument's scale.

The quantities 2.5 cm and 2.54 cm have two and three significant figures, respectively. This is rather evident. However, some confusion may arise when a quantity contains one or more zeros. For example, how many significant figures does the quantity 0.0254 m have? What about 104.6 m? 2705.0 m? In such cases, we will use the following rules:

1. Zeros at the beginning of a number are not significant. They merely locate the decimal point. For example,

    0.0254 m has three significant figures (2, 5, 4)

2. Zeros within a number are significant. For example,

    104.6 m has four significant figures (1, 0, 4, 6)

3. Zeros at the end of a number after the decimal point are significant. For example,

    2705.0 m has five significant figures (2, 7, 0, 5, 0).

4. In whole numbers without a decimal point that end in one or more zeros (trailing zeros)—for example, 500 kg—the zeros may or may not be significant. In such cases, it is not clear which zeros serve only to locate the decimal point and which are actually part of the measurement. That is, if the first zero from the left (5$\underline{0}$0 kg) is the estimated digit in the measurement, then only two digits are reliably known, and there are only two significant figures. Similarly, if the last zero is the estimated digit (50$\underline{0}$ kg), then there are

▲ **FIGURE 1.10 Significant figures and insignificant figures** For the division operation 5.3/1.67, a calculator with a floating decimal point gives many digits. A calculated quantity can be no more accurate than the least accurate quantity involved in the calculation, so this result should be rounded off to two significant figures—that is, 3.2.

Demonstration/activity: Many general-purpose electric meters have more than one scale, for different sensitivities. Display one such meter (an overhead-projection meter, if available), and discuss the difference in significant figures.

three significant figures. This ambiguity may be removed by using scientific (powers-of-ten) notation:

$$5.0 \times 10^2 \text{ kg has two significant figures}$$
$$5.00 \times 10^2 \text{ kg has three significant figures}$$

This notation is helpful in expressing the results of calculations with the proper numbers of significant figures, as we shall see shortly. (Appendix I includes a review of scientific notation.)

(*Note*: To avoid confusion regarding numbers having trailing zeros used as given quantities in text examples and exercises, we will consider the trailing zeros to be significant. For example, assume that a time of 20 s has two significant figures, even if it is not written out as 2.0×10^1 s.)

It is important to report the results of mathematical operations with the proper number of significant figures. This is accomplished by using rules for (1) multiplication and division and (2) addition and subtraction. To obtain the proper number of significant figures, the results are rounded off. Here are some general rules that will be used for mathematical operations and rounding.

Significant Figures in Calculations

1. When multiplying and dividing quantities, leave as many significant figures in the answer as there are in the quantity with the least number of significant figures.
2. When adding or subtracting quantities, leave the same number of decimal places (rounded) in the answer as there are in the quantity with the least number of decimal places.

Rules for Rounding*

1. If the first digit to be dropped is less than 5, leave the preceding digit as is.
2. If the first digit to be dropped is 5 or greater, increase the preceding digit by one.

The rules for significant figures mean that the result of a calculation can be no more accurate than the least accurate quantity used. That is, you cannot gain accuracy performing mathematical operations. Thus, the result that should be reported for the division operation discussed at the beginning of this section is

$$\frac{\overset{(2\,sf)}{5.3 \text{ m}}}{\underset{(3\,sf)}{1.67 \text{ m/s}}} = 3.2 \text{ s} \quad (2\,sf)$$

The result is rounded off to two significant figures. (See Fig. 1.10.)

Applications of these rules are shown in the following Examples.

Example 1.7 ■ Using Significant Figures in Multiplication and Division: Rounding Applications

The following operations are performed and the results rounded off to the proper number of significant figures:

Multiplication

$$\underset{(2\,sf)}{2.4 \text{ m}} \times \underset{(3\,sf)}{3.65 \text{ m}} = 8.76 \text{ m}^2 = 8.8 \text{ m}^2 \quad (rounded\ to\ two\ sf)$$

Division

$$\frac{\overset{(4\,sf)}{725.0 \text{ m}}}{\underset{(3\,sf)}{0.125 \text{ s}}} = 5800 \text{ m/s} = 5.80 \times 10^3 \text{ m/s} \quad (represented\ with\ three\ sf;\ why?)$$

*It should be noted that these rounding rules give an approximation of accuracy, as opposed to the results provided by more advanced statistical methods.

Follow-Up Exercise. Perform the following operations, and express the answers in the standard powers-of-ten notation (one digit to the left of the decimal point) with the proper number of significant figures: (a) $(2.0 \times 10^5 \text{ kg})(0.035 \times 10^2 \text{ kg})$ and (b) $(148 \times 10^{-6} \text{ m})/(0.4906 \times 10^{-6} \text{ m})$. *(Answers to all Follow-Up Exercises are at the back of the text.)*

Example 1.8 ■ Using Significant Figures in Addition and Subtraction: Application of Rules

The following operations are performed by finding the number that has the least number of decimal places. (Units have been omitted for convenience.)

Addition
In the numbers to be added, note that 23.1 has the least number of decimal places (one):

$$
\begin{array}{r}
23.1 \\
0.546 \\
\underline{1.45} \\
25.096
\end{array}
\xrightarrow{\textit{(rounding off)}} 25.1
$$

Subtraction
The same rounding procedure is used. Here, 157 has the least number of decimal places (none).

$$
\begin{array}{r}
157 \\
\underline{-5.5} \\
151.5
\end{array}
\xrightarrow{\textit{(rounding off)}} 152
$$

Follow-Up Exercise. Given the numbers 23.15, 0.546, and 1.058, (a) add the first two numbers and (b) subtract the last number from the first. *(Answers to all Follow-Up Exercises are at the back of the text.)*

Teaching tip: Encourage students to practice this method in the end-of-chapter exercises.

Suppose that you must deal with mixed operations—multiplication and/or division *and* addition and/or subtraction. What do you do in this case? Just follow the regular rules for order of algebraic operations, and observe significant figures as you go.

The number of digits reported in a result depends on the number of digits in the given data. The rules for rounding will generally be observed in this book. However, there will be exceptions that may make a difference, as explained in the following Problem-Solving Hint.

Problem-Solving Hint: The "Correct" Answer

When working problems, you naturally strive to get the correct answer and will probably want to check your answers against those listed in the Answers to Odd-Numbered Exercises section in the back of the book. However, on occasion, you may find that your answer differs slightly from that given, even though you have solved the problem correctly. There are several reasons why this could happen.

As stated previously, it is best to round off only the final result of a multipart calculation, but this practice is not always convenient in elaborate calculations. Sometimes, the results of intermediate steps are important in themselves and need to be rounded off to the appropriate number of digits as if each were a final answer. Similarly, Examples in this book are often worked in steps to show the stages in the *reasoning* of the solution. The results obtained when the results of intermediate steps are rounded off may differ slightly from those obtained when only the final answer is rounded.

Rounding differences may also occur when using conversion factors. For example, in changing 5.0 mi to kilometers using the conversion factor listed in the front of this book in different forms,

$$ 5.0 \text{ mi} \left(\frac{1.609 \text{ km}}{1 \text{ mi}} \right) = (8.045 \text{ km}) = 8.0 \text{ km} \quad \textit{(two significant figures)} $$

and

$$ 5.0 \text{ mi} \left(\frac{1 \text{ km}}{0.621 \text{ mi}} \right) = (8.051 \text{ km}) = 8.1 \text{ km} \quad \textit{(two significant figures)} $$

(continues on next page)

PHYSLET®

Exploration 1.1 Click-Drag to Get Position

The difference arises because of rounding of the conversion factors. Actually, 1 km = 0.6214 mi, so 1 mi = (1/0.6214) km = 1.609 269 km ≈ 1.609 km. (Try repeating these conversions with the unrounded factors, and see what you get.) To avoid rounding differences in conversions, we will generally use the multiplication form of a conversion factor, as in the first of the foregoing equations, unless there is a convenient exact factor, such as 1 min/60 s.

Slight differences in answers may occur when different methods are used to solve a problem, because of rounding differences. Keep in mind that when solving a problem (a general procedure for which is given in Section 1.7), *if your answer differs from that in the text in only the last digit, the disparity is most likely the result of a rounding difference for an alternative method of solution being used.*

1.7 Problem Solving

OBJECTIVES: To (a) establish a general problem-solving procedure, and (b) apply it to typical problems.

An important aspect of physics is problem solving. In general, this involves the application of physical principles and equations to data from a particular situation in order to find some unknown or wanted quantity. There is no universal method for approaching problem solving that will automatically produce a solution. However, although there is no magic formula for problem solving, there are some sound practices that can be very useful. The steps in the following procedure are intended to provide you with a framework that can be applied to solving most of the problems you will encounter during your course of study. (You may wish to make modifications to suit your own style.)

We will generally use these steps in dealing with the Example problems throughout the text. Additional Problem-Solving Hints will be given where appropriate.

General Problem-Solving Steps

1. *Read the problem carefully, and analyze it.* What is wanted, and what is given?
2. *Where appropriate, draw a diagram as an aid in visualizing and analyzing the physical situation of the problem.* This step may not be necessary in every case, but it is often useful.
3. *Write down the given data and what is to be found. Make sure the data is expressed in the same system of units (usually SI).* If necessary, use the unit conversion procedure learned earlier in the chapter. Some data may not be given explicitly. For example, if a car "starts from rest," its initial speed is zero ($v_o = 0$); in some instances, you may be expected to know certain quantities, such as the acceleration due to gravity, g, or to look them up in tables.
4. *Determine which principle(s) and equation(s) are applicable to the situation, and how they can be used to get from the information given to what is to be found.* You may have to devise a strategy that involves several steps. Also, try to simplify equations as much as possible through algebraic manipulation. The fewer calculations you do, the less likely you are to make a mistake—*so don't put in numbers until you have to.*
5. *Substitute the given quantities (data) into the equation(s) and perform calculations.* Report the result with the proper units and proper number of significant figures.
6. *Consider whether the results are reasonable.* Does the answer have an appropriate magnitude? (This means, is it in the right ballpark?) For example, if a person's calculated mass turns out to be 4.60×10^2 kg, the result should be questioned, since a mass of 460 kg has a weight of 1010 lb. ◄Fig. 1.11 summarizes the main steps in the form of a flowchart.

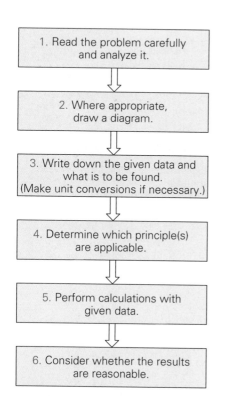

▲ **FIGURE 1.11** A flowchart for the suggested problem-solving procedure

In general, there are three types of examples in this text as listed in Table 1.4. The preceding steps would be applicable to the first two types, because they include calculations. Conceptual Examples, in general, do not follow these steps, being primarily conceptual in nature.

In reading the worked Examples and Integrated Examples, you should be able to recognize the general application or flow of the preceding steps. This format will be used throughout the text. Let's take an Example and an Integrated Example as illustrations. Comments will be made in these examples to point out the problem-solving approach and steps that will not be made in the text Examples, but should be understood. Since no physical principles have really been covered, we will use math and trig problems, which should serve as a good review.

Example 1.9 ■ Finding the Outside Surface Area of a Cylindrical Container

A closed cylindrical container used to store material from a manufacturing process has an outside radius of 50.0 cm and a height of 1.30 m. What is the total outside surface area of the container?

Thinking It Through. (In this type of Example, the Thinking It Through section generally combines problem-solving steps 1 and 2 given previously.)

It should be noted immediately that the length units are given in mixed units, so a unit conversion will be in order. To visualize and analyze the cylinder, drawing a diagram is helpful (▶Fig. 1.12). With this information in mind, proceed to finding the solution, using the expression for the area of a cylinder (the combined areas of the circular ends and the cylinder's side).

Solution. Writing what is given and what is to be found (step 3 in our procedure):

Given: $r = 50.0$ cm   *Find:* A (the outside surface area of the cylinder)
$h = 1.30$ m

First, let's tend to the mixed units. You should be able in this case to immediately write $r = 50.0$ cm $= 0.500$ m. But often conversions are not obvious, so going through the unit conversion for illustration:

$$r = 50.0 \text{ cm} \left(\frac{1 \text{ m}}{100 \text{ cm}} \right) = 0.500 \text{ m}$$

There are general equations for areas (and volumes) of commonly shaped objects. The area of a cylinder can be easily looked up (given in Appendix I), but suppose you didn't have such a source. In this case, you can figure it out for yourself. Looking at Fig. 1.12, note that the outside surface area of a cylinder consists of that of two circular ends and that of a rectangle (the body of the cylinder laid out flat). Equations for the areas of these common shapes are generally remembered. So the area of the two ends would be

$$2A_e = 2 \times \pi r^2 \quad \begin{array}{l} \textit{(2 times the area of the circular end;} \\ \textit{area of a circle} = \pi r^2) \end{array}$$

and the area of the body of the cylinder is

$$A_b = 2\pi r \times h \quad \begin{array}{l} \textit{(circumference of circular end} \\ \textit{times height)} \end{array}$$

Then the total area is

$$A = 2A_e + A_b = 2\pi r^2 + 2\pi rh$$

The data could be put into the equation, but sometimes an equation may be simplified to save some calculation steps.

$$A = 2\pi r(r + h) = 2\pi(0.500 \text{ m})(0.500 \text{ m} + 1.30 \text{ m})$$
$$= \pi(1.80 \text{ m}^2) = 5.65 \text{ m}^2$$

and the result appears reasonable considering the cylinder's dimensions.

Follow-Up Exercise. If the wall thickness of the cylinder's side and ends is 1.00 cm, what is the inside volume of the cylinder? *(Answers to all Follow-Up Exercises are at the back of the text.)*

TABLE 1.4	Types of Examples

Example—primarily mathematical in nature

Sections: **Thinking It Through Solution**

Integrated Example—(a) conceptual multiple choice, (b) mathematical follow-up

Sections: **(a) Conceptual Reasoning (b) Quantitative Reasoning and Solution**

Conceptual Example—in general, needs only reasoning to obtain the answer, although some simple math may be required at times to justify the reasoning

Sections: **Reasoning and Answer**

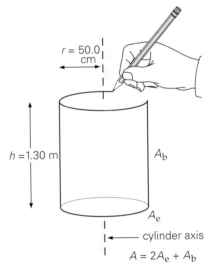

▲ **FIGURE 1.12 A helpful step in problem solving** Drawing a diagram helps you visualize and better understand the situation. See Example 1.9.

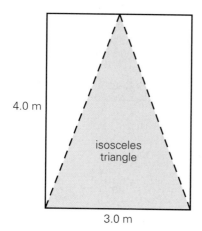

▲ **FIGURE 1.13 A flower bed project** Two types of triangles for a new flower bed. See Example 1.10.

Teaching tip: This Example gives the instructor a great opportunity to have the students review some trigonometry.

Basic trigonometric functions:

$$\cos\theta = \frac{x}{r}\left(\frac{\text{side adjacent}}{\text{hypotenuse}}\right)$$

$$\sin\theta = \frac{y}{r}\left(\frac{\text{side opposite}}{\text{hypotenuse}}\right)$$

$$\tan\theta = \frac{\sin\theta}{\cos\theta} = \frac{y}{x}\left(\frac{\text{side opposite}}{\text{side adjacent}}\right)$$

Integrated Example 1.10 ■ Sides and Angles

(a) A gardener has a rectangular plot measuring 3.0 m × 4.0 m. She wishes to use half of this area to make a triangular flower bed. Of the two types of triangles shown in ◄Fig. 1.13, which should she use to do this? (1) The one right triangle, (2) the isosceles triangle—two sides equal, or (3) either one? (b) In laying out the flower bed, the gardener decides to use a right triangle. Wishing to line the sides with rows of stone, she wants to know the total length (L) of the triangle sides. She would also like to know the values of the acute angles of the triangle. Can you help her so she doesn't have to do physical measurements?

(a) Conceptual Reasoning. The rectangular plot has a total area of 3.0 m × 4.0 m = 12 m². It is obvious that the right triangle divides the plot in half (Fig. 1.13). This is not as obvious for the isosceles triangle. But with a little study you should see that the white areas could be arranged such that their combined area would be the same as that of the shaded isosceles triangle. So the isosceles triangle also divides the plot in half and the answer is (3). [This could be proven mathematically by computing the areas of the triangles. Area = $\frac{1}{2}$(altitude × base).]

(b) Quantitative Reasoning and Solution. To find the total length of the sides, we need to find the length of the hypotenuse of the triangle. This can be done using the Pythagorean theorem, $x^2 + y^2 = r^2$, and

$$r = \sqrt{x^2 + y^2} = \sqrt{(3.0\text{ m})^2 + (4.0\text{ m})^2} = \sqrt{25\text{ m}^2} = 5.0\text{ m}$$

(Or directly, you may have noticed that this is a 3-4-5 right triangle.) Then,

$$L = 3.0\text{ m} + 4.0\text{ m} + 5.0\text{ m} = 12\text{ m}$$

The acute angles of the triangle can be found by using trigonometry. Referring to the angles in Fig. 1.13,

$$\tan\theta_1 = \frac{\text{side opposite}}{\text{side adjacent}} = \frac{4.0\text{ m}}{3.0\text{ m}}$$

and

$$\theta_1 = \tan^{-1}\left(\frac{4.0\text{ m}}{3.0\text{ m}}\right) = 53°$$

Similarly,

$$\theta_2 = \tan^{-1}\left(\frac{3.0\text{ m}}{4.0\text{ m}}\right) = 37°$$

which add to 90° as would be expected with the right angle (90° + 90° = 180°).

Follow-Up Exercise. What are the total length of the sides and the interior angles for the isosceles triangle in Fig. 1.13? *(Answers to all Follow-Up Exercises are at the back of the text.)*

These examples illustrate how the problem-solving steps are woven into finding the solution of a problem. You will see this pattern throughout the solved examples in the text, although not as explicitly explained. Try to develop your problem-solving skills in a similar manner.

Finally, let's take a Conceptual Example that has conceptual reasoning and some simple calculations.

Conceptual Example 1.11 ■ Climbing at an Angle

A jet pilot puts his plane through two straight, sharp inclined climbs at different angles. On the first climb, the plane travels 40.0 km at an angle of 15 degrees relative to the horizontal. On the second inclined climb, the plane travels 20.0 km at an angle of 30 degrees relative to the horizontal. How do the vertical distances of the two climbes compare? (a) That of the first incline is larger, (b) that of the second incline is larger, or (c) they both are the same.

Reasoning and Answer. At first glance it may appear that the vertical distances are the same. After all, the first incline's angle is half that of the second incline. And, the hypotenuse (distance) of the first climb is twice that of the second, so won't the two effects cancel out making the correct answer (c)? No. The flaw here is that the vertical distance is based on the sine of the angle (make a sketch) and the sine of an angle is *not* proportional to the angle. Check it out your calculator. 2 × sin 15° = 0.518 and

$\sin 30° = 0.500$. So they don't offset. Half the distance at double the angle produces a smaller vertical distance, and the correct answer is (a).

Follow-Up Exercise. In this Example, would the second climb have to be steeper than 30 degrees or less steep to make the climb distances the same? What should the angle be in this case?

Approximation and Order-of-Magnitude Calculations

At times when solving a problem, you may not be interested in an exact answer, but want only an estimate or a "ballpark" figure. Approximations can be made by rounding off quantities so as to make the calculations easier and, perhaps, obtainable without the use of a calculator. For example, suppose you want to get an idea of the area of a circle with radius $r = 9.5$ cm. Then, rounding 9.5 cm ≈ 10 cm, and $\pi \approx 3$ instead of 3.14,

$$A = \pi r^2 \approx 3(10 \text{ cm})^2 = 300 \text{ cm}^2$$

(Note that significant figures are not a concern in calculations involving approximations.) The answer is not exact, but it is a good approximation. Compute the exact answer and see.

Powers-of-ten, or scientific, notation is particularly convenient in making estimates or approximations in what are called **order-of-magnitude calculations**. *Order of magnitude* means that we express a quantity to the power of 10 closest to the actual value. For example, in the foregoing calculation, approximating 9.5 cm ≈ 10 cm is expressing 9.5 as 10^1, and we say that the radius is *on the order of* 10 cm. Expressing a distance of 75 km $\approx 10^2$ km indicates that the distance is on the order of 10^2 km. The radius of the Earth is 6.4×10^3 km $\approx 10^4$ km, or on the order of 10^4 km. A nanostructure with a width of 8.2×10^{-9} m is on the order of 10^{-8} m, or 10 nm. (Why an exponent of -8?)

An order-of-magnitude calculation gives only an estimate, of course. But this estimate may be enough to provide you with a better grasp or understanding of a physical situation. Usually, the result of an order-of-magnitude calculation is precise within a power of 10, or *within an order of magnitude*. That is, the prefix to the power of 10 is somewhere between 1 and 10. For example, if we got a time result of 10^5 s, we would expect the exact answer to be somewhere between 1×10^5 s and 10×10^5 s.

Example 1.12 ■ Order-of-Magnitude Calculation: Drawing Blood

A medical technologist draws 15 cc of blood from a patient's vein. Back in the lab, it is determined that this volume of blood has a mass of 16 g. Estimate the density of the blood, in standard SI units.

Thinking It Through. The data are given in cgs (centimeter–gram–second) units, which are often used for practicality when dealing with small, whole-number quantities in some situations. The cc abbreviation is commonly used in the medical and chemistry fields for cm^3. Density (ρ) is mass per unit volume, where $\rho = m/V$ (Section 1.4).

Solution.

Given: $\quad m = 16 \text{ g} \left(\dfrac{1 \text{ kg}}{1000 \text{ g}} \right) = 1.6 \times 10^{-2} \text{ kg} \approx 10^{-2} \text{ kg}$ *Find:*   estimate of ρ (density)

$$V = 15 \text{ cm}^3 \left(\frac{1 \text{ m}}{10^2 \text{ cm}} \right)^3 = 1.5 \times 10^{-5} \text{ m}^3 \approx 10^{-5} \text{ m}^3$$

So, we have

$$\rho = \frac{m}{V} \approx \frac{10^{-2} \text{ kg}}{10^{-5} \text{ m}^3} \approx 10^3 \text{ kg/m}^3$$

This result is quite close to the average density of whole blood, 1.05×10^3 kg/m^3.

Follow-Up Exercise. A patient receives 750 cc of whole blood. Estimate the mass of the blood, in standard units. (*Answers to all Follow-Up Exercises are at the back of the text.*)

Example 1.13 ■ How Many Red Cells in Your Blood?

The blood volume in the human body varies with a person's age, body size, and sex. On average, this volume is about 5 L. A typical value of red blood cells (erythrocytes) per volume is 5 000 000 (5×10^6) cells per cubic millimeter. Estimate how many red blood cells you have in your body.

Thinking It Through. The red blood cell count in cells per cubic millimeter is sort of a red blood cell "number density." Multiplying this figure by the total volume of blood [(cells/volume) × total volume] will give the total number of cells. But note that we must have the volumes in the same units.

Solution.

Given:

$$V = 5\,L$$
$$= 5\,L\left(10^{-3}\,\frac{m^3}{L}\right)$$
$$= 5 \times 10^{-3}\,m^3 \approx 10^{-2}\,m^3$$
$$\text{cells/volume} = 5 \times 10^6\,\frac{\text{cells}}{mm^3} \approx 10^7\,\frac{\text{cells}}{mm^3}$$

Find: the approximate number of red cells in the body

Then, changing to cubic meters,

$$\frac{\text{cells}}{\text{volume}} \approx 10^7\,\frac{\text{cells}}{mm^3}\left(\frac{10^3\,mm}{1\,m}\right)^3 \approx 10^{16}\,\frac{\text{cells}}{m^3}$$

(*Note*: The conversion factor for liters to cubic meters was obtained directly from the conversion tables, but there is no conversion factor given for converting cubic millimeters to cubic meters, so we just use a known conversion and cube it.) Then,

$$\left(\frac{\text{cells}}{\text{volume}}\right)(\text{total volume}) \approx \left(10^{16}\,\frac{\text{cells}}{m^3}\right)(10^{-2}\,m^3) = 10^{14} \text{ red blood cells}$$

Red blood cells (erythrocytes) are one of the most abundant cells in the human body.

Follow-Up Exercise. The average number of white blood cells (leukocytes) in human blood is normally 5000 to 10 000 cells per cubic millimeter. Estimate the number of white blood cells you have in your body. (*Answers to all Follow-Up Exercises are at the back of the text.*)

Chapter Review

- **SI units of length, mass, and time.** The meter (m), the kilogram (kg), and the second (s), respectively.

LENGTH: METER

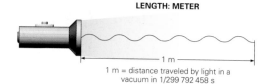

1 m

1 m = distance traveled by light in a
vacuum in 1/299 792 458 s

MASS: KILOGRAM

0.10 m

water

0.10 m

0.10 m

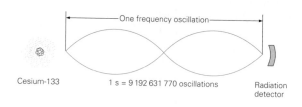

One frequency oscillation

Cesium-133   1 s = 9 192 631 770 oscillations   Radiation detector

- **Liter (L).** A volume of 1000 mL, or 1000 cm³. To a good approximation, a liter of water has a mass of 1 kg.

Volume

1 L

1 qt

1 L = 1.06 qt

1 qt = 0.947 L

- **Unit analysis.** Unit analysis can be used to determine the consistency of an equation, that is, if the equation is dimensionally correct. Unit analysis can also be used to find the unit of a quantity.

- **Significant figures (digits).** The digits that are known with certainty, plus one digit that is uncertain, in a measured value.

- **Problem solving.** Problems should be worked using a consistent procedure. Order-of-magnitude calculations may be done only when an estimated value is desired.

Suggested Procedure for Problem Solving:

1. Read the problem carefully and analyze it.
2. Where appropriate draw a diagram.
3. Write down the given data and what is to be found. (Make unit conversions if necessary.)
4. Determine which principle(s) are applicable.
5. Perform calculations with given data.
6. Consider if the results are reasonable.

- **Density (ρ).** The mass per unit volume of an object or substance, which is a measure of the compactness of the material it contains:

$$\rho = \frac{m}{V} \left(\frac{\text{mass}}{\text{volume}} \right) \tag{1.1}$$

Exercises*

MC = *Multiple Choice Question,* **CQ** = *Conceptual Question, and* **IE** = *Integrated Exercise. Throughout the text, many exercise sections will include "paired" exercises. These exercise pairs, identified with* **red numbers,** *are intended to assist you in problem solving and learning. In a pair, the first exercise (even numbered) is worked out in the Study Guide so that you can consult it should you need assistance in solving it. The second exercise (odd numbered) is similar in nature, and its answer is given at the back of the book.*

1.2 SI Units of Length, Mass, and Time

1. **MC** How many base units are there in the SI: (a) 3, (b) 5, (c) 7, or (d) 9? (c)

2. **MC** The only SI standard represented by material standard is the (a) meter, (b) kilogram, (c) second, (d) electric charge. (b)

3. **MC** Which of the following is *not* an SI base quantity: (a) mass; (b) weight; (c) length; or (d) time? (b)

4. **MC** Which of the following is the SI base unit for mass: (a) pound; (b) gram; (c) kilogram; (d) ton? (c)

5. **CQ** Why are there not more SI base units? no more fundamental quantities

6. **CQ** Why is weight not a base quantity? weight changes depending on gravity of locations

7. **CQ** What replaced the original definition of the second and why? Is the replacement still used? mean solar day, no, atomic clocks now used

8. **CQ** Give a couple of major differences between the SI and the British system. decimal vs. duodecimal bases

1.3 More about the Metric System

9. **MC** The prefix *giga-* means (a) 10^{-9}, (b) 10^9, (c) 10^{-6}, (d) 10^6. (b)

10. **MC** The prefix *micro-* means (a) 10^6, (b) 10^{-6}, (c) 10^3, (d) 10^{-3}. (b)

11. **MC** A new technology is concerned with objects the size of what metric prefix? (a) *nano-* (b) *micro-* (c) *mega-* (d) *giga-* (a)

12. **MC** One liter of water has a volume of (a) 1 m³, (b) 1 qt, (c) 1000 cm³, (d) 10^4 mm³. (c)

13. **CQ** If a fellow student tells you he saw a 3-cm-long ladybug, would you believe him? How about another student saying she caught a 10-kg salmon? no, yes

14. **CQ** Explain why 1 mL is equivalent to 1 cm³. 1 L = 1000 mL and 1 L = 1000 cm³

15. **CQ** Explain why a metric ton is equivalent to 1000 kg. see ISM

16. ● The metric system is a decimal (base-10) system, and the British system is, in part, a duodecimal (base-12) system. Discuss the ramifications if our monetary system had a duodecimal base. What would be the possible values of our coins if this were the case? see ISM

17. ● (a) In the British system, 16 oz = 1 pt and 16 oz = 1 lb. Is something wrong here? Explain. (b) Here's an old one: A pound of feathers weighs more than a pound of gold. How can that be? [*Hint*: Look up *ounce* in the dictionary.] see ISM

*Keep in mind here and throughout the text that your answer to an odd-numbered exercise may differ slightly from that given at the back of the book because of rounding. See the Problem-Solving Hint: The "Correct" Answer in this chapter.

18. ●● A sailor tells you that if his ship is traveling at 25 knots (nautical miles per hour), it is moving faster than the 25 mi/h your car travels. How can that be? 1 nautical mi = 6076 ft

1.4 Unit Analysis*

19. **MC** Both sides of an equation are equal in (a) numerical value, (b) units, (c) dimensions, (d) all of the preceding. (d)

20. **MC** Unit analysis of an equation cannot tell you if (a) the equation is dimensionally correct, (b) the equation is physically correct, (c) the numerical value is correct, (d) both b and c. (d)

21. **MC** Which of the following is true for the quantity $\frac{x}{t}$: (a) It may have the same dimensions but different units; (b) it may have the same units but different dimensions; or (c) both a and b are true. (a)

22. CQ Can unit analysis tell you whether you have used the correct equation in solving a problem? Explain. no

23. CQ The equation for the area of a circle from two sources is given as $A = \pi r^2$ and $A = \pi d^2/2$. Can unit analysis tell you which is correct? Explain. no

24. CQ How might unit analysis help determine the units of a quantity? by putting in units and solving for those of unknown quantity

25. ● Show that the equation $x = x_o + vt$ is dimensionally correct, where v is velocity and x and x_o are lengths, and t is time. (length) = (length) + (length)

26. ● If x refers to distance, v_o and v to speeds, a to acceleration, and t to time, which of the following equations is dimensionally correct: (a) $x = v_o t + at^3$; (b) $v^2 = v_o^2 + 2at$; (c) $x = at + vt^2$; or (d) $v^2 = v_o^2 + 2ax$? (d)

27. ●● Use SI unit analysis to show that the equation $A = 4\pi r^2$, where A is the area and r is the radius of a sphere, is dimensionally correct. $m^2 = m^2$

28. ●● You are told that the volume of a sphere is given by $V = \pi d^3/4$, where V is the volume and d is the diameter of the sphere. Is this equation dimensionally correct? (Use SI unit analysis to find out.) yes; $m^3 = m^3$

29. ●● The correct equation for the volume of a sphere is $V = 4\pi r^3/3$, where r is the radius of the sphere. Is the equation in Exercise 28 correct? If not, what should it be when expressed in terms of d? no; $V = \pi d^3/6$

30. ●● The kinetic energy (K) of an object of mass m moving with speed v is given by $K = \frac{1}{2}mv^2$. The name for the unit of kinetic energy in the SI system is the joule (J). What are the units of the joule in terms of SI base units? $kg \cdot m^2/s^2$

31. ●● The general equation for a parabola is $y = ax^2 + bx + c$, where a, b, and c are constants. What are the units of each constant if y and x are in meters? a, 1/m; b, dimensionless; c, m

32. ●● The units for pressure (p) in terms of SI base units are known to be $\dfrac{kg}{m \cdot s^2}$. For a physics class assignment, a student derives an expression for the pressure exerted by the wind on a wall in terms of the air density (ρ) and wind speed (v) and her result is $p = \rho v^2$. Use SI unit analysis to show that her result is dimensionally consistent. Does this prove that this relationship is physically correct? no

33. ●● Density is defined as the mass of an object divided by the volume of the object. Using SI unit analysis, determine the SI unit for density. (See Section 1.4 for units of mass and volume.) kg/m^3

34. ●● Is the equation for the area of a trapezoid, $A = \frac{1}{2}a(b_1 + b_2)$, where a is the height and b_1 and b_2 are the bases, dimensionally correct? (▼Fig. 1.14.) yes

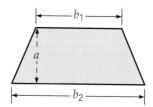

▲ **FIGURE 1.14 The area of a trapezoid** See Exercise 34.

35. ●● One student, using unit analysis, says that the equation $v = \sqrt{2ax}$ is dimensionally correct. Another says it isn't. With whom do you agree, and why? the first student; see ISM

36. ●●● Newton's second law of motion (Chapter 4) is expressed by the equation $F = ma$, where F represents force, m is mass, and a is acceleration. (a) The SI unit of force is, appropriately, called the newton (N). What are the units of the newton in terms of base quantities? (b) An equation for force associated with uniform circular motion (Chapter 7) is $F = mv^2/r$, where v is speed and r is the radius of the circular path. Does this equation give the same units for the newton? (a) $kg \cdot m/s^2$ (b) yes

37. ●●● The angular momentum (L) of a particle of mass m moving at a constant speed v in a circle of radius r is given by $L = mvr$. (a) What are the units of angular momentum in terms of SI base units? (b) The units of kinetic energy in terms of SI base units are $\dfrac{kg \cdot m^2}{s^2}$. Using SI

*Units of velocity and acceleration are given in Table 1.3.

unit analysis, show that the expression for the kinetic energy of this particle in terms of its angular momentum, $K = \dfrac{L^2}{2mr^2}$, is dimensionally correct. (c) In the previous equation, the term mr^2 is called the *moment of inertia* of the particle in the circle. What are the units of moment of inertia in terms of SI base units? (a) kg · m²/s (b) see ISM (c) kg · m²

38. ●●● Einstein's famous mass–energy equivalence is expressed by the equation $E = mc^2$, where E is energy, m is mass, and c is the speed of light. (a) What are the SI base units of energy? (b) Another equation for energy is $E = mgh$, where m is mass, g is the acceleration due to gravity, and h is height. Does this equation give the same units as in part (a)? (a) kg · m²/s² (b) yes

1.5 Unit Conversions*

39. **MC** A good way to ensure proper unit conversion is to (a) use another measurement instrument, (b) always work in the same system of units, (c) use unit analysis, (d) have someone check your math. (c)

40. **MC** You often see 1 kg = 2.2 lb. This expression means that (a) 1 kg is equivalent to 2.2 lb, (b) this is a true equation, (c) 1 lb = 2.2 kg, (d) none of the preceding. (a)

41. **MC** You have a quantity of water and wish to express this in volume units that give the largest number. Should you use (a) in³; (b) mL; (c) μL; or (d) cm³? (c)

42. **CQ** Are an equation and an equivalence statement the same? Explain. no

43. **CQ** Does it make any difference whether you multiply or divide by a conversion factor? Explain. yes

44. **CQ** Does unit analysis apply to unit conversions? Explain. yes

45. ● Figure 1.8 (top) shows the elevation of a location in both feet and meters. If a town is 130 ft above sea level, what is the elevation in meters? 39.6 m

46. **IE** ● (a) If you wanted to express your height with the largest number, you would use (1) meters, (2) feet, (3) inches, (4) centimeters. Why? (b) If you are 6.00 ft tall, what is your height in centimeters?
(a) (4) cm (b) 183 cm

47. ● If the capillaries of an average adult were unwound and spread out end to end, they would extend to a length over 40 000 mi (Fig. 1.9). If you are 1.75 m tall, how many times your height would the capillary length equal? 37 000 000 times

*Conversion factors are listed inside the front cover of the text.

48. **IE** ● (a) Compared with a 2-L soda bottle, a half-gallon soda bottle holds (1) more, (2) the same amount of, (3) less soda. (b) Verify your answer for part (a).
(a) (3) less soda (b) 2 L by 0.11 L more

49. ● (a) A football field is 300 ft long and 160 ft wide. What are the field's dimensions in meters? (b) A football is 11.0 to $11\frac{1}{4}$ in. long. What is its length in centimeters?
(a) 91.5 m by 48.8 m (b) 27.9 cm to 28.6 cm

50. ● Suppose that when the United States goes completely metric, the dimensions of a football field are established as 100 m by 54 m. Which would be larger, the metric football field or a current football field (see Exercise 49a), and what would be the difference between the areas? metric; 9.4 × 10² m²

51. ●● If blood flows with an average speed of 0.35 m/s in the human circulatory system, how many miles does a blood cell travel in 1.0 h? 0.78 mi

52. ●● Driving a jet-powered car, Royal Air Force pilot Andy Green broke the sound barrier on land for the first time and achieved a record land speed of more than 763 mi/h in Black Rock Desert, Nevada, on October 15, 1997 (▾Fig. 1.15). (a) What is this speed expressed in m/s? (b) How long would it take the jet-powered car to travel the length of a 300-ft football field at this speed? (a) 341 m/s (b) 0.268 s

▲ **FIGURE 1.15 Record run** See Exercise 52.

53. **IE** ●● (a) Which of the following represents the greatest speed: (1) 1 m/s; (2) 1 km/h; (3) 1 ft/s; or (4) 1 mi/h? (b) Express the speed 15.0 m/s in mi/h.
(a) (1) 1 m/s (b) 33.6 mi/h

54. ●● An automobile speedometer is shown in ▾Fig. 1.16. (a) What would be the equivalent scale readings (for each empty box) in kilometers per hour? (b) What would be the 70-mi/h speed limit in kilometers per hour?
(a) 16 km/h for each 10 mi/h (b) 1.1 × 10² km/h

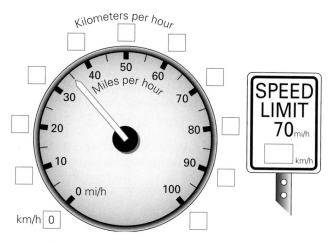

▲ **FIGURE 1.16 Speedometer readings** See Exercise 54.

55. ●● A person weighs 170 lb. (a) What is her mass in kilograms? (b) Assuming the density of the average human body is about that of water (which is true), estimate her body's volume in both cubic meters and liters. Explain why the smaller unit of the liter is more appropriate (convenient) for describing this size volume. (a) 77.3 kg (b) 0.0773 m³ or about 77.3 L

56. ●● If the components of the human circulatory system (arteries, veins, and capillaries) were completely extended and placed end to end, the length would be on the order of 100 000 km. Would the length of the circulatory system reach around the circumference of the Moon? If so, how many times? yes, 9.1 times

57. ●● The human heartbeat, as determined by the pulse rate, is normally about 60 beats/min. If the heart pumps 75 mL of blood per beat, what volume of blood is pumped in one day in liters? 6.5 × 10³ L/day

58. ●● A typical wide receiver in American football can run the 40-yd dash in about 4.5 s starting from rest. (a) What is his average speed in m/s? (b) What is his average speed in mi/h? (a) 8.1 m/s (b) 18 mi/h

59. ●● Some common product labels are shown in ▼Fig. 1.17. From the units on the labels, find (a) the number of milliliters in 2 fl. oz and (b) the number of ounces in 100 g. (a) 59.1 mL (b) 3.53 oz

▲ **FIGURE 1.17 Conversion factors** See Exercise 59.

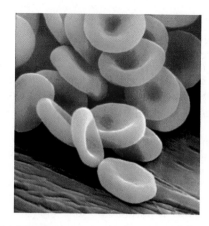

▲ **FIGURE 1.18 Red blood cells** See Exercise 60.

60. ●● ▲Fig. 1.18 is a picture of red blood cells seen under a scanning electron microscope. Normally, women possess about 4.5 million of these cells in each cubic millimeter of blood. If the blood flow to the heart organ is 250 mL/min, how many red blood cells does a woman's heart receive each second? 1.9 × 10¹⁰/s

61. ●●● A student was 18 in. long when she was born. She is now 5 ft 6 in. tall and 20 years old. How many centimeters a year did she grow on average? 6.1 cm

62. ●●● The density of metal mercury is 13.6 g/cm³. (a) What is this density as expressed in kilograms per cubic meter? (b) How many kilograms of mercury would be required to fill a 0.250-L container? (a) 1.36 × 10⁴ kg/m³ (b) 3.40 kg

63. ●●● The Roman Coliseum used to be flooded with water to recreate ancient naval battles. Assuming the floor of the Coliseum to be 250 m in diameter and the water to have a depth of 10 ft, (a) how many cubic meters of water are required? (b) How much mass would this water have in kilograms? (c) How much would the water weigh in pounds? (a) 1.5 × 10⁵ m³ (b) 1.5 × 10⁸ kg (c) 3.3 × 10⁸ lb

64. ●●● In the Bible, Noah is instructed to build an ark 300 cubits long, 50.0 cubits wide, and 30.0 cubits high (▼Fig. 1.19). Historical records indicate a cubit is equal to half a yard. (a) What would the dimensions of the ark be in meters? (b) What would the ark's volume be in cubic meters? To approximate, assume that the ark is to be rectangular. (a) 137 m × 22.9 m × 13.7 m (b) 4.30 × 10⁴ m³

▲ **FIGURE 1.19 Noah and his ark** See Exercise 64.

1.6 Significant Figures

65. **MC** Which of the following has the greatest number of significant figures: (a) 103.07; (b) 124.5; (c) 0.09916; or (d) 5.408×10^5? (a)

66. **MC** Which of the following numbers has four significant figures: (a) 140.05; (b) 276.02; (c) 0.004 006; or (d) 0.073 004? (c)

67. **MC** In a multiplication and/or division operation involving the numbers 15 437, 201.08, and 408.0×10^5, the result should have how many significant figures? (a) 3 (b) 4 (c) 5 (d) any number (b)

68. **CQ** What is the purpose of significant figures? to provide an estimation of accuracy

69. **CQ** Are all the significant figures reported for a measured value accurately known? Explain. no

70. **CQ** How are the number of significant figures determined for the results of calculations involving (a) multiplication, (b) division, (c) addition, and (d) subtraction? see ISM

71. ● Express the length $50 500\ \mu m$ (micrometers) in centimeters, decimeters, and meters, to three significant figures. 5.05 cm; 5.05×10^{-1} dm; 5.05×10^{-2} m

72. ● Using a meterstick, a student measures a length and reports it to be 0.8755 m. What is the smallest division on the meterstick scale? 0.001 m, or 1 mm

73. ● Determine the number of significant figures in the following measured numbers: (a) 1.007 m; (b) 8.03 cm; (c) 16.272 kg; (d) $0.015\ \mu s$ (microseconds). (a) 4 (b) 3 (c) 5 (d) 2

74. ● Express each of the numbers in Exercise 73 with two significant figures.
(a) 1.0 m (b) 8.0 cm (c) 16 kg (d) $1.5 \times 10^{-2}\ \mu s$

75. ● Which of the following quantities has three significant figures: (a) 305.0 cm; (b) 0.0500 mm; (c) 1.000 81 kg; or (d) 8.06×10^4 m²? (b) and (d); (a) has four and (c) has six

76. ●● The cover of your physics book measures 0.274 m long and 0.222 m wide. What is its area in square meters? 6.08×10^{-2} m²

77. ●● The interior storage compartment of a restaurant refrigerator measures 1.3 m high, 1.05 m wide, and 67 cm deep. Determine its volume in cubic feet. 32 ft³

78. **IE** ●● The top of a rectangular table measures 1.245 m by 0.760 m. (a) The smallest division on the scale of the measurement instrument is (1) m, (2) cm, (3) mm. Why? (b) What is the area of the tabletop? (a) (2) cm (b) 0.946 m²

79. **IE** ●● The outside dimensions of a cylindrical soda can are reported as 12.559 cm for the diameter and 5.62 cm for the height. (a) How many significant figures will the total outside area have: (1) two; (2) three; (3) four; or (4) five? Why? (b) What is the total outside surface area of the can in cubic centimeters? (a) (2) three (b) 469 cm²

80. ●● Express the following calculations to the proper number of significant figures: (a) $12.634 + 2.1$; (b) $13.5 - 2.134$; (c) $\pi(0.25\ m)^2$; (d) $\sqrt{2.37/3.5}$.
(a) 14.7 (b) 11.4 (c) 0.20 m² (d) 0.82

81. **IE** ●●● In doing a problem, a student adds 46.9 m and 5.72 m and then subtracts 38 m from the result. (a) How many decimal places will the final answer have: (1) zero; (2) one; or (3) two? Why? (b) What is the final answer? (a) (1) zero (b) 15 m

82. ●●● Work this exercise by the two given procedures as directed, commenting on and explaining any difference in the answers. Use your calculator for the calculations. Compute $p = mv$, where $v = x/t$, given $x = 8.5$ m, $t = 2.7$ s, and $m = 0.66$ kg. (a) First compute v and then p. (b) Compute $p = mx/t$ without an intermediate step. (c) Are the results the same? If not, why?
(a) 2.0 kg·m/s (b) 2.1 kg·m/s (c) no, rounding difference

1.7 Problem Solving

83. **MC** An important step in problem solving before mathematically solving an equation is (a) checking units, (b) checking significant figures, (c) checking with a friend, (d) checking to see if the result will be reasonable. (a)

84. **MC** An important final step in problem solving before reporting an answer is (a) saving your calculations, (b) reading the problem again, (c) seeing if the answer is reasonable, (d) checking your results with another student. (c)

85. **MC** In order-of-magnitude calculations, you should (a) pay close attention to significant figures, (b) work primarily in the British system, (c) get results within a factor of 100, (d) express a quantity to the power of 10 closest to the actual value. (d)

86. **CQ** How many steps are in a good problem-solving procedure as suggested in this chapter? six

87. **CQ** What are the main steps in a problem-solving procedure? all six steps as listed in the chapter

88. **CQ** When you do order-of-magnitude calculations, should you be concerned about significant figures? Explain. no

89. **CQ** When doing an order-of-magnitude calculation, how accurate can you expect the answer to be? Explain. within an order of 10

90. ● A corner construction lot has the shape of a right triangle. If the two sides perpendicular to each other are 37 m long and 42.3 m long, respectively, what is the length of the hypotenuse? 56 m

91. ● The lightest solid material is silica aerogel, which has a typical density of only about 0.10 g/cm³. The molecular structure of silica aerogel is typically 95% empty space. What is the mass of 1 m³ of silica aerogel? 100 kg

92. ●● Nutrition Facts labels now appear on most foods. An abbreviated label concerned with fat is shown in ▼Fig. 1.20. When burned in the body, each gram of fat supplies 9 Calories. (A food Calorie is really a kilocalorie, as we shall see in Chapter 11.) (a) What percentage of the Calories in one serving is supplied by fat? (b) You may notice that our answer doesn't agree with the listed Total Fat percentage in Fig. 1.20. This is because the given Percent Daily Values are the percentages of the maximum recommended amounts of

nutrients (in grams) contained in a 2000-Calorie diet. What are the maximum recommended amounts of total fat and saturated fat for a 2000-Calorie diet? (a) 52% (b) 64 g, 20 g

Nutrition Facts
Serving Size: 1 can
Calories: 310

Amount Per Serving	% Daily Value*
Total Fat 18 g	28%
Saturated Fat 7g	35%

* Percent Daily Values are based on a 2,000 Calorie diet.

▲ **FIGURE 1.20 Nutrition Facts** See Exercise 92.

93. ●● The thickness of the total of numbered pages of a textbook is measured to be 3.75 cm. (a) If the last page of the book is numbered 860, what is the average thickness of a page? (b) Repeat the calculation by using order-of-magnitude calculations.
(a) 8.72×10^{-3} cm (b) about 10^{-2} cm

94. **IE** ●● To go to a football stadium from your house, you first drive 1000 m north, then 500 m west, and finally 1500 m south. (a) Relative to your home, the football stadium is (1) north of west, (2) south of east, (3) north of east, (4) south of west. (b) What is the straight-line distance from your house to the stadium?
(a) (4) south of west (b) 707 m

95. ●● Two chains of length 1.0 m are used to support a lamp, as shown in ▼Fig. 1.21. The distance between the two chains is 1.0 m along the ceiling. What is the vertical distance from the lamp to the ceiling? 0.87 m

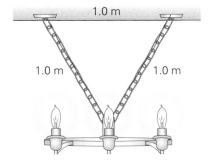

▲ **FIGURE 1.21 Support the lamp** See Exercise 95.

96. ●● Tony's Pizza Palace sells a medium 9.0-in. (diameter) pizza for $7.95, and a large 12-in. pizza for $13.50. Which pizza is the better buy?
12-in. better buy; 9.0-in.: 8.0 in.²/$, 12-in.: 8.4 in.²/$

97. ●● In ▶Fig. 1.22, which black region has the greater area, the center circle or the outer ring?
same area for both, 1.3 cm²

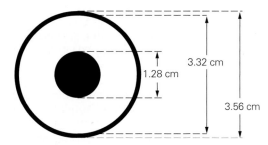

▲ **FIGURE 1.22 Which black area is greater?** See Exercise 97.

98. ●● The Channel Tunnel, or "Chunnel," which runs under the English Channel between Great Britain and France, is 31 mi long. (There are actually three separate tunnels.) A shuttle train that carries passengers through the tunnel travels with an average speed of 75 mi/h. On average, how long, in minutes, does the shuttle take to make a one-way trip through the Chunnel? 25 min

99. ●● Human adult blood contains on the average $7000/\text{mm}^3$ white blood cells (leukocytes) and $250\,000/\text{mm}^3$ platelets (thrombocytes). If a person has a blood volume of 5.0 L, estimate the total number of white cells and platelets in the blood. 3.5×10^{10} white cells, 1.3×10^{12} platlets

100. ●● A 10-ft by 20-ft lawn area is to be made into an outdoor patio by placing 2.0-ft-diameter circular concrete "pads" in an array so they touch each other. The existing grass will fill the spaces. (a) How many of these pads are required to do the job? (b) When the project is completed, what percentage of the original grass will still remain? (a) 50 (b) 21.5%

101. ●● Experimentally, the force felt on an auto due to its movement through the (still) air varies approximately as the *square* of the car's speed. (This force is sometimes called "air resistance.") Assume this force varies exactly as the square of the speed. Around town at 30 mi/h, measurements indicate that a certain car experiences an air resistance force of 100 lb. What size force would you expect to the car to experience traveling on highway at 65 mi/h?
4.7×10^2 lb

102. ●● The average number of hairs on the normal human scalp is 125 000. A healthy person loses about 65 hairs per day. (New hair from the hair follicle pushes the old hair out.) (a) How many hairs are lost in one month? (b) Pattern baldness (top-of-the-head hair loss) affects about 35 million men in the United States. With an average of 15% of the scalp bald, how many hairs are lost per year by one of these "bald is beautiful" people?
(a) 1950 hairs (b) 2.0×10^4 hairs

103. ●●● Approximately 118 mi wide and 307 mi long and averaging 279 ft in depth, Lake Michigan is the second-largest Great Lake by volume. Estimate its volume of water in cubic meters. about 10^{12} m³

104. **IE** ●●● In the Tour de France, a rider races up two successive (straight) hills of different slope and length. The first is 2.00 km long at an angle of 5° above the horizontal. This is immediately followed by one 3.00 km long at 7°. (a) What will be the overall (net) angle from start to finish: (1) smaller than 5°; (2) between 5° and 7°; or (3) greater than 7°? (b) Calculate the actual overall (net) angle of rise experienced by this racer from start to finish, to corroborate your reasoning in part (a). (a) (2) between 5° and 7° (b) 6.2°

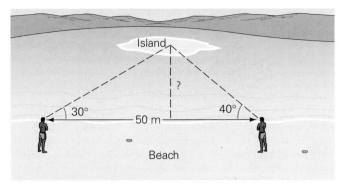

▲ **FIGURE 1.23 Measuring with lines of sight** See Exercise 105.

105. ●●● A student wants to determine the distance of a small island from the lakeshore (▲Fig. 1.23). He first draws a 50-m line parallel to the shore. Then, he goes to the ends of the line and measures the angles of the lines of sight from the island relative to the line he has drawn. The angles are 30° and 40°. How far is the island from the shore? 17 m

Comprehensive Exercises

106. **IE** A car is driven 13 mi east and then a certain distance due north and ends up at a position 25° north of east of its initial position. (a) The distance traveled by the car due north is (1) less than, (2) equal to, (3) greater than 13 mi. Why? (b) What distance does the car go due north?
(a) (1) less than (b) 6.1 mi

107. An airplane flies 100 mi south from city A to city B, 200 mi east from city B to city C, and then 300 mi north from city C to city D. (a) What is the straight-line distance from city A to city D? (b) What is the direction of city D relative to city A? (a) 283 mi (b) 45° north of east

108. In a radioactivity experiment, a solid lead brick (same measurements as a patio brick, 2.00″ × 4.00″ × 8.00″ except with a density that is 11.4 times that of water) is to be modified to hold a solid cylindrical piece of plastic. To accomplish this, the machinists are told to drill a cylin-

drical hole 2.0 cm in diameter through the center of the brick parallel to the longest side of the brick. (a) What is the mass of lead (in kilograms) removed from the brick? (b) What percentage of the original lead remains in the brick? (c) Assuming the cylindrical hole is completely filled with plastic (with a density twice that of water), determine the overall (average) density of the brick/plastic combination after fabrication is complete.
(a) 0.727 kg (b) 93.9% (c) 1.08×10^4 kg/m³

109. On a certain night, an observer on the Earth determines that the angle between the direction to Mars and the direction to the Sun is 50°. On that night, assuming circular orbits, determine the distance to Mars from the Earth using the known radii of the orbits of both planets. 3.0×10^8 km

110. Estimate the number of water molecules in a cup (8 oz. exactly) of water 1 (fluid) = 0.0296 L. [*Hint:* You may find it helpful to recall that the mass of a hydrogen atom is about 1.67×10^{-27} kg and that an oxygen atom's mass is about sixteen times that value.]
on the order of 10^{25} mole cules

111. **IE** At the Indianapolis 500 time trials, each car gets a chance to make four consecutive laps, with its overall or average speed determining that car's place on race day. Each lap covers 2.5 mi (exact). During a practice run, cautiously and gradually taking his car faster and faster, a driver records the following average speed for each successive lap: 160 mi/h, 180 mi/h, 200 mi/h, and 220 mi/h. (a) Will his average speed be (1) exactly the average of these speeds (190 mi/h), (2) greater than 190 mi/h, or (3) less than 190 mi/h? Explain. (b) To corroborate your conceptual reasoning, calculate the car's average speed.
(a) (3) less than 190 mi/h (b) 187 mi/h

112. A student doing a lab experiment drops a small solid cube into a cylindrical gas of water. The inner diameter of the glass is 6.00 cm. The cube sinks to the bottom, and the water level in the glass rises 1.00 cm. If the mass of the cube is 73.6 g, (a) determine the length of one side of the cube, and (b) calculate the cube's density. (Work the exercise in cgs units for convenience).
(a) 3.05 cm (b) 2.60 g/cm³

 The following Physlet Physics Problem can be used with this chapter.
1.1

KINEMATICS: DESCRIPTION OF MOTION

PHYSICS FACTS

- "Give me matter and motion and I will construct the universe." Rene Descartes (1640)

- Nothing can exceed the speed of light (in vacuum), 3.0×10^8 m/s (186 000 mi/s).

- NASA's X-43A uncrewed jet flew at a speed of 7700 km/h (4800 mi/h)—faster than a speeding bullet.

- A bullet from a high-powered rifle travels at a speed of about 2900 km/h (1800 mi/h).

- Electrical signals between your brain and muscles travel at about 435 km/h (270 mi/h).

- A person at the equator is traveling at a speed of 1600 km/h (1000 mi/h) due to the Earth's rotation.

- Fast and slow (approximate maximum speeds):
  - Cheetah 113 km/h (70 mi/h)
  - Horse 76 km/h (47 mi/h)
  - Greyhound 63 km/h (39 mi/h)
  - Rabbit 56 km/h (35 mi/h)
  - Cat 48 km/h (30 mi/h)
  - Human 45 km/h (28 mi/h)
  - Chicken 14 km/h (9 mi/h)
  - Snail 0.05 km/h (0.03 mi/h)

- Aristotle thought heavy objects fall faster than lighter ones. Galileo wrote, "Aristotle says that an iron ball falling from a height of one hundred cubits reaches the ground before a one-pound ball has fallen a single cubit. I say they arrive at the same time."

The cheetah is running at full stride. This fastest of all land animals is capable of attaining speeds up to 113 km/h, or 70 mi/h. The sense of motion in the chapter opening photograph is so strong that you can almost feel the air rushing by you. And yet this sense of motion is an illusion. Motion takes place in time, but the photo can "freeze" only a single instant. You'll find that without the dimension of time, you can hardly describe motion at all.

The description of motion involves the representation of a restless world. Nothing is ever perfectly still. You may sit, apparently at rest, but your blood flows, and air moves into and out of your lungs. The air is composed of gas molecules moving at different speeds and in different directions. And while you experience stillness, you, your chair, the building you are in, and the air you breathe are all revolving through space with the Earth, part of a solar system in a spiraling galaxy in an expanding universe.

The branch of physics concerned with the study of motion, and what produces and affects motion, is called **mechanics**. The roots of mechanics and of human interest in motion go back to early civilizations. The study of the motions of heavenly bodies, or *celestial mechanics*, grew out of the need to measure time and location. Several early Greek scientists, notably Aristotle, put forth theories of motion that were useful descriptions, but were later proved to be incomplete or incorrect. Our currently accepted concepts of motion were formulated in large part by Galileo (1564–1642) and Isaac Newton (1642–1727).

Mechanics is usually divided into two parts: (1) kinematics and (2) dynamics. **Kinematics** deals with the *description* of the motion of objects, without consideration of what causes the motion. **Dynamics** analyzes the *causes* of motion. This chapter covers kinematics and reduces the description of motion to its simplest terms by considering motion in a straight line. You'll learn to analyze changes in motion—speeding up, slowing down, and stopping. Along the way, we'll deal with a particularly interesting case of accelerated motion: free fall under the influence only of gravity.

2.1 Distance and Speed: Scalar Quantities

OBJECTIVES: To (a) define distance and calculate speed, and (b) explain what is meant by a scalar quantity.

Distance

We observe motion all around us. But what is motion? This question seems simple; however, you might have some difficulty giving an immediate answer (and it's not fair to use forms of the verb *to move* to describe motion). After a little thought, you should be able to conclude that **motion** (or moving) involves changing position. Motion can be described in part by specifying *how far* something travels in changing position—that is, the distance it travels. **Distance** is simply the *total path length* traversed in moving from one location to another. For example, you may drive to school from your hometown and express the distance traveled in miles or kilometers. In general, the distance between two points depends on the path traveled (▸Fig. 2.1).

Along with many other quantities in physics, distance is a scalar quantity. A **scalar quantity** is a quantity with only magnitude, or size. That is, a *scalar* has only a numerical value, such as 160 km or 100 mi. (Note that the magnitude includes units.) Distance tells you the magnitude only—how far, but not the direction. Other examples of scalars are quantities such as 10 s (time), 3.0 kg (mass), and 20°C (temperature). Some scalars may have negative values; for example, −10°F.

Speed

When something is in motion, its position changes with time. That is, it moves a certain distance in a given amount of time. Both length and time are therefore important quantities in describing motion. For example, imagine a car and a pedestrian moving down a street and traveling a distance of one block. You would expect the car to travel faster, and thus to cover the same distance in a shorter time than the person. A length-time relationship can be expressed by using the *rate* at which distance is traveled, or what we call **speed**.

Average speed ($\bar{s}$) is the distance d traveled, that is, the actual length of the path, divided by the total time Δt elapsed in traveling that distance:

$$\text{average speed} = \frac{\text{distance traveled}}{\text{total time to travel that distance}} \quad (2.1)$$

$$\bar{s} = \frac{d}{\Delta t} = \frac{d}{t_2 - t_1}$$

SI unit of speed: meters per second (m/s)

A symbol with a bar over it is commonly used to denote an average. The Greek letter delta, Δ, is used to represent a change or difference in a quantity, in this case the time difference between the beginning (t_1) and end (t_2) of a trip, or the elapsed time.

The SI standard unit of speed is meters per second (m/s, length/time), although kilometers per hour (km/h) is used in many everyday applications. The British standard unit is feet per second (ft/s), but we often use miles per hour (mi/h). Often, the initial time is taken to be zero, $t_1 = 0$, as in resetting a stopwatch, and thus the equation is written $\bar{s} = d/t$, where it is understood that t is the total time.

Since distance is a scalar (as is time), speed is also a scalar. The distance does *not* have to be in a straight line. (See Fig. 2.1.) For example, you probably have computed the average speed of an automobile trip by using the distance obtained from the

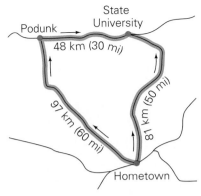

▲ **FIGURE 2.1 Distance—total path length** In driving to State University from Hometown, one student may take the shortest route and travel a distance of 81 km (50 mi). Another student takes a longer route in order to visit a friend in Podunk before returning to school. The longer trip is in two segments, but the distance traveled is the total length, 97 km + 48 km = 145 km (90 mi).

Note: A scalar quantity has magnitude, but no direction.

Teaching tip: Emphasize throughout this chapter and in future chapters which quantities are scalars and which quantities are vectors.

Teaching tip: Point out that any distance can be traveled in a very short period of time at a very high speed, but it is *not* possible to cover any distance in *zero* time.

▲ **FIGURE 2.2** **Instantaneous speed** The speedometer of a car gives the speed over a very short interval of time, so its reading approaches the instantaneous speed.

starting and ending odometer readings. Suppose these readings were 17 455 km and 17 775 km, respectively, for a 4.0-h trip. (We'll assume that you have a car with odometer readings in kilometers.) Subtracting the readings gives a total traveled distance d of 320 km, so the average speed of the trip is $d/t = 320$ km/4.0 h $= 80$ km/h (or about 50 mi/h).

Average speed gives a general description of motion over a time interval Δt. In the case of the auto trip with an average speed of 80 km/h, the car's speed wasn't *always* 80 km/h. With various stops and starts on the trip, the car must have been moving more slowly than the average speed at various times. It therefore had to be moving more rapidly than the average speed another part of the time. With an average speed, you don't know how fast the car was moving at any particular instant of time during the trip. In analogy, the average test score of a class doesn't tell you the score of any particular student.

The **instantaneous speed** is a quantity that tells how fast something is moving *at a particular instant of time*. The speedometer of a car gives an approximate instantaneous speed. For example, the speedometer shown in ◄Fig. 2.2 indicates a speed of about 44 mi/h, or 70 km/h. If the car travels with constant speed (so the speedometer reading does not change), then the average and instantaneous speeds will be equal. (Do you agree? Think of the previous average test score analogy. What if all of the students got the same score?)

Example 2.1 ■ Slow Motion: Rover Moves Along

In January 2004, the Mars Exploration Rover touched down on the surface of Mars and rolled out for exploration (◄Fig. 2.3). The average speed of a rover on flat, hard ground is 5.0 cm/s. (a) Assuming the rover traveled continuously over this terrain at its average speed, how much time would it take to travel 2.0 m in a straight line? (b) However, in order to ensure a safe drive, the rover was equipped with hazard avoidance software that cause it to stop and assess its location every few seconds. It was programmed to drive at average speed for 10 s, then stop and observe the terrain for 20 s before moving onward for another 10 s and repeating the cycle. Taking its programming into account, what would be the rover's average speed in traveling the 2.0 m?

Thinking It Through. (a) Knowing the average speed and the distance, the time can be computed from the equation for average speed (Eq. 2.1). (b) Here, to calculate the average speed, the total time, which includes stops, must be used.

Solution. Listing the data in symbol form: (cm/s is converted directly to m/s)

Given:
(a) $\bar{s} = 5.0$ cm/s $= 0.050$ m/s
     $d = 2.0$ m
(b) cycles of 10 s travel, 20 s stops

Find:
(a) Δt (time to travel distance)
(b) $\bar{s}$ (average speed)

(a) From Eq. 2.1, we have $\bar{s} = \dfrac{d}{\Delta t}$

Rearranging,

$$\Delta t = \frac{d}{\bar{s}} = \frac{2.0 \text{ m}}{0.050 \text{ m/s}} = 40 \text{ s}$$

(b) Here we need the total time for the 2.0-m distance. In each 10-s interval, a distance of 0.050 m/s × 10 s = 0.50 m would be traveled. So, the total time would be four 10-s intervals for actual travel, and three 20-s intervals of stopping, giving $\Delta t = 4 \times 10$ s $+ 3 \times 20$ s $= 100$ s. Then

$$\bar{s} = \frac{d}{\Delta t} = \frac{d}{t_2 - t_1} = \frac{2.0 \text{ m}}{100 \text{ s}} = 0.020 \text{ m/s}$$

Follow-Up Exercise. Suppose the Rover's programming was for 5.0 s of travel and for 10-s stops. How long would it take to travel the 2.0 m in this case? (*Answers to all Follow-Up Exercises are at the back of the text.*)

▲ **FIGURE 2.3** **Mars Exploration Rovers** Twin Rover landed on opposite sides of the Martian planet in search of answers about the history of water on Mars.

2.2 One-Dimensional Displacement and Velocity: Vector Quantities

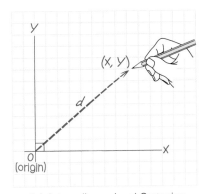

OBJECTIVES: To (a) define displacement and calculate velocity, and (b) explain the difference between scalar and vector quantities.

Displacement

For straight-line, or linear, motion, it is convenient to specify position by using the familiar two-dimensional Cartesian coordinate system, with x- and y-axes at right angles. A straight-line path can be in any direction relative to the axes, but for convenience, we usually orient the coordinate axes so that the motion is along one of them. (See the accompanying Learn by Drawing.)

As we have seen, distance is a scalar quantity with only magnitude (and units). However, to more completely describe motion, more information should be given by adding a *direction*. This information is particularly easy to convey for a change of position in a straight line. We define **displacement** as the straight-line distance between two points, along with the *direction* from the starting point to the final position. Unlike distance (a scalar), displacement can have either positive or negative values, with the signs indicating the directions along a coordinate axis.

As such, displacement is a **vector quantity**. In other words, a *vector* has both magnitude and direction. For example, when we describe the displacement of an airplane as 25 km north, we are giving a *vector* description (magnitude and direction). Other vector quantities include velocity and acceleration, as will be seen.

There is an algebra that applies to vectors, but we have to know how to specify and deal with the direction part of the vector. This process is relatively simple in one dimension by using + and − signs to indicate directions. To illustrate this with displacements, consider the situation shown in ▼Fig. 2.4, where x_1 and x_2 indicate the initial and final positions, respectively, on the x-axis as a student moves in a straight line from his locker to the physics lab. As can be seen in Fig. 2.4a, the scalar distance he travels is 8.0 m. To specify displacement (a vector) between x_1 and x_2, we use the expression

$$\Delta x = x_2 - x_1 \qquad (2.2)$$

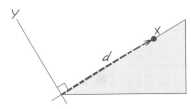

(a) A two-dimensional Cartesian coordinate system. A displacement vector d locates a point (x, y).

(b) For one-dimensional, or straight-line, motion, it is convenient to orient one of the coordinate axes along the direction of motion.

◀ **FIGURE 2.4 Distance (scalar) and displacement (vector)** (a) The distance (straight-line path) between the student and the physics lab is 8.0 m and is a scalar quantity. (b) To indicate displacement, x_1 and x_2 specify the initial and final positions, respectively. The displacement is then $\Delta x = x_2 - x_1 = 9.0\text{ m} - 1.0\text{ m} = +8.0\text{ m}$—that is, 8.0 m in the positive x-direction.

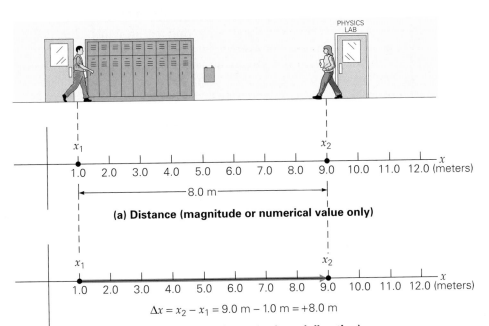

(a) Distance (magnitude or numerical value only)

$\Delta x = x_2 - x_1 = 9.0\text{ m} - 1.0\text{ m} = +8.0\text{ m}$

(b) Displacement (magnitude and direction)

where Δ is again used to represent a change in a quantity. Then, as in Fig. 2.4b, we have

$$\Delta x = x_2 - x_1 = +9.0 \text{ m} - (+1.0 \text{ m}) = +8.0 \text{ m}$$

Note: Δ always means *final minus initial*, just as a *change* in your bank account is the final balance minus the initial balance.

where the $+$ signs indicate the positions on the axis. Hence, the student's displacement (magnitude and direction) is 8.0 m in the positive x-direction, as indicated by the positive ($+$) result in Fig. 2.4b. (As in "regular" mathematics, the plus sign is often omitted, so it is understood, so this displacement can be written as $\Delta x = 8.0$ m instead of $\Delta x = +8.0$ m.)

Vector quantities in this book are usually indicated by boldface type with an overarrow; for example, a displacement vector is indicated by $\vec{d}$ or $\vec{x}$, and a velocity vector is indicated by $\vec{v}$. However, when working in one dimension, this notation is not needed and can be simplified by using plus and minus signs to indicate the only two possible directions. The x-axis is commonly used for horizontal motions, and a plus ($+$) sign is taken to indicate the direction to the right, or in the "positive x-direction," and a minus ($-$) sign indicates the direction to the left, or in the "negative x-direction."

Note: For displacement in *one direction*, the distance is the magnitude of the displacement.

Keep in mind that these signs only "point" in *particular directions*. An object moving along the negative x-axis toward the origin would be moving in the positive x-direction, even though its value is negative. How about an object moving along the positive x-axis toward the origin? If you said in the negative x-direction, you are correct. (In art figures, vector arrows indicate directions with the associated magnitudes alongside.)

Suppose the other student in Fig. 2.4 walks from the physics lab (her initial position is different, $x_1 = +9.0$ m) to the end of the lockers (the final position is now $x_2 = +1.0$ m). Her displacement would be

$$\Delta x = x_2 - x_1 = +1.0 \text{ m} - (+9.0 \text{ m}) = -8.0 \text{ m}$$

The minus sign indicates that the direction of the displacement was in the negative x-direction or to the left in the figure. In this case, we say that the two students' displacements are equal (in magnitude) and opposite (in direction).

Velocity

Note: Don't confuse velocity (a vector) with speed (a scalar).

As we have seen, speed, like the distance it incorporates, is a scalar quantity—it has magnitude only. Another more descriptive quantity used to describe motion is *velocity*. Speed and velocity are often used synonymously in everyday conversation, but the terms have different meanings in physics. Speed is a scalar, and velocity is a vector—velocity has both magnitude and direction. Unlike speed (but like displacement), one-dimensional velocities can have both positive and negative values, indicating the only two possible directions.

Velocity tells you how fast something is moving *and* in which direction it is moving. And just as we can speak of average and instantaneous speeds, there are average and instantaneous velocities involving vector displacements. The **average velocity** is the displacement divided by the total travel time. In one dimension, this involves just motion along one axis, which we take to be the x-axis. In this case,

$$\text{average velocity} = \frac{\text{displacement}}{\text{total travel time}} \qquad (2.3)^*$$

$$\bar{v} = \frac{\Delta x}{\Delta t} = \frac{x_2 - x_1}{t_2 - t_1}$$

SI unit of velocity: meters per second (m/s)*

---

*Another common form of this equation is

$$\bar{v} = \frac{\Delta x}{\Delta t} = \frac{(x_2 - x_1)}{(t_2 - t_1)} = \frac{(x - x_o)}{(t - t_o)} = \frac{(x - x_o)}{t}$$

or, after rearranging,

$$x = x_o + \bar{v}t, \qquad (2.3)$$

where x_o is the initial position, x is the final position, and $\Delta t = t$ with $t_o = 0$. See Section 2.3 for more on this notation.

In the case of more than one displacement (such as for successive displacements), the average velocity is equal to the total or net displacement divided by the total time. The total displacement is found by adding the displacements algebraically according to the directional signs.

You might be wondering whether there is a relationship between average speed and average velocity. A quick look at Fig. 2.4 will show you that if all the motion is in one direction, that is, there is no reversal of direction, the distance is equal to the magnitude of the displacement. Then the average speed is equal to the magnitude of the average velocity. *However, be careful.* This set of relationships is not true if there is a reversal of direction, as Example 2.2 shows.

Teaching tip: Point out to students that average speed depends on path, whereas average velocity depends only on initial and final positions.

Example 2.2 ■ There and Back: Average Velocities

A jogger jogs from one end to the other of a straight 300-m track in 2.50 min and then jogs back to the starting point in 3.30 min. What was the jogger's average velocity (a) in jogging to the far end of the track, (b) coming back to the starting point, and (c) for the total jog?

Note: For displacements in both the + and − directions (reversal of direction), the distance is not the magnitude of the total displacement.

Thinking It Through. The average velocities are computed from the defining equation. Note that the times given are the Δt's associated with the particular displacements.

Solution. From the problem, we have:

Given: $\Delta x_1 = 300$ m (taking the initial direction as positive)
$\Delta x_2 = -300$ m (taking the direction of the return trip as negative)
$\Delta t_1 = 2.50$ min $(60 \text{ s/min}) = 150$ s
$\Delta t_2 = 3.30$ min $(60 \text{ s/min}) = 198$ s

Find: Average velocities for
(a) the first leg of the jog,
(b) the return jog,
(c) the total jog

(a) The jogger's average velocity for the trip down the track is found from Eq. 2.3:

$$\bar{v}_1 = \frac{\Delta x_1}{\Delta t_1} = \frac{+300 \text{ m}}{150 \text{ s}} = +2.00 \text{ m/s}$$

(b) Similarly, for the return trip, we have

$$\bar{v}_2 = \frac{\Delta x_2}{\Delta t_2} = \frac{-300 \text{ m}}{198 \text{ s}} = -1.52 \text{ m/s}$$

(c) For the total trip, there are two displacements to consider, down and back, so these are added together to get the total displacement, and then divided by the total time:

$$\bar{v}_3 = \frac{\Delta x_1 + \Delta x_2}{\Delta t_1 + \Delta t_2} = \frac{300 \text{ m} + (-300 \text{ m})}{150 \text{ s} + 198 \text{ s}} = 0 \text{ m/s}$$

The average velocity for the total trip is zero! Do you see why? Recall from the definition of displacement that the magnitude of displacement is the straight-line distance between two points. The displacement from one point back to the same point is zero; hence the average velocity is zero. (See ▸Fig. 2.5.)

The total or net displacement could have been found by simply taking $\Delta x = x_{\text{final}} - x_{\text{initial}} = 0 - 0 = 0$, where the initial and final positions are taken to be the origin, but it was done in parts here for illustration purposes.

Follow-Up Exercise. Find the jogger's average speed for each of the cases in this Example, and compare it with the magnitudes of the respective average velocities. [Will the average speed for part (c) be zero?] (*Answers to all Follow-Up Exercises are at the back of the text.*)

▲ **FIGURE 2.5** Back home again! Despite having covered nearly 110 m on the base paths, at the moment the runner slides through the batter's box (his original position) into home plate, his displacement is zero—at least, if he is a right-handed batter. No matter how fast he ran the bases, his average velocity for the round trip is also zero.

As Example 2.2 shows, average velocity provides only an overall description of motion. One way to take a closer look at motion is to take smaller time intervals; that is, to let the observation time interval (Δt) become smaller and smaller. As with speed, when Δt approaches zero, we obtain the **instantaneous velocity,** which describes how fast something is moving and in which direction *at a particular instant of time.*

Instantaneous velocity is defined mathematically as

$$v = \lim_{\Delta t \to 0} \frac{\Delta x}{\Delta t} \qquad (2.4)$$

This expression is read as "the instantaneous velocity is equal to the limit of $\Delta x / \Delta t$ as Δt goes to zero." The time interval does not ever equal zero (why?), but *approaches* zero. Instantaneous velocity is technically still an average velocity, but over such a small Δt that it is essentially an average "at an instant in time," which is why we call it the *instantaneous* velocity.

Note: The word *uniform* means "constant."

Uniform motion means motion with a constant velocity (constant magnitude *and* constant direction). As a one-dimensional example of this, the car in ▼Fig. 2.6 has a uniform velocity. It travels the same distance and experiences the same displacement in equal time intervals (50 km each hour), and the direction of its motion does not change.

Illustration 1.3 Getting Data Out

Graphical Analysis

Graphical analysis is often helpful in understanding motion and its related quantities. For example, the motion of the car in Fig. 2.6a may be represented on a plot of position versus time, or x versus t. As can be seen from Fig. 2.6b, a straight line is obtained for a uniform, or constant, velocity on such a graph.

▶ **FIGURE 2.6** Uniform linear motion—constant velocity In uniform linear motion, an object travels at a constant velocity, covering the same distance in equal time intervals. **(a)** Here, a car travels 50 km each hour. **(b)** An x-versus-t plot is a straight line, since equal displacements are covered in equal times. The numerical value of the slope of the line is equal to the magnitude of the velocity, and the sign of the slope gives its direction. (The average velocity equals the instantaneous velocity in this case. Why?)

Δx (km)	Δt (h)	$\Delta x / \Delta t$
50	1.0	50 km/1.0 h = 50 km/h
100	2.0	100 km/2.0 h = 50 km/h
150	3.0	150 km/3.0 h = 50 km/h

(a)

Illustration 2.1 Position and Displacement

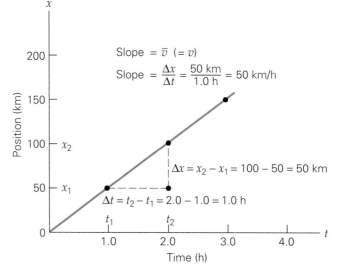

Uniform velocity

(b)

Exploration 2.1 Compare Position vs. Time and Velocity vs. Time Graphs

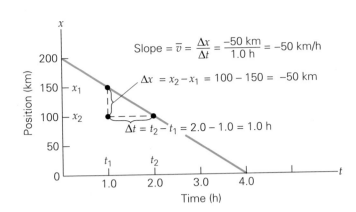

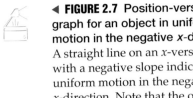

◀ **FIGURE 2.7 Position-versus-time graph for an object in uniform motion in the negative *x*-direction** A straight line on an *x*-versus-*t* plot with a negative slope indicates uniform motion in the negative *x*-direction. Note that the object's location changes at a constant rate. At *t* = 4.0 h, the object is at *x* = 0. How would the graph look if the motion continues for *t* > 4.0 h?

Recall from Cartesian graphs of *y* versus *x* that the slope of a straight line is given by $\Delta y/\Delta x$. Here, with a plot of *x* versus *t*, the slope of the line, $\Delta x/\Delta t$, is therefore equal to the average velocity $\bar{v} = \Delta x/\Delta t$. For uniform motion, this value is equal to the instantaneous velocity. That is, $\bar{v} = v$. (Why?) The numerical value of the slope is the magnitude of the velocity, and the sign of the slope gives the direction. A positive slope indicates that *x* increases with time, so the motion is in the positive *x*-direction. (The plus sign is often omitted as being understood, which we will do in general from here on.)

Suppose that a plot of position versus time for a car's motion is a straight line with a negative slope, as in ▲ Fig. 2.7. What does this indicate? As the figure shows, the position (*x*) values get smaller with time at a constant rate, indicating that the car is traveling in uniform motion, but now in the negative *x*-direction which correlates with the negative value of the slope.

In most instances, the motion of an object is *nonuniform*, meaning that different distances are covered in equal intervals of time. An *x*-versus-*t* plot for such motion in one dimension is a curved line, as illustrated in ▼Fig. 2.8. The average velocity of the object during any interval of time is the slope of a straight line between the two points on the curve that correspond to the starting and ending times of the interval. In the figure, since $\bar{v} = \Delta x/\Delta t$, the average velocity of the total trip is the slope of the straight line joining the beginning and ending points of the curve.

PHYSLET®

Exploration 2.2 Determine the Correct Graph

Teaching tip: Point out that the slope of a curve at a point is $\Delta x/\Delta t$, with units—not the angle of the tangent in degrees (a common error).

PHYSLET®

Illustration 2.2 Average Velocity

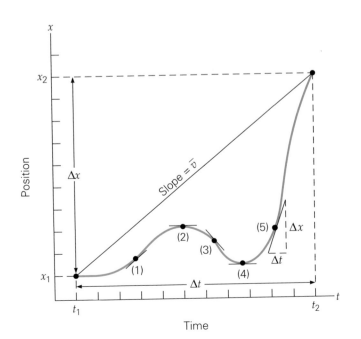

◀ **FIGURE 2.8 Position-versus-time graph for an object in nonuniform linear motion** For a nonuniform velocity, an *x*-versus-*t* plot is a curved line. The slope of the line between two points is the average velocity between those positions, and the instantaneous velocity is the slope of a line tangent to the curve at any point. Five tangent lines are shown, with the intervals for $\Delta x/\Delta t$ in the fifth. Can you describe the object's motion in words?

Illustration 2.3 Average and Instantaneous Velocity

The instantaneous velocity is equal to the slope of the tangent line to the curve at the time of interest. Five typical tangent lines are shown in Fig. 2.8. At (1), the slope is positive, and the motion is therefore in the positive x-direction. At (2), the slope of a horizontal tangent line is zero, so there is no motion for an instant. That is, the object has instantaneously stopped ($v = 0$) at that time. At (3), the slope is negative, so the object is moving in the negative x-direction. Thus, the object stopped in the process of changing direction at point (2). What is happening at points (4) and (5)?

Drawing various tangent lines along the curve, we see that their slopes vary, in magnitude and direction (sign), indicating that the instantaneous velocity is changing with time. An object in nonuniform motion can speed up, slow down, or change direction. How we describe motion with changing velocity is the topic of Section 2.3.

Demonstration/activity: An inclined air track or ramp and cart with a strobe light can be used to demonstrate accelerated motion.

2.3 Acceleration

OBJECTIVES: To (a) explain the relationship between velocity and acceleration, and (b) perform graphical analyses of acceleration.

The basic description of motion involves the time rate of change of position, which we call *velocity*. Going one step further, we can consider how this *rate of change* itself changes. Suppose that something is moving at a constant velocity and then the velocity changes. Such a change in velocity is called an *acceleration*. The gas pedal on an automobile is commonly called the *accelerator*. When you press down on the accelerator, the car speeds up; when you let up on the accelerator, the car slows down. In either case, there is a change in velocity with time. We define **acceleration** as the time rate of change of velocity.

Analogous to average velocity, the **average acceleration** is defined as the change in velocity divided by the time taken to make the change:

$$\text{average acceleration} = \frac{\text{change in velocity}}{\text{time to make the change}} \qquad (2.5)$$

$$\bar{a} = \frac{\Delta v}{\Delta t}$$

$$= \frac{v_2 - v_1}{t_2 - t_1} = \frac{v - v_o}{t - t_o}$$

SI unit of acceleration: meters per second squared (m/s^2)

Note that we have changed the initial and final variables to a more commonly used notation. That is, v_o and t_o are the initial or original velocity and time, respectively, and v and t are the general velocity and time at some time in the future, such as when you want to know the velocity v at a particular time t. (This may or may not be the final velocity of a particular situation.)

From $\Delta v/\Delta t$, the SI units of acceleration can be seen to be meters per second (Δv) per second (Δt), that is, (m/s)/s or m/(s·s), which is commonly expressed as meters per second squared (m/s^2). In the British system, the units are feet per second squared (ft/s^2).

Note: In compound units, multiplication is indicated by a dot.

Because velocity is a vector quantity, so is acceleration, since acceleration represents a change in velocity. Since velocity has both magnitude and direction, a change in velocity may involve changes in either or both of these factors. Thus an acceleration may result from a change in *speed* (magnitude), a change in *direction*, or a change in *both*, as illustrated in ▶Fig. 2.9.

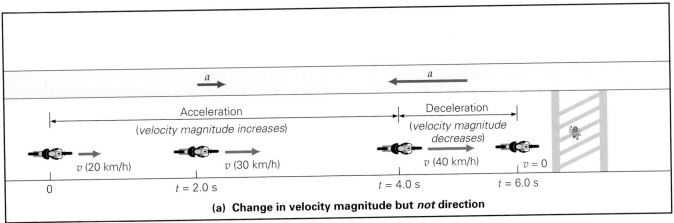

(a) Change in velocity magnitude but *not* direction

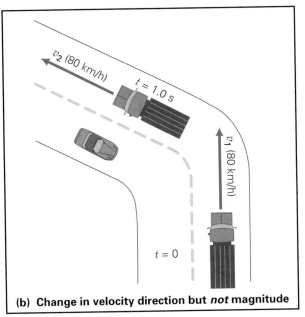

(b) Change in velocity direction but *not* magnitude

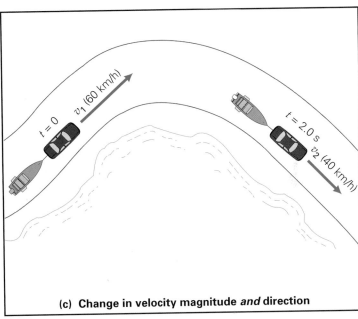

(c) Change in velocity magnitude *and* direction

▲ **FIGURE 2.9 Acceleration—the time rate of change of velocity** Since velocity is a vector quantity, with magnitude and direction, an acceleration can occur when there is **(a)** a change in magnitude, but not direction; **(b)** a change in direction, but not magnitude; or **(c)** a change in both magnitude and direction.

For straight-line, linear motion, plus and minus signs will be used to indicate the directions of velocity and acceleration, as was done for linear displacements. Eq. 2.5 is commonly simplified and written as

$$\bar{a} = \frac{v - v_{\text{o}}}{t} \tag{2.6}$$

where t_{o} is taken to be zero. (v_{o} may not be zero, so it cannot generally be omitted.)

Analogous to instantaneous velocity, **instantaneous acceleration** is the acceleration at a particular instant of time. This quantity is expressed mathematically as

$$a = \lim_{\Delta t \to 0} \frac{\Delta v}{\Delta t} \tag{2.7}$$

The conditions of the time interval approaching zero are the same here as described for instantaneous velocity.

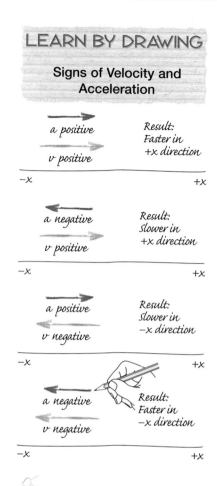

LEARN BY DRAWING

Signs of Velocity and Acceleration

a positive

v positive

Result:
Faster in
+x direction

−x   +x

a negative

v positive

Result:
Slower in
+x direction

−x   +x

a positive

v negative

Result:
Slower in
−x direction

−x   +x

a negative

v negative

Result:
Faster in
−x direction

−x   +x

Example 2.3 ■ Slowing It Down: Average Acceleration

A couple in a sport-utility vehicle (SUV) are traveling at 90 km/h on a straight highway. The driver sees an accident in the distance and slows down to 40 km/h in 5.0 s. What is the average acceleration of the SUV?

Thinking It Through. To find the average acceleration, the variables as defined in Eq. 2.6 must be given, and they are.

Solution. From the statement of the problem, we have the following data:

Given: $v_o = (90 \text{ km/h})\left(\dfrac{0.278 \text{ m/s}}{1 \text{ km/h}}\right)$ *Find:* $\bar{a}$ (average acceleration)

$\qquad\quad = 25 \text{ m/s}$

$\qquad v = (40 \text{ km/h})\left(\dfrac{0.278 \text{ m/s}}{1 \text{ km/h}}\right)$

$\qquad\quad = 11 \text{ m/s}$

$\qquad t = 5.0 \text{ s}$

[Here, the instantaneous velocities are assumed to be in the positive direction, and conversions to standard units (meters per second) are made right away, since it is noted that the time is given in seconds. In general, we always work with standard units.]

Given the initial and final velocities and the time interval, the average acceleration can be found by using Eq. 2.6:

$$\bar{a} = \frac{v - v_o}{t} = \frac{11 \text{ m/s} - (25 \text{ m/s})}{5.0 \text{ s}} = -2.8 \text{ m/s}^2$$

The minus sign indicates the direction of the (vector) acceleration. In this case, the acceleration is opposite to the direction of the motion ($v > 0$) and the car slows. Such an acceleration is sometimes called a *deceleration*, since the car is slowing.

Follow-Up Exercise. Does a negative acceleration necessarily mean that a moving object is slowing down (decelerating) or that its speed is decreasing? [*Hint*: See the accompanying Learn by Drawing.] (*Answers to all Follow-Up Exercises are at the back of the text.*)

Constant Acceleration

Although acceleration can vary with time, our study of motion will generally be restricted to constant accelerations for simplicity. (An important constant acceleration is the acceleration due to gravity near the Earth's surface, which will be considered in the next section.) Since for a constant acceleration, the average is equal to the constant value ($\bar{a} = a$), the bar over the acceleration in Eq. 2.6 may be omitted. Thus, for a constant acceleration, the equation relating velocity, acceleration, and time is commonly written (rearranging Eq. 2.6) as follows:

$$v = v_o + at \quad \textit{(constant acceleration only)} \tag{2.8}$$

(Note that the *at* term represents the *change* in velocity, since $at = v - v_o = \Delta v$.)

Example 2.4 ■ Fast Start, Slow Stop: Motion with Constant Acceleration

A drag racer starting from rest accelerates in a straight line at a constant rate of 5.5 m/s² for 6.0 s. (a) What is the racer's velocity at the end of this time? (b) If a parachute deployed at this time causes the racer to slow down uniformly at 2.4 m/s², how long will the racer take to come to a stop?

Thinking It Through. The racer first speeds up and then slows down, so close attention must be given to the directional signs of the vector quantities. Choose a coordinate system with the positive direction in the direction of the initial velocity. (Draw a sketch of the situation for yourself.) The answers can then be found by using the appropriate equations. Note that there are two different phases to the motion, and thus two different accelerations. Let's distinguish these phases with 1 and 2 subscripts.

Solution. Taking the initial motion to be in the positive direction, we have the following data:

Given: (a) $v_0 = 0$ (at rest)   **Find:** (a) v_1 (final velocity for first phase of motion)
    $a_1 = 5.5 \text{ m/s}^2$ (b) t_2 (time for second phase of motion)
    $t_1 = 6.0 \text{ s}$
  (b) $v_0 = v_1$ [from part(a)]
    $v_2 = 0$ (comes to stop)
    $a_2 = -2.4 \text{ m/s}^2$ (opposite direction of v_0)

The data have been listed in two parts. This practice helps avoid confusion with symbols. Note that the final velocity v_1 that is to be found in part (a) becomes the initial velocity v_0 for part (b).

(a) To find the final velocity v_1, we use Eq. 2.8 directly:

$$v_1 = v_0 + a_1 t_1 = 0 + (5.5 \text{ m/s}^2)(6.0 \text{ s}) = 33 \text{ m/s}$$

(b) Here we want to find time, so solving Eq. 2.6 for t_2 and using $v_0 = v_1 = 33 \text{ m/s}$ from part (a),

$$t_2 = \frac{v_2 - v_0}{a_2} = \frac{0 - (33 \text{ m/s})}{-2.4 \text{ m/s}^2} = 14 \text{ s}$$

Note that the time comes out positive, as it should.

Follow-Up Exercise. What is the racer's instantaneous velocity 10 s after the parachute is deployed? (*Answers to all Follow-Up Exercises are at the back of the text.*)

Teaching tip: Students may continue to confuse velocity with acceleration for some time. Consider a scalar analog of money and the rate at which money is lost or gained. Is it possible to be "broke," yet be losing or making money?

    Motions with constant accelerations are easy to represent graphically by plotting instantaneous velocity versus time. In this case, the v-versus-t plot is a straight line, the slope of which is equal to the acceleration, as illustrated in ▼Fig. 2.10. Note that Eq. 2.8 can be written as $v = at + v_0$, which, as you may recognize, has the form of an equation of a straight line, $y = mx + b$ (slope m and intercept b). In Fig. 2.10a, the motion

PHYSLET®

Illustration 2.5 Motion on a Hill or Ramp

▼ **FIGURE 2.10 Velocity-versus-time graphs for motions with constant accelerations** The slope of a v-versus-t plot is the acceleration. (a) A positive slope indicates an increase in the velocity in the positive direction. The vertical arrows to the right indicate how the acceleration adds velocity to the initial velocity v_0. (b) A negative slope indicates a decrease in the initial velocity v_0, or a deceleration. (c) Here a negative slope indicates a negative acceleration, but the initial velocity is in the negative direction, $-v_0$, so the speed of the object increases in that direction. (d) The situation here is initially similar to that of part (b) but ends up resembling that in part (c). Can you explain what happened at time t_1?

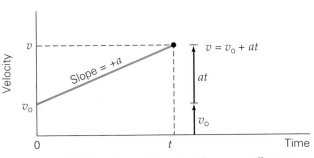

(a) Motion in positive direction—speeding up

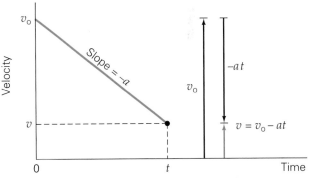

(b) Motion in positive direction—slowing down

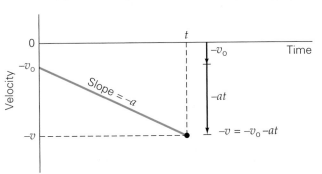

(c) Motion in negative direction—speeding up

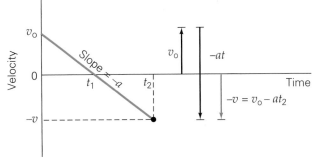

(d) Changing direction

Exploration 2.3 A Curtain Blocks Your View of a Golf Ball

is in the positive direction, and the acceleration term adds to the velocity after $t = 0$, as illustrated by the vertical arrows at the right of the graph. Here, the slope is positive ($a > 0$). In Fig. 2.10b, the negative slope ($a < 0$) indicates a negative acceleration that produces a slowing down, or deceleration. However, Fig. 2.10c illustrates how a negative acceleration can speed things up (for motion in the negative direction). The situation in Fig. 2.10d is slightly more complex. Can you explain what is happening there?

When an object moves at a constant acceleration, its velocity changes by the same amount in each equal time interval. For example, if the acceleration is 10 m/s^2 in the same direction as that of the initial velocity, the object's velocity increases by 10 m/s each second. As an example of this, suppose that the object has an initial velocity v_o of 20 m/s in a particular direction at $t_\text{o} = 0$. Then, for $t = 0$, 1.0, 2.0, 3.0, and 4.0 s, the magnitudes of the velocities are 20, 30, 40, 50, and 60 m/s, respectively.

The average velocity may be computed in the regular manner (Eq. 2.3), or you may immediately recognize that the uniformly increasing series of numbers 20, 30, 40, 50, and 60 has an average value of 40 (the midway value of the series) and $\bar{v} = 40$ m/s. Note that the average of the initial and final values also gives the average of the series—that is, $(20 + 60)/2 = 40$. Only when the velocity changes at a uniform rate because of a constant acceleration is $\bar{v}$ then the average of the initial and final velocities:

$$\bar{v} = \frac{v + v_\text{o}}{2} \quad \text{(constant acceleration only)} \qquad (2.9)$$

Example 2.5 ■ On the Water: Using Multiple Equations

A motorboat starting from rest on a lake accelerates in a straight line at a constant rate of 3.0 m/s^2 for 8.0 s. How far does the boat travel during this time?

Thinking It Through. We have only one equation for distance (Eq. 2.3, $x = x_\text{o} + \bar{v}t$), but this equation cannot be used directly. The average velocity must first be found, so multiple equations and steps are involved.

Solution. Reading the problem, summarizing the given data, and identifying what is to be found, we have the following (assuming the boat to accelerate in the $+x$-direction):

Given: $x_\text{o} = 0$ *Find:* x (distance)
        $v_\text{o} = 0$
        $a = 3.0 \text{ m/s}^2$
        $t = 8.0 \text{ s}$

(Note that all of the units are standard.)

In analyzing the problem, we might reason: To find x, we will need to use Eq. 2.3 in the form $x = x_\text{o} + \bar{v}t$. (The average velocity $\bar{v}$ must be used because the velocity is changing and thus not constant.) With time given, the solution to the problem then involves finding $\bar{v}$. By Eq. 2.9, $\bar{v} = (v + v_\text{o})/2$, and with $v_\text{o} = 0$, we need only find the final velocity v to solve the problem. Equation 2.8, $v = v_\text{o} + at$, enables us to calculate v from the given data. So we have the following:

The velocity of the boat at the end of 8.0 s is

$$v = v_\text{o} + at = 0 + (3.0 \text{ m/s}^2)(8.0 \text{ s}) = 24 \text{ m/s}$$

The average velocity over that time interval is

$$\bar{v} = \frac{v + v_\text{o}}{2} = \frac{24 \text{ m/s} + 0}{2} = 12 \text{ m/s}$$

Finally, the magnitude of the displacement, which in this case is the same as the distance traveled, is given by Eq. 2.3 (choosing the boat's initial location at the origin, so $x_\text{o} = 0$):

$$x = \bar{v}t = (12 \text{ m/s})(8.0 \text{ s}) = 96 \text{ m}$$

Follow-Up Exercise. (Sneak preview.) In Section 2.4, the following equation will be derived: $x = v_\text{o}t + \frac{1}{2}at^2$. Use the data in this Example to see if this equation gives the distance traveled. (*Answers to all Follow-Up Exercises are at the back of the text.*)

2.4 Kinematic Equations (Constant Acceleration)

<u>OBJECTIVES:</u> To (a) explain the kinematic equations of constant acceleration and (b) apply them to physical situations.

The description of motion in one dimension with constant acceleration requires only three basic equations. From previous sections, these equations are

$$x = x_o + \bar{v}t \qquad (2.3)$$

$$\bar{v} = \frac{v + v_o}{2} \qquad \text{(constant acceleration only)} \qquad (2.9)$$

$$v = v_o + at \qquad \text{(constant acceleration only)} \qquad (2.8)$$

(Keep in mind that the first equation, Eq. 2.3, is general and is not limited to situations in which there is constant acceleration, as the latter two equations are.)

However, as Example 2.5 showed, the description of motion in some instances requires multiple applications of these equations, which may not be obvious at first. It would be helpful if there were a way to reduce the number of operations in solving kinematic problems, and there is—by combining equations algebraically.

For instance, suppose we want an expression that gives location x in terms of time and acceleration rather than in terms of time and average velocity (as in Eq. 2.3). We can eliminate v from Eq. 2.3 by substituting for v from Eq. 2.9 into Eq. 2.3:

$$x = x_o + \bar{v}t$$

and

$$x = x_o + \tfrac{1}{2}(v + v_o)t \qquad \text{(constant acceleration only)} \qquad (2.10)$$

Then, substituting for v from Eq. 2.8 gives

$$x = x_o + \tfrac{1}{2}(v_o + at + v_o)t$$

Simplifying,

$$x = x_o + v_o t + \tfrac{1}{2}at^2 \qquad \text{(constant acceleration only)} \qquad (2.11)$$

Essentially, this series of steps was done in Example 2.5. The combined equation allows the displacement of the motorboat in that Example to be computed directly:

$$x - x_o = \Delta x = v_o t + \tfrac{1}{2}at^2 = 0 + \tfrac{1}{2}(3.0 \text{ m/s}^2)(8.0 \text{ s})^2 = 96 \text{ m}$$

Much easier, isn't it?

We may want an expression that gives velocity as a function of position x rather than time (as in Eq. 2.8). We can eliminate t from Eq. 2.8 by using Eq. 2.10 in the form

$$v + v_o = 2\frac{(x - x_o)}{t}$$

Then, multiplying this equation by Eq. 2.8 in the form $(v - v_o) = at$ yields

$$(v + v_o)(v - v_o) = 2a(x - x_o)$$

and using the relationship $v^2 - v_o^2 = (v + v_o)(v - v_o)$, we have

$$v^2 = v_o^2 + 2a(x - x_o) \qquad \text{(constant acceleration only)} \qquad (2.12)$$

Teaching tip: Point out which variables are scalars and which are vectors.

Exploration 2.4 Set the x(t) of a Monster Truck

Note: $\Delta x = x - x_o$ is displacement, but with $x_o = 0$, as it often is, then $\Delta x = x$, and the value of the x position is the same as that of the displacement, which saves writing $\Delta x = x - x_o$ every time.

Illustration 2.4 Constant
Acceleration and Measurement

Problem-Solving Hint

Students in introductory physics courses are sometimes overwhelmed by the various kinematic equations. Keep in mind that equations and mathematics are the tools of physics. As any mechanic or carpenter will tell you, tools make your work easier as long as you are familiar with them and know how to use them. The same is true for physics tools.

Summarizing the equations for linear motion with *constant* acceleration, we have:

$$v = v_o + at \tag{2.8}$$

$$x = x_o + \tfrac{1}{2}(v + v_o)t \tag{2.10}$$

$$x = x_o + v_o t + \tfrac{1}{2}at^2 \tag{2.11}$$

$$v^2 = v_o^2 + 2a(x - x_o) \tag{2.12}$$

This set of equations is used to solve the majority of kinematic problems. (Occasionally we are interested in average speed or velocity, but as noted earlier, averages generally don't tell you a great deal.)

Note that each of the equations in the list has four or five variables. All but one of the variables in an equation must be known in order to be able to solve for what you are trying to find. Generally, we choose an equation with the unknown or wanted quantity. But, as pointed out, the other variables in the equation must be known. If they are not, then the wrong equation was chosen or another equation must be used to find the variable. (Another possibility is that not enough data is given to solve the problem, but that is not the case in this textbook.)

Always try to understand and visualize a problem. Listing the data as described in the suggested problem-solving procedure in Chapter 1 may help you decide which equation to use, by determining the known and unknown variables. Remember this approach as you work through the remaining Examples in the chapter. Also, don't overlook any *implied data*, an error illustrated by Example 2.6.

Exploration 2.5 Determine x(t) and
v(t) of the Lamborghini

Conceptual Example 2.6 ■ Something Is Wrong!

A student working a problem with a constantly accelerating object wants to find v, and is given that $v_o = 0$ and $t = 3.0$ s, but is not given the acceleration a. He observes the kinematic equations and decides, using $v = at$ and $x = \tfrac{1}{2}at^2$ (with $x_o = v_o = 0$), that the unknown a can be eliminated. With $a = v/t$ and $a = 2x/t^2$ and equating,

$$v/t = 2x/t^2$$

but x is not known, so he decides to use $x = vt$ to eliminate it, and

$$v/t = 2vt/t^2$$

Simplifying,

$$v = 2v \qquad \text{or} \qquad 1 = 2!$$

What's wrong here?

Reasoning and Answer. Obviously something is big-time wrong, and it goes back to the problem-solving procedure given in Section 1.7. Step 4 there states: *Determine which principle(s) and equation(s) are applicable to the situation.* Since only equations were used, one equation must not apply to the situation. On inspection and analyzing, this can be seen to be $x = vt$, which applies only to nonaccelerated motion, and hence doesn't apply to the problem.

Follow-Up Exercise. Given only v_o and t, is there any way to find v using the given kinematic equations? Explain. *(Answers to all Follow-Up Exercises are at the back of the text.)*

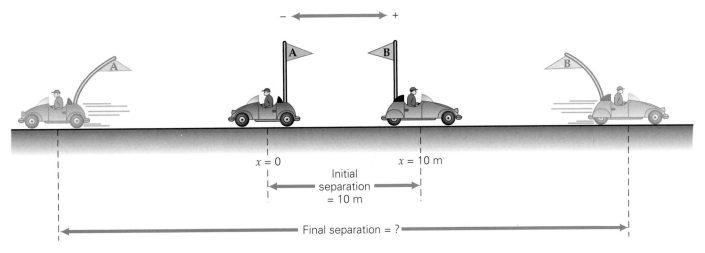

▲ **FIGURE 2.11** **Away they go!** Two dune buggies accelerate away from each other. How far are they apart at a later time? See Example 2.7.

Example 2.7 ■ Moving Apart: Where Are They Now?

Two riders on dune buggies sit 10 m apart on a long, straight track, facing in opposite directions. Starting at the same time, both riders accelerate at a constant rate of 2.0 m/s². How far apart will the dune buggies be at the end of 3.0 s?

Thinking It Through. We know only that the dune buggies are initially 10 m apart, so they can be positioned anywhere on the x-axis. It is convenient to place one at the origin so that one initial position (x_o) is zero. A sketch of the situation is shown in ▲Fig. 2.11.

Solution. Referring to the sketch, we have the following data:

Given: $x_{o_A} = 0$ *Find:* separation distance at $t = 3.0$ s
  $a_A = -2.0$ m/s²
  $t = 3.0$ s
  $x_{o_B} = 10$ m
  $a_B = 2.0$ m/s²

The displacement of each vehicle is given by Eq. 2.11 [the only displacement (Δx) equation with acceleration (a)]: $x = x_o + v_o t + \frac{1}{2}at^2$. But wait, we don't have v_o in the Given list. Some implied data may have been missed. We quickly note that $v_o = 0$ for both vehicles, so

$$x_A = x_{o_A} + v_{o_A}t + \tfrac{1}{2}a_At^2$$
$$= 0 + 0 + \tfrac{1}{2}(-2.0 \text{ m/s}^2)(3.0 \text{ s})^2 = -9 \text{ m}$$

and

$$x_B = x_{o_B} + v_{o_B}t + \tfrac{1}{2}a_Bt^2$$
$$= 10 \text{ m} + 0 + \tfrac{1}{2}(2.0 \text{ m/s}^2)(3.0 \text{ s})^2 = 19 \text{ m}$$

Hence vehicle A is 9 m to the left of the origin on the $-x$-axis, whereas vehicle B is at a position of 19 m to the right of the origin on the $+x$-axis. And so, the separation distance between the two dune buggies is 28 m.

Follow-Up Exercise. Would it make any difference in the separation distance if vehicle B had been initially put at the origin instead of vehicle A? Try it and find out. (*Answers to all Follow-Up Exercises are at the back of the text.*)

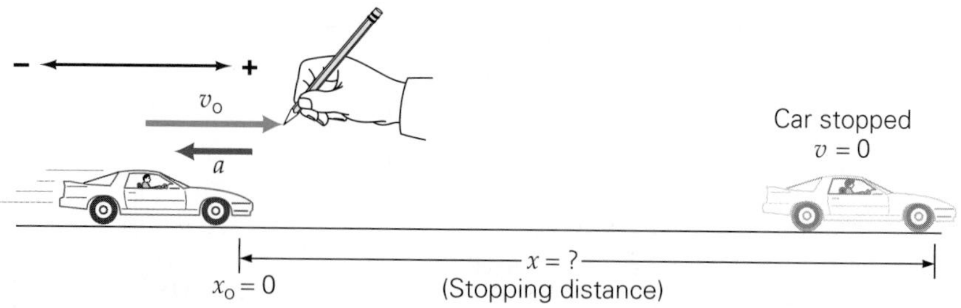

▲ **FIGURE 2.12 Vehicle stopping distance** A sketch to help visualize the situation in Example 2.8.

Example 2.8 ■ Putting On the Brakes: Vehicle Stopping Distance

The stopping distance of a vehicle is an important factor in road safety. This distance depends on the initial speed (v_0) and the braking capacity, which produces the deceleration, a, assumed to be constant. (We will take the sign of acceleration to be negative since it is opposite that of the velocity, which we will take to be positive. Thus, the car slows to a stop.) Express the stopping distance x in terms of these quantities.

Thinking It Through. Again, a kinematic equation is required, and the appropriate one is determined by listing what is given and what is to be found. Notice that the distance x is wanted, and time is not involved.

Solution. Here we are working with variables, so we can represent quantities only in symbolic form.

Given: v_0 (positive x direction)   *Find:*   stopping distance x
         a (< 0, opposite direction of v_0)   (in terms of the given variables)
         $v = 0$ (car comes to stop)
         $x_0 = 0$ (car taken to be initially at the origin)

Again, it is helpful to make a sketch of the situation, particularly when vector quantities are involved (▲Fig. 2.12). Since Eq. 2.12 has the variables we want, it should allow us to find the stopping distance x. Expressing the negative acceleration explicitly ($-a$) and assuming $x_0 = 0$ gives

$$v^2 = v_0^2 - 2ax$$

Since the vehicle comes to a stop ($v = 0$), we can solve for x:

$$x = \frac{v_0^2}{2a}$$

This equation gives us x expressed in terms of the vehicle's initial speed and stopping acceleration. Notice that the stopping distance x is proportional to the *square* of the initial speed. Doubling the initial speed therefore increases the stopping distance by a factor of 4 (for the same deceleration). That is, if the stopping distance is x_1 for an initial speed of v_1, then for a twofold increase in the initial speed ($v_2 = 2v_1$), the stopping distance would increase fourfold:

$$x_1 = \frac{v_1^2}{2a}$$

$$x_2 = \frac{v_2^2}{2a} = \frac{(2v_1)^2}{2a} = 4\left(\frac{v_1^2}{2a}\right) = 4x_1$$

We can get the same result by using ratios:

$$\frac{x_2}{x_1} = \frac{v_2^2}{v_1^2} = \left(\frac{v_2}{v_1}\right)^2 = 2^2 = 4$$

Do you think this consideration is important in setting speed limits, for example, in school zones? (The driver's reaction time should also be considered. A method for approximating a person's reaction time is given in Section 2.5.)

Follow-Up Exercise. Tests have shown that the Chevy Blazer has an average braking deceleration of 7.5 m/s², while that of a Toyota Celica is 9.2 m/s². Suppose these two vehicles are being driven down a straight, level road at 97 km/h (60 mi/h), with the Celica in front of the Blazer. A cat runs across the road ahead of them, and both drivers apply their brakes at the same time and come to safe stops (not hitting the cat). Assuming constant acceleration and the same reaction times for both drivers, what is the minimum safe tailgating distance for the Blazer so that there won't be a rear-end collision with the Celica when the two vehicles come to a stop? *(Answers to all Follow-Up Exercises are at the back of the text.)*

Graphical Analysis of Kinematic Equations

As was shown in Fig. 2.10, plots of v versus t gave straight-line graphs where the slopes were values of the constant accelerations. There is another interesting aspect of v-versus-t graphs. Consider the one shown in ►Fig. 2.13a, particularly the shaded area under the curve. Suppose we calculate the area of the shaded triangle, where, in general, $A = \frac{1}{2}ab$ $\left[\text{Area} = \frac{1}{2}(\text{altitude})(\text{base})\right]$.

For the graph in Fig. 2.13a, the altitude is v and the base is t, so $A = \frac{1}{2}vt$. But from the equation $v = v_0 + at$, we have $v = at$, where $v_0 = 0$ (zero intercept on graph). Therefore,

$$A = \tfrac{1}{2}vt = \tfrac{1}{2}(at)t = \tfrac{1}{2}at^2 = \Delta x$$

Hence, Δx, the displacement, is equal to the area under a v-versus-t curve.

Now look at Fig. 2.13b. Here, there is a nonzero value of v_0 at $t = 0$, so the object is initially moving. Consider the two shaded areas. We know that the area of the triangle is $A_2 = \frac{1}{2}at^2$, and the area of the rectangle can be seen (with $x_0 = 0$) to be $A_1 = v_0 t$. Adding these areas to get the total area yields

$$A_1 + A_2 = v_0 t + \tfrac{1}{2}at^2 = \Delta x$$

This is just Eq. 2.11 which is equal to the area under the v-versus-t curve.

2.5 Free Fall

To use the kinematic equations to analyze free fall.

One of the more common cases of constant acceleration is the acceleration due to gravity near the Earth's surface. When an object is dropped, its initial velocity (at the instant it is released) is zero. At a later time while falling, it has a nonzero velocity. There has been a change in velocity and thus, by definition, an acceleration. This **acceleration due to gravity (g)** near the Earth's surface has an approximate magnitude of

$$g = 9.80 \text{ m/s}^2 \quad \textit{(acceleration due to gravity)}$$

(or 980 cm/s²) and is directed downward (toward the center of the Earth). In British units, the value of g is about 32.2 ft/s².

The values given here for g are only approximate because the acceleration due to gravity varies slightly at different locations as a result of differences in elevation and regional average mass densities of the Earth. These small variations will be ignored in this book unless otherwise noted. (Gravitation is studied in more detail in Chapter 7.) Air resistance is another factor that affects (reduces) the acceleration of a falling object, but it too will be ignored here for simplicity. (The frictional effect of air resistance will be considered in Chapter 4.)

Objects in motion solely under the influence of gravity are said to be in **free fall**. The words *free fall* may bring to mind dropped objects. However, the term applies to any motion under the *sole* influence of gravity. Objects released from rest or

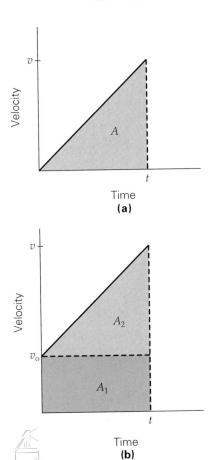

▲ **FIGURE 2.13** *v*-versus-*t* graphs, one more time (a) In the straight-line plot for a constant acceleration, the area under the curve is equal to *x*, the distance covered. (b) If v_0 is not zero, the distance is still given by the area under the curve, here divided into two parts, areas A_1 and A_2.

PHYSLET®

Exploration 2.8 Determine Area Under a(t) and v(t)

Demonstration/activity: Stand on top of a desk and drop objects of different masses at the same time. Air resistance has little effect on such things as coins, marbles, and stones, so they hit the floor at the same time. Even a tightly crumpled sheet of paper works.

thrown upward or downward are all in free fall once they are released. That is, after $t = 0$ (the time of release), only gravity is influencing the motion. (Even when an object projected upward is traveling upward, *it is still accelerating downward*.) Thus, the set of equations for motion in one dimension with constant acceleration can be used to describe free fall.

The acceleration due to gravity, g, has the same value for all free-falling objects, regardless of their mass or weight. It was once thought that heavier bodies accelerate faster than lighter bodies. This concept was part of Aristotle's theory of motion. You can easily observe that a coin accelerates faster than a sheet of paper when dropped simultaneously from the same height. But in this case, air resistance plays a noticeable role. If the paper is crumpled into a compact ball, it gives the coin a much better race. Similarly, a feather "floats" down much more slowly than a coin falls. However, in a near-vacuum, where there is negligible air resistance, the feather and the coin have the same acceleration—the acceleration due to gravity (▼ Fig. 2.14).

Astronaut David Scott performed a similar experiment on the Moon in 1971 by simultaneously dropping a feather and a hammer from the same height. He did not need a vacuum pump: The Moon has no atmosphere and therefore no air resistance. The hammer and the feather reached the lunar surface together, but both had a much smaller acceleration and fell at a slower rate than on Earth. The acceleration due to gravity near the Moon's surface is only about one sixth of that near the Earth's surface ($g_M \approx g/6$).

Currently accepted ideas about the motion of falling bodies are due in large part to Galileo. He challenged Aristotle's theory and experimentally investigated the motion of such objects. Legend has it that Galileo studied the accelerations of falling bodies by dropping objects of different weights from the top of the Leaning Tower of Pisa. (See the accompanying Insight on Galileo.)

Demonstration/activity: Drop a book and a piece of paper to show the effect of air friction. Place a sheet of paper on top of the book and another sheet on the bottom of the book, and drop the book flat side down. Explain your observation.

▼ **FIGURE 2.14 Free fall and air resistance (a)** When dropped simultaneously from the same height, a feather falls more slowly than a coin, because of air resistance. But when both objects are dropped in an evacuated container with a good partial vacuum, where air resistance is negligible, the feather and the coin both have the same constant acceleration. **(b)** An actual demonstration with multiflash photography: An apple and a feather are released simultaneously through a trap door into a large vacuum chamber, and they fall together—almost. Because the chamber has only a partial vacuum, there is still some air resistance. (Can you tell?)

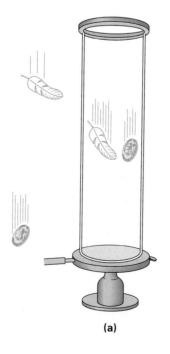

(a)

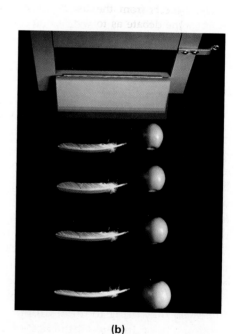

(b)

INSIGHT   2.1 GALILEO GALILEI AND THE LEANING TOWER OF PISA

FIGURE 1 Galileo Galileo is alleged to have performed free-fall experiments by dropping objects off the Leaning Tower of Pisa.

FIGURE 2 The Leaning Tower of Pisa The tower, constructed as a belfry for a nearby cathedral, was built on shifting subsoil. Construction began in 1173, and the tower started to shift one way and then the other before inclining to its present direction. Today, the tower leans about 5 m (16 ft) from the vertical at the top. It was closed in 1990, and efforts were made to stabilize and correct the leaning. Some improvement was made and the tower is now open to the public.

Galileo Galilei (▲Fig. 1) was born in Pisa, Italy, in 1564 during the Renaissance. Today, he is known throughout the world by his first name and is often referred to as the father of modern science and experimental physics, which attests to the magnitude of his scientific contributions.

One of Galileo's greatest contributions to science was the establishment of the scientific method—that is, investigation through experiment. In contrast, Aristotle's approach was based on logical deduction. By the scientific method, for a theory to be valid, it must correctly predict, or agree, with experimental results. If it doesn't, it is invalid or requires modification. Galileo said, "I think that in the discussion of natural problems we ought not to begin at the authority of places of Scripture, but at sensible experiments and necessary demonstrations."*

Probably the most popular and well-known legend about Galileo is that he performed experiments with falling bodies by dropping objects from the Leaning Tower of Pisa (▶Fig. 2). There is some debate as to whether Galileo actually did this, but there is little doubt that he questioned Aristotle's view on the motion of falling objects. In 1638, Galileo wrote,

> Aristotle says that an iron ball of one hundred pounds falling from a height of one hundred cubits reaches the ground before a one-pound ball has fallen a single cubit. I say that they arrive at the same time. You find, on making the experiment, that the larger outstrips the smaller by two finger-breadths, that is, when the larger has reached the ground, the other is short of it by two finger-breadths; now you would not hide behind these two fingers the ninety-nine cubits of Aristotle.[†]

*From *Growth of Biological Thought: Diversity, Evolution & Inheritance*, by F. Meyr (Cambridge, MA: Harvard University Press, 1982).

[†]From *Aristotle, Galileo, and the Tower of Pisa*, by L. Cooper (Ithaca, NY: Cornell University Press, 1935).

This and other writings show that Galileo was aware of the effect of air resistance.

The experiments at the Tower of Pisa were supposed to have taken place around 1590. In his writings of about that time, Galileo mentions dropping objects from a high tower, but never specifically names the Tower of Pisa. A letter written to Galileo from another scientist in 1641 describes the dropping of a cannonball and a musket ball from the Tower of Pisa. The first account of Galileo doing a similar experiment was written a dozen years after his death by Vincenzo Viviani, his last pupil and first biographer. It is not known whether Galileo told this story to Viviani in his declining years or Viviani created this picture of his former teacher.

The important point is that Galileo recognized (and probably experimentally showed) that free-falling objects fall with the same acceleration regardless of their mass or weight. (See Fig. 2.14.) Galileo gave no reason as to why all objects in free fall have the same acceleration, but Newton did, as you will learn in a later chapter.

It is customary to use y to represent the vertical direction and to take upward as positive (as with the vertical y-axis of Cartesian coordinates). Because the acceleration due to gravity is always downward, it is in the negative y-direction. This negative acceleration, $a = -g = -9.80 \text{ m/s}^2$, should be substituted into the equations of motion. However, the relationship $a = -g$ may be expressed explicitly in the equations for linear motion for convenience:

$$v = v_\text{o} - gt \tag{2.8'}$$

$$y = y_\text{o} + v_\text{o}t - \tfrac{1}{2}gt^2 \qquad \text{\textit{(Free-fall equations with} } a_y = -g \text{ \textit{expressed explicity)}} \tag{2.11'}$$

$$v^2 = v_\text{o}^2 - 2g(y - y_\text{o}) \tag{2.12'}$$

illustration 2.6 Free Fall

Equation 2.10 applies as well, but it does not contain g:

$$y = y_\text{o} + \tfrac{1}{2}(v + v_\text{o})t \tag{2.10'}$$

The origin ($y = 0$) of the frame of reference is usually taken to be at the initial position of the object. Writing $-g$ explicitly in the equations reminds you of its direction. Then the value of g is simply inserted as 9.80 m/s^2.

The equations can also be written with $a = g$, for example, $v = v_\text{o} + gt$, with the directional minus sign associated directly with g. In this case, a value of $g = -9.80 \text{ m/s}^2$ must be substituted for g each time. However, either method works, and *the choice is arbitrary.* Your instructor may prefer one method over the other.

Note that you must be explicit about the directions of vector quantities. The location y and the velocities v and v_o may be positive (up) or negative (down), but the acceleration due to gravity is always downward.

The use of these equations and the sign convention (with $-g$ explicitly expressed in the equations) are illustrated in the following Examples. (This convention will be used throughout the text.)

Example 2.9 ■ A Stone Thrown Downward: The Kinematic Equations Revisited

A boy on a bridge throws a stone vertically downward with an initial speed of 14.7 m/s toward the river below. If the stone hits the water 2.00 s later, what is the height of the bridge above the water?

Thinking It Through. This is a free-fall problem, but note that the initial velocity is downward, which is taken as negative. It is important to express this factor explicitly. Draw a sketch to help you analyze the situation if needed.

Solution. As usual, we first write what is given and what is to be found:

Given: $v_\text{o} = -14.7 \text{ m/s}$ (downward taken as the   *Find:* y (bridge height above water)
$t = 2.00 \text{ s}$ negative direction)
$g\ (= 9.80 \text{ m/s}^2)$

Notice that g is a positive number, since by our convention the directional minus sign has already been put into the previous equations of motion.

Which equation(s) will provide the solution using the given data? It should be evident that the distance the stone travels in an amount of time t is given directly by Eq. 2.11'. Taking $y_\text{o} = 0$:

$$y = v_\text{o}t - \tfrac{1}{2}gt^2 = (-14.7 \text{ m/s})(2.00 \text{ s}) - \tfrac{1}{2}(9.80 \text{ m/s}^2)(2.00 \text{ s})^2$$
$$= -29.4 \text{ m} - 19.6 \text{ m} = -49.0 \text{ m}$$

The minus sign indicates that the rock's displacement is downward, as it should be. Thus the height bridge is 49.0 m.

Follow-Up Exercise. How much longer would it take for the stone to reach the river if the boy in this Example had dropped the ball rather than thrown it? (*Answers to all Follow-Up Exercises are at the back of the text.*)

Reaction time is the time it takes a person to notice, think, and act in response to a situation—for example, the time between first observing and then responding to an obstruction on the road ahead by applying the brakes. Reaction time varies with the complexity of the situation (and with the individual). In general, the largest part of a person's reaction time is spent thinking, but practice in dealing with a given situation can reduce this time. The following Example gives a method for measuring reaction time.

Example 2.10 ■ Measuring Reaction Time: Free Fall

A person's reaction time can be measured by having another person drop a ruler (without warning) even with and through the first person's thumb and forefinger, as shown in ▶Fig. 2.15. After observing the unexpected release, the first person grasps the falling ruler as quickly as possible, and the length of the ruler below the top of the finger is noted. Suppose the ruler descends 18.0 cm before it is caught. What is the person's reaction time?

Thinking It Through. Both distance and time are involved. This observation indicates which kinematic equation should be used.

Solution. Notice that only the distance of fall is given. However, we know a couple of other things, such as v_o and g, so, taking $y_o = 0$:

Given: $y = -18.0 \text{ cm} = -0.180 \text{ m}$ *Find:* t (reaction time)
$v_o = 0$
$g \ (= 9.80 \text{ m/s}^2)$

(Note that the distance y has been converted to meters. Why?) We can see that Eq. 2.11′ applies here (with $v_o = 0$), giving us

$$y = -\tfrac{1}{2}gt^2$$

Solving for t gives

$$t = \sqrt{\frac{2y}{-g}} = \sqrt{\frac{2(-0.180 \text{ m})}{-9.80 \text{ m/s}^2}} = 0.192 \text{ s}$$

Try this experiment with a fellow student and measure your reaction time. Why do you think another person besides you should drop the ruler?

Follow-Up Exercise. A popular trick is to substitute a crisp dollar bill lengthwise for the ruler in Fig. 2.15, telling the person that he or she can have the dollar if able to catch it. Is this proposal a good deal? (The length of a dollar is 15.7 cm.) *(Answers to all Follow-Up Exercises are at the back of the text.)*

▲ **FIGURE 2.15 Reaction time** A person's reaction time can be measured by having the person grasp a dropped ruler. See Example 2.10.

Example 2.11 ■ Free Fall Up and Down: Using Implicit Data

A worker on a scaffold in front of a billboard throws a ball straight up. The ball has an initial speed of 11.2 m/s when it leaves the worker's hand at the top of the billboard (▼Fig. 2.16). (a) What is the maximum height the ball reaches relative to the top of the billboard? (b) How long does it take the ball to reach this height? (c) What is the position of the ball at $t = 2.00$ s?

Thinking It Through. In part (a), only the upward part of the motion has to be considered. Note that the ball stops (instantaneous zero velocity) at the maximum height, which allows us to determine this height. In part (b), knowing the maximum height, we can then determine the upward time of flight. In part (c), the distance–time equation (Eq. 2.11′) applies for any time and therefore allows us to calculate the position (y) of the ball relative to the launch point at $t = 2.00$ s.

(continues on next page)

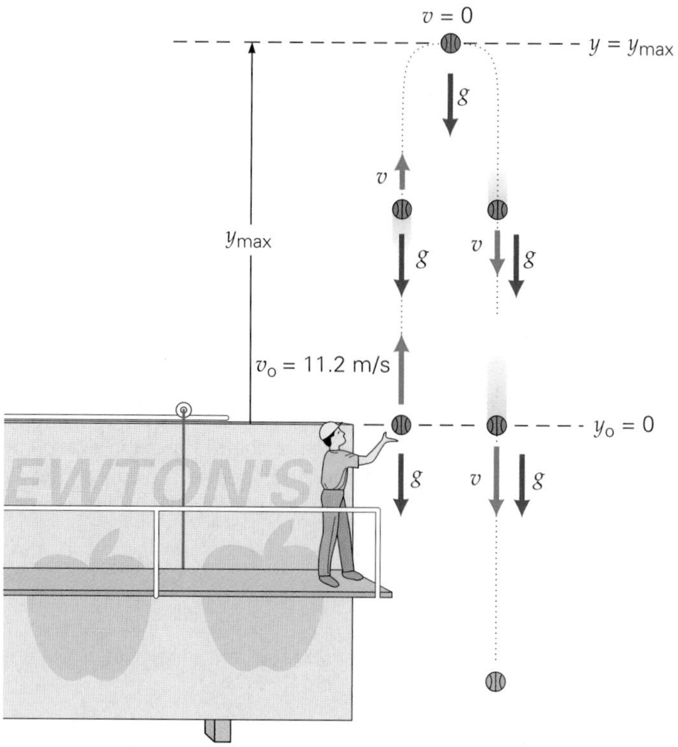

$v = 0$

$y = y_{max}$

g

v

g

v g

y_{max}

$v_o = 11.2$ m/s

$y_o = 0$

g v g

EWTON'S

▲ **FIGURE 2.16** **Free fall up and down** Note the lengths of the velocity and acceleration vectors at different times. (The upward and downward paths of the ball are horizontally displaced for illustration purposes.) See Example 2.11.

Solution. It might appear that all that is given is the initial velocity v_o at time t_o. However, a couple of other pieces of information are implied that should be understood. One, of course, is the acceleration g, and the other is the velocity at the maximum height where the ball stops. Here, in changing direction, the velocity of the ball is momentarily zero, so we have (again taking $y_o = 0$):

Given: $v_o = 11.2$ m/s   **Find:** (a) y_{max} (maximum height above launch point)
$g \, (= 9.80$ m/s^2)   (b) t_u (time upward)
$v = 0$ (at y_{max})   (c) y (at $t = 2.00$ s)
$t = 2.00$ s [for part (c)]

(a) Notice that we reference the height ($y = 0$) to the top of the billboard. For this part of the problem, we need be concerned with only the upward motion—a ball is thrown upward and stops at its maximum height y_{max}. With $v = 0$ at this height, y_{max} may be found directly from Eq. 2.12′ as

$$v^2 = 0 = v_o^2 - 2gy_{max}$$

So,

$$y_{max} = \frac{v_o^2}{2g} = \frac{(11.2 \text{ m/s})^2}{2(9.80 \text{ m/s}^2)} = 6.40 \text{ m}$$

relative to the top of the billboard ($y_o = 0$; see Fig. 2.16).

(b) The time the ball travels upward to its maximum height is designated t_u, that is, the time it takes for the ball to reach y_{max}, where $v = 0$. Since we know v_o and v, we can find the time t_u directly from Eq. 2.8′ as

$$v = 0 = v_o - gt_u$$

So,

$$t_u = \frac{v_o}{g} = \frac{11.2 \text{ m/s}}{9.80 \text{ m/s}^2} = 1.14 \text{ s}$$

Teaching tip: A graph of velocity vs. time for the motion in Example 2.11 can be used to show that the acceleration (that is, the slope) at the high point is ($-g$) even though the velocity is zero.

(c) The height of the ball at $t = 2.00$ s is given directly by Eq. 2.11′:

$$y = v_o t - \tfrac{1}{2}gt^2$$
$$= (11.2 \text{ m/s})(2.00 \text{ s}) - \tfrac{1}{2}(9.80 \text{ m/s}^2)(2.00 \text{ s})^2 = 22.4 \text{ m} - 19.6 \text{ m} = 2.8 \text{ m}$$

Note that this height is 2.8 m above, or measured upward from, the reference point ($y_o = 0$). The ball has reached its maximum height and is on its way back down.

Considered from another reference point, the situation in part (c) can be analyzed by imagining dropping a ball from a height of y_{max} above the top of the billboard with $v_o = 0$ and asking how far it falls in a time $t = 2.00$ s $- t_u = 2.00$ s $- 1.14$ s $= 0.86$ s. The answer is (with $y_o = 0$ at the maximum height)

$$y = v_o t - \tfrac{1}{2}gt^2 = 0 - \tfrac{1}{2}(9.80 \text{ m/s}^2)(0.86 \text{ s})^2 = -3.6 \text{ m}$$

This height is the same as the position found previously, but is measured with respect to the maximum height as the reference point; that is,

$$y_{max} - 3.6 \text{ m} = 6.4 \text{ m} - 3.6 \text{ m} = 2.8 \text{ m}$$

above the starting point.

Follow-Up Exercise. At what height does the ball in this Example have a speed of 5.00 m/s? [*Hint*: The ball attains this height twice—once on the way up, and once on the way down.] *(Answers to all Follow-Up Exercises are at the back of the text.)*

Here are some interesting facts about free-fall motion of an object thrown upward in the absence of air resistance. First, if the object returns to its launch elevation, then the times of flight upward and downward are the same. Similarly, note that at the very top of the trajectory, the object's velocity is zero for an instant, but the acceleration (even at the top) remains a constant 9.8 m/s² downward. If the acceleration went to zero, the object would remain there, as if gravity had been turned off!

Finally, the object returns to the starting point with the same speed as that at which it was launched. (The velocities have the same magnitude, but are opposite in direction.)

Teaching tip: Because of the confusion between velocity and acceleration, students tend to think that an object thrown upward has an acceleration of zero at its maximum height. Explain that gravity still exists at that point and the velocity is still changing. At that instant, the velocity is zero merely because it is changing from a positive to a negative value as the object changes direction. If it were zero, it would stop at the top.

Problem-Solving Hint

When working projectile problems involving motions up and down, it is often convenient to divide the problem into two parts and consider each part separately. As seen in Example 2.11, for the upward part of the motion, the velocity is zero at the maximum height. A quantity of zero usually simplifies the calculations. Similarly, the downward part of the motion is analogous to that of an object dropped from a height where the initial velocity can be taken as zero.

However, as Example 2.11 shows, the appropriate equations may be used directly for any position or time of the motion. For instance, note in part (c) that the height was found directly for a time *after* the ball had reached the maximum height. The velocity of the ball at that time could also have been found directly from Eq. 2.8′, $v = v_o - gt$.

Also, note that the initial position was consistently taken as $y_o = 0$. This assumption is taken for convenience when the situation involves only one object (then $y_o = 0$ at $t_o = 0$). Using this convention can save a lot of time in writing and solving equations.

The same is true with only one object in horizontal motion: You can usually take $x_o = 0$ at $t_o = 0$. There are a couple of exceptions to this case, however: first, if the problem specifies the object to be initially located at a position other than $x_o = 0$; and second, if the problem involves two objects, as in Example 2.7. In the latter case, if one object is taken to be initially at the origin, the other's initial position is not zero.

Example 2.12 ■ Free Fall on Mars

Exploration 2.7 Drop Two Balls; One
with a Delayed Drop

The Mars Polar Lander was launched in January 1999 and was lost near the Martian surface in December 1999. What became of the spacecraft is unknown. (See the Chapter 1 Insight, Is Unit Conversion Important?) Let's assume the retrorockets were fired early with the Lander coming to a stop and falling to the surface from a height of 40 m. (Highly unlikely, but let's assume this is correct.) Considering the spacecraft to be in free fall, with what speed did it hit the surface?

Thinking It Through. This appears to be analogous to a simple problem of dropping an object from a height—and it is, but this situation takes place on Mars. It was noted in this section that the acceleration due to gravity on the surface of the Moon is one sixth of that on the Earth. Acceleration due to gravity also varies for other planets. So we need to know g_{Mars}. Try looking in Appendix III. (The appendices contain a lot of useful information, so don't forget to check them out.)

Solution.

Given: $y = -40 \text{ m } (y_o = 0 \text{ again})$ *Find:* v (magnitude, speed)
$v_o = 0$
$g_{Mars} = (0.379)g = (0.379)(9.8 \text{ m/s}^2)$
$= 3.7 \text{ m/s}^2 \text{ (from Appendix III)}$

Then Eq. 2.12′ can be used:

$$v^2 = v_o^2 - 2g_{Mars}y = 0 - 2(3.7 \text{ m/s}^2)(-40 \text{ m})$$

So,

$$v^2 = 296 \text{ m}^2/\text{s}^2 \quad \text{and} \quad v = \sqrt{296 \text{ m}^2/\text{s}^2} = \pm 17 \text{ m/s}$$

This is the velocity, which we know is downward, so the negative root is selected, and $v = -17$ m/s. Since the speed is the magnitude of the velocity, it is 17 m/s.

Follow-Up Exercise. From the 40-m height, how long did the lander's descent take? Compute this using two different kinematic equations and compare the answers. (*Answers to all Follow-Up Exercises are at the back of the text.*)

Exploration 2.6 Toss the Ball to
Barely Touch the Ceiling

Chapter Review

- **Motion** involves a change of position; it can be described in terms of the distance moved (a scalar) or the displacement (a vector).

- A **scalar** quantity has magnitude (value and units) only; a **vector** quantity has magnitude *and* direction.

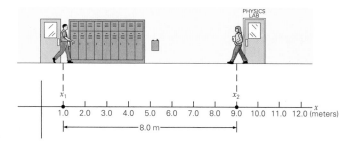

- **Average speed** $\bar{s}$ (a scalar) is the distance traveled divided by the total time:

$$\text{average speed} = \frac{\text{distance traveled}}{\text{total time to travel that distance}}$$

$$\bar{s} = \frac{d}{\Delta t} = \frac{d}{t_2 - t_1} \tag{2.1}$$

- **Average velocity** (a vector) is the displacement divided by the total travel time:

$$\text{average velocity} = \frac{\text{displacement}}{\text{total travel time}}$$

$$\bar{v} = \frac{\Delta x}{\Delta t} = \frac{x_2 - x_1}{t_2 - t_1} \quad \text{or} \quad x = x_o + \bar{v}t \tag{2.3}$$

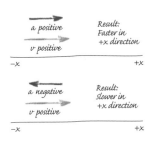

- **Instantaneous velocity** (a vector) describes how fast something is moving and in what direction at a particular instant of time.

- **Acceleration** is the time rate of change of velocity and hence is a vector quantity:

$$\text{average acceleration} = \frac{\text{change in velocity}}{\text{time to make the change}}$$

$$\bar{a} = \frac{\Delta v}{\Delta t} = \frac{v_2 - v_1}{t_2 - t_1} \qquad (2.5)$$

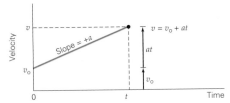

Motion in positive direction — speeding up

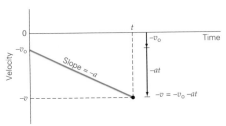

Motion in negative direction — speeding up

- **The kinematic equations for *constant* acceleration:**

$$\bar{v} = \frac{v + v_o}{2} \qquad (2.9)$$

$$v = v_o + at \qquad (2.8)$$

$$x = x_o + \frac{1}{2}(v + v_o)t \qquad (2.10)$$

$$x = x_o + v_o t + \frac{1}{2}at^2 \qquad (2.11)$$

$$v^2 = v_o^2 + 2a(x - x_o) \qquad (2.12)$$

- An object in **free fall** has a constant acceleration of magnitude $g = 9.80 \text{ m/s}^2$ (acceleration due to gravity) near the surface of the Earth.

- Expressing $a = -g$ in the kinematic equations for constant acceleration in the y-direction yields the following:

$$v = v_o - gt \qquad (2.8')$$

$$y = y_o + \frac{1}{2}(v + v_o)t \qquad (2.10')$$

$$y = y_o + v_o t - \frac{1}{2}gt^2 \qquad (2.11')$$

$$v^2 = v_o^2 - 2g(y - y_o) \qquad (2.12')$$

Exercises

MC = *Multiple Choice Question,* **CQ** = *Conceptual Question, and* **IE** = *Integrated Exercise. Throughout the text, many exercise sections will include "paired" exercises. These exercise pairs, identified with* **red numbers***, are intended to assist you in problem solving and learning. In a pair, the first exercise (even numbered) is worked out in the Study Guide so that you can consult it should you need assistance in solving it. The second exercise (odd numbered) is similar in nature, and its answer is given at the back of the book.*

**2.1 Distance and Speed: Scalar Quantities *and*
2.2 One-Dimensional Displacement and Velocity: Vector Quantities**

1. **MC** A scalar quantity has (a) only magnitude, (b) only direction, (c) both magnitude and direction. (a)

2. **MC** What can be said about distance traveled relative to the magnitude of displacement? (a) greater than, (b) equal to, (c) both a and b. (c)

3. **MC** A vector quantity has (a) only magnitude, (b) only direction, (c) both direction and magnitude. (c)

4. **MC** What can be said about average speed relative to the magnitude of the average velocity? (a) greater than, (b) equal to, (c) both a and b. (c)

5. **CQ** Can the displacement of a person's trip be zero, yet the distance involved in the trip be nonzero? How about the reverse situation? Explain. see ISM

6. **CQ** You are told that a person has walked 750 m. What can you safely say about the person's final position relative to the starting point? no final position can be given; may be 0 to 750 m from start

7. **CQ** If the displacement of an object is 300 m north, what can you say about the distance traveled by the object?
    the distance traveled is greater than or equal to 300 m

8. **CQ** Speed is the magnitude of velocity. Is average speed the magnitude of average velocity? Explain. no, see ISM

9. **CQ** The average velocity of a jogger on a straight track is computed to be +5 km/h. Is it possible for the jogger's instantaneous velocity to be negative at any time during the jog? Explain. yes, see ISM

10. • What is the magnitude of the displacement of a car that travels half a lap along a circle that has a radius of 150 m? How about when the car travels a full lap?
    300 m; zero

11. • A student throws a rock straight upward at shoulder level, which is 1.65 m above the ground. What is the displacement of the rock when it hits the ground?
    1.65 m down

12. • In 1999, the Moroccan runner Hicham El Guerrouj ran the 1-mi race in 3 min, 43.13 s. What was his average speed during the race in m/s? 7.2 m/s

13. • A senior citizen walks 0.30 km in 10 min, going around a shopping mall. (a) What is her average speed in meters per second? (b) If she wants to increase her average speed by 20% in walking a second lap, what would her travel time in minutes have to be?
    (a) 0.50 m/s (b) 8.3 min

14. •• A hospital patient is given 500 cc of saline by IV. If the saline is received at a rate of 4.0 mL/min, how long will it take for the half liter to run out? 125 min

15. •• A hospital nurse walks 25 m to a patient's room at the end of the hall in 0.50 min. She talks with the patient for 4.0 min, and then walks back to the nursing station at the same rate she came. What was the nurse's average speed? 0.17 m/s

16. •• On a cross-country trip, a couple drives 500 mi in 10 h on the first day, 380 mi in 8.0 h on the second day, and 600 mi in 15 h on the third day. What was the average speed for the whole trip? 45 mi/h

17. **IE** •• A car travels three quarters of a lap on a circular track of radius R. (a) The magnitude of the displacement is (1) less than R, (2) greater than R, but less than $2R$,

(3) greater than $2R$. (b) If $R = 50$ m, what is the magnitude of the displacement?
    (a) (2) greater than R, but less than $2R$ (b) 71 m

18. **IE** •• A race car travels a complete lap on a circular track of radius 500 m in 50 s. (a) The average velocity of the race car is (1) zero, (2) 100 m/s, (3) 200 m/s, (4) none of the preceding. Why? (b) What is the average speed of the race car? (a) (1) zero (b) 63 m/s

19. **IE** •• A student runs 30 m east, 40 m north, and 50 m west. (a) The magnitude of the student's net displacement is (1) between 0 and 20 m, (2) between 20 m and 40 m, (3) between 40 m and 60 m. (b) What is his net displacement? (a) (3) between 40 m and 60 m (b) 45 m at 27° west of north

20. •• A student throws a ball vertically upward such that it travels 7.1 m to its maximum height. If the ball is caught at the initial height 2.4 s after being thrown, (a) what is the ball's average speed, and (b) what is its average velocity?
    (a) 5.9 m/s (b) zero, displacement is zero

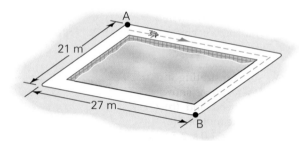

▲ **FIGURE 2.17 Speed versus velocity** See Exercise 21. (Not drawn to scale; insect is displaced for clarity.)

21. •• An insect crawls along the edge of a rectangular swimming pool of length 27 m and width 21 m (▼Fig. 2.17). If it crawls from corner A to corner B in 30 min, (a) what is its average speed, and (b) what is the magnitude of its average velocity? (a) 2.7 cm/s (b) 1.9 cm/s

22. •• Consider motion on the Earth's surface during one complete day. (a) What is the average velocity of a person on the Earth's equator? (b) What is the average speed of a person on the Earth's equator? (c) Compare these two results to a person located exactly at the Earth's North Pole. (a) zero (b) 4.7×10^3 m/s (c) both zero

23. •• A high school kicker makes a 30.0-yd field goal attempt (in American football) and hits the crossbar at a height of 10.0 ft. (a) What is the net displacement of the football from the time it leaves the ground until it hits the crossbar? (b) Assuming the football took 2.5 s to hit the crossbar, what was its average velocity? (c) Explain why you *cannot* determine its average speed from this data. (a) 90.6 ft, 6.3° above horizontal (b) 36.2 ft/s at 6.3° (c) see ISM

24. •• A plot of position versus time is shown in ▶Fig. 2.18 for an object in linear motion. (a) What are the average velocities for the segments AB, BC, CD, DE, EF, FG,

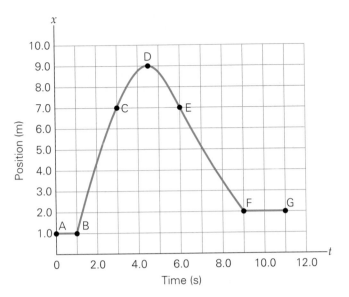

▲ **FIGURE 2.18 Position versus time** See Exercise 24.

and BG? (b) State whether the motion is uniform or nonuniform in each case. (c) What is the instantaneous velocity at point D? see ISM

25. ●● In demonstrating a dance step, a person moves in one dimension, as shown in ▼Fig. 2.19. What are (a) the average speed and (b) the average velocity for each phase of the motion? (c) What are the instantaneous velocities at $t = 1.0$ s, 2.5 s, 4.5 s, and 6.0 s? (d) What is the average velocity for the interval between $t = 4.5$ s and $t = 9.0$ s? [*Hint*: Recall that the overall displacement is the displacement between the starting point and the ending point.] see ISM

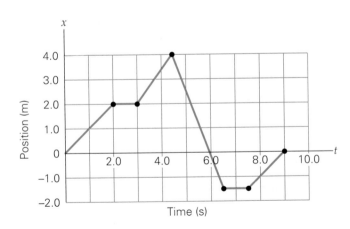

▲ **FIGURE 2.19 Position versus time** See Exercise 25.

26. ●● You can determine the speed of a car by measuring the time it takes to travel between mile markers on a highway. (a) How many seconds should elapse between two consecutive mile markers if the car's average speed is 70 mi/h? (b) What is the car's average speed if it takes 65 s to travel between the mile markers?
(a) 51 s (b) 55 mi/h

27. ●● Short hair grows at a rate of about 2.0 cm/month. A college student has his hair cut to a length of 1.5 cm. He will have it cut again when the length is 3.5 cm. How long will it be until his next trip to the barber shop? 1 month

28. ●●● A student driving home for the holidays starts at 8:00 AM to make the 675-km trip, practically all of which is on nonurban interstate highway. If she wants to arrive home no later than 3:00 PM, what must be her minimum average speed? Will she have to exceed the 65-mi/h speed limit? 59.9 mi/h; no

29. ●●● A regional airline flight consists of two legs with an intermediate stop. The airplane flies 400 km due north from airport A to airport B. From there, it flies 300 km due east to its final destination at airport C. (a) What is the plane's displacement from its starting point? (b) If the first leg takes 45 min and the second leg 30 min, what is the average velocity for the trip? (c) What is the average speed for the trip? (d) Why is the average speed not the same as the magnitude for the average velocity? see ISM

30. ●●● Two runners approaching each other on a straight track have constant speeds of 4.50 m/s and 3.50 m/s, respectively, when they are 100 m apart (▼Fig. 2.20). How long will it take for the runners to meet, and at what position will they meet if they maintain these speeds? 12.5 s, 56.3 m (relative to runner on left)

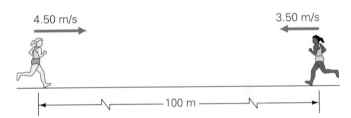

▲ **FIGURE 2.20 When and where do they meet?** See Exercise 30.

2.3 Acceleration

31. **MC** On a position-versus-time plot for an object that has a constant acceleration, the graph is (a) a horizontal line, (b) a nonhorizontal and nonvertical straight line, (c) a vertical line, (d) a curve. (d)

32. **MC** An acceleration may result from (a) an increase in speed, (b) a decrease in speed, (c) a change of direction, (d) all of the preceding. (d)

33. **MC** A negative acceleration can cause (a) an increase in speed, (b) a decrease in speed, (c) either a or b. (c)

34. **MC** The gas pedal of an automobile is commonly referred to as the *accelerator*. Which of the following might also be called an accelerator: (a) the brakes; (b) the steering wheel; (c) the gear shift; or (d) all of the preceding? Explain. (d)

35. **CQ** A car is traveling at a constant speed of 60 mi/h on a circular track. Is the car accelerating? Explain. yes, see ISM

36. **CQ** Does a fast-moving object always have higher acceleration than a slower object? Give a few examples, and explain. not necessarily, see ISM

37. **CQ** A classmate states that a negative acceleration always means that a moving object is decelerating. Is this statement true? Explain. not necessarily, see ISM

38. **CQ** Describe the motions of the two objects that have the velocity-versus-time plots shown in ▼Fig. 2.21. See ISM

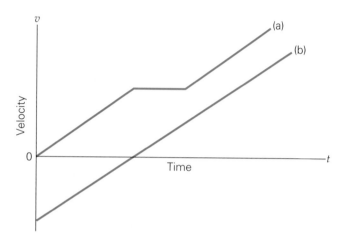

▲ **FIGURE 2.21** Description of motion
See Exercise 38.

39. **CQ** An object traveling at a constant velocity v_o experiences a constant acceleration in the same direction for a period of time t. Then an acceleration of equal magnitude is experienced in the opposite direction of v_o for the same period of time t. What is the object's final velocity? v_o

40. ● An automobile traveling at 15.0 km/h along a straight, level road accelerates to 65.0 km/h in 6.00 s. What is the magnitude of the auto's average acceleration? 2.32 m/s²

41. ● A sports car can accelerate from 0 to 60 mi/h in 3.9 s. What is the magnitude of the average acceleration of the car in meters per second squared? 6.9 m/s²

42. ● If the sports car in Exercise 41 can accelerate at a rate of 7.2 m/s², how long does the car take to accelerate from 0 to 60 mi/h? 3.7 s

43. **IE** ●● A couple is traveling by car 40 km/h down a straight highway. They see an accident in the distance, so the driver applies the brakes, and in 5.0 s the car slows down uniformly to rest. (a) The direction of the acceleration vector is (1) in the same direction as, (2) opposite to, (3) at 90° relative to the velocity vector. Why? (b) By how much must the velocity change

each second from the start of braking to the car's complete stop? (a) (2) opposite to velocity (b) −2.2 m/s each second, opposite direction of velocity

44. ●● A paramedic drives an ambulance at a constant speed of 75 km/h on a straight street for ten city blocks. Because of heavy traffic, the driver slows to 30 km/h in 6.0 s and travels two more blocks. What was the average acceleration of the vehicle? −2.1 m/s²

45. ●● With good tires and brakes, a car traveling 50 mi/h on dry pavement can travel 400 ft when the driver reacts to something he sees and brings the vehicle to a stop. If this is done uniformly, what is the car's acceleration? (These are actual conditions and 400 ft is about the length of a city block.) −2.0 m/s²

46. ●● A ball is thrown straight up at an initial speed of 9.8 m/s and, on returning to your hand, hits it moving downward at that same speed. If the whole trip took 2.0 s, determine the ball's (a) average acceleration and (b) average velocity. (a) −9.80 m/s² (b) 0

47. ●● After landing, a jetliner on a straight runway taxis to a stop at an average velocity of −35.0 km/h. If the plane takes 7.00 s to come to rest, what are the plane's initial velocity and acceleration?
−70.0 km/h or −19.4 m/s, +2.78 m/s²

48. ●● A train on a straight, level track has an initial speed of 35.0 km/h. A uniform acceleration of 1.50 m/s² is applied while the train travels 200 m. (a) What is the speed of the train at the end of this distance? (b) How long did it take for the train to travel the 200 m? (a) 26.3 m/s (b) 11.1 s

49. ●● A hockey puck sliding along the ice hits the boards head-on moving to the left with a speed of 35 m/s. As it reverses direction, it is in contact with the boards for 0.095 s, before rebounding at a slower speed of 11 m/s. Determine the average acceleration the puck experienced while hitting the boards. Typical car accelerations are 5 m/s². Comment on the size of your answer, and why it is so different from this value especially when the puck speeds are similar to car speeds. 4.8 × 10² m/s², see ISM

50. ●● What is the acceleration for each graph segment in ▼Fig. 2.22? Describe the motion of the object over the total time interval. $\bar{a}_{0-4} = 2.0\,\text{m/s}^2$; $\bar{a}_{4-10} = 0$; $\bar{a}_{10-18} = -1.0\,\text{m/s}^2$

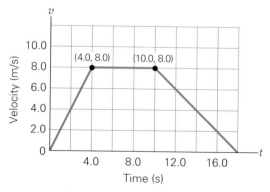

▲ **FIGURE 2.22** Velocity versus time
See Exercises 50 and 75.

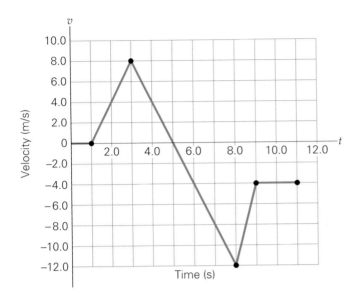

▲ **FIGURE 2.23 Velocity versus time** See Exercises 51 and 79.

51. ●● ▲Figure 2.23 shows a plot of velocity versus time for an object in linear motion. (a) Compute the acceleration for each phase of motion. (b) Describe how the object moves during the last time segment. see ISM

52. ●● A car initially traveling to the right at a steady speed of 25 m/s for 5.0 s applies its brakes and slows at a constant rate of 5 m/s² for 3.0 s. It then continues traveling to the right at a steady but slower speed with no additional braking for another 6.0 s. (a) To help with the calculations, make a sketch of the car's velocity versus time, being sure to show all three time intervals. (b) What is its velocity after the 3.0 s of braking? (c) What was its displacement during the total 14.0 s of its motion? (d) What was its average speed for the 14.0 s?
(a) see ISM (b) 10 m/s (c) 2.4 × 10² m/s (d) 17 m/s

53. ●●● A train normally travels at a uniform speed of 72 km/h on a long stretch of straight, level track. On a particular day, the train must make a 2.0-min stop at a station along this track. If the train decelerates at a uniform rate of 1.0 m/s² and, after the stop, accelerates at a rate of 0.50 m/s², how much time is lost because of stopping at the station? 150 s

2.4 Kinematic Equations (Constant Acceleration)

54. **MC** For a constant linear acceleration, the velocity-versus-time graph is (a) a horizontal line, (b) a vertical line, (c) a nonhorizontal and nonvertical straight line, (d) a curved line. (c)

55. **MC** For a constant linear acceleration, the position-versus-time graph would be (a) a horizontal line, (b) a vertical line, (c) a nonhorizontal and nonvertical straight line, (d) a curve. (d)

56. **MC** An object accelerates uniformly from rest for t seconds. The object's average speed for this time interval is (a) $\frac{1}{2}at$, (b) $\frac{1}{2}at^2$, (c) $2at$, (d) $2at^2$. (a)

57. **CQ** If an object's velocity-versus-time graph is a horizontal line, what can you say about the object's acceleration? it is zero

58. **CQ** In solving a kinematic equation for x, which has a negative acceleration, is x necessarily negative? not necessarily, see ISM

59. **CQ** How many variables must be known to solve a kinematic equation? all but one

60. **CQ** A classmate states that a negative acceleration always means that a moving object is decelerating. Is this statement true? Explain. no; negative acceleration would speed up an object having $-v$, or $v_o = 0$

61. ● At a sports-car rally, a car starting from rest accelerates uniformly at a rate of 9.0 m/s² over a straight-line distance of 100 m. The time to beat in this event is 4.5 s. Does the driver do it? If not, what must the minimum acceleration be to do so? no, 9.9 m/s²

62. ● A car accelerates from rest at a constant rate of 2.0 m/s² for 5.0 s. (a) What is the speed of the car at the end of that time? (b) How far does the car travel in this time? (a) 10 m/s (b) 25 m

63. ● A car traveling at 25 mi/h is to stop on a 35-m-long shoulder of the road. (a) What is the required magnitude of the minimum acceleration? (b) How much time will elapse during this minimum deceleration until the car stops?
(a) 1.8 m/s² (b) 6.3 s

64. ● A motorboat traveling on a straight course slows uniformly from 60 km/h to 40 km/h in a distance of 50 m. What is the boat's acceleration? −1.6 m/s²

65. ●● The driver of a pickup truck going 100 km/h applies the brakes, giving the truck a uniform deceleration of 6.50 m/s² while it travels 20.0 m. (a) What is the speed of the truck in kilometers per hour at the end of this distance? (b) How much time has elapsed?
(a) 81.4 km/h (b) 0.794 s

66. ●● An experimental rocket car starting from rest reaches a speed of 560 km/h after a straight 400-m run on a level salt flat. Assuming that acceleration is constant, (a) what was the time of the run, and (b) what is the magnitude of the acceleration? (a) 5.14 s (b) 30.3 m/s²

67. ●● A rocket car is traveling at a constant speed of 250 km/h on a salt flat. The driver gives the car a reverse thrust, and the car experiences a continuous and constant deceleration of 8.25 m/s². How much time elapses until the car is 175 m from the point where the reverse thrust is applied? Describe the situation(s) for your answer(s). 3.09 s and 13.7 s, see ISM

68. ●● Two identical cars capable of accelerating at 3.00 m/s² are racing on a straight track with running starts.

Car A has an initial speed of 2.50 m/s; car B starts with an initial speed of 5.00 m/s. (a) What is the separation of the two cars after 10 s? (b) Which car is moving faster after 10 s? (a) 25 m (b) car B

69. ●● According to Newton's laws of motion (which we will meet in Chapter 4) a frictionless 30° incline should provide an acceleration of 4.90 m/s² down the incline. A student with a stopwatch measures an object, starting from rest, to slide down a 15.00-m very smooth incline in exactly 3.00 s. Is the incline frictionless? no, see ISM

70. IE ●● An object moves in the +x-direction at a speed of 40 m/s. As it passes through the origin, it starts to experience a constant acceleration of 3.5 m/s² in the −x-direction. (a) What will happen next? (1) The object will reverse its direction of travel at the origin; (2) the object will keep traveling in the +x-direction; (3) the object will travel in the +x-direction and then reverses its direction. Why? (b) How much time elapses before the object returns to the origin? (c) What is the velocity of the object when it returns to the origin? see ISM

71. ●● A rifle bullet with a muzzle speed of 330 m/s is fired directly into a special dense material that stops the bullet in 25 cm. Assuming the bullet's deceleration to be constant, what is its magnitude? 2.2×10^5 m/s²

72. ●● The speed limit in a school zone is 40 km/h (about 25 mi/h). A driver traveling at this speed sees a child run onto the road 13 m ahead of his car. He applies the brakes, and the car decelerates at a uniform rate of 8.0 m/s². If the driver's reaction time is 0.25 s, will the car stop before hitting the child? yes; $x = 10.5$ m

73. ●● Assuming a reaction time of 0.50 s for the driver in Exercise 72, will the car stop before hitting the child? no, $x = 13.3$ m

74. ●● A bullet traveling horizontally at a speed of 350 m/s hits a board perpendicular to the surface, passes through it, and emerges on the other side at a speed of 210 m/s. If the board is 4.00 cm thick, how long does the bullet take to pass through it? 1.43×10^{-4} s

75. ●● (a) Show that the area under the curve of a velocity-versus-time plot for a constant acceleration is equal to the displacement. [*Hint*: The area of a triangle is $ab/2$, or one half the altitude times the base.] (b) Compute the distance traveled for the motion represented by Fig. 2.22. (a) $A = v_0 t + \frac{1}{2}at^2$ (b) 96 m

76. IE ●● An object initially at rest experiences an acceleration of 2.00 m/s² on a level surface. Under these conditions, it travels 6.00 m. Let's designate the first 3.00 m as phase 1 with a subscript 1 for those quantities, and the second 3.00 m as phase 2 with subscript 2. (a) The times for traveling each phase should be related by

which condition: (1) $t_1 < t_2$; (2) $t_1 = t_2$; or (3) $t_1 > t_2$? (b) Now calculate the two travel times and compare them quantitatively. (a) (3) $t_1 > t_2$ (b) 1.73 s, 0,718 s

77. IE ●● A car initially at rest experiences loss of its parking brake as it rolls down a straight hill with a constant acceleration of 0.850 m/s², and travels a total of 100 m. Let's designate the first half of the distance as phase 1 with subscript 1 for those quantities, and the second half as phase 2 with subscript 2. (a) The car's speeds at the end of each phase should be related by which condition: (1) $v_1 < \frac{1}{2}v_2$; (2) $v_1 = \frac{1}{2}v_2$; or (3) $v_1 > \frac{1}{2}v_2$? (b) Now calculate the two speeds and compare them quantitatively.
(a) (3) $v_1 > \frac{1}{2}v_2$ (b) 9.22 m/s, 13.0 m/s

78. ●● An object initially at rest experiences an acceleration of 1.5 m/s² for 6.0 s and then travels at that constant velocity for another 8.0 s. What is the object's average velocity over the 14-s interval? 7.1 m/s

79. ●●● Figure 2.23 shows a plot of velocity versus time for an object in linear motion. (a) What are the instantaneous velocities at $t = 8.0$ s and $t = 11.0$ s? (b) Compute the final displacement of the object. (c) Compute the total distance the object travels.
(a) −12 m/s; −4.0 m/s (b) −18 m (c) 50 m

80. IE ●●● (a) A car traveling at a speed of v can brake to an emergency stop in a distance x. Assuming all other driving conditions are all similar, if the traveling speed of the car doubles, the stopping distance will be (1) $\sqrt{2}x$, (2) $2x$, (3) $4x$. (b) A driver traveling at 40.0 km/h in a school zone can brake to an emergency stop in 3.00 m. What would be the braking distance if the car were traveling at 60.0 km/h? (a) (3) 4x (b) 6.75 m

81. ●●● A car accelerates horizontally from rest on a level road at a constant acceleration of 3.00 m/s². Down the road, it passes through two photocells ("electric eyes" designated by 1 for the first one and 2 for the second one) that are separated by 20.0 m. The time interval to travel this 20.0-m distance as measured by the electric eyes is 1.40 s. (a) Calculate the speed of the car as it passes *each* electric eye. (b) How far is it from the start to the first electric eye? (c) How long did it take the car to get to the first electric eye?
(a) 12.2 m/s, 16.4 m/s (b) 24.8 (c) 4.07 s

82. ●●● An automobile is traveling on a long straight highway at a steady 75.0 mi/h when the driver sees a wreck 150 m ahead. At that instant, she applies the brakes (ignore reaction time). Between her and the wreck are two different surfaces. First there is 100 m of ice (this is the Midwest!), where her deceleration is only 1.00 m/s². From then on she is on dry concrete, where her deceleration is a more normal 7.00 m/s². (a) What was her speed just after leaving the icy portion of the road? (b) How much total distance does it take her to stop? (c) What is the total time it took her to stop? (a) 30.4 m/s (b) 166 m (c) 7.44 s

2.5 Free Fall

Neglect air resistance in the following Exercises.

83. **MC** An object is thrown vertically upward. Which of the following statements is true: (a) Its velocity changes non-uniformly; (b) its maximum height is independent of the initial velocity; (c) its travel time upward is slightly greater than its travel time downward; (d) the speed on returning to its starting point is the same as its initial speed? (d)

84. **MC** The free-fall motion described in this section applies to (a) an object dropped from rest, (b) an object thrown vertically downward, (c) an object thrown vertically upward, (d) all of the preceding. (d)

85. **MC** A dropped object in free fall (a) falls 9.8 m each second, (b) falls 9.8 m during the first second, (c) has an increase in speed of 9.8 m/s each second, (d) has an increase in acceleration of 9.8 m/s² each second. (c)

86. **MC** An object is thrown straight upward. At its maximum height, (a) its velocity is zero, (b) its acceleration is zero, (c) both a and b. (a)

87. **MC** When an object is thrown vertically upward, it is accelerating on (a) the way up, (b) on the way down, (c) both a and b. (c)

88. **CQ** When a ball is thrown upward, what are its velocity and acceleration at its highest point?
zero and 9.8 m/s² downward

89. **CQ** Imagine you are in space far away from any planet, and you throw a ball as you would on the Earth. Describe the ball's motion.
The ball moves with a constant velocity

90. **CQ** A person drops a stone from the window of a building. One second later, she drops another stone. How does the distance between the stones vary with time?
increases

91. **CQ** How would free fall on the Moon differ from that on the Earth? $g_M = g_E/6$

92. ● A student drops a ball from the top of a tall building; the ball takes 2.8 s to reach the ground. (a) What was the ball's speed just before hitting the ground? (b) What is the height of the building?
(a) 27 m/s (b) 38 m

93. **IE** ● The time it takes for an object dropped from the top of cliff A to hit the water in the lake below is twice the time it takes for another object dropped from the top of cliff B to reach the lake. (a) The height of cliff A is (1) one half, (2) two, (3) four times that of cliff B. (b) If it takes 1.80 s for the object to fall from cliff A to the water, what are the heights of cliffs A and B?
(a) (3) four times (b) 15.9 m and 3.97 m

94. ● For the motion of a dropped object in free fall, sketch the general forms of the graphs of (a) v versus t and (b) y versus t. (a) linear, slope $-g$ (b) parabola

95. ● You can perform a popular trick by dropping a dollar bill (lengthwise) through the thumb and forefinger of a fellow student. Tell your fellow student to grab the dollar bill as fast as possible, and he or she can have the dollar if able to catch it. (The length of a dollar is 15.7 cm, and the average human reaction time is about 0.2 s. See Fig. 2.15.) Is this proposal a good deal? Justify your answer.
no, not a good deal

96. ● A boy throws a stone straight upward with an initial speed of 15 m/s. What maximum height will the stone reach before falling back down? 11 m

97. ● In Exercise 96, what would be the maximum height of the stone if the boy and the stone were on the surface of the Moon, where the acceleration due to gravity is only 1.67 m/s² ? 67 m

98. ●● The ceiling of a classroom is 3.75 m above the floor. A student tosses an apple vertically upward, releasing it 0.50 m above the floor. What is the maximum initial speed that can be given to the apple if it is not to touch the ceiling?
slightly less than 8.0 m/s

99. ●● The Petronas Twin Towers in Malaysia and the Chicago Sears Tower have heights of about 452 m and 443 m, respectively. If objects were dropped from the top of each, what would be the difference in the time it takes the objects to reach the ground? $\Delta t = 0.096$ s

100. ●● You throw a stone vertically upward with an initial speed of 6.0 m/s from a third-story office window. If the window is 12 m above the ground, find (a) the time the stone is in flight and (b) the speed of the stone just before it hits the ground. (a) 2.3 s (b) 16 m/s

101. **IE** ●● A Super Ball is dropped from a height of 4.00 m. Assuming the ball rebounds with 95% of its impact speed, (a) would the ball bounce to (1) less than 95%, (2) equal to 95.0%, or (3) more than 95% of the initial height? (b) How high will the ball go? (a) (1) less than 95% (b) 3.61 m

102. ●● In an indoor domed baseball stadium the ceiling is designed so that a batted ball could not possibly strike the ceiling. Assume a major league fastball maximum speed of 95.0 mi/h and that the wooden bat might decrease this to 80.0 mi/h. Assume that the ball leaves the bat at a height of 1.00 m above the playing field. (a) Determine the minimum ceiling height so that a major league "pop-up" (that is, a ball that leaves the bat moving straight up) does not hit the ceiling. (b) In a real game, a pop-up comes within 10.0 m of this ceiling height. What was the ball's speed as it left the bat? (a) 66.2 m (b) 32.9 m/s

103. ●● During the experiment mentioned in the text involving dropping a feather and a hammer on the Moon, both objects were released from a height of 1.30 m. According to the video of the experiment, they both took 1.26 s to hit the lunar surface. (a) What is the local value for acceleration due to gravity at that place on the Moon? (b) What speed did the two objects have just before hitting the surface?
(a) 1.64 m/s² (b) 2.07 m/s

104. •• In ▼Fig. 2.24, a student at a window on the second floor of a dorm sees his math professor walking on the sidewalk beside the building. He drops a water balloon from 18.0 m above the ground when the professor is 1.00 m from the point directly beneath the window. If the professor is 170 cm tall and walks at a rate of 0.450 m/s, does the balloon hit her? If not, how close does it come? hits 14 cm in front of the professor

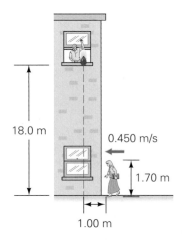

▲ **FIGURE 2.24 Hit the professor** See Exercise 104. (This figure is not drawn to scale.)

105. •• A photographer in a helicopter ascending vertically at a constant rate of 12.5 m/s accidentally drops a camera out the window when the helicopter is 60.0 m above the ground. (a) How long will the camera take to reach the ground? (b) What will its speed be when it hits? (a) 5.00 s (b) 36.5 m/s

106. IE •• The acceleration due to gravity on the Moon is about one sixth of that on the Earth. (a) If an object were dropped from the same height on the Moon and on the Earth, the time it would take to reach the surface on the Moon is (1) $\sqrt{6}$, (2) 6, (3) 36 times that it would take on the Earth. (b) For a projectile with an initial velocity of 18.0 m/s upward, what would be the maximum height and the total time of flight on the Moon and on the Earth? (a) (1) $\sqrt{6}$ (b) see ISM

107. ••• It takes 0.210 s for a dropped object to pass a window that is 1.35 m tall. From what height above the top of the window was the object released? (See ▶Fig. 2.25.) 1.49 m above the top of the window

108. ••• A tennis ball is dropped from a height of 10.0 m. It rebounds off the floor and comes up to a height of only 4.00 m on its first rebound. (Ignore the small amount of time the ball is in contact with the floor.) (a) Determine the ball's speed just before it hits the floor on the way down. (b) Determine the ball's speed as it leaves the floor on its way up to its first rebound height. (c) How

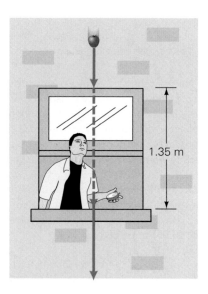

▲ **FIGURE 2.25 From where did it come?** See Exercise 107.

long is the ball in the air from the time it is dropped until the time it reaches its maximum height on the first rebound? (a) 14.0 m/s (b) 8.85 m/s (c) 2.33 s

109. ••• A pollution-sampling rocket is launched straight upward with rockets providing a constant acceleration of 12.0 m/s² for the first 1000 m of flight. At that point the rocket motors cut off and the rocket itself is in free fall. Ignore air resistance. (a) What is the rocket's speed when the engines cut off? (b) What is the maximum altitude reached by this rocket? (c) What is the time it takes to get to its maximum altitude?
(a) 155 m/s (b) 2.22 × 10³ m (c) 28.7 s

110. ••• A test rocket containing a probe to determine the composition of the upper atmosphere is fired vertically upward from an initial position at ground level. During the time t while its fuel lasts, the rocket ascends with a constant upward acceleration of magnitude $2g$. Assume that the rocket travels a small enough height that the Earth's gravitational force can be considered constant. (a) What are the speed and height of the rocket when its fuel runs out? (b) What is the maximum height the rocket reaches? (c) If $t = 30.0$ s, calculate the rocket's maximum height.
(a) $2gt$, gt^2 (b) $3gt^2$ (c) 2.65 × 10⁴ m

111. ••• A car and a motorcycle start from rest at the same time on a straight track, but the motorcycle is 25.0 m behind the car (▶Fig. 2.26). The car accelerates at a uniform rate of 3.70 m/s² and the motorcycle at a uniform rate of 4.40 m/s². (a) How much time elapses before the motorcycle overtakes the car? (b) How far will each have traveled during that time? (c) How far ahead of the car will the motorcycle be 2.00 s later? (Both vehicles are still accelerating.)
(a) 8.45 s (b) x_M = 157 m; x_C = 132 m (c) 13 m

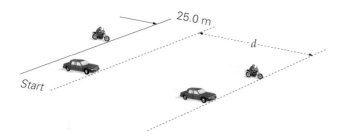

▲ **FIGURE 2.26** A tie race See Exercise 111. (This figure is not drawn to scale.)

Comprehensive Exercises

112. Two joggers run at the same average speed. Jogger A cuts directly north across the diameter of the circular track, while jogger B takes the full semicircle to meet his partner on the opposite side of the track. Assume their common average speed is 2.70 m/s and the track has a diameter of 150 m. (a) How many seconds ahead of jogger B does jogger A arrive? (b) How do their travel distances compare? (c) How do their displacements compare? (d) How do their average velocities compare? (a) 32.7 s (b) see ISM (c) 150 m north (d) 2.70 m/s north, 1.72 m/s north

113. Many highways with steep downhill areas have "runaway truck" inclined paths just off the main roadbed. These paths are designed so that if a vehicle's braking system gives out, the driver can steer it onto this incline (usually composed of loose gravel). The idea is that the vehicle can then roll up the incline and come permanently and safely to rest (in the gravel) with no need of a braking system. In one region of Hawaii the incline distance is 300 m and provides a (constant) deceleration of 2.50 m/s². (a) What is the maximum speed that a runaway vehicle can have as it enters the incline? (b) How long would such a vehicle take to come to rest? (c) Suppose another vehicle moving 10 mi/h (4.47 m/s) faster than the maximum value enters the incline. What speed will it have as it leaves the gravel-filled area? (a) 38.7 m/s (b) 15.5 s (c) 19.2 m/s

114. The world's tallest building is the Taipei 101 Tower in Taipei, Taiwan—a 509 m (1667-ft), 101-story building (▶Fig. 2.27). The outdoor observation deck is on the 89th floor, and two high-speed elevators that service it reach a peak speed of 1008 m/min on the way up and 610 m/min on the way down. Assuming these peak speeds are reached at the midpoint of the run and that the accelerations are constant for each leg of the runs, (a) what are the accelerations for the up and down runs? (b) How much longer is the trip down than the trip up? (a) 0.629 m/s², 0.232 m/s² (b) 34.5 s

115. From street level, Superman spots Lois Lane in trouble as the evil villain, Lex Luthor, drops her from near the top of the Empire State Building. At that very in-

▲ **FIGURE 2.27** A tall one–the tallest The Taipei 101 Tower in Taipei, Taiwan is the world's tallest building. With 101 stories, it has a height of 509 m (1671 ft). The tower was completed in 2004.

stant, the Man of Steel starts upward at a constant acceleration to attempt a midair rescue of Lois. Assuming she was dropped from a height of 300 m and that Superman can accelerate straight upward at 15 m/s², determine (a) how far Lois falls before he catches her, (b) how long Superman takes to reach her, and (c) their speeds at the instant he reaches her. Comment on whether these speeds might be a danger to Lois, who, being a mere mortal, might get hurt running into the impervious Man of Steel if the speeds are too great. (a) 119 m (b) 4.92 s (c) Lois: 48.2 m/s, Superman: 73.8 m/s

116. In the 1960s there was a contest for the car that could do the following two maneuvers (one right after the other) in the shortest *total* time: First, accelerate from rest to 100 mi/h (45.0 m/s), and then brake to a complete stop. (Ignore the reaction time correction that occurs between the speeding-up and slowing-down phases and assume that all accelerations are constant.) For several years, the winner was the "James Bond car," the Aston Martin. One year it won the contest when it took only a *total* of 15.0 seconds to perform these two tasks! Its braking acceleration (deceleration) was known to be an excellent 9.00 m/s². (a) Calculate the time it took during the braking phase. (b) Calculate the distance it took during the braking phase. (c) Calculate the car's acceleration during the speeding-up phase. (d) Calculate the distance it took to reach 100 mi/h. (a) 5.00 s (b) 113 m (c) 4.50 m/s² (d) 225 m

117. Let's investigate a possible vertical landing on Mars that includes two segments: free fall followed by a parachute deployment. Assume the probe is close to the surface, so the Martian acceleration due to gravity is constant at 3.00 m/s^2. Suppose the lander is initially moving vertically downward at 200 m/s at a height of 20 000 m above the surface. Neglect air resistance during the free-fall phase. Assume it first free-falls for 8 000 m. (The parachutes don't open until it is 12 000 m from the surface. See ▸Fig. 2.28.) (a) Determine the lander's speed at the end of the 8000-m free-fall drop. (b) At 12 000 m above the surface, the parachute deploys and the lander *immediately* begins to slow. If it can survive hitting the surface at up to 20.0 m/s, determine the minimum constant deceleration needed during this phase. (c) What is the total time taken to land from the original height of 20 000 m ?

(a) -297 m/s (b) 3.66 m/s^2 (c) 108 s

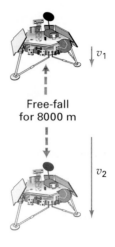

Free-fall
for 8000 m

Parachute
slowdown for
the last 12,000 m

Just above the Martian surface

▲ **FIGURE 2.28 Down she comes** See Exercise 117.

The following Physlet Physics Problems can be used with this chapter.
1.2, 1.3, 2.1, 2.2, 2.3, 2.4, 2.5, 2.6, 2.7, 2.8, 2.9, 2.10, 2.11, 2.12, 2.13, 2.14, 2.18

3

MOTION IN TWO DIMENSIONS

PHYSICS FACTS

- Word origins:
  - *kinematics*: from the Greek *kinema*, meaning "motion."
  - *velocity*: from the Latin *velocitas*, meaning "swiftness."
  - *acceleration*: from the Latin *accelerare*, meaning "hasten."
- Projectiles:
  - "Big Bertha," a gun used by the Germans in World War I, with a barrel length of 6.7 m (22 ft), could project an 820-kg (1800-lb) shell 15 km (9.3 mi).
  - The "Paris Gun", also used by the Germans in World War I, with a barrel length of 34 m (112 ft), could project a 120-kg (264-lb) shell 131 km (81 mi). Designed to bombard Paris, France, the shell reached a maximum height of 40 km (25 mi) during its 170-s trajectory.
  - To get maximum range on level ground, a projectile ideally should be projected at an angle of 45°. With air resistance, the speed of the projectile is reduced, and so is the range. The angle of projection for maximum range in this case is less than 45°, which gives a greater initial horizontal velocity component to help compensate for the air resistance.
  - In discus throwing, the discus is aerodynamic and some lift is provided. Therefore, for maximum range, a greater initial horizontal velocity component is wanted—traveling more distance horizontally while being vertically lifted.
- Discus distance records:
  - Women: 76.80 m (252 ft)
  - Men: 74.08 m (243 ft)
  - The men's discus has a mass of 2 kg (4.4 lb), and the women's discus a mass of 1 kg (2.2 lb).

You *can* get there from here! It's just a matter of knowing which way to head at the crossroads. But did you ever wonder why so many roads meet at right angles? There's a good reason. Living on the Earth's surface, we are used to describing locations in two dimensions, and one of the easiest ways to do this is by referring to a pair of mutually perpendicular axes. When you want to tell someone how to get to a particular place in the city, you might say, "Go uptown for four blocks and then cross town for three more blocks." In the country, it's "Go south for five miles and then another half mile east." Either way, though, you need to know how far to go in each of *two* directions that are 90° apart.

You can use the same approach to describe motion—and the motion doesn't have to be in a straight line. As you will shortly see, we can use a generalized version of vectors, introduced in Chapter 2, to describe motion in curved paths as well. Such analysis of *curvilinear* motion will eventually allow you to analyze the behavior of batted balls, planets circling the Sun, and even electrons in atoms.

Curvilinear motion can be analyzed by using rectangular components of motion. Essentially, you break down, or *resolve*, the curved motion into rectangular (*x* and *y*) components and look at the motion in both dimensions simultaneously. You can apply the kinematic equations introduced in Chapter 2 to these components. For an object moving in a curved path, for example, the *x*- and *y*-coordinates of its motion will then give the object's position at any time.

3.1 Components of Motion

OBJECTIVES: To (a) analyze motion in terms of its components, and (b) apply the kinematic equations to components of motion.

In Chapter 1, an object moving in a straight line was considered to be moving along one of the Cartesian axes (x or y). But what if the motion is not along an axis? For example, consider the situation illustrated in ▼Fig. 3.1. Here, three balls are moving uniformly across a tabletop. The ball rolling in a straight line along the side of the table, designated as the x-direction, is moving in one dimension. That is, its motion can be described with a single coordinate, x, as was

▼ **FIGURE 3.1 Components of motion** **(a)** The velocity (and displacement) for uniform, straight-line motion—that of the dark purple ball—may have x- and y-components (v_x and v_y as shown in the pencil drawing), because of the chosen orientation of the coordinate axes. Note that the velocity and displacement of the ball in the x-direction are exactly the same as those that a ball rolling along the x-axis with a uniform velocity of v_x would have. A comparable relationship holds true for the ball's motion in the y-direction. Since the motion is uniform, the ratio v_y/v_x (and therefore θ) is constant. **(b)** The coordinates (x, y) of the ball's position and the distance d the ball has traveled from the origin can be found at any time t.

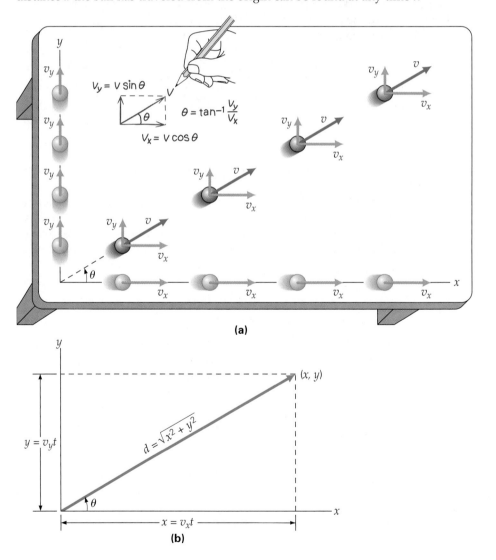

(a)

(b)

done for motions in Chapter 2. Similarly, the motion of the ball rolling in the y-direction can be described by a single y-coordinate. However, for this coordinate choice, both x- and y-coordinates are needed to describe the motion of the ball rolling diagonally across the table. We say that this motion is described *in two dimensions*.

You might observe that if the diagonally moving ball were the only object you had to consider, the x-axis could be chosen to be in the direction of that ball's motion, and the motion would thereby be reduced to one dimension. This observation is true, but once the coordinate axes are fixed, motions not along the axes must be described with two coordinates (x, y), or in two dimensions. Also, keep in mind that not all motions in a plane (two dimensions) are in straight lines. Think about the path of a ball you toss to another person. The path is curved for such projectile motion. (This motion will be considered in Section 3.3.) In the general case, both coordinates are needed.

In considering the motion of the ball moving diagonally across the table in Fig. 3.1a, we can think of it as moving in the x- and y-directions simultaneously. That is, it has a velocity in the x-direction (v_x) *and* a velocity in the y-direction (v_y) at the same time. The combined velocity components describe the actual motion of the ball. If the ball has a constant velocity v in a direction at an angle θ relative to the x-axis, then the velocities in the x- and y-directions are obtained by resolving, or breaking down, the velocity vector into **components of motion** in these directions, as shown in the pencil drawing in Fig. 3.1a. As this drawing shows, the v_x and v_y components have magnitudes of

Illustration 3.1 *Vector Decomposition*

$$v_x = v \cos \theta \qquad (3.1a)$$

and

$$v_y = v \sin \theta \qquad (3.1b)$$

respectively. (Notice that $v = \sqrt{v_x^2 + v_y^2}$, so v is a combination of the velocities in the x- and y-directions.)

You are familiar with the use of two-dimensional length components in finding the x- and y-coordinates in a Cartesian system. For the ball rolling on the table, its position (x, y), or the distance traveled from the origin in each of the component directions at time t, is given by (Eq. 2.11 with $a = 0$)

$$x = x_\text{o} + v_x t \quad \textit{Magnitudes of displacement components} \qquad (3.2a)$$
$$\textit{(under condition of constant velocity}$$
$$y = y_\text{o} + v_y t \quad \textit{and zero acceleration)} \qquad (3.2b)$$

respectively. (Here, the x_o and y_o are the ball's coordinates at $a = 0$, which may be other than zero.) The ball's straight-line distance from the origin is then $d = \sqrt{x^2 + y^2}$. (Fig. 3.1b).

Note that $\tan \theta = v_y/v_x$. (See the hand-drawn sketch in Fig. 3.1a.) So the direction of the motion relative to the x-axis is given by $\theta = \tan^{-1}(v_y/v_x)$. Also, $\theta = \tan^{-1}(y/x)$. Why?

In this introduction to components of motion, the velocity vector has been taken to be in the first quadrant ($0 < \theta < 90°$), where both the x- and y-components are positive. But, as will be shown in more detail in the next section, vectors may be in any quadrant, and one or both of their components can be negative. Can you tell in which quadrants the v_x or v_y components would be negative?

Example 3.1 ■ On a Roll: Using Components of Motion

If the diagonally moving ball in Fig. 3.1a has a constant velocity of 0.50 m/s at an angle of 37° relative to the x-axis, find how far it travels in 3.0 s by using x- and y-components of its motion.

Thinking It Through. Given the magnitude and direction (angle) of the velocity of the ball, we can find the x- and y-components of the velocity. Then we can compute the distance in each direction. Since the x- and y-axes are at right angles to each other, the Pythagorean theorem gives the distance of the straight-line path of the ball, as shown in Fig. 3.1b. (Note the procedure: Separate the motion into components, calculate what is needed in each direction, and recombine if necessary.)

Solution. Organizing the data, we have

Given: $v = 0.50$ m/s   *Find:* d (distance traveled)
$\qquad\quad\theta = 37°$
$\qquad\quad t = 3.0$ s

The distance traveled by the ball in terms of its x- and y-components is given by $d = \sqrt{x^2 + y^2}$. To find x and y as given by Eq. 3.2, we must first compute the velocity components v_x and v_y (Eq. 3.1):

$$v_x = v \cos 37° = (0.50 \text{ m/s})(0.80) = 0.40 \text{ m/s}$$
$$v_y = v \sin 37° = (0.50 \text{ m/s})(0.60) = 0.30 \text{ m/s}$$

Then, taking $x_0 = 0$ and $y_0 = 0$, the component distances are

$$x = v_x t = (0.40 \text{ m/s})(3.0 \text{ s}) = 1.2 \text{ m}$$

and

$$y = v_y t = (0.30 \text{ m/s})(3.0 \text{ s}) = 0.90 \text{ m}$$

and the distance of the path is

$$d = \sqrt{x^2 + y^2} = \sqrt{(1.2 \text{ m})^2 + (0.90 \text{ m})^2} = 1.5 \text{ m}$$

Follow-Up Exercise. Suppose that a ball is rolling diagonally across a table with the same speed as in this Example, but from the lower right corner, which is taken as the origin of the coordinate system, toward the upper left corner at an angle of 37° relative to the $-x$-axis. What would be the velocity components in this case? (Would the distance change?) *(Answers to all Follow-Up Exercises are at the back of the text.)*

Problem-Solving Hint

Note that for this simple case, the distance can also be obtained directly from $d = vt = (0.50 \text{ m/s})(3.0 \text{ s}) = 1.5$ m. However, we have solved this Example in a more general way to illustrate the use of components of motion. The direct solution would have been evident if the equations had been combined algebraically before calculation, that is, as

$$x = v_x t = (v \cos \theta)t$$

and

$$y = v_y t = (v \sin \theta)t$$

from which it follows that

$$d = \sqrt{x^2 + y^2} = \sqrt{(v \cos \theta)^2 t^2 + (v \sin \theta)^2 t^2} = \sqrt{v^2 t^2 (\cos^2 \theta + \sin^2 \theta)} = vt$$

Before embarking on the first solution strategy that occurs to you, pause for a moment to see whether there might be an easier or more direct way of approaching the problem.

Kinematic Equations for Components of Motion

Example 3.1 involved two-dimensional motion in a plane. With a constant velocity (constant components v_x and v_y), the motion is in a straight line. The motion may also be accelerated. For motion in a plane with a *constant acceleration* that has components a_x and a_y, the displacement and velocity components are given by the kinematic equations of Chapter 2 written separately for the x- and y-directions:

$$x = x_o + v_{x_o}t + \tfrac{1}{2}a_xt^2 \qquad (3.3a)$$

$$y = y_o + v_{y_o}t + \tfrac{1}{2}a_yt^2 \qquad (3.3b)$$

(constant acceleration only)

$$v_x = v_{x_o} + a_xt \qquad (3.3c)$$

$$v_y = v_{y_o} + a_yt \qquad (3.3d)$$

Kinematic equations for displacement and velocity components

If an object is initially moving with a constant velocity and suddenly experiences an acceleration in the direction of the velocity or opposite to it, it will continue in a straight-line path, either speeding up or slowing down, respectively.

If, however, the acceleration is at some angle other than 0° or 180° to the velocity vector, the motion is along a curved path. For the motion of an object to be *curvilinear*—that is, to vary from a straight-line path—an acceleration not parallel to the velocity is required. For such a curved path, the ratio of the velocity components varies with time. That is, the direction of the motion, $\theta = \tan^{-1}(v_y/v_x)$, varies with time, because one or both of the velocity components do.

Consider a ball initially moving along the x-axis, as illustrated in ▼Fig. 3.2. Assume that, starting at a time $t_o = 0$, the ball receives a constant acceleration a_y in

PHYSLET

Exploration 3.2 **Run the Gauntlet, Controlling x, v, and a**

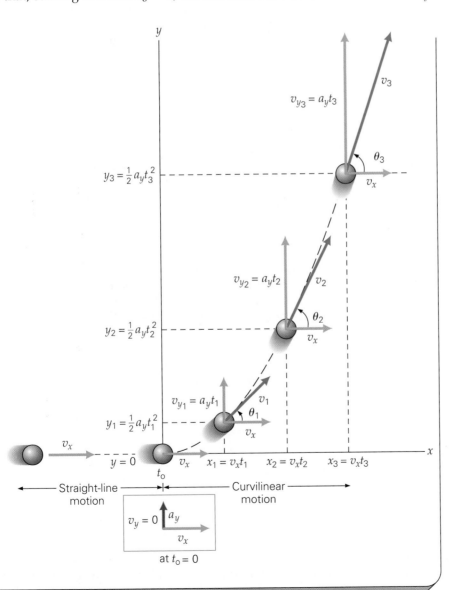

◄ **FIGURE 3.2** **Curvilinear motion** An acceleration not parallel to the instantaneous velocity produces a curved path. Here, an acceleration a_y is applied at $t_o = 0$ to a ball initially moving with a constant velocity v_x. The result is a curved path with the velocity components as shown. Notice how v_y increases with time, while v_x remains constant.

Illustration 3.3 The Direction of
Velocity and Acceleration Vectors

the y-direction. The magnitude of the x-component of the ball's displacement is given by $x = v_x t$; where the $\frac{1}{2}a_x t^2$ term of Eq. 3.3a drops out because there is no acceleration in the x-direction. Prior to t_o, the motion was in a straight line along the x-axis. But at any time after t_o, the y-coordinate is not zero, but is given by $y = \frac{1}{2}a_y t^2$ (Eq. 3.3b with $y_o = 0$ and $v_{y_o} = 0$). The result is a curved path for the ball.

Note that the length (magnitude) of the velocity component v_y changes with time, while that of the v_x component remains constant. The total velocity vector *at any time* is tangent to the curved path of the ball. It is at an angle θ relative to the positive x-axis, given by $\theta = \tan^{-1}(v_y/v_x)$, which now changes with time, as can be seen in Fig. 3.2 and in Example 3.2.

Example 3.2 ■ A Curving Path: Vector Components

Suppose that the ball in Fig. 3.2 has an initial velocity of 1.50 m/s along the x-axis. Starting at $t_o = 0$, the ball receives an acceleration of 2.80 m/s^2 in the y-direction. (a) What is the position of the ball 3.00 s after t_o? (b) What is the velocity of the ball at that time?

Thinking It Through. Keep in mind that the motions in the x- and y-directions can be analyzed independently. For part (a), simply compute the x- and y-positions at the given time, taking into account the acceleration in the y-direction. For part (b), find the component velocities, and vectorially combine them to get the total velocity.

Solution. Referring to Fig. 3.2, we have the following:

Given: $v_{x_o} = v_x = 1.50$ m/s *Find:* (a) (x, y) (position coordinates)
$v_{y_o} = 0$ (b) v (velocity, magnitude and direction)
$a_x = 0$
$a_y = 2.80$ m/s^2
$t = 3.00$ s

(a) At 3.00 s after $t_o = 0$, Eqs. 3.3a and 3.3b tell us that the ball has traveled the following distances from the origin ($x_o = y_o = 0$) in the x- and y-directions, respectively:

$$x = v_{x_o}t + \tfrac{1}{2}a_x t^2 = (1.50 \text{ m/s})(3.00 \text{ s}) + 0 = 4.50 \text{ m}$$
$$y = v_{y_o}t + \tfrac{1}{2}a_y t^2 = 0 + \tfrac{1}{2}(2.80 \text{ m/s}^2)(3.00 \text{ s})^2 = 12.6 \text{ m}$$

Thus, the position of the ball is $(x, y) = (4.50 \text{ m}, 12.6 \text{ m})$. If you had computed the distance $d = \sqrt{x^2 + y^2}$, what would you have gotten? (Note that this quantity is not the actual distance the ball has traveled in 3.00 s, but rather the magnitude of the *displacement*, or straight-line distance, from the origin to the position at $t = 3.00$ s.)

(b) The x-component of the velocity is given by Eq. 3.3c:

$$v_x = v_{x_o} + a_x t = 1.50 \text{ m/s} + 0 = 1.50 \text{ m/s}$$

(This component is constant, since there is no acceleration in the x-direction.) Similarly, the y-component of the velocity is given by Eq. 3.3d:

$$v_y = v_{y_o} + a_y t = 0 + (2.80 \text{ m/s}^2)(3.00 \text{ s}) = 8.40 \text{ m/s}$$

The velocity therefore has a magnitude of

$$v = \sqrt{v_x^2 + v_y^2} = \sqrt{(1.50 \text{ m/s})^2 + (8.40 \text{ m/s})^2} = 8.53 \text{ m/s}$$

and its direction relative to the $+x$-axis is

$$\theta = \tan^{-1}\left(\frac{v_y}{v_x}\right) = \tan^{-1}\left(\frac{8.40 \text{ m/s}}{1.50 \text{ m/s}}\right) = 79.9°$$

Follow-Up Exercise. Suppose that the ball in this Example also received an acceleration of 1.00 m/s^2 in the $+x$-direction starting at t_o. What would be the position of the ball 3.00 s after t_o in this case?

Note: Don't confuse the direction of the velocity with the direction of the displacement from the origin. The direction of the velocity is always tangent to the path.

Problem-Solving Hint

When using the kinematic equations, it is important to note that motion in the x- and y-directions can be analyzed independently—the factor connecting them being time t. That is, you can find (x, y) and/or (v_x, v_y) at a given time t. Also, keep in mind that we often set $x_o = 0$ and $y_o = 0$, which means that the object is located at the origin at $t_o = 0$. If the object is actually elsewhere at $t_o = 0$, then the values of x_o and/or y_o would have to be used in the appropriate equations. (See Eqs. 3.3a and b.)

3.2 Vector Addition and Subtraction

OBJECTIVES: To (a) learn vector notation, (b) be able to add and subtract vectors graphically and analytically, and (c) use vectors to describe motion in two dimensions.

Many physical quantities, including those describing motion, have a direction associated with them—that is, they are vectors. You have already worked with a few such quantities related to motion (displacement, velocity, and acceleration) and will encounter more during this course of study. A very important technique in the analysis of many physical situations is the addition (and subtraction) of vectors. By adding or combining such quantities (**vector addition**), you can obtain the overall, or net vector. This *resultant* vector is called the *vector sum.*

You have already been adding vectors. In Chapter 2, displacements in one dimension were added to get the net displacement. In this chapter, vector components of motion in two dimensions will be added to get net effects. Notice that in Example 3.2, the velocity components v_x and v_y were combined to get the resultant velocity.

In this section, we will look at vector addition and subtraction in general, along with common vector notation. As you will learn, these operations are not the same as scalar or numerical addition and subtraction, with which you are already familiar. Vectors have magnitudes *and* directions, so different rules apply.

In general, there are geometrical (graphical) methods and analytical (computational) methods of vector addition. The geometrical methods are useful in helping you visualize the concepts of vector addition, particularly with a quick sketch. Analytical methods are more commonly used, however, because they are faster and more precise.

In Section 3.1, we were concerned chiefly about vector components. The notation for the magnitudes of components was, for example, v_x and v_y. To represent vectors, the notation $\vec{A}$ and $\vec{B}$—a boldface symbol with an overarrow—will be used.

Note: In vector notation, vectors are generally represented with overarrows by boldface symbols, such as $\vec{A}$ and $\vec{B}$, and their magnitudes by italic symbols, A and B. In most figures, vectors will be represented by arrows (for *direction*) with their *magnitude* displayed next to them.

Vector Addition: Geometric Methods

Triangle Method   To add two vectors—say, to add $\vec{B}$ to $\vec{A}$ (that is, to find $\vec{A} + \vec{B}$) by the **triangle method**—you first draw $\vec{A}$ on a sheet of graph paper to some scale (▼Fig. 3.3a). For example, if $\vec{A}$ represents a displacement in meters, a convenient scale is 1 cm : 1 m, or 1 cm of vector length on the graph corresponds to 1 m of displacement. As shown in Fig. 3.3b, the direction of the $\vec{A}$ vector is specified as being at an angle θ_A relative to a coordinate axis, usually the x-axis.

Next, draw $\vec{B}$ with its tail starting at the tip of $\vec{A}$. (Thus, this method is also called the *tip-to-tail method*.) The vector from the tail of $\vec{A}$ to the tip of $\vec{B}$ is then the vector sum $\vec{R}$, or the resultant of the two vectors: $\vec{R} = \vec{A} + \vec{B}$.

If the vectors are drawn to scale, the magnitude of $\vec{R}$ can be found by measuring its length and using the scale conversion. In such a graphical approach, the direction angle θ_R is measured with a protractor. If we know the magnitudes and directions (angles θ) of $\vec{A}$ and $\vec{B}$, the magnitude and direction of $\vec{R}$ can be found analytically by using trigonometric methods. For the nonright triangle in Fig. 3.3b, the laws of sines and cosines can be used. (See Appendix I.) This tip-to-tail method

Note: A vector (arrow) can be moved around in vector addition methods—as long as you don't change its length (magnitude) or direction, you don't change the vector.

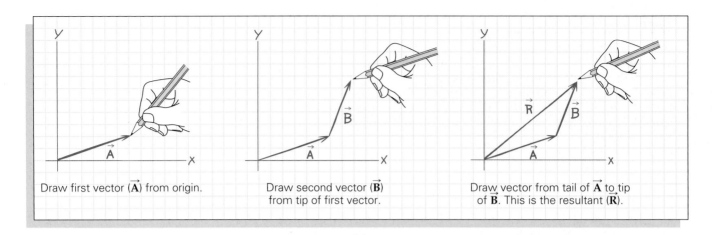

Draw first vector ($\vec{\mathbf{A}}$) from origin.

Draw second vector ($\vec{\mathbf{B}}$) from tip of first vector.

Draw vector from tail of $\vec{\mathbf{A}}$ to tip of $\vec{\mathbf{B}}$. This is the resultant ($\vec{\mathbf{R}}$).

(a)

Exploration 3.1 Addition of Displacement Vectors

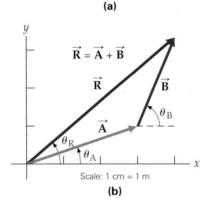

Scale: 1 cm = 1 m

(b)

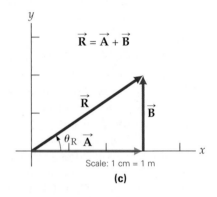

Scale: 1 cm = 1 m

(c)

▲ **FIGURE 3.3 Triangle method of vector addition** **(a)** The vectors $\vec{\mathbf{A}}$ and $\vec{\mathbf{B}}$ are placed tip to tail. The vector that extends from the tail of $\vec{\mathbf{A}}$ to the tip of $\vec{\mathbf{B}}$, forming the third side of the triangle, is the resultant or sum $\vec{\mathbf{R}} = \vec{\mathbf{A}} + \vec{\mathbf{B}}$. **(b)** When the vectors are drawn to scale, the magnitude of $\vec{\mathbf{R}}$ can be found by measuring the length of $\vec{\mathbf{R}}$ and using the scale conversion, and the direction angle θ_R can be measured with a protractor. Analytical methods can also be used. For a nonright triangle, as in part (b), the laws of sines and cosines can be used to determine the magnitude of $\vec{\mathbf{R}}$ and θ_R (Appendix I). **(c)** If the vector triangle is a right triangle, $\vec{\mathbf{R}}$ is easily obtained via the Pythagorean theorem, and the direction angle is given by an inverse trigonometric function.

can be extended to any number of vectors. The vector from the tail of the first vector to the tip of the last vector is the resultant or vector sum. For more than two vectors, it is called the polygon method.

The resultant of the vector right triangle in Fig. 3.3c would be much easier to find using the Pythagorean theorem for the magnitude and an inverse trigonometric function to find the direction angle. Notice that $\vec{\mathbf{R}}$ is made up of x- and y-vectors $\vec{\mathbf{A}}$ and $\vec{\mathbf{B}}$. Such x- and y-components are the basis of the convenient analytical component method, which will be discussed shortly.

Vector Subtraction   Vector subtraction is a special case of vector addition:

$$\vec{\mathbf{A}} - \vec{\mathbf{B}} = \vec{\mathbf{A}} + (-\vec{\mathbf{B}})$$

That is, to subtract $\vec{\mathbf{B}}$ from $\vec{\mathbf{A}}$, a *negative* $\vec{\mathbf{B}}$ is added to $\vec{\mathbf{A}}$. In Chapter 2, you learned that a minus sign simply means that the direction of a vector is opposite that of one with a plus sign (for example, $+x$ and $-x$). The same is true with vectors represented by boldface notation. The vector $-\vec{\mathbf{B}}$ has the same magnitude as the vector $\vec{\mathbf{B}}$, but is in the opposite direction (▶Fig. 3.4). The vector diagram in Fig. 3.4 provides a graphical representation of $\vec{\mathbf{A}} - \vec{\mathbf{B}}$.

Vector Components and the Analytical Component Method

Probably the most widely used analytical method for adding multiple vectors is the **component method**. It will be used again and again throughout the course of our study, so a basic understanding of the method is *essential*. Learn this section well.

Adding Rectangular Vector Components   By *rectangular components*, we mean vector components at right (90°) angles to each other, usually taken in the rectangular-coordinate *x*- and *y*-directions. You have already had an introduction to the addition of such components in the discussion of the velocity components of motion in Section 3.1. For the general case, suppose that $\vec{A}$ and $\vec{B}$, two vectors at right angles, are added, as illustrated in ▼Fig. 3.5a. The right angle makes the math easy. The magnitude of $\vec{C}$ is given by the Pythagorean theorem:

$$C = \sqrt{A^2 + B^2} \tag{3.4a}$$

The orientation of $\vec{C}$ relative to the *x*-axis is given by the angle

$$\theta = \tan^{-1}\left(\frac{B}{A}\right) \tag{3.4b}$$

This notation is how a resultant is expressed in **magnitude–angle form**.

Resolving a Vector into Rectangular Components; Unit Vectors   Resolving a vector into rectangular components is essentially the reverse of adding the rectangular components of the vector. Given a vector $\vec{C}$, Fig. 3.5b illustrates how it may be resolved into *x* and *y* vector components $\vec{C}_x$ and $\vec{C}_y$. Simply complete the vector triangle with *x*- and *y*-components. As the diagram shows, the magnitudes, or vector lengths, of these components are given by

$$C_x = C \cos \theta \tag{3.5a}$$
$$C_y = C \sin \theta \qquad \textit{(vector components)} \tag{3.5b}$$

respectively (similar to $v_x = v \cos \theta$ and $v_y = v \sin \theta$ in Example 3.1).* The angle of direction of $\vec{C}$ can also be expressed in terms of the components, since $\tan \theta = C_y/C_x$, or

$$\theta = \tan^{-1}\left(\frac{C_y}{C_x}\right) \qquad \begin{array}{l}\textit{(direction of vector} \\ \textit{from magnitudes of components)}\end{array} \tag{3.6}$$

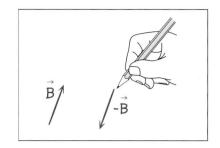

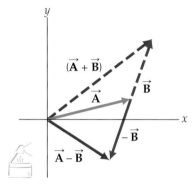

▲ **FIGURE 3.4** Vector subtraction Vector subtraction is a special case of vector addition; that is, $\vec{A} - \vec{B} = \vec{A} + (-\vec{B})$, where $-\vec{B}$ has the same magnitude as $\vec{B}$, but is in the opposite direction. (See the sketch.) Thus, $\vec{A} + \vec{B}$ is not the same as $\vec{B} - \vec{A}$, in either length or direction. Can you show that $\vec{B} - \vec{A} = -(\vec{A} - \vec{B})$ geometrically?

Magnitude–angle form of a vector

◄ **FIGURE 3.5** Vector components **(a)** The vectors $\vec{A}$ and $\vec{B}$ along the *x*- and *y*-axes, respectively, add to give $\vec{C}$. **(b)** A vector $\vec{C}$ may be resolved into rectangular components $\vec{C}_x$ and $\vec{C}_y$.

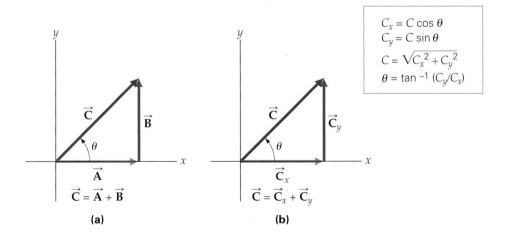

$$C_x = C \cos \theta$$
$$C_y = C \sin \theta$$
$$C = \sqrt{C_x{}^2 + C_y{}^2}$$
$$\theta = \tan^{-1}(C_y/C_x)$$

(a)   **(b)**

*Figure 3.5b illustrates only a vector in the first quadrant, but the equations hold for all quadrants when vectors are referenced to either the positive or negative *x*-axis. The directions of the components are indicated by + and − signs, as will be shown shortly.

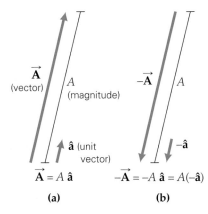

▲ **FIGURE 3.6 Unit vectors (a)** A unit vector $\hat{a}$ has a magnitude of unity, or one, and thereby simply indicates a vector's direction. Written with the magnitude A, it represents the vector $\vec{A}$, and $\vec{A} = A\hat{a}$. **(b)** For the vector $-\vec{A}$, the unit vector is $-\hat{a}$, and $-\vec{A} = -A\hat{a} = A(-\hat{a})$.

Another way of expressing the magnitude and direction of a vector involves the use of unit vectors. For example, as illustrated in ◄Fig. 3.6, a vector $\vec{A}$ can be written as $\vec{A} = A\hat{a}$. The numerical magnitude is represented by A, and $\hat{a}$ is called a **unit vector**. That is, it has a magnitude of unity, or one, but no units and thus simply indicates the vector's direction. For example, a velocity along the x-axis can be written $\vec{v} = (4.0 \text{ m/s}) \hat{x}$ (that is, 4.0 m/s magnitude in the $+x$-direction).

Note in Fig. 3.6 how $-\vec{A}$ would be represented in this notation. Although the minus sign is sometimes put in front of the numerical magnitude, this quantity is an absolute number; the minus actually goes with the unit vector: $-\vec{A} = -A\hat{a} = A(-\hat{a})$.* That is, the unit vector is in the $-\hat{a}$ direction (opposite $\hat{a}$). A velocity of $\vec{v} = (-4.0 \text{ m/s}) \hat{x}$ has a magnitude of 4.0 m/s in the $-x$ direction; that is, $\vec{v} = (4.0 \text{ m/s})(-\hat{x})$.

This notation can be used to express explicitly the rectangular components of a vector. For example, the ball's displacement from the origin in Example 3.2 could be written $\vec{d} = (4.50 \text{ m}) \hat{x} + (12.6 \text{ m}) \hat{y}$, where $\hat{x}$ and $\hat{y}$ are unit vectors in the x- and y-directions, respectively. In some instances, it may be more convenient to express a general vector in this unit-vector **component form**:

$$\vec{C} = C_x\hat{x} + C_y\hat{y} \qquad (3.7)$$

Vector Addition Using Components

The **analytical component method of vector addition** involves resolving the vectors into rectangular vector components and adding the components for each axis independently.

This method is illustrated graphically in ▼Fig. 3.7 for two vectors $\vec{F}_1$ and $\vec{F}_2$.† *The sums of the x- and y-component vectors being added are then equal to the corresponding vector components of the resultant vector.*

► **FIGURE 3.7 Component addition (a)** In adding vectors by the component method, each vector is first resolved into its x- and y-component vectors. **(b)** The sums of the x- and y-components of vectors $\vec{F}_1$ and $\vec{F}_2$ are $\vec{F}_x = \vec{F}_{x_1} + \vec{F}_{x_2}$ and $\vec{F}_y = \vec{F}_{y_1} + \vec{F}_{y_2}$, respectively.

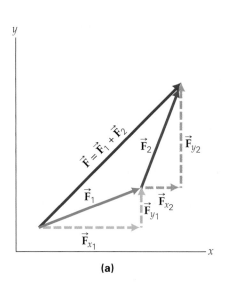

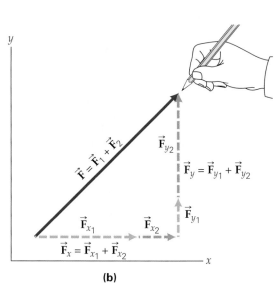

(a)

(b)

*The notation is sometimes written with an absolute value, $\vec{A} = |A|\hat{a}$, or $-\vec{A} = -|A|\hat{a}$, so as to clearly show that the magnitude of $\vec{A}$ is a positive quantity.

†The symbol $\vec{F}$ is commonly used to denote force, a very important vector quantity that you will study in Chapter 4. Here, $\vec{F}$ is employed as a general vector, but its use provides familiarity with the notation used in the next chapter, where knowledge of the addition of forces is essential.

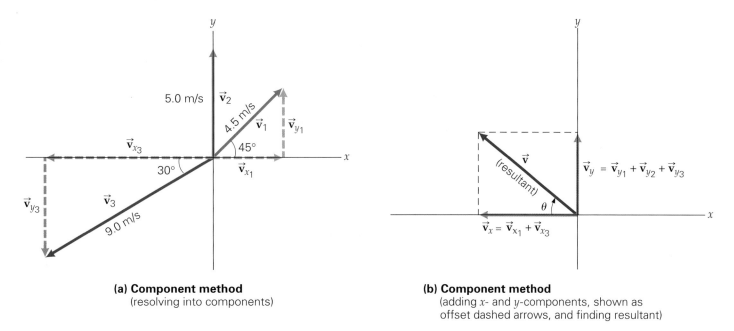

(a) Component method
(resolving into components)

(b) Component method
(adding x- and y-components, shown as
offset dashed arrows, and finding resultant)

▲ **FIGURE 3.8** **Component method of vector addition** **(a)** In the analytical component method, all the vectors to be added ($\vec{\mathbf{v}}_1$, $\vec{\mathbf{v}}_2$, and $\vec{\mathbf{v}}_3$) are first placed with their tails at the origin so that they may be easily resolved into rectangular components. **(b)** The respective summations of all the x-components and all the y-components are then added to give the components of the resultant $\vec{\mathbf{v}}$.

The same principle applies if you are given three (or more) vectors to add. You could find the resultant by applying the graphical tip-to-tail method. However, this technique involves drawing the vectors to scale and using a protractor to measure angles, which can be time consuming. But if you use the component method, you do not have to draw the vectors tip to tail. In fact, it is usually more convenient to put all of the tails together at the origin, as shown in ▲Fig. 3.8a. Also, the vectors do not have to be drawn to scale, since the approximate sketch is just a visual aid in applying the analytical method.

In the component method, you resolve the vectors to be added into their x- and y-components, add the respective components, and recombine to find the resultant. The resultant is shown in Fig. 3.8b. By looking at the x-components, it can be seen that the vector sum of these components is in the $-x$-direction. Similarly, the sum of the y-components is in the $+y$-direction. (Note that $\vec{\mathbf{v}}_2$ is in the y-direction and has a zero x-component, just as a vector in the x-direction would have a zero y-component.)

In using the plus and minus notation to indicate directions, we write the x- and y-components of the resultant: $v_x = v_{x_1} - v_{x_3}$ and $v_y = v_{y_1} + v_{y_2} - v_{y_3}$. When the numerical values of the vector components are computed and put into these equations, you will have values for $v_x < 0$ and $v_y > 0$ as shown in Fig. 3.8b.

Notice also in Fig. 3.8b that the directional angle θ of the resultant is referenced to the x-axis, as are the individual vectors in Fig. 3.8a. *In adding vectors by the component method, we will reference all vectors to the nearest x-axis—that is, the +x-axis or −x-axis.* This policy eliminates angles greater than 90° (as occurs when we customarily measure angles counterclockwise from the $+x$-axis) and the use of double-angle formulas, such as $\cos(\theta + 90°)$, and greatly simplifies calculations. The recommended procedures for adding vectors analytically by the component method can be summarized as follows:

Teaching tip: Show students the *nearest x*-axis from each of the four quadrants.

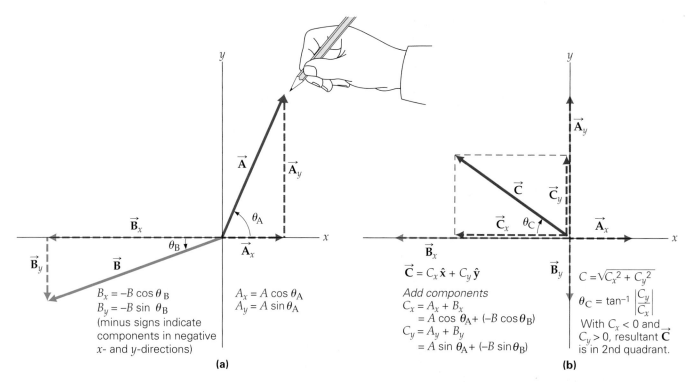

$$\vec{C} = C_x\hat{x} + C_y\hat{y}$$

Add components
$$C_x = A_x + B_x$$
$$= A\cos\theta_A + (-B\cos\theta_B)$$
$$C_y = A_y + B_y$$
$$= A\sin\theta_A + (-B\sin\theta_B)$$

$$C = \sqrt{C_x^2 + C_y^2}$$

$$\theta_C = \tan^{-1}\left|\frac{C_y}{C_x}\right|$$

With $C_x < 0$ and $C_y > 0$, resultant $\vec{C}$ is in 2nd quadrant.

$B_x = -B\cos\theta_B$
$B_y = -B\sin\theta_B$
(minus signs indicate components in negative x- and y-directions)

$A_x = A\cos\theta_A$
$A_y = A\sin\theta_A$

(a)

(b)

▲ **FIGURE 3.9 Vector addition by the analytical component method (a)** Resolve the vectors into their x- and y-components. **(b)** Add all of the x-components and all of the y-components together vectorially to obtain the x- and y-components $\vec{C}_x$ and $\vec{C}_y$, respectively, of the resultant. Express the resultant in either component form or magnitude–angle form. All angles are referenced to the $+x$- or $-x$-axis to keep them less than $90°$.

Procedures for Adding Vectors by the Component Method

1. Resolve the vectors to be added into their x- and y-components. Use the acute angles (angles less than $90°$) between the vectors and the x-axis, and indicate the directions of the components by plus and minus signs (▲Fig. 3.9).
2. Add all of the x-components together, and all of the y-components together vectorially to obtain the x- and y-components of the resultant, or vector sum.
3. Express the resultant vector, using:
   (a) the unit vector component form—for example, $\vec{C} = C_x\hat{x} + C_y\hat{y}$—or
   (b) the magnitude–angle form.

For the latter notation, find the magnitude of the resultant by using the summed x- and y-components and the Pythagorean theorem:

$$C = \sqrt{C_x^2 + C_y^2}$$

Find the angle of direction (relative to the x-axis) by taking the inverse tangent ($\tan^{-1}$) of the *absolute value* (that is, the positive value, ignoring any minus signs) of the ratio of the magnitudes of y- and x-components:

$$\theta = \tan^{-1}\left|\frac{C_y}{C_x}\right|$$

Note: The absolute value indicates that minus signs are ignored (for example, $|-3| = 3$). This operation is done to avoid negative values and angles greater than $90°$.

Designate the quadrant in which the resultant lies. This information is obtained from the signs of the summed components or from a sketch of their addition via the triangle method. (See Fig. 3.9.) The angle θ is the angle between the resultant and the x-axis in that quadrant.

Example 3.3 ■ Applying the Analytical Component Method: Separating and Combining x- and y-Components

Let's apply the procedural steps of the component method to the addition of the vectors in Fig. 3.8b. The vectors with units of meters per second represent velocities.

Thinking It Through. Follow and learn the steps of the procedure. Basically, you resolve the vectors into components and add the respective components to get the components of the resultant, which then may be expressed in component form or magnitude–angle form.

Solution. The rectangular components of the vectors are shown in Fig. 3.8b. Summing these components and taking the values from Fig. 3.8a,

$$\vec{\mathbf{v}} = v_x\hat{\mathbf{x}} + v_y\hat{\mathbf{y}} = (v_{x_1} + v_{x_2} + v_{x_3})\,\hat{\mathbf{x}} + (v_{y_1} + v_{y_2} + v_{y_3})\,\hat{\mathbf{y}}$$

where

$$v_x = v_{x_1} + v_{x_2} + v_{x_3} = v_1\cos 45° + 0 - v_3\cos 30°$$
$$= (4.5\text{ m/s})(0.707) - (9.0\text{ m/s})(0.866) = -4.6\text{ m/s}$$

and

$$v_y = v_{y_1} + v_{y_2} + v_{y_3} = v_1\sin 45° + v_2 - v_3\sin 30°$$
$$= (4.5\text{ m/s})(0.707) + (5.0\text{ m/s}) - (9.0\text{ m/s})(0.50) = 3.7\text{ m/s}$$

In tabular form, the components are provided as follows:

	x-Components		y-Components
v_{x_1}	$+v_1\cos 45° = +3.2$ m/s	v_{y_1}	$+v_1\sin 45° = +3.2$ m/s
v_{x_2}	$= 0$ m/s	v_{y_2}	$= +5.0$ m/s
v_{x_3}	$-v_3\cos 30° = -7.8$ m/s	v_{y_3}	$-v_3\sin 30° = -4.5$ m/s
Sums:	$v_x = -4.6$ m/s		$v_y = +3.7$ m/s

The directions of the components are indicated by signs. (The + sign is sometimes omitted as being understood.) In this case, v_2 has no x-component. Note that in general, for the analytical component method, the x-components are cosine functions and the y-components are sine functions, as long as we reference to the nearest part of the x-axis.

In component form, the resultant vector is

$$\vec{\mathbf{v}} = (-4.6\text{ m/s})\,\hat{\mathbf{x}} + (3.7\text{ m/s})\,\hat{\mathbf{y}}$$

In magnitude–angle form, the resultant velocity has a magnitude of

$$v = \sqrt{v_x^2 + v_y^2} = \sqrt{(-4.6\text{ m/s})^2 + (3.7\text{ m/s})^2} = 5.9\text{ m/s}$$

Since the x-component is negative and the y-component is positive, the resultant lies in the *second quadrant* at an angle of

$$\theta = \tan^{-1}\left|\frac{v_y}{v_x}\right| = \tan^{-1}\left(\frac{3.7\text{ m/s}}{4.6\text{ m/s}}\right) = 39°$$

above the −x-axis (see Fig. 3.8b).

Follow-Up Exercise. Suppose in this Example that there were an additional velocity vector $\vec{\mathbf{v}}_4 = (+4.6\text{ m/s})\,\hat{\mathbf{x}}$. What would be the resultant of all four vectors in this case?

Although our discussion is limited to motion in two dimensions (in a plane), the component method is easily extended to three dimensions. For a velocity in three dimensions, the vector has x-, y-, and z-components: $\vec{\mathbf{v}} = v_x\hat{\mathbf{x}} + v_y\hat{\mathbf{y}} + v_z\hat{\mathbf{z}}$ and magnitude $v = \sqrt{v_x^2 + v_y^2 + v_z^2}$.

LEARN BY DRAWING

Make a Sketch and Add Them Up

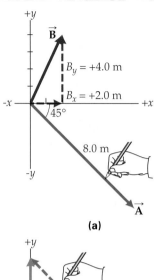

(a)

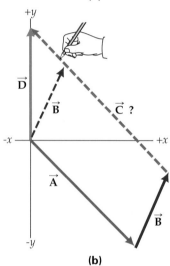

(b)

(a) A sketch is made for the vectors $\vec{A}$ and $\vec{B}$. In a vector drawing, the vector lengths are usually set to some scale—for example, 1 cm : 1 m—but in a quick sketch, the vector lengths are estimated. (b) By shifting $\vec{B}$ to the tip of $\vec{A}$ and putting in $\vec{D}$, the vector $\vec{C}$ can be found from $\vec{A} + \vec{B} + \vec{C} = \vec{D}$.

PHYSLET®

Exploration 3.3 Acceleration of a Golf Ball that Rims the Hole

Example 3.4 ■ Find the Vector: Add Them Up

You are given two displacement vectors: $\vec{A}$, with a magnitude of 8.0 m in a direction 45° below the +x-axis, and $\vec{B}$, which has an x-component of +2.0 m and a y-component of +4.0 m. Find a vector $\vec{C}$ so that $\vec{A} + \vec{B} + \vec{C}$ equals a vector $\vec{D}$ that has a magnitude of 6.0 m in the +y-direction.

Thinking It Through. Here again, a sketch helps to understand the situation and gives a general idea of the attributes of $\vec{C}$. This would be something like the accompanying Learn By Drawing. Note that in part (a) both $\vec{A}$ and $\vec{B}$ have +x-components, so $\vec{C}$ would have to have a −x-component to cancel these components. (We are given that the resultant $\vec{D}$ points only in the +y-direction.) $\vec{B}_y$ and $\vec{D}$ are in the +y-direction, but the $\vec{A}_y$-component is larger in the −y-direction, so $\vec{C}$ would have to have a +y-component. With this information, we see that $\vec{C}$ lies in the second quadrant. A polygon sketch (shown in part (b) of Learn by Drawing) confirms this observation.

So we know that $\vec{C}$ has second-quadrant components and that it has a relatively large magnitude (from the lengths of the vectors in the polygon drawing). This information gives us an idea of what we are looking for, making it easer to see if the results from the analytic solution are reasonable.

Solution.

Given: $\vec{A}$: 8.0 m, 45° below the −x-axis   *Find:* $\vec{C}$ such that
         (fourth quadrant) $\vec{A} + \vec{B} + \vec{C} = \vec{D} = (+6.0\text{ m})\,\hat{y}$
         $\vec{B}_x = (2.0\text{ m})\,\hat{x}$
         $\vec{B}_y = (4.0\text{ m})\,\hat{y}$

Let's set up the components in tabular form again so they can be easily seen:

x-Components	y-Components
$A_x = A\cos 45° = (8.0\text{ m})(0.707) = +5.7\text{ m}$	$A_y = -A\sin 45° = -(8.0\text{ m})(0.707)$
$B_x = +2.0\text{ m}$	$ = -5.7\text{ m}$
$C_x = ?$	$B_y = +4.0\text{ m}$
$D_x = 0$	$C_y = ?$
	$D_y = +6.0\text{ m}$

To find the components of $\vec{C}$, where $\vec{A} + \vec{B} + \vec{C} = \vec{D}$, we sum the x- and y-components separately:

$$x:\quad \vec{A}_x + \vec{B}_x + \vec{C}_x = \vec{D}_x$$

or

$$+5.7\text{ m} + 2.0\text{ m} + C_x = 0 \quad \text{and} \quad C_x = -7.7\text{ m}$$

$$y:\quad \vec{A}_y + \vec{B}_y + \vec{C}_y = \vec{D}_y$$

or

$$-5.7\text{ m} + 4.0\text{ m} + C_y = 6.0\text{ m} \quad \text{and} \quad C_y = +7.7\text{ m}$$

So,

$$\vec{C} = (-7.7\text{ m})\,\hat{x} + (7.7\text{ m})\,\hat{y}$$

We can also express the result in magnitude–angle form:

$$C = \sqrt{C_x^2 + C_y^2} = \sqrt{(-7.7\text{ m})^2 + (7.7\text{ m})^2} = 11\text{ m}$$

and

$$\theta = \tan^{-1}\left|\frac{C_y}{C_x}\right| = \tan^{-1}\left|\frac{7.7\text{ m}}{-7.7\text{ m}}\right| = 45° \quad (\textit{above the −x-axis, why?})$$

Follow-Up Example. Suppose $\vec{D}$ pointed in the opposite direction $[\vec{D} = (-6.0\text{ m})\,\hat{y}]$. What would $\vec{C}$ be in this case?

3.3 Projectile Motion

OBJECTIVES: To analyze projectile motion to find (a) position, (b) time of flight, and (c) range.

A familiar example of two-dimensional, curvilinear motion is that of objects that are thrown or projected by some means. The motion of a stone thrown across a stream or a golf ball driven off a tee is **projectile motion**. A special case of projectile motion in one dimension occurs when an object is projected vertically upward (or downward). This case was treated in Chapter 2 in terms of free fall (air resistance neglected). Projectile motion is free fall too, because the only acceleration of a projectile is that due to gravity. Vector components can be used to analyze projectile motion by simply breaking up the motion into its x- and y-components and treating them separately.

Horizontal Projections

It is instructive to first analyze the motion of an object projected horizontally, or parallel to a level surface. Suppose that you throw an object horizontally with an initial velocity v_{x_o} as in ▾Fig. 3.10. Projectile motion is analyzed beginning at the instant of release ($t = 0$). Once the object is released, there is no longer a horizontal acceleration ($a_x = 0$), so throughout the object's path, the horizontal velocity remains constant: $v_x = v_{x_o}$.

According to the equation $x = x_o + v_x t$ (Eq. 3.2a), the projected object would continue to travel in the horizontal direction indefinitely. However, you know that this is not what happens. As soon as the object is projected, it is in free fall in the vertical direction, with $v_{y_o} = 0$ (vertically it behaves as though it had been dropped) and $a_y = -g$. In other words, the projected object travels at a uniform velocity in the horizontal direction while *at the same time* undergoing acceleration in the downward direction under the influence of gravity. The result is a curved path, as illustrated in Fig. 3.10. (Compare the motions in Fig. 3.10 and Fig. 3.2. Do you see any similarities?) If there were no horizontal motion, the object would simply drop to the ground in a straight line. In fact, the time of flight of the projected object is *exactly the same as if it were a dropped object falling vertically.*

Note: Review Section 2.5, one-dimensional free fall.

Demonstration/activity: While standing on a desk, throw an object such as a marble horizontally outward while dropping an identical object vertically at the same time. Listen for the sound of the objects striking the floor. They will tend to do so at the same time.

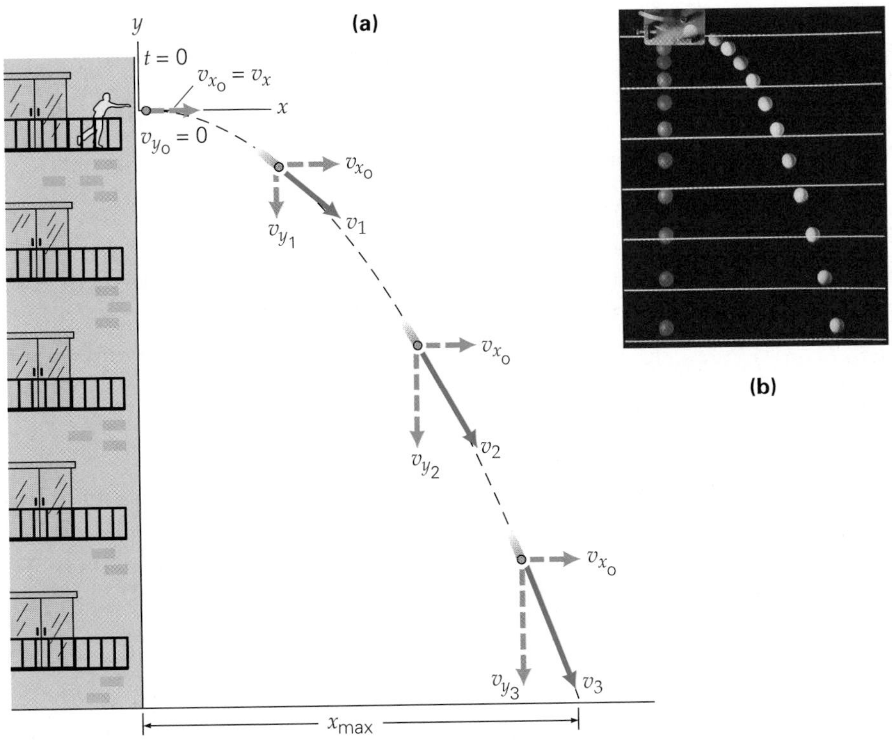

◀ **FIGURE 3.10** Horizontal projection **(a)** The velocity components of a projectile launched horizontally show that the projectile travels to the right as it falls downward, as indicated by the minus sign. **(b)** A multiflash photograph shows the paths of two golf balls. One was projected horizontally at the same time that the other was dropped straight down. The horizontal lines are 15 cm apart, and the interval between flashes was $\frac{1}{30}$ s. The vertical motions of the balls are the same. Why? Can you describe the horizontal motion of the yellow ball?

Demonstration/activity: The constant horizontal speed in Fig. 3.10a can be demonstrated by placing a blank transparency over the photograph and marking the position of the balls. Then place the transparency on the overhead projector and construct straight vertical lines through the balls, observing their equal spacing.

Note the components of the velocity vector in Fig. 3.10a. The length of the horizontal component of the velocity vector remains the same, but the length of the vertical component increases with time. What is the instantaneous velocity at any point along the path? (Think in terms of vector addition, covered in Section 3.3.) The photo in Fig. 3.10b shows the actual motions of a horizontally projected golf ball and one that is simultaneously dropped from rest. The horizontal reference lines show that the balls fall vertically at the same rate. The only difference is that the horizontally projected ball also travels to the right as it falls.

Example 3.5 ■ Starting at the Top: Horizontal Projection

Suppose that the ball in Fig. 3.10a is projected from a height of 25.0 m above the ground and is thrown with an initial horizontal velocity of 8.25 m/s. (a) How long is the ball in flight before striking the ground? (b) How far from the building does the ball strike the ground?

Thinking It Through. In looking at the components of motion, we find that part (a) involves the time it takes the ball to fall vertically, analogous to a ball dropped from that height. This time is also the time the ball travels in the horizontal direction. The horizontal speed is constant, so the horizontal distance requested in part (b) can be found.

Solution. Writing the data with the origin chosen as the point from which the ball is thrown and downward taken as the negative direction, we have

Given: $y = -25.0$ m $\qquad$ *Find:* (a) t (time of flight)
$\qquad v_{x_o} = 8.25$ m/s $\qquad\qquad\quad$ (b) x (horizontal distance)
$\qquad a_x = 0$
$\qquad v_{y_o} = 0$
$\qquad a_y = -g$
$\qquad$ ($x_o = 0$ and $y_o = 0$ because of our choice of axes location.)

(a) As noted previously, the time of flight is the same as the time it takes for the ball to fall vertically to the ground. To find this time, we can use the equation $y = y_o + v_{y_o}t - \frac{1}{2}gt^2$, in which the negative direction of g is expressed explicitly, as was done in Chapter 2. With $v_{y_o} = 0$, we have

$$y = -\tfrac{1}{2}gt^2$$

So,

$$t = \sqrt{\frac{2y}{-g}} = \sqrt{\frac{2(-25.0\ \text{m})}{-9.80\ \text{m/s}^2}} = 2.26\ \text{s}$$

(b) The ball travels in the x-direction for the same amount of time it travels in the y-direction (that is, 2.26 s). Since there is no acceleration in the horizontal direction, the ball travels in this direction with a uniform velocity. Thus, with $x_o = 0$ and $a_x = 0$,

$$x = v_{x_o}t = (8.25\ \text{m/s})(2.26\ \text{s}) = 18.6\ \text{m}$$

Follow-Up Exercise. (a) Choose the axes to be at the base of the building, and show that the resulting equation is the same as in the Example. (b) What is the velocity (in component form) of the ball just before it strikes the ground?

Teaching tip: Point out how the expression for time is first obtained algebraically from the fundamental expression prior to substituting in the values and calculating. Encourage students to practice this analytical technique.

PHYSLET®

Illustration 3.4 Projectile Motion

Projections at Arbitrary Angles

The general case of projectile motion involves an object projected at an arbitrary angle θ relative to the horizontal—for example, a golf ball hit by a club (▶Fig. 3.11). During projectile motion, the object travels up and down while traveling horizontally with a constant velocity. (Does the ball have acceleration? Yes. At each point of the motion, gravity acts, and $\vec{a} = -g\,\hat{y}$.)

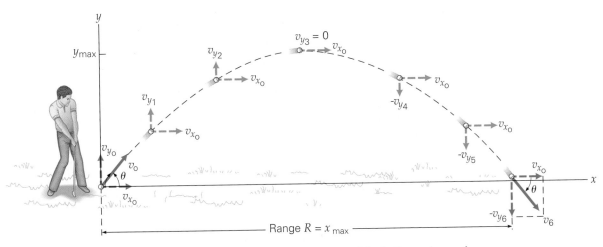

▲ FIGURE 3.11 Projection at an angle The velocity components of the ball are shown for various times. (Directions are indicated by signs, with the + sign being omitted as conventionally understood.) Note that $v_y = 0$ at the top of the arc, or at y_{max}. The range R is the maximum horizontal distance, or x_{max}. (Notice that $v_0 = v_6$. Why?)

This motion is also analyzed by using its components. As before, upward is taken as the positive direction and downward as the negative direction. The initial velocity v_o is first resolved into rectangular components:

$$v_{x_o} = v_o \cos \theta \tag{3.8a}$$
$$\text{\textit{(initial velocity components)}}$$
$$v_{y_o} = v_o \sin \theta \tag{3.8b}$$

Teaching tip: Remind students to treat the x- and y-components of motion independently.

There is no horizontal acceleration and the acceleration due to gravity acts in the negative y-direction. Thus, the x-component of the velocity is constant and the y-component varies with time (see Eq. 3.3d):

$$v_x = v_{x_o} = v_o \cos \theta \tag{3.9a}$$
$$\text{\textit{(projectile motion velocity components)}}$$
$$v_y = v_{y_o} - gt = v_o \sin \theta - gt \tag{3.9b}$$

▼ FIGURE 3.12 Parabolic arcs
Sparks of hot metal from welding describe parabolic arcs.

The components of the instantaneous velocity at various times are illustrated in Fig. 3.11. The instantaneous velocity is the sum of these components and is tangent to the curved path of the ball at any point. Notice that the ball strikes the ground at the same speed as it was launched (but with $-v_{y_o}$) and at the same angle below the horizontal.

Similarly, the displacement components are given by ($x_o = y_o = 0$):

$$x = v_{x_o}t = (v_o \cos \theta)t \tag{3.10a}$$
$$y = v_{y_o}t - \tfrac{1}{2}gt^2 = (v_o \sin \theta)t - \tfrac{1}{2}gt^2 \tag{3.10b}$$
$$\text{\textit{(projectile motion displacement components)}}$$

The curve produced by these equations, or the path of motion of the projectile, is called a **parabola**. The path of projectile motion is often referred to as a *parabolic arc*. Such arcs are commonly observed (▶Fig. 3.12).

Note that, as in the case of horizontal projection, *time is the common feature shared by the components of motion.* Aspects of projectile motion that may be of interest in various situations include the time of flight, the maximum height reached, and the **range** (R), which is the maximum horizontal distance traveled.

Note: A projectile follows a parabolic path.

Example 3.6 ■ Teeing Off: Projection at an Angle

Suppose a golf ball is hit off the tee with an initial velocity of 30.0 m/s at an angle of 35° to the horizontal, as in Fig. 3.11. (a) What is the maximum height reached by the ball? (b) What is its range?

Thinking It Through. The maximum height involves the y-component; the procedure for finding it is like that for finding the maximum height of a ball projected vertically upward. The ball travels in the x-direction for the same amount of time it would take for the ball to go up and down.

Solution.

Given: $v_o = 30.0$ m/s   *Find:* (a) y_{max}
    $\theta = 35°$ (b) $R = x_{max}$
    $a_y = -g$
    (x_o and $y_o = 0$ *and* final $y = 0$)

Let us compute v_{x_o} and v_{y_o} explicitly so that we can use simplified kinematic equations:

$$v_{x_o} = v_o \cos 35° = (30.0 \text{ m/s})(0.819) = 24.6 \text{ m/s}$$

$$v_{y_o} = v_o \sin 35° = (30.0 \text{ m/s})(0.574) = 17.2 \text{ m/s}$$

(a) Just as for an object thrown vertically upward, $v_y = 0$ at the maximum height (y_{max}). Thus, we can find the time to reach the maximum height (t_u) by using Eq. 3.9b with v_y set equal to zero:

$$v_y = 0 = v_{y_o} - gt_u$$

Solving for t_u, we have

$$t_u = \frac{v_{y_o}}{g} = \frac{17.2 \text{ m/s}}{9.80 \text{ m/s}^2} = 1.76 \text{ s}$$

(Note that t_u represents the amount of time the ball moves upward.)

The maximum height y_{max} is then obtained by substituting t_u into Eq. 3.10b:

$$y_{max} = v_{y_o}t_u - \tfrac{1}{2}gt_u^2 = (17.2 \text{ m/s})(1.76 \text{ s}) - \tfrac{1}{2}(9.80 \text{ m/s}^2)(1.76 \text{ s})^2 = 15.1 \text{ m}$$

The maximum height could also be obtained directly from Eq. 2.11′, $v_y^2 = v_{y_o}^2 - 2gy$, with $y = y_{max}$ and $v_y = 0$. However, the method of solution used here illustrates how the time of flight is obtained.

(b) As in the case of vertical projection, the time in going up is equal to the time in coming down, so the total time of flight is $t = 2t_u$ (to return to the elevation from which the object was projected, $y = y_o = 0$, as can be seen from $y - y_o = v_{y_o}t - \tfrac{1}{2}gt^2 = 0$, and $t = 2v_{y_o}/g = 2t_u$.)

The range R is equal to the horizontal distance traveled (x_{max}), which is easily found by substituting the total time of flight $t = 2t_u = 2(1.76 \text{ s}) = 3.52 \text{ s}$ into Eq. 3.10a:

$$R = x_{max} = v_x t = v_{x_o}(2t_u) = (24.6 \text{ m/s})(3.52 \text{ s}) = 86.6 \text{ m}$$

Follow-Up Exercise. How would the values of maximum height (y_{max}) and the range (x_{max}) compare with those found in this Example if the golf ball had been similarly teed off on the surface of the Moon? [*Hint:* $g_M = g/6$; that is, acceleration due to gravity on the Moon is one sixth of that on the Earth.] Do not do any numerical calculations. Find the answers by "sight reading" the equations.

Demonstration/activity: A water hose directing a steady stream of water into the air can be used to demonstrate the parabolic path of a projectile. Using a chalkboard protractor and an aquarium or sink to catch the water, show how the range can vary with the angle. Try illuminating the water in a dark room with a strobe light!

The range of a projectile is an important consideration in various applications. This factor is particularly important in sports in which a maximum range is desired, such as golf and javelin throwing.

In general, what is the range of a projectile launched with velocity v_o at an angle θ? In order to answer this question, we consider the equation used in Example 3.6 to calculate the range, $R = v_x t$. First let's look at the expressions for v_x and t. Since there is no acceleration in the horizontal direction, we know that

$$v_x = v_{x_o} = v_o \cos \theta$$

and the total time t (as shown in Example 3.6) is

$$t = \frac{2v_{y_o}}{g} = \frac{2v_o \sin \theta}{g}$$

Then,

$$R = v_x t = (v_o \cos \theta)\left(\frac{2v_o \sin \theta}{g} \right) = \frac{2v_o^2 \sin \theta \cos \theta}{g}$$

Using the trigonometric identity $\sin 2\theta = 2 \cos \theta \sin \theta$ (see Appendix I), we have

$$R = \frac{v_o^2 \sin 2\theta}{g} \quad \begin{array}{l} \textit{projectile range } x_{max} \\ \textit{(only for } y_{initial} = y_{final}) \end{array} \qquad (3.11)$$

Note that the range depends on the magnitude of the initial velocity (or speed), v_o, and the angle of projection, θ, and g are assumed to be constant. Keep in mind that this equation applies only to the *special*, but common, case of $y_{initial} = y_{final}$—that is, when the landing point is at the same height as the launch point.

Teaching tip: Make sure students understand that Eq. 3.11 should be used only when the initial launch point and the final return point are at the same height ($y_{initial} = y_{final}$).

Example 3.7 ■ A Throw from the Bridge

A young girl standing on a bridge throws a stone with an initial velocity of 12 m/s at a downward angle of 45° to the horizontal, in an attempt to hit a block of wood floating in the river below (▼Fig. 3.13). If the stone is thrown from a height of 20 m and it just reaches the water when the block is 13 m from the bridge, does the stone hit the block? (Assume that the block does not move appreciably and that it is in the plane of the throw.)

Thinking It Through. The question is, what is the range of the stone? If this range is the same as the distance between the block and the bridge, then the stone hits the block. To find the range of the stone, we need to find the time of descent (from the y-component of motion) and then use this time to find the distance x_{max}. (Time is the connecting factor.)

Solution.

Given: $v_o = 12$ m/s
$\theta = 45°$ $\quad v_{x_o} = v_o \cos 45° = 8.5$ m/s
$y = -20$ m $\quad v_{y_o} = -v_o \sin 45° = -8.5$ m/s
$x_{block} = 13$ m
$(x_o = y_o = 0)$

Find: Range or x_{max} of stone from bridge. (Is it the same as the block's distance from the bridge?)

To find the time for upward travel, we have previously used $v_y = v_{y_o} - gt$, where $v_y = 0$ at the top of the arc. However, in this case, v_y is not zero when the stone reaches the river, so to use this equation, we need to find v_y. This value may be found from the kinematic equation Eq. 2.11',

$$v_y^2 = v_{y_o}^2 - 2gy$$

(continues on next page)

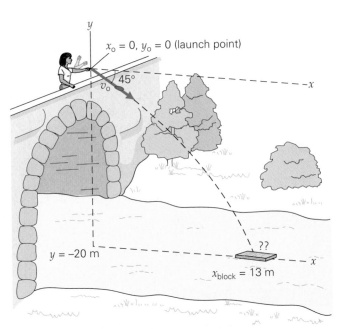

▲ **FIGURE 3.13** A throw from the bridge—hit or miss? See Example 3.7.

as

$$v_y = \sqrt{(-8.5 \text{ m/s})^2 - 2(9.8 \text{ m/s}^2)(-20 \text{ m})} = -22 \text{ m/s}$$

(negative root because v_y is downward).
   Then solving $v_y = v_{y_o} - gt$ for t,

$$t = \frac{v_{y_o} - v_y}{g} = \frac{-8.5 \text{ m/s} - (-22 \text{ m/s})}{9.8 \text{ m/s}^2} = 1.4 \text{ s}$$

The stone's horizontal distance from the bridge at this time is

$$x_{max} = v_{x_o} t = (8.5 \text{ m/s})(1.4 \text{ s}) = 12 \text{ m}$$

So the girl's throw falls short by a meter (block at 13 m).
   Note that Eq. 3.10b, $y = y_o + v_{y_o}t - \frac{1}{2}gt^2$, could have been used to find the time, but this calculation would have involved solving a quadratic equation.

Follow-Up Exercise. (a) Why was it assumed that the block was in the plane of the throw? (b) Why wasn't Eq. 3.11 used in this Example to find the range? Show that Eq. 3.11 works in Example 3.6, but not in Example 3.7, by computing the range in each case and comparing your results with the answers found in the Examples.

Conceptual Example 3.8 ■ Which Has the Greater Speed?

Consider two balls, both thrown with the same initial speed v_o, but one at an angle of 45° above the horizontal and the other at an angle of 45° below the horizontal (▼ Fig. 3.14). Determine whether, upon reaching the ground, (a) the ball projected upward will have the greater speed, (b) the ball projected downward will have the greater speed, or (c) both balls will have the same speed. *Clearly establish the reasoning and physical principle(s) used in determining your answer before checking it next. That is, **why** did you select your answer?*

Reasoning and Answer. At first, you might think the answer is (b), because this ball is projected downward. But the ball projected upward falls from a greater maximum height, so perhaps the answer is (a). To solve this dilemma, look at the horizontal line in Fig. 3.14 between the two velocity vectors that extends beyond the upper trajectory. From this diagram, you should be able to see that the trajectories for both balls are the same below this

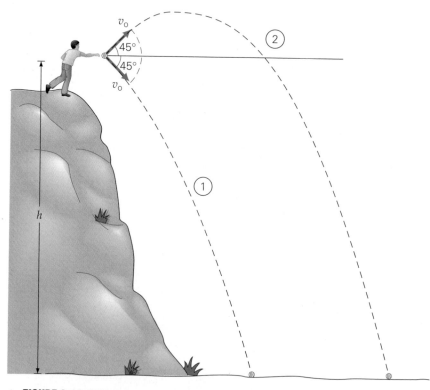

▲ **FIGURE 3.14 Which has the greater speed?** See Example 3.8.

line. Moreover, the downward velocity of the upper ball on reaching this line is v_o at an angle of 45° below the horizontal. (See Fig. 3.11.) Therefore, relative to the horizontal line and below, the conditions are identical, with the same y-component and the same constant x-component. So the answer is (c).

Follow-Up Exercise. Suppose the ball thrown downward was thrown at an angle of −40°. Which ball would hit the ground with the greater speed in this case?

Problem-Solving Hint

The range of a downward projectile, as in Fig. 3.14, is found as illustrated in Example 3.7. But what about the range of the upward projectile? This task might be thought of as an "extended-range" problem. One way to solve it is to divide the trajectory into two parts—(1) the arc above the horizontal line and (2) the downward part below the horizontal line—such that $x_{max} = x_1 + x_2$. You know how to find x_1 (Example 3.6) and x_2 (Example 3.7). Another way to solve the problem is to use $y = y_o + v_{y_o}t - \frac{1}{2}gt^2$, where y is the final position of the projectile, and solve for t, the total time of flight. You would then use that value in the equation $x = v_{x_o}t$.

Exploration 3.5 Uphill and Downhill Projectile Motion

Equation 3.11, $R = \dfrac{v_o^2 \sin 2\theta}{g}$, allows us to compute the range for a particular projection angle and initial velocity. However, we are sometimes interested in the maximum range for a given initial velocity—for example, the maximum range of an artillery piece that fires a projectile with a particular muzzle velocity. Is there an optimum angle that gives the maximum range? Under ideal conditions, the answer is yes.

For a particular v_o, the range is a maximum (R_{max}) when $\sin 2\theta = 1$, since this value of θ yields the maximum value of the sine function (which varies from 0 to 1). Thus,

$$R_{max} = \frac{v_o^2}{g} \quad (y_{initial} = y_{final}) \qquad (3.12)$$

Because this maximum range is obtained when $\sin 2\theta = 1$ and because $\sin 90° = 1$, we have

$$2\theta = 90° \quad \text{or} \quad \theta = 45°$$

for the maximum range for a given initial speed when the projectile returns to the elevation from which it was projected. At a greater or smaller angle, for a projectile with the same initial speed, the range will be less as illustrated in ▼ Fig. 3.15. Also, the range is the same for angles equally above and below 45°, such as 30° and 60°.

Thus, to get the maximum range, a projectile *ideally* should be projected at an angle of 45°. However, up to now, we have neglected air resistance. In actual situations, such as when a ball or object is thrown or hit hard, this factor may have a significant effect. Air resistance reduces the speed of the projectile, thereby reducing the

Teaching tip: Special equations such as for the range and maximum height are fun to derive and use, but discourage students from memorizing them. Understanding how to derive them is much more important than recognizing them.

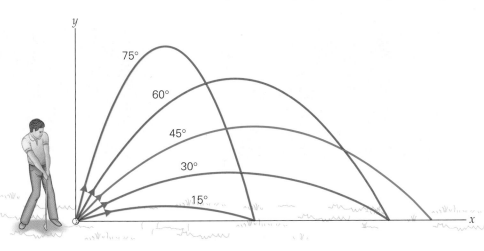

◀ **FIGURE 3.15 Range** For a projectile with a given initial speed, the maximum range is ideally attained with a projection of 45° (no air resistance). For projection angles above and below 45°, the range is shorter, and it is equal for angles equally different from 45° (for example, 30° and 60°).

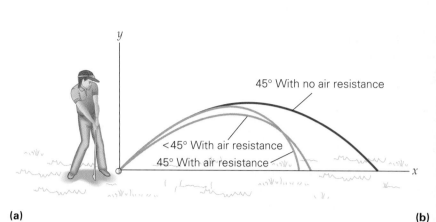

(a)

(b)

▲ **FIGURE 3.16 Air resistance and range** **(a)** When air resistance is a factor, the angle of projection for maximum range is less than 45°. **(b)** Javelin throw. Because of air resistance, the javelin is thrown at an angle less than 45° in order to achieve maximum range.

range. As a result, when air resistance is a factor, the angle of projection for maximum range is less than 45°, which gives a greater initial horizontal velocity (▲Fig. 3.16). Other factors, such as spin and wind, may also affect the range of a projectile. For example, backspin on a driven golf ball provides lift and the projection angle for the maximum range may be considerably less than 45°.

Keep in mind that for the maximum range to occur at a projection angle of 45°, the components of initial velocity must be equal—that is, $\tan^{-1}(v_{y_0}/v_{x_0}) = 45°$ and $\tan 45° = 1$, so that $v_{y_0} = v_{x_0}$. However, this condition may not always be physically possible, as Conceptual Example 3.9 shows.

Conceptual Example 3.9 ■ The Longest Jump: Theory and Practice

In a long-jump event, does the jumper normally have a launch angle of (a) less than 45°, (b) exactly 45°, or (c) greater than 45°? *Clearly establish the reasoning and physical principle(s) used in determining your answer before checking it next. That is, **why** did you select your answer?*

Reasoning and Answer. Air resistance is not a major factor here (although wind speed is taken into account for record setting in track-and-field events). Therefore, it would seem that in order to achieve maximum range, the jumper would take off at an angle of 45°. But there is another physical consideration. Let's look more closely at the jumper's initial velocity components (▼Fig. 3.17a).

To maximize a long jump, the jumper runs as fast as possible and then pushes upward as strongly as he can to maximize both velocity components. The initial vertical velocity component v_{y_0} depends on the upward push of the jumper's legs, whereas the initial horizontal velocity component v_{x_0} depends mostly on the running speed toward the jump point. In general, a greater velocity can be achieved by running than by jumping, so $v_{x_0} > v_{y_0}$. Then, since $\theta = \tan^{-1}(v_{y_0}/v_{x_0})$, we have $\theta < 45°$, where $v_{y_0}/v_{x_0} < 1$ in this case.

(a)

(b)

▲ **FIGURE 3.17 Athletes in action** **(a)** To maximize a long jump, a jumper runs as fast as possible and then pushes upward as strongly as he can to maximize the velocity components (v_x and v_y). **(b)** When driving in toward the basket and jumping to score, basketball players seem to be suspended momentarily, or "hang" in the air.

Hence, the answer is (a)—it certainly could not be (c). A typical launch angle for a long jump is 20° to 25°. (If a jumper increased her launch angle to be closer to the ideal 45°, then her running speed would have to decrease, resulting in a decrease in range.)

Follow-Up Exercise. When driving in and jumping to score, basketball players seem to be suspended momentarily, or to "hang" in the air (Fig. 3.17b). Explain the physics of this effect.

Example 3.10 ■ A "Slap Shot": Is It Good?

A hockey player hits a "slap shot" in practice (with no goalie present) when he is 15.0 m directly in front of the net. The net is 1.20 m high, and the puck is initially hit at an angle of 5.00° above the ice with a speed of 35.0 m/s. Determine whether the puck makes it into the net. If it does, determine whether the puck is rising or falling vertically as it crosses the front plane of the net.

Thinking It Through. First let's make a sketch of the situation using x–y coordinates, assuming that the puck is at the origin at the time it is hit and showing the net and its height as in ▼ Fig. 3.18. Note that the launch angle is exaggerated. An angle of 5.00° is quite small, but then again, the top of the net is not overly high (1.20 m).

To determine whether the shot is of goal quality, we need to know whether the puck's trajectory takes it above the net or into the net. That is, what is the puck's height (y) when its horizontal distance is $x = 15.0$ m? Whether the puck is rising or falling at this horizontal distance depends on when the puck reaches its maximum height. The appropriate equation(s) should tell us this information; we must keep mind that time is the connecting factor between the x- and y-components.

Solution. Listing the data as usual,

Given: $x = 15.0$ m, $x_o = 0$ *Find:*   Whether the puck goes
         $y_{net} = 1.20$ m, $y_o = 0$ into the net, and if so,
         $\theta = 5.00°$ if it is rising or falling
         $v_o = 35.0$ m/s
         $v_{x_o} = v_o \cos 5.00° = 34.9$ m/s
         $v_{y_o} = v_o \sin 5.00° = 3.05$ m/s

The vertical location of the puck at any time t is given by $y = v_{y_o}t - \frac{1}{2}gt^2$, so we need to know how long the puck takes to travel the 15.0 m to the net. The connecting factor of the components is time, so this time can be found from the x motion:

$$x = v_{x_o}t \quad \text{or} \quad t = \frac{x}{v_{x_o}} = \frac{15.0 \text{ m}}{34.9 \text{ m/s}} = 0.430 \text{ s}$$

So on reaching the front of the net, the puck is at a height of

$$y = v_{y_o}t - \frac{1}{2}gt^2 = (3.05 \text{ m/s})(0.430 \text{ s}) - \frac{1}{2}(9.80 \text{ m/s}^2)(0.430 \text{ s})^2$$
$$= 1.31 \text{ m} - 0.906 \text{ m} = 0.40 \text{ m}$$

Goal!

The time (t_u) for the puck to reach its maximum height is given by $v_y = v_{y_o} - gt_u$, where $v_y = 0$ and

$$t_u = \frac{v_{y_o}}{g} = \frac{3.05 \text{ m/s}}{9.80 \text{ m/s}^2} = 0.311 \text{ s}$$

and with the puck reaching the net in 0.430 s, it is descending.

Follow-Up Exercise. At what distance from the net did the puck start to descend?

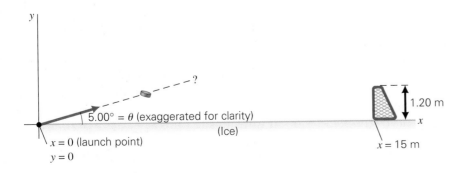

◀ **FIGURE 3.18 Slap shot** Is it a goal? See Example 3.10.

3.4 Relative Velocity

OBJECTIVE: To understand and determine relative velocities through vector addition and subtraction.

Velocity is not absolute, but is dependent on the observer. That is, it is *relative* to the observer's state of motion. If we observe an object moving with a certain velocity, then that velocity must be relative to something else. For example, a bowling ball moves down the alley with a certain velocity, and then its velocity is relative to the alley. The motions of objects are often described as relative to the Earth or ground, which we commonly think of as a *stationary* frame of reference. In other instances it is convenient to use a *moving* frame of reference.

Measurements must be made with respect to some reference. This reference is usually taken to be the origin of a coordinate system. The point you designate as the origin of a set of coordinate axes is arbitrary and entirely a matter of choice. For example, you may "attach" the coordinate system to the road or the ground and then measure the displacement or velocity of a car relative to these axes. For a "moving" frame of reference, the coordinate axes may be attached to a car moving along a highway. In analyzing motion from another reference frame, you do not change the physical situation or what is taking place, only the point of view from which you describe it. Hence, we say that motion is *relative* (to some reference frame), and we refer to **relative velocity**. Since velocity is a vector, vector addition and subtraction are helpful in determining relative velocities.

Teaching tip: Emphasize the importance of drawing the axes.

Relative Velocities in One Dimension

When the velocities are linear (along a straight line) in the same or opposite directions and all have the same reference (such as the ground), we can find relative velocities by using vector subtraction. As an illustration, consider cars moving with constant velocities along a straight, level highway, as in ▶Fig. 3.19. The velocities of the cars shown in the figure are *relative to the Earth, or the ground,* as indicated by the reference set of coordinate axes in Fig. 3.19a, with motions along the *x*-axis. They are also relative to the stationary observers standing by the highway and sitting in the parked car A. That is, these observers see the cars as moving with velocities $\vec{v}_B = +90$ km/h and $\vec{v}_C = -60$ km/h. The relative velocity of two objects is given by the velocity (vector) difference between them. For example, the velocity of car B *relative to car A* is given by

Note: Use subscripts carefully! $\vec{v}_{AB}$ = velocity of *A* relative to *B*.

$$\vec{v}_{BA} = \vec{v}_B - \vec{v}_A = (+90 \text{ km/h})\,\hat{x} - 0 = (+90 \text{ km/h})\,\hat{x}$$

Thus, a person sitting in car A would see car B move away (in the positive *x*-direction) with a speed of 90 km/h. For this linear case, the directions of the velocities are indicated by plus and minus signs (in addition to the minus sign in the equation).

Similarly, the velocity of car C relative to an observer in car A is

$$\vec{v}_{CA} = \vec{v}_C - \vec{v}_A = (-60 \text{ km/h})\,\hat{x} - 0 = (-60 \text{ km/h})\,\hat{x}$$

The person in car A would see car C approaching (in the negative *x*-direction) with a speed of 60 km/h.

But suppose that you want to know the velocities of the other cars *relative to car B* (that is, from the point of view of an observer in car B) or relative to a set of coordinate axes with the origin fixed to car B (Fig. 3.19b). Relative to these axes, car B is not moving; it acts as the fixed reference point. The other cars are moving relative to car B. The velocity of car C relative to car B is

$$\vec{v}_{CB} = \vec{v}_C - \vec{v}_B = (-60 \text{ km/h})\,\hat{x} - (+90 \text{ km/h})\,\hat{x} = (-150 \text{ km/h})\,\hat{x}$$

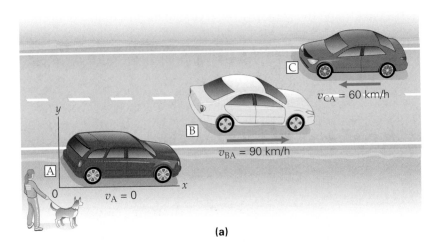

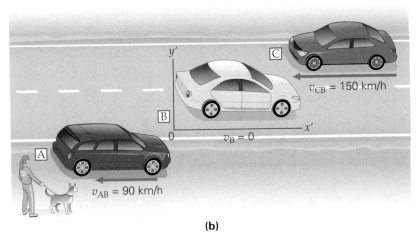

(a)

(b)

◀ **FIGURE 3.19 Relative velocity** The observed velocity of a car depends on, or is relative to, the frame of reference. The velocities shown in **(a)** are relative to the ground or to the parked car. In **(b)**, the frame of reference is with respect to car B, and the velocities are those that a driver of car B would observe. (See text for description.) **(c)** These aircraft, performing air-to-air refueling, are normally described as traveling at hundreds of kilometers per hour. To what frame of reference do these velocities refer? What is their velocity relative to each other?

(c)

Similarly, car A has a velocity relative to car B of

$$\vec{\mathbf{v}}_{AB} = \vec{\mathbf{v}}_A - \vec{\mathbf{v}}_B = 0 - (+90 \text{ km/h})\,\hat{\mathbf{x}} = (-90 \text{ km/h})\,\hat{\mathbf{x}}$$

Notice that relative to B, the other cars are both moving in the negative x-direction. That is, C is approaching B with a velocity of 150 km/h in the negative x-direction, and A appears to be receding from B with a velocity of 90 km/h in the negative x-direction. (Imagine yourself in car B, and take that position as stationary. Car C would appear to be coming toward you at a high speed, and car A would be getting farther and farther away, as though it were moving backward relative to you.) Note that, in general,

$$\vec{\mathbf{v}}_{AB} = -\vec{\mathbf{v}}_{BA}$$

What about the velocities of cars A and B relative to car C? From the point of view (or reference point) of car C, cars A and B would both appear to be approaching or moving in the positive x-direction. For the velocity of B relative to C, we have

$$\vec{\mathbf{v}}_{BC} = \vec{\mathbf{v}}_B - \vec{\mathbf{v}}_C = (90 \text{ km/h})\,\hat{\mathbf{x}} - (-60 \text{ km/h})\,\hat{\mathbf{x}} = (+150 \text{ km/h})\,\hat{\mathbf{x}}$$

Can you show that $\vec{\mathbf{v}}_{AC} = (+60 \text{ km/h})\,\hat{\mathbf{x}}$? Also note the situation in Fig. 3.19c.

In some instances, we may need to work with velocities that do not all have the same reference point. In such cases, relative velocities can be found by means of vector addition. To solve problems of this kind, *it is essential to identify the velocity references with care.*

Let's look first at a one-dimensional (linear) example. Suppose that a straight moving walkway in a major airport moves with a velocity of $\vec{\mathbf{v}}_{wg} = (+1.0 \text{ m/s})\,\hat{\mathbf{x}}$, where the subscripts indicate the velocity of the walkway (w) relative to the

Exploration 9.5 Two Airplanes with Different Land Speeds

ground (g). A passenger (p) on the walkway (w) trying to make a flight connection walks with a velocity of $\vec{v}_{pw} = (+2.0 \text{ m/s})\,\hat{x}$ relative to the walkway. What is the passenger's velocity relative to an observer standing next to the walkway (that is, relative to the ground)?

The velocity we are seeking, $\vec{v}_{pg}$, is given by

$$\vec{v}_{pg} = \vec{v}_{pw} + \vec{v}_{wg} = (2.0 \text{ m/s})\,\hat{x} + (1.0 \text{ m/s})\,\hat{x} = (3.0 \text{ m/s})\,\hat{x}$$

Thus, the stationary observer sees the passenger as traveling with speed of 3.0 m/s down the walkway. (Make a sketch, and show how the vectors add.) An explanation of the indicator line on the w symbols follows.

Problem-Solving Hint

> Notice the pattern of the subscripts in this example. On the right side of the equation, the two inner subscripts out of the four total subscripts are the same (w). Basically, the walkway (w) is used as an intermediate reference frame. The outer subscripts (p and g) are sequentially the same as those for the relative velocity on the left side of the equation. When adding relative velocities, always check to make sure that the subscripts have this relationship—it indicates that you have set up the equation correctly.

What if a passenger got on the walkway going in the opposite direction and walked with the same speed as that of the walkway? Now it is essential to indicate the direction in which the passenger is walking by means of a minus sign: $\vec{v}_{pw} = (-1.0 \text{ m/s})\,\hat{x}$. In this case, relative to the stationary observer,

$$\vec{v}_{pg} = \vec{v}_{pw} + \vec{v}_{wg} = (-1.0 \text{ m/s})\,\hat{x} + (1.0 \text{ m/s})\,\hat{x} = 0$$

so the passenger is stationary with respect to the ground, and the walkway acts as a treadmill. (Excellent physical exercise!)

Relative Velocities in Two Dimensions

Of course, velocities are not always in the same or opposite directions. However, knowing how to use rectangular components to add or subtract vectors, we can solve problems involving relative velocities in two dimensions, as Examples 3.11 and 3.12 show.

Example 3.11 ■ Across and Down the River: Relative Velocity and Components of Motion

Exploration 9.3 *Compare Relative Motion in Different Frames*

The current of a 500-m-wide straight river has a flow rate of 2.55 km/h. A motorboat that travels with a constant speed of 8.00 km/h in still water crosses the river (▶Fig. 3.20). (a) If the boat's bow points directly across the river toward the opposite shore, what is the velocity of the boat relative to the stationary observer sitting at the corner of the bridge? (b) How far downstream will the boat's landing point be from the point directly opposite its starting point?

Thinking It Through. Careful designation of the given quantities is very important—the velocity of what, relative to what? Once this is done, part (a) should be straightforward. (See the previous Problem-Solving Hint.) For part (b), we use kinematics, where the time it takes the boat to cross the river is the key.

Solution. As indicated in Fig. 3.20, we take the river's flow velocity ($\vec{v}_{rs}$, river to shore) to be in the x-direction and the boat's velocity ($\vec{v}_{br}$, boat to river) to be in the y-direction. Note that the river's flow velocity is *relative to the shore* and that the boat's velocity is *relative to the river*, as indicated by the subscripts. Listing the data, we have:

Given: $y_{max} = 500 \text{ m}$ (river width)

$\vec{v}_{rs} = (2.55 \text{ km/h})\,\hat{x}$ (velocity of river
$\phantom{\vec{v}_{rs}} = (0.709 \text{ m/s})\,\hat{x}$ *relative to shore*)

$\vec{v}_{br} = (8.00 \text{ km/h})\,\hat{y}$
$\phantom{\vec{v}_{br}} = (2.22 \text{ m/s})\,\hat{y}$ (velocity of boat
relative to river)

Find: (a) $\vec{v}_{bs}$ (velocity of boat
relative to shore)

(b) x (distance downstream)

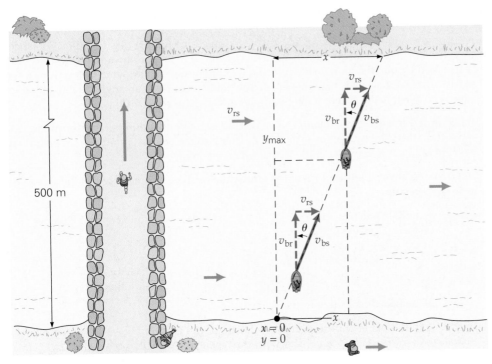

▲ **FIGURE 3.20 Relative velocity and components of motion** As the boat moves across the river, it is carried downstream by the current. See Example 3.11.

Notice that as the boat moves toward the opposite shore, it is also carried downstream by the current. These velocity components would be clearly apparent relative to the jogger crossing the bridge and to the person sauntering downstream in Fig. 3.20. If both observers stay even with the boat, the velocity of each will match one of the components of the boat's velocity. Since the velocity components are constant, the boat travels in a straight line diagonally across the river (much like the ball rolling across the table in Example 3.1).

(a) The velocity of the boat relative to the shore ($\vec{v}_{bs}$) is given by vector addition. In this case, we have

$$\vec{v}_{bs} = \vec{v}_{br} + \vec{v}_{rs}$$

Since the velocities are not along one axis, their magnitudes cannot be added directly. Notice in Fig. 3.20 that the vectors form a right triangle, so we can apply the Pythagorean theorem to find the magnitude of v_{bs}:

$$v_{bs} = \sqrt{v_{br}^2 + v_{rs}^2} = \sqrt{(2.22 \text{ m/s})^2 + (0.709 \text{ m/s})^2}$$
$$= 2.33 \text{ m/s}$$

The direction of this velocity is defined by

$$\theta = \tan^{-1}\left(\frac{v_{rs}}{v_{br}}\right) = \tan^{-1}\left(\frac{0.709 \text{ m/s}}{2.22 \text{ m/s}}\right) = 17.7°$$

(b) To find the distance x that the current carries the boat downstream, we use components. Note that in the y-direction, $y_{max} = v_{br}t$, and

$$t = \frac{y_{max}}{v_{br}} = \frac{500 \text{ m}}{2.22 \text{ m/s}} = 225 \text{ s}$$

which is the time it takes the boat to cross the river.

During this time, the boat is carried downstream by the current a distance of

$$x = v_{rs}t = (0.709 \text{ m/s})(225 \text{ s}) = 160 \text{ m}$$

Follow-Up Exercise. What is the distance traveled by the boat in crossing the river?

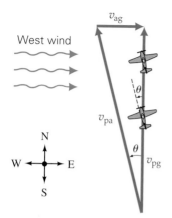

▲ FIGURE 3.21 Flying into the wind To fly directly north, the plane's heading (θ-direction) must be west of north. See Example 3.12.

Example 3.12 ■ Flying into the Wind: Relative Velocity

An airplane with an airspeed of 200 km/h (its speed in still air) flies in a direction such that with a west wind of 50.0 km/h, it travels in a straight line northward. (Wind direction is specified by the direction *from* which the wind blows, so a west wind blows from west to east.) To maintain its course due north, the plane must fly at an angle, as illustrated in ◄Fig. 3.21. What is the speed of the plane along its northward path?

Thinking It Through. Here again, the velocity designations are important, but Fig. 3.21 shows that the velocity vectors form a right triangle, and the magnitude of the unknown velocity can be found by using the Pythagorean theorem.

Solution. As always, it is important to identify the reference frame to which the given velocities are relative.

Given: $\vec{\mathbf{v}}_{pg}$ = 200 km/h at angle θ *Find:* v_{pg} (ground speed of plane)
(velocity of plane with respect to still air = air speed)
$\vec{\mathbf{v}}_{ag}$ = 50.0 km/h east (velocity of air with respect to the Earth, or ground = wind speed)
Plane flies due north with velocity $\vec{\mathbf{v}}_{pg}$

The speed of the plane with respect to the Earth, or the ground, v_{pg}, is called the plane's *ground speed*; v_{pa} is its airspeed. Vectorially, the respective velocities are related by

$$\vec{\mathbf{v}}_{pg} = \vec{\mathbf{v}}_{pa} + \vec{\mathbf{v}}_{ag}$$

If no wind were blowing (v_{ag} = 0), the ground speed and airspeed would be equal. However, a headwind (a wind blowing directly toward the plane) would cause a slower ground speed, and a tailwind would cause a faster ground speed. The situation is analogous to that of a boat going upstream versus downstream.

Here, $\vec{\mathbf{v}}_{pg}$ is the resultant of the other two vectors, which can be added by the triangle method. We use the Pythagorean theorem to find v_{pg}, noting that v_{pa} is the hypotenuse of the triangle:

$$v_{pg} = \sqrt{v_{pa}^2 - v_{ag}^2} = \sqrt{(200 \text{ km/h})^2 - (50.0 \text{ km/h})^2} = 194 \text{ km/h}$$

(Note that it was convenient to use the units of kilometers per hour, since the calculation did not involve any other units.)

Follow-Up Exercise. What must be the plane's heading (θ-direction) in this Example for the plane to fly directly north?

Teaching tip: Explain that an airspeed indicator in a light plane gives the plane's speed relative to the air. A fully instrumented plane will also have a ground-speed indicator.

Chapter Review

• Motion in two dimensions is analyzed by considering the motion of linear components. The connecting factor between components is time.

Components of Initial Velocity:

$$v_x = v_o \cos \theta \qquad (3.1a)$$
$$v_y = v_o \sin \theta \qquad (3.1b)$$

Components of Displacement (constant acceleration only):

$$x = x_o + v_{x_o} t + \tfrac{1}{2} a_x t^2 \qquad (3.3a)$$
$$y = y_o + v_{y_o} t + \tfrac{1}{2} a_y t^2 \qquad (3.3b)$$

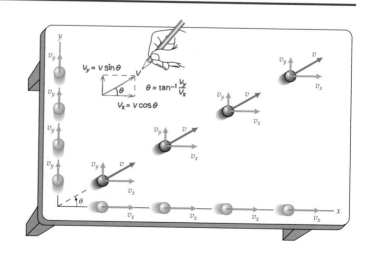

Components of Velocity (constant acceleration only):

$$v_x = v_{x_o} + a_x t \qquad (3.3c)$$
$$v_y = v_{y_o} + a_y t \qquad (3.3d)$$

- Of the various methods of vector addition, the component method is most useful. A resultant vector can be expressed in **magnitude–angle form** or in **unit-vector component form**.

Vector Representation:

$$\left.\begin{aligned} C &= \sqrt{C_x^2 + C_y^2} \\ \theta &= \tan^{-1}\left|\frac{C_y}{C_x}\right| \end{aligned}\right\} \quad \textit{(magnitude–angle form)} \qquad (3.4a)$$

$$\vec{C} = C_x \hat{x} + C_y \hat{y} \quad \textit{(component form)} \qquad (3.7)$$

- **Projectile motion** is analyzed by considering horizontal and vertical components separately—constant velocity in the horizontal direction and an acceleration due to gravity, g, in the downward vertical direction. (The foregoing equations for constant acceleration then have an acceleration of $a = -g$ instead of a.)

- **Range (R)** is the maximum horizontal distance traveled.

$$R = \frac{v_o^2 \sin 2\theta}{g} \quad \begin{array}{l} \textit{projectile range, } x_{max} \\ \textit{(only for } y_{initial} = y_{final}) \end{array} \qquad (3.11)$$

$$R_{max} = \frac{v_o^2}{g} \quad (y_{initial} = y_{final}) \qquad (3.12)$$

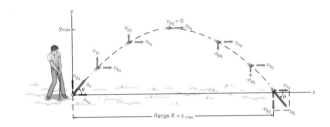

- **Relative velocity** is expressed *relative* to a particular reference frame.

Exercises

MC = *Multiple Choice Question,* **CQ** = *Conceptual Question, and* **IE** = *Integrated Exercise. Throughout the text, many exercise sections will include "paired" exercises. These exercise pairs, identified with* **red numbers**, *are intended to assist you in problem solving and learning. In a pair, the first exercise (even numbered) is worked out in the Study Guide so that you can consult it should you need assistance in solving it. The second exercise (odd numbered) is similar in nature, and its answer is given at the back of the book.*

3.1 Components of Motion

1. **MC** On Cartesian axes, the x-component of a vector is generally associated with a (a) cosine, (b) sine, (c) tangent, (d) none of the foregoing. (a)

2. **MC** The equation $x = x_o + v_{x_o}t + \frac{1}{2}a_x t^2$ applies (a) to all kinematic problems, (b) only if v_{y_o} is zero, (c) to constant accelerations, (d) to negative times. (c)

3. **MC** For an object in curvilinear motion, (a) the object's velocity components are constant, (b) the y-velocity component is necessarily greater than the x-velocity component, (c) there is an acceleration nonparallel to the object's path, (d) the velocity and acceleration vectors must be at right angles (90°). (c)

4. **CQ** Can the x-component of a vector be greater than the magnitude of the vector? How about the y-component? Explain. no, no

5. **CQ** Is it possible for an object's velocity to be perpendicular to the object's acceleration? If so, describe the motion. yes, e.g., circular motion

6. **CQ** Describe the motion of an object that is initially traveling with a constant velocity and then receives an acceleration of constant magnitude (a) in a direction parallel to the initial velocity, (b) in a direction perpendicular to the initial velocity, and (c) that is always perpendicular to the instantaneous velocity or direction of motion. see ISM

7. **IE** ● A golf ball is hit with an initial speed of 35 m/s at an angle less than 45° above the horizontal. (a) The horizontal velocity component is (1) greater than, (2) equal to, (3) less than the vertical velocity component. Why? (b) If the ball is hit at an angle of 37°, what are the initial horizontal- and vertical-velocity components? (a) (1) greater than (b) 28 m/s, 21 m/s

8. **IE** ● The x- and y-components of an acceleration vector are 3.0 m/s² and 4.0 m/s², respectively. (a) The magnitude of the acceleration vector is (1) less than 3.0 m/s², (2) between 3.0 m/s² and 4.0 m/s², (3) between 4.0 m/s² and 7.0 m/s², (4) equal to 7.0 m/s². (b) What are the magnitude and direction of the acceleration vector? (a) (3) between 4.0 m/s² and 7.0 m/s² (b) 5.0 m/s², 53° above +x-axis

9. ● If the magnitude of a velocity vector is 7.0 m/s and the x-component is 3.0 m/s, what is the y-component? ±6.3 m/s

10. ●● The x-component of a velocity vector that has an angle of 37° to the +x-axis has a magnitude of 4.8 m/s. (a) What is the magnitude of the velocity? (b) What is the magnitude of the y-component of the velocity? (a) 6.0 m/s (b) 3.6 m/s

11. **IE** ●● A student walks 100 m west and 50 m south. (a) To get back to the starting point, the student must walk in a general direction of (1) south of west, (2) north of east, (3) south of east, (4) north of west. (b) What displacement will bring the student back to the starting point? (a) (2) north of east (b) 1.1×10^2 m, 27° north of east

12. ●● A student strolls diagonally across a level rectangular campus plaza, covering the 50-m distance in 1.0 min (▼Fig. 3.22). (a) If the diagonal route makes a 37° angle with the long side of the plaza, what would be the distance if the student had walked halfway around the outside of the plaza instead of along the diagonal route? (b) If the student had walked the outside route in 1.0 min at a constant speed, how much time would she have spent on each side? (a) 70 m (b) 0.57 min, 0.43 min

▲ **FIGURE 3.22 Which way?** See Exercise 12.

13. ●● A ball rolls at a constant velocity of 1.50 m/s at an angle of 45° below the +x-axis in the fourth quadrant. If we take the ball to be at the origin at $t = 0$, what are its coordinates (x, y) 1.65 s later? $x = 1.75$ m; $y = -1.75$ m

14. ●● A ball rolling on a table has a velocity with rectangular components $v_x = 0.60$ m/s and $v_y = 0.80$ m/s. What is the displacement of the ball in an interval of 2.5 s? 2.5 m at 53° above +x-axis

15. ●● A small plane takes off at a constant velocity of 150 km/h at an angle of 37°. At 3.00 s, (a) how high is the plane above the ground, and (b) what horizontal distance has the plane traveled from the liftoff point? (a) 75.2 m (b) 99.8 m

16. IE ●● During part of its trajectory (which lasts exactly 1 min) a missile travels at a constant speed of 2000 mi/h while maintaining a constant orientation angle of 20° from the vertical. (a) During this phase, what is true about its velocity components: (1) $v_y > v_x$, (2) $v_y = v_x$, or (3) $v_y < v_x$? [Hint: Make a sketch and be careful of the angle.] (b) Determine the two velocity components analytically to confirm your choice in part (a) and also calculate how far the missile will rise during this time. (a) (1) $v_y > v_x$ (b) $v_x = 306$ m/s, $v_y = 840$ m/s, $y = 5.04 \times 10^4$ m = 50.4 km

17. ●● At the instant a ball rolls off a rooftop it has a horizontal velocity component of +10.0 m/s and a vertical component (downward) of 15.0 m/s. (a) Determine the angle of the roof. (b) What is the ball's speed as it leaves the roof? (a) $\theta = 56.3°$ below the horizontal (b) 18.0 m/s

18. ●● A particle moves at a speed of 3.0 m/s in the +x-direction. Upon reaching the origin, the particle receives a continuous constant acceleration of 0.75 m/s² in the −y-direction. What is the position of the particle 4.0 s later? $(x, y) = (12$ m, -6.0 m)

19. ●●● At a constant speed of 60 km/h, an automobile travels 700 m along a straight highway that is inclined 4.0° to the horizontal. An observer notes only the vertical motion of the car. What is the car's (a) vertical velocity magnitude and (b) vertical travel distance? (a) 1.2 m/s (b) 49 m

20. ●●● A baseball player hits a home run into the right field upper deck. The ball lands in a row that is 135 m horizontally from home plate and 25.0 m above the playing field. An avid fan measures its time of flight to be 4.10 s. (a) Determine the ball's average velocity components. (b) Determine the

magnitude and angle of its average velocity. (c) Explain why you cannot determine its average *speed* from the data given. see ISM

3.2 Vector Addition and Subtraction*

21. MC Two linear vectors of magnitudes 3 and 4, respectively, are added. The magnitude of the resultant vector is (a) 1, (b) 7, (c) between 1 and 7. (c)

22. MC The resultant of $\vec{A} - \vec{B}$ is the same as (a) $\vec{B} - \vec{A}$, (b) $-\vec{A} + \vec{B}$, or (c) $-(\vec{A} + \vec{B})$, (d) $-(\vec{B} - \vec{A})$. (d)

23. MC A unit vector has (a) magnitude, (b) direction, (c) neither of these, (d) both of these. (d)

24. CQ In Exercise 21, under what condition would the magnitude of the resultant equal 1? How about 7? How about 5? opposite direction, same direction, at right angle

25. CQ Can a nonzero vector have a zero x-component? Explain. yes, when the vector is in the y-direction

26. CQ Is it possible to add a vector quantity to a scalar quantity? no

27. CQ Can $\vec{A} + \vec{B}$ equal zero, when $\vec{A}$ and $\vec{B}$ have nonzero magnitudes? Explain. yes, if equal and opposite

28. CQ Are any of the vectors in ▼Fig. 3.23 equal? yes, all are equal

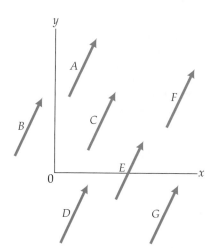

▲ **FIGURE 3.23 Different vectors?** See Exercise 28.

29. ● Using the triangle method, show graphically that (a) $\vec{A} + \vec{B} = \vec{B} + \vec{A}$ and (b) if $\vec{A} - \vec{B} = \vec{C}$, then $\vec{A} = \vec{B} + \vec{C}$. see ISM

30. IE ● (a) Is vector addition associative? That is, does $(\vec{A} + \vec{B}) + \vec{C} = \vec{A} + (\vec{B} + \vec{C})$? (b) Justify your answer graphically. (a) yes (b) see ISM

---

*A few exercises in this section use force vectors ($\vec{F}$). These vectors should be treated as vectors to be added, just as you would treat velocity vectors. The SI unit of force is the newton (N). A force vector might be written as $\vec{F} = 50$ N at an angle of 20°. Some familiarity with $\vec{F}$ vectors will be helpful in Chapter 4.

31. ● A vector has an x-component of -2.5 m and a y-component of 4.2 m. Express the vector in magnitude–angle form.
4.9 m, 59° above $-x$-axis

32. ● For the two vectors $\vec{x}_1 = (20\ \text{m})\ \hat{x}$ and $\vec{x}_2 = (15\ \text{m})\ \hat{x}$, compute and show graphically (a) $\vec{x}_1 + \vec{x}_2$, (b) $\vec{x}_1 - \vec{x}_2$, and (c) $\vec{x}_2 - \vec{x}_1$. (a) (35 m) $\hat{x}$ (b) (5 m) $\hat{x}$ (c) $(-5\ \text{m})\ \hat{x}$

33. ● During a takeoff (in still air), an airplane moves at a speed of 120 mi/h at an angle of 20° above the ground. What is the horizontal speed of the plane? 113 mi/h

34. ●● Two boys are pulling a box across a horizontal floor as shown in ▼Fig 3.24. If $F_1 = 50.0$ N and $F_2 = 100$ N, find the resultant (or sum) force by (a) the graphical method and (b) the component method.
(a) see ISM (b) 145 N at 50.1° north of east

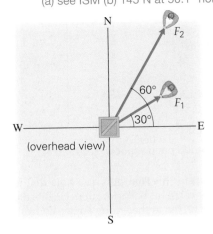

◀ **FIGURE 3.24**
Adding force vectors
See Exercises 34 and 54.

35. ●● For each of the given vectors, give a vector that, when added to it, yields a *null vector* (a vector with a magnitude of zero). Express the vector in the form other than that in which it is given (component or magnitude–angle). (a) $\vec{A} = 4.5$ cm, 40° above the $+x$-axis; (b) $\vec{B} = (2.0\ \text{cm})\ \hat{x} - (4.0\ \text{cm})\ \hat{y}$; (c) $\vec{C} = 8.0$ cm at an angle of 60° above the $-x$-axis. see ISM

36. **IE** ●● (a) If each of the two components (x and y) of a vector are doubled, (1) the vector's magnitude doubles, but the direction remains unchanged; (2) the vector's magnitude remains unchanged, but the direction angle doubles; or (3) both the vector's magnitude and direction angle double. (b) If the x- and y-components of a vector of 10 m at 45° are tripled, what is the new vector? (a) (1) magnitude doubles, but direction remains unchanged (b) 30 m at 45°

37. ●● Two brothers are pulling their other brother on a sled (▼Fig 3.25). (a) Find the resultant (or sum) of the vectors

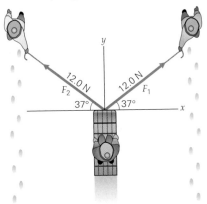

◀ **FIGURE 3.25**
Vector addition
See Exercise 37.

$\vec{F}_1$ and $\vec{F}_2$. (b) If $\vec{F}_1$ in the figure were at an angle of 27° instead of 37° with the $+x$-axis, what would be the resultant (or sum) of $\vec{F}_1$ and $\vec{F}_2$?
(a) (14.4 N) $\hat{y}$ (b) 12.7 N at 85.0° above $+x$-axis

38. ●● Given two vectors $\vec{A}$, which has a length of 10.0 and makes an angle of 45° below the $-x$-axis, and $\vec{B}$, which has an x-component of $+2.0$ and a y-component of $+4.0$, (a) sketch the vectors on x–y axes, with all their "tails" starting at the origin, and (b) calculate $\vec{A} + \vec{B}$. (a) see ISM (b) $(-5.1)\ \hat{x} + (-3.1)\ \hat{y}$ or 5.9 at 31° below $-x$-axis

39. ●● The velocity of object 1 in component form is $\vec{v}_1 = (+2.0\ \text{m/s})\ \hat{x} + (-4.0\ \text{m/s})\ \hat{y}$. Object 2 has twice the speed of object 1 but moves in the opposite direction. (a) Determine the velocity of object 2 in component notation. (b) What is the speed of object 2?
(a) $\vec{v}_2 = (-4.0\ \text{m/s})\ \hat{x} + (8.0\ \text{m/s})\ \hat{y}$ (b) 8.9 m/s

40. ●● For the vectors shown in ▼Fig. 3.26, determine $\vec{A} + \vec{B} + \vec{C}$. 16 m/s at 79° above the $-x$-axis

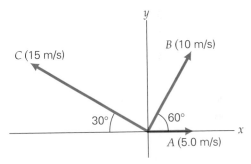

▲ **FIGURE 3.26 Adding vectors** See Exercises 40 and 41.

41. ●● For the velocity vectors shown in Fig. 3.26, determine $\vec{A} - \vec{B} - \vec{C}$. 21 m/s at 51° below the $+x$-axis

42. ●● Given two vectors $\vec{A}$ and $\vec{B}$ with magnitude A and B, respectively, you can subtract $\vec{B}$ from $\vec{A}$ to get a third vector $\vec{C} = \vec{A} - \vec{B}$. If the magnitude of $\vec{C}$ is equal to $C = A + B$, what is the relative orientation of vectors $\vec{A}$ and $\vec{B}$?
opposite

43. ●● In two successive chess moves, a player first moves his queen two squares forward, then moves the queen three steps to the left (from the player's view). Assume each square is 3.0 cm on a side. (a) Using forward (toward his opponent) as the positive y-axis and right as the positive x-axis, write the queen's net displacement in component form. (b) At what net angle was the queen moved relative to the leftward direction? (a) $(-9.0\ \text{cm})\ \hat{x} + (6.0\ \text{cm})\ \hat{y}$ (b) 33.7°, relative to $-x$ axis

44. ●● Two force vectors $\vec{F}_1 = (3.0\ \text{N})\ \hat{x} - (4.0\ \text{N})\ \hat{y}$ and $\vec{F}_2 = (-6.0\ \text{N})\ \hat{x} + (4.5\ \text{N})\ \hat{y}$ are applied to a particle. What third force $\vec{F}_3$ would make the net, or resultant, force on the particle zero? 3.4 N at 27° below $+x$-axis

45. ●● Two force vectors $\vec{F}_1 = 8.0$ N at an angle of 60° above the $+x$-axis and $\vec{F}_2 = 5.5$ N at an angle of 45° below the $+x$-axis are applied to a particle at the origin. What third force $\vec{F}_3$ would make the net, or resultant, force on the particle zero?
8.5 N at 21° below $-x$-axis

46. ●● A student works three problems involving the addition of two different vectors $\vec{F}_1$ and $\vec{F}_2$. He states that the magnitudes of the three resultants are given by (a) $F_1 + F_2$, (b) $F_1 - F_2$, and (c) $\sqrt{F_1^2 + F_2^2}$. Are these results possible? If so, describe the vectors in each case.
(a) same direction (b) opposite directions (c) at 90°

47. ●● A block weighing 50 N rests on an inclined plane. Its weight is a force directed vertically downward, as illustrated in ▼Fig. 3.27. Find the components of the force parallel to the surface of the plane and perpendicular to it.
parallel 30 N, perpendicular 40 N

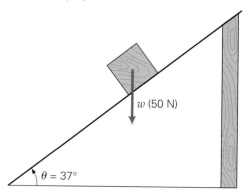

▲ **FIGURE 3.27 Block on an inclined plane** See Exercise 47.

48. ●● Two displacements, one with a magnitude of 15.0 m and a second with a magnitude of 20.0 m, can have any angle you want. (a) How would you create the sum of these two vectors so it has the largest magnitude possible? What is that magnitude? (b) How would you orient them so the magnitude of the sum was at its minimum? What value would that be? (c) Generalize the result to any two vectors.
see ISM

49. ●●● A person walks from point A to point B as shown in ▼Fig. 3.28. What is the person's displacement relative to point A? 27 m at 72° above −x-axis

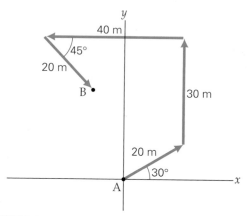

▲ **FIGURE 3.28 Adding displacement vectors**
See Exercise 49.

50. IE ●●● A meteorologist tracks the movement of a thunderstorm with Doppler radar. At 8:00 PM, the storm was 60 mi northeast of her station. At 10:00 PM, the storm is at 75 mi north. (a) The general direction of the thunderstorm's velocity is (1) south of east, (2) north of west, (3) north of east, (4) south of west. (b) What is the average velocity of the storm?
(a) (2) north of west (b) 26.7 mi/h at 37.6° north of west

51. IE ●●● A flight controller determines that an airplane is 20.0 mi south of him. Half an hour later, the same plane is 35.0 mi northwest of him. (a) The general direction of the airplane's velocity is (1) east of south, (2) north of west, (3) north of east, (4) west of south. (b) If the plane is flying with constant velocity, what is its velocity during this time?
(a) (2) north of west (b) 102 mi/h at 61.1° north of west

52. IE ●●● ▼Fig. 3.29 depicts a decorative window (the thick inner square) weighing 100 N suspended in a patio opening (the thin outer square). The upper two corner cables are each at 45° and the left one exerts a force (F_1) of 100 N on the window. (a) How does the magnitude of the force exerted by the upper right cable (F_2) compare to that exerted by the upper left cable: (1) $F_2 > F_1$, (2) $F_2 = F_1$, or $F_2 < F_1$? (b) Use your result from part (a) to help determine the force exerted by the bottom cable (F_3).
(a) (2) $F_2 = F_1$ (b) 41 N, down

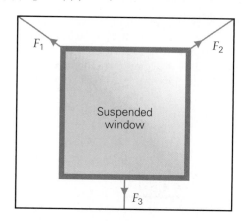

▲ **FIGURE 3.29 A suspended patio window** See Exercise 52.

53. ●●● A golfer lines up for his first putt at a hole that is 10.5 m exactly northwest of his ball's location. He hits the ball 10.5 m and straight, but at the wrong angle, 40° from due north. In order for the golfer to have a "two-putt green," determine (a) the angle of the second putt and (b) the magnitude of the second putt's displacement. (c) Determine why you cannot determine the length of travel of the second putt. (a) 42.8 south of west (b) 0.91 m
(c) might travel in a curve.

54. ●●● Two students are pulling a box as shown in Fig. 3.24. If $F_1 = 100$ N and $F_2 = 150$ N and a third student wants to stop the box, what force should he apply?
242 N at 48° below −x-axis

3.3 Projectile Motion*

55. **MC** If air resistance is neglected, the motion of an object projected at an angle consists of a uniform downward acceleration combined with (a) an equal horizontal acceleration, (b) a uniform horizontal velocity, (c) a constant upward velocity, (d) an acceleration that is always perpendicular to the path of motion. (b)

56. **MC** A football is thrown on a long pass. Compared to the ball's initial horizontal velocity, the horizontal component of its velocity at the highest point is (a) greater, (b) less, (c) the same. (c)

57. **MC** A football is thrown on a long pass. Compared to the ball's initial vertical velocity, the vertical component of its velocity at the highest point is (a) greater, (b) less, (c) the same. (b)

58. **CQ** A golf ball is hit on a level fairway. When it lands, its velocity vector has rotated through an angle of 90°. What was the launch angle of the golf ball? [*Hint*: See Fig. 3.11.] 45°

---

*Assume angles to be exact for significant figure purposes.

59. CQ Figure 3.10b shows a multiflash photograph of one ball dropping from rest and, at the same time, another ball projected horizontally at the same height. The two balls hit the ground at the same time. Why? Explain. the horizontal motion does not affect the vertical motion

60. CQ In ▼Fig. 3.30, a spring-loaded "cannon" on a wheeled car fires a metal ball vertically. The car is given a push and set in motion horizontally with constant velocity. A pin is pulled with a string to launch the ball, which travels upward and then falls back into the moving cannon every time. Why does the ball always fall back into the cannon? Explain. the vertical motion does not affect the horizontal motion

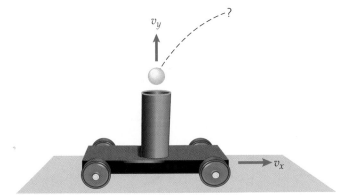

▲ **FIGURE 3.30 A ballistics car** See Exercises 60 and 69.

61. ● A ball with a horizontal speed of 1.0 m/s rolls off a bench 2.0 m high. (a) How long will the ball take to reach the floor? (b) How far from a point on the floor directly below the edge of the bench will the ball land? (a) 0.64 s (b) 0.64 m

62. ● An electron is ejected horizontally at a speed of 1.5×10^6 m/s from the electron gun of a computer monitor. If the viewing screen is 35 cm from the end of the gun, how far will the electron travel in the vertical direction before hitting the screen? Based on your answer, do you think designers need to worry about this gravitational effect? 2.7×10^{-13} m; no

63. ● A ball rolls horizontally with a speed of 7.6 m/s off the edge of a tall platform. If the ball lands 8.7 m from the point on the ground directly below the edge of the platform, what is the height of the platform? 6.4 m

64. ● A ball is thrown horizontally, with a speed of 15 m/s, from the top of a 6.0-m-tall hill. How far from the point on the ground directly below the launch point does the ball strike the ground? 17 m

65. ● If Exercise 64 were to take place on the surface of the Moon, where the acceleration due to gravity is only 1.67 m/s², what would be the answer? 40 m

66. ●● A pitcher throws a fastball horizontally at a speed of 140 km/h toward home plate, 18.4 m away. (a) If the batter's combined reaction and swing times total 0.350 s, how long can the batter watch the ball after it has left the pitcher's hand before swinging? (b) In traveling to the plate, how far does the ball drop from its original horizontal line? (a) 0.123 s (b) 1.10 m

67. IE ●● Ball A rolls at a constant speed of 0.25 m/s on a table 0.95 m above the floor, and ball B rolls on the floor directly under the first ball and with the same speed and direction. (a) When ball A rolls off the table and hits the floor, (1) ball B is ahead of ball A, (2) ball B collides with ball A, (3) ball A is ahead of ball B. Why? (b) When ball A hits the floor, how far from the point directly below the edge of the table will each ball be? (a) (2) ball B collides with ball A (b) 0.11 m, 0.11 m

68. ●● A package of supplies is to be dropped from an airplane so that it hits the ground at a designated spot near some campers. The airplane, moving horizontally at a constant velocity of 140 km/h, approaches the spot at an altitude of 0.500 km above level ground. Having the designated point in sight, the pilot prepares to drop the package. (a) What should the angle be between the horizontal and the pilot's line of sight when the package is released? (b) What is the location of the plane when the package hits the ground? (a) 51.8° (b) directly over impact point

69. ●● A wheeled car with a spring-loaded cannon fires a metal ball vertically (Fig. 3.30). If the vertical initial speed of the ball is 5.0 m/s as the cannon moves horizontally at a speed of 0.75 m/s, (a) how far from the launch point does the ball fall back into the cannon, and (b) what would happen if the cannon were accelerating? (a) 0.77 m (b) ball would not fall back in

70. ●● A soccer player kicks a stationary ball, giving it a speed of 15.0 m/s at an angle of 15.0° to the horizontal. (a) What is the maximum height reached by the ball? (b) What is the ball's range? (c) How could the range be increased? (a) 0.768 m (b) 11.5 m (c) increase v_o and/or increase the angle

71. ●● An arrow has an initial launch speed of 18 m/s. If it must strike a target 31 m away at the same elevation, what should be the projection angle? 35° or 55°

72. ●● An astronaut on the Moon fires a projectile from a launcher on a level surface so as to get the maximum range. If the launcher gives the projectile a muzzle velocity of 25 m/s, what is the range of the projectile? [*Hint:* The acceleration due to gravity on the Moon is only one sixth of that on the Earth.] 3.8×10^2 m

73. ●● In 2004 two Martian probes successfully landed on the Red Planet. The final phase of the landing involved bouncing the probes until they came to rest (they were surrounded by protective inflated "balloons"). During one of the bounces, the telemetry (electronic data sent back to Earth) indicated that the probe took off at 25.0 m/s at an angle of 20° and landed 110 m away (and then bounced again). Assuming the landing region was level, determine the acceleration due to gravity near the Martian surface. 3.65 m/s²

74. ●● In laboratory situations, a projectile's range can be used to determine its speed. To see how this is done, suppose a ball rolls off a horizontal table and lands 1.5 m out from the edge of the table. If the tabletop is 90 cm above the floor, determine (a) the time the ball is in the air, and (b) the ball's speed as it left the table top. (a) 0.43 s (b) 3.5 m/s

75. ●● A stone thrown off a bridge 20 m above a river has an initial velocity of 12 m/s at an angle of 45° above the horizontal (▼Fig. 3.31). (a) What is the range of the stone? (b) At what velocity does the stone strike the water? (a) 26 m (b) 23 m/s at 68° below horizontal

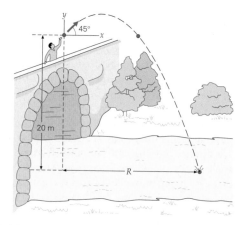

▲ **FIGURE 3.31 A view from the bridge** See Exercise 75.

76. ●● If the maximum height reach by a projectile launched on level ground is equal to half the projectile's range, what is the launch angle? 63°

77. ●● William Tell is said to have shot an apple off his son's head with an arrow. If the arrow was shot with an initial speed of 55 m/s and the boy was 15 m away, at what launch angle did Bill aim the arrow? (Assume that the arrow and apple are initially at the same height above the ground.) 1.4°

78. ●●● This time, William Tell is shooting at an apple that hangs on a tree (▼Fig. 3.32). The apple is a horizontal distance of 20.0 m away and at a height of 4.00 m above the ground. If the arrow is released from a height of 1.00 m above the ground and hits the apple 0.500 s later, what is the arrow's initial velocity? 40.9 m/s, 11.9° above horizontal

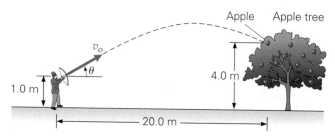

▲ **FIGURE 3.32 Hit the apple** See Exercise 78. (Not drawn to scale.)

79. ●●● A hockey player hits a "slap shot" in practice at a horizontal distance of 15 m from the net (with no goalie present). The net is 1.2 m high, and the puck is initially hit at an angle of 5.0° above the horizontal with a speed of 50 m/s. Does the puck make it into the net? yes, see ISM

80. ●●● In two attempts, a javelin is thrown at angles of 35° and 60°, respectively, to the horizontal from the same height and at the same speed in each case. For which throw does the javelin go farther, and how many times farther? (Assume that the landing place is at the same height as the launching place.) $R_{35} = 1.1R_{60}$

81. ●●● A ditch 2.5 m wide crosses a trail bike path (▼Fig 3.33). An upward incline of 15° has been built up on the approach so that the top of the incline is level with the top of the ditch. What is the minimum speed a trail bike must be moving to clear the ditch? (Add 1.4 m to the range for the back of the bike to clear the ditch safely.) 8.7 m/s

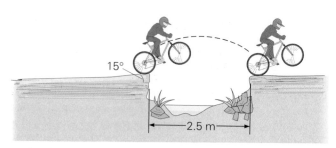

▲ **FIGURE 3.33 Clear the ditch** See Exercise 81. (Not drawn to scale.)

82. ●●● A ball rolls down a roof that makes an angle of 30° to the horizontal (▼Fig. 3.34). It rolls off the edge with a speed of 5.00 m/s. The distance to the ground from that point is two stories or 7.00 m. (a) How long is the ball in the air? (b) How far from the base of the house does it land? (c) What is its speed just before landing?
(a) 0.967 s (b) 4.18 m (c) 12.8 m/s

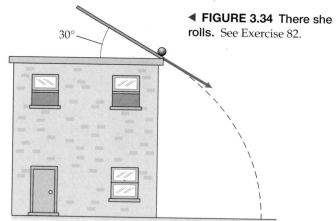

◄ **FIGURE 3.34 There she rolls.** See Exercise 82.

83. ●●● A quarterback passes a football—at a velocity of 50 ft/s at an angle of 40° to the horizontal—toward an intended receiver 30 yd downfield. The pass is released 5.0 ft above the ground. Assume that the receiver is stationary and that he will catch the ball if it comes to him. Will the pass be completed? If not, will the throw be long or short? the pass is short

84. ●●● A 2.05-m-tall basketball player takes a shot when he is 6.02 m from the basket (at the three-point line). If the launch angle is 25° and the ball was launched at the level of the player's head, what must be the release speed of the ball for the player to make the shot? The basket is 3.05 m above the floor. 10.9 m/s

85. ●●● The hole on a level, elevated golf green is a horizontal distance of 150 m from the tee and at an elevation of 12.0 m above the tee. A golfer hits a ball at an angle 10.0° greater than that of the hole above the tee and makes a hole in one! (a) Sketch the situation. (b) What was the initial speed of the ball? (c) Suppose the next golfer hits her ball toward the hole at the same speed, but at an angle of 10.5° greater than that of the hole above the tee. Does the ball go in the hole, or is the shot too long or too short?
(a) see ISM (b) 66.0 m/s (c) too long

3.4 Relative Velocity

86. **MC** You are traveling in a car on a straight, level road going 70 km/h. A car coming toward you appears to be traveling 130 km/h. How fast is the other car going: (a) 130 km/h, (b) 60 km/h, (c) 70 km/h, or (d) 80 km/h? (b)

87. **MC** Two cars approach each other on a straight, level high-way. Car A travels at 60 km/h and car B at 80 km/h. The driver of car B sees car A approaching at a speed of (a) 60 km/h, (b) 80 km/h, (c) 20 km/h, (d) greater than 100 km/h. (d)

88. **MC** For the situation in Exercise 87, at what speed does the driver of car A see car B approaching: (a) 60 km/h. (b) 80 km/h, (c) 20 km/h, or (d) greater than 100 km/h? (d)

89. **CQ** We often take the Earth or ground as a stationary frame of reference. Is this really true? Explain.
No. Earth undergoes several movements

90. **CQ** A student walks on a treadmill moving at 4.0 m/s and remains at the same place in the gym. (a) What is the student's velocity relative to the gym floor? (b) What is the student's speed relative to the treadmill?
(a) zero (b) 4.0 m/s

91. **CQ** You are running in the rain along a straight sidewalk to your dorm. If the rain is falling vertically downward relative to the ground, how should you hold your umbrella so as to minimize the rain landing on you? Explain. tilted forward

92. **CQ** When driving to the basket for a layup, a basketball player usually tosses the ball gently upward relative to herself. Explain why? ball already has a relative velocity to the ground as the player jumps up

93. **CQ** When you are riding in a fast-moving car, in what direction would you throw an object up so it will return to your hand? Explain. straight up, see ISM

94. ● While you are traveling in a car on a straight, level interstate highway at 90 km/h, another car passes you in the same direction; its speedometer reads 120 km/h. (a) What is your velocity relative to the other driver? (b) What is the other car's velocity relative to you?
(a) −30 km/h (b) +30 km/h

95. ● A shopper is in a hurry to catch a bargain in a department store. She walks up the escalator, rather than letting it carry her, at a speed of 1.0 m/s relative to the escalator. If the escalator is 10 m long and moves at a speed of 0.50 m/s, how long does it take for the shopper to get to the next floor?
6.7 s

96. ● A person riding in the back of a pickup truck traveling at 70 km/h on a straight, level road throws a ball with a speed of 15 km/h relative to the truck in the direction opposite to the truck's motion. What is the velocity of the ball (a) relative to a stationary observer by the side of the road, and (b) relative to the driver of a car moving in the same direction as the truck at a speed of 90 km/h?
(a) +55 km/h (b) −35 km/h

97. ● In Exercise 96, what are the relative velocities if the ball is thrown in the direction of the truck?
(a) +85 km/h (b) −5 km/h

98. ●● In a 500-m stretch of a river, the speed of the current is a steady 5.0 m/s. How long does a boat take to finish a round trip (upstream and downstream) if the speed of the boat is 7.5 m/s relative to still water? 4.0 min

99. ●● A moving walkway in an airport is 75 m long and moves at a speed of 0.30 m/s. A passenger, after traveling 25 m while standing on the walkway, starts to walk at a speed of 0.50 m/s relative to the surface of the walkway. How long does she take to travel the total distance of the walkway?
146 s = 2.43 min

100. **IE** ●● A swimmer swims north at 0.15 m/s relative to still water across a river that flows at a rate of 0.20 m/s from west to east. (a) The general direction of the swimmer's velocity, relative to the riverbank, is (1) north of east, (2) south of west, (3) north of west, (4) south of east. (b) Calculate the swimmer's velocity relative to the riverbank.
(a) (1) north of east (b) 0.25 m/s at 37° north of east

101. ●● A boat that travels at a speed of 6.75 m/s in still water is to go directly across a river and back (▼ Fig. 3.35). The current flows at 0.50 m/s. (a) At what angle(s) must the boat be steered? (b) How long does it take to make the round trip? (Assume that the boat's speed is constant at all times, and neglect turnaround time.)
(a) same both ways, 4.25° up stream (b) 44.6 s

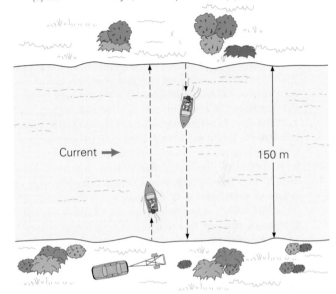

▲ **FIGURE 3.35 Over and back** See Exercise 101. (Not drawn to scale.)

102. **IE** ●● It is raining, and there is no wind. When you are sitting in a stationary car, the rain falls straight down relative to the car and the ground. But when you're driving, the rain appears to hit the windshield at an angle. (a) As the velocity of the car increases, this angle (1) also increases, (2) remains the same, (3) decreases. Why? (b) If the raindrops fall straight down at a speed of 10 m/s, but appear to make an angle of 25° to the vertical, what is the speed of the car?
(a) (1) also increases (b) 4.7 m/s

103. ●● If the flow rate of the current in a straight river is greater than the speed of a boat in the water, the boat cannot make a trip *directly across* the river. Prove this statement. see ISM

104. **IE** ●● You are in a fast powerboat that is capable of a sustained steady speed of 20.0 m/s in still water. On a swift straight section of a river you travel parallel to the bank of the river. You note that you take 15.0 s to go between two trees on the riverbank that are 400 m apart. (a) Are you traveling (1) with the current, (2) against the current, or (3) is there no current? (b) If there is a current [reasoned in part (a)], determine its speed. (a) (1) with the current (b) 6.7 m/s

105. ●● A motorboat is capable of traveling at a steady 5.00 m/s in still water. The motorboat heads across a small river (200 m wide) at an angle of 25° upstream from directly across the river. The boat ends up 40.0 m upstream from the "straight across" direction when it reaches the far bank. Determine the speed of the river current. 1.21 m/s

106. ●●● A shopper in a mall is on an escalator that is moving downward at an angle of 41.8° below the horizontal at a constant speed of 0.75 m/s. At the same time a little boy drops a toy parachute from a floor above the escalator and it descends at a steady vertical speed of 0.50 m/s. Determine the speed of the parachute toy as observed from the moving escalator. 0.56 m/s

107. ●●● An airplane is flying at 150 mi/h (its speed in still air) in a direction such that with a wind of 60.0 mi/h blowing from east to west, the airplane travels in a straight line southward. (a) What must be the plane's heading (direction) for it to fly directly south? (b) If the plane has to go 200 mi in the southward direction, how long does it take?
(a) 24° east of south (b) 1.5 h

Comprehensive Exercises

108. A field goal is attempted when the football is at the center of the field, 40 yd from the goalposts. If the kicker gives the ball a velocity of 70 ft/s toward the goalposts at an angle of 45° to the horizontal, will the kick be good? (The crossbar of the goalposts is 10 ft above the ground, and the ball must be higher than the crossbar when it reaches the goalposts for the field goal to be good.)
the kick is good

109. The apparatus for a popular lecture demonstration is shown in ▼Fig. 3.36. A gun is aimed directly at a can, which is simultaneously released when the gun is fired. This gun won't miss as long as the initial speed of the bullet is sufficient to reach the falling target before the target hits the floor. Verify this statement, using the figure. [*Hint*: Note that $y_0 = x \tan \theta$.] see ISM

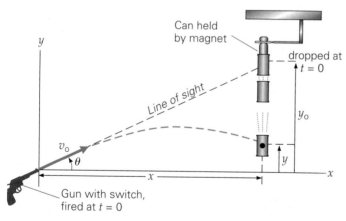

▲ **FIGURE 3.36 A sure shot** See Exercise 109. (Not drawn to scale.)

110. **IE** A shot-putter launches the shot from a vertical distance of 2.0 m off the ground (from just above her ear) at a speed of 12.0 m/s. The initial velocity is at an angle of 20° above the horizontal. Assume the ground is flat. (a) Compared to a projectile launched at the same angle and speed at ground level, would the shot be in the air (1) a longer time, (2) a shorter time, or (3) the same amount of time? (b) Justify your answer explicitly, determine the shot's range and velocity just before impact in unit-vector (component) notation. (a) (1) a longer time (b) see ISM, 13.3 m, $(11.3 \text{ m/s}) \, \hat{\mathbf{x}} + (-7.46 \text{ m/s}) \, \hat{\mathbf{y}}$

111. An early technique for "dropping" a nuclear bomb was not to drop it, but to let it go while the plane was climbing at a rapid rate. The idea was to "toss" it upward at a steep angle, thus giving the plane time to turn and get away before the bomb exploded. Assume the plane is traveling at 600 km/h when it releases the bomb at an angle of 75° above the horizontal. Also assume that the plane releases the bomb at an altitude of 4000 m above the ground, and the bomb is set to detonate at an altitude of 500 m above the ground. Ignoring air resistance, (a) how long does the plane have to get away before the bomb detonates? (b) What maximum height above ground level does the bomb reach? (c) What is the bomb's speed just as it detonates?
(a) 47.8 s (b) 5.32×10^3 m (c) 310 m/s

112. At a merging on-ramp of a busy Los Angeles freeway, car A is moving directly east on the freeway at a steady speed of 35.0 m/s. Car B is merging onto the freeway from the on-ramp, which points 10° north of due east, moving at 30.0 m/s. (See ▼Fig. 3.37.) If they collide, it will be at the point marked x in the figure, which initially is 350 m down the road from the position of car A. Use the x–y coordinate system to signify E–W versus N–S directions. (a) What is the velocity of car B relative to car A? (b) Show that they do *not* collide at point x. (c) Determine how far apart the cars are (and which is ahead) when car B reaches point x.
(a) $(-5.5 \text{ m/s}) \, \hat{\mathbf{x}} + (5.2 \text{ m/s}) \, \hat{\mathbf{y}}$ (b) no collision (c) 63 m

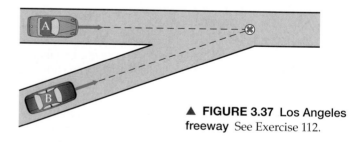

▲ **FIGURE 3.37 Los Angeles freeway** See Exercise 112.

The following Physlet Physics Problems can be used with this chapter.
PHYSLET® 3.1, 3.2, 3.3, 3.4, 3.5, 3.6, 3.7, 3.8, 3.9, 3.10, 3.11, 3.12, 9.1, 9.2, 9.7, 9.9

FORCE AND MOTION

PHYSICS FACTS

- Isaac Newton was born on Christmas Day, 1642, the same day that Galileo died. (By our current Gregorian calendar, Newton's birthdate is January 4, 1643. England did not begin using the Gregorian calendar until 1752.)

- Newton
  - found white light to be a mixture of colors, theorized that light was made up of particles, which he called corpuscles, rather than waves. We now have the dual nature of light, with light both behaving as a wave and made up of particles called photons.
  - developed the fundamentals of calculus. Gottfried Leibniz, a German mathematician, independently developed a similar version of calculus. There was a life-long, bitter dispute between Newton and Leibniz over who should receive the credit for doing so first.
  - built the first reflecting telescope with a power of 40X

- The astronomer Edmond Halley used Newton's work on gravitation and orbits to predict that a comet he had observed in 1682 would return in 1758. The comet returned as predicted and was named Halley's Comet in his honor. Contrary to popular belief, Halley did not discover the comet. Its periodic appearance had been recorded since 263 BCE, when it was first seen been by Chinese astronomers. Halley died in 1742 and did not get to see the return of his comet.

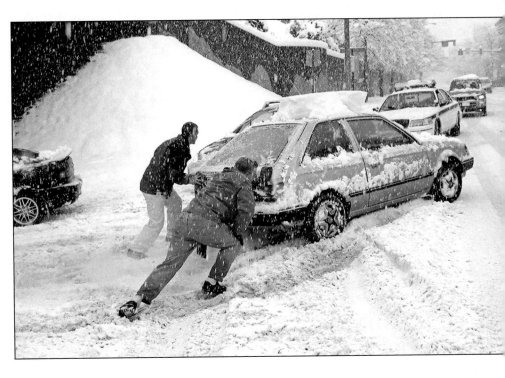

Y ou don't have to understand any physics to know that what's needed to get the car in the picture (or anything else) moving is a push or a pull. If the frustrated men (or the tow truck that may soon be called) can apply enough *force*, the car will move.

But what's keeping the car stuck in the snow? A car's engine can generate plenty of force—so why doesn't the driver just put the car into reverse and back out? For a car to move, another force is needed besides that exerted by the engine: *friction*. Here, the problem is most likely that there is not enough friction between the tires and the snow.

In Chapters 2 and 3, we learned how to analyze motion in terms of kinematics. Now our attention turns to the study of *dynamics*—that is, what *causes* motion and changes in motion. This will lead us to the concept of force and inertia.

The study of force and motion occupied many early scientists. The English scientist Isaac Newton (1642–1727 ▶Fig. 4.1) summarized the various relationships and principles of those early scientists into three statements, or laws, which not surprisingly are known as *Newton's laws of motion*. These laws sum up the concepts of dynamics. In this chapter, you'll learn what Newton had to say about force and motion.

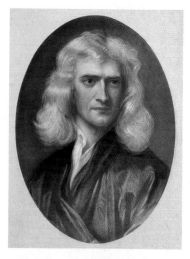

▲ **FIGURE 4.1** Isaac Newton
Newton (1642–1727), one of the greatest scientific minds of all time, made fundamental contributions to mathematics, astronomy, and several branches of physics, including optics and mechanics. He formulated the laws of motion and universal gravitation (Chapter 7) and was one of the inventors of calculus. He did some of his most profound work when he was in his mid-twenties.

Note: In the notation $\sum \vec{F}_i$, the Greek letter sigma means the "sum of" the individual forces, as indicated by the i subscript: $\sum \vec{F}_i = \vec{F}_1 + \vec{F}_2 + \vec{F}_3 + \cdots$, that is, a vector sum. The i subscripts are sometimes omitted as being understood, and we write $\sum \vec{F}$.

Teaching tip: Explain that balanced and unbalanced forces correspond to zero and nonzero net forces, respectively; they are not merely another classification of single forces.

▶ **FIGURE 4.2** Net force
(a) Opposite forces are applied to a crate. **(b)** If the forces are of equal magnitude, the vector resultant, or the net force acting on the crate is zero. The forces acting on the crate are said to be balanced. **(c)** If the forces are unequal in magnitude, the resultant is not zero. A nonzero net force (F_{net}), or unbalanced force, then acts on the crate, producing an acceleration (for example, setting the crate in motion if it was initially at rest).

4.1 The Concepts of Force and Net Force

OBJECTIVES: To (a) relate force and motion, and (b) explain what is meant by a net or unbalanced force.

Let's first take a closer look at the meaning of force. It is easy to give examples of forces, but how would you generally define this concept? An operational definition of force is based on observed effects. That is, a force is described in terms of what it does. From your own experience, you know that *forces can produce changes in motion.* A force can set a stationary object into motion. It can also speed up or slow down a moving object or change the direction of its motion. In other words, a force can produce a change in velocity (speed and/or direction)—that is, an acceleration. Therefore, an observed *change* in motion, including motion starting from rest, is evidence of a force. This concept leads to a common definition of **force**:

> A force is something that is capable of changing an object's state of motion, that is, changing its velocity.

The word *capable* is very significant here. It takes into account the fact that a force may be acting on an object, but its capability to produce a change in motion may be balanced, or canceled, by one or more other forces. The net effect is then zero. Thus, a single force *may not necessarily* produce a change in motion. However, it follows that if a force acts *alone*, the object on which it acts *will* accelerate.

Since a force can produce an acceleration—a vector quantity—force must itself be a vector quantity, with both magnitude and direction. When several forces act on an object, you will often be interested in their combined effect—the net force. The **net force**, $\vec{F}_{net}$, is the vector sum $\sum \vec{F}_i$, or resultant, of all the forces acting on an object or system. (See the note in the margin.) Consider the opposite forces illustrated in ▼Fig. 4.2a. The net force is zero when forces of equal magnitude act in opposite directions (Fig. 4.2b, where signs are used for directions). Such forces are said to be balanced forces. A nonzero net force is referred to as an unbalanced force (Fig. 4.2c). In this case, the situation can be analyzed as though only one force equal to the net force were acting. An unbalanced, or nonzero, net force always produces an acceleration. In some instances, an applied unbalanced force may also deform

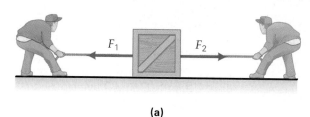

(a)

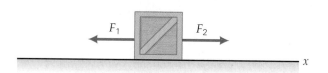

(b) Zero net force (balanced forces)

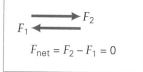

$F_{net} = F_2 - F_1 = 0$

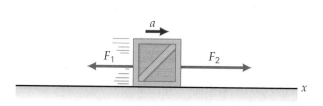

(c) Nonzero net force (unbalanced forces)

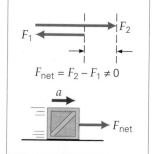

$F_{net} = F_2 - F_1 \neq 0$

an object, that is, change its size and/or shape (as we shall see in Chapter 9). A deformation involves a change in motion for some part of an object; hence, there is an acceleration.

Forces are sometimes divided into two types or classes. The more familiar of these classes is *contact forces*. Such forces arise because of physical contact between objects. For example, when you push on a door to open it or throw or kick a ball, you exert a contact force on the door or ball.

The other class of forces is called *action-at-a-distance forces*. Examples of these forces include gravity, the electrical force between two charges, and the magnetic force between two magnets. The Moon is attracted to the Earth and maintained in orbit by a gravitational force, but there seems to be nothing physically transmitting that force. In Chapter 30, you will learn the modern view of how such action-at-a-distance forces are thought to be transmitted.

Now, with a better understanding of the concept of force, let's see how force and motion are related through Newton's laws.

4.2 Inertia and Newton's First Law of Motion

OBJECTIVES: To (a) state and explain Newton's first law of motion, and (b) describe inertia and its relationship to mass.

The groundwork for Newton's first law of motion was laid by Galileo. In his experimental investigations, Galileo dropped objects to observe motion under the influence of gravity. (See the related Insight in Chapter 2.) However, the relatively large acceleration due to gravity causes dropped objects to move quite fast and quite far in a short time. From the kinematic equations in Chapter 2, you can see that 3.0 s after being dropped, an object in free fall has a speed of about 29 m/s (64 mi/h) and has fallen a distance of 44 m (about 48 yd, or almost half the length of a football field). Thus, experimental measurements of free-fall distance versus time were particularly difficult to make with the instrumentation available in Galileo's time.

To slow things down so that he could study motion, Galileo used balls rolling on inclined planes. He allowed a ball to roll down one inclined plane and then up another with a different degree of incline (▼Fig. 4.3). Galileo noted that the ball rolled to approximately the same height in every case, but it rolled farther in the horizontal direction when the angle of incline was smaller. When allowed to roll onto a horizontal surface, the ball traveled a considerable distance and went even farther when the surface was made smoother. Galileo wondered how far the ball would travel if the horizontal surface could be made perfectly smooth (frictionless). Although this situation is impossible to attain experimentally, Galileo reasoned that in this ideal case with an infinitely long surface, the ball would continue to travel indefinitely with straight-line, uniform motion, since there would be nothing (no net force) to cause its motion to change.

According to Aristotle's theory of motion, which had been accepted for about 1500 years prior to Galileo's time, the normal state of a body was to be at rest (with the exception of celestial bodies, which were thought to be naturally in motion). Aristotle probably observed that objects moving on a surface tend to slow down and come to rest, so this conclusion would have seemed logical to him. However, from his experiments, Galileo concluded that bodies in motion exhibit the behavior of maintaining that motion and that if an object is initially at rest, it will remain so, unless something causes it to move.

Exploration 4.2 Change the Two Forces Applied

Exploration 4.3 Change the Force Applied to Get to the Goal

Teaching tip: Have students name different types of forces. Always try to identify what is exerting the force and what the force is acting on.

Illsutration 3.2 Motion on an Incline

◄ **FIGURE 4.3 Galileo's experiment** A ball rolls farther along the upward incline as the angle of incline is decreased. On a smooth, horizontal surface, the ball rolls a greater distance before coming to rest. How far would the ball travel on an ideal, perfectly smooth surface? (The ball would slide in this case because of the absence of friction.)

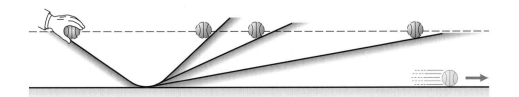

Galileo called this tendency of an object to maintain its initial state of motion **inertia.** That is,

> Inertia is the natural tendency of an object to maintain a state of rest or to remain in uniform motion in a straight line (constant velocity).

For example, if you've ever tried to stop a slowly rolling automobile by pushing on it, you felt its resistance to a change in motion, to slowing down. Physicists describe the property of inertia in terms of observed behavior. A comparative example of inertia is illustrated in ◄Fig. 4.4. If the two punching bags have the same density (mass per unit volume; see Chapter 1), the larger one has more mass and therefore more inertia, as you would quickly notice when you try to punch both bags.

Newton related the concept of inertia to mass. Originally, he called mass a quantity of matter, but he later redefined it as follows:

Note: Inertia is *not* a force.

> Mass is a quantitative measure of inertia.

That is, a massive object has more inertia, or more resistance to a change in motion, than does a less massive object. For example, a car has more inertia than a bicycle.

Newton's first law of motion, sometimes called the *law of inertia*, summarizes these observations:

Newton's first law—the law of inertia

> In the absence of an unbalanced applied force ($\vec{\mathbf{F}}_{net} = 0$), a body at rest remains at rest, and a body already in motion remains in motion with a constant velocity (constant speed and direction).

That is, if the net force acting on an object is zero, then its acceleration is zero. It may be moving with a constant velocity, or be at rest—in both cases, $\Delta\vec{\mathbf{v}} = 0$ or $\vec{\mathbf{v}} = $ constant.

4.3 Newton's Second Law of Motion

OBJECTIVES: To (a) state and explain Newton's second law of motion, (b) apply it to physical situations, and (c) distinguish between weight and mass.

A change in motion, or an acceleration (that is, a change in velocity—speed and/or direction), is evidence of a net force. All experiments indicate that the acceleration of an object is directly proportional to, and in the direction of, the applied net force; that is,

$$\vec{\mathbf{a}} \propto \vec{\mathbf{F}}_{net}$$

where the boldface symbols with arrows indicate vector quantities. For example, suppose you hit two identical balls. If you hit the second ball twice as hard as the first one (that is, you applied twice as much force), you would expect the acceleration of the second ball to be twice as great as that of the first ball (and still in the direction of the force).

However, as Newton recognized, the inertia or mass of the object also plays a role. For a given net force, the more massive the object, the less its acceleration will be. For example, if you hit two balls of different masses with the same force, the less massive ball would experience a greater acceleration. That is, the acceleration is inversely proportional to mass.

Then we have

$$\vec{\mathbf{a}} \propto \frac{\vec{\mathbf{F}}_{net}}{m}$$

or, in words,

> The acceleration of an object is directly proportional to the net force acting on it and inversely proportional to its mass. The direction of the acceleration is in the direction of the applied net force.

▲ **FIGURE 4.4 A difference in inertia** The larger punching bag has more mass and hence more inertia, or resistance to a change in motion.

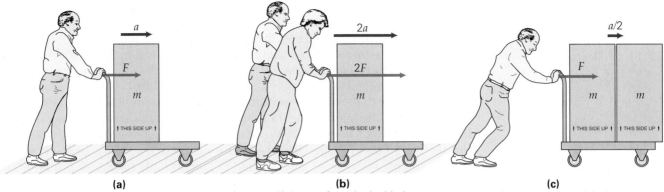

(a)
A nonzero net force accelerates the crate: $a \propto F/m$

(b)
If the net force is doubled, the acceleration is doubled.

(c)
If the mass is doubled, the acceleration is halved.

 ▲ **FIGURE 4.5 Newton's second law** The relationships among force, acceleration, and mass shown here are expressed by Newton's second law of motion (assuming no friction).

▲ Figure 4.5 presents some illustrations of this principle.

Rewritten as $\vec{F}_{net} \propto m\vec{a}$, **Newton's second law of motion** is commonly expressed in equation form as

$$\vec{F}_{net} = m\vec{a} \quad \textit{Newton's second law} \qquad (4.1)$$

SI unit of force: newton (N) or kilogram-meter per second squared ($\text{kg} \cdot \text{m/s}^2$)

where $\vec{F}_{net} = \Sigma \vec{F}_i$. Equation 4.1 defines the SI unit of force, which is appropriately called the **newton (N)**.

By unit analysis, Eq. 4.1 shows that a newton in base units is defined as $1\ \text{N} = 1\ \text{kg} \cdot \text{m/s}^2$. That is, a net force of 1 N gives a mass of 1 kg an acceleration of $1\ \text{m/s}^2$ (▶Fig. 4.6). The British-system unit of force is the pound (lb). One pound is equivalent to about 4.5 N (actually, 4.448 N). An average apple weighs about 1 N.

Newton's second law, $\vec{F}_{net} = m\vec{a}$, allows the quantitative analysis of force and motion. We might think of it as a cause-and-effect relationship, with the force being the cause and acceleration being the motional effect.

Notice that if the net force acting on an object is zero, the object's acceleration is zero, and it remains at rest or in uniform motion, which is consistent with the first law. For a nonzero net force (an unbalanced force), the resulting acceleration is in the same direction as the net force.*

Weight

Equation 4.1 can be used to relate mass and weight. Recall from Chapter 1 that weight is the gravitational force of attraction that a celestial body exerts on an object. For us, this force is the gravitational attraction of the Earth. Its effects are easily demonstrated: When you drop an object, it falls (accelerates) toward the Earth. Since only one force is acting on the object, its **weight** ($\vec{w}$) is the net force $\vec{F}_{net}$, and the acceleration due to gravity ($\vec{g}$) can be substituted for $\vec{a}$ in Eq. 4.1. We can therefore write, in terms of magnitude,

$$w = mg \qquad (4.2)$$
$$(F_{net} = ma)$$

Thus the weight of an object with 1.0 kg of mass is $w = mg = (1.0\,\text{kg})(9.8\,\text{m/s}^2) = 9.8\,\text{N}$.

That is, 1.0 kg of mass has a weight of approximately 9.8 N, or 2.2 lb near the Earth's surface. Although weight and mass are simply related through Eq. 4.2, keep

*It may appear that Newton's first law is a special case of Newton's second law, but this is not so. The first law *defines* what is called an *inertial reference system* (as we shall see in Chapter 26): a system in which there is no net force, that is, not accelerating, or in which an isolated object is stationary or moves with a constant velocity. If Newton's first law holds, then the second law in the form $\vec{F}_{net} = m\vec{a}$ applies to the system.

Illustration 4.3 Newton's Second Law and Force

Newton's second law—force and acceleration

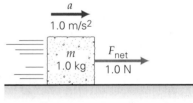

$$F_{net} = ma$$
$$1.0\ \text{N} = (1.0\ \text{kg})\ (1.0\ \text{m/s}^2)$$

▲ **FIGURE 4.6 The newton (N)** A net force of 1.0 N acting on a mass of 1.0 kg produces an acceleration of $1.0\ \text{m/s}^2$ (on a frictionless surface).

Teaching tip: Students find it difficult to distinguish between the net force and the individual forces. Writing Newton's second law as

    sum of forces = $m\vec{a}$

and

    net force = $m\vec{a}$

may help. Use of different colors of chalk when constructing vectors may also help.

Illustration 4.1 Newton's First Law and Reference Frames

in mind that *mass is the fundamental property*. Mass doesn't depend on the value of g, but weight does. As pointed out previously, the acceleration due to gravity on the Moon is about one sixth that on the Earth. The weight of an object on the Moon would thus be one sixth of its weight on the Earth, but its mass, which reflects the quantity of matter it contains and its inertia, would be the same in both places.

Newton's second law, along with the fact that $w \propto m$, explains why all objects in free fall have the same acceleration. Consider, for example, two falling objects, one with twice the mass of the other. The object with twice as much mass would have twice as much weight, or two times as much gravitational force acting on it. But the more massive object also has twice the inertia, so twice as much force is needed to give it the same acceleration. Expressing this relationship mathematically, for the smaller mass (m), we can write $a = F_{net}/m = mg/m = g$, and for the larger mass ($2m$), we have the same acceleration: $a = F_{net}/m = 2mg/(2m) = g$ (▸Fig. 4.7). Some other effects of g, which you may have experienced, are discussed in Insight 4.1.

INSIGHT · 4.1 *g's* OF FORCE AND EFFECTS ON THE HUMAN BODY

The value of g at the Earth's surface is referred to as the *standard acceleration* and is used as a nonstandard unit. For example, when a spacecraft lifts off, astronauts are said to experience an acceleration of "several g's." This expression means that the astronauts' acceleration is several times the standard acceleration g. Since $g = w/m$, we can also think of g as the (weight) *force per unit mass*. Thus, the term **g's of force** is used to express force in terms of multiples of the standard acceleration.

To help understand this nonstandard unit of force, let's look at some examples. During the takeoff of a jet airliner, passengers experience an average horizontal force of about $0.20g$. This means that as the plane accelerates down the runway, the seat back exerts a horizontal force on you of about one fifth of your weight (to accelerate you along with the plane), but you experience a feeling of being pushed back into the seat. On takeoff at an angle of 30°, the force increases to about $0.70g$.

When a person is subjected to several g's vertically, blood can begin to pool in the lower extremities, which may cause blood vessels to distend or capillaries to rupture. Under such conditions, the heart has a difficult time pumping blood throughout the body. At a force of about $4g$, the pooling of blood in the lower body deprives the head of sufficient oxygen. Lack of blood circulation to the eyes can cause temporary blindness, and if the brain is deprived of oxygen, a person becomes disoriented and quickly "blacks out" or loses consciousness. The average person can withstand several g's of force for only a short period of time.

The maximum force on astronauts in a space shuttle on blastoff is about $3g$. But jet fighter pilots are subjected to as much as $9g$ when pulling out of a downward dive. These pilots wear "g-suits," which are designed to prevent blood pooling. The common g-suit is inflated by compressed air and applies pressure to the pilot's lower body to prevent blood from accumulating there. Work is being done on the development of a hydrostatic g-suit that contains liquid, which is less restrictive than air. When the number of g's increases, the liquid, like the blood in the body, flows into the lower part of the suit and applies pressure to the legs.

On the Earth, where only $1g$ is experienced, a partial "g-suit" of sorts is being used to prevent blood clots in patients who

have undergone hip replacement surgery. Each year 400 to 800 people die in the first three months after such surgery, primarily because blood clots form in a leg, break off into the bloodstream, and lodge in the lungs—giving rise to a condition called *pulmonary embolism*. In other cases, a blood clot in the leg may slow the flow of blood to the heart. These complications arise more often after hip replacement surgery than after almost any other surgery and occur after the patient has left the hospital.

Studies have shown that pneumatic (operated by air) compression of the legs during the hospital stay reduces these risks. A plastic leg cuff inflates every few minutes, forcing blood from the lower leg (Fig. 1). This mechanical massaging prevents blood from pooling in the veins and clotting. By using both this technique and anticlotting drug therapy, it is hoped that many of the postoperative deaths can be prevented.

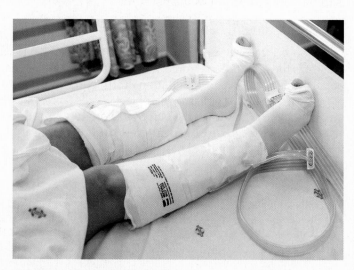

FIGURE 1 Pneumatic massage The leg cuffs inflate periodically, forcing the blood from the lower legs and preventing it from pooling in the veins.

Newton's second law allows us to analyze dynamic situations. In using this law, you should keep in mind that F_{net} is the *magnitude of the net force* and *m* is the *total mass of the system*. The boundaries defining a system may be real or imaginary. For example, a system might consist of all the gas molecules in a particular sealed vessel. But you might also define a system to be all the gas molecules in an arbitrary cubic meter of air. In studying dynamics, we often have occasion to work with systems made up of one or more discrete masses—the Earth and Moon, for instance, or a series of blocks on a tabletop, or a tractor and wagon, as in Example 4.1.

Example 4.1 ■ Newton's Second Law: Finding Acceleration

A tractor pulls a loaded wagon on a level road with a constant horizontal force of 440 N (▼ Fig. 4.8). If the total mass of the wagon and its contents is 275 kg, what is the magnitude of the wagon's acceleration? (Ignore any frictional forces.)

Thinking It Through. This problem is a direct application of Newton's second law. Note that the total mass is given; we treat the two separate masses (wagon and contents) as one and look at the whole system.

Solution. Listing the data, we have

Given: $F = 440$ N   *Find:*   *a* (acceleration)
        $m = 275$ kg

In this case, F is the net force, and the acceleration is given by Eq. 4.2, $F_{net} = ma$. Solving for the magnitude of *a*,

$$a = \frac{F_{net}}{m} = \frac{440 \text{ N}}{275 \text{ kg}} = 1.60 \text{ m/s}^2$$

and the direction of *a* is that in which the tractor is pulling.

Note that *m* is the *total* mass of the wagon and its contents. If the masses of the wagon and its contents had been given separately—say, $m_1 = 75$ kg and $m_2 = 200$ kg, respectively—they would have been added together in Newton's law: $F_{net} = (m_1 + m_2)a$. Also, in reality, there would be an opposing force of friction f. Suppose there were an effective frictional force of $f = 140$ N. In this case, the net force would be the vector sum of the force exerted by the tractor and the frictional force. Then the acceleration would be (using directional signs)

$$a = \frac{F_{net}}{m} = \frac{F - f}{m_1 + m_2} = \frac{440 \text{ N} - 140 \text{ N}}{275 \text{ kg}} = 1.09 \text{ m/s}^2$$

Again, the direction of *a* would be the direction in which the tractor is pulling.

With a constant net force, the acceleration is also constant, so the kinematic equations of Chapter 2 can be applied. Suppose the wagon started from rest ($v_0 = 0$). Could you find how far it traveled in 4.00 s? Using the appropriate kinematic equation (Eq. 2.11, with $x_0 = 0$) for the case with friction, we have

$$x = v_0 t + \tfrac{1}{2}at^2 = 0 + \tfrac{1}{2}(1.09 \text{ m/s}^2)(4.00 \text{ s})^2 = 8.72 \text{ m}$$

Follow-Up Exercise. Suppose the applied force on the wagon is 550 N. With the same frictional force, what would be the wagon's velocity 4.00 s after starting from rest? *(Answers to all Follow-Up Exercises are at the back of the text.)*

Note: The *m* in $F_{net} = ma$, is the total mass of the system.

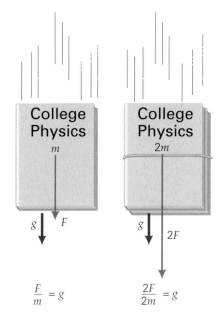

$$\frac{F}{m} = g \qquad \frac{2F}{2m} = g$$

▲ **FIGURE 4.7 Newton's second law and free fall** In free fall, all objects fall with the same constant acceleration *g*. An object with twice the mass of another has twice as much gravitational force acting on it. But with twice the mass, the object also has twice as much inertia, so twice as much force is needed to give it the same acceleration.

◄ **FIGURE 4.8 Force and acceleration** See Example 4.1.

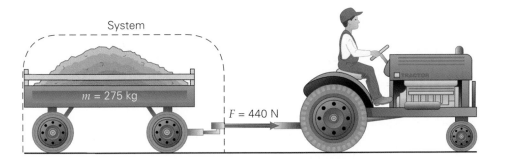

Example 4.2 ■ Newton's Second Law: Finding Mass

A student weighs 588 N. What is her mass?

Thinking It Through. Newton's second law allows us to determine an object's mass if we know the object's weight (force), since g is known.

Solution.

Given: $w = 588$ N   *Find:* m (mass)

Recall that weight is a (gravitational) force and it is related to the mass of an object by $w = mg$ (Eq. 4.2), where g is the acceleration due to gravity (9.80 m/s²). Rearranging the equation, we have

$$m = \frac{w}{g} = \frac{588 \text{ N}}{9.80 \text{ m/s}^2} = 60.0 \text{ kg}$$

In countries that use the metric system, the kilogram unit of mass is used to express "weight" rather than a force unit. It would be said that this student weighs 60.0 "kilos."

Recall that 1 kg of mass has a weight of 2.2 lb on the Earth's surface. Then in British units, she would weigh 60.0 kg (2.2 lb/kg) = 132 lb.

Follow-Up Exercise. (a) A person in Europe is a bit overweight and would like to lose 5.0 "kilos." What would be the equivalent loss in pounds? (b) What is your "weight" in kilos?

As we have seen, a dynamic system may consist of more than one object. In applications of Newton's second law, it is often advantageous, and sometimes necessary, to isolate a given object within a system. This isolation is possible because *the motion of any part of a system is also described by Newton's second law*, as Example 4.3 shows.

Example 4.3 ■ Newton's Second Law: All or Part of the System?

Illustration 4.5 *Pull Your Wagons*

Two blocks with masses $m_1 = 2.5$ kg and $m_2 = 3.5$ kg rest on a frictionless surface and are connected by a light string (▼Fig. 4.9).* A horizontal force (F) of 12.0 N is applied to m_1, as shown in the figure. (a) What is the magnitude of the acceleration of the masses (that is, of the total system)? (b) What is the magnitude of the force (T) in the string? [When a rope or string is stretched taut, it is said to be under tension. For a very light string, the force at the right end of the string has the same magnitude (T) as the force at the left end.]

Thinking It Through. It is important to remember that Newton's second law may be applied to a total system or any part of it (a subsystem, so to speak). This capability allows for the analysis of a particular component of a system, if desired. Identification of all of

▼ **FIGURE 4.9 An accelerated system** See Example 4.3.

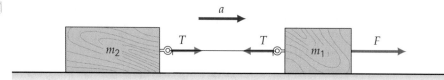

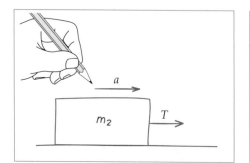
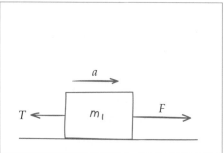

Isolating the masses

*When an object is described as being "light," its mass can be ignored in analyzing the situation given in the problem. That is, its mass is negligible relative to the other masses.

the acting forces is critical, as this Example shows. We then apply $F_{net} = ma$ to each subsystem or component.

Solution. Carefully listing the data and what is to be found:

Given: $m_1 = 2.5$ kg   *Find:*   (a) a (acceleration)
           $m_2 = 3.5$ kg   (b) T (tension, a force)
           $F = 12.0$ N

Given an applied force, the acceleration of the masses can be found from Newton's second law. It is important to keep in mind that Newton's second law applies to the total system *or to any part of it*—that is, to the total mass $(m_1 + m_2)$, to m_1 individually, or to m_2 individually. However, *you must be sure to identify correctly the appropriate force or forces in each case.* The net force acting on the combined masses, for example, is not the same as the magnitude of the net force acting on m_2 considered separately, as will be seen.

(a) First, taking the system as a whole (that is, considering both m_1 and m_2), the net force acting on this system is F. Note that in considering the total system, we are concerned only about the net external force acting on it. The *internal* equal and opposite T forces are not a consideration in this case, since they cancel. Representing the total mass as M, in Newton's second law equation:

$$a = \frac{F_{net}}{M} = \frac{F}{m_1 + m_2} = \frac{12.0 \text{ N}}{2.5 \text{ kg} + 3.5 \text{ kg}} = 2.0 \text{ m/s}^2$$

The acceleration is in the direction of the applied force, as indicated in the figure.

(b) Under tension, a force is exerted on an object by strings (or ropes or wires) and is directed along the string. Note in the figure that we are assuming the tension to be transmitted *undiminished* through the string. That is, the tension is the same everywhere in the string. Thus, the magnitude of T acting on m_2 is the same as that acting on m_1. This is actually true only if the string has zero mass. Only such idealized *light* (that is, of negligible mass) strings or ropes will be considered in this book.

So there is a force of magnitude T on each of the masses, because of tension in the connecting string. To find the value of T, we must consider a *part* of the system that is affected by this force.

Each block may be considered as a separate system to which Newton's second law applies. In these subsystems, the tension comes into play explicitly. Looking at the sketch of the isolated m_2 in Fig. 4.9, we see that the only force acting to accelerate this mass is T. From the values of m_2 and a, the magnitude of this force is given directly by

$$F_{net} = T = m_2 a = (3.5 \text{ kg})(2.0 \text{ m/s}^2) = 7.0 \text{ N}$$

An isolated sketch of m_1 is also shown in Fig. 4.9, and Newton's second law can equally well be applied to this block to find T. The forces must be added vectorially to get the net force on m_1 that produces its acceleration. Recalling that vectors in one dimension can be written with directional signs and magnitudes, we have

$$F_{net} = F - T = m_1 a \quad \text{(direction of F taken to be positive)}$$

Then, solving for T,

$$T = F - m_1 a$$
$$= 12.0 \text{ N} - (2.5 \text{ kg})(2.0 \text{ m/s}^2) = 12.0 \text{ N} - 5.0 \text{ N} = 7.0 \text{ N}$$

Follow-Up Exercise. Suppose that an additional horizontal force to the left of 3.0 N is applied to m_2 in Fig. 4.9. What would be the tension in the connecting string in this case?

The Second Law in Component Form

Not only does Newton's second law hold for any part of a system, but it also applies to each component of the acceleration. For example, a force may be expressed in component notation in two dimensions as follows:

$$\sum \vec{F}_i = m\vec{a}$$

and

$$\sum (F_x \hat{x} + F_y \hat{y}) = m(a_x \hat{x} + a_y \hat{y}) = ma_x \hat{x} + ma_y \hat{y} \qquad (4.3a)$$

Hence, to satisfy both x and y directions independently, we have

$$\Sigma F_x = ma_x \quad \text{and} \quad \Sigma F_y = ma_y \qquad (4.3b)$$

and Newton's second law applies separately to each component of motion. Note that *both* equations must be true. (Also, $\Sigma F_z = ma_z$ in three dimensions.) Example 4.4 demonstrates how the second law is applied using components.

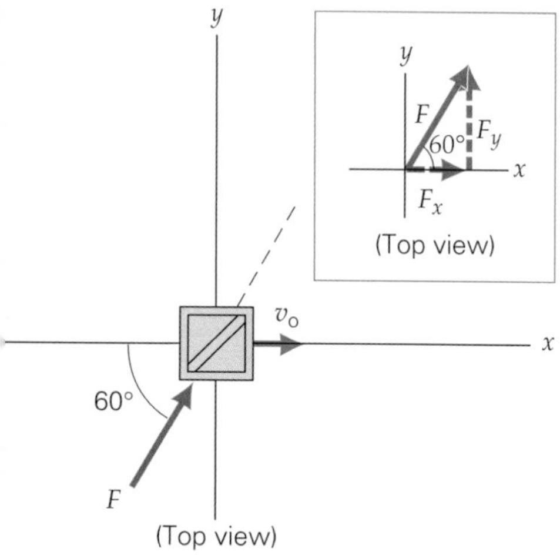

Example 4.4 ■ Newton's Second Law: Components of Force

A block of mass 0.50 kg travels with a speed of 2.0 m/s in the positive x-direction on a flat, frictionless surface. On passing through the origin, the block experiences a constant force of 3.0 N at an angle of 60° relative to the x-axis for 1.5 s (◀Fig. 4.10). What is the velocity of the block at the end of this time?

Thinking It Through. With the force at an angle to the initial motion, it would appear that the solution is complicated. But note in the insert in Fig. 4.10 that the force can be resolved into components. The motion can then be analyzed in each component direction.

Solution. First, listing the given data and what is to be found:

Given: $m = 0.50$ kg   *Find:* $\vec{v}$ (velocity at the end of 1.5 s)
$v_{x_o} = 2.0$ m/s
$v_{y_o} = 0$
$F = 3.0$ N, $\theta = 60°$
$t = 1.5$ s

Let's find the magnitudes of the force components:

$$F_x = F\cos 60° = (3.0\,\text{N})(0.500) = 1.5\,\text{N}$$
$$F_y = F\sin 60° = (3.0\,\text{N})(0.866) = 2.6\,\text{N}$$

Then, applying Newton's second law to each direction to find the components of acceleration, we get

$$a_x = \frac{F_x}{m} = \frac{1.5\,\text{N}}{0.50\,\text{kg}} = 3.0\,\text{m/s}^2$$

$$a_y = \frac{F_y}{m} = \frac{2.6\,\text{N}}{0.50\,\text{kg}} = 5.2\,\text{m/s}^2$$

Next, from the kinematic equation relating velocity and acceleration (Eq. 2.8), the velocity components of the block are given by

$$v_x = v_{x_o} + a_x t = 2.0\,\text{m/s} + (3.0\,\text{m/s}^2)(1.5\,\text{s}) = 6.5\,\text{m/s}$$
$$v_y = v_{y_o} + a_y t = 0 + (5.2\,\text{m/s}^2)(1.5\,\text{s}) = 7.8\,\text{m/s}$$

Then, at the end of the 1.5 s, the velocity of the block is

$$\vec{v} = v_x\hat{x} + v_y\hat{y} = (6.5\,\text{m/s})\hat{x} + (7.8\,\text{m/s})\hat{y}$$

Follow-Up Exercise. (a) What is the direction of the velocity at the end of the 1.5 s? (b) If the force were applied at an angle of 30° (rather than 60°) relative to the x-axis, how would the results of this Example be different?

▲ **FIGURE 4.10 Off the straight and narrow** A force is applied to a moving block when it reaches the origin, and the block then begins to deviate from its straight-line path. See Example 4.4.

PHYSLET®

Exploration 4.4 Set the Force on a Hockey Puck

PHYSLET®

Exploration 4.5 Space Probe with Multiple Engines

PHYSLET®

Exploration 4.6 Pulled Golf Ball Breaks Toward the Hole

4.4 Newton's Third Law of Motion

OBJECTIVES: To (a) state and explain Newton's third law of motion, and (b) identify action–reaction force pairs.

Newton formulated a third law that is as far-reaching in its physical significance as the first two laws. For a simple introduction to the third law, consider the forces involved in seatbelt safety. When the brakes are suddenly applied when you are riding in a moving car, because of your inertia you continue to move forward as the car slows. (The frictional force on the seat of your pants is not enough to stop you.)

Teaching tip: Students have difficulty with contact forces between bodies. Slightly separating the objects on diagrams and constructing the two vectors with different colors may help. In discussions, have students state *what is exerting the force* and *which object the force is acting on.*

In doing so, you exert forward forces on the seatbelt and shoulder strap. The belt and strap exert corresponding backward reaction forces on you, causing you to slow down with the car. If you haven't buckled up, you keep going (Newton's first law) until another backward force, such as that applied by the dashboard or windshield, slows you down.

We commonly think of forces as occurring singly. However, Newton recognized that it is impossible to have a single force. He observed that in any application of force, there is always a mutual interaction; therefore, forces always occur in pairs. An example given by Newton was the following: If you press on a stone with a finger, then the finger is also pressed by, or receives a force from, the stone.

Newton termed the paired forces *action* and *reaction*, and **Newton's third law of motion** is as follows:

| For every force (action), there is an equal and opposite force (reaction). |

In symbol notation, Newton's third law is

$$\vec{\mathbf{F}}_{12} = -\vec{\mathbf{F}}_{21}$$

That is, $\vec{\mathbf{F}}_{12}$ is the force exerted *on* object 1 *by* object 2, and $-\vec{\mathbf{F}}_{21}$ is the equal and opposite force exerted *on* object 2 *by* object 1. (The minus sign indicates the opposite direction.) *Which force is considered the action or the reaction is arbitrary;* $\vec{\mathbf{F}}_{21}$ may be the reaction to $\vec{\mathbf{F}}_{12}$ or vice versa.

Newton's third law at first glance may seem to contradict Newton's second law: If there are always equal and opposite forces, how can there be a nonzero net force? An important thing to remember about the force pair of the third law is that *the action–reaction forces do not act on the same object.* The second law is concerned with force(s) acting on a particular object (or system). The opposing forces of the third law act on *different* objects. Hence, these forces cannot cancel each other out or have a vector sum of zero when we apply the second law to the individual objects.

To illustrate this distinction, consider the situations shown in ▸Fig. 4.11. We often tend to forget the reaction force. For example, in the left portion of Fig. 4.11a, the obvious force that acts on a block sitting on a table is the Earth's gravitational attraction, which is expressed by the weight mg. But, *there has to be another force* acting on the block. For the block not to accelerate, the table must exert an upward force $\vec{\mathbf{N}}$ whose magnitude is equal to the block's weight. Thus, $\Sigma F_y = +N - mg = ma_y = 0$, where the directions of the vectors are indicated by plus and minus signs.

In reaction to $\vec{\mathbf{N}}$, the block exerts a downward force on the table, $-\vec{\mathbf{N}}$, whose magnitude is the same as the block's weight, mg. However, $-\vec{\mathbf{N}}$ is *not* the object's weight. Weight and $-\vec{\mathbf{N}}$ have two different origins: Weight is the action-at-a-distance gravitational force, and $-\vec{\mathbf{N}}$ is a contact force between the two surfaces.

You can easily demonstrate that this upward force on the block is there by placing the block on your hand and holding it stationary—you exert an upward force on the block, and you would feel a reaction force of $-\vec{\mathbf{N}}$ on your hand. If you applied a greater force, that is, $N > mg$, then the block would accelerate upward.

We call the force that a surface exerts on an object a *normal* force and use the symbol N to denote the force. *Normal* means *perpendicular*. The **normal force** that a surface exerts on an object is always perpendicular to the surface. In Fig. 4.11a, the normal force is equal and opposite to the weight of the block. However, the normal force is not always equal and opposite to an object's weight. The normal force is a "reaction" force; it reacts to the situation. Examples of this are given in Figs. 4.11a, b, c, and d, described here with the summation of the vertical components (ΣF_y).

(Fig. 4.11b) Applied force at a downward angle.

$$\Sigma F_y: \quad N - mg - F_y = ma_y = 0, \quad \text{and} \quad N = mg + F_y \quad (N > mg)$$

(Fig. 4.11c) Applied force at an upward angle.

$$\Sigma F_y: \quad N - mg + F_y = ma_y = 0, \quad \text{and} \quad N = mg - F_y \quad (N < mg)$$

Newton's third law—action and reaction

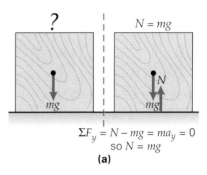

$$\Sigma F_y = N - mg = ma_y = 0$$
$$\text{so } N = mg$$
(a)

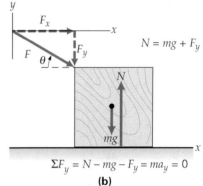

$$\Sigma F_y = N - mg - F_y = ma_y = 0$$
(b)

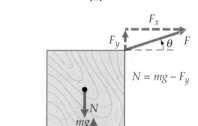

$$\Sigma F_y = N - mg + F_y = ma_y = 0$$
(c)

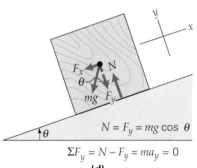

$$N = F_y = mg \cos \theta$$
$$\Sigma F_y = N - F_y = ma_y = 0$$
(d)

▲ **FIGURE 4.11** Distinctions between Newton's second and third laws Newton's second law deals with the forces acting on a particular object (or system). Newton's third law deals with the force pair that acts on different objects. See Conceptual Example 4.5.

(Fig. 4.11d) Block on an inclined plane. (Normal force perpendicular to the surface of the plane.)

$$\Sigma F_y: \quad N - F_y = ma_y = 0, \quad \text{and} \quad N = F_y = mg \cos \theta$$

In this case, the weight component down the plane, F_x, would accelerate the block down the plane in the absence of an equal opposing frictional force between the block and surface of the plane.

Conceptual Example 4.5 ■ Where Are the Newton's Third-Law Force Pairs?

A woman waiting to cross the street holds a briefcase in her hand as shown in ▼Fig. 4.12a. Identify all of the third-law force pairs involving the briefcase in this situation.

Reasoning and Answer. Being held motionless, the briefcase's acceleration is zero, and $\Sigma F_y = 0$. Focusing only on the case, two equal and opposite forces acting on it can be identified—the downward weight of the case and the upward applied force by the hand. However, these two forces *cannot* be a third-law force pair because they act on the *same* object.

On an overall inspection, you should realize that the reaction force to the upward force of the hand on the briefcase is a downward force on the hand. Then how about the reaction force to the weight of the case? Since weight is the attractive gravitational force on the case by the Earth, the corresponding force on the Earth by the case makes up the third-law force pair.

Follow-Up Exercise. The woman inadvertently drops her briefcase as illustrated in Fig. 4.12b. Are there any third-law force pairs in this situation? Explain.

Teaching tip: As an example of a third law force pair, explain that the Earth gravitationally attracts you while you exert an attractive force of equal magnitude on the Earth. Emphasize that these two gravitational forces act on *different objects*. Make sure students don't confuse the foregoing force pair with the following force pair: The Earth exerts a downward gravitational force on you, while the surface of the Earth exerts an upward normal force on you. Point out that theses two forces act on the *same object*.

▶ **FIGURE 4.12 Force pairs of Newton's third law (a)** When a person holds a briefcase, there are two force pairs: a contact pair ($\vec{F}_1$ and $-\vec{F}_1$) and an action-at-a-distance (gravity) pair ($\vec{F}_2$ and $-\vec{F}_2$). The net force acting on the briefcase is zero: The upward contact force ($\vec{F}_1$) balances the downward weight force. Note, however, that the upward contact force and downward weight force are *not* a third-law pair. **(b)** Any third-law force pairs? See the Example Follow-Up Exercise.

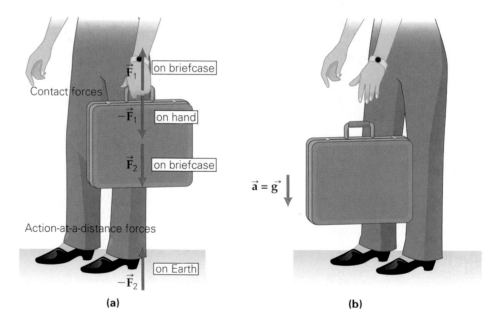

(a)   (b)

Jet propulsion is yet another example of Newton's third law in action. In the case of a rocket, the rocket and exhaust gases exert equal and opposite forces on each other. As a result, the exhaust gases are accelerated away from the rocket, and the rocket is accelerated in the opposite direction. When a big rocket "blasts off," as in a space shuttle launch, it produces a fiery release of exhaust. A common misconception is that the exhaust gases "push" against the launch pad to accelerate the rocket. If this interpretation were true, there would be no space travel, since there is nothing to "push" against in space. The correct explanation is one of action (rocket exerting a force on the gases) and reaction (gases exerting an opposite force on the rocket).

Another action–reaction pair is given in Insight 4.2 on p. 115.

INSIGHT 4.2 SAILING INTO THE WIND—TACKING

A sailboat can easily sail in the direction of the wind (which fills the sails). However, after sailing some distance in the windward direction, the skipper usually wants to return to home port—which involves somehow "sailing into the wind." This may sound rather impossible, but it isn't. It's called *tacking*, and it can be explained and understood by using force vectors and Newton's laws.

A sailboat cannot sail directly upwind, since the wind force on the sail would accelerate the boat backward, opposite the desired direction. The wind filling the sail exerts a force F_s perpendicular to the sail (Fig. 1a). If the boat is steered at an angle relative to the wind direction, there is a component of force parallel to the boat's heading ($F_\parallel$). On this course, some distance upwind is gained, but it would never get the boat back to port. The perpendicular component ($F_\perp$) acts sideways and would put the boat way off course.

So, being an old salt, the skipper "tacks" or maneuvers the boat so that the parallel force component is changed by 90°

(Fig. 1b). The skipper continually repeats the maneuver, and using this zigzag course, the boat gets back to port (Fig. 2a).

What about the perpendicular force component? You might think that this would take the boat way off course. It would, and does a little, but most of the perpendicular force is balanced by the keel of the boat, which is underneath (Fig. 2b). The water resistance exerts an opposite force on the keel, which cancels out most of the sideways perpendicular force, producing little, if any, acceleration in that direction.

(a)

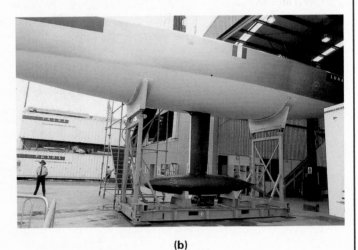

(b)

FIGURE 2 Into the wind (a) As the skipper turns the boat into the wind, the tacking begins. **(b)** The perpendicular force component in tacking would take the boat off course sideways. But, the water resistance on the keel under the boat exerts an opposite force and cancels out most of the sideways force.

F_s (force perpendicular to sail)

$F_\perp$ $F_\parallel$

Wind velocity

(a)

$F_\perp$ $F_\parallel$

Wind velocity

$F_\parallel$

$F_\perp$

(b)

FIGURE 1 Let's go tacking (a) The wind filling the sail exerts a force perpendicular to the sail (F_s). We can resolve this force vector into components. The one parallel to the motion of the boat ($F_\parallel$) has an upwind component. **(b)** By changing the direction of the sail, the skipper can "tack" the boat upwind.

Illustration 4.2 Free-Body Diagrams

LEARN BY DRAWING

Forces on an Object on an Inclined Plane and Free-body Diagrams

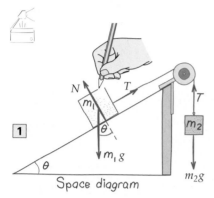

1

Space diagram

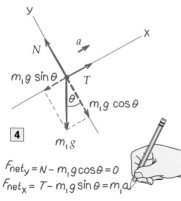

2

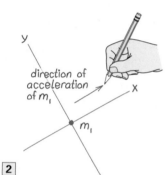

3

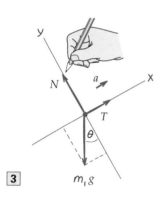

4

$$F_{net_y} = N - m_1 g \cos\theta = 0$$
$$F_{net_x} = T - m_1 g \sin\theta = m_1 a$$

4.5 More on Newton's Laws: Free-Body Diagrams and Translational Equilibrium

OBJECTIVES: To (a) apply Newton's laws in analyzing various situations using free-body diagrams, and (b) understand the concept of translational equilibrium.

Now that you have been introduced to Newton's laws and some applications in analyzing motion, the importance of these laws should be evident. They are so simply stated, yet so far-reaching. The second law is probably the most often applied, because of its mathematical relationship. However, the first and third laws are often used in qualitative analysis, as our continuing study of the different areas of physics will reveal.

In general, we will be concerned with applications that involve constant forces. Constant forces result in constant accelerations and allow the use of the kinematic equations from Chapter 2 in analyzing the motion. When there is a variable force, Newton's second law holds for the *instantaneous* force and acceleration, but the acceleration will vary with time, requiring advanced mathematics to analyze. We will generally limit ourselves to constant accelerations and forces. Several examples of applications of Newton's second law are presented in this section so that you can become familiar with its use. This small but powerful equation will be used again and again throughout this text.

There is still one more item in the problem-solving arsenal that is a great help with force applications—free-body diagrams. These are explained in the following Problem-Solving Strategy.

Problem-Solving Strategy: Free-Body Diagrams

In illustrations of physical situations, sometimes called *space diagrams*, force vectors may be drawn at different locations to indicate their points of application. However, because we are presently concerned only with linear motions, vectors in *free-body diagrams* (FBD) may be shown as emanating from a common point, which is usually chosen as the origin of the *x–y* axes. One of the axes is generally chosen along the direction of the net force acting on a body, since that is the direction in which the body will accelerate. Also, it is often important to resolve force vectors into components, and properly chosen *x–y* axes simplify this task.

In a free-body diagram, the vector arrows do not have to be drawn exactly to scale. However, the diagram should clearly show whether there is a net force and whether forces balance each other in a particular direction. When the forces aren't balanced, we know from Newton's second law that there must be an acceleration.

In summary, the general steps in constructing and using free-body diagrams are as follows (refer to the accompanying Learn by Drawing as you read):

1. Make a sketch, or space diagram, of the situation (if one is not already available) and identify the forces acting on each body of the system. A space diagram is an illustration of the physical situation that identifies the force vectors.

2. Isolate the body for which the free-body diagram is to be constructed. Draw a set of Cartesian axes, with the origin at a point through which the forces act and with one of the axes along the direction of the body's acceleration. (The acceleration will be in the direction of the net force, if there is one.)

3. Draw properly oriented force vectors (including angles) on the diagram, emanating from the origin of the axes. If there is an unbalanced force, assume a direction of acceleration and indicate it with an acceleration vector. Be sure to include only those forces that act on the isolated body of interest.

4. Resolve any forces that are not directed along the *x*- or *y*-axis into *x*- or *y*-components (use plus and minus signs to indicate direction). Use the free-body diagram and force components to analyze the situation in terms of Newton's second law of motion. (*Note:* If you assume that the acceleration is in one direction, and in the solution it comes out with the opposite sign, then the acceleration is actually in the opposite direction from that assumed. For example, if you assume that $\vec{a}$ is in the +*x*-direction, but you get a negative answer, then $\vec{a}$ is in the −*x*-direction.)

Free-body diagrams are a particularly useful way of following one of the Suggested Problem-Solving Procedures in Chapter 1: Draw a diagram as an aid in visualizing and analyzing the physical situation of the problem. *Make it a practice to draw free-body diagrams for force problems, as is done in the following Examples.*

PHYSLET®

Illustration 4.4 Mass on an Incline

PHYSLET®

Exploration 4.1 Vectors for a Box on an Incline

Example 4.6 ■ Up or Down? Motion on a Frictionless Inclined Plane

Two masses are connected by a light string running over a light pulley of negligible friction as illustrated in a previous Learn by Drawing diagram. One mass ($m_1 = 5.0$ kg) is on a frictionless 20° inclined plane, and the other ($m_2 = 1.5$ kg) is freely suspended. What is the acceleration of the masses? (Only the free-body diagram for m_1 is shown in the Learn by Drawing (LBD) diagram. You need to draw the free-body diagram for m_2.)

Thinking It Through. Apply the preceding Problem-Solving Strategy.

Solution. Following our usual procedure, we write

Given: $m_1 = 5.0$ kg   *Find:* $\vec{a}$ (acceleration)
      $m_2 = 1.5$ kg
      $\theta = 20°$

To help visualize the forces involved, we isolate m_1 and m_2 and draw free-body diagrams for each mass. For mass m_1, there are three concurrent forces (forces acting through a common point). These forces are T, its weight m_1g, and N, where T is the tension force of the string on m_1 and N is the normal force of the plane on the block (LBD-3). The forces are shown as emanating from their common point of action. (Recall that a vector can be moved as long as its direction and magnitude are not changed.)

We will start by assuming that m_1 accelerates up the plane, which is taken to be in the $+x$-direction. (It makes no difference whether we assume that m_1 accelerates up or down the plane, as we shall see shortly.) Notice that m_1g (the weight) is broken down into components. The x-component is opposite to the assumed direction of acceleration, and the y-component acts perpendicularly to the plane and is balanced by the normal force N. (There is no acceleration in the y-direction, so there is no net force in this direction.)

Then, applying Newton's second law in component form (Eq. 4.3b) to m_1, we have

$$\Sigma F_{x_1} = T - m_1g \sin\theta = m_1a$$
$$\Sigma F_{y_1} = N - m_1g \cos\theta = m_1a_y = 0 \quad (a_y = 0, \text{ no net force, so the forces cancel})$$

And for m_2,
$$\Sigma F_{y_2} = m_2g - T = m_2a_y = m_2a$$

where the masses of the string and pulley have been neglected. Since they are connected by a string, the accelerations of m_1 and m_2 have the same magnitudes, we can use $a_x = a_y = a$.

Adding the first and last equations to eliminate T, we have

$$m_2g - m_1g \sin\theta = (m_1 + m_2)a$$
$$\text{(net force} = total \text{ mass} \times \text{acceleration)}$$

(Note that this is the equation that would be obtained by applying Newton's second law to the system as a whole, because in the system of both blocks, the $\pm T$ forces are internal forces and cancel.)

Then, solving for a:

$$a = \frac{m_2g - m_1g \sin 20°}{m_1 + m_2}$$
$$= \frac{(1.5 \text{ kg})(9.8 \text{ m/s}^2) - (5.0 \text{ kg})(9.8 \text{ m/s}^2)(0.342)}{5.0 \text{ kg} + 1.5 \text{ kg}}$$
$$= -0.32 \text{ m/s}^2$$

The minus sign indicates that the acceleration is opposite to the assumed direction. That is, m_1 actually accelerates down the plane, and m_2 accelerates upward. As this example shows, if you assume the acceleration to be in the wrong direction, the sign on the result will give you the correct direction anyway.

Could you find the tension force T in the string if you were asked to do so? How this task could be done should be quite evident from the free-body diagram.

Follow-Up Exercise. (a) In this Example, what is the minimum amount of mass for m_2 that would cause m_1 not to accelerate up or down the plane? (b) Keeping the masses the same as in the Example, how should the angle of incline be adjusted so that m_1 would not accelerate up or down the plane?

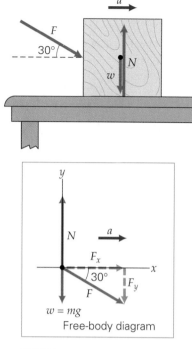

▲ **FIGURE 4.13 Finding force from motional effects** See Example 4.7.

Example 4.7 ■ Components of Force and Free-Body Diagrams

A force of 10.0 N is applied at an angle of 30° to the horizontal on a 1.25-kg block initially at rest on a frictionless surface, as illustrated in ◄Fig. 4.13. (a) What is the magnitude of the block's acceleration? (b) What is the magnitude of the normal force?

Thinking It Through. The applied force may be resolved into components. The horizontal component accelerates the block. The vertical component affects the normal force. (Review Fig. 4.11 if necessary.)

Solution. First writing the given data and what is to be found:

Given: $F = 10.0$ N   *Find:* (a) a (acceleration)
$m = 1.25$ kg   (b) N (normal force)
$\theta = 30°$
$v_o = 0$

Then drawing a free-body diagram for the block, as in Fig. 4.13.

(a) The acceleration of the block can be calculated using Newton's second law. We choose our axes so that a is in the +x-direction. As the free-body diagram shows, only a component (F_x) of the applied force F acts in this direction. The component of F in the direction of motion is $F_x = F \cos \theta$. We apply Newton's second law in the x-direction to calculate the acceleration:

$$F_x = F \cos 30° = ma_x$$

$$a_x = \frac{F \cos 30°}{m} = \frac{(10.0 \text{ N})(0.866)}{1.25 \text{ kg}} = 6.93 \text{ m/s}^2$$

(b) The acceleration found in part (a) is the acceleration of the block, since the block accelerates only in the x-direction (that is, it does not accelerate in the y-direction). Since $a_y = 0$, the sum of the forces in the y-direction must be zero. That is, the downward component of F acting on the block, F_y, and its downward weight force w must be balanced by the upward normal force N that the surface exerts on the block. If this were not the case, then there would be a net force and an acceleration in the y-direction.

Summing the forces in the y-direction with upward taken as positive and

$$\Sigma F_y = N - F_y - w = 0$$

or

$$N - F \sin 30° - mg = 0$$

and

$$N = F \sin 30° + mg = (10.0 \text{ N})(0.500) + (1.25 \text{ kg})(9.80 \text{ m/s}^2) = 17.3 \text{ N}$$

The surface then exerts a force of 17.3 N upward on the block, which balances the sum of the downward forces acting on it.

Follow-Up Exercise. (a) Suppose the applied force on the block is applied for only a short time. What is the magnitude of the normal force after the applied force is removed? (b) If the block slides off the edge of the table, what would be the net force on the block just after it leaves the table (with the applied force removed)?

Problem-Solving Hint

There is no single fixed way to go about solving a problem. However, some general strategies or procedures are helpful in solving problems involving Newton's second law. When using the Suggested Problem-Solving Procedures introduced in Chapter 1, you might include the following steps when solving problems involving force applications:

- Draw a free-body diagram for each individual body, showing all of the forces acting on that body.

- Depending on what is to be found, apply Newton's second law either to the system as a whole (in which case internal forces cancel) or to a part of the system. Basically, *you want to obtain an equation (or set of equations) containing the quantity for which you want to solve.* Review Example 4.3. (If there are two unknown quantities, application of Newton's second law to two parts of the system may give you two equations and two unknowns. See Example 4.6.)

- Keep in mind that Newton's second law may be applied to components of acceleration and that forces may have to be resolved into components to do this. Review Example 4.7.

Exploration 4.7 Atwood's Machine

Translational Equilibrium

Several forces may act on an object without producing an acceleration. In such a case, with $\vec{a} = 0$, we know from Newton's second law that

$$\sum \vec{F}_i = 0 \qquad (4.4)$$

That is, the vector sum of the forces, or the net force, is zero, so the object either remains at rest (as in ▸Fig. 4.14) *or* moves with a constant velocity. In such cases, objects are said to be in **translational equilibrium**. When remaining at rest, an object is said to be in *static translational equilibrium*.

It follows that the sums of the rectangular components of the forces for an object in translational equilibrium are also zero (why?):

$$\begin{aligned} \sum F_x &= 0 \\ \sum F_y &= 0 \end{aligned} \quad \textit{(translational equilibrium only)} \qquad (4.5)$$

For three-dimensional problems, it is noted that $\sum F_z = 0$. However, we will restrict our discussion of forces to two dimensions.

Equations 4.5 give what is often referred to as the **condition for translational equilibrium**. (Conditions for rotational considerations will be given in Chapter 8.) Let's apply this translational-equilibrium condition to a case involving static equilibrium.

Example 4.8 ■ Keep It Straight: In Static Equilibrium

To keep a broken leg bone straight while it is healing sometimes requires *traction*, which is the procedure in which the bone is held under stretching tension forces at both ends to keep it aligned. Consider a leg under tractional tension as shown in ▾Fig. 4.15. The cord is attached to a suspended mass of 5.0 kg and runs over a pulley. The attached cord above the pulley makes an angle of $\theta = 40°$ with the vertical. Neglecting the mass of the lower leg and the pulley and assuming all the strings are ideal, determine the magnitude of the tension in the horizontal cord.

Thinking It Through. The pulley is in a static equilibrium and thus has no net force on it. If the forces are summed both vertically and horizontally, they independently should add to zero. This should enable us to find the tension in the horizontal string.

Given: Listing the data: *Find:* T in the horizontal cord
   $m = 5.0 \text{ kg}$
   $\theta = 40°$

Solution. Draw the free-body diagrams for the pulley and suspended mass (shown in Fig. 4.15). It should be clear that the horizontal string must exert a force to the left on the pulley as shown. Summing the vertical forces on m, we see that $T_1 = mg$. Summing the vertical forces on the pulley, we have

$$\sum F_y = +T_2 \cos \theta - T_1 = 0$$

and summing the horizontal forces:

$$\sum F_x = +T_2 \sin \theta - T = 0$$

Solving the latter equation for T, and substituting T_2 from the first:

$$T = T_2 \sin \theta = \frac{T_1}{\cos \theta} \sin \theta = mg \tan \theta$$

where $T_1 = mg$. Putting in the numbers,

$$T = mg \tan \theta = (5.0 \text{ kg})(9.8 \text{ m/s}^2) \tan 40° = 41 \text{ N}$$

Follow-Up Exercise. Suppose the attending physician requires a tractional force on the bottom of the foot of 55 N. Keeping the suspended mass the same, would you increase or decrease the angle of the upper string? Prove your answer by calculating the required angle.

(a)

(b)

▲ **FIGURE 4.14 Many forces, no acceleration** (a) At least five different external forces act on this physics professor. (Here, f is the force of friction.) Nevertheless, she experiences no acceleration. Why? (b) Adding the force vectors by the polygon method reveals that the vector sum of the forces is zero. The professor is in static translational equilibrium. (She is also in static rotational equilibrium; we'll see why in Chapter 8.)

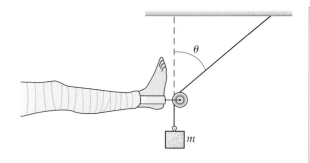

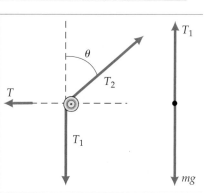

◀ **FIGURE 4.15 Static translational equilibrium** See Example 4.8.

Example 4.9 ■ On Your Toes: In Static Equilibrium

An 80-kg person stands on one foot with the heel elevated (▼ Fig. 4.16a). This gives rise to a tibia force F_1 and an Achilles-tendon "pull" force F_2, as illustrated in Fig. 4.16b. Typical angles are $\theta_1 = 15°$ and $\theta_2 = 21°$, respectively. (a) Find general equations for F_1 and F_2, and show that θ_2 must be greater than θ_1 to prevent damage to the Achilles tendon. (b) Compare the force of the Achilles tendon with the weight of the person.

Thinking It Through. This is a case of static translational equilibrium, so we can sum the F_2 x- and y-components to get equations for F_1 and F_2.

Solution. Listing what is given and what is to be found,

Given: $m = 80$ kg
$\quad\quad\quad F_1 = $ tibia force
$\quad\quad\quad F_2 = $ tendon "pull"
$\quad\quad\quad \theta_1 = 15°, \theta_2 = 21°$

$\quad\quad$ (The mass of the foot m_f is not given.)

Find: (a) general equations for F_1 and F_2
$\quad\quad\quad$ (b) comparison of tendon force F_2 and the person's weight

(a) It is assumed that the person of mass m is at rest, standing on one foot. Then, summing the force components on the foot (Fig. 4.16b),

$$\Sigma F_x = +F_1 \sin \theta_1 - F_2 \sin \theta_2 = 0$$
$$\Sigma F_y = +N - F_1 \cos \theta_1 + F_2 \cos \theta_2 - m_f g = 0$$

where m_f is the mass of the foot. From the F_x equation, we have

$$F_1 = F_2 \left(\frac{\sin \theta_2}{\sin \theta_1} \right) \tag{1}$$

Substituting into the F_y equation,

$$N - F_2 \left(\frac{\sin \theta_2}{\sin \theta_1} \right) \cos \theta_1 + F_2 \cos \theta_2 - m_f g = 0$$

Solving for F_2 with $N = mg$, yields,

$$F_2 = \frac{N - m_f g}{\left(\dfrac{\sin \theta_2}{\tan \theta_1} \right) - \cos \theta_2} = \frac{mg - m_f g}{\cos \theta_2 \left(\dfrac{\tan \theta_2}{\tan \theta_1} - 1 \right)} \tag{2}$$

▼ **FIGURE 4.16 On your toes** **(a)** A person stands on one foot with the heel elevated. **(b)** The foot forces involved for this position (not to scale). See Example 4.9.

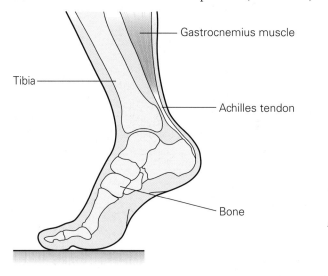
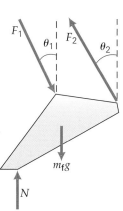

(a)   **(b)**

Examining the F_2 in Eq. 2, we see that if $\theta_2 = \theta_1$, or $\tan \theta_2 = \tan \theta_1$, then F_2 is very large. (Why?) So to have a finite force, we must have $\tan \theta_2 > \tan \theta_1$ or $\theta_2 > \theta_1$, and $21° > 15°$, so Nature obviously knows her physics.

Then, substituting F_2 into Eq. 1 to find F_1,

$$F_1 = F_2 \left(\frac{\sin \theta_2}{\sin \theta_1} \right) = \left[\frac{(m - m_f)g}{\cos \theta_2 \left(\dfrac{\tan \theta_2}{\tan \theta_1} \right) - 1} \right] \left(\frac{\sin \theta_2}{\sin \theta_1} \right)$$

$$= \frac{(m - m_f)g \tan \theta_2}{\left(\dfrac{\tan \theta_2}{\tan \theta_1} - 1 \right) \sin \theta_1} = \frac{\tan \theta_2 \,(m - m_f)g}{\cos \theta_1 \tan \theta_2 - \sin \theta_1}$$

(Check the trig manipulation on this last step)

(b) The person's weight is $w = mg$, where m is the mass of the person's body. This is to be compared with F_2. Then, with $m \gg m_f$ (total body mass much greater than mass of the foot), to a good approximation, m_f may be assumed negligible compared to m, that is, $w - m_f g = mg - m_f g \approx w$. So for F_2 we have

$$F_2 = \frac{w - m_f g}{\cos \theta_2 \left(\dfrac{\tan \theta_2}{\tan \theta_1} - 1 \right)} \approx \frac{w}{\cos 21° \left(\dfrac{\tan 21°}{\tan 15°} - 1 \right)} = 2.5w$$

The Achilles tendon force is thus approximately 2.5 times the person's weight. No wonder folks stretch or tear this tendon, even without jumping!

Follow-Up Exercise. (a) Compare the tibia force with the weight of the person. (b) Suppose the person jumped upward from the one-foot toe position (as in taking a running jump shot in basketball). How would this jump affect F_1 and F_2?

4.6 Friction

OBJECTIVES: To explain (a) the causes of friction, and (b) how friction is described by using coefficients of friction.

Friction refers to the ever-present resistance to motion that occurs whenever two materials, or media, are in contact with each other. This resistance occurs for all types of media—solids, liquids, and gases—and is characterized as the **force of friction (f)**. For simplicity, up to now we have generally ignored all kinds of friction (including air resistance) in examples and problems. Now that you know how to describe motion, you are ready to consider situations that are more realistic, in that the effects of friction are included.

In some real situations, we want to increase friction—for example, by putting sand on an icy road or sidewalk to improve traction. This might seem contradictory, since an increase in friction presumably would increase the resistance to motion. We commonly say that friction opposes motion, and think that the force of friction is in the opposite direction of motion. However, consider the forces involved in walking, as illustrated in ▶Fig. 4.17. The force of friction does resist motion (that of the foot), but is in the direction of the (walking) motion. Without friction, the foot would slip backward. (Think about walking on a slippery surface.) As another example, consider a worker standing in the center of the bed of a flatbed truck that is accelerating in the forward direction. If there were no friction between the worker's shoes and the truck bed, he would slide backwards. Obviously, there is friction between the shoes and the bed, which keeps him from sliding backward, and it is in the forward direction.

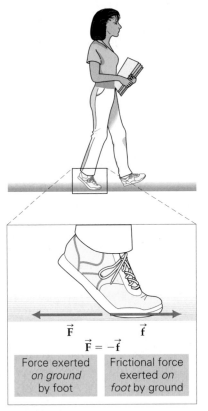

▲ **FIGURE 4.17 Friction and walking** The force of friction, $\vec{f}$, is shown in the direction of the walking motion. The force of friction prevents the foot from slipping backward while the other foot is brought forward. If you walk on a deep-pile rug, $\vec{F}$ is evident in that the pile will be bent backward.

(a)

(b)

▲ **FIGURE 4.18** Increasing and decreasing friction **(a)** To get a fast start, drag racers need to make sure that their wheels don't slip when the starting light goes on. Just before the start of the race, they floor the accelerator to maximize the friction between their tires and the track by "burning in" the tires. This "burn in" is done by spinning the wheels with the brakes on until the tires are extremely hot. The rubber becomes so sticky that it almost welds itself to the surface of the road. **(b)** Water serves as a good lubricant to reduce friction in rides such as this one.

So there are situations where friction is desired (◄Fig. 4.18a), and situations where reduced friction is needed (Fig. 4.18b). Another situation where we try to reduce friction, we lubricate moving machine parts to allow them to move more freely, thereby lessening wear and reducing the expenditure of energy. Automobiles would not run without friction-reducing oils and greases.

This section is concerned chiefly with friction between solid surfaces. All surfaces are microscopically rough, no matter how smooth they appear or feel. It was originally thought that friction was due primarily to the mechanical interlocking of surface irregularities, or *asperities* (high spots). However, research has shown that friction between the contacting surfaces of ordinary solids (metals in particular) is due mostly to local adhesion. When surfaces are pressed together, local welding or bonding occurs in a few small patches where the largest asperities make contact. To overcome this local adhesion, a force great enough to pull apart the bonded regions must be applied.

Friction between solids is generally classified into three types: static, sliding (kinetic), and rolling. **Static friction** includes all cases in which the frictional force is sufficient to prevent relative motion between surfaces. Suppose you want to move a large desk. You push it, but the desk doesn't move. The force of static friction between the desk's legs and the floor opposes and equals the horizontal force you are applying, so there is no motion—a static condition.

Sliding friction, or **kinetic friction**, occurs when there is relative (sliding) motion at the interface of the surfaces in contact. When pushing the desk, you can eventually get it sliding, but there is still a great deal of resistance between the desk's legs and the floor—kinetic friction.

Rolling friction occurs when one surface rotates as it moves over another surface, but does not slip or slide at the point or area of contact. Rolling friction, such as occurs between a train wheel and a rail, is attributed to small local deformations in the contact region. This type of friction is difficult to analyze and will not be discussed.

Frictional Forces and Coefficients of Friction

In this subsection, we will consider the forces of friction on stationary and sliding objects. These forces are called the *force of static friction* and the *force of kinetic (or sliding) friction*, respectively. Experimentally, it has been found that the force of friction depends on both the nature of the two surfaces and the *load*, or the normal force that presses the surfaces together. Thus we can write $f \propto N$. For an object on a horizontal surface, this force is equal in magnitude to the object's weight. (Why?) However, as was shown in the LBD figure on p. 116, on an inclined plane, only a component of the weight force contributes to the load.

The force of static friction (f_s) between surfaces in contact acts in the direction that opposes the initiation of relative motion between the surfaces. The magnitude takes on a range of values given by

$$f_s \leq \mu_s N \quad \text{(static conditions)} \tag{4.6}$$

where μ_s is a constant of proportionality called the **coefficient of static friction**. ("μ" is the Greek letter mu. Note that it is dimensionless. How do you know this from the equation?)

The less-than-or-equal-to sign ($\leq$) indicates that the force of static friction may have different values from zero up to some maximum value. To understand this concept, look at ►Fig. 4.19. In Fig. 4.19a, one person pushes on a file cabinet, but it doesn't move. With no acceleration, the net force on the cabinet is zero, and $F - f_s = 0$, or $F = f_s$. Suppose that a second person also pushes, and the file cabinet still doesn't budge. Then f_s must now be larger, since the applied force has been increased. Finally, if the applied force is made large enough to overcome the static friction, motion occurs (Fig. 4.19c). The greatest, or maximum, force of static friction is exerted just before the cabinet starts to slide (Fig. 4.19b), and for this case, Eq. 4.6 gives the maximum value of static friction:

$$f_{s_{max}} = \mu_s N \qquad (4.7)$$

Once an object is sliding, the force of friction changes to kinetic friction (f_k). This force acts in the direction opposite to the direction of the object's motion and has a magnitude of

$$f_k = \mu_k N \quad \textit{(sliding conditions)} \qquad (4.8)$$

where μ_k is the **coefficient of kinetic friction** (sometimes called the *coefficient of sliding friction*). Note that Eqs. 4.7 and 4.8 are *not* vector equations, since f and N are in different directions. Generally, the coefficient of kinetic friction is less than

▼ **FIGURE 4.19 Force of friction versus applied force** (a) In the static region of the graph, as the applied force F increases, so does f_s; that is, $f_s = F$ and $f_s < \mu_s N$. **(b)** When the applied force F exceeds $f_{s_{max}} = \mu_s N$, the heavy file cabinet is set into motion. **(c)** Once the cabinet is moving, the frictional force decreases, since kinetic friction is less than static friction ($f_k < f_{s_{max}}$). Thus, if the applied force is maintained, there is a net force, and the cabinet is accelerated. For the cabinet to move with constant velocity, the applied force must be reduced to equal the kinetic friction force: $f_k = \mu_k N$.

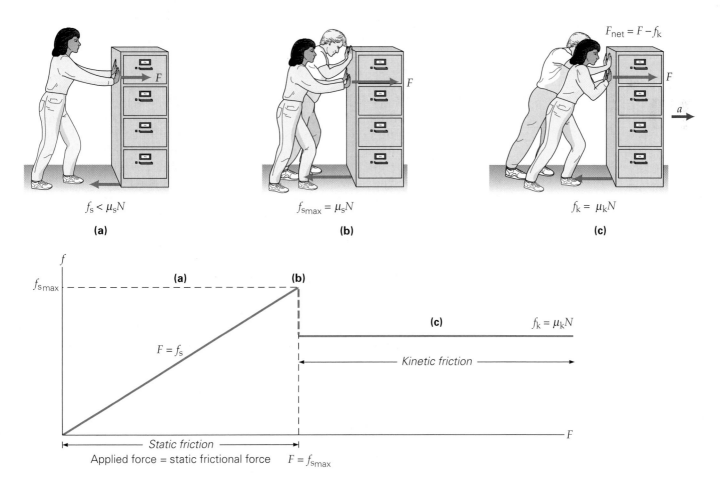

TABLE 4.1	Approximate Values for Coefficients of Static and Kinetic Friction between Certain Surfaces		
Friction between Materials		μ_s	μ_k
Aluminum on aluminum		1.90	1.40
Glass on glass		0.94	0.35
Rubber on concrete			
dry		1.20	0.85
wet		0.80	0.60
Steel on aluminum		0.61	0.47
Steel on steel			
dry		0.75	0.48
lubricated		0.12	0.07
Teflon on steel		0.04	0.04
Teflon on Teflon		0.04	0.04
Waxed wood on snow		0.05	0.03
Wood on wood		0.58	0.40
Lubricated ball bearings		<0.01	<0.01
Synovial joints (at the ends of most long bones—for example, elbows and hips)		0.01	0.01

Illustration 5.1 Static and Kinetic Friction

Demonstration/activity: Pull a weighted block of wood on the demonstration desk with a large demonstration-type spring scale. Then turn the block on edge with the same weight and show how the changed surface area has little effect on the pulling (frictional) force. Be careful not to extrapolate this demonstration to tires on roadways.

the coefficient of static friction ($\mu_k < \mu_s$), which means that the force of kinetic friction is less than $f_{s_{max}}$. The coefficients of friction between some common materials are listed in Table 4.1.

Note that the force of static friction (f_s) exists in response to an applied force. The magnitude of f_s and its direction depend on the magnitude and direction of the applied force. Up to its maximum value, the force of static friction is equal in magnitude and opposite in direction to the applied force (F), since there is no acceleration ($F - f_s = ma = 0$). Thus, if the person in Fig. 4.19a were to push on the cabinet in the opposite direction, f_s would also change direction to oppose the new push. If there were no applied force F, then f_s would be zero. When the magnitude of F exceeds that of $f_{s_{max}}$, the cabinet begins moving (accelerates), and kinetic friction comes into effect, with $f_k = \mu_k N$. If the magnitude of F is reduced to that of f_k, the cabinet will slide with a constant velocity; if the magnitude of F is maintained greater than that of f_k, the cabinet will continue to accelerate.

It has been experimentally determined that the coefficients of friction (and therefore the forces of friction) are nearly independent of the contact area between metal surfaces. This means that the force of friction between a brick-shaped metal block and a metal surface is the same regardless of whether the block is lying on a larger side or a smaller side.

Finally, keep in mind that although the equation $f = \mu N$ holds in general for frictional forces, it may not remain linear. That is, μ is not always constant. For example, the coefficient of kinetic friction varies somewhat with the relative speed of the surfaces. However, for speeds up to several meters per second, the coefficients are relatively constant. For simplicity, our discussion will neglect any variations due to speed (or area), and the forces of static and kinetic friction will be assumed to depend only on the load (N) and the nature of the two surfaces as expressed by the given coefficients of friction.

Example 4.10 ■ Pulling a Crate: Static and Kinetic Forces of Friction

(a) In ▼Fig. 4.20, if the coefficient of static friction between the 40.0-kg crate and the floor is 0.650. What is the magnitude of the minimum horizontal force the worker must pull to get the crate moving? (b) If the worker maintains that force once the crate starts to move and the coefficient of kinetic friction between the surfaces is 0.500, what is the magnitude of the acceleration of the crate?

Thinking It Through. This scenario involves applications of the forces of friction. In (a), the maximum force of static friction must be calculated. In (b), if the worker maintains an applied force of this magnitude after the crate is in motion, there will be an acceleration, since $f_k < f_{s_{max}}$.

Solution. Listing the given data and what is to be found,

Given: $m = 40.0$ kg   *Find:*   (a) F (minimum force necessary to move crate)
        $\mu_s = 0.650$ (b) a (acceleration)
        $\mu_k = 0.500$

(a) The crate will not move until the magnitude of the applied force F slightly exceeds that of the maximum static frictional force $f_{s_{max}}$. So $f_{s_{max}}$ must be found to see what force the worker needs to apply. The weight of the crate and the normal force are equal in magnitude in this case (see the free-body diagram in Fig. 4.20), so the magnitude of the maximum force of static friction is

$$f_{s_{max}} = \mu_s N = \mu_s (mg)$$
$$= (0.650)(40.0 \text{ kg})(9.80 \text{ m/s}^2) = 255 \text{ N}$$

So the crate will begin to move when the applied force F exceeds 255 N.

(b) Now the crate is in motion, and the worker maintains a constant applied force of $F = f_{s_{max}} = 255$ N. The force of kinetic friction f_k acts on the crate. However, this force is smaller than the applied force F, because $\mu_k < \mu_s$. Hence, there is a net force on the crate, and the acceleration of the crate may be found by using Newton's second law in the x-direction:

$$\Sigma F_x = +F - f_k = F - \mu_k N = ma_x$$

Solving for a_x, we obtain

$$a_x = \frac{F - \mu_k N}{m} = \frac{F - \mu_k (mg)}{m}$$
$$= \frac{255 \text{ N} - (0.500)(40.0 \text{ kg})(9.80 \text{ m/s}^2)}{40.0 \text{ kg}} = 1.48 \text{ m/s}^2$$

Follow-Up Exercise. On the average, by what factor does μ_s exceed μ_k for nonlubricated, metal-on-metal surfaces? (See Table 4.1.)

▼ **FIGURE 4.20** Forces of static and kinetic friction See Example 4.10.

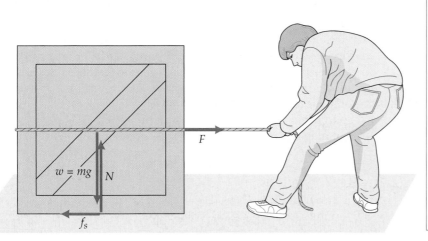

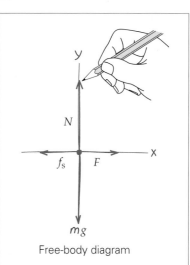

Free-body diagram

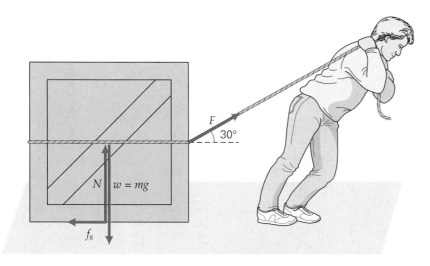

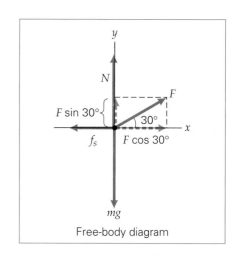

Free-body diagram

▲ **FIGURE 4.21** Pulling at an angle: a closer look at the normal force See Example 4.11.

Let's look at another worker with the same crate, but this time assume that the worker applies the force at an angle (▲ Fig. 4.21).

Example 4.11 ■ Pulling at an Angle: A Closer Look at the Normal Force

A worker pulling a crate applies a force at an angle of 30° to the horizontal, as shown in Fig. 4.21. What is the magnitude of the minimum force he must apply to move the crate? (Before looking at the solution, would you expect that the force needed in this case would be greater or less than that in Example 4.10?)

Thinking It Through. Since the applied force is at an angle to the horizontal surface, the vertical component will affect the normal force. (See Fig. 4.11.) This change in the normal force will, in turn, affect the maximum force of static friction.

Solution. The data are the same as in Example 4.10, except that the force is applied at an angle.

Given: $\theta = 30°$ *Find:* F (minimum force necessary to move the crate)

In this case, the crate will begin to move when the *horizontal component* of the applied force, $F \cos 30°$, slightly exceeds the maximum static friction force. So we may write the following for the maximum friction:

$$F \cos 30° = f_{s_{max}} = \mu_s N$$

However, the magnitude of the normal force is not equal to the weight of the crate here, because of the upward component of the applied force. (See the free-body diagram in Fig. 4.21.) Then by Newton's second law, since $a_y = 0$, we have

$$\Sigma F_y = +N + F \sin 30° - mg = 0$$

or

$$N = mg - F \sin 30°$$

In effect, the applied force partially supports the weight of the crate. Substituting this expression for N into the first equation gives

$$F \cos 30° = \mu_s (mg - F \sin 30°)$$

Solving for F gives

$$F = \frac{mg}{(\cos 30°/\mu_s) + \sin 30°}$$

$$= \frac{(40.0 \text{ kg})(9.80 \text{ m/s}^2)}{(0.866/0.650) + 0.500} = 214 \text{ N}$$

Teaching tip: Knowing how to break down forces into components is of utmost importance. Problems with this skill become evident when you see students drawing the component vectors at the wrong size. Use dashed lines to complete the vector rectangle and similar elements.

Thus, less applied force is needed in this case, reflecting the fact that the frictional force is less, because of the reduced normal force.

Follow-Up Exercise. Note that in this Example, applying the force at an angle produces two effects. As the angle between the applied force and the horizontal increases, the horizontal component of the applied force is reduced. However, the normal force also gets smaller, resulting in a lower $f_{s_{max}}$. Does one effect always outweigh the other? That is, does the applied force F necessary to move the crate always decrease with increasing angle? (*Hint*: Investigate F for different angles. For example, compute F for 20° and 50°. You already have a value for 30°. What do the results tell you?)

Example 4.12 ■ No Slip, No Slide: Static Friction

A crate sits in the middle of the bed on a flatbed truck that is traveling at 80 km/h on a straight, level road. The coefficient of static friction between the crate and the truck bed is 0.40. When the truck comes uniformly to a stop, the crate does not slide, but remains stationary on the truck. What is the minimum stopping distance for the truck so the crate does not slide on the truck bed?

Thinking It Through. There are three forces on the crate, as shown in the free-body diagram in ▶Fig. 4.22 (assuming that the truck is initially traveling in the $+x$-direction). But wait. There is a net force in the $-x$-direction, and hence there should be an acceleration in that direction ($a_x < 0$). What does this mean? It means that relative to the ground, the crate is decelerating at the same rate as the truck, which is necessary for the crate not to slide—the crate and the truck slow down uniformly together.

The force creating this acceleration for the crate is the static force of friction. The acceleration is found using Newton's second law, and then used in one of the kinematic equations to find the distance.

Solution.

Given: $v_{x_o} = 80 \text{ km/h} = 22 \text{ m/s}$ *Find:* minimum stopping distance
        $\mu_s = 0.40$

Applying Newton's second law to the crate using the maximum f_s to find the minimum stopping distance,

$$\Sigma F_x = -f_{s_{max}} = -\mu_s N = -\mu_s mg = ma_x$$

Solving for a_x,

$$a_x = -\mu_s g = -(0.40)(9.8 \text{ m/s}^2) = -3.9 \text{ m/s}^2$$

which is the maximum deceleration of the truck so the crate does not slide.

Hence, the minimum stopping distance (x) for the truck is based on this acceleration and given by Eq. 2.12, where $v_x = 0$ and x_o is taken to be zero. So,

$$v_x^2 = 0 = v_{x_o}^2 + 2(a_x)x$$

Solving for x, we obtain

$$x = \frac{v_{x_o}^2}{-2a_x} = \frac{(22 \text{ m/s})^2}{-2(-3.9 \text{ m/s}^2)} = 62 \text{ m}$$

Is the answer reasonable? This distance is about two thirds of a football field.

Follow-Up Exercise. Draw a free-body diagram and describe what happens in terms of accelerations and coefficients of friction if the crate starts to slide forward on the truck bed when the truck is braking to a stop (in other words, if a_x exceeds -3.9 m/s^2).

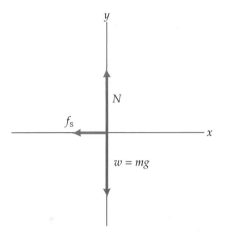

▲ **FIGURE 4.22** Free-body diagram
See Example 4.12.

Air Resistance

Air resistance refers to the resistance force acting on an object as it moves through air. In other words, air resistance is a type of frictional force. In analyses of falling objects, you can usually ignore the effect of air resistance and still get good approximations for falling relatively short distances. However, for longer distances, air resistance cannot be ignored.

▲ **FIGURE 4.23 Airfoil** The airfoil at the top of the truck's cab makes the truck more streamlined and therefore reduces air resistance.

Illustration 5.5 Air Friction

Air resistance occurs when a moving object collides with air molecules. Therefore, air resistance depends on the object's shape and size (which determine the area of the object that is exposed to collisions) as well as its speed. The larger the object and the faster it moves, the more collisions there will be with air molecules. (Air density is also a factor, but this quantity can be assumed to be constant near the Earth's surface.) To reduce air resistance (and fuel consumption), automobiles are made more "streamlined," and airfoils are used on trucks and campers (◄Fig. 4.23).

Consider a falling object. Since air resistance depends on speed, as a falling object accelerates under the influence of gravity, the retarding force of air resistance increases (▼Fig. 4.24a). Air resistance for human-sized objects is proportional to the square of the speed, v^2, so the resistance builds up rather rapidly. Thus when the speed doubles, the air resistance increases by a factor of 4. Eventually, the magnitude of the retarding force equals that of the object's weight force (Fig. 4.24b), so the net force on it is zero. The object then falls with a maximum constant velocity, which is called the **terminal velocity**, with magnitude v_t.

This can be easily seen from Newton's second law. For the falling object, we have

$$F_{net} = ma$$

or

$$mg - f = ma$$

where downward has been taken as positive for convenience. Solving for a, we obtain

$$a = g - \frac{f}{m}$$

where a is the magnitude of the instantaneous downward acceleration.

Notice that the acceleration for a falling object when air resistance is included is less than g; that is, $a < g$. As the object continues to fall, its speed increases, and the force of air resistance, f, increases (since it is speed dependent) until $a = 0$, when $f = mg$ and $f - mg = 0$. The object then falls at its constant terminal velocity.

For a skydiver with an unopened parachute, terminal velocity is about 200 km/h (about 125 mi/h). To reduce the terminal velocity so that it can be reached sooner and the time of fall extended, a skydiver will try to increase exposed body area to a maximum by assuming a spread-eagle position (►Fig. 4.25). This position takes advantage of the dependence of air resistance on the size and shape of the falling object. Once the parachute is open (giving a larger exposed area and a shape that catches the air), the additional air resistance slows the diver down to about 40 km/h (25 mi/h), which is preferable for landing.

▼ **FIGURE 4.24 Air resistance and terminal velocity** (a) As the speed of a falling object increases, so does the frictional force of air resistance. (b) When this force of friction equals the weight of the object, the net force is zero, and the object falls with a constant (terminal) velocity. (c) A plot of speed versus time, showing these relationships.

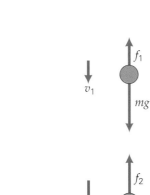

(a) As v increases, so does f.

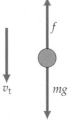

(b) When $f = mg$, the object falls with a constant (terminal) velocity.

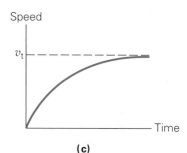

(c)

◀ **FIGURE 4.25 Terminal velocity**
Skydivers assume a spread-eagle position to maximize air resistance. This causes them to reach terminal velocity more quickly and prolongs the time of fall. Shown here is a formation of sky divers viewed from below.

Conceptual Example 4.13 ▪ Race You Down: Air Resistance and Terminal Velocity

From a high altitude, a balloonist simultaneously drops two balls of identical size, but appreciably different in weight. Assuming that both balls reach terminal velocity during the fall, which of the following is true: (a) The heavier ball reaches terminal velocity first; (b) the balls reach terminal velocity at the same time; (c) the heavier ball hits the ground first; (d) the balls hit the ground at the same time? *Clearly establish the reasoning and physical principle(s) used in determining your answer before checking it next. That is, **why** did you select your answer?*

Reasoning and Answer. Terminal velocity is reached when the weight of a ball is balanced by the frictional air resistance. Both balls initially experience the same acceleration, g, and their speeds and the retarding forces of air resistance increase at the same rate. The weight of the lighter ball will be balanced first, so (a) and (b) are incorrect. The lighter ball reaches terminal velocity ($a = 0$) first, but the heavier ball continues to accelerate, speeding up, and pulls ahead of the lighter ball. Hence, the heavier ball hits the ground first, and the answer is (c), and (d) is incorrect.

Follow-Up Exercise. Suppose the heavier ball were much larger in size than the lighter ball. How might this difference affect the outcome?

You see an example of terminal velocity quite often. Why do clouds stay seemingly suspended in the sky? Certainly the water droplets or ice crystals (high clouds) should fall—and they do. However, they are so small that their terminal velocity is reached quickly, and the very slow rate of their descent goes unnoticed. Buoyancy in the air is also a factor (see Chapter 9). In addition, there may be some helpful updrafts that keep the water and ice from reaching the ground.

An extraterrestrial use of "air" resistance is called *aerobraking*. This spaceflight technique uses a planetary atmosphere to slow down an orbiting spacecraft. As the craft passes through the top layer of the planetary atmosphere, the atmospheric "drag" slows and lowers the craft's speed so as to put it in the desired orbit. Many passes may be needed, with the spacecraft passing in and out of the atmosphere to achieve the proper final orbit.

Aerobraking is a worthwhile technique because it eliminates the need for a heavy load of chemical propellants that would otherwise be needed to place the spacecraft in orbit. This allows a greater payload of scientific instruments for investigations. Aerobraking was used to adjust the spacecraft *Odyssey*'s orbit around Mars in 2001.

Demonstration/activity: Have students sketch graphs of acceleration, velocity, and distance versus time for an object falling in air and reaching a terminal velocity. Compare these graphs with ones for free fall without friction. Discuss how denser air at lower altitudes changes your graphs.

Chapter Review

- A **force** is something that is capable of changing an object's state of motion. To produce a change in motion, there must be a nonzero net, or unbalanced, force:

$$\vec{F}_{net} = \Sigma \vec{F}_i$$

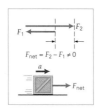

- **Newton's first law of motion** is also called the *law of inertia*, where inertia is the natural tendency of an object to maintain its state of motion. It states that in the absence of a net applied force, a body at rest remains at rest, and a body in motion remains in motion with constant velocity.

- **Newton's second law** relates the net force acting on an object or system to the (total) mass and the resulting acceleration. It defines the cause-and-effect relationship between force and acceleration:

$$\Sigma \vec{F}_i = \vec{F}_{net} = m\vec{a} \qquad (4.1)$$

A nonzero net force accelerates the crate: $a \propto F/m$

The equation for **weight** in terms of mass is a form of Newton's second law:

$$w = mg \qquad (4.2)$$

The component form of Newton's second law:

$$\Sigma(F_x\hat{\mathbf{x}} + F_y\hat{\mathbf{y}}) = m(a_x\hat{\mathbf{x}} + a_y\hat{\mathbf{y}}) = ma_x\hat{\mathbf{x}} + ma_y\hat{\mathbf{y}} \qquad (4.3a)$$

and

$$\Sigma F_x = ma_x \qquad \text{and} \qquad \Sigma F_y = ma_y \qquad (4.3b)$$

- **Newton's third law** states that for every force, there is an equal and opposite reaction force. The opposing forces of a third-law force pair always act on different objects.

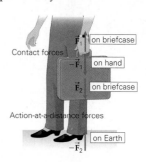

- An object is said to be in **translational equilibrium** when it either is at rest or moves with a constant velocity. When remaining at rest, an object is said to be in *static translational equilibrium*. The condition for translational equilibrium is represented as

$$\Sigma \vec{F}_i = 0 \qquad (4.4)$$

or

$$\Sigma F_x = 0 \qquad \text{and} \qquad \Sigma F_y = 0 \qquad (4.5)$$

- **Friction** is the resistance to motion that occurs between contacting surfaces. (In general, friction occurs for all types of media—solids, liquids, and gases.)

- The frictional force between surfaces is characterized by coefficients of friction (μ), one for the static case and one for the kinetic (moving) case. In many cases, $f = \mu N$, where N is the normal force—the force perpendicular to the surface (that is, the force exerted *by* the surface *on* the object). As a ratio of forces (f/N), μ is unitless.

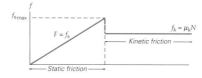

Force of Static Friction:

$$f_s \leq \mu_s N \qquad (4.6)$$

$$f_{s_{max}} = \mu_s N \qquad (4.7)$$

Force of Kinetic (Sliding) Friction:

$$f_k = \mu_k N \qquad (4.8)$$

- The force of air resistance on a falling object increases with increasing speed. It eventually attains a constant velocity, called the *terminal velocity*.

Exercises*

MC = *Multiple Choice Question,* **CQ** = *Conceptual Question, and* **IE** = *Integrated Exercise. Throughout the text, many exercise sections will include "paired" exercises. These exercise pairs, identified with* **red numbers**, *are intended to assist you in problem solving and learning. In a pair, the first exercise (even numbered) is worked out in the Study Guide so that you can consult it should you need assistance in solving it. The second exercise (odd numbered) is similar in nature, and its answer is given at the back of the book.*

4.1 The Concepts of Force and Net Force
and 4.2 Inertia and Newton's First Law of Motion

1. **MC** Mass is related to an object's (a) weight, (b) inertia, (c) density, (d) all of the preceding. (d)

2. **MC** A force (a) always produces motion, (b) is a scalar quantity, (c) is capable of producing a change in motion, (d) both a and b. (c)

3. **MC** If an object is moving at constant velocity, (a) there must be a force in the direction of the velocity, (b) there must be no force in the direction of the velocity, (c) there must be no net force, (d) there must be a net force in the direction of the velocity. (c)

4. **MC** If the net force on an object is zero, the object could (a) be at rest, (b) be in motion at a constant velocity, (c) have zero acceleration, (d) all of the preceding. (d)

5. **MC** The force required to keep a rocket ship moving at a constant velocity in deep space is (a) equal to the weight of the ship, (b) dependent on how fast the ship is moving, (c) equal to that generated by the rocket's engines at half power, (d) zero. (d)

6. **CQ** (a) If an object is at rest, there must be no forces acting on it. Is this statement correct? Explain. (b) If the net force on an object is zero, can you conclude the object is at rest? Explain. (a) no (b) no

7. **CQ** When on a jet airliner that is taking off, you feel that you are being "pushed" back into the seat. Use Newton's first law to explain why. see ISM

8. **CQ** An object weighs 300 N on Earth and 50 N on the Moon. Does the object also have less inertia on the Moon? no, same mass, same inertia

9. **CQ** Consider an air-bubble level that is sitting on a horizontal surface (▼Fig. 4.26). Initially, the air bubble is in the middle of the horizontal glass tube. (a) If the level is pushed and a force is applied to accelerate it, which way would the bubble move? Which way would the bubble move if the force is then removed and the level slows down, due to friction? (b) Such a level is sometimes used as an "accelerometer" to indicate the direction of the acceleration. Explain the principle involved. [*Hint*: Think about pushing a pan of water.] see ISM

10. **CQ** As a follow-up to Exercise 9, consider a child holding a helium balloon in a closed car at rest. What would the child observe when the car (a) accelerates from rest and (b) brakes to a stop? (The balloon does not touch the roof of the car.) balloon moves (a) forward (b) backward

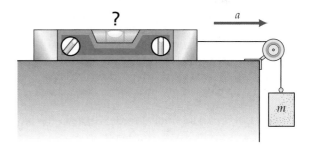

▲ **FIGURE 4.26** An air-bubble level/accelerometer
See Exercise 9.

11. **CQ** The following is an old trick (▼Fig. 4.27). If a tablecloth is yanked out very quickly, the dishes on it will barely move. Why? see ISM

▼ **FIGURE 4.27 Magic or physics?** See Exercise 11.

*Unless otherwise stated, all objects are located near the Earth's surface, where $g = 9.80 \text{ m/s}^2$.

12. ● Which has more inertia, 20 cm² of water or 10 cm³ of aluminum, and how many times more? (See Table 9.2.) $m_{Al} = 1.4m_{water}$

13. ● A net force of 4.0 N gives an object an acceleration of 10 m/s². What is the mass of the object? 0.40 kg

14. ● Two forces act on a 5.0-kg object sitting on a frictionless horizontal surface. One force is 30 N in the +x-direction, and the other 35 N in the −x-direction. What is the acceleration of the object? −1.0 m/s²

15. ● In Exercise 14, if the 35-N force acted downward at an angle of 40° relative to the horizontal, what would be the acceleration in this case? 0.64 m/s²

16. ● Consider a 2.0-kg ball and a 6.0-kg ball in free fall. (a) What is the force acting on each? (b) What is the acceleration of each? (a) 2.0 kg: 20 N; 6.0 kg: 59 N (b) 9.80 m/s²

17. **IE ●●** A hockey puck with a weight of 0.50 lb is sliding freely across a section of very smooth (frictionless) horizontal ice. (a) When it is sliding freely, how does the upward force of the ice on the puck (the normal force) compare with the upward force when the puck is sitting permanently at rest: (1) The upward force is greater when the puck is sliding; (2) the upward force is less when it is sliding; or (3) the upward force is the same in both situations? (b) Calculate the upward force on the puck in both situations. (a) (3) the upward force is the same in both situations (b) 0.50 lb

18. **●●** A 5.0-kg block at rest on a frictionless surface is acted on by forces $F_1 = 5.5$ N and $F_2 = 3.5$ N, as illustrated in ▼Fig. 4.28. What additional force will keep the block at rest? $\vec{F}_3 = (-7.6\ \text{N})\hat{x} + (0.64\ \text{N})\hat{y}$

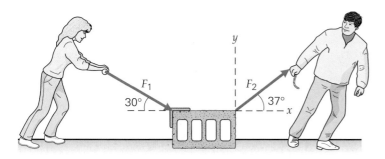

▲ **FIGURE 4.28 Two applied forces** See Exercise 18.

19. **IE ●●** (a) You are told that an object has zero acceleration. Which of the following is true: (1) The object is at rest; (2) the object is moving with constant velocity; (3) either (1) or (2) is possible; or (4) neither 1 nor 2 is possible. (b) Two forces on the object are $F_1 = 3.6$ N at 74° below the +x-axis and $F_2 = 3.6$ N at 34° above the −x-axis. Is there a third force on the object, and why? If yes, what is it? (a) (3) either 1 or 2 is possible (b) see ISM

20. **IE ●●** A fish weighing 25 lb is caught and hauled onto the boat. (a) Compare the tension in the fishing line when the fish is brought up vertically at a constant speed to the tension when the fish is held vertically at rest for the picture-taking ceremony on the wharf. In which case is the tension largest: (1) When the fish is moving up; (2) when the fish is being held steady; or (3) the tension is the same in both situations? (b) Calculate the tension in the fishing line. (a) (3) tension is the same for both situations (b) 25 lb

21. **●●●** A 1.5-kg object moves up the y-axis at a constant speed. When it reaches the origin, the forces $F_1 = 5.0$ N at 37° above the +x-axis, $F_2 = 2.5$ N in the +x-direction, $F_3 = 3.5$ N at 45° below the −x-axis, and $F_4 = 1.5$ N in the −y-direction are applied to it. (a) Will the object continue to move along the y-axis? (b) If not, what simultaneously applied force will keep it moving along the y-axis at a constant speed? (a) no (b) $F_5 = 4.1$ N at 13° above the −x-axis

22. **IE ●●●** Three horizontal forces (the only horizontal ones) act on a box sitting on a floor. One (call it F_1) acts due east and has a magnitude of 150 lb. A second force (call it F_2) has an easterly component of 30.0 lb and a southerly component of 40.0 lb. The box remains at rest. (Neglect friction.) (a) Sketch the two known forces on the box. In which quadrant is the unknown third force: (1) the first quadrant; (2) the second quadrant; (3) the third quadrant; or (4) the fourth quadrant? (b) Find the unknown third force in newtons and compare your answer to the sketched estimate. (a) (2) the second quadrant (b) $F_3 = 184$ N at 12.5° above −x axis

4.3 Newton's Second Law of Motion

23. **MC** The newton unit of force is equivalent to (a) kg · m/s, (b) kg · m/s², (c) kg · m²/s, (d) none of the preceding. (b)

24. **MC** The acceleration of an object is (a) inversely proportional to the acting net force, (b) directly proportional to its mass, (c) directly proportional to the net force and inversely proportional to its mass, (d) none of these. (c)

25. **MC** The weight of an object is directly proportional to (a) its mass, (b) its inertia, (c) the acceleration due to gravity, (d) all of the preceding. (d)

26. **CQ** An astronaut has a mass of 70 kg when measured on Earth. What is his weight in deep space, far from any celestial body? What is his mass there? zero; 70 kg

27. **CQ** In general, this chapter has considered forces that are applied to objects of constant mass. What would be the situation if mass were added to or lost from a system while a constant force was being applied to the system? Give examples of situations in which this set of events might happen. see ISM

28. **CQ** The engines of most rockets produce a constant thrust (forward force). However, when a rocket is fired, its acceleration increases with time as the engine continues to operate. Is this situation a violation of Newton's second law? Explain. no, both mass and gravity decrease

29. **CQ** In football, good wide receivers usually have "soft" hands for catching balls (▶Fig. 4.29). How would you interpret this description on the basis of Newton's second law? see ISM

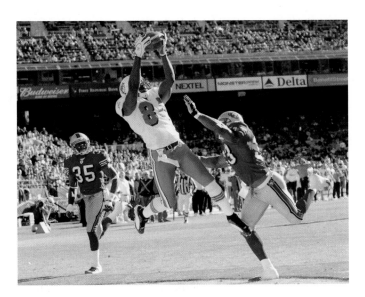

▲ **FIGURE 4.29 Soft hands** See Exercise 29.

30. ● A 6.0-N net force is applied to a 1.5-kg mass. What is the object's acceleration?
    4.0 m/s² in the direction of the net force
31. ● What is the mass of an object that accelerates at 3.0 m/s² under the influence of a 5.0-N net force? 1.7 kg
32. ● A loaded Boeing 747 jumbo jet has a mass of 2.0×10^5 kg. What net force is required to give the plane an acceleration of 3.5 m/s² down the runway for take-offs? 7.0×10^5 N
33. IE ● A 6.0-kg object is brought to the Moon, where the acceleration due to gravity is only one sixth of that on the Earth. (a) The mass of the object on the Moon is (1) zero, (2) 1.0 kg, (3) 6.0 kg, (4) 36 kg. Why? (b) What is the weight of the object on the Moon? (a) (3) 6.0 kg (b) 9.8 N
34. ● What is the weight of a 150-lb person in newtons? What is his mass in kilograms? 668 N; 68.1 kg
35. IE ●● ▼Fig. 4.30 shows a product label. (a) This label is correct (1) on the Earth; (2) on the Moon, where the acceleration due to gravity is only one sixth of that on the Earth; (3) in deep space, where there is little gravity;

▲ **FIGURE 4.30 Correct label?** See Exercise 35.

(4) all of the preceding. (b) What mass of lasagne would a label show for an amount that weighs 2 lb on the Moon? (a) (1) on the Earth (b) 5.4 kg
36. ●● In a college homecoming competition, eighteen students lift a sports car. While holding the car off the ground, each student exerts an upward force of 400 N. (a) What is the mass of the car in kilograms? (b) What is its weight in pounds? (a) 735 kg (b) 1.62×10^3 lb
37. ●● (a) A horizontal force acts on an object on a frictionless horizontal surface. If the force is halved and the mass of the object is doubled, the acceleration will be (1) four times, (2) two times, (3) one half, (4) one fourth as great. (b) If the acceleration of the object is 1.0 m/s², and the force on it is doubled and its mass is halved, what is the new acceleration? (a) (4) one fourth as great (b) 4.0 m/s²
38. ●● The engine of a 1.0-kg toy plane exerts a 15-N forward force. If the air exerts an 8.0-N resistive force on the plane, what is the magnitude of the acceleration of the plane? 7.0 m/s²
39. ●● When a horizontal force of 300 N is applied to a 75.0-kg box, the box slides on a level floor, opposed by a force of kinetic friction of 120 N. What is the magnitude of the acceleration of the box? 2.40 m/s²
40. IE ●● A rocket is far away from all planets and stars, so gravity is not a consideration. It is using its rocket engines to accelerate upward with an acceleration $a = 9.80$ m/s². On the floor of the main deck is a crate (object with brick pattern) with a mass of 75.0 kg (▼Fig. 4.31). (a) How many forces are acting on the crate: (1) zero; (2) one; (3) two; (4) three? (b) Determine the normal force on the crate and compare it to the normal force the crate would experience if it were at rest on the surface of the Earth. (a) (2) one (b) 735 N, same

▲ **FIGURE 4.31 Away we go** See Exercise 40.

41. ●● An object (mass 10.0 kg) slides *upward* on a slippery vertical wall. A force F of 60 N acts at an angle of 60° as shown in ▼ Fig. 4.32. (a) Determine the normal force exerted on the object by the wall. (b) Determine the object's acceleration. (a) 30 N (b) −4.60 m/s²

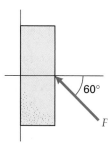

▲ **FIGURE 4.32 Up a wall** See Exercise 41.

42. ●● In an emergency stop to avoid an accident, a shoulder-strap seatbelt holds a 60-kg passenger in place. If the car was initially traveling at 90 km/h and came to a stop in 5.5 s along a straight, level road, what was the average force applied to the passenger by the seatbelt? 2.7 × 10² N

43. ●● A jet catapult on an aircraft carrier accelerates a 2000-kg plane uniformly from rest to a launch speed of 320 km/h in 2.0 s. What is the magnitude of the net force on the plane? 8.9 × 10⁴ N

44. ●●● In serving, a tennis player accelerates a 56-g tennis ball horizontally from rest to a speed of 30 m/s. Assuming that the acceleration is uniform when the racquet force is applied over a distance of 0.50 m, what is the magnitude of the force exerted on the ball by the racquet? 50 N

45. ●●● A car is skidding out of control on a horizontal icy (frictionless) road. It has a mass of 2000 kg and it is headed directly at Lois Lane at a speed of 45.0 m/s. When the car is 200 m from her, Superman begins to exert a steady force on the car, F, relative to the horizontal with a magnitude of 1.30 × 10⁴ N (he's a strong dude) at a downward angle of 30°. Was Superman correct? Was this force enough to stop the car before hitting Lois? Lois is saved, see ISM

4.4 Newton's Third Law of Motion

46. **MC** The action and reaction forces of Newton's third law (a) are in the same direction, (b) have different magnitudes, (c) act on different objects, (d) can be the same force. (c)

47. **MC** A brick hits a glass window. The brick breaks the glass, so (a) the magnitude of the force of the brick on the glass is greater than the magnitude of the force of the glass on the brick, (b) the magnitude of the force of the brick on the glass is smaller than the magnitude of the force of the glass on the brick, (c) the magnitude of the force of the brick on the glass is equal to the magnitude of the force of the glass on the brick, (d) none of the preceding. (c)

48. **MC** A freight truck collides head-on with a passenger car, causing a lot more damage to the car than to the truck. From this condition, we can say that (a) the magnitude of the force of the truck on the car is greater than the magnitude of the force of the car on the truck, (b) the magnitude of the force of the truck on the car is smaller than the magnitude of the force of the car on the truck, (c) the magnitude of the force of the truck on the car is equal to the magnitude of the force of the car on the truck, (d) none of the preceding. (c)

49. **CQ** Here is a story of a horse and a farmer: One day, the farmer attaches a heavy cart to the horse and demands that the horse pull the cart. "Well," says the horse, "I cannot pull the cart, because, according to Newton's third law, if I apply a force to the cart, the cart will apply an equal and opposite force on me. The net result will be that I cannot pull the cart, since all the forces will cancel. Therefore, it is impossible for me to pull this cart." The farmer was very upset! What could he say to persuade the horse to move? the forces on different objects (one on horse, one on cart) cannot cancel

50. **CQ** Is something wrong with the following statements? When a baseball is hit with a bat, there are equal and opposite forces on the bat and baseball. The forces then cancel, and there is no motion. yes; the forces on different objects (one on ball, one on bat) cannot cancel

51. **IE** ● A book is sitting on a horizontal surface. (a) There is (are) (1) one. (2) two, or (3) three force(s) acting on the book. (b) Identify the reaction force to each force on the book. (a) (2) two (b) see ISM

52. ●● In an Olympic figure-skating event, a 65-kg male skater pushes a 45-kg female skater, causing her to accelerate at a rate of 2.0 m/s². At what rate will the male skater accelerate? What is the direction of his acceleration? 1.4 m/s² opposite to her's

53. **IE** ●● A sprinter of mass 65.0 kg starts his race by pushing horizontally backward on the starting blocks with a force of 200 N. (a) What force causes him to accelerate out of the blocks: (1) His push on the blocks; (2) the downward force of gravity; or (3) the force the blocks exert forward on him? (b) Determine his initial acceleration as he leaves the blocks. (a) (3) the force the blocks exert forward on him (b) 3.08 m/s²

54. ●● Jane and John, with masses of 50 kg and 60 kg, respectively, stand on a frictionless surface 10 m apart. John pulls on a rope that connects him to Jane, giving Jane an acceleration of 0.92 m/s² toward him. (a) What is John's acceleration? (b) If the pulling force is applied constantly, where will Jane and John meet? (a) 0.77 m/s² toward Jane (b) 4.5 m from John's original position

55. **IE** ●●● During a daring rescue, a helicopter rescue squad initially accelerates a little girl (mass 25.0 kg) vertically off the roof of a burning building. They do this by dropping a rope down to her, which she holds onto as they pull her up. Neglect the mass of the rope. (a) What force causes the girl to accelerate vertically upward: (1) Her weight; (2) the pull of the helicopter on the rope; (3) the pull of the girl on the rope; or (4) the pull of the rope on the girl? (b) Determine the pull of the rope (the tension) if she initially accelerates at $a_y = +0.750$ m/s². (a) (4) the pull of the rope on the girl (b) 264 N

4.5 More on Newton's Laws: Free-Body Diagrams and Translational Equilibrium

56. **MC** The kinematic equations of Chapter 2 can be used (a) only with constant forces, (b) only with constant velocities, (c) with variable accelerations, (d) all of the preceding. (a)

57. **MC** The condition(s) for translational equilibrium is (are) (a) $\sum F_x = 0$, (b) $\sum F_y = 0$, (c) $\sum \vec{F}_i = 0$, (d) all of the preceding. (d)

58. **CQ** Draw the free-body diagram for a person sitting in the seat of an aircraft that is (a) accelerating down the runway for takeoff, and (b) after takeoff at a 20° angle to the ground. see ISM

59. **CQ** A person pushes perpendicularly on a block of wood that has been placed against a wall. Draw a free-body diagram and identify the reaction forces to all the forces on the block. see ISM

60. **CQ** A person on a bathroom scale (not the digital type) stands on the scale with his arms at his side. He then quickly raise his arms over his head, and notices that the scale reading increases as he brings his arms upward. Similarly, there is a decrease as he brings his arms downward. Why does the scale reading change? (Try this yourself.) see ISM

61. **IE ●** (a) When an object is on an inclined plane, the normal force exerted by the inclined plane on the object is (1) less than, (2) equal to, (3) more than the weight of the object. Why? (b) For a 10-kg object on a 30° inclined plane, what are the object's weight and the normal force exerted on the object by the inclined place? (a) (1) less than (b) 98 N and 85 N

62. **●●** A 75.0-kg person is standing on a scale in an elevator. What is the reading of the scale in newtons if the elevator is (a) at rest, (b) moving up at a constant velocity of 2.00 m/s, and (c) accelerating up at 2.00 m/s²? (a) 735 N (b) 735 N (c) 885 N

63. **●●** In Exercise 62, what if the elevator is accelerating down at 2.00 m/s²? 585 N

64. **IE ●●** The weight of a 500-kg object is 4900 N. (a) When the object is on a moving elevator, its measured weight could be (1) zero, (2) between zero and 4900 N, (3) more than 4900 N, (4) all of the preceding. Why? (b) Describe the motion if the object's measured weight is only 4000 N in a moving elevator. (a) (4) all of the preceding, see ISM (b) $a = 1.8$ m/s² downward

65. **●●** (a) A 75-kg water skier is pulled by a boat with a horizontal force of 400 N due east with a water drag on the skis of 300 N. A sudden gust of wind supplies another horizontal force of 50 N on the skier at an angle of 60° north of east. At that instant, what is the skier's acceleration? (b) What would be the skier's acceleration if the wind force were in the opposite direction to that in part (a)? (a) 1.7 m/s² at 19° north of east (b) 1.2 m/s² at 30° south of east

66. **●●** A boy pulls a box of mass 30 kg with a force of 25 N in the direction shown in ▼Fig. 4.33. (a) Ignoring friction, what is the acceleration of the box? (b) What is the normal force exerted on the box by the ground? (a) 0.72 m/s² (b) 2.8 × 10² N

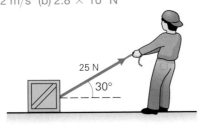

▲ **FIGURE 4.33** Pulling a box See Exercise 66.

67. **●●** A girl pushes a 25-kg lawn mower as shown in ▼Fig. 4.34. If $F = 30$ N and $\theta = 37°$, (a) what is the acceleration of the mower, and (b) what is the normal force exerted on the mower by the lawn? Ignore friction. (a) 0.96 m/s² (b) 2.6 × 10² N

▲ **FIGURE 4.34** Mowing the lawn See Exercise 67.

68. **●●** A 3000-kg truck tows a 1500-kg car by a chain. If the net forward force on the truck by the ground is 3200 N, (a) what is the acceleration of the car, and (b) what is the tension in the connecting chain? (a) 0.711 m/s² (b) 1067 N

69. **●●** A block of mass 25.0 kg slides down a frictionless surface inclined at 30°. To ensure that the block does not accelerate, what is the smallest force that you must exert on it and what is its direction? 123 N up the incline

70. **IE ●●** (a) An Olympic skier coasts down a slope with an angle of inclination of 37°. Neglecting friction, there is (are) (1), one, (2) two, (3) three force(s) acting on the skier. (b) What is the acceleration of the skier? (c) If the skier has a speed of 5.0 m/s at the top of the slope, what is his speed when he reaches the bottom of the 35-m-long slope? (a) (2) two (b) 5.9 m/s² (c) 21 m/s

71. **●●** A car coasts (engine off) up a 30° grade. If the speed of the car is 25 m/s at the bottom of the grade, what is the distance traveled by the car before it comes to rest? 64 m

72. **●●** Assuming ideal frictionless conditions for the apparatus shown in ▼Fig. 4.35, what is the acceleration of the system if (a) $m_1 = 0.25$ kg, $m_2 = 0.50$ kg, and $m_3 = 0.25$ kg, and (b) $m_1 = 0.35$ kg, $m_2 = 0.15$ kg, and $m_3 = 0.50$ kg? see ISM

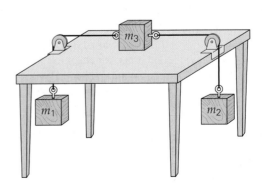

▲ **FIGURE 4.35** Which way will they accelerate? See Exercises 72, 110, and 111.

73. **IE ●●** A rope is fixed at both ends on two trees and a bag is hung in the middle of the rope (causing the rope to sag vertically). (a) The tension in the rope depends on (1) only the tree separation, (2) only the sag, (3) both the tree separation and sag, (4) neither the tree separation nor the sag. (b) If the tree separation is 10 m, the mass of the bag is 5.0 kg, and the sag is 0.20 m, what is the tension in the line? (a) (3) both the tree separation and sag (b) 6.1×10^2 N

74. **●●** A 55-kg gymnast hangs vertically from a pair of parallel rings. (a) If the ropes supporting the rings are attached to the ceiling directly above, what is the tension in each rope? (b) If the ropes are supported so that they make an angle of 45° with the ceiling, what is the tension in each rope? (a) 2.7×10^2 N (b) 3.8×10^2 N

75. **●●** A physicist's car has a small lead weight suspended from a string attached to the interior ceiling. Starting from rest, after a fraction of a second the car accelerates at a steady rate for about 10 s. During that time, the string (with the weight on the end of it) makes a backward (opposite the acceleration) angle of 15.0° from the vertical. Determine the car's (and the weight's) acceleration during the 10-s interval. 2.63 m/s²

76. **●●** A child attaches a 50.0-g mass (*m*) by a string to a toy car (mass *M* = 350 g). The string is run over the edge of a table by a frictionless pulley (neglect its mass and the string's mass) and the string is horizontal. Assuming the car has wheels with negligible friction, calculate (a) the acceleration of the car and (b) the tension in the string. (a) 1.23 m/s² (b) 0.429 N

77. **●●** At the end of most landing runways in airports, an extension of the runway is constructed using a special substance called *formcrete*. Formcrete can support the weight of cars, but crumbles under the weight of airplanes to slow them down if they run off the end of a runway. If a plane of mass 2.00×10^5 kg is to stop from a speed of 25.0 m/s on a 100-m-long stretch of formcrete, what is the average force exerted on the plane by the formcrete? 6.25×10^5 N

78. **●●** A rifle weighs 50.0 N and its barrel is 0.750 m long. It shoots a 25.0-g bullet, which leaves the barrel at a speed (muzzle velocity) of 300 m/s after being uniformly accelerated. What is the magnitude of the force exerted on the rifle by the bullet? 1.50×10^3 N

79. **●●** A horizontal force of 40 N acting on a block on a frictionless, level surface produces an acceleration of 2.5 m/s². A second block, with a mass of 4.0 kg, is dropped onto the first. What is the magnitude of the acceleration of the combination of blocks if the same force continues to act? (Assume that the second block does not slide on the first block.) 2.0 m/s²

80. **●●** The *Atwood machine* consists of two masses suspended from a fixed pulley, as shown in ▸Fig. 4.36. It is named after the British scientist George Atwood (1746–1807), who used it to study motion and to measure the value of *g*. If m_1 = 0.55 kg and m_2 = 0.80 kg, (a) what is the acceleration of the system, and (b) what is the magnitude of the tension in the string? (a) 1.8 m/s² (b) 6.4 N

81. **●●** An Atwood machine (see Fig. 4.36) has suspended masses of 0.25 kg and 0.20 kg. Under ideal conditions, what will be the acceleration of the smaller mass? 1.1 m/s² up

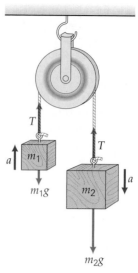

▲ **FIGURE 4.36 Atwood machine** See Exercises 80, 81, and 82.

82. **●●●** One mass, m_1 = 0.215 kg, of an ideal Atwood machine (see Fig. 4.36) rests on the floor 1.10 m below the other mass, m_2 = 0.255 kg. (a) If the masses are released from rest, how long does it take m_2 to reach the floor? (b) How high will mass m_1 ascend from the floor? [*Hint:* When m_2 hits the floor, m_1 continues to move upward.] (a) 1.62 s (b) 1.19 m

83. **IE ●●●** Two blocks are connected by a light string and accelerated upward by a pulling force *F*. The mass of the upper block is 50.0 kg and that of the lower block is 100 kg. The upward acceleration of the system as a whole is 1.50 m/s². Neglect the mass of the string. (a) Draw the free-body diagram of each block. Use the diagrams to determine which of the following is true for the magnitude of the string tension *T* compared to other forces: (1) $T > w_2$ and $T < F$; (2) $T > w_2$ and $T > F$; (3) $T < w_2$ and $T < F$; or (4) $T = w_2$ and $T < F$? (b) Apply Newton's laws to find the required pull, *F*. (c) Find the tension in the string, *T*. (a) (1) $T > w_2$ and $T < F$ (b) 1.70×10^3 N (c) 1.13×10^3 N

84. **●●●** In the frictionless apparatus shown in ▾Fig. 4.37, m_1 = 2.0 kg. What is m_2 if both masses are at rest? How about if both masses are moving at constant velocity? 1.2 kg; 1.2 kg

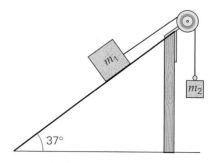

▲ **FIGURE 4.37 Inclined Atwood machine** See Exercises 84, 85, and 112.

85. ●●● In the ideal setup shown in Fig. 4.37, $m_1 = 3.0$ kg and $m_2 = 2.5$ kg. (a) What is the acceleration of the masses? (b) What is the tension in the string? (a) 1.2 m/s², m_1 up and m_2 down (b) 21 N

86. ●●● Two blocks on a level frictionless table are in contact. The mass of the left block is 5.00 kg and the mass of the right block is 10.0 kg and they accelerate to the left at 1.50 m/s². A person on the left exerts a force (F_1) of 75.0 N to the right. Another person exerts an unknown force (F_2) to the left. (a) Determine the force F_2. (b) Calculate the force of contact N between the two blocks (that is, the normal force at their vertical touching surfaces). (a) 97.5 N (b) 82.5 N

4.6 Friction

87. **MC** In general, the frictional force (a) is greater for smooth than rough surfaces, (b) depends on sliding speeds, (c) is proportional to the normal force, (d) depends significantly on the surface area of contact. (c)

88. **MC** The coefficient of kinetic friction, μ_k, (a) is usually greater than the coefficient of static friction, μ_s; (b) usually equals μ_s; (c) is usually smaller than μ_s; (d) equals the applied force that exceeds the maximum static force. (c)

89. **MC** A crate sits in the middle of the bed of a flatbed truck. The driver accelerates the truck gradually from rest to a normal speed, but then has to make a sudden stop to avoid hitting a car. If the crate slides as the truck stops, the frictional force would be (a) in the forward direction, (b) in the backward direction, (c) zero. (b)

90. **CQ** Identify the direction of the friction force in the following cases: (a) a book sitting on a table; (b) a box sliding on a horizontal surface; (c) a car making a turn on a flat road; (d) the initial motion of a machine part delivered on a conveyor belt in an assembly line. (a) no friction (b) opposite its direction of velocity (c) sideways (d) forward, in direction of velocity

91. **CQ** The purpose of a car's antilock brakes is to prevent the wheels from locking up so as to keep the car rolling rather than sliding. Why would rolling decrease the stopping distance as compared with sliding? see ISM

92. **CQ** Shown in ▼Fig. 4.38 are the front and rear wings of an Indy racing car. These wings generate *down force*, which is the vertical downward force produced by the air moving over the car. Why is such a down force desired? An Indy car can create a down force equal to twice its weight. Why not simply make the cars heavier? see ISM

▲ **FIGURE 4.38 Down force** See Exercise 92.

93. **CQ** (a) We commonly say that friction opposes motion. Yet when we walk, the frictional force is in the direction of our motion (Fig. 4.17). Is there an inconsistency in terms of Newton's second law? Explain. (b) What effects would wind have on air resistance? [*Hint*: The wind can blow in different directions.] see ISM

94. **CQ** Why are drag-racing tires wide and smooth, whereas passenger-car tires are narrower and have tread (▼Fig. 4.39)? Are there frictional and/or safety considerations? Does this difference between the tires contradict the fact that friction is independent of surface area? see ISM

▲ **FIGURE 4.39 Racing tires versus passenger-car tires: safety** See Exercise 94.

95. **IE** ● A 20-kg box sits on a rough horizontal surface. When a horizontal force of 120 N is applied, the object accelerates at 1.0 m/s². (a) If the applied force is doubled, the acceleration will (1) increase, but less than double; (2) also double; (3) increase, but more than double. Why? (b) Calculate the acceleration to prove your answer to part (a). (a) (3) increase, but more than double (b) 7.0 m/s²

96. ● In moving a 35.0-kg desk from one side of a classroom to the other, a professor finds that a horizontal force of 275 N is necessary to set the desk in motion, and a force of 195 N is necessary to keep it in motion at a constant speed. What are the coefficients of (a) static and (b) kinetic friction between the desk and the floor? (a) 0.802 (b) 0.569

97. ● A 40-kg crate is at rest on a level surface. If the coefficient of static friction between the crate and the surface is 0.69, what horizontal force is required to get the crate moving? 2.7×10^2 N

98. ● The coefficients of static and kinetic friction between a 50.0-kg box and a horizontal surface are 0.500 and 0.400 respectively. (a) What is the acceleration of the object if a 250-N horizontal force is applied to the box? (b) What is the acceleration if the applied force is 235 N? (a) 1.1 m/s² (b) 0

99. ●● A packing crate is placed on a 20° inclined plane. If the coefficient of static friction between the crate and the plane is 0.65, will the crate slide down the plane if released from rest? Justify your answer. $\theta_{min} = \tan^{-1} 0.65 = 33° > 20°$, so it will not move

100. ●● A 1500-kg automobile travels at 90 km/h along a straight concrete highway. Faced with an emergency situation, the driver jams on the brakes, and the car skids to a stop. What will be the car's stopping distance for (a) dry pavement and (b) wet pavement? (a) 38 m (b) 53 m

101. ●● A hockey player hits a puck with his stick, giving the puck an initial speed of 5.0 m/s. If the puck slows uniformly and comes to rest in a distance of 20 m, what is the coefficient of kinetic friction between the ice and the puck? 0.064

102. ●● In attempting to move a heavy couch (mass 200 kg) across a rug-covered floor, a man finds he must exert a horizontal force of 700 N to get the couch to barely move. Once the couch starts moving, the man continues to push with 700 N and his daughter (a physics major) estimates that it then accelerates at 1.10 m/s². Determine (a) the coefficient of static friction and (b) the coefficient of kinetic friction between the couch and the rug. (a) 0.357 (b) 0.245

103. IE ●● A person has a choice while trying to push a crate across a horizontal pad of concrete: push it at a downward angle of 30°, or pull it at an upward angle of 30°. (a) Which choice is most likely to require less force on the part of the person: (1) Pushing at a downward angle; (2) pulling at the same angle, but upward; or (3) pushing or pulling shouldn't matter? (b) If the crate has a mass of 50.0 kg and the coefficient of kinetic friction between it and the concrete is 0.750, calculate the required force to move it across the concrete at a steady speed for both situations. (a) (2) pulling at the same angle (b) 296 N; 748 N

104. ●● Suppose the slope conditions for the skier shown in ▼Fig. 4.40 are such that the skier travels at a constant velocity. From the photo, could you find the coefficient of kinetic friction between the snowy surface and the skis? If so, describe how this would be done. yes; $\tan \theta = \mu_k$

▲ **FIGURE 4.40 A downslope run** See Exercise 104.

105. ●● A 5.0-kg wooden block is placed on an adjustable wooden inclined plane. (a) What is the angle of incline above which the block will *start* to slide down the plane? (b) At what angle of incline will the block then slide down the plane at a constant speed? (a) 30° (b) 22°

106. ●● A block that has a mass of 2.0 kg and is 10 cm wide on each side just begins to slide down an inclined plane with a 30° angle of incline (▼Fig. 4.41). Another block of the same height and same material has base dimensions of 20 cm × 10 cm and thus a mass of 4.0 kg. (a) At what critical angle will the more massive block start to slide down the plane? Why? (b) Estimate the coefficient of static friction between the block and the plane. (a) 30° (b) 0.58

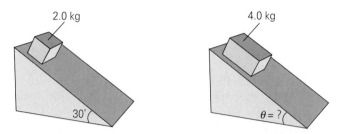

▲ **FIGURE 4.41 At what angle will it begin to slide?** See Exercise 106.

107. ●● In the apparatus shown in ▼Fig. 4.42, m_1 = 10 kg and the coefficients of static and kinetic friction between m_1 and the table are 0.60 and 0.40, respectively. (a) What mass of m_2 will just barely set the system in motion? (b) After the system begins to move, what is the acceleration? (a) 6.0 kg (b) 1.2 m/s²

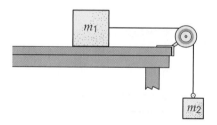

▲ **FIGURE 4.42 Friction and motion** See Exercise 107.

108. ●● In loading a fish delivery truck, a person pushes a block of ice up a 20° incline at constant speed. The push is 150 N in magnitude and parallel to the incline. The block has a mass of 35.0 kg. (a) Is the incline frictionless? (b) If not, what is the force of kinetic friction on the block of ice? (a) no (b) 33 N

109. ●●● An object (mass 3.0 kg) slides *upward* on a vertical wall *at constant velocity* when a force F of 60 N acts on it at an angle of 60° to the horizontal. (a) Draw the free-body diagram of the object. (b) Using Newton's laws find the normal force on the object. (c) Determine the force of kinetic friction on the object. (a) see ISM (b) 30 N (c) 23 N

110. ●●● For the apparatus in Fig. 4.35, what is the minimum value of the coefficient of static friction between the block (m_3) and the table that would keep the system at rest if m_1 = 0.25 kg, m_2 = 0.50 kg, and m_3 = 0.75 kg? 0.33

111. ●●● If the coefficient of kinetic friction between the block and the table in Fig. 4.35 is 0.560, and m_1 = 0.150 kg and m_2 = 0.250 kg, (a) what should m_3 be if the system is to move with a constant speed? (b) If m_3 = 0.100 kg, what is the magnitude of the acceleration of the system? (a) 0.179 kg (b) 0.862 m/s²

112. ●●● In the apparatus shown in Fig. 4.37, $m_1 = 2.0$ kg and the coefficients of static and kinetic friction between m_1 and the inclined plane are 0.30 and 0.20, respectively. (a) What is m_2 if both masses are at rest? (b) What is m_2 if both masses are moving at constant velocity? (a) between 0.72 kg and 1.7 kg (b) between 0.88 kg and 1.5 kg

Comprehensive Exercises

113. **IE** One block (A, mass 2.00 kg) rests atop another (B, mass 5.00 kg) on a horizontal surface. The surface is a powered walkway accelerating to the right at 2.50 m/s². B does not slip on the walkway surface, nor does A slip on B's top surface. (a) Sketch the free-body diagram of each block. Use these to determine the force responsible for A's acceleration. Is it (1) the pull of the walkway, (2) the normal force on A by the top surface of B, (3) the force of static friction on the bottom surface of B, or (4) the force of static friction acting on A due to the top surface of B? (b) Determine the forces of static friction on each block. (a) (4) the force of static friction on A due to top surface of B (b) 5.00 N, 17.5 N

114. In moving a box of mass 75.0 kg down a slippery (but not frictionless) incline (angle of 20°), a mover finds he must exert a horizontal force of 200 N to keep the box from accelerating down the incline. (a) Find the normal force N on the box. (b) Find the coefficient of kinetic friction between the box and the moving ramp. (a) 759 N (b) 0.0835

115. Two blocks (A and B) remain stuck together as they are pulled to the right by a force $F = 200$ N (▼Fig. 4.43). B is on a rough horizontal tabletop (coefficient of kinetic friction of 0.800). (a) What is the acceleration of the system? (b) What is the force of friction between the two objects? (a) 5.5 m/s² (b) 173 N

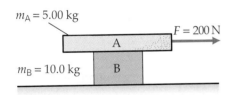

$m_A = 5.00$ kg

$F = 200$ N

A

$m_B = 10.0$ kg   B

▲ **FIGURE 4.43 Take it away** See Exercise 115.

116. A two-stage toy rocket is launched vertically from rest. While the stages are together, the rocket engines that are attached to the lower stage exert an upward force of 500 N. The top stage (nose cone) has a mass (m) of 2.00 kg and the bottom stage has a mass (M) of 8.00 kg. (a) Draw the free-body diagram for each stage and determine whether any forces in the two diagrams are an action–reaction pair. If so, tell which they are. (There may be more than one pair.) (b) Calculate the rocket's acceleration. (c) Determine and the contact force between the two stages (that is, the upward normal force exerted on the upper stage by the lower stage). (a) see ISM (b) 40.2 m/s² (c) 100 N

117. A 5.00-kg block (M) on a 30° incline is connected by a light string over a frictionless pulley to an unknown mass, m. The coefficient of kinetic friction between the block and the incline is 0.100. When the system is released from rest, mass m accelerates upward at 2.00 m/s². Determine (a) the string tension and (b) the value of m. (a) 10.3 N (b) 0.954 kg

118. **IE** In hauling a boat out of the water for the winter, the storage facility uses a wide strap with cables operating at the same angle (measured from the horizontal) on either side of the boat (▼Fig. 4.44). (a) As the boat comes up vertically and θ decreases, the tension in the cables (1) increases, (2) decreases, (3) stays the same. (b) Determine the tension in each cable if the boat has a mass of 500 kg and the angle of each cable is 45° from the horizontal and the boat is being momentarily held at rest. Compare this to the tension when the boat is raised and held at rest so the angle becomes 30°. (a) (1) increases (b) 45°: 3.46 × 10³ N; 30°: 4.90 × 10³ N

T_2 T_1

θ θ

▲ **FIGURE 4.44 Hoist it up** See Exercise 118.

The following Physlet Physics Problems can be used with this chapter.

PHYSLET® 4.1, 4.2, 4.3, 4.4, 4.5, 4.6, 4.7, 4.8, 4.9, 4.10, 4.11, 4.13, 5.1, 5.2, 5.3, 5.4, 5.5, 5.6, 5.7

5

WORK AND ENERGY

PHYSICS FACTS

- *Kinetic* comes from the Greek *kinein*, meaning "to move."
- *Energy* comes from the Greek *energeia*, meaning "activity."
- The United States has 5% of the world's population, yet consumes 26% of its energy supply.
- Recycling aluminum takes 95% less energy than making aluminum from raw materials.
- The human body operates within the limits imposed by the law of conservation of total energy, needing dietary energy equal to the energy expended in the external work of daily activities, internal activities, and system heat losses.
- The basal metabolic rate (BMR) is a measurement of the rate at which the human body expends energy. An average 70-kg man requires about 7.5×10^6 J of basal energy per day for simply living or maintaining basic functions—for example, breathing, blood circulation, and digestion. Any type of exercise requires more energy—for example, an additional 4.6×10^6 J per hour to climb stairs.
- The human body uses muscles to propel itself, turning stored energy into motion. There are 630 active muscles in your body and they act in groups.

Source: Harold E. Edgerton/©Harold & Esther Edgerton Foundation, 2002, courtesy of Palm Press, Inc.

A description of pole vaulting, as shown in the photo, might be as follows: The athlete runs with a pole, plants it into the ground, and tries to vault his body over a bar set at a certain height. However, a physicist might give a different description: The athlete has chemical potential energy stored in his body. He uses this potential energy to do work in running down the path to gain speed, or kinetic energy. When he plants the pole, most of his kinetic energy goes into elastic potential energy of the bent pole. This potential energy is used to lift the vaulter or to do work against gravity, and is partially converted into gravitational potential energy. At the top, there is just enough kinetic energy left to carry the vaulter over the bar. On the way down, the gravitational potential energy is converted back to kinetic energy, which is absorbed by the mat in doing work to stop the fall. The pole vaulter participates in a game of work–energy, a game of give and take.

This chapter centers on these two concepts that are important in both science and everyday life—*work* and *energy*. We commonly think of work as being associated with doing or accomplishing something. Because work makes us physically (and sometimes mentally) tired, machines have been invented to decrease the amount of effort we expend personally. Thinking about energy tends to bring to mind the cost of fuel for transportation and heating, or perhaps the food that supplies the energy our bodies need to carry out life processes and to do work.

Although these notions do not really define work and energy, they point us in the right direction. As you might have guessed, work and energy are closely related. In physics, as in everyday life, when something possesses energy, it has the ability to do work. For example, water rushing through the sluices of a dam has energy of motion, and this energy allows the water to do the work of driving a turbine or dynamo to generate electricity. Conversely, no work can be performed without energy.

Energy exists in various forms: mechanical energy, chemical energy, electrical energy, heat energy, nuclear energy, and so on. A transformation from one form to another may take place, but the total amount of energy is *conserved*, meaning it always remains the same. This point makes the concept of energy very useful. When a physically measurable quantity is conserved, it not only gives us an insight that leads to a better understanding of nature, but also usually provides another approach to practical problems. (You will be introduced to other conserved quantities during the course of our study of physics.)

5.1 Work Done by a Constant Force

OBJECTIVES: To (a) define mechanical work, and (b) compute the work done in various situations.

The word *work* is commonly used in a variety of ways: We go to work; we work on projects; we work at our desks or on computers; we work problems. In physics, however, *work* has a very specific meaning. Mechanically, work involves force and displacement, and the word *work* is used to describe quantitatively what is accomplished when a force acts on an object as it moves through a distance. In the simplest case of a *constant* force acting on an object, work that the force does is defined as follows:

> The **work** done by a constant force acting on an object is equal to the product of the magnitudes of the displacement and the component of the force parallel to that displacement.

Work then involves a force acting on an object moving through a distance. A force may be applied, as in ▼Fig. 5.1a, but *if there is no motion (no displacement), then no work is done*. For a constant force F acting *in the same direction* as the displacement d (Fig. 5.1b), the work (W) done is defined as the product of their magnitudes:

$$W = Fd \qquad (5.1)$$

and work is a scalar. (As you might expect, when work is done as in Fig. 5.1b, energy is expended. We shall discuss the relationship between work and energy in Section 5.3.)

In general, work is done on an object only by a force, or force *component*, parallel to the line of motion or displacement of the object (Fig. 5.1c). That is, if the force acts at an angle θ to the object's displacement, then $F_{\parallel} = F \cos \theta$ is the

Work involves force and displacement

Teaching tip: Discuss whether carrying an object horizontally across the room involves work according to the definition given here.

Note: The product of two vectors (force and displacement) in this case is a special type of vector multiplication and yields a scalar quantity equal to $(F \cos \theta)d$. Thus, work is a scalar—it does not have direction. It can, however, be positive, zero, or negative, depending on the angle.

▼ **FIGURE 5.1 Work done by a constant force—the product of the magnitudes of the parallel component of force and the displacement** **(a)** If there is no displacement, no work is done: $W = 0$. **(b)** For a constant force in the same direction as the displacement, $W = Fd$. **(c)** For a constant force at an angle to the displacement, $W = (F \cos \theta)d$.

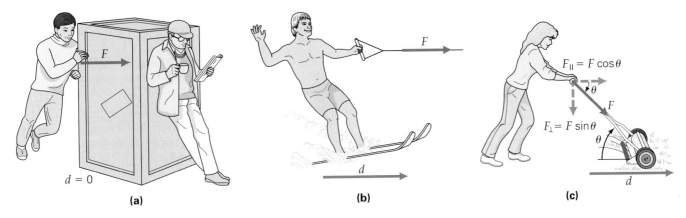

component of the force parallel to the displacement. Thus, a more general equation for work done by a constant force is

$$W = F_\parallel d = (F \cos \theta)d \quad \text{(work done by a constant force)} \quad (5.2)$$

Notice that θ is the angle *between* the force and the displacement vectors. To remind yourself of this factor, you can write $\cos \theta$ between the magnitudes of the force and displacement, $W = F(\cos \theta)d$. If $\theta = 0°$ (that is, force and displacement are in the same direction as in Fig. 5.1b), then $W = F(\cos 0°)d = Fd$, so Eq. 5.2 reduces to Eq. 5.1. The perpendicular component of the force, $F_\perp = F \sin \theta$, does no work, since there is no displacement in this direction.

The units of work can be determined from the equation $W = Fd$. With force in newtons and displacement in meters, work has the SI unit of newton-meter (N·m). This unit is called a **joule** (J):

$$Fd = W$$
$$1\,\text{N·m} = 1\,\text{J}$$

For example, the work done by a force of 25 N on an object as the object moves through a parallel displacement of 2.0 m is $W = Fd = (25\,\text{N})(2.0\,\text{m}) = 50\,\text{N·m}$, or 50 J.

From the previous displayed equation, we also see that in the British system, work would have the unit pound-foot. However, this name is commonly written in reverse: The British standard unit of work is the **foot-pound (ft·lb)**. One ft·lb is equal to 1.36 J.

We can analyze work graphically. Suppose a constant force F in the x direction acts on an object as it moves a distance x. Then $W = Fx$, and if F versus x is plotted, a straight-line graph is obtained such as shown in the accompanying Learn by Drawing. The area under the line is Fx, so this area is equal to the work done by the force over the given distance. We will consider a nonconstant, or variable, force later.*

Remember that *work is a scalar quantity* and, as such, may have a positive or negative value. In Fig. 5.1b, the work is positive, because the force acts in the same direction as the displacement (and $\cos 0°$ is positive). The work is also positive in Fig. 5.1c, because a force component acts in the direction of the displacement (and $\cos \theta$ is positive).

However, if the force, or a force component, acts in the opposite direction of the displacement, the work is negative, since the cosine term is negative. For example, for $\theta = 180°$ (force opposite to the displacement), $\cos 180° = -1$, so the work is negative: $W = F_\parallel d = F(\cos 180°)d = -Fd$. An example is a braking force that slows down or decelerates an object. See Learn by Drawing on p. 143.

Example 5.1 ■ Applied Psychology: Mechanical Work

A student holds her psychology textbook, which has a mass of 1.5 kg, out a second-story dormitory window until her arm is tired; then she releases it (◄Fig. 5.2). (a) How much work is done on the book by the student in simply holding it out the window? (b) How much work is done by the force of gravity during the time in which the book falls 3.0 m?

Thinking It Through. Analyze the situations in terms of the definition of work, keeping in mind that force and displacement are the key factors.

Solution. Listing the data, we have

Given: $v_o = 0$ (initially at rest)
$m = 1.5\,\text{kg}$
$d = 3.0\,\text{m}$

Find: (a) W (work done by student in holding)
(b) W (work done by gravity in falling)

(a) Even though the student gets tired (because work is performed within the body to maintain muscles in a state of tension), she does *no work on the book* in merely holding it stationary. She exerts an upward force on the book (equal in magnitude to its weight), but the displacement is zero in this case ($d = 0$). Thus, $W = Fd = F \times 0 = 0\,\text{J}$.

The joule (J), pronounced "jool," was named in honor of James Prescott Joule (1818–1889), a British scientist who investigated work and energy.

LEARN BY DRAWING

Work: Area under the F-versus-x Curve

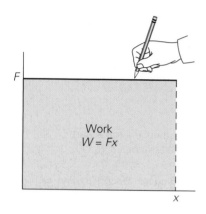

Work
$W = Fx$

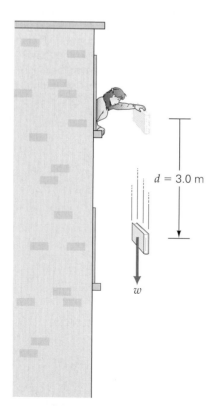

$d = 3.0\,\text{m}$

w

▲ **FIGURE 5.2** Mechanical work requires motion See Example 5.1.

*Work is the area under the F-versus-x curve even if the curve is not a straight line. Finding the work in such cases generally requires advanced mathematics.

(b) While the book is falling, the only force acting on it is the force of gravity (neglecting air resistance), which is equal in magnitude to the weight of the book: $F = w = mg$. The displacement is in the same direction as the force ($\theta = 0°$) and has a magnitude of $d = 3.0$ m, so the work done by gravity is

$$W = F(\cos 0°)d = (mg)d = (1.5\text{ kg})(9.8\text{ m/s}^2)(3.0\text{ m}) = +44\text{ J}$$

(+ because the force and displacement are in the same direction.)

Follow-Up Exercise. A 0.20-kg ball is thrown upward. How much work is done on the ball by gravity as the ball rises between heights of 2.0 m and 3.0 m? (*Answers to all Follow-Up Exercises are at the back of the text.*)

Example 5.2 ■ Hard Work

A worker pulls a 40.0-kg crate with a rope, as illustrated in ▼Fig. 5.3. The coefficient of kinetic (sliding) friction between the crate and the floor is 0.550. If he moves the crate with a constant velocity a distance of 7.00 m, how much work is done?

Thinking It Through. A good first thing to do in problems such as this is to draw a free-body diagram. This is shown in the figure. To find the work, the force F must be known. As usual in such cases, this is done by summing the forces.

Solution.

Given: $\quad m = 40.0$ kg $\qquad$ *Find:* $\quad W$ (the work done in moving the
$\qquad \mu_k = 0.550 \qquad\qquad\qquad\qquad$ crate 7.00 m)
$\qquad d = 7.00$ m
$\qquad \theta = 30°$ (from figure)

Then, summing the forces in the x and y directions:

$$\Sigma F_x = F\cos 30° - f_k = F\cos 30° - \mu_k N = ma_x = 0$$
$$\Sigma F_y = N + F\sin 30° - mg = ma_y = 0$$

To find F, the second equation may be solved for N, which is then substituted in the first equation.

$$N = mg - F\sin 30°$$

(Notice that N is not equal to the weight of the crate. Why?) And, substituting into the first equation,

$$F\cos 30° - \mu_k(mg - F\sin 30°) = 0 \qquad \text{(continues on next page)}$$

(continues on next page)

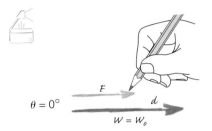

LEARN BY DRAWING

Determining the Sign of Work

$\theta = 0°$
$W = W_o$

$\theta < 90°$
$W > 0$ but $< W_o$

$\theta = 90°$
$W = 0$ (why?)

$\theta > 90°$
$W < 0$

$\theta = 180°$
$W = -W_o$

▼ **FIGURE 5.3 Doing some work.** See Example 5.2.

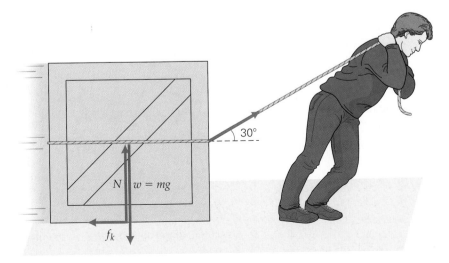

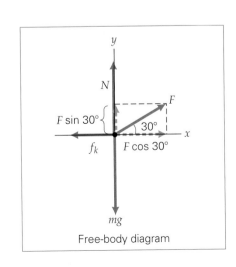

Free-body diagram

Solving for F and putting in values:

$$F = \frac{\mu_k mg}{(\cos 30° + \mu_k \sin 30°)} = \frac{(0.550)(40.0 \text{ kg})(9.80 \text{ m/s}^2)}{(0.866) + (0.550)(0.500)]} = 189 \text{ N}$$

Then, $W = F(\cos 30°)d = (189 \text{ N})(0.866)(7.00 \text{ m}) = 1.15 \times 10^3 \text{ J}$

Follow-Up Exercise. It takes about 3.80×10^4 of work to lose 1.00 g of body fat. What distance would the worker have to pull the crate to lose 1 g of fat? (Assume all the work goes into fat reduction.) Make an estimate before solving and see how close you come.

Illustration 6.2 *Constant Forces*

We commonly specify which force is doing work *on* which object. For example, the force of gravity does work on a falling object, such as the book in Example 5.1. Also, when you lift an object, *you* do work *on* the object. We sometimes describe this as doing work *against* gravity, because the force of gravity acts in the direction opposite that of the applied lift force and opposes it. For example, an average-sized apple has a weight of about 1 N. So if you lifted such an apple a distance of 1 m with a force equal to its weight, you would have done 1 J of work against gravity [$W = Fd = (1 \text{ N})(1 \text{ m}) = 1$ J]. This example gives you an idea of how much work 1 J represents.

In both Examples 5.1 and 5.2, work was done by a single constant force. If more than one force acts on an object, the work done by each can be calculated separately:

> The *total*, or *net*, *work* is defined as the work done by all the forces acting on the object, or the scalar sum of all those quantities of work.

This concept is illustrated in Example 5.3.

Example 5.3 ■ Total or Net Work

A 0.75-kg block slides with a uniform velocity down a 20° inclined plane (▼Fig. 5.4). (a) How much work is done by the force of friction on the block as it slides the total length of the plane? (b) What is the net work done on the block? (c) Discuss the net work done if the angle of incline is adjusted so that the block accelerates down the plane.

Thinking It Through. (a) The length of the plane can be found using trigonometry, so this part boils down to finding the force of friction. (b) The net work is the sum of all the work done by the individual forces. (*Note*: Since the block has a uniform, or constant, velocity, the net force on it is zero. This observation should tell you the answer, but it will be shown explicitly in the solution.) (c) If there is acceleration, Newton's second law applies, which involves a net force, so there will be net work.

Solution. We list the information that is given. In addition, it is equally important to list specifically what is to be found.

▶ **FIGURE 5.4 Total or net work** See Example 5.3.

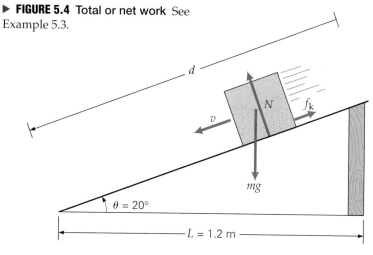

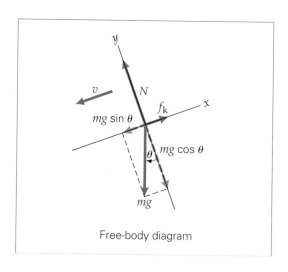

Free-body diagram

Given: $m = 0.75$ kg   *Find:*   (a) W_f (work done on the block by friction)
          $\theta = 20°$ (b) W_{net} (net work on the block)
          $L = 1.2$ m (from Fig. 5.4)   (c) W (discuss net work with block accelerating)

(a) Note from Fig. 5.4 that only two forces do work, because there are only two forces parallel to the motion: f_k, the force of kinetic friction, and $mg \sin \theta$, the component of the block's weight acting down the plane. The normal force N and $mg \cos \theta$, the components of the block's weight acting perpendicular to the plane, do no work on the block. (Why?)
    We first find the work done by the frictional force:

$$W_f = f_k(\cos 180°)d = -f_k d = -\mu_k N d$$

Note: Recall the discussion of friction in Section 4.6.

The angle 180° indicates that the force and displacement are in opposite directions. (It is common in such cases to write $W_f = -f_k d$ directly, since kinetic friction typically opposes motion.) The distance d the block slides down the plane can be found by using trigonometry. Note that $\cos \theta = L/d$, so

$$d = \frac{L}{\cos \theta}$$

We know that $N = mg \cos \theta$, but what is μ_k? It would appear that some information is lacking. When this situation occurs, look for another approach to solve the problem. As noted earlier, there are only two forces parallel to the motion, and they are opposite, so with a constant velocity their magnitudes are equal, $f_k = mg \sin \theta$. Thus,

$$W_f = -f_k d = -(mg \sin \theta)\left(\frac{L}{\cos \theta}\right) = -mgL \tan 20°$$
$$= -(0.75 \text{ kg})(9.8 \text{ m/s}^2)(1.2 \text{ m})(0.364) = -3.2 \text{ J}$$

(b) To find the net work, we need to calculate the work done by gravity and then add it to our result in part (a). Since $F_\parallel$ for gravity is just $mg \sin \theta$,

$$W_g = F_\parallel d = (mg \sin \theta)\left(\frac{L}{\cos \theta}\right) = mgL \tan 20° = +3.2 \text{ J}$$

where the calculation is the same as in part (a) except for the sign. Then

$$W_{net} = W_g + W_f = +3.2 \text{ J} + (-3.2 \text{ J}) = 0$$

Remember that work is a scalar quantity, so scalar addition is used to find net work.

(c) If the block accelerates down the plane, then from Newton's second law, we have $F_{net} = mg \sin \theta - f_k = ma$. The component of the gravitational force ($mg \sin \theta$) is greater than the opposing frictional force (f_k), so net work is done on the block, because now $|W_g| > |W_f|$. You may be wondering what the effect of nonzero net work is. As you will learn shortly, nonzero net work causes a change in the amount of energy an object has.

Teaching tip: Point out that kinetic friction does negative work on the object, while the component of the weight down the plane does positive work on the object. The net work is zero only because the forces are equal and opposite.

Follow-Up Exercise. In part (c) of this Example, is it possible for the frictional work to be greater in magnitude than the gravitational work? What would this condition mean in terms of the block's speed?

Problem-Solving Hint

Note that in part (a) of Example 5.3, the equation for W_f was simplified by using the algebraic expressions for N and d instead of by computing these quantities initially. It is a good rule of thumb not to plug numbers into an equation until you have to. Simplifying an equation through cancellation is easier with symbols and saves computation time.

5.2 Work Done by a Variable Force

OBJECTIVES: To (a) differentiate between work done by constant and variable forces, and (b) compute the work done by a spring force.

The discussion in the preceding section was limited to work done by constant forces. In general, however, forces are variable; that is, they change in magnitude and/or angle with time and/or position. For example, someone might push harder

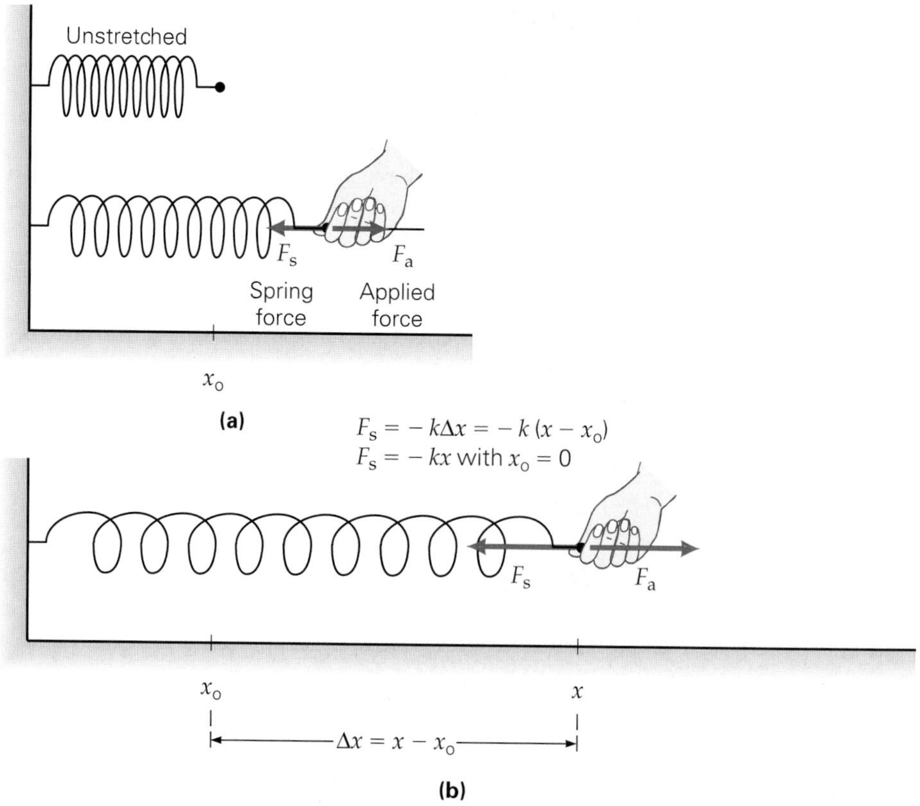

▶ **FIGURE 5.5 Spring force (a)** An applied force F_a stretches the spring, and the spring exerts an equal and opposite force F_s on the hand. **(b)** The magnitude of the force depends on the change Δx in the spring's length. This change is measured from to the end of the unstretched spring at x_o.

Unstretched

F_s F_a

Spring force   Applied force

x_o

(a)

$$F_s = -k\Delta x = -k(x - x_o)$$
$$F_s = -kx \text{ with } x_o = 0$$

F_s F_a

x_o x

$$\Delta x = x - x_o$$

(b)

Note: In Fig. 5.5, the hand applies a variable force F_a in stretching the spring. At the same time, the spring exerts an equal and opposite force F_s on the hand.

and harder on an object to overcome the force of static friction, until the applied force exceeds $f_{s_{max}}$. However, the force of static friction does no work, because there is no motion or displacement.

An example of a variable force that does work is illustrated in ▲Fig. 5.5, which depicts a spring being stretched by an applied force F_a. As the spring is stretched (or compressed) farther and farther, its restoring force (the force that opposes the stretching or compression) becomes greater, and an increased applied force is required. For most springs, the spring force is directly proportional to the change in length of the spring from its unstretched length. In equation form, this relationship is expressed as

$$F_s = -k\Delta x = -k(x - x_o)$$

or, if $x_o = 0$,

$$F_s = -kx \quad \textit{(ideal spring force)} \quad (5.3)$$

where x now represents the distance the spring is stretched (or compressed) from its unstretched length. As can be seen, the force varies with x. We describe this relationship by saying that the *force is a function of position*.

The k in this equation is a constant of proportionality and is commonly called the **spring constant**, or **force constant**. The greater the value of k, the stiffer or stronger is the spring. As you should be able to prove to yourself, the SI unit of k is newton per meter (N/m). The minus sign indicates that the spring force acts in the direction opposite to the displacement when the spring is either stretched or compressed. Equation 5.3 is a form of what is known as *Hooke's law*, named after Robert Hooke, a contemporary of Newton.

The relationship expressed by the spring force equation holds only for ideal springs. Real springs approximate this linear relationship between force and displacement within certain limits. If a spring is stretched beyond a certain point, called its *elastic limit*, the spring will be permanently deformed, and the linear relationship no longer applies.

To compute the work done by variable forces generally requires calculus. But we are fortunate in that the spring force is a special case that can be computed by using a graph. A plot of F (the applied force) versus x is shown in ▶Fig. 5.6. The graph has a straight-line slope of k, with $F = kx$, where F is the applied force doing work in stretching the spring.

Teaching tip: Point out that expressions for constant acceleration are not applicable when the force varies.

PHYSLET

Illustration 6.3 *Force and Displacement*

As described earlier, work is the area under an F-versus-x curve, and here it is in the form of a triangle as indicated by the shaded area in the figure. Then, computing this area,

$$\text{area} = W = \tfrac{1}{2}(\text{altitude} \times \text{base})$$

or

$$W = \tfrac{1}{2}Fx = \tfrac{1}{2}(kx)x = \tfrac{1}{2}kx^2$$

where $F = kx$. Thus we have

$$W = \tfrac{1}{2}kx^2 \qquad \begin{array}{l}\textit{work done in stretching (or compressing)}\\ \textit{a spring from } x_o = 0\end{array} \qquad (5.4)$$

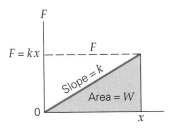

▲ **FIGURE 5.6** Work done by a uniformly variable spring force A graph of F versus x, where F is the applied force doing work in stretching a spring, is a straight line with a slope of k. The work is equal to the area under the line, which is that of a triangle with area $= \tfrac{1}{2}(\text{altitude} \times \text{base})$. Then $W = \tfrac{1}{2}Fx = \tfrac{1}{2}(kx)x = \tfrac{1}{2}kx^2$.

Example 5.4 ■ Determining the Spring Constant

A 0.15-kg mass is attached to a vertical spring and hangs at rest a distance of 4.6 cm below its original position (▶Fig. 5.7). An additional 0.50-kg mass is then suspended from the first mass and allowed to descend to a new equilibrium position. What is the total extension of the spring? (Neglect the mass of the spring.)

Thinking It Through. The spring constant k appears in Eq. 5.3. Therefore, to find the value of k for a particular instance, the spring force and distance the spring is stretched (or compressed) must be known.

Solution. The data given are as follows:

Given: $m_1 = 0.15$ kg   *Find:* x (total stretch distance)
         $x_1 = 4.6$ cm $= 0.046$ m
         $m_2 = 0.50$ kg

The total stretch distance is given by $x = F/k$, where F is the applied force, which in this case is the weight of the mass suspended on the spring. However, the spring constant k is not given. This quantity may be found from the data pertaining to the suspension of m_1 and resulting displacement x_1. (This method is commonly used to determine spring constants.) As seen in Fig. 5.7a, the magnitudes of the weight force and the restoring spring force are equal, since $a = 0$, so we may equate their magnitudes:

$$F_s = kx_1 = m_1g$$

Solving for k, we obtain

$$k = \frac{m_1g}{x_1} = \frac{(0.15 \text{ kg})(9.8 \text{ m/s}^2)}{0.046 \text{ m}} = 32 \text{ N/m}$$

Then, knowing k, we find the total extension of the spring from the balanced-force situation shown in Fig. 5.7b:

$$F_s = (m_1 + m_2)g = kx$$

Thus,

$$x = \frac{(m_1 + m_2)g}{k} = \frac{(0.15 \text{ kg} + 0.50 \text{ kg})(9.8 \text{ m/s}^2)}{32 \text{ N/m}} = 0.20 \text{ m (or 20 cm)}$$

Follow-Up Exercise. How much work is done by gravity in stretching the spring through both displacements in Example 5.4?

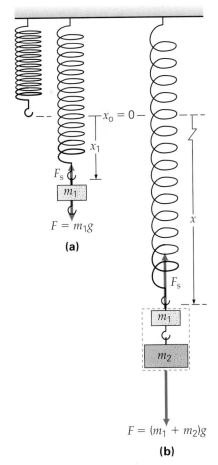

▲ **FIGURE 5.7** Determining the spring constant and the work done in stretching a spring See Example 5.4.

Problem-Solving Hint

The reference position x_o used to determine the change in length of a spring is arbitrary and is usually chosen as zero for convenience. *The important quantity in computing work is the difference in position, Δx, or the net change in the length of the spring from its unstretched length.* As shown in ▼Fig. 5.8 for a mass suspended on a spring, x_o can be referenced to the unloaded length of the spring or to the loaded position, which may be taken as the zero position for convenience. In Example 5.4, x_o was referenced to the end of the unloaded spring.

When the net force on the suspended mass is zero, the mass is said to be at its *equilibrium position* (as in Fig. 5.7a with m_1 suspended). This position, rather than the unloaded length, may be taken as a zero reference ($x_o = 0$; see Fig. 5.8b). The equilibrium position is a convenient reference point for cases in which the mass oscillates up and down on the spring. Also, since the displacement is in the vertical direction, the x's are often replaced by y's.

▶ **FIGURE 5.8 Displacement reference** The reference position x_o is arbitrary and is usually chosen for convenience. It may be **(a)** at the end of the spring at its unloaded position or **(b)** at the equilibrium position when a mass is suspended on the spring. The latter is particularly convenient in cases in which the mass oscillates up and down on the spring.

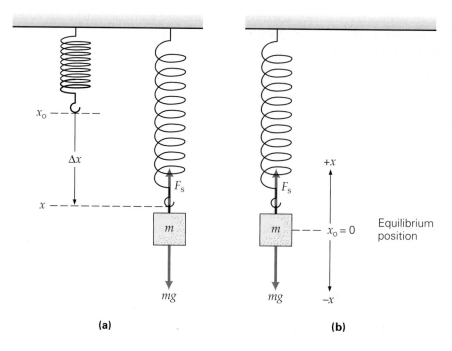

(a)   (b)

Illustration 6.4 Springs

Exploration 6.1 An Operational Definition of Work

5.3 The Work–Energy Theorem: Kinetic Energy

<u>**OBJECTIVES:**</u> To (a) study the work–energy theorem, and (b) apply it in solving problems.

Now that we have an operational definition of work, let's take a look at how work is related to energy. Energy is one of the most important concepts in science. We describe it as a quantity that objects or systems possess. Basically, work is something that is *done on* objects, whereas energy is something that objects *have*—the ability to do work.

One form of energy that is closely associated with work is *kinetic energy*. (Another form of energy, *potential energy*, will be described in Section 5.4.) Consider an object at rest on a frictionless surface. Let a horizontal force act on the object and set it in motion. Work is done *on* the object, but where does the work "go," so to speak? It goes into setting the object into motion, or changing its *kinetic* conditions. Because of its motion, we say the object has gained energy—kinetic energy, which gives it the capability to do work.

For a constant force doing work on a moving object, as illustrated in ▼Fig. 5.9, the force does an amount of work $W = Fx$. But what are the kinematic effects? The force gives the object a constant acceleration, and from Eq. 2.12, $v^2 = v_o^2 + 2ax$ (with $x_o = 0$),

$$a = \frac{v^2 - v_o^2}{2x}$$

▶ **FIGURE 5.9 The relationship of work and kinetic energy** The work done on a block by a constant force in moving it along a horizontal frictionless surface is equal to the change in the block's kinetic energy: $W = \Delta K$.

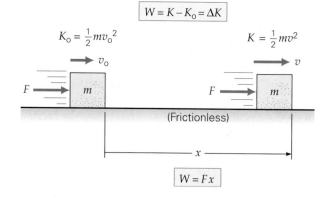

where v_0 may or may not be zero. Writing the magnitude of the force in the form of Newton's second law and substituting in the expression for a from the previous equation gives

$$F = ma = m\left(\frac{v^2 - v_0^2}{2x}\right)$$

Using this expression in the equation for work, we have

$$W = Fx = m\left(\frac{v^2 - v_0^2}{2x}\right)x$$

$$= \tfrac{1}{2}mv^2 - \tfrac{1}{2}mv_0^2$$

It is convenient to define $\tfrac{1}{2}mv^2$ as the **kinetic energy (K)** of the moving object:

$$K = \tfrac{1}{2}mv^2 \quad \text{(kinetic energy)} \tag{5.5}$$

SI unit of energy: joule (J)

Kinetic energy is often called the *energy of motion*. Note that it is directly proportional to the square of the (instantaneous) speed of a moving object, and therefore cannot be negative.

Then, in terms of kinetic energy, the previous expression for work can be written as

$$W = \tfrac{1}{2}mv^2 - \tfrac{1}{2}mv_0^2 = K - K_0 = \Delta K$$

or

$$W = \Delta K \tag{5.6}$$

where it is understood that *W is the net work if more than one force acts on the object*, as shown in Example 5.3. This equation is called the **work–energy theorem**, and it relates the work done on an object to the change in the object's kinetic energy. That is, *the net work done on a body by all the forces acting on it is equal to the change in kinetic energy of the body*. Both work and energy have units of joules, and both are *scalar* quantities. Keep in mind that the work–energy theorem is true in general for variable forces and not just for the special case considered in deriving Eq. 5.6.

To illustrate that net work is equal to the change in kinetic energy, recall that in Example 5.1 the force of gravity did +44 J of work on a book that fell from rest through a distance of $y = 3.0$ m. At that position and instant, the falling book had 44 J of kinetic energy. Since $v_0 = 0$ in this case, $\tfrac{1}{2}mv^2 = mgy$. Substituting this expression into the equation for the work done on the falling book by gravity, we get

$$W = Fd = mgy = \frac{mv^2}{2} = K = \Delta K$$

where $K_0 = 0$. Thus the kinetic energy gained by the book is equal to the net work done on it: 44 J in this case. (As an exercise, confirm this fact by calculating the speed of the book and evaluating its kinetic energy.)

The work–energy theorem tells us that when work is done on an object, there is a change in or a transfer of energy. In general, then, we might say that *work is a measure of the transfer of kinetic energy* to the object. For example, a force doing work on an object that causes the object to speed up gives rise to an increase in the object's kinetic energy. Conversely, (negative) work done by the force of kinetic friction may cause a moving object to slow down and decrease its kinetic energy. So for an object to have a change in its kinetic energy, net work must be done on the object, as Eq. 5.6 tells us.

When an object is in motion, it possesses kinetic energy and has the capability to do work. For example, a moving automobile has kinetic energy and can do work in crumpling a fender in a fenderbender—not *useful* work in that case, but still work. Another example of work done by kinetic energy is shown in ▸Fig. 5.10.

Teaching tip: Emphasize that net work done causes a change in kinetic energy. Discuss the process of pitching and catching a baseball. Catching a baseball involves negative work, which decreases kinetic energy. Toss a ball to a student to initiate the discussion.

Work–energy theorem

▲ **FIGURE 5.10 Kinetic energy and work** A moving object, such as a wrecking ball, processes kinetic energy and can do work. A massive ball is used in demolishing buildings.

PHYSLET®

Exploration 6.4 Change the Direction
of the Force Applied

Example 5.5 ■ A Game of Shuffleboard: The Work–Energy Theorem

A shuffleboard player (▼Fig. 5.11) pushes a 0.25-kg puck that is initially at rest such that a constant horizontal force of 6.0 N acts on it through a distance of 0.50 m. (Neglect friction.) (a) What are the kinetic energy and the speed of the puck when the force is removed? (b) How much work would be required to bring the puck to rest?

Thinking It Through. Apply the work–energy theorem. If you can find the amount of work done, you know the change in kinetic energy, and vice versa.

Solution. Listing the given data as usual, we have

Given: $m = 0.25$ kg   *Find:* (a) K (kinetic energy)
$F = 6.0$ N v (speed)
$d = 0.50$ m   (b) W (work done in stopping puck)
$v_o = 0$

(a) Since the speed is not known, we cannot compute the kinetic energy $\left(K = \frac{1}{2}mv^2\right)$ directly. However, kinetic energy is related to work by the work–energy theorem. The work done on the puck by the player's applied force F is

$$W = Fd = (6.0\,\text{N})(0.50\,\text{m}) = +3.0\,\text{J}$$

Then, by the work–energy theorem, we obtain

$$W = \Delta K = K - K_o = +3.0\,\text{J}$$

But $K_o = \frac{1}{2}mv_o^2 = 0$, because $v_o = 0$, so

$$K = 3.0\,\text{J}$$

The speed can be found from the kinetic energy. Since $K = \frac{1}{2}mv^2$,

$$v = \sqrt{\frac{2K}{m}} = \sqrt{\frac{2(3.0\,\text{J})}{0.25\,\text{kg}}} = 4.9\,\text{m/s}$$

(b) As you might guess, the work required to bring the puck to rest is equal to the puck's kinetic energy (that is, the amount of energy that the puck must lose to stop its motion). To confirm this equality, we essentially perform the reverse of the previous calculation, with $v_o = 4.9$ m/s and $v = 0$:

$$W = K - K_o = 0 - K_o = -\tfrac{1}{2}mv_o^2 = -\tfrac{1}{2}(0.25\,\text{kg})(4.9\,\text{m/s})^2 = -3.0\,\text{J}$$

The minus sign indicates that the puck loses energy as it slows down. The work is done *against* the motion of the puck; that is, the opposing force is in a direction opposite that of the motion. (In a real-life situation, the opposing force would be friction.)

Follow-Up Exercise. Suppose the puck in this Example had twice the final speed when released. Would it then take twice as much work to stop the puck? Justify your answer numerically.

▶ **FIGURE 5.11** Work and kinetic energy See Example 5.5.

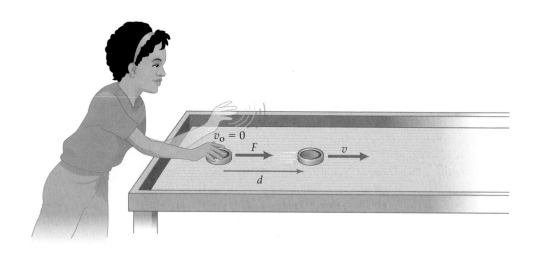

Problem-Solving Hint

Notice how work–energy considerations were used to find speed in Example 5.5. This operation can be done in another way as well. First, the acceleration could be found from $a = F/m$, and then the kinematic equation $v^2 = v_0^2 + 2ax$ could be used to find v (where $x = d = 0.50$ m). The point is that many problems can be solved in different ways, and finding the fastest and most efficient way is often the key to success. As our discussion of energy progresses, you will see how useful and powerful the notions of work and energy are, both as theoretical concepts and as practical tools for solving many kinds of problems.

Conceptual Example 5.6 ■ Kinetic Energy: Mass versus Speed

In a football game, a 140-kg guard runs at a speed of 4.0 m/s, and a 70-kg free safety moves at 8.0 m/s. Which of the following is a correct statement? (a) Both players have the same kinetic energy. (b) The safety has twice as much kinetic energy as the guard. (c) The guard has twice as much kinetic energy as the safety. (d) The safety has four times as much kinetic energy as the guard?

PHYSLET®

Exploration 6.2 The Two-Block Push

Reasoning and Answer. The kinetic energy of a body depends on both its mass and its speed. You might think that, with half the mass but twice the speed, the safety would have the same kinetic energy as the guard, but this is not the case. As observed from the relationship $K = \frac{1}{2}mv^2$, kinetic energy is directly proportional to the mass, but it is also proportional to the *square* of the speed. Thus, having half the mass decreases the kinetic energy by a factor of two; so if the two athletes had equal speeds, the safety would have half as much kinetic energy as the guard.

However, doubling the speed increases the kinetic energy, not by a factor of 2 but by a factor of 2^2, or 4. Thus, the safety, with half the mass but twice the speed, would have $\frac{1}{2} \times 4 = 2$ times as much kinetic energy as the guard, and so the answer is (b).

Note that to answer this question, it was not necessary to calculate the kinetic energy of each player. But this can be done to verify our conclusions:

$$K_{\text{safety}} = \tfrac{1}{2}m_s v_s^2 = \tfrac{1}{2}(70 \text{ kg})(8.0 \text{ m/s})^2 = 2.2 \times 10^3 \text{ J}$$

$$K_{\text{guard}} = \tfrac{1}{2}m_g v_g^2 = \tfrac{1}{2}(140 \text{ kg})(4.0 \text{ m/s})^2 = 1.1 \times 10^3 \text{ J}$$

Thus, we see explicitly that our answer was correct.

Follow-Up Exercise. Suppose that the safety's speed were only 50 percent greater than the guard's, or 6.0 m/s. Which athlete would then have the greater kinetic energy, and how much greater?

Problem-Solving Hint

Note that the work–energy theorem relates the work done to the *change* in the kinetic energy. Often, we have $v_0 = 0$ and $K_0 = 0$, so $W = \Delta K = K$. But take care! You *cannot* simply use the square of the change in speed, $(\Delta v)^2$, to calculate ΔK, as you might at first think. In terms of speed, we have

$$W = \Delta K = K - K_0 = \tfrac{1}{2}mv^2 - \tfrac{1}{2}mv_0^2 = \tfrac{1}{2}m(v^2 - v_0^2)$$

But $v^2 - v_0^2$ is not the same as $(v - v_0)^2 = (\Delta v)^2$, because $(v - v_0)^2 = v^2 - 2vv_0 + v_0^2$. Hence, the change in kinetic energy, is *not* equal to $\frac{1}{2}m(v - v_0)^2 = \frac{1}{2}m(\Delta v)^2 \neq \Delta K$.

This observation means that to calculate work, or the change in kinetic energy, you must compute the kinetic energy of an object at one point or time (using the instantaneous speed to get the instantaneous kinetic energy) and also at another location or time. Then subtract the quantities to find the change in kinetic energy, or the work. Alternatively, you can find the difference of the *squares* of the speeds $(v^2 - v_0^2)$ first in computing the change, but remember never to use the square of the difference of the speeds. To see this hint in action, look at Conceptual Example 5.7.

Conceptual Example 5.7 ■ An Accelerating Car: Speed and Kinetic Energy

A car traveling at 5.0 m/s speeds up to 10 m/s, with an increase in kinetic energy that requires work W_1. Then the car's speed increases from 10 m/s to 15 m/s, requiring additional work W_2. Which of the following relationships accurately compares the two amounts of work: (a) $W_1 > W_2$, (b) $W_1 = W_2$, or (c) $W_2 > W_1$?

Reasoning and Answer. As noted previously, the work–energy theorem relates the work done on the car to the *change* in its kinetic energy. Since the speeds have the same increment in each case ($\Delta v = 5.0$ m/s), it might appear that (b) would be the answer. However, keep in mind that the work is equal to the *change* in kinetic energy and involves $v_2^2 - v_1^2$, not $(\Delta v)^2 = (v_2 - v_1)^2$.

So the greater the speed of an object, the greater its kinetic energy, and we would expect the *difference* in kinetic energy in changing speeds (or the work required to change speed) to be greater for higher speeds for the same Δv. So, (c) is the answer.

The main point is that the Δv values are the same, but more work is required to increase the kinetic energy of an object at higher speeds.

Follow-Up Exercise. Suppose the car speeds up a third time, from 15 m/s to 20 m/s, a change requiring work W_3. How does the work done in this increment compare with W_2? Justify your answer numerically. [*Hint*: Use a ratio.]

5.4 Potential Energy

OBJECTIVES: To (a) define and understand potential energy, and (b) learn about gravitational potential energy.

An object in motion has kinetic energy. However, whether an object is in motion or not, it may have another form of energy—potential energy. As the name implies, an object having potential energy has the *potential* to do work. You can probably think of many examples: a compressed spring, a drawn bow, water held back by a dam, a wrecking ball poised to drop. In all such cases, the potential to do work derives from the *position* or *configuration* of bodies. The spring has energy because it is compressed, the bow because it is drawn, the water and the ball because they have been lifted above the surface of the Earth (◄Fig. 5.12). Consequently, **potential energy, (U)**, is often called *the energy of position* (and/or configuration).

In a sense, potential energy can be thought of as stored work, just as kinetic energy can. You have already seen an example of potential energy in Section 5.2 when work was done in compressing a spring from its equilibrium position. Recall that the work done in such a case is $W = \frac{1}{2}kx^2$ (with $x_o = 0$). Note that the amount of work done depends on the amount of compression (x). Because work is done, there is a *change* in the spring's potential energy (ΔU), which is equal to the work done *by the applied force* in compressing (or stretching) the spring:

$$W = \Delta U = U - U_o = \frac{1}{2}kx^2 - \frac{1}{2}kx_o^2$$

Thus, with $x_o = 0$ and $U_o = 0$, as they are commonly taken for convenience, the *potential energy of a spring* is

$$U = \frac{1}{2}kx^2 \quad \text{(potential energy of a spring)} \tag{5.7}$$

SI unit of energy: joule (J)

[*Note*: Since the potential energy varies as x^2, the previous Problem-Solving Hint also applies. That is, when $x_o \neq 0$, then $x^2 - x_o^2 \neq (x - x_o)^2$.]

(a)

(b)

▲ **FIGURE 5.12 Potential energy** Potential energy has many forms. **(a)** Work must be done to bend the bow, giving it potential energy. That energy is converted into kinetic energy when the arrow is released. **(b)** Gravitational potential energy is converted into kinetic energy when an object falls. (Where did the gravitational potential energy of the water and the diver come from?)

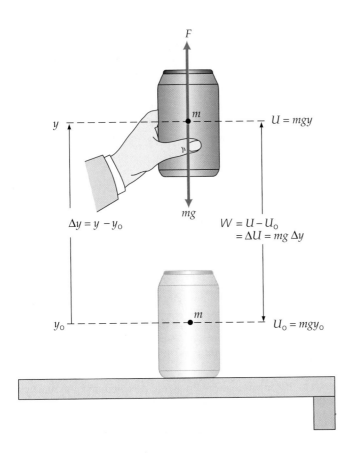

▲ **FIGURE 5.13 Gravitational potential energy** The work done in lifting an object is equal to the change in gravitational potential energy: $W = F\Delta y = mg(y - y_0)$.

Perhaps the most common type of potential energy is gravitational potential energy. In this case, position refers to the height of an object above some reference point, such as the floor or the ground. Suppose that an object of mass m is lifted a distance Δy (▲Fig. 5.13). Work is done against the force of gravity, and an applied force at least equal to the object's weight is necessary to lift the object: $F = w = mg$. The work done in lifting is then equal to the change in potential energy. Expressing this relationship in equation form, since there is no overall change in kinetic energy, we have

work done by external force = change in gravitational potential energy

or

$$W = F\Delta y = mg(y - y_0) = mgy - mgy_0 = \Delta U = U - U_0$$

where y is used as the vertical coordinate. With the common choice of $y_0 = 0$, such that $U_0 = 0$, the **gravitational potential energy** is

$$U = mgy \tag{5.8}$$

SI unit of energy: joule (J)

(Eq. 5.8 represents the gravitational potential energy on or near the Earth's surface, where g is considered to be constant. A more general form of gravitational potential energy will be given in Section 7.5.)

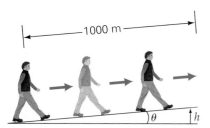

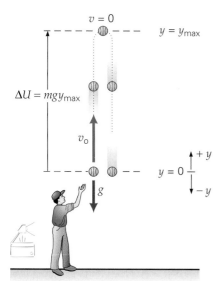

▲ **FIGURE 5.14** Adding potential energy See Example 5.8.

Example 5.8 ■ More Energy Needed

To walk 1000 m on level ground, a 60-kg person requires an expenditure of about 1.0×10^5 J of energy. What is the total energy required if the walk is extended another 1000 m along a 5.0° incline as shown in ◀Fig. 5.14?

Thinking It Through. To walk an additional 1000 m would require the 1.0×10^5 J *plus* the additional energy for doing work against gravity in walking up the incline. From the figure, the increase in height can be seen to be $h = d \sin \theta$.

Solution. Listing the given data:

Given: $m = 60$ kg   *Find:* Total expended energy
$E_o = 1.0 \times 10^5$ J (for 1000 m)
$\theta = 5.0°$
$d = 1000$ m (for each part of the work)

The additional expended energy in going up the incline is equal to gravitational potential energy gained. So,

$$\Delta U = mgh = (60 \text{ kg})(9.8 \text{ m/s}^2)(1000 \text{ m}) \sin 5.0° = 5.1 \times 10^4 \text{ J}$$

Then, the total energy expended for the 2000 m walk is

$$\text{Total } E = 2E_o + \Delta U = 2(1.0 \times 10^5 \text{ J}) + 0.51 \times 10^5 \text{ J} = 2.5 \times 10^5 \text{ J}$$

Notice that the value of ΔU was expressed as a multiple of 10^5 in the last equation so it could be added to the E_o term, and the result was rounded to two significant figures per our rules from Chapter 1.

Follow-Up Exercise. If the angle of incline were doubled and the walk *just* up the incline is repeated, will the additional energy expended by the person in doing work against gravity be doubled? Justify your answer.

Example 5.9 ■ A Thrown Ball: Kinetic Energy and Gravitational Potential Energy

A 0.50-kg ball is thrown vertically upward with an initial velocity of 10 m/s (◀Fig. 5.15). (a) What is the change in the ball's kinetic energy between the starting point and the ball's maximum height? (b) What is the change in the ball's potential energy between the starting point and the ball's maximum height? (Neglect air resistance.)

Thinking It Through. Kinetic energy is lost and gravitational potential energy is gained as the ball travels upward.

Solution. Studying Fig. 5.15 and listing the data,

Given: $m = 0.50$ kg   *Find:* (a) ΔK (the change in kinetic energy)
$v_o = 10$ m/s   (b) ΔU (the change in potential energy between y_o
$a = g$ and y_{max})

(a) To find the *change* in kinetic energy, the kinetic energy is computed at each point. We know the initial velocity v_o, and at the maximum height $v = 0$, so $K = 0$. Thus,

$$\Delta K = K - K_o = 0 - K_o = -\tfrac{1}{2}mv_o^2 = -\tfrac{1}{2}(0.50 \text{ kg})(10 \text{ m/s})^2 = -25 \text{ J}$$

That is, the ball loses 25 J of kinetic energy as negative work is done on it by the force of gravity. (The gravitational force and the ball's displacement are in opposite directions.)

(b) To find the change in potential energy, we need to know the ball's height above its starting point when $v = 0$. Using Eq. 2.11', $v^2 = v_o^2 - 2gy$ (with $y_o = 0$ and $v = 0$) to find y_{max},

$$y_{max} = \frac{v_o^2}{2g} = \frac{(10 \text{ m/s})^2}{2(9.8 \text{ m/s}^2)} = 5.1 \text{ m}$$

Then, with $y_o = 0$ and $U_o = 0$,

$$\Delta U = U = mgy_{max} = (0.50 \text{ kg})(9.8 \text{ m/s}^2)(5.1 \text{ m}) = +25 \text{ J}$$

The potential energy increases by 25 J, as might be expected.

Follow-Up Exercise. In this Example, what are the overall changes in the ball's kinetic and potential energies when the ball returns to the starting point?

▲ **FIGURE 5.15** Kinetic and potential energies See Example 5.9. (The ball is displaced sideways for clarity.)

Exploration 7.2 *Choice of Zero for Potential Energy*

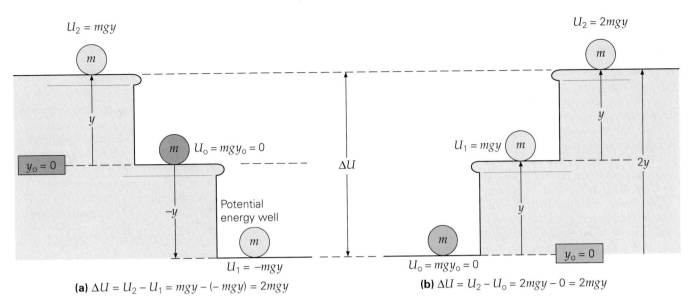

(a) $\Delta U = U_2 - U_1 = mgy - (-mgy) = 2mgy$ **(b)** $\Delta U = U_2 - U_o = 2mgy - 0 = 2mgy$

▲ **FIGURE 5.16** Reference point and change in potential energy **(a)** The choice of a reference point (zero height) is arbitrary and may give rise to a negative potential energy. An object is said to be in a potential-energy well in this case. **(b)** The well may be avoided by selecting a new zero reference. Note that the difference, or *change*, in potential energy (ΔU) associated with the two positions is the same, regardless of the reference point. There is no physical difference, even though there are two coordinate systems and two different zero reference points.

Zero Reference Point

An important point is illustrated in Example 5.9, namely, the choice of a zero reference point. Potential energy is the energy of *position*, and the potential energy at a particular position (U) is meaningful only when referenced to the potential energy at some other position (U_o). The reference position or point is arbitrary, as is the origin of a set of coordinate axes for analyzing a system. Reference points are usually chosen with convenience in mind—for example, $y_o = 0$. The value of the potential energy at a particular position depends on the reference point used. However, *the difference, or change, in potential energy associated with two positions is the same regardless of the reference position.*

If, in Example 5.9, ground level had been taken as the zero reference point, then U_o at the release point would not have been zero. However, U at the maximum height would have been greater, and $\Delta U = U - U_o$ would have been the same. This concept is illustrated in ▲Fig. 5.16. Note in Fig. 5.16a that the potential energy can be negative. When an object has a negative potential energy, it is said to be in a potential-energy *well*, which is analogous to being in an actual well: work is needed to raise the object to a higher position in the well or to get it out of the well.

It is also said that gravitational potential energy is *independent of path*. This means that only the change in height Δh (or Δy) is the consideration, not the path that leads to the change in height. An object could travel many paths leading to the same Δh.

Teaching tip: Make sure students understand that potential energy is a scalar quantity; however, unlike kinetic energy, it can be either positive or negative, depending on the choice of the zero position.

5.5 Conservation of Energy

OBJECTIVES: To (a) distinguish between conservative and nonconservative forces, and (b) explain their effects on the conservation of energy.

Conservation laws are the cornerstones of physics, both theoretically and practically. Most scientists would probably name conservation of energy as the most profound and far-reaching of these important laws. When we say that a physical quantity is *conserved*, we mean that it is constant, or has a constant value. Because so many things continually change in physical processes, conserved quantities are extremely helpful in our attempts to understand and describe the universe. Keep in mind, though, that many quantities are conserved only under special conditions.

One of the most important conservation laws is that concerning conservation of energy. (You may have seen this topic coming in Example 5.9.) A familiar statement is that the total energy of the universe is conserved. This statement is true, because

Demonstration/activity: Consider using this demonstration as an introduction to the conservation of energy. Suspend a 1-kg mass on a string 1–2 m long so the mass is about chest high. Pull the mass back through an angle of several degrees to a point right underneath your chin or just in front of your nose. Then release it from rest. The mass, conserving energy, will not hit you on return. (Don't give it a push, and don't move!)

INSIGHT   5.1 PEOPLE POWER: USING BODY ENERGY

The human body is energy inefficient. That is, a lot of energy doesn't go into doing useful work and is wasted. It would be advantageous to convert some of this energy into useful work. Attempts are being made to do this through "energy harvesting" from the human body. Normal body activities produce motion, flexing and stretching, compression, and body heat—this is energy there for the taking. Harvesting the energy is a difficult job, but using advances in nanotechnology (Chapter 1) and materials science, the effort is being made.

One older example of using body energy is the self-winding wristwatch, which is wound mechanically from the wearer's arm movements. (Today, batteries have all but taken over.) An ultimate goal in "energy harvesting" is to convert some of the body's energy into electricity—even if only a small amount. How might this be done? Here are a couple of ways:

- Piezoelectric devices. When mechanically stressed, piezoelectric substances, which like some ceramics, can generate electrical energy.

- Thermoelectric materials, which convert heat resulting from some temperature difference into electrical energy.

These methods have severe limitations and produce only small amounts of electricity. But with miniaturization and nanotechnology, the results could be enough. Researchers have already developed boots that use the compression of walking on a compound to produce enough energy to power a radio.

A more recent application is the "backpack generator." The mounted backpack's load is suspended by springs. The up and down hip motion of a person wearing the backpack makes the suspended load bounce up and down. This motion turns a gear connected to a simple magnetic coil generator, similar to those used in flashlights that are energized by a rhythmic shaking (see Insight 20.1, page 664). The body's mechanical energy with this device can generate up to 7 watts of electrical energy. A typical cell phone operates on about 1 watt. (The watt is a unit of power, J/s, energy/second, see Section 5.6).

Who knows what the future of technology may hold? Reflect on how many advances have occurred in your own lifetime.

Note: A system is a physical arrangement with real or imaginary boundaries. A classroom might be considered a system, and so might an arbitrary cubic meter of air.

the whole universe is taken to be a system. A *system* is defined as a definite quantity of matter enclosed by boundaries, either real or imaginary. In effect, the universe is the largest possible closed, or isolated, system we can imagine. Within a *closed system*, particles can interact with each other, but have absolutely no interaction with anything outside. In general, then, the amount of energy in a system remains constant when no mechanical work is done on or by the system, and no energy is transferred to or from the system (including thermal energy and radiation).

Thus, the **law of conservation of total energy** may be stated as follows:

Conservation of total energy

| The total energy of an isolated system is always conserved. |

Within such a system, energy may be converted from one form to another, but the total amount of all forms of energy is constant, or unchanged. Total energy can never be created or destroyed. The use of body energy is discussed in the accompanying Insight 5.1.

PHYSLET®

Illustration 7.1 *Choice of System*

Conceptual Example 5.10 ■ Violation of the Conservation of Energy?

A static, uniform liquid is in one side of a double container as shown in ▶Fig. 5.17a. If the valve is open, the level will fall, because the liquid has (gravitational) potential energy. This may be computed by assuming all the mass of the liquid to be concentrated at its center of mass, which is at a height $h/2$. (More on the center of mass in Chapter 6.) When the valve is open, the liquid flows into the container on the right, and when static equilibrium is reached, each container has liquid to a height of $h/2$, with centers of mass at $h/4$. This being the case, the potential energy of the liquid before opening the valve was $U_o = (mg)h/2$, and afterward, with half the total mass in each container (Fig. 5.16b), $U = (m/2)g(h/4) + (m/2)g(h/4) = 2(m/2)g(h/4) = (mg)h/4$. Whoa. Was half of the energy lost?

Reasoning and Answer. No; by the conservation of total energy, it must be around somewhere. Where might it have gone? When the liquid flows from one container to the other, because of internal friction and friction against the walls, half of the potential energy is converted to heat (thermal energy), which is transferred to the surroundings as the liquid comes to equilibrium. (This means the same temperature and no internal fluctuations.)

Follow-Up Exercise. What would happen in this Example in the absence of friction?

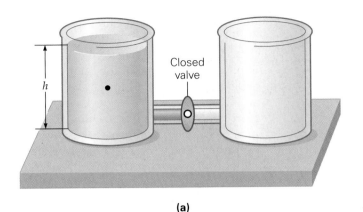

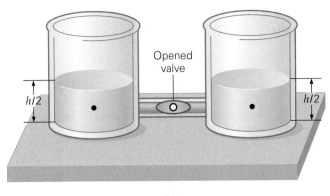

(a)   (b)

▲ **FIGURE 5.17** Is energy lost? See Conceptual Example 5.10.

Conservative and Nonconservative Forces

We can make a general distinction among systems by considering two categories of forces that may act within or on them: conservative and nonconservative forces. You have already been introduced to a couple of conservative forces: the force due to gravity and the spring force. We considered a classic nonconservative force, friction, in Chapter 4. A conservative force is defined as follows:

Note: Friction is discussed in Section 4.6.

> A force is said to be conservative if the work done by it in moving an object is independent of the object's path.

Conservative force—work independent of path

This definition means that the work done by a **conservative force** depends only on the initial and final positions of an object.

The concept of conservative and nonconservative forces is sometimes difficult to comprehend at first. Because this concept is so important in the conservation of energy, let's consider some illustrative examples to increase understanding.

First, what does *independent of path* mean? As an example of path independence, consider picking an object up from the floor and placing it on a table. This is doing work against the *conservative force of gravity*. The work done is equal to the potential energy gained, $mg\Delta h$, where Δh is the *vertical* distance between the object's position on the floor and its position on the table. This is the important point. You may have carried the object over to the sink before putting it on the table, or walked around to the other side of the table. But only the vertical displacement makes a difference in the work done because that is in the direction of the vertical force. For any horizontal displacement no work is done, since the displacement and force are at right angles. The magnitude of the work done is equal to the change in potential energy (under frictionless conditions only), and in fact, *the concept of potential energy is associated only with conservative forces.* A change in potential energy can be defined in terms of the work done by a conservative force.

Conversely, a **nonconservative force** *does* depend on path.

Demonstration/activity: To demonstrate the independence of path, use a demo cart of 2–5 kg on good wheels. First pull the cart up a ramp, using a spring scale, and record the distance and force. Then pull the cart horizontally (slowly, so that the force is minimal) for the length of the ramp, and lift it vertically to the same position, with the scale showing its full weight. Compare the work done and the final position of the cart in each case.

> A force is said to be nonconservative if the work done by it in moving an object does depend on the object's path.

Nonconservative force—work dependent on path

Friction is a nonconservative force. A longer path would produce more work done by friction than a shorter one, and more energy would be lost to heat on the longer path. So the work done against friction certainly depends on the path. Hence, in a sense, a conservative force allows you to conserve or store energy as potential energy, whereas a nonconservative force does not.

Another approach to explain the distinction between conservative and nonconservative forces is through an equivalent statement of the previous definition of conservative force:

> A force is conservative if the work done by it in moving an object through a round trip is zero.

Another way of describing a conservative force

Notice that for the *conservative* gravitational force, the force and displacement are sometimes in the same direction (in which case positive work is done by the force) and sometimes in opposite directions (in which case negative work is done by the force) during a round trip. Think of the simple case of the book falling to the floor and being placed back on the table. With positive and negative work, the total work done by gravity is zero.

However, for only a *nonconservative* force like that of kinetic friction, which always opposes the motion or is in the opposite direction to the displacement, the total work done in a round trip can *never* be zero and is always negative (that is, energy is lost). But don't get the idea that nonconservative forces only take energy away from a system. On the contrary, we often supply nonconservative pushes and pulls (forces) that add to the energy of a system, such as when you push a stalled car.

Conservation of Total Mechanical Energy

The idea of a conservative force allows us to extend the conservation of energy to the special case of mechanical energy, which greatly helps us better analyze many physical situations. The sum of the kinetic and potential energies is called the **total mechanical energy**:

$$E = K + U \tag{5.9}$$
$$\text{total mechanical energy} = \text{kinetic energy} + \text{potential energy}$$

For a **conservative system** (that is, a system in which only conservative forces do work) the total mechanical energy is constant, or conserved:

$$E = E_o$$

Substituting for E and E_o from Eq. 5.9,

$$K + U = K_o + U_o \tag{5.10a}$$

or

$$\tfrac{1}{2}mv^2 + U = \tfrac{1}{2}mv_o^2 + U_o \tag{5.10b}$$

Equation 5.10b is a mathematical statement of the **law of the conservation of mechanical energy**:

Conservation of mechanical energy

In a conservative system, the sum of all types of kinetic energy and potential energy is constant and equals the total mechanical energy of the system.

PHYSLET®

Illustration 7.2 Representations of Energy

Note: while the kinetic and potential energies in a conservative system may change, their sum is always constant. For a conservative system when work is done and energy is transferred within a system, we can write Eq. 5.10a as

$$(K - K_o) + (U - U_o) = 0 \tag{5.11a}$$

or as

$$\Delta K + \Delta U = 0 \quad \textit{(for a conservative system)} \tag{5.11b}$$

Demonstration/activity: A Hot Wheels track or any similar loop-the-loop apparatus can be used to demonstrate the conservation of energy. What occurs when the speed of the mass at the top of the circle is less than the critical speed to go around the loop? What is the critical speed? Also consider swinging a mass on a string in a vertical circle.

This expression tells us that these quantities are related in a seesaw fashion: If there is a decrease in potential energy, then the kinetic energy must increase by an equal amount to keep the sum of the changes equal to zero. However, in a nonconservative system, mechanical energy is usually lost (for example, to the heat of friction), and thus $\Delta K + \Delta U < 0$. But keep in mind, as pointed out previously, that a nonconservative force may instead add energy to a system (or have no effect at all).

Example 5.11 ■ Look Out Below! Conservation of Mechanical Energy

PHYSLET®

Exploration 7.3 Elastic Collision

A painter on a scaffold drops a 1.50-kg can of paint from a height of 6.00 m. (a) What is the kinetic energy of the can when the can is at a height of 4.00 m? (b) With what speed will the can hit the ground? (Neglect air resistance.)

Thinking It Through. Total mechanical energy is conserved, since only the conservative force of gravity acts on the system (the can). The initial total mechanical energy can be found, and the potential energy decreases as the kinetic energy (as well as speed) increases.

Solution. Listing what is given and what we are to find, we have:

Given: $m = 1.50$ kg
$y_o = 6.00$ m
$y = 4.00$ m
$v_o = 0$

Find: (a) K (kinetic energy at $y = 4.00$ m)
(b) v (speed just before hitting the ground)

(a) First, it is convenient to find the can's total mechanical energy, since this quantity is conserved while the can is falling. Initially, with $v_o = 0$, the can's total mechanical energy is all potential energy. Taking the ground as the zero reference point,

$$E = K_o + U_o = 0 + mgy_o = (1.50 \text{ kg})(9.80 \text{ m/s}^2)(6.00 \text{ m}) = 88.2 \text{ J}$$

The relation $E = K + U$ continues to hold while the can is falling, but now we know what E is. Rearranging the equation, $K = E - U$ and U can be found at $y = 4.00$ m:

$$K = E - U = E - mgy = 88.2 \text{ J} - (1.50 \text{ kg})(9.80 \text{ m/s}^2)(4.00 \text{ m}) = 29.4 \text{ J}$$

Alternatively, we could have computed the change in (in this case, the loss of) potential energy, ΔU. Whatever potential energy was lost must have been gained as kinetic energy (Eq. 5.11). Then,

$$\Delta K + \Delta U = 0$$
$$(K - K_o) + (U - U_o) = (K - K_o) + (mgy - mgy_o) = 0$$

With $K_o = 0$ (because $v_o = 0$), we obtain

$$K = mg(y_o - y) = (1.50 \text{ kg})(9.8 \text{ m/s}^2)(6.00 \text{ m} - 4.00 \text{ m}) = 29.4 \text{ J}$$

(b) Just before the can strikes the ground ($y = 0, U = 0$), the total mechanical energy is all kinetic energy, or

$$E = K = \tfrac{1}{2}mv^2$$

Thus,

$$v = \sqrt{\frac{2E}{m}} = \sqrt{\frac{2(88.2 \text{ J})}{1.50 \text{ kg}}} = 10.8 \text{ m/s}$$

Basically, all of the potential energy of a free-falling object released from some height y is converted into kinetic energy just before the object hits the ground, so

$$|\Delta K| = |\Delta U|$$

Thus,

$$\tfrac{1}{2}mv^2 = mgy$$

or

$$v = \sqrt{2gy}$$

Note that the mass cancels and is not a consideration. This result is also obtained from the kinematic equation $v^2 = 2gy$ (Eq. 2.11′), with $v_o = 0$ and $y_o = 0$.

Follow-Up Exercise. A painter on the ground wishes to toss a paintbrush vertically upward a distance of 5.0 m to his partner on the scaffold. Use methods of conservation of mechanical energy to determine the minimum speed that he must give to the brush.

Demonstration/activity: This demonstration is a graphing exercise. Drop a 1-kg mass from rest from a height of 20 m. Have students graph the potential, kinetic, and total energies as functions of the displacement and the time. For a more challenging exercise, give the mass an initial horizontal or vertical velocity.

► **FIGURE 5.18** Speed and energy
See Conceptual Example 5.12.

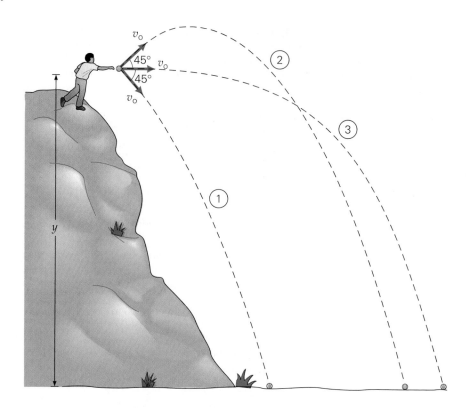

Conceptual Example 5.12 ■ A Matter of Direction? Speed and Conservation of Energy

Three balls of equal mass m are projected with the same speed in different directions, as shown in ▲Fig. 5.18. If air resistance is neglected, which ball would you expect to strike the ground with the greatest speed: (a) ball 1; (b) ball 2; (c) ball 3; or (d) all balls strike with the same speed?

Reasoning and Answer. All of the balls have the same initial kinetic energy, $K_o = \frac{1}{2}mv_o^2$. (Recall that energy is a scalar quantity, and the different directions of projection do not produce any difference in the kinetic energies.) Regardless of their trajectories, all of the balls ultimately descend a distance y relative to their common starting point, so they all lose the same amount of potential energy. (Recall that U is energy of *position* and is *independent* of path.)

By the law of conservation of mechanical energy, the amount of potential energy each ball loses is equal to the amount of kinetic energy it gains. Since all of the balls start with the same amount of kinetic energy and gain the same amount of kinetic energy, all three will have equal kinetic energies just before striking the ground. This means that their speeds must be equal, so the answer is (d).

Note that although balls 1 and 2 are projected at 45° angles, this factor is not relevant. Since the change in potential energy is independent of path, it is independent of the projection angle. The vertical distance between the starting point and the ground is the same (for y) for projectiles at any angle. (*Note*: Although the strike speeds are equal, the *times* the balls take to reach the ground are different. Refer to Conceptual Example 3.11 for another approach.)

Follow-Up Exercise. Would the balls strike the ground with different speeds if their masses were different? (Neglect air resistance.)

Example 5.13 ■ Conservative Forces: Mechanical Energy of a Spring

A 0.30-kg block sliding on a horizontal frictionless surface with a speed of 2.5 m/s, as depicted in ►Fig. 5.19, strikes a light spring that has a spring constant of 3.0×10^3 N/m. (a) What is the total mechanical energy of the system? (b) What is the kinetic energy K_1 of the block when the spring is compressed a distance $x_1 = 1.0$ cm? (Assume that no energy is lost in the collision.)

Thinking It Through. (a) Initially, the total mechanical energy is all kinetic energy. (b) The total energy is the same as in part (a), but it is now divided between kinetic energy and spring potential energy (assuming the spring is not fully compressed).

Solution.

Given: $m = 0.30$ kg
$v_o = 2.5$ m/s
$k = 3.0 \times 10^3$ N/m
$x_1 = 1.0$ cm $= 0.010$ m

Find: (a) E (total mechanical energy)
(b) K_1 (kinetic energy)

(a) Before the block makes contact with the spring, the total mechanical energy of the system is all in the form of kinetic energy; therefore,

$$E = K_o = \tfrac{1}{2}mv_o^2 = \tfrac{1}{2}(0.30 \text{ kg})(2.5 \text{ m/s})^2 = 0.94 \text{ J}$$

Since the system is conservative (that is, no mechanical energy is lost), this quantity is the total mechanical energy at any time.

(b) When the spring is compressed a distance x_1, it has gained potential energy $U_1 = \tfrac{1}{2}kx_1^2$, and

$$E = K_1 + U_1 = K_1 + \tfrac{1}{2}kx_1^2$$

Solving for K_1, we have

$$K_1 = E - \tfrac{1}{2}kx_1^2$$
$$= 0.94 \text{ J} - \tfrac{1}{2}(3.0 \times 10^3 \text{ N/m})(0.010 \text{ m})^2 = 0.94 \text{ J} - 0.15 \text{ J} = 0.79 \text{ J}$$

Follow-Up Exercise. How far will the spring in this Example be compressed when the block comes to a stop? (Solve using energy principles.)

See the accompanying Learn by Drawing for another example of energy exchange.

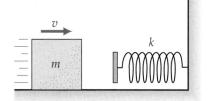

▲ **FIGURE 5.19 Conservative force and the mechanical energy of a spring** See Example 5.13.

Illustration 7.5 A Block on an Incline

Exploration 7.4 A Ball Hits a Mass Attached to a Spring

LEARN BY DRAWING ENERGY EXCHANGES: A FALLING BALL

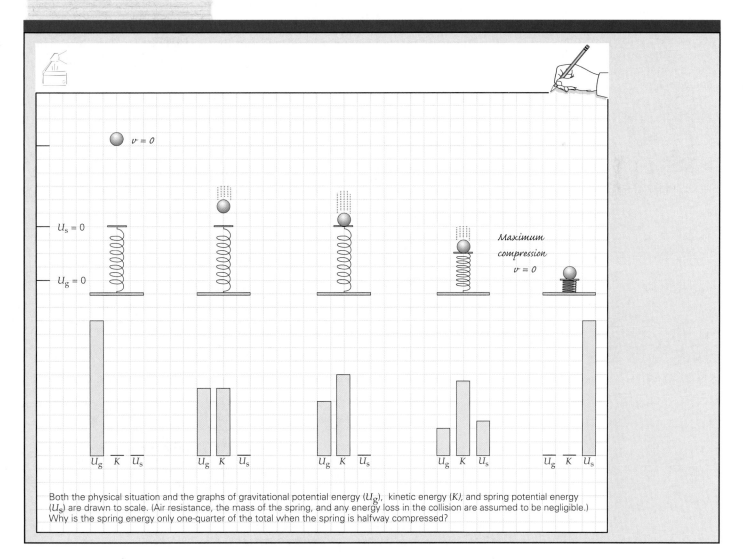

Both the physical situation and the graphs of gravitational potential energy (U_g), kinetic energy (K), and spring potential energy (U_s) are drawn to scale. (Air resistance, the mass of the spring, and any energy loss in the collision are assumed to be negligible.) Why is the spring energy only one-quarter of the total when the spring is halfway compressed?

▲ FIGURE 5.20 Nonconservative force and energy loss Friction is a nonconservative force—when friction is present and does work, mechanical energy is not conserved. Can you tell from the photo what is happening to the work being done by the motor on the grinding wheel after the work is converted into rotational kinetic energy? (Note that the worker is wisely wearing a face shield rather than just goggles as the sign in the background suggests.)

Illustration 7.4 External Forces

Total Energy and Nonconservative Forces

In the preceding examples, we ignored the force of friction, which is probably the most common nonconservative force. In general, both conservative and nonconservative forces can do work on objects. However, when some nonconservative forces do work, the total mechanical energy is not conserved. Mechanical energy is "lost" through the work done by nonconservative forces, such as friction.

You might think that we can no longer use an energy approach to analyze problems involving such nonconservative forces, since mechanical energy can be lost or dissipated (◄Fig. 5.20). However, in some instances, we can use the total energy to find out how much energy was lost to the work done by a nonconservative force. Suppose an object initially has mechanical energy and that nonconservative forces do an amount of work W_{nc} on it. Starting with the work–energy theorem, we have

$$W = \Delta K = K - K_o$$

In general, the net work (W) may be done by both conservative forces (W_c) and nonconservative forces (W_{nc}), so

$$W_c + W_{nc} = K - K_o \qquad (5.12)$$

But recall that the work done by conservative forces is equal to $-\Delta U$, or $W_{nc} = U_o - U$, and Eq. 5.12 then becomes

$$W_{nc} = K - K_o - (U_o - U)$$
$$= (K + U) - (K_o + U_o)$$

Therefore,

$$W_{nc} = E - E_o = \Delta E \qquad (5.13)$$

Hence, the work done by the nonconservative forces acting on a system is equal to the change in mechanical energy. Notice that for dissipative forces, $E_o > E$. Thus, the change is negative, indicating a decrease in mechanical energy. This condition agrees in sign with W_{nc}, which, for friction, would also be negative. Example 5.14 illustrates this concept.

Example 5.14 ■ Nonconservative Force: Downhill Racer

A skier with a mass of 80 kg starts from rest at the top of a slope and skis down from an elevation of 110 m (▼Fig. 5.21). The speed of the skier at the bottom of the slope is 20 m/s. (a) Show that the system is nonconservative. (b) How much work is done by the nonconservative force of friction?

Thinking It Through. (a) If the system is nonconservative, then $E_o \neq E$, and these quantities can be computed. (b) We cannot determine the work from force–distance considerations, but W_{nc} is equal to the difference in total energies (Eq. 5.13).

Solution.

Given: $m = 80$ kg
$v_o = 0$
$v = 20$ m/s
$y_o = 110$ m

Find: (a) Show that E is not conserved.
(b) W_{nc} (work done by friction)

▶ FIGURE 5.21 Work done by a nonconservative force See Example 5.14.

$v = 20$ m/s

110 m

(a) If the system is conservative, the total mechanical energy is constant. Taking $U_o = 0$ at the bottom of the hill, the initial energy at the top of the hill is

$$E_o = U = mgy_o = (80\text{ kg})(9.8\text{ m/s}^2)(110\text{ m}) = 8.6 \times 10^4\text{ J}$$

And the energy at the bottom of the slope is

$$E = K = \tfrac{1}{2}mv^2 = \tfrac{1}{2}(80\text{ kg})(20\text{ m/s})^2 = 1.6 \times 10^4\text{ J}$$

Therefore, $E_o \neq E$, so this system is not conservative.

(b) The amount of work done by the nonconservative force of friction is equal to the change in the mechanical energy, or to the amount of mechanical energy lost (Eq. 5.13):

$$W_{nc} = E - E_o = (1.6 \times 10^4\text{ J}) - (8.6 \times 10^4\text{ J}) = -7.0 \times 10^4\text{ J}$$

This quantity is more than 80% of the initial energy. (Where did this energy actually go?)

Follow-Up Exercise. In free fall, air resistance is sometimes negligible, but for skydivers, air resistance has a very practical effect. Typically, a skydiver descends about 450 m before reaching a terminal velocity (Section 4.6) of 60 m/s. (a) What is the percentage of energy loss to nonconservative forces during this descent? (b) Show that after terminal velocity is reached, the rate of energy loss in J/s is given by (60 mg), where m is the mass of the skydiver.

Example 5.15 ■ Nonconservative Force: One More Time

A 0.75-kg block slides on a frictionless surface with a speed of 20 m/s. It then slides over a rough area 1.0 m in length and onto another frictionless surface. The coefficient of kinetic friction between the block and the rough surface is 0.17. What is the speed of the block after it passes across the rough surface?

Thinking It Through. The task of finding the final speed implies that we use equations involving kinetic energy, where the final kinetic energy can be found by using the conservation of *total* energy. Note that the initial and final energies are kinetic energies, since there is no change in gravitational potential energy. It is always good to make a sketch of the situation for clarity and understanding. See ▼Fig. 5.22.

Solution. Listing the data as usual,

Given: $m = 0.75$ kg   *Find:* v (final speed of block)
          $x = 1.0$ m
          $\mu_k = 0.17$
          $v_o = 2.0$ m/s

For this nonconservative system, we have from Eq. 5.13

$$W_{nc} = E - E_o = K - K_o$$

In the rough area, the block loses energy, because of the work done against friction (W_{nc}), and thus

$$W_{nc} = -f_k x = -\mu_k N x = -\mu_k mgx$$

[negative because f_k and the displacement x are in opposite directions; that is, $(f_k \cos 180°)x = -f_k x$].
     Then, rearranging the energy equation and writing the terms out in detail,

$$K = K_o + W_{nc}$$

or

$$\tfrac{1}{2}mv^2 = \tfrac{1}{2}mv_o^2 - \mu_k mgx$$

(continues on next page)

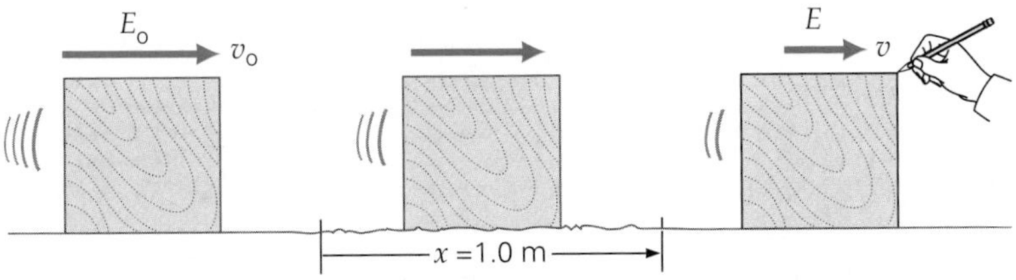

◀ **FIGURE 5.22** A nonconservative rough spot See Example 5.15.

Simplifying yields

$$v = \sqrt{v_0^2 - 2\mu_k gx} = \sqrt{(2.0 \text{ m/s})^2 - 2(0.17)(9.8 \text{ m/s}^2)(1.0 \text{ m})} = 0.82 \text{ m/s}$$

Note that the mass of the block was not needed. And, it can be easily shown that the block lost more than 80% of its energy to friction.

Follow-Up Exercise. Suppose the coefficient of kinetic friction between the block and the rough surface were 0.25. What would happen to the block in this case?

Note that in a nonconservative system, the *total energy* (*not* the total mechanical energy) is conserved (including nonmechanical forms of energy, such as heat), but not all of it is available for mechanical work. For a conservative system, you get back what you put in, so to speak. That is, if you do work on the system, the transferred energy is available to do work. But keep in mind that conservative systems are idealizations, because all real systems are nonconservative to some degree. However, working with ideal conservative systems gives us an insight into the conservation of energy.

Total energy is always conserved. During this course of study, you will learn about other forms of energy, such as thermal, electrical, nuclear, and chemical energies. In general, on the microscopic and submicroscopic levels, these forms of energy can be described in terms of kinetic energy and potential energy. Also, you will learn that mass is a form of energy and that the law of conservation of energy must take this form into account in order to be applied to the analysis of nuclear reactions.

An example of energy conversion is given in Insight 5.2.

5.6 Power

OBJECTIVES: To (a) define power, and (b) describe mechanical efficiency.

A particular task may require a certain amount of work, but that work might be done over different lengths of time or at different rates. For example, suppose that you have to mow a lawn. This task takes a certain amount of work, but you might do the job in a half hour, or you might take an hour or two. There's a practical distinction to be made here. That is, there is usually not only an interest in the amount of work done, but also an interest in how fast it is done–that is, the rate at which it is done. *The time rate of doing work* is called **power**.

The average power is the work done divided by the time it takes to do the work, or work per unit of time:

$$\overline{P} = \frac{W}{t} \tag{5.14}$$

INSIGHT 5.2 HYBRID ENERGY CONVERSION

As you have learned, energy may be transformed from one form to another. An interesting example is the conversion that takes place in the new hybrid automobiles. A hybrid car has both a gasoline (internal combustion) engine and a battery-driven electric motor, both of which may be used to power the vehicle.

A moving car has kinetic energy, and when you step on the brake pedal to slow the car down, kinetic energy is lost. Normally, the brakes of a car accomplish this slowing by friction, and energy is dissipated as heat (conservation of energy). However, with the braking of a hybrid car, some of the energy is converted to electrical energy and stored in the battery of the electric motor. This is called *regenerative braking*. That is, instead of using regular friction brakes to slow the car, the electric motor is used. In this mode, the motor runs in reverse and acts as a generator, converting the lost kinetic energy into electrical energy. (See Section 20.2 for generator operation.) The energy is stored in the battery for later use (Fig. 1).

Hybrid cars must also have regular friction brakes to be used when rapid braking is needed. (See Insight 20.2, page 666 for a more detailed discussion on hybrids.)

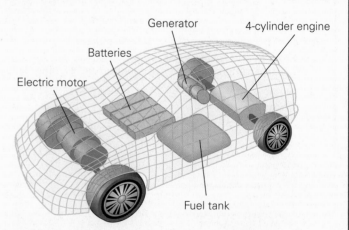

FIGURE 1 Hybrid car A diagram showing the major components. See text for description.

If we are interested in the work (and power) done by a constant force of magnitude F acting while an object moves through a parallel displacement of magnitude d, then

$$\overline{P} = \frac{W}{t} = \frac{Fd}{t} = F\left(\frac{d}{t}\right) = F\overline{v} \qquad (5.15)$$

SI unit of power: J/s or watt (W)

where it is assumed that the force is in the direction of the displacement. Here, $\overline{v}$ is the magnitude of the average velocity. If the velocity is constant, then $\overline{P} = P = Fv$. If the force and displacement are not in the same direction, then we can write

$$\overline{P} = \frac{F(\cos\theta)d}{t} = F\overline{v}\cos\theta \qquad (5.16)$$

where θ is the angle between the force and the displacement.

As you can see from Eq. 5.15, the SI unit of power is joules per second (J/s), but this unit is given another name, the **watt (W)**:

$$1\,\text{J/s} = 1\,\text{watt (W)}$$

The SI unit of power is named in honor of James Watt (1736–1819), a Scottish engineer who developed one of the first practical steam engines. A common unit of electrical power is the *kilowatt* (kW).

The British unit of power is foot-pound per second (ft · lb/s). However, a larger unit, the **horsepower (hp)**, is more commonly used:

$$1\,\text{hp} = 550\,\text{ft}\cdot\text{lb/s} = 746\,\text{W}$$

Power tells you how fast work is being done *or* how fast energy is transferred. For example, motors have power ratings commonly given in horsepower. A 2-hp motor can do a given amount of work in half the time that a 1-hp motor would take, or twice the work in the same amount of time. That is, a 2-hp motor is twice as "powerful" as a 1-hp motor.

Note: In Watt's time, steam engines were replacing horses for work in mines and mills. To characterize the performance of his new engine, which was more efficient than existing ones, Watt used the average rate at which a horse could do work as a unit—a horsepower.

Example 5.16 ■ A Crane Hoist: Work and Power

A crane hoist like the one shown in ▶Fig. 5.23 lifts a load of 1.0 metric ton a vertical distance of 25 m in 9.0 s at a constant velocity. How much useful work is done by the hoist each second?

Thinking It Through. The useful work done each second (that is, per second) is the power output, so this quantity is what is to be found.

Solution.

Given: $m = 1.0$ metric ton   *Find:*   Work per second ($=$ power, P)
            $= 1.0 \times 10^3$ kg
        $y = 25$ m
        $t = 9.0$ s

Keep in mind that the work per unit time (work per second) is power, so this quantity is what we need to compute. Since the load moves with a constant velocity, $\overline{P} = P$. (Why?) The work is done against gravity, so $F = mg$, and

$$P = \frac{W}{t} = \frac{Fd}{t} = \frac{mgy}{t}$$

$$= \frac{(1.0 \times 10^3\,\text{kg})(9.8\,\text{m/s}^2)(25\,\text{m})}{9.0\,\text{s}} = 2.7 \times 10^4\,\text{W (or 27 kW)}$$

Thus, since a watt (W) is a joule per second (J/s), the hoist did 2.7×10^4 J of work each second. Note that the velocity has a magnitude of $v = d/t = 25$ m/9.0 s $= 2.8$ m/s, and the power could be found using $P = Fv$.

Follow-Up Exercise. If the hoist motor of the crane in this Example is rated at 70 hp, what percentage of this power output goes into useful work?

Demonstration/activity: Power is the rate of doing work. Pull a loaded lab cart slowly across the room with a large demonstration spring scale. Have students record the work done (that is, force times distance) as a function of time, and graph the relationship on the board. What is the physical significance of the slope of the line? Does the product of the magnitude of the force and velocity give the same value? Warn students not to use this product if the velocity and the force are not constant, unless they want to obtain the instantaneous power.

▲ **FIGURE 5.23 Power delivery** See Example 5.16.

Example 5.17 ■ Cleaning Up: Work and Time

The motors of two vacuum cleaners have net power outputs of 1.00 hp and 0.500 hp, respectively. (a) How much work in joules can each motor do in 3.00 min? (b) How long does each motor take to do 97.0 kJ of work?

Thinking It Through. (a) Since power is work/time ($P = W/t$), the work can be computed. Note that power is given in horsepower units which is converted to watts. (b) This part of the problem is another application of Eq. 5.15.

Solution.

Given:
$$P_1 = 1.00 \text{ hp} = 746 \text{ W}$$
$$P_2 = 0.500 \text{ hp} = 373 \text{ W}$$
$$t = 3.00 \text{ min} = 180 \text{ s}$$
$$W = 97.0 \text{ kJ} = 97.0 \times 10^3 \text{ J}$$

Find: (a) W (work for each)
(b) t (time for each)

(a) Since $P = W/t$,

$$W_1 = P_1 t = (746 \text{ W})(180 \text{ s}) = 1.34 \times 10^5 \text{ J}$$

and

$$W_2 = P_2 t = (373 \text{ W})(180 \text{ s}) = 0.67 \times 10^5 \text{ J}$$

Note that in the same amount of time, the smaller motor does half as much work as the larger one, as you would expect.

(b) The times are given by $t = W/P$, and for the same amount of work,

$$t_1 = \frac{W}{P_1} = \frac{97.0 \times 10^3 \text{ J}}{746 \text{ W}} = 130 \text{ s}$$

and

$$t_2 = \frac{W}{P_2} = \frac{97.0 \times 10^3 \text{ J}}{373 \text{ W}} = 260 \text{ s}$$

Note that the smaller motor takes twice as long as the larger one to do the same amount of work.

Follow-Up Exercise. (a) A 10-hp motor breaks down and is temporarily replaced with a 5-hp motor. What can you say about the rate of work output? (b) Suppose the situation were reversed—a 5-hp motor is replaced with a 10-hp motor. What can you say about the rate of work output for this case?

Efficiency

Machines and motors are commonly used items in our daily lives, and we often talk about their efficiency. Efficiency involves work, energy, and/or power. Both simple and complex machines that do work have mechanical parts that move, so some input energy is always lost because of friction or some other cause (perhaps in the form of sound). Thus, not all of the input energy goes into doing useful work.

Mechanical efficiency is essentially a measure of what you get out for what you put in—that is, the *useful* work output compared with the energy input. **Efficiency**, ε, is given as a fraction (or percentage):

$$\varepsilon = \frac{\text{work output}}{\text{energy input}} (\times 100\%) = \frac{W_{\text{out}}}{E_{\text{in}}} (\times 100\%) \qquad (5.17)$$

Efficiency is a unitless quantity

For example, if a machine has a 100-J (energy) input and a 40-J (work) output, then its efficiency is

$$\varepsilon = \frac{W_{\text{out}}}{E_{\text{in}}} = \frac{40 \text{ J}}{100 \text{ J}} = 0.40 (\times 100\%) = 40\%$$

An efficiency of 0.40, or 40%, means that 60% of the energy input is lost because of friction or some other cause and doesn't serve its intended purpose. Note that if both terms of the ratio in Eq. 5.17 are divided by time t, we obtain $W_{out}/t = P_{out}$ and $E_{in}/t = P_{in}$. So, we can also write efficiency in terms of power, P:

$$\varepsilon = \frac{P_{out}}{P_{in}} \ (\times 100\%) \qquad (5.18)$$

Example 5.18 ■ Home Improvement: Mechanical Efficiency and Work Output

The motor of an electric drill with an efficiency of 80% has a power input of 600 W. How much useful work is done by the drill in 30 s?

Thinking It Through. This example is an application of Eq. 5.18 and the definition of power.

Solution.

Given: $\varepsilon = 80\% = 0.80$ *Find:* W_{out} (work output)
$P_{in} = 600 \text{ W}$
$t = 30 \text{ s}$

Given the efficiency and power input, we can readily find the power output P_{out} from Eq. 5.18, and this quantity is related to the work output ($P_{out} = W_{out}/t$). First, we rearrange Eq. 5.18:

$$P_{out} = \varepsilon P_{in} = (0.80)(600 \text{ W}) = 4.8 \times 10^2 \text{ W}$$

Then, substituting this value into the equation relating power output and work output, we obtain

$$W_{out} = P_{out}t = (4.8 \times 10^2 \text{ W})(30 \text{ s}) = 1.4 \times 10^4 \text{ J}$$

Follow-Up Exercise. (a) Is it possible to have a mechanical efficiency of 100%? (b) What would an efficiency of greater than 100% imply?

Table 5.1 lists the typical efficiencies of some machines. You may be surprised by the relatively low efficiency of the automobile. Much of the energy input (from gasoline combustion) is lost as exhaust heat and through the cooling system (more than 60%), and friction accounts for a great deal more. About 20% of the input energy is converted to useful work that goes into propelling the vehicle. Air conditioning, power steering, radio, and tape and CD players are nice, but they also use energy and contribute to the car's decrease in efficiency.

TABLE 5.1	Typical Efficiencies of Some Machines
Machine	*Efficiency (approximate %)*
Compressor	85
Electric motor	70–95
Automobile (hybrid cars increase fuel efficiency by 25%)	20
Human muscle*	20–25
Steam locomotive	5–10

*Technically not a machine, but used to perform work.

Chapter Review

- **Work done by a constant force** is the product of the magnitude of the displacement and the component of the force parallel to the displacement:

$$W = (F \cos \theta)d \qquad (5.2)$$

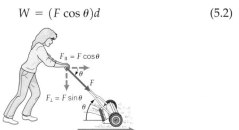

- Calculating work done by a variable force requires advanced mathematics. An example of a variable force is the **spring force**, given by *Hooke's law*:

$$F_s = -kx \qquad (5.3)$$

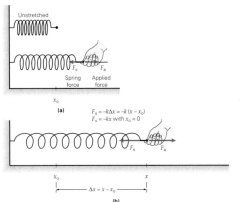

The **work done by a spring force** is given by

$$W = \tfrac{1}{2}kx^2 \qquad (5.4)$$

- **Kinetic energy** is the energy of motion and is given by

$$K = \tfrac{1}{2}mv^2 \qquad (5.5)$$

- By the **work–energy theorem**, the net work done on an object is equal to the change in the kinetic energy of the object:

$$W = K - K_o = \Delta K \qquad (5.6)$$

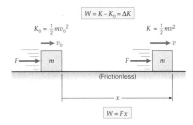

- **Potential energy** is the energy of position and/or configuration. The elastic **potential energy of a spring** is given by

$$U = \tfrac{1}{2}kx^2 \qquad (\text{with } x_o = 0) \qquad (5.7)$$

The most common type of potential energy is **gravitational potential energy**, associated with the gravitational attraction near the Earth's surface.

$$U = mgy \qquad (\text{with } y_o = 0) \qquad (5.8)$$

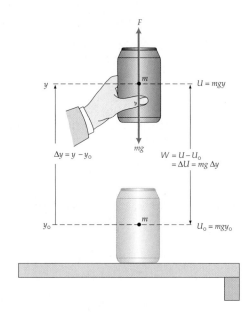

- **Conservation of energy**: The total energy of the universe or of an isolated system is always conserved.

   Conservation of mechanical energy: The total mechanical energy (kinetic plus potential) is constant in a conservative system:

$$\tfrac{1}{2}mv^2 + U = \tfrac{1}{2}mv_o^2 + U_o \qquad (5.10b)$$

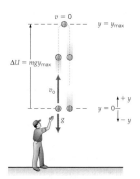

- In systems with **nonconservative forces**, where mechanical energy is lost, the work done by a nonconservative force is given by

$$W_{nc} = E - E_o = \Delta E \qquad (5.13)$$

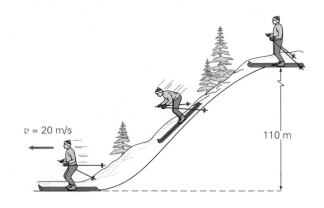

• **Power** is the time rate of doing work (or expending energy). **Average power** is given by

$$\bar{P} = \frac{W}{t} = \frac{Fd}{t} = F\bar{v} \qquad (5.15)$$

(constant force in direction of d and v)

$$\bar{P} = \frac{F(\cos\theta)d}{t} = F\bar{v}\cos\theta \qquad (5.16)$$

(constant force acts at an angle θ between d and v)

• **Efficiency** relates work output to energy (work) input as a percent:

$$\varepsilon = \frac{W_{out}}{E_{in}} \; (\times 100\%) \qquad (5.17)$$

$$\varepsilon = \frac{P_{out}}{P_{in}} \; (\times 100\%) \qquad (5.18)$$

Exercises

MC = *Multiple Choice Question,* **CQ** = *Conceptual Question, and* **IE** = *Integrated Exercise. Throughout the text, many exercise sections will include "paired" exercises. These exercise pairs, identified with* **red numbers***, are intended to assist you in problem solving and learning. In a pair, the first exercise (even numbered) is worked out in the Study Guide so that you can consult it should you need assistance in solving it. The second exercise (odd numbered) is similar in nature, and its answer is given at the back of the book.*

5.1 Work Done by a Constant Force

1. **MC** The units of work are (a) N·m, (b) kg·m²/s², (c) J, (d) all of the preceding. (d)

2. **MC** For a particular force and displacement, the most work is done when the angle between them is (a) 30°, (b) 60°, (c) 90°, (d) 180°. (d)

3. **MC** A pitcher throws a fastball. When the catcher catches it, (a) positive work is done, (b) negative work is done, (c) the net work is zero. (b)

4. **MC** Work done in free fall is (a) only positive, (b) only negative or (c) can be either positive or negative. (c)

5. **CQ** (a) As a weightlifter strains to lift a barbell from the floor (▼Fig. 5.24a), is he doing work? Why or why not? (b) In raising the barbell above his head, is he doing work? Explain. (c) In holding the barbell above his head (Fig. 5.24b), is he doing more work, less work, or the same amount of work as in lifting the barbell? Explain. (d) If the weightlifter drops the barbell, is work done on the barbell? Explain what happens in this situation. see ISM

(a)   (b)

▲ **FIGURE 5.24 Man at work?** See Exercise 5.

6. **CQ** You are carrying a backpack across campus. What is the work done by your vertical carrying force on the backpack? Explain. zero, $\theta = 90°$ so $\cos\theta = 0$

7. **CQ** A jet plane flies in a vertical circular loop. In what regions of the loop is the work done by the plane's weight positive and/or negative? Is the work constant? If not, are there maximum and minimum instantaneous values? Explain. see ISM

8. ● If a person does 50 J of work in moving a 30-kg box over a 10-m distance on a horizontal surface, what is the minimum force required? 5.0 N

9. ● A 5.0-kg box slides a 10-m distance on ice. If the coefficient of kinetic friction is 0.20, what is the work done by the friction force? −98 J

10. ● A passenger at an airport pulls a rolling suitcase by its handle. If the force used is 10 N and the handle makes an angle of 25° to the horizontal, what is the work done by the pulling force while the passenger walks 200 m? 1.8×10^3 J

11. ● A college student earning some summer money pushes a lawn mower on a level lawn with a constant force of 200 N at an angle of 30° downward from the horizontal. How far does the student push the mower in doing 1.44×10^3 J of work? 8.31 m

12. ●● A 3.00-kg block slides down a frictionless plane inclined 20° to the horizontal. If the length of the plane's surface is 1.50 m, how much work is done, and by what force? 15.1 J, gravitational force

13. ●● Suppose the coefficient of kinetic friction between the block and the plane in Exercise 12 is 0.275. What would be the net work done in this case? 3.7 J

14. ●● A father pulls his young daughter on a sled with a constant velocity on a level surface through a distance of 10 m, as illustrated in ▼Fig. 5.25a. If the total mass of the sled and the girl is 35 kg and the coefficient of kinetic friction between the sled runners and the snow is 0.20, how much work does the father do? 6.1×10^2 J

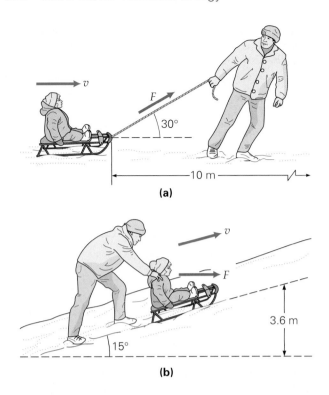

(a)

(b)

▲ **FIGURE 5.25 Fun and work** See Exercises 14 and 15.

15. ●● A father pushes horizontally on his daughter's sled to move it up a snowy incline, as illustrated in Fig. 5.25b. If the sled moves up the hill with a constant velocity, how much work is done by the father in moving it from the bottom to the top of the hill? (Some necessary data are given in Exercise 14.) 2.3×10^3 J

16. **IE** ●● A hot-air balloon ascends at a constant rate. (a) The weight of the balloon does (1) positive work, (2) negative work, (3) no work. Why? (b) A hot-air balloon with a mass of 500 kg ascends at a constant rate of 1.50 m/s for 20.0 s. How much work is done by the upward buoyant force? (Neglect air resistance.)
(a) (2) negative work (b) 1.47×10^5 J

17. **IE** ●● A hockey puck with a mass of 200 g and an initial speed of 25.0 m/s slides freely to rest in the space of 100 m on a sheet of horizontal ice. How many forces do nonzero work on it as it slows: (a) (1) none, (2) one, (3) two, or (4) three? Explain. (b) Determine the work done by all the individual forces on the puck as it slows.
(a) (2) one (b) −62.5 J

18. **IE** ●● An eraser with a mass of 100 g sits on a book at rest. The eraser is initially 10.0 cm from any edge of the book. The book is suddenly yanked very hard and slides out from under the eraser. In doing so, it partially drags the eraser with it, although not enough to stay on the book. The coefficient of kinetic friction between the book and the eraser is 0.150. (a) The sign of the work done by the force of kinetic friction of the book on the eraser is (1) positive, (2) negative, or (3) zero work is done by kinetic friction. Explain. (b) How much work is done by the book's frictional force on the eraser by the time it falls off the edge of the book? (a) (1) positive (b) 0.0147 J

19. ●●● A 500-kg, ligh-weight helicopter ascends from the ground with an acceleration of 2.00 m/s². Over a 5.00-s

interval, what is (a) the work done by the lifting force, (b) the work done by the gravitational force, and (c) the net work done on the helicopter?
(a) 1.48×10^5 J (b) -1.23×10^5 J (c) 2.50×10^4 J

20. ●●● A man pushes horizontally on a desk that rests on a rough wooden floor. The coefficient of static friction between the desk and floor is 0.750 and the coefficient of kinetic friction is 0.600. The desk's mass is 100 kg. He pushes just hard enough to get the desk moving and continues pushing with that force for 5.00 s. How much work does he do on the desk? 1.35×10^4 J

21. **IE** ●●● A student could either pull or push, at an angle of 30° from the horizontal, a 50-kg crate on a horizontal surface, where the coefficient of kinetic friction between the crate and surface is 0.20. The crate is to be moved a horizontal distance of 15 m. (a) Compared with pushing, pulling requires the student to do (1) less, (2) the same, or (3) more work. (b) Calculate the minimum work required for both pulling and pushing.
(a) (1) less (b) pulling: 1.3×10^3 J; pushing: 1.7×10^3 J

5.2 Work Done by a Variable Force

22. **MC** The work done by a variable force of the form $F = kx$ is equal to (a) kx^2, (b) kx, (c) $\frac{1}{2}kx^2$, (d) none of the preceding. (c)

23. **CQ** Does it take twice the work to stretch a spring 2 cm from its equilibrium position as it does to stretch it 1 cm from its equilibrium position? no, it takes more than twice the work, force increases as spring stretches

24. **CQ** If a spring is compressed 2.0 cm from its equilibrium position and then compressed an additional 2.0 cm, how much more work is done in the second compression than in the first? Explain. three times as much

25. ● To measure the spring constant of a certain spring, a student applies a 4.0-N force, and the spring stretches by 5.0 cm. What is the spring constant? 80 N/m

26. ● A spring has a spring constant of 30 N/m. How much work is required to stretch the spring 2.0 cm from its equilibrium position? 6.0×10^{-3} J

27. ● If it takes 400 J of work to stretch a spring 8.00 cm, what is the spring constant? 1.25×10^5 N/m

28. ● If a 10-N force is used to compress a spring with a spring constant of 4.0×10^2 N/m, what is the resulting spring compression? 2.5 cm

29. **IE** ● A certain amount of work is required to stretch a spring from its equilibrium position. (a) If twice the work is performed on the spring, the spring will stretch more by a factor of (1) $\sqrt{2}$, (2) 2, (3) $1/\sqrt{2}$, (4) $\frac{1}{2}$. Why? (b) If 100 J of work is done to pull a spring 1.0 cm, what work is required to stretch it 3.0 cm? (a) (1) $\sqrt{2}$ (b) 900 J

30. ●● Compute the work done by the variable force in the graph of F versus x in ►Fig. 5.26. [*Hint:* The area of a triangle is $A = \frac{1}{2}$altitude × base.] −3.0 J

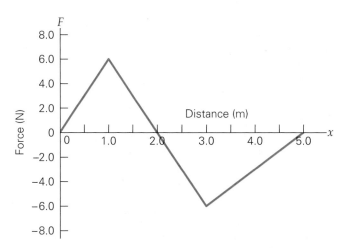

▲ **FIGURE 5.26 How much work is done?** See Exercise 30.

31. **IE ●●** A spring with a force constant of 50 N/m is to be stretched from 0 to 20 cm. (a) The work required to stretch the spring from 10 cm to 20 cm is (1) more than, (2) the same as, (3) less than that required to stretch it from 0 to 10 cm. (b) Compare the two work values to prove your answer to part (a).
(a) (1) more than (b) 0–10 cm: 0.25 J; 10–20 cm: 0.75 J

32. **IE ●●** In gravity-free interstellar space, a spaceship fires its rockets to speed up. The rockets are programmed to increase thrust from zero to 1.00×10^4 N with a linear increase over the course of 18.0 km. Then the thrust decreases linearly back to zero over the next 18.0 km. Assuming the rocket was stationary to start, (a) during which segment will more work (magnitude) be done: (1) the first 60 s, (2) the second 60 s, (3) the work done is the same in both segments? Explain your reasoning. (b) Determine quantitatively how much work is done in each segment.
(a) (3) the work done is the same (b) see ISM

33. **●●** A particular spring has a force constant of 2.5×10^3 N/m. (a) How much work is done in stretching the relaxed spring by 6.0 cm? (b) How much more work is done in stretching the spring an additional 2.0 cm?
(a) 4.5 J (b) 3.5 J

34. **●●** For the spring in Exercise 33, how much mass would have to be suspended from the vertical spring to stretch it (a) the first 6.0 cm and (b) the additional 2.0 cm?
(a) 15 kg (b) 5.1 kg more

35. **●●●** In stretching a spring in an experiment, a student inadvertently stretches it past its elastic limit; the force-versus-stretch graph is shown in ▼Fig. 5.27. Basically, after it reaches its limit, the spring begins to behave as if it were considerably stiffer. How much work was done on the spring? Assume that on the force axis, the tick marks are every 10 N, and on the x-axis, they are every 10 cm or 0.10 m. 6.0 J

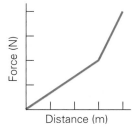

◀ **FIGURE 5.27 Past the limit** See Exercise 35

36. **●●●** A spring (spring 1) with a spring constant of 500 N/m is attached to a wall and connected to another weaker spring (spring 2) with a spring constant of 250 N/m on a horizontal surface. Then an external force of 100 N is applied to the end of the weaker spring (#2). How much potential energy is stored in each spring?
$U_1 = 10$ J, $U_2 = 20$ J

5.3 The Work–Energy Theorem: Kinetic Energy

37. **MC** Which of the following is a scalar quantity: (a) work, (b) force, (c) kinetic energy, or (d) both a and c? (d)

38. **MC** If the angle between the net force and the displacement of an object is greater than 90°, (a) kinetic energy increases, (b) kinetic energy decreases, (c) kinetic energy remains the same, (d) the object stops. (b)

39. **MC** Two identical cars, A and B, traveling at 55 mi/h collide head-on. A third identical car, C, crashes into a brick wall going 55 mi/h. Which car has the least damage: (a) car A, (b) car B, (c) car C, or (d) all the same? (c)

40. **MC** Which of the following objects has the least kinetic energy: (a) an object of mass $4m$ and speed v; (b) an object of mass $3m$ and speed $2v$; (c) an object of mass $2m$ and speed $3v$; or (d) an object of mass m and speed $4v$? (a)

41. **CQ** You want to decrease the kinetic energy of an object as much as you can. You can do so by either reducing the mass by half or reducing the speed by half. Which option should you pick, and why? reducing speed

42. **CQ** A certain amount of work W is required to accelerate a car from rest to a speed v. How much work is required to accelerate the car from rest to a speed of $v/2$? $W/4$

43. **CQ** A certain amount of work W is required to accelerate a car from rest to a speed v. If instead an amount of work equal to $2W$ is done on the car, what is the car's speed? $\sqrt{2}v$

44. **IE ●** A 0.20-kg object with a horizontal speed of 10 m/s hits a wall and bounces directly back with only half the original speed. (a) What percentage of the objects' initial kinetic energy is lost? (1) 25%, (2) 50%, (3) 75%. (b) How much kinetic energy is lost in the ball's collision with the wall? (a) (3) 75% (b) 7.5 J

45. **●** A 2.5-g bullet traveling at 350 m/s hits a tree and slows uniformly to a stop while penetrating a distance of 12 cm into the tree's trunk. What force was exerted on the bullet in bringing it to rest? -1.3×10^3 N

46. **●** A 1200-kg automobile travels at 90 km/h. (a) What is its kinetic energy? (b) What net work would be required to bring it to a stop? (a) 3.8×10^5 J (b) -3.8×10^5 J

47. **●** A constant net force of 75 N acts on an object initially at rest through a parallel distance of 0.60 m. (a) What is the final kinetic energy of the object? (b) If the object has a mass of 0.20 kg, what is its final speed?
(a) 45 J (b) 21 m/s

48. **IE ●●** A 2.00-kg mass is attached to a vertical spring with a spring constant of 250 N/m. A student pushes on the mass vertically upward with his hand while slowly lowering it to its equilibrium position. (a) How many forces do nonzero work on the object: (1) one, (2) two, or (3) three? Explain your reasoning. (b) Calculate the work done on the object by each of the forces acting on it as it is lowered into position. (a) (3) three (b) $W_g = 1.54$ J, $W_{sp} = -0.768$ J, $W_h = -0.77$ J

49. **●●** The stopping distance of a vehicle is an important safety factor. Assuming a constant braking force, use the work–energy theorem to show that a vehicle's stopping distance is proportional to the square of its initial speed. If an automobile traveling at 45 km/h is brought to a stop in 50 m, what would be the stopping distance for an initial speed of 90 km/h? see ISM; 200 m

50. **IE ●●** A large car of mass $2m$ travels at speed v. A small car of mass m travels with a speed $2v$. Both skid to a stop with the same coefficient of friction. (a) The small car will have (1) a longer, (2) the same, (3) a shorter stopping distance. (b) Calculate the ratio of the stopping distance of the small car to that of the large car. (Use the work–energy theorem, not Newton's laws.) (a) (1) a longer (b) 4

51. **●●●** An out-of-control truck with a mass of 5000 kg is traveling at 35.0 m/s (about 80 mi/h) when it starts descending a steep (15°) incline. The incline is icy, so the coefficient of friction is only 0.30. Use the work–energy theorem to determine how far the truck will skid (assuming it locks its brakes and skids the whole way) before it comes to rest. 2.0×10^3 m

52. **●●●** If the work required to speed up a car from 10 km/h to 20 km/h is 5.0×10^3 J, what would be the work required to increase the car's speed from 20 km/h to 30 km/h? 8.3×10^3 J

5.4 Potential Energy

53. **MC** A change in gravitational potential energy (a) is always positive, (b) depends on the reference point, (c) depends on the path, (d) depends only on the initial and final positions. (d)

54. **MC** The change in gravitational potential energy can be found by calculating $mg\Delta h$ and subtracting the reference point potential energy. (a) True, (b) false. (b)

55. **MC** The reference point for gravitational potential energy may be (a) zero, (b) negative, (c) positive, (d) all of the preceding. (d)

56. **CQ** If a spring changes its position from x_o to x, the change in potential energy is then proportional to what? (Express the answer in terms of x_o and x.) $x^2 - x_o^2$

57. **CQ** Two cars travel from the bottom to the top of a hill by different routes, one of which has more twists and turns. At the top, which has more potential energy?
Same potential energy

58. **●** How much more gravitational potential energy does a 1.0-kg hammer have when it is on a shelf 1.2 m high than when it is on a shelf 0.90 m high? 2.9 J

59. **IE ●** You are told that the gravitational potential energy of a 2.0-kg object has decreased by 10 J. (a) With this information, you can determine (1) the object's initial height, (2) the object's final height, (3) both the initial and the final height, (4) only the difference between the two heights. Why? (b) What can you say has physically happened to the object? (a) (4) only the difference between the two heights (b) position is lowered 0.51 m

60. **●●** A 0.20-kg stone is thrown vertically upward with an initial velocity of 7.5 m/s from a starting point 1.2 m above the ground. (a) What is the potential energy of the stone at its maximum height relative to the ground? (b) What is the change in the potential energy of the stone between its launch point and its maximum height? (a) 8.0 J (b) 5.6 J

61. **IE ●●** The floor of the basement of a house is 3.0 m below ground level, and the floor of the attic is 4.5 m above ground level. (a) If an object in the attic were brought to the basement, the change in potential energy will be greatest relative to which floor: (1) attic, (2) ground, (3) basement, or (4) all the same? Why? (b) What are the respective potential energies of 1.5-kg objects in the basement and attic, relative to ground level? (c) What is the change in potential energy if the object in the attic is brought to the basement? (a) (4) all the same (b) $U_b = -44$ J, $U_a = 66$ J (c) -1.1×10^2 J

62. **●●** A 0.50-kg mass is placed on the end of a vertical spring that has a spring constant of 75 N/m and eased down into its equilibrium position. (a) Determine the change in spring (elastic) potential energy of the system. (b) Determine the system's change in gravitational potential energy. (a) 0.16 J (b) −0.32 J

63. **●●** A horizontal spring, resting on a frictionless tabletop, is stretched 15 cm from its unstretched configuration and a 1.00-kg mass is attached to it. The system is released from rest. A fraction of a second later, the spring finds itself compressed 3.0 cm from its unstretched configuration. How does its final potential energy compare to its initial potential energy? (Give your answer as a ratio, final to initial.) $^1/_{25}$

64. **●●●** A student has six textbooks, each with a thickness of 4.0 cm and a weight of 30 N. What is the minimum work the student would have to do to place all the books in a single vertical stack, starting with all the books on the surface of the table? 18 J

65. **●●●** A 1.50-kg mass is placed on the end of a spring that has a spring constant of 175 N/m. The mass-spring system rests on a frictionless incline that is at an angle of 30° from the horizontal (▼Fig. 5.28). The system is eased into its equilibrium position, where it stays. (a) Determine the change in elastic potential energy of the system. (b) Determine the system's change in gravitational potential energy. (a) 0.154 J (b) −0.309 J

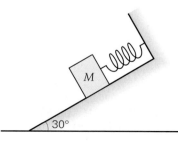

◀ **FIGURE 5.28**
Changes in potential energy See Exercise 65.

5.5 Conservation of Energy

66. **MC** Energy cannot be (a) transferred, (b) conserved, (c) created, (d) in different forms. (c)

67. **MC** If a nonconservative force acts on an object, and does work, then (a) the object's kinetic energy is conserved, (b) the object's potential is conserved, (c) the mechanical energy is conserved, (d) the mechanical energy is not conserved. (d)

68. **MC** The speed of a pendulum is greatest (a) when the pendulum's kinetic energy is a minimum, (b) when the pendulum's acceleration is a maximum, (c) when the pendulum's potential energy is a minimum, (d) none of the preceding. (c)

69. **CQ** For a classroom demonstration, a bowling ball suspended from a ceiling is displaced from the vertical position to one side and released from rest just in front of the nose of a student (▼Fig. 5.29). If the student doesn't move, why won't the bowling ball hit his nose? final height = initial height

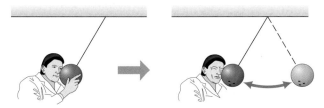

▲ **FIGURE 5.29 In your face?** See Exercise 69.

70. **CQ** When you throw an object into the air, is its initial speed the same as its speed just before it returns to your hand? Explain by applying the concept of conservation of mechanical energy. see ISM

71. **CQ** A student throws a ball vertically upward so it just reaches the height of a window on the second floor of a dormitory. At the same time the ball is thrown upward, a student at the window drops a ball. Are the mechanical energies of the balls the same at half the height of the window? Explain. yes, see ISM

72. ● A 0.300-kg ball is thrown vertically upward with an initial speed of 10.0 m/s. If the initial potential energy is taken as zero, find the ball's kinetic, potential, and mechanical energies (a) at its initial position, (b) at 2.50 m above the initial position, and (c) at its maximum height. (a) 15.0 J, zero, 15.0 J (b) 7.65 J, 7.35 J, 15.0 J (c) see ISM

73. ● What is the maximum height reached by the ball in Exercise 72? 5.10 m

74. ●● A 0.50-kg ball thrown vertically upward has an initial kinetic energy of 80 J. (a) What are its kinetic and potential energies when it has traveled three fourths of the distance to its maximum height? (b) What is the ball's speed at this point? (c) What is its potential energy at the maximum height? (Assume a reference is chosen to be zero at the launch point.) (a) 20 J, 60 J (b) 8.9 m/s (c) 80 J

75. **IE** ●● A girl swings back and forth on a swing with ropes that are 4.00 m long. The maximum height she reaches is 2.00 m above the ground. At the lowest point of the swing, she is 0.500 m above the ground. (a) The girl attains the maximum speed (1) at the top, (2) in the middle, (3) at the bottom of the swing. Why? (b) What is the girl's maximum speed? (a) (3) at the bottom of the swing (b) 5.42 m/s

76. ●● A block M (1.00 kg) on a frictionless 5° incline is connected by a light string running over a frictionless pulley to a suspended block m (200 g). The blocks are released from rest and the suspended mass falls 1.00 m before hitting the floor. Determine the speed of the blocks just before m hits the floor. 1.36 m/s

77. ●● A 1.00-kg block (M) is on a flat frictionless surface (▼Fig. 5.30). This block is attached to a spring initially at its relaxed length (spring constant is 50.0 N/m). A light string is attached to the block and runs over a frictionless pulley to a 450-g dangling mass (m). If the dangling mass is released from rest, how far does it fall before stopping? 0.176 m

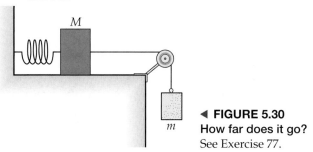

◀ **FIGURE 5.30**
How far does it go?
See Exercise 77.

78. **IE** ●● A 500-g (small) mass on the end of a 1.50-m-long string is pulled aside 15° from the vertical and shoved downward (toward the bottom of its motion) with a speed of 2.00 m/s. (a) Is the angle on the other side (1) greater than, (2) less than, or (3) the same as the angle on the initial side (15°)? Explain in terms of energy. (b) Calculate the angle it goes to on the other side, neglecting air resistance. (a) (1) greater than (b) 33.9°

79. ●● When a certain rubber ball is dropped from a height of 1.25 m onto a hard surface, it loses 18.0% of its mechanical energy on each bounce. (a) How high will the ball bounce on the first bounce? (b) How high will it bounce on the second bounce? (c) With what speed would the ball have to be thrown downward to make it reach its original height on the first bounce? (a) 1.03 m (b) 0.841 m (c) 2.32 m/s

80. ●● A skier coasts down a very smooth, 10-m-high slope similar to the one shown in Fig. 5.21. If the speed of the skier on the top of the slope is 5.0 m/s, what is his speed at the bottom of the slope? 15 m/s

81. ●● A roller coaster travels on a frictionless track as shown in ▼Fig. 5.31. (a) If the speed of the roller coaster at point A is 5.0 m/s, what is its speed at point B? (b) Will it reach point C? (c) What minimum speed at point A is required for the roller coaster to reach point C? (a) 11 m/s (b) no (c) 7.7 m/s

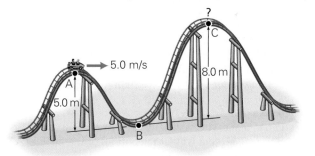

▲ **FIGURE 5.31 Energy conversion(s)** See Exercise 81.

82. •• A simple pendulum has a length of 0.75 m and a bob with a mass of 0.15 kg. The bob is released from an angle of 25° relative to a vertical reference line (▼Fig. 5.32). (a) Show that the vertical height of the bob when it is released is $h = L(1 - \cos 25°)$. (b) What is the kinetic energy of the bob when the string is at an angle of 9.0°? (c) What is the speed of the bob at the bottom of the swing? (Neglect friction and the mass of the string.)
(a) see ISM (b) 9.0×10^{-2} J (c) 1.2 m/s

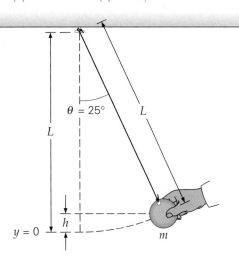

▲ **FIGURE 5.32 A pendulum swings** See Exercise 82.

83. •• Suppose the simple pendulum in Exercise 82 were released from an angle of 60°. (a) What would be the speed of the bob at the bottom of the swing? (b) To what height would the bob swing on the other side? (c) What angle of release would give half the speed of that for the 60° release angle at the bottom of the swing?
(a) 2.7 m/s (b) 0.38 m (c) 29°

84. •• A 1.5-kg box that is sliding on a frictionless surface with a speed of 12 m/s approaches a horizontal spring. (See Fig. 5.19.) The spring has a spring constant of 2000 N/m. (a) How far will the spring be compressed in stopping the box? (b) How far will the spring be compressed when the box's speed is reduced to half of its initial speed?
(a) 0.33 m (b) 0.28 m

85. •• A 28-kg child initially at rest slides down a playground slide from a height of 3.0 m above the bottom of the slide. If her speed at the bottom is 2.5 m/s, what is the work done by nonconservative forces? -7.4×10^2 J

86. ••• A hiker plans to swing on a rope across a ravine in the mountains, as illustrated in ▶Fig. 5.33, and to drop when she is just above the far edge. (a) At what horizontal speed should she be moving when she starts to swing? (b) Below what speed would she be in danger of falling into the ravine? Explain.
(a) at least 2.9 m/s (b) starting with $v_0 < 2.9$ m/s

87. ••• In Exercise 80, if the skier has a mass of 60 kg and the force of friction retards his motion by doing 2500 J of work, what is his speed at the bottom of the slope? 12 m/s

88. ••• A 1.00-kg block (M) is on a frictionless, 20° degree inclined plane. The block is attached to a spring ($k = 25$ N/m) that is fixed to a wall at the bottom of the incline. A light string attached to the block runs over a frictionless pulley

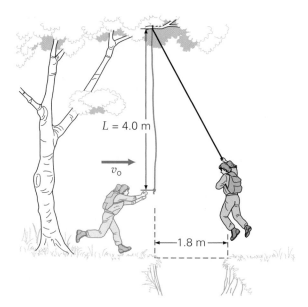

▲ **FIGURE 5.33 Can she make it?** See Exercise 86.

to a 40.0-g suspended mass. The suspended mass is given an initial downward speed of 1.50 m/s How far does it drop before coming to rest? (Assume the spring is unlimited in how far it can stretch) 0.21 m

5.6 Power

89. **MC** Which of the following is not a unit of power: (a) J/s; (b) W · s; (c) W; (d) hp? (b)

90. **MC** Consider a 2.0-hp motor and a 1.0-hp motor. Compared to the 2.0-hp motor, for a given amount of work, the 1.0-hp motor can (a) do twice as much work in half the time, (b) half the work in the same time, (c) one quarter of the work in three quarters of the time, (d) none of the preceding. (b)

91. **CQ** If you check your electricity bill, you will note that you are paying the power company for so many kilowatt-hours (kWh). Are you really paying for power? Explain. Also, convert 2.5 kWh to J. no, energy; 9.0×10^6 J

92. **CQ** (a) Does efficiency describe how fast work is done? Explain. (b) Does a more powerful machine always perform more work than a less powerful one? Explain. See ISM

93. **CQ** Two students who weigh the same start at the same ground-floor location at the same time to go to the same classroom on the third floor by different routes. If they arrive at different times, which student will have expended more power? Explain. see ISM

94. • What is the power in watts of a motor rated at $\frac{1}{2}$ hp? 373 W

95. • A girl consumes 8.4×10^6 J (2000 food calories) of energy per day while maintaining a constant weight. What is the average power she produces in a day? 97 W

96. • A 1500-kg race car can go from 0 to 90 km/h in 5.0 s. What average power is required to do this?
9.4×10^4 W = 1.3×10^2 hp

97. • The two 0.50-kg weights of a cuckoo clock descend 1.5 m in a three-day period. At what rate is their total gravitational potential energy decreased? 5.7×10^{-5} W

98. • A 60-kg woman runs up a staircase 15 m high (vertically) in 20 s. (a) How much power does she expend? (b) What is her horsepower rating? (a) 4.4×10^{2} W (b) 0.59 hp

99. •• An electric motor with a 2.0-hp output drives a machine with an efficiency of 40%. What is the energy output of the machine per second? 6.0×10^{2} J

100. •• Water is lifted out of a well 30.0 m deep by a motor rated at 1.00 hp. Assuming 90% efficiency, how many kilograms of water can be lifted in 1 min? 1.37×10^{2} kg

101. •• In a time of 10 s, a 70-kg student runs up two flights of stairs whose combined vertical height is 8.0 m. Compute the student's power output in doing work against gravity in (a) watts and (b) horsepower.
(a) 5.5×10^{2} W (b) 0.74 hp

102. •• How much power must you exert to horizontally drag a 25.0-kg table 10.0 m across a brick floor in 30.0 s at constant velocity, assuming the coefficient of kinetic friction between the table and floor is 0.550? 44.9 W

103. ••• A 3250-kg aircraft takes 12.5 min to achieve its cruising altitude of 10.0 km and cruising speed of 850 km/h. If the plane's engines deliver, on average, 1500 hp during this time, what is the efficiency of the engines? 48.7%

104. ••• A sleigh and driver with a total mass of 120 kg are pulled up a hill with a 15° incline by a horse, as illustrated in ▼Fig. 5.34. (a) If the overall retarding frictional force is 950 N and the sled moves up the hill with a constant velocity of 5.0 km/h, what is the power output of the horse? (Express in horsepower, of course. Note the magnitude of your answer, and explain.) (b) Suppose that in a spurt of energy, the horse accelerates the sled uniformly from 5.0 km/h to 20 km/h in 5.0 s. What is the horse's maximum instantaneous power output? Assume the same force of friction. (a) 2.3 hp (b) 10 hp

105. ••• A construction hoist exerts an upward force of 500 N on an object with a mass of 50 kg. Starting from rest, determine the power the hoist exerted to lift the object vertically for 10 s under these conditions. 5.0×10^{2} W

Comprehensive Exercises

106. A spring with a spring constant of 2000 N/m is compressed 10.0 cm on a horizontal surface (▼Fig. 5.35). Then a 1.00-kg object is attached to it and released. At the relaxed-length position of the spring, the mass leaves the spring and the table goes from very smooth to rough, with a coefficient of friction of 0.500. There is a wall 50.0 cm away from the release point. (a) Determine whether the mass will make it back to the spring after one bounce off the wall, assuming it rebounds elastically (no speed loss) off the wall. (b) If it does make it to the spring, how far does it compress it? If it doesn't make it to the spring, where is its final location? (a) Hits spring (b) 0.0714 m

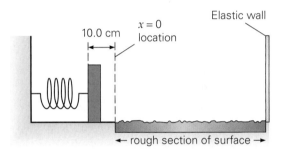

▲ **FIGURE 5.35 Come back** See Exercise 106

107. A block slides from rest down a frictionless incline. The incline is 2.50 m long and its angle is 40°. At the bottom is a small smooth curved section that joins into a rough section of horizontal floor. The block slides an additional horizontal distance of 3.00 m before stopping. Determine the coefficient of kinetic friction between the block and the floor. 0.536

108. Two identical springs (neglect their masses) are used to "play catch" with a small block of mass 100 g (▼Fig. 5.36). Spring A is attached to the floor and compressed 10.0 cm

▲ **FIGURE 5.34 A one-horse open sleigh** See Exercise 104.

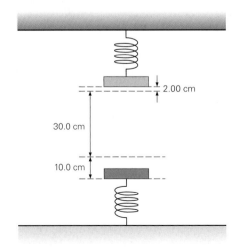

▲ **FIGURE 5.36 Playing catch** See Exercise 108.

with the mass on the end of it (loosely). Spring A is released from rest and the mass is accelerated upward. It impacts the spring attached to the ceiling, compresses it 2.00 cm, and stops after traveling a distance of 30.0 cm from the relaxed position of spring A to the relaxed position of spring B as shown. Determine the spring constant of the two springs (same since they are identical). 86.8 N/m

109. George of the Jungle grabs a vine that is 15.0 m long and swings down to the jungle floor. He starts from rest with the vine at 60°, lets go at the bottom of the swing, and slides along level dirt ground to a stop. If George has a mass of 100 kg, and the coefficient of kinetic friction between him and the jungle floor is 0.75, determine how far he slides before coming to rest. 10 m

110. A light spring initially stretched 20.0 cm has a 300-g mass on the end of it. The system is on a rough horizontal tabletop. The coefficient of kinetic friction is 0.60. The mass is initially shoved inward with a speed of 1.50 m/s and compresses the spring 5.00 cm before stopping. Calculate the spring constant. 5.5 N/m

The following Physlet Physics Problems can be used with this chapter.
6.3, 6.4, 6.5, 6.6, 6.7, 6.8, 6.9, 6.10, 6.11, 6.12, 6.14, 7.1, 7.2, 7.3, 7.4, 7.5, 7.6, 7.7, 7.8, 7.10

PHYSICS FACTS

- *Momentum* is the Latin word for "motion".

- Newton called momentum a "quantity of motion." From the *Principia*: *The quantity of motion is the measure of the same, arising from the velocity and quantity of matter, conjointly.*

- Newton called impulse a "motive force."

- A collision is the meeting or interaction of particles or objects, causing an exchange of energy and/or momentum.

- Before takeoff, the Space Shuttle, its external fuel tank (for the shuttle's engines), and its two solid booster rockets have a total weight on the order of 20 million N (4.4 million lb). To put the Space Shuttle into space, on launch the two solid rockets generate an average of 24 million N (5.3 million lb) of thrust over a 2-min burn, and the shuttle's three engines add 1.7 million N (375 000 lb) of thrust over an 8-min burn. The rockets and external fuel tank are later jettisoned.

- It is a common misconception that on rocket blastoff, the fiery engine exhaust striking and "pushing" against the launch pad propels the rocket upward. If this were the case, how could rocket engines be used in space where there is nothing to push against?

Tomorrow, sportscasters may say that the momentum of the entire game changed as a result of the clutch hit shown in the photograph. One team is said to have gained momentum and went on to win the game. But regardless of the effect on the team, it's clear that the momentum of the *ball* in the photograph must have changed dramatically. The ball was traveling toward the plate at a good rate of speed—with a lot of momentum. But a collision with a hardwood bat—with plenty of momentum of its own—changed the ball's direction in a fraction of a second. A fan might say that the batter turned the ball around. After studying Chapter 4, you might say that the force the bat applied to the ball gave it a large acceleration, reversing its velocity vector. Yet if you summed the momentum of the ball and bat just before the collision and just afterward, you'd discover that although both the ball and the bat had momentum changes, the total momentum never changed!

If you were bowling and the ball bounced off the pins and rolled toward you, you would probably be very surprised. But why? What leads us to expect that the ball will send the pins flying and continue on its way, rather than rebounding? You might say that the momentum of the ball carries it onward even after the collision (and you would be right)—but what does that really mean? In this chapter, you will study the concept of *momentum* and learn how it is particularly useful in analyzing motion and collisions.

6.1 Linear Momentum

OBJECTIVE: To define and compute linear momentum and the components of momentum.

The term *momentum* may bring to mind a football player running down the field, knocking down players who are trying to stop him. Or you might have heard someone say that a team lost its momentum (and so lost the game). Such everyday usages give some insight into the meaning of momentum. They suggest the idea of mass in motion and therefore of inertia. We tend to think of heavy or massive objects in motion as having a great deal of momentum, even if they move very slowly. However, according to the technical definition of momentum, a light object can have just as much momentum as a heavier one, and sometimes more.

Newton referred to what modern physicists term **linear momentum (*p*)** as "the quantity of motion . . . arising from velocity and the quantity of matter conjointly." In other words, the momentum of a body is proportional to *both* its mass and its velocity. By definition,

Note: The momentum vector of a single object is in the direction of the object's velocity.

| the linear momentum of an object is the product of its mass and velocity: |

$$\vec{\mathbf{p}} = m\vec{\mathbf{v}} \qquad (6.1)$$

SI unit of momentum: kilogram-meter per second ($\text{kg} \cdot \text{m/s}$)

Teaching tip: Stress that linear momentum is a vector quantity.

It is common to refer to linear momentum as simply *momentum*. Momentum is a vector quantity that has the same direction as the velocity, and *x–y* components with magnitudes of $p_x = mv_x$ and $p_y = mv_y$, respectively.

Equation 6.1 expresses the momentum of a single object or particle. For a system of more than one particle, the **total linear momentum ($\vec{\mathbf{P}}$)** of the system is the vector sum of the *momenta* (plural of *momentum*) of the individual particles:

Note: Total linear momentum—a vector sum

$$\vec{\mathbf{P}} = \vec{\mathbf{p}}_1 + \vec{\mathbf{p}}_2 + \vec{\mathbf{p}}_3 + \cdots = \sum \vec{\mathbf{p}}_i \qquad (6.2)$$

(*Note:* $\vec{\mathbf{P}}$ signifies the *total* momentum, while $\vec{\mathbf{p}}$ signifies an *individual* momentum.)

Example 6.1 ■ Momentum: Mass and Velocity

A 100-kg football player runs with a velocity of 4.0 m/s straight down the field. A 1.0-kg artillery shell leaves the barrel of a gun with a muzzle velocity of 500 m/s. Which has the greater momentum (magnitude), the football player or the shell?

Thinking It Through. Given the mass and velocity of an object, the magnitude of its momentum can be calculated from Eq. 6.1.

Solution. As usual, we first list the given data and what is to be found, using the subscripts p and s to refer to the player and shell, respectively.

Given: $m_\text{p} = 100 \text{ kg}$ *Find:* p_p and p_s (magnitudes of the momenta)
$\qquad\quad v_\text{p} = 4.0 \text{ m/s}$
$\qquad\quad m_\text{s} = 1.0 \text{ kg}$
$\qquad\quad v_\text{s} = 500 \text{ m/s}$

The magnitude of the momentum of the football player is

$$p_\text{p} = m_\text{p}v_\text{p} = (100 \text{ kg})(4.0 \text{ m/s}) = 4.0 \times 10^2 \text{ kg} \cdot \text{m/s}$$

and that of the shell is

$$p_\text{s} = m_\text{s}v_\text{s} = (1.0 \text{ kg})(500 \text{ m/s}) = 5.0 \times 10^2 \text{ kg} \cdot \text{m/s}$$

Thus, the less massive shell has the greater momentum. Remember, the magnitude of momentum depends on *both* the mass *and* the magnitude of the velocity.

Follow-Up Exercise. What would the football player's speed have to be for his momentum to have the same magnitude as the artillery shell's momentum? Would this speed be realistic? (*Answers to all Follow-Up Exercises are at the back of the text.*)

Integrated Example 6.2 ■ Linear Momentum: Some Ballpark Comparisons

Consider the three objects shown in ▶Fig. 6.1—a .22-caliber bullet, a cruise ship, and a glacier. Assuming each to be moving at its normal speed, (a) which would you expect to have the greatest linear momentum: (1) the bullet, (2) the ship, or (3) the glacier? (b) Estimate the masses and speeds and compute order-of-magnitude values of the linear momentum of the objects.

(a) Conceptual Reasoning. Certainly the bullet travels the fastest and the glacier the slowest, with the cruise ship in between. But momentum, $p = mv$, is equally dependent on mass as well as speed. The fast bullet has a tiny mass compared with that of the ship and the glacier. The slow glacier has a huge mass that greatly overshadows that of the bullet, and to some extent that of the ship. The cruise ship weighs a great deal and has considerable mass. Which object has the greater momentum also depends on the relative speeds. The glacier "creeps" along compared with the ship, so the very slow speed of the glacier counterbalances its huge mass to make its momentum less than might be expected. Assuming the speed difference to be greater than the mass difference for the ship and glacier, the ship would have the larger momentum. Similarly, because of the fast bullet's relatively tiny mass, it would be expected to have the least momentum. So with this reasoning, the largest momentum goes to the ship and the smallest momentum to the bullet, and the answer would be (2).

(b) Quantitative Reasoning and Solution. With no physical data given, you are asked to estimate the masses and velocities (speeds) of the objects so as to be able to compute their momenta [which will verify the reasoning in part (a)]. As is often the case in real-life problems, you may have difficulty estimating the values, so you would try to look up approximate values for the various quantities. For this example, the estimates will be provided. (Note that the units given in references vary, and it is important to convert units correctly.)

Given: Estimates (given shortly) of weight (mass) and speed for the bullet, cruise ship, and glacier.

Find: The approximate magnitudes of the momenta for the bullet (p_b), cruise ship (p_s), and glacier (p_g).

Bullet: A typical .22-caliber bullet has a weight of about 30 grains and a muzzle velocity of about 1300 ft/s. (A grain, abbreviated gr, is an old British unit. It was once commonly used for pharmaceuticals, such as 5-gr aspirin tablets; 1 lb = 7000 gr.)

Ship: A ship like the one shown in Fig. 6.1b would have a weight of about 70 000 tons and a speed of about 20 knots. (A knot is another old unit, still commonly used in nautical contexts; 1 knot = 1.15 mi/h.)

Glacier: The glacier might be 1 km wide, 10 km long, and 250 m deep and move at a rate of 1.0 m per day. (There is much variation among glaciers. Therefore, these figures must involve more assumptions and rougher estimates than those for the bullet or ship. For example, we are assuming a uniform, rectangular cross-sectional area for the glacier. The depth is particularly difficult to estimate from a photograph; a minimum value is given by the fact that glaciers must be at least 50–60 m thick before they can "flow." Observed speeds range from a few centimeters to as much as 40 m a day for valley glaciers such as the one shown in Fig. 6.1c. The value chosen here is considered a typical one.)

Then, converting the data to metric units and giving orders of magnitude yields the following:

Bullet:

$$m_b = 30 \text{ gr}\left(\frac{1 \text{ lb}}{7000 \text{ gr}}\right)\left(\frac{1 \text{ kg}}{2.2 \text{ lb}}\right) = 0.0019 \text{ kg} \approx 10^{-3} \text{ kg}$$

$$v_b = (1.3 \times 10^3 \text{ ft/s})\left(\frac{0.305 \text{ m/s}}{\text{ft/s}}\right) = 4.0 \times 10^2 \text{ m/s} \approx 10^2 \text{ m/s}$$

Ship:

$$m_s = 7.0 \times 10^4 \text{ ton}\left(\frac{2.0 \times 10^3 \text{ lb}}{\text{ton}}\right)\left(\frac{1 \text{ kg}}{2.2 \text{ lb}}\right) = 6.4 \times 10^7 \text{ kg} \approx 10^8 \text{ kg}$$

$$v_s = 20 \text{ knots}\left(\frac{1.15 \text{ mi/h}}{\text{knot}}\right)\left(\frac{0.447 \text{ m/s}}{\text{mi/h}}\right) = 10 \text{ m/s} = 10^1 \text{ m/s}$$

(a)

(b)

(c)

▲ **FIGURE 6.1** Three moving objects: a comparison of momenta and kinetic energies **(a)** A .22-caliber bullet shattering a ballpoint pen; **(b)** a cruise ship; **(c)** a glacier, Glacier Bay, Alaska. See Example 6.2.

(continues on next page)

Glacier:

$$\text{width} \approx 10^3 \, \text{m}, \text{ length} \approx 10^4 \, \text{m}, \text{ depth} \approx 10^2 \, \text{m}$$

$$v_g = (1.0 \, \text{m/day})\left(\frac{1 \, \text{day}}{86 \, 400 \, \text{s}}\right) = 1.2 \times 10^{-5} \, \text{m/s} \approx 10^{-5} \, \text{m/s}$$

We have all the speeds and masses except for m_g, the mass of the glacier. To compute this value, the density of ice is needed, since $m = \rho V$ (Eq. 1.1). The density of ice is less than that of water (ice floats in water), but the two are not very different, so we will use the density of water, $1.0 \times 10^3 \, \text{kg/m}^3$, to simplify the calculations.

Thus, the mass of the glacier is approximated as

$$m_g = \rho V = \rho(l \times w \times d)$$

$$\approx (10^3 \, \text{kg/m}^3)[(10^4 \, \text{m})(10^3 \, \text{m})(10^2 \, \text{m})] = 10^{12} \, \text{kg}$$

Then, calculating the magnitudes of the momenta of the objects, we have

Bullet:

$$p_b = m_b v_b \approx (10^{-3} \, \text{kg})(10^2 \, \text{m/s}) = 10^{-1} \, \text{kg} \cdot \text{m/s}$$

Ship:

$$p_s = m_s v_s \approx (10^8 \, \text{kg})(10^1 \, \text{m/s}) = 10^9 \, \text{kg} \cdot \text{m/s}$$

Glacier:

$$p_g = m_g v_g \approx (10^{12} \, \text{kg})(10^{-5} \, \text{m/s}) = 10^7 \, \text{kg} \cdot \text{m/s}$$

So the ship does have the largest momentum, and the bullet has the smallest.

Follow-Up Exercise. Which of the objects in this Example has (1) the greatest kinetic energy and (2) the least kinetic energy? Justify your choices using order-of-magnitude calculations. (Notice here that the dependence is on the square of the speed, $K = \frac{1}{2}mv^2$.)

Example 6.3 ■ Total Momentum: A Vector Sum

What is the total momentum for each of the systems of particles illustrated in ▶Fig. 6.2a and b?

Thinking It Through. The total momentum is the vector sum of the individual momenta (Eq. 6.2). This quantity can be computed using the components of each vector.

Solution.

Given: Magnitudes and directions of momenta from Fig. 6.2

Find: (a) Total momentum ($\vec{P}$) for Fig. 6.2a
(b) Total momentum ($\vec{P}$) for Fig. 6.2b

(a) The total momentum of a system is the vector sum of the momenta of the individual particles, so

$$\vec{P} = \vec{p}_1 + \vec{p}_2 = (2.0 \, \text{kg} \cdot \text{m/s})\hat{x} + (3.0 \, \text{kg} \cdot \text{m/s})\hat{x} = (5.0 \, \text{kg} \cdot \text{m/s})\hat{x} \quad (\textit{+x-direction})$$

(b) Computing the total momenta in the *x*- and *y*-directions gives

$$\vec{P}_x = \vec{p}_1 + \vec{p}_2 = (5.0 \, \text{kg} \cdot \text{m/s})\hat{x} + (-8.0 \, \text{kg} \cdot \text{m/s})\hat{x}$$

$$= -(3.0 \, \text{kg} \cdot \text{m/s})\hat{x} \quad (\textit{-x-direction})$$

$$\vec{P}_y = \vec{p}_3 = (4.0 \, \text{kg} \cdot \text{m/s})\hat{y} \quad (\textit{+y-direction})$$

Then

$$\vec{P} = \vec{P}_x + \vec{P}_y = (-3.0 \, \text{kg} \cdot \text{m/s})\hat{x} + (4.0 \, \text{kg} \cdot \text{m/s})\hat{y}$$

or

$$P = 5.0 \, \text{kg} \cdot \text{m/s at } 53° \text{ relative to the negative } x \text{ axis.}$$

Follow-Up Exercise. In this Example, if $\vec{p}_1$ and $\vec{p}_2$ in part (a) were added to $\vec{p}_2$ and $\vec{p}_3$ in part (b), what would be the total momentum?

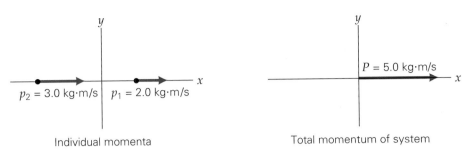

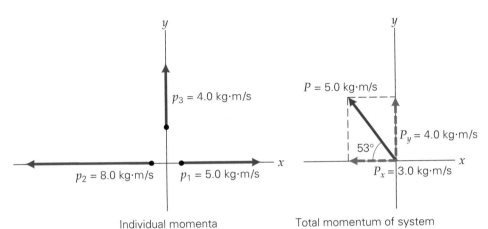

(a) $\vec{\mathbf{P}} = \vec{\mathbf{p}}_1 + \vec{\mathbf{p}}_2$

(b) $\vec{\mathbf{P}} = \vec{\mathbf{p}}_1 + \vec{\mathbf{p}}_2 + \vec{\mathbf{p}}_3$

◀ **FIGURE 6.2** Total momentum The total momentum of a system of particles is the vector sum of the particles' individual momenta. See Example 6.3.

In Example 6.3a, the momenta were along the coordinate axes and thus were added straightforwardly. If the motion of one (or more) of the particles is not along an axis, its momentum vector may be broken up, or resolved, into rectangular components, and individual components can then be added to find the components of the total momentum, just as you learned to do with force components in Chapter 4.

Since momentum is a vector, a change in momentum can result from a change in magnitude and/or direction. Examples of changes in the momenta of particles because of changes of direction on collision are illustrated in ▼Fig. 6.3. In the figure, the magnitude of a particle's momentum is taken to be the same both before and after collision (as indicated by the arrows of equal length). Figure 6.3a illustrates a direct rebound—a 180° change in direction. Note that the change in momentum ($\Delta\vec{\mathbf{p}}$) is the vector difference and that directional signs for the vectors are important. Figure 6.3b shows a glancing collision, for which the change in momentum is given by analyzing changes in the x- and y-components.

▼ **FIGURE 6.3** Change in momentum The change in momentum is given by the *difference* in the momentum vectors. **(a)** Here, the vector sum is zero, but the vector *difference*, or change in momentum, is not. (The particles are displaced for convenience.) **(b)** The change in momentum is found by computing the change in the components.

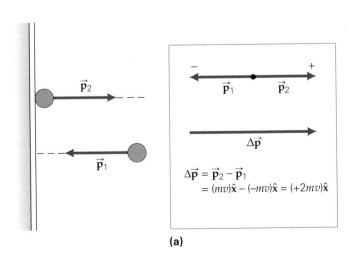

(a)

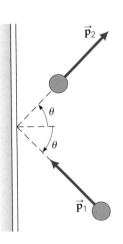

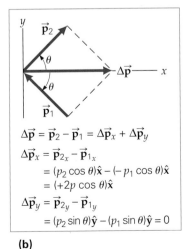

(b)

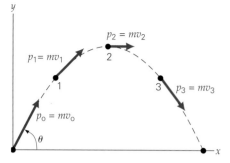

▲ **FIGURE 6.4 Change in the momentum of a projectile** The total momentum vector of a projectile is tangential to the projectile's path (as is its velocity); this vector changes in magnitude and direction, because of the action of an external force (gravity). The *x*-component of the momentum is constant. (Why?)

Demonstration/activity: Drop a good rubber ball (Super Ball) and an equivalent clay mass sequentially from equal heights onto a steel plate (ring-stand base) held by a student. The student holding the plate can feel which one gives the larger "push." Which one? Why?

(a)

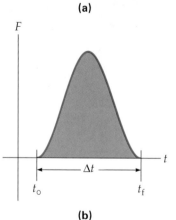

(b)

▲ **FIGURE 6.5 Collision impulse**
(a) Collision impulse causes the football to be deformed. **(b)** The impulse is the area under the curve of an *F*-versus-*t* graph. Note that the impulse force on the ball is not constant, but rises to a maximum.

Force and Momentum

As you know from Chapter 4, if an object has a change in velocity (an acceleration), a net force must be acting on it. Similarly, since momentum is directly related to velocity, a change in momentum also requires a net force. In fact, Newton originally expressed his second law of motion in terms of momentum rather than acceleration. The force–momentum relationship may be seen by starting with $\vec{F}_{net} = m\vec{a}$ and using $\vec{a} = (\vec{v} - \vec{v}_o)/\Delta t$, where the mass is assumed to be constant. Thus,

$$\vec{F}_{net} = m\vec{a} = \frac{m(\vec{v} - \vec{v}_o)}{\Delta t} = \frac{m\vec{v} - m\vec{v}_o}{\Delta t} = \frac{\vec{p} - \vec{p}_o}{\Delta t} = \frac{\Delta\vec{p}}{\Delta t}$$

or

$$\vec{F}_{net} = \frac{\Delta\vec{p}}{\Delta t} \quad \begin{array}{l} \textit{Newton's second law of motion} \\ \textit{in terms of momentum} \end{array} \quad (6.3)$$

where $\vec{F}_{net}$ is the *average* net force on the object if the acceleration is not constant (or the *instantaneous* net force if Δt goes to zero).

Expressed in this form, Newton's second law states that *the net external force acting on an object is equal to the time rate of change of the object's momentum*. It is easily seen from the development of Eq. 6.3 that the equations $\vec{F}_{net} = m\vec{a}$ and $\vec{F}_{net} = \Delta\vec{p}/\Delta t$ are equivalent if the mass is constant. In some situations, however, the mass may vary. This factor will not be a consideration here in our discussion of particle collisions, but a special case will be given later in the chapter. The more general form of Newton's second law, Eq. 6.3, is true even if the mass varies.

Just as the equation $\vec{F}_{net} = m\vec{a}$ indicates that an acceleration is evidence of a net force, the equation $\vec{F}_{net} = \Delta\vec{p}/\Delta t$ indicates that *a change in momentum is evidence of a net force*. For example, as illustrated in ◄Fig. 6.4, the momentum of a projectile is tangential to the projectile's parabolic path and changes in both magnitude and direction. The change in momentum indicates that there is a net force acting on the projectile, which you know is the force of gravity. Changes in momentum were illustrated in Fig. 6.3. Can you identify the forces in these two cases? Think in terms of Newton's third law.

6.2 Impulse

OBJECTIVES: To relate (a) impulse and momentum, and (b) kinetic energy and momentum.

When two objects—such as a hammer and a nail, a golf club and a golf ball, or even two cars—collide, they can exert large forces on one another for a short period of time (◄Fig. 6.5a). The force is not constant in this situation. However, Newton's second law in momentum form is still useful for analyzing such situations by using average values. Written in this form, the law states that the *average* force is equal to the time rate of change of momentum: $\vec{F}_{avg} = \Delta\vec{p}/\Delta t$ (Eq. 6.3). Rewriting the equation to express the change in momentum, we have (with only one force acting on the object)

$$\vec{F}_{avg}\Delta t = \Delta\vec{p} = \vec{p} - \vec{p}_o \quad (6.4)$$

The term $\vec{F}_{avg}\Delta t$ is known as the **impulse** ($\vec{I}$) of the force:

$$\vec{I} = \vec{F}_{avg}\Delta t = \Delta\vec{p} = m\vec{v} - m\vec{v}_o \quad (6.5)$$

SI unit of impulse and momentum: newton-second (N·s)

Thus, *the impulse exerted on an object is equal to the change in the object's momentum*. This statement is referred to as the **impulse–momentum theorem**. Impulse

TABLE 6.1	Some Typical Contact Times (Δt)
	Δt (milliseconds)
Golf ball (hit by a driver)	1.0
Baseball (hit off tee)	1.3
Tennis (forehand)	5.0
Football (kick)	8.0
Soccer (header)	23.0

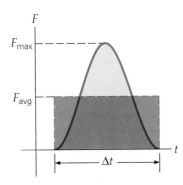

▲ **FIGURE 6.6** Average impulse force The area under the average-force curve ($F_{avg}\Delta t$, within the dashed red lines) is the same as the area under the F-versus-t curve, which is usually difficult to evaluate.

Illustration 8.1 Force and Impulse

has units of newton-second (N·s), which are also units of momentum (1 N·s = 1 kg·m/s²·s = 1 kg·m/s).

In Chapter 5, you learned that by the work–energy theorem ($W_{net} = F_{net}\Delta x = \Delta K$), the area under an F_{net}-versus-x curve is equal to the net work, or change in kinetic energy. Similarly, the area under an F_{net}-versus-t curve is equal to the impulse, or the change in momentum (Fig. 6.5b). An impulse force usually varies with time and is therefore not a constant force. However, in general, it is convenient to talk about the equivalent *constant* average force $\vec{F}_{avg}$ acting over a time interval Δt to give the same impulse (same area under the force-versus-time curve), as shown in ▶Fig. 6.6. Some typical contact times in sports are given in Table 6.1.

Example 6.4 ■ Teeing Off: The Impulse–Momentum Theorem

A golfer drives a 0.046-kg ball from an elevated tee, giving the ball an initial horizontal speed of 40 m/s (about 90 mi/h). What is the magnitude of the average force exerted by the club on the ball during this time?

Thinking It Through. The average force on the ball is equal to the time rate of change of its momentum, and this can be computed (Eq. 6.5).

Solution.

Given: $m = 0.046$ kg   *Find:* F_{avg} (average force)
         $v = 40$ m/s
         $v_0 = 0$
         $\Delta t = 1.0$ ms $= 1.0 \times 10^{-3}$ s (Table 6.1)

The mass and the initial and final velocities are given, so the change in momentum can be easily found. Then the magnitude of the average force can be computed from the impulse–momentum theorem:

$$F_{avg}\Delta t = p - p_0 = mv - mv_0$$

Thus,

$$F_{avg} = \frac{mv - mv_0}{\Delta t} = \frac{(0.046 \text{ kg})(40 \text{ m/s}) - 0}{1.0 \times 10^{-3} \text{ s}} = 1800 \text{ N (or about 410 lb)}$$

[This a very large force compared with the weight of the ball, $w = mg = (0.046 \text{ kg})(9.8 \text{ m/s}^2) = 0.45$ N (or about 0.22 lb).] The force is in the direction of the acceleration and is the *average* force. The instantaneous force is even greater than this value near the midpoint of the time interval of the collision (Δt in Fig. 6.6).

Follow-Up Exercise. Suppose the golfer in this Example drives the ball with the same average force, but "follows through" on the swing so as to increase the contact time to 1.5 ms. What effect would this change have on the initial horizontal speed of the drive?

$$F_{\text{avg}} \Delta t = mv_0$$

(a)

$$F_{\text{avg}} \Delta t = mv_0$$

(b)

▲ **FIGURE 6.7** Adjust the impulse
(a) The change in momentum in catching the ball is a constant mv_0. If the ball is stopped quickly (small Δt), the impulse force is large (big F_{avg}) and stings the catcher's bare hands. **(b)** Increasing the contact time (large Δt) by moving the hands with the ball reduces the impulse force and makes catching more enjoyable.

Example 6.4 illustrates the large forces that colliding objects can exert on one another during short contact times. In some cases, we shorten the contact time to maximize the impulse force—for example, in a karate chop. However, in other instances, the Δt may be manipulated to reduce the force. Suppose there is a fixed change in momentum in a given situation. Then, since $\Delta p = F_{\text{avg}} \Delta t$, if Δt could be made longer, the average impulse force F_{avg} would be reduced.

You have probably tried to minimize the impulse force on occasion. For example, in catching a hard, fast-moving ball, you quickly learn not to catch it with your arms rigid, but rather to move your hands with the ball. This movement increases the contact time and reduces the impulse force and the "sting" (◄Fig. 6.7).

When jumping from a height onto a hard surface, you try not to land stiff-legged. The abrupt stop (small Δt) would apply a large impulse force to your leg bones and joints and could cause injury. If you bend your knees as you land, the impulse is vertically upward, opposite your velocity ($F_{\text{avg}} \Delta t = \Delta p = -mv_0$, with the final velocity being zero). Thus, increasing the time interval Δt makes the impulse force smaller. Another example in which the contact time is increased to decrease the impulse force is given in Insight 6.1, The Automobile Air Bags and Martian Air Bags, on page 186.

Example 6.5 ■ Impulse Force and Body Injury

A 70.0-kg worker jumps stiff-legged from a height of 1.00 m onto a concrete floor. What is the magnitude of the impulse he feels on landing, assuming a sudden stop in 8.00 ms?

Thinking It Through. The impulse is $F_{\text{avg}} \Delta t$, which cannot be calculated directly from the given data. But impulse is equal to the change in momentum, $F_{\text{avg}} \Delta t = \Delta p = mv - mv_0$. So the impulse can be calculated from the difference in momenta.

Solution.

Given: $m = 70.0$ kg   *Find:* Impulse (I) on the worker
$\quad\quad\quad h = 1.00$ m
$\quad\quad\quad \Delta t = 8.00$ ms $= 8.00 \times 10^{-3}$ s

There are two different parts here: (a) the worker descending after jumping and (b) the sudden stop after hitting the floor. So we must be careful with notation.

(a) Here, $v_{0_1} = 0$, and the final velocity may be found using $v^2 = v_0^2 - 2gh$ (Eq. 2.12′), with the result of

$$v_1 = -\sqrt{2gh}$$

(b) The v_1 of the first process is then the initial velocity with which the stiff-legged worker hits the floor, that is, $v_{0_2} = v_1 = -\sqrt{2gh}$, and the final velocity at the second phase is $v_2 = 0$. Then,

$$I = F_{\text{avg}} \Delta t = \Delta p = mv_2 - mv_{0_2} = 0 - m(-\sqrt{2gh}) = +m\sqrt{2gh}$$

$$= (70.0 \text{ kg})\sqrt{2(9.80 \text{ m/s}^2)(1.00 \text{ m})} = 310 \text{ kg} \cdot \text{m/s}$$

where the impulse is in the upward direction.

With a Δt of 6.0×10^{-3} s for the sudden stop on impact, this would give force of

$$F_{\text{avg}} = \frac{\Delta p}{\Delta t} = \frac{310 \text{ kg} \cdot \text{m/s}}{8.00 \times 10^{-3} \text{ s}} = 3.88 \times 10^4 \text{ N} \quad (about\ 8.73 \times 10^3\ lb\ of\ force!)$$

and the force is upward on the stiff legs.

Follow-Up Exercise. Suppose the worker bent his knees and increased the contact time to 0.60 s on landing. What would be the impulse force on him in this case?

In some instances, the applied impulse force may be relatively constant and the contact time (Δt) deliberately increased to produce a greater impulse, and thus a greater change in momentum ($F_{\text{avg}} \Delta t = \Delta p$). This is the principle of "following through" in sports, for example, when hitting a ball with a bat or racquet, or driving

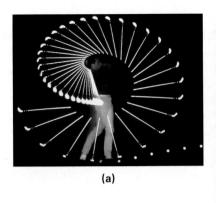

(a)

(b)

◄ **FIGURE 6.8 Increasing the contact time (a)** A golfer follows through on a drive. One reason he does so is to increase the contact time so that the ball receives greater impulse and momentum. **(b)** The follow-through on a long putt increases the contact time for greater momentum, but the main reason here is for directional control.

a golf ball. In the latter case (▲Fig. 6.8a), assuming that the golfer supplies the same average force with each swing, the longer the contact time, the greater will be the impulse or change in momentum the ball receives. That is, with $F_{avg}\Delta t = mv$ (since $v_o = 0$), the greater the value of Δt, the greater will be the final speed of the ball. (This principle was illustrated in the Follow-Up Exercise in Example 6.4.) In some instances, a long follow-through that may primarily be used to improve control of the ball's direction (Fig. 6.8b).

The word *impulse* implies that the impulse force acts only briefly (like an "impulsive" person), and this is true in many instances. However, the definition of *impulse* places no limit on the time interval over which the force may act. Technically, a comet at its closest approach to the Sun is involved in a collision, because in physics, collision forces do *not* have to be contact forces. Basically, a **collision** is any interaction between objects in which there is an exchange of momentum and/or energy.

As you might expect from the work–energy theorem and the impulse–momentum theorem, momentum and kinetic energy are directly related. A little algebraic manipulation of the equation for kinetic energy (Eq. 5.5) allows us to express kinetic energy (K) in terms of the *magnitude* of momentum (p):

$$K = \tfrac{1}{2}mv^2 = \frac{(mv)^2}{2m} = \frac{p^2}{2m} \tag{6.6}$$

Thus, kinetic energy and momentum are intimately related, but they are different quantities.

PHYSLET®

Illustration 8.2 The Difference between Impulse and Work

6.3 Conservation of Linear Momentum

OBJECTIVES: To (a) explain the conditions for the conservation of linear momentum, and (b) apply them to physical situations.

Like total mechanical energy, the total momentum of a system is a conserved quantity under certain conditions. This fact allows us to analyze a wide range of situations and solve many problems readily. Conservation of momentum is one of the most important principles in physics. In particular, it is used to analyze collisions of objects ranging from subatomic particles to automobiles in traffic accidents.

For the linear momentum of a single object to be conserved (that is, to remain constant with time), one condition must hold that is apparent from the momentum form of Newton's second law (Eq. 6.3). If the net force acting on a particle is zero, that is,

$$\vec{F}_{net} = \frac{\Delta\vec{p}}{\Delta t} = 0$$

INSIGHT 6.1 THE AUTOMOBILE AIR BAG AND MARTIAN AIR BAGS

A dark, rainy night—a car goes out of control and hits a big tree head-on! But the driver walks away with only minor injuries, because he had his seatbelt buckled and his car's air bags deployed. Air bags, along with seatbelts, are safety devices designed to prevent (or lessen) injuries to passengers in automobile collisions.

When a car collides with something basically immovable, such as a tree or a bridge abutment, or has a head-on collision with another vehicle, the car stops almost instantaneously. If the front-seat passengers have not buckled up (and there are no air bags), they keep moving until acted on by an external force (by Newton's first law). For the driver, this force is supplied by the steering wheel and column, and for the passenger by the dashboard and/or windshield.

Even when everyone has buckled up, there can be injuries. Seatbelts absorb energy by stretching, and they widen the area over which the force is exerted. However, if a car is going fast enough and hits something truly immovable, there may be too much energy for the belts to absorb. This is where the air bag comes in. The bag inflates automatically on hard impact (Fig. 1), cushioning the driver (and front-seat passenger if both sides are equipped with air bags). In terms of impulse, the air bag increases the stopping contact time—the fraction of a second it takes your head to sink into the inflated bag is many times longer than the instant in which you would have stopped otherwise by hitting a solid surface such as the windshield. A longer contact time means a reduced average impact force and thus less likelihood of an injury. (Because the bag is large, the total impact force is also spread over a greater area of the body, so the force on any one part of the body is also less.)

How does an air bag inflate during the little time that elapses between a front-end impact and the instant the driver would hit the steering column? An air bag is equipped with sensors that detect the sharp deceleration associated with a head-on collision the instant it begins. If the deceleration exceeds the sensors' threshold settings, a control unit sends an electric current to an igniter in the air bag, which sets off a chemical explosion that generates gas to inflate the bag at an explosive rate. The complete process from sensing to full inflation takes only on the order of 25 *thousandths* of a second (0.025 s).

Air bags have saved many lives. However, in some cases, the deployment of air bags has caused problems. An air bag is not a soft, fluffy pillow. When activated, it is ejected out of its compartment at speeds up to 320 km/h (200 mi/h) and could hit a person with enough force to cause severe injury and even death. Adults are advised to sit at least 13 cm (6 in.) from the air bag compartment and to buckle up—always. Children should sit in the rear seat, out of the reach of air bags.*

Martian Air Bags

Air bags on Mars? There was one in 1997 when the spacecraft *Pathfinder* was landing a rover on Mars. And in 2004, more air bags

FIGURE 1 Impulse and safety An automobile air bag increases the contact time that a person in a crash would experience with the dashboard or windshield, thereby decreasing the impulse force that could cause injury.

*Guidelines from the National Highway Traffic Safety Administration (www.nhtsa.dot.gov).

then

$$\Delta\vec{\mathbf{p}} = 0 = \vec{\mathbf{p}} - \vec{\mathbf{p}}_o$$

where $\vec{\mathbf{p}}_o$ is the initial momentum and $\vec{\mathbf{p}}$ is the momentum at some later time. Since these two values are equal, the momentum is conserved, and

$$\vec{\mathbf{p}} = \vec{\mathbf{p}}_o \quad \text{or} \quad m\vec{\mathbf{v}} = m\vec{\mathbf{v}}_o$$
final momentum = initial momentum

Note that this conservation is consistent with Newton's first law: An object remains at rest ($\vec{\mathbf{p}} = 0$), or in motion with a *uniform* velocity (constant $\vec{\mathbf{p}} \neq 0$), unless acted on by a net external force.

The conservation of momentum can be extended to a system of particles if we write Newton's second law in terms of the net force acting on the system and of the momenta of the particles: $\vec{\mathbf{F}}_{net} = \sum\vec{\mathbf{F}}_i$ and $\vec{\mathbf{P}} = \sum\vec{\mathbf{p}}_i = \sum m\vec{\mathbf{v}}_i$.

Because $\vec{\mathbf{F}}_{net} = \Delta\vec{\mathbf{P}}/\Delta t$, and if there is no net external force acting *on the system*, then $\vec{\mathbf{F}}_{net} = 0$, and $\Delta\vec{\mathbf{F}} = 0$; so $\vec{\mathbf{P}} = \vec{\mathbf{P}}_o$, and the *total* momentum is conserved. This generalized condition is referred to as the law of **conservation of linear momentum**:

$$\vec{\mathbf{P}} = \vec{\mathbf{P}}_o \tag{6.7}$$

Note: A lowercase $\vec{\mathbf{p}}$ means individual momentum. A capital $\vec{\mathbf{P}}$ means total momentum of the system. Both are vectors. ($\vec{\mathbf{P}} = \sum\vec{\mathbf{p}}_i$)

Conservation of momentum—no net external force

Thus, the total linear momentum of a system, $\vec{\mathbf{P}} = \sum\vec{\mathbf{p}}_i$, is conserved if the net external force acting on the system is zero.

(a)

(b)

arrived with the Mars Exploration Rover Mission. Spacecraft landings are usually softened by retrorockets fired intermittently toward the planet surface. However, firing retrorockets very near the Martian surface would have left trace amounts of foreign combustion chemicals on the surface. Since one objective of the Mars missions was to analyze the chemical composition of Martian rocks and soil, another method of landing had to be developed.

The solution? Probably the most expensive air bag system ever created, costing approximately $5 million to develop and install. The Rovers were surrounded by 4.6 m-(15-ft)-diameter "beach balls" for an air bag landing (Fig. 2a).

On entering the Martian atmosphere, the spacecraft was traveling at about 27 000 km/h (17 000 mi/h). A high-altitude rocket system and parachute slowed it down to about 80–100 km/h (50–60 mi/h). At an altitude of about 200 m (660 ft), gas generators inflated the air bags, which allowed the bag-covered Rovers to bounce and roll a bit on landing (Fig. 2b). The air bags then deflated, and out rolled the Rover (Fig. 2c).

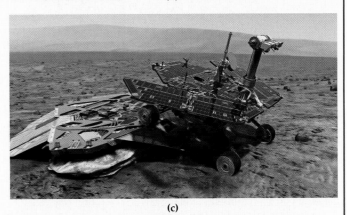

(c)

FIGURE 2 More bounce to the ounce (a) "Beach ball" air bags were used to protect *Pathfinder* and Mars Rovers. **(b)** Artist's conception of a Mars rover bouncing in its air bags. **(c)** A rover coming out safely.

There are many ways to achieve this condition. For example, recall from Chapter 5 that a *closed*, or *isolated*, system is one on which no net external force acts, so the total linear momentum of an isolated system is conserved.

Within a system, internal forces may act—for example, when particles collide. These are force pairs of Newton's third law, and there is a good reason that such forces are not explicitly referred to in the condition for the conservation of momentum. By Newton's third law, these internal forces are equal and opposite and vectorially cancel each other. Thus, *the net internal force of a system is always zero.*

An important point to understand, however, is that the momenta of *individual* particles or objects within a system may change. But in the absence of a net external force, the *vector sum* of all the momenta (the total system momentum $\vec{P}$) remains the same. If the objects are initially at rest (that is, the total momentum is zero) and then are set in motion as the result of internal forces, the total momentum must still add to zero. This principle is illustrated in ▼ Fig. 6.9 and analyzed in Example 6.6. Objects in an isolated system may transfer momentum among themselves, but the total momentum after the changes must add up to the initial value, assuming the net external force on the system is zero.

The conservation of momentum is often a powerful and convenient tool for analyzing situations involving motion and collisions. Its application is illustrated in the following Examples. (Notice that conservation of momentum, in many cases, bypasses the need to know the forces involved.)

Note: Third-law force pairs were discussed in Section 4.4.

Illustration 8.3 Hard and Soft Collisions and the Third Law

▶ **FIGURE 6.9** An internal force and the conservation of momentum The spring force is an internal force, so the momentum of the system is conserved. See Example 6.6.

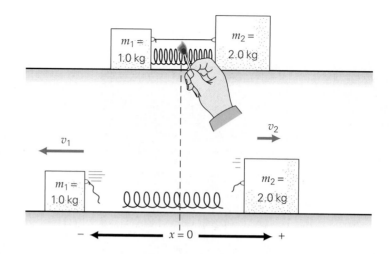

Exploration 8.6 An Explosive Collision

Demonstration/activity: Other demonstrations of the conservation of linear momentum:
- Suspend from strings a loosely stoppered test tube containing a small amount of water. Heat the water to produce steam pressure and a small explosion. Have students explain this effect in terms of momentum.
- Suspend a block of wood from strings. Attach a glob of clay to one side. Then, using a BB gun or other spring gun, fire a projectile at the clay and observe the motion of the block. Fire at the other, hard side of the block so that the projectile rebounds, and discuss the observed differences. Students are often surprised. The air track can be used to provide quantitative data for analysis.

Example 6.6 ■ Before and After: Conservation of Momentum

Two masses, $m_1 = 1.0$ kg and $m_2 = 2.0$ kg, are held on either side of a light compressed spring by a light string joining them, as shown in Fig. 6.9. The string is burned (negligible external force), and the masses move apart on the frictionless surface, with m_1 having a velocity of 1.8 m/s to the left. What is the velocity of m_2?

Thinking It Through. With no net external force (the weights are each canceled by a normal force), the total momentum of the system is conserved. It is initially zero, so after the string is burned, the momentum of m_2 must be *equal to and opposite* that of m_1. (Vector addition gives zero total momentum. Also, note that the term *light* indicates that the masses of the spring and string can be ignored.)

Solution. Listing the masses and speed given, we have

Given: $m_1 = 1.0$ kg   *Find:* v_2 (velocity—speed and direction)
         $m_2 = 2.0$ kg
         $v_1 = -1.8$ m/s (left)

Here, the system consists of the two masses and the spring. Since the spring force is internal to the system, the momentum of the system is conserved. It should be apparent that the initial total momentum of the system $(\vec{P}_o)$ is zero, and therefore the final momentum must also be zero. Thus, we may write

$$\vec{P}_o = \vec{P} = 0 \quad \text{and} \quad \vec{P} = \vec{p}_1 + \vec{p}_2 = 0$$

(The momentum of the "light" spring does not come into the equations, because its mass is negligible.) Then,

$$\vec{p}_2 = -\vec{p}_1$$

which means that the momenta of m_1 and m_2 are equal and opposite. Using directional signs (with + indicating the direction to the right in the figure), we have

$$m_2 v_2 = -m_1 v_1$$

and

$$v_2 = -\left(\frac{m_1}{m_2}\right)v_1 = -\left(\frac{1.0 \text{ kg}}{2.0 \text{ kg}}\right)(-1.8 \text{ m/s}) = +0.90 \text{ m/s}$$

Thus, the velocity of m_2 is 0.90 m/s in the positive x-direction, or to the right in the figure. This value is half of v_1, as you might have expected, since m_2 has twice the mass of m_1.

Follow-Up Exercise. (a) Suppose that the large block in Fig. 6.9 were attached to the Earth's surface so that the block could not move when the string was burned. Would momentum be conserved in this case? Explain. (b) Two girls, each having a mass of 50 kg, stand at rest on skateboards with negligible friction. The first girl tosses a 2.5-kg ball to the second. If the speed of the ball is 10 m/s, what is the speed of each girl after the ball is caught, and what is the momentum of the ball before it is tossed, while it is in the air, and after it is caught?

Integrated Example 6.7 ■ Conservation of Linear Momentum: Fragments and Components

Teaching tip: Remind students that this law is the second conservation law of mechanics.

A 30-g bullet with a speed of 400 m/s strikes a glancing blow to a target brick of mass 1.0 kg. The brick breaks into two fragments. The bullet deflects at an angle of 30° above the +x-axis and has a reduced speed of 100 m/s. One piece of the brick (with mass 0.75 kg) goes off to the right, or in the initial direction of the bullet, with a speed of 5.0 m/s. (a) Taking the x-axis to the right, will the other piece of the brick move in the (1) second quadrant, (2) third quadrant, or (3) fourth quadrant? (b) Determine the speed and direction of the other piece of the brick immediately after collision (where gravity can be neglected).

(a) Conceptual Reasoning. The conservation of linear momentum can be applied because there is no net external force on the system—bullet + brick. Initially, all of the momentum is in the forward +x-direction, (▼Fig. 6.10). Afterward, one piece of the brick flies off in the +x-direction, and the bullet at an angle of 30° to the x-axis. The bullet's momentum has a positive y-component, so the other piece of the brick must have a negative y-component because there was no initial momentum in the y-direction. Hence, with the total momentum in the +x-direction (before and after), the answer is (3) or fourth quadrant.

(b) Quantitative Reasoning and Solution. There is one object with momentum before collision (the bullet), and three with momenta afterward (the bullet and the two fragments). By the conservation of linear momentum, the total (vector) momentum after collision equals that before collision. As is often the case, a sketch of the situation is helpful, with the vectors resolved in component form (Fig. 6.10). Applying the conservation of linear momentum should allow the velocity (speed and direction) of the second fragment to be determined.

Given: $m_b = 30$ g $= 0.030$ kg
$v_{b_0} = 400$ m/s (initial bullet speed)
$v_b = 100$ m/s (final bullet speed)
$\theta_b = 30°$ (final bullet angle)
$M = 1.0$ kg (brick mass)
$m_1 = 0.75$ kg and $\theta_1 = 0°$ (mass and angle of the large fragment)
$v_1 = 5.0$ m/s
$m_2 = 0.25$ kg (mass of small fragment)

Find: v_2 (speed of the smaller brick fragment)
θ_2 (direction of the fragment relative to the original direction of the bullet)

With no external forces (gravity neglected), the total linear momentum is conserved. Therefore, we can equate both the x- and y-components of the total momentum, before and after, as follows (see Fig. 6.10):

$$x: \quad \overset{before}{m_b v_{b_0}} = \overset{after}{m_b v_b \cos\theta_b + m_1 v_1 + m_2 v_2 \cos\theta_2}$$
$$y: \quad 0 = m_b v_b \sin\theta_b - m_2 v_2 \sin\theta_2$$

(continues on next page)

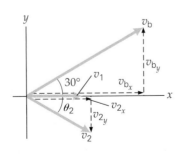

◄ **FIGURE 6.10** A glancing collision Momentum is conserved in an isolated system. The motion in two dimensions may be analyzed in terms of the components of momentum, which are also conserved. See Integrated Example 6.7.

The x-equation can be rearranged to solve for the magnitude of the x-velocity of the smaller fragment:

$$v_2 \cos \theta_2 = \frac{m_b v_{b_o} - m_b v_b \cos \theta_b - m_1 v_1}{m_2}$$

$$= \frac{(3.0 \times 10^{-2}\,\text{kg})(4.0 \times 10^2\,\text{m/s}) - (3.0 \times 10^{-2}\,\text{kg})(10^2\,\text{m/s})(0.866) - (0.75\,\text{kg})(5.0\,\text{m/s})}{0.25\,\text{kg}}$$

$$= 23\,\text{m/s}$$

Similarly, the y-equation can be solved for the magnitude of the y-velocity component of the smaller fragment:

$$v_2 \sin \theta_2 = \frac{m_b v_b \sin \theta_b}{m_2} = \frac{(3.0 \times 10^{-2}\,\text{kg})(10^2\,\text{m/s})(0.50)}{0.25\,\text{kg}} = 6.0\,\text{m/s}$$

Forming a ratio,

$$\frac{v_2 \sin \theta_2}{v_2 \cos \theta_2} = \frac{6.0\,\text{m/s}}{23\,\text{m/s}} = 0.26 = \tan \theta_2$$

(where the v_2 terms cancel, and $\dfrac{\sin \theta_2}{\cos \theta_2} = \tan \theta_2$). Then,

$$\theta_2 = \tan^{-1}(0.26) = 15°$$

and from the x-equation,

$$v_2 = \frac{23\,\text{m/s}}{\cos 15°} = \frac{23\,\text{m/s}}{0.97} = 24\,\text{m/s}$$

Follow-Up Exercise. Is the kinetic energy conserved for the collision in this Example? If not, where did the energy go?

Example 6.8 ■ Physics on Ice

A physicist is lowered from a helicopter to the middle of a smooth, level, frozen lake, the surface of which has negligible friction, and challenged to make her way off the ice. Walking is out of the question. (Why?) As she stands there pondering her predicament, she decides to use the conservation of momentum by throwing her heavy, identical mittens, which will provide her with the momentum to get herself to shore. To get to the shore more quickly, which should this sly physicist do: throw both mittens at once or throw them separately, one after the other with the same speed?

Thinking It Through. The initial momentum of the system (physicist and mittens) is zero. With no net external force, by the conservation of momentum, the total momentum *remains* zero, so if the physicist throws the mittens in one direction, she will go in the opposite direction (because momenta vectors in opposite directions can add to zero). So which way of throwing gives greater speed? If both the mittens were thrown together, the magnitude of their momentum would be $2mv$, where v is relative to the ice and m is the mass of one mitten.

When thrown separately, the first mitten would have a momentum of mv. The physicist and the second mitten would then be in motion, and throwing the second mitten would add some more momentum to the physicist and increase her speed, but would the speed now be greater than that if both mittens were thrown simultaneously? Let's analyze the conditions of the second throw. After she throws the first mitten, the physicist "system" would have less mass. With less mass, the second throw would produce a greater acceleration and speed things up. But on the other hand, after the first throw, the second mitten is moving with the person, and when thrown in the opposite direction, the mitten would have a velocity less than v relative to the ice (or to a stationary observer). So which effect would be greater? What do you think? Sometimes situations are difficult to analyze intuitively, and you must apply scientific principles to figure them out.

Solution.

Given: m = mass of single mitten
M = mass of physicist
$-v$ = velocity of thrown mitten(s), in the negative direction
V_p = velocity of physicist in the positive direction

Find: Which method of mitten throwing gives the physicist the greater speed

When the mittens are thrown together, by the conservation of momentum,

$$0 = 2m(-v) + MV_\mathrm{p} \quad \text{and} \quad V_\mathrm{p} = \frac{2mv}{M} \quad \textit{(thrown together)} \tag{1}$$

When they are thrown separately,

First throw: $0 = m(-v) + (M + m)V_{\mathrm{p}_1}$ and $V_{\mathrm{p}_1} = \dfrac{mv}{M + m}$ *(thrown separately)* (2)

Second throw: $(M + m)V_{\mathrm{p}_1} = m(V_{\mathrm{p}_1} - v) + MV_{\mathrm{p}_2}$

Note that in the last m term, the quantities in the parentheses represent the velocity of the mitten is that relative to the ice. With an initial velocity of $+V_{\mathrm{p}_1}$ when the first mitten is thrown in the negative direction, we have $V_{\mathrm{p}_1} - v$. (Recall relative velocities from Chapter 3.)
Solving for V_{p_2}:

$$V_{\mathrm{p}_2} = V_{\mathrm{p}_1} + \left(\frac{m}{M}\right)v = \frac{mv}{M + m} + \left(\frac{m}{M}\right)v = \left(\frac{m}{M + m} + \frac{m}{M}\right)v \tag{3}$$

where (2) was substituted for V_{p_1} after the first throw.
Now, when the mittens are thrown together (Eq. 1),

$$V_\mathrm{p} = \left(\frac{2m}{M}\right)v$$

so the question is whether the result of (3) is greater or less than that of (1). Notice that with a greater denominator for the $m/(M + m)$ term in (3) it is less than the m/M term. So,

$$\left(\frac{m}{M + m} + \frac{m}{M}\right) < \frac{2m}{M}$$

and therefore, $V_\mathrm{p} > V_{\mathrm{p}_2}$, or (thrown together) > (thrown separately).

Follow-Up Exercise. Suppose the second throw were in the direction of the physicist's velocity from the first throw. Would this throw bring her to a stop?

As mentioned previously, the conservation of momentum is used to analyze the collisions of objects ranging from subatomic particles to automobiles in traffic accidents. In many instances, however, external forces may be acting on the objects, which means that the momentum is not conserved.

But as you will learn in the next section, the conservation of momentum often allows a good approximation *over the short time of a collision*, during which the internal forces (which conserve system momentum) are much greater than the external forces. For example, external forces such as gravity and friction also act on colliding objects, but are often relatively small compared with the internal forces. (This concept was implied in Example 6.7.) Therefore, if the objects interact for only a brief time, the effects of the external forces may be negligible compared with those of the large internal forces during that time and we may correctly use the conservation of linear momentum.

6.4 Elastic and Inelastic Collisions

OBJECTIVE: To describe the conditions on kinetic energy and momentum in elastic and inelastic collisions.

In general, a *collision* maybe defined as a meeting or interaction of particles or objects that causes an exchange of energy and/or momentum. Taking a closer look at collisions in terms of the conservation of momentum is simpler if we consider an isolated system, such as a system of particles (or balls) involved in head-on collisions. For simplicity, we will consider only collisions in one dimension. Such collisions can also be analyzed in terms of the conservation of energy. On the basis of what happens to the total kinetic energy, we can define two types of collisions: *elastic* and *inelastic*.

▶ **FIGURE 6.11** Collisions
(a) Approximate elastic collisions.
(b) An inelastic collision.

(a)   (b)

Elastic collision—total kinetic energy conserved, as is momentum

In an **elastic collision**, the total kinetic energy is conserved. That is, the *total* kinetic energy of all the objects of the system after the collision is the same as the *total* kinetic energy before the collision (▲Fig. 6.11a). Kinetic energy may be traded between objects of a system, but the total kinetic energy in the system remains constant. That is,

$$\text{total } K \text{ after} = \text{total } K \text{ before}$$
$$K_f = K_i$$

(condition for an elastic collision) (6.8)

Exploration 8.4 Elastic and Inelastic Collisions and $\Delta \vec{p}$

During such a collision, some or all of the initial kinetic energy is temporarily converted to potential energy as the objects are deformed. But after the maximum deformations occur, the objects *elastically* "spring" back to their original shapes, and the system regains all of its original kinetic energy. For example, two steel balls or two billiard balls may have a nearly elastic collision, with each ball having the same shape afterward as before; that is, there is no permanent deformation.

Inelastic collision—total kinetic energy not conserved, but momentum is

In an **inelastic collision** (Fig. 6.11b), total kinetic energy is *not* conserved. For example, one or more of the colliding objects may not regain the original shapes, and/or sound or frictional heat may be generated and some kinetic energy is lost. Then,

$$\text{total } K \text{ after} < \text{total } K \text{ before}$$
$$K_f < K_i$$

(condition for an inelastic collision) (6.9)

Note: In reality, only atoms and subatomic particles have truly elastic collisions, but some larger hard objects have nearly elastic collisions in which the kinetic energy is approximately conserved.

For example, a hollow aluminum ball that collides with a solid steel ball may be dented. Permanent deformation of the ball takes work, and that work is done at the expense of the original kinetic energy of the system. Everyday collisions are inelastic.

For isolated systems, momentum is conserved in both elastic and inelastic collisions. *For an inelastic collision, only an amount of kinetic energy consistent with the conservation of momentum may be lost.* It may seem strange that kinetic energy can be lost and momentum still conserved, but this fact provides insight into the difference between scalar and vector quantities and the differences in their conservation requirements.

Momentum and Energy in Inelastic Collisions

To see how momentum can remain constant while the kinetic energy changes (decreases) in inelastic collisions, consider the examples illustrated in ▼Fig. 6.12. In Fig. 6.12a, two balls of equal mass ($m_1 = m_2$) approach each other with equal and opposite velocities ($v_{1_o} = -v_{2_o}$). Hence, the total momentum before the collision is (vectorially) zero, but the (scalar) total kinetic energy is *not* zero. After the collision, the balls are stuck together and stationary, so the total momentum is unchanged—still zero. Momentum is conserved because the forces of collision are internal to the system of the two balls—there is no net external force on the system. The total kinetic energy, however, has decreased to zero. In this case, some of the kinetic energy went into the work done in permanently deforming the balls.

Before   *Collision*   *After*

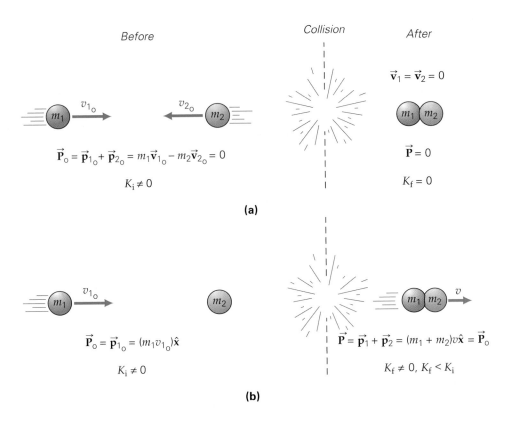

$$\vec{v}_1 = \vec{v}_2 = 0$$

$$\vec{P}_o = \vec{p}_{1_o} + \vec{p}_{2_o} = m_1\vec{v}_{1_o} - m_2\vec{v}_{2_o} = 0$$

$$\vec{P} = 0$$

$$K_i \neq 0$$

$$K_f = 0$$

(a)

$$\vec{P}_o = \vec{p}_{1_o} = (m_1 v_{1_o})\hat{\mathbf{x}}$$

$$\vec{P} = \vec{p}_1 + \vec{p}_2 = (m_1 + m_2)v\hat{\mathbf{x}} = \vec{P}_o$$

$$K_i \neq 0$$

$$K_f \neq 0, K_f < K_i$$

(b)

◀ **FIGURE 6.12** Inelastic collisions In inelastic collisions, momentum is conserved, but kinetic energy is not. Collisions like the ones shown here, in which the objects stick together, are called *completely* or *totally inelastic collisions*. The maximum amount of kinetic energy lost is consistent with the law of conservation of momentum.

Some energy may also have gone into doing work against friction (producing heat) or may have been lost in some other way (for example, in producing sound).

It should be noted that the balls need not stick together after collision. In a less inelastic collision, the balls may recoil in opposite directions at reduced, but equal, speeds. The momentum would still be conserved (still equal to zero—why?), but the kinetic energy would again not be conserved. Under all conditions, the amount of kinetic energy that can be lost must be consistent with the conservation of momentum.

In Fig. 6.12b, one ball is initially at rest as the other approaches. The balls stick together after collision, but are still in motion. Both of these cases are examples of a **completely inelastic collision**, in which the objects stick together, and hence both objects have the same velocity after colliding. The coupling of colliding railroad cars is a practical example of a completely inelastic collision.

Assume that the balls in Fig. 6.12b have different masses. Since the momentum is conserved even in inelastic collisions,

$$\underset{before}{m_1 v_{1_o}} = \underset{after}{(m_1 + m_2)v}$$

and

$$v = \left(\frac{m_1}{m_1 + m_2}\right)v_{1_o} \quad \begin{array}{l} \textit{(m$_2$ initially at rest,} \\ \textit{completely inelastic collision only)} \end{array} \quad (6.10)$$

Thus, v is less than v_{1_o}, since $m_1/(m_1 + m_2)$ must be less than 1. Now consider how much kinetic energy has been lost. Initially, $K_i = \frac{1}{2}m_1 v_o^2$, and after collision the final kinetic energy is:

$$K_f = \tfrac{1}{2}(m_1 + m_2)v^2$$

Substituting for v from Eq. 6.10 and simplifying the result, we have

$$K_f = \tfrac{1}{2}(m_1 + m_2)\left(\frac{m_1 v_{1_o}}{m_1 + m_2}\right)^2 = \frac{\frac{1}{2}m_1^2 v_{1_o}^2}{m_1 + m_2}$$

$$= \left(\frac{m_1}{m_1 + m_2}\right)\tfrac{1}{2}m_1 v_{1_o}^2 = \left(\frac{m_1}{m_1 + m_2}\right)K_i$$

and

$$\frac{K_f}{K_i} = \frac{m_1}{m_1 + m_2} \qquad \begin{array}{l}(m_2 \text{ initially at rest,} \\ \text{completely inelastic collision only})\end{array} \qquad (6.11)$$

Equation 6.11 gives the fractional amount of the initial kinetic energy that remains with the system after a completely inelastic collision. For example, if the masses of the balls are equal ($m_1 = m_2$), then $m_1/(m_1 + m_2) = \frac{1}{2}$, and $K_f/K_i = \frac{1}{2}$, or $K_f = K_i/2$. That is, half of the initial kinetic energy is lost.

Note that not all of the kinetic energy can be lost in this case, no matter what the masses of the balls are. The total momentum after collision cannot be zero, since it was not zero initially. Thus, after the collision, the balls must be moving and must have some kinetic energy ($K_f \neq 0$). *In a completely inelastic collision, the maximum amount of kinetic energy lost must be consistent with the conservation of momentum.*

PHYSLET®

Exploration 8.3 An Inelastic Collision with Unknown Masses

Example 6.9 ■ Stuck Together: Completely Inelastic Collision

A 1.0-kg ball with a speed of 4.5 m/s strikes a 2.0-kg stationary ball. If the collision is completely inelastic, (a) what are the speeds of the balls after the collision? (b) What percentage of the initial kinetic energy do the balls have after the collision? (c) What is the total momentum after the collision?

Thinking It Through. Recall the definition of a *completely inelastic collision*. The balls stick together after collision; kinetic energy is *not* conserved, but total momentum is.

Solution. Using the labeling as in the preceding discussion, we have

Given: $m_1 = 1.0$ kg   *Find:* (a) v (speed after collision)
$m_2 = 2.0$ kg
$v_o = 4.5$ m/s   (b) $\dfrac{K_f}{K_i}$ ($\times 100\%$)

                                      (c) $\vec{P}_f$ (total momentum after collision)

(a) The momentum is conserved and

$$\vec{P}_f = \vec{P}_o \qquad \text{or} \qquad (m_1 + m_2)v = m_1 v_o$$

The balls stick together and have the same speed after collision. This speed is then

$$v = \left(\frac{m_1}{m_1 + m_2}\right)v_o = \left(\frac{1.0 \text{ kg}}{1.0 \text{ kg} + 2.0 \text{ kg}}\right)(4.5 \text{ m/s}) = 1.5 \text{ m/s}$$

(b) The fractional part of the initial kinetic energy that the balls have after the completely inelastic collision is given by Eq. 6.11. Notice that this fraction, as given by the masses, is the same as that for the speeds (Eq. 6.10) in this special case. By inspection, we can write

$$\frac{K_f}{K_i} = \frac{m_1}{m_1 + m_2} = \frac{1.0 \text{ kg}}{1.0 \text{ kg} + 2.0 \text{ kg}} = \frac{1}{3} = 0.33 (\times 100\%) = 33\%$$

Let's show this relationship explicitly:

$$\frac{K_f}{K_i} = \frac{\frac{1}{2}(m_1 + m_2)v^2}{\frac{1}{2}m_1 v_o^2} = \frac{\frac{1}{2}(1.0 \text{ kg} + 2.0 \text{ kg})(1.5 \text{ m/s})^2}{\frac{1}{2}(1.0 \text{ kg})(4.5 \text{ m/s})^2} = 0.33 \ (= 33\%)$$

Keep in mind that Eq. 6.11 applies *only* to *completely* inelastic collisions in which m_2 is initially at rest. For other types of collisions, the initial and final values of the kinetic energy must be computed explicitly.

(c) The total momentum is conserved in all collisions (in the absence of external forces), so the total momentum after collision is the same as before collision. That value is the momentum of the incident ball, with a magnitude of

$$P_f = p_{1_o} = m_1 v_o = (1.0 \text{ kg})(4.5 \text{ m/s}) = 4.5 \text{ kg} \cdot \text{m/s}$$

and the same direction as that of the incoming ball. Also, as a double check,

$$P_f = (m_1 + m_2)v = 4.5 \text{ kg} \cdot \text{m/s}.$$

Follow-Up Exercise. A small hard-metal ball of mass m collides with a larger, stationary, soft-metal ball of mass M. A *minimum* amount of work W is required to make a dent in the larger ball. If the smaller ball initially has kinetic energy $K = W$, will the larger ball be dented in a completely inelastic collision between the two balls?

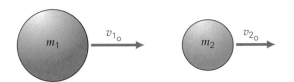

◀ **FIGURE 6.13** Elastic collision coming up Two objects traveling prior to collision with $v_{1_0} > v_{2_0}$. See text for description

Momentum and Energy in Elastic Collisions

For *elastic* collisions, there are two conservation criteria: conservation of momentum (which holds for both elastic and inelastic collisions) and conservation of kinetic energy (for elastic collisions only). That is, for the elastic collision of two objects:

$$\text{Conservation of momentum } \vec{\mathbf{P}}: \qquad \overset{before}{m_1\vec{\mathbf{v}}_{1_0} + m_2\vec{\mathbf{v}}_{2_0}} = \overset{after}{m_1\vec{\mathbf{v}}_1 + m_2\vec{\mathbf{v}}_2} \qquad (6.12)$$

$$\text{Conservation of kinetic energy } K: \quad \tfrac{1}{2}m_1v_{1_0}^2 + \tfrac{1}{2}m_2v_{2_0}^2 = \tfrac{1}{2}m_1v_1^2 + \tfrac{1}{2}m_2v_2^2 \qquad (6.13)$$

▲Figure 6.13 illustrates two objects traveling prior to a one-dimensional, head-on collision with $v_{1_0} > v_{2_0}$ (both in the positive x-direction). For this two-object situation we may write

$$\text{Total momentum:} \qquad \overset{before}{m_1v_{1_0} + m_2v_{2_0}} = \overset{after}{m_1v_1 + m_2v_2} \qquad (1)$$

(where signs are used to indicate directions and the v's indicate magnitudes).

$$\text{Kinetic energy:} \qquad \tfrac{1}{2}m_1v_{1_0}^2 + \tfrac{1}{2}m_2v_{2_0}^2 = \tfrac{1}{2}m_1v_1^2 + \tfrac{1}{2}m_2v_2^2 \qquad (2)$$

If the masses and the initial velocities of the objects are known (which we usually do), then there are two unknown quantities, the final velocities after collision. To find them, equations (1) and (2) are solved simultaneously. First the equation for momentum conservation is written as follows:

$$m_1(v_{1_0} - v_1) = -m_2(v_{2_0} - v_2) \qquad (3)$$

Then, canceling the $\frac{1}{2}$ terms in (2), rearranging, and factoring $[a^2 - b^2 = (a - b)(a + b)]$:

$$m_1(v_{1_0} - v_1)(v_{1_0} + v_1) = -m_2(v_{2_0} - v_2)(v_{2_0} + v_2) \qquad (4)$$

Dividing equation (4) by (3) are rearranging yields

$$v_{1_0} - v_{2_0} = -(v_1 - v_2) \qquad (5)$$

This equation shows that the magnitudes of the relative velocities before and after collision are equal. That is, the relative speed of approach of object m_1 to object m_2 before collision is the same as their relative speed of separation after collision. (See Section 3.4.) Notice that this relation is independent of the values of the masses of the objects, and holds for any mass combination as long as the collision is elastic and *one-dimensional*.

Then, combining equation (5) with (3) to eliminate v_2 and get v_1 in terms of the two initial velocities,

$$v_1 = \left(\frac{m_1 - m_2}{m_1 + m_2}\right)v_{1_0} + \left(\frac{2m_2}{m_1 + m_2}\right)v_{2_0} \qquad (6.14)$$

Similarly, eliminating v_1 to find v_2,

$$v_2 = \left(\frac{2m_1}{m_1 + m_2}\right)v_{1_0} - \left(\frac{m_1 - m_2}{m_1 + m_2}\right)v_{2_0} \qquad (6.15)$$

PHYSLET®

Illustration 8.4 Relative Velocity in Collisions

One Object Initially at Rest

For this common, special case, say with $v_{2_0} = 0$, we have only the first terms in Eqs. 6.14 and 6.15. In addition, if $m_1 = m_2$, then $v_1 = 0$ and $v_2 = v_{1_0}$. That is, the objects *completely* exchange momentum and kinetic energy. The incoming object is stopped on collision, and the originally stationary object moves off with the same velocity as the incoming ball, obviously conserving the system's momentum and kinetic energy. (A real-world example that comes close to these conditions is the head-on collision of billiard balls.)

You can also get some approximates for special cases from the equations for one object initially at rest (taken to be m_2):

For $m_1 \gg m_2$ (massive incoming ball): $\quad v_1 \approx v_{1_o} \quad$ and $\quad v_2 \approx 2v_{1_o}$

That is, the massive incoming object is slowed down only slightly and the light (less massive) object is knocked away with a velocity almost twice that of the initial velocity of the massive object. (Think of a bowling ball hitting a pin.)

For $m_1 \ll m_2$ (light incoming ball): $\quad v_1 \approx -v_{1_o} \quad$ and $\quad v_2 \approx 0$

That is, if a light (small mass) object elastically collides with a massive stationary one, the massive object remains *almost* stationary and the light object recoils backward with approximately the same speed that it had before collision.

Example 6.10 ■ Elastic Collision: Conservation of Momentum and Kinetic Energy

A 0.30-kg billiard ball with a speed of 2.0 m/s in the positive x-direction has a head-on elastic collision with a stationary 0.70-kg billiard ball. What are the velocities of the balls after collision?

Thinking It Through. The incoming ball is less massive than the stationary one, so we might expect the objects to separate in opposite directions after collision, the less massive ball recoiling from the more massive one. Equations 6.14 and 6.15 can be used to find the velocities with $v_{2_o} = 0$.

Solution. Using the notation used previously, we have

Given: $\quad m_1 = 0.30$ kg and $v_{1_o} = 2.0$ m/s $\qquad$ *Find:* $\quad v_1$ and v_2
$\qquad\qquad m_2 = 0.70$ kg and $v_{2_o} = 0$

Directly from Eqs. 6.13 and 6.14, the velocities after collision are

$$v_1 = \left(\frac{m_1 - m_2}{m_1 + m_2}\right)v_{1_o} = \left(\frac{0.30 \text{ kg} - 0.70 \text{ kg}}{0.30 \text{ kg} + 0.70 \text{ kg}}\right)(2.0 \text{ m/s}) = -0.80 \text{ m/s}$$

$$v_2 = \left(\frac{2m_1}{m_1 + m_2}\right)v_{1_o} = \left[\frac{2(0.30 \text{ kg})}{0.30 \text{ kg} + 0.70 \text{ kg}}\right](2.0 \text{ m/s}) = 1.2 \text{ m/s}$$

Follow-Up Exercise. What would be the separation distance of the two objects 2.5 s after collision?

PHYSLET®

Exploration 8.2 An Elastic Collision

Two Colliding Objects, Both Initially Moving

Now let's look at some examples where both terms in Eq. 6.14 and 6.15 are needed.

Example 6.11 ■ Collisions: Overtaking and Coming Together

The precollision conditions for two elastic collisions are shown in ▶Fig. 6.14. What are the final velocities in each case?

Thinking It Through. These collisions are direct applications of Eq. 6.14 and 6.15. Notice that in (a) the 4.0-kg object will overtake and collide with the 1.0-kg object.

Solution. Listing the data from the figure with the +x-direction taken to the right.

Given: (a) $m_1 = 4.0$ kg $\qquad v_{1_o} = 10$ m/s $\qquad$ *Find:* $\quad v_1$ and v_2 (velocities after collision)
$\qquad\qquad m_2 = 1.0$ kg $\qquad v_{2_o} = 5.0$ m/s
$\qquad$ (b) $m_1 = 2.0$ kg $\qquad v_{1_o} = 6.0$ m/s
$\qquad\qquad m_2 = 4.0$ kg $\qquad v_{2_o} = -6.0$ m/s

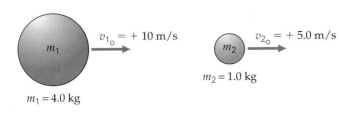

◀ **FIGURE 6.14** Collisions: (a) Overtaking and (b) coming together See Example 6.11.

Then, substituting into the collision equations,

(a) Eq. 6.14:

$$v_1 = \left(\frac{m_1 - m_2}{m_1 + m_2}\right)v_{1_0} + \left(\frac{2m_2}{m_1 + m_2}\right)v_{2_0}$$

$$= \left(\frac{4.0\ \text{kg} - 1.0\ \text{kg}}{4.0\ \text{kg} + 1.0\ \text{kg}}\right)10\ \text{m/s} + \left(\frac{2[1.0\ \text{kg}]}{4.0\ \text{kg} + 1.0\ \text{kg}}\right)5.0\ \text{m/s}$$

$$= \tfrac{3}{5}(10\ \text{m/s}) + \tfrac{2}{5}(5.0\ \text{m/s}) = 8.0\ \text{m/s}$$

Similarly, Eq. 6.15 gives:

$$v_2 = 13\ \text{m/s}$$

So the more massive object overtakes and collides with the less massive object, transferring momentum (increasing velocity).

(b) Applying the collision equations for this situation, we have (Eq. 6.14):

$$v_1 = \left(\frac{2.0\ \text{kg} - 4.0\ \text{kg}}{2.0\ \text{kg} + 4.0\ \text{kg}}\right)6.0\ \text{m/s} + \left(\frac{2[4.0\ \text{kg}]}{2.0\ \text{kg} + 4.0\ \text{kg}}\right)(-6.0\ \text{kg})$$

$$= -\left(\tfrac{1}{3}\right)6.0\ \text{m/s} + \left(\tfrac{4}{3}\right)(-6.0\ \text{m/s}) = -10\ \text{m/s}$$

Similarly, Eq. 6.15 gives

$$v_2 = 2.0\ \text{m/s}$$

Here, the less massive object goes in the opposite (negative) direction after collision, with a greater momentum obtained from the more massive object.

Follow-Up Exercise. Show that in parts (a) and (b) of this Exercise, the amount momentum gained by one object is the same as that lost by the other.

Integrated Example 6.12 ■ Equal and Opposite

Two balls of equal mass with equal but opposite velocities approach each other for a head-on elastic collision. (a) After collision, the balls will (1) move off stuck together, (2) both be at rest, (3) move off in the same direction, or (4) recoil in opposite directions. (b) Prove your answer explicitly.

(a) Conceptual Reasoning. Make a sketch of the situation. Then, looking at the choices, (1) is eliminated because if they stuck together it would be an inelastic collision. If they come to rest after collision, momentum would be conserved (why?), but not kinetic energy, so (2) is not applicable for an elastic collision. If they both moved off in the same direction after collision, the momentum would not be conserved (zero before, nonzero after). The answer must be (4). This is the only option by which momentum and kinetic energy could be conserved. To maintain the zero momentum before collision, the objects would have to recoil in opposite directions with the same speeds as before collision.

(continues on next page)

(b) Quantitative Reasoning and Solution. To explicitly show that (4) is correct, Eqs. 6.13 and 6.14 may be used. Since no numerical values are given, we work with symbols.

Given: $m_1 = m_2 = m$ (taking m_1 to be initially traveling in the $+x$ direction:)
v_{1_o} and $-v_{2_o}$ (with equal speeds)

Find: v_1 and v_2

Then, substituting into Eqs. 6.14 and 6.15, without writing the equations out [see part (a) in Example 6.11],

$$v_1 = \left(\frac{0}{2m}\right)v_{1_o} + \left(\frac{2m}{2m}\right)(-v_{2_o}) = -v_{2_o}$$

and

$$v_2 = \left(\frac{2m}{2m}\right)v_{1_o} + \left(\frac{0}{2m}\right)(-v_{2_o}) = v_{1_o}$$

From the results, it can be seen that after collision the balls recoil in opposite directions.

Follow-Up Exercise. Show that momentum and kinetic energy are conserved in this Example.

Exploration 8.5 Two and Three Ball Collisions

Conceptual Example 6.13 ■ Two In, One Out?

A novelty collision device, as shown in ◄Fig. 6.15, consists of five identical metal balls. When one ball swings in, after multiple collisions, one ball swings out at the other end of the row of balls. When two balls swing in, two swing out; when three swing in, three swing out, and so on—always the same number out as in.

Suppose that two balls, each of mass m, swing in at velocity v and collide with the next ball. Why doesn't one ball swing out at the other end with a velocity $2v$?

Reasoning and Answer. The collisions along the horizontal row of balls are approximately elastic. The case of two balls swinging in and one ball swinging out with twice the velocity wouldn't violate the conservation of momentum: $(2m)v = m(2v)$. However, another condition applies if we assume elastic collisions—the conservation of kinetic energy. Let's check to see if this condition is upheld for this case:

$$\begin{array}{cc} before & after \\ K_i & = K_f \end{array}$$

$$\tfrac{1}{2}(2m)v^2 \overset{?}{=} \tfrac{1}{2}m(2v)^2$$

$$mv^2 \neq 2mv^2$$

Hence, the kinetic energy would *not* be conserved if this happened, and the equation is telling us that this situation violates established physical principles and does not occur. Note that there's a big violation—more energy out than in.

Follow-Up Exercise. Suppose the first ball of mass m were replaced with a ball of mass $2m$. When this ball is pulled back and allowed to swing in, how many balls will swing out? [*Hint*: Think about the analogous situation for the first two balls as in Fig. 6.14a, and remember that the balls in the row are actually colliding. It may help to think of them as being separated.]

▲ **FIGURE 6.15** One in, one out See Conceptual Example 6.13.

6.5 Center of Mass

OBJECTIVES: To (a) explain the concept of the center of mass and compute its location for simple systems, and (b) describe how the center of mass and center of gravity are related.

The conservation of total momentum gives us a method of analyzing a "system of particles." Such a system may be virtually anything—for example, a volume of gas, water in a container, or a baseball. Another important concept, the center of mass, allows us to analyze the overall motion of a system of particles. It involves representing the whole system as a single particle or point mass. This concept will be introduced here and applied in more detail in the upcoming chapters.

We have seen that if no net external force acts on a particle, the particle's linear momentum is constant. Similarly, if no net external force acts on a *system* of particles, the linear momentum of the system is constant. This similarity implies that a system of particles might be represented by an *equivalent* single particle. Moving rigid objects, such as balls, automobiles, and so forth, are essentially systems of particles and can be effectively represented by equivalent single particles when we analyze motion. Such representation is done through the concept of the **center of mass (CM)**:

> The center of mass is the point at which all of the mass of an object or system may be considered to be concentrated, for the purposes of linear or translational motion only.

Even if a rigid object is rotating, an important result (beyond the scope of this text to derive) is that the center of mass still moves as though it were a particle (▼Fig. 6.16). The center of mass is sometimes described as the *balance point* of a solid object. For example, if you balance a meterstick on your finger, the center of mass of the stick is located directly above your finger, and all of the mass (or weight) seems to be concentrated there.

An expression similar to Newton's second law for a single particle applies to a *system* when the center of mass is used:

$$\vec{F}_{net} = M\vec{A}_{CM} \tag{6.16}$$

Here, $\vec{F}_{net}$ is the net *external* force on the system, M is the total mass of the system or the sum of the masses of the particles of the system ($M = m_1 + m_2 + m_3 + \cdots + m_n$, where the system has n particles), and $\vec{A}_{CM}$ is the acceleration of the center of mass of the system. In words, Eq. 6.16 says that the *center of mass* of a system of particles moves as though all the mass of the system were concentrated there and acted on by the resultant of the external forces. Note that the movement of the individual parts of the system is *not* predicted by Eq. 6.16.

It follows that *if the net external force on a system is zero*, the total linear momentum of the center of mass is conserved (that is, it stays constant), because

$$\vec{F}_{net} = M\vec{A}_{CM} = M\left(\frac{\Delta\vec{V}_{CM}}{\Delta t}\right) = \frac{\Delta(M\vec{V}_{CM})}{\Delta t} = \frac{\Delta\vec{P}}{\Delta t} = 0 \tag{6.17}$$

Then, $\Delta\vec{P}/\Delta t = 0$, which means that there is no change in $\vec{P}$ during a time Δt, or the total momentum of the system, $\vec{P} = M\vec{V}_{CM}$, is constant (but not necessarily zero). Since M is constant (why?), $\vec{V}_{CM}$ is a constant in this case. Thus, the center of mass either moves with a constant velocity or remains at rest.

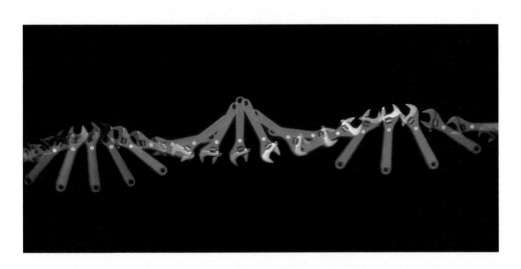

◄ **FIGURE 6.16 Center of mass** The center of mass of this sliding wrench moves in a straight line as though it were a particle. Note the white dot on the wrench that marks the center of mass.

PHYSLET®

Illustration 8.7 *Center of Mass and Gravity*

Although you may more readily visualize the center of mass of a solid object, the concept of the center of mass applies to any system of particles or objects, even a quantity of gas. For a system of n particles arranged in one dimension, along the x-axis (▾Fig. 6.17), the location of the center of mass is given by

$$\vec{X}_{CM} = \frac{m_1\vec{x}_1 + m_2\vec{x}_2 + m_3\vec{x}_3 + \cdots + m_n\vec{x}_n}{m_1 + m_2 + m_3 + \cdots + m_n} \tag{6.18}$$

That is, X_{CM} is the x-coordinate of the center of mass of a system of particles. In shorthand notation (using signs to indicate vector directions in one dimension), this relationship is expressed as

$$X_{CM} = \frac{\sum m_i x_i}{M} \tag{6.19}$$

where $\sum$ is the summation of the products $m_i x_i$ for n particles ($i = 1, 2, 3, \ldots, n$). If $\sum m_i x_i = 0$, then $X_{CM} = 0$, and the center of mass of the one-dimensional system is located at the origin.

Other coordinates of the center of mass for systems of particles are similarly defined. For a two-dimensional distribution of masses, the coordinates of the center of mass are (X_{CM}, Y_{CM}).

Example 6.14 ■ Finding the Center of Mass: A Summation Process

Three masses—2.0 kg, 3.0 kg, and 6.0 kg—are located at positions (3.0, 0), (6.0, 0), and (−4.0, 0), respectively, in meters from the origin (Fig. 6.17). Where is the center of mass of this system?

Thinking It Through. Since all $y_i = 0$, obviously $Y_{CM} = 0$, and the CM lies somewhere on the x-axis. The masses and the positions are given, so we can use Eq. 6.19 to calculate X_{CM} directly. However, keep in mind that the positions are located by vector displacements from the origin and are indicated in one dimension by the appropriate signs (+ or −).

Solution. Listing the data,

Given: $m_1 = 2.0 \text{ kg}$ *Find:* X_{CM} (CM coordinate)
$m_2 = 3.0 \text{ kg}$
$m_3 = 6.0 \text{ kg}$
$x_1 = 3.0 \text{ m}$
$x_2 = 6.0 \text{ m}$
$x_3 = -4.0 \text{ m}$

Then, performing the summation as indicated in Eq. 6.19 yields

$$X_{CM} = \frac{\sum m_i x_i}{M}$$

$$= \frac{(2.0 \text{ kg})(3.0 \text{ m}) + (3.0 \text{ kg})(6.0 \text{ m}) + (6.0 \text{ kg})(-4.0 \text{ m})}{2.0 \text{ kg} + 3.0 \text{ kg} + 6.0 \text{ kg}} = 0$$

The center of mass is at the origin.

Follow-Up Exercise. At what position should a fourth mass of 8.0 kg be added so the CM is at $x = +1.0 \text{ m}$?

▶ **FIGURE 6.17 System of particles in one dimension** Where is the system's center of mass? See Example 6.14.

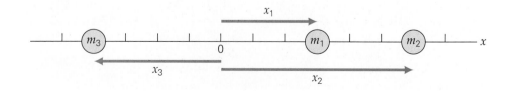

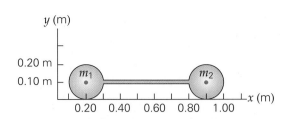

◀ **FIGURE 6.18** Location of the
center of mass See Example 6.15.

Example 6.15 ■ A Dumbbell: Center of Mass Revisited

A dumbbell (▲ Fig. 6.18) has a connecting bar of negligible mass. Find the location of the center of mass (a) if m_1 and m_2 are each 5.0 kg and (b) if m_1 is 5.0 kg and m_2 is 10.0 kg.

Thinking It Through. This Example shows how the location of the center of mass depends on the distribution of mass. In part (b), you might expect the center of mass to be located closer to the more massive end of the dumbbell.

Solution. Listing the data, with the coordinates to be used in Eq. 6.19,

Given: $x_1 = 0.20$ m   *Find:* (a) (X_{CM}, Y_{CM}) (CM coordinates), with $m_1 = m_2$
             $x_2 = 0.90$ m   (b) (X_{CM}, Y_{CM}), with $m_1 \neq m_2$
             $y_1 = y_2 = 0.10$ m
     (a) $m_1 = m_2 = 5.0$ kg
     (b) $m_1 = 5.0$ kg
         $m_2 = 10.0$ kg

Note that each mass is considered to be a particle located at the center of the sphere (its center of mass).

(a) X_{CM} is given by a two-term sum.

$$X_{CM} = \frac{m_1 x_1 + m_2 x_2}{m_1 + m_2}$$

$$= \frac{(5.0 \text{ kg})(0.20 \text{ m}) + (5.0 \text{ kg})(0.90 \text{ m})}{5.0 \text{ kg} + 5.0 \text{ kg}} = 0.55 \text{ m}$$

Similarly, we find that $Y_{CM} = 0.10$ m. (You might have seen this right away, since each center of mass is at this height.) The center of mass of the dumbbell is then located at $(X_{CM}, Y_{CM}) = (0.55 \text{ m}, 0.10 \text{ m})$, or midway between the end masses.

(b) With $m_2 = 10.0$ kg,

$$X_{CM} = \frac{m_1 x_1 + m_2 x_2}{m_1 + m_2}$$

$$= \frac{(5.0 \text{ kg})(0.20 \text{ m}) + (10.0 \text{ kg})(0.90 \text{ m})}{5.0 \text{ kg} + 10.0 \text{ kg}} = 0.67 \text{ m}$$

which is two thirds of the way between the masses. (Note that the distance of the CM from the center of m_1 is $\Delta x = 0.67$ m $-$ 0.20 m $=$ 0.47 m. With the distance $L = 0.70$ m between the centers of the masses, $\Delta x/L = 0.47$ m/0.70 m $=$ 0.67, or $\frac{2}{3}$.) You might expect the balance point of the dumbbell in this case to be closer to m_2, and it is. The y-coordinate of the center of mass is again $Y_{CM} = 0.10$ m, as you can prove for yourself.

Follow-Up Exercise. In part (b) of this Example, take the origin of the coordinate axes to be at the point where m_1 touches the x-axis. What are the coordinates of the CM in this case, and how does its location compare with that found in the Example?

In Example 6.15, when the value of one of the masses changed, the x-coordinate of the center of mass changed. You might have expected the y-coordinate to change also. However, the centers of the end masses were still at the same height, and Y_{CM} remained the same. To increase Y_{CM}, one or both of the end masses would have to be in a higher position.

Now let's see how the concept of the center of mass can be applied to a realistic situation.

▶ **FIGURE 6.19** Walking toward shore See Example 6.16.

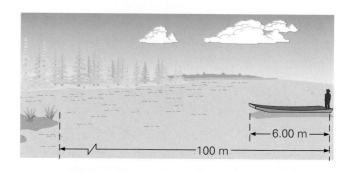

Illustration 8.8 Moving Objects and Center of Mass

Integrated Example 6.16 ■ Internal Motion: Where's the Center of Mass and the Man?

A 75.0-kg man stands in the far end of a 50.0-kg boat 100 m from the shore, as illustrated in ▲ Fig. 6.19. If he walks to the other end of the 6.00-m-long boat, (a) does the CM (1) move to the right, (2) move to the left, or (3) remain stationary? Neglect friction and assume the CM of the boat is at its midpoint. (b) After walking to the other end of the boat, how far is he from the shore?

(a) Conceptual Reasoning. With no net external force, the acceleration of the center of mass of the man–boat system is zero (Eq. 6.18), and so is the total momentum by Eq. 6.17 ($\vec{P} = M\vec{V}_{CM} = 0$). Hence, the velocity of the center of mass of the system is zero, or the center of mass is stationary and remains so to conserve system momentum; that is, X_{CM_i} (initial) $= X_{CM_f}$ (final), so the answer is (3).

(b) Quantitative Reasoning and Solution. The answer is *not* 100 m − 6.00 m = 94.0 m, because the boat moves as the man walks. Why? The positions of the masses of the man and the boat determine the location of the CM of the system, both before and after the man walks. Since the CM does not move, we know that $X_{CM_i} = X_{CM_f}$. Using this fact and finding the value of X_{CM_i}, this value can be used in the calculation of X_{CM_f}, which will contain the unknown we are looking for.

Taking the shore as the origin ($x = 0$),

Given: $m_m = 75.0$ kg   **Find:** x_{m_f} (distance of man from shore)
$\quad\quad x_{m_i} = 100$ m
$\quad\quad m_b = 50.0$ kg
$\quad\quad x_{b_i} = 94.0$ m + 3.00 m = 97.0 m (CM position of the boat)

Note that if we take the man's final position to be a distance x_{m_f} from the shore, then the final position of the boat's center of mass will be $x_{b_f} = x_{m_f} + 3.00$ m, since the man will be at the front of the boat, 3.00 m from its CM, but on the other side.

Then initially,

$$X_{CM_i} = \frac{m_m x_{m_i} + m_b x_{b_i}}{m_m + m_b}$$

$$= \frac{(75.0\text{ kg})(100\text{ m}) + (50.0\text{ kg})(97.0\text{ m})}{75.0\text{ kg} + 50.0\text{ kg}} = 98.8\text{ m}$$

Finally, the CM must be at the same location, since $V_{CM} = 0$. From Eq. 6.19 we have

$$X_{CM_f} = \frac{m_m x_{m_f} + m_b x_{b_f}}{m_m + m_b}$$

$$= \frac{(75.0\text{ kg})x_{m_f} + (50.0\text{ kg})(x_{m_f} + 3.00\text{ m})}{75.0\text{ kg} + 50.0\text{ kg}} = 98.8\text{ m}$$

Here, $X_{CM_f} = 98.8$ m $= X_{CM_i}$, since the CM does not move. Then, solving for x_{m_f}, we have

$$(125\text{ kg})(98.8\text{ m}) = (125\text{ kg})x_{m_f} + (50.0\text{ kg})(3.00\text{ m})$$

and

$$x_{m_f} = 97.6\text{ m}$$

from the shore.

Follow-Up Exercise. Suppose the man then walks back to his original position at the opposite end of the boat. Would he then be 100 m from shore again?

Center of Gravity

As you know, mass and weight are related. Closely associated with the concept of the center of mass is the concept of the **center of gravity (CG)**, the point where all of the *weight* of an object may be considered to be concentrated when the object is represented as a particle. If the acceleration due to gravity is constant both in magnitude and direction over the extent of the object, Eq. 6.20 can be rewritten (with all $g_i = g$) as

$$MgX_{CM} = \sum m_i g x_i \qquad (6.20)$$

Then, the object's weight Mg acts as though its mass were concentrated at X_{CM}, and the center of mass and the center of gravity coincide. As you may have noticed, the location of the center of gravity was implied in some previous figures in Chapter 4, where the vector arrows for weight ($w = mg$) were drawn from a point at or near the center of an object. For practical purposes, the center of gravity is considered to coincide with the center of mass. That is, the acceleration due to gravity is constant for all parts of the object. (Note the constant g in Eq. 6.20.) There would be a difference in the locations of the two points if an object were so large that the acceleration due to gravity were different at different parts of the object.

In some cases, the center of mass or the center of gravity of an object may be located by symmetry. For example, for a spherical object that is homogeneous (that is, the mass is distributed evenly throughout), the center of mass is at the geometrical center (or center of symmetry). In Example 6.15a, where the end masses of the dumbbell were equal, it was probably apparent that the center of mass was midway between them.

The location of the center of mass or center of gravity of an irregularly shaped object is not so evident and is usually difficult to calculate (even with advanced mathematical methods that are beyond the scope of this book). In some instances, the center of mass may be located experimentally. For example, the center of mass of a flat, irregularly shaped object can be determined experimentally by suspending it freely from different points (▼Fig. 6.20). A moment's thought should convince

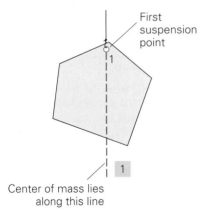

First suspension point

Center of mass lies along this line

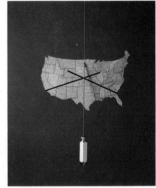

(b)

Second suspension point

CM

Center of mass also lies along this line

(a)

▲ **FIGURE 6.20 Location of the center of mass by suspension (a)** The center of mass of a flat, irregularly shaped object can be found by suspending the object from two or more points. The CM (and CG) lies on a vertical line under any point of suspension, so the intersection of two such lines marks its location midway through the thickness of the body. The sheet could be balanced horizontally at this point. Why? **(b)** The process is illustrated with a cutout map of the United States. Note that a plumb line dropped from any other point (third photo) does in fact pass through the CM as located in the first two photos.

(a)

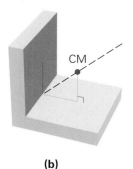

(b)

▲ **FIGURE 6.21 The center of mass may be located outside a body** The center of mass (and center of gravity) may lie either inside or outside a body, depending on the distribution of that object's mass. **(a)** For a uniform ring, the center of mass is at the center of the ring. **(b)** For an L-shaped object, if the mass distribution is uniform and the legs are of equal length, the center of mass lies on the diagonal between the legs.

▼ **FIGURE 6.22 Center of gravity** By arching his body backwards over the bar, the high jumper lowers his center of gravity. See text for description.

▶ **FIGURE 6.23 Jet propulsion** Squid and octopi propel themselves by squirting jets of water. Shown here is a Giant Octopus jetting away.

you that the center of mass (or center of gravity) always lies vertically below the point of suspension. Since the center of mass is defined as the point at which all the mass of a body can be considered to be concentrated, this is analogous to a particle of mass suspended from a string. Suspending the object from two or more points and marking the vertical lines on which the center of mass must lie locates the center of mass as the intersection of the lines.

The center of mass (or center of gravity) of an object may lie outside the body of the object (◀Fig. 6.21). For example, the center of mass of a homogeneous ring is at the ring's center. The mass in any section of the ring is compensated for by the mass in an equivalent section directly across the ring, and by symmetry, the center of mass is at the center of the ring. For an L-shaped object with uniform legs, equal in mass and length, the center of mass lies on a line that makes a 45° angle with both legs. Its location can easily be determined by suspending the L from a point on one of the legs and noting where a vertical line from that point intersects the diagonal line.

In the high jump, the location of center of gravity is very important. Jumping raises the CG. It takes energy to do this, and the higher the jump, the more energy it takes. Therefore a high jumper wants to clear the bar while keeping his CG low. A jumper will try to keep his CG as close to the bar as possible when passing over it. In the "Fosbury flop" style, made famous by Dick Fosbury in the 1968 Olympics, the jumper arches his body backwards over the bar (◀Fig. 6.22). With the legs, head, and arms below the bar, the CG is lower than in the "layout" style where the body is nearly parallel to the ground when going over the bar. With the "flop," a jumper may be able to make his CG (which is outside the body) pass underneath the bar.

6.6 Jet Propulsion and Rockets

OBJECTIVE: To apply the conservation of momentum in the explanation of jet propulsion and the operation of rockets.

The word *jet* is sometimes used to refer to a stream of liquid or gas emitted at a high speed—for example, a jet of water from a fountain or a jet of air from an automobile tire. **Jet propulsion** is the application of such jets to the production of motion. This concept usually brings to mind jet planes and rockets, but squid and octopi propel themselves by squirting jets of water (▼Fig. 6.23).

You have probably tried the simple application of blowing up a balloon and releasing it. Lacking any guidance or rigid exhaust system, the balloon zigzags around, driven by the escaping air. In terms of Newton's third law, the air is forced out by the contraction of the stretched balloon—that is, the balloon exerts a force on the air. Thus, there must be an equal and opposite reaction force exerted by the air on the balloon. It is this force that propels the balloon on its erratic path.

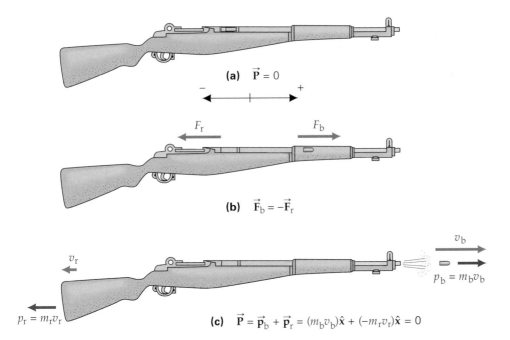

(a) $\vec{P} = 0$

(b) $\vec{F}_b = -\vec{F}_r$

(c) $\vec{P} = \vec{p}_b + \vec{p}_r = (m_b v_b)\hat{x} + (-m_r v_r)\hat{x} = 0$

◀ **FIGURE 6.24 Conservation of momentum** (a) Before the rifle is fired, the total momentum of the rifle and bullet (as an isolated system) is zero. (b) During firing, there are equal and opposite internal forces, and the instantaneous total momentum of the rifle–bullet system remains zero (neglecting external forces, such as arise when a rifle is being held). (c) When the bullet leaves the barrel, the total momentum of the system is still zero. (The vector equation is written in boldface (vector) notation and then in sign–magnitude notation so as to indicate directions.)

Jet propulsion is explained by Newton's third law, and in the absence of external forces, the conservation of momentum also applies. You may understand this concept better by considering the recoil of a rifle, taking the rifle and the bullet as an isolated system (▲Fig. 6.24). Initially, the total momentum of this system is zero. When the rifle is fired (by remote control to avoid external forces), the expansion of the gases from the exploding charge accelerates the bullet down the barrel. These gases push backward on the rifle as well, producing a recoil force (the "kick" experienced by a person firing a weapon). Since the initial momentum of the system is zero and the force of the expanding gas is an internal force, the momenta of the bullet and of the rifle must be exactly equal and opposite at any instant. After the bullet leaves the barrel, there is no propelling force, so the bullet and the rifle move with constant velocities (unless acted on by a net external force such as gravity or air resistance).

Similarly, the thrust of a rocket is created by exhausting the gas from burning fuel out the rear of the rocket. The expanding gas exerts a force on the rocket that propels the rocket in the forward direction (▼Fig. 6.25). The rocket exerts

(b)

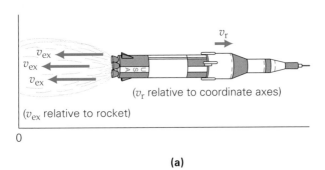

(v_{ex} relative to rocket)

(v_r relative to coordinate axes)

(a)

▲ **FIGURE 6.25 Jet propulsion and mass reduction** (a) A rocket burning fuel is continuously losing mass and thus becomes easier to accelerate. The resulting force on the rocket (the thrust) depends on the product of the rate of change of its mass with time and the velocity of the exhaust gases: $(\Delta m/\Delta t)\vec{v}_{ex}$. Since the mass is decreasing, $\Delta m/\Delta t$ is negative, and the thrust is opposite $\vec{v}_{ex}$. (b) The space shuttle uses a multistage rocket. Both the two booster rockets and the huge external fuel tank are jettisoned in flight. (c) The first and second stages of a *Saturn V* rocket separating after 148 s of burn time.

(c)

Teaching tip: Packages of model-rocket engines typically contain graphs of *F* versus *t*. Obtain one of these graphs and discuss the physical significance of the area under the curve. Also point out how to calculate the change in velocity produced by one of those engines.

a reaction force on the gas, so all of the gas is directed out the exhaust nozzle. If the rocket is at rest when the engines are turned on and there are no external forces (as in deep space, where friction is zero and gravitational forces are negligible), then the instantaneous momentum of the exhaust gas is equal and opposite to that of the rocket. The numerous exhaust-gas molecules have small masses and high velocities, and the rocket has a much larger mass and a smaller velocity.

Unlike a rifle firing a single shot, which has negligible mass, a rocket continuously loses mass when burning fuel. (The rocket is more like a machine gun.) Thus, the rocket is a system for which the mass is not constant. As the mass of the rocket decreases, it accelerates more easily. Multistage rockets take advantage of this fact. The hull of a burnt-out stage is jettisoned to give a further in-flight reduction in mass (Fig. 6.25c). The payload (cargo) is typically a very small part of the initial mass of rockets for space flights.

Suppose that the purpose of a spaceflight is to land a payload on the Moon. At some point on the journey, the gravitational attraction of the Moon will become greater than that of the Earth, and the spacecraft will accelerate toward the Moon. A soft landing is desirable, so the spacecraft must be slowed down enough to go into orbit around the Moon or land on it. This slowing down is accomplished by using the rocket engines to apply a *reverse thrust*, or braking thrust. The spacecraft is maneuvered through a 180° angle, or turned around, which is quite easy to do in space. The rocket engines are then fired, expelling the exhaust gas toward the Moon and supplying a braking action. That is, the force on the rocket is opposite its velocity.

You have experienced a reverse-thrust effect if you have flown in a commercial jet. In this instance, however, the craft is not turned around. Instead, after touchdown, the jet engines are revved up, and a braking action can be felt. Ordinarily, revving up the engines accelerates the plane forward. The reverse thrust is accomplished by activating thrust reversers in the engines that deflect the exhaust gases forward (▼Fig. 6.26). The gas experiences an impulse force and a change in momentum in the forward direction (see Fig. 6.3b), and the engine and the aircraft have an equal and opposite momentum change and thus experience a braking impulse force.

Question: There are no end-of-chapter exercises on the material covered in this section, so test your knowledge with this one: Astronauts use handheld maneuvering devices (small rockets) to move around on space walks. Describe how these rockets are used. Is there any danger on an untethered space walk?

▼ **FIGURE 6.26 Reverse thrust** Thrust reversers are activated on jet engines during landing to help slow the plane. The gas experiences an impulse force and a change in momentum in the forward direction, and the plane experiences an equal and opposite momentum change and a braking impulse force.

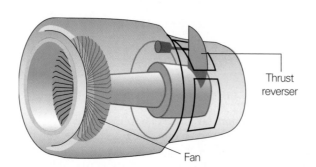

Thrust reverser

Fan

Normal operation

Thrust reverser activated

Chapter Review

- The **linear momentum** ($\vec{p}$) of a particle is a vector and is defined as the product of mass and velocity.

$$\vec{p} = m\vec{v} \tag{6.1}$$

- The **total linear momentum** ($\vec{P}$) of a system is the vector sum of the momenta of the individual particles:

$$\vec{P} = \vec{p}_1 + \vec{p}_2 + \vec{p}_3 + \cdots = \Sigma\vec{p}_i \tag{6.2}$$

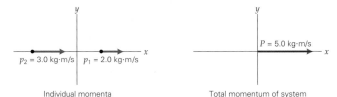

Individual momenta   Total momentum of system

- **Newton's second law in terms of momentum (for a particle):**

$$\vec{F}_{net} = \frac{\Delta\vec{p}}{\Delta t} \tag{6.3}$$

- The **impulse–momentum theorem** relates the impulse acting on an object to its change in momentum:

$$\text{Impulse} = \vec{F}_{avg}\Delta t = \Delta\vec{p}_o = m\vec{v} - m\vec{v}_o \tag{6.5}$$

- **Conservation of linear momentum:** In the absence of a net external force, the total linear momentum of a system is conserved.

$$\vec{P} = \vec{P}_o \tag{6.7}$$

- **In an elastic collision, the total kinetic energy of the system is conserved.**

- **Momentum is conserved in both elastic and inelastic collisions.** In a completely inelastic collision, objects stick together after impact.

- **Conditions for an elastic collision:**

$$\vec{P}_f = \vec{P}_i$$
$$K_f = K_i \tag{6.8}$$

- **Conditions for an inelastic collision:**

$$\vec{P}_f = \vec{P}_i$$
$$K_f < K_i \tag{6.9}$$

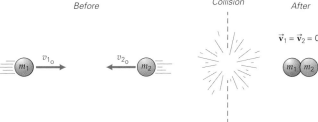

- **Final velocity in head-on, two-body completely inelastic collision** ($v_{2_o} = 0$)

$$v = \left(\frac{m_1}{m_1 + m_2}\right)v_{1_o} \tag{6.10}$$

- **Ratio of kinetic energies in head-on, two-body completely inelastic collision** ($v_{2_o} = 0$)

$$\frac{K_f}{K_i} = \frac{m_1}{m_1 + m_2} \tag{6.11}$$

- **Final velocities in head-on, two-body elastic collions**

$$v_1 = \left(\frac{m_1 - m_2}{m_1 + m_2}\right)v_{1_o} + \left(\frac{2m_2}{m_1 + m_2}\right)v_{2_o} \tag{6.14}$$

$$v_2 = \left(\frac{2m_1}{m_1 + m_2}\right)v_{1_o} - \left(\frac{m_1 - m_2}{m_1 + m_2}\right)v_{2_o} \tag{6.15}$$

- The **center of mass** is the point at which all of the mass of an object or system may be considered to be concentrated. The center of mass does not necessarily lie within an object. (The **center of gravity** is the point at which all the weight may be considered to be concentrated.)

- **Coordinates of the center of mass** (using signs for directions):

$$X_{CM} = \frac{\Sigma m_i x_i}{M} \tag{6.19}$$

Exercises

MC = *Multiple Choice Question,* **CQ** = *Conceptual Question, and* **IE** = *Integrated Exercise. Throughout the text, many exercise sections will include "paired" exercises. These exercise pairs, identified with* **red numbers***, are intended to assist you in problem solving and learning. In a pair, the first exercise (even numbered) is worked out in the Study Guide so that you can consult it should you need assistance in solving it. The second exercise (odd numbered) is similar in nature, and its answer is given at the back of the book.*

6.1 Linear Momentum

1. **MC** Linear momentum has units of (a) N/m, (b) kg·m/s, (c) N/s, (d) all of the preceding. (b)

2. **MC** Linear momentum is (a) always conserved, (b) a scalar quantity, (c) a vector quantity, (d) unrelated to force. (c)

3. **MC** A net force on an object can cause (a) an acceleration, (b) a change in momentum, (c) a change in velocity, (d) all of the preceding. (d)

4. **CQ** Does a fast-running running back always have more linear momentum than a slow-moving, more massive lineman? Explain. no, mass is also a factor

5. **CQ** Two objects have the same momentum. Do they necessarily have the same kinetic energy? Explain.
not necessarily; see ISM

6. **CQ** Two objects have the same kinetic energy. Do they necessarily have the same momentum? Explain.
no, mass is a factor

7. ● If a 60-kg woman is riding in a car traveling at 90 km/h, what is her linear momentum relative to (a) the ground and (b) the car? (a) 1.5×10^3 kg·m/s (b) zero

8. ● The linear momentum of a runner in a 100-m dash is 7.5×10^2 kg·m/s. If the runner's speed is 10 m/s, what is his mass? 75 kg

9. ● Find the magnitude of the linear momentum of (a) a 7.1-kg bowling ball traveling at 12 m/s and (b) a 1200-kg automobile traveling at 90 km/h.
(a) 85 kg·m/s (b) 3.0×10^4 kg·m/s

10. ● In a football game, a lineman usually has more mass than a running back. (a) Will a lineman always have greater linear momentum than a running back? Why? (b) Who has greater linear momentum, a 75-kg running back running at 8.5 m/s or a 120-kg lineman moving at 5.0 m/s?
(a) no (b) the running back, 38 kg·m/s more

11. ● How fast would a 1200-kg car travel if it had the same linear momentum as a 1500-kg truck traveling at 90 km/h?
31 m/s

12. ●● A 0.150-kg baseball traveling with a horizontal speed of 4.50 m/s is hit by a bat and then moves with a speed of 34.7 m/s in the opposite direction. What is the change in the ball's momentum? 5.88 kg·m/s in direction opposite v_o

13. ●● A 15.0-g rubber bullet hits a wall with a speed of 150 m/s. If the bullet bounces straight back with a speed of 120 m/s, what is the change in momentum of the bullet?
4.05 kg·m/s in direction opposite v_o

14. **IE** ●● Two protons approach each other with different speeds. (a) Will the magnitude of the total momentum of the two-proton system be (1) greater than the magnitude of the momentum of either proton, (2) equal to the difference between the magnitudes of momenta of the two protons, or (3) equal to the sum of the magnitudes of momenta of the two protons? Why? (b) If the speeds of the two protons are 340 m/s and 450 m/s, respectively, what is the total momentum of the two-proton system? [*Hint*: Find the mass of a proton in one of the tables inside the backcover.] (a) (2) equal to the difference
(b) 1.8×10^{-25} kg·m/s in direction of the faster proton

15. ●● Taking the density of air to be 1.29 kg/m³, what is the magnitude of the linear momentum of a cubic meter of air moving with a wind speed of (a) 36 km/h and (b) 74 mi/h (the wind speed at which a tropical storm becomes a hurricane)? (a) 13 kg·m/s (b) 43 kg·m/s

16. ●● Two runners of mass 70 kg and 60 kg, respectively, have a total linear momentum of 350 kg·m/s. The heavier runner is running at 2.0 m/s. Determine the possible velocities of the lighter runner. 3.5 m/s in same direction or 8.2 m/s in opposite direction

17. ●● A 0.20-kg billiard ball traveling at a speed of 15 m/s strikes the side rail of a pool table at an angle of 60° (►Fig. 6.27). If the ball rebounds at the same speed and angle, what is the change in its momentum?
$\Delta \vec{p} = (-3.0 \, \text{kg·m/s}) \hat{y}$

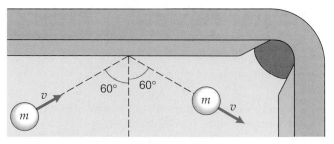

▲ **FIGURE 6.27 Glancing collision** See Exercises 17, 18, and 44.

18. ●● Suppose the billiard ball in Fig. 6.27 approaches the rail at a speed of 15 m/s and an angle of 60°, as shown, but rebounds at a speed of 10 m/s and an angle of 50°. What is the change in momentum in this case? [*Hint*: Use components.] $(-1.1 \, \text{kg·m/s}) \hat{x} - (2.8 \, \text{kg·m/s}) \hat{y}$

19. ●● A person pushes a 10-kg box from rest and accelerates it to a speed of 4.0 m/s with a constant force. If the box is pushed for a time of 2.5 s, what force is exerted by the person? 16 N

20. ●● A loaded tractor-trailer with a total mass of 5000 kg traveling at 3.0 km/h hits a loading dock and comes to a stop in 0.64 s. What is the magnitude of the average force exerted on the truck by the dock? 6.5×10^3 N

21. ●● A 2.0-kg mud ball drops from rest at a height of 15 m. If the impact between the ball and the ground lasts 0.50 s, what is the average net force exerted by the ball on the ground? 68 N

22. **IE** ●● In football practice, two wide receivers run different pass receiving patterns. One with a mass of 80.0 kg runs at 45° northeast at a speed of 5.00 m/s. The second receiver (mass of 90.0 kg) runs straight down the field (due east) at 6.00 m/s. (a) What is the direction of their total momentum: (1) Exactly northeast; (2) to the north of northeast; (3) exactly east; or (4) to the east of northeast? (b) Justify your answer in part (a) by actually computing their total momentum. (a) (4) to east of northeast (b) $(823 \, \text{kg·m/s}) \hat{x} + (283 \, \text{kg·m/s}) \hat{y}$

23. ●● A major league catcher catches a fastball moving at 95.0 mi/h and his hand and glove recoil 10.0 cm in bringing the ball to rest. If it took 0.00470 seconds to bring the ball (with a mass of 250 g) to rest in the glove, (a) what is the magnitude and direction of the change in momentum of the ball? (b) Find the average force the ball exerts on the hand and glove.
(a) 10.6 kg·m/s opposite v_o (b) 2.26×10^3 N

24. ●●● At a basketball game, a 120-lb cheerleader is tossed vertically upward with a speed of 4.50 m/s by a male cheerleader. (a) What is the cheerleader's change in momentum from the time she is released to just before being caught if she is caught at the height at which she was released? (b) Would there be any difference if she were caught 0.30 m below the point of release? If so, what is the change then? (a) 491 kg·m/s downward (b) yes; 524 kg·m/s downward

25. ●●● A ball of mass 200 g is released from rest at a height of 2.00 m above the floor and it rebounds straight up to a height of 0.900 m. (a) Determine the ball's change in momentum due to its contact with the floor. (b) If the contact time with the floor was 0.0950 seconds, what was the average force the floor exerted on the ball, and in what direction? (a) 2.09 kg·m/s (b) 24.0 N upward

6.2 Impulse

26. **MC** Impulse has units (a) kg·m/s, (b) N·s, (c) the same as momentum, (d) all of the preceding. (d)

27. **MC** Impulse is equal to (a) $F\Delta x$, (b) the change in kinetic energy, (c) the change in momentum, (d) $\Delta p/\Delta t$. (c)

28. **CQ** "Follow-through" is very important in many sports, such as in serving a tennis ball. Explain how follow-through can increase the speed of the tennis ball when it is served. increase contact time so to increase impulse

29. **CQ** A karate student tries *not* to follow through in order to break a board, as shown in ▼Fig. 6.28. How can the abrupt stop of the hand (with no follow-through) generate so much force? see ISM

▲ **FIGURE 6.28 A karate chop** See Exercises 29 and 35.

30. **CQ** Explain the difference for each of the following pairs of actions in terms of impulse: (a) a golfer's long drive and a short chip shot; (b) a boxer's jab and a knockout punch; (c) a baseball player's bunting action and a home-run swing. see ISM

31. **CQ** Explain the principle behind (a) the use of Styrofoam as packing material to prevent objects from breaking, (b) the use of shoulder pads by football players to prevent injuries, and (c) the thicker glove used by a baseball catcher than by his teammates in the field. greater contact time—less average force

32. ● When tossed upward and hit horizontally by a batter, a 0.20-kg softball receives an impulse of 3.0 N·s. With what horizontal speed does the ball move away from the bat? 15 m/s

33. ● An automobile with a linear momentum of 3.0×10^4 kg·m/s is brought to a stop in 5.0 s. What is the magnitude of the average braking force? 6.0×10^3 N

34. ● A pool player imparts an impulse of 3.2 N·s to a stationary 0.25-kg cue ball with a cue stick. What is the speed of the ball just after impact? 13 m/s

35. ●● For the karate chop in Exercise 29, assume that the hand has a mass of 0.35 kg and that the speeds of the hand just before and just after hitting the board are 10 m/s and 0, respectively. What is the average force exerted by the fist on the board if (a) the fist follows through, so the contact time is 3.0 ms, and (b) the fist stops abruptly, so the contact time is only 0.30 ms? (a) 1.2×10^3 N (b) 1.2×10^4 N

36. **IE** ●● When bunting, a baseball player uses the bat to change both the speed and direction of the baseball. (a) Will the magnitude of the change in momentum of the baseball before and after the bunt be (1) greater than the magnitude of the momentum of the baseball either before or after the bunt, (2) equal to the difference between the magnitudes of momenta of the baseball before and after the bunt, or (3) equal to the sum of the magnitudes of momenta of the baseball before and after the bunt? Why? (b) The baseball has a mass of 0.16 kg; its speeds before and after the bunt are 15 m/s and 10 m/s, respectively; the bunt lasts 0.025 s. What is the change in momentum of the baseball? (c) What is the average force on the ball by the bat? see ISM

37. **IE** ●● A car with a mass of 1500 kg is rolling on a level road at 30.0 m/s. It receives an impulse with a magnitude of 2000 N·m and its speed is reduced as much as possible by an impulse of this size. (a) Was this impulse caused by (1) the driver hitting the accelerator, (2) the driver putting on the brakes, or (3) the driver turning the steering wheel? (b) What was the car's speed after the impulse was applied?
(a) (2) the driver putting on the brakes (b) 28.7 m/s

38. ●● An astronaut (mass of 100 kg, with equipment) is headed back to her space station at a speed of 0.750 m/s but at the wrong angle. To correct her direction, she fires rockets from her backpack at right angles to her motion for a brief time. These directional rockets exert a constant force of 100.0 N for only 0.200 s. [Neglect the small loss of mass due to burning fuel and assume the impulse is at right angles to her initial momentum.] (a) What is the magnitude of the impulse delivered to the astronaut? (b) What is her new direction (relative to the initial direction)? (c) What is her new speed?
(a) 20.0 N·s, (b) 14.9°, (c) 0.776 m/s

39. ●● A volleyball is traveling toward you. (a) Which action will require a greater force on the volleyball, your catching the ball or your hitting the ball back? Why? (b) A 0.45-kg volleyball travels with a horizontal velocity of 4.0 m/s over the net. You jump up and hit the ball back with a horizontal velocity of 7.0 m/s. If the contact time is 0.040 s, what was the average force on the ball?
(a) hitting it back (b) 1.2×10^2 N in direction opposite v_o

40. ●● A 1.0-kg ball is thrown horizontally with a velocity of 15 m/s against a wall. If the ball rebounds horizontally with a velocity of 13 m/s and the contact time is 0.020 s, what force is exerted on the ball by the wall?
1.4×10^3 N in direction opposite v_o

41. ●● A boy catches—with bare hands and his arms rigidly extended—a 0.16-kg baseball coming directly toward him at a speed of 25 m/s. He emits an audible "Ouch!" because the ball stings his hands. He learns quickly to move his hands with the ball as he catches it. If the contact time of the collision is increased from 3.5 ms to 8.5 ms in this way, how do the magnitudes of the average impulse forces compare? 1.1×10^3 N and 4.7×10^2 N

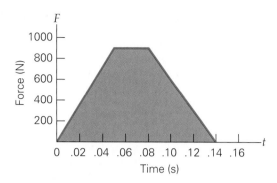

▲ **FIGURE 6.29 Force-versus-time graph** See Exercise 42.

42. ●● A one-dimensional impulse force acts on a 3.0-kg object as diagrammed in ▲ Fig. 6.29. Find (a) the magnitude of the impulse given to the object, (b) the magnitude of the average force, and (c) the final speed if the object had an initial speed of 6.0 m/s. (a) 77 N·s (b) 5.5 × 10² N (c) 32 m/s

43. ●● A 0.45-kg piece of putty is dropped from a height of 2.5 m above a flat surface. When it hits the surface, the putty comes to rest in 0.30 s. What is the average force exerted on the putty by the surface? 15 N upward

44. ●● If the billiard ball in Fig. 6.27 is in contact with the rail for 0.010 s, what is the magnitude of the average force exerted on the ball? (See Exercise 17.) 3.0 × 10² N

45. ●● A 15000-N automobile travels at a speed of 45 km/h northward along a street, and a 7500-N sports car travels at a speed of 60 km/h eastward along an intersecting street. (a) If neither driver brakes and the cars collide at the intersection and lock bumpers, what will the velocity of the cars be immediately after the collision? (b) What percentage of the initial kinetic energy will be lost in the collision? (a) 36 km/h, 56° north of east (b) 49% lost

46. ●●● In a simulated head-on crash test, a car impacts a wall at 25 mi/h (40 km/h) and comes abruptly to rest. A 120-lb passenger dummy (with a mass of 55 kg), without a seatbelt, is stopped by an air bag, which exerts a force on the dummy of 2400 lb. How long was the dummy in contact with the air bag while coming to a stop? 0.057 s

47. ●●● A baseball player pops a pitch straight up. The ball (mass 200 g) was traveling horizontally at 35.0 m/s just before contact with the bat, and 20.0 m/s just after contact. Determine the direction and magnitude of the impulse delivered to the ball by the bat. 8.06 kg·m/s, 29.7° above −x axis

6.3 Conservation of Linear Momentum

48. **MC** The conservation of linear momentum is described by the (a) momentum-impulse theorem, (b) work-energy theorem, (c) Newton's first law, (d) conservation of energy. (a)

49. **MC** The linear momentum of an object is conserved if (a) the force acting on the object is conservative, (b) a single, unbalanced internal force is acting on the object, (c) the mechanical energy is conserved, (d) none of the preceding. (d)

50. **MC** Internal forces do not affect the conservation of momentum because (a) they cancel each other, (b) their effects are canceled by external forces, (c) they can never produce a change in velocity, (d) Newton's second law is not applicable to them. (a)

51. **CQ** An airboat of the type used in swampy and marshy areas is shown in ▼Fig. 6.30. Explain the principle of its propulsion. Using the concept of conservation of linear momentum, determine what would happen to the boat if a sail were installed behind the fan. see ISM

▲ **FIGURE 6.30 Fan propulsion** See Exercise 51.

52. **CQ** Imagine yourself standing in the middle of a frozen lake. The ice is so smooth that it is frictionless. How could you get to shore? (You couldn't walk. Why?) throw something, or even blow breath

53. **CQ** A stationary object receives a direct hit by another object moving toward it. Is it possible for both objects to be at rest after the collision? Explain. no, because there is an initial momentum

54. **CQ** When a golf ball is driven off the tee, its speed is often much greater than the speed of the golf club. Explain how this situation can happen. golf ball has much smaller mass

55. ● A 60-kg astronaut floating at rest in space outside a space capsule throws his 0.50-kg hammer such that it moves with a speed of 10 m/s relative to the capsule. What happens to the astronaut? moves 0.083 m/s in opposite direction

56. ● In a pairs figure-skating competition, a 65-kg man and his 45-kg female partner stand facing each other on skates on the ice. If they push apart and the woman has a velocity of 1.5 m/s eastward, what is the velocity of her partner? (Neglect friction.) 1.0 m/s westward

57. ●● To get off a frozen, frictionless lake, a 65.0-kg person takes off a 0.150-kg shoe and throws it horizontally, directly away from the shore with a speed of 2.00 m/s. If the person is 5.00 m from the shore, how long does he take to reach it? 1.08 × 10³ s = 18.1 min

58. **IE** ●● An object initially at rest explodes and splits into three fragments. The first fragment flies off to the west, and the second fragment flies off to the south. The third fragment will fly off toward a general direction of (1) southwest, (2) north of east, (3) either due north or due east. Why? (b) If the object has a mass of 3.0 kg, the first fragment has a mass of 0.50 kg and a speed of 2.8 m/s, and the second fragment has a mass of 1.3 kg and a speed of 1.5 m/s, what are the speed and direction of the third fragment? (a) (2) north of east (b) 2.0 m/s, 54° north of east

59. ●● Suppose that the 3.0-kg object in Exercise 58 is initially traveling at a speed of 2.5 m/s in the positive x-direction. What will be the speed and direction of the third fragment in this case? 7.6 m/s, 12° above the +x-axis

60. ●● Two balls of equal mass (0.50 kg) approach the origin along the positive x- and y-axes at the same speed (3.3 m/s). (a) What is the total momentum of the system? (b) Will the balls necessarily collide at the origin? What is the total momentum of the system after both balls have passed through the origin? (a) 2.3 kg·m/s, 45° below the −x axis (b) no; same as in (a)

61. ●● Two identical cars hit each other and lock bumpers. In each of the following cases, what are the speeds of the cars immediately after coupling bumpers? (a) A car moving with a speed of 90 km/h approaches a stationary car; (b) two cars approach each other with speeds of 90 km/h and 120 km/h, respectively; (c) two cars travel in the same direction with speeds of 90 km/h and 120 km/h, respectively.
(a) 45 km/h (b) 15 km/h (c) 105 km/h

62. ●● A 1200-kg car moving to the right with a speed of 25 m/s collides with a 1500-kg truck and locks bumpers with the truck. Calculate the velocity of the combination after the collision if the truck is initially (a) at rest, (b) moving to the right with a speed of 20 m/s, and (c) moving to the left with a speed of 20 m/s.
(a) 11 m/s to the right (b) 22 m/s to the right (c) at rest

63. ●● A 10-g bullet moving horizontally at 400 m/s penetrates a 3.0-kg wood block resting on a horizontal surface. If the bullet slows down to 300 m/s after emerging from the block, what is the speed of the block immediately after the bullet emerges (▼Fig. 6.31)? 0.33 m/s

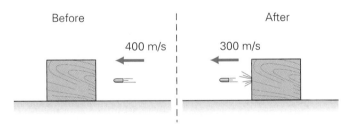

▲ **FIGURE 6.31 Momentum transfer?** See Exercise 63.

64. ●● An explosion of a 10.0-kg bomb releases only two separate pieces. The bomb was initially at rest and a 4.00-kg piece travels westward at 100 m/s immediately after the explosion. (a) What is the speed and direction of the other piece immediately after the explosion? (b) How much kinetic energy was released in this explosion?
(a) −66.7 m/s, east; (b) 3.33 × 10⁴ J

65. ●● A 1600-kg (empty) truck rolls with a speed of 2.5 m/s under a loading bin, and a mass of 3500 kg is deposited in the truck. What is the truck's speed immediately after loading? 0.78 m/s

66. **IE** ●● A new crowd control method utilizes "rubber" bullets instead of real ones. Suppose that, in a test, one of these "bullets" with a mass of 500 g is traveling at 250 m/s to the right. It hits a stationary target head-on. The target's mass is 25.0 kg and it rests on a smooth surface. The bullet bounces backward (to the left) off the target at 100 m/s. (a) Which way must the target move after the collision: (1) right, (2) left, (3) it could be stationary, or (4) you can't tell from the data given? (b) Determine the recoil speed of the target after the collision.
(a) (1) right (b) 7.00 m/s

67. ●● For a movie scene, a 75-kg stuntman drops from a tree onto a 50-kg sled that is moving on a frozen lake with a velocity of 10 m/s toward the shore. (a) What is the speed of the sled after the stuntman is on board? (b) If the sled hits the bank and stops, but the stuntman keeps on going, with what speed does he leave the sled? (Neglect friction.) (a) 4.0 m/s (b) 4.0 m/s

68. ●● A 90-kg astronaut is stranded in space at a point 6.0 m from his spaceship, and he needs to get back in 4.0 min to control the spaceship. To get back, he throws a 0.50-kg piece of equipment so that it moves at a speed of 4.0 m/s directly away from the spaceship. (a) Does he get back in time? (b) How fast must he throw the piece of equipment so he gets back in time? (a) no (b) 4.5 m/s

69. ●●● A projectile that is fired from a gun has an initial velocity of 90.0 km/h at an angle of 60.0° above the horizontal. When the projectile is at the top of its trajectory, an internal explosion causes it to separate into two fragments of equal mass. One of the fragments falls straight downward as though it had been released from rest. How far from the gun does the other fragment land? 82.8 m

70. ●●● A moving shuffleboard puck has a glancing collision with a stationary puck of the same mass, as shown in ▼Fig. 6.32. If friction is negligible, what are the speeds of the pucks after the collision? 0.61 m/s, 0.73 m/s

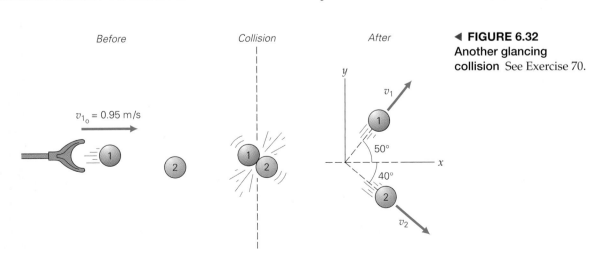

◄ **FIGURE 6.32
Another glancing
collision** See Exercise 70.

71. ●●● A small asteroid (mass of 10 g) strikes a glancing blow at a satellite in empty space. The satellite was initially at rest and the asteroid traveling at 2000 m/s. The satellite's mass is 100 kg. The asteroid is deflected 10° from its original direction and its speed decreases to 1000 m/s, but neither object loses mass. Determine the (a) direction and (b) speed of the satellite after the collision.
(a) 9.7° (b) 0.10 m/s

72. ●●● A *ballistic pendulum* is a device used to measure the velocity of a projectile—for example, the muzzle velocity of a rifle bullet. The projectile is shot horizontally into, and becomes embedded in, the bob of a pendulum, as illustrated in ▼ Fig. 6.33. The pendulum swings upward to some height h, which is measured. The masses of the block and the bullet are known. Using the laws of momentum and energy, show that the initial velocity of the projectile is given by $v_o = [(m + M)/m]\sqrt{2gh}$. see ISM

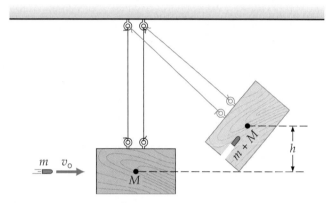

▲ **FIGURE 6.33 A ballistic pendulum** See Exercises 72 and 98.

6.4 Elastic and Inelastic Collisions

73. **MC** Which of the following is *not* conserved in an inelastic collision: (a) momentum, (b) mass, (c) kinetic energy, or (d) total energy? (c)

74. **MC** A rubber ball of mass m traveling horizontally with a speed v hits a wall and bounces back with the same speed. The change in momentum is (a) mv, (b) $-mv$, (c) $-mv/2$, (d) $+2mv$. (d)

75. **MC** In a head-on elastic collision, mass m_1 strikes a stationary mass m_2. There is a complete transfer of energy if (a) $m_1 = m_2$, (b) $m_1 \gg m_2$, (c) $m_1 \ll m_2$, (d) the masses stick together. (a)

76. **MC** The condition for a two-object inelastic collision is (a) $K_f < K_i$, (b) $p_i \neq p_f$, (c) $m_1 > m_2$, (d) $v_1 < v_2$. (a)

77. **CQ** Since $K = p^2/2m$, how can kinetic energy be lost in an inelastic collision while the total momentum is still conserved? Explain. see ISM

78. **CQ** Discuss the common and different characteristics of an elastic collision and an inelastic collision. see ISM

79. **CQ** Can all of the kinetic energy be lost in the collision of two objects? Explain. yes, see ISM

80. ●● For the apparatus in Fig. 6.15, one ball swinging in at a speed of $2v_o$ will not cause two balls to swing out with speeds v_o. (a) Which law of physics precludes this situation from happening, the law of conservation of momentum or the law of conservation of mechanical energy? (b) Prove this law mathematically.
(a) conservation of mechanical energy (b) see ISM

81. ●● A proton of mass m moving with a speed of 3.0×10^6 m/s undergoes a head-on elastic collision with an alpha particle of mass $4m$, which is initially at rest. What are the velocities of the two particles after the collision? $v_p = -1.8 \times 10^6$ m/s; $v_a = 1.2 \times 10^6$ m/s

82. ●● A 4.0-kg ball with a velocity of 4.0 m/s in the $+x$-direction collides head-on elastically with a stationary 2.0-kg ball. What are the velocities of the balls after the collision? $v_1 = +1.3$ m/s; $v_2 = +5.3$ m/s

83. ●● A ball with a mass of 0.10 kg is traveling with a velocity of 0.50 m/s in the $+x$-direction and collides head-on with a 5.0-kg ball that is at rest. Find the velocities of the balls after the collision. Assume that the collision is elastic. $v_1 = -0.48$ m/s; $v_2 = +0.020$ m/s

84. ●● At a county fair, two children ram each other head-on while riding on the bumper car attraction. Jill and her car, traveling left at 3.50 m/s, have a total mass of 325 kg. Jack and his car, traveling to the right at 2.00 m/s, have a total mass of 290 kg. Assuming the collision to be elastic, determine their velocities after the collision. 1.69 m/s, right; −3.81 m/s, left

85. ●● In a high-speed chase, a policeman's car bumps a criminal's car directly from behind to get his attention. The policeman's car is moving at 40.0 m/s to the right and has a total mass of 1800 kg. The criminal's car is initially moving in the same direction at 38.0 m/s. His car has a total mass of 1500 kg. Assuming an elastic collision, determine their two velocities immediately after the bump. $v_p = 38.2$ m/s, $v_c = 40.2$ m/s

86. **IE** ●● ▼Fig. 6.34 shows a bird catching a fish. Assume that initially the fish jumps up and that the bird coasts horizontally and does not touch water with its feet or flap its wings. (a) Is this kind of collision (1) elastic, (2) inelastic, or (3) completely inelastic? Why? (b) If the mass of the bird is 5.0 kg, the mass of the fish is 0.80 kg, and the bird coasts with a speed of 6.5 m/s before grabbing, what is the speed of the bird after grabbing the fish?
(a) (3) completely inelastic (b) 5.6 m/s

▶ **FIGURE 6.34 Elastic or inelastic?** See Exercise 86.

87. ●● A 1.0-kg object moving at 10 m/s collides with a stationary 2.0-kg object as shown in ▼Fig. 6.35. If the collision is perfectly inelastic, how far along the inclined plane will the combined system travel? Neglect friction. 0.94 m

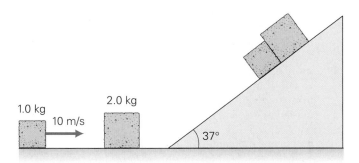

▲ **FIGURE 6.35 How far is up?** See Exercises 87 and 92.

88. ●● In a pool game, a cue ball traveling at 0.75 m/s hits the stationary eight ball. The eight ball moves off with a velocity of 0.25 m/s at an angle of 37° relative to the cue ball's initial direction. Assuming that the collision is inelastic, at what angle will the cue ball be deflected, and what will be its speed? 15°, 0.57 m/s

89. ●● Two balls with masses of 2.0 kg and 6.0 kg travel toward each other at speeds of 12 m/s and 4.0 m/s, respectively. If the balls have a head-on, inelastic collision and the 2.0-kg ball recoils with a speed of 8.0 m/s, how much kinetic energy is lost in the collision? 1.1×10^2 J

90. ●● Two balls approach each other as shown in ▼Fig. 6.36, where $m = 2.0$ kg, $v = 3.0$ m/s, $M = 4.0$ kg, and $V = 5.0$ m/s. If the balls collide and stick together at the origin, (a) what are the components of the velocity v of the balls after collision, and (b) what is the angle θ? (a) $v_x = 1.0$ m/s; $v_y = 3.3$ m/s (b) 73°

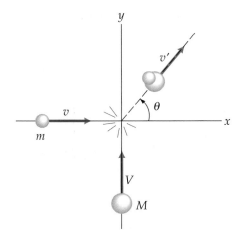

▲ **FIGURE 6.36 A completely inelastic collision**
See Exercise 90.

91. IE ●● A car traveling east and a minivan traveling south collide in a completely inelastic collision at a perpendicular intersection. (a) Right after the collision, will the car and minivan move toward a general direction (1) south of east, (2) north of west, or (3) either due south or due east? Why? (b) If the initial speed of the 1500-kg car was 90.0 km/h and the initial speed of the 3000-kg minivan was 60.0 km/h, what is the velocity of the vehicles immediately after collision?
(a) (1) south of east (b) 13.9 m/s, 53.1° south of east

92. ●● A 1.0-kg object moving at 2.0 m/s collides elastically with a stationary 1.0-kg object, similar to the situation shown in Fig. 6.35. How far will the initially stationary object travel along a 37° inclined plane? Neglect friction. 0.34 m

93. ●● A fellow student states that the total momentum of a three-particle system ($m_1 = 0.25$ kg, $m_2 = 0.20$ kg, and $m_3 = 0.33$ kg) is initially zero. He calculates that after an inelastic triple collision the particles have velocities of 4.0 m/s at 0°, 6.0 m at 120°, and 2.5 m/s at 230°, respectively, with angles measured from the +x axis. Do you agree with his calculations? If not, assuming the first two answers to be correct, what should be the momentum of the third particle so the total momentum is zero? no; 1.1 kg·m/s, 249°

94. ●● A freight car with a mass of 25 000 kg rolls down an inclined track through a vertical distance of 1.5 m. At the bottom of the incline, on a level track, the car collides and couples with an identical freight car that was at rest. What percentage of the initial kinetic energy is lost in the collision? 50%

95. ●●● In nuclear reactors, subatomic particles called *neutrons* are slowed down by allowing them to collide with the atoms of a moderator material, such as carbon atoms, which are 12 times as massive as neutrons. (a) In a head-on elastic collision with a carbon atom, what percentage of a neutron's energy is lost? (b) If the neutron has an initial speed of 1.5×10^7 m/s, what will be its speed after collision?
(a) 28% (b) 1.3×10^7 m/s

96. ●●● In a noninjury chain-reaction accident on a foggy freeway, car 1 (mass of 2000 kg) moving at 15.0 m/s to the right elastically collides with car 2, initially at rest. The mass of car 2 is 1500 kg. In turn, car 2 then goes on to lock bumpers (that is, it is a completely inelastic collision) with car 3, which has a mass of 2500 kg and was also at rest. Determine the speed of all cars immediately after this unfortunate accident.
$v_1 = 2.14$ m/s, $v_2 = 1.71$ m/s, $v_{23} = 6.41$ m/s

97. ●●● Pendulum 1 is made of a 1.50-m string with a small Super Ball attached as a bob. It is pulled aside 30° and released. At the bottom of its arc, it collides with another pendulum bob of the same length, but the second Pendulum 2 has a bob made from a Super Ball whose mass is twice that of the bob of Pendulum 1. Determine the angles to which both pendulums rebound (when they come to rest) after they collide and bounce back.
$\theta_1 = 9.94°$, $\theta_2 = 19.8°$

98. ●●● Show that the fraction of kinetic energy lost in a ballistic-pendulum collision (as in Fig. 6.33) is equal to $M/(m + M)$. see ISM

6.5 Center of Mass

99. **MC** The center of mass of an object (a) always lies at the center of the object, (b) is at the location of the most massive particle in the object, (c) always lies within the object, (d) none of the preceding. (d)

100. **MC** The center of mass and center of gravity coincide (a) if the acceleration due to gravity is constant, (b) if momentum is conserved, (c) if momentum is not conserved, (d) only for irregularly shaped objects. (a)

101. **CQ** ▼ Figure 6.37 shows a flamingo standing on one of its two legs, with its other leg lifted. What can you say about the location of the flamingo's center of mass?
directly above the foot on the ground

▶ **FIGURE 6.37 Delicate balance** See Exercise 101.

102. **CQ** Two identical objects are located a distance d apart. If one of the objects remains at rest and the other moves away with a constant velocity what is the effect on the CM of the system? moves with $v/2$ in v direction

103. ●● (a) The center of mass of a system consisting of two 0.10-kg particles is located at the origin. If one of the particles is at $(0, 0.45 \text{ m})$, where is the other? (b) If the masses are moved so their center of mass is located at $(0.25 \text{ m}, 0.15 \text{ m})$, can you tell where the particles are located? (a) $(0, -0.45 \text{ m})$ (b) no, only that they are equidistant from CM

104. ● The centers of a 4.0-kg sphere and a 7.5-kg sphere are separated by a distance of 1.5 m. Where is the center of mass of the two-sphere system?
0.98 m from less massive sphere

105. ● (a) Find the center of mass of the Earth–Moon system. [*Hint*: Use data from the tables on the inside cover of the book, and consider the distance between the Earth and Moon to be measured from their centers.] (b) Where is that center of mass relative to the surface of the Earth? (a) 4.6×10^6 m from centers (b) 1.8×10^6 m below

106. ●● Find the center of mass of a system composed of three spherical objects with masses of 3.0 kg, 2.0 kg, and 4.0 kg and centers located at $(-6.0 \text{ m}, 0)$, $(1.0 \text{ m}, 0)$, and $(3.0 \text{ m}, 0)$, respectively. $(-0.44 \text{ m}, 0)$

107. ●● Rework Exercise 69, using the concept of the center of mass, and compute the distance the other fragment landed from the gun. 82.8 m

108. **IE** ●● A 3.0-kg rod of length 5.0 m has at opposite ends point masses of 4.0 kg and 6.0 kg. (a) Will the center of mass of this system be (1) nearer to the 4.0-kg mass, (2) nearer to the 6.0-kg mass, or (3) at the center of the rod? Why? (b) Where is the center of mass of the system? see ISM

109. ●● A piece of uniform sheet metal measures 25 cm by 25 cm. If a circular piece with a radius of 5.0 cm is cut from the center of the sheet, where is the sheet's center of mass now? still at center of sheet

110. ●● Locate the center of mass of the system shown in ▼ Fig. 6.38 (a) if all of the masses are equal; (b) if $m_2 = m_4 = 2m_1 = 2m_3$; (c) if $m_1 = 1.0$ kg, $m_2 = 2.0$ kg, $m_3 = 3.0$ kg, and $m_4 = 4.0$ kg.
(a) (2.0 m, 2.0 m) (b) (2.0 m, 2.0 m) (c) (2.8 m, 2.0 m)

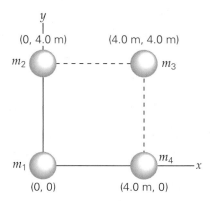

▲ **FIGURE 6.38 Where's the center of mass?** See Exercise 110.

111. ●● Two cups are placed on a uniform board that is balanced on a cylinder (▼Fig. 6.39). The board has a mass of 2.00 kg and is 2.00 m long. The mass of cup 1 is 200 g and it is placed 1.05 m to the left of the balance point. The mass of cup 2 is 400 g. Where should Cup 2 be placed for balance (relative to the right end of the board)? 0.175 m

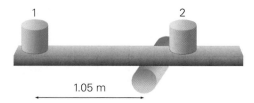

▲ **FIGURE 6.39 Don't let it roll** See Exercise 111.

112. ●● A 100-kg astronaut (mass includes space gear) on a space walk is 5.0 m from a 3000-kg space capsule and at the full length of her safety cord. To return to the capsule, she pulls herself along the cord. Where do the astronaut and capsule meet?
0.16 m from capsule's original position

113. •• Two skaters with masses of 65 kg and 45 kg, respectively, stand 8.0 m apart, each holding one end of a piece of rope. (a) If they pull themselves along the rope until they meet, how far does each skater travel? (Neglect friction.) (b) If only the 45-kg skater pulls along the rope until she meets her friend (who just holds onto the rope), how far does each skater travel? see ISM

114. ••• Three particles, each with a mass of 0.25 kg, are located at $(-4.0 \text{ m}, 0)$, $(2.0 \text{ m}, 0)$, and $(0, 3.0 \text{ m})$ and are acted on by forces $\vec{F}_1 = (-3.0 \text{ N})\hat{y}$, $\vec{F}_2 = (5.0 \text{ N})\hat{y}$, and $\vec{F}_3 = (4.0 \text{ N})\hat{x}$, respectively. Find the acceleration (magnitude and direction) of the center of mass of the system. [*Hint:* Consider the components of the acceleration.] see ISM

Comprehensive Exercises

115. An unbalanced baton consists of a very light aluminum rod 60.0 cm long. The mass at one end is three times the mass at the "light" end. Initially the baton is launched with a speed of 15.0 m/s at an angle of 45 degrees from ground level. When the baton reaches the apex of its flight, it is oriented vertically, with the "light end" above the "heavy end." Determine the height of each end of the baton above the ground at the apex. massive end, 5.59 m; less massive end, 6.19 m

116. A 20.0-g bullet traveling at 300 m/s passes completely through a wooden block, initially at rest on a smooth table. The block has a mass of 1000 g. The bullet emerges traveling in the same direction, but at 50.0 m/s. (a) What is the speed of the block afterward? (b) What fraction (or percentage) of the total initial kinetic energy is lost in this process? (a) 5.00 m/s, (b) 95.8%

117. A small asteroid (a rock with a mass of 100 g) strikes a glancing blow with a space probe with a mass of 1000 kg. Assume all planets are far away so gravitational forces can be neglected. Because of the collision, the asteroid is deflected at 40° from its original direction and its speed reduced to 12 000 m/s from its initial speed of 20 000 m/s. The probe was originally moving at 200 m/s in the +x direction. (a) Determine the angle at which the probe is deflected from its initial direction. (b) What is the speed of the probe after the collision? (a) 0.222° (b) 199 m/s

118. **IE** In the radioactive decay of a nucleus of an atom called americium-241 (symbol ^{241}Am, mass of 4.03×10^{-25} kg), it emits an alpha particle (designated as α) with a mass of 6.68×10^{-27} kg to the right with a kinetic energy of 8.64×10^{-13} J. (This is typical of nuclear energies, small on the everyday scale.) The remaining nucleus is neptunium-237 (^{237}Np) and has a mass of 3.96×10^{-25} kg. Assume the initial nucleus was at rest. (a) Will the neptunium nucleus have (1) more, (2) less, or (3) the same amount of kinetic energy compared to the alpha particle? (b) Determine the kinetic energy of the ^{237}Np nucleus afterward. (a) (2) less (b) 1.46×10^{-14} J

119. A hockey player is initially moving at 5.00 m/s to the east. Her mass is 50.0 kg. She intercepts and catches on her stick a puck initially moving at 35.0 m/s at an angle of 30 degrees (▼Fig. 6.40). Assume that the puck's mass is 0.50 kg and the two form a single object for a few seconds. (a) Determine the direction angle and speed of the puck and skater after the collision. (b) Was this collision elastic or inelastic? Prove your answer with numbers. (a) 4.65 m/s $\theta = 2.13°$ (b) $K < K_o$ inelastic

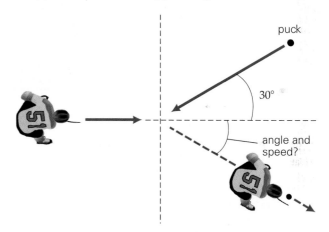

▲ **FIGURE 6.40 Catching on** See Exercise 119.

CIRCULAR MOTION AND GRAVITATION

PHYSICS **FACTS**

- An ultracentrifuge can spin samples with a force of 15 000 *g*. Such force is needed for harvesting protein precipitates, bacteria, and other cells.

- Newton coined the word *gravity* from *gravitas*, the Latin word for "weight" or "heaviness."

- If you want to "lose" weight, go to the Earth's equator. Because the Earth bulges slightly, the acceleration due to gravity is slightly less there and you would weigh less (but your mass would still be the same).

- The centripetal force that keeps planets in orbit is supplied by gravity. The centripetal force that keeps atomic electrons in orbit about the nuclear proton(s) is supplied by the electrical force. The electrical force between an electron and a proton is about 10^{40} times greater than the gravitational force between them (Chapter 15).

- Kepler's laws that describe planetary orbital motion were empirical—that is, derived from observations, with no theoretical foundation. Working primarily with observational data on the planet Mars, Kepler took years and an immense number of calculations to formulate his laws. There are nearly a thousand surviving sheets of calculations. Kepler referred to this as "my war with Mars." Kepler's laws follow directly from analyses using Newton's law of gravitation and calculus. The laws can be derived theoretically on a page or two of calculations.

People often say that rides like the spirling circular one in the photograph "defy gravity." Of course, you know that in reality, gravity cannot be defied; it commands respect. There is nothing that will shield you from it and no place in the universe where you can go to be entirely free of it.

Circular motion is everywhere, from atoms to galaxies, from flagella of bacteria to Ferris wheels. Two terms are frequently used to describe such motion. In general, we say that an object *rotates* when the axis of rotation lies within the body and that it *revolves* when the axis is outside the body. Thus, the Earth rotates on its axis and revolves about the Sun.

Circular motion is motion in two dimensions, and so it can be described by rectangular components as used in Chapter 3. However, it is usually more convenient to describe circular motion in terms of angular quantities that will be introduced in this chapter. Being familiar with the description of circular motion will make the study of rotating rigid bodies much easier, as you will find in Chapter 8.

Gravity plays a large role in determining the motions of the planets, since it supplies the force necessary to maintain their orbits. This chapter will consider Newton's law of gravitation, which describes this fundamental force, and will analyze planetary motion in terms of this and related basic laws. The same considerations will help you understand the motions of the Earth's satellites, which include one natural satellite (the Moon) and many artificial ones.

7.1 Angular Measure

OBJECTIVES: To (a) define units of angular measure, and (b) show how angular measure is related to circular arc length.

Motion is described as a time rate of change of position (Section 2.1). As you might guess, *angular speed* and *angular velocity* also involve a time rate of change of position, which is expressed by an *angular change*. Consider a particle traveling in a circular path, as shown in ▶Fig. 7.1. At a particular instant, the particle's position (P) may be designated by the Cartesian coordinates x and y. However, the position may also be designated by the polar coordinates r and θ. The distance r extends from the origin, and the angle θ is commonly measured counterclockwise from the positive x-axis. The transformation equations that relate one set of coordinates to the other are

$$x = r \cos \theta \tag{7.1a}$$

$$y = r \sin \theta \tag{7.1b}$$

as can be seen from the x- and y-coordinates of point P in Fig. 7.1.

Note that r is the same for any point on a given circle. As a particle travels in a circle, the value of r is constant, and only θ changes with time. Thus, circular motion can be described by using one polar coordinate (θ) that changes with time, instead of two Cartesian coordinates (x) and (y), both of which change with time.

Analogous to linear displacement is **angular displacement**, the magnitude of which is given by

$$\Delta\theta = \theta - \theta_{o} \tag{7.2}$$

or simply $\Delta\theta = \theta$ when we choose $\theta_{o} = 0°$. (The direction of the angular displacement will be explained in the next section on angular velocity.) A unit commonly used to express angular displacement is the degree (°); there are 360° in one complete circle, or revolution.*

It is important to be able to relate the angular description of circular motion to the orbital or tangential description—that is, to relate the angular displacement to the arc length s. The *arc length* is the distance traveled along the circular path, and the angle θ is said to *subtend* (define) the arc length. A quantity that is very convenient for relating angle to arc length is the **radian (rad)**. The angle in radians is given by the ratio of the arc length (s) and the radius (r)—that is, θ (in radians) $= s/r$. When $s = r$, the angle is equal to one radian, $\theta = s/r = r/r = 1$ rad (▶Fig. 7.2).

Thus, we can write (with the angle in radians),

$$s = r\theta \tag{7.3}$$

which is an important relationship between the circular arc length s, and the radius of the circle r. (Notice that since $\theta = s/r$, the angle in radians is the ratio of two lengths. This means that a radian measure is a pure number—that is, it is dimensionless and has no units.)

To get a general relationship between radians and degrees, let's consider the distance around a complete circle (360°). For one full circle, with $s = 2\pi r$ (the circumference), there are a total of $\theta = s/r = 2\pi r/r = 2\pi$ rad in 360° or

$$2\pi \text{ rad} = 360°$$

This relationship can be used to obtain convenient conversions of common angles (Table 7.1). Also, dividing both sides of this relationship by 2π, we have

$$1 \text{ rad} = 360°/2\pi = 57.3°$$

Notice in Table 7.1 that the angles in radians are expressed in terms of π explicitly, for convenience.

---

*A degree may be divided into the smaller units of minutes (1 degree = 60 minutes) and seconds (1 minute = 60 seconds). These divisions have nothing to do with time units.

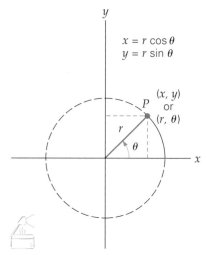

▲ FIGURE 7.1 Polar coordinates A point may be described by polar coordinates instead of Cartesian coordinates—that is, by (r, θ) instead of (x, y). For a circle, θ is the angular distance and r is the radial distance. The two types of coordinates are related by the transformation equations $x = r \cos \theta$ and $y = r \sin \theta$.

Teaching tip: Does it seem odd that a circle is divided into 360°? Ask students if they know why that value was selected.

Illustration 10.1 *Coordinates for Circular Motion*

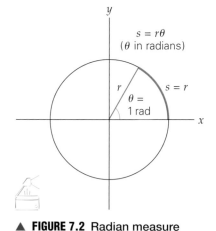

▲ FIGURE 7.2 Radian measure Angular displacement may be measured either in degrees or in radians (rad). An angle θ is subtended by an arc length s. When $s = r$, the angle subtending s is defined to be 1 rad. More generally, $\theta = s/r$, where θ is in radians. One radian is equal to 57.3°.

Finish

N
W ← → E
S

256 m

Start

▲ **FIGURE 7.3 Arc length—found by means of radians** See Example 7.1.

TABLE 7.1

Equivalent Degree and Radian Measures

Degrees	Radians
360°	2π
180°	π
90°	$\pi/2$
60°	$\pi/3$
57.3°	1
45°	$\pi/4$
30°	$\pi/6$

Teaching tip: Warn students not to measure angular quantities in revolutions or degrees when using the expression $s = r\theta$. The angle θ must be in radians.

Example 7.1 ■ Finding Arc Length: Using Radian Measure

A spectator standing at the center of a circular running track observes a runner start a practice race 256 m due east of her own position (◄Fig. 7.3). The runner runs on the track to the finish line, which is located due north of the observer's position. What is the distance of the run?

Thinking It Through. Note that the subtending angle of the section of circular track is $\theta = 90°$. The arc length (s) can be found, since the radius r of the circle is known.

Solution. Listing what is given and what is to be found,

Given: $r = 256$ m   *Find:* s (arc length)
           $\theta = 90° = \pi/2$ rad

Simply using Eq. 7.3 to find the arc length, we get

$$s = r\theta = (256 \text{ m})\left(\frac{\pi}{2}\right) = 402 \text{ m}$$

Note that the rad is omitted, and the equation is dimensionally correct. Why?

Follow-Up Exercise. What would be the distance of one complete lap around the track in this Example? *(Answers to all Follow-Up Exercises are at the back of the text.)*

Example 7.2 ■ How Far Away? A Useful Approximation

A sailor sights a distant ship and finds that it subtends an angle of 1.15° as illustrated in ▼Fig. 7.4a. He knows from the shipping charts that the tanker is 150 m in length. Approximately how far away is the tanker?

Thinking It Through. Note in the accompanying Learn by Drawing (p. 219) that for small angles, the arc length approximates the y-length of the triangle (the opposite side from θ), or $s \approx y$. Hence, if we know the length and the angle, the radial distance can be found, which is approximately equal to the tanker's distance from the sailor.

Solution. To approximate the distance, we take the ship's length to be nearly equal to the arc length subtended by the measured angle. This approximation is good for small angles. The data are then as follows:

Given: $\theta = 1.15°$ (1 rad/57.3°) = 0.0201 rad   *Find:* r (radial distance)
           $s = 150$ m

▶ **FIGURE 7.4 Angular distance** For small angles, the arc length is approximately a straight line, or the chord length. Knowing the length of the tanker, we can find how far away it is by measuring its angular size. See Example 7.2. (Drawing not to scale for clarity.)

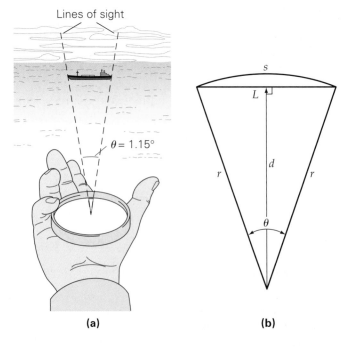

Lines of sight

$\theta = 1.15°$

(a)

s

L

r d r

θ

(b)

Knowing the arc length and angle, Eq. 7.3 can be used to find r (note that the unitless rad is omitted):

$$r = \frac{s}{\theta} = \frac{150 \text{ m}}{0.0201} = 7.46 \times 10^3 \text{ m} = 7.46 \text{ km}$$

As noted, the distance r is an approximation, obtained by assuming that for small angles, the arc length s and the straight-line chord length L are very nearly the same length (Fig. 7.4b). How good is this approximation? To see, let's compute the actual perpendicular distance d to the middle of the ship. From the geometry, we have $\tan(\theta/2) = (L/2)/d$, so

$$d = \frac{L}{2\tan(\theta/2)} = \frac{150 \text{ m}}{2\tan(1.15°/2)} = 7.47 \times 10^3 \text{ m} = 7.47 \text{ km}$$

The first calculation is a pretty good approximation—the values derived by the two methods are nearly equal.

Follow-Up Exercise. As pointed out, the approximation used in this Example is for *small* angles. You might wonder what is small. To investigate this question, what would be the percentage error of the approximated distance to the tanker for angles of 10° and 20°?

Problem-Solving Hint

In computing trigonometric functions such as $\tan\theta$ or $\sin\theta$, the angle may be expressed in degrees or radians; for example, $\sin 30° = \sin[(\pi/6) \text{ rad}] = \sin(0.524 \text{ rad}) = 0.500$. When finding trig functions with a calculator, note that there is usually a way to change the angle entry between *deg* and *rad* modes. Hand calculators commonly are set in *deg* (degree) mode, so if you want to find the value of, say, $\sin(1.22 \text{ rad})$, first change to *rad* mode and enter sin 1.22, and $\sin(1.22 \text{ rad}) = 0.939$. (Or you could convert rads to degrees first and use *deg* mode.)

Your calculator may have a third mode, *grad*. The grad is a little-used angular unit. A grad is 1/100 of a right (90°) angle; that is, there are 100 grads in a right angle.

7.2 Angular Speed and Velocity

OBJECTIVES: To (a) describe and compute angular speed and velocity, and (b) explain their relationship to tangential speed.

The description of circular motion in angular form is analogous to the description of linear motion. In fact, you'll notice that the equations are almost identical mathematically, with different symbols being used to indicate that the quantities have different meanings. The lowercase Greek letter omega with a bar over it ($\bar{\omega}$) is used to represent **average angular speed**, the magnitude of the angular displacement divided by the total time to travel the distance:

$$\bar{\omega} = \frac{\Delta\theta}{\Delta t} = \frac{\theta - \theta_o}{t - t_o} \quad \textit{average angular speed} \tag{7.4}$$

We say that the units of angular speed are radians per second. Technically, this is $1/s$ or s^{-1}, since the radian is unitless. But, it is useful to keep the rad in there to indicate that the quantity is angular speed. The **instantaneous angular speed (ω)** is given by considering a very small time interval—that is, as Δt approaches zero.

As in the linear case, if the angular speed is *constant*, then $\bar{\omega} = \omega$. Taking θ_o and t_o to be zero in Eq. 7.4,

$$\omega = \frac{\theta}{t} \quad \text{or} \quad \theta = \omega t \quad \textit{istantaneous angular speed} \tag{7.5}$$

SI unit of angular speed: radians per second (rad/s or s^{-1})

Another common descriptive unit for angular speed is revolutions per minute (rpm); for example, a CD (compact disc) rotates at a speed of 200–500 rpm (depending on the location of the track). This nonstandard unit of revolutions per

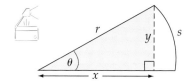

Demonstration/activity: Have students hold a transparent ruler at arm's length and determine the angle subtended by such things as a window in a nearby building, a car, or the height of a person one block away. Then have the students calculate either the size of the object or its distance from them.

Teaching tip: Point out the similarities between linear motion and rotational motion. Emphasize the vector properties of both types of motion.

Exploration 10.1 *Constant Angular Velocity Equation*

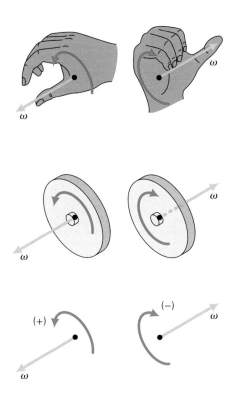

Illustration 10.2 Motion about a Fixed Axis

▲ **FIGURE 7.5 Angular velocity**
The direction of the angular-velocity vector for an object in rotational motion is given by the right-hand rule: When the fingers of the right hand are curled in the direction of the rotation, the extended thumb points in the direction of the angular-velocity vector. Circular senses or directions are commonly indicated by (a) plus and (b) minus signs.

minute is readily converted to radians per second, since 1 revolution = 2π rad. For example, $(150 \text{ rev/min})(2\pi \text{ rad/rev})(1 \text{ min/60 s}) = 5.0\pi \text{ rad/s} (= 16 \text{ rad/s}).^*$

The **average and instantaneous angular velocities** are analogous to their linear counterparts. Angular velocity is associated with angular displacement. Both are vectors and thus have direction; however, this directionality is, by convention, specified in a special way. In one-dimensional, or linear, motion, a particle can go only in one direction or the other (+ or −), so the displacement and velocity vectors can have only these two directions. In the angular case, a particle moves one way or the other, but the motion is along its *circular path*. Thus, the angular-displacement and angular-velocity vectors of a particle in circular motion can have only two directions, which correspond to going around the circular path with either increasing or decreasing angular displacement from θ_o—that is, counterclockwise or clockwise. Let's focus on the angular-velocity vector $\vec{\omega}$. (The direction of the angular displacement will be the same as that of the angular velocity. Why?)

The *direction* of the angular-velocity vector is given by a *right-hand rule*, as illustrated in ◂Fig. 7.5a. When the fingers of your right hand are curled in the direction of the circular motion, your extended thumb points in the direction of $\vec{\omega}$. Note since circular motion can be in only one of two circular *senses*, clockwise or counterclockwise, plus and minus signs can be used to distinguish circular rotation directions. It is customary to take a counterclockwise rotation as positive (+), since positive angular distance (and displacement) is conventionally measured counterclockwise from the positive x-axis.

Why not just designate the direction of the angular-velocity vector to be either clockwise or counterclockwise? This designation is not used because clockwise (cw) and counterclockwise (ccw) are directional senses or indications rather than actual directions. These rotational senses are like right and left. If you faced another person and each of you were asked whether something was on the right or left, your answers would disagree. Similarly, if you held this book up toward a person facing you and rotated it, would it be rotating cw or ccw for both of you?

We can use cw and ccw to indicate rotational "directions" when they are specified relative to a reference—for example, the positive x-axis as in the preceding discussion. Referring to Fig. 7.5, imagine yourself being first on one side of one of the rotating disks and then on the other. Then apply the right-hand rule on both sides. You should find that the direction of the angular velocity vector is the same for both locations (because it is referenced to the right hand). Relative to this vector—for example, looking at the tip—there is no ambiguity in using + and − to indicate rotational senses or directions.

Relationship between Tangential and Angular Speeds

A particle moving in a circle has an instantaneous velocity tangential to its circular path. For a constant angular velocity and speed, the particle's orbital speed, or **tangential speed**, v (the magnitude of the tangential velocity), is also constant. How the angular and tangential speeds are related is revealed by starting with Eq. 7.3 ($s = r\theta$) and Eq. 7.5 ($\theta = \omega t$):

$$s = r\theta = r(\omega t)$$

The arc length, or distance, is also given by

$$s = vt$$

Combining the equations for s gives the relationship between the tangential speed (v) and the angular speed (ω),

$$v = r\omega \qquad \text{(7.6)}$$

tangential speed relation to angular speed for circular motion

*It is often convenient to leave π in symbol form.

where ω is in radians per second. Equation 7.6 holds in general for instantaneous tangential and angular speeds for solid- or rigid-body rotation about a fixed axis, even when ω might vary with time.

Note that all the particles of a solid object rotating with constant angular velocity have the same angular speed, but the tangential speeds are different at different distances from the axis of rotation (►Fig. 7.6a).

Example 7.3 ■ Merry-Go-Rounds: Do Some Go Faster Than Others?

An amusement-park merry-go-round at its constant operational speed makes one complete rotation in 45 s. Two children are on horses, one at 3.0 m from the center of the ride and the other farther out, 6.0 m from the center. What are (a) the angular speed and (b) the tangential speed of each child?

Thinking It Through. The angular speed of each child is the same, since both children make a complete rotation in the same time. However, the tangential speeds will be different, because the radii are different. That is, the child at the greater radius travels in a larger circle during the rotation time and thus must travel faster.

Solution.

Given: $\theta = 2\pi$ rad (one rotation)   *Find:* (a) ω_1 and ω_2 (angular speeds)
$t = 45$ s   (b) v_1 and v_2 (tangential speeds)
$r_1 = 3.0$ m
$r_2 = 6.0$ m

(a) As noted, $\omega_1 = \omega_2$, that is, both riders rotate at the same angular speed. All points on the merry-go-round travel through 2π rad in the time it takes to make one rotation. The angular speed can be found from Eq. 7.5 (constant ω) as

$$\omega = \frac{\theta}{t} = \frac{2\pi \text{ rad}}{45 \text{ s}} = 0.14 \text{ rad/s}$$

Hence, $\omega = \omega_1 = \omega_2 = 0.14$ rad/s.

(b) The tangential speed is different at different radial locations on the merry-go-round. All of the "particles" making up the merry-go-round go through one rotation in the same amount of time. Therefore, the farther a particle is from the center, the longer its circular path will be, and the greater is its tangential speed, as Eq. 7.6 indicates. (See also Fig. 7.6a.) Thus,

$$v_1 = r_1\omega = (3.0 \text{ m})(0.14 \text{ rad/s}) = 0.42 \text{ m/s}$$

and

$$v_2 = r_2\omega = (6.0 \text{ m})(0.14 \text{ rad/s}) = 0.84 \text{ m/s}$$

(Note that the unitless radian has been dropped from the answer.)

Then, a rider on the outer part of the ride has a greater tangential speed than a rider closer to the center. Here, rider 2 has a radius twice that of rider 1 and therefore goes twice as fast.

Follow-Up Exercise. (a) On an old 45-rpm record, the beginning track is 8.0 cm from the center, and the end track is 5.0 cm from the center. What are the angular speeds and the tangential speeds at these distances when the record is spinning at 45 rpm? (b) Why, on oval racetracks, do inside and outside runners have different starting points (called a "staggered" start), such that some runners start "ahead" of others?

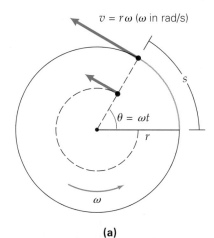

$v = r\omega$ (ω in rad/s)

$\theta = \omega t$

(a)

(b)

▲ **FIGURE 7.6 Tangential and angular speeds** (a) Tangential and angular speeds are related by $v = r\omega$, where ω is in radians per second. Note that all of the particles of an object rotating about a fixed axis travel in circles. All the particles have the same angular speed ω, but particles at different distances from the axis of rotation have different tangential speeds. (b) Sparks from a grinding wheel provide a graphic illustration of instantaneous tangential velocity. (Why do the paths curve slightly?)

Note: Whenever a tangential or linear quantity is calculated from an angular quantity, the angular unit radian is dropped in the final answer. When an angular quantity is asked for, the unit radian is usually included in the final answer for clarity.

Period and Frequency

Some other quantities commonly used to describe circular motion are period and frequency. The time it takes an object in circular motion to make one complete revolution, or *cycle*, is called the **period** (T). For example, the period of revolution of the Earth about the Sun is one year, and the period of the Earth's axial rotation is 24 h. The standard unit of period is the second (s). The period is sometimes given in seconds per revolution (s/rev) or seconds per cycle (s/cycle).

Note: Frequency (f) and period (T) are inversely related.

Demonstration/activity: Use a strobe light to view a fan or other rotating object, and explain how to measure the frequency and angular frequency.

Closely related to the period is the **frequency** (f), which is the number of revolutions, or cycles, made in a given time, generally a second. For example, if a particle traveling uniformly in a circular orbit makes 5.0 revolutions in 2.0 s, the frequency (of revolution) is $f = 5.0$ rev/2.0 s $= 2.5$ rev/s, or 2.5 cycles/s (cps, or cycles per second).

Revolution and *cycle* are merely descriptive terms used for convenience and are *not* units. Without these descriptive terms, we see that the unit of frequency is inverse seconds (1/s, or s^{-1}), which is called the **hertz (Hz)** in the SI.

The two quantities are inversely related by

$$f = \frac{1}{T} \quad \text{frequency and period} \tag{7.7}$$

SI unit of frequency: hertz (Hz, 1/s or s^{-1})

where the period is in seconds and the frequency is in hertz, or inverse seconds.

For uniform circular motion, the tangential orbital speed is related to the period T by $v = 2\pi r/T$—that is, the distance traveled in one revolution divided by the time for one revolution (one period). The frequency can also be related to the angular speed.

Since an angular distance of 2π rad is traveled in one period (by definition of the period), we have

$$\omega = \frac{2\pi}{T} = 2\pi f \quad \begin{array}{l} \text{angular speed} \\ \text{in terms of period and frequency} \end{array} \tag{7.8}$$

Notice that ω and f have the same units, $\omega =$ rad/s $= 1/s$ and $2\pi f =$ rad/s $= 1/s$. This notation can easily cause confusion, which is why the unitless radian term is often added.

Example 7.4 ■ Frequency and Period: An Inverse Relationship

A compact disc (CD) rotates in a player at a constant speed of 200 rpm. What are the CD's (a) frequency and (b) period of revolution?

Thinking It Through. We can use the relationships for the frequency (f), the period (T), and the angular frequency ω, expressed in Eqs. 7.7 and 7.8.

Solution. The angular speed is not in standard units and so must be converted. Revolutions per minute (rpm) can be converted to radians per second (rad/s).

Given: $\omega = \left(\dfrac{200 \text{ rev}}{\text{min}}\right)\left(\dfrac{1 \text{ min}}{60 \text{ s}}\right)\left(\dfrac{2\pi \text{ rad}}{\text{rev}}\right) = 20.9$ rad/s   **Find:** (a) f (frequency)
(b) T (period)

[Note the above unit conversion could be done using one convenient factor: $1(\text{rev/min}) = (\pi/30)$ rad/s.]

(a) Rearranging Eq. 7.8 and solving for f, we get

$$f = \frac{\omega}{2\pi} = \frac{20.9 \text{ rad/s}}{2\pi \text{ rad/cycle}} = 3.33 \text{ Hz}$$

The units of 2π are rad/cycle or revolution, so the result is in cycles/second or inverse seconds, which is the hertz.

(b) Equation 7.8 could be used to find T, but Eq. 7.7 is a bit simpler:

$$T = \frac{1}{f} = \frac{1}{3.33 \text{ Hz}} = 0.300 \text{ s}$$

Thus, the CD takes 0.300 s to make one revolution. (Notice that since Hz $= 1/s$, the equation is dimensionally correct.)

Follow-Up Exercise. If the period of a particular CD is 0.500 s, what is the CD's angular speed in revolutions per minute?

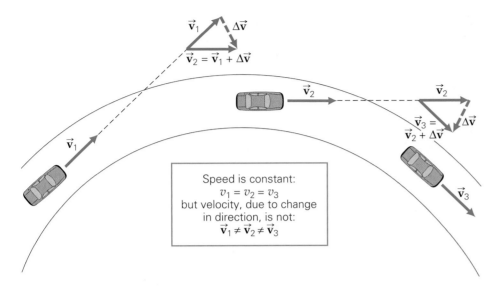

Illustration 3.5 Uniform Circular Motion and Acceleration

Note: Review the discussion of curvilinear motion in Section 3.1.

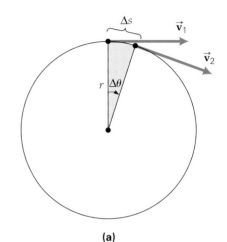

(a)

$$\vec{v}_2 = \vec{v}_1 + \Delta\vec{v} \text{ or}$$
$$\vec{v}_2 - \vec{v}_1 = \Delta\vec{v}$$

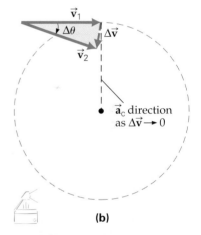

(b)

▲ **FIGURE 7.7 Uniform circular motion** The speed of an object in uniform circular motion is constant, but the object's velocity changes in the direction of motion. Thus, there is an acceleration.

7.3 Uniform Circular Motion and Centripetal Acceleration

OBJECTIVES: To (a) explain the causes of centripetal acceleration in constant or uniform circular motion, and (b) compute centripetal acceleration.

A simple, but important, type of circular motion is **uniform circular motion**, which occurs when an object moves at a constant speed in a circular path. An example of this movement is a car going around a circular track (▲Fig. 7.7). The motion of the Moon around the Earth is approximated by uniform circular motion. Such motion is curvilinear, so you know from the discussion in Chapter 3 that there must be an acceleration. But what are its magnitude and direction?

Centripetal Acceleration

The acceleration of uniform circular motion is not in the same direction as the instantaneous velocity (which is tangent to the circular path at any point). If it were, the object would speed up, and the circular motion wouldn't be uniform. Recall that acceleration is the time rate of change of velocity and that velocity has both *magnitude* and *direction*. In uniform circular motion, the direction of the velocity is continuously changing, which is a clue to the direction of the acceleration. (See Fig. 7.7.)

The velocity vectors at the beginning and end of a time interval give the change in velocity, or $\Delta\vec{v}$, via vector subtraction. All of the instantaneous-velocity vectors have the same magnitude or length (constant speed), but they differ in direction. Note that because $\Delta\vec{v}$ is not zero, there must be an acceleration ($\vec{a} = \Delta\vec{v}/\Delta t$).

As illustrated in ▶Fig. 7.8, as Δt (or $\Delta\theta$) becomes smaller, $\Delta\vec{v}$ points more toward the center of the circular path. As Δt approaches zero, the instantaneous change in the velocity, and therefore the acceleration, points exactly toward the center of the circle. As a result, the acceleration in uniform circular motion is called **centripetal acceleration**, which means "center-seeking" acceleration (from the Latin *centri*, "center," and *petere*, "to fall toward" or "to seek").

The centripetal acceleration must be directed radially inward, that is, with no component in the direction of the perpendicular (tangential) velocity, or else the magnitude of that velocity would change (▼Fig. 7.9). Note that for an object in uniform circular motion, the direction of the centripetal acceleration is continuously changing. In terms of x- and y-components, a_x and a_y are not constant. Can you describe how this differs from the acceleration in projectile motion?

▲ **FIGURE 7.8 Analysis of centripetal acceleration (a)** The velocity vector of an object in uniform circular motion is constantly changing direction. **(b)** As Δt, the time interval for $\Delta\theta$, is taken to be smaller and smaller and approaches zero, $\Delta\vec{v}$ (the change in the velocity, and therefore an acceleration) is directed toward the center of the circle. The result is a centripetal, or center-seeking, acceleration which has a magnitude of $a_c = v^2/r$.

Teaching tip: Make sure students understand that, although the speed is constant, the velocity is changing, because of the changing direction.

The magnitude of the centripetal acceleration can be deduced from the small shaded triangles in Fig. 7.8. (For very short time intervals, the arc length Δs is almost a straight line—the chord.) These two triangles are similar, because each has a pair of equal sides surrounding the same angle $\Delta\theta$. (Note that the velocity vectors have the same magnitude.) Thus, Δv is to v as Δs is to r, which we can write as

$$\frac{\Delta v}{v} \approx \frac{\Delta s}{r}$$

The arc length Δs is the distance traveled in time Δt; thus, $\Delta s = v\Delta t$, so

$$\frac{\Delta v}{v} \approx \frac{\Delta s}{r} = \frac{v\Delta t}{r}$$

and

$$\frac{\Delta v}{\Delta t} \approx \frac{v^2}{r}$$

Then, as Δt approaches zero, this approximation becomes exact. The instantaneous centripetal acceleration, $a_c = \Delta v/\Delta t$, thus has a magnitude of

$$a_c = \frac{v^2}{r} \quad \begin{array}{l}\textit{magnitude of cenripetal acceleration}\\ \textit{in terms of tangential speed}\end{array} \quad (7.9)$$

▲ FIGURE 7.9 Centripetal acceleration For an object in uniform circular motion, the centripetal acceleration is directed radially inward. There is no acceleration component in the tangential direction; if there were, the magnitude of the velocity (tangential *speed*) would change.

Note: Centripetal acceleration depends on tangential speed (v) and radius (r).

PHYSLET®

Exploration 3.6 Uniform Circular Motion

Using Eq. 7.6 ($v = r\omega$), we can also write the equation for centripetal acceleration in terms of the angular speed:

$$a_c = \frac{v^2}{r} = \frac{(r\omega)^2}{r} = r\omega^2 \quad \begin{array}{l}\textit{magnitude of cenripetal acceleration}\\ \textit{in terms of angular speed}\end{array} \quad (7.10)$$

Orbiting satellites have centripetal accelerations, and a down-to-Earth medical application of centripetal acceleration is discussed in Insight 7.2.

Example 7.5 ■ A Centrifuge: Centripetal Acceleration

A laboratory centrifuge like that shown in ◄Fig. 7.10 operates at a rotational speed of 12 000 rpm. (a) What is the magnitude of the centripetal acceleration of a red blood cell at a radial distance of 8.00 cm from the centrifuge's axis of rotation? (b) How does this acceleration compare with g?

Thinking It Through. Here, the angular speed and the radius are given, so the magnitude of the centripetal acceleration can be computed directly from Eq. 7.10. The result can be compared with g by using $g = 9.80$ m/s^2.

Solution. The data are as follows:

Given: $\omega = (1.20 \times 10^4 \text{ rpm})\left[\dfrac{(\pi/30)\text{rad/s}}{\text{rpm}}\right]$ *Find:* (a) a_c
           $= 1.26 \times 10^3$ rad/s   (b) how a_c compares with g
           $r = 8.00$ cm $= 0.0800$ m

(a) The centripetal acceleration is found from Eq. 7.10:

$$a_c = r\omega^2 = (0.0800 \text{ m})(1.26 \times 10^3 \text{ rad/s})^2 = 1.27 \times 10^5 \text{ m/s}^2$$

(b) Using the relationship $1\,g = 9.80$ m/s^2 to express a_c in terms of g, we have

$$a_c = (1.27 \times 10^5 \text{ m/s}^2)\left(\frac{1\,g}{9.80 \text{ m/s}^2}\right) = 1.30 \times 10^4\,g \; (= 13\,000\,g!)$$

Follow-Up Exercise. (a) What angular speed in revolutions per minute would give a centripetal acceleration of $1\,g$ at the radial distance in this Example, and, taking gravity into account, what would be the resultant acceleration? (b) Compare the effect of gravity on the spinning tubes at the rotational speeds in the Example and in part (a) of this Follow-Up Exercise.

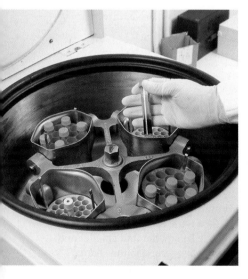

▲ FIGURE 7.10 Centrifuge Centrifuges are used to separate particles of different sizes and densities suspended in liquids. For example, red and white blood cells can be separated from each other and from the plasma that makes up the liquid portion of the blood in the centrifuge tube.

INSIGHT 7.1 THE CENTRIFUGE: SEPARATING BLOOD COMPONENTS

The centrifuge is a machine with rotating parts that is used to separate particles of different sizes and densities suspended in a liquid (or a gas). For example, cream is separated from milk by centrifuging, and blood components are separated in centrifuges in medical laboratories (see Fig. 7.10).

There is a much slower process to separate blood components. They will eventually settle in layers in a vertical test tube—a process called *sedimentation*—under the influence of normal gravity alone. The viscous drag of the plasma on the particles is analogous to (but much greater than) the air resistance that determines the terminal velocity of falling objects (Section 4.6). Red blood cells settle in the bottom layer of the tube, because they have a greater terminal velocity than do the white blood cells and platelets and reach the bottom sooner. The white cells settle in the next layer and the platelets settle on top. However, gravitational sedimentation is generally a very slow process.

The erythrocyte sedimentation rate (ESR) has some diagnostic value, but clinicians usually do not want to wait a long time to see the fractional volume of red cells (erythrocites) in the blood or separate them from the plasma. And so, centrifugation is used to speed up the process. Centrifuge tubes are pivoted and spin horizontally. The resistance of the fluid medium on the particles supplies the centripetal acceleration that keeps them moving in slowly widening circles as they toward the bottom of the tube. The bottom of the tube itself must exert a strong force on the contents as a whole and must be strong enough not to break.

Laboratory centrifuges commonly operate at speeds sufficient to produce centripetal accelerations thousands of times larger than g. (See Example 7.5.) Since the principle of the centrifuge involves centripetal acceleration, perhaps "centripuge" would be a more descriptive name.

Centripetal Force

For an acceleration to exist, there must be a net force. Thus, for a centripetal (inward) acceleration to exist, there must be a **centripetal force** (net inward force). Expressing the magnitude of this force in terms of Newton's second law ($\vec{F}_{net} = m\vec{a}$) and inserting the expression for centripetal acceleration from Eq. 7.9, we can write

PHYSLET®

Exploration 5.1 Circular Motion

$$F_c = ma_c = \frac{mv^2}{r} \quad \textit{magnitude of centripetal force} \quad (7.11)$$

The centripetal force, like the centripetal acceleration, is directed radially toward the center of the circular path.

PHYSLET®

Illustration 5.3 The Ferris Wheel

Conceptual Example 7.6 ■ Breaking Away

A ball attached to a string is swung with uniform motion in a horizontal circle above a person's head (▼Fig. 7.11a). If the string breaks, which of the trajectories shown in Fig. 7.11b (viewed from above) would the ball follow?

Reasoning and Answer. When the string breaks, the centripetal force goes to zero. There is no force in the outward direction, so the ball could not follow trajectory (a). Newton's first law states that if no force acts on an object in motion, the object will continue to move in a straight line. This factor rules out trajectories *b*, *d*, and *e*.

It should be evident from the previous discussion that at any instant (including the instant when the string breaks), the isolated ball has a horizontal, tangential velocity. The downward force of gravity acts on it, but this force affects only its vertical motion, which is not visible in Fig. 7.11b. The ball thus flies off tangentially and is essentially a horizontal projectile (with $v_{x_o} = v$, $v_{y_o} = 0$, and $a_y = -g$). Viewed from above, the ball would appear to follow the path labeled *c*.

Follow-Up Exercise. If you swing a ball in a horizontal circle about your head, can the string be exactly horizontal? (See Fig. 7.11a.) Explain your answer. [*Hint:* Analyze the forces acting on the ball.]

Keep in mind that, in general, a net force applied at an angle to the direction of motion of an object produces changes in the magnitude *and* direction of the velocity. However, when a net force of constant magnitude is continuously applied at an angle of 90° to the direction of motion (as is centripetal force), only the direction of the velocity changes. This is because there is no force component parallel to the velocity. Also

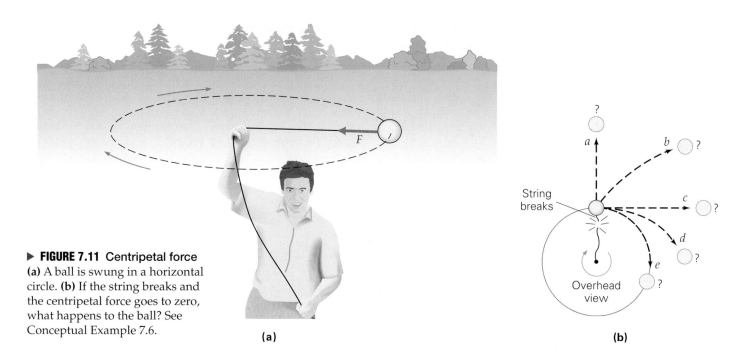

▶ **FIGURE 7.11** Centripetal force **(a)** A ball is swung in a horizontal circle. **(b)** If the string breaks and the centripetal force goes to zero, what happens to the ball? See Conceptual Example 7.6.

(a)

(b)

notice that because the centripetal force is always perpendicular to the direction of motion, this force does no work. (Why?) Therefore, by the work–energy theorem (Section 5.3), a centripetal force does not change the kinetic energy or speed of the object.

Note that the centripetal force in the form $F_c = mv^2/r$ is *not* a new individual force, but rather the cause of the centripetal acceleration and supplied by either a real force or the vector sum of several forces.

The force supplying the centripetal acceleration for satellites is gravity. In Conceptual Example 7.6, it was the tension in the string. Another force that often supplies centripetal acceleration is friction. Suppose that an automobile moves into a level, circular curve. To negotiate the curve, the car must have a centripetal acceleration, which is supplied by the force of friction between the tires and the road.

However, this (static; why?) friction has a maximum limiting value. If the speed of the car is high enough or the curve is sharp enough, the friction will not be sufficient to supply the necessary centripetal acceleration, and the car will skid outward from the center of the curve. If the car moves onto a wet or icy spot, the friction between the tires and the road may be reduced, allowing the car to skid at an even lower speed. (Banking a curve helps vehicles negotiate the curve without slipping.)

Example 7.7 ■ Where the Rubber Meets the Road: Friction and Centripetal Force

A car approaches a level, circular curve with a radius of 45.0 m. If the concrete pavement is dry, what is the maximum speed at which the car can negotiate the curve at a constant speed?

Thinking It Through. The car is in uniform circular motion on the curve, so there must be a centripetal force. This force is supplied by static friction, so the maximum static frictional force provides the centripetal force when the car is at its maximum tangential speed.

Solution. Writing down what is given and what is to be found,

Given: $r = 45.0$ m   *Find:* v (maximum speed)

To go around the curve at a particular speed, the car must have a centripetal acceleration, and therefore a centripetal force must act on it. This inward force is supplied by static friction between the tires and the road. (The tires are not slipping or skidding relative to the road.)

Recall from Chapter 4 that the maximum frictional force is given by $f_{s_{max}} = \mu_s N$ (Eq. 4.7), where N is the magnitude of the normal force on the car and is equal in magnitude to the weight of the car, mg, on the level road (why?). Thus the magnitude of the maximum static frictional force is equal to the magnitude of the centripetal force ($F_c = mv^2/r$).

From this we can find the maximum speed. To find $f_{s_{max}}$, the coefficient of friction between rubber and concrete is needed and from Table 4.1, it is $\mu_s = 1.20$. Then,

$$f_{s_{max}} = F_c$$

$$\mu_s N = \mu_s mg = \frac{mv^2}{r}$$

So

$$v = \sqrt{\mu_s rg} = \sqrt{(1.20)(45.0 \text{ m})(9.80 \text{ m/s}^2)} = 23.0 \text{ m/s}$$

(about 83 km/h, or 52 mi/h).

Follow-Up Exercise. Would the centripetal force be the same for all types of vehicles in this Example?

The proper safe speed for driving on a highway curve is an important consideration. The coefficient of friction between tires and the road may vary, depending on weather, road conditions, the design of the tires, the amount of tread wear, and so on. When a curved road is designed, safety may be promoted by banking, or inclining, the roadway. This design reduces the chances of skidding because the normal force exerted on the car by the road then has a component toward the center of the curve that reduces the need for friction. In fact, for a circular curve with a given banking angle and radius, there is one speed for which no friction is required at all. This condition is used in banking design. (See Exercise 45.)

Let's look at one more example of centripetal force, this time with two objects in uniform circular motion. Example 7.8 will help you better understand the motions of satellites in circular orbits, discussed in a later section.

Example 7.8 ■ Strung Out: Centripetal Force and Newton's Second Law

Suppose that two masses, $m_1 = 2.5$ kg and $m_2 = 3.5$ kg, are connected by light strings and are in uniform circular motion on a horizontal frictionless surface as illustrated in ►Fig. 7.12, where $r_1 = 1.0$ m and $r_2 = 1.3$ m. The forces acting on the masses are $T_1 = 4.5$ N and $T_2 = 2.9$ N, which are the tensions in the strings, respectively. Find the magnitude of the centripetal acceleration and the tangential speed of (a) mass m_2 and (b) mass m_1.

Thinking It Through. The centripetal forces on the masses are supplied by the tensions (T_1 and T_2) in the strings. By isolating the masses, we can find a_c for each mass, because the net force on a mass is equal to the mass's centripetal force ($F_c = ma_c$). The tangential speeds can then be found, since the radii are known ($a_c = v^2/r$).

Solution.

Given: $r_1 = 1.0$ m and $r_2 = 1.3$ m
$m_1 = 2.5$ kg and $m_2 = 3.5$ kg
$T_1 = 4.5$ N
$T_2 = 2.9$ N

Find: centripetal acceleration (a_c) and tangential speed (v) of
(a) m_2
(b) m_1

(a) By isolating m_2 in the figure, it can seen that the centripetal force is provided by the tension in the string. (T_2 is the only force acting on m_2 toward the center of its circular path.) Thus,

$$T_2 = m_2 a_{c_2}$$

and

$$a_{c_2} = \frac{T_2}{m_2} = \frac{2.9 \text{ N}}{3.5 \text{ kg}} = 0.83 \text{ m/s}^2$$

where the acceleration is toward the center of the circle.

The tangential speed of m_2 can be found from $a_c = v^2/r$:

$$v_2 = \sqrt{a_{c_2} r_2} = \sqrt{(0.83 \text{ m/s}^2)(1.3 \text{ m})} = 1.0 \text{ m/s}$$

(b) The situation is a bit different for m_1. In this case, two radial forces are acting on m_1: the string tensions T_1 (inward) and $-T_2$ (outward). By Newton's second law, in order to have a centripetal acceleration, there must be a net force, which is given by the difference in the two tensions, so we expect $T_1 > T_2$, and

$$F_{net_1} = +T_1 + (-T_2) = m_1 a_{c_1} = \frac{m_1 v_1^2}{r_1}$$

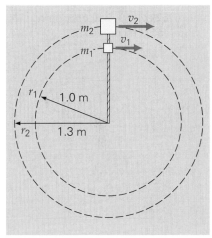

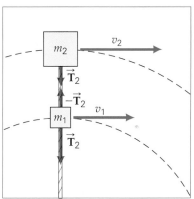

▲ **FIGURE 7.12** Centripetal force and Newton's second law See Example 7.8.

(continues on next page)

where the radial direction (toward the center of the circular path) is taken to be positive. Then

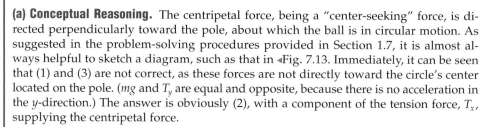

$$a_{c_1} = \frac{T_1 - T_2}{m_1} = \frac{4.5 \text{ N} + (-2.9 \text{ N})}{2.5 \text{ kg}} = 0.64 \text{ m/s}^2$$

and

$$v_1 = \sqrt{a_{c_1} r_1} = \sqrt{(0.64 \text{ m/s}^2)(1.0 \text{ m})} = 0.80 \text{ m/s}$$

Follow-Up Exercise. Notice in this Example that the centripetal acceleration of m_2 is greater than that of m_1, yet $r_2 > r_1$, and $a_c \propto 1/r$. Is something wrong here? Explain.

Integrated Example 7.9 ■ Center-Seeking Force: One More Time

A 1.0-m cord is used to suspend a 0.50-kg tetherball from the top of the pole. After being hit several times, the ball goes around the pole in uniform circular motion with a tangential speed of 1.1 m/s at an angle of 20° relative to the pole. (a) The force that supplies the centripetal acceleration is (1) the weight of the ball, (2) a component of the tension force in the string, (3) the total tension in the string. (b) What is the magnitude of the centripetal force?

(a) Conceptual Reasoning. The centripetal force, being a "center-seeking" force, is directed perpendicularly toward the pole, about which the ball is in circular motion. As suggested in the problem-solving procedures provided in Section 1.7, it is almost always helpful to sketch a diagram, such as that in ◂Fig. 7.13. Immediately, it can be seen that (1) and (3) are not correct, as these forces are not directly toward the circle's center located on the pole. (mg and T_y are equal and opposite, because there is no acceleration in the y-direction.) The answer is obviously (2), with a component of the tension force, T_x, supplying the centripetal force.

(b) Quantitative Reasoning and Solution. T_x supplies the centripetal force, and the given data are for the dynamical form of the centripetal force, that is, $T_x = F_c = mv^2/r$ (Eq. 7.11).

Given: $L = 1.0 \text{ m}$ *Find:* F_c (magnitude of the centripetal force)
$v = 1.1 \text{ m/s}$
$m = 0.50 \text{ kg}$
$\theta = 20°$

As pointed out previously, the magnitude of the centripetal force may be found using Eq. 7.11:

$$F_c = T_x = \frac{mv^2}{r}$$

But, we need the radial distance r. From the figure, this quantity can be seen to be $r = L \sin 20°$, so

$$F_c = \frac{mv^2}{L \sin 20°} = \frac{(0.50 \text{ kg})(1.1 \text{ m/s})^2}{(1.0 \text{ m})(0.342)} = 1.8 \text{ N}$$

Follow-Up Exercise. What is the magnitude of the tension T in the string?

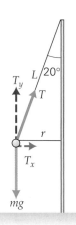

▲ **FIGURE 7.13 Ball on a string** See Integrated Example 7.9.

Exploration 5.2 Force an Object Around a Circle

7.4 Angular Acceleration

OBJECTIVES: To (a) define angular acceleration, and (b) analyze rotational kinematics.

As you might have guessed, there is another type of acceleration besides linear, and that is *angular acceleration*. This quantity is the time rate of change of angular velocity. In circular motion, if there is an angular acceleration, the motion is not uniform, because the speed and/or direction would be changing. Analogous to the linear case, the magnitude of the **average angular acceleration ($\overline{\alpha}$)** is given by

$$\overline{\alpha} = \frac{\Delta \omega}{\Delta t}$$

where the bar over the alpha indicates that it is an average value, as usual. Taking $t_o = 0$, and if the angular acceleration is constant, so that $\bar{\alpha} = \alpha$, we have

$$\alpha = \frac{\omega - \omega_o}{t} \quad \text{(constant angular acceleration)}$$

SI unit of angular acceleration:
radians per second squared (rad/s^2)

or

$$\omega = \omega_o + \alpha t \quad \begin{array}{l}\text{(constant angular}\\ \text{acceleration only)}\end{array} \qquad (7.12)$$

No boldface vector symbols with over arrows are used in Eq. 7.12, because, plus and minus signs will be used to indicate angular directions, as described earlier. As in the case of linear motion, if the angular acceleration increases the angular velocity, both quantities have the same sign, meaning that their vector directions are the same (that is, α is in the same direction as ω as given by the right-hand rule). If the angular acceleration decreases the angular velocity, then the two quantities have opposite signs, meaning that their vectors are opposed (that is, α is in the direction opposite to ω as given by the right-hand rule, or is an angular deceleration, so to speak).

Teaching tip: Students often confuse tangential and centripetal accelerations. It may be helpful to discuss the following question: Is it possible for a particle in circular motion to have a tangential acceleration and a centripetal acceleration at the same time?

Example 7.10 ■ A Rotating CD: Angular Acceleration

A CD accelerates uniformly from rest to its operational speed of 500 rpm in 3.50 s. (a) What is the angular acceleration of the CD during this time? (b) What is the angular acceleration of the CD after this time? (c) If the CD comes uniformly to a stop in 4.50 s, what is its angular acceleration during that part of the motion?

Thinking It Through. (a) We are given the initial and final angular velocities; the constant (uniform) angular acceleration can be calculated (Eq. 7.12), since the amount of time during which the CD accelerates is known. (b) Keep in mind that the operational angular speed is constant. (c) Everything is given for Eq. 7.12, but a negative result should be expected. Why?

Solution.

Given: $\omega_o = 0$

$\omega = (500 \text{ rpm})\left[\dfrac{(\pi/30) \text{ rad/s}}{\text{rpm}}\right] = 52.4 \text{ rad/s}$

$t = 3.50$ s (starting up)
$t = 4.50$ s (coming to a stop)

Find:
(a) α (during start-up)
(b) α (in operation)
(c) α (in coming to a stop)

(a) Using Eq. 7.12,

$$\alpha = \frac{\omega - \omega_o}{t} = \frac{52.4 \text{ rad/s} - 0}{3.50 \text{ s}} = 15.0 \text{ rad/s}^2$$

in the direction of the angular velocity.

(b) After the CD reaches its operational speed, the angular velocity remains constant, so $\alpha = 0$.

(c) Again using Eq. 7.12, but this time with $\omega_o = 500$ rpm and $\omega = 0$,

$$\alpha = \frac{\omega - \omega_o}{t} = \frac{0 - 52.4 \text{ rad/s}}{4.50 \text{ s}} = -11.6 \text{ rad/s}^2$$

where the minus sign indicates that the angular acceleration is in the direction opposite that of the angular velocity (which is taken as +).

Follow-Up Exercise. (a) What are the directions of the $\vec{\omega}$ and $\vec{\alpha}$ vectors in part (a) of this Example if the CD rotates clockwise when viewed from above? (b) Do the directions of these vectors change in part (c)?

As with arc length and angle ($s = r\theta$) and tangential and angular speeds ($v = r\omega$), there is a relationship between the magnitudes of the tangential acceleration and the angular acceleration. The **tangential acceleration** (a_t) is associated with changes in tangential speed and hence continuously changes direction. The magnitudes of the

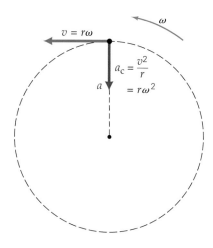

(a) Uniform circular motion
$(a_t = r\,\alpha = 0)$

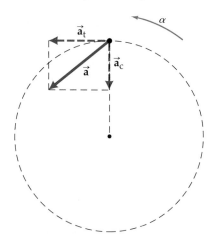

(b) Nonuniform circular motion
$(\vec{a} = \vec{a}_t + \vec{a}_c)$

▲ **FIGURE 7.14 Acceleration and circular motion (a)** In uniform circular motion, there is centripetal acceleration, but no angular acceleration ($\alpha = 0$) or tangential acceleration ($a_t = r\alpha = 0$). **(b)** In nonuniform circular motion, there are angular and tangential accelerations, and the total acceleration is the vector sum of the tangential and centripetal accelerations.

Exploration 10.2 *Constant Angular Acceleration Equation*

tangential and angular accelerations are related by a factor of r. For circular motion with a constant radius r,

$$a_t = \frac{\Delta v}{\Delta t} = \frac{\Delta(r\omega)}{\Delta t} = \frac{r\Delta\omega}{\Delta t} = r\alpha$$

so

$$a_t = r\alpha \quad \textit{magnitude of tangential acceleration} \qquad (7.13)$$

The tangential acceleration (a_t) is written with a subscript t to distinguish it from the radial, or centripetal, acceleration (a_c). Centripetal acceleration is necessary for circular motion, but tangential acceleration is not. For uniform circular motion, there is no angular acceleration ($\alpha = 0$) or tangential acceleration, as can be seen from Eq. 7.13. There is only centripetal acceleration (◄Fig. 7.14a).

However, when there is an angular acceleration α (and therefore a tangential acceleration of magnitude $a_t = r\alpha$), there is a change in *both* the angular *and* tangential velocities. As a result, the centripetal acceleration $a_c = v^2/r = r\omega^2$ must increase or decrease if the object is to maintain the same circular orbit (that is, if r is to stay the same). When there are both tangential and centripetal accelerations, the total instantaneous acceleration is their vector sum (Fig. 7.14b). The tangential-acceleration vector and the centripetal-acceleration vector are perpendicular to each other at any instant, and the total acceleration is $\vec{a} = a_t\hat{t} + a_c\hat{r}$, where $\hat{t}$ and $\hat{r}$ are unit vectors directed tangentially and radially inward, respectively. You should be able to find the magnitude of $\vec{a}$ and the angle it makes relative to $\vec{a}_t$ by using trigonometry (Fig. 7.14b).

Other equations for angular kinematics can be derived, as was done for the linear equations in Chapter 2. That development will not be shown here; the set of angular equations with their linear counterparts for constant accelerations are listed in Table 7.2. A quick review of Chapter 2 (with a change of symbols) will show you how the angular equations are derived.

Example 7.11 ■ Even Cooking: Rotational Kinematics

A microwave oven has a 30-cm diameter rotating plate for even cooking. The plate accelerates from rest at a uniform rate of 0.87 rad/s² for 0.50 s before reaching its constant operational speed. (a) How many revolutions does the plate make before reaching its operational speed? (b) What are the operational angular speed of the plate and the operational tangential speed at its rim?

Thinking It Through. This Example involves the use of the angular kinematic equations (Table 7.2). In (a), the angular distance θ will give the number of revolutions. For (b), first find ω and then $v = r\omega$.

Solution. Listing the given data and what is to be found:

Given: $d = 30$ cm, $r = 15$ cm $= 0.15$ m (radius) *Find:* (a) θ (in revolutions)
$\quad\quad\quad\omega_o = 0$ (at rest) (b) ω and v (angular and
$\quad\quad\quad\alpha = 0.87$ rad/s² tangential speeds,
$\quad\quad\quad t = 0.50$ s respectively)

(a) To find the angular distance θ *in radians*, use Eq. 4 from Table 7.2 with $\theta_o = 0$:

$$\theta = \omega_o t + \tfrac{1}{2}\alpha t^2 = 0 + \tfrac{1}{2}(0.87 \text{ rad/s}^2)(0.50 \text{ s})^2 = 0.11 \text{ rad}$$

Since 2π rad $= 1$ rev,

$$\theta = (0.11 \text{ rad})\left(\frac{1 \text{ rev}}{2\pi \text{ rad}}\right) = 0.018 \text{ rev}$$

so the plate reaches its operational speed in only a small fraction of a revolution.

(b) From Table 7.2, we see that Eq. 3 gives the angular speed, and

$$\omega = \omega_o + \alpha t = 0 + (0.87 \text{ rad/s}^2)(0.50 \text{ s}) = 0.44 \text{ rad/s}$$

Then, Eq. 7.6 gives the tangential speed at the rim radius:

$$v = r\omega = (0.15 \text{ m})(0.44 \text{ rad/s}) = 0.066 \text{ m/s}$$

Follow-Up Exercise. When the oven is turned off, the plate makes half a revolution before stopping. What is the plate's angular acceleration during this period?

TABLE 7.2	Equations for Linear and Angular Motion with Constant Acceleration*	
Linear	**Angular**	
$x = \bar{v}t$	$\theta = \bar{\omega}t$	(1)
$\bar{v} = \dfrac{v + v_o}{2}$	$\bar{\omega} = \dfrac{\omega + \omega_o}{2}$	(2)
$v = v_o + at$	$\omega = \omega_o + \alpha t$	(3)
$x = x_o + v_o t + \frac{1}{2}at^2$	$\theta = \theta_o + \omega_o t + \frac{1}{2}\alpha t^2$	(4)
$v^2 = v_o^2 + 2a(x - x_o)$	$\omega^2 = \omega_o^2 + 2\alpha(\theta - \theta_o)$	(5)

*The first equation in each column is general, that is, not limited to situations where the acceleration is constant.

7.5 Newton's Law of Gravitation

OBJECTIVES: To (a) describe Newton's law of gravitation and how it relates to the acceleration due to gravity, and (b) investigate how the law applies to gravitational potential energy.

Another of Isaac Newton's many accomplishments was the formulation of the **universal law of gravitation**. This law is very powerful and fundamental. Without it, for example, we would not understand the cause of tides or know how to put satellites into particular orbits around the Earth. This law allows us to analyze the motions of planets, comets, stars, and even galaxies. The word *universal* in the name indicates that it is believed to apply everywhere in the universe. (This term highlights the importance of the law, but for brevity, it is common to refer simply to *Newton's law of gravitation* or the *law of gravitation*.)

Newton's law of gravitation in mathematical form gives a simple relationship for the gravitational interaction between two particles, or point masses, m_1 and m_2, separated by a distance r (▶Fig. 7.15a). Basically, every particle in the universe has an attractive gravitational interaction with every other particle because of their masses. The forces of mutual interaction are equal and opposite, forming a force pair as described by Newton's third law (Chapter 4) that is, $\vec{F}_{12} = -\vec{F}_{21}$ in Fig. 7.15a.

The gravitational attraction, or force (F), decreases as the square of the distance (r^2) between two point masses increases; that is, the magnitude of the gravitational force and the distance separating the two particles are related as follows:

$$F \propto \frac{1}{r^2}$$

(This type of relationship is called an *inverse-square law*, that is, F is inversely proportional to r^2.)

Newton's law also correctly postulates that the gravitational force, or attraction of a body, depends on the body's mass—the greater the mass, the greater the attraction. However, because gravity is a mutual interaction between masses, it should be directly proportional to both masses, that is, to their product ($F \propto m_1 m_2$).

Hence, **Newton's law of gravitation** has the form $F \propto m_1 m_2 / r^2$. Expressed as an equation with a constant of proportionality, the magnitude of the mutually attractive gravitational force (F_g) between two masses is given by

$$F_g = \frac{Gm_1 m_2}{r^2} \qquad (7.14)$$

where G is a constant called the **universal gravitational constant** and has a value of

$$G = 6.67 \times 10^{-11}\,\text{N} \cdot \text{m}^2/\text{kg}^2$$

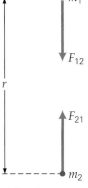

(a) Point masses

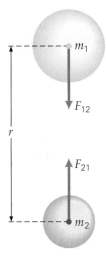

(b) Homogeneous spheres

$$F_{12} = F_{21} = \frac{Gm_1 m_2}{r^2}$$

▲ **FIGURE 7.15 Universal law of gravitation** **(a)** Any two particles, or point masses, are gravitationally attracted to each other with a force that has a magnitude given by Newton's universal law of gravitation. **(b)** For homogeneous spheres, the masses may be considered to be concentrated at their centers.

Newton's universal law of gravitation

Teaching tip: A graph of F versus r could be shown at this time to help students understand the inverse-square law.

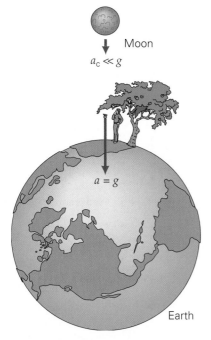

▲ FIGURE 7.16 Gravitational insight? Newton developed his law of gravitation while studying the orbital motion of the Moon. According to legend, his thinking was spurred when he observed an apple falling from a tree. He supposedly wondered whether the force causing the apple to accelerate toward the ground could extend to the Moon and cause it to "fall" or accelerate toward Earth. That is, supply its orbital centripetal acceleration.

Illustration 12.1 Projectile and Satellite Orbits

This constant is often referred to as "big G" to distinguish it from "little g," the acceleration due to gravity. Note from Eq. 7.14 that F_g approaches zero only when r becomes infinitely large. That is, the gravitational force has, or acts over, an *infinite range*.

How did Newton come to his conclusions about the force of gravity? Legend has it that his insight came after he observed an apple fall from a tree to the ground. Newton had been wondering what supplied the centripetal force to keep the Moon in orbit and might have had this thought: "If gravity attracts an apple toward the Earth, perhaps it also attracts the Moon, and the Moon is falling, or accelerating toward the Earth, under the influence of gravity" (◄Fig. 7.16).

Whether or not the legendary apple did the trick, Newton assumed that the Moon and the Earth were attracted to each other and could be treated as point masses, with their total masses concentrated at their centers (Fig 7.15b). The inverse-square relationship had been speculated on by some of his contemporaries. Newton's achievement was demonstrating that the relationship could be deduced from one of Johannes Kepler's laws of planetary motion (Section 7.6).

Newton expressed Eq. 7.14 as a proportion ($F_g \propto m_1 m_2/r^2$) because he did not know the value of G. It was not until 1798 (seventy-one years after Newton's death) that the value of the universal gravitational constant was experimentally determined by an English physicist, Henry Cavendish. Cavendish used a sensitive balance to measure the gravitational force between separated spherical masses (as illustrated in Fig. 7.15b). If F, r, and the m's are known, G can be computed from Eq. 7.14.

As mentioned earlier, Newton considered the nearly spherical Earth and Moon to be point masses located at their respective centers. It took him some years, using mathematical methods he developed, to prove that this is the case only for spherical, *homogeneous* objects.* The concept is illustrated in ►Fig. 7.17.

Example 7.12 ■ Gravitational Attraction between the Earth and the Moon: A Centripetal Force

Compute the magnitude of the mutual gravitational force between the Earth and the Moon. (Assume that the Earth and the Moon are homogeneous spheres.)

Thinking It Through. This Example is an application of Eq. 7.14 for which the masses and the distance may be looked up. (See inside back cover.)

Solution. No data are given, so they must be available from references.

Given: (from tables inside the back cover of this text)
$M_E = 6.0 \times 10^{24}$ kg (mass of the Earth)
$m_M = 7.4 \times 10^{22}$ kg (mass of the Moon)
$r_{EM} = 3.8 \times 10^8$ m (distance between)

Find: F_g (gravitational force)

The average distance from the Earth to the Moon (r_{EM}) is taken to be the distance from the center of one to the center of the other. Using Eq. 7.14, we get

$$F_g = \frac{Gm_1 m_2}{r^2} = \frac{GM_E m_M}{r_{EM}^2}$$

$$= \frac{(6.67 \times 10^{-11}\,\text{N} \cdot \text{m}^2/\text{kg}^2)(6.0 \times 10^{24}\,\text{kg})(7.4 \times 10^{22}\,\text{kg})}{(3.8 \times 10^8\,\text{m})^2}$$

$$= 2.1 \times 10^{20}\,\text{N}$$

This is the magnitude of the centripetal force that keeps the Moon revolving in its orbit around the Earth. It is a very large force, but the Moon is a very massive object, with a correspondingly large inertia. Because of the inward or radial acceleration, it is sometimes said that the Moon is "falling" toward the Earth. This motion, combined with the tangential motion, results in the Moon's nearly circular orbit about the Earth.

Follow-Up Exercise. With what acceleration is the Moon "falling" toward the Earth?

---

*For a homogeneous sphere, the equivalent point mass is located at the center of mass. However, this is a special case. The center of gravitational force and the center of mass of a configuration of particles or an object do not generally coincide.

(a)   (b)

◄ **FIGURE 7.17 Uniform spherical masses (a)** Gravity acts between any two particles. The resultant gravitational force exerted on an object outside a homogeneous sphere by two particles at symmetric locations within the sphere is directed toward the center of the sphere. **(b)** Because of the sphere's symmetry and uniform distribution of mass, the net effect is as though all the mass of the sphere were concentrated as a particle at its center. For this special case, the gravitational center of force and center of mass coincide, but this is generally not true for other objects. (Only a few of the red force arrows are shown because of space considerations.)

The acceleration due to gravity at a particular distance from a planet can also be investigated by using Newton's second law of motion and his law of gravitation. The magnitude of the acceleration due to gravity, which we will generally write as a_g, at a distance r from the center of a spherical mass M is found by setting the force of gravitational attraction due to that spherical mass equal to ma_g. This is the net force on an object of mass m at a distance r:

$$ma_g = \frac{GmM}{r^2}$$

Note: The symbol g is reserved for the acceleration due to gravity at the Earth's surface; a_g is more generally the acceleration due to gravity at some greater radial distance.

Then, the acceleration due to gravity at any distance r from the planet's center is

$$a_g = \frac{GM}{r^2} \tag{7.15}$$

Notice that a_g is proportional to $1/r^2$, so the farther away an object is from the planet, the smaller its acceleration due to gravity and the smaller the attractive force (ma_g) on the object. The force is directed toward the center of the planet.

Equation 7.15 can be applied to the Moon or any planet. For example, taking the Earth to be a point mass M_E located at its center and R_E as its radius, we obtain the acceleration due to gravity at the Earth's surface ($a_{g_E} = g$) by setting the distance r to be equal to R_E.

$$a_{g_E} = g = \frac{GM_E}{R_E^2} \tag{7.16}$$

This equation has several interesting implications. First, it reveals that taking g to be constant everywhere on the surface of the Earth involves the assumption that the Earth has a homogeneous distribution of mass and that the distance from the center of the Earth to any location on its surface is the same. These two assumptions are not exactly true. Therefore, taking g to be a constant is only an approximation, but one that works pretty well for most situations.

Teaching tip: Remind students that r is measured from the Earth's center, not from the surface.

Also, you can see why the acceleration due to gravity is the same for all free-falling objects—that is, independent of the mass of the object. The mass of the object doesn't appear in Eq. 7.16, so all objects in free fall accelerate at the same rate.

Finally, if you're observant, you'll notice that Eq. 7.16 can be used to compute the mass of the Earth. All of the other quantities in the equation are measurable, and their values are known, so M_E can readily be calculated. This is what Cavendish did after he determined the value of G experimentally.

The acceleration due to gravity does vary with altitude. At a distance h above the Earth's surface, we have $r = R_E + h$. The acceleration is then given by

$$a_g = \frac{GM_E}{(R_E + h)^2} \tag{7.17}$$

Problem-Solving Hint

When comparing accelerations due to gravity or gravitational forces, you will often find it convenient to work with ratios. For example, comparing a_g with g (Eqs. 7.15 and 7.16) for the Earth gives

$$\frac{a_g}{g} = \frac{GM_E/r^2}{GM_E/R_E^2} = \frac{R_E^2}{r^2} = \left(\frac{R_E}{r}\right)^2 \quad \text{or} \quad \frac{a_g}{g} = \left(\frac{R_E}{r}\right)^2$$

Note how the constants cancel out. Taking $r = R_E + h$, you can easily compute a_g/g, or the acceleration due to gravity at some altitude h above the Earth compared with g on the Earth's surface (9.80 m/s^2).

Because R_E is very large compared with everyday altitudes above the Earth's surface, the acceleration due to gravity does not decrease very rapidly with height. At an altitude of 16 km (10 mi, about twice as high as modern jet airliners fly), $a_g/g = 0.99$, and thus a_g is still 99% of the value of g at the Earth's surface. At an altitude of 320 km (200 mi), a_g is 91% of g. This is the approximate altitude of an orbiting space shuttle. (Astronauts in orbit do have weight. The so-called weightless condition is discussed in Section 7.6.)

Example 7.13 ■ Geosynchronous Satellite Orbit

Some communication and weather satellites are launched into circular orbits above the Earth's equator so they are *synchronous* (from the Greek *syn-*, same, and *chronos*, time) with the Earth's rotation. That is, they remain "fixed" or "hover" over one point on the equator. At what altitude are these geosynchronous satellites?

Thinking It Through. To remain above one location at the equator, the period of the satellite's revolution must be the same as the Earth's period of rotation—24 h. Also, the centripetal force keeping the satellite in orbit is supplied by the gravitational force of the Earth, $F_g = F_c$. The distance between the center of the Earth and the satellite is $r = R_E + h$. (R_E is the radius of the Earth and h is the height or altitude of the satellite above the Earth's surface.)

Solution. Listing the known data,

Given: T (period) $= 24 \text{ h} = 8.64 \times 10^4 \text{ s}$ *Find:* h (altitude)
$r = R_E + h$
From solar system data inside back cover:
$R_E = 6.4 \times 10^3 \text{ km} = 6.4 \times 10^6 \text{ m}$
$M_E = 6.0 \times 10^{24} \text{ kg}$

Setting the magnitudes of gravitational force and the motional centripetal force equal ($F_g = F_c$), where m is the mass of the satellite, and putting the values in terms of angular speed,

$$F_g = F_c$$

$$\frac{GmM_E}{r^2} = \frac{mv^2}{r} = \frac{m(r\omega)^2}{r} = mr\omega^2$$

and

$$r^3 = \frac{GM_E}{\omega^2} = GM_E\left(\frac{T}{2\pi}\right)^2 = \left(\frac{GM_E}{4\pi^2}\right)T^2$$

using the relationship $\omega = 2\pi/T$. Then substituting values:

$$r^3 = \frac{(6.7 \times 10^{-11} \text{ N}\cdot\text{m}^2/\text{kg}^2)(6.4 \times 10^{24} \text{ kg})(8.64 \times 10^4 \text{ s})^2}{4\pi^2} = 81 \times 10^{21} \text{ m}^3$$

and

$$r = 4.3 \times 10^7 \text{ m}$$

So,

$$h = r - R_E = 4.3 \times 10^7 \text{ m} - 0.64 \times 10^7 = 3.7 \times 10^7 \text{ m}$$
$$= 3.7 \times 10^4 \text{ km} \,(= 23\,000 \text{ mi})$$

Follow-Up Exercise. Show that the period of a satellite in orbit close ($h \ll R_E$) to the Earth's surface (neglecting air resistance) may be approximated by $T^2 \approx 4R_E$ and compute T.

Another aspect of the decrease of g with altitude concerns potential energy. In Chapter 5, you learned that $U = mgh$ for an object at a height h above some zero reference point, since g is essentially constant near the Earth's surface. This potential energy is equal to the work done in raising the object a distance h above the Earth's surface in a *uniform* gravitational field. But what if the change in altitude is so large that g cannot be considered constant while work is done in moving an object, such as a satellite? In this case, the equation $U = mgh$ doesn't apply. In general, it can be shown (using mathematical methods beyond the scope of this book) that the **gravitational potential energy** of two point masses separated by a distance r is given by

$$U = -\frac{Gm_1m_2}{r} \tag{7.18}$$

The minus sign in Eq. 7.18 arises from the choice of the zero reference point (the point where $U = 0$), which is $r = \infty$.

In terms of the Earth and a mass m at an altitude h above the Earth's surface,

$$U = -\frac{Gm_1m_2}{r} = -\frac{GmM_E}{R_E + h} \tag{7.19}$$

where r is the distance separating the Earth's center and the mass. This means that on the Earth we are in a negative gravitational potential energy well (▼Fig. 7.18) that extends to infinity, because the force of gravity has an infinite range. As h increases, so does U. That is, U becomes *less negative*, or gets closer to zero (that is, more positive), corresponding to a higher position in the potential energy well.

▼ **FIGURE 7.18 Gravitational potential energy well** On the Earth, we are in a negative gravitational potential energy well. As with an actual well or hole in the ground, work must be done against gravity to get higher in the well. The potential energy of an object increases as the object moves higher in the well. This means that the value of U becomes less negative. The top of the Earth's gravitational well is at infinity, where the gravitational potential energy is, by choice, zero.

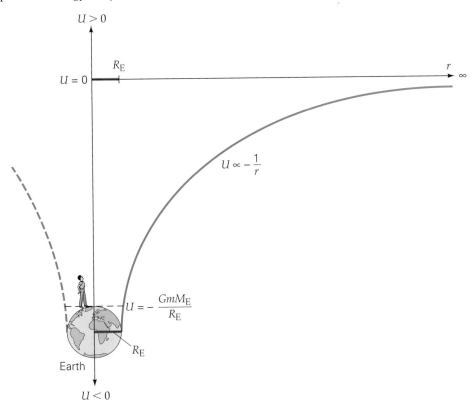

Thus, when gravity does negative work (an object moves higher in the well) or gravity does positive work (an object falls lower in the well), there is a *change* in potential energy. As with finite potential energy wells, this *change* in energy is one of the most important things in analyzing situations such as these.

Example 7.14 ■ Different Orbits: Change in Gravitational Potential Energy

Two 50-kg satellites move in circular orbits about the Earth at altitudes of 1000 km (about 620 mi) and 37000 km (about 23000 mi), respectively. The lower one monitors particles about to enter the atmosphere, and the higher, geosynchronous one takes weather pictures from its stationary position with respect to the Earth's surface over the equator (see Example 7.13). What is the difference in the gravitational potential energies of the two satellites in their respective orbits?

Thinking It Through. The potential energies of the satellites are given by Eq. 7.19. Since an increase in altitude (h) results in a less negative value of U, the satellite with the greater h is higher in the gravitational-potential-energy well and has more gravitational potential energy.

Solution. Listing the data so we can better see what's given (with two significant figures),

Given: $m = 50$ kg
$h_1 = 1000$ km $= 1.0 \times 10^6$ m
$h_2 = 37000$ km $= 37 \times 10^6$ m
$M_E = 6.0 \times 10^{24}$ kg (from the inside the back cover of the book)
$R_E = 6.4 \times 10^6$ m

Find: ΔU (difference in potential energy)

The difference in the gravitational potential energy can be computed directly from Eq. 7.19. Keep in mind that the potential energy is the energy of position, so we compute the potential energies for each position or altitude and subtract one from the other. Thus, we have

$$\Delta U = U_2 - U_1 = -\frac{GmM_E}{R_E + h_2} - \left(-\frac{GmM_E}{R_E + h_1}\right) = GmM_E\left(\frac{1}{R_E + h_1} - \frac{1}{R_E + h_2}\right)$$

$$= (6.67 \times 10^{-11} \text{ N} \cdot \text{m}^2/\text{kg}^2)(50 \text{ kg})(6.0 \times 10^{24} \text{ kg})$$

$$\times \left[\frac{1}{6.4 \times 10^6 \text{ m} + 1.0 \times 10^6 \text{ m}} - \frac{1}{6.4 \times 10^6 \text{ m} + 37 \times 10^6 \text{ m}}\right]$$

$$= +2.2 \times 10^9 \text{ J}$$

Because ΔU is positive, m_2 is higher in the gravitational potential energy well than m_1. Note that even though both U_1 and U_2 are negative, U_2 is "more positive," or "less negative," and closer to zero. Thus, it takes more energy to get a satellite farther from the Earth.

Follow-Up Exercise. Suppose that the altitude of the higher satellite in this Example were doubled, to 72000 km. Would the difference in the gravitational potential energies of the two satellites then be twice as great? Justify your answer.

Note: For communications purposes, many satellites are launched into a circular orbit above the equator at an altitude of about 37000 km. Satellites there are synchronous to the Earth's rotation. That is, they remain "fixed" over one point on the equator; to an observer on the Earth, they are always seen in the same position in the sky.

Substituting the gravitational potential energy (Eq. 7.18) into the equation for the total mechanical energy gives the equation a different form than it had in Chapter 5. For example, the total mechanical energy of a mass m_1 moving near mass m_2 is

$$E = K + U = \tfrac{1}{2}m_1v^2 - \frac{Gm_1m_2}{r} \tag{7.20}$$

This equation and the principle of the conservation of energy can be applied to the Earth's motion about the Sun, by neglecting other gravitational forces. The Earth's orbit is not quite circular, but slightly elliptical. At *perihelion* (the point of the Earth's closest approach to the Sun), the mutual gravitational potential energy is less (a larger negative number) than it is at *aphelion* (the point farthest from the Sun). Therefore, as can be seen from Eq. 7.20 in the form $\tfrac{1}{2}m_1v^2 = E + Gm_1m_2/r$, where E is constant, the Earth's kinetic energy and orbital

Note: the potential energy $U = -Gm_1m_2/r$ is *not* written as *mgh*.

speed are greatest at perihelion (the smallest value of r) and least at aphelion (the greatest value of r). Or, in general, the Earth's orbital speed is greater when it is nearer the Sun than when it is farther away.

Mutual gravitational potential energy also applies to a group, or *configuration*, of more than two masses. That is, there is gravitational potential energy due to the several masses in a configuration, because work was needed to be done in bringing the masses together. Suppose that there is a single fixed mass m_1, and another mass m_2 is brought close to m_1 from an infinite distance (where $U = 0$). The work done against the attractive force of gravity is negative (why?) and equal to the change in the mutual potential energy of the masses, which are now separated by a distance r_{12}; that is, $U_{12} = -Gm_1m_2/r_{12}$.

If a third mass m_3 is brought close to the other two fixed masses, there are then two forces of gravity acting on m_3, so $U_{13} = -Gm_1m_3/r_{13}$ and $U_{23} = -Gm_2m_3/r_{23}$. The total gravitational potential energy of the configuration is therefore

$$U = U_{12} + U_{13} + U_{23}$$
$$= -\frac{Gm_1m_2}{r_{12}} - \frac{Gm_1m_3}{r_{13}} - \frac{Gm_2m_3}{r_{23}} \qquad (7.21)$$

A fourth mass could be brought in to further prove the point, but this development should be sufficient to suggest that the total gravitational potential energy of a configuration of particles is equal to the sum of the individual potential energies for all pairs of particles.

Teaching tip: Explain the significance of the minus sign in the equation for total mechanical energy. Under what conditions can E be negative? What does it mean?

Example 7.15 ■ Total Gravitational Potential Energy: Energy of Configuration

Three masses are in a configuration as shown in ▶Fig. 7.19. What is their total gravitational potential energy?

Thinking It Through. Equation 7.21 applies, but be sure to keep your masses and their distances distinct.

Solution. From the figure, we have

Given: $m_1 = 1.0\ \text{kg}$ *Find:* U (total gravitational potential energy)
$m_2 = 2.0\ \text{kg}$
$m_3 = 2.0\ \text{kg}$
$r_{12} = 3.0\ \text{m}; r_{13} = 4.0\ \text{m}; r_{23} = 5.0\ \text{m}$ (3–4–5 right triangle)

We can use Eq. 7.21 directly, since only three masses are used in this example. (Note that Eq. 7.21 can be extended to any number of masses.) Then,

$$U = U_{12} + U_{13} + U_{23}$$
$$= -\frac{Gm_1m_2}{r_{12}} - \frac{Gm_1m_3}{r_{13}} - \frac{Gm_2m_3}{r_{23}}$$
$$= (6.67 \times 10^{-11}\ \text{N}\cdot\text{m}^2/\text{kg}^2)$$
$$\times \left[-\frac{(1.0\ \text{kg})(2.0\ \text{kg})}{3.0\ \text{m}} - \frac{(1.0\ \text{kg})(2.0\ \text{kg})}{4.0\ \text{m}} - \frac{(2.0\ \text{kg})(2.0\ \text{kg})}{5.0\ \text{m}} \right]$$
$$= -1.3 \times 10^{-10}\ \text{J}$$

Follow-Up Exercise. Explain what the *negative* potential energy in this Example means in physical terms.

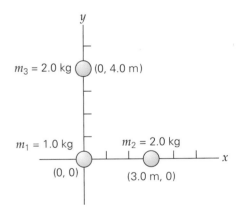

▲ **FIGURE 7.19** Total gravitational potential energy See Example 7.15.

Many of the effects of gravity are familiar to us. When lifting an object, we may think of it as being heavy, but we are working against gravity. Gravity causes rocks to tumble down and causes mudslides. But we often put gravity to use. For example, fluids from bottles for intravenous infusions flow because of gravity. An extraterrestrial application of how we put gravity to work is given in Insight 7.2 on space exploration on page 238.

INSIGHT 7.2 SPACE EXPLORATION: GRAVITY ASSISTS

After a seven-year, 3.5-billion-km (2.2-billion-mi) journey, the *Cassini-Huygens* spacecraft arrived at Saturn in July 2004, having made two Venus flybys, a Jupiter flyby, and one Earth flyby (Fig. 1).* Why was the spacecraft launched toward Venus, an inner planet, in order to go to Saturn, an outer planet?

Although space probes can be launched from the Earth with current rocket technology, there are limitations—namely, fuel versus payload: the more fuel, the smaller the payload. Using rockets alone, planetary spacecraft are realistically limited to visiting Venus, Mars, and Jupiter. The other planets could not be reached by a spacecraft of reasonable size without taking decades to get there.

So how did *Cassini* get to Saturn in 2004, almost seven years after its 1997 launch? By using gravity in a clever scheme called *gravity assist*. By using gravity assists, missions to all of the planets in our solar system are possible. Rocket energy is needed to get a spacecraft to the first planet, and after that, the energy is more or less "free." Basically, during a planetary fly-by (or swing-by), there is an exchange of energy between the planet and the spacecraft, which enables the spacecraft to increase its speed relative to the Sun. (This phenomenon is sometimes called a *slingshot effect*.)

Let's take a brief look at the physics of this ingenious use of gravity. Imagine the *Cassini* spacecraft making a swing-by of Jupiter. Recall from Chapter 6 that a collision is an interaction of objects in which there is an exchange of momentum and energy. Technically, in a swing-by, a spacecraft is having a "collision" with a planet.

When the spacecraft approaches from "behind" the planet and leaves in "front" (relative to the planet's motional direction), the gravitational interaction gives rise to a change in momentum—that is, a greater magnitude afterward and the direction is different. Then there is a $\Delta \vec{p}$ in the general "forward" direction of the spacecraft. Since $\Delta \vec{p} \propto \vec{F}$, a force acting on the craft gives it a "kick" of energy in that direction. So positive net work is done and there is an increase in kinetic energy ($W_{net} = \Delta K > 0$, by the

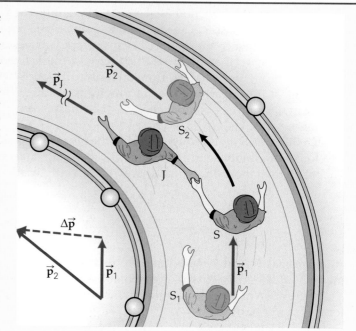

FIGURE 2 Skating swing-by Anaglous to a planetary swing-by is a roller derby "slingshot maneuver." Skater J slings skater S, who comes out of the "flyby" with greater speed than she had before (S_1, S, and S_2 sequence). In this case, the change in momentum on skater J, the slinger, would probably be noticeable when it is not for planets. (Why?)

work–energy theorem). The spacecraft leaves with more energy, a greater speed, and a new direction. (If the swing-by occurred in the opposite direction, the spacecraft would slow down.)

Momentum and energy are conserved in this elastic collision, and the planet gets an equal and opposite change in momentum, giving a retarding effect. But because the planet's mass is so much larger than that of the spacecraft, the effect on the planet is negligible.

To help you grasp the idea of a gravity assist, consider the analogous roller derby "slingshot maneuver" illustrated in Fig. 2. The skaters interact, and skater S comes out of the "flyby" with increased speed. Here, the change in momentum of the "slinger," skater J, would probably be noticeable, but that would not be so for Jupiter or any other planet.

*Cassini was the French–Italian astronomer who studied Saturn, discovering four of its moons and that Saturn's rings are separated into two parts by a narrow gap, now called the *Cassini Division*. The *Cassini-Huygens* spacecraft released a Huygens probe to Saturn's moon Titan, which was discovered by the Dutch scientist Christiaan Huygens.

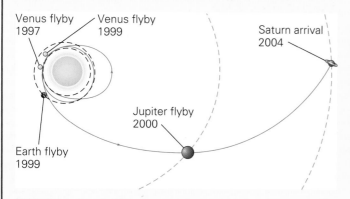

FIGURE 1 *Cassini-Huygens* **spacecraft trajectory** See text for description.

7.6 Kepler's Laws and Earth Satellites

OBJECTIVES: To (a) state and explain Kepler's laws of planetary motion, and (b) describe the orbits and motions of satellites.

The force of gravity determines the motions of the planets and satellites and holds the solar system (and galaxy) together. A general description of planetary motion had been set forth shortly before Newton's time by the German astronomer and mathematician Johannes Kepler (1571–1630). Kepler formulated three *empirical* laws from observational data gathered during a twenty-year period by the Danish astronomer Tycho Brahe (1546–1601).

Kepler went to Prague to assist Brahe, who was the official mathematician at the court of the Holy Roman Emperor. Brahe died the next year, and Kepler succeeded him, inheriting his records of the positions of the planets. Analyzing these data, Kepler announced the first two of his three laws in 1609 (the year Galileo built his first telescope). These laws were applied initially only to Mars. Kepler's third law came ten years later.

Interestingly enough, Kepler's laws of planetary motion, which took him about fifteen years to deduce from observed data, can now be derived theoretically with a page or two of calculations. These three laws apply not only to planets, but also to any system composed of a body revolving about a much more massive body, to which the inverse-square law of gravitation applies (such as the Moon, artificial Earth satellites, and solar-bound comets).

Kepler's first law (the law of orbits):

Planets move in elliptical orbits, with the Sun at one of the focal points.

An ellipse, shown in ▼Fig. 7.20a, has, in general, an oval shape, resembling a flattened circle. In fact, a circle is a special case of an ellipse in which the focal points, or *foci* (plural of *focus*), are at the same point (the center of the circle). Although the orbits of the planets are elliptical, most do not deviate very much from circles (Mercury and Pluto are notable exceptions; see Appendix III, "Eccentricity"). For example, the difference between the perihelion and aphelion of the Earth (its closest and farthest distances from the Sun, respectively) is about 5 million km. This distance may sound like a lot, but it is only a little more than 3% of 150 million km, which is the average distance between the Earth and the Sun.

Kepler's second law (the law of areas):

A line from the Sun to a planet sweeps out equal areas in equal lengths of time.

This law is illustrated in Fig. 7.20b. Since the time to travel the different orbital distances (s_1 and s_2) is the same such that the areas swept out (A_1 and A_2) are equal, this law tells you that the orbital speed of a planet varies in different parts of its orbit. Because a planet's orbit is elliptical, its orbital speed is greater when it is

Kepler's first law

Illustration 12.3 Circular and Noncircular Motion

Kepler's second law

Illustration 12.5 Kepler's Second Law

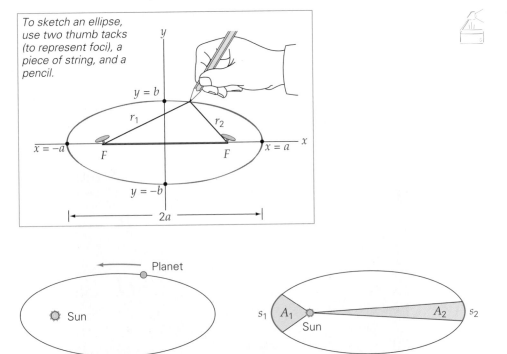

◀ **FIGURE 7.20** Kepler's first and second laws of planetary motion **(a)** In general, an ellipse has an oval shape. The sum of the distances from the focal points F to any point on the ellipse is constant: $r_1 + r_2 = 2a$. Here, $2a$ is the length of the line joining the two points on the ellipse at the greatest distance from its center, called the *major axis*. (The line joining the two points closest to the center is b, the *minor axis*.) Planets revolve about the Sun in elliptical orbits for which the Sun is at one of the focal points and nothing is at the other. **(b)** A line joining the Sun and a planet sweeps out equal areas in equal times. Since $A_1 = A_2$, a planet travels faster along s_1 than along s_2.

To sketch an ellipse, use two thumb tacks (to represent foci), a piece of string, and a pencil.

(a)

(b)

closer to the Sun than when it is farther away. The conservation of energy was used in Section 7.5 (Eq. 7.20) to deduce this relationship for the Earth.

Kepler's third law

Kepler's third law (the law of periods):

The square of the orbital period of a planet is directly proportional to the cube of the average distance of the planet from the Sun; that is, $T^2 \propto r^3$.

Kepler's third law is easily derived for the special case of a planet with a circular orbit, using Newton's law of gravitation. Since the centripetal force is supplied by the force of gravity, the expressions for these forces can be set equal:

Teaching tip: Emphasize this relationship for orbiting bodies. Also show this expression to be a statement of Newton's second law, $\vec{F}_{net} = m\vec{a}$.

PHYSLET®

Exploration 12.1 Different x_o or v_o for Planetary Orbits

$$\underset{\substack{\text{centripetal}\\\text{force}}}{\frac{m_p v^2}{r}} = \underset{\substack{\text{gravitational}\\\text{force}}}{\frac{Gm_p M_S}{r^2}}$$

or

$$v = \sqrt{\frac{GM_S}{r}}$$

In these equations, m_p and M_S are the masses of the planet and the Sun, respectively, and v is the planet's orbital speed. But $v = 2\pi r/T$ (circumference/period = distance/time), so

$$\frac{2\pi r}{T} = \sqrt{\frac{GM_S}{r}}$$

Squaring both sides and solving for T^2 gives

$$T^2 = \left(\frac{4\pi^2}{GM_S}\right)r^3$$

or

$$T^2 = Kr^3 \tag{7.22}$$

PHYSLET®

Exploration 12.3 Properties of Elliptical Orbits

The constant K for solar-system planetary orbits is easily evaluated from orbital data (for T and r) for the Earth: $K = 2.97 \times 10^{-19}$ s^2/m^3. (As an exercise, you might wish to convert K to the more useful units of y^2/km^3.) *Note*: This value of K applies to all the planets in our solar system, but does *not* apply to planet satellites as Example 7.16 will show.

If you look inside the back cover and in Appendix III, you will find the masses of the Sun and the planets of the solar system. How were these masses determined? The following Example shows how Kepler's third law can be used to do this.

Example 7.16 ■ By Jove!

The planet Jupiter (Roman name Jove) is the largest in the solar system, both in volume and mass. Jupiter has 62 known moons, the four largest being discovered by Galileo in 1610. Two of these moons, Io and Europa, are shown in ▼Fig. 7.21. Given that Io is an average distance of 4.22×10^5 km from Jupiter and has an orbital period of 1.77 days, compute the mass of Jupiter.

▶ **FIGURE 7.21** Jupiter and moons Two of Jupiter's moons, Europa and Io, discovered by Galileo are shown here. Europa is on the left, and Io on the right over the Great Red Spot. Io and Europa are comparable in size to our Moon. The Great Red Spot, roughly twice the size of the Earth, is believed to be a huge storm, similar to a hurricane on the Earth.

Thinking It Through. Given the values for Io's distance from planet (r) and period (T), this would appear to be an application of Kepler's third law, and it is. However, keep in mind that the M_S in Eq. 7.22 is the mass of the Sun, which the planets orbit. The third law can be applied to any satellite, as long as the M is that of the body being orbited by the satellite. In this case, it will be M_J, the mass of Jupiter.

Solution.

Given: $r = 4.22 \times 10^5$ km $= 4.22 \times 10^8$ m   *Find:* M_J (mass of Jupiter)
$T = 1.77$ days $(8.64 \times 10^4$ s/day$) = 1.53 \times 10^5$ s

With r and T known, K can be found in Eq. 7.22 (written K_J, indicating it is for Jupiter)

$$K_J = \frac{T^2}{r^3} = \frac{(1.53 \times 10^5 \text{ s})^2}{(4.22 \times 10^8 \text{ m})^3} = 3.11 \times 10^{-16} \text{ s}^2/\text{m}^3$$

Then, writing K_J explicitly, $K_J = \dfrac{4\pi^2}{GM_J}$, and

$$M_J = \frac{4\pi^2}{GK_J} = \frac{4\pi^2}{(6.67 \times 10^{-11} \text{ N}\cdot\text{m}^2/\text{kg}^2)(3.11 \times 10^{-16} \text{ s}^2/\text{m}^3)} = 1.90 \times 10^{27} \text{ kg}$$

Follow-Up Exercise. Compute the mass of the Sun from Earth's orbital data.

Earth's Satellites

We are only a little more than half a century into the space age. Since the 1950s, numerous uncrewed satellites have been put into orbit about the Earth, and now astronauts regularly spend weeks or months in orbiting space laboratories.

Putting a spacecraft into orbit about the Earth (or any planet) is an extremely complex task. However, you can get a basic understanding of the method from fundamental principles of physics. First, suppose that a projectile could be given the initial speed required to take it just to the top of the Earth's potential-energy well. At the exact top of the well, which is an infinite distance away ($r = \infty$), the potential energy is zero. By the conservation of energy and Eq. 7.18,

$$\overset{initial}{K_o + U_o} = \overset{final}{K + U}$$

or

$$\overset{initial}{\tfrac{1}{2}mv_{esc}^2 - \frac{GmM_E}{R_E}} = \overset{final}{0 + 0}$$

where v_{esc} is the **escape speed**—that is, the initial speed needed to escape from the surface of the Earth. The final energy is zero, since the projectile stops at the top of the well (at very large distances, and it is barely moving, $K \approx 0$), and $U = 0$ there. Solving for v_{esc} gives

$$v_{esc} = \sqrt{\frac{2GM_E}{R_E}} \tag{7.23}$$

Since $g = GM_E/R_E^2$ (Eq. 7.17), it is convenient to write

$$v_{esc} = \sqrt{2gR_E} \tag{7.24}$$

Although derived here for the Earth, this equation may be used generally to find the escape speeds for other planets and our Moon (using their accelerations due to gravity). The escape speed for Earth turns out to be 11 km/s, or about 7 mi/s.

A tangential speed less than the escape speed is required for a satellite to orbit. Consider the centripetal force on a satellite in circular orbit about the Earth. Since the centripetal force on the satellite is supplied by the gravitational attraction between the satellite and the Earth, we may again write

$$F_c = \frac{mv^2}{r} = \frac{GmM_E}{r^2}$$

Teaching tip: Newton's law of gravitation suggests that gravitational force diminishes with distance, but never goes to zero. Nevertheless, we describe the ability to "escape the Earth's gravitational attraction." Be prepared to discuss this concept in terms of the conservation of energy.

Then

$$v = \sqrt{\frac{GM_E}{r}} \tag{7.25}$$

where $r = R_E + h$. For example, suppose that a satellite is in a circular orbit at an altitude of 500 km (about 300 mi); its tangential speed must be

$$v = \sqrt{\frac{GM_E}{r}} = \sqrt{\frac{GM_E}{R_E + h}} = \sqrt{\frac{(6.67 \times 10^{-11}\, \text{N} \cdot \text{m}^2/\text{kg}^2)(6.0 \times 10^{24}\, \text{kg})}{(6.4 \times 10^6\, \text{m} + 5.0 \times 10^5\, \text{m})}}$$

$$= 7.6 \times 10^3\, \text{m/s} = 7.6\, \text{km/s (about 4.7 mi/s)}$$

This speed is about 27 000 km/h, or 17 000 mi/h. As can be seen from Eq. 7.25, the required circular orbital speed *decreases* with altitude.

In practice, a satellite is given a tangential speed by a component of the thrust from a rocket stage (▼Fig. 7.22a). The inverse-square relationship of Newton's law of gravitation means that the satellite orbits that are possible about a massive planet or star are ellipses, of which a circular orbit is a special case. This condition is illustrated

▶ **FIGURE 7.22 Satellite orbits (a)** A satellite is put into orbit by giving it a tangential speed sufficient for maintaining an orbit at a particular altitude. The higher the orbit, the smaller the tangential speed. **(b)** At an altitude of 500 km, a tangential speed of 7.6 km/s is required for a circular orbit. With a tangential speed between 7.6 km/s and 11 km/s (the escape speed), the satellite would move out of the circular orbit. Since it would not have the escape speed, it would "fall" around the Earth in an elliptical orbit, with the Earth's center at one focal point. A tangential speed less than 7.6 km/s would also give an elliptical path about the center of the Earth, but because the Earth is not a point mass, a certain minimum speed would be needed to keep the satellite from striking the Earth's surface.

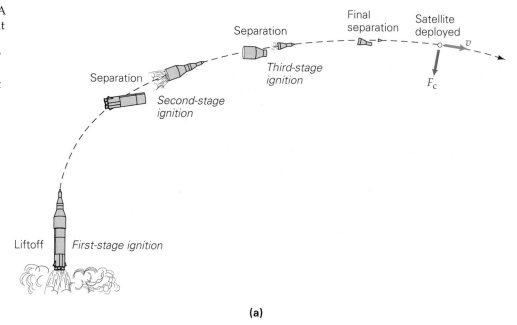

(a)

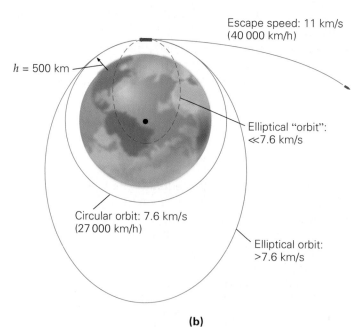

(b)

in Fig. 7.22b for the Earth, using the previously calculated values. If a satellite is not given a sufficient tangential speed, it will fall back to the Earth (and possibly be burned up while falling through the atmosphere). If the tangential speed reaches the escape speed, the satellite will leave its orbit and go off into space.

Finally, the total energy of an orbiting satellite in circular orbit is

$$E = K + U = \tfrac{1}{2}mv^2 - \frac{GmM_E}{r} \tag{7.26}$$

Substituting the expression for v from Eq. 7.25 into the kinetic energy term in Eq. (7.26) gives

$$E = \frac{GmM_E}{2r} - \frac{GmM_E}{r}$$

Thus,

$$E = -\frac{GmM_E}{2r} \qquad \text{\textit{total energy of an}} \atop \text{\textit{Earth-orbiting satellite}} \tag{7.27}$$

Note that the total energy of the satellite is negative: More work is required to put a satellite into a higher orbit, where it has more potential and total energy. The total energy E increases as its *numerical value* becomes smaller—that is, less negative—as the satellite goes to a higher orbit toward the zero potential at the top of the well. That is, the farther a satellite is from Earth, the greater its total energy. The relationship of speed and energy to orbital radius is summarized in Table 7.3.

To help understand why the total energy increases when its value becomes less negative, think of a change in energy from, say, 5.0 J to 10 J. This change would be considered an increase in energy. Similarly, a change from -10 J to -5.0 J would also be an increase in energy, even though the *absolute* value has decreased:

$$\Delta U = U - U_o = -5.0\,\text{J} - (-10\,\text{J}) = +5.0\,\text{J}$$

Also note from the development of Eq. 7.27 that the kinetic energy of an orbiting satellite is equal to the absolute value of the satellite's total energy:

$$K = \frac{GmM_E}{2r} = |E| \tag{7.28}$$

Adjustments in satellite altitude (r) are made by applying forward or reverse thrusts. For example, reverse thrust, provided by the engines of docked cargo ships, was used to put the Russian space station *Mir* into lower orbits and ultimately led to its final destruction in March 2001. A final thrust sent the station into a decaying orbit and into our atmosphere. Most of the 120-ton *Mir* burned up in the atmosphere; however, some pieces did fall into the Pacific Ocean.

The advent of the space age and the use of orbiting satellites have brought us the terms *weightlessness* and *zero gravity*, because astronauts appear to "float" about in orbiting spacecraft (▼ Fig. 7.23a). However, these terms are misnomers. As mentioned earlier in the chapter, gravity is an infinite-range force, and the Earth's

TABLE 7.3	Relationship of Radius, Speed, and Energy for Circular Orbital Motion	
	Increasing r (larger orbit)	*Decreasing r (smaller orbit)*
ω	decreases	increases
v	decreases	increases
K	decreases	increases
U	increases (smaller negative value)	decreases (larger negative value)
$E\,(= K + U)$	increases (smaller negative value)	decreases (larger negative value)

gravity acts on a spacecraft and astronauts, supplying the centripetal force necessary to keep them in orbit. Gravity there is not zero, so there must be weight.*

A better term to describe the floating effect of astronauts in orbiting spacecraft would be *apparent weightlessness*. The astronauts "float" because both the astronauts and the spacecraft are centripetally accelerating (or falling) toward the Earth at the same rate. To help you understand this effect, consider the analogous situation of a person standing on a scale in an elevator (Fig. 7.23b). The "weight" measurement that the scale registers is actually the normal force N of the scale on the person. In a nonaccelerating elevator ($a = 0$), we have $N = mg = w$, and N is equal to the true weight of the individual. However, suppose the elevator is descending with an acceleration a, where $a < g$. As the vector diagram in the figure shows,

$$mg - N = ma$$

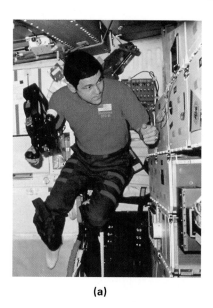

(a)

▲ **FIGURE 7.23 Apparent weightlessness** **(a)** An astronaut "floats" in a spacecraft, seemingly in a weightless condition. **(b)** In a stationary elevator (top), a scale reads the passenger's true weight. The weight reading is the reaction force N of the scale on the person. If the elevator is descending with an acceleration $a < g$ (middle), the reaction force and apparent weight are less than the true weight. If the elevator were in free fall ($a = g$; bottom), the reaction force and indicated weight would be zero, since the scale would be falling as fast as the person.

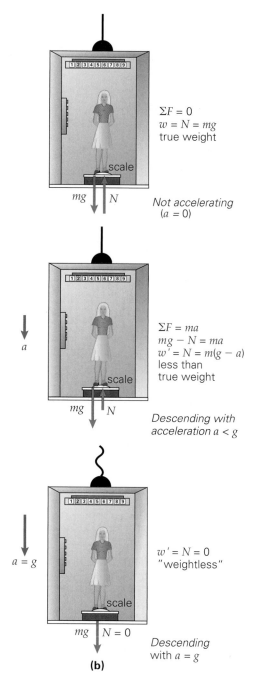

$\Sigma F = 0$
$w = N = mg$
true weight

mg $\quad N$

Not accelerating
$(a = 0)$

a

$\Sigma F = ma$
$mg - N = ma$
$w' = N = m(g - a)$
less than
true weight

mg $\quad N$

Descending with acceleration $a < g$

$a = g$

$w' = N = 0$
"weightless"

mg $\quad N = 0$

Descending with $a = g$

(b)

*Another term used to describe astronaut "floating" is *microgravity*, implying that it is caused by an apparent large reduction in gravity. This too is a misnomer. Using Eq. 17.17, at a typical satellite altitude of 300 km, it can be shown that the reduction in the acceleration due to gravity is about 10%.

and the *apparent* weight w' is

$$w' = N = m(g - a) < mg$$

where the downward direction is taken as positive in this instance. With a downward acceleration a, we see that N is less than mg, hence the scale indicates that the person weighs less than his or her true weight. Note that the *apparent acceleration due to gravity* is $g' = g - a$.

Now suppose the elevator were in free fall, with $a = g$. As you can see, N (and thus the apparent weight w') would be zero. Essentially, the scale is accelerating, or falling, at the same rate as the person. The scale may indicate a "weightless" condition ($N = 0$), but gravity still acts, as would be noted by the sudden stop at the bottom of the shaft. (See Insight 7.3 on the effects of "weightlessness" on the human body in space.)

Space has been called the *final frontier*. Someday, instead of brief stays in Earth-orbiting spacecraft, there may be permanent space colonies with "artificial"

INSIGHT   7.3   "WEIGHTLESSNESS": EFFECTS ON THE HUMAN BODY

Astronauts spend weeks and months in orbiting spacecraft and space stations. Although gravity acts on them, the astronauts experience long durations of "zero gravity" (zero-g),* due to the centripetal motion. On Earth, gravity provides the force that causes our muscles and bones to develop to the proper strength so we may function in our environment. That is, our muscles and bones must be strong enough for us to be able to walk, lift objects, and so on. And we exercise and eat properly to maintain our ability to function against the pull of gravity.

However, in a zero-g environment, muscle atrophy occurs quickly, because the body perceives no need for muscles. That is, muscles lose mass if there is no need for them to respond to gravity. In zero-g, muscle mass may deplete as much as 5% per week. Bone loss also occurs at a rate of about 1% per month. Models show that the total bone loss could reach 40–60%. The bone loss raises the calcium level in the blood, which may lead to kidney stones.

The circulatory system is affected too. On Earth, gravity causes blood to pool in our feet. When we stand, the blood pressure in our feet (about 200 mm Hg) is much greater than that in our heads (60–80 mm Hg), because of the downward force of gravity. (See Section 9.2 for a discussion of the measurement of blood pressure.) In the zero-g experienced by astronauts, this force is not there, and the blood pressure equalizes throughout the body at about 100 mm Hg. This condition causes fluid to flow from the legs to the head, and gives rise to the so-called puffy-face and bird-leg syndromes. The veins in the neck and face stand out more than usual and the eyes become red and swollen. An astronaut's legs become thinner, because the blood flow to them is no longer gravity-assisted, and it is difficult for the heart to pump blood to them (Fig. 1).

Even more serious, the condition of above-normal blood pressure in the head is interpreted by the brain as indicating that there is too much blood in the body, and blood production is signaled to slow down. Astronauts can lose up to 22% of their blood as a result of the equalized body pressure in zero-g. Also, with equalized blood pressure, the heart doesn't work as hard, and the heart muscles may atrophy.

All of these phenomena explain why astronauts undergo rigorous physical-fitness programs before going into space and exercise in space using elastic restraints. On returning to Earth, their bodies have to readjust to a normal "9.8 m/s^2 g" environment.

*This term will be used here for description, with the understanding that it is *apparent* zero-g.

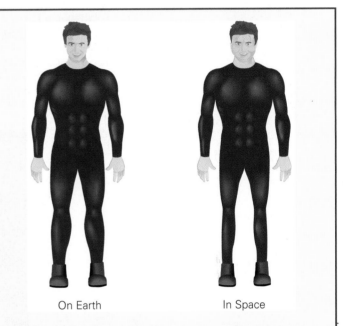

On Earth   In Space

FIGURE 1 Puffy-face syndrome In zero-g, without a gravity gradient the blood pressure equalizes throughout the body and fluid flows from the legs to the head, giving rise to the so-called puffy-face syndrome. An astronaut's legs become thinner (bird-leg syndrome) because the blood flow to them is no longer gravity-assisted and it is difficult for the heart to pump blood to them.

Each of the bodily losses requires different recovery times. Blood volume is typically restored in a few days, with astronauts drinking lots of liquids. Most muscles are regenerated in a month or so, depending on the length of stay in zero-g. Bone recovery takes much longer. Astronauts spending three to six months in space may require two or three years to regain the lost bone, if it is regained at all. Exercise and nutrition are very important in all the recovery processes.

There is much to learn about the effects of zero-g—or even reduced-g. Uncrewed spacecraft have visited Mars, with the aim of one day sending astronauts to the Red Planet. This task would involve perhaps a six-month trip in zero-g and, on arrival, a Martian surface gravity that is only 38% of the Earth's gravity. No one yet understands completely the effects that such a space journey would have on an astronaut's body.

gravity in the future. One proposal is to have a huge, rotating space colony in the form of a wheel—somewhat like an automobile tire, with the inhabitants living inside the tire. As you know, centripetal force is necessary to keep an object in rotational circular motion. On the rotating Earth, that force is supplied by gravity, and we refer to it as *weight*. We exert a force on the ground, and the normal force (by Newton's third law) upward on our feet is actually what we sense and gives us the feeling of "having our feet on solid ground."

In a rotating space colony, the situation is somewhat reversed. The rotating colony would supply the centripetal force on the inhabitants, and the centripetal force would be perceived as a normal force acting on the soles of our feet, provided artificial gravity. Rotation at the proper speed would produce a simulation of "normal" gravity $(a_c \approx g = 9.80 \text{ m/s}^2)$ within the colony wheel. Note that in the colonists' world, "down" would be outward, toward the periphery of the space station, and "up" would always be inward, toward the axis of rotation (▼Fig. 7.24).

▼ **FIGURE 7.24 Space colony and artificial gravity** (top) It has been suggested that a space colony could be housed in a huge, rotating wheel as in this artist's conception. The rotation would supply the "artificial gravity" for the colonists. (bottom) **(a)** In the frame of reference of someone in a rotating space colony, centripetal force, coming from the normal force N of the floor, would be perceived as weight sensation or artificial gravity. We are used to feeling N upward on our feet to balance gravity. Rotation at the proper speed would simulate normal gravity. To an outside observer, a dropped ball would follow a tangential straight-line path, as shown in figure (a). **(b)** A colonist on board the space colony would observe the ball to fall downward as in a normal gravitational situation.

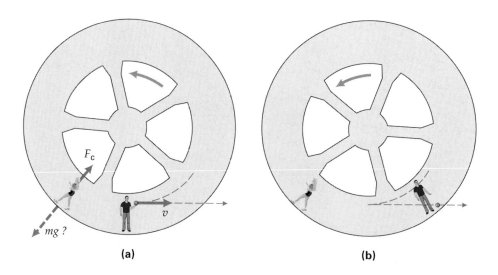

(a)   (b)

Chapter Review

- The **radian (rad)** is a measure of angle; 1 rad is the angle of a circle subtended by an arc length (s) equal to the radius (r):

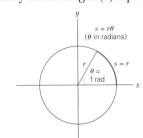

Arc Length (angle in radians):

$$s = r\theta \tag{7.3}$$

Angular Kinematic Equations for $\theta_o = 0$ and $t_o = 0$; (see Table 7.2 for linear analogues):

$$\theta = \bar{\omega}t \quad \text{(general, not limited to constant acceleration)} \tag{7.5}$$

$$\left.\begin{array}{l} \bar{\omega} = \dfrac{\omega + \omega_o}{2} \\[2mm] \omega = \omega_o + \alpha t \\[2mm] \theta = \theta_o + \omega_o t + \frac{1}{2}\alpha t^2 \\[2mm] \omega = \omega_o^2 + 2\alpha\,(\theta - \theta_o) \end{array}\right\} \begin{array}{l} \\ constant \\ acceleration \\ only \end{array}$$

(2, Table 7.2)

(7.12)

(4, Table 7.2)

(5, Table 7.2)

- **Tangential speed (v)** and **angular speed (ω)** for circular motion are directly proportional, with the radius r being the constant of proportionality:

$$v = r\omega \tag{7.6}$$

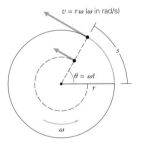

- The **frequency (f)** and **period (T)** are inversely related:

$$f = \frac{1}{T} \tag{7.7}$$

- **Angular speed** (*with uniform circular motion*) in terms of period (T) and frequency (f):

$$\omega = \frac{2\pi}{T} = 2\pi f \tag{7.8}$$

- In uniform circular motion, a **centripetal acceleration (a_c)** is required and is always directed toward the center of the circular path, and its magnitude is given by:

$$a_c = \frac{v^2}{r} = r\omega^2 \tag{7.10}$$

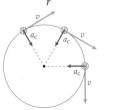

- A **centripetal force**, F_c, (the net force directed toward the center of a circle) is a requirement for circular motion, the magnitude of which is

$$F_c = ma_c = \frac{mv^2}{r} \tag{7.11}$$

- **Angular acceleration (α)** is the time rate of change of angular velocity and is related to the **tangential acceleration (a_t)** by

$$a_t = r\alpha \tag{7.13}$$

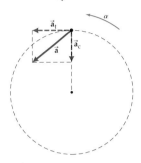

Nonuniform circular motion
($\vec{a} = \vec{a}_t + \vec{a}_c$)

- According to **Newton's law of gravitation**, every particle attracts every other particle in the universe with a force that is proportional to the masses of both particles and inversely proportional to the square of the distance between them:

$$F_g = \frac{Gm_1m_2}{r^2}$$
$$(G = 6.67 \times 10^{-11} \text{ N}\cdot\text{m}^2/\text{kg}^2) \tag{7.14}$$

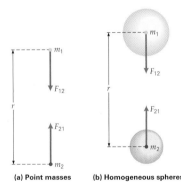

(a) Point masses   (b) Homogeneous spheres

- **Acceleration due to gravity at an altitude h:**

$$a_g = \frac{GM_E}{(R_E + h)^2} \tag{7.17}$$

• **Gravitational potential energy of two particles**:

$$U = -\frac{Gm_1m_2}{r} \qquad (7.18)$$

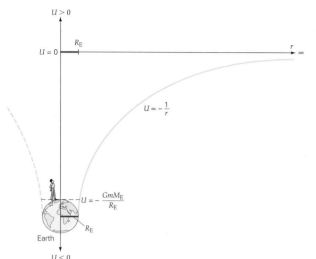

• **Kepler's first law (law of orbits):** Planets move in elliptical orbits, with the Sun at one of the focal points.

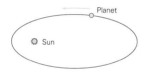

• **Kepler's second law (law of areas):** A line from the Sun to a planet sweeps out equal areas in equal lengths of time.

• **Kepler's third law (law of periods):**

$$T^2 = Kr^3 \qquad (7.22)$$

(K depends on the mass of the object orbited; for objects orbiting the Sun, $K = 2.97 \times 10^{-19}$ s²/m³)

• **Escape speed** (*from the Earth*) is

$$v_{\text{esc}} = \sqrt{\frac{2GM_E}{R_E}} = \sqrt{2gR_E} \qquad (7.23, 7.24)$$

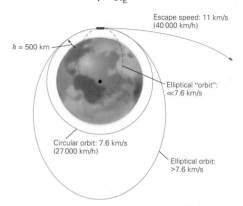

• Earth's satellites are in a negative potential-energy well; the higher the object is in the well, the greater is the object's potential energy and the less is its kinetic energy.

• **Energy of a satellite orbiting Earth:**

$$E = -\frac{GmM_E}{2r} \qquad (7.27)$$

$$K = |E| \qquad (7.28)$$

Exercises

MC = *Multiple Choice Question,* **CQ** = *Conceptual Question, and* **IE** = *Integrated Exercise. Throughout the text, many exercise sections will include "paired" exercises. These exercise pairs, identified with* **red numbers***, are intended to assist you in problem solving and learning. In a pair, the first exercise (even numbered) is worked out in the Study Guide so that you can consult it should you need assistance in solving it. The second exercise (odd numbered) is similar in nature, and its answer is given at the back of the book.*

7.1 Angular Measure

1. **MC** The radian unit is equivalent to a ratio of (a) degree/time, (b) length, (c) length/length, (d) length/time. (c)

2. **MC** For the polar coordinates of a particle traveling in a circle, the variables are (a) both *r* and *θ*, (b) only *r*, (c) only *θ*, (d) none of these. (c)

3. **CQ** Why does 1 rad equal 57.3°? Wouldn't it be more convenient to have an even number of degrees?
   2π rad = 360°, so 1 rad = 57.3°

4. **CQ** A wheel rotates about a rigid axis through its center. Do all points on the wheel travel the same distance? How about the same *angular* distance? no, yes

5. • The Cartesian coordinates of a point on a circle are (1.5 m, 2.0 m). What are the polar coordinates (*r, θ*) of this point? (2.5 m, 53°)

6. • The polar coordinates of a point are (5.3 m, 32°). What are the point's Cartesian coordinates? (4.5 m, 2.8 m)

7. • You can determine the diameter of the Sun by measuring the angle it subtends. If the angle is 0.535° and the average distance from the Earth to the Sun is 1.5 × 10¹¹ m, what is the diameter of the Sun? 1.4 × 10⁹ m

8. • Convert the following angles from degrees to radians, to two significant figures: (a) 15°, (b) 45°, (c) 90°, and (d) 120°.
   (a) 0.26 rad (b) 0.79 rad (c) 1.6 rad (d) 2.1 rad

9. ● Convert the following angles from radians to degrees: (a) $\pi/6$ rad, (b) $5\pi/12$ rad, (c) $3\pi/4$ rad, and (d) π rad. (a) 30° (b) 75° (c) 135° (d) 180°

10. ● You measure the length of a distant car to be subtended by an angular distance of 1.5°. If the car is actually 5.0 m long, approximately how far away is the car? 1.4×10^2 m

11. ● A jogger on a circular track that has a radius of 0.250 km runs a distance of 1.00 km. What angular distance does the jogger cover in (a) radians and (b) degrees? (a) 4.00 rad (b) 229°

12. ●● The hour, minute, and second hands on a clock are 0.25 m, 0.30 m, and 0.35 m long, respectively. What are the distances traveled by the tips of the hands in a 30-min interval? 0.065 m, 0.94 m, 66 m, respectively

13. ●● In Europe, a large circular walking track with a diameter of 0.900 km is marked in angular distances in radians. An American tourist who walks 3.00 mi daily goes to the track. How many radians should he walk per day to maintain his daily routine? 10.7 rad

14. ●● You ordered a 12-in. pizza for a party of five. For the pizza to be distributed evenly, how should it be cut in triangular pieces? (▼ Fig. 7.25)? 0.19 m arc length, or 72°

▶ **FIGURE 7.25 Tough pizza to cut** See Exercise 14.

15. **IE** ●● To attend the 2000 Summer Olympics, a fan flew from Mosselbaai, South Africa (34°S, 22°E) to Sydney, Australia (34°S, 151°E). (a) What is the smallest angular distance the fan has to travel: (1) 34°, (2) 12°, (3) 117°, or (4) 129°? Why? (b) Determine the approximate shortest flight distance, in kilometers. (a) (4) 129° (b) about 1.4×10^4 km

16. **IE** ●● A bicycle wheel has a small pebble embedded in its tread. The bike is set upside down, and the rider accidentally bumps the wheel and the pebble moves through an arc length of 25.0 cm before coming to rest. In that time, the wheel spins 35°. (a) The radius of the wheel is therefore (1) more than 25.0 cm, (2) less than 25.0 cm, or (3) equal to 25.0 cm. (b) Determine the radius of the wheel. (a) (1) more than 25.0 cm (b) 40.9 cm

17. ●● At the end of her routine, an ice skater spins through 7.50 revolutions with her arms always fully outstretched at right angles to her body. If her arms are 60.0 cm long, through what linear arc length distance do the tips of her fingers move during her finish? 28.3 m

18. ●●● (a) Could a circular pie be cut such that all of the wedge-shaped pieces have an arc length along the outer crust equal to the pie's radius? (b) If not, how many such pieces could you cut, and what would be the angular dimension of the final piece? (a) no (b) six such pieces and one 0.28-rad piece

19. ●●● Electrical wire with a diameter of 0.50 cm is wound on a spool with a radius of 30 cm and a height of 24 cm. (a) Through how many radians must the spool be turned to wrap one even layer of wire? (b) What is the length of this wound wire? (a) 3.0×10^2 rad (b) 91 m

20. ●●● A yo-yo with an axle diameter of 1.00 cm has a 90.0-cm length of string wrapped around it many times, but in such a way that the string completely covers the surface of its axle, but there are no double layers of string. The outermost portion of the yo-yo is 5.00 cm from the center of the axle. (a) If the yo-yo is dropped with the string fully wound, through what angle does it rotate by the time it reaches the bottom of its fall? (b) How much arc length distance has a piece of the yo-yo on its outer edge traveled by the time it bottoms out? (a) 180 rad, (b) 9.00 m

7.2 Angular Speed and Velocity

21. **MC** Viewed from above, a turntable rotates counterclockwise. The angular velocity vector is then (a) tangential to the turntable's rim, (b) out of the plane of the turntable, (c) counterclockwise, (d) none of the preceding. (b)

22. **MC** The frequency unit of hertz is equivalent to (a) that of the period, (b) that of the cycle, (c) radian/s, (d) s^{-1}. (d)

23. **MC** The particles in a uniformly rotating object all have the same (a) angular acceleration, (b) angular speed, (c) tangential velocity, (d) both (a) and (b). (d)

24. **CQ** Do all points on a wheel rotating about a fixed axis through its center have the same angular velocity? The same tangential speed? Explain. yes, no; see ISM

25. **CQ** When "clockwise" or "counterclockwise" is used to describe rotational motion, why is a phrase such as "viewed from above" added? views from opposite sides would give different circular senses

26. **CQ** Imagine yourself standing on the edge of an operating merry-go-round. How would your tangential motion be affected if you walked toward the center? (Watch out for the horses going up and down.) your tangential speed decreases

27. ● A computer DVD-ROM has a variable angular speed from 200 rpm to 450 rpm. Express this range of angular speed in radians per second. 21 rad/s to 47 rad/s

28. ● A race car makes two and a half laps around a circular track in 3.0 min. What is the car's average angular speed? 0.087 rad/s

29. ● If a particle is rotating with an angular speed of 3.5 rad/s, how long does the particle take to go through one revolution? 1.8 s

30. ● What is the period of revolution for (a) a 9 500-rpm centrifuge and (b) a 9 500-rpm computer hard-disk drive? (a) 4.80×10^{-3} s (b) 6.32×10^{-3} s

31. ●● Determine which has the greater angular speed: particle A, which travels 160° in 2.00 s, or particle B, which travels 4π rad in 8.00 s. particle B

32. ●● The tangential speed of a particle on a rotating wheel is 3.0 m/s. If the particle is 0.20 m from the axis of rotation, how long will the particle take to go through one revolution? 0.42 s

33. ●● A merry-go-round makes 24 revolutions in a 3.0-min ride. (a) What is its average angular speed in rad/s? (b) What are the tangential speeds of two people 4.0 m and 5.0 m from the center, or axis of rotation? (a) 0.84 rad/s (b) 3.4 m/s, 4.2 m/s

34. ●● In Exercise 16, suppose the wheel took 1.20 s to stop after it was bumped. Assume as you face the plane of the wheel, it was rotating counterclockwise. During this time, determine (a) the average angular speed and tangential speed of the pebble, (b) the average angular speed and tangential speed of a piece of grease on the wheel's axle (radius 1.50 cm), and (c) the direction of their respective angular velocities. see ISM

35. IE ●● The Earth rotates on its own axis once a day and revolves around the Sun once a year. (a) Which is greater, the rotating angular speed or the revolving angular speed? Why? (b) Calculate both angular speeds in rad/s. see ISM

36. ●● A little boy jumps onto a small merry-go-round (radius of 2.00 m) in a park and rotates for 2.30 s through an arc length distance of 2.55 m before coming to rest. If he landed (and stayed) at a distance of 1.75 m from the central axis of rotation of the merry-go-round, what was his average angular speed and average tangential speed? 0.634 rad/s, 1.11 m/s

37. ●●● The driver of a car sets the cruise control and ties the steering wheel so that the car travels at a uniform speed of 15 m/s in a circle with a diameter of 120 m. (a) Through what angular distance does the car move in 4.00 min? (b) What linear distance does it travel in this time? (a) 60 rad (b) 3.6 × 10³ m

38. ●●● In a noninjury noncontact skid on icy pavement on an empty road, a car spins 1.75 revolutions while it skids to a halt. It was initially moving at 15.0 m/s, and because of the ice it was able to decelerate at a rate of only 1.50 m/s². Viewed from above, the car spun clockwise. Determine its average angular velocity as it spun and slid to a halt. 1.10 rad/s, down

7.3 Uniform Circular Motion and Centripetal Acceleration

39. **MC** Uniform circular motion requires (a) centripetal acceleration, (b) angular speed, (c) tangential velocity, (d) all of the preceding. (d)

40. **MC** In uniform circular motion, there is a (a) constant velocity, (b) constant angular velocity, (c) zero acceleration, (d) nonzero tangential acceleration. (b)

41. **MC** If the centripetal force on a particle in uniform circular motion is increased, (a) the tangential speed will remain constant, (b) the tangential speed will decrease, (c) the radius of the circular path will increase, (d) the tangential speed will increase and/or the radius will decrease. (d)

42. **CQ** The spin cycle of a washing machine is used to extract water from recently washed clothes. Explain the physical principle(s) involved. see ISM

43. **CQ** The apparatus illustrated in ▼Fig. 7.26 is used to demonstrate forces in a rotating system. The floats are in jars of water. When the arm is rotated, which way will the floats move? Does it make a difference which way the arm is rotated? in direction of acceleration, inward; no

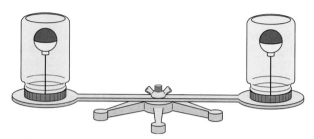

▲ **FIGURE 7.26** **When set in motion, a rotating system** See Exercise 43.

44. **CQ** When rounding a curve in a fast-moving car, we experience a feeling of being thrown outward (▼Fig. 7.27). It is sometimes said that this effect occurs because of an outward centrifugal (center-fleeing) force. However, in terms of Newton's laws, this pseudo, or false, force doesn't really exist. Analyze the situation in the figure to show that this is the case (that is, that the force does not exist). [*Hint*: Start with Newton's first law.] see ISM

▲ **FIGURE 7.27** **A center-fleeing force?** See Exercise 44.

45. **CQ** Many racetracks have banked turns, which allow the cars to travel faster around the curves than if the curves were flat. Actually, cars could also make turns on these banked curves if there were no friction at all. Explain this statement using the free-body diagram shown in ▼Fig. 7.28. see ISM

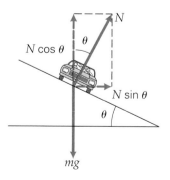

▲ **FIGURE 7.28** **Banking safety** See Exercise 45.

46. ● An Indy car with a speed of 120 km/h goes around a level, circular track with a radius of 1.00 km. What is the centripetal acceleration of the car? 1.11 m/s²

47. ● A wheel of radius 1.5 m rotates at a uniform speed. If a point on the rim of the wheel has a centripetal acceleration of 1.2 m/s², what is the point's tangential speed? 1.3 m/s

48. ● A rotating cylinder about 16 km long and 7.0 km in diameter is designed to be used as a space colony. With what angular speed must it rotate so that the residents on it will experience the same acceleration due to gravity on Earth? 0.053 rad/s or about 730 rev/day

49. ●● The Moon revolves around the Earth in 27.3 days in a nearly circular orbit with a radius of 3.8×10^5 km. Assuming that the Moon's orbital motion is a uniform circular motion, what is the Moon's acceleration as it "falls" toward the Earth? 2.69×10^{-3} m/s²

50. ●● Imagine that you swing about your head a ball attached to the end of a string. The ball moves at a constant speed in a horizontal circle. (a) Can the string be exactly horizontal? Why? (b) If the mass of the ball is 0.250 kg, the radius of the circle is 1.50 m, and it takes 1.20 s for the ball to make one revolution, what is the ball's tangential speed? (c) What centripetal force are you imparting to the ball via the string? (a) no (b) 7.85 m/s (c) 10.3 N

51. ●● In Exercise 50, if you supplied a tension force of 12.5 N to the string, what angle would the string make relative to the horizontal? 11.3°

52. ●● A car with a constant speed of 83.0 km/h enters a circular flat curve with a radius of curvature of 0.400 km. If the friction between the road and the car's tires can supply a centripetal acceleration of 1.25 m/s², does the car negotiate the curve safely? Justify your answer. no; 1.33 m/s² > 1.25 m/s²

53. **IE** ●● A student is to swing a bucket of water in a vertical circle without spilling any (▼Fig. 7.29). (a) Explain how this task is possible. (b) If the distance from his shoulder to the center of mass of the bucket of water is 1.0 m, what is the minimum speed required to keep the water from coming out of the bucket at the top of the swing? (a) weight is supplying centripetal force (b) 3.1 m/s

54. ●● In performing a "figure 8" maneuver, a figure skater wants to make the top part of the 8 approximately a circle of radius 2.20 m. He needs to glide through this part of the figure at approximately a constant speed, taking 4.50 s. His skates digging into the ice are capable of providing a maximum centripetal acceleration of 3.25 m/s². Will he be able to do this as planned? If not, what adjustment can he make if he wants this part of the figure to remain the same size (assume the ice conditions and skates don't change)? a_c = 4.28 m/s, exceeds maximum; reduce speed and/or increase time

55. ●● A light string of length of 56.0 cm connects two small square blocks, each with a mass of 1.50 kg. The system is placed on a slippery (frictionless) sheet of horizontal ice and spun so the two blocks rotate uniformly about their common center of mass, which itself does not move. They are supposed to rotate with a period of 0.750 s. If the string can exert a force of only 100 N before it breaks, determine whether this string will work. 29.5 N, string will work

56. **IE** ●● A jet pilot puts an aircraft with a constant speed into a vertical circular loop. (a) Which is greater, the normal force exerted on the seat by the pilot at the bottom of the loop or that at the top of the loop? Why? (b) If the speed of the aircraft is 700 km/h and the radius of the circle is 2.0 km, calculate the normal forces exerted on the seat by the pilot at the bottom and top of the loop. Express your answer in terms of the pilot's weight, *mg*. (a) at the bottom (b) (2.9)*mg* (bottom), (0.93)*mg* (top)

57. ●●● A block of mass *m* slides down an inclined plane into a loop-the-loop of radius *r* (▼Fig. 7.30). (a) Neglecting friction, what is the minimum speed the block must have at the highest point of the loop in order to stay in the loop? [*Hint:* What force must act on the block at the top of the loop to keep the block on a circular path?] (b) At what vertical height on the inclined plane (in terms of the radius of the loop) must the block be released if it is to have the required minimum speed at the top of the loop? (a) $v = \sqrt{rg}$ (b) $h = (5/2)r$

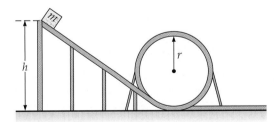

▲ **FIGURE 7.30 Loop-the-loop** See Exercise 57.

58. ●●● For a scene in a movie, a stunt driver drives a 1.50×10^3-kg SUV with a length of 4.25 m around a circular curve with a radius of curvature of 0.333 km (▼Fig. 7.31). The vehicle is to be driven off the edge a gully 10.0 m wide, and land on the other side 2.96 m below the initial side. What is the minimum centripetal acceleration the truck must have in going around the circular curve to clear the gully and land on the other side? 1.01 m/s²

▲ **FIGURE 7.29 Weightless water?** See Exercise 53.

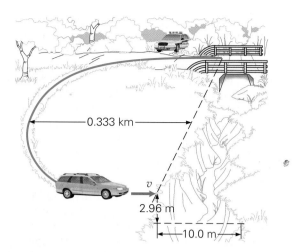

▲ **FIGURE 7.31 Over the gully** See Exercise 58.

59. ●●● Consider a simple pendulum of length L that has a small mass (the bob) of mass m attached to the end of its string. If the pendulum starts out horizontally and is released from rest, show (a) that the speed at the bottom of the swing is $v_{max} = \sqrt{2gL}$ and (b) that the tension in the string at that point is three times the weight of the bob, or $T_{max} = 3mg$. [*Hint:* Use conservation of energy to determine the speed at the bottom and centripetal-force ideas and a free-body diagram to determine the tension at the bottom.] see ISM

7.4 Angular Acceleration

60. **MC** The angular acceleration in circular motion (a) is equal in magnitude to the tangential acceleration divided by the radius, (b) increases the angular velocity if both angular velocity and angular acceleration are in the same direction, (c) has units of s^{-2}, (d) all of the preceding. (d)

61. **MC** In circular motion, the tangential acceleration (a) does not depend on the angular acceleration, (b) is constant, (c) has units of s^{-2}, (d) none of these. (d)

62. **MC** For uniform circular motion, (a) $\alpha = 0$, (b) $\omega = 0$, (c) $r = 0$, (d) none of the preceding. (a)

63. **CQ** A car increases its speed when it is on a circular track. Does the car have centripetal acceleration? How about angular acceleration? Explain. yes, yes, ISM

64. **CQ** Is it possible for a car traveling on a circular track to have angular acceleration, but not centripetal acceleration? Explain. no; see ISM

65. **CQ** Is it possible for a car traveling on a circular track to have an increase in tangential acceleration and not centripetal acceleration? no

66. ● During an acceleration, the angular speed of an engine increases from 600 rpm to 2500 rpm in 3.0 s. What is the average angular acceleration of the engine? 80 rad/s²

67. ● A merry-go-round accelerating uniformly from rest achieves its operating speed of 2.5 rpm in five revolutions. What is the magnitude of its angular acceleration? 1.1×10^{-3} rad/s²

68. ●● To maintain their bone density and other vital bodily signs, the inhabitants of a cylindrically shaped living space want to generate a "1-g environment" on their way to a distant destination. Assume the cylinder has a diameter of 250 m (they live on the inner surface) and is initially *not* rotating about its long axis. For minimal disruption, the (constant) angular acceleration is a very gentle 0.00010 rad/s². Determine how long it takes for them to reach their goal of a "1-g environment." 2.8×10^{3} s

69. **IE** ●● A car on a circular track accelerates from rest. (a) The car experiences (1) only angular acceleration, (2) only centripetal acceleration, (3) both angular and centripetal accelerations. Why? (b) If the radius of the track is 0.30 km and the magnitude of the constant angular acceleration is 4.5×10^{-3} rad/s², how long does the car take to make one lap around the track? (c) What is the total (vector) acceleration of the car when it has completed half of a lap? see ISM

70. ●● The blades of a fan running at low speed turn at 250 rpm. When the fan is switched to high speed, the rotation rate increases uniformly to 350 rpm in 5.75 s. (a) What is the magnitude of the angular acceleration of the blades? (b) How many revolutions do the blades go through while the fan is accelerating? (a) 1.82 rad/s² (b) 28.7 rev

71. ●● In the spin-dry cycle of a modern washing machine, a wet towel with a mass of 1.50 kg is "stuck to" the inside surface of the perforated (to allow the water out) washing cylinder. To have decent removal of water, damp/wet clothes need to experience a centripetal acceleration of at least $10g$. Assuming this value, and that the cylinder has a radius of 35.0 cm, determine the constant angular acceleration of the towel required if the washing machine takes 2.50 s to achieve its final angular speed. 6.69 rad/s²

72. ●●● A pendulum swinging in a circular arc under the influence of gravity, as shown in ▼ Fig. 7.32, has both centripetal and tangential components of acceleration. (a) If the pendulum bob has a speed of 2.7 m/s when the cord makes an angle of $\theta = 15°$ with the vertical, what are the magnitudes of the components at this time? (b) Where is the centripetal acceleration a maximum? What is the value of the tangential acceleration at that location? (a) $a_t = 2.5$ m/s²; $a_c = 9.7$ m/s² (b) lowest point; $a_t = 0$

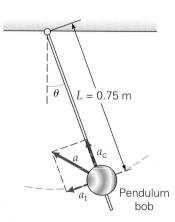

▲ **FIGURE 7.32 A swinging pendulum** See Exercise 72.

73. ●●● A simple pendulum of length 2.00 m is released from a horizontal position. When it makes an angle of 30° from the vertical, determine (a) its angular acceleration, (b) its centripetal acceleration, and (c) the tension in the string. Assume the bob's mass is 1.50 kg.
(a) 2.45 rad/s² (b) 17.0 m/s² (c) 38.2 N

7.5 Newton's Law of Gravitation

74. **MC** The gravitational force is (a) a linear function of distance, (b) an inverse function of distance, (c) an inverse function of distance squared, (d) sometimes repulsive. (c)

75. **MC** The acceleration due to gravity of an object on the Earth's surface (a) is a universal constant, like G; (b) does not depend on the Earth's mass; (c) is directly proportional to the Earth's radius; (d) does not depend on the object's mass. (d)

76. **MC** Compared with its value on the Earth's surface, the value of the acceleration due to gravity at an altitude of one Earth radius is (a) the same, (b) two times as great, (c) one half as great, (d) one fourth as great. (d)

77. **CQ** Astronauts in a spacecraft orbiting the Earth or out for a "space walk" (▾Fig. 7.33) are seen to "float" in midair. This phenomenon is sometimes referred to as *weightlessness* or *zero gravity* (zero-*g*). Are these terms correct? Explain why an astronaut appears to float in or near an orbiting spacecraft. no, see ISM

▲ **FIGURE 7.33 Out for a walk** Why does this astronaut seem to "float"? See Exercise 77.

78. **CQ** If the cup in ▾Fig. 7.34 were dropped, no water would run out. Explain. see ISM

▲ **FIGURE 7.34 Let it go** See Exercise 78.

79. **CQ** Can you determine the mass of the Earth simply by measuring the gravitational acceleration near the Earth's surface? If yes, give the details. yes, see ISM

80. ● From the known mass and radius of the Moon (see the tables inside the back cover of the book), compute the value of the acceleration due to gravity, g_M, at the surface of the Moon. 1.6 m/s²

81. ● Calculate the gravitational force between the Earth and the Moon. 2.0 × 10²⁰ N

82. ●● For a spacecraft going directly from the Earth to the Moon, beyond what point will lunar gravity begin to dominate? That is, where will the lunar gravitational force be equal in magnitude to the Earth's gravitational force? Are the astronauts onboard a spacecraft at this point truly weightless? 3.4 × 10⁸ m from Earth; no

83. ●● Four identical masses of 2.5 kg each are located at the corners of a square with 1.0-m sides. What is the net force on any one of the masses?
8.0 × 10⁻¹⁰ N, toward opposite corner

84. ●● What is the acceleration due to gravity on the top of Mt. Everest? The summit is about 8.80 km above sea level. (Report to three significant figures.) 9.77 m/s²

85. ●● A man has a mass of 75 kg on the Earth's surface. How far above the surface of the Earth would he have to go to "lose" 10% of his body weight? 3.4 × 10⁵ m

86. **IE** ●● Two objects are attracting each other with a certain gravitational force. (a) If the distance between the objects is reduced to half, the new gravitational force will (1) increase by a factor of 2, (2) increase by a factor of 4, (3) decrease by a factor of 2, (4) decrease by a factor of 4. Why? (b) If the original force between the two objects is 0.90 N, and the distance is tripled, what is the new gravitational force between the objects?
(a) (2) increase by a factor of 4 (b) 0.10 N

87. ●● During the Apollo lunar explorations of the late 1960s and early 1970s, the main section of the spaceship remained in orbit about the Moon with one astronaut in it while the other two descended to the surface in the landing module. If the main section orbited about 50 mi above the lunar surface, determine that section's centripetal acceleration. 1.5 m/s²

88. ●● Referring to Exercise 87, determine the (a) the gravitational potential energy, (b) the total energy, and (c) the energy needed to "escape" the Moon for the main section of the lunar exploration mission in orbit. Assume the mass of this section is 5000 kg.
(a) −1.3 × 10¹⁰ J (b) −6.8 × 10⁹ J (c) 6.8 × 10⁹ J

89. ●●● (a) What is the mutual gravitational potential energy of the configuration shown in ▾Fig. 7.35 if all the masses are 1.0 kg? (b) What is the gravitational force per unit mass at the center of the configuration? (a) −2.5 × 10⁻¹⁰ J (b) 0

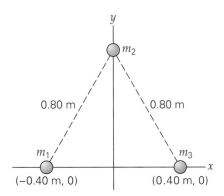

▲ **FIGURE 7.35 Gravitational potential, gravitational force, and center of mass** See Exercise 89.

90. **IE ●●●** A deep space probe mission is planned to explore the composition of interstellar space. Assuming the three most important objects in the solar system for this project are the Sun, the Earth, and Jupiter, (a) what would be the distance of the Earth relative to Jupiter that would result in the lowest escape speed needed if the probe is to be launched from the Earth: (1) the Earth should be as close as possible to Jupiter; (2) the Earth should be as far as possible from Jupiter; or (3) the distance of the Earth relative to Jupiter doesn't matter? (b) Estimate the least escape speed for this probe, assuming planetary circular orbits, and only the Earth, Sun, and Jupiter are important. (See data in Appendix III.) Comment on which of the three objects, if any, determines most of the escape speed. (a) (2) the Earth should be as far as possible from Jupiter (b) 4.3 × 10⁴ m/s, Sun

7.6 Kepler's Laws and Earth Satellites

91. **MC** A new planet is discovered and its period determined. The new planet's distance from the Sun could then be found by using Kepler's (a) first law, (b) second law, (c) third law. (c)

92. **MC** As a planet moves in its elliptical orbit, (a) its speed is constant. (b) its distance from the Sun is constant, (c) it moves faster when it is closer to the Sun, (d) it moves slower when it is closer to the Sun. (c)

93. **MC** If a satellite near the Earth's surface does not have a minimum tangential speed of 11 km/s, it could (a) go into an elliptical orbit, (b) go into a circular orbit, (c) crash into the Earth, (d) all of the preceding. (d)

94. **CQ** (a) In one revolution, how much work does the centripetal force do on a satellite in circular orbit about the Earth? (b) A person in a freely falling elevator thinks he can avoid injury by jumping upward just before the elevator strikes the floor. Would this strategy work? (a) zero (b) no, see ISM

95. **CQ** (a) In putting a satellite into orbit over the equator, should the rocket be launched eastward or westward? Why? (b) In the United States, such satellites are launched from Florida. Why not from California, which generally has better weather conditions? see ISM

96. **CQ** As you pilot your spaceship in orbit, you see a piece of equipment ahead of you in the same orbit. (a) Can you speed up your spaceship with one rocket burst in order to catch the equipment? Explain. (b) What do you have to do to catch the equipment? (a) no; see ISM (b) decrease the orbital radius, see ISM

97. **●** An instrument package is projected vertically upward to collect data near the top of the Earth's atmosphere (at an altitude of about 900 km). (a) What initial speed is required at the Earth's surface for the package to reach this height? (b) What percentage of the escape speed is this initial speed? (a) 3.7 × 10³ m/s (b) 34%

98. **●●** In the year 2056, Martian Colony I wants to put a Mars-synchronous communication satellite in orbit about Mars to facilitate communications with the new bases being planned on the Red Planet. At what distance above the Martian equator would this satellite be placed? (To a good approximation, the Martian day is the same length as that of the Earth's.) 2.0 × 10⁷ m

99. **●●** The asteroid belt that lies between Mars and Jupiter may be the debris of a planet that broke apart or that was not able to form as a result of Jupiter's strong gravitation. The asteroid belt has a period of about 5.0 yr. Approximately how far from the Sun would this "fifth" planet have been? 4.4 × 10¹¹ m

100. **●●** Using a development similar to Kepler's law of periods for planets orbiting the Sun, find the required altitude of geosynchronous satellites above the Earth. [*Hint:* The period of such satellites is the same as that of the Earth.] 3.6 × 10⁷ m

101. **●●** Venus has a rotational period of 243 days. What would be the altitude of a synchronous satellite for this planet (similar to geosynchronous satellite on the Earth)? 1.53 × 10⁹ m

102. **●●●** A small space probe is put into circular orbit about a newly discovered moon of Saturn. The moon's radius is known to be 550 km. If the probe orbits at a height of 1500 km above the moon's surface and takes 2.00 Earth days to make one orbit, determine the moon's mass. 1.7 × 10²⁰ kg

Comprehensive Exercises

103. Just an instant before reaching the very bottom of a semicircular section of a roller coaster ride, the automatic emergency brake inadvertently goes on. Assume the car has a total mass of 750 kg, the radius of that section of the track is 55.0 m, and the car entered the bottom after descending vertically (from rest) 25.0 m on a frictionless straight incline. If the braking force is a steady 1700 N, determine (a) the car's centripetal acceleration (including direction), (b) the normal force of the track on the car, (c) the tangential acceleration of the car (including direction), and (d) the total acceleration of the car. see ISM

104. A 60.0-kg passenger is riding in a car traveling at a steady 25.0 m/s. The car takes a left turn on a flat curve with a radius of 75.0 m at this constant speed. The passenger is not wearing her seatbelt, but there is friction between her and the seat. The coefficient of static friction is 0.600. (a) Will she slide to the right on the seat during this turn? (b) If she does slip, when she contacts the door, what is the normal force exerted sideways on her? (a) she will slide (b) 147 N

105. A horizontal disk of radius 5.00 cm accelerates angularly from rest with an angular acceleration of 0.250 rad/s². A small pebble located halfway out from the center has a coefficient of static friction with the disk of 0.15. (a) How long will the pebble take to slip on the disk surface? (b) How many revolutions will the disk have rotated through at that time. (a) 31 s (b) 19 rev

106. A simple pendulum consists of a light string 1.50 m long with a small 0.500-kg mass attached. The pendulum starts out at 45° below the horizontal and is given an initial downward speed of 1.50 m/s. At the bottom of the arc, determine (a) the centripetal acceleration of the bob and (b) the tension in the string. (a) 7.24 m/s² (b) 8.52 N

107. To see in principle how astronomers determine stellar masses, consider the following. Unlike our solar system, many systems have two or more stars. If there are two, it is a *binary* star system. The simplest possible case is that of two identical stars in a circular orbit about their common center of mass midway between them (small black dot in ▼Fig. 7.36). Using telescopic measurements, it is sometimes possible to measure the distance, D, between the star's centers and the time (orbital period), T, for one orbit. Assume uniform circular motion and the following data. The stars have the same mass, the distance between them is one billion km (1.0×10^9 km), and the time each takes for one orbit is 10.0 Earth years. Determine the mass of each star. 2.97×10^{30} kg

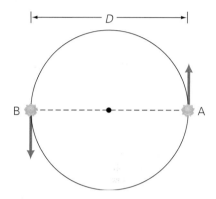

▲ **FIGURE 7.36 Binary stars** See Exercise 107.

ROTATIONAL MOTION AND EQUILIBRIUM

PHYSICS FACTS

- If it were not for torques supplied by our muscles, we would be without body mobility.

- Antilock brakes are used on cars because the rolling stopping distance is less than that of a locked-brake stopping distance.

- The Earth's rotational axis, which is tilted $23\frac{1}{2}°$, precesses (rotates) with a period of 26 000 years. As a result, Polaris, toward which the axis currently points, has not always been, nor will it always be, the North Star.

- The December 2004 "megathrust" earthquake, which measured 9.0 on the Richter scale and spawned a disastrous tsunami, wobbled the Earth by a few centimeters. This caused the Earth's rotation to slightly accelerate and it was calculated that the days were shortened by a couple of microseconds, which is not measurable.

- Some figure skaters in jumps reach rotational speeds on the order of 7 rev/s, or 420 rpm (revolutions per minute). Some automobiles engines have idle speeds of 600–800 rpm.

It's always a good idea to keep your equilibrium—but it's more important in some situations than in others. When you look at a photo like this, your first reaction is probably to wonder how does this tightrope walker traversing the Niagara River at Horseshoe Falls keep from falling. Presumably, the pole must help—but in what way? You'll find out in this chapter.

We might say that the tightrope walker is in equilibrium. Translational equilibrium ($\sum \vec{F}_i = 0$) was discussed in Chapter 4, but here there is another consideration: rotation. Should he start to fall (and we hope he doesn't), there would initially be a sideways rotation about the wire. (Wires are generally used now.) To avoid this calamity, another condition must be met: rotational equilibrium, which we will discuss in this chapter.

The tightrope walker is striving to avoid rotational motion. But rotational motion is very important in physics, because rotating objects are all around us: wheels on vehicles, gears and pulleys in machinery, planets in our solar system, and even many bones in the human body. (Can you think of a few bones that rotate in sockets?)

Fortunately, the equations describing rotational motion can be written as almost direct analogues of those for translational (linear) motion. In Chapter 7, this similarity was pointed out with respect to the linear and angular kinematic equations. With the addition of equations describing rotational dynamics, you will be able to analyze the general motions of real objects that can rotate as well as translate.

8.1 Rigid Bodies, Translations, and Rotations

OBJECTIVES: To (a) distinguish between pure translational and pure rotational motions of a rigid body, and (b) state the condition(s) for rolling without slipping.

In previous chapters, it was convenient to consider motion with the understanding that an object can be represented by a particle located at the center of mass of the object. Rotation, or spinning, was not a consideration, because a particle, a point mass, has no physical dimensions. Rotational motion becomes relevant when we analyze the motion of solid extended objects or *rigid bodies*, the focus of this chapter.

> A **rigid body** is an object or a system of particles in which the distances between particles are fixed (remain constant).

A quantity of liquid water is not a rigid body, but the ice that would form if the water were frozen would be. The discussion of rigid-body rotation is therefore restricted to solids. Actually, the concept of a rigid body is an idealization. In reality, the particles (atoms and molecules) of a solid vibrate constantly. Also, solids can undergo elastic (and inelastic) deformations (Chapter 6). Even so, most solids can be considered rigid bodies for purposes of analyzing rotational motion.

A rigid body may be subject to either or both of two types of motions: translational and rotational. Translational motion is basically the linear motion studied in previous chapters. If an object has only (pure) **translational motion** every particle in it has the same instantaneous velocity, which means that the object is not rotating (▼Fig. 8.1a).

An object may have only (pure) **rotational motion**—motion about a fixed axis (Fig. 8.1b). In this case, all of the particles of the object have the same instantaneous *angular* velocity and travel in circles about the axis of rotation (Fig. 8.1b).

Note: The words *rotation* and *revolution* are commonly used synonymously. In general, this book uses *rotation* when the axis of rotation goes through the body (for example, the Earth's rotation on its axis, in a period of 24 h) and *revolution* when the axis is outside the body (for example, the revolution of the Earth about the Sun, in a period of 365 days).

Demonstration/activity: To distinguish between translational and rotational motions, mount an object such as a wheel or a stick so that it can rotate about an axis. Then set the support on which it is mounted on a lab cart and translate it across the room.

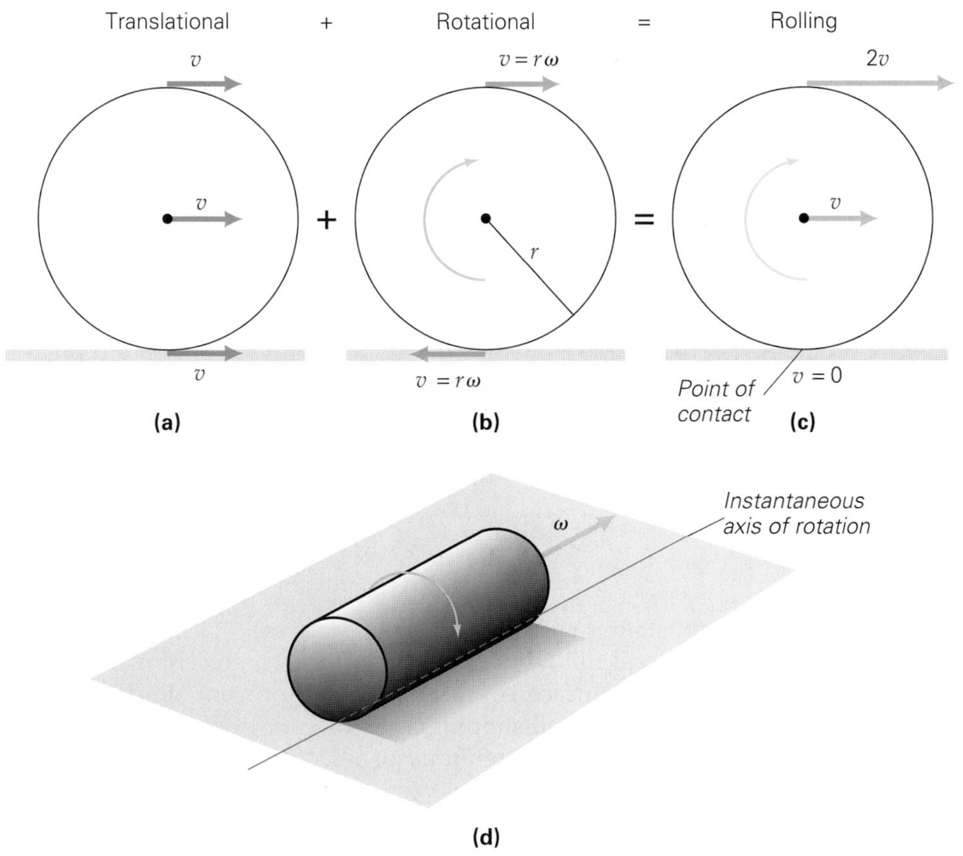

(a)   **(b)**   **(c)**

(d)

◀ **FIGURE 8.1 Rolling—a combination of translational and rotational motions** **(a)** In pure translational motion, all the particles of an object have the same instantaneous velocity. **(b)** In pure rotational motion, all the particles of an object have the same instantaneous angular velocity. **(c)** Rolling is a combination of translational and rotational motions. Summing the velocity vectors for these two motions shows that the point of contact (for a sphere) or the line of contact (for a cylinder) is instantaneously at rest. **(d)** The line of contact for a cylinder (or, for a sphere, a line through the point of contact) is called the *instantaneous axis of rotation*. Note that the center of mass of a rolling object on a level surface moves linearly and remains over the point or line of contact.

General rigid-body motion is a combination of translational and rotational motions. When you throw a ball, the translational motion is described by the motion of its center of mass (as in projectile motion). But the ball may also spin, or rotate, and usually does. A common example of rigid-body motion involving both translation and rotation is rolling, as illustrated in Fig. 8.1c. The combined motion of any point or particle is given by the vector sum of the particle's instantaneous velocity vectors. (Three points or particles are shown in the figure—one at the top, one in the middle, and one at the bottom, of the object.)

At each instant, a rolling object rotates about an **instantaneous axis of rotation** through the point of contact of the object with the surface it is rolling on (for a sphere) or along the line of contact of the object with the surface (for a cylinder, Fig. 8.1d). The location of this axis changes with time. However, note in Fig. 8.1c that the point or line of contact of the body with the surface is instantaneously at rest (and thus has zero velocity), as can be seen from the vector addition of the combined motions at that point. Also, the point on the top has twice the tangential speed ($2v$) of the middle (center-of-mass) point (v), because the top point is twice as far away from the instantaneous axis of rotation as the middle point is. (With a radius r, for the middle point, $r\omega = v$, and for the top point, $2r\omega = 2v$.)

When an object rolls without slipping, its translational and rotational motions are simply related as discussed in Chapter 7. For example, when a uniform ball (or cylinder) rolls in a straight line on a flat surface, it turns through an angle θ, and a point (or line) on the object that was initially in contact with the surface moves through an arc distance s (◄Fig. 8.2). And from Chapter 7, $s = r\theta$ (Eq. 7.3). The center of mass of the ball is directly over the point of contact and moves a linear distance s. Then

$$v_{CM} = \frac{s}{t} = \frac{r\theta}{t} = r\omega$$

where $\omega = \theta/t$. In terms of the speed of the center of mass and the angular speed ω, the **condition for rolling without slipping** is

$$v_{CM} = r\omega \quad \text{(rolling, no slipping)} \tag{8.1}$$

The condition for rolling without slipping is also expressed by

$$s = r\theta \tag{8.1a}$$

where s is the distance the object rolls (the distance the center of mass moves).

Carrying Eq. 8.1 one step further, we can write an expression for the time rate of change of the velocity. Assuming it started from rest, $\Delta v_{CM} = v_{CM}/t = (r\omega)/t$. This yields an equation for *accelerated rolling without slipping*:

$$a_{CM} = \frac{v_{CM}}{t} = \frac{r\omega}{t} = r\alpha \tag{8.1b}$$

where $\alpha = \omega/t$ (for a constant α with ω_o assumed to be zero).

Essentially, an object will roll without slipping if the coefficient of static friction between the object and surface is great enough to prevent slippage. There may be a combination of rolling and slipping motions—for example, the slipping of a car's wheels when traveling through mud or ice. If there is rolling and slipping, then there is no clear relationship between the translational and rotational motions, and $v_{CM} = r\omega$ does not hold.

Illustration 11.2 Rolling Motion

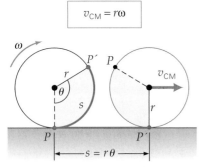

▲ **FIGURE 8.2 Rolling without slipping** As an object rolls without slipping, the length of the arc between two points of contact on the circumference is equal to the linear distance traveled. (Think of paint coming off a roller.) This distance is $s = r\theta$. The speed of the center of mass is $v_{CM} = r\omega$.

Integrated Example 8.1 ■ Rolling without Slipping

A cylinder rolls on a horizontal surface without slipping. (a) At any point in time, the tangential speed of the top of the cylinder is (1) v, (2) $r\omega$, (3) $v + r\omega$, or (4) zero. (b) The cylinder has a radius of 12 cm and a center-of-mass speed of 0.10 m/s, as it rolls without slipping. If it continues to travel at this speed for 2.0 s, through what angle does the cylinder rotate during this time?

(a) Conceptual Reasoning. Since the cylinder rolls without slipping, the relationship $v_{CM} = r\omega$ applies. As was shown in Fig. 8.1, the speed at the point of contact is zero, v for the center (of mass), and $2v$ at the top. With $v = r\omega$, the answer is (3), $v + r\omega = v + v = 2v$.

(b) Quantitative Reasoning and Solution. Knowing the radius and the translational speed, the angular speed can be calculated from the nonslipping condition, $v_{CM} = r\omega$. With this and the time, the angle of rotation may be calculated.

We list the data as usual:

Given: $r = 12$ cm $= 0.12$ m   *Find:* θ (angle of rotation)
     $v_{CM} = 0.10$ m/s
       $t = 2.0$ s

Using $v_{CM} = r\omega$ to find the angular speed,

$$\omega = \frac{v_{CM}}{r} = \frac{0.10 \text{ m/s}}{0.12 \text{ m}} = 0.083 \text{ rad/s}$$

Then,

$$\theta = \omega t = (0.83 \text{ m/s})(2.0 \text{ s}) = 1.7 \text{ rad}$$

The cylinder makes a little over one quarter of a rotation. (Right? Check it yourself.)

Follow-Up Exercise. How far does the CM of the cylinder travel linearly in part (b) of this Example? Find the distance by using two different methods: translational and rotational. *(Answers to all Follow-Up Exercises are at the back of the text.)*

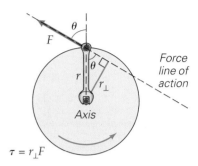

$\tau = r_\perp F$

(a) Counterclockwise torque

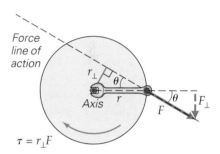

$\tau = r_\perp F$

(b) Smaller clockwise torque

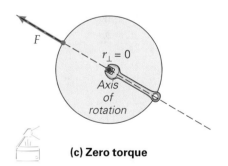

(c) Zero torque

▲ **FIGURE 8.3** Torque and moment arm **(a)** The perpendicular distance $r_\perp$ from the axis of rotation to the line of action of a force is called the *moment arm* (or *lever arm*) and is equal to $r\sin\theta$. The torque, or twisting force, that produces rotational motion is $\tau = r_\perp F$. **(b)** The same force in the opposite direction with a smaller moment arm produces a smaller torque in the opposite direction. Note that $r_\perp F = rF_\perp$, or $(r\sin\theta)F = r(F\sin\theta)$. **(c)** When a force acts through the axis of rotation, $r_\perp = 0$ and $\tau = 0$.

8.2 Torque, Equilibrium, and Stability

OBJECTIVES: To (a) define torque, (b) apply the conditions for mechanical equilibrium, and (c) describe the relationship between the location of the center of gravity and stability.

Torque

As with translational motion, a force is necessary to produce a change in rotational motion. The rate of change of motion depends not only on the magnitude of the force, but also on the perpendicular distance of its line of action from the axis of rotation, $r_\perp$ (▸Fig. 8.3a, b). The line of action of a force is an imaginary line extending through the force vector arrow—that is, the line along which the force acts.

Figure 8.3 shows that $r_\perp = r\sin\theta$, where r is the straight-line distance between the axis of rotation and the point at which the force acts and θ is the angle between the line of r and the force $\vec{F}$. This perpendicular distance $r_\perp$ is called the **moment arm** or **lever arm**.

The product of the force and the lever arm is called **torque** $\vec{\tau}$ (from the Latin *torquere*, meaning "to twist"). The magnitude of the torque provided by the force is

$$\tau = r_\perp F = rF\sin\theta \tag{8.2}$$

SI unit of torque: meter-newton (m·N)

(The symbolism $r_\perp F$ is commonly used to denote torque, but note from Fig. 8.3 that $r_\perp F = rF_\perp$.) The SI units of torque are meters × newtons (m·N), the same as the units of work, $W = Fd$ (N·m, or J). However, we will write the units of torque in reverse order as m·N to avoid confusion. But keep in mind, that torque is *not* work, and its unit is *not* the joule.

Rotational acceleration is not *always* produced when a force acts on a stationary rigid body. From Eq. 8.2, we see that when the force acts through the axis of rotation such that $\theta = 0$, then $\tau = 0$ (Fig. 8.3c). Also, when $\theta = 90°$, the torque is maximum and the force acts perpendicularly to r. Thus, the angular acceleration depends on *where* a perpendicular force is applied (and therefore on the length of the lever arm). As a practical example, think of applying a force to a heavy glass door that swings in and out. Where you apply the force makes a great difference in how easily the door opens or rotates (through the hinge axis). Have you ever tried to open such a door and inadvertently pushed on the side near the hinges? This force produces a small torque and thus little or no rotational acceleration.

Torque in rotational motion can be thought of as the analogue of force in translational motion. An unbalanced or net force changes translational motion, and an unbalanced or net torque changes rotational motion. Torque is a vector. Its direction is always perpendicular to the plane of the force and moment arm vectors and is given by a *right-hand rule* similar to that for angular velocity given in Section 7.2. If the fingers of the right hand are curled around the axis of rotation in the direction that the torque would produce a rotational (angular) acceleration, the extended thumb points in the direction of the torque. A sign convention, as in the case of linear motion, can be used to represent torque directions, as will be discussed shortly.

Demonstration/activity: Clamp a typical mechanic's torque wrench to the table and pull at various angles to the lever with a large demonstration spring scale. Show how the same force can produce different amounts of torque, depending on the angle between the line of force and the lever.

Example 8.2 ■ Lifting and Holding: Muscle Torque at Work

In our bodies, torques produced by the contraction of our muscles cause some bones to rotate at joints. For example, when you lift something with your forearm, a torque is applied on the lower arm by the biceps muscle (▼Fig. 8.4). With the axis of rotation through the elbow joint and the muscle attached 4.0 cm from the joint, what are the magnitudes of the muscle torques for cases (a) and (b) in Fig. 8.4 if the muscle exerts a force of 600 N?

Thinking It Through. As in many rotational situations, it is important to know the orientations of the $\vec{r}$ and $\vec{F}$ vectors so that we can find the angle *between* them to determine the lever arm. Note in the inset in Fig. 8.4a that if the tails of the $\vec{r}$ and $\vec{F}$ vectors were put together, the angle between them would be greater than 90°, that is, $30° + 90° = 120°$. In Fig. 8.4b, the angle is 90°.

Solution. First we list the data given here and in the figure. This Example demonstrates an important point, namely that θ is the angle *between* the radial vector $\vec{r}$ and the force $\vec{F}$

Given: $r = 4.0 \text{ cm} = 0.040 \text{ m}$ *Find:*   (a) τ (muscle torque magnitude) for Fig. 8.4a
          $F = 600 \text{ N}$ (b) τ (muscle torque magnitude) for Fig. 8.4b
          $\theta_a = 30° + 90° = 120°$
          $\theta_b = 90°$

▼ **FIGURE 8.4 Human torque** See Example 8.2.

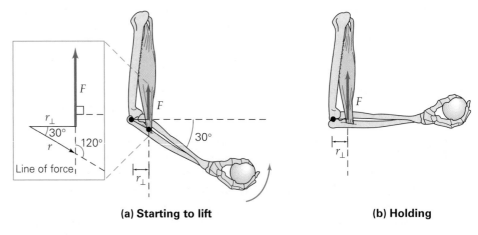

(a) Starting to lift   **(b) Holding**

(a) In this case, $\vec{r}$ is directed along the forearm, so the angle between the $\vec{r}$ and $\vec{F}$ vectors is $\theta_a = 120°$. Using Eq. 8.2, we have

$$\tau = rF \sin(120°) = (0.040 \text{ m})(600 \text{ N})(0.866) = 21 \text{ m} \cdot \text{N}$$

at the instant in question.

(b) Here, the distance r and the line of action of the force are perpendicular $(\theta_b = 90°)$, and $r_\perp = r \sin 90° = r$. Then,

$$\tau = r_\perp F = rF = (0.040 \text{ m})(600 \text{ N}) = 24 \text{ m} \cdot \text{N}$$

The torque is greater in (b). This is to be expected because the maximum value of the torque (τ_{max}) occurs when $\theta = 90°$.

Follow-Up Exercise. In part (a) of this Example, there must have been a net torque, since the ball was accelerated upward by a rotation of the forearm. In part (b), the ball is just being held and there is no rotational acceleration, so there is no net torque on the system. Identify the other torque(s) in each case.

Conceptual Example 8.3 ■ My Aching Back

A person bends over as shown in ▶Fig. 8.5a. For most of us, the center of gravity of the human body is in or near the chest region. When we bend over, this gives rise to a torque that tends to produce rotation about an axis at the base of the spine and cause us to fall over. Why don't we fall when we bend over like this? (Consider only the upper torso.)

Reasoning and Answer. Indeed, if this were the only torque acting, we would fall when bending over. But since we don't fall, another force must be producing a torque such that the net torque is zero. Where does this other torque come from? Obviously from inside the body through a complicated combination of back muscles.

Representing the vector sum as all the back muscle forces as the net force F_b (as shown in Fig. 8.5b), it can be seen that the back muscles exert a force that counterbalances the torque of the center of gravity.

Follow-Up Exercise. Suppose the person was bent over holding a heavy object he had just picked up. How would this affect the back muscle force?

(a)

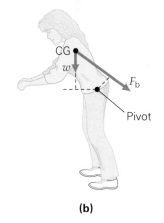

(b)

▲ **FIGURE 8.5** Torque but no rotation **(a)** When bending over, a person's weight, acting through their center of gravity, gives rise to a counterclockwise torque that tends to produce rotation about an axis at the base of the spine. **(b)** However, the back muscles combine to produce a force, F_b, and the resulting clockwise torque counterbalances that of gravity.

Before considering rotational dynamics with net torques and rotational motions, let's look at a situation in which the forces and torques acting on an object are balanced, or in equilibrium.

Equilibrium

In general, equilibrium means that things are in balance or are stable. This definition applies in the mechanical sense to forces and torques. Unbalanced forces produce translational accelerations, but *balanced* forces produce the condition we call *translational equilibrium*. Similarly, unbalanced torques produce rotational accelerations, and *balanced* torques produce *rotational equilibrium*.

According to Newton's first law of motion, when the sum of the forces acting on a body is zero, the body remains either at rest (static) or in motion with a constant velocity. In either case, the body is said to be in **translational equilibrium**. Stated another way, the *condition for translational equilibrium* is that the net force on a body is zero; that is, $\vec{F}_{net} = \Sigma \vec{F}_i = 0$. It should be apparent that this condition is satisfied for the situations illustrated in ▼Fig. 8.6a and b. Forces with lines of action through the same point are called **concurrent forces**. When these forces vectorially add to zero, as in Figs. 8.6a and b, the body is in translational equilibrium.

But what about the situation pictured in Fig. 8.6c? Here, $\Sigma \vec{F}_i = 0$, but the opposing forces will cause the object to rotate, and it will clearly not be in a state of static equilibrium. (Such a pair of equal and opposite forces that do not have the same line of action is called a *couple*.) Thus, the condition $\Sigma \vec{F}_i = 0$ is a necessary, but *not sufficient*, condition for static equilibrium.

Since $\vec{F}_{net} = \Sigma \vec{F}_i = 0$ is the condition for translational equilibrium, you might predict (and correctly so) that $\vec{\tau}_{net} = \Sigma \vec{\tau}_i = 0$ is the *condition for rotational equilibrium*.

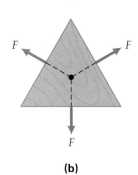

(a)

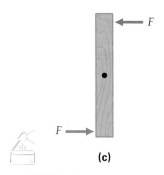

(b)

(c)

▲ **FIGURE 8.6** Equilibrium and forces Forces with lines of action through the same point are said to be *concurrent*. The resultants of the concurrent forces acting on the objects in **(a)** and **(b)** are zero, and the objects are in equilibrium, because the net torque *and* net force are zero. In **(c)**, the object is in *translational* equilibrium, but it will undergo angular acceleration; thus, the object is *not* in rotational equilibrium.

Teaching tip: A review of free-body diagrams may be necessary.

Exploration 13.2 Static Friction on a Horizontal Beam

That is, if the sum of the *torques* acting on an object is zero, then the object is in **rotational equilibrium**—it remains rotationally at rest or rotates with a constant angular velocity.

Thus, we see that there are actually *two* equilibrium conditions. Taken together, they define **mechanical equilibrium**. A body is said to be in mechanical equilibrium when the conditions for both translational and rotational equilibrium are satisfied:

$$\vec{F}_{net} = \Sigma \vec{F}_i = 0 \quad \text{(for translational equilibrium)} \tag{8.3}$$
$$\vec{\tau}_{net} = \Sigma \vec{\tau}_i = 0 \quad \text{(for rotational equilibrium)}$$

A rigid body in mechanical equilibrium may be either at rest or moving with a constant linear and/or angular velocity. An example of the latter is an object rolling without slipping on a level surface, with the center of mass of the object having a constant velocity. However, this is an ideal condition. Of greater practical interest is **static equilibrium**, the condition that exists when a rigid body remains at rest—that is, a body for which $v = 0$ and $\omega = 0$. There are many instances in which we do not want things to move, and this absence of motion can occur only if the equilibrium conditions are satisfied. It is particularly comforting to know, for example, that a bridge you are crossing is in static equilibrium and not subject to translational or rotational motion.

Let's consider examples of static translational equilibrium and static rotational equilibrium separately and then an example in which both apply.

Example 8.4 ■ Translational Static Equilibrium: No Translational Acceleration or Motion

A picture hangs motionless on a wall as shown in ▶Fig. 8.7a. If the picture has a mass of 3.0 kg, what are the magnitudes of the tension forces in the wires?

Thinking It Through. Since the picture remains motionless, it must be in static equilibrium, so applying the conditions for mechanical equilibrium should give equations that yield the tensions. Note that all the forces (tension and weight force) are concurrent; that is, their lines of action pass through a common point, the nail. Because of this, the condition for rotational equilibrium ($\Sigma \vec{\tau}_i = 0$) is automatically satisfied: With respect to an axis of rotation at the nail, the moment arms ($r_\perp$) of the forces are zero, and therefore the torques are zero. Thus, we need consider only translational equilibrium.

Solution.

Given: $\theta_1 = 45°, \theta_2 = 50°$ *Find:* T_1 and T_2
        $m = 3.00$ kg

You will find it helpful to isolate the forces acting on the picture in a free-body diagram, as was done in Chapter 4 for force problems (Fig. 8.7b). The diagram shows the concurrent forces acting through their common point. Note that we have moved all the force vectors to that point, which is taken as the origin of the coordinate axes. The weight force mg acts downward.

With the system in static equilibrium, the net force on the picture is zero; that is, $\Sigma \vec{F}_i = 0$. Thus, the sums of the rectangular components are also zero: $\Sigma F_x = 0$ and $\Sigma F_y = 0$. Then (using ± for direction), we have

$$\Sigma F_x: \quad +T_1 \cos \theta_1 - T_2 \cos \theta_2 = 0 \tag{1}$$

$$\Sigma F_y: \quad +T_1 \sin \theta_1 + T_2 \sin \theta_2 - mg = 0 \tag{2}$$

Then, solving for T_2 in Eq. 1 (or T_1 if you like),

$$T_2 = T_1 \left(\frac{\cos \theta_1}{\cos \theta_2} \right) \tag{3}$$

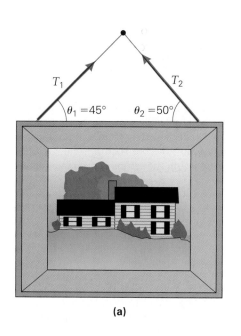

(a)

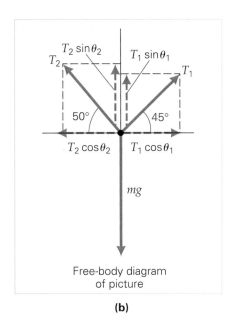

Free-body diagram
of picture

(b)

◀ **FIGURE 8.7 Translational static equilibrium (a)** Since the picture hangs motionless on the wall, the sum of the forces acting on it must be zero. The forces are concurrent, with their lines of action passing through a common point at the nail. **(b)** In the free-body diagram, all the forces are represented as acting at the common point at the nail. T_1 and T_2 have been moved to this point for convenience. Note, however, that the forces shown are acting on the *picture*, not on the nail. See Example 8.4.

and substituting into Eq. 2, with a little algebra, we have

$$T_1\left[\sin 45° + \left(\frac{\cos 45°}{\cos 50°}\right)\sin 50°\right] - mg$$

$$= T_1\left[0.707 + \left(\frac{0.707}{0.643}\right)(0.766)\right] - (3.00\text{ kg})(9.80\text{ m/s}^2) = 0$$

and

$$T_1 = \frac{29.4\text{ N}}{1.55} = 19.0\text{ N}$$

Then, from Eq. (2),

$$T_2 = T_1\left(\frac{\cos\theta_1}{\cos\theta_2}\right) = 19.0\text{ N}\left(\frac{0.707}{0.643}\right) = 20.9\text{ N}$$

Follow-Up Exercise. Analyze the situation in Fig. 8.7 that would result if the wires were at equal angles and were shortened such that the angles were decreased, but kept equal. Carry your analysis to the limit where the angles approach zero. Is the answer realistic?

As pointed out earlier, torque is a vector and therefore has direction. Similar to the way we treated linear motion (Chapter 2), in which we used plus and minus signs to express opposite directions (for example, $+x$ and $-x$), we can designate torque directions as being plus and minus, depending on the rotational acceleration they tend to produce. The rotational "directions" are taken as clockwise or counterclockwise around the axis of rotation. A torque that tends to produce a counterclockwise rotation will be taken as positive $(+)$, and a torque that tends to produce a clockwise rotation will be taken as negative $(-)$. (See the right-hand rule in Section 7.2.) To illustrate, let's apply our convention to the situation in Example 8.5.

(Since we will consider only fixed axis rotations, about which there are just two possible torque directions, the sign-magnitude notation will be used for torque vectors in examples, similar to the motion in one-dimension in Chapter 2.)

Illustration 13.4 The Diving Board Problem

Illustration 13.3 The Force and Torque for Equilibrium

Exploration 13.1 Balance a Mobile

Exploration 13.3 Distributed Load

Example 8.5 ■ Rotational Static Equilibrium: No Rotational Motion

Three masses are suspended from a meterstick as shown in ▼ Fig. 8.8a. How much mass must be suspended on the right side for the system to be in static equilibrium? (Neglect the mass of the meterstick.)

Thinking It Through. As the free-body diagram (Fig. 8.8b) shows, the translational equilibrium condition will be satisfied with the upward normal force $\vec{\mathbf{N}}$ balancing the downward weight forces, so long as the stick is horizontal. But $\vec{\mathbf{N}}$ is not known if m_3 is unknown, so applying the condition for rotational equilibrium should give the required value of m_3. (Note that the lever arms are measured from the pivot point, the center of the meterstick.)

Solution. From the figure, we have

Given: $m_1 = 25$ g   **Find:** m_3 (unknown mass)
$\qquad\quad r_1 = 50$ cm
$\qquad\quad m_2 = 75$ g
$\qquad\quad r_2 = 30$ cm
$\qquad\quad r_3 = 35$ cm

Because the condition for translational equilibrium ($\Sigma\vec{\mathbf{F}}_i = 0$) is satisfied (there is no $\vec{\mathbf{F}}_{net}$ in the y-direction), $N - Mg = 0$, or $N = Mg$, where M is the total mass. This is true no matter what the total mass may be—that is, regardless of how much mass we add for m_3. However, unless the proper mass for m_3 is placed on the right side, the stick will experience a net torque and begin to rotate.

Notice that the masses on the left side produce torques that would tend to rotate the stick counterclockwise, and the mass on the right side produces a torque that would tend to rotate the stick clockwise. We apply the condition for rotational equilibrium by summing the torques about an axis. Let's take this to be through the center of the stick at the 50-cm position, or point A in Fig. 8.8b. Then, noting that N passes through the axis of rotation ($r_\perp = 0$) and produces no torque, we have

$$\Sigma\tau_i:\quad \tau_1 + \tau_2 + \tau_3 = +r_1F_1 + r_2F_2 - r_3F_3 \quad (\textit{using our sign convention for torque vectors})$$
$$= r_1(m_1g) + r_2(m_2g) - r_3(m_3g) = 0$$

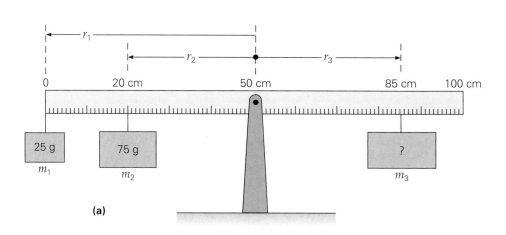

(a)

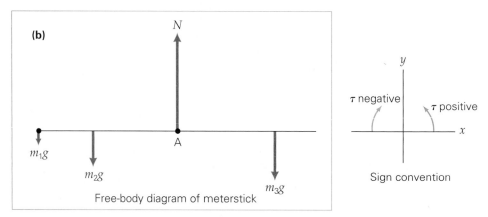

(b)

Free-body diagram of meterstick

Sign convention

▶ **FIGURE 8.8 Rotational static equilibrium** For the meterstick to be in rotational equilibrium, the sum of the torques acting about any selected axis must be zero. (The mass of the meterstick is considered negligible.) See Example 8.5.

Noting that the g's cancel and solving for m_3, we have

$$m_3 = \frac{m_1r_1 + m_2r_2}{r_3} = \frac{(25\text{ g})(50\text{ cm}) + (75\text{ g})(30\text{ cm})}{35\text{ cm}} = 100\text{ g}$$

where it was convenient not to convert to standard units. (The mass of the stick was neglected. If the stick is uniform, however, its mass will not affect the equilibrium, as long as the pivot point is at the 50-cm mark. Why?)

Follow-Up Exercise. The axis of rotation could have been taken through any point along the stick. That is, if a system is in static rotational equilibrium, the condition $\sum \tau_i = 0$ holds for *any* axis of rotation. Show that the preceding statement is true for the system in this Example by taking the axis of rotation through the left end of the stick ($x = 0$).

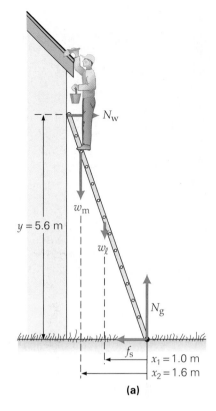

In general, the conditions for both translational and rotational equilibrium need to be written explicitly to solve a statics problem. Example 8.6 is one such case.

Example 8.6 ■ Static Equilibrium: No Translation, No Rotation

A ladder with a mass of 15 kg rests against a smooth wall (▸Fig. 8.9a). A painter who has a mass of 78 kg stands on the ladder as shown in the figure. What is the magnitude of frictional force that must act on the bottom of the ladder to keep it from slipping?

Thinking It Through. Here we have a variety of forces and torques. However, the ladder will not slip as long as the conditions for static equilibrium are satisfied. Summing both the forces and torques to zero should enable us to solve for the necessary frictional force. Also, as we will see, choosing a convenient axis of rotation, such that one or more τ's are zero in the summation of the torques, can simplify the torque equation.

Solution.

Given: $m_\ell = 15\text{ kg}$ *Find:* f_s (force of static friction)
            $m_m = 78\text{ kg}$
            Distances given in figure

Because the wall is smooth, there is negligible friction between it and the ladder, and only the normal reaction force of the wall (N_w) acts on the ladder at this point (Fig. 8.9b).

In applying the conditions for static equilibrium, you are free to choose any axis of rotation for the rotational condition. (The conditions must hold for all parts of a system that is in static equilibrium; that is, there can't be motion in any part of the system.) Note that choosing an axis at the end of the ladder where it touches the ground eliminates the torques due to f_s and N_g, since the moment arms are zero. Then you can write three equations (using mg for w):

$$\sum F_x: \quad N_w - f_s = 0$$
$$\sum F_y: \quad N_g - m_m g - m_\ell g = 0$$

and

$$\sum \tau_i: \quad (m_\ell g)x_1 + (m_m g)x_2 + (-N_w y) = 0$$

The weight of the ladder is considered to be concentrated at its center of gravity. Solving the third equation for N_w and substituting the given values for the masses and distances gives

$$N_w = \frac{(m_\ell g)x_1 + (m_m g)x_2}{y}$$

$$= \frac{(15\text{ kg})(9.8\text{ m/s}^2)(1.0\text{ m}) + (78\text{ kg})(9.8\text{ m/s}^2)(1.6\text{ m})}{5.6\text{ m}} = 2.4 \times 10^2\text{ N}$$

From the first equation, then,

$$f_s = N_w = 2.4 \times 10^2\text{ N}$$

Follow-Up Exercise. In this Example, would the frictional force between the ladder and the ground (call it f_{s_1}) remain the same if there were friction between the wall and the ladder (call it f_{s_2})? Justify your answer.

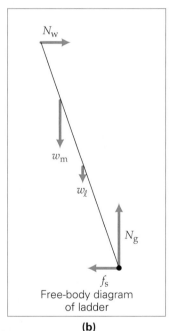

▲ **FIGURE 8.9 Static equilibrium** For the painter's sake, the ladder has to be in static equilibrium; that is, both the sum of the forces and the sum of the torques must be zero. See Example 8.6.

Teaching tip: Discuss the problem of a ladder sliding down because of too little friction. Does the angle at which sliding tends to occur depend on how high the painter is working?

(a)

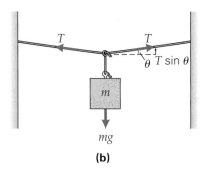

(b)

▲ **FIGURE 8.10** The Iron Cross
(a) The static iron cross gymnastic
position is one of the most strenuous
and difficult to perform. **(b)** An
analogous situation of a weight
suspended on a rope tied at both
ends. See Conceptual Example 8.7.

Note: Stable equilibrium—a
restoring torque.

Note: Unstable equilibrium—a
torque that topples the condition of
stable equilibrium.

Problem-Solving Hint

As the preceding Examples have shown, a good procedure to follow in working
problems involving static equilibrium is as follows:

1. Sketch a space diagram of the problem.
2. Draw a free-body diagram, showing and labeling all external forces and, if necessary, resolving the forces into x- and y-components.
3. Apply the equilibrium conditions. Sum the forces: $\Sigma \vec{F}_i = 0$, usually in component form; $\Sigma F_x = 0$ and $\Sigma F_y = 0$. Sum the torques: $\Sigma \vec{\tau}_i = 0$. Remember to select an appropriate axis of rotation to reduce the number of terms as much as possible. Use $\pm$ sign conventions for both $\vec{F}$ and $\vec{\tau}$.
4. Solve for the unknown quantities.

Conceptual Example 8.7 ■ No Net Torque: The Iron Cross

The static "iron cross" gymnastic position is one of the most strenuous and difficult to perform (◄Fig. 8.10a). What makes it so difficult?

Reasoning and Answer. The gymnast must be extremely strong in order to achieve and maintain such a static position because it requires a huge muscle force to suspend his body from the rings. This can be shown by considering an analogous situation of a weight suspended on a rope tied at each end (Fig. 8.10b). The closer the rope (or gymnast's arms) get to the horizontal position, the more force is needed to keep the weight suspended.

From the figure, it can be seen that the vertical components of the tension force (T) in the rope must balance the downward weight force. (T is analogous to the arm muscle forces.) That is,

$$2T \sin \theta = mg$$

Note that for the rope (or gymnast's arms) to become horizontal, the angle must approach zero $(\theta \to 0)$. Then, what happens to the tension force $T \sin \theta$? As $\theta \to 0$, then $T \to \infty$, making it very difficult to achieve a small θ, and impossible for the rope (and gymnast's arms) to be exactly horizontal.

Follow-Up Exercise. What is the least stressful position for a gymnast on the rings?

Stability and Center of Gravity

The equilibrium of a particle or a rigid body can be stable or unstable in a gravitational field. For rigid bodies, these categories of equilibria are conveniently analyzed in terms of the center of gravity of the body. Recall from Chapter 6 that the **center of gravity** is the point at which all the weight of an object may be considered to be acting as if the object were a particle. When the acceleration due to gravity is constant, the center of gravity and the center of mass coincide.

If an object is in **stable equilibrium**, any small displacement results in a restoring force or torque, which tends to return the object to its original equilibrium position. As illustrated in ►Fig. 8.11a, a ball in a bowl is in stable equilibrium. Analogously, the center of gravity of an extended body on the right is in stable equilibrium. Any slight displacement raises its center of gravity, and a restoring gravitational force tends to return it to the position of minimum potential energy. This force actually produces a restoring torque that is due to a component of the weight force and that tends to rotate the object about a pivot point back to its original position.

For an object in **unstable equilibrium**, any small displacement from equilibrium results in a torque that tends to rotate the object farther away from its equilibrium position. This situation is illustrated in Fig. 8.11b. Note that the center of gravity of the object is at the top of an overturned, or inverted, potential-energy bowl; that is, the potential energy is at a maximum in this case. Small displacements or slight disturbances have profound effects on objects that are in unstable equilibrium: It doesn't take much to cause such an object to change its position.

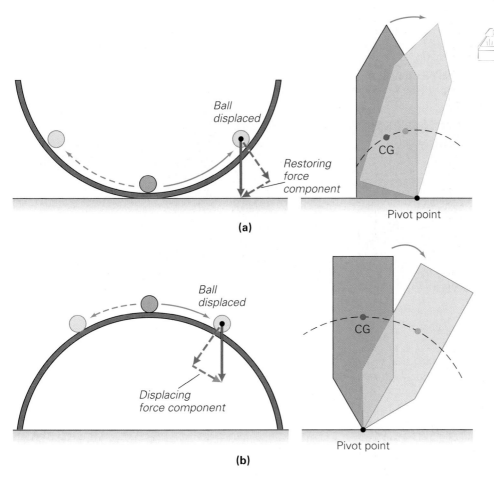

(a)

(b)

◄ FIGURE 8.11 Stable and unstable equilibria (a) When an object is in stable equilibrium, any small displacement from an equilibrium position results in a force or torque that tends to return the object to that position. A ball in a bowl (left) returns to the bottom after being displaced. Analogously, the center of gravity (CG) of an extended object (right) can be thought of as being in a potential-energy "bowl": A small displacement raises the CG, increasing the object's potential energy. **(b)** For an object in unstable equilibrium, any small displacement from its equilibrium position results in a force or torque that tends to take the object farther away from that position. The ball on top of an overturned bowl (left) is in unstable equilibrium. For an extended object (right), the CG can be thought of as being on an inverted potential-energy bowl. A small displacement lowers the CG, decreasing the object's potential energy.

Yet even if the angular displacement of an object in stable equilibrium is quite substantial, the object will still be restored to its equilibrium position. As you might have summarized, the **condition of stable equilibrium** is:

> An object is in stable equilibrium as long as its center of gravity after a small displacement still lies above and inside the object's original base of support. That is, the line of action of the weight force through the center of gravity intersects the original base of support.

When this is the case, there will always be a restoring gravitational torque (▼Fig. 8.12a). However, when the center of gravity or center of mass falls outside the base of support, over goes the object—because of a gravitational torque that rotates it away from its equilibrium position (Fig. 8.12b).

Demonstration/activity: Try to stand a meterstick on its end on the desk. If it stands, it is in stable equilibrium. Why is it considered stable even though many students will think it is unstable?

Teaching tip: Ask students to consider the center of gravity of humans. What happens to the center of gravity of women when they are pregnant? The resulting torque can cause back problems.

▼ FIGURE 8.12 Examples of stable and unstable equilibria (a) When the center of gravity is above and inside an object's base of support, the object is in stable equilibrium. (There is a restoring torque.) Note how the line of action of the weight intersects the center of gravity (CG) intersects the original base of support after the displacement. **(b)** When the center of gravity lies outside the base of support, or the line of action of the weight does not intersect the original base of support, the object is unstable. (There is a displacing torque.)

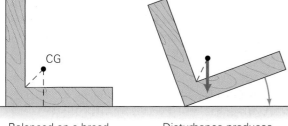
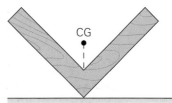
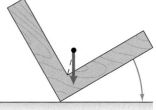

| Balanced on a broad base of support | Disturbance produces restoring torque | Balanced carefully on narrow base of support (point) | Disturbance produces displacing torque |

(a) Stable Equilibrium   **(b) Unstable Equilibrium**

▶ **FIGURE 8.13 Stable and unstable** (a) Race cars are very stable because of their wide wheel bases and low center of gravity. (b) The acrobat's base of support is very narrow: the small area of head-to-head contact. As long as his center of gravity remains above this area, he is in equilibrium, but a displacement of only a few centimeters would probably be enough to topple him. (Why he is in a spread-eagle position will become clearer in Section 8.3.)

(a)

(b)

▼ **FIGURE 8.14 The challenge**
(a) The male student leans forward with his head on the wall. He is to lift the chair and stand up—but he can't. Yet, the female student can easily perform this simple feat. (b) But wait. He applies physics and swings the chair back, and he can stand. Why?

(a)

(b)

Exploration 13.4 The Stacking of Bricks

Rigid bodies with wide bases and low centers of gravity are therefore most stable and least likely to tip over. This relationship is evident in the design of high-speed race cars, which have wide wheel bases and centers of gravity close to the ground (▲Fig. 8.13). SUVs, on the other hand, can roll over more easily. Why?

The location of the center of gravity of the human body has an effect on certain physical abilities. For example, women can generally bend over and touch their toes or touch their palms to the floor more easily than can men, who often fall over trying. On the average, men have higher centers of gravity (larger shoulders) than do women (larger pelvises), so it is more likely that a man's center of gravity will be outside his base of support when he bends over. Conceptual Example 8.8 gives another real-life example of equilibrium and stability.

Conceptual Example 8.8 ■ The Center-of-Gravity Challenge

A female student issues a challenge to a male student. She states she can perform a simple physical feat that he can't. She places a straight-back chair (like most kitchen chairs) with its back against a wall. He is to face the wall and stand next to the chair with his toes touching the wall, then step two foot-lengths backward. (That is, he is to bring the toe of one foot behind the heel of the other foot twice and end up with his feet together, away from the wall.) Next, he is to lean forward and place the top of his head against the wall, reach over to bring the chair directly in front of him, and place one hand on each side of the chair (◀Fig. 8.14a). Finally, without moving his feet, he is to stand up while lifting the chair. The female student demonstrates this and easily stands up.

Most males can't perform this feat, but most females can. Why?

Reasoning and Answer. When the male student bends over and tries to lift the chair, he is in unstable equilibrium (but fortunately he doesn't fall over). That is, the center of gravity of the male student/chair system falls outside (in front of) the system's base of support—his feet. Males tend to have a higher center of gravity (larger shoulders and narrower pelvis) than do females (larger pelvis). When the female student bends over and lifts the chair, the center of gravity of the female student/chair system does not fall outside the system's base of support (her feet). She is in stable equilibrium and so is able to stand up from the bent position while she lifts the chair.

But wait! The male student applies physics and swings the chair back (Fig. 8.14b). The combined center of gravity is now over his base of support, and he can stand while holding the chair.

Follow-Up Exercise. Why might some males be able to stand while lifting the chair, and some females not be able to do this?

Another classic example of equilibrium is the Leaning Tower of Pisa (▸Fig. 8.15a), from which Galileo allegedly performed his "free-fall" experiments. (See Insight 2.1 in Chapter 2 on page 51.) The tower started leaning before its completion in 1350 CE, because of the soft subsoil beneath it. In 1990, the lean was about 5.5° from the vertical (about 5 m, or 17 ft, at the top) with an average increase in the lean of about 1.2 mm a year.

Attempts have been made to stop the lean increase. In 1930, cement was injected under the base, but the lean continued to increase. In the 1990s, major actions were taken. The tower was cabled back and counterweights added to the high side (Fig. 8.15b). Drilling was done diagonally below the foundation on the high side so as to create cavities from which soil could be removed. The tower settled back to about a 5° lean, or a shift of about 40 cm at the top. Moral of the story: Keep that center of gravity above the base of support.

(a)

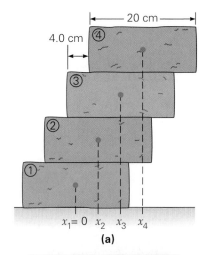

(b)

▲ **FIGURE 8.15 Hold it stable!**
(a) The Leaning Tower of Pisa. Although leaning, it is still in stable equilibrium. Why? **(b)** Tons of lead counterweight being used to help correct the tower's lean.

Example 8.9 ■ Stack Them Up: Center of Gravity

Uniform, identical bricks 20 cm long are stacked so that 4.0 cm of each brick extends beyond the brick beneath, as shown in ▸Fig. 8.16a. How many bricks can be stacked in this way before the stack falls over?

Thinking It Through. As each brick is added, the center of mass (or center of gravity) of the stack moves to the right. The stack will be stable as long as the combined center of mass (CM) is over the base of support—the bottom brick. All of the bricks have the same mass, and the center of mass of each is located at its midpoint. So the horizontal location of the stack's CM must be computed as bricks are added, until the CM extends beyond the base. The location of the CM was discussed in Chapter 6 (see Eq. 6.19).

Solution.

Given: brick length = 20 cm

Find: maximum number of bricks that yields stability
displacement of each brick = 4.0 cm

Taking the origin to be at the center of the bottom brick, we find that the horizontal coordinate of the center of mass (or center of gravity) for the first two bricks in the stack is given by Eq. 6.19, where $m_1 = m_2 = m$ and x_2 is the displacement of the second brick:

$$X_{CM_2} = \frac{mx_1 + mx_2}{m + m} = \frac{m(x_1 + x_2)}{2m} = \frac{x_1 + x_2}{2} = \frac{0 + 4.0 \text{ cm}}{2} = 2.0 \text{ cm}$$

The masses of the bricks cancel out (since they are all the same). For three bricks,

$$X_{CM_3} = \frac{m(x_1 + x_2 + x_3)}{3m} = \frac{0 + 4.0 \text{ cm} + 8.0 \text{ cm}}{3} = 4.0 \text{ cm}$$

For four bricks,

$$X_{CM_4} = \frac{m(x_1 + x_2 + x_3 + x_4)}{4m} = \frac{0 + 4.0 \text{ cm} + 8.0 \text{ cm} + 12.0 \text{ cm}}{4} = 6.0 \text{ cm}$$

and so on.

This series of results shows that the center of mass of the stack moves horizontally 2.0 cm for each brick added to the bottom one. For a stack of six bricks, the center of mass is 10 cm from the origin and directly over the edge of the bottom brick (2.0 cm × 5 *added* bricks = 10 cm, which is half the length of the bottom brick), so the stack is just at unstable equilibrium. The stack may not topple if the sixth brick is positioned very carefully, but it is doubtful that this could be done in practice. A seventh brick would definitely cause the stack to fall off the bottom brick. Why? (As shown in Fig. 8.16b, you can try it yourself and stack them up using books. Don't let the librarian catch you.)

Follow-Up Exercise. If the bricks in this Example were stacked so that, alternately, 4.0 cm and 6.0 cm extended beyond the brick beneath, how many bricks could be stacked before the stack toppled?

(a)

(b)

▲ **FIGURE 8.16 Stack them up!**
(a) How many bricks can be stacked like this before the stack falls? See Example 8.7. **(b)** Try a similar experiment with books.

For another case of stability, see Insight 8.1: Stability in Action on page 271.

$$\tau_{net} = r_\perp F_{net} = rF_\perp = mr^2\alpha$$

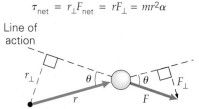

▲ **FIGURE 8.17 Torque on a particle** The magnitude of the torque on a particle of mass m is $\tau = mr^2\alpha$.

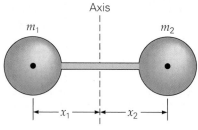

(a) $m_1 = m_2 = 30$ kg
$x_1 = x_2 = 0.50$ m

(b) $m_1 = 40$ kg, $m_2 = 10$ kg
$x_1 = x_2 = 0.50$ m

(c) $m_1 = m_2 = 30$ kg
$x_1 = x_2 = 1.5$ m

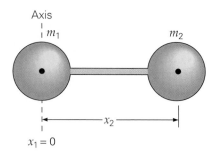

(d) $m_1 = m_2 = 30$ kg
$x_1 = 0$, $x_2 = 3.0$ m

(e) $m_1 = 40$ kg, $m_2 = 10$ kg
$x_1 = 0$, $x_2 = 3.0$ m

▲ **FIGURE 8.18 Moment of inertia** The moment of inertia depends on the distribution of mass relative to a particular axis of rotation and, in general, has a different value for each axis. This difference reflects the fact that objects are easier or more difficult to rotate about certain axes. See Example 8.10.

8.3 Rotational Dynamics

OBJECTIVES: To (a) describe the moment of inertia of a rigid body, and (b) apply the rotational form of Newton's second law to physical situations.

Moment of Inertia

Torque is the rotational analogue of force in linear motion, and a net torque produces rotational motion. To analyze this relationship, consider a constant net force acting on a particle of mass m about the given axis (◄Fig. 8.17). The magnitude of the torque on the particle is

$$\tau_{net} = r_\perp F_{net} = rF_\perp = rma_\perp = mr^2\alpha \quad \textit{torque on a particle} \quad (8.4)$$

where $a_\perp = a_t = r\alpha$ is the tangential acceleration (a_t, Eq. 7.13). For the rotation of a rigid body about a fixed axis, this equation can be applied to each particle in the object and the results summed over the entire body (n particles) to find the total torque. Since all the particles of a rotating rigid body have the same angular acceleration, we can simply add the individual torque magnitudes:

$$\tau_{net} = \Sigma\tau_i = \tau_1 + \tau_2 + \tau_3 + \cdots + \tau_n$$
$$= m_1 r_1^2 \alpha + m_2^2 r_2 \alpha + m_3 r_3^2 \alpha + \cdots + m_n r_n^2 \alpha$$
$$= (m_1 r_1^2 + m_2 r_2^2 + m_3 r_3^2 + \cdots + m_n r_n^2)\alpha$$

$$\Sigma\tau_{net} = \left(\Sigma m_i r_i^2\right)\alpha \quad (8.5)$$

But for a rigid body, the masses (m_i's) and the distances from the axis of rotation (r_i's) do not change. Therefore, the quantity in the parentheses in Eq. 8.5 is constant, and it is called the **moment of inertia**, I (for a given axis):

$$I = \Sigma m_i r_i^2 \quad \textit{moment of inertia} \quad (8.6)$$

SI unit of moment of inertia: kilogram-meters squared ($kg \cdot m^2$)

The magnitude of the net torque can be conveniently written as

$$\tau_{net} = I\alpha \quad \textit{net torque on a rigid body} \quad (8.7)$$

This is the *rotational form of Newton's second law* ($\vec{\tau}_{net} = I\vec{\alpha}$, in vector form). Keep in mind that, like *net* forces, *net* torques (τ_{net}) are necessary to produce angular accelerations.

As you might infer by comparing the rotational form of Newton's second law with the translational form ($\vec{F}_{net} = m\vec{a}$), the moment of inertia I is a measure of *rotational inertia*, or a body's tendency to resist change in its rotational motion. Although I is constant for a rigid body and is the rotational analogue of mass, you must keep in mind that, unlike the mass of a particle, the moment of inertia of a body is referenced to a particular axis and can have different values for different axes.

The moment of inertia also depends on the mass distribution of the body *relative* to its axis of rotation. It is easier (that is, it takes less torque) to give an object an angular acceleration about some axes than about others. The following Example illustrates this point.

Example 8.10 ■ Rotational Inertia: Mass Distribution and Axis of Rotation

Find the moment of inertia about the axis indicated for each of the one-dimensional dumbbell configurations in ◄Fig. 8.18. (Neglect the mass of the connecting bar, and give your answers to three significant figures for comparison.)

Thinking It Through. This is a direct application of Eq. 8.6 for cases with different masses and distances. It will show that the moment of inertia of an object depends on the axis of rotation and on the mass distribution relative to the axis of rotation. The sum for I will include only two terms (two masses).

Solution.

Given: Values of m and r from the figure   *Find:* $I = \Sigma m_i r_i^2$

With $I = m_1 r_1^2 + m_2 r_2^2$:

(a) $I = (30 \text{ kg})(0.50 \text{ m})^2 + (30 \text{ kg})(0.50 \text{ m})^2 = 15.0 \text{ kg} \cdot \text{m}^2$

(b) $I = (40 \text{ kg})(0.50 \text{ m})^2 + (10 \text{ kg})(0.50 \text{ m})^2 = 12.5 \text{ kg} \cdot \text{m}^2$

(c) $I = (30 \text{ kg})(1.5 \text{ m})^2 + (30 \text{ kg})(1.5 \text{ m})^2 = 135 \text{ kg} \cdot \text{m}^2$

(d) $I = (30 \text{ kg})(0 \text{ m})^2 + (30 \text{ kg})(3.0 \text{ m})^2 = 270 \text{ kg} \cdot \text{m}^2$

(e) $I = (40 \text{ kg})(0 \text{ m})^2 + (10 \text{ kg})(3.0 \text{ m})^2 = 90.0 \text{ kg} \cdot \text{m}^2$

This Example clearly shows how the moment of inertia depends on mass *and* its distribution relative to a particular axis of rotation. In general, the moment of inertia is larger the farther the mass is from the axis of rotation. This principle is important in the design of flywheels, which are used in automobiles to keep the engine running smoothly between cylinder firings. The mass of a flywheel is concentrated near the rim, giving a large moment of inertia, which resists changes in motion.

Follow-Up Exercise. In parts (d) and (e) of this Example, would the moments of inertia be different if the axis of rotation went through m_2? Explain.

INSIGHT 8.1 STABILITY IN ACTION

FIGURE 1 Leaning into a curve When rounding a curve or making a turn, a bicycle rider must lean into the curve. (This rider could have told you why.)

When riding a bicycle and going around a curve or making a turn on a level surface, the rider instinctively leans into the curve (Fig. 1). Why? You might think that leaning over, rather than remaining upright, is more likely to cause a spill. However, leaning really does increase stability—it's all a matter of torques.

When a vehicle goes around a level circular curve, a centripetal force is needed to keep the vehicle on the road, as we learned in Chapter 7. This force is generally supplied by the force of static friction between the tires and the road. As illustrated in ▶Fig. 2a, the reaction force $\vec{\mathbf{R}}$ of the ground on the bicycle provides the required centripetal force $(\vec{\mathbf{R}}_x = \vec{\mathbf{F}}_c = \vec{\mathbf{f}}_s)$ to round the curve, and the normal force $(\vec{\mathbf{R}}_y = \vec{\mathbf{N}})$.

Suppose the rider tried to remain upright while going around the curve with these forces operative, as shown in Fig. 2a. Note that the line of action of $\vec{\mathbf{R}}$ does not go through the system's center of gravity (indicated by a dot). With an axis of rotation

through the center of gravity, a counterclockwise torque would tend to rotate the bicycle in such a way that the wheels would slide inward underneath the rider. However, if the rider leans inward at the proper angle (Fig. 2b), both the line of action of $\vec{\mathbf{R}}$ and the weight force act through the center of gravity, and there is no rotational instability (as the gentleman on the bicycle well knew).

There is still a torque on the leaning rider, however. Indeed, when the rider leans into the curve, the weight force gives rise to a torque about an axis through the point of contact with the ground. This torque, along with the turning of the handlebars, causes the bicycle to turn. If the bicycle were not moving, there would be a rotation about this axis, and the bicycle and rider would fall over.

The need to lean into a curve is readily apparent in bicycle and motorcycle races on level tracks. Things can be made easier for the riders if tracks or roadways are banked to provide a natural lean (Section 7.3).

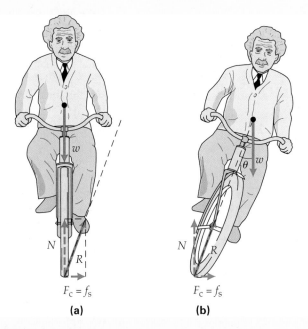

FIGURE 2 Make that turn See text for description.

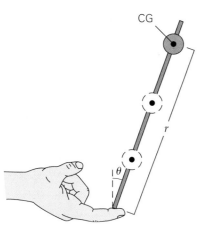

CG

r

θ

▲ **FIGURE 8.19** Greater stability with a higher center of gravity? See Integrated Example 8.11.

Integrated Example 8.11 ■ Balancing Act: Locating the Center of Gravity

(a) A rod with a movable ball, like that shown in ◀Fig. 8.19, is more easily balanced if the ball is in a higher position. Is this because, when the ball in a higher position, (1) the system has a higher center of gravity and more stability; (2) the center of gravity is off the vertical, and there is less torque and a smaller angular acceleration; (3) the center of gravity is closer to the axis of rotation; or (4) the moment of inertia about the axis of rotation is larger? (b) Suppose the distance of the ball from the finger for the farthest position in Fig. 8.19 is 60 cm and the distance to the closest position is 20 cm. When the rod rotates, how many times greater is the angular acceleration of the rod with the ball at the closest position than that with the ball at the farthest position? (Neglect the mass of the rod.)

(a) Conceptual Reasoning. With the ball at any position and the rod vertical, the system is in unstable equilibrium. We saw in Section 8.2 that rigid bodies with wide bases and *low* centers of gravity are more stable, so answer (1) isn't correct. Any slight movement would cause the rod to rotate about an axis through its point of contact. With the center of gravity (CG) at a higher position and off the vertical, there would be a greater lever arm (and thus a *greater* torque), so (2) is also incorrect. With the ball in a higher position, the center of gravity is *farther* from the axis of rotation, which makes (3) incorrect. This leaves (4) by a process of elimination, but let's justify it as the correct answer.

Moving the CG farther from the axis of rotation has an interesting consequence: a greater moment of inertia, or resistance to change in rotational motion, and hence a smaller angular acceleration. However, with the ball in a higher position, as the rod starts to rotate there is a greater torque. The net result is that the increased moment of inertia produces an even greater resistance to rotational motion and hence a smaller angular acceleration. [Note that the torque ($\tau = rF \sin \theta$) varies as r, whereas the moment of inertia ($I = mr^2$) varies as r^2 and so has a larger increase with increasing r. What effect does $\sin \theta$ have?] Then, the smaller the angular acceleration, the more time there is available to adjust your hand under the rod to balance it by bringing the finger and the axis of rotation under the center of gravity. The torque is then zero and the rod is again in equilibrium, albeit unstable. And so, the answer is (4).

(b) Quantitative Reasoning and Solution. Being asked how many times greater or less something is compared to something else usually implies the use of a ratio in which some quantity (or quantities) that is not given cancels. Note that the mass of the ball is not given, which would be needed to compute the gravitational torque (τ). Also, the angle θ is not given. So it is best to start with basic equations and see what happens.

Given: $r_1 = 20$ cm   *Find:* How many times greater the rod's angular acceleration
　　　　$r_2 = 60$ cm   is with the ball at r_1 compared to when it is at r_2

The magnitude of the angular acceleration is given by Eq. 8.7, $\alpha = \tau_{net}/I$. So attention turns to the torque τ_{net} and the moment of inertia I. From the basic equations of the chapter, $\tau_{net} = r_{\perp}F = rF \sin \theta$ (Eq. 8.2), or $\tau_{net} = rmg \sin \theta$, where $F = mg$ in this case, with m being the mass of the ball. Similarly, $I = mr^2$ (Eq. 8.6). Thus,

$$\alpha = \frac{\tau_{net}}{I} = \frac{rmg \sin \theta}{mr^2} = \frac{g \sin \theta}{r}$$

(Note that the angular acceleration α is inversely proportional to the lever arm r; that is, the longer the lever arm, the smaller the angular acceleration.) The $\sin \theta$ is still there, but note what happens when the ratio of the angular accelerations is formed:

$$\frac{\alpha_1}{\alpha_2} = \frac{g \sin \theta / r_1}{g \sin \theta / r_2} = \frac{r_2}{r_1} = \frac{60 \text{ cm}}{20 \text{ cm}} = 3 \quad \text{or} \quad \alpha_2 = \frac{\alpha_1}{3}$$

Hence, the angular acceleration of the rod with the ball at the upper position is one-third that with the ball at the lower position.

Follow-Up Exercise. When walking on a thin bar or rail, such as a railroad rail, you have probably found that it helps to hold your arms outstretched. Similarly, tightrope walkers often carry long poles, as in the chapter-opening photo. How does this posture help a performer maintain balance?

As Integrated Example 8.11 shows, the moment of inertia is an important consideration in rotational motion. By changing the axis of rotation and the relative mass distribution, the value of I can be changed and the motion affected. You were probably told to do this when playing softball or baseball as a child. When at bat, children are often instructed to "choke up" on the bat—to move their hands farther up on the handle.

Now you know why. In doing so, the child moves the axis of rotation of the bat closer to the more massive end of the bat (or its center of mass). Hence, the moment of inertia of the bat is decreased (smaller r in the mr^2 term). Then, when a swing is taken, the angular acceleration is greater. The bat gets around quicker, and the chance of hitting the ball before it goes past is greater. A batter has only a fraction of a second to swing, and with $\theta = \frac{1}{2}\alpha t^2$, a larger α allows the bat to rotate more quickly (swing faster).

Parallel-Axis Theorem

Calculations of the moments of inertia of most extended rigid bodies require math that is beyond the scope of this book. The results for some common shapes are given in ▼Fig. 8.20. The rotational axes are generally taken along axes of symmetry—that is, axes running through the center of mass so as to

Demonstration/activity: Use two $\frac{1}{2}$-in. wooden dowels 36 in. long and four 500-g masses. Tape a 500-g mass to each end of one dowel, then tape a 500-g mass about 12 in. in from each end of the other dowel. Have a student grab the sticks in the centers, one in each hand, and attempt to twist them back and forth at the same rate. Can you calculate how much "harder" it is to twist one than the other?

▼ **FIGURE 8.20** Moments of inertia of some uniform-density objects with common shapes

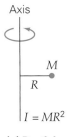

$I = MR^2$

(a) Particle

$I = \frac{1}{12}ML^2$

(b) Thin rod

$I = \frac{1}{3}ML^2$

(c) Thin rod

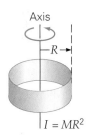

$I = MR^2$

(d) Thin cylindrical shell, hoop, or ring

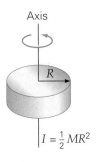

$I = \frac{1}{2}MR^2$

(e) Solid cylinder or disk

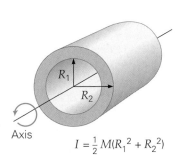

$I = \frac{1}{2}M(R_1{}^2 + R_2{}^2)$

(f) Annular cylinder

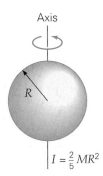

$I = \frac{2}{5}MR^2$

(g) Solid sphere about any diameter

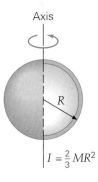

$I = \frac{2}{3}MR^2$

(h) Thin spherical shell

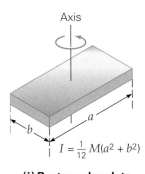

$I = \frac{1}{12}M(a^2 + b^2)$

(i) Rectangular plate

$I = \frac{1}{12}ML^2$

(j) Thin rectangular sheet

$I = \frac{1}{3}ML^2$

(k) Thin rectangular sheet

give a symmetrical mass distribution. One exception is the rod with an axis of rotation through one end (Fig. 8.20c). This axis is parallel to an axis of rotation through the center of mass of the rod (Fig. 8.20b). The moment of inertia about such a parallel axis is given by a useful theorem called the **parallel-axis theorem**, namely,

$$I = I_{CM} + Md^2 \tag{8.8}$$

where I is the moment of inertia about an axis that is parallel to one through the center of mass and at a distance d from it, I_{CM} is the moment of inertia about an axis through the center of mass, and M is the total mass of the body (◀Fig. 8.21). For the axis through the end of the rod (Fig. 8.20c), the moment of inertia is obtained by applying the parallel-axis theorem to the thin rod in Fig. 8.20b:

$$I = I_{CM} + Md^2 = \tfrac{1}{12}ML^2 + M\left(\frac{L}{2}\right)^2 = \tfrac{1}{12}ML^2 + \tfrac{1}{4}ML^2 = \tfrac{1}{3}ML^2$$

Note: $I = I_{CM}$, the minimum I value, when $d = 0$.

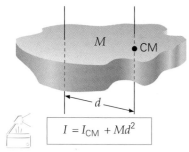

$$I = I_{CM} + Md^2$$

▲ **FIGURE 8.21** Parallel-axis theorem The moment of inertia about an axis parallel to another through the center of mass of a body is $I = I_{CM} + Md^2$, where M is the total mass of the body and d is the distance between the two axes.

Applications of Rotational Dynamics

The rotational form of Newton's second law allows us to analyze dynamic rotational situations. Examples 8.12 and 8.13 illustrate how this is done. In such situations, it is very important to make certain that all the data are properly listed to help with the increasing number of variables.

Example 8.12 ■ Opening the Door: Torque in Action

A student opens a uniform 12-kg door by applying a constant force of 40 N at a perpendicular distance of 0.90 m from the hinges (◀Fig. 8.22). If the door is 2.0 m in height and 1.0 m wide, what is the magnitude of its angular acceleration? (Assume that the door rotates freely on its hinges.)

Thinking It Through. From the given information, we can calculate the applied net torque. To find the angular acceleration of the door, the moment of inertia is needed. This can be calculated, since the door's mass and dimensions are known.

Solution. From the information set forth in the problem, we can list the following:

Given: $M = 12$ kg   *Find:* α (magnitude of angular acceleration)
         $F = 40$ N
         $r_\perp = r = 0.90$ m
         $h = 2.0$ m   (door height)
         $w = 1.0$ m   (door width)

We need to apply the rotational form of Newton's second law (Eq. 8.7), $\tau_{net} = I\alpha$, where I is about the hinge axis. τ_{net} can be found from the given data, so the problem boils down to determining the moment of inertia of the door.

Looking at Fig. 8.20, it can be seen that case (k) applies to a door (treated as a uniform rectangle) rotating on hinges, so $I = \tfrac{1}{3}ML^2$, where $L = w$, the width of the door. Then,

$$\tau_{net} = I\alpha$$

or

$$\alpha = \frac{\tau_{net}}{I} = \frac{r_\perp F}{\tfrac{1}{3}ML^2} = \frac{3rF}{Mw^2} = \frac{3(0.90 \text{ m})(40 \text{ N})}{(12 \text{ kg})(1.0 \text{ m})^2} = 9.0 \text{ rad/s}^2$$

Follow-Up Exercise. In this Example, if the constant torque were applied through an angular distance of 45° and then removed, how long would the door take to swing completely open (90°)?

▲ **FIGURE 8.22** Torque in action See Example 8.12.

In problems involving pulleys in Chapter 4, the mass (as well as the inertia) of the pulley was always neglected in order to simplify things. Now we know how to include those quantities and can treat pulleys more realistically as seen in the next Example.

Example 8.13 ■ Pulleys Have Mass, Too: Taking Account of Pulley Inertia

A block of mass m hangs from a string wrapped around a frictionless, disk-shaped pulley of mass M and radius R, as shown in ▶Fig. 8.23. If the block descends from rest under the influence of gravity, what is the magnitude of its linear acceleration? (Neglect the mass of the string.)

Thinking It Through. Real pulleys have mass and rotational inertia, which affect their motion. The suspended mass (via the string) applies a torque to the pulley. Here we use the rotational form of Newton's second law to find the angular acceleration of the pulley and then its tangential acceleration, which is the same in magnitude as the linear acceleration of the block. (Why?) No numerical values are given so the answer will be in symbol form.

Solution. The linear acceleration of the block depends on the angular acceleration of the pulley, so we look at the pulley system first. The pulley is treated as a disk and thus has a moment of inertia $I = \frac{1}{2}MR^2$ (Fig. 8.20e). A torque due to the tension force in the string (T) acts on the pulley. With $\tau = I\alpha$ (considering only the upper dashed box in Fig. 8.23), we obtain

$$\tau_{net} = r_\perp F = RT = I\alpha = \left(\tfrac{1}{2}MR^2\right)\alpha$$

so that

$$\alpha = \frac{2T}{MR}$$

The linear acceleration of the block and the angular acceleration of the pulley are related by $a = R\alpha$, where a is the tangential acceleration, and

$$a = R\alpha = \frac{2T}{M} \tag{1}$$

But T is unknown. Looking at the descending mass (the lower dashed box) and summing the forces in the vertical direction (positive in the direction of motion) gives

$$mg - T = ma$$

or

$$T = mg - ma \tag{2}$$

Using Eq. 2 to eliminate T from Eq. 1 yields

$$a = \frac{2T}{M} = \frac{2(mg - ma)}{M}$$

And solving for a,

$$a = \frac{2mg}{(2m + M)} \tag{3}$$

Note that if $M \rightarrow 0$ (as in the case of ideal, massless pulleys in previous chapters), then $I \rightarrow 0$ and $a = g$ (from Eq. 3). Here, however, $M \neq 0$, so we have $a < g$. (Why?)

Follow-Up Exercise. Pulleys can be analyzed even more realistically. In this Example, friction was neglected, but practically, a frictional torque (τ_f) exists and should be included. What would be the form, as in Eq. 3, of the angular acceleration in this case? Show that your result is dimensionally correct.

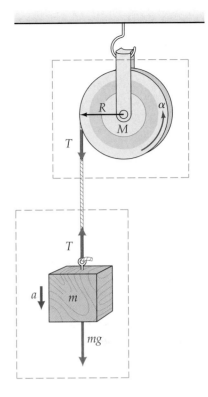

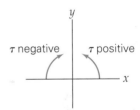

▲ **FIGURE 8.23 Pulley with inertia** Taking the mass, or rotational inertia, of a pulley into account allows a more realistic description of the motion. The directional sign convention for torque is shown. See Example 8.13.

Exploration 10.3 Torque and Moment of Inertia

Exploration 10.4 Torque on Pulley Due to the Tension of Two Strings

In pulley problems, as before, the mass of the string will be neglected—an approach that still gives a good approximation if the string is relatively light. Taking the mass of the string into account would give a continuously varying mass hanging on the pulley, thus producing a variable torque. Such a problem is beyond the scope of this book.

Suppose you had masses suspended from each side of a pulley. Here, you'd have to compute the net torque. If you didn't know the values of the masses or which way the pulley would rotate, then you could simply assume a direction. As in the linear case, if the result came out with the opposite sign, it would indicate that you had assumed the wrong direction.

Problem-Solving Hint

For problems such as those of Examples 8.13 and 8.14, dealing with coupled rotational and translational motions, keep in mind that with no rope slippage, the magnitudes of the accelerations are usually related by $a = r\alpha$, while $v = r\omega$ relates the magnitudes of the velocities at any instant of time. Applying Newton's second law (in rotational or linear form) to different parts of the system gives equations that can be combined by using such relationships. Also, for rolling without slipping, $a = r\alpha$ and $v = r\omega$ relate the angular quantities to the linear motion of the center of mass.

Another application of rotational dynamics is the analysis of motion of objects that can roll.

Conceptual Example 8.14 ■ Applying a Torque One More Time: Which Way Does the Yo-Yo Roll?

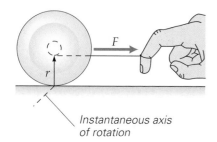

▲ **FIGURE 8.24 Pulling the yo-yo's string** See Conceptual Example 8.14.

The string of a yo-yo sitting on a level surface is pulled as shown in ◄Fig. 8.24. Will the yo-yo roll (a) toward the person or (b) away from the person?

Reasoning and Answer. Apply the physics we have just studied to the situation. Note that the instantaneous axis of rotation is along the line of contact of the yo-yo with the surface. If you had a stick standing vertically in place of the $\vec{r}$ vector and pulled on a string attached to the top of the stick in the direction of $\vec{F}$, which way would the stick rotate? Of course, it would rotate clockwise (about its instantaneous axis of rotation). The yo-yo reacts similarly; that is, it rolls in the direction of the pull, so the answer is (a). (Get a yo-yo and try it if you're a nonbeliever.)

There is more interesting physics in our yo-yo situation. The pull force is not the only force acting on the yo-yo; there are three others. Do they contribute torques? Let's identify these forces. There's the weight of the yo-yo and the normal force from the surface. Also, there is a horizontal force of static friction between the yo-yo and the surface. (Otherwise the yo-yo would slide rather than roll.) But these three forces act through the line of contact or through the instantaneous axis of rotation, so they produce no torques here. (Why?)

What would happen if we increased the angle of the string or pull force (relative to the horizontal) as illustrated in ▼Fig. 8.25a? The yo-yo would still roll to the right. As can be seen in Fig. 8.25b, at some critical angle θ_c the line of force goes through the axis of rotation, and the net torque on the yo-yo becomes zero, so the yo-yo does not roll.

If this critical angle is exceded (Fig. 8.25c), the yo-yo will begin to roll counterclockwise, or to the left. Note that the line of action of the force is on the other side of the axis of rotation from that in Fig. 8.26a and that the lever arm ($r_\perp$) has changed directions, resulting in a reversed net torque direction.

Follow-Up Exercise. Suppose you set the yo-yo string at the critical angle, with the string over a round, horizontal bar at the appropriate height, and you suspend a weight on the end of the string to supply the force for the equilibrium condition. What will happen if you then pull the yo-yo toward you, away from its equilibrium position, and release it?

Demonstration/activity: For a classroom demonstration, construct a big "yo-yo" by cutting two circular pieces (about 20 cm in diameter) of aluminum sheet or plywood and attaching them to a round, relatively large wooden spindle. Using this yo-yo configuration on the lecture desk, demonstrate the physics of Conceptual Example 8.14. It is helpful to paint a radial design on the circular pieces so that the rotations will be more easily observed.

▶ **FIGURE 8.25 The angle makes a difference (a)** With the line of action to the left of the instantaneous axis of rotation, the yo-yo rolls to the right. **(b)** At a critical angle θ_c, the line of action passes through the axis, and the yo-yo is in equilibrium. **(c)** When the line of action is to the right of the axis, the yo-yo rolls to the left. See Conceptual Example 8.14.

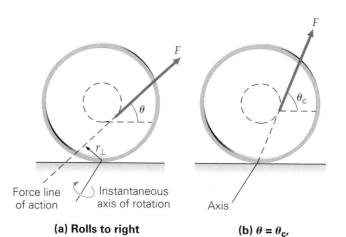

(a) Rolls to right

(b) $\theta = \theta_c$, in rotational equilibrium does not roll

(c) $\theta > \theta_c$, rolls to left ($r_\perp$ to the right)

8.4 Rotational Work and Kinetic Energy

OBJECTIVES: To discuss, explain, and use the rotational forms of (a) work, (b) kinetic energy, and (c) power.

This section gives the rotational analogues of various equations of linear motion associated with work and kinetic energy for constant torques. Because their development is similar to that given for their linear counterparts, detailed discussion is not needed. As in Chapter 5, it is understood that W is the net work if more than one force or torque acts on an object.

Rotational Work   We can go directly from work done by a force to work done by a torque, since the two are related ($\tau = r_\perp F$). For rotational motion, the **rotational work** $W = Fs$ done by a single force F acting tangentially along an arc length s is

Rotational work

$$W = Fs = F(r_\perp \theta) = \tau\theta$$

where θ is in radians. Thus, for a single torque acting through an angle of rotation, θ,

$$W = \tau\theta \quad (single\ force) \tag{8.9}$$

In this book, both the torque (τ) and angular displacement (θ) vectors are almost always along the fixed axis of rotation, so you will not need to be concerned about parallel components, as you were for translational work. The torque and angular displacement may be in opposite directions, in which case the torque does negative work and slows the rotation of the body. Negative rotational work is analogous to F and d being in opposite directions for translational motion.

Rotational Power   An expression for the instantaneous **rotational power**, the rotational analogue of power (the time rate of doing work), is easily obtained from Eq. 8.9:

Rotational power

$$P = \frac{W}{t} = \tau\left(\frac{\theta}{t}\right) = \tau\omega \tag{8.10}$$

The Work–Energy Theorem and Kinetic Energy

The relationship between the net rotational work done on a rigid body (more than one force acting) and the change in rotational kinetic energy of the body can be derived as follows, starting with the equation for rotational work:

$$W_{net} = \tau\theta = I\alpha\theta$$

Since we assume the torques are due only to constant forces, α is constant. But from rotational kinematics in Chapter 7, we know that for a constant angular acceleration, $\omega^2 = \omega_0^2 + 2\alpha\theta$, and

$$W_{net} = I\left(\frac{\omega^2 - \omega_0^2}{2}\right) = \tfrac{1}{2}I\omega^2 - \tfrac{1}{2}I\omega_0^2$$

From Eq. 5.6 (work–energy), we know that $W_{net} = \Delta K$. Therefore,

$$W_{net} = \tfrac{1}{2}I\omega^2 - \tfrac{1}{2}I\omega_0^2 = K - K_o = \Delta K \tag{8.11}$$

Rotational analogue of work–energy theorem

Then the expression for **rotational kinetic energy**, K, is

$$K = \tfrac{1}{2}I\omega^2 \tag{8.12}$$

Rotational kinetic energy

Thus, *the net rotational work done on an object is equal to the change in the rotational kinetic energy of the object* (with zero linear kinetic energy). Consequently, to change the rotational kinetic energy of an object, a net torque must be applied.

It is possible to derive the expression for the kinetic energy of a rotating rigid body (on a fixed axis) directly. Summing the instantaneous kinetic energies of the body's individual particles relative to the fixed axis gives

$$K = \tfrac{1}{2}\sum m_i v_i^2 = \tfrac{1}{2}\left(\sum m_i r_i^2\right)\omega^2 = \tfrac{1}{2}I\omega^2$$

where, for each particle of the body, $v_i = r_i\omega$. So, Eq. 8.12 doesn't represent a new form of energy; rather, it is simply another expression for kinetic energy, in a form that is more convenient for rigid-body rotation.

TABLE 8.1	Translational and Rotational Quantities and Equations		
Translational		*Rotational*	
Force:	$\vec{\mathbf{F}}$	Torque (magnitude):	$\tau = rF \sin\theta$
Mass (inertia):	m	Moment of inertia:	$I = \sum m_i r_i^2$
Newton's second law:	$\vec{\mathbf{F}}_{net} = m\vec{\mathbf{a}}$	Newton's second law:	$\vec{\boldsymbol{\tau}}_{net} = I\vec{\boldsymbol{\alpha}}$
Work:	$W = Fd$	Work:	$W = \tau\theta$
Power:	$P = Fv$	Power:	$P = \tau\omega$
Kinetic energy:	$K = \frac{1}{2}mv^2$	Kinetic energy:	$K = \frac{1}{2}I\omega^2$
Work–energy theorem:	$W_{net} = \frac{1}{2}mv^2 - \frac{1}{2}mv_0^2 = \Delta K$	Work–energy theorem:	$W_{net} = \frac{1}{2}I\omega^2 - \frac{1}{2}I\omega_0^2 = \Delta K$
Linear momentum:	$\vec{\mathbf{p}} = m\vec{\mathbf{v}}$	Angular momentum:	$\vec{\mathbf{L}} = I\vec{\boldsymbol{\omega}}$

A summary of translational and rotational analogues is given in Table 8.1. (The table also contains angular momentum, which we shall discuss in Section 8.5.)

When an object has both translational and rotational motion, its total kinetic energy may be divided into parts to reflect the two kinds of motion. For example, for a cylinder rolling without slipping on a level surface, the motion is purely rotational relative to the instantaneous axis of rotation (the point or line of contact), which is instantaneously at rest. The total kinetic energy of the rolling cylinder is

$$K = \tfrac{1}{2}I_i\omega^2$$

where I_i is the moment of inertia about the instantaneous axis. This moment of inertia about the point of contact (our axis) is given by the parallel-axis theorem (Eq. 8.8), $I_i = I_{CM} + MR^2$, where R is the radius of the cylinder. Then

$$K = \tfrac{1}{2}I_i\omega^2 = \tfrac{1}{2}(I_{CM} + MR^2)\omega^2 = \tfrac{1}{2}I_{CM}\omega^2 + \tfrac{1}{2}MR^2\omega^2$$

But since there is no slipping, $v_{CM} = R\omega$, and

$$K = \tfrac{1}{2}I_{CM}\omega^2 + \tfrac{1}{2}Mv_{CM}^2 \quad \text{(rolling, no slipping)} \tag{8.13}$$

$$\underset{KE}{\text{total}} = \underset{KE}{\text{rotational}} + \underset{KE}{\text{translational}}$$

Note that although a cylinder was used as an example here, this is a general result and applies to any object that is rolling without slipping.

Thus, *the total kinetic energy of such an object is the sum of two contributions: the translational kinetic energy of the object's center of mass and the rotational kinetic energy of the object relative to a horizontal axis through its center of mass.*

Note: A rolling body has both translational kinetic energy and rotational kinetic energy.

PHYSLET®

Illustration 11.3 Translational and Rotational Kinetic Energy

Example 8.15 ■ Division of Energy: Rotational and Translational

A uniform, solid 1.0-kg cylinder rolls without slipping at a speed of 1.8 m/s on a flat surface. (a) What is the total kinetic energy of the cylinder? (b) What percentage of this total is rotational kinetic energy?

Thinking It Through. The cylinder has both rotational and translational kinetic energies, so Eq. 8.13 applies, and its terms are related by the condition of rolling without slipping.

Solution.

Given: $M = 1.0$ kg
$v_{CM} = 1.8$ m/s
$I_{CM} = \frac{1}{2}MR^2$ (from Fig. 8.20e)

Find: (a) K (total kinetic energy)
(b) $\dfrac{K_r}{K}$ ($\times 100\%$) (percentage of rotational energy)

(a) The cylinder rolls without slipping, so the condition $v_{CM} = R\omega$ applies. Then the total kinetic energy is the sum of the rotational kinetic energy K_r and the translational kinetic energy of the center of mass, K_{CM} (Eq. 8.13):

$$K = \tfrac{1}{2}I_{CM}\omega^2 + \tfrac{1}{2}Mv_{CM}^2 = \tfrac{1}{2}\left(\tfrac{1}{2}MR^2\right)\left(\frac{v_{CM}}{R}\right)^2 + \tfrac{1}{2}Mv_{CM}^2 = \tfrac{1}{4}Mv_{CM}^2 + \tfrac{1}{2}Mv_{CM}^2$$

$$= \tfrac{3}{4}Mv_{CM}^2 = \tfrac{3}{4}(1.0\text{ kg})(1.8\text{ m/s})^2 = 2.4\text{ J}$$

(b) The rotational kinetic energy K_r of the cylinder is the first term of the preceding equation, so, forming a ratio in symbol form, we get

$$\frac{K_r}{K} = \frac{\frac{1}{4}Mv_{CM}^2}{\frac{3}{4}Mv_{CM}^2} = \frac{1}{3}(\times 100\%) = 33\%$$

Thus, the total kinetic energy of the cylinder is made up of rotational and translational parts, with one third being rotational.

Note that in part (b) the radius of the cylinder was not needed and neither was the mass. Because we used a ratio, these quantities canceled out. *Don't* think that this exact division of energy is a general result, however. It is easy to show that the percentage is different for objects with different moments of inertia. For example, you should expect a rolling sphere to have a smaller percentage of rotational kinetic energy than a cylinder has, because the sphere has a smaller moment of inertia $\left(I = \frac{2}{5}MR^2\right)$.

Follow-Up Exercise. Potential energy can be brought into the act by applying the conservation of energy to an object rolling up or down an inclined plane. In this Example, suppose that the cylinder rolled up a 20° inclined plane without slipping. (a) At what vertical height (measured by the vertical distance of its CM) on the plane does the cylinder stop? (b) To find the height in part (a), you probably equated the initial total kinetic energy to the final gravitational potential energy. That is, the total kinetic energy was reduced by the work done by gravity. However, a frictional force also acts (to prevent slipping). Is there not work done here, too?

Example 8.16 ■ Rolling Down versus Sliding Down: Which Is Faster?

A uniform cylindrical hoop is released from rest at a height of 0.25 m near the top of an inclined plane (▶Fig. 8.26). If the hoop rolls down the plane without slipping and no energy is lost due to friction, what is the linear speed of the cylinder's center of mass at the bottom of the incline?

Thinking It Through. Here, gravitational potential energy is converted into kinetic energy—both rotational and translational. The conservation of (mechanical) energy applies, since W_f is zero.

Solution.

Given: $\quad h = 0.25$ m $\qquad\qquad$ *Find:* $\quad v_{CM}$ (speed of CM)
$\qquad\quad I_{CM} = MR^2 \quad$ (from Fig. 8.20d)

Because the total mechanical energy of the cylinder is conserved, you can write

$$E_o = E$$

or, since $v_o = 0$ at the top of the incline and assuming that $U = 0$ at the bottom,

$$U_o = K$$

$$\underbrace{Mgh}_{\text{initially at rest}} = \underbrace{\tfrac{1}{2}I_{CM}\omega^2 + \tfrac{1}{2}Mv_{CM}^2}_{\text{at bottom of incline}}$$

Using the rolling condition $v_{CM} = R\omega$ gives

$$Mgh = \tfrac{1}{2}(MR^2)\left(\frac{v_{CM}}{R}\right)^2 + \tfrac{1}{2}Mv_{CM}^2 = Mv_{CM}^2$$

Solving for v_{CM},

$$v_{CM} = \sqrt{gh} = \sqrt{(9.8 \text{ m/s}^2)(0.25 \text{ m})} = 1.6 \text{ m/s}$$

Again, not much numerical information was needed here. Note that the hoop rolls down from the same height that the solid cylinder rolled up the incline in the Example 8.15 Follow-Up Exercise, yet the speed of the hoop is less than that of the cylinder at the bottom of the incline. Why? Because of differences in the moment of inertia.

Follow-Up Exercise. Suppose the inclined plane in this Example were frictionless and the hoop slid down the plane instead of rolling. How would the speed at the bottom compare in this case? Why are the speeds different?

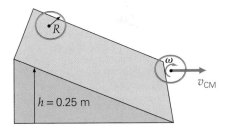

▲ **FIGURE 8.26 Rolling motion and energy** When an object rolls down an inclined plane, potential energy is converted to translational *and* rotational kinetic energy. This makes the rolling slower than frictionless sliding. See Example 8.16.

Exploration 11.3 Rolling Down an Incline

A Fixed Race

As Example 8.16 shows for an object rolling down an incline without slipping, v_{CM} is independent of M and R. The masses and radii cancel out, so all objects of a particular shape (with the same equation for the moment of inertia) roll with the same speed, regardless of their size or density. But the rolling speed does vary with the moment of inertia, which varies with the shape. Therefore, rigid bodies with different shapes roll with different speeds. For example, if you released a cylindrical hoop, a solid cylinder, and a uniform sphere at the same time from the top of an inclined plane, the sphere would win the race to the bottom, followed by the cylinder, with the hoop coming in last—every time!

You can try this as an experiment with a couple of cans of food or other cylindrical containers—one full of some solid material (in effect, a rigid body) and one empty and with the ends cut out—and a smooth, solid ball. Remember that the masses and the radii make no difference. You might think that an annular cylinder (a hollow cylinder with inner and outer radii that vary appreciably—Fig. 8.20f) would be a possible front-runner, or "front-roller," in such a race, but it wouldn't win. The rolling race down an incline is fixed even when you vary the masses and the radii.

Another aspect of rolling is discussed in Insight 8.2: Slide or Roll to a Stop.

Demonstration/activity: Place a cylinder, a hoop, and a solid sphere of equal mass and radius at the top of an inclined plane. Release the three at the same time and find out which object wins the race. Analyze this race in terms of the conservation of energy. (See Exercise 92.)

Suppose the three objects just described were given the same speed on a horizontal surface and were allowed to roll up a ramp. Would they all reach the same height?

8.5 Angular Momentum

OBJECTIVES: To (a) define angular momentum, and (b) apply the conservation of angular momentum to physical situations.

Another important quantity in rotational motion is angular momentum. Recall from Section 6.1 how the linear momentum of an object is changed by a force. Analogously, changes in angular momentum are associated with torques. As we have seen, torque is the product of a moment arm and a force. In a similar manner, **angular momentum (L)** is the product of a moment arm and a linear momentum. For a particle of mass m, the magnitude of the linear momentum is $p = mv$, where $v = r\omega$. The magnitude of the angular momentum is

$$L = r_\perp p = mr_\perp v = mr_\perp^2 \omega \quad \textit{single-particle angular momentum} \quad (8.14)$$

SI unit of angular momentum: kilogram-meters squared per second $(\text{kg} \cdot \text{m}^2/\text{s})$ where v is the speed of the particle, $r_\perp$ is the moment arm, and ω is the angular speed.

For circular motion, $r_\perp = r$, since $\vec{v}$ is perpendicular to $\vec{r}$. For a system of particles making up a rigid body, all the particles travel in circles, and the magnitude of the total angular momentum is

$$L = (\Sigma m_i r_i^2)\omega = I\omega \quad \textit{rigid-body angular momentum} \quad (8.15)$$

which, for rotation about a fixed axis, is (in vector notation)

$$\vec{L} = I\vec{\omega} \quad (8.16)$$

Thus, $\vec{L}$ is in the direction of the angular velocity vector $(\vec{\omega})$. This direction is given by the right-hand rule.

For linear motion, the change in the total linear momentum of a system is related to the net external force by $\vec{F}_{net} = \Delta\vec{P}/\Delta t$. Angular momentum is analogously related to net torque (in magnitude form):

$$\tau_{net} = I\alpha = \frac{I\Delta\omega}{\Delta t} = \frac{\Delta(I\omega)}{\Delta t} = \frac{\Delta L}{\Delta t}$$

That is,

$$\tau_{net} = \frac{\Delta L}{\Delta t} \quad (8.17)$$

Thus, the net torque is equal to *the time rate of change of angular momentum*. In other words, a net torque causes a *change* in angular momentum.

PHYSLET®

Illustration 10.3 Moment of Inertia, Rotational Energy, and Angular Momentum

PHYSLET®

Exploration 11.4 Moment of Inertia and Angular Momentum

Note: Review Eq. 6.3, Section 6.1.

INSIGHT   8.2 SLIDE OR ROLL TO A STOP? ANTILOCK BRAKES

While driving, in an emergency you may instinctively jam on the brakes, trying to come to a quick stop—that is, to stop in the shortest distance. But with the wheels locked, the car skids or slides, often out of control. In this case, the force of sliding friction is acting on the wheels.

To prevent skidding, you may have learned to pump the brakes in order to roll rather than slide to a stop, particularly on wet or icy roads. Most newer automobiles have a computerized antilock braking system (ABS) that pumps the brakes automatically. When the brakes are applied firmly and the car begins to slide, sensors in the wheels note the sliding motion, and a computer takes control of the braking system. It momentarily releases the brakes and then varies the brake-fluid pressure with a pumping action (up to thirteen times per second!) so that the wheels will continue to roll without slipping.

In the absence of sliding, both rolling friction and static friction act. In many cases, however, the force of rolling friction is small, and only static friction need be taken into account. The ABS works to keep static friction near the maximum, $f_s \approx f_{s_{max}}$, which you can't do easily by foot.

Does sliding instead of rolling make a big difference in an automobile's stopping distance? We can calculate the difference by assuming that rolling friction is negligible. Although the external force of static friction does no work to dissipate energy in slowing a car (this is done internally by friction on the brake pads), it does determine whether the wheels roll or slide.

In Example 2.8, a vehicle's stopping distance was given by

$$x = \frac{v_o^2}{2a}$$

By Newton's second law, the net force in the horizontal direction is $F = f = \mu N = \mu mg = ma$, and the stopping acceleration is then $a = \mu g$. Thus,

$$x = \frac{v_o^2}{2\mu g} \qquad (1)$$

But, as was noted in Chapter 4, the coefficient of sliding (kinetic) friction is generally less than that of static friction; that is, $\mu_k < \mu_s$. The general difference between rolling stops and sliding stops can be seen by using the same initial velocity v_o for both cases. Then, using Eq. 1 to form a ratio, we obtain

$$\frac{x_{roll}}{x_{slide}} = \frac{\mu_k}{\mu_s} \qquad \text{or} \qquad x_{roll} = \left(\frac{\mu_k}{\mu_s}\right)x_{slide}$$

From Table 4.1, the value of μ_k for rubber on wet concrete is 0.60, and the value of μ_s for these surfaces is 0.80. Using these values for a comparison of the stopping distances gives

$$x_{roll} = \left(\frac{0.60}{0.80}\right)x_{slide} = (0.75)x_{slide}$$

Thus, the car comes to a rolling stop in 75% of the distance required for a sliding stop—for example, 15 m instead of 20 m. Although this distance may vary for different conditions, it could be an important, perhaps lifesaving, difference.

Conservation of Angular Momentum

Equation 8.17 was derived by using $\tau_{net} = I\alpha$, which applies to a rigid system of particles or a rigid body having a constant moment of inertia. However, Eq. 8.17 is a general equation that also applies to even nonrigid systems of particles. In such a system, there may be a change in the mass distribution and a change in the moment of inertia. As a result, there may be an angular acceleration even in the absence of a net torque. How can this be?

If the net torque on a system is zero, then, by Eq. 8.17, $\vec{\tau}_{net} = \Delta\vec{L}/\Delta t = 0$, and

$$\Delta\vec{L} = \vec{L} - \vec{L}_o = I\vec{\omega} - I_o\vec{\omega}_o = 0$$

or

$$I\omega = I_o\omega_o \qquad (8.18)$$

Thus, the condition for the **conservation of angular momentum** is as follows:

In the absence of an external, unbalanced torque, the total (vector) angular momentum of a system is conserved (remains constant).

Conservation of angular momentum

As with total linear momentum, the internal torques arising from internal forces cancel out.

For a rigid body with a constant moment of inertia (that is, $I = I_o$), the angular speed remains constant ($\omega = \omega_o$) in the absence of a net torque. But it is possible for the moment of inertia to change in some systems, giving rise to a change in the angular speed, as the following Example illustrates.

Note: Angular momentum is conserved when the net torque is zero. ($\vec{L}$ stays fixed.) This is the third conservation law in mechanics.

▶ **FIGURE 8.27 Conservation of angular momentum** When the string is pulled downward through the tube, the revolving ball speeds up. See Example 8.17.

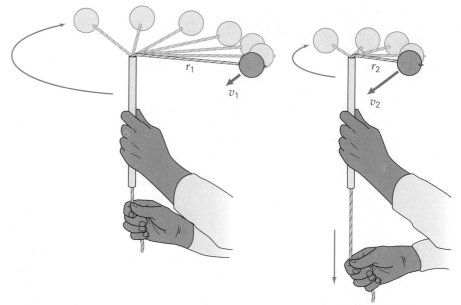

Example 8.17 ■ Pull It Down: Conservation of Angular Momentum

A small ball at the end of a string that passes through a tube is swung in a circle, as illustrated in ▲Fig. 8.27. When the string is pulled downward through the tube, the angular speed of the ball increases. (a) Is the increase in angular speed caused by a torque due to the pulling force? (b) If the ball is initially moving at a 2.8 m/s in a circle with a radius of 0.30 m, what will be its tangential speed if the string is pulled down to reduce the radius of the circle to 0.15 m? (Neglect the mass of the string.)

Thinking It Through. (a) A force is applied to the ball via the string, but consider the axis of rotation. (b) In the absence of a net torque, the angular momentum is conserved (Eq. 8.18), and the tangential speed is related to the angular speed by $v = r\omega$.

Solution.

Given: $r_1 = 0.30$ m   *Find:* (a) Cause of the increase in angular speed
$r_2 = 0.15$ m   (b) v_2 (final tangential speed)
$v_1 = 2.8$ m/s

(a) The change in the angular velocity, or an angular acceleration, is not caused by a torque due to the pulling force. The force on the ball, as transmitted by the string (tension), acts through the axis of rotation, and therefore the torque is zero. Because the rotating portion of the string is shortened, the moment of inertia of the ball ($I = mr^2$, from Fig. 8.20a) decreases. Because in the absence of an external torque, the angular momentum ($I\omega$) of the ball is conserved, and if I is reduced, ω must increase.

(b) Because the angular momentum is conserved, we can equate the magnitudes of the angular momenta:

$$I_o\omega_o = I\omega$$

Then, using $I = mr^2$ and $\omega = v/r$ gives

$$mr_1v_1 = mr_2v_2$$

and

$$v_2 = \left(\frac{r_1}{r_2}\right)v_1 = \left(\frac{0.30 \text{ m}}{0.15 \text{ m}}\right)2.8 \text{ m/s} = 5.6 \text{ m/s}$$

When the radial distance is shortened, the ball speeds up.

Follow-Up Exercise. Let's look at the situation in this Example in terms of work and energy. If the initial speed is the same and the vertical pulling force is 7.8 N, what is the final speed of the 0.10-kg ball?

Example 8.17 should help you understand Kepler's law of equal areas (Chapter 7) from another viewpoint. A planet's angular momentum is conserved to a good approximation by neglecting the weak gravitational torques from other planets. (The Sun's gravitational force on a planet produces little or no torque. Why?) When a planet is closer to the Sun in its elliptical orbit and so has a shorter moment arm, its speed is greater, by the conservation of angular momentum. [This is the basis of Kepler's second law (the law of areas), Section 7.6.] Similarly, when an orbiting satellite's altitude varies during the course of an elliptical orbit about a planet, the satellite speeds up or slows down in accordance with the same principle.

Real-Life Angular Momentum

A popular demonstration of the conservation of angular momentum is shown in ▼Fig. 8.28a. A person sitting on a stool that rotates holds weights with his arms outstretched and is started slowly rotating. An external torque to start this rotation must be supplied by someone else, because the person on the stool cannot initiate the motion by himself. (Why not?) Once rotating, if the person brings his arms inward, the angular speed increases and he spins much faster. Extending his arms again slows him down. Can you explain this phenomenon?

If L is constant, what happens to ω when I is made smaller by reducing r? The angular speed must increase to compensate and keep L constant. Ice skaters do dizzying spins by pulling in and raising their arms to reduce their moment of inertia (Fig. 8.28b). Similarly, a diver spins during a high dive by tucking in the body and limbs, greatly decreasing his or her moment of inertia. The enormous wind speeds of tornadoes and hurricanes represent another example of the same effect (Fig. 8.28c).

Angular momentum also plays a role in ice-skating jumps in which the skater spins in the air, such as a triple axel or triple lutz. A torque applied on the jump gives the skater angular momentum, and the arms and legs are drawn into the body, which, as in spinning on one's toes, decreases the moment of inertia and increases the angular speed so that multiple spins can be made during the jump. To land with a smaller rate of spin, the skater opens the arms and projects the nonlanding leg. You may have noticed that most jump landings proceed in a curved arc, which allows the skater to gain control.

PHYSLET®

Exploration 11.5 *Conservation of Angular Momentum*

(a)

(b)

(c)

◄ **FIGURE 8.28 Change in moment of inertia (a)** When the student spins slowly with masses in outstretched arms, his moment of inertia is relatively large. (The masses are farther from the axis of rotation.) Note that he is isolated, with no external torques (neglecting friction) acting on him, so his angular momentum, $L = I\omega$, is conserved. Pulling his arms inward decreases his moment of inertia. (Why?) Consequently, ω must increase, and he goes into a dizzying spin. **(b)** Ice skaters change their moment of inertia to increase ω in doing spins. **(c)** The same principle helps explain the violence of the winds that spiral around the center of a hurricane. As air rushes in toward the center of the storm, where the pressure is low, its rotational speed must increase for angular momentum to be conserved.

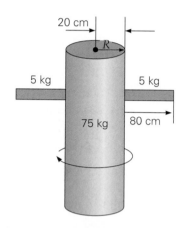

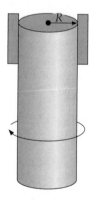

20 cm

R

5 kg   5 kg

75 kg   80 cm

(a) Arms extended (not to scale)

R

(b) Arms overhead

▲ **FIGURE 8.29** Skater model.
Change in moment of inertia and
spin. See Example 8.18.

Example 8.18 ■ A Skater Model

Real-life situations are generally complicated, but some can be approximately analyzed
by using simple models. Such a model for a skater's spin is shown in ◄Fig. 8.29, with a
cylinder and rods representing the skater. In (a), the skater goes into the spin with the
"arms" out, and in (b) the "arms" are over the head to achieve a faster spin by the conservation of angular momentum. If the initial spin rate is 1 revolution per 1.5 s, what (angular speed) is the angular speed when the arms are tucked in?

Thinking It Through. The body and arms of a skater are approximated by the cylinder and
rods, for which we know the moments of inertia (Fig. 8.20). Special attention must be
given to finding the moment of inertia of the arms around the axis of rotation (through
the cylinder). This can be done by applying the parallel axis theorem (Eq. 8.8).

 With the angular momentum conserved, $L = L_o$ or $I\omega = I_o\omega_o$, knowing the initial
angular speed, and given quantities to evaluate the moments of inertia (Fig. 8.29), the
final angular speed can be found.

Solution. Listing the given data (see Fig. 8.29):

Given: $\omega_o = (1 \text{ rev}/1.5 \text{ s})(2\pi \text{ rad/rev}) = 4.2 \text{ rad/s}$ *Find:* ω (final angular speed)
     $M_c = 75 \text{ kg}$ (cylinder or body)
     $M_r = 5.0 \text{ kg}$ (one rod or arm)
     $R = 20 \text{ cm} = 0.20 \text{ m}$
     $L = 80 \text{ cm} = 0.80 \text{ m}$

Moment of inertia (from Fig. 8.20).

cylinder: $I_c = \frac{1}{2}M_cR^2$ rod: $I_r = \frac{1}{12}M_rL^2$

Let's first compute the moments of inertia of the system using the parallel axis theorem,
$I = I_{cm} + Md^2$ (Eq. 8.8).

Before: The I_c of the cylinder is straight forward (Fig. 8.20e):

$$I_c = \tfrac{1}{2}M_cR^2 = \tfrac{1}{2}(75 \text{ kg})(0.20 \text{ m})^2 = 1.5 \text{ kg} \cdot \text{m}^2$$

Referencing the moment of inertia of a horizontal rod (Fig. 8.29a) to the cylinder's axis of
rotation using the parallel axis theorem:

$$I_r = I_{cm(rod)} + Md^2$$
$$= \tfrac{1}{12}M_rL^2 + M_r(R + L/2)^2 \text{ where the parallel axis through the CM of the rod is a dis-}$$
$$\text{tance of } R + L/2 \text{ from the axis of rotation.}$$
$$= \tfrac{1}{12}(5.0 \text{ kg})(0.80 \text{ m})^2 + (5.0 \text{ kg})(0.20 \text{ m} + 0.40 \text{ m})^2 = 2.1 \text{ kg} \cdot \text{m}^2$$

And, $I_o = I_c + 2I_r = 1.5 \text{ kg} \cdot \text{m}^2 + 2(2.1 \text{ kg} \cdot \text{m}^2) = 5.7 \text{ kg} \cdot \text{m}^2$

After: In Fig. 8.29b, treating an arm mass as if its center of mass in now only about 20 cm
from the axis of rotation, the moment of inertia of each arm is $I = M_rR^2$ (Fig. 8.20b), and

$$I = I_c + 2(M_rR^2) = 1.5 \text{ kg} \cdot \text{m}^2 + 2(5.0 \text{ kg} \cdot \text{m}^2)(0.20 \text{ m})^2 = 1.9 \text{ kg} \cdot \text{m}^2$$

Then with the conservation of angular momentum, $L = L_o$ or $I\omega = I_o\omega_o$ and

$$\omega = \left(\frac{I_a}{I_b}\right)\omega_o = \left(\frac{5.7 \text{ kg} \cdot \text{m}^2}{1.9 \text{ kg} \cdot \text{m}^2}\right)(4.2 \text{ rad/s}) = 13 \text{ rad/s}$$

So the angular speed increases by a factor of 3.

Follow-Up Exercise. Suppose a skater with 75% of the mass of the skater in the Exercise
did a spin. What would be the spin rate ω in this case? (Consider all masses to be reduced
by 75%.)

 Angular momentum, $\vec{\mathbf{L}}$, is a vector, and when it is conserved or constant, its
magnitude *and* direction must remain unchanged. Thus, when no external torques
act, the direction of $\vec{\mathbf{L}}$ is fixed in space. This is the principle behind passing a football
accurately, as well as that behind the movement of a gyrocompass (►Fig. 8.30). A football is normally passed with a spiraling rotation. This spin, or gyroscopic action, sta-
bilizes the ball's spin axis in the direction of motion. Similarly, rifle bullets are set
spinning by the rifling in the barrel for directional stability.

 The $\vec{\mathbf{L}}$ vector of a spinning gyroscope in the compass is set in a particular direction (usually north). In the absence of external torques, the compass direction
remains fixed, even though its carrier (an airplane or ship, for example) changes
directions. You may have played with a toy gyroscope that is set spinning and
placed on a pedestal. In a "sleeping" condition, the gyro stands straight up with

▶ **FIGURE 8.30 Constant direction of angular momentum** When angular momentum is conserved, its direction is constant in space. **(a)** This principle can be demonstrated by a passed football. **(b)** Gyroscopic action also occurs in a gyroscope, a rotating wheel that is universally mounted on gimbals (rings) so that it is free to turn about any axis. When the frame moves, the wheel maintains its direction. This is the principle of the gyrocompass.

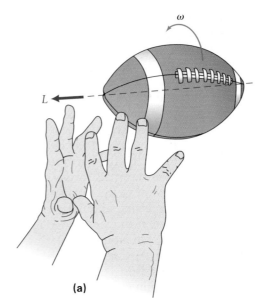

(a)

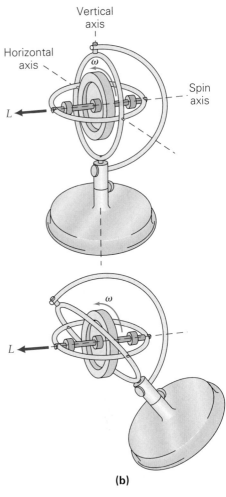

(b)

its angular-momentum vector fixed in space for some time. The gyro's center of gravity is on the axis of rotation, so there is no net torque due to its weight.

However, the gyroscope eventually slows down because of friction, causing $\vec{L}$ to tilt. In watching this motion, you may have noticed that the spin axis revolves, or *precesses*, about the vertical axis. It revolves tilted over, so to speak (Fig. 8.30b). Since the gyroscope precesses, the angular-momentum vector $\vec{L}$ is no longer constant in direction, indicating that a torque must be acting to produce a change ($\Delta\vec{L}$) with time. As can be seen from the figure, the torque arises from the vertical component of the weight force, since the center of gravity no longer lies directly above the point of support or on the vertical axis of rotation. The instantaneous torque is such that the gyroscope's axis moves or precesses about the vertical axis.

In a similar manner, the Earth's rotational axis precesses. The Earth's spin axis is tilted $23\frac{1}{2}°$ with respect to a line perpendicular to the plane of its revolution about the Sun; the axis precesses about this line (▼Fig. 8.31). The precession is due to slight gravitational torques exerted on the Earth by the Sun and the Moon.

The period of the precession of the Earth's axis is about 26 000 years, so the precession has little day-to-day effect. However, it does have an interesting long-term effect. Polaris will not always be (nor has it always been) the North Star—that is, the star toward which the Earth's axis of rotation points. About 5000 years ago, Alpha Draconis was the North Star, and 5000 years from now it will be Alpha Cephei, which is at an angular distance of about 68° away from Polaris on the circle described by the precession of the Earth's axis.

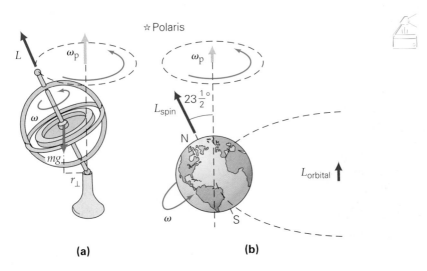

(a)

(b)

◀ **FIGURE 8.31 Precession** An external torque causes a change in angular momentum. **(a)** For a spinning gyroscope, this change is directional, and the axis of rotation precesses at angular acceleration ω_p about a vertical line. (The torque due to the weight force would point out of the page as drawn here, as would $\Delta\vec{L}$.) Note that although there is a torque that would topple a nonspinning gyroscope, a spinning gyroscope doesn't fall. **(b)** Similarly, the Earth's axis precesses because of gravitational torques caused by the Sun and the Moon. We don't notice this motion because the period of precession is about 26 000 years.

There are some other long-term torque effects on the Earth and the Moon. Did you know that the Earth's daily spin rate is slowing down and hence the days are getting longer? Also, that the Moon is receding, or getting farther away, from the Earth? This is due primarily to ocean tidal friction, which gives rise to a torque. As a result, the Earth's spin angular momentum, and therefore its rate of rotation, is changing. The slowing rate of rotation causes the average day to be longer; this century will be about 25 seconds longer than the previous one.

But the rate we are talking about is an average rate. At times, the Earth's rotation speeds up for relatively short periods. This increase is thought to be associated with the rotational inertia of the liquid layer of the Earth's core. (See the Insight 13.1 on page 453.)

The tidal torque on the Earth results chiefly from the Moon's gravitational attraction, which is the main cause of ocean tides. This torque is *internal* to the Earth–Moon system, so the total angular momentum of that system is conserved. Since the Earth is losing angular momentum, the Moon must be gaining angular momentum to keep the total angular momentum of the system constant. The Earth loses rotational (spin) angular momentum, and the Moon gains orbital angular momentum. As a result, the Moon drifts slightly farther from Earth and its orbital speed decreases. The Moon moves away from the Earth at about 4 cm per year. Thus, the Moon moves in a slowly widening spiral.

Finally, a common example in which angular momentum is an important consideration is the helicopter. What would happen if a helicopter had a single rotor? Since the motor supplying the torque is internal, the angular momentum would be conserved. Initially, $\vec{L} = 0$; hence, to conserve the total angular momentum of the system (rotor plus body), the separate angular momenta of the rotor and body would have to be in opposite directions to cancel. Thus, on takeoff, the rotor would rotate one way and the helicopter body the other, which is not a desirable situation.

To prevent this situation, helicopters have two rotors. Large helicopters have two overlapping rotors (▼Fig. 8.32a). The oppositely rotating rotors cancel each other's angular momenta, so the helicopter body does not have to rotate to provide canceling angular momentum. The rotors are offset at different heights so that they do not collide.

Small helicopters with a single overhead rotor have small "antitorque" tail rotors (Fig. 8.32b). The tail rotor produces a thrust like a propeller and supplies the torque to counterbalance the torque produced by the overhead rotor. The tail rotor also helps in steering the craft. By increasing or decreasing the tail rotor's thrust, the helicopter turns (rotates) one way or the other.

▼ **FIGURE 8.32 Different rotors** See text for description.

Front rotor   Rear rotor

$\vec{L}$ $-\vec{L}$

(top view)

(a) $\Sigma\vec{L} = 0$

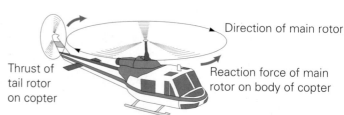

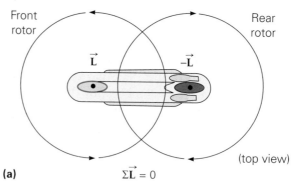

Direction of main rotor

Thrust of tail rotor on copter

Reaction force of main rotor on body of copter

(b)

Chapter Review

- In **pure translational motion**, all of the particles that make up a rigid object have the same instantaneous velocity.

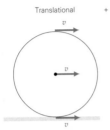

Translational   +

- In **pure rotational motion (about a fixed axis)**, all of the particles that make up a rigid object have the same instantaneous angular velocity.

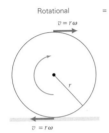

Rotational   =

Condition for rolling without slipping:

$$v_{CM} = r\omega \tag{8.1}$$
$$(\text{or } s = r\theta \quad \text{or} \quad a_{CM} = r\alpha)$$

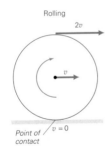

Rolling

- **Torque** ($\vec{\tau}$), the rotational analogue of force, is the product of a force and a moment arm, or lever arm.

  Torque (magnitude):
  $$\tau = r_{\perp}F = rF \sin\theta \tag{8.2}$$
  (Direction is given by right-hand rule.)

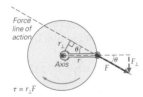

- **Mechanical equilibrium** requires that the net force, or summation of the forces, be zero (translational equilibrium) and that the net torque, or summation of the torques, be zero (rotational equilibrium).

  Conditions for translational and rotational mechanical equilibrium, respectively:
  $$\vec{F}_{net} = \Sigma\vec{F}_i = 0 \quad \text{and} \quad \vec{\tau}_{net} = \Sigma\vec{\tau}_i = 0 \tag{8.3}$$

- An object is in **stable equilibrium** as long as its center of gravity, upon small displacement, lies above and inside the object's original base of support.

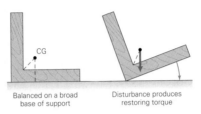

Balanced on a broad   Disturbance produces
base of support   restoring torque

Stable Equilibrium

- **Moment of inertia (*I*)** is the rotational analogue of mass and is given by
  $$I = \Sigma m_i r_i^2 \tag{8.6}$$

  Rotational form of Newton's second law:
  $$\vec{\tau}_{net} = I\vec{\alpha} \tag{8.7}$$

  Parallel-axis theorem:
  $$I = I_{CM} + Md^2 \tag{8.8}$$

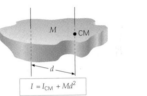

$$I = I_{CM} + Md^2$$

  Rotational work:
  $$W = \tau\theta \tag{8.9}$$

  Rotational power:
  $$P = \tau\omega \tag{8.10}$$

  Work–energy theorem (rotational):
  $$W_{net} = \tfrac{1}{2}I\omega^2 - \tfrac{1}{2}I\omega_o^2 = \Delta K \tag{8.11}$$

  Rotational kinetic energy:
  $$K = \tfrac{1}{2}I\omega^2 \tag{8.12}$$

  Kinetic energy of a rolling object (no slipping):
  $$K = \tfrac{1}{2}I_{CM}\omega^2 + \tfrac{1}{2}Mv_{CM}^2 \tag{8.13}$$

- **Angular momentum**: The product of a moment arm and linear momentum or the product of a moment of inertia and angular velocity.

  Angular momentum of a particle in circular motion (magnitude):
  $$L = r_{\perp}p = mr_{\perp}v = mr_{\perp}^2\omega \tag{8.14}$$

  Angular momentum of a rigid body:
  $$\vec{L} = I\vec{\omega} \tag{8.16}$$

  Torque as change in angular momentum (magnitude form):
  $$\vec{\tau}_{net} = \frac{\Delta\vec{L}}{\Delta t} \tag{8.17}$$

  Conservation of angular momentum (with $\vec{\tau}_{net} = 0$):
  $$L = L_o \quad \text{or} \quad I\omega = I_o\omega_o \tag{8.18}$$

  The angular momentum is conserved in the absence of an external, unbalanced torque.

Exercises

MC = *Multiple Choice Question,* **CQ** = *Conceptual Question, and* **IE** = *Integrated Exercise. Throughout the text, many exercise sections will include "paired" exercises. These exercise pairs, identified with* **red numbers,** *are intended to assist you in problem solving and learning. In a pair, the first exercise (even numbered) is worked out in the Study Guide so that you can consult it should you need assistance in solving it. The second exercise (odd numbered) is similar in nature, and its answer is given at the back of the book.*

8.1 Rigid Bodies, Translations, and Rotations

1. **MC** In pure rotational motion of a rigid body, (a) all the particles of the body have the same angular velocity, (b) all the particles of the body have the same tangential velocity, (c) acceleration is always zero, (d) there are always two simultaneous axes of rotation. (a)

2. **MC** For an object with only rotational motion, all particles of the object have the same (a) instantaneous velocity, (b) average velocity, (c) distance from the axis of rotation, (d) instantaneous angular velocity. (d)

3. **MC** The condition for rolling without slipping is (a) $a_c = r\omega^2$, (b) $v_{CM} = r\omega$, (c) $F = ma$, (d) $a_c = v^2/r$. (b)

4. **MC** A rolling object (a) has an axis of rotation through the axis of symmetry, (b) has a zero velocity at the point or line of contact, (c) will slip if $s = r\theta$, (d) all of the preceding. (b)

5. **MC** For the tires on your skidding car, (a) $v_{CM} = r\omega$, (b) $v_{CM} > r\omega$, (c) $v_{CM} < r\omega$, (d) none of the preceding. (b)

6. **CQ** Suppose someone in your physics class says that it is possible for a rigid body to have translational motion and rotational motion at the same time. Would you agree? If so, give an example. yes; rolling motion

7. **CQ** For a rolling cylinder, what would happen if the tangential speed v were less than $r\omega$? Is it possible for v to be greater than $r\omega$? Explain. slipping; yes, sliding

8. **CQ** If the top of your automobile tire is moving with a speed of v, what is the reading of your speedometer? $v/2$

9. ● A wheel rolls uniformly on level ground without slipping. A piece of mud on the wheel flies off when it is at the 9 o'clock position (rear of wheel). Describe the subsequent motion of the mud. see ISM

10. ● A rope goes over a circular pulley with a radius of 6.5 cm. If the pulley makes four revolutions without the rope slipping, what length of rope passes over the pulley? 1.6 m

11. ● A wheel rolls five revolutions on a horizontal surface without slipping. If the center of the wheel moves 3.2 m, what is the radius of the wheel? 0.10 m

12. ●● A bowling ball with a radius of 15.0 cm travels down the alley so that its center of mass is moving at 3.60 m/s. The bowler estimates that it makes about 7.50 complete revolutions in 2.00 seconds. Is it rolling without slipping? Prove your answer, assuming that the bowler's quick observation limits answers to two significant figures. $\omega \approx 24$ rad/s, no slipping

13. ●● A ball with a radius of 15 cm rolls on a level surface, and the translational speed of the center of mass is 0.25 m/s. What is the angular speed about the center of mass if the ball rolls without slipping? 1.7 rad/s

14. **IE** ●● (a) When a disk rolls without slipping, should the product $r\omega$ be (1) greater than, (2) equal to, or (3) less than v_{CM}? (b) A disk with a radius of 0.15 m rotates through 270° as it travels 0.71 m. Does the disk roll without slipping? Prove your answer. (a) (2) equal to (b) yes

15. ●●● A bocce ball with a diameter of 6.00 cm rolls without slipping on a level lawn. It has an initial angular speed of 2.35 rad/s and comes to rest after 2.50 m. Assuming constant deceleration, (a) determine the magnitude of its angular deceleration and (b) the magnitude of the maximum tangential acceleration of the ball's surface (tell where that part is located). (a) 0.0331 rad/s² (b) 1.99 × 10⁻³ m/s²

16. ●●● A cylinder with a diameter of 20 cm rolls with an angular speed of 0.50 rad/s on a level surface. If the cylinder experiences a uniform tangential acceleration of 0.018 m/s² without slipping until its angular speed is 1.25 rad/s, through how many complete revolutions does the cylinder rotate during the time it accelerates? 0.58 rotations

8.2 Torque, Equilibrium, and Stability

17. **MC** It is possible to have a net torque when (a) all forces act through the axis of rotation, (b) $\Sigma \vec{F}_i = 0$, (c) an object is in rotational equilibrium, (d) an object remains in unstable equilibrium. (b)

18. **MC** If an object in unstable equilibrium is displaced slightly, (a) its potential energy will decrease, (b) the center of gravity is directly above the axis of rotation, (c) no gravitational work is done, (d) stable equilibrium follows. (a)

19. **MC** Torque has the same units as (a) work, (b) force, (c) angular velocity, (d) angular acceleration. (a)

20. **CQ** When lifting an object with the use of the back rather than the legs, we often experience back pain. Why? back muscles have to exert greater torque

21. **CQ** A gymnast on the balance beam will squat when she feels she is losing her balance. Why? to lower center of gravity

22. **CQ** Explain the balancing acts in ►Fig. 8.33. Where are the centers of gravity? the centers of gravity must be directly below the base of support

▲ **FIGURE 8.33 Balancing acts** See Exercise 22. *Left*: A toothpick on the rim of the glass supports a fork and spoon. *Right*: toy birds balance on their beaks.

23. **CQ** "Popping a wheelie" is a motorcycle stunt in which the front end of the cycle raises up off the ground on a fast start, and can remain there for some distance. Explain the physics involved in this stunt. frictional torque causes cycle to rotate upward, balanced by weight torque

24. **CQ** In the cases of both stable and unstable equilibrium, a small displacement of the center of gravity causes gravitational work to be done. (See the balls and bowls in Fig. 8.11.) However, there is another type of equilibrium in which the displacement of the center of mass involves no gravitational work. This is called *neutral equilibrium*, and the displaced center of gravity essentially moves in a straight line. Give an example of an object in neutral equilibrium. a cylinder or sphere on a level surface

25. ● In Fig. 8.4a, if the arm makes a 37° angle with the horizontal and a torque of 18 m·N is to be produced, what force must the biceps muscle supply? 5.6×10^2 N

26. ● The drain plug on a car's engine has been tightened to a torque of 25 m·N. If a 0.15-m-long wrench is used to change the oil, what is the minimum force needed to loosen the plug? 1.7×10^2 N

27. ● In Exercise 26, due to limited work space, you must crawl under the car. The force thus cannot be applied perpendicularly to the length of the wrench. If the applied force makes a 30° angle with the length of the wrench, what is the force required to loosen the drain plug? 3.3×10^2 N

28. ● How many different positions of stable equilibrium and unstable equilibrium are there for a cube? Consider each surface, edge, and corner to be a different position. 6 stable (faces), 20 unstable (12 edges and 8 corners)

29. **IE** ● Two children are sitting on opposite ends of a uniform seesaw of negligible mass. (a) Can the seesaw be balanced if the masses of the children are different? How? (b) If a 35-kg child is 2.0 m from the pivot point (or fulcrum), how far from the pivot point will her 30-kg playmate have to sit on the other side for the seesaw to be in equilibrium? (a) yes (b) 2.3 m

30. ● A uniform meterstick pivoted at its center, as in Example 8.5, has a 100-g mass suspended at the 25.0-cm position. (a) At what position should a 75.0-g mass be suspended to put the system in equilibrium? (b) What mass would have to be suspended at the 90.0-cm position for the system to be in equilibrium? (a) 83.3 cm (b) 62.5 g

31. ●● Show that the balanced meterstick in Example 8.5 is in static rotational equilibrium about a horizontal axis through the 100-cm end of the stick. see ISM

32. **IE** ●● Telephone and electrical lines are allowed to sag between poles so that the tension will not be too great when something hits or sits on the line. (a) Is it possible to have the lines perfectly horizontal? Why or why not? (b) Suppose that a line were stretched almost perfectly horizontally between two poles that are 30 m apart. If a 0.25-kg bird perches on the wire midway between the poles and the wire sags 1.0 cm, what would be the tension in the wire? (a) no (b) 1.8×10^3 N (>400 lb!)

33. ●● In ▼Fig. 8.34, what is the force F_m supplied by the deltoid muscle so as to hold up the outstretched arm if the mass of the arm is 3.0 kg? (F_j is the joint force on the bone of the upper arm—the humerus.) 1.6×10^2 N

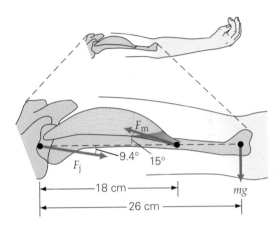

▲ **FIGURE 8.34 Arm in static equilibrium** See Exercise 33.

34. ●● In Figure 8.4b, determine the force exerted by the bicep muscle, assuming that the hand is holding a ball with a mass of 5.00 kg. Assume that the mass of the forearm is 8.50 kg with its center of mass located 20.0 cm away from the elbow joint (the black dot in the figure). Assume also that the center of mass of the ball in the hand is 30.0 cm away from the elbow joint. (The muscle contact is 4.00 cm from the elbow joint, Example 8.2.) 784 N

35. ●● A bowling ball (mass 7.00 kg and radius 17.0 cm) is released so fast that it skids without rotating down the alley (at least for a while). Assume the ball skids to your right and the coefficient of sliding friction between the ball and the lane surface is 0.400. (a) What is the direction of the torque exerted by the friction on the ball about the center of mass of the ball? (b) Determine the magnitude of this torque (again about the ball's center of mass). (a) clockwise (b) 4.66 m·N

36. ●● A variation of Russell traction (▼Fig. 8.35) supports the lower leg in a cast. Suppose that the patient's leg and cast have a combined mass of 15.0 kg and m_1 is 4.50 kg. (a) What is the reaction force of the leg muscles to the traction? (b) What must m_2 be to keep the leg horizontal? (a) 88.2 N (b) 10.5 kg

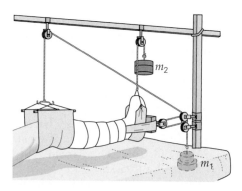

▲ **FIGURE 8.35** Static traction See Exercise 36.

37. ●● In doing physical therapy for an injured knee joint, a person raises a 5.0-kg weighted boot as shown in ▼Fig. 8.36. Compute the torque due to the boot for each position shown. 0, 9.80 m·N, 17.0 m·N, 19.6 m·N

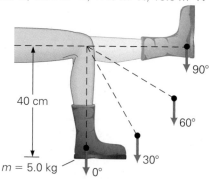

▲ **FIGURE 8.36** Torque in physical therapy See Exercise 37.

38. ●● An artist wishes to construct a birds-and-bees mobile, as shown in ▼Fig. 8.37. If the mass of the bee on the lower left is 0.10 kg and each vertical support string has a length of 30 cm, what are the masses of the other birds and bees? (Neglect the masses of the bars and strings.) $m_2 = 0.20$ kg, $m_3 = 0.50$ kg, $m_4 = 0.40$ kg

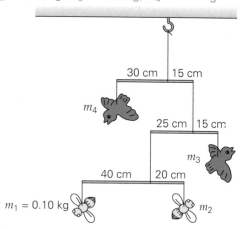

▲ **FIGURE 8.37** Birds and bees See Exercise 38.

39. **IE** ●● The location of a person's center of gravity relative to his or her height can be found by using the arrangement shown in ▶Fig. 8.38. The scales are initially adjusted to zero with the board alone. (a) Would you expect the location of the center of gravity to be (1) midway between the scales, (2) toward the scale at the person's head, or (3) toward the scale at the person's feet? Why? (b) Locate the center of gravity of the person relative to the horizontal dimension. (a) (2) toward the scale at the person's head (b) 0.87 m from the feet

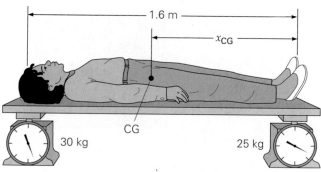

▲ **FIGURE 8.38** Locating the center of gravity See Exercise 39.

40. ●● (a) How many uniform, identical textbooks of width 25.0 cm can be stacked on top of each other on a level surface without the stack falling over if each successive book is displaced 3.00 cm in width relative to the book below it? (b) If the books are 5.00 cm thick, what will be the height of the center of mass of the stack above the level surface? (a) 9 books (b) 22.5 cm

41. ●● If four metersticks were stacked on a table with 10 cm, 15 cm, 30 cm, and 50 cm, respectively, hanging over the edge, as shown in ▼Fig. 8.39, would the top meterstick remain on the table? yes; see ISM

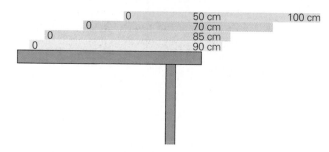

▲ **FIGURE 8.39** Will they fall off? See Exercise 41.

42. ●● A 10.0-kg solid uniform cube with 0.500-m sides rests on a level surface. What is the minimum amount of work necessary to put the cube into an unstable equilibrium position? 10.2 J

43. ●● While standing on a long board resting on a scaffold, a 70-kg painter paints the side of a house, as shown in ▼Fig. 8.40. If the mass of the board is 15 kg, how close to the end can the painter stand without tipping the board over? 1.2 m from left end of board

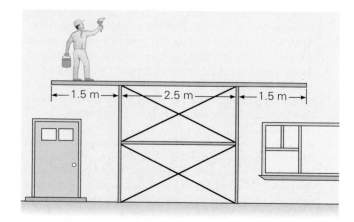

▲ **FIGURE 8.40** Not too far! See Exercises 43 and 46.

44. ●● A mass is suspended by two cords as shown in ▼Fig. 8.41. What are the tensions in the cords? $T_1 = 49$ N; $T_2 = 40$ N

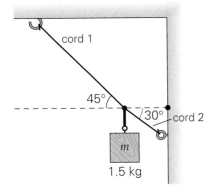

◀ **FIGURE 8.41 A lot of tension** See Exercises 44 and 45.

45. ●● If the cord attached to the vertical wall in Fig. 8.41 were horizontal (instead of at a 30° angle), what would the tensions in the cords be? $T_1 = 21$ N; $T_2 = 15$ N

46. ●●● Suppose that the board in Fig. 8.40 were suspended from vertical ropes attached to each end instead of resting on scaffolding. If the painter stood 1.5 m from one end of the board, what would the tensions in the ropes be? (See Exercise 43 for additional data.) 2.6×10^2 N; 5.7×10^2 N

47. IE ●●● In a circus act, a uniform board (length 3.00 m, mass 35.0 kg) is suspended from a rope at one end, and the other end rests on a concrete pillar. When a clown (mass 75.0 kg) steps out halfway onto the board, the board tilts so the rope end is 30° from the horizontal and the rope stays vertical. (a) In which situation will the rope tension be larger: (1) the board without the clown on it, (2) the board with the clown on it, or (3) you can't tell from the data given? (b) Calculate the force exerted by the rope in both situations. see ISM

48. IE ●●● The forces acting on Einstein and the bicycle (Fig. 2 of the Insight on p. 271) are the total weight of Einstein and the bicycle (mg) at the center of gravity of the system, the normal force (N) exerted by the road, and the force of static friction (f_s) acting on the tires due to the road. (a) If Einstein is to maintain balance, should the tangent of the lean angle θ ($\tan\theta$) be (1) greater than, (2) equal to, or (3) less than f_s/N? (b) The angle θ in the picture is about 11°. What is the minimum coefficient of static friction μ_s between the road and the tires? (c) If the radius of the circle is 6.5 m, what is the maximum speed of Einstein's bicycle? [*Hint:* The net torque about the center of gravity must be zero for rotational equilibrium.]
(a) (2) equal to (b) 0.19 (c) 3.5 m/s

8.3 Rotational Dynamics

49. **MC** The moment of inertia of a rigid body (a) depends on the axis of rotation, (b) cannot be zero, (c) depends on mass distribution, (d) all of the preceding. (d)

50. **MC** Which of the following best describes the physical quantity called torque: (a) rotational analogue of force, (b) energy due to rotation, (c) rate of change of linear momentum, or (d) force that is tangent to a circle? (a)

51. **MC** In general, the moment of inertia is greater when (a) more mass is farther from the axis of rotation, (b) more mass is closer to the axis of rotation, (c) it makes no difference. (a)

52. **MC** The moment of inertia about an axis parallel to the axis through the center of mass depends on (a) the mass of the rigid body, (b) the distance between the axes, (c) the moment of inertia about the axis through the center of mass, or (d) all of the preceding. (d)

53. **CQ** (a) Does the moment of inertia of a rigid body depend in any way on the center of mass of the body? Explain. (b) Can a moment of inertia have a negative value? If so, explain what this would mean. (a) yes (b) no; see ISM

54. **CQ** Why does the moment of inertia of a rigid body have different values for different axes of rotation? What does this mean physically?
depends on how mass is distributed about an axis

55. **CQ** When a hard-boiled egg lying on a table is given a quick torque (spin), it will rise up and spin on one end like a top. A raw egg will not. Why the difference?
hard-boiled egg is a rigid body

56. **CQ** Why does jerking a paper towel from a roll cause the paper to tear better than pulling it smoothly? Will the amount of paper on the roll affect the results? see ISM

57. **CQ** Tight-rope walkers are continually in danger of falling (unstable equilibrium). Commonly, a performer carries a long pole while walking the tight rope, as shown in the chapter-opening photo. What is the purpose of the pole? (In walking along a narrow board or rail, you probably extend your arms for the same reason.) see ISM

58. ● A fixed 0.15-kg solid-disk pulley with a radius of 0.075 m is acted on by a net torque of 6.4 m · N. What is the angular acceleration of the pulley? 1.5×10^4 rad/s²

59. ● What net torque is required to give a uniform 20-kg solid ball with a radius of 0.20 m an angular acceleration of 20 rad/s²? 0.64 m · N

60. ● For the system of masses shown in ▼Fig. 8.42, find the moment of inertia about (a) the x-axis, (b) the y-axis, and (c) an axis through the origin and perpendicular to the page (z-axis). Neglect the masses of the connecting rods.
(a) 22.5 kg · m² (b) 62.5 kg · m² (c) 85.0 kg · m²

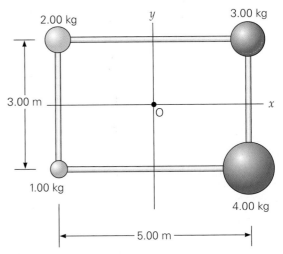

▲ **FIGURE 8.42 Moments of inertia about different axes** See Exercise 60.

61. ● A light meterstick is loaded with masses of 2.0 kg and 4.0 kg at the 30-cm and 75-cm positions, respectively. (a) What is the moment of inertia about an axis through the 0-cm end of the meterstick? (b) What is the moment of inertia about an axis through the center of mass of the system? (c) Use the parallel-axis theorem to find the moment of inertia about the axis through the 0-cm end of the stick, and compare the result with the result of part (a). (a) $2.4 \ kg \cdot m^2$ (b) $0.27 \ kg \cdot m^2$ (c) $2.4 \ kg \cdot m^2$ (same)

62. ●● A 2000-kg Ferris wheel accelerates from rest to an angular speed of 2.0 rad/s in 12 s. Approximate the Ferris wheel as a circular disk with a radius of 30 m. What is the net torque on the wheel? $1.5 \times 10^5 \ m \cdot N$

63. ●● A 15-kg uniform sphere with a radius of 15 cm rotates about an axis tangent to its surface at 3.0 rad/s. A constant torque of $10 \ m \cdot N$ then increases the rotational speed to 7.5 rad/s. Through what angle does the sphere rotate while accelerating? 1.1 rad

64. **IE** ●● Two objects of different masses are joined by a light rod. (a) Is the moment of inertia about the center of mass the minimum or the maximum? Why? (b) If the two masses are 3.0 kg and 5.0 kg and the length of the rod is 2.0 m, find the moments of inertia of the system about an axis perpendicular to the rod, through the center of the rod and the center of mass. (a) minimum (b) $8.0 \ kg \cdot m^2$; $7.5 \ kg \cdot m^2$

65. ●● Two masses are suspended from a pulley as shown in ▼Fig. 8.43 (the Atwood machine revisited; see Chapter 4, Exercise 68). The pulley itself has a mass of 0.20 kg, a radius of 0.15 m, and a constant torque of $0.35 \ m \cdot N$ due to the friction between the rotating pulley and its axle. What is the magnitude of the acceleration of the suspended masses if $m_1 = 0.40$ kg and $m_2 = 0.80$ kg? (Neglect the mass of the string.) $1.2 \ m/s^2$

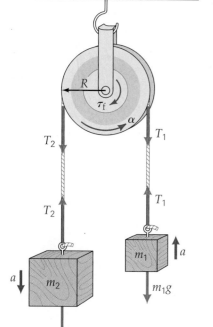

◀ **FIGURE 8.43 The Atwood machine revisited** See Exercise 65.

66. ●● A submarine door is designed so it has a rectangular plate that rotates on two rectangular shafts as shown in ▼Fig. 8.44. Each shaft has a mass of 50.0 kg and a length of 25.0 cm. The door has a mass of 200 kg and measures 50.0 cm by 1.00 m. Determine the moment of inertia of this door/hatch system about the hinge line (shown as the dashed vertical line in the figure). $56.3 \ kg \cdot m^2$

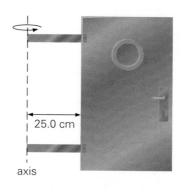

25.0 cm

axis

◀ **FIGURE 8.44 Submarine door** (not to scale) See Exercise 66.

67. ●● To start her lawn mower, Julie pulls on a cord that is wrapped around a pulley. The pulley has a moment of inertia about its central axis of $I = 0.550 \ kg \cdot m^2$ and a radius of 5.00 cm. There is an equivalent frictional torque impeding her pull of $\tau_f = 0.430 \ m \cdot N$. To accelerate the pulley at $\alpha = 4.55 \ rad/s^2$, (a) how much torque does Julie need to apply to the pulley? (b) How much tension must the rope exert? (a) $2.93 \ m \cdot N$ (b) 58.7 N

68. ●● For the system shown in ▼Fig. 8.45, $m_1 = 8.0$ kg, $m_2 = 3.0$ kg, $\theta = 30°$, and the radius and mass of the pulley are 0.10 m and 0.10 kg, respectively. (a) What is the acceleration of the masses? (Neglect friction and the string's mass.) (b) If the pulley has a constant frictional torque of $0.050 \ m \cdot N$ when the system is in motion, what is the acceleration of the masses? [*Hint*: Isolate the forces. The tensions in the strings are different. Why?] (a) $0.89 \ m/s^2$ (b) $0.84 \ m/s^2$

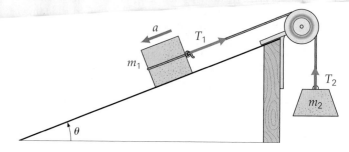

▲ **FIGURE 8.45 Inclined plane and pulley** See Exercise 68.

69. ●● A meterstick pivoted about a horizontal axis through the 0-cm end is held in a horizontal position and let go. (a) What is the tangential acceleration of the 100-cm position? Are you surprised by this result? (b) Which position has a tangential acceleration equal to the acceleration due to gravity? (a) 1.5g (b) 67-cm position

70. ●● Pennies are placed every 10 cm on a meterstick. One end of the stick is put on a table and the other end is held horizontally with a finger, as shown in ▶Fig. 8.46. If the finger is pulled away, what happens to the pennies? see ISM

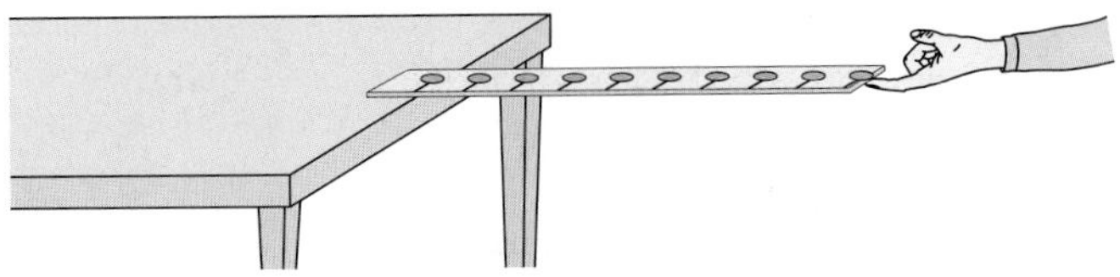

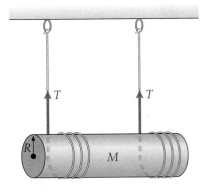

▲ **FIGURE 8.46 Money left behind?** See Exercise 70.

71. ●●● A uniform 2.0-kg cylinder of radius 0.15 m is suspended by two strings wrapped around it (▼Fig. 8.47). As the cylinder descends, the strings unwind from it. What is the acceleration of the center of mass of the cylinder? (Neglect the mass of the string.) 6.5 m/s²

◀ **FIGURE 8.47 Unwinding with gravity** See Exercise 71.

72. ●●● A planetary space probe is in the shape of a cylinder. To protect it from heat on one side (from the Sun's rays), operators on the Earth put it into a "barbecue mode," that is, they set it rotating about its long axis. To do this, they fire four small rockets mounted tangentially as shown in ▼Fig. 8.48 (the probe is shown coming toward you). The object is to get the probe to rotate completely once every 30 seconds, starting from no rotation at all. They wish to do this by firing all four rockets for a certain length of time. Each rocket can exert a thrust of 50.0 N. Assume the probe is a uniform solid cylinder with a radius of 2.50 m and a mass of 1000 kg and neglect the mass of each rocket engine. Determine the amount of time the rockets need to be fired. 1.31 s

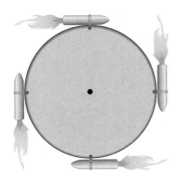

▲ **FIGURE 8.48 Space probe in the "barbecue mode"** See Exercise 72.

73. IE ●●● A ball of radius R and mass M rolls down an incline of angle θ. (a) For the ball to roll without slipping, should the tangent of the maximum angle of incline ($\tan \theta$) be equal to (1) $3 \mu_s/2$, (2) $5 \mu_s/2$, (3) $7 \mu_s/2$, or (4) $9 \mu_s/2$? Here, μ_s is the coefficient of static friction. (b) If the ball is made of wood and the surface is also wood, what is the maximum angle of incline? [*Hint:* See Table 4.1.] (a) (3) $7 \mu_s/2$ (b) 63.8°

8.4 Rotational Work and Kinetic Energy

74. **MC** From $W = \tau\theta$, the unit of rotational work is the (a) watt, (b) N · m, (c) kg · rad/s², (d) N · rad. (b)

75. **MC** A bowling ball rolls without slipping on a flat surface. The ball has (a) rotational kinetic energy, (b) translational kinetic energy, (c) both translational and rotational kinetic energy, (d) neither translational nor rotational kinetic energy. (c)

76. **MC** A rolling cylinder on a level surface has (a) rotational kinetic energy, (b) translational kinetic energy, (c) both translational and rotational kinetic energies. (c)

77. **CQ** Can you increase the rotational kinetic energy of a wheel without changing its translational kinetic energy? Explain. yes, see ISM

78. **CQ** In order to produce fuel-efficient vehicles, automobile manufacturers want to minimize rotational kinetic energy and maximize translational kinetic energy when a car is traveling. If you were the designer of wheels of a certain diameter, how would you design them? small total mass with more mass near the center

79. **CQ** What is required to produce a change in rotational kinetic energy? rotational work

80. ● A constant retarding torque of 12 m · N stops a rolling wheel of diameter 0.80 m in a distance of 15 m. How much work is done by the torque? 4.5×10^2 J

81. ● A person opens a door by applying a 15-N force perpendicular to it at a distance 0.90 m from the hinges. The door is pushed wide open (to 120°) in 2.0 s. (a) How much work was done? (b) What was the average power delivered? (a) 28 J (b) 14 W

82. ● A constant torque of 10 m · N is applied to a 10-kg uniform disk of radius 0.20 m. What is the angular speed of the disk about an axis through its center after it rotates 2.0 revolutions from rest? 35 rad/s

83. ● A 2.5-kg pulley of radius 0.15 m is pivoted about an axis through its center. What constant torque is required for the pulley to reach an angular speed of 25 rad/s after rotating 3.0 revolutions, starting from rest? 0.47 m · N

84. IE ● In Fig. 8.23, a mass m descends a vertical distance from rest. (Neglect friction and the mass of the string.) (a) From the conservation of mechanical energy, will the linear speed of the descending mass be (1) greater than, (2) equal to, or (3) less than $\sqrt{2gh}$? Why? (b) If $m = 1.0$ kg, $M = 0.30$ kg, and $R = 0.15$ m, what is the linear speed of the mass after it has descended a vertical distance of 2.0 m from rest? (a) (3) less than $\sqrt{2gh}$ (b) 5.8 m/s

85. ●● A sphere with a radius of 15 cm rolls on a level surface with a constant angular speed of 10 rad/s. To what height on a 30° inclined plane will the sphere roll before coming to rest? (Neglect frictional losses.) 0.16 m

86. ●● Estimate the ratio of the translational kinetic energy of the Earth as it orbits the Sun to the rotational kinetic energy it has about its N–S axis. 1.0×10^4

87. •• You wish to accelerate a small merry-go-round from rest to a rotational speed of one third of a revolution per second by pushing tangentially on it. Assume the merry-go-round is a disk with a mass of 250 kg and a radius of 1.50 m. Ignoring friction, how hard do you have to push tangentially to accomplish this in 5.00 s? (Use energy methods and assume a constant push on your part.) 78.5 N

88. •• A thin 1.0-m-long rod pivoted at one end falls (rotates) from a horizontal position, starting from rest and with no friction. What is the angular speed of the rod when it is vertical? [*Hint:* Consider the center of mass and use the conservation of mechanical energy.] 5.4 rad/s

89. •• A uniform sphere and a uniform cylinder with the same mass and radius roll at the same velocity side by side on a level surface without slipping. If the sphere and the cylinder approach an inclined plane and roll up it without slipping, will they be at the same height on the plane when they come to a stop? If not, what will be the percentage difference of the heights? cylinder goes higher by 7.1%

90. •• A hoop starts from rest at a height 1.2 m above the base of an inclined plane and rolls down under the influence of gravity. What is the linear speed of the hoop's center of mass just as the hoop leaves the incline and rolls onto a horizontal surface? (Neglect friction.) 3.4 m/s

91. •• An industrial flywheel with a moment of inertia of 4.25×10^2 kg·m² rotates with a speed of 7500 rpm. (a) How much work is required to bring the flywheel to rest? (b) If this work is done uniformly in 1.5 min, how much power is expended? (a) 1.31×10^8 J (b) 1.46×10^6 W

92. •• A cylindrical hoop, a cylinder, and a sphere of equal radius and mass are released at the same time from the top of an inclined plane. Using the conservation of mechanical energy, show that the sphere always gets to the bottom of the incline first with the fastest speed and that the hoop always arrives last with the slowest speed. see ISM

93. •• For the following objects, which all roll without slipping, determine the rotational kinetic energy about the center of mass as a percentage of the total kinetic energy: (a) a solid sphere, (b) a thin spherical shell, and (c) a thin cylindrical shell. (a) 29% (b) 40% (c) 50%

94. ••• In a tumbling clothes dryer, the cylindrical drum (radius 50.0 cm and mass 35.0 kg) rotates once every second. (a) Determine its rotational kinetic energy about its central axis. (b) If it started from rest and reached that speed in 2.50 s, determine the average net torque on the dryer drum. (a) 173 J (b) 22.0 m·N

95. ••• A steel ball rolls down an incline into a loop-the-loop of radius R (▶Fig. 8.49a). (a) What minimum speed must the ball have at the top of the loop in order to stay on the track? (b) At what vertical height (h) on the incline, in terms of the radius of the loop, must the ball be released in order for it to have the required minimum speed at the top of the loop? (Neglect frictional losses.) (c) Figure 8.49b shows the loop-the-loop of a roller coaster. What are the sensations of the riders if the roller coaster has the minimum or a greater speed at the top of the loop? [*Hint:* In case the speed is below the minimum, seat and shoulder straps hold the riders in.] (a) $v = \sqrt{gR}$ (b) $h = 2.7R$ (c) weightlessness

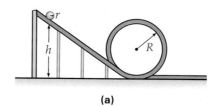

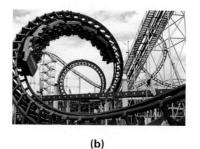

◀ **FIGURE 8.49**
Loop-the-loop and rotational speed
See Exercise 95.

8.5 Angular Momentum

96. **MC** The units of angular momentum are (a) N·m, (b) kg·m/s², (c) kg·m²/s, (d) J·m. (c)

97. **MC** The Earth's orbital speed is greatest about (a) March 21, (b) June 21, (c) Sept. 21, (d) Dec. 21. (d)

98. **MC** The angular momentum may be increased by (a) decreasing the moment of inertia, (b) decreasing the angular velocity, (c) increasing the product of the angular momentum and moment of inertia, (d) none of these. (c)

99. **CQ** A child stands on the edge of a rotating playground merry-go-round (the hand-driven type). He then starts to walk toward the center of the merry-go-round. This results in a dangerous situation. Why? see ISM

100. **CQ** The release of vast amounts of carbon dioxide may result in an increase in the Earth's average temperature through the so-called greenhouse effect and cause melting of the polar ice caps. If this occurred and the ocean level rose substantially, what effect would it have on the Earth's rotation and on the length of the day? longer day; see ISM

101. **CQ** In the classroom demonstration illustrated in ▼Fig. 8.50, a person on a rotating stool holds a rotating bicycle wheel by handles attached to the wheel. When the wheel is held horizontally, she rotates one way (clockwise as viewed from above). When the wheel is turned over, she rotates in the opposite direction. Explain why this occurs. [*Hint:* Consider angular-momentum vectors.] see ISM

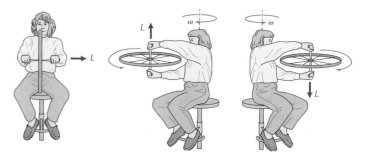

▲ **FIGURE 8.50 Faster rotation** See Exercise 101.

102. **CQ** Cats usually land on their feet when they fall, even if held upside down when dropped (▼Fig. 8.51). While a cat is falling, there is no external torque and its center of mass falls as a particle. How can cats turn themselves over while falling? see ISM

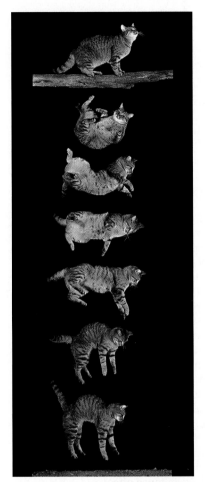

◄ **FIGURE 8.51 A double rotation** See Exercise 102.

103. **CQ** Two ice skaters that weigh the same skate towards each other with the same speed on parallel paths. As they pass each other, they link arms. (a) What is the velocity of their center of mass after they link arms? (b) What happens to their initial, linear kinetic energies? see ISM

104. ● What is the angular momentum of a 2.0-g particle moving counterclockwise (as viewed from above) with an angular speed of 5π rad/s in a horizontal circle of radius 15 cm? (Give the magnitude and direction.)
7.1×10^{-4} kg·m²/s toward you

105. ● A 10-kg rotating disk of radius 0.25 m has an angular momentum of 0.45 kg·m²/s. What is the angular speed of the disk? 1.4 rad/s

106. ●● Compute the ratio of the magnitudes of the Earth's orbital angular momentum and its rotational angular momentum. Are these momenta in the same direction?
$L_{orbital}/L_{rotation} = 3.8 \times 10^6$; no

107. ●● The period of the Moon's rotation is the same as the period of its revolution: 27.3 days (sidereal). What is the angular momentum for each rotation and revolution? (Because the periods are equal, we see only one side of the Moon from Earth.)
$L_{rot} = 2.4 \times 10^{29}$ kg·m²/s; $L_{rev} = 2.8 \times 10^{34}$ kg·m²/s

108. **IE** ●● Circular disks are used in automobile clutches and transmissions. When a rotating disk couples to a stationary one through frictional force, the energy from the rotating disk can transfer to the stationary one. (a) Is the angular speed of the coupled disks (1) greater than, (2) less than, or (3) the same as the angular speed of the original rotating disk? Why? (b) If a disk rotating at 800 rpm couples to a stationary disk with three times the moment of inertia, what is the angular speed of the combination? (a) (2) less than (b) 200 rpm

109. ●● A father gently places his small son on a rotating merry-go-round. The merry-go-round is essentially a disk with a mass of 250 kg and a radius of 2.50 m initially rotating at one revolution every 5.00 seconds. Assume the boy has a mass of 15.0 kg and is placed (without slipping) near the edge of the merry-go-round. Determine the final angular speed of the merry-go-round/boy combination. 1.18 rad/s

110. ●● A skater has a moment of inertia of 100 kg·m² when his arms are outstretched and a moment of inertia of 75 kg·m² when his arms are tucked in close to his chest. If he starts to spin at an angular speed of 2.0 rps (revolutions per second) with his arms outstretched, what will his angular speed be when they are tucked in? 2.7 rps

111. ●● An ice skater spinning with outstretched arms has an angular speed of 4.0 rad/s. She tucks in her arms, decreasing her moment of inertia by 7.5%. (a) What is the resulting angular speed? (b) By what factor does the skater's kinetic energy change? (Neglect any frictional effects.) (c) Where does the extra kinetic energy come from?
(a) 4.3 rad/s (b) $K = 1.1K_o$ (c) work done by skater

112. ●● A billiard ball at rest is struck (bold arrow in ▼Fig. 8.52) by a cue with an average force of 5.50 N lasting for 0.050 s. The cue contacts the ball's surface so that the lever arm is half the radius of the ball as shown. If the cue ball has a mass of 200 g and a radius of 2.50 cm, determine the angular speed of the ball immediately after the blow. 69 rad/s

1.25 cm

▶ **FIGURE 8.52 Cueing low** See Exercise 112.

113. ●●● A comet approaches the Sun as illustrated in ▼Fig. 8.53 and is deflected by the Sun's gravitational attraction. This event is considered a collision, and b is called the *impact parameter*. Find the distance of closest approach (d) in terms of the impact parameter and the velocities (v_o at large distances and v at closest approach). Assume that the radius of the Sun is negligible compared to d. (As the figure shows, the tail of a comet always "points" away from the Sun.) $d = b(v_o/v)$

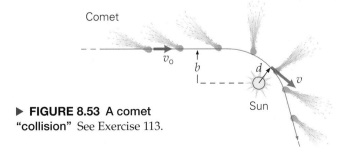

Comet
v_o
b
d
v
Sun

▶ **FIGURE 8.53 A comet "collision"** See Exercise 113.

114. ●●● While repairing his bicycle, a student has it upside down and has the front wheel spinning at 2.00 rev/s. Assume the wheel has a mass of 3.25 kg and all of the mass is located on the rim, which has a radius of 41.0 cm. To slow the wheel, he places his hand on the tire, thereby exerting a tangential force of friction on the wheel. It takes 3.50 s to come to rest. Use the change in angular momentum to determine the force he exerts on the wheel. Assume the frictional force of the axle is negligible. 4.78 N

115. IE ●●● A kitten stands on the edge of a lazy Susan (a turntable). Assume that the lazy Susan has frictionless bearings and is initially at rest. (a) If the kitten starts to walk around the edge of the lazy Susan, the lazy Susan will (1) remain lazy and stationary, (2) rotate in the direction opposite that in which the kitten is walking, or (3) rotate in the direction the kitten is walking. Explain. (b) The mass of the kitten is 0.50 kg, and the lazy Susan has a mass of 1.5 kg and a radius of 0.30 m. If the kitten walks at a speed of 0.25 m/s, relative to the ground, what will be the angular speed of the lazy Susan? (c) When the kitten has walked completely around the edge and is back at its starting point, will that point be above the same point on the ground as it was at the start? If not, where is the kitten relative to the starting point? (Speculate on what might happen if everyone on the Earth suddenly started to run eastward. What effect might this have on the length of a day?) see ISM

Comprehensive Exercises

116. In a "modern art" exhibit, a multicolored empty industrial wire spool is suspended from two light wires as shown in ▶Fig. 8.54. The spool has a mass of 50.0 kg, with an outer diameter of 75.0 cm and an inner axle diameter of 18.0 cm. One wire (#1) is attached tangentially to the axle and makes a 10° angle with the vertical. The other wire (#2) is attached tangentially to the outer edge and makes an unknown angle, θ, with the vertical. Determine the tension in each wire and the angle θ. 46.3°, $T_1 = 426$ N, $T_2 = 102$ N

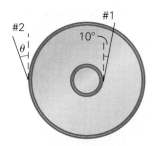

◀ **FIGURE 8.54 Modern art** See Exercise 116.

117. Modern bowling alleys have automatic ball returns. The ball is lifted to a height of 2.00 m at the end of the alley, and starting from rest rolls down a ramp. It continues to roll horizontally and eventually rolls up a ramp at the other end that is 0.500 m off the ground. Assuming the mass of the bowling ball is 7.00 kg and its radius is 16.0 cm, (a) determine the rotation rate of the ball during the middle horizontal travel, (b) its linear speed during the middle horizontal travel, and (c) the final rotation rate and linear speed. (a) 33.1 rad/s (b) 5.29 m/s (c) 28.6 rad/s, 4.58 m/s

118. An ice skater with a mass of 80.0 kg and a moment of inertia (about her central vertical axis) of 3.00 kg · m² catches a baseball with her outstretched arm. The catch is made at a distance of 1.00 m from the central axis. The ball has a mass of 300 g and is traveling at 20.0 m/s before the catch. (a) What linear speed does the system (skater + ball) have after the catch? (b) What is the angular speed of the system (skater + ball) after the catch? (c) What percentage of the ball's initial kinetic energy is lost during the catch? Neglect friction with the ice. 0.0747 m/s, 1.82 rad/s, 91%

119. A spring (spring constant 500 N/m) is stretched 10.0 cm by pulling on a string that passes over a pulley (moment of inertia about its axis is 0.550 kg · m² and a radius of 5.40 cm). The string is then connected to a mass (1.50 kg) on the other side. The dangling mass is released from rest and rises. Determine the speed of the mass when the spring is in its relaxed position (unstretched). Neglect friction. 0.104 m/s

The following Physlet Physics Problems can be used with this chapter.
PHYSLET® 10.7, 10.9, 10.10, 10.11, 10.12, 10.13, 10.14, 11.1, 11.2, 11.3, 11.4, 11.5, 11.6, 11.7, 11.8, 11.9, 11.10, 13.2, 13.3, 13.6, 13.7, 13.8, 13.13

SOLIDS AND FLUIDS

PHYSICS FACTS

- The Mariana Trench in the Pacific Ocean is the deepest known point on Earth. It is about 11 km (6.8 mi) below sea level. At this depth, the ocean water exerts a pressure of about 108 MPa (15 900 lb/in^2), or more than 1000 atmospheres of pressure.

- The German zeppelin, *Hindenburg*, had a hydrogen gas volume of 20 000 m^3 (7 062 000 ft^3). It went down in a fiery crash in 1937 in Lakehurst, NJ. (Hydrogen is very flammable.) The ship was originally designed for nonflammable helium. But, the majority of helium was produced by the United States and a law was enacted that forbade the sale of helium to Nazi Germany.

- Although Archimedes is credited with the principle of buoyancy, it is questionable whether this occurred to him in his bathtub while pondering a way to see if the king's crown was pure gold and contained no silver, as the story goes. According to a Roman account, the solution came to him when he got into a bathtub and it overflowed. Supposedly, quantities of pure gold and silver equal in weight to the king's crown were each put into bowls filled with water, and the silver caused more water to overflow. Testing the crown, more water overflowed than for the pure gold, which implied some silver content. A crown of pure gold? Pure gold is soft and malleable (can be beaten into thin sheets) and ductile (can be drawn into thin wire).

S hown in the photo are solid mountains and the unseen fluid air that makes gliding possible. We walk on the solid surface of the Earth and in our daily lives use solid objects of all sorts, from scissors to computers. But we are surrounded by fluids—liquids and gases—on which we depend. Without the water that we drink, we could survive only for a few days at most; without the oxygen in the air we breathe, we could not live for more than a few minutes. Indeed, we ourselves are not nearly as solid as we think. By far the most abundant substance in our bodies is water, and it is in the watery environment of our cells that all chemical processes on which life depends take place.

On the basis of general physical distinctions, matter is commonly divided into three phases: solid, liquid, and gas. A *solid* has a definite shape and volume. A *liquid* has a fairly definite volume, but assumes the shape of its container. A *gas* takes on the shape and volume of its container. Solids and liquids are sometimes called *condensed matter*. We will use a different classification scheme and consider matter in terms of solids and fluids. Liquids and gases are referred to collectively as fluids. A **fluid** is a substance that can flow; liquids and gases qualify, but solids do not.

A simplistic description of solids is that they are made up of particles called atoms that are held rigidly together by interatomic forces. In Chapter 8, the concept of an ideal rigid body was used to describe rotational motion. Real solid bodies are not absolutely rigid and can be elastically deformed by external forces. Elasticity usually brings to mind a rubber band or spring that will resume its original dimensions even after being greatly deformed. In fact, all materials—even very hard steel—are elastic to some degree. But, as you will learn, such deformation has an *elastic limit*.

Fluids, however, have little or no elastic response to a force. Instead, the force merely causes an unconfined fluid to flow. This chapter pays particular attention to the behavior of fluids, shedding light on such questions as how hydraulic lifts work, why icebergs and ocean liners float, and what "10W-30" on a can of motor oil means. You'll also discover why the person in the photo can neither float like a helium balloon nor fly like a hummingbird, yet, with the aid of a suitably shaped piece of plastic, can soar like an eagle.

Because of their fluidity, liquids and gases have many properties in common, and it is convenient to study them together. There are important differences as well. For example, liquids are not very compressible, whereas gases are easily compressed.

9.1 Solids and Elastic Moduli

OBJECTIVES: To (a) distinguish between stress and strain, and (b) use elastic moduli to compute dimensional changes.

As stated previously, all solid materials are elastic to some degree. That is, a body that is slightly deformed by an applied force will return to its original dimensions or shape when the force is removed. The deformation may not be noticeable for many materials, but it's there.

You may be able to visualize why materials are elastic if you think in terms of the simplistic model of a solid in ◄Fig. 9.1. The atoms of the solid substance are imagined to be held together by springs. The elasticity of the springs represents the resilient nature of the interatomic forces. The springs resist permanent deformation, as do the forces between atoms. The elastic properties of solids are commonly discussed in terms of stress and strain. **Stress** is a measure of the force causing a deformation. **Strain** is a relative measure of the deformation a stress causes. Quantitatively, *stress is the applied force per unit cross-sectional area*:

$$\text{stress} = \frac{F}{A} \tag{9.1}$$

SI unit of stress: newton per square meter (N/m²)

Here, F is the magnitude of the applied force normal (perpendicular) to the cross-sectional area. Equation 9.1 shows that the SI units for stress are newtons per square meter (N/m²).

As illustrated in ▼Fig. 9.2, a force applied to the ends of a rod gives rise to either a *tensile stress* (an elongating tension, $\Delta L > 0$) or a *compressional stress* (a shortening tension, $\Delta L < 0$), depending on the direction of the force. In both these cases, the *tensile strain* is the ratio of the change in length ($\Delta L = L - L_o$) to the original length (L_o), without regard to the sign, so we use the absolute value, $|\Delta L|$:

$$\text{strain} = \frac{|\text{change in length}|}{\text{original length}} = \frac{|\Delta L|}{L_o} = \frac{|L - L_o|}{L_o} \tag{9.2}$$

Strain is a positive unitless quantity

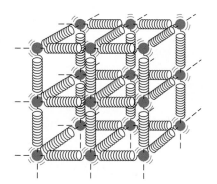

▲ FIGURE 9.1 A springy solid The elastic nature of interatomic forces is indicated by simplistically representing them as springs, which, like the forces, resist deformation.

Demonstration/activity: A student's weight of approximately 150 lb (68 kg) can be supported by an aluminum can. Have a student stand vertically on the narrow circular end of an empty can. Describe the mechanical properties present. Poke the side of the can lightly, and it will collapse. Discuss the mechanical properties this experiment demonstrates. Use a micrometer to measure the thickness of the can, and calculate the stress applied by the 150-lb student.

► FIGURE 9.2 Tensile and compressional stresses Tensile and compressional stresses are due to forces applied normally to the surface area of the ends of bodies.
(a) A tension, or tensile stress, tends to increase the length of an object.
(b) A compressional stress tends to shorten the length. $\Delta L = L - L_o$ can be positive, as in (a), or negative, as in (b). The sign is not needed in Eq. 9.2, so the absolute value, $|\Delta L|$, is used.

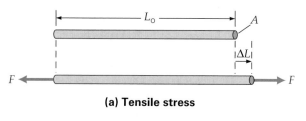

(a) Tensile stress

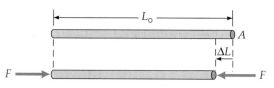

(b) Compressional stress

Thus the strain is the *fractional change* in length. For example, if the strain is 0.05, the length of the material has changed by 5% of the original length.

As might be expected, the resulting strain depends on the applied stress. For relatively small stresses, this is a direct (or linear) proportion, that is, stress ∝ strain. For relatively small stresses, this is a direct (or linear) proportion. The constant of proportionality, which depends on the nature of the material, is called the **elastic modulus**, that is,

stress = elastic modulus × strain

or

$$\text{elastic modulus} = \frac{\text{stress}}{\text{strain}} \qquad (9.3)$$

SI unit of elastic modulus: newton per square meter (N/m^2)

The elastic modulus is the stress divided by the strain, and the elastic modulus has the same units as stress. (Why?)

Three general types of elastic moduli (plural of *modulus*) are associated with stresses that produce changes in length, shape, or volume. These are called *Young's modulus*, the *shear modulus*, and the *bulk modulus*, respectively.

Change in Length: Young's Modulus

▼Figure 9.3 is a typical graph of the tensile stress versus the strain for a metal rod. The curve is a straight line up to a point called the *proportional limit*. Beyond this point, the strain begins to increase more rapidly to another critical point called the **elastic limit**. If the tension is removed at this point, the material will return to its original length. If the tension is applied beyond the elastic limit and then removed, the material will recover somewhat, but will retain some permanent deformation.

The straight-line part of the graph shows a direct proportionality between stress and strain. This relationship, first formalized by the English physicist Robert Hooke in 1678, is now known as *Hooke's law*. (It is the same general relationship as that given for a spring in Section 5.2—see Fig. 5.5.) The elastic modulus for a tension or a compression is called **Young's modulus (Y):**[*]

$$\frac{F}{A} = Y\left(\frac{\Delta L}{L_o}\right) \quad \text{or} \quad Y = \frac{F/A}{\Delta L/L_o} \qquad (9.4)$$

stress   strain

SI unit of Young's modulus: newton per square meter (N/m^2)

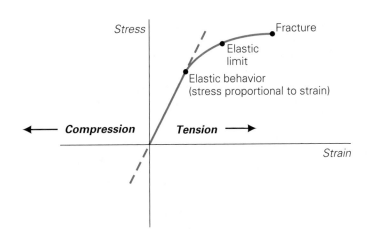

◀ **FIGURE 9.3 Stress versus strain** A plot of stress versus strain for a typical metal rod is a straight line up to the proportional limit. Then elastic deformation continues until the elastic limit is reached. Beyond that, the rod will be permanently deformed and will eventually fracture or break.

[*]Thomas Young (1773–1829) was an English physician and physicist who also demonstrated the wave nature of light. See Young's double-slit experiment, Section 24.1.

TABLE 9.1	Elastic Moduli for Various Materials (in N/m²)		
Substance	Young's modulus (Y)	Shear modulus (S)	Bulk modulus (B)
Solids			
Aluminum	7.0×10^{10}	2.5×10^{10}	7.0×10^{10}
Bone (limb)	Tension: 1.5×10^{10}	1.2×10^{10}	
	Compression: 9.3×10^{9}		
Brass	9.0×10^{10}	3.5×10^{10}	7.5×10^{10}
Copper	11×10^{10}	3.8×10^{10}	12×10^{10}
Glass	5.7×10^{10}	2.4×10^{10}	4.0×10^{10}
Iron	15×10^{10}	6.0×10^{10}	12×10^{10}
Nylon	5.0×10^{8}	8.0×10^{8}	
Steel	20×10^{10}	8.2×10^{10}	15×10^{10}
Liquids			
Alcohol, ethyl			1.0×10^{9}
Glycerin			4.5×10^{9}
Mercury			26×10^{9}
Water			2.2×10^{9}

The units of Young's modulus are the same as those of stress, newtons per square meter (N/m^2), since the strain is unitless. Some typical values of Young's modulus are given in Table 9.1.

To obtain a conceptual or physical understanding of Young's modulus, let's solve Eq. 9.4 for ΔL:

$$\Delta L = \left(\frac{FL_\text{o}}{A}\right)\frac{1}{Y} \quad \text{or} \quad \Delta L \propto \frac{1}{Y}$$

Hence, the larger Young's the modulus of a material, the smaller its change in length (other parameters being equal).

Example 9.1 ■ Pulling My Leg: Under a Lot of Stress

The femur (upper leg bone) is the longest and strongest bone in the body. Taking a typical femur to be approximately circular in cross-section with a radius of 2.0 cm, how much force would be required to extend a patient's femur by 0.010% while in horizontal traction?

Thinking It Through. Equation 9.4 should apply, but where does the percentage increase fit in? This question can be answered as soon as it is recognized that the $\Delta L/L_\text{o}$ term is the *fractional* increase in length. For example, if you had a spring with a length of 10 cm (L_o), and you stretched it 1.0 cm (ΔL), then $\Delta L/L_\text{o} = 1.0 \text{ cm}/10 \text{ cm} = 0.10$. This ratio can readily be changed to a percentage, and we would say the spring's length was increased by 10%. So the percentage increase is really just the value of the $\Delta L/L_\text{o}$ term (multiplied by 100%).

Solution. Listing the data,

Given: $r = 2.0 \text{ cm} = 0.020 \text{ m}$ *Find:* F (tensile force)
          $\Delta L/L_\text{o} = 0.010\% = 1.0 \times 10^{-4}$
             $Y = 1.5 \times 10^{10} \text{ N/m}^2$ (for bone from Table 9.1)

Using Eq. 9.4, we have

$$F = Y(\Delta L/L_\text{o})A = Y(\Delta L/L_\text{o})\pi r^2$$
$$= (1.5 \times 10^{10} \text{ N/m}^2)(1.0 \times 10^{-4})\pi(0.020 \text{ m})^2 = 1.9 \times 10^{3} \text{ N}$$

How much force is this? Quite a bit—in fact, more than 400 lb. The femur is a pretty strong bone.

Follow-Up Exercise. A total mass of 16 kg is suspended from a 0.10-cm-diameter steel wire. (a) By what percentage does the length of the wire increase? (b) The tensile or ultimate strength of a material is the maximum stress the material can support before breaking or fracturing. If the tensile strength of the steel wire in (a) is $4.9 \times 10^8 \, \text{N/m}^2$, how much mass could be suspended before the wire would break? *(Answers to all Follow-Up Exercises are at the back of the text.)*

Most types of bone are composed of protein collagen fibers that are tightly bound together and overlapping. Collagen has great tensile strength, and the calcium salts within the collagen give bone great compressional strength. Collagen also makes up cartilage, tendons, and skin, which have good tensile strength.

Change in Shape: Shear Modulus

Another way an elastic body can be deformed is by a *shear stress*. In this case, the deformation is due to an applied force that is *tangential* to the surface area (▶Fig. 9.4a). A change in shape results without a change in volume. The *shear strain* is given by x/h, where x is the relative displacement of the faces and h is the distance between them.

The shear strain is sometimes defined in terms of the *shear angle* ϕ. As Fig. 9.4b shows, $\tan \phi = x/h$. But the shear angle is usually quite small, so a good approximation is $\tan \phi \approx \phi \approx x/h$, where ϕ is in radians.* (If $\phi = 10°$, for example, there is only 1.0% difference between ϕ and $\tan \phi$.) The **shear modulus (S)**, sometimes called the *modulus of rigidity*, is then

$$S = \frac{F/A}{x/h} \approx \frac{F/A}{\phi} \qquad (9.5)$$

SI unit of shear modulus: newton per square meter (N/m^2)

Note in Table 9.1 that the shear modulus is generally less than Young's modulus. In fact, S is approximately $Y/3$ for many materials, which indicates a greater response to a shear stress than to a tensile stress. Note also the inverse relationship $\phi \approx 1/S$, similar to that pointed out previously for Young's modulus.

A shear stress may be of the torsional type, resulting from the twisting action of a torque. For example, a torsional shear stress may shear off the head of a bolt that is being tightened.

Liquids do not have shear moduli (or Young's moduli)—hence the gaps in Table 9.1. A shear stress cannot be effectively applied to a liquid or a gas because fluids deform continuously in response. It is often said that *fluids cannot support a shear.*

Change in Volume: Bulk Modulus

Suppose that a force directed inward acts over the entire surface of a body (▼Fig. 9.5). Such a *volume stress* is often applied by pressure transmitted by a fluid. An elastic material will be compressed by a volume stress; that is, the material will show a change in volume, but not in general shape, in response to a pressure change Δp. (Pressure is force per unit area, as we shall see in Section 9.2.) The change in pressure is equal to

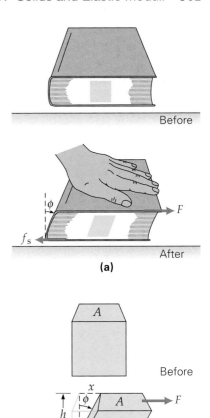

▲ **FIGURE 9.4 Shear stress and strain** (a) A shear stress is produced when a force is applied tangentially to a surface area. (b) The strain is measured in terms of the relative displacement of the object's faces, or the shear angle ϕ.

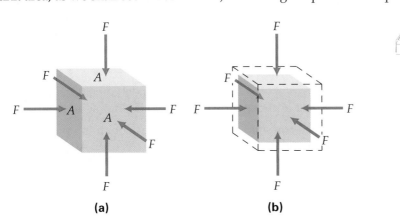

◀ **FIGURE 9.5 Volume stress and strain** (a) A volume stress is applied when a normal force acts over an entire surface area, as shown here for a cube. This type of stress most commonly occurs in gases. (b) The resulting strain is a change in volume.

*See Learn by Drawing on page 219.

the volume stress, or $\Delta p = F/A$. The *volume strain* is the ratio of the volume change (ΔV) to the original volume (V_o). The **bulk modulus (B)** is then

$$B = \frac{F/A}{-\Delta V/V_o} = -\frac{\Delta p}{\Delta V/V_o} \qquad (9.6)$$

SI unit of bulk modulus: newton per square meter (N/m^2)

The minus sign is introduced to make B a positive quantity, since $\Delta V = V - V_o$ is negative for an increase in external pressure (when Δp is positive). Similarly to the previous moduli relationships, $\Delta V \propto 1/B$.

Bulk moduli of some selected solids and liquids are listed in Table 9.1. Gases also have bulk moduli, since they can be compressed. For a gas, it is common to talk about the reciprocal of the bulk modulus, which is called the **compressibility (k)**:

$$k = \frac{1}{B} \quad \textit{(compressibility for gases)} \qquad (9.7)$$

The change in volume ΔV is thus directly proportional to the compressibility k.

Solids and liquids are relatively incompressible and thus have small values of compressibility. Conversely, gases are easily compressed and have large compressibilities, which vary with pressure and temperature.

Example 9.2 ■ Compressing a Liquid: Volume Stress and Bulk Modulus

By how much should the pressure on a liter of water be changed to compress it by 0.10%?

Thinking It Through. Similarly to the fractional change in length, $\Delta L/L_o$, the fractional change in volume is given by $-\Delta V/V_o$, which may be expressed as a percentage. The pressure change can then be found from Eq. 9.6. Compression implies a negative ΔV.

Solution.

Given: $\quad -\Delta V/V_o = 0.0010$ (or 0.10%) $\qquad\qquad$ *Find:* $\quad \Delta p$
$\qquad\qquad V_o = 1.0\,\text{L} = 1000\,\text{cm}^3$
$\qquad\qquad B_{H_2O} = 2.2 \times 10^9\,\text{N/m}^2$ (from Table 9.1)

Note that $-\Delta V/V_o$ is the *fractional* change in the volume. With $V_o = 1000\,\text{cm}^3$, the change (reduction) in volume is

$$-\Delta V = 0.0010\,V_o = 0.0010(1000\,\text{cm}^3) = 1.0\,\text{cm}^3$$

However, the change in volume is not needed. The fractional change, as listed in the given data, can be used directly in Eq. 9.6 to find the increase in pressure:

$$\Delta p = B\left(\frac{-\Delta V}{V_o}\right) = (2.2 \times 10^9\,\text{N/m}^2)(0.0010) = 2.2 \times 10^6\,\text{N/m}^2$$

(This increase is about twenty-two times normal atmospheric pressure. Not too compressible.)

Follow-Up Exercise. If an extra $1.0 \times 10^6\,\text{N/m}^2$ of pressure above normal atmospheric pressure is applied to a half liter of water, what is the change in the water's volume?

9.2 Fluids: Pressure and Pascal's Principle

OBJECTIVES: To (a) explain the pressure–depth relationship, and (b) state Pascal's principle and describe how it is used in practical applications.

A force can be applied to a solid at a point of contact, but this won't work with a fluid, since a fluid cannot support a shear. With fluids, a force must be applied over an area. Such an application of force is expressed in terms of **pressure**, or the *force per unit area*:

$$p = \frac{F}{A} \qquad (9.8a)$$

SI unit of pressure: newton per square meter (N/m^2), or pascal (Pa)

The force in this equation is understood to be acting normally (perpendicularly) to the surface area. F may be the perpendicular component of a force that acts at an angle to a surface (▸Fig. 9.6).

As Figure 9.6 shows, in the more general case we should write

$$p = \frac{F_\perp}{A} = \frac{F \cos \theta}{A} \qquad (9.8b)$$

Pressure is a scalar quantity (with magnitude only), even though the force producing it is a vector.

Pressure has SI units of newton per square meter (N/m^2), or **pascal (Pa)**, in honor of the French scientist and philosopher Blaise Pascal (1623–1662), who studied fluids and pressure. By definition,*

$$1 \, Pa = 1 \, N/m^2$$

In the British system, a common unit of pressure is pound per square inch $(lb/in^2$, or psi). Other units, some of which will be introduced later, are used in special applications. Before going on, here's a "solid" example of the relationship between force and pressure.

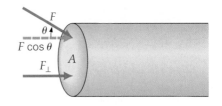

$$p = \frac{F_\perp}{A} = \frac{F \cos \theta}{A}$$

▲ **FIGURE 9.6 Pressure** Pressure is usually written $p = F/A$, where it is understood that F is the force or component of force normal to the surface. In general, then, $p = (F \cos \theta)/A$.

Conceptual Example 9.3 ■ Force and Pressure: Taking a Nap on a Bed of Nails

Suppose you are getting ready to take a nap, and you have a choice of lying stretched out on your back on (a) a bed of nails, (b) a hardwood floor, or (c) a couch. Which one would you choose for the most comfort, and *why*?

Reasoning and Answer. The comfortable choice is quite apparent—the couch. But here, the conceptual question is *why*.

First let's look at the prospect of lying on a bed of nails, an old trick that originated in India and used to be demonstrated in carnival sideshows (See Fig. 9.27). There is really no trick here, just physics—namely, force and pressure. It is the force per unit area, or pressure $(p = F/A)$, that determines whether a nail will pierce the skin. The force is determined by the weight of the person lying on the nails. The area is determined by the *effective* area of the nails in contact with the skin (neglecting one's clothes).

If there were only one nail, the person's weight would not be supported by the nail, and with such a small area, the pressure would be very great—a situation in which the lone nail would pierce the skin. However, when a bed of nails is used, the same force (weight) is distributed over hundreds of nails, which gives a relatively large effective area of contact. The pressure is then reduced to a level at which the nails do not pierce the skin.

When you are lying on a hardwood floor, the area in contact with your body is appreciable and the pressure is reduced, but it still may be a bit uncomfortable. Parts of your body, such as your neck and the small of your back, are *not* in contact with a surface, but they would be on a soft couch, making for a comfortable pressure—the lower the pressure, the more comfort (same force over a larger area). So (c) is the answer.

Follow-Up Exercise. What are a couple of important considerations in constructing a bed of nails to lie on?

Now, let's take a quick review of density, which is an important consideration in the study of fluids. Recall from Chapter 1 that the density (ρ) of a substance is defined as mass per *unit* volume (Eq. 1.1):

$$density = \frac{mass}{volume}$$

$$\rho = \frac{m}{V}$$

SI unit of density: kilogram per cubic meter (kg/m^3)
(common cgs unit: gram per cubic centimeter, or g/cm^3)

The densities of some common substances are given in Table 9.2.

*Notice that the unit of pressure is equivalent to energy per volume, $N/m^2 = N \cdot m/m^3 = J/m^3$, an energy density.

INSIGHT 9.1 OSTEOPOROSIS AND BONE MINERAL DENSITY (BMD)

Bone is a living, growing tissue. Your body is continuously taking up old bone (resorption) and making new bone tissue. In the early years of life, bone growth is greater than bone loss. This continues until a peak bone mass is reached as a young adult. After this, bone growth is slowly outpaced by bone loss. Bones naturally become less dense and weaker with age. Osteoporosis ("porous bone") occurs when bones deteriorate to the point where they are easily fractured (Fig. 1).

Osteoporosis and low bone mass affect an estimated 24 million Americans, most of whom are women. Osteoporosis results in an increased risk of bone fractures, particularly of the hip and the spine. Many women take calcium supplements to help prevent this.

To understand how bone density is measured, let's first distinguish between *bone* and *bone tissue*. Bone is the solid material composed of a protein matrix, most of which has calcified. Bone tissue includes the marrow spaces within the matrix.

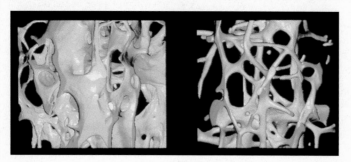

FIGURE 1 **Bone mass loss** An X-ray micrograph of the bone structure of the vertebrae of a 50-year old (left) and a 70-year old (right). Osteoporosis, a condition characterized by bone weakening caused by loss of bone mass, is evident for the vertebrae on the right.

(Marrow is the soft, fatty, vascular tissue in the interior cavities of bones and is a major site of blood cell production.) The marrow volume varies with the bone type.

If the volume of an intact bone is measured (for example, by water displacement), then, the *bone tissue density* can be computed, commonly in grams per cubic centimeter after the bone is weighed to determine mass. If you burn the bone, weigh the remaining ash, and divide by the volume of the overall bone (bone tissue), you get the *bone tissue mineral density*, which is commonly called the **bone mineral density (BMD)**.

To measure the BMD of bones *in vivo*, types of radiation transmission through the bone is measured, and this is related to the amount of bone mineral present. Also, a "projected" area of the bone is measured. Using these measurements, an areal or projected BMD is computed in units of mg/cm^2. Figure 2 illustrates the magnitude of the effect of bone density loss with aging.

The diagnosis of osteoporosis relies primarily on the measurement of BMD. The mass of a bone, measured by a BMD test (also called a *bone densitometry test*), generally correlates to the bone strength. It is possible to predict fracture risk, much as blood pressure measurements can help predict stroke risk. Bone density testing is recommended for all women age 65 and older, and for younger women at an increased risk of osteoporosis. This also applies to men. Osteoporosis is often thought to be a woman's disease, but 20% of osteoporosis cases occur in men. A BMD test cannot predict the certainty of developing a fracture, but only predicts the degree of risk.

So how is BMD measured? This is where the physics comes in. Various instruments, divided into *central devices* and *peripheral devices*, are used. Central devices are used primarily to measure the bone density of the hip and spine. Peripheral devices are smaller, portable machines that are used to measure the bone density in such places as the heel or finger.

TABLE 9.2	Densities of Some Common Substances (in kg/m^3)				
Solids	*Density (ρ)*	*Liquids*	*Density (ρ)*	*Gases**	*Density (ρ)*
Aluminum	2.7×10^3	Alcohol, ethyl	0.79×10^3	Air	1.29
Brass	8.7×10^3	Alcohol, methyl	0.82×10^3	Helium	0.18
Copper	8.9×10^3	Blood, whole	1.05×10^3	Hydrogen	0.090
Glass	2.6×10^3	Blood plasma	1.03×10^3	Oxygen	1.43
Gold	19.3×10^3	Gasoline	0.68×10^3	Water vapor (100°C)	0.63
Ice	0.92×10^3	Kerosene	0.82×10^3		
Iron (and steel)	7.8×10^3 (general value)	Mercury	13.6×10^3		
Lead	11.4×10^3	Seawater (4°C)	1.03×10^3		
Silver	10.5×10^3	Water, fresh (4°C)	1.00×10^3		
Wood, oak	0.81×10^3				

*At 0°C and 1 atm, unless otherwise specified.

A common peripheral device uses *quantitative ultrasound* (QUS). Instead of X-rays, a bone density screening is made using high-frequency sound waves (ultrasound). (See Section 14.1 for a discussion of ultrasound.) QUS measurements are usually done on the heel. The test takes only a minute or two; and the devices are now found in some pharmacies or drugstores. Its purpose is to tell you if you are "at risk," and may need a more thorough DXA test.

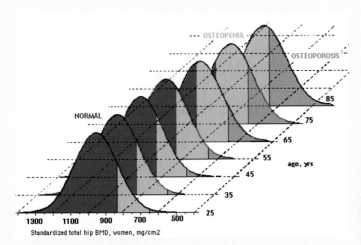

FIGURE 2 Bone density loss with aging An illustration of how normal bone density loss for a female hip bone increases with age (scale on right). Osteopenia refers to decreased calcification or bone density. A person with osteopenia is at risk for developing osteoporosis, a condition that causes bones to become brittle and prone to fracture.

The most widely used central device relies on *dual energy X-ray absorptiometry* (DXA), which uses X-ray imaging to measure bone density. (See Section 20.4 for a discussion of X-rays.) The DXA scanner produces two X-ray beams of different energy levels. The amount of X-rays that pass through a bone is measured for each beam and the amounts vary with the density of bone. The calculated bone density is based on the difference between the two beams. The procedure is nonintrusive and takes 10–20 min, and the X-ray exposure is usually about one tenth of that of a chest X-ray (Fig. 3).

FIGURE 3 Osteoporosis bone scanning A technician runs an X-ray bone scan on an elderly patient for osteoporosis. X-ray images are displayed on the monitor. Such images can confirm the presence of osteoporosis. Also, such bone densitometry tests can be used to diagnosis rickets, a children's disease characterized by softening of the bones.

Water has a density of 1.00×10^3 kg/m³ (or 1.00 g/cm³), from the original definition of the kilogram (Chapter 1). Mercury has a density of 13.6×10^3 kg/m³ (or 13.6 g/cm³). Hence, mercury is 13.6 times as dense as water. Gasoline, however, is less dense than water. See Table 9.2. (*Note*: Be careful not to confuse the symbol for density, ρ (Greek rho) with that for pressure, p.)

Note: $p \neq \rho$

We say that density is a measure of the compactness of the matter of a substance: The greater the density, the more matter or mass in a given volume. Notice that density quantifies the amount or mass per unit volume. For an important density consideration, see Insight 9.1 on Osteoporosis and Bone Mineral Density (BMD).

Pressure and Depth

If you have gone scuba diving, you well know that pressure increases with depth, having felt the increased pressure on your eardrums. An opposite effect is commonly felt when you fly in a plane or ride in a car going up a mountain: with increasing altitude, your ears may "pop" because of *reduced* external air pressure.

How the pressure in a fluid varies with depth can be demonstrated by considering a container of liquid at rest. Imagine that you can isolate a rectangular column of water,

▶ **FIGURE 9.7** Pressure and depth
The extra pressure at a depth h in a liquid is due to the weight of the liquid above: $p = \rho gh$, where ρ is the density of the liquid (assumed to be constant). This is shown for an imaginary rectangular column of liquid.

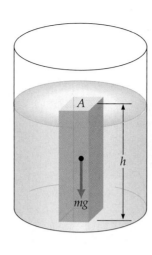

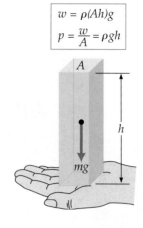

$$w = \rho(Ah)g$$
$$p = \frac{w}{A} = \rho gh$$

PHYSLET®

Illustration 14.1 Pressure in a Liquid

Pressure–depth relationship

as shown in ▲Fig. 9.7. Then the force on the bottom of the container below the column (or the hand) is equal to the weight of the liquid making up the column: $F = w = mg$. Since density is $\rho = m/V$, the mass in the column is equal to the density times the volume; that is, $m = \rho V$. (The liquid is assumed incompressible, so ρ is constant.)

The volume of the isolated liquid column is equal to the height of the column times the area of its base, or $V = hA$. Thus, we can write

$$F = w = mg = \rho Vg = \rho ghA$$

With $p = F/A$, the pressure at a depth h due to the weight of the column is

$$p = \rho gh \qquad (9.9)$$

This is a general result for incompressible liquids. The pressure is the same everywhere on a horizontal plane at a depth h (with ρ and g constant). Note that Eq. 9.9 is independent of the base area of the rectangular column: We could have taken the whole cylindrical column of the liquid in the container in Fig. 9.7 and gotten the same result.

The derivation of Eq. 9.9 did not take into account pressure being applied to the open surface of the liquid. This factor adds to the pressure at a depth h to give a *total* pressure of

$$p = p_{\text{o}} + \rho gh \qquad \begin{array}{l}\textit{(incompressible liquid}\\\textit{at constant density)}\end{array} \qquad (9.10)$$

where p_{o} is the pressure applied to the liquid surface (that is, the pressure at $h = 0$). For an open container, $p_{\text{o}} = p_{\text{a}}$, atmospheric pressure, or the weight (force) per unit area due to the gases in the atmosphere above the liquid's surface. The average atmospheric pressure at sea level is sometimes used as a unit, called an **atmosphere (atm)**:

$$1\ \text{atm} = 101.325\ \text{kPa} = 1.01325 \times 10^5\ \text{N/m}^2 \approx 14.7\ \text{lb/in}^2$$

The measurement of atmospheric pressure will be described shortly.

Demonstration/activity: Demonstrate the effect of atmospheric pressure by using Magdeburg spheres. Recall that the area involved in this demonstration is the cross-section, not the surface area, of the sphere. Why?

Example 9.4 ■ A Scuba Diver: Pressure and Force

(a) What is the total pressure on the back of a scuba diver in a lake at a depth of 8.00 m? (b) What is the force on the diver's back due to the water alone, taking the surface of the back to be a rectangle 60.0 cm by 50.0 cm?

Thinking It Through. (a) This is a direct application of Eq. 9.10 in which p_{o} is taken as the atmospheric pressure p_{a}. (b) Knowing the area and the pressure due to the water, the force can be found from the definition of pressure, $p = F/A$.

Solution.

Given: $h = 8.00\ \text{m}$
$A = 60.0\ \text{cm} \times 50.0\ \text{cm}$
$\quad = 0.600\ \text{m} \times 0.500\ \text{m} = 0.300\ \text{m}^2$
$\rho_{\text{H}_2\text{O}} = 1.00 \times 10^3\ \text{kg/m}^3$ (from Table 9.2)
$p_{\text{a}} = 1.01 \times 10^5\ \text{N/m}^2$

Find: (a) p (total pressure)
(b) F (force due to water)

(a) The total pressure is the sum of the pressure due to the water and the atmospheric pressure (p_a). By Eq. 9.10, this is

$$p = p_a + \rho g h$$
$$= (1.01 \times 10^5 \text{ N/m}^2) + (1.00 \times 10^3 \text{ kg/m}^3)(9.80 \text{ m/s}^2)(8.00 \text{ m})$$
$$= (1.01 \times 10^5 \text{ N/m}^2) + (0.784 \times 10^5 \text{ N/m}^2) = 1.79 \times 10^5 \text{ N/m}^2 \text{ (or Pa)}$$
$$\text{(expressed in atmospheres)} \approx 1.8 \text{ atm}$$

This is also the pressure on the diver's cardrums.

(b) The pressure p_{H_2O} due to the water alone is the $\rho g h$ portion of the preceding equation, so $p_{H_2O} = 0.784 \times 10^5 \text{ N/m}^2$.
Then, $p_{H_2O} = F/A$, and

$$F = p_{H_2O}A = (0.784 \times 10^5 \text{ N/m}^2)(0.300 \text{ m}^2)$$
$$= 2.35 \times 10^4 \text{ N (or } 5.29 \times 10^3 \text{ lb—about 2.6 tons!)}$$

Follow-Up Exercise. You might question the answer to part (b) of this Example—how could the diver support such a force? To get a better idea of the forces our bodies can support, what would be the force on the diver's back at the water surface from atmospheric pressure alone? How do you suppose our bodies can support such forces or pressures?

Pascal's Principle

When the pressure (for example, air pressure) is increased on the entire open surface of an incompressible liquid at rest, the pressure at any point in the liquid or on the boundary surfaces increases by the same amount. The effect is the same if pressure is applied to any surface of an enclosed fluid by means of a piston (▶Fig. 9.8). The transmission of pressure in fluids was studied by Pascal, and the observed effect is called **Pascal's principle**:

Pressure applied to an enclosed fluid is transmitted undiminished to every point in the fluid and to the walls of the container.

For an incompressible liquid, the change in pressure is transmitted essentially instantaneously. For a gas, a change in pressure will generally be accompanied by a change in volume or temperature (or both), but after equilibrium has been reestablished, Pascal's principle remains valid.

Common practical applications of Pascal's principle include the hydraulic braking systems used on automobiles. Through tubes filled with brake fluid, a force on the brake pedal transmits a force to the wheel brake cylinder. Similarly, hydraulic lifts and jacks are used to raise automobiles and other heavy objects (▼Fig. 9.9).

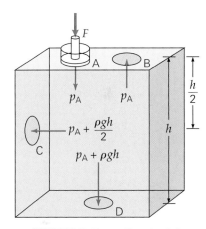

▲ **FIGURE 9.8 Pascal's principle** The pressure applied at point A is fully transmitted to all parts of the fluid and to the walls of the container. There is also pressure due to the weight of the fluid above at different depths (for instance, $\rho g h/2$ at C and $\rho g h$ at D).

▼ **FIGURE 9.9 The hydraulic lift and shock absorbers** **(a)** Because the input and output pressures are equal (Pascal's principle), a small input force gives a large output force proportional to the ratio of the piston areas. **(b)** A simplified exposed view of one type of shock absorber. (See Follow-Up Exercise 9.5 for description.)

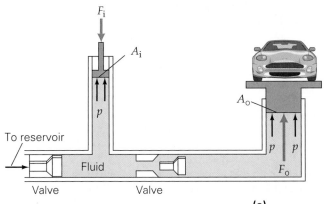

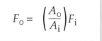

$$F_o = \left(\frac{A_o}{A_i}\right)F_i$$

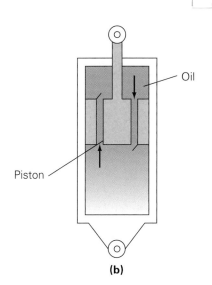

(a)

(b)

Illustration 14.2 Pascal's Principle

Using Pascal's principle, we can show how such systems allow us not only to transmit force from one place to another, but also to multiply that force. The input pressure p_i supplied by compressed air for a garage lift, for example, gives an input force F_i on a small piston area A_i (Fig. 9.9). The full magnitude of the pressure is transmitted to the output piston, which has an area A_o. Since $p_i = p_o$, it follows that

$$\frac{F_i}{A_i} = \frac{F_o}{A_o}$$

and

$$F_o = \left(\frac{A_o}{A_i}\right) F_i \quad \textit{hydraulic force multiplication} \qquad (9.11)$$

With A_o larger than A_i, F_o will be larger than F_i. The input force is greatly multiplied if the input piston has a relatively small area.

Example 9.5 ■ The Hydraulic Lift: Pascal's Principle

A garage lift has input and lift (output) pistons with diameters of 10 cm and 30 cm, respectively. The lift is used to hold up a car with a weight of 1.4×10^4 N. (a) What is the magnitude of the force on the input piston? (b) What pressure is applied to the input piston?

Thinking It Through. (a) Pascal's principle, as expressed in the hydraulic Eq. 9.11, has four variables, and three are given (areas via diameters). (b) The pressure is simply $p = F/A$.

Solution.

Given: $d_i = 10$ cm $= 0.10$ m   Find: (a) F_i (input force)
$\quad\quad\quad d_o = 30$ cm $= 0.30$ m   (b) p_i (input pressure)
$\quad\quad\quad F_o = 1.4 \times 10^4$ N

(a) Rearranging Eq. 9.11 and using $A = \pi r^2 = \pi d^2/4$ for the circular piston ($r = d/2$) gives

$$F_i = \left(\frac{A_i}{A_o}\right) F_o = \left(\frac{\pi d_i^2/4}{\pi d_o^2/4}\right) F_o = \left(\frac{d_i}{d_o}\right)^2 F_o$$

or

$$F_i = \left(\frac{0.10 \text{ m}}{0.30 \text{ m}}\right)^2 F_o = \frac{F_o}{9} = \frac{1.4 \times 10^4 \text{ N}}{9} = 1.6 \times 10^3 \text{ N}$$

The input force is one ninth of the output force; in other words, the force was multiplied by 9 (that is, $F_o = 9F_i$).

(Note that we didn't really need to write the complete expressions for the areas. We know that the area of a circle is proportional to the square of the diameter of the circle. If the ratio of the piston diameters is 3 to 1, the ratio of their areas must therefore be 9 to 1, and we could have used this ratio directly in Eq. 9.11.)

(b) Then we apply Eq. 9.8a:

$$p_i = \frac{F_i}{A_i} = \frac{F_i}{\pi r_i^2} = \frac{F_i}{\pi (d_i/2)^2} = \frac{1.6 \times 10^3 \text{ N}}{\pi (0.10 \text{ m})^2/4}$$

$$= 2.0 \times 10^5 \text{ N/m}^2 \;(= 200 \text{ kPa})$$

This pressure is about 30 lb/in^2, a common pressure used in automobile tires and about twice atmospheric pressure (which is approximately 100 kPa, or 15 lb/in^2.)

Follow-Up Exercise. Pascal's principle is used in shock absorbers on automobiles and on the landing gear of airplanes. (The polished steel piston rods can be seen above the wheels on aircraft.) In these devices, a large force (the shock produced on hitting a bump in the road or an airport runway at high speed) must be reduced to a safe level by removing energy. Basically, fluid is forced by the motion of a large-diameter piston through small channels in the piston on each stroke cycle (Fig. 9.9b).

Note that the valves allow for fluid through the channel, which creates resistance to the motion of the piston (effectively the reverse of the situation in Fig. 9.9a). The piston goes up and down, dissipating the energy of the shock. This is called *damping* (Section 13.2). Suppose that the input piston of a shock absorber on a jet plane has a diameter of 8.0 cm. What would be the diameter of an output channel that would reduce the force by a factor of 10?

As Example 9.5 shows, we can relate forces produced by pistons directly to their diameters: $F_i = (d_i/d_o)^2 F_o$ or $F_o = (d_o/d_i)^2 F_i$. By making $d_o \gg d_i$, we can get huge factors of force multiplication, as is typical for hydraulic presses, jacks, and earth-moving equipment. (The shiny input piston rods are often visible on front loaders and backhoes.) Inversely, we can get a force reduction by making $d_i > d_o$, as in Follow-Up Exercise 9.5.

However, don't think that you are getting something for nothing with large force multiplications: Energy is still a factor, and it can never be multiplied by a machine. (Why not?) Looking at the work involved and assuming that the work output is equal to the work input, $W_o = W_i$ (an ideal condition—why?), we have, from Eq. 5.1,

$$F_o x_o = F_i x_i$$

or

$$F_o = \left(\frac{x_i}{x_o}\right) F_i$$

where x_o and x_i are the output and input distances moved by the respective pistons.

Thus, the output force can be much greater than the input force only if the input distance is much greater than the output distance. For example, if $F_o = 10 F_i$, then $x_i = 10 x_o$, and the input piston must travel 10 times the distance of the output piston. We say that *force is multiplied at the expense of distance.*

Pressure Measurement

Pressure can be measured by mechanical devices that are often spring loaded (such as a tire gauge). Another type of instrument, called a manometer, uses a liquid—usually mercury—to measure pressure. An *open-tube manometer* is illustrated in ▼Fig. 9.10a.

▼ **FIGURE 9.10 Pressure measurement (a)** For an open-tube manometer, the pressure of the gas in the container is balanced by the pressure of the liquid column and atmospheric pressure acting on the open surface of the liquid. The absolute pressure of the gas equals the sum of the atmospheric pressure (p_a) and ρgh, the gauge pressure. **(b)** A tire gauge measures gauge pressure, the difference between the pressure in the tire and atmospheric pressure: $p_{gauge} = p - p_a$. Thus, if a tire gauge reads 200 kPa (30 lb/in²), the actual pressure within the tire is 1 atm higher, or 300 kPa. **(c)** A barometer is a closed-tube manometer that is exposed to the atmosphere and thus reads only atmospheric pressure.

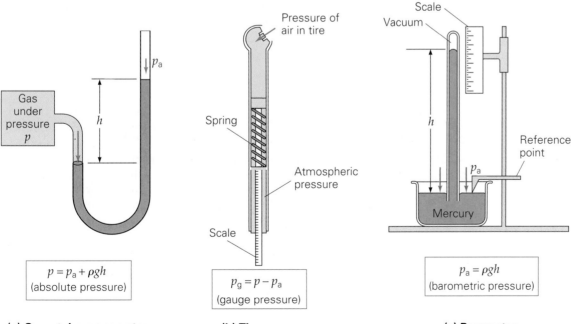

$p = p_a + \rho gh$
(absolute pressure)

(a) Open-tube manometer

$p_g = p - p_a$
(gauge pressure)

(b) Tire gauge

$p_a = \rho gh$
(barometric pressure)

(c) Barometer

One end of the U-shaped tube is open to the atmosphere, and the other is connected to the container of gas whose pressure is to be measured. The liquid in the U-tube acts as a reservoir through which pressure is transmitted according to Pascal's principle.

The pressure of the gas (p) is balanced by the weight of the column of liquid (of height h, the difference in the heights of the columns) and the atmospheric pressure (p_a) on the open liquid surface:

$$p = p_a + \rho g h \tag{9.12}$$

The pressure p is called the **absolute pressure**.

You may have measured pressure using pressure gauges; a tire gauge used to measure air pressure in automobile tires is a common example (Fig. 9.10b). Such gauges, quite appropriately, measure **gauge pressure**: A pressure gauge registers only the pressure *above* (*or below*) atmospheric pressure. Hence, to get the absolute pressure (p), you have to add the atmospheric pressure (p_a) to the gauge pressure (p_g):

$$p = p_a + p_g$$

For example, suppose your tire gauge reads a pressure of 200 kPa ($\approx$ 30 lb/in^2). The absolute pressure within the tire is then $p = p_a + p_g = 101$ kPa $+ 200$ kPa $= 301$ kPa, where normal atmospheric pressure is about 101 kPa (14.7 lb/in^2), as will be shown shortly.

Teaching tip: Adding atmospheric pressure is confusing to some students. Point out that a tire pressure of 35 psi, as measured with a standard gauge, does *not* take account of atmospheric pressure.

The gauge pressure of a tire keeps the tire rigid or operational. In terms of the more familiar pounds per square inch (psi, or lb/in^2), a tire with a gauge pressure of 30 psi has an absolute pressure of about 45 psi (30 + 15, with atmospheric pressure $\approx$ 15 psi). Hence, the pressure on the inside of the tire is 45 psi, and that on the outside is 15 psi. The Δp of 30 psi keeps the tire inflated. If you open the valve or get a puncture, the internal and external pressures equalize and you have a flat!

Atmospheric pressure itself can be measured with a *barometer*. The principle of a mercury barometer is illustrated in Fig. 9.10c. The device was invented by Evangelista Torricelli (1608–1647), Galileo's successor as professor of mathematics at an academy in Florence. A simple barometer consists of a tube filled with mercury that is inverted into a reservoir. Some mercury runs from the tube into the reservoir, but a column supported by the air pressure on the surface of the reservoir remains in the tube. This device can be considered a *closed-tube manometer*, and the pressure it measures is just the atmospheric pressure, since the gauge pressure (the pressure *above* atmospheric pressure) is zero.

The atmospheric pressure is then equal to the pressure due to the weight of the column of mercury, or

$$p_a = \rho g h \tag{9.13}$$

A *standard atmosphere* is defined as the pressure supporting a column of mercury exactly 76 cm in height at sea level and at 0°C. (For a common biological atmospheric effect because of pressure changes, see Insight 9.2 on Possible Earaches.)

Changes in atmospheric pressure can be observed as changes in the height of a column of mercury. These changes are due primarily to high- and low-pressure air masses that travel across the country. Atmospheric pressure is commonly reported in terms of the height of the barometer column, and weather forecasters say that the barometer is rising or falling. That is,

$$1 \text{ atm (about 101 kPa)} = 76 \text{ cm Hg} = 760 \text{ mm Hg}$$
$$= 29.92 \text{ in. Hg (about 30 in. Hg)}$$

INSIGHT 9.2 AN ATMOSPHERIC EFFECT: POSSIBLE EARACHES

Variations in atmospheric pressure can have a common physiological effect: changes in pressure in the ears with changes in altitude. This "plugging up" and "popping" of the ears is frequently experienced in ascents and descents on mountain roads or on airplanes. The eardrum, so important to your hearing, is a membrane that separates the middle ear from the outer ear. [See Fig. 1 in the Chapter 14 (Sound) Insight 14.2 on Hearing, p. 475, to view the anatomy of the ear.] The middle ear is connected to the throat by the Eustachian tube, the end of which is normally closed. The tube opens during swallowing or yawning to permit air to escape, so the internal and external pressures are equalized.

However, when climbing relatively quickly in an airplane or in a car in a mountainous region, the air pressure outside the ear may be less than that in the middle ear. This difference in pressure forces the eardrum outward. If the outward pressure is not relieved, you will soon have an earache. The pressure is relieved by a "pushing" of air through the Eustachian tube into the throat, which produces a popping sound. We often swallow or yawn to assist this process. Similarly, when we descend, the higher outside pressure at lower altitudes needs to be equalized with the lower pressure in the middle ear. Swallowing allows air to flow into the middle ear in this case.

Nature takes care of us, but it is important to understand what is going on. If you have a throat infection, the opening of the Eustachian tube to the throat might be swollen, partially blocking the tube. You may be tempted to hold your nose and blow with your mouth closed in order to clear your ears. Don't do it! You could blow mucus into the inner ear and cause a painful inner-ear infection. Instead, swallow hard several times and give some big yawns to help open the Eustachian tube and equalize the pressure.

In honor of Torricelli, a pressure supporting 1 mm of mercury is given the name *torr*:

$$1 \text{ mm Hg} \equiv 1 \text{ torr}$$

and

$$1 \text{ atm} = 760 \text{ torr*}$$

Because mercury is highly toxic, it is sealed inside a barometer. A safer and less expensive device that is widely used to measure atmospheric pressure is the *aneroid* ("without fluid") *barometer*. In an aneroid barometer, a sensitive metal diaphragm on an evacuated container (something like a drumhead) responds to pressure changes, which are indicated on a dial. This is the kind of barometer you frequently find in homes in decorative wall mountings.

Since air is compressible, the atmospheric density and pressure are greatest at the Earth's surface and decrease with altitude. We live at the bottom of the atmosphere, but don't notice its pressure very much in our daily activities. Remember that our bodies are composed largely of fluids, which exert a matching outward pressure. Indeed, the external pressure of the atmosphere is so important to our normal functioning that we take it with us wherever we can. The pressurized suits worn by astronauts in space or on the Moon are needed not only to supply oxygen, but also to provide an external pressure similar to that on the Earth's surface.

A very important gauge pressure reading is discussed in Insight 9.3: Blood Pressure and Its Measurement on page 312. Read it before going on to Example 9.6.

---

*In the SI, one atmosphere has a pressure of 1.013×10^5 N/m^2, or about 10^5 N/m^2. Meteorologists use yet another unit of pressure called the *millibar* (mb). A *bar* is defined to be 10^5 N/m^2, and because a bar = 1000 mb, we have 1 atm $\approx$ 1 bar = 1000 mb. With 1000 mb, the small changes in atmospheric pressure are more easily reported.

INSIGHT 9.3 BLOOD PRESSURE AND ITS MEASUREMENT

Basically, a pump is a machine that transfers mechanical energy to a fluid, thereby increasing the pressure and causing the fluid to flow. One pump that is of interest to everyone is the heart, a muscular pump that drives blood throughout the body's circulatory network of arteries, capillaries, and veins. With each pumping cycle, the human heart's interior chambers enlarge and fill with freshly oxygenated blood from the lungs (Fig. 1).

The human heart contains two pairs of chambers: two ventricles and two atria. When the ventricles contract, blood is forced out through the arteries. Smaller and smaller arteries branch off from the main ones, until the very small capillaries are reached. There, oxygen and nutrients being carried by the blood are exchanged with the surrounding tissues, and carbon-dioxide (a waste gas) is picked up. The blood then flows into the veins to the lungs to expel carbon dioxide, back to the heart to complete the circuit.

When the ventricles contract, forcing blood into the arterial system, the pressure in the arteries increases sharply. The maximum pressure achieved during ventricular contraction is called the *systolic pressure*. When the ventricles relax, the arterial pressure drops, and the lowest pressure before the next contraction, called *diastolic pressure*, is reached. (These pressures are named after two parts of the pumping cycle, *systole* and *diastole*.)

The walls of the arteries have considerable elasticity and expand and contract with each pumping cycle. This alternating

FIGURE 1 The heart as a pump The human heart is analogous to a mechanical force pump. Its pumping action, consisting of **(a)** intake and **(b)** output, gives rise to variations in blood pressure.

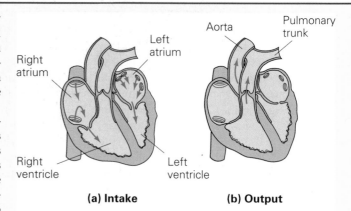

(a) Intake **(b) Output**

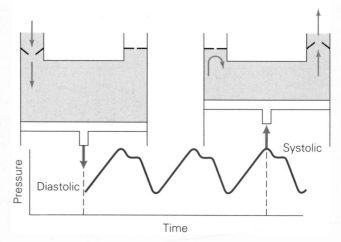

Illustration 14.4 Pumping Water Up from a Well

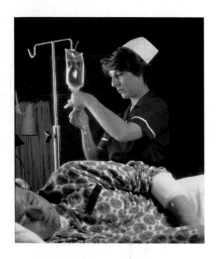

▲ **FIGURE 9.11 What height is needed?** See Example 9.6.

Example 9.6 ■ An IV: A Gravity Assist

An IV (*intravenous injection*) is quite a different type of gravity assist from that discussed for space probes in Chapter 7. Consider a hospital patient who receives an IV under gravity flow, as shown in ◄Fig. 9.11. If the blood gauge pressure in the vein is 20.0 mm Hg, above what height should the bottle be placed for the IV blood transfusion to function properly?

Thinking It Through. The fluid gauge pressure at the bottom of the IV tube must be greater than the pressure in the vein and can be computed from Eq. 9.9. (The liquid is assumed to be incompressible.)

Solution.

Given: $p_v = 20.0$ mm Hg (vein gauge pressure)   *Find:* h (height for $p_v > 20$ mm Hg)
  $\rho = 1.05 \times 10^3$ kg/m^3 (whole blood density from Table 9.2)

First, the common medical unit of mm Hg (or torr) needs to be changed to the SI unit of pascal (Pa, or N/m^2):

$$p_v = (20.0 \text{ mm Hg})[133 \text{ Pa}/(\text{mm Hg})] = 2.66 \times 10^3 \text{ Pa}$$

Then, for $p > p_v$,

$$p = \rho g h > p_v$$

or

$$h > \frac{p_v}{\rho g} = \frac{2.66 \times 10^3 \text{ Pa}}{(1.05 \times 10^3 \text{ kg/m}^3)(9.80 \text{ m/s}^2)} = 0.259 \text{ m} (\approx 26 \text{ cm})$$

expansion and contraction can be felt as a *pulse* in an artery near the surface of the body. For example, the radial artery near the surface of the wrist is commonly used to measure a person's pulse. The pulse rate is equal to the ventricular contraction rate, and hence the pulse rate indicates the heart rate.

Taking a person's blood pressure involves measuring the pressure of the blood on the arterial walls. This is done with a *sphygmomanometer*. (The Greek word *sphygmo* means "pulse.") An inflatable cuff is inflated to shut off the blood flow temporarily. The cuff pressure is slowly released, and the artery is monitored with a stethoscope (Fig. 2). Soon blood is just forced through the constricted artery. This flow is turbulent and gives rise to a specific sound with each heartbeat. When the sound is first heard, the systolic pressure is noted on the gauge. When the turbulent beats disappear because blood begins to flow smoothly, the diastolic pressure is taken.

Blood pressure is commonly reported by giving the systolic and diastolic pressures, separated by a slash—for example, 120/80 (mm Hg, read as "120 over 80"). (The gauge in Fig. 2 is an aneroid type; older types of sphygmomanometers used a mercury column to measure blood pressure.) Normal blood pressure ranges are 120–139 for systolic and 80–89 for diastolic. (Blood pressure is a gauge pressure. Why?)

Away from the heart, blood vessels branch to smaller and smaller diameters. The pressure in the blood vessels decreases as their diameter decreases. In small arteries, such as those in the arm, the blood pressure is on the order of 10 to 20 mm Hg, and there is no systolic–diastolic variation.

High blood pressure is a common health problem. The elastic walls of the arteries expand under the hydraulic force of the blood

pumped from the heart. Their elasticity may diminish with age, however. Cholesterol deposits can narrow and roughen the arterial passageways, impeding the blood flow and giving rise to a form of arteriosclerosis, or hardening of the arteries. Because of these defects, the driving pressure must increase to maintain a normal blood flow. The heart must work harder, which places a greater demand on the heart muscles. A relatively slight decrease in the effective cross-sectional area of a blood vessel has a rather large effect (an increase) on the flow rate, as will be shown in Section 9.4.

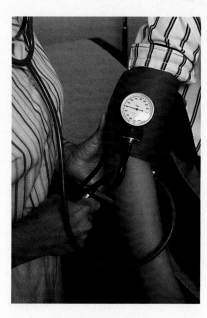

FIGURE 2 Measuring blood pressure The pressure is indicated on the gauge in millimeters Hg.

The IV bottle needs to be at least 26 cm above the injection site.

Follow-Up Exercise. The normal (gauge) blood pressure range is commonly reported as 120/80 (in millimeters Hg). Why is the blood pressure of 20 mm Hg in this Example so low?

9.3 Buoyancy and Archimedes' Principle

OBJECTIVES: To (a) relate the buoyant force to Archimedes' principle, and (b) tell whether an object will float in a fluid, on the basis of relative densities.

When placed in a fluid, an object will either sink or float. This is most commonly observed with liquids; for example, objects float or sink in water. But the same effect occurs in gases: A falling object sinks in the atmosphere, and other bodies float (▼Fig. 9.12).

Things float because they are buoyant, or are buoyed up. For example, if you immerse a cork in water and release it, the cork will be buoyed up to the surface and float there. From your knowledge of forces, you know that such motion requires an upward net force on the object. That is, there must be an upward force acting on the object that is greater than the downward force of its weight. The forces are equal when the object floats in equilibrium. The upward force resulting from an object being wholly or partially immersed in a fluid is called the **buoyant force**.

▲ **FIGURE 9.12 Fluid buoyancy**
The air is a fluid in which objects such as this dirigible floats. The helium inside the blimps is less dense than the surrounding air and the blimp is supported by the resulting buoyant force.

Illustration 14.3 Buoyant Force

How the buoyant force comes about can be seen by considering a buoyant object being held under the surface of a fluid (▶Fig. 9.13a). The pressures on the upper and lower surfaces of the block are $p_1 = \rho_f g h_1$ and $p_2 = \rho_f g h_2$, respectively, where ρ_f is the density of the fluid. Thus, there is a pressure difference $\Delta p = p_2 - p_1 = \rho_f g(h_2 - h_1)$ between the top and bottom of the block, which gives an upward force (the buoyant force) F_b. This force is balanced by the applied force and the weight of the block.

It is not difficult to derive an expression for the magnitude of the buoyant force. We know that pressure is force per unit area. Thus, if both the top and bottom areas of the block are A, the magnitude of the net buoyant force in terms of the pressure difference is

$$F_b = p_2 A - p_1 A = (\Delta p)A = \rho_f g(h_2 - h_1)A$$

Since $(h_2 - h_1)A$ is the volume of the block, and hence the volume of fluid displaced by the block, V_f, we can write the expression for F_b as

$$F_b = \rho_f g V_f$$

But $\rho_f V_f$ is simply the mass of the fluid displaced by the block, m_f. Thus, we can write the expression for the buoyant force as $F_b = m_f g$: The magnitude of the buoyant force is equal to the weight of the fluid displaced by the block (Fig. 9.13b). This general result is known as **Archimedes' principle**:

A body immersed wholly or partially in a fluid experiences a buoyant force equal in magnitude to the weight of the *volume of fluid* that is displaced:

$$F_b = m_f g = \rho_f g V_f \tag{9.14}$$

Archimedes (287 BCE–212 BCE) was given the task of determining whether a gold crown made for the king was pure gold or contained a quantity of silver. Legend has it that the solution to the problem came to him when he was in the bath tub. (See the Physics Facts at the beginning of the chapter.) It is said that he was so excited he jumped out of the tub and ran home through the streets of the city (unclothed) shouting "Eureka! Eureka!" (Greek for "I have found it"). Although Archimedes' solution to the problem involved density and volume, it presumably got him thinking about buoyancy.

Integrated Example 9.7 ■ Lighter Than Air: Buoyant Force

A spherical helium-filled weather balloon has a radius of 1.10 m. (a) Does the buoyant force on the balloon depend on the density of (1) helium, (2) air, or (3) the weight of the rubber "skin"? [$\rho_{air} = 1.29$ kg/m^3 and $\rho_{He} = 0.18$ kg/m^3.] (b) Compute the magnitude of the buoyant force on the balloon. (c) The balloon's rubber skin has a mass of 1.20 kg. When released, what is the magnitude of the balloon's initial acceleration if it carries a payload with a mass of 3.52 kg?

(a) Conceptual Reasoning. The buoyant force has nothing to do with the helium or rubber skin and is equal to the weight of the displaced air, which can be found from the balloon's volume and the density of air. So the answer is (2).

(b, c) Quantitative Reasoning and Solution.

Given: $\rho_{air} = 1.29$ kg/m^3 *Find:* (b) F_b (buoyant force)
$\rho_{He} = 0.18$ kg/m^3 (c) a (initial acceleration)
$m_s = 1.20$ kg
$m_p = 3.52$ kg
$r = 1.10$ m

(b) The volume of the balloon is

$$V = (4/3)\pi r^3 = (4/3)\pi(1.10 \text{ m})^3 = 5.58 \text{ m}^3$$

Then the buoyant force is equal to the weight of the air displaced:

$$F_b = m_{air}g = (\rho_{air}V)g = (1.29 \text{ kg/m}^3)(5.58 \text{ m}^3)(9.80 \text{ m/s}^2) = 70.5 \text{ N}$$

(c) Draw a free-body diagram. There are three weight forces downward—those of the helium, the rubber skin, and the payload—and the upward buoyant force. Sum these forces to find the net force, and then use Newton's second law to find the acceleration. The weights of the helium, rubber skin, and payload are as follows:

$$w_{He} = m_{He}g = (\rho_{He}V)g = (0.18 \text{ kg/m}^3)(5.58 \text{ m}^3)(9.80 \text{ m/s}^2) = 9.84 \text{ N}$$
$$w_s = m_s g = (1.20 \text{ kg})(9.80 \text{ m/s}^2) = 11.8 \text{ N}$$
$$w_p = m_p g = (3.52 \text{ kg})(9.80 \text{ m/s}^2) = 35.5 \text{ N}$$

Summing the forces (taking upward as positive),

$$F_{net} = F_b - w_{He} - w_s - w_p = 70.5 \text{ N} - 9.84 \text{ N} - 11.8 \text{ N} - 35.5 \text{ N} = 13.4 \text{ N}$$

and

$$a = \frac{F_{net}}{m_{total}} = \frac{F_{net}}{m_{He} + m_s + m_p} = \frac{13.4 \text{ N}}{0.994 \text{ kg} + 1.20 \text{ kg} + 3.52 \text{ kg}} = 2.35 \text{ m/s}^2$$

Follow-Up Exercise. As the balloon rises, it eventually stops accelerating and rises at a constant velocity for a short time, then starts sinking toward the ground. Explain this behavior in terms of atmospheric density and temperature. (*Hint*: Temperature and air density decreases with altitude. The pressure of a quantity of gas is directly proportional to temperature.)

Example 9.8 ■ Your Buoyancy in Air

Air is a fluid and our bodies displace air. And so, a buoyant force is acting on each of us. Estimate the magnitude of the buoyant force on a 75-kg person due to the air displaced.

Thinking It Through. The key word here is *estimate*, because not much data is given. We know that the buoyant force is $F_b = \rho_a gV$, where ρ_a is the density of air (which can be found in Table 9.2), and V is the volume of the air displaced, which is the same as the volume of the person. The question is, how do we find the volume of the person?

The mass is given, and if the density of the person were known, the volume could be found ($\rho = m/V$ or $V = m/\rho$). Here is where the estimate comes in. Most people can barely float in water, so the density of the human body is about that of water, $\rho = 1000 \text{ kg/m}^3$. Using this estimate, the buoyant force can also be estimated.

Solution.

Given: $m = 75 \text{ kg}$
$\quad\quad \rho_a = 1.29 \text{ kg/m}^3$ (Table 9.2)
$\quad\quad \rho_p = 1000 \text{ kg/m}^3$ (estimated density
$\quad\quad\quad\quad\quad\quad$ of person)

Find: F_b (buoyant force)

First, let's find the volume of the person,

$$V_p = \frac{m}{\rho_p} = \frac{75 \text{ kg}}{1000 \text{ kg/m}^3} = 0.075 \text{ m}^3$$

Then,

$$F_b = \rho_a gV_p = \rho_a g\left(\frac{m}{\rho_p}\right) = (1.29 \text{ kg/m}^3)(9.8 \text{ m/s}^2)(0.075 \text{ m}^3)$$
$$= 0.95 \text{ N} (\approx 1.0 \text{ N or } 0.225 \text{ lb})$$

Not much when you weigh yourself. But it does mean that your weight is ≈ 0.2 lb more than the scale reading.

Follow-Up Exercise. Estimate the buoyant force on a helium-filled weather balloon that has a diameter on the order of a meteorologist's arm span (arms held horizontally), and compare with the result in the Example.

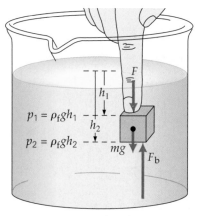

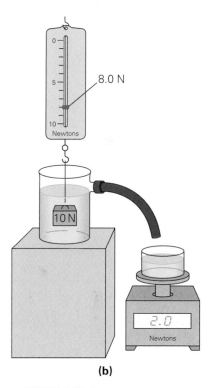

$$\Delta p = \rho_f g(h_2 - h_1)$$

(a)

8.0 N

(b)

▲ **FIGURE 9.13** Buoyancy and Archimedes' principle **(a)** A buoyant force arises from the difference in pressure at different depths. The pressure on the bottom of the submerged block (p_2) is greater than that on the top (p_1), so there is a (buoyant) force directed upward. (It is shifted for clarity.) **(b)** Archimedes' principle: The buoyant force on the object is equal to the weight of the volume of fluid displaced. (The scale is set to read zero when the container is empty.)

Exploration 14.2 Buoyant Force

Integrated Example 9.9 ■ Weight and Buoyant Force: Archimedes' Principle

A container of water with an overflow tube, similar to that shown in Fig. 9.13b, sits on a scale that reads 40 N. The water level is just below the exit tube in the side of the container. (a) An 8.0-N cube of wood is placed in the container. The water displaced by the floating cube runs out the exit tube into another container that is not on the scale. Will the scale reading then be (1) exactly 48 N, (2) between 40 N and 48 N, (3) exactly 40 N, or (4) less than 40 N? (b) Suppose you pushed down on the wooden cube with your finger such that the top surface of the cube was even with the water level. How much force would have to be applied if the wooden cube measured 10 cm on a side?

(a) Conceptual Reasoning. By Archimedes' principle, the block is buoyed upward with a force equal in magnitude to the weight of the water displaced. Since the block floats, the upward buoyant force must balance the weight of the cube and so has a magnitude of 8.0 N. Thus, a volume of water weighing 8.0 N is displaced from the container as 8.0 N of weight is added to the container. The scale still reads 40 N, so the answer is (3).

Note that the upward buoyant force and the block's weight act *on the block*. The reaction force (pressure) of the block *on the water* is transmitted to the bottom of the container (Pascal's principle) and is registered on the scale. (Make a sketch showing the forces on the cube.)

(b) Quantitative Reasoning and Solution. Here three forces are acting on the stationary cube: the buoyant force upward and the weight and the force applied by the finger downward. The weight of the cube is known, so to find the applied finger force, we need to determine the buoyant force on the cube.

Given: $\ell = 10$ cm $= 0.10$ m (side length of cube)   *Find:* downward applied force
 $w = 8.0$ N (weight of cube)   necessary to put cube even with water level

The summation of the forces acting on the cube is $\Sigma F_i = +F_b - w - F_f = 0$, where F_b is the upward buoyant force and F_f is the downward force applied by the finger. Hence, $F_f = F_b - w$. As we know, the magnitude of the buoyant force is equal to the weight of the water the cube displaces, which is given by $F_b = \rho_f g V_f$ (Eq. 9.14). The density of the fluid is that of water, which is known (1.0×10^3 kg/m^3, Table 9.2), so

$$F_b = \rho_f g V_f = (1.0 \times 10^3 \text{ kg/m}^3)(9.8 \text{ m/s}^2)(0.10 \text{ m})^3 = 9.8 \text{ N}$$

Thus,

$$F_f = F_b - w = 9.8 \text{ N} - 8.0 \text{ N} = 1.8 \text{ N}$$

Follow-Up Exercise. In part (a), would the scale still read 40 N if the object had a density greater than that of water? In part (b), what would the scale read?

Demonstration/activity: Push an object, such as a piece of wood, that is lighter than water, into a container of water. (Do not submerge the object.) As the object is pushed deeper, the reaction force of the water becomes greater.

Demonstration/activity: An indoor hot-air balloon can be constructed with a very thin plastic bag of the kind used by dry cleaners to protect clothing. Several small birthday candles mounted on a light frame suspended by threads from the balloon will power it. Be careful with the fire.

Demonstration/activity: Use a demonstration scale to weigh a 1-kg mass both in the air and in water. Explain the difference. Do this experiment with both a steel mass and a brass mass if available. Is there a difference? Does a 1-kg mass *really* weigh 9.8 N in air?

Buoyancy and Density

We commonly say that helium and hot-air balloons float because they are lighter than air. To be technically correct, we should say they are *less dense than air*. An object's density will tell you whether it will sink or float in a fluid, as long as you also know the density of the fluid. Consider a solid uniform object that is totally immersed in a fluid. The weight of the object is

$$w_o = m_o g = \rho_o V_o g$$

The weight of the volume of fluid displaced, or the magnitude of the buoyant force, is

$$F_b = w_f = m_f g = \rho_f V_f g$$

If the object is *completely submerged*, $V_f = V_o$. Dividing the second equation by the first gives

$$\frac{F_b}{w_o} = \frac{\rho_f}{\rho_o} \quad \text{or} \quad F_b = \left(\frac{\rho_f}{\rho_o}\right)w_o \quad \text{(object completely submerged)} \quad (9.15)$$

Thus, if ρ_o is less than ρ_f, then F_b will be greater than w_o, and the object will be buoyed to the surface and float. If ρ_o is greater than ρ_f, then F_b will be less than w_o, and the object will sink. If ρ_o equals ρ_f, then F_b will be equal to w_o, and the object will remain in equilibrium at any submerged depth (as long as the density of the fluid is constant). If the object is not uniform, so that its density varies over its volume, then the density of the object in Eq. 9.15 means average density.

Expressed in words, these three conditions are as follows:

> An object will float in a fluid if the average density of the object is less than the density of the fluid ($\rho_o < \rho_f$).
> An object will sink in a fluid if the average density of the object is greater than the density of the fluid ($\rho_o > \rho_f$).
> An object will be in equilibrium at any submerged depth in a fluid if the average density of the object and the density of the fluid are equal ($\rho_o = \rho_f$).

See ▸Fig. 9.14 for an example of the last condition.

A quick look at Table 9.2 will tell you whether an object will float in a fluid, regardless of the shape or volume of the object. The three conditions just stated also apply to a fluid in a fluid, provided that the two are immiscible (do not mix). For example, you might think that cream is "heavier" than skim milk, but that's not so: Since cream floats on milk, it is less dense than milk.

In general, the densities of objects or fluids will be assumed to be uniform and constant in this book. (The density of the atmosphere does vary with altitude, but is relatively constant near the surface of the Earth.) In any event, in practical applications it is the *average* density of an object that often matters with regard to floating and sinking. For example, an ocean liner is, on average, less dense than water, even though it is made of steel. Most of its volume is occupied by air, so the liner's average density is less than that of water. Similarly, the human body has air-filled spaces, so most of us float in water. The surface depth at which a person floats depends on his or her density. (Why?)

In some instances, the overall density of an object is purposefully varied. For example, a submarine submerges by flooding its tanks with seawater (called "taking on ballast"), which increases its average density. When the sub is to surface, the water is pumped out of the tanks, so the average density of the sub becomes less than that of the surrounding seawater.

Similarly, many fish control their depths by using their *swim bladders* or *gas bladders*. A fish changes or maintains buoyancy by regulating the volume of gas in the gas bladder. Maintaining neutral buoyancy (neither rising or sinking) is important because it allows the fish to stay at a particular depth for feeding. Some fish may move up and down in the water in search of food. Instead of using up energy to swim up and down, the fish alters its buoyancy so as to go up and down.

This is accomplished by adjusting the quantities of gas in the gas bladder. Gas is transferred from the gas bladder to the adjoining blood vessels and back again. Deflating the bladder decreases the volume and increases the average density, and the fish sinks. Gas is forced into the surrounding blood vessels and carried away.

Conversely, to inflate the bladder, gases are forced into the bladder from the blood vessels, thereby increasing the volume and decreasing the average density, and the fish rises. These processes are complex, but Archimedes' principle is being applied in a biological setting.

PHYSLET®

Exploration 14.1 Floating and Density

▲ **FIGURE 9.14 Equal densities and buoyancy** This soft drink contains colored gelatin beads that remain suspended for months with virtually no change. What is the density of the beads compared to the density of the drink?

Demonstration/activity: Place ice in a beaker of water filled to the rim. Ask students to predict what will occur when the ice melts. Start this demo at the beginning of a class period.

Exploration 14.3 Buoyancy and Oil on Water

▲ **FIGURE 9.15 The tip of the iceberg** The vast majority of an iceberg's bulk is underneath the water as illustrated here.

Example 9.10 ■ Float or Sink? Comparison of Densities

A uniform solid cube of material 10 cm on each side has a mass of 700 g. (a) Will the cube float in water? (b) If so, how much of its volume would be submerged?

Thinking It Through. (a) The question is whether the density of the material the cube is made of is greater or less than that of water, so we compute the cube's density. (b) If the cube floats, then the buoyant force and the cube's weight are equal. Both of these forces are related to the cube's volume, so we can write them in terms of that volume and equate them.

Solution. It is sometimes convenient to work in cgs units in comparing small quantities. For densities in grams per cubic centimeter divide the values in Table 9.2 by 10^3, or drop the "$\times 10^3$" from the values given for solids and liquids, and replace with "$\times 10^{-3}$" for gases.

Given: $m = 700$ g
$L = 10$ cm
$\rho_{H_2O} = 1.00 \times 10^3$ kg/m^3
$= 1.00$ g/cm^3 (Table 9.2)

Find: (a) Whether the cube will float in water
(b) The percentage of the volume submerged if the cube does float

(a) The density of the cube is

$$\rho_c = \frac{m}{V_c} = \frac{m}{L^3} = \frac{700 \text{ g}}{(10 \text{ cm})^3} = 0.70 \text{ g/cm}^3 < \rho_{H_2O} = 1.00 \text{ g/cm}^3$$

Since ρ_c is less than ρ_{H_2O} the cube will float.

(b) The weight of the cube is $w_c = \rho_c g V_c$. When the cube is floating, it is in equilibrium, which means that its weight is balanced by the buoyant force. That is, $F_b = \rho_{H_2O} g V_{H_2O}$, where V_{H_2O} is the volume of water the submerged part of the cube displaces. Equating the expressions for weight and buoyant force gives

$$\rho_{H_2O} g V_{H_2O} = \rho_c g V_c$$

or

$$\frac{V_{H_2O}}{V_c} = \frac{\rho_c}{\rho_{H_2O}} = \frac{0.70 \text{ g/cm}^3}{1.00 \text{ g/cm}^3} = 0.70$$

Thus, $V_{H_2O} = 0.70 V_c$, and 70% of the cube is submerged.

Follow-Up Exercise. Most of an iceberg floating in the ocean (◄Fig. 9.15) is submerged. What is seen is the proverbial "tip of the iceberg." What percentage of an iceberg's volume is seen above the surface? [*Note:* Icebergs are frozen *fresh* water floating in salty sea water.]

A quantity called specific gravity is related to density. It is commonly used for liquids, but also applies to solids. The **specific gravity (sp. gr.)** of a substance is equal to the ratio of the density of the substance (ρ_s) to the density of water (ρ_{H_2O}) at 4°C, the temperature for maximum density:

$$sp. \ gr. = \frac{\rho_s}{\rho_{H_2O}}$$

Because it is a ratio of densities, specific gravity has no units. In cgs units, $\rho_{H_2O} = 1.00$ g/cm^3, so

$$sp. \ gr. = \frac{\rho_s}{1.00} = \rho_s \quad (\rho_s \text{ in g/cm}^3 \text{ only})$$

That is, the specific gravity of a substance is equal to the numerical value of its density *in cgs units*. For example, if a liquid has a density of 1.5 g/cm^3, its specific gravity is 1.5, which tells you that it is 1.5 times as dense as water. (As pointed out earlier, to get density values for solids and liquids in grams per cubic centimeter, divide the value in Table 9.2 by 10^3.)

9.4 Fluid Dynamics and Bernoulli's Equation

<u>OBJECTIVES:</u> To (a) identify the simplifications used in describing ideal fluid flow, and (b) use the continuity equation and Bernoulli's equation to explain common effects of ideal fluid flow.

In general, fluid motion is difficult to analyze. For example, think of trying to describe the motion of a particle (a molecule, as an approximation) of water in a rushing stream. The overall motion of the stream may be apparent, but a mathematical description of the motion of any one particle of it may be virtually impossible because of eddy currents (small whirlpool motions), the gushing of water over rocks, frictional drag on the stream bottom, and so on. A basic description of fluid flow is conveniently obtained by ignoring such complications and considering an ideal fluid. Actual fluid flow can then be approximated with reference to this simpler theoretical model.

In this simplified approach to fluid dynamics, it is customary to consider four characteristics of an **ideal fluid**. In such a fluid, flow is (1) *steady*, (2) *irrotational*, (3) *nonviscous*, and (4) *incompressible*.

> Condition 1: *Steady flow* means that all the particles of a fluid have the same velocity as they pass a given point.

Steady flow might be called smooth or regular flow. The path of steady flow can be depicted in the form of **streamlines** (▶Fig. 9.16a). Every particle that passes a particular point moves along a streamline. That is, every particle moves along the same path (streamline) as particles that passed by earlier. Streamlines never cross; if they did, a particle would have alternative paths and abrupt changes in its velocity, in which case the flow would not be steady.

Steady flow requires low velocities. For example, steady flow is approximated by the flow relative to a canoe that is gliding slowly through still water. When the flow velocity is high, eddies tend to appear, especially near boundaries, and the flow becomes turbulent as in Fig. 9.16b.

Streamlines also indicate the relative magnitude of the velocity of a fluid. The velocity is greater where the streamlines are closer together. Notice this effect in Fig. 9.16a. The reason for it will be explained shortly.

> Condition 2: *Irrotational flow* means that a fluid element (a small volume of the fluid) has no net angular velocity. This condition eliminates the possibility of whirlpools and eddy currents. (The flow is nonturbulent.)

Consider the small paddle wheel in Fig. 9.16a. With a zero net torque, the wheel does not rotate. Thus, the flow is irrotational.

> Condition 3: *Nonviscous flow* means that viscosity is negligible.

Viscosity refers to a fluid's internal friction, or resistance to flow. (For example, honey has a much greater viscosity than water.) A truly nonviscous fluid would flow freely with no internal energy loss within it. Also, there would be no frictional drag between the fluid and the walls containing it. In reality, when a liquid flows through a pipe, the speed is lower near the walls because of frictional drag and is higher toward the center of the pipe. (Viscosity is discussed in more detail in Section 9.5.)

> Condition 4: *Incompressible flow* means that the fluid's density is constant.

Liquids can usually be considered incompressible. Gases, by contrast, are quite compressible. Sometimes, however, gases approximate incompressible flow—for example, air flowing relative to the wings of an airplane traveling at low speeds. Theoretical or ideal fluid flow is not characteristic of most real situations, but the analysis of ideal flow provides results that approximate, or generally describe, a variety of applications. Usually, this analysis is derived, not from Newton's laws, but instead from two basic principles: conservation of mass and conservation of energy.

Streamlines

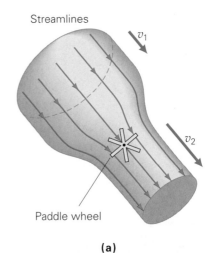

(a)

(b)

▲ **FIGURE 9.16 Streamline flow**
(a) Streamlines never cross and are closer together in regions of greater fluid velocity. The stationary paddle wheel indicates that the flow is irrotational, or without whirlpools and eddy currents. (b) The smoke from an extinguished candle begins to rise in nearly streamline flow, but quickly becomes rotational and turbulent.

▶ **FIGURE 9.17 Flow continuity**
Ideal fluid flow can be described in
terms of the conservation of mass
by the equation of continuity.

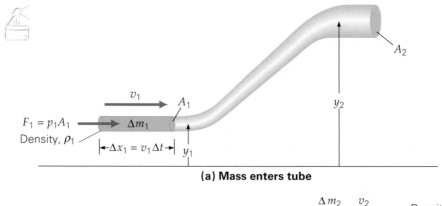

(a) Mass enters tube

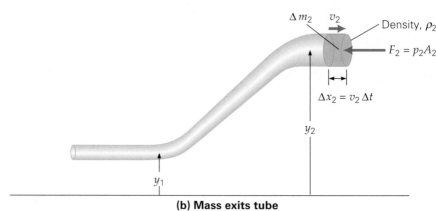

(b) Mass exits tube

Illustration 15.1 *The Continuity Equation*

▲ **FIGURE 9.18 Flow rate** By the
flow rate equation, the speed of a
fluid is greater when the cross-
sectional area of the tube through
which the fluid is flowing is
smaller. Think of a hose that is
equipped with a nozzle such that
the cross-sectional area of the hose
is made smaller.

Equation of Continuity

If there are no losses of fluid within a uniform tube, the mass of fluid flowing into
the tube in a given time must be equal to the mass flowing out of the tube in the
same time (by the conservation of mass). For example, in ▲Fig. 9.17a, the mass
(Δm_1) entering the tube during a short time (Δt) is

$$\Delta m_1 = \rho_1 \Delta V_1 = \rho_1 (A_1 \Delta x_1) = \rho_1 (A_1 v_1 \Delta t)$$

where A_1 is the cross-sectional area of the tube at the entrance and, in a time Δt, a
fluid particle moves a distance equal to $v_1 \Delta t$. Similarly, the mass leaving the tube
in the same interval is (Fig. 9.17b)

$$\Delta m_2 = \rho_2 \Delta V_2 = \rho_2 (A_2 \Delta x_2) = \rho_2 (A_2 v_2 \Delta t)$$

Since the mass is conserved, $\Delta m_1 = \Delta m_2$, and it follows that

$$\rho_1 A_1 v_1 = \rho_2 A_2 v_2 \quad \text{or} \quad \rho A v = \text{constant} \quad (9.16)$$

This general result is called the **equation of continuity**.
   For an incompressible fluid, the density ρ is constant, so

$$A_1 v_1 = A_2 v_2 \quad \text{or} \quad Av = \text{constant} \quad \text{(for an incompressible fluid)} \quad (9.17)$$

This is sometimes called the **flow rate equation**. Av is called the *volume rate of flow*,
and is the volume of fluid that passes by a point in the tube per unit time.
(Av: m² · m/s = m³/s, volume per time.)
   Note that the flow rate equation shows that the fluid velocity is greater where
the cross-sectional area of the tube is smaller. That is,

$$v_2 = \left(\frac{A_1}{A_2}\right) v_1$$

and v_2 is greater than v_1 if A_2 is less than A_1. This effect is evident in the common
experience that the speed of water is greater from a hose fitted with a nozzle than
from the same hose without a nozzle (◀Fig. 9.18).

The flow rate equation can be applied to the flow of blood in your body. Blood flows from the heart into the aorta. It then makes a circuit through the circulatory system, passing through arteries, arterioles (small arteries), capillaries, and venules (small veins) and back to the heart through veins. The speed is lowest in the capillaries. Is this a contradiction? No: The *total* area of the capillaries is much larger than that of the arteries or veins, so the flow rate equation is still valid.

Example 9.11 ■ Blood Flow: Cholesterol and Plaque

High cholesterol in the blood can cause fatty deposits called plaques to form on the walls of blood vessels. Suppose a plaque reduces the effective radius of an artery by 25%. How does this partial blockage affect the speed of blood through the artery?

Thinking It Through. The flow rate equation (Eq. 9.17) applies, but note that no values of area or speed are given. This indicates that we should use ratios.

Solution. Taking the unclogged artery to have a radius r_1, we can say that the plaque then reduces the effective radius to r_2.

Given: $r_2 = 0.75r_1$ (for a 25% reduction)   *Find:* v_2

Writing the flow rate equation in terms of the radii, we have

$$A_1 v_1 = A_2 v_2$$
$$(\pi r_1^2)v_1 = (\pi r_2^2)v_2$$

Rearranging and canceling,

$$v_2 = \left(\frac{r_1}{r_2}\right)^2 v_1$$

From the given information, $r_1/r_2 = 1/0.75$, so

$$v_2 = (1/0.75)^2 v_1 = 1.8v_1$$

Hence, the speed through the clogged artery increases by 80%.

Follow-Up Exercise. By how much would the effective radius of an artery have to be reduced to have a 50% increase in the speed of the blood flowing through it?

Example 9.12 ■ Speed of Blood in the Aorta

Blood flows at a rate of 5.00 L/min through an aorta with a radius of 1.00 cm. What is the speed of blood flow in the aorta?

Thinking It Through. It is noted that the flow rate is a volume flow rate, which implies the use of the flow rate equation (Eq. 9.17), $Av = $ constant. Since the constant is in terms of volume/time, the given flow rate is the constant.

Solution. Listing the data:

Given: Flow rate $= 5.00$ L/min   *Find:* v (blood speed)
          $r = 1.00$ cm $= 10^{-2}$ m

Let's first find the cross-sectional area of the circular aorta.

$$A = \pi r^2 = (3.14)(10^{-2}\,\text{m})^2 = 3.14 \times 10^{-4}\,\text{m}^2$$

Then the (volume) flow rate needs to be put into standard units.

$$5.00\,\text{L/min} = (5.00\,\text{L/m})(10^{-3}\,\text{m}^3/\text{L})(1\,\text{min}/60\,\text{s}) = 8.33 \times 10^{-5}\,\text{m}^3/\text{s}$$

Using the flow rate equation, we have

$$v = \frac{\text{constant}}{A} = \frac{8.33 \times 10^{-5}\,\text{m}^3/\text{s}}{3.14 \times 10^{-4}\,\text{m}^2} = 0.265\,\text{m/s}$$

Follow-Up Exercise. Constrictions of the arteries occur with hardening of the arteries. If the radius of the aorta in this Example were constricted to 0.900 cm, what would be the percentage change in blood flow?

PHYSLET®

Exploration 15.1 *Blood Flow and the Continuity Equation*

▶ **FIGURE 9.19 Flow rate and pressure** Taking the horizontal difference in flow heights to be negligible in a constricted pipe, we obtain, for Bernoulli's equation, $p + \frac{1}{2}\rho v^2 = $ constant. In a region of smaller cross-sectional area, the flow speed is greater (see flow rate equation); from Bernoulli's equation, the pressure in that region is lower than in other regions.

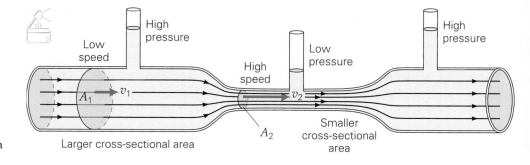

PHYSLET®

Exploration 15.2 Bernoulli's Equation

Bernoulli's Equation

The conservation of energy or the general work–energy theorem leads to another relationship that has great generality for fluid flow. This relationship was first derived in 1738 by the Swiss mathematician Daniel Bernoulli (1700–1782) and is named for him. Bernoulli's result was

$$W_{\text{net}} = \Delta K + \Delta U$$

$$\frac{\Delta m}{\rho}(p_1 - p_2) = \frac{1}{2}\Delta m(v_2^2 - v_1^2) + \Delta mg(y_2 - y_1)$$

where Δm is a mass increment as in the derivation of the continuity equation.

Teaching tip: Emphasize that Bernoulli's equation is simply an application of the law of conservation of energy.

Note that in working with a fluid, the terms in Bernoulli's equation are work or energy per unit volume (J/m^3). That is, $W = F\Delta x = p(A\Delta x) = p\Delta V$ and therefore $p = W/\Delta V$ (work/volume). Similarly, with $\rho = m/V$, we have $\frac{1}{2}\rho v^2 = \frac{1}{2}mv^2/V$ (energy/volume) and $\rho gy = mgy/V$ (energy/volume).

Canceling each Δm and rearranging gives the common form of **Bernoulli's equation**:

$$p_1 + \frac{1}{2}\rho v_1^2 + \rho gy_1 = p_2 + \frac{1}{2}\rho v_2^2 + \rho gy_2 \qquad (9.18)$$

or

$$p + \frac{1}{2}\rho v^2 + \rho gy = \text{constant}$$

Note: Compare the derivation of Eq. 5.10 in Section 5.5.

Bernoulli's equation, or principle, can be applied to many situations. For example, for a fluid at rest ($v_2 = v_1 = 0$), Bernoulli's equation becomes

$$p_2 - p_1 = \rho g(y_1 - y_2)$$

This is the pressure–depth relationship derived earlier (Eq. 9.10). Also, if there is horizontal flow ($y_1 = y_2$), then $p + \frac{1}{2}\rho v^2 = $ constant, which indicates that the pressure decreases if the speed of the fluid increases (and vice versa). This effect is illustrated in ▲Fig. 9.19, where the difference in flow heights through the pipe is considered negligible (so the ρgy term drops out).

Chimneys and smokestacks are tall in order to take advantage of the more consistent and higher wind speeds at greater heights. The faster the wind blows over the top of a chimney, the lower the pressure, and the greater the pressure difference between the bottom and top of the chimney. Thus, the chimney draws exhaust out better. Bernoulli's equation and the continuity equation ($Av = $ constant) also tell you that if the cross-sectional area of a pipe is reduced, so that the velocity of the fluid passing through it is increased, then the pressure is reduced.

The Bernoulli effect (as it is sometimes called) gives a *simplistic* explanation for the lift of an airplane. Ideal airflow over an airfoil or wing is shown in ◀Fig. 9.20. (Turbulence is neglected.) The wing is curved on the top side and is angled relative to the incident streamlines. As a result, the streamlines above the wing are closer together than those below, which cause a higher air speed and lower pressure above the wing. With a higher pressure on the bottom of the wing, there is a net upward force, or *lift*.

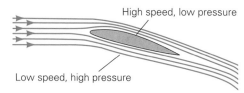

▲ **FIGURE 9.20 Airplane lift—Bernoulli's principle in action** Because of the shape and orientation of an airfoil or airplane wing, the air streamlines are closer together, and the air speed is greater above the wing than below it. By Bernoulli's principle, the resulting pressure difference supplies part of the upward force called the lift.

This rather common explanation of lift is termed simplistic because Bernoulli's effect does not apply to the situation. Bernoulli's principle requires ideal fluid flow and energy conservation within the system, neither of which are satisfied in aircraft flying conditions. It is perhaps better to rely on Newton's laws, which always must be satisfied. Basically overall, the wing deflects the airflow downward, giving rise to a downward change in the airflow momentum and a downward force (Newton's second law). This results in an upward reaction force on the wing (Newton's third law). When this upward force exceeds the weight of the plane, you have enough lift for take off and flight.

Teaching tip: Explain how lift affects an airplane.

Illustration 15.4 Airplane Lift

Example 9.13 ■ Flow Rate from a Tank: Bernoulli's Equation

A cylindrical tank containing water has a small hole punched in its side below the water level, and water runs out (▼Fig. 9.21). What is the approximate initial flow rate of water out of the tank?

Thinking It Through. Equation 9.17 ($A_1v_1 = A_2v_2$) is the flow rate equation, where Av has units of m³/s, or volume/time. The v terms can be related by Bernoulli's equation, which also contains y, which can be used to find differences in height. The areas are not given, so relating the v terms might require some sort of approximation, as will be seen. (Note that the *approximate* initial flow rate is wanted.)

Solution.

Given: No specific values are given, so symbols will be used.

Find An expression for the approximate initial water flow rate from the hole

Bernoulli's equation,

$$p_1 + \tfrac{1}{2}\rho v_1^2 + \rho g y_1 = p_2 + \tfrac{1}{2}\rho v_2^2 + \rho g y_2$$

can be used. Recall that $y_2 - y_1$ is just the height of the surface of the liquid above the hole. The atmospheric pressures acting on the open surface and at the hole, p_1 and p_2, respectively, are essentially equal and cancel from the equation, as does the density, so we are left with

$$v_1^2 - v_2^2 = 2g(y_2 - y_1)$$

By the equation of continuity (the flow rate equation, Eq. 9.17), $A_1v_1 = A_2v_2$, where A_2 is the cross-sectional area of the tank and A_1 is that of the hole. Since A_2 is much greater than A_1, v_1 is much greater than v_2 (initially, $v_2 \approx 0$). So, to a good approximation,

$$v_1^2 = 2g(y_2 - y_1) \quad \text{or} \quad v_1 = \sqrt{2g(y_2 - y_1)}$$

The flow rate (volume/time) is then

$$\text{flow rate} = A_1v_1 = A_1\sqrt{2g(y_2 - y_1)}$$

Given the area of the hole and the height of the liquid above it, you can find the initial speed of the water coming from the hole and the flow rate. (What happens as the water level falls?)

Follow-Up Exercise. What would be the percentage change in the initial flow rate from the tank in this Example if the diameter of the small circular hole were increased by 30.0%?

Illustration 15.2 Bernoulli's Principle at Work

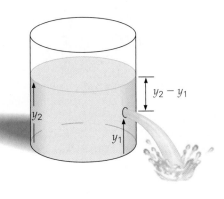

◀ **FIGURE 9.21 Fluid flow from a tank** The flow rate is given by Bernoulli's equation. See Example 9.13.

Exploration 15.3 Application of Bernoulli's Equation

Conceptual Example 9.14 ■ A Stream of Water: Smaller and Smaller

You have probably observed that a small steady stream of water flowing out of a kitchen faucet gets smaller the farther the water falls from the faucet. Why does that happen?

Reasoning and Answer. This effect can be explained by Bernoulli's principle. As the water falls, it accelerates and its speed increases. Then, by Bernoulli's principle, the liquid pressure inside the stream decreases. (See Fig. 9.19.) A pressure difference between that inside stream and the atmospheric pressure on the outside is thus created. As a result, there is an increasing inward force as the stream falls, so it becomes smaller. Eventually, the stream may get so thin that it breaks up into individual droplets.

Follow-Up Exercise. The equation of continuity can also be used to explain this stream effect. Give this explanation.

*9.5 Surface Tension, Viscosity, and Poiseuille's Law

OBJECTIVES: To (a) describe the source of surface tension, and (b) discuss fluid viscosity.

Surface Tension

The molecules of a liquid exert small attractive forces on each other. Even though molecules are electrically neutral overall, there is often some slight asymmetry of charge that gives rise to attractive forces between them (called *van der Waals forces*). Within a liquid, any molecule is completely surrounded by other molecules, and the net force is zero (▼Fig. 9.22a). However, for molecules at the surface of the liquid, there is no attractive force acting from above the surface. (The effect of air molecules is small and considered negligible.) As a result, net forces act upon the molecules of the surface layer, due to the attraction of neighboring molecules just below the surface. This inward pull on the surface molecules causes the surface of the liquid to contract and to resist being stretched or broken, a property called **surface tension**.

If a sewing needle is carefully placed on the surface of a bowl of water, the surface acts like an elastic membrane under tension. There is a slight depression in the surface, and molecular forces along the depression act at an angle to the surface (Fig. 9.22b). The vertical components of these forces balance the weight (mg) of the needle, and the needle "floats" on the surface. Similarly, surface tension supports the weight of a water strider (Fig. 9.22c).

The net effect of surface tension is to make the surface area of a liquid as small as possible. That is, a given volume of liquid tends to assume the shape that has the least surface area. As a result, drops of water and soap bubbles have spherical shapes, because a sphere has the smallest surface area for a given volume (▼Fig. 9.23). In forming a drop or bubble, surface tension pulls the molecules together to minimize the surface area. (See Insight 9.4 on The Lungs and Baby's First Breath for an example of surface tension in respiration.)

Demonstration/activity: The following are some demonstrations of the Bernoulli effect:
- Suspend two lightbulbs from strings, with the bulbs 2 cm apart. Blow air between the bulbs.
- Place an index card on top of a lab table and have students try to move the card by blowing.
- Use an air track blower to blow a stream of air vertically. Place a light ball in the stream, where it will levitate in a stable position.
- Throw a spinning Ping-Pong ball, plastic beach ball, and so on.

▼ FIGURE 9.22 Surface tension
(a) The net force on a molecule in the interior of a liquid is zero, because the molecule is surrounded by other molecules. However, a nonzero net force acts on a molecule at the surface, due to the attractive forces of the neighboring molecules just below the surface. **(b)** For an object such as a needle to form a depression on the surface, work must be done, since more interior molecules must be brought to the surface to increase its area. As a result, the surface area acts like a stretched elastic membrane, and the weight of the object is supported by the upward components of the surface tension. **(c)** Insects such as this water strider can walk on water because of the upward components of the surface tension, much as you might walk on a large trampoline. Note the depressions in the surface of the liquid where the legs touch it.

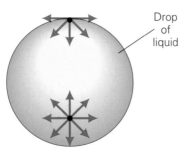

(a)

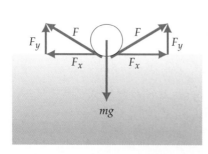

(b)

(c)

INSIGHT   9.4 THE LUNGS AND BABY'S FIRST BREATH

Respiration, or breathing, is vital to life. It is a fascinating procedure that supplies oxygen to the blood and carries away carbon dioxide—and a lot of physics is involved.

The process of respiration involves the lowering of the diaphragm to increase the volume of the thoracic cavity. Figure 1 shows a bell-jar model of respiration. By the ideal gas law (Section 10.3), the lowering of the diaphragm and increasing the volume of the thoracic cavity lowers the pressure ($p \propto 1/V$), and air is inhaled. The inhalation process inflates the alveoli—small balloonlike structures in the lungs, as illustrated in Fig. 2a. (Figure 2b shows an illustration of a damaged lung, the cause and effects of which will be discussed shortly.)

The oxygen exchange with the blood takes place across the membrane surfaces of the alveoli. The total membrane surface in the lungs is on the order of 100 m², with a thickness of less than a millionth of a meter ($< 1 \ \mu m$, micrometer), making the gas exchange is very efficient. The behavior of the alveoli may be described by Laplace's law and surface tension.*

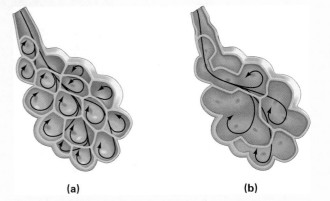

(a)   **(b)**

FIGURE 2 Alveoli (a) Inhalation inflates the alveoli, balloonlike structures of the lungs. There are between 300 million and 400 million alveoli in each lung. **(b)** Pulmonary disease can cause the enlargement of the alveoli as some are destroyed and others enlarge or combine. As a result, there is less oxygen exchange and a shortness of breath.

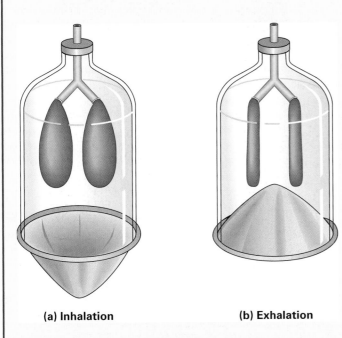

(a) Inhalation   **(b) Exhalation**

FIGURE 1 Bell-jar model of respiration (a) Lowering the diaphragm (rubber sheet) and increasing the volume of the thoracic cavity lowers the pressure and air is inhaled into the lungs (balloons). **(b)** When the diaphragm moves upward, the process is reversed and air is exhaled.

*Pierre-Simon de Laplace (1749–1827) was a French astronomer and mathematician.

Laplace's law states that the larger a spherical membrane, the greater the wall tension required to withstand the pressure of an internal fluid. That is, the wall tension is directly proportional to the spherical radius. So when the alveoli inflate, there is greater tension. Once they are inflated, exhalation is accomplished when the diaphragm relaxes and the wall tension of the alveoli acts to force the air out. Also, there is a fluid coating on the alveoli, which is a *surfactant*—a substance that lowers surface tension. A reduction in surface tension makes it easier to inflate the alveoli on inhalation.

The pulmonary disease *emphysema*, most common in long-term smokers, results from an enlargement of the alveoli as some are destroyed and others either enlarge or combine (Fig. 2b). Normally it would take twice the pressure to inflate a membrane with twice the radius. The enlarged alveoli provide less recoil on exhalation, and a person with emphysema has difficulty breathing as well as reduced oxygen exchange.

Now, about the baby's first breath. Most everyone knows that it is much more difficult to blow up a balloon for the first time than to blow it up again. This is because the applied pressure does not create much tension in the balloon to start the stretching process. According to Laplace's law, it would take a greater tension increase to expand a small balloon than to expand a large balloon. Consider the tension ratios for a 3-cm expansion in radius—say, from 1 cm to 4 cm $\left(\frac{4}{1} = 4\right)$ and 10 cm to 13 cm $\left(\frac{13}{10} = 1.3\right)$.

In a newborn baby, the alveoli are small and collapsed and must be inflated with an initial inhalation. The traditional practice to accomplish this are slaps on the baby's bottom to make the newborn cry and inhale.

Viscosity

All real fluids have an internal resistance to flow, or **viscosity**, which can be considered to be friction between the molecules of a fluid. In liquids, viscosity is caused by short-range cohesive forces, and in gases, it is caused by collisions between molecules. (See the discussion of air resistance in Section 4.6.) The viscous

▶ **FIGURE 9.23 Surface tension at work** Because of surface tension, **(a)** water droplets and **(b)** soap bubbles tend to assume the shape that minimizes their surface area—that of a sphere.

(a)

(b)

drag for both liquids and gases depends on their velocity and may be directly proportional to it in some cases. However, the relationship varies with the conditions; for example, the drag is approximately proportional to either v^2 or v^3 in turbulent flow.

Internal friction causes the layers of a fluid to move relative to each other in response to a shear stress. This layered motion, called *laminar flow*, is characteristic of steady flow for viscous liquids at low velocities (▼Fig. 9.24a). At higher velocities, the flow becomes rotational, or *turbulent*, and difficult to analyze.

Since there are shear stresses and shear strains (deformation) in laminar flow, the viscous property of a fluid can be described by a coefficient, like the elastic moduli discussed in Section 9.1. Viscosity is characterized by a *coefficient of viscosity*, η (the Greek letter eta), commonly referred to as simply the viscosity.

The coefficient of viscosity is, in effect, the ratio of the shear stress to the rate of change of the shear strain (since motion is involved). Unit analysis shows that the SI unit of viscosity is the pascal-second (Pa·s). This combined unit is called the *poiseuille* (Pl), in honor of the French scientist Jean Poiseuille (1797–1869), who studied the flow of liquids, particularly blood. (Poiseuille's law on flow rate will be presented shortly.) The cgs unit of viscosity is the *poise* (P). A smaller multiple, the *centipoise* (cP), is widely used because of its convenient size; $1\ P = 10^2\ cP$.

The viscosities of some fluids are listed in Table 9.3. The greater the viscosity of a liquid, which is easier to visualize than that of a gas, the greater is the shear stress required to get the layers of the liquid to slide along each other. Note, for example, the large viscosity of glycerin compared to that of water.*

▶ **FIGURE 9.24 Laminar flow** **(a)** A shear stress causes layers of a fluid to move over each other in laminar flow. The shear force and the flow rate depend on the viscosity of the fluid. **(b)** For laminar flow through a pipe, the speed of the fluid is less near the walls of the pipe than near the center because of frictional drag between the walls and the fluid.

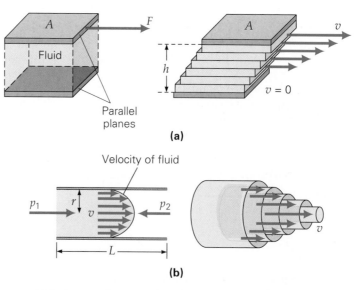

*If you want to think about a substance with a very large viscosity, consider that of glass. It has been said that the glass in the stained-glass windows of medieval churches has "flowed" over time, such that the panes are now thicker at the bottom than at the top. A recent analysis indicates that window glass may even flow over incredibly long periods that exceed the limits of human history. On human time scales, such a flow would not be evident. [See E. D. Zanotto, *American Journal of Physics*, 66 (May 1998), 392–395.]

TABLE 9.3	Viscosities of Various Fluids*
	Viscosity (η)
Fluid	Poiseuille (Pl)
Liquids	
Alcohol, ethyl	1.2×10^{-3}
Blood, whole (37°C)	1.7×10^{-3}
Blood plasma (37°C)	2.5×10^{-3}
Glycerin	1.5×10^{-3}
Mercury	1.55×10^{-3}
Oil, light machine	1.1
Water	1.00×10^{-3}
Gases	
Air	1.9×10^{-5}
Oxygen	2.2×10^{-5}

*At 20°C unless otherwise indicated.

As you might expect, viscosity, and thus fluid flow, varies with temperature, which is evident from the old saying, "slow as molasses in January." A familiar application is the viscosity grading of motor oil used in automobiles. In winter, a low-viscosity, or relatively thin, oil should be used (such as SAE grade 10W or 20W), because it will flow more readily, particularly when the engine is cold at start-up. In summer, a higher-viscosity, or thicker, oil is used (SAE 30, 40, or even 50). Seasonal changes in the grade of motor oil are not necessary if you use the multigrade, year-round oils. These oils contain additives called viscosity improvers, which are polymers whose molecules are long, coiled chains. An increase in temperature causes the molecules to uncoil and intertwine. Thus, the normal decrease in viscosity is counteracted. The action is reversed on cooling, and the oil maintains a relatively small viscosity range over a large temperature range. Such motor oils are graded, for example, as SAE 10W-30 (or 10W-30, for short).

Note: *SAE* stands for *Society of Automotive Engineers*, an organization that designates the grades of motor oils based on their viscosity.

Poiseuille's Law

Viscosity makes analyzing fluid flow difficult. For example, when a fluid flows through a pipe, there is frictional drag between the liquid and the walls, and the fluid velocity is greater toward the center of the pipe (Fig. 9.24b). In practice, this effect makes a difference in a fluid's *average flow rate* $Q = A\bar{v} = \Delta V/\Delta t$ (see Eq. 9.17), which describes the volume (ΔV) of fluid flowing past a given point during a time Δt. The SI unit of flow rate is cubic meters per second (m³/s). The flow rate depends on the properties of the fluid and the dimensions of the pipe, as well as on the pressure difference (Δp) between the ends of the pipe.

Jean Poiseuille studied flow in pipes and tubes, assuming a constant viscosity and steady or laminar flow. He derived the following relationship, known as **Poiseuille's law**, for the flow rate:

Illustration 15.3   Ideal and Viscous Fluid Flow

$$Q = \frac{\Delta V}{\Delta t} = \frac{\pi r^4 \Delta p}{8\eta L} \qquad (9.19)$$

Here, r is the radius of the pipe and L is its length.

As expected, the flow rate is inversely proportional to the viscosity (η) and the length of the pipe. Also as expected, the flow rate is directly proportional to the pressure difference Δp between the ends of the pipe. Somewhat surprisingly, however, the flow rate is proportional to r^4, which makes it more highly dependent on the radius of the tube than we might have thought.

An application of fluid flow in a medical IV was examined in Example 9.6. However, Poiseuille's law, which incorporates the flow rate, affords more reality to this application, as the next Example shows.

Example 9.15 ■ Poiseuille's Law: A Blood Transfusion

A hospital patient needs a blood IV (intravenous) transfusion, which will be administered through a vein in the arm via a gravity IV. The physician wishes to have 500 cc of whole blood delivered over a period of 10 min by an 18-gauge needle with a length of 50 mm and an inner diameter of 1.0 mm. At what height above the arm should the bag of blood be hung? (Assume a venous blood pressure of 15 mm Hg.)

Thinking It Through. This is an application of Poiseuille's law (Eq. 9.19) to find the pressure needed at the inlet of the needle that will provide the required flow rate (Q). Note that $\Delta p = p_{in} - p_{out}$ (inlet pressure minus outlet pressure). Knowing the inlet pressure, we can find the required height of the bag as in Example 9.6. (*Caution:* There are a lot of nonstandard units here, and some quantities are assumed to be known from tables.)

Solution. First we write the given (and known) quantities, converting to standard SI units as we go:

Given: $\Delta V = 500 \text{ cc} = 500 \text{ cm}^3 \ (1 \text{ m}^3/10^6 \text{ cm}^3)$ *Find:* h (height of bag)
$\qquad\qquad = 5.00 \times 10^{-4} \text{ m}^3$
$\qquad \Delta t = 10 \text{ min} = 600 \text{ s} = 6.00 \times 10^2 \text{ s}$
$\qquad\quad L = 50 \text{ mm} = 5.0 \times 10^{-2} \text{ m}$
$\qquad\quad d = 1.0 \text{ mm, or } r = 0.50 \text{ mm} = 5.0 \times 10^{-4} \text{ m}$
$\qquad p_{out} = 15 \text{ mm Hg} = 15 \text{ torr } (133 \text{ Pa/torr}) = 2.0 \times 10^3 \text{ Pa}$
$\qquad\quad \eta = 1.7 \times 10^{-3} \text{ Pl (whole blood, from Table 9.3)}$

The flow rate is

$$Q = \frac{\Delta V}{\Delta t} = \frac{5.00 \times 10^{-4} \text{ m}^3}{6.00 \times 10^2 \text{ s}} = 8.33 \times 10^{-7} \text{ m}^3/\text{s}$$

We insert this number into Eq. 9.19 and solve for Δp:

$$\Delta p = \frac{8\eta L Q}{\pi r^4} = \frac{8(1.7 \times 10^{-3} \text{ Pl})(5.0 \times 10^{-2} \text{ m})(8.33 \times 10^{-7} \text{ m}^3/\text{s})}{\pi(5.0 \times 10^{-4} \text{ m})^4} = 2.9 \times 10^3 \text{ Pa}$$

With $\Delta p = p_{in} - p_{out}$, we have

$$p_{in} = \Delta p + p_{out} = (2.9 \times 10^3 \text{ Pa}) + (2.0 \times 10^3 \text{ Pa}) = 4.9 \times 10^3 \text{ Pa}$$

Then, to find the height of the bag that will deliver this amount of pressure, we use $p_{in} = \rho g h$ (where $\rho_{whole \ blood} = 1.05 \times 10^3 \text{ kg/m}^3$, from Table 9.2). Thus,

$$h = \frac{p_{in}}{\rho g} = \frac{4.9 \times 10^3 \text{ Pa}}{(1.05 \times 10^3 \text{ kg/m}^3)(9.80 \text{ m/s}^2)} = 0.48 \text{ m}$$

Hence, for the prescribed flow rate, the bag of blood should be hung about 48 cm above the needle in the arm.

Follow-Up Exercise. Suppose the physician wants to follow up the blood transfusion with 500 cc of saline solution at the same rate of flow. At what height should the saline bag be placed? (The *isotonic* saline solution administered by IV is a 0.85% aqueous salt solution, which has the same salt concentration as do body cells. To a good approximation, saline has the same density as water.)

Gravity-flow IVs are still used, but with modern technology, the flow rates of IVs are now often controlled and monitored by machines (◄Fig. 9.25).

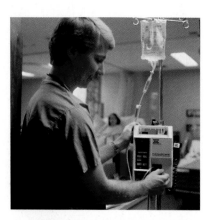

▲ **FIGURE 9.25 IV technology** The mechanism of intravenous injection is still a gravity assist, but IV flow rates are now commonly controlled and monitored by machines.

Chapter Review

• In the deformation of elastic solids, **stress** is a measure of the force causing the deformation:

$$\text{stress} = \frac{F}{A} \qquad (9.1)$$

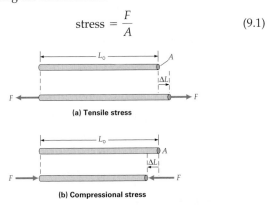

(a) Tensile stress

(b) Compressional stress

Strain is a relative measure of the deformation a stress causes:

$$\text{strain} = \frac{\text{change in length}}{\text{original length}} = \frac{|\Delta L|}{L_o} = \frac{|L - L_o|}{L_o} \qquad (9.2)$$

• An **elastic modulus** is the ratio of stress to strain.

Young's modulus:

$$Y = \frac{F/A}{\Delta L/L_o} \qquad (9.4)$$

Shear modulus:

$$S = \frac{F/A}{x/h} \approx \frac{F/A}{\phi} \qquad (9.5)$$

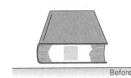

Before

Before

After

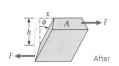

After

Bulk modulus:

$$B = \frac{F/A}{-\Delta V/V_o} = -\frac{\Delta p}{\Delta V/V_o} \qquad (9.6)$$

• Pressure is the force per unit area.

$$p = \frac{F}{A} \qquad (9.8a)$$

Pressure–depth relationship (for an incompressible fluid at constant density):

$$p = p_o + \rho g h \qquad (9.10)$$

• **Pascal's principle**. Pressure applied to an enclosed fluid is transmitted undiminished to every point in the fluid and to the walls of the container.

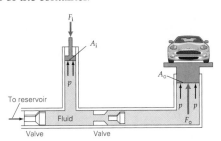

• **Archimedes' principle**. A body immersed wholly or partially in a fluid is buoyed up by a force equal in magnitude to the weight of the volume of fluid displaced.

Buoyant force:

$$F_b = m_f g = \rho_f g V_f \qquad (9.14)$$

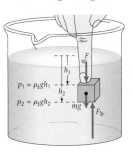

• An object will float in a fluid if the average density of the object is less than the density of the fluid. If the average density of the object is greater than the density of the fluid, the object will sink.

• For an ideal fluid, the flow is (1) steady, (2) irrotational, (3) nonviscous, and (4) incompressible. The following equations describe such a flow:

Equation of continuity:

$$\rho_1 A_1 v_1 = \rho_2 A_2 v_2 \qquad \text{or} \qquad \rho A v = \text{constant} \qquad (9.16)$$

Flow rate equation (for an incompressible fluid):

$$A_1 v_1 = A_2 v_2 \qquad \text{or} \qquad A v = \text{constant} \qquad (9.17)$$

Bernoulli's equation (for an incompressible fluid):

$$p_1 + \tfrac{1}{2}\rho v_1^2 + \rho g y_1 = p_2 + \tfrac{1}{2}\rho v_2^2 + \rho g y_2$$

or

$$p + \tfrac{1}{2}\rho v^2 + \rho g y = \text{constant} \qquad (9.18)$$

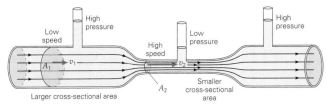

• Bernoulli's equation is a statement of the conservation of energy for a fluid.

• **Surface tension**: The inward pull on the surface molecules of a liquid that causes the surface to contract and resist being stretched or broken.

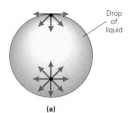

(a)

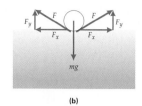

(b)

• **Viscosity**: a fluid's internal resistance to flow. All real fluids have a nonzero viscosity.

Poiseuille's law (flow rate in pipes and tubes for fluids with constant viscosity and steady or laminar flow):

$$Q = \frac{\Delta V}{\Delta t} = \frac{\pi r^4 \Delta p}{8\eta L} \qquad (9.19)$$

Exercises

MC = *Multiple Choice Question,* **CQ** = *Conceptual Question, and* **IE** = *Integrated Exercise. Throughout the text, many exercise sections will include "paired" exercises. These exercise pairs, identified with* **red numbers**, *are intended to assist you in problem solving and learning. In a pair, the first exercise (even numbered) is worked out in the Study Guide so that you can consult it should you need assistance in solving it. The second exercise (odd numbered) is similar in nature, and its answer is given at the back of the book.*

9.1 Solids and Elastic Moduli

Use as many significant figures as you need to show small changes.

1. **MC** The pressure on an elastic body is described by (a) a modulus, (b) work, (c) stress, (d) strain. (c)

2. **MC** Shear moduli are not zero for (a) solids, (b) liquids, (c) gases, (d) all of these. (a)

3. **MC** A relative measure of deformation is (a) a modulus, (b) work, (c) stress, (d) strain. (d)

4. **MC** The volume stress for the bulk modulus is (a) Δp, (b) ΔV, (c) V_0, (d) $\Delta V/V_0$. (a)

5. **CQ** Which has a greater Young's modulus, a steel wire or a rubber band? Explain. steel wire

6. **CQ** Why are scissors sometimes called shears? Is this a descriptive name in the physical sense? see ISM

7. **CQ** Ancient stonemasons sometimes split huge blocks of rock by inserting wooden pegs into holes drilled in the rock and then pouring water on the pegs. Can you explain the physics that underlies this technique? [*Hint:* Think about sponges and paper towels.] see ISM

8. ● A tennis racket has nylon strings. If one of the strings with a diameter of 1.0 mm is under a tension of 15 N, how much is it lengthened from its original length of 40 cm? 0.0015 m

9. ● Suppose you use the tip of one finger to support a 1.0-kg object. If your finger has a diameter of 2.0 cm, what is the stress on your finger? $3.1 \times 10^4 \text{ N/m}^2$

10. ● A 5.0-m-long rod is stretched 0.10 m by a force. What is the strain in the rod? 0.020

11. ● A 250-N force is applied at a 37° angle to the surface of the end of a square bar. The surface is 4.0 cm on a side. What are (a) the compressional stress and (b) the shear stress on the bar? (a) $9.4 \times 10^4 \text{ N/m}^2$ (b) $1.2 \times 10^5 \text{ N/m}^2$

12. ●● A 4.0-kg object is supported by an aluminum wire of length 2.0 m and diameter 2.0 mm. How much will the wire stretch? 0.36 mm

13. ●● A copper wire has a length of 5.0 m and a diameter of 3.0 mm. Under what load will its length increase by 0.3 mm? 47 N

14. ●● A metal wire 1.0 mm in diameter and 2.0 m long hangs vertically with a 6.0-kg object suspended from it. If the wire stretches 1.4 mm under the tension, what is the value of Young's modulus for the metal? $1.1 \times 10^{11} \text{ N/m}^2$

15. **IE** ●● When railroad tracks are installed, gaps are left between the rails. (a) Should a greater gap be used if the rails are installed on (1) a cold day or (2) a hot day? Or (3) Does it make any difference? Why? (b) Each steel rail is 8.0 m long and has a cross-sectional area of 0.0025 m². On a hot day, each rail thermally expands as much as 3.0×10^{-3} m. If there were no gaps between the rails, what would be the force on the ends of each rail? (a) (1) a cold day (b) 1.9×10^5 N

16. ●● A rectangular steel column (20.0 cm × 15.0 cm) supports a load of 12.0 metric tons. If the column is 2.00 m in length before being stressed, what is the decrease in length? 3.92×10^{-5} m

17. **IE** ●● A bimetallic rod as illustrated in ▶Fig 9.26 is composed of brass and copper. (a) If the rod is subjected to a compressive force, will the rod bend toward the brass or the copper? Why? (b) Justify your answer mathematically if the compressive force is 5.00×10^4 N. (a) bends toward brass (b) see ISM

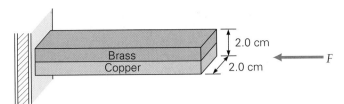

▲ **FIGURE 9.26** Bimetallic rod and mechanical stress
See Exercise 17.

18. **IE ●●** Two same-size metal posts, one aluminum and one copper, are subjected to equal shear stresses. (a) Which post will show the larger deformation angle, (1) the copper post or (2) the aluminum post? Or (3) Is the angle the same for both? Why? (b) By what factor is the deformation angle of one post greater than the other?
(a) (2) the aluminium post (b) $\phi_{Al} = 1.5\phi_{Cu}$

19. **●●** A 85.0-kg person stands on one leg and 90% of the weight is supported by the upper leg connecting the knee and hip joint—the femur. Assuming the femur is 0.650 m long and has a radius of 2.00 cm, by how much is the bone compressed? 4.2×10^{-7} m

20. **●●** Two metal plates are held together by two steel rivets, each of diameter 0.20 cm and length 1.0 cm. How much force must be applied parallel to the plates to shear off both rivets? 1.0×10^5 N

21. **IE ●●** (a) Which of the liquids in Table 9.1 has the greatest compressibility? Why? (b) For equal volumes of ethyl alcohol and water, which would require more pressure to be compressed by 0.10%, and how many times more? (a) ethyl alcohol (b) $\Delta p_w / \Delta p_{ea} = 2.2$

22. **●●●** A brass cube 6.0 cm on each side is placed in a pressure chamber and subjected to a pressure of 1.2×10^7 N/m² on all of its surfaces. By how much will each side be compressed under this pressure? 3.2×10^{-6} m

23. **●●●** A cylindrical eraser of negligible mass is dragged across a paper at a constant velocity to the right by its pencil. The coefficient of kinetic friction between eraser and paper is 0.650. The pencil pushes down with 4.20 N. The height of the eraser is 1.10 cm and its diameter is 0.760 cm. Its top surface is displaced horizontally 0.910 mm relative to the bottom. Determine the shear modulus of the eraser material. 7.28×10^5 N/m²

24. **●●●** A 45-kg traffic light is suspended from two steel cables of equal length and radii 0.50 cm. If each cable makes a 15° angle with the horizontal, what is the fractional increase in their length due to the weight of the light? 5.4×10^{-5}

9.2 Fluids: Pressure and Pascal's Principle

25. **MC** For a liquid in an open container, the total pressure at any depth depends on (a) atmospheric pressure, (b) liquid density, (c) acceleration due to gravity, (d) all of the preceding. (d)

26. **MC** For the pressure–depth relationship for a fluid ($p = \rho g h$), it is assumed that (a) the pressure decreases with depth, (b) a pressure difference depends on the reference point, (c) the fluid density is constant, (d) the relationship applies only to liquids. (c)

27. **MC** When measuring automobile tire pressure, what type of pressure is this: (a) gauge, (b) absolute, (c) relative, or (d) all of the preceding? (a)

28. **CQ** ▼Figure 9.27 shows a famous "bed of nails" trick. The woman lies on a bed of nails with a cinder block on her chest. A person hits the anvil with a sledgehammer. The nails do not pierce the woman's skin. Explain why. see ISM

▲ **FIGURE 9.27** A bed of nails See Exercise 28.

29. **CQ** Automobile tires are inflated to about 30 lb/in.², whereas thin bicycle tires are inflated to 90 to 115 lb/in.²—at least three times as much pressure! Why? see ISM

30. **CQ** (a) Why is blood pressure usually measured at the arm? (b) Suppose the pressure reading were taken on the calf of the leg of a standing person. Would there be a difference, in principle? Explain. see ISM

31. **CQ** (a) Two dams form artificial lakes of equal depth. However, one lake backs up 15 km behind the dam, and the other backs up 50 km behind. What effect does the difference in length have on the pressures on the dams? (b) Dams are usually thicker at the bottom. Why? see ISM

32. **CQ** Water towers (storage tanks) are generally bulb shaped as shown in ▼Fig. 9.28. Wouldn't it be better to have a cylindrical storage tank of the same height? Explain. see ISM

▶ **FIGURE 9.28** Why a bulb-shaped water tower? See Exercise 32.

33. **CQ** (a) Liquid storage cans, such as gasoline cans, generally have capped vents. What is the purpose of the vents, and what happens if you forget to remove the cap before you pour the liquid? (b) Explain how a medicine dropper works. (c) Explain how we breathe (inhalation and exhalation). *see ISM*

34. **CQ** A water dispenser for pets has an inverted plastic bottle, as shown in ▼Fig. 9.29. (The water is dyed blue for contrast.) When a certain amount of water is drunk from the bowl, more water flows automatically from the bottle into the bowl. The bowl never overflows. Explain the operation of the dispenser. Does the height of the water in the bottle depend on the surface area of the water in the bowl? *see ISM*

◄ **FIGURE 9.29 Pet barometer** See Exercise 34.

35. **IE** ● In his original barometer, Pascal used water instead of mercury. (a) Water is less dense than mercury, so the water barometer would have (1) a higher height than, (2) a lower height than, or (3) the same height as the mercury barometer. Why? (b) How high would the water column have been? *(a) (1) a higher height than (b) 10 m*

36. ● If you dive to 10 m below the surface of a lake, (a) what is the pressure due to the water alone? (b) What is the total or absolute pressure at that depth? *(a) 9.8×10^4 Pa (b) 2.0×10^5 Pa*

37. **IE** ● In an open U-tube, the pressure of a water column on one side is balanced by the pressure of a column of gasoline on the other side. (a) Compared to the height of the water column, the gasoline column will have (1) a higher, (2) a lower, or (3) the same height. Why? (b) If the height of the water column is 15 cm, what is the height of the gasoline column? *(a) (1) a higher (b) 22 cm*

38. ● A 75-kg athlete does a single-hand handstand. If the area of the hand in contact with the floor is 125 cm², what pressure is exerted on the floor? *5.9×10^4 Pa*

39. ● The gauge pressure in both tires of a bicycle is 690 kPa. If the bicycle and the rider have a combined mass of 90.0 kg, what is the area of contact of *each* tire with the ground? (Assume that each tire supports half the total weight of the bicycle.) *6.39×10^{-4} m²*

40. ●● In a sample of seawater taken from an oil spill, an oil layer 4.0 cm thick floats on 55 cm of water. If the density of the oil is 0.75×10^3 kg/m³, what is the absolute pressure on the bottom of the container? *1.07×10^5 Pa*

41. **IE** ●● In a lecture demonstration, an empty can is used to demonstrate the force exerted by air pressure (▼Fig. 9.30). A small quantity of water is poured into the can, and the water is brought to a boil. Then the can is sealed with a rubber stopper. As you watch, the can is slowly crushed with sounds of metal bending. (Why is a rubber stopper used as a safety precaution?) (a) This is because of (1) thermal expansion and contraction, (2) a higher steam pressure inside the can, or (3) a lower pressure inside the can as steam condenses. Why? (b) Assuming the dimensions of the can are 0.24 m × 0.16 m × 0.10 m and the inside of the can is in a perfect vacuum, what is the total force exerted on the can by the air pressure? *(a) (3) a lower pressure inside the can as steam condenses (b) 1.6×10^4 N = 3600 lb*

▲ **FIGURE 9.30 Air pressure** See Exercise 41.

42. ●● What is the fractional decrease in pressure when a barometer is raised 40.0 m to the top of a building? (Assume that the density of air is constant over that distance.) *0.50%*

43. ●● A student decides to compute the standard barometric reading on top of Mt. Everest (29 028 ft) by assuming that the density of air has the same constant density as at sea level. Try this yourself. What does the result tell you? *air density decreases rapidly with altitude; see ISM*

44. ●● To drink a soda (assume same density as water) through a straw requires you to lower the pressure at the top of the straw. What does the pressure need to be at the top of a straw that is 15.0 cm above the surface of the soda in order to get soda to your lips? *9.98×10^4 Pa*

45. ●● During a plane flight, a passenger experiences ear pain due to a head cold that has clogged his Eustachian tubes. Assuming the pressure in his tubes remained at 1.00 atm (from sea level) and the cabin pressure is maintained at 0.900 atm, determine the air pressure force (including its direction) on one eardrum, assuming it has a diameter of 0.800 cm. *0.51 N*

46. ●● Here is a demonstration Pascal used to show the importance of a fluid's pressure on the fluid's depth (▶Fig. 9.31): An oak barrel with a lid of area 0.20 m² is filled with water. A long, thin tube of cross-sectional area 5.0×10^{-5} m² is inserted into a hole at the center of the lid, and water is poured into the tube. When the water reaches 12 m high, the barrel bursts. (a) What was the weight of the water in the tube? (b) What was the pressure of the water on the lid of the barrel? (c) What was the net force on the lid due to the water pressure? *(a) 5.9 N (b) 1.2×10^5 Pa (c) 2.4×10^4 N*

◀ **FIGURE 9.31**
Pascal and the
bursting barrel See
Exercise 46.

47. ●● The door and the seals on an aircraft are subject to a tremendous amount of force during flight. At an altitude of 10 000 m (about 33 000 ft), the air pressure outside the airplane is only $2.7 \times 10^4 \text{ N/m}^2$, while the inside is still at normal atmospheric pressure, due to pressurization of the cabin. Calculate the net force due to the air pressure on a door of area 3.0 m². $2.2 \times 10^5 \text{ N (about 50 000 lb)}$

48. ●● The pressure exerted by a person's lungs can be measured by having the person blow as hard as possible into one side of a manometer. If a person blowing into one side of an open-tube manometer produces an 80-cm difference between the heights of the columns of water in the manometer arms, what is the gauge pressure of the lungs? $7.8 \times 10^3 \text{ Pa}$

49. ●● In a head-on auto collision, the driver had his air bags disconnected and his head hits the windshield, fracturing his skull. Assuming the driver's head has a mass of 4.0 kg, the area of the head to hit the windshield to be 25 cm², and an impact time of 3.0 ms, with what speed does the head hit the windshield? (Take the compressive fracture strength of the cranial bone to be $1.0 \times 10^8 \text{ Pa}$.) $1.9 \times 10^2 \text{ m/s}$

50. ●● A cylinder has a diameter of 15 cm (▼Fig. 9.32). The water level in the cylinder is maintained at a constant height of 0.45 m. If the diameter of the spout pipe is 0.50 cm, how high is h, the vertical stream of water? (Assume the water to be an ideal fluid.) 0.45 m

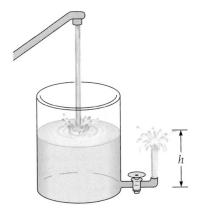

◀ **FIGURE 9.32** How
high a fountain? See
Exercise 50.

51. ●● In 1960, the U.S. Navy's bathyscaphe *Trieste* (a submersible) descended to a depth of 10 912 m (about 35 000 ft) into the Mariana Trench in the Pacific Ocean. (a) What was the pressure at that depth? (Assume that seawater is incompressible.) (b) What was the force on a circular observation window with a diameter of 15 cm? (a) $1.1 \times 10^8 \text{ Pa}$ (b) $1.9 \times 10^6 \text{ N}$

52. ●● The output piston of a hydraulic press has a cross-sectional area of 0.25 m². (a) How much pressure on the input piston is required for the press to generate a force of $1.5 \times 10^6 \text{ N}$? (b) What force is applied to the input piston if it has a diameter of 5.0 cm? (a) $6.0 \times 10^6 \text{ Pa}$ (b) $1.2 \times 10^4 \text{ N}$

53. ●● A hydraulic lift in a garage has two pistons: a small one of cross-sectional area 4.00 cm² and a large one of cross-sectional area 250 cm². (a) If this lift is designed to raise a 3500-kg car, what minimum force must be applied to the small piston? (b) If the force is applied through compressed air, what must be the minimum air pressure applied to the small piston? (a) 549 N (b) $1.37 \times 10^6 \text{ Pa}$

54. ●●● A hypodermic syringe has a plunger of area 2.5 cm² and a 5.0×10^{-3}-cm² needle. (a) If a 1.0-N force is applied to the plunger, what is the gauge pressure in the syringe's chamber? (b) If a small obstruction is at the end of the needle, what force does the fluid exert on it? (c) If the blood pressure in a vein is 50 mm Hg, what force must be applied on the plunger so that fluid can be injected into the vein? (a) $4.0 \times 10^3 \text{ N/m}^2$ (b) $2.0 \times 10^{-3} \text{ N}$ (c) 1.7 N

55. ●●● A funnel has a cork blocking its drain tube. The cork has a diameter of 1.50 cm and is held in place by static friction with the sides of the drain tube. When water is filled to 10.0 cm above the cork, it comes flying out. Determine the maximum force of static friction between the cork and drain tube. Neglect the weight of the cork. 0.173 N

56. ●●● A hydraulic balance used to detect small changes in mass is shown in ▼Fig. 9.33. If a mass m of 0.25 g is placed on the balance platform, by how much will the height of the water in the smaller, 1.0-cm-diameter cylinder have changed when the balance comes to equilibrium? 2.6 mm

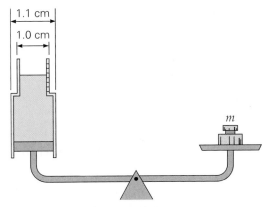

▲ **FIGURE 9.33 A hydraulic balance** See Exercise 56.

9.3 Buoyancy and Archimedes' Principle

57. **MC** A wood block floats in a swimming pool. The buoyant force exerted on the block by water depends on (a) the volume of water in the pool, (b) the volume of the wood block, (c) the volume of the wood block under water, (d) all of the preceding. (c)

58. **MC** If a submerged object displaces an amount of liquid of greater weight than its own and is then released, the object will (a) rise to the surface and float, (b) sink, (c) remain in equilibrium at its submerged position. (a)

59. **MC** Comparing an object's density (ρ_o) to that of a fluid (ρ_f), what is the condition for the object to float: (a) $\rho_o < \rho_f$, or (b) $\rho_f < \rho_o$? (a)

60. **CQ** (a) What is the most important factor in constructing a life jacket that will keep a person afloat? (b) Why is it so easy to float in Utah's Great Salt Lake? see ISM

61. **CQ** An ice cube floats in a glass of water. As the ice melts, how does the level of the water in the glass change? Would it make any difference if the ice cube were hollow? Explain. see ISM

62. **CQ** Oceangoing ships in port are loaded to the so-called *Plimsoll mark*, which is a line indicating the maximum safe loading depth. However, in New Orleans, located at the mouth of the Mississippi River, where the water is brackish (partly salty and partly fresh), ships are loaded until the Plimsoll mark is somewhat below the water line. Why? port water (partly fresh) is less dense than seawater

63. **CQ** Two blocks of equal volume, one iron and one aluminum, are dropped into a body of water. Which block will experience the greater buoyant force? Why? same

64. **CQ** An inventor comes up with an idea for a perpetual motion machine, as illustrated in ▼Fig. 9.34. It contains a sealed chamber with mercury (Hg) in one half and water (H_2O) in the other. A cylinder is mounted in the center and is free to rotate. He reasons that since mercury is much denser than water (13.6 g/cm³ to 1.00 g/cm³), the weight of the mercury displaced by half the cylinder is much greater than the water displaced by the other half. Then, the buoyant force on the mercury side is greater than that on the water side—more than thirteen times greater. The difference in forces and torques should cause the cylinder to rotate—perpetually. Would you invest any money in this invention? Why or why not? see ISM

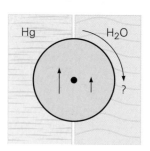

◀ **FIGURE 9.34** Perpetual motion? See Exercise 64.

65. **IE ●** (a) If the density of an object is exactly equal to the density of a fluid, the object will (1) float, (2) sink, (3) stay at any height in the fluid, as long as it is totally immersed. (b) A cube 8.5 cm on each side has a mass of 0.65 kg. Will the cube float or sink in water? Prove your answer.
(a) (3) stay at any height (b) sink; see ISM

66. **●** Suppose that Archimedes found that the king's crown had a mass of 0.750 kg and a volume of 3.980×10^{-5} m³. (a) What simple approach did Archimedes use to determine the crown's volume? (b) Was the crown pure gold?
(a) water displacement (b) no; see ISM

67. **●** A rectangular boat, as illustrated in ▼Fig. 9.35, is overloaded such that the water level is just 1.0 cm below the top of the boat. What is the combined mass of the people and the boat? 2.6×10^3 kg

▲ **FIGURE 9.35** An overloaded boat See Exercise 67.

68. **●●** An object has a weight of 8.0 N in air. However, it apparently weighs only 4.0 N when it is completely submerged in water. What is the density of the object? 2.0×10^3 kg/m³

69. **●●** When a 0.80-kg crown is submerged in water, its apparent weight is measured to be 7.3 N. Is the crown pure gold? no; see ISM

70. **●●** A steel cube 0.30 m on each side is suspended from a scale and immersed in water. What will the scale read? 1.8×10^3 N

71. **●●** A wood cube 0.30 m on each side has a density of 700 kg/m³ and floats levelly in water. (a) What is the distance from the top of the wood to the water surface? (b) What mass has to be placed on top of the wood so that its top is just at the water level? (a) 0.09 m (b) 8.1 kg

72. **●●** (a) Given a piece of metal with a light string attached, a scale, and a container of water in which the piece of metal can be submersed, how could you find the volume of the piece without using the variation in the water level? (b) An object has a weight of 0.882 N. It is suspended from a scale, which reads 0.735 N when the piece is submerged in water. What are the volume and density of the piece of metal?
(a) see ISM (b) 1.50×10^{-5} m³, 6.00×10^3 kg/m³

73. **●●** An aquarium is filled with a liquid. A cork cube, 10.0 cm on a side, is pushed and held at rest completely submerged in the liquid. It takes a force of 7.84 N to hold it under the liquid. If the density of cork is 200 kg/m³, find the density of the liquid. 1.00×10^3 kg/m³ (probably H_2O)

74. **●●** A block of iron quickly sinks in water, but ships constructed of iron float. A solid cube of iron 1.0 m on each side is made into sheets. To make these sheets into a hollow cube that will not sink, what should be the minimum length of the sides of the sheets? 2.0 m

75. **●●** Plans are being made to bring back the zeppelin, a lighter-than-air airship like the Goodyear blimp that carries passengers and cargo, but is filled with helium, not flammable hydrogen as was used in the ill-fated *Hindenburg*. (See opening Physics Facts.) One design calls for the ship to be 110 m long and to have a total mass (without helium) of 30.0 metric tons. Assuming the ship's "envelope" to be cylindrical, what would its diameter have to be so as to lift the total weight of the ship and the helium? 17.7 m

76. ●●● A girl floats in a lake with 97% of her body beneath the water. What are (a) her mass density and (b) her weight density? (a) 9.7×10^2 kg/m³ (b) 9.5×10^3 N/m³

77. ●●● A spherical navigation buoy is tethered to the lake floor by a vertical cable (▼Fig. 9.36). The outside diameter of the buoy is 1.00 m. The interior of the buoy consists of an aluminum shell 1.0 cm thick and the rest is solid plastic. The density of aluminum is 2700 kg/m³ and the density of the plastic is 200 kg/m³. The buoy is set to float exactly halfway out of the water. Determine the tension in the cable. 8.1×10^2 N

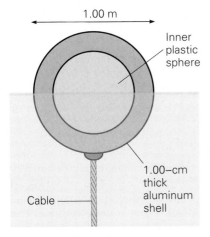

1.00 m

Inner plastic sphere

Cable

1.00–cm thick aluminum shell

◀ FIGURE 9.36 It's a buoy (Not to scale.) See Exercise 77.

78. ●●● ▼Figure 9.37 shows a simple laboratory experiment. Calculate (a) the volume and (b) the density of the suspended sphere. (Assume that the density of the sphere is uniform and that the liquid in the beaker is water.) (c) Would you be able to make the same determinations if the liquid in the beaker were mercury? (See Table 9.2.) Explain. (a) 9.8×10^{-4} m³ (b) 1.5×10^3 kg/m³ (c) see ISM

▲ FIGURE 9.37 Dunking a sphere See Exercise 78.

9.4 Fluid Dynamics and Bernoulli's Equation

79. **MC** If the speed at some point in a fluid changes with time, the fluid flow is *not* (a) steady, (b) irrotational, (c) incompressible, (d) nonviscous. (a)

80. **MC** An ideal fluid is not (a) steady, (b) compressible, (c) irrotational, or (d) nonviscous. (b)

81. **MC** Bernoulli's equation is based primarily on (a) Newton's laws, (b) conservation of momentum, (c) a nonideal fluid, (d) conservation of energy. (d)

82. **MC** According to Bernoulli's equation, if the pressure on the liquid in Fig. 9.19 is increased, (a) the flow speed always increases, (b) the height of the liquid always increases, (c) both the flow speed and the height of the liquid may increase, (d) none of the preceding. (c)

83. **CQ** The speed of blood flow is greater in arteries than in capillaries. However, the flow rate equation (Av = constant) seems to predict that the speed should be greater in the smaller capillaries. Can you resolve this apparent inconsistency? there are many capillaries

84. **CQ** (a) Explain why water shoots out farther from a hose if you put your finger over the tip of the hose. (b) Give a human analogy of constricted flow and greater speed. (a) smaller area, greater speed (b) arteries to capillaries

85. **CQ** (a) If an Indy racer had a flat bottom, it would be highly unstable (like an airplane wing) due to the lift it gets when it moves at a high speed. To increase friction and stability, the bottom has a concave section called the *Venturi tunnel* (▼Fig. 9.38). (a) In terms of Bernoulli's equation, explain how this concavity supplies extra downward force to the car in addition to that supplied by the front and rear wings. (b) What is the purpose of the "spoiler" on the back of the racer? see ISM

▲ FIGURE 9.38 Venturi tunnel and spoiler See Exercise 85.

86. **CQ** Here are two common demonstrations of Bernoulli effects: (a) If you hold a narrow strip of paper in front of your mouth and blow over the top surface, the strip will rise (▼Fig. 9.39a). (Try it.) Why? (b) A plastic egg is supported vertically by a stream of air from a tube (Fig. 9.39b). The egg will not move away from the midstream position. Why not? see ISM

(a)   (b)

▲ FIGURE 9.39 Bernoulli effects See Exercise 86.

87. • An ideal fluid is moving at 3.0 m/s in a section of a pipe of radius 0.20 m. If the radius in another section is 0.35 m, what is the flow speed there? 0.98 m/s

88. **IE** • (a) If the radius of a pipe narrows to half of its original size, will the flow speed in the narrow section (1) increase by a factor of 2, (2) increase by a factor of 4, (3) decrease by a factor of 2, or (4) decrease by a factor of 4? Why? (b) If the radius widens to three times its original size, what is the ratio of the flow speed in the wider section to that in the narrow section? (a) (2) increase by a factor of 4 (b) decrease by a factor of 9

89. •• The speed of blood in a major artery of diameter 1.0 cm is 4.5 cm/s. (a) What is the flow rate in the artery? (b) If the capillary system has a total cross-sectional area of 2500 cm², the average speed of blood through the capillaries is what percentage of that through the major artery? (c) Why must blood flow at low speed through the capillaries? (a) 3.5 cm³/s (b) 0.031% (c) see ISM

90. •• The blood flow speed through an aorta with a radius of 1.00 cm is 0.265 m/s. If hardening of the arteries causes the aorta to be constricted to a radius of 0.800 cm, by how much would the blood flow speed increase? 0.149 m/s

91. •• Using the data and result of Exercise 90, calculate the pressure difference between the two areas of the aorta. (Blood density: $\rho = 1.06 \times 10^3 \text{ kg/m}^3$.) 53.6 Pa

92. •• In a dramatic lecture demonstration, a physics professor blows hard across the top of a copper penny that is at rest on a level desk. By doing this at the right speed, he can get the penny to accelerate vertically, into the airstream, and then deflect it into a tray, as shown in ▼Fig. 9.40. Assuming the diameter of a penny is 1.80 cm and it has a mass of 3.50 g, what is the minimum airspeed needed to lift the penny off the tabletop? Assume the air under the penny remains at rest. 14.5 m/s

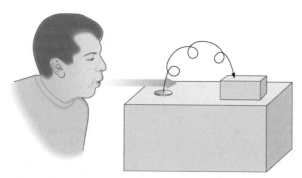

▲ **FIGURE 9.40 A big blow** See Exercise 92.

93. •• A room measures 3.5 m by 4.5 m by 6.0 m. If the heating and air-conditioning ducts to and from the room are circular with diameter 0.30 m and all the air in the room is to be exchanged every 12 min, (a) what is the average flow rate? (b) What is the necessary flow speed in the duct? (Assume that the density of the air is constant.) (a) 0.13 m³/s (b) 1.8 m/s

94. •• The spout heights in the container in ▼Fig. 9.41 are 10 cm, 20 cm, 30 cm, and 40 cm. The water level is maintained at a 45-cm height by an outside supply. (a) What is the speed of the water out of each hole? (b) Which water stream has the greatest range relative to the base of the container? Justify your answer. (a) 0.99 m/s; 1.7 m/s; 2.2 m/s; 2.6 m/s (b) 0.45 m, from $y = 20$ cm

▲ **FIGURE 9.41 Streams as projectiles** See Exercise 94.

95. ••• Water flows at a rate of 25 L/min through a horizontal 7.0-cm-diameter pipe under a pressure of 6.0 Pa. At one point, calcium deposits reduce the cross-sectional area of the pipe to 30 cm². What is the pressure at this point? (Consider the water to be an ideal fluid.) 2.2 Pa

96. ••• As a fire-fighting method, a homeowner in the deep woods rigs up a water pump to bring water from a lake that is 10.0 m below the level of the house. If the pump is capable of producing a gauge pressure of 140 kPa, at what rate (in L/s) can water be pumped to the house assuming the hose has a radius of 5.00 cm? 71.9 L/s

97. ••• A Venturi meter can be used to measure the flow speed of a liquid. A simple such device is shown in ▼Fig. 9.42. Show that the flow speed of an ideal fluid is given by

$$v_1 = \sqrt{\frac{2g\Delta h}{(A_1^2/A_2^2) - 1}}.$$ see ISM

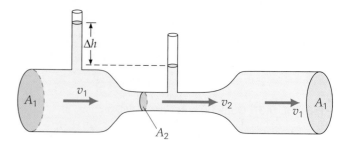

▲ **FIGURE 9.42 A flow speed meter** See Exercise 97.

*9.5 Surface Tension, Viscosity, and Poiseuille's Law

98. **MC** Water droplets and soap bubbles tend to assume the shape of a sphere. This effect is due to (a) viscosity, (b) surface tension, (c) laminar flow, (d) none of the preceding. (b)

99. **MC** Some insects can walk on water because (a) the density of water is greater than that of the insect, (b) water is viscous, (c) water has surface tension, (d) none of the preceding. (c)

100. **MC** The viscosity of a fluid is due to (a) forces causing friction between the molecules, (b) surface tension, (c) density, or (d) none of the above. (a)

101. **CQ** A motor oil is labeled 10W-40. What do the numbers 10 and 40 measure? How about the W? viscosity; winter

102. **CQ** Why are clothes washed in hot water and a detergent added? to reduce surface tension

103. ●● The pulmonary artery, which connects the heart to the lungs, is about 8.0 cm long and has an inside diameter of 5.0 mm. If the flow rate in it is to be 25 mL/s, what is the required pressure difference over its length? 3.5×10^2 Pa

104. ●● A hospital patient receives a quick 500-cc blood transfusion through a needle with a length of 5.0 cm and an inner diameter of 1.0 mm. If the blood bag is suspended 0.85 m above the needle, how long does the transfusion take? (Neglect the viscosity of the blood flowing in the plastic tube between the bag and the needle.) 2.0×10^2 s

105. ●● A nurse needs to draw 20.0 cc of blood from a patient and deposit it into a small plastic container whose interior is at atmospheric pressure. He inserts the needle end of a long tube into a vein where the average gauge pressure is 30.0 mm Hg. This allows the internal pressure in the vein to push the blood into the collection container. The needle is 0.900 mm in diameter and 2.54 cm long. The long tube is wide and smooth enough that we can assume its resistance is negligible, and that all the resistance to blood flow occurs in the narrow needle. How long does it take him to collect the sample? 13.5 s

Comprehensive Exercises

106. Show that specific gravity is equivalent to a ratio of densities, given that its strict definition is the ratio of the weight of a given volume of a substance to the weight of an equal volume of water. see ISM

107. A rock is suspended from a string in air. The tension in the string is 2.94 N. When the rock is then dunked into a liquid, and the rope let go slack, it sinks and comes to rest on a spring whose spring constant is 200 N/m. The spring's final compression is 1.00 cm. If the density of the rock is known to be 2500 kg/m³, what is the density of the liquid? 8.0×10^2 kg/m³

108. An unevenly weighted baton (cylindrical in shape) consists of two sections: a denser (lower) section and a less dense (upper) section. When placed in water, it is upright and barely floats. The baton has a diameter of 2.00 cm; its lower part is made of steel with a density of 7800 kg/m³, and the upper part is made of wood with a density of 810 kg/m³. The steel part has a length of 5.00 cm. Find the length of the wooden section. 1.79 m

109. A crude shower is rigged up by a team of campers. It consists of a large (top open) cylindrical container hung from a tree. Its bottom area is punctured by a lot of small holes, each 1.00 mm in diameter, and the container is 30.0 cm in diameter and 75.0 cm high. (a) Initially, what is the speed the water comes out of the holes? (b) How many holes are needed if you want a 1.20 L/s total flow rate? (a) 3.83 m/s (b) 399

110. What is the difference in volume (due only to pressure changes, not temperature or other factors) between 1000 kg of water at the surface (assume 4°C) of the ocean and the same mass at the deepest known depth, 8.00 km? (Mariana trench; assume also 4°C.) −0.0356 m³

 The following Physlet Physics Problems can be used with this chapter.

PHYSLET 14.1, 14.2, 14.3, 14.4, 14.5, 14.6, 14.7, 14.8, 14.9, 14.10, 15.1, 15.2, 15.3, 15.4, 15.5, 15.6, 15.7, 15.8, 15.9, 15.10

TEMPERATURE AND KINETIC THEORY

PHYSICS FACTS

- Daniel Gabriel Fahrenheit (1686–1736), a German instrument maker, constructed the first alcohol thermometer (1709) and mercury thermometer (1714). Fahrenheit used temperatures of 0° and 96° for reference points. The freezing and boiling points of water were then measured to be 32°F and 212°F.

- Anders Celsius (1701–1744), a Swedish astronomer, invented the Celsius temperature scale with a 100-degree interval between the freezing and boiling point of water (0°C and 100°C). Celsius' original scale was reversed, 100°C (freezing) and 0°C (boiling). This was later changed.

- The Celsius and Fahrenheit temperature scales have equal readings at −40°, that is −40°C = −40°F.

- The lowest possible temperature is absolute zero (−273.15°C). There is no known upper limit on temperature.

- The Golden Gate Bridge over San Francisco Bay varies in length by almost 1 m between summer and winter (thermal expansion).

- Almost all substances have positive coefficients of thermal expansion (expanding on heating). A few have negative coefficients (contraction on heating). Water does over a particular temperature range. The volume of a quantity of water decreases (contracts) on heating from 0°C to 4°C.

Like sailboats, hot-air balloons are low-tech devices in a high-tech world. You can equip a balloon with the latest satellite-linked, computerized navigational system and attempt to fly across the Pacific. However, the basic principles that keep you aloft were known and understood centuries ago. As seen in the photo, air is heated and with an increase in temperature, the heated, less dense air rises. When enough hot air is in the balloon, it becomes buoyant and up you go.

Temperature and heat are frequent subjects of conversation, but if you had to tell what the words really mean, you might find yourself at a loss. We use various types of thermometers to measure temperatures, which provide an objective equivalent for our sensory experience of hot and cold. A temperature change generally results from the application or removal of heat. Temperature, therefore, is related to heat. But how? And what is heat? In this chapter, you'll find that the answers to such questions lead to an understanding of some far-reaching physical principles.

An early theory of heat considered it to be a fluidlike substance called *caloric* (from the Latin word *calor*, meaning "heat") that could be made to flow into and out of a body. Even though this theory has been abandoned, we still speak of heat as flowing from one object to another. Heat is now known to be energy in transit, and temperature and thermal properties are explained by considering the atomic and molecular behavior of substances. This and the next two chapters examine the nature of temperature and heat in terms of microscopic (molecular) theory and macroscopic observations. Here, you'll explore the nature of heat and the ways temperature is measured. You'll also encounter the gas laws, which explain not only the behavior of hot-air balloons, but also more important phenomena, such as how our lungs supply us with the oxygen we need to live.

10.1 Temperature and Heat

OBJECTIVE: To distinguish between temperature and heat.

A good way to begin studying thermal physics is with definitions of temperature and heat. **Temperature** is a relative measure, or indication, of hotness or coldness. A hot stove is said to have a high temperature and an ice cube to have a low temperature. An object that has a higher temperature than another object is said to be hotter or the other object colder. Note that *hot* and *cold* are relative terms, like *tall* and *short*. We can perceive temperature by touch. However, this temperature sense is somewhat unreliable, and its range is too limited to be useful for scientific purposes.

Heat is related to temperature and describes the process of energy transfer from one object to another. That is, **heat** is *the net energy transferred from one object to another because of a temperature difference.* Heat is energy in transit, so to speak. Once transferred, the energy becomes part of the total energy of the molecules of the object or system, that is, its **internal energy**. So heat (energy) transfers between objects can result in internal energy changes.*

On a microscopic level, temperature is associated with molecular motion. In kinetic theory (Section 10.5), which treats gas molecules as point particles, temperature is a measure of the average random *translational* kinetic energy of the molecules. However, diatomic molecules and other real substances, besides having such translational "temperature" energy, also may have kinetic energy due to vibrations and rotations, as well as potential energy due to the attractive forces between molecules. These energies do not contribute to the temperature of the gas but are part of its internal energy, which is the sum of all such energies (▾Fig. 10.1).

Teaching tip: Average random kinetic energy is sometimes referred to as thermal energy, but the latter term is used to mean different things. When thermal energy is associated with random translational motion, it is the "temperature energy." Explain that an object moving as a whole, such as a thrown ball, is in translational motion, but the motion is ordered, not random.

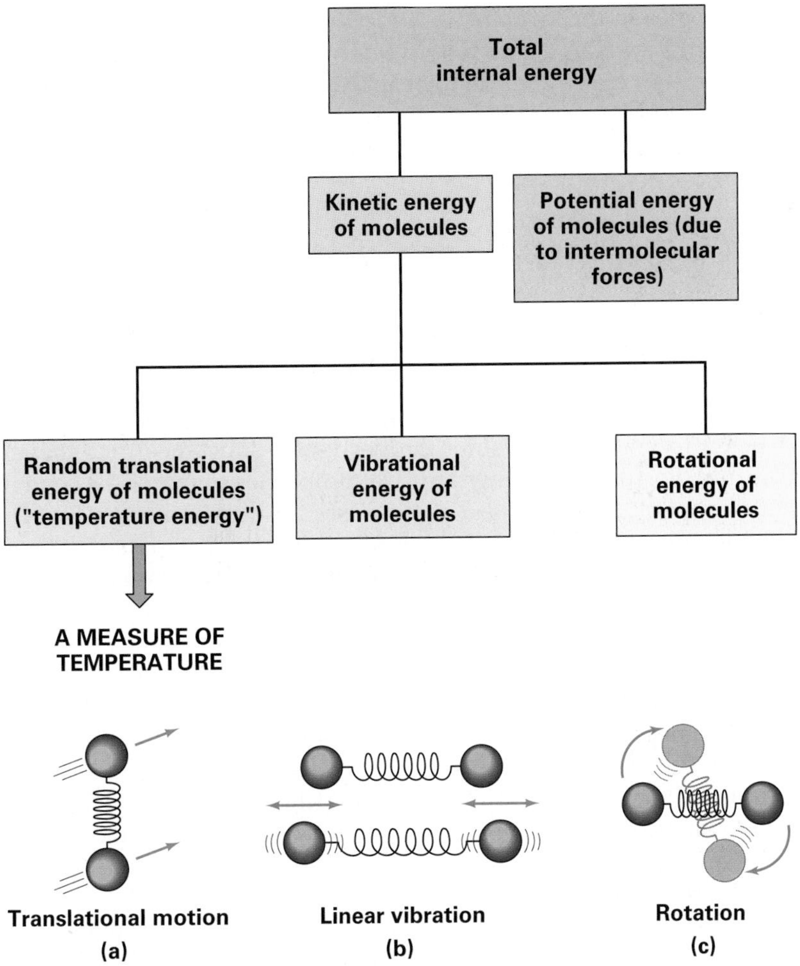

◀ **FIGURE 10.1** Molecular motions The total internal energy is made up of kinetic and potential energies. The kinetic energy has the following forms: **(a)** Temperature is associated with random translational motion of molecules. Neither the **(b)** linear vibrational motion and **(c)** rotation motion contribute to the temperature, nor does the intermolecular potential energy.

```
                    ┌─────────────────────┐
                    │       Total          │
                    │   internal energy    │
                    └─────────────────────┘
                       │                │
          ┌────────────────────┐  ┌────────────────────┐
          │   Kinetic energy   │  │  Potential energy  │
          │    of molecules    │  │  of molecules (due │
          └────────────────────┘  │  to intermolecular │
                                   │      forces)       │
                                   └────────────────────┘
```

| Random translational energy of molecules ("temperature energy") | Vibrational energy of molecules | Rotational energy of molecules |

A MEASURE OF TEMPERATURE

Translational motion
(a)

Linear vibration
(b)

Rotation
(c)

*Note: Some of the energy may go into doing work and not into internal energy (Section 12.2).

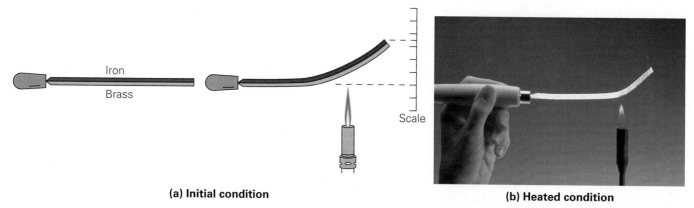

(a) Initial condition

(b) Heated condition

▲ **FIGURE 10.2** Thermal expansion (a) A bimetallic strip is made of two strips of different metals bonded together. (b) When such a strip is heated, it bends because of unequal expansions of the two metals. Here, brass expands more than iron, so the deflection is toward the iron. The deflection of the end of a strip could be used to measure temperature.

Note that a higher temperature does not necessarily mean that one system has a greater internal energy than another. For example, in a classroom on a cold day, the air temperature is relatively high compared to that of the outdoor air. But all that cold air outside the classroom has far more internal energy than does the warm air inside, simply because there is so much *more* of it. If this were not the case, heat pumps would not be practical (Chapter 12). In other words, the internal energy of a system also depends on its mass, or the number of molecules in the system.

When heat is transferred between two objects, regardless of whether they are touching, the objects are said to be in *thermal contact*. When there is no longer a net heat transfer between objects in thermal contact, they have come to the same temperature and are said to be in *thermal equilibrium*.

10.2 The Celsius and Fahrenheit Temperature Scales

OBJECTIVES: To (a) explain how a temperature scale is constructed, and (b) convert temperatures from one scale to another.

A measure of temperature is obtained by using a **thermometer**, a device constructed to make use of some property of a substance that changes with temperature. Fortunately, many physical properties of materials change sufficiently with temperature to be used as the bases for thermometers. By far the most obvious and commonly used property is **thermal expansion** (Section 10.4), a change in the dimensions or volume of a substance that occurs when the temperature changes.

Almost all substances expand with increasing temperature, and do so to different extents. Most substances also contract with decreasing temperature. (Thermal expansion refers to both expansion and contraction; contraction is considered a negative expansion.) Because some metals expand more than others, a bimetallic strip (a strip made of two different metals bonded together) can be used to measure temperature changes. As heat is added, the composite strip will bend away from the side made of the metal that expands more (▲Fig. 10.2). Coils formed from such strips are used in dial thermometers and in common household thermostats (◄Fig. 10.3).

A common thermometer is the liquid-in-glass type, which is based on the thermal expansion of a liquid. A liquid in a glass bulb expands into a glass stem, rising in a capillary bore (a thin tube). Mercury and alcohol (usually dyed

(a)

(b)

▲ **FIGURE 10.3** Bimetallic coil Bimetallic coils are used in (a) dial thermometers (the coil is in the center) and (b) household thermostats (the coil is to the right). Thermostats are used to regulate a heating or cooling system, turning off and on as the temperature of the room changes. The expansion and contraction of the coil causes the tilting of a glass vial containing mercury, which makes and breaks electrical contact.

red to make it more visible) are the liquids used in most liquid-in-glass thermometers. These substances are chosen because of their relatively large thermal expansion and because they remain liquids over normal temperature ranges.

Thermometers are calibrated so that a numerical value can be assigned to a given temperature. For the definition of any standard scale or unit, two fixed reference points are needed. The ice point and the steam point of water at standard atmospheric pressure are two convenient fixed points. More commonly known as the freezing and boiling points, these are the temperatures at which pure water freezes and boils, respectively, under a pressure of 1 atm (standard pressure).

The two most familiar temperature scales are the **Fahrenheit temperature scale** (used in the United States) and the **Celsius temperature scale** (used in the rest of the world). As shown in ▶Fig. 10.4, the ice and steam points have values of 32°F and 212°F, respectively, on the Fahrenheit scale and 0°C and 100°C, respectively, on the Celsius scale. On the Fahrenheit scale, there are 180 equal intervals, or degrees (F°), between the two reference points; on the Celsius scale, there are 100 degrees (C°). Therefore, since 180/100 = 9/5 = 1.8, a Celsius degree is almost twice as large as a Fahrenheit degree. (See margin note for the difference between °C and C°.)

A relationship for converting between the two scales can be obtained from a graph of Fahrenheit temperature (T_F) versus Celsius temperature (T_C), such as the one in ▼Fig. 10.5. The equation of the straight line (in slope–intercept form, $y = mx + b$) is $T_F = (180/100)T_C + 32$, and

$$T_F = \tfrac{9}{5}T_C + 32$$

or

Celsius-to-Fahrenheit conversion   (10.1)

$$T_F = 1.8T_C + 32$$

where $\tfrac{9}{5}$ or 1.8 is the slope of the line and 32 is the intercept on the vertical axis. Thus, to change from a Celsius temperature (T_C) to its equivalent Fahrenheit temperature (T_F), you simply multiply the Celsius reading by $\tfrac{9}{5}$ and add 32.

The equation can be solved for T_C to convert from Fahrenheit to Celsius:

$$T_C = \tfrac{5}{9}(T_F - 32) \quad \textit{Fahrenheit-to-Celsius conversion} \qquad (10.2)$$

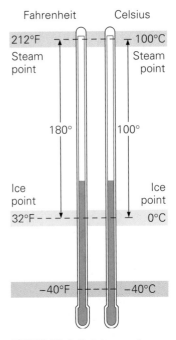

▲ **FIGURE 10.4** Celsius and Fahrenheit temperature scales Between the ice and steam fixed points, there are 100 degrees on the Celsius scale and 180 degrees on the Fahrenheit scale. Thus, a Celsius degree is 1.8 times as large as a Fahrenheit degree.

Note: For distinction, a particular temperature measurement, such as $T = 20°C$, is written with °C (pronounced 20 degrees Celsius), whereas a temperature interval, such as $\Delta T = 80°C - 60°C = 20\ C°$, is written with C° (pronounced 20 Celsius degrees).

 ▼ **FIGURE 10.5** Fahrenheit versus Celsius A plot of Fahrenheit temperature versus Celsius temperature gives a straight line of the general form $y = mx + b$, where $T_F = \tfrac{9}{5}T_C + 32$.

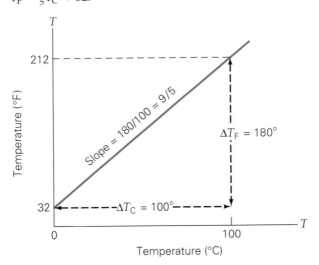

Example 10.1 ■ Converting Temperature Scale Readings: Fahrenheit and Celsius

What are (a) the typical room temperature of 20°C and a cold temperature of −18°C on the Fahrenheit scale; and (b) another cold temperature of −10°F and normal body temperature, 98.6°F, on the Celsius scale?

Thinking It Through. This is a direct application of Eqs. 10.1 and 10.2.

Solution.

Given: (a) $T_C = 20°C$ and $T_C = −18°C$ *Find:* for each temperature,
(b) $T_F = −10°F$ and $T_F = 98.6°F$ (a) T_F
(b) T_C

(a) Equation 10.1 is for changing Celsius readings to Fahrenheit:

$$20°C: \quad T_F = \tfrac{9}{5}T_C + 32 = \tfrac{9}{5}(20) + 32 = 68°F$$

$$−18°C: \quad T_F = \tfrac{9}{5}T_C + 32 = \tfrac{9}{5}(−18) + 32 = 0°F$$

(This typical room temperature of 20°C is a good one to remember.)

(b) Equation 10.2 changes Fahrenheit to Celsius:

$$−10°F: \quad T_C = \tfrac{5}{9}(T_F − 32) = \tfrac{5}{9}(−10 − 32) = −23°C$$

$$98.6°F: \quad T_C = \tfrac{5}{9}(T_F − 32) = \tfrac{5}{9}(98.6 − 32) = 37.0°C$$

From the last calculation, note that normal body temperature has a whole-number value on the Celsius scale. Keep in mind that a Celsius degree is 1.8 times (almost twice) as large as a Fahrenheit degree, so a temperature elevation of several degrees on the Celsius scale makes a big difference. For example, a temperature of 40.0°C represents an elevation of 3.0 C° over normal body temperature. However, on the Fahrenheit scale, this is an increase of 3.0 × 1.8 = 5.4 F°, or a temperature of 98.6 + 5.4 = 104.0°F.

Follow-Up Exercise. Convert the following temperatures: (a) −40°F to Celsius and (b) −40°C to Fahrenheit. (*Answers to all Follow-Up Exercises are at the back of the text.*)

Problem-Solving Hint

Because Eqs. 10.1 and 10.2 are so similar, it is easy to miswrite them. Since they are equivalent, you need to know only one of them—say, Celsius to Fahrenheit (Eq. 10.1, $T_F = \tfrac{9}{5}T_C + 32$). Solving this equation for T_C algebraically gives Eq. 10.2. A good way to make sure that you have written the conversion equation correctly is to test it with a known temperature, such as the boiling point of water. For example, $T_C = 100°C$, so

$$T_F = \tfrac{9}{5}T_C + 32 = \tfrac{9}{5}(100) + 32 = 212°F$$

Thus, we know the equation is correct.

Liquid-in-glass thermometers are adequate for many temperature measurements, but problems arise when highly accurate determinations are needed. A material may not expand uniformly over a wide temperature range. When calibrated to the ice and steam points, an alcohol thermometer and a mercury thermometer have the same readings at those points, but because alcohol and mercury have different expansion properties, the thermometers will not have exactly the same reading at an intermediate temperature, such as room temperature. For very sensitive temperature measurements and to define intermediate temperatures precisely, some other type of thermometer must be used. One such thermometer, a *gas thermometer*, is discussed next. But first, a couple of Insights on body temperatures—Insights 10.1 and 10.2.

INSIGHT 10.1 HUMAN BODY TEMPERATURE

We commonly take "normal" human body temperature to be 98.6°F (or 37.0°C). The source of this value is a study of human temperature readings done in 1868—more than 135 years ago. A more recent study, conducted in 1992, notes that the 1868 study used thermometers that were not as accurate as modern electronic (digital) thermometers. The new study has some interesting results.

The normal human body temperature from oral measurements varies among individuals over a range of about 96°F to 101°F, with an average temperature of 98.2°F. After strenuous exercise, the oral temperature can rise as high as 103°F. When the body is exposed to cold, oral temperatures can fall below 96°F. A rapid drop in temperature of 2 to 3 F° produces uncontrollable shivering. The skeletal muscles contract and so do the tiny muscles attached to the hair follicles. The result is "goose bumps."

Your body temperature is typically lowest in the morning, after you have slept and your digestive processes are at a low point. Normal body temperature generally rises during the day to a peak and then recedes. The 1992 study also indicated that women have a slightly higher average body temperature than do men (98.4°F versus 98.1°F).

What about the extremes? A fever temperature is typically between 102°F and 104°F. A body temperature above 106°F is extremely dangerous. At such temperatures, the enzymes that take part in certain chemical reactions in the body begin to be inactivated, and a total breakdown of body chemistry can result. On the cold side, decreased body temperature results in memory lapses and slurred speech, muscular rigidity, erratic heartbeats, and loss of consciousness. Below 78°F, death occurs due to heart failure. However, mild hypothermia (lower-than-normal body

temperature) can be beneficial. A decrease in body temperature slows down the body's chemical reactions, and cells use less oxygen than they normally do. This effect is applied in some surgeries (Fig. 1). A patient's body temperature may be lowered significantly to avoid damage to the brain and to the heart, which must be stopped during some procedures.

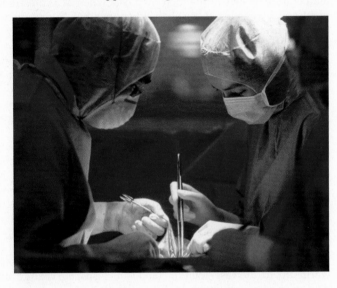

FIGURE 1 Lower than normal During some surgeries, the patient's body temperature is lowered to slow down the body's chemical reactions and to reduce the need for blood to supply oxygen to the tissues.

10.3 Gas Laws, Absolute Temperature, and the Kelvin Temperature Scale

OBJECTIVES: To (a) describe the ideal gas law, (b) explain how it is used to determine absolute zero, and (c) understand the Kelvin temperature scale.

Whereas different liquid-in-glass thermometers show slightly different readings for temperatures other than fixed points because of the liquids' different expansion properties, a thermometer that uses a gas gives the same readings regardless of the gas used. The reason is that at very low densities all gases exhibit the same expansion behavior.

The variables that describe the behavior of a given quantity (mass) of gas are pressure, volume, and temperature (p, V, and T). When temperature is held constant, the pressure and volume of a quantity of gas are related as follows:

$$pV = \text{constant} \quad \text{or} \quad p_1V_1 = p_2V_2 \quad \textit{(at constant temperature)} \quad (10.3)$$

That is, the product of pressure and volume is a constant. This relationship is known as *Boyle's law*, after Robert Boyle (1627–1691), the English chemist who discovered it.

When the pressure is held constant, the volume of a quantity of gas is related to the *absolute* temperature (to be defined shortly):

$$\frac{V}{T} = \text{constant} \quad \text{or} \quad \frac{V_1}{T_1} = \frac{V_2}{T_2} \quad \textit{(at constant pressure)} \quad (10.4)$$

Demonstration/activity: (1) Place marshmallows or a partially filled balloon in a vacuum chamber, and pump out the air. (2) Ask students to explain why their ears "pop" when they drive in the mountains or ascend and descend in an airplane.

INSIGHT 10.2 WARM-BLOODED VERSUS COLD-BLOODED

With few exceptions, all mammals and birds are warm-blooded and all fish, reptiles, amphibians, and insects are cold-blooded. The difference is that warm-blooded creatures try to maintain their bodies at a relatively constant temperature, while cold-blooded creatures take on the temperature of their surroundings (Fig. 1).

Warm-blooded creatures maintain a relatively constant body temperature by generating their own heat when in a cold environment and by cooling themselves when in a hot environment. To generate heat, warm-blooded animals convert food into energy. To stay cool on hot days, they sweat, pant, or get wet and thereby remove heat by water evaporation. Primates (humans, apes, monkeys, and so on) have sweat glands all over their bodies. Dogs and cats have sweat glands only in their feet. Pigs and whales have no sweat glands. Pigs generally rely on wallowing in mud for cooling, and whales can change water depths for temperature changes or seasonally migrate.

Also, some animals have fur coats for warmth in the winter and shed them to cool off. Warm-blooded animals can shiver to activate certain muscles to increase metabolism and thereby generate heat. Birds (and some people) migrate between colder and warmer regions.

The body temperature of cold-blooded creatures changes with the temperature of their environment. They are very active in warm environments and are sluggish when it is cold. This is because their muscle activity depends on chemical reactions that vary with temperature. Cold-blooded creatures often bask in the sun to warm up to increase their metabolism. Fish can change water depths or seasonally migrate. Frogs, toads, and lizards hibernate during winter. To stay warm, honeybees crowd together and rapidly flap their wings to generate heat.

Some animals do not fall into the strict definitions of being warm-blooded or cold-blooded. Bats, for example, are mammals that cannot maintain a constant body temperature, and they cool off when not active. Some warm-blooded animals, such as bears, groundhogs, and gophers, hibernate in winter. During the hibernation period, they live off stored body fat; their body temperatures may drop as much as 10 C° (18 F°).

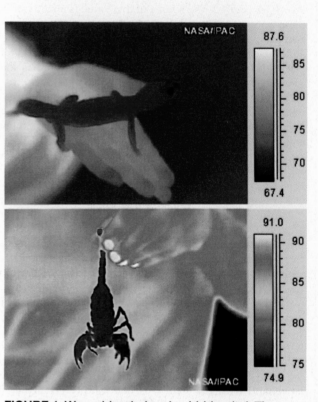

FIGURE 1 Warm-blooded and cold-blooded The infrared images show that cold-blooded creatures take on the temperature of their surroundings. Both the gecko and the scorpion are at the same temperature (color) as the air surrounding them. Notice the difference between these cold-blooded creatures and the warm-blooded humans holding them.

That is, the ratio of the volume to the temperature is a constant. This relationship is known as *Charles's law*, named for the French scientist Jacques Charles (1746–1823), who made early hot-air balloon flights and was therefore quite interested in the relationship between the volume and temperature of a gas. A popular demonstration of Charles's law is shown in ►Fig. 10.6.

Low-density gases obey these laws, which may be combined into a single relationship. Since pV = constant and V/T = constant for a given quantity of gas, pV/T must also equal a constant. This relationship is the **ideal gas law**:

$$\frac{pV}{T} = \text{constant} \qquad \text{or} \qquad \frac{p_1V_1}{T_1} = \frac{p_2V_2}{T_2} \qquad \textit{ideal gas law (ratio form)} \quad (10.5)$$

That is, the ratio pV/T at one time (t_1) is the same as at another time (t_2), or at any other time, as long as the quantity (number of molecules or mass) of gas does not change.

This relationship can be written in a more general form that applies not just to a given quantity of a single gas, but to any quantity of any low-pressure, dilute gas. With a quantity of gas determined by the number of molecules (N) in the gas (that is, $pV/T \propto N$), it follows that

$$\frac{pV}{T} = Nk_B \qquad \text{or} \qquad pV = Nk_BT \qquad \textit{ideal gas law} \quad (10.6)$$

PHYSLET®

Exploration 20.3 Ideal Gas Law

where k_B is a constant of proportionality known as *Boltzmann's constant*:

$$k_B = 1.38 \times 10^{-23} \, J/K$$

The K stands for temperature on the Kelvin scale, discussed shortly. (Can you show that the units are correct?) Note that the mass of the sample does not appear explicitly in Eq. 10.6. However, the number of molecules N in a sample of a gas is proportional to the total mass of the gas. The ideal gas law, sometimes called the *perfect gas law*, applies to real gases with low pressures and densities, and de-scribes the behavior of most gases fairly accurately at normal densities.

Macroscopic Form of the Ideal Gas Law

Equation 10.6 is a *microscopic (micro* means extremely small) form of the ideal gas law in that it refers specifically to the number of molecules, N. However, the law can be rewritten in a *macroscopic (macro* means large) form, which involves quantities that can be measured with everyday laboratory equipment. In this form, we have

$$pV = nRT \quad \textit{ideal gas law} \quad (10.7)$$

using nR rather than Nk_B for convenience since $n \propto N$. Here, n is the number of moles (mol) of the gas, a quantity defined next, and R is called the *universal gas constant*: $R = 8.31 \, J/(mol \cdot K)$.

In chemistry, a **mole** (abbreviated mol) of a substance is defined as the quanti-ty that contains **Avogadro's number** (N_A) of molecules:

$$N_A = 6.02 \times 10^{23} \, molecules/mol$$

Thus, n and N in the two forms of the ideal gas law are related by $N = nN_A$. From Eq. 10.7, it can be shown that 1 mol of *any* gas occupies 22.4 L at 0°C and 1 atm. These conditions, 0°C and 1 atm, are known as *standard temperature and pressure (STP)*.

It is important to note what these equations for the macroscopic (Eq. 10.7) and microscopic (Eq. 10.6) forms of the ideal gas law represent. For the macro-scopic form of the ideal gas law, the constant $R = pV/(nT)$ has units of $J/(mol \cdot K)$. For the microscopic form of the law, $k_B = pV/(NT)$, with units of $J/(molecule \cdot K)$. Note that the difference between the macroscopic and micro-scopic forms of the ideal gas law is moles versus molecules, and we usually measure gas quantities in moles.

Equation 10.7 is a practical form of the ideal gas law, because (macroscopic or laboratory) quantities are measured in moles (n) of gases rather than the number of molecules (N). To use Eq. 10.7, we need to know the number of moles of a quan-tity of gas. This is done by finding the *formula mass* of a compound or element, which is the sum of the atomic masses given in the formula (for example, H_2O) of the substance. Because the masses are so small in relation to the SI standard kilo-gram, another unit, the *atomic mass unit* (u), is used:

$$1 \text{ atomic mass unit (u)} = 1.66054 \times 10^{-27} \, kg*$$

The formula mass is determined from the chemical formula and the atomic mass-es of the atoms. (The latter are listed in Appendix IV and are commonly rounded to the nearest one half.) For example, water, H_2O, with two hydrogen atoms and one oxygen atom, has a formula mass of $2m_H + 1m_O = 2(1.0 \, u) + 1(16.0 \, u) = 18.0 \, u$, because the atomic mass of each hydrogen atom is 1.0 u and that of an oxygen atom is 16.0 u. Then, 1 mol of water has a formula mass of 18.0. Similarly, the oxygen we breathe, O_2, has a formula mass of $2 \times 16.0 \, u = 32.0 \, u$. Hence, a mole of oxygen has a mass of 32.0 u. The mass of 1 mol of any substance is its formula mass expressed in grams. For example, 32.0 g of oxygen is one mole and would occupy 22.4 L at STP.

Note: The temperature T in the ideal gas law is absolute (Kelvin) temperature.

Note: N is the total number of molecules; N_A is Avogadro's number; $n = N/N_A$ is the number of moles.

(a)   **(b)**

▲ **FIGURE 10.6 Charles's law in action** Demonstrations of the relationship between the volume and the temperature of a quantity of gas. A weighted balloon, initially at room temperature, is placed in a beaker of water. **(a)** When ice is placed in the beaker and the temperature falls, the balloon's volume is reduced. **(b)** When the water is heated and the temperature rises, the balloon's volume increases.

---

*The atomic mass unit is based on assigning a carbon atom the value of exactly 12 u.

▶ **FIGURE 10.7** Constant-volume gas thermometer Such a thermometer indicates temperature as a function of pressure, since, for a low-density gas, $p \propto T$. **(a)** At some initial temperature, the pressure reading has a certain value. **(b)** When the gas thermometer is heated, the pressure (and temperature) reading is higher, because, on average, the gas molecules are moving faster.

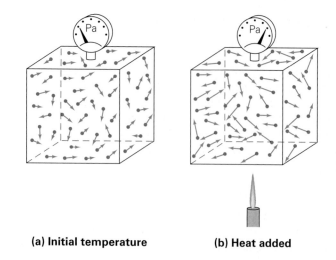

(a) Initial temperature   **(b) Heat added**

It is interesting to note that Avogadro's number allows you to compute the mass of a particular type of molecule. For example, suppose you want to know the mass of a water molecule (H_2O). As we have just seen, the formula mass of 1 mol of water is 18.0 g, or 18.0 g/mol. The *molecular mass* (m) is then given by

$$m = \frac{\text{formula mass (in kilograms)}}{N_A}$$

and, converting grams to kilograms, we have

$$m_{H_2O} = \frac{(18.0 \text{ g/mol})(10^{-3} \text{ kg/g})}{6.02 \times 10^{23} \text{ molecules/mol}} = 2.99 \times 10^{-26} \text{ kg/molecule}$$

Absolute Zero and the Kelvin Temperature Scale

The product of the pressure and the volume of a sample of ideal gas is directly proportional to the temperature of the gas: $pV \propto T$. This relationship allows a gas to be used to measure temperature in a *constant-volume gas thermometer*. Holding the volume of the gas constant, which can be done easily in a rigid container, means that $p \propto T$ (▲Fig. 10.7). Then using a constant-volume gas thermometer, one reads the temperature in terms of pressure. A plot of pressure versus temperature gives a straight line in this case (▼Fig. 10.8a).

▼ **FIGURE 10.8** Pressure versus temperature **(a)** A low-density gas kept at a constant volume gives a straight line on a graph of p versus T, that is, $p = (Nk_B/V)T$. When the line is extended to the zero pressure value, a temperature of $-273.15°C$ is obtained, which is taken to be absolute zero. **(b)** Extrapolation of lines for all low-density gases indicates the same absolute zero temperature. The actual behavior of gases deviates from this straight-line relationship at low temperatures because the gases start to liquefy.

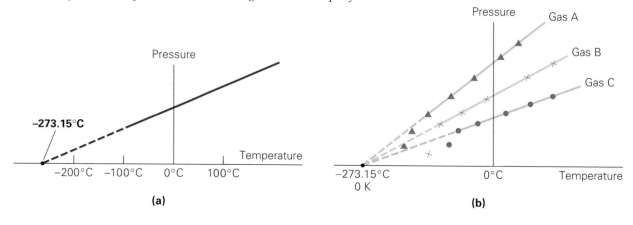

As can be seen in Fig. 10.8b, measurements of real gases (plotted data points) deviate from the values predicted by the ideal gas law at very low temperatures. This is because the gases liquefy at such temperatures. However, the relationship is linear over a large temperature range, and it looks as though the pressure might reach zero with decreasing temperature if the gas were to continue to be gaseous (ideal or perfect).

The absolute minimum temperature for an ideal gas is therefore inferred by extrapolating, or extending the straight line to the axis, as in Fig. 10.8b. This temperature is found to be −273.15°C and is designated as **absolute zero**. Absolute zero is believed to be the lower limit of temperature, but it has never been attained. In fact, there is a law of thermodynamics that says it never can be achieved (Section 12.5).* There is no known upper limit to temperature. For example, the temperatures at the centers of some stars are estimated to be greater than 100 million degrees (K or °C, take your choice).

Absolute zero is the foundation of the **Kelvin temperature scale**, named after the British scientist Lord Kelvin who proposed it in 1848.[†] On this scale, −273.15°C is taken as the zero point—that is, as 0 K (▼Fig. 10.9). The size of a single unit of Kelvin temperature is the same as that of the Celsius degree, so temperatures on these scales are related by

$$T_K = T_C + 273.15 \quad \textit{Celsius-to-Kelvin conversion} \tag{10.8}$$

where T_K is the temperature in **kelvins** (*not* degrees Kelvin; for example, 300 kelvins). The kelvin is abbreviated as K (*not* °K). For general calculations, it is common to round the 273.15 in Eq. 10.8 to 273, that is,

$$T_K = T_C + 273 \quad \textit{(for general calculations)} \tag{10.8a}$$

The absolute Kelvin scale is the official SI temperature scale; however, the Celsius scale is used in most parts of the world for everyday temperature readings. The absolute temperature in kelvins is used primarily in scientific applications.

◀ **FIGURE 10.9 The Kelvin temperature scale** The lowest temperature on the Kelvin scale (corresponding to −273.15°C) is absolute zero. A unit interval on the Kelvin scale, called a kelvin and abbreviated K, is equivalent to a temperature change of 1 C°; thus, $T_K = T_C + 273.15$. (The constant is usually rounded to 273 for convenience.) For example, a temperature of 0°C is equal to 273 kelvins.

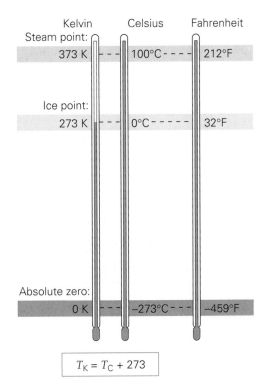

Kelvin	Celsius	Fahrenheit
Steam point:		
373 K	100°C	212°F
Ice point:		
273 K	0°C	32°F
Absolute zero:		
0 K	−273°C	−459°F

$$T_K = T_C + 273$$

*At the time of this writing, the lowest temperature scientists have been able to attain is 250×10^{-12} K, that is, 250 pK (picokelvins) above absolute zero.

[†]Lord Kelvin, born William Thomson (1824–1907), developed devices to improve telegraphy and the compass and was involved in the laying of the first transatlantic cable. When he received his title, it is said that he considered choosing Lord Cable or Lord Compass as the title, but decided on Lord Kelvin, after a river that runs near the University of Glasgow in Scotland, where he was a professor of physics for fifty years.

Problem-Solving Hint

Keep in mind that Kelvin temperatures *must* be used with the ideal gas law. It is a common mistake to use Celsius or Fahrenheit temperatures in that equation. Suppose you used a Celsius temperature of $T = 0°C$ in the gas law. You would have $pV = 0$, which makes no sense, since neither p nor V is zero at the freezing point of water.

Note that there can be no negative temperatures on the Kelvin scale if absolute zero is the lowest possible temperature. That is, the Kelvin scale doesn't have an arbitrary zero temperature somewhere within the scale as on the Fahrenheit and Celsius scales—zero K is absolute zero, period.

Example 10.2 ■ Deepest Freeze: Absolute Zero on the Fahrenheit Scale

What is absolute zero on the Fahrenheit scale?

Thinking It Through. This requires the conversion of 0 K to the Fahrenheit scale. But first a conversion to the Celsius scale is in order. (Why?)

Solution.

Given: $T_K = 0\,K$ *Find:* T_F

Temperatures on the Kelvin scale are related directly to Celsius temperatures by $T_K = T_C + 273.15$ (Eq. 10.8), so first we convert 0 K to a Celsius value:

$$T_C = T_K - 273.15 = 0 - 273.15 = -273.15°C$$

(We use $-273.15°C$ for absolute zero to give a more accurate value of absolute zero on the Fahrenheit scale.) Then, converting to Fahrenheit (Eq. 10.1) gives

$$T_F = \tfrac{9}{5}T_C + 32 = \tfrac{9}{5}(-273.15) + 32 = -459.67°F$$

Thus, absolute zero is about $-460°F$.

Follow-Up Exercise. There is an absolute temperature scale associated with the Fahrenheit temperature scale called the Rankine scale. A Rankine degree is the same size as a Fahrenheit degree, and absolute zero is taken as 0°R (zero degrees Rankine). Write the conversion equations between (a) the Rankine and the Fahrenheit scales, (b) the Rankine and the Celsius scales, and (c) the Rankine and the Kelvin scales.

Initially, gas thermometers were calibrated by using the ice and steam points. The Kelvin scale uses absolute zero and a second fixed point adopted in 1954 by the International Committee on Weights and Measures. This second fixed point is the **triple point of water**, at which water coexists simultaneously in equilibrium as a solid (ice), liquid (water), and gas (water vapor). The triple point occurs at a unique set of values for temperature and pressure—a temperature of 0.01°C and a pressure of 4.58 mm Hg—and provides a reproducible reference temperature for the Kelvin scale. The temperature of the triple point on the Kelvin scale was assigned a value of 273.16 K. The SI kelvin unit is then defined as 1/273.16 of the temperature at the triple point of water.*

Now let's use the ideal gas law, which requires absolute temperatures.

Example 10.3 ■ The Ideal Gas Law: Using Absolute Temperatures

A quantity of low-density gas in a rigid container is initially at room temperature (20°C) and a particular pressure (p_1). If the gas is heated to a temperature of 60°C, by what factor does the pressure change?

Thinking It Through. A "factor" of change implies a ratio (p_2/p_1), so Eq. 10.5 should apply. Note that the container is rigid, which means that $V_1 = V_2$.

Solution.

Given: $T_1 = 20°C$ *Find:* p_2/p_1 (pressure ratio or factor)
$T_2 = 60°C$
$V_1 = V_2$

*The 273.16 value given here for the triple point temperature and the -273.15 value, as determined in Fig. 10.8, indicate different things. The $-273.15°C$ is taken as 0 K. The 273.16 K (or 0.01°C) is a different reading on a different temperature scale.

Since the factor by which the pressure changes is wanted, we write p_2/p_1 as a ratio. For example, if $p_2/p_1 = 2$, then $p_2 = 2p_1$, or the pressure would change (increase) by a factor of 2. The ratio also indicates that we should use the ideal gas law in ratio form. The law requires *absolute* temperatures, so we first change the Celsius temperatures to kelvins:

$$T_1 = 20°C + 273 = 293 \text{ K}$$
$$T_2 = 60°C + 273 = 333 \text{ K}$$

Note: *Always* use Kelvin (absolute) temperatures with the ideal gas law.

Observe that a rounded value of 273 was used in Eq. 10.8 for convenience. Then, using the ideal gas law (Eq. 10.5) in the form $p_2V_2/T_2 = p_1V_1/T_1$, we have, with $V_1 = V_2$,

$$p_2 = \left(\frac{T_2}{T_1}\right)p_1 = \left(\frac{333 \text{ K}}{293 \text{ K}}\right)p_1 = 1.14p_1$$

So, p_2 is 1.14 times p_1; that is, the pressure increases by a factor of 1.14, or 14 percent. (What would the factor be if the Celsius temperatures were *incorrectly* used? It would be much larger: $60°C/20°C = 3$, or $p_2 = 3p_1$.)

Follow-Up Exercise. If the gas in this Example is heated from an initial temperature of 20°C (room temperature) so that the pressure increases by a factor of 1.26, what is the final Celsius temperature?

Because of its absolute nature, the Kelvin temperature scale has special significance. As will be seen in Section 10.5, the absolute temperature is directly proportional to the internal energy of an ideal gas and so can be used as an indication of that energy. There are no negative values on the absolute scale. Negative absolute temperatures would imply negative internal energy for the gas, a meaningless concept. Suppose you were asked to double the temperatures of, say, −10°C and 0°C. What would you do? The following Integrated Example should help.

Integrated Example 10.4 ■ Some Like It Hot: Doubling the Temperature

The evening weather report gives the day's high temperature as 10°C and predicts the next day's high to be 20°C. (a) A father tells his son that this means it will be twice as warm tomorrow, but the son says it does not. Do you agree with (1) the father or (2) the son? (b) Prove your result by using the absolute (Kelvin) temperature scale. Use a ratio.

(a) Conceptual Reasoning. Keep in mind that temperature gives a relative *indication* of hotness or coldness. Certainly, 20°C would be warmer than 10°C. But just because the numerical value of the higher temperature is twice as great (or greater by a factor of 2, because $20°C/10°C = 2$) does not necessarily mean it is twice as warm or there is twice the energy. It means that the air temperature is 10 degrees higher and therefore relatively warmer. So the son wins and the answer is (2).

(b) Quantitative Reasoning and Solution. The Kelvin temperatures can be computed directly from Eq. 10.8a, and a ratio of these temperatures will give the factor of increase based on internal energy.

Given: $T_{C_1} = 10°C$ *Find:* T_{K_2}/T_{K_1}
          $T_{C_2} = 20°C$

The equivalent absolute temperatures are

$$T_{K_1} = T_{C_1} + 273 = 10°C + 273 = 283 \text{ K}$$
$$T_{K_2} = T_{C_2} + 273 = 20°C + 273 = 293 \text{ K}$$

and

$$\frac{T_{K_2}}{T_{K_1}} = \frac{293 \text{ K}}{283 \text{ K}} = 1.04$$

So there is an increase of 0.04, or 4% in temperature.

Follow-Up Exercise. The weather report gives the day's high temperature as 0°C. If the next day's temperature were double that, what would the temperature be in degrees Celsius? Would this be environmentally possible?

10.4 Thermal Expansion

OBJECTIVE: To understand and be able to calculate the thermal expansions of solids and liquids.

Changes in the dimensions and volumes of materials are common thermal effects. As you learned earlier, thermal expansion provides a means of measuring temperature. The thermal expansion of gases is generally described by the ideal gas law and is very obvious. Less dramatic, but by no means less important, is the thermal expansion of solids and liquids.

Note: Solids are discussed in Section 9.1.

Thermal expansion results from a change in the average distance separating the atoms of a substance as it is heated. The atoms are held together by bonding forces, which can be simplistically represented as springs in a simple model of a solid. (See Fig. 9.1.) The atoms vibrate back and forth; with increased temperature (that is, more internal energy), they become increasingly active and vibrate over greater distances. With wider vibrations in all dimensions, the solid expands as a whole.

The change in one dimension of a solid (length, width, or thickness) is called *linear* expansion. For small temperature changes, linear expansion (or contraction) is approximately proportional to ΔT, or $T - T_o$ (▼Fig. 10.10a). The *fractional* change in length is $(L - L_o)/L_o$, or $\Delta L/L_o$, where L_o is the original length of the solid at the initial temperature.* This ratio is related to the change in temperature by

Demonstration/activity: Use the standard ring-and-ball apparatus to demonstrate linear expansion.

Exploration 19.2 Expansion of Materials

$$\frac{\Delta L}{L_o} = \alpha \Delta T \qquad \text{or} \qquad \Delta L = \alpha L_o \Delta T \qquad (10.9)$$

where α is the **thermal coefficient of linear expansion**. Note that the unit of α is inverse temperature: inverse Celsius degrees ($1/C°$, or $C°^{-1}$). Values of α for some materials are given in Table 10.1.

A solid may have different coefficients of linear expansion for different directions, but for simplicity we will assume that the same coefficient applies to all directions (in other words, that solids show *isotropic* expansion). Also, the coefficient of expansion may vary slightly for different temperature ranges. Since this variation is negligible for most common applications, α will be considered to be constant and independent of temperature.

▼ **FIGURE 10.10 Thermal expansion** **(a)** Linear expansion is proportional to the temperature change; that is, the change in length ΔL is proportional to ΔT, and $\Delta L/L_o = \alpha \Delta T$, where α is the thermal coefficient of linear expansion. **(b)** For isotropic expansion, the thermal coefficient of area expansion is approximately 2α. **(c)** The thermal coefficient of volume expansion for solids is about 3α.

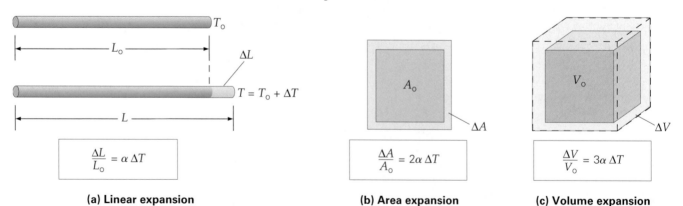

(a) Linear expansion   **(b) Area expansion**   **(c) Volume expansion**

*A fractional change may also be expressed as a percent change. For example, by analogy, if you invested \$100 (\$_o) and made \$10 (Δ\$), then the fractional change would be Δ\$/\$_o = 10/100 = 0.10$, or an increase (percent change) of 10%.

TABLE 10.1	Values of Thermal Expansion Coefficients (in $C^{\circ -1}$) for Some Materials at 20°C		
Material	Coefficient of linear expansion (α)	Material	Coefficient of volume expansion (β)
Aluminum	24×10^{-6}	Alcohol, ethyl	1.1×10^{-4}
Brass	19×10^{-6}	Gasoline	9.5×10^{-4}
Brick or concrete	12×10^{-6}	Glycerin	4.9×10^{-4}
Copper	17×10^{-6}	Mercury	1.8×10^{-4}
Glass, window	9.0×10^{-6}	Water	2.1×10^{-4}
Glass, Pyrex	3.3×10^{-6}		
Gold	14×10^{-6}	Air (and most other gases at 1 atm)	3.5×10^{-3}
Ice	52×10^{-6}		
Iron and steel	12×10^{-6}		

Equation 10.9 can be rewritten to give the final length (L) after a change in temperature:

$$\Delta L = \alpha L_o \Delta T$$
$$L - L_o = \alpha L_o \Delta T$$
$$L = L_o + \alpha L_o \Delta T$$

or

$$L = L_o(1 + \alpha \Delta T) \qquad (10.10)$$

Equation 10.10 can be used to compute the thermal expansion of *areas* of flat objects. Since area (A) is length squared (L^2) for a square,

$$A = L^2 = L_o^2(1 + \alpha \Delta T)^2 = A_o(1 + 2\alpha \Delta T + \alpha^2 \Delta T^2)$$

where A_o is the original area. Because the values of α for solids are much less than 1 ($\sim 10^{-5}$, as shown in Table 10.1), the second-order term (containing $\alpha^2 \simeq (10^{-5})^2 = 10^{-10} \ll 10^{-5}$) can be dropped with negligible error. As a first-order approximation, then, and with the understanding that the change in area, $\Delta A = A - A_o$, we have

$$A = A_o(1 + 2\alpha \Delta T) \qquad \text{or} \qquad \frac{\Delta A}{A_o} = 2\alpha \Delta T \qquad (10.11)$$

Thus, the **thermal coefficient of area expansion** (Fig. 10.10b) is twice as large as the coefficient of linear expansion. (That is, it is equal to 2α.) This relationship is valid for all flat shapes. (See the accompanying Learn by Drawing.)

Similarly, a first-order expression for thermal *volume* expansion is

$$V = V_o(1 + 3\alpha \Delta T) \qquad \text{or} \qquad \frac{\Delta V}{V_o} = 3\alpha \Delta T \qquad (10.12)$$

The **thermal coefficient of volume expansion** (Fig. 10.10c) is equal to 3α (for isotropic solids and liquids).

The equations for thermal expansions are approximations. (Why?) Even though an equation is a description of a physical relationship, always keep in mind that it may be only an approximation of physical reality or may apply only in certain situations.

The thermal expansion of materials is an important consideration in construction. Seams are put in concrete highways and sidewalks to allow room for expansion

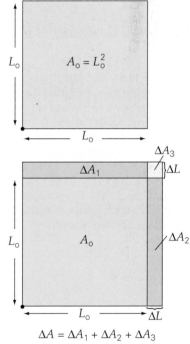

LEARN BY DRAWING

Thermal Area Expansion

$\Delta A = \Delta A_1 + \Delta A_2 + \Delta A_3$
$\Delta A_1 = \Delta A_2 = L_o \, \Delta L$
$\quad = L_o \, (\alpha L_o \, \Delta T) = \alpha A_o \, \Delta T$
Since ΔA_3 is very small
compared to ΔA_1 and ΔA_2,
$\quad \Delta A \approx 2\alpha A_o \, \Delta T$

(a)

(b)

▲ **FIGURE 10.11 Expansion gaps**
(a) Expansion gaps are built into bridge roadways to prevent contact stresses produced by thermal expansion. **(b)** These loops in oil pipelines serve a similar purpose. As hot oil passes through them, the pipes expand, and the loops take up the extra length. The loops also accommodate expansions resulting from day–night temperature variations.

and to prevent cracking. Expansion gaps in large bridges and between railroad rails are necessary to prevent damage (◄Fig. 10.11a). The Golden Gate Bridge across San Francisco Bay varies in length by about 1 m between summer and winter. Similarly, expansion loops are found in oil pipelines (Fig. 10.11b). The height of the Eiffel Tower in Paris varies 0.36 cm for each Celsius degree change.

The thermal expansion of steel beams and girders can produce tremendous pressures, as the following Example shows.

Example 10.5 ■ Temperature Rising: Thermal Expansion and Stress

A steel beam is 5.0 m long at a temperature of 20°C (68°F). On a hot day, the temperature rises to 40°C (104°F). (a) What is the change in the beam's length due to thermal expansion? (b) Suppose that the ends of the beam are initially in contact with rigid vertical supports. How much force will the expanded beam exert on the supports if the beam has a cross-sectional area of 60 cm²?

Thinking It Through. (a) This is a direct application of Eq. 10.9. (b) As the constricted beam expands, it applies a stress, and hence a force, to the supports. For linear expansion, Young's modulus (Section 9.1) should come into play.

Solution.

Given: $L_0 = 5.0$ m   *Find:* (a) ΔL (change in length)
$T_0 = 20°C$ (b) F (force)
$T = 40°C$
$\alpha = 12 \times 10^{-6}$ C$^{\circ -1}$ (from Table 10.1)
$A = 60 \text{ cm}^2\left(\dfrac{1 \text{ m}}{100 \text{ cm}}\right)^2 = 6.0 \times 10^{-3} \text{ m}^2$

(a) Using Eq. 10.9 to find the change in length with $\Delta T = T - T_0 = 40°C - 20°C = 20$ C°, we have

$$\Delta L = \alpha L_0 \Delta T = (12 \times 10^{-6} \text{ C}^{\circ -1})(5.0 \text{ m})(20 \text{ C}°) = 1.2 \times 10^{-3} \text{ m} = 1.2 \text{ mm}$$

This may not seem like much of an expansion, but it can give rise to a great deal of force if the beam is constrained and kept from expanding, as part (b) will show.

(b) By Newton's third law, if the beam is kept from expanding, the force the beam exerts on its constraint supports is equal to the force exerted by the supports to prevent the beam from expanding by a length ΔL. This is the same as the force that would be required to compress the beam by that length. Using the Young's modulus form of Hooke's law (Section 9.1) with $Y = 20 \times 10^{10}$ N/m² (Table 9.1), the stress on the beam is

$$\frac{F}{A} = \frac{Y\Delta L}{L_0} = \frac{(20 \times 10^{10} \text{ N/m}^2)(1.2 \times 10^{-3} \text{ m})}{5.0 \text{ m}} = 4.8 \times 10^7 \text{ N/m}^2$$

The force is then

$$F = (4.8 \times 10^7 \text{ N/m}^2)A = (4.8 \times 10^7 \text{ N/m}^2)(6.0 \times 10^{-3} \text{ m}^2)$$
$$= 2.9 \times 10^5 \text{ N (about 65 000 lb, or 32.5 tons!)}$$

Follow-Up Exercise. Expansion gaps between identical steel beams laid end to end are specified to be 0.060% of the length of a beam at the installation temperature. With this specification, what is the temperature range for noncontact expansion?

Conceptual Example 10.6 ■ Larger or Smaller? Area Expansion

A circular piece is cut from a flat metal sheet (▶Fig. 10.12a). If the sheet is then heated in an oven, the size of the hole will (a) become larger, (b) become smaller, (c) remain unchanged.

Reasoning and Answer. It is a common misconception to think that the area of the hole will shrink because the metal expands inwardly around it. To counter this misconception, think of the piece of metal removed from the hole rather than of the hole itself. This piece would expand with increasing temperature. The metal in the heated sheet reacts as if the piece that is removed were still part of it. (Think of putting the piece of metal back into the hole after heating, as in Fig. 10.12b, or consider drawing a circle on an uncut metal sheet and heating it.) So the answer is (a).

Follow-Up Exercise. A circular ring of iron has a tight-fitting metal bar inside it, across its diameter. If the ring is heated in an oven to a high temperature, would it be distorted or bent out of shape, or would it remain circular?

Fluids (liquids and gases), like solids, normally expand with increasing temperature. Because fluids have no definite shape, only volume expansion (and not linear or area expansion) is meaningful. The expression is

$$\frac{\Delta V}{V_0} = \beta \Delta T \quad \textit{fluid volume expansion} \qquad (10.13)$$

where β is the coefficient of volume expansion for fluids. Note in Table 10.1 that the values of β for fluids are typically larger than the values of 3α for solids.

Unlike most liquids, water exhibits an anomalous expansion in volume near its freezing point. The volume of a given amount of water decreases as it is cooled from room temperature, until its temperature reaches 4°C (▼Fig. 10.13a). Below 4°C, the volume increases, and therefore the density decreases (Fig. 10.13b). This means that water has its maximum density ($\rho = m/V$) at 4°C (actually, 3.98°C).

When water freezes, its molecules form a hexagonal (six-sided) lattice pattern. (This is why snowflakes have hexagonal shapes.) It is the open structure of this lattice that gives water its almost unique property of expanding on freezing and being less dense as a solid than as a liquid. (This is why ice floats in water and frozen water pipes burst—water expands by about 9% on freezing.) The variation in the density of water over the temperature range from 4°C to 0°C indicates that the open lattice structure is beginning to form at about 4°C rather than exactly at the freezing point.

This property has an important environmental effect: Bodies of water such as lakes and ponds freeze at the top first, and the ice that forms floats. As a lake cools toward 4°C, water near the surface loses energy to the atmosphere, becomes denser, and sinks. The warmer, less dense water near the bottom rises. However, once the colder water on top reaches temperatures below 4°C, it becomes less dense and remains at the surface, where it freezes. If water did not have this property, lakes and ponds would freeze from the bottom up, which would destroy much of their animal and plant life (and would make ice skating a lot less popular). There would also be no oceanic ice caps at the polar regions. Instead, there would be a thick layer of ice at the bottom of the ocean, covered by a layer of water.

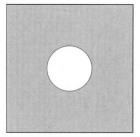

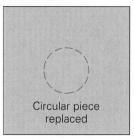

(a) Metal plate with hole

(b) Metal plate without hole

▲ **FIGURE 10.12** A larger or smaller hole? See Conceptual Example 10.6.

Teaching tip: Students commonly assume that the diameter of a hole in a large plate will decrease when the plate is heated, yet they feel that the diameter of a thin ring will increase. After discussing this example, ask students what would happen to a letter of the alphabet if it were heated?

Demonstration/activity: Have students watch the reading on a mercury thermometer when a hot flame is passed near the bulb. The reading will drop slightly while the glass bulb heats up; then the mercury will start expanding, as expected.

Demonstration/activity: Fill a flask with water to the brim. Heat the flask. The water level will rise, and water will spill over as the temperature increases.

◀ **FIGURE 10.13** Thermal expansion of water Water exhibits nonlinear expansion behavior near its freezing point. **(a)** Above 4°C (actually, 3.98°C), water expands with increasing temperature, but from 4°C down to 0°C, it expands with decreasing temperature. **(b)** As a result, water has its maximum density near 4°C.

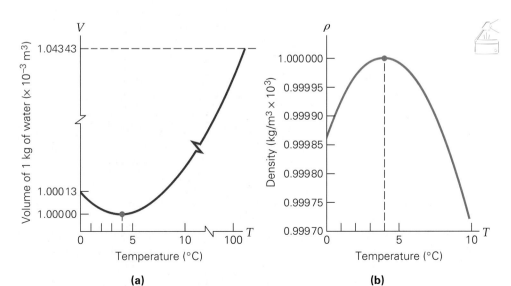

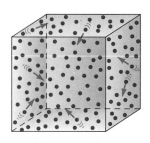

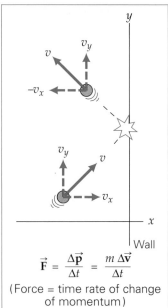

$$\vec{F} = \frac{\Delta \vec{p}}{\Delta t} = \frac{m\,\Delta \vec{v}}{\Delta t}$$

(Force = time rate of change of momentum)

▲ **FIGURE 10.14 Kinetic theory of gases** The pressure a gas exerts on the walls of a container is due to the force resulting from the change in momentum of the gas molecules that collide with the wall. The force exerted by an individual molecule is equal to the time rate of change of momentum; that is, $\vec{F} = \Delta\vec{p}/\Delta t = m\Delta\vec{v}/\Delta t$, where $\vec{p} = m\vec{v}$. The sum of the instantaneous normal components of the collision forces gives rise to the average pressure on the wall.

Teaching tip: The average kinetic energy of gas molecules is dependent only on the temperature. Compare the rms speeds of molecules of different masses at the same temperature. Also, compare the temperatures of molecules having different masses, but the same speed.

Note: Elastic collisions are discussed in Section 6.4.

Conceptual Example 10.7 ■ Quick Chill: Temperature and Density

Ice is put into a container of water at room temperature. For faster cooling, the ice should (a) be allowed to float naturally in the water (b) be pushed to the bottom of the container with a stick and held there.

Reasoning and Answer. When the ice melts, the water in its vicinity is cooled and hence becomes denser (Fig. 10.13b). If the ice is permitted to float at the top, the denser water sinks and the warmer, less dense water rises. This mixing causes the water to cool quickly. If the ice were at the bottom of the container, however, the colder, denser water would remain there, and cooling of the top layer of warm water would be slowed, so the answer is (a).

Follow-Up Exercise. Suppose that the density-versus-temperature curve for water (Fig. 10.13b) were inverted, so that it dipped downward. What would this imply for the situation in this Example and for the freezing of lakes? Explain.

10.5 The Kinetic Theory of Gases

OBJECTIVES: To (a) relate kinetic theory and temperature, and (b) explain the process of diffusion.

If the molecules of a sample of gas are viewed as colliding particles, the laws of mechanics can be applied to each molecule of the gas. We should then be able to describe the gas's microscopic characteristics, such as pressure, internal energy, and so on, in terms of molecular motion. Because of the large number of particles involved, however, a statistical approach is employed for such a microscopic description.

One of the major accomplishments of theoretical physics was to do exactly that—derive the ideal gas law from mechanical principles. This derivation led to a new interpretation of temperature in terms of the translational kinetic energy of the gas molecules. As a theoretical starting point, the molecules of an ideal gas are viewed as point masses in random motion with relatively large distances separating them.

In this section, we consider primarily the kinetic theory of *monatomic* (single-atom) gases, such as He, and learn about the internal energy of such a gas. In the next section, the internal energy of *diatomic* (two-atom molecules) gases, such as O_2, will be considered. In either case, the vibrational and rotational motions can be ignored with regard to temperature and pressure, since these quantities depend only on *linear* motion.

According to the **kinetic theory of gases**, the molecules of an ideal gas undergo perfectly elastic collisions with the walls of its container. (With the molecules taken to be point particles, molecular collisions can be neglected.) From Newton's laws of motion, the force on the walls of the container can be calculated from the change in momentum of the gas molecules when they collide with the walls (◄Fig. 10.14). If this force is expressed in terms of pressure (force/area), the following equation is obtained (see Appendix II for derivation):

$$pV = \tfrac{1}{3}Nmv_{\text{rms}}^2 \qquad (10.14)$$

Here, V is the volume of the container or gas, N is the number of gas molecules in the closed container, m is the mass of a gas molecule, and the speed v_{rms} is the average speed of the molecules, but a special kind of average. It is obtained by averaging the squares of the speeds and then taking the square root of the average—that is, $\sqrt{\overline{v^2}} = v_{\text{rms}}$. As a result, v_{rms} is called the *root-mean-square* (*rms*) speed.

Solving Eq. 10.6 for pV and equating the resulting expression with Eq. 10.14 shows how temperature came to be interpreted as a measure of translational kinetic energy:

$$pV = Nk_{\text{B}}T = \tfrac{1}{3}Nmv_{\text{rms}}^2 \quad \text{or} \quad \tfrac{1}{2}mv_{\text{rms}}^2 = \tfrac{3}{2}k_{\text{B}}T \quad \textit{(for all ideal gases)} \qquad (10.15)$$

Thus, the temperature of a gas (and that of the walls of the container or a thermometer bulb in thermal equilibrium with the gas) is directly proportional to its average random kinetic energy (per molecule), since $\overline{K} = \tfrac{1}{2}mv_{\text{rms}}^2 = \tfrac{3}{2}k_{\text{B}}T$. (Don't forget that T is the absolute temperature in kelvins.)

Example 10.8 ■ Molecular Speed: Relation to Absolute Temperature

What is the average (rms) speed of a helium atom (He) in a helium balloon at room temperature? (Take the mass of the helium atom to be 6.65×10^{-27} kg.)

Thinking It Through. All the data we need to solve for the average speed in Eq. 10.15 are known.

Solution.

Given: $m = 6.65 \times 10^{-27}$ kg   *Find:* v_{rms} (rms speed)
$\quad\quad\quad T = 20°C$ (room temperature)
$\quad\quad\quad k_B = 1.38 \times 10^{-23}$ J/K (known)

Eq. 10.15 will be used, so we list k_B among the given quantities.
    The Celsius temperature must be changed to kelvins, and note that k_B has units of J/K, so

$$T_K = T_C + 273 = 20°C + 273 = 293 \text{ K}$$

Rearranging Eq. 10.15,

$$v_{rms} = \sqrt{\frac{3k_B T}{m}} = \sqrt{\frac{3(1.38 \times 10^{-23} \text{ J/K})(293 \text{ K})}{6.65 \times 10^{-27} \text{ kg}}} = 1.35 \times 10^3 \text{ m/s} = 1.35 \text{ km/s}$$

This is more than 3000 mi/h—pretty fast!

Follow-Up Exercise. In this Example, if the temperature of the gas were increased by 10 C°, what would be the corresponding percentage increases in the average (rms) speed and in the average kinetic energy?

Interestingly, Eq. 10.15 predicts that at absolute zero ($T = 0$ K), all translational molecular motion of a gas would cease. According to classical theory, this would correspond to absolute zero energy. However, modern quantum theory says that there would still be some zero-point motion, and a corresponding minimum *zero-point energy*. Basically, absolute zero is the temperature at which all the energy that *can* be removed from an object has been removed.

Internal Energy of Monatomic Gases

Because the "particles" in an ideal monatomic gas do not vibrate or rotate, as explained previously, the total translational kinetic energy of all the molecules is equal to the total internal energy of the gas. That is, the gas's internal energy is all "temperature" energy (Section 10.1). With N molecules in a system, we can use Eq. 10.15, expressing the energy per molecule, to write an equation for the total internal energy U:

$$U = N\left(\tfrac{1}{2}mv_{rms}^2\right) = \tfrac{3}{2}Nk_B T = \tfrac{3}{2}nRT \quad \textit{(for ideal monatomic gases only)} \quad (10.16)$$

Thus, we see that the internal energy of an ideal monatomic gas is directly proportional to its absolute temperature. (In Section 10.6, we will see that this is true regardless of the molecular structure of the gas. However, the expression for U will be a bit different for gases that are not monatomic.) This means that if the absolute temperature of a gas is doubled (by heat transfer), for example, from 200 K to 400 K, then the internal energy of the gas is also doubled.

Diffusion

We depend on our sense of smell to detect odors, such as the smell of smoke from something burning. That you can smell something from a distance implies that molecules get from one place to another in the air—from the source to your nose. This process of random molecular mixing in which particular molecules move from a region where they are present in higher concentration to one where they are in lower concentration is called **diffusion**. Diffusion also occurs readily in

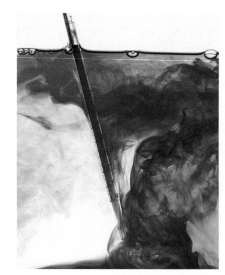

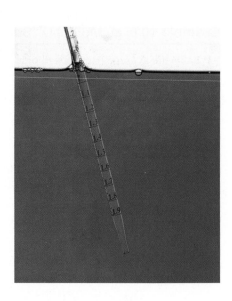

▲ **FIGURE 10.15 Diffusion in liquids** Random molecular motion would eventually distribute the dye throughout the water. Here there is some distribution due to mixing, and the ink colors the water after a few minutes. The distribution would take more time by diffusion only.

liquids; think about what happens to a drop of ink in a glass of water (▲Fig. 10.15). It even occurs to some degree in solids.

The rate of diffusion for a particular gas depends on the rms speed of its molecules. Even though gas molecules have large average speeds (Example 10.8), their average positions change slowly, and the molecules do not fly from one side of a room to the other. Instead, there are frequent collisions, and as a result, the molecules "drift" rather slowly. For example, suppose someone opened a bottle of ammonia on the other side of a closed room. It would take some time for the ammonia to diffuse across the room until you could smell it. (Much of the movement that people commonly attribute to diffusion is actually due to air currents.)

Gases can also diffuse through porous materials or permeable membranes. (This process is sometimes referred to as *effusion*.) Energetic molecules enter the material through the pores (openings), and, colliding with the pore walls, they slowly meander through the material. Such gaseous diffusion can be used to physically separate the different gases making up a mixture.

The kinetic theory of gases says that the average translational kinetic energy (per molecule) of a gas is proportional to the absolute temperature of the gas: $\frac{1}{2}mv_{rms}^2 = \frac{3}{2}k_BT$. So on the average, the molecules of different gases (having different masses) move at different speeds at a given temperature. As you might expect, because they move faster, lighter gas molecules diffuse through the tiny openings of a porous material faster than do heavier gas molecules.

For instance, at a particular temperature, molecules of oxygen (O_2) move faster on the average than do the more massive molecules of carbon dioxide (CO_2). Because of this difference in molecular speed, oxygen can diffuse through a barrier faster than carbon dioxide can. Suppose that a mixture of equal volumes of oxygen and carbon dioxide is contained on one side of a porous barrier (▼Fig. 10.16). After a while, some O_2 molecules and some CO_2 molecules will have diffused through the barrier, but more oxygen than carbon dioxide. Repeating the process on this diffused gas mixture would cause the oxygen concentration to become even greater

▶ **FIGURE 10.16 Separation by gaseous diffusion** The molecules of both gases diffuse (or effuse) through the porous barrier, but because oxygen molecules have the greater average speed, more of them pass through. Thus, over time, there is a greater concentration of oxygen molecules on the other side of the barrier.

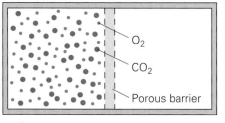

Equal volumes of O_2 and CO_2

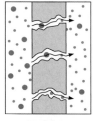

Diffusion through barrier

INSIGHT 10.3 PHYSIOLOGICAL DIFFUSION IN LIFE PROCESSES

Diffusion plays a central role in many life processes. For example, consider a cell membrane in the lung. Such a membrane is permeable to a number of substances, any of which will diffuse through the membrane from a region where its concentration is high to a region where its concentration is low. Most important, the lung membrane is permeable to oxygen (O_2), and the transfer of O_2 across the membrane occurs because of a concentration gradient.

The blood carried to the lungs is low in O_2, having given up the oxygen during its circulation through the body to tissues requiring O_2 for metabolism. Conversely, the air in the lungs is high in O_2, because there is a continuous exchange of fresh air in the breathing process. As a result of this concentration difference, or gradient, O_2 diffuses from the space within the lungs into the blood that flows through the lung tissue, and the blood leaving the lungs is high in O_2.

Exchanges between the blood and the tissues occur across capillary walls, and diffusion again is a major factor. The chemical composition of arterial blood is regulated to maintain the proper concentrations of particular solutes (substances dissolved in the blood solution), so diffusion takes place in the appropriate directions across capillary walls. For example, as cells take up O_2 and nutrients, including glucose (blood sugar), the blood continuously brings in fresh supplies of the substances to maintain the concentration gradient needed for diffusion to the cells. The continuous production of carbon dioxide (CO_2) and metabolic wastes in the cells produces concentration gradients in the opposite direction for these substances. They therefore diffuse out of the cells into the blood, to be carried away from the tissues by the circulatory system.

During periods of physical exertion, cellular activity increases. More O_2 is used up and more CO_2 is produced, thereby increasing the concentration gradients and the diffusion rates. How do the lungs respond to an increased demand for O_2 to the blood? As you might expect, the rate of diffusion depends on the surface area and thickness of the lung membrane. Deeper breathing during exercise causes the alveoli (small air sacs in the lungs) to increase in volume. The alveolar surface area increases accordingly, and the thickness of the membrane wall decreases, allowing more rapid diffusion.

Also, the heart works harder during exercise, and the blood pressure is raised. The increased pressure forces open capillaries that are normally closed during rest or mild activity. As a result, the total exchange area between the blood and cells is increased. Each of these changes helps expedite the exchange of gases during exercise.

on the far side of the barrier. Almost pure oxygen can be obtained by repeating the separation process many times. Separation by gaseous diffusion is a key process in obtaining enriched uranium, which was used in the first atomic bomb and in early nuclear reactors that generate electricity (Chapter 30).

Fluid diffusion is very important to organisms. In plant photosynthesis, carbon dioxide from the air diffuses into leaves, and oxygen and water vapor diffuse out. The diffusion of liquid water across a permeable membrane down a concentration gradient (a concentration difference) is called **osmosis**, a process that is vital in living cells. Osmotic diffusion is also important to kidney functioning: Tubules in the kidneys concentrate waste matter from the blood in much the same way that oxygen is removed from mixtures. (See the accompanying Insight 10.3 for other examples of diffusion.)

Osmosis is the tendency for the solvent of a solution, such as water, to diffuse across a semipermeable membrane from the side where the solvent is at higher concentration to the side where it is at lower concentration. When pressure is applied to the side with the lower concentration, the diffusion is reversed—a process called *reverse osmosis*. Reverse osmosis is used in desalination plants to provide freshwater from seawater in dry coastal regions.

Reverse osmosis is also used in water purification. You may have drunk such purified water. One of the popular bottled waters is purified "using state of the art treatment by reverse osmosis," according to the label.

*10.6 Kinetic Theory, Diatomic Gases, and the Equipartition Theorem

OBJECTIVES: To understand (a) the difference between monatomic and diatomic gases, (b) the meaning of the equipartition theorem, and (c) the expression for the internal energy of a diatomic gas.

In the real world, most of the gases we deal with are *not* monatomic gases. Recall from chemistry that monatomic gases are elements known as *noble* or *inert* gases, because they do not readily combine with other atoms. These elements are found on the far right side of the periodic chart: helium, neon, argon, krypton, xenon, and radon.

However, the mixture of gases we breathe (collectively known as "air") consists mainly of diatomic molecules of nitrogen (N_2, 78% by volume) and oxygen (O_2, 21% by volume). Each of these gases has two identical atoms chemically bonded together to form a single molecule. How do we deal with these realistic, more complicated molecules in terms of the kinetic theory of gases? [There are even more complicated gas molecules consisting of more than two atoms, such as carbon dioxide (CO_2). However, because of the complexity of such gas molecules, we will limit our discussion to diatomic molecules.]

The Equipartition Theorem

As we saw in Section 10.5, only the translational kinetic energy of a gas is determined by the gas's temperature. Thus, for any type of gas, regardless of how many atoms make up its molecules, it is *always true* that the average *translational* kinetic energy per molecule is still proportional to the temperature of the gas (Eq. 10.15): $\frac{1}{2}mv_{rms}^2 = \frac{3}{2}k_BT$ (for all gases).

Recall that for monatomic gases, the total internal energy U consists solely of translational kinetic energy. For diatomic molecules, this is not true, because a diatomic molecule is free to rotate and vibrate in addition to moving linearly. Therefore, these extra forms of energy must be taken into account. The expression given in Eq. 10.16 $\left(U = \frac{3}{2}Nk_BT\right)$ for monatomic gases, which assumes that the total energy is due only to translational kinetic energy, therefore does *not* hold for diatomic gases.

Scientists wondered exactly how the expression for the internal energy of a diatomic gas might differ from that for a monatomic gas. In looking at the derivation of Eq. 10.16 from the kinetic theory, they realized that the factor of 3 in that equation was due to the fact that the gas molecules had three independent linear ways (dimensions) of moving. Thus, for each molecule, there were three independent ways of possessing kinetic energy: with x, y, and z linear motion. Each independent way a molecule has for possessing energy is called a **degree of freedom**.

According to this scheme, a monatomic gas has only three degrees of freedom, since its molecules can move only linearly and can possess kinetic energy in three dimensions. Scientists reasoned that, quite possibly, a diatomic gas could vibrate (see Fig. 10.1), thus having vibrational kinetic and potential energies (two additional degrees of freedom). A diatomic molecule might also rotate.

Consider a symmetric diatomic molecule—for example, O_2. A classical model describes such a diatomic molecule as though the molecules were particles connected by a rigid rod (◄Fig. 10.17). The rotational moment of inertia, I, has the same value about each of the axes (x and y) that pass perpendicularly through the center of the rod. The moment of inertia about the z-axis is essentially zero. (Why?) Thus, only two degrees of freedom are associated with the kinetic energies of diatomic molecule rotation.

On the basis of the understanding of monatomic gases and their three degrees of freedom, the **equipartition theorem** was proposed. (As the name implies, the total energy of a gas or molecule is "partitioned," or divided, equally for each degree of freedom.) That is,

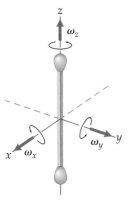

▲ **FIGURE 10.17 Model of a diatomic gas molecule** A dumbbell-like molecule can rotate about three axes. The moment of inertia, I, about the x- and y-axes is the same. The masses (molecules) on the ends of the rod are pointlike particles, so the moment of inertia about the z-axis is I_z is negligible compared to I_x and I_y.

> On average, the total internal energy U of an ideal gas is divided equally among each degree of freedom its molecules possess. Furthermore, each degree of freedom contributes $\frac{1}{2}Nk_BT$ (or $\frac{1}{2}nRT$) to the total internal energy of the gas.

The equipartition theorem fits the special case of a monatomic gas, since the theorem predicts that $U = \frac{3}{2}Nk_BT$, which we know to be true. With three degrees of freedom, $U = 3\left(\frac{1}{2}Nk_BT\right)$, in agreement with the monatomic result given earlier (Eq. 10.16).

The Internal Energy of a Diatomic Gas

Exactly how does the equipartition theorem enable us to calculate the internal energy of a diatomic gas such as oxygen? To perform such a calculation, we must realize that U now includes all the available degrees of freedom. In addition to the translational degrees of freedom, what other motions are the molecules undergoing? The analysis

Exploration 20.4 Equipartition Theorem

is complicated and beyond the scope of this text, so we give only the general results. *Typically, for normal (room) temperatures, quantum theory predicts (and experiment verifies) that only the rotational motions are important in the degrees of freedom.*

So the total internal energy of a diatomic gas is composed of the internal energies due to the three linear degrees of freedom and the two rotational degrees of freedom, for a total of five degrees of freedom. Hence, we may write

$$U = K_{\text{trans}} + K_{\text{rot}} = 3\left(\tfrac{1}{2}nRT\right) + 2\left(\tfrac{1}{2}nRT\right) \qquad \begin{array}{l}\textit{(for diatomic gases}\\ \textit{near room temperature)}\end{array} \qquad (10.17)$$
$$= \tfrac{5}{2}nRT = \tfrac{5}{2}Nk_BT$$

Thus, a given monatomic sample of gas at normal room temperature has 40% less internal energy than a similar diatomic sample at the same temperature. Or, equivalently, the monatomic sample possesses only 60% of the internal energy of the diatomic sample.

Example 10.9 ■ Monatomic versus Diatomic: Are Two Atoms Better Than One?

More than 99% of the air we breathe consists of diatomic gases, mainly nitrogen (N_2, 78%) and oxygen (O_2, 21%). There are traces of other gases, one of which is radon (Ra), a monatomic gas arising from radioactive decay of uranium in the ground. [Radon itself is radioactive, which is irrelevant here, but this fact does make radon a possible health danger when concentrated inside a house (as we shall see in Chapter 29).] (a) Calculate the total internal energy of 1.00-mol samples each of oxygen and radon at room temperature (20°C). (b) For each sample, calculate the amount of internal energy associated with molecular *translational* kinetic energy.

Thinking It Through. (a) We have to consider the number of degrees of freedom in a monatomic gas and a diatomic gas in computing the internal energy U. (b) Only three linear degrees of freedom contribute to the translational kinetic energy portion (U_{trans}) of the internal energy.

Solution. We list the data and convert to kelvins right away because we know that the internal energy is expressed in terms of absolute temperature:

Given: $n = 1.00$ mol   *Find:* (a) U for O_2 and Ra samples
            $T = 20°C + 273 = 293$ K   (b) U_{trans} for O_2 and Ra samples at
              at room temperature   room temperature

(a) Let's compute the total internal energy of the (monatomic) radon sample first, using Eq. 10.16:

$$U_{\text{Ra}} = \tfrac{3}{2}nRT = \tfrac{3}{2}(1.00 \text{ mol})[8.31 \text{ J}/(\text{mol} \cdot \text{K})](293 \text{ K}) = 3.65 \times 10^3 \text{ J}$$

Since this sample is at room temperature, the (diatomic) oxygen will also include internal energy stored as two extra degrees of freedom, due to rotation. Thus, we have

$$U_{O_2} = \tfrac{5}{2}nRT = \tfrac{5}{2}(1.00 \text{ mol})[8.31 \text{ J}/(\text{mol} \cdot \text{K})](293 \text{ K}) = 6.09 \times 10^3 \text{ J}$$

As we have seen, even though each sample has the same number of molecules and the same temperature, the oxygen sample has about 67% more total internal energy.

(b) For (monatomic) radon, all the internal energy is in the form of translational kinetic energy; hence, the answer is the same as in part (a):

$$U_{\text{trans}} = U_{\text{Ra}} = 3.65 \times 10^3 \text{ J}$$

For (diatomic) oxygen, only $\tfrac{3}{2}nRT$ of the total internal energy $\left(\tfrac{5}{2}nRT\right)$ is in the form of translational kinetic energy, so the answer is the same as for radon; that is, $U_{\text{trans}} = 3.65 \times 10^3$ J for both gas samples.

Follow-Up Exercise. (a) In this Example, how much energy is associated with the rotational motion of the oxygen molecules? (b) Which sample has the highest rms speed? (*Note:* The mass of one radon atom is about seven times the mass of an oxygen molecule.) Explain your reasoning.

Chapter Review

Celsius–Fahrenheit conversions:

$$T_F = \tfrac{9}{5}T_C + 32 \quad \text{or} \quad T_F = 1.8T_C + 32 \qquad (10.1)$$

$$T_C = \tfrac{5}{9}(T_F - 32) \qquad (10.2)$$

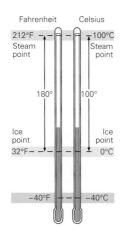

- **Heat** is the net energy transferred from one object to another because of temperature differences. Once transferred, the energy becomes part of the internal energy of the object (or system).

- The **ideal (or perfect) gas law** relates the pressure, volume, and absolute temperature of an ideal or dilute gas.

Ideal (or perfect) gas law (always use absolute temperatures):

$$\frac{p_1 V_1}{T_1} = \frac{p_2 V_2}{T_2} \quad \text{or} \quad pV = Nk_B T \qquad (10.5\text{–}6)$$

or

$$pV = nRT \qquad (10.7)$$

where $k_B = 1.38 \times 10^{-23}$ J/K and $R = 8.31$ J/(mol·K)

- **Absolute zero (0 K)** corresponds to $-273.15°C$.

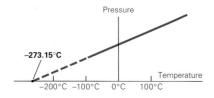

Celsius–Kelvin conversion:

$$T_K = T_C + 273.15 \qquad (10.8)$$

$$T_K = T_C + 273 \quad \textit{(for general calculations)} \qquad (10.8\text{a})$$

- **Thermal coefficients of expansion** relate the fractional change in dimension(s) to a change in temperature.

Thermal expansion of solids:

linear: $\quad \dfrac{\Delta L}{L_o} = \alpha \Delta T \ \text{ or } \ L = L_o(1 + \alpha \Delta T) \qquad (10.9, 10.10)$

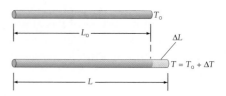

area: $\quad \dfrac{\Delta A}{A_o} = 2\alpha \Delta T \ \text{ or } \ A = A_o(1 + 2\alpha \Delta T) \qquad (10.11)$

volume: $\quad \dfrac{\Delta V}{V_o} = 3\alpha \Delta T \ \text{ or } \ V = V_o(1 + 3\alpha \Delta T) \qquad (10.12)$

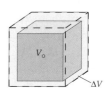

Thermal volume expansion of fluids:

$$\frac{\Delta V}{V_o} = \beta \Delta T \qquad (10.13)$$

- According to the **kinetic theory of gases**, the absolute temperature of a gas is directly proportional to the average random kinetic energy per molecule.

Results of kinetic theory of gases:

$$pV = \tfrac{1}{3}Nmv_{rms}^2 \qquad (10.14)$$

$$\tfrac{1}{2}mv_{rms}^2 = \tfrac{3}{2}k_B T \qquad \textit{(for all ideal gases)} \qquad (10.15)$$

$$U = \tfrac{3}{2}Nk_B T = \tfrac{3}{2}nRT \qquad \textit{(ideal monatomic gases only)} \qquad (10.16)$$

$$U = \tfrac{5}{2}Nk_B T = \tfrac{5}{2}nRT \qquad \textit{(for diatomic gases near room temperature)} \qquad (10.17)$$

Exercises*

MC = *Multiple Choice Question,* CQ = *Conceptual Question, and* IE = *Integrated Exercise. Throughout the text, many exercise sections will include "paired" exercises. These exercise pairs, identified with* **red numbers**, *are intended to assist you in problem solving and learning. In a pair, the first exercise (even numbered) is worked out in the Study Guide so that you can consult it should you need assistance in solving it. The second exercise (odd numbered) is similar in nature, and its answer is given at the back of the book.*

10.1 Temperature and Heat *and* 10.2 The Celsius and Fahrenheit Temperature Scales

1. **MC** Temperature is associated with molecular (a) rotational energy, (b) random translational energy, (c) vibrational energy, (d) all of the preceding. (b)

2. **MC** A typical room temperature of 68°F on the Celsius scale is (a) 10°C, (b) 20°C, (c) 30°C. (b)

3. **MC** A particular temperature interval, as opposed to a particular temperature value is written (a) C°, (b) °C, (c) C° − C°, (d) it makes no difference. (a)

4. **CQ** Heat flows spontaneously from a body at a higher temperature to one at a lower temperature that is in thermal contact with it. Does heat always flow from a body with more internal energy to one with less internal energy? Explain. not necessarily

5. **CQ** What is the hottest (highest-temperature) item in a home? [*Hint:* Think about this one, and maybe a light will come on.] incandescent lamp filament, up to 3000°C

6. **CQ** The tires of commercial jumbo jets are inflated with nitrogen, not air. Why?
air has water and it may freeze at high altitudes

7. **CQ** When temperature changes during the day, which scale, Celsius or Fahrenheit, will read a smaller change? Explain. Celsius

8. ● Convert the following to Celsius readings: (a) 500°F, (b) 0°F, (c) −20°F, and (d) −40°F.
(a) 260°C (b) −18°C (c) −29°C (d) −40°C

9. ● Convert the following to Fahrenheit readings: (a) 150°C, (b) 32°C, (c) −25°C, and (d) −273°C.
(a) 302°F (b) 90°F (c) −13°F (d) −459°F

10. ● The coldest inhabited village in the world is Oymyakon, a town located in eastern Siberia, where it gets as cold as −94°F. What is this temperature on the Celsius scale?
−70°C

11. ● Which is the lower temperature: (a) 245°C or 245°F? (b) 200°C or 375°F? (a) 245°F (b) 375°F

12. ● A person running a fever has a body temperature of 39.4°C. What is this temperature on the Fahrenheit scale?
103°F

13. ● The highest and lowest recorded air temperatures in the United States are, respectively, 134°F (Death Valley, California, 1913) and −80°F (Prospect Creek, Alaska, 1971). What are these temperatures on the Celsius scale?
56.7°C and −62°C

14. ● The highest and lowest recorded air temperatures in the world are, respectively, 58°C (Libya, 1922) and −89°C (Antarctica, 1983). What are these temperatures on the Fahrenheit scale? 136°F; −128°F

15. **IE** ●● There is one temperature at which the Celsius and Fahrenheit scales have the same reading. (a) To find that temperature, would you set (1) $5T_F = 9T_C$, (2) $9T_F = 5T_C$, or (3) $T_F = T_C$? Why? (b) Find the temperature.
(a) (3) $T_F = T_C$ (b) −40°F = −40°C

16. ●● During open-heart surgery it is common to cool the patient's body down to slow body processes and gain an extra margin of safety. A drop of 8.5 C° is typical in these types of operations. If a patient's normal body temperature is 98.2°F, what is her final temperature in both Celsius and Fahrenheit? 28.3°C, 82.9°F

17. ●●● (a) The largest temperature drop recorded in the United States in one day occurred in Browning, Montana, in 1916, when the temperature went from 7°C to −49°C. What is the corresponding change on the Fahrenheit scale? (b) On the Moon, the average surface temperature is 127°C during the day and −183°C during the night. What is the corresponding change on the Fahrenheit scale?
(a) −101 F° (b) 558 F°

18. ●●● Astronomers know that the temperatures of stellar interiors are "extremely high." By this they mean they can convert from Fahrenheit to Celsius temperature using a rough rule of thumb:

$$T(\text{in °C}) \approx \tfrac{1}{2}T(\text{in °F}).$$

(a) Determine the exact fraction (it isn't $\tfrac{1}{2}$) and (b) the percentage error astronomers make by using $\tfrac{1}{2}$ at high temperatures. (a) $0.555T_F$ (b) 11%

19. **IE** ●●● Fig. 10.5 is a plot of Fahrenheit temperature versus Celsius temperature. (a) Is the value of the y-intercept found by setting (1) $T_F = T_C$, (2) $T_C = 0$, or (3) $T_F = 0$? Why? (b) Compute the value of the y-intercept. (c) What would be the slope and y-intercept if the graph were plotted the opposite way (Celsius versus Fahrenheit)?
(a) (2) $T_C = 0$ (b) 32°F (c) 5/9; −18°C

10.3 Gas Laws, Absolute Temperature, and the Kelvin Temperature Scale

20. **MC** The temperature used in the ideal gas law must be expressed on which scale: (a) Celsius, (b) Fahrenheit, (c) Kelvin, or (d) any of the preceding? (c)

21. **MC** Which of the following has the smallest degree interval: (a) Fahrenheit, (b) Celsius, or (c) Kelvin? (a)

*Assume all temperatures to be exact, and neglect significant figures for small changes in dimension.

22. **MC** When the temperature of a quantity of gas is increased, (a) the pressure must increase, (b) the volume must increase, (c) both the pressure and volume must increase, (d) none of the preceding. (d)

23. **CQ** A type of constant-volume gas thermometer is shown in ▼Fig. 10.18. Describe how it operates. see ISM

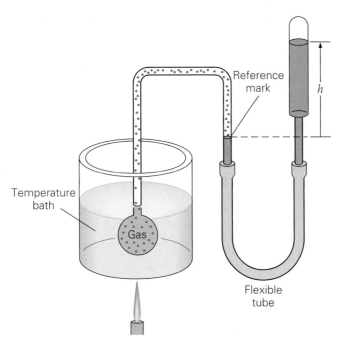

▲ **FIGURE 10.18 A type of constant-volume gas thermometer** See Exercise 23.

24. **CQ** Describe how a constant-pressure gas thermometer might be constructed. see ISM

25. **CQ** In terms of the ideal gas law, what would a temperature of absolute zero imply? How about a negative absolute temperature?
   zero pressure or volume; negative pressure or volume

26. **CQ** Excited about a New Year's Eve party in Times Square, you pump up ten balloons in your warm apartment and take them to the cold square. However, you are very disappointed with your decorations. Why? balloons collapsed; see ISM

27. **CQ** Which has more molecules, 1 mole of oxygen or 1 mole of nitrogen? Explain. same

28. ● Convert the following temperatures to absolute temperatures in kelvins: (a) 0°C, (b) 100°C, (c) 20°C, and (d) −35°C. (a) 273 K (b) 373 K (c) 293 K (d) 238 K

29. ● Convert the following temperatures to degrees Celsius: (a) 0 K, (b) 250 K, (c) 273 K, and (d) 325 K.
   (a) −273°C (b) −23°C (c) 0°C (d) 52°C

30. ● (a) Derive an equation for converting Fahrenheit temperatures directly to absolute temperatures in kelvins. (b) Which is the lower temperature, 300°F or 300 K?
   (a) $T_K = \frac{5}{9} T_F + 255{,}37$, (b) 300 K

31. ● When lightning strikes, it can heat the air around it to more than 30 000 K, five times the surface temperature of the Sun. (a) What is this temperature on the Fahrenheit and Celsius scales? (b) The temperature is sometimes reported to be 30 000°C. Assuming that 30 000 K is correct, what is the percentage error of this Celsius value?
   (a) 53 541°F; 29 727°C (b) 0.910%

32. ● How many moles are in (a) 40 g of water, (b) 245 g of H_2SO_4 (sulfuric acid), (c) 138 g of NO_2 (nitrogen dioxide), and (d) 56 L of SO_2 (sulfur dioxide) at STP (standard temperature of exactly 0°C and pressure of exactly 1 atm)?
   (a) 2.2 mol (b) 2.5 mol (c) 3.0 mol (d) 2.5 mol

33. **IE** ● (a) In a constant-volume gas thermometer, if the pressure of the gas decreases, will the temperature of the gas (1) increase, (2) decrease, or (3) remain the same? Why? (b) The initial absolute pressure of a gas is 1000 Pa at room temperature (20°C). If the pressure increases to 1500 Pa, what is the new Celsius temperature?
   (a) (2) decrease (b) 167°C

34. ●● In the troposphere (the lowest part of the atmosphere), the temperature decreases rather uniformly with altitude at a so-called "lapse" rate of about 6.5 C°/km. What are the temperatures (a) near the top of the troposphere (which has an average thickness of 11 km) and (b) outside a commercial aircraft flying at a cruising altitude of 34 000 ft? (Assume that the ground temperature is normal room temperature.) (a) −52°C (b) −47°C

35. ●● An athlete has a large lung capacity, 7.0 L. Assuming air to be an ideal gas, how many molecules of air are in the athlete's lungs when the air temperature in the lungs is 37°C under normal atmospheric pressure? 1.7×10^{23}

36. ●● Show that 1.00 mol of ideal gas under STP occupies a volume of 0.0224 m^3 = 22.4 L. see ISM

37. ●● What volume is occupied by 6.0 g of hydrogen under a pressure of 2.0 atm and a temperature of 300 K? 0.037 m^3

38. ●● Is there a temperature that has the same numerical value on the Kelvin and Fahrenheit scales? Justify your answer. 574.58 K; see ISM

39. ●● A husband buys a helium-filled anniversary balloon for his wife. The balloon has a volume of 3.5 L in the warm store at 74°F. When he takes it outside, where the temperature is 48°F, he finds it has shrunk. By how much has the volume decreased? 0.16 L

40. ●● On a warm day (92°F), an air-filled balloon occupies a volume of 0.20 m^3 and is under a pressure of 20.0 lb/in². If the balloon is cooled to 32°F in a refrigerator while its pressure is reduced to 14.7 lb/in², what is the volume of the air in the container? (Assume that the air behaves as an ideal gas.) 0.24 m^3

41. ●● A steel-belted radial automobile tire is inflated to a gauge pressure of 30.0 lb/in² when the temperature is 61°F. Later in the day, the temperature rises to 100°F. Assuming that the volume of the tire remains constant, what is the tire's pressure at the elevated temperature? [*Hint:* Remember that the ideal gas law uses absolute pressure.]
   33.4 lb/in²

42. **IE ●●** (a) If the temperature of an ideal gas increases and its volume decreases, will the pressure of the gas (1) increase, (2) remain the same, or (3) decrease? Why? (b) The Kelvin temperature of an ideal gas is doubled and its volume is halved. How is the pressure affected? (a) (1) increase (b) $p_2 = 4p_1$

43. **●●** A scuba diver takes a steel tank of air on a deep dive. The tank's volume is 5.35 L and it is completely filled with air at a total pressure of 2.45 atm at the start of the dive. The air temperature at the surface is 94°F and he ends up in deep water at 60°F. Assuming thermal equilibrium and neglecting air loss, determine the total internal pressure of the air when it is cold. 2.31 atm

44. **●●** If 2.4 m³ of a gas initially at STP is compressed to 1.6 m³ and its temperature is raised to 30°C, what is the final pressure? 1.7 atm

45. **IE ●●** The pressure on a low-density gas in a cylinder is kept constant as its temperature is increased. (a) Does the volume of the gas (1) increase, (2) decrease, or (3) remain the same? Why? (b) If the temperature is increased from 10°C to 40°C, what is the percentage change in the volume of the gas? (a) (1) increase (b) 10.6%

46. **●●●** (a) Show that for the Kelvin temperature range

$$T \gg 273\ \text{K}, \quad T \approx T_{\text{C}} \approx \frac{5}{9}T_{\text{F}}.$$

(b) For room temperature, what percentage error would result in using this to determine the Kelvin temperature? (c) For a typical stellar interior temperature of 10 million °F, what is the percentage error in the Kelvin temperature? (Carry as many significant figures as needed.) (a) see ISM (b) 87% (c) 4.6×10^{-3}%

47. **●●●** A diver releases an air bubble of volume 2.0 cm³ from a depth of 15 m below the surface of a lake, where the temperature is 7.0°C. What is the volume of the bubble when it reaches just below the surface of the lake, where the temperature is 20°C? 5.1 cm³

10.4 Thermal Expansion

48. **MC** Are the units of the thermal coefficient of linear expansion (a) m/C°, (b) m²/C°, (c) m·C°, or (d) 1/C°? (d)

49. **MC** Is the thermal coefficient of volume expansion for a solid (a) 2α, (b) $2\alpha^2$, (c) 3α, or (d) α^3? (c)

50. **MC** Which of the following describes the behavior of water density in the temperature range of 0°C to 4°C: (a) increases with increasing temperature, (b) remains constant, (c) decreases with decreasing temperature, or (d) both a and c. (d)

51. **CQ** A cube of ice sits on a bimetallic strip at room temperature (▼ Fig. 10.19). What will happen if (a) the upper strip is aluminum and the lower strip is brass, (b) the upper strip is iron and the lower strip is copper? (c) If the cube is made of a hot metal rather than ice and the two strips are brass and copper, should the brass or copper be on top to keep the cube from falling off? (a) ice moves upward (b) ice moves downward (c) copper

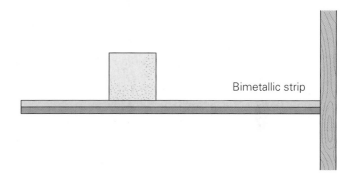

Bimetallic strip

▲ FIGURE 10.19 Which way will the cube go? See Exercise 51.

52. **CQ** A solid metal disk rotates freely, so the conservation of angular momentum applies (Chapter 8). If the disk is heated while it is rotating, will there be any effect on the rate of rotation (the angular speed)? decreases; see ISM

53. **CQ** A demonstration of thermal expansion is shown in ▼ Fig. 10.20. (a) Initially, the ball goes through the ring made of the same metal. When the ball is heated (b), it does not go through the ring (c). If both the ball and the ring are heated, the ball again goes through the ring. Explain what is being demonstrated. see ISM

◄ FIGURE 10.20 Ball-and-ring expansion See Exercises 53 and 63.

(a)

(b)

(c)

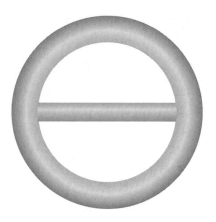

◄ **FIGURE 10.21**
Stress out of shape?
See Exercise 54.

54. **CQ** A circular ring of iron has a tight-fitting iron bar across its diameter, as illustrated in ▲Fig. 10.21. If the arrangement is heated in an oven to a high temperature, will the circular ring be distorted? What if the bar is made of aluminum? no; yes; see ISM

55. **CQ** We often use hot water to loosen tightly sealed metal lids on glass jars. Explain why this works.
lid expands more than glass

56. ● A steel beam 10 m long is installed in a structure at 20°C. What are the beam's changes in length at the temperature extremes of −30°C to 45°C? −6.0 mm to 3.0 mm

57. **IE** ● An aluminum tape measure is accurate at 20°C. (a) If the tape measure is placed in a freezer, would it read (1) high, (2) low, or (3) the same? Why? (b) If the temperature of the freezer is −5.0°C, what would be the stick's percentage error because of thermal contraction?
(a) (1) high (b) 0.060%

58. ● Concrete highway slabs are poured in lengths of 5.00 m. How wide should the expansion gaps between the slabs be at a temperature of 20°C to ensure that there will be no contact between adjacent slabs over a temperature range of −25°C to 45°C? 1.5 mm

59. ● A man's gold wedding ring has an inner diameter of 2.4 cm at 20°C. If the ring is dropped into boiling water, what will be the change in the inner diameter of the ring? 0.0027 cm

60. ●● What temperature change would cause a 0.10% increase in the volume of a quantity of water that was initially at 20°C? 4.8 C°

61. ●● A piece of copper tubing used in plumbing has a length of 60.0 cm and an inner diameter of 1.50 cm at 20°C. When hot water at 85°C flows through the tube, what are (a) the tube's new length and (b) the change in its cross-sectional area? Does the latter affect the flow speed? (a) 60.1 cm (b) 3.91 × 10⁻³ cm²; yes

62. **IE** ●● A circular piece is cut from an aluminum sheet at room temperature. (a) When the sheet is then placed in an oven, will the hole (1) get larger, (2) get smaller, or (3) remain the same? Why? (b) If the diameter of the hole is 8.00 cm at 20°C and the temperature of the oven is 150°C, what will be the new area of the hole? (a) (1) get larger (b) 50.6 cm²

63. **IE** ●● In Fig. 10.20, the steel ring of diameter 2.5 cm is 0.10 mm smaller in diameter than the steel ball at 20°C. (a) For the ball to go through the ring, should you heat (1) the ring, (2) the ball, or (3) both so that the ball will go through the ring? Why? (b) What is the minimum required temperature? (a) (1) the ring (b) 353°C

64. ●● A circular steel plate of radius 15 cm is cooled from 350°C to 20°C. By what percentage does the plate's area decrease? 0.79%

65. ●● A pumpkin pie has filling up to the brim. The pie plate is made of Pyrex and its expansion can be neglected. It is a cylinder with an inside depth of 2.10 cm and an inside diameter of 30.0 cm. It is prepared at a room temperature of 68°F and placed in an oven at 400°F. When it taken out, 151 cc of the pie filling has flowed out and over the rim. Determine the coefficient of volume expansion of the pie filling, assuming it is a fluid. 5.52 × 10⁻⁴/C°

66. ●● One morning, an employee at a rental car company fills a car's steel gas tank to the top and then parks the car a short distance away. (a) That afternoon, when the temperature increases, will there be any gas spill or not? Why? (b) If the temperatures in the morning and afternoon are, respectively, 10°C and 30°C and the gas tank can hold 25 gal in the morning, how much gas will be lost? (Neglect the expansion of the tank.) (a) spill (b) 0.48 gal

67. ●● A copper block has an internal spherical cavity with a 10-cm diameter (▼Fig. 10.22). The block is heated in an oven from 20°C to 500 K. (a) Does the cavity get larger or smaller? (b) What is the change in the cavity's volume?
(a) larger (b) 5.5 × 10⁻⁶ m³

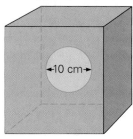

◄ **FIGURE 10.22 A hole in a block** See Exercise 67.

68. ●● When exposed to sunlight, a hole in a sheet of copper expands in diameter by 0.153% compared to its diameter at 68°F. What is the temperature of the copper sheet in the sun? 110°C

69. ●●● A brass rod has a circular cross-section of radius 5.00 cm. The rod fits into a circular hole in a copper sheet with a clearance of 0.010 mm completely around it when both the rod and the sheet are at 20°C. (a) At what temperature will the clearance be zero? (b) Would such a tight fit be possible if the sheet were brass and the rod were copper? (a) 116°C (b) no

70. ●●● Table 10.1 states that the (experimental) coefficient of volume expansion β for air (and most other ideal gases at 1 atm and 20°C) is 3.5 × 10⁻³/C°. Use the definition of the volume expansion coefficient to show that this value can, to a very good approximation, be predicted from the ideal gas law and that the result holds for all ideal gases, not just air. see ISM

71. ••• A Pyrex beaker that has a capacity of 1000 cm³ at 20°C contains 990 cm³ of mercury at that temperature. Is there some temperature at which the mercury will completely fill the beaker? Justify your answer. (Assume that no mass is lost by vaporization.) yes, 79°C

10.5 The Kinetic Theory of Gases

72. **MC** If the average kinetic energy of the molecules in an ideal gas initially at 20°C doubles, what is the final temperature of the gas: (a) 10°C, (b) 40°C, (c) 313°C, or (d) 586°C? (c)

73. **MC** If the temperature of a quantity of ideal gas is raised from 100 K to 200 K, is the internal energy of the gas (a) doubled, (b) halved, (c) unchanged, or (d) none of the preceding? (a)

74. **MC** The smelling of odors is generally the result of (a) effusion, (b) diffusion, (c) osmosis, (d) reverse osmosis. (b)

75. **CQ** Equal volumes of helium gas (He) and neon gas (Ne) at the same temperature (and pressure) are on opposite sides of a porous membrane (▼ Fig. 10.23). Describe what happens after a period of time, and why. see ISM

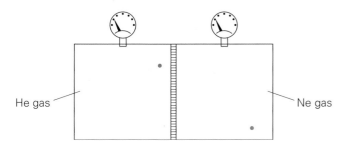

▲ **FIGURE 10.23 What happens as time passes?**
See Exercise 75.

76. **CQ** Natural gas is odorless; to alert people to gas leaks, the gas company inserts an additive that has a distinctive scent. When there is a gas leak, the additive reaches your nose before the gas does. What can you conclude about the masses of the additive molecules and gas molecules?
$m_{additive} < m_{gas}$

77. • What is the average kinetic energy per molecule in an ideal gas at (a) 20°C and (b) 100°C?
(a) 6.07×10^{-21} J (b) 7.72×10^{-21} J

78. **IE** • If the Celsius temperature of an ideal gas is doubled, (a) will the internal energy of the gas (1) double, (2) increase by less than a factor of 2, (3) be half as much, or (4) decrease by less than a factor of 2? Why? (b) If the temperature is raised from 20°C to 40°C, by how much will the internal energy of 2.0 mol of an ideal gas change?
(a) (2) increase by less than a factor of 2 (b) 5.0×10^2 J

79. • (a) What is the average kinetic energy per molecule of an ideal gas at a temperature of 25°C? (b) What is the average (rms) speed of the molecules if the gas is helium? (A helium molecule consists of a single atom of mass 6.65×10^{-27} kg.) (a) 6.21×10^{-21} J (b) 1.37×10^3 m/s

80. • What is the average speed of the molecules in low-density oxygen gas at 0°C? (The mass of an oxygen molecule, O_2, is 5.31×10^{-26} kg.) 4.61×10^2 m/s

81. •• If the temperature of an ideal gas increases from 300 K to 600 K, what happens to the rms speed of the gas molecules? increases by a factor of $\sqrt{2}$

82. •• At a given temperature, which would be greater, the rms speed of oxygen (O_2) or of ozone (O_3), and how much greater? $v_{oxygen} = 1.22 v_{ozone}$

83. •• (a) Estimate the total amount of translational kinetic energy in a classroom at normal room temperature. Assume the room measures 4.00 m by 10.0 m by 3.00 m. (b) If this energy were all harnessed, how high would it be able to lift an elephant with a mass of 1200 kg?
(a) 1.82×10^7 J (b) 1.55×10^3 m

84. •• If the temperature of an ideal gas is raised from 25°C to 100°C, how much faster is the new average (rms) speed of the gas molecules? 1.12 times as fast

85. •• A quantity of an ideal gas is at 0°C. An equal quantity of another ideal gas is twice as hot. What is its temperature?
273°C

86. •• If 2.0 mol of oxygen gas is confined in a 10-L bottle under a pressure of 6.0 atm, what is the average kinetic energy of an oxygen molecule? 7.5×10^{-21} J

87. •• If the rms speed of the molecules in an ideal gas at 20°C increases by a factor of two, what is the new temperature?
899°C

88. •• Calculate the number of gas molecules in a container of volume 0.10 m³ filled with gas under a partial vacuum of pressure 20 Pa at 20°C. 4.9×10^{20}

89. **IE** ••• During the race to the atomic bomb in World War II, it was necessary to separate a lighter isotope of uranium (U-235 was the fissionable one needed for bomb material) from a heavier variety (U-238). The uranium was converted into a gas, uranium hexafluoride (UF₆) and the differences in their average speeds was used to separate the two uranium isotopes by *gaseous diffusion*. As a two-component molecular mixture at room temperature, which of the two types of molecules would be moving faster, on average: (1) ²³⁵UF₆, (2) ²³⁸UF₆, or (3) they had the same average speed? (b) Determine the ratio of their speeds, light molecule to heavy molecule. Treat the molecules as ideal gases and neglect rotations and/or vibrations of the molecules. The masses of the three atoms in atomic mass units are 238, 235, and 19 for fluorine.
(a) (1) ²³⁵UF₆ (b) 1.00429

*10.6 Kinetic Theory, Diatomic Gases, and the Equipartition Theorem

90. **MC** Is the temperature of a diatomic molecule like O_2 a measure of its (a) translational kinetic energy, (b) rotational kinetic energy, (c) vibrational kinetic energy, or (d) all of the preceding? (a)

91. **MC** On average, is the total internal energy of a gas divided equally among (a) each atom, (b) each degree of freedom, (c) linear motion, rotational motion, and vibrational motion, or (d) none of the preceding? (b)

92. **CQ** Why does a sample of gas with diatomic molecules have more internal energy than a similar sample with monatomic molecules at the same temperature?
it has more degrees of freedom

93. CQ What is the difference in the internal energies of monatomic and diatomic molecules? *nRT*

94. ● If 1.0 mol of ideal monoatomic gas has a total internal energy of 5.0×10^3 J at a certain temperature, what is the total internal energy of 1.0 mol of a diatomic gas at the same temperature? 8.3×10^3 J

95. ● What is the total internal energy of 1.0 mol of O_2 gas at 20°C? 6.1×10^3 J

96. ●● For an average molecule of N_2 gas at 10°C, what are its (a) translational kinetic energy, (b) rotational kinetic energy, and (c) total energy?
(a) 5.65×10^{-21} J (b) 3.77×10^{-21} J (c) 9.42×10^{-21} J

97. ●● (a) In the classroom in Exercise 83, how much of the energy is in the form of rotational kinetic energy? (b) how much total kinetic energy is there in the air?
(a) 1.21×10^7 J (b) 3.03×10^7 J

Comprehensive Exercises

98. IE (a) As they are cooled, the densities of most objects (1) increases, (2) decrease, (3) stay the same. (b) By what percentage does the density of a bowling ball change (assuming it is a uniform sphere) when it is taken from room temperature (68°F) into the cold night air in Nome, Alaska (−40°F). Assume the ball is made out of material that has a linear coefficient of expansion α of 75.2×10^{-6}/C°.
(a) (1) increases (b) 1.37%

99. When a full copper kettle is tipped vertically at room temperature (68°F), water initially pours out of its spout at 100 cc/s. By what percentage will this change if the kettle instead contains boiling water at 212°F. Assume that the only significant change is due to the change in size of the spout. 0.272%

100. An ideal gas occupies a container of volume 0.75 L at STP. Find (a) the number of moles and (b) the number of molecules of the gas. (c) If the gas is carbon monoxide (CO), what is its mass?
(a) 3.3×10^{-2} mol (b) 2.0×10^{22} molecules (c) 0.94 g

101. IE The escape speed from the Earth is about 11 000 m/s. Assume that for a given type of gas to eventually escape the Earth's atmosphere requires that its average molecular speed be about 10% of the escape speed. (a) Which gas would be more likely to escape the Earth? (1) oxygen, (2) nitrogen, or (3) helium? (b) Assuming a temperature of −40°F in the upper atmosphere, determine the average translational speed of a molecule of oxygen. Is it enough to escape the Earth? (Data: The mass of an oxygen molecule is 5.34×10^{-26} kg, the mass of a nitrogen molecule is 4.68×10^{-26} kg, and that of helium is 6.68×10^{-27} kg.
(a) (3) helium (b) 425 m/s ≪ 1100 m/s

> The following Physlet Physics Problems can be used with this chapter.
>
> PHYSLET® 19.3, 19.4, 19.5, 20.1, 20.2, 20.4, 20.5, 20.6, 20.7

PHYSICS FACTS

- With a skin temperature of 34°C (93.2°F), a person sitting in a room at 23°C (73.4°F) will lose about 100 J of heat per second, which is a power output approximately that which is a 100-W lightbulb. This is why a closed room full of people tends to get very warm.

- A couple of inches of fiberglass in the attic can cut heat loss by as much as 90% (see Example 11.7).

- If the Earth did not have an atmosphere (hence no greenhouse effect), its average surface temperature would be 30 C° (86 F°) lower than it is now. That would freeze liquid water and basically eliminate life as we know it.

- Most metals are excellent thermal conductors. However, iron and steel are relatively poor conductors; they conduct only about 12% as much as copper.

- Styrofoam is one of the best thermal insulators. It conducts only 25% as much as wool under similar conditions.

- During a race on a hot day, a professional cyclist can evaporate as much as seven liters of water in three hours in getting rid of the heat generated by this vigorous activity.

Heat is crucial to our existence. Our bodies must balance heat loss and gain to stay within the narrow temperature range necessary for life. This thermal balance is delicate and any disturbance can have serious consequences. Sickness can disrupt the balance and our bodies produce a chill or fever.

To maintain our health, we exercise by doing mechanical work such as lifting weights, riding bicycles, and so on. Our bodies convert food energy (chemical potential) to mechanical work; however, this process is not perfect. That is, the body cannot convert all the food energy into mechanical work—in fact, less than 20%, depending on which muscle groups are doing the work. The rest goes into heat. The leg muscles are the largest and most efficient in performing mechanical work; for example, cycling and running are relatively efficient processes. The arm and shoulder muscles are less efficient; hence, snow shoveling is a low-efficiency exercise. The body must have special cooling mechanisms to get rid of excess heat generated during intense exercise. The most efficient mechanism is through perspiring, or the evaporation of water. The Olympic marathon champion Gezahgne Abera tries to promote cooling and evaporation by pouring water over his head, as shown in the photograph.

On a larger scale, heat is important to our planet's ecosystem. The average temperature of the Earth, so critical to our environment and to the survival of the organisms that inhabit it, is maintained through a heat-exchange balance. Each day, a vast quantity of solar energy reaches our planet's atmosphere and surface. Scientists are concerned that a buildup of atmospheric "greenhouse" gases, a product of our industrial society, could significantly raise the Earth's average temperature. This change would undoubtedly have a negative effect on life on the Earth.

At a more practical level, most of us know to be very careful while handling anything that has recently been in contact with a flame or other source of heat. Yet while the copper bottom of a steel pot on the stove can be very hot, the steel pot handle is only warm to the touch. Sometimes direct contact isn't necessary for heat to be transmitted, but how was heat transferred? And why was the steel handle not nearly as hot as the pot? It has to do with thermal conduction as you will learn.

In this chapter, you'll learn what heat is and how it is measured. You will also study the various mechanisms by which heat is transferred from one object to another. This knowledge will allow you to explain many everyday phenomena, as well as provide a basis for understanding the conversion of thermal energy into useful mechanical work.

11.1 Definition and Units of Heat

OBJECTIVES: To (a) define heat, (b) distinguish the various units of heat, and (c) define the mechanical equivalent of heat.

Like work, heat involves a transfer of energy. In the 1800s, it was thought that heat described the amount of energy an object possessed, but this is not true. Rather, **heat** is the name used to describe a type of energy *transfer*. When we refer to "heat" or "heat energy," this is the amount of energy added to, or removed from, the total internal energy of an object due to temperature differences.

Because heat is energy *in transit*, it is measured in standard energy units. As usual, we use SI units (the joule), but for completeness, and define other commonly used units of heat. An important one is the **kilocalorie (kcal)** (▼Fig. 11.1a):

> One kilocalorie (kcal) is defined as the amount of heat needed to raise the temperature of 1 kg of water by 1 C° (from 14.5°C to 15.5°C).

(This kilocalorie is technically known as the "15° kilocalorie.") The temperature range is specified because the energy needed varies slightly with temperature—a variation so small that it can be ignored at the temperature in question.

For smaller quantities, the **calorie (cal)** is sometimes used (1 kcal = 1000 cal). One calorie is the amount of heat needed to raise the temperature of 1 g of water by 1 C° (from 14.5°C to 15.5°C) (Fig. 11.1b).

A familiar use of the larger unit, the kilocalorie, is for specifying the energy values of foods. In this context, the word is usually shortened to *Calorie* (Cal). That is, people on diets really count kilocalories. This quantity refers to the food energy that is available for conversion to heat to be used, for mechanical movement, to maintain body temperature, or to increase body mass. (See Example 11.1.) The capital C distinguishes the larger kilogram-Calorie, or kilocalorie, from the smaller gram-calorie, or calorie. They are sometimes referred to as "big Calorie" and "little calorie." (In some countries, the joule is used for food values—see ▶Fig. 11.2.)

▶ **FIGURE 11.1 Units of heat (a)** A kilocalorie raises the temperature of 1 kg of water by 1 C°. **(b)** A calorie raises the temperature of 1 g of water by 1 C°. **(c)** A Btu raises the temperature of 1 lb of water by 1 F°. (Not drawn to scale.)

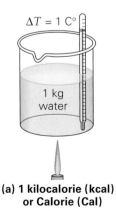

$\Delta T = 1\ \text{C}°$

1 kg water

(a) 1 kilocalorie (kcal) or Calorie (Cal)

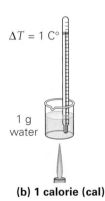

$\Delta T = 1\ \text{C}°$

1 g water

(b) 1 calorie (cal)

$\Delta T = 1\ \text{F}°$

1 lb water

(c) 1 British thermal unit (Btu)

A unit of heat sometimes used in American industry is the **British thermal unit (Btu)**. One Btu is the amount of heat needed to raise the temperature of 1 lb of water by 1 F° (from 63°F to 64°F; Fig. 11.1c), and 1 Btu = 252 cal = 0.252 kcal. If you buy an air conditioner or an electric heater, you will find that it is rated in Btu, which is really Btu per hour—in other words, a power rating. For example, window air conditioners range from 4000 to 25 000 Btu/h. This specifies the rate at which the appliance can remove heat.

The Mechanical Equivalent of Heat

The idea that heat is actually a transfer of energy is the result of work by many scientists. Some early observations were made by the American Benjamin Thompson (Count Rumford), 1753–1814, while he was supervising the boring of cannon barrels in Germany. Rumford noticed that water put into the bore of the cannon (to prevent overheating during drilling) boiled away and had to be replenished. The theory of heat at that time pictured it as a "caloric fluid," which flowed from hot objects to colder ones. Rumford did several experiments to detect "caloric fluid" by measuring changes in the weights of heated substances. Since no weight change was ever detected, he concluded that the mechanical work done by friction was actually responsible for the heating of the water.

This conclusion was later proven quantitatively by the English scientist James Joule (after whom the unit of energy is named, see Section 5.6). Using the apparatus illustrated in ▸Fig. 11.3, Joule demonstrated that when a given amount of mechanical work was done, the water was heated, as indicated by an increase in its temperature. He found that for every 4186 J of work done, the temperature of the water rose 1 C° per kg, or that 4186 J was equivalent to 1 kcal:

$$1 \text{ kcal} = 4186 \text{ J} = 4.186 \text{ kJ} \qquad \text{or} \qquad 1 \text{ cal} = 4.186 \text{ J}$$

This relationship is called the **mechanical equivalent of heat**. Example 11.1 illustrates an everyday use of these conversion factors.

▲ **FIGURE 11.2 It's a joule!** In Australia, diet drinks are labeled as being "low joule." In Germany, the labeling is a bit more specific: "Less than 4 kilojoules (1 kcal) in 0.3 Liter." How does this labeling compare to that for diet drinks in the United States?

Teaching tip: Make sure students understand the difference between a calorie and a Calorie.

Exploration 19.1 Mechanical Equivalent of Heat

Example 11.1 ■ Working Off That Birthday Cake: Mechanical Equivalent of Heat to the Rescue

At a birthday party, a student eats a piece of cake (food energy value of 400 Cal). To prevent this energy from being stored as fat, she takes a stationary-bicycle workout class right after the party. This exercise requires the body to do work at an average rate of 200 watts. How long must the student bicycle to achieve her goal of "working off" the cake's energy?

Thinking It Through. Power is the rate at which the student does work, and the watt (W) is its SI unit (1 W = 1 J/s, Section 5.6). To find the time to do this work, we express the food energy content in joules and use the definition of average power, $\overline{P} = W/t$ (work/time).

Solution. The work required to "burn up" the energy content of the cake is at least 400 Cal. Listing the data given and converting to SI units. (Remember that Cal means kcal.)

Given: $W = (400 \text{ kcal})\left(\dfrac{4186 \text{ J}}{\text{kcal}}\right) = 1.67 \times 10^6 \text{ J}$ *Find:* Time t to "burn up" 400 Cal
$\overline{P} = 200 \text{ W} = 200 \text{ J/s}$

Rearranging the equation for average power,

$$t = \frac{W}{\overline{P}} = \frac{1.67 \times 10^6 \text{ J}}{200 \text{ J/s}} = 8.35 \times 10^3 \text{ s} = 139 \text{ min} = 2.32 \text{ h}$$

Follow-Up Exercise. If the 400 Cal in this Example were used to increase the student's gravitational potential energy, how high would she rise? (Assume her mass is 60 kg.) *(Answers to all Follow-Up Exercises are at the back of the text.)*

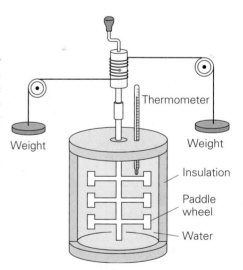

▲ **FIGURE 11.3 Joule's apparatus for determining the mechanical equivalent of heat** As the weights descend, the paddle wheels churn the water, and the mechanical energy, or work, is converted into heat energy, raising the temperature of the water. For every 4186 J of work done, the temperature of the water rises 1 C° per kilogram. Thus, 4186 J is equivalent to 1 kcal.

11.2 Specific Heat and Calorimetry

OBJECTIVES: **OBJECTIVES:** To (a) define specific heat, and (b) explain how the specific heats of materials are measured using the technique of calorimetry.

Specific Heats of Solids and Liquids

Illustration 19.1 Specific Heat

Teaching tip: Students should be able to state the specific heat for water from the definition of a calorie or a kilocalorie.

Demonstration/activity: Heat equal masses of lead, steel, and aluminum to the same temperature in a hot water bath. Then place them on a block of paraffin. What do the results of this activity explain about the specific heats of these elements?

Teaching tip: Point out that water, because of its high specific heat, is used to store heat energy in solar homes. It is also used as a coolant in cars, because it can store, and then transfer, a great deal of heat energy away from the engine—and it is inexpensive.

Recall from Chapter 10 that when heat is added to a solid or liquid, the energy may go toward increasing the average molecular kinetic energy (temperature change) and also toward increasing the potential energy associated with the molecular bonds. Different substances have different molecular configurations and bonding patterns. Thus, if equal amounts of heat are added to equal masses of different substances, the resulting temperature changes will *not* generally be the same.

The amount of heat (Q) required to change the temperature of a substance is proportional to the mass (m) of the substance and to the change in its temperature (ΔT). That is, $Q \propto m\Delta T$, or, in equation form, with a proportionality constant c,

$$Q = cm\Delta T \quad \text{or} \quad c = \frac{Q}{m\Delta T} \quad \textit{specific heat} \tag{11.1}$$

Here, $\Delta T = T_f - T_i$ is the temperature change of the object and c is the *specific heat capacity*, or simply its **specific heat**. The SI units of specific heat are $J/(kg \cdot K)$ or $J/(kg \cdot C°)$, because $1 K = 1 C°$. Specific heat is characteristic of the substance type. Thus, the specific heat gives us an indication of a material's internal molecular configuration and bonding.

Note that the specific heat physically means the heat (transfer) required to raise (or lower) the temperature of 1 kg of a substance by 1 C°. The specific heats of some common substances are given in Table 11.1. Specific heats vary slightly with temperature, but at everyday temperatures they can be considered constant.

The larger the specific heat of a substance, the more energy must be transferred to or taken from it (per kilogram of mass) to change its temperature by a given amount. That is, a substance with a higher specific heat requires more heat

TABLE 11.1	Specific Heats of Various Substances (Solids and Liquids) at 20°C and 1 atm	
	Specific Heat (c)	
Substance	J/(kg·C°)	kcal/(kg·C°) or cal/(g·C°)
Solids		
Aluminum	920	0.220
Copper	390	0.0932
Glass	840	0.201
Ice (−10°C)	2100	0.500
Iron or steel	460	0.110
Lead	130	0.0311
Soil (average)	1050	0.251
Wood (average)	1680	0.401
Human body (average)	3500	0.84
Liquids		
Ethyl alcohol	2450	0.585
Glycerin	2410	0.576
Mercury	139	0.0332
Water (15°C)	4186	1.000
Gas		
Water vapor (H_2O)	2000	0.48

for a given temperature change and mass than one with a lower specific heat. Table 11.1 shows that metals have specific heats considerably lower than that of water. Thus it takes only a small amount of heat to produce a relatively large temperature increase in a metal object, compared to the same mass of water.

Compared to most common materials, water has a large specific heat of $4186 \, \text{J/(kg} \cdot \text{C}°)$, or $1.00 \, \text{kcal/(kg} \cdot \text{C}°)$. You have been the victim of the high specific heat of water if you have ever burned your mouth on a baked potato or the hot cheese on a pizza. These foods have high water content and with its high specific heat, this means they don't cool off as quickly as some other drier foods do. The large specific heat of water is also responsible for the mild climate of places near large bodies of water. (See Section 11.4 for more details.)

Note from Eq. 11.1 that when there is a temperature increase, ΔT is positive ($T_f > T_i$) then Q is positive. This condition corresponds to energy being *added* to a system or object. Conversely, ΔT and Q are negative when energy is *removed from* a system or object. This sign convention will be used throughout this book.

Teaching tip: Emphasize that c and ΔT are inversely related for a given m. Therefore, if equal amounts of heat are added to equal masses of two different materials, the mass with the higher specific heat will have the *smaller* change in temperature.

Example 11.2 ■ Drinking Off That Birthday Cake: Specific Heat to the Rescue?

At a birthday party in Example 11.1, another student ate a piece of cake (400 Cal). To prevent this energy from being stored as fat, she decides to drink ice water at 0°C. She reasons that the ingested ice water will be warmed up to her normal body temperature of 37°C and absorb the energy. How much ice water would she have to drink to absorb the energy generated by metabolizing the birthday cake?

Thinking It Through. The heat to warm up a certain mass of ice water from 0°C to 37°C is equal to the 400 Cal of heat energy metabolized from the birthday cake. Since the heat, the specific heat, and the temperature change of the ice water are known, we can find the required mass of ice water using Eq. 11.1.

Solution. The heat required to warm up the ice water is 400 Cal. Listing the data given and converting to SI units. (Remember that Cal means kcal.)

Given: $Q = (400 \, \text{kcal}) \left(\dfrac{4186 \, \text{J}}{\text{kcal}} \right) = 1.67 \times 10^6 \, \text{J}$ *Find:* Mass m of water to "drink off" 400 Cal

$\quad\quad T_i = 0°\text{C}$
$\quad\quad T_f = 37°\text{C}$
$\quad\quad c = 4.186 \, \text{kJ/(kg} \cdot \text{C}°)$ (from Table 11.1)

From Eq. 11.1, $Q = cm\Delta T = cm(T_f - T_i)$. Solving for m gives

$$m = \frac{Q}{c\Delta T} = \frac{1.67 \times 10^6 \, \text{J}}{[4186 \, \text{J/(kg} \cdot \text{C}°)](37°\text{C} - 0°\text{C})} = 10.8 \, \text{kg}$$

This mass of water occupies almost 3 gal or 12 L—quite a bit to drink. (Can you show this?)

Follow-Up Exercise. In this Example, how would the answer change if she drinks ice water at a temperature of 5°C rather than 0°C?

Integrated Example 11.3 ■ Cooking Class 101: Studying Specific Heats While Learning How to Boil Water

To prepare pasta, you bring a pot of water from room temperature (20°C) to its boiling point (100°C). The pot itself has a mass of 0.900 kg, is made of steel, and holds 3.00 kg of water. (a) Which of the following is true: (1) The pot requires more heat than the water, (2) the water requires more heat than the pot, or (3) they both require the same amount of heat? (b) Determine the required heat for both the water and the pot, and the ratio Q_w/Q_{pot}.

(a) Conceptual Reasoning. The temperature increase is the same for the water and the pot. Thus, the required heat is affected by the product of mass and specific heat. There is 3.0 kg of water to heat. This is more than three times the mass of the pot. From Table 11.1, the specific heat of water is about nine times larger than that of steel. Both factors together indicate that the water will require significantly more heat than the pot, so the answer is (2).

(continues on next page)

(b) Quantitative Reasoning and Solution. The heat needed can be found using Eq. 11.1, after looking up the specific heats. The temperature change is easily determined from the initial and final values.

Listing the data given:

Given: $m_{pot} = 0.900$ kg $\qquad$ *Find:* The required heat for the water and
$\qquad m_w = 3.00$ kg $\qquad\qquad\qquad$ the pot, and the heat ratio Q_w/Q_{pot}
$\qquad c_{pot} = 460$ J/kg·C° (from Table 11.1)
$\qquad c_w = 4186$ J/kg·C° (from Table 11.1)
$\qquad \Delta T = T_f - T_i = 100°C - 20°C = 80$ C°

In general, the amount of heat is given by $Q = cm \, \Delta T$. The temperature increase (ΔT) for both objects is 80 C°. Thus, the heat for the water is

$$Q_w = c_w m_w \Delta T_w$$
$$= [4186 \text{ J}/(\text{kg·C}°)](3.00 \text{ kg})(80 \text{ C}°) = 1.00 \times 10^6 \text{ J}$$

and the heat required for the pot is

$$Q_{pot} = c_{pot} m_{pot} \Delta T_{pot}$$
$$= [460 \text{ J}/(\text{kg·C}°)](0.900 \text{ kg})(80 \text{ C}°) = 3.31 \times 10^4 \text{ J}$$

Because

$$\frac{Q_w}{Q_{pot}} = \frac{1.00 \times 10^6 \text{ J}}{3.31 \times 10^4 \text{ J}} = 30.2$$

water requires more than 30 times the heat, because it has more mass and a greater specific heat.

Follow-Up Exercise. (a) In this Example, if the pot were the same mass but instead made out of aluminum, would the heat ratio (water to pot) be smaller or larger than the answer for the steel pot? Explain. (b) Verify your answer.

Calorimetry

Teaching tip: Calorimetry problems involve the concept of conservation of energy. Discuss the general nature of this concept, which is illustrated by the following questions: What lost heat? How much? What gained heat? How much?

Calorimetry is a technique that quantitatively measures heat exchanges. Such measurements are made by using an instrument called a *calorimeter* (cal-oh-RIM-i-ter), usually an insulated container that allows little heat exchange with the environment (ideally none). A simple laboratory calorimeter is shown in ◄Fig. 11.4.

The specific heat of a substance can be determined by measuring the masses and temperature changes of the objects involved and using Eq. 11.1.* Usually the unknown is the unknown specific heat, *c*. Typically a substance of known mass and temperature is put into a quantity of water in a calorimeter. The water is at a different temperature from that of the substance, usually a lower one. The principle of the conservation of energy is then applied to determine the substance's specific heat, *c*. This procedure is called the *method of mixtures*. Example 11.4 illustrates the use of this procedure. Such heat-exchange problems are just a matter of "thermal accounting," involving the conservation of energy. The total of all the heat losses ($Q < 0$) must equal the total of all the heat gains ($Q > 0$). This means the algebraic sum of all the heat transfers must equal zero, or $\Sigma Q_i = 0$, assuming negligible heat exchange with the environment.

▲ FIGURE 11.4 Calorimetry **apparatus** The calorimetry cup (center, with black insulating ring), goes into the larger container. The cover with the thermometer and stirrer is seen at the right. Metal shot or pieces of metal are heated in the small cup (with the handle) in the steam generator on the tripod.

Example 11.4 ■ Calorimetry Using the Method of Mixtures

Students in a physics lab are to determine the specific heat of copper experimentally. They place 0.150 kg of copper shot into boiling water and let it stay for a while, so as to reach a temperature of 100°C. Then they carefully pour the hot shot into a calorimeter cup (Fig. 11.4) containing 0.200 kg of water at 20.0°C. The final temperature of the mixture in the cup is measured to be 25.0°C. If the aluminum cup has a mass of 0.0450 kg, what is the specific heat of copper? (Assume that there is no heat exchange with the surroundings.)

*In this section, calorimetry will *not* involve phase changes, such as ice melting or water boiling. These effects are discussed in Section 11.3.

Thinking It Through. The conservation of heat energy is involved: $\Sigma Q_i = 0$, taking into account the correct positive and negative signs. In calorimetry problems, it is important to identify and label all of the quantities with proper signs. Identification of the heat gains and losses is crucial. You will probably use this method in the laboratory.

Solution. The subscripts Cu, w, and Al will be used to refer to the copper, water, and aluminum calorimeter cup, respectively. The subscripts h, i, and f to refer to the temperature of the *hot* metal shot, the water (and cup) *initially* at room temperature, and the *final* temperature of the system, respectively. With this notation,

Given: $m_{Cu} = 0.150 \text{ kg}$ *Find:* c_{Cu} (specific heat)
         $m_w = 0.200 \text{ kg}$
         $c_w = 4186 \text{ J/(kg} \cdot \text{C}°)$ (from Table 11.1)
         $m_{Al} = 0.0450 \text{ kg}$
         $c_{Al} = 920 \text{ J/(kg} \cdot \text{C}°)$ (from Table 11.1)
         $T_h = 100°\text{C}$, $T_i = 20.0°\text{C}$, and $T_f = 25.0°\text{C}$

If there is no heat exchange with the surroundings, the system's total energy is conserved, $\Sigma Q_i = 0$, and

$$\Sigma Q_i = Q_w + Q_{Al} + Q_{Cu} = 0$$

Substituting the relationship in Eq. 11.1 for these heats,

$$c_w m_w \Delta T_w + c_{Al} m_{Al} \Delta T_{Al} + c_{Cu} m_{Cu} \Delta T_{Cu} = 0$$

or

$$c_w m_w (T_f - T_i) + c_{Al} m_{Al} (T_f - T_i) + c_{Cu} m_{Cu} (T_f - T_h) = 0$$

Here, the water and aluminum cup, initially at T_i, are heated to T_f, so $\Delta T_w = \Delta T_{Al} = (T_f - T_i)$. The copper initially at T_h is cooled to T_f, so $\Delta T_{Cu} = (T_f - T_h)$ and this is a negative quantity, indicating a temperature drop for the copper. Solving for c_{Cu} gives

$$c_{Cu} = -\frac{(c_w m_w + c_{Al} m_{Al})(T_f - T_i)}{m_{Cu}(T_f - T_h)}$$

$$= -\frac{\{[4186 \text{ J/(kg} \cdot \text{C}°)](0.200 \text{ kg}) + [920 \text{ J/(kg} \cdot \text{C}°)](0.0450 \text{ kg})\}(25.0°\text{C} - 20.0°\text{C})}{(0.150 \text{ kg})(25.0°\text{C} - 100°\text{C})}$$

$$= 390 \text{ J/(kg} \cdot \text{C}°)$$

Notice that the proper use of signs resulted in a positive answer for c_{Cu}, as required. If, for example, the Q_{Cu} term had not had the correct sign, the answer would have been negative—a big clue that you had an initial sign error.

Follow-Up Exercise. In this example, what would the final equilibrium temperature be if the calorimeter (water and cup) initially had been at a warmer 30°C?

Specific Heat of Gases

When heat is added to or taken from most materials, they expand or contract. During expansion, for example, the materials would then do work on the atmosphere. For most solids and liquids, this work is negligible, because the volume changes are very small (Chapter 10). This is why this effect wasn't included in our discussion of specific heat of solids and liquids.

However, for gases, expansion and contraction *can* be significant. It is therefore important to specify the *conditions* under which heat is transferred when referring to a gas. If heat is added to a gas at constant volume (a *rigid* container), the gas does no work. (Why?) In this case, all of the heat goes into increasing the gas's internal energy and, therefore, to increasing its temperature. However, if the same amount of heat is added at constant pressure (a *nonrigid* container allowing a volume change), a portion of the heat is converted to work as the gas expands. Thus, not all of the heat will go into the gas's internal energy. This process results in a *smaller* temperature change than occurred during the constant-volume process.

To designate the physical quantities that are held constant while heat is added to or removed from a gas, we use a subscript notation: c_p means specific heat under conditions of constant pressure (p), and c_v means specific heat under conditions

Demonstration/activity: Perform a standard calorimetry experiment using different amounts of hot and cold water. Have students predict the final temperature of each water sample when given the initial masses and temperatures. Reasonable results can be obtained using a 2-lb coffee can; neglect the mass of the can and perform the demonstration quickly.

PHYSLET

Exploration 19.3   Calorimetry

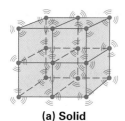

(a) Solid

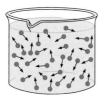

(b) Liquid

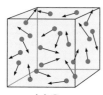

(c) Gas

▲ **FIGURE 11.5 Three phases of matter** **(a)** The molecules of a solid are held together by bonds; consequently, a solid has a definite shape and volume. **(b)** The molecules of a liquid can move more freely, so a liquid has a definite volume and assumes the shape of its container. **(c)** The molecules of a gas interact weakly and are separated by relatively large distances; thus, a gas has no definite shape or volume, unless it is confined in a container.

Note: Solid, liquid, and gas are sometimes referred to as states of matter rather than phases of matter, but the *state* of a system has a different meaning in physics, as you will learn in Chapter 12.

of constant volume (v). The specific heat for water vapor (H_2O) given in Table 11.1 is the specific heat under constant pressure (c_p).

An important result is that for a particular gas, c_p is always greater than c_v. This is true because for a specific mass of gas, $c \propto Q/\Delta T$. Since for a given Q, ΔT_v is as large as it can be, and c_v will be less than c_p. In other words, $\Delta T_v > \Delta T_p$. Specific heats of gases play an important role in adiabatic thermodynamic processes. (See Section 12.3.)

11.3 Phase Changes and Latent Heat

OBJECTIVES: To **(a)** compare and contrast the three common phases of matter, and **(b)** relate latent heat to phase changes.

Matter normally exists in one of three *phases*: solid, liquid, or gas (◄Fig. 11.5). However, this division into three common phases is only approximate since there are other phases, such as a plasma phase and a superconducting phase. The phase that a substance is in depends on the substance's internal energy (as indicated by its temperature) and the pressure on it. However, it is likely that you think of adding or removing heat as the way to change the phase of a substance.

In the **solid phase**, molecules are held together by attractive forces, or bonds (Fig. 11.5a). Adding heat causes increased motion about the molecular-equilibrium positions. If enough heat is added to provide sufficient energy to break the intermolecular bonds, most solids undergo a phase change and become liquids. The temperature at which this phase change occurs is called the **melting point**. The temperature at which a liquid becomes a solid is called the **freezing point**. In general, these temperatures are the same for a given substance, but they can differ slightly.

In the **liquid phase**, molecules of a substance are relatively free to move and a liquid assumes the shape of its container (Fig. 11.5b). In certain liquids, there may be some locally ordered structure, giving rise to liquid crystals, such as those used in LCDs (liquid crystal displays) of calculators and computer displays (Chapter 24).

Adding even more heat increases the motion of the molecules of a liquid. When they have enough energy to become separated, the liquid changes to the **gaseous (or vapor) phase*.** This change may occur slowly, by *evaporation* (p. 379), or rapidly, at a particular temperature called the **boiling point**. The temperature at which a gas condenses into a liquid is the **condensation point**.

Some solids, such as dry ice (solid carbon dioxide), mothballs, and certain air fresheners, change directly from the solid to the gaseous phase at standard pressure. This process is called **sublimation**. Like the rate of evaporation, the rate of sublimation increases with temperature. A phase change from a gas to a solid is called *deposition*. Frost, for example, is solidified water vapor (gas) deposited on grass, car windows, and other objects. Frost is *not* frozen dew (liquid water), as is sometimes mistakenly assumed.

Latent Heat

In general, when heat is transferred to a substance, its temperature increases as the average kinetic energy per molecule increases. However, when heat is added (or removed) during a phase change, the temperature of the substance does *not* change. For example, if heat is added to a quantity of ice at −10°C, the temperature of the ice increases until it reaches its melting point of 0°C. At this point, the addition of more heat does not increase the ice's temperature, but causes it to melt, or change phase. (The heat must be added slowly so that the ice and melted water remain in thermal equilibrium, otherwise, the ice water can warm above 0°C even though the ice remains at 0°C.) Only after the ice is completely melted does adding more heat cause

---

*The terms *vapor* and *vapor phase* are sometimes used interchangeably with the term *gaseous phase*. Strictly speaking, *vapor* refers to the gaseous phase of a substance in contact with its liquid phase.

the temperature of the water to rise. A similar situation occurs during the liquid–gas phase change at the boiling point. Adding more heat to boiling water only causes more vaporization. A temperature increase occurs only *after* the water is completely boiled, resulting in *superheated steam*.

During a phase change, the heat goes into breaking the attractive bonds and separating molecules (increasing their potential, rather than kinetic, energies) rather than into increasing the temperature. The heat required for a phase change is called the **latent heat (L)***, which is defined as the magnitude of the heat needed per unit mass to induce a phase change:

$$L = \frac{|Q|}{m} \quad \text{latent heat} \tag{11.2}$$

where m is the mass of the substance. Latent heat has SI units of joules per kilogram (J/kg), or kilocalories per kilogram (kcal/kg).

The latent heat for a solid–liquid phase change is called the **latent heat of fusion** (L_f), and that for a liquid–gas phase change is called the **latent heat of vaporization** $(L_v.)$ These quantities are often referred to as simply the *heat of fusion* and the *heat of vaporization*. The latent heats of some substances, along with their melting and boiling points, are given in Table 11.2. (The latent heat for the less common solid–gas phase change is called the *latent heat of sublimation* and is symbolized by L_s.) As you might expect, the latent heat (in joules per kilogram) is the amount of energy per kilogram *given up* when the phase change is in the opposite direction, that is, from liquid to solid or gas to liquid.

A more useful form of Eq. 11.3 is given by solving for Q and including a positive/negative sign for the two possible directions of heat flow:

$$Q = \pm mL \quad \text{(signs with latent heat)} \tag{11.3}$$

This equation is more practical for problem solving because in calorimetry problems, you are typically interested in applying conservation of energy in the form of $\Sigma Q_i = 0$. The positive/negative sign $(\pm)$ must be explicitly expressed because heat can flow either into $(+)$ or out of $(-)$ the object or system of interest.

When solving calorimetry problems involving phase changes, you must be careful to use the correct sign for those terms, in agreement with our sign conventions (▼Fig. 11.6). For example, if water is condensing from steam into liquid droplets, *removal* of heat is involved, necessitating the choice of the *negative* sign.

Note: Keep in mind that ice can be colder than 0°C and steam can be hotter than 100°C.

Teaching tip: Emphasize that there is no change in temperature in the equation $Q = \pm mL$. In other words, while ice is melting, its temperature does not change. The water that is present immediately after melting has the same temperature as that of the ice immediately prior to melting.

| **TABLE 11.2** | Temperatures of Phase Changes and Latent Heats for Various Substances (at 1 atm) |

Substance	Melting Point	L_f		Boiling Point	L_v	
		J/kg	kcal/kg		J/kg	kcal/kg
Alcohol, ethyl	−114°C	1.0×10^5	25	78°C	8.5×10^5	204
Gold	1063°C	0.645×10^5	15.4	2660°C	15.8×10^5	377
Helium†	—	—	—	−269°C	0.21×10^5	5
Lead	328°C	0.25×10^5	5.9	1744°C	8.67×10^5	207
Mercury	−39°C	0.12×10^5	2.8	357°C	2.7×10^5	65
Nitrogen	−210°C	0.26×10^5	6.1	−196°C	2.0×10^5	48
Oxygen	−219°C	0.14×10^5	3.3	−183°C	2.1×10^5	51
Tungsten	3410°C	1.8×10^5	44	5900°C	48.2×10^5	1150
Water	0°C	3.33×10^5	80	100°C	22.6×10^5	540

†Not a solid at a pressure of 1 atm; melting point is −272°C at 26 atm.

*Latent is Latin for "hidden".

▶ **FIGURE 11.6** Phase changes and latent heats **(a)** At 0°C, 3.33×10^5 J must be added to 1 kg of ice or removed from 1 kg of liquid water to change its phase. **(b)** At 100°C, 22.6×10^5 J must be added to 1 kg of liquid water or removed from 1 kg of steam to change its phase.

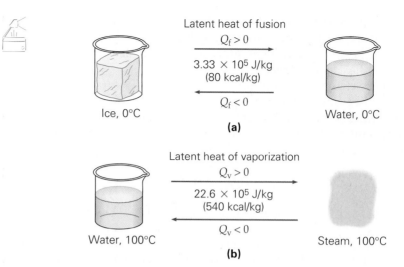

Latent heat of fusion
$Q_f > 0$
3.33×10^5 J/kg
(80 kcal/kg)
$Q_f < 0$

Ice, 0°C   Water, 0°C

(a)

Latent heat of vaporization
$Q_v > 0$
22.6×10^5 J/kg
(540 kcal/kg)
$Q_v < 0$

Water, 100°C   Steam, 100°C

(b)

Problem-Solving Hint

Recall in Section 11.2 that where there were no phase changes, the expression for heat $Q = cm\Delta T$ automatically gave the correct sign for Q from the sign of ΔT. But there is no ΔT during a phase change. *Choosing the correct sign is up to you.*

For water, the latent heats of fusion and vaporization are

$$L_f = 3.33 \times 10^5 \text{ J/kg}$$
$$L_v = 22.6 \times 10^5 \text{ J/kg}$$

The accompanying Learn by Drawing, numerically expressed in Example 11.5, shows explicitly the two types of heat terms (specific heat and latent heat) that must be employed in the general situation when any of the objects undergo a temperature change *and* a phase change.

Example 11.5 ■ From Cold Ice to Hot Steam

Heat is added to 1.00 kg of cold ice at −10°C. How much heat is required to change the cold ice to hot steam at 110°C?

Thinking It Through. Five steps are involved: (1) heating ice to its melting point (specific heat of ice), (2) melting ice to ice water at 0°C (latent heat, a phase change), (3) heating liquid water (specific heat of liquid water), (4) vaporizing water at 100°C (latent heat, a phase change), and (5) heating steam (specific heat of water vapor). The key idea here is that the temperature does not change during a phase change. [Refer to Learn by Drawing on page 377.]

Solution.

Given: $m = 1.00$ kg   *Find:* Q_{total} (total heat)
$T_i = -10°C$
$T_f = 110°C$
$L_f = 3.33 \times 10^5$ J/kg (from Table 11.2)
$L_v = 22.6 \times 10^5$ J/kg (from Table 11.2)
$c_{ice} = 2100$ J/(kg·C°) (from Table 11.1)
$c_{water} = 4186$ J/(kg·C°) (from Table 11.1)
$c_{steam} = 2000$ J/(kg·C°) (from Table 11.1)

(1) $Q_1 = c_{ice}m\Delta T_1 = [2100 \text{ J/(kg·C°)}](1.00 \text{ kg})[0°C - (-10°C)]$ *(heating ice)*
    $= +2.10 \times 10^4$ J

(2) $Q_2 = +mL_v = (1.00 \text{ kg})(3.33 \times 10^5 \text{ J/kg}) = +3.33 \times 10^5$ J   *(melting ice)*

(3) $Q_3 = c_w m\Delta T_2 = [4186 \text{ J/(kg·C°)}](1.00 \text{ kg})(100°C - 0°C)$ *(heating water)*
    $= +4.19 \times 10^5$ J

(4) $Q_4 = +mL_v = (1.00 \text{ kg})(22.6 \times 10^5 \text{ J/kg}) = +2.26 \times 10^6$ J   *(vaporizing water)*

(5) $Q_5 = c_{steam}m\Delta T_3 = [2000 \text{ J/(kg·C°)}](1.00 \text{ kg})(110°C - 100°C)$ *(heating steam)*
    $= +2.00 \times 10^4$ J

Teaching tip: When solving phase-change problems, students often forget to take into account the changes in specific heat. For example, if steam condenses and there is a change in the temperature of water, they will sometimes still use the specific heat of steam rather than the specific heat of water.

The total heat required is

$$Q_{\text{total}} = \Sigma Q_i = 2.1 \times 10^4 \text{ J} + 3.33 \times 10^5 \text{ J} + 4.19 \times 10^5 \text{ J} + 2.26 \times 10^6 \text{ J} + 2.00 \times 10^4 \text{ J}$$
$$= 3.05 \times 10^6 \text{ J}$$

The latent heat of vaporization is, by far, the largest. It is actually greater than the sum of the other four terms.

Follow-Up Exercise. How much heat must a freezer remove from liquid water (initially at 20°C) to create 0.250 kg of ice at −10°C?

Problem-Solving Hint

Note that you must compute the latent heat at each phase change. It is a common error to use the specific-heat equation with a temperature interval *that includes* a phase change. Also, you cannot assume a complete phase change until you have checked for it numerically. (See Example 11.6.)

Teaching tip: Ask students what will be the final result if equal masses of ice at 0°C and steam at 100°C are added together.

LEARN BY DRAWING FROM COLD ICE TO HOT STEAM

It can be helpful to focus on the fusion and vaporization of water graphically. To heat up a piece of cold ice at −10°C all the way to hot steam at 110°C, five separate specific heat and latent heat calculations are necessary. (Most freezers are at a temperature of about −10°C.) At the phase change (0°C and 100°C) heat is added without a temperature change. Once each phase change is complete, adding more heat causes the temperature to increase. The slopes of the lines in the drawings are not all the same, which indicates that the specific heats of the various phases are not the same. (Why do different slopes mean different specific heats?) The numbers come from Example 11.5

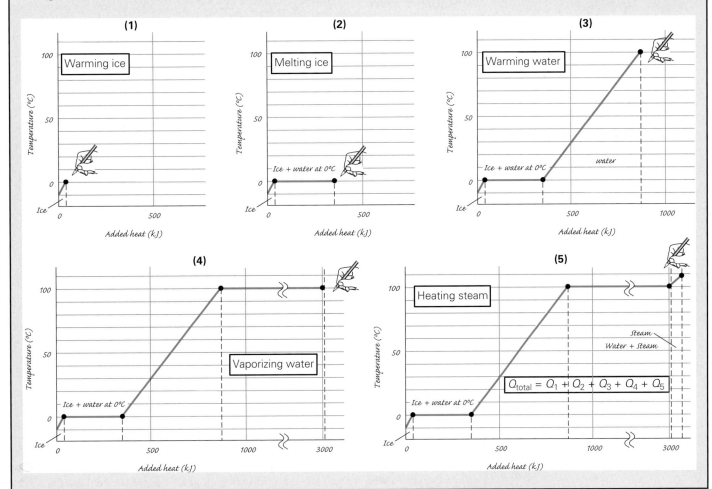

Teaching tip: Explain how a change in pressure can affect the phases of matter—for example, in a pressure cooker. For a thought question, ask students what should happen to the boiling and freezing points of water at high altitudes. Do eggs really have to be boiled longer in Denver than in Detroit?

Point out that as a general rule, the increase of pressure favors the liquid phase of matter—molecules are less likely to vaporize.

Technically the freezing and boiling points of water (0°C and 100°C, respectively) apply only at 1 atm of pressure. Phase-change temperatures generally vary with pressure. For example, the boiling point of water decreases with decreasing pressure. At high altitudes, where there is lower atmospheric pressure, the boiling point of water is lowered. For example, at Pikes Peak, Colorado, at an elevation of about 4300 m, the atmospheric pressure is about 0.79 atm and water boils at about 94°C rather than at 100°C. The lower temperature lengthens the cooking time of food. Conversely, some cooks use a pressure cooker to *reduce* cooking time—by increasing the pressure, a pressure cooker raises the boiling point.

The freezing point of water actually *decreases* with increasing pressure. This inverse relationship is characteristic of only a very few substances, including water (Section 10.4), that expand when they freeze.

Example 11.6 ■ Practical Calorimetry: Using Phase Changes to Save a Life

Organ transplants are becoming commonplace. Many times, the procedure involves removing a healthy organ from a deceased person and flying it to the recipient. During that time, to prevent its deterioration, the organ is packed in ice in an insulated container. Assume that a liver has a mass of 0.500 kg and is initially at 29°C. The specific heat of the human liver is 3500 J/(kg·C°). The liver is surrounded by 2.00 kg of ice initially at −10°C. Calculate the final equilibrium temperature.

Thinking It Through. Clearly, the liver will cool, and the ice will warm. However, it is not clear what temperature the ice will reach. If it gets to the freezing point, it will begin to melt, and a phase change must be considered. If all of it melts, then additional heat required to warm that water to a temperature above 0°C must be considered. Thus, care must be taken, since it *cannot* be assumed that all the ice melts, or even that the ice reaches its melting point. Hence, the calorimetry equation (conservation of energy) cannot be written down until the terms in it are determined. First we need to review the *possible* heat transfers. Only then can the final temperature be determined.

Solution. Listing the data given and the information obtained from tables,

Given: $m_l = 0.500$ kg   *Find:* The final temperature of the system
       $m_{ice} = 2.00$ kg
        $c_l = 3500$ J/(kg·C°)
        $c_{ice} = 2100$ J/(kg·C°)   (from Table 11.1)
        $L_f = 3.33 \times 10^5$ J/(kg·C°) (from Table 11.2)

The amount of heat required to bring the ice to 0°C from −10°C is

$$Q_{ice} = c_{ice} m_{ice} \Delta T_{ice} = [2100 \text{ J/(kg·C°)}](2.00 \text{ kg})(+10 \text{ C°}) = +4.20 \times 10^4 \text{ J}$$

Since this heat must come from the liver, the *maximum* heat available from the liver needs to be calculated; that is, if its temperature drops all the way from 29°C to 0°C:

$$Q_{l,max} = c_l m_l \Delta T_{l,max} = [3500 \text{ J/(kg·C°)}](0.500 \text{ kg})(-29 \text{ C°}) = -5.08 \times 10^4 \text{ J}$$

This *is* enough to bring the ice to 0°C. If 4.20×10^4 J of heat flows into the ice (bringing it to 0°C), the liver is still not at 0°C. Then how much ice melts? This depends on how much more heat can be transferred from the liver.

How much more heat Q' would be extracted from the liver if its temperature were to drop to 0°C? This value is just the maximum amount minus the heat that went into warming the ice, or

$$Q' = |Q_{l,max}| - 4.20 \times 10^4 \text{ J}$$
$$= 5.08 \times 10^4 \text{ J} - 4.20 \times 10^4 \text{ J} = 8.8 \times 10^3 \text{ J}$$

Compare this with the magnitude of the heat needed to melt the ice completely ($|Q_{melt}|$) to decide whether this can, in fact, happen. The heat required to melt *all* the ice is

$$|Q_{melt}| = +m_{ice} L_{ice} = +(2.00 \text{ kg})(3.33 \times 10^5 \text{ J/kg}) = +6.66 \times 10^5 \text{ J}$$

Since this amount of heat is much larger than the amount available from the liver, only part of the ice melts. In the process, the liver has dropped to 0°C, and the remainder of the ice is at 0°C. Since everything in the "calorimeter" is at the same temperature, heat flow stops, and the final system temperature is 0°C. Thus the final result is that the liver is in a container with ice and some liquid water, all at 0°C. Since the container is a very good insulator, it will prevent any inward heat flow that might raise the liver's temperature. It is therefore expected that the liver will arrive at its destination in good shape.

Follow-Up Exercise. (a) In this Example, how much of the ice melts? (b) If the ice originally had been at its melting point (0°C), what would the equilibrium temperature have been?

Problem-Solving Hint

Notice in Example 11.6 that numbers were *not* plugged directly into the $\Sigma Q_i = 0$ equation, which is equivalent to assuming that all the ice melts. In fact, if this step had been done, we would have been on the wrong track. For calorimetry problems *involving phase changes*, a careful step-by-step numerical "accounting" procedure should be followed until all of the pieces of the system are at the same temperature. At that point, the problem is over, because no more heat exchanges can happen.

Evaporation

The **evaporation** of water from an open container becomes evident only after a relatively long period of time. This phenomenon can be explained in terms of the kinetic theory (Section 10.5). The molecules in a liquid are in motion at different speeds. A faster-moving molecule near the surface may momentarily leave the liquid. If its speed is not too large, the molecule will return to the liquid, because of the attractive forces exerted by the other molecules. Occasionally, however, a molecule has a large enough speed to leave the liquid entirely. The higher the temperature of the liquid, the more likely this phenomenon is to occur.

The escaping molecules take their energy with them. Since those molecules with greater-than-average energy are the ones most likely to escape, the average molecular energy, and thus the temperature of the remaining liquid, will be reduced. That is, *evaporation is a cooling process* for the object from which the molecules escape. You have probably noticed this phenomenon when drying off after a bath or shower. You can read more about this in Insight 11.1 on page 380 on Physiological Regulation of Body Temperature.

Illustration 20.4 Evaporative Cooling

Demonstration/activity: Take a fan to class, and let the students explain its purpose. Have students dip their hands in water and then hold them in the air flow in front of the fan.

11.4 Heat Transfer

OBJECTIVES: To (a) describe the three mechanisms of heat transfer, and (b) give practical applications of each.

The transfer of heat is an important topic and has many practical applications. Heat can move from place to place by three different mechanisms: conduction, convection, and radiation.

Conduction

You can keep a pot of coffee hot on an electric stove because heat is conducted through the bottom of the coffeepot from the hot metal burner. The process of **conduction** results from molecular interactions. Molecules at a higher-temperature region on an object move relatively rapidly. They collide with, and transfer some of their energy to, the less energetic molecules in a nearby cooler part of the object. In this way, energy is conductively transferred from a higher-temperature region to a lower-temperature region—transfer as a result of a temperature difference.

Solids can be divided into two general categories: metals and nonmetals. Metals are generally good conductors of heat, or **thermal conductors**. Metals have a large number of electrons that are free to move around (not permanently bound to a particular molecule or atom). These free electrons (rather than the interaction between adjacent atoms) are primarily responsible for the good heat conduction in metals. Nonmetals, such as wood and cloth, have relatively few free electrons. The absence of this transfer mechanism makes them poor heat conductors relative to metals. A poor heat conductor is called a **thermal insulator**.

Illustration 19.2 Heat Transfer, Conduction

INSIGHT   11.1 PHYSIOLOGICAL REGULATION OF BODY TEMPERATURE

Being warm-blooded, humans must maintain a narrow range of body temperature. (See Insight 10.1, page 343.) The generally accepted value for the average normal body temperature is 37.0°C (98.6°F). However, it can be as low as 35.5°C (96°F) in the early-morning hours on a cold day and as high as 39.5°C (103°F) during intense exercise on a hot day. For females, the body temperature at rest rises very slightly after ovulation as a result of a rise in the hormone progesterone. This can be used to predict on which day ovulation will occur in the next cycle.

When the ambient temperature is lower than the body temperature, the body loses heat. If the body loses too much heat, a circulatory mechanism causes a reduction of blood flow to the skin in order to reduce heat loss. A physiological response to this mechanism through shivering is to increase heat generation (and thus warm the body) by "burning" the body's reserves of carbohydrate or fat. If the body temperature drops below 33°C (91.4°F), *hypothermia* may result, which can cause severe thermal injuries to organs and even death.

At the other extreme, if the body is subjected to ambient temperatures higher than the body temperature, along with intense exercise, the body can overheat. *Heat stroke* is a prolonged elevation of body temperature above 40°C (104°F). If the body becomes overheated, the blood vessels to the skin dilate, carrying more warm blood to the skin, enabling the interior of the body and organs to remain cooler. (The person's face may turn red.)

Usually, radiation, conduction, natural convection (discussed later in Section 11.4), and possibly slight evaporation of perspiration on the skin are sufficient to maintain a heat loss at a rate that keeps our body temperature in the safe range. However, when the ambient temperature becomes too high, these mechanisms cannot do the job completely. To avoid heat stroke, as a last resort (and the most efficient one), the body produces heavy perspiration. The evaporation of water from our skin removes a lot of heat, thanks to the large value of latent heat of vaporization of water.

Evaporation lowers the temperature of the perspiration on our bodies, which can then draw heat from our skin and thus cool our bodies. The removal of a minimum of 2.26×10^6 J of heat from the body is required to evaporate each kilogram (liter) of water.* For the body of a 75-kg person, which is mainly composed of water, the heat loss due to evaporation of 1 kg of water could lower the body temperature as much as

$$\Delta T = \frac{Q}{cm} = \frac{2.26 \times 10^6 \text{ J}}{[4186 \text{ J}/(\text{kg} \cdot \text{C}°)](75 \text{ kg})} = 7.2 \text{ C}°$$

On a hot day's race, a professional cyclist can evaporate as much as 7.0 kg of water in 3.5 h. This heat loss through sweat is the mechanism that enables the body to keep its temperature in the safe range.

On a summer day, a person may stand in front of a fan and remark how "cool" the blowing air feels. But the fan is merely blowing hot air from one place to another. The air *feels* cool because it is relatively drier (has low humidity compared to the sweaty body) and therefore its flow promotes evaporation, which removes latent-heat energy.

*The actual latent heat of vaporization of perspiration is about 2.42×10^6 J (greater than the 2.26×10^6 J value used here for temperature at 100°C).

In general, the ability of a substance to conduct heat depends on the substance's phase. Gases are poor thermal conductors; their molecules are relatively far apart, and collisions are therefore infrequent. Liquids and solids are better thermal conductors than gases, because their molecules are closer together and can interact more readily.

Heat conduction is usually described using the *time rate* of heat flow ($\Delta Q / \Delta t$) in a material for a given temperature difference (ΔT), as illustrated in ▸Fig. 11.7. Experiment has established that the rate of heat flow through a substance depends on the temperature difference between its boundaries. Heat conduction also depends on the size and shape of the object as well as its composition. In our analysis of heat flow, we will use uniform slabs for simplicity.

Experimentally, it is found that the heat flow rate ($\Delta Q / \Delta t$ in J/s or W) through a slab of material is directly proportional to the material's surface area (A) and the temperature difference across its ends (ΔT), and inversely proportional to its thickness (d). That is,

$$\frac{\Delta Q}{\Delta t} \propto \frac{A \Delta T}{d}$$

Using a constant of proportionality k allows us to write the relation as an equation:

$$\frac{\Delta Q}{\Delta t} = \frac{kA\Delta T}{d} \quad \textit{(conduction only)} \quad (11.4)$$

The constant k, called the **thermal conductivity**, characterizes the heat-conducting ability of a material and depends only on the type of material. The

TABLE 11.3   Thermal Conductivities of Some Substances

Substance	Thermal Conductivity, k	
	J/(m·s·C°) or W/(m·C°)	kcal/(m·s·C°)
Metals		
Aluminum	240	5.73×10^{-2}
Copper	390	9.32×10^{-2}
Iron and steel	46	1.1×10^{-2}
Silver	420	10×10^{-2}
Liquids		
Transformer oil	0.18	4.3×10^{-5}
Water	0.57	14×10^{-5}
Gases		
Air	0.024	0.57×10^{-5}
Hydrogen	0.17	4.1×10^{-5}
Oxygen	0.024	0.57×10^{-5}
Other Materials		
Brick	0.71	17×10^{-5}
Concrete	1.3	31×10^{-5}
Cotton	0.075	1.8×10^{-5}
Fiberboard	0.059	1.4×10^{-5}
Floor tile	0.67	16×10^{-5}
Glass (typical)	0.84	20×10^{-5}
Glass wool	0.042	1.0×10^{-5}
Goose down	0.025	0.59×10^{-5}
Human tissue (average)	0.20	4.8×10^{-5}
Ice	2.2	53×10^{-5}
Styrofoam	0.042	1.0×10^{-5}
Wood, oak	0.15	3.6×10^{-5}
Wood, pine	0.12	2.9×10^{-5}
Vacuum	0	0

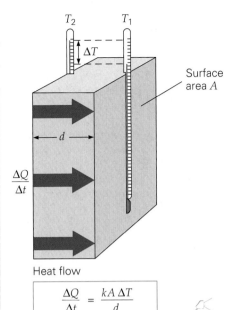

$$\frac{\Delta Q}{\Delta t} = \frac{kA \, \Delta T}{d}$$

▲ **FIGURE 11.7 Thermal conduction** Heat conduction is characterized by the time rate of heat flow ($\Delta Q/\Delta t$) in a material with a temperature difference across it of ΔT. For a slab of material, $\Delta Q/\Delta t$ is directly proportional to the cross-sectional area (A) and the thermal conductivity (k) of the material; it is inversely proportional to the thickness of the slab (d).

▼ **FIGURE 11.8 Copper-bottomed pots** Copper is used on the bottoms of some stainless steel pots and saucepans. The high thermal conductivity of copper ensures the rapid and even spread of heat from the burner; the low thermal conductivity of stainless steel retains the heat in the pot and keeps the handle not too hot to touch. (The thermal conductivity of stainless steel is only 12% that of copper.)

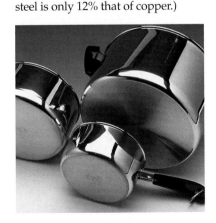

greater the value of k for a material, the better it will conduct heat, all other factors being equal. The units of k are J/(m·s·C°) = W/(m·C°). The thermal conductivities of various substances are listed in Table 11.3. These values vary slightly with temperature, but can be considered constant over normal temperature ranges.

Compare the relatively large thermal conductivities of the good thermal conductors, the metals, with the relatively small thermal conductivities of some good thermal insulators, such as Styrofoam and wood. Some stainless steel cooking pots have copper bottoms (▶Fig. 11.8). Being a good conductor of heat, the copper conducts heat faster to the food being cooked and also promotes the distribution of heat over the bottom of a pot for even cooking. Conversely, Styrofoam is a good insulator, mainly because it contains small, trapped pockets of air, thus reducing conduction and convection losses (p. 383). When you step on a tile floor with one bare foot and on an adjacent rug with the other bare foot, you feel that the tile is "colder" than the rug. However, both the tile and rug are actually at the same temperature. The reason is that the tile is a much better thermal conductor, so it removes or conducts heat from your foot better than the rug, making your foot on tile feel colder.

▶ **FIGURE 11.9** Insulation and thermal conductivity **(a)**, **(b)** Attics should be insulated to prevent loss of heat by the mechanism of conduction. See Example 11.7 and Insight 11.2 (page 384) on Physics, the Construction Industry, and Energy Conservation. **(c)** This thermogram of a house allows us to visualize the house's heat loss. Blue represents the areas that have the lowest rate of heat leaking; white, pink, and red indicate areas with increasingly larger heat losses. (Red areas have the most loss). What recommendations would you make to the owner of this house to save both money and energy? (Compare this figure with Fig. 11.15.)

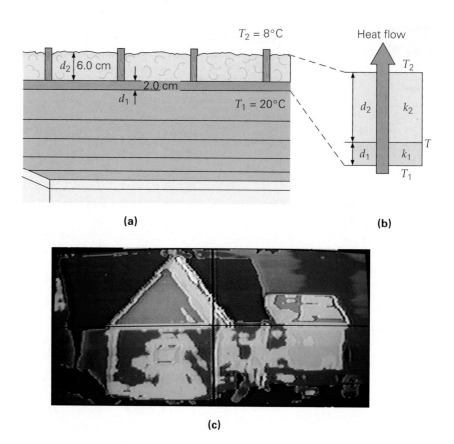

(a)

(b)

(c)

Exploration 19.4 Heat Balance

Example 11.7 ■ Thermal Insulation: Helping Prevent Heat Loss

A room with a pine ceiling that measures 3.0 m by 5.0 m and is 2.0 cm thick has a layer of glass-wool insulation above it that is 6.0 cm thick (▲Fig. 11.9a). On a cold day, the temperature inside the room at ceiling height is 20°C, and the temperature in the attic above the insulation layer is 8°C. Assuming that the temperatures remain constant with a steady heat flow, how much energy does the layer of insulation save in 1.0 h? Assume that losses are due to conduction only.

Thinking It Through. Here we have two materials, so we must consider Eq. 11.4 for two different thermal conductivities (k). We want to find $\Delta Q/\Delta t$ for the combination so that we can get ΔQ for $\Delta t = 1.0$ h. The situation is a bit complicated, because the heat flows through two materials. But we know that at a steady rate, *the heat flows must be the same through both* (why?). To find the energy saved in 1.0 h, we need to compute how much heat is conducted in this time with and without the layer of insulation.

Solution. Computing some of the quantities in Eq. 11.4 and making conversions as we list the data,

Given: $A = 3.0 \text{ m} \times 5.0 \text{ m} = 15 \text{ m}^2$ *Find:* Energy saved in one hour
$d_1 = 2.0 \text{ cm} = 0.020 \text{ m}$
$d_2 = 6.0 \text{ cm} = 0.060 \text{ m}$
$\Delta T = T_1 - T_2 = 20°\text{C} - 8.0°\text{C} = 12 \text{ C}°$
$\Delta t = 1.0 \text{ h} = 3.6 \times 10^3 \text{ s}$
$k_1 = 0.12 \text{ J}/(\text{m} \cdot \text{s} \cdot \text{C}°) (\text{wood, pine})$
$k_2 = 0.042 \text{ J}/(\text{m} \cdot \text{s} \cdot \text{C}°) (\text{glass wool})$ } (from Table 11.3)

(In working such problems with several given quantities, it is especially important to label all the data correctly.)

First, let's consider how much heat would be conducted in one hour through the wooden ceiling with no insulation present. Since we know Δt, we can rearrange Eq. 11.4*

---

*Equation 11.4 can be extended to any number of layers or slabs of materials: $\Delta Q/\Delta t = A(T_2 - T_1)/\Sigma(d_i/k_i)$. (See Insight 11.2 involving insulation in building construction, p. 384).

to find ΔQ_c (heat conducted through the wooden ceiling alone, assuming the same ΔT):

$$\Delta Q_c = \left(\frac{k_1 A \Delta T}{d_1}\right)\Delta t = \left\{\frac{[0.12 \text{ J}/(\text{m}\cdot\text{s}\cdot\text{C}°)](15 \text{ m}^2)(12 \text{ C}°)}{0.020 \text{ m}}\right\}(3.6 \times 10^3 \text{ s}) = 3.9 \times 10^6 \text{ J}$$

Now we need to find the heat conducted through the ceiling *and* the insulation layer together. Let T be the temperature at the interface of the materials and T_1 and T_2 be the warmer and cooler temperatures, respectively (Fig. 11.9b). Then

$$\frac{\Delta Q_1}{\Delta t} = \frac{k_1 A(T_1 - T)}{d_1} \quad \text{and} \quad \frac{\Delta Q_2}{\Delta t} = \frac{k_2 A(T - T_2)}{d_2}$$

We don't know T, but when the conduction is steady, the flow rates are the same for both materials; that is, $\Delta Q_1/\Delta t = \Delta Q_2/\Delta t$, or

$$\frac{k_1 A(T_1 - T)}{d_1} = \frac{k_2 A(T - T_2)}{d_2}$$

The A's cancel, and solving for T gives

$$\begin{aligned}
T &= \frac{k_1 d_2 T_1 + k_2 d_1 T_2}{k_1 d_2 + k_2 d_1} \\
&= \frac{[0.12 \text{ J}/(\text{m}\cdot\text{s}\cdot\text{C}°)](0.060 \text{ m})(20.0°\text{C}) + [0.042 \text{ J}/(\text{m}\cdot\text{s}\cdot\text{C}°)](0.020 \text{ m})(8.0°\text{C})}{[0.12 \text{ J}/(\text{m}\cdot\text{s}\cdot\text{C}°)](0.060 \text{ m}) + [0.042 \text{ J}/(\text{m}\cdot\text{s}\cdot\text{C}°)](0.020 \text{ m})} \\
&= 18.7°\text{C}
\end{aligned}$$

Since the flow rate is the same through the wood and the insulation, we can use the expression for either material to calculate it. Let's use the expression for the wood ceiling. Here, care must be taken to use the correct ΔT. The temperature at the wood–insulation interface is 18.7°C; thus,

$$\Delta T_{\text{wood}} = |T_1 - T| = |20°\text{C} - 18.7°\text{C}| = 1.3°\text{C}$$

Therefore, the heat flow rate is

$$\frac{\Delta Q_1}{\Delta t} = \frac{k_1 A |\Delta T_{\text{wood}}|}{d_1} = \frac{[0.12 \text{ J}/(\text{m}\cdot\text{s}\cdot\text{C}°)](15 \text{ m}^2)(1.3 \text{ C}°)}{0.020 \text{ m}} = 1.2 \times 10^2 \text{ J/s (or W)}$$

In 1.0 h, the heat loss with insulation in place is

$$\Delta Q_1 = \frac{\Delta Q_1}{\Delta t} \times \Delta t = (1.2 \times 10^2 \text{ J/s})(3600 \text{ s}) = 4.3 \times 10^5 \text{ J}$$

This value represents a decreased heat loss of

$$\Delta Q_c - \Delta Q_1 = 3.9 \times 10^6 \text{ J} - 4.3 \times 10^5 \text{ J} = 3.5 \times 10^6 \text{ J}$$

This amount represents a savings of $\dfrac{3.5 \times 10^6 \text{ J}}{3.9 \times 10^6 \text{ J}} \times (100\%) = 90\%$.

Follow-Up Exercise. Verify that the heat flow rate is the same through the insulation as through the wood (1.2×10^2 J/s) in this Example.

Convection

In general, compared with solids, liquids and gases are not good thermal conductors. However, the mobility of molecules in fluids permits heat transfer by another process—convection. (A fluid is a substance that can flow, and hence includes both liquids and gases.) **Convection** is heat transfer as a result of mass transfer, which can be natural or forced.

Natural convection cycles occur in liquids and gases. For example, when cold water is in contact with a hot object, such as the bottom of a pot on a stove, the object transfers heat to the water adjacent to the pot by conduction. But the water carries the heat away with it by natural convection, and a cycle is set up in which upper, colder (more dense) water replaces the rising warm (less dense) water. Such cycles are also

INSIGHT 11.2 PHYSICS, THE CONSTRUCTION INDUSTRY, AND ENERGY CONSERVATION

In the last few decades, many homeowners have found it cost-effective to provide their homes with better insulation. To quantify the insulating properties of various materials, the insulation and construction industries do not use thermal conductivity k. Rather, they use a quantity called *thermal resistance*, which is related to the *inverse* of k.

To see how these two quantities are related, consider Eq. 11.4 rewritten as

$$\frac{\Delta Q}{\Delta t} = \left(\frac{k}{d}\right)A\Delta T = \left(\frac{1}{R_t}\right)A\Delta T$$

where the *thermal resistance* is $R_t = d/k$. Note that R_t depends not only on the material's properties (expressed in the thermal conductivity k), but also on its thickness d. R_t is a measure of how "resistant" to heat flow the slab of material is.

The heat flow rate is proportional to the area of the material and to the temperature difference. More area means more heat conduction, and temperature differences are the fundamental cause of the heat flow in the first place. But note that the heat flow rate $\Delta Q/\Delta t$ is inversely related to the thermal resistance: More thermal resistance results in less heat flow. More resistance is attained using *thicker* material with a *low* conductivity.

For homeowners, the lesson is clear. To reduce heat flow (and thus minimize heat loss in the winter and heat gain in the summer), they should reduce areas of low thermal resistance, such as windows, or at least increase the window's resistance by switching to double or triple panes. Similarly for walls, increasing the thermal resistance by adding or upgrading insulation is the way to go. Lastly, changing interior temperature requirements (changing $\Delta T = |T_{\text{exterior}} - T_{\text{interior}}|$) can make a big difference. In summer, homeowners should raise the ther-mostat setting on air conditioning (lowering ΔT by increasing T_{interior}), and in winter, they should lower the thermostat setting on the heating system (lowering ΔT by decreasing T_{interior}).

Insulation and building materials are classified according to their *R-values*—that is, their thermal resistance values. In the United States, the units of R_t are $\text{ft}^2 \cdot \text{h} \cdot \text{F}°/\text{Btu}$. While these units may seem awkward, the important point is that they are proportional to the thermal resistance of the material. Thus, wall insulation with a value of R-31 (meaning $R_t = 31 \text{ ft}^2 \cdot \text{h} \cdot \text{F}°/\text{Btu}$) is about 1.6 times (or 31/19) as resistive (0.6 times, or 19/31, as conductive) as insulation with a value of R-19. A comparison photo of various types of insulation is shown in Fig. 1.

FIGURE 1 Differences in R-values For insulation blankets made of identical materials, the R-values are proportional to the materials' thickness.

important in atmospheric processes, as illustrated in ▼Fig. 11.10. During the day, the ground heats up more quickly than do large bodies of water, as you may have noticed if you have been to the beach. This phenomenon occurs both because the water has a higher specific heat than land and because convection currents disperse the absorbed heat throughout the larger volume of water. The air in contact with the warm ground is heated and expands, becoming less dense. As a result, the warm air rises (air currents) and, to fill the space, other air moves horizontally (winds)—creating a sea breeze near a large body of water. Cooler air descends, and a thermal convection cycle is set up, which transfers heat away from the land. At night, the ground loses its heat more quickly than the water, and the surface of the water is warmer than the land. As a result, the cycle is reversed. Since the prevailing jet streams over the Northern

▶ **FIGURE 11.10 Convection cycles** During the day, natural convection cycles give rise to sea breezes near large bodies of water. At night, the pattern of circulation is reversed, and the land breezes blow. The temperature differences between land and water are the result of their specific-heat differences. Water has a much larger specific heat, so the land warms up more quickly during the day. At night, the land cools more quickly, while the water remains warmer, because of its larger specific heat.

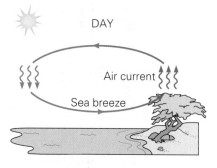

Land warmer than water

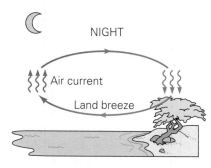

Water warmer than land

Hemisphere are mostly from west to east, west coasts usually have milder climate than east coasts. The winds move the ocean air with more constant temperature toward the west coasts. On a smaller scale, there is usually a smaller temperature fluctuation on the west coasts than a few miles inland, where desert conditions prevail.

In *forced convection*, the fluid is moved mechanically. This condition results in transfer without a temperature difference. In fact, we can transfer heat energy from a low-temperature region to a high-temperature region this way, as in the case of the forced convection of a refrigerator coolant removing energy from the cooler interior of the refrigerator. (The circulating coolant carries heat energy from the inside of the refrigerator, and this heat is given up to the environment, as we will see in Section 12.5.)

Common examples of forced convection systems are forced-air heating systems in homes (▸Fig. 11.11), the human circulatory system, and the cooling system of an automobile engine. The human body does not use all of the energy obtained from food; a great deal is lost in the form of heat. (There's usually a temperature difference between your body and your surroundings.) So that body temperature will stay normal, the internally generated heat is transferred close to the surface of the skin by blood circulation. From the skin, the energy is conducted to the air or lost by radiation (the other heat-transfer mechanism, to be discussed shortly). This circulatory system is highly adjustable; blood flow can be increased or decreased to specific areas depending on needs.

Water or some other coolant is circulated (pumped) through most automobile cooling systems. (Some smaller engines are air-cooled.) The coolant carries engine heat to the radiator (a form of *heat exchanger*), where forced-air flow produced by the fan and car movement carries it away. The *radiator* of an automobile is actually misnamed—most of the heat is transferred from it by forced convection rather than by radiation.

Conceptual Example 11.8 ■ Foam Insulation: Better Than Air?

Polymer foam insulation is sometimes blown into the space between the inner and outer walls of a house. Since air is a better thermal insulator than foam, why is the foam insulation needed: (a) To prevent loss of heat by conduction, (b) to prevent loss of heat by convection, or (c) for fireproofing?

Reasoning and Answer. Polymer foams will generally burn, so (c) isn't likely to be the answer. Air is a poor thermal conductor, even poorer than plastic foam (Styrofoam—see Table 11.3), so the answer can't be (a). However, as a gas, the air is subject to convection *within the wall space*. In the winter, the air near the warm inner wall is heated and rises, thus setting up a convection cycle in the space and transferring heat to the cold outer wall. In the summer, with air conditioning, the heat-loss cycle is reversed. Foam blocks the movement of air and thus stops such convection cycles. Hence, the answer is (b).

Follow-Up Exercise. Thermal underwear and thermal blankets are loosely knit with lots of small holes. Wouldn't they be more effective if the material were closely knit?

Radiation

Conduction and convection require some material as a transport medium. The third mechanism of heat transfer needs no medium; it is called **radiation**, which refers to energy transfer by electromagnetic waves (Chapter 20). Heat is transferred to the Earth from the Sun through empty space by radiation. Visible light and other forms of electromagnetic radiation are commonly referred to as *radiant energy*.

You have experienced heat transfer by radiation if you've ever stood near an open fire (▾Fig. 11.12). You can feel the heat on your exposed hands and face. This heat transfer is not due to convection or conduction, since heated air rises and air is a poor conductor. Visible radiation is emitted from the burning material, but most of the heating effect comes from the invisible **infrared radiation** emitted by the glowing embers or coals. You feel this radiation because it is absorbed by water molecules in your skin. (Body tissue is about 85% water.) The water molecule has an internal vibration whose frequency coincides with that of infrared radiation, which is therefore readily absorbed. (This effect is called *resonance absorption*. The electromagnetic wave

Teaching tip: Stress that heat spontaneously flows from a higher-temperature region to a lower-temperature region.

Teaching tip: Explain that convection is the only form of heat transfer that involves transfer of mass.

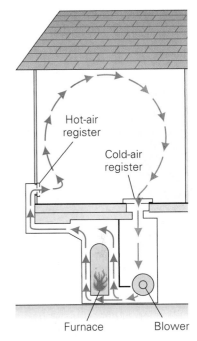

▲ **FIGURE 11.11 Forced convection** Houses are commonly heated by forced convection. Registers or gratings in the floors or walls allow heated air to enter and cooler air to return to the heat source. (Can you explain why the registers are located near the floor?) In older homes, hot water runs through pipes along the wall baseboards, and natural convection distributes the heat vertically upward.

Note: Resonance is discussed in Section 13.5.

▶ **FIGURE 11.12 Heating by conduction, convection, and radiation** The hands on top of the flame are warmed by the convection of rising hot air (and some radiation). The gloved hand is warmed by conduction. The hands to the right of the flame are warmed by radiation.

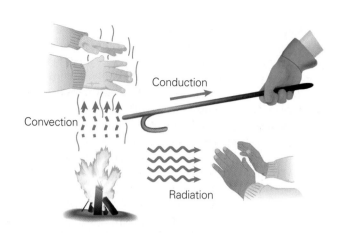

▲ **FIGURE 11.13 A practical application of heat transfer by radiation** A Tibetan teakettle is heated by focusing sunlight, using a metal reflector.

Illustration 19.3 Heat Transfer, Radiation

drives the molecular vibration, and energy is transferred to the molecule, somewhat like pushing a swing. See Chapter 13 on oscillations for more details.) Heat transfer by radiation can play a practical role in daily living (◀Fig. 11.13).

Infrared radiation is sometimes referred to as "heat radiation" or thermal radiation. You may have noticed the reddish infrared lamps used to keep food warm in cafeterias. Heat transfer by infrared radiation is also important in maintaining our planet's warmth by a mechanism known as the *greenhouse effect*. This important environmental topic is discussed in Insight 11.3 on page 388 on The Greenhouse Effect.

Although infrared radiation is invisible to the human eye, it can be detected by other means. Infrared detectors can measure temperature remotely (▼Fig. 11.14). Also, some cameras can use special infrared film. A picture taken with this film is an image consisting of contrasting light and dark areas, corresponding to regions of higher and lower temperatures, respectively. Special instruments that apply such *thermography* are used in industry and medicine; the images they produce are called *thermograms* (▶Fig. 11.15).

A new application of thermograms is for security. The system consists of an infrared camera and a computer that identifies an individual by means of the unique heat pattern emitted by the facial blood vessels. The camera takes a picture of the radiation from a person's face, which is compared with an earlier image stored in the computer memory.

The rate at which an object radiates energy has been found to be proportional to the fourth power of the object's absolute temperature (T^4). This relationship is expressed in an equation known as **Stefan's law**, expressed as

$$P = \frac{\Delta Q}{\Delta t} = \sigma A e T^4 \quad \textit{(radiation only)} \tag{11.5}$$

▶ **FIGURE 11.14 Detecting SARS** Infrared thermometers were used to measure body temperature during the Severe Acute Respiratory Syndrome (SARS) outbreak in 2003.

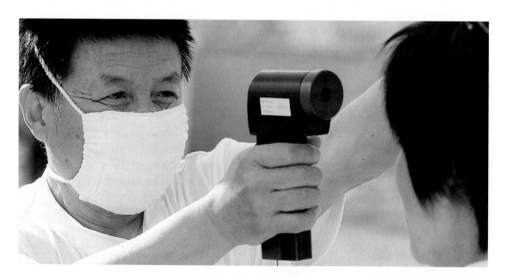

where P is the power radiated in watts (W), or joules per second (J/s). A is the object's surface area and T is its temperature in Kelvin. The symbol σ (the Greek letter sigma) is the *Stefan–Boltzmann constant*: $\sigma = 5.67 \times 10^{-8} \, W/(m^2 \cdot K^4)$. The **emissivity** ($e$) is a unitless number between 0 and 1 that is characteristic of the material. Dark surfaces have emissivities close to 1, and shiny surfaces have emissivities close to 0. The emissivity of human skin is about 0.70.

Dark surfaces not only are better emitters of radiation, but they are also good absorbers. This must be the case because to maintain a constant temperature, the incident energy absorbed must equal the emitted energy. *Thus, a good absorber is also a good emitter.* An ideal, or perfect, absorber (and emitter) is referred to as a **black body** ($e = 1.0$). Shiny surfaces are poor absorbers, since most of the incident radiation is reflected. This fact can be demonstrated easily, as shown in ▼Fig. 11.16. (Can you see why it is better to wear light-colored clothes in the summer and dark-colored clothes in the winter?)

When an object is in thermal equilibrium with its surroundings, its temperature is constant; thus, it must be emitting and absorbing radiation at the same rate. However, if the temperatures of the object and its surroundings are different, there will be a net flow of radiant energy. If an object is at a temperature T and its surroundings are at a temperature T_s, the net rate of energy loss or gain per unit time (power) is given by

$$P_{net} = \sigma Ae(T_s^4 - T^4) \qquad (11.6)$$

Note that if T_s is less than T, then P (or $\Delta Q/\Delta t$) will be negative, indicating a net heat energy loss, in keeping with our heat-flow sign convention. Keep in mind that the temperatures used in calculating radiated power are the absolute temperatures in kelvins.

You may have noticed in Chapter 10 that heat was defined as the net thermal energy transfer due to temperature differences. The word *net* here is important. It is possible to have energy transfer between an object and its surroundings, or between objects, at the same temperature. Note that if $T_s = T$ (that is, there is no temperature difference), there is a continuous exchange of radiant energy (Eq. 11.6 still holds), but there is no *net* change of the internal energy of the object.

Example 11.9 ■ Body Heat: Radiant-Heat Transfer

Suppose that your skin has an emissivity of 0.70, a temperature of 34°C, and a total area of 1.5 m². How much net energy per second will be radiated from your skin if the ambient room temperature is 20°C?

Thinking It Through. Everything is given for us to find P_{net} from Eq. 11.6. The net radiant-energy transfer is between the skin and the surroundings. We must remember to work with temperatures in kelvins.

Solution.

Given: $T_s = 20°C + 273 = 293 \, K$ *Find:* P_{net} (net power)
          $T = 34°C + 273 = 307 \, K$
          $e = 0.70$
          $A = 1.5 \, m^2$
          $\sigma = 5.67 \times 10^{-8} \, W/(m^2 \cdot K^4)$ (known)

Using Eq. 11.6 directly,

$$\begin{aligned} P_{net} &= \sigma Ae(T_s^4 - T^4) \\ &= [5.67 \times 10^{-8} \, W/(m^2 \cdot K^4)](1.5 \, m^2)(0.70)[(293 \, K)^4 - (307 \, K)^4] \\ &= -90 \, W \text{ (or } -90 \, J/s) \end{aligned}$$

Thus, 90 J of energy is radiated, or *lost* (as indicated by the minus sign), each second. That is, the human body loses heat at a rate that is close to that of a 100-W lightbulb! No wonder a room full of people can get warm!

Follow-Up Exercise. (a) In this Example, suppose the skin had been exposed to an ambient room temperature of only 10°C. What would the rate of heat loss be? (b) Elephants have huge body masses and large daily caloric food intakes. Can you explain how their huge ear flaps (large surface area) might help stabilize their body temperature?

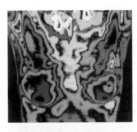

▲ **FIGURE 11.15 Applied thermography** Thermograms can be used to detect breast cancer by detecting tumor regions that are higher in temperature than normal. The upper photo shows a thermogram scan of a woman without breast cancer. The lower photo shows the result for a woman with breast cancer. The "hot spots" in this scan tells the physician where the cancer resides.

Demonstration/activity: Wrap one of two identical cans with masking tape, and leave the surface of the other can shiny. Fill both with boiling water at the start of a lecture period, and measure their temperatures about every 5 min. The can wrapped with tape will cool faster. Explain why.

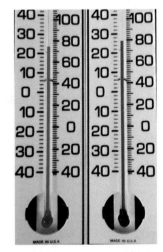

▲ **FIGURE 11.16 Good absorber** Black objects are generally good absorbers of radiation. The bulb of the thermometer on the right has been painted black. Note the difference in temperature readings.

INSIGHT 11.3 THE GREENHOUSE EFFECT

The *greenhouse effect* helps regulate the Earth's long-term average temperature, which has been fairly constant for some centuries. By this natural phenomenon, a portion of the solar radiation (mostly visible) reaches and warms the Earth's surface. The Earth, in turn, reradiates energy in the form of infrared (IR) radiation. The balance between absorption and radiation is a major factor in stabilizing the Earth's temperature.

This balance is affected by the concentration of atmospheric *greenhouse* gases—primarily water vapor, carbon dioxide (CO_2), and methane. As the reradiated infrared radiation passes back through the atmosphere, some of the radiation is absorbed by the greenhouse gases. These gases are selective absorbers: They absorb radiation at certain IR wavelengths but not at others (Fig. 1a).

If IR radiation is absorbed, the atmosphere warms, and in turn the Earth is warmed. Without the atmosphere absorbing the IR radiation, life on the Earth would probably not exist because, the average surface temperature would be a cold $-18°C$, rather than the present $15°C$.

Why is this phenomenon called the *greenhouse effect*? The reason is that the atmosphere functions somewhat like the glass in a greenhouse. That is, the absorption and transmission properties of glass are similar to those of the atmospheric greenhouse gases—in general, visible radiation is transmitted, but infrared radiation is selectively absorbed (Fig. 1b). We have all observed the warming effect of sunlight passing through glass—for example, in a closed car on a sunny but cold day. Similarly, a greenhouse heats up by trapping the reradiated infrared radiation inside. Thus, it is quite warm in a greenhouse on a sunny day, even in winter. (The glass enclosure also keeps warm air from escaping upward. In practice, this elimination of heat loss by convection is the chief factor in maintaining an elevated temperature in a greenhouse.)

The problem on Earth is that human activities since the beginning of the industrial age have been accentuating greenhouse warming. With the combustion of hydrocarbon fuels (gas, oil, coal, and so on) for heating and industrial processes, vast amounts of CO_2 and other greenhouse gases are vented into the atmosphere, where they trap increasingly more IR radiation. There is grave concern that the result of this trend will be—and in fact already is—increased *global warming*, an increase in the Earth's average temperature. Such an increase could dramatically affect the environment. For example, some detailed calculations indicate that the climate in many parts of the globe would be altered, with agricultural production and world food supplies affected in ways that are very difficult to predict. A general rise in temperature would cause partial melting of the polar ice caps. Sea levels would rise, flooding low-lying regions and endangering coastal ports and population centers.

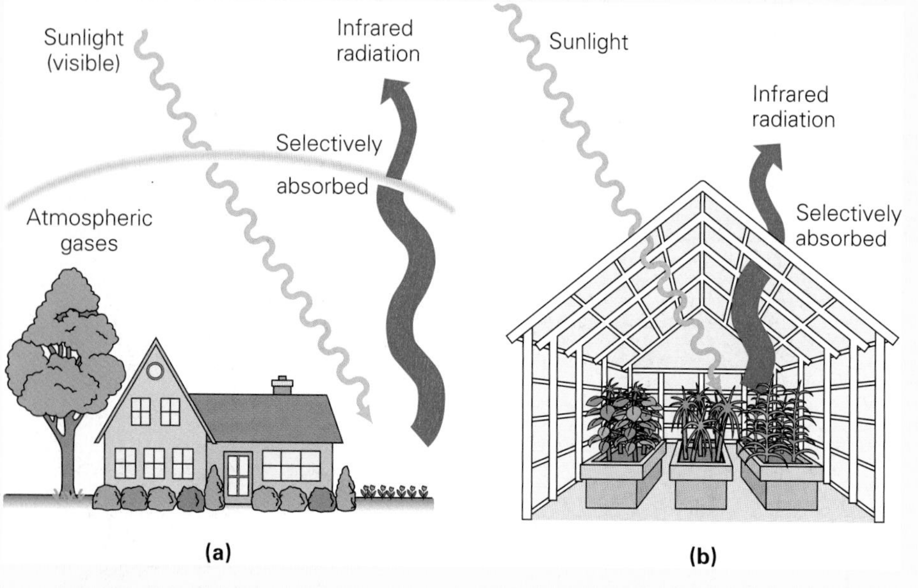

FIGURE 1 The greenhouse effect **(a)** The greenhouse gases of the atmosphere, particularly water vapor, methane, and carbon dioxide, are selective absorbers with absorption properties similar to those of the glass used in greenhouses. Visible light is transmitted and heats the Earth's surface, while some of the infrared radiation that is re-emitted is absorbed and trapped in the Earth's atmosphere. **(b)** A greenhouse operates in a similar way.

Problem-Solving Hint

Note that in Example 11.9, the fourth powers of the temperatures were found first, and then their difference was found. It is *not* correct to find the temperature difference and then raise it to the fourth power: $T_s^4 - T^4 \neq (T_s - T)^4$.

Let's consider a practical example of heat transfer that is becoming more commonplace as energy costs rise—solar panels.

Conceptual Example 11.10 ■ Solar Panels: Reducing the Heat Transfer

Solar panels are used to collect solar energy to heat water, which may then be used directly to heat a home at night. The panel boxes have black interiors (why?) through which the piping runs to carry the water, and the top is covered with glass. However, ordinary glass absorbs a lot of the Sun's ultraviolet radiation. This absorption reduces the heating effect. Wouldn't it be better to leave the glass off the panel boxes?

Reasoning and Answer. Some energy is absorbed by the glass, but the glass serves a useful purpose and saves a lot more energy than it absorbs. As the black interior and piping of the panel box heat up, there could be heat loss by radiation (infrared) and convection. The glass prevents this loss from occurring via the greenhouse effect. It absorbs much of the infrared radiation and also keeps the convection currents *inside* the solar-panel box. (See the Insight on The Greenhouse Effect.)

Follow-Up Exercise. Is there any practical reason for window drapes (other than privacy)?

Let's look at a few more real-life examples of heat transfer. In the spring, a late frost could kill the buds on fruit trees in an orchard. To reduce heat transfer, some growers spray water on the trees to form ice before a hard frost occurs. Using ice to save buds? Ice is a relatively poor (and inexpensive) thermal conductor, so it has an insulating effect. It will maintain the buds' temperature at 0°C, not going below that value, and therefore protects the buds.

Another method to protect orchards from freezing is the use of smudge pots, containers in which material is burned to create a dense cloud of smoke. At night, when the Sun-warmed ground cools off by radiation, the cloud absorbs this heat and reradiates it back to the ground. Thus, the ground takes longer to cool, hopefully without reaching freezing temperatures before the Sun comes up. (Recall that frost is the direct condensation of water vapor in the air to ice—not frozen dew.)

A Thermos bottle (▶Fig. 11.17) keeps cold beverages cold and hot ones hot. It consists of a double-walled, partially evacuated container with silvered walls (mirrored interior). The bottle is constructed to minimize all three mechanisms of heat transfer. The double-walled and partially evacuated container counteracts conduction and convection because both processes depend on a medium to transfer the heat (the double walls are more for holding the partially evacuated region than for reducing conduction and convection). The mirrored interior minimizes loss by radiation. The stopper on top of the thermos stops convection off the top of the liquid as well.

Look at ▼Fig. 11.18. Why would anyone wear a dark robe in the desert? We have learned that dark objects absorb radiation (Fig. 11.16). Wouldn't a white robe be better? A dark robe definitely absorbs more radiant energy and warms the air

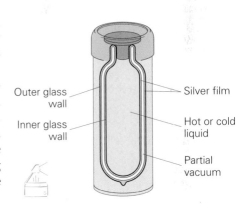

▲ **FIGURE 11.17 Thermal insulation** The Thermos bottle minimizes all three mechanisms of heat transfer.

◀ **FIGURE 11.18 A dark robe in the desert?** Dark objects absorb more radiation than do lighter ones, and they become hotter. What's going on here? See the text for an explanation.

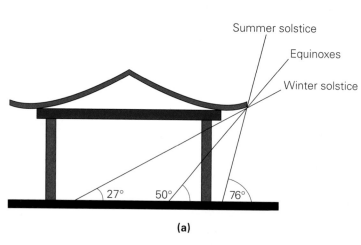

(a)

(b)

▲ **FIGURE 11.19 Aspects of passive solar design in ancient China** (a) In summer, with the sun angle high, the overhangs provide shade to the building. The brick and mud walls are thick to reduce conductive heat flow to the interior. In winter, the sun angle is low so the sunlight streams into the building, especially with the help of the upward curved overhangs. The leaves of nearby deciduous trees provide additional shade in the summer but allow sunlight in when they have dropped their leaves in the winter. (b) A photo of such a building in Beijing, China, in December.

inside near the body. But note that the robe is open at the bottom. The warm air rises (since it is less dense) and exits at the neck area, and outside cooler air enters the robe at the bottom—a natural-convection air circulation.

Finally, consider some of the thermal factors involved in "passive" solar house design used as far back as in ancient China (▲Fig. 11.19). The term *passive* means that the design elements require no active use of energy to conserve energy. In Beijing, China, for example, the angles of the sunlight are 76°, 50°, and 27° above the horizon at the summer solstice, the spring and fall equinoxes, and the winter solstice, respectively. With a proper combination of column height and roof overhang length, maximum sunlight is allowed *into* the building in the winter, but most of the sunlight will *not* reach the inside of the building in the summer. The overhangs of the roofs are also curved upward, not just for good looks, but also for letting the maximum amount of light into the building in the winter. Trees planted on the south side of the building can also play important roles in both summer and winter. In the summer, the leaves block and filter the sunlight; in the winter, the dropped leaves will let sunlight through.

Chapter Review

- **Heat** (Q) is the energy exchanged between objects, commonly because they are at different temperatures.

- The **specific heat** (c) tells how much heat is needed to raise the temperature of 1 kg of a particular material by 1 C°. It is characteristic of the type of material and is defined by

$$Q = cm\Delta T \quad \text{or} \quad c = \frac{Q}{m\Delta T} \quad (11.1)$$

- **Calorimetry** is a technique that uses heat transfer between objects, most commonly to measure specific heats of materials. It is based on conservation of energy, written as $\Sigma Q_i = 0$, assuming no heat losses or gains to the environment.

- **Latent heat** (L) is the heat required to change the phase of an object *per kilogram* of mass. During the phase change, the temperature of the system does not change. Its general definition is

$$L = \frac{|Q|}{m} \quad \text{or} \quad Q = \pm mL \quad (11.2, 11.3)$$

• Heat transfer due to direct contact of objects that have different temperatures is called **conduction**. The rate of heat flow by conduction through a slab of material is given by

$$\frac{\Delta Q}{\Delta t} = \frac{kA\Delta T}{d} \qquad (11.4)$$

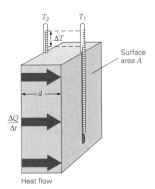

• **Convection** refers to heat transfer due to mass movement of gas or liquid molecules. *Natural convection* is driven by density differences caused by temperature differences. In *forced convection*, the movement is driven by mechanical means.

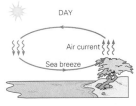

• **Radiation** refers to heat transferred by electromagnetic radiation between objects that have different temperatures, usually an object and its surroundings. The rate of transfer is given by

$$P_{net} = \sigma Ae(T_s^4 - T^4) \qquad (11.6)$$

where σ is the Stefan–Boltzmann constant, whose value is $5.67 \times 10^{-8} \, \text{W}/(\text{m}^2 \cdot \text{K}^4)$.

Exercises*

MC = *Multiple Choice Question,* **CQ** = *Conceptual Question, and* **IE** = *Integrated Exercise. Throughout the text, many exercise sections will include "paired" exercises. These exercise pairs, identified with* **red numbers**, *are intended to assist you in problem solving and learning. In a pair, the first exercise (even numbered) is worked out in the Study Guide so that you can consult it should you need assistance in solving it. The second exercise (odd numbered) is similar in nature, and its answer is given at the back of the book.*

11.1 Definition and Units of Heat

1. **MC** The SI unit of heat energy is the (a) calorie, (b) kilocalorie, (c) Btu, (d) joule. (d)

2. **MC** Which of the following is the largest unit of heat energy: (a) calorie, (b) Btu, (c) joule, or (d) kilojoule. (b)

3. **CQ** Discuss the difference between a calorie and a Calorie. 1 Cal = 1000 cal

4. **CQ** If someone says that a hot object contains more heat than a cold one, would you agree? Why? see ISM

5. ● A person goes on a 1500-Cal-per-day diet to lose weight. What is his daily allowance expressed in joules? $6.279 \times 10^6 \, \text{J}$

6. ● A window air conditioner has a rating of 20 000 Btu/h. What is this rating in watts? $5.86 \times 10^3 \, \text{W}$

7. ●● A typical person's normal metabolic rate (the rate at which it converts food/stored energy into heat, movement, and so on) is about $4 \times 10^5 \, \text{J/h}$ and the average food energy in a Big Mac is 600 Calories. If a person lived on nothing but Big Macs, how many per day would he or she have to eat to maintain a constant body weight? 4

8. ●● A student ate a Thanksgiving dinner that totaled 2800 Cal. He wants to use up all that energy by lifting a 20-kg mass a distance of 1.0 m. Assume that he lifts the mass with constant velocity and no work is required in lowering the mass. (a) How many times must he lift the mass? (b) If he can lift and lower the mass once every 5.0 s, how long does this exercise take? (a) 60 000 times (b) 83 h

11.2 Specific Heat and Calorimetry

9. **MC** The amount of heat necessary to change the temperature of 1 kg of a substance by 1 C° is called the substance's (a) specific heat, (b) latent heat, (c) heat of combustion, (d) mechanical equivalent of heat. (a)

10. **MC** For gases, which of the following is true about the specific heat under constant pressure, c_p, and specific heat under constant volume, c_v: (a) $c_p > c_v$, (b) $c_p = c_v$, or (c) $c_p < c_v$? (a)

11. **MC** The same amount of heat Q is added to two objects of the same mass. If object 1 experienced a greater temperature change than object 2, $\Delta T_1 > \Delta T_2$, then (a) $c_1 > c_2$, (b) $c_1 < c_2$, (c) $c_1 = c_2$. (b)

12. **CQ** If you live near a lake, does the lake water or the lake beach get hotter during a summer day? Which gets colder during a winter night? Explain. beach during summer day and also beach during winter night

*Neglect heat losses to the external environment in the exercises unless instructed otherwise, and consider all temperatures to be exact.

13. **CQ** Equal amounts of heat are added to two different objects at the same initial temperature. What factors can cause the final temperature of the two objects to be different? specific heat and mass

14. **CQ** Thousands of people have performed firewalking. (You should not try this at home!) In firewalking, people walk on a bed of red-hot coals (temperature over 2000°F) with bare feet. How is this possible? [*Hint*: Human tissues largely consist of water.] see ISM

15. **IE ●** The temperature of a lead block and a copper block, both 1.0 kg and at 20°C, are to be raised to 100°C. (a) The copper will require (1) more heat, (2) the same heat, (3) less heat than the lead. Why? (b) Calculate the difference between the heat required for the two blocks to prove your answer to part (a). (a) (1) more heat (b) copper requires 2.1×10^4 J more

16. **●** A 5.0-g pellet of aluminum at 20°C gains 200 J of heat. What is its final temperature? 63°C

17. **●** How many joules of heat must be added to 5.0 kg of water at 20°C to bring it to the boiling point? 1.7×10^6 J

18. **●** Blood can carry excess heat from the interior to the surface of the body, where the heat is dispersed. If 0.250 kg of blood at a temperature of 37.0°C flows to the surface and loses 1500 J of heat, what is the temperature of the blood when it flows back into the interior? Assume blood has the same specific heat as water. 35.6°C

19. **IE ●●** Equal amounts of heat are added to an aluminum block and a copper block of different masses to achieve the same temperature increase. (a) The mass of the aluminum block is (1) more, (2) the same, (3) less than the mass of the copper block. Why? (b) If the mass of the copper block is 3.00 kg, what is the mass of the aluminum block? (a) (3) less than (b) 1.27 kg

20. **●●** A modern engine of alloy construction consists of 25 kg of aluminum and 80 kg of iron. How much heat does the engine absorb as its temperature increases from 20°C to 120°C as it warms up to operating temperature? 6.0×10^6 J

21. **●●** A 0.200-kg glass cup at 20°C is filled with 0.40 kg of hot water at 90°C. Neglecting any heat losses to the environment, what is the equilibrium temperature of the water? 84°C

22. **●●** A 0.250-kg coffee cup at 20°C is filled with 0.250 kg of brewed coffee at 100°C. The cup and the coffee come to thermal equilibrium at 80°C. If no heat is lost to the environment, what is the specific heat of the cup material? [*Hint*: Consider the coffee essentially to be water.] 1.4×10^3 J/(kg·C°)

23. **●●** An aluminum spoon at 100°C is placed in a Styrofoam cup containing 0.200 kg of water at 20°C. If the final equilibrium temperature is 30°C and no heat is lost to the cup itself or environment, what is the mass of the aluminum spoon? 0.13 kg

24. **IE ●●** Equal amounts of heat are added to different quantities of copper and lead. The temperature of the copper increases by 5.0 C° and the temperature of the lead by 10 C°. (a) The lead has (1) a greater mass than the copper, (2) the same amount of mass as the copper, (3) less mass than the copper. (b) Calculate the mass ratio of the lead to the copper to prove your answer to part (a). (a) (1) a greater mass than the copper (b) 1.5

25. **IE ●●** Initially at 20°C, 0.50 kg of aluminum and 0.50 kg of iron are heated to 100°C. (a) The aluminum gains (1) more heat than the iron, (2) the same amount of heat as the iron, (3) less heat than the iron. Why? (b) Calculate the difference in heat required to prove your answer to part (a). (a) (1) more heat than the iron (b) Al by 1.8×10^4 J more

26. **●●** To determine the specific heat of a new metal alloy, 0.150 kg of the substance is heated to 400°C and then placed in a 0.200-kg aluminum calorimeter cup containing 0.400 kg of water at 10.0°C. If the final temperature of the mixture is 30.5°C, what is the specific heat of the alloy? (Ignore the calorimeter stirrer and thermometer.) 687 J/(kg·C°)

27. **IE ●●** In a calorimetry experiment, 0.50 kg of a metal at 100°C is added to 0.50 kg of water at 20°C in an aluminum calorimeter cup. The cup has a mass of 0.250 kg. (a) If some water splashed out of the cup when the metal was added, the measured specific heat will appear to be (1) higher, (2) the same, (3) lower than the value calculated for the case in which the water does not splash out. Why? (b) If the final temperature of the mixture is 25°C, and no water splashed out, what is the specific heat of the metal? (a) (1) higher (b) 3.1×10^2 J/(kg·C°)

28. **●●** A student doing an experiment pours 0.150 kg of heated copper shot into a 0.375-kg aluminum calorimeter cup containing 0.200 kg of water at 25°C. The mixture (and the cup) comes to thermal equilibrium at 28°C. What was the initial temperature of the shot? 89°C

29. **●●** At what average rate would heat have to be removed from 1.5 L of (a) water and (b) mercury to reduce the liquid's temperature from 20°C to its freezing point in 3.0 min? (a) 7.0×10^2 W (b) 9.4×10^2 W

30. **●●** When resting, a person gives off heat at a rate of about 100 W. If the person is submerged in a tub containing 500 kg of water at 27°C and the heat from the person goes only into the water, how many hours will it take for the water temperature to rise to 28°C? 5.8 h

31. **●●●** Lead pellets of total mass 0.60 kg are heated to 100°C and then placed in a well insulated aluminum cup of mass 0.20 kg that contains 0.50 kg of water initially at 17.3°C. What is the equilibrium temperature of the mixture? 20.0°C

32. **●●●** A student mixes 1.0 L of water at 40°C with 1.0 L of ethyl alcohol at 20°C. Assuming that no heat is lost to the container or the surroundings, what is the final temperature of the mixture? [*Hint*: See Table 11.1.] 34°C

11.3 Phase Changes and Latent Heat

33. **MC** The SI units of latent heat are (a) $1/C°$, (b) $J/(kg \cdot C°)$, (c) $J/C°$, (d) J/kg. (d)

34. **MC** Latent heat is always (a) part of the specific heat, (b) related to the specific heat, (c) the same as the mechanical equivalent of heat, (d) involved in a phase change. (d)

35. **MC** When a substance undergoes a phase change, the added heat changes (a) the temperature, (b) the kinetic energy, (c) the potential energy, (d) the mass of the substance. (c)

36. **CQ** You are monitoring the temperature of some cold ice cubes ($-5.0°C$) in a cup as the ice and cup are heated. Initially, the temperature rises, but it stops at $0°C$. After a while, it begins rising again. Is anything wrong with the thermometer? Explain. see ISM

37. **CQ** In general, you would get a more severe burn from steam at $100°C$ than from the same mass of hot water at $100°C$. Why? see ISM

38. **CQ** When you breathe out in the winter, you can see your "breath" like fog. Explain. the water molecules in your breath condense to water droplets

39. ● How much heat is required to boil away 0.500 kg of water that is initially at $100°C$? 1.13×10^6 J

40. **IE** ● (a) Converting 1.0 kg of water at $100°C$ to steam at $100°C$ requires (1) more heat, (2) the same amount of heat, (3) less heat than converting 1.0 kg of ice at $0°C$ to water at $0°C$. Explain. (b) Calculate the difference in heat required to prove your answer to part (a). (a) (1) more heat (b) vaporization by 1.93×10^6 J more

41. ● First calculate the heat that needs to be removed to convert 1.0 kg of steam at $100°C$ to water at $40°C$ and then compute the heat that needs to be removed to lower the temperature of water at $100°C$ to water at $40°C$. Compare the two results. Are you surprised by the result? 2.5×10^6 J and 2.5×10^5 J; yes

42. ● An artist wants to melt some lead to make a statue. How much heat must be added to 0.75 kg of lead at $20°C$ to cause it to melt completely? 4.9×10^4 J

43. ● Water is boiled to add moisture to the air in the winter to help a congested person breathe better. Calculate the heat required to boil away 0.50 L of water that is initially at $50°C$. 1.2×10^6 J

44. ● How much heat is required to completely boil away 0.50 L of liquid nitrogen at $-196°C$? (Take the density of liquid nitrogen to be 0.80×10^3 kg/m³.) 8.0×10^4 J

45. **IE** ●● Heat has to be removed to condense mercury vapor at a temperature of 630 K into liquid mercury. (a) This heat involves (1) only specific heat, (2) only latent heat, or (3) both specific and latent heats. Explain. (b) If the mass of the mercury vapor is 15 g, how much heat would have to be removed? (a) (2) only latent heat (b) 4.1×10^3 J

46. ●● How much ice (at $0°C$) must be added to 1.0 kg of water at $100°C$ to end up with all liquid at $20°C$? 0.81 kg

47. ●● If 0.050 kg of ice at $0°C$ is added to 0.300 kg of water at $25°C$ in a 0.100-kg aluminum calorimeter cup, what is the final temperature of the water? $11°C$

48. **IE** ●● An alcohol rub can rapidly decrease body (skin) temperature. (a) This is because of (1) the cooler temperature of the alcohol, (2) the evaporation of alcohol, (3) the high specific heat of the human body. (b) For a 65-kg person to decrease his body temperature by 1.0 C°, what mass of alcohol must be evaporated from his skin? Ignore the heat involved in raising the temperature of alcohol to its boiling point (why?) and approximate the human body as water. (a) (2) the evaporation of alcohol (b) 0.32 kg

49. ●● Steam at $100°C$ is bubbled into 0.250 kg of water at $20°C$ in a calorimeter cup, where it condenses into liquid form. How much steam will have been added when the water in the cup reaches $60°C$? (Ignore the effect of the cup.) 1.8×10^{-2} kg

50. ●● Ice (initially at $0°C$) is added to 0.75 L of tea at $20°C$ to make the coldest possible iced tea. If enough ice is added so the final mixture is all liquid, how much liquid is in the pitcher when this condition occurs? 0.94 kg or 0.94 L

51. ●● 0.50 L of water at $16°C$ is put into an aluminum ice-cube tray of mass 0.250 kg at the same temperature. How much energy must the freezer remove from this system to turn the water into ice at $-8.0°C$? 2.1×10^5 J

52. **IE** ●● Evaporation of water from our skin is a very important mechanism for controlling body temperature. (a) This is because (1) water has a high specific heat, (2) water has a high latent heat of vaporization, (3) water contains more heat when hot, (4) water is a good heat conductor. (b) In a 3.5-h intense cycling race, a cyclist could lose as much as 7.0 kg of water through sweat. Estimate how much heat the cyclist loses in the process. (a) (2) water has a high latent heat of vaporization (b) 1.6×10^7 J

53. **IE** ●●● A 0.50-kg piece of ice at $-10°C$ is placed in an equal mass of water at $10°C$. (a) When thermal equilibrium is reached between the two, (1) all the ice will melt, (2) some of the ice will melt, (3) none of the ice will melt. (b) How much ice melts? (a) (2) some of the ice will melt (b) 0.5032 kg

54. ●●● One kilogram of a substance experimentally shows the T-versus-Q graph in ▼Fig. 11.20. (a) What are its

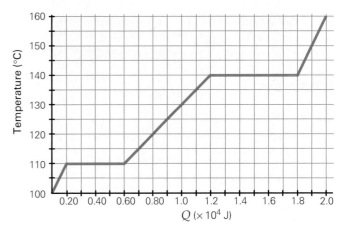

▲ **FIGURE 11.20 Temperature versus heat input** See Exercise 54.

melting and boiling points? In SI units, what are (b) the specific heats of the substance during its various phases and (c) the latent heats of the substance at the various phase changes? see ISM

55. ●●● Some ceramic materials become superconducting when immersed in liquid nitrogen. In such an experiment, a 0.150-kg piece of such a material at 20°C is placed in liquid nitrogen at its boiling point to cool in a perfectly insulated flask, which allows the gaseous N_2 to immediately escape. How many liters of liquid nitrogen will be boiled away during this operation? (Take the specific heat of the ceramic material to be that of glass, and the density of liquid nitrogen to be $0.80 \times 10^3 \, \text{kg/m}^3$.) 0.17 L

11.4 Heat Transfer

56. **MC** The warming of the Earth's atmosphere involves (a) conduction, (b) convection, (c) radiation, (d) all of the preceding. (d)

57. **MC** Which of the following is the dominant heat transfer mechanism for the Earth to receive energy from the Sun: (a) conduction, (b) convection, (c) radiation, (d) all of the preceding? (c)

58. **MC** Water is a poor heat conductor but a pot of water can be heated more quickly than you might think at first glance. This decrease in time is mainly due to heat (a) conduction, (b) convection, (c) radiation, (d) all of the preceding. (b)

59. **CQ** A plastic ice-cube tray and a metal ice-cube tray are removed from the same freezer, at the same initial temperature. However, when your hands touch both, the metal one feels cooler. Why?
metal conducts heat away from your hand more quickly

60. **CQ** Why is the warning shown on the highway road sign in ▼Fig. 11.21 necessary? see ISM

▲ **FIGURE 11.21 A cold warning** See Exercise 60.

61. **CQ** Polar bears have an excellent heat insulation system. (Sometimes infrared cameras cannot even detect them.) Polar-bear hairs are actually hollow inside. Explain how this helps the bears maintain their body temperature in the cold winter. air is a poor heat conductor

62. **IE** ● Assume that a tile floor and an oak floor each have the same temperature and thickness. (a) Compared with the oak floor, the tile floor will conduct heat away your

bare feet (1) faster, (2) at the same rate, or (3) slower. Why? (b) Calculate the ratio of the rate of heat flow of the tile floor to that of the oak floor. (a) faster (b) 4.5

63. ● The single glass pane in a window has dimensions of 2.00 m by 1.50 m and is 4.00 mm thick. How much heat will flow through the glass in 1.00 h if there is a temperature difference of 2 C° between the inner and outer surfaces? (Consider conduction only.) 4.54×10^6 J

64. ● Assume a goose has a 2.0-cm-thick layer of down (on average) and a body surface area of 0.15 m². What is the rate of heat loss (conduction only) if the goose with a body temperature of 41°C is outside on a winter day when the air temperature is 2°C? 7.3 J/s

65. ● Assume that your skin has an emissivity of 0.70, a normal temperature of 34°C, and a total exposed area of 0.25 m². How much heat energy per second do you lose due to radiation if the outside temperature is 22°C? 13 J

66. ● The U.S. five-cent coin, the nickel, has a mass of 5.1 g, a volume of 0.719 cm³, and a total surface area of 8.54 cm². Assume that a nickel is an ideal radiator; how much radiant energy per second comes from the nickel, if it is at 20°C? 0.36 J/s

67. **IE** ●● An aluminum bar and a copper bar of identical cross-sectional area have the same temperature difference between their ends and conduct heat at the same rate. (a) The copper bar is (1) longer, (2) of the same length, (3) shorter than the aluminum bar. Why? (b) Calculate the ratio of the length of the copper bar to that of the aluminum bar. (a) (1) longer, because it has a higher thermal conductivity (b) 1.63

68. ●● Assuming that the human body has a 1.0-cm-thick layer of skin tissue and a surface area of 1.5 m², estimate the rate at which heat is conducted from inside the body to the surface if the skin temperature is 34°C. (Assume a normal body temperature of 37°C for the temperature of the interior.) 90 J/s

69. ●● A copper teakettle has a circular bottom 30.0 cm in diameter that has a uniform thickness of 2.50 mm. It sits on a burner whose temperature is 150°C. (a) If the teakettle is full of boiling water, what is the rate of heat conduction through its bottom? (b) Assuming that the heat from the burner is the only heat input, how much water is boiled away in 5.0 min? Is your answer unreasonably large? If yes, explain why.
(a) 5.5×10^5 J/s (b) 73 kg; yes, see ISM

70. **IE** ●● The emissivity of an object is 0.60. (a) Compared with a perfect blackbody at the same temperature, this object would radiate (1) more power, (2) the same amount of power, (3) less power. Why? (b) Calculate the ratio of the power radiated by the blackbody to that radiated by the object. (a) (3) less power (b) 1.7

71. ●● A lamp filament radiates energy at a rate of 100 W when the temperature of the surroundings is 20°C, and only 99.5 W when the surroundings are at 30°C. If the temperature of the filament is the same in each case, what is its temperature in Celsius? 411°C

72. **IE** ●● The thermal insulation used in building is commonly rated in terms of its *R-value*, defined as d/k, where d is the thickness of the insulation in inches and k is its thermal conductivity. (See Insight 11.2 on p. 384.) In the United States, R-values are expressed in British units. For example, 3.0 in. of foam plastic would have an R-value of $3.0/0.30 = 10$, where, $k = 0.30$ Btu·in./(ft²·h·F°). This value is expressed as R-10. (a) Better insulation has a (a) (1) high, (2) low, or (3) zero R-value. Explain. (b) What thicknesses of (1) styrofoam and (2) brick would give an R-value of R-10? (a) (1) a high R-value (b) 4.2 in.; 51 in.

73. **IE** ●● A piece of pine 14 in. thick has an R-value of 19. (a) If glass wool is to have the same R-value, its thickness should be (a) thicker than, (2) the same as, or (3) thinner than 14 in. Why? (b) Calculate the required thickness of such a piece of glass wool. (See Exercise 72 and Insight 11.2 on p. 384.) (a) (3) thinner than (b) 4.9 in.

74. **IE** ●● (a) If the Kelvin temperature of an object is doubled, its radiated power increases by (1) 2, (2) 4, (3) 8, (4) 16 times. Explain. (b) If its temperature is increased from 20°C to 40°C, by how much does the radiated power change? (a) (4) 16 (b) increased by a factor of 1.3

75. ●● Solar heating takes advantage of solar collectors such as the type shown in ▼Fig. 11.22. During daylight hours, the average intensity of solar radiation at the top of the atmosphere is about 1400 W/m². About 50% of this radiation reaches the Earth during daylight hours. (The rest is reflected, scattered, absorbed, and so on.) How much heat energy would be received, on average, by the cylindrical collector shown in the figure during 10 h of daylight? 1.0×10^8 J

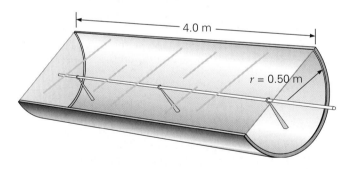

▲ **FIGURE 11.22** Solar collector and solar heating
See Exercise 75.

For Exercises 76–81, read Example 11.7 and the footnote on p. 382.

76. ●●● A large picture window measures 2.0 m by 3.0 m. At what rate will heat be conducted through the window when the room temperature is 20°C and the outside temperature is 0°C (a) if the window consists of a single pane of glass 4.0 mm thick and (b) if the window instead has a double pane of glass (a "thermopane"), in which each pane is 2.0 mm thick, with an intervening air space of 1.0 mm? (Assume that there is a constant temperature difference and consider conduction only.) (a) 2.5×10^4 J/s (b) 2.6×10^3 J/s

77. ●●● The lowest natural temperature ever recorded on the Earth was at Vostok, a Russian Antarctic station, when a temperature of −89.4°C (−129°F) was recorded on July 21, 1983. A typical person has a body temperature of 37.0°C, skin tissue 0.0250 m thick, and a total skin surface area of 1.50 m². Suppose such a person has a 0.100-m-thick goose down jacket and pants capable of covering the whole body. (a) What would be the rate of heat loss of a naked human? (b) What would be the rate of heat loss of a human wearing the down jacket and pants? (a) 1.5×10^3 J/s (b) 46 J/s

78. ●●● The wall of a house is composed of solid concrete block with an outside brick veneer and is faced on the inside with fiberboard, as illustrated in ▼Fig. 11.23. If the outside temperature on a cold day is −10°C and the inside temperature is 20°C, how much energy is conducted through it in 1.0 h if the wall measures 3.5 m by 5.0 m? 3.4×10^6 J

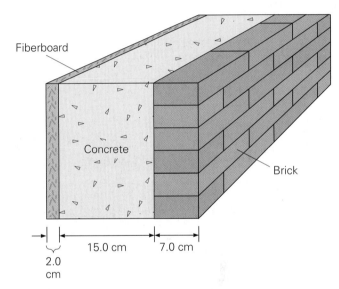

▲ **FIGURE 11.23** Thermal conductivity and heat loss
See Exercise 78.

79. ●●● Suppose you wished to cut the heat loss through the wall in Exercise 78 in half by installing insulation. What thickness of Styrofoam should be placed between the fiberboard and concrete block to accomplish this goal? 2.3 cm

80. ●●● A steel cylinder of radius 5.0 cm and length 4.0 cm is placed in end-to-end thermal contact with a copper cylinder of the same dimensions. If the free ends of the two cylinders are maintained at constant temperatures of 95°C (steel) and 15°C (copper), how much heat will flow through the cylinders in 20 min? 7.8×10^5 J

81. ●●● In Exercise 80, what is the temperature at the interface of the cylinders? 23°C

Comprehensive Exercises

82. 0.60 kg of ice at −10°C is placed in 0.30 kg of water at 50°C. How much liquid is left when the system reaches thermal equilibrium? 0.45 kg

83. A large Styrofoam cooler has a surface area of 1.0 m² and a thickness of 2.5 cm. If 5.0 kg of ice at 0°C is stored inside and the outside temperature is a constant 35°C, how long does it take for all the ice to melt? (Consider conduction only.) 7.8 h

84. After performing a barrel jump, a 65-kg ice skater traveling at 25 km/h comes to a stop. If 40% of the frictional heat generated by the skate blades goes into melting the ice (assumed to be at 0°C), how much ice is melted? Where does the other 60% of the energy go?
    1.9 g; into heating skates, generating noise, etc.

85. A 0.030-kg lead bullet hits a steel plate, both initially at 20°C. The bullet melts and splatters on impact. (This action has been photographed.) Assuming that 80% of the bullet's kinetic energy is converted to heat to melt it, what is the minimum speed it must have to melt on impact? 4.0×10^2 m/s

86. A waterfall is 75 m high. If 30% of the gravitational potential energy of the water were converted into heat energy, by how much would the temperature of the water increase in going from the top to the bottom of the falls? [*Hint*: Consider a kilogram of water going over the falls.] 0.053C°

87. A cyclist with a total skin area of 1.5 m² is riding a bicycle on a day when the air temperature is 20°C and her skin temperature is 34°C. The cyclist does work at about 100 W (moving the pedals) but her efficiency is only about 20% in terms of converting energy into mechanical work. Estimate the amount of water this cyclist must evaporate per hour (through sweat) to get rid of the excess body heat she produces. Assume a skin emissivity of 0.70. 0.49 kg

The following Physlet Physics Problems can be used with this chapter.
PHYSLET® 19.1, 19.2, 19.6, 19.7, 19.8, 19.9, 19.10, 19.11

THERMODYNAMICS

PHYSICS FACTS

- An automobile with a typical thermodynamic efficiency of twenty percent will lose about one third of the energy from the burned gasoline through the exhaust, another one third to the coolant, and about one-tenth to the surroundings.

- In Europe, 35% of passenger cars sold are diesel-powered. As of 2004 the percentage increased to 60% for passenger cars with engine sizes between 2.5 L and 3.3 L. This is mainly due to the higher efficiency of diesel engines and lower diesel fuel price. In North America, only 2% to 3% of passenger vehicles sold annually are diesel-powered.

- The efficiency of the human body, as measured by work output versus energy consumed, can be as high as 20% when large muscle groups, such as leg muscles are used, but as low as 3% to 5% when only the small muscle groups, such as arm muscles, are used.

- Professional cyclists can do work at a rate of up to 2 hp (about 1500 W) in short bursts.

At certain locations on the Earth, water heated deep in the interior rises to the surface as hot springs. In Yellowstone National Park, this produces boiling pools and geysers such as Old Faithful, shown in the photograph. In Iceland, the hot water warms the ocean and can create warm lagoons surrounded by glaciers. Such intriguing settings are popular vacation spots.

But the uses of such hot springs extend beyond recreation. Iceland's capital city of Reykjavik is heated by piping hot-spring water to homes and businesses. Furthermore, whenever there is a temperature difference, the potential exists for obtaining useful work. For instance, geothermal power plants draw on the energy of geysers as a renewable resource to generate electric energy while producing virtually no pollution. In this chapter, you'll learn under what conditions, and with what efficiency, heat can be exploited to perform work in the human body and in machines as different as automobile engines and home freezers. You'll find that the laws governing such energy conversions include some of the most general and far-reaching laws in all of physics.

As the word implies, **thermodynamics** deals with the transfer (dynamics) of heat (the Greek word for "heat" is *therme*). The development of thermodynamics started about 200 years ago and grew out of efforts to develop heat engines. The steam engine was one of the first such devices, designed to convert heat to mechanical work. Steam engines in factories and locomotives powered the Industrial Revolution, which changed the world. Although our study is primarily concerned with heat and work, thermodynamics is a broad and comprehensive science that includes a great deal more than heat-engine theory. In this chapter, the laws on which thermodynamics is based, as well as the concept of entropy will be presented.

12.1 Thermodynamic Systems, States, and Processes

OBJECTIVES: To (a) define thermodynamic systems and states of systems, and (b) explain how thermal processes affect such systems.

Thermodynamics is a field that describes systems with very many particles— think of the number of molecules in a gas sample—that using ordinary dynamics (Newton's laws) to keep track of them is impossible. Therefore, even though the underlying physics is the same as for other systems, we generally use alternative (macroscopic) variables, such as pressure and temperature, to describe thermodynamic systems as a whole. Because of this difference in language, it is important to become familiar with the terms and definitions at the outset.

The term **system**, as used in thermodynamics, refers to a definite quantity of matter enclosed by boundaries or surfaces, either real or imaginary. For example, a quantity of gas in the piston cylinder of an engine has real boundaries, and imaginary boundaries enclose a cubic meter of air in a room. The boundaries of a system need not have a definite shape or enclose a fixed volume. For example, a cylinder of gas experiences a volume change when the piston is moved.

Occasionally, it is necessary to consider systems between which matter is transferred. However, for the most part, we will consider systems of constant mass. More important is the interchange of energy between a system and its surroundings. This exchange may occur through a transfer of heat and/or the performance of mechanical work. For example, if a balloon is warmed (meaning that heat is transferred into it), it can expand and do work on its containing surface (its latex "skin") and the atmosphere by exerting a force through a distance, as discussed in Chapter 5.

If heat transfer into or out of the system is impossible, the system is said to be a **thermally isolated system**. However, work may be done on a thermally isolated system, thus transferring energy to it. For example, a thermally isolated cylinder (perhaps surrounded by heavy insulation) filled with gas can be compressed by an external force exerted on a piston cover. As such, work is done on the system, and as we know, work is a way of transferring energy.

When heat does enter or leave a system, it is usually taken in from or given up to the surroundings or to what we call heat reservoirs. A **heat reservoir** is a large separate system assumed to have unlimited heat capacity. Any amount of heat can be withdrawn from or added to a heat reservoir without appreciably changing its temperature. For example, pouring a bottle of warm water into a cold lake does not noticeably raise the lake's temperature. This cold lake is an example of a low-temperature heat reservoir.

State of a System

Just as there are kinematic equations to describe the motion of an object, there are **equations of state** to describe the conditions of thermodynamic systems. Such an equation expresses a mathematical relationship between the thermodynamic variables that describe the system. The ideal gas law, $pV = nRT$ (Section 10.3), is an example of an equation of state. This expression establishes a relationship among the pressure (p), volume (V), absolute temperature (T), and number of moles (n, or equivalently, N, the number of molecules, since from Section 10.3, we know that $N = nN_A$) of a gas. These ideal gas quantities are examples of *state variables*. Clearly, then, different states have different sets of values for these variables.

For a quantity of ideal gas, a set of these three variables (p, V, and T) that satisfies the ideal gas law specifies its state completely as long as the system is in thermal equilibrium and has a uniform temperature. Such a system is said to be in a definite *state*. It is convenient to plot the states according to the thermodynamic coordinates (p, V, T), much as we plot graphs using Cartesian coordinates (x, y, z). A general two-dimensional illustration of such a plot is shown in ◄Fig. 12.1.

Just as the coordinates (x, y) specify individual points on a Cartesian graph, the coordinates (V, p) specify individual *states* on the p–V graph or diagram. This

Teaching tip: Students will be able to identify with the concept of a thermally isolated house. Heavy use of insulation is an attempt to reduce heat transfer to minimum.

Teaching tip: The ideal gas law will be used extensively in this chapter. Have students review it if necessary.

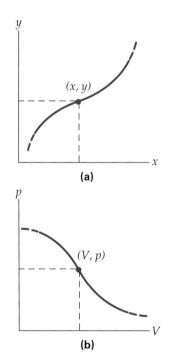

▲ **FIGURE 12.1 Graphing states**
(a) On a Cartesian graph, the coordinates (x, y) represent an individual point. **(b)** Similarly, on a p–V graph or diagram, the coordinates (V, p) represent a particular state of a system. [It is common to say p–V, rather than V–p, because the plot is a p vs. V graph.]

is because the ideal gas law, $pV = nRT$, can be solved for the unique temperature of a gas if we know the gas's pressure, volume, and the number of molecules or moles in the sample. In other words, on a p–V diagram, each "coordinate" gives the pressure and volume of a gas directly, and the temperature of the gas indirectly. Thus, to describe a gas completely, only a p–V plot is necessary. In some cases, however, it can be instructive to refer to other plots, such as p–T or T–V plots. (Notice that Fig. 12.1b could illustrate a phenomenon that you might be familiar with—reduction of the pressure of a gas, resulting in its expansion.)

Note: Because p, V, and T are connected by the ideal gas law, specifying the values of any two of these variables automatically tells you the value of the third variable.

Processes

A **process** is any *change* in the state, or the thermodynamic coordinates, of a system. For instance, when an ideal gas undergoes a process, its state variables p, V, and T will, in general, all change. Suppose a gas initially in state 1, described by state variables (p_1, V_1, T_1), changes to a second state, state 2. State 2 will, in general, be described by a different set of state variables (p_2, V_2, T_2). A system that has undergone a change of state has been subjected to a *thermodynamic process*.

Processes are classified as either reversible or irreversible. Suppose that a system of gas in equilibrium (with known p, V, and T values) is allowed to expand quickly when the pressure on it is reduced. The state of the system will change rapidly and unpredictably, but eventually the system will reach a different state of equilibrium, with another set of thermodynamic coordinates. On a p–V diagram (▶Fig. 12.2), we would be able to show the initial and final states (labeled 1 and 2, respectively), but *not* what happened in between them. This type of process is called an **irreversible process**—a process for which the intermediate steps are nonequilibrium states. "Irreversible" does not mean that the system can't be taken back to the initial state; it means only that the process path can't be retraced, because of the nonequilibrium conditions that existed. An explosion is an example of an irreversible process.

If, however, the gas changes state very, very slowly, passing from one equilibrium state to a neighboring one and eventually arriving at the final state (see Fig. 12.2, initial and final states 3 and 4, respectively), then the process path is known. In such a situation, the system could be brought back to its initial conditions by "traveling" the path in the opposite direction, re-creating every intermediate state (again, in many small steps) along the way. Such a process is called a **reversible process**. In practice, a perfectly reversible process cannot be achieved. All real thermodynamic processes are irreversible to some degree, because they follow complicated paths with many intermediate nonequilibrium states. However, the concept of an ideal reversible process is useful and will be our primary tool in discussing the thermodynamics of an ideal gas.

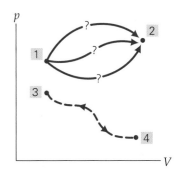

▲ **FIGURE 12.2 Paths of reversible and irreversible processes** If a gas quickly goes from state 1 to state 2, the process is irreversible, since we do not know the "path." If, however, the gas is taken through many closely spaced equilibrium states (as in going from state 3 to state 4), the process is reversible (from state 4 back to state 3), in principle. Reversible means "exactly retraceable."

12.2 The First Law of Thermodynamics

OBJECTIVES: To (a) explain the relationship among internal energy, heat, and work as expressed by the first law of thermodynamics, and (b) learn the technique for calculating work done by gases.

Recall from mechanics (Chapter 5) that work describes the transfer of energy from one object to another by application of a force. For example, when you push on a chair initially at rest, some of the work you do (exerting a force through a distance) on the chair goes into increasing its kinetic energy. At the same time, you lose stored (chemical) energy in your body in doing so. This type of work is done in an *orderly* way, that is, by applying various forces, in well-defined directions, on objects of interest. For example, when a gas (enclosed in a cylinder and fitted with a piston) is allowed to expand, it does work on the piston at the expense of some of its internal energy. From Chapters 10 and 11, we now know there is a second way to change the energy of a system—by adding or removing heat energy. Thus, internal energy is lost by a hot object when the heat is transferred to a cold object, which then gains internal energy. This process changes both objects' internal energies, but in opposite ways.

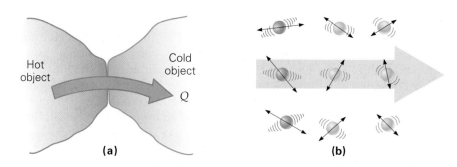

▶ **FIGURE 12.3** Heat flow (via conduction) on the atomic scale **(a)** Macroscopically, heat is transferred by conduction from the hot object to the cold one. **(b)** On the atomic scale, heat conduction is explained as the energy transfer from the more energetic atoms (in the hotter object) to the less energetic atoms (in the colder object). This transfer of energy from an atom to its neighbor results in the heat transfer observed in part (a).

Although we can't see the actual process, heat transfer is really the same concept as the work we know from mechanics, but on a microscopic (atomic) level. During a conduction process, for example, energy is transferred from a hot solid object to a colder solid object, because the faster-vibrating atoms of the warmer object do work on the slower atoms of the colder object (▲ Fig. 12.3). This energy is then transferred farther into the volume of the cold object as more work is done on the neighboring (slower-vibrating) atoms. This ongoing process is the "flow" or "transfer" of energy we observe macroscopically as heat transfer.

The **first law of thermodynamics** describes how work and heat are related to a system's internal energy. This law is a restatement of *energy conservation* in terms of thermodynamic variables. It relates the change in internal energy (ΔU) *of a system* to the work (W) done *by or on that system* and the heat energy transferred (Q) *to or from that system*. Depending on the conditions, heat transfer Q can result in a change in that system's internal energy, ΔU. However, because of the heat transfer, the system might do work on the environment. Thus, heat transferred to a system can end up in two places: a change in the internal energy of the system and/or work done by the system. This is just a statement of energy conservation. Therefore, the first law of thermodynamics can be written as

$$Q = \Delta U + W \tag{12.1}$$

As always, it is important to remember what the symbols mean and what their sign conventions denote (shown in ▼ Fig. 12.4). Q is the net heat *added to or removed from the system*, ΔU is the change in internal energy *of the system* and W is the work done *by the system* (on the environment).* For example, a gas may absorb 1000 J of heat and do 400 J of work on the environment, thus leaving 600 J as the increase in the gas's internal energy. If the gas were to do more than 400 J of work, less energy would go to the internal energy of the gas. The first law does *not* tell you the values

▶ **FIGURE 12.4** Sign conventions for **Q, W,** and **ΔU** **(a)** If heat flows into a system, Q is positive. Heat flowing out is designated as negative. **(b)** The experimental way to tell if a gas's internal energy changes is to take its temperature, assuming there is no phase change. Since internal energy is determined by temperature, a rise or fall in one of these quantities implies a similar rise or fall in the other. **(c)** If a gas expands, the work W it does is positive. If the gas is compressed, the work done is designated as negative.

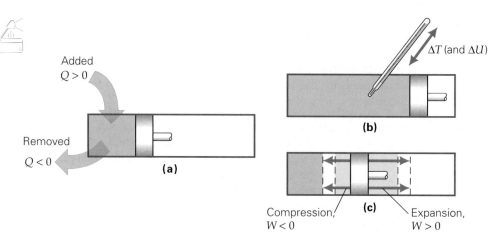

*In some chemistry and engineering books, the first law of thermodynamics is written as $Q = \Delta U - W'$. The two equations are the same but have a different emphasis. In this expression, W' means the work done *by the environment on the system* and is thus the negative of our work W (why?), or $W = -W'$. The first law was discovered by researchers interested in building heat engines (Sections 12.5 and 12.6). Their emphasis was on finding the work done *by* the system, W, not W'. Since we want to understand heat engines, the historical definition is adopted: W means the work done by the system.

of ΔU or W in processes. These amounts depend, as we shall see, on the system's conditions or the specific process involved (constant pressure, constant volume, and so on) as the heat energy is transferred (Section 12.3).

It is important to note that heat flow is *not* necessary to change temperature. As a soda bottle is opened, as shown in ▶Fig. 12.5, the gas inside the bottle expanded because it was at a higher pressure than the atmosphere. In doing so it did (positive) work on the surroundings (the atmospheric gases) so its internal energy decreased. This is because the net heat flow is zero in this process. Since $\Delta U = Q - W$, then ΔU is negative (U decreases) if $Q = 0$ and W is positive. This reduction in internal energy will cause the water vapor in the bottled gas to condense into a cloud of tiny liquid water droplets.

Consider the application of the first law of thermodynamics to exercise and weight loss.

Example 12.1 ■ Energy Balancing: Exercising Using Physics

A 65-kg worker shovels coal for 3.0 h. During the shoveling, the worker did work at an average rate of 20 W and lost heat to the environment at an average rate of 480 W. Ignoring the loss of water by the evaporation of perspiration from his skin, how much fat will the worker lose? The energy value of fat (E_f) is 9.3 kcal/g.

Thinking It Through. Since the time duration of the shoveling, the rate of work being done (power), and the rate of heat lost is known, we can calculate the total work done and the heat. Then the change in internal energy can be found using the first law of thermodynamics. This change in internal energy (a decrease) results in a loss of fat.

Solution. Listing the given values, and converting power to work and heat:

Given: $W = Pt = (20 \text{ J/s})(3.0 \text{ h})(3600 \text{ s/h}) = 2.16 \times 10^5 \text{ J}$ *Find:*   mass of fat burned
$Q = -(480 \text{ J/s})(3.0 \text{ h})(3600 \text{ s/h}) = -5.18 \times 10^6 \text{ J}$
     (Q is negative because heat is lost)
$E_f = 9.3 \text{ kcal/g}$
     $= 9.3 \times 10^3 \text{ kcal/kg} = (9.3 \times 10^3 \text{ kcal/kg})(4186 \text{ J/kcal})$
     $= 3.89 \times 10^7 \text{ J/kg}$

From the first law of thermodynamics, $Q = \Delta U + W$, we have

$$\Delta U = Q - W = -5.18 \times 10^6 \text{ J} - 2.16 \times 10^5 \text{ J} = -5.40 \times 10^6 \text{ J}$$

Thus the mass of fat loss is

$$m = \frac{|\Delta U|}{E_f} = \frac{5.40 \times 10^6 \text{ J}}{3.89 \times 10^7 \text{ J/kg}} = 0.14 \text{ kg}$$

That is about a third of a pound, or about 5 ounces.

Follow-Up Exercise. How much fat would be lost if the worker were playing basketball, doing work at a rate of 120 W and generating heat at a rate of 600 W? (*Answers to all Follow-Up Exercises are at the back of the text.*)

When applying the first law of thermodynamics, the proper use of signs (shown in Fig. 12.4) cannot be overemphasized. The signs for work are easy to remember if you keep in mind that positive work is done by a force that acts generally in the direction of the displacement, such as when a gas expands. Similarly, negative work means that the force acts generally opposite to the direction of the displacement, as when a gas contracts.

But how do you compute the work done by the gas? To answer this question, consider a cylindrical piston with end area A, containing a known sample of gas (▶Fig. 12.6). Let us imagine that the gas is allowed to expand over a very small distance Δx. If the volume of the gas does not change appreciably, then the pressure remains constant. In moving the piston slowly and steadily outward, the gas does positive work on the piston. Thus from the definition of work,

$$W = F\Delta x \cos \theta = F\Delta x \cos 0° = F\Delta x$$

▲ **FIGURE 12.5 Temperature decrease without removing heat** The gas does positive work on the outside air upon the opening of the bottle. This results in a decrease of both its internal energy and temperature.

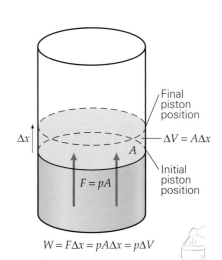

$$W = F\Delta x = pA\Delta x = p\Delta V$$

▲ **FIGURE 12.6 Work in thermodynamic terms** If a gas expands by a very small amount and slowly, its pressure remains constant. The small amount of work done by the gas is $p\Delta V$.

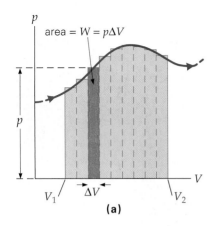

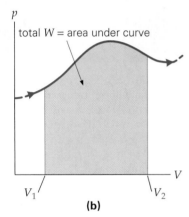

▲ **FIGURE 12.7** Thermodynamic work as the area under the process curve (a) If a gas expands by a significant amount, the work done can be computed by treating the expansion in little steps, each one yielding a small amount of work. The total work is determined (approximately) by adding up the many rectangular strips. (b) If the number of rectangular strips becomes large, and each one becomes very thin, the calculation of the area becomes exact. The work done is equal to the area between the process curve and the V-axis.

In terms of pressure, $p = F/A$, or $F = pA$. Substituting for F, we have

$$W = pA\Delta x$$

But $A\Delta x$ is the volume of a right cylinder with end area A and height Δx. Here, that volume represents the change in volume of the gas, or $\Delta V = A\Delta x$, and

$$W = p\Delta V$$

Note that the work done in Fig. 12.6 is positive because ΔV is positive. If the gas contracts, the work is negative because the volume change is negative ($\Delta V = V_2 - V_1 < 0$).

Of course, gases don't always change their volumes by small amounts and aren't usually subject to constant pressure. In fact, changes in volume and pressure can be significant. How do we handle the calculation of work under these circumstances? The answer is seen in ◄Fig. 12.7. Here, we have a reversible path on a p–V diagram. Notice that during each small step, the pressure remains approximately constant. Therefore, for each step, we approximate the work done by $p\Delta V$. Graphically, this quantity is just the area of a small narrow rectangle, extending from the process curve to the V-axis. However, as the volume changes, so does the pressure. Therefore, to approximate the *total* work, we add up these small amounts of work or $W \approx \Sigma(p\Delta V)$. To get an exact value, think of the area as made up of a large number of very thin rectangles. As the number of rectangles becomes infinitely large, each rectangle's thickness approaches zero. This process involves calculus and is beyond the scope of this text. However, it should be clear that the following is true:

> The work done by a system is equal to the area under the process curve on a p–V diagram.

Before we discuss specific types of processes, note that there is a fundamental difference between U and both Q and W. Any system "contains" a certain amount of internal energy U. However, it is wrong to say that a system "possesses" certain "amounts" of heat or work, as these quantities represent energy *transfers*, not total energies. A further distinction is that both Q and W depend on how the gas gets from its initial to its final state, whereas ΔU does not. The heat added to, or removed from, a system *depends on the conditions* under which this transfer is done (Section 11.2). Similarly, from the area-under-the-curve representation, the work *depends on the path* (▼Fig. 12.8). For example, more work is done if the process takes place at higher pressures. This situation is represented as a larger area, with more work done by a larger force with the same volume change.

Contrast these properties with those of ΔU for an ideal gas, for example. To find ΔU, we need know only the internal energies at the ends of the path. This is because for an ideal gas (with a fixed number of moles), the internal energy depends only on the absolute temperature of the gas. For that case (see Chapter 10), $U \propto nRT$; thus, $\Delta U = U_2 - U_1$ depends only on ΔT. To summarize, the change in the internal energy, ΔU, is *independent* of the process path for an ideal gas, whereas Q and W both *depend* on the path.

▶ **FIGURE 12.8** Thermodynamic work depends on the process path This graph shows the work done by a gas as it expands the same amount, but by three different processes. The work done during process I is larger than process II's work, which in turn is larger than process III's work. Fundamentally, applying a larger force (pressure) through the same distance (volume change) requires more work. Process I includes the blue, green, and pink areas; process II includes just the green and pink areas; and process III includes just the pink area.

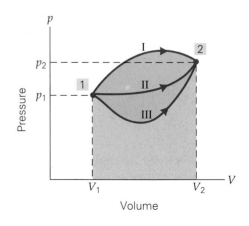

12.3 Thermodynamic Processes for an Ideal Gas

OBJECTIVES: To (a) describe and understand the four fundamental thermodynamic processes using an ideal gas, and (b) analyze the work done, heat flow, and change in internal energy that occurs during each of these processes.

The first law of thermodynamics can be applied to several processes for a system consisting of an ideal gas. Note that in three of the processes, one thermodynamic variable is kept constant. Such processes have names that begin with *iso-* (from the Greek *isos*, meaning "equal").

Isothermal Process   An **isothermal process** is a constant-temperature process (*iso* for equal, *thermal* for temperature). In this case, the process path is called an *isotherm*, or a curve of constant temperature. (See ▼Fig. 12.9.) The ideal gas law may be rewritten as $p = nRT/V$. Since the gas remains at constant temperature, nRT is a constant. Therefore, p is inversely proportional to V—that is, $p \propto 1/V$, which is a hyperbola. (Recall that a hyperbola is written as $y = a/x$ or $y \propto 1/x$, and it plots as a downward-curving line on a set of x–y axes.)

In the expansion from state 1 (initial) to state 2 (final) in Fig. 12.9, heat is added to the system, while both the pressure and volume vary in such a way as to keep the temperature constant. Positive work is done by the expanding gas. On an isotherm, $\Delta T = 0$; therefore, $\Delta U = 0$. The heat added to the gas is exactly equal to the amount of work done by the gas, and none of the heat goes into increasing the gas's internal energy. See the Learn by Drawing on Leaning on Isotherms on page 409.

In terms of the first law of thermodynamics, we can write

$$Q = \Delta U + W = 0 + W$$

or

$$Q = W \quad \textit{(ideal gas isothermal process)} \qquad (12.2)$$

The magnitude of the work done on the gas is equal to the area under the curve (requiring calculus to compute). We simply state it as follows:

$$W_{\text{isothermal}} = nRT \ln\!\left(\frac{V_2}{V_1}\right) \quad \textit{(ideal gas isothermal process)} \qquad (12.3)$$

Since the product nRT is a constant along a given isotherm, the work done depends on the ratio of the endpoint volumes.

Note: *Isothermal* means "constant temperature."

Teaching tip: *pV* is a constant for an isothermal process. Discuss the problems you might have in attempting to demonstrate this process in class.

PHYSLET®

Illustration 20.3 Thermodynamic Processes

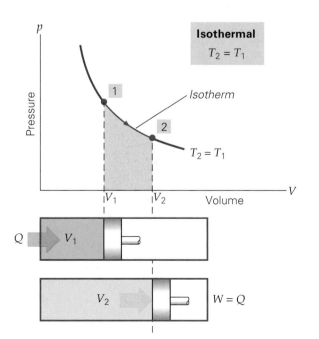

◀ **FIGURE 12.9** Isothermal (constant-temperature) process All of the heat added to the gas goes into doing work (the expanding gas moves the piston): Because $\Delta T = 0$, then $\Delta U = 0$, and from the first law of thermodynamics, $Q = W$. As always, the work is equal to the area (shaded) under the isotherm on the p–V diagram.

Problem-Solving Hint

In Eq. 12.3, the function "ln" stands for *natural logarithm*. Recall that *common logarithms* ("log") are referenced to the base 10 (see Appendix I). For this type, the exponent of the base 10 is the logarithm of the number in question. For example, $100 = 10^2$, the logarithm of 100 is 2, or, in equation terms, $\log 100 = 2$. In general, if $y = 10^x$, then x is the logarithm of y, or $x = \log y$. The natural logarithm is similar, except it uses a different base, e, which is an irrational number ($e \approx 2.7183$). As a check, find the natural logarithm of 100 on your calculator. (The answer is $\ln 100 = 4.605$.)

Note: *Isobaric* means "constant pressure."

Demonstration/activity: Fill a plastic garbage bag partially full of air and close the end. Then heat the bag over an electric heater to observe the change in volume with little or no change in pressure. (Be careful not to melt the plastic!)

Isobaric Process A constant-pressure process is called an **isobaric process** (*iso* for equal, and *bar* for pressure)*. An isobaric process for an ideal gas is illustrated in ▾Fig. 12.10. On a p–V diagram, an isobaric process is represented by a horizontal line called an *isobar*. When heat is added to or removed from an ideal gas at constant pressure, the ratio V/T remains constant (since $V/T = nR/p =$ constant). As the heated gas expands, its temperature must increase, and the gas crosses to higher temperature isotherms. This temperature increase means that the internal energy of the gas increases, since $\Delta U \propto \Delta T$.

As you can see from the isobar in Fig. 12.10, the area representing the work is rectangular. Thus, the work is relatively easy to compute (length times width):

$$W_{\text{isobaric}} = p(V_2 - V_1) = p\Delta V \quad \text{(ideal gas isobaric process)} \quad (12.4)$$

For example, when heat is added to or removed from a gas under isobaric conditions, the gas's internal energy changes *and* the gas expands or contracts, doing positive or negative work respectively. (See Integrated Example 12.2 for the signs.) We can write this relationship, using the first law of thermodynamics, with the work expression appropriate for isobaric conditions (Eq. 12.4):

$$Q = \Delta U + W = \Delta U + p\Delta V \quad \text{(ideal gas isobaric process)} \quad (12.5)$$

To see a detailed comparison of an isobaric process and an isothermal process, consider the following Integrated Example.

▶ **FIGURE 12.10** Isobaric (constant-pressure) process The heat added to the gas in the frictionless piston goes into work done by the gas and into changing the internal energy of the gas: $Q = \Delta U + W$. The work is equal to the area under the isobar (from state 1 to state 2 here, shown as shaded) on the p–V diagram. Note the two isotherms. They are not part of the isobaric process, but they help show us that the temperature rises during the isobaric expansion.

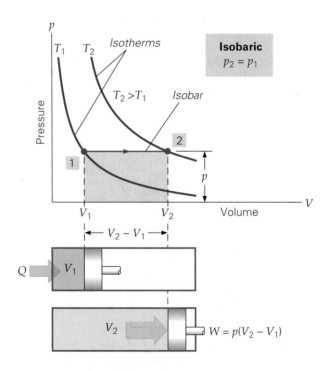

*Pressure is measured in bars or millibars.

Integrated Example 12.2 ■ Isotherms versus Isobars: Which Area?

Two moles of a monatomic ideal gas, initially at 0°C and 1.00 atm, are expanded to twice their original volume, using two different processes. They are expanded first isothermally and then, starting in the same initial state, isobarically. (a) During which process does the gas do more work: (1) the isothermal process, (2) the isobaric process, or (3) it does the same amount of work during both processes? Explain. (b) To prove your answer, determine the work done by the gas in each case.

(a) Conceptual Reasoning. As shown in ►Fig. 12.11, both processes involve expansion. The isobar is horizontal, and the isotherm is a decreasing hyperbola. Thus, the gas does more work during the isobaric expansion (more area under the curve). Fundamentally, this is because the isobaric process is done at higher (constant) pressure than is the isothermal process (where the pressure drops as the gas expands). In both cases, the work is positive. (How do we know this?) Thus, the correct answer to part (a) is that the isobaric process does more work.

(b) Quantitative Reasoning and Solution. If we know the volumes, Eqs. 12.3 and 12.4 can be used. These quantities can be calculated from the ideal gas law.

Listing the data,

Given: $p_1 = 1.00 \text{ atm} = 1.01 \times 10^5 \text{ N/m}^2$ *Find:* The work done during the isothermal
$T_1 = 0°C = 273 \text{ K}$ and isobaric processes
$n = 2.00 \text{ mol}$ (see Section 10.3)
$V_2 = 2V_1$

For the isotherm, use Eq. 12.3 (the natural logarithm of the volume ratio is ln 2 = 0.693):

$$W_{\text{isothermal}} = nRT \ln\left(\frac{V_2}{V_1}\right) = (2.00 \text{ mol})[8.31 \text{ J}/(\text{mol} \cdot \text{K})](273 \text{ K})(\ln 2)$$
$$= +3.14 \times 10^3 \text{ J}$$

For the isobar, we need to know the two volumes. Using the ideal gas law, we obtain

$$V_1 = \frac{nRT_1}{p_1} = \frac{(2.00 \text{ mol})[8.31 \text{ J}/(\text{mol} \cdot \text{K})](273 \text{ K})}{1.01 \times 10^5 \text{ N/m}^2}$$
$$= 4.49 \times 10^{-2} \text{ m}^3$$

and therefore
$$V_2 = 2V_1 = 8.98 \times 10^{-2} \text{ m}^3$$

The work is given by Eq. 12.4 as
$$W_{\text{isobar}} = p(V_2 - V_1)$$
$$= (1.01 \times 10^5 \text{ N/m}^2)(8.98 \times 10^{-2} \text{ m}^3 - 4.49 \times 10^{-2} \text{ m}^3) = +4.53 \times 10^3 \text{ J}$$

This amount is larger than the isothermal work, as expected from part (a).

Follow-Up Exercise. In this Example, what is the heat flow in each process?

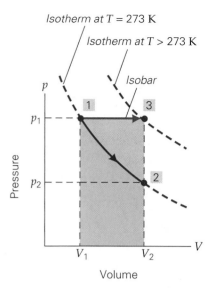

▲ **FIGURE 12.11 Comparing work** In Integrated Example 12.2, the gas does positive work while expanding. It does more work under isobaric conditions (from state 1 to state 3) than under isothermal conditions (from state 1 to state 2), because the pressure remains constant on the isobar but decreases along the isotherm. (Compare areas under the curves.)

Isometric Process An **isometric process** (short for *isovolumetric*, or constant-volume, process), sometimes called an *isochoric process*, is a constant-volume process. As illustrated in ▼Fig. 12.12, the process path on a *p–V* diagram is a vertical line,

Note: *Isometric* means "constant volume."

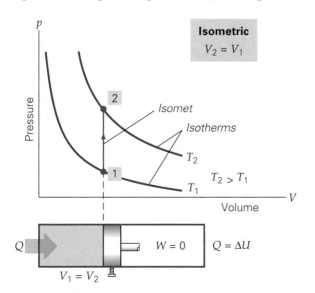

◄ **FIGURE 12.12 Isometric (constant-volume) process** All of the heat added to the gas goes into increasing the gas's internal energy, because there is no work done ($W = 0$); thus, $Q = \Delta U$. (Notice the locking nut on the piston, which prevents any movement.) Again, the isotherms, although not part of the isometric process, tell us visually that the temperature of the gas rises.

called an *isomet*. No work is done, since the area under such a curve is zero. (There is no displacement, so there is no change in volume.) Because the gas can do no work, if heat is added it must go completely into increasing the gas' internal energy and, therefore, its temperature. In terms of the first law of thermodynamics,

$$Q = \Delta U + W = \Delta U + 0 = \Delta U$$

and thus

$$Q = \Delta U \quad \text{(ideal gas isometric process)} \tag{12.6}$$

Consider the following example of an isometric process in action.

Example 12.3 ■ A Practical Isometric Exercise: How *Not* to Recycle a Spray Can

Many "empty" aerosol cans contain remnant propellant gases under approximately 1 atm of pressure (assume 1.00 atm) at 20°C. They display the warning "Do not dispose of this can in an incinerator or open fire." (a) Explain why it is dangerous to throw such a can into a fire. (b) What is the change in internal energy of such a gas if 500 J of heat is added to it, raising its temperature to 2000°F. (c) What is the final pressure of the gas?

Thinking It Through. This is an isovolumetric process; hence, all the heat goes into increasing the gas' internal energy. A pressure rise is expected, which is where the danger lies. The final pressure can be obtained using the ideal gas law.

Solution. Listing the data and converting given temperatures into kelvins. (Again, for qualitative reasoning, refer to the Learn by Drawing on page 409.)

Given: $p_1 = 1.00 \text{ atm} = 1.01 \times 10^5 \text{ N/m}^2$ *Find:* (a) Explain the danger in
        $V_1 = V_2$ heating the can.
        $T_1 = 20°C = 293 \text{ K}$ (b) ΔU (change in internal energy)
        $T_2 = 2000°F = 1.09 \times 10^3 \,°C$ (c) p_2 (final pressure of gas)
            $= 1.37 \times 10^3 \text{ K}$
        $Q = +500 \text{ J}$

(a) When heat is added, it all goes into increasing the gas's internal energy. Because at constant volume, pressure is proportional to temperature, the final pressure will be greater than 1.00 atm. The danger is that the container could explode into metallic fragments like a grenade if the maximum design pressure of the container is exceeded.

(b) To calculate the change in internal energy, we use the first law of thermodynamics. Recall that the work done in an isometric process is zero (why?).

$$\Delta U = Q - W = Q - 0 = Q = +500 \text{ J}$$

(c) The final pressure of the gas is determined directly from the ideal gas law:

$$\frac{p_2 V_2}{T_2} = \frac{p_1 V_1}{T_1} \quad \text{or} \quad p_2 = p_1 \left(\frac{V_1}{V_2}\right)\left(\frac{T_2}{T_1}\right) = (1 \text{ atm})\left(\frac{V_1}{V_1}\right)\left(\frac{1.37 \times 10^3 \text{ K}}{293 \text{ K}}\right) = 4.68 \text{ atm}$$

Follow-Up Exercise. Suppose the can were designed to withstand pressures up to 3.5 atm. What would be the highest temperature it could reach without exploding?

Adiabatic Process In an **adiabatic process**, no heat is transferred into or out of the system. That is, $Q = 0$ (▶Fig. 12.13). (The Greek word *adiabatos* means "impassable.") This condition is satisfied for a thermally isolated system, one completely surrounded by "perfect" insulation. This is an ideal situation, since there is some heat transfer even with the best of materials, if you wait long enough. Thus, for real-life conditions, we can only approximate adiabatic processes. For example, nearly adiabatic processes can take place if the changes occur rapidly enough so that there isn't time for significant heat to flow into or out of the system. In other words, *quick* processes can approximate adiabatic conditions.

The curve for this process is called an *adiabat*. During an adiabatic process, all three thermodynamic coordinates (p, V, T) change. For example, if the pressure on a gas is reduced, the gas expands. However, no heat flows into the gas. Without a compensating input of heat, this work is done at the expense of the gas' internal energy. Therefore, ΔU must be negative. Since the internal energy, and thus the temperature, both decrease, such an expansion is a cooling process. Similarly, an adiabatic compression is a warming process (temperature increase).

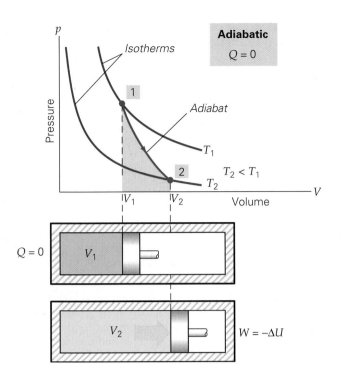

◄ **FIGURE 12.13** Adiabatic (no heat transfer) process In an adiabatic process (shown here for a cylinder with heavy insulation), no heat is added to or removed from the system; thus, $Q = 0$. During expansion (shown here), positive work is done by the gas at the expense of its internal energy: $W = -\Delta U$. The pressure, volume, and temperature all change in the process. The work done by the gas is the shaded area between the adiabat and the V-axis.

From the first law of thermodynamics, an adiabatic process can be described by

$$Q = 0 = \Delta U + W$$

or

$$\Delta U = -W \quad \text{(adiabatic process)} \quad (12.7)$$

For completeness, we state some other relationships in an adiabatic process. In such a process, an important factor is the ratio of the gas's molar specific heats, defined by a dimensionless quantity $\gamma = c_p/c_v$, where c_p and c_v are the specific heats at constant pressure and volume, respectively. For the two common types of gas molecules, monatomic and diatomic, the values of γ are about 1.67 and 1.40, respectively. The volume and pressure at any two points on an adiabat are related by

$$p_1 V_1^\gamma = p_2 V_2^\gamma \quad \text{(ideal gas adiabatic process)} \quad (12.8)$$

The work done by an ideal gas during an adiabatic process can be shown to be

$$W_{\text{adiabatic}} = \frac{p_1 V_1 - p_2 V_2}{\gamma - 1} \quad \text{(ideal gas adiabatic process)} \quad (12.9)$$

Demonstration/activity: Examples of approximately adiabatic processes abound. The syringe fire (Exercise 10) works well in lecture, as does feeling the cold sides of a spray can after a 10-second release of gas from the can.

Conceptual Example 12.4 ■ Exhaling: Blowing Hot and Cold?

The air in your lungs is warm. This can be demonstrated by putting near your mouth your forearm and blow air with your mouth opened wide. If you repeat this with your lips puckered, the air will feel (a) warmer, (b) cooler, (c) the same.

Reasoning and Answer. We'll go from answer to reasoning on this one, since it is easy to determine the correct answer experimentally. Try it, and you might be surprised to find that the answer is (b).

The interesting thing is *why* this situation occurs. When you blow with mouth open, you get a gush of warm air (roughly at body temperature). However, when you blow with your lips puckered, the stream of air is compressed. Then, as it emerges, the air expands, doing positive work against the atmosphere. The process is approximately adiabatic, since it takes place quickly. From the first law, since $Q = 0$, then $\Delta U = -W$; therefore, ΔU is negative, and the temperature drops. The work is done at the expense of the air's internal energy.

Follow-Up Exercise. Even during winter days with snow on the slopes, it is not uncommon in the Rocky Mountains to experience blasts of warm air coming down the slopes. (These gusts are called *chinooks*.) Explain how these winds could experience a significant rise in temperature while there is still snow and ice on the ground.

To clear up confusion that often occurs between isotherms and abiabats, see the following Integrated Example.

Integrated Example 12.5 ■ Adiabats versus Isotherms: Two Different Processes That Are Often Confused

A sample of helium expands to triple its initial volume adiabatically in one case and isothermally in another. In both cases, it starts from the same initial state. The sample contains 2.00 mol of helium, initially at 20°C and 1.00 atm. (a) During which process does the gas do more work: (1) the adiabatic process, (2) the isothermal process, or (3) the work done is the same during both processes? (b) Calculate the work done during each process to verify your reasoning in part (a). The specific heat ratio, or the γ value, for helium is 1.67.

(a) Conceptual Reasoning. To determine graphically which process involves more work, we look at the areas under the process curves (see Fig. 12.13). The area under the isothermal process curve is larger; thus, the gas does more work during its isothermal expansion, and the correct answer is (2). Physically, the isothermal expansion involves more work because the pressures are always higher during the isothermal expansion than during the adiabatic one.

(b) Quantitative Reasoning and Solution. To determine the isothermal work, we need the initial and final volumes, which can be determined using the ideal gas law. For the adiabatic work, the ratio of specific heats, γ, is important, as are the final pressure and volume. The final pressure can be found using Eq. 12.8, and the ideal gas law enables us to determine the final volume.

Listing the given values and converting the given temperature into kelvins:

Given: $p_1 = 1.00 \text{ atm} = 1.01 \times 10^5 \text{ N/m}^2$ *Find:*   Work done during each process
$n = 2.00 \text{ mol}$
$T_1 = 20°C + 273 = 293 \text{ K}$
$V_2 = 3V_1$
$\gamma = 1.67$

The data necessary to calculate the isothermal work from Eq. 12.3 are given. Since the volume ratio is 3 and $\ln 3 = 1.10$, we have

$$W_{\text{isothermal}} = nRT \ln\left(\frac{V_2}{V_1}\right)$$

$$= (2.00 \text{ mol})[8.31 \text{ J/(mol} \cdot \text{K)}](293 \text{ K})(\ln 3) = +5.35 \times 10^3 \text{ J}$$

For the adiabatic process, we can determine the work from Eq. 12.9, but first we need the final pressure and volume. The final pressure can be determined from a ratio form of Eq. 12.8:

$$p_2 = p_1\left(\frac{V_1}{V_2}\right)^\gamma = p_1\left(\frac{V_1}{3V_1}\right)^\gamma = p_1\left(\frac{1}{3}\right)^{1.67} = 0.160p_1$$

$$= (0.160)(1.01 \times 10^5 \text{ N/m}^2) = 1.62 \times 10^4 \text{ N/m}^2$$

The initial volume is determined from the ideal gas law:

$$V_1 = \frac{nRT_1}{p_1} = \frac{(2.00 \text{ mol})[8.31 \text{ J/(mol} \cdot \text{K)}](293 \text{ K})}{1.01 \times 10^5 \text{ N/m}^2}$$

$$= 4.82 \times 10^{-2} \text{ m}^3$$

Therefore, $V_2 = 3V_1 = 1.45 \times 10^{-1} \text{ m}^3$. Now, applying Eq. 12.9, we have

$$W_{\text{adiabatic}} = \frac{p_1V_1 - p_2V_2}{\gamma - 1}$$

$$= \frac{(1.01 \times 10^5 \text{ N/m}^2)(4.82 \times 10^{-2} \text{ m}^3) - (1.62 \times 10^4 \text{ N/m}^2)(1.45 \times 10^{-1} \text{ m}^3)}{1.67 - 1}$$

$$= +3.76 \times 10^3 \text{ J}$$

As expected, this result is less than the isothermal work.

Follow-Up Exercise. In this Example, (a) calculate the final temperature of the gas in the adiabatic expansion. (b) During the adiabatic expansion, determine the gas' change in internal energy by using the expression for the internal energy of a monatomic gas. Does it equal the negative of the work done (as calculated in the Example)? Explain.

TABLE 12.1	Important Thermodynamic Processes		
Process	Characteristic	Result	The First Law of Thermodynamics
Isothermal	T = constant	$\Delta U = 0$	$Q = W$
Isobaric	p = constant	$W = p\Delta V$	$Q = \Delta U + p\Delta V$
Isometric	V = constant	$W = 0$	$Q = \Delta U$
Adiabatic	$Q = 0$		$\Delta U = -W$

As a final summary for these thermodynamic processes, their characteristics and consequences are listed in Table 12.1.

LEARN BY DRAWING   LEANING ON ISOTHERMS

When you are analyzing thermodynamic processes, it is sometimes hard to keep track of the signs of heat flow (Q), work (W), and internal-energy change (ΔU). One method that can help with this bookkeeping is to superimpose a series of isotherms on the p–V diagram you are working with (as in Figs. 12.9 through 12.13). This method is useful even if the situation you are studying does not involve isothermal processes.

Before starting, recall that an isothermal process is one in which the temperature remains constant:

1. In an isothermal process for an ideal gas, ΔU is zero. (Why?)

2. Since T is constant, pV must also be constant, since, from the ideal gas law (Eq. 10.3), $pV = nRT$ = constant. You may recall from algebra that $p = k/V$ is the equation of a hyperbola. Thus, on a p–V diagram, an isothermal process is described by a hyperbola. The farther from the axes the hyperbola is, the higher is the temperature it represents (Fig. 1).

To take advantage of these properties, follow these steps:

- Sketch a set of isotherms for a series of increasing temperatures on the p–V diagram (Fig. 1).

- Then sketch the process you are analyzing—for example, the isobar shown in Fig. 2. isomet = constant volume (vertical line), isobar = constant pressure

(horizontal line), and adiabat = no heat flow (downward sloping curve, steeper than an isotherm)].

- Next, use the graphs to determine the signs of W and ΔU. W is represented by the area under the p–V curve for the process represented, and its sign is determined by whether the gas expanded (positive) or was compressed (negative). The sign of ΔT will be clear from the isotherms, since they serve as a temperature scale. For example, a rise in T implies an increase in U.

- Last, determine the sign of Q from the first law of thermodynamics, $Q = \Delta U + W$. From the sign of Q, it should be clear whether heat was transferred into or out of the system.

The example in Fig. 2 shows the power of this visual approach. Here, we are to decide whether heat flows into or out of the gas during an isobaric expansion. Expansion implies positive work done by the gas. But what is the direction of the heat flow (or is it zero)? Sketching the isobar, it can be seen that it crosses from lower-temperature isotherms to higher-temperature ones. Hence, there is a temperature increase, and ΔU is positive. From $Q = \Delta U + W$, we see that Q is the sum of two positive quantities, ΔU and W. Therefore, Q must be positive, which means that heat enters the gas.

As an exercise, try analyzing an isometric process, using this graphical approach. See also Integrated Examples 12.2 and 12.5.

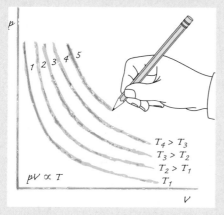

FIGURE 1 Isotherms on a p–V diagram

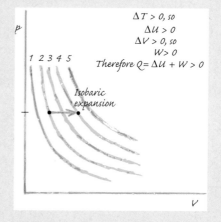

FIGURE 2 An isobaric expansion

12.4 The Second Law of Thermodynamics and Entropy

OBJECTIVES: To (a) state and explain the second law of thermodynamics in several forms, and (b) explain the concept of entropy.

Suppose that a piece of hot metal is placed in an insulated container of cool water. Heat will be transferred from the metal to the water, and the two will eventually come to thermal equilibrium at some intermediate temperature. For a thermally isolated system, the system's total energy remains constant. Could heat have been transferred from the cooler water to the hot metal instead? This process would not happen naturally. But if it did, the total energy of the system would remain constant, and this "impossible" inverse process would *not* violate energy conservation or the first law of thermodynamics.

Clearly, there must be another principle that specifies the *direction* in which a process can take place. This principle is embodied in the **second law of thermodynamics**, which says that certain processes do not take place, or have never been observed to take place, even though they may be consistent with the first law.

There are several equivalent statements of the second law, which are worded according to their application. One applicable to the aforementioned situation is as follows:

| Heat will not flow spontaneously from a colder body to a warmer body.

An equivalent alternative statement of the second law involves thermal cycles. A *thermal cycle* typically consists of several separate thermal processes after which the system ends up back at its starting conditions. If the system is a gas, this means the same p–V–T state from which it started. The second law, stated in terms of a thermal cycle (operating as a heat engine; see Section 12.5), is as follows:

| In a thermal cycle, heat energy cannot be completely transformed into mechanical work.

Note: Real machines can be made to be virtually frictionless, but their efficiency is still less than 100%. The second-law limit does not refer to frictional losses.

In general, the second law of thermodynamics applies to all forms of energy. It is considered true because no one has ever found an exception to it. If it were not valid, a perpetual-motion machine could be built. Such a machine could first transform heat completely into work and motion (mechanical energy), with no energy loss. The mechanical energy could then be transformed back into heat and be used to reheat the reservoir from where the heat came originally (again with no loss). Since the processes could be repeated indefinitely, the machine would run perpetually, just shifting energy back and forth. All of the energy is accounted for, so this situation does *not* violate the first law. However, it is obvious that real machines are always less than 100% efficient—that is, the work output is always less than the energy input. Another statement of the second law is therefore as follows:

| It is impossible to construct an operational perpetual-motion machine.

Attempts have been made to construct such machines, with no success.*

It would be convenient to have some way of expressing the *direction* of a process in terms of the thermodynamic properties of a system. One such property is temperature. In analyzing a conductive heat-transfer process, you need to know

---

*Although perpetual-motion *machines* cannot exist, (very nearly) perpetual motion is known to exist—for example, the planets have been in motion around the Sun for about 5 billion years.

the temperatures of the system and its surroundings. Knowing the temperature difference between the two processes allows you to state the direction in which the heat transfer will spontaneously take place. Another useful quantity, particularly during the discussion of heat engines, is entropy.

Entropy

A property that indicates the *natural direction* of a process was first described by Rudolf Clausius (1822–1888), a German physicist. This property is called **entropy**. Entropy is a multifaceted concept, with various different physical interpretations:

- Entropy is a measure of a system's ability to do useful work. As a system loses the ability to do work, its entropy increases.
- Entropy determines the direction of time. It is "time's arrow" that points out the forward flow of events, distinguishing past events from future ones.
- Entropy is a measure of disorder. A system naturally moves toward greater disorder, or disarray. The more order there is, the lower is the system's entropy.
- Entropy of the universe is increasing.

All of these statements (and others) turn out to be equally valid interpretations of entropy and are physically equivalent, as you will see in the upcoming discussions. First, however, we introduce the definition of the change in entropy. The change in a system's entropy (ΔS) when an amount of heat (Q) is added or removed by a reversible process at a constant temperature is

$$\Delta S = \frac{Q}{T} \quad \begin{array}{l} \text{(change in entropy} \\ \text{at constant temperature)} \end{array} \qquad (12.10)$$

SI unit of entropy: joule per kelvin (J/K)

If the temperature changes during the process, calculating the change in entropy requires advanced mathematics. Our discussions will be limited to isothermal processes or those involving small temperature changes. For the latter, we can approximate entropy changes by average temperatures, as in Example 12.7. But first, let's look at an example of a change in entropy and how it is interpreted.

Example 12.6 ■ Change in Entropy: An Isothermal Process

While doing physical exercise at 34°C, an athlete loses 0.400 kg of water per hour by the evaporation of persperation from his skin. Estimate the change in entropy of the water as it vaporizes. The latent heat of vaporization of persperation is about 24.2×10^5 J/kg.

Thinking It Through. A phase change occurs at constant temperature; hence, we can apply Eq. 12.10 ($\Delta S = Q/T$) after converting to kelvins. From Eq. 11.2 ($Q = mL_v$) we can compute the amount of heat added.

Solution. From the statement of the problem, we have

Given: $m = 0.400$ kg   *Find:* ΔS (change in entropy)
$\quad\quad\quad T = 34°C + 273 = 307$ K
$\quad\quad\quad L_v = 24.2 \times 10^5$ J/kg

Since a phase change occurs, latent heat is absorbed by water:

$$Q = mL_v = (0.400 \text{ kg})(24.2 \times 10^5 \text{ J/kg}) = 9.68 \times 10^5 \text{ J}$$

Then

$$\Delta S = \frac{Q}{T} = \frac{+9.68 \times 10^5 \text{ J}}{307 \text{ K}} = +3.15 \times 10^3 \text{ J/K}$$

Q is positive, because heat is added to water. The change in entropy, then, is also positive, and the entropy of the water increases. This outcome is reasonable, because a gaseous state is more random (disordered) than a liquid state.

Follow-Up Exercise. What is the change in entropy of a 1.00-kg water sample when it freezes to form ice at 0°C?

Teaching tip: Entropy can be viewed as a measure of disorder.

Illustration 21.2 Entropy and Reversible/Irreversible Processes

Note: ΔS is positive if a system absorbs heat ($Q > 0$) and negative if heat leaves ($Q < 0$). The sign of ΔS is determined by the sign of Q.

Note: Always use temperature in kelvins when computing entropy. What would happen if you used the Celsius temperature to calculate ΔS for a phase change from ice to water at 0°C?

Teaching tip: Remind students that Q can be either positive or negative; therefore, the entropy change may be either positive or negative.

Example 12.7 ■ A Warm Spoon into Cool Water: System Entropy Increase or Decrease?

PHYSLET®

Illustration 21.3 Entropy and Heat Exchange

A metal spoon at 24°C is immersed in 1.00 kg of water at 18°C. The system (spoon and water) is thermally isolated and comes to equilibrium at a temperature of 20°C. (a) Find the approximate change in the entropy of the system. (b) Repeat the calculation, assuming, although this can't happen, that the water temperature dropped to 16°C and the spoon's temperature increased to 28°C. Comment on how entropy shows that the situation in part (b) cannot happen.

Thinking It Through. The system is thermally isolated, so there is only heat exchange between the spoon and the water, that is, $Q_m + Q_w = 0$, where the subscripts m and w stand for metal and water, respectively. Q_w can be determined from the known water mass, specific heat, and temperature change. Therefore, we can determine both Q values (equal but opposite signs). Strictly speaking, Eq. 12.10 cannot be used because it is applicable only for constant-temperature processes. However, here the temperature changes are small, so a good approximation for ΔS can be obtained by using each object's *average* temperature $\overline{T}$.

Solution. Using i and f to stand for *initial* and *final*, respectively,

Given: $T_{m,i} = 24°C$ Find: (a) ΔS (change in entropy of the
$T_{w,i} = 18°C$ system in a realistic situation)
$m_w = 1.00$ kg   (b) ΔS (change in entropy of
$c_w = 4186$ J/(kg·C°) (from Table 11.1)   the system in an unrealistic
(a) $T_f = 20°C$ situation)
(b) $T_{m,f} = 28°C$; $T_{w,f} = 16°C$

(a) The amount of heat transferred (Q) needs to be determined in order to solve for ΔS. With $\Delta T_w = T_f - T_{w,i} = 20°C - °18°C = +2.0$ C°, the heat gained by the water is

$$Q_w = c_w m_w \Delta T = [4186 \text{ J}/(\text{kg} \cdot \text{C}°)](1.00 \text{ kg})(2.0 \text{ C}°) = +8.37 \times 10^3 \text{ J}$$

from Eq. 11.1. This quantity is also the magnitude of the heat *lost* by the metal. Therefore,

$$Q_m = -8.37 \times 10^3 \text{ J}$$

The average temperatures are

$$\overline{T}_w = \frac{T_{w,i} + T_f}{2} = \frac{18°C + 20°C}{2} = 19°C = 292 \text{ K}$$

$$\overline{T}_m = \frac{T_{m,i} + T_f}{2} = \frac{24°C + 20°C}{2} = 22°C = 295 \text{ K}$$

We then use these average temperatures and Eq. 12.10 to compute the approximate entropy changes for the water and the metal:

$$\Delta S_w \approx \frac{Q_w}{\overline{T}_w} = \frac{+8.37 \times 10^3 \text{ J}}{292 \text{ K}} = +28.7 \text{ J/K}$$

$$\Delta S_m \approx \frac{Q_m}{\overline{T}_m} = \frac{-8.37 \times 10^3 \text{ J}}{295 \text{ K}} = -28.4 \text{ J/K}$$

The change in the entropy of the *system* is the sum of these, or

$$\Delta S = \Delta S_w + \Delta S_m \approx +28.7 \text{ J/K} - 28.4 \text{ J/K} = +0.3 \text{ J/K}$$

The entropy of the metal decreased, because heat was lost. The entropy of the water increased by a greater amount, so overall, the system's entropy increased.

(b) Although this situation conserves energy, it violates the second law of thermodynamics. To see this violation in terms of entropy, let's repeat the foregoing calculation, using the second set of numbers. With $\Delta T_w = T_f - T_{w,i} = 16°C - 18°C = -2.0$ C°, the heat lost by the water is

$$Q_w = c_w m_w \Delta T = [4186 \text{ J}/(\text{kg} \cdot \text{C}°)](1.00 \text{ kg})(-2.0 \text{ C}°) = -8.37 \times 10^3 \text{ J}$$

Again using the average temperatures, $\overline{T}_w = 17°C = 290 \text{ K}$ and $\overline{T}_m = 26°C = 299 \text{ K}$, to compute the approximate entropy changes for the water and the metal spoon:

$$\Delta S_w \approx \frac{Q_w}{\overline{T}_w} = \frac{-8.37 \times 10^3 \text{ J}}{290 \text{ K}} = -28.9 \text{ J/K}$$

$$\Delta S_m \approx \frac{Q_m}{\overline{T}_m} = \frac{+8.37 \times 10^3 \text{ J}}{299 \text{ K}} = +28.0 \text{ J/K}$$

The change in the entropy of the *system* is:

$$\Delta S = \Delta S_w + \Delta S_m \approx -28.9 \text{ J/K} + 28.0 \text{ J/K} = -0.9 \text{ J/K}$$

In this unrealistic scenario, the entropy of the metal increased, but the entropy of the water decreased by a greater amount, and the total system entropy decreased.

Follow-Up Exercise. What should the initial temperatures in this Example be to make the overall system entropy change zero? Explain in terms of heat transfers.

Note how the entropy change of the system in Example 12.7a is positive, because the process is a *natural* one. That is, it is a process that is always observed to occur. In general, the direction of any process is toward an increase in total system entropy. That is, *the entropy of an isolated system never decreases.* Another way to state this observation is to say that *the entropy of an isolated system increases for every natural process* ($\Delta S > 0$). In coming to an intermediate temperature, the water and spoon in Example 12.7a are undergoing a natural process. In 12.7b the process would never be observed, and the decrease in system entropy indicates this. Similarly, water at room temperature in an isolated ice-cube tray will not naturally (spontaneously) turn into ice.

However, if a system is *not* isolated, it may undergo a decrease in entropy. For example, if the ice-cube tray filled with water is instead put into a freezer compartment, the water will freeze, undergoing a decrease in entropy. But there will be a *larger increase in entropy somewhere else* in the universe. In this case, the refrigerator freezer warms the kitchen as it freezes the ice, and the total entropy of the system (ice plus kitchen) actually increases.

Note: For every natural process, the entropy of a closed system increases.

Thus, a statement of the second law of thermodynamics in terms of entropy (for naturally occurring processes) is as follows:

| The total entropy of the universe increases in every *natural* process. |

Processes exist for which the entropy is constant. One obvious such process is any adiabatic process, since $Q = 0$. In this case, $\Delta S = Q/T = 0$. Similarly, any reversible isothermal expansion that is followed immediately by an isothermal compression along the same path has zero net entropy change. This last example is true because the two heat flows are the same, but opposite in sign, and the temperatures are also the same; thus, $\Delta S = Q/T + (-Q/T) = 0$. With the realization that, under some circumstances, it *is* possible to have $\Delta S = 0$, the previous statement of the second law of thermodynamics is generalized to include all possible processes. This is as follows:

Note: Total energy can be neither created nor destroyed; total entropy can be created but not destroyed.

| During *any* process, the entropy of the universe can only increase or remain constant ($\Delta S \geq 0$). |

To appreciate one of the many alternative (and equivalent) interpretations of entropy, consider the foregoing statements rewritten in terms of order and disorder. Here, entropy is interpreted as a measure of the disorder of a system. Thus, a larger value for entropy means more disorder (or, equivalently, less order):

| All naturally occurring processes move toward a state of greater disorder or disarray. |

A working definition of order and disorder may be extracted from everyday observations. Suppose you are making a pasta salad and have chopped tomatoes ready to toss into the cooked pasta. Before you mix the pasta and tomatoes, there is a relative amount of order; that is, the ingredients are separate and unmixed. Upon mixing, the separate ingredients become one dish, and there is less order (or more disorder, if you prefer). The pasta salad, once mixed, will never separate into the individual ingredients on its own (that is, by a natural process—one that happens on its own). Of course, you could go in and pick out the individual tomato pieces, but that would not be a natural process. Similarly, broken eyeglasses do

INSIGHT 12.1 LIFE, ORDER, AND THE SECOND LAW

The second law of thermodynamics is one of the foundations of physics and operates universally—the total entropy of the universe increases in every natural process. Certainly, life is a natural process. Yet clearly there is an increase in order during the development of an embryo into a mature adult. Moreover, in virtually all life processes, larger, more complex molecules and cellular structures are continuously being synthesized from simpler components. Does this mean that living things are exempt from the second law?

To answer this question, it must be realized that cells are *not* closed systems. They are involved in continuous exchanges of matter and energy with their surroundings. Living cells are constantly occupied in the exchange and conversion of energy from one form to another, which is governed by the first law of thermodynamics. For example, cells convert chemical potential energy into kinetic energy (movement) and electrical energy (the basis of nerve impulses). Many plant cells convert sunlight energy into chemical energy through photosynthesis (Fig. 1a). We have taken a hint from nature in attempting to harness the energy in sunlight (Fig. 1b).

In all biological processes, living systems maintain or even increase their organization while causing a larger decrease in the order of their surroundings. Living organisms are in a state of low entropy and require the addition of energy in the form of food and oxygen to keep them in this state. The second law is not violated by a *local* decrease in entropy as long as the *total* entropy of the universe increases in the process.

(a)   (b)

FIGURE 1 Solar collectors **(a)** Green plants capture huge amounts of solar energy every day and have been doing so for hundreds of million of years. The chemical energy they store in the form of sugars and other complex molecules is used by nearly all organisms on the Earth to maintain and increase the highly organized state that we call life. **(b)** Humans have developed devices (such as these reflective solar-energy concentrators) to capture some of the energy of sunlight.

not, on their own, go back together. Nor do mixed gases suddenly separate back into their separate species. Disorder is sometimes described as a measure of the randomness in a system. In the pasta dish after the ingredients are thoroughly mixed, the pieces are distributed randomly throughout the salad.

To better understand this interpretation of entropy, consider Insight 12.1 on Life, Order, and the Second Law.

12.5 Heat Engines and Thermal Pumps

OBJECTIVES: To (a) explain the concept of a heat engine and compute thermal efficiency, and (b) explain the concept of a thermal pump and compute the coefficient of performance.

A **heat engine** is any device that converts heat energy into work. Since the second law of thermodynamics prohibits perpetual-motion machines, some of the heat supplied to a heat engine will necessarily be lost and not go into work output. (In our study, we will not be concerned with the details of the mechanical components of an engine, such as pistons and cylinders.)

For our purposes, a heat engine is any device that takes heat from a high-temperature source (a hot reservoir), converts some of it to useful work, and transfers the rest to its surroundings (a cold, or low-temperature, reservoir). For example, most turbines that generate electricity (Chapter 20) are heat engines, using input heat from various sources. These sources include the burning of chemical ("fossil") fuels (oil, gas, or coal); nuclear reactions (Chapter 30); and the thermal energy already present beneath the Earth's surface (see the chapter-opening photo). They might be cooled by river water, for example, thus losing heat to this low-temperature reservoir. A generalized heat engine is represented in ▸Fig. 12.14a.

A few reminders about our sign convention are in order before starting analyzing heat engines. For engines, we are interested primarily in the work W done

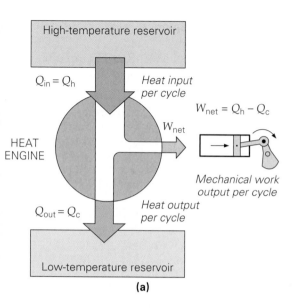

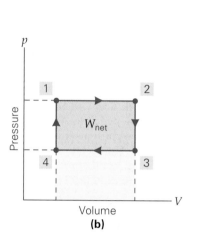

(a)

(b)

◀ **FIGURE 12.14 Heat engine
(a)** Energy flow for a generalized cyclic heat engine. Note that the width of the arrow representing Q_h (heat flow out of hot reservoir) is equal to the combined widths of the arrows representing W_{net} and Q_c (heat flow into cold reservoir), reflecting the conservation of energy: $Q_h = Q_c + W_{net}$. **(b)** This specific cyclic process consists of two isobars and two isomets. The net work output per cycle is the area of the rectangle formed by the process paths. (See Example 12.10 for the analysis of this particular cycle.)

by the gas on the environment. During expansion, the gas does positive work. Similarly, during a compression, the work done *by the gas* is negative. Also, it will be assumed that the "working substance" (the material that absorbs the heat and does the work) behaves like an ideal gas. The fundamental physics on which heat engines are based is the same regardless of the working substance. However, using ideal gases makes the mathematics easier.

Adding heat to a gas can produce work. But since a *continuous* output is usually wanted, practical heat engines operate in a **thermal cycle**, or a series of processes that brings the system back to its original condition. Cyclic heat engines include steam engines and internal combustion engines, such as automobile engines.

An idealized, rectangular thermodynamic cycle is shown in Fig. 12.14b. It consists of two isobars and two isomets. When these processes occur in the sequence indicated, the system goes through a cycle (1–2–3–4–1), returning to its original condition. When the gas expands (during 1 to 2), it does (positive) work equal to the area under the isobar. Doing positive work is exactly the desired output of an engine. (Think of a leaf blower pushing air out, or a car engine piston moving the crankshaft.) However, there must be a compression (during 3 to 4) of the gas to bring it back to its initial conditions. During this phase, the work done by the gas is negative, which is *not* the object of an engine. In a sense, a portion of the positive work done by the gas is "cancelled" by the negative work done during the compression.

From this discussion, you can see that the important quantity in engine design is not the amount of expansional work, but rather the *net work* W_{net} per cycle. This quantity is represented graphically as the area enclosed by the process curves that make up the cycle in the accompanying Learn By Drawing. (In Fig 12.14b, the area

LEARN BY DRAWING   **REPRESENTING WORK IN THERMAL CYCLES**

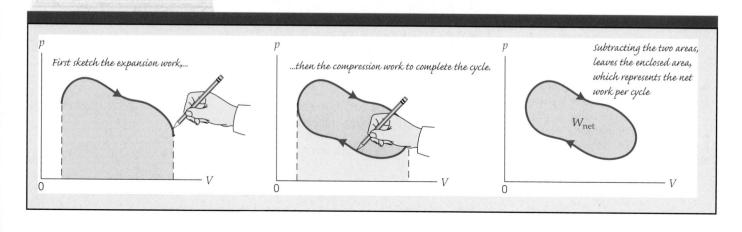

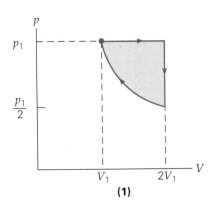

(1)

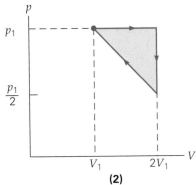

(2)

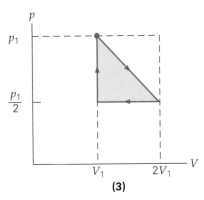

(3)

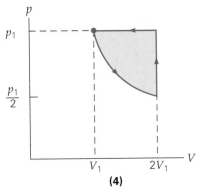

(4)

▲ **FIGURE 12.15 Comparing heat-engine cycles** Comparing the net work done per cycle, using graphical (visual) methods. (See Learn By Drawing on p. 415 and Conceptual Example 12.8 for details.)

is rectangular.) When the paths are not straight lines, numerical calculations of the areas may be difficult, but the concept is the same.

Before discussing engine efficiency, let's look at a Conceptual Example that uses these graphical techniques.

Conceptual Example 12.8 ■ Net Work: A Matter of Comparing Areas

A group of engineers is charged with designing a heat engine. They have narrowed their choices to four cycles (◀Fig. 12.15), and must decide which will generate the most net work. Can you help them decide? Order the given cycles from *least* net work to *most* net work per cycle. Assume in all cases that the gas starts in the same initial condition and that all expansions involve a doubling of the initial volume.

The cycles under consideration are as follows: (1) The gas expands isobarically and then has its pressure halved isometrically. Finally, it is isothermally compressed to its original state. (2) The gas expands (again) isobarically and then has its pressure halved isometrically. Lastly, it is compressed along a straight-line path (on a *p–V* diagram) to its original state. (3) The gas has its pressure halved (while doubling its volume) along a straight-line path (on a *p–V* diagram). Then it is isobarically compressed to its original volume. Finally, its pressure is raised isometrically to bring it back to its original state. (4) A gas is isothermally expanded, then its pressure is doubled isometrically, and last it is brought back to its original state isobarically.

Reasoning and Answer. The key here is to translate the words into process lines and graphical areas, as illustrated in Fig. 12.15. There each cycle is sketched (to scale) and then the enclosed areas are compared.

1. An isobaric expansion is represented by a horizontal line to the right and represents positive work done by the gas. Isometric compression is represented by a vertical line and does no work. At this point, the temperature is the same as the initial value, because the pressure is halved and the volume is doubled: $p_f V_f = (p_i/2)(2V_i) = p_i V_i = nRT_i$. An isothermal process then takes the gas back to its initial state.

2. This is the same as (1) for the first two processes, but the return is *not* isothermal.

3. During the first process, the gas does positive work. It expands along a straight line that ends up at the initial temperature. (How can you tell?) The isobaric compression is represented by a horizontal line to the left. The final process is a vertical isometric pressure increase back to initial conditions, during which zero work is done.

4. First there is an isothermal expansion to halve the initial pressure and double the initial volume. Here, the gas does positive work. Then a pressure increase brings the gas back to its initial pressure, but requires no work. Finally, the isobaric compression restores the gas to its initial conditions.

First, it should be clear that cycle 4 is not a heat engine, since it takes more work to compress it than the work it does on the environment, and the person who suggested this cycle should go back to school.

Cycle 1 does more net work than cycle 2. It is also clear that cycle 2 and cycle 3 do the same amount of net work, although they do it in a different way. Thus, the winner is cycle 1, with cycles 2 and 3 tied for second place.

Follow-Up Exercise. (a) In this Example, describe what changes you would make to cycle 2 to make it the winner. [*Hint:* There are many ways to achieve this goal.]

Thermal Efficiency

Thermal efficiency is an important parameter used to rate heat engines. The **thermal efficiency** (ε) of a heat engine is defined as

$$\varepsilon = \frac{\text{net work out}}{\text{heat in}} = \frac{W_{net}}{Q_{in}} \quad \text{\textit{thermal efficiency}} \atop \text{\textit{of a heat engine}} \qquad (12.11)$$

Efficiency tells us how much useful work (W_{net}) the engine does in comparison with the input heat it receives (Q_{in}). For example, modern automobile engines have an efficiency of about 20% to 25%. This means that only about one fourth of the heat generated by igniting the air–gas mixture is actually converted into mechanical work, which turns the car wheels and so on. Alternatively, you could say that the engine wastes about three fourths of the heat, eventually transferring it to the atmosphere through the hot exhaust system, radiator system, and metal engine.

For one cycle of an ideal gas heat engine, W_{net} is determined by applying the first law of thermodynamics to the complete cycle. Recall that our heat sign convention designates Q_{out} as negative. For our discussion of heat engines and pumps, *all heat symbols (all Q's) will represent magnitude only*. Therefore, Q_{out} is written as $-Q_c$ (the negative of a positive quantity Q_c to indicate flow *out* of the engine into a cold reservoir). Q_{in} is positive by our sign convention and is shown as $+Q_h$ (to indicate flow *to* the engine from the hot gas ignition).

Applying the first law of thermodynamics to the expansional part of the cycle and showing the work done by the gas as $W = +W_{exp}$, we have $\Delta U_h = +Q_h - W_{exp}$. For the compression part of the cycle, the work done by the gas is shown explicitly as being negative ($W = -W_{comp}$ and $\Delta U_c = -Q_c + W_{comp}$). Adding these equations and realizing that for an ideal gas, $\Delta U_{cycle} = \Delta U_h + \Delta U_c = 0$ (why?),

$$0 = (Q_h - Q_c) + (W_{comp} - W_{exp})$$

or

$$W_{exp} - W_{comp} = Q_h - Q_c$$

However, $W_{net} = W_{exp} - W_{comp}$, so the final result is (remember Q represents magnitude here)

$$W_{net} = Q_h - Q_c$$

So the thermal efficiency of a heat engine can be rewritten in terms of the heat flows as

$$\varepsilon = \frac{W_{net}}{Q_h} = \frac{Q_h - Q_c}{Q_h} = 1 - \frac{Q_c}{Q_h} \qquad \begin{array}{l} \textit{efficiency of} \\ \textit{an ideal gas heat engine} \end{array} \qquad (12.12)$$

Like mechanical efficiency, thermal efficiency is a dimensionless fraction and is commonly expressed as a percentage. Equation 12.12 indicates that a heat engine could have 100% efficiency if Q_c were zero. This condition would mean that no heat energy would be lost and all the input (hot reservoir) heat would be converted to useful work. However, this situation is impossible according to the second law of thermodynamics. In 1851, this observation led Lord Kelvin (who developed the kelvin temperature scale, Section 10.3) to state the second law in yet another, but physically equivalent, manner:

No cyclic heat engine can convert its heat input completely to work.

From Eq. 12.12 it can be seen that to maximize the work output per cycle of a heat engine, we must minimize Q_c/Q_h, which increases the efficiency.

Almost all automobile gasoline engines use a *four-stroke cycle*. An approximation to this important cycle involves the steps shown in ▼Fig. 12.16, along with a *p–V* diagram of the thermodynamic processes that make up the cycle. This theoretical cycle is called the *Otto cycle*, for the German engineer Nikolaus Otto (1832–1891), who built one of the first successful gasoline engines.

During the intake stroke (1–2), an isobaric expansion, the air–fuel mixture is admitted at atmospheric pressure through the open intake valve as the piston drops. This mixture is compressed adiabatically (quickly) on the compression stroke (2–3). This step is followed by fuel ignition (3–4, when the spark plug fires,

Note: Representing Work in Thermal Cycles. Here, all Q's are positive; heat flow in or out will be shown by a + or − sign, respectively.

PHYSLET®

Exploration 21.1 Engine Efficiency

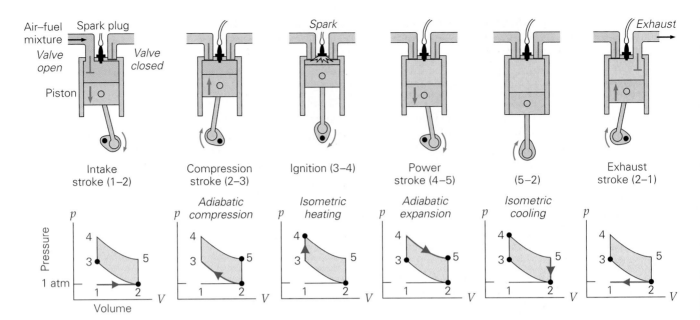

▲ **FIGURE 12.16 The four-stroke cycle of a heat engine** The process steps for the four-stroke Otto cycle. The piston moves up and down twice each cycle, making a total of four strokes per cycle.

Exploration 21.2 Internal Combustion Engine

giving an isometric pressure rise). Next, an adiabatic expansion occurs during the power stroke (4–5). Following this step is an isometric cooling of the system when the piston is at its lowest position (5–2). The final, exhaust stroke is along the isobaric leg of the Otto cycle (2–1). Notice that it takes two up and down motions of the piston to produce one power stroke.

Example 12.9 ■ Thermal Efficiency: What You Get Out of What You Put In

The small, gasoline-powered engine of a leaf blower absorbs 800 J of heat energy from a high-temperature reservoir (the ignited gas–air mixture) and exhausts 700 J to a low-temperature reservoir (the outside air, through its cooling fins). What is the engine's thermal efficiency?

Thinking It Through. The definition of thermal efficiency of a heat engine (Eq. 12.11) can be used if W_{net} is determined. (Keep in mind that the Qs mean heat magnitudes.)

Solution.

Given: $Q_h = 800$ J   *Find:* ε (thermal efficiency)
        $Q_c = 700$ J

The net work done by the engine per cycle is

$$W_{net} = Q_h - Q_c = 800 \text{ J} - 700 \text{ J} = 100 \text{ J}$$

Therefore, the thermal efficiency is

$$\varepsilon = \frac{W_{net}}{Q_h} = \frac{100 \text{ J}}{800 \text{ J}} = 0.125 \text{ (or 12.5\%)}$$

Follow-Up Exercise. (a) What would be the net work per cycle of the engine in this Example if the efficiency were raised to 15% and the input heat per cycle were raised to 1000? (b) How much heat would be exhausted in this case?

To see how the efficiency of an ideal gas cycle is calculated using the law of thermodynamics, consider the following Example.

Example 12.10 ■ Thermal Efficiency: Applying the Basic Definition

Assume you have 0.100 mol of an ideal monatomic gas that follows the cycle given in Fig. 12.14b and that the pressure and temperature at the lower left-hand corner of that figure are 1.00 atm and 20°C, respectively. Further assume that the pressure doubles during the isometric pressure increase and the volume doubles during the isobaric expansion. What would be the thermal efficiency of this cycle?

Thinking It Through. The definition of thermal efficiency of a heat engine applies (Eq. 12.11). However, we have to be careful, because heat exchanges can occur during more than one of the processes in the cycle. To determine heat input during the isobaric expansion, we need to know the change in internal energy and thus the change in temperature. So it seems likely that the temperatures at all four corners of the cycle will be needed.

Solution. Let's label the four corners with numbers, as shown in Figure 12.14b. Listing the data given and converting to SI units,

Given: $p_4 = p_3 = 1.00 \text{ atm} = 1.01 \times 10^5 \text{ N/m}^2$ *Find:* ε (thermal efficiency)
$n = 0.100 \text{ mol}$
$T_4 = 20°C = 293 \text{ K}$
$p_1 = p_2 = 2.00 \text{ atm} = 2.02 \times 10^5 \text{ N/m}^2$
$V_2 = V_3 = 2V_4 = 2V_1$

First, the volumes and temperatures at the corners is computed, using the ideal gas law:

$$V_4 = V_1 = \frac{nRT_1}{p_1} = \frac{(0.100 \text{ mol})[8.31 \text{ J/(mol} \cdot \text{K)}](293 \text{ K})}{1.01 \times 10^5 \text{ N/m}^2} = 2.41 \times 10^{-3} \text{ m}^3$$

Therefore,

$$V_2 = V_3 = 2V_1 = 4.82 \times 10^{-3} \text{ m}^3$$

During isometric processes, temperature (absolute in kelvins) is directly proportional to pressure, and during isobaric processes, temperature is directly proportional to volume. Therefore,

$$T_1 = 2T_4 = 586 \text{ K}$$
$$T_2 = 2T_1 = 1172 \text{ K}$$
$$T_3 = \tfrac{1}{2}T_2 = 586 \text{ K}$$

Now the heat transfers can be calculated. $W = 0$ during the 4–1 process, and for a monatomic gas, $\Delta U = \tfrac{3}{2}nR\Delta T$. Therefore

$$Q_{41} = \Delta U_{41} = \tfrac{3}{2}nR\Delta T_{41} = \tfrac{3}{2}(0.100 \text{ mol})[8.31 \text{ J/(mol} \cdot \text{K)}](586 \text{ K} - 293 \text{ K}) = +365 \text{ J}$$

During the 1–2 process, the gas expands and its internal energy increases. The work done by the gas is

$$W_{12} = p_1 \Delta V_{12} = (2.02 \times 10^5 \text{ N/m}^2)(4.82 \times 10^{-3} \text{ m}^3 - 2.41 \times 10^{-3} \text{ m}^3) = +487 \text{ J}$$

Since work was done *and* the internal energy went up, we have

$$Q_{12} = \Delta U_{12} + W_{12} = \tfrac{3}{2}nR\Delta T_{12} + 487 \text{ J}$$
$$= \tfrac{3}{2}(0.100 \text{ mol})[8.31 \text{ J/(mol} \cdot \text{K)}](1172 \text{ K} - 586 \text{ K}) + 487 \text{ J}$$
$$= +730 \text{ J} + 487 \text{ J} = +1.22 \times 10^3 \text{ J}$$

Thus the total heat input per cycle Q_h is

$$Q_h = Q_{41} + Q_{12} = 1.59 \times 10^3 \text{ J}$$

To find the net work, we need the area enclosed by the cycle. Therefore,

$$W_{\text{net}} = (\Delta p_{23})(\Delta V_{12}) = (1.01 \times 10^5 \text{ N/m}^2)(2.41 \times 10^{-3} \text{ m}^3) = +243 \text{ J}$$

and the efficiency is

$$\varepsilon = \frac{W_{\text{net}}}{Q_h} = \frac{243 \text{ J}}{1.59 \times 10^3 \text{ J}} = 0.153 \text{ or } 15.3\%$$

Follow-Up Exercise. Calculate the total output heat (Q_c) in this Example by calculating the Qs involved in processes 2–3 and 3–4 and adding them. Your answer should agree with the value of Q_c found from simple subtraction, using the results in the Example. Do you get such agreement?

INSIGHT 12.2 THERMODYNAMICS AND THE HUMAN BODY

Like the bodies of all other organisms, the human body is not a closed system. We must consume food and oxygen to survive. Both the first and second laws of thermodynamics have interesting implications for our bodies.

The human body metabolizes the chemical energy stored in food and/or the fatty tissues of the body. This is quite an efficient process, because typically 95% of the energy content in food is eventually metabolized. Some of this metabolized energy is converted into work, W, to circulate blood, perform daily tasks, and so on. The rest is lost to the environment in the form of heat, Q. For a typical 65-kg male, about 80 J of work per second is needed just to keep the body parts, such as liver, brain, and skeletal muscles, performing their functions.

The first law of thermodynamics, or the law of conservation of energy, can be written as $\Delta U = Q - W$.

Here ΔU is the change in internal energy of the body, which could come from two contributions. One is from the consumed food, and the other from the body's stored fat in fatty tissues. Thus we can write $\Delta U = \Delta U_{food} + \Delta U_{fat}$. Hence ΔU is a negative quantity, because as the energy stored in food and fat are converted to heat and work, our bodies have less energy stored (until more food is consumed again). Since Q is heat lost to the environment, it is also a negative quantity.

The human body is an example of a biological heat engine. The energy source of this engine is the energy metabolized from food and fatty tissues. Some of this energy is converted into work, and the rest is expelled to the environment in the form of heat. This situation is directly analogous to a heat engine taking in heat from the hot reservoir, doing mechanical work, and exhausting waste heat into the environment. Thus the efficiency of the human body is

$$\varepsilon = \frac{\text{work ouput}}{\text{energy input}} = \frac{W}{|\Delta U|}$$

Since W, Q, and ΔU vary widely from one activity to another, efficiency is often determined by using the time rate of these quantities—that is, the work per unit time (power P), $W/\Delta t$, and the energy consumed per unit time (metabolic rate), $|\Delta U|/\Delta t$:

$$\varepsilon = \frac{W}{|\Delta U|} = \frac{W/\Delta t}{(|\Delta U|/\Delta t)} = \frac{P}{(|\Delta U|/\Delta t)}$$

The power exerted during a particular activity such as running or cycling, can be measured by a device called a *dynamometer*. The metabolic rate has been found to be directly proportional to the rate of oxygen consumption, so this rate ($|\Delta U|/\Delta t$) can be measured by using breathing devices, as in Fig. 1. Thus, the efficiency of the body performing different activities can be measured by measuring the rate of oxygen consumption associated with each separate activity.

The efficiency of the human body, for the most part, depends on muscle activity and which muscles are used. The largest muscles in the body are leg muscles, so if an activity uses these muscles, the efficiency associated with that activity is relatively high. For example, some professional bicycle racers can achieve an efficiency as high as 20%, generating more than 2 hp of power in short bursts. Arm muscles, conversely, are relatively small, so activities such as bench pressing have efficiencies of less than 5%. Like any other heat engine, the human body can never achieve 100% efficiency. When people exercise, a lot of waste heat is generated; they must get rid of it through processes such as perspering, to avoid overheating. Read more about this in the Insight 11.1 (page 380) on Physiological Regulation of Body Temperature in Chapter 11.

FIGURE 1 Measuring energy consumed and work performed A cyclist is on a breathing device and a dynamometer so that both the power and the metabolic rate can be measured.

Thermal Pumps: Refrigerators, Air Conditioners, and Heat Pumps

The function performed by a thermal pump is basically the reverse of that of a heat engine. The name **thermal pump** is a generic term for *any* device, including refrigerators, air conditioners, and heat pumps, that transfers heat energy from a low-temperature reservoir to a high-temperature reservoir (▶Fig. 12.17a). For such a transfer to occur, there must be work input. Since the second law of thermodynamics says that heat will not *spontaneously* flow from a cold body to a hot body, we must provide the means for this process to happen, that is, using work from the environment.

Note: As used here, the term *thermal pump* is general and does not refer specifically to the devices that are commonly used for heating buildings. These devices are called *heat pumps*.

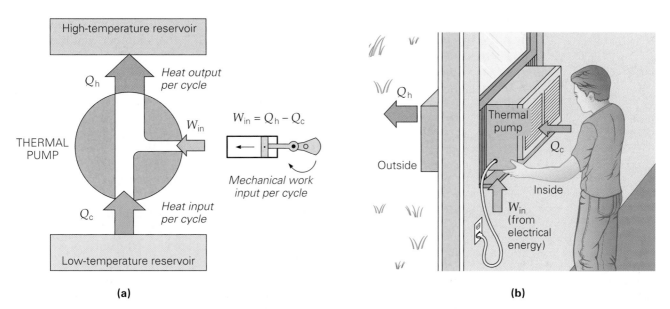

▲ **FIGURE 12.17 Thermal pumps** **(a)** An energy-flow diagram for a generalized cyclic thermal pump. The width of the arrow representing Q_h, the heat transferred into the high-temperature reservoir, is equal to the combined widths of the arrows representing W_{in} and Q_c, reflecting the conservation of energy: $Q_h = W_{in} + Q_c$. **(b)** An air conditioner is an example of a thermal pump. Using the input work, it transfers heat (Q_c) from a low-temperature reservoir (inside the house) to a high-temperature reservoir (outside).

A familiar example of a thermal pump is an air conditioner. By using input work from electrical energy, heat is transferred from the inside of the house (low-temperature reservoir) to the outside of the house (high-temperature reservoir), as shown in Fig. 12.17b. A refrigerator (▶Fig. 12.18) works on the exact same principles and processes. With the work performed by the compressor (W_{in}), heat (Q_c) is transferred to the evaporator coils inside of the refrigerator. The combination of this heat and work (Q_h) is then expelled to the outside of the refrigerator through the condenser.

In essence, a refrigerator or air conditioner pumps heat *up* a temperature gradient, or "hill." (Think of pumping water up an actual hill against the force of gravity.) The cooling efficiency of this operation is based on the amount of heat *extracted* from the cold-temperature reservoir (the refrigerator, the freezer, or the inside of a house), Q_c, compared with the work W_{in} needed to do so. Since a practical refrigerator operates in a cycle to provide continuous removal of heat, $\Delta U = 0$ for the cycle. Then, by the conservation of energy (the first law of thermodynamics), $Q_c + W_{in} = Q_h$, where Q_h is the heat ejected to the high-temperature reservoir or the outside.

The measure of air conditioner's or a refrigerator's performance is defined differently from that for a heat engine, because of the difference in their functions. For the cooling appliances, the efficiency is expressed as a **coefficient of performance (COP)**. Since the purpose is to extract the most heat (Q_c to make or keep things cold) per unit of work input (W_{in}), the coefficient of performance for a refrigerator or air conditioner (COP_{ref}) is the ratio of these two quantities:

$$COP_{ref} = \frac{Q_c}{W_{in}} = \frac{Q_c}{Q_h - Q_c} \quad \text{(refrigerator or air conditioner)} \quad (12.13)$$

Thus, the greater the COP, the better the performance—that is, more heat is extracted for each unit of work done. For normal operation, the work input is less than the heat removed, so the COP is greater than 1. The COPs of typical refrigerators and air conditioners range from 3 to 5, depending on operating conditions and mechanical design details. This range means that the amount of heat removed from the cold reservoir (the refrigerator, freezer, or house interior) is three to five times the amount of work needed to remove it.

Any machine that transfers heat in the opposite direction to that in which it would naturally flow is called a *thermal pump*. The term **heat pump** is specifically

Teaching tip: Ask students if they can find where the condenser coils are located on a refrigerator or on a car air conditioner.

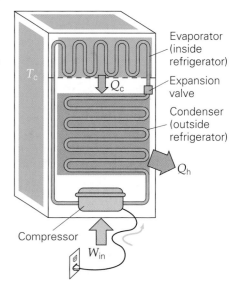

▲ **FIGURE 12.18 Refrigerator** Heat (Q_c) is carried away from the interior by the refrigerant as latent heat. This heat energy and that of the work input (W_{in}) are discharged from the condenser to the surroundings (Q_h). A refrigerator can be thought of as a remover of heat (Q_c) from an already cold region (its interior) *or* as a heat pump that adds heat (Q_h) to an already warm area (the kitchen).

applied to commercial devices used to cool homes and offices in the summer and to heat them in the winter. The summer operation is that of an air conditioner. In this mode, it cools the interior of the house and heats the outdoors. Operating in its winter heating mode, a heat pump heats the interior and cools the outdoors, usually by taking heat energy from the cold air or ground.

For a heat pump in its heating mode, the heat *output* (to warm something up or keep something warm) is the item of interest, so the COP for this case is defined differently than that of a refrigerator or air conditioner. As you might guess, it is the ratio of Q_h to W_{in} (the heating you get for the work put in), or

$$\text{COP}_{hp} = \frac{Q_h}{W_{in}} = \frac{Q_h}{Q_h - Q_c} \quad \text{(heat pump in heating mode)} \quad (12.14)$$

where, again, $Q_c + W_{in} = Q_h$ is used. Typical COPs for heat pumps range between 2 and 4, again depending on the operating conditions and design.

Compared with electrical heating, heat pumps are very efficient. For each unit of electric energy consumed, a heat pump typically pumps in from 1.5 to 3 times as much heat as direct electric heating systems provide. Some heat pumps use water from underground reservoirs, wells, or buried loops of pipe as a low-temperature heat reservoir. These heat pumps are more efficient than the ones that use the outside air, because water has a larger specific heat than air, and the average temperature difference between the water and the inside air is usually smaller.

Example 12.11 ■ Air Conditioner/Heat Pump: Thermal Switch Hitting

An air conditioner operating in summer extracts 100 J of heat from the interior of the house for every 40 J of electric energy required to operate it. Determine (a) the air conditioner's COP and (b) its COP if it runs as a heat pump in the winter. Assume it is capable of moving the same amount of heat for the same amount of electric energy, regardless of the direction in which it runs.

Thinking It Through. The input work and input heat in part (a) is known, so the definition of COP for a refrigerator (Eq. 12.13) can be applied. For the reverse operation, it is the output heat that is important, so we find this quantity from the conservation of energy.

Solution.

Given: $Q_c = 100$ J   *Find:* (a) COP_{ref}
         $W_{in} = 40$ J   (b) COP_{hp}

(a) From Eq. 12.13, the COP for this engine operating as an air conditioner is

$$\text{COP}_{ref} = \frac{Q_c}{W_{in}} = \frac{100 \text{ J}}{40 \text{ J}} = 2.5$$

(b) When the device operates as a heat pump, the relevant heat is the output heat, which can be calculated from the conservation of energy:

$$Q_h = Q_c + W_{in} = 100 \text{ J} + 40 \text{ J} = 140 \text{ J}$$

Thus, the COP for this engine operating as a heat pump in winter is, from Eq. 12.14,

$$\text{COP}_{hp} = \frac{Q_h}{W_{in}} = \frac{140 \text{ J}}{40 \text{ J}} = 3.5$$

Follow-Up Exercise. (a) Suppose you redesigned the device in this Example to perform the same operation, but with 25% less work input. What would be the new values of the two COPs? (b) Which COP would have the larger percentage increase?

12.6 The Carnot Cycle and Ideal Heat Engines

OBJECTIVES: To (a) explain how the Carnot cycle applies to heat engines, (b) compute the ideal Carnot efficiency, and (c) state the third law of thermodynamics.

Lord Kelvin's statement of the second law of thermodynamics says that any *cyclic* heat engine, regardless of its design, must always exhaust some heat energy (Section 12.5). But how much heat must be lost in the process? In other words,

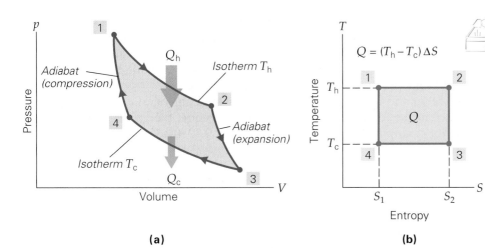

(a)

◀ **FIGURE 12.19 The Carnot cycle** **(a)** The Carnot cycle consists of two isotherms and two adiabats. Heat is absorbed during the isothermal expansion and exhausted during the isothermal compression. **(b)** On a T–S diagram, the Carnot cycle forms a rectangle, the area of which is equal to Q.

what is the *maximum* possible efficiency of a heat engine? In designing heat engines, engineers strive to make them as efficient as possible, but there must be some theoretical limit, and, according to the second law, it must be less than 100%.

Sadi Carnot (1796–1832), a French engineer, studied this limit. The first thing he sought was the thermodynamic cycle an *ideal* heat engine would use, that is, the most efficient cycle. Carnot found that the ideal heat engine absorbs heat from a *constant* high-temperature reservoir (T_h) and exhausts it to a *constant* low-temperature reservoir (T_c). These processes are ideally reversible isothermal processes and may be represented as two isotherms on a p–V diagram. But what are the processes that complete the cycle? Carnot showed that these processes are reversible adiabatic processes. As we saw in Section 12.3, the curves on a p–V diagram are called adiabats and are steeper than isotherms (▲Fig. 12.19a). An irreversible heat engine operating between two heat reservoirs at constant temperatures cannot have an efficiency greater than that of a reversible heat engine operating between the same two temperatures.

Thus, the ideal **Carnot cycle** consists of two isotherms and two adiabats and is conveniently represented on a T–S diagram, where it forms a rectangle (Fig. 12.19b). The area under the upper isotherm (1–2) is the heat added to the system from the high-temperature reservoir: $Q_h = T_h \Delta S$. Similarly, the area under the lower isotherm (3–4) is the heat exhausted: $Q_c = T_c \Delta S$. Here, Q_h and Q_c are the heat transfers at *constant* temperatures (T_h and T_c, respectively). There is no heat transfer ($Q = 0$) during the adiabatic legs of the cycle. (Why?)

The difference between these heat transfers is the work output, which is equal to the area enclosed by the process paths (the shaded areas on the diagrams):

$$W_{net} = Q_h - Q_c = (T_h - T_c)\Delta S$$

Since ΔS is the same for both isotherms (see Fig. 12.19b, processes 1–2 and 3–4), we can use the expressions to relate the temperatures and heats. That is, since

$$\Delta S = \frac{Q_h}{T_h} \quad \text{and} \quad \Delta S = \frac{Q_c}{T_c}$$

we have

$$\frac{Q_h}{T_h} = \frac{Q_c}{T_c} \quad \text{or} \quad \frac{Q_c}{Q_h} = \frac{T_c}{T_h}$$

This equation can be used to express the efficiency of an ideal heat engine in terms of temperature. From Eq. 12.12, this ideal **Carnot efficiency** (ε_C) is

$$\varepsilon_C = 1 - \frac{Q_c}{Q_h} = 1 - \frac{T_c}{T_h}$$

or

$$\varepsilon_C = 1 - \frac{T_c}{T_h} \quad \textit{Carnot efficiency (ideal heat engine)} \qquad (12.15)$$

where, as usual, the fractional efficiency is often expressed as a percentage. Note that T_c and T_h must be expressed in kelvins.

PHYSLET®

Illustration 21.4 Engines and Entropy

PHYSLET®

Illustration 21.1 Carnot Engine

Note: The temperatures in the Carnot efficiency are the absolute, or Kelvin, temperatures.

The Carnot efficiency expresses the theoretical upper limit on the thermodynamic efficiency of a cyclic heat engine operating between two known temperature extremes. In practice, this limit can never be achieved because no real engine processes are reversible. A true Carnot engine cannot be built, because the necessary reversible processes can only be approximated.

However, the Carnot efficiency does illustrate a general idea: The greater the difference in the temperatures of the heat reservoirs, the greater the Carnot efficiency. For example, if T_h is twice T_c, or $T_c/T_h = 0.50$, the Carnot efficiency is

Note: The Carnot efficiency, which can never be attained, is an ideal upper limit.

$$\varepsilon_C = 1 - \frac{T_c}{T_h} = 1 - 0.50 = 0.50 \ (\times \ 100\%) = 50\%$$

However, if T_h is four times T_c, or $T_c/T_h = 0.25$, then

$$\varepsilon_C = 1 - \frac{T_c}{T_h} = 1 - 0.25 = 0.75 \ (\times \ 100\%) = 75\%$$

Since a heat engine can never attain 100% thermal efficiency, it is useful to compare its actual efficiency ε with its theoretical maximum efficiency, that of a Carnot cycle, ε_C. To see the importance of this concept in more detail, study the next Example carefully.

There are also Carnot COPs for refrigerators and heat pumps. (See Exercise 98.)

Example 12.12 ■ Carnot Efficiency: The True Measure of Efficiency for Any Real Engine

An engineer is designing a cyclic heat engine to operate between the temperatures of 150°C and 27°C. (a) What is the maximum theoretical efficiency that can be achieved? (b) Suppose the engine, when built, does 100 J of work per cycle for every 500 J of input heat per cycle. What is its efficiency, and how close is it to the Carnot efficiency?

Thinking It Through. The maximum efficiency for specific high and low temperatures is given by Eq. 12.15. Remember, we must convert absolute temperatures. In part (b), we calculate the actual efficiency and compare it with our answer in part (a).

Solution.

Given: $T_h = 150°C + 273 = 423 \ K$ *Find:* (a) ε_C (Carnot efficiency)
$T_c = 27°C + 273 = 300 \ K$ (b) ε (actual efficiency) and compare it
$W_{net} = 100 \ J$ with ε_C
$Q_h = 500 \ J$

(a) Using Eq. 12.15 to find the maximum theoretical efficiency, we get

$$\varepsilon_C = 1 - \frac{T_c}{T_h} = 1 - \frac{300 \ K}{423 \ K} = 0.291 \ (\times \ 100\%) = 29.1\%$$

(b) The actual efficiency is, from Eq. 12.12,

$$\varepsilon = \frac{W_{net}}{Q_h} = \frac{100 \ J}{500 \ J} = 0.200 \ (\text{or } 20.0\%)$$

Thus,

$$\frac{\varepsilon}{\varepsilon_C} = \frac{0.200}{0.291} = 0.687 \ (\text{or } 68.7\%)$$

In other words, the heat engine is operating at 68.7% of its theoretical maximum. That's pretty good.

Follow-Up Exercise. If the operating high temperature of the engine in this Example were increased to 200°C, what would be the change in the theoretical efficiency?

The Third Law of Thermodynamics

Another inference might be drawn from the expression for the Carnot efficiency (Eq. 12.15). It would seem possible to have ε_C equal to 100%, if only T_c could be absolute zero. (See Section 10.3.) However, absolute zero has never been achieved, although ultralow-temperature (cryogenic) experiments have gotten within 20 nK (2×10^{-8} K) of it. Apparently, reducing the temperature of a system already close to absolute zero in a finite number of steps is impossible. This is embodied in the **third law of thermodynamics** which, simply stated, reads:

| It is impossible to reach absolute zero in a finite number of thermal processes. |

In other words, it appears to be impossible to ever reach absolute zero experimentally, althrough attempts have been made. (See Chapter 10 footnote on page 347).

Chapter Review

- The **first law of thermodynamics** is a statement of the conservation of energy for a thermodynamic system. Expressed in equation form, it relates the change in a system's internal energy to the heat flow and the work done on it and is written as

$$Q = \Delta U + W \qquad (12.1)$$

- Some **thermodynamic processes** (for gases) are

 isothermal: a process that occurs at constant temperature
 isobaric: a process that occurs at constant pressure
 isometric: a process that occurs at constant volume
 adiabatic: a process involving no heat flow

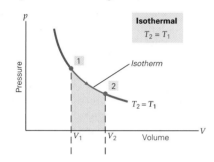

- The expressions for **thermodynamic work** done by an ideal gas during various processes are

$$W_{isothermal} = nRT \ln\left(\frac{V_2}{V_1}\right) \quad \text{(ideal gas isothermal process)} \qquad (12.3)$$

$$W_{isobaric} = p(V_2 - V_1) = p\Delta V \quad \text{(ideal gas isobaric process)} \qquad (12.4)$$

$$W_{adiabatic} = \frac{p_1 V_1 - p_2 V_2}{\gamma - 1} \quad \text{(ideal gas adiabatic process)} \qquad (12.9)$$

(In the adiabatic process, $\gamma = c_p/c_v$ is the ratio of specific heats at constant pressure and volume, respectively.)

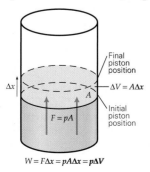

$$W = F\Delta x = pA\Delta x = p\Delta V$$

- The **second law of thermodynamics** states whether a process can take place naturally or, alternatively, specifies the direction a process can take.

- **Entropy** (S) is a measure of the disorder of a system. The **change in entropy** of an object at constant temperature is given by

$$\Delta S = \frac{Q}{T} \qquad (12.10)$$

The total entropy of the universe increases in every natural process.

- A **heat engine** is a device that converts heat into work. Its **thermal efficiency** ε is the ratio of work output to heat input, or

$$\varepsilon = \frac{W_{net}}{Q_h} = \frac{Q_h - Q_c}{Q_h} = 1 - \frac{Q_c}{Q_h} \qquad (12.12)$$

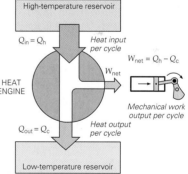

- A **thermal pump** is a device that transfers heat energy from a low-temperature reservoir to a high-temperature reservoir. The coefficient of performance (COP) is the ratio of heat transferred to the input work. The COP differs depending on whether the thermal pump is used as a heat pump or an air conditioner/refrigerator.

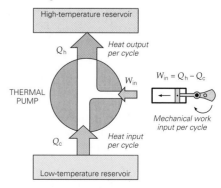

• A **Carnot cycle** is a theoretical heat-engine cycle consisting of two isotherms and two adiabats. Its efficiency is the highest possible efficiency that any heat engine could have, operating between two temperature extremes. The efficiency of a Carnot cycle is

$$\varepsilon_C = 1 - \frac{T_c}{T_h} \qquad (12.15)$$

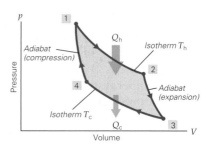

Exercises*

MC = *Multiple Choice Question*, CQ = *Conceptual Question, and* IE = *Integrated Exercise. Throughout the text, many exercise sections will include "paired" exercises. These exercise pairs, identified with* **red numbers**, *are intended to assist you in problem solving and learning. In a pair, the first exercise (even numbered) is worked out in the Study Guide so that you can consult it should you need assistance in solving it. The second exercise (odd numbered) is similar in nature, and its answer is given at the back of the book.*

12.1 Thermodynamic Systems, States, and Processes

1. **MC** On a *p–V* diagram, a reversible process is a process (a) whose path is known, (b) whose path is unknown, (c) for which the intermediate steps are nonequilibrium states, (d) none of the preceding. (a)

2. **MC** There may be an exchange of heat with the surroundings for (a) a thermally isolated system, (b) a completely isolated system, (c) a heat reservoir, (d) none of the preceding. (c)

3. **MC** Only initial and final states are known for irreversible processes on (a) *p–V* diagrams, (b) *p–T* diagrams, (c) *V–T* diagrams, (d) all of the preceding. (d)

4. **CQ** Explain why the process shown in Fig. 12.1b is *not* that for an ideal gas at constant temperature. see ISM

5. **CQ** Does an irreversible process mean the system cannot return to its original state? Explain. no, all it means is that the intermediate steps are nonequilibrium states.

12.2 The First Law of Thermodynamics
and
12.3 Thermodynamic Processes for an Ideal Gas

6. **MC** There is no heat flow into or out of the system in an (a) isothermal process, (b) adiabatic process, (c) isobaric process, (d) isometric process. (b)

7. **MC** According to the first law of thermodynamics, if work is done on a system, then (a) the internal energy of the system must change, (b) heat must be transferred from the system, (c) the internal energy of the system must change and/or heat must be transferred from the system, (d) heat must be transferred to the system. (c)

8. **MC** When heat is added to a system of ideal gas during an isothermal expansion process, (a) work is done on the system, (b) the internal energy decreases, (c) the effect is the same as for an isometric process, (d) none of the preceding. (d)

9. **CQ** On a *p–V* diagram, sketch a cyclic process that consists of an isothermal expansion followed by an isobaric compression, and lastly followed by an isometric process—in that order. see ISM

10. **CQ** In ▾Fig. 12.20, the plunger of a syringe is pushed in quickly, and the small pieces of paper in the syringe catch fire. Explain this phenomenon, using the first law of thermodynamics. (Similarly, in a diesel engine, there are no spark plugs. How can the air–fuel mixture ignite?) see ISM

◀ **FIGURE 12.20 Syringe fire** See Exercise 10.

11. **CQ** Discuss heat, work, and the change in internal energy of your body when you play basketball. lost heat, did work, decreased internal energy

12. **CQ** In an adiabatic process, there is no heat exchange between the system and the environment, but the temperature of an ideal gas changes. How can that be? Explain. through work

*Take temperatures to be exact.

13. **CQ** An ideal gas initially at temperature T_o, pressure p_o, and volume V_o is compressed to one half its initial volume. As shown in ▼Fig. 12.21, process 1 is adiabatic, 2 is isothermal, and 3 is isobaric. Rank the work done on the gas and the final temperatures of the gas, from high to low, for all three processes, and explain how you made your ranking.
work: 1, 2, 3; final temperature 1, 2, 3

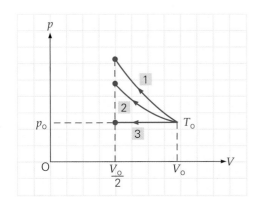

▲ **FIGURE 12.21 Thermodynamic processes**
See Exercise 13.

14. **IE ●** A rigid container contains 1.0 mol of an ideal gas that slowly receives 2.0×10^4 J of heat. (a) The work done by the gas is (1) positive, (2) zero, (3) negative. Why? (b) What is the change in the internal energy of the gas?
(a) (2) zero, because V = constant (b) 2.0×10^4 J

15. **IE ●** A quantity of ideal gas goes through a cyclic process and does 400 J of net work. (a) The temperature of the gas at the end of the cycle is (1) higher than, (2) the same as, (3) less than when it started. Why? (b) Is a net amount of heat added to or removed from the system, and how much is involved?
(a) (2) the same as, because ΔU = 0 (b) added, 400 J

16. **●** While playing in a tennis match, you lost 6.5×10^5 J of heat, and your internal energy also decreased by 1.2×10^6 J. How much work did you do in the match?
5.5×10^5 J

17. **IE ●** While doing 500 J of work, an ideal gas expands adiabatically to 1.5 times its initial volume. (a) The temperature of the gas (1) increases, (2) remains the same, (3) decreases. Why? (b) What is the change in the internal energy of the gas?
(a) (3) decreases, as Q = 0 and W is positive (b) −500 J

18. **IE ●** An ideal gas expands from 1.0 m³ to 3.0 m³ at atmospheric pressure while absorbing 5.0×10^5 J of heat in the process. (a) The temperature of the system, (1) increases, (2) stays the same, or (3) decreases. Explain. (b) What is the change in internal energy of the system?
(a) (1) increases (b) 3.0×10^5 J

19. **●●** A gas at low density (that is, it behaves as an ideal gas) is under an initial pressure of 1.65×10^4 Pa and occupies a volume of 0.20 m³. The slow addition of 8.4×10^3 J of heat to this gas causes it to expand isobarically to a volume of 0.40 m³. (a) How much work is done by the gas in the process? (b) Does the internal energy of the gas change? If so, by how much? (a) 3.3×10^3 J (b) yes, 5.1×10^3 J

20. **●●** An Olympic weight lifter lifts 145 kg a vertical distance of 2.1 m. When he does so, his internal energy decreases by 6.0×10^4 J. For his body, how much heat flows, and in what direction? −5.7×10^4 J out (lost)

21. **IE ●●** An ideal gas is taken through the reversible processes shown in ▼Fig. 12.22. (a) Is the overall change in the internal energy of the gas (1) positive, (2) zero, or (3) negative? Explain. (b) In terms of state variables p and V, how much work is done by or on the gas, and (c) what is the net heat transfer in the overall process? (a) (2) zero (b) −p_1V_1 (on the gas) (c) −p_1V_1 (out of the gas)

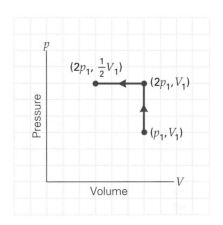

▲ **FIGURE 12.22 A p–V diagram for an ideal gas**
See Exercise 21.

22. **●●** A fixed quantity of gas undergoes the reversible changes illustrated in the p–V diagram in ▼Fig. 12.23. How much work is done in each process?
1–2: 0; 2–3: 2.5×10^4 J; 3–4: 0; 4–5: 2.5×10^4 J

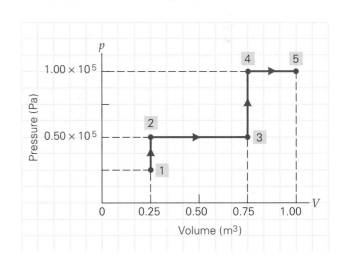

▲ **FIGURE 12.23 A p–V diagram and work** See Exercises 22 and 23.

23. **●●** Suppose that after the final process in Fig. 12.23 (see Exercise 22), the pressure of the gas is decreased isometrically from 1.0×10^5 Pa to 0.70×10^5 Pa, and then the gas is compressed isobarically from 1.0 m³ to 0.80 m³. What is the total work done in all of these processes, including 1 through 5? 3.6×10^4 J

24. **IE ●●** 2.0 mol of an ideal gas expands isothermally from a volume of 20 L to 40 L while its temperature remains at 300 K. (a) The work done by the gas is (1) positive, (2) negative, (3) zero. Explain. (b) What is the magnitude of the work? (a) (1) positive (b) 3.5×10^3 J

25. **IE ●●** A gas is enclosed in a cylindrical piston with a 12.0-cm radius. Heat is slowly added to the gas while the pressure is maintained at 1.00 atm. During the process, the piston moves 6.00 cm. (a) This is an (1) isothermal, (2) isobaric, (3) adiabatic process. Explain. (b) If the heat transferred to the gas during the expansion is 420 J, what is the change in the internal energy of the gas? (a) (2) isobaric (b) 146 J

26. **●●** A monatomic ideal gas ($\gamma = 1.67$) is compressed adiabatically from a pressure of 1.00×10^5 Pa and volume of 240 L to a volume of 40.0 L. (a) What is the final pressure of the gas? (b) How much work is done on the gas? (a) 1.99×10^6 Pa (b) -8.30×10^4 J

27. **IE ●●●** The temperature of 2.0 mol of ideal gas is increased from 150°C to 250°C by two different processes. In process 1, 2500 J of heat is added to the gas; in process 2, 3000 J of heat is added. (a) In which case is more work done, (1) process 1, (2) process 2, or (3) the same? Explain. [*Hint:* See Eq. 10.16.] (b) Calculate the change in internal energy and work done for each process. (a) (2) process 2 (b) $\Delta U = 2.5 \times 10^3$ J for both; $W_1 = 0$, $W_2 = 5.0 \times 10^2$ J

28. **IE ●●●** One mole of ideal gas is compressed as shown on the *p*–*V* diagram in ▼Fig. 12.24. (a) Is the work done by the gas (1) positive, (2) zero, or (3) negative? Why? (b) What is the work done by the gas? (c) What is the change in temperature of the gas? (a) (3) negative (b) -1.8×10^5 J (c) -4.8×10^4 K

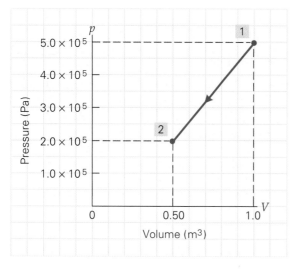

▲ **FIGURE 12.24 A variable *p*–*V* process and work** See Exercise 28.

29. **●●●** One mole of ideal gas is taken through the cyclic process shown in ▶Fig. 12.25. (a) Compute the work involved for each of the four processes. (b) Find ΔU, W, and Q for the complete cycle. (c) What is T_3? see ISM

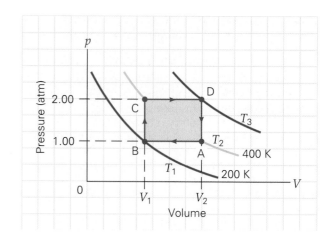

▲ **FIGURE 12.25 A cyclic process** See Exercise 29.

12.4 The Second Law of Thermodynamics and Entropy

30. **MC** In any natural process, the overall change in the entropy of the universe could not be (a) negative, (b) zero, (c) positive. (a)

31. **MC** The second law of thermodynamics (a) describes the state of a system, (b) applies only when the first law is satisfied, (c) precludes perpetual motion machines, (d) does not apply to an isolated system. (c)

32. **MC** An ideal gas is compressed isothermally. The change in entropy for this process is (a) positive, (b) negative, (c) zero, (d) none of the preceding. (b)

33. **CQ** Does the entropy of each of these objects increase or decrease? (a) *Ice* as it melts; (b) *water vapor* as it condenses; (c) *water* as it is heated on a stove; (d) *food* as it is cooled in a refrigerator. (a) increases (b) decreases (c) increases (d) decreases

34. **CQ** When a quantity of hot water is mixed with a quantity of cold water, the combined system comes to thermal equilibrium at some intermediate temperature. How does the entropy of the system (both liquids) change? increases

35. **CQ** A student challenges the second law of thermodynamics by saying that entropy does not have to increase in all situations, such as when water freezes to ice. Is this challenge valid? Why or why not? no, ice or water itself is not an isolated system

36. **CQ** Is a living organism an open system or an isolated system? Explain. open system because it has to obtain energy outside itself

37. **IE ●** 1.0 kg of ice melts completely into liquid water at 0°C. (a) The change in entropy of the ice (water) in this process is (1) positive, (2) zero, (3) negative. Explain. (b) What *is* the change in entropy of the ice (water)? (a) (1) positive (b) $+1.2 \times 10^3$ J/K

38. **IE ●** A process involves 0.50 kg of steam condensing to water at 100°C. (a) The change in entropy of the steam (water) is (1) positive, (2) zero, (3) negative. Why? (b) What *is* the change in entropy of the steam (water)? (a) (3) negative (b) -3.0×10^3 J/K

39. **●** What is the change in entropy of mercury vapor ($L_v = 2.7 \times 10^5$ J/kg) when 0.50 kg of it condenses to a liquid at its boiling point of 357°C? -2.1×10^2 J/K

40. ●● In an isothermal expansion at 27°C, an ideal gas does 30 J of work. What is the change in entropy of the gas? 0.10 J/K

41. ●● During a liquid-to-solid phase change of a substance, its change in entropy is -4.19×10^3 J/K. If 1.67×10^6 J of heat is removed in the process, what is the freezing point of the substance in degrees Celsius?   126°C

42. IE ●● One mole of an ideal gas undergoes an isothermal compression at 20°C, and 7.5×10^3 J of work is done in compressing the gas. (a) Will the entropy of the gas (1) increase, (2) remain the same, or (3) decrease? Why? (b) What is the change in entropy of the gas? (a) (3) decreases, $Q < 0$ (b) -26 J/K

43. IE ●● A quantity of ideal gas undergoes a reversible isothermal expansion at 0°C and does 3.0×10^3 J of work on its surroundings in the process. (a) Will the entropy of the gas (1) increase, (2) remain the same, or (3) decrease? Explain. (b) What is the change in the entropy of the gas? (a) (1) increases, $Q > 0$ (b) $+11$ J/K

44. IE ●● An isolated system consists of two very large thermal reservoirs at constant temperatures of 373 K and 273 K. Assume 1000 J of heat were to flow from the cold reservoir to the hot reservoir spontaneously. (a) The total change in entropy of the isolated system (both reservoirs) would be (1) positive, (2) zero, (3) negative. Explain. (b) Calculate the total change in entropy of this isolated system. (a) (3) negative (b) -0.98 J/K

45. IE ●● Two heat reservoirs at temperatures 200°C and 60°C, respectively, are brought into thermal contact, and 1.50×10^3 J of heat spontaneously flows from one to the other with no significant temperature change. (a) The change in the entropy of the two-reservoir system is (1) positive, (2) zero, (3) negative. Explain. (b) Calculate the change in the entropy of the two-reservoir system. (a) (1) positive (b) $+1.33$ J/K

46. ●● In the winter, heat from a house with an inside temperature of 18°C leaks out at a rate of 2.0×10^4 J per second. The outside temperature is 0°C. (a) What is the change in entropy per second of the house? (b) What is the total change in entropy per second of the house–outside system? (a) -69 J/K (b) 4.5 J/K

47. IE ●●● A system goes from state 1 to state 3 as shown on the *T–S* diagram in ▼Fig. 12.26. (a) The heat transfer for the process going from state 2 to state 3 is (1) positive, (2) zero, (3) negative. Explain. (b) Calculate the total heat transferred in going from state 1 to state 3. (a) (2) zero, $\Delta S = 0$ (b) 2.73×10^4 J

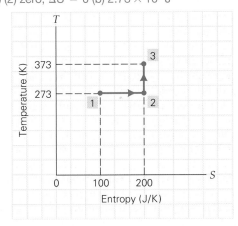

▲ **FIGURE 12.26 Entropy and heat** See Exercises 47 and 48.

48. IE ●●● Suppose that the system described by the *T–S* diagram in Fig. 12.26 is returned to its original state, state 1, by a reversible process depicted by a straight line from state 3 to state 1. (a) The change in entropy of system for this overall cyclic process is (1) positive, (2) zero, (3) negative. Explain. (b) How much heat is transferred in the cyclic process? [*Hint*: See Example 12.7.] (a) (2) zero, $\Delta S = 0$ (b) -5.0×10^3 J

49. ●●● A 50.0-g ice cube at 0°C is placed in 500 mL of water at 20°C. *Estimate* the change in entropy (after all the ice has melted) (a) for the ice, (b) for the water, and (c) for the ice–water system. (a) 61.0 J/K (b) -57.8 J/K (c) 3.2 J/K

12.5 Heat Engines and Thermal Pumps*

50. **MC** For a cyclic heat engine, (a) $\varepsilon > 1$, (b) $Q_h = W_{net}$, (c) $\Delta U = W_{net}$, (d) $Q_h > Q_c$. (d)

51. **MC** A thermal pump (a) is rated by thermal efficiency, (b) requires work input, (c) is not consistent with the second law of thermodynamics, (d) violates the first law of thermodynamics. (b)

52. **MC** Which of the following determines the thermal efficiency of a heat engine? (a) $Q_c \times Q_h$; (b) Q_c/Q_h; (c) $Q_h - Q_c$; (d) $Q_h + Q_c$. (b)

53. **CQ** What happens to the internal energy of a cyclic heat engine after a complete cycle? it is unchanged, because it returns to its original value

54. **CQ** Is leaving a refrigerator door open a practical way to air-condition a room? Explain. no, see ISM

55. **CQ** Lord Kelvin's statement of the second law of thermodynamics as applied to heat engines ("No heat engine operating in a cycle can convert its heat input completely to work") refers to their operation *in a cycle*. Why is the phrase "in a cycle" included? see ISM

56. **CQ** The heat output of a thermal pump is greater than the energy used to operate the pump. Is this device a violation of the first law of thermodynamics? no, see ISM

57. **CQ** In normal atmospheric convection cycles, colder air from a higher altitude is transferred to a lower, warmer level. Does this process violate the second law of thermodynamics? Explain. no, see ISM

58. ● A gasoline engine has a thermal efficiency of 28%. If the engine absorbs 2000 J of heat per cycle, (a) what is the net work output per cycle? (b) How much heat is exhausted per cycle? (a) 5.6×10^2 J (b) 1.4×10^3 J

59. ● If an engine does 200 J of net work and exhausts 600 J of heat per cycle, what is its thermal efficiency? 25%

60. ● A heat engine with a thermal efficiency of 20% does 800 J of net work each cycle. How much heat per cycle is lost to the surroundings (the low-temperature reservoir)? 3.2×10^3 J

61. ● An internal combustion engine with a thermal efficiency of 15.0% does 2.60×10^4 J of net work each cycle. How much heat is lost by the engine in each cycle? 1.47×10^5 J

*Consider efficiencies to be exact.

62. IE ● The heat output of a particular engine is 7.5×10^3 J per cycle, and the net work out is 4.0×10^3 J per cycle. (a) The heat input is (1) less than 4.0×10^3 J, (2) between 4.0×10^3 J and 7.5×10^3 J, (3) greater than 7.5×10^3 J. Explain. (b) What is the heat input and thermal efficiency of the engine?
(a) (3) greater than 7.5×10^3 J (b) 11.5×10^3 J, 35%

63. ●● A gasoline engine burns fuel that releases 3.3×10^8 J of heat per hour. (a) What is the energy input during a 2.0-h period? (b) If the engine delivers 25 kW of power during this time, what is its thermal efficiency?
(a) 6.6×10^8 J (b) 27%

64. IE ●● An engineer redesigns a heat engine and improves its thermal efficiency from 20% to 25%. (a) Does the ratio of the heat output to heat input (1) increase, (2) remain the same, or (3) decrease? Why? (b) What is the *change* in Q_c/Q_h in this example? (a) (3) decreases (b) -0.05

65. IE ●● A steam engine is to have its thermal efficiency improved from 8.00% to 10.0% while continuing to produce 4500 J of useful work each cycle. (a) Does the ratio of the heat input to heat output (1) increase, (2) remain the same, or (3) decrease? Explain. (b) What is its change in Q_h/Q_c?
(a) (1) increases (b) $+0.024$

66. ●● When running, a refrigerator removes heat from its cold interior at a rate of 7.5 kW when the required input work is done at a rate of 2.5 kW. At what rate is heat exhausted to the kitchen? 10 kW

67. ●● A refrigerator with a COP of 2.2 removes 4.2×10^5 J of heat from its interior each cycle. (a) How much heat is exhausted each cycle? (b) What is the total work input in joules for 10 cycles? (a) 6.1×10^5 J (b) 1.9×10^6 J

68. ●● A heat pump removes 2.0×10^3 J of heat from the outdoors and delivers 3.5×10^3 J of heat to the inside of a house each cycle. (a) How much work is required per cycle? (b) What is the COP of this pump? (a) 1.5×10^3 J (b) 2.3

69. ●● An air conditioner has a COP of 2.75. What is the power rating of the unit if it is to remove 1.00×10^7 J of heat from a house interior in 20 min? 3.0 kW

70. ●● A steam engine has a thermal efficiency of 30.0%. If its heat input for each cycle is supplied by the condensation of 8.00 kg of steam at 100°C, (a) what is the net work output per cycle, and (b) how much heat is lost to the surroundings in each cycle? (a) 5.42×10^6 J (b) 1.27×10^7 J

71. ●●● A gasoline engine has a thermal efficiency of 25.0%. If heat is expelled from the engine at a rate of 1.50×10^6 J per hour, how long does the engine take to perform a task that requires an amount of work of 3.0×10^6 J? 6.0 h

72. ●●● A coal-fired power plant produces 900 MW of electric power and operates at a thermal efficiency of 25%. (a) What is the rate of heat input to the plant? (b) What is the rate of heat discharge from the plant? (c) Water at 15°C from a nearby river is used to cool the discharged heat. If the cooling water is not to exceed a temperature of 40°C, how many gallons per minute of the cooling water is required? (a) 3.6×10^3 MW (b) 2.7×10^3 MW (c) 4.1×10^5 gal/min

73. ●●● A four-stroke engine runs on the Otto cycle. It delivers 150 hp at 3600 rpm. (a) How many cycles are in 1 min? (b) If the thermal efficiency of the engine is 20%, what is the heat input per minute? (c) How much heat is wasted (per minute) to the environment? (a) 1800 (b) 3.4×10^7 J (c) 2.7×10^7 J

12.6 The Carnot Cycle and Ideal Heat Engines

74. MC The Carnot cycle consists of (a) two isobaric and two isothermal processes, (b) two isometric and two adiabatic processes, (c) two adiabatic and two isothermal processes, (d) four arbitrary processes that return the system to its initial state. (c)

75. MC Which of the following temperature-reservoir relationships would yield the highest efficiency for a Carnot engine: (a) $T_c = 0.15T_h$, (b) $T_c = 0.25T_h$, (c) $T_c = 0.50T_h$, or (d) $T_c = 0.90T_h$? (a)

76. MC For a heat engine that operates between two reservoirs of temperatures T_c and T_h, the Carnot efficiency is the (a) highest possible value, (b) lowest possible value, (c) average value, (d) none of the preceding. (a)

77. CQ Automobile engines can be either air cooled or water cooled. Which type of engine would you expect to be more efficient, and why? water cooled, because efficiency depends on ΔT and water can maintain a large ΔT

78. CQ If you have the choice of running your heat engine between the following two sets of temperatures for the cold and hot reservoirs, which would you choose, and why: between 100°C and 300°C, or between 50°C and 250°C? 50°C and 250°C; higher efficiency

79. CQ Diesel engines are more efficient than gasoline engines. Which type of engine runs hotter? Why? diesel; see ISM

80. ● A steam engine operates between 100°C and 30°C. What is the Carnot efficiency of the ideal engine that operates between these temperatures? 19%

81. ● A Carnot engine has an efficiency of 35% and takes in heat from a high-temperature reservoir at 147°C. What is the Celsius temperature of the engine's low-temperature reservoir? 0°C

82. ● What is the temperature of the hot reservoir of a Carnot engine that is 30% efficient and has a 20°C cold reservoir? 146°C

83. ● It has been proposed that temperature differences in the ocean could be used to run a heat engine to generate electricity. In tropical regions, the water temperature is about 25°C at the surface and about 5°C at very large depths. (a) What would be the maximum theoretical efficiency of such an engine? (b) Would a heat engine with such a low efficiency be practical? Explain. (a) 6.7% (b) probably not; see ISM

84. ● An engineer wants to run a heat engine with an efficiency of 40% between a high-temperature reservoir at 350°C and a low-temperature reservoir. Below what temperature must the low-temperature reservoir be for practical operation of the engine? 101°C

85. ●● A Carnot engine takes 2.7×10^4 J of heat per cycle from a high-temperature reservoir at 320°C and exhausts some of it to a low-temperature reservoir at 120°C. How much net work is done by the engine per cycle? 9.1 × 10³ J

86. ●● A Carnot engine with an efficiency of 40% operates with a low-temperature reservoir at 50°C and exhausts 1200 J of heat each cycle. What are (a) the heat input per cycle and (b) the Celsius temperature of the high-temperature reservoir? (a) 2000 J (b) 265°C

87. IE ●● A Carnot engine takes in heat from a reservoir at 327°C and has an efficiency of 30%. The exhaust temperature is not changed and the efficiency is increased to 40%. (a) The temperature of the hot reservoir is (1) lower than (2) equal to, (3) higher than 327°C. Explain. (b) What is the new temperature of the hot reservoir?
(a) (3) higher than 327°C (b) 427°C

88. ●● An inventor claims to have developed a heat engine that, each cycle, takes in 5.0×10^5 J of heat from a high-temperature reservoir at 400°C and exhausts 2.0×10^5 J to the surroundings at 125°C. Would you invest your money in the production of this engine? Explain.
no, such an engine is not possible; see ISM

89. ●● An inventor claims to have created a heat engine that produces 10.0 kW of power for a 15.0-kW heat input while operating between reservoirs at 27°C and 427°C. (a) Is this claim valid? (b) To produce 10.0 kW of power, what is the minimum heat input required?
(a) no; ε_C = 57% while ε = 67% (b) 17.5 kW

90. ●● A heat engine operates at a thermal efficiency that is 45% of the Carnot efficiency. If the temperatures of the high-temperature and low-temperature reservoirs are 400°C and 100°C, respectively, what are the Carnot efficiency and the thermal efficiency of the engine?
45% and 20%

91. ●● A heat engine has half the thermal efficiency of a Carnot engine operating between temperatures of 100°C and 375°C. (a) What is the Carnot efficiency of the heat engine? (b) If the heat engine absorbs heat at a rate of 50 kW, at what rate is heat exhausted? (a) 42% (b) 39 kW

92. IE ●● Equation 12.15 shows that the greater the temperature difference between the reservoirs of a heat engine, the greater the engine's Carnot efficiency. Suppose you had the choice of raising the temperature of the high-temperature reservoir by a certain number of kelvins or lowering the temperature of the low-temperature reservoir by the same number of kelvins. (a) To produce the largest increase in efficiency, you should choose (1) to raise the high-temperature reservoir, (2) to lower the low-temperature reservoir, (3) both 1 and 2 produce the same change in efficiency, so it does not matter which you choose. Explain. (b) Prove your answer to part (a) numerically. (a) (2) to lower the low-temperature reservoir (b) see ISM

93. ●● The working substance of a cyclic heat engine is 0.75 kg of an ideal gas. The cycle consists of two isobaric processes and two isometric processes, as shown in ▼Fig. 12.27. What would be the efficiency of a Carnot engine operating with the same high temperature and low temperature reservoirs? 53%

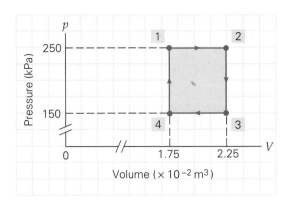

▲ **FIGURE 12.27 Thermal efficiency** See Exercise 93.

94. ●● In each cycle, a Carnot engine takes 800 J of heat from a high-temperature reservoir and discharges 600 J to a low-temperature reservoir. What is the ratio of the temperature of the high-temperature reservoir to that of the low-temperature reservoir? 1.3

95. IE●● A Carnot engine operating between reservoirs at 27°C and 227°C does 1500 J of work in each cycle. (a) The change in entropy for the engine for each cycle is (1) negative, (2) zero, (3) positive. Why? (b) What is the heat input of the engine?
(a) (2) zero; returns to original value (b) 3750 J

96. ●● The *autoignition temperature* of a fuel is defined as the temperature at which the fuel-air mixture would self-explode and ignite. Thus, it sets an upper limit on the temperature of the hot reservoir in a modern automobile engine. The autoignition temperatures for commonly available gasoline and diesel fuel are about 500°F and 600°F, respectively. What are the maximum Carnot efficiencies of a gasoline engine and a diesel engine if the cold-reservoir temperature is 27°C? 44% and 49%

97. ●● Because of limitations on materials, the maximum temperature of the superheated steam used in a turbine for the generation of electricity is about 540°C. (a) If the steam condenser operates at 20°C, what is the maximum Carnot efficiency of a steam turbine generator? (b) The actual efficiency of such generators is about 35 to 40%. What does this range tell you?
(a) 64% (b) see ISM

98. ●●● There is a Carnot coefficient of performance (COP_C) for an ideal, or Carnot, refrigerator. (a) Show that this quantity is given by

$$COP_C = \frac{T_c}{T_h - T_c}$$

(b) What does this tell you about adjusting the temperatures for maximum efficiency of a refrigerator? (Can you guess the equation for the COP_C for a heat pump?) see ISM

99. ●●● A salesperson tells you that a new refrigerator with a high COP removes 2.6×10^3 J (each cycle) from the inside of the refrigerator at 5.0°C and expels 2.8×10^3 J into the 30°C kitchen. (a) What is the refrigerator's COP? (b) Is this scenario possible? Justify your answer. (a) 13 (b) no, COP_C = 11

100. ●●● An ideal heat pump is equivalent to a Carnot engine running in reverse. (a) Show that the Carnot COP of the heat pump is

$$COP_C = \frac{1}{\varepsilon_C},$$

where ε_C is the Carnot efficiency of the heat engine. (b) If a Carnot engine has an efficiency of 40%, what would be the COP_C when it runs in reverse as a heat pump? (See Exercise 98.) (a) see ISM (b) 2.5

Comprehensive Exercises

101. When cruising at 75 mi/h on a highway, a car's engine develops 45 hp. If this engine has a thermodynamic efficiency of 25% and 1 gal of gasoline has an energy content of 1.3×10^8 J, what is the fuel efficiency (in miles per gallon) of this car? 20 mi/gal

102. A gram of water (volume of 1.00 cm^3) at 100°C is converted to 1.67×10^3 cm^3 of steam at atmospheric pressure. What is the change in the internal energy of the water (steam)? 2.09×10^3 J

103. In a highly competitive game, a basketball player can produce 300 W of power. Assuming the efficiency of the player's "engine" is 15% and the dissipated heat is primarily through the evaporation of sweat, what mass of sweat is evaporated per hour? 2.7 kg

104. A Carnot engine is to produce 400 J of work per cycle. If 600 J of heat is exhausted to a 27°C cold reservoir per cycle, what is the change in entropy of the hot reservoir per cycle? −2.0 J/K

105. A quantity of an ideal gas at an initial pressure of 2.00 atm undergoes an adiabatic expansion to atmospheric pressure. What is the ratio of the final temperature to the initial temperature of the gas? 0.157

106. A 100-MW power generating plant has an efficiency of 40%. If water is used to carry off the wasted heat and the temperature of the water is not to increase by more than 12 C°, what mass of water must flow through the plant each second? 3.0×10^3 kg

The following Physlet Physics Problems can be used with this chapter.

PHYSLET® 20.8, 20.9, 20.10, 20.11, 21.1, 21.2, 21.3, 21.4, 21.5, 21.6, 21.7, 21.8.

13

VIBRATIONS AND WAVES

PHYSICS FACTS

- Waves (of different types) can travel through solids, liquids, gases, and vacuum.

- Disturbances set up waves. Soldiers marching across older wooden bridges are told to break step and not march at a periodic cadence. This might correspond to a natural frequency of the bridge, resulting in resonance and large oscillations that could damage the bridge and even cause it to collapse.

- *Brain waves* are tiny oscillating electrical voltages in the brain. These are measured by attaching electrodes on the scalp hooked to an EEG (*electro*encephalo*graph*) to get a recording (*graph*) of electrical signals (*electro*) from the brain (*encephalo*). Electrical signals from the brain are displayed in the form of brain waves, the frequency of which depends on the brain's activity.

- *Tidal waves* are not related to tides. More appropriate is the Japanese name *tsunami*, which means "harbor wave." The effects of a tsunami are intensified in the confined spaces of bays and harbors. The waves are generated by subterranean earthquakes and can race across the ocean at speeds up to 960 km/h, with little surface evidence. When a tsunami reaches the shallow coast, friction slows the wave down, at the same time causing it to roll up into a 5- to 30-m-high wall of water that crashes down on the shoreline.

The photograph depicts what a lot of people probably first think of when they hear the word *wave*. We're all familiar with ocean waves or their smaller relatives, the ripples that form on the surface of a lake or pond when something disturbs the surface. Yet in many ways, the waves that are most important to us, as well as most interesting to physicists, either are invisible or don't look like waves. Sound, for example, is a wave. Perhaps most surprisingly, light is a wave. In fact, all electromagnetic radiations are waves—radio waves, microwaves, X-rays, and so on. Whenever you peer through a microscope, put on a pair of glasses, or look at a rainbow, you are experiencing wave energy in the form of light. In Chapter 28, you'll learn how even moving particles have wavelike properties. But first we need to look at the basic description of waves.

In general, waves are related to vibrations or oscillations—back-and-forth motion—such as that of a mass on a spring or a swinging pendulum, and fundamental to such motions are restoring forces or torques. In a material medium, the restoring force is provided by intermolecular forces. If a molecule is disturbed, restoring forces exerted by its neighbors tend to return the molecule to its original position, and it begins to oscillate. In so doing, it affects adjacent molecules, which are in turn set into oscillation. This is referred to as *propagation*. This raises the question, "What is propagated by the molecules in a material?" The answer is *energy*. A single disturbance, which happens when you give the end of a stretched rope a quick shake, gives rise to a *wave pulse*. A continuous, repetitive disturbance gives rise to a continuous propagation of energy that we call *wave motion*. But before looking at waves in media, it is helpful to analyze the oscillations of a single mass.

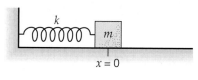

(a) Equilibrium

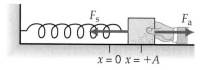

(b) $t = 0$ Just before release

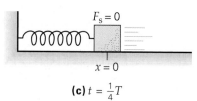

(c) $t = \frac{1}{4}T$

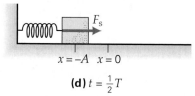

(d) $t = \frac{1}{2}T$

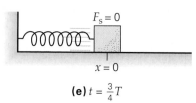

(e) $t = \frac{3}{4}T$

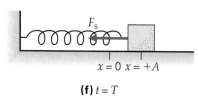

(f) $t = T$

▲ **FIGURE 13.1** Simple harmonic motion (SHM) When an object on a spring **(a)** is displaced from its equilibrium position, $x = 0$, and **(b)** is released, the object undergoes SHM (assuming no frictional losses). The time it takes to complete one cycle is the period of oscillation (T). (Here, F_s is the spring force and F_a is the applied force.) **(c)** At $t = T/4$, the object is back at its equilibrium position; **(d)** at $t = T/2$, it is at $x = -A$. **(e)** During the next half cycle, the motion is to the right; **(f)** at $t = T$, the object is back at its initial $(t = 0)$ starting position as in **(b)**.

13.1 Simple Harmonic Motion

OBJECTIVES: To (a) describe simple harmonic motion, and (b) describe how energy and speed vary in such motion.

The motion of an oscillating object depends on the restoring force that makes the object go back and forth. It is convenient to begin to study such motion by considering the simplest type of force acting along the x-axis: a force that is directly proportional to the object's displacement from equilibrium. A common example is the (ideal) spring force, described by **Hooke's law** (Section 5.2),

$$F_s = -kx \tag{13.1}$$

where k is the spring constant. Recall from Chapter 5 that the minus sign indicates that the force is always in the direction opposite that of the displacement. That is, the force always tends to *restore* the object to the spring's equilibrium position.

Suppose that an object on a horizontal frictionless surface is connected to a spring as shown in ◀Fig. 13.1. When the object is displaced to one side of its equilibrium position and released, it will move back and forth—that is, it will vibrate, or oscillate. Here, an oscillation or a vibration is clearly a *periodic motion*—a motion that repeats itself again and again along the same path. For linear oscillations, like those of an object attached to a spring, the path may be back and forth or up and down. For the angular oscillation of a pendulum, the path is back and forth along a circular arc.

Motion under the influence of the type of force described by Hooke's law is called **simple harmonic motion (SHM)**, because the force is the simplest restoring force and because the motion can be described by harmonic functions (sines and cosines), as you will see later in the chapter. The directed distance of an object in SHM from its equilibrium position is the object's **displacement**. Note in Fig. 13.1 that the displacement can be either positive or negative, which indicates direction. The maximum displacements are $+A$ and $-A$ (Fig. 13.1b, d). The magnitude of the maximum displacement, or the maximum distance of an object from its equilibrium position, is called the object's **amplitude (A)**, a scalar quantity that expresses the distance of both extreme displacements from the equilibrium position.

Besides the amplitude, two other important quantities used in describing an oscillation are its period and frequency. The **period (T)** is the time it takes the object to complete one cycle of motion. A cycle is a *complete* round trip, or motion through a complete oscillation. For example, if an object starts at $x = A$ (Fig. 13.1b), then when it returns to $x = A$ (as in Fig. 13.1f), it will have completed one cycle in a time we call one period. If an object were initially at $x = 0$ when disturbed, then its second return to this point would mark a cycle. (Why a *second* return?) In either case, the object would travel a distance of $4A$ during one cycle. Can you show this?

The **frequency (f)** is the number of cycles per second. The frequency and the period are related by

$$f = \frac{1}{T} \quad \text{frequency and period} \tag{13.2}$$

SI unit of frequency: hertz (Hz), or cycle per second (cycle/s)

The inverse relationship is reflected in the units. The period is the number of seconds per cycle, and the frequency is the number of cycles per second. For example, if $T = \frac{1}{2}$ s/cycle, then it completes 2 cycles each second or $f = 2$ cycles/s.

The standard unit of frequency is the **hertz (Hz)**, which is one cycle per second.* From Eq. 13.2, frequency has the unit inverse seconds (1/s, or s^{-1}),

*The unit is named for Heinrich Hertz (1857–1894), a German physicist and early investigator of electromagnetic waves.

TABLE 13.1	Terms Used to Describe Simple Harmonic Motion
displacement—the directed distance of an object ($\pm x$) from its equilibrium position.	
amplitude (A)—the magnitude of the maximum displacement, or the maximum distance, of an object from its equilibrium position.	
period (T)—the time for one complete cycle of motion.	
frequency (f)— the number of cycles per second (in hertz or inverse seconds, where $f = 1/T$).	

since the period is a measure of time. Although cycle is not really a unit, you might find it convenient at times to express frequency in cycles per second to help with unit analysis. This is similar to the way the radian (rad) is used in the description of circular motion in Sections 7.1 and 7.2.

The preceding terms used to describe SHM are summarized in Table 13.1.

Energy and Speed of a Mass–Spring System in SHM

Recall from Chapter 5 that the potential energy stored in a spring that is stretched or compressed a distance $\pm x$ from equilibrium (chosen to be $x = 0$) is

$$U = \tfrac{1}{2}kx^2 \tag{13.3}$$

The *change* in potential energy of an object oscillating on a spring is related to the work done by the spring force. An object with mass m oscillating on a spring also has kinetic energy. The kinetic and potential energies together give the total mechanical energy E of the system:

$$E = K + U = \tfrac{1}{2}mv^2 + \tfrac{1}{2}kx^2 \tag{13.4}$$

When the object is at one of its maximum displacements, $+A$ or $-A$, it is instantaneously at rest, $v = 0$ (▼Fig. 13.2). Thus, all the energy is in the form of potential energy (U_{max}) at this location; that is,

$$E = \tfrac{1}{2}m(0)^2 + \tfrac{1}{2}k(\pm A)^2 = \tfrac{1}{2}kA^2$$

or

$$E = \tfrac{1}{2}kA^2 \qquad \text{\textit{total energy of an object} \atop \textit{in SHM on a spring}} \tag{13.5}$$

This outcome is a general result for SHM:

The total energy of an object in simple harmonic motion is directly proportional to the square of the amplitude.

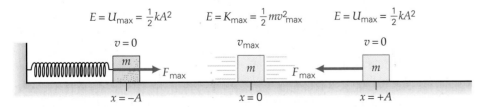

▲ **FIGURE 13.2 Oscillations and energy** For a mass oscillating in SHM on a spring (on a frictionless surface), the total energy at the amplitude positions ($\pm A$) is all potential energy (U_{max}), and $E = \tfrac{1}{2}kx^2 = \tfrac{1}{2}kA^2$, which is the total energy of the system. At the center position ($x = 0$), the total energy is all kinetic energy ($E = \tfrac{1}{2}mv_{max}^2$, where m is the mass of the block). How is the total energy divided at locations somewhere between $x = 0$ and $x = \pm A$?

PHYSLET®

Illustration 16.3 Energy and Simple Harmonic Motion

Note: This discussion will be limited to light springs, the mass of which can be considered negligible.

Equation 13.5 allows us to express the velocity of an object oscillating on a spring as a function of position:

$$E = K + U \qquad \text{or} \qquad \tfrac{1}{2}kA^2 = \tfrac{1}{2}mv^2 + \tfrac{1}{2}kx^2$$

Solving for v^2 and taking the square root:

$$v = \pm\sqrt{\frac{k}{m}(A^2 - x^2)} \quad \textit{velocity of an object in SHM} \qquad (13.6)$$

where the positive and negative signs indicate the direction of the velocity. Note that at $x = \pm A$ the velocity is zero, since the object is instantaneously at rest at its maximum displacement from equilibrium.

Note also that when the oscillating object passes through its equilibrium position ($x = 0$), its potential energy is zero. At that instant, the energy is all kinetic, and the object is traveling at its maximum speed v_{max}. The expression for the energy in this case is

$$E = \tfrac{1}{2}kA^2 = \tfrac{1}{2}mv_{max}^2$$

and,

$$v_{max} = \sqrt{\frac{k}{m}}\,A \quad \begin{array}{l}\textit{maximum speed}\\ \textit{of mass on a spring}\end{array} \qquad (13.7)$$

In the next Example, as well as in the accompanying Learn by Drawing, you can visualize the continuous trade-off between kinetic and potential energy.

Example 13.1 ■ A Block and a Spring: Simple Harmonic Motion

A block with a mass of 0.25 kg sitting on a frictionless surface is connected to a light spring that has a spring constant of 180 N/m (see Fig. 13.1). If the block is displaced 15 cm from its equilibrium position and released, what are (a) the total energy of the system and (b) the speed of the block when it is 10 cm from its equilibrium position?

Thinking It Through. The total energy depends on the spring constant (k) and the amplitude (A), which are given. At $x = 10$ cm, the speed should be less than the maximum speed. (Why?)

Solution. First we list the given data, as usual, and what is to be found. The initial displacement is the amplitude. (Why?)

Given: $m = 0.25$ kg
  $k = 180$ N/m
  $A = 15$ cm $= 0.15$ m
  $x = 10$ cm $= 0.10$ m

Find: (a) E (total energy)
  (b) v (speed)

(a) The total energy is given by Eq. 13.5:

$$E = \tfrac{1}{2}kA^2 = \tfrac{1}{2}(180 \text{ N/m})(0.15 \text{ m})^2 = 2.0 \text{ J}$$

(b) The instantaneous speed of the block at a distance 10 cm from the equilibrium position is given by Eq. 13.6 without directional signs:

$$v = \sqrt{\frac{k}{m}(A^2 - x^2)} = \sqrt{\frac{180 \text{ N/m}}{0.25 \text{ kg}}[(0.15 \text{ m})^2 - (0.10 \text{ m})^2]} = \sqrt{9.0 \text{ m}^2/\text{s}^2} = 3.0 \text{ m/s}$$

What would the speed be at $x = -10$ cm?

Follow-Up Exercise. In part (b) of this Example, the block at $x = 10$ cm is at two thirds, or 67%, of its maximum displacement. Is its speed at that position therefore 67% of its maximum speed? Justify your answer mathematically. (*Answers to all Follow-Up Exercises are at the back of the text.*)

LEARN BY DRAWING OSCILLATING IN A PARABOLIC POTENTIAL WELL

A way to visualize the conservation of energy in simple harmonic motion is shown in Fig. 1. The potential energy of a mass–spring system can be sketched on a plot of energy (E) versus position (x). Since $U = \frac{1}{2}kx^2 \propto x^2$, the graph is a *parabola*.

In the absence of nonconservative forces, the total energy of the system, E, is constant. But E is the sum of the kinetic and potential energies. During the oscillations, there is a continuous trade-off between the two types of energies, but their sum remains constant. Mathematically, this relationship is written as $E = K + U$. In Fig. 2, U (indicated by a blue arrow) is represented by the vertical distance from the x-axis.

Since E is constant and independent of x, it is plotted as a horizontal line (shown in green). The kinetic energy is the part of the total energy that is *not* potential energy; that is, $K = E - U$. It can be graphically interpreted (purple arrow) as the vertical distance between the potential-energy parabola and the horizontal green total-energy line. As the object oscillates on the x-axis, the energy trade-offs can be visualized as the lengths of the two arrows change.

A general location, x_1, is shown in Fig. 2. Neither the kinetic energy nor the potential energy is at its maximum value of E there. These maximum values occur instead at $x = 0$ and $x = \pm A$, respectively. The motion cannot exceed $x = \pm A$, because that would imply a negative kinetic energy, which is physically impossible. (Why?) The amplitude positions are sometimes called the *endpoints* of the motion, because they are the locations where the speed is instantaneously zero and the object reverses direction.

Try using the graphical approach to answer the following questions (and make up some of your own): What do you have to do to E to increase the amplitude of oscillation, and how might you do this? What happens to the amplitude of a real-life system in SHM in the presence of a force such as friction when E decays with time?

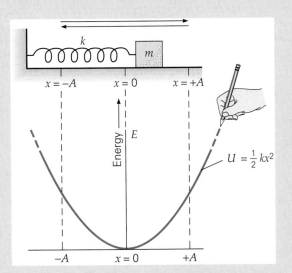

FIGURE 1 The potential-energy "well" of a spring–mass system The potential energy of a spring that is stretched or compressed from its equilibrium position ($x = 0$) is a parabola, since $U \propto x^2$. At $x = \pm A$, all of the system's energy is potential.

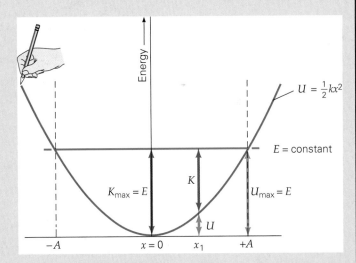

FIGURE 2 Energy transfers as the spring–mass system oscillates The vertical distance from the x-axis to the parabola is the system's potential energy. The remainder—the vertical distance between the parabola and the horizontal line representing the system's constant total energy E—is the system's kinetic energy (K).

The spring constant is commonly determined by placing an object of known mass on the end of a spring and letting it settle vertically to a new equilibrium position. The next Example shows some typical results.

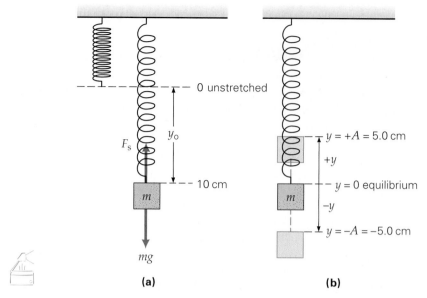

▲ **FIGURE 13.3 Determination of the spring constant** (a) When an object suspended on a spring is in equilibrium, the two forces on the object cancel, so that $F_s = w$, or $ky_o = mg$. Thus, the spring constant k can be computed: $k = mg/y_o$. (b) The zero reference point of an object in SHM and suspended on a spring is conveniently taken to be the new equilibrium position, as the motion is symmetric about that point. (See Example 13.2.)

Example 13.2 ■ The Spring Constant: Experimental Determination

When a 0.50-kg mass is suspended from a spring, the spring stretches a distance of 10 cm to a new equilibrium position (▲Fig. 13.3a). (a) What is the spring constant of the spring? (b) The mass is then pulled down another 5.0 cm and released. What is the highest position of the oscillating mass?

Thinking It Through. At the equilibrium position, the net force on the mass is zero because $a = 0$. In part (b), negative y will be used to designate "downward," as is customary in problems involving vertical motion.

Solution.

Given: $m = 0.50$ kg   *Find:* (a) k (spring constant)
$\quad\quad\quad y_o = 10$ cm $= 0.10$ m   (b) A (amplitude)
$\quad\quad\quad y = -5.0$ cm $= -0.050$ m
$\quad\quad\quad$ (new reference point)

(a) When the suspended mass is in equilibrium (Fig. 13.3a), the net force on the mass is zero. Thus, the weight of the mass and the spring force are equal and opposite. Then, equating their magnitudes,

$$F_s = w$$

or

$$ky_o = mg$$

Hence,

$$k = \frac{mg}{y_o} = \frac{(0.50 \text{ kg})(9.8 \text{ m/s}^2)}{0.10 \text{ m}} = 49 \text{ N/m}$$

(b) Once set into motion, the mass oscillates up and down through the equilibrium position. Since the motion is symmetric about this point, it is convenient to designate it as the zero reference point of the oscillation (Fig. 13.3b). The initial displacement is $-A$, so the highest position of the mass will be 5.0 cm above the equilibrium position $(+A)$.

Follow-Up Exercise. How much more potential energy does the spring in this Example have at the bottom of its oscillation than at the top?

13.2 Equations of Motion

OBJECTIVES: To (a) understand the equation of SHM, and (b) explain what is meant
by phase and phase differences.

The equation that gives the object's position as a function of time is referred to as the
equation of motion. For example, the equation of motion with a constant linear
acceleration is $x = x_o + v_o t + \frac{1}{2}at^2$, where v_o is the initial velocity (Chapter 2). However, the acceleration is not constant in simple harmonic motion, so the kinematic
equations of Chapter 2 do not apply to this case.

The equation of motion for an object in SHM can be found from a relationship
between simple harmonic and uniform circular motions. SHM can be simulated
by a component of uniform circular motion, as illustrated in ▼Fig. 13.4. As the illuminated object moves in uniform circular motion (with constant angular
speed ω) in a vertical plane, its shadow moves back and forth vertically, following the same path as the object on the spring, which is in simple harmonic motion.
Since the shadow and the object have the same position at any time, it follows that
the equation of motion for the shadow of the object in circular motion is the same
as the equation of motion for the oscillating object on the spring.

From the reference circle in Fig. 13.4b, the y-coordinate (position) of the object
is given by

$$y = A \sin \theta$$

But the object moves with a constant angular velocity of magnitude ω. In terms of
the angular distance θ, assuming that $\theta = 0°$ at $t = 0$, we have $\theta = \omega t$, so

$$y = A \sin \omega t \quad \begin{array}{l}(SHM \ for \ y_o = 0, \\ initial \ upward \ motion)\end{array} \qquad (13.8)$$

Note that as t increases from zero, y increases in the positive direction, so the
equation describes initial upward motion.

With Eq. 13.8 as the equation of motion, the mass *must* always be initially at
$y_o = 0$. But what if the mass on the spring were initially at the amplitude position
$+A$? In that case, the sine equation would not describe the motion, because it does

Teaching tip: Have students move an
index finger in a horizontal circle in front
of one eye and then close the other eye.
When you lack depth perception, the
motion appears oscillatory!

Illustration 16.1 Representations of
Simple Harmonic Motion

▼ **FIGURE 13.4 Reference circle for vertical motion** **(a)** The shadow of an object in
uniform circular motion has the same vertical motion as an object oscillating on a spring
in simple harmonic motion. **(b)** The motion can be described by $y = A \sin \theta = A \sin \omega t$
(assuming that $y = 0$ at $t = 0$).

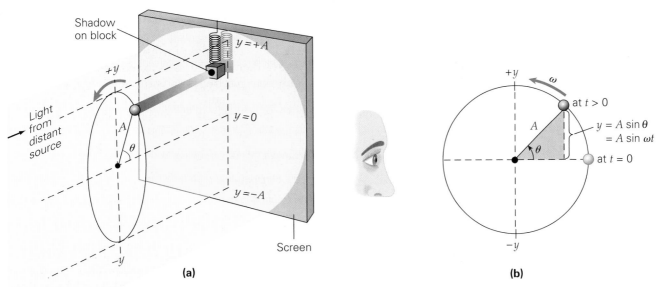

▶ **FIGURE 13.5** Sinusoidal equation of motion As time passes, the oscillating object traces out a sinusoidal curve on the moving paper. In this case, $y = A \cos \omega t$, because the object's initial displacement is $y_o = +A$.

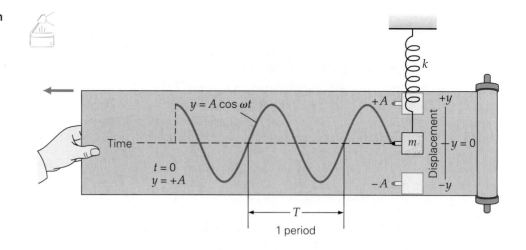

not describe the *initial condition*—that is, $y_o = +A$ at $t_o = 0$. So another equation of motion is needed, and $y = A \cos \omega t$ applies. By this equation, at $t_o = 0$, the mass is at $y_o = A \cos \omega t = A \cos \omega(0) = +A$, and the cosine equation correctly describes the initial conditions (▲ Fig. 13.5):

$$y = A \cos \omega t \qquad \text{(initial downward motion with } y_o = +A) \qquad (13.9)$$

Here, the initial motion is downward, because, for times shortly after $t_o = 0$, the value of y decreases. If the amplitude were $-A$, the mass would initially be at the bottom and the initial motion would be upward.

Thus, the equation of motion for an oscillating object may be either a sine or a cosine function. Both of these functions are referred to as being *sinusoidal*. That is, simple harmonic motion is described by a sinusoidal function of time.

The angular speed ω (in rad/s) of the *reference circle object* (Fig. 13.4) is called the *angular frequency* of the oscillating object, since $\omega = 2\pi f$, where f is the frequency of revolution or rotation of the object (Section 7.2). Figure 13.4 shows that the frequency of the "orbiting" object is the same as the frequency of oscillation of the object on the spring. Thus, using $f = 1/T$, we can write Eq. 13.8 as

$$y = A \sin(2\pi f t) = A \sin\left(\frac{2\pi t}{T}\right) \qquad \text{(SHM for } y_o = 0, \text{ initial upward motion)} \qquad (13.10)$$

Note that this equation is for initial upward motion, because after $t_o = 0$, the value of y increases positively. For initial downward motion, the amplitude term would be $-A$.

Equations 13.8 and 13.10 give three equivalent forms of the equation of motion for an object in SHM. Any one of them can be used for convenience, depending on the known parameters. For example, suppose you are given the time t in terms of the period T—say, $t_o = 0$, $t_1 = T/4$, and $t_2 = 3T/4$—and are asked to find the position of an object in SHM at these times. In this case, it is convenient to use Eq. 13.10, and

$$t_o = 0 \qquad y_o = A \sin[2\pi(0)/T] = A \sin 0 = 0$$

$$t_1 = \frac{T}{4} \qquad y_1 = A \sin[2\pi(T/4)/T] = A \sin \pi/2 = A$$

$$t_2 = \frac{3T}{4} \qquad y_2 = A \sin[2\pi(3T/4)/T] = A \sin 3\pi/2 = -A$$

The results tell us that the object was initially at $y = 0$ (equilibrium), as we knew. One quarter of a period later, it was at $y = A$, or the amplitude of its oscillation; and at a time of three quarters of a period ($3T/4$), it was at the $-A$ position, which is to be expected, since the motion is periodic. (Where would the object be at $T/2$ and T?)

Hence, we may write in general,

$$y = \pm A \sin \omega t = \pm A \sin(2\pi f t) = \pm A \sin\left(\frac{2\pi t}{T}\right) \quad \begin{array}{l}(+ \text{ for initial motion} \\ \text{upward with } y_o = 0; \\ - \text{ for initial motion} \\ \text{downward with } y_o = 0)\end{array} \quad (13.8a)$$

By a similar development, Eq. 13.9 has the general form

$$y = \pm A \cos \omega t = \pm A \cos(2\pi f t) = \pm A \cos\left(\frac{2\pi t}{T}\right) \quad \begin{array}{l}(+ \text{ for initial motion} \\ \text{downward with } y_o = +A; \\ - \text{ for initial motion} \\ \text{upward with } y_o = -A)\end{array} \quad (13.9b)$$

To show the convenience of the reference circle, let us use it to compute the period of the spring–object system. Note that the time for the object in the reference circle to make one complete "orbit" is exactly the time it takes for the oscillating object to make one complete cycle. (See Fig. 13.4.) Thus, all we need is the time for one orbit around the reference circle, and we have the period of oscillation. Because the object "orbiting" the reference circle is in uniform circular motion at a constant speed equal to the maximum speed of oscillation v_{max}, the object travels a distance of one circumference in one period. Then $t = d/v$, where $t = T$, the circumference is d, and v is v_{max} given by Equation 13.7; that is,

$$T = \frac{d}{v_{max}} = \frac{2\pi A}{\sqrt{k/m}\, A}$$

or

$$T = 2\pi\sqrt{\frac{m}{k}} \quad \begin{array}{l}\textit{period of object} \\ \textit{oscillating on a spring}\end{array} \quad (13.11)$$

Because the amplitudes canceled out in Eq. 13.11, *the period (and frequency) are independent of the amplitude of the motion.* This statement is a general feature of simple harmonic oscillators—that is, oscillators driven by a linear restoring force, such as a spring obeying Hooke's law.

We see from Eq. 13.11 that the greater the mass, the longer the period, and the greater the spring constant (or the stiffer the spring), the shorter the period. It is the *ratio* of mass to stiffness that determines the period. Thus, you can offset an increase in mass by using a stiffer spring.

Since $f = 1/T$,

$$f = \frac{1}{2\pi}\sqrt{\frac{k}{m}} \quad \begin{array}{l}\textit{frequency of mass} \\ \textit{oscillating on a spring}\end{array} \quad (13.12)$$

Thus, the greater the spring constant (the stiffer the spring), the more frequently the system vibrates, as you might expect.

Also, note that since $\omega = 2\pi f$, we may write

$$\omega = \sqrt{\frac{k}{m}} \quad \begin{array}{l}\textit{angular frequency of mass} \\ \textit{oscillating on a spring}\end{array} \quad (13.13)$$

As another example, a simple pendulum (a small, heavy object on a string) will undergo simple harmonic motion for small angles of oscillation. The period of a simple pendulum oscillating through a small angle $\theta \lesssim 10°$ is given, to a good approximation, by

$$T = 2\pi\sqrt{\frac{L}{g}} \quad \textit{period of a simple pendulum} \quad (13.14)$$

where L is the length of the pendulum and g is the acceleration due to gravity. A pendulum-driven clock that is not properly rewound and is running down would still keep correct time, because the period would remain unchanged as the amplitude decreased. As shown by Eq. 13.14, the period is independent of amplitude.

An important difference between the period of the mass–spring system and that of the pendulum is that the latter is independent of the mass of the bob. (See Eq. 13.11 and Eq. 13.14.) Can you explain why? Think about what supplies the restoring force for the pendulum's oscillations. It is gravity. Hence, the acceleration (along with the velocity

Demonstration/activity: The deflection of most materials is approximately linearly proportional to the force for small deflections. Clamp one end of a meterstick to the desk and put weights of 20 g, 40 g, etc., on the end of the stick to observe the linear deflection. Measure the "spring constant" of the apparatus. Then tape a larger weight of about 500 g to the end of the stick, and start it oscillating up and down. Measure the meterstick's period and compare it with the calculated period. Are the periods close?

Note: Period and frequency are independent of amplitude in SHM.

PHYSLET®

Illustration 16.2 The Simple Pendulum and Spring Motion

PHYSLET®

Exploration 16.1 Spring and Pendulum Motion

Demonstration/activity: Set up several pendulums to show how the period is independent of the mass and amplitude for small amplitudes.

and period) is expected to be independent of mass. That is, the gravitational force automatically provides the same acceleration to different bob masses on pendulums with the same length. We have seen that similar effects occur in free fall (Chapter 2) and with blocks sliding and cylinders rolling down inclines (Chapters 4 and 8, respectively). The next Example demonstrates the usage of the equation of motion for SHM.

Example 13.3 ■ An Oscillating Mass: Applying the Equation of Motion

A mass on a spring oscillates vertically with an amplitude of 15 cm, a frequency of 0.20 Hz, and an equation of motion given by Eq. 13.8, with $y_o = 0$ at $t_o = 0$ and initial upward motion. (a) What are the position and direction of motion of the mass at $t = 3.1$ s? (b) How many oscillations (cycles) does the mass make in a time of 12 s?

Thinking It Through. Part (a) is a straightforward application of Eq. 13.8. In part (b), the number of oscillations means the number of cycles, and recall that frequency is sometimes expressed in cycles per second. Hence, multiplying the frequency by the time would give the number of cycles or oscillations.

Solution.

Given: $A = 15$ cm $= 0.15$ m
$f = 0.20$ Hz
$y = A \sin \omega t$ (Eq. 13.8)
(a) $t = 3.1$ s (b) $t = 12$ s

Find: (a) y (position and direction of motion)
(b) n (number of oscillations or cycles)

(a) First, since the frequency f is given, it is convenient to use the equation of motion in the form $y = A \sin 2\pi f t$ (Eq. 13.10). As can be seen from the equation, at $t_o = 0$, $y_o = 0$, so initially the mass is at the zero (equilibrium) position. Then, at $t = 3.1$ s,

$$y = A \sin 2\pi f t$$
$$= (0.15 \text{ m}) \sin[2\pi(0.20 \text{ s}^{-1})(3.1 \text{ s})]$$
$$= (0.15 \text{ m}) \sin(3.9 \text{ rad}) = -0.10 \text{ m}$$

So the mass is at $y = -0.10$ m at $t = 3.1$ s. But what is its direction of motion? Let's look at the period (T) and see what part of its cycle the mass is in. By Eq. 13.2,

$$T = \frac{1}{f} = \frac{1}{0.20 \text{ Hz}} = 5.0 \text{ s}$$

In $t = 3.1$ s, the mass has gone through 3.1 s/5.0 s = 0.62, or 62%, of a period or cycle, so it is moving downward. [The motion is up ($\frac{1}{4}$cycle) and back ($\frac{1}{4}$ cycle) to $y_o = 0$ in $\frac{1}{2}$, or 50%, of the cycle, and therefore downward during the next $\frac{1}{4}$ cycle].

(b) The number of oscillations (cycles) is equal to the product of the frequency (cycles/s) and the elapsed time (s), both of which are given:

$$n = ft = (0.20 \text{ cycles/s})(12 \text{ s}) = 2.4 \text{ cycles}$$

or with $f = 1/T$,

$$n = \frac{t}{T} = \frac{12 \text{ s}}{5.0 \text{ s}} = 2.4 \text{ cycles}$$

(Note that *cycle* is not a unit and is used only for convenience.)

Thus, the mass has gone through two complete cycles and 0.4 of another, which means that it is on its way back to $y_o = 0$ from its amplitude position of $+A$. (Why?)

Follow-Up Exercise. Find what is asked for in this Example at times (1) $t = 4.5$ s and (2) $t = 7.5$ s.

Problem-Solving Hint

Note that in the calculation in part (a) of Example 13.3, where we have sin 3.9, the angle is in radians, *not* degrees. Don't forget to set your calculator to radians (rather than degrees) when finding the value of a trigonometric function in equations for simple harmonic or circular motion.

Example 13.4 ■ Fun with a Pendulum: Frequency and Period

A helpful older brother takes his sister to play on the swings in the park. He pushes her from behind on each return. Assuming that the swing behaves as a simple pendulum with a length of 2.50 m, (a) what would be the frequency of the oscillations, and (b) what would be the interval between the brother's pushes?

Thinking It Through. (a) The period is given by Eq. 13.14, and the frequency and period are inversely related: $f = 1/T$. (b) Since the brother pushes from one side on each return, he must push once every cycle that is completed, so the time between his pushes is equal to the swing's period.

Solution.

Given: $L = 2.50$ m   *Find:* (a) f (frequency)
                                        (b) T (period)

(a) We can take the reciprocal of Eq. 13.14 to solve directly for the frequency:

$$f = \frac{1}{T} = \frac{1}{2\pi}\sqrt{\frac{g}{L}} = \frac{1}{2\pi}\sqrt{\frac{9.80 \text{ m/s}^2}{2.50 \text{ m}}} = 0.315 \text{ Hz}$$

(b) The period is then found from the frequency:

$$T = \frac{1}{f} = \frac{1}{0.315 \text{ Hz}} = 3.17 \text{ s}$$

The brother must push every 3.17 s to maintain a steady swing (and to keep his sister from complaining).

Follow-Up Exercise. In this Example, the older brother, a physics buff, carefully measures the period of the swing to be 3.18 s, not 3.17 s. If the length of 2.50 m is accurate, what is the acceleration due to gravity at the location of the park? Considering this accurate value of g, do you think the park is at sea level?

Initial Conditions and Phase

You may be wondering how to decide whether to use a sine or cosine function to describe a particular case of simple harmonic motion. In general, the form of the function is determined by the initial displacement and velocity of the object: the *initial conditions* of the system. These initial conditions are the values of the displacement and velocity at $t = 0$; taken together, they tell how the system is initially set into motion.

Let's look at four special cases. If an object in vertical SHM has an initial displacement of $y = 0$ at $t = 0$ and moves initially upward, the equation of motion is $y = A \sin \omega t$ (▼Fig. 13.6a). Note that $y = A \cos \omega t$ does not satisfy the initial condition, because $y_o = A \cos \omega t = A \cos \omega(0) = A$, since $\cos 0 = 1$.

Suppose that the object is initially released ($t = 0$) from its positive amplitude position ($+A$), as in the case of the object on a spring shown in Fig. 13.5. Here, the equation of motion is $y = A \cos \omega t$ (Fig. 13.6b). This expression satisfies the initial condition: $y_o = A \cos \omega(0) = A$.

The other two cases are (1) $y = 0$ at $t = 0$, with motion initially downward (for an object on a spring) or in the negative direction (for horizontal SHM), and (2) $y = -A$ at $t = 0$, meaning that the object is initially at its negative-amplitude position. These motions are described by $y = -A \sin \omega t$ and $y = -A \cos \omega t$, respectively, as illustrated in Figs. 13.6c and d.

Only these four initial conditions will be considered in our study. Should y_o have a value other than 0 or $\pm A$, the equation of motion is somewhat complicated. Note in Fig. 13.6 that if the curves are extended in the negative direction of the horizontal axis (dashed purple lines), they all have the same shape, but have been "shifted," so to speak. In (a) and (b), one curve is ahead of the other by 90°, or $\frac{1}{4}$ cycle. That is, the two curves are shifted by a quarter cycle with respect to one another. The oscillations are then said to have a *phase difference* of 90°. In (a) and (c), the curves are shifted 180° and are 180° out of phase. (Note in this case that the oscillations are opposite: When one mass is going up, the other is going down.) What about the oscillations in (a) and (d)?

Note: Initial conditions include both the displacement y_o and velocity v_o at $t = 0$.

▶ **FIGURE 13.6 Initial conditions and equations of motion** The initial conditions (y_0 and t_0) determine the form of the equation of motion—for the cases shown here, either a sine or a cosine. For $t_0 = 0$, the initial displacements are **(a)** $y_0 = 0$, **(b)** $y_0 = +A$, **(c)** $y_0 = 0$, and **(d)** $y_0 = -A$. The equations of motion must match the initial conditions. (See text for description.)

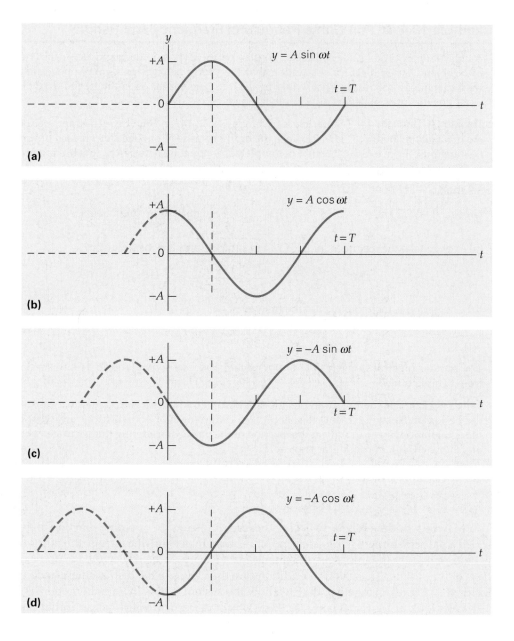

A figure with a 360° (or 0°) phase shift is not shown, because this would be the same as that in (a). When two objects in SHM have the same equation of motion, they are said to be oscillating *in phase*, which means that they are oscillating together with identical motions. Objects with a 180° phase shift or difference are said to be *completely out of phase* and will always be going in opposite directions and be at opposite amplitudes at the same time.

Velocity and Acceleration in SHM

Expressions for the velocity and acceleration of an object in SHM can also be obtained. Using calculus, one can show that $v = \Delta y / \Delta t = \Delta(A \sin \omega t)/\Delta t$ in the limit as Δt goes to zero gives the following expression for the instantaneous velocity:

Note: Maximum speed $v = \omega A$.

$$v = \omega A \cos \omega t \qquad \begin{array}{l} \textit{(vertical velocity if } v_0 \textit{ is upward} \\ t_0 = 0, y_0 = 0) \end{array} \qquad (13.15)$$

Here, the signs indicating direction are given by the cosine function.

The acceleration can be found by using Newton's second law with the spring force $F_s = -ky$:

$$a = \frac{F_s}{m} = \frac{-ky}{m} = -\frac{k}{m}A \sin \omega t$$

Since $\omega = \sqrt{k/m}$,

$$a = -\omega^2 A \sin \omega t = -\omega^2 y \quad \begin{array}{l} \text{(vertical acceleration if } v_o \text{ is} \\ \text{upward at } t_o = 0, \, y_o = 0) \end{array} \quad (13.16)$$

Note that the functions for the velocity and acceleration are out of phase with that for the displacement. Since the velocity is 90° out of phase with the displacement, the speed is greatest when $\cos \omega t = \pm 1$ at $y = 0$, that is, when the oscillating object is passing through its equilibrium position. The acceleration is 180° out of phase with the displacement (as indicated by the minus sign on the right-hand side of Eq. 13.16). Therefore, the magnitude of the acceleration is a maximum when $\sin \omega t = \pm 1$ at $y = \pm A$, that is, when the displacement is a maximum, or when the object is at an amplitude position. At any position except the equilibrium position, the directional sign of the acceleration is opposite that of the displacement, as it should be for an acceleration resulting from a restoring force. At the equilibrium position, both the displacement and acceleration are zero. (Can you see why?)

Note also that the acceleration in SHM is not constant with time. Hence, the kinematic equations for acceleration (Chapter 2) *cannot* be used, since they describe constant acceleration.

Note: Maximum acceleration magnitude $a = \omega^2 A$.

Damped Harmonic Motion

Simple harmonic motion with constant amplitude implies that there are no losses of energy, but in practical applications there are always some frictional losses. Therefore, to maintain a constant amplitude motion, energy must be added to a system by some external driving force, such as someone pushing a swing. Without a driving force, the amplitude and the energy of an oscillator decrease with time, giving rise to **damped harmonic motion** (▼Fig. 13.7a). The time required for the

▼ **FIGURE 13.7 Damped harmonic motion** **(a)** When a driving force adds energy to a system in an amount equal to the energy losses of the system, the oscillation is steady with a constant amplitude. When the driving force is removed, the oscillations decay (that is, they are damped), and the amplitude decreases nonlinearly with time. **(b)** In some applications, damping is desirable and even promoted, as with shock absorbers in automobile suspension systems. Otherwise, the passengers would be in for a bouncy ride.

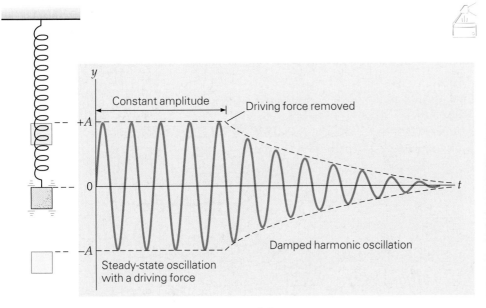

Constant amplitude · Driving force removed

Steady-state oscillation with a driving force

Damped harmonic oscillation

(a)

(b)

▲ **FIGURE 13.8** Energy transfer The propagation of a disturbance, or a transfer of energy through space, is seen in a row of falling dominoes.

▶ **FIGURE 13.9 Wave pulse** The hand disturbs the stretched rope in a quick up-and-down motion, and a wave pulse propagates along the rope. (The red arrows represent the velocities of the hand and of pieces of the rope at different times and locations.) The rope "particles" move up and down as the pulse passes. The energy in the pulse is thus *both* kinetic (elastic) and potential (gravitational).

oscillations to cease, or damp out, depends on the magnitude and type of the damping force (such as air resistance).

In many applications involving continuous periodic motion, damping is unwanted and necessitates an energy input. However, in some instances, damping is desirable. For example, the dial in a spring-operated bathroom scale oscillates briefly before stopping at a weight reading. If not properly damped, these oscillations would continue for some time, and you would have to wait before you could read your weight. Shock absorbers provide damping in the suspension systems of automobiles (Fig. 13.7b; also see Fig. 9.9b). Without "shocks" to dissipate energy after hitting a bump, the ride would be bouncy. In California, many new buildings incorporate damping mechanisms (giant shock absorbers) to dampen their oscillatory motion after they are set in motion by earthquake waves.

13.3 Wave Motion

OBJECTIVES: To (a) describe wave motion in terms of various parameters, and (b) identify different types of waves.

The world is full of waves of various types; some examples are water waves, sound waves, waves generated by earthquakes, and light waves. All waves result from a disturbance, the source of the wave. In this chapter, we will be concerned with mechanical waves—those that are propagated in some medium. (Light waves, which do not require a propagating medium, will be considered in more detail in later chapters.)

When a medium is disturbed, energy is imparted to it. Suppose that energy is added to a material mechanically, such as by impact or (in the case of a gas) by compression. The addition of the energy sets some of the particles in the medium vibrating. Because the particles are linked by intermolecular forces, the oscillation of each particle affects that of its neighbors. The added energy propagates, or spreads, by means of interactions among the particles of the medium. An analogy to this process is shown in ◀Fig. 13.8, where the "particles" are dominoes. As each domino falls, it topples the one next to it. Thus, energy is transferred from domino to domino, and the disturbance propagates through the medium—the energy travels, not the medium.

In this case, there is no restoring force between the dominoes, so they do not oscillate, as do particles in a continuous material medium. Therefore, the disturbance moves in space, but it does not repeat itself in time at any one location.

Similarly, if the end of a stretched rope is given a quick shake, the disturbance transfers energy from the hand to the rope, as illustrated in ▼Fig. 13.9. The forces acting

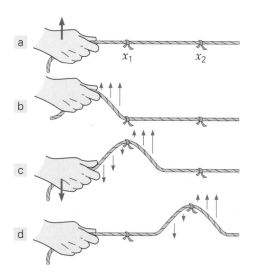

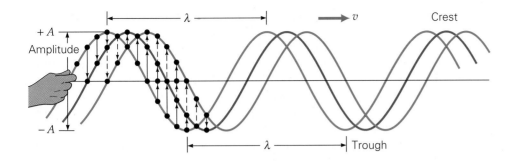

◀ **FIGURE 13.10** Periodic wave A continuous harmonic disturbance can set up a sinusoidal wave in a stretched rope, and the wave travels down the rope with wave speed v. Note that the "particles" in the rope oscillate vertically in simple harmonic motion. The distance between two successive points that are in phase (for example, at two crests) on the waveform is the wavelength λ of the wave. Can you tell how much time has elapsed, as a fraction of the period T, between the first (red) and last (blue) waves?

between the "particles" in the rope cause them to move in response to the motion of the hand, and a *wave pulse* travels down the rope. Each "particle" goes up and then back down as the pulse passes by. This motion of individual particles and the propagation of the wave pulse as a whole can be observed by tying pieces of ribbon onto the rope (at x_1 and x_2 in the figure). As the disturbance passes point x_1, the ribbon rises and falls, as do the rope's "particles." Later, the same happens to the ribbon at x_2, which indicates that the energy disturbance is propagating, or traveling, along the rope.

In a continuous material medium, particles interact with their neighbors, and restoring forces cause them to oscillate when they are disturbed. Thus, a disturbance not only propagates through space, but may be repeated over and over in time at each position. Such a regular, rhythmic disturbance in both time and space is called a **wave**, and the transfer of energy is said to take place by means of **wave motion**.

A continuous wave motion, or *periodic wave*, requires a disturbance from an oscillating source (▲Fig. 13.10). In this case, the particles move up and down continuously. If the driving source is such that a constant amplitude is maintained (the source oscillates in simple harmonic motion), the resulting particle motion is also simple harmonic.

Such periodic wave motion will have sinusoidal forms (sine or cosine) in both time and space. Being *sinusoidal in space* means that if you took a photograph of the wave at any instant ("freezing" it in time), you would see a sinusoidal waveform (such as one of the curves in Fig. 13.10). However, if you looked at a single point in space as a wave passed by, you would see a particle of the medium oscillating up and down *sinusoidally with time*, like the mass on a spring discussed in Section 13.2. (For example, imagine looking through a thin slit at a fixed location on the moving paper in Fig. 13.5. The wave trace would be seen rising and falling like a particle.)

Note: A wave is a combination of oscillations in space and time.

Wave Characteristics

Specific quantities are used to describe sinusoidal waves. As with a particle in simple harmonic motion, the **amplitude (A)** of a wave is the magnitude of the maximum displacement, or the maximum distance, from the particle's equilibrium position (Fig. 13.10). This quantity corresponds to the height of a wave crest or the depth of a trough. Recall from Section 13.2 that, in SHM, the total energy of the oscillator is proportional to the square of the amplitude. Similarly, the energy *transported* by a wave is proportional to the square of its amplitude ($E \propto A^2$). Note the difference, though: A wave is one way of *transmitting* energy through space, whereas an oscillator's energy is localized in space.

For a periodic wave, the distance between two successive crests (or troughs) is called the **wavelength (λ)** (Fig. 13.10). Actually, it is the distance between any two successive parts of the wave that are in phase (that is, that are at identical points on the waveform). The crest and trough positions are usually used for convenience. Note that a wavelength corresponds spatially to one cycle. Keep in mind that the wave, not the medium or material, is traveling.

The **frequency (f)** of a periodic wave is the number of cycles per second—that is, the number of complete waveforms, or wavelengths, that pass by a given point during each second. The frequency of the wave is the same as the frequency of the SHM source that created it.

Illustration 17.2 Wave Functions

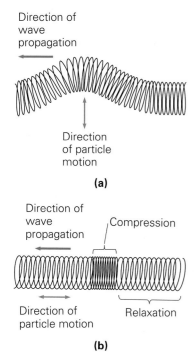

Direction of wave propagation

Direction of particle motion

(a)

Direction of wave propagation

Compression

Direction of particle motion

Relaxation

(b)

▲ **FIGURE 13.11 Transverse and longitudinal waves** (Wave pulses shown here for simplicity) **(a)** In a transverse wave, the motion of the particles is perpendicular to the direction of the wave velocity, as shown here in a spring for a wave moving to the left. A transverse wave is sometimes called a *shear wave*, because it supplies a force that tends to shear the medium. Transverse shear waves can propagate only in solids. (Why?) **(b)** In a longitudinal wave, the particle motion is parallel to (or *along*) the direction of the wave velocity. Here, a wave pulse also moves to the left. A longitudinal wave is sometimes called a *compressional wave*, because the force tends to compress the medium. Longitudinal compressional waves can propagate in all media—solid, liquid, and gas. Can you explain the motion of the wave *source* for both types of waves?

A periodic wave is said to possess a **period (*T*)**. The period $T = 1/f$ is the time for one complete waveform (a wavelength) to pass by a given point. Since a wave moves, it also has a **wave speed (*v*)** (or velocity if the wave's direction is specified). Any particular point on the wave (such as a crest) travels a distance of one wavelength λ in a time of one period *T*. Then, since $v = d/t$ and $f = 1/T$, we have

$$v = \frac{\lambda}{T} = \lambda f \quad \textit{wave speed} \tag{13.17}$$

Note that the dimensions of *v* are correct (length/time). In general, the wave speed depends on the nature of the medium, in addition to the source frequency *f*.

Example 13.5 ■ Dock of the Bay: Finding Wave Speed

A person on a pier observes a set of incoming waves that have a sinusoidal form with a distance of 1.6 m between the crests. If a wave laps against the pier every 4.0 s, what are (a) the frequency and (b) the speed of the waves?

Thinking It Through. Knowing the period and wavelength, the definition of frequency and Eq. 13.17 for wave speed can be used.

Solution. The distance between crests is the wavelength, so we have the following information:

Given: $\lambda = 1.6$ m   *Find:* (a) *f* (frequency)
$\quad\quad\quad T = 4.0$ s $\quad\quad\quad\quad\quad\quad$ (b) *v* (wave speed)

(a) The lapping indicates the arrival of a wave crest; hence, 4.0 s is the wave period—the time it takes to travel one wavelength (the crest-to-crest distance). Then

$$f = \frac{1}{T} = \frac{1}{4.0 \text{ s}} = 0.25 \text{ s}^{-1} = 0.25 \text{ Hz}$$

(b) The frequency or the period can be used in Eq. 13.17 to find the wave speed:

$$v = \lambda f = (1.6 \text{ m})(0.25 \text{ s}^{-1}) = 0.40 \text{ m/s}$$

Alternatively,

$$v = \frac{\lambda}{T} = \frac{1.6 \text{ m}}{4.0 \text{ s}} = 0.40 \text{ m/s}$$

Follow-Up Exercise. On another day, the person measures the speed of sinusoidal water waves at 0.25 m/s. (a) How far does a wave crest travel in 2.0 s? (b) If the distance between successive crests is 2.5 m, what is the frequency of these waves?

Types of Waves

In general, waves may be divided into two types, based on the direction of the particles' oscillations relative to that of the wave velocity. In a **transverse wave**, the particle motion is perpendicular to the direction of the wave velocity. The wave produced in a stretched string (Fig. 13.10) is an example of a transverse wave, as is the wave shown in ◄Fig. 13.11a. A transverse wave is sometimes called a *shear wave*, because the disturbance supplies a force that tends to shear the medium—to separate layers of that medium at a right angle to the direction of the wave velocity. Shear waves can propagate only in solids, since a liquid or a gas cannot support a shear. That is, a liquid or a gas does not have sufficient restoring forces between its particles to propagate a transverse wave.

In a **longitudinal wave**, the particle oscillation is parallel to the direction of the wave velocity. A longitudinal wave can be produced in a stretched spring by moving the coils back and forth along the spring axis (Fig. 13.11b). Alternating pulses of compression and relaxation travel along the spring. A longitudinal wave is sometimes called a *compressional wave*.

Sound waves in air are another example of longitudinal waves. A periodic disturbance produces compressions in the air. Between the compressions are *rarefactions*—regions where the density of the air is reduced, or rarefied. A loudspeaker oscillating

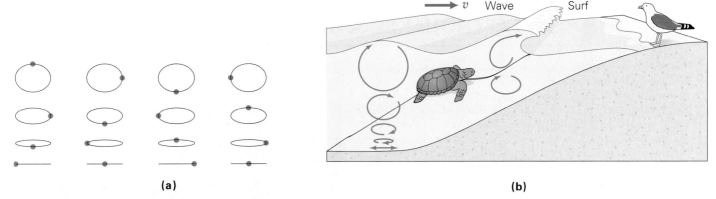

(a)   **(b)**

▲ **FIGURE 13.12 Water waves** Water waves are a combination of longitudinal and transverse motions. **(a)** At the surface, the water particles move in circles, but their motions become more longitudinal with depth. **(b)** When a wave approaches the shore, the lower particles are forced into steeper paths until, finally, the wave breaks or falls over to form surf.

back and forth, for example, can create these compressions and rarefactions, which travel out into the air as sound waves. (Sound is discussed in detail in Chapter 14.)

Longitudinal waves can propagate in solids, liquids, and gases, because all phases of matter can be compressed to some extent. The propagations of transverse and longitudinal waves in different media give information about the Earth's interior structure, as discussed in Insight 13.1 on Earthquakes, Seismic Waves, and Seismology, page 450.

The sinusoidal profile of water waves might make you think that they are transverse waves. Actually, they reflect a combination of longitudinal and transverse motions (▲Fig. 13.12). The particle motion may be nearly circular at the surface and becomes more elliptical with depth, eventually longitudinal. A hundred meters or so below the surface of a large body of water, the wave disturbances have little effect. For example, a submarine at these depths is undisturbed by large waves on the ocean's surface. As a wave approaches shallower water near shore, the water particles have difficulty completing their elliptical paths. When the water becomes too shallow, the particles can no longer move through the bottom parts of their paths, and the wave breaks. Its crest falls forward to form breaking surf as the waves' kinetic energy is transformed into potential energy—a water "hill" that eventually topples over.

PHYSLET®

Illustration 17.1 Wave Types

13.4 Wave Properties

OBJECTIVE: To explain various wave properties and resulting phenomena.

Among the properties exhibited by all waves are superposition, interference, reflection, refraction, dispersion, and diffraction.

Superposition and Interference

When two or more waves meet or pass through the same region of a medium, they pass through each other and proceed without being altered. While they are in the same region, the waves are said to be interfering.

What happens during interference? That is, what does the combined waveform look like? The relatively simple answer is given by the **principle of superposition**:

> At any time, the combined waveform of two or more interfering waves is given by the sum of the displacements of the individual waves at each point in the medium.

Using the principle of superposition, **interference** is illustrated in ▼Fig. 13.13. The displacement of the combined waveform at any point is $y = y_1 + y_2$, where y_1 and y_2 are the displacements of the individual pulses at that point. (Directions are indicated by plus and minus signs.) Interference, then, is the physical addition of waves. In adding waves, we must take into account the possibility that they are producing disturbances in opposite directions. In other words, we must treat the disturbances in terms of vector addition.

INSIGHT 13.1 EARTHQUAKES, SEISMIC WAVES, AND SEISMOLOGY

The structure of the Earth's interior is something of a mystery. The deepest mine shafts and drillings extend only a few kilometers into the Earth, compared with a depth of about 6400 km to the Earth's center. Using waves to probe the Earth's structure is one way to investigate it further. Waves generated by earthquakes have proved to be especially useful for this purpose. Seismology is the study of these waves, called *seismic waves*.

Earthquakes are caused by the sudden release of built-up stress along cracks and faults, such as the famous San Andreas Fault in California (Fig. 1). According to the geological theory of plate tectonics, the outer layer of the Earth consists of rigid plates—huge slabs of rock that move very slowly relative to one another. Stresses continuously build up, particularly along boundaries between plates.

When slippage of plates finally occurs, the energy from this stress-relieving event propagates outward as (seismic) waves from a site below the surface called the *focus*. The point on the Earth's surface directly above the focus is called the *epicenter*, and receives the greatest impact of a quake. Seismic waves are of two general types: surface waves and body waves. *Surface waves*, which move along the Earth's surface, account for most earthquake damage (Fig. 2). *Body waves* travel through the Earth and are both longitudinal and transverse waves. The compressional (longitudinal) waves are called *P waves*, and the shear (transverse) waves are called *S waves* (Fig. 3).

P and S stand for *primary* and *secondary* and indicate the waves' relative speeds (actually, their arrival times at monitoring stations). In general, primary waves travel through materials faster than do secondary waves and are detected first. An earthquake's rating on the Richter scale is related to the energy released in the form of seismic waves.

Seismic stations around the world monitor P and S waves with sensitive detecting instruments called *seismographs*. From the data gathered, we can map the paths of the waves through the Earth and thereby learn about the structure of the interior of our planet. The Earth's interior seems to be divided into three general regions: the crust, the mantle, and the core, which itself has a solid inner region and a liquid outer region.*

The locations of these regions' boundaries are determined in part by *shadow zones*, regions where no waves of a particular type are detected. These zones appear because, although longitudinal

*In most places, the crust is about 24–30 km (15–20 mi) thick; the mantle is 2900 km (1800 mi) thick; and the core has a radius of 3450 km (2150 mi). The solid inner core has a radius of about 1200 km (750 mi).

FIGURE 2 **Bad vibrations** Earthquake damage caused by the surface waves of a major shock that struck Kobe, Japan, in January 1995.

waves can travel through solids *or* liquids, transverse waves can travel only through solids. When an earthquake occurs, P waves are detected on the side of the Earth opposite the focus, but S waves are not. (See Fig. 3.) The absence of S waves in a shadow zone leads to the conclusion that the Earth must have a region near its center that is in the liquid phase.

When the transmitted P waves enter and leave the liquid region, they are refracted (bent). This refraction gives rise to a P-wave shadow zone, which indicates that only the outer part of the core is liquid. As you will learn in Chapter 19, the combination of a liquid outer core and the Earth's rotation may be responsible for the Earth's magnetic field.

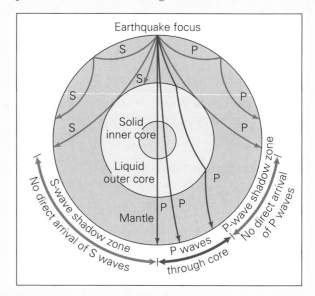

FIGURE 3 **Compressional and shear waves** Earthquakes produce waves that travel through the Earth. Because transverse (S) waves are not detected on the opposite side of the Earth, scientists believe that at least part of the Earth's core is a viscous liquid under high pressures and temperatures. The waves bend continuously, or refract, because their speed varies with depth.

FIGURE 1 **The San Andreas Fault** A small section of the fault, which runs through the San Francisco Bay area as well as across the more rural regions of California, is shown here.

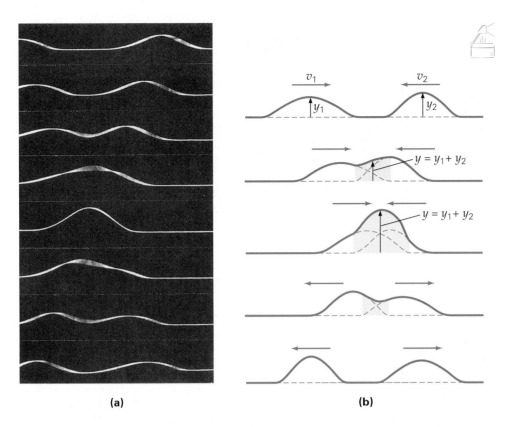

(a)

◀ **FIGURE 13.13 Principle of superposition (a)** When two waves meet, they interfere as shown in the photograph. **(b)** The beige tint marks the area where the two waves, moving in opposite directions, overlap and combine. The displacement at any point on the combined wave is equal to the sum of the displacements on the individual waves: $y = y_1 + y_2$.

PHYSLET®

Illustration 17.3 Superposition of Pulses

(b)

In the figure, the vertical displacements of the two pulses are in the same direction, and the amplitude of the combined waveform is greater than that of either pulse. This situation is called **constructive interference**. Conversely, if one pulse has a negative displacement, the two pulses tend to cancel each other when they overlap, and the amplitude of the combined waveform is smaller than that of either pulse. This situation is called **destructive interference**.

The special cases of total constructive and total destructive interference for traveling wave pulses of the same width and amplitude are shown in ▾Fig. 13.14. At the instant these interfering waves exactly overlap (crest coinciding with crest), the amplitude of the combined waveform is twice that of either individual wave. This case is referred to as **total constructive interference**. When the interfering pulses have opposite displacements and are exactly superimposed (crest coinciding with trough), the waveforms momentarily disappear; that is, the amplitude of the combined wave is zero. This case is called **total destructive interference**.

The word *destructive* unfortunately tends to imply that the energy, as well as the form of the waves, is destroyed. This is not the case. At the point of total

Demonstration/activity: Use a long Slinky, or tape two or more short Slinkies end to end. Have a student on each end shake a pulse at the same time. Observe how the pulses pass through each other. Have each student shake a pulse of about the same amplitude, but on opposite sides of the Slinky. At one instant, as the pulses pass through each other, the Slinky is perfectly straight. Where is the energy at that instant?

PHYSLET®

Exploration 17.1 Superposition of Two Pulses

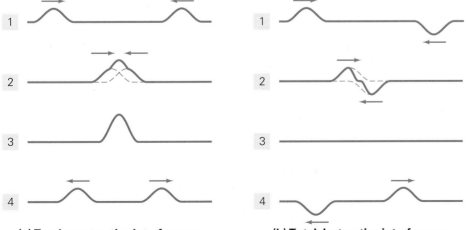

(a) Total constructive interference   **(b) Total destructive interference**

◀ **FIGURE 13.14 Interference (a)** When two wave pulses of the same amplitude meet and are in phase, they interfere constructively. When the pulses are exactly superimposed (3), total constructive interference occurs. **(b)** When the interfering pulses have opposite amplitudes and are exactly superimposed (3), total destructive interference occurs.

▶ **FIGURE 13.15** Destructive interference in action (a) Pilots use headphones mounted with a microphone that picks up low-frequency engine noise. (b) A wave is generated that is inverse that of the engine noise. When played back through the headphones, destructive interference produces a quieter engine noise. This process is called "active noise cancellation."

Exploration 17.3 Traveling Pulses and Barriers

(a)

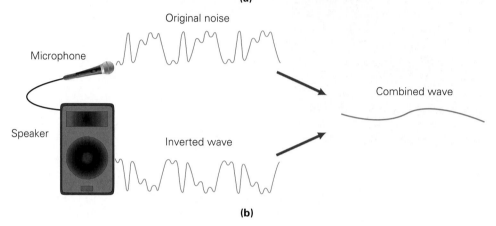

(b)

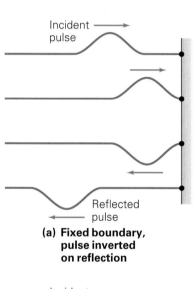

(a) Fixed boundary, pulse inverted on reflection

(b) Free (movable) boundary, pulse not inverted on reflection

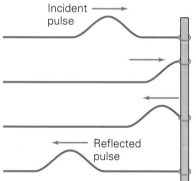

▲ **FIGURE 13.16** Reflection (a) When a wave (pulse) on a string is reflected from a fixed boundary, the reflected wave is inverted. (b) If the string is free to move at the boundary, the phase of the reflected wave is not shifted from that of the incident wave.

destructive interference, when the net wave shape and, hence potential energy, are zero, the wave energy is stored in the medium completely in the form of kinetic energy. That is, the straight string has instantaneous velocity.

There are several practical applications of destructive interference. One of these is automobile mufflers. Exhaust gases from the engine passing from a high pressure in the cylinders to normal atmospheric pressure would produce loud noises. These are reduced by mufflers. Typically, a muffler consists of a metal jacket containing perforated pipes and chambers. The pipes and chambers are arranged so that the pressure waves of the exhaust gases are reflected back and forth, giving rise to destructive interference. This greatly reduces the noise of the exhaust coming from the tail pipe.

Other applications are termed "active noise control" or "active noise cancellation." This involves sound modification, particularly sound cancellation by electro-acoustical means. A particularly important application is for airline or helicopter pilots who need to hear what's going on around them over the engine noise. Pilots use special headphones mounted with a microphone that picks up the low-frequency engine noise. A component in the headphone then creates a wave that is the inverse of the engine noise. This is played back through the headphones and destructive interference produces a quieter background (▲Fig. 13.15). A pilot can then better hear the mid- and high-frequency sounds, such as conversation and instrument warning sounds.

Reflection, Refraction, Dispersion, and Diffraction

Besides meeting other waves, waves can (and do) meet objects or a boundary with another medium. In such cases, several things may occur. One of these is **reflection**, which occurs when a wave strikes an object or comes to a boundary with another medium and is at least partly diverted back into the original medium. An echo is the reflection of sound waves, and mirrors reflect light waves.

Two cases of reflection are illustrated in ◀Fig. 13.16. If the end of the string is fixed, the reflected pulse is inverted (Fig. 13.16a). This is because the pulse causes the string to exert an upward force on the wall, and the wall exerts an equal and opposite downward force on the string (by Newton's third law). The downward force creates the downward, or inverted, reflected pulse. If the end of the string is free to move, then the reflected pulse is not inverted. (There is no phase shift.) This is illustrated in Fig. 13.16b, which shows the string attached to a light ring that can move freely on a smooth pole. The ring is accelerated upward by the front portion of the incoming pulse and then comes back down, thus creating a noninverted reflected pulse.

More generally, when a wave strikes a boundary, the wave is not completely reflected. Instead, some of the wave's energy is reflected and some is transmitted or absorbed. When a wave crosses a boundary into another medium, its speed generally changes because the new material has different characteristics. Entering the medium obliquely (at an angle), the transmitted wave moves in a direction different from that of the incident wave. This phenomenon is called **refraction** (▸Fig. 13.17).

Since refraction depends on changes in the speed of the wave, you might be wondering which physical parameters determine the wave speed. Generally, there are two types of situations. The simplest kind of wave is one whose speed does *not* depend on its wavelength (or frequency). All such waves travel at the same speed, determined solely by the properties of the medium. These waves are called *nondispersive waves,* because they do not disperse, or spread apart from one another. An example of a nondispersive transverse wave is a wave on a string, whose speed, as we shall see, is determined only by the tension and mass density of the string (Section 13.5). Sound is a nondispersive longitudinal wave; the speed of sound (in air) is determined only by the compressibility and density of the air. Indeed, if the speed of sound did depend on the frequency, at the back of the symphony hall you might hear the violins well before the clarinets, even though the two sound waves were in perfect synchronization when they left the orchestra pit.

When the wave speed *does* depend on wavelength (or frequency), the waves are said to exhibit **dispersion**: Waves of different frequencies spread apart from one another. Although waves of light are nondispersive in a vacuum, when they enter some media, they are spread out or dispersed. This is the basis for prisms separating sunlight into a color spectrum and for the formation of a rainbow, as we shall see in Chapter 22. Dispersion will be most important for us in our study of light, but you should remember that waves other than light can also be dispersive under the right conditions.

Diffraction refers to the bending of waves around an edge of an object and is not related to refraction. For example, if you stand along an outside wall of a building near the corner of the street, you can hear people talking around the corner. Assuming that there are no reflections or air motion (wind), this would not be possible if the sound waves traveled in a straight line. As the sound waves pass the corner, instead of being sharply cut off, they "wrap around" the edge, and you can hear the sound.

In general, the effects of diffraction are evident only when the size of the diffracting object or opening is about the same as or smaller than the wavelength of the waves. The dependence of diffraction on the wavelength and size of the object or opening is illustrated in ▾Fig. 13.18. For many waves, diffraction is negligible under normal circumstances. For instance, visible light has wavelengths on the order of 10^{-6} m. Such wavelengths are much too small to exhibit diffraction when they pass through common-sized openings, such as an eyeglass lens.

Reflection, refraction, dispersion, and diffraction will be considered in more detail when we study light waves in Chapters 22 and 24.

(a)   (b)

▲ **FIGURE 13.17 Refraction** The refraction of water waves is shown from overhead. As the crests approach the beach, their left edge slows because it enters shallow water first. Thus, the whole crest rotates, approaching the beach more or less head-on.

◂ **FIGURE 13.18 Diffraction** Diffraction effects are greatest when the opening (or object) is about the same size as or smaller than the wavelength of the waves. **(a)** With an opening much larger than the wavelength of these plane water waves, diffraction is noticeable only near the edges. **(b)** With an opening about the same size as the wavelength of the waves, diffraction produces nearly semicircular waves.

Illustration 17.4 Superposition of
Traveling Waves

Demonstration/activity: A $\frac{1}{4}$-in. rubber
rope several feet long works well for
demonstrating standing waves. Stretch
the rope the length of the lecture room.
Shake a pulse in the rope, and measure
the time it takes for the pulse to make
three or four round trips. Calculate the
speed of the pulse. Then measure the
mass per unit length of the rope and the
tension, and calculate the speed from
the expression given in the text. Do the
two figures agree?

13.5 Standing Waves and Resonance

OBJECTIVES: To (a) describe the formation and characteristics of standing waves, and (b) explain the phenomenon of resonance.

If you shake one end of a stretched rope, waves travel down it to the fixed end and are reflected back. The waves going down and back interfere. In most cases, the combined waveforms have a changing, jumbled appearance. But if the rope is shaken at just the right frequency, a steady waveform, or series of uniform loops, appears to stand in place along the rope. Appropriately, this phenomenon is called a **standing wave** (▼Fig. 13.19). It arises because of interference with the reflected waves, which have the same wavelength, amplitude, and speed as the incident waves. Since the two identical waves travel in opposite directions, the net energy flow down the rope is zero. In effect, the energy is standing in the loops.

Some points on the rope remain stationary at all times and are called **nodes**. At these points, the displacements of the interfering waves are *always* equal and opposite. Thus, by the principle of superposition, the interferring waves cancel each other completely at these points, and the rope does not undergo displacement there. At all other points, the rope oscillates back and forth at the same frequency. The points of maximum amplitude, where constructive interference is greatest, are called **antinodes**. As you can see in Fig. 13.19a, adjacent antinodes are separated by a half-wavelength ($\lambda/2$), or one loop; adjacent nodes are also separated by a half-wavelength.

Standing waves can be generated in a rope by more than one driving frequency; the higher the frequency, the more oscillating half-wavelength loops are in the rope. The only requirement is that an integer number of half-wavelengths "fit" the length of the rope. The frequencies at which large-amplitude standing waves are produced are called **natural frequencies**, or **resonant frequencies**. The resulting standing-wave patterns are called *normal*, or *resonant, modes of vibration*. In general, all systems that oscillate have one or more natural frequencies, which depend on such factors as mass, elasticity or restoring force, and geometry (boundary conditions). The natural frequencies of a system are sometimes called its *characteristic* frequencies.

A stretched string or rope can be analyzed to determine its natural frequencies. The boundary condition is that the ends are fixed; thus, there must be a node at each end. The number of closed segments or loops of a standing wave that will fit between the nodes at the ends (along the length of the string) is

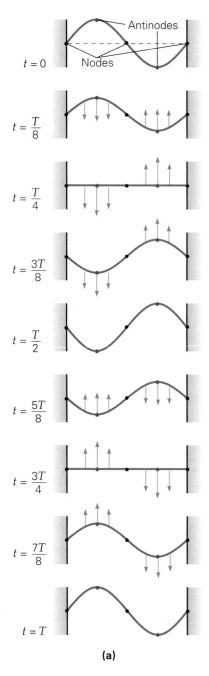

(a)

▼ **FIGURE 13.19 Standing waves** (a) Conditions of destructive and constructive interference recur as each wave travels a distance of $\lambda/4$ in a time $t = T/4$. The velocities of the rope's particles are indicated by the arrows. This motion gives rise to standing waves with stationary nodes and maximum-amplitude antinodes (b) Standing waves are formed by interfering waves traveling in opposite directions..

(b)

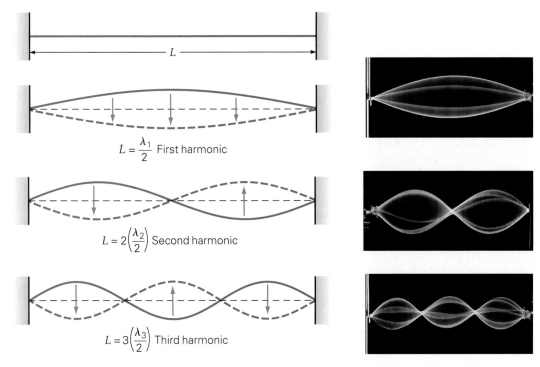

▲ **FIGURE 13.20** **Natural frequencies** A stretched string can have standing waves only at certain frequencies. These correspond to the numbers of half-wavelength loops that will fit along the length of string between the nodes at the fixed ends.

equal to an integral number of *half*-wavelengths (▲Fig. 13.20). Note that $L = \lambda_1/2$, $L = 2(\lambda_2/2)$, $L = 3(\lambda_3/2)$, $L = 4(\lambda_4/2)$, and so on. In general,

$$L = n\left(\frac{\lambda_n}{2}\right) \quad \text{or} \quad \lambda_n = \frac{2L}{n} \quad (\text{for } n = 1, 2, 3, \dots)$$

The natural frequencies of oscillation are therefore

$$f_n = \frac{v}{\lambda_n} = n\left(\frac{v}{2L}\right) = nf_1 \quad \text{for } n = 1, 2, 3, \dots \quad \begin{array}{l} \textit{natural frequencies} \\ \textit{for a stretched string} \end{array} \quad (13.18)$$

where v is the speed of waves on a string. The lowest natural frequency ($f_1 = v/2L$ for $n = 1$) is called the **fundamental frequency**. All of the other natural frequencies are integral multiples of the fundamental frequency: $f_n = nf_1$ (for $n = 1, 2, 3, \dots$). The set of frequencies f_1, $f_2 = 2f_1$, $f_3 = 3f_1, \dots$ is called a **harmonic series**: f_1 (the fundamental frequency) is the *first harmonic*, f_2 the *second harmonic*, and so on.

Strings that are fixed at each end are found in stringed musical instruments such as violins, pianos, and guitars. When such a string is excited, the resulting vibration generally includes several harmonics in addition to the fundamental frequency. The number of harmonics depends on how and where the string is excited—that is, whether it is plucked, struck, or bowed. It is the combination of harmonic frequencies that gives a particular instrument its characteristic sound quality (more on this in Chapter 14). As Eq. 13.18 shows, the fundamental frequency of a stretched string, as well as the other harmonics, depends on the length of the string. Think of how different notes are obtained on a particular string of a violin or a guitar (▶Fig. 13.21).

▲ **FIGURE 13.21** **Fundamental frequencies** Performers of a stringed instrument such as the violin or guitar use their fingers to stop or fret the strings. By pressing a string against the fingerboard, the player reduces the amount of its length that is free to vibrate. This reduction changes the resonant frequency of the string and the pitch of the tone it produces.

Natural frequencies also depend on other parameters, such as mass and force, which affect the wave speed in the string. For a stretched string, the wave speed (v) can be shown to be

$$v = \sqrt{\frac{F_T}{\mu}} \quad \begin{array}{l} \text{wave speed} \\ \text{in a stretched string} \end{array} \tag{13.19}$$

where F_T is the tension in the string and μ is the linear mass density (mass per unit length, $\mu = m/L$). (We use F_T, rather than the T of previous chapters, so as not to confuse the tension with the period T.) Thus, Eq. 13.18 can be written as

$$f_n = n\left(\frac{v}{2L}\right) = \frac{n}{2L}\sqrt{\frac{F_T}{\mu}} = nf_1 \quad (\text{for } n = 1, 2, 3, \dots) \tag{13.20}$$

Note that the greater the linear mass density of a string, the lower its natural frequencies. As you may know, the low-note strings on a violin or guitar are thicker, or more massive, than the high-note strings. By tightening a string, we increase all frequencies of that string. Changing the tension in the string is how violinists, for example, tune their instruments before a performance.

Example 13.6 ■ A Piano String: Fundamental Frequency and Harmonics

A piano string with a length of 1.15 m and a mass of 20.0 g is under a tension of 6.30×10^3 N. (a) What is the fundamental frequency of the string when it is struck? (b) What are the frequencies of the next two harmonics?

Thinking It Through. We have the tension and can calculate the linear mass density from the data. This will allow us to find the fundamental frequency, and from that, the harmonics can be calculated.

Solution.

Given: $L = 1.15$ m
$m = 20.0$ g $= 0.0200$ kg
$F_T = 6.30 \times 10^3$ N

Find: (a) f_1 (fundamental frequency)
(b) f_2 and f_3 (frequencies of next two harmonics)

(a) The linear mass density of the string is

$$\mu = \frac{m}{L} = \frac{0.0200 \text{ kg}}{1.15 \text{ m}} = 0.0174 \text{ kg/m}$$

Then, using Eq. 13.20, we have

$$f_1 = \frac{1}{2L}\sqrt{\frac{F_T}{\mu}} = \frac{1}{2(1.15 \text{ m})}\sqrt{\frac{6.30 \times 10^3 \text{ N}}{0.0174 \text{ kg/m}}} = 262 \text{ Hz}$$

This is approximately the frequency of middle C (C_4) on a piano.

(b) Since $f_2 = 2f_1$ and $f_3 = 3f_1$, it follows that

$$f_2 = 2f_1 = 2(262 \text{ Hz}) = 524 \text{ Hz}$$

and

$$f_3 = 3f_1 = 3(262 \text{ Hz}) = 786 \text{ Hz}$$

The second harmonic corresponds approximately to C_5 on a piano, since, by definition, the frequency doubles with each octave (every eighth white key).

Follow-Up Exercise. A musical note is referenced to the fundamental vibrational frequency, or first harmonic. In musical terms, the second harmonic is the first overtone, the third harmonic is the second overtone, and so on. If an instrument has a third overtone with a frequency of 880 Hz, what is the frequency of the first overtone?

Note: Don't be confused by the language: *First* overtone means the first frequency above the fundamental frequency—that is, the *second* harmonic.

Integrated Example 13.7 ■ Tuning Up: Increasing the Frequency of a Guitar String

Suppose you want to increase the fundamental frequency of a guitar string. (a) Would you (1) loosen the string to halve its tension, (2) tighten the string to double its tension, (3) use another string of the same material with half the diameter at the same tension, or (4) use another string of the same material with twice the diameter at the same tension? (b) In actuality, you want to go from the A note (220 Hz) below middle C to the A note (440 Hz) above middle C. If the guitar strings are made of steel ($\rho = 7.8 \times 10^3$ kg/m^3, Table 9.2), and an initial thicker string has a diameter of 0.30 cm, show that a string with half the diameter will have double the fundamental frequency.

(a) Conceptual Reasoning. The fundamental frequency of a stretched string is given by Eq. 13.20:

$$f = \frac{1}{2L}\sqrt{\frac{F_T}{\mu}} \qquad (for\ n = 1)$$

The frequency of the string is proportional to the *square root* of the tension force F_T, so loosening the string—that is, decreasing F_T—would not increase the frequency. Nor would doubling the tension double the frequency (because $\sqrt{2F_T} \neq 2\sqrt{F_T}$). So, neither (1) nor (2) is the correct answer.

In using another string, the question is, then, how does the frequency vary with the linear mass density μ of the string? If the strings are of the same material (same density ρ), then the greater the diameter of a string, the greater is its mass per unit length (greater μ). Hence, a thinner string, with a smaller μ, will vibrate at a higher frequency, and the answer is (3).

(b) Quantitative Reasoning and Solution. At first, one might think of computing the frequencies directly, using Eq. 13.20. But this can't be done, because not enough data are given, which usually implies the use of a ratio. To show the difference in frequencies with different-diameter strings, the frequency equation needs to be expressed in terms of the diameter of a string. Inspecting Eq. 13.20, we see that the diameter of the string does not depend on the length L (assumed constant between the bridge and neck of the guitar) or on the constant tension force F_T (assuming negligible stretching). This leads to a consideration of the linear mass density $\mu = m/L$.

Recall that the mass of a string depends on its density and volume; that is, $\rho = m/V$, or $m = \rho V$. Then, the volume V of a length (L) of wire, which may be thought of as that of a long cylinder with circular cross-section A, can be determined by $V = AL$. The circular area is proportional to the square of the diameter of the wire, so this is the key to our proof.

Given: $f = \dfrac{1}{2L}\sqrt{\dfrac{F_T}{\mu}}$ (for $n = 1$, Eq. 13.20)

$\rho = 7.8 \times 10^3$ kg/m^3 (steel)

$d_1 = 0.30$ cm

$d_2 = d_1/2 = 0.15$ cm

Find: Prove that a string with a diameter d_2 that is half that of a thicker string (diameter d_1) will have twice the fundamental frequency of the thicker string.

As noted in the Quantative Reasoning and Solution above, the linear mass density of the wire string can be expressed in terms of its density and volume, the latter of which is proportional to the square of the string's diameter:

$$\mu = \frac{m}{L} = \frac{\rho V}{L} = \frac{\rho AL}{L} = \rho\left(\frac{\pi d^2}{4}\right)$$

Putting this into Eq. 13.20 yields

$$f = \frac{1}{2L}\sqrt{\frac{F_T}{\mu}} = \frac{1}{2L}\sqrt{\frac{4F_T}{\rho\pi d^2}} = \left(\frac{1}{L}\sqrt{\frac{F_T}{\rho\pi}}\right)\frac{1}{d}$$

(continues on next page)

The quantities in the brackets are constant, and $f \propto 1/d$, so, in ratio form,

$$\frac{f_2}{f_1} = \frac{d_1}{d_2} = \frac{0.30 \text{ cm}}{0.15 \text{ cm}} = 2 \quad \text{and} \quad f_2 = 2f_1$$

Follow-Up Exercise. The fundamental frequency of a violin string is A below middle C (220 Hz). How could you tune this string to middle C (264 Hz) without changing strings as was done in this Example?

Illustration 17.5 Resonant Behavior on a String

When an oscillating system is driven at one of its natural, or resonant, frequencies, the maximum amount of energy is transferred to the system. The natural frequencies of a system are the frequencies at which the system "wants" to vibrate, so to speak. The condition of driving a system at a natural frequency is referred to as **resonance**.

A common example of a system in mechanical resonance is someone being pushed on a swing. Basically, a swing is a simple pendulum and has only one resonant

INSIGHT 13.2 DESIRABLE AND UNDESIRABLE RESONANCES

When we hold a large seashell to our ear, sounds like that of the ocean are heard. The cause of this sounds is a resonance effect. Ambient sounds enter the shell, which acts as a resonant cavity. The air in the shell resonates at the shell's natural frequencies. Standing sound waves are set up that can be heard when the shell is held close to the ear. (More about sound in Chapter 14.)

The air in the shell resonates at the shell's natural frequencies. The variations in sound arise from the different ambient sounds, or the coming and going of different resonant frequencies. "The coming and going of these resonant frequencies gives the listener the illusion of hearing the ocean waves come and go."* Basically, the brain processes the sound, looking for a pattern that has been previously experienced. Most of us have heard the sound of ocean waves, so this is what we associate with when we listen to a seashell. You can also hear a similar "ocean" sound by placing an empty glass or cupping your hand over your ear.

When a large number of soldiers march over a small bridge, they are generally ordered to break step. The reason is that the marching frequency may correspond to one of the natural frequencies of the bridge and set it into resonant vibration, which could cause it to collapse. This actually occurred on a suspension bridge in England in 1831. The bridge was weak and in need of repair, but it collapsed as a direct result of the resonance vibrations induced by the marching soldiers—with some injuries.

Another incident of a bridge vibrating was due not to marching soldiers but to the driving force of the wind. On November 7, 1940, winds with speeds of 65 to 72 km/h (40 to 45 mi/h) started the main span of the Tacoma Narrows Bridge (in Washington State) vibrating. The bridge, 855 m (2800 ft) long and 12 m (39 ft) wide, had been opened to traffic only four months earlier.

During the first month that the bridge was used, small transverse modes of vibration had been observed. But on November 7, special wind effects drove the bridge in near resonance, and the main span vibrated at a frequency of 36 vib/min

*From *The Flying Circus of Physics*, by J. Walker. (New York: Wiley, 1977).

and an amplitude of 1.5 ft. At 10 A.M., the main span began to vibrate in a torsional (twisting) mode in two segments at a frequency of 14 vib/min. The wind continued to drive the bridge in resonance, and the vibrational amplitude increased. Shortly after 11 A.M., the main span collapsed (Fig. 1).†

"Galloping Gertie" (the nickname given to the bridge) was rebuilt on the same tower foundations. However, the new design made the structure stiffer, to increase its resonant frequency so that high winds could not produce unwanted resonance.

†It is doubtful that the gusting of the wind set the bridge into vibration. The wind velocity was moderately steady, and gust fluctuations are normally random. One explanation for the driving source of the oscillations involves the formation of vortices as the wind blew past the bridge. Vortices are like the eddies that form in water at the end of an oar when a boat is rowed. The wind blowing over and under the bridge formed vortices that rotated in opposite directions. The formation and "shedding" of the vortices (like eddies coming off oars) would have imparted energy to the bridge. If the frequency of this action approximated a natural frequency, a standing wave would have been set up.

FIGURE 1 Galloping Gertie The collapse of the Tacoma Narrows Bridge on November 7, 1940, is captured in this frame from a movie camera.

frequency for a given length $\left[f = 1/T = 1/(2\pi)\sqrt{g/L} \right]$. If you push the swing with this frequency and in phase with its motion, its amplitude and energy increase (▶Fig. 13.22). If you push at a slightly different frequency, the energy transfer is no longer a maximum. (What do you think happens if you push with the resonant frequency, but 180° out of phase with the swing's motion?)

Unlike a simple pendulum, a stretched string has many natural frequencies. Almost any driving frequency will cause a disturbance in the string. However, if the frequency of the driving force is not equal to one of the natural frequencies, the resulting wave will be relatively small and jumbled. By contrast, when the frequency of the driving force matches one of the natural frequencies, the maximum amount of energy is transferred to the string. A steady standing-wave pattern results, with the amplitude at the antinodes becoming relatively large.

Mechanical resonance is not the only type of resonance. When you tune a radio, you are changing the resonant frequency of an electrical circuit (Chapter 21) so that it will be driven by, or will pick up, a signal at the frequency of the station you want. Other examples of resonance are described in Insight 13.2, Desirable and Undesirable Resonances.

▲ **FIGURE 13.22 Resonance in the playground** The swing behaves like a pendulum in SHM. To transfer energy efficiently, the man must time his pushes to the natural frequency of the swing.

Chapter Review

• **Simple harmonic motion (SHM)** requires a restoring force directly proportional to the displacement, such as an ideal spring force, which is given by Hooke's law.

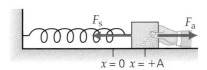

Hooke's law:

$$F_s = -kx \qquad (13.1)$$

• The **frequency** (f) and **period** (T) for SMH are the inverse of each other.

Frequency and period for SHM:

$$f = \frac{1}{T} \qquad (13.2)$$

• In general, the total energy of an object in SHM is directly proportional to the square of the amplitude.

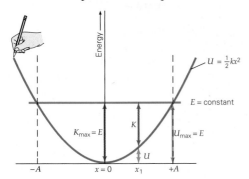

Total energy of a spring and mass in SHM:

$$E = \tfrac{1}{2}kA^2 = \tfrac{1}{2}mv^2 + \tfrac{1}{2}kx^2 \qquad (13.4\text{--}5)$$

• The form of an equation of motion for an object in SHM depends on the object's initial (y_o) displacement.

Equations of motion for SHM:

$$y = \pm A \sin \omega t = \pm A \sin(2\pi ft) = \pm A \sin\left(\frac{2\pi t}{T}\right) \qquad (13.8a)$$

+ *for initial motion upward with* $y_o = 0$

− *for initial motion downward with* $y_o = 0$

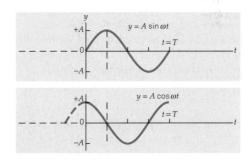

$$y = \pm A \cos \omega t = \pm A \cos(2\pi ft) = \pm A \cos\left(\frac{2\pi t}{T}\right) \qquad (13.9a)$$

for initial motion downward with $y_o = +A$

for initial motion upward with $y_o = -A$

Velocity of a mass oscillating on a spring:

$$v = \pm\sqrt{\frac{k}{m}(A^2 - x^2)} \qquad (13.6)$$

Period of a mass oscillating on a spring:

$$T = 2\pi\sqrt{\frac{m}{k}} \qquad (13.11)$$

Angular frequency of a mass oscillating on a spring:

$$\omega = 2\pi f = \sqrt{\frac{k}{m}} \qquad (13.13)$$

Period of a simple pendulum (small-angle approximation):

$$T = 2\pi\sqrt{\frac{L}{g}} \qquad (13.14)$$

Velocity of a mass in SHM:

$$v = \omega A \cos \omega t \quad \begin{array}{l}\text{(vertical velocity if } v_o \\ \text{is upward at } t_o = 0, y_o = 0)\end{array} \qquad (13.15)$$

Acceleration of a mass in SHM:

$$a = -\omega^2 A \sin \omega t = -\omega^2 y \quad \begin{array}{l}\text{(vertical acceleration if } v_o \\ \text{is upward at } t_o = 0, y_o = 0)\end{array} \qquad (13.16)$$

• A wave is a disturbance in time and space; energy is transferred or propagated by wave motion.

Wave speed:

$$v = \frac{\lambda}{T} = \lambda f \qquad (13.17)$$

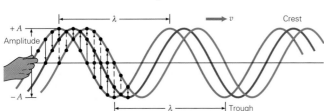

• At any time, the combined waveform of two or more interfering waves is given by the sum of the displacements of the individual waves at each point in the medium.

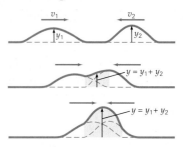

• At natural frequencies, **standing waves** can form on a string as a result of the **interference** of two waves of identical wavelength, amplitude, and speed traveling in opposite directions on a string.

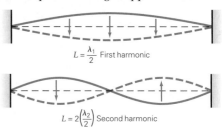

Natural frequencies for a stretched string:

$$f_n = n\left(\frac{v}{2L}\right) = \frac{n}{2L}\sqrt{\frac{F_T}{\mu}} = nf_1 \quad \text{(for } n = 1, 2, 3, \ldots) \qquad (13.20)$$

Exercises

MC = *Multiple Choice Question,* **CQ** = *Conceptual Question, and* **IE** = *Integrated Exercise. Throughout the text, many exercise sections will include "paired" exercises. These exercise pairs, identified with **red numbers**, are intended to assist you in problem solving and learning. In a pair, the first exercise (even numbered) is worked out in the Study Guide so that you can consult it should you need assistance in solving it. The second exercise (odd numbered) is similar in nature, and its answer is given at the back of the book.*

13.1 Simple Harmonic Motion

1. **MC** A particle in SHM has (a) variable amplitude, (b) a restoring force in the form of Hooke's law, (c) a frequency directly proportional to its period, (d) a position that is represented graphically by $x(t) = at + b$. (b)

2. **MC** The maximum kinetic energy of a mass–spring system in SHM is equal to (a) A, (b) A^2, (c) kA, (d) $kA^2/2$. (d)

3. **MC** If the period of a system in SHM is doubled, the frequency of the system is (a) doubled, (b) halved, (c) four times as large, (d) one-quarter as large. (b)

4. **MC** When a particle in a horizontal SHM is at the equilibrium position, the potential energy of the system is (a) zero, (b) maximum, (c) negative, (d) none of the preceding. (a)

5. **CQ** If the amplitude of a mass in SHM is doubled, how are (a) the energy and (b) the maximum speed affected? (a) four times as large (b) twice as large

6. **CQ** How does the speed of a mass in SHM change as the mass approaches its equilibrium position? Explain. increases

7. **CQ** A mass–spring system in SHM has an amplitude A and period T. How long does the mass take to travel a distance A? How about $2A$? $T/4$, $T/2$

8. **CQ** A tennis player uses a racket to bounce a ball up and down with a constant period. Is this a simple harmonic motion? Explain.
no, restoring force does not obey Hooke's law

9. ● A particle oscillates in SHM with an amplitude A. What is the total *distance* the particle travels in one period? $4A$

10. ● A 0.75-kg toy oscillating on a spring completes a cycle every 0.60 s. What is the frequency of this oscillation? 1.7 Hz

11. ● A particle in simple harmonic motion has a frequency of 40 Hz. What is the period of this oscillation? 0.025 s

12. ● The frequency of a simple harmonic oscillator is doubled from 0.25 Hz to 0.50 Hz. What is the change in its period? decrease of 2.0 s

13. ● What is the spring constant of a spring scale that stretches 6.0 cm when a basket of vegetables of mass 0.25 kg is suspended from it? 41 N/m

14. ● An object of mass 0.50 kg is attached to a spring with spring constant 10 N/m. If the object is pulled down 0.050 m from the equilibrium position and released, what is its maximum speed? 0.22 m/s

15. ●● Atoms in a solid are in continuous vibrational motion due to thermal energy. At room temperature, the amplitude of these atomic vibrations is typically about 10^{-9} cm, and their frequency is on the order of 10^{12} Hz. (a) What is the approximate period of oscillation of a typical atom? (b) What is the maximum speed of such an atom? (a) 10^{-12} s (b) 63 m/s

16. IE ●● (a) At what position is the magnitude of the force on a mass in a mass–spring system minimum: (1) $x = 0$, (2) $x = -A$, or (3) $x = +A$? Why? (b) If $m = 0.500$ kg, $k = 150$ N/m, and $A = 0.150$ m, what are the magnitude of the force on the mass and the acceleration of the mass at $x = 0$, 0.050 m, and 0.150 m? (a) (1) $x = 0$ (b) 0 N, 0 m/s²; 7.50 N, 15.0 m/s²; 22.5 N, 45.0 m/s²

17. IE ●● (a) At what position is the speed of a mass in a mass–spring system maximum: (1) $x = 0$, (2) $x = -A$, or (3) $x = +A$? Why? (b) If $m = 0.250$ kg, $k = 100$ N/m, and $A = 0.10$ m, what is the maximum speed? (a) (1) $x = 0$ (b) 2.0 m/s

18. ●● A mass–spring system is in SHM in the horizontal direction. If the mass is 0.25 kg, the spring constant is 12 N/m, and the amplitude is 15 cm, (a) what would be the maximum speed of the mass, and (b) where would this occur? (c) What would be the speed at a half-amplitude position? (a) 1.0 m/s (b) at the equilibrium position (c) 0.90 m/s

19. ●● In Exercise 18, (a) what is the speed of the mass if it is at $x = 10$ cm? (b) What is the magnitude of the force exerted by the spring on the mass? (a) 0.77 m/s (b) 1.2 N

20. ●● A horizontal spring on a frictionless level air track has a 150-g object attached to it and it is stretched 6.50 cm. Then the object is given an outward initial velocity of 2.20 m/s. If the spring constant is 35.2 N/m, determine how much farther the spring stretches. 9.26 cm

21. ●● A 350-g block moving vertically upward collides with a light vertical spring and compresses it 4.50 cm before coming to rest. If the spring constant is 50.0 N/m, what was the initial speed of the block? (Ignore energy losses to sound and other factors during the collision.) 1.08 m/s

22. ●● A 0.25-kg object suspended on a light spring is released from a position 15 cm above the stretched equilibrium position. The spring has a spring constant of 80 N/m. (a) What is the total energy of the system? (Neglect gravitational potential energy.) (b) Does this energy depend on the mass of the object? Explain. (a) 0.90 J (b) no

23. ●● What is the speed of the object in Exercise 22 when the object is (a) 5.0 cm above its equilibrium position

and (b) 5.0 cm below its equilibrium position? (c) What is the object's maximum speed, and where does this occur? (a) 2.5 m/s (b) 2.5 m/s (c) 2.7 m/s, equilibrium position

24. ●●● A 75-kg circus performer jumps from a 5.0-m height onto a trampoline and stretches it downward 0.30 m. Assuming that the trampoline obeys Hooke's law, (a) how far will it stretch if the performer jumps from a height of 8.0 m? (b) How far will the trampoline stretch if the performer stands still on it while taking a bow? (a) 0.38 m (b) 8.5×10^{-3} m

25. ●●● A vertical spring has a 200-g mass attached to it. The mass is released from rest and falls 22.3 cm before stopping. (a) Determine the spring constant. (b) Determine the speed of the mass when it has fallen only 10.0 cm. (a) 17.6 N/m (b) 1.04 m/s

26. ●●● A 0.250-kg ball is dropped from a height of 10.0 cm onto a spring, as illustrated in ▼Fig. 13.23. If the spring has a spring constant of 60.0 N/m, (a) what distance will the spring be compressed? (Neglect energy loss during collision.) (b) On recoiling upward, how high will the ball go? (a) 0.140 m (b) 10.0 cm (original position)

▲ **FIGURE 13.23 How far down?** See Exercise 26.

13.2 Equations of Motion

27. **MC** The equation of motion for a particle in SHM (a) is always a cosine function, (b) reflects damping action, (c) is independent of the initial conditions, or (d) gives the position of the particle as a function of time. (d)

28. **MC** For the SHM equation $y = A \sin \omega t$, the initial position y_0 is (a) $+A$, (b) $-A$, (c) 0, or (d) any of the preceding. (c)

29. **MC** For the SHM equation $y = A \sin(2\pi t/T)$, the y-position of the object in three quarters of the period is (a) $+A$, (b) $-A$, (c) $A/2$, (d) 0. (b)

30. **CQ** If the length of a pendulum is doubled, what is the ratio of the new period to the old one? $\sqrt{2}$ times as large

31. CQ The apparatus in Fig. 13.5 demonstrates that the motion of a mass on a spring can be described by a sinusoidal function of time. How could this same relationship be demonstrated for a pendulum?
tracing out path of the object on scrolling horizontal paper

32. CQ Could simple harmonic motion be described by a tangent function? Explain. no; tan 90° goes to ∞

33. CQ Would the period of a pendulum in an upward-accelerating elevator be increased or decreased compared with its period in a nonaccelerating elevator? Explain. decreased, see ISM

34. CQ If a mass–spring system is taken to the Moon, will the period of the system change? How about the period of a pendulum taken to the Moon? Explain. no, yes

35. ● What mass on a spring with a spring constant of 100 N/m will oscillate with a period of 2.0 s? 10 kg

36. ● A 0.50-kg mass oscillates in simple harmonic motion on a spring with a spring constant of 200 N/m. What are (a) the period and (b) the frequency of the oscillation? (a) 0.31 s (b) 3.2 Hz

37. ● The simple pendulum in a tall clock is 0.75 m long. What are (a) the period and (b) the frequency of this pendulum? (a) 1.7 s (b) 0.57 Hz

38. ● A breeze sets a suspended lamp into oscillation. If the period is 1.0 s, what is the distance from the ceiling to the lamp at the lowest point? Assume that the lamp acts as a simple pendulum. 0.25 m

39. ● Write the general equation of motion for a mass that is on a horizontal frictionless surface and is connected to a spring at equilibrium (a) if the mass is initially given a quick push away from the spring and (b) if the mass is pulled away from the spring and released. (a) $x = A \sin \omega t$ (b) $x = A \cos \omega t$

40. ● The equation of motion for an oscillator in vertical SHM is given by $y = (0.10 \text{ m}) \sin(100)t$. What are the (a) amplitude, (b) frequency, and (c) period of this motion? (a) 0.10 m (b) 16 Hz (c) 0.063 s

41. ● The displacement of an object is given by $y = (5.0 \text{ cm}) \sin(20\pi)t$. What are the object's (a) amplitude, (b) frequency, and (c) period of oscillation? (a) 5.0 cm (b) 10 Hz (c) 0.10 s

42. ● If the displacement of an oscillator in SHM is described by the equation $y = (0.25 \text{ m}) \cos(314)t$, where y is in meters and t is in seconds, what is the position of the oscillator at (a) $t = 0$, (b) $t = 5.0$ s, and (c) $t = 15$ s? (a) 0.25 m (b) 0.17 m (c) −0.18 m

43. IE ●● The oscillations of two oscillating mass–spring systems are graphed in ▶Fig. 13.24. The mass in System A is four times that in System B. (a) Compared with System B, System A has (1) more, (2) the same, or (3) less energy. Why? (b) Calculate the ratio of energy between System B and System A. (a) (3) less (b) 1.8×10^2

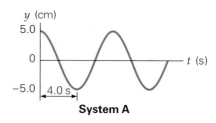

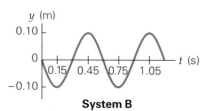

▲ **FIGURE 13.24 Wave energy and equation of motion** See Exercises 43, 56, and 57.

44. ●● Show that the total energy of a mass–spring system in simple harmonic motion is given by $\frac{1}{2}m\omega^2 A^2$. see ISM

45. ●● The velocity of a vertically oscillating mass–spring system is given by $v = (0.750 \text{ m/s}) \sin(4t)$. Determine (a) the amplitude and (b) maximum acceleration of this oscillator. (a) 0.188 m (b) 3.00 m/s²

46. IE ●● (a) If the mass in a mass–spring system is doubled, the new period is (1) 2, (2) $\sqrt{2}$, (3) $1/\sqrt{2}$ times the old period. Why? (b) If the initial period is 3.0 s and the mass is reduced to 1/3 of its initial value, what is the new period? (a) (2) $\sqrt{2}$ (b) 1.7 s

47. IE ●● (a) If the spring constant in a mass–spring system is tripled, the new period is (1) 3, (2) $\sqrt{3}$, (3) $1/\sqrt{3}$ times the old period. Why? (b) If the initial period is 2.0 s and the spring constant is halved, what is the new period? (a) (3) $1/\sqrt{3}$ (b) 2.8 s

48. ●● Show that for a pendulum to oscillate at the same frequency as a mass on a spring, the pendulum's length must be $L = mg/k$. see ISM

49. ●● Since there is little gravity in outer space, astronauts cannot measure their mass on a spring scale as we do on the Earth. (a) Can you design a method whereby they could measure their mass on a spring scale in space? (b) If the period of oscillation of an astronaut on a 3000-N/m spring is 1.0 s, what is the astronaut's mass? (a) using the period of vibration (b) 76 kg

50. ●● Students use a simple pendulum with a length of 36.90 cm to measure the acceleration of gravity at the location of their school. If the period of the pendulum is 1.220 s, what is the experimental value of g at the school? 9.787 m/s²

51. ●● What is the maximum kinetic energy of a simple horizontal mass–spring oscillator whose equation of motion is given by $x = (0.350 \text{ m}) \sin(7t)$? The mass on the end of the spring is 900 g. 2.70 J

52. ●● The equation of motion of a particle in vertical SHM is given by $y = (10\text{ cm}) \sin(0.50)t$. What are the particle's (a) displacement, (b) velocity, and (c) acceleration at $t = 1.0$ s? (a) 4.8 cm (b) 4.4 cm/s (c) -1.2 cm/s²

53. ●● For a few seconds during an earthquake, the floor of an apartment building is measured to oscillate in approximately simple harmonic motion with a period of 1.95 seconds and an amplitude of 8.65 cm. Determine the maximum speed and acceleration of the floor during this motion. Express the acceleration as a fraction of g. 0.279 m/s, 0.897 m/s² = (0.00915) g

54. ●● Two equal masses oscillate on light springs, the second with a spring constant twice that of the first. Which system will have the greater frequency and how much greater? the second system, $f_2 = \sqrt{2}f_1$

55. **IE**●● (a) If a pendulum clock were taken to the Moon, where the acceleration due to gravity is only one sixth (assume the figure to be exact) that on the Earth, will the period of vibration (1) increase, (2) remain the same, or (3) decrease? Why? (b) If the period on the Earth is 2.0 s, what is the period on the Moon? (a) (1) increase (b) 4.9 s

56. ●● The motion of a particle is described by the curve for System A in Fig. 13.24. Write the equation of motion in terms of a cosine function. (5.0 cm) $\cos(\pi t/4)$

57. ●● The motion of a 0.25-kg mass oscillating on a light spring is described by the curve for System B in Fig. 13.24. (a) Write the equation for the displacement of the mass as a function of time. (b) What is the spring constant of the spring? (a) $y = -(0.10\text{ m}) \sin(10\pi/3)t$ (b) $k = 27$ N/m

58. ●●● The forces acting on a simple pendulum are shown in ▼ Fig. 13.25. (a) Show that, for the small-angle approximation ($\sin \theta \approx \theta$), the force producing the motion has the same form as Hooke's law. (b) Show by analogy with a mass on a spring that the period of a simple pendulum is given by $T = 2\pi\sqrt{L/g}$. [*Hint:* Think of the effective spring constant.] see ISM

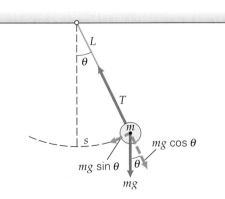

▲ **FIGURE 13.25 SHM of a pendulum** See Exercise 58.

59. ●●● A simple pendulum is set into small-angle motion making a maximum angle with the vertical of 5°. Its period is 2.21 s. (a) Determine its length. (b) Determine its maximum speed. (c) What is its maximum angular speed? (a) 1.21 m (b) 0.301 m/s (c) 0.248 rad/s

60. ●●● A clock uses a pendulum that is 75 cm long. The clock is accidentally broken, and when it is repaired, the length of the pendulum is shortened by 2.0 mm. Consider the pendulum to be a simple pendulum. (a) Will the repaired clock gain or lose time? (b) By how much will the time indicated by the repaired clock differ from the correct time (taken to be the time determined by the original pendulum in 24 h)? (c) If the pendulum string were metal, would the surrounding temperature make a difference in the timekeeping of the clock? Explain.
(a) gain time (b) 1.9 min (c) yes, because of linear expansion

13.3 Wave Motion

61. **MC** Wave motion in a material medium involves (a) the propagation of a disturbance, (b) interparticle interactions, (c) the transfer of energy, (d) all of the preceding. (d)

62. **MC** For a periodic wave propagating at a speed v, the following relationship holds: (a) $\lambda = v/f$, (b) $v = \lambda/f$, (c) $v = \lambda f^2$, or (d) $f = \lambda/v$. (a)

63. **MC** A water wave is (a) transverse, (b) longitudinal, (c) a combination of transverse and longitudinal, (d) none of the preceding. (c)

64. **CQ** What type(s) of wave(s), transverse or longitudinal, will propagate through (a) solids, (b) liquids, and (c) gases? (a) transverse and longitudinal (b) longitudinal (c) longitudinal

65. ● ▼ Figure 13.26 shows pictures of two mechanical waves. Identify each as transverse or longitudinal. see ISM

▲ **FIGURE 13.26 Transverse or longitudinal?** See Exercise 65.

66. Standing on a hill and looking at a tall wheat field, you see a beautiful wave traveling across the field when there is a breeze. What type of wave is this? Explain. longitudinal

67. ● A longitudinal sound wave has a speed of 340 m/s in air. If this wave produces a tone with a frequency of 1000 Hz, what is its wavelength? 0.340 m

68. ● A transverse wave has a wavelength of 0.50 m and a frequency of 20 Hz. What is the wave speed? 10 m/s

69. ● A student reading his physics book on a lake dock notices that the distance between two incoming wave crests is about 0.75 m, and he then measures the time of arrival between the crests to be 1.6 s. What is the approximate speed of the waves? 0.47 m/s

70. ● Light waves travel in a vacuum at a speed of 300 000 km/s. The frequency of visible light is about 5×10^{14} Hz. What is the approximate wavelength of the light? 6×10^{-7} m

71. ●● The range of sound frequencies audible to the human ear extends from about 20 Hz to 20 kHz. If the speed of sound in air is 345 m/s, what are the limits of this audible range, expressed in wavelengths? 1.7 cm to 17 m

72. ● A certain laser emits light of wavelength 633×10^{-9} m. What is the frequency of this light in a vacuum? 4.74×10^{-14} Hz

73. ●● A laser gives off a light wave with a wavelength of 500 nm at a frequency of 4.00×10^{14} Hz. Is this light traveling in a vacuum? Prove your answer. no, see ISM

74. IE ●● The AM frequencies on a radio dial range from 550 kHz to 1600 kHz, and the FM frequencies range from 88.0 MHz to 108 MHz. All of these radio waves travel at a speed of 3.00×10^8 m/s (speed of light). (a) Compared with the FM frequencies, the AM frequencies have (1) longer, (2) the same, or (3) shorter wavelengths. Why? (b) What are the wavelength ranges of the AM band and the FM band? (a) (1) longer (b) AM: 188 m to 545 m; FM: 2.78 m to 3.41 m

75. ●● A sonar generator on a submarine produces periodic ultrasonic waves at a frequency of 2.50 MHz. The wavelength of the waves in seawater is 4.80×10^{-4} m. When the generator is directed downward, an echo reflected from the ocean floor is received 10.0 s later. How deep is the ocean at that point? (Assume that wavelength is constant at all depths.) 6.00 km

76. ●● A wave traveling in the +x-direction is shown in ▶Fig. 13.27a. The particle displacement at a particular location in the medium through which the wave travels is shown in Fig. 13.27b. (a) What is the amplitude of the traveling wave? (b) What is the wave speed? (a) 0.15 m (b) 0.15 m/s

77. ● Assume that P and S (primary and secondary) waves from an earthquake with a focus near the Earth's surface travel through the Earth at nearly constant average speeds of 8.0 km/s and 6.0 km/s, respectively. Assume that there is no deflection or refraction of the waves. (a) How long is the delay between the arrivals of successive waves at a seismic monitoring station located 90° in latitude from the epicenter (the spot on the surface directly above the

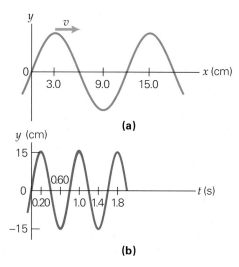

▲ **FIGURE 13.27 How high and how fast?** See Exercise 76.

focus) of the quake? (b) Do the waves cross the boundary of the mantle? (c) How long do the waves take to arrive at a monitoring station on the opposite side of the Earth?
(a) 3.8×10^2 s (b) yes, see ISM (c) 1.6×10^3 s

78. ●●● The speed of longitudinal waves traveling in a long, solid rod is given by $v = \sqrt{Y/\rho}$, where Y is Young's modulus and ρ is the density of the solid. If a disturbance has a frequency of 40 Hz, what is the wavelength of the waves it produces in (a) an aluminum rod and (b) a copper rod? [*Hint:* See Tables 9.1 and 9.2.]
(a) 1.3×10^2 m (b) 88 m

79. ●●● Fred strikes a steel train rail with a hammer at a frequency of 2.50 Hz, and Wilma puts her ear to the rail 1.0 km away. (a) How long after the first strike does Wilma hear the sound? (b) What is the time interval between the successive sound pulses she hears? [*Hint:* See Tables 9.1 and 9.2 and Exercise 78.] (a) 0.20 s (b) 0.40 s

80. ●●● A sinusoidal traveling transverse string wave has a frequency of 10.0 Hz and travels at 25.0 m/s along the *x*-axis. (a) Locate the points on the string that have maximum speed at any given time. (b) Determine the maximum speed, and (c) the distance between successive high and low spots on the string.
(a) x = 0 (crossing axis) (b) 3.45 m/s (c) 1.25 m

13.4 Wave Properties

81. **MC** When two waves meet each other and interfere, the resultant waveform is determined by (a) reflection, (b) refraction, (c) diffraction, (d) superposition. (d)

82. **MC** Refraction (a) involves constructive interference, (b) refers to a change in direction at media interfaces, (c) is identical to diffraction, (d) occurs only for media or mechanical waves. (b)

83. **MC** You can often hear people talking from around a corner of a building. This is due primarily to (a) reflection, (b) refraction, (c) interference, (d) diffraction. (d)

84. CQ What is destroyed when destructive interference occurs? What happens to energy? Explain. only waveform; energy is not destroyed, but redistributed

85. CQ Dolphins and bats determine the location of their prey by emitting ultrasonic sound waves. Which wave phenomenon is involved? reflection (this is called echolocation)

86. CQ If sound waves were dispersive (that is, if the speed of sound depended on its frequency), what consequences would you experience if you were listening to an orchestra in a concert hall? sounds from different instruments would arrive at different times

13.5 Standing Waves and Resonance

87. MC For two traveling waves to form standing waves, the waves must have the same (a) frequency, (b) amplitude, (c) speed, (d) all of the preceding. (d)

88. MC The points of maximum amplitude on a rope with a standing wave waveform are called (a) nodes, (b) antinodes, (c) fundamentals, (d) resonance points. (b)

89. MC When a stretched violin string oscillates in its third harmonic mode, the standing wave in the string will exhibit (a) 3 wavelengths, (b) 1/3 wavelength, (c) 3/2 wavelengths, (d) 2 wavelengths. (c)

90. CQ A child's swing (a pendulum) has only one natural frequency f_1, yet it can be driven or pushed smoothly at frequencies of $f_1/2, f_1/3, 2f_1$, and $3f_1$. How is this possible? see ISM

91. CQ By rubbing the circular lip of a thin wine glass with a moist finger, you can make the glass "sing." (Try it.) (a) What causes this? (b) What would happen to the frequency of the sound if you added water to the glass? see ISM

92. CQ A wave travels in a string with a fixed tension. How does the wavelength change if the frequency is increased? How does the wave speed change? decreases; no change

93. CQ Given the same tension and length, will a thicker or thinner guitar string sound higher in frequency? Why? thinner; speed is higher and so is frequency

94. • The fundamental frequency of a stretched string is 150 Hz. What are the frequencies of (a) the second harmonic and (b) the third harmonic? (a) 300 Hz (b) 450 Hz

95. • If the frequency of the third harmonic of a vibrating string is 450 Hz, what is the frequency of the first harmonic? 150 Hz

96. • A standing wave is formed in a stretched string that is 3.0 m long. What are the wavelengths of (a) the first harmonic and (b) the third harmonic? (a) 6.0 m (b) 2.0 m

97. IE •• A piece of rubber tubing is stretched by a force. (a) If the force doubles, the transverse wave speed (1) doubles, (2) halves, (3) increases by $\sqrt{2}$, (4) decreases by $\sqrt{2}$. Why? (b) If the linear mass density of a 10.0-m length of tubing is 0.125 kg/m and it is stretched by a force of 9.00 N, what is the transverse wave speed in the tubing? (c) What are its waves' natural frequencies? see ISM

98. •• Will a standing wave be formed in a 4.0-m length of stretched string that transmits waves at a speed of 12 m/s if it is driven at a frequency of (a) 15 Hz or (b) 20 Hz? (a) yes (b) no, see ISM

99. •• Two waves of equal amplitude and wavelength of 0.80 m travel in opposite directions at a speed of 250 m/s in a string. If the string is 2.0 m long, for which harmonic mode is the standing wave set up in the string? $n = 5$

100. •• On a violin, a correctly tuned A string has a frequency of 440 Hz. If an A string produces sound at 450 Hz under a tension of 500 N, what should the tension be to produce the correct frequency? 478 N

101. •• A university physics department buys 1000 m of string and determines its total mass to be 1.50 kg. This string is used to set up a standing-wave laboratory demonstration between two posts 3.0 m apart. If the desired first overtone (second harmonic) frequency is 35 Hz, what is the required string tension? 16.5 N

102. •• Two stretched strings A and B have the same tension and linear mass density. Are any of the first six harmonics of the strings equal if the string lengths are (a) 1.0 m and 3.0 m or (b) 1.5 m and 2.0 m, respectively? see ISM

103. •• You are setting up two standing string waves. You have a length of uniform piano wire that is 3.0 m long and has a mass of 150 g. You cut this into two lengths, one of 1.0 m and the other of 2.0 m, and place each length under tension. What should be the ratio of tensions (expressed as short to long) so that their fundamental frequencies are the same? $\frac{1}{4}$

104. IE •• A violin string is tuned to a certain frequency (the fundamental frequency, or first harmonic). (a) If a violinist wants a higher frequency, should the string be (1) lengthened, (2) kept the same length, or (3) shortened? Why? (b) If the string is tuned to 520 Hz and the violinist puts a finger down on the string one eighth of the string length from the neck end, what is the frequency of the string when the instrument is played this way? (a) (3) shortened (b) 594 Hz

105. ••• A tight uniform string with a length of 2.50 m is tied down at both ends and placed under a tension of 100 N. When it vibrates in its second-overtone mode (draw a sketch) the sound given off has a frequency of 75.0 Hz. What is the mass of the string? 0.016 kg

106. ••• In a common laboratory experiment on standing waves, the waves are produced in a stretched string by an electrical vibrator that oscillates at 60 Hz (▼Fig. 13.28). The string runs over a pulley, and a hanger is suspended from the end. The tension in the string is varied by adding weights to the hanger. If the active length of the string (the part that vibrates) is 1.5 m and this length of the string has a mass of 0.10 g, what masses must be suspended to produce the first four harmonics in that length? 0.22 kg; 0.055 kg; 0.024 kg; 0.014 kg

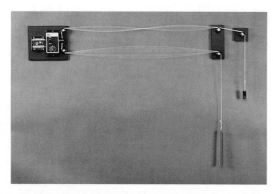

▲ **FIGURE 13.28 Standing waves in strings** Twin vibrating strings with standing waves. This demonstration model allows you to vary the string's tension, length, and type (linear mass density). Also, the vibration frequency can be adjusted. See Exercise 106.

Comprehensive Exercises

107. The velocity of a vertically oscillating mass–spring system is given by $v = (-0.600 \text{ m/s}) \sin(6t)$. (a) Where does the motion start and in what direction does the object move initially? Describe the initial phase of the motion. (b) Determine the period of the motion. (c) Determine the equation of motion (y). (d) Determine the maximum acceleration. (a) Released from rest, downward (b) 1.05 s (c) $y = (0.010 \text{ m}) \cos(6t)$ (d) 3.6 m/s²

108. A vertical spring has a 500-g mass attached to it and the mass is given an initial downward speed of 1.50 m/s. The mass travels downward 25.3 cm before stopping and returning. (a) Determine the spring constant. (b) What is its speed after it falls 5.00 cm? (c) What is the acceleration of the mass at the very bottom of the motion? (a) 56.3 N (b) 1.72 m/s (c) 18.7 m/s² (up)

109. A student cuts 5.00 m of string from a spool and determines that its mass is 10.0 g. 2.00 m of this string is then stretched and "tweaked." The first overtone or second harmonic of this string vibrates at 25.0 Hz. (a) Calculate the tension the string is under. (b) Calculate the fundamental frequency for this string. (c) If you wanted to increase the fundamental frequency by 25%, where would you grasp the string and which part would you "tweak"? (a) 5.00 N (b) 12.5 Hz (c) 0.40 m from one end

110. During an earthquake, the corner of a tall building oscillates with an amplitude of 20 cm at 0.50 Hz. What are the magnitudes of (a) the maximum displacement, (b) the maximum velocity, and (c) the maximum acceleration of the corner of the building? (Assume SHM.) (a) 20 cm from equilibrium (b) 0.63 m/s (c) 2.0 m/s²

111. A mass resting on a horizontal frictionless surface is connected to a fixed spring. The mass is displaced 16 cm from its equilibrium position and released. At $t = 0.50$ s, the mass is 8.0 cm from its equilibrium position (and has not passed through it yet). What is the period of oscillation of the mass? 3.0 s

PHYSLET® The following Physlet Physics Problems can be used with this chapter. 16.2, 16.3, 16.5, 16.6, 16.7, 16.8, 17.1, 17.2, 17.3, 17.4, 17.5, 17.6, 17.8

SOUND

PHYSICS FACTS

- Sound is (a) the physical propagation of a disturbance (energy) in a medium, and (b) the physiological and psychological response generally to pressure waves in air. (For example, see the discussion of a tree falling in the forest in the chapter introduction.)

- Humans cannot hear sounds with frequencies below 20 Hz—infrasound. Both elephants and rhinoceroses communicate by infrasound. Infrasound is produced by avalanches, meteors, tornadoes, earthquakes, and ocean waves. Some migratory birds are able to hear infrasounds produced when ocean waves break, which allows them to orient themselves with the coastline.

- The normal audible frequency range of human hearing is between 20 Hz and 20 kHz.

- The visible part of the outer ear is called the *pinna* or ear flap. Many animals move the ear flap in order to focus their hearing in a certain direction. Humans cannot do so.

- Ultrasound (frequency > 20 kHz) is used to make fetal images—"baby's first picture."

- Loud noise exposure—for example, from rock bands—is a common cause of tinnitus, or ringing in the ears.

Good vibrations, clearly! We owe a lot to sound waves. Not only do they provide us with one of our main sources of enjoyment in the form of music, but they also bring us a wealth of vital information about our environment, from the chime of a doorbell to the warning shrill of a police siren to the song of a mockingbird. Indeed, sound waves are the basis for our major form of communication: speech. These waves can also constitute a highly irritating distraction (noise). But sound waves become music, speech, or noise only when our ears perceive them. Physically, sound is simply waves that propagate in solids, liquids, and gases. Without a medium, there can be no sound; in a vacuum such as outer space, there is utter silence.

This distinction between the sensory and physical meanings of sound gives you a way to answer the old philosophical question: If a tree falls in the forest where there is no one to hear it, does it make a sound? The answer depends on how sound is defined—it is no if we are thinking in terms of sensory hearing, but yes if we are considering only the physical waves. Since sound waves are all around us most of the time, we are exposed to many interesting sound phenomena. You'll explore some of the most important of these in this chapter.

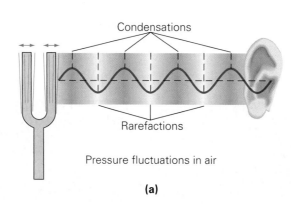

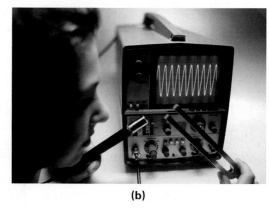

(a) **(b)**

▲ **FIGURE 14.1 Vibrations make waves (a)** A vibrating tuning fork disturbs the air, producing alternating high-pressure regions (condensations) and low-pressure regions (rarefactions), which form sound waves. **(b)** After being picked up by a microphone, the pressure variations are converted to electrical signals. When these signals are displayed on an oscilloscope, the sinusoidal waveform is evident.

Illustration 18.2 *Molecular View of a Sound Wave*

Demonstration/activity: Show how sound is produced by striking the desk, activating a tuning fork, blowing across a bottle, whistling, talking, swinging a flexible corrugated pipe around in circles, using an oscillator and small speaker, and so on.

Demonstration/activity: Show that air is necessary for sound to be produced and transmitted by placing a sound source in a bell jar and removing the air.

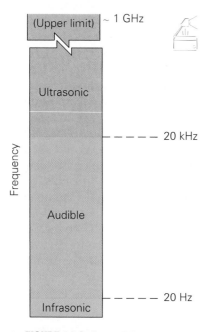

▲ **FIGURE 14.2 Sound frequency spectrum** The audible region of sound for humans lies between about 20 Hz and 20 kHz. Below this is the infrasonic region, and above it is the ultrasonic region. The upper limit is about 1 GHz, because of the elastic limitations of materials.

14.1 Sound Waves

OBJECTIVES: To (a) define sound, and (b) explain the sound frequency spectrum.

For sound waves to exist there must be a disturbance or vibrations in some medium. This disturbance may be the clapping of hands or the skidding of tires as a car comes to a sudden stop. Under water, you can hear the click of rocks against one another. If you put your ear to a thin wall, you can hear sounds from the other side of the wall. **Sound waves** in gases and liquids (both are fluids) are primarily longitudinal waves. However, sound disturbances moving through solids can have both longitudinal and transverse components. The intermolecular interactions in solids are much stronger than in fluids and allow transverse components to propagate.

The characteristics of sound waves can be visualized by considering those produced by a tuning fork, essentially a metal bar bent into a U shape (▲ Fig. 14.1). The prongs, or tines, vibrate when struck. The fork vibrates at its fundamental frequency (with an antinode at the end of each tine), so a single tone is heard. (A *tone* is sound with a definite frequency.) The vibrations disturb the air, producing alternating high-pressure regions called *condensations* and low-pressure regions called *rarefactions*. Assuming the fork steadily vibrates, the disturbances propagate outward, and a series of them can be described by a sinusoidal wave.

When the disturbances traveling through the air reach the ear, the eardrum (a thin membrane) is set into vibration by the pressure variations. On the other side of the eardrum, tiny bones (the hammer, anvil, and stirrup) carry the vibrations to the inner ear, where they are picked up by the auditory nerve. (See Insight 14.2 on the ear on page 475.)

Characteristics of the ear limit the perception of sound. Only sound waves with frequencies between about 20 Hz and 20 kHz (kilohertz) initiate nerve impulses that are interpreted by the human brain as sound. This frequency range is called the **audible region** of the **sound frequency spectrum** (◀Fig. 14.2). Hearing is most acute in the 1000–10 000 Hz range, with speech mainly in the 1000–4000 Hz range.

Infrasound

Frequencies lower than 20 Hz are in the **infrasonic region**. Waves in this region, which humans are unable to hear, are found in nature. Longitudinal waves generated by earthquakes have infrasonic frequencies, and these waves are used to study the Earth's interior (see Insight 13.1, p. 450). Infrasonic waves, or *infrasound*, are also generated by wind and weather patterns. Elephants and cattle have hearing responses in the infrasonic region and may even give early warnings of earthquakes and weather disturbances, such as tornadoes. (Elephants can detect sounds with frequencies as low as 1 Hz, but the pigeon takes the infrasound hearing prize, being able to detect sound frequencies as low as 0.1 Hz.) It has been found that the vortex of a tornado produces infrasound. Also, the frequency changes, with low frequencies when the vortex is small and higher frequencies when the vortex is large. Infrasound can be detected miles away from a tornado, and so may be a method for gaining increased warning times for tornado approaches.

There are infrasound listening stations. Nuclear explosions produce infrasound, and after the Nuclear Test Ban Treaty of 1963, infrasound listening stations were set up to detect possible violations. Now these stations can be used to detect other sources such as earthquakes and tornadoes.

Ultrasound

Above 20 kHz is the **ultrasonic region**. Ultrasonic waves can be generated by high-frequency vibrations in crystals. Ultrasonic waves, or *ultrasound*, cannot be detected by humans, but can be by other animals. The audible region for dogs extends to about 40 kHz, so ultrasonic or "silent" whistles can be used to call dogs without disturbing people. Cats and bats have even higher audible ranges, up to about 70 kHz and 100 kHz, respectively.

There are many practical applications of ultrasound. Since ultrasound can travel for kilometers in water, it is used in sonar to detect underwater objects and their ranges, much like radar uses radio waves. Sound pulses generated by the sonar apparatus are reflected by underwater objects, and the resulting echoes are picked up by a detector. The time required for a sound pulse to make one round trip, together with the speed of sound in water, gives the distance or range of the object. Sonar also is widely used by fishermen to detect schools of fish, and in a similar manner, ultrasound is used in autofocus cameras. Distance measurement allows focal adjustments to be made.

There are applications of ultrasonic sonar in nature. Sonar appeared in the animal kingdom long before it was developed by human engineers. On their nocturnal hunting flights, bats use a kind of natural sonar to navigate in and out of their caves and to locate and catch flying insects (▼Fig. 14.3a). The bats emit pulses of ultrasound and track their prey by means of the reflected echoes. The technique is known as *echolocation*. The auditory system and data-processing capabilities of bats are truly amazing. (Note the size of the bat's ears in Fig. 14.3b.)

On the basis of the intensity of the echo, a bat can tell how big an insect is—the smaller the insect, the less intense is the echo. The direction of motion of an insect is sensed by the frequency of the echo. If an insect is moving away from the bat, the returning echo will have a lower frequency. If the insect is moving toward the bat, the

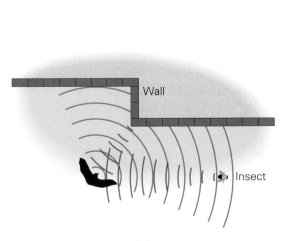

(a)　　　　　　　　　(b)

◀ **FIGURE 14.3** Echolocation **(a)** With the aid of their own natural sonar systems, bats hunt flying insects. The bats emit pulses of ultrasonic waves, which lie within their audible region, and use the echoes reflected from their prey to guide their attack. **(b)** Note the size of the ears—good for ultrasonic hearing. Do you know why bats roost hanging upside down? See text for answer.

Wall

Insect

INSIGHT   14.1 ULTRASOUND IN MEDICINE

Probably the best known applications of ultrasound are in medicine. For instance, ultrasound is used to obtain an image of a fetus, avoiding potentially dangerous X-rays. Ultrasonic generators (transducers) made of piezoelectric materials produce high-frequency pulses that are used to scan the designated region of the body.* When the pulses encounter a boundary between two tissues that have different densities, the pulses are reflected. These reflections are monitored by a receiving transducer, and a computer constructs an image from the reflected signals. Images of the fetus are recorded several times each second as the transducer is scanned across the mother's abdomen. A still shot or "echogram" of a fetus is shown in Fig. 1. A developing fetus,

*When an electric field is applied to a piezoelectric material, it undergoes mechanical distortion. Periodic applications allow the generation of ultrasonic waves. Conversely, when the material experiences mechanical pressure, an electric voltage develops. This allows the detection of ultrasonic waves.

which is surrounded by a sac containing the amniotic fluid, can be distinguished from other anatomical features, and the position, size, sex, and possible abnormalities may be detected.

Ultrasound can be used to assess stroke risk. Plaque deposits may accumulated on the inner walls of blood vessels and restrict blood flow. One of the major causes of stroke is the obstruction of the carotid artery in the neck, which directly affects the blood supply to the brain. The presence and severity of such obstructions may be detected by using ultrasound (Fig. 2). An ultrasonic generator is placed on the neck, and the reflections from blood cells moving through the artery are monitored to determine the rate of blood flow, thereby providing an indication of the severity of any blockage. This procedure involves shifting the frequency of the reflected waves, as described by the *Doppler effect*. (There will be more on this effect in Section 14.5, along with more on Doppler "flow meters.")

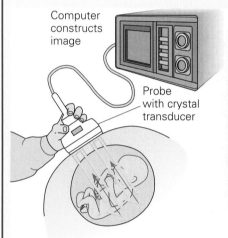

Computer
constructs
image

Probe
with crystal
transducer

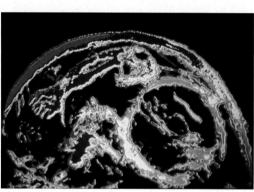

FIGURE 1 Ultrasound in use
Ultrasound generated by transducers, which convert electrical oscillations into mechanical vibrations and vice versa, is transmitted through tissue and is reflected from internal structures. The reflected waves are detected by the transducers, and the signals are used to construct an image, or echogram, shown here for a well-developed fetus.

echo will have a higher frequency. The change in frequency is known as the *Doppler effect*, which is presented in more detail in Section 14.5. Dolphins also use ultrasonic sonar to locate objects. This is very efficient since sound travels almost five times as fast in water as in air.

The bat, the only mammal to have evolved true flight, is a much maligned and feared creature. However, because they feed on tons of insects yearly, bats save the environment from a lot of insecticides. "Blind as a bat" is a common expression, yet bats have fairly good vision, which complements their use of echolocation. Finally, do you know why bats roost and hang upside down (Fig. 14.3b)? That is their take-off position. Unlike birds, bats can't launch themselves from the ground. Their wings don't produce enough lift to allow takeoff directly from the ground, and their legs are so small and underdeveloped that they can't run to build up takeoff speed. So they use their claws to hang, and fall into flight when they are ready to fly.

Ultrasound is used to clean teeth with ultrasonic toothbrushes. In industrial and home applications, ultrasonic baths are used to clean metal machine parts, dentures, and jewelry. The high-frequency (short-wavelength) ultrasound vibrations loosen particles in otherwise inaccessible places. Perhaps the best known medical application of ultrasound is to view a fetus without exposing it to harmful X-rays. (See Insight 14.1 on Ultrasound in Medicine.) Also, ultrasound is used to diagnose gallstones and kidney stones, and can be used to break these up by a technique called *lithotripsy* (Greek, "stone breaking").

Another widely used ultrasonic device is the ultrasonic scalpel, which uses ultrasonic energy for both precise cutting and coagulation. Vibrating at about 55 kHz, the scalpel makes small incisions, at the same time causing a protein clot to form that seals blood vessels—"bloodless" surgery, so to speak. The ultrasonic scalpel has been used in gynecological procedures such as the removal of fibroid tumors, in tonsillectomies, and many other types of surgical procedures.[†]

In cases of uncontrolled bleeding, such as blunt trauma resulting from a car accident or severe wounds received in com-

[†]One of the most remarkable and complicated inventions of nature is the blood clot. It can be life-saving–when it magically forms and stops a site of bleeding, or it can be life-threatening–when it blocks an artery in the heart or the brain.

bat, rapid hemostasis (termination of bleeding) is essential. Solutions being investigated and developed include the use of diagnostic ultrasound to detect the site of bleeding and high-intensity focused ultrasound (HIFU) to induce hemostasis by ultrasonic cauterization. In China, ultrasound-guided HIFU has been used successfully for several years and is becoming the treatment of choice for many forms of cancer.

Note: Adapted from the plenary lecture given by Dr. Lawrence A. Crum at the 18th International Congress on Acoustics in Kyoto, Japan, in the summer of 2004. Professor Crum is at the Applied Physics Laboratory at the University of Washington in Seattle, Washington.

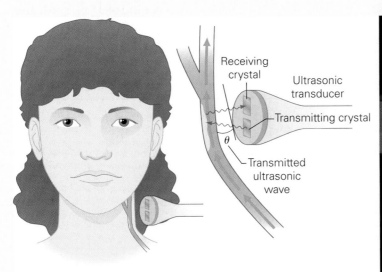 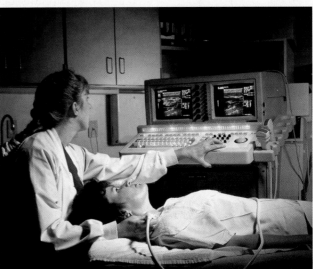

FIGURE 2 Carotid artery blockage? Ultrasound is used to measure the blood flow through the carotid artery to see if there is a blockage. See text for description.

Ultrasonic frequencies extend into the megahertz (MHz) range, but the sound frequency spectrum does not continue indefinitely. There is an upper limit of about 10^9 Hz, or 1 GHz (gigahertz), which is determined by the upper limit of the elasticity of the materials through which the sound propagates.

14.2 The Speed of Sound

OBJECTIVES: To (a) tell how the speed of sound differs in different media, and (b) describe the temperature dependence of the speed of sound in air.

In general, the speed at which a disturbance moves through a medium depends on the elasticity and density of the medium. For example, as you learned in Chapter 13, the wave speed in a stretched string is given by $v = \sqrt{F_T/\mu}$, where F_T is the tension in the string and μ is the linear mass density of the string.

Similar expressions describe wave speeds in solids and liquids, for which the elasticity is expressed in terms of moduli (Chapter 9). In general, the speed of sound in a solid and in a liquid is given by $v = \sqrt{Y/\rho}$ and $v = \sqrt{B/\rho}$, respectively, where Y is Young's modulus, B is the bulk modulus, and ρ is the density. The speed of sound in a gas is inversely proportional to the square root of the molecular mass, but the equation will not be presented here.

Demonstration/activity: Measure the speed of sound in aluminum, steel, brass, and other materials. Hold a lab equipment-support rod lightly at its center and strike the end with a small hammer to produce a longitudinal standing wave with a node in the center. Many students can quickly match the resulting frequency with a variable oscillator and speaker. The speed is then the product of the frequency and twice the length of the rod. Don't confuse the longitudinal wave with the lower-frequency transverse wave that is also produced.

TABLE 14.1

Speed of Sound in Various Media (typical values)

Medium	Speed (m/s)
Solids	
Aluminum	5100
Copper	3500
Iron	4500
Glass	5200
Polystyrene	1850
Zinc	3200
Liquids	
Alcohol, ethyl	1125
Mercury	1400
Water	1500
Gases	
Air (0°C)	331
Air (100°C)	387
Helium (0°C)	965
Hydrogen (0°C)	1284
Oxygen (0°C)	316

Solids are generally more elastic than liquids, which in turn are more elastic than gases. In a highly elastic material, the restoring forces between the atoms or molecules cause a disturbance to propagate faster. Thus, the speed of sound is generally about two to four times as fast in solids as in liquids and about ten to fifteen times as fast in solids as in gases such as air (Table 14.1).

Although not expressed explicitly in the preceding equations, the speed of sound generally depends on the temperature of the medium. In dry air, for example, the speed of sound is 331 m/s (about 740 mi/h) at 0°C. As the temperature increases, so does the speed of sound. For *normal environmental temperatures*, the speed of sound in air increases by about 0.6 m/s for each degree Celsius above 0°C. Thus, a good approximation of the speed of sound in air for a particular (environmental) temperature is given by

$$v = (331 + 0.6T_C) \text{ m/s} \quad \textit{speed of sound in dry air} \quad (14.1)$$

where T_C is the air temperature in degrees Celsius.* Although not written explicitly, the units associated with the factor 0.6 are meters per second per Celsius degree $[\text{m}/(\text{s} \cdot \text{C}°)]$.

Let's take a comparative look at the speed of sound in different media.

Example 14.1 ■ Solid, Liquid, Gas: Speed of Sound in Different Media

From their material properties, find the speed of sound in (a) a solid copper rod, (b) liquid water, and (c) air at room temperature (20°C).

Thinking It Through. We know that the speed of sound in a solid or a liquid depends on the elastic modulus and the density of the solid or liquid. These values are available in Tables 9.1 and 9.2. The speed of sound in air is given by Eq. 14.1.

Solution.

Given: $Y_{Cu} = 11 \times 10^{10} \text{ N/m}^2$ *Find:* (a) v_{Cu} (speed in copper)
$B_{H_2O} = 2.2 \times 10^9 \text{ N/m}^2$ (b) v_{H_2O} (speed in water)
$\rho_{Cu} = 8.9 \times 10^3 \text{ kg/m}^3$ (c) v_{air} (speed in air)
$\rho_{H_2O} = 1.0 \times 10^3 \text{ kg/m}^3$
(all values from Tables 9.1 and 9.2)
$T_C = 20°C$ (for air)

(a) To find the speed of sound in a copper rod, we use the expression $v = \sqrt{Y/\rho}$:

$$v_{Cu} = \sqrt{\frac{Y}{\rho}} = \sqrt{\frac{11 \times 10^{10} \text{ N/m}^2}{8.9 \times 10^3 \text{ kg/m}^3}} = 3.5 \times 10^3 \text{ m/s}$$

(b) For water, $v = \sqrt{B/\rho}$:

$$v_{H_2O} = \sqrt{\frac{B}{\rho}} = \sqrt{\frac{2.2 \times 10^9 \text{ N/m}^2}{1.0 \times 10^3 \text{ kg/m}^3}} = 1.5 \times 10^3 \text{ m/s}$$

(c) For air at 20°C, by Eq. 14.1, we have

$$v_{air} = (331 + 0.6T_C) \text{ m/s} = [331 + 0.6(20)] \text{ m/s} = 343 \text{ m/s} = 3.43 \times 10^2 \text{ m/s}$$

Follow-Up Exercise. In this Example, how many times faster is the speed of sound in copper (a) than in water and (b) than in air (at room temperature)? Compare your results with the values given at the beginning of the section. (*Answers to all Follow-Up Exercises are at the back of the text.*)

---

*A better approximation of these and higher temperatures is given by the expression

$$v = \left(331\sqrt{1 + \frac{T_C}{273}}\right) \text{ m/s}$$

In Table 14.1, see v for air at 100°C, which is outside the normal environmental temperature range.

A generally useful approximate value for the speed of sound in air is $\frac{1}{3}$ km/s (or $\frac{1}{5}$ mi/s). Using this value, you can, for example, estimate how far away lightning is by counting the number of seconds between the time you observe the flash and the time you hear the associated thunder. Because the speed of light is so fast, you see the lightning flash almost instantaneously. The sound waves of the thunder travel relatively slowly, at about $\frac{1}{3}$ km/s. For example, if the interval between the two events is measured to be 6 s (often by counting "one thousand one, one thousand two, ..."), the lightning stroke was about 2 km away $\left(\frac{1}{3} \text{ km/s} \times 6 \text{ s} = 2 \text{ km, or } \frac{1}{5} \text{ mi/s} \times 6 \text{ s} = 1.2 \text{ mi}\right)$.

You may also have noticed the delay in the arrival of sound relative to that of light at a baseball game. If you're sitting in the outfield stands, you see the batter hit the ball before you hear the crack of the bat.

Example 14.2 ■ Good Approximations?

(a) Show how good the approximations of $\frac{1}{3}$ km/s and $\frac{1}{5}$ mi/s are for the speed of sound. Use room temperature and dry air conditions. (b) Find the percent error of each.

Thinking It Through. Taking the actual speed of sound to be given by Eq. 14.1, and converting $\frac{1}{3}$ km/s and $\frac{1}{5}$ mi/s to m/s, comparisons can be made.

Solution. Listing what is given, along with the calculation of the speed of sound:

Given: $T_C = 20°C$ (room temperature)   *Find:* (a) How approximations compare to
$\qquad v = (331 + 0.6T_C) \text{ m/s}$ actual value
$\qquad\quad = [331 + 0.6(20)] \text{ m/s} = 343 \text{ m/s}$ (b) Percent errors
$\qquad v_{km} = \frac{1}{3} \text{ km/s}$
$\qquad v_{mi} = \frac{1}{5} \text{ mi/s}$

(a) Then, doing the conversions:

$$v_{km} = \tfrac{1}{3} \text{ km/s} (10^3 \text{ m/km}) = 333 \text{ m/s}$$

$$v_{mi} = \tfrac{1}{5} \text{ mi/s} (1609 \text{ m/mi}) = 322 \text{ m/s}$$

The approximations are somewhat reasonable, with v_{km} being the better.

(b) The percent error is given by the absolute difference of the values, divided by the accepted value times 100%. So, (where units cancel)

$$v_{km} = \tfrac{1}{3} \text{ km/s} \qquad \% \text{ error} = \frac{|343 - 333|}{343} \times 100\% = \frac{10}{343} \times 100\% = 2.9\%$$

$$v_{mi} = \tfrac{1}{5} \text{ mi/s} \qquad \% \text{ error} = \frac{|343 - 322|}{343} \times 100\% = \frac{21}{343} \times 100\% = 6.1\%$$

so the kilometers per second approximation is considerably better.

Follow-Up Exercise. Suppose the thunderstorm and lightning occurs on a very hot day with a dry-air temperature of 38°C. Would the percent errors in the Example increase or decrease? Justify your answer.

The speed of sound in air depends on various factors. Temperature is the most important, but there are other considerations, such as the homogeneity and composition of the air. For example, the air composition may not be "normal" in a polluted area. These effects are relatively small and will not be considered, except conceptually in the next Example.

Conceptual Example 14.3 ■ Speed of Sound: Sound Traveling Far and Wide

Note that the speed of sound in *dry* air for a given temperature is given to a good approximation by Eq. 14.1. However, the moisture content of the air (humidity) varies, and this variation affects the speed of sound. At the same temperature, would sound travel faster in (a) dry air or (b) moist air?

Reasoning and Answer. According to an old folklore saying, "Sound traveling far and wide, a stormy day will betide." This saying implies that sound travels faster on a highly humid day, when a storm or precipitation is likely. But is the saying true?

Near the beginning of this section, we learned that the speed of sound in a gas is inversely proportional to the square root of the molecular mass of the gas. So at constant pressure, is moist air more or less dense than dry air?

In a volume of moist air, a large number of water (H_2O) molecules occupy the space normally occupied by either nitrogen (N_2) or oxygen (O_2) molecules, which make up 98% of the air. Water molecules are less massive than both nitrogen and oxygen molecules. [From Section 10.3, the molecular (formula) masses are H_2O, 18 g; N_2, 28 g; and O_2, 32 g.] Thus, the average molecular mass of a volume of moist air is less than that of dry air, and the speed of sound is greater in moist air.

We can look at this situation another way: Since water molecules are less massive, they have less inertia and respond to the sound wave faster than nitrogen or oxygen molecules do. The water molecules therefore propagate the disturbance faster.

Follow-Up Exercise. Considering only molecular masses, would you expect the speed of sound to be greatest in nitrogen, oxygen, or helium (at the same temperature and pressure)? Explain.

Note: Humidity was included here as an interesting consideration for the speed of sound in air. However, henceforth, in computing the speed of sound in air at a certain temperature, we will consider only dry air (Eq. 14.1) unless otherwise stated.

Always keep in mind that our discussion generally assumes ideal conditions for the propagation of sound. Actually, the speed of sound depends on many things, one of which is humidity, as the preceding Conceptual Example shows. A variety of other properties affect the propagation of sound. As an example, let's ask the question, "Why do ships' foghorns have such a low pitch or frequency?" The answer is that low-frequency sound waves travel farther than high-frequency ones under identical conditions. This effect is explained by a couple of characteristics of sound waves. First, sound waves are attenuated (that is, they lose energy) because of the viscosity of the air (Section 9.5). Second, sound waves tend to interact with oxygen and water molecules in the air. The combined result of these two properties is that the total attenuation of sound in air depends on the frequency of the sound: the higher the frequency, the more the attenuation and the shorter the distance traveled. It turns out that the attenuation increases as the *square* of the frequency. For example, a 200-Hz sound will travel 16 times as far as an 800-Hz sound to obtain the same attenuation. So, low-frequency foghorns are used. With a shock wave dependence on frequency, you might notice that when a storm's lightning is farther away, the thunder you generally hear is a low-frequency rumble. (See Insight 14.2 for more on the ear and hearing.)

14.3 Sound Intensity and Sound Intensity Level

OBJECTIVES: To (a) define sound intensity and explain how it varies with distance from a point source, and (b) calculate sound intensity levels on the decibel scale.

Wave motion involves the propagation of energy. The rate of energy transfer is expressed in terms of **intensity**, which is the energy transported per unit time across a unit area. Since energy divided by time is power, intensity is power divided by area:

$$\text{intensity} = \frac{\text{energy/time}}{\text{area}} = \frac{\text{power}}{\text{area}} \qquad \left[I = \frac{E/t}{A} = \frac{P}{A} \right]$$

The standard units of intensity (power/area) are watts per square meter (W/m^2).

INSIGHT 14.2 THE PHYSIOLOGY AND PHYSICS OF THE EAR AND HEARING

The ear consists of three basic parts: the outer ear, the middle ear, and the inner ear (Fig. 1). The visible part of the ear is the *pinna* (or ear flap), and it collects and focuses sound waves. Many animals can move the ear flap in order to focus their hearing in a particular direction; humans have generally lost this ability and must turn the head. The sound enters the ear and travels through the *ear canal* to the *eardrum* of the middle ear.

The eardrum is a membrane that vibrates in response to the pressure of impinging sound waves. The vibrations are transmitted through the middle ear by an intricate set of three bones called the *malleus*, or hammer; the *incus*, or anvil; and the *stapes*, or stirrup. These bones form a linkage to the *oval window*, the opening to the inner ear. The eardrum transmits sound vibrations to the bones of the middle ear, which in turn transmits the vibrations through the oval window to the fluid of the inner ear.

The inner ear consists of the *semicircular canals*, the *cochlea*, and the *auditory nerve*. The semicircular canals and the cochlea are filled with a water-like liquid. The liquid and the nerve cells in the

semicircular canal play no role in the process of hearing but serve to detect rapid movements and assist in maintaining balance.

The inner surface of the cochlea, a snail-shaped organ, is lined with more than 25 000 hairlike nerve cells. These nerve cells differ from each other slightly in length and have different degrees of resiliency to the fluid waves passing through the cochlea. Different hair cells have sensitivity to particular frequencies of waves. When the frequency of a compressional wave matches the natural frequency of hair cells, the cells resonate (Section 13.5) with a larger amplitude of vibration. This causes the release of electrical impulses from the nerve cells, which are transmitted to the auditory nerve. The auditory nerve carries the signals to the brain, where they are interpreted as sound.

The hair cells of the cochlea are very critical to hearing. Damage to those cells can give rise to *tinnitus*, or "ringing in the ears." Exposure to loud noises is a common cause of tinnitus and often leads to hearing loss as well. After a loud rock concert in an enclosed room, people often hear a temporary ringing in the ears and some slight loss of hearing. Hair cells can be damaged by loud noises temporarily or permanently. Over time, loud sounds can cause permanent injury because hair cells are lost. Because the hair cells are (resonance) frequency specific, a person may be unable to hear sounds at particular frequencies.

In a quiet room, put both thumbs in your ears firmly and listen. Do you hear a low pulsating sound? You are hearing the sound, at about 25 Hz, made by the contracting and relaxing of the muscle fibers in your hands and arms. Although in the audible range, these sounds are not normally heard, because the human ear is relatively insensitive to low-frequency sounds.

The middle ear is connected to the throat by the Eustachian tube, the end of which is normally closed. It opens during swallowing and yawning to permit air to enter and leave, so that internal and external pressures are equalized. You have probably experienced a "stopping up" of your ears with a sudden change in atmospheric pressure (e.g., during rapid ascents or descents in elevators or airplanes). Swallowing opens the Eustachian tubes and relieves the excess pressure difference on the middle ear. (See Insight 9.2, An Atmospheric Effect: Possible Earaches, p. 311.)

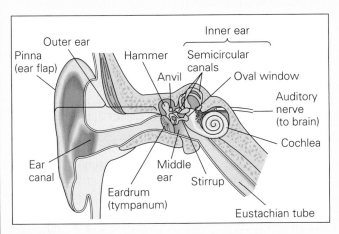

FIGURE 1 Anatomy of the human ear The ear converts pressure waves in the air into electrical nerve impulses that are interpreted as sounds by the brain.

Consider a point source that sends out spherical sound waves, as shown in ▼Fig. 14.4. If there are no losses, the sound intensity at a distance R from the source is

$$I = \frac{P}{A} = \frac{P}{4\pi R^2} \quad \text{(point source only)} \tag{14.2}$$

where P is the power of the source and $4\pi R^2$ is the area of a sphere of radius R, through which the sound energy passes perpendicularly.

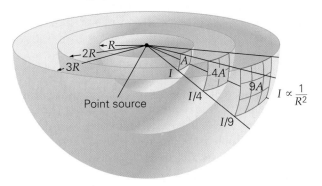

◀ **FIGURE 14.4 Intensity of a point source** The energy emitted from a point source spreads out equally in all directions. Since intensity is power divided by area, $I = P/A = P/(4\pi R^2)$, where the area is that of a spherical surface. The intensity then decreases with the distance from the source as $1/R^2$ (figure not to scale).

The intensity of a point source of sound is therefore *inversely proportional to the square of the distance from the source* (an inverse-square relationship). Two intensities at different distances from a point source of constant power can be compared as a ratio:

$$\frac{I_2}{I_1} = \frac{P/(4\pi R_2^2)}{P/(4\pi R_1^2)} = \frac{R_1^2}{R_2^2}$$

or

$$\frac{I_2}{I_1} = \left(\frac{R_1}{R_2}\right)^2 \quad \text{(point source only)} \tag{14.3}$$

Suppose that the distance from a point source is doubled; that is, $R_2 = 2R_1$, or $R_1/R_2 = \frac{1}{2}$. Then

$$\frac{I_2}{I_1} = \left(\frac{R_1}{R_2}\right)^2 = \left(\frac{1}{2}\right)^2 = \frac{1}{4}$$

and

$$I_2 = \frac{I_1}{4}$$

Since the intensity decreases by a factor of $1/R^2$, doubling the distance decreases the intensity to a quarter of its original value.

A good way to understand this inverse-square relationship intuitively is to look at the geometry of the situation. As Fig. 14.4 shows, the greater the distance from the source, the larger the area over which a given amount of sound energy is spread, and thus the lower its intensity. (Imagine having to paint two walls of different areas. If you had the same amount of paint to use on each, you'd have to spread it more thinly over the larger wall.) Since this area increases as the square of the radius R, the intensity decreases accordingly—that is, as $1/R^2$.

Sound intensity is perceived by the ear as **loudness**. On the average, the human ear can detect sound waves (at 1 kHz) with an intensity as low as 10^{-12} W/m². This intensity (I_o) is referred to as the *threshold of hearing*. Thus, for us to hear a sound, it must not only have a frequency in the audible range, but also be of sufficient intensity. As the intensity is increased, the perceived sound becomes louder. At an intensity of 1.0 W/m², the sound is uncomfortably loud and may be painful to the ear. This intensity (I_p) is called the *threshold of pain*.

Note that the thresholds of pain and hearing differ by a factor of 10^{12}:

$$\frac{I_p}{I_o} = \frac{1.0 \text{ W/m}^2}{10^{-12} \text{ W/m}^2} = 10^{12}$$

That is, the intensity at the threshold of pain is a *trillion* times that at the threshold of hearing. Within this enormous range, the perceived loudness is not directly proportional to the intensity. That is, if the intensity is doubled, the perceived loudness does not double. In fact, a doubling of perceived loudness corresponds approximately to an increase in intensity by a factor of 10. For example, a sound with an intensity of 10^{-5} W/m² would be perceived to be twice as loud as one with an intensity of 10^{-6} W/m². (The smaller the negative exponent, the larger the intensity.)

Sound Intensity Level: The Bel and the Decibel

It is convenient to compress the large range of sound intensities by using a logarithmic scale (base 10) to express *intensity levels* (not to be confused with sound intensity in W/m²). The intensity level of a sound must be referenced to a standard intensity, which is taken to be that of the threshold of hearing, $I_o = 10^{-12}$ W/m². Then, for any intensity I, the intensity level is the logarithm

Threshold of hearing: $I_o = 10^{-12}$ W/m²

Threshold of pain: $I_p = 1.0$ W/m²

(or log) of the ratio of I to I_o, that is, $\log I/I_o$. For example, if a sound has an intensity of $I = 10^{-6}$ W/m²,

$$\log \frac{I}{I_o} = \log \frac{10^{-6} \text{ W/m}^2}{10^{-12} \text{ W/m}^2} = \log 10^6 = 6 \text{ B}$$

(Recall that $\log_{10} 10^x = x$.)* The exponent of the power of 10 in the final log term is taken to have a unit called the **bel (B)**. Thus, a sound with an intensity of 10^{-6} W/m² has an intensity level of 6 B on this scale. That way, the intensity range from 10^{-12} W/m² to 1.0 W/m² is compressed into a scale of intensity levels ranging from 0 B to 12 B.

A finer intensity scale is obtained by using a smaller unit, the **decibel (dB)**, which is a tenth of a bel. The range from 0 to 12 B corresponds to 0 to 120 dB. In this case, the equation for the relative **sound intensity level**, or **decibel level (β)**, is

$$\beta = 10 \log \frac{I}{I_o} \qquad \text{(where } I_o = 10^{-12} \text{ W/m}^2\text{)} \tag{14.4}$$

Note that the sound intensity level (in decibels, which are dimensionless) is *not* the same as the sound intensity (in watts per square meter).

The decibel intensity scale and familiar sounds at some intensity levels are shown in ▼Fig. 14.5. Sound intensities can have detrimental effects on hearing, and because of this, the U.S. government has set occupational noise-exposure limits.

Note: The bel was named in honor of Alexander Graham Bell, the inventor of the telephone.

Sound intensity level in decibels

▼ **FIGURE 14.5 Sound intensity levels and the decibel scale** The intensity levels of some common sounds on the decibel (dB) scale.

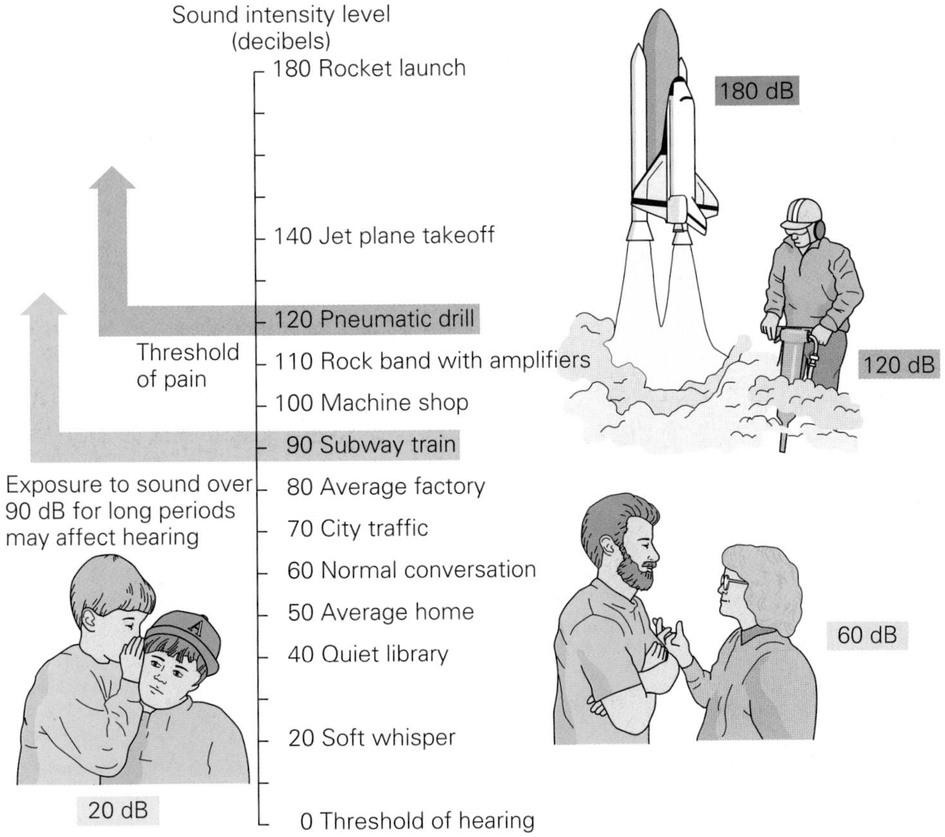

Sound intensity level (decibels)

- 180 Rocket launch — 180 dB
- 140 Jet plane takeoff
- 120 Pneumatic drill — 120 dB
- 110 Rock band with amplifiers
- 100 Machine shop
- 90 Subway train
- 80 Average factory
- 70 City traffic
- 60 Normal conversation — 60 dB
- 50 Average home
- 40 Quiet library
- 20 Soft whisper — 20 dB
- 0 Threshold of hearing

Threshold of pain

Exposure to sound over 90 dB for long periods may affect hearing

*A review of logarithms is given in Appendix I.

Example 14.4 ■ Sound Intensity Levels: Using Logarithms

What are the intensity levels of sounds with intensities of (a) 10^{-12} W/m^2 and (b) 5.0×10^{-6} W/m^2?

Thinking It Through. The sound intensity levels can be found by using Eq. 14.4.

Solution.

Given: (a) $I = 10^{-12}$ W/m^2 *Find:* (a) β (sound intensity level)
     (b) $I = 5.0 \times 10^{-6}$ W/m^2 (b) β

(a) Using Eq. 14.4, we have

$$\beta = 10 \log \frac{I}{I_o} = 10 \log\left(\frac{10^{-12} \text{ W/m}^2}{10^{-12} \text{ W/m}^2}\right) = 10 \log 1 = 0 \text{ dB}$$

The intensity 10^{-12} W/m^2 is the same as that at the threshold of hearing. (Recall that $\log 1 = 0$, since $1 = 10^0$ and $\log 10^0 = 0$.) Note that an intensity level of 0 dB does not mean that there is no sound.

(b) $\beta = 10 \log \frac{I}{I_o} = 10 \log\left(\frac{5.0 \times 10^{-6} \text{ W/m}^2}{10^{-12} \text{ W/m}^2}\right)$

$$= 10 \log(5.0 \times 10^6) = 10(\log 5.0 + \log 10^6) = 10(0.70 + 6.0) = 67 \text{ dB}$$

Follow-Up Exercise. Note in this Example that the intensity of 5.0×10^{-6} W/m^2 is halfway between 10^{-6} and 10^{-5} (or 60 and 70 dB), yet this intensity does not correspond to a midway value of 65 dB. (a) Why? (b) What intensity *does* correspond to 65 dB? (Compute it to three significant figures.)

Example 14.5 ■ Intensity Level Differences: Using Ratios

(a) What is the difference in the intensity levels if the intensity of a sound is doubled? (b) By what factors does the intensity increase for intensity level *differences* of 10 dB and 20 dB?

Thinking It Through. (a) If the intensity is doubled, then $I_2 = 2I_1$, or $I_2/I_1 = 2$. We can then use Eq. 14.4 to find the intensity difference. Recall that $\log a - \log b = \log a/b$. (b) Here, it is important to note that these values are intensity level *differences*, $\Delta\beta = \beta_2 - \beta_1$, *not* intensity *levels*. The equation developed in part (a) will work. (Why?)

Solution. Listing the data,

Given: (a) $I_2 = 2I_1$ *Find:* (a) $\Delta\beta$ (intensity level difference)
     (b) $\Delta\beta = 10$ dB   (b) I_2/I_1 (factors of increase)
        $\Delta\beta = 20$ dB

(a) Using Eq. 14.4 and the relationship $\log a - \log b = \log a/b$, we have, for the difference, $\Delta\beta = \beta_2 - \beta_1 = 10[\log(I_2/I_o) - \log(I_1/I_o)] = 10 \log[(I_2/I_o)/(I_1/I_o)] = 10 \log I_2/I_1$. Then,

$$\Delta\beta = 10 \log \frac{I_2}{I_1} = 10 \log 2 = 3 \text{ dB}$$

Thus, doubling the intensity increases the intensity level by 3 dB (such as an increase from 55 dB to 58 dB).

(b) For a 10-dB difference,

$$\Delta\beta = 10 \text{ dB} = 10 \log \frac{I_2}{I_1} \quad \text{and} \quad \log \frac{I_2}{I_1} = 1.0$$

Since $\log 10^1 = 1$, the intensity ratio is 10:1 because

$$\frac{I_2}{I_1} = 10^1 \quad \text{and} \quad I_2 = 10\, I_1$$

Similarly, for a 20-dB difference,

$$\Delta\beta = 20\ \text{dB} = 10\log\frac{I_2}{I_1} \quad \text{and} \quad \log\frac{I_2}{I_1} = 2.0$$

Since $\log 10^2 = 2$,

$$\frac{I_2}{I_1} = 10^2 \quad \text{and} \quad I_2 = 100\ I_1$$

Thus, an intensity-level difference of 10 dB corresponds to changing (increasing or decreasing) the intensity by a factor of 10. An intensity-level difference of 20 dB corresponds to changing the intensity by a factor of 100.

You should be able to guess the factor that corresponds to an intensity-level difference of 30 dB. In general, the factor of the intensity change is $10^{\Delta\beta}$, where $\Delta\beta$ is the difference in levels of bels. Since 30 dB = 3 B and $10^3 = 1000$, the intensity changes by a factor of 1000 for an intensity level difference of 30 dB.

Follow-Up Exercise. A $\Delta\beta$ of 20 dB and a $\Delta\beta$ of 30 dB correspond to factors of 100 and 1000, respectively, in intensity changes. Does a $\Delta\beta$ of 25 dB correspond to an intensity-change factor of 500? Explain.

Example 14.6 ■ Combined Sound Levels: Adding Intensities

Sitting at a sidewalk restaurant table, a friend talks to you in normal conversation (60 dB). At the same time, the intensity level of the street traffic reaching you is also 60 dB. What is the total intensity level of the combined sounds?

Thinking It Through. It is tempting simply to add the two sound intensity levels together and say that the total is 120 dB. But intensity levels in decibels are logarithmic, so you can't add them in the normal way. However, intensities (I) can be added arithmetically, since energy and power are scalar quantities. Then the combined intensity level can be found from the sum of the intensities.

Solution. We have the following information:

Given: $\beta_1 = 60\ \text{dB}$ *Find:* Total β
$\beta_2 = 60\ \text{dB}$

Let's find the intensities associated with the intensity levels:

$$\beta_1 = 60\ \text{dB} = 10\log\frac{I_1}{I_o} = 10\log\left(\frac{I_1}{10^{-12}\ \text{W/m}^2}\right)$$

By inspection, we have

$$I_1 = 10^{-6}\ \text{W/m}^2$$

Similarly, $I_2 = 10^{-6}\ \text{W/m}^2$, since both intensity levels are 60 dB. So the total intensity is

$$I_{\text{total}} = I_1 + I_2 = 1.0 \times 10^{-6}\ \text{W/m}^2 + 1.0 \times 10^{-6}\ \text{W/m}^2 = 2.0 \times 10^{-6}\ \text{W/m}^2$$

Then, converting back to intensity level, we get

$$\beta = 10\log\frac{I_{\text{total}}}{I_o} = 10\log\left(\frac{2.0 \times 10^{-6}\ \text{W/m}^2}{10^{-12}\ \text{W/m}^2}\right) = 10\log(2.0 \times 10^6)$$

$$= 10(\log 2.0 + \log 10^6) = 10(0.30 + 6.0) = 63\ \text{dB}$$

This value is a long way from 120 dB! Notice that the combined intensities doubled the intensity value, and the intensity level increased by 3 dB, in agreement with our finding in part (a) of Example 14.5.

Follow-Up Exercise. In this Example, suppose the added noise gave a total that *tripled* the sound intensity level of the conversation. What would be the total combined intensity level in this case?

▲ **FIGURE 14.6** **Protect your hearing** Continuous loud sounds can damage hearing, so our ears may need protection as shown here. Note in Table 14.2 that the intensity level of lawn mowers is on the order of 90 dB.

Protect Your Hearing

Hearing may be damaged by excessive noise, so our ears sometimes need protection from continuous loud sounds (▲Fig. 14.6). Hearing damage depends on the sound intensity level (decibel level) and the exposure time. The exact combinations vary for different people, but a general guide to noise levels is given in Table 14.2. Studies have shown that sound levels of 90 dB and above will damage receptor nerves in the ear, resulting in a loss of hearing. At 90 dB, it takes eight hours or less for damage to occur. In general, if the sound level is increased by 5 decibels, the safe exposure time is cut in half. For example, if a

TABLE 14.2		Sound Intensity Levels and Ear Damage Exposure Times	
	Decibels (dB)	Examples	Damage Can Occur with Nonstop Exposure
Faint	30	Quiet library, whispering	
Moderate	60	Normal conversation, sewing machine	
Very loud	80	Heavy traffic, noisy restaurant, screaming child	10 hours
	90	Lawn mower, motorcycle, loud party	Less than 8 hours
	100	Chainsaw, subway train, snowmobile	Less than 2 hours
Extremely loud	110	Stereo headset at full blast, rock concert	30 minutes
	120	Dance clubs, car stereos, action movies, some musical toys	15 minutes
	130	Jackhammer, loud computer games, loud sporting events	Less than 15 minutes
Painful	140	Boom stereos, gunshot blast, firecrackers	Any length (for example, hearing loss can occur from a few shots of a high-powered gun if protection is not worn)

Courtesy of The EAR Foundation.

sound level of 95 dB (that of a very loud lawn mower or motorcycle) takes four hours to damage your hearing, then a sound level of 105 dB takes only one hour to do damage.

14.4 Sound Phenomena

OBJECTIVES: To (a) explain the reflection, refraction, and diffraction of sound, and (b) distinguish between constructive and destructive interference.

Reflection, Refraction, and Diffraction

An echo is a familiar example of the *reflection* of sound—sound "bouncing" off a surface. Sound *refraction* is less obvious than reflection, but you may have experienced it on a calm summer evening, when it is possible to hear distant voices or other sounds that ordinarily would not be audible. This effect is due to the refraction, or bending (change in direction), of the sound waves as they pass from one region into a region where the air density is different. The effect is similar to what would happen if the sound passed into another medium.

The required conditions for sound to be refracted downward are a layer of cooler air near the ground or water and a layer of warmer air above it. These conditions occur frequently over bodies of water, which cool after sunset (▼Fig. 14.7). As a result of the cooling, the waves are refracted in an arc that may allow a distant person to receive an increased intensity of sound.

Another bending phenomenon is *diffraction*, described in Section 13.4. Sound may be diffracted, or bent, around corners or around an object. We usually think of waves as traveling in straight lines. However, you can hear someone you cannot see standing around a corner. This bending is different from that of refraction, in which no obstacle causes the bending.

Reflection, refraction, and diffraction are described in a general sense here for sound. These phenomena are important considerations for light waves as well and will be discussed more fully in Chapters 22 and 24.

▼ **FIGURE 14.7 Sound refraction** Sound travels more slowly in the cool air near the water surface than in the upper, warmer air. As a result, the waves are refracted, or bent downward. This bending increases the intensity of the sound at a distance where it otherwise might not be heard.

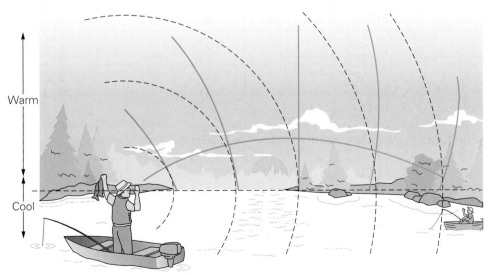

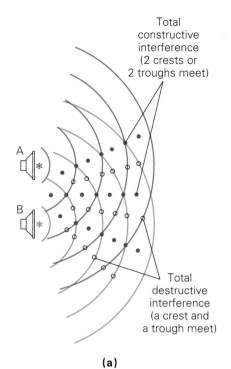

Total constructive interference (2 crests or 2 troughs meet)

Total destructive interference (a crest and a trough meet)

(a)

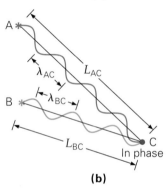

In phase

(b)

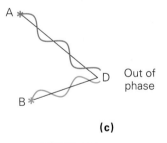

Out of phase

(c)

▲ **FIGURE 14.8 Interference**
(a) Sound waves from two point sources spread out and interfere. **(b)** At points where the waves arrive in phase (with a zero phase difference), such as point C, constructive interference occurs. **(c)** At points where the waves arrive completely out of phase (with a phase difference of 180°), such as point D, destructive interference occurs. The phase difference at a particular point depends on the path lengths the waves travel to reach that point.

Interference

Like waves of any kind, sound waves *interfere* when they meet. Suppose that two loudspeakers separated by some distance emit sound waves in phase at the same frequency. If we consider the speakers to be point sources, then the waves will spread out spherically and interfere (◄Fig. 14.8a). The lines from a particular speaker represent wave crests (or condensations), and the troughs (or rarefactions) lie in the intervening white areas.

In particular regions of space, there will be constructive or destructive interference. For example, if two waves meet in a region where they are exactly in phase (two crests or two troughs coincide), there will be total **constructive interference** (Fig. 14.8b). Notice that the waves have the same motion at point C in the figure. If, instead, the waves meet such that the crest of one coincides with the trough of the other (at point D), the two waves will cancel each other out (Fig. 14.8c). The result will be total **destructive interference**.

It is convenient to describe the path lengths traveled by the waves in terms of wavelength (λ) to determine whether they arrive in phase. Consider the waves arriving at point C in Fig. 14.8b. The path lengths in this case are $L_{AC} = 4\lambda$ and $L_{BC} = 3\lambda$. The **phase difference** ($\Delta\theta$) is related to the **path-length difference** (ΔL) by the simple relationship

$$\Delta\theta = \frac{2\pi}{\lambda}(\Delta L) \quad \begin{array}{l} \textit{phase difference} \\ \textit{and path-length difference} \end{array} \quad (14.5)$$

Since 2π rad is equivalent, in angular terms, to a full wave cycle or wavelength, multiplying the path-length difference by $2\pi/\lambda$ gives the phase difference in radians. For the example illustrated in Fig. 14.8b, we have

$$\Delta\theta = \frac{2\pi}{\lambda}(L_{AC} - L_{BC}) = \frac{2\pi}{\lambda}(4\lambda - 3\lambda) = 2\pi \text{ rad}$$

When $\Delta\theta = 2\pi$ rad, the waves are shifted by one wavelength. This is the same as $\Delta\theta = 0°$, so the waves are in phase. Thus, the waves interfere constructively in the region of point C, increasing the intensity, or loudness, of the sound detected there.

From Eq. 14.5, we see that the sound waves are in phase at any point where the path-length difference is zero or an integral multiple of the wavelength. That is,

$$\Delta L = n\lambda \quad (n = 0, 1, 2, 3, \dots) \quad \begin{array}{l} \textit{condition for} \\ \textit{constructive interference} \end{array} \quad (14.6)$$

A similar analysis of the situation in Fig. 14.8c, where $L_{AD} = 2\frac{3}{4}\lambda$ and $L_{BD} = 2\frac{1}{4}\lambda$, gives

$$\Delta\theta = \frac{2\pi}{\lambda}\left(2\tfrac{3}{4}\lambda - 2\tfrac{1}{4}\lambda\right) = \pi \text{ rad}$$

or $\Delta\theta = 180°$. At point D, the waves are completely out of phase, and destructive interference occurs in this region.

Sound waves will be out of phase at any point where the path-length difference is an odd number of half-wavelengths ($\lambda/2$), or

$$\Delta L = m\left(\frac{\lambda}{2}\right) \quad (m = 1, 3, 5, \dots) \quad \begin{array}{l} \textit{condition for} \\ \textit{destructive interference} \end{array} \quad (14.7)$$

At these points, a softer, or less intense, sound will be heard or detected. If the amplitudes of the waves are exactly equal, the destructive interference is total and no sound is heard.

Destructive interference of sound waves provides a way to reduce loud noises, which can be distracting and cause hearing discomfort. The procedure is to have a reflected wave or an introduced wave with a phase difference that cancels out the original sound as much as possible. Ideally, this would be 180° out of phase with the undesirable noise. A couple of such applications, automobile mufflers and pilot headphones, were discussed in Section 13.4.

Example 14.7 ■ Pump Up the Volume: Sound Interference

At an open-air concert on a hot day (with an air temperature of 25°C), you sit 7.00 m and 9.10 m, respectively, from a pair of speakers, one at each side of the stage. A musician, warming up, plays a single 494-Hz tone. What do you hear in terms of intensity? (Consider the speakers to be point sources.)

Thinking It Through. The sound waves from the speakers interfere. Is the interference, on which the intensity depends constructive, destructive, or something in between? This depends on the path-length difference, which can be computed from the given distances.

Solution.

Given: $d_1 = 7.00$ m and $d_2 = 9.10$ m   *Find:* ΔL (path-length difference
        $f = 494$ Hz   in wavelength units)
        $T = 25$°C

The path-length difference (2.10 m) between the waves arriving at your location must be expressed in terms of the wavelength of the sound. To do this, we first need to know the wavelength. Given the frequency, the wavelength can be found from the relationship $\lambda = v/f$, provided that the speed of sound, v, at the given temperature is known. The speed v can be found by using Eq. 14.1:

$$v = (331 + 0.6T_C) \text{ m/s} = [331 + 0.6(25)] \text{ m/s} = 346 \text{ m/s}$$

The wavelength of the sound waves is then

$$\lambda = \frac{v}{f} = \frac{346 \text{ m/s}}{494 \text{ Hz}} = 0.700 \text{ m}$$

Thus, the distances in terms of wavelength are

$$d_1 = (7.00 \text{ m})\left(\frac{\lambda}{0.700 \text{ m}}\right) = 10.0\lambda \quad \text{and} \quad d_2 = (9.10 \text{ m})\left(\frac{\lambda}{0.700 \text{ m}}\right) = 13.0\lambda$$

The path-length difference in terms of wavelengths is

$$\Delta L = d_2 - d_1 = 13.0\lambda - 10.0\lambda = 3.0\lambda$$

This is an integral number of wavelengths ($n = 3$), so constructive interference occurs. The sounds of the two speakers reinforce each other, and you hear an intense (loud) tone at 494 Hz.

Follow-Up Exercise. Suppose that in this Example the tone traveled to a person sitting 7.00 m and 8.75 m, respectively, from the two speakers. What would be the situation in that case?

Another interesting interference effect occurs when two tones of nearly the same frequency ($f_1 \approx f_2$) are sounded simultaneously. The ear senses pulsations in loudness known as **beats**. The human ear can detect as many as seven beats per second before they sound "smooth" (continuous, without any pulsations).

Suppose that two sinusoidal waves with the same amplitude, but slightly different frequencies, interfere (▼Fig. 14.9a). Figure 14.9b represents the resulting sound wave. The amplitude of the combined wave varies sinusoidally, as shown by the black curves (known as *envelopes*) that outline the wave.

What does this variation in amplitude mean in terms of what the listener perceives? A listener will hear a pulsating sound (beats), as determined by the envelopes. The maximum amplitude is 2A (at the point where the maxima of the two original waves interfere constructively). Detailed mathematics shows

Illustration 18.3 Interference in Time and Beats

(a)

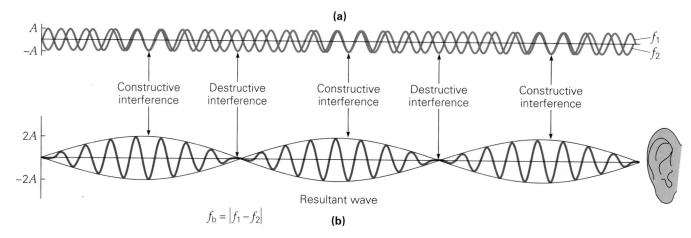

Constructive interference Destructive interference Constructive interference Destructive interference Constructive interference

Resultant wave

$f_b = |f_1 - f_2|$

(b)

▲ **FIGURE 14.9 Beats** Two traveling waves of equal amplitude and slightly different frequencies interfere and give rise to pulsating tones called beats. The beat frequency is given by $f_b = |f_1 - f_2|$.

that a listener will hear the beats at a frequency called the **beat frequency** (f_b), given by

$$f_b = |f_1 - f_2| \tag{14.8}$$

The absolute value is taken because the frequency f_b cannot be negative, even if $f_2 > f_1$. A negative beat frequency would be meaningless.

Beats can be produced when tuning forks of nearly the same frequency are vibrating at the same time. For example, using forks with frequencies of 516 Hz and 513 Hz, one can generate a beat frequency of $f_b = 516$ Hz $- 513$ Hz $= 3$ Hz, and three beats are heard each second. Musicians tune two stringed instruments to the same note by adjusting the tensions in the strings until the beats disappear ($f_1 = f_2$).

Exploration 18.3 A Microphone between Two Loudspeakers

14.5 The Doppler Effect

OBJECTIVES: To (a) describe and explain the Doppler effect, and (b) give some examples of its occurrences and applications.

If you stand along a highway and a car or truck approaches you with its horn blowing, the **pitch** (the perceived frequency) of the sound is higher as the vehicle approaches and lower as it recedes. You can also hear variations in the frequency of the motor noise when you watch a race car going around a track. A variation in the perceived sound frequency due to the motion of the source is an example of the **Doppler effect**. (The Austrian physicist Christian Doppler (1803–1853) first described this effect.)

As ▶Fig. 14.10 shows, the sound waves emitted by a moving source tend to bunch up in front of the source and spread out in back. The Doppler shift in frequency can be found by assuming that the air is at rest in a reference frame such as that depicted in ▶Fig. 14.11. The speed of sound in air is v, and the speed of the moving source is v_s. The frequency of the sound produced by the source is f_s. In one period, $T = 1/f_s$, a wave crest moves a distance $d = vT = \lambda$. (The sound wave would travel this distance in still air in any case, regardless of whether the source is moving.) But in one period, the source travels a distance $d_s = v_s T$ before emitting another wave crest. The distance between the successive wave crests is thus shortened to a wavelength λ':

$$\lambda' = d - d_s = vT - v_s T = (v - v_s)T = \frac{v - v_s}{f_s}$$

Demonstration/activity: Use an oscillator and a small speaker connected to a cord. Swing the speaker in a circular path parallel to the floor and around your head to demonstrate the Doppler effect. Use small wires taped to the cord to transmit the oscillator signal to the speaker. Try swinging the speaker as a simple pendulum.

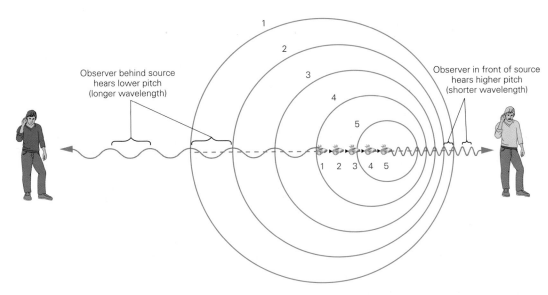

▲ **FIGURE 14.10 The Doppler effect for a moving source** The sound waves bunch up in front of a moving source—the whistle—giving a higher frequency there. The waves trail out behind the source, giving a lower frequency there. (The figure is not drawn to scale. Why?)

The frequency heard by the observer (f_o) is related to the shortened wavelength by $f_o = v/\lambda'$, and substituting λ' gives

$$f_o = \frac{v}{\lambda'} = \left(\frac{v}{v - v_s} \right) f_s$$

or

$$f_o = \left(\frac{1}{1 - \dfrac{v_s}{v}} \right) f_s \qquad \begin{array}{l} \textit{(source moving toward} \\ \textit{a stationary observer} \\ \textit{where } v_s = \textit{speed of source} \\ \textit{and } v = \textit{speed of sound)} \end{array} \qquad (14.9)$$

Since $1 - (v_s/v)$ is less than 1, f_o is greater than f_s in this situation. For example, suppose that the speed of the source is a tenth of the speed of sound; that is, $v_s = v/10$, or $v_s/v = \frac{1}{10}$. Then, by Eq. 14.9, $f_o = \frac{10}{9} f_s$.

PHYSLET®

Illustration 18.4 The Doppler Effect

▼ **FIGURE 14.11 The Doppler effect and wavelength** Sound from a moving car's horn travels a distance d in a time T. During this time, the car (the source) travels a distance d_s before putting out a second pulse, thereby shortening the observed wavelength of the sound in the approaching direction.

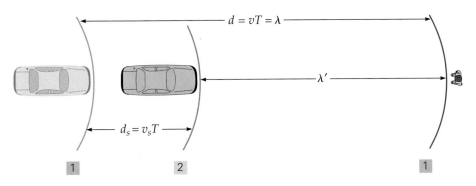

Similarly, when the source is moving away from the observer ($\lambda' = d + d_s$), the observed frequency is given by

$$f_o = \left(\frac{v}{v + v_s}\right)f_s = \left(\frac{1}{1 + \dfrac{v_s}{v}}\right)f_s \qquad \begin{array}{l}\textit{(source moving away}\\ \textit{from a stationary}\\ \textit{observer)}\end{array} \qquad (14.10)$$

Here, f_o is less than f_s. (Why?)

Combining Eqs. 14.9 and 14.10 yields a general equation for the observed frequency with a moving source and a stationary observer:

$$f_o = \left(\frac{v}{v \pm v_s}\right)f_s = \left(\frac{1}{1 \pm \dfrac{v_s}{v}}\right)f_s \qquad \left\{\begin{array}{l}- \textit{for source moving toward}\\ \quad\textit{stationary observer}\\ + \textit{for source moving away}\\ \quad\textit{from stationary observer}\end{array}\right. \qquad (14.11)$$

As you might expect, the Doppler effect also occurs with a moving observer and a stationary source, although this situation is a bit different. As the observer moves toward the source, the distance between successive wave crests is the normal wavelength (or $\lambda = v/f_s$), but the measured wave speed is different. Relative to the approaching observer, the sound from the stationary source has a wave speed of $v' = v + v_o$, where v_o is the speed of the observer and v is the speed of sound in still air. (The observer moving toward the source is moving in a direction opposite that of the propagating waves and thus meets more wave crests in a given time.)

With $\lambda = v/f_s$, the observed frequency is then

$$f_o = \frac{v'}{\lambda} = \left(\frac{v + v_o}{v}\right)f_s$$

or

$$f_o = \left(1 + \frac{v_o}{v}\right)f_s \qquad \begin{array}{l}\textit{(observer moving toward}\\ \textit{a stationary source}\\ \textit{where } v_o = \textit{speed of observer}\\ \textit{and } v = \textit{speed of sound})\end{array} \qquad (14.12)$$

Similarly, for an observer moving away from a stationary source, the perceived wave speed is $v' = v - v_o$, and

$$f_o = \frac{v'}{\lambda} = \left(\frac{v - v_o}{v}\right)f_s$$

or

$$f_o = \left(1 - \frac{v_o}{v}\right)f_s \qquad \begin{array}{l}\textit{(observer moving away}\\ \textit{from a stationary source)}\end{array} \qquad (14.13)$$

Equations 14.12 and 14.13 can be combined into a general equation for a moving observer and a stationary source:

$$f_o = \left(\frac{v \pm v_o}{v}\right)f_s = \left(1 \pm \frac{v_o}{v}\right)f_s \qquad \left\{\begin{array}{l}+ \textit{for observer moving}\\ \quad\textit{toward stationary source}\\ - \textit{for observer moving away}\\ \quad\textit{from stationary source}\end{array}\right. \qquad (14.14)$$

Problem-Solving Hint

You may find it difficult to remember whether a plus or minus sign is used in the general equations for the Doppler effect. Let your experience help you. For the common case of a stationary observer, the frequency of the sound increases when the source approaches, so the denominator in Eq. 14.11 must be smaller than the numerator. Accordingly, in this case you use the minus sign. When the source is receding, the frequency is lower. The denominator in Eq. 14.11 must then be larger than the numerator, and you use the plus sign. Similar reasoning will help you choose a plus or minus sign for the numerator in Eq. 14.14. See Eq. 14.14a in footnote.

Example 14.8 ■ On the Road Again: The Doppler Effect

Exploration 18.5   An Ambulance
Drives by with Its Siren On

As a truck traveling at 96 km/h approaches and passes a person standing along the highway, the driver sounds the horn. If the horn has a frequency of 400 Hz, what are the frequencies of the sound waves heard by the person (a) as the truck approaches and (b) after it has passed? (Assume that the speed of sound is 346 m/s.)

Thinking It Through. This situation is an application of the Doppler effect, Eq. 14.11, with a moving source and a stationary observer. In such problems, it is important to identify the data correctly.

Solution. In Doppler situations, it is important to clearly list what is to be found.

Given: $v_s = 96$ km/h $= 27$ m/s   *Find:* (a) f_o (observed frequency while
$\quad\quad\quad f_s = 400$ Hz $\quad\quad\quad\quad\quad$ truck is approaching)
$\quad\quad\quad v = 346$ m/s $\quad\quad\quad\quad$ (b) f_o (observed frequency while truck is
$\quad\quad\quad\quad\quad\quad\quad\quad\quad\quad\quad\quad\quad\quad\quad$ moving away)

(a) From Eq. 14.11 with a minus sign (source approaching stationary observer),

$$f_o = \left(\frac{v}{v - v_s}\right)f_s = \left(\frac{346 \text{ m/s}}{346 \text{ m/s} - 27 \text{ m/s}}\right)(400 \text{ Hz}) = 434 \text{ Hz}$$

(b) A plus sign is used in Eq. 14.11 when the source is moving away:

$$f_o = \left(\frac{v}{v + v_s}\right)f_s = \left(\frac{346 \text{ m/s}}{346 \text{ m/s} + 27 \text{ m/s}}\right)(400 \text{ Hz}) = 371 \text{ Hz}$$

Follow-Up Exercise. Suppose that the observer in this Example were initially moving toward and then past a stationary 400-Hz source at a speed of 96 km/h. What would be the observed frequencies? (Would they differ from those for the moving source?)

There are also cases in which both the source and the observer are moving, either toward or away from one another. We will not consider them mathematically here, but will do so conceptually in the next Conceptual Example.*

Conceptual Example 14.9 ■ It's All Relative: Moving Source and Moving Observer

Suppose a sound source and an observer are moving away from each other in opposite directions, each at half the speed of sound in air. In this case, the observer would (a) receive sound with a frequency higher than the source frequency, (b) receive sound with a frequency lower than the source frequency, (c) receive sound with the same frequency as the source frequency, or (d) receive no sound from the source.

(continues on next page)

*In the case of both the observer and source moving,

$$f_o = \left(\frac{v \pm v_o}{v \mp v_s}\right)f_s \quad\quad\quad\quad (14.14a)$$

From experience, you should know there would be a frequency increase when the observer and source approach each other, and vice versa. That is, the upper signs in the numerator and denominator apply if the observer and source move toward each other, and the lower signs apply if they are moving away. (See Exercise 71.)

Reasoning and Answer. As we know, when a source moves away from a stationary observer, the observed frequency is lower (Eq. 14.10). Similarly, when an observer moves away from a stationary source, the observed frequency is also lower (Eq. 14.13). With both source and observer moving away from each other in opposite directions, the combined effect would make the observed frequency even less, so neither (a) nor (c) is the answer.

It would appear that (b) is the correct answer, but we must logically eliminate (d) for completeness. Remember that the speed of sound relative to the air is constant. Therefore, (d) would be correct *only if the observer is moving faster than the speed of sound* relative to the air. Since the observer is moving at only half the speed of sound, (b) is the correct answer.

Follow-Up Exercise. In this Example, what would be the result if both the source and the observer were traveling in the same direction with the same subsonic speed? (*Subsonic*, as opposed to *supersonic*, refers to a speed that is less than the speed of sound in air.)

The Doppler effect also applies to light waves, although the equations describing the effect are different from those just given. When a distant light source such as a star moves away from us, the frequency of the light we receive from it is lowered. That is, the light is shifted toward the red (long-wavelength) end of the spectrum, an effect known as a *Doppler red shift*. Similarly, the frequency of light from an object approaching us is increased—the light is shifted toward the blue (short-wavelength) end of the spectrum, producing a *Doppler blue shift*. The magnitude of the shift is related to the speed of the source.

The Doppler shift of light from astronomical objects is very useful to astronomers. The rotation of a planet, a star, or some other body can be established by looking at the Doppler shifts of light from opposite sides of the object; because of the rotation, one side is receding (and hence is red-shifted) and the other is approaching (and thus is blue-shifted). Similarly, the Doppler shifts of light from stars in different regions of our galaxy, the Milky Way, indicate that the galaxy is rotating.

You have been subjected to a practical application of the Doppler effect if you have ever been caught speeding in your car by police radar, which uses reflected radio waves. (*Radar* stands for *ra*dio *d*etecting *a*nd *r*anging and is similar to underwater sonar, which uses ultrasound.) If radio waves are reflected from a parked car, the reflected waves return to the source with the same frequency. But for a car that is moving toward a patrol car, the reflected waves have a higher frequency, or are Doppler-shifted. Actually, there is a double Doppler shift: In receiving the wave, the moving car acts like a moving observer (the first Doppler shift), and in reflecting the wave, the car acts like a moving source emitting a wave (the second Doppler shift). The magnitudes of the shifts depend on the speed of the car. A computer quickly calculates this speed and displays it for the police officer.

For other important medical and weather applications of the Doppler effect, see Insight 4.3 on Doppler Applications: Blood Cells and Raindrops on page 490.

Sonic Booms

Consider a jet plane that can travel at supersonic speeds. As the speed of a moving source of sound approaches the speed of sound, the waves ahead of the source come close together (▸Fig. 14.12a). When a plane is traveling at the speed of sound, the waves can't outrun it, and they pile up in front. At supersonic speeds, the waves overlap. This overlapping of a large number of waves produces many points of constructive interference, forming a large pressure ridge, or *shock wave*. This kind of wave is sometimes called a *bow wave* because it is analogous to the wave produced by the bow of a boat moving through water at a speed greater than the speed of the water waves. Figure 14.12b shows the shock wave of a bullet traveling at 500 m/s.

From aircraft traveling at supersonic speed, the shock wave trails out to the sides and downward. When this pressure ridge passes over an observer on the

PHYSLET®

Exploration 18.4 Doppler Effect and the Velocity of the Source

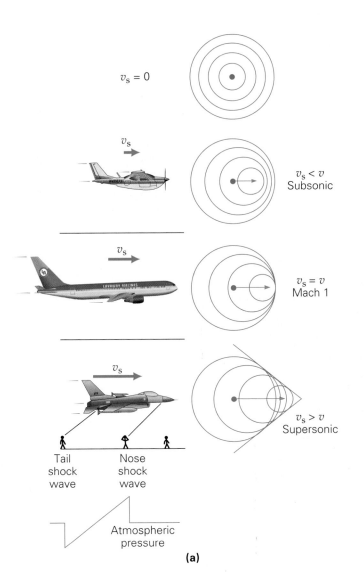

$v_s = 0$

v_s

$v_s < v$
Subsonic

v_s

$v_s = v$
Mach 1

v_s

$v_s > v$
Supersonic

Tail
shock
wave

Nose
shock
wave

Atmospheric
pressure

(a)

(b)

◀ **FIGURE 14.12 Bow waves and sonic booms (a)** When an aircraft exceeds the speed of sound in air, v_s, the sound waves form a pressure ridge, or shock wave. As the trailing shock wave passes over the ground, observers hear a sonic boom (actually, two booms, because shock waves are formed at the front and tail of the plane). **(b)** A bullet traveling at a speed of 500 m/s. Note the shock waves produced (and the turbulence behind the bullet). The image was made by using interferometry with polarized light and a pulsed laser, with an exposure time of 20 ns.

ground, the large concentration of energy produces what is known as a **sonic boom**. There is really a double boom, because shock waves are formed at both ends of the aircraft. Under certain conditions, the shock waves can break windows and cause other damage to structures on the ground. (Sonic booms are no longer heard as frequently as in the past. Pilots are now instructed to fly supersonically only at high altitudes and away from populated areas.)

On a smaller scale, you have probably heard a "mini" sonic boom—the "crack" of a whip. This must mean that the whip's tip has somehow attained supersonic speed. How does this happen? Whips generally taper down from the handle to the tip, which may have several frayed strands. When the whip is given a flick of the wrist, a wave pulse is sent down the length of the whip. Treating the whip pulse as a string wave pulse, recall that the speed of the pulse depends inversely on the linear mass density which decreases toward the tip. Thus the speed of the pulse increases to the point that at the tip, it is greater than the speed of sound. The "crack" is made by the air rushing back into the region of reduced pressure created by the final flip of the whip's tip, much as the sonic from a supersonic jet trails behind the jet.

A common misconception is that a sonic boom is heard only when a plane breaks the sound barrier. As an aircraft approaches the speed of sound, the pressure ridge in front of it is essentially a barrier that must be overcome with extra power. However, once supersonic speed is reached, the barrier is no longer there, and the shock waves, continuously created, trail behind the plane, producing booms along its ground path.

Illustration 18.5 *The Location of a Supersonic Airplane*

14.3 DOPPLER APPLICATIONS: BLOOD CELLS AND RAINDROPS

Blood Cells

As we saw in Insight 14.1, "Ultrasound in Medicine", on page 470, ultrasound provides a variety of uses in the medical field. Since the Doppler effect can be used to detect and provide information on moving objects, it is used to examine blood flow in the major arteries and veins of the arms and legs (Fig. 1a). In this application, the Doppler effect is used to measure the blood flow speed. Ultrasound reflects from red blood cells with a change in frequency according to the speed of the cells. The overall flow speed helps physicians diagnose such things as blood clots, arterial occlusion (closing), and venous insufficiency. Ultrasound procedures offer a less invasive alternative to other diagnostic procedures, such as arteriography (X-ray pictures of an artery after the injection of a dye).

Another medical use of ultrasound is the echocardiogram, which is an examination of the heart. On a monitor, this ultrasonic technique can display the beating movements of the heart, and the physician can see the heart's chambers, valves, and blood flow into and out of the organ (Fig. 1b).

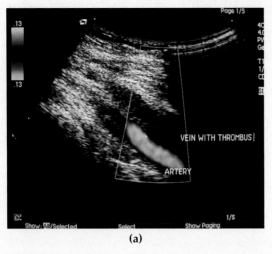

(a)

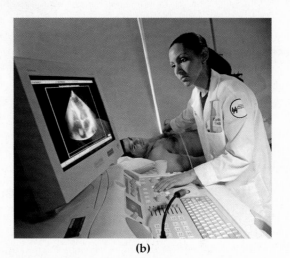

(b)

FIGURE 1 (a) **Blood flow and blockage** This Doppler ultrasound scan shows a deep vein thrombosis in a patient's leg. The thrombus (clot) blocking the vein is the dark area right of center. Blood flow in an adjacent artery (orange) is slowed due to the clot. In extreme cases, a clot can break away and be carried to the lungs, where it can block an artery and cause a potentially fatal pulmonary embolism (blockage of a blood vessel). (b) **Echocardiogram** This ultrasonic procedure can display the beating movements of the heart, the heart chambers, valves, and blood flow as it makes it way in and out of the organ.

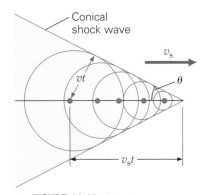

▲ **FIGURE 14.13 Shock wave cone and Mach number** When the speed of the source (v_s) is greater than the speed of sound in air (v), the interfering spherical sound waves form a conical shock wave that appears as a V-shaped pressure ridge when viewed in two dimensions. The angle θ is given by $\sin \theta = v/v_s$, and the inverse ratio v_s/v is called the Mach number.

Ideally, the sound waves produced by a supersonic aircraft form a cone-shaped shock wave (◄Fig. 14.13). The waves travel outward with a speed v, and the speed of the source (plane) is v_s. Note from the figure that the angle between a line tangent to the spherical waves and the line along which the plane is moving is given by

$$\sin \theta = \frac{vt}{v_s t} = \frac{v}{v_s} = \frac{1}{M} \qquad (14.15)$$

The inverse ratio of the speeds is called the **Mach number** (M), named after Ernst Mach (1838–1916), an Austrian physicist who used it in studying supersonics, and is given by

$$M = \frac{v_s}{v} = \frac{1}{\sin \theta} \qquad (14.16)$$

If v equals v_s, the plane is flying at the speed of sound, and the Mach number is 1 (that is, $v_s/v = 1$). Therefore, a Mach number less than 1 indicates a sub-

Raindrops

Radar has been used since the early 1940s to provide information about rainstorms and other forms of precipitation. This information is obtained from the intensity of the reflected signal. Such conventional radars can also detect the hooked (rotational) "signature" of a tornado, but only after the storm is well developed.

A major improvement in weather forecasting came about with the development of a radar system that could measure the Doppler frequency shift in addition to the magnitude of the echo signal reflected from precipitation (usually raindrops). The Doppler shift is related to the velocity of the precipitation blown by the wind.

A Doppler-based radar system (Fig. 2a) can penetrate a storm and monitor its wind speeds. The direction of a storm's wind-driven rain gives a wind "field" map of the affected region. Such maps provide strong clues of developing tornadoes, so meteorologists can detect tornadoes much earlier than was ever before possible (Fig. 2b). With Doppler radar, forecasters have been able to predict tornadoes as much as 20 min before they touch down, compared with just over 2 min for conventional radar. Doppler radar has saved many lives with this increased warning time. The National Weather Service has a network of Doppler radars around the United States, and Doppler radar scans are now common on both TV weather forecasts and the Internet.

Doppler radars installed at major airports have another use: to detect wind shears. Several airplane crashes and near-crashes have been attributed to downward wind bursts (also known as microbursts or downbursts). Such strong downdrafts cause wind shears capable of forcing landing aircraft to crash. Wind bursts generally result from high-speed downdrafts in the turbulence of thunderstorms, but they can also occur in clear air when rain evaporates high above ground. Since Doppler radar can detect the wind speed and the direction of raindrops in clouds, as well as dust and other objects floating in the air, it can provide an early warning against dangerous wind shear conditions. Two or three radar sites are needed to detect motions in two or three directions (dimensions), respectively.

(a)

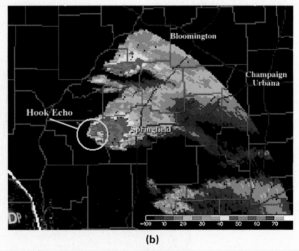

(b)

FIGURE 2 Doppler radar **(a)** A Doppler radar installation. **(b)** Doppler radar depicts the precipitation inside a thunderstorm. A hook echo is a signature of a possible tornado.

sonic speed, and a Mach number greater than 1 indicates a supersonic speed. In the latter case, the Mach number tells the speed of the aircraft in terms of a multiple of the speed of sound. A Mach number of 2, for instance, indicates a speed twice the speed of sound. Note that since $\sin \theta \leq 1$, no shock wave can exist unless $M \geq 1$.

14.6 Musical Instruments and Sound Characteristics

OBJECTIVE: To explain some of the sound characteristics of musical instruments in physical terms.

Musical instruments provide good examples of standing waves and boundary conditions. On some stringed instruments, different notes are produced by using finger pressure to vary the lengths of the strings (▸Fig. 14.14). As was learned in Chapter 13, the natural frequencies of a stretched string (fixed at each end, as is the case for the strings on an instrument) are $f_n = n(v/2L)$,

▲ **FIGURE 14.14 A shorter vibrating string, a higher frequency** Different notes are produced on stringed instruments such as guitars, violins, and cellos by placing a finger on a string to change its effective, or vibrating, length.

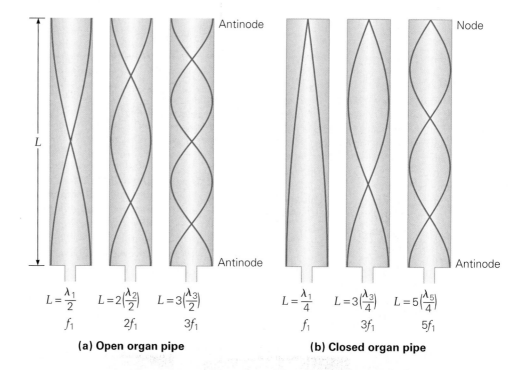

$$L = \frac{\lambda_1}{2} \qquad L = 2\left(\frac{\lambda_2}{2}\right) \qquad L = 3\left(\frac{\lambda_3}{2}\right)$$

$$f_1 \qquad\qquad 2f_1 \qquad\qquad 3f_1$$

(a) Open organ pipe

$$L = \frac{\lambda_1}{4} \qquad L = 3\left(\frac{\lambda_3}{4}\right) \qquad L = 5\left(\frac{\lambda_5}{4}\right)$$

$$f_1 \qquad\qquad 3f_1 \qquad\qquad 5f_1$$

(b) Closed organ pipe

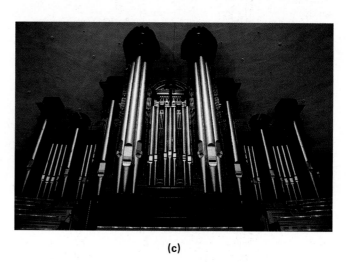

(c)

▲ **FIGURE 14.15 Organ pipes** Longitudinal standing waves (illustrated here as sinusoidal curves) are formed in vibrating air columns in pipes. **(a)** An open pipe has antinodes at both ends. **(b)** A closed pipe has a closed (node) end and an open (antinode) end. **(c)** A modern pipe organ. The pipes can be open or closed.

from Eq. 13.20, where the speed of the wave in the string is given by $v = \sqrt{F_T/\mu}$. Initially adjusting the tension in a string tunes it to a particular (fundamental) frequency. Then the effective length of the string is varied by finger location and pressure.

Standing waves can also be set up in wind instruments. For example, consider a pipe organ with fixed lengths of pipe, which may be open or closed (▲Fig. 14.15). An open pipe is open at both ends, and a closed pipe is closed at one end and open at the other (the end with the antinode). Analysis similar to that done in Chapter 13 for a stretched string with the proper boundary conditions shows that the natural frequencies of the pipes are

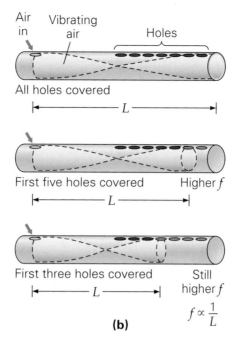

(a)   **(b)**

▲ **FIGURE 14.16 Wind instruments (a)** Wind instruments are essentially open tubes. (Here, Kenny G performs on a soprano saxophone at a Macy's Thanksgiving Day parade.) **(b)** The effective length of the air column, and hence the pitch of the sound, is varied by opening and closing holes along the tube. The frequency f is inversely proportional to the effective length L of the air column.

$$f_n = \frac{v}{\lambda_n} = n\left(\frac{v}{2L}\right) = nf_1 \qquad (n = 1, 2, 3, \dots)$$

(natural frequencies for an open pipe—open on both ends) (14.17)

and

$$f_m = \frac{v}{\lambda_m} = m\left(\frac{v}{4L}\right) = mf_1 \qquad (m = 1, 3, 5, \dots)$$

(natural frequencies for a closed pipe— closed on one end) (14.18)

where v is the speed of sound in air. Note that the natural frequencies depend on the length of the pipe. This is an important consideration in a pipe organ (Fig. 14.15c), particularly in selecting the dominant or fundamental frequency. (The diameter of the pipe is also a factor, but is not considered in this simple analysis.)

The same physical principles apply to wind and brass instruments. In all of these, human breath is used to create standing waves in an open tube. Most such instruments allow the player to vary the effective length of the tube and thus the pitch produced—either with the help of slides or valves that vary the actual length of tubing in which the air can resonate, as in most brasses, or by opening and closing holes in the tube, as in woodwinds (▲ Fig. 14.16).

Recall from Chapter 13 that a musical note or tone is referenced to the fundamental vibrational frequency of an instrument. In musical terms, the first overtone is the second harmonic, the second overtone is the third harmonic, and so on. Note that for a closed organ pipe (Eq. 14.18), the even harmonics are missing.

Example 14.10 ■ Pipe Dreams: Fundamental Frequency

A particular open organ pipe has a length of 0.653 m. Taking the speed of sound in air to be 345 m/s, what is the fundamental frequency of this pipe?

Thinking It Through. The fundamental frequency ($n = 1$) of an open pipe is given directly by Eq. 14.17. Physically, there is a half-wavelength ($\lambda/2$) in the length of the pipe, so $\lambda = 2L$.

Solution.

Given: $L = 0.653$ m 　　　　　　*Find:* f_1 (fundamental frequency)
　　　　$v = 345$ m/s (speed of sound)

With $n = 1$,

$$f_1 = \frac{v}{2L} = \frac{345 \text{ m/s}}{2(0.653 \text{ m})} = 264 \text{ Hz}$$

This frequency is middle C (C_4).

Follow-Up Exercise. A closed organ pipe has a fundamental frequency of 256 Hz. What would be the frequency of its first overtone? Is this frequency audible?

Perceived sounds are described by terms whose meanings are similar to those used to describe the physical properties of sound waves. Physically, a wave is generally characterized by intensity, frequency, and waveform (harmonics). The corresponding terms used to describe the sensations of the ear are loudness, pitch, and quality (or timbre). These general correlations are shown in Table 14.3. However, the correspondence is not perfect. The physical properties are objective and can be measured directly. The sensory effects are subjective and vary from person to person. (Think of temperature as measured by a thermometer and by the sense of touch.)

Sound intensity and its measurement on the decibel scale were covered in Section 14.3. Loudness is related to intensity, but the human ear responds differently to sounds of different frequencies. For example, two tones with the same intensity (in watts per square meter) but different frequencies might be judged by the ear to be different in loudness.

Frequency and *pitch* are often used synonymously, but again there is an objective–subjective difference: If the same low-frequency tone is sounded at two intensity levels, most people say that the more intense sound has a lower pitch, or perceived frequency.

The curves in the graph of intensity level versus frequency shown in ▼Fig. 14.17 are called *equal-loudness contours* (or Fletcher–Munson curves, after the researchers who generated them). These contours join points representing intensity–frequency combinations that a person with average hearing judges to be equally loud. The top curve shows that the decibel level of the threshold of pain (120 dB) does not vary a great deal over the normal hearing range, regardless of the frequency of the sound. In contrast, the threshold of hearing, represented by the lowest contour, varies widely

▶ **FIGURE 14.17 Equal-loudness contours** The curves indicate tones that are judged to be equally loud, although they have different frequencies and intensity levels. For example, on the lowest contour, a 1000-Hz tone at 0 dB sounds as loud as a 50-Hz tone at 40 dB. Note that the frequency scale is logarithmic to compress the large frequency range.

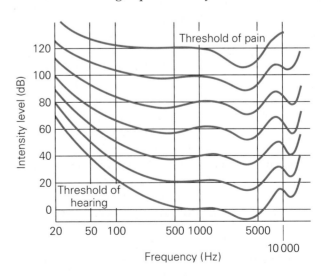

TABLE 14.3	General Correlation between Perceptual and Physical Characteristics of Sound	
Sensory Effect	*Physical Wave Property*	
Loudness	Intensity	
Pitch	Frequency	
Quality (timbre)	Waveform (harmonics)	

with frequency. For a tone with a frequency of 2000 Hz, the threshold of hearing is 0 dB, but a 20-Hz tone would have to have an intensity level of over 70 dB just to be heard (the extrapolated *y*-intercept of the lowest curve).

It is interesting to note the dips (or minima) in the curves. The hearing curves show a significant dip in the 2000–5000 Hz range, the ear being most sensitive around 4000 Hz. A tone with a frequency of 4000 Hz can be heard at intensity levels *below* 0 dB. The high sensitivity in the 2000–5000 Hz region is very important for the understanding of speech. (Why?) Another dip in the curves, or region of sensitivity, occurs at about 12 000 Hz.

The minima occur as a result of resonance in a closed cavity in the auditory canal (similar to a closed pipe). The length of the cavity is such that it has a fundamental resonance frequency of about 4000 Hz, resulting in extra sensitivity. As in a closed cavity, the next natural frequency is the third harmonic (see Eq. 14.18), which is three times the fundamental frequency, or about 12 000 Hz.

Example 14.11 ■ The Human Ear Canal: Standing Waves

Consider the human ear canal to be a cylindrical tube of length 2.54 cm (1.0 in.; see Fig. 1 in Insight 14.2 on p. 475). What would be the lowest sound resonance frequency? Take the air temperature in the canal to be 37°C.

Thinking It Through. The auditory ear canal as described is essentially a closed pipe—open at one end (outer ear canal) and closed at the other (eardrum). The lowest-frequency resonance standing wave that will fit in the pipe is $L = \lambda/4$ (Fig. 14.15), and $\lambda = 4L$. Then, the frequency is given by $f_1 = v/\lambda_1 = v/4L$ (Eq. 14.18), where v is the speed of sound in air.

Solution.

Given: $L = 2.54 \text{ cm} = 0.0254 \text{ m}$ *Find:*   lowest resonance frequency, f_1
             $T = 37°C$ (normal body temperature)

First finding the speed of sound at 37°C,

$$v = (331 + 0.6T_C) \text{ m/s} = [331 + 0.6(37.0)] \text{ m/s} = 353 \text{ m/s}$$

and

$$f_1 = \frac{v}{4L} = \frac{353 \text{ m/s}}{4(0.0254 \text{ m})} = 3.47 \times 10^3 \text{ Hz} = 3.47 \text{ kHz}$$

Compare with the curves in Fig. 14.17, and note the dip in the curves at about this frequency. How about the other dip just above 10 kHz? Check out the next natural frequency for the ear canal, f_3.

Follow-Up Exercise. Children have smaller ear canals than adults, on the order of 1.30 cm in length. What is the lowest fundamental frequency for a child's ear canal? Use the same air temperature as in the Example. (Note: With growth the ear canal lengthens, and it has been experimentally determined that the "adult" ear canal length and lowest fundamental frequency is reached about age 7.)

The **quality** of a tone is the characteristic that enables it to be distinguished from another tone of basically the same intensity and frequency. Tone quality depends on the waveform—specifically, the number of harmonics (overtones) present and their relative intensities (▼Fig. 14.18). The tone of a voice depends in large

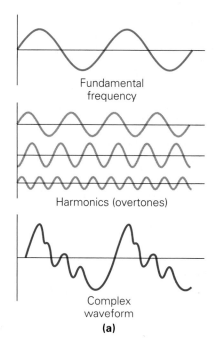

Fundamental frequency

Harmonics (overtones)

Complex waveform

(a)

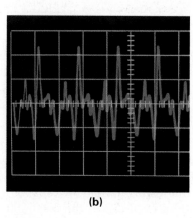

(b)

▲ **FIGURE 14.18 Waveform and quality (a)** The superposition of sounds of different frequencies and amplitudes gives a complex waveform. The harmonics, or overtones, determine the quality of the sound. **(b)** The waveform of a violin tone is displayed on an oscilloscope.

Exploration 18.2 *Creating Sounds by Adding Harmonics*

part on the vocal resonance cavities. One person can sing a tone with the same basic frequency and intensity as another, but different combinations of overtones give the two voices different qualities.

The notes of a musical scale correspond to certain frequencies; as we saw in Example 14.10, middle C (C_4) has a frequency of 264 Hz. When a note is played on an instrument, its assigned frequency is that of the first harmonic, which is the fundamental frequency. (The second harmonic is the first overtone, the third harmonic is the second overtone, and so on.) The fundamental frequency is dominant over the accompanying overtones that determine the sound quality of the instrument. Recall from Chapter 13 that the overtones that are produced depend on how an instrument is played. Whether a violin string is plucked or bowed, for example, can be discerned from the quality of identical notes.

Chapter Review

- The sound frequency spectrum is divided into infrasonic ($f < 20$ Hz), audible (20 Hz $< f <$ 20 kHz), and ultrasonic ($f > 20$ kHz) frequency regions.

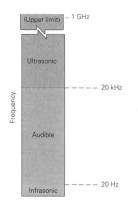

- The speed of sound in a medium depends on the elasticity of the medium and its density. In general, $v_{solids} > v_{liquids} > v_{gases}$.

  Speed of sound in dry air (meters per second):

  $$v = (331 + 0.6T_C) \text{ m/s} \qquad (14.1)$$

- The intensity of a point source is inversely proportional to the square of the distance from the source.

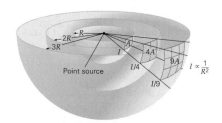

Intensity of a point source:

$$I = \frac{P}{4\pi R^2} \quad \text{and} \quad \frac{I_2}{I_1} = \left(\frac{R_1}{R_2}\right)^2 \quad (14.2, 14.3)$$

• The sound intensity level is a logarithmic function of the sound intensity and is expressed in decibels (dB).

Intensity level (in decibels, dB):

$$\beta = 10 \log \frac{I}{I_o} \quad \text{where} \quad I_o = 10^{-12} \text{ W/m}^2 \quad (14.4)$$

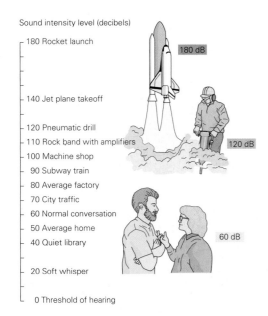

Sound intensity level (decibels)

- 180 Rocket launch
- 140 Jet plane takeoff
- 120 Pneumatic drill
- 110 Rock band with amplifiers
- 100 Machine shop
- 90 Subway train
- 80 Average factory
- 70 City traffic
- 60 Normal conversation
- 50 Average home
- 40 Quiet library
- 20 Soft whisper
- 0 Threshold of hearing

• Sound wave interference of two point sources depends on phase difference as related to path-length difference. Sound waves that arrive at a point in phase reinforce each other (constructive interference); sound waves that arrive at a point out of phase cancel each other (destructive interference).

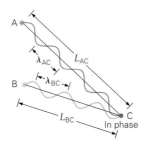

Phase difference (where ΔL is the path-length difference):

$$\Delta\theta = \frac{2\pi}{\lambda}(\Delta L) \quad (14.5)$$

Condition for constructive interference:

$$\Delta L = n\lambda \quad (n = 0, 1, 2, 3, \dots) \quad (14.6)$$

Condition for destructive interference:

$$\Delta L = m\left(\frac{\lambda}{2}\right) \quad (m = 1, 3, 5, \dots) \quad (14.7)$$

Beat frequency:

$$f_b = |f_1 - f_2| \quad (14.8)$$

• The Doppler effect depends on the velocities of the sound source and observer relative to still air. When the relative motion of the source and observer is toward each other, the observed pitch increases; when the relative motion of source and observer is away from each other, the observed pitch decreases.

Doppler effect:

Moving source, stationary observer

$$f_o = \left(\frac{v}{v \pm v_s}\right)f_s = \left(\frac{1}{1 \pm \frac{v_s}{v}}\right)f_s \quad (14.11)$$

$$\begin{cases} - \text{ for source moving toward stationary observer} \\ + \text{ for source moving away from stationary observer} \end{cases}$$

where v_s = speed of source
and v = speed of sound

Moving observer, stationary source

$$f_o = \left(\frac{v \pm v_o}{v}\right)f_s = \left(1 \pm \frac{v_o}{v}\right)f_s \quad (14.14)$$

$$\begin{cases} + \text{ for observer moving toward stationary source} \\ - \text{ for observer moving away from stationary source} \end{cases}$$

where v_o = speed of observer
and v = speed of sound

Moving observer and moving source

$$f_o = \left(\frac{v \pm v_o}{v \mp v_s}\right)f_s \quad (14.14a)$$

Upper signs if moving toward each other
Lower signs if moving away from each other

Angle for conical shock wave:

$$\sin\theta = \frac{vt}{v_s t} = \frac{v}{v_s} = \frac{1}{M} \quad (14.15)$$

Mach number:

$$M = \frac{v_s}{v} = \frac{1}{\sin\theta} \quad (14.16)$$

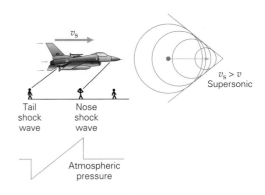

Natural frequencies of an open organ pipe—open on both ends:

$$f_n = n\left(\frac{v}{2L}\right) = nf_1 \qquad (n = 1, 2, 3, \dots) \qquad (14.17)$$

Natural frequencies of a closed organ pipe—closed on one end:

$$f_m = m\left(\frac{v}{4L}\right) = mf_1 \qquad (m = 1, 3, 5, \dots) \qquad (14.18)$$

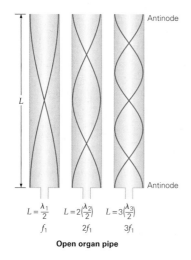

Open organ pipe

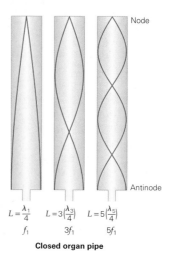

Closed organ pipe

Exercises

MC = *Multiple Choice Question,* **CQ** = *Conceptual Question, and* **IE** = *Integrated Exercise. Throughout the text, many exercise sections will include "paired" exercises. These exercise pairs, identified with* **red numbers***, are intended to assist you in problem solving and learning. In a pair, the first exercise (even numbered) is worked out in the Study Guide so that you can consult it should you need assistance in solving it. The second exercise (odd numbered) is similar in nature, and its answer is given at the back of the book.*

14.1 Sound Waves *and* 14.2 The Speed of Sound

1. **MC** A sound wave with a frequency of 15 Hz is in what region of the sound spectrum: (a) audible, (b) infrasonic, (c) ultrasonic, or (d) supersonic? (b)

2. **MC** A sound wave in air (a) is longitudinal, (b) is transverse, (c) has longitudinal and transverse components, (d) travels faster than a sound wave through a liquid. (a)

3. **MC** The speed of sound is generally greatest in (a) solids, (b) liquids, (c) gases, (d) a vacuum. (a)

4. **MC** The speed of sound in air (a) is about 1/3 km/s, (b) is about 1/5 mi/s, (c) depends on temperature, (d) all of the preceding. (d)

5. **CQ** Suggest a possible explanation of why some flying insects produce buzzing sounds and some do not. sounds not all in our audible range

6. **CQ** Explain why sound travels faster in warmer than in colder air. warmer air molecules vibrate faster

7. **CQ** Two sounds that differ in frequency are emitted from a single loudspeaker. Which sound will reach your ear first, the one with the lower or the higher frequency? same time

8. **CQ** The speed of sound in air depends on temperature. What effect, if any, should humidity have? speed increases with increasing humidity

9. ● The thunder from a lightning flash is heard by an observer 3.0 s after she sees the flash. What is the approximate distance to the lightning strike in (a) kilometers and (b) miles? (a) 1.0 km (b) 0.60 mi

10. ● What is the speed of sound in air at (a) 10°C and (b) 20°C? (a) 337 m/s (b) 343 m/s

11. ● The speed of sound in air on a summer day is 350 m/s. What is the air temperature? 32°C

12. ● Sonar is used to map the ocean floor. If an ultrasonic signal is received 2.0 s after it is emitted, how deep is the ocean floor at that location? 1.5×10^3 m

13. ● The wave speed in a liquid is given by $v = \sqrt{B/\rho}$, where B is the bulk modulus of the liquid and ρ is its density. Show that this equation is dimensionally correct. What about $v = \sqrt{Y/\rho}$ for a solid? (Y is Young's modulus.) see ISM

14. ●● A 1.00-m-long string with a mass of 15.0 g is placed under a tension of 35.0 N. It is plucked so it vibrates in its fundamental standing-wave mode. What are the (a) frequency and (b) wavelength of the sound? Assume normal room temperature. (a) 24.2 Hz (b) 14.2 m

15. IE ●● A tuning fork vibrates at a frequency of 256 Hz. (a) When the air temperature increases, the wavelength of the sound from the tuning fork (1) increases, (2) remains the same, (3) decreases. Why? (b) If the temperature rises from 0°C to 20°C, what is the change in the wavelength? (a) (1) increases (b) +0.047 m

16. ●● Particles approximately 3.0×10^{-2} cm in diameter are to be scrubbed loose from machine parts in an aqueous ultrasonic cleaning bath. Above what frequency should the bath be operated to produce wavelengths of this size and smaller? 5.0×10^6 Hz

17. ●● Medical ultrasound uses a frequency of around 20 MHz to diagnose human conditions and ailments. (a) If the speed of sound in tissue is 1500 m/s, what is the smallest detectable object? (b) If the penetration depth is about 200 wavelengths, how deep can this instrument penetrate? (a) 7.5×10^{-5} m (b) 1.5×10^{-2} m

18. ●● Brass is an alloy of copper and zinc. Does the addition of zinc to copper cause an increase or decrease in the speed of sound in brass rods compared to copper? Explain. decrease, see ISM.

19. ●● The speed of sound in steel is about 4.50 km/s. A steel rail is struck with a hammer, and an observer 0.400 km away has one ear to the rail. (a) How much time will elapse from the time the sound is heard through the rail until the time it is heard through the air? Assume that the air temperature is 20°C and that no wind is blowing. (b) How much time would elapse if the wind were blowing toward the observer at 36.0 km/h from where the rail was struck? (a) 1.08 s (b) 1.04 s

20. ●● A person holds a rifle horizontally and fires at a target. The bullet has a muzzle speed of 200 m/s, and the person hears the bullet strike the target 1.00 s after firing it. The air temperature is 72° F. What is the distance to the target? 127 m

21. ●● A freshwater dolphin sends ultrasonic sound to locate a prey. If the echo off the prey is received by the dolphin 0.12 s after being sent, how far is the prey from the dolphin? 90 m

22. ●● A submarine on the ocean surface gets a sonar echo indicating an underwater object. The echo comes back at an angle of 20° above the horizontal and the echo took 2.32 s to get back to the submarine. What is the object's depth? 595 m

23. ●● The speed of sound in human tissue is on the order of 1500 m/s. A 3.50-MHz probe is used for an ultrasonic procedure. (a) If the effective physical depth of the ultrasound is 250 wavelengths, what is the physical depth in meters? (b) What is the time lapse for the ultrasound to make a round trip if reflected from an object at the effective depth? (c) The smallest detail capable of being detected is on the order of one wavelength of the ultrasound. What would this be? (a) 0.107 m (b) 1.43×10^{-4} s (c) 4.29×10^{-4} m

24. ●● The size of your eardrum (the tympanum; see Fig. 1 in Insight 14.2 on p. 475) partially determines the upper frequency limit of your audible region, usually between 16 000 Hz and 20 000 Hz. If the wavelength is on the order of twice the diameter of the eardrum and the air temperature is 20° C, how wide is your eardrum? Is your answer reasonable? 8.6×10^{-3} m to 1.1×10^{-2} m; yes

25. IE ●●● On hiking up a mountain that has several overhanging cliffs, a climber drops a stone at the first cliff to determine its height by measuring the time it takes to hear the stone hit the ground. (a) At a second cliff that is twice the height of the first, the measured time of the sound from the dropped stone is (1) less than double, (2) double, or (3) more than double that of the first. Why? (b) If the measured time is 4.8 s for the stone dropping from the first cliff, and the air temperature is 20°C, how high is the cliff? (c) If the height of a third cliff is three times that of the first one, what would be the measured time for a stone dropped there to reach the ground? (a) (1) less than double (b) 1.0×10^2 m (c) 8.7 s

26. ●●● A bat moving at 15.0 m/s emits a high-frequency sound as it approaches a wall that is 25.0 m away. Assuming the bat continues straight toward the wall, how far away is he when he receives the echo? The temperature in the bat cave is a cool 8.0°C. 22.9 m

27. ●●● Sound propagating through air at 30°C passes through a vertical cold front into air that is 4.0°C. If the sound has a frequency of 2500 Hz, by what percentage does its wavelength change in crossing the boundary? 4.5%

14.3 Sound Intensity and Sound Intensity Level

28. **MC** If the air temperature increases, would the sound intensity from a constant-output point source (a) increase, (b) decrease, or (c) remain unchanged? (c)

29. **MC** The decibel scale is referenced to a standard intensity of (a) 1.0 W/m², (b) 10^{-12} W/m², (c) normal conversation, (d) the threshold of pain. (b)

30. **MC** If the intensity level of a sound at 20 dB is increased to 40 dB, the intensity would increase by a factor of (a) 10, (b) 20, (c) 40, (d) 100. (d)

31. **CQ** Where is the intensity greater and by what factor? (1) At a point a distance R from a power source P, or (2) at a point a distance $2R$ from a power source of $2P$. Explain. (1) by a factor of 2

32. **CQ** The Richter scale, used to measure the intensity level of earthquakes, is a logarithmic scale, as is the decibel scale. Why are such scales used? to compress a large physical range into a smaller numerical scale

33. **CQ** Can there be negative decibel levels, such as −10 dB? If so, what would these mean? yes, an intensity below the intensity of threshold of hearing

34. ● Calculate the intensity generated by a 1.0-W point source of sound at a location (a) 3.0 m and (b) 6.0 m from it. (a) 8.8×10^{-3} W/m^2 (b) 2.2×10^{-3} W/m^2

35. **IE** ● (a) If the distance from a point sound source triples, the sound intensity will be (1) 3, (2) 1/3, (3) 9, (4) 1/9 times the original value. Why? (b) By how much must the distance from a point source be increased to reduce the sound intensity by half? (a) (4) 1/9 (b) 1.4 times

36. ● Assuming that the diameter of your eardrum is 1 cm (see Exercise 24), what is the sound power received by the eardrum at the threshold of (a) hearing and (b) pain? (a) 8×10^{-17} W (b) 8×10^{-5} W

37. ● A middle C note (262 Hz) is sounded on a piano to help tune a violin string. When the string is sounded, nine beats are heard in 3.0 s. (a) How much is the violin string off tune? (b) Should the string be tightened or loosened to sound middle C?
(a) 3.0 Hz (b) not enough information given

38. ● Calculate the intensity level for (a) the threshold of hearing and (b) the threshold of pain. (a) 0 dB (b) 120 dB

39. ● Find the intensity levels in decibels for sounds with intensities of (a) 10^{-2} W/m^2, (b) 10^{-6} W/m^2, and (c) 10^{-15} W/m^2. (a) 100 dB (b) 60 dB (c) −30 dB

40. **IE** ●● (a) If the power of a sound source doubles, the intensity level at a certain distance from the source (1) increases, (2) exactly doubles, or (3) decreases. Why? (b) What are the intensity levels at a distance of 10 m from a 5.0-W and a 10-W source, respectively?
(a) (1) increases (b) 96 dB; 99 dB

41. ●● The intensity levels of two people holding a conversation are 60 dB and 70 dB, respectively. What is the intensity of the combined sounds? 1.1×10^{-5} W/m^2

42. ●● A person has a hearing loss of 30 dB for a particular frequency. What is the sound intensity that is heard at this frequency that has an intensity of the threshold of pain? 1.0×10^{-3} W/m^2

43. ●● Noise levels for some common aircraft are given in Table 14.4. What are the lowest and highest intensities for (a) takeoff and (b) landing? (a) 3.72×10^{-4} W/m^2; 1.00×10^{-1} W/m^2 (b) 9.55×10^{-3} W/m^2; 6.03×10^{-2} W/m^2

44. **IE** ●● If the distance to a sound source is halved, (a) will the sound intensity level change by a factor of (1) 2, (2) 1/2, (3) 4, (4) 1/4, or (5) none of the preceding? Why? (b) What is the change in the sound intensity level?
(a) (5) none of the preceding (b) increases by 6 dB

45. ●● A compact speaker puts out 100 W of sound power. (a) Neglecting losses to the air, at what distance would the sound intensity be at the pain threshold? (b) Neglecting losses to the air, at what distance would the sound intensity be that of normal speech? Does your answer seem reasonable? Explain.
(a) 2.82 m (b) 2.82×10^3 m, unreasonable

46. ●● What is the intensity level of a 23-dB sound after being amplified (a) 10 thousand times, (b) a million times, (c) a billion times? (a) 63 dB (b) 83 dB (c) 113 dB

47. ●● In a neighborhood challenge to see who can climb a tree the fastest, you are ready to climb. Your friends have surrounded you in a circle as a cheering section; each individual alone would cause a sound intensity level of 80 dB at your location. If the actual sound level at your location is 87 dB, how many people are rooting for you? five

48. **IE** ●● A dog's bark has a sound intensity level of 40 dB. (a) If two of the same dogs are barking, the intensity level is (1) less than 40 dB, (2) between 40 dB and 80 dB, (3) 80 dB. (b) What would be the intensity level?
(a) (2) between 40 dB and 80 dB (b) 43 dB

49. ●● At a rock concert, the average sound intensity level for a person in a front-row seat is 110 dB for a single band. If all the bands scheduled to play produce sound of that same intensity, how many of them would have to play simultaneously for the sound level to be at or above the threshold of pain? 10 bands

TABLE 14.4	Takeoff and Landing Noise Levels for Some Common Commercial Jet Aircraft* (See Exercise 43.)	
Aircraft	*Takeoff Noise (dB)*	*Landing Noise (dB)*
737	85.7–97.7	99.8–105.3
747	89.5–110.0	103.8–107.8
DC-10	98.4–103.0	103.8–106.6
L-1011	95.9–99.3	101.4–102.8

*Noise level readings are taken from 198 m (650 ft). The range depends on the aircraft model and the type of engine used.

50. ●● At a distance of 12.0 m from a point source, the intensity level is measured to be 70 dB. At what distance from the source will the intensity level be 40 dB? 379 m

51. ●● At a Fourth of July celebration, a firecracker explodes (▼Fig. 14.19). Considering the firecracker to be a point source, what are the intensities heard by observers at points B, C, and D, relative to that heard by the observer at A? $I_B = 0.563I_A$, $I_C = 0.250I_A$, $I_D = 0.173I_A$

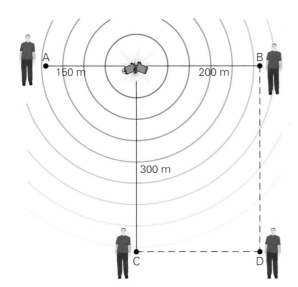

▲ **FIGURE 14.19 A big bang** See Exercise 51.

52. ●● An office in an e-commerce company has fifty computers, which generate a sound intensity level of 40 dB (from the keyboards). The office manager tries to cut the noise to half as loud by removing twenty-five computers. Does he achieve his goal? What is the intensity level generated by twenty-five computers? no; 37 dB

53. ●●● A 1000-Hz tone from a loudspeaker has an intensity level of 100 dB at a distance of 2.5 m. If the speaker is assumed to be a point source, how far from the speaker will the sound have intensity levels (a) of 60 dB and (b) barely high enough to be heard? (a) 2.5×10^2 m (b) 2.5×10^5 m

54. ●●● During practice in a huddle, a quarterback shouts the play in anticipation of crowd noise during the actual game. To a receiver 0.750 m away from the quarterback in the huddle, it seems as loud as the noise from a screaming child. When they get into practice formation, the quarterback yells at twice the output power, yet the instructions seem only about as loud as normal conversation. Use typical values in Table 14.2 to estimate how far from the quarterback the receiver is in the formation. 10.6 m

55. ●●● A bee produces a buzzing sound that is barely audible to a person 3.0 m away. How many bees would have to be buzzing at that distance to produce a sound with an intensity level of 50 dB? 10^5 bees

14.4 Sound Phenomena *and* 14.5 The Doppler Effect

56. **MC** Constructive and destructive interference of sound waves depends on (a) the speed of sound, (b) diffraction, (c) phase difference, (d) all of the preceding. (c)

57. **MC** Beats are the direct result of (a) interference, (b) refraction, (c) diffraction, (d) the Doppler effect. (a)

58. **MC** Police radar makes use of (a) refraction, (b) the Doppler effect, (c) interference, (d) sonic boom. (b)

59. **CQ** Do interference beats have anything to do with the "beat" of music? Explain. no, see ISM

60. **CQ** (a) Is there a Doppler effect if a sound source and an observer are moving with the same velocity? (b) What would be the effect if a moving source accelerated toward a stationary observer? (a) no (b) increasing frequency

61. **CQ** As a person walks in between a pair of loudspeakers that produce tones of the same amplitude and frequency, he hears a varying sound intensity. Explain. see ISM

62. **CQ** How can Doppler radar used in weather forecasting measure both the location and motion of the clouds? echolocation and Doppler effect

63. ● Two adjacent point sources, A and B, are directly in front of an observer and emit identical 1000-Hz tones. To what closest distance behind source B would source A have to be moved for the observer to hear no sound? (Assume that the air temperature is 20°C and ignore the falling off of intensity with distance.) 0.172 m

64. ● A violinist and a pianist simultaneously sound notes with frequencies of 436 Hz and 440 Hz, respectively. What beat frequency will the musicians hear? 4 Hz

65. **IE** ● A violinist tuning her instrument to a piano note of 264 Hz detects three beats per second. (a) The frequency of the violin could be (1) less than 264 Hz, (2) equal to 264 Hz, (3) greater than 264 Hz, (4) both (1) and (3). Why? (b) What are the possible frequencies of the violin tone? (a) (4) both (1) and (3) (b) 267 Hz or 261 Hz

66. ● What is the frequency heard by a person driving 60 km/h directly toward a factory whistle ($f = 800$ Hz) if the air temperature is 0°C? 840 Hz

67. **IE** ● On a day with a temperature of 20°C and no wind blowing, the frequency heard by a moving person from a 500-Hz stationary siren is 520 Hz. (a) The person is (1) moving toward, (2) moving away from, or (3) stationary relative to the siren. Why? (b) What is the person's speed? (a) (1) moving toward (b) 14 m/s

68. ●● While standing near a railroad crossing, you hear a train horn. The frequency emitted by the horn is 400 Hz. If the train is traveling at 90.0 km/h and the air temperature is 25°C, what is the frequency you hear (a) when the train is approaching and (b) after it has passed? (a) 431 Hz (b) 373 Hz

69. ●● Two identical strings on different cellos are tuned to the 440-Hz A note. The peg holding one of the strings slips, so its tension is decreased by 1.5%. What is the beat frequency heard when the strings are then played together? 3.3 Hz

70. ●● How fast, in kilometers per hour, must a sound source be moving toward you to make the observed frequency 5.0% greater than the true frequency? (Assume that the speed of sound is 340 m/s.) 58 km/h

71. **IE** ●● You are driving east at 25.0 m/s as you notice an ambulance traveling west toward you at 35.0 m/s. The sound you detect from the sirens has a frequency of 300 Hz. (a) Is the true frequency of the sirens (1) greater than 300 Hz, (2) less than 300 Hz, or (3) exactly 300 Hz? (b) Determine the true frequency of the sirens. Assume normal room temperature. (a) (2) less than 300 Hz (b) 251 Hz

72. ●● The frequency of an ambulance siren is 700 Hz. What are the frequencies heard by a stationary pedestrian as the ambulance approaches and moves away from her at a speed of 90.0 km/h. (Assume that the air temperature is 20°C.) 755 Hz approaching, 652 Hz moving away

73. ●● A jet flies at a speed of Mach 2.0. What is the half-angle of the conical shock wave formed by the aircraft? Can you tell the speed of the shock wave? 30°, yes

74. **IE** ●● A fighter jet flies at a speed of Mach 1.5. (a) If the jet were to fly faster than Mach 1.5, the half-angle of the conical shock wave would (1) increase, (2) remain the same, or (3) decrease. Why? (b) What is the half-angle of the conical shock wave formed by the jet plane at Mach 1.5? (a) (3) decrease (b) 42°

75. ●● The half-angle of the conical shock wave formed by a supersonic jet is 30°. What are (a) the Mach number of the aircraft and (b) the actual speed of the aircraft if the air temperature is −20°C? (a) 2.0 (b) 638 m/s

76. ●●● Two point-source loudspeakers are a certain distance apart, and a person stands 12.0 m in front of one of them on a line perpendicular to the baseline of the speakers. If the speakers emit identical 1000-Hz tones, what is their minimum nonzero separation so that the observer hears little or no sound? (Take the speed of sound to be exactly 340 m/s.) 2.03 m

77. ●● An observer is traveling between two identical sources of sound (frequency 100 Hz). His speed is 10.0 m/s as he approaches one and recedes from the other. (a) What frequency tone does he hear from both sources combined? (b) How many beats per second does he hear? Assume normal room temperature. (a) 103 Hz approaching, 97.0 Hz receding (b) 6 Hz

78. ●●● A bystander hears a siren vary in frequency from 476 Hz to 404 Hz as a fire truck approaches, passes by, and moves away on a straight street (▶Fig. 14.20). What is the speed of the truck? (Take the speed of sound in air to be 343 m/s.) 28 m/s

476 Hz   404 Hz

▲ **FIGURE 14.20 The siren's wail** See Exercise 78.

79. ●●● Bats emit sounds of frequencies around 35.0 kHz and use echolocation to find their prey. If a bat is moving with a speed of 12.0 m/s toward a hovering, stationary insect, (a) what is the frequency received by the insect if the air temperature is 20°C? (b) What frequency of the reflected sound heard by the bat? (c) If the insect were initially moving directly away from the bat, would this affect the frequencies? Explain. (a) 36.3 kHz (b) 37.6 kHz (c) yes

80. ●●● A supersonic jet flies directly overhead relative to an observer, at an altitude of 2.0 km (▼Fig. 14.21). When the observer hears the first sonic boom, the plane has flown a horizontal distance of 2.5 km at a constant speed. (a) What is the angle of the shock wave cone? (b) At what Mach number is the plane flying? (Assume that the speed of sound is at an average constant temperature of 15°C.) (a) 39° (b) 1.6

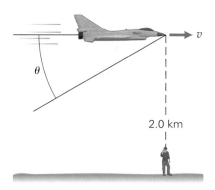

▲ **FIGURE 14.21 Faster than a speeding bullet**
See Exercise 80.

14.6 Musical Instruments and Sound Characteristics

81. **MC** Given open and closed pipes of the same length, which would have the lowest natural frequency: (a) the open pipe, (b) the closed pipe, or (c) both have the same low frequency? (b)

82. **MC** The human ear can hear tones best at (a) 1000 Hz, (b) 4000 Hz, (c) 6000 Hz, (d) all frequencies. (b)

83. **MC** Equal loudness curves vary with sound (a) quality, (b) harmonics, (c) waveform, (d) pitch. (d)

84. **MC** The quality of sound depends on its (a) waveform, (b) frequency, (c) speed, (d) intensity. (a)

85. CQ (a) After a snowfall, why does it seem particularly quiet? (b) Why do empty rooms sound hollow? (c) Why do people's voices sound fuller or richer when they sing in the shower? see ISM

86. CQ Why aren't the frets on a guitar evenly spaced? see ISM

87. CQ Is it possible for an open organ pipe and an organ pipe closed at one end, each of the same length, to produce notes of the same frequency? Justify your answer. no, see ISM

88. CQ When you blow across the mouth of an empty bottle, a particular tone is produced. If the bottle is filled to one third its height, how would the tone be affected? Explain. How about if it were filled to one half? see ISM

89. CQ Why are there no even harmonics in a pipe that is closed on one end? the closed end must be a node

90. ● The first three natural frequencies of an organ pipe are 126 Hz, 378 Hz, and 630 Hz. (a) Is the pipe an open or a closed pipe? (b) Taking the speed of sound in air to be 340 m/s, find the length of the pipe. (a) closed pipe (b) 0.675 m

91. ● A closed organ pipe has a fundamental frequency of 528 Hz (a C note) at 20°C. What is the fundamental frequency of the pipe when the temperature is 0°C? 510 Hz

92. ● The human ear canal is about 2.5 cm long. It is open at one end and closed at the other. (See Fig. 1 in Insight 14.2 on p. 475.) (a) What is the fundamental frequency of the ear canal at 20°C? (b) To what frequency is the ear most sensitive? (c) If a person's ear canal is longer than 2.5 cm, is the fundamental frequency higher or lower than that in part (a)? Explain. (a) 3.4 kHz (b) 3.4 kHz (c) lower

93. ●● An organ pipe that is closed at one end has a length of 0.80 m. At 20°C, what is the distance between a node and an adjacent antinode for (a) the second harmonic and (b) the third harmonic? (a) does not exist, only odd harmonics (b) 0.30 m

94. ●● An open organ pipe and an organ pipe that is closed at one end both have lengths of 0.52 m at 20°C. What is the fundamental frequency of each pipe? open, 330 Hz; closed, 165 Hz

95. ●● (a) To have an open organ pipe with a first-overtone frequency of 512 Hz in the cold outdoors ($-10°C$), what would be the required length of the pipe? (b) What would be the fundamental frequency of this pipe if it were brought inside a hockey rink, with a temperature of $+10°C$? Neglect length changes to the pipe due to thermal expansion. (a) 0.635 m (b) 265 Hz

96. ●● An organ pipe that is closed at one end is 1.10 m long. It is oriented vertically and filled with carbon dioxide gas (which is denser than air and thus will stay in the pipe). A tuning fork with a frequency of 60.0 Hz

can be used to set up a standing wave in the fundamental mode. What is the speed of sound in carbon dioxide? 264 m/s

97. ●● An open organ pipe 0.750 m long is found to have its first overtone at a frequency of 441 Hz. What is the temperature of the air in the pipe? 0°C

98. IE ●● When all of its holes are closed, a flute is essentially a tube that is open at both ends, with the length being from the mouthpiece to the far end (as in Fig. 14.16b). If a hole is open, then the length of the tube is effectively from the mouthpiece to the hole. (a) Is the position at the mouthpiece (1) a node, (2) an antinode, or (3) neither a node nor an antinode? Why? (b) If the lowest fundamental frequency on a flute is 262 Hz, what is the minimum length of the flute at 20°C? (c) If a note of frequency 440 Hz is to be played, which hole should be open? Express your answer as a distance from the hole to the mouthpiece. (a) (2) antinode (b) 0.655 m (c) 0.390 m

99. ●●● In a classroom demonstration, the instructor breathes some helium gas, and when he talks his speech has a "Donald Duck" sound. (This also occurs for deep-sea divers, who breathe a mixture of helium, nitrogen, and oxygen.) Considering a simplistic vocal tract to be a 15.0-cm closed pipe (glottis at one end), and taking the speeds of sound for air and helium to be 331 m/s and 965 m/s, respectively, show why the "Donald Duck" effect occurs. f_{air} = 552 Hz, f_{He} = 1610 Hz

100. ●●● An organ pipe that is closed at one end is filled with helium. The pipe has a fundamental frequency of 660 Hz in air at 20°C. What is the pipe's fundamental frequency with the helium in it? 1.92×10^3 Hz

101. ●●● An open organ pipe has a length of 50.0 cm. While in its fundamental mode, a second pipe, closed at one end, is also in its fundamental mode. A beat frequency of 2.00 Hz is heard. Determine the possible lengths of the closed pipe. Assume normal room temperature. 0.249 m and 0.251 m

Comprehensive Exercises

102. At a rock concert there are two main speakers, each putting out 500 W of sound power. You are 5.00 m from one and 10.0 m from the other. (a) What are the sound intensities at your location from each speaker and the total sound intensity? (b) What are the sound intensity levels at your location due to each speaker and what is the total sound intensity level? (c) About how long can you sit there without suffering permanent hearing damage? see ISM

103. Bats typically give off an ultrahigh-frequency sound at about 50 000 Hz. If the bat is approaching a stationary object at 18.0 m/s, what will be the reflected frequency it detects? (Assume the air in the cave is at 5°C.) [Hint: You will need to apply the Doppler equations twice. Why?] 5.55×10^4 Hz

104. You hear sound from two organ pipes that are equidistant from you. Pipe A is open at one end and closed at the other, while pipe B is open at both ends. When both are oscillating in their first-overtone mode, you hear a beat frequency of 5.0 Hz. Assume normal room temperature. (a) If the length of pipe A is 1.00 m, calculate the possible lengths of pipe B. (b) Assuming your shortest length for pipe B, what would the beat frequency be (assuming both are still in their first-overtone modes) on a hot desert summer day with a temperature of 40 °C?
(a) 1.36 m or 1.31 m (b) 5.1 Hz

105. IE An open organ pipe with a length of 50.0 cm is oscillating in its second-overtone or third-harmonic mode. Assume the air to be at room temperature and the pipe to be at rest in still air. A person moves toward this pipe at 2.00 m/s and, at the same time, away from a highly reflective wall. (a) Will the observer hear beats? (1) yes, (2) no, or (3) can't tell from the data given. (b) Calculate the frequency of sound emitted. (c) Calculate the beat frequency she would hear. [*Hint*: There are two frequencies, one directly from the pipe and one from the wall.]
(a) (1) yes (b) 1.03×10^3 Hz (c) 12 Hz

The following Physlet Physics Problems can be used with this chapter.
18.1, 18.2, 18.3, 18.4, 18.5, 18.7, 18.8, 18.9, 18.10, 18.11, 18.12, 18.13, 18.14, 18.15, 18.16

15

ELECTRIC CHARGE, FORCES, AND FIELDS

PHYSICS FACTS

- Charles Augustin de Coulomb (1736–1806), a French scientist and the discoverer of the force law between charged objects, had a diverse career. In addition, he made significant contributions in hospital reform, the cleanup of the Parisian water supply, Earth magnetism, soils engineering, and the construction of forts, the latter two while he served in the military.

- The Taser stun gun, as used by law enforcement agencies, works by generating a large electric charge separation and applying it to parts of the body, disrupting the normal electrical signals and causing temporary incapacity. The stun gun needs to physically contact the body with its two electrodes, and the shock can be delivered even through thick clothing. A long-distance version of the Taser works by firing barbed electrodes with trailing wires.

- The electric eel (which can grow up to 6 feet in length and is actually a fish) acts electrically in a similar way to a Taser. More than 80% of the eel's body is tail, with its vital organs located behind its small head. It uses the electric field it creates for both locating prey and stunning them before eating.

- Home air purifiers use the electric force to reduce dust, bacteria, and other particulates in the air. The electric force removes electrons from the pollutants, making them positively charged. These particles are attracted to negatively charged plates, where they stay until manually removed. When working properly, these purifiers can reduce the particulate level by more than 99%.

Few natural processes deliver such an enormous amount of energy in a fraction of a second as a lightning bolt. Yet most people have never experienced its power at close range; luckily, only a few hundred people per year are struck by lightning in the United States.

It might surprise you to realize that you have almost certainly had a similar experience, at least in a physics context. Have you ever walked across a carpeted room and gotten a shock when you reached for a metallic doorknob? Although the scale is dramatically different, the physical process involved (static electricity discharge) is much the same as being struck by lightning—mini-lightning, so to speak.

Electricity sometimes gives rise to dramatic effects such as sparking electrical outlets or lightning strikes. We know that electricity can sometimes be dangerous, but we also know that electricity can be "domesticated." In the home or office, its usefulness is taken for granted. Indeed, our dependence on electric energy becomes evident only when the power goes off unexpectedly, providing a dramatic reminder of the role that it plays in our daily lives. Yet less than a century ago there were no power lines crossing the country, no electric lights or appliances—none of the electrical applications that are all around us today.

Physicists now know that the electric force is related to the magnetic force (see Chapter 20). Together they are called the "electromagnetic force," which is one of the four fundamental forces in nature. (Gravity [Chapter 7] and two types of short-range nuclear forces discussed in Chapters 29 and 30 are the other three.) We begin here by studying the electric force and its properties. Eventually (Chapter 20) the magnetic and electric forces will be interconnected.

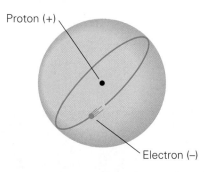

(a) Hydrogen atom

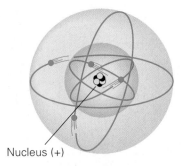

(b) Beryllium atom

▲ **FIGURE 15.1 Simplistic model of atoms** The so-called solar system model of **(a)** a hydrogen atom and **(b)** a beryllium atom views the electrons (negatively charged) as orbiting the nucleus (positively charged), analogously to the planets orbiting the Sun. The electronic structure of atoms is actually much more complicated than this.

Note: Recall the discussion of Newton's third law in Section 4.4.

▶ **FIGURE 15.2 The charge–force law, or law of charges (a)** Like charges repel. **(b)** Unlike charges attract.

15.1 Electric Charge

OBJECTIVES: To (a) distinguish between the two types of electric charge, (b) state the charge–force law that operates between charged objects, and (c) understand and use the law of charge conservation.

What is *electricity*? One simple answer is that it is a term describing phenomena associated with the electricity we have in our homes. But fundamentally it really involves the study of the interaction between *electrically charged* objects. To demonstrate this, our study will start with the simplest situation, electro*statics*, when electrically charged objects are *at rest*.

Like mass, **electric charge** is a fundamental property of matter (Chapter 1). Electric charge is associated with particles that make up the atom: the electron and the proton. The simplistic solar system model of the atom, as illustrated in ◀Fig. 15.1, likens its structure to that of the planets orbiting the Sun. The *electrons* are viewed as orbiting a nucleus, a core containing most of the atom's mass in the form of *protons* and electrically neutral particles called *neutrons*. As seen in Section 7.5, the centripetal force that keeps the planets in orbit about the Sun is supplied by gravity. Similarly, the force that keeps the electrons in orbit around the nucleus is the electrical force. However, there are important distinctions between gravitational and electrical forces.

One difference is that there is only one type of mass and gravitational forces are only attractive. Electric charge, however, comes in two types, distinguished by the labels positive (+) and negative (−). Protons carry a positive charge, and electrons carry a negative charge. Different combinations of the two types of charge can produce *either* attractive *or* repulsive electrical forces.

The directions of the electric forces when charges interact with one another are given by the following principle, called the **law of charges** or the **charge–force law**:

| Like charges repel, and unlike charges attract. |

That is, two negatively charged particles or two positively charged particles repel each other, whereas particles with opposite charges attract each other (▼Fig. 15.2). The repulsive and attractive forces are equal and opposite, and act on different objects, in keeping with Newton's third law (action–reaction).

The charge on an electron and that on a proton are equal in magnitude, but opposite in sign. The magnitude of the charge on an electron is abbreviated as *e* and is the fundamental unit of charge, because it is the smallest charge observed in nature.* The SI unit of charge is the **coulomb (C)**, named for the French physicist/engineer Charles A. de Coulomb (1736–1806), who discovered a relationship between electric force and charge (Section 15.3). The charges and masses of the electron, proton, and

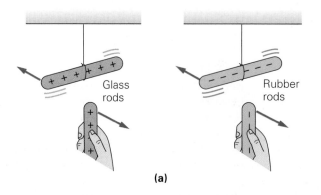

Glass rods

Rubber rods

(a)

(b)

*Protons, as well as neutrons and other particles, are now known to be made up of more fundamental particles called *quarks*, which carry charges of $\pm\frac{1}{3}$ and $\pm\frac{2}{3}$ of the electronic charge. There is experimental evidence of the existence of quarks within the nucleus, but free quarks have not been detected. Current theory implies that direct detection of quarks may, in principle, be impossible (Chapter 30).

TABLE 15.1	Sub-atomic Particles and Their Electric Charge	
Particle	Electric Charge*	Mass*
Electron	-1.602×10^{-19} C	$m_e = 9.109 \times 10^{-31}$ kg
Proton	$+1.602 \times 10^{-19}$ C	$m_p = 1.673 \times 10^{-27}$ kg
Neutron	0	$m_n = 1.675 \times 10^{-27}$ kg

*Even though the values are displayed to four significant figures, we will usually use only two or three in our calculations.

neutron are given in Table 15.1, where we see that $e = 1.602 \times 10^{-19}$ C. Our general symbol for charge will be q or Q. The charge on the electron is written as $q_e = -e = -1.602 \times 10^{-19}$ C, and that on the proton as $q_p = +e = +1.602 \times 10^{-19}$ C.

Other terms are frequently used when discussing charged objects. Saying that an object has a **net charge** means that the object has an excess of either positive or negative charges. (It is common, however, to ask about the "charge" of an object when we really mean the net charge.) As you will see in Section 15.2, excess charge is most commonly produced by a transfer of electrons, *not* protons. (Protons are bound in the nucleus and, under most common situations, do not leave.) For example, if an object has a (net) charge of $+1.6 \times 10^{-18}$ C, then it has had electrons removed from it. Specifically it has a deficiency of *ten* electrons, because $10 \times 1.6 \times 10^{-19}$ C $= 1.6 \times 10^{-18}$ C. That is, the total number of electrons on the object no longer completely cancels the positive charge of all the protons—resulting in a net positive charge. On an atomic level, some of the atoms that compose the object are deficient in electrons. Such positively charged atoms are termed *positive ions*. Atoms with an excess of electrons are *negative ions*.

Since the charge of the electron is such a tiny fraction of a coulomb, an object having a net charge on the order of one coulomb is rarely seen in everyday situations. Therefore, it is common to express amounts of charge using *microcoulombs* (μC, or 10^{-6} C), *nanocoulombs* (nC, or 10^{-9} C), and *picocoulombs* (pC, or 10^{-12} C).

Because the (net) electric charge on an object is caused by either a deficiency or an excess of electrons, it must always be an integer multiple of the charge on an electron. A plus sign or a minus sign will indicate whether the object has a deficiency or an excess of electrons respectively. Thus, for the (net) charge of an object, we may write

$$q = \pm ne \qquad (15.1)$$

SI unit of charge: coulomb (C)

where $n = 1, 2, 3, \ldots$. It is sometimes said that charge is "quantized," which means that it occurs only in integral multiples of the fundamental electronic charge.

In dealing with any electrical phenomena, another important principle is **conservation of charge**:

| The net charge of an isolated system remains constant. |

That is, the net charge remains constant even though it may not be zero. Suppose, for example, that a system consists initially of two electrically neutral objects, and one million electrons are transferred from one to the other. The object with the added electrons will then have a net negative charge, and the object with the reduced number of electrons will have a net positive charge of equal magnitude. (See Example 15.1.) But the net charge of the *system* remains zero. If the universe is considered as a whole, conservation of charge means that the net charge *of the universe* is constant.

Note that this principle doesn't prohibit the creation or destruction of charged particles. In fact, physicists have known for a long time that charged particles can be created and destroyed on the atomic and nuclear levels. However, because of charge conservation, charged particles are created or destroyed only in pairs with equal and opposite charges.

Integrated Example 15.1 ■ On the Carpet: Conservation of Quantized Charge

You shuffle across a carpeted floor on a dry day and the carpet acquires a net positive charge (for details on this mechanism, see Section 15.2). (a) Will you have a (1) deficiency or (2) an excess of electrons? (b) If the charge the carpet acquired has a magnitude of 2.15 nC, how many electrons were transferred?

(a) Conceptual Reasoning. (a) Since the carpet has a net positive charge, it must have lost electrons and you must have gained them. Thus, your charge is negative, indicating an excess of electrons, and the correct answer is (2).

(b) Quantitative Reasoning and Solution. Because the charge of one electron is known, we can quantify the excess of electrons. Express the charge in coulombs, and state what is to be found.

Given: $q_c = +(2.15 \text{ nC})\left(\dfrac{10^{-9} \text{ C}}{1 \text{ nC}}\right)$ *Find:* n, number of transferred electrons

$\qquad\qquad = +2.15 \times 10^{-9} \text{ C}$

$\qquad q_e = -1.60 \times 10^{-19} \text{ C}$ (from Table 15.1)

The net charge on you is

$$q = -q_c = -2.15 \times 10^{-9} \text{ C}$$

Thus

$$n = \frac{q}{q_e} = \frac{-2.15 \times 10^{-9} \text{ C}}{-1.60 \times 10^{-19} \text{ C/electron}} = 1.34 \times 10^{10} \text{ electrons}$$

As can be seen, net charges, even in everyday situations, can involve huge numbers of electrons (here, more than 13 billion), because the charge of any one electron is very small.

Follow-Up Exercise. In this Example, if your mass is 80 kg, by what percentage has your mass increased due to the excess electrons? *(Answers to all Follow-Up Exercises are at the back of the text.)*

15.2 Electrostatic Charging

OBJECTIVES: To (a) distinguish between conductors and insulators, (b) explain the operation of the electroscope, and (c) distinguish among charging by friction, conduction, induction, and polarization.

The existence of two types of electric charge along with the attractive and repulsive electrical forces can be easily demonstrated. Before learning how this is done, let's distinguish between electrical conductors and insulators. What distinguishes these broad groups of substances is their ability to conduct, or transmit, electric charge. Some materials, particularly metals, are good **conductors** of electric charge. Others, such as glass, rubber, and most plastics, are **insulators**, or poor electrical conductors. A comparison of the relative magnitudes of the conductivities of some materials is given in ▸Fig. 15.3.

In conductors, the *valence* electrons of the atoms—that is, the electrons in the outermost orbits—are loosely bound. As a result, they can be easily removed from the atom and moved about in the conductor, or they can leave the conductor altogether. That is, the valence electrons are not permanently bound to a particular atom. In insulators, however, even the loosest bound electrons are too tightly bound to be easily removed from their atoms. Thus, charge does not readily move through, nor is it readily removed from, an insulator.

As Fig. 15.3 shows, there is also a "middle" class of materials called **semiconductors**. Their ability to conduct charge is intermediate between that of insulators and conductors. The movement of electrons in semiconductors is much more difficult to describe than the simple valence electron approach used for insulators and conductors. In fact, the details of semiconductor properties can be understood only with the aid of quantum mechanics, which is beyond the scope of this book.

However, it is interesting to note that the conductivity of semiconductors can be adjusted by adding atomic impurities in varying concentrations. Beginning in the 1940s, scientists undertook research into the properties of semiconductors to create applications for such materials. Scientists used semiconductors to create transistors, then solid-

Teaching tip: Make sure that students can distinguish among conductors, insulators, and semiconductors.

Teaching tip: To connect with an earlier chapter, discuss why good thermal conductors are generally good electrical conductors.

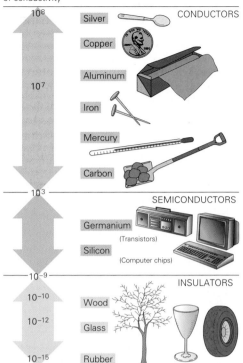

Relative magnitude   Material
of conductivity

◀ **FIGURE 15.3** Conductors, semiconductors, and insulators A comparison of the relative magnitudes of the electrical conductivities of various materials (not drawn to scale).

state circuits, and, eventually, modern computer microchips. The microchip is one of the major developments responsible for the high-speed computer technology of today.

Now that we know a bit about conductors and insulators, let's investigate a way of determining the sign of the charge on an object. The *electroscope* is one of the simplest devices used to determine electric charge (▼Fig. 15.4). In its simplest form, it consists of a metal rod with a metallic bulb at one end. The rod is attached to a solid, rectangular piece of metal that has an attached foil "leaf," usually made of gold or aluminum. This arrangement is insulated from its protective glass container by a nonconducting frame. When charged objects are brought close to the bulb, electrons in the bulb are either attracted to or repelled by the charged objects. For example, if a negatively charged rod is brought near the bulb, electrons in the bulb are repelled, and the bulb is left with a positive charge. The electrons are conducted down to the metal rectangle and its attached foil leaf, which then will swing away, because they have like charges (Fig. 15.4b). Similarly, if a positively charged rod is brought near the bulb, the leaf also swings away. (Can you explain why?)

Notice that the net charge on the electroscope remains zero in these instances. Because the device is isolated, only the *distribution* of charge is altered. However, it

Note: An uncharged electroscope can detect only whether an object is electrically charged. If the electroscope is previously charged with a known sign, it can then also determine the sign of the charge on the object.

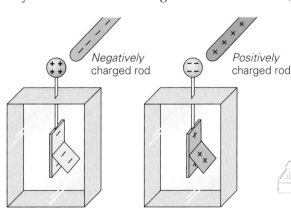

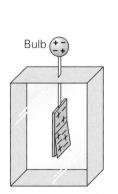

(a) Neutral electroscope has charges evenly distributed; leaf is vertical.

(b) Electrostatic forces cause leaf to separate away. (only excess or net charge is shown)

◀ **FIGURE 15.4** The electroscope An electroscope can be used to determine whether an object is electrically charged. When a charged object is brought near the bulb, the leaf moves away from the metal piece.

is possible to give an electroscope (and other objects) a net charge by different methods, all of which are said to involve **electrostatic charging**. Consider the following types of processes that produce electrostatic charging.

Charging by Friction

In the frictional charging process, certain insulator materials are rubbed with cloth or fur, and they become electrically charged by a transfer of charge. For example, if a hard rubber rod is rubbed with fur, the rod will acquire a net negative charge; rubbing a glass rod with silk will give the rod a net positive charge. This process is called **charging by friction**. The transfer of charge is due to the contact between the materials, and the amount of charge transferred depends, as you might expect, on the nature of those materials.

Example 15.1 was an example of frictional charging, in which a net charge was picked up from the carpet. If you had reached for a metal object such as a doorknob, you might have been "zapped" by a spark. As your hand approaches, the knob becomes positively changed, thus attracting the electrons from your hand. As they travel, they collide with, and excite, the atoms of the air, which give off light as they de-excite (lose energy). This light is seen as the spark of "mini-lightning" between your hand and the knob.

Charging by Conduction (Contact)

Bringing a charged rod close to an electroscope will reveal that the rod is charged, but it does not tell you what type of charge the rod has (positive or negative). The sign of the charge can be determined, however, if the electroscope is first given a known type of (net) charge. For example, electrons can be transferred to the electroscope from a negatively charged object, as illustrated in ▾Fig. 15.5a. The electrons in

Note: From an external viewpoint, you cannot tell whether the rubber rod gained negative charges or the fur gained positive charges. In other words, moving electrons to the rubber rod results in the same physical situation as moving positive charges to the fur. However, because the rubber is an insulator and its electrons are therefore tightly bound, we might suspect that the fur lost electrons and the rubber gained them. In solids, the protons, being in the nuclei of the atoms, do not move; only electrons move. It is just a question of which material most easily loses electrons.

▶ **FIGURE 15.5 Charging by conduction (a)** The electroscope is initially neutral (but the charges are separated), as a charged rod touches the bulb. **(b)** Charge is transferred to the electroscope. **(c)** When a rod of the same charge is brought near the bulb, the leaf moves farther apart. **(d)** When an oppositely charged rod is brought nearby, the leaf collapses.

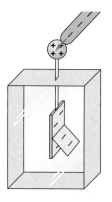

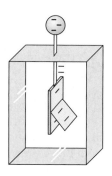

(a) Neutral electroscope is touched with negatively charged rod.

(b) Charges are transferred to bulb; electroscope has net negative charge.

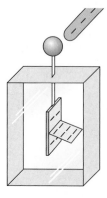

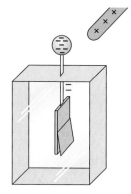

(c) Negatively charged rod repels electrons; leaf moves further.

(d) Positively charged rod attracts electrons; leaf collapses.

the rod repel one another, and some will transfer onto the electroscope. Notice that the leaf is now permanently diverged from the metal. In this case, we say that the electroscope has been **charged by contact** or by **conduction** (Fig. 15.5b). "Conduction" in this case refers to the flow of charge during the short period of time the electrons are transferred.

If a negatively charged rod is brought close to the now negatively charged electroscope, the leaf will diverge even further as more electrons are repelled down from the bulb (Fig. 15.5c). A positively charged rod will cause the leaf to collapse by attracting electrons up to the bulb and away from the leaf area (Fig. 15.5d).

Charging by Induction

Using a negatively charged rubber rod (already charged by friction), you might ask whether it is possible to create an electroscope that is positively charged. The answer is yes. This can be accomplished by **charging by induction**. Starting with an uncharged electroscope, you touch the bulb with a finger, which *grounds* the electroscope—that is, provides a path by which electrons can escape from the bulb (▼Fig. 15.6). Then, when a negatively charged rod is brought close to (but not touching) the bulb, the rod repels electrons from the bulb through your finger and body and down into the Earth (hence the term *ground*). Removing your finger *while the charged rod is kept nearby* leaves the electroscope with a net positive charge. This is because when the rod is removed, the electrons that moved to the Earth have no way back because the return path is gone.

Charge Separation by Polarization

Charging by contact and charging by induction create a net charge through the removal of charge from an object. However, charge can be moved *within the object* while keeping its net charge zero. For example, the induction process described previously initially causes **polarization**, or separation of positive and negative charge. If the object is not grounded, it will remain electrically neutral, but have equal and opposite amounts of charge at its ends. In this situation, we say that it has become an *electric dipole* (Section 15.4). On the molecular level, electric dipoles can be permanent; that is, they don't need a nearby charged object to retain their charge separation. A good example of this is the water molecule. Examples of both permanent and nonpermanent electric dipoles and forces

▼ **FIGURE 15.6 Charging by induction** **(a)** Touching the bulb with a finger provides a path to ground for charge transfer. The symbol e⁻ stands for "electron." **(b)** When the finger is removed, the electroscope has a net positive charge, opposite that of the rod.

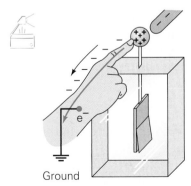

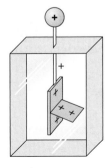

(a) Repelled by the nearby negatively charged rod, electrons are transferred to ground through hand.

(b) After removing the finger first, then later the rod, the electroscope is left positively charged.

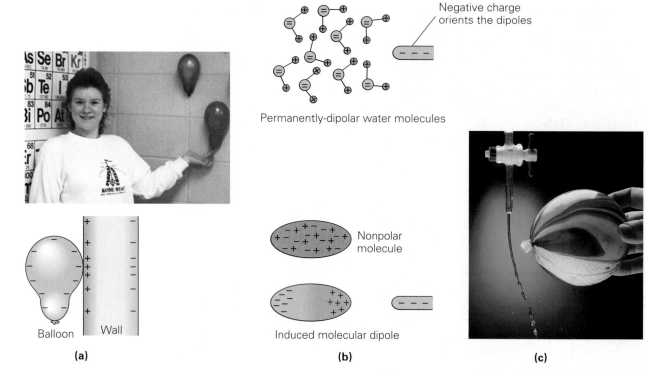

▲ FIGURE 15.7 Polarization (a) When the balloons are charged by friction and placed in contact with the wall, the wall is polarized. That is, an opposite charge is induced on the wall's surface, to which the balloons then stick by the force of electrostatic attraction. The electrons on the balloon do not leave the balloon because its material (rubber) is a poor conductor. **(b)** Some molecules, such as those of water, are polar by nature; that is, they have permanently separated regions of positive and negative charge. But even some molecules that are not normally dipolar can be polarized temporarily by the presence of a nearby charged object. The electric force induces a separation of charge and, consequently, temporary molecular dipoles. **(c)** A stream of water bends toward a charged balloon. The negatively charged balloon attracts the positive ends of the water molecules, thus causing the stream to bend.

Illustration 22.4 *Charging Objects and Static Cling*

Teaching tip: Students may be rusty on operations with vectors. Electric forces are noncontact forces and can be more difficult for some students to comprehend. Exercises at the end of the chapter allow students to practice finding resultant forces.

that can act on them are shown in ▲Fig. 15.7. Now you can understand why, when you rub a balloon on your sweater, it can stick to the wall. The balloon is charged by friction, and bringing it near to the wall polarizes the wall. The opposite sign charge on the wall's nearest surface creates a net attractive force.

Electrostatic charge can be annoying, as when static cling causes clothes and papers to stick together, or even dangerous, such as when electrostatic spark discharges start a fire or cause an explosion in the presence of a flammable gas. To discharge electric charge, many large trucks have dangling metal chains in contact with the ground. At gas stations, there are warnings to fill your gas cans while they are on the ground, not on the truck bed or car trunk surface (why?).

However, electrostatic forces can also be beneficial. For example, the air we breathe is cleaner because of electrostatic precipitators used in smokestacks. In these devices, electrical discharges cause the particles (by-products of fuel combustion) to acquire a net charge. The charged particles can then be removed from the flue gases by attracting them to electrically charged surfaces. On a smaller scale, electrostatic air cleaners are available for the home (see opening Physics Fact).

15.3 Electric Force

OBJECTIVES: To (a) understand Coulomb's law, and (b) use it to calculate the electric force between charged particles.

We know that the *directions* of electric forces on interacting charges are given by the charge–force law. However, what about their *magnitudes*? This was investigated by Coulomb, who found that the magnitude of the electric force between two "point" (very small) charges q_1 and q_2 depended directly on the product of the magnitude of the charges and inversely on the square of the distance between

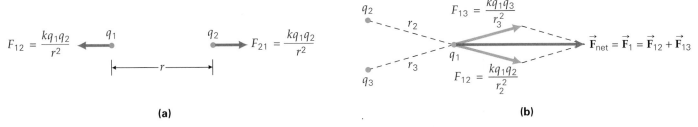

(a)   **(b)**

▲ **FIGURE 15.8 Coulomb's law (a)** The mutual electrostatic forces on two point charges are equal and opposite. **(b)** For a configuration of two or more point charges, the force on a particular charge is the vector sum of the forces on it due to all the other charges. (*Note:* In each of these situations, all of the charges are of the same sign. How can we tell that this is true? Can you tell their *sign*? What is the direction of the force on q_2 due to q_3?)

them. That is, $F_e \propto q_1q_2/r^2$. (q is the charge *magnitude*; thus, q_1 means the magnitude of q_1. This relationship is mathematically similar to that for the force of gravity between two point masses ($F_g \propto m_1m_2/r^2$); see Chapter 7.

Like Cavendish's measurements to determine the universal gravitational constant G (Section 7.5), Coulomb's measurements provided a constant of proportionality, k, so that the electric force could be written in equation form. Thus, the magnitude of the electric force between two point charges is described by an equation called **Coulomb's law**:

$$F_e = \frac{kq_1q_2}{r^2} \quad \begin{array}{l}\textit{(point charges only,}\\ \textit{q means charge magnitude)}\end{array} \quad (15.2)$$

Here, r is the distance between the charges (▲ Fig. 15.8a) and k a constant whose experimental value is

$$k = 8.988 \times 10^9 \, \text{N} \cdot \text{m}^2/\text{C}^2 \approx 9.00 \times 10^9 \, \text{N} \cdot \text{m}^2/\text{C}^2$$

Equation 15.2 gives the force between any two charged particles, but in many instances, we are concerned with the forces between more than two charges. In this situation, the net electric force on any particular charge is the vector sum of the forces on that charge due to all the other charges (Fig. 15.8b). For a review of vector addition, using electric forces, see the next two Examples.

Conceptual Example 15.2 ■ Free of Charge: Electric Forces

You may have done this. A rubber comb combed through dry hair can acquire a net negative charge. That comb will then attract small pieces of *uncharged* paper. This would seem to violate Coulomb's force law. Since the paper has no net charge, you might expect there to be no electric force on it. Which charging mechanism explains this phenomenon, and how does it explain it: (a) conduction, (b) friction, or (c) polarization?

Reasoning and Answer. Because the comb doesn't touch the paper, the paper cannot be charged by either conduction or friction, because both of these require contact. Thus, the answer must be (c). When the charged comb is near the paper, the paper becomes polarized (▶Fig. 15.9). The key to understanding the attraction is to observe that the charged ends of the paper are *not* the same distance from the comb. The positive end of the paper is closer to the comb than the negative end. Since the electric force decreases with distance, the attraction ($\vec{\mathbf{F}}_1$) between the comb and the positive end of the paper is greater than the repulsion ($\vec{\mathbf{F}}_2$) between the comb and the paper's negative end. Therefore, after adding these two forces vectorially, we find that the net force on the paper points toward the comb, and if it is light enough, the paper will accelerate in that direction.

Follow-Up Exercise. Does the phenomenon described in this Example tell you the sign of the charge on the comb? Why or why not?

Note: Coulomb's law gives the electric force only between point charges, not objects with extended charged areas.

Note: In calculations, we will take k to be exact at $9.00 \times 10^9 \, \text{N} \cdot \text{m}^2/\text{C}^2$ for significant-figure purposes.

Demonstration/activity: Paint two Ping-Pong balls with metallic paint and suspend them from 50–100 cm of silk thread. Charge them with an electrostatic generator and observe that they stand apart due to the repulsive force. Measure their weight and angle of separation, and use Coulomb's law to calculate the magnitude of charge on the balls. (This activity requires dry weather, but it provides a nice way to revisit Newton's laws, forces, static equilibrium, and free-body diagrams.)

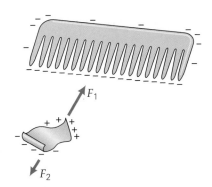

▲ **FIGURE 15.9 Comb and paper** See Conceptual Example 15.2.

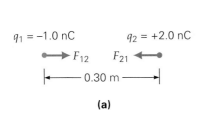

$q_1 = -1.0$ nC $\qquad q_2 = +2.0$ nC

(a)

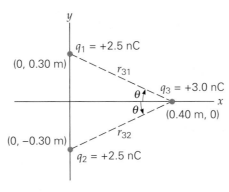

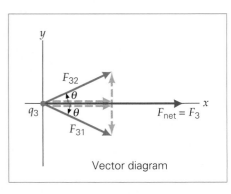

Vector diagram

(b)

▲ **FIGURE 15.10** Coulomb's law and electrostatic forces See Example 15.3.

Exploration 22.4 Dipole Symmetry

Exploration 22.1 Equilibrium

Example 15.3 ■ Coulomb's Law: Vector Addition Involving Trigonometry

(a) Two point charges of -1.0 nC and $+2.0$ nC are separated by a distance of 0.30 m (▲Fig. 15.10a). What is the electric force on each particle? (b) A configuration of three charges is shown in Fig. 15.10b. What is the net electric force on q_3?

Thinking It Through. Adding electric forces is no different from adding any other type of force. The only difference here is that we first must use Coulomb's law to calculate the force *magnitudes*. Then it is just a matter of computing components. (a) For the two point charges, we use Coulomb's law (Eq. 15.2), noting that the forces are attractive. (Why?) (b) Here we must use components to vectorially add the two forces acting on q_3 due to q_1 and q_2. We can find θ from the distances between charges. This angle is necessary to calculate the x and y force components. (See the Problem-Solving Hint on p. 515.)

Solution. Listing the data and converting nanocoulombs to coulombs, we have

Given: (a) $q_1 = -(1.0 \text{ nC})\left(\dfrac{10^{-9} \text{ C}}{1 \text{ nC}}\right) = -1.0 \times 10^{-9} \text{ C}$ *Find:* (a) $\vec{F}_{12}$ and $\vec{F}_{21}$
(b) $\vec{F}_3$

$q_2 = +(2.0 \text{ nC})\left(\dfrac{10^{-9} \text{ C}}{1 \text{ nC}}\right) = +2.0 \times 10^{-9} \text{ C}$

$r = 0.30$ m

(b) Data given in Figure 15.10b. Convert charges to coulombs as in part (a).

(a) Equation 15.2 gives the magnitude of the force acting on each charge using the charge magnitudes and distance between them:

$$F_{12} = F_{21} = \frac{kq_1q_2}{r^2} = \frac{(9.00 \times 10^9 \text{ N} \cdot \text{m}^2/\text{C}^2)(1.0 \times 10^{-9} \text{ C})(2.0 \times 10^{-9} \text{ C})}{(0.30 \text{ m})^2}$$

$$= 0.20 \times 10^{-6} \text{ N} = 0.20 \ \mu\text{N}$$

Note that Coulomb's law gives only the force's magnitude. However, because the charges are of opposite sign, the forces must be mutually attractive as shown in Fig. 15.10a.

(b) The forces $\vec{F}_{31}$ and $\vec{F}_{32}$ must be added vectorially, using trigonometry and components, to find the net force. Since all the charges are positive, the forces are repulsive, as shown in the vector diagram in Fig. 15.10b. Since $q_1 = q_2$ and the charges are equidistant from q_3, it follows that $\vec{F}_{31}$ and $\vec{F}_{32}$ have the same magnitude.

Also from the figure, we can see that $r_{31} = r_{32} = 0.50$ m. (Why?) With data from the figure, using Eq. 15.2:

$$F_{32} = \frac{kq_2q_3}{r_{32}^2} = \frac{(9.00 \times 10^9 \text{ N} \cdot \text{m}^2/\text{C}^2)(2.5 \times 10^{-9} \text{ C})(3.0 \times 10^{-9} \text{ C})}{(0.50 \text{ m})^2}$$

$$= 0.27 \times 10^{-6} \text{ N} = 0.27 \ \mu\text{N}$$

Taking into account the directions of $\vec{F}_{31}$ and $\vec{F}_{32}$ we see by symmetry that the y-components cancel. Thus, $\vec{F}_3$ (the net force on q_3) acts along the positive x-axis and has a magnitude of

$$F_3 = F_{31_x} + F_{32_x} = 2 F_{31_x}$$

because $F_{31} = F_{32}$.

The angle θ can be determined from the triangles; that is, $\theta = \tan^{-1}\left(\dfrac{0.30 \text{ m}}{0.40 \text{ m}}\right) = 37°$. Thus $\vec{F}_3$ has a magnitude of

$$F_3 = 2F_{31_x} = 2F_{32} \cos\theta$$
$$= 2(0.27 \ \mu N) \cos 37° = 0.43 \ \mu N$$

and acts in the positive x-direction (to the right).

Follow-Up Exercise. In part (b) of this Example, calculate the net force $\vec{F}_1$ on q_1.

The magnitudes of the charges in Example 15.3 are typical of static charges produced by frictional rubbing; that is, they are tiny. Thus, the forces involved are very small by everyday standards, much smaller than any force we have studied so far. However, on the atomic scale, even tiny forces can produce huge accelerations, because the particles (such as electrons and protons) have extremely small mass. Consider the answers in Example 15.4 compared with the answers in Example 15.3.

Problem-Solving Hint

The signs of the charges can be used explicitly in Eq. 15.2 with a positive value for F meaning a repulsive force and a negative value an attractive force. *However, such an approach is not recommended*, because this sign convention is useful only for one-dimensional forces, that is, those that have only one component, as in Example 15.3a. When forces are two-dimensional, thus requiring components, Eq. 15.2 should instead be used to calculate the *magnitude* of the force, using only the *magnitude* of the charges (as in Example 15.3b). Then the charge–force law determines the direction of the force between each pair of charges. (Draw a sketch and put in the angles.) Lastly, use trigonometry to calculate each force's components and then combine them appropriately. This latter approach is recommended and the one that will be used in this text.

Example 15.4 ■ Inside the Nucleus: Repulsive Electrostatic Forces

(a) What is the magnitude of the repulsive electrostatic force between two protons in a nucleus? Take the distance from center to center of these protons to be 3.00×10^{-15} m. (b) If the protons were released from rest, how would the magnitude of their initial acceleration compare with that of the acceleration due to gravity on the Earth's surface, g?

Thinking It Through. (a) Coulomb's law must be applied to find the repulsive force. (b) To find the initial acceleration, we use Newton's second law ($F_{net} = ma$).

Solution. Listing the known quantities, we have the following:

Given: $r = 3.00 \times 10^{-15}$ m   *Find:* (a) F_e (magnitude of force)

   $q_1 = q_2 = +1.60 \times 10^{-19}$ C (from Table 15.1)   (b) $\dfrac{a}{g}$ (magnitude of acceleration compared with g)

   $m_p = 1.67 \times 10^{-27}$ kg (from Table 15.1)

(a) Using Coulomb's law (Eq. 15.2), we have

$$F_e = \frac{kq_1q_2}{r^2} = \frac{(9.00 \times 10^9 \ \text{N} \cdot \text{m}^2/\text{C}^2)(1.60 \times 10^{-19} \ \text{C})(1.60 \times 10^{-19} \ \text{C})}{(3.00 \times 10^{-15} \ \text{m})^2} = 25.6 \ \text{N}$$

This force is much larger than that in the previous Example and is equivalent to the weight of an object with a mass of about 2.5 kg. Thus, with its small mass, we expect the proton to experience a huge acceleration.

(b) If it acted alone on a proton, this force would produce an acceleration of

$$a = \frac{F_e}{m_p} = \frac{25.6 \ \text{N}}{1.67 \times 10^{-27} \ \text{kg}} = 1.53 \times 10^{28} \ \text{m/s}^2$$

Then

$$\frac{a}{g} = \frac{1.53 \times 10^{28} \ \text{m/s}^2}{9.8 \ \text{m/s}^2} = 1.56 \times 10^{27}$$

(continues on next page)

That is, $a \approx 10^{27} g$. The factor of 10^{27} is enormous. To help see how large it is, if a uranium atom were subject to this acceleration, the net force required would be about the same as the weight of a polar bear (a thousand pounds or so)!

Most atoms contain more than two protons in their nuclei. With these enormous repulsive forces, you would expect nuclei to fly apart. Because this doesn't generally happen, there must be a stronger attractive force holding the nucleus together. This is called the nuclear (or strong) force, and will be discussed in Chapters 29 and 30.

Follow-Up Exercise. Suppose you could anchor a proton to the ground and you wished to place a second one directly above the first so that the second proton was in equilibrium (that is, so the electrical repulsion force acting on the second proton balanced its weight force). How far apart would the protons be?

Although there is a striking similarity between the mathematical form of the expressions for the electric and gravitational forces, there is a huge difference in the relative strengths of the two forces, as is shown in the next Example.

Example 15.5 ■ Inside the Atom: Electric Force versus Gravitational Force

Determine the ratio of the electric to gravitational force between a proton and an electron. In other words, how many times larger is the electric force than the gravitational force?

Thinking It Through. The distance between the proton and electron is not given. However, both the electrical force and the gravitational force vary as the inverse square of the distance, so the distance will cancel out in a ratio. By using Coulomb's law and Newton's law of gravitation (Chapter 7), the ratio can be determined if one knows the charges, masses, and appropriate electric and gravitational constants.

Solution. The charges and masses of the particles are known (Table 15.1), as are the electrical constant k and the universal gravitational constant G.

Given: $q_e = -1.60 \times 10^{-19}$ C $\qquad\qquad$ **Find:** $\dfrac{F_e}{F_g}$ (ratio of forces)
$\quad\quad\quad q_p = +1.60 \times 10^{-19}$ C
$\quad\quad\quad m_e = 9.11 \times 10^{-31}$ kg
$\quad\quad\quad m_p = 1.67 \times 10^{-27}$ kg

The expressions for the forces are

$$F_e = \frac{kq_e q_p}{r^2} \quad \text{and} \quad F_g = \frac{Gm_e m_p}{r^2}$$

Forming a ratio of magnitudes for comparison purposes (and to cancel r) gives

$$\frac{F_e}{F_g} = \frac{kq_e q_p}{Gm_e m_p}$$

$$= \frac{(9.00 \times 10^9 \text{ N} \cdot \text{m}^2/\text{C}^2)(1.60 \times 10^{-19} \text{ C})^2}{(6.67 \times 10^{-11} \text{ N} \cdot \text{m}^2/\text{kg}^2)(9.11 \times 10^{-31} \text{ kg})(1.67 \times 10^{-27} \text{ kg})} = 2.27 \times 10^{39}$$

or

$$F_e = (2.27 \times 10^{39})F_g$$

The magnitude of the electrostatic force between a proton and an electron is more than 10^{39} times the magnitude of the gravitational force. While a factor of 10^{39} is incomprehensible to most, it should be perfectly clear that because of this large value, the gravitational force between charged particles can generally be neglected in our study of electrostatics.

Follow-Up Exercise. With respect to this Example, show that gravity is even more negligible compared with the electric force between two electrons. Explain why this is so.

15.4 Electric Field

<u>**OBJECTIVES:**</u> To (a) understand the definition of the electric field, and (b) plot electric field lines and calculate electric fields for simple charge distributions.

The electric force, like the gravitational force, is an "action-at-a-distance" force. Since the range of the electric force is infinite ($F_e \propto 1/r^2$ and approaches zero only as r approaches infinity), a particular configuration of charges can have an effect on an additional charge placed anywhere nearby.

The idea of a force acting across space was difficult for early investigators to accept, and the modern concept of a *force field*, or simply a field, was introduced. An *electric field* is envisioned as surrounding every arrangement of charges. Thus, the electric field represents the *physical effect* of a particular configuration of charges on the nearby space. The field represents what is different about the nearby space because those charges are there. The concept treats charges as interacting with the electric field created by other charges, not directly with the charges "at a distance." The main idea of the electric field concept is as follows: A configuration of charges creates an electric field in the space nearby. When another charge is placed in this field, *the field* will exert an electric force on that charge. Thus:

> Charges create electric fields, and these fields in turn exert electric forces on other charges.

An electric field is actually a *vector field* (it has direction as well as magnitude). It enables us to determine the force (including direction) exerted on a charge at a location. *However, the electric field is not that force.* Instead, the magnitude (or strength) of the field is defined as the electric force exerted per unit charge. Determining an electric field's strength may be theoretically imagined using the following procedure. Place a very small charge (called a *test charge*) at the location of interest. Measure the force acting on that test charge and divide by the amount of its charge, thus determining the force that would be exerted *per coulomb*. Next imagine removing the test charge. The force disappears (why?), but the field remains, because it is created by the nearby charges, which remain. When the electric field is determined in many locations, we have a "map" of the electric field strength but no direction. Thus the "mapping" is incomplete.

Because the field direction is specified by the direction of the force on the test charge, it depends on whether the test charge is chosen as positive or negative. The convention is that a *positive test charge* (q_+) is used for measuring electric field direction (see ▸Fig. 15.11). That is:

> The electric field direction is in the direction of the force experienced by a positive test charge.

Once the electric field's magnitude and direction due to a charge configuration is known, you can ignore the "source" charge configuration and talk in terms of the field they produce. This way of visualizing electric interactions between charges often facilitates calculations.

The **electric field $\vec{E}$** at any location is defined as follows

$$\vec{E} = \frac{\vec{F}_{\text{on } q_+}}{q_+} \tag{15.3}$$

SI unit of electric field: newton/coulomb (N/C)

The direction of $\vec{E}$ is in the direction of the force on a small *positive* test charge at that location.

For the special case of a point charge, we can use Coulomb's force law. To determine the magnitude of the electric field due to a point charge at a distance r from that point charge, Eq. 15.3 can be used:

$$E = \frac{F_{\text{on } q_+}}{q_+} = \frac{(kqq_+/r^2)}{q_+} = \frac{kq}{r^2}$$

Note: Charges set up an electric field, and this field then acts on other charges placed in it.

Demonstration/activity: Bring a charged pith ball near a Van de Graaff generator to illustrate electric fields. The direction will be obvious. Using a charged Ping-Pong ball as described previously, you can actually get an approximate value of the electric field. Students are surprised at the large number, indicating a field strength in thousands of newtons per coulomb.

Exploration 23.1 Fields and Test Charges

Note: A *test charge* (q_+) is small and positive.

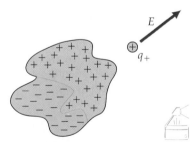

▲ **FIGURE 15.11 Electric field direction** By convention, the direction of the electric field $\vec{E}$ is in the direction of the force experienced by an imaginary (positive) test charge. To see the direction, ask which way the test charge would accelerate if released. Here the "system of charges" produces a (net) electric field upward and to the right at the location of the test charge. In this particular arrangement, can you explain this direction by observing the signs and locations of charges in the system?

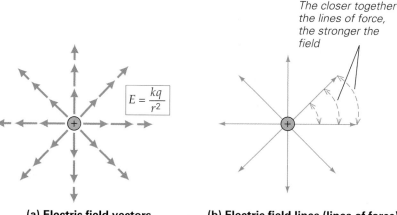

(a) Electric field vectors   **(b) Electric field lines (lines of force)**

▲ **FIGURE 15.12 Electric field** **(a)** The electric field points away from a positive point charge, in the direction a force would be exerted on a small positive test charge. The field's magnitude (the lengths of the vectors) decreases as the distance from the charge increases, reflecting the inverse-square distance relationship characteristic of the field produced by a point charge. **(b)** In this simple case, the vectors are easily connected to give the electric field line pattern due to a positive point charge.

That is,

$$E = \frac{kq}{r^2} \quad \begin{array}{l}\textit{(magnitude of electric field}\\ \textit{due to point charge q)}\end{array} \qquad (15.4)$$

Notice that in deriving Eq. 15.4, q_+ canceled out. *This must always happen*, because the field is produced by the other charges, *not* the test charge q_+.

Some electric field vectors in the vicinity of a positive charge are illustrated in ▲Fig. 15.12a. Note that their directions are *away from the positive charge* because a positive test charge would feel a force in this direction. Notice also that the magnitude of the field (the arrow length) decreases with increasing distance r.

If there is more than one charge creating an electric field, then the total, or net, electric field at any point is found using the **superposition principle for electric fields**, which can be stated as follows.

> For a configuration of charges, the total, or net, electric field at any point is the vector sum of the electric fields due to the individual charges.

This principle is demonstrated in the next two Examples, and a way to qualitatively determine the direction of the electric field when more than one charge is involved is shown in the accompanying Learn by Drawing on Using the Superposition Principle to Determine the Electric Field Direction.

Illustration 23.2 Electric Fields from Point Charges

Note: Think of the electric field definition as being useful in the same way that the price per pound is for food items. Knowing how much you want of an item, you can compute how much it will cost if you know the price per pound. Similarly, given the magnitude of a charge placed in an electric field, you can compute the force on it if you know the field strength in newtons per coulomb.

LEARN BY DRAWING

USING THE SUPERPOSITION PRINCIPLE TO DETERMINE THE ELECTRIC FIELD DIRECTION

To estimate the direction of the electric field at any point P, draw the individual electric fields vectors and add them, taking into account the relative field magnitudes if you can. In this situation, $\vec{\mathbf{E}}_1$ is much smaller than $\vec{\mathbf{E}}_2$ because of both distance and charge factors. Can you explain why the vector representing $\vec{\mathbf{E}}_2$, if drawn accurately, would be about eight times as long as that of $\vec{\mathbf{E}}_1$? The final step is to complete the vector addition.

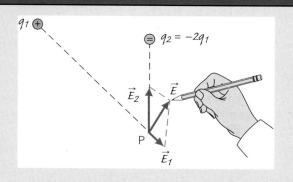

Example 15.6 ■ Electric Fields in One Dimension: Zero Field by Superposition

Two point charges are placed on the x-axis as in ▶Fig. 15.13. Find all locations on the axis where the electric field is zero.

Thinking It Through. Each point charge produces its own field. By the superposition principle, the electric field is the vector sum of the two fields. We are looking for locations where these fields are equal and opposite, so as to cancel and give no (*total* or *net*) electric field.

Solution. Let us specify the location as a distance x from q_1 (located at $x = 0$) and convert charges from microcoulombs to coulombs as usual.

Given: $d = 0.60$ m (distance between charges) *Find:* x [the location(s) of zero E]
$q_1 = +1.5\ \mu C = +1.5 \times 10^{-6}$ C
$q_2 = +6.0\ \mu C = +6.0 \times 10^{-6}$ C

Since both charges are positive, their fields point to the right at all locations to the right of q_2. Therefore, the fields cannot cancel in that region. Similarly, to the left of q_1, both fields point to the left and cannot cancel. The only possibility of cancellation is *between* the charges. In that region, the two fields will cancel if their magnitudes are equal, because they are oppositely directed. Setting the magnitudes equal and solving for x:

$$E_1 = E_2 \quad \text{or} \quad \frac{kq_1}{x^2} = \frac{kq_2}{(d-x)^2}$$

Rearranging this expression and canceling the constant k yields

$$\frac{1}{x^2} = \frac{(q_2/q_1)}{(d-x)^2}$$

With $q_2/q_1 = 4$, taking the square root of both sides:

$$\sqrt{\frac{1}{x^2}} = \sqrt{\frac{q_2/q_1}{(d-x)^2}} = \sqrt{\frac{4}{(d-x)^2}} \quad \text{or} \quad \frac{1}{x} = \frac{2}{d-x}$$

Solving, $x = d/3 = 0.60$ m$/3 = 0.20$ m. (Why don't we use the negative square root? Try it.) The result being closer to q_1 makes sense physically. Because q_2 is the larger charge, for the two fields to be equal in magnitude, the location must be closer to q_1.

Follow-Up Exercise. Repeat this Example, changing the only sign of the right-hand charge.

Integrated Example 15.7 ■ Electric Fields in Two Dimensions: Using Vector Components and Superposition

▼Fig. 15.14a shows a configuration of three point charges. **(a)** In what quadrant is the electric field at the origin: (1) the first quadrant, (2) the second quadrant, or (3) the third quadrant? Explain your reasoning, using the superposition principle. **(b)** Calculate the magnitude and direction of the electric field at the origin due to these charges.

Where is $E = 0$?

$q_1 = +1.5\ \mu C$ $q_2 = +6.0\ \mu C$

0   0.10 0.20 0.30 0.40 0.50 0.60 x (m)

▲ **FIGURE 15.13 Electric field in one dimension** See Example 15.6.

Note: Total electric field: $\vec{E} = \Sigma \vec{E}_i$.

Teaching tip: The concept of an electric field is difficult for most students. Emphasize that an electric field exists in a region where a charge would be acted on by an electric force. The field strength is the ratio of the force to the charge. The direction of the field is in the direction of the force on a positive charge.

(continues on next page)

◀ **FIGURE 15.14 Finding the electric field** See Integrated Example 15.7.

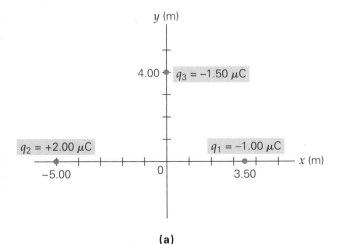

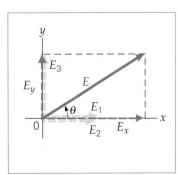

(a)   (b)

(a) Conceptual Reasoning. The electric field points towards negative point charges and away from positive point charges. Therefore, $\vec{E}_1$ and $\vec{E}_2$ point in the positive x-direction and $\vec{E}_3$ points along the positive y-axis. Because the electric field is the sum of these three fields, both of its components are positive. Therefore $\vec{E}$ is in the first quadrant (Fig. 15.14b). Thus, the correct answer is (1).

(b) Quantitative Reasoning and Solution. The directions of the individual electric fields are shown in the sketch in part (a). According to the superposition principle, we need to add the fields vectorially to find the electric field $(\vec{E} = \vec{E}_1 + \vec{E}_2 + \vec{E}_3)$.

Listing the data given and converting the charges into coulombs, we have:

Given: $q_1 = -1.00\ \mu C = -1.00 \times 10^{-6}\ C$ **Find:** $\vec{E}$ (electric field at origin)
$q_2 = +2.00\ \mu C = +2.00 \times 10^{-6}\ C$
$q_3 = -1.50\ \mu C = -1.50 \times 10^{-6}\ C$
$r_1 = 3.50\ m$
$r_2 = 5.00\ m$
$r_3 = 4.00\ m$

From the sketch E_y is due to $\vec{E}_3$ and E_x is the sum of the magnitudes of $\vec{E}_1$ and $\vec{E}_2$. The magnitudes of the three fields are determined from Eq. 15.4. These magnitudes are

$$E_1 = \frac{kq_1}{r_1^2} = \frac{(9.00 \times 10^9\ N \cdot m^2/C^2)(1.00 \times 10^{-6}\ C)}{(3.50\ m)^2} = 7.35 \times 10^2\ N/C$$

$$E_2 = \frac{kq_2}{r_2^2} = \frac{(9.00 \times 10^9\ N \cdot m^2/C^2)(2.00 \times 10^{-6}\ C)}{(5.00\ m)^2} = 7.20 \times 10^2\ N/C$$

$$E_3 = \frac{kq_3}{r_3^2} = \frac{(9.00 \times 10^9\ N \cdot m^2/C^2)(1.50 \times 10^{-6}\ C)}{(4.00\ m)^2} = 8.44 \times 10^2\ N/C$$

The x- and y-components of the field are

$$E_x = E_1 + E_2 = +7.35 \times 10^2\ N/C + 7.20 \times 10^2\ N/C = +1.46 \times 10^3\ N/C$$

and

$$E_y = E_3 = +8.44 \times 10^2\ N/C$$

In component form,

$$\vec{E} = E_x\hat{x} + E_y\hat{y} = (1.46 \times 10^3\ N/C)\hat{x} + (8.44 \times 10^2\ N/C)\hat{y}$$

You should be able to show that in magnitude–angle form this is

$$E = 1.69 \times 10^3\ N/C\ \text{at}\ \theta = 30.0°\quad (\theta\ \text{is in the first quadrant, relative to the}\ +x\text{-axis})$$

Follow-Up Exercise. In this Example, suppose q_1 was moved to the origin. Find the electric field at its former location.

Electric Lines of Force

A convenient way of *graphically* representing the electric field is by use of *electric lines of force*, or **electric field lines**. To start, consider the electric field vectors near a positive point charge, as in Fig. 15.12a. In Fig. 15.12b these vectors have been "connected." This constructs the *electric field line pattern* due to a positive point charge. Notice that the field lines come closer together (their spacing decreases) as we near the charge, because the field increases in strength. Also note that at any location on a field line, the electric field *direction* is tangent to the line. (The lines usually have arrows attached to them that indicate the general field direction.) It should be clear that electric field lines can't cross. If they did, it would mean that at the crossing spot there would be two directions for the force on a charge placed there—a physically unreasonable result.

PHYSLET®

Illustration 23.3 Field-Line
Representation of Vector Fields

The general rules for sketching and interpreting electric field lines are as follows:

1. The closer together the field lines, the stronger the electric field.
2. At any point, the direction of the electric field is tangent to the field lines.
3. The electric field lines start at positive charges and end at negative charges.
4. The number of lines leaving or entering a charge is proportional to the magnitude of that charge.
5. Electric field lines can never cross.

These rules enable us to "map" the pattern of electric lines of force due to various charge configurations. (See the accompanying Learn by Drawing on Sketching Electric Lines of Force.)

Let's now apply these rules and the superposition principle to map out the electric field line pattern due to an *electric dipole* in Example 15.8. An **electric dipole** consists of two equal, but opposite, electric charges (or "poles," as they were known historically). Even though the net charge on the dipole is zero, it creates an electric field because the charges are separated. If the charges were at the same location, their fields would cancel everywhere.

In addition to using dipoles to learn about electric field sketching, dipoles are important in themselves, because they occur in nature. For example, electric dipoles can serve as a model for important polarized molecules, such as the water molecule. (See Fig. 15.7.) Also see Insight 15.2 on Electric Fields in Law Enforcement and Nature: Stun Guns and Electric Fish on page 524.

Example 15.8 ■ Constructing the Electric Field Pattern Due to a Dipole

Use the superposition principle and the electric field line rules to construct a typical electric field line due to an electric dipole.

Thinking It Through. The construction involves vector addition of the individual electric fields from the two opposite ends of the dipole.

Solution.

Given: an electric dipole of two equal and opposite charges separated by a distance, d

Find: a typical electric field line

An electric dipole is shown in ▼Fig. 15.15a. To keep track of the two fields, let's label the positive charge q_+ and the negative charge q_-. Their individual fields, $\vec{E}_+$ and $\vec{E}_-$, will be designated by the same subscripts.

Because electric fields (and field lines) start at positive charges, let's begin at location A, near charge q_+. Because this is much closer to q_+ it follows that $E_+ > E_-$. We know that $\vec{E}_+$ will *always* point away from q_+ and $\vec{E}_-$ will *always* point toward q_-. Putting these two facts together enables us to qualitatively draw the two fields at A. The parallelogram method determines their vector sum: the electric field at A.

To map the electric field line, the general direction of the electric field at A points us approximately to our next location, B. At B, there is a reduced magnitude (why?) and slight directional change for both $\vec{E}_+$ and $\vec{E}_-$. You should now be able to see how the fields at C and D are determined. Location D is special because it is on the perpendicular bisector of the dipole axis (the line that connects the two charges). The electric field points downward anywhere on this line. You should be able to continue the construction at points E, F, and G.

Lastly, to construct the electric field line, start at the positive end of the dipole, because the field lines leave that end. Because the electric field vectors are tangent to the field lines, we draw the line to fulfill this requirement. [You should be able to sketch in the other lines and understand the complete dipole field pattern shown in Fig. 15.15b.]

Follow-Up Exercise. Using the techniques in this Example, construct the field lines that start (a) just above the positive charge, (b) just below the negative charge, and (c) just below the positive charge.

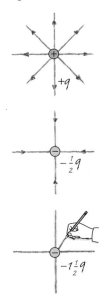

▶ **FIGURE 15.15** Mapping the electric field due to a dipole (a) The construction of one electric field line from a dipole is shown. The electric field is the vector sum of the two fields produced by the two ends of the dipole. (See Example 15.8 for details.) (b) The full electric dipole field is determined by following the procedure in part (a) at other locations near the dipole.

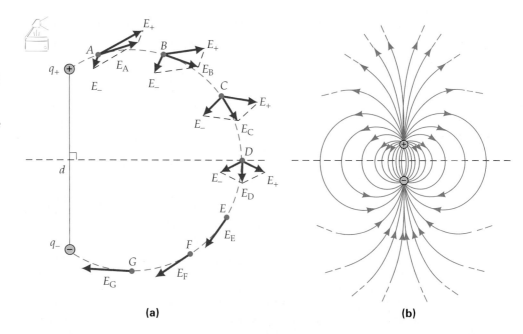

(a)

(b)

▼ Figure 15.16a shows the use of the superposition principle to construct the electric field line pattern due to a large positively charged plate. Notice that the field points perpendicularly away from the plate on both sides. Figure 15.16b shows the result if the plate is negatively charged, the only difference being the field direction. Putting these two together, we can find the field between two closely spaced and oppositely charged plates. The result is the pattern in Fig. 15.16c. Due to the cancellation of the horizontal field components (as long as we stay away from the plate edges), the electric field is uniform and points from the positive to negative. (Think of the direction of the force acting on a positive test charge placed between the plates.)

The derivation of the expression for the electric field magnitude between two closely spaced plates is beyond the scope of this text. However, the result is

$$E = \frac{4\pi kQ}{A} \quad \textit{(electric field between parallel plates)} \tag{15.5}$$

where Q is the magnitude of the total charge on *one* of the plates and A is the area of *one* plate. Parallel plates are common in electronic applications. For example, in Chapter 16 we will study an important circuit element called a *capacitor*, which, in its simplest form, is just a set of parallel plates. Capacitors play a crucial role in life-saving devices such as *heart defibrillators*, as we shall also see in Chapter 16.

Cloud-to-ground lightning can be approximated by closely spaced parallel plates as in the next Example. (See Insight 15.1 on Lightning and Lightning Rods on accompanying page.)

▶ **FIGURE 15.16** Electric field due to very large parallel plates (a) Above a positively charged plate, the net electric field points upward. Here, the horizontal components of the electric fields from various locations on the plate cancel out. Below the plate, $\vec{E}$ points downward. (b) For a negatively charged plate, the electric field directions (shown on both sides of the plate) are reversed. (c) Superimposing the fields from both plates results in cancellation outside the plates and an approximately uniform field between them.

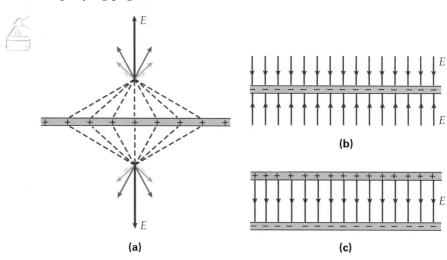

(a)

(b)

(c)

INSIGHT   15.1 LIGHTNING AND LIGHTNING RODS

Although the violent release of electrical energy in the form of lightning is common, we have a lot to learn about its formation. It is known that during the development of a cumulonimbus (storm) cloud, a separation of charge occurs. How the separation of charge takes place in a cloud is not fully understood, but it must be associated with the rapid vertical movement of air and moisture within storm clouds. Whatever the mechanism, the cloud acquires regions of different charge, with the bottom usually negatively charged.

As a result, an opposite charge is induced on the Earth's surface (Fig. 1a). Eventually, lightning may reduce this charge difference by ionizing the air, allowing a flow of charge between the cloud and the ground. However, air is a good insulator, so the electric field must be quite strong for ionization to occur. (See Example 15.9 for a quantitative estimate of the charge on a cloud.)

Most lightning occurs entirely within a cloud (intracloud discharges), where it cannot be seen. However, visible discharges do take place between clouds (cloud-to-cloud discharges) and between a cloud and the Earth (cloud-to-ground discharges). Special high-speed-camera photographs of cloud-to-ground discharges reveal a nearly invisible downward ionization path. The lightning discharges in a series of jumps or steps and so is called a *stepped leader*. As the leader nears the ground, positively charged ions in the form of a *streamer* rise from trees, tall buildings, or the ground to meet it.

When a streamer and a leader make contact, the electrons along the leader channel begin to flow downward. The initial flow is near the ground. As it continues, electrons positioned successively higher begin to migrate downward. Hence, the path of electron flow is extended upward in a *return stroke*. The surge of charge in the return stroke causes the conductive path to be illuminated, producing the bright flash seen by the eye and recorded in time-exposure photographs (Fig. 1b). Most lightning flashes have a duration of less than 0.50 s. Usually, after the initial discharge, ionization again takes place along the original channel, and another return stroke occurs. Typical lightning events have three or four return strokes.

Ben Franklin is often said to have been the first to demonstrate the electrical nature of lightning. In 1750, he suggested an experiment using a metal rod on a tall building. However, a Frenchman named Thomas François d'Alibard was the one who first set up the experiment using a rod during a thunderstorm (Fig. 1c). Franklin later performed a similar experiment with a kite, also during a thunderstorm.

A practical outcome of Franklin's work was the *lightning rod*, a pointed metal rod connected by a wire to a metal rod driven into the Earth, or "grounded." The elevated rod's tip, with its dense accumulation of induced positive charge and large electric field (see Fig. 15.19b), intercepts the downward, ionized stepped leader, discharging it harmlessly to the ground before the leader reaches a structure or makes contact with an upward streamer. This prevents the formation of a damaging electrical surge associated with a return stroke.

(a)

(b)

(c)

FIGURE 1 Lightning and lightning rods (a) Cloud polarization induces a charge on the Earth's surface. (b) When the electric field becomes large enough, an electrical discharge results, which we call lightning. (c) A lightning rod mounted atop a tall structure provides a path to ground so as to prevent damage.

INSIGHT

15.2 ELECTRIC FIELDS IN LAW ENFORCEMENT AND NATURE: STUN GUNS AND ELECTRIC FISH

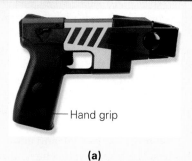

— Hand grip

(a)

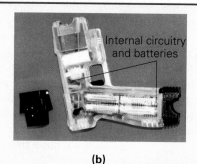

Internal circuitry and batteries

(b)

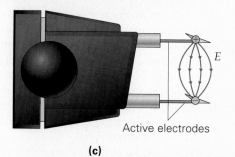

E

Active electrodes

(c)

FIGURE 1 The Taser stun gun (a) The exterior of a stun gun; notice the grip and two electrodes. **(b)** The interior: the circuitry necessary to increase the electric field and charge separation to the strength required to disrupt nerve communication. **(c)** A sketch of the electric field between the electrodes. (The charges change sign periodically, producing an oscillating electric field.)

Stun guns and electric fish have similar electric field properties. Stun guns (a generic name for several types, the most familiar being the hand-held Taser) generate a charge separation by using batteries and internal circuitry. This circuitry can produce a large charge polarization—that is, equal and opposite charge on the electrodes. Figures 1a and 1b show a typical Taser. The charges on the electrodes oscillate in sign, but at any instant, the field is close to that of a dipole (Fig. 1c). Tasers are used for subduing a criminal, theoretically without permanent harm. A law enforcement officer, holding the grip, applies the electrodes to the body—say, the thigh. The electric field disrupts the electrical signals in the nerves that control the large thigh muscle, rendering the muscle inoperative and making the criminal more easily subdued.

The phrase *electric fish* conjures an image of an electric eel (which is actually an eel-shaped fish). However, there are other fish that are "electric." The electric eel and others such as the electric catfish are *strongly electric fish*. They can generate large electric fields to stun prey but can also use the fields for location and communication. *Weakly electric fish*, such as the elephant nose (Fig. 2a), use their fields (Fig. 2b) for location and communication only. Fish that actively produce electric fields are called *electrogenic fish*.

In electrogenic fish, the charge separation is accomplished by the *electric organ* (shown for the elephant nose fish in Fig. 2b), which is a specialized stack of *electroplates*. Each electroplate is a disklike structure that is normally uncharged. When the brain sends a signal, the disks become polarized through a chemical process similar to that of nerve action, creating the fish's field.

Weakly electric fish are capable of producing electric fields about the same as those produced by batteries. This is good only for *electrocommunication* and *electrolocation*. Strongly electric fish produce fields hundreds of times stronger and can kill prey if the fish touches them simultaneously with the oppositely charged areas. The electric eel has thousands of electroplates stacked in the electric organ, which typically extends from behind its head well into its tail and may take up to 50% of its body length (Fig. 2c).

As an example how these fields are used for electrolocation, consider the change to the elephant nose fish's normal electric field pattern (Fig. 2b) if it approaches a small conducting object (Fig. 3). Notice that the field lines change to bend toward the object; because the object is conducting, the field lines must be oriented at right angles to its surface. This results in a stronger field at the part of the fish's skin surface near the object. Skin sensors detect this increase and send a signal to the brain to that effect. A nonconducting object, such as a rock, would have the opposite effect. In reality electrolocation and electrocommunication are determined by an interplay of the electric field and the sensory organs. However, the basic properties of electrostatic fields provide us with the general idea of how such fish operate.

Example 15.9 ■ Parallel Plates: Estimating the Charge on Storm Clouds

The electric field (magnitude) E required to ionize moist air is about 1.0×10^6 N/C. When the field reaches this value, the least bound electrons are pulled off their molecules (ionization of the molecules), which can lead to a lightning stroke. Assume that the value for E between the negatively charged lower cloud surface and the positively charged ground is 1.00% of this, or 1.0×10^4 N/C. (See Fig. 1a of Insight 15.1 on page 523.) Take the clouds to be squares 10 miles on each side. Estimate the magnitude of the total negative charge on the lower surface.

Thinking It Through. The electric field is given, so Eq. 15.5 can be used to estimate Q. The cloud area A (one of the "plates") must be expressed in square meters.

Solution.

Given: $E = 1.0 \times 10^4$ N/C
$d = 10$ mi $\approx 1.6 \times 10^4$ m

Find: Q (the magnitude of the charge on the lower cloud surface)

(a)

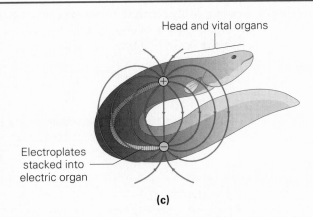

Head and vital organs

Electroplates
stacked into
electric organ

(c)

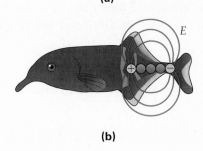

E

(b)

FIGURE 2 Electric fish (a) The elephant nose fish, which is a weakly electric fish—a fish that uses its electric field for electrolocation and communication. **(b)** At an instant in time, the approximate electric field produced by the elephant nose fish's electric organ, located near its tail. (The field actually oscillates.) **(c)** At an instant in time, the approximate electric field produced by an electric eel. The electric organ in the eel is capable of producing fields that can kill and stun as well as locate and communicate.

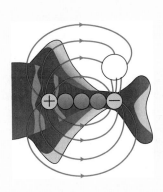

FIGURE 3 Electrolocation The field from an elephant nose fish with a conducting object nearby. Note the decrease in spacing of the field lines as they enter the skin surface. This increase in field strength is picked up by sensory organs on the skin, which send a signal to the fish's brain.

Using $A = d^2$ for the area of a square and solving Eq. 15.5 for the magnitude of the charge (the cloud surface is negative):

$$Q = \frac{EA}{4\pi k} = \frac{(1.0 \times 10^4 \text{ N/C})(1.6 \times 10^4 \text{ m})^2}{4\pi(9.0 \times 10^9 \text{ N} \cdot \text{m}^2/\text{C}^2)} = 23 \text{ C}$$

This expression is justified only if the distance between the clouds and the ground is much less than their size. (Why?) Such an assumption is equivalent to assuming that the 10-mile-long clouds are less than several miles from the Earth's surface.

This amount of charge is huge compared with the frictional static charges developed when shuffling on a carpet. However, because the cloud charge is spread out over a large area, any region of the cloud does not contain a lot of charge.

Follow-Up Exercise. In this Example, (a) what is the direction of the electric field between the cloud and the Earth? (b) How much charge would be required to ionize moist air?

Illustration 23.4 *Practical Uses of Charges and Electric Fields*

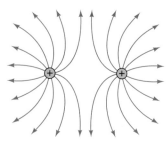

(a) Like point charges

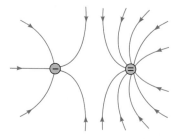

(b) Unequal like point charges

▲ **FIGURE 15.17** Electric fields
Electric fields for **(a)** like point
charges and **(b)** unequal like point
charges.

The electric field patterns for some other common charge configurations are shown in ◄Fig. 15.17. You should be able to see how they are sketched qualitatively. Note that the electric field lines begin on positive charges and end on negative ones (or at infinity when there is no nearby negative charge). Choose the number of lines emanating from or ending at a charge in proportion to the magnitude of that charge. (See the Learn by Drawing on Sketching Electric Lines of Force on page 521.)

15.5 Conductors and Electric Fields

OBJECTIVES: To (a) describe the electric field near the surface and in the interior of a conductor, (b) determine where charge accumulates on a charged conductor, and (c) sketch the electric field line outside a charged conductor.

The electric fields associated with charged conductors have several interesting properties. By definition in electrostatics, the charges are at rest. Because conductors possess electrons that are free to move, and they don't, the electrons must experience no electric force and thus no electric field. Hence we conclude that:

| The electric field is zero *inside* a charged conductor.

Excess charges on a conductor tend to get as far away from each other as possible, because they are highly mobile. Then:

| Any *excess* charge on an isolated conductor resides entirely on the surface of the conductor.

Another property of static electric fields and conductors is that there cannot be a tangential component of the field at the surface of the conductor. If this were not true, charges would move *along* the surface, contrary to our assumption of a static situation. Thus:

| The electric field at the surface of a charged conductor is perpendicular to the surface.

Lastly, the excess charge on a conductor of irregular shape is most closely packed where the surface is highly curved (at the sharpest points). Since the charge is densest there, the electric field will also be the largest at these locations. That is:

| Excess charge tends to accumulate at sharp points, or locations of highest curvature, on charged conductors. As a result, the electric field is greatest at such locations.

These last two results are summarized in ►Fig. 15.18. *Note that they are true only for conductors under static conditions.* Electric fields *can* exist inside nonconducting materials and inside conductors when conditions vary with time.

To understand *why* most of the charge accumulates in the highly curved surface regions, consider the forces acting *between* charges on the surface of the conductor. (See ►Fig. 15.19a.) Where the surface is fairly flat, these forces will be directed nearly parallel to the surface. The charges will spread out until the parallel forces from neighboring charges in opposite directions cancel out. At a sharp end, the forces between charges will be directed more nearly perpendicular to the surface, and so there will be little tendency for the charges to move parallel to the surface. Therefore one would expect highly curved regions of the surface to accumulate the highest concentration of charge.

An interesting situation occurs if there is a large concentration of charge on a conductor with a sharp point (Fig. 15.19b). The electric field above the point may be high enough to ionize air molecules (to pull or push electrons off the molecules). The freed electrons are then further accelerated by the field and can cause secondary ionizations by striking other molecules. This results in an "avalanche" of electrons, visible as a spark discharge. More charge can be placed

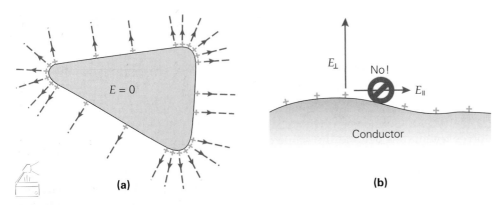

(a)   **(b)**

▲ **FIGURE 15.18** **Electric fields and conductors** **(a)** Under static conditions, the electric field is zero inside a conductor. Any excess charge resides on the conductor's surface. For an irregularly shaped conductor, the excess charge accumulates in the regions of highest curvature (the sharpest points), as shown. The electric field near the surface is perpendicular to that surface and strongest where the charge is densest. **(b)** Under static conditions, the electric field must *not* have a component tangential to the conductor's surface.

on a gently curved conductor, such as a sphere, before a spark discharge will occur. The concentration of charge at the sharp point of a conductor is one reason for the effectiveness of lightning rods. (See Insight 15.1 on Lightning and Lightning Rods on page 523.)

For some law enforcement and biological applications of electric fields and conductors, refer to Insight 15.2 on Stun Guns and Electric Fish on page 524.

As an illustration of an early experiment done on conductors with excess charge, consider the following Example.

Conceptual Example 15.10 ▦ The Classic Ice Pail Experiment

A positively charged rod is held inside an isolated metal container that has uncharged electroscopes conductively attached to its inside surface and to its outside surface (▾Fig. 15.20). What will happen to the leaves of the electroscopes? (Justify your answer.) (a) Neither electroscope's leaf will show a deflection. (b) Only the outside-connected electroscope's leaf will show a deflection. (c) Only the inside-connected electroscope's leaf will show a deflection. (d) The leaves of both electroscopes will show deflections.

Reasoning and Answer. The positively charged rod will attract negative charges, causing the inside of the metal container to become negatively charged. The outside electroscope will thus acquire a positive charge. Hence, both electroscopes will be charged (though with opposite signs) and show deflections, so the answer is (d). This experiment was performed by the nineteenth-century English physicist Michael Faraday using ice pails, so this setup is often called *Faraday's ice pail experiment.*

Follow-Up Exercise. Suppose in this Example that the positively charged rod actually *touched* the metal container. How would the electroscopes react now?

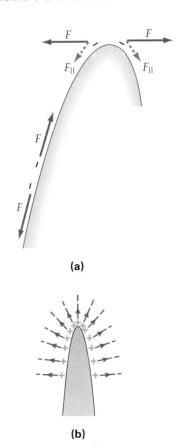

(a)

(b)

▲ **FIGURE 15.19** **Concentration of charge on a curved surface** **(a)** On a flat surface, the repulsive forces between excess charges are parallel to the surface and tend to push the charges apart. On a sharply curved surface, in contrast, these forces are directed at an angle to the surface. Their components parallel to the surface are smaller, allowing charge to concentrate in such areas. **(b)** Taken to the extreme, a sharply pointed metallic needle has a dense concentration of charge at the tip. This produces a large electric field in the region above the tip, which is the principle of the lightning rod.

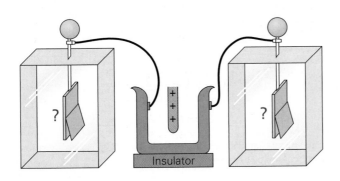

◀ **FIGURE 15.20** **An ice pail experiment** See Conceptual Example 15.10.

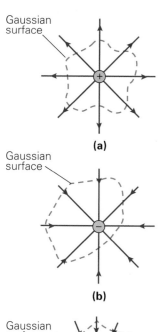

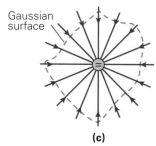

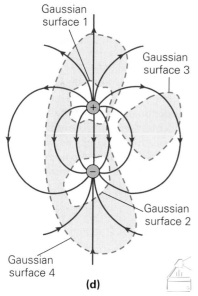

▲ **FIGURE 15.21** Various Gaussian surfaces and lines of force (a) Surrounding a single positive point charge, **(b)** surrounding a single negative point charge, and **(c)** surrounding a larger negative point charge. **(d)** Four different surfaces surrounding various parts of an electric dipole.

*15.6 Gauss's Law for Electric Fields: A Qualitative Approach

OBJECTIVES: To (a) state the physical basis of Gauss's law, and (b) use the law to make qualitative predictions.

One of the fundamental laws of electricity was discovered by Karl Friedrich Gauss (1777–1855), a German mathematician. Using it for quantitative calculations involves techniques beyond the scope of this book. However, a conceptual look at this law can teach us some interesting physics.

Consider the single positive charge in ◄Fig. 15.21a. Now picture an *imaginary closed surface* surrounding this charge. Such a surface is called a **Gaussian surface**. Let us designate electric field lines that pass through the surface outwardly as positive and inward pointing ones as negative. If the lines of both types are counted and totaled (that is, subtract the number of negative lines from the number of positive ones), we find that the total is positive, because in this case there are *only* positive lines. This result reflects the fact that there is a net number of outward-pointing electric field lines through the surface. Similarly, for a negative charge (Fig. 15.21b), the count would yield a negative total, indicating a net number of inward-pointing lines passing through the surface. Note that these results would be true for *any* closed surface surrounding the charge, regardless of its shape or size. If we double the magnitude of the negative charge (Fig. 15.21c), our negative field line count would also double. (Why?)

Figure 15.21d shows a dipole with four different imaginary Gaussian surfaces. surface 1 encloses a net positive charge and therefore has a positive line count. Similarly, surface 2 has a negative line count. The more interesting cases are surfaces 3 and 4. Note that both include zero net charge—surface 3 because it includes no charges at all and Surface 4 because it includes equal and opposite charges. Note that both surfaces 3 and 4 have a net line count of zero, correlating with no net charge enclosed.

These situations can be generalized (conceptually) to give us the underlying physical principle of **Gauss's law:***

> The net number of electric field lines passing through an imaginary closed surface is proportional to the amount of net charge enclosed within that surface.

An everyday analogy illustrated in ▼Fig. 15.22 may help you understand this principle. If you surround a lawn sprinkler with an imaginary surface (surface 1), you find that there is a net flow of water out through that surface—because inside is a "source" of water (disregarding the pipe's bringing water into the sprinkler). In an analogous way, a net outward-pointing electric field indicates the presence of a net positive charge inside the surface, because positive charges are "sources" of the electric field. Similarly, a puddle would form inside our imaginary surface 2 because there would be a net inward water flow through the surface. The following Example illustrates the power of Gauss's law in its qualitative form.

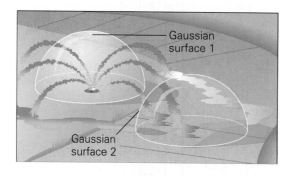

◄ **FIGURE 15.22** Water analogy to Gauss's law A net outward flow of water indicates a source of water inside closed surface 1. A net inward flow of water indicates a water drain inside closed surface 2.

---

*Strictly speaking, this is Gauss's law for electric fields. There is also a version of Gauss's law for magnetic fields, which will not be discussed.

Conceptual Example 15.11 ■ Charged Conductors Revisited: Gauss's Law

A net charge Q is placed on a conductor of arbitrary shape (►Fig. 15.23). Use the qualitative version of Gauss's law to prove that all the charge must lie on the conductor's surface under electrostatic conditions.

Reasoning and Answer. Because the situation is static equilibrium, there can be no electric field inside the volume of the conductor; otherwise, the almost-free electrons would move around. Take a Gaussian surface that follows the shape of the conductor, but is *just barely* inside the actual surface. Because there are no electric field lines inside the conductor, there are also no electric field lines passing through our imaginary surface. Thus zero electric field lines penetrate the Gaussian surface. But by Gauss's law, the net number of field lines is proportional to the amount of charge inside the surface. Therefore, there must be no net charge within the surface.

   Because our surface can be made as close as we wish to the conductor surface, it follows that the excess charge, if it cannot be inside the volume of the conductor, must be on the surface.

Follow-Up Exercise. In this Example, if the net charge on the conductor is negative, what is the sign of the net number of lines through a Gaussian surface that completely encloses the conductor? Explain your reasoning.

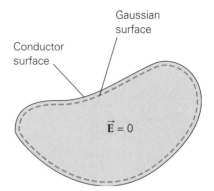

▲ **FIGURE 15.23** Gauss's law: **Excess charge on a conductor** See Conceptual Example 15.12.

Chapter Review

- The **law of charges**, or **charge–force law**, states that like charges repel and opposite charges attract.

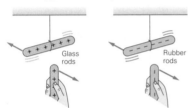

- **Charge conservation** means that the net charge of an isolated system remains constant.
- **Conductors** are materials that conduct electric charge readily because their atoms have one or more loosely bound electrons.
- **Insulators** are materials that do not easily gain, lose, or conduct electric charge.
- **Electrostatic charging** involves processes that enable an object to gain a net charge. Among these processes are charging by friction, contact (conduction), and induction.

- The **electric polarization** of an object involves creating separate and equal amounts of positive and negative charge in different locations on that object.

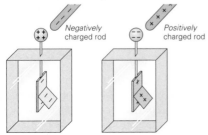

- **Coulomb's law** expresses the magnitude of the force between two point charges:

$$F_e = \frac{kq_1q_2}{r^2} \quad \text{(two point charges)} \tag{15.2}$$

where $k \approx 9.00 \times 10^9 \text{ N} \cdot \text{m}^2/\text{C}^2$.

$$F_{12} = \frac{kq_1q_2}{r^2} \longleftarrow \overset{q_1}{\bullet} \qquad \overset{q_2}{\bullet} \longrightarrow F_{21} = \frac{kq_1q_2}{r^2}$$

- The **electric field** is a vector field that describes how charges modify the space around them. It is defined as the electric force per unit positive charge, or

$$\vec{\mathbf{E}} = \frac{\vec{\mathbf{F}}_{\text{on } q_+}}{q_+} \tag{15.3}$$

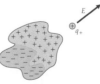

- According to the **superposition principle for electric fields**, the (net) electric field at any location due to a configuration of charges is the vector sum of the individual electric fields from individual charges of that configuration.

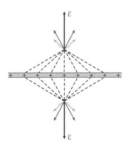

• **Electric field lines** are a visualization of the electric field. The line spacing is inversely related to the field strength, and tangents to the lines give the field direction.

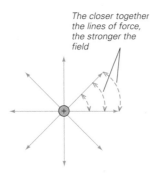

The closer together the lines of force, the stronger the field

• Under electrostatic conditions, the **electric fields associated with conductors** have the following properties:

The electric field is zero inside a charged conductor.

Excess charge on a conductor resides entirely on its surface.

The electric field near the surface of a charged conductor is perpendicular to that surface.

Excess charge on the surface of a conductor is most dense at locations of highest surface curvature.

The electric field near the surface of a charged conductor is greatest at locations of highest surface curvature.

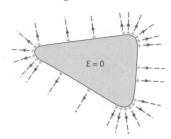

Exercises*

MC = *Multiple Choice Question,* CQ = *Conceptual Question, and* IE = *Integrated Exercise. Throughout the text, many exercise sections will include "paired" exercises. These exercise pairs, identified with* **red numbers,** *are intended to assist you in problem solving and learning. In a pair, the first exercise (even numbered) is worked out in the Study Guide so that you can consult it should you need assistance in solving it. The second exercise (odd numbered) is similar in nature, and its answer is given at the back of the book.*

15.1 Electric Charge

1. **MC** A combination of two electrons and three protons would have a net charge of (a) +1, (b) −1, (c) $+1.6 \times 10^{-19}$ C, (d) -1.6×10^{-19} C. (c)

2. **MC** An electron is just above a fixed proton. The direction of the force on the electric proton is (a) up, (b) down, (c) zero. (a)

3. **MC** In Exercise 2, which one feels the bigger size force: (a) The electron, (b) proton, or (c) both feel the same size force? (c)

4. **CQ** (a) How do we know that there are two types of electric charge? (b) What would be the effect of designating the charge on the electron as positive and the charge on the proton as negative? see ISM

5. **CQ** An electrically neutral object can be given a net charge by several means. Does this violate the conservation of charge? Explain. no, see ISM

6. **CQ** If a solid neutral object becomes positively charged, does its mass increase or decrease? What if it becomes negatively charged? decrease; increase

7. **CQ** How can you determine the type of charge on an object using an electroscope that has a net charge of a known sign? Explain. see ISM

*Take *k* to be exact at 9.00×10^9 N·m^2/C^2 and *e* to be exact at 1.60×10^{-19} C for significant-figure purposes.

8. **CQ** If two objects electrically repel each other, are both necessarily charged? What if they attract each other? yes; not necessarily (polarization)

9. ● What is the net charge of an object that has 1.0 million excess electrons? -1.6×10^{-13} C

10. ● In walking across a carpet, you acquire a net negative charge of 50 μC. How many excess electrons do you have? 3.1×10^{14} electrons

11. ●● An alpha particle is the nucleus of a helium atom with no electrons. What would be the charge on two alpha particles? $+6.40 \times 10^{-19}$ C

12. **IE** ●● A glass rod rubbed with silk acquires a charge of $+8.0 \times 10^{-10}$ C. (a) Is the charge on the silk (1) positive, (2) zero, or (3) negative? Why? (b) What is the charge on the silk, and how many electrons have been transferred to the silk? (c) How much mass has the glass rod lost? see ISM

13. **IE** ●● A rubber rod rubbed with fur acquires a charge of -4.8×10^{-9} C. (a) Is the charge on the fur (1) positive, (2) zero, or (3) negative? Why? (b) What is the charge on the fur, and how much mass is transferred to the rod? (c) How much mass has the rubber rod gained? (a) (1) positive (b) $+4.8 \times 10^{-9}$ C, 2.7×10^{-20} kg, (c) 2.7×10^{-20} kg

15.2 Electrostatic Charging

14. **MC** A rubber rod is rubbed with fur. The fur is then quickly brought near the bulb of an uncharged electroscope. The sign of the charge on the leaves of the electroscope is (a) positive, (b) negative, (c) zero. (a)

15. **MC** A stream of water is deflected toward a nearby electrically charged object that is brought close to it. The sign of the charge on the object (a) is positive, (b) is negative, (c) is zero, (d) can't be determined by the data given. (d)

16. **MC** A balloon is charged and then clings to a wall. The sign of the charge on the balloon (a) is positive, (b) is negative, (c) is zero, (d) can't be determined by the data given. (d)

17. **CQ** Fuel trucks often have metal chains reaching from their frames to the ground. Why is this important?
to remove excess charge due to friction of rubber on road

18. **CQ** Is there a gain or loss of electrons when an object is electrically polarized? Explain.
no, charges simply reorient themselves

19. **CQ** Explain carefully the steps you would use to create an electroscope that is positively charged by induction. After you are done, how can you verify that the electroscope is positively (and thus not negatively) charged?
see ISM

20. **CQ** Two metal spheres mounted on insulated supports are in contact. Bringing a negatively charged object close to the right-hand sphere would enable you to temporarily charge both spheres by induction. Explain clearly how this would work and what the sign of the charge on each sphere would be. see ISM

15.3 Electric Force

21. **MC** How does the magnitude of the electric force between two point charges change as the distance between them is increased? The force (a) decreases, (b) increases, (c) stays the same. (a)

22. **MC** Compared with the electric force, the gravitational force between two protons is (a) about the same, (b) somewhat larger, (c) very much larger, (d) very much smaller. (d)

23. **CQ** The Earth attracts us by its gravitational force, but we have seen that the electric force is much greater than the gravitational force. Why don't we experience an electric force from the Earth? objects are electrically neutral

24. **CQ** Two nearby electrons would fly apart if released. How could you prevent this by placing a single charge in their neighborhood? Explain clearly what the sign of the charge and its location would have to be. see ISM

25. **CQ** Coulomb's law is an example of an inverse-square law. Use this inverse-square idea to determine the ratio of electric force (final divided by initial) between two charges when the distance between them is cut to one third of its initial value. 9

26. **IE ●** An electron that is a certain distance from a proton is acted on by an electrical force. (a) If the electron were moved twice that distance away from the proton, would the electrical force be (1) 2, (2) $\frac{1}{2}$, (3) 4, or (4) $\frac{1}{4}$ times the original force? Why? (b) If the initial electric force is F, and the electron were moved to one third the original distance toward the proton, what would be the new electrical force? (a) (4) $\frac{1}{4}$ (b) $9F$

27. **●** Two identical point charges are a fixed distance apart. By what factor would the magnitude of the electric force between them change if (a) one of their charges were doubled and the other were halved, (b) both their charges were halved, and (c) one charge were halved and the other were left unchanged? (a) 1 (b) $\frac{1}{4}$ (c) $\frac{1}{2}$

28. **●** In a certain organic molecule, the nuclei of two carbon atoms are separated by a distance of 0.25 nm. What is the magnitude of the electric force between them?
1.3×10^{-7} N

29. **●** An electron and a proton are separated by 2.0 nm. (a) What is the magnitude of the force on the electron? (b) What is the net force on the system?
(a) 5.8×10^{-11} N (b) zero

30. **IE ●** Two charges originally separated by a certain distance are moved farther apart until the force between them has decreased by a factor of 10. (a) Is the new distance (1) less than 10, (2) equal to 10, or (3) greater than 10 times the original distance? Why? (b) If the original distance was 30 cm, how far apart are the charges?
(a) (1) less than 10 times (b) 95 cm

31. **●** Two charges are brought together until they are 100 cm apart, causing the electric force between them to increase by a factor of exactly 5. What was their initial separation distance? 2.24 m

32. **●** The distance between neighboring singly charged sodium and chlorine ions in crystals of table salt (NaCl) is 2.82×10^{-10} m. What is the attractive electric force between the ions? 2.90×10^{-9} N

33. **●●** Two point charges of -2.0 μC are fixed at opposite ends of a meterstick. Where on the meterstick could (a) a free electron and (b) a free proton be in electrostatic equilibrium? (a) 50 cm (b) 50 cm

34. **●●** Two point charges of -1.0 μC and $+1.0$ μC are fixed at opposite ends of a meterstick. Where could (a) a free electron and (b) a free proton be in electrostatic equilibrium? (a) nowhere (b) nowhere

35. **●●** Two charges, q_1 and q_2, are located at the origin and at (0.50 m, 0), respectively. Where on the x-axis must a third charge, q_3, of arbitrary sign be placed to be in electrostatic equilibrium if (a) q_1 and q_2 are like charges of equal magnitude, (b) q_1 and q_2 are unlike charges of equal magnitude, and (c) $q_1 = +3.0$ μC and $q_2 = -7.0$ μC? (a) $x = 0.25$ m (b) nowhere (c) $x = -0.94$ m for $\pm q_3$

36. **●●** Compute the gravitational force and electrical force between the electron and proton in the hydrogen atom (▼Fig. 15.24) assuming they are 5.3×10^{-11} m apart. Then calculate the ratio of the magnitudes of the electric force to the gravitational force.
$F_g = 3.6 \times 10^{-47}$ N, $F_e = 8.2 \times 10^{-8}$ N, 2.3×10^{39}

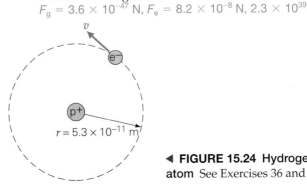

◄ **FIGURE 15.24 Hydrogen atom** See Exercises 36 and 37.

37. ●●● On average, the electron and proton in a hydrogen atom are separated by a distance of 5.3×10^{-11} m (Fig. 15.24). Assuming the orbit of the electron to be circular, (a) what is the electric force on the electron? (b) What is the electron's orbital speed? (c) What is the magnitude of the electron's centripetal acceleration in units of g? (a) 8.2×10^{-8} N (b) 2.2×10^{6} m/s (c) 9.2×10^{21} g

38. ●●● Three charges are located at the corners of an equilateral triangle, as depicted in ▼ Fig. 15.25. What are the magnitude and the direction of the force on q_1? 3.6 N in the positive x-direction

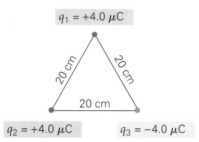

$q_1 = +4.0 \ \mu C$

20 cm   20 cm

20 cm

$q_2 = +4.0 \ \mu C$ $q_3 = -4.0 \ \mu C$

▲ **FIGURE 15.25 Charge triangle** See Exercises 38, 59, and 60.

39. ●●● Four charges are located at the corners of a square, as illustrated in ▼ Fig. 15.26. What are the magnitude and the direction of the force (a) on charge q_2 and (b) on charge q_4? (a) 96 N, 39° below positive x-axis (b) 61 N, 84° above negative x-axis

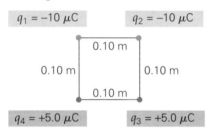

$q_1 = -10 \ \mu C$ $q_2 = -10 \ \mu C$

0.10 m

0.10 m   0.10 m

0.10 m

$q_4 = +5.0 \ \mu C$ $q_3 = +5.0 \ \mu C$

▲ **FIGURE 15.26 Charge rectangle** See Exercises 39, 61, and 65.

40. ●●● Two 0.10-g pith balls are suspended from the same point by threads 30 cm long. (Pith is a light insulating material once used to make helmets worn in tropical climates.) When the balls are given equal charges, they come to rest 18 cm apart, as shown in ▼ Fig. 15.27. What is the magnitude of the charge on each ball? (Neglect the mass of the thread.) 3.3×10^{-8} C

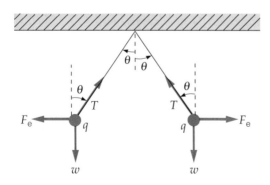

▲ **FIGURE 15.27 Repelling pith balls** See Exercise 40.

15.4 Electric Field

41. **MC** How is the magnitude of the electric field due to a point charge reduced when the distance from that charge is tripled: (a) It stays the same, (b) it is reduced to one third of its original value charge, (c) it is reduced to one ninth of its original value charge, or (d) it is reduced to one twenty-seventh of its original value charge? (c)

42. **MC** The SI units of electric field are (a) C, (b) N/C, (c) N, (d) J. (b)

43. **MC** At a point in space, an electric force acts vertically upward on an electron. The direction of the electric field at that point is (a) down, (b) up, (c) zero, (d) undetermined by the data. (a)

44. **CQ** How is the relative magnitude of the electric field in different regions determined from a field vector diagram? from the relative lengths of the electric field vectors

45. **CQ** How can the relative magnitudes of the field in different regions be determined from an electric field line diagram? by the relative density or spacing of the field lines

46. **CQ** Explain clearly why electric field lines can never cross. see ISM

47. **CQ** A positive charge is inside an isolated metal spherical shell, as shown in ▼ Fig. 15.28. Describe the electric field in the three regions: between the charge and the inside surface of the shell, inside the shell itself, and outside the outer shell surface. What is the sign of the charge on the two shell surfaces? How would your answers change if the charge were negative? see ISM

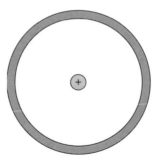

Metal conductor

◀ **FIGURE 15.28 A point charge inside a thick metal spherical shell** See Exercise 47.

48. **CQ** At a certain location, the electric field due to the excess charge on the Earth's surface points downward. What is the sign of the charge on the Earth's surface at that location? Why? negative; see ISM

49. **CQ** (a) Could the electric field due to two identical negative charges ever be zero at some location(s) nearby? Explain. If yes, describe and sketch the situation. (b) How would your answer change if the charges were equal but oppositely charged? Explain. (a) yes, see ISM (b) no, see ISM

50. **IE** ● (a) If the distance from a charge is doubled, is the magnitude of the electric field (1) increased, (2) decreased, or (3) the same compared to the initial value? (b) If the original electric field due to a charge is 1.0×10^{-4} N/C, what is the magnitude of the new electric field at twice the distance from the charge? (a) (2) decreased (b) 2.5×10^{-5} N/C

51. ● An electron is acted on by an electric force of 3.2×10^{-14} N. What is the magnitude of the electric field at the electron's location? 2.0×10^5 N/C

52. ● What is the magnitude and direction of the electric field at a point 0.75 cm away from a point charge of $+2.0$ pC? 3.2×10^2 N/C away from the charge

53. ● At what distance from a proton is the magnitude of its electric field 1.0×10^5 N/C?
    1.2×10^{-7} m away form the charge

54. IE ●● Two fixed charges, $-4.0\,\mu$C and $-5.0\,\mu$C, are separated by a certain distance. (a) Is the net electric field at a location halfway between the two charges (1) directed toward the $-4.0\,\mu$C charge, (2) zero, or (3) directed toward the $-5.0\,\mu$C charge? Why? (b) If the charges are separated by 20 cm, calculate the magnitude of the net electric field halfway between the charges?
    (a) (3) toward the $-5.0\,\mu$C charge (b) 9.0×10^5 N/C

55. ●● What would be the magnitude and the direction of an electric field that would just support the weight of a proton near the surface of the Earth? What about an electron?
    1.0×10^{-7} N/C upward; 5.6×10^{-11} N/C downward

56. IE ●● Two charges, $-3.0\,\mu$C and $-4.0\,\mu$C, are located at $(-0.50$ m, 0) and $(0.50$ m, 0), respectively. There is a point on the x-axis between the two charges where the electric field is zero. (a) Is that point (1) left of the origin, (2) at the origin, or (3) right of the origin? (b) Find the location of the point where the electric field is zero.
    (a) (1) left of the origin (b) $(-0.036$ m, 0)

57. ●● Three charges, $+2.5\,\mu$C, $-4.8\,\mu$C, and $-6.3\,\mu$C, are located at $(-0.20$ m, 0.15 m), $(0.50$ m, -0.35 m), and $(-0.42$ m, -0.32 m), respectively. What is the electric field at the origin?
    $\vec{E} = (2.2 \times 10^5$ N/C$)\hat{x} + (-4.1 \times 10^5$ N/C$)\hat{y}$

58. ●● Two charges of $+4.0\,\mu$C and $+9.0\,\mu$C are 30 cm apart. Where on the line joining the charges is the electric field zero? 12 cm from the charge of $4.0\,\mu$C (between the charges)

59. ●●● What is the electric field at the center of the triangle in Fig. 15.25? 5.4×10^6 N/C toward the charge of $-4.0\,\mu$C

60. ●●● Compute the electric field at a point midway between charges q_1 and q_2 in Fig. 15.25.
    1.2×10^6 N/C toward the charge of $-4.0\,\mu$C

61. ●●● What is the electric field at the center of the square in Fig. 15.26? 3.8×10^7 N/C in the $+y$-direction

62. ●●● A particle with a mass of 2.0×10^{-5} kg and a charge of $+2.0\,\mu$C is released in a (parallel-plate) uniform horizontal electric field of 12 N/C. (a) How far horizontally does the particle travel in 0.50 s? (b) What is the horizontal component of its velocity at that point? (c) If the plates are 5.0 cm on each side, how much charge is on each?
    (a) 0.15 m (b) 0.60 m/s (c) ±0.27 pC

63. ●●● Two very large parallel plates are oppositely and uniformly charged. If the field between the plates is 1.7×10^6 N/C, how dense is the charge on each plate (in μC/m²)? 15 μC/m²

64. ●●● Two square, oppositely charged conducting plates measure 20 cm on each side. The plates are close together and parallel to each other. Their charges are $+4.0$ nC and -4.0 nC, respectively. (a) What is the electric field between the plates? (b) What force is exerted on an electron between the plates? (a) 1.1×10^4 N/C toward negative plate (b) 1.8×10^{-15} N toward positive plate

65. ●●● Compute the electric field at a point 4.0 cm from q_2 along a line running toward q_3 in Fig. 15.26.
    $\vec{E} = (-4.4 \times 10^6$ N/C$)\hat{x} + (7.3 \times 10^7$ N/C$)\hat{y}$

66. ●●● Two equal and opposite point charges form a dipole, as shown in ▼Fig. 15.29. (a) Add the electric fields due to each end at point P, thus graphically determining the direction of the field there. (b) Derive a symbolic expression for the magnitude of the electric field at point P, in terms of k, q, d, and x. (c) If point P is very far away, use the exact result to show that $E \approx kqd/x^3$. (d) Why is it an inverse-*cube* fall-off instead of inverse-*square*? Explain. see ISM

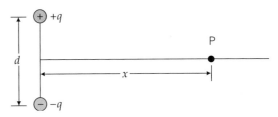

▲ **FIGURE 15.29 Electric dipole field** See Exercise 66.

15.5 Conductors and Electric Fields

67. **MC** In electrostatic equilibrium, is the electric field just below the surface of a charged conductor (a) the same value as the field just above the surface (b) zero, (c) dependent on the amount of charge on the conductor, or (d) given by kq/R^2? (b)

68. **MC** An uncharged thin metal slab is placed in an external electric field that points horizontally to the left. What is the electric field inside the slab: (a) zero, (b) the same value as the original external field but oppositely directed, (c) less than the original external field value but not zero, or (d) depends on the magnitude of the external field? (a)

69. **MC** The direction of the electric field at the surface of a charged conductor under electrostatic conditions (a) is parallel to the surface, (b) is perpendicular to the surface, (c) is at a 45° angle to the surface, or (d) depends on the charge on the conductor. (b)

70. **CQ** Is it safe to stay in a car in a lightning storm (▼Fig. 15.30)? Explain. yes, see ISM

▲ **FIGURE 15.30 Safe inside a car?** See Exercise 70.

71. **CQ** Under electrostatic conditions, the excess charge on a conductor is uniformly spread over its surface. What is the shape of the surface? spherical

72. **CQ** Tall buildings have lightning rods to protect them from lightning strikes. Explain why the rods are pointed in shape and taller than the buildings. charges accumulate at sharp points and lightning hits the tall rods first

73. **IE ●** A solid conducting sphere is surrounded by a thick, spherical conducting shell. Assume that a total charge $+Q$ is placed at the center of the sphere and released. (a) After equilibrium is reached, the inner surface of the shell will have (1) negative, (2) zero, (3) positive charge. (b) In terms of Q, how much charge is on the interior of the sphere? (c) The surface of the sphere? (d) The inner surface of the shell? (e) The outer surface of the shell? (a) (1) negative (b) zero (c) $+Q$ (d) $-Q$ (e) $+Q$

74. **●** In Exercise 73, what is the electric field direction (a) in the interior of the solid sphere, (b) between the sphere and the shell, (c) inside the shell, and (d) outside the shell? (a) none (b) outward (c) none (d) outward

75. **●●** In Exercise 73, write expressions for the electric field magnitude (a) in the interior of the solid sphere, (b) between the sphere and the shell, (c) inside the shell, and (d) outside the shell. Your answer should be in terms of Q, r (the distance from the center of the sphere), and k. (a) zero (b) kQ/r^2 (c) zero (d) kQ/r^2

76. **●●** A flat, triangular piece of metal with rounded corners has a net positive charge on it. Sketch the charge distribution on the surface and the electric field lines near the surface of the metal (including their direction). see ISM

77. **●●●** Approximate a metal needle as a long cylinder with a very pointed, but slightly rounded, end. Sketch the charge distribution and outside electric field lines if the needle has an excess of electrons on it. see ISM

*15.6 Gauss's Law for Electric Fields: A Qualitative Approach

78. **MC** A Gaussian surface surrounds an object with a net charge of $-5.0\,\mu C$. Which of the following is true? (a) More electric field lines will point outward than inward. (b) More electric field lines will point inward than outward. (c) The net number of field lines through the surface is zero. (d) There must be only field lines passing inward through the surface. (b)

79. **MC** What can you say about the net number of electric field lines passing through a Gaussian surface located completely within the region between a set of oppositely charged parallel plates? (a) The net number point outward. (b) The net number point inward. (c) The net number is zero. (d) The net number depends on the amount of charge on each plate. (c)

80. **MC** Two concentric spherical surfaces enclose a charged particle. The radius of the outer sphere is twice that of the inner one. Which sphere will have more electric field lines passing through its surface? (a) The larger one. (b) The smaller one. (c) Both spheres would have the same number of field lines passing through them. (d) The answer depends on the amount of charge on the particle. (c)

81. **CQ** The same Gaussian surface is used to surround two charged objects separately. The net number of field lines penetrating the surface is the same in both cases, but the lines are oppositely directed. What can you say about the net charges on the two objects? the net charges are equal and opposite in sign

82. **CQ** If a net number of electric field lines point outward from a Gaussian surface, does that necessarily mean there are no negative charges in the interior? Explain with an example. no, just more positives than negatives: net positive

83. **●** Suppose a Gaussian surface encloses both a positive point charge that has six field lines leaving it and a negative point charge with twice the magnitude of charge of the positive one. What is the net number of field lines passing through the Gaussian surface? -6 lines, or net of 6 lines entering it

84. **IE ●●** A Gaussian surface has 16 field lines leaving it when it surrounds a point charge of $+10.0\,\mu C$ and 75 field lines entering it when it surrounds an unknown point charge. (a) The magnitude of the unknown charge is (1) greater than $10.0\,\mu C$, (2) equal to $10.0\,\mu C$, or (3) less than $10.0\,\mu C$. Why? (b) What is the unknown charge? (a) (1) greater than $10.0\,\mu C$ (b) $-46.9\,\mu C$

85. **●●** If 10 field lines leave a Gaussian surface when it completely surrounds the positive end of an electric dipole, what would the count be if the surface surrounded just the other end? 10 field lines entering (negative)

Comprehensive Exercises

86. A negatively charged pith ball (mass 6.00×10^{-3} g, charge -1.50 nC) is suspended vertically from a light nonconducting string of length 15.5 cm. This apparatus is then placed in a horizontal uniform electric field. After being released, the pith ball comes to a stable position at an angle of $12.3°$ to the left of the vertical. (a) What is the direction of the external electric field? (b) Determine magnitude of the electric field. (a) right (b) 8.55×10^3 N/C

87. A positively charged particle with a charge of 9.35 pC is suspended in equilibrium in the electric field between two oppositely charged horizontal parallel plates. The square plates each have a charge 5.50×10^{-5} C, are separated by 6.25 mm, and have an edge length of 11.0 cm. (a) Which plate must be positively charged? (b) Determine the mass of the particle. (a) bottom (b) 4.90×10^{-4} kg

88. **CQ** Use the superposition principle and/or symmetry arguments to determine the direction of the electric field (a) at the center of a uniformly positively charged semicircular wire, (b) in the plane of a negatively charged flat plate, just off one of the edges, and (c) on the perpendicular bisector axis of a long thin insulator with more negative charge on one end than positive charge on the other end. see ISM

89. An electron starts from one plate of a charged closely spaced (vertical) parallel plate arrangement with a velocity of 1.63×10^4 m/s to the right. Its speed on reaching the other plate, 2.10 cm away, is 4.15×10^4 m/s. (a) What type of charge is on each plate? (b) What is the direction of the electric field between the plates? (c) If the plates are square with an edge length of 25.4 cm, determine the charge on each. see ISM

90. Two fixed charges, $-3.0\,\mu C$ and $-5.0\,\mu C$, are 0.40 m apart. (a) Where should a third charge of $-1.0\,\mu C$ be placed to put the system of three charges in electrostatic equilibrium? (b) What if the third charge was $+1.0\,\mu C$ instead?

    (a) 0.17 m from $-3.0\,\mu C$ in between charges (b) same

91. Find the electric field at point O for the charge configuration shown in ▼Fig. 15.31.

    5.5×10^3 N/C at 66° below positive x-axis

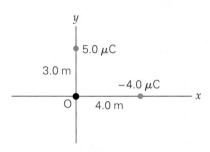

▲ **FIGURE 15.31 Electric field** See Exercise 91.

92. **CQ** A uniform metal slab (less thick than the plate separation distance) is inserted between and parallel to a pair of oppositely charged parallel plates. Sketch the resulting electric field everywhere between the plates including the slab. see ISM

93. An electron in a computer monitor enters midway between two parallel oppositely charged plates, as shown in ▼Fig. 15.32. The initial speed of the electron is 6.15×10^7 m/s and its vertical deflection (d) is 4.70 mm. (a) What is the magnitude of the electric field between the plates? (b) Determine the magnitude of the surface charge density on the plates in C/m².

    (a) 2.02×10^4 N/C (b) 1.79×10^{-7} C/m²

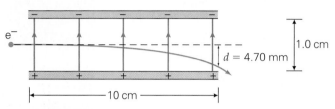

▲ **FIGURE 15.32 Electron in a computer monitor** See Exercise 93.

94. **CQ** For an electric dipole, the product qd is called the *dipole moment* and given the symbol p. The dipole moment is actually a vector $\vec{p}$ that points from the negative to the positive end. Assuming an electric dipole is free to move and rotate and starts from rest, (a) use a sketch to show that if it is placed in a uniform field it will rotate as it tries to "line up" with the field direction. (b) What is different about the motion of the dipole if the field is not uniform? see ISM

The following Physlet Physics Problems can be used with this chapter.

22.1, 22.2, 22.3, 22.4, 22.5, 22.7, 22.8, 22.9, 23.1, 23.2, 23.4, 23.6, 23.7, 23.8

PHYSICS FACTS

- The unit of electrical capacitance, the farad, is named for the British scientist Michael Faraday (1791–1867). With little formal education he was appointed (at age 21), a laboratory assistant at the Royal Institution (London). Eventually he became director of the laboratory. He discovered electromagnetic induction, which is the principle behind modern electric generating plants.

- In electrochemistry, an important amount of charge called the faraday is equal to 96 485.3415 coulombs. The name honors Michael Faraday for his electrochemistry experiments that showed that 1 faraday of charge is required to deposit 1 mole of silver on the negatively charged cathode of his apparatus.

- Count Alessandro Volta was born in Como, Italy, in 1745. Because he did not speak until age 4, his family was convinced he was mentally retarded. However, in 1778 he was the first to isolate methane (the main component of natural gas). Like many chemists of his time, he did significant work with electricity in relationship to chemical reactions. He constructed the first electric battery, and the unit of electromotive force, the volt (V), was named in his honor.

- Electric eels can kill or stun prey by producing potential differences (or voltages) up to 650 volts, more than fifty times as large as that of a car battery. Other electric fish, such as the elephant nose, generate only about 1 volt, useful for electrolocation but not hunting.

The girl in the photo is experiencing some electrical effects as she is electrically charged to several thousand volts. Household circuits operate at 120 volts and can give you a potentially dangerous shock. Yet this girl doesn't seem to be having a problem. What's going on? You'll find the explanation of this and other electrical phenomena in this and the two following chapters. Here the concept of electric potential will be introduced and its properties and usefulness examined.

Although this chapter concentrates on the study of fundamental electrical concepts such as voltage and capacitance, there are discussions involving practical applications. For example, your dentist's X-ray machine works by using high voltage to accelerate electrons. Heart defibrillators use capacitors to temporarily store the electrical energy required to stimulate the heart into its correct rhythm. Capacitors are used to store the energy that triggers the flash unit in your camera. Our body's nervous system, its communication network, is capable of sending thousands of electrical voltages per second shuttling back and forth along "cables" we call nerves. These signals are generated by chemical activity. The body uses them to enable us to do many things we take for granted such as muscle movement, thought processes, vision, and hearing. And, in following chapters, we will see even more practical uses of electricity, such as electric appliances, computers, medical instruments, electrical energy distribution systems, and household wiring.

16.1 Electric Potential Energy and Electric Potential Difference

OBJECTIVES: To (a) understand the concept of electric potential difference (voltage) and its relationship to electric potential energy, and (b) calculate electric potential differences.

In Chapter 15, electrical effects were analyzed in terms of electric field vectors and lines of force. Recall that the study of mechanics in early chapters began similarly through the use of Newton's laws, free-body diagrams, and forces (*vectors*). A search for a simpler approach led to *scalar* quantities such as work, kinetic energy, and potential energy. Using these quantities, energy methods were employed to solve problems that are much more difficult when using the vector (force) approach. It turns out to be extremely useful, both conceptually and for problem solving, to extend the energy methods to the study of electric fields.

Electric Potential Energy

To investigate electric potential energy, let's start with one of the simplest electric field patterns: the field between two large, oppositely charged parallel plates. As was learned in Chapter 15, near the center of the plates the field is uniform in magnitude and direction (▾Fig. 16.1a). Suppose a small positive charge q_+ is moved at constant velocity against the electric field, $\vec{E}$, in a straight line from the negative plate (A) to the positive one (B). An external force ($\vec{F}_{ext}$) with the same magnitude as the electric force is required (why?), and so we have $F_{ext} = q_+E$. The work done by this external force is positive, because the force and displacement are in the same direction. Then the work done by the external force is $W_{ext} = F_{ext}(\cos 0°)d = q_+Ed$.

Suppose the positive charge is now released from the positive plate. It will accelerate toward the negative plate, gaining kinetic energy. This kinetic energy is a result of the work done on the charge, and the initial energy (when there is no kinetic energy at B) must be some type of potential energy. In moving from A to B, the charge's **electric potential energy, U_e,** has increased ($U_B > U_A$) by an amount equal to the external work done on it. So the *change* in electric potential energy of the charge is

$$\Delta U_e = U_B - U_A = q_+Ed$$

The gravitational analogy to the parallel-plate electric field is the gravitational field near the Earth's surface, where it is uniform. When an object is raised a vertical distance h at constant velocity, the change in its potential energy is positive ($U_B > U_A$) and equal to the work done by the external (lifting) force. If it is assumed that there is no acceleration, this force must equal the object's weight, or $F_{ext} = w = mg$ (Fig. 16.1b). The increase in gravitational potential energy is then

$$\Delta U_g = U_B - U_A = F_{ext}h = mgh$$

(*Note:* Different distance symbols (h and d) are used to distinguish between the two different situations.)

Note: From Eq. 15.3, $\vec{E} = \vec{F}/q_+$, and $\vec{F} = q_+\vec{E}$.

PHYSLET

Illustration 25.1 Energy and Voltage

▼ **FIGURE 16.1 Changes in potential energy in uniform electric and gravitational fields (a)** Moving a positive charge q_+ against the electric field requires positive work and increases the electric potential energy. **(b)** Moving a mass m against the gravitational field requires positive work and increases the gravitational potential energy.

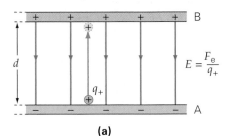

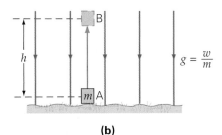

(a)   **(b)**

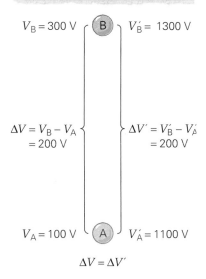

LEARN BY DRAWING

ΔV Is Independent of Reference Point

$V_B = 300$ V

$V'_B = 1300$ V

$\Delta V = V_B - V_A$
$\quad = 200$ V

$\Delta V' = V'_B - V'_A$
$\quad = 200$ V

$V_A = 100$ V

$V'_A = 1100$ V

$\Delta V = \Delta V'$

Note: We commonly assign a value of zero for the electric potential of the negative plate, but this is arbitrary. Only potential *differences* are meaningful.

Electric Potential Difference

Recall that defining the electric field as the electric force *per unit positive test charge* eliminated the dependence on the test charge. Thus knowing the electric field, the force on *any* charge placed could be determined from $F_e = q_+E$. Similarly, the **electric potential difference**, ΔV, between any two points in space is defined as the change in potential energy *per unit positive test charge*:

$$\Delta V = \frac{\Delta U_e}{q_+} \quad \text{(electric potential difference)} \tag{16.1}$$

SI unit of electric potential difference:
joule/coulomb (J/C), or volt (V)

The SI unit of electric potential difference is the joule per coulomb. This unit is named the **volt (V)** in honor of Alessandro Volta (1745–1827), an Italian scientist who constructed the first battery (Chapter 17), and 1 V = 1 J/C. Potential difference is commonly called **voltage**, and the symbol for potential difference is routinely changed from ΔV to just V, as is done later in this chapter.

Notice a crucial point: Electric potential difference, although based on electric potential-*energy* difference, is *not* the same. Electric potential difference is defined as electric potential energy difference *per unit charge*, and therefore does *not* depend on the amount of charge moved. Like the electric field, potential difference is a very useful quantity. If ΔV is known, we can then calculate ΔU_e for *any* amount of charge moved. To illustrate this, let's calculate the potential difference associated with the uniform field between two parallel plates:

$$\Delta V = \frac{\Delta U_e}{q_+} = \frac{q_+Ed}{q_+} = Ed \quad \begin{array}{l} \textit{potential difference} \\ \textit{(parallel plates only)} \end{array} \tag{16.2}$$

Notice that the amount of charge moved, q_+, cancels out. The potential difference ΔV depends only on the characteristics of the charged plates—that is, the field produced (E) and the separation (d). The language used is as follows:

> For a pair of oppositely charged parallel plates, the positively charged plate is at a *higher electric potential* than the negatively charged one by an amount ΔV.

Notice that electric potential *difference* is defined without defining electric potential itself (V). Although this may seem backward, there is a good reason for it. Of the two, electric potential *difference* is the physically meaningful quantity; that is, the quantity we actually measure. (Electric potential differences, or voltages, are measured with voltmeters, see Chapter 18.). The electric potential V, in contrast, isn't definable in an absolute way—it depends entirely on the choice of a reference point. This means that an arbitrary constant can be added to, or subtracted from, potentials, changing them. However this has no affect on the meaningful quantity of potential *difference*.

We encountered this idea during the study of potential energy associated with springs and gravitation (Sections 5.2 and 5.4). Recall that only *changes* in potential energies were important. Specific values of potential energy could be determined, but only after the zero reference point was defined. For example, in the case of gravity, zero gravitational potential energy is sometimes chosen at the Earth's surface. However, it is just as correct (and sometimes more convenient) to define zero at an infinite distance from Earth (Section 7.5).

These ideas also hold for electric potential energy and potential. The electric potential may be chosen as zero at the negative plate of a pair of parallel plates. However, it is sometimes convenient to locate the zero value at infinity, as will be seen in the case of a point charge. Either way, *differences* are unaffected. For a visualization of this, refer to the Learn by Drawing feature on this page. In this figure, with a certain choice of zero electric potential, point A is a potential of $+100$ V and B at $+300$ V. With a different zero, the potential at A might be $+1100$ V, in which case, the electric potential at B would then be $+1300$ V. Regardless of the zero, B will *always* be 200 V higher in potential than A.

The following Example illustrates the relationship between electric potential energy and electric potential.

Example 16.1 ■ Energy Methods in Moving a Proton: Potential Energy versus Potential

Imagine moving a proton from the negative plate to the positive plate of a parallel-plate arrangement (►Fig. 16.2a). The plates are 1.50 cm apart, and the field is uniform with a magnitude of 1500 N/C. (a) What is the change in the proton's electric potential energy? (b) What is the electric potential difference (voltage) between the plates? (c) If the proton is released from rest at the positive plate (Fig. 16.2b), what speed will it have just before it hits the negative plate?

Thinking It Through. (a) The change in potential energy can be computed from the work required to move the charge. (b) The electric potential difference between the plates can then be found by dividing the work by the charge moved. (c) When the proton is released, its electric potential energy is converted into kinetic energy. Knowing the proton's mass, we can calculate its speed.

Solution. The magnitude of the electric field, E, is given. Because a proton is involved, we can find its mass and charge (Table 15.1).

Given: $E = 1500$ N/C
$q_p = +1.60 \times 10^{-19}$ C
$m_p = 1.67 \times 10^{-27}$ kg
$d = 1.50$ cm $= 1.50 \times 10^{-2}$ m

Find: (a) ΔU_e (potential-energy change)
(b) ΔV (potential difference between plates)
(c) v (speed of released proton just before it reaches negative plate)

(a) Th electric potential energy is increased, so positive work is done to move the proton against the field, toward the positive plate:

$$\Delta U_e = q_p Ed = (+1.60 \times 10^{-19} \text{ C})(1500 \text{ N/C})(1.50 \times 10^{-2} \text{ m})$$
$$= +3.60 \times 10^{-18} \text{ J}$$

(b) The potential difference, or voltage, is the potential energy *change* per unit charge (defined by Eq. 16.1):

$$\Delta V = \frac{\Delta U_e}{q_p} = \frac{+3.60 \times 10^{-18} \text{ J}}{+1.60 \times 10^{-19} \text{ C}} = +22.5 \text{ V}$$

We would say that the positive plate is 22.5 V higher in electric potential than the negative one.

(c) The total energy of the proton is constant; therefore, $\Delta K + \Delta U_e = 0$. The proton has no initial kinetic energy ($K_o = 0$). Hence, $\Delta K = K - K_o = K$. From this, the speed of the proton can be calculated:

$$\Delta K = K = -\Delta U_e$$

or

$$\tfrac{1}{2} m_p v^2 = -\Delta U_e$$

But on its return to the negative plate, the proton's potential energy change is negative (why?), hence $\Delta U_e = -3.60 \times 10^{-18}$ J. So its speed is

$$v = \sqrt{\frac{2(-\Delta U_e)}{m_p}} = \sqrt{\frac{2[-(-3.60 \times 10^{-18} \text{ J})]}{1.67 \times 10^{-27} \text{ kg}}} = 6.57 \times 10^4 \text{ m/s}$$

Notice that even though the kinetic energy gained is very small, the proton acquires a high speed because its mass is extremely small.

Follow-Up Exercise. In this Example, how would your answers change if an alpha particle were moved instead of a proton? (An alpha particle is the nucleus of a helium atom and has a charge of $+2e$ and a mass approximately four times that of a proton.) (*Answers to all Follow-Up Exercises are at the back of the text.*)

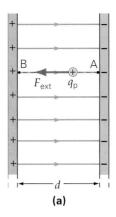

(a)

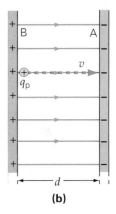

(b)

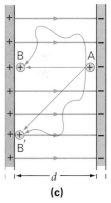

(c)

▲ **FIGURE 16.2 Accelerating a charge** **(a)** Moving a proton from the negative to the positive plate increases the proton's potential energy. (See Example 16.1.) **(b)** When it is released from the positive plate, the proton accelerates toward the negative plate, gaining kinetic energy at the expense of electric potential energy. **(c)** The work done to move a proton between any two points in an electric field, such as A and B or A and B′, is independent of the path.

The principles in Example 16.1 can be used to show another interesting property of electric potential energy (and potential) changes: Both are *independent of the path* on which the charged particle is taken. Recall from Section 5.5 that this means the *electrostatic force is conservative*. As shown in Fig. 16.2c, the work done in moving the proton from A to B is the same, regardless of the route. The alternative wiggly paths from A to B and A to B′ require the same work as do the straight-line paths. This is fundamentally because movement at right angles to the field requires no work. (Why?)

Note: Potential differences, like electric fields, are defined in terms of positive charges. Negative charges are subject to the same potential difference, but the opposite potential energy change. To determine whether the potential increases or decreases, decide whether an external force does positive or negative work on a *positive* test charge.

The gravitational analogy to Example 16.1 is that of raising an object in a uniform gravitational field. When the object is raised, its gravitational potential energy increases, because the force of gravity acts downward. However, with electricity we know that there are two types of charge, and the force between them can be repulsive or attractive. At this point, the analogy to gravity breaks down.

To understand why the analogy fails, consider how the discussion in Example 16.1 would differ if instead an electron had been moved. Because an electron is negatively charged, it would be attracted to plate B, and the external force would be *opposite* the electron's displacement (to prevent the electron from accelerating). For an electron, this force would do *negative* work, thereby *decreasing* the electric potential energy. Unlike the proton, the electron would be attracted to the positive plate (the plate with the higher electric potential). If allowed to move freely, electrons would "fall" (accelerate) toward regions of higher potential. Recall that the proton "fell" (accelerated) toward the region of lower potential. Regardless, both the proton and electron ended up losing electric potential energy and gaining kinetic energy. Thus the behavior of charged particles in electric fields can be summarized, in "potential language" as follows:

> Positive charges, when released, accelerate toward regions of lower electric potential.

> Negative charges, when released, accelerate toward regions of higher electric potential.

Consider the following medical application involving the creation of X-rays from fast-moving electrons, accelerated by large electric potential differences (voltages).

Example 16.2 ■ Creating X-Rays: Accelerating Electrons

Modern dental offices use X-ray machines for diagnosing hidden dental problems (◄Fig. 16.3a). Typically, electrons are accelerated through electric potential differences (voltages) of 25 000 V. When the electrons hit the positive plate, their kinetic energy is converted into high-energy particles called *X-ray photons* (Fig. 16.3b). (Photons are particles of light discussed in Chapter 27.) Suppose a single electron's kinetic energy is distributed equally among five X-ray photons. How much energy would one photon have?

Thinking It Through. From energy conservation, the kinetic energy gained by one electron is equal in magnitude to the electric potential energy it loses. From the kinetic energy lost by one electron, the energy of one X-ray photon can be calculated.

Solution. The charge of an electron is known (from Table 15.1), and the accelerating voltage is given.

Given: $q = -1.60 \times 10^{-19}$ C   *Find:*   energy (E) of one X-ray photon
        $\Delta V = 2.50 \times 10^4$ V

The electron leaves the negatively charged plate and moves toward the region of highest electric potential ("uphill"). Thus, the change in its electric potential energy is

$$\Delta U_e = q\Delta V = (-1.60 \times 10^{-19}\,\text{C})(+2.50 \times 10^4\,\text{V}) = -4.00 \times 10^{-15}\,\text{J}$$

The gain in kinetic energy comes from this loss in electric potential energy. Because the electrons have no appreciable kinetic energy when they start,

$$K = \left|\Delta U_e\right| = 4.00 \times 10^{-15}\,\text{J}$$

Therefore, if equally shared, one photon will have an energy of

$$E = \frac{K}{5} = 8.00 \times 10^{-16}\,\text{J}$$

Follow-Up Exercise. In this Example, use energy methods to determine the speed of one electron when it is halfway to the positive plate.

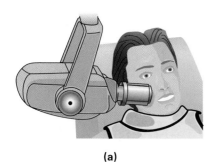

(a)

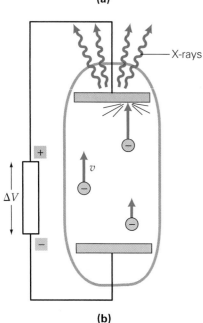

(b)

▲ **FIGURE 16.3** X-ray production
(a) An illustration of a dental X-ray machine. **(b)** A schematic diagram of the X-ray production.

Electric Potential Difference Due to a Point Charge

In nonuniform electric fields, the potential difference between two points is determined by applying the fundamental definition (Eq. 16.1). However, in this case the field strength (and thus work done) varies, making the calculation beyond the scope of this text. The only nonuniform field we will consider in any detail is that due to a point charge (▶Fig. 16.4). Here we will simply state the result for the potential difference (voltage) between two points at distances r_A and r_B from a point charge q:

$$\Delta V = \frac{kq}{r_B} - \frac{kq}{r_A} \qquad \begin{array}{l}\textit{electric potential difference}\\ \textit{(point charge only)}\end{array} \qquad (16.3)$$

In Fig. 16.4, the point charge is positive. Since point B is closer to the charge than A, the potential difference is positive, that is $V_B - V_A > 0$ or $V_B > V_A$. Thus B is at a higher potential than A. This is fundamentally because changes in potential are determined by visualizing the movement of a positive test charge. Here it takes *positive* work to move such a charge from A to B.

From this we see that electric potential increases as we move nearer to a positive charge. Notice also (in Fig. 16.4) that the work done on path II is the same as that for path I. Because the electric force is conservative, the potential difference is also the same, regardless of path.

Consider what would happen if the point charge were negative. In this case, B would be at a *lower* potential than A because the work required to move a positive test charge closer would be *negative* (why?).

Electric potential values change according to the following:

| Electric potential increases when moving nearer to positive charges *or* farther from negative charges

and

| Electric potential decreases when moving farther from positive charges *or* nearer to negative charges.

The potential at a very large distance from a point charge is usually chosen to be zero (as was done for the gravitational case of a point mass in Chapter 7). With this choice, the *electric potential V* at a distance r from a point charge is

$$V = \frac{kq}{r} \qquad \begin{array}{l}\textit{electric potential}\\ \textit{(point charge only,}\\ \textit{zero at infinity)}\end{array} \qquad (16.4)$$

Even though this expression is for the electric potential, V, keep in mind that only electric potential *differences* (ΔV) are important, as the next Example illustrates.

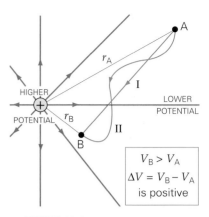

▲ **FIGURE 16.4 Electric field and potential due to a point charge** Electric potential increases as you move closer to a positive charge. Thus, B is at a higher potential than A.

Integrated Example 16.3 ■ Describing the Hydrogen Atom: Potential Differences Near a Proton

According to the Bohr model of the hydrogen atom (Chapter 27), the electron in orbit around the proton can exist only in certain sized circular orbits. The smallest has a radius of 0.0529 nm, and the next largest has a radius of 0.212 nm. (a) How do the values of electric potential compare between each orbit: (1) The smaller one is at a higher potential, (2) the larger is at a higher potential, or (3) they both have the same potential? Explain your reasoning. (b) Verify your answer to part (a) by calculating the values of the electric potential at the locations of the two orbits.

(a) Conceptual Reasoning. The electron orbits in the field of a proton, whose charge is positive. Because electric potential increases with decreasing distance from a positive charge, the answer must be (1).

(b) Quantitative Reasoning and Solution. We know the charge on the proton, so Eq. 16.4 can be used to find the potential values. Listing the values,

Given: $q_p = +1.60 \times 10^{-19}$ C   *Find:* The value of the electric
$\qquad r_1 = 0.0529$ nm $= 5.29 \times 10^{-11}$ m   potential (V) for each orbit
$\qquad r_2 = 0.212$ nm $= 2.12 \times 10^{-10}$ m (Note: 1 nm $= 10^{-9}$ m)

Applying Equation 16.4 we find, for the smaller orbit,

$$V_1 = \frac{kq_p}{r_1} = \frac{(9.00 \times 10^9 \ \text{N} \cdot \text{m}^2/\text{C}^2)(+1.60 \times 10^{-19} \ \text{C})}{5.29 \times 10^{-11} \ \text{m}} = +27.2 \ \text{V}$$

(continues on next page)

and for the larger orbit, we have

$$V_2 = \frac{kq_p}{r_2} = \frac{(9.00 \times 10^9 \text{ N} \cdot \text{m}^2/\text{C}^2)(+1.60 \times 10^{-19} \text{ C})}{2.12 \times 10^{-10} \text{ m}} = +6.79 \text{ V}$$

Follow-Up Exercise. In this Example, suppose the electron were moved from the smallest to the next orbit. (a) Has it moved to a region of higher or lower electrical potential? (b) What would be the change in the electric potential *energy* of the electron?

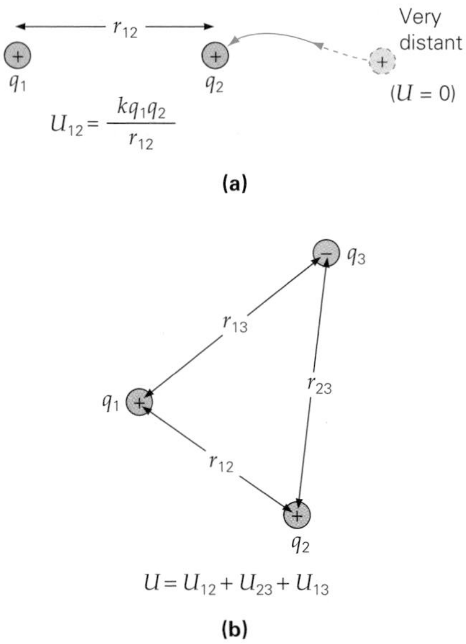

Electric Potential Energy of Various Charge Configurations

In Chapter 7 the gravitational potential energy of *systems* of masses was considered in some detail. The expressions for electric force and gravitational force are mathematically similar, and so are those for potential energy, except that charge takes the place of mass (remembering that charge comes in two signs). In the gravitational case of two masses, the mutual gravitational potential energy is negative, because the force is always *attractive*. For electric potential energy, the result can be positive *or* negative, because the electric force can be *repulsive* or *attractive*.

For example, consider a positive point charge, q_1, fixed in space. Suppose a second positive charge q_2 is brought toward it from a very large distance (that is, let its initial location $r \to \infty$) to a distance r_{12} (◄Fig. 16.5a). In this case, the work required is positive (why?). Therefore, this particular system gains electric potential energy. The potential at a large distance (V_∞) is, as is usual for point charges and masses, chosen as zero. (The zero point is arbitrary.) Thus, from Eq. 16.3, the change in potential energy is

$$\Delta U_e = q_2 \Delta V = q_2(V_1 - V_\infty) = q_2\left(\frac{kq_1}{r_{12}} - 0\right) = \frac{kq_1 q_2}{r_{12}}$$

Because the large-distance value of the electric potential energy is chosen as zero, it follows that $\Delta U_e = U_{12} - U_\infty = U_{12}$. With this choice of reference, the potential energy of *any* two-charge system is

$$U_{12} = \frac{kq_1 q_2}{r_{12}} \qquad \begin{array}{l}\textit{mutual electric potential}\\ \textit{energy (two charges)}\end{array} \tag{16.5}$$

Notice that for *unlike* charges the electric potential energy is negative and for *like* charges, the value is positive. So if the two charges are of the same sign, when released, they will move apart, gaining kinetic energy as they lose potential energy. Conversely, it would take positive work to increase the separation of two opposite charges, such as the proton and the electron, much like stretching a spring. (See the Follow-Up Exercise related to Integrated Example 16.3.)

Because energy is a scalar, for a configuration of any number of point charges, the *total* potential energy (U) is the algebraic sum of the mutual potential energies of all pairs of charges:

$$U = U_{12} + U_{23} + U_{13} + U_{14} \cdots \tag{16.6}$$

Only the first three terms of Eq. 16.6 would be needed for the configuration shown in Fig. 16.5b. Note that the signs of the charges keep things straight mathematically, as the biomolecular situation in Example 16.4 shows.

$$U_{12} = \frac{kq_1 q_2}{r_{12}}$$

(a)

$$U = U_{12} + U_{23} + U_{13}$$

(b)

▲ **FIGURE 16.5** Mutual electric potential energy of point charges **(a)** If a positive charge is moved from a large distance to a distance r_{12} from another positive charge, there is an increase in potential energy because positive work must be done to bring the mutually repelling charges closer. **(b)** For more than two charges, the system's electric potential energy is the sum of the mutual potential energies of each pair.

Teaching tip: Warn students not to "double count" the terms in Eq. 16.6. For two charges, there is one term; for three charges, three terms; for four charges, six terms; and so on.

Example 16.4 ■ Molecule of Life: The Electric Potential Energy of a Water Molecule

The water molecule is the foundation of life as we know it. Many of its properties (such as the reason it is a liquid on the Earth's surface) are related to the fact that it is a permanent polar molecule (see Section 15.4 on electric dipoles). A simple picture of the water molecule, including the charges, is shown in ▶Fig. 16.6. The distance from each hydrogen atom to the oxygen atom is 9.60×10^{-11} m, and the angle (θ) between the two hydrogen–oxygen bond directions is 104°. What is the total electrostatic energy of the water molecule?

Thinking It Through. The model of this molecule involves three charges. The charges are given, but the distance between the hydrogen atoms must be calculated using trigonometry. The total electrostatic potential energy is the algebraic sum of the potential energies of the three pairs of charges (that is, Eq. 16.6 will have three terms).

Solution. The following data are taken from Fig. 16.6.

Given: $q_1 = q_2 = +5.20 \times 10^{-20}$ C $\qquad$ **Find:** U (total electrostatic potential
$\qquad q_3 = -10.4 \times 10^{-20}$ C $\qquad\qquad\qquad$ of energy water molecule)
$\qquad r_{13} = r_{23} = 9.60 \times 10^{-11}$ m
$\qquad \theta = 104°$

Notice that $(r_{12}/2)/r_{13} = \sin(\theta/2)$. Hence, we can solve for r_{12}:

$$r_{12} = 2r_{13}\left(\sin\frac{\theta}{2}\right) = 2(9.60 \times 10^{-11} \text{ m})(\sin 52°) = 1.51 \times 10^{-10} \text{ m}$$

Before determining the total potential energy of this system, let's calculate each pair's contribution separately. Note that $U_{13} = U_{23}$. (Why?) Applying Eq. 16.5,

$$U_{12} = \frac{kq_1q_2}{r_{12}} = \frac{(9.00 \times 10^9 \text{ N} \cdot \text{m}^2/\text{C}^2)(+5.20 \times 10^{-20} \text{ C})(+5.20 \times 10^{-20} \text{ C})}{1.51 \times 10^{-10} \text{ m}}$$
$$= +1.61 \times 10^{-19} \text{ J}$$

and

$$U_{13} = U_{23} = \frac{kq_2q_3}{r_{23}} = \frac{(9.00 \times 10^9 \text{ N} \cdot \text{m}^2/\text{C}^2)(+5.20 \times 10^{-20} \text{ C})(-10.4 \times 10^{-20} \text{ C})}{9.60 \times 10^{-11} \text{ m}}$$
$$= -5.07 \times 10^{-19} \text{ J}$$

Thus the total electrostatic potential energy is

$$U = U_{12} + U_{13} + U_{23} = (+1.61 \times 10^{-19} \text{ J}) + (-5.07 \times 10^{-19} \text{ J}) + (-5.07 \times 10^{-19} \text{ J})$$
$$= -8.53 \times 10^{-19} \text{ J}$$

The negative result indicates that the molecule requires positive work to break it apart. (That is, it must be pulled apart.)

Follow-Up Exercise. Another common polar molecule is carbon monoxide (CO), a toxic gas commonly produced by automobiles when fuel combustion is incomplete. The carbon atom is positively charged and the oxygen atom is negative. The distance between the carbon and oxygen atoms is 1.20×10^{-10} m, and the (average) charge on each is 6.60×10^{-20} C. Determine the electrostatic energy of this molecule. Is it more or less electrically stable than a water molecule in this Example?

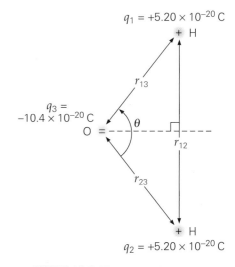

▲ **FIGURE 16.6** Electrostatic potential energy of a water molecule A charge configuration has electric potential energy, because work is required to bring the charges together from large distances. The charges shown on the water molecule are net average charges because the atoms within the molecule share electrons. So the charges on the ends of the water molecule can be smaller than the charge on the electron or proton. (See Example 16.4 for details.)

16.2 Equipotential Surfaces and the Electric Field

OBJECTIVES: To (a) explain what is meant by an equipotential surface, (b) sketch equipotential surfaces for simple charge configurations, and (c) explain the relationship between equipotential surfaces and electric fields.

Equipotential Surfaces

Suppose a positive charge is moved perpendicularly to an electric field (such as path I of ▼Fig. 16.7a). As the charge moves from A to A′, *no work* is done by the electric field (why?). If no work is done, then the potential energy of the charge does not change, so $\Delta U_{AA'} = 0$. From this, it can be concluded that these two points (A and A′)—and *all* other points on path I—are at the same potential V; that is,

$$\Delta V_{AA'} = V_{A'} - V_A = \frac{\Delta U_{AA'}}{q} = 0 \qquad \text{or} \qquad V_A = V_{A'}$$

This result actually holds for all points on the *plane* parallel to the plates and containing path I. A surface like this plane, on which the potential is constant, is called an **equipotential surface** (or simply an *equipotential*). The word *equipotential*

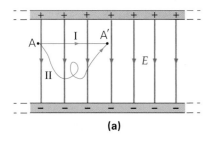

(a)

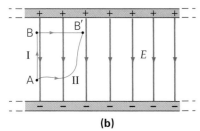

(b)

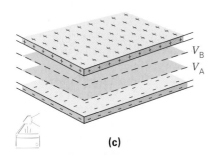

(c)

▲ **FIGURE 16.7 Construction of equipotential surfaces between parallel plates** (a) The work done in moving a charge is zero as long as you start and stop on the same equipotential surface. (Compare paths I and II.) (b) Once the charge moves to a higher potential (for example, from point A to point B), it can stay on that new equipotential surface by moving perpendicularly to the electric field (B to B'). The change in potential is independent of the path, since the same change occurs whether path I or path II is used. (Why?) (c) The actual equipotential surfaces within the parallel plates are planes parallel to those plates. Two such plates are shown, with $V_B > V_A$.

means "same potential." Note that, unlike this special case, an equipotential need not be a flat plane.

Since no work is required to move a charge along an equipotential surface, it must be generally true that

| Equipotential surfaces are always at right angles to the electric field. |

Moreover, because the electric field is conservative, the work is the same whether path I, path II, or *any other* path from A to A' is taken (Fig. 16.7a). As long as the charge returns to the same equipotential surface from which it started, the work done on it is zero and the value of the electric potential is the same.

If the positive charge is moved opposite to $\vec{E}$ (path I in Fig. 16.7b)—at right angles to the equipotentials—the electric potential energy, and hence the electric potential, increases. (Why?) When B is reached, the charge is on a different equipotential—one of a higher potential than A. If, instead, the charge had been moved from A to B', the work would be the same as that in moving from A to B. Hence, B and B' are on the same equipotential surface. For parallel plates, the equipotentials are planes parallel to the plates (Fig. 16.7c).

To help understand the concept of an electric equipotential surface, consider a gravitational analogy. If the gravitational potential energy is designated as zero at ground level and an object is raised a height $h = h_B - h_A$ (from A to B in ▼Fig. 16.8), then the work done by an external force is mgh and is positive. For horizontal movement, the potential energy does not change. This means that the dashed plane at height h_B is a gravitational equipotential surface—and so is the plane at h_A, but it has a lower potential value than the plane at h_B. Therefore, surfaces of constant gravitational potential energy are planes parallel to the Earth's surface. Topographic maps, which display land contours by plotting lines of constant elevation (usually relative to sea level), are actually maps of constant gravitational potential (▶Fig. 16.9a, b). Note how the equipotentials near a point charge (Fig. 16.9c, d) are qualitatively similar to the gravitational contours due to a hill.

It is useful to know how to sketch equipotential surfaces, because they are intimately related to the electric field and to practical aspects such as voltage. The Learn by Drawing on Graphical Relationship between Electric Field Lines and Equipotentials (page 547) summarizes a qualitative method useful for sketching equipotential surfaces, given an electric field line pattern. As this feature shows, the method is also useful for the converse problem: sketching the electric field lines if the equipotential surfaces are given. Can you see how these ideas were used to construct the equipotentials of an electric dipole in ▶Fig. 16.10?

To determine the relationship between the electric field (E) and the electric potential (V), consider the special case of a uniform electric field (▶Fig. 16.11). The potential difference (ΔV) between any two equipotential planes (labeled V_1 and V_2 in the figure) can be calculated with the same technique used to derive Eq. 16.2. The result is

$$\Delta V = V_3 - V_1 = E\,\Delta x \qquad (16.7)$$

Thus, if you start on equipotential surface 1 and move *perpendicularly away* from it and *opposite to* the electric field to equipotential surface 3, there is a potential *increase* (ΔV) that depends on the electric field strength (E) and the distance (Δx).

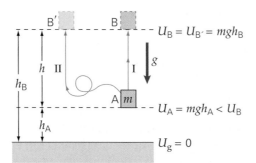

◀ **FIGURE 16.8 Gravitational potential energy analogy** Raising an object in a uniform gravitational field results in an increase in gravitational potential energy, and, $U_B > U_A$. At a given height, the object's potential energy is constant as long as it remains on that (gravitational) equipotential surface. Here, $\vec{g}$ points downward, like $\vec{E}$ in Fig. 16.7.

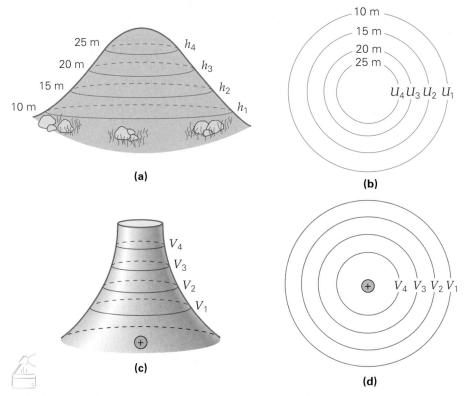

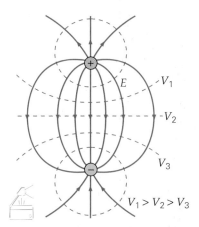

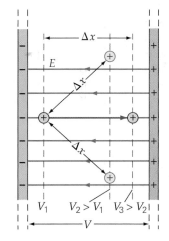

▲ **FIGURE 16.9** Topographic maps—a gravitational analogy to equipotential surfaces **(a)** A symmetrical hill with slices at different elevations. Each slice is a plane of constant gravitational potential. **(b)** A topographic map of the slices in (a). The contours, where the planes intersect the surface, represent increasingly larger values of gravitational potential as one goes up the hill. **(c)** The electric potential V near a point charge q forms a similar symmetrical hill. V is constant at fixed distances from q. **(d)** Electrical equipotentials around a point charge are spherical (in two dimensions they are circles) centered on the charge. The closer the equipotential to the positive charge, the larger its electric potential.

▲ **FIGURE 16.10** Equipotentials of an electric dipole Equipotentials are perpendicular to electric field lines. $V_1 > V_2$ because equipotential surface 1 is closer to the positive charge than is surface 2. To understand how equipotentials are constructed, see the accompanying Learn by Drawing. (See page 547.)

For a given distance Δx, this perpendicular movement yields the maximum possible gain in potential. Think of taking one step of length Δx in *any* direction, starting from surface 1. The way to maximize the increase would be to step onto surface 3. A step in any direction not perpendicular to surface 1 (for example, ending on surface 2) yields a smaller increase in potential.

By finding the direction of the *maximum* potential increase, we are finding the direction opposite that of $\vec{E}$. Thus, as a general rule, we can state that:

The direction of the electric field $\vec{E}$ is that in which the electric potential decreases the most rapidly.

Then, at any location, the magnitude of the electric field is the maximum rate of change of the potential with distance, or

$$E = \left| \frac{\Delta V}{\Delta x} \right|_{max} \qquad (16.8)$$

The unit of electric field is volts per meter (V/m). Previously, E was expressed in newtons per coulomb (N/C; see Section 15.4). You should show, through dimensional analysis, that 1 V/m = 1 N/C. A graphical interpretation of the relationship between $\vec{E}$ and V is shown in the Learn by Drawing on page 547.

In most practical situations, it is the potential difference (*voltage*), rather than the electric field, that is specified. For example, a D-cell flashlight battery has a terminal voltage of 1.5 V, meaning that it can maintain a potential difference of 1.5 V between its terminals. Most automotive batteries have a terminal voltage of about 12 V. Some of the common potential differences, or voltages, are listed in Table 16.1.

▲ **FIGURE 16.11** Relationship between the potential change (ΔV) and the electric field ($\vec{E}$) The electric field direction is that of maximum decrease in potential, or opposite the direction of maximum increase in potential (here, this maximum is in the direction of the solid blue arrow, not the angled ones; why?). The electric field magnitude is given by the maximum rate at which the potential changes over distance (usually in volts per meter).

Illustration 25.2 Work and
Equipotentials

Illustration 25.3 Electric Potential of
Charged Spheres

TABLE 16.1	Common Electric Potential Differences (Voltages)
Source	*Approximate Voltage (ΔV)*
Across nerve membranes	100 mV
Small-appliance batteries	1.5 to 9.0 V
Automotive batteries	12 V
Household outlet (United States)	110 to 120 V
Houschold outlets (Europe)	220 to 240 V
Automotive ignitions (spark plug firing)	10 000 V
Laboratory generators	25 000 V
High-voltage electric power delivery lines	300 kV or more
Cloud-to-Earth surface during thunderstorm	100 MV or more

Whether you know it or not, you live in an electric field near the Earth's surface. This field varies with weather conditions and, consequently, can be an indicator of approaching storms. Example 16.5 applies the equipotential surface concept to help in understanding the Earth's electric field.

Example 16.5 ■ The Earth's Electric Field and Equipotential Surfaces: Electric Barometers?

Under normal atmospheric conditions, the Earth's surface is electrically charged. This creates an approximately constant electric field of about 150 V/m pointing *down* near the surface. (a) Under these conditions, what is the shape of the equipotential surfaces, and in what direction does the electric potential decrease the most rapidly? (b) How far apart are two equipotential surfaces that have a 1000-V difference between them? Which has a higher potential, the one farther from the Earth or the one closer?

Thinking It Through. (a) Near the Earth's surface, the electric field is approximately uniform, so the equipotentials are similar to those of parallel plates. The discussion of Eqs. 16.7 and 16.8 enables us to determine which way the potential increases. (b) Equation 16.8 can then be used to determine how far apart the equipotential surfaces are.

Solution. Listing the data,

Given: $E = 150$ V/m, downward   *Find:* (a) shape of equipotential surfaces and
     $\Delta V = 1000$ V   direction of decrease in potential
                 (b) Δx (distance between equipotentials)

(a) Uniform electric fields are associated with plane equipotentials; in this case, the planes are parallel to the Earth's surface. The electric field points downward. This is the direction in which the potential decreases most rapidly.

(b) To determine the distance between the two equipotentials, think of moving vertically so that $\Delta V/\Delta x$ has its maximum value. Solving Eq. 16.8 for Δx yields

$$\Delta x = \frac{\Delta V}{E} = \frac{1000 \text{ V}}{150 \text{ V/m}} = 6.67 \text{ m}$$

Because the potential decreases as we move downward (in the direction of $\vec{E}$), the higher potential is associated with the surface that is 6.67 m *farther* from the ground.

Follow-Up Exercise. Re-examine this Example under storm conditions. During a lightning storm, the electric field can rise to many times the normal value. (a) Under these conditions, if the field is 900 V/m and points *upward*, how far apart are two equipotential surfaces that differ by 2000 V? (b) Which surface is at a higher potential, the one closer to the Earth or the one farther away? (c) Can you tell how far the two surfaces are from the ground? Why or why not?

Exploration 25.1 Investigate
Equipotential Lines

Exploration 25.2 Electric Field Lines
and Equipotentials

LEARN BY DRAWING

GRAPHICAL RELATIONSHIP BETWEEN ELECTRIC FIELD LINES AND EQUIPOTENTIALS

Because it takes no work to move a charge along an equipotential surface, such surfaces must be perpendicular to the electric field lines. Also, the electric field has a magnitude equal to the change in potential per unit distance (V/m) and points in the direction in which the potential decreases most rapidly. These facts can be used to construct equipotentials if we know the field pattern. The reverse is also true: Given the equipotentials, the electric field lines can be constructed. Furthermore, if the potential (in volts) associated with each equipotential is known, the strength and direction of the field can be estimated from the rate at which the potential changes with distance (Eq. 16.8).

A couple of examples should provide a graphical insight into the connection between equipotential surfaces and their associated electric fields. Consider Fig. 1, in which you are given the electric field lines and want to determine the shape of the equipotentials. Pick any point, such as A, and begin moving at right angles to the field lines. Keep moving so as to maintain this perpendicular orientation to the lines. Between lines you may have to approximate, but plan ahead to the next field line so it is crossed at a right angle. To find another equipotential, start at another point, such as B, and proceed the same way. Sketch as many equipotentials as you need to map the area of interest. The figure shows the result of sketching four equipotentials, from A (at the highest potential—can you tell why?) to D (at the lowest potential).

Now suppose you are given the equipotentials instead of the field lines (Fig. 2). The electric field lines point in the direction of decreasing V and are perpendicular to

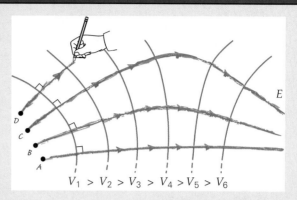

$$V_1 > V_2 > V_3 > V_4 > V_5 > V_6$$

FIGURE 2 Mapping the electric field from equipotentials Start at a convenient point, and trace a line that crosses each equipotential at a right angle. Repeat the process as often as needed to reveal the field pattern, adding arrows to indicate the direction of the field lines from high to low potential. In going from one potential to the next, plan ahead so that each succeeding equipotential is also crossed at right angles.

the equipotential surfaces. Thus, to map the field, start at any point, and move in such a way that your path intersects each equipotential surface at a right angle. The resulting field line is shown in Fig. 2, beginning at point A. Starting at points B, C, and D provides additional field lines that suggest the complete electric field pattern; you need only add the arrows in the direction of decreasing potential.

Lastly, suppose you want to estimate the magnitude of $\vec{E}$ at some point P (Fig. 3), knowing the values of the equipotentials 1.0 cm on either side of it. From this, you know that the field points roughly from A to B (why?) and its approximate magnitude would be

$$E = \left| \frac{\Delta V}{\Delta x} \right|_{max} = \frac{(1000 \text{ V} - 950 \text{ V})}{2.0 \times 10^{-2} \text{ m}}$$
$$= 2.5 \times 10^3 \text{ V/m}$$

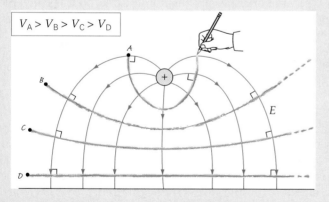

$$V_A > V_B > V_C > V_D$$

FIGURE 1 Sketching equipotentials from electric field lines If you know the electric field pattern, pick a point in the region of interest and move so that your path is always perpendicular to the next field line. Keep your path as smooth as possible, planning ahead so that each succeeding field line is also crossed at right angles. To map a surface with a higher (or lower) potential, move in the opposite (or the same) direction as the electric field and repeat the process. Here, $V_A > V_B$, and so on.

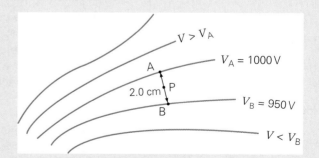

FIGURE 3 Estimating the magnitude of the electric field The magnitude of the potential change per meter at any point gives the strength of the electric field at that point.

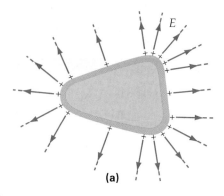

(a)

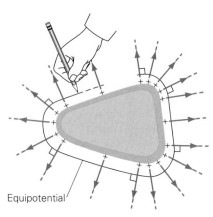

Equipotential

(b)

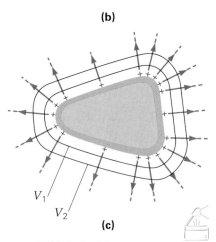

V_1
V_2
(c)

▲ **FIGURE 16.12** Equipotential surfaces near a charged conductor See Conceptual Example 16.6.

PHYSLET®

Exploration 25.3 Electric Potential around Conductors

Note: The electron-volt is a unit of energy. The volt is not. Do not confuse the two!

Equipotential surfaces can be useful for describing the field near a charged conductor, as the following Conceptual Example shows.

Conceptual Example 16.6 ■ The Equipotential Surfaces Outside a Charged Conductor

A solid conductor with an excess positive charge is shown in ◄Fig. 16.12a. Which of the following best describes the shape of the equipotential surfaces just outside the conductor's surface: (a) flat planes, (b) spheres, or (c) approximately the shape of the conductor's surface? Explain your reasoning.

Reasoning and Answer. Choice (a) can be eliminated immediately, because flat (plane) equipotential surfaces are associated with flat plates. While it might be tempting to pick answer (b), a quick look at the electric field near the surface (Chapter 15), in conjunction with the Learn by Drawing on p. 547, shows that the correct answer is (c). To verify that (c) is the correct answer, recall that near the surface the electric field is perpendicular to that surface. Since the equipotential surfaces are perpendicular to the field lines, they must follow the contour of the conductor's surface (Fig. 16.12b).

Follow-Up Exercise. In this Example, (a) which of the two equipotentials (1 or 2) shown in Fig. 16.12c is at a higher potential? (b) What is the approximate shape of the equipotential surfaces *very far from* this conductor? Explain your reasoning. [*Hint:* What does the conductor look like when you are very far from it?]

The Electron-Volt

The concept of electric potential provides a unit of energy that is particularly useful in molecular, atomic, nuclear, and elementary particle physics. The **electron-volt (eV)** is defined as the kinetic energy acquired by an electron (or proton) accelerated through a potential difference, or voltage, of exactly 1 V. The gain in kinetic energy is equal (but opposite) to the change in electric potential energy. For an electron, its gain in kinetic energy in joules is:

$$\Delta K = -\Delta U_e = -(e\,\Delta V) = -(-1.60 \times 10^{-19}\,\text{C})(1.00\,\text{V}) = +1.60 \times 10^{-19}\,\text{J}$$

Since this is what is meant by 1 electron-volt, the conversion factor between the electron volt and the joule (to three significant figures) is

$$1\,\text{eV} = 1.60 \times 10^{-19}\,\text{J}$$

The electron-volt is typical of energies on the atomic scale, so it is convenient to express atomic energies in terms of electron-volts instead of joules. The energy of *any* charged particle accelerated through *any* potential difference can be expressed in electron-volts. For example, if an electron is accelerated through a potential difference of 1000 V, its gain in kinetic energy (ΔK) is one thousand times that of a 1-eV electron, or

$$\Delta K = e\,\Delta V = (1\,e)(1000\,\text{V}) = 1000\,\text{eV} = 1\,\text{keV}$$

The abbreviation *keV* stands for *kiloelectron-volt*.

The electron-volt is defined in terms of a particle with the minimum charge (the electron or proton). However, the energy of a particle with *any* charge can also be expressed in electron-volts. Thus, if a particle with a charge of $+2e$, such as an alpha particle, were accelerated through a potential difference of 1000 volts, it would gain a kinetic energy of $\Delta K = e\,\Delta V = (2\,e)(1000\,\text{V}) = 2000\,\text{eV} = 2\,\text{keV}$. Note how easy it is to compute the kinetic energy if you work in electron-volts.

Occasionally, larger units than the electron-volt are needed. For example, in nuclear and elementary particle physics, it is not uncommon to find particles with energies of *megaelectron-volts* (MeV) and *gigaelectron-volts* (GeV); 1 MeV = 10^6 eV and 1 GeV = 10^9 eV.*

---

*At one time, a billion electron-volts was referred to as BeV, but this usage was abandoned because confusion arose. In some countries, such as Great Britain and Germany, a billion means 10^{12} (which is called a trillion in the United States).

In working problems, it is important to be aware that the electron-volt (eV) is *not* an SI unit. Hence, when using energies, you must convert from electron-volts to joules. For example, to calculate the speed of an electron accelerated from rest through 10.0 V, first convert the kinetic energy (10.0 eV) to joules:

$$K = (10.0 \text{ eV})(1.60 \times 10^{-19} \text{ J/eV}) = 1.60 \times 10^{-18} \text{ J}$$

Continuing in the SI system, the mass of the electron must be in kilograms. Then the speed is

$$v = \sqrt{2K/m} = \sqrt{2(1.60 \times 10^{-18} \text{ J})/(9.11 \times 10^{-31} \text{ kg})} = 1.87 \times 10^6 \text{ m/s}$$

16.3 Capacitance

OBJECTIVES: To (a) define capacitance and explain what it means physically, and (b) calculate the charge, voltage, electric field, and energy storage for parallel-plate capacitors.

A pair of parallel plates, if charged, stores electrical energy (▼Fig. 16.13). Such an arrangement of conductors is an example of a **capacitor**. (Any pair of conductors qualifies as a capacitor.) The energy storage occurs because it takes work to transfer the charge from one plate to the other. Imagine that one electron is moved between a pair of initially uncharged plates. Once that is done, transferring a *second* electron would be more difficult, because it is not only repelled by the first electron on the negative plate, but also attracted by a double positive charge on the positive plate. Separating the charges requires more and more work as more and more charge accumulates on the plates. (This is analogous to stretching a spring. The more you stretch it, the harder it is to stretch it further.)

The work needed to charge parallel plates can be done quickly (usually in a few microseconds) by a battery. Although we won't discuss battery action in detail until the next chapter, all you need to know now is that a battery removes electrons from the positive plate and transfers, or "pumps," them through a wire to the negative plate. In the process of doing work, the battery loses some of its internal chemical potential energy. Of primary interest here is the result: a separation of charge and the creation of an electric field in the capacitor. The battery will continue to charge the capacitor until the potential difference between the plates is equal to the terminal voltage of the battery—for example, 12 V if you use a standard automotive battery. When the capacitor is disconnected from the battery, it becomes a storage "reservoir" of electrical energy.

For a capacitor, the potential difference across the plates is proportional to the charge Q on the plates, or $Q \propto V$.* (Here, Q denotes the magnitude of the charge

Note: Capacitors store energy in their electric fields.

Note: Recall that our notation for potential difference, or voltage (ΔV), will be replaced by V for convenience.

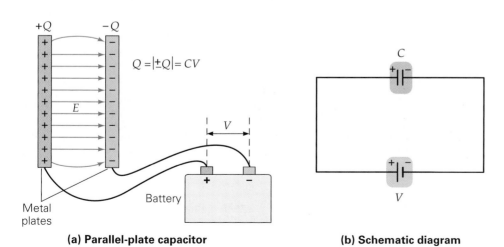

(a) Parallel-plate capacitor   (b) Schematic diagram

◀ **FIGURE 16.13 Capacitor and circuit diagram (a)** Two parallel metal plates are charged by a battery that moves electrons from the positive plate to the negative one through the wire. Work is done while charging the capacitor, and energy is stored in the electric field. **(b)** This diagram represents the charging situation shown in part (a). It also shows the symbols commonly used for a battery (V) and a capacitor (C). The longer line of the battery symbol is the positive terminal, and the shorter line represents the negative terminal. The symbol for a capacitor is similar, but the lines are of equal length.

*At this point, we will begin using V to denote potential differences instead of ΔV. This is a common practice. Always remember that the important quantity is potential difference, ΔV.

Note: The charges on the plates are $+Q$ and $-Q$, but it is customary to refer in general to the magnitude of these charges, Q (meaning $|\pm Q|$), on a capacitor.

on *either* plate, *not* the net charge on the whole capacitor, which is zero.) This proportionality can be made into an equation by using a constant, C, called *capacitance*:

$$Q = CV \quad \text{or} \quad C = \frac{Q}{V} \quad (16.9)$$

SI unit of capacitance: coulomb per volt (C/V), or farad (F)

Note: The farad was named for the English scientist Michael Faraday (1791–1867), an early investigator of electrical phenomena who first introduced the concept of the electric field.

The coulomb per volt equals a **farad**, 1 C/V = 1 F. The farad is a large unit (see Example 16.7), so the *microfarad* (1 μF = 10^{-6} F), the *nanofarad* (1 nF = 10^{-9} F), and the *picofarad* (1 pF = 10^{-12} F) are commonly used.

Capacitance represents the charge stored *per volt*. When a capacitor has a large capacitance, it holds a large amount of charge *per volt* compared with one of smaller capacitance. If you connected the same battery to two different capacitors, the one with the larger capacitance would store more charge and more energy.

Note: Generally, we will use the lowercase letter q to represent charges on single particles and the uppercase letter Q for the larger amounts of charges on capacitor plates.

Capacitance depends *only* on the geometry (size, shape, and spacing) of the plates (and the material between the plates, Section 16.5) and *not* the charge on the plates. Consider the parallel-plate capacitor, which has an electric field given by Eq. 15.5:

$$E = \frac{4\pi k Q}{A}$$

The voltage across the plates can be computed from Eq. 16.2:

$$V = Ed = \frac{4\pi k Q d}{A}$$

The capacitance of a parallel-plate arrangement is then

$$C = \frac{Q}{V} = \left(\frac{1}{4\pi k}\right)\frac{A}{d} \quad \text{(parallel plates only)} \quad (16.10)$$

It is common to replace the expression in the parentheses in Eq. 16.10 with a single quantity called the **permittivity of free space (ε_o)**. The value of this constant (to three significant figures) is

PHYSLET®

Illustration 26.2 A Capacitor
Connected to a Battery

$$\varepsilon_o = \frac{1}{4\pi k} = 8.85 \times 10^{-12} \text{ C}^2/(\text{N} \cdot \text{m}^2) \quad \text{permittivity of free space} \quad (16.11)$$

ε_o describes the electrical properties of free space (vacuum), but its value in air is only 0.05% larger. In our calculations, they will be taken to be the same.

It is common to rewrite Eq. 16.10 in terms of ε_o:

$$C = \frac{\varepsilon_o A}{d} \quad \text{(parallel plates only)} \quad (16.12)$$

Let's use Eq. 16.12 in the next Example to show just how unrealistically large an air-filled capacitor with a capacitance of 1.0 F would be.

Example 16.7 ■ Parallel-Plate Capacitors: How Large Is a Farad?

What would be the plate area of an air-filled 1.0-F parallel-plate capacitor if the plate separation were 1.0 mm? Would it be realistic to consider building such a capacitor?

Thinking It Through. The area can be calculated directly from Eq. 16.12. Remember to keep all quantities in SI units, so that the answer will be in square meters. The vacuum value of ε_o for air can be used without creating a significant error.

Solution.

Given: $\quad C = 1.0$ F $\qquad\qquad\qquad$ *Find:* $\quad A$ (area of one of the plates)
$\qquad\qquad d = 1.0$ mm $= 1.0 \times 10^{-3}$ m

Solving Eq. 16.12 for the area gives

PHYSLET®

Exploration 26.1 Energy

$$A = \frac{Cd}{\varepsilon_o} = \frac{(1.0 \text{ F})(1.0 \times 10^{-3} \text{ m})}{8.85 \times 10^{-12} \text{ C}^2/(\text{N} \cdot \text{m}^2)} = 1.1 \times 10^8 \text{ m}^2$$

This is more than 100 km² (40 mi²), that is, a square more than 10 km (6 mi) on a side. It is unrealistic to build a capacitor that big; 1.0 F is therefore a very large value of capacitance. There are ways, however, to make high-capacity capacitors (Section 16.4).

Follow-Up Exercise. In this Example, what would the plate spacing have to be if you wanted the capacitor to have a plate area of 1 cm²? Compare your answer with a typical atomic diameter of 10^{-9} to 10^{-10} m. Is it feasible to build this capacitor?

The expression for the energy stored in a capacitor can be obtained by graphical analysis, since both Q and V vary during charging—for example, as the charge is separated by a battery. A plot of voltage versus charge for charging a capacitor is a straight line with a slope of $1/C$, because $V = (1/C)Q$ (►Fig. 16.14). The graph represents the charging of an initially uncharged capacitor ($V_o = 0$) to a final voltage (V). The work done is equivalent to transferring the total charge, using an average voltage $\overline{V}$. Because the voltage varies linearly with charge, the average voltage is half the final voltage V:

$$\overline{V} = \frac{V_{\text{final}} + V_{\text{initial}}}{2} = \frac{V + 0}{2} = \frac{V}{2}$$

The energy stored in the capacitor (equal to the work done by the battery) is then

$$U_C = W = Q\overline{V} = \tfrac{1}{2}QV$$

Because $Q = CV$, this equation can be written in several equivalent forms:

$$U_C = \tfrac{1}{2}QV = \frac{Q^2}{2C} = \tfrac{1}{2}CV^2 \quad \textit{energy storage in a capacitor} \tag{16.13}$$

Typically, the form $U_C = \tfrac{1}{2}CV^2$ is the most practical, since the capacitance and the voltage are usually the known quantities. A very important medical application of the capacitor is in the *cardiac defibrillator* discussed in the next Example.

Example 16.8 ■ Capacitors to the Rescue: Energy Storage in a Cardiac Defibrillator

During a heart attack, the heart beats in an erratic fashion, called *fibrillation*. One way to get it back to normal rhythm is to shock it with electrical energy supplied by a *cardiac defibrillator* (►Fig. 16.15). About 300 J of energy is required to produce the desired effect. Typically, a defibrillator stores this energy in a capacitor charged by a 5000-V power supply. (a) What capacitance is required? (b) What is the charge on the capacitor's plates?

Thinking It Through. (a) To find the capacitance, solve for C in Eq. 16.13. (b) The charge then follows from the definition of capacitance (Eq. 16.9).

Solution. We list the given data:

Given: $U_C = 300$ J   *Find:* (a) C (the capacitance)
          $V = 5000$ V   (b) Q (charge on capacitor)

(a) The most useful form of Eq. 16.13 is $U_C = \tfrac{1}{2}CV^2$. Solving for C,

$$C = \frac{2U_C}{V^2} = \frac{2(300 \text{ J})}{(5000 \text{ V})^2} = 2.40 \times 10^{-5} \text{ F} = 24.0 \ \mu\text{F}$$

(b) The charge (magnitude) on either plate is then

$$Q = CV = (2.40 \times 10^{-5} \text{ F})(5000 \text{ V}) = 0.120 \text{ C}$$

Follow-Up Exercise. For the capacitor in this Example, if the maximum allowable energy for any single defibrillation attempt is 750 J, what is the maximum voltage that should be used?

Sometimes capacitors can successfully model real-life phenomena. For example, a lightning storm can be considered to be the discharge of a negatively charged cloud to the positively charged ground—in effect, a "cloud-ground" capacitor. Another interesting application of electric potential treats nerve membranes as cylindrical capacitors to help explain nerve signal transmission. (See Insight 16.1 on Electric Potential and Nerve Signal Transmission on page 552.)

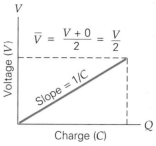

▲ **FIGURE 16.14 Capacitor voltage versus charge** A plot of voltage (V) versus charge (Q) for a capacitor is a straight line with slope $1/C$ (because $V = (1/C)Q$). The average voltage is $\overline{V} = \tfrac{1}{2}V$, and the total work done is equivalent to transferring the charge through $\overline{V}$. Thus, $U_C = W = Q\overline{V} = \tfrac{1}{2}QV$, the area under the curve (a triangle).

Note: Do not confuse U_C, the energy stored in a capacitor, with ΔU_e, the change in electric potential energy of a charged particle. (See Section 16.1.)

Note: Practice using the various forms of capacitor energy. In a pinch, you need recall only one, together with the definition of capacitance, $C = Q/V$.

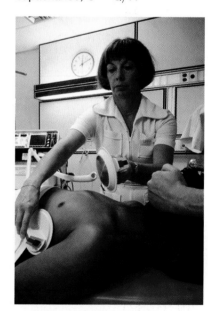

▲ **FIGURE 16.15 Defibrillator** A burst of electric current (flow of charge) from a defibrillator may restore a normal heartbeat in people in cardiac arrest. Capacitors store the electrical energy on which the device depends.

INSIGHT 16.1 ELECTRIC POTENTIAL AND NERVE SIGNAL TRANSMISSION

The human body's nervous system is responsible for the reception of external stimuli through our senses (such as touch) as well as communication between the brain and our organs and muscles. If you touch something hot, nerves in your hand detect the problem and send a signal to your brain; your brain then sends the signal "Pull back!" through other parts of the nervous system to your hand. But what are these signals, and how do they work?

A typical nerve consists of a bundle of nerve cells called *neurons*, much like individual telephone wires bundled into a single cable. The structure of a typical neuron is shown in Fig. 1a. The cell body, or *soma*, has long branchlike extensions called *dendrites*, which receive the input signal. The soma is responsible for processing the signal and transmitting it down a long extension called the *axon*. At the other end of the axon are projections with knobs called *synaptic terminals*. At these knobs, the electrical signal is transmitted to another neuron across a gap called the *synapse*. The human body contains on the order of 100 billion neurons, and each neuron can have several hundred synapses! Running the nervous system costs the body about 25% of its energy intake.

To understand the electrical nature of nerve signal transmission, let us focus on the axon. A vital component of the axon is its cell membrane, which is typically about 10 nm thick and consists of *phospholipids* (electrically polarized hydrocarbon molecules) and embedded protein molecules (Fig. 1b). The membrane has proteins called *ion channels*, which form pores where large protein molecules regulate the flow of ions (primarily sodium) across the membrane. The key to nerve signal transmission is the fact that these ion channels are selective: They allow only certain types of ions to cross the membrane; others cannot.

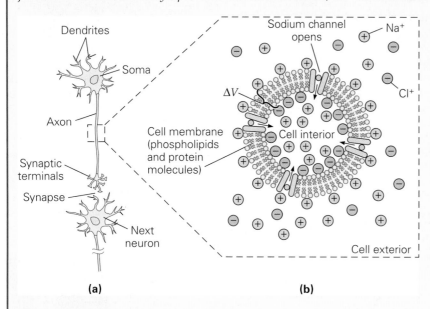

FIGURE 1 **(a)** The structure of a typical neuron. **(b)** An enlargement of the axon membrane, showing the membrane (about 10 nm thick) and the concentration of ions inside and outside the cell. The charge polarization across the membrane leads to a voltage, or membrane potential. When triggered by an external stimulus, the sodium ion channels open, allowing sodium ions into the cell. This influx changes the membrane potential.

(a)

(b)

16.4 Dielectrics

OBJECTIVES: To (a) understand what a dielectric is, and (b) understand how it affects the physical properties of a capacitor.

In most capacitors, a sheet of insulating material, such as paper or plastic, is between the plates. Such an insulating material, called a **dielectric**, serves several purposes. For one, it keeps the plates from coming into contact. Contact would allow the electrons to flow back onto the positive plate, neutralizing the charge on the capacitor and the energy stored. A dielectric also allows flexible plates of metallic foil to be rolled into a cylinder, giving the capacitor a more compact (more practical) size. Finally, a dielectric increases the charge storage capacity of the capacitor and therefore, under the right conditions, the energy stored in the capacitor. This capability depends on the type of material and is characterized by the **dielectric constant** (κ). Values of the dielectric constant for some common materials are given in Table 16.2.

Illustration 26.1 Microscopic View of a Capacitor

The fluid outside the axon, although electrically neutral, contains sodium ions (Na^+) and chlorine ions (Cl^-) in solution. In contrast, the axon's internal fluid is rich in potassium ions (K^+) and negatively charged protein molecules. If it were not for the selective nature of the cell membrane, the Na^+ concentration would be equal on both sides of the membrane. Under normal (or *resting*) conditions, it is difficult for Na^+ to penetrate the interior of the nerve cell. This ion selectivity gives rise to a polarization of charge across the membrane. The exterior is positive (with Na^+ trying to enter the region of lower concentration), attracting the negative proteins to the inner surface of the membrane (Fig. 1b). Thus a cylindrical capacitor-like charge-storage system exists across an axon membrane at rest. The *resting membrane potential* (the voltage across the membrane) is defined as $\Delta V = V_{in} - V_{out}$. Because the outside is positively charged, as defined, the resting potential is a negative quantity; it ranges from about -40 to -90 mV (millivolts), with a typical value of -70 mV in humans.

Signal conduction occurs when the cell membrane receives a stimulus from the dendrites. Only then does the membrane potential change, and this change is propagated down the axon. The stimulus triggers Na^+ channels in the membrane (which are closed while resting, like a gate) to open and temporarily allows sodium ions to enter the cell (Fig. 1b). These positive ions are attracted to the negative charge layer on the interior and are driven by the difference in concentration. In about 0.001 s, enough sodium ions have passed through the gated channel to cause a reversal of polarity, and the membrane potential rises, typically to $+30$ mV in humans. The time sequence for this change in membrane potential is shown in Fig. 2. When the difference in Na^+ concentration causes the membrane voltage to become positive, the Na^+ channels close. A chemical process involving proteins known as the *Na/K–ATPase molecular pump* then re-establishes the resting potential at about -70 mV by selectively transporting the excess Na^+ back to the cell's exterior.

This temporary change in membrane potential (a total of 100 mV, from -70 mV to $+30$ mV) is called the cell's *action potential*. The action potential is the signal that is transmitted down the axon. This "voltage wave" travels at speeds of 1 to 100 m/s on its way to triggering another such pulse in the adjacent neuron. This speed, along with other factors such as time delays in the synapse region, is responsible for typical human reaction times totaling a few tenths of a second.

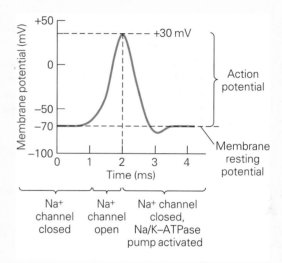

FIGURE 2 As the sodium channels open and sodium ions rush to the cell's interior, the membrane potential changes quickly from its resting value of -70 mV to about $+30$ mV. The resting potential is restored (about 4 ms later) by a protein "pumping" process that chemically removes the excess Na^+ after the sodium channels have closed (at 2 ms).

TABLE 16.2 Dielectric Constants for Some Materials

Material	Dielectric Constant (κ)	Material	Dielectric Constant (κ)
Vacuum	1.0000	Glass (range)	3–7
Air	1.00059	Pyrex glass	5.6
Paper	3.7	Bakelite	4.9
Polyethylene	2.3	Silicon oil	2.6
Polystyrene	2.6	Water	80
Teflon	2.1	Strontium titanate	233

How a dielectric affects the electrical properties of a capacitor is illustrated in ▾Fig. 16.16. The capacitor is fully charged (creating a field $\vec{E}_o$) and disconnected from the battery, after which a dielectric is inserted (Fig. 16.16a). In the dielectric material,

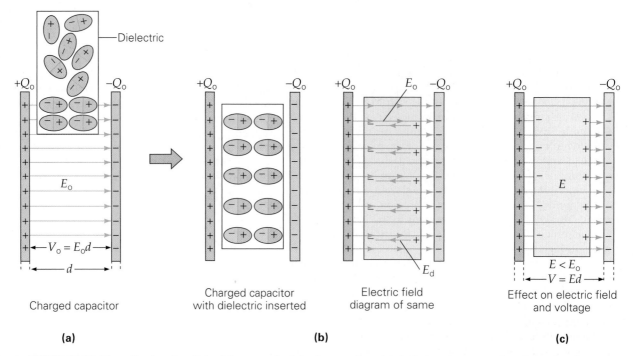

Charged capacitor

(a)

Charged capacitor
with dielectric inserted

Electric field
diagram of same

Effect on electric field
and voltage

(b)

(c)

 ▲ **FIGURE 16.16 The effects of a dielectric on an isolated capacitor** **(a)** A dielectric material with randomly oriented permanent molecular dipoles (or dipoles induced by the electric field) is inserted between the plates of an isolated charged capacitor. As the dielectric is inserted, the capacitor tends to pull it in, thus doing work on it. (Note the attractive forces between the plate charges and those induced on the dielectric surfaces.) **(b)** When the material is in the capacitor's electric field, the dipoles orient themselves with the field, giving rise to an opposing electric field $\vec{E}_d$. **(c)** The dipole field partially cancels the field due to the plate charges. The net effect is a decrease in both the electric field and the voltage. Because the stored charge remains the same, the capacitance increases.

Illustration 26.3 Capacitor with a
Dielectric

Note: Equation 16.14 holds only if
the battery is disconnected.

work is done on molecular dipoles by the existing electric field, aligning them with that field (Fig. 16.16b). (The molecular polarization may be permanent or temporarily induced by the electric field. In either case, the effect is the same.) Work is also done on the dielectric sheet as a whole, because the charged plates pull it into them.

The result is that the dielectric creates a "reverse" electric field ($\vec{E}_d$ in Fig. 16.16c) that partially cancels the field between the plates. This means that the *net* field ($\vec{E}$) between the plates is reduced, and so is the voltage across the plates (because $V = Ed$). The dielectric constant κ of the material is defined as the ratio of the voltage with the material in place (V) to the vacuum voltage (V_o). Because V is proportional to E, this ratio is the same as the electric field ratio:

$$\kappa = \frac{V_o}{V} = \frac{E_o}{E} \quad \begin{array}{l} \text{(only when the capacitor} \\ \text{charge is constant)} \end{array} \tag{16.14}$$

Note that κ is dimensionless and is greater than 1, because $V < V_o$. Equation 16.14 shows that the dielectric constant can be determined by measuring the two voltages. (Voltmeters are discussed in detail in Chapter 18.) Because the battery was disconnected and the capacitor isolated, the charge on the plates, Q_o, is unaffected. Because $V = V_o/\kappa$, the value of the capacitance with the dielectric inserted is larger than the vacuum value by a factor of κ. In effect, the same amount of charge is now being stored at a lower voltage, and the result is an increase in capacitance. To understand this effect, apply the definition of capacitance:

$$C = \frac{Q}{V} = \frac{Q_o}{(V_o/\kappa)} = \kappa\left(\frac{Q_o}{V_o}\right) \quad \text{or} \quad C = \kappa C_o \tag{16.15}$$

So inserting a dielectric into an isolated capacitor results in a larger capacitance. But what about energy storage? Because there is no energy input (the battery is disconnected) and the capacitor does work on the dielectric by pulling it into the region

Teaching tip: Point out that with the
battery removed, Q must stay the same.
The voltage drops when the dielectric is
inserted, which is why C is increased by
a factor of κ.

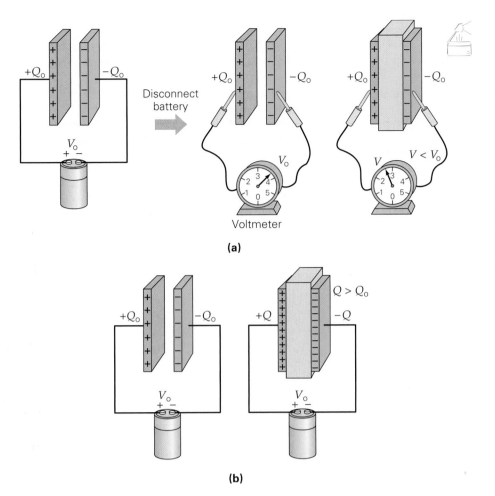

◀ **FIGURE 16.17** Dielectrics and capacitance (a) A parallel-plate capacitor in air (no dielectric) is charged by a battery to a charge Q_o and a voltage V_o (left). If the battery is disconnected and the potential across the capacitor is measured by a voltmeter, a reading of V_o is obtained (center). But if a dielectric is now inserted between the capacitor plates, the voltage drops to $V = V_o/\kappa$ (right), so the stored energy decreases. (Can you estimate the dielectric constant from the voltage readings?) (b) A capacitor is charged as in part (a), but the battery is left connected. When a dielectric is inserted into the capacitor, the voltage is maintained at V_o. (Why?) However, the charge on the plates increases to $Q = \kappa Q_o$. Therefore, more energy is now stored in the capacitor. In both cases, the capacitance increases by a factor of κ.

between the plates, the stored energy *drops* by a factor of κ (▲Fig. 16.17a), as the following calculation shows:

$$U_C = \frac{Q^2}{2C} = \frac{Q_o^2}{2\kappa C_o} = \frac{Q_o^2/2C_o}{\kappa} = \frac{U_o}{\kappa} < U_o \quad \text{(battery disconnected)}$$

A different situation occurs, however, if the dielectric is inserted *and the battery remains connected*. In this case, the voltage stays constant and the battery supplies (pumps) more charge—and therefore does work (Fig. 16.17b). Because the battery does work, we expect the energy stored in the capacitor to *increase*. With the battery remaining connected, the charge on the plates increases by a factor κ, or $Q = \kappa Q_o$. Once again the capacitance increases, but now it is because more charge is stored at the same voltage. From the definition of capacitance, the result is the same as Eq. 16.15, because $C = Q/V = \kappa Q_o/V_o = \kappa(Q_o/V_o) = \kappa C_o$. Thus,

> the effect of a dielectric is to increase the capacitance by a factor of κ, regardless of the conditions under which the dielectric is inserted.

In the case of a capacitor kept at constant voltage, the energy storage of the capacitor increases at the expense of the battery. To see this, let's calculate the energy with the dielectric in place under these conditions:

$$U_C = \tfrac{1}{2}CV^2 = \tfrac{1}{2}\kappa C_o V_o^2 = \kappa\left(\tfrac{1}{2}C_o V_o^2\right) = \kappa U_o > U_o \quad \text{(battery connected)}$$

For a parallel-plate capacitor with a dielectric, the capacitance is increased over its (air) value in Eq. 16.12 by a factor of κ:

$$C = \kappa C_o = \frac{\kappa \varepsilon_o A}{d} \quad \text{(parallel plates only)} \tag{16.16}$$

This relationship is sometimes written as $C = \varepsilon A/d$, where $\varepsilon = \kappa \varepsilon_o$ is called the **dielectric permittivity** of the material, which is always greater than ε_o. (How do you know this?)

Teaching tip: Make sure that students distinguish carefully between the cases of constant Q and constant V. The results are different! However, C increases on the insertion of a dielectric, regardless.

Demonstration/activity: There are now very compact capacitors with values of about 1 F. A good demonstration is to charge them with a hand-cranked generator, storing several joules of energy. Then connect the capacitor across a small lightbulb, and the bulb will light for several seconds. Later, in the section on dielectrics, you can point out how such a large capacitance can be made.

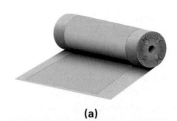

(a)

(b)

▲ **FIGURE 16.18 Capacitors in use**
(a) The dielectric material between the capacitor plates enables the plates to be constructed so that they are close together, thus increasing the capacitance. In addition, the plates can then be rolled up into a compact, more practical capacitor.
(b) Capacitors (flat, brown circles and purple cylinders) among other circuit elements in a microcomputer.

A sketch of the inside of a typical cylindrical capacitor and an assortment of real capacitors is shown in ◄Fig. 16.18. Changes in capacitance can be used to monitor motion in our technological world, as the next Example shows.

Example 16.9 ■ The Capacitor as a Motion Detector: Computer Keyboards

Consider a capacitor (with dielectric) underneath a computer key (▼Fig. 16.19). The capacitor is connected to a 12.0-V battery and has a normal (uncompressed—without a keystroke) plate separation of 3.00 mm and a plate area of 0.750 cm². (a) What is the required dielectric constant if the capacitance is 1.10 pF? (b) How much charge is stored on the plates under normal conditions? (c) How much charge flows onto the plates (that is, what is the change in their charge) if they are compressed to a separation of 2.00 mm?

Thinking It Through. (a) The capacitance of air-filled plates can be found from Eq. 16.12, and then the dielectric constant can be determined from Eq. 16.15. (b) The charge follows from Eq. 16.9. (c) The compressed-plate separation distance must be used to recompute the capacitance. Then the new charge can be found as in (b).

Solution. The given data are as follows:

Given: $V = 12.0$ V
$d = 3.00$ mm $= 3.00 \times 10^{-3}$ m
$A = 0.750$ cm² $= 7.50 \times 10^{-5}$ m²
$C = 1.10$ pF $= 1.10 \times 10^{-12}$ F
$d' = 2.00$ mm $= 2.00 \times 10^{-3}$ m

Find: (a) κ (dielectric constant)
(b) Q (initial capacitor charge)
(c) ΔQ (change in capacitor charge)

(a) From Eq. 16.12, the capacitance if the plates were separated by air would be

$$C_o = \frac{\varepsilon_o A}{d} = \frac{(8.85 \times 10^{-12} \text{ C}^2/\text{N} \cdot \text{m}^2)(7.50 \times 10^{-5} \text{ m}^2)}{3.00 \times 10^{-3} \text{ m}} = 2.21 \times 10^{-13} \text{ F}$$

Because the dielectric increases the capacitance, its value is

$$\kappa = \frac{C}{C_o} = \frac{1.10 \times 10^{-12} \text{ F}}{2.21 \times 10^{-13} \text{ F}} = 4.98$$

(b) The initial charge is then

$$Q = CV = (1.10 \times 10^{-12} \text{ F})(12.0 \text{ V}) = 1.32 \times 10^{-11} \text{ C}$$

(c) Under compressed conditions, the capacitance is

$$C' = \frac{\kappa \varepsilon_o A}{d'} = \frac{(4.98)(8.85 \times 10^{-12} \text{ C}^2/\text{N} \cdot \text{m}^2)(7.50 \times 10^{-5} \text{ m}^2)}{2.00 \times 10^{-3} \text{ m}} = 1.65 \times 10^{-12} \text{ F}$$

The voltage remains the same, $Q' = C'V = (1.65 \times 10^{-12} \text{ F})(12.0 \text{ V}) = 1.98 \times 10^{-11}$ C. Because the capacitance increased, the charge increased by

$$\Delta Q = Q' - Q = (1.98 \times 10^{-11} \text{ C}) - (1.32 \times 10^{-11} \text{ C}) = +6.60 \times 10^{-12} \text{ C}$$

As the key is depressed, a charge, whose magnitude is related to the displacement, flows onto the capacitor, providing a way of measuring the movement electrically.

Follow-Up Exercise. In this Example, suppose instead that the spacing between the plates were *increased* by 1.00 mm from the normal value of 3.00 mm. Would charge flow onto or away from the capacitor? How much charge would this be?

▶ **FIGURE 16.19 Capacitors in use**
Capacitors can be used to convert movement into electrical signals that can be measured and analyzed by computer. As the distance between the plates changes, so does the capacitance, which causes a change in the charge on the capacitor. Some computer keyboards operate in this way, as do other instruments such as seismographs (Chapter 13). See Example 16.9.

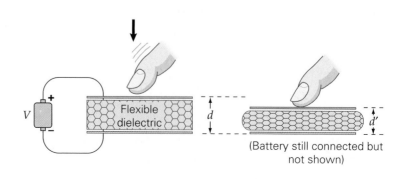

(Battery still connected but not shown)

16.5 Capacitors in Series and in Parallel

OBJECTIVES: To (a) find the equivalent capacitance of capacitors in series and in parallel, (b) calculate the charge, voltage, and energy storage of individual capacitors in series and parallel configurations, and (c) analyze capacitor networks that include both series and parallel arrangements.

Capacitors can be connected in two basic ways: *in series* or *in parallel*. In series, the capacitors are connected head to tail (▼Fig. 16.20a). When connected in parallel, all the leads on one side of the capacitors have a common connection. (Think of all the "tails" connected together and all the "heads" connected together; Fig. 16.20b.)

Note: For so-called sandwiched dielectric parallel-plate capacitors, there is no head or tail distinction between the leads. Some types of capacitors do have particular positive and negative sides, and then the distinction must be made.

Illustration 26.4 Microscopic View of Capacitors in Series and Parallel

▼ **FIGURE 16.20 Capacitors in series and in parallel** **(a)** All capacitors connected in series have the same charge, and the sum of the voltage drops is equal to the voltage of the battery. The total series capacitance is equivalent to the value of C_s. **(b)** When capacitors are connected in parallel, the voltage drops across the capacitors are the same, and the total charge is equal to the sum of the charges on the individual capacitors. The total parallel capacitance is equivalent to the value of C_p. **(c)** In a parallel connection, thinking of the plates makes it easier to see why the total charge is the sum of the individual charges. In effect, this arrangement represents a capacitor with two large plates.

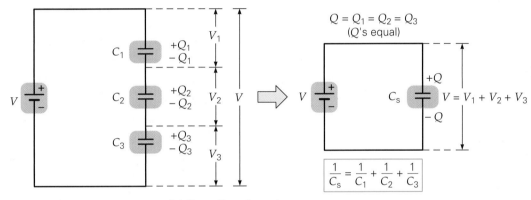

$$\frac{1}{C_s} = \frac{1}{C_1} + \frac{1}{C_2} + \frac{1}{C_3}$$

(a) Capacitors in series

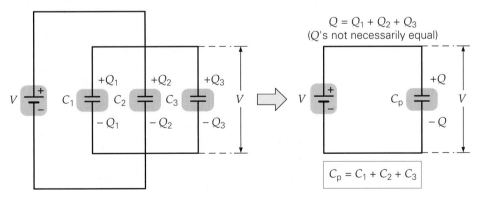

$$C_p = C_1 + C_2 + C_3$$

(b) Capacitors in parallel

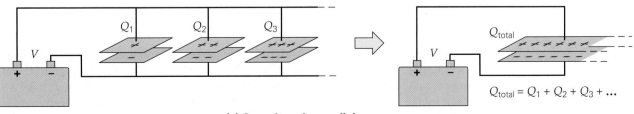

$$Q_{total} = Q_1 + Q_2 + Q_3 + \cdots$$

(c) Capacitors in parallel

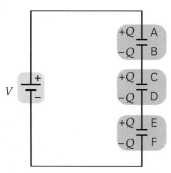

▲ **FIGURE 16.21 Charges on capacitors in series** Plates B and C together had zero net charge to start. When the battery placed $+Q$ on plate A, charge $-Q$ was induced on B; thus, C must have acquired $+Q$ for the BC combination to remain neutral. Continuing this way through the string, we see that all the charges must be the same in magnitude.

Capacitors in Series

When capacitors are wired in series, the charge Q must be the same on all the plates:

$$Q = Q_1 = Q_2 = Q_3 = \cdots$$

To see why this must be true, examine ◀Fig. 16.21. Note that only plates A and F are actually connected to the battery. Because the plates labeled B and C are isolated, the total charge on them must always be zero. So if the battery puts a charge of $+Q$ on plate A, then $-Q$ is induced on B at the expense of plate C, which acquires a charge of $+Q$. This charge in turn induces $-Q$ on D, and so on down the line.

As we have seen, "voltage drop" is just another name for "change in electrical potential energy per unit charge." When we add up all the series capacitor voltage drops (see Fig 16.20a), we must get the same value as the voltage across the battery terminals. The sum of the individual voltage drops across all the capacitors is equal to the voltage of the source:

$$V = V_1 + V_2 + V_3 + \cdots$$

The **equivalent series capacitance**, C_s, is defined as the value of a single capacitor that could replace the series combination and store the same charge at the same voltage. Because the combination of capacitors stores a charge of Q at a voltage of V, it follows that $C_s = Q/V$, or $V = Q/C_s$. However, the individual voltages are related to the individual charges by $V_1 = Q/C_1$, $V_2 = Q/C_2$, $V_3 = Q/C_3$, and so on.

Substituting these expressions into the voltage equation, we have

$$\frac{Q}{C_s} = \frac{Q}{C_1} + \frac{Q}{C_2} + \frac{Q}{C_3} + \cdots$$

Canceling the common Q's, we get

$$\frac{1}{C_s} = \frac{1}{C_1} + \frac{1}{C_2} + \frac{1}{C_3} + \cdots \quad \textit{equivalent series capacitance} \qquad (16.17)$$

This relationship means that C_s is always smaller than the smallest capacitance in the series combination. For example, try Eq. 16.17 with $C_1 = 1.0\ \mu\text{F}$ and $C_2 = 2.0\ \mu\text{F}$. You should be able to show that $C_s = 0.67\ \mu\text{F}$, which is less than $1.0\ \mu\text{F}$ (the general proof will be left to you). Physically, the reasoning goes like this. In series, all the capacitors have the same charge, so the charge stored by this arrangement is $Q = C_iV_i$ (where the subscript i refers to *any* of the individual capacitors in the string). Because $V_i < V$, the series arrangement stores *less* charge than any individual capacitor connected by itself to the same battery.

It makes sense that in series the smallest capacitance receives the largest voltage. A small value of C means less charge stored per volt. In order for the charge on all the capacitors to be the same, the smaller the value of capacitance, the larger the fraction of the total voltage required ($Q = CV$).

Capacitors in Parallel

With a parallel arrangement (Fig. 16.20b), the voltages across the capacitors are the same (why?), and each individual voltage is equal to that of the battery:

$$V = V_1 = V_2 = V_3 = \cdots$$

The total charge is the sum of the charges on each capacitor (Fig 16.20c):

$$Q_{\text{total}} = Q_1 + Q_2 + Q_3 + \cdots$$

The equivalent capacitance in parallel is expected to be larger than the largest capacitance, because more charge per volt can be stored in this way than if any one capacitor were connected to the battery by itself. The individual charges are given by $Q_1 = C_1V$, $Q_2 = C_2V$, and so on. A capacitor with the **equivalent parallel capacitance**, C_p, would hold this same total charge when connected to the battery, so $C_p = Q_{total}/V$, or $Q_{total} = C_pV$. Substituting these expressions into the previous equation gives

$$C_pV = C_1V + C_2V + C_3V + \cdots$$

and canceling the common V we obtain

$$C_p = C_1 + C_2 + C_3 + \cdots \quad \textit{equivalent parallel capacitance} \qquad (16.18)$$

In the parallel case, the equivalent capacitance C_p is the sum of the individual capacitances. In this case, the equivalent capacitance is larger than the largest individual capacitance. Because capacitors in parallel have the same voltage, the largest capacitance will store the most charge. For a comparison of capacitors in series and in parallel, consider the next Example.

Example 16.10 ■ Charging without Credit Cards: Capacitors in Series and in Parallel?

Given two capacitors, one with a capacitance of 2.50 μF and the other of 5.00 μF, what are the charge on each and the total charge stored if they are connected across a 12.0-V battery (a) in series and (b) in parallel?

Thinking It Through. (a) Capacitors in series have the same charge. From Eq. 16.17 we can find the equivalent capacitance and then the charge on each capacitor. (b) Capacitors in parallel have the same voltage; from that the charge on each can be easily determined because their individual capacitances are known.

Solution. Listing the data, we have the following:

Given: $C_1 = 2.50\ \mu\text{F} = 2.50 \times 10^{-6}\ \text{F}$ *Find:* (a) Q on each capacitor in series and
$C_2 = 5.00\ \mu\text{F} = 5.00 \times 10^{-6}\ \text{F}$ Q_{total} (total charge)
$V = 12.0\ \text{V}$ (b) Q on each capacitor in parallel and
                              Q_{total} (total charge)

(a) In series, the total (equivalent) capacitance is determined as follows:

$$\frac{1}{C_s} = \frac{1}{2.50 \times 10^{-6}\ \text{F}} + \frac{1}{5.00 \times 10^{-6}\ \text{F}} = \frac{3}{5.00 \times 10^{-6}\ \text{F}}$$

so

$$C_s = 1.67 \times 10^{-6}\ \text{F}$$

(*Note:* C_s is less than the smallest capacitance in the series, as expected.)

Because the charge on each capacitor is the same in series (and the same as the total), we have

$$Q_{total} = Q_1 = Q_2 = C_sV = (1.67 \times 10^{-6}\ \text{F})(12.0\ \text{V}) = 2.00 \times 10^{-5}\ \text{C}$$

(b) Here, the parallel equivalent capacitance relationship is used:

$$C_p = C_1 + C_2 = 2.50 \times 10^{-6}\ \text{F} + 5.00 \times 10^{-6}\ \text{F} = 7.50 \times 10^{-6}\ \text{F}$$

(This result is reasonable because it is greater than the largest individual value in the parallel arrangement.)

Therefore,

$$Q_{total} = C_pV = (7.50 \times 10^{-6}\ \text{F})(12.0\ \text{V}) = 9.00 \times 10^{-5}\ \text{C}$$

In parallel, each capacitor has the full 12.0 V across it; hence,

$$Q_1 = C_1V = (2.50 \times 10^{-6}\ \text{F})(12.0\ \text{V}) = 3.00 \times 10^{-5}\ \text{C}$$
$$Q_2 = C_2V = (5.00 \times 10^{-6}\ \text{F})(12.0\ \text{V}) = 6.00 \times 10^{-5}\ \text{C}$$

As a final double check, notice that the total stored charge is equal to the sum of the charges on both capacitors.

Follow-Up Exercise. In this Example, determine which combination, series or parallel, stores the most energy.

Exploration 26.4 Equivalent Capacitance

Capacitor arrangements generally can involve *both* series and parallel connections, as shown in the next Example. In this situation, you simplify the circuit, using the equivalent parallel and series capacitance expressions, until you end up with one single, overall equivalent capacitance. To find the results for each individual capacitor, you work backward until you get to the original arrangement.

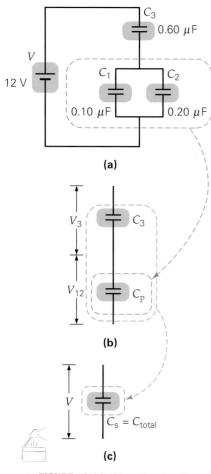

(a)

(b)

(c)

▲ **FIGURE 16.22 Circuit reduction**
When capacitances are combined, the combination of capacitors is reduced to a single equivalent capacitance. See Example 16.11.

Example 16.11 ■ The Electrical Two-Step—Forward, Then Backward: A Series–Parallel Combination of Capacitors

Three capacitors are connected in a circuit as shown in ◄Fig. 16.22a. What is the voltage across each capacitor?

Thinking It Through. The voltage across each capacitor could be found from $V = Q/C$ if the charge on each capacitor were known. The total charge on the capacitors is found by reducing the series–parallel combination to a single equivalent capacitance. Two of the capacitors are in parallel. Their single equivalent capacitance (C_p) is itself in series with the last capacitor—a fact that enables the total capacitance to be found. Working backward will allow the voltage across each capacitor to be found.

Solution.

Given: Values of capacitance and voltage from the figure

Find: V_1, V_2, and V_3 (voltages across capacitors)

Starting with the parallel combination, we have

$$C_p = C_1 + C_2 = 0.10 \ \mu F + 0.20 \ \mu F = 0.30 \ \mu F$$

Now the arrangement is partially reduced, as shown in Fig. 16.22b. Next, considering C_p in series with C_3, we can find the total, or overall, equivalent capacitance of the original arrangement:

$$\frac{1}{C_s} = \frac{1}{C_3} + \frac{1}{C_p} = \frac{1}{0.60 \ \mu F} + \frac{1}{0.30 \ \mu F} = \frac{1}{0.60 \ \mu F} + \frac{2}{0.60 \ \mu F} = \frac{1}{0.20 \ \mu F}$$

Therefore,

$$C_s = 0.20 \ \mu F = 2.0 \times 10^{-7} \ F$$

This is the total equivalent capacitance of the arrangement (Fig. 16.22c). Treating the problem as one single capacitor, we find the charge on that equivalent capacitance:

$$Q = C_s V = (2.0 \times 10^{-7} \ F)(12 \ V) = 2.4 \times 10^{-6} \ C$$

This is the charge on C_3 and C_p, because they are in series. We can use this to calculate the voltage across C_3:

$$V_3 = \frac{Q}{C_3} = \frac{2.4 \times 10^{-6} \ C}{6.0 \times 10^{-7} \ F} = 4.0 \ V$$

The sum of the voltages across the capacitors equals the voltage across the battery terminals. The voltages across C_1 and C_2 are the same because they are in parallel. Because the voltage across C_1 (or C_2) plus the voltage across C_3 equals the total voltage (the battery voltage), we can write $V = V_{12} + V_3 = 12 \ V$. (See Fig. 16.22a.) Here, V_{12} represents the voltage across either C_1 or C_2. Solving for V_{12},

$$V_{12} = V - V_3 = 12 \ V - 4.0 \ V = 8.0 \ V$$

Notice that C_p is less than C_3. Because C_p and C_3 are in series, it follows that C_p (and therefore C_1 and C_2) have most of the voltage.

Follow-Up Exercise. In this Example, find (a) the charge stored on each capacitor and (b) the energy stored in each.

Chapter Review

- The **electric potential difference** (or **voltage**) between two points is the work done per unit positive charge between those two points, or the change in electric potential energy per unit positive charge. Expressed in equation form, this relationship is

$$\Delta V = \frac{\Delta U_e}{q_+} = \frac{W}{q_+} \tag{16.1}$$

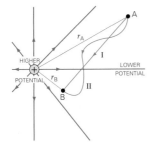

- **Equipotential surfaces** (surfaces of constant electric potential, also called **equipotentials**) are surfaces on which a charge has a constant electric potential energy. These surfaces are everywhere perpendicular to the electric field.

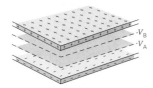

- The expression for the **electric potential due to a point charge** (choosing $V = 0$ at $r = \infty$) is

$$V = \frac{kq}{r} \tag{16.4}$$

- The **electric potential energy for a pair of point charges** is (choosing $U = 0$ at $r = \infty$)

$$U_{12} = \frac{kq_1q_2}{r_{12}} \tag{16.5}$$

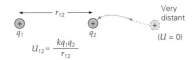

- The **electric potential energy of a configuration of more than two point charges** is the sum of point-charge pair terms from Eq. 16.5:

$$U_{total} = U_{12} + U_{23} + U_{13} + \cdots \tag{16.6}$$

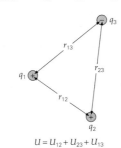

$$U = U_{12} + U_{23} + U_{13}$$

- The electric field is related to the rate of change of electric potential with distance. The electric field ($\vec{E}$) is in the direction of the most rapid decrease in electric potential (V). The electric field magnitude (E) is the rate of change of the potential with distance, or

$$E = \left| \frac{\Delta V}{\Delta x} \right|_{max} \tag{16.8}$$

- The **electron-volt (eV)** is the kinetic energy gained by an electron or a proton accelerated through a potential difference of 1 volt.

- A **capacitor** is any arrangement of two metallic plates. Capacitors store charge on their plates, and therefore electric energy.

- **Capacitance** is a quantitative measure of how effective a capacitor is in storing charge. It is the magnitude of the charge stored on either plate per volt, or

$$Q = CV \quad \text{or} \quad C = \frac{Q}{V} \tag{16.9}$$

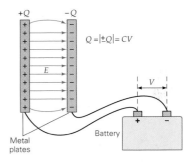

- The **capacitance of a parallel-plate capacitor** (in air) is

$$C = \frac{\varepsilon_o A}{d} \tag{16.12}$$

where $\varepsilon_o = 8.85 \times 10^{-12} \, C^2/(N \cdot m^2)$ is called the **permittivity of free space**.

- The **energy stored in a capacitor** depends on its capacitance and the charge the capacitor stores (or, equivalently, the voltage across its plates). There are three equivalent expressions for this energy:

$$U_C = \tfrac{1}{2}QV = \frac{Q^2}{2C} = \tfrac{1}{2}CV^2 \tag{16.13}$$

- A **dielectric** is a nonconducting material that increases capacitance.

- The **dielectric constant κ** describes the effect of a dielectric on capacitance. A dielectric increases the capacitor's capacitance over its value with air between the plates by a factor of κ

$$C = \kappa C_o \tag{16.15}$$

• Capacitors in series are equivalent to one capacitor, with a capacitance called the **equivalent series capacitance** C_s. The equivalent series capacitance is

$$\frac{1}{C_s} = \frac{1}{C_1} + \frac{1}{C_2} + \frac{1}{C_3} + \cdots \qquad (16.17)$$

• Capacitors in parallel are equivalent to one capacitor, with a capacitance called the **equivalent parallel capacitance** C_p. In parallel, all the capacitors have the same voltage. The equivalent parallel capacitance is

$$C_p = C_1 + C_2 + C_3 + \cdots \qquad (16.18)$$

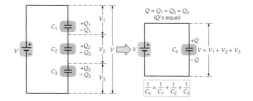

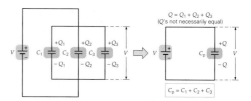

Exercises

MC = *Multiple Choice Question*, **CQ** = *Conceptual Question, and* **IE** = *Integrated Exercise. Throughout the text, many exercise sections will include "paired" exercises. These exercise pairs, identified with* **red numbers**, *are intended to assist you in problem solving and learning. In a pair, the first exercise (even numbered) is worked out in the Study Guide so that you can consult it should you need assistance in solving it. The second exercise (odd numbered) is similar in nature, and its answer is given at the back of the book.*

16.1 Electric Potential Energy and Electric Potential Difference

1. **MC** The SI unit of electric potential difference is the (a) joule, (b) newton per coulomb, (c) newton-meter, (d) joule per coulomb. (d)

2. **MC** How does the electrostatic potential energy of two positive point charges change when the distance between them is tripled: (a) It is reduced to one third its original value, (b) it is reduced to one ninth its original value, (c) it is unchanged, or (d) it is increased to three times its original value? (a)

3. **MC** An electron is moved from the positive to negative plate of a charged parallel plate arrangement. How does the sign of the change d in its electrostatic potential energy compare to the sign of the change in electrostatic potential it experiences: (a) Both are positive, (b) the energy change is positive, the potential change is negative, (c) the energy change is negative, the potential change is positive, or (d) both are negative? (b)

4. **CQ** What is the difference (a) between electrostatic potential energy and electric potential and (b) between electric potential difference and voltage?
(a) see ISM (b) no difference

5. **CQ** When a proton approaches another fixed proton, what happens to (a) the kinetic energy of the approaching proton, (b) the electric potential energy of the system, and (c) the total energy of the system?
(a) decreases (b) increases (c) remains the same

6. **CQ** Using the language of electrical potential and energy (not forces), explain why positive charges speed up as they approach negative charges. see ISM

7. **CQ** An electron is released in a region where the electric potential decreases to the left. Which way will the electron begin to move? Explain. to the right, see ISM

8. **CQ** An electron is released in a region where the electric potential is constant. Which way will the electron accelerate? Explain. it doesn't, see ISM

9. **CQ** If two locations are at the same electrical potential, how much work does it take to move a charge from the first location to the second? Explain. zero, see ISM

10. ● A pair of parallel plates is charged by a 12-V battery. How much work is required to move a particle with a charge of $-4.0~\mu$C from the positive to the negative plate?
$+4.8 \times 10^{-5}$ J

11. ● If it takes $+1.6 \times 10^{-5}$ J to move a positively charged particle between two charged parallel plates, (a) what is the charge on the particle if the plates are connected to a 6.0-V battery? (b) Was it moved from the negative to the positive plate or from the positive to the negative plate?
(a) 2.7 μC (b) negative to positive

12. ● What are the magnitude and direction of the electric field between the two charged parallel plates in Exercise 11 if the plates are separated by 4.0 mm?
1.5×10^3 V/m pointing from positive to negative

13. ● In a dental X-ray machine, a beam of electrons is accelerated by a potential difference of 10 kV. At the end of the acceleration, how much kinetic energy does each electron have if they start from rest? 1.6×10^{-15} J

14. ● An electron is accelerated by a uniform electric field (1000 V/m) pointing vertically upward. Use Newton's laws to determine the electron's velocity after it moves 0.10 cm from rest. 5.9×10^5 m/s, down

15. ● (a) Repeat Exercise 14, but find the speed by using energy methods. Get the direction in which the electron is moving by considering electric potential-energy changes. (b) Does the electron gain or lose potential energy? (a) 5.9×10^5 m/s, down (b) lose potential energy

16. IE ● Consider two points at different distances from a positive point charge. (a) The point closer to the charge is at a (1) higher, (2) equal, (3) lower potential than the point farther away. Why? (b) How much different is the electric potential 20 cm from a charge of 5.5 μC compared to 40 cm from the same charge? (a) (1) higher (b) $+1.2 \times 10^5$ V

17. IE ●● (a) At one third the original distance from a positive point charge, by what factor is the electric potential changed: (1) 1/3, (2) 3, (3) 1/9, or (4) 9? Why? (b) How far from a $+1.0$-μC charge is a point with an electric potential value of 10 kV? (c) How much of a change in potential would occur if the point were moved to three times that distance? (a) (2) 3 (b) 0.90 m (c) -6.7 kV

18. IE ●● In the Bohr model of the hydrogen atom (see Chapter 27), we will learn that the electron can exist only in circular orbits of certain radii about a proton. (a) Will a larger orbit have a (1) a higher, (2) an equal, or (3) a lower electric potential than a smaller orbit? Why? (b) Determine the potential difference between two orbits of radii 0.21 nm and 0.48 nm. (a) (3) a lower potential (b) 3.9 V

19. ●● In Exercise 18, by how much does the potential energy of the atom change if the electron goes (a) from the lower to the higher orbit, (b) from the higher to the lower orbit, and (c) from the larger orbit to a very large distance? (a) gains 6.2×10^{-19} J (b) loses 6.2×10^{-19} J (c) gains 4.8×10^{-19} J

20. ●● How much work is required to completely separate two charges (each -1.4 μC) and leave them at rest if they were initially 8.0 mm apart? -2.2 J

21. ●● In Exercise 20, if the two charges are released at their initial separation distance, how much kinetic energy would each have when they are very distant from one another? 1.1 J

22. ●● It takes $+6.0$ J of work to move two charges from a large distance apart to 1.0 cm from one another. If the charges have the same magnitude, (a) how large is each charge, and (b) what can you tell about their signs? (a) 2.5 μC (b) same sign

23. ●● A $+2.0$-μC charge is initially 0.20 m from a fixed -5.0-μC charge and is then moved to a position 0.50 m from the fixed charge. (a) How much work is required to move the charge? (b) Does the work depend on the path through which the charge is moved? (a) $+0.27$ J (b) no

24. ●● An electron is moved from point A to point B and then to point C along two legs of an equilateral triangle with sides of length 0.25 m (▼Fig. 16.23). If the horizontal electric field is 15 V/m, (a) what is the magnitude of the work required? (b) What is the potential difference between points A and C? (c) Which point is at a higher potential? (a) 6.0×10^{-19} J (b) 3.8 V (c) point C

25. ●● Compute the energy necessary to bring together (from a very large distance) the charges in the configuration shown in ▼Fig. 16.24. -0.72 J

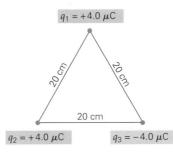

$q_1 = +4.0$ μC

20 cm   20 cm

20 cm

$q_2 = +4.0$ μC $q_3 = -4.0$ μC

◀ **FIGURE 16.24** A charge triangle See Exercises 25 and 27.

26. ●● Compute the energy necessary to bring together (from a very large distance) the charges in the configuration shown in ▼Fig. 16.25. -4.1 J

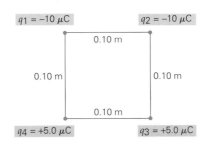

$q_1 = -10$ μC $q_2 = -10$ μC

0.10 m

0.10 m   0.10 m

0.10 m

$q_4 = +5.0$ μC $q_3 = +5.0$ μC

◀ **FIGURE 16.25** A charge rectangle See Exercises 26 and 28.

27. ●●● What is the value of the electric potential at (a) the center of the triangle and (b) a point midway between q_2 and q_3 in Fig. 16.24? (a) 3.1×10^5 V (b) 2.1×10^5 V

28. ●●● What is the value of electric potential at (a) the center of the square and (b) a point midway between q_1 and q_4 in Fig. 16.25? (a) -1.3×10^6 V (b) -1.3×10^6 V

29. IE ●●● In a computer monitor, electrons are accelerated from rest through a potential difference in an "electron gun" arrangement (▼Fig. 16.26). (a) Should the left side of the gun be at (1) a higher, (2) an equal, or (3) a lower potential than the right side? Why? (b) If the potential difference in the gun is 5.0 kV, what is the "muzzle speed" of the electrons emerging from the gun? (c) If the gun is directed at a screen 25 cm away, how long do the electrons take to reach the screen? (a) (3) a lower (b) 4.2×10^7 m/s, (c) 6.0×10^{-9} sec

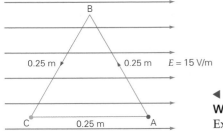

B

0.25 m   0.25 m $E = 15$ V/m

C   0.25 m   A

◀ **FIGURE 16.23** Work and energy See Exercise 24.

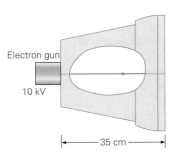

Electron gun

10 kV

⟵ 35 cm ⟶

◀ **FIGURE 16.26** Electron speed See Exercise 29.

16.2 Equipotential Surfaces and the Electric Field

30. **MC** On an equipotential surface (a) the electric potential is constant, (b) the electric field is zero, (c) the electric potential is zero, (d) there must be equal amounts of negative and positive charge. (a)

31. **MC** Equipotential surfaces (a) are parallel to the electric field, (b) are perpendicular to the electric field, (c) can be at any angle with respect to the electric field. (b)

32. **MC** An electron is moved from an equipotential surface at +5.0 V to one at +10.0 V. It is moving generally in a direction (a) parallel to the electric field, (b) opposite to the electric field, (c) in the same direction as the electric field. (b)

33. **CQ** Sketch the topographic map you would expect as you walk away from the ocean up a gently sloping uniform beach. Label the gravitational equipotentials as to relative height and potential value. Show how to predict, from the map, which way a ball would accelerate if it is initially rolled up the beach away from the water. see ISM

34. **CQ** Explain why two equipotential surfaces cannot intersect. see ISM

35. **CQ** Suppose a charge starts at rest on an equipotential surface, is moved off that surface, and then is eventually returned to the same surface at rest after a round trip. How much work did it take to do this? Explain. zero, see ISM

36. **CQ** What geometrical shape are the equipotential surfaces between two charged parallel plates? planes, parallel to the plates

37. **CQ** (a) What is the approximate shape of the equipotential surfaces inside the axon cell membrane? (See Fig. 1, p. 552.) (b) Under resting-potential conditions, where is the region of highest electric potential inside the membrane? (c) What about during reversed polarity conditions? (a) cylindrical (b) near the outer surface (c) near the inner surface

38. **CQ** Near a fixed positive point charge, if you go from one equipotential surface to another one with a smaller radius, (a) what happens to the value of the potential? (b) What was your general direction relative to the electric field? (a) Increases (b) opposite $\vec{E}$

39. **CQ** (a) If a proton is accelerated from rest by a potential difference of 1 million volts, how much kinetic energy does it gain? (b) How would your answer to part (a) change if the accelerated particle had twice the charge of the proton (same sign) and four times the mass? (a) 1.60×10^{-13} J (b) it would double

40. **CQ** (a) Can the electric field at a point be zero while there is also a nonzero electric potential at that point? (b) Can the electric potential at a point be zero while there is also a nonzero electric field at that point? Explain. If the answer to either part is yes, give an example. (a) yes, see ISM (b) yes, see ISM

41. ● For a +3.50-μC point charge, what is the radius of the equipotential surface that is at a potential of 2.50 kV? 12.6 m

42. ● A uniform electric field of 10 kV/m points vertically upward. How far apart are the equipotential planes that differ by 100 V? 1.0 cm

43. ● In Exercise 42, if the ground is designated as zero potential, how far above the ground is the equipotential surface corresponding to 7.0 kV? 70 cm

44. ● Determine the potential 2.5 mm from the negative plate of a pair of parallel plates separated by 10 mm and connected to a 24-V battery. +6.0 V

45. ● Relative to the positive plate in Exercise 44, where is the point with a potential of 20 V? 1.7 mm away from the positive plate, toward the negative plate

46. ● If the radius of the equipotential surface of point charge is 14.3 m at a potential of 2.20 kV, what is the magnitude of the point charge creating the potential? +3.50 μC

47. **IE** ● (a) The equipotential surfaces in the neighborhood of a point charge are (1) concentric spheres, (2) concentric cylinders, (3) planes. (b) Calculate the amount of work (in electron-volts) it would take to move an electron from 12.6 m to 14.3 m away from a +3.50-μC point charge. (a) (1) concentric spheres (b) +298 eV

48. ● The potential difference involved in a typical lightning discharge may be up to 100 MV (million volts). What is the gain in kinetic energy of an electron accelerated through this potential difference? Give your answer in both electron-volts and joules. (Assume that there are no collisions.) 1.00×10^8 eV or 1.60×10^{-11} J

49. ● In a typical Van de Graaff linear accelerator, protons are accelerated through a potential difference of 20 MV. What is their kinetic energy if they started from rest? Give your answer in (a) eV, (b) keV, (c) MeV, (d) GeV, and (e) joules. see ISM

50. ● In Exercise 49, how do your answers change if a doubly charged (+2e) alpha particle is accelerated instead? (An alpha particle consists of two neutrons and two protons.) all are doubled, see ISM

51. ●● In Exercises 49 and 50, compute the speed of the proton and alpha particle after being accelerated. see ISM

52. ●● Calculate the voltage required to accelerate a beam of protons initially at rest, and calculate their speed if they have a kinetic energy of (a) 3.5 eV, (b) 4.1 keV, and (c) 8.0×10^{-16} J. see ISM

53. ●● Repeat the calculation in Exercise 52 for electrons instead of protons. see ISM

54. ●●● Two large parallel plates are separated by 3.0 cm and connected to a 12-V battery. Starting at the negative plate and moving 1.0 cm toward the positive plate at a 45° angle (▼Fig. 16.27), (a) what value of potential would be reached,

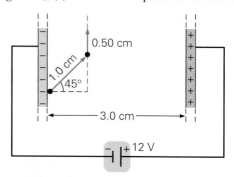

▲ **FIGURE 16.27 Reaching our potential** See Exercises 54 and 55.

assuming the negative plate were defined as zero potential? (b) What would be the value of the potential if you then moved 0.50 cm parallel to the plates? +2.8 V, +2.8 V

55. ●●● Consider a point midway between the two large charged plates in Fig. 16.27. Compute the change in electric potential if from there you moved (a) 1.0 mm toward the positive plate, (b) 1.0 mm toward the negative plate, and (c) 1.0 mm parallel to the plates.
(a) +0.40 V (b) −0.40 V (c) zero

56. ●●● Using the results of Exercise 55, determine the electric field (direction and magnitude) at the midway point between the plates. 400 V/m toward negative plate

16.3 Capacitance

57. **MC** A capacitor is first connected to a 6.0-V battery and then disconnected and connected to a 12.0-V battery. How does its capacitance change: (a) It increases, (b) it decreases, or (c) it stays the same? (c)

58. **MC** A capacitor is first connected to a 6.0-V battery and then disconnected and connected to a 12.0-V battery. How does the charge on one of its plates change: (a) It increases, (b) it decreases, or (c) it stays the same? (a)

59. **MC** A capacitor is first connected to a 6.0-V battery and then disconnected and connected to a 12.0-V battery. By how much does the electric field strength between its plates change: (a) two times, (b) four times, or (c) it stays the same. (a)

60. **MC** A capacitor has the distance between its plates cut in half. By what factor does its capacitance change: (a) It is cut in half, (b) it is reduced to one fourth its original value, (c) it is doubled, or (d) it is quadrupled? (c)

61. **MC** A capacitor has the area of its plates reduced. How would you adjust the distance between those plates to keep the capacitance constant: (a) increase it, (b) decrease it, or (c) changing the distance cannot ever make up for the plate area change? (b)

62. **CQ** If the plates of an isolated parallel-plate capacitor are moved closer to each other, does the energy storage increase, decrease, or remain the same? Explain.
it decreases; see ISM

63. **CQ** If the potential difference across a capacitor is doubled, what happens to (a) the charge on the capacitor and (b) the energy stored in the capacitor?
(a) it doubles (b) it quadruples

64. **CQ** A capacitor is connected to a 12-V battery. If the plate separation is tripled and the capacitor remains connected to the battery, by what factor does the charge on the capacitor change? decreases to 1/3 of its original value

65. ● How much charge flows through a 12-V battery when a 2.0-μF capacitor is connected across its terminals?
2.4 × 10⁻⁵ C

66. ● A parallel-plate capacitor has a plate area of 0.50 m² and a plate separation of 2.0 mm. What is its capacitance?
2.2 × 10⁻⁹ F

67. ● What plate separation is required for a parallel-plate capacitor to have a capacitance of 5.0 × 10⁻⁹ F if the plate area is 0.40 m²? 0.71 mm

68. **IE** ● (a) For a parallel-plate capacitor, a larger plate area results in (1) a larger, (2) an equal, (3) a smaller capacitance. (b) A 2.5 × 10⁻⁹ F parallel-plate capacitor has a plate area of 0.425 m². If the capacitance is to double, what is the required plate area?
(a) (1) a larger (b) 0.850 m²

69. ●● A 12-V battery is connected to a parallel-plate capacitor with a plate area of 0.20 m² and a plate separation of 5.0 mm. (a) What is the charge on the capacitor? (b) How much energy is stored in the capacitor?
(a) 4.2 × 10⁻⁹ C (b) 2.5 × 10⁻⁸ J

70. ●● If the plate separation of the capacitor in Exercise 69 changed to 10 mm after the capacitor is disconnected from the battery, how do your answers change?
(a) same (b) doubles to 5.0 × 10⁻⁸ J

71. ●●● Current state-of-the-art capacitors are capable of storing many times the energy of older ones. Such a capacitor, with a capacitance of 1.0 F, is able to light a small 0.50-W bulb at steady full power for 5.0 s before it quits. What was the terminal voltage of the battery that charged the capacitor?
2.2 V

72. ●●● A 1.50-F capacitor is connected to a 12.0-V battery for a long time, and then is disconnected. The capacitor briefly runs a 1.00-W toy motor for 2.00 s. After this time, (a) by how much has the energy stored in the capacitor decreased? (b) What is the voltage across the plates? (c) How much charge is stored on the capacitor? (d) How much longer could the capacitor run the motor, assuming the motor ran at full power until the end? see ISM

73. ●●● Two parallel plates have a capacitance value of 0.17 μF when they are 1.5 mm apart. They are connected permanently to a 100-V power supply. If you pull the plates out to a distance of 4.5 mm, (a) what is the electric field between them? (b) By how much has the capacitor's charge changed? (c) By how much has its energy storage changed? (d) Repeat these calculations assuming the power supply is disconnected before you pull the plates further apart. see ISM

16.4 Dielectrics

74. **MC** Putting a dielectric in a charged parallel-plate capacitor that is not connected to a battery (a) decreases the capacitance, (b) decreases the voltage, (c) increases the charge, (d) causes a discharge because the dielectric is a conductor. (b)

75. **MC** A parallel-plate capacitor is connected to a battery. If a dielectric is inserted between the plates, (a) the capacitance decreases, (b) the voltage increases, (c) the voltage decreases, (d) the charge increases. (d)

76. **MC** A parallel-plate capacitor is connected to a battery and then disconnected. If a dielectric is then inserted between the plates, what happens to the charge on its plates: (a) The charge decreases, (b) the charge increases, or (c) the charge stays the same? (c)

77. **CQ** Give several reasons why a conductor would not be a good choice as a dielectric for a capacitor. see ISM

78. **CQ** A parallel-plate capacitor is connected to a battery and then disconnected. If a dielectric is inserted between the plates, what happens to (a) the capacitance and (b) the voltage? (a) increases (b) decreases

79. **CQ** Explain clearly why the electric field between two parallel plates of a capacitor decreases when a dielectric is inserted if the capacitor is not connected to a power supply, but remains the same when it is connected to a power supply. see ISM

80. ● A capacitor has a capacitance of 50 pF, which increases to 150 pF when a dielectric material is between its plates. What is the dielectric constant of the material? $\kappa = 3.0$

81. ● A 50-pF capacitor is immersed in silicone oil ($\kappa = 2.6$). When the capacitor is connected to a 24-V battery, what will be the charge on the capacitor and the amount of stored energy? 3.1×10^{-9} C; 3.7×10^{-8} J

82. ●● The dielectric of a parallel-plate capacitor is to be constructed from glass that completely fills the volume between the plates. The area of each plate is 0.50 m². (a) What thickness should the glass have if the capacitance is to be 0.10 μF? (b) What is the charge on the capacitor if it is connected to a 12-V battery?
(a) 0.20 mm (b) 1.2 μC

83. ●●● A parallel-plate capacitor has a capacitance of 1.5 μF with air between the plates. The capacitor is connected to a 12-V battery and charged. The battery is then removed. When a dielectric is placed between the plates, a potential difference of 5.0 V is measured across the plates. (a) What is the dielectric constant of the material? (b) Did the energy stored in the capacitor increase, decrease, or stay the same? (c) By how much did the energy storage of this capacitor change when the dielectric was inserted?
(a) $\kappa = 2.4$ (b) decreased (c) -6.3×10^{-5} J

84. **IE** ●●● An air-filled parallel-plate capacitor has rectangular plates with dimensions of 6.0 cm × 8.0 cm. It is connected to a 12-V battery. While the battery remains connected, a sheet of 1.5-mm-thick Teflon ($\kappa = 2.1$) is inserted and completely fills the space between the plates. (a) While the dielectric was being inserted, (a) charge flowed onto the capacitor, (2) charge flowed off the capacitor, (3) no charge flowed. (b) Determine the change in the charge storage of this capacitor because of the dielectric insertion.
(a) (1) charge flowed on to the capacitor (b) 3.7×10^{-10} C

16.5 Capacitors in Series and in Parallel

85. **MC** Capacitors in series have the same (a) voltage, (b) charge, (c) energy storage. (b)

86. **MC** Capacitors in parallel have the same (a) voltage, (b) charge, (c) energy storage. (a)

87. **MC** Capacitors 1, 2, and 3 have the same capacitance value C. 1 and 2 are in series and their combination is in parallel with 3. What is their effective total capacitance. (a) C, (b) 1.5C, (c) 3C, or (d) C/3? (b)

88. **CQ** Under what conditions would two capacitors in series have the same voltage? equal capacitance

89. **CQ** Under what conditions would two capacitors in parallel have the same charge? equal capacitance

90. **CQ** If you are given two capacitors, how should you connect them to get (a) maximum equivalent capacitance and (b) minimum equivalent capacitance? (a) parallel (b) series

91. **CQ** You have N (an even number ≥ 2) identical capacitors, each with a capacitance of C. In terms of N and C, what is their total effective capacitance if (a) they are all connected in series? (b) they are all connected in parallel? (c) two halves (N/2 each) are connected in series and these two sets are connected in parallel?
(a) C/N (b) NC (c) 4C/N

92. ● What is the equivalent capacitance of two capacitors with capacitances of 0.40 μF and 0.60 μF when they are connected (a) in series and (b) in parallel?
(a) 0.24 μF (b) 1.0 μF

93. **IE** ● (a) Two capacitors can be connected to a battery in either a series or parallel combination. The parallel combination will draw (1) more, (2) equal, or (3) less energy from a battery than the series combination. Why? (b) When a series combination of two uncharged capacitors is connected to a 12-V battery, 173 μJ of energy is drawn from the battery. If one of the capacitors has a capacitance of 4.0 μF, what is the capacitance of the other? (a) (1) more (b) 6.0 μF

94. ●● For the arrangement of three capacitors in ▼Fig. 16.28, what value of C_1 will give a total equivalent capacitance of 1.7 μF? 1.2 μF

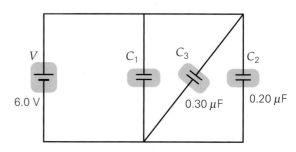

▲ **FIGURE 16.28 A capacitor triad** See Exercises 94 and 98.

95. **IE** ●● (a) Three capacitors of equal capacitance are connected in parallel to a battery, and together they draw a certain amount of charge Q from that battery. Will the charge on each capacitor be (1) Q, (2) 3Q, or (3) Q/3? (b) Three capacitors of 0.25 μF each are connected in parallel to a 12-V battery. What is the charge on each capacitor? (c) How much charge is drawn from the battery? (a) (3) Q/3 (b) 3.0 μC (c) 9.0 μC

96. **IE** ●● (a) If you are given three identical capacitors, you can obtain (1) three, (2) five, (3) seven different capacitance values. (b) If the three capacitors each have a capacitance of 1.0 μF, what are the different values of equivalent capacitance? (a) (3) seven values
(b) 0.33, 0.50, 0.67, 1.0, 1.5, 2.0, and 3.0 μF

97. ●● What are the maximum and minimum equivalent capacitances that can be obtained by combinations of three capacitors of 1.5 μF, 2.0 μF, and 3.0 μF?
max. 6.5 μF; min. 0.67 μF

98. ●●● If the capacitance $C_1 = 0.10$ μF, what is the charge on each of the capacitors in the circuit in Fig. 16.28?
C_1: 0.60 μC; C_2: 1.2 μC; C_3: 1.8 μC

99. ●●● Four capacitors are connected in a circuit as illustrated in ▶Fig. 16.29. Find the charge on, and the voltage difference across, each of the capacitors. see ISM

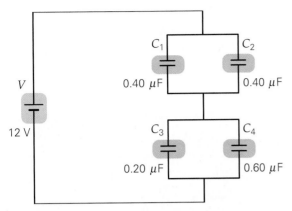

▲ **FIGURE 16.29 Double parallel in series** See Exercise 99.

Comprehensive Exercises

100. **IE** A tiny dust particle in the form of a long thin needle has charges of ±7.14 pC on its end. The length of the particle is 3.75 μm. (a) Which location is at a higher potential: (1) 7.65 μm above the positive end, (2) 5.15 μm above the positive end, or (3) both locations are at the same potential? (b) Compute the potential at the two the points in part (a). (c) Use your answer from part (b) to determine the work needed to move an electron from the near point to the far point. (a) (2) 5.15 μm (b) 2.76 $\times$ 10³ V, 5.26 $\times$ 10³ V (c) 4.00 $\times$ 10⁻¹⁶ J

101. A vacuum tube has a vertical height of 50.0 cm. An electron leaves from the top at a speed of 3.2×10^6 m/s downward and is subjected to a "typical" Earth field of 150 V/m downward. (a) Use energy methods to determine if it reaches the bottom surface of the tube. (b) If it does, with what speed does it hit, if not, how close does it come to the bottom surface? (a) $K_o = 29.2$ eV, total $\Delta U = 75$ eV, so it can't (b) 30.6 cm from bottom

102. **CQ** Sketch the equipotential surfaces and the electric field line pattern outside a uniformly (negatively) charged long wire. Label the surfaces with relative potential value and indicate the electric field direction. see ISM

103. A helium atom with one electron already removed (a positive helium ion) consists of a single orbiting electron and a nucleus of two protons. The electron is in its minimum orbital radius of 0.027 nm. (a) What is the potential energy of the system? (b) What is the centripetal acceleration of the electron? (c) What is the total energy of the system? (d) What is the minimum energy required to ionize this atom so the electron leaves completely? see ISM

104. Suppose that the three capacitors in Figure 16.22 have the following values: $C_1 = 0.15\ \mu$F, $C_2 = 0.25\ \mu$F, and $C_3 = 0.30\ \mu$F. (a) What is the equivalent capacitance of this arrangement? (b) How much charge will be drawn from the battery? (c) What is the voltage across each capacitor? (a) 0.17 μF (b) 2.1 μC (c) $V_1 = V_2 = 5.1$ V, $V_3 = 6.9$ V

105. **IE** Two very large horizontal parallel plates are separated by 1.50 cm. An electron is to be suspended in midair between them. (a) The top plate should be at (1) a higher potential, (2) an equal potential, (3) a lower potential compared with the bottom plate. Explain. (b) What voltage across the plates is required? (c) Does the electron have to be positioned midway between the plates, or is any location between the plates just as good? (a) (1) a higher potential (b) 8.37 $\times$ 10⁻¹³ V (c) any location

106. (See the Insight on Electric Potential and Nerve Signal Transmission on p. 552 and the Learn by Drawing on graphical relationships between $\vec{E}$ and V on p. 547.) Suppose an (axon) cell membrane is experiencing the end of a stimulus event and the voltage across the cell membrane is instantaneously at 30 mV. Assume the membrane is 10 nm thick. At this point the Na/K-ATPase molecular pump starts to move the excess Na⁺ ions back to the exterior. (a) How much work does it take for the pump to move the first sodium ion? (b) Estimate the electric field (including direction) in the membrane under these conditions? (c) What is the electric field (including direction) under normal conditions when the voltage across the membrane is −70 mV? (a) 4.8 $\times$ 10⁻²¹ J (b) 3.0 MV/m outward (c) 7.0 MV/m inward

107. In exercise 106, assume that the inside and outside surfaces of the axon membrane act like a parallel plate capacitor with an area of 1.1×10^{-9} m². (a) Estimate the capacitance of an axon's membrane, assuming it is filled with lipids with a dielectric constant of 3.0. (b) How much charge would be on each surface under resting potential conditions? (a) 2.9 pF (b) 0.20 pC

108. Two parallel plates, 9.25 cm on a side, are separated by 5.12 mm. (a) Determine their capacitance if the volume from one plate to midplane is filled with a material of dielectric constant 2.55 and the rest is filled with a different material (dielectric constant 4.10). See ▼Fig. 16.30a. [*Hint*: Do you see two capacitors in series?] (b) Repeat part (a), except fill the volume from one edge to the middle with the same two materials. See Figure 16.30b. (Do you see two capacitors in parallel?) (a) 4.65 $\times$ 10⁻¹¹ F (b) 4.92 $\times$ 10⁻¹¹ F

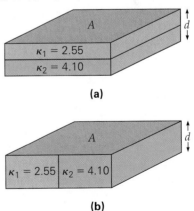

▲ **FIGURE 16.30 Double-stuffed capacitor** See Exercise 108.

The following Physlet Physics Problems can be used with this chapter.
PHYSLET® 25.1, 25.2, 25.3, 25.4, 25.5, 25.6, 25.7, 26.1, 26.2, 26.3, 26.5, 26.9, 26.11

ELECTRIC CURRENT AND RESISTANCE

PHYSICS FACTS

- André Marie Ampère (1775–1836) was a French physicist/mathematician known for his work with electric currents. His name is used for the SI unit of current, the ampere (usually shortened to *amp*). He also worked in chemistry, being involved in the classification of the elements and the discovery of fluorine. In physics, Ampère is famous for being one of the first to attempt a combined theory of electricity and magnetism. Ampere's law, which describes the *magnetic* field created by a flow of *electric* charge, is one of the four fundamental equations of classical electromagnetism.

- In a metal wire, the electric *energy* travels at the speed of light (in the wire), which is much faster than the speed of the charge carriers themselves. The speed of the latter is only several millimeters per second.

- The SI unit of electrical resistance, the ohm (Ω), is named after Georg Simon Ohm (1789–1854), a German mathematician and physicist. A quantity called electrical conductivity, proportional to the *inverse* of resistance, is named, appropriately enough, the mho—his last name spelled backwards.

- Using a voltage of up to 600 volts, electric eels and rays can, for brief times, discharge as much as 1 ampere of current through flesh. The energy is delivered at a rate of 600 J/s, or about three-fourths of a horsepower.

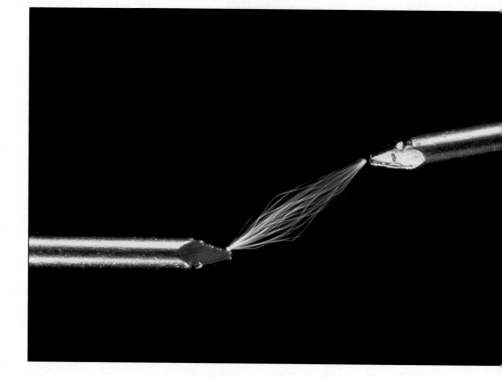

If you were asked to think of electricity and its uses, many favorable images would probably come to mind, including such diverse applications as lamps, television remote controls, computers, and electric leaf blowers. You might also think of some unfavorable images, such as dangerous lightning, a shock, or sparks you may have experienced from an overloaded electric outlet.

Common to all of these images is the concept of electric energy. For an electric appliance, energy is supplied by electric current in wires; for lightning or a spark, it is conducted through the air. In either case, the light, heat, or mechanical energy given off is simply electric energy converted to a different form. In the photograph, for example, the light given off by the spark is emitted by air molecules.

In this chapter, we are concerned with the fundamental principles governing electric circuits. These principles will enable us to answer questions such as the following: What is electric current and how does it travel? What causes an electric current to move through an appliance when we flip a switch? Why does the electric current cause the filament in a bulb to glow brightly, but not affect the connecting wires in the same way? We can apply electrical principles to gain an understanding of a wide range of phenomena, from the operation of household appliances to the workings of Nature's spectacular fireworks—lightning.

17.1 Batteries and Direct Current

OBJECTIVES: To (a) introduce the properties of a battery, (b) explain how a battery produces a direct current in a circuit, and (c) learn various circuit symbols for sketching schematic circuit diagrams.

After studying electric force and energy in Chapters 15 and 16, you can probably guess what is required to produce an *electric current*, or a flow of charge. Here are some analogies to help. Water naturally flows downhill, from higher to lower gravitational potential energy—that is, because there is a *difference* in gravitational potential energy. Heat flows naturally because of temperature *differences*. In electricity, a flow of electric charge is caused by an *electric* potential *difference*—what we call "voltage."

In solid conductors, particularly metals, some of the outer electrons of atoms are relatively free to move. (In liquid conductors and charged gases called *plasmas*, positive and negative ions as well as electrons can move.) Energy is required to move electric charge. Electric energy is generated through the conversion of other forms of energy, giving rise to a potential difference, or voltage. Any device that can produce and maintain a potential difference is called by the general name of a *power supply*.

Battery Action

One common type of power supply is the battery. A **battery** converts stored *chemical* potential energy into electrical energy. The Italian scientist Alessandro Volta constructed one of the first practical batteries. A simple battery consists of two unlike metal *electrodes* in an *electrolyte*, a solution that conducts electricity. With the appropriate electrodes and electrolyte, a potential difference develops across the electrodes as a result of chemical action (▶Fig. 17.1).

When a complete circuit is formed, for example, by connecting a lightbulb and wires (Fig. 17.1), electrons from the more negative electrode (B) will move through the wire and bulb to the less negative electrode (A).* The result is a flow of electrons in the wire. As electrons move through the bulb's filament, colliding with and transferring energy to its atoms (typically tungsten), the filament reaches a sufficient temperature to give off visible light (glow). Since electrons move to regions of higher potential, electrode A is at a higher potential than B. Thus the battery action has created a potential *difference* (V) across its terminals. Electrode A is the **anode** and labeled with a plus ($+$) sign. Electrode B is the **cathode** and labeled as negative ($-$). It is easy to keep track of this sign convention because the negatively charged electrons will move through the wire from B ($-$) to A ($+$).

For the study of circuits, we can just picture a battery as a "black box" that maintains a constant potential difference across its terminals. Inserted into a circuit, a battery can do work on, and transfer energy to, electrons in the wire (at the expense of its own internal chemical energy), which in turn delivers that energy to external circuit elements. In these elements, the energy is converted into other forms, such as mechanical motion (as in electric fans), heat (as in immersion heaters), and light (as in flashlights). Other sources of voltage, such as generators and photocells, will be considered later.

To help better visualize the role of a battery, consider the gravitational analogy in ▶Fig. 17.2. A gasoline-fueled pump (analogous to the battery) does work on the water as it lifts it. The increase in the water's gravitational potential energy comes at the expense of the chemical potential energy of the gasoline molecules. The water then returns to the pump by flowing down the trough (analogous to the wire) into the pond. On the way down, the water does work on the wheel, resulting in rotational kinetic energy, analogous to the electrons transferring energy to a lightbulb.

*As we shall see soon, a *complete circuit* is any complete loop consisting of wires and electrical devices (such as batteries and lightbulbs).

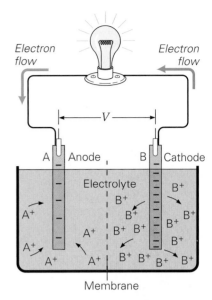

▲ **FIGURE 17.1 Battery action in a chemical battery or cell** Chemical processes involving an electrolyte and two unlike metal electrodes cause ions of both metals to dissolve into the solution at different rates. Thus, one electrode (the cathode) becomes more negatively charged than the other (the anode). The anode is at a higher potential than the cathode. By convention, the anode is designated the positive terminal and the cathode the negative. This potential difference (V) can cause a current, or a flow of charge (electrons), in the wire. The positive ions migrate as shown. (A membrane is necessary to prevent mixing of the two types of ion; why?)

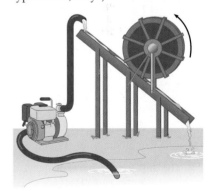

▲ **FIGURE 17.2 Gravitational analogy to a battery and lightbulb** A gasoline-powered pump lifts water from the pond, increasing the potential energy of the water. As the water flows downhill, it transfers energy to (or does work on) a waterwheel, causing the wheel to spin. This action is analogous to the delivery of energy to a lightbulb by an electrical current (for example, as in Fig. 17.1).

▶ **FIGURE 17.3 Electromotive force (emf) and terminal voltage (a)** The emf ($\mathscr{E}$) of a battery is the maximum potential difference across its terminals. This maximum occurs when the battery is not connected to an external circuit. **(b)** Because of internal resistance (r), the terminal voltage V when the battery is in operation is less than the emf $\mathscr{E}$. Here, R is the resistance of the lightbulb.

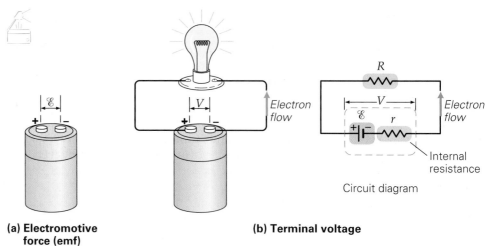

(a) **Electromotive force (emf)**

(b) **Terminal voltage**

Circuit diagram

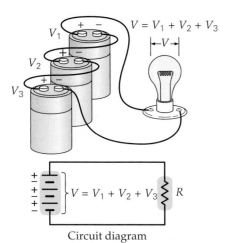

$$V = V_1 + V_2 + V_3$$

$$V = V_1 + V_2 + V_3$$

Circuit diagram

(a) Batteries in series

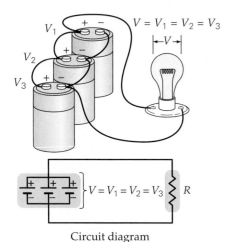

$$V = V_1 = V_2 = V_3$$

$$V = V_1 = V_2 = V_3$$

Circuit diagram

(b) Batteries in parallel (equal voltages)

▲ **FIGURE 17.4 Batteries in series and in parallel (a)** When batteries are connected in series, their voltages add, and the voltage across the resistance R is the sum of the voltages. **(b)** When batteries of the same voltage are connected in parallel, the voltage across the resistance is the same, as if only a single battery were present. In this case, each battery supplies part of the total current.

Battery EMF and Terminal Voltage

The potential difference across the terminals of a battery *when it is not connected* to a circuit is called the battery's **electromotive force (emf)**, symbolized by $\mathscr{E}$. The name is misleading, because emf is *not* a force, but a potential difference, or voltage. To avoid confusion with force, we will call electromotive force just *emf*. Thus a battery's emf represents the work done by the battery *per coulomb* of charge that passes through it. If a battery does 1 joule of work on 1 coulomb of charge, then its emf is 1 joule per coulomb (1 J/C), or 1 volt (1 V).

The emf actually represents the maximum potential difference across the terminals (▲Fig. 17.3a). Under practical circumstances, when a battery is in a circuit and charge flows, the voltage across the terminals is always slightly *less* than the emf. This "operating voltage" (V) of a battery (the battery symbol is the pair of unequal-length parallel lines in Fig. 17.3b) is called its **terminal voltage**. Because batteries in actual operation are of most interest, it is the terminal voltage that is important.

Under many conditions, the emf and terminal voltage are essentially the same. Any difference is due to the battery's *internal resistance* (r), shown explicitly in the circuit diagram in Fig. 17.3b. (Resistance, defined in Section 17.3, is a quantitative measure of the opposition to charge flow.) Internal resistances are typically small, so the terminal voltage of a battery is essentially the same as the emf, that is, $V \approx \mathscr{E}$. However, when a battery supplies a large current or when its internal resistance is high (older batteries), the terminal voltage may drop appreciably below the emf. The reason is that it takes some voltage just to produce a current in the internal resistance itself. Mathematically, the terminal voltage is related to the emf, current, and internal resistance by $V = \mathscr{E} - Ir$ where I is the *electric current* (Section 17.2) in the battery.

For example, most modern cars have a battery "voltage readout." Upon startup, the 12-V battery's voltage typically reads only 10 V (this value is normal). Because of the enormous current required at startup, the Ir term (2 V) reduces the emf by about 2 V to the measured terminal voltage of 10 V. When the engine is running and supplying most of the electric energy to run the car's functions, the current required from the battery is essentially zero and the battery readout rises back to normal voltage levels. Thus, the terminal voltage, and not the emf, is a true indication of the state of the battery. Unless otherwise specified, we will assume negligible internal resistance, so that $V \approx \mathscr{E}$.

There is a wide variety of batteries in use. One of the most common is the 12-V automobile battery, consisting of six 2-V cells connected in *series*.* That is, the positive terminal of each cell is connected to the negative terminal of the next (see the three cells in ◀Fig. 17.4a). When batteries or cells are connected in this fashion, their voltages add. If cells are connected in *parallel*, their positive terminals are connected to each other, as are the negative ones (Fig. 17.4b). When identical batteries are connected this way, the potential difference or terminal

*Chemical energy is converted to electrical energy in a chemical *cell*. The term *battery* generally refers to a collection, or "battery," of cells.

voltage is the same for all of them. However, each supplies only a fraction of the current to the circuit. For example, if you have three batteries with equal voltages connected in parallel, each supplies one third of the total current. A parallel connection of two batteries is the main method for "jump-starting" a car. For such a start, the weak (high r) battery is connected in parallel to a normal (low r) battery, which delivers most of the current to start the car.

Circuit Diagrams and Symbols

To help analyze and visualize circuits, it is common to draw *circuit diagrams* that are schematic representations of the wires, batteries, appliances, and so on. Each element in the circuit is represented by its own symbol in such a diagram. As in Fig. 17.3b and Fig. 17.4, the battery symbol is two parallel lines, the longer representing the positive terminal and the shorter the negative terminal. Any element (such as a lightbulb or appliance) that *opposes* the flow of charge is represented by the symbol —⋀⋀—. (Electrical resistance is defined in Section 17.3; here we merely introduce the symbol.) Connecting wires are unbroken lines and assumed, unless stated otherwise, to have negligible resistance. Where lines cross, it is assumed that they do *not* contact one another, unless they have a dot at their intersection. Lastly, switches are shown as "drawbridges," capable of going up (to open the circuit and stop the current) and down (closed to complete the circuit and allow current). These symbols, along with that of the capacitor (from Chapter 16), are summarized in the accompanying Learn by Drawing. An example of the use of these symbols and circuit diagrams to understand circuits conceptually is shown in the next example.

Conceptual Example 17.1 ■ Asleep at the Switch?

▸Fig. 17.5 shows a circuit diagram that represents two identical batteries (each with a terminal voltage V) connected in parallel to a lightbulb (represented by a resistor). Because it is assumed that the wires have no resistance, we know that before switch S_1 is opened, the voltage across the lightbulb equals V (that is, $V_{AB} = V$). What happens to the voltage across the lightbulb when S_1 is opened? (a) The voltage remains the same (V) as before the switch was opened. (b) The voltage drops to $V/2$, because only one battery is now connected to the bulb. (c) The voltage drops to zero.

Reasoning and Answer. It might be tempting to choose answer (b), because there is now just one battery. But look again. The remaining battery is still connected to the lightbulb. This means that there must be *some* voltage across the lightbulb, so the answer certainly cannot be (c). But it also means that the answer cannot be (b), because the remaining battery itself will maintain a voltage of V across the bulb. Hence, the answer is (a).

Follow-Up Exercise. In this Example, what would the correct answer be if, in addition to opening S_1, switch S_2 were also opened? Explain your answer and reasoning. (*Answers to all Follow-Up Exercises are at the back of the text.*)

17.2 Current and Drift Velocity

OBJECTIVES: To (a) define electric current, (b) distinguish between electron flow and conventional current, and (c) explain the concept of drift velocity and electric energy transmission.

As we have just seen, to sustain an electric current requires a voltage source and a **complete circuit**—the name given to a continuous conducting path. Most practical circuits include a switch to "open" or "close" the circuit. An open switch eliminates the continuous part of the path, thereby stopping the flow of charge in the wires.

Electric Current

Because it is the electrons that move in any circuit's wires, the charge flow is away from the negative terminal of the battery. Historically, however, circuit analysis has been done in terms of **conventional current**. The conventional current's direction is

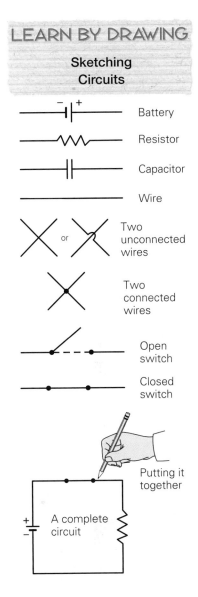

LEARN BY DRAWING

Sketching Circuits

— Battery
— Resistor
— Capacitor
— Wire
Two unconnected wires
Two connected wires
Open switch
Closed switch

Putting it together

A complete circuit

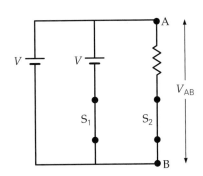

▲ **FIGURE 17.5 What happens to the voltage?** See Example 17.1.

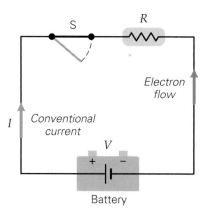

▲ **FIGURE 17.6 Conventional current** For historical reasons, circuit analysis is usually done with conventional current. Conventional current is in the direction in which positive charges would flow, or opposite to the electron flow.

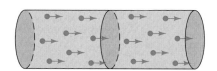

▲ **FIGURE 17.7 Electric current** Electric current (I) in a wire is defined as the rate at which the net charge (q) passes through the wire's cross-sectional area: $I = q/t$. The units of I are amperes (A), or *amps* for short.

that in which positive charges would flow, that is, *opposite* the actual electron flow (◄Fig. 17.6). (Some situations do exist in which a positive charge flow *is* responsible for the current—for example, in semiconductors.)

The battery is said to *deliver* current to a circuit or a component of that circuit (a circuit element). Alternatively, we sometimes say that a circuit (or its components) *draws* current from the battery. The current then returns to the battery. A battery can produce a current in only one direction. One-directional charge flow is called **direct current (dc)**. (Note if the current changes direction and/or magnitude, it is *alternating current*. We will study this type of situation in detail in Chapter 21.)

Quantitatively, the **electric current (I)** is the time rate of flow of net charge. We will be concerned primarily with steady charge flow. In this case, if a net charge q passes through a cross-sectional area in a time interval t (◄Fig. 17.7), the electric current is defined as

$$I = \frac{q}{t} \quad \textit{electric current} \tag{17.1}$$

SI unit of current: coulomb per second (C/s) or ampere (A)

The coulomb per second is designated the **ampere (A)** in honor of the French physicist André Ampère (1775–1836), an early investigator of electrical and magnetic phenomena. In everyday usage, the ampere is commonly shortened to *amp*. Thus, a current of 10 A is read as "ten amps." Small currents are routinely expressed in *milliamperes* (mA, or 10^{-3} A), *microamperes* (μA, or 10^{-6} A), or *nanoamperes* (nA, or 10^{-9} A). These are usually shortened to *milliamps, microamps,* and *nanoamps,* respectively. In a typical household circuit, it is not unusual for the wires to carry several amps of current. To understand the relationship between charge and current, consider the next example.

Example 17.2 ■ Counting Electrons: Current and Charge

Suppose there is a steady current of 0.50 A in a flashlight bulb lasting for 2.0 min. How much charge passes through the bulb during this time? How many electrons does this represent?

Thinking It Through. The current and time elapsed are given; therefore, the definition of current (Eq. 17.1) allows us to find the charge q. Since each electron carries a charge of magnitude 1.6×10^{-19} C, then q can be converted into the number of electrons.

Solution. Listing the data given and converting the time into seconds:

Given: $I = 0.50$ A   *Find:* q (amount of charge)
        $t = 2.0$ min $= 1.2 \times 10^2$ s n (number of electrons)

By Eq. 17.1, $I = q/t$, so the magnitude of the charge is
$$q = It = (0.50 \text{ A})(1.2 \times 10^2 \text{ s}) = (0.50 \text{ C/s})(1.2 \times 10^2 \text{ s}) = 60 \text{ C}$$

Solving for the number of electrons (n), we have

$$n = \frac{q}{e} = \frac{60 \text{ C}}{1.6 \times 10^{-19} \text{ C/electron}} = 3.8 \times 10^{20} \text{ electrons}$$

(It takes a lot of electrons.)

Follow-Up Exercise. Many sensitive laboratory instruments can measure currents in the nanoamp range or smaller. How long, in years, would it take for 1.0 C of charge to flow past a given point in a wire that carries a current of 1.0 nA?

Drift Velocity, Electron Flow, and Electric Energy Transmission

Although we frequently mention charge flow in analogy to water flow, electric charge traveling in a conductor does not flow the same way that water flows in a pipe. In the absence of a potential difference across the ends of a metal wire, the free electrons move randomly at high speeds, colliding many times per second with the metal atoms. As a result, there is no net flow of charge, because equal amounts of charge pass through a given cross-sectional area in opposite directions during a specific time interval.

However, when a potential difference (voltage) *is* applied across the wire (such as by a battery), an electric field appears in the wire in one direction. A flow of electrons then begins *opposite* that direction (why?). This does *not* mean the electrons move directly from one end of the wire to the other. They still move in all directions as they collide with the atoms of the conductor but there is now an *added* component (in one direction) to their velocities (►Fig. 17.8). The result is that their velocities are now, on average, more toward the positive terminal of the battery than away.

This net electron flow is characterized by an average velocity called the **drift velocity**. The drift velocity is much smaller than the random (thermal) velocities of the electrons themselves. Typically the magnitude of the drift velocity is on the order of 1 mm/s. At that speed, it would take an electron about 17 min to travel 1 m along a wire. Yet a lamp comes on almost instantaneously when the switch is closed (completing the circuit), and the electronic signals carrying telephone conversations travel almost instantaneously over miles of wire. How can that be?

Evidently, *something* must be moving faster than the "drifting" electrons. Indeed, this something is the electric field. When a potential difference is applied, the associated electric field in the conductor travels at a speed close to that of light (in the material, roughly 10^8 m/s). Thus the electric field influences the electrons *throughout the conductor* almost instantaneously. This means that the current starts everywhere in the circuit essentially simultaneously. You don't have to wait for electrons to "get there" from a distant place (say, near the switch). Thus in the light bulb, the electrons that are *already* in its filament begin to move almost immediately, delivering energy and creating light with no noticeable delay.

This effect is analogous to toppling a row of standing dominos. When you tip a domino at one end, that *signal* (or energy) is transmitted rapidly down the row. Very quickly, at the other end, the last domino topples (and delivers energy). Note that the domino delivering the signal or energy is *not* the one you pushed. It was the energy, not the dominos, that traveled down the row.

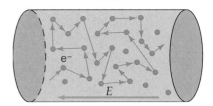

Drift velocity v_d

▲ **FIGURE 17.8 Drift velocity** Because of collisions with the atoms of the conductor, electron motion is random. However, when the conductor is connected, for example, to a battery to form a complete circuit, there is a small net motion in the direction opposite the electric field [toward the high-potential (positive) terminal, or anode]. The speed and direction of this net motion form the drift velocity of the electrons.

17.3 Resistance and Ohm's Law

OBJECTIVES: To (a) define electrical resistance and explain what is meant by an ohmic resistor, (b) summarize the factors that determine resistance, and (c) calculate the effect of these factors in simple situations.

If you place a voltage (potential difference) across the ends of any conducting material, what factors determine the current? As might be expected, usually the greater the voltage, the greater the current. However, another factor also influences current. Just as internal friction (viscosity; see Chapter 9) affects fluid flow in pipes, the resistance of the wire's material will affect the flow of charge. Any object that offers significant resistance to electrical current is called a *resistor* and is represented by the zigzag symbol (Section 17.1). This symbol is used to represent all types of resistors, from the cylindrical color-coded ones on printed circuit boards to electrical devices and appliances such as hair dryers and lightbulbs (►Fig. 17.9).

But how is resistance quantified? We know, for example, that if a large voltage applied across an object produces only a small current, then that object has a high resistance. Thus, the **resistance (R)** of any object should be related to the ratio of the voltage across the object to the resulting current through that object. Resistance is therefore defined as

$$R = \frac{V}{I} \quad \text{electrical resistance} \quad (17.2a)$$

SI unit of resistance: volt per ampere (V/A), or ohm (Ω)

The units of resistance are volts per ampere (V/A), called the **ohm** (Ω) in honor of the German physicist Georg Ohm (1789–1854), who investigated the relationship between current and voltage. Large values of resistance are commonly expressed as kilohms (kΩ) and megohms (MΩ). A schematic circuit diagram

Note: *Resistor* is a generic term for any object that possesses significant electrical resistance.

Teaching tip: Use the units to give the meaning: 1 Ω means that 1 V is required to produce 1 amp; 2 Ω means that 2 V are required to produce 1 amp, and so on.

Note: Remember, *V* stands for ΔV.

▲ **FIGURE 17.9 Resistors in use** A printed circuit board, typically used in computers, includes resistors of different values. The large, striped cylinders are resistors; their four-band color code indicates their resistance in ohms.

Note: *Ohmic* means "having a constant resistance."

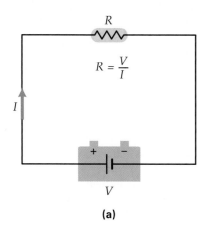

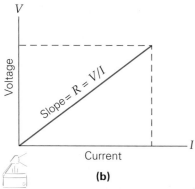

▲ **FIGURE 17.10 Resistance and Ohm's law** (a) In principle, any object's electrical resistance can be determined by dividing the voltage across it by the current through it. (b) If the object obeys Ohm's law (applicable only to a constant resistance), then a plot of voltage versus current is a straight line with a slope equal to R, the element's resistance. (Its resistance does not change with voltage.)

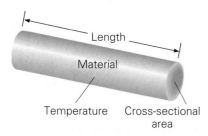

▲ **FIGURE 17.11 Resistance factors** Factors directly affecting the electrical resistance of a cylindrical conductor are the type of material it is made of, its length (L), its cross-sectional area (A), and its temperature (T).

showing how, in principle, resistance is determined is illustrated in ◄Fig. 17.10a. (In Chapter 18, we will study the instruments that are used to measure electrical currents and voltages, called *ammeters* and *voltmeters*, respectively.)

For some materials, the resistance may be constant over a range of voltages. A resistor that exhibits constant resistance is said to obey **Ohm's law**, or to be *ohmic*. The law was named after Ohm, who found materials possessing this property. A plot of voltage versus current for a material with an ohmic resistance gives a straight line with a slope equal to its resistance R (Fig. 17.10b). A common and practical form of Ohm's law is

$$V = IR \quad (Ohm's\ law) \tag{17.2b}$$

(or $I \propto V$, only when R = constant)

Ohm's law is *not* a fundamental law in the same sense as, for example, the law of conservation of energy. There is no "law" that states that materials *must* have constant resistance. Indeed, many of our advances in electronics are based on materials such as semiconductors, which have *nonlinear* (nonohmic) voltage–current relationships.

Unless specified otherwise, we will assume resistor to be ohmic. Always remember, however, that many materials are nonohmic. For instance, the tungsten filaments in lightbulbs have resistance that increases with temperature, being much larger at their operating temperature than at room temperature. The following Example shows how the resistance of the human body can make the difference between life and death.

Example 17.3 ■ Danger in the House: Human Resistance

Any room in the house that is exposed to water and electrical voltage can present hazards. (See the discussion of electrical safety in Section 18.5.) For example, suppose a person steps out of a shower and inadvertently touches an exposed 120-V wire (perhaps a frayed cord on a hair dryer) with a finger. The human body, when wet, can have an electrical resistance as low as 300 Ω. Using this value, estimate the current in that person's body.

Thinking It Through. The wire is at an electrical potential of 120 V above the floor, which is "ground" and taken to be at 0 V. Therefore the voltage (or potential difference) across the body is 120 V. To determine the current, we can use Eq. 17.2, the definition of resistance.

Solution. Listing the data,

Given: $V = 120\ V$ *Find:* I (current in the body)
$R = 300\ \Omega$

From Eq. 17.2, we have

$$I = \frac{V}{R} = \frac{120\ V}{300\ \Omega} = 0.400\ A = 400\ mA$$

While this is a small current by everyday standards, it is a large current for the human body. A current over 10 mA can cause severe muscle contractions, and currents on the order of 100 mA can stop the heart. So this current is potentially deadly. (See Insight 18.2 on Electricity and Personal Safety, Table 1, in Chapter 18 on page 614.)

Follow-Up Exercise. When the human body is dry, its resistance (over its length) can be as high as 100 kΩ. What voltage would be required to produce a current of 1.0 mA (the value that a person can barely feel)?

Factors That Influence Resistance

On the atomic level, resistance arises when electrons collide with the atoms that make up a material. Thus, resistance partially depends on the type of material of which an object is composed. However, geometrical factors also influence resistance. In summary, the resistance of an object of uniform cross-section, such as a length of wire, depends on four properties: (1) the type of material, (2) its length, (3) its cross-sectional area, and (4) its temperature (◄Fig. 17.11).

As you might expect, the resistance of an object (such as a piece of wire) is *inversely* proportional to its cross-sectional area (A), and *directly* proportional to its length (L); that is, $R \propto L/A$. For example, a uniform metal wire 4 m long offers twice as much resistance as a similar wire 2.0 m long, but a wire with a cross-sectional area of 0.50 mm^2 has only half the resistance of one with an area of 0.25 mm^2. These geometrical resistance conditions are analogous to those for liquid flow in a pipe. The longer the pipe, the more is its resistance (drag), but the larger its cross-sectional area, the more liquid it can carry per second. To see an interesting use of the dependence of resistance on length and area by living organisms, read Insight 17.1 on The "Bio-Generation" of High Voltage on this page.

Illustration 30.5 Ohm's "Law"

INSIGHT | 17.1 THE "BIO-GENERATION" OF HIGH VOLTAGE

From the discussion of battery action in Section 17.1, you know that two different metals in acid can generate a constant separation of charge (a voltage) and thus can produce electric current. However, living organisms can also create voltages by a process sometimes called "bio-generation." Electric eels (see Insight 15.2 on page 527), in particular, can generate 600 V, more than enough to kill humans. But how do they accomplish this? As we shall see, the process has similarities both to regular "dry cells" and to nerve signal transmission.

Eels have three organs related to its electrical activities. The Sachs' organ generates low-voltage pulsations for navigation. The others, named the Hunter organ and the Main organ, are sources of high voltage (Fig. 1). In these organs, cells called *electrocytes*, or *electroplates*, are arranged in a stack. Each cell has a flat, disklike shape. The electroplate stack is a series connection similar to that in a car battery, in which there are six cells at 2 V each, producing a total of 12 V. Each electroplate is capable of producing a voltage of only about 0.15 V, but four or five thousand in series can add up to a voltage of 600 V. The electroplates are electrically similar to muscle cells in that they, like muscle cells, receive nerve impulses by synaptic connection. However, these nerve impulses do not cause movement. Instead they trigger voltage generation by the following mechanism.

Each electroplate has the same structure. The top and bottom membranes behave similarly to nerve membranes (see Insight 16.1 on page 552). Under resting conditions, the Na$^+$ ions cannot penetrate the membrane. To equilibrate their concentrations on both sides, the ions reside near the outside surface. This, in turn, attracts the (interior) negatively charged proteins to the interior surface. As a result, the interior is at a potential of 0.08 V lower than the outside. Therefore, under resting conditions, the outside

top (toward the head, or anterior) surface and the outside bottom (posterior) surface of *all* the electroplates are positive (one is shown in Fig. 2a) and exhibit no voltage ($\Delta V_1 = 0$). Thus, under resting conditions a series stack has no voltage $\left(\Delta V_{total} = \sum \Delta V_i = 0\right)$ from top to bottom (Fig. 2b).

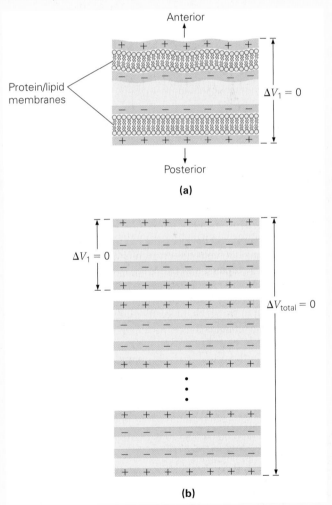

FIGURE 2 (a) A single, resting electroplate One of the thousands of electroplates in the eel's electric organs has, under resting conditions, equal amounts of positive charge at its top and bottom, resulting in no voltage. **(b) Resting electroplates in series** Several thousand electroplates in series under resting conditions have a total voltage of zero.

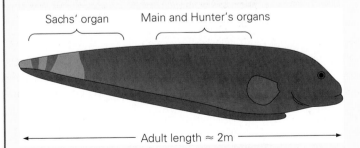

FIGURE 1 Anatomy of an electric eel 80% of an electric eel's body is devoted to voltage generation. Most of that portion contains the two organs (Main and Hunter's) responsible for the high voltage associated with killing of prey. The Sachs' organ produces a lower pulsating voltage used for navigation.

(continues on next page)

However, when an eel locates prey, the eel's brain sends a signal along a neuron *to only the bottom membrane* of each electroplate (one cell is shown in Fig. 3a). A chemical (*acetylcholine*) diffuses across the synapse onto the membrane, briefly opening the ion channels and allowing in Na^+. *For a few milliseconds the lower membrane polarity is reversed*, creating a voltage across one cell of $\Delta V_1 \approx 0.15$ V. The whole stack does this simultaneously, causing a large voltage across the ends of the stack ($\Delta V_{total} \approx 4000 \, \Delta V_1 = 600$ V; see Fig. 3b). When the eel touches the prey with the stack ends, the resulting current pulse through the prey (about 0.5 A) delivers enough energy to kill or at least stun.

An interesting biological "wiring" arrangement enables all electroplates to be triggered simultaneously—a requirement crucial for generation of the maximum voltage. Since each electroplate is a different distance from the brain, the action potential traveling down the neurons must be carefully timed. To do this, the neurons attached to the top of the stack (closest to the brain) are longer and thinner than those attached to the bottom. From what you know about resistance (see, for example, the discussion of Eq. 17.3 and $R \propto L/A$), it should be clear that both a reduction in area and increase in length of the neurons serve to increase neuron resistance compared with those attached to more distant electroplates. Increased resistance means that the action potential travels slower through the closer neurons, thus enabling the closer electroplates to receive their signal at the same time as the more distant ones—a very interesting and practical use of physics (from the eel's perspective, not the prey's).

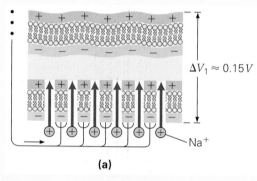

(a)

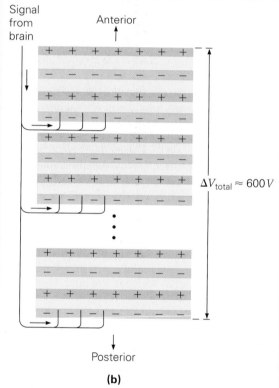

(b)

FIGURE 3 (a) **An electroplate in action** Upon location of prey, a signal is sent from the eel's brain to each electroplate along a neuron attached only to the bottom of the plate. This triggers a brief opening of the ion channel allowing Na^+ ions to the interior, temporarily reversing the polarity of the lower membrane. This creates a temporary electric potential difference (voltage) between the top and bottom membranes. Each electroplate voltage is typically a few tenths of a volt. (b) **A series stack of electroplates in action** When each electroplate in the stack is triggered into action by the lower neuron signal, this results in a large voltage between the top and bottom of the stack, typically on the order of 600 V. This large voltage enables the eel to deliver a pulse of current on the order of a few tenths of an ampere through the prey. The energy deposited in the prey is usually enough to stun or kill it.

Resistivity

Note: Do not confuse resistivity with mass density, which has the same symbol (ρ).

The resistance of an object is partly determined by its material's atomic properties, quantitatively described by that material's **resistivity (ρ)**. The resistance of uniform cross-section object is given by

$$R = \rho\left(\frac{L}{A}\right) \qquad (17.3)$$

SI unit of resistivity: ohm-meter ($\Omega \cdot m$)

The units of resistivity (ρ) are ohm-meters ($\Omega \cdot m$). (You should show this.) Thus, knowing its resistivity (the material type) and using Eq. 17.3, the resistance of any

TABLE 17.1	Resistivities (at 20°C) and Temperature Coefficients of Resistivity for Various Materials*					
	$\rho\ (\Omega \cdot m)$	$\alpha\ (1/C°)$			$\rho\ (\Omega \cdot m)$	$\alpha\ (1/C°)$
Conductors				*Semiconductors*		
Aluminum	2.82×10^{-8}	4.29×10^{-3}		Carbon	3.6×10^{-5}	-5.0×10^{-4}
Copper	1.70×10^{-8}	6.80×10^{-3}		Germanium	4.6×10^{-1}	-5.0×10^{-2}
Iron	10×10^{-8}	6.51×10^{-3}		Silicon	2.5×10^{2}	-7.0×10^{-2}
Mercury	98.4×10^{-8}	0.89×10^{-3}				
Nichrome (alloy of nickel and chromium)	100×10^{-8}	0.40×10^{-3}		*Insulators*		
Nickel	7.8×10^{-8}	6.0×10^{-3}		Glass	10^{12}	
Platinum	10×10^{-8}	3.93×10^{-3}		Rubber	10^{15}	
Silver	1.59×10^{-8}	4.1×10^{-3}		Wood	10^{10}	
Tungsten	5.6×10^{-8}	4.5×10^{-3}				

*Values for semiconductors are general ones, and resistivities for insulators are typical orders of magnitude.

constant-area object can be calculated, as long as its length and cross-sectional area is known.

The values of the resistivities of some conductors, semiconductors, and insulators are given in Table 17.1. The values apply at 20°C, because resistivity can depend on temperature. Most common wires are composed of copper or aluminum with cross-sectional areas on the order of 10^{-6} m^2 or 1 mm^2. For a length of 1.5 m, you should be able to show that the resistance of a copper wire with this area is about 0.025 Ω (= 25 mΩ). This explains why wire resistances are neglected in circuits—their values are much less than most household devices.

An interesting and potentially important medical application involves the measurement of human body resistance and its relationship to body fat. (See Insight 17.2 on Bioelectrical Impedance Analysis on page 578.) To get a feeling for the magnitudes of these quantities in living tissue, consider the following Example.

Example 17.4 ■ Electric Eels: Cooking with Bio-Electricity?

Suppose an electric eel touches the head and tail of a cylindrically shaped fish and applies a voltage of 600 V across it. (See Insight 17.1 on page 575.) If a current of 0.80 A results (likely killing the prey), estimate the average resistivity of the fish's flesh, assuming it is 20 cm long and 4.0 cm in diameter.

Thinking It Through. With cylindrical geometry, we know its length and can find its cross-sectional area from the given dimensions. From the voltage and current, its resistance can be determined. Lastly, its resistivity can be estimated using Eq. 17.3.

Solution. Listing the data:

Given: $L = 20$ cm $= 0.20$ m   *Find:* f (resistivity)
$\quad\quad\quad d = 4.0$ cm $= 4.0 \times 10^{-2}$ m
$\quad\quad\quad V = 600$ V
$\quad\quad\quad I = 0.80$ A

The cross-sectional area of the fish is

$$A = \pi r^2 = \pi \left(\frac{d}{2}\right)^2 = \frac{\pi(2.0 \times 10^{-2}\ \text{m})^2}{4} = 3.1 \times 10^{-4}\ \text{m}^2$$

(continues on next page)

We also know that the fish's overall resistance is $R = \dfrac{V}{I} = \dfrac{600 \text{ V}}{0.80 \text{ A}} = 7.5 \times 10^2 \ \Omega$. From Eq. 17.3, we have

$$\rho = \frac{RA}{L} = \frac{(7.5 \times 10^2 \ \Omega)(3.1 \times 10^{-4} \text{ m}^2)}{0.20 \text{ m}} = 1.2 \ \Omega \cdot \text{m} \text{ or about } 120 \ \Omega \cdot \text{cm}$$

Comparing this to the values in Table 17.1, you can see that, as expected, the fish's flesh is much more resistive than metals, but certainly not a great insulator. The value is on the order of the resistivities measured for different human tissues; for example, cardiac muscle has a resistivity of about 175 $\Omega \cdot$ cm and liver on the order of 200 $\Omega \cdot$ cm. Clearly our answer is an average over the whole fish and tells us nothing about different regions of the fish.

Follow-Up Exercise. Suppose for its next meal, the eel in this Example chooses a different species of fish. The next fish has twice the average resistivity, half the length, and half the diameter of the first fish. What current would be expected in this fish if the eel applied 400 V across its body?

For many materials, especially metals, the temperature dependence of resistivity is nearly linear if the temperature change is not too great. That is, the resistivity at a temperature T after a temperature change $\Delta T = T - T_o$ is given by

$$\rho = \rho_o(1 + \alpha \Delta T) \qquad \begin{array}{l}\textit{temperature variation} \\ \textit{of resistivity}\end{array} \qquad (17.4)$$

where α is a constant (usually only over a certain temperature range) called the **temperature coefficient of resistivity** and ρ_o is a reference resistivity at T_o (usually 20°C). Equation 17.4 can be rewritten as

$$\Delta \rho = \rho_o \alpha \Delta T \qquad (17.5)$$

Note: Compare the form of Eq. 17.5 with Eq. 10.10 for the linear expansion of a solid.

where $\Delta \rho = \rho - \rho_o$ is the change in resistivity that occurs when the temperature changes by ΔT. The ratio $\Delta \rho / \rho_o$ is dimensionless, so α has units of inverse Celsius degrees, written as $1/\text{C}°$. Physically, α represents the fractional change in resistivity $(\Delta \rho / \rho_o)$ per Celsius degree. The temperature coefficients of resistivity for some materials are listed in Table 17.1. These coefficients are assumed to be constant over normal temperature ranges. Notice that for semiconductors and insulators the coefficients are generally orders of magnitude and are usually not constant.

INSIGHT 17.2 BIOELECTRICAL IMPEDANCE ANALYSIS (BIA)

Traditional methods for estimating body-fat percentages involve the use of buoyancy tanks (for density measurements, see Chapter 9) or calipers to pinch the flesh. However, in recent years, electrical resistance experiments have been designed to measure the body fat of the human body.* In theory, these measurements—termed *bioelectrical impedance analysis* (BIA)—have the potential to determine, with more accuracy than traditional methods, a patient's total water content, fat-free mass, and body fat (*adipose tissue*).

The principle of BIA is based on the water content of the human body. Water in the human body is a relatively good conductor of electric current, due to the presence of ions such as potassium (K^+) and sodium (Na^+). Because muscle tissue holds more water per kilogram than fat holds, it is a better conductor than fat. Thus for a given voltage, the difference in currents should be a good indicator of the fat-to-muscle percentage.

In practice, one electrode of a low-voltage power supply is connected to the wrist and the other to the opposite ankle during a BIA test. The current is kept below 1 mA for safety, with typical currents being about 800 μA. The subject cannot feel this small current. Typical resistance values are about 250 Ω. From Ohm's law, the required voltage is $V = IR = (8 \times 10^{-4} \text{ A})(250 \ \Omega) = 0.200 \text{ V}$, or about 200 mV. In actuality, the voltage alternates in polarity at a frequency of 50 kHz, because this frequency is known *not* to trigger electrically excitable tissues, such as nerves and cardiac muscle.

From what has been presented in this chapter (for example, Eq. 17.3), you should be able to understand some of the factors involved in interpreting the results of human resistance measurements. The measured resistance is the total resistance. However, the current travels not through a uniform material but rather through the arm, trunk, and leg. Not only does each of these body parts have a different fat-to-muscle ratio, which affects resistivity (ρ), but also they differ widely in length (L) and cross-sectional area (A). The arm and the leg, usually dominated by muscle and with a small cross-sectional area, offer the most resistance. The trunk, which usually contains a relatively high percentage of fat and has a large cross-sectional area, has low resistance.

By subjecting BIAs to statistical analysis, researchers hope to understand how the wide range of physical and genetic parameters present in humans affects resistance measurements. Among these parameters are height, weight, body type, and ethnicity. Once the correlations are understood, BIA may well become a valuable medical tool in routine physicals and in the diagnosis of various diseases.

*Technically, this technique measures the body's *impedance*, which includes effects of capacitance and magnetic effects as well as resistance. (See Chapter 21.) However, these contributions are about 10% of the total. Hence, the word *resistance* is used here.

Resistance is directly proportional to resistivity (Eq. 17.3). This means that an object's resistance has the same dependence on temperature as its resistivity (Eqs. 17.3 and 17.4). The resistance of an object of uniform cross section varies with temperature:

$$R = R_o(1 + \alpha\Delta T) \quad \text{or} \quad \Delta R = R_o\alpha\Delta T \qquad \text{\textit{temperature variation of resistance}} \qquad (17.6)$$

Here, $\Delta R = R - R_o$, the change in resistance relative to its reference value R_o, usually taken to be at 20°C. The variation of resistance with temperature provides a means of measuring temperature in the form of an *electrical resistance thermometer*, as illustrated in the next Example.

Example 17.5 ■ An Electrical Thermometer: Variation of Resistance with Temperature

A platinum wire has a resistance of 0.50 Ω at 0°C. It is placed in a water bath, where its resistance rises to a final value of 0.60 Ω. What is the temperature of the bath?

Thinking It Through. From the temperature coefficient of resistivity for platinum from Table 17.1, ΔT can be found from Eq. 17.6 and added to 0°C, the initial temperature, to find the temperature of the bath.

Solution.

Given: $T_o = 0°C$ *Find:* T (temperature of the bath)
       $R_o = 0.50 \ \Omega$
       $R = 0.60 \ \Omega$
       $\alpha = 3.93 \times 10^{-3}/C° \ \text{(Table 17.1)}$

The ratio $\Delta R/R_o$ is the fractional change in the initial resistance R_o (at 0°C). We solve Eq. 17.6 for ΔT, using the given values:

$$\Delta T = \frac{\Delta R}{\alpha R_o} = \frac{R - R_o}{\alpha R_o} = \frac{0.60 \ \Omega - 0.50 \ \Omega}{(3.93 \times 10^{-3}/C°)(0.50 \ \Omega)} = 51 \ C°$$

Thus, the bath is at $T = T_o + \Delta T = 0°C + 51 \ C° = 51°C$.

Follow-Up Exercise. In this Example, if the material had been copper with $R_o = 0.50 \ \Omega$, rather than platinum, what would its resistance be at 51°C? From this, you should be able to explain which material makes the more "sensitive" thermometer, one with a high temperature coefficient of resistivity or one with a low value.

Superconductivity

Carbon and other semiconductors have negative temperature coefficients of resistivity. However, many materials, including most metals, have positive temperature coefficients, which means that their resistances increase as temperature increases. You might wonder how far electrical resistance can be reduced by lowering the temperature. In certain cases, the resistance can reach zero—not just close to zero, but, as accurately as can be measured, *exactly* zero. This phenomenon is called **superconductivity** (first discovered in 1911 by Heike Kamerlingh Onnes, a Dutch physicist). Currently the required temperatures are about 100 K or below. Thus its current usage is restricted to high-tech laboratory apparatus and research equipment.

However, it does have the potential for important new everyday applications, especially if materials can be found whose superconducting temperature is near room temperature. Among the applications are superconducting magnets (already in use in labs and small-scale naval propulsion units). In the absence of resistance, high currents and very high magnetic fields are possible (Chapter 19). Used in motors or engines, superconducting electromagnets would be more efficient, providing more power for the same energy input. Superconductors might also be used as electrical transmission lines with no resistive losses. Some envision superfast superconducting computer memories. The absence of electrical resistance opens almost endless possibilities. You're likely to hear more about superconductor applications in the future as new materials are developed.

17.4 Electric Power

OBJECTIVES: To (a) define electric power, (b) calculate the power delivery of simple electric circuits, and (c) explain joule heating and its significance.

When a sustained current exists in a circuit, the electrons are given energy by the voltage source, such as a battery. As these charge carriers pass through circuit components, they collide with the atoms of the material (that is, they encounter resistance) and lose energy. The energy transferred in the collisions can result in an increase in the temperature of the components. In this way, electrical energy can be transformed, at least partially, into thermal energy.

However, electric energy can also be converted into other forms of energy such as light (as in lightbulbs) and mechanical motion (as in electric drills). According to conservation of energy, whatever forms the energy may take, the *total* energy delivered to the charge carriers by the battery must be *completely* transferred to the circuit elements (neglecting losses in the wires). That is, on return to the voltage source or battery, a charge carrier loses all the electric potential energy it gained from that source, and is ready to repeat the process.

The energy gained by an amount of charge q from a voltage source (voltage V) is qV [by units, $C(J/C) = J$]. Over a time interval t, the *rate* at which energy is delivered may not be constant. Thus the average rate of energy delivery, called the average **electric power, $\overline{P}$,** is given by

$$\overline{P} = \frac{W}{t} = \frac{qV}{t}$$

In the special case when the current and voltage are steady with time (as with a battery), then the average power is the same as the power at all times. For steady (dc) currents, $I = q/t$ (Eq. 17.1). Thus, we can rewrite the preceding equation as:

$$P = IV \quad \textit{electric power} \tag{17.7a}$$

Recall from Chapter 5 that the SI unit of power is the watt (W). The ampere (the unit of current I) times the volt (the unit of voltage V) gives the joule per second (J/s), or watt. (You should check this.)

A visual mechanical analogy to help explain Eq. 17.7a is given in ▼Fig. 17.12. The figure depicts a simple electric circuit as a system for transferring energy, in analogy to a conveyor belt delivery system.

Because $R = V/I$, power can be written in three equivalent forms:

$$P = IV = \frac{V^2}{R} = I^2R \quad \textit{electric power} \tag{17.7b}$$

Joule Heat

The thermal energy expended in a current-carrying resistor is referred to as **joule heat,** or **I^2R losses** (pronounced "I squared R" losses). In many instances (such as in electrical transmission lines), joule heating is an undesirable side effect. However, in

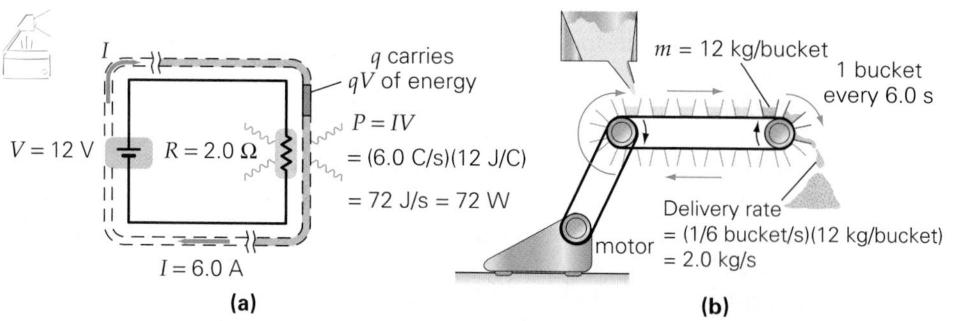

Teaching tip: Once again, the units tell the tale: $(C/s)(J/C) = J/s = W$.

▶ **FIGURE 17.12 Electric power analogy** Electric circuits can be thought of as energy delivery systems much like a conveyor belt. **(a)** Imagine the current as being made up of consecutive segments of charge $q = 1.0$ C, each carrying $qV = 12$ J of energy supplied by the battery. The current is $I = V/R = 6.0$ A, or 6.0 C/s. Then the power (or energy delivery rate) to the resistor is $(6.0$ C/s$)(12$ J/C$) = 72$ J/s $= 72$ W. **(b)** The conveyor belt comprises a series of buckets, each carrying 12 kg of sand (analogous to the energy carried by each charge q), arriving at a rate of one bucket every 6.0 s (analogous to current I). The delivery rate in kg/s is analogous to the power in J/s in part (a).

(a)
$V = 12$ V
$R = 2.0\ \Omega$
$I = 6.0$ A

q carries qV of energy

$P = IV$
$= (6.0$ C/s$)(12$ J/C$)$
$= 72$ J/s $= 72$ W

$m = 12$ kg/bucket
1 bucket every 6.0 s

motor

Delivery rate
$= (1/6$ bucket/s$)(12$ kg/bucket$)$
$= 2.0$ kg/s

(b)

other situations, the conversion of electrical energy to thermal energy is the main purpose. Heating applications include the heating elements (burners) of electric stoves, hair dryers, immersion heaters, and toasters.

Electric lightbulbs are rated in watts (power)—for example, 60 W (►Fig. 17.13a). Incandescent lamps are relatively inefficient as light sources. Typically, less than 5% of the electrical energy is converted to visible light; most of the energy produced is invisible infrared radiation and heat.

Electrical appliances are tagged or stamped with their power ratings. Either the voltage and power requirements or the voltage and current requirements are given (Fig. 17.13b). In either case, the current, power, and effective resistance can be calculated. Typical power requirements of some household appliances are given in Table 17.2. Even though most common appliances specify a nominal operating voltage of 120 V, it should be noted that household voltage can vary from 110 V to 120 V and still be considered in the "normal" range.

(a)

Integrated Example 17.6 ■ A Modern Appliance Dilemma: Computing or Eating

(a) Consider two appliances that operate at the same voltage. Appliance A has a higher power rating than appliance B. (a) How does the resistance of A compare with that of B: (1) larger, (2) smaller, or (3) the same? (b) A computer system includes a color monitor with a power requirement of 200 W, whereas a countertop broiler/toaster oven is rated at 1500 W. What is the resistance of each if both are designed to run at 120 V?

(a) Conceptual Reasoning. Power depends on current and voltage. Because the two appliances operate at the same voltage, they can't carry the same current and yet have different power requirements. Therefore, answer (3) cannot be correct. Because both appliances operate at the same voltage, the one with higher power (A) must carry more current. For A to carry more current at the same voltage as B, it must have less resistance than B. Therefore, the correct answer is (2); A has less resistance than B.

(b) Quantitative Reasoning and Solution. The definition of resistance is $R = V/I$ (Eq. 17.2). To use this definition, we need the current, which can be determined from Eq. 17.7 ($P = IV$). This will be done twice, once for the monitor and then for the broiler/toaster. We list the data, using the subscript m for monitor and b for broiler/toaster:

Given: $P_m = 200$ W   *Find:* R (resistance of each appliance)
    $P_b = 1500$ W
    $V = 120$ V

The monitor current is (using Eq. 17.7)

$$I_m = \frac{P_m}{V} = \frac{200\text{ W}}{120\text{ V}} = 1.67\text{ A}$$

and that in the broiler/toaster is

$$I_b = \frac{P_b}{V} = \frac{1500\text{ W}}{120\text{ V}} = 12.5\text{ A}$$

Thus their resistances are

$$R_m = \frac{V}{I_m} = \frac{120\text{ V}}{1.67\text{ A}} = 71.9\ \Omega$$

and

$$R_b = \frac{V}{I_b} = \frac{120\text{ V}}{12.5\text{ A}} = 9.60\ \Omega$$

Because they all operate at the same voltage, an appliance's power output is controlled by its resistance. The appliance's resistance is *inversely* related to the appliance's power requirement.

Follow-Up Exercise. An immersion heater is a common "appliance" in most college dorms, useful for heating water for tea, coffee, or soup. Assuming 100% of the heat goes into the water, what must be the heater's resistance (operating at 120 V) to heat a cup of water (mass 250 g) from room temperature (20°C) to boiling in 3.00 minutes?

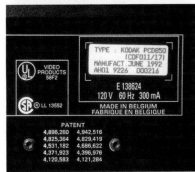

(b)

▲ **FIGURE 17.13** Power ratings (a) Lightbulbs are rated in watts. Operated at 120 V, this 60-W bulb uses 60 J of energy each second. (b) Appliance ratings list either voltage and power or voltage and current. From either, the current, power, and effective resistance can be found. Here, one appliance is rated at 120 V and 18 W and the other at 120 V and 300 mA. Can you compute the current and resistance for the former and the power required and resistance for the latter?

TABLE 17.2	Typical Power and Current Requirements for Various Household Appliances (120 V)				
Appliance	Power	Current	Appliance	Power	Current
Air conditioner, room	1500 W	12.5 A	Heater, portable	1500 W	12.5 A
Air-conditioning, central	5000 W	41.7 A*	Microwave oven	900 W	5.2 A
Blender	800 W	6.7 A	Radio–cassette player	14 W	0.12 A
Clothes dryer	6000 W	50 A*	Refrigerator, frost-free	500 W	4.2 A
Clothes washer	840 W	7.0 A	Stove, top burners	6000 W	50.0 A*
Coffeemaker	1625 W	13.5 A	Stove, oven	4500 W	37.5 A*
Dishwasher	1200 W	10.0 A	Television, color	100 W	0.83 A
Electric blanket	180 W	1.5 A	Toaster	950 W	7.9 A
Hair dryer	1200 W	10.0 A	Water heater	4500 W	37.5 A*

*A high-power appliance such as this is typically wired to a 240-V house supply to reduce the current to half these values (Section 18.5).

Example 17.7 ■ A Potentially Dangerous Repair: Don't Do It Yourself!

A hair dryer is rated at 1200 W for a 115-V operating voltage. The uniform wire filament breaks near one end, and the owner repairs it by removing the section near the break and simply reconnecting it. The filament is then 10.0% shorter than its original length. What will be the heater's power output after this "repair"?

Thinking It Through. The wire always operates at 115 V. Thus shortening the wire, which decreases its resistance, will result in a larger current. With this increase in current, one would expect the power output to go up.

Solution. Let's use a subscript 1 to indicate "before breakage" and a subscript 2 to mean "after the repair." Listing the given values,

Given: $P_1 = 1200$ W   *Find:* P_2 (power output after the repair)
$V_1 = V_2 = 115$ V
$L_2 = 0.900L_1$

After the repair, the wire has 90.0% of its original resistance, because (see Eq. 17.3) a wire's resistance is directly proportional to its length. To show the reduction to 90% explicitly, let's express the resistance after repair ($R_2 = \rho L_2/A$) in terms of the original resistance ($R_1 = \rho L_1/A$):

$$R_2 = \rho \frac{L_2}{A} = \rho \frac{0.900L_1}{A} = 0.900 \left(\rho \frac{L_1}{A} \right) = 0.900R_1$$

as expected.

The current will increase, because the voltage is the same ($V_2 = V_1$). This requirement can be written as $V_2 = I_2 R_2 = V_1 = I_1 R_1$. So the new current in terms of the original current is

$$I_2 = \left(\frac{R_1}{R_2} \right) I_1 = \left(\frac{R_1}{0.900R_1} \right) I_1 = (1.11)I_1$$

which means that the current after the repair is about 11% larger than before.

The original power is $P_1 = I_1 V = 1200$ W. The power after the repair is $P_2 = I_2 V$ (note that the voltages have no subscripts because they remained the same and will cancel out). Forming a ratio gives

$$\frac{P_2}{P_1} = \frac{I_2 V}{I_1 V} = \frac{I_2}{I_1} = 1.11$$

which can be solved for P_2:

$$P_2 = 1.11P_1 = 1.11(1200 \text{ W}) = 1.33 \times 10^3 \text{ W}$$

The power output of the dryer has been increased by about 120 W. *Do not perform such a repair job!*

Follow-Up Exercise. In this Example, determine the (a) initial and final resistances and (b) initial and final currents.

We often complain about our electric bills, but what do we actually pay for? What is bought is electric *energy* measured in units of the **kilowatt-hour (kWh)**. Power is the rate at which work is done ($P = W/t$ or $W = Pt$), so work has units of watt-seconds (power × time). Converting this unit to the larger unit of kilowatt-hours (kWh), it can be seen that the kilowatt-hour is a unit of work (or energy), equivalent to 3.6 million joules, because

$$1 \text{ kWh} = (1000 \text{ W})(3600 \text{ s}) = (1000 \text{ J/s})(3600 \text{ s}) = 3.6 \times 10^6 \text{ J}$$

Thus, we pay the "power" company for electrical energy that we use to do work with our appliances.

The cost of electric energy varies with location. In the United States, it ranges from a low of several cents (per kilowatt-hour) to several times that value. Recently, electric energy rates have been deregulated. Coupled with increasing demand (without a corresponding increase in supply), deregulation has given rise to skyrocketing rates in some areas of the country. Do you know the price of electricity in your locality? Check an electric bill to find out, especially if you are in one of the areas affected by dramatically rising rates. Let's look at the cost of electricity to run a typical appliance in the next Example.

Example 17.8 ■ Electric Energy Cost: The Price of Coolness

If the motor of a frost-free refrigerator runs 15% of the time, how much does it cost to operate per month (to the nearest cent) if the power company charges 11¢ per kilowatt-hour? (Assume 30 days in a month.)

Thinking It Through. From the power and the time the motor is on per day, we can compute the electrical energy the refrigerator requires *daily* and project that to a 30-day month.

Solution. Practical energy amounts are usually written in kilowatt-hours because the joule is a relatively tiny unit. Listing the data,

Given: $P = 500$ W (Table 17.2)   *Find:*   operating cost per month
        Cost = $0.11/kWh

The refrigerator motor operates 15% of the time, so in one day it runs $t = (0.15)(24 \text{ h}) = 3.60$ h. Because $P = W/t$, the electrical energy required *per day* is

$$W = Pt = (500 \text{ W})(3.60 \text{ h/day}) = 1.80 \times 10^3 \text{ Wh} = 1.80 \text{ kWh/day}$$

So the cost per day is

$$\left(\frac{1.80 \text{ kWh}}{\text{day}}\right)\left(\frac{\$0.11}{\text{kWh}}\right) = \frac{\$0.20}{\text{day}}$$

or 20¢ per day. For a 30-day month, the cost would be

$$\left(\frac{\$0.20}{\text{day}}\right)\left(\frac{30 \text{ day}}{\text{month}}\right) \approx \$6 \text{ per month}$$

Follow-Up Exercise. How long would you have to leave a 60-W lightbulb on to use the same amount of electrical energy that the refrigerator motor in this Example uses each hour it is on?

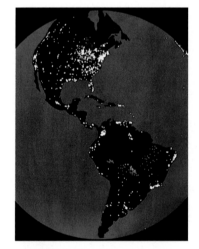

▲ **FIGURE 17.14 All lit up** A satellite image of the Americas at night. Can you identify the major population centers in the United States and elsewhere? The red spots across part of South America indicate large-scale burning of vegetation. The small yellow spot in Central America shows burning gas flares at oil production sites. At the top right edge, you can just glimpse a few of the white city lights of Europe. This image was recorded by a visible–infrared system.

Electrical Efficiency and Natural Resources

About 25% of the electric energy generated in the United States goes to lighting (▶Fig. 17.14). This percentage is roughly equivalent to the output of 100 electric-generating (power) plants. Refrigerators consume about 7% of the electric energy produced in the United States (the output of about twenty-eight such plants).

This huge (and growing) consumption of electric energy has prompted the federal government and many state governments to set minimum efficiency limits for such appliances as refrigerators, freezers, air conditioners, and water heaters

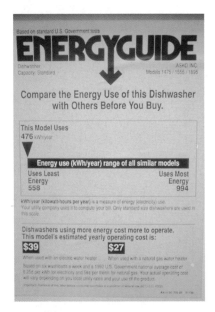

▲ FIGURE 17.15 Energy guide
Consumers are made aware of the efficiencies of appliances in terms of the average yearly cost of their operation. Sometimes the yearly cost is given for different kilowatt-hour (kWh) rates, which vary around the United States.

(◄Fig. 17.15). Also, efficient fluorescent lighting has been developed and put into more common use. The most efficient fluorescent lamp now in use consumes about 25–30% less energy than the average fluorescent lamp and roughly 75% less energy than incandescent lamps with an equivalent light output.

The result of all these measures has been significant energy savings as new, more efficient appliances gradually replace inefficient models. Energy saved translates directly into savings of fuels and other natural resources, as well as a reduction in environmental hazards such as chemical pollution and global warming. To see what kind of results can be achieved by applying just one energy efficiency standard, consider the next Example.

Example 17.9 ■ What We Can Save: Increasing Electrical Efficiency

Many modern power plants produce electric energy at a rate of about 1.0 GW (gigawatt electric power output). Estimate how many fewer such power plants California would need if all its households switched from the 500-W refrigerators of Example 17.8 to more-efficient 400-W refrigerators. (Assume that there are about 10 million homes in California with an average of 1.2 refrigerators operating per home.)

Thinking It Through. The results from Example 17.8 can be used to calculate the overall effect.

Solution.

Given: Plant rate = 1.0 GW = 1.0×10^6 kW   *Find:*   how many fewer power plants
Energy requirement,   are required by switching to
    500-watt model = 1.80 kWh/day   more efficient refrigerators
    (Example 17.8)
Number of homes = 10×10^6
Number of refrigerators per home = 1.2

For the entire state, the energy usage per day with the less-efficient refrigerators is

$$\left(\frac{1.80 \text{ kWh/day}}{\text{refrig}}\right)(10 \times 10^6 \text{ homes})\left(\frac{1.2 \text{ refrigerators}}{\text{home}}\right) = 2.2 \times 10^7 \frac{\text{kWh}}{\text{day}}$$

The more-efficient refrigerators would use only 80% (400 W/500 W = 0.80) of this, or 1.7×10^7 kWh/day. The difference, 5.0×10^6 kWh/day, is the rate at which electric energy is saved. One 1.0-GW power plant produces

$$\left(1.0 \times 10^6 \frac{\text{kW}}{\text{plant}}\right)\left(24\frac{\text{h}}{\text{day}}\right) = 2.4 \times 10^7 \frac{(\text{kWh/plant})}{\text{day}}$$

So the replacement refrigerators would save about

$$\frac{5.0 \times 10^6 \text{ kWh/day}}{2.4 \times 10^7 \text{ kWh/(plant-day)}} = 0.21 \text{ plant}$$

or about 20% of the output of a typical plant. Note that this saving results from a change in a *single* appliance. Imagine what could be done if all appliances, including lighting, were made more efficient. Developing and using more efficient electrical appliances is one way to avoid having to build new electric energy generating plants.

Follow-Up Exercise. Electric and gas water heaters are often said to be equally efficient—typically, about 95%. In reality, while gas water heaters are capable of 95% efficiency, it might be more accurate to describe electric water heaters as only about 30% efficient, even though approximately 95% of the *electrical* energy they require is transferred to the water in the form of heat. Explain. [*Hint*: What is the source of energy for an electric water heater? Compare this to the energy delivery of natural gas. Recall the discussion of electrical generation in Section 12.4 and Carnot efficiency in Section 12.5.]

Chapter Review

- A **battery** produces an **electromotive force (emf)**, or voltage, across its terminals. The high-voltage terminal is the **anode,** and the low-voltage one is the **cathode**.

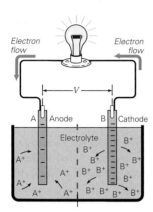

- **Electromotive force (emf $\mathscr{E}$)** is measured in volts and represents the number of joules of energy that a battery (power supply) gives to 1 coulomb of charge passing through it; that is, $1\,\text{J/C} = 1\,\text{V}$.

- **Electric current (I)** is the rate of charge flow. Its direction is that of **conventional current**, which is the direction in which positive charge actually flows or appears to flow. In metals, because the charge flow is electrons, the current direction is opposite the electron flow direction. Current is measured in **amperes** ($1\,\text{A} = 1\,\text{C/s}$) and defined as

$$I = \frac{q}{t} \qquad (17.1)$$

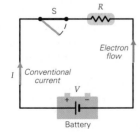

- For an electric current to exist in a circuit, it must be a **complete circuit**—that is, a circuit (set of circuit elements and wires) that connects both terminals of a battery or power supply with no break.

- The electrical **resistance (R)** of an object is the voltage across the object divided by the current in it, or

$$R = \frac{V}{I} \qquad \text{or} \qquad V = IR \qquad (17.2)$$

The units of resistance are the **ohm** or the volt per ampere.

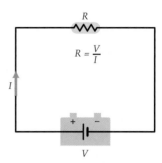

- A circuit element obeys **Ohm's law** if it exhibits constant electrical resistance. Ohm's law is commonly written $V = IR$, where R is constant.

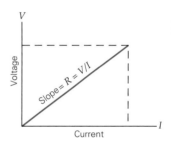

- The resistance of an object depends on the **resistivity (ρ)** of the material (based on its atomic properties), the cross-sectional area A, and the length L. For objects of uniform cross section,

$$R = \rho\left(\frac{L}{A}\right) \qquad (17.3)$$

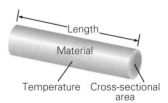

- **Electric power** (P) is the rate at which work is done by a battery (power supply), or the rate at which energy is transferred to a circuit element. The power delivered to a circuit element depends on the element's resistance, the current in it, and the voltage across it. Electrical power can be written in three equivalent ways:

$$P = IV = \frac{V^2}{R} = I^2 R \qquad (17.7b)$$

Exercises

MC = *Multiple Choice Question*, **CQ** = *Conceptual Question, and* **IE** = *Integrated Exercise. Throughout the text, many exercise sections will include "paired" exercises. These exercise pairs, identified with* **red numbers**, *are intended to assist you in problem solving and learning. In a pair, the first exercise (even numbered) is worked out in the Study Guide so that you can consult it should you need assistance in solving it. The second exercise (odd numbered) is similar in nature, and its answer is given at the back of the book.*

In this chapter, assume that all batteries have negligible internal resistance unless otherwise indicated.

17.1 Batteries and Direct Current

1. **MC** When a battery is part of a complete circuit, the voltage across its terminals is its (a) emf, (b) terminal voltage, (c) power output, (d) all of the preceding. (b)

2. **MC** As a battery ages, its (a) emf increases, (b) emf decreases, (c) terminal voltage increases, (d) terminal voltage decreases. (d)

3. **MC** When four 1.5-V batteries are connected, the output voltage of the combination is 1.5 V. These batteries are connected in (a) series, (b) parallel, (c) a pair in series connected in parallel to the other pair in series, (d) none of the preceding. (b)

4. **MC** When helping someone whose car has a "dead" battery, how should your car's battery be connected in relation to the "dead" battery: (a) series, (b) parallel, or (c) either series or parallel would work fine? (b)

5. **MC** When several 1.5-V batteries are connected in series, the overall output voltage of the combination is measured to be 12 V. How many batteries are needed to achieve this voltage: (a) two, (b) ten, (c) eight, or (d) six? (c)

6. **CQ** Why does the battery design shown in Fig. 17.1 require a chemical membrane? to prevent the two ions from each electrode from mixing

7. **CQ** A battery has its voltage measured while sitting on the workbench and the technician reads its manufacturer's rating, which is 12 V. Does this mean it will perform as expected when placed in a complete circuit? Explain. no; terminal voltage is lower than emf

8. **CQ** Sketch the following *complete* circuits, using the symbols show in the Learn by Drawing on page 571. (a) Two ideal 6.0-V batteries in series wired to a capacitor followed by a resistor. (b) Two ideal 12.0-V batteries in parallel, connected as a unit to two identical resistors in series with one another. (c) A nonideal battery (one with internal resistance) wired to two identical capacitors that are in parallel with each other, followed by two resistors in series with one another. see ISM

9. ● (a) Three 1.5-V dry cells are connected in series. What is the total voltage of the combination? (b) What would be the total voltage if the cells were connected in parallel? (a) 4.5 V (b) 1.5 V

10. ● What is the voltage across six 1.5-V batteries when they are connected (a) in series and (b) in parallel? (a) 9.0 V (b) 1.5 V

11. ●● Two 6.0-V batteries and one 12-V battery are connected in series. (a) What is the voltage across the whole arrangement? (b) What arrangement of these three batteries would give a total voltage of 12 V? (a) 24 V (b) two 6.0-V in series, together in parallel with the 12-V

12. ●● Given three batteries with voltages of 1.0 V, 3.0 V, and 12 V, respectively, how many different voltages could be obtained by connecting one or more of the batteries in series or parallel, and what are these voltages? seven different voltages; see ISM

13. **IE** ●● You are given four AA batteries that are rated at 1.5 V each. The batteries are grouped in pairs. In arrangement A, the two batteries in each pair are in series, and then the pairs are connected in parallel. In arrangement B, the two batteries in each pair are in parallel, and then the pairs are connected in series. (a) Compared with arrangement B, will arrangement A have (1) a higher, (2) the same, or (3) a lower total voltage? (b) What are the total voltages for each arrangement? (a) (2) the same (b) 3.0 V, 3.0 V

17.2 Current and Drift Velocity

14. **MC** In which of these situations does more charge flow past a given point on a wire: when the wire has (a) a current of 2.0 A for 1.0 min, (b) 4.0 A for 0.5 min, (c) 1.0 A for 2.0 min, or (d) all are the same charge? (d)

15. **MC** Which of these situations involves the least current: a wire that has (a) 1.5 C passing a given point in 1.5 min, (b) 3.0 C passing a given point in 1.0 min, or (c) 0.5 C passing a given point in 0.10 min? (a)

16. **MC** In a dental X-ray machine, the movement of accelerated electrons is east. The current associated with these electrons is in what direction: (a) east, (b) west, or (c) zero? (b)

17. **CQ** In the circuit shown in Fig. 17.4a, what is the direction of (a) the electron flow in the resistor, (b) the current in the resistor, and (c) the current in the battery? (a) upward (b) downward (c) upward

18. **CQ** Drift velocity, the average speed of electrons in a complete circuit, is typically a few mm per second. Yet a lamp 3.0 m away from a light switch comes on instantaneously when you flip the switch. Explain this apparent paradox. electric field travels close to the speed of light

19. ● A net charge of 30 C passes through the cross-sectional area of a wire in 2.0 min. What is the current in the wire? 0.25 A

20. ● How long would it take for a net charge of 2.5 C to pass a location in a wire if it is to carry a steady current of 5.0 mA? 8.3 min

21. ● A small toy car draws a 0.50-mA current from a 3.0-V nicad (nickel–cadmium) battery. In 10 min of operation, (a) how much charge flows through the toy car, and (b) how much energy is lost by the battery? (a) 0.30 C (b) 0.90 J

22. ●● A car's starter motor draws 50 A from the car's battery during startup. If the startup time is 1.5 s, how many electrons pass a given location in the circuit during that time? 4.7×10^{20} electrons

23. ●● A net charge of 20 C passes a location in a wire in 1.25 min. How long does it take for a net 30-C charge to pass that location if the current in the wire is doubled? 56 s

24. ●●● Car batteries are often rated in "ampere-hours" or A·h. (a) Show that A·h has units of charge and that the value of 1 A·h is 3600 C. (b) A fully charged, heavy-duty battery is rated at 100 A·h and can deliver a current of 5.0 A steadily until depleted. What is the maximum time this battery can deliver current, assuming it isn't being recharged? (c) How much charge will the battery deliver in this time? (a) see ISM (b) 20 h (c) 3.60×10^5 C

25. IE ●●● Imagine that some protons are moving to the left at the same time that some electrons are moving to the right past the same location. (a) Will the net current be (1) to the right, (2) to the left, (3) zero, or (4) none of the preceding? (b) In 4.5 s, 6.7 C of electrons flow to the right at the same time that 8.3 C of protons flow to the left. What is the magnitude of the total current? (a) (2) to the left (b) 3.3 A

26. ●●● In a proton linear accelerator, a 9.5-mA proton current hits a target. (a) How many protons hit the target each second? (b) How much energy is delivered to the target each second if the protons each have a kinetic energy of 20 MeV and lose all their energy in the target? (a) 5.9×10^{16} protons (b) 1.9×10^5 J each second

17.3 Resistance and Ohm's Law*

27. **MC** The ohm is just another name for the (a) volt per ampere, (b) ampere per volt (c) watt, or (d) volt. (a)

28. **MC** Two ohmic resistors are placed across a 12-V battery one at a time. The resulting current in resistor A is measured to be twice as that in B. What can you say about their resistance values? (a) $R_A = 2R_B$, (b) $R_A = R_B$, (c) $R_A = R_B/2$, or (d) none of the preceding. (c)

29. **MC** An ohmic resistor is placed across two different batteries. When connected to battery A, the resulting current is measured to be three times the current when the resistor is attached to battery B. What can you say about the battery voltages: (a) $V_A = 3V_B$ (b) $V_A = V_B$ (c) $V_B = 3V_A$ or (d) none of the preceding. (a)

30. **MC** If you double the voltage across an ohmic resistor while at the same time cutting its resistance to one-third its original value, what happens to the current in the resistor: (a) it doubles, (b) it triples, (c) it goes up by six times, or (d) can't tell from the data given? (c)

31. **CQ** If the voltage (V) were plotted on the same graph versus current (I) for two ohmic conductors with different resistances, how could you tell which is less resistive? the one with the shallower slope

32. **CQ** Filaments in lightbulbs usually fail just after the bulbs are turned on rather than when they have already been on for a while. Why? resistance is low and current is high at turn-on

*Assume that the temperature coefficients of resistivity given in Table 17.1 apply over large temperature ranges.

33. **CQ** A wire is connected across a steady voltage source. (a) If that wire is replaced with one of the same material that is twice as long and has twice the cross-sectional area, how will the current in the wire be affected? (b) How will the current be affected if, instead, the new wire has the same length as the old one but half the diameter? (a) same (b) one-quarter the current

34. **CQ** A real battery always has some internal resistance r (▼Fig. 17.16) that increases with the battery's age. Explain why the terminal voltage drops as the internal resistance increases. see ISM

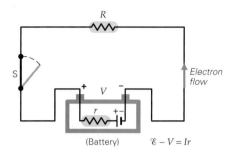

▲ **FIGURE 17.16 Emf and terminal voltage** See Exercises 34 and 35.

35. ● A battery labeled 12.0 V supplies 1.90 A to a 6.00-Ω resistor (Fig. 17.16). (a) What is the terminal voltage of the battery? (b) What is its internal resistance? (a) 11.4 V (b) 0.32 Ω

36. ● What is the emf of a battery with a 0.15-Ω internal resistance if the battery delivers 1.5 A to an externally connected 5.0-Ω resistor? 7.7 V

37. IE ● Some states allow the use of aluminum wire in houses in place of copper. (a) If you wanted the resistance of your aluminum wires to be the same as that of copper, would the aluminum wire have to have (1) a greater diameter than, (2) a smaller diameter than, or (3) the same diameter as the copper wire? (b) Calculate the thickness ratio of aluminum to that of copper. (a) (1) a greater diameter (b) 1.29

38. ● How much current is drawn from a 12-V battery when a 15-Ω resistor is connected across its terminals? 0.80 A

39. ● What voltage must a battery have to produce a 0.50-A current through a 2.0-Ω resistor? 1.0 V

40. ● During a research experiment on the conduction of current in the human body, a medical technician attaches one electrode to the wrist and a second to the shoulder. If 100 mV are applied across the two electrodes and the resulting current is 12.5 mA, what is the overall resistance of the patient's arm? 8.00 Ω

41. ●● A 0.60-m-long copper wire has a diameter of 0.10 cm. What is the resistance of the wire? 1.3×10^{-2} Ω

42. ●● A material is formed into a long rod with a square cross-section 0.50 cm on each side. When a 100-V voltage is applied across a 20-m length of the rod, a 5.0-A current is carried. (a) What is the resistivity of the material? (b) Is the material a conductor, an insulator, or a semiconductor? (a) 2.5×10^{-5} Ω·m (b) likely a semiconductor

43. ●● Two copper wires have equal cross-sectional areas and lengths of 2.0 m and 0.50 m, respectively. (a) What is the ratio of the current in the shorter wire to that in the longer one if they are connected to the same power supply? (b) If you wanted the two wires to carry the same current, what would the ratio of their cross-sectional areas have to be? (Give your answer as a ratio of longer to shorter.) (a) 4 (b) 4

44. IE ●● Two copper wires have equal lengths, but the diameter of one is three times that of the other. (a) The resistance of the thinner wire is (1) 3, (2) $\frac{1}{3}$, (3) 9, (4) $\frac{1}{9}$ times that of the resistance of the thicker wire. Why? (b) If the thicker wire has a resistance of 1.0 Ω, what is the resistance of the thinner wire? (a) (3) 9 times (b) 9.0 Ω

45. ●● The wire in a heating element of an electric stove burner has a 0.75-m effective length and a 2.0×10^{-6}-m² cross-sectional area. (a) If the wire is made of iron and operates at a 380°C temperature, what is its operating resistance? (b) What is its resistance when the stove is "off"? (a) 0.13 Ω (b) 0.038 Ω

46. ●● (a) What is the percentage variation of the resistivity of copper over the temperature range from room temperature (20°C) to 100°C? (b) Assume a copper wire's resistance changes only due to resistivity changes over this temperature range. Further assume it is connected to the same power supply. By what percentage would its current change? Would it be an increase or decrease? (a) 54% increase (b) 35% decrease

47. ●● A copper wire has a 25-mΩ resistance at 20°C. When the wire is carrying a current, heat produced by the current causes the temperature of the wire to increase by 27 C°. (a) What is the change in the wire's resistance? (b) If its original current was 10.0 mA, what is its final current? (a) 4.6 mΩ (b) 8.5 mA

48. ●● When a resistor is connected to a 12-V source, it draws a 185-mA current. The same resistor connected to a 90-V source draws a 1.25-A current. Is the resistor ohmic? Justify your answer mathematically. not ohmic, see ISM

49. ●● A particular application requires a 20-m length of aluminum wire to have a 0.25-mΩ resistance at 20°C. What must be the wire's diameter? 5.4×10^{-2} m

50. ●● If the resistance of the wire in Exercise 49 cannot vary by more than ±5.0%, what is the wire's operating temperature range? 8.2°C to 32°C

51. IE ●●● As a wire is stretched out so that its length increases, its cross-sectional area decreases, while the total volume of the wire remains constant. (a) Will the resistance after the stretch be (1) greater than, (2) the same as, or (3) less than that before the stretch? (b) A 1.0-m length of copper wire with a 2.0-mm diameter is stretched out; its length increases by 25% while its cross-sectional area decreases, but remains uniform. Compute the resistance ratio (final to initial). (a) (1) greater than (b) 1.6

52. ●●● ▼ Figure 17.17 shows data on the dependence of the current through a resistor on the voltage across that resistor. (a) Is the resistor ohmic? Explain. (b) What is its resistance? (c) Use the data to predict what voltage would be needed to produce a 4.0-A current in the resistor. (a) yes (b) 2.0 Ω (c) 8.0 V

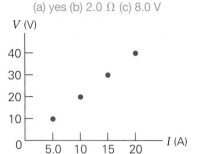

◀ FIGURE 17.17 An ohmic resistor? See Exercise 52.

53. ●●● At 20°C, a silicon rod is connected to a battery with a terminal voltage of 6.0 V and a 0.50-A current results. The temperature of the rod is then increased to 25°C. Assume its temperature coefficient of resistance is constant. (a) What is its new resistance? (b) How much current does it carry? (c) If you wanted to cut the current from its room temperature value of 0.50 A to 0.40 A, at what temperature would the sample have to be? (a) 7.8 Ω (b) 0.77 A (c) 16.4°C

54. IE ●●● A platinum wire is connected to a battery. (a) If the temperature increases, will the current in the wire (1) increase, (2) remain the same, or (3) decrease? Why? (b) An electrical resistance thermometer is made of platinum wire that has a 5.0-Ω resistance at 20°C. The wire is connected to a 1.5-V battery. When the thermometer is heated to 2020°C, by how much does the current change? (a) (3) decrease (b) decrease by 0.27 A

17.4 Electric Power

55. **MC** The electric power unit, the watt, is equivalent to what combination of SI units: (a) $A^2 \cdot \Omega$, (b) J/s, (c) V^2/Ω, or (d) all of the preceding? (d)

56. **MC** If the voltage across an ohmic resistor is doubled, the power expended in the resistor (a) increases by a factor of 2, (b) increases by a factor of 4, (c) decreases by half, or (d) none of the preceding. (b)

57. **MC** If the current through an ohmic resistor is halved, the power expended in the resistor (a) increases by a factor of 2, (b) increases by a factor of 4, (c) decreases by half, (d) decreases by a factor of 4. (d)

58. **CQ** Assuming that the resistance of your hair dryer obeys Ohm's law, what would happen to its power output if you plugged it directly into a 240-V outlet in Europe if it is designed to be used in the 120-V outlets of the United States? power output would quadruple, and it would overheat and burn out

59. **CQ** Most lightbulb filaments are made of tungsten and are about the same length. What would be different about the filament in a 60-W bulb compared with that in a 40-W bulb? the wire in the 60-W bulb would be thicker

60. **CQ** Which one consumes more power from a 12-V battery, a 5.0-Ω resistor or a 10-Ω resistor? Why? 5.0 Ω; same voltage, more current

61. ● A digital video disk (DVD) player is rated at 100 W at 120 V. What is its resistance? 144 Ω

62. ● A freezer of resistance 10 Ω is connected to a 110-V source. What is the power delivered when this freezer is on? 1.2×10^3 W

63. ● The current through a refrigerator with a resistance of 12 Ω is 13 A (when the refrigerator is on). What is the power delivered to the refrigerator? 2.0×10^3 W

64. ● Show that the quantity volts squared per ohm (V^2/Ω) has SI units of power. see ISM

65. ● An electric water heater is designed to produce 50 kW of heat when it is connected to a 240-V source. What must be the resistance of the heater? 1.2 Ω

66. ●● If the heater in Exercise 65 is 90% efficient, how long would it take to heat 50 gal of water from 20°C to 80°C? 18 min

67. IE ●● An ohmic resistor in a circuit is designed to operate at 120 V. (a) If you connect the resistor to a 60-V power source, will the resistor dissipate heat at (1) 2, (2) 4, (3) $\frac{1}{2}$, or (4) $\frac{1}{4}$ times the designed power? Why? (b) If the designed power is 90 W at 120 V, but the resistor is connected to 40 V, what is the power delivered to the resistor at the lower voltage? (a) (4) $\frac{1}{4}$ (b) 10 W

68. ●● An electric toy with a resistance of 2.50 Ω is operated by a 1.50-V battery. (a) What current does the toy draw? (b) Assuming that the battery delivers a steady current for its lifetime of 6.00 h, how much charge passed through the toy? (c) How much energy was delivered to the toy? (a) 0.600 A (b) 1.30×10^4 C (c) 1.94×10^4 J

69. ●● A welding machine draws 18 A of current at 240 V. (a) What is its power rating? (b) What is its resistance? (a) 4.3×10^3 W (b) 13 Ω

70. ●● On average, an electric water heater operates for 2.0 h each day. (a) If the cost of electricity is $0.15/kWh, what is the cost of operating the heater during a 30-day month? (b) What is the resistance of a typical water heater? [*Hint:* See Table 17.2.] (a) $40.50 (b) 6.4 Ω

71. ●● (a) What is the resistance of a heating coil if it is to generate 15 kJ of heat per minute when it is connected to a 120-V source? (b) How would you change the resistance if instead you wanted 10 kJ of heat per minute? (a) 58 Ω (b) 86 Ω

72. ●● A 200-W computer power supply is on 10 h per day. If the cost of electricity is $0.15/kWh, what is the cost (to the nearest dollar) of using the computer for a year (365 days)? $110

73. ●● A 120-V air conditioner unit draws 15 A of current. If it operates for 20 min, (a) how much energy in kilowatt-hours does it use? (b) If the cost of electricity is $0.15/kWh, what is the cost (to the nearest penny) of operating the unit for 20 min? (a) 0.60 kWh (b) $0.09

74. ●● Two resistors, 100 Ω and 25 kΩ, are rated for a maximum power output of 1.5 W and 0.25 W, respectively. What is the maximum voltage that can be safely applied to each resistor? 100 Ω: 12 V; 25 kΩ: 79 V

75. ●● A wire 5.0 m long and 3.0 mm in diameter has a resistance of 100 Ω. A 15-V potential difference is applied across the wire. Find (a) the current in the wire, (b) the resistivity of its material, and (c) the rate at which heat is being produced in the wire. (a) 0.15 A (b) 1.4×10^{-4} Ω · m (c) 2.3 W

76. IE ●● When connected to a voltage source, a coil of tungsten wire initially dissipates 500 W of power. In a short time, the temperature of the coil increases by 150 C° because of joule heating. (a) Will the dissipated power (1) increase, (2) remain the same, or (3) decrease? Why? (b) What is the corresponding change in the power? (a) (3) decrease (b) decrease by 202 W

77. ●● A 20-Ω resistor is connected to four 1.5-V batteries. What is the joule heat loss per minute in the resistor if the batteries are connected (a) in series and (b) in parallel? (a) 1.1×10^2 J (b) 6.8 J

78. ●● A 5.5-kW water heater operates at 240 V. (a) Should the heater circuit have a 20-A or a 30-A circuit breaker? (A circuit breaker is a safety device that opens the circuit at its rated current.) (b) Assuming 85% efficiency, how long will the heater take to heat the water in a 55-gal tank from 20° to 80°C? (a) 30-A breaker (b) 3.1 h

79. ●● A student uses an immersion heater to heat 0.30 kg of water from 20°C to 80°C for tea. If the heater is 75% efficient and takes 2.5 min, what is its resistance? (Assume 120-V household voltage.) 21 Ω

80. ●● An ohmic appliance is rated at 100 W when it is connected to a 120-V source. If the power company cuts the voltage by 5.0% to conserve energy, what is (a) the current in the appliance and (b) the power consumed by the appliance after the voltage drop? (a) 0.79 A (b) 90 W

81. ●● A lightbulb's output is 60 W when it operates at 120 V. If the voltage is cut in half and the power dropped to 20 W during a brownout, what is the ratio of the bulb's resistance at full power to its resistance during the brownout? $R_{120}/R_{60} = 4/3$

82. ●● To empty a flooded basement, a water pump must do work (lift the water) at a rate of 2.00 kW. If the pump is wired to a 240-V source and is 84% efficient, (a) how much current does it draw and (b) what is its resistance? (a) 9.9 A (b) 24 Ω

83. ●●● Find the total monthly (30-day) electric bill (to the nearest dollar) for the following household appliance usage if the utility rate is $0.12/kWh: Central air conditioning runs 30% of the time; a blender is used 0.50 h/month; a dishwasher is used 8.0 h/month; a microwave oven is used 15 min/day; the motor of a frost-free refrigerator runs 15% of the time; a stove (burners plus oven) is used a total of 10 h/month; and a color television is operated 120 h/month. (Use the information given in Table 17.2.) $152

Comprehensive Exercises

84. IE A piece of carbon and a piece of copper have the same resistance at room temperature. (a) If the temperature of each piece is increased by 10.0 C°, will the copper piece have (1) a higher resistance than, (2) the same resistance as, or (3) a lower resistance than the carbon piece? Why? (b) Calculate the ratio of the resistance of copper to that of carbon at the raised temperature. (a) (1) a higher resistance than (b) 1.07

85. Two pieces of aluminum and copper wire are identical in length and diameter. At some temperature, one of the wires will have the same resistance that the other has at 20°C. What is that temperature? (Is there more than one temperature?) 117°C for Cu or −72.6°C for Al

86. A battery delivers 2.54 A to an ohmic resistor rated at 4.52 Ω. When it is connected to a 2.21-Ω resistor, it delivers 4.98 A. Determine the battery's (a) internal resistance (assumed constant), (b) emf , and (c) terminal voltage (in both cases).
(a) 0.195 Ω (b) 12.0 V (c) 11.5 V; 11.0 V

87. An external resistor is connected to a battery with a variable emf but constant internal resistance. At an emf of 3.00 V, the resistor draws a current of 0.500 A, and at 6.00 V, the resistor draws a current of 1.00 A. Is the external resistor ohmic? Prove your answer. yes, see ISM

88. An electric eel delivers a current of 0.75 A to a small pencil-thin prey 15 cm long. If the eel's "bio-battery" was charged to 500 V, and it was constant for 20 ms before dropping to zero, estimate (a) the resistance of the fish, (b) the energy delivered to the fish, and (c) the average electric field (magnitude) in the fish's flesh.
(a) 6.7×10^2 Ω (b) 7.5 J (c) 3.3×10^3 V/m

89. Most modern TVs have an "instant warm-up" feature. Even though the set appears to be off, it is "off" only in that there is no picture and audio. To provide a "quick on" feature, the TV's electronics are kept ready. This takes about 10 W of electric power, constantly. Assume that there is one TV with this feature for every two households and estimate how many electric power plants this feature takes to run in the United States. about one-half, see ISM

90. ▼Figure 17.18 shows free-charge carriers that each have a charge q and move with a speed v_d (drift speed) in a conductor of cross-sectional area A. Let n be the number of free charge carriers per unit volume. (a) Prove that the total charge (ΔQ) free to move in the volume element shown is given by $\Delta Q = (nAx)q$. (b) Prove that the current in the conductor is given by $I = nqv_d A$.
see ISM

91. A copper wire with a cross-sectional area of 13.3 mm^2 (AWG No. 6) carries a 1.2-A current. If the wire contains 8.5×10^{22} free electrons per cubic centimeter, what is the drift velocity of the electrons? [*Hint*: See Exercise 90 and Fig. 17.18.] 6.6×10^{-6} m/s

92. A computer CD-ROM drive that operates on 120 V is rated at 40 W when it is operating. (a) How much current does the drive draw? (b) What is the drive's resistance?
(a) 0.33 A (b) 3.6×10^2 Ω

93. The tungsten filament of an incandescent lamp has a resistance of 200 Ω at room temperature. What would the resistance be at an operating temperature of 1600°C?
1.6×10^3 Ω

94. A common sight in our modern world is high-voltage lines carrying electric energy over long distances from the power plant to populated areas. The delivery voltage of these lines is typically 500 kV, whereas by the time the energy reaches our households it is down to 120 V (see Chapter 20 for how this is done). (a) Explain clearly why electric power is delivered over long distances at high voltages when we know that high voltages can be dangerous. (b) Calculate the ratio of heating loss in a given length of wire (assumed ohmic) carrying current at 500 kV to when it operates at 120 V.
(a) to reduce I^2R losses (b) 5.76×10^{-8}

95. In a country setting it is common to see hawks sitting on a single high-voltage electric power line searching for a road-kill meal (▼Fig. 17.19). To understand why this bird isn't electrocuted, let's do a ballpark estimate of the voltage between her feet. Assume dc conditions in a power line that is 1.0 km long, has a resistance of 30 Ω, and is at an electric potential of 250 kV above the other wire (the one the bird is not on), which is at ground or zero volts. (a) If the wires are carrying energy at the rate of 100 MW, what is the current in them? (b) Assuming the bird's feet are 15 cm apart, what is the resistance of that segment of the hot wire? (c) What is the voltage *difference* between the bird's feet? Comment on the size of your answer and whether you think this might be dangerous. (d) What is the voltage difference between her feet if she places one on the ground wire while continuing to hold onto the hot wire? Comment on the size of your answer and whether you think this might be dangerous.
(a) 400 A (b) 4.5×10^{-3} Ω (c) 1.8 V (d) 250 kV

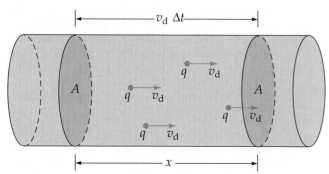

▲ **FIGURE 17.18 Total charge and current** See Exercises 90 and 91.

◄ **FIGURE 17.19 Bird on a wire** See Exercise 95.

 The following Physlet Physics Problem can be used with this chapter.
30.5

18

BASIC ELECTRIC CIRCUITS

PHYSICS FACTS

- Physics for parents of teenagers: Usually more than one hair dryer/blower cannot be on the same household circuit without tripping the circuit breaker. Either two separate circuits are needed in the bathroom, or someone has to go to another room and use a different circuit.

- An ammeter wired *incorrectly* in parallel with a circuit element not only measures the wrong current, but risks being burned out. Hence all ammeters have protective fuses. On the other hand, if a voltmeter is *incorrectly* connected in series with a circuit element, the circuit current drops to zero, and although the voltage measurement is incorrect, there is no damage.

- Less than 0.01 of an amp of current through the human body can trigger muscle paralysis. If the person cannot then let go of the exposed wiring, death could result if the current passes through a vital organ, such as the heart.

- Specialized *pacemaker cells* located in a small region of the heart trigger your heartbeat. Their electrical signals travel across the heart in about 50 ms. If these fail, other parts of the heart's electrical system can take over as a backup. The pacemaker cells can be influenced by the body's nervous system, so the rate at which they tell the heart to beat can vary dramatically—for example, from a calm 60 beats per minute when asleep to more than 100 beats per minute during physical exertion.

We usually think of metallic wires as the "connectors" between resistors in a circuit. However, wires are not the only conductors of electricity, as the photo shows. Because the bulb is lit, the circuit must be complete. It can be concluded, therefore, that the "lead" in a pencil (actually a form of carbon called *graphite*) conducts electricity. The same must be true for the liquid in the beaker—in this case, a solution of water and ordinary table salt.

Electric circuits are of many kinds and can be designed for many purposes, from boiling water to lighting a Christmas tree. Circuits containing "liquid" conductors (as in the photo), have practical applications in the laboratory and in industry; for example, they can be used to synthesize or purify chemical substances and to *electroplate* metals. (Electroplating means to chemically attach metals to surfaces using electrical techniques, such as in making silver plate.) Armed with the principles learned in Chapters 15, 16, and 17, you are now ready to analyze some electric circuits. This analysis will give you an appreciation of how electricity actually works.

Circuit analysis most often deals with voltage, current, and power requirements. A circuit may be analyzed theoretically before being assembled. The analysis might show that the circuit would not function properly as designed or that there could be a safety problem (such as overheating due to joule heat). To help in this chapter's analysis, we will rely heavily on diagrams of circuits to visualize and understand their function. A few of these diagrams were included in Chapter 17.

Let's begin our analysis of circuits by looking at arrangements of resistive elements, such as lightbulbs, toasters, and immersion heaters.

18.1 Resistances in Series, Parallel, and Series–Parallel Combinations

OBJECTIVES: To (a) determine the equivalent resistance of resistors in series, parallel, and series–parallel combinations, and (b) use equivalent resistances to analyze simple circuits.

The resistance symbol ─⋀⋀─ can represent *any* type of circuit element such as a lightbulb or toaster. Here it is assumed that all the elements are ohmic (constant resistance) unless otherwise stated. (Note that lightbulbs, in particular, are not ohmic because their resistance increases significantly as they warm up.) In addition, as usual, wire resistance will be neglected.

Resistors in Series

In analyzing a circuit, because voltage represents energy per unit charge, to conserve energy, the *sum of the voltages around a complete circuit loop is zero.* Remember that *voltage* means "change in electrical potential," so voltage gains and losses are represented by $+$ and $-$ signs, respectively. For the circuit in ▼ Fig. 18.1a, by conservation of energy (per coulomb) the individual voltages (V_i, where $i = 1, 2,$ or 3) across the resistors add to equal the voltage (V) across the battery terminals. Each resistor in series must carry the same current (I) because charge can't "pile up" or "leak out" at any location in the circuit. Summing the voltage gains and losses, we have $V - \Sigma V_i = 0$. Finally, we know how the voltage is related to the resistance for each resistor, namely $V_i = IR_i$. Substituting this into the previous equation,

$$V - \Sigma(IR_i) = 0 \quad \text{or} \quad V = \Sigma(IR_i) \tag{18.1}$$

The elements in Fig. 18.1a are said to be connected in **series**, or end to end. *When resistors are in series, the current must be the same through all the resistors,* as required by the conservation of charge. If this were not true, then charge would build up or disappear, which cannot happen. ▶Figure 18.2 shows the analogous flow of water in a smooth streambed punctuated by a series of rapids (representing "resistance").

Labeling the common current in the resistors as I, then Eq. 18.1 can be written explicitly for three resistors (such as in Fig. 18.1a):

$$V = V_1 + V_2 + V_3$$
$$= IR_1 + IR_2 + IR_3 = I(R_1 + R_2 + R_3)$$

To **equivalent series resistance, (R_s)** is the value of a single resistor that could replace the three resistors by one resistor R_s and maintain the same current means that $V = IR_s$, or $R_s = V/I$. Hence, the resistors in series have an equivalent resistance

$$R_s = \frac{V}{I} = R_1 + R_2 + R_3$$

▶ **FIGURE 18.1 Resistors in series (a)** When resistors (representing the resistances of lightbulbs here) are in series, the current in each is the same. ΣV_i, the sum of the voltage drops across the resistors, is equal to V, the battery voltage. **(b)** The equivalent resistance R_s of the resistors in series is the sum of the resistances.

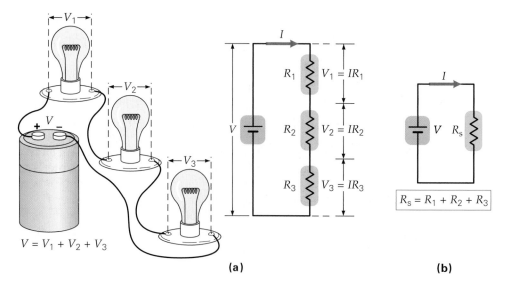

$V = V_1 + V_2 + V_3$

$$R_s = R_1 + R_2 + R_3$$

(a)   **(b)**

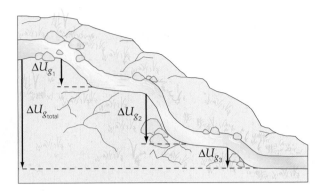

◀ **FIGURE 18.2** Water-flow analogy to resistors in series Even though, in general, a different amount of gravitational potential energy (per kilogram) is lost as the water flows down each set of rapids, the *current* of water is the same everywhere. The *total* loss of gravitational potential energy (per kilogram) is the sum of the losses. (To make this a "complete" water circuit, some external agent, such as a pump, would need to continuously do work on the water by returning it to the top of the hill, restoring its original gravitational potential energy.)

That is, the equivalent resistance of resistors in series is the sum of the individual resistances. This means that the three resistors (bulbs) in Fig. 18.1a could be replaced by a single resistor of resistance R_s (Fig. 18.1b) without affecting the current. For example, if each resistor in Fig. 18.1a had a value of 10 Ω, then R_s would be 30 Ω. *Note that the equivalent series resistance is larger than the resistance of the largest resistor in the series.*

This result can be extended to any number of **resistors in series**:

$$R_s = R_1 + R_2 + R_3 + \cdots = \Sigma R_i \quad \begin{array}{l} \textit{equivalent series} \\ \textit{resistance} \end{array} \quad (18.2)$$

Note: R_s is larger than the largest resistance in a series arrangement.

Series connections are not common in some circuits such as house wiring, because there are two major disadvantages compared to parallel wiring. The first is clear considering what happens if one of the bulbs in Fig. 18.1a burns out (or you wish to turn only that one bulb off). In this case, *all* of the bulbs go out, because the circuit would no longer be complete, or continuous. In this situation, the circuit is said to be *open*. An *open circuit* has an infinite equivalent resistance, because the current in it is zero, even though the battery voltage is not.

A second disadvantage of series connections is that each resistor operates at less than the battery voltage (V). Consider what would happen if a fourth resistor were added. The result would be that the voltage across each of the original bulbs (and the current) would decrease, resulting in reduced power delivered to all the bulbs. That is, the bulbs would not glow at the rated brightness or light output. Clearly, this situation is *not* acceptable in a household setting.

Resistors in Parallel

Resistors can also be connected to a battery in **parallel** (▼Fig. 18.3a). In this case, all the resistors have common connections—that is, all the leads on one side of the resistors are attached together to one terminal of the battery. All the leads on the other side are attached to the other terminal. *When resistors are connected in parallel to a source of emf, the voltage drop across each resistor is the same.* It may not surprise

Demonstration/activity: Build a parallel circuit and show how the addition of each successive element in parallel causes the current from the power supply to increase. Note that the resistance of elements in parallel is less than the resistance of the individual elements. Also, discuss how the power changes when resistors are placed in series and in parallel.

◀ **FIGURE 18.3** Resistors in parallel (a) When resistors are connected in parallel, the voltage drop across each resistor is the same. The current from the battery divides (generally unequally) among the resistors. (b) The equivalent resistance, R_p, of resistors in parallel is given by a reciprocal relationship.

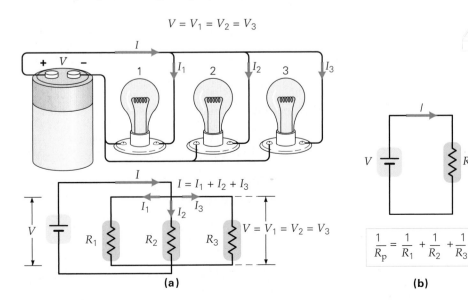

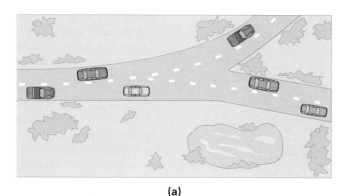

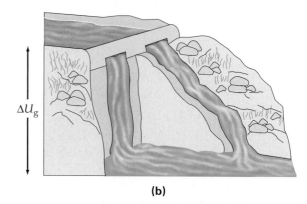

ΔU_g

(a)

(b)

▲ **FIGURE 18.4** Analogies for resistors in parallel (a) When a road forks, the total number of cars entering the two branches each minute is equal to the number of cars arriving at the fork each minute. Movement of charge into and then out of a junction can be considered in the same way. **(b)** When water flows from a dam, the gravitational potential energy lost (per kilogram of water) in falling to the stream below is the same regardless of the path. This is analogous to voltages across parallel resistors.

Note: In actuality, the wires of a circuit are not ordinarily arranged in the neat rectangular pattern of a circuit diagram. The rectangular form is a convention that provides a neater and consistent presentation along with an easier visualization of the actual circuit.

Teaching tip: Simple series and parallel combinations can be made without much reference to Kirchhoff's rules, but the derivation of equivalent resistances really depends on Kirchhoff's rules, which are not introduced until the next section. You can make the voltage drop seem plausible with a simple energy-conservation argument. The division of current makes more sense if explained in terms of charge conservation.

Teaching tip: Encourage students to do the necessary algebra, and show the best way of using the calculator for these types of problems. Some students will use the 1/x function on their calculators and then round to one significant figure prior to adding the next term. In many of the simpler cases, the fractions can be added directly without a calculator, using common denominators. Always carry units so that the need to invert will be obvious at the end of the calculation.

Note: R_p is smaller than the smallest resistance in a parallel arrangement.

you to learn that household circuits are wired in parallel. (See Section 18.5.) This is because when wired in parallel, each appliance operates at full voltage, and turning one appliance off or on does not affect the others.

Unlike resistors in series, the current in a parallel circuit divides into the different paths (Fig. 18.3a). This occurs whenever we have a *junction* (a location where several wires come together), much as traffic divides when it reaches a fork in the road (▲Fig. 18.4a). The total current out of the battery is equal to the sum of these currents. Specifically, for three resistors in parallel, $I = I_1 + I_2 + I_3$. Notice that if the resistances are equal, the current will divide so that each resistor has the same current. However, in general, the resistances will not be equal and the current will divide among the resistors in inverse proportion to their resistances. This means that the largest current will take the path of least resistance. Remember, however, that no one resistor carries the total current.

The **equivalent parallel resistance (R_p)** is the value of a single resistor that could replace all the resistors and maintain the same current. Thus, $R_p = V/I$, or $I = V/R_p$. In addition, the voltage drop (V) must be the same across each resistor. To visualize this situation, imagine a water analogy. Consider two separate water paths, each leading from the top of a dam to the bottom. The water loses the same amount of gravitational potential energy (analogous to V) regardless of the path (Fig. 18.4b). For electricity, a given amount of charge loses the same amount of electrical potential energy, regardless of which parallel resistor it passes through.

The current through each resistor is $I_i = V/R_i$. (The subscript i represents *any* of the resistors: 1, 2, 3, ...) Substituting for each current, we obtain

$$I = I_1 + I_2 + I_3 = \frac{V}{R_1} + \frac{V}{R_2} + \frac{V}{R_3}$$

Therefore,

$$\frac{V}{R_p} = V\left(\frac{1}{R_p}\right) = V\left(\frac{1}{R_1} + \frac{1}{R_2} + \frac{1}{R_3}\right)$$

By equating the two expressions in the parentheses, we see that R_p is related to the individual resistances by a reciprocal equation

$$\frac{1}{R_p} = \frac{1}{R_1} + \frac{1}{R_2} + \frac{1}{R_3}$$

This result can be generalized to include any number of **resistors in parallel**:

$$\frac{1}{R_p} = \frac{1}{R_1} + \frac{1}{R_2} + \frac{1}{R_3} + \cdots = \Sigma\left(\frac{1}{R_i}\right) \qquad \begin{array}{l}\textit{equivalent parallel}\\ \textit{resistance}\end{array} \qquad (18.3)$$

For the special case when there are just two resistors, the equivalent resistance can be rewritten (using a common denominator) as

$$\frac{1}{R_p} = \frac{1}{R_1} + \frac{1}{R_2} = \frac{R_1 + R_2}{R_1 R_2}$$

or

$$R_p = \frac{R_1 R_2}{R_1 + R_2} \qquad \begin{array}{l}\textit{(only for two}\\ \textit{resistors in parallel)}\end{array} \qquad (18.3a)$$

Problem-Solving Hint

Note that Eq. 18.3 gives $1/R_p$, *not* R_p. At the end of the calculation, the reciprocal must be taken to find R_p. Unit analysis will show that the units are not ohms until inverted. As usual, carrying units along with calculations makes errors of this type less likely to occur.

The equivalent resistance of resistors in parallel is always less than the smallest resistance in the arrangement. For example, two parallel resistors—say, of resistances 6.0 Ω and 12.0 Ω—are equivalent to a single with a resistance of 4.0 Ω (you should show this). But why should we expect this seemingly strange answer?

Physically, the reason can be seen by first considering a 12-V battery in a circuit with a *single* 6.0-Ω resistor. The current in the circuit is 2.0 A ($I = V/R$). Now imagine connecting a 12.0-Ω resistor in parallel to the 6.0-Ω resistor. The current through the 6.0-Ω resistor will be unaffected—it will remain at 2.0 A. (Why?) However, the new resistor will have a current of 1.0 A (using $I = V/R$ again). Thus the *total* current in the circuit is 1.0 A + 2.0 A = 3.0 A. Now look at the overall result. When the second resistor is attached to the first *in parallel,* the total current delivered by the battery increases. Since the voltage did not increase, the equivalent resistance of the circuit *must have decreased* (below its initial value of 6.0 Ω) when the 12-Ω resistor was attached. In other words, every time an extra parallel path is added, the result is more total current. Thus the circuit behaves as if its equivalent resistance *decreased.*

Notice that this argument does not depend on the value of the added resistor. All that matters is that another path with some resistance is added. (Try this using a 2-Ω or a 2-MΩ resistor in place of the 12-Ω resistor. A decrease in equivalent resistance happens again. Note, however, that the *value* of the equivalent resistance will be different.)

In general, then, series connections provide a way to increase total resistance and parallel connections provide a way to decrease total resistance. To see how these ideas work, consider Example 18.1.

Example 18.1 ■ Connections Count: Resistors in Series and in Parallel

What is the equivalent resistance of three resistors (1.0 Ω, 2.0 Ω, and 3.0 Ω) when connected (a) in series (Fig. 18.1a) and (b) in parallel (Fig. 18.3a)? (c) How much current will be delivered by a 12-V battery in each of these arrangements?

Thinking It Through. To find the equivalent resistances for parts (a) and (b), apply Eqs. 18.2 and 18.3, respectively. For the series current in part (c), calculate the current through the battery by treating the battery as if it were connected to a single resistor—the series equivalent resistance. For the parallel arrangement, the total current can be determined by using the parallel equivalent resistance. From the knowledge that each resistor in parallel has the same voltage across it, the individual currents can be calculated.

Solution. Listing the data

Given: $R_1 = 1.0$ Ω   *Find:* (a) R_s (series resistance)
        $R_2 = 2.0$ Ω   (b) R_p (parallel resistance)
        $R_3 = 3.0$ Ω   (c) I (total current for each case)
        $V = 12$ V

(a) The equivalent series resistance (Eq. 18.2) is

$$R_s = R_1 + R_2 + R_3 = 1.0 \text{ Ω} + 2.0 \text{ Ω} + 3.0 \text{ Ω} = 6.0 \text{ Ω}$$

Our result is larger than the largest resistance, as expected.

(b) The equivalent parallel resistance is determined from Eq. 18.3

$$\frac{1}{R_p} = \frac{1}{R_1} + \frac{1}{R_2} + \frac{1}{R_3} = \frac{1}{1.0 \text{ Ω}} + \frac{1}{2.0 \text{ Ω}} + \frac{1}{3.0 \text{ Ω}}$$

$$= \frac{6.0}{6.0 \text{ Ω}} + \frac{3.0}{6.0 \text{ Ω}} + \frac{2.0}{6.0 \text{ Ω}} = \frac{11}{6.0 \text{ Ω}}$$

or, after inverting,

$$R_p = \frac{6.0 \text{ Ω}}{11} = 0.55 \text{ Ω}$$

which is less than the least resistance, also as expected.

PHYSLET®

Illustration 30.3 Current and Voltage Dividers

PHYSLET®

Illustration 30.4 Batteries and Switches

Demonstration/activity: An effective lecture demo involves two lightbulbs of different power ratings, wired first in parallel and then in series, in both cases using 120-V wall voltage. In the parallel arrangement, students are pleased to see the lightbulbs both operating at normal brightness and independently of one another. (Switch either one off, and the other is unaffected.) However, in series, the bulb with the higher wattage, which was previously the brighter of the two, will be very dim or apparently not lit at all (especially if you choose two very different wattages). In this case, point out that if one bulb is lit, the other must also be "on." Why? Ask an adventuresome student in the front row to come up and feel the bulb that seems not to be working. It will be warm, but it is giving off only infrared, not visible, light. Turn off one bulb, and they both go off. Turn them both back on, and ask students which way they would prefer the appliances in their house to be wired!

(continues on next page)

(c) From the equivalent series resistance and the battery voltage:

$$I = \frac{V}{R_s} = \frac{12 \text{ V}}{6.0 \text{ }\Omega} = 2.0 \text{ A}$$

Let's calculate the voltage drop across each resistor:

$$V_1 = IR_1 = (2.0 \text{ A})(1.0 \text{ }\Omega) = 2.0 \text{ V}$$
$$V_2 = IR_2 = (2.0 \text{ A})(2.0 \text{ }\Omega) = 4.0 \text{ V}$$
$$V_3 = IR_3 = (2.0 \text{ A})(3.0 \text{ }\Omega) = 6.0 \text{ V}$$

Notice that to ensure that the current through each resistor is the same, it must be that *in series, the larger resistors require more voltage*. As a check, note that the sum of the resistor voltage drops $(V_1 + V_2 + V_3)$ equals the battery voltage.

For the parallel arrangement, the total current is:

$$I = \frac{V}{R_p} = \frac{12 \text{ V}}{0.55 \text{ }\Omega} = 22 \text{ A}$$

Note that the current for the parallel combination is much larger than that for the series combination. (Why?) Now the current through each resistor can be determined, because each has a voltage of 12 V across it. Therefore,

$$I_1 = \frac{V}{R_1} = \frac{12 \text{ V}}{1.0 \text{ }\Omega} = 12 \text{ A}$$

$$I_2 = \frac{V}{R_2} = \frac{12 \text{ V}}{2.0 \text{ }\Omega} = 6.0 \text{ A}$$

$$I_3 = \frac{V}{R_3} = \frac{12 \text{ V}}{3.0 \text{ }\Omega} = 4.0 \text{ A}$$

As a check, note that the sum of the currents is equal to the current through the battery.

As can be seen, for resistors in parallel, the resistor with the smallest resistance gets most of the total current because resistors in parallel experience the same voltage. (Note that for parallel arrangements the least resistance never has *all* the current, just the largest.)

Follow-Up Exercise. (a) Calculate the power delivered to each resistor for both arrangements in this Example. (b) What generalizations can you make? For instance, which resistor gets the most power in series? in parallel? (c) For each arrangement, does the total power delivered to all the resistors equal the power output of the battery? *(Answers to all Follow-Up Exercises are at the back of the text.)*

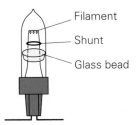

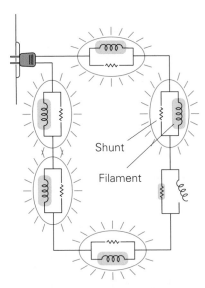

▲ **FIGURE 18.5 Shunt-wired Christmas tree lights** A shunt, or "jumper," in parallel with the bulb filament reestablishes a complete circuit when one of the filaments burns out (lower right bulb). Without the shunt, if one were to burn out, all the bulbs would go out.

As a wiring application, consider strings of Christmas tree lights. In the past, such strings had large bulbs connected in series. When one bulb burned out, all the others in the string went out, leaving you to hunt for the faulty bulb. Now, with newer strings that have smaller bulbs, one or more bulbs may burn out, but the others remain lit. Does this mean that the bulbs are now wired in parallel? No; parallel wiring would give a small resistance and large current, which could be dangerous.

Instead, an insulated jumper, or "shunt," is wired in parallel with each bulb's filament (◄Fig. 18.5). In normal operation, the shunt is insulated from the filament wires and does not carry current. When the filament breaks or "burns out," there is *momentarily* an open circuit, and there is no current anywhere in the string. Thus, the voltage across the open circuit at the broken filament will be the full 120-V household voltage. This voltage causes sparking that burns off the shunt's insulating material. Now the shunt is in electrical contact with the other filament wires, again completing the circuit, and the rest of the lights in the string continue to glow. (The shunt, a wire with little resistance, is indicated by the small resistance symbol in the circuit diagram of Fig. 18.5. Under normal operation, there is a gap—the insulation—between the shunt and the filament wire.) To understand what happens to the remaining bulbs in a string with a burnt-out bulb, consider the following Example.

Conceptual Example 18.2 ■ Oh, Tannenbaum! Christmas Tree Lights Burning Brightly

Consider a string of Christmas tree lights composed of bulbs with jumper shunts. If the filament of one bulb burns out and the shunt completes the circuit, will the other bulbs each (a) glow a little more brightly, (b) glow a little more dimly, or (c) be unaffected?

Reasoning and Answer. If one bulb filament burns out and its shunt completes the circuit, there will be less total resistance in the circuit, because the shunt's resistance is much less than the filament's resistance. (Note that the filaments of the good bulbs and the shunt of the burnt-out bulb are in now series, so the resistances add.)

With less total resistance, there will be more current in the circuit, and the remaining good bulbs will glow a little brighter because the light output of a bulb is directly related to the power delivered to that bulb. (Recall that electrical power is related to the current by $P = I^2 R$.) So the answer is (a). For example, suppose the string initially has 18 identical bulbs. Because the total voltage across the string is 120 V, the voltage drop in any one bulb is $(120 \text{ V})/18 = 6.7$ V. If one bulb is out (and thus shunted), the voltage across each of the remaining lighted bulbs becomes $(120 \text{ V})/17 = 7.1$ V. This increased voltage causes the current to increase. Both increases contribute to more power delivered to each bulb, and brighter lights (recall the alternative expression for electric power, $P = IV$).

Follow-Up Exercise. In this Example, if you were to remove one bulb, what would be the voltage across (a) the empty socket and (b) any of the remaining bulbs? Explain.

PHYSLET®

Illustration 30.1 Complete Circuits

Series–Parallel Resistor Combinations

Resistors may be connected in a circuit in a variety of series–parallel combinations. As shown in ▼Fig. 18.6, circuits with only one voltage source can sometimes be reduced to a single equivalent loop, containing just the voltage source and one equivalent resistance, by applying the series and parallel results.

A procedure for analyzing circuits (determining voltage and current for each circuit element) for such combinations is as follows:

1. Determine which groups of resistors are in series and which are in parallel, and reduce all groups to equivalent resistances, using Eqs. 18.2 and 18.3.

PHYSLET®

Illustration 30.2 Switches, Voltages, and Complete Circuits

Teaching tip: Students must master series and parallel circuits before beginning series–parallel combinations.

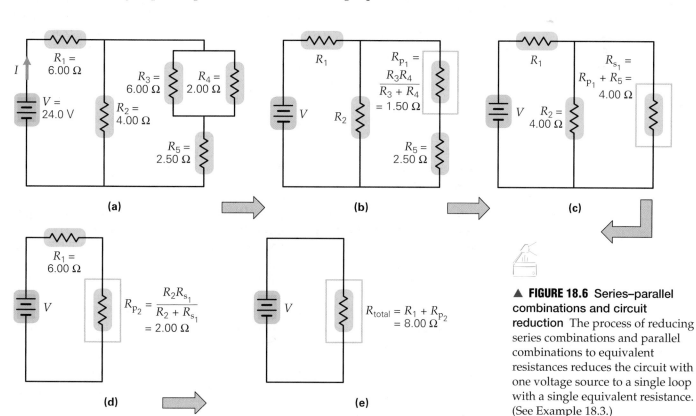

▲ **FIGURE 18.6** Series–parallel combinations and circuit reduction The process of reducing series combinations and parallel combinations to equivalent resistances reduces the circuit with one voltage source to a single loop with a single equivalent resistance. (See Example 18.3.)

2. Reduce the circuit further by treating the separate equivalent resistances (from Step 1) as individual resistors. Proceed until you get to a single loop with one total (overall or equivalent) resistance value.

3. Find the current delivered to the reduced circuit using $I = V/R_{\text{total}}$.

4. Expand the reduced circuit back to the actual circuit by reversing the reduction steps, one at a time. Use the current of the reduced circuit to find the currents and voltages for the resistors in each step.

To see this procedure in use, consider the next Example.

Example 18.3 ■ Series–Parallel Combination of Resistors: Same Voltage or Same Current?

What are the voltages across and the currents in each of the resistors R_1 through R_5 in Fig. 18.6a?

Thinking It Through. Applying the steps described previously, it is important to identify parallel and series combinations before starting. It should be clear that R_3 is in parallel with R_4 (written $R_3\|R_4$). This parallel combination is itself in series with R_5. Furthermore, the $(R_3\|R_4) + R_5$ leg is in parallel with R_2. Lastly, this parallel combination is in series with R_1. Combining the resistors step by step should enable us to determine the total equivalent circuit resistance (Step 2). From that value, the total current can be calculated. Then, working backward, we can find the current in, and voltage across, each resistor.

Solution. To avoid rounding errors, the results will be carried to three significant figures.

Given: Values in Fig. 18.6a   *Find:* Current and voltage for each resistor (Fig. 18.6a)

The parallel combination at the right-hand side of the circuit diagram can be reduced to the equivalent resistance R_{P_1} (see Fig. 18.6b), using Eq. 18.3:

$$\frac{1}{R_{\text{P}_1}} = \frac{1}{R_3} + \frac{1}{R_4} = \frac{1}{6.00\ \Omega} + \frac{1}{2.00\ \Omega} = \frac{4}{6.00\ \Omega}$$

This expression is equivalent to

$$R_{\text{P}_1} = 1.50\ \Omega$$

This operation leaves a series combination of R_{P_1} and R_5 along that side, which is reduced to R_{S_1}, using Eq. 18.2 (Fig. 18.6c):

$$R_{\text{S}_1} = R_{\text{P}_1} + R_5 = 1.50\ \Omega + 2.50\ \Omega = 4.00\ \Omega$$

Then, R_2 and R_{S_1} are in parallel and can be reduced (again using Eq. 18.3) to R_{P_2} (Fig. 18.6d):

$$\frac{1}{R_{\text{P}_2}} = \frac{1}{R_2} + \frac{1}{R_{\text{S}_1}} = \frac{1}{4.00\ \Omega} + \frac{1}{4.00\ \Omega} = \frac{2}{4.00\ \Omega}$$

This expression is equivalent to

$$R_{\text{P}_2} = 2.00\ \Omega$$

This operation leaves two resistances (R_1 and R_{P_2}) in series. These resistances combine to give the total equivalent resistance (R_{total}) of the circuit (Fig. 18.6e):

$$R_{\text{total}} = R_1 + R_{\text{P}_2} = 6.00\ \Omega + 2.00\ \Omega = 8.00\ \Omega$$

Thus, the battery delivers a current of

$$I = \frac{V}{R_{\text{total}}} = \frac{24.0\ \text{V}}{8.00\ \Omega} = 3.00\ \text{A}$$

Now let's work backward and "rebuild" the actual circuit. Note that the battery current is the same as the current through R_1 and R_{P_2}, because they are all in series. (In Fig. 18.6d, $I = I_1 = 3.00$ A and $I = I_{\text{P}_2} = 3.00$ A.) Therefore, the voltages across these resistors are

$$V_1 = I_1 R_1 = (3.00\ \text{A})(6.00\ \Omega) = 18.0\ \text{V}$$

and

$$V_{\text{P}_2} = I_{\text{P}_2} R_{\text{P}_2} = (3.00\ \text{A})(2.00\ \Omega) = 6.00\ \text{V}$$

Because R_{P_2} is made up of R_2 and R_{S_1} (Fig. 18.6c and d), there must be a 6.00-V drop across both of these resistors. We can use this to calculate the current in each.

$$I_2 = \frac{V_2}{R_2} = \frac{6.00\ \text{V}}{4.00\ \Omega} = 1.50\ \text{A} \qquad \text{and} \qquad I_{\text{S}_1} = \frac{V_{\text{S}_1}}{R_{\text{S}_1}} = \frac{6.00\ \text{V}}{4.00\ \Omega} = 1.50\ \text{A}$$

Next, notice that I_{S_1} is also the current in R_{P_1} and R_5, because they are in series. (In Fig. 18.6b, $I_{\text{S}_1} = I_{\text{P}_1} = I_5 = 1.50$ A.)

The resistors' individual voltages are therefore

$$V_{P_1} = I_{s_1}R_{P_1} = (1.50\text{ A})(1.50\ \Omega) = 2.25\text{ V}$$

and

$$V_5 = I_{s_1}R_5 = (1.50\text{ A})(2.50\ \Omega) = 3.75\text{ V}$$

respectively. (As a check, note that the voltages do, in fact, add to 6.00 V.)

Finally, the voltage across R_3 and R_4 is the same as V_{P_1} (why?), and

$$V_{P_1} = V_3 = V_4 = 2.25\text{ V}$$

With these voltages and known resistances, the last two currents, I_3 and I_4, are

$$I_3 = \frac{V_3}{R_3} = \frac{2.25\text{ V}}{6.00\ \Omega} = 0.38\text{ A}$$

and

$$I_4 = \frac{V_4}{R_4} = \frac{2.25\text{ V}}{2.00\ \Omega} = 1.13\text{ A}$$

The current (I_{s_1}) is expected to divide at the R_3–R_4 junction. Thus, a check is available: $I_3 + I_4$ does equal I_{s_1}, within rounding errors.

Follow-Up Exercise. In this Example, verify that the total power delivered to all of the resistors is the same as the power output of the battery.

18.2 Multiloop Circuits and Kirchhoff's Rules

OBJECTIVES: To (a) understand the physical principles that underlie Kirchhoff's circuit rules, and (b) apply these rules in the analysis of actual circuits.

Series–parallel circuits with a single voltage source can always be reduced to a single loop, as we have seen in Example 18.3. However, circuits may contain several loops, each one having several voltage sources, resistances, or both. In many cases, resistors may not be connected either in series or in parallel. A multiloop circuit, which does not lend itself to the methods of Section 18.1, is shown in ▼Fig. 18.7a. Even though some groups of resistors may be replaced by their equivalent resistances (Fig. 18.7b), this circuit can be reduced only so far by using parallel and series procedures.

Analyzing these types of circuits requires a more general approach—that is, the application of **Kirchhoff's rules**.* These embody conservation of charge and energy. (Although not stated specifically, Kirchhoff's rules were applied to the parallel and series arrangements in Section 18.1.) First, it is useful to introduce some terminology that will help us describe more complex circuits:

- A point where three or more wires are joined is called a **junction**, or **node**—for example, point A in Fig. 18.7b.
- A path connecting two junctions is called a **branch**. A branch may contain one or more circuit elements and there may be more than two branches between two junctions.

Teaching tip: Try not to give the impression that Kirchhoff's rules apply only to complex circuits. Show how they also apply to simple series and/or parallel circuits with only one voltage source.

Teaching tip: The junction theorem is generally accepted by most students. However, the concept that the current into a lightbulb equals the current out of the bulb is more difficult for some students. Students tend to think that current gets "used up" on its way through an element. Placing two identical ammeters such that one is on each side of a circuit element helps illustrate the important point that current is never "lost." Stress that in electric circuits, current is simply the vehicle for delivering energy and converting it from one form into another.

Note: Kirchhoff's rules were developed by the German physicist Gustav Kirchhoff (1824–1887).

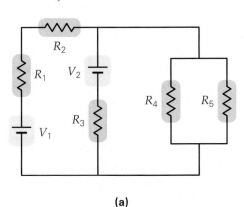
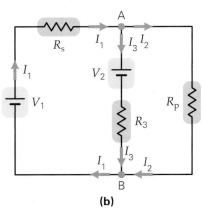

(a)   **(b)**

◀ **FIGURE 18.7 Multiloop circuit** In general, a circuit that contains voltage sources in more than one loop cannot be completely reduced by series and parallel methods alone. However, some reductions within each loop may be possible, such as from part **(a)** to part **(b)**. At a junction the current divides or comes together, as at junctions A and B in part **(b)**, respectively. Any path between two junctions is called a *branch*. In part (b), there are three branches—that is, there are three different ways to get from junction A to junction B.

*Gustav Robert Kirchhoff (1824–1887) was a German scientist who made important contributions to electrical circuit theory and light spectroscopy. He invented the spectroscope, a device that separated light in its colors and studied the emitted light "signature" of various elements (see Chapter 27).

Illustration 30.7 The Loop Rule

Exploration 30.1 Circuit Analysis

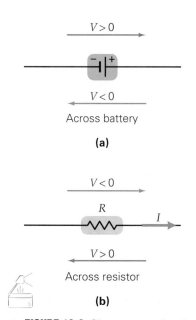

▲ **FIGURE 18.8** Sign convention for Kirchhoff's loop theorem **(a)** The battery voltage is taken as positive if it is traversed from the negative to the positive terminal. It is assigned a negative value if traversed from the positive to the negative. **(b)** The voltage across a resistor is taken as negative if the resistor is traversed in the direction of the assigned current ("downstream"). It is taken as positive if the resistance is traversed opposite that of the assigned branch current ("upstream").

Kirchhoff's Junction Theorem

Kirchhoff's first rule, or **junction theorem**, states that the algebraic sum of the currents at any junction is zero:

$$\Sigma I_i = 0 \qquad \text{sum of currents at a junction} \qquad (18.4)$$

This means that the sum of the currents entering at a junction (taken as positive) and the currents leaving the junction (taken as negative) is zero. This rule is just a statement of charge conservation—no charge can pile up at a junction (why?). For the junction at point A in Fig. 18.7b, for example, the algebraic sum of the currents is $I_1 - I_2 - I_3 = 0$; equivalently,

$$I_1 = I_2 + I_3$$
current in = current out

(This rule was applied in analyzing parallel resistances in Section 18.1.)

Problem-Solving Hint

> Sometimes it is not evident whether a particular current is directed into or out of a junction just by looking at a circuit diagram. In this case, a direction is simply *assumed*. Then the currents are calculated, without worry about their directions. If some of the assumptions turn out to be opposite to the actual directions, then negative answers for these currents will result. This outcome means that the directions of these currents are opposite to the directions initially chosen (or guessed).

Kirchhoff's Loop Theorem

Kirchhoff's second rule, or **loop theorem**, states that the algebraic sum of the potential differences (voltages) across all of the elements of any *closed loop* is zero:

$$\Sigma V_i = 0 \qquad \text{sum of voltages around a closed loop} \qquad (18.5)$$

This expression means that the sum of the voltage rises (an increase in potential) equals the sum of the voltage drops (a decrease in potential) around a closed loop, which must be true if energy is conserved. (This rule was used in analyzing series resistances in Section 18.1.)

Notice that traversing a circuit loop in different directions will yield either a voltage rise or a voltage drop across each circuit element. Thus, it is important to establish a sign convention for voltages. We will use the convention illustrated in ◄Fig. 18.8. The voltage across a battery is taken as positive (a voltage rise) if it is traversed from the negative terminal to the positive (Fig. 18.8a). Thus voltage across a battery will be negative if it is traversed in the opposite direction, from the positive to negative. (Note that the direction of the *current* through the battery has *nothing* to do with the sign of the battery voltage. The sign of this voltage depends only on the direction we choose to cross the battery.)

The voltage across a resistor is taken to be negative (a decrease) if the resistor is traversed in the same direction as the assigned current, in essence going "downhill" potential-wise (Fig. 18.8b). Clearly the voltage will be positive if the resistor is traversed in the opposite direction (going counter to the current direction, gaining electrical potential). Used together, these sign conventions allow the summation of the voltages around a closed loop, regardless of the direction chosen to do that sum. It should be clear that Eq. 18.5 is the same in either case. To see this, note that reversing the chosen loop direction simply amounts to multiplying Eq. 18.5 (from the original direction) by −1. This operation, of course, does not change the equation.

Problem-Solving Hint

> In applying Kirchhoff's loop theorem, the sign of a voltage across a resistor is determined by the direction of the current in that resistor. However, there can be situations in which the current direction is not obvious. How do you handle the voltage signs in such cases? The answer is simple: After assuming a direction for the current, follow the voltage sign convention *based on this assumed direction*. This guarantees that the two sign conventions are mathematically consistent. Thus, if it turns out that the actual current direction is opposite your choice, the voltage drops will automatically reflect that.

A graphical interpretation of Kirchhoff's loop theorem is presented in the Learn by Drawing future on page 602. Integrated Example 18.4, in which a simple parallel circuit is reexamined using Kirchhoff's rules, shows that our previous series–parallel considerations were consistent with these circuit rules. In this Example, take care to notice how important it is draw a correct circuit diagram—it can guide you as to how to proceed.

Teaching tip: It is extremely important for students to get the correct signs in their expressions when applying Kirchhoff's rules.

Integrated Example 18.4 ■ A Simple Circuit: Using Kirchhoff's Rules

Two resistors R_1 and R_2 are connected in parallel. This combination is in series with a third resistor R_3 that has the largest resistance of the three. A battery completes the circuit, with one electrode connected to the beginning and the other to the end of this network. (a) Which resistor will carry the most current, (1) R_1, (2) R_2, or (3) R_3? Explain. (b) In the actual circuit, assume $R_1 = 6.0\ \Omega$, $R_2 = 3.0\ \Omega$, $R_3 = 10.0\ \Omega$, and the battery's terminal voltage is 12.0 V. Apply Kirchhoff's rules to determine the current in each resistor and the voltage across each resistor.

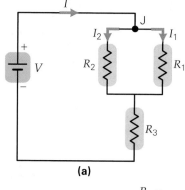

(a) Conceptual Reasoning. It is best to first look at a schematic circuit diagram based on the word description of the network (◄Fig. 18.9). You might think that the resistor with the least resistance would carry the most current. But be careful; this holds only if *all* the resistors are in parallel. This is *not* true here. The two parallel resistors each carry only a portion of the total current. However, because the total of their two currents is in R_3, that resistor carries the total, and therefore the most, current. Thus, the correct answer is (3).

(b) Quantitative Reasoning and Solution.

Given: $R_1 = 6.0\ \Omega$
$\qquad\quad R_2 = 3.0\ \Omega$
$\qquad\quad R_3 = 10.0\ \Omega$
$\qquad\quad V = 12.0$ V

Find: the current in each resistor and the voltage across each resistor

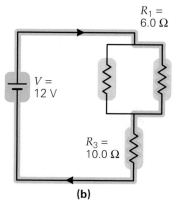

There are three unknown currents: the total current (I) and the currents in each of the parallel resistors (labeled as I_1 and I_2). Since there is only one battery, the current must be clockwise (shown in the figure). Applying Kirchhoff's junction theorem to the first junction (J in Fig.18.9a), we have

$$\Sigma I_i = 0 \qquad \text{or} \qquad I - I_1 - I_2 = 0 \tag{1}$$

Using the loop theorem in the clockwise direction in Fig. 18.9b, we cross the battery from the negative to the positive terminal and then traverse R_1 and R_3 to complete the loop. The resulting equation (showing the voltage signs explicitly) is

$$\Sigma V_i = 0 \qquad \text{or} \qquad +V + (-I_1R_1) + (-IR_3) = 0 \tag{2}$$

A third equation can be obtained by applying the loop theorem but this time going through R_2 instead of R_1 (Fig. 18.9c). This yields

$$\Sigma V_i = 0 \qquad \text{or} \qquad +V + (-I_2R_2) + (-IR_3) = 0 \tag{3}$$

Putting in the battery voltage (in volts) and resistances (in ohms) and rearranging:

$$I = I_1 + I_2 \tag{1a}$$
$$12 - 6I_1 - 10I = 0 \qquad \text{or} \qquad 6 - 3I_1 - 5I = 0 \tag{2a}$$
$$12 - 3I_2 - 10I = 0 \tag{3a}$$

Adding Eqs. (2a) and (3a) yields $18 - 3(I_1 + I_2) - 15I = 0$. However, from Eq. (1a), $I = I_1 + I_2$. Therefore, this equation becomes

$$18 - 3I - 15I = 0 \qquad \text{or} \qquad 18I = 18$$

and solving for the total current yields $I = 1.00$ A.

Eqs. (3a) and (1a) can then be solved for the remaining currents:

$$I_2 = \tfrac{2}{3}\text{A} \qquad \text{and} \qquad I_1 = \tfrac{1}{3}\text{A}$$

These answers are consistent with our circuit-diagram reasoning in part (a).

Because the currents are now known, the voltages can be obtained from Ohm's law, $V = IR$. Thus,

$$V_1 = I_1R_1 = \left(\tfrac{1}{3}\text{A}\right)(6.0\ \Omega) = 2.0 \text{ V}$$
$$V_2 = I_2R_2 = \left(\tfrac{2}{3}\text{A}\right)(3.0\ \Omega) = 2.0 \text{ V}$$
$$V_3 = I_3R_3 = (1.0\text{ A})(10.0\ \Omega) = 10.0 \text{ V}$$

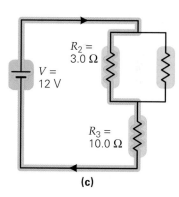

▲ **FIGURE 18.9** Sketching circuit diagrams and Kirchhoff's rules **(a)** The circuit diagram resulting from the written description in Integrated Example 18.4. **(b)** and **(c)** The two loops used in the analysis of Integrated Example 18.4.

Take a quick look at the results to see if they are reasonable. As expected, the voltage drops across the parallel resistors are equal. Because of that, two thirds of the total current is in the resistor with the least resistance. Also, the total voltage across the network is 12.0 V, as it must be.

Exploration 30.2 Lightbulbs

(continues on page 603)

LEARN BY DRAWING

KIRCHHOFF PLOTS: A GRAPHICAL INTERPRETATION OF KIRCHHOFF'S LOOP THEOREM

The equation form of Kirchhoff's loop theorem has a geometrical visualization that may help you develop better insight into its meaning. This graphical approach allows us to visualize the potential changes in a circuit, either to anticipate the results of mathematical analysis or to qualitatively confirm the results. (Don't forget that a complete analysis usually also includes the junction theorem—see Example 18.5.)

The idea is to make a three-dimensional plot based on the circuit diagram. The wires and elements of the circuit form the basis for the x–y plane, or the diagram's "floor." Plotted perpendicularly to this plane, along the z-axis, is the electric potential (V), with an appropriate choice for zero. Such a diagram is called a *Kirchhoff plot* (Fig. 1).

The rules for constructing a Kirchhoff plot are simple: Start at a known potential value, and go around a complete loop, finishing where you started. Because you come back to the same location, the sum of all the rises in potential (positive voltages) must be balanced by the sum of the drops (negative voltages). This requirement is the geometrical expression of energy conservation, embodied mathematically by Kirchhoff's loop theorem.

Thus, if the potential increases (say, in traversing a battery from negative terminal to positive terminal), draw a rise in the z-direction. In this instance, the rise represents the terminal voltage of the battery. Similarly, if the potential decreases (for example, in traversing a resistor in the direction of the current), make sure the potential drops. If possible, try to draw the rises and drops (the voltages) to scale. That is, if there is a large rise in potential (such as across a high-voltage battery), then draw that rise to be large in proportion to the others on the diagram.

For elaborate circuits, this graphical method may prove to be too complicated for practical use. Nevertheless, it is always good to keep this concept in mind, as it illustrates the fundamental physics behind the loop theorem.

As an example of the power of this method, consider the circuit in Fig. 1: a battery with internal resistance r wired to a single external resistor R. The direction of the current is from the anode to the cathode through the external resistor. The potential of the battery's cathode is chosen as zero. Starting there and traversing the circuit in the direction of the current, there is a rise in potential from the battery cathode to the anode. From there the potential remains constant as the current travels through the wires to the external resistor. That is, no significant voltage *drop* should be indicated along connecting wires (why?).

At the resistor, there must be a drop in potential. However, it must not drop to zero, because there must be some voltage left to produce current through the internal resistance. Thus it can be reasoned visually why the terminal voltage of the battery, V, must be less than its emf (the rise between a and b).

Figure 2 shows two resistors in series, and that combination in parallel with a third resistor. For simplicity, all three resistors have the same resistance (R) and the battery's internal resistance is assumed to be zero. Starting at point a, there is a rise in potential corresponding to the battery voltage. Then, as the loop is traced, it goes through the single resistor, so there must be one drop in potential equal in magnitude to $\mathcal{E}$.

Following the loop that includes the two resistors, each must have half the total drop (why?). So each will carry half the current of the single resistor. Recall that in parallel circuits, the largest resistance carries the least current. Notice how nicely the geometrical approach helps develop your intuition and allows anticipation of quantitative results.

As an exercise, try redrawing Fig. 2 if, instead, the series resistors had resistance values of R and 2R. Which of these two now has the largest voltage? How do the currents in the resistors compare to the previous situation? Lastly, analyze the circuit mathematically, to see whether your expectations are confirmed.

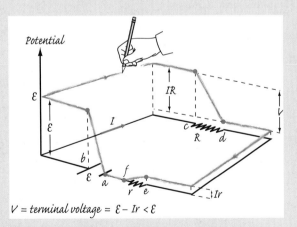

$V = terminal\ voltage = \mathcal{E} - Ir < \mathcal{E}$

FIGURE 1 Kirchhoff plots: A graphical problem-solving strategy The schematic of the circuit is laid out in the x–y plane, and the electric potential is plotted perpendicularly along the z-axis. Usually, the zero of the potential is taken to be the negative terminal of the battery. A direction for current is assigned, and the value of the potential is plotted around the circuit, following the rules for gains and losses. This particular plot shows a rise in potential when the battery is traversed from cathode to anode, followed by a drop in potential across the external resistor, and a smaller drop in potential across the battery's internal resistance.

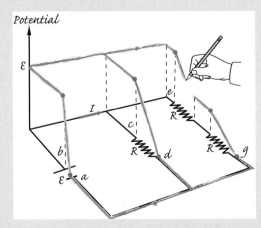

FIGURE 2 Kirchhoff plot of a more complex circuit Imagine how the plot would change if you were to vary the values of the three resistors. Then analyze the circuit mathematically to see whether your plot allowed you to anticipate the voltages and currents.

A special note before leaving this Example: We know that the answers must be in amperes and volts because we used amps, volts, and ohms were used consistently throughout. If you stay within this system (that is, express quantities in volts, amps and ohms), you don't *need* to carry units; the answers will automatically be in these units. (Of course, it is always a good idea to check your units if there is a question.)

Follow-Up Exercise. (a) In this Example, predict what will happen to each of the currents if R_2 is increased. Explain your reasoning. (b) Rework part (b) of this Example, changing R_2 to 8.0 Ω, and see if your reasoning is correct.

Application of Kirchhoff's Rules

Integrated Example 18.4 could have been worked using the expressions for equivalent resistances. However, more complicated, multiloop circuits (which may have resistors neither in parallel or series) require a more structured approach. In this book, the following general steps will be used when applying Kirchhoff's rules:

1. Assign a current and direction of current for each branch in the circuit. This assignment is done most conveniently at junctions.
2. Indicate the loops and the directions in which they are to be traversed (▸Fig. 18.10). Every branch *must* be in at least one loop.
3. Apply Kirchhoff's first rule (junction rule) at each junction that gives a unique equation. (This step gives a set of equations that includes *all* currents, but you may have redundant equations from two different junctions.)
4. Traverse the number of loops necessary to include all branches. In traversing a loop, apply Kirchhoff's second rule, the loop theorem (using $V = IR$ for each resistor), and write the equations, using our sign conventions.

If this procedure is applied properly, Steps 3 and 4 give a set of N equations if there are N unknown currents. These equations may then be solved for the currents. If more loops are traversed than necessary, you might also have redundant loop equations. Only the number of loops that includes each branch *once* is needed.

This procedure may seem complicated, but it's generally straightforward, as the following Example shows.

Example 18.5 ■ Branch Currents: Using Kirchhoff's Rules

For the circuit diagrammed in Fig. 18.10, find the current in each branch.

Thinking It Through. Series or parallel calculations cannot be used here. (Why?) Instead, the solution is begun by assigning current directions ("best guesses") in each loop, and then the junction theorem and the loop theorem are used (twice—once for each inner loop) to generate three equations, because there are three currents.

Solution.

Given: Values in Fig. 18.10   *Find:* The current in each of the three branches

The chosen current directions and loop traversal directions are shown in the figure. (Remember, these directions are not unique; choose them, work the problem, and check the final current signs to see if your choices were correct.) There is a current in every branch, and every branch is in at least one loop. (Some branches are in more than one loop, which is acceptable.)
Applying Kirchhoff's first rule at the left-hand junction gives

$$I_1 - I_2 - I_3 = 0$$

or, after rearranging,

$$I_1 = I_2 + I_3 \qquad (1)$$

(For the other junction, $I_2 + I_3 - I_1 = 0$ but this is equivalent to Eq. (1), so we are done with the junctions.)
Going around loop 1 as in Fig. 18.10 and applying Kirchhoff's loop theorem with the sign conventions gives

$$\Sigma V_i = +V_1 + (-I_1 R_1) + (-V_2) + (-I_3 R_3) = 0 \qquad (2)$$

Putting in the numerical values gives

$$+6 - 6I_1 - 12 - 2I_3 = 0$$

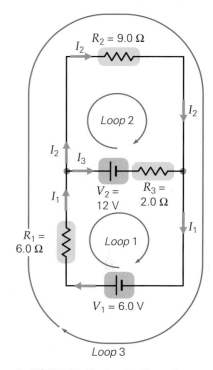

▲ **FIGURE 18.10 Application of Kirchhoff's rules** To analyze a circuit such as the one shown for Example 18.5, assign a current and its direction for each branch in the circuit (most conveniently done at junctions). Identify each loop and the direction of traversal. Then write current equations for each independent junction (using Kirchhoff's junction theorem). Also write voltage equations for as many loops as needed to include every branch (using Kirchhoff's loop theorem). Be careful to observe sign conventions.

(continues on next page)

Rearranging this equation and dividing both sides by 2, we have

$$3I_1 + I_3 = -3$$

For convenience, units are omitted (they are all in amps and ohms, and thus are self-consistent).
For loop 2, the loop theorem yields

$$\Sigma V_i = +V_2 + (-I_2 R_2) + (+I_3 R_3) = 0 \qquad (3)$$

Again, after substituting in the values and rearranging, we have

$$9I_2 - 2I_3 = 12 \qquad (3a)$$

Equations (1), (2a), and (3a) form a set of three equations with three unknowns. You can solve for the currents in many ways. For example, substitute Eq. (1) into Eq. (2a) to eliminate I_1:

$$3(I_2 + I_3) + I_3 = -3$$

which, after rearranging and dividing by 3, simplifies to

$$I_2 = -1 - \tfrac{4}{3}I_3 \qquad (4)$$

Then, substituting Eq. (4) into Eq. (3a) eliminates I_2:

$$9\left(-1 - \tfrac{4}{3}I_3\right) - 2I_3 = 12$$

Finishing the algebra and solving for I_3, we have

$$-14I_3 = 21 \qquad \text{or} \qquad I_3 = -1.5 \text{ A}$$

The minus sign tells us that the wrong direction was assumed for I_3.
Putting the value of I_3 into Eq. (4) gives I_2:

$$I_2 = -1 - \tfrac{4}{3}(-1.5 \text{ A}) = 1.0 \text{ A}$$

Then, from Eq. (1),

$$I_1 = I_2 + I_3 = 1.0 \text{ A} - 1.5 \text{ A} = -0.5 \text{ A}$$

Again, the minus sign indicates that the direction of I_1 was also initially chosen incorrectly.
Note that this analysis did not have to use loop 3. The equation for this loop would be redundant, containing no new information (can you show this?).

Follow-Up Exercise. Rework this Example, using the junction theorem and loops 3 and 1 instead of loops 1 and 2.

18.3 RC Circuits

OBJECTIVES: To (a) understand the charging and discharging of a capacitor through a resistor, and (b) calculate the current and voltage at specific times during these processes.

Until now, only circuits that have constant currents have been considered. In some direct-current (dc) circuits, the current can *vary with time* while maintaining a constant direction (it still is "dc"). Such is the case in **RC circuits**, which most generally consist of several resistors and capacitors.

Charging a Capacitor through a Resistor

The charging of an uncharged capacitor by a battery is depicted in ◀Fig. 18.11. After the switch is closed, even though there is a gap (the capacitor plates), charge *must* flow while the capacitor is charging.

The maximum charge (Q_o) that the capacitor can attain depends on its capacitance (C) and the battery voltage (V_o). To determine the value of Q_o and understand how both the current and capacitor charge vary with time, consider the following argument. At $t = 0$, there is no charge on the capacitor and thus no voltage across it. By Kirchhoff's loop theorem, this means that the full battery voltage must appear across the resistor, resulting in an initial (maximum) current $I_o = V_o/R$. As charge on the capacitor increases, so must the voltage across its plates, thereby reducing the resistor's voltage and current. Eventually, when the capacitor is charged to its maximum, the current becomes zero. At this time, the resistor's voltage is zero and the capacitor's voltage must be V_o. Because of the relationship between the charge on a capacitor and its voltage (Ch. 16, Eq 16.19), the maximum capacitor charge is given by $Q_o = CV_o$. (This sequence is depicted in Fig. 18.11.)

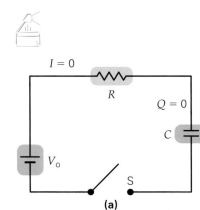
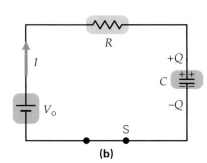

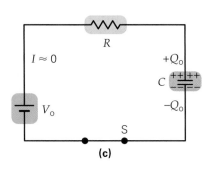

▲ **FIGURE 18.11** Charging a capacitor in a series RC circuit **(a)** Initially there is no current and no charge on the capacitor. **(b)** When the switch is closed, there is a current in the circuit until the capacitor is charged to its maximum value. The rate of charging depends on the circuit's time constant, τ ($= RC$). **(c)** For times much larger than τ, the current is very close to zero, and the capacitor is said to be fully charged.

The resistance is one of two factors that determines how fast the capacitor is charged, because the larger its value, the greater the resistance to charge flow. The capacitance is the other factor that influences the charging speed—it simply takes longer to charge a larger capacitor. Analysis of this type of circuit requires mathematics beyond the level of this book. However, it can be shown that the voltage across the capacitor increases exponentially with time according to

$$V_C = V_o[1 - e^{-t/(RC)}] \quad \begin{array}{l}\textit{(charging capacitor}\\ \textit{voltage in an RC circuit)}\end{array} \quad (18.6)$$

where e has an approximate value of 2.718. (Recall that the irrational number e is the base of the system of *natural logarithms*.) A graph of V_C versus t is shown in ▸Fig. 18.12a. As expected, V_C approaches V_o, the capacitor's maximum voltage, after a "long" time.

A graph of I versus t is given in Fig. 18.12b. The current varies with time according to

$$I = I_o e^{-t/(RC)} \quad (18.7)$$

Thus the current decreases exponentially with time and has its largest value initially, as expected.

According to Eq. 18.6, it would take an infinite time for the capacitor to become fully charged. However, in practice, we know that most capacitors become essentially completely charged in relatively short times. It is therefore customary to use a special value to express the "charging time." This value, called the **time constant (τ)**, is

$$\tau = RC \quad \textit{time constant for RC circuits} \quad (18.8)$$

(You should be able to show that RC has units of seconds.) After an elapsed time of one time constant, that is $t = \tau = RC$, the voltage across the charging capacitor has risen to 63% of the maximum possible. This can be seen by evaluating V_C (Eq. 18.6), replacing t with τ ($= RC$):

$$V_C = V_o(1 - e^{-\tau/\tau}) = V_o(1 - e^{-1})$$
$$\approx V_o\left(1 - \frac{1}{2.718}\right) = 0.63V_o$$

Because $Q \propto V_C$, this result means that the capacitor has 63% of its maximum possible charge after one time constant has elapsed. You should be able to show that after one time constant, the current has dropped to 37% of its initial (maximum) value, I_o.

After a time of two time constants has elapsed ($t = 2\tau = 2RC$), you should be able to show that the capacitor is charged to more than 86% of its maximum value; at $t = 3\tau = 3RC$, the capacitor is charged to 95% of its maximum value; and so on. As a general rule of thumb, a capacitor is considered to be "fully charged" after "several time constants" have elapsed.

Discharging a Capacitor through a Resistor

▸Figure 18.13a shows a capacitor being *discharged* through a resistor. In this case, the voltage across the capacitor *decreases* exponentially with time, as does the current. The expression for the decay of the capacitor's voltage (from its maximum voltage of V_o) is

$$V_C = V_o e^{-t/(RC)} = V_o e^{-t/\tau} \quad \begin{array}{l}\textit{(discharging capacitor}\\ \textit{voltage in an RC circuit)}\end{array} \quad (18.9)$$

Then after one time constant, the capacitor voltage is at 37% of its original value (Fig. 18.13b). The current in the circuit decays exponentially also, following Eq. 18.7. This is also the behavior of a capacitor in a heart defibrillator as it discharges its stored energy (as a flow of charge or current) through the heart (resistance R) in a discharge time of about 0.1 sec. RC circuits are also an integral part of cardiac pacemakers, which alternately charge a capacitor, transfer the energy to the heart, and repeat this at a rate determined by the time constant. For details on these interesting and important instruments, refer to Insight 18.11 on Applications of RC Circuits to Cardiac Medicine on page 608. Aspects of RC circuits in medical settings are also covered in Exercises 107 and 108. As a practical application, consider their use in modern cameras in Example 18.6.

Demonstration/activity: Use a square-wave oscillator and oscilloscope to demonstrate the charging and discharging of an RC circuit.

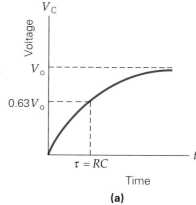

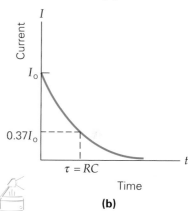

▲ **FIGURE 18.12** Capacitor charging in a series RC circuit
(a) In a series RC circuit, as the capacitor charges, the voltage across it increases nonlinearly, reaching 63% of its maximum voltage (V_o) in one time constant, τ. **(b)** The current in this circuit is initially a maximum ($I_o = V_o/R$) and decays exponentially, falling to 37% of its initial value in one time constant, τ.

Illustration 30.6 RC Circuit

Exploration 30.6 RC Time Constant

Note: Most calculators now have an e^x button. For exponential calculations, practice using it. For example, make sure your calculator gives you $e^{-1} \approx 0.37$.

Teaching tip: Show students how to use their calculators for operations involving e^x and ln x to calculate exponentially varying current and voltage.

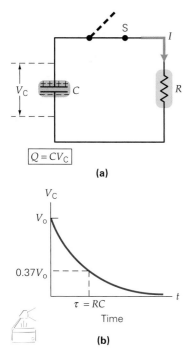

$Q = CV_C$

(a)

$0.37V_0$

$\tau = RC$

Time

(b)

▲ FIGURE 18.13 Capacitor discharging in a series RC circuit **(a)** The capacitor is initially fully charged. When the switch is closed, current appears in the circuit as the capacitor begins to discharge. **(b)** In this case, the voltage across the capacitor (and the current in the circuit) decays exponentially with time, falling to 37% of its initial value in one time constant, τ.

Teaching tip: Capacitors with values of 1.0 F or larger are readily available nowadays. Charge a 1.0-F capacitor with a 12-V storage cell, and 72 J of energy is stored. Next, discharge it through a small lightbulb rated at, say, 6 W (J/s; at a full 12 V). Make the argument that during its decay, it will operate, on average, at about 3 W and therefore should stay on for about 20–25 seconds. Make this prediction and try to show it experimentally.

Teaching tip: Placing an ammeter in series with the element through which the current is being measured increases the resistance of the circuit. Explain why this increase should be kept to a minimum. (This phenomenon is one reason that ammeters are expensive.)

▶ FIGURE 18.14 Blinker circuit **(a)** When a neon tube is connected across the capacitor in a series RC circuit that has the proper voltage source, the voltage across the tube will oscillate with time. As a result, the tube periodically flashes or blinks. **(b)** A graph of tube voltage versus time shows the voltage oscillating between V_b, the "breakdown" voltage, and V_m, the "maintaining" voltage. See text for detailed discussion.

Example 18.6 ■ RC Circuits in Cameras: Flash Photography Is as Easy as Falling Off a Log(arithm)

In many cameras, the built-in flash gets its energy from that stored in a capacitor. The capacitor is charged using long-life batteries with voltages of typically 9.00 V. Once the bulb is fired, the capacitor must recharge quickly through an internal RC circuit. If the capacitor has a value of 0.100 F, what must the resistance be so the capacitor is charged to 80% of its maximum charge (the minimum charge to fire the bulb again) in 5.00 s?

Thinking It Through. After one time constant, the capacitor will be charged to 63% of its maximum voltage and charge. Because the capacitor needs 80%, the time constant must be *less* than 5.00 s. Eq. 18.6 can be used (along with a calculator) to determine the time constant. From that the required value of resistance can be determined.

Solution. The data given include the final voltage across the capacitor, V_C, which is 80% of the battery's voltage, which means that Q is 80% of the maximum charge.

Given: $C = 0.100$ F
$V_B = V_o = 9.00$ V
$V_C = 0.80V_o = 7.20$ V
$t = 5.00$ s

Find: R (the resistance required so that the capacitor is 80% charged in 5.00 s)

Putting the data into Eq. 18.6, $V_C = V_o(1 - e^{-t/\tau})$, we have

$$7.20 = 9.00(1 - e^{-5.00/\tau})$$

Rearranging this equation yields $e^{-5.00/\tau} = 0.20$, and the reciprocal of this expression (to make the exponent positive) is

$$e^{5.00/\tau} = 5.00$$

To solve for the time constant, recall that if $e^a = b$, then a is the *natural logarithm* (ln) of b. Thus, in our case, $5.00/\tau$ is the natural logarithm of 5.00. Using a calculator, we find that $\ln 5.00 = 1.61$. Therefore

$$\frac{5.00}{\tau} = \ln 5.00 = 1.61$$

or

$$\tau = RC = \frac{5.00}{1.61} = 3.11 \text{ s}$$

Solving for R yields

$$R = \frac{3.11 \text{ s}}{C} = \frac{3.11 \text{ s}}{0.10 \text{ F}} = 31 \ \Omega$$

As expected, the time constant is less than 5.0 s, because achieving 80% of the maximum voltage requires a time interval longer than one time constant.

Follow-Up Exercise. (a) In this Example, how does the energy stored in the capacitor (after 5.00 s) compare with the maximum energy storage? Explain why it isn't 80%. (b) If you waited 10.00 s to charge the capacitor, what would its voltage be? Why isn't it twice the voltage that exists across the capacitor after 5.00 s?

An application of an RC circuit is diagrammed in ▼Fig. 18.14a. This circuit is called a *blinker circuit* (or a *neon-tube relaxation oscillator*). The resistor and capacitor are initially wired in series, and then a miniature neon tube is connected in parallel with the capacitor.

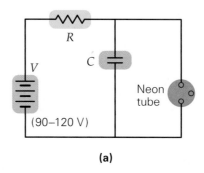

(a)

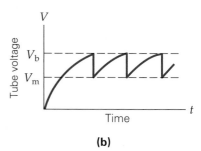

(b)

When the circuit is closed, the voltage across the capacitor (and the neon tube) rises from 0 to V_b, which is the *breakdown voltage* of the neon gas in the tube (about 80 V). At that voltage, the gas becomes ionized (that is, electrons are freed from atoms, creating positive and negative charges that are free to move). Thus the gas begins to conduct electricity, and the tube lights. When the tube is in this conducting state, the capacitor discharges through it, and the voltage falls rapidly (Figure 18.14b). When the voltage drops below V_m, called the *maintaining voltage*, the ionization in the tube cannot be sustained, and the tube stops conducting. The capacitor begins charging again, the voltage rises from V_m to V_b, and the cycle repeats. This continual repetition causes the tube to blink on and off.

18.4 Ammeters and Voltmeters

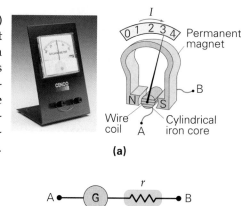

(a)

(b)

▲ **FIGURE 18.15 The galvanometer**
(a) A galvanometer is a current-sensitive device whose needle deflection is proportional to the current in its coil. **(b)** The circuit symbol for a galvanometer is a circle containing a G. The internal resistance (r) of the meter is indicated explicitly as r.

<u>OBJECTIVES:</u> To understand (a) how galvanometers are used to make ammeters and voltmeters, (b) how multirange versions of these devices are constructed, and (c) how they are connected to measure current and voltage in real circuits.

As the names imply, an **ammeter** measures current *through* circuit elements and a **voltmeter** measures voltages *across* circuit elements. The basic component common to both of these meters is a **galvanometer** (▶Fig. 18.15a). The galvanometer operates on magnetic principles covered in Chapter 19. In this chapter, it will be treated simply as a circuit element with an internal resistance r (typically about 50 Ω) whose needle deflection is proportional to the current in it (Fig. 18.15b).

The Ammeter

A galvanometer measures current, but because of its small resistance, only currents in the microampere range can be measured without burning out its wires. However, there is a way to construct an ammeter to measure larger currents with a galvanometer. To do this, a small *shunt resistor* (with a resistance of R_s) is employed in parallel with a galvanometer. The job of the shunt resistor (or "shunt" for short) is to take most of the current (▼Fig. 18.16). This requires the shunt to have much less resistance than the galvanometer ($R_s \ll r$). The following Example illustrates how the resistance of the shunt is determined in the design of an ammeter.

Note: Ammeters are connected in series with the element whose current they are measuring (Fig. 18.16b).

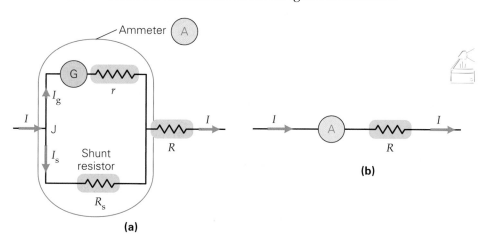

(a)

(b)

◀ **FIGURE 18.16 A dc ammeter**
Here, R is the resistance of the resistor whose current is being measured. **(a)** A galvanometer in parallel with a shunt resistor (R_s) creates an ammeter capable of measuring various ranges of current, depending on the value of R_s. **(b)** The circuit symbol for an ammeter is a circle with an A inside it. (See Example 18.7 for a detailed discussion of ammeter design.)

Exploration 30.4 *Galvanometers and Ammeters*

Example 18.7 ■ Ammeter Design Using Kirchhoff's Rules: Choosing a Shunt Resistor

Suppose you have a galvanometer that can safely carry a maximum coil current of 200 μA (called its *full-scale sensitivity*) and that has a coil resistance of 50 Ω. It is to be used in an ammeter designed to measure currents up to 3.0 A (at full scale). What is the required shunt resistance? (See Fig. 18.16a).

Thinking It Through. The galvanometer itself can carry only a small current, so most of the current will have to be diverted, or "shunted," through the shunt. Thus, the shunt resistance

(continues on next page)

will have to be much less than the galvanometer's internal resistance. Because the shunt and coil resistance are really two resistors in parallel, they have the same voltage across them. This reasoning along with Kirchhoff's laws should enable us to determine the value of R_s.

Solution. Listing the data, we have

Given: $I_g = 200\,\mu A = 2.00 \times 10^{-4}$ A   *Find:* R_s (shunt resistance)
$r = 50\,\Omega$
$I_{max} = 3.0$ A

Because the voltages across the galvanometer and the shunt resistor are equal, we can write (using subscripts "g" for galvanometer and "s" for shunt—see Fig. 18.16a)

$$V_g = V_s \quad \text{or} \quad I_g r = I_s R_s$$

Using Kirchhoff's junction rule at J, the current I in the external circuit is $I = I_g + I_s$, or $I_s = I - I_g$. Substituting this into the previous equation, we have

$$I_g r = (I - I_g) R_s$$

INSIGHT

18.1 APPLICATIONS OF RC CIRCUITS TO CARDIAC MEDICINE

The normal human heart beats between 60 and 70 times per minute, with each beat delivering about 70 mL of blood—about a gallon per minute. Your heart is essentially a pump composed of specialized muscle cells. The cells are triggered to beat when they receive electrical signals (Fig. 1). These signal (see Insight 16.1 on Nerve Transmission in Chapter 16) are sent by special *pacemaker cells* located in the *sinotrial node* (*SA node* for short) in one of the upper chambers of the heart.

During a heart attack or after an electrical shock, the heart may go into an unregulated beat pattern. If left untreated, this condition would be fatal in minutes. Fortunately, it is possible to return the heart to its normal pattern by passing an electrical current through it. The instrument that does this is called a *cardiac defibrillator*. The main component of a defibrillator is, for our purposes, a capacitor charged to a high voltage.*

Several hundred joules of electric energy are needed to restart the heart. The high-voltage and low-voltage plates of the capacitor are attached to the patient's skin by two "paddles" placed just above the two sides of the heart (Fig. 2a and Fig. 2b). When a

*Because portable batteries aren't capable of high voltages, the charging uses a phenomenon called electromagnetic induction, which will be studied in Chapter 20.

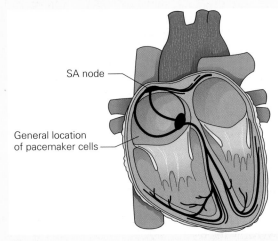

FIGURE 1 The heart The pacemaker cells are located primarily in the SA node. Electrical signals that trigger a heartbeat reach the lower areas of the heart in about 50 milliseconds.

switch is thrown, current flows through the heart, thus transferring the capacitor's energy to the heart in an attempt to cause it to beat correctly.

This discharge through the heart is essentially that of an RC circuit. Typically the capacitor has a value of 10 μF and is charged to 1000 V. (Recent advances in dielectrics have produced compact capacitors at 1 F or more. This reduces the need for high voltage because the stored energy is proportional to the capacitance, $U_C = CV^2/2$.) The resistance of the heart (R_h) is typically about 1000 Ω, giving a time constant (for discharge) of $\tau = R_h C = 10^{-2}$ s = 10 ms.

Because of this 10 ms discharge time constant, the capacitor is essentially fully discharged after 50 ms. If needed, a second charge can be applied. To be able to do this, the capacitor should be able to recharge in about 5 s (Fig. 2c). Thus the *charging* time constant should be about 1 s. This means the charging resistor should have an approximate value (R_c) of $R_c = \tau/C \approx 10^5\,\Omega$.

In some forms of heart disease, the heartbeats are irregular due to problems with the pacemaker cells. The heart can be triggered to beat correctly by using an electronic (implanted) *cardiac pacemaker*. These units are the size of a book of matches, powered by a long-life battery, and usually inserted surgically near the SA node.

Most pacemakers are controlled by a sophisticated triggering circuit that allows the pacemaker to send signals to the heart only if they are needed ("on demand" pacemakers; see Figs. 3a and b). The triggering circuit sends a signal to the pacemaker to "fire" only if the heart does not beat. If the heart is beating normally, the capacitor switch is left in the fully charged position, waiting for the signal to fire.

For our purposes, the pacemaker is simply an RC circuit. The capacitor (typically 10 μF) is kept charged by the battery, and must be ready to release its energy as rapidly as 70 times per minute—the worst-case scenario if the heart's own pacemaker cells are not operating at all. Typically, the resistance of the heart muscle between the pacemaker leads is about 100 Ω meaning the pacemaker discharge time constant is $\tau \approx 1$ ms. Thus it is effectively fully discharged in 5 ms.

To operate at 70 times per second, the capacitor has to charge, fire, and recharge in 1/70 $\approx$ 14 ms. Since it takes about 5 ms to discharge, it has about 9 ms to recharge, meaning a recharge time constant of about 2 ms. This requires a recharge resistor R_C (the one in the circuit through which the capacitor is charged) to be, at most, about 200 Ω (Fig. 3c).

Thus the shunt's resistance R_s is

$$R_s = \frac{I_g r}{I_{max} - I_g}$$

$$= \frac{(2.00 \times 10^{-4}\,\text{A})(50\,\Omega)}{3.0\,\text{A} - 2.00 \times 10^{-4}\,\text{A}}$$

$$= 3.3 \times 10^{-3}\,\Omega = 3.3\,\text{m}\Omega$$

The shunt resistance is very small compared with the coil's resistance. This allows most of the current (2.9998 A at full scale) to pass through the shunt. This ammeter will be able to read currents linearly up to 3.0 A. For example, if a current of 1.5 A were to flow into the ammeter, there would be 100 μA (half the maximum) in the coil, which would give a half-scale reading, or 1.5 A.

Follow-Up Exercise. In this Example, if we had used a shunt resistance of 1.0 mΩ, what would be the full-scale reading (maximum current reading) of the ammeter?

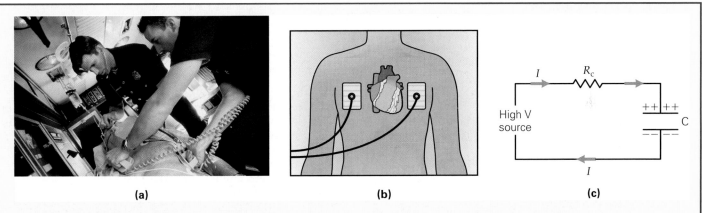

FIGURE 2 Restart the heart! (a) Paddles are placed externally to either side of the heart, and energy from a charged capacitor passes through it, hopefully triggering it into a normal beating pattern. **(b)** This shows the schematic diagram for correct defibrillator use. The discharge is that of an RC circuit. **(c)** Recharging the defibrillator's capacitor, getting it ready to go again, through a (charging) resistor $R_C \approx 10^5\,\Omega$.

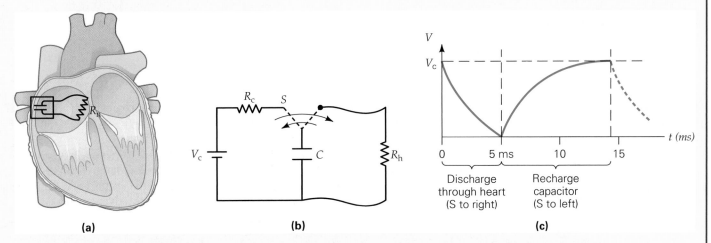

FIGURE 3 Cardiac pacemaker (a) The typical pacemaker (shown as a capacitor in a box) is implanted surgically on or near the heart surface, with its leads attached to the heart muscle (resistance R_n). (The capacitor's charging circuit is not shown.) Other leads (not shown) receive signals from the heart to determine whether the pacemaker needs to "fire." **(b)** The sensing circuit determines the position of the capacitor's "switch." If the heart is not beating, the sensing circuit flips the switch to the right, initiating energy discharge through the heart muscle. If the heart is beating properly, the sensing circuit sets the switch to the left, keeping the capacitor fully charged. **(c)** If the pacemaker is in operation, one full cycle requires about 15 ms. About 5 ms is for discharge through the heart muscle, and 10 ms is required to recharge the capacitor. The recharge is accomplished using a long-life battery, V_c.

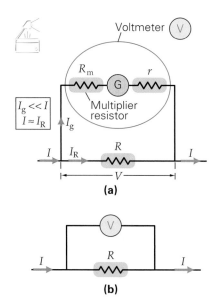

Voltmeter (V)

$I_g \ll I$
$I \approx I_R$

(a)

(b)

▲ **FIGURE 18.17 A dc voltmeter**
Here, R is the resistance of the resistor whose voltage is being measured. **(a)** A galvanometer in series with a multiplier resistor (R_m) is a voltmeter capable of measuring various ranges of voltage, depending on the value of R_m. **(b)** The circuit symbol for a voltmeter is a circle with a V inside it. (See Example 18.8 for a detailed discussion of voltmeter design.)

Teaching tip: Connecting a voltmeter across (in parallel with) the element to measure the difference in potential reduces the resistance of the circuit. Explain why the resistance of the voltmeter should be as large as possible.

Teaching tip: Warn students that ammeters are connected in series and voltmeters are wired in parallel. An ammeter connected in parallel will burn up, and a voltmeter connected in series will stop the current.

Note: Voltmeters are connected in parallel with, or across, the element whose voltage they are measuring (Fig. 18.17b).

Exploration 30.5 Voltmeters

The Voltmeter

A voltmeter that is capable of reading voltages higher than the microvolt range (anything higher than a microvolt would burn out the galvanometer if it were alone) is constructed by connecting a large *multiplier resistor* in *series* with a galvanometer (◄Fig. 18.17). Because the voltmeter has a large resistance, due to the multiplier resistor, it draws little current from the circuit element whose voltage it measures. However, the current that exists in the voltmeter is proportional to the voltage across the circuit element. Thus, the voltmeter can be calibrated in volts. To better understand this configuration, consider Example 18.8.

Example 18.8 ■ Voltmeter Design: Using Kirchhoff's Rules to Choose a Multiplier Resistor

Suppose that the galvanometer in Example 18.7 is to be used instead in a voltmeter with a full-scale reading of 3.0 V. What is the required multiplier resistance?

Thinking It Through. To turn a galvanometer into a voltmeter, we need a reduction in current—accomplished by adding a large "multiplier resistor" in series. All the data necessary to calculate the multiplier resistance are given here and in Example 18.7.

Solution. First, we list the data:

Given: $I_g = 200 \, \mu A$ *Find:* R_m (multiplier resistance)
              $= 2.00 \times 10^{-4} \, A$ (from Example 18.7)
        $r = 50 \, \Omega$ (from Example 18.7)
     $V_{max} = 3.0 \, V$

The resistances of the galvanometer and multiplier are in series. This combination is itself in parallel with the external circuit element (R). Therefore, the voltage across the external circuit element is the sum of the voltages across the galvanometer and multiplier (Fig. 18.17):

$$V = V_g + V_m$$

The voltages across the galvanometer and multiplier resistors are

$$V_g = I_g r \quad \text{and} \quad V_m = I_g R_m$$

Combining these three equations, we have

$$V = V_g + V_m = I_g r + I_g R_m = I_g(r + R_m)$$

Solving for the resistance of the multiplier, we have

$$R_m = \frac{V - I_g r}{I_g}$$

$$= \frac{3.0 \, V - (2.00 \times 10^{-4} \, A)(50 \, \Omega)}{2.00 \times 10^{-4} \, A}$$

$$= 1.5 \times 10^4 \, \Omega = 15 \, k\Omega$$

Notice that the second term in the numerator ($I_g r$) is negligible compared with the full-scale reading of 3.0 V. Thus, to a good approximation, $R_m \approx V/I_g$, or $V \propto I_g$, and the measured voltage is proportional to the current in the galvanometer.

Follow-Up Exercise. The voltmeter in this Example is used to measure the voltage of a resistor in a circuit. A current of 3.00 A flows through the resistor (1.00 Ω) *before* the voltmeter is connected. Assuming that the total *incoming* current (I in Fig. 18.17b) remains the same after the voltmeter is connected, calculate the current in the galvanometer.

For versatility, ammeters and voltmeters may be constructed with a multi-range feature. This is accomplished by providing a choice of shunt or multiplier resistors (▶Fig. 18.18a and b). Combinations of these meters are manufactured and sold as *multimeters*, which are capable of measuring voltage, current, and often resistance. Electronic digital multimeters are now commonplace (Fig. 18.18c). In place of mechanical galvanometers, these use electronic circuits to analyze digital signals and calculate voltages, currents, and resistances, which are then displayed.

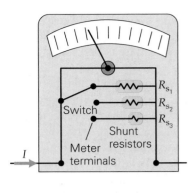

(a) Multirange ammeter

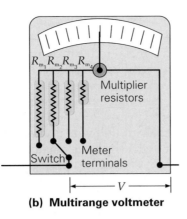

(b) Multirange voltmeter

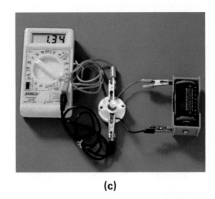

(c)

▲ **FIGURE 18.18 Multirange meters** **(a)** An ammeter or **(b)** a voltmeter can measure different ranges of current and voltage by switching among different shunt or multiplier resistors, respectively. (Instead of a switch, there may be an exterior terminal for each range.) **(c)** Both functions can be combined in a multimeter, shown here on the left measuring the voltage across a lightbulb. (How can you tell it is not measuring the current?).

18.5 Household Circuits and Electrical Safety

OBJECTIVES: To understand (a) how household circuits are wired, and (b) the underlying principles that govern electrical safety devices.

Although household circuits use alternating current, which has not yet been discussed, you can understand their operation (and many practical applications) using the circuit principles just studied.

For example, would you expect the elements (lamps, appliances, and so on) in a household circuit to be in series or parallel? From the discussion of Christmas tree lights (Section 18.1), it should be apparent that they must be connected in parallel. For example, when a bulb in a lamp in your kitchen burns out, other appliances on that circuit, such as the coffee maker, must continue to work. Moreover, household appliances and lamps are generally rated to work at 120 V. If these elements were in series, none of the individual appliances would have a voltage of 120 V across it.

Electrical power is supplied to a house by a three-wire system (▼Fig. 18.19). There is a difference in potential of 240 V between the two "hot," or high-potential,

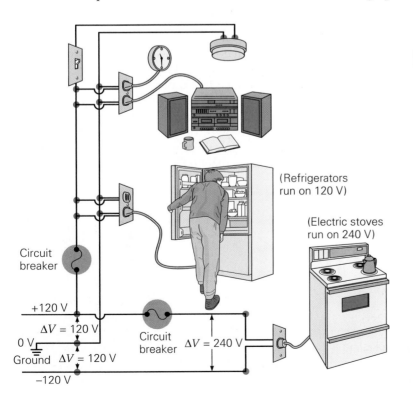

(Refrigerators run on 120 V)

(Electric stoves run on 240 V)

Circuit breaker

+120 V

$\Delta V = 120$ V

0 V

Ground $\Delta V = 120$ V

Circuit breaker $\Delta V = 240$ V

−120 V

◀ **FIGURE 18.19 Household wiring schematic** A 120-V circuit is obtained by connecting between either of the "hot" lines and the ground line. A voltage of 240 V (for appliances that require a lot of power such as electric stoves) can be obtained by connecting between the two "hot" lines of opposite polarity. (*Note:* For clarity, the dedicated ground wire [the third line that takes the rounded prong] is not shown.)

Note: Household voltage can fluctuate, under normal conditions, between 110 and 120 V. Similarly, 240-V connections can be as low as 220 V and still be considered normal.

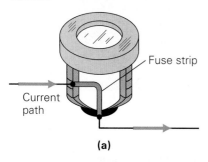

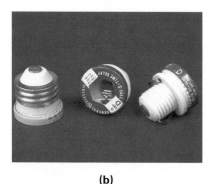

(a)

(b)

▲ **FIGURE 18.20 Fuses** **(a)** A fuse contains a metallic strip that melts when the current exceeds a rated value. This action opens the circuit and prevents overheating. **(b)** Edison-base fuses (left) have threads similar to lightbulbs. The threads are identical in this type of fuse; thus, different ampere-rated fuses can be interchanged—something that is not desirable (why?). Type-S fuses (right) have different threads for different ratings and thus cannot be interchanged.

wires. Each of these "hot" wires has a 120-V difference in potential with respect to the ground. The third wire is grounded at the point where the wires enter the house, usually by a metal rod driven into the ground. This wire is defined to be at zero potential and is called the *ground*, or *neutral, wire*.

The 120 V needed for most appliances is obtained by connecting them between the ground and either high-potential wire. The result is the same in either case, because $\Delta V = 120\ \text{V} - 0\ \text{V} = 120\ \text{V}$ or $\Delta V = 0\ \text{V} - (-120\ \text{V}) = 120\ \text{V}$. (See Fig. 18.19.)

Even though the ground wire is at zero potential, it *is* a current-carrying wire, because it is part of the complete circuit. Large appliances such as central air conditioners, ovens, and water heaters operate at 240 V. This value is obtained by connecting them between the two hot wires: $\Delta V = 120\ \text{V} - (-120\ \text{V}) = 240\ \text{V}$. The current in an appliance (under 120-V operating conditions) may be given on a rating tag. If not, it can usually be determined from the power rating on the tag (from $I = P/V$). Thus a stereo rated at 180 W would draw a current of 1.50 A (because $I = P/V = 180\ \text{W}/120\ \text{V} = 1.50\ \text{A}$).

There are limitations on the number of appliances in a circuit because of a limitation on the *total* current in the wires of that circuit. Specifically, joule heating (or I^2R loss) of the wires must be carefully considered. Remember that the more elements in parallel, the smaller their equivalent resistance. Thus adding appliances (in their "on" position) increases the total current. Real wires have resistance and could be subject to significant joule heating if the current is large. Therefore, by adding too many appliances, it is possible to overload a household circuit and produce too much heat *in the wires*. This could melt insulation and perhaps even start a fire.

This potential overloading is prevented by limiting the current. Two types of devices are commonly employed as limiters: fuses and circuit breakers. **Fuses** are still fairly common in older homes (◄Fig. 18.20). An Edison-base fuse has threads like those on the base of a lightbulb. (See Fig. 18.20b.) Inside the fuse is a metal strip that melts when the current is larger than the rated value (typically 15 A for a 120-V circuit). The melting of the strip opens the circuit, and the current drops to zero.

Circuit breakers are used exclusively in newer homes. One type (▼Fig. 18.21) uses a bimetallic strip (see Chapter 10). As the current in the strip increases, the strip becomes warmer and bends. At the rated value of current, the strip will bend sufficiently to open the circuit. The strip then cools, so the breaker can be reset. However, a blown fuse or a tripped circuit breaker indicates that the circuit is attempting to draw too much current! *Find and correct the problem before replacing the fuse or resetting the circuit breaker.* Also, under no circumstances should the blown fuse be even temporarily replaced by one of a higher current rating (why?). If a fuse of the correct current rating is not available, for safety purposes it is better to leave that circuit open (unless it controls items needed for emergency or crucial for living) until the correct fuse is found.

Switches, fuses, and circuit breakers are placed in the "hot" (high-potential) side of the circuit. They would, of course, also work if placed in the grounded side. To see why they aren't, consider the following. If they were placed there, even if the switch were open, the fuse blown, or the breaker tripped, the appliances would all remain connected to a high voltage—which could be potentially dangerous if a person made electrical contact (▶Fig. 18.22a).

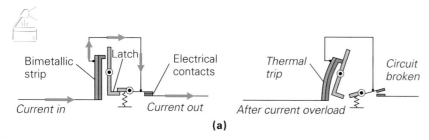

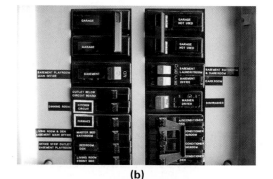

▲ **FIGURE 18.21 Circuit breakers** **(a)** A diagram of a thermal trip element. With increased current and joule heating, the element bends until it opens the circuit at some preset current value. Trip elements using magnetic principles also exist. **(b)** A typical bank of household circuit breakers.

(b)

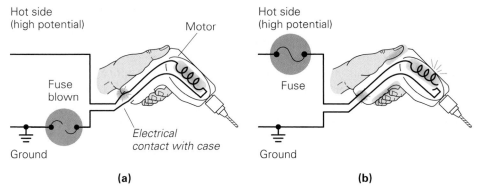

▲ **FIGURE 18.22** Electrical safety **(a)** Switches and fuses or circuit breakers should always be wired in the hot side of the line, *not* in the grounded side as shown. If these elements are wired in the grounded side, the line (and potentially the metallic case of the appliance) remains at a high voltage even when the fuse is blown or a switch is open. **(b)** Even if the fuse or circuit breaker is wired in the hot side, a potentially dangerous situation exists. If an internal wire comes in contact with the metal casing of an appliance or power tool, a person touching the casing, which is at high voltage, can get a shock. To prevent this possibility, a third dedicated ground line runs from the case to ground (see Fig. 18.23).

▲ **FIGURE 18.23** Dedicated grounding For safety, a third wire is connected from an appliance or power tool to ground. This dedicated grounding wire normally carries no current (as opposed to the grounded wire of the circuit). If the hot wire should come in contact with the metal case, the current will follow the ground wire (path of least resistance) rather than go through the body of the operator holding the case. The plug used for this is shown in Fig. 18.24.

Even with fuses or circuit breakers wired correctly into the "hot" side of the circuit, there is still a possibility of an electrical shock from a defective appliance that has a metal casing, such as a hand drill. For example, if a wire comes loose inside, it could make contact with the casing, which would then be at a high voltage (Fig. 18.22b). A person's body could then provide a path for current to ground, thus receiving a shock. For a discussion of the effects of electric shock, see Insight 18.2 on Electricity and Personal Safety on page 614.

To prevent a shock, a third, dedicated grounding wire is added to the circuit that grounds the metal casing of appliances or power tools (▶Fig. 18.23). This wire provides a path of very low resistance, bypassing the tool. This wire does not normally carry current. If a hot wire comes in contact with the casing, the circuit is completed through this grounding wire. Then the fuse is blown or the circuit breaker tripped. Most of the current would be in the ground wire rather than in you. Most likely you would not be harmed. Remember, however, that the breaker, if reset, will continue to trip unless you find the source of the problem and fix it.

On **three-prong grounded plugs**, the large, round prong connects with the grounding wire. Adapters can be used between a three-prong plug and a two-prong socket. Such an adapter has a grounding lug or grounding wire (▼Fig. 18.24a) that should be fastened to the receptacle box by the plate-fastening screw. The receptacle box is itself attached to the grounding wire. If the adapter lug or wire is *not* connected, the system is left unprotected, which defeats the purpose of the dedicated grounding safety feature.

You may have noticed another type of plug, a two-prong plug that fits in the socket only in one orientation, because one prong is wider than the other and one

Teaching tip: Ask students to imagine the consequences of having their kitchen wired in series. Point out that they would have to have everything on (mixer, dishwasher, lights, etc.). just to make toast. Even then, they would need special appliances, each designed with just the proper resistance to operate at a specific fraction of 120 V. If any appliance were added or disconnected, all the remaining appliances would have to be changed!

◀ **FIGURE 18.24** Plugging into ground **(a)** To accommodate the dedicated ground wire (Fig. 18.23) a three-prong plug is used. The adapter shown enables a three-prong plug to be used in a two-prong socket. The grounding lug (loop) on the adapter should be connected to the plate-fastening screw on the grounded receptacle box—otherwise, this safety feature is lost. **(b)** A polarized plug. The differently sized prongs permit prewired identification of the high and ground sides of the line. See text for details.

(a)

(b)

of the slits of the receptacle is also larger (Fig. 18.24b). This type is called a **polarized plug**. *Polarizing* in the electrical circuit sense is a method of identifying the hot and grounded sides of the line so that particular connections can be made.

Such polarized plugs and sockets are now a common safety feature. Wall receptacles are wired so that the small slit connects to the hot side and the large slit connects to the neutral, or ground, side. Having the hot side identified in this way makes two safeguards possible. First, the manufacturer of an electrical appliance can design it so that the switch is always in the hot side of the line. Thus, all of the wiring of the appliance beyond the switch is safely neutral when the switch is open and the appliance off. Moreover, the casing of an appliance is connected by the manufacturer to the ground side by means of a polarized plug. Should a hot wire inside the appliance come loose and contact the metal casing, the effect would be similar to that with a dedicated grounding system. The hot side of the line would be shorted to the ground, which would blow a fuse or trip a circuit breaker. Once again, you would be spared.

Another type of electrical safety device, the ground fault circuit interrupter, or GFCI, is discussed in Chapter 20.

INSIGHT 18.2 ELECTRICITY AND PERSONAL SAFETY

Safety precautions are necessary to prevent injuries when people work with electrical devices or wiring. Electrical conductors (wires) are coated with insulating materials so they can be handled safely. However, if a person comes in contact with a charged conductor, a difference in potential could exist across part of the person's body. A bird can sit on a high-voltage line without any problem because both of its feet are at the same potential—thus there is *no difference in potential* to generate a current in the bird. But if a person carrying an aluminum (conducting) ladder touches it to a bare electrical line, a difference in potential exists between the line and the ground. Thus the ladder and person become part of a current-carrying circuit.

The extent of personal injury in such cases depends on the size of the current in the body and on its path through the body. We know that the current in the body is given by $I = V/R_{body}$. Clearly, the current depends on the body's resistance.

The body's resistance can vary. If the skin is dry, the total body resistance can be $0.50\,M\Omega$ $(0.50 \times 10^6\,\Omega)$ or more. For a voltage of 120 V, there would be a current of about one-quarter of a milliamp, because

$$I = \frac{V}{R_{body}} = \frac{120\,V}{0.50 \times 10^6\,\Omega} = 0.24 \times 10^{-3}\,A = 0.24\,mA$$

This current is almost too weak to be felt (see Table 1). But if the skin is wet, then R_{body} can drop to as low as $5.0\,k\Omega\,(5.0 \times 10^3\,\Omega)$ and the current will be 24 mA (you should show this), a value which could potentially be dangerous. (See Table 1 again.)

A basic precaution is to avoid contact with any exposed electrical conductor that might cause a voltage across any part of your body. The physical damage resulting from such contact depends on the current path through the body. If that path is from a finger to the thumb on one hand, a large current probably would result only in a burn. However, if the path is from hand to hand through the chest (and therefore likely through the heart), the effect could be much worse. Some of the possible effects of this type of path are also given in Table 1.

Injury results because the current interferes with muscle function and/or causes burns. Muscle function is regulated by electrical nerve impulses (see Chapter 16), which can be influenced by external currents. Muscle reaction and pain can occur from a current of just a few milliamperes. At about 10 mA, muscle paralysis can prevent a person from releasing the conductor. At about 20 mA, contraction of the chest muscles occurs, which can cause impairment or stoppage of breathing. Death can occur in a few minutes. At 100 mA, rapid uncoordinated movements of the heart muscles (called *ventricular fibrillation*) prevent the proper heart pumping action and can be fatal in seconds. Working safely with electricity requires a knowledge of fundamental electrical principles *and* common sense. Electricity must be treated with respect.

Related Exercises: 94 and 95

TABLE 1	Effects of Electric Current on the Human Body*
Current (approximate)	*Effect*
2.0 mA (0.002 A)	Mild shock or heating
10 mA (0.01 A)	Paralysis of motor muscles
20 mA (0.02 A)	Paralysis of chest muscles, causing respiratory arrest; fatal in a few minutes
100 mA (0.1 A)	Ventricular fibrillation, preventing coordination of the heart's beating; fatal in a few seconds
1000 mA (1 A)	Serious burns; fatal almost instantly

*The effect of a given amount of current depends on a variety of conditions. This table gives only general and relative descriptions and assumes a circuit path that includes the upper chest.

Chapter Review

- When resistors are wired in **series**, the current through each of them is the same. The **equivalent resistance** of resistors in series is

$$R_s = R_1 + R_2 + R_3 + \cdots = \Sigma R_i \qquad (18.2)$$

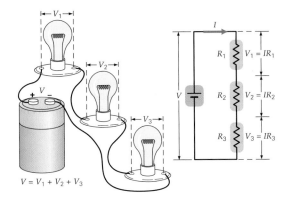

- When resistors are wired in **parallel**, the voltage across each of them is the same. The **equivalent resistance** is

$$\frac{1}{R_p} = \frac{1}{R_1} + \frac{1}{R_2} + \frac{1}{R_3} + \cdots = \Sigma \frac{1}{R_i} \qquad (18.3)$$

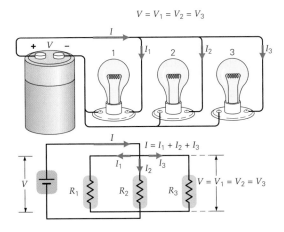

- **Kirchhoff's junction theorem** states that the total current into any **junction** equals the total current out of that junction (conservation of electric charge).

$$\Sigma I_i = 0 \quad \textit{sum of currents at a junction} \qquad (18.4)$$

- **Kirchhoff's loop theorem** states that in traversing a complete circuit loop, the algebraic sum of the voltage gains and losses is zero, or the sum of the voltage gains equals the sum of the voltage losses (conservation of energy in an electric circuit). In terms of voltages, this can be written as

$$\Sigma V_i = 0 \quad \begin{array}{l}\textit{sum of voltage around}\\ \textit{a closed loop}\end{array} \qquad (18.5)$$

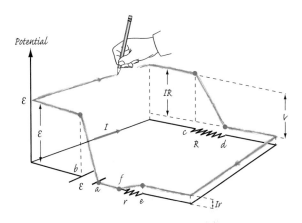

- The **time constant (τ)** for an RC circuit is a characteristic time by which we measure the capacitor's charging and discharging rate. τ is given by

$$\tau = RC \qquad (18.8)$$

- An **ammeter** is a device for measuring current; it consists of a galvanometer and a shunt resistor in parallel. Ammeters are connected in series with the circuit element carrying the current to be measured, and have very little resistance.

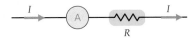

- A **voltmeter** is a device for measuring voltage; it consists of a galvanometer and a multiplier resistor wired in series. Voltmeters are connected in parallel, with the circuit element experiencing the voltage to be measured, and have large resistance.

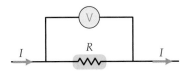

Exercises

*MC = Multiple Choice Question, CQ = Conceptual Question, and IE = Integrated Exercise. Throughout the text, many exercise sections will include "paired" exercises. These exercise pairs, identified with **red numbers**, are intended to assist you in problem solving and learning. In a pair, the first exercise (even numbered) is worked out in the Study Guide so that you can consult it should you need assistance in solving it. The second exercise (odd numbered) is similar in nature, and its answer is given at the back of the book.*

Assume that all resistors are ohmic unless otherwise stated.

18.1 Resistances in Series, Parallel, and Series–Parallel Combinations

1. **MC** Which of the following quantities must be the same for resistors in series? (a) voltage, (b) current, (c) power, (d) energy. (b)

2. **MC** Which of the following quantities must be the same for resistors in parallel? (a) voltage, (b) current, (c) power, (d) energy. (a)

3. **MC** Two resistors (A and B) are connected in series to a 12-V battery. Resistor A has 9 V across it. Which resistor has the least resistance? (a) A, (b) B, (c) both have the same, (d) can't tell from the data given. (b)

4. **MC** Two resistors (A and B) are connected in parallel to a 12-V battery. Resistor A has 2.0 A in it and the total current in the battery is 3.0 A. Which resistor has the most resistance? (a) A, (b) B, (c) both have the same, (d) can't tell from the data given. (b)

5. **MC** Two resistors (one with a resistance of 2.0 Ω and the other with 6.0 Ω resistance) are connected in parallel to a battery. Which one produces the most joule heating? (a) 2.0 Ω, (b) 6.0 Ω, (c) both produce the same, (d) can't tell from the data given. (a)

6. **MC** Two lightbulbs (bulb A has a rating of 100 W *at 120 V* and B has a rating of 60 W *at 120 V*) are connected in series to a wall socket at 120 V. Which one produces the most light? (a) A, (b) B, (c) both produce the same, (d) can't tell from the data given. (b)

7. **CQ** Are the voltage drops across resistors in series generally the same? If not, under what circumstance(s) could they be the same? no; only if all resistances are equal

8. **CQ** Are the currents in resistors in parallel generally the same? If not, under what circumstance(s) could they be the same? no; only if all resistances are equal

9. **CQ** If a large resistor and a small resistor are connected in series, will the effective resistance be closer in value to that of the large resistance or the small one? What if they are connected in parallel? large; small

10. **CQ** Lightbulbs have a power output marked on them by the manufacturer. For example, a 60-W light bulb assumes the bulb is connected to a 120-V source. Suppose you have two bulbs. One labeled 60 W is followed by a 40 W bulb in series to a 120-V source. Which one glows the brightest? Why? What happens if you switch the order of the bulbs? Are either of them at full power rating? Explain. see ISM

11. **CQ** Three identical resistors are connected to a battery. Two are wired in parallel, and that combination is followed in series by the third resistor. Which resistor(s) has (a) the largest current, (b) the largest voltage, and (c) the largest power output? (a) third (b) third (c) third

12. **CQ** Three resistors have values of 5 Ω, 2 Ω, and 1 Ω. The first one is followed in series by the last two wired in parallel. When this arrangement is connected to a battery, which resistor(s) has (a) the largest current, (b) the largest voltage, and (c) the largest power output? (a) 5 Ω (b) 5 Ω (c) 5 Ω

13. ● Three resistors that have values of 10 Ω, 20 Ω, and 30 Ω, respectively, are to be connected. (a) How should you connect them to get the maximum equivalent resistance, and what is this maximum value? (b) How should you connect them to get the minimum equivalent resistance, and what is this minimum value? (a) in series, 60 Ω (b) in parallel, 5.5 Ω

14. ● Two identical resistors (R) are connected in series and then wired in parallel to a 20-Ω resistor. If the total equivalent resistance is 10 Ω, what is the value of R? 10 Ω

15. ● Two identical resistors (R) are connected in parallel and then wired in series to a 40-Ω resistor. If the total equivalent resistance is 55 Ω, what is the value of R? 30 Ω

16. **IE** ● (a) In how many different ways can three 4.0-Ω resistors, be wired: (1) three, (2) five, or (3) seven? (b) Sketch the different ways you found in part (a) and determine the equivalent resistance for each. (a) (3) seven (b) 1.3 Ω, 2.0 Ω, 2.7 Ω, 4.0 Ω, 6.0 Ω, 8.0 Ω, and 12 Ω; see ISM

17. ● Three resistors with values of 5.0 Ω, 10 Ω, and 15 Ω, respectively, are connected in series in a circuit with a 9.0-V battery. (a) What is the total equivalent resistance? (b) What is the current in each resistor? (c) At what rate is energy delivered to the 15-Ω resistor? (a) 30 Ω (b) 0.30 A (c) 1.4 W

18. ● Find the equivalent resistances for all possible combinations of two or more of the three resistors in Exercise 17. see ISM

19. ● Three resistors with values 1.0 Ω, 2.0 Ω, and 4.0 Ω, respectively, are connected in parallel in a circuit with a 6.0-V battery. What are (a) the total equivalent resistance, (b) the voltage across each resistor, and (c) the power delivered to the 4.0-Ω resistor? (a) 0.57 Ω (b) 6.0 V (c) 9.0 W

20. **IE** ●● (a) If you had only an infinite supply of 1.0-Ω resistors, what is the minimum number of resistors required to make an equivalent resistance of 1.5 Ω: (1) two, (2) three,

or (3) four? (b) Describe or show by a sketch how the resistors should be connected. (a) (2) three (b) two in parallel, connected in series to the third; see ISM

21. **IE ●●** A length of wire with a resistance R is cut into two equal segments. The segments are then twisted together to form a conductor half as long as the original wire. (a) The resistance of the shortened conductor is (1) $R/4$, (2) $R/2$, (3) R. (b) If the resistance of the original wire is $27 \mu\Omega$ and the wire is cut into three equal segments, what is the resistance of the shortened conductor? (a) (1) $R/4$ (b) 3.0 $\mu\Omega$

22. **IE ●●** You have four 5.00-Ω resistors. (a) Can you connect all the resistors to produce an effective total resistance of 3.75 Ω? (b) Describe how you would connect them. (a) yes (b) three in series, connected in parallel to the fourth

23. **●●** Three resistors with values of 2.0 Ω, 4.0 Ω, and 6.0 Ω, respectively, are connected in series in a circuit with a 12-V battery. (a) How much current is delivered to the circuit by the battery? (b) What is the current in each resistor? (c) How much power is delivered to each resistor? (d) How does this power compare with the power delivered to the total equivalent resistance? see ISM

24. **●●** Suppose that the resistors in Exercise 23 are connected in parallel. (a) How much current is delivered to the circuit by the battery? (b) What is the current in each resistor? (c) How much power is delivered to each resistor? (d) How does this power compare with the power delivered to the total equivalent resistance? see ISM

25. **●●** Two 8.0-Ω resistors are connected in parallel, as are two 4.0-Ω resistors. These two combinations are then connected in series in a circuit with a 12-V battery. What is the current in each resistor and the voltage across each resistor? 1.0 A (for all); $V_{8.0}$ = 8.0 V; $V_{4.0}$ = 4.0 V

26. **●●** What is the equivalent resistance of the resistors in ▼Fig. 18.25? 0.80 Ω

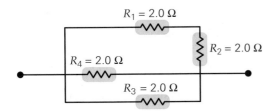

▲ **FIGURE 18.25 Series–parallel combination** See Exercises 26 and 34.

27. **●●** What is the equivalent resistance between points A and B in ▼Fig. 18.26? 2.7 Ω

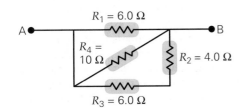

▲ **FIGURE 18.26 Series–parallel combination** See Exercises 27 and 36.

28. **●●** What is the equivalent resistance of the arrangement of resistors shown in ▼Fig. 18.27? 7.5 Ω

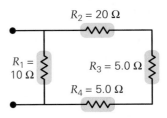

▲ **FIGURE 18.27 Series–parallel combination** See Exercise 28.

29. **●●** Several 60-W lightbulbs are connected in parallel with a 120-V source. The very last bulb blows a 15-A fuse in the circuit. (a) Sketch a schematic circuit diagram to show the fuse in relation to the bulbs. (b) How many lightbulbs are in the circuit (including the very last bulb)? (a) see ISM (b) 31

30. **●●** Find the current in and voltage across the 10-Ω resistor shown in ● ▼Fig. 18.28. 0.25 A; 2.5 V

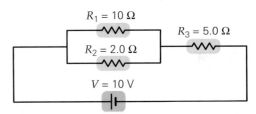

▲ **FIGURE 18.28 Current and voltage drop of a resistor** See Exercises 30 and 52.

31. **●●** For the circuit shown in ▼Fig. 18.29, find (a) the current in each resistor, (b) the voltage across each resistor, and (c) the total power delivered. (a) 1.0 A; 0.50 A; 0.50 A (b) 20 V; 10 V; 10 V (c) 30 W

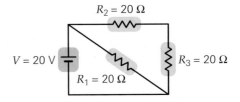

▲ **FIGURE 18.29 Circuit reduction** See Exercises 31 and 53.

32. **●●** A 120-V circuit has a circuit breaker rated to trip (to create an open circuit) at 15 A. How many 300-Ω resistors could be connected in parallel without tripping the breaker? 37

33. **●●** In your dorm room, you have two 100-W lights, a 150-W color TV, a 300-W refrigerator, a 900-W hairdryer, and a 200-W computer (including the monitor). If there is a 15-A circuit breaker in the 120-V power line, will the breaker trip (create an open circuit)? no, since I = 14.6 A < 15 A

34. **●●** Suppose that the resistor arrangement in Fig. 18.25 is connected to a 12-V battery. What will be (a) the current in each resistor, (b) the voltage drop across each resistor, and (c) the total power delivered? see ISM

35. ●● To make hot tea, you use a 500-W heater connected to a 120-V line to heat 0.20 kg of water from 20°C to 80°C. Assuming that there is no heat loss other than that delivered to the water, how long does this process take? 100 s = 1.7 min

36. ●●● The terminals of a 6.0-V battery are connected to points A and B in Fig. 18.26. (a) How much current is in each resistor? (b) How much power is delivered to each? (c) Compare the sum of the individual powers with the power delivered to the equivalent resistance for the circuit. see ISM

37. ●●● Lightbulbs with the power ratings (expressed in watts) given in ▼Fig. 18.30 are connected in a circuit as shown. (a) What current does the voltage source deliver to the circuit? (b) Find the power delivered to each bulb. (Take the bulbs' resistances to be the same as at their normal operating voltage.) see ISM

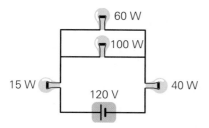

▲ **FIGURE 18.30 Watt's up?** See Exercise 37.

38. ●●● Two resistors R_1 and R_2 are in series with a 7.0-V battery. If R_1 has a resistance of 2.0 Ω and R_2 receives energy at the rate of 6.0 W, what is (are) the value(s) for the circuit's current(s)? (There may be more than one answer.) 1.5 A or 2.0 A

39. ●●● For the circuit in ▼Fig. 18.31, find (a) the current in each resistor and (b) the voltage across each resistor. see ISM

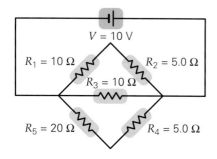

▲ **FIGURE 18.31 Resistors and currents** See Exercise 39.

40. ●●● What is the total power delivered to the circuit shown in ▼Fig. 18.32? 35 W

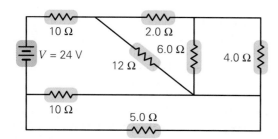

▲ **FIGURE 18.32 Power dissipation** See Exercise 40.

41. ●●● What is the equivalent resistance of the arrangement shown in ▼Fig. 18.33? 8.1 Ω

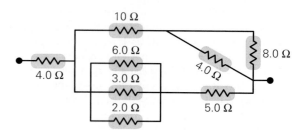

▲ **FIGURE 18.33 Equivalent resistance** See Exercise 41.

42. ●●● The circuit in ▼Fig. 18.34, named the *Wheatstone bridge* after Sir Charles Wheatstone (1802–1875), is used to measure resistance without the corrections sometimes necessary when using ammeter–voltmeter measurements. (See, for example, Exercises 88 and 89.) The resistances R_1, R_2, and R_s are known, and R_x is the unknown. R_s is variable and is adjusted until the bridge circuit is balanced—that is, until the galvanometer (G) reads zero (no current). Show that when the bridge is balanced, R_x can be determined from the following relationship:

$$R_x = \left(\frac{R_2}{R_1}\right)R_s.$$ see ISM

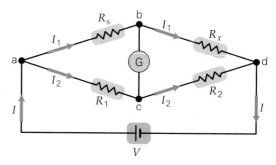

▲ **FIGURE 18.34 Wheatstone bridge** See Exercise 42.

18.2 Multiloop Circuits and Kirchhoff's Rules

43. **MC** You have a multiloop circuit with one battery. After leaving the battery, the current encounters a junction into two wires. One wire carries 1.5 A and the other 1.0 A. What is the current in the battery? (a) 2.5 A, (b) 1.5 A, (c) 1.0 A, (d) 5.0 A, (e) can't be determined from the given data. (a)

44. **MC** By our sign convention, if a resistor is traversed in the actual direction of the current in it, what can you say about the sign of the change in electric potential (the voltage): (a) it is negative, (b) it is positive, (c) it is zero, or (d) you can't tell from the data given? (a)

45. **MC** By our sign convention, if a battery is traversed in the actual direction of the current in it, what can you say about the sign of the change in electric potential (the battery's terminal voltage): (a) It is negative, (b) it is positive, (c) it is zero, or (d) you can't tell from the data given. (d)

46. **MC** You have a multiloop circuit with one battery that has a terminal voltage of 12 V. After leaving the positive terminal

of the battery, a short wire takes you to a junction where the current splits into three wires. From that point until you return to the negative terminal of the battery, what can you say about the sum of the voltages in each wire: (a) they total $+12$ V, (b) they total -12 V, (c) their magnitude is less than 12 V, or (d) their magnitude is greater than 12 V. (b)

47. **CQ** Must the current in a battery (in a complete circuit) always travel from its negative terminal to its positive terminal? Explain. If not, give an example. no, see ISM

48. **CQ** Use Kirchhoff's junction theorem to explain why the total equivalent resistance of a circuit is reduced by connecting a second resistor in parallel to another resistor. see ISM

49. **CQ** Use Kirchhoff's loop theorem to explain why a 60-W lightbulb produces more light than one rated at 100 W when they are connected in series to a 120-V source. [*Hint*: Recall that the power ratings are meaningful only at 120 V.] see ISM

50. ● Traverse loop 3 *opposite* to the direction shown in Fig. 18.10, and demonstrate that the resulting equation is the same as if you had followed the direction of the arrows. see ISM

51. ● For the circuit shown in Fig. 18.10, reverse the directions of loops 1 and 2, and demonstrate that equations equivalent to those in Example 18.5 are obtained. see ISM

52. ●● Use Kirchhoff's loop theorem to find the current in each resistor in Fig. 18.28. $I_1 = 0.25$ A; $I_2 = 1.25$ A; $I_3 = 1.50$ A

53. ●● Apply Kirchhoff's rules to the circuit in Fig. 18.29 to find the current in each resistor. $I_1 = 1.0$ A; $I_2 = I_3 = 0.50$ A

54. **IE** ●● Two batteries with terminal voltages of 10 V and 4 V, respectively, are connected with their positive terminals together. A 12-Ω resistor is wired between their negative terminals. (a) The current in the resistor is (1) 0 A, (2) between 0 A and 1.0 A, (3) greater than 1.0 A. Why? (b) Use Kirchhoff's loop theorem to find the current in the circuit and the power delivered to the resistor. (c) Compare this result with the power output of each battery.
(a) (2) between 0 A and 1.0 A (b) and (c) see ISM

55. ●● Using Kirchhoff's rules, find the current in each resistor in ▼Fig. 18.35. $I_1 = 0.33$ A (left); $I_2 = 0.33$ A (right)

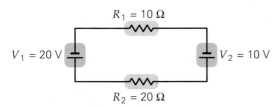

▲ **FIGURE 18.35 Single-loop circuit** See Exercise 55.

56. ●● Apply Kirchhoff's rules to the circuit in ▶Fig. 18.36, and find (a) the current in each resistor and (b) the rate at which energy is being delivered to the 8.0-Ω resistor. see ISM

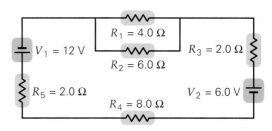

▲ **FIGURE 18.36 A loop in a loop** See Exercise 56.

57. ●●● Find the current in each resistor in the circuit shown in ▼Fig. 18.37.
$I_1 = 3.75$ A (up); $I_2 = 1.25$ A (left); $I_3 = 1.25$ A (right)

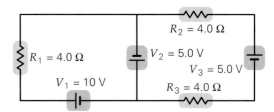

▲ **FIGURE 18.37 Double-loop circuit** See Exercise 57.

58. ●●● Find the currents in the circuit branches in ▼Fig. 18.38.
$I_1 = 3.23$ A (down); $I_2 = 1.54$ A (down); $I_3 = 1.02$ A (left); $I_4 = I_5 = I_6 = 1.93$ A (left)

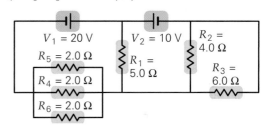

▲ **FIGURE 18.38 How many loops?** See Exercise 58.

59. ●●● For the multiloop circuit shown in ▼Fig. 18.39, what is the current in each branch? see ISM

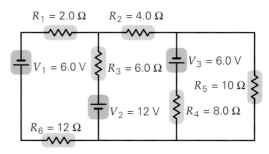

▲ **FIGURE 18.39 Triple-loop circuit** See Exercise 59.

18.3 RC Circuits

60. **MC** As a capacitor discharges through a resistor, the voltage across the resistor is a maximum (a) at the beginning of the process, (b) near the middle of the process, (c) at the end of the process, (d) after one time constant. (a)

61. **MC** When a capacitor discharges through a resistor, the current in the circuit is a minimum (a) at the beginning of the process, (b) near the middle of the process, (c) at the end of the process, (d) after one time constant. (c)

62. **MC** A charged capacitor discharges through a resistor (call this #1). If the value of the resistor is then doubled and the identical capacitor allowed to discharge again (call this #2), how do the time constants compare? (a) $\tau_1 = 2\tau_2$, (b) $\tau_1 = \tau_2$, (c) $\tau_1 = \frac{1}{2}\tau_2$, (d) $\tau_2 = 4\tau_1$. (c)

63. **MC** A capacitor discharges through a resistor (call this #1). The capacitor is then recharged to twice the initial charge in #1, and the discharge occurs through the same resistor (call this #2). How do the time constants compare? (a) $\tau_1 = 2\tau_2$, (b) $\tau_1 = \tau_2$, (c) $\tau_1 = \frac{1}{2}\tau_2$, (d) can't tell from the data given. (b)

64. **CQ** Another way to describe the discharge time of an RC circuit is to use a time interval called the *half-life*, which is defined as the time for the capacitor to lose half its initial charge. Is the time constant longer or shorter than the half-life? Explain your reasoning. longer, see ISM

65. **CQ** Does charging a capacitor in an RC circuit to 25% its maximum value take longer or shorter than one time constant? Explain. shorter, see ISM

66. **CQ** Explain why the current in a charging RC circuit decreases as the capacitor is being charged. charges on capacitor resist more charges to be transferred

67. ● In Fig. 18.11b, the switch is closed at $t = 0$, and the capacitor begins to charge. What is the voltage across the resistor and across the capacitor, expressed as fractions of V_o (to two significant figures), (a) just after the switch is closed, (b) after two time constants have elapsed, and (c) after many time constants have elapsed? see ISM

68. ● A capacitor in a single-loop RC circuit is charged to 63% of its final voltage in 1.5 s. Find (a) the time constant for the circuit and (b) the percentage of the circuit's final voltage after 3.5 s. (a) 1.5 s (b) 90%

69. **IE** ● In a flashing neon sign display, a certain time constant is desired. (a) To increase this time constant, you should (1) increase the capacitance, (2) decrease the capacitance, or (3) eliminate the capacitor. Why? (b) If a 2.0-s time constant is desired and you have a 1.0-μF capacitor, what resistance should you use in the circuit? (a) (1) increase the capacitance (b) 2.0 MΩ

70. ●● How many time constants will it take for an initially charged capacitor to be discharged to half of its initial voltage? 0.693τ

71. ●● A 1.00-μF capacitor, initially charged to 12 V, discharges when it is connected in series with a resistor. (a) What resistance is necessary to cause the capacitor to have only 37% of its initial charge 1.50 s after starting? (b) What is the voltage across the capacitor at $t = 3\tau$ if the capacitor is instead *charged* by the same battery through the same resistor? (a) 1.50 MΩ (b) 11.4 V

72. ●● A series RC circuit with $C = 40\ \mu$F and $R = 6.0\ \Omega$ has a 24-V source in it. With the capacitor initially uncharged, an open switch in the circuit is closed. (a) What is the voltage across the resistor immediately afterward?

(b) What is the voltage across the capacitor at that time? (c) What is the current in the resistor at that time? (a) 24 V (b) 0 (c) 4.0 A

73. ●● (a) For the circuit in Exercise 72, after the switch has been closed for $t = 4\tau$, what is the charge on the capacitor? (b) After a long time has passed, what are the voltages across the capacitor and the resistor? (a) 9.4×10^{-4} C (b) $V_C = 24$ V; $V_R = 0$

74. ●●● An RC circuit with a 5.0-MΩ resistor and a 0.40-μF capacitor is connected to a 12-V source. If the capacitor is initially uncharged, what is the change in voltage across it between $t = 2\tau$ and $t = 4\tau$? 1.4 V

75. ●●● A 3.0-MΩ resistor is connected in series with a 0.28-μF capacitor. This arrangement is then connected across four 1.5-V batteries (also in series). (a) What is the maximum current in the circuit and when does it occur? (b) What percentage of the maximum current is in the circuit after 4.0 s? (c) What is the maximum charge on the capacitor and when does it occur? (d) What percentage of the maximum charge is on the capacitor after 4.0 s? see ISM

18.4 Ammeters and Voltmeters

76. **MC** To accurately measure the voltage across a 1-kΩ resistor, the voltmeter should have a resistance that is (a) much larger than 1 kΩ, (b) much smaller than 1 kΩ, (c) about the same as 1 kΩ, (d) zero. (a)

77. **MC** To accurately measure the current in a 1-kΩ resistor, the ammeter should have a resistance that is (a) much larger than 1 kΩ, (b) much smaller than 1 kΩ, (c) about the same as 1 kΩ, (d) as large as possible—essentially infinite if possible. (b)

78. **MC** To correctly measure the voltage across a circuit element, a voltmeter should be connected (a) in series with it, (b) in parallel with it, (c) between the high potential side of the element and ground, (d) none of the preceding. (b)

79. **CQ** (a) What would happen if an ammeter were connected in parallel with a current-carrying circuit element? (b) What would happen if a voltmeter were connected in series with a current-carrying circuit element? see ISM

80. **CQ** Explain clearly, using Kirchhoff's laws, why the resistance of an ideal voltmeter is infinite. see ISM

81. **CQ** If designed properly, a good ammeter should have a very small resistance. Why? Explain clearly, using Kirchhoff's laws. see ISM

82. **IE** ● A galvanometer with a full-scale sensitivity of 2000 μA has a coil resistance of 100 Ω. It is to be used in an ammeter with a full-scale reading of 30 A. (a) Should you use (1) a shunt resistor, (2) a zero-resistor, or (3) a multiplier resistor? Why? (b) What is the necessary resistance? (a) (1) shunt resistor (b) 6.7 mΩ

83. **IE** ● The galvanometer in Exercise 82 is to be used in a voltmeter with a full-scale reading of 15 V. (a) Should you use (1) a shunt resistor, (2) a zero-resistor, or (3) a multiplier resistor? Why? (b) What is the required resistance? (a) (3) multiplier resistor (b) 7.4 kΩ

84. ● A galvanometer with a full-scale sensitivity of 600 μA and a coil resistance of 50 Ω is to be used to build an am-

meter designed to read 5.0 A at full scale. What is the required shunt resistance? 6.0 mΩ

85. •• A galvanometer has a coil resistance of 20 Ω. A current of 200 μA deflects the needle through 10 divisions at full scale. What resistance is needed to convert the galvanometer to a full-scale 10-V voltmeter? 50 kΩ

86. •• An ammeter has a resistance of 1.0 mΩ. Find the current in the ammeter when it is properly connected to a 10-Ω resistor and a 6.0-V source. (Express your answer to five significant figures to show how it differs from 0.60 A.) 0.59994 A

87. •• A voltmeter has a resistance of 30 kΩ. What is the current in the meter when it is properly connected across a 10-Ω resistor that is hooked to a 6.0-V source? 0.20 mA

88. IE ••• An ammeter and a voltmeter can measure the value of a resistor. Suppose that the ammeter is connected in series with the resistor and that the voltmeter is placed across the resistor only. (a) For accurate measurement, the internal resistance of the voltmeter should be (1) zero, (2) equal to the resistance to be measured, (3) infinite. Why? (b) Explain why the correct resistance is *not* given by $R = \dfrac{V}{I}$. (c) Show that the correct resistance is actually larger than the result in part (b) and is given by $R = \dfrac{V}{1 - (V/R_V)}$ where V is the voltage measured by the voltmeter, I is the current measured by the ammeter, and R_V is the resistance of the voltmeter. (d) Show that the result in part (c) reduces to $R = \dfrac{V}{I}$ for an ideal voltmeter.
(a) (3) infinite (b) see ISM

89. IE ••• An ammeter and a voltmeter can measure the value of a resistor. Suppose that the ammeter is connected in series with the resistor and that the voltmeter is placed across both the ammeter and the resistor. (a) For accurate measurement, the internal resistance of the ammeter should be (1) zero, (2) equal to the resistance to be measured, (3) infinite. Why? (b) Explain why the correct resistance is *not* given by $R = \dfrac{V}{I}$. (c) Show that the correct resistance is actually smaller than the result in part (b) and is given by $R = (V/I) - R_A$ where V is the voltage measured by the voltmeter, I is the current measured by the ammeter, and R_A is the resistance of the ammeter. (d) Show that the result in part (c) reduces to $R = \dfrac{V}{I}$ for an ideal ammeter. (a) (1) zero (b), (c) and (d) see ISM

18.5 Household Circuits and Electrical Safety

90. **MC** The ground wire in household wiring (a) is a current-carrying wire, (b) is at a voltage of 240 V from one of the "hot" wires, (c) carries no current, (d) none of the preceding.
(a)

91. **MC** A dedicated grounding wire (a) is the basis for the polarized plug, (b) is necessary for a circuit breaker, (c) normally carries no current, (d) none of the preceding.
(c)

92. **CQ** In terms of electrical safety, explain clearly what is wrong with the circuit in ▶Fig. 18.40, and why? the fuse and switch are on the ground side of the circuit

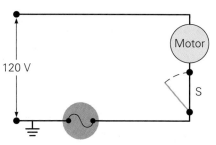

▲ **FIGURE 18.40 A safety problem?** See Exercise 92.

93. **CQ** The severity of bodily injury from electrocution depends on the magnitude of the current and its path, yet you commonly see signs that warn "Danger: High Voltage" (▼Fig. 18.41). Shouldn't such signs be changed to refer to high current? Explain. no, high voltage could produce harmful current even if resistance is high

▲ **FIGURE 18.41 Danger—high voltage** Shouldn't it be "high current" instead of "high voltage?" See Exercise 93.

94. **CQ** Explain why it is perfectly safe for birds to perch with both feet on the same high-voltage wire, even if the insulation is worn through.
the voltage between the feet is small

95. **CQ** After a collision with a power pole, you are trapped in your car, with a high-voltage line (frayed insulation) in contact with the hood of the car. Is it safer to step out of the car one foot at a time or to jump with both feet leaving the car at the same time? Explain your reasoning.
safer to jump, see ISM

96. **CQ** Most electrical codes require the metal case of an electric clothes dryer to have a wire running from the case to a nearby faucet (or any metal plumbing). Explain why.
the case is grounded so that its potential is always zero

Comprehensive Exercises

97. Find the current in each resistor in the circuit in ▼Fig. 18.42.
$I_1 = 2.6$ A (right); $I_2 = 1.7$ A (left); $I_3 = 0.86$ A (down)

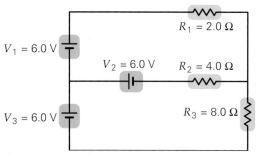

▲ **FIGURE 18.42 Kirchhoff's rules** See Exercise 97.

98. Four resistors are connected to a 90-V source, as shown in ▼Fig. 18.43. (a) Which resistor(s) receive(s) the most power and how much is that? (b) What is the total power delivered to the circuit by the power supply?
(a) R_1 and R_2; 8.1×10^2 W (b) 2.2×10^3 W

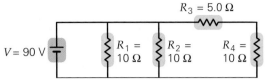

▲ **FIGURE 18.43 How much power is delivered?** See Exercise 98.

99. Four resistors are connected in a circuit with a 110-V source, as shown in ▼ Fig. 18.44. (a) What is the current in each resistor? (b) How much power is delivered to each resistor? see ISM

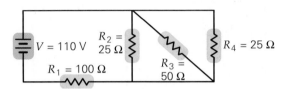

▲ **FIGURE 18.44 Joule-heat losses** See Exercise 99.

100. Nine resistors, each of value R, are connected in a "ladder" fashion, as shown in ▼Fig. 18.45. (a) What is the effective resistance of this network between points A and B? (b) If $R = 10\ \Omega$ and a 12.0-V battery is connected from point A to point B, how much current is in each resistor? see ISM

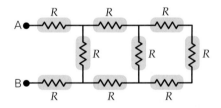

▲ **FIGURE 18.45 A resistance ladder** See Exercise 100.

101. A 4.0-Ω resistor and a 6.0-Ω resistor are connected in series. A third resistor is connected in parallel with the 6.0-Ω resistor. The whole arrangement gives a total equivalent resistance of 7.0 Ω. What is the value of the third resistor?
6.0 Ω

102. A galvanometer with an internal resistance of 50 Ω and a full-scale sensitivity of 200 μA is used to construct a multirange voltmeter. What values of multiplier resistors allow for three full-scale voltage readings of 20 V, 100 V, and 200 V? (See Fig. 18.18b.) 100 kΩ, 500 kΩ, 1.00 MΩ

103. A galvanometer with an internal resistance of 100 Ω and a full-scale sensitivity of 100 μA is used to construct a multirange ammeter. What values of shunt resistors allow for three full-scale current readings of 1.0 A, 5.0 A and 10 A? (See Fig. 18.18a.) 10 mΩ, 2.0 mΩ, and 1.0 mΩ

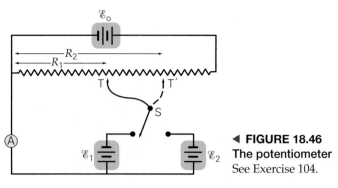

◀ **FIGURE 18.46**
The potentiometer
See Exercise 104.

104. ▲Fig. 18.46 shows a schematic circuit of an instrument called a *potentiometer* that is a very accurate device for determining emf values of power supplies. It consists of three batteries, an ammeter, several resistors, and a uniform wire that can be "tapped" for a specific fraction of its total resistance. $\mathscr{E}_0$ is the emf of a working battery, $\mathscr{E}_1$ designates a battery with a precisely known emf, and $\mathscr{E}_2$ designates a battery whose emf is unknown. The switch S is thrown toward battery 1, and the point T (for "tapped") is moved along the resistor until the ammeter reads zero. Let's call resistance of this arrangement R_1. This procedure is repeated with the switch thrown toward battery 2, and the point T is moved to T' until the ammeter again reads zero. Let's designate the resistance of this arrangement as R_2. Show that the unknown emf can be determined from the following relationship: $\mathscr{E}_2 = \dfrac{R_2}{R_1}\mathscr{E}_1$. see ISM

105. If a combination of three 30-Ω resistors receives energy at the rate of 3.2 W when connected to a 12-V battery, how are the resistors connected? two in parallel with each other and in series with the other resistor

106. A battery has three cells, each with an internal resistance of 0.020 Ω and an emf of 1.50 V. The battery is connected in parallel with a 10.0-Ω resistor. (a) Determine the voltage across the resistor. (b) How much current is in each cell? (The cells in a battery are in series.) (a) 4.47 V (b) 0.447 A

107. A 10.0-μF capacitor in a heart defibrillator unit is charged fully by a 10 000-V power supply. Each capacitor plate is connected to the chest of a patient by wires and flat "paddles," one on either side of the heart. The energy stored in the capacitor is delivered through an RC circuit, where R is the resistance of the body between the two paddles. Data indicates that it takes 75.1 ms for the voltage to drop to 20.0 V. (a) Find the time constant. (b) Determine the resistance, R. (c) How much time does it take for the capacitor to lose 90% of its stored energy? (a) 12.1 ms (b) 1.21 kΩ (c) 13.9 ms

108. During an operation, one of the electrical instruments has its metal case shorted to the 120-V "hot" wire that powers it. The attending physician is isolated from ground because of rubber-soled shoes, and inadvertently touches the case with his elbow, while simultaneously having his opposite hand in contact with the patient's chest. The patient, lying on a metal table, is well grounded. If the patient's head-to-ground resistance is 2200 Ω, what is the minimum resistance for the physician so that they both feel, at most, a "mild shock"?
58 kΩ

The following Physlet Physics Problems can be used with this chapter.
PHYSLET® 30.1, 30.2, 30.3, 30.4, 30.6, 30.7, 30.8, 30.11, 30.12

PHYSICS FACTS

- The SI unit of current, the ampere or the coulomb per second, is officially defined in terms of the magnetic field it creates and the magnetic force that field can exert on another current.

- Nikola Tesla (1856–1943) was a Serbian-American researcher perhaps best known for the Tesla coil, which is capable of producing high voltages (see Chapter 20) and is a common sight at high school science fairs. Tesla's name became the SI unit of magnetic field. When Westinghouse secured the patent rights to his alternating-current designs, this resulted in a battle between Edison's direct-current system and the Tesla-Westinghouse alternating-current system. The latter eventually won out and has become the primary means of delivering electric energy throughout the world.

- Pierre Curie (1859–1906) pioneered in widely varying areas ranging from magnetism to radioactivity. He discovered that ferromagnetic substances exhibited a temperature transition above which they lost their ferromagnetic behavior. This is now known as the Curie temperature.

When thinking about magnetism, most people tend to think of an attraction because it is well known that certain things can be picked up with a magnet. You have probably encountered magnetic latches that hold cabinet doors shut, or used magnets to stick notes to the refrigerator. It is less likely that one thinks of repulsion. Yet repulsive magnetic forces exist—and they can be just as useful as attractive ones.

In this regard, the photo shows an interesting example. At first glance, the vehicle looks like an ordinary train. But where are the wheels? In fact, it isn't a conventional train at all, but a high-speed, *magnetically levitated* one. It doesn't physically touch the rails. Rather it "floats" above them, supported by repulsive forces produced by powerful magnets. The advantages are obvious: with no wheels, there is no rolling friction and no bearings to lubricate—in fact, there are few moving parts of any kind.

Where do these useful magnetic forces come from? For centuries, the properties of magnets were attributed to the supernatural. The original "natural" magnets were called lodestone. Magnetism is now associated with electricity, because physicists discovered that both are really different aspects of a single force: the electromagnetic force. Electromagnetism is used in motors, generators, radios, and many other familiar applications. In the future, the development of high-temperature superconductors (Chapter 17) may open the way for the practical application of many devices now found only in the laboratory.

Although electricity and magnetism are fundamentally different manifestations of the same force, it is instructive to consider them individually and then put them together, so to speak, as electromagnetism. This chapter and the next will investigate magnetism and its intimate relationship to electricity.

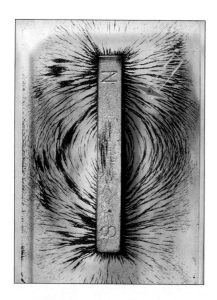

▲ **FIGURE 19.1 Bar magnet** The iron filings indicate the poles, or centers of force, of a common bar magnet. The compass direction designates these poles as north (N) and south (S). (See Fig. 19.3.)

Illustration 27.2 Earth's Magnetic Field

Demonstration/activity: Break or cut a small bar magnet in half to demonstrate that the magnet will remain a dipole.

19.1 Magnets, Magnetic Poles, and Magnetic Field Direction

OBJECTIVES: To (a) learn the force rule between magnetic poles, and (b) explain how the direction of a magnetic field is determined with a compass.

One of the features of a common bar magnet is that it has two "centers" of force, called *poles*, near each end (◄Fig. 19.1). To avoid confusion with the plus-minus designation used for electric charge, these poles are instead called north (N) and south (S). This terminology stems from the early use of the magnetic compass to determine direction. The north pole of a compass magnet was historically defined as the *north-seeking* end—that is, the end that points *north* on the Earth. The other end of the compass was labeled as south, or a south pole.

By using two bar magnets, the nature of the forces acting between the magnetic poles can be determined. Each pole of a bar magnet is attracted to the opposite pole of the other and repelled by the same pole of the other. Thus we have the **pole–force law**, or **law of poles**:

> Like magnetic poles repel each other, and unlike magnetic poles attract each other (▼Fig. 19.2).

One immediate (and sometimes confusing) result of the historical definition of a magnet's north pole has to do with the Earth's magnetic field. Because the north pole of a bar magnet is attracted to the Earth's north *polar* region (that is, *geographic* north), that area must be acting, or magnetically speaking, as a south (magnetic) pole. (See Section 19.8 for more details on the geophysics of the Earth's magnetic field.) So the Earth's south magnetic pole is in the general vicinity of its north geographic pole.

Two opposite magnetic poles, such as those of a bar magnet, form a *magnetic dipole*. At first glance, a bar magnet 's field might appear to be the magnetic analog of the electric dipole. There are, however, fundamental differences between the two. For example, permanent magnets always have two poles, never one. You might think that breaking a bar magnet in half would yield two isolated poles. However, the resulting pieces of the magnet always turn out to be two shorter magnets, *each with its own set of north and south poles*. While a single magnetic pole (a *magnetic monopole*) could exist in theory, it has yet to be found experimentally.

The fact that there is no magnetic analog to electric charge provides a strong hint about the differences between electric and magnetic fields. For example, the source of magnetism is electric charge, just like the electric field. However, as we will see in Sections 19.6 and 19.7, magnetic fields are produced only when electric charges are *in motion*, such as electric currents in circuits and orbiting (or spinning) atomic electrons. The latter is actually the source of the bar magnet's field.

Magnetic Field Direction

The historical approach to analyzing a bar magnet's field was to try to express the magnetic force between poles in a mathematical form similar to Coulomb's law for the electric force (Chapter 15). In fact, Coulomb developed such a law, using magnetic pole strengths in place of electric charge. However, this approach is rarely used, because it does not fit our modern understanding—that is, single magnetic poles do not exist. Instead, the modern description uses the concept of the *magnetic field*.

Recall that electric charges produce electric fields, which can be represented by electric field lines. The electric field (vector) is defined as the force per unit charge at

▶ **FIGURE 19.2 The pole–force law, or law of poles** Like poles (N–N or S–S) repel, and unlike poles (N–S) attract.

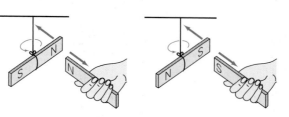

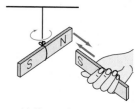

Like poles repel   Unlike poles attract

any location in space, or $\vec{E} = \vec{F}_e / q_0$. Similarly, magnetic interactions can be described in terms of the **magnetic field**, a vector quantity represented by the symbol $\vec{B}$. Just as electric fields exist near electric charges, we know that magnetic fields occur near permanent magnets. The magnetic field pattern surrounding a magnet can be made visible by sprinkling iron filings over a magnet covered with paper or glass (Fig. 19.1). Because of the magnetic field, the iron filings themselves become magnetized into little magnets (basically compass needles) and, behaving like compasses, line up in the direction of $\vec{B}$.

Because the magnetic field is a vector field, both magnitude (or "strength") and direction must be specified. The direction of a magnetic field (usually called a "B field") is defined in terms of a compass that has been calibrated (for direction) using the Earth's magnetic field:

> The direction of a magnetic field ($\vec{B}$) at any location is the direction that the north pole of a compass would point if placed at that location.

This provides a method for mapping a magnetic field by moving a small compass to various locations in the field. At any location, the compass needle will line up in the direction of the B field that exists there. If the compass is then moved in the direction in which its needle (the north end) points, the path of the needle traces out a *magnetic field line*, as illustrated in ▼Fig. 19.3a.

Because the north end of a compass points away from the north pole of a bar magnet, the field lines of a bar magnet point away from that pole and point toward its south pole. The rules that govern the interpretation of the magnetic field lines are the same that apply to electric field lines:

> The closer together (that is, the denser) the B field lines, the stronger the magnetic field. At any location, the direction of the magnetic field is tangent to the field line, or equivalently, the way the north end of a compass points.

Note the concentration of iron filings in the pole regions (Figs. 19.3b and c). This indicates closely spaced field lines and therefore a relatively strong magnetic field compared with other locations. As for field direction, note that just outside the middle of the magnet the field points vertically downward, tangent to the field line at that point (Fig. 19.3a, point P).

You might think that the magnitude of $\vec{B}$ would be defined as the magnetic force per unit pole strength, in analogy to $\vec{E}$. However, because magnetic monopoles don't exist, the magnitude of $\vec{B}$ is defined in terms of the magnetic force exerted on a moving electric charge, which is discussed next.

PHYSLET

Illustration 27.1 **Magnets and Compass Needles**

Demonstration/activity: Place magnets on an overhead projector and overlay them with a clear plastic sheet. Then sprinkle iron filings over the plastic to observe the magnetic fields. Look at various types of magnets. Can you tell which poles are which?

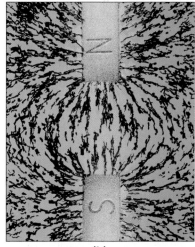

(b)

▼ **FIGURE 19.3 Magnetic fields** (a) Magnetic field lines can be traced and outlined by using iron filings or a compass, as shown in the case of the magnetic field due to a bar magnet. The filings behave like tiny compasses and line up with the field. The closer together the field lines, the stronger the magnetic field. (b) Iron filing pattern for the magnetic field between unlike poles; the field lines converge. (c) Iron filing pattern for the magnetic field between like poles; the field lines diverge.

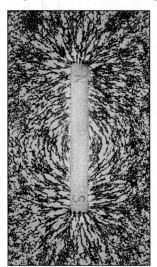

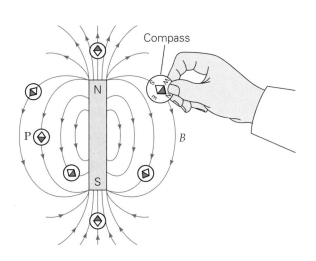

(a)

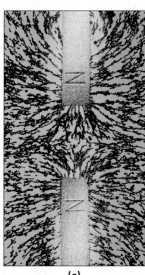

(c)

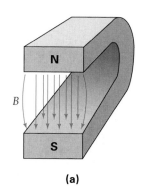

(a)

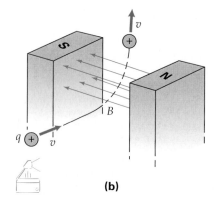

(b)

▲ **FIGURE 19.4** Force on a moving charged particle **(a)** A horseshoe magnet, created by bending a permanent bar magnet, produces a fairly uniform field between its poles. **(b)** When a charged particle enters a magnetic field, the particle is acted on by a force whose direction is obvious by the deflection of the particle from its original path.

Note: The magnetic field plays a vital role in magnetic resonance imaging (MRI), a technique widely used in medical diagnostics. (See Chapter 28.)

19.2 Magnetic Field Strength and Magnetic Force

OBJECTIVES: To (a) define magnetic field strength, and (b) determine the magnetic force exerted by a magnetic field on a moving charged particle.

Experiments indicate that one quantity that determines the *magnetic* force on a particle is its *electric* charge. That is, there is a connection between *electrical* properties of objects and how they respond to *magnetic* fields. The study of these interactions is called **electromagnetism**. Consider the following electromagnetic interaction. Suppose a positively charged particle is moving at a constant velocity as it enters a uniform magnetic field. For simplicity, let us also assume that its velocity is perpendicular to the field. (A fairly uniform B field exists between the poles of a "horseshoe" magnet, as shown in ◄Fig. 19.4a.) When the charged particle enters the field, it is *deflected* into an upward curved path, which is actually part of a circular path (if the B field is uniform), as shown in Fig. 19.4b.

From our study of circular motion (Section 7.3), for a particle to move in a circular arc, a centripetal force must exist that is always perpendicular to its velocity. But what provides this force here? No electric field is present. The gravitational force, besides being too weak to cause such a deflection, would deflect the particle into a downward parabolic arc, not an upward circular one. Evidently, the force is a magnetic one, due to the interaction of the moving charge and the magnetic field. *Thus a magnetic field can exert a force on a moving charged particle.*

From detailed measurements, the magnitude of this force is found to be proportional to the particle's charge and its speed. When the particle's velocity ($\vec{v}$) is perpendicular to the magnetic field ($\vec{B}$), the magnitude of the field, or the field strength B, is defined as

$$B = \frac{F}{qv} \quad \text{(valid only when } \vec{v} \text{ is perpendicular to } \vec{B}\text{)} \tag{19.1}$$

SI unit of magnetic field:
newton per ampere-meter [N/(A·m), or tesla (T)]

Physically, B means the magnetic force exerted on a charged particle *per unit charge* (coulomb) and *per unit speed* (m/s). From this relationship, the units of B are N/(C·m/s) or N/(A·m), because 1 A = 1 C/s. This combination of units is named the **tesla (T)** after Nikola Tesla (1856–1943), an early researcher in magnetism, and 1 T = 1 N/(A·m). Most everyday magnetic field strengths, such as those from permanent magnets, are much smaller than 1 T. In such situations, it is common to express magnetic field strengths in milliteslas (1 mT = 10^{-3} T) or microteslas (1 μT = 10^{-6} T). A non-SI unit commonly used by geologists and geophysicists, called the *gauss* (G), is defined as one ten-thousandth of a tesla (1 G = 10^{-4} T = 0.1 mT). For example, the Earth's magnetic field is on the order of several tenths of a gauss or several hundredths of a millitesla. On the other hand, conventional laboratory magnets can produce fields as high as 3 T, and superconducting magnets up to 25 T or higher.

Once the magnetic field strength has been determined (Eq. 19.1), the force on any charged particle moving at any speed can be found.* Equation 19.1 can be solved to find the force's magnitude:

$$F = qvB \quad \text{(valid only if } \vec{v} \text{ is perpendicular to } \vec{B}\text{)} \tag{19.2}$$

In the more general case, a particle's velocity may *not* be perpendicular to the field. Then the magnitude of the force depends on the sine of the angle (θ) between the velocity vector and the magnetic field vector. In general, the magnitude of the magnetic force is

$$F = qvB \sin \theta \quad \begin{array}{l} \text{magnetic force on} \\ \text{a charged particle} \end{array} \tag{19.3}$$

Note this means that the magnetic force is zero when $\vec{v}$ and $\vec{B}$ are parallel ($\theta = 0°$) or oppositely directed ($\theta = 180°$), because $\sin 0° = \sin 180° = 0$. The force has maximum value when these two vectors are perpendicular. With $\theta = 90°$ ($\sin 90° = 1$), this maximum value is $F = qvB \sin 90° = qvB$.

*Strictly speaking, the speeds must be considerably less than the speed of light to avoid relativistic complications (Chapter 26).

The Right-Hand Force Rule for Moving Charges

The *direction* of the magnetic force on any moving charged particle is determined by the orientation of the particle's velocity relative to the magnetic field. Experiment shows that the magnetic force direction is given by the **right-hand force rule** (▾Fig. 19.5a):

> When the fingers of the right hand are pointed in the direction of a charged particle's velocity $\vec{v}$ and then curled (through the smallest angle) toward the vector $\vec{B}$, the extended thumb points in the direction of the magnetic force $\vec{F}$ that acts on a *positive* charge. If the particle has a negative charge, the magnetic force is in the direction opposite to that of the thumb.

You might imagine the fingers of the right hand to be physically turning or rotating the vector $\vec{v}$ into $\vec{B}$ so that $\vec{v}$ and $\vec{B}$ are aligned.

Notice that the magnetic force is always *perpendicular to the plane formed by $\vec{v}$ and $\vec{B}$* (Fig. 19.5b). Because the force is perpendicular to the particle's direction of motion ($\vec{v}$), it cannot do any work on the particle. (This follows from the definition of work in Chapter 5, with a right angle between the force and displacement, $W = Fd \cos 90° = 0$.) Therefore, a magnetic field does not change the speed (that is, kinetic energy) of a particle—only its direction.

Several common alternative (and physically equivalent) right-hand force rules are shown in Fig. 19.5c. For negative charges, it is suggested that you start by assuming that the charge is positive. Next, determine the force direction, using the right-hand force rule. Lastly, *reverse* this direction to find the actual force direction on the negative charge. To see how this rule is applied to both charge signs, consider the following Conceptual Example.

Note: *B* fields that point into the plane of the page are designated by ✕. *B* fields that point out of the plane are indicated by •. Visualize these symbols as the feathered end and the tip of an arrow, respectively.

Teaching tip: Remind students that they don't have the correct magnetic force direction if **F** isn't perpendicular to both $\vec{v}$ and **B**.

Teaching tip: For negative charges, the magnetic force is in the direction opposite to that for positive charges. Suggest that students first treat negative charges as if they were positive, then determine the force direction from the right-hand force rule, and, when they are done, reverse the answer.

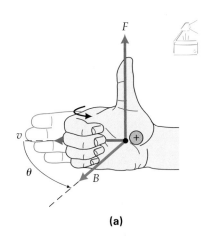

(a)

◀ **FIGURE 19.5 Right-hand rules for magnetic force** **(a)** When the fingers of the right hand are pointed in the direction of $\vec{v}$ and then curled toward the direction of $\vec{B}$, the extended thumb points in the direction of the force $\vec{F}$ on a *positive* charge. **(b)** The magnetic force is always perpendicular to the plane of $\vec{B}$ and $\vec{v}$ and thus is always perpendicular to the direction of the particle's motion. **(c)** When the extended forefinger of the right hand points in the direction of $\vec{v}$ and the middle finger points in the direction of $\vec{B}$, the extended right thumb points in the direction of $\vec{F}$ on a *positive* charge. **(d)** When the fingers of the right hand are pointed in the direction of $\vec{B}$ and the thumb in the direction of $\vec{v}$, the palm points in the direction of the force $\vec{F}$ on a *positive* charge. (Regardless of the rule you employ, remember to use your right hand and reverse the direction for a negative charge.)

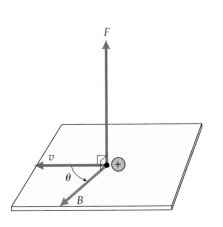

(b)

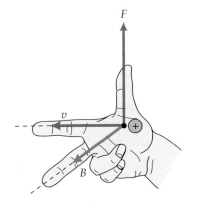

(c)

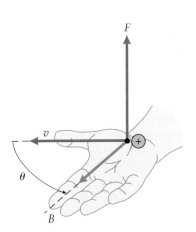

(d)

Conceptual Example 19.1 ■ Even "Lefties" Use the Right-Hand Rule

In a linear particle accelerator, a beam of protons travels horizontally northward. To deflect the protons eastward with a uniform magnetic field, the field should point in which direction? (a) vertically downward, (b) west, (c) vertically upward, or (d) south.

Reasoning and Answer. Because the force is perpendicular to the plane formed by $\vec{v}$ and $\vec{B}$, the field *cannot* be horizontal. If it were, it would deflect the protons down or up. Thus, we can eliminate choices (b) and (d). Use the right-hand force rule (assuming a positive charge) to see if $\vec{B}$ could be downward (answer a). You should verify (make a sketch) that for a downward magnetic field, the force would be to the west. Hence, the answer must be (c). The magnetic field must point upward to deflect the protons to the east. You should verify that this answer is correct, using the right-hand force rule.

Follow-Up Exercise. What direction would the particles in this Example deflect if they were electrons moving southward? (*Answers to all Follow-Up Exercises are at the back of the text.*)

According to our previous discussion, charged particles traveling in uniform magnetic fields have circular arcs as their trajectory. To see this in more detail, consider Example 19.2 carefully.

Example 19.2 ■ Going Around in Circles: Force on a Moving Charge

A particle with a charge of -5.0×10^{-4} C and a mass of 2.0×10^{-9} kg moves at 1.0×10^3 m/s in the $+x$-direction. It enters a uniform magnetic field of 0.20 T that points in the $+y$-direction (see ◀Fig. 19.6a). (a) Which way will the particle deflect just as it enters the field? (b) What is the magnitude of the force on the particle just as it enters the field? (c) What is the radius of the circular arc that the particle will travel while in the field?

Thinking It Through. The initial deflection is, of course, in the direction of the initial magnetic force. Since the particle is negative, care must be taken in applying the right-hand force rule. A circular arc is expected, as the magnetic force is always perpendicular to the particle's velocity. The magnitude of the magnetic force on a single charge is given by Eq. 19.3. This force is the only significant force on the electron; therefore it is also the net force. From this, Newton's second law should enable us to determine the circular orbit radius.

Solution. We list the given data.

Given:
$q = -5.0 \times 10^{-4}$ C
$v = 1.0 \times 10^3$ m/s ($+x$-direction)
$m = 2.0 \times 10^{-9}$ kg
$B = 0.20$ T ($+y$-direction)

Find: (a) Initial deflection direction
(b) Magnitude of initial magnetic force F
(c) Radius r of the orbit

(a) By the right-hand force rule, the force on a positive charge would be in the $+z$-direction (see direction of palm, Fig 19.6a). Because the charge is negative, the force is actually opposite this; thus, the particle will begin to deflect in the $-z$-direction.

(b) The force's magnitude can be determined from Eq. 19.3. Because only the magnitude is of interest, the sign of q can be dropped.

$$F = qvB \sin \theta$$
$$= (5.0 \times 10^{-4} \text{ C})(1.0 \times 10^3 \text{ m/s})(0.20 \text{ T})(\sin 90°) = 0.10 \text{ N}$$

(c) Because the magnetic force is the only force acting on the particle, it is also the net force (Fig. 19.6b). This net force points toward the circle's center and is called the centripetal force ($\vec{F}_{net} = \vec{F}_c$; see Chapter 7). Therefore, in describing circular motion, Newton's second law becomes

$$\vec{F}_c = m\vec{a}_c$$

Now substitute the magnetic force (from Eq. 19.2, because $\theta = 90°$) for the net force and the expression for centripetal acceleration ($a_c = v^2/r$; see Section 7.3) to obtain:

$$qvB = \frac{mv^2}{r} \quad \text{or} \quad r = \frac{mv}{qB}$$

Finally, inserting the numerical values,

$$r = \frac{mv}{qB} = \frac{(2.0 \times 10^{-9} \text{ kg})(1.0 \times 10^3 \text{ m/s})}{(5.0 \times 10^{-4} \text{ C})(0.20 \text{ T})} = 2.0 \times 10^{-2} \text{ m} = 2.0 \text{ cm}$$

Follow-Up Exercise. In this Example, if the particle had been a proton traveling initially in the $+z$-direction, (a) in what direction would it be initially deflected? (b) If the radius of its circular path were 10 cm and its speed were 1.0×10^6 m/s, what would be the magnetic field strength?

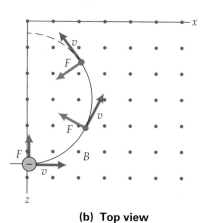

PHYSLET®

Illustration 27.3 A Mass Spectrometer

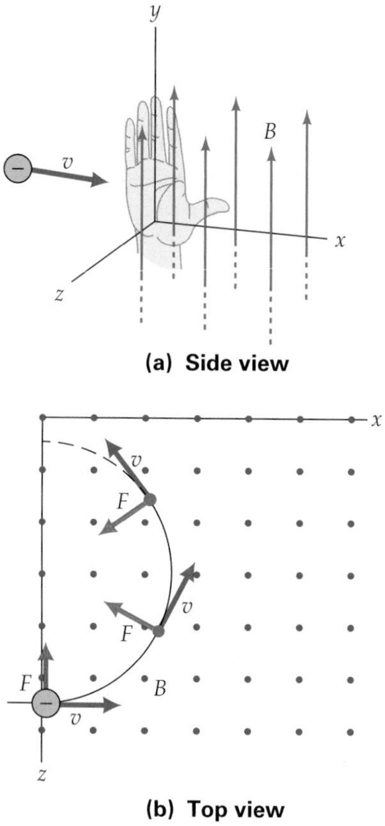

(a) Side view

(b) Top view

▲ **FIGURE 19.6 Path of a charged particle in a magnetic field (a)** A charged particle entering a uniform magnetic field will be deflected— here toward the *xy*-plane by the right-hand rule, because the charge is negative. **(b)** In the field, the force is always perpendicular to the particle's velocity. The particle moves in a circular path if the field is constant and the particle enters the field perpendicularly to its direction. (See Example 19.2.)

19.3 Applications: Charged Particles in Magnetic Fields

OBJECTIVE: To understand how the magnetic force on charged particles is employed in various practical applications.

We have seen that a charged particle moving in a magnetic field usually experiences a magnetic force. This force deflects the particle by an amount that depends on its mass, charge, and velocity (speed and direction), as well as the field strength. Let's take a look at the important role this force plays in some common appliances, machines, and instruments.

The Cathode-Ray Tube (CRT): Oscilloscope Screens, Television Sets, and Computer Monitors*

The **cathode-ray tube (CRT)** is a vacuum tube that is commonly used as a display screen for a laboratory instrument called an *oscilloscope* (▶Fig. 19.7). The basic operation of the oscilloscope and television picture tube is similar, and is shown in ▼Fig. 19.8. Electrons are emitted from a hot metallic filament and accelerated by a voltage applied between the cathode (−) and the anode (+) in an "electron gun" arrangement. In one kind of design, these tubes use current-carrying coils to produce a magnetic field (Section 19.6), which in turn controls the deflection of the electron beam. As the field strength is quickly varied, the electron beam scans the fluorescent screen in a fraction of a second. When the electrons hit the fluorescent material, they cause the atoms to emit light (Section 27.4). For a black-and-white TV, the signals reproduce an image on the screen as a mosaic of light and dark dots, depending on whether the beam is on or off at a particular instant.

   Producing the images on a color TV or a color computer monitor is somewhat more involved. A common color picture tube has three beams, one for each of the primary colors (red, green, and blue; Chapter 25). Phosphor dots on the screen are arranged in groups of three (triads), with one dot for each primary color. The excitation of the appropriate dots and the resulting emission (fluorescence) of a combination of colors produce a color picture.

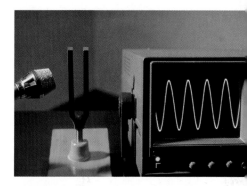

▲ **FIGURE 19.7 Cathode-ray tube (CRT)** The motion of the deflected beam traces a pattern on a fluorescent screen.

The Velocity Selector and the Mass Spectrometer

Have you ever thought about how the mass of an atom or molecule is measured? Electric and magnetic fields provide a way in a **mass spectrometer** ("mass spec" for short). Mass spectrometers perform many functions in modern laboratories. For example, they can be used to track short-lived molecules in studies of the biochemistry of living organisms. They can also determine the structure of large organic molecules and to analyze the composition of complex mixtures, such as a sample of smog-laden air.

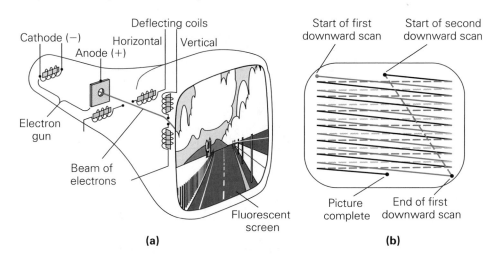

(a)   (b)

◀ **FIGURE 19.8 Television tube**
(a) A television picture tube is a cathode-ray (electron) tube, or CRT. The electrons are accelerated between the cathode and anode and are then deflected to the proper location on a fluorescent screen by magnetic fields produced by current-carrying coils. **(b)** In this design the beam scans every other line on the screen in one downward pass that takes $\frac{1}{60}$ s and then scans the lines in between in a second pass of $\frac{1}{60}$ s. This yields a complete picture of 525 lines in $\frac{1}{30}$ s.

*More and more, vacuum tube displays are being replaced by flat-screen displays that employ materials such as liquid crystals (LCD TVs and flat-screen computer monitors). These do not use magnetic forces in their operation.

In criminal cases, forensic chemists use the mass spectrometer to identify traces of materials, such as a streak of paint from a car accident. In other fields such as archaeology and paleontology, these instruments can be used to separate atoms to establish the age of ancient rocks and human artifacts. In modern hospitals, mass spectrometers are essential for measuring and maintaining the proper balance of gaseous medications, such as anesthetic gases administered during an operation.

In actuality, it is the masses of *ions*, or charged molecules, that are measured in mass spectrometer.* Ions with a known charge $(+q)$ are produced by removing electrons from atoms and molecules. At this point, the ion beam would have a distribution of speeds, rather than a single speed. If these entered the mass spectrometer, then different-speed ions would take different paths in the mass spectrometer. Thus, before they enter the spectrometer, a specific ion velocity must be selected for analysis. This can be accomplished by using a *velocity selector*. This instrument consists of an electric field and magnetic field at right angles.

This arrangement allows particles traveling only at a unique velocity to pass through undeflected. To see this, consider a positive ion approaching the crossed-field arrangement at right angles to both fields. The electric field produces a downward force $(F_e = qE)$, and the magnetic field produces an upward force $(F_m = qvB_1)$. (You should verify each force's direction in ▼Fig. 19.9.)

If the beam is not to be deflected, the resultant, or net, force on each particle must be zero. In other words, these two forces cancel. Therefore they are equal in magnitude and oppositely directed. Equating the two force magnitudes,

$$F_e = F_m \quad \text{or} \quad qE = qvB_1$$

which can be solved for the "selected" speed:

$$v = \frac{E}{B_1}$$

If the plates are parallel, the electric field between them is given by $E = V/d$, where V is the voltage across the plates and d is the distance between them. A more practical version of the previous equation is

$$v = \frac{V}{B_1 d} \quad \begin{array}{l} \textit{speed selected by} \\ \textit{a velocity selector} \end{array} \tag{19.4}$$

The desired speed can be selected by varying V, as usually B_1 and d are difficult to change.

Once through the velocity selector, the beam passes through a slit into another magnetic field $(\vec{B}_2)$ that is perpendicular to the beam direction. At this point, the particles are bent into a circular arc. The analysis is identical to that in Example 19.2. Therefore,

$$F_c = ma_c \quad \text{or} \quad qvB_2 = m\frac{v^2}{r}$$

▶ **FIGURE 19.9 Principle of the mass spectrometer** Ions pass through the velocity selector; only those with a particular velocity $(v = E/B_1)$ then enter a magnetic field (B_2). These ions are deflected, with the radius of the circular path depending on the mass and charge of the ion. Paths of two different radii indicate that the beam contains ions of two different masses (assuming that they have the same charge).

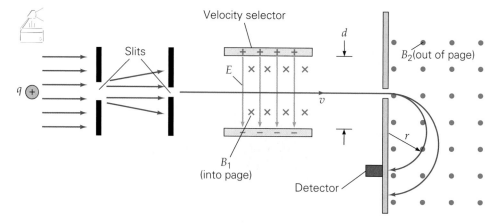

Top view

*Recall that removing electrons from or adding electrons to an atom or molecule produces an ion. However, an ion's mass is negligibly different from that of its neutral atom, because the electron's mass is very small compared to the masses of the protons and neutrons in the atomic nuclei.

Using Eq. 19.4, the mass of the particle is given by

$$m = \left(\frac{qdB_1B_2}{V}\right)r \quad \begin{array}{l}\textit{(mass determined with}\\\textit{a mass spectrometer)}\end{array} \quad (19.5)$$

The quantity inside the parentheses is a constant (assuming all ions have the same charge). Hence, the greater the mass of an ion, the greater its circular-path radius. Two paths of different radii are shown in Fig. 19.9. This indicates that the beam actually contains ions of two different masses. If the radius is measured (say, by recording the position where the ions hit a detector), the ion mass can be determined using Eq. 19.5.

In a mass spectrometer of slightly different design, the detector is kept at a fixed position. This design employs a time-varying magnetic field (B_2), and a computer records and stores the detector reading as a function of time. Notice that in this design, m is proportional to B_2. To see this, rewrite Eq. 19.5 as $m = (qdB_1r/V)B_2$. Because the quantity in the parentheses is a constant, then $m \propto B_2$. Thus as B_2 is varied, the detector data in connection with a high-speed computer enables us to determine the masses and relative number (that is, percentage) of ion of each mass. Regardless of design, the result—called a *mass spectrum* (the number of ions plotted against their mass)—is typically displayed on an oscilloscope or computer screen and digitized for storage and analysis (▶Fig. 19.10). Consider the following Example of a mass spectrometer arrangement.

Example 19.3 ■ The Mass of a Molecule: a Mass Spectrometer

One electron is removed from a methane molecule before it enters the mass spectrometer in Fig. 19.9. After passing through the velocity selector, the ion has a speed of 1.00×10^3 m/s. It then enters the main magnetic field region, in which the field strength is 6.70×10^{-3} T. From there it follows a circular path and lands 5.00 cm from the field entrance. Determine the mass of this molecule. (Neglect the mass of the electron that is removed.)

Thinking It Through. The centripetal force is provided by the magnetic force on the ion. Because the velocity and magnetic field are at right angles, the magnetic force is given by Eq. 19.2. Applying Newton's second law to circular motion, we can solve for the molecule's mass.

Solution. First, list the given data.

Given: $q = 1.60 \times 10^{-19}$ C (electron)
$r = d/2 = (5.00\ \text{cm})/2 = 0.0250$ m
$B_2 = 6.70 \times 10^{-3}$ T
$v = 1.00 \times 10^3$ m/s

Find: m (mass of a methane molecule)

The centripetal force on the ion ($F_c = mv^2/r$) is provided by the magnetic force ($F_m = qvB_2$):

$$\frac{mv^2}{r} = qvB_2$$

Solving this equation for m and putting in the numerical values, we have:

$$m = \frac{qB_2r}{v} = \frac{(1.60 \times 10^{-19}\ \text{C})(6.70 \times 10^{-3}\ \text{T})(0.0250\ \text{m})}{1.00 \times 10^3\ \text{m/s}} = 2.68 \times 10^{-26}\ \text{kg}$$

Follow-Up Exercise. In this Example, if the magnetic field between the velocity selector's parallel plates, which are 10.0 mm apart, is 5.00×10^{-2} T, what voltage must be applied to the plates?

Silent Propulsion: Magnetohydrodynamics

In a search for quiet and efficient methods of propulsion at sea, engineers have invented a system based on *magnetohydrodynamics*—the study of the interactions of moving fluids and magnetic fields. This method of propulsion relies on the magnetic force and does not require moving parts, such as motors, bearings, and shafts, that are common to most ships and submarines. To avoid detection, this "silent-running" feature is of particular importance to modern submarine design.

Basically, seawater enters the front of the unit and is expelled at high speeds out the rear (▶Fig. 19.11). A superconducting electromagnet (see Section 19.7) is used to produce a large magnetic field. At the same time, an electric generator produces a large dc voltage, sending a current through the seawater. [Recall that seawater is a

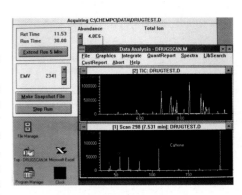

▲ **FIGURE 19.10 Mass spectrometer** Display of a mass spectrometer, with the number of molecules plotted vertically and the molecular mass horizontally. The molecule being analyzed is myoglobin, a protein that stores oxygen in muscle tissue. For myoglobin, each peak on the display represents the mass of an ionized fragment. Such patterns, *mass spectra*, help determine the composition and structure of large molecules. The mass spectrometer can also be used to identify tiny amounts of a molecule in a complex mixture.

Exploration 27.3 Mass Spectrometer

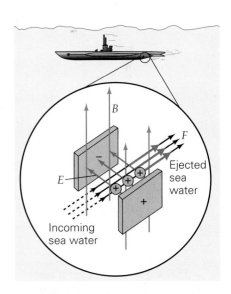

▲ **FIGURE 19.11 Propulsion via magnetohydrodynamics** In magnetohydrodynamic propulsion, seawater has an electric current passed through it by a dc voltage. A magnetic field exerts a force on the current, pushing the water out of the submarine or boat. The reaction force pushes the vessel in the opposite direction.

good conductor because it has a large concentration of sodium (Na^+) and chlorine (Cl^-) ions.] The magnetic force on the electric current pushes the water backward, expelling it as a jet of water. By Newton's third law, a reaction force acts forward on the submarine, enabling it to accelerate silently.

19.4 Magnetic Forces on Current-Carrying Wires

<u>OBJECTIVES:</u> To (a) calculate the magnetic force on a current-carrying wire and the torque on a current-carrying loop, and (b) explain the concept of the magnetic moment of a loop or coil.

Any charged particle moving in a magnetic field will generally experience a magnetic force. Because an electric current is composed of moving charges, we should expect a current-carrying wire, when placed in a magnetic field, to experience such a force as well. The sum of the individual magnetic forces on the charges that make up the current yield the total magnetic force on the wire.

Recall that the direction of the "conventional current" assumes that electric current in a wire is due to the motion of positive charges, as depicted in ▼Fig. 19.12.* In this orientation, the magnetic force is a maximum, because $\theta = 90°$. In a time t, any one charge q_i (that is, an electron) would move on the average a length $L = vt$, where v is the average drift speed. Because all the moving charges (total charge $= \Sigma q_i$) in this length of wire are acted on by a magnetic force in the same direction, the magnitude of the total force on this length of wire is (assuming the field to be uniform), from Eq. 19.2,

$$F = (\Sigma q_i)vB$$

Substituting L/t for v and rearranging gives

$$F = (\Sigma q_i)\left(\frac{L}{t}\right)B = \left(\frac{\Sigma q_i}{t}\right)LB$$

But $\Sigma q_i/t$ is just the current (I). Therefore, in terms of the circuit current, we can write

$$F = ILB \quad \text{(valid only if current and magnetic field are perpendicular)} \tag{19.6}$$

This result yields the maximum force on the wire. If the current makes an angle θ with respect to the field direction, then the force on the wire will be less. In general, the force on a length of current-carrying wire in a uniform magnetic field is

$$F = ILB \sin \theta \quad \text{magnetic force on a current–carrying wire} \tag{19.7}$$

If the current is parallel to or directly opposite to the field, then the force on the wire is zero.

The direction of the magnetic force on a current-carrying wire is also given by a right-hand rule. As was the case for individual charged particles, there are several equivalent versions of the **right-hand force rule for a current-carrying wire**, the most common being:

When the fingers of the right hand are pointed in the direction of the conventional current I and then curled toward the vector $\vec{B}$, the extended thumb points in the direction of the magnetic force on the wire (see ▶Figs. 19.13a and b).

Illustration 27.4 Magnetic Forces on Currents

Exploration 27.1 Map Field Lines and Determine Forces

Note: Remember that thinking of a current in terms of a positive charge is simply a useful convention. In reality, negative electrons are the charge carriers in ordinary electric current.

▶ **FIGURE 19.12 Force on a wire segment** Magnetic fields exert forces on wires that carry currents, because electric current is composed of moving charged particles. The maximum magnetic force on a current-carrying wire is shown, because the angle between the charge velocity and the field is 90°.

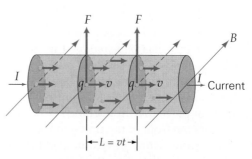

*Use the right-hand force rule to convince yourself that electrons traveling to the left would give the same magnetic force direction.

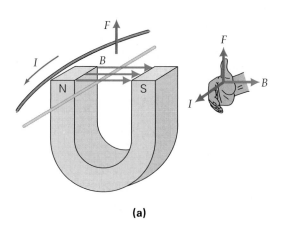

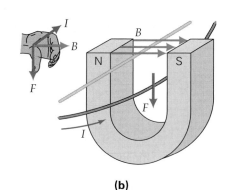

◀ **FIGURE 19.13** A right-hand force rule for current-carrying wires The direction of the force is given by pointing the fingers of the right hand in the direction of the conventional current I and then curling them toward $\vec{B}$. The extended thumb points in the direction of $\vec{F}$. The force is **(a)** upward and **(b)** downward.

An equivalent alternative is shown in ▸Fig. 19.14.

> When the fingers of the right hand are extended in the direction of the magnetic field and the thumb pointed in the direction of the conventional current I carried by the wire, the palm of the right hand points in the direction of the magnetic force on the wire.

Both rules give the same direction, because they are extensions of the right-hand force rules for individual charges. To see how two current-carrying wires interact magnetically, consider the next example.

Integrated Example 19.4 ■ Magnetic Forces on Wires Suspended at the Equator

Because a current-carrying wire is acted on by a magnetic force, it would seem possible to suspend such a wire at rest above the ground using the Earth's magnetic field. (a) A long, straight wire is located at the equator. In what direction would the current in the wire have to be to perform such a feat: (1) up, (2) down, (3) east, or (4) west? Draw a sketch to help you decide. (b) Calculate the current required to suspend the wire, assuming that the Earth's magnetic field is 0.40 G (gauss) at the equator and the wire is 1.0 m long with a mass of 30 g.

(a) Conceptual Reasoning. The magnetic force direction must be upward, because gravity acts downward (▸Fig. 19.15). The Earth's magnetic field at the equator is parallel to the ground and points north. Because the magnetic force is perpendicular to both the current and the field, the current cannot be up or down, which eliminates the first two choices. To decide between east and west, simply choose one and see if it works (or doesn't work). Suppose the current is to the west. Using the right-hand force rule shows that the magnetic force acts downward. Because this is incorrect, the only correct answer is (3), eastward. You should verify that this is correct by using the force right-hand rule directly.

(b) Quantitative Reasoning and Solution. The mass of the wire is known, and therefore its weight can be calculated. This must be equal and opposite to the magnetic force. The current and the field are at right angles to each other; hence, the magnetic force is given by Eq. 19.6, and from that, the current can be determined.

First we list the data (and convert to SI units at the same time):

Given: $m = 30\,\text{g} = 3.0 \times 10^{-2}\,\text{kg}$ *Find:* I (current required
$B = (0.40\,\text{G})(10^{-4}\,\text{T/G}) = 4.0 \times 10^{-5}\,\text{T}$ to suspend the wire)
$L = 1.0\,\text{m}$

The wire's weight is $w = mg = (3.0 \times 10^{-2}\,\text{kg})(9.8\,\text{m/s}^2) = 0.29\,\text{N}$. With the wire suspended, this would equal the magnetic force, or

$$w = ILB$$

Therefore,

$$I = \frac{w}{LB} = \frac{0.29\,\text{N}}{(1.0\,\text{m})(4.0 \times 10^{-5}\,\text{T})} = 7.4 \times 10^3\,\text{A}$$

This is a huge current, so suspending the wire in this manner is probably not a practical idea.

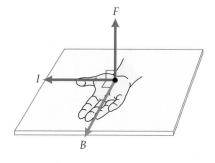

▲ **FIGURE 19.14** An alternative right-hand force rule When the fingers of the right hand are extended in the direction of the magnetic field $\vec{B}$ and the thumb is pointed in the direction of the conventional current I, then the palm points in the direction of $\vec{F}$. You should check that this gives the same direction as the equivalent rule shown in Fig. 19.13.

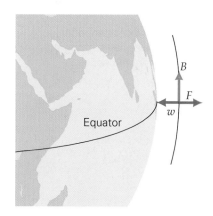

▲ **FIGURE 19.15** Defying gravity by using a magnetic field? Near the Earth's equator it is theoretically possible to cancel the pull of gravity with an upward magnetic force on a wire. What must be the direction and magnitude of the current? (See Integrated Example 19.4.)

(continues on next page)

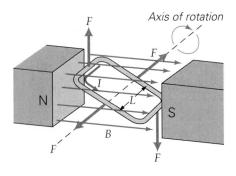

(a)

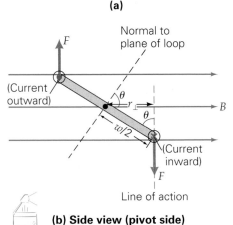

(b) Side view (pivot side)

▲ **FIGURE 19.16** Force and torque on a current-carrying pivoted loop **(a)** A current-carrying rectangular loop oriented in a magnetic field as shown is acted on by a force on each of its sides. Only the forces on the sides parallel to the axis of rotation produce a torque that causes the loop to rotate. **(b)** A side view shows the geometry for determining the torque. (See text for details.)

Note: A *coil* consists of N loops of the same size, all carrying the same current I in series.

Follow-Up Exercise. (a) Using the right-hand force rule, show that the idea of suspending a wire, as in this Example, could not work at either the south or north magnetic poles of the Earth. (b) In this Example, what would the wire's mass have to be for it to be suspended when carrying a more reasonable current of 10 A? Does this seem like a reasonable mass for a 1-m length of wire?

Torque on a Current-Carrying Loop

Another important use of magnetism is to exert forces and torques on current-carrying loops, such as the rectangular loop in a uniform field shown in ◄Fig. 19.16a. (The wires connecting the loop to a voltage source are not shown.) Suppose the loop is free to rotate about an axis passing through opposite sides, as shown. There is no net force or torque due to the forces acting on its pivoted sides (the sides through which the rotation axis passes). The forces on these sides are equal and opposite and in the plane of the loop. Therefore they produce no net torque or force. However, the equal and opposite forces on the two sides of the loop *parallel* to the axis of rotation, while not creating a net force, *do* create a net torque (see Chapter 8).

To see how this works, consider Fig. 19.16b, which is a side view of Fig.19.16a. The magnitude of the magnetic force F on each of the non-pivoted sides (length L) is given by $F = ILB$. The torque produced by this force (Section 8.2) is $\tau = r_\perp F$, where $r_\perp$ is the perpendicular distance (lever arm) from the axis of rotation to the line of action of the force. From Fig. 19.16b, $r_\perp = \frac{1}{2}w \sin\theta$, where w is the width of the loop and θ is the angle between the normal to the loop's plane and the direction of the field. The net torque τ is due to the torques from both forces and is their sum, or twice one of them (why?)

$$\tau = 2r_\perp F = 2\left(\tfrac{1}{2}w \sin\theta\right)F = wF \sin\theta$$
$$= w(ILB)\sin\theta$$

Then, because wL is the area (A) of the loop, the torque on a single pivoted, current-carrying loop can be rewritten as

$$\tau = IAB \sin\theta \quad \textit{torque on one current-carrying loop} \quad (19.8)$$

Although derived for a rectangular loop, Eq. 19.8 is valid for a flat loop of *any* shape and area. A *coil* is composed of N loops connected in series (where $N = 2, 3, \ldots$). Thus, for a coil, the torque is N times that on one loop (because each loop carries the same current). Therefore the torque on a coil is

$$\tau = NIAB \sin\theta \quad \begin{array}{l}\textit{torque on current-carrying}\\ \textit{coil of N loops}\end{array} \quad (19.9)$$

The magnitude of a coil's **magnetic moment** vector, m, is defined as

$$m = NIA \quad \textit{magnetic moment of a coil} \quad (19.10)$$

(SI units of magnetic moment: ampere · meter² or A · m²)

The direction of the magnetic moment vector $\vec{\mathbf{m}}$ is determined by circling the fingers of the right hand in the direction of the (conventional) current. The thumb gives the direction of this vector. Note that $\vec{\mathbf{m}}$ is always perpendicular to the plane of the coil (▶Fig. 19.17a). Equation 19.10 can be rewritten in terms of the magnetic moment:

$$\tau = mB \sin\theta \quad (19.11)$$

The magnetic torque tends to align the magnetic moment vector ($\vec{\mathbf{m}}$) with the magnetic field direction. To see this, notice that a loop or coil in a magnetic field is subject to a torque until $\sin\theta = 0$ (that is, $\theta = 0°$), at which point the forces producing the torque are parallel to the plane of the loop (see Fig. 19.17b). This situation exists when the plane of the loop is perpendicular to the field. If the loop is started from rest with its magnetic moment making some angle with the field, the loop will undergo an angular acceleration that will rotate it to the zero angle position. Rotational inertia will carry it through the equilibrium position (zero angle, Fig. 19.17c) to the other side. On that side, the torque will slow the loop, stop it, and then reaccelerate it back toward equilibrium. In other words, the torque on the loop is *restoring* and tends to cause the magnetic moment to oscillate about the field direction, much like a compass needle as it settles down to point north.

Example 19.5 ■ Magnetic Torque: Doing the Twist?

A laboratory technician makes a circular coil out of 100 loops of thin copper wire with a resistance of 0.50 Ω. The coil diameter is 10 cm and the coil is connected to a 6.0-V battery. (a) Determine the magnetic moment (magnitude) of the coil. (b) Determine the maximum torque (magnitude) on the coil if it were placed between the pole faces of a magnet where the magnetic field strength was 0.40 T.

Thinking It Through. The magnetic moment includes not only the number of loops and the area of the coil, but also the current in the wires. Ohm's law can be used to find the current. The maximum torque occurs when the angle between the magnetic moment vector and the B field is 90°, as given by Eq. 19.11.

Solution. Listing the given data with the radius of the circle expressed in SI units:

Given: $N = 100$ loops
$r = d/2 = 5.0$ cm $= 5.0 \times 10^{-2}$ m
$R = 0.50\ \Omega$
$V = 6.0$ V

Find: (a) m (coil magnetic moment)
(b) τ (maximum torque on the coil)

(a) The magnetic moment is given by Eq. 19.10, so the area and current are needed:

$$A = \pi r^2 = (3.14)(5.0 \times 10^{-2}\ \text{m})^2 = 7.9 \times 10^{-3}\ \text{m}^2$$

and

$$I = \frac{V}{R} = \frac{6.0\ \text{V}}{0.50\ \Omega} = 12\ \text{A}$$

Therefore the magnetic moment magnitude is

$$m = NIA = (100)(12\ \text{A})(7.9 \times 10^{-3}\ \text{m}^2) = 9.5\ \text{A} \cdot \text{m}^2$$

(b) The magnitude of the maximum torque (using $\theta = 90°$ in Eq. 19.11) is:

$$\tau = mB \sin \theta = (9.5\ \text{A} \cdot \text{m}^2)(0.40\ \text{T})(\sin 90°) = 3.8\ \text{m} \cdot \text{N}$$

Follow-Up Exercise. In this Example, (a) show that if the coil were rotated so that its magnetic moment vector was at 45°, the torque would *not* be half the maximum torque. (b) At what angle would the torque be half the maximum torque?

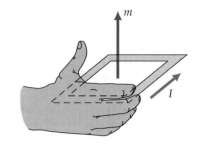

(a)

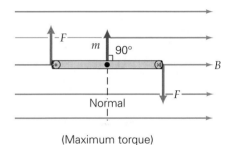

(Maximum torque)

(b)

19.5 Applications: Current-Carrying Wires in Magnetic Fields

OBJECTIVE: To explain the operation of various instruments whose functions depend on electromagnetic interactions between currents and magnetic fields.

With the principles of electromagnetic interactions learned so far, you can understand the operation of some familiar applications such as the ones that follow in this section.

The Galvanometer: The Foundation of the Ammeter and Voltmeter

Recall that ammeters and voltmeters use the *galvanometer* as the heart of their design.* Now you can understand exactly how the galvanometer works. As ▼Fig. 19.18a shows, a galvanometer consists of a coil of wire loops on an iron core that pivots between the pole faces of a permanent magnet. When a current is in the coil, a torque is exerted on the coil. A small spring supplies a countertorque, and when the two torques cancel (equilibrium) a pointer indicates a deflection angle ϕ that is proportional to the coil's current.

A problem arises if the galvanometer's magnetic field is not shaped correctly. If the coil rotated from its position of maximum torque ($\theta = 90°$), the torque would become less, and the pointer deflection ϕ would *not* then be proportional to the current. This problem is avoided by making the pole faces curved and by wrapping the coil on a cylindrical iron core. The core tends to concentrate the field lines such that $\vec{B}$ is always perpendicular to the nonpivoted side of the coil (Fig. 19.18b). With this design, the deflection angle is proportional to the current through the galvanometer ($\phi \propto I$), as required.

*Even though most voltmeters and ammeters are now digital, it is useful to understand how the mechanical versions use magnetic forces to make electrical measurements.

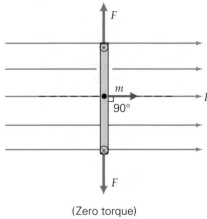

(Zero torque)

(c)

▲ **FIGURE 19.17 Magnetic moment of a current-carrying loop (a)** A right-hand rule determines the direction of the loop's magnetic moment vector $\vec{m}$. The fingers wrap around the loop in the direction of the current, and the thumb gives the direction of $\vec{m}$. **(b)** Condition of maximum torque. **(c)** Condition of zero torque. If the loop is free to rotate, the magnetic moment vector of the loop will tend to align with the direction of the external magnetic field.

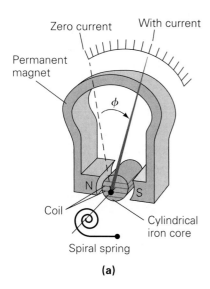

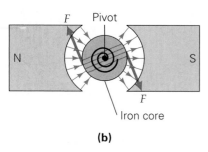

▲ **FIGURE 19.18 The galvanometer**
(a) The deflection (ϕ) of the needle from its zero-current position is proportional to the current in the coil. A galvanometer can therefore detect and measure currents.
(b) A magnet with curved pole faces is used so that the field lines are always perpendicular to the core surface and the torque does not vary with ϕ.

The dc Motor

In general, *an electric motor is a device that converts electrical energy into mechanical energy.* Such a conversion actually occurs during the movement of a galvanometer needle. However, a galvanometer is not considered a motor, because a practical **dc motor** must have continuous rotation for continuous energy output.

A pivoted, current-carrying coil in a magnetic field will rotate, but for only a half-cycle. When the magnetic field and the coil's magnetic moment are lined up ($\sin \theta = 0$), the torque on the coil is zero and it is in equilibrium.

To provide continuous rotation, the current is reversed every half-turn so that the torque-producing forces are reversed. This is done by using a *split-ring commutator*, which is an arrangement of two metal half-rings insulated from each other (▼Fig. 19.19a). The ends of the wire of the coil are fixed to the half-rings so that they rotate together. The current is supplied to the coil through the commutator by means of contact brushes. Then, with one half-ring electrically positive and the other negative, the coil and ring will rotate. When they have gone through half a rotation, the half-rings come in contact with the opposite brushes. Because their polarity is now reversed, the current in the coil also reverses. In turn, this action reverses the directions of the magnetic forces, keeping the torque in the same direction (clockwise in Fig. 19.19b). Even though the torque is zero at the equilibrium position, the coil is in unstable equilibrium and has enough rotational momentum to continue through the equilibrium point, whereupon the torque takes over and the coil rotates another half-cycle. The process repeats in continuous operation. In a real motor, the rotating shaft is called the *armature*.

The Electronic Balance

Traditional laboratory balances measure mass by balancing the weight of an unknown mass against a known one. Digital electronic balances (▶Fig. 19.20a) work on a different principle. In one type of design, there is still a suspended beam with a pan on one end that holds the object to be weighed, but no known mass is needed. Instead the balancing downward force is supplied by current-carrying coils of wire in the field of a permanent magnet (Fig. 19.20b). The coils move up and down in the cylindrical gap of the magnet, and the downward force is proportional to the current in the coils. The weight of the object in the pan is determined from the coil current, which produces a force just sufficient to balance the beam. From the weight, the balance determines the object's mass using the local value for g from $m = w/g$.

▼ **FIGURE 19.19 A dc motor** **(a)** A split-ring commutator reverses the polarity and current each half-cycle, so the coil rotates continuously. **(b)** An end view shows the forces on the coil and its orientation during a half-cycle. [For simplicity, we depict a single loop, but the coil actually has many (N) loops.] Notice the current reversal (shown by the dot and cross notation) between situations (3) and (4).

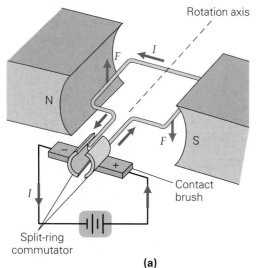

(a)

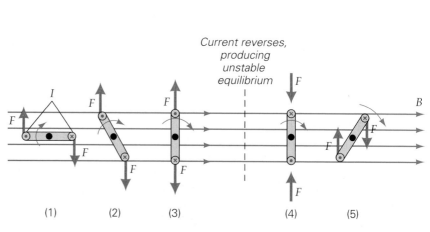

(b) Side view of loop, showing clockwise loop rotation sequence

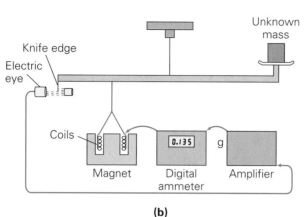

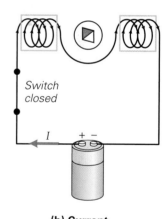

◀ FIGURE 19.20 Electronic balance **(a)** A digital electronic balance. **(b)** Diagram of the principle of an electronic balance. The balance force is supplied by electromagnetism.

(a)

(b)

The current required to produce balance is controlled automatically by photo-sensors and an electronic feedback loop. When the beam is balanced and horizontal, a knife-edge obstruction cuts off part of the light from a source that falls on a photo-sensitive "electric eye," the resistance of which depends on the amount of light falling on it. This resistance controls the current that an amplifier sends through the coil. For example, if the beam tilts so that the knife edge rises and more light strikes the eye, the current in the coil is increased to counterbalance the tilting. In this man-ner, the beam is electronically maintained in nearly horizontal equilibrium. Lastly, the current that keeps the beam in the horizontal position is read out on a digital am-meter calibrated in grams or milligrams instead of amperes.

19.6 Electromagnetism: The Source of Magnetic Fields

OBJECTIVES: To (a) understand the production of a magnetic field by electric currents, (b) calculate the strength of the magnetic field in simple cases, and (c) use the right-hand source rule to determine the direction of the magnetic field from the direction of the current that produces it.

Electric and magnetic phenomena, although clearly different, are closely and fun-damentally related. As we have seen, the *magnetic* force on a particle depends on the particle's *electric* charge. But what is the source of the magnetic field? Danish physicist Hans Christian Oersted discovered the answer in 1820, when he found that *electric currents produce magnetic fields*. His studies marked the beginnings of the discipline called **electromagnetism**, which involves the relationship between electric currents and magnetic fields.

In particular, Oersted first noted that an electric current could produce a deflection of a compass needle. This property can be demonstrated with an arrangement like that in ▶Fig. 19.21. When the circuit is open and there is no current, the compass needle points, as usual, in the northerly direction. However, when the switch is closed and there is current in the circuit, the compass needle points in a different direction, indicat-ing that an additional magnetic field (due to the current) must be affecting the needle.

Developing expressions for the magnetic field created by various configura-tions of current-carrying wires requires mathematics beyond the scope of this book. This section will present the results for the magnetic fields for several com-mon current configurations.

Magnetic Field Near a Long, Straight, Current-Carrying Wire

At a perpendicular distance d from a long, straight wire carrying a current I (▼Fig. 19.22), the magnitude of $\vec{B}$ is given by

$$B = \frac{\mu_o I}{2\pi d} \quad \begin{array}{l}\textit{magnetic field due}\\ \textit{to a long, straight wire}\end{array} \quad (19.12)$$

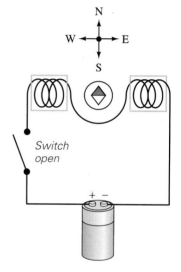

(a) No current

(b) Current

▲ FIGURE 19.21 Electric current and magnetic field (a) With no current in the wire, the compass needle points north. **(b)** With a current in the wire, the compass needle is deflected, indicating the presence of an additional magnetic field superimposed on that of the Earth. In this case, the strength of the additional field is roughly equal in magnitude to that of the Earth. How can you tell?

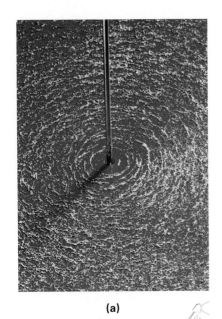

(a)

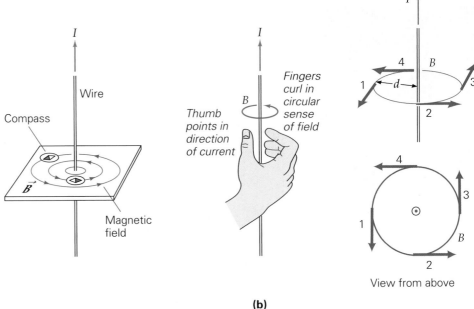

(b)

View from above

▲ **FIGURE 19.22 Magnetic field around a long, straight current-carrying wire (a)** The field lines form concentric circles around the wire, as revealed by the pattern of iron filings. **(b)** The circular sense of the field lines is given by the right-hand source rule, and the magnetic field vector is tangent to the circular field line at any point.

Demonstration/activity: Iron filings can be used to demonstrate the magnetic field around a long, straight wire if a high-current source is available. Otherwise use several small magnetic compasses to show the magnetic field. Connect several meters of insulated wire to a dc source, and thread the wire among the students in class. Then pass out magnetic compasses to the students and have them hold the compass above, below, and alongside the wire to observe the magnetic field.

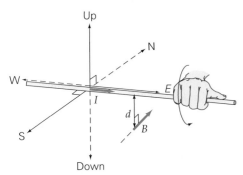

▲ **FIGURE 19.23 Magnetic field** Finding the magnitude and direction of the magnetic field produced by a straight current-carrying wire. (See Example 19.6.)

where $\mu_o = 4\pi \times 10^{-7}$ T·m/A is a proportionality constant called the **magnetic permeability of free space**. For long, straight wires, the field lines are closed circles centered on the wire (Fig. 19.22a). Note in Fig. 19.22b that the direction of $\vec{B}$ due to a current in a long straight wire is given by a **right-hand source rule**:

> If a long straight current-carrying wire is grasped with the right hand and the extended thumb points in the direction of the current (I), then the curled fingers indicate the circular sense of the magnetic field lines.

Example 19.6 ■ Common Fields: Magnetic Field from a Current-Carrying Wire

The maximum household current in a wire is about 15 A. Assume that this current exists in a long straight wire in a west-to-east direction (◄Fig. 19.23). What are the magnitude and direction of the magnetic field the current produces 1.0 cm directly below the wire?

Thinking It Through. To find the magnitude of the field, we should use Eq. 19.12 for a long, straight wire. The direction of the field is given by the right-hand source rule.

Solution. Listing the data:

Given: $I = 15$ A $\qquad$ *Find:* $\vec{B}$ (magnitude and direction)
$\qquad$ $d = 1.0$ cm $= 0.010$ m

From Eq. 19.12, the magnitude of the field 1.0 cm below the wire is

$$B = \frac{\mu_o I}{2\pi d} = \frac{(4\pi \times 10^{-7} \text{ T·m/A})(15 \text{ A})}{2\pi(0.010 \text{ m})} = 3.0 \times 10^{-4} \text{ T}$$

By the source rule (Fig. 19.23), the field direction directly below the wire is north.

Follow-Up Exercise. (a) In this Example, what is the field direction 5.0 cm above the wire? (b) What current is needed to produce a magnetic field at this location with one-half the strength of the field in the Example?

Magnetic Field at the Center of a Circular Current-Carrying Wire Loop

At the *center* of a circular coil of wire consisting of N loops, each of radius r and each carrying the same current I (▶Fig. 19.24a shows one such loop), the magnitude of $\vec{B}$ is

$$B = \frac{\mu_o N I}{2r} \qquad \text{\textit{magnetic field at center of circular coil of N loops}} \qquad (19.13)$$

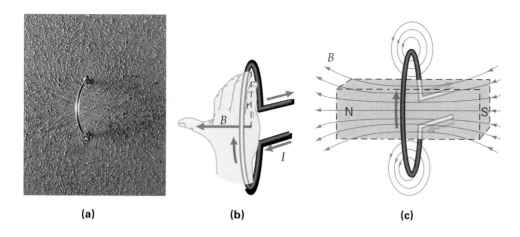

(a)　　　(b)　　　(c)

Illustration 28.1 Field from Wires and Loops

In this case (and all circular current arrangements, such as solenoids, discussed next), it is convenient to determine the magnetic field direction using the right-hand source rule that is slightly different from (but equivalent to) the one for straight wires:

> If a circular loop of current-carrying wires is grasped with the right hand so the fingers are curled in the direction of the current, the magnetic field direction *inside* the circular area formed by the loop is the direction in which the extended thumb points. (see Fig. 19.24b)

In all cases, the magnetic field lines form closed loops, the direction of which is determined by the right-hand source rule. Recall however, the direction of $\vec{B}$ is tangent to the field line, and therefore depends on the location (Fig. 19.24c). Notice that the overall field pattern of the loop is geometrically similar to that of a bar magnet. More about this later.

Magnetic Field in a Current-Carrying Solenoid

A *solenoid* is constructed by winding a long wire into a coil, or *helix*, with many circular loops, as shown in ▼Fig. 19.25. If the solenoid's radius is small compared with its length (L), the interior magnetic field is parallel to the solenoid's longitudinal axis and constant in magnitude. The longer the solenoid, the more uniform the internal field. Notice how the solenoid's field (Fig. 19.25) closely resembles that of a permanent bar magnet.

As usual, the direction of the interior field is given by the right-hand source rule for circular geometry. If the solenoid has N turns and each carries a current I, the magnitude of the magnetic field near its center is given by

$$B = \frac{\mu_o NI}{L} \quad \begin{array}{l}\textit{magnetic field near}\\\textit{the center of a solenoid}\end{array} \quad (19.14)$$

Notice that the solenoid field depends on how closely packed (note the N/L) or how densely the turns are wound. Thus n is defined as $n = N/L$ to quantify this. Its units are turns per meter, and it is called the *linear turn density*. In terms of this, Eq. 19.14 is sometimes rewritten as $B = \mu_o nI$.

To see why the solenoid might be best suited for magnetic applications requiring a large magnetic field, consider the next Example.

▼ **FIGURE 19.25** Magnetic field of a solenoid **(a)** The magnetic field of a current-carrying solenoid is fairly uniform near the central axis of the solenoid, as seen in this pattern of iron filings. **(b)** The direction of the field in the interior can be determined by applying the right-hand source rule to any of the loops. Notice its resemblance to the field near a bar magnet.

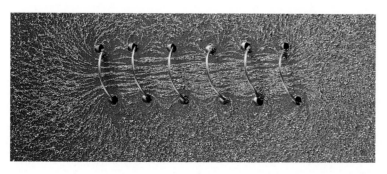

(a)

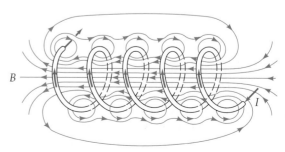

(b)

Illustration 28.2 Forces Between Wires

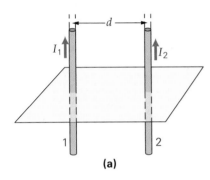

(a)

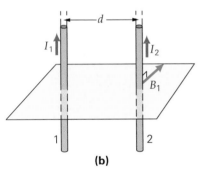

(b)

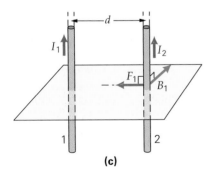

(c)

▲ **FIGURE 19.26** Mutual interaction of parallel current-carrying wires **(a)** Two parallel wires carry current in the same direction. **(b)** Wire 1 creates a magnetic field at the site of wire 2. **(c)** Wire 2 is pulled toward wire 1 by a force. (See Integrated Example 19.8 for details.)

Example 19.7 ■ Wire versus Solenoid: Concentrating the Magnetic Field

A solenoid is 0.30 m long with 300 turns and carries a current of 15.0 A. (a) What is the magnitude of the magnetic field at the center of this solenoid? (b) Compare your result with the field near the single wire in Example 19.6, which carries the same current, and comment.

Thinking It Through. The B field depends on the number of turns (N), the solenoid length (L), and the current (I). This is a direct application of Eq. 19.14.

Solution.

Given: $I = 15.0$ A   *Find:* (a) B (magnitude of magnetic field near the
$N = 300$ turns   solenoid center)
$L = 0.30$ m   (b) Compare the answer in part (a) with that
                          from a long straight wire in Example 19.6

(a) From Eq. 19.14,

$$B = \frac{\mu_0 N I}{L} = \frac{(4\pi \times 10^{-7}\ \text{T}\cdot\text{m/A})(300)(15.0\ \text{A})}{0.30\ \text{m}} = 6\pi \times 10^{-3}\ \text{T} \approx 18.8\ \text{mT}$$

(b) Notice that this is more than *60 times as large as* the field near the wire in Example 19.6. Winding many loops close together in a helix fashion increases the field while enabling the use of the same current. The fundamental reason is that the solenoid's field is the vector sum of the fields from 300 loops, and the individual magnetic field directions are all approximately the same.

Follow-Up Exercise. In this Example, if the current were reduced to 1.0 A and the solenoid shortened to 0.10 m, how many turns would be needed to create the same magnetic field?

In the next Integrated Example, all aspects of electromagnetism are involved: forces on electric currents and the production of magnetic fields by electric currents. Study the example carefully, especially the use of the appropriate right-hand rule.

Integrated Example 19.8 ■ Attraction or Repulsion: Magnetic Force between Two Parallel Wires

Two long, parallel wires carry currents in the same direction, as illustrated in ◄Fig. 19.26a. (a) Is the magnetic force between these wires (1) attractive or (2) repulsive? Make a sketch to show how you obtained your result. (b) If both wires carry the same current of 5.0 A, have a length of 50 cm, and are separated by 3.0 mm, determine the magnitude of the force on each wire.

(a) Conceptual Reasoning. Choose one wire and determine the direction of the magnetic field it produces at the location of the other wire. (Note that the field of interest is determined first using the right-hand source rule.) In Fig. 19.26b, wire 1 was chosen. The field produced by the current in wire 1 is the field in which wire 2 is located. Next, use the right-hand force rule on wire 2 to determine the direction of the force on it. The result, shown in Fig. 19.26c, is an attractive force, so (1) is the correct answer. You should be able to show that, at the same time, wire 2 exerts an attractive force on wire 1, in keeping with Newton's third law.

(b) Quantitative Reasoning and Solution. To find the magnetic field strength produced by wire 1, use Eq. 19.12. Because the field is at right angles to the current in wire 2, the magnitude of the force on it is ILB. Be careful to use the appropriate field and current. The symbols are shown in Fig. 19.26. We list the given data and convert to SI units.

Given: $I_1 = I_2 = 5.0$ A   *Find:* F (the magnitude of the magnetic
$d = 3.0$ mm $= 3.0 \times 10^{-3}$ m   force between the wires)
$L = 50$ cm $= 5.0 \times 10^{-1}$ m

The magnetic field due to I_1 at the location of wire 2 is

$$B_1 = \frac{\mu_0 I_1}{2\pi d} = \frac{(4\pi \times 10^{-7}\ \text{T}\cdot\text{m/A})(5.0\ \text{A})}{2\pi(3.0 \times 10^{-3}\ \text{m})} = 3.3 \times 10^{-4}\ \text{T}$$

The magnitude of the magnetic force on wire 2 due to the field created by wire 1 is

$$F_2 = I_2 L B_1 = (5.0\ \text{A})(0.50\ \text{m})(3.3 \times 10^{-4}\ \text{T}) = 8.3 \times 10^{-4}\ \text{N}$$

Follow-Up Exercise. (a) In this Example, determine the force direction if the current in either one of them is reversed. (b) If the magnitude of the force between the wires is kept the same as in the Example, but the current tripled, how far apart are the wires?

The magnetic force between parallel wires in an arrangement like that analyzed in Integrated Example 19.8 provides the modern basis for the definition of the ampere. The National Institute of Standards and Technology (NIST) defines the ampere as

> that current which, if maintained in each of two long parallel wires separated by a distance of exactly 1 m in free space, would produce a magnetic force between the wires of exactly 2×10^{-7} N for each meter of wire.

This definition was chosen as a universal standard, in part because it is easier to measure forces than to count electrons (that is, measure charge) over time.

19.7 Magnetic Materials

OBJECTIVES: To (a) explain how ferromagnetic materials enhance external magnetic fields, (b) understand the concept of a material's magnetic permeability, (c) explain how "permanent" magnets are produced, and (d) explain how "permanent" magnetism can be destroyed.

Why are some materials magnetic or easily magnetized, whereas others are not? How can a bar magnet create a magnetic field when it carries no obvious current? To answer these questions, let's start with some basics. It is known that a current is needed to produce a magnetic field. If the magnetic fields of a bar magnet and a long solenoid are compared (see Figs. 19.1 and 19.25), it seems that the magnetic field of the bar magnet might be due to *internal* currents. Perhaps these "invisible" currents are due to electrons orbiting the atomic nucleus or electron spin. However, detailed analysis of atomic structure shows that the net magnetic field produced by orbital motion is zero or very small.

What then *is* the source of the magnetism produced by magnetic materials? Modern atomic quantum theory tells us that the permanent type of magnetism, like that exhibited by an iron bar magnet, is produced by *electron spin*. Classical physics likens a spinning electron to the Earth rotating on its axis. However, this mechanical analog is *not* actually the case. Electron spin is a quantum mechanical effect with no direct classical analog. Nonetheless, the picture of spinning electrons creating magnetic fields is useful for qualitative thinking and reasoning. In effect, each "spinning" electron produces a field similar to a current loop (Fig. 19.24c). This pattern, resembling that from a small bar magnet, enables us to treat electrons, magnetically speaking, as tiny compass needles.

In multielectron atoms, the electrons *usually* are arranged in pairs with their spins oppositely aligned (that is, one with "spin up" and one with "spin down" in chemistry parlance). In this case, their magnetic fields will effectively cancel, and the material cannot be magnetic. Aluminum is such a material.

However, in certain materials, known as **ferromagnetic materials**, the fields due to electron spins in individual atoms do not cancel. Thus each atom possesses a magnetic moment. There is a strong interaction between these neighboring moments that leads to the formation of regions called **magnetic domains**. In a given domain, the electron spin moments are aligned in approximately the same direction, producing a relatively strong (net) magnetic field. Not many ferromagnetic materials occur naturally. The most common are iron, nickel, and cobalt. Gadolinium and certain manufactured alloys, such as neodymium and other rare earth alloys, are also ferromagnetic.

In an unmagnetized ferromagnetic material, the domains are randomly oriented and there is no net magnetization (▼Fig. 19.27a). But when a ferromagnetic material

▼ **FIGURE 19.27 Magnetic domains** **(a)** With no external magnetic field, the magnetic domains of a ferromagnetic material are randomly oriented and the material is unmagnetized. **(b)** In an external magnetic field, domains with orientations parallel to the field may grow at the expense of other domains, and the orientations of some domains may become more aligned with the field. **(c)** As a result, the material becomes magnetized, or exhibits magnetic properties.

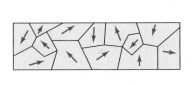

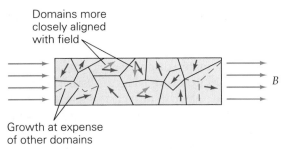

Domains more closely aligned with field

Growth at expense of other domains

(a) No external magnetic field

(b) With external magnetic field

(c) Resulting bar magnet

INSIGHT 19.1 THE MAGNETIC FORCE IN FUTURE MEDICINE

From ancient times to present, humans have sought healing power in magnetism. Claims that magnetism can heal bunions, eliminate tennis elbow, and cure cancer have been made, but none have ever been substantiated. However, there are many magnetic applications in modern medicine, such as the well-known magnetic resonance imaging (MRI) system (see Chapter 28).

Many new ideas are on the horizon as well. For example, certain types of bacteria can create nanometer-sized permanent magnets inside themselves (see Insight 19.2 on Magnetism in Nature on page 645). Scientists have proposed harvesting these tiny magnets, which are just small enough to fit through a hypodermic needle. These magnets could then be attached to drug molecules. By placing a magnetic field near the site of interest, the molecules could be attracted and held there. Holding drug molecules in place would increase their effectiveness and reduce side effects that can occur when drugs circulate into other parts of the body.

A major problem with this proposal is the need to develop techniques for extracting the tiny bacterial magnets and growing them in large enough quantities. Alternative proposals include creating nano-sized unmagnetized pieces of iron by chemical means, attaching them to the drug molecules, and moving them around with magnetic fields in a microscopic version of iron filings. Both proposals present dangers, such as the drug molecules attracting each other, thus clumping and potentially blocking blood flow.

Instead, perhaps, tiny magnetic microspheres could be filled with drugs or radioactive material and steered to the site and held in place by external magnetic fields. One application would be to the treatment of diabetic foot ulcers—open lesions that are difficult to heal due to diabetes-related circulation problems. The wound would be covered by thin, strong magnets in a bandage. An injection of microspheres filled with slow-release drugs, such as an antibiotic, would follow. The magnets would attract the microspheres to the ulcer site and hold them in place. As the microspheres would break down over the course of several weeks, they would release the drugs slowly, helping the body repair the wound. Microspheres filled with radioactive material could potentially treat liver, lung, brain, and other deep tumors.

Another experimental therapy uses magnetically induced heating (*hyperthermic* techniques) to target breast cancer. This therapy could be especially important in destroying the smaller tumors now being found by modern imaging techniques. For these tumors, fluid magnetite (Fe_3O_4) would be injected directly into the tumor. For tumors larger than a few cubic millimeters, the iron particles could be delivered via the circulatory system if they were first attached to biomolecules that specifically target cancer cells. Regardless of the method, the treatment would be the same.

In the presence of an external alternating magnetic field, the iron particles would heat up due to induced currents (see Chapter 20 on electromagnetic induction). A rise in temperature of only a few degrees Celsius above normal body temperature has been shown to kill cancer cells. In theory, this heating would occur locally only at the tumor sites and be minimally invasive. Initial experiments have had positive results, and the outlook is promising.

Illustration 27.5 Permanent Magnets and Ferromagnetism

(such as an iron bar) is placed in an external magnetic field, the domains change their orientation and size (Fig. 19.27b). Remember the picture of the electron as a small compass; the electrons begin to "line up" in the external field. As the external field and the iron bar begin to interact, the iron exhibits the following two effects:

1. Domain boundaries change, and the domains with magnetic orientations in the direction of the external field grow at the expense of the others.

2. The magnetic orientation of some domains may change slightly so as to be more aligned with the field.

On removal of the external fields, the iron domains remain more or less aligned in the original external field direction, thus creating an overall "permanent" magnetic field of their own.

Now you can also understand why an unmagnetized piece of iron is attracted to a magnet and why iron filings line up with a magnetic field. Essentially, the pieces of iron become induced magnets (Fig. 19.27c). Some uses of permanent magnets and magnetic forces in modern medicine are discussed in Insight 19.1 on The Magnetic Force in Future Medicine on this page.

Electromagnets and Magnetic Permeability

Ferromagnetic materials are used to make electromagnets, usually by wrapping a wire around an iron core (▶Fig. 19.28a). The current in the coil creates a magnetic field in the iron, which in turn creates its own field that is typically many times larger than the coil's field. By turning the current on and off, we can turn the magnetic field on and off at will. When the current is on, it induces magnetism in ferromagnetic materials (for example, the iron sliver in Fig. 19.28b) and, if the forces are large enough, it can be used to pick up large amounts of scrap iron (Fig. 19.28c).

The iron used in an electromagnet is called *soft iron*. In this type of iron, when the external field is removed, the magnetic domains become unaligned and the iron reverts to its demagnetized state. The adjective *soft* refers not to the metal's mechanical hardness, but to its magnetic properties.

When an electromagnet is on (lower drawing in Fig. 19.28a), the iron core is magnetized and adds to the field of the solenoid. The total field is expressed as

$$B = \frac{\mu NI}{L} \qquad \text{\textit{magnetic field at the center}} \atop \text{\textit{of the iron--core solenoid}} \qquad (19.15)$$

Notice that this equation is identical to that for the magnetic field of an air-core solenoid (Eq. 19.14), except that it contains μ instead of μ_o, the permeability of free space. Here, μ represents the **magnetic permeability** *of the core material*, not free space. The role permeability plays in magnetism is similar to that of the permittivity ε in electricity (Chapter 16). For magnetic materials, the magnetic permeability is defined in terms of its value in free space; thus,

$$\mu = \kappa_m \mu_o \qquad (19.16)$$

where κ_m is called the *relative* permeability (dimensionless) and is the magnetic analog of the dielectric constant κ.

The value of κ_m for a vacuum is equal to unity. (Why?) Because, for ferromagnetic materials, the total magnetic field far exceeds that from the wire wrapping, it follows that $\mu \gg \mu_o$ and $\kappa_m \gg 1$. A core of a ferromagnetic material with a large permeability in an electromagnet can enhance its field thousands of times compared with an air core. In other words, ferromagnetic materials have values of κ_m on the order of thousands. To see the effect of ferromagnetic materials, refer to the next Example.

Example 19.9 ■ Magnetic Advantage: Using Ferromagnetic Materials

A laboratory solenoid with 200 turns in a length of 30 cm is limited to carrying a maximum current of 2.0 A. The scientists need an interior magnetic field strength of at least 2.0 T and are debating whether they need to employ a ferromagnetic core. (a) Is their field possible if no material fills its core? (b) If not, determine the minimum magnetic permeability of the ferromagnetic material that would comprise the core.

Thinking It Through. The *B* field depends on the number of turns (*N*), the solenoid length (*L*), the current (*I*), and the permeability of the core material (μ). This is a direct application of Eqs. 19.14 and 19.15.

Solution.

Given: $I_{max} = 2.0$ A   *Find:* (a) Is $B = 2.0$ T possible with no core material?
          $N = 200$ turns   (b) Magnetic permeability required to attain $B = 2.0$ T
          $L = 0.30$ m

(a) From Eq. 19.14, without any core material, the interior field clearly would not be large enough:

$$B = \frac{\mu_o NI}{L} = \frac{(4\pi \times 10^{-7}\,\text{T} \cdot \text{m/A})(200)(2.0\,\text{A})}{0.30\,\text{m}} = 1.7 \times 10^{-3}\,\text{T} = 1.7\,\text{mT}$$

(b) The required field is about $2.0\,\text{T}/1.7 \times 10^{-3}\,\text{T}$ or about 1200 times as strong as the answer to part (a). Thus, because $B \propto \mu$ if everything else is constant, to attain a value of 2.0 T requires a permeability $\mu \geq 1200\,\mu_o$ or $\mu \geq 1.5 \times 10^{-3}\,\text{T} \cdot \text{m/A}$.

Follow-Up Exercise. In this example, if the scientists found a way for the solenoid to handle up to 5.0 A, what would be the new required permeability?

From Eq. 19.15, the electromagnet's field strength also depends on its current. Large currents produce large fields, but this is accompanied by much greater joule heating ($I^2 R$ losses) in the wires, which then require water cooling. The problem can be alleviated by using a superconducting wire. This is because superconducting materials have zero resistance and there is no joule heating. This technology currently exists in laboratories. For commercial use, superconducting magnets are not yet practical because of the energy required in the form of cooling to keep the conductors at their low temperatures and thus in their superconducting state. If

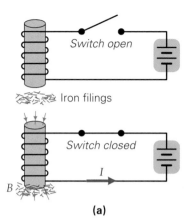
(a)

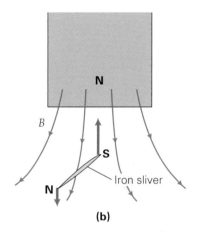
(b)

(c)

▲ **FIGURE 19.28 Electromagnet**
(a) *(top)* With no current in the circuit, there is no magnetic force. *(bottom)* However, with a current in the coil, there is a magnetic field and the iron core becomes magnetized. **(b)** Detail of the lower end of the electromagnet in part (a). The sliver of iron is attracted to the end of the electromagnet. **(c)** An electromagnet picking up scrap metal.

(and when) near–room temperature superconductors are found, high-strength magnetic fields will become commonplace in many appliances and applications.

The type of iron that retains some of its magnetism after being in an external magnetic field is called *hard iron* and is used to make so-called permanent magnets. You may have noticed that a paper clip or a screwdriver blade becomes slightly magnetized after being near a magnet. Permanent magnets are produced by heating pieces of some ferromagnetic material in an oven and then cooling them in a strong magnetic field to get the maximum effect. In permanent magnets, the domains do *not* become unaligned when the external magnetic field is removed.

A *permanent* magnet is not necessarily truly permanent, because its magnetism can be destroyed. Hitting such a magnet with a hard object or dropping it on the floor can cause a loss of some domain alignment, reducing the magnet's magnetic field. Also, it can cause a loss of magnetism, because the increase in random (thermal) motions of atoms tends to disrupt the domain alignment. One of the worst things you can do to a magnetic VHS or audiocassette tape is to leave it on the dashboard of a car on a hot day. The increased thermal motion of the electron spins can partially destroy the magnetic voice or video signal imprinted on the tape. Above a certain critical temperature, called the **Curie temperature** (or Curie point), domain coupling is destroyed by these increased thermal oscillations, and a ferromagnetic material loses its ferromagnetism. This effect was discovered by the French physicist Pierre Curie (1859–1906), husband of Madame Marie Curie. The Curie temperature for iron is 770°C.

Ferromagnetic domain alignment plays an important role in geology and geophysics. For example, it is well known that when cooled, lava flows that initially contain iron above its Curie temperature can retain some magnetism due to the Earth's field as it existed when the lava cooled below the Curie temperature and hardened. Measuring the strength and orientation of older lava flows at various locations has enabled geophysicists to map the changes in the Earth's magnetic field and polarity over time.

Some of the first evidence in support of plate tectonic motion came from measuring the direction of the magnetic polarity of seafloor samples containing iron.* The seafloor near the mid-Atlantic ridge, for example, is composed of lava flows from underwater volcanoes. These solidified flows were found to exhibit permanent magnetism, but the polarity varied with time (older samples are farther out from the ridge) as the Earth's magnetic polarity changed.

*19.8 Geomagnetism: The Earth's Magnetic Field

OBJECTIVES: To (a) state some of the general characteristics of the Earth's magnetic field, (b) explain some theories about its possible source, and (c) discuss some of the ways in which the Earth's magnetic field affects our planet's local environment.

The magnetic field of the Earth was used for centuries before people had any clues about its origin. In ancient times, navigators used lodestones or magnetized needles to locate north. Some other forms of life, including certain bacteria and homing pigeons, also use the Earth's magnetic field for navigation. (See Insight 19.2 on Magnetism in Nature.)

An early study of magnetism was carried out by the English scientist Sir William Gilbert about 1600. In investigating the magnetic field of a specially cut *lodestone* (the name for naturally occurring magnetized rocks) that simulated that of the Earth, he concluded that the Earth as a whole acts as a magnet. Gilbert thought that a large body of permanently magnetized material within the Earth might produce its field.

In fact, the Earth's external magnetic field, or the *geomagnetic field*, as it is called, does have a configuration similar to that which would be produced by a large interior bar magnet with the south pole of the magnet pointing north (◄Fig. 19.29).

*The outermost solid layer of the Earth's crust is known to be composed of sections or "plates." These plates are in constant, but very slow, motion—a centimeter per year is a typical speed. At certain intersections, such as that where the Pacific plate meets the North American plate along the southern coastline of Alaska, the plates collide, giving rise to volcanism and earthquakes. At other intersections, such as the mid-Atlantic ridge, the plates are receding from one another, with new material coming up from the Earth's interior in the form of hot lava.

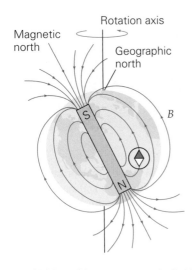

▲ **FIGURE 19.29 Geomagnetic field** The Earth's magnetic field is similar to that of a bar magnet. However, a permanent solid magnet could not exist within the Earth because of the high temperatures there. The Earth's magnetic field is believed to be associated with motions in the liquid outer core deep within the planet.

INSIGHT   19.2 MAGNETISM IN NATURE

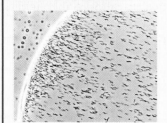

FIGURE 1 Magnetotactic bacteria in migration Bacteria in a drop of muddy water, as viewed under a microscope, align along the direction of the applied magnetic field (north to the left) and accumulate at the edge. When the field is reversed, so is the direction of migration.

FIGURE 2 A magnetotactic bacterium A freshwater magnetotactic bacterium, shown in an electron micrograph. Two whiplike appendages, or flagella, are clearly visible, along with a chain of magnetite particles.

For centuries, humans have relied on compasses to provide directional information (Fig. 19.29). Studies indicate that certain organisms seem to have their own built-in directional sensors. For example, some species of bacteria are *magnetotactic*—that is, able to sense the presence and direction of the Earth's magnetic field.

In the 1980s, experiments were performed on bacteria commonly found in bogs, swamps, and ponds.* In a laboratory magnetic field, when a droplet of muddy water was viewed under a microscope, one species of bacteria always aligned and migrated in the direction of the field (Fig. 1)—just as these bacteria do in their natural environment with the Earth's field. Furthermore, when these bacteria died and therefore could no longer migrate, they maintained their alignment with the field even when its direction changed. It became apparent that members of this species act like biological magnetic dipoles, or biological compasses. Once aligned with the field, they migrate along magnetic field lines by moving their *flagella* (whiplike appendages), as shown in Fig. 2.

What makes these bacteria act as living compasses? Even among known magnetotactic species, "new" bacteria (formed by cell division) do not initially have this magnetotactic sense. However, if they live in a solution containing a minimum concentration of iron, they are able to synthesize a chain of small magnetic particles (Fig. 2). Oddly enough, these internal com-

passes have the same chemical composition as the original slivers of naturally occurring ore used as compasses by ancient sailors: magnetite (chemical symbol Fe_3O_4). The individual particles in the chain are approximately 50 nm across, and the chain of a mature bacterium typically contains about 20 such particles, each of which is a single magnetic domain.

In essence, these bacteria are passively steered by their internal compasses. But why is it biologically important for these bacteria to follow the Earth's magnetic field? A piece of the puzzle was found while investigators were studying the same species from Southern Hemisphere waters. These bacteria migrate *opposite* the direction of the Earth's field, unlike their Northern Hemisphere counterparts. Recall that in the Northern Hemisphere the Earth's magnetic field inclines downward and that the reverse is true in the Southern Hemisphere. This discovery lead scientists to believe that the bacteria are using the field direction for survival. Oxygen is toxic to them, so they are most likely to survive in the muddy, oxygen-poor depths of their bog, swamp, or pond, and following the Earth's magnetic field direction enables them to head that way (Fig. 3). This directional sense also aids them near the equator. There it does not direct them downward but instead keeps them at a constant depth, thus avoiding an upward migration to the deadly oxygen-rich surface waters.

Evidence of magnetic field navigation has been found not only in bacteria, but also in such diverse organisms as bees, butterflies, homing pigeons, and dolphins.

*See, for example, R. P. Blakemore and R. B. Frankel, "Magnetic Navigation in Bacteria," *Scientific American*, December 1981. We are indebted to Professor Frankel for several interesting discussions of this topic.

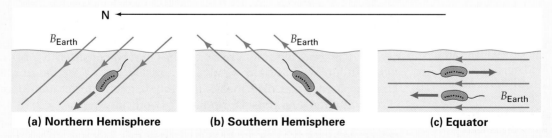

FIGURE 3 Survival of the fittest? (a) In the Northern Hemisphere, where the Earth's magnetic field inclines downward, magnetotactic bacteria follow the field to the oxygen-poor depths. **(b)** In the Southern Hemisphere, the Earth's field is inclined upward, but the bacteria migrate opposite the field and therefore are also able to head for deep waters, like their Northern Hemisphere cousins. **(c)** Near the equator, the bacteria move parallel to the water surface and thus are kept away from shallow, oxygen-rich (and hence toxic to them) waters.

The magnitude of the horizontal component of the Earth's magnetic field at the magnetic equator is on the order of 10^{-5} T (about 0.4 G), and the vertical component at the geomagnetic poles is about 10^{-4} T (roughly, 1 G). It has been calculated that for a ferromagnetic material of maximum magnetization to produce this field, it would have to occupy only about 0.01% of the Earth's volume.

The idea of a ferromagnet of this size within the Earth may not seem unreasonable at first, but this cannot be a correct model. It is known that the interior temperatures deep inside the Earth are well above the Curie temperatures of iron and nickel, the ferromagnetic materials believed to be the most abundant in the Earth's interior. For example, the Curie temperature for iron is attained at a depth of only 100 km below the Earth's surface. The temperatures are even higher at greater depths. So the existence of a permanent internal Earth magnet is not possible.

Knowing that electric currents produce magnetic fields has led scientists to speculate that the Earth's field is associated with motions in the liquid outer core, which, in turn, may be connected some way with the Earth's rotation. We know that Jupiter, a planet that is largely gaseous and rotates very rapidly, has a magnetic field much larger than that of Earth. Mercury and Venus have very weak magnetic fields; these planets are more like Earth and rotate relatively slowly.

Several theoretical models have been proposed to explain the Earth's magnetic field. For example, it has been suggested that the field arises from currents associated with thermal convection cycles in the liquid outer core caused by heat from the inner core. But the details of this mechanism are still not clear.

It is well known that the axis of the Earth's magnetic field does *not* coincide with the planet's rotational axis, which defines the geographic poles. Hence, the Earth's (south) magnetic pole and the geographic North Pole do not coincide (see Fig. 19.29). The magnetic pole is several thousand kilometers south of the geographic North Pole (true north). The Earth's north magnetic pole is displaced even more from its south geographic pole, meaning that the magnetic axis does not even pass through the center of the Earth.

A compass indicates the direction of *magnetic north*, not "true," or geographic, north. The angular difference in these two directions is called the *magnetic declination* (◂Fig. 19.30). The magnetic declination varies with location. Knowing these variations has historically been particularly important in the accurate navigation of airplanes and ships, as you can imagine. Most recently, with the advent of super-accurate GPSs (Global Positioning Systems), high-tech travelers no longer depend on compasses as much as before.

The Earth's magnetic field also exhibits a variety of fluctuations with time. As we have discussed, the permanent magnetism created in iron-rich rocks as they cooled in the Earth's magnetic field has provided us with much evidence of these fluctuations over long time scales. For example, the Earth's magnetic poles have switched polarity at various times in the past, most recently about 700 000 years ago. During a period of reversed polarity, the south magnetic pole is near the south geographic pole—the opposite of today's polarity. The mechanism for this periodic magnetic polarity reversal is not clearly understood and scientists are still investigating it.

On a shorter time scale, the magnetic poles also tend to "wander," or change location. For example, the Earth's south magnetic pole (near the north geographic pole) has recently been moving about 1° in latitude (roughly 110 km or 70 mi) per decade. For some unknown reason, it has moved consistently northward from its 1904 latitude of 69°N, and westward, crossing the 100°W longitudinal meridian. This long-term polar drift means that the magnetic declination map (Fig. 19.30b) varies with time and must be updated periodically.

On a still shorter time scale, there are sometimes dramatic daily shifts of magnetic pole location by as much as 80 km (50 mi), followed by a return to the starting position. These shifts are thought to be caused by charged particles from the Sun that reach the Earth's upper atmosphere and set up currents that change the planet's overall magnetic field.

Charged particles from the Sun entering the Earth's magnetic field give rise to another phenomenon. A charged particle that enters a uniform magnetic field at an angle that is *not* perpendicular to the field spirals in a helix (▸Fig. 19.31a). This

Demonstration/activity: To illustrate the sometimes subtle effects of the Earth's magnetic field, turn on a CRT such as a computer display and mark the location of something on the screen with a piece of tape. Then turn the CRT upside down and notice that the image on the screen has moved to one side. Explain.

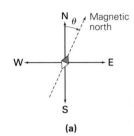

(a)

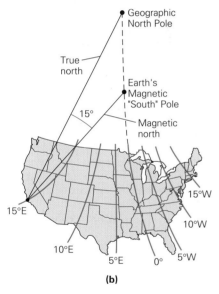

(b)

▲ **FIGURE 19.30** Magnetic declination **(a)** The angular difference between magnetic north and "true," or geographic, north is called the magnetic declination. **(b)** The magnetic declination varies with location and time. The map shows *isogonic* lines (lines with the same magnetic declination) for the continental United States. For locations on the 0° line, magnetic north is in the same direction as true (geographic) north. On either side of this line, a compass has an easterly or westerly variation. For example, on a 15°E line, a compass has an easterly declination of 15°. (Magnetic north is 15° east of true north.)

▲ **FIGURE 19.31 Magnetic confinement** **(a)** A charged particle entering a uniform magnetic field at an angle other than 90° moves in a spiraling path. **(b)** In a nonuniform, bulging magnetic field, particles spiral back and forth as though confined in a magnetic bottle. **(c)** Charged particles are trapped in the Earth's magnetic field, and the regions where they are concentrated are called Van Allen belts.

is because the component of the particle's velocity parallel to the field does not change. (Recall that a magnetic field acts only on the perpendicular component of the velocity.) The motions of charged particles in a nonuniform field are quite complex. However, for a bulging field such as that depicted in Fig. 19.31b, the particles spiral back and forth as though in a "magnetic bottle."

An analogous phenomenon occurs in the Earth's magnetic field, giving rise to regions with concentrations of charged particles. Two large donut-shaped regions at altitudes of several thousand kilometers are called the *Van Allen radiation belts* (Fig. 19.31c). In the lower Van Allen belt, light emissions called *auroras* occur—the aurora borealis, or northern lights, in the Northern Hemisphere and the aurora australis, or southern lights, in the Southern Hemisphere. These eerie, flickering lights are most commonly observed in the Earth's polar regions, but have been seen at lower latitudes (▶Fig. 19.32).

An aurora is created when charged solar particles become trapped in the Earth's magnetic field. Maximum aurora activity occurs after a solar disturbance, such as a solar flare—a violent magnetic storm on the Sun that spews out enormous quantities of charged particles. Trapped in the Earth's magnetic field, these particles are guided toward the polar regions, where they excite or ionize oxygen and nitrogen atoms in the atmosphere. When the excited atoms return to their normal state and the ions regain their normal number of electrons, light is emitted (Chapter 27), producing the glow of the aurora.

▲ **FIGURE 19.32 Aurora borealis: the northern lights** This spectacular display is caused by energetic solar particles trapped in the Earth's magnetic field. The particles excite or ionize air atoms; on de-excitation (or recombination) of the atoms, light is emitted.

Chapter Review

- The **pole–force law,** or **law of poles:** opposite magnetic poles attract and like poles repel.

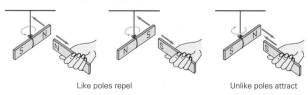

Like poles repel   Unlike poles attract

- The **magnetic field ($\vec{B}$)** has SI units of the **tesla (T)**, where $1\ \text{T} = 1\ \text{N}/(\text{A}\cdot\text{m})$. Magnetic fields can exert forces on moving charged particles and electric currents. The magnitude of the magnetic force on a charged particle is

$$F = qvB \sin \theta \qquad (19.3)$$

The magnitude of the magnetic force on a current-carrying wire is

$$F = ILB \sin \theta \qquad (19.7)$$

- **Right-hand force rules** determine the direction of a *magnetic force* on moving charged particles and current-carrying wires.

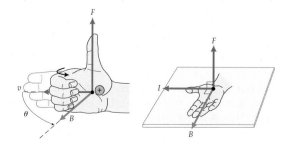

- A series of N current-carrying circular loops, each with a plane area A and carrying a current I, can experience a **magnetic torque** when placed in a magnetic field. The magnitude of the torque on such an arrangement is

$$\tau = NIAB \sin \theta \qquad (19.9)$$

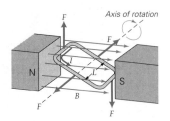

- The magnitude of the **magnetic field** produced by a long, straight wire is

$$B = \frac{\mu_{\text{o}} I}{2\pi d} \qquad (19.12)$$

where $\mu_{\text{o}} = 4\pi \times 10^{-7} \text{ T} \cdot \text{m/A}$ is the **magnetic permeability of free space**. For long, straight wires, the field lines are closed circles centered on the wire.

- The magnitude of the magnetic field produced at the center of an arrangement of N circular current-carrying loops of radius r is

$$B = \frac{\mu_{\text{o}} NI}{2r} \qquad (19.13)$$

- The magnitude of the magnetic field produced near the center of the interior of a *solenoid* with N windings and a length L is

$$B = \frac{\mu_{\text{o}} NI}{L} \qquad (19.14)$$

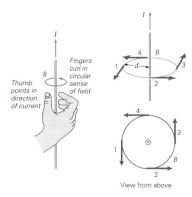

- **Right-hand source rules** are used to determine the direction of the magnetic field from various current configurations.

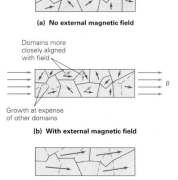

View from above

- In **ferromagnetic materials,** the electron spins align, creating **domains.** When an external field is applied, the effect is to increase the size of those domains that already point in the direction of the field at the expense of the others. When the external magnetic field is removed, a **permanent magnet** remains.

(a) No external magnetic field

Domains more closely aligned with field

Growth at expense of other domains

(b) With external magnetic field

(c) Resulting bar magnet

Exercises

MC = *Multiple Choice Question*, **CQ** = *Conceptual Question, and* **IE** = *Integrated Exercise. Throughout the text, many exercise sections will include "paired" exercises. These exercise pairs, identified with* **red numbers,** *are intended to assist you in problem solving and learning. In a pair, the first exercise (even numbered) is worked out in the Study Guide so that you can consult it should you need assistance in solving it. The second exercise (odd numbered) is similar in nature, and its answer is given at the back of the book.*

19.1 Magnets, Magnetic Poles, and Magnetic Field Direction

1. **MC** When the ends of two bar magnets are near each other, they attract one another. The ends must be (a) one north, the other south, (b) both north, (c) both south, (d) either a or b. (a)

2. **MC** A compass that has been calibrated in the Earth's magnetic field is placed near the end of a permanent bar magnet, and points away from that end of the magnet. It can be concluded that this end of the magnet (a) acts as a north magnetic pole, (b) acts as a south magnetic pole, (c) you can't conclude anything about the magnetic properties of the permanent magnet. (a)

3. **MC** If you look directly at the south pole of a bar magnet, its magnetic field points (a) to the right, (b) to the left, (c) away from you, or (d) toward you. (c)

4. **CQ** Given two identical iron bars, one of which is a permanent magnet and the other unmagnetized, how could you tell which is which by using only the two bars? see ISM

5. **CQ** The direction of any magnetic field is taken to be in the direction that an Earth-calibrated compass points. Explain why this means that magnetic field lines must leave from the north pole of a permanent bar magnet and enter its south pole. see ISM

19.2 Magnetic Field Strength and Magnetic Force

6. **MC** A proton moves vertically upward perpendicularly to a uniform magnetic field and deflects to the right as you watch it. What is the magnetic field direction: (a) directly away from you, (b) directly toward you, (c) to the right, or (d) to the left? (b)

7. **MC** An electron is moving horizontally to the east in a uniform magnetic field that is vertical. It is found to deflect north. What direction is the magnetic field? (a) up (b) down or (c) the direction can't be determined from the given data. (a)

8. **MC** If a negatively charged particle were moving downward along the right edge of this page, which way should a magnetic field (known to be perpendicular to the plane of the paper) be oriented so that the particle would initially be deflected to the left: (a) out of the page, (b) in the plane of the page, or (c) into the page? (c)

9. **MC** An electron passes through a magnetic field without being deflected. What do you conclude about the angle between the magnetic field direction and that of the electron's velocity, assuming that no other forces act: (a) They could be in the same direction, (b) they could be perpendicular, (c) they could be opposite, or (d) both a and c are possible? (d)

10. **CQ** A proton and an electron are moving at the same velocity perpendicularly to a constant magnetic field. (a) How do the magnitudes of the magnetic forces on them compare? (b) What about the magnitudes of their accelerations? (a) same (b) electron experiences greater acceleration

11. **CQ** If a charged particle moves in a straight line and there are no other forces on it except possibly from a magnetic field, can you say with certainty that no magnetic field is present? Explain. not necessarily

12. **CQ** Three particles enter a uniform magnetic field as shown in ▶Fig. 19.33a. Particles 1 and 3 have equal speeds and charges of the same magnitude. What can you say about (a) the charges of the particles and (b) their masses? see ISM

13. **CQ** You want to deflect a positively charged particle in an S-shaped path, as shown in Fig. 19.33b, using only magnetic fields. (a) Explain how this could be done by using magnetic fields perpendicular to the plane of the page. (b) How does the magnitude of an emerging particle's velocity compare with the particle's initial velocity? see ISM

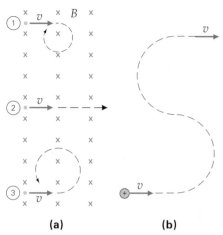

▲ **FIGURE 19.33 Charges in motion** See Exercises 12 and 13.

14. **IE** ● A positive charge moves horizontally to the right across this page and enters a magnetic field directed vertically downward in the plane of the page. (a) What is the direction of the magnetic force on the charge: (1) into the page, (2) out of the page, (3) downward in the plane of the page, or (4) upward in the plane of the page? Explain. (b) If the charge is 0.25 C, its speed is 2.0×10^2 m/s, and it is acted on by a force of 20 N, what is the magnetic field strength?
(a) (1) into the page (b) 0.40 T

15. ● A charge of 0.050 C moves vertically in a field of 0.080 T that is oriented 45° from the vertical. What speed must the charge have such that the force acting on it is 10 N? 3.5×10^3 m/s

16. ●● A magnetic field can be used to determine the sign of charge carriers in a current-carrying wire. Consider a wide conducting strip in a magnetic field oriented as shown in ▼Fig. 19.34. The charge carriers are deflected by the magnetic force and accumulate on one side of the strip, giving rise to a measurable voltage across it. (This phenomenon is known as the *Hall effect*.) If the sign of the charge carriers is unknown (they are either positive charges moving as indicated by the arrows in the figure or negative charges moving in the opposite direction), how does polarity or sign of the measured voltage allow the sign of the charge to be determined? Assume that only one type of charge carrier is responsible for the current. see ISM

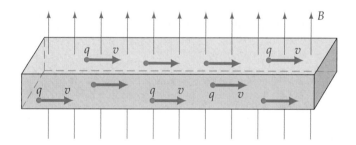

▲ **FIGURE 19.34 The Hall effect** See Exercise 16.

17. ●● A beam of protons is accelerated to a speed of 5.0×10^6 m/s in a particle accelerator and emerges horizontally from the accelerator into a uniform magnetic field. What $\vec{B}$ field perpendicular to the velocity of the proton would cancel the force of gravity and keep the beam moving exactly horizontally? 2.0×10^{-14} T, left, looking in the direction of the velocity

18. **IE ●●** An electron travels in the $+x$-direction in a magnetic field and is acted on by a magnetic force in the $-y$-direction. (a) In which of the following directions could the magnetic field be oriented: (1) $-x$-, (2) $+y$-, (3) $+z$-, or (4) $-z$-direction? Explain. (b) If the electron speed is 3.0×10^6 m/s and the magnitude of the force is 5.0×10^{-19} N, what is the magnetic field strength? (a) (4) $-z$- (b) 1.0×10^{-6} T

19. **●●** An electron travels at a speed of 2.0×10^4 m/s through a uniform magnetic field whose magnitude is 1.2×10^{-3} T. What is the magnitude of the magnetic force on the electron if its velocity and the magnetic field (a) are perpendicular, (b) make an angle of $45°$, (c) are parallel, and (d) are exactly opposite? (a) 3.8×10^{-18} N (b) 2.7×10^{-18} N (c) zero (d) zero

20. **●●** What angle(s) does a particle's velocity have to make with the magnetic field direction for the particle to be subjected to half the maximum possible magnetic force? $30°$ or $150°$

21. **●●●** A proton beam is first accelerated due east to a speed of 3.0×10^5 m/s in a particle accelerator. The beam then enters a uniform magnetic field of 0.50 T, which is oriented at an angle of $37°$ above the horizontal relative to the direction of the beam. (a) What is the initial acceleration of a proton in the accelerated beam? (b) What if the magnetic field were angled at $37°$ below the horizontal instead? (c) If the beam were made of electrons rather than protons and the field angle upward at $37°$, what would be the difference in the force on the particles as the beam entered the magnetic field? see ISM

19.3 Applications: Charged Particles in Magnetic Fields

22. **MC** In a mass spectrometer two ions with identical charge and speed are accelerated into two different semicircular arcs. Ion A's arc has a radius of 25.0 cm and ion B's arc's radius is 50.0 cm. What can you say about their relative masses? (a) $m_A = m_B$ (b) $m_A = 2m_B$ (c) $m_A = \frac{1}{2}m_B$ (d) you can't say anything given just this data. (c)

23. **MC** In a mass spectrometer two ions with identical mass and speed are accelerated into two different semicircular arcs. Ion A's arc has a radius of 25.0 cm and ion B's arc's radius is 50.0 cm. What can you say about their net charges? (a) $q_A = q_B$ (b) $q_A = 2q_B$ (c) $q_A = \frac{1}{2}q_B$ (d) you can't say anything given just this data. (b)

24. **MC** In the velocity selector shown in Fig. 19.9, which way will an ion be deflected if its velocity is larger than E/B_1? (a) up (b) down (c) There will be no deflection. (a)

25. **CQ** Explain how a nearby magnet can distort the display of a computer monitor or television picture tube (▼Fig. 19.35). magnetic force on electron beam

◀ **FIGURE 19.35 Magnetic disturbance** See Exercise 25.

26. **CQ** The enlarged circular inset in Fig. 19.11 shows how the positive (Na^+) ions in seawater are accelerated out the rear of the submarine to provide a propulsive force. But what about the negative (Cl^-) ions in the seawater? Because they have charge of the opposite sign, aren't they accelerated *forward* resulting in a net force of zero on the submarine? Explain. see ISM

27. **CQ** Explain clearly why the speed selected in a velocity selector setup (similar to that in Fig. 19.9) does not depend on the charges of any of the ions passing through. see ISM

28. **●** An ionized deuteron (a particle with a $+e$ charge) passes through a velocity selector whose perpendicular magnetic and electric fields have magnitudes of 40 mT and 8.0 kV/m, respectively. Find the speed of the ion. 2.0×10^5 m/s

29. **●** In a velocity selector, the uniform magnetic field of 1.5 T is produced by a large magnet. Two parallel plates with a separation of 1.5 cm produce the perpendicular electric field. What voltage should be applied across the plates so that (a) a singly charged ion traveling at 8.0×10^4 m/s will pass through undeflected or (b) a doubly charged ion traveling at the same speed will pass through undeflected? (a) 1.8×10^3 V (b) same voltage, independent of the charge

30. **●** A charged particle travels undeflected through perpendicular electric and magnetic fields whose magnitudes are 3000 N/C and 30 mT, respectively. Find the speed of the particle if it is (a) a proton or (b) an alpha particle. (An alpha particle is a helium nucleus—a positive ion with a double positive charge.) (a) 1.0×10^5 m/s (b) same speed, independent of charge

31. **●●** In an experimental technique for treating deep tumors, unstable positively charged pions (π^+, elementary particles with a mass of 2.25×10^{-28} kg) are aimed to penetrate the flesh and to disintegrate at the tumor site, releasing energy to kill cancer cells. If pions with a kinetic energy of 10 keV are required and if a velocity selector with an electric field strength of 2.0×10^3 V/m is used, what must be the magnetic field strength? 5.3×10^{-4} T

32. **●●** In a mass spectrometer, a singly charged ion having a particular velocity is selected by using a magnetic field of 0.10 T perpendicular to an electric field of 1.0×10^3 V/m. This same magnetic field is then used to deflect the ion, which moves in a circular path with a radius of 1.2 cm. What is the mass of the ion? 1.9×10^{-26} kg

33. **●●** In a mass spectrometer, a doubly charged ion having a particular velocity is selected by using a magnetic field of 100 mT perpendicular to an electric field of 1.0 kV/m. This same magnetic field is then used to deflect the ion in a circular path with a radius of 15 mm. Find (a) the mass of the ion and (b) the kinetic energy of the ion. (c) Does the kinetic energy of the ion increase in the circular path? Explain. (a) 4.8×10^{-26} kg (b) 2.4×10^{-18} J (c) no, work equals zero

34. **●●●** In a mass spectrometer, a beam of protons enters a magnetic field. Some protons make exactly a one-quarter circular arc of radius 0.50 m. If the field is always perpendicular to the proton's velocity, what is the field's magnitude if exiting protons have a kinetic energy of 10 keV? 2.9×10^{-2} T

19.4 Magnetic Forces on Current-Carrying Wires
and
19.5 Applications: Current-Carrying Wires in Magnetic Fields

35. **MC** A long, straight, horizontal wire located on the equator carries a current directed toward the east. What is the direction of the force on the wire due to the Earth's magnetic field: (a) east, (b) west, (c) south, or (d) upward? (d)

36. **MC** A long, straight, horizontal wire located on the equator carries a current. In what direction should the current be if the object is to attempt to balance the wire's weight with the magnetic force on it: (a) east, (b) west, (c) south, or (d) upward? (a)

37. **MC** You are looking horizontally due west directly at the circular plane of a current-carrying coil. The coil is in a uniform vertically upward magnetic field. When released, the top of the coil starts to rotate away from you as the bottom rotates toward you. Which direction is the current in the coil: (a) clockwise, (b) counterclockwise, or (c) can't tell from the data given? (b)

38. **CQ** Two straight wires are parallel to each other and carry different currents in the same direction. Do they attract or repel each other? How do the magnitudes of these forces on each wire compare?
    attract; same by Newton's third law, see ISM

39. **CQ** Predict what should happen to the length of a spring when a large current passes through it. [*Hint*: Consider the current direction in the neighboring spring coils.]
    it shortens

40. **CQ** Is it possible to orient a current loop in a uniform magnetic field so that there is no torque on the loop? If yes, describe the possible orientation(s). yes, with magnetic field perpendicular to plane of loop

41. **CQ** Explain the operation of the doorbell and door chimes illustrated in ▼Fig. 19.36. see ISM

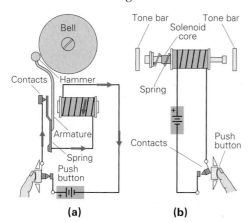

(a)   **(b)**

▲ **FIGURE 19.36 Electromagnetic applications** Both **(a)** a doorbell and **(b)** door chimes have electromagnets. See Exercise 41.

42. **IE ●** A straight, horizontal segment of wire carries a current in the +x-direction in a magnetic field that is directed in the −z-direction. (a) Is the magnetic force on the wire directed in the (1) −x-, (2) +z-, (3) +y-, or (4) −y-direction? Explain. (b) If the wire is 1.0 m long and carries a current of 5.0 A and the magnitude of the magnetic field is 0.30 T, what is the magnitude of the force on the wire?
    (a) (3) +y- (b) 1.5 N

43. **●** A 2.0-m length of straight wire carries a current of 20 A in a uniform magnetic field of 50 mT whose direction is at an angle of 37° from the direction of the current. Find the force on the wire.
    1.2 N perpendicular to the plane of $\vec{B}$ and *I*

44. **●●** Show how you can use a right-hand force rule to find the direction of the current in a wire in a magnetic field if you know the force on the wire. The forces on some specific wires are shown in ▼Fig. 19.37. (a) right (b) toward top of the page (c) into the page (d) left (e) into or out of the page

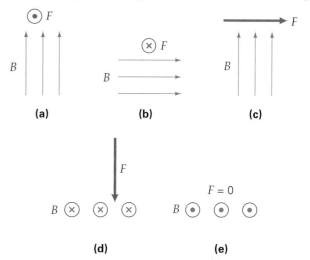

▲ **FIGURE 19.37 The right-hand force rule** See Exercise 44.

45. **●●** A straight wire 50 cm long conducts a current of 4.0 A directed vertically upward. If the wire is acted on by a force of 1.0×10^{-2} N in the eastward direction due to a magnetic field at right angles to the length of the wire, what are the magnitude and direction of the magnetic field?
    5.0×10^{-3} T north to south

46. **●●** A horizontal magnetic field of 1.0×10^{-4} T is at an angle of 30° to the direction of the current in a straight, horizontal wire 75 cm long. If the wire carries a current of 15 A, what is the magnitude of the force on the wire?
    5.6×10^{-4} N

47. **●●** A wire carries a current of 10 A in the +x-direction in a uniform magnetic field of 0.40 T. Find the magnitude of the force per unit length and the direction of the force on the wire if the magnetic field points in (a) the +x-direction, (b) the +y-direction, (c) the +z-direction, (d) the −y-direction, and (e) the −z-direction. see ISM

48. **●●** A straight wire 25 cm long is oriented vertically in a uniform horizontal magnetic field of 0.30 T pointing in the −x-direction. What current (including direction) would cause the wire to be subject to a force of 0.050 N in the +y-direction? 0.67 A; −z

49. **●●** A wire carries a current of 10 A in the +x-direction. Find the force per unit length on the wire if it is in a magnetic field that has components of $B_x = 0.020$ T, $B_y = 0.040$ T, and $B_z = 0$ T. 0.40 N/m; +z

50. **●●** A set of jumper cables used to start a car from another car's battery is connected to the terminals of both batteries. If 15 A of current exists in the cables during the starting procedure and the cables are parallel and 15 cm apart, what is the force per unit length on the cables?
    3.0×10^{-4} N/m, repulsive

51. IE ●● Two long, straight, parallel wires carry current in the same direction. (a) Use the right-hand source and force rules to determine whether the forces on the wires are (1) attractive or (2) repulsive. (b) If the wires are 24 cm apart and carry currents of 2.0 A and 4.0 A, respectively, find the force per unit length on each wire.
(a) (1) attractive (b) 6.7×10^{-6} N/m

52. IE ●● Two long, straight, parallel wires 10 cm apart carry currents in opposite directions. (a) Use the right-hand source and force rules to determine whether the forces on the wires are (1) attractive or (2) repulsive. (b) If the wires carry equal currents of 3.0 A, what is the force per unit length on the wires? (a) (2) repulsive (b) 1.8×10^{-5} N/m

53. ●● A nearly horizontal dc power line in the midlatitudes of North America carries a current of 1000 A directly eastward. If the Earth's magnetic field at the location of the power line is northward with a magnitude of 5.0×10^{-5} T at an angle of 45° below the horizontal, what are the magnitude and direction of the magnetic force on a 15-m section of the line? 0.53 N northward at an angle of 45° above the horizontal

54. ●● What is the force (including direction) per unit length on wire 1 in ▼Fig. 19.38? 2.7×10^{-5} N/m toward wire 2

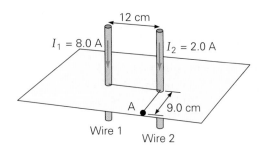

▲ **FIGURE 19.38 Parallel current-carrying wires** See Exercises 54, 55, 73, and 76.

55. ●● What is the force (including direction) per unit length on wire 2 in Fig. 19.38? 2.7×10^{-5} N/m toward wire 1

56. IE ●● A long wire is placed 2.0 cm directly below a rigidly mounted second wire (▼Fig. 19.39). (a) Use the right-hand source and force rules to determine whether the currents in the wires should be in (1) the same or (2) the opposite direction so that the lower wire is in equilibrium. (It "floats.") (b) If the lower wire has a linear mass density of 1.5×10^{-3} kg/m and the wires carry the same current, what should be the current?
(a) (1) the same (b) 38 A

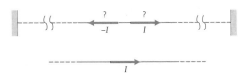

▲ **FIGURE 19.39 Magnetic suspension** The bottom wire is magnetically attracted to the top (rigidly fixed) wire. See Exercise 56.

57. ●● A wire is bent as shown in ▼ Fig. 19.40 and placed in a magnetic field with a magnitude of 1.0 T in the indicated direction. Find the net force on the wire if $x = 50$ cm and it carries a current of 5.0 A in the direction shown.
7.5 N upward in the plane of the paper

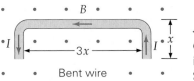

◄ **FIGURE 19.40** Current-carrying wire in a magnetic field See Exercise 57.

58. IE ●●● A loop of current-carrying wire is in a 1.6-T magnetic field. (a) For the magnetic torque on the loop to be at maximum, should the plane of the coil be (1) parallel, (2) perpendicular, or (3) at a 45° angle to the magnetic field? Explain. (b) If the loop is rectangular with dimensions 20 cm by 30 cm and carries a current of 1.5 A, what is the magnitude of the magnetic moment of the loop, and what is the maximum torque? (c) What would be the angle(s) between the magnetic moment vector and the magnetic field direction if the loop felt only 20% of its maximum torque? (a) (1) parallel (b) 9.0×10^{-2} A·m², 0.14 m·N (c) 12° or 168°

59. ●●● Two straight wires are positioned at right angles to each other as in ▼ Fig. 19.41. What is the net force on each wire? Is there a net torque on each wire? zero, yes

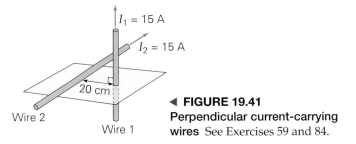

◄ **FIGURE 19.41** Perpendicular current-carrying wires See Exercises 59 and 84.

60. ●●● A rectangular wire loop with a cross-sectional area of 0.20 m² carries a current of 0.25 A. The loop is free to rotate about an axis that is perpendicular to a uniform magnetic field with strength 0.30 T. The plane of the loop is at an angle of 30° to the direction of the magnetic field. (a) What is the magnitude of the torque on the loop? (b) How would you change the magnetic field to double the magnitude of the torque in part (a)? (c) Could you double the torque in part (a) only by changing the angle? Explain. If so, find that angle.
(a) 0.013 m·N (b) double (c) no, see ISM

19.6 Electromagnetism: The Source of Magnetic Fields

61. MC A long, straight wire is parallel to the ground and carries a steady current to the east. At a point directly below the wire, what is the direction of the magnetic field the wire produces: (a) north, (b) east, (c) south, or (d) west? (a)

62. MC You are looking directly into one end of a long solenoid. The magnetic field at its center points at you. What is the direction of the current in the solenoid, as viewed by you: (a) clockwise, (b) counterclockwise, (c) directly toward you, or (d) directly away from you? (b)

63. **MC** A current-carrying loop of wire is in the plane of this paper. Outside the loop, its magnetic field points into the paper. What is the direction of the current in the loop: (a) clockwise, (b) counterclockwise, or (c) can't tell from the data given? (b)

64. **CQ** A circular current-carrying loop is lying flat on a table and creates a field at the loop's center. A calibrated compass, when placed at the center of the loop, points downward. If you look straight down on the loop, what is the direction of the current? Explain your reasoning. clockwise, see ISM

65. **CQ** If you doubled your distance from a long current-carrying wire, what would you have to do to the current to keep the magnetic field strength the same as at the close position but reverse its direction? Explain. double the current and reverse the direction

66. **CQ** There are two solenoids, one with 100 turns and the other with 200 turns. If both carry the same current, will the one with more turns necessarily produce a stronger magnetic field at its center? Explain. not necessarily

67. **CQ** To minimize the effects of the magnetic field, most appliance wires are placed close together. Explain how this works to reduce the external field from the current in the wire. see ISM

68. **CQ** Two circular wire loops are coplanar (that is, their areas are in the same plane) and have a common center. The outer one carries a current of 10 A in the clockwise direction. To create a zero magnetic field at their center, what should be the direction of the current in the inner loop? Should its current be 10 A, larger than 10 A, or smaller than 10 A? Explain your reasoning. counterclockwise, smaller, see ISM

69. ● The magnetic field at the center of a 50-turn coil of radius 15 cm is 0.80 mT. Find the current in the coil. 3.8 A

70. ● A long, straight wire carries a current of 2.5 A. Find the magnitude of the magnetic field 25 cm from the wire. 2.0×10^{-6} T

71. ● In a physics lab, a student discovers that the magnitude of the magnetic field at a certain distance from a long wire is 4.0 μT. If the wire carries a current of 5.0 A, what is the distance of the magnetic field from the wire? 0.25 m

72. ● A solenoid is 0.20 m long and consists of 100 turns of wire. At its center, the solenoid produces a magnetic field with a strength of 1.5 mT. Find the current in the coil. 2.4 A

73. ●● Two long, parallel wires carry currents of 8.0 A and 2.0 A (Fig. 19.38). (a) What is the magnitude of the magnetic field midway between the wires? (b) Where on a line perpendicular to and joining the wires is the magnetic field zero? (a) 2.0×10^{-5} T (b) 9.6 cm from wire 1

74. ●● Two long, parallel wires separated by 50 cm each carry currents of 4.0 A in a horizontal direction. Find the magnetic field midway between the wires if the currents are (a) in the same direction and (b) in opposite directions. (a) 0 (b) 6.4×10^{-6} T

75. ●● Two long, parallel wires separated by 0.20 m carry equal currents of 1.5 A in the same direction. Find the magnitude of the magnetic field 0.15 m away from each wire on the side opposite the other wire (▼Fig. 19.42). both 2.9×10^{-6} T

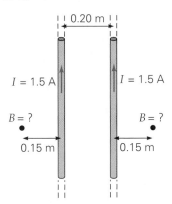

▲ **FIGURE 19.42 Magnetic field summation** See Exercise 75.

76. ●● In Fig. 19.38, find the magnetic field (magnitude and direction) at point A, which is located 9.0 cm away from wire 2 on a line perpendicular to the line joining the wires. 1.4×10^{-5} T at 38° below a horizontal line to the left

77. ●● Suppose that the current in wire 1 in Fig. 19.38 were in the opposite direction. What would be the magnetic field midway between the wires? 3.3×10^{-5} T

78. ●● How much current must flow in a circular loop of radius 10 cm to produce a magnetic field at the center of the loop that is the same magnitude as the horizontal component of the Earth's magnetic field at the equator (about 0.40 G)? 6.4 A

79. ●● A coil of four circular loops of radius 5.0 cm carries a current of 2.0 A clockwise, as viewed from above the coil's plane. What is the magnetic field at the center of the coil? 1.0×10^{-4} T, away from the observer

80. **IE** ●● A circular loop of wire in the horizontal plane carries a counterclockwise current, as viewed from above. (a) Use the right-hand source rule to determine whether the direction of the magnetic field at the center of the loop is (1) toward or (2) away from the observer. (b) If the diameter of the loop is 12 cm and the current is 1.8 A, what is the magnitude of the magnetic field at the center of the loop? (a) (1) toward (b) 1.9×10^{-5} T

81. ●● A circular loop of wire with a radius of 5.0 cm carries a current of 1.0 A. Another circular loop of wire is concentric with (that is, has a common center with) the first and has a radius of 10 cm. The magnetic field at the center of the loops is double what the field would be from the first one alone, but oppositely directed. What is the current in the second loop? 4.0 A

82. ●●● A current-carrying solenoid is 10 cm long and is wound with 1000 turns of wire. It produces a magnetic field of 4.0×10^{-4} T at the solenoid's center. (a) How long would you make the solenoid in order to produce a field of 6.0×10^{-4} T at its center? (b) Adjusting only the windings, what number would be needed to produce a field of 8.0×10^{-4} T at the center? (c) What current in the solenoid would be needed to produce a field of 9.0×10^{-4} T but in the opposite direction? (a) 6.7 cm (b) 2000 turns (c) opposite the original current direction; 0.072 A

83. ●●● A solenoid is wound with 200 turns per centimeter. An outer layer of insulated wire with 180 turns per centimeter is wound over the solenoid's first layer of wire. When the solenoid is operating, the inner coil carries a current of 10 A and the outer coil carries a current of 15 A in the direction opposite to that of the current in the inner coil (▼Fig. 19.43). (a) What is the magnitude of the magnetic field at the center of the doubly wound solenoid? (b) What is the direction of the magnetic field at the center for this configuration? (a) 8.8×10^{-2} T (b) to the right

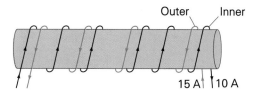

Outer   Inner

15 A   10 A

▲ **FIGURE 19.43 Double it up?** See Exercise 83.

84. ●●● Two long, perpendicular wires carry currents of 15 A, as illustrated in Fig. 19.41. What is the magnitude of the magnetic field at the midpoint of the line joining the wires? 4.2×10^{-5} T

85. ●●● Four wires running through the corners of a square with sides of length a, as shown in ▼Fig. 19.44, carry equal currents I. Calculate the magnetic field at the center of the square in terms of these parameters. $B = \sqrt{2}\mu_o I/(\pi a)$, at 45° toward the lower left wire

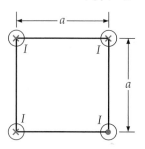

◀ **FIGURE 19.44 Current-carrying wires in a square array** See Exercise 85.

86. ●●● A particle with charge q and mass m moves in a horizontal plane at right angles to a uniform vertical magnetic field B. (a) What is the frequency f of the particle's circular motion in terms of q, B, and m? (This frequency is called the *cyclotron frequency*.) (b) Show that the time required for any charged particle to make one complete revolution is independent of its speed and radius. (c) Compute the path radius and the cyclotron frequency if the particle is an electron with speed $v = 1.0 \times 10^5$ m/s and the field strength is $B = 1.0 \times 10^{-4}$ T. see ISM

19.7 Magnetic Materials

87. **MC** The main source of magnetism in magnetic materials is from (a) electron orbits, (b) electron spin, (c) magnetic poles, (d) nuclear properties. (b)

88. **MC** When a ferromagnetic material is placed in an external magnetic field, (a) the domain orientation may change, (b) the domain boundaries may change, (c) new domains are created, (d) both a and b. (d)

89. **CQ** If you are looking down on the orbital plane of the electron in a hydrogen atom and the electron orbits counterclockwise, what is the direction of the magnetic field the electron produces at the proton? away from you

90. **CQ** What is the purpose of the iron core often used at the center of a solenoid? to increase the magnetic permeability and the magnetic field

91. **CQ** Discuss several ways you destroy or reduce the magnetic field of a permanent magnet. by hitting or heating it

92. ●● A solenoid with 100 turns per centimeter has an iron core with a relative permeability of 2000. The solenoid carries a current of 0.040 A. (a) What is the magnetic field at the center of the solenoid? (b) How much greater is the magnetic field with the iron core than it would be without it? (a) 1.0 T (b) $B = (2.0 \times 10^3)B_o$

93. ●●● What is the magnetic field (due to the electron only) at the center of the circular orbit of the electron in a hydrogen atom? The orbital radius is 0.0529 nm. [*Hint*: Find the electron's period by considering the centripetal force.] 12 T

*19.8 Geomagnetism: The Earth's Magnetic Field

94. **MC** The Earth's magnetic field (a) has poles that coincide with the geographic poles, (b) only exists at the poles, (c) reverses polarity every few hundred years, (d) none of these. (d)

95. **MC** Auroras (see Fig. 19.32) (a) occur only in the Northern Hemisphere, (b) are related to the lower Van Allen belt, (c) occur because of Earth's magnetic pole reversals, (d) happen predominantly when there are no solar disturbances. (b)

96. **MC** If the direction of your calibrated compass pointed straight down, where would you be: (a) near the Earth's north geographic pole, (b) near the equator, or (c) near the Earth's south geographic pole? (a)

97. **MC** If a proton was orbiting above the Earth's equator in the Van Allen belt, which way would it have to be orbiting: (a) to the west, (b) to the east, or (c) either direction would work? (a)

98. **CQ** Determine the direction of the force on an electron due to the Earth's magnetic field on an electron near the equator for each of the following situations. The electron's velocity is directed (a) due south, (b) northwest, or (c) upward. (a) zero (b) up (c) east

99. **CQ** In a relatively short time geologically speaking, it is conjectured that the Earth's magnetic field direction will reverse. After that, what would be the polarity of the magnetic pole near the Earth's geographic North Pole? north magnetic pole

Comprehensive Exercises

100. A beam of protons is accelerated from rest through a potential difference of 3.0 kV. It enters a region where its velocity is initially perpendicular to an electric field. The field is created by two parallel plates separated by 10 cm with a potential difference of 250 V across them. Find the magnitude of the magnetic field (perpendicular to $\vec{E}$) needed so the beam passes undeflected through the plates. 3.3×10^{-3} T

101. A solenoid 10 cm long has 3000 turns of wire and carries a current of 5.0 A. A 2000-turn coil of wire of the same length as the solenoid surrounds it and is concentric (shares a common central axis) with it. The outer coil carries a current of 10 A in the same direction as that of the current in the solenoid. Find the magnetic field at the common center. 0.44 T

102. IE A horizontal beam of electrons travels from north to south in a discharge tube located in the Northern Hemisphere. (a) Is the magnetic force on the electron directed (1) west, (2) east, (3) south, or (4) north? Explain. (b) If their speed is 1.0×10^3 m/s and the vertical component of the Earth's magnetic field at that location is known to be 5.0×10^{-5} T, what is the magnitude of the force on each electron? (a) (1) west (b) 8.0×10^{-21} N

103. A proton enters a uniform magnetic field that is at a right angles to its velocity. The field strength is 0.80 T and the proton follows a circular path with a radius of 4.6 cm. What is its (a) momentum and (b) kinetic energy? (a) 5.9×10^{-21} kg·m/s (b) 1.0×10^{-14} J

104. Exiting a linear accelerator, a narrow horizontal beam of protons travels due north. If 1.75×10^{13} protons pass a given point per second, determine the magnetic field direction and strength at a location of 2.40 m east of the beam. Does it seem likely this would interfere with a magnetic strip on an ATM card, in comparison to the Earth's field? 2.3×10^{-13} T down, no

105. A 200-turn circular coil of wire has a radius of 10.0 cm and a total resistance of 0.115 Ω. At its center the magnetic field strength is 7.45 mT. Determine the voltage of the power supply creating the current in the coil. 0.682 V

106. A 100-turn circular coil of wire has a radius of 20.0 cm and carries a current of 0.400 A. The normal to the coil area points due east. A compass, when placed at the center of the coil, does not point east, but instead makes an angle of 60° north of east. Using this data, determine (a) the magnitude of the horizontal component of the Earth's field at that location and (b) the magnitude of the Earth's field at that location if it makes an angle of 55° below the horizontal. (a) 2.18×10^{-4} T (b) 3.80×10^{-4} T

107. IE Two long straight wires are oriented perpendicularly to the page. Wire 1 carries current of 20.0 A into the page and 15.0 cm to its left, wire 2 carries a 5.00 A current. Somewhere on the line joining the two wires there is to be a zero magnetic field. (a) What is the direction of the current in wire 2: (1) out of the paper, (2) into the paper, or (3) can't tell from the data given? (b) Find the location where the zero magnetic field exists. (a) (2) into the page (b) 0.030 m from left wire

108. IE A circular coil of wire has the normal to its area pointing upward. A second smaller concentric coil carries a current in the opposite direction. (a) Where, in the plane of these coils, could the magnetic field be zero: (1) only inside the smaller one, (2) only between the inner and outer one, (3) only outside the larger one, (4) or inside the smaller one and outside the larger one? (b) The larger one is a 200-turn coil of wire with a radius of 9.50 cm and carries a current of 11.5 A. The second one is a 100-turn coil with a radius of 2.50 cm. Determine the current in the inner coil so the magnetic field at their common center is zero. Neglect the Earth's field. (a) (4) inside the smaller one and outside the larger one (b) 6.05 A

109. A 50-cm-long solenoid has 100 turns of wire and carries a current of 0.95 A. It has a ferromagnetic core completely filling its interior where the field is 0.71 T. Determine the (a) magnetic permeability and (b) relative magnetic permeability of the material. (a) 3.74×10^{-3} T·m/A (b) 3.0×10^3

The following Physlet Physics Problems can be used with this chapter. 27.1, 27.2, 27.3, 27.4, 27.5, 27.6, 27.7, 27.8, 27.9, 27.10, 28.1, 28.6

ELECTROMAGNETIC INDUCTION AND WAVES

PHYSICS FACTS

- Nikola Tesla (1856–1943), the Serbian-American scientist-inventor whose last name is the SI unit of magnetic field strength, invented ac dynamos, transformers, and motors. He sold the patent rights to these to George Westinghouse. This eventually led to a struggle between Thomas Edison's dc systems and Westinghouse's ac version of power generation and distribution. The latter eventually won, with the installation of the first large-scale electric generator at Niagara Falls.

- To prove the safety of electric energy to a skeptical public at the turn of the twentieth century, Tesla gave exhibitions of lighting lamps by allowing electricity to flow through his body. Westinghouse used his system to light the World's Columbian Exposition at Chicago in 1893. Tesla proved that the Earth could be used as a conductor and lighted 200 lamps without wires at a distance of 25 miles. With his giant transformer (a Tesla coil) Westinghouse created artificial lightning, producing bolts measuring over 100 feet long.

- Radio waves, radar, visible light, and X-rays are all electromagnetic waves. Better known as light, they all obey the same mathematical relationships. The only difference is their frequency and wavelength. In a vacuum, they all travel at exactly the same speed, c (3.00×10^8 m/s).

- The Scottish physicist James Clerk Maxwell (1831–1879) fully developed and integrated the equations of electricity and magnetism. This set became known as Maxwell's equations, and his interpretation of them was one of the great achievements of nineteenth-century physics.

As we saw in Chapter 19, an electric current produces a magnetic field. But the relationship between electricity and magnetism does not stop there. In this chapter, you will learn that under the right conditions, a magnetic field can produce an electric current. How is this done? Chapter 19 considered only *constant* magnetic fields. No current is produced in a loop of wire that is stationary in a constant magnetic field. However, if the magnetic field changes with time, or if the wire loop moves into or out of, or is rotated in, the field, a current *is* produced in the wire.

The uses of this interrelationship of electricity and magnetism are many. One example happens during the playing of a videotape, which is actually a magnetic tape that has information encoded on it as variations in its magnetism. These variations can be used to produce electrical currents, which, in turn, are amplified and the signal sent for replay to the television set. Similar processes are involved when information is stored on or retrieved from a magnetic disk in your computer.

On a larger scale, consider the generation of the electric energy that provides the basis for our modern civilization. At hydroelectric plants such as that in this photo, one of the oldest and simplest energy sources on Earth—falling water—is used to generate electric energy. The gravitational potential energy of the water is converted into kinetic energy, and some of this kinetic energy is transformed, eventually, into electric energy. But how does this last step take place? Regardless of the ultimate source of the energy—the burning of oil, coal, or gas; a nuclear reactor; or falling water—the actual conversion to electric energy is accomplished by means of magnetic fields and electromagnetic induction. This chapter not only examines the underlying principles that make such conversion possible, but also discusses several practical applications. Moreover, we will also see that the creation and propagation of electromagnetic radiation is intimately related to electromagnetic induction.

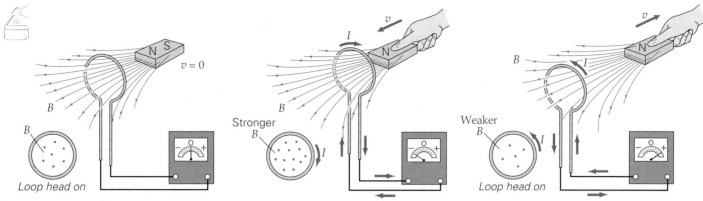

(a) No motion between magnet and loop **(b) Magnet is moved toward loop** **(c) Magnet is moved away from loop**

20.1 Induced emf: Faraday's Law and Lenz's Law

OBJECTIVES: To (a) define magnetic flux and explain how an induced emf is created, and (b) determine induced emfs and currents.

Recall from Chapter 17 that the term *emf* stands for *electromotive force*, which is a voltage or electric potential difference capable of creating an electric current. It is observed experimentally that a magnet held stationary near a conducting wire loop does *not* induce an emf (and therefore produces no current) in that loop (▲Fig. 20.1a). If the magnet is moved toward the loop, however, as shown in Fig. 20.1b, the deflection of the galvanometer needle indicates that current exists in the loop, but only during the motion. Furthermore, if the magnet is moved away from the loop, as shown in Fig. 20.1c, the galvanometer needle is deflected in the opposite direction, which indicates a reversal of the current's direction, but, again, only during the motion.

Deflections of the galvanometer needle, indicating the presence of *induced currents*, also occur if the loop is moved toward or away from the stationary magnet. The effect thus depends on *relative* motion of the loop and magnet. It also turns out that the magnitude of the induced current depends on the speed of that motion. However, experimentally, there is a noteworthy exception. If a loop is moved (but not rotated) in a *uniform* magnetic field, as shown in ▶Fig. 20.2, no current is induced. We will see why this is so later in this section.

Yet another way to induce a current in a stationary wire loop is to vary the current in another, nearby loop. When the switch in the battery-powered circuit in ▼Fig. 20.3a is closed, the current in the loop on the right goes from zero to some constant value in a short time. Only during the buildup time does the magnetic field caused by the current in this loop increase in the region of the loop on the left. During the buildup, the galvanometer needle deflects, indicating current in the left loop. When the current in the right loop attains its steady value, the field it produces becomes constant, and the current in the left loop drops to zero. Similarly, when the switch in the right loop is opened (Fig. 20.3b), its current and field decrease to zero, and the galvanometer deflects in the opposite direction, indicating a reversal in direction of the current induced in the left loop. The important fact to note is that *induced current in a loop occurs only when the magnetic field through that loop changes.*

In Fig. 20.1, moving the magnet changed the magnetic environment in a loop, caused an induced emf that, in turn, caused an induced current. For the case of *two* stationary loops (Fig. 20.3), a changing current in the right loop produced a changing magnetic environment in the left loop, thereby inducing an emf and a current in the left loop.* There is a convenient way of summarizing what is happening in both Fig. 20.1 and Fig. 20.3: To induce currents in a loop or complete circuit, a process called **electromagnetic induction**, all that matters is whether the magnetic field through the loop or circuit is changing.

---

*The term *mutual induction* is used to describe the situation in which emfs and currents are induced between two (or more) loops.

▲ **FIGURE 20.1 Electromagnetic induction (a)** When there is no relative motion between the magnet and the wire loop, the number of field lines through the loop (in this case, 7) is constant, and the galvanometer shows no deflection. **(b)** Moving the magnet toward the loop increases the number of field lines passing through the loop (now 12), and an induced current is detected. **(c)** Moving the magnet away from the loop decreases the number of field lines passing through the loop (to 5). The induced current is now in the opposite direction. (Note the needle deflection.)

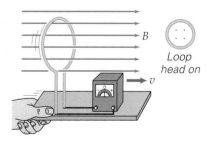

▲ **FIGURE 20.2 Relative motion and no induction** When a loop is moved parallel to a uniform magnetic field, there is no change in the number of field lines passing through the loop, and there is no induced current.

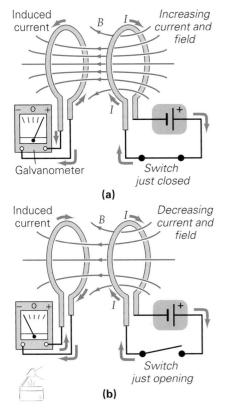

▲ **FIGURE 20.3** Mutual induction
(a) When the switch is closing in the right-loop circuit, the buildup of current produces a changing magnetic field in the other loop, inducing a current in it. **(b)** When the switch is opened, the magnetic field collapses, and the magnetic field in the left loop decreases. The induced current in this loop is then in the opposite direction. The induced currents occur only when the magnetic field passing through a loop changes and vanish when the field reaches a constant value.

Detailed experiments on electromagnetic induction were done independently by Michael Faraday in England and Joseph Henry in the United States around 1830. Faraday found that the important factor in electromagnetic induction was the time rate of change of the number of magnetic field lines passing through the loop or circuit area. That is, he discovered that

> an induced emf is produced in a loop or complete circuit whenever the number of magnetic field lines passing through the plane of the loop or circuit changes.

Magnetic Flux

Because the induced emf in a loop depends on the rate of change of the number of magnetic field lines passing through it, determining induced emf demands that the number of field lines through the loop be quantified. Consider a loop of wire in a uniform magnetic field (▼Fig. 20.4a). The number of field lines through the loop depends on the loop's area, its orientation relative to the field, and the strength of that field. To describe the loop's orientation, the concept of an *area vector* ($\vec{A}$) is employed. Its direction is normal to the loop's plane, and its magnitude is equal to the loop area. θ, the angle between the magnetic field ($\vec{B}$) and the area vector ($\vec{A}$), is a measure of their relative orientation. For example, in Fig. 20.4a, $\theta = 0°$, meaning that the vectors are in the same direction, or, alternatively, the area plane is perpendicular to the field.

For the case of a magnetic field that does not vary over the area, the number of magnetic field lines passing through a particular area (the area inside a loop in our case) is proportional to the **magnetic flux (Φ)**, which is defined as

$$\Phi = BA \cos \theta \qquad \text{\textit{magnetic flux}} \atop \text{\textit{(in a constant magnetic field)}} \qquad (20.1)$$

SI unit of magnetic flux: tesla-meter squared ($T \cdot m^2$), or weber (Wb)*

The SI unit of the magnetic field is the tesla, and magnetic flux has SI units of $T \cdot m^2$. This combination is sometimes expressed as the weber, defined as $1 \text{ Wb} = 1 \text{ T} \cdot m^2$. The orientation of the loop with respect to the magnetic field affects the number of field lines passing through it, and this factor is accounted for by the cosine term in Eq. 20.1. Let us consider several possible orientations:

- If $\vec{B}$ and $\vec{A}$ are parallel ($\theta = 0°$), then the magnetic flux is positive and has a maximum value of $\Phi_{max} = BA \cos 0° = +BA$. The maximum possible number of magnetic field lines pass through the loop in this orientation (Fig. 20.4b).
- If $\vec{B}$ and $\vec{A}$ are oppositely directed ($\theta = 180°$), then the magnitude of the magnetic flux is a maximum again, but of opposite sign: $\Phi_{180°} = BA \cos 180° = -BA = -\Phi_{max}$ (Fig. 20.4c).

▼ **FIGURE 20.4** Magnetic flux **(a)** Magnetic flux (Φ) is a measure of the number of field lines passing through an area (A). The area can be represented by a vector $\vec{A}$ perpendicular to the plane of the area. **(b)** When the plane of a loop is perpendicular to the field and $\theta = 0°$, then $\Phi = \Phi_{max} = +BA$. **(c)** When $\theta = 180°$, the magnetic flux has the same magnitude, but is opposite in direction: $\Phi = -\Phi_{max} = -BA$. **(d)** When $\theta = 90°$, then $\Phi = 0$. **(e)** As the loop's plane is changed from being perpendicular to the field to one more parallel to the field, less area is available to the field lines, and therefore the flux decreases. In general, $\Phi = BA \cos \theta$.

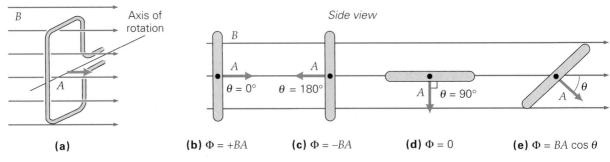

(a) **(b)** $\Phi = +BA$ **(c)** $\Phi = -BA$ **(d)** $\Phi = 0$ **(e)** $\Phi = BA \cos \theta$

*Wilhelm Eduard Weber (1804–1891), a German physicist, was noted for his work in magnetism and electricity, particularly terrestrial magnetism. The unit name *weber* was introduced as the SI unit of magnetic flux in 1935.

- If $\vec{B}$ and $\vec{A}$ are perpendicular, then there are no field lines passing through the plane of the loop, and the flux is zero: $\Phi_{90°} = BA \cos 90° = 0$ (Fig. 20.4d).
- For situations at intermediate angles, the flux is less than the maximum value, but nonzero (Fig. 20.4e). $A \cos \theta$ can be interpreted as the effective area of the loop perpendicular to the field lines (▸Fig. 20.5a). Alternatively, $B \cos \theta$ can be viewed as the perpendicular component of the field through the full area of the loop, A, as shown in Fig. 20.5b. Thus, Eq. 20.1 can be thought of as either $\Phi = (B \cos \theta)A$ or $\Phi = B(A \cos \theta)$, depending on the interpretation. In either case, the answer is the same.

Faraday's Law of Induction and Lenz's Law

From quantitative experiments, Faraday determined that the emf ($\mathcal{E}$) induced in a coil (a coil, by definition, consists of a series connection of N individual loops) depends on the time rate of change of the number of magnetic field lines through all the loops, or the *time rate of change of the magnetic flux through all the loops (total flux)*. This dependence, known as **Faraday's law of induction**, is expressed mathematically as

$$\mathcal{E} = -N \frac{\Delta \Phi}{\Delta t} = -\frac{\Delta(N\Phi)}{\Delta t} \quad \textit{Faraday's law for induced emf} \quad (20.2)$$

where $\Delta \Phi$ is the change in flux through one loop. In a coil consisting of N loops, the total change in flux is $N\Delta\Phi$. Note that the induced emf in Eq. 20.2 is an average value over the time interval Δt (why?).

The minus sign is included in Eq. 20.2 to give an indication of the *direction* of the induced emf, which we haven't discussed as yet. The Russian physicist, Heinrich Lenz (1804–1865), discovered the law that governs the direction of the induced emf. **Lenz's law** is stated as follows:

> An induced emf in a wire loop or coil has a direction such that the current it creates produces its own magnetic field that opposes the *change* in magnetic flux through that loop or coil.

This law means that the magnetic field *due to the induced current* is in a direction that tries to keep the flux through the loop from changing. For example, if the flux increases in the +x-direction, the magnetic field due to the induced current will be in the −x-direction (▾Fig. 20.6a). This effect tends to cancel the increase in the flux, or *oppose the change*. Essentially, the magnetic field due to the induced current tries to maintain the existing magnetic flux. This effect is sometimes called "electromagnetic inertia," by analogy to the tendency of objects to resist changes in their velocity. In the long run, the induced current cannot prevent the magnetic flux from changing. However, during the time that the flux is changing, the induced magnetic field will oppose that change.

The direction of the induced current is given by the **induced-current right-hand rule**:

> With the thumb of the right hand pointing in the direction of the induced field, the fingers curl in the direction of the induced current.

(See Fig. 20.6b and Integrated Example 20.1.) You might recognize this rule as a version of the right-hand rules used to find the direction of a magnetic field produced

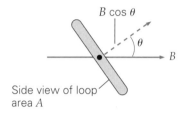

$$\Phi = B (A \cos \theta)$$

(a)

$$\Phi = (B \cos \theta) A$$

(b)

▲ **FIGURE 20.5 Magnetic flux through a loop: An alternative interpretation** Instead of defining the flux (Φ) **(a)** in terms of the magnetic field magnitude (B) passing through a reduced area ($A \cos \theta$), we can define it **(b)** in terms of the perpendicular component of the magnetic field ($B \cos \theta$) passing through A. Either way, Φ is a measure of the number of field lines passing through A and is given by $\Phi = BA \cos \theta$ (Eq. 20.1).

Illustration 29.2 Loop in a Changing Magnetic Field

Exploration 29.1 Lenz's Law

◀ **FIGURE 20.6 Finding the direction of the induced current** **(a)** An external magnetic field is shown increasing to the right. The induced current creates its own magnetic field to try to counteract the flux change that is occurring. **(b)** The (induced) current right-hand (source) rule determines the direction of the induced current. Here the direction of the induced field must be to the left. With the thumb of the right hand pointing left, the fingers give the induced current direction.

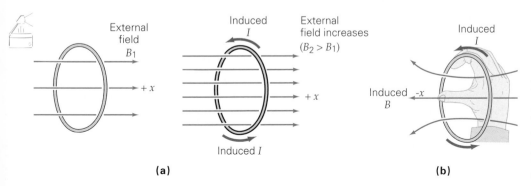

(a)

(b)

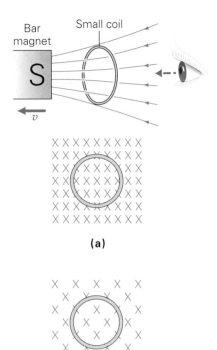

(a)

(b)

Induced I

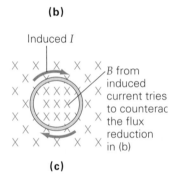

B from induced current tries to counterac the flux reduction in (b)

(c)

▲ **FIGURE 20.7** Using a bar magnet to induce currents **(a)** The south end of a bar magnet is pulled away from a wire loop. **(b)** The view from the right of the loop shows the magnetic field pointing away from the observer, or into the page, and decreasing. **(c)** To counteract this loss of flux into the page, current is induced in the clockwise direction, so as to provide its own field into the page. See Integrated Example 20.1.

Exploration 29.3 Loop Near a Wire

Demonstration/activity: Place an aluminum ring on a large electromagnet operating under ac conditions. A repulsive force should move the ring upward. Then take an identical ring with a small section cut out of it. Ask the students to explain why the second ring doesn't move. This demonstration emphasizes that even though there is an induced emf, an induced current may not always result.

by a current (Chapter 19). Here it is used in reverse. Typically, you know the induced field direction (for example, $-x$ in Fig. 20.6b) and want the direction of the current that produces it. An application of Lenz's law is illustrated in Integrated Example 20.1.

Integrated Example 20.1 ■ Lenz's Law and Induced Currents

(a) The south end of a bar magnet is pulled far away from a small wire coil. (See ◄Fig. 20.7a.) Looking from behind the coil toward the south end of the magnet (Fig. 20.7b), what is the direction of the induced current: (1) counterclockwise, (2) clockwise, or (3) there is no induced current? **(b)** Suppose that the magnetic field over the area of the coil is initially constant at 40 mT, the coil's radius is 2.0 mm, and there are 100 loops in the coil. Determine the magnitude of the average induced emf in the coil if the bar magnet is removed in 0.75 s.

(a) Conceptual Reasoning. There is initially magnetic flux *into* the plane of the coil (Fig. 20.7b), and later, when the magnet is far away from the coil, there is no flux; and the flux has changed. Therefore, there must be an induced emf, so answer (3) cannot be correct. As the bar magnet is pulled away, the field weakens, but maintains the same direction. The induced emf will produce an (induced) current that, in turn, will produce a magnetic field into the page so as to try to prevent this decrease in flux. Therefore, the induced emf and current are in the clockwise direction, as found using the induced-current right-hand rule (Fig. 20.7c) and the correct answer is (2), clockwise.

(b) Quantitative Reasoning and Solution. This example is a straightforward application of Eq. 20.2. The initial flux is the maximum possible. The following data are given and converted to SI units:

Given: $B_i = 40 \text{ mT} = 0.040 \text{ T}$ *Find:* Average induced emf $\mathscr{E}$ (magnitude)
$r = 2.00 \text{ mm} = 2.00 \times 10^{-3} \text{ m}$
$N = 100 \text{ loops}$
$\Delta t = 0.75 \text{ s}$

To find the initial magnetic flux through one loop of the coil, use Eq. 20.1 with an angle of $\theta = 0°$. (Why?) The area is $A = \pi r^2 = \pi (2.00 \times 10^{-3} \text{ m})^2 = 1.26 \times 10^{-5} \text{ m}^2$. Therefore, the initial flux, Φ_i, through one loop is positive (why?) and given by

$$\Phi_i = B_i A \cos \theta = (0.040 \text{ T})(1.26 \times 10^{-5} \text{ m}^2) \cos 0° = +5.03 \times 10^{-7} \text{ T} \cdot \text{m}^2$$

Because the final flux is zero, $\Delta \Phi = \Phi_f - \Phi_i = 0 - \Phi_i = -\Phi_i$. Therefore, the absolute value of the average induced emf is

$$|\mathscr{E}| = N \frac{|\Delta \Phi|}{\Delta t} = (100 \text{ loops}) \frac{(5.03 \times 10^{-7} \text{ T} \cdot \text{m}^2 \text{ loop})}{(0.75 \text{ s})} = 6.70 \times 10^{-5} \text{ V}$$

Follow-Up Exercise. In this Example, (a) in which direction is the induced current if instead a north magnetic pole approaches the coil quickly? Explain. (b) In this example, what would be the average induced current if the coil had a total resistance of 0.20 Ω? *(Answers to all Follow-Up Exercises are at the back of the text.)*

Lenz's law incorporates the principle of energy conservation. Consider a situation in which a wire loop has an increasing magnetic flux through its area. Contrary to Lenz's law, suppose instead that the magnetic field from the induced current *added* to the flux instead of keeping it at its original value. This increased flux would then lead to an even greater induced current. In turn, this greater induced current would produce a still greater magnetic flux, which in turn would give a greater induced current, and so on. Such a something-for-nothing energy situation would violate conservation of energy.

To understand the direction of the induced emf in a loop in terms of forces, consider the case of the moving magnet (for example, Fig. 20.1b). Recall that a current-carrying loop creates its own magnetic field similar to that of a bar magnet. (See Figs. 19.3 and 19.25.) The induced current sets up a magnetic field in the loop, and that loop acts like a bar magnet with a polarity that will oppose the motion of the real bar magnet (▶Fig. 20.8). You should be able to show that if the bar magnet is pulled away from the loop, the loop exerts a magnetic *attraction* to try to keep the magnet from leaving—electromagnetic inertia in action.

Substituting the expression for the magnetic flux (Φ) given by Eq. 20.1 into Eq. 20.2, we have

$$\mathscr{E} = -N \frac{\Delta \Phi}{\Delta t} = -\frac{N \Delta (BA \cos \theta)}{\Delta t} \qquad (20.3)$$

Thus, an induced emf results if

1. the strength of the magnetic field changes,
2. the loop area changes, and/or
3. the orientation between the loop area and the field direction changes.

In situation (1), a flux change is created by a time-varying field, such as that from a time-varying current in a nearby circuit or created by moving a magnet near a coil, as in Fig. 20.1 (or by moving the coil near the magnet).

In situation (2), a flux change results because of a varying loop area. This situation might occur if a loop had an adjustable circumference (such as the loop around an inflatable balloon, as in Exercise 23 at the end of the chapter).

Finally, in situation (3), a change in flux can result from a *change in orientation of the loop*. This situation can occur when a coil is rotated in a magnetic field. The change in the number of field lines through a single loop is evident in the sequential views in Fig. 20.4. Rotating a coil in a field is a common way of inducing an emf and will be considered separately in Section 20.2. The emfs that result from changing the field strength and loop area are analyzed in the next two Examples. Also, see Insight 20.1 on Electromagnetic Induction at Work: Flashlights and Antiterrorism on page 664 for ways in which electromagnetic induction helps make our everyday lives safer and easier.

Conceptual Example 20.2 ▪ Fields in the Fields: Electromagnetic Induction

In rural areas where electric power lines carry electricity to big cities, it is possible to generate small electric currents by means of induction in a conducting loop. The overhead power lines carry alternating currents that periodically reverse direction 60 times per second. How would you orient the plane of the loop to maximize the induced current if the power lines run north to south: (a) parallel to the Earth's surface, (b) perpendicular to the Earth's surface in the north–south direction, or (c) perpendicular to the Earth's surface in the east–west direction? (See ▸Fig. 20.9a.)

Reasoning and Answer. Magnetic field lines from long wires are circular. (See Fig. 19.23.) By the source right-hand rule, the magnetic field direction at ground level is parallel to the Earth's surface and alternates in direction. The orientation choices are shown in Fig. 20.9b. Neither answer (a) nor (c) can be correct, because in these orientations there would *never* be any magnetic flux passing through the loop. In this situation, the flux would be constant and there would be no induced emf. Hence, the answer is (b). If the loop is oriented perpendicular to the Earth's surface with its plane in the north–south direction, the flux through it would vary from zero to its maximum value and back sixty times per second, and this would maximize the induced emf and current in the loop.

Follow-Up Exercise. Suggest possible ways of increasing the induced current in this Example by changing only properties of the loop and not of the overhead wires.

Example 20.3 ■ Induced Currents: A Potential Hazard to Equipment?

Electrical instruments can be damaged or destroyed if they are in a rapidly changing magnetic field. This can occur if an instrument is located near an electromagnet operating under ac conditions; the electromagnet's external field could produce a changing flux within a nearby instrument. If the induced currents are large enough, they could damage the instrument. Consider a computer speaker that is near such an electromagnet (▾Fig. 20.10, p. 662). Suppose an electromagnet exposes the speaker to a maximum magnetic field of 1.00 mT that reverses direction every 1/120 s.

Assume that the speaker's coil consists of 100 circular loops (each with a radius of 3.00 cm) and has a total resistance of 1.00 Ω. According to the manufacturer of the speaker, the current in the coil should not exceed 25.0 mA. (a) Calculate the magnitude of the average induced emf in the coil during the 1/120 s interval. (b) Is the induced current likely to damage the speaker coil?

Thinking It Through. (a) The flux goes from a (maximum) positive to a (maximum) negative value in 1/120 s. The magnetic flux change can be determined from Eq. 20.1 with $\theta = 0°$ and $\theta = 180°$. The average induced emf can then be calculated from Eq. 20.2. (b) Once we know the emf, the induced current can be calculated from $I = \mathcal{E}/R$.

(continues on next page)

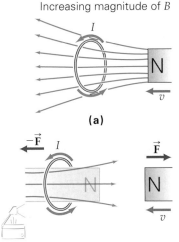

Increasing magnitude of B

(a)

▲ **FIGURE 20.8** Lenz's law in terms of forces **(a)** If the north end of a bar magnet is moved rapidly toward a wire loop, current is induced in the direction shown. **(b)** While the induced current exists, the loop then acts like a bar magnet with its "north end" close to the north end of the real bar magnet. Thus there is a magnetic repulsion. This is an alternative way of viewing Lenz's law: Induce a current so as to try to keep the flux from changing—in this case, to try to keep the bar magnet away and maintain the initial value of flux, zero.

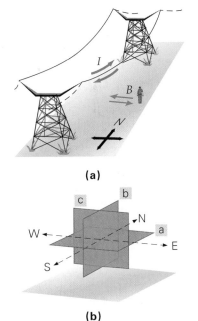

(a)

(b)

▲ **FIGURE 20.9** Induced emfs below power lines **(a)** If current-carrying wires run in the north–south direction, then directly below the alternating current produces a magnetic field that oscillates between pointing east and west. **(b)** These are the three choices for loop orientation in Conceptual Example 20.2.

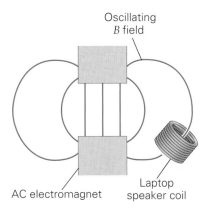

▲ **FIGURE 20.10** Instrument hazard? The coil of a computer speaker system is close to an alternating-current electromagnet. The changing flux in the coil produces an induced emf and, thus, an induced current that depends on the resistance of the coil. See Example 20.3.

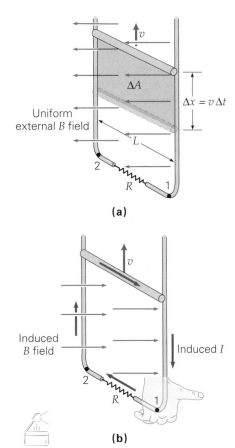

▲ **FIGURE 20.11** Motional emf (a) As the metal rod is pulled on the metal frame, the area of the rectangular loop varies with time. A current is induced in the loop as a result of the changing flux. (b) To counteract the increase of flux to the left, an induced current creates its own magnetic field to the right. See Integrated Example 20.4.

Solution. Listing the data and converting to SI units,

Given: $B_i = +1.00 \text{ mT} = +1.00 \times 10^{-3} \text{ T}$
   (+ pointing one way)
   $B_f = -1.00 \text{ mT} = -1.00 \times 10^{-3} \text{ T}$
   (pointing the opposite way)
   $\Delta t = 1/120 \text{ s} = 8.33 \times 10^{-3} \text{ s}$
   $N = 100 \text{ loops}$
   $R = 1.00 \ \Omega$
   $r = 3.00 \text{ cm} = 3.00 \times 10^{-2} \text{ m}$
   $I_{max} = 25.0 \text{ mA} = 2.50 \times 10^{-2} \text{ A}$

Find: (a) $\mathcal{E}$ (magnitude of average induced emf)
(b) I (magnitude of average induced current)

(a) The circular loop area is $A = \pi r^2 = \pi(3.00 \times 10^{-2} \text{ m})^2 = 2.83 \times 10^{-3} \text{ m}^2$. Thus the initial flux through *one* loop is (see Eq. 20.1):

$$\Phi_i = B_i A \cos \theta = (1.00 \times 10^{-3} \text{ T})(2.83 \times 10^{-3} \text{ m}^2/\text{loop})(\cos 0°) = 2.83 \times 10^{-6} \text{ T} \cdot \text{m}^2/\text{loop}$$

Because the final flux is the negative of this, the change in flux through one loop is

$$\Delta\Phi = \Phi_f - \Phi_i = -\Phi_i - \Phi_i = -2\Phi_i = -5.66 \times 10^{-6} \text{ T} \cdot \text{m}^2/\text{loop}$$

Therefore, the average induced emf is (using Eq. 20.2)

$$\mathcal{E} = N\frac{|\Delta\Phi|}{\Delta t} = (100 \text{ loops})\left(\frac{5.66 \times 10^{-6} \text{ T} \cdot \text{m}^2/\text{loop}}{8.33 \times 10^{-3} \text{ s}}\right) = 6.79 \times 10^{-2} \text{ V}$$

(b) This voltage is small by everyday standards, but keep in mind that the speaker coil's resistance is also small. To determine the induced current in the coil, use the relationship between voltage, resistance, and current:

$$I = \frac{\mathcal{E}}{R} = \frac{6.79 \times 10^{-2} \text{ V}}{1.00 \ \Omega} = 6.79 \times 10^{-2} \text{ A} = 67.9 \text{ mA}$$

This value exceeds the allowed speaker current of 25.0 mA and therefore the speaker coil is possibly subject to damage.

Follow-Up Exercise. In this example, if the speaker coil were moved farther from the electromagnet, it could reach a point where the induced average current would be below the "dangerous" level of 25 mA. Determine the magnetic field strength B_{max} at this point.

As a special case, emfs and currents can be induced in conductors as they are moved through a magnetic field. In this situation, the induced emf is called a *motional emf*. To see how this works, consider the situation in ◀Fig. 20.11a. As the bar moves upward, the circuit area increases by $\Delta A = L\Delta x$ (Fig. 20.11a.) At constant speed, the distance traveled by the bar in a time Δt is $\Delta x = v\Delta t$. Therefore $\Delta A = Lv\Delta t$. The angle between the magnetic field and the normal to the area (θ) is always 0°. However, the area is changing, so the flux varies. However, we know that $\Phi = BA \cos 0° = BA$; hence we can write $\Delta\Phi = B\Delta A$, or $\Delta\Phi = BLv\Delta t$. Therefore, from Faraday's law, the magnitude of this "motional" (induced) emf, $\mathcal{E}$, is $|\mathcal{E}| = |\Delta\Phi|/\Delta t = BLv\Delta t/\Delta t = BLv$. This is the fundamental idea behind electric energy generation: Move a conductor in a magnetic field, and convert the work done on it into electrical energy. To see some of the details, consider the following Integrated Example.

Integrated Example 20.4 ■ The Essence of Electric-Energy Generation: Mechanical Work into Electrical Current

Consider the situation in Fig. 20.11a. An external force does work as the movable bar moves, and this work is converted to electrical energy. Because the "circuit" (wires, resistor, and bar) is in a magnetic field, the flux through it changes with time, inducing a current. (a) What is the direction of the induced current in the resistor, (1) from 1 to 2 or (2) from 2 to 1? (b) If the bar is 20 cm long and is pulled at a steady speed of 10 cm/s, what is the induced current if the resistor has a value of 5.0 Ω and the circuit is in a uniform magnetic field of 0.25 T?

(a) Conceptual Reasoning. In Fig. 20.11a, the magnetic flux points left and increases. According to Lenz's law, the field due to the induced current must then be to the right. Using the induced-current right-hand rule, we find that the direction of the induced current is from 1 to 2 (Fig. 20.11b), and the correct answer is (1).

(b) Quantitative Reasoning and Solution. The flux change is due to an area change as the bar is pulled upward. The analysis for motional emfs has been done in the preceding text. Lastly, once the motional emf has been found, the induced current can be determined using Ohm's law.

Listing the data and converting to SI units, we have:

Given: $B = 0.25$ T   *Find:* Induced current in the resistor
$L = 20$ cm $= 0.20$ m
$v = 10$ cm/s $= 0.10$ m/s
$R = 5.0$ Ω

In the preceding text, it was shown that the magnitude of the induced emf $\mathscr{E}$ is given by BLv, so numerically we have:

$$|\mathscr{E}| = BLv = (0.25 \text{ T})(0.20 \text{ m})(0.10 \text{ m/s}) = 5.0 \times 10^{-3} \text{ V}$$

Hence the induced current is

$$I = \frac{\mathscr{E}}{R} = \frac{5.0 \times 10^{-3} \text{ V}}{5.0 \text{ Ω}} = 1.0 \times 10^{-3} \text{ A}$$

Clearly this arrangement isn't a practical way to generate large amounts of electrical energy. Here the power dissipated in the resistor is only 5.0×10^{-6} W. (You should verify this.)

Follow-Up Exercise. In this Example, if the field were increased by three times and the bar's width changed to 45 cm, what would the bar's speed be to induce a current of 0.1 A?

20.2 Electric Generators and Back emf

OBJECTIVES: To (a) understand the operation of electrical generators and calculate the emf produced by an ac generator, and (b) explain the origin of back emf and its effect on the behavior of motors.

One way to induce an emf in a loop is through a change in the loop's orientation in its magnetic field (Fig. 20.4). This is the operational principle behind electric generators.

Electric Generators

An *electric generator* is a device that converts mechanical energy into electrical energy. Basically, the function of a generator is the reverse of that of a motor.

Recall that a battery supplies direct current (dc). That is, the voltage polarity (and therefore the current direction) do not change. However, most generators produce *alternating current* (ac), named because the polarity of the voltage (and therefore the current direction) change periodically. Thus, the electric energy used in homes and industry is delivered in the form of alternating voltage and current. (See Chapter 21 for analysis of ac circuits and Chapter 18 for household wiring diagrams.)

An **ac generator** is sometimes called an *alternator*. The elements of a simple ac generator are shown in ▸Fig. 20.12. A wire loop called an *armature* is mechanically rotated in a magnetic field by some external means, such as water flow or steam hitting turbine blades. The rotation of the blades in turn causes a rotation of the loop. This results in a change in the loop's magnetic flux and an induced emf in the loop. The ends of the loop are connected to an external circuit by means of slip rings and brushes. In this case, the induced currents will be delivered to that circuit. In practice, generators have many loops, or windings, on their armatures.

When the loop is rotated at a constant angular speed (ω), the angle (θ) between the magnetic-field vector and the area vector of the loop changes with time: $\theta = \omega t$ (assuming that $\theta = 0°$ at $t = 0$). As a result, the number of field lines through the loop changes with time, causing an induced emf. From Eq. 20.1, the flux (for one loop) varies as

$$\Phi = BA \cos \theta = BA \cos \omega t$$

From this it can be seen that the induced emf will also vary with time. For a rotating coil of N loops, Faraday's law yields

$$\mathscr{E} = -N\frac{\Delta \Phi}{\Delta t} = -NBA\left(\frac{\Delta(\cos \omega t)}{\Delta t}\right)$$

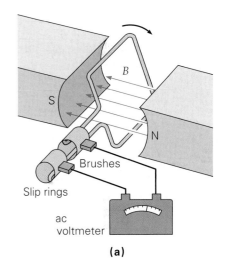

(a)

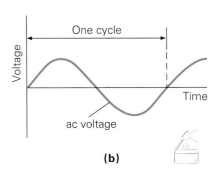

(b)

▲ **FIGURE 20.12** A simple ac generator **(a)** The rotation of a wire loop in a magnetic field produces **(b)** a voltage output whose polarity reverses with each half-cycle. This alternating voltage is picked up by a brush/slip ring arrangement as shown.

INSIGHT	# 20.1 ELECTROMAGNETIC INDUCTION AT WORK: FLASHLIGHTS AND ANTITERRORISM

We use electromagnetic induction in many ways in our daily lives, in most cases without realizing it. One recent invention is a flashlight that works without a battery (Fig. 1a). As the flashlight is shaken, a strong permanent magnet in it oscillates through induction coils, inducing an oscillating emf and current. To charge a capacitor, the ac current must be *rectified* into dc current and not change direction. The schematic of this flashlight is shown in Fig. 1b. Here a solid-state *rectifier circuit* (triangular symbol) acts as a "one-way current valve." In this diagram, only clockwise dc current goes to the capacitor and charges it. After about a minute,

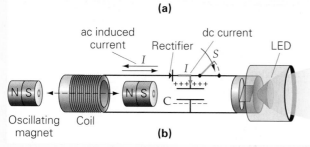

(a)

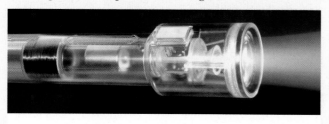

(b)

FIGURE 1 A batteryless flashlight (a) A photo of a relatively new type of flashlight that produces light using electric energy generated by shaking (induction). **(b)** A schematic diagram of the flashlight shown in part (a). As the flashlight is shaken, its internal permanent magnet passes through a coil, inducing a current. This current alternates in direction (why?) and thus needs to be turned into dc ("rectified") before it can charge a capacitor. Once the capacitor is fully charged, it can be used to create a current through a light-emitting diode (LED), which in turn gives off light, typically for several minutes.

the capacitor is fully charged. When the switch S is thrown, the capacitor discharges through an efficient *light-emitting diode* (LED). The resulting beam of light lasts for several minutes before the flashlight needs to be reshaken. This device could, at the least, play an important backup role to the more traditional flashlights that rely on batteries.

In air travel safety, induction is used to prevent dangerous metallic objects (such as knives and guns) from being carried onto airplanes. As a passenger walks through the arch of an airport metal detector (see Fig. 2), a series of large "spiked" currents is periodically delivered to a coil (solenoid) in one of the nonmagnetic sides. In the most common system, called PI (for *pulsed induction*), these current spikes occur hundreds of times per second. As the current rises and falls, a changing magnetic field is created in the passenger. If the passenger is carrying nothing metallic, there will be no significant induced current and no induced magnetic field. However, if the passenger has a metal object, a current will be induced in that object, which in turn will produce its own (induced) magnetic field that can be sensed by the emitting coil—that is, a "magnetic echo." Sophisticated electronics measure the echo-induced emf and trigger a warning light to suggest that further inspection of that passenger is warranted.

FIGURE 2 Screening at the airport As passengers walk through the arch, they are subjected to a series of magnetic field pulses. If they have a metal object on their person, the currents induced in that object create their own magnetic field "echo" that, when detected, gives the safety inspectors reason to check the passenger more closely.

Here, we have removed B and A from the time rate of change, because they are constant. By using methods beyond the scope of this book, it can be shown that the induced emf expression can be rewritten as

$$\mathcal{E} = (NBA\omega) \sin \omega t$$

Notice that the product of terms, $NBA\omega$, represents the magnitude of the maximum emf, which occurs whenever $\sin \omega t = \pm 1$. If $NBA\omega$ is called $\mathcal{E}_o$, the maximum value of the emf, then the previous equation can be rewritten compactly as

$$\mathcal{E} = \mathcal{E}_o \sin \omega t \qquad (20.4)$$

Because the sine function varies between ± 1, the polarity of the emf changes with time (▸Fig. 20.13). Note that the emf has its maximum value $\mathcal{E}_o$ when $\theta = 90°$ or $\theta = 270°$. That is, at the instants when the plane of the loop is parallel to the field, and the magnetic flux is zero, the emf will be at its largest (magnitude). The

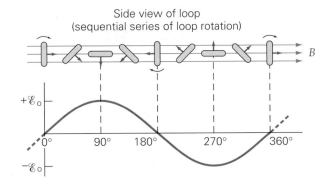

Side view of loop
(sequential series of loop rotation)

▶ **FIGURE 20.13** An ac generator
output A graph of the sinusoidal
output of a generator, with a side
view of the corresponding loop
orientations during a cycle,
showing the flux variation with
time. Note that the emf is a
maximum when the flux changes
most rapidly, as it passes through
zero and changes in sign.

PHYSLET®

Illustration 29.3 Electric Generator

change in flux is greatest at these angles, because although the flux is momentarily zero, it is changing rapidly due to a *sign* change. Near the angles that produce the flux's largest value ($\theta = 0°$ and $\theta = 180°$), the flux is approximately constant and thus the induced emf is zero at those angles.

Because the induced current is produced by this alternating induced emf, the current also changes direction periodically. In everyday applications, it is common to refer to the frequency (f) of the armature [in hertz (Hz) or rotations per second], rather than the angular frequency (ω). Because they are related by $\omega = 2\pi f$, Eq. 20.4 can be rewritten as

$$\mathcal{E} = \mathcal{E}_o \sin(2\pi ft) \quad \text{alternator emf} \qquad (20.5)$$

The ac frequency in the United States and most of the western hemisphere is 60 Hz. A frequency of 50 Hz is common in Europe and other areas.

Keep in mind that Eqs. 20.4 and 20.5 give the instantaneous value of the emf and that $\mathcal{E}$ varies between $+\mathcal{E}_o$ and $-\mathcal{E}_o$ over half of an armature rotational period (1/120 of a second in the United States). For practical ac electrical circuits, time-averaged values for ac voltage and current are more important. This concept will be developed in Chapter 21. To see how various factors influence the generator's output, examine the next Example closely. Also, see Insight 20.2 on Electromagnetic Induction at Play on page 666 for ways in which electromagnetic induction makes for an interesting hobby and helps generate the electric energy needed to power hybrid automobiles for more fuel-efficient transportation.

Demonstration/activity: Light a bulb using a hand-cranked generator. Show how the generator is easier to turn when the bulb is not connected and is harder to turn when the bulb is connected. Is there an emf across the ends of the wire when the crank is turned while the bulb is disconnected?

Example 20.5 ■ An ac Generator: Renewable Electric Energy

A farmer decides to use a waterfall to create a small hydroelectric power plant for his farm. He builds a coil consisting of 1500 circular loops of wire with a radius of 20 cm, which rotates on the generator's armature at 60 Hz in a magnetic field. To generate an rms voltage of 120 V, he needs to generate a maximum emf of 170 V (we will learn more about ac voltages in Chapter 21). What is the magnitude of the generator's magnetic field necessary for this to happen?

Thinking It Through. We can determine the magnetic field from the expression for $\mathcal{E}_o$.

Solution.

Given: $\mathcal{E}_o = 170$ V

$\quad\quad\quad N = 1500$ loops

$\quad\quad\quad r = 20$ cm $= 0.20$ m

$\quad\quad\quad f = 60$ Hz

Find: magnitude of the magnetic field (B)

The generator's maximum (or peak) emf is given by $\mathcal{E}_o = NBA\omega$. Because $\omega = 2\pi f$ and, for a circle, $A = \pi r^2$, this can be rewritten as

$$\mathcal{E}_o = NB(\pi r^2)(2\pi f) = 2\pi^2 NBr^2 f$$

Solving for B,

$$B = \frac{\mathcal{E}_o}{2\pi^2 Nr^2 f} = \frac{170 \text{ V}}{2\pi^2 (1500)(0.20 \text{ m})^2 (60 \text{ Hz})} = 2.4 \times 10^{-3} \text{ T}$$

Follow-Up Exercise. In this Example, suppose that the farmer wanted to generate an emf with an rms value of 240 V, which requires a maximum emf of 340 V. If he chose to do so by changing the size of the coils, what would their new radius have to be?

INSIGHT 20.2 ELECTROMAGNETIC INDUCTION AT PLAY: HOBBIES AND TRANSPORTATION

FIGURE 1 A two-coil metal detector Note both the transmitter (larger outer) and receiver (smaller inner) coils.

Electromagnetic induction plays an important part in our leisure and transportation activities. For example, some hobbyists use metal detectors to hunt for "buried treasure" of a metallic kind. A common design consists of two coils of wire at the end of a shaft used for sweeping just above the ground (see Fig. 1). At the hand-held end are electronics for displaying information about any detected items. The outer, or *transmitter*, coil contains a current oscillating at several thousand hertz, creating an ever-changing magnetic field in the ground below it. (Usually it can penetrate a foot or more below the surface, depending on soil type and condition.) If no metallic objects are within range of this oscillating field, then no significant currents will be induced. Therefore no induced magnetic field "echo" will be detected by the inner, or *receiver*, coil. However, if a metallic object is present, the current induced in it will create a magnetic echo (field) that the receiver will detect as an induced emf and current. By means of sophisticated computer software to evaluate the strength of the induced signal, the object's depth and chemical makeup can be estimated.

The increasing price of gasoline has many drivers turning to *gas-electric hybrid* automobiles, in which the gasoline engine is considerably smaller than conventional ones. In addition to helping power the car, at least part of the hybrid engine's job is to supply electric energy (through induction in a generator) to batteries and an electric motor, which in turn supply power to the wheels. In this way, more work can be extracted from a gallon of gasoline then in a conventional engine.

A cutaway of a typical hybrid car is shown in Fig. 2a. Hybrid cars currently come in two basic designs: parallel and series. In the *parallel hybrid* arrangement (Fig. 2b), the gasoline engine is connected to the wheels via a standard transmission. However, it also turns a generator that, through induction, creates and supplies electric energy to charge the batteries and/or to operate the electric motor. Sophisticated power electronics monitor the charge on the batteries and divert current to where it is needed. The electric motor is connected to the wheels through its own separate transmission, hence the name *parallel hybrid*—as the gasoline and electric motor work together, in parallel. *Fully hybrid* models are capable of moving the car with either engine alone (for maximum fuel economy—say, while cruising on a freeway) or both simultaneously (when more power is needed— say, while accelerating on to a freeway).

Alternatively the engines/motors can be connected in series—in the *series hybrid* automobile. Here the electric motor is what actually powers the wheels (Fig. 2c). The job of the gasoline engine is to supply electric energy (through induction in its generator) to the batteries and electric motor. If the batteries are fully charged and the motor is running well, the gasoline engine can idle down or power off. With frequent accelerations, when the electric motor is called on for a high-power output, the batteries may drain quickly. Under these conditions, the power electronics direct the gasoline engine to begin generating electric energy to recharge the batteries.

Regardless of the design, the object is the same: higher efficiency—that is, more miles per gallon. Hybrids, unlike purely electric cars, are never "plugged in"; they derive all their energy from burning gasoline. However, they are much more efficient, and thus considerably less polluting, than conventional cars. Some recent car models employ hybrid engines that are capable of more horsepower than their gasoline counterparts. For these reasons, the hybrid engine is increasingly likely to be the engine of choice for many drivers in the near future.

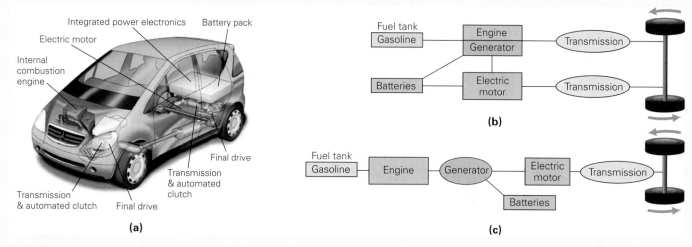

FIGURE 2 Hybrid automobiles **(a)** A cutaway of a typical modern hybrid vehicle. **(b)** A schematic of the main systems in a parallel hybrid. **(c)** A schematic of the main systems in a series hybrid.

(a)

(b)

◀ **FIGURE 20.14** Electrical generation **(a)** Turbines such as those depicted here generate electric energy in much larger quantities than can the "small hydro" plant in the chapter-opening photo. **(b)** Gravitational potential energy of water, here trapped behind the Glen Canyon dam on the Colorado River in Arizona, is converted into electric energy.

In most large-scale ac generators (power plants), the armature is actually stationary, and magnets revolve about it. The revolving magnetic field produces a time-varying flux through the coils of the armature and thus an ac output. A turbine supplies the mechanical energy required to spin the magnets in the generator (▲ Fig. 20.14a). Turbines are typically powered by steam generated from the heat of combustion of fossil fuels or by heat generated from nuclear fission (see Chapters 29 and 30), but they can also be rotated by falling water (*hydroelectricity*), as in Fig. 20.14b. Thus the basic difference between the various types of power plants is the source of the energy that turns the turbines.

Back emf

Although their main job is to convert electric energy into mechanical energy, motors also generate (induced) emfs at the same time. Like a generator, a motor has a rotating armature in a magnetic field. For motors, the induced emf is called a **back emf** (or *counter emf*), $\mathscr{E}_b$, because its direction is opposite that of the line voltage and tends to reduce the current in the armature coils.

If V is the line voltage, then the net voltage driving the motor is less than V (because the line voltage and the back emf are of opposite polarity). Therefore the net voltage is $V_{net} = V - \mathscr{E}_b$. If the motor's armature has a resistance of R, the currentthe motor draws while in operation is $I = V_{net}/R = (V - \mathscr{E}_b)/R$ or, solving for the back emf,

$$\mathscr{E}_b = V - IR \quad \text{(back emf of a motor)} \quad (20.6)$$

where V is the line voltage.

The back emf of a motor depends on the rotational speed of the armature and increases from zero to some maximum value as the armature goes from rest to its normal operating speed. On startup, the back emf is zero (why?). Therefore the starting current is a maximum (Eq. 20.6 with $\mathscr{E}_b = 0$). Ordinarily, a motor turns something, such as a drill bit; that is, it has a mechanical load. Without a load, the armature speed will increase until the back emf almost equals the line voltage. The result is a small current in the coils, just enough to overcome friction and joule heat losses. Under normal load conditions, the back emf is less than the line voltage. The larger the load, the slower the motor rotates and the smaller the back emf. If a motor is overloaded and turns very slowly, the back emf may be reduced so much that the current becomes very large (note that V_{net} increases as $\mathscr{E}_b$ decreases) and may burn out the coils. The back emf plays a vital role in the regulation of a motor's operation by limiting the current in it.

Schematically, a back emf in a dc motor circuit can be represented as an "induced battery" with polarity opposite that of the driving voltage (▶ Fig. 20.15). To see how the back emf affects the current in a motor, consider the next Example.

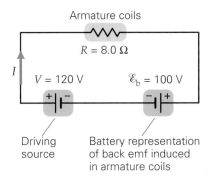

▲ **FIGURE 20.15** Back emf The back emf in the armature of a dc motor can be represented as a battery with polarity opposite that of the driving voltage.

Example 20.6 ■ Getting up to Speed: Back emf in a dc Motor

A dc motor is built with windings that have a resistance of 8.00 Ω and operates at a line voltage of 120 V. With a normal load, there is a back emf of 100 V when the motor reaches full speed. (See Fig. 20.15.) Determine (a) the starting current drawn by the motor and (b) the armature current at operating speed under a normal load.

Thinking It Through. (a) The only difference between startup and full speed is that there is no back emf at startup. The net voltage and resistance determine the current, so Eq. 20.6 can be applied. (b) At operating speed, the back emf increases and is opposite in polarity to the line voltage. Equation 20.6 can again be used to determine the current.

Solution. Let's list the data as usual:

Given: $R = 8.00\ \Omega$
$V = 120$ V
$\mathcal{E}_b = 100$ V

Find: (a) I_s (starting current)
(b) I (operating current)

(a) From Eq. 20.6, the current in the windings is

$$I_s = \frac{V}{R} = \frac{120\ \text{V}}{8.00\ \Omega} = 15.0\ \text{A}$$

(b) When the motor is at full speed, the back emf is 100 V; thus, the current is less.

$$I = \frac{V - \mathcal{E}_b}{R} = \frac{120\ \text{V} - 100\ \text{V}}{8.00\ \Omega} = 2.50\ \text{A}$$

With little or no back emf, the *starting* current is relatively large. When a big motor, such as that of a central air-conditioning unit, starts up, the lights in the building might momentarily dim, because of the large starting current that the motor draws. In some designs, resistors are temporarily connected in series with a motor's coil to protect the windings from burning out as a result of large starting currents.

Follow-Up Exercise. In this Example, (a) how much energy is required to bring the motor to operating speed if it takes 10 s and the back emf averages 50 V during that time? (b) Compare this amount with the amount of energy required to keep the motor running for 10 s once it reaches its operating conditions.

Because motors and generators are opposites, so to speak, and a back emf develops in a motor, you may be wondering whether a back force develops in a generator. The answer is yes. When an operating generator is not connected to an external circuit, no current exists, and therefore there is no magnetic force on the armature coils. However, when the generator delivers energy to an external circuit and current *is* in the coils, the magnetic force on the armature coils produces a *countertorque* that opposes the rotation of the armature. As more current is drawn, the countertorque increases and a greater driving force is needed to turn the armature. Therefore, the higher the generator's current output, the greater the energy expended (that is, fuel consumed) in overcoming the countertorque.

20.3 Transformers and Power Transmission

OBJECTIVES: To (a) explain transformer action in terms of Faraday's law, (b) calculate the output of step-up and step-down transformers, and (c) understand the importance of transformers in electric energy delivery systems.

Electric energy is transmitted by power lines over long distances. It is desirable to minimize I^2R losses (joule heat) that can occur in these transmission lines. Because the resistance of a line is fixed, reducing I^2R losses means reducing current. However, the power output of a generator is determined by its outputs of current and voltage ($P = IV$), and for a fixed voltage, such as 120 V, a reduction in current would mean a reduced power output. It might appear that there is no way to reduce the current while maintaining the power level. Fortunately, electromagnetic induction enables us to reduce power-transmission losses by increasing voltage while simultaneously reducing current in such a way that the delivered *power* is essentially unchanged. This is done using a device called a **transformer**.

A simple transformer consists of two coils of insulated wire wound on the same iron core (▶Fig. 20.16a). When ac voltage is applied to the input coil, or *primary coil*, the alternating current produces an alternating magnetic flux concentrated in the iron core, without any significant leakage of flux outside the core. Under these conditions, the same changing flux also passes through the output coil, or *secondary coil*, inducing an alternating voltage and current in it. (Note that it is common in transformer design to refer to emfs as "voltages," as was done in Chapter 18. We will use this language here also.)

The ratio of the induced voltage in the secondary coil to that of the voltage in the primary coil depends on the ratio of the numbers of turns in the two coils. By Faraday's law, the induced voltage in the secondary coil is

$$V_s = -N_s \frac{\Delta \Phi}{\Delta t},$$

where N_s is the number of turns in the secondary coil. The changing flux in the primary coil produces a back emf of

$$V_p = -N_p \frac{\Delta \Phi}{\Delta t},$$

where N_p is the number of turns in the primary coil. If the resistance of the primary coil is neglected, this back emf is equal in magnitude to the external voltage applied to the primary coil (why?). Forming a ratio of output voltage (secondary) to input voltage (primary) yields

$$\frac{V_s}{V_p} = \frac{-N_s(\Delta \Phi / \Delta t)}{-N_p(\Delta \Phi / \Delta t)}$$

or

$$\frac{V_s}{V_p} = \frac{N_s}{N_p} \quad \text{(voltage ratio in a transformer)} \tag{20.7}$$

If the transformer is 100% efficient (that is, there are no energy losses), then the power input is equal to the power output. Because $P = IV$, we have

$$I_p V_p = I_s V_s \tag{20.8}$$

Although some energy is always lost, this equation is a good approximation, as a well-designed transformer will have an efficiency greater than 95%. (The sources of energy losses will be discussed shortly.) Assuming this ideal case, from Eq. 20.8, the transformer currents and voltages are related to the turn ratio by

$$\frac{I_p}{I_s} = \frac{V_s}{V_p} = \frac{N_s}{N_p} \tag{20.9}$$

To summarize the transformer action in terms of voltage and current output, we have

$$V_s = \left(\frac{N_s}{N_p}\right) V_p \tag{20.10a}$$

(ideal relationship between transformer voltage and current)

and

$$I_s = \left(\frac{N_p}{N_s}\right) I_p \tag{20.10b}$$

If the secondary coil has more windings than the primary coil does (that is, $N_s/N_p > 1$), as in Fig. 20.16a, the voltage is "stepped up," because $V_s > V_p$. This is called a *step-up transformer*. Notice that because of this there is *less* current in the secondary than in the primary ($N_p/N_s < 1$ and $I_s < I_p$).

If the secondary coil has fewer turns than the primary does, we have a *step-down transformer* (Fig. 20.16b). In the usual transformer language, this means that the voltage is "stepped down," and the current, therefore, is increased. Depending on the design details, a step-up transformer may be used as a step-down transformer by simply reversing output and input connections.

Demonstration/activity: Near a large electromagnet that is operating under ac conditions, place a circuit consisting of just a coil of wire and a small lightbulb. In spite of having no battery in its circuit, the bulb lights because of the induced emf in the coil. This demonstration simulates transformer action.

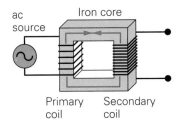

(a) Step-up transformer: high-voltage (low-current) output

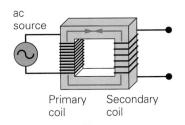

(b) Step-down transformer: low-voltage (high-current) output

▲ **FIGURE 20.16** Transformers **(a)** A step-up transformer has more turns in the secondary coil than in the primary coil. **(b)** A step-down transformer has more turns in the primary coil than in the secondary coil.

Teaching tip: Some students tend to associate high voltage with large amounts of energy, and they may even view transformers as devices that give more energy out than what is put in. Always refer to voltage as a measure of energy per unit charge.

Integrated Example 20.7 ■ Transformer Orientation: Step-Up or Step-Down Configuration?

An ideal 600-W transformer has 50 turns on its primary coil and 100 turns on its secondary coil. (a) Is this transformer a (1) a step-up or (2) step-down arrangement? (b) If the primary coil is connected to a 120-V source, what are the output voltage and current of this transformer?

(a) Conceptual Reasoning. Step-up or step-down refers to what happens to the voltage, not the current. Because the voltage is proportional to the number of turns, in this case the secondary voltage is greater than the primary voltage. Thus the correct answer is (1) a step-up transformer.

(b) Quantitative Reasoning and Solution. The output voltage can be determined from Eq. 20.10a, once the turn ratio is established. From the power, the current can be determined.

Given: $N_p = 50$
$\qquad\quad N_s = 100$
$\qquad\quad V_p = 120$ V

Find: V_s and I_s (secondary voltage and current)

The secondary voltage can be found using Eq. 20.10a with a turn ratio of 2, because $N_s = 2N_p$:

$$V_s = \left(\frac{N_s}{N_p}\right)V_p = (2)(120 \text{ V}) = 240 \text{ V}$$

If the transformer is ideal, then the input power equals the output power. On the primary side, the input power is $P_p = I_p V_p = 600$ W, so the input current must be

$$I_p = \frac{600 \text{ W}}{V_p} = \frac{600 \text{ W}}{120 \text{ V}} = 5.00 \text{ A}$$

Because the voltage is stepped up by a factor of two, the output current should be stepped down by a factor of two. From Eq. 20.10b,

$$I_s = \left(\frac{N_p}{N_s}\right)I_p = \left(\frac{1}{2}\right)(5.00 \text{ A}) = 2.50 \text{ A}$$

Follow-Up Exercise. (a) When a European visitor (the average ac voltages are 240 V in Europe) visits the United States, what type of transformer should be used to enable her hair dryer to work properly? Explain. (b) For a 1500-W hair dryer (assumed ohmic), what would be the transformer's input current in the United States, assuming it to be ideal?

The preceding relationships strictly apply only to ideal (or "lossless") transformers; actual transformers have energy losses. Well-designed transformers generally have a loss of less than 5%. Consequently there is no such thing as an ideal transformer. Many factors combine to determine how close a real transformer comes to performing like an ideal one.

First, there is flux leakage; that is, not all of the flux passes through the secondary coil. In some transformer designs, one of the insulated coils is wound directly on top of the other (interlocking) rather than having two separate coils. This configuration helps minimize flux leakage while reducing transformer size.

Second, the ac current in the primary means there is a changing magnetic flux through those coils. In turn this gives rise to an induced emf in the primary. This is called *self-induction*. By Lenz's law, the self-induced emf will oppose the change in current and thus limit the primary current (this is a similar effect to that of the back emf in a motor).

A third reason that transformers are less than ideal is joule heating (I^2R losses) due to the resistance of the wires. Usually this loss is small because the wires have little resistance.

Lastly, consider the effect of induction in the core material. To increase magnetic flux, the core is made of a highly permeable material (such as iron), but such materials are also good conductors. The changing magnetic flux in the core induces emfs there, which in turn create *eddy (or "swirling") currents* in the core material. These eddy currents can cause energy loss between the primary and secondary by heating the core (I^2R losses again).

To reduce the loss of energy due to eddy currents, transformer cores are made of thin sheets of material (usually iron) laminated with an insulating glue between them. The insulating layers between the sheets break up the eddy currents or confine them to the thin sheets, greatly reducing energy loss.

The effects of eddy currents can be demonstrated by allowing a plate made of a conductive, but nonmagnetic, metal, such as aluminum, to swing through a magnetic

Demonstration/activity: Drop a magnet and an identical-looking nonmagnet through a long copper tube. Copper is conducting, but nonmagnetic. Explain why the time taken to fall through the tube is different for the two items. Try the same experiment with a PVC tube. "Cow magnets" make excellent objects, as they are relatively small with large magnetic fields and can be purchased from scientific supply houses.

Teaching tip: The setup in Fig. 20.17 is easy to use as a demonstration. Bring in a large permanent magnet and suspend aluminum sheets above the magnet in a pendulum fashion. Pull the sheets aside and release them, allowing them to swing into the field. The one with the slices cut into it will swing through to the other side, whereas the solid panel will stop abruptly.

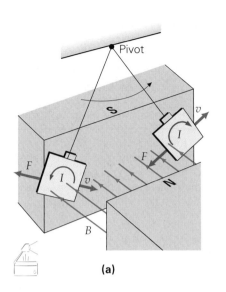
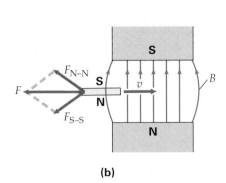
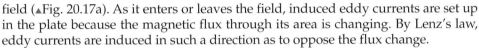

(a) **(b)** **(c)**

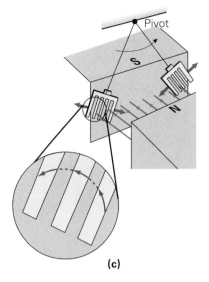

▲ **FIGURE 20.17 Eddy currents**
(a) Eddy currents are induced in a metal plate moving in a magnetic field. The induced currents oppose the change in flux. These currents then experience a retarding magnetic force to oppose the motion first into, then out of, the field region. To see this, note that the currents reverse direction as the plate leaves the field. **(b)** An overhead view as the plate swings toward the field from the left. The retarding force $\vec{F}$ (to slow it from entering the field) results from the two repulsive forces ($\vec{F}_{N-N}$ and $\vec{F}_{S-S}$) acting between magnetic poles. This is because the side of the plate closest to the north pole of the permanent magnet acts as a north pole, and the other side acts as a south pole. **(c)** If the plate has slits, the eddy currents, and thus magnetic forces, are drastically reduced and the plate will swing more freely.

field (▲Fig. 20.17a). As it enters or leaves the field, induced eddy currents are set up in the plate because the magnetic flux through its area is changing. By Lenz's law, eddy currents are induced in such a direction as to oppose the flux change.

When the plate enters the field (the position of the left-hand plate in Fig. 20.17a), a counterclockwise current is induced. (You should apply Lenz's law to show this.) The induced current produces its own magnetic field, which means that, in effect, the plate has a north magnetic pole near the permanent magnet's north pole and a south magnetic pole near the permanent magnet's south pole (Fig. 20.17b). Two repulsive magnetic forces act on the plate. The effect of the net force is to slow the plate down as it enters the field. The plate's eddy currents are reversed in direction as it leaves the field, producing a net attractive magnetic force, thus tending to slow the plate from leaving the field. In both cases, the induced emfs act to slow the plate's motion.

The reduction of eddy currents (similar to how the laminated layers in a transformer work) can be demonstrated by using a plate with slits cut into it (Fig. 20.17c). When this plate swings between the magnet's poles, it swings relatively freely, because the eddy currents are greatly reduced by the air gaps (slits). Consequently, the magnetic force on the plate is also reduced.

The damping effect of eddy currents has been applied in the braking systems of rapid-transit railcars. When an electromagnet (housed in the car) is turned on, it applies a magnetic field to a rail. The repulsive force due to the induced eddy currents in the rail acts as a braking force (▾Fig. 20.18). As the car slows, the eddy currents in the rail decrease, allowing a smooth braking action.

Power Transmission and Transformers

For power transmission over long distances, transformers provide a way to increase the voltage and reduce the current of an electric generator, thus cutting down the joule heating (I^2R) losses in the transmission wires that are carrying current. A schematic diagram of an ac power distribution system is shown in ▾Fig. 20.19. The voltage output of the generator is stepped up, reducing the current. The energy is transmitted over long distances to an area substation near the consumers. There,

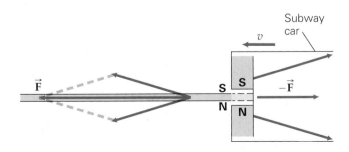

◀ **FIGURE 20.18 Electromagnetic braking and mass transit** When braking, a train energizes an electromagnet onboard. This electromagnet straddles a long metal rail. The induced currents in the rail produce a mutually repulsive force between the rail and the train, thereby slowing the train.

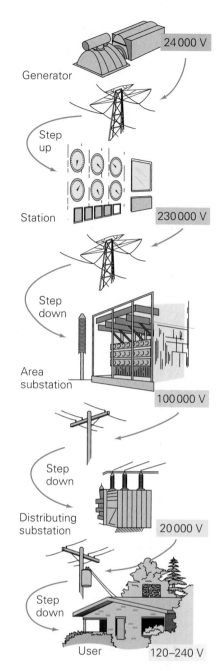

▲ **FIGURE 20.19** Power transmission A diagram of a typical electrical-power distribution system.

Illustration 31.3 Transformers

the voltage is stepped down, increasing the current. There are further step-downs at distributing substations and utility poles before the electricity is supplied to homes and businesses at the normal voltage and current.

The following Example illustrates the benefits of being able to step up the voltage (and step down the current) for electrical power transmission.

Example 20.8 ■ Cutting Your Losses: Power Transmission at High Voltage

A small hydroelectric power plant produces energy in the form of electric current at 10 A and a voltage of 440 V. The voltage is stepped up to 4400 V (by an ideal transformer) for transmission over 40 km of power line, which has a total resistance of 20 Ω. (a) What percentage of the original energy would have been lost in transmission if the voltage had not been stepped up? (b) What percentage of the original energy is actually lost when the voltage is stepped up?

Thinking It Through. (a) The power output can be computed from $P = IV$ and compared with the power lost in the wire, $P = I^2R$. (b) Equations 20.10a and 20.10b should be used to determine the stepped-up voltage and stepped-down currents, respectively. Then the calculation is repeated, and the results are compared with those of part (a).

Solution.

Given: $I_p = 10$ A   *Find:* (a) percentage energy loss without voltage step-up
   $V_p = 440$ V   (b) percentage energy loss with voltage step-up
   $V_s = 4400$ V
   $R = 20$ Ω

(a) The power output by the generator is

$$P = I_pV_p = (10 \text{ A})(440 \text{ V}) = 4400 \text{ W}$$

The rate of energy loss of the wire (joules per second, or watts) in transmitting a current of 10 A is very high, because

$$P_{loss} = I^2R = (10 \text{ A})^2(20 \text{ Ω}) = 2000 \text{ W}$$

Thus, the percentage of the produced energy lost to joule heat in the wires is nearly 50% because

$$\% \text{ loss} = \frac{P_{loss}}{P} \times 100\% = \frac{2000 \text{ W}}{4400 \text{ W}} \times 100\% = 45\%$$

(b) When the voltage is stepped up to 4400 V, this allows for transmission of energy at a current that is reduced by a factor of 10 from its value in part (a). We have

$$I_s = \left(\frac{V_p}{V_s}\right)I_p = \left(\frac{440 \text{ V}}{4400 \text{ V}}\right)(10 \text{ A}) = 1.0 \text{ A}$$

The power is thus reduced by a factor of 100, because it varies as the square of the current:

$$P_{loss} = I^2R = (1.0 \text{ A})^2(20 \text{ Ω}) = 20 \text{ W}$$

Therefore, the percentage of power lost is also reduced by a factor of 100 to a much more acceptable level:

$$\% \text{ loss} = \frac{P_{loss}}{P} \times 100\% = \frac{20 \text{ W}}{4400 \text{ W}} \times 100\% = 0.45\%$$

Follow-Up Exercise. Some heavy-duty electrical appliances, such as water pumps, can be wired to 240 V or 120 V. Their power rating is the same regardless of the voltage at which they run. (a) Explain the efficiency advantage of operating such appliances at the higher voltage. (b) For a 1.00-hp pump (746 W), estimate the ratio of the power lost in the wires at 240 V to the power lost at 120 V (assuming that all resistances are ohmic and the connecting wires are the same).

20.4 Electromagnetic Waves

OBJECTIVES: To (a) explain the nature, origin, and propagation of electromagnetic waves, and (b) describe some of the properties and uses of the different types of electromagnetic waves.

Electromagnetic waves (or *electromagnetic radiation*) were considered as a means of heat transfer in Section 11.4. You can now understand the production and characteristics of electromagnetic radiation, because these waves are composed of electric and magnetic fields.

Scottish physicist James Clerk Maxwell (1831–1879) is credited with first bringing together, or *unifying*, electric and magnetic phenomena. Using mathematics beyond the scope of this book, he took the equations that governed each field and predicted the existence of electromagnetic waves. In fact, he went further and calculated their speed in a vacuum, and his prediction agreed with experiment. Because of these contributions, the set of equations is known as *Maxwell's equations*, although they were, for the most part, developed by others (for example, Faraday's law of induction).

Essentially, Maxwell showed how the electric field and the magnetic field could be thought of as a single electromagnetic field. The apparently separate fields are symmetrically related in the sense that either one can create the other under the proper conditions. This symmetry is evident by looking at the equations (not shown). A qualitative summary of the results is sufficient:

A time-varying magnetic field produces a time-varying electric field.
A time-varying electric field produces a time-varying magnetic field.

The first statement summarizes our observations in Section 20.1: A changing magnetic flux gives rise to an induced emf, which in turn can cause a current. The second statement (which we will not study in detail) is crucial to the self-propagating characteristic of electromagnetic waves. Together, these two phenomena enable these waves to travel through a vacuum, whereas all other waves, such as string waves, require a supporting medium.

According to Maxwell's theory, *accelerating* electric charges, such as an oscillating electron, produce electromagnetic waves. The electron in question could, for example, be one of the many electrons in the metal antenna of a radio transmitter, driven by an electrical (voltage) oscillator at a frequency of 10^6 Hz (1 MHz). As each electron oscillates, it continually accelerates and decelerates and thus radiates an electromagnetic wave (▾Fig. 20.20a). The driven oscillations of many electrons produce time-varying electric and magnetic fields in the vicinity of the antenna. The electric field, shown in red in Fig. 20.20a, is in the plane of the page and continually changes direction, as does the magnetic field (shown in blue and pointing into and out of the paper).

Both the electric and the magnetic fields carry energy and propagate outward at the speed of light. This speed is symbolized by the letter c. To three significant figures, $c = 3.00 \times 10^8$ m/s. Maxwell's results showed that at large distances from the source, these electromagnetic waves become plane waves. (Figure 20.20b shows a wave at an instant in time.) Here, the electric field ($\vec{E}$) is perpendicular to the magnetic field ($\vec{B}$), and each varies sinusoidally with time. Both $\vec{E}$ and $\vec{B}$ are perpendicular to the direction of wave propagation. Thus, electromagnetic waves are *transverse* waves, with the *fields* oscillating perpendicularly to the direction of propagation. According to Maxwell's theory, as one field changes, it creates the other. This process, repeated again and again, gives rise to the traveling electromagnetic wave we call light. An important result of all this is as follows:

In a vacuum, all electromagnetic waves, regardless of frequency or wavelength, travel at the same speed, $c = 3.00 \times 10^8$ m/s.

Illustration 32.1 Creation of Electromagnetic Waves

▾ **FIGURE 20.20** Source of electromagnetic waves Electromagnetic waves are produced, fundamentally, by accelerating electric charges. **(a)** Here charges (electrons) in a metal antenna are driven by an oscillating voltage source. As the antenna polarity and current direction periodically change, alternating electric and magnetic fields propagate outward. The electric and magnetic fields are perpendicular to the direction of wave propagation. Thus, electromagnetic waves are transverse waves. **(b)** At large distances from the source, the initially curved wavefronts become planer.

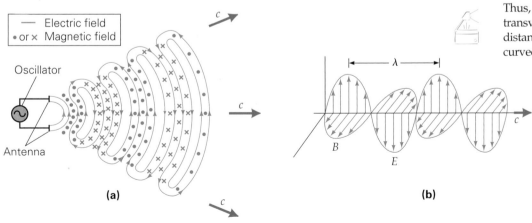

For everyday distances, the time delay due to the speed of light can usually be neglected. However, for interplanetary trips, this delay can be a problem. Consider the following Example.

Example 20.9 ■ Long-Distance Guidance: The Speed of Electromagnetic Waves in a Vacuum

The first successful Mars landings were the *Viking* probes in 1976. They sent radio and TV signals (both are electromagnetic waves) back to the Earth. How much longer would it take for a signal to reach us when Mars was farthest from the Earth than when it was closest to us? The average distances of Mars and the Earth from the Sun are 229 million km (d_M) and 150 million km (d_E), respectively. Assume that both planets have circular orbits, and use the average distances as the radii of the circles.

Thinking It Through. This situation calls for a time–distance calculation. The planets are farthest apart when they are on opposite sides of the Sun and separated by a distance of $d_M + d_E$. (This arrangement requires signals to be sent through the Sun, which, of course, is not possible. However, it does serve to determine the upper limit on transmission times.) The planets are closest when they are aligned on the same side of the Sun. In this case, their separation distance is at a minimum value of $d_M - d_E$. (Draw a diagram to help visualize this.) Because the speed of electromagnetic waves in a vacuum is known, the times can be found from $t = d/c$.

Solution. Listing the data and converting the distances to meters,

Given: $d_M = 229 \times 10^6$ km $= 2.29 \times 10^{11}$ m   *Find:* Δt (difference in time for light to
$d_E = 150 \times 10^6$ km $= 1.50 \times 10^{11}$ m   travel the longest and shortest
                                                              distances)

Radio and TV waves travel at speed c. The longest travel time t_L is

$$t_L = \frac{d_M + d_E}{c} = \frac{3.79 \times 10^{11} \text{ m}}{3.00 \times 10^8 \text{ m/s}} = 1.26 \times 10^3 \text{ s} \quad \text{(or 21.1 min)}$$

For the shortest distance the shortest travel time t_s is

$$t_S = \frac{d_M - d_E}{c} = \frac{7.90 \times 10^{10} \text{ m}}{3.00 \times 10^8 \text{ m/s}} = 2.63 \times 10^2 \text{ s} \quad \text{(or 4.39 min)}$$

The difference in the times is thus $\Delta t = t_L - t_s = 1.00 \times 10^3$ s (or 16.7 min).

Follow-Up Exercise. Assume that a *Rover* Martian vehicle (see Example 2.1 on page 34) is heading for a collision with a rock 2.0 m ahead of it. When it is at that distance, the vehicle sends a picture of the rock to controllers on Earth. If Mars is at the closest point to the Earth, what is the maximum speed that the *Rover* could have and still avoid a collision? Assume that the video signal from the *Rover* reaches the Earth and that the signal for it to stop is sent back immediately.

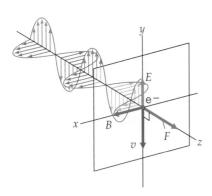

▲ **FIGURE 20.21 Radiation pressure** The electric field of an electromagnetic wave that strikes a surface acts on an electron, giving it a velocity. The magnetic field then exerts a force on that moving charge in the direction of propagation of the incident light. (Verify this direction, using the magnetic right-hand force rule.)

Radiation Pressure

An electromagnetic wave carries energy. Consequently, it can do work and can exert a force on a material it strikes. Consider light striking an electron at rest on a surface (◄Fig. 20.21). The electric field of the wave exerts a force on the electron, giving it a downward velocity ($\vec{v}$) as shown in the figure. Because a charged particle moving in a magnetic field experiences a force, there is a magnetic force on the electron, due to the magnetic-field component of the light wave. By the force right-hand rule, this force is in the direction that the wave is propagating (Fig. 20.21a). Therefore, because the electromagnetic wave will produce the same force on many electrons in a material, it exerts a force on the surface as a whole, and that force is in the direction in which it is traveling.

The radiation force per area is called **radiation pressure**. Radiation pressure is negligibly small for most everyday situations, but it can be important in atmospheric and astronomical phenomena, as well as in atomic and nuclear physics, where masses are small and there is no friction. For example, radiation pressure plays a key role in determining the direction in which the tail of a comet points. Sunlight delivers energy to the comet's "head," which consists of ice and dust. Some of this material evaporates as the comet nears the Sun, and the evaporated gases are pushed away from the Sun by radiation pressure. Thus, the tail generally points away from the Sun, no matter whether the comet is approaching or leaving the Sun's vicinity.

Another potential use of radiation pressure from sunlight is to propel interplanetary "sailing" satellites outward from the Sun toward the outer planets in an ever enlarging, spiraling orbit (▶Fig. 20.22a). To create enough force, given the extremely low pressure of the sunlight, the sails would have to be very large in area, and the satellite would have to have as little mass as possible. The payoff is that no fuel (except for small amounts for course corrections) would be needed once the satellite was launched. Consider the following Conceptual Example, which concerns radiation pressure and space travel.

Conceptual Example 20.10 ■ Sailing the Sea of Space: Radiation Pressure in Action

Consider the design of a relatively light spacecraft with a huge "sail" to be used as an interplanetary probe. With little or no power of its own, it would be designed to use the pressure of sunlight to propel it to the outer planets. To get the maximum propulsive force, what kind of surface should the sail have: (a) shiny and reflective, (b) dark and absorptive, or (c) surface characteristics would not matter?

Reasoning and Answer. At first glance, you might think that the answer is (c). However, as we have seen, radiation is capable of exerting force and can transfer momentum to whatever it strikes. The interaction between the radiation and the sail can be described in terms of conservation of momentum, as shown in Fig. 20.22b. (See Section 6.3.) If the radiation is absorbed, the situation is analogous to a completely inelastic collision (such as a putty wad sticking to a door), and the sail would acquire all of the momentum ($\vec{\mathbf{p}}$) originally possessed by the radiation.

However, if the radiation is *reflected*, the situation is analogous to a completely elastic collision, like a Superball bouncing off a wall (see Section 6.1). Because the momentum of the radiation after the collision would be equal to its original momentum in magnitude, but opposite in direction, its momentum would thus be reversed (from $\vec{\mathbf{p}}$ to $-\vec{\mathbf{p}}$). To conserve momentum, the momentum transferred to the shiny sail would be twice as great ($2\vec{\mathbf{p}}$) as that for the dark sail. Because force is the rate of change of momentum, reflective sails would experience, on average, twice as much force as absorptive ones. So the answer is (a).

Follow-Up Exercise. (a) Would the sail in this Example provide less or more acceleration as the interplanetary sailing ship moves farther from the Sun? (b) Explain how a change in sail area could counteract this change.

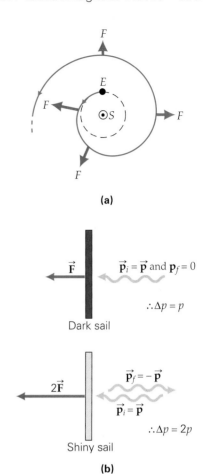

▲ **FIGURE 20.22** "Sailing" the solar system (a) A space probe launched from the Earth (E) equipped with a large sail would be acted on by radiation pressure from sunlight (Sun is at S). This cost-free force would cause the satellite to spiral outward. With proper planning, the craft could get to outer planets with little or no extra fuel. Note the reduction in force with distance. (b) Is it better for the sail to be dark or shiny? See Conceptual Example 20.10 and momentum conservation.

Types of Electromagnetic Waves

Electromagnetic waves are classified by the range of frequencies or wavelengths they encompass. Recall from Chapter 13 that frequency and wavelength are inversely related by the traveling-wave relationship $\lambda = c/f$, where the speed of light, c, has been substituted for the general wave speed v. The higher the frequency, the shorter the wavelength, and vice versa. The electromagnetic spectrum is continuous, so the limits of the various types of radiation are approximate (▼Fig. 20.23). Table 20.1 (p. 676) lists these ranges for the general types of electromagnetic waves.

Power Waves Electromagnetic waves with a frequency of 60 Hz result from alternating currents in electric power lines. These power waves have a wavelength of 5.0×10^6 m, or 5000 km (more than 3000 mi). Waves of such low frequency are of little practical use. They may occasionally produce a so-called 60-Hz hum on your stereo or introduce, via induction, unwanted electrical noise in delicate instruments. More serious concerns have been expressed about the possible effects of these waves

▶ **FIGURE 20.23 The electromagnetic spectrum** The spectrum of frequencies or wavelengths is divided into various regions, or ranges. The visible-light region is a very small part of the total spectrum. For visible light, wavelengths are usually expressed in nanometers (1 nm = 10^{-9} m). (The relative sizes of the wavelengths at the top of the figure are not to scale.)

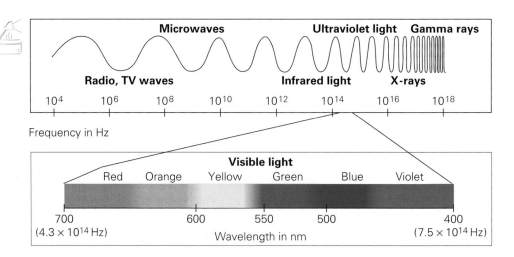

TABLE 20.1	Classification of Electromagnetic Waves		
Type of Wave	*Approximate Frequency Range (Hz)*	*Approximate Wavelength Range (m)*	*Some Typical Sources*
Power waves	60	5.0×10^6	Electric currents
Radio waves—AM	$(0.53 \times 10^6)–(1.7 \times 10^6)$	570–186	Electric circuits/antennae
Radio waves—FM	$(88 \times 10^6)–(108 \times 10^6)$	3.4–2.8	Electric circuits/antennae
TV	$(54 \times 10^6)–(890 \times 10^6)$	5.6–0.34	Electric circuits/antennae
Microwaves	$10^9–10^{11}$	$10^{-1}–10^{-3}$	Special vacuum tubes
Infrared radiation	$10^{11}–10^{14}$	$10^{-3}–10^{-7}$	Warm and hot bodies, stars
Visible light	$(4.0 \times 10^{14})–(7.0 \times 10^{14})$	10^{-7}	The Sun and other stars, and lamps
Ultraviolet radiation	$10^{14}–10^{17}$	$10^{-7}–10^{-10}$	Very hot bodies, stars, and special lamps
X-rays	$10^{17}–10^{19}$	$10^{-10}–10^{-12}$	High-speed electron collisions and atomic processes
Gamma rays	above 10^{19}	below 10^{-12}	Nuclear reactions and nuclear decay processes

on health. Some early research tended to suggest that very low-frequency fields may have potentially harmful biological effects on cells and tissues. However, recent surveys indicate that this is not the case.

Radio and TV Waves Radio and TV waves are generally in the frequency range from 500 kHz to about 1000 MHz. The AM (*amplitude-modulated*) band runs from 530 to 1710 kHz (1.71 MHz). Higher frequencies, up to 54 MHz, are used for "shortwave" bands. TV bands range from 54 MHz to 890 MHz. The FM (*frequency-modulated*) radio band runs from 88 to 108 MHz, which lies in a gap between channels 6 and 7 of the range of TV bands. Cellular phones use radio waves to transmit voice communication in the ultrahigh-frequency (UHF) band, with frequencies similar to those used for TV channels 13 and higher.

Early global communications used the "shortwave" bands, as do amateur (ham) radio operators today. But how are the normally straight-line radio waves transmitted around the curvature of the Earth? This feat is accomplished by reflection off ionic layers in the upper atmosphere. Energetic particles from the Sun ionize gas molecules, giving rise to several ion layers. Certain of these layers reflect radio waves. By "bouncing" radio waves off these layers, radio transmissions can be sent beyond the horizon, to any region of the Earth.

It turns out that such reflection requires the ionic layers to have fairly uniform density. When, from time to time, a solar disturbance produces a larger-than-average shower of energetic charged particles that upsets this uniformity, a communications "blackout" can occur as the radio waves are scattered in various directions rather than reflected in straight lines. To avoid such disruptions, global communications have, in the past, relied largely on transoceanic cables. Now we also have communications satellites, which can provide line-of-sight transmission to any point on the globe.

Microwaves   Microwaves, with frequencies in the gigahertz (GHz) range, are produced by special vacuum tubes (called *klystrons* and *magnetrons*). Microwaves are commonly used in communications and radar applications. In addition to its roles in navigation and guidance, radar provides the basis for the speed guns used to time such things as baseball pitches and motorists through the use of the Doppler effect (see Section 14.5). When the waves are reflected off an object of interest, the magnitude and sign of the frequency shift allow determination of the object's velocity.

Infrared (IR) Radiation   The infrared region of the electromagnetic spectrum lies adjacent to the low-frequency, or long-wavelength, end of the visible spectrum. A warm body emits IR radiation, which depends on that body's temperature. (See Chapter 27.) An object at or near room temperature emits radiation in the far infrared region. ("Far" means relative to the visible region.)

Recall from Section 11.4 that infrared radiation is sometimes referred to as "heat rays." This is because water molecules, which are present in most materials and which possess electrical permanent polarization, readily absorb electromagnetic radiation at frequencies in the infrared-wavelength region. When they do, their random thermal motion is increased—and the molecules "heat up," as well as their surroundings. Infrared lamps are used in therapeutic applications, such as easing pain in strained muscles, and to keep food warm. IR is also associated with maintaining the Earth's temperature through the *greenhouse effect*. In this effect, incoming visible light (which passes relatively easily through the atmosphere) is absorbed by the Earth's surface and reradiated as infrared radiation. This IR radiation is in turn trapped by "greenhouse gases," such as carbon dioxide and water vapor, which are opaque to IR radiation. Its name comes from actual glass-enclosed greenhouse, where the glass rather than atmospheric gases traps the IR energy.

Visible Light   The region of visible light occupies only a small portion of the electromagnetic spectrum and covers a frequency range from 4×10^{14} Hz to about 7×10^{14} Hz. In terms of wavelengths, the range is from about 700 to 400 nm (Fig. 20.23). Recall that 1 nanometer (nm) $= 10^{-9}$ m. Only the radiation in this region activates the receptors on the retina of human eyes. Visible light emitted or reflected from objects around us provides us with visual information about our world. Visible light and optics will be discussed in Chapters 22 to 25.

It is interesting to note that not all animals are sensitive to the same range of wavelengths. For example, snakes can visually detect infrared radiation, and the visible range of many insects extends well into the ultraviolet range. The sensitivity range of the human eye conforms closely to the spectrum of wavelengths emitted by the Sun. The human eye's maximum sensitivity is in the same yellow-green region where the Sun's energy output is at its maximum (wavelengths of about 550 nm).

Ultraviolet (UV) Radiation   The Sun's spectrum has a small component of ultraviolet (UV) light, whose frequency range lies beyond the violet end of the visible region. UV is also produced artificially by special lamps and very hot objects. In addition to causing tanning of the skin, UV radiation can cause sunburn and/or skin cancer if exposure to it is too high.

Upon its arrival at the Earth, most of the Sun's ultraviolet emission is absorbed in the ozone (O_3) layer in the atmosphere, at an altitude of about 30 to 50 km. Because the ozone layer plays a protective role, there is concern about its depletion by chlorofluorocarbon gases (such as Freon, once commonly used in refrigerators) that drift upward and react with the ozone.

Most UV radiation is absorbed by ordinary glass. Therefore, you cannot get much of a tan through glass windows. Sunglasses are labeled to indicate which UV protection standards they meet in shielding the eyes from this potentially harmful radiation. Certain types of high-tech glass (called "photogray" glass) darken when exposed to UV. These materials are used to create "transition" sunglasses that darken when

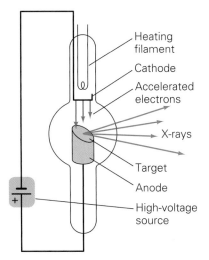

▲ **FIGURE 20.24 The X-ray tube**
Electrons accelerated through a large voltage strike a target electrode. There they slow down and interact with the electrons of the target material. Energy is emitted in the form of X-rays during this "braking" (deceleration) process.

exposed to sunlight. Of course, these sunglasses aren't very useful while driving a car. (Why?) Welders wear special glass goggles to protect their eyes from large amounts of UV produced by the arcs of welding torches. Similarly, it is important to shield your eyes from sunlamps or snow-covered surfaces. The ultraviolet component of sunlight reflected from snow-covered surfaces can produce snowblindness in unprotected eyes.

X-Rays Beyond the ultraviolet region of the electromagnetic spectrum is the important X-ray region. We are familiar with X-rays primarily through medical applications. X-rays were discovered accidentally in 1895 by the German physicist Wilhelm Roentgen (1845–1923) when he noted the glow of a piece of fluorescent paper, evidently caused by some mysterious radiation coming from a cathode-ray tube. Because of the apparent mystery involved, this radiation was named *x-radiation* or *X-rays* for short.

The basic elements of an X-ray tube are shown in ◄Fig. 20.24. An accelerating voltage, typically several thousand volts, is applied across the electrodes in a sealed, evacuated tube. Electrons emitted from the heated negative electrode (cathode) are accelerated toward the positive electrode (anode). When they strike the anode, some of their kinetic-energy loss is converted to electromagnetic energy in the form of X-rays.

A similar process takes place in color TV picture tubes, which use high voltages and electron beams. When the high-speed electrons hit the screen, they can emit X-rays. Fortunately, all modern televisions have the shielding necessary to protect viewers from exposure to this radiation. In the early days of color television, this was not always the case—hence the warning that came with the set: "Do not sit too close to the screen."

As you will learn in Chapter 27, the energy carried by electromagnetic radiation depends on its frequency. High-frequency X-rays have very high energies and can cause cancer, skin burns, and other harmful effects. However, at low intensities, X-rays can be used with relative safety to view the internal structure of the human body and other opaque objects.* X-rays can pass through materials that are opaque to other types of radiation. The denser the material, the greater is its absorption of X-rays and the less intense the transmitted radiation will be. For example, as X-rays pass through the human body, many more are absorbed or scattered by bone than by tissue. If the transmitted radiation is directed onto a photographic plate or film, the exposed areas show variations in intensity—a picture of internal structures.

The combination of the computer with modern X-ray machines permits the formation of three-dimensional images by means of a technique called *computerized tomography*, or *CT* (▼Fig. 20.25).

Gamma Rays The electromagnetic waves of the uppermost frequency range of the known electromagnetic spectrum are called *gamma rays* (γ-rays). This high-frequency radiation is produced in nuclear reactions, in particle accelerators, and in certain types of nuclear decay (radioactivity). Gamma rays will be discussed in more detail in Chapter 29.

▶ **FIGURE 20.25 CT scan** In an ordinary X-ray image, the entire thickness of the body is projected onto the film and internal structures often overlap, making details hard to distinguish. In CT—computerized tomography (from the Greek *tomo*, meaning "slice," and *graph*, meaning "picture")—X-ray beams scan a slice of the body. **(a)** The transmitted radiation is recorded by a series of detectors and processed by a computer. Using information from multiple slices, the computer constructs a three-dimensional image. Any single slice can be displayed for further study. **(b)** CT image of a brain with a benign tumor.

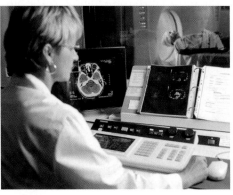

(a)

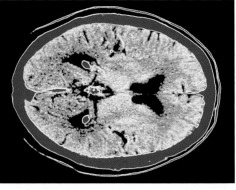

(b)

*Many health scientists believe that there is no safe "threshold" level for X-rays or other energetic radiation—that is, no level of exposure that is completely risk-free—and that some of the dangerous effects are cumulative over a lifetime. People should therefore avoid unnecessary medical X-rays or any other unwarranted exposure to "hard" radiation (Chapter 29). However, when properly used, X-rays can be an extremely useful diagnostic tool capable of saving lives.

Chapter Review

- **Magnetic flux (Φ)** is a measure of the number of magnetic field lines that pass through an area. For a single wire loop of area A, it is defined as

$$\Phi = BA \cos \theta \qquad (20.1)$$

where B is the magnetic field strength (assumed constant), A is the loop area, and θ is the angle between the direction of the magnetic field and the normal to the area's plane.

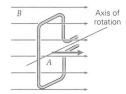

- **Faraday's law of induction** relates the induced emf in a loop (or coil composed of N loops in series) to the time rate of change of the magnetic flux through that loop (or coil).

$$\mathcal{E} = -N \frac{\Delta \Phi}{\Delta t} \qquad (20.2)$$

where $\Delta \Phi$ is the change in flux through *one loop* and there are N total loops.

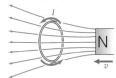

- **Lenz's law** states that when a change in magnetic flux induces an emf in a coil, loop, or circuit, the resulting, or induced, current direction is such as to create a magnetic field to oppose the change in flux.

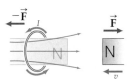

- An **ac generator** converts mechanical energy into electrical energy. The generator's emf as a function of time is

$$\mathcal{E} = \mathcal{E}_0 \sin \omega t \qquad (20.4)$$

where $\mathcal{E}_0$ is the maximum emf.

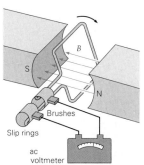

- A **transformer** is a device that changes the voltage supplied to it by means of induction. The voltage applied to the input, or primary (p), side of the transformer is changed into the output, or secondary (s), voltage. The current and voltage relationships for a transformer are

$$V_s = \left(\frac{N_s}{N_p} \right) V_p \qquad (20.10a)$$

$$I_s = \left(\frac{N_p}{N_s} \right) I_p \qquad (20.10b)$$

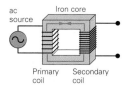

- An **electromagnetic wave** (light) consists of time-varying electric and magnetic fields that propagate at a speed of c (3.00×10^8 m/s) in a vacuum. The different types of electromagnetic radiation (such as UV, radio waves, and visible light) differ in frequency and wavelength.

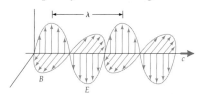

Exercises

MC = *Multiple Choice Question,* **CQ** = *Conceptual Question, and* **IE** = *Integrated Exercise. Throughout the text, many exercise sections will include "paired" exercises. These exercise pairs, identified with* **red numbers**, *are intended to assist you in problem solving and learning. In a pair, the first exercise (even numbered) is worked out in the Study Guide so that you can consult it should you need assistance in solving it. The second exercise (odd numbered) is similar in nature, and its answer is given at the back of the book.*

20.1 Induced emf: Faraday's Law and Lenz's Law

1. **MC** A unit of magnetic flux is (a) Wb, (b) T · m², (c) T · m/A, or (d) both a and b. (d)

2. **MC** The magnetic flux through a loop can change due to a change in (a) the area of the coil, (b) the strength of the magnetic field, (c) the orientation of the loop, or (d) all of the preceding. (d)

3. **MC** For a current to be induced in a wire loop, (a) there must be a large magnetic flux through the loop, (b) the loop's plane must be parallel to the magnetic field, (c) the loop's plane must be perpendicular to the magnetic field, or (d) the magnetic flux through the loop must vary with time. (d)

4. **MC** Identical single loops A and B are oriented so they initially have the maximum amount of flux in a magnetic field. Loop A is then quickly rotated so its normal is perpendicular to the magnetic field, and in the same time, B is rotated so its normal makes an angle of only 45° with the field. How do their induced emfs compare? (a) They are the same (b) A's is larger than B's (c) B's is larger than A's or (d) you can't tell the relative emf magnitudes from the data given. (b)

5. **MC** Identical single loops A and B are oriented so they have the maximum amount of flux when placed in a magnetic field. Both loops maintain their orientation relative to the field, but in the same amount of time A is moved to a region of stronger field, while B is moved to a region of weaker field. How do their induced emfs compare? (a) They are the same (b) A's is larger than B's (c) B's is larger than A's or (d) you can't tell the relative emf magnitudes from the data given. (d)

6. **CQ** A bar magnet is dropped through a coil of wire as shown in ▼Fig. 20.26. (a) Describe what is observed on the galvanometer by sketching a graph of induced emf versus *t*. (b) Does the magnet fall freely? Explain. see ISM

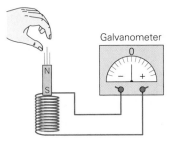

◀ **FIGURE 20.26**
A time-varying magnetic field
What will the galvanometer measure? See Exercise 6.

7. **CQ** In Fig. 20.1b, what would be the direction of the induced current in the loop if the south pole of the magnet were approaching instead of the north pole?
counterclockwise (in head-on view)

8. **CQ** In Fig. 20.7a, how would you move the coil so to prevent any current from being induced in it? Explain.
with the same velocity as the bar magnet, see ISM

9. **CQ** Does the induced emf in a closed loop depend on the value of the magnetic flux in the loop? Explain.
no, it depends on the rate of the flux change with time

10. **CQ** Two identical strong magnets are dropped simultaneously by two students into two vertical tubes of the same dimensions (▼Fig. 20.27). One tube is made of copper, and the other is made of plastic. From which tube will the magnet emerge first? Why? from the plastic tube; see ISM

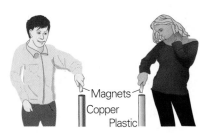

◀ **FIGURE 20.27**
Free fall?
See Exercise 10.

11. **CQ** A basic telephone has both a speaker-transmitter and a receiver (▼Fig. 20.28). Until the advent of digital phones in the 1990s, the transmitter had a diaphragm coupled to a carbon chamber (called the *button*), which contained loosely packed granules of carbon. As the diaphragm vibrated because of incident sound waves, the pressure on the granules varied, causing them to be more or less closely packed. As a result, the resistance of the button changed. The receiver converted these electrical impulses to sound. Applying the principles of electricity and magnetism that you have learned, explain the basic operation of this type of telephone.
see ISM

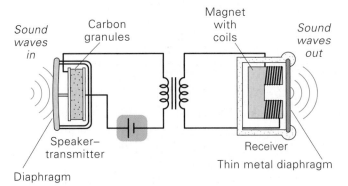

▲ **FIGURE 20.28** **Telephone operation** See Exercise 11.

12. • A circular loop with an area of 0.015 m² is in a uniform magnetic field of 0.30 T. What is the flux through the loop's plane if it is (a) parallel to the field, (b) at an angle of 37° to the field, and (c) perpendicular to the field?
(a) 0 (b) 2.7×10^{-3} T·m² (c) 4.5×10^{-3} T·m²

13. • A circular loop (radius of 20 cm) is in a uniform magnetic field of 0.15 T. What angle(s) between the normal to the plane of the loop and the field would result in a flux with a magnitude of 1.4×10^{-2} T·m²? 42° or 138°

14. • The plane of a conductive loop with an area of 0.020 m² is perpendicular to a uniform magnetic field of 0.30 T. If the field drops to zero in 0.0045 s, what is the magnitude of the average emf induced in the loop? 1.3 V

15. • A loop in the form of a right triangle with one side of 40.0 cm and a hypotenuse of 50.0 cm lies in a plane perpendicular to a uniform magnetic field of 550 mT. What is the flux through the loop? 3.3×10^{-2} T·m²

16. • A square coil of wire with 10 turns is in a magnetic field of 0.25 T. The total flux through the coil is 0.50 T·m². Find the area of one turn if the field (a) is perpendicular to the plane of the coil and (b) makes an angle of 60° with the plane of the coil. (a) 0.20 m² (b) 0.23 m²

17. •• An ideal solenoid with a current of 1.5 A has a radius of 3.0 cm and a turn density of 250 turns/m. What is the magnetic flux (due to its own field) through only one of its loops at its center? 1.3×10^{-6} T·m²

18. •• A uniform magnetic field is at right angles to the plane of a wire loop. If the field decreases by 0.20 T in 1.0×10^{-3} s and the magnitude of the average emf induced in the loop is 80 V, what is the area of the loop? 0.40 m²

19. ●● A square loop of wire with sides of length 40 cm is in a uniform magnetic field perpendicular to its area. If the field's strength is initially 100 mT and it decays to zero in 0.010 s, what is the magnitude of the average emf induced in the loop? 1.6 V

20. ●● The magnetic flux through one loop of wire is reduced from 0.35 Wb to 0.15 Wb in 0.20 s. The average induced current in the coil is 10 A. Find the resistance of the wire. 0.10 Ω

21. ●● When the magnetic flux through a single loop of wire increases by 30 T · m², an average current of 40 A is induced in the wire. Assuming that the wire has a resistance of 2.5 Ω, over what period of time did the flux increase? 0.30 s

22. ●● In 0.20 s, a coil of wire with 50 loops experiences an average induced emf of 9.0 V, due to a changing magnetic field perpendicular to the plane of the coil. The radius of the coil is 10 cm, and the initial strength of the magnetic field is 1.5 T. Assuming that the strength of the field decreased with time, what is the final strength of the field? 0.35 T

23. IE ●● A single strand of wire of adjustable length is wound around the circumference of a round balloon. A uniform magnetic field is perpendicular to the plane of the loop (▼Fig. 20.29). (a) If the balloon is inflated, in which direction is the induced current, looking down from above: (1) counterclockwise, (2) clockwise, or (3) there is no induced current? (b) If the magnitude of the magnetic field is 0.15 T and the wire's diameter increases from 20 cm to 40 cm in 0.040 s, what is the magnitude of the average value of the emf induced in the loop? (a) (1) counterclockwise (b) 0.35 V

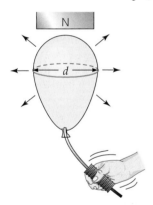

◄ **FIGURE 20.29 Pumping energy** See Exercise 23.

24. ●● The magnetic field perpendicular to the plane of a wire loop with an area of 0.10 m² changes with time as shown in ▶Fig. 20.30. What is the magnitude of the average emf induced in the loop for each segment of the graph (for example, from 0 to 2.0 ms)?
from left to right: 50 V, 13 V, 0 V, 12 V

25. IE ●● A boy is traveling due north at a constant speed while carrying a metal rod. The rod's length is oriented in the east–west direction and is parallel to the ground. (a) There will be no induced emf when the rod is (1) at the equator, (2) near the Earth's magnetic poles, (3) somewhere between the equator and the poles. Why? (b) Assume that the Earth's magnetic field is 1.0×10^{-4} T near the North Pole and 1.0×10^{-5} T near the Equator. If the boy runs with a speed of 5.0 m/s northward near each location, and the rod is 1.0 m long, calculate the induced emf in the rod in each location. (a) (1) at the equator (b) 0.50 mV at the pole, zero at the equator

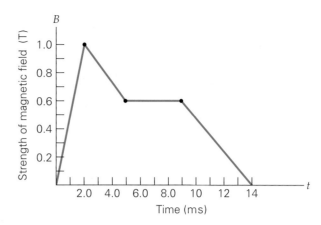

▲ **FIGURE 20.30 Magnetic field versus time** See Exercise 24.

26. ●● A metal airplane with a wingspan of 30 m flies horizontally at a constant speed of 320 km/h in a region where the vertical component of the Earth's magnetic field is 5.0×10^{-5} T. What is the induced emf between the tips of the plane's wings? 0.13 V

27. ●● Suppose that the metal rod in Fig. 20.11 is 20 cm long and is moving at a speed of 10 m/s in a magnetic field of 0.30 T and that the metal frame is covered with an insulating material. Find (a) the magnitude of the induced emf across the rod and (b) the current in the rod.
(a) 0.60 V (b) 0 A

28. ●●● The flux through a loop of wire changes uniformly from +40 Wb to −20 Wb in 1.5 ms. (a) What is the significance of the negative flux? (b) What is the average induced emf in the loop? (c) If you wanted to double the average induced emf by changing only the time, what would the new time interval be? (d) If you wanted to double the average induced emf by changing only the final flux value, what would it be? see ISM

29. ●●● A coil of wire with 10 turns and an area of 0.055 m² is placed in a magnetic field of 1.8 T and oriented so that the area is perpendicular to the field. The coil is then flipped by 90° in 0.25 s and ends up with the area parallel to the field (▼Fig. 20.31). What is the magnitude of the average emf induced in the coil? 4.0 V

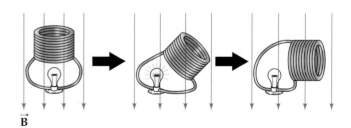

▲ **FIGURE 20.31 Flipping the coil** See Exercises 29 and 30.

30. IE ●●● In Fig 20.31, the coil is flipped by 180° in the same time interval as in Exercise 29. (a) How does the magnitude of the average emf compare with that in Exercise 29, where the coil was flipped only 90°: (1) higher, (2) the same, or (3) lower? Why? (b) What is the magnitude of the average emf in this case? (a) (1) higher (b) 8.0 V

31. ●●● A uniform magnetic field of 0.50 T penetrates a double-incline block as shown in ▼Fig. 20.32. (a) Determine the magnetic flux through each inclined surface of the block. (b) Determine the flux through the vertical back surface of the block. (c) Determine the flux through the flat horizontal surface of the block. (d) What is the total flux through all the outside surfaces? Explain the meaning of your answer. see ISM

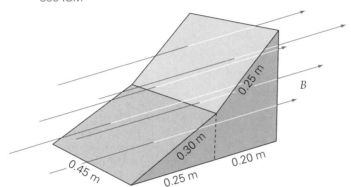

▲ **FIGURE 20.32 Magnetic flux** See Exercise 31. (Not drawn to scale.)

20.2 Electric Generators and Back emf

32. **MC** Doing nothing but increasing the coil area in an ac generator would result in (a) an increase in the frequency of rotation, (b) a decrease in the maximum induced emf, (c) an increase in the maximum induced emf. (c)

33. **MC** The back emf of a motor depends on (a) the input voltage, (b) the input current, (c) the armature's rotational speed, or (d) none of the preceding. (c)

34. **CQ** What is the orientation of the armature loop in a simple ac generator when the value of (a) the emf is a maximum and (b) the magnetic flux is a maximum? Explain why the maximum emf does not occur when the flux is a maximum. see ISM

35. **CQ** A student has a bright idea for a generator: For the arrangement shown in ▼Fig. 20.33, the magnet is pulled down and released. With a highly elastic spring, the student thinks that there should be a relatively continuous electrical output. What is wrong with this idea? see ISM

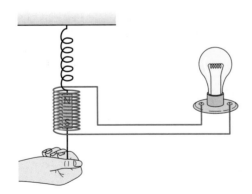

▲ **FIGURE 20.33 Inventive genius?** See Exercise 35.

36. **CQ** In a dc motor, if the armature is jammed or turns very slowly under a heavy load, the coils in the motor may burn out. Why?
there is no back emf, and thus there is a large current

37. **CQ** If you wanted to make a more compact ac generator by reducing the area of the coils, how would could you compensate with other factors in order to maintain the same output as before? increase N, B or ω, see ISM

38. ● A hospital emergency room ac generator operates at a rotation frequency of 60 Hz. If the output voltage is a maximum (in magnitude) at $t = 0$, when is it next (a) a maximum (in magnitude), (b) zero, and (c) at its initial value? (a) 1/120 s (b) 1/240 s (c) 1/60 s

39. ● A student makes a simple generator by using a single square loop 10 cm on each side. The loop is then rotated at a frequency of 60 Hz in a magnetic field of 0.015 T. (a) What is the maximum emf output? (b) What would be the maximum emf output if 10 such loops were used instead? (a) 0.057 V (b) 0.57 V

40. ●● A simple ac generator consists of a coil with 10 turns (each turn has an area of 50 cm²). The coil rotates in a uniform magnetic field of 350 mT with a frequency of 60 Hz. (a) Write an expression in the form of Eq. 20.5 for the generator's emf variation with time. (b) Compute the maximum emf. (a) $\mathscr{E} = \mathscr{E}_o \sin 120\pi t$ (b) 6.6 V

41. **IE** ●● A 60-Hz ac voltage source has a maximum voltage of 120 V. A student wants to determine the voltage 1/180 s after it is zero. (a) How many possible voltages are there: (1) one, (2) two, or (3) three? Why? (b) Determine all possible voltages. (a) (2) two (b) ±104 V

42. ●● An ac generator with a maximum emf of 400 V is to be constructed from wire loops (radius 0.15 m). It will operate at a frequency of 60 Hz and use a magnetic field of 200 mT. How many loops will be needed? 75 loops

43. ●● An ac generator operates at a rotational frequency of 60 Hz and has a maximum emf of 100 V. Assume that it has zero emf at startup. What is the instantaneous emf (a) 1/240 s after startup and (b) 1/120 s after the emf passes through zero as it begins to reverse polarity? (a) 100 V (b) 0 V

44. ●● The armature of a simple ac generator has 20 circular loops of wire, each with a radius of 10 cm. It is rotated with a frequency of 60 Hz in a uniform magnetic field of 800 mT. What is the maximum emf induced in the loops, and how often is this value attained? 1.9×10^2 V, every 1/120 s

45. ●● The armature of an ac generator has 100 turns. Each turn is a rectangular loop measuring 8.0 cm by 12 cm. The generator has a sinusoidal voltage output with an amplitude of 24 V. If the magnetic field of the generator is 250 mT, with what frequency does the armature turn? 16 Hz

46. **IE** ●● (a) To increase the output of an ac generator, a student has the choice of doubling *either* the generator's magnetic field *or* its frequency. To maximize the increase in emf output, (1) he should double the magnetic field, (2) he should double the frequency, (3) it does not matter which one he doubles. Explain. (b) Two students display their ac generators at a science fair. The generator made by student A has a loop area of 100 cm² rotating in a magnetic field of 20 mT at 60 Hz. The one made by student B has a loop area of 75 cm² rotating in a magnetic field of 200 mT at 120 Hz. Which one generates the largest maximum emf? Justify your answer mathematically. (a) (3) it does not matter which one he doubles (b) student B's, see ISM

47. **IE ●●** A motor has a resistance of 2.50 Ω and is connected to a 110-V line. (a) Is the operating current of the motor (1) higher than 44 A, (2) 44 A, or (3) lower than 44 A? Why? (b) If the back emf of the motor at operating speed is 100 V, what is its operating current?
(a) (3) lower than 44 A (b) 4.00 A

48. **●●●** The starter motor in an automobile has a resistance of 0.40 Ω in its armature windings. The motor operates on 12 V and has a back emf of 10 V when running at normal operating speed. How much current does the motor draw (a) when running at its operating speed, (b) when the motor is running at half its final rotational speed, and (c) when starting up?
(a) 5.0 A (b) 18 A (c) 30 A

49. **●●●** A 240-V dc motor has an armature whose resistance is 1.50 Ω. When running at its operating speed, it draws a current of 16.0 A. (a) What is the back emf of the motor when it is operating normally? (b) What is the starting current? (Assume that there is no additional resistance in the circuit.) (c) What series resistance would be required to limit the starting current to 25 A? (a) 216 V (b) 160 A (c) 8.1 Ω

20.3 Transformers and Power Transmission

50. **MC** A transformer in the electrical energy delivery system just before your house has (a) more windings in the primary coil, (b) more windings in the secondary coil, or (c) the same number of windings in the primary and secondary coils. (a)

51. **MC** The output power delivered by a realistic step-down transformer is (a) greater than the input power, (b) less than the input power, or (c) the same as the input power. (b)

52. **CQ** Explain why electric energy delivery systems operate at such high voltages when such voltages can be dangerous. high voltage and thus low current in order to reduce joule heat

53. **CQ** In your automotive workshop emergency repair, you need a step-down transformer, but have only step-up transformers on the shelves. Can you use a step-up transformer as a step-down one? If so, explain how you would wire it. yes, reverse the primary and secondary coils

54. **IE ●** The secondary coil of an ideal transformer has 450 turns, and the primary coil has 75 turns. (a) Is this transformer a (1) step-up or (2) step-down transformer? Why? (b) What is the ratio of the current in the primary coil to the current in the secondary coil? (c) What is the ratio of the voltage across the primary coil to the voltage in the secondary coil?
(a) (1) step-up (b) 6:1 (c) 1:6

55. **●** An ideal transformer steps 8.0 V up to 2000 V, and the 4000-turn secondary coil carries 2.0 A. (a) Find the number of turns in the primary coil. (b) Find the current in the primary coil. (a) 16 (b) 5.0×10^2 A

56. **●** The primary coil of an ideal transformer has 720 turns, and the secondary coil has 180 turns. If the primary coil carries 15 A at a voltage of 120 V, what are (a) the voltage and (b) the output current of the secondary coil?
(a) 30 V (b) 60 A

57. **●** The transformer in the power supply for a computer's 250-MB Zip drive changes a 120-V input to a 5.0-V output. Find the ratio of the number of turns in the primary coil to the number of turns in the secondary coil. 24:1

58. **●** The primary coil of an ideal transformer is connected to a 120-V source and draws 10 A. The secondary coil has 800 turns and a current of 4.0 A. (a) What is the voltage across the secondary coil? (b) How many turns are in the primary coil?
(a) 3.0×10^2 V (b) 3.2×10^2 turns

59. **●** An ideal transformer has 840 turns in its primary coil and 120 turns in its secondary coil. If the primary coil draws 2.50 A at 110 V, what are (a) the current and (b) the output voltage of the secondary coil? (a) 17.5 A (b) 15.7 V

60. **●●** The efficiency e of a transformer is defined as the ratio of the power output to the power input:

$$e = \frac{P_s}{P_p} = \frac{I_s V_s}{I_p V_p}.$$

(a) Show that in terms of the ratios of currents and voltages given in Eq. 20.10 for an ideal transformer, an efficiency of 100% is obtained for that situation. (b) Suppose a step-up transformer increased the line voltage from 120 to 240 V, while at the same time the output current was reduced to 5.0 A from 12 A. What is the transformer's efficiency? Is it ideal? see ISM

61. **IE ●●** The specifications of a transformer used with a small appliance read as follows: Input, 120 V, 6.0 W; Output, 9.0 V, 300 mA. (a) Is this transformer (1) an ideal or (2) a nonideal transformer? Why? (b) What is its efficiency? (See Exercise 60.)
(a) (2) a nonideal, because $P_s < P_p$ (b) 45%

62. **●●** A circuit component operates at 20 V and 0.50 A. A transformer with 300 turns in its primary coil is used to convert 120-V household electricity to the proper voltage. (a) How many turns must the secondary coil have? (b) How much current is in the primary coil? (a) 50 (b) 8.3×10^{-2} A

63. **●●** A transformer in a door chime steps down the voltage from 120 V to 6.0 V and supplies a current of 0.50 A to the chime mechanism. (a) What is the turn ratio of the transformer? (b) What is the current input to the transformer?
(a) N_s/N_p is 1:20 (b) 2.5×10^{-2} A

64. **●●** An ac generator supplies 20 A at 440 V to a 10 000-V power line. If the step-up transformer has 150 turns in its primary coil, how many turns are in the secondary coil?
3.4×10^3 turns

65. **●●** The electricity supplied in Exercise 64 is transmitted over a line 80.0 km long with a resistance of 0.80 Ω/km. (a) How many kilowatt-hours are saved in 5.00 h by stepping up the voltage? (b) At $0.10/kWh, how much of a savings (to the nearest $10) is this to the consumer in a 30-day month, assuming that the energy is supplied continuously?
(a) 128 kWh (b) $1840

66. **●●** At an area substation, the power-line voltage is stepped down from 100 000 V to 20 000 V. If 10 MW of power is delivered to the 20 000-V circuit, what are the primary and secondary currents in the transformer?
100 A, 500 A

67. **●●** A voltage of 200 000 V in a transmission line is reduced to 100 000 V at an area substation, to 7200 V at a distributing substation, and finally to 240 V at a utility pole outside a house. (a) What turn ratio N_s/N_p is required for each reduction step? (b) By what factor is the transmission-line current stepped up in each voltage step-down? (c) What is the overall factor by which the current is stepped up from transmission line to utility pole?
(a) 1:2, 1:14, 1:30 (b) 2.0, 14, 30 (c) 833

68. ●● A plant produces electric energy at 50 A and 20 kV. The energy is transmitted 25 km over transmission lines whose resistance is 1.2 Ω/km. (a) What is the power loss in the lines if the energy is transmitted at 20 kV? (b) What should be the output voltage of the generator to decrease the power loss by a factor of 15? (a) 75 kW (b) 77 kV

69. ●● Electrical power is transmitted through a power line 175 km long with a resistance of 1.2 Ω/km. The generator's output is 50 A at its operating voltage of 440 V. This voltage has a single step-up for transmission at 44 kV. (a) How much power is lost as joule heat during the transmission? (b) What must be the turn ratio of a transformer at the delivery point in order to provide an output voltage of 220 V? (Neglect the voltage drop in the line.) (a) 53 W (b) $N_p/N_s = 200$

20.4 Electromagnetic Waves

70. **MC** Relative to the blue end of the visible spectrum, the yellow and green regions have (a) higher frequencies, (b) longer wavelengths, (c) shorter wavelengths, or (d) both a and c. (b)

71. **MC** Which of the following electromagnetic waves has the lowest frequency? (a) UV (b) IR (c) X-ray or (d) microwave (d)

72. **MC** Which of the following electromagnetic waves travels fastest in a vacuum? (a) green light (b) infrared light (c) gamma rays (d) radiowaves or (e) they all have the same speed (e)

73. **MC** If you doubled the frequency of a blue light source, what kind of light would it then put out? (a) red (b) blue (c) violet or (d) UV (e) X-ray (d)

74. **CQ** An antenna is connected to a car battery. Will the antenna emit electromagnetic radiation? Why or why not? Explain.
no, dc current cannot generate a varying magnetic field

75. **CQ** On a cloudy summer day, you work outside and feel cool, yet that evening you find that you are sunburned. Explain how this is possible. UV radiation causes sunburn and it can still get through the clouds, see ISM

76. **CQ** Radiation exerts pressure on surfaces on which it falls (*radiation pressure*). Will this pressure be greater on a shiny surface or a dark surface? Will it be greater using a bright source, or one of the same color but fainter? Explain both of your answers.
shiny and more if brighter, see ISM

77. **CQ** Radar operates at wavelengths on the order of centimeters, whereas FM radio operates at wavelengths of several meters. How do radar frequencies compare with the frequencies of the FM band on a radio? How do their speeds in a vacuum compare? see ISM

78. ● Find the frequencies of electromagnetic waves with wavelengths of (a) 3.0 cm, (b) 650 nm, and (c) 1.2 fm. (d) Classify the type of light in each case. see ISM

79. ● In a small town there are only two AM radio stations, one at 920 kHz and one at 1280 kHz. What are the wavelengths of the radio waves transmitted by each station? 326 m and 234 m

80. ● A meteorologist in a TV station is using radar to determine the distance to a cloud. He notes that a time of 0.24 ms elapses between the sending and the return of a radar pulse. How far away is the cloud? 36 km

81. ● How long does a laser beam take to travel from the Earth to a reflector on the Moon and back? Take the distance from the Earth to the Moon to be 2.4×10^5 mi. (This experiment was done when the *Apollo* flights of the early 1970s left laser reflectors on the lunar surface.) 2.6 s

82. ●● Orange light has a wavelength of 600 nm, and green light has a wavelength of 510 nm. (a) What is the difference in frequency between the two types of light? (b) If you doubled the wavelength of both of these, what type of light would they become? (a) 8.8×10^{13} Hz (b) IR

83. ●● A certain type of radio antenna is called a *quarter-wavelength antenna*, because its length is equal to one quarter of the wavelength to be received. If you were going to make such antennae for the AM and FM radio bands by using the middle frequencies of each band, what lengths of wire would you use? AM: 67 m; FM: 0.77 m

84. **IE** ●●● Microwave ovens can have cold spots and hot spots, due to standing electromagnetic waves, analogous to standing wave nodes and antinodes in strings (▼Fig. 20.34). (a) The longer the distance between the cold spots, (1) the higher the frequency, (2) the lower the frequency, (3) the frequency is independent of this distance. Why? (b) In your microwave the cold spots (nodes) are approximately every 5.0 cm, and your neighbor's microwave produces them every 6.0 cm. Which microwave operates at a higher frequency and by how much? (a) (2) the lower the frequency (b) your microwave, 5.0×10^8 Hz

▲ **FIGURE 20.34 Cold spots?** See Exercise 84.

Comprehensive Exercises

85. **IE** In ▼Fig. 20.35, a metal bar of length L moves in a region of constant magnetic field. That field is directed into the page. (a) The direction of the induced current through the resistor is (1) up, (2) down, (3) there is no current. Why? (b) If the magnitude of the magnetic field is 250 mT, what is the current? (a) (1) up (b) 25 mA

86. Suppose a modern power plant has a 1.00-GW electric power output at 500 V. It is stepped up to a transmission voltage of 750 kV in a series of five identical transformers. (a) What is the output current at the plant? (b) What is the turn ratio in each of the transformers? (c) What is the current in the high voltage delivery lines?
(a) 2.00×10^6 A (b) 1:4.32 (c) 1.33×10^3 A

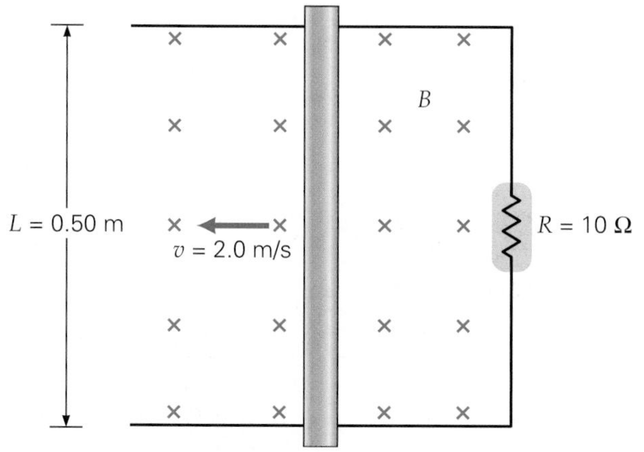

▲ **FIGURE 20.35 Motional emf** See Exercise 85.

87. A transformer is used by a European traveler while she is visiting the United States. Primarily she uses it to run a 1200-watt hair dryer/blower she brought with her. When plugged in to her hotel room outlet in Los Angeles, she notices it runs *exactly* as it does at home. The input voltage and current are 120 V and 11.0 A, respectively. (a) Is this an ideal transformer? Explain how you came to your conclusion. (b) If it is not an ideal transformer, what is its efficiency? (a) no, input power is greater than output power (b) 90.9%

88. A solenoid of length 20.0 cm is made of 5000 circular coils. It carries a steady current of 10.0 A. Near its center is placed a very flat and small coil made of 100 circular loops, each with a radius of 3.00 mm. This small coil is oriented so that its area receives the maximum magnetic flux. A switch is opened in the solenoid circuit and its current drops to zero in 15.0 ms. (a) What was the initial magnetic flux through the inner coil? (b) Determine the average induced emf in the small coil during the 15.0 ms. (c) If you look along the long axis of the solenoid so the initial 10.0 A current is clockwise, determine the direction of the induced current in the small inner coil during the time the current drops to zero. (a) 3.55×10^{-7} T·m² (b) 2.37×10^{-3} V (c) clockwise

89. **IE** A flat coil of copper wire consists of 100 loops and has a total resistance of 0.500 Ω. The coil diameter is 4.00 cm and it is in a uniform magnetic field pointing away from you (into the page). Its orientation is in the plane of the page. It is then pulled to the right and completely out of the field. (a) What is the direction of the induced current in the coil: (a) clockwise, (b) counterclockwise, or (c) there is no induced current? (b) During the time the coil leaves the field, an average induced current of 10.0 mA is measured. What is the average induced emf in the coil? (c) If the field strength is 3.50 mT, how much time did it take to pull the coil out? (a) (1) clockwise (b) 5.00×10^{-3} V (c) 0.0879 s

90. The transformer on a utility pole steps the voltage down from 20 000 V to 220 V for use in a college science building. During the day, the transformer delivers electric energy at the rate of 6.60 kW. (a) Assuming the transformer to be ideal, during that time what are the primary and secondary currents in the transformer? (b) If the transformer is only 95% efficient (but still delivers energy at a rate of 6.60 kW at 220 V), how does its input current compare to the ideal case? (c) At what rate is heat lost in the nonideal transformer? (a) 0.330 A, 30.0 A (b) it is larger, 0.348 A (c) 350 W

91. Suppose you wanted to build an electric generator using the Earth's magnetic field. Assume it has a strength of 0.040 mT at your location. Your generator design calls for a coil of 1000 windings rotated at exactly 60 Hz. The coil is oriented so that the normal to the area lines up with the Earth's field at the end of each cycle. What must the coil diameter be to generate a maximum voltage of 170 V (required in order to average 120 V)? Does this seem like a practical way to generate electric energy? 3.79 m, no

92. A radio signal is sent to a deep space probe traveling in the plane of the solar system. 3.5 days later, Earth receives a response. Assuming the probe computers took 4.5 hours to process the signal instructions and to send out the return, is the probe within the Solar System? (Assume a Solar System radius of about 40 times the Earth–Sun distance.) no, see ISM

93. A 100-loop coil of wire with a diameter of 2.50 cm is oriented in a 0.250-T constant magnetic field so that it initially has no magnetic flux. In 0.115 sec, the coil is turned so that its normal makes an angle of 45° with the field direction. If a 4.75-mA average current is induced in the coil during this rotation, what is the coil's resistance? 0.159 Ω

The following Physlet Physics Problems can be used with this chapter.

PHYSLET® 29.1, 29.2, 29.5, 29.6, 29.7, 29.8

AC CIRCUITS

PHYSICS FACTS

- Under ac conditions (alternating voltage direction), a capacitor, even with the gap between the plates, allows current in the circuit during the charging and discharging stages. Under dc conditions (steady voltage across the plates), there is no current.

- Under dc voltages, an inductor offers no impedance to the flow of charge and thus can readily conduct current. However, under ac conditions, an inductor impedes the change in current by producing a reverse emf in accordance with Faraday's law of induction.

- A capacitor, inductor, and resistor connected in series to an ac power supply is analogous to a mechanical damped, driven spring–mass system. Driven at its natural frequency, the circuit "resonates," that is, exhibits a current maximum, just as the mechanical system has its largest amplitude under the same condition. The capacitor stores electric potential energy in analogy to the spring's elastic potential energy. The inductor stores magnetic energy (associated with moving charges), in analogy to the mass on the spring having kinetic (moving) energy. The resistor will dissipate the system's energy, as air resistance might do to the mechanical system.

Direct-current (dc) circuits have many uses, but the nuclear reactor control panel in the photo operates many devices that use alternating current (ac). The electric power delivered to our homes and offices is also ac, and most everyday devices and appliances require alternating current.

There are several reasons for our reliance on alternating current. For one thing, almost all electric energy generators produce electric energy using electromagnetic induction, and thus produce ac outputs (Chapter 20). Furthermore, electrical energy produced in ac fashion can be transmitted economically over long distances through the use of transformers. But perhaps the most important reason is that ac currents produce electromagnetic effects that can be exploited in a variety of devices. For example, when you tune a radio to a station, you take advantage of a special *resonance* property of ac circuits (studied in this chapter).

To determine currents in dc circuits, resistance values were of main concern. There is, of course, resistance present in ac circuits as well, but additional factors can affect the flow of charge. For instance, a capacitor in a dc circuit is equivalent to an infinite resistance (an open circuit). However, in an ac circuit the alternating voltage continually charges and discharges a capacitor. Under such conditions, current can exist in a circuit even if it contains a capacitor. Moreover, wrapped coils of wire can oppose an ac current through the principles electromagnetic induction (Lenz's law; Section 20.1).

In this chapter, the principles of ac circuits will be studied. More generalized forms of Ohm's law and expressions for power, applicable to ac circuits, will be developed. Finally, the phenomenon and uses of *circuit resonance* will be explored.

21.1 Resistance in an AC Circuit

OBJECTIVES: To (a) specify how voltage, current, and power vary with time in an ac circuit, (b) understand the concepts of rms and peak values, and (c) learn how resistors respond under ac conditions.

An ac circuit contains an ac voltage source (such as a small generator or simply a household outlet) and one or more elements. An ac circuit with a single resistive element is shown in ▶Fig. 21.1. If the source's output voltage varies sinusoidally, as that from a generator (see Section 20.2), the voltage across the resistor varies with time in accordance with the equation

$$V = V_o \sin \omega t = V_o \sin 2\pi ft \tag{21.1}$$

where ω is the angular frequency of the voltage (in rad/s) and is related to its frequency f (in Hz) by $\omega = 2\pi f$. The voltage oscillates between $+V_o$ and $-V_o$ as $\sin 2\pi ft$ oscillates between ± 1. The voltage V_o, called the **peak** (or *maximum*) **voltage**, represents the amplitude of the voltage oscillations.

AC Current and Power

Under ac conditions, the current through the resistor oscillates in direction and magnitude. From Ohm's law, the ac current in the resistor, as a function of time, is given by

$$I = \frac{V}{R} = \left(\frac{V_o}{R}\right) \sin 2\pi ft$$

Because V_o represents the peak voltage across the resistor, the expression in parentheses represents the maximum current in the resistor. Thus, this expression can be rewritten as

$$I = I_o \sin 2\pi ft \tag{21.2}$$

where the amplitude of the current is $I_o = V_o/R$ and is called the **peak** (or *maximum*) **current**.

▶Figure 21.2 shows both current and voltage as functions of time for a resistor. Note that they are in step, or *in phase*. That is, both reach their zero, minimum, and maximum values at the same time. The current oscillates and takes on both positive and negative values, indicating its directional changes during each cycle. Because the current spends equal time in both directions, *the average current is zero*. Mathematically, this is because the time-averaged value of the sine function over one or more *complete* (360°) cycles is zero. Using overbars to denote a time-averaged value, we have $\overline{\sin \theta} = \overline{\sin 2\pi ft} = 0$. Similarly, $\overline{\cos \theta} = 0$.

Even though the *average* current is zero, this does not mean that there is no joule heating (I^2R losses). This is because the dissipation of electrical energy in a resistor does not depend on the current's direction. The instantaneous power as a function of time is obtained from the instantaneous current (Eq. 21.2). Thus,

$$P = I^2R = (I_o^2 R) \sin^2 2\pi ft \tag{21.3}$$

Even though the current changes sign, the *square* of the current, I^2, is always positive. Thus the average value of I^2R is *not* zero. The average, or mean, value of I^2 is

$$\overline{I^2} = \overline{I_o^2 \sin^2 2\pi ft} = I_o^2 \overline{\sin^2 2\pi ft}$$

Using the trigonometric identity $\sin^2 \theta = \frac{1}{2}(1 - \cos 2\theta)$, we obtain $\overline{\sin^2 \theta} = \frac{1}{2}(1 - \overline{\cos 2\theta})$. Because $\overline{\cos 2\theta} = 0$ (just as $\overline{\cos \theta} = 0$), it follows that $\overline{\sin^2 \theta} = \frac{1}{2}$. Thus the previous expression for $\overline{I^2}$ can be rewritten as

$$\overline{I^2} = I_o^2 \overline{\sin^2 2\pi ft} = \frac{1}{2}I_o^2 \tag{21.4}$$

The average power is therefore

$$\overline{P} = \overline{I^2}R = \frac{1}{2}I_o^2 R \tag{21.5}$$

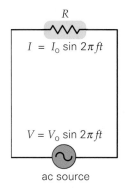

▲ **FIGURE 21.1 A purely resistive circuit** The ac source delivers a sinusoidal voltage to a circuit consisting of a single resistor. The voltage across, and current in, the resistor vary sinusoidally at the frequency of the applied ac voltage.

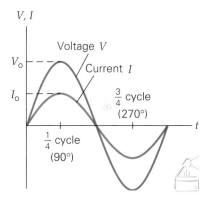

▲ **FIGURE 21.2 Voltage and current in phase** In a purely resistive ac circuit, the voltage and current are in step, or in phase.

Demonstration/activity: Watching the bulb of a lamp plugged into a wall socket, we cannot see current, voltage, or power oscillate, because the frequency is too high for our eyes to detect and the filament does not cool off significantly during current fluctuations. However, small LED bulbs will show the power fluctuation if you swing them in a circle in a dark room. The faster they move, the longer is the length of arc between the pulses of light, and you can estimate the frequency of the pulses (120 Hz; why?) from estimates of the distance between pulses, the radius of the circle, and the period of the orbit.

Teaching tip: An alternative argument for determining the average of the square of the sine or cosine function is as follows: $\sin^2 \theta + \cos^2 \theta = 1$, so the averaging process yields $\overline{\sin^2 \theta} + \overline{\cos^2 \theta} = 1$. But because the sine and cosine functions are identical except for a shift of 90°, you expect identical results when you average each over a cycle. In other words, we have $\overline{\sin^2 \theta} = \overline{\cos^2 \theta}$, so $2\overline{\sin^2 \theta} = 1$ and $\overline{\sin^2 \theta} = \frac{1}{2}$.

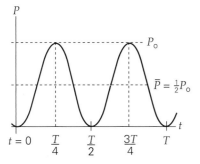

▲ **FIGURE 21.3** Power variation with time in a resistor Although both current and voltage oscillate in direction (sign), their product (power) is always a positive oscillating quantity. The average power is one-half the peak power.

Demonstration/activity: Connect a 10-V bulb (Christmas tree bulbs work well) to a dc power supply and adjust the output until the bulb glows. Connect an identical bulb to a high-power audio oscillator and adjust its output until that bulb glows with the same brightness. Then change the frequency of the ac oscillator to see if the brightness of the bulb changes. What happens when the frequency is very low? Use an oscilloscope to measure the potential across each bulb. How does the maximum potential across the ac bulb compare with the constant potential across the dc bulb?

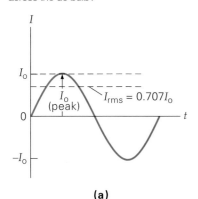

(a)

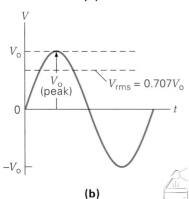

(b)

▲ **FIGURE 21.4** Root-mean-square (rms) current and voltage The rms values of (a) current and (b) voltage are 0.707, or $1/\sqrt{2}$, times the peak (maximum) values.

It should be emphasized that ac power has the same form as dc power ($P = I^2 R$) and is *valid at all times*. It is customary, however, to work with average power and a special kind of "average" current defined as follows:

$$I_{rms} = \sqrt{\overline{I^2}} = \sqrt{\tfrac{1}{2}I_o^2} = \frac{I_o}{\sqrt{2}} = \frac{\sqrt{2}}{2}I_o = 0.707 I_o \qquad (21.6)$$

I_{rms} is called the **rms current**, or **effective current**. (Here, *rms* stands for *root-mean-square*, indicating the square *root* of the *mean* value of the *square* of the current.) The rms current represents the value of a steady (dc) current required to produce the same power as its ac current counterpart, hence the name *effective* current.

Using $I_{rms}^2 = \left(I_o/\sqrt{2}\right)^2 = \tfrac{1}{2}I_o^2$, we can rewrite the average power (Eq. 21.5) as

$$\overline{P} = \tfrac{1}{2}I_o^2 R = I_{rms}^2 R \quad \text{(time-averaged power)} \qquad (21.7)$$

The average power is just the time-varying (oscillating) power averaged over time (◄Fig. 21.3).

AC Voltage

The peak values of voltage and current for a resistor are related by $V_o = I_o R$. Using a development similar to that for rms current, we define the **rms voltage**, or **effective voltage**, as

$$V_{rms} = \frac{V_o}{\sqrt{2}} = \frac{\sqrt{2}}{2}V_o = 0.707 V_o \qquad (21.8)$$

For resistors under ac conditions, then, dc relationships can be used—as long as we realize that the quantities represent rms values. Thus for ac situations involving only a resistor, the relationship between rms values of current and voltage is

$$V_{rms} = I_{rms}R \quad \text{voltage across a resistor} \qquad (21.9)$$

Combining Eqs. 21.9 and 21.7, we have several physically equivalent expressions for ac power:

$$\overline{P} = I_{rms}^2 R = I_{rms}V_{rms} = \frac{V_{rms}^2}{R} \quad \text{ac power} \qquad (21.10)$$

It is customary to measure and specify rms values when dealing with ac quantities. For example, the household line voltage of 120 V, as you may have guessed, is really the rms value of the voltage. It actually has a peak value of

$$V_o = \sqrt{2}\,V_{rms} = 1.414(120 \text{ V}) = 170 \text{ V}$$

Visual interpretations of peak and rms values of current and voltage are shown in ◄Fig. 21.4.

Example 21.1 ■ A Bright Lightbulb: Its RMS and Peak Values

A lamp with a 60-W bulb is plugged into a 120-V outlet. (a) What are the rms and peak currents through the lamp? (b) What is the resistance of the bulb under these conditions?

Thinking It Through. (a) Because the average power and rms voltage are known, we can find the rms current from Eq. 21.10. From the rms current, Eq. 21.6 can be used to calculate the peak current. (b) The resistance is found from Eq. 21.9.

Solution. The average power and the rms voltage of the source are given.

Given: $\overline{P} = 60$ W $\qquad$ *Find:* (a) I_{rms} and I_o (rms and peak currents)
$\qquad\quad V_{rms} = 120$ V $\qquad\qquad\qquad$ (b) R (bulb resistance)

(a) The rms current is

$$I_{rms} = \frac{\overline{P}}{V_{rms}} = \frac{60\ W}{120\ V} = 0.50\ A$$

and the peak current is determined by rearranging Eq. 21.6:

$$I_o = \sqrt{2}I_{rms} = \sqrt{2}(0.50\ A) = 0.71\ A$$

(b) The resistance of the bulb is

$$R = \frac{V_{rms}}{I_{rms}} = \frac{120\ V}{0.50\ A} = 240\ \Omega$$

Follow-Up Exercise. What would be the (a) rms current and (b) peak current in a 60-W light-bulb in Great Britain, where the house rms voltage is 240 V at 50 Hz? (c) How would the resistance of a 60-W bulb in Great Britain compare with one designed for operation at 120 V? Why are the two resistances different? (*Answers to all Follow-Up Exercises are at the back of the text.*)

Conceptual Example 21.2 ■ Across the Pond: British versus American Electrical Systems

In many countries, the line voltage is 240 V. If a British tourist visiting in the United States plugged in a hair dryer from home (where the voltage is 240 V), you would expect it (a) not to operate, (b) to operate normally, (c) to operate poorly, or (d) to burn out.

Reasoning and Answer. British appliances operate at 240 V. At a decreased voltage (namely 120 V), there would be decreased current ($I = V/R$) and reduced joule heating (because $P = IV$). If the resistance of the appliance were constant, then, at half the voltage, there would be only one-fourth the power output. Thus the heating element of the hair dryer might get warm, but it would not work as expected, so the answer is (c). In addition, the decreased current could cause the motor to run slower than normal.

When on foreign travel, most people do not make this mistake. This is because plugs and sockets vary from country to country. If you are traveling with appliances from home, a converter/adapter kit can be useful (▸Fig. 21.5). This kit contains a selection of plugs for adapting to the foreign sockets, as well as a voltage converter. The converter is a device that converts 240 V to 120 V for U.S. travelers and vice versa for tourists visiting the United States who have 240 V electrical supplies at home. (There are appliances that can be switched between 120 V and 240 V.)

Follow-Up Exercise. What happens if an American tourist inadvertently plugs a 120-V appliance into a British 240-V outlet without a converter? Explain.

21.2 Capacitive Reactance

OBJECTIVES: To (a) explain the behavior of capacitors in ac circuits, and (b) calculate the effect of a capacitor on ac current (capacitive reactance).

In Chapter 16, we learned that when a capacitor is connected to a dc voltage source, current exists only for the short time required to charge the capacitor. As charge accumulates on the capacitor's plates, the voltage across them increases, opposing the external voltage and reducing the current. When the capacitor is fully charged, the current drops to zero.

Things are different when a capacitor is driven by an ac voltage source (▸Fig. 21.6a). Under these conditions, the capacitor limits the current, but doesn't completely prevent the flow of charge. This is because the capacitor is alternately charged and discharged as the current and voltage reverse each half-cycle.

Plots of ac current and voltage versus time for a circuit with just a capacitor are shown in Fig. 21.6b. Let's look at the changing conditions of the capacitor with time (▾Fig. 21.7).

- In Fig. 21.7a, $t = 0$ is arbitrarily chosen as the time of maximum voltage ($V = V_o$)*. At the start, the capacitor is assumed fully charged ($Q_o = CV_o$) with the polarity shown. Because the plates cannot accommodate more charge, there is no current in the circuit.

*We have arbitrarily chosen the polarity of initial capacitor voltage as positive (Figs. 21.6b and 21.7a).

Illustration 31.2 AC Voltage and Current

▲ **FIGURE 21.5** Converter and adapters In countries that have 240-V line voltages, U.S. tourists need conversion to 120 V to operate normal U.S. appliances properly. Note the different types of plugs for different countries. The small plugs go into the foreign sockets, and the converter prongs fit into the back of a socket. A U.S. standard two-prong plug fits into the converter, which has a 120-V output. (See Conceptual Example 21.2.)

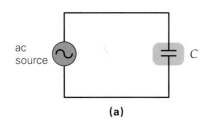

(a)

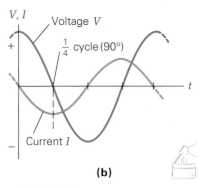

(b)

▲ **FIGURE 21.6** A purely capacitive circuit **(a)** In a circuit with only capacitance, **(b)** the current leads the voltage by 90°, or one quarter cycle. Half of a cycle of voltage and current, shown, corresponds to Fig. 21.7.

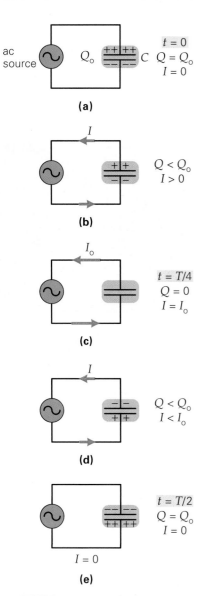

▲ **FIGURE 21.7 A capacitor under ac conditions** This sequence shows the voltage, charge, and current in a circuit containing only a capacitor and an ac voltage source. All five circuit diagrams taken together represent physically what is plotted in the first half of the cycle (from $t = 0$ to $t = T/2$) in the graph shown in Fig. 21.6b.

Teaching tip: Ask students to use unit analysis to show that capacitive reactance is measured in ohms.

- As the voltage decreases so that $0 < V < V_o$, the capacitor begins to discharge, giving rise to a counterclockwise current (labeled negative, compare Fig. 21.6b to Fig. 21.7b).
- The current reaches its maximum value when the voltage drops to zero and the capacitor plates are completely discharged (Fig. 21.7c). This occurs one quarter of the way through the cycle ($t = T/4$).
- The ac voltage source now reverses polarity and starts to increase in magnitude, so that $-V_o < V < 0$. The capacitor begins to charge, this time with the opposite polarity (Fig. 21.7d). With the plates uncharged, there is no opposition to the current, so the current is at its maximum value. However, as the plates accumulate charge, they begin to inhibit the current, and the current decreases in magnitude.
- Halfway through the cycle ($t = T/2$), the capacitor is fully charged, but opposite in polarity to its starting condition (Fig. 21.7e). The current is zero, and the voltage is at its maximum magnitude, but opposite the initial polarity ($V = -V_o$).

During the next half-cycle (not shown in Fig. 21.7), the process is reversed and the circuit returns to its initial condition.

Note that the current and voltage are *not* in step (that is, *not* in phase). The current reaches its maximum a quarter-cycle *ahead* of the voltage. The phase relationship between the current and the voltage for a capacitor is commonly stated this way:

In a purely capacitive ac circuit, the *current* leads the *voltage* by 90°, or one-quarter $\left(\frac{1}{4}\right)$ cycle.

Thus in an ac situation, a capacitor provides opposition to the charging process, but it is not totally limiting. (Recall that under dc conditions it behaves as an open circuit). The quantitative measure of this "capacitive opposition" to current is called the capacitor's **capacitive reactance (X_C)**. In an ac circuit, capacitive reactance is given by

$$X_C = \frac{1}{\omega C} = \frac{1}{2\pi f C} \quad \textit{capacitive reactance} \quad (21.11)$$

SI unit of capacitive reactance:
ohm (Ω), or second per farad (s/F)

where, as usual, $\omega = 2\pi f$, C is the capacitance (in farads), and f is the frequency (in Hz). Like resistance, reactance is measured in ohms (Ω). Using unit analysis, you should be able to show that the ohm is equivalent to the second per farad.

Equation 21.11 shows that the reactance is inversely proportional to both the capacitance (C) and the voltage frequency (f). Both of these dependencies can be understood physically as follows.

Recall that *capacitance* means "charge stored per volt" ($C = Q/V$). Therefore, for a particular voltage, the greater the capacitance, the more charge the capacitor can accommodate. This requires a larger charge flow rate, or current. Increasing the capacitance offers less opposition to charge flow (that is, a reduced capacitive reactance) at a given frequency.

To see the frequency dependence, consider the fact that the greater the frequency of the voltage, the *shorter* the time for charging each cycle. A shorter charging time means less charge will be able to accumulate on the plates and there will be less opposition to the current. Increasing the frequency results in a decrease in capacitive reactance. Hence capacitive reactance is inversely proportional to *both* frequency and capacitance.

It is always good to check a general relationship to see if it gives a result you know to be true in a special case [or several special case(s)]. As a special case for the capacitor, note that if $f = 0$ (that is, nonoscillating dc conditions), the capacitive reactance is infinite. As expected under such conditions, there would be no current.

Capacitive reactance is related to the voltage across the capacitor and the current by an equation that has the same form as $V = IR$ for pure resistances:

$$V_{rms} = I_{rms} X_C \quad \textit{voltage across a capacitor} \quad (21.12)$$

Consider the next Example: a capacitor connected to an ac voltage source.

Example 21.3 ■ Current under ac Conditions: Capacitive Reactance

A 15.0-μF capacitor is connected to a 120-V, 60-Hz source. What are (a) the capacitive reactance and (b) the current (rms and peak) in the circuit?

Thinking It Through. The capacitive reactance can be obtained from the capacitance and the frequency by using Eq. 21.11. (b) We can then calculate the rms current from the reactance and rms voltage via Eq. 21.12. Finally, Eq. 21.6 gives the peak current.

Solution. Assuming the 60-Hz frequency to be exact, our answers are to three significant figures. We list the data:

Given: $C = 15.0\ \mu\text{F} = 15.0 \times 10^{-6}\ \text{F}$ *Find:* (a) X_C (capacitive reactance)
$V_{\text{rms}} = 120\ \text{V}$ (b) I_o (peak current), I_{rms} (rms current)
$f = 60\ \text{Hz}$

(a) The capacitive reactance is

$$X_C = \frac{1}{2\pi f C} = \frac{1}{2\pi (60\ \text{Hz})(15.0 \times 10^{-6}\ \text{F})} = 177\ \Omega$$

(b) Then, the rms current is

$$I_{\text{rms}} = \frac{V_{\text{rms}}}{X_C} = \frac{120\ \text{V}}{177\ \Omega} = 0.678\ \text{A}$$

and therefore, the peak current is

$$I_o = \sqrt{2} I_{\text{rms}} = \sqrt{2}(0.678\ \text{A}) = 0.959\ \text{A}$$

The current oscillates at 60 cycles per second with a magnitude of 0.959 A.

Follow-Up Exercise. In this Example, (a) what is the peak voltage and (b) what frequency would give the same current if the capacitance were reduced by half?

21.3 Inductive Reactance

OBJECTIVES: To (a) explain what an inductor is, (b) explain the behavior of inductors in ac circuits, and (c) calculate the effect of inductors on ac current (inductive reactance).

Inductance is a measure of the opposition a circuit element presents to a time-varying current (by Lenz's law). In principle, all circuit elements (even resistors) have some inductance. However, a coil of wire with negligible resistance has, in effect, only inductance. When placed in a circuit with a time-varying current, such a coil, called an **inductor**, exhibits a reverse voltage, or back emf, in opposition to the changing current. The changing current through the coil produces a changing magnetic field and flux. The back emf is the induced emf in opposition to this changing flux. Because the back emf is induced in the inductor as a result of its own changing magnetic field, this phenomenon is called *self-induction*.

The self-induced emf (for a coil consisting of N loops) is given by Faraday's law (Eq. 20.2): $\mathcal{E} = -N\Delta\Phi/\Delta t$. The time rate of change of the total flux through the coil, $N\Delta\Phi/\Delta t$, is proportional to the rate of change of the current in the coil, $\Delta I/\Delta t$. This is because the current produces the magnetic field responsible for the changing flux. Thus the back emf is proportional to, and oppositely directed to, the rate of current change. This relationship is expressed using a proportionality constant, L:

$$\mathcal{E} = -L\left(\frac{\Delta I}{\Delta t}\right) \tag{21.13}$$

where L is the *inductance* of the coil (more properly, its *self*-inductance). You should be able to show, using unit analysis, that the units of inductance are volt-seconds per ampere (V·s/A). This combination is called a **henry (H**, 1 H = 1 V·s/A), in honor of Joseph Henry (1797–1878), an American physicist and early investigator of electromagnetic induction. Smaller units, such as the millihenry (mH), are commonly used (1 mH = 10^{-3} H).

Note: Lenz's law is given in Section 20.1.

Teaching tip: Explain that Henry discovered electromagnetic induction independently at about the same time as Michael Faraday did, but Faraday published his findings first.

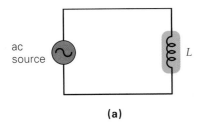

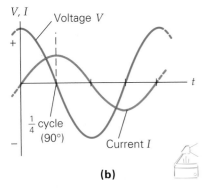

▲ **FIGURE 21.8** A purely inductive circuit **(a)** In a circuit with only inductance, **(b)** the voltage leads the current by 90°, or one quarter cycle.

Illustration 31.4 *Phase Shift*

Exploration 31.2 *Reactance*

The opposition presented to current by an inductor under ac conditions depends on the inductance and the voltage frequency. This is expressed quantitatively by the circuit's **inductive reactance (X_L)**, which is

$$X_L = \omega L = 2\pi f L \quad \textit{inductive reactance} \tag{21.14}$$

SI unit of inductive reactance:
ohm (Ω), or henry per second (H/s)

where f is the frequency of the driving voltage, $\omega = 2\pi f$, and L is the inductance. Like capacitive reactance, inductive reactance is measured in ohms (Ω), which you should be able to show are equivalent to henrys per second.

Note that the inductive reactance is proportional to both the coil inductance (L) and the voltage frequency (f). The inductance is a property of the coil that depends on the number of turns, the coil's diameter and length, and the material of the core (if any). The frequency of the voltage plays a role because the more rapidly the current in the coil changes, the greater the rate of change of its magnetic flux. This implies a larger self-induced (back or reverse) emf to oppose the changes in current.

In terms of X_L, the voltage across an inductor is related to the current and inductive reactance by:

$$V_{rms} = I_{rms}X_L \quad \textit{voltage across an inductor} \tag{21.15}$$

The circuit symbol for an inductor and the graphs of the voltage across the inductor and the current in the circuit are shown in ◄Fig. 21.8. When an inductor is connected to an ac voltage source, maximum voltage corresponds to zero current. When the voltage drops to zero, the current is maximum. This happens because, as the voltage changes polarity (causing the magnetic flux through the inductor to drop to zero), the inductor acts to prevent the change in accordance with Lenz's law, so the induced emf creates a current. In an inductor, the current *lags* one quarter cycle behind the voltage, a relationship commonly expressed as follows:

In a purely inductive ac circuit, the *voltage* leads the *current* by 90°, or one quarter ($\frac{1}{4}$) cycle.

Because the phase relationships between current and voltage for purely inductive and purely capacitive circuits are opposite, there is a phrase that may help you remember the difference: *ELI* the *ICE* man. Here E represents voltage (for *e*mf) and I represents current. The three letters *ELI* indicate that for inductance (L), the voltage leads the current (I)—reading the acronym from left to right. Similarly, *ICE* means that for capacitance (C), the current leads the voltage.

Example 21.4 ■ Current Opposition without Resistance: Inductive Reactance

A 125-mH inductor is connected to a 120-V, 60-Hz source. What are (a) the inductive reactance and (b) the rms current in the circuit?

Thinking It Through. Because the inductance and frequency are known, we can compute the inductive reactance from Eq. 21.14 and the current from Eq. 21.15.

Solution. Listing the given data,

Given: $L = 125$ mH $= 0.125$ H   *Find:* (a) X_L (inductive reactance)
        $V_{rms} = 120$ V   (b) I_{rms}
            $f = 60$ Hz

(a) The inductive reactance is

$$X_L = 2\pi f L = 2\pi(60 \text{ Hz})(0.125 \text{ H}) = 47.1 \ \Omega$$

(b) The rms current is then

$$I_{rms} = \frac{V_{rms}}{X_L} = \frac{120 \text{ V}}{47.1 \ \Omega} = 2.55 \text{ A}$$

Follow-Up Exercise. In this Example, (a) what is the peak current? (b) What voltage frequency would yield the same current if the inductance were reduced to one third the value in this Example?

21.4 Impedance: RLC Circuits

OBJECTIVES: To (a) calculate currents and voltages when a combination of reactive and resistive circuit elements are present in ac circuits, (b) use phase diagrams to calculate overall impedance and rms currents, and (c) understand and use the concept of the power factor in ac circuits.

In the previous sections purely capacitive or purely inductive circuits were considered separately and without resistance present. However, in the real world, it is impossible to have purely reactive circuits, because there is always some resistance—at a minimum, that from the connecting wires. Thus resistances, capacitive reactances, and inductive reactances *combine* to impede the current in ac circuits. An analysis of some combination circuits illustrates these effects.

Series RC Circuit

Suppose an ac circuit consists of a voltage source, a resistor, and a capacitor connected in series (►Fig. 21.9a). The phase relationship between the current and the voltage is different for each circuit element. As a result, a special graphical method is needed to find the overall opposition to the current in the circuit. This method employs a *phase diagram*.

In a phase diagram, such as in Fig. 21.9b for an RC circuit, the resistance and reactance in the circuit are endowed with vectorlike properties and their magnitudes represented by arrows called *phasors*. On a set of x–y coordinate axes, the resistance is plotted on the positive x-axis (that is, at $0°$), because the voltage–current phase difference for a resistor is zero. The capacitive reactance is plotted along the negative y-axis, to reflect a phase difference (ϕ) of $-90°$ because for a capacitor, the voltage lags behind the current by one quarter of a cycle.

The phasor sum is the effective, or net, opposition to the current, which we call the **impedance (Z)**. Phasors must be added in the same way as vectors because the effects of the resistor and capacitor are not in phase. For the series RC circuit,

$$Z = \sqrt{R^2 + X_C^2} \quad \textit{series RC circuit impedance} \tag{21.16}$$

The unit of impedance is the ohm.

The generalization of Ohm's law to circuits containing capacitors and inductors along with resistors is

$$V_{rms} = I_{rms}Z \quad \textit{Ohm's law for ac circuits} \tag{21.17}$$

To illustrate how phasors can be used to analyze an RC circuit, consider the next Example. Take particular note of part (b), in which there is an *apparent* violation of Kirchhoff's loop theorem—explained by phase differences between the voltages across the two elements in the circuit.

Example 21.5 ■ RC Impedance and Kirchhoff's Loop Theorem

A series RC circuit has a resistance of 100 Ω and a capacitance of 15.0 μF. (a) What is the (rms) current in the circuit when it is driven by a 120-V, 60-Hz source? (b) Compute the (rms) voltage across each circuit element and the two elements combined. Compare it with that of the voltage source. Is Kirchhoff's loop theorem satisfied? Comment and explain your reasoning.

Thinking It Through. (a) Note that the voltage and capacitor values are the same as those in Example 21.3 and a resistor has been added in series. This will help in reducing the necessary calculation. Then, using phasors, the capacitive reactance and the resistance can be combined to determine the overall impedance (Eq. 21.16). From Eq. 21.17, the impedance and voltage can be used to find the current. (b) Because the current is the same everywhere at any given time in a series circuit, the result of part (a) can be used to calculate the voltages. The rms voltage across both elements together is found by recalling that the individual voltages are out of phase by 90°. What this means physically is that they reach their peak values *not* at the same time, but rather one fourth of a period apart. Thus, we *cannot* simply add the voltages.

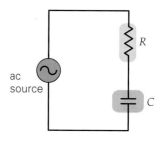

(a) RC circuit diagram

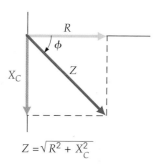

$$Z = \sqrt{R^2 + X_C^2}$$

(b) Phase diagram

▲ **FIGURE 21.9** A series RC circuit **(a)** In a series RC circuit, **(b)** the impedance Z is the phasor sum of the resistance R and the capacitive reactance X_C.

Note: *Impedance* (denoted by Z) refers to the overall circuit opposition to current. We reserve *reactance* and *resistance* for opposition in individual elements.

Note: $V_{rms} = I_{rms}Z$ can be applied to any circuit, as long as the impedance Z is properly calculated, using phasors.

(continues on next page)

Solution.

Given: $R = 100\ \Omega$
$\quad\quad\quad\ \ C = 15.0\ \mu\text{F} = 15.0 \times 10^{-6}\ \text{F}$
$\quad\quad\quad\ \ V_{\text{rms}} = 120\ \text{V}$
$\quad\quad\quad\ \ f = 60\ \text{Hz}$

Find: (a) I (rms current)
$\quad\quad$ (b) V_C (rms voltage across capacitor)
$\quad\quad\quad\ \ V_R$ (rms voltage across resistor)
$\quad\quad\quad\ \ V_{(R+C)}$ (combined rms voltage)

(a) In Example 21.3, we found that the reactance for this capacitor at this frequency was $X_C = 177\ \Omega$. Now Eq. 21.16 can be used to calculate the circuit impedance:

$$Z = \sqrt{R^2 + X_C^2} = \sqrt{(100\ \Omega)^2 + (177\ \Omega)^2} = 203\ \Omega$$

Because $V_{\text{rms}} = I_{\text{rms}}Z$, the rms current is

$$I_{\text{rms}} = \frac{V_{\text{rms}}}{Z} = \frac{120\ \text{V}}{203\ \Omega} = 0.591\ \text{A}$$

(b) Using Eq. 21.17 first for the rms voltage across the resistor alone ($Z = R$),

$$V_R = I_{\text{rms}}R = (0.591\ \text{A})(100\ \Omega) = 59.1\ \text{V}$$

For the capacitor alone ($Z = X_C$), the rms voltage across the capacitor is

$$V_C = I_{\text{rms}}X_C = (0.591\ \text{A})(177\ \Omega) = 105\ \text{V}$$

The algebraic sum of these two rms voltages is 164 V, which is *not* the same as the rms value of the voltage source (120 V). This does *not* mean that Kirchhoff's loop theorem has been violated. In fact, the source voltage does equal the combined voltages across the capacitor and resistor *if you account for phase differences*. The combined voltage must be calculated properly to take into account the 90° phase difference between the two voltages. Using the Pythagorean theorem to get the total voltage, we have

$$V_{(R+C)} = \sqrt{V_R^2 + V_C^2} = \sqrt{(59.1\ \text{V})^2 + (105\ \text{V})^2} = 120\ \text{V}$$

Thus when the individual voltages are combined properly (taking into account that the voltages do not peak at the same time), Kirchhoff's law are still valid. Here it has been shown that the total rms voltage across both elements is equal to the rms voltage of the source. We must take care to add voltages this way because they are out of phase in general. *Thus, Kirchhoff's laws are valid at any instant of time, not just for rms values, but care must be taken to account for phase differences.*

Follow-Up Exercise. (a) How would the result in part (a) of this Example change if the circuit were driven by a voltage source with the same rms voltage, but oscillating at 120 Hz? (b) Is the resistor or the capacitor responsible for the change?

Series RL Circuit

The analysis of a series RL circuit (◄Fig. 21.10) is similar to that of a series RC circuit. However, the inductive reactance is plotted along the *positive y*-axis in the phase diagram, to reflect a phase difference of $+90°$ with respect to the resistance. Remember that a positive phase angle means that the voltage *leads* the current, as is true for an inductor.

Thus the impedance in an RL series circuit is

$$Z = \sqrt{R^2 + X_L^2} \quad \textit{series RL circuit impedance} \quad (21.18)$$

Series RLC Circuit

More generally, an ac circuit may contain all three circuit elements—a resistor, an inductor, and a capacitor—as shown in series in ▶Fig. 21.11. Again, phasor addition must be used to determine the overall circuit impedance. Combining the vertical components (that is, inductive and capacitive reactances) gives the *total reactance*, $X_L - X_C$. Subtraction is used because the phase difference between X_L and X_C is 180°. The overall circuit impedance is the phasor sum of the resistance and the total reactance. Employing the Pythagorean theorem once more on the phasor diagram, we have

$$Z = \sqrt{R^2 + (X_L - X_C)^2} \quad \textit{series RLC circuit impedance} \quad (21.19)$$

The **phase angle (ϕ)** between the source voltage and the current in the circuit is the angle between the overall impedance phasor (Z) and the $+x$-axis (Fig. 21.11b), or

$$\tan \phi = \frac{X_L - X_C}{R} \quad \textit{phase angle in series RLC circuit} \quad (21.20)$$

PHYSLET®

Illustration 31.6 Voltage and Current Phasors

PHYSLET®

Illustration 31.7 RC Circuits and Phasors

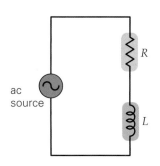

(a) RL circuit diagram

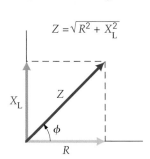

(b) Phase diagram

▲ **FIGURE 21.10** A series RL circuit (a) In a series RL circuit, (b) the impedance Z is the phasor sum of the resistance R and the inductive reactance X_L.

TABLE 21.1	Impedances and Phase Angles for Series Circuits	
Circuit Element(s)	Impedance Z (in Ω)	Phase Angle ϕ
R	R	$0°$
C	X_C	$-90°$
L	X_L	$+90°$
RC	$\sqrt{R^2 + X_C^2}$	negative (meaning that ϕ is between $0°$ and $-90°$)
RL	$\sqrt{R^2 + X_L^2}$	positive (meaning that ϕ is between $0°$ and $+90°$)
RLC	$\sqrt{R^2 + (X_L - X_C)^2}$	positive if $X_L > X_C$ negative if $X_C > X_L$

Notice that if X_L is greater than X_C (as in Fig. 21.11b), the phase angle is positive ($+\phi$), and the circuit is said to be *inductive*, because the nonresistive part of the impedance (that is, the reactance) is dominated by the inductor. If X_C is greater than X_L, the phase angle is negative ($-\phi$), and the circuit is said to be *capacitive*, because capacitive reactance dominates over inductive reactance.

A summary of impedances and phase angles for the three circuit elements and various combinations is given in Table 21.1. Example 21.6 analyzes an RLC circuit.

Example 21.6 ■ All Together Now: Impedance in an RLC Circuit

A series RLC circuit has a resistance of 25.0 Ω, a capacitance of 50.0 μF, and an inductance of 0.300 H. If the circuit is driven by a 120-V, 60-Hz source, what are (a) the total impedance of the circuit, (b) the rms current in the circuit, and (c) the phase angle between the current and the voltage?

Thinking It Through. (a) To calculate the overall impedance from Eq. 21.19, the individual reactances must first be determined. (b) The current is computed from the generalization of Ohm's law, $V_{rms} = I_{rms}Z$ (Eq. 21.17). (c) The phase angle is calculated from Eq. 21.20.

Solution. We are given all the necessary data:

Given: $R = 25.0$ Ω
$C = 50.0\ \mu F = 5.00 \times 10^{-5}$ F
$L = 0.300$ H
$V_{rms} = 120$ V
$f = 60$ Hz

Find: (a) Z (overall circuit impedance)
(b) I_{rms}
(c) ϕ (phase angle)

(a) The individual reactances are

$$X_C = \frac{1}{2\pi f C} = \frac{1}{2\pi(60\ \text{Hz})(5.00 \times 10^{-5}\ \text{F})} = 53.1\ \Omega$$

and

$$X_L = 2\pi f L = 2\pi(60\ \text{Hz})(0.300\ \text{H}) = 113\ \Omega$$

Then,

$$Z = \sqrt{R^2 + (X_L - X_C)^2} = \sqrt{(25.0\ \Omega)^2 + (113\ \Omega - 53.1\ \Omega)^2} = 64.9\ \Omega$$

(b) Because $V_{rms} = I_{rms}Z$, we have

$$I_{rms} = \frac{V_{rms}}{Z} = \frac{120\ \text{V}}{64.9\ \Omega} = 1.85\ \text{A}$$

(c) Solving $\tan\phi = (X_L - X_C)/R$ for the phase angle gives

$$\phi = \tan^{-1}\left(\frac{X_L - X_C}{R}\right) = \tan^{-1}\left(\frac{113\ \Omega - 53.1\ \Omega}{25.0\ \Omega}\right) = +67.3°$$

We should have expected a positive phase angle, because the inductive reactance is greater than the capacitive reactance [see part (a)]. Thus this circuit is *inductive* in nature.

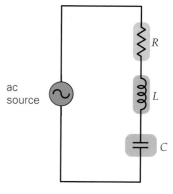

(a) RLC circuit diagram

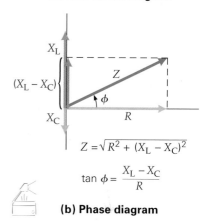

$$Z = \sqrt{R^2 + (X_L - X_C)^2}$$

$$\tan\phi = \frac{X_L - X_C}{R}$$

(b) Phase diagram

▲ **FIGURE 21.11** A series RLC circuit **(a)** In a series RLC circuit, **(b)** the impedance Z is the phasor sum of the resistance R and the total (or net) reactance ($X_L - X_C$). Note that the phasor diagram is drawn for the case of $X_L > X_C$.

(continues on next page)

Follow-Up Exercise. (a) Consider the RLC circuit in this Example, but with the driving frequency doubled. Reasoning conceptually, should the phase angle ϕ be greater or less than the $+67.3°$ after the increase? (b) Compute the new phase angle to show that your reasoning is correct.

By now, you should appreciate the usefulness of phasor diagrams in determining impedances, voltages, and currents in ac circuits. However, you might still be wondering what the use and meaning are of the phase angle ϕ. To illustrate its importance, let's examine the power loss in an RLC circuit. Note that this power analysis also depends on the use of phasor diagrams.

Power Factor for a Series RLC Circuit

In considering an RLC circuit, a crucial thing to realize is that any circuit power loss (joule heating) can take place only in the resistor. *There are no power losses associated with capacitors and inductors.* Capacitors and inductors simply store energy and give it back, without loss. Ideally, neither has any resistance, and thus any joule heating attributed to them is zero.

The average (rms) power dissipated by a resistor is $P_{rms} = I_{rms}^2 R$. This rms power can also be expressed in terms of the rms current and voltage, but the voltage *must be that across the resistor* (V_R), because it is the only dissipative element. The average power dissipated in a series RLC circuit can be alternatively expressed as

$$\overline{P} = P_R = I_{rms} V_R$$

The voltage across the resistor can be found from a voltage triangle that corresponds to the phasor triangle (◄Fig. 21.12). The rms voltages across the individual components in an RLC circuit are $V_R = I_{rms}R$, $V_L = I_{rms}X_L$, and $V_C = I_{rms}X_C$. Combining the last two voltages, we can write $(V_L - V_C) = I_{rms}(X_L - X_C)$. If each leg of the phasor triangle (Fig. 21.12a) is multiplied by the rms current, an equivalent voltage triangle results (Fig. 21.12b). As this figure shows, the voltage across the resistor is

$$V_R = V_{rms} \cos\phi \qquad (21.21)$$

The term $\cos\phi$ is called the **power factor**. From Fig. 21.11,

$$\cos\phi = \frac{R}{Z} \qquad \textit{series RLC power factor} \qquad (21.22)$$

The average power, rewritten in terms of the power factor, is

$$\overline{P} = I_{rms}V_{rms}\cos\phi \qquad \textit{series RLC power} \qquad (21.23)$$

Because power is dissipated only in the resistance ($\overline{P} = I_{rms}^2 R$), Eq. 21.22 enables us to express the average power as

$$\overline{P} = I_{rms}^2 Z \cos\phi \qquad \textit{series RLC power} \qquad (21.24)$$

Note that $\cos\phi$ varies from a maximum of $+1$ (when $\phi = 0°$) to a minimum of zero (when $\phi = \pm90°$). When $\phi = 0°$, the circuit is said to be *completely resistive*. That is, there is maximum power dissipation (as though the circuit contained only a resistor). The power factor decreases as the phase angle increases in either direction [because $\cos(-\phi) = \cos\phi$]—in other words, as the circuit becomes inductive or capacitive. At $\phi = +90°$, the circuit is *completely inductive*; at $\phi = -90°$, it is *completely capacitive*. In these cases, the circuit contains only an inductor or a capacitor, respectively, so no power is dissipated. In practice, because there is always some resistance, a circuit can never be completely inductive or capacitive. It is possible, however, for an RLC circuit to *appear* to be completely resistive even if it contains a capacitor and an inductor, as we shall see in Section 21.5. Let's look at our previous RLC example with an emphasis on power.

Teaching tip: Point out to students that in an LC circuit the energy oscillates between the capacitor and the inductor. The behavior is similar to that of a mass on a spring, in which the energy oscillates between the spring and the mass.

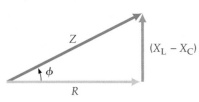

(a) Phasor triangle

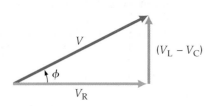

(b) Equivalent voltage triangle

▲ **FIGURE 21.12 Phasor and voltage triangles** The rms voltages across the components of a series RLC circuit are given by $V_R = I_{rms}R$, $V_L = I_{rms}X_L$, and $V_C = I_{rms}X_C$. Because the current is the same through each, **(a)** the phasor triangle can be converted to **(b)** a voltage triangle. Note that $V_R = V\cos\phi$. Both phasor diagrams are drawn for the case of $X_L > X_C$.

Illustration 31.5 *Power and Reactance*

Exploration 31.4 *Phase Angle and Power*

Example 21.7 ■ Power Factor Revisited

What is the average power dissipated in the circuit described in Example 21.6?

Thinking It Through. The power factor can be determined because the resistance (R) and impedance (Z) are known. Once the power factor is known, the actual power can be calculated.

Solution.

Given: See Example 21.6   *Find:* $\overline{P}$ (average power)

In Example 21.6, it was determined that the circuit had an impedance of $Z = 64.9\ \Omega$, and its resistance was $R = 25.0\ \Omega$. Therefore its power factor is

$$\cos\phi = \frac{R}{Z} = \frac{25.0\ \Omega}{64.9\ \Omega} = 0.385$$

Using the other data from Example 21.6 and Eq. 21.23 gives

$$\overline{P} = I_{rms}V_{rms}\cos\phi = (1.85\ \text{A})(120\ \text{V})(0.385) = 85.5\ \text{W}$$

This is less than the power that would be dissipated without a capacitor and an inductor. (Can you show this to be true? Why is it true?)

Follow-Up Exercise. If the frequency were doubled and the capacitor removed from this Example, what would be the rms power?

21.5 Circuit Resonance

OBJECTIVES: To (a) understand the concept of resonance in ac circuits, and (b) calculate the resonance frequency of an RLC circuit.

From the previous discussion, it can be seen that when the power factor ($\cos\phi$) of an RLC series circuit is equal to unity, maximum power is transferred to the circuit. In this situation, the current in the circuit must be maximum, because the impedance is at its minimum. This occurs because at this unique frequency, the inductive and capacitive reactances *effectively cancel*—that is, they are equal in magnitude and 180° out of phase, or opposite. This situation can happen in any RLC circuit if the appropriate source frequency is chosen.

The key to finding this frequency is to realize that because inductive and capacitive reactances are frequency dependent, so is the overall impedance. From the expression for the RLC series impedance, $Z = \sqrt{R^2 + (X_L - X_C)^2}$, it can be seen that the impedance is a minimum when $X_L - X_C = 0$. This occurs at a frequency f_o, found by setting $X_L = X_C$. Using the expressions for the reactances, this means that $2\pi f_o L = 1/2\pi f_o C$. Solving for f_o yields

$$f_o = \frac{1}{2\pi\sqrt{LC}} \quad \text{series RLC resonance frequency} \quad (21.25)$$

This frequency satisfies the condition of minimum impedance—and therefore maximizes the current in the circuit. In analogy to pumping a swing at just the right frequency or having a violin string in one of its normal modes, f_o is called the circuit's **resonance frequency**. A plot of capacitive and inductive reactances versus frequency is shown in ▼ Fig. 21.13a. The curves X_C and X_L intersect at f_o, the frequency at which their values are equal.

A Physical Explanation of Resonance

The physical explanation of resonance in a series RLC circuit is worth exploring. We have seen that the capacitor and inductor voltages are *always* 180° out of phase, or have opposite polarity. In other words, they tend to cancel out but usually don't completely do so because their values are not equal. If this is the case, then the voltage across the resistor is less than that of the source voltage because there is a net voltage across the

Illustration 31.8 Impedance and Resonance, RLC Circuit

Exploration 31.7 RLC Circuit

Demonstration/activity: Bring in a spring–mass system, tweak it to get its natural frequency, and then smoothly move the top of the spring up and down at that same frequency to show that the amplitude gets very large. Demonstrate that extremely high and low frequencies produce very small amplitudes. Bring in a rope tied at one end and remind students of the standing-wave resonance idea.

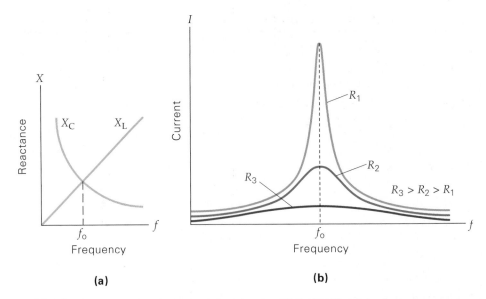

(a)

(b)

▲ **FIGURE 21.13 Resonance frequency for a series RLC circuit** **(a)** At the resonance frequency (f_o), the capacitive and inductive reactances are equal ($X_L = X_C$). On a graph of X versus f, this is the frequency at which the curves of X_C and X_L intersect. **(b)** On a graph of I versus f, the current is a maximum at f_o. The curve becomes sharper and narrower as the resistance in the circuit decreases.

Demonstration/activity: Following the demonstration of spring–mass resonance, bring in a circuit with an obvious capacitor, inductor, and lightbulb (for resistance) connected to a variable-frequency power supply. Show how adjusting the frequency dramatically changes the light output of the bulb. Because the light output is due to joule heating, its large change is a resonance effect, maximizing current and power dissipation in the circuit. If you know the capacitance and inductance, you can estimate the resonance frequency beforehand and compare it with the experimental value.

▲ **FIGURE 21.14 Variable air capacitor** Rotating the movable plates between the fixed plates changes the overlap area and thus the capacitance. Such capacitors were common in tuning circuits in older radios.

combination of capacitor and inductor. This means that the power dissipated in the resistor is less than its maximum value. However, in the special situation when the capacitive and inductive voltages do cancel, the full source voltage appears across the resistor, the power factor becomes 1, and the resistor dissipates the maximum possible power. This is what we mean by the circuit being driven "at resonance."

Applications of Resonance

As we have seen, when a series RLC circuit is driven at its resonance frequency, both the current in the circuit and the power transfer to the circuit are at a maximum. A graph of rms current versus driving frequency is shown in Fig. 21.13b for several different values of resistance. As expected, the maximum current occurs at frequency f_o. Notice also that the curve becomes sharper and narrower as the resistance decreases.

Resonant circuits have a variety of applications. One common application is in the tuning mechanism of a radio. Each radio station has an assigned broadcast frequency at which its radio waves are transmitted (see Insight 21.1 on Oscillator Circuits: Broadcasters of Electromagnetic Radiation). When the waves are received at the antenna, their oscillating electric and magnetic fields set the electrons in the antenna into regular back-and-forth motion. In other words, they produce an alternating current in the receiver circuit, just as a regular ac voltage source would do.

In a given area, each radio station is assigned its own broadcast frequency. Usually several different radio signals reach an antenna together, but a good receiver circuit selectively picks up only the one with a frequency at or near its resonance frequency. Most radios allow you to alter this resonance frequency to "tune in" different stations. In the early days of radio, variable air capacitors were used for this purpose (▶Fig. 21.14). Today, more compact variable capacitors in smaller radios have a polymer dielectric between thin plates. The polymer sheets help maintain the plate separation and increase the capacitance, thus allowing manufacturers to use plates of a much smaller area. (Recall from Chapter 16 that $C = \kappa \varepsilon_o A/d$.) In most modern radios, solid-state devices replace variable capacitors.

INSIGHT 21.1 OSCILLATOR CIRCUITS: BROADCASTERS OF ELECTROMAGNETIC RADIATION

To generate the high-frequency electromagnetic waves used in radio communications and television (Fig. 1), electric current must be made to oscillate at high frequencies in electronic circuits. This can be accomplished with RLC circuits. Such circuits are called *oscillator circuits*, because the current in them oscillates at a frequency determined by their inductive and capacitive elements.

When the resistance in an RLC circuit is very small, the circuit is essentially an LC circuit. The current in such a circuit oscillates at a frequency f, which is the circuit's "natural" frequency and also its resonance frequency (Eq. 21.25). Any small resistance in the circuit would dissipate energy. However, in an ideal LC circuit (which we are considering here) with no resistance, the oscillation would continue indefinitely.

To understand this oscillation, consider the energy oscillations in an ideal (resistanceless) parallel LC circuit, shown in Fig. 2a. Let's assume that the capacitor is initially charged and the switch is then closed ($t = 0$). The following sequence of events occurs:

1. The capacitor would discharge instantaneously (because $RC = 0$) if it were not for the current having to pass through the coil. At $t = 0$, the current in the coil is zero (Fig. 2a). As the current builds, so does the magnetic field in the coil. By Lenz's law, the increasing magnetic field and change of flux in the coil induce a back emf to oppose this current increase. Because of this back emf, the capacitor does takes time to discharge.

2. When the capacitor is fully discharged (Fig. 2b), all of its energy (in its electric field) has been transferred to the inductor in the form of its magnetic field. (Because in this circuit it is assumed that $R = 0$, no energy is lost to joule heating.) At this time (one-quarter of a period; $T/4$), the magnetic field and the current in the coil are maximum, and all the energy is stored in the inductor. (Refer to the energy "histograms" accompanying the circuit diagrams in Figs. 2a, 2b, and 2c to visualize the energy trades as the cycle proceeds.)

3. As the magnetic field collapses from its maximum value, an emf that opposes the collapse is induced in the coil. This

emf acts in a direction that tends to continue the current in the coil even as it is decreasing (Lenz's law again). The polarity of the emf is now opposite that in step 1. Thus current continues to charge to the capacitor, but the result is a polarity reversed from its initial polarity.

4. When the capacitor is again fully charged (but at reverse polarity), it has its initial energy again (Fig. 2c). This occurs halfway through the cycle, or half a period from the start ($T/2$). The magnetic field in the coil is zero, as is the circuit current.

5. The capacitor again begins to discharge, and these four steps are repeated over and over. Thus we have a current and energy oscillation in the circuit. In an ideal case of a circuit without resistance, the oscillations would continue indefinitely.

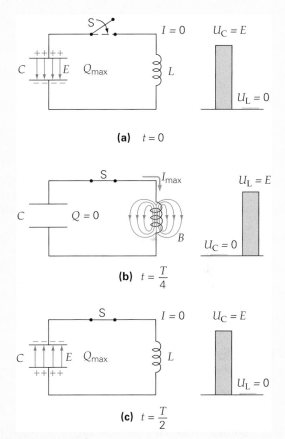

(a) $t = 0$

(b) $t = \dfrac{T}{4}$

(c) $t = \dfrac{T}{2}$

FIGURE 2 An oscillating LC circuit If the resistance is negligible, this circuit will oscillate indefinitely. Half a full cycle is shown between $t = 0$ and $t = T/2$. Energy is transferred back and forth between magnetic and electric types of energy (as shown by the energy histograms to the right). The oscillating electrons in the wire will give off electromagnetic radiation at the circuit's oscillating frequency.

FIGURE 1 A broadcast antenna

Integrated Example 21.8 ■ AM versus FM: Resonance in Radio Reception

(a) When you switch from an AM station (on the "AM band"—the word *band* refers to a specific range of frequencies) to one on the FM band, you are effectively changing the capacitance of the receiving circuit, assuming constant inductance. Is the capacitance (1) increased or (2) decreased when you make this change? (b) Suppose you were listening to news on an AM station at 920 kHz and switched to a music on the FM band at 99.7 MHz. By what factor would you have changed the capacitance of the receiving circuit in the radio, assuming constant inductance?

(a) Conceptual Reasoning. Because FM stations broadcast at significantly higher frequencies than do AM stations (see Table 20.1, p. 676), the resonance frequency of the receiver must be increased to receive signals on the FM band. An increase of the resonance frequency requires reducing the capacitance since the inductance is fixed. Thus the correct answer is (2).

(b) Quantitative Reasoning and Solution. The resonance frequency (Eq. 21.25) depends on the inductance and the capacitance. Because the question asks for a "factor," it is clearly asking for a ratio of the new capacitance to the original capacitance. The frequencies have to be expressed in the same units, so convert MHz into kHz and use unprimed quantities for AM and primed quantities for FM.

Given: $f_o = 920$ kHz   *Find:* C'/C (ratio of FM capacitance to
$f'_o = 99.7$ MHz $= 99.7 \times 10^3$ kHz   AM capacitance)

From Eq. 21.25, the two resonant frequencies are given by

$$f_o = \frac{1}{2\pi\sqrt{LC}} \quad \text{and} \quad f'_o = \frac{1}{2\pi\sqrt{LC'}}$$

Dividing the first of these equations by the second gives

$$\frac{f_o}{f'_o} = \frac{2\pi\sqrt{LC'}}{2\pi\sqrt{LC}} = \sqrt{\frac{C'}{C}}$$

Solving for the capacitance ratio by squaring, and substituting the numbers, we have

$$\frac{C'}{C} = \left(\frac{f_o}{f'_o}\right)^2 = \left(\frac{920 \text{ kHz}}{99.7 \times 10^3 \text{ kHz}}\right)^2 = 8.51 \times 10^{-5}$$

Thus, $C' = 8.51 \times 10^{-5}\,C$ and the capacitance was decreased by a factor of almost one ten-thousandth ($8.51 \times 10^{-5} \approx 10^{-4}$).

Follow-Up Exercise. (a) Based on the resonance curves shown in Fig. 21.13b, can you explain how it is possible to pick up two radio stations *simultaneously* on your radio? (You may have encountered this phenomenon, particularly between two cities located far apart from one another. Two stations are sometimes granted licenses for broadcasting at closely spaced frequencies under the assumption that they won't both be received by the same radio. However, under certain atmospheric conditions, this may not be true.) (b) In part (b) of this Example, if you next increased the capacitance by a factor of two (starting with the news at 920 kHz) to listen to a hockey game, to what new frequency on the AM band would you now be tuned?

Chapter Review

- An **ac voltage** is described by

$$V = V_o \sin \omega t = V_o \sin 2\pi f t \qquad (21.1)$$

- For a sinusoidally varying current, called **ac current,** the **peak current** I_o and the **rms** (root-mean-square or effective) **current** I_{rms} are related by

$$I_{rms} = \frac{I_o}{\sqrt{2}} = 0.707 I_o \qquad (21.6)$$

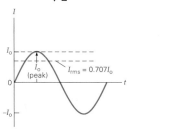

- For an ac voltage, the **peak voltage** V_o is related to its the **rms** (root-mean-square) **voltage** V_{rms} by

$$V_{rms} = \frac{V_o}{\sqrt{2}} = 0.707 V_o \qquad (21.8)$$

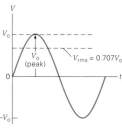

- The current in a resistor is in phase with the voltage across it. For a capacitor, the current is 90° (one quarter of a cycle) ahead of the voltage. For an inductor, the current lags the voltage by 90°.

- In ac circuits, joule heating is due entirely to the resistive elements, and the time-averaged **power dissipation** is

$$\overline{P} = I_{rms}^2 R \qquad (21.10)$$

- In an ac circuit, capacitors and inductors allow current and create opposition to current. This opposition is characterized by **capacitive reactance (X_C)** and **inductive reactance (X_L)**, respectively. The capacitive reactance is given by

$$X_C = \frac{1}{\omega C} = \frac{1}{2\pi f C} \qquad (21.11)$$

The inductive reactance is given by

$$X_L = \omega L = 2\pi f L \qquad (21.14)$$

- Ohm's law, as applied to each type of circuit element, is a generalization of the version from dc circuits. The relationship between rms current and the rms voltage for a resistor is:

$$V_{rms} = I_{rms} R \qquad (21.9)$$

The relationship between rms current and the rms voltage for a capacitor is

$$V_{rms} = I_{rms} X_C \qquad (21.12)$$

The relationship between rms current and the rms voltage for an inductor is

$$V_{rms} = I_{rms} X_L \qquad (21.15)$$

- **Phasors** are vectorlike quantities that allow resistances and reactances to be represented graphically.

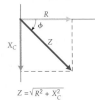

$$Z = \sqrt{R^2 + X_C^2}$$

- **Impedance (Z)** is the total, or effective, opposition to current that takes into account both resistances and reactances. Impedance is related to current and circuit voltage by a generalization of Ohm's law:

$$V_{rms} = I_{rms} Z \qquad (21.17)$$

- The **impedance for a series RLC circuit** is

$$Z = \sqrt{R^2 + (X_L - X_C)^2} \qquad (21.19)$$

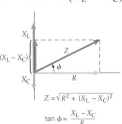

$$Z = \sqrt{R^2 + (X_L - X_C)^2}$$

$$\tan\phi = \frac{X_L - X_C}{R}$$

- The **phase angle (ϕ)** between the rms voltage and the rms current in a series RLC circuit is

$$\tan\phi = \frac{X_L - X_C}{R} \qquad (21.20)$$

- The **power factor (cos ϕ)** for a series RLC circuit is a measure of how close to the maximum power dissipation the circuit is. The power factor is

$$\cos\phi = \frac{R}{Z} \qquad (21.22)$$

The average power dissipated (joule heating in the resistor) is

$$\overline{P} = I_{rms} V_{rms} \cos\phi \qquad (21.23)$$

or

$$\overline{P} = I_{rms}^2 Z \cos\phi \qquad (21.24)$$

- The **resonance frequency (f_o)** of an RLC circuit is the frequency at which the circuit dissipates maximum power. This frequency is

$$f_o = \frac{1}{2\pi\sqrt{LC}} \qquad (21.25)$$

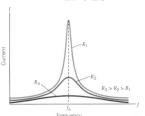

Exercises

MC = *Multiple Choice Question*, **CQ** = *Conceptual Question, and* **IE** = *Integrated Exercise. Throughout the text, many exercise sections will include "paired" exercises. These exercise pairs, identified with* **red numbers**, *are intended to assist you in problem solving and learning. In a pair, the first exercise (even numbered) is worked out in the Study Guide so that you can consult it should you need assistance in solving it. The second exercise (odd numbered) is similar in nature, and its answer is given at the back of the book.*

21.1 Resistance in an AC Circuit

1. **MC** Which of the following voltages is larger for a sinusoidally varying ac voltage: (a) V_o, (b) V_{rms}, or (c) they have the same value? (a)

2. **MC** During the course of one ac voltage cycle (United States) how long does the direction of the current stay constant in a resistor: (a) 1/60 s, (b) 1/120 s, or (c) 1/30 s? (b)

3. **MC** During seven complete ac voltage cycles (United States) what is the average voltage: (a) 0 V, (b) 60 V, (c) 120 V, or (d) 170 V? (a)

4. **CQ** The average current in a resistor in an ac circuit is zero. Explain why the average power delivered to a resistor isn't zero. average current is zero due to direction change; power is delivered regardless of current direction.

5. **CQ** The voltage and current associated with a resistor in an ac circuit are *in phase*. What does that mean? both reach maximum or minimum at the same time

6. **CQ** A 60-W lightbulb designed to work at 240 V in England is instead connected to a 120-V source. Discuss the changes in the bulb's rms current and power when it is at 120 V compared with 240 V. Assume the bulb is ohmic. current is cut in half and power drops to 25%

7. **CQ** If the ac voltage and current for a particular circuit element are given respectively by $V = 120\sin(120\pi t)$ and $I = 30\sin(120\pi t + \pi/2)$, could the circuit element be a resistor? Is the frequency 60 Hz? Explain. see ISM

8. ● What are the peak voltages of a 120-V ac line and a 240-V ac line? 170 V; 339 V

9. ● An ac circuit has an rms current of 5.0 A. What is the peak current? 7.1 A

10. ● The maximum voltage across a resistor in an ac circuit is 156 V. Find the resistor's rms voltage. 110 V

11. ● How much ac rms current must be in a 10-Ω resistor to produce an average power of 15 W? 1.2 A

12. ● An ac circuit contains a resistor with a resistance of 5.0 Ω. The resistor has an rms current of 0.75 A. (a) Find its rms voltage and peak voltage. (b) Find the average power delivered to the resistor. (a) V_{rms} = 3.8 V; V_0 = 5.3 V (b) 2.8 W

13. ● A hair dryer is rated at 1200 W when plugged into a 120-V outlet. Find (a) its rms current, (b) its peak current, and (c) its resistance. (a) 10.0 A (b) 14.1 A (c) 12.0 Ω

14. **IE** ●● The voltage across a 10-Ω resistor varies as $V = (170$ V$) \sin(100\pi t)$. (a) Will the current in the resistor (1) be in phase with the voltage, (2) lead the voltage by 90°, or (3) lag the voltage by 90°? (b) Write the expression for the current in the resistor as a function of time and determine the voltage frequency. (a) (1) be in phase with the voltage (b) $I = (17$ A$) \sin(100\pi t)$; 50 Hz

15. ●● An ac voltage is applied to a 25.0-Ω resistor so that it dissipates 500 W of power. Find the resistor's (a) rms and peak currents and (b) rms and peak voltages. (a) 4.47 A, 6.32 A (b) 112 V, 158 V

16. **IE** ●● An ac voltage source has a peak voltage of 85 V and a frequency of 60 Hz. The voltage at $t = 0$ is zero. (a) A student wants to calculate the voltage at $t = 1/240$ s. How many possible answers are there: (1) one, (2) two, or (3) three? Why? (b) Determine all possible answers. (a) (2) two (b) ±85 V

17. ●● An ac voltage source has an rms voltage of 120 V. Its voltage goes from zero to its maximum value in 4.20 ms. Write an expression for the voltage as a function of time. $V = (170$ V$) \sin(119\pi t)$

18. ●● What are the resistance and rms current of a 100-W, 120-V computer monitor? 144 Ω; 0.833 A

19. ●● Find the rms and peak currents in a 40-W, 120-V lightbulb. 0.33 A; 0.47 A

20. ●● A 50-kW heater is designed to run using a 240-V ac source. Find its (a) peak current and (b) peak voltage. (a) 2.9×10^2 A (b) 3.4×10^2 V

21. ●● The current in a resistor is given by $I = (8.0$ A$) \sin(40\pi t)$ when a voltage given by $V = (60$ V$) \sin(40\pi t)$ is applied to it. (a) What is the frequency and period of the voltage source? (b) What is the average power delivered to the resistor? (a) 20 Hz, 0.050 s (b) 2.4×10^2 W

22. ●● The current and voltage outputs of an operating ac generator have peak values of 2.5 A and 16 V, respectively. (a) What is the average power output of the generator? (b) What is the effective resistance of the circuit it is in? (a) 20 W (b) 6.4 Ω

23. ●●● The current in a 60-Ω resistor is given by $I = (2.0$ A$) \sin(380t)$. (a) What is the frequency of the current? (b) What is the rms current? (c) How much average power is delivered to the resistor? (d) Write an equation for the voltage across the resistor as a function of time.

(e) Write an equation for the power delivered to the resistor as a function of time. (f) Show that the rms power obtained in part (e) is the same as your answer to part (c). see ISM

21.2 Capacitive Reactance
and
21.3 Inductive Reactance

24. **MC** In a purely capacitive ac circuit, (a) the current and voltage are in phase, (b) the current leads the voltage, (c) the current lags the voltage, or (d) none of the preceding. (b)

25. **MC** A single capacitor is connected to an ac voltage source. When the voltage across the capacitor is at a maximum, then the charge on it is (a) zero (b) at a maximum or (c) neither of the preceding, but somewhere in between. (b)

26. **MC** A single inductor is connected to an ac voltage source. When the voltage across the inductor is at a maximum, then the current in it is not changing. (a) true (b) false or (c) cannot be determined from the given information. (b)

27. **CQ** Explain why, under very low frequency ac conditions, a capacitor acts almost as an open circuit while an inductor acts almost as a short circuit. see ISM

28. **CQ** Can an inductor oppose dc current? What about a capacitor? Explain each and why they are different. inductor, no; capacitor, yes; see ISM

29. **CQ** If the current on a 10-μF capacitor is described by $I = (120$ A$) \sin(120\pi t + \pi/2)$, explain why the instantaneous voltage across it at $t = 0$ is zero whereas the current at that time is not. current leads voltage by 90°, they are out of phase, see ISM

30. ● Find the frequency at which a 25-μF capacitor has a reactance of 25 Ω. 2.5×10^2 Hz

31. ● A single 2.0-μF capacitor is connected across the terminals of a 60-Hz voltage source, and a current of 2.0 mA is measured on an ac ammeter. What is the capacitive reactance of the capacitor? 1.3×10^3 Ω

32. ● What capacitance would have a reactance of 100 Ω in a 60-Hz ac circuit? 2.7×10^{-5} F

33. ● A single 50-mH inductor forms a complete circuit when connected to an ac voltage source at 120 V and 60 Hz. (a) What is the inductive reactance of the circuit? (b) How much current is in the circuit? (c) What is the phase angle between the current and the applied voltage? (Assume negligible resistance.) (a) 19 Ω (b) 6.4 A (c) voltage leads current by 90°

34. ● How much current is in a circuit containing only a 50-μF capacitor connected to an ac generator with an output of 120 V and 60 Hz? 2.3 A

35. ●● A variable capacitor in a circuit with a 120-V, 60-Hz source initially has a capacitance of 0.25 μF. The capacitance is then increased to 0.40 μF. What is the percentage change in the current in the circuit? an increase of 60%

36. ●● An inductor has a reactance of 90 Ω in a 60-Hz ac circuit. What is its inductance? 0.24 H

37. ●● Find the frequency at which a 250-mH inductor has a reactance of 400 Ω. 255 Hz

38. **IE ●●** A capacitor is connected to a variable-frequency ac voltage source. (a) If the frequency increases by a factor of 3, the capacitive reactance will be (1) 3, (2) $\frac{1}{3}$, (3) 9, (4) $\frac{1}{9}$ times the original reactance. Why? (b) If the capacitive reactance of a capacitor at 120 Hz is 100 Ω, what is its reactance if the frequency is changed to 60 Hz?
    (a) (2) $\frac{1}{3}$ (b) 200 Ω

39. **●●** With a single 150-mH inductor in a circuit with a 60-Hz voltage source, a current of 1.6 A is measured on an ac ammeter. (a) What is the rms voltage of the source? (b) What is the phase angle between the current and that voltage?
    (a) 90 V (b) voltage leads current by 90°

40. **●●** What inductance has the same reactance in a 120-V, 60-Hz circuit as a capacitance of 10 μF? 0.70 H

41. **●●** A circuit with a single capacitor is connected to a 120-V, 60-Hz source. What is its capacitance if there is a current of 0.20 A in the circuit? 4.4 μF

42. **IE ●●** An inductor is connected to a variable-frequency ac voltage source. (a) If the frequency decreases by a factor of 2, the rms current will be (1) 2, (2) $\frac{1}{2}$, (3) 4, (4) $\frac{1}{4}$ times the original rms current. Why? (b) If the rms current in an inductor at 40 Hz is 9.0 A, what is its rms current if the frequency is changed to 120 Hz? (a) (1) 2 (b) 3.0 A

21.4 Impedance: RLC Circuits
and
21.5 Circuit Resonance

43. **MC** The impedance of an RLC circuit depends on (a) frequency, (b) inductance, (c) capacitance, or (d) all of the preceding. (d)

44. **MC** If the capacitance of a series RLC circuit is decreased, (a) the capacitive reactance increases, (b) the inductive reactance increases, (c) the current remains constant, or (d) the power factor remains constant. (a)

45. **MC** When a series RLC circuit is driven at its resonance frequency, (a) energy is dissipated only by the resistive element, (b) the power factor has a value of one, (c) there is maximum power delivered to the circuit, or (d) all of the preceding. (d)

46. **CQ** What is the impedance of an RLC circuit at resonance and why? $Z = R$, impedance is at minimum

47. **CQ** Is any power ever delivered to capacitors or inductors in ac circuits? Why or why not?
    no, $\phi = 90°$, so $\cos\phi = 0$

48. **CQ** What are the factors that determine the resonant frequency of an RLC circuit? Is resistance a factor? Explain?
    capacitance and inductance, no, see ISM

49. **●** A coil in a 60-Hz circuit has a resistance of 100 Ω and an inductance of 0.45 H. Calculate (a) the coil's reactance and (b) the circuit's impedance.
    (a) 1.7×10^2 Ω (b) 2.0×10^2 Ω

50. **●** A series RC circuit has a resistance of 200 Ω and a capacitance of 25 μF and is driven by a 120-V, 60-Hz source. (a) Find the capacitive reactance and impedance of the circuit. (b) How much current is drawn from the source? (a) 1.1×10^2 Ω; 2.3×10^2 Ω (b) 0.53 A

51. **●** A series RL circuit has a resistance of 100 Ω and an inductance of 100 mH and is driven by a 120-V, 60-Hz source. (a) Find the inductive reactance and the impedance of the circuit. (b) How much current is drawn from the source?
    (a) 38 Ω; 1.1×10^2 Ω (b) 1.1 A

52. **●** An RC circuit has a resistance of 250 Ω and a capacitance of 6.0 μF. If the circuit is driven by a 60-Hz source, find (a) the capacitive reactance and (b) the impedance of the circuit. (a) 4.4×10^2 Ω (b) 5.1×10^2 Ω

53. **IE ●** An RC circuit has a resistance of 100 Ω and a capacitive reactance of 50 Ω. (a) Will the phase angle be (1) positive, (2) zero, or (3) negative? Why? (b) What is the phase angle of this circuit? (a) (3) negative (b) −27°

54. **●●** A series RLC circuit has a resistance of 25 Ω, an inductance of 0.30 H, and a capacitance of 8.0 μF. (a) At what frequency should the circuit be driven for the maximum power to be transferred from the driving source? (b) What is the impedance at that frequency?
    (a) (1) 1.0×10^2 Hz (b) 25 Ω

55. **IE ●●** In a series RLC circuit, $R = X_C = X_L = 40$ Ω for a particular driving frequency. (a) This circuit is (1) inductive, (2) capacitive, (3) in resonance. Why? (b) If the driving frequency is doubled, what will be the impedance of the circuit?
    (a) (3) in resonance, $X_L = X_C$ so $Z = R$ (b) 72 Ω

56. **IE ●●** (a) An RLC series circuit is in resonance. Which one of the following can you change without upsetting the resonance: (1) resistance, (2) capacitance, (3) inductance, or (4) frequency? Why? (b) A resistor, an inductor, and a capacitor have values of 500 Ω, 500 mH, and 3.5 μF, respectively. They are connected in series to a power supply of 240 V with a frequency of 60 Hz. What values of resistance and inductance would be required for this circuit to be in resonance (without changing the capacitor)?
    (a) (1) resistance (b) 2.0 H for any resistance

57. **●●** How much power is dissipated in the circuit described in Exercise 56b using the initial values of resistance, inductance and capacitance? 50 W

58. **●●** What is the resonant frequency of an RLC circuit with a resistance of 100 Ω, an inductance of 100 mH, and a capacitance of 5.00 μF? 225 Hz

59. **●●** A tuning circuit in a radio receiver has a fixed inductance of 0.50 mH and a variable capacitor. If the circuit is tuned to a radio station broadcasting at 980 kHz on the AM dial, what is the capacitance of the capacitor?
    5.3×10^{-11} F

60. **●●** What would be the range of the variable capacitor in Exercise 59 required for tuning over the complete AM band? [*Hint*: See Table 20.1.]
    1.7×10^{-11} F to 1.8×10^{-10} F

61. **●●** Find the currents supplied by the ac source for all possible connections in ▼Fig. 21.15. see ISM

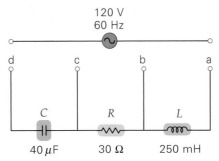

◄ FIGURE 21.15
A series RLC circuit
See Exercise 61.

62. IE ●● A coil with a resistance of 30 Ω and an inductance of 0.15 H is connected to a 120-V, 60-Hz source. (a) Is the phase angle of this circuit (1) positive, (2) zero, or (3) negative? Why? (b) What is the phase angle of the circuit? (c) How much rms current is in the circuit? (d) What is the average power delivered to the circuit? (a) (1) positive (b) 62° (c) 1.9 A (d) 1.1 × 10² W

63. ●● A small welder uses a voltage source of 120 V at 60 Hz. When the source is operating, it requires 1200 W of power, and the power factor is 0.75. Find the rms current in the welder. 13 A

64. ●● A series circuit is connected to a 220-V, 60-Hz power supply. The circuit has the following components: a 10-Ω resistor, a coil with an inductive reactance of 120 Ω, and a capacitor with a reactance of 120 Ω. Compute the rms voltage across (a) the resistor, (b) the inductor, and (c) the capacitor. (a) 220 V (b) 2.64 × 10³ V (c) 2.64 × 10³ V

65. ●● A series RLC circuit has a resistance of 25 Ω, a capacitance of 0.80 μF, and an inductance of 250 mH. The circuit is connected to a variable-frequency source with a fixed rms voltage output of 12 V. If the frequency that is supplied is set at the circuit's resonance frequency, what is the rms voltage across each of the circuit elements? $(V_{rms})_R = 12$ V; $(V_{rms})_L = 2.7 \times 10^3$ V; $(V_{rms})_C = 2.7 \times 10^3$ V

66. ●● (a) In Exercises 64 and 65, determine the numerical (scalar) sum of the rms voltages across the three circuit elements and explain why it is much larger than the source voltage. (b) Determine the sum of these voltages using the proper phasor techniques and show that your result is equal to the source voltage. see ISM

67. IE ●● (a) If the circuit in▼Fig. 21.16 is in resonance, the impedance of the circuit is (1) greater than 25 Ω, (2) equal to 25 Ω, (3) less than 25 Ω. Why? (b) If the driving frequency is 60 Hz, what is the circuit's impedance? (a) (2) equal to 25 Ω (b) 362 Ω

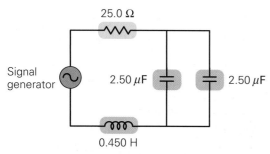

25.0 Ω

Signal generator

2.50 μF   2.50 μF

0.450 H

▲ **FIGURE 21.16 Tune to resonance** See Exercise 67.

68. ●●● A series RLC circuit with a resistance of 400 Ω has capacitive and inductive reactances of 300 Ω and 500 Ω, respectively. (a) What is the power factor of the circuit? (b) If the circuit operates at 60 Hz, what additional capacitance should be connected to the original capacitance to give a power factor of unity, and how should the capacitors be connected? (a) 0.894 (b) 13.3 μF; series

69. ●●● A series RLC circuit has components with R = 50 Ω, L = 0.15 H, and C = 20 μF. The circuit is driven by a 120-V, 60-Hz source. What is the power delivered to the circuit, expressed as a percentage of the power delivered when the circuit is in resonance? 30%

Comprehensive Exercises

70. A series RLC radio receiver circuit with an inductance of 1.50 μH is tuned to an FM station at 98.9 MHz by adjusting a variable capacitor. When the circuit is tuned to this station, (a) what is its inductive reactance? (b) What is its capacitive reactance? (c) What is its capacitance? (a) 932 Ω (b) 932 Ω (c) 1.70 × 10⁻¹² F

71. A circuit connected to a 110-V, 60-Hz source contains a 50-Ω resistor and a coil with an inductance of 100 mH. Find (a) the reactance of the coil, (b) the impedance of the circuit, (c) the current in the circuit, and (d) the power dissipated by the coil, and (e) calculate the phase angle between the current and the applied voltage. (a) 38 Ω (b) 63 Ω (c) 1.8 A (d) zero (e) 37°

72. A 1.0-μF capacitor is connected to a 120-V, 60-Hz source. (a) What is the capacitive reactance of the circuit? (b) How much current is in the circuit? (c) What is the phase angle between the current and the applied voltage? (a) 2.7 × 10³ Ω (b) 4.5 × 10⁻² A (c) −90°

73. IE (a) If an RLC circuit is in resonance, the phase angle of the circuit is (1) positive, (2) zero, (3) negative. Why? (b) A circuit has an inductive reactance of 280 Ω at 60 Hz. What value of capacitance would set this circuit into resonance? (a) (2) zero as $X_L = X_C$ (b) 9.4 μF

74. The circuit in ▼Fig. 21.17a is called a *low-pass filter* because a large current and voltage (and thus a lot of power) is delivered to the load resistor (R_L) only by a low-frequency source. The circuit in Fig. 21.17b is called a *high-pass filter* because a large current and voltage (and thus a lot of power) is delivered to the load only by a high-frequency source. Describe conceptually why the circuits have these characteristics. see ISM

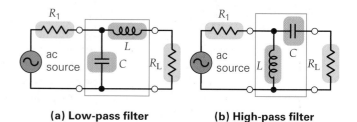

(a) Low-pass filter   **(b) High-pass filter**

▲ **FIGURE 21.17 Low-pass and high-pass filters** See Exercise 74.

The following Physlet Physics Problems can be used with this chapter.
31.1, 31.3, 31.4, 31.5, 31.8, 31.9, 31.11, 31.12, 31.14

22

REFLECTION AND REFRACTION OF LIGHT

PHYSICS FACTS

- Due to total internal reflection, optical fibers allow signals to travel for long distances without repeaters (amplifiers), to compensate for reductions in signal strength. Fiber-optic repeaters are currently about 100 km (about 62 mi) apart, compared to about 1.5 km (about 1 mi) for electrical (wire-based) systems.

- Every day installers lay enough new fiber-optic cables for computer networks to circle the Earth three times. Optical fibers can be drawn to smaller diameters than copper wire. Fibers can be as small as 10 microns in diameter. In comparison, the average human hair is about 25 microns in diameter.

- Most camera lenses are coated with a thin film to reduce light loss due to reflection. For a typical seven-element camera lens, about 50% of the light would be lost due to reflection if the lens were not coated with thin films.

- In 1998, scientists at MIT made a perfect mirror, a mirror with 100% reflection. A tube lined with this type of mirror would transmit light over long distances better than optical fibers.

We live in a visual world, surrounded by eye-catching images such as that refractive image of the turtle shown in the photo. How these images are formed is something taken largely for granted—until we see something that can't be easily explained. *Optics* is the study of light and vision. Human vision requires *visible light* of wavelength from 400 nm to 700 nm (see Fig. 20.23). Optical properties, such as reflection and refraction, are shared by all electromagnetic waves. Light acts as a wave in its propagation (Chapter 24) and as a particle (photon) when it interacts with matter (Chapter 27–30).

In this chapter, we will investigate the basic optical phenomena of reflection, refraction, total internal reflection, and dispersion. The principles that govern reflection explain the behavior of mirrors, while those that govern refraction explain the properties of lenses. With the aid of these and other optical principles, we can understand many optical phenomena experienced every day—why a glass prism spreads light into a spectrum of colors, what causes mirages, how rainbows are formed, and why the legs of a person standing in a lake or swimming pool seem to shorten. Some less familiar but increasingly useful territory, including the fascinating field of fiber optics will also be explored.

A simple geometrical approach involving straight lines and angles can be used to investigate many aspects of the properties of light, especially how light propagates. For these purposes, we need not be concerned with the physical (wave) nature of electromagnetic waves described in Chapter 20. The principles of geometrical optics will be introduced here and applied in greater detail in the study of mirrors and lenses in Chapter 23.

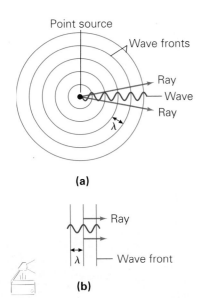

(a)

(b)

▲ **FIGURE 22.1 Wave fronts and rays** A wave front is defined by adjacent points on a wave that are in phase, such as those along wave crests or troughs. A line perpendicular to a wave front in the direction of the wave's propagation is called a ray. **(a)** Near a point source, the wave fronts are circular in two dimensions and spherical in three dimensions. **(b)** Very far from a point source, the wave fronts are approximately linear or planar and the rays nearly parallel.

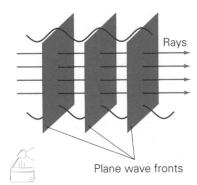

▲ **FIGURE 22.2 Light rays** A plane wave travels in a direction perpendicular to its wave fronts. A beam of light can be represented by a group of parallel rays (or by a single ray).

22.1 Wave Fronts and Rays

OBJECTIVE: To define and explain the concepts of wave fronts and rays.

Waves, electromagnetic or otherwise, are conveniently described in terms of wave fronts. A **wave front** is the line or surface defined by adjacent portions of a wave that are in phase. If an arc is drawn along one of the crests of a circular water wave moving out from a point source, all the particles on the arc will be in phase (◄Fig. 22.1a). An arc along a wave trough would work equally well. For a three-dimensional spherical wave, such as a sound or light wave emitted from a point source, the wave front is a spherical surface rather than a circle.

Very far from the source, the curvature of a short segment of a circular or spherical wave front is extremely small. Such a segment may be approximated as a *linear wave front* (in two dimensions) or a **plane wave front** (in three dimensions), just as we take the surface of the Earth to be locally flat (Fig. 22.1b). A plane wave front can also be produced directly by a large luminous flat surface. In a uniform medium, wave fronts propagate outward from the source at a speed characteristic of the medium. This was seen for sound waves in Chapter 14, and the same occurs for light, although at a much faster speed. The speed of light is greatest in a vacuum: $c = 3.00 \times 10^8$ m/s. The speed of light in air, for all practical purposes, is the same as that in vacuum.

The geometrical description of a wave in terms of wave fronts tends to neglect the fact that the wave is actually oscillating, like those studied in Chapter 13. This simplification is carried a step further with the concept of a ray. As illustrated in Fig. 22.1, a line drawn perpendicular to a series of wave fronts and pointing in the direction of propagation is called a **ray**. Note that a ray points in the direction of the energy flow of a wave. A plane wave is assumed to travel in a straight line in a medium in the direction of its rays, perpendicular to its plane wave fronts. A beam of light can be represented by a group of rays or simply as a single ray (◄Fig. 22.2). The representation of light as rays is adequate and convenient for describing many optical phenomena.

How do we see things and objects around us? We see them because rays from the objects, or rays that appear to come from the objects, enter our eyes (▼Fig. 22.3). In the eyes rays form images of the objects on the retina. The rays could be coming directly from the objects as in the case of light sources or could be reflected or refracted by the objects or other optical systems. Our eyes and brain working together, however, cannot tell whether the rays actually come from the objects or only *appear* to come from the objects. This is one way magicians can fool our eyes with seemingly impossible illusions.

The use of the geometrical representations of wave fronts and rays to explain phenomena such as the reflection and refraction of light is called **geometrical optics**. However, certain other phenomena, such as the interference of light, cannot be treated in this manner and must be explained in terms of actual wave characteristics. These phenomena will be considered in Chapter 24.

▶ **FIGURE 22.3 How we see things** We see things because **(a)** rays from the objects or **(b)** rays appearing to come from the objects enter our eyes.

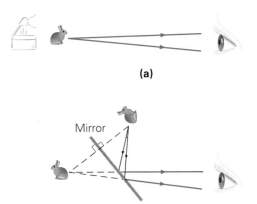

(a)

(b)

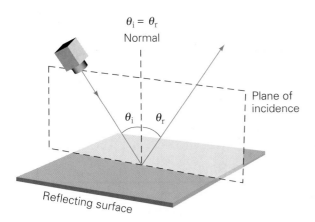

$\theta_i = \theta_r$
Normal

Plane of incidence

θ_i | θ_r

Reflecting surface

◀ **FIGURE 22.4** The law of reflection According to the law of reflection, the angle of incidence (θ_i) is equal to the angle of reflection (θ_r). Note that the angles are measured relative to a normal (a line perpendicular to the reflecting surface). The normal and the incident and reflected rays always lie in the same plane.

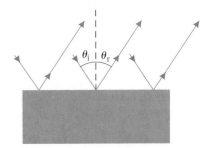

θ_i | θ_r

(a) Specular (regular) reflection (diagram)

22.2 Reflection

<u>OBJECTIVES:</u> To (a) explain the law of reflection, and (b) distinguish between regular (specular) and irregular (diffuse) reflection.

The reflection of light is an optical phenomenon of enormous importance: If light were not reflected to our eyes by objects around us, we wouldn't see the objects at all. **Reflection** involves the absorption and re-emission of light by means of complex electromagnetic vibrations in the atoms of the reflecting medium. However, the phenomenon is easily described by using rays.

A light ray incident on a surface is described by an **angle of incidence (θ_i)**. This angle is measured relative to a *normal*—a line perpendicular to the reflecting surface (▲ Fig. 22.4). Similarly, the reflected ray is described by an **angle of reflection (θ_r)**, also measured from the normal. The relationship between these angles is given by the **law of reflection**: The angle of incidence is equal to the angle of reflection, or

$$\theta_i = \theta_r \quad \textit{law of reflection} \qquad (22.1)$$

Two other attributes of reflection are that the incident ray, the reflected ray, and the normal all lie in the same plane, which is sometimes called the plane of incidence, and that the incident and the reflected rays are on opposite sides of the normal.

When the reflecting surface is smooth and flat, the reflected rays from parallel incident rays are also parallel (▶Fig. 22.5a). This type of reflection is called **specular**, or **regular**, **reflection**. The reflection from a highly polished flat mirror is an example of specular (regular) reflection (Fig. 22.5b). If the reflecting surface is rough, however, the reflected rays are not parallel, because of the irregular nature of the surface (▶Fig. 22.6). This type of reflection is termed **diffuse**, or **irregular**, **reflection**. The reflection of light from this page is an example of diffuse reflection because the paper is microscopically rough. Insight 22.1 on A Dark, Rainy Night on page 709, discusses more about the difference between specular and diffuse reflection in a real-life situation.

Note in Fig. 22.5a and Fig. 22.6 that the law of reflection still applies locally to both specular and diffuse reflection. However, the type of reflection involved determines whether we see images from a reflecting surface. In specular reflection, the reflected, parallel rays produce an image when they are viewed by an optical system such as an eye or a camera. Diffuse reflection does not produce an image, because the light is reflected in various directions.

Experience with friction and direct investigations show that all surfaces are rough on a microscopic scale. What, then, determines whether reflection is specular or diffuse? In general, if the dimensions of the surface irregularities are greater than the wavelength of the light, the reflection is diffuse. Therefore, to make a good mirror, glass (with a metal coating) or metal must be polished at least until the surface

(b) Specular (regular) reflection (photo)

▲ **FIGURE 22.5** Specular (regular) reflection **(a)** When a light beam is reflected from a smooth surface and the reflected rays are parallel, the reflection is said to be specular or regular. **(b)** Specular (regular) reflection from a smooth water surface produces an almost perfect mirror image of salt mounds at this Australian salt mine.

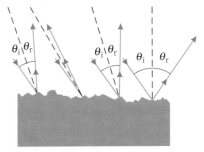

θ_i θ_r θ_i θ_r θ_i | θ_r

▲ **FIGURE 22.6** Diffuse (irregular) reflection Reflected rays from a relatively rough surface, such as this page, are not parallel; the reflection is said to be diffuse or irregular. (Note that the law of reflection still applies locally to each individual ray.)

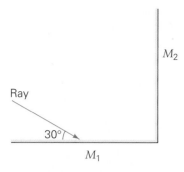

▲ **FIGURE 22.7 Trace the ray** See Example 22.1.

Note: Drawing diagrams like these is extremely important in the study of geometrical optics.

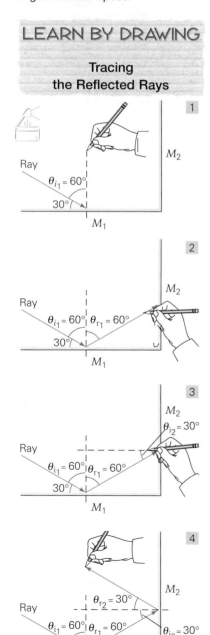

LEARN BY DRAWING

Tracing the Reflected Rays

irregularities are about the same size as the wavelength of light. Recall from Chapter 20 that the wavelength of visible light is on the order of 10^{-7} m. (You will learn more about reflection from a mirror in the Learn by Drawing presented in Example 22.1.)

Diffuse reflection enables us to see illuminated objects, such as the Moon. If the Moon's spherical surface were smooth, only the reflected sunlight from a small region would come to an observer on the Earth, and only that small illuminated area would be seen. Also, you can see the beam of light from a flashlight or spotlight because of diffuse reflection from dust and particles in the air.

Example 22.1 ■ Tracing the Reflected Ray

Two mirrors, M_1 and M_2, are perpendicular to each other, with a light ray incident on one of the mirrors as shown in ◄Fig. 22.7. (a) Sketch a diagram to trace the path of the reflected light ray. (b) Find the direction of the ray after it is reflected by M_2.

Thinking It Through. The law of reflection can be used to determine the direction of the ray after it leaves the first and then the second mirror.

Solution.

Given: $\theta = 30°$ (angle relative to M_1)   **Find:** (a) Sketch a diagram tracing the light ray
                                                    (b) θ_{r_2} (angle of reflection from M_2)

Follow steps 1–4 in Learn by Drawing:

(a) 1. Since the incident and reflected rays are measured from the normal (a line perpendicular to the reflecting surface), we draw the normal to mirror M_1 at the point the incident ray hits M_1. From geometry, it can be seen that the angle of incidence on M_1 is $\theta_{i_1} = 60°$.

2. According to the law of reflection, the angle of reflection from M_1 is also $\theta_{r_1} = 60°$. Next, draw this reflected ray with an angle of reflection of 60°, and extend it until it hits M_2.

3. Draw another normal to M_2 at the point where the ray hits M_2. Also from geometry (focus on the triangle in the diagram), the angle of incidence on M_2 is $\theta_{i_2} = 30°$. (Why?)

(b) 4. The angle of reflection off M_2 is $\theta_{r_2} = \theta_{i_2} = 30°$. This is the final direction of the ray reflected after both mirrors.

What if the directions of the rays are reversed? In other words, if a ray is first incident on M_2, in the direction opposite that of the one drawn for part (b), will all the rays reverse their directions? Draw another diagram to prove that this is indeed the case. Light rays are reversible.

Follow-Up Exercise. When following an eighteen-wheel truck, you may see a sign on the back stating, "If you can't see my mirror, I can't see you." What does this mean? (*Answers to all Follow-Up Exercises are at the back of the text.*)

22.3 Refraction

OBJECTIVES: To (a) explain refraction in terms of Snell's law and the index of refraction, and (b) give examples of refractive phenomena.

Refraction refers to the change in direction of a wave at a boundary where the wave passes from one transparent medium into another. In general, when a wave is incident on a boundary between media, some of the wave's energy is reflected and some is transmitted. For example, when light traveling in air is incident on a transparent material such as glass, it is partially reflected and partially transmitted (▶Fig. 22.8). But the direction of the transmitted light is different from the direction of the incident light, so the light is said to have been refracted; in other words, it has changed direction.

This change in direction is caused by the fact that light travels with different speeds in different media. Intuitively, you might expect the passage of light to take longer through a medium with more atoms per volume, and the

INSIGHT   22.1 A DARK, RAINY NIGHT

When you drive on a dry night, you can clearly see the road and the street signs ahead of you from your headlights. However, here is a familiar scene on a dark, rainy night: Even with headlights, you can hardly see the road ahead. When a car approaches, the situation becomes even worse. You see the reflections of the approaching car's headlights from the surface of the road, and they appear brighter than usual. Often nothing can be seen except the reflective glare of the oncoming headlights.

What causes these conditions? When the road surface is dry, the reflection of light off the road is diffuse (irregular), because the surface is rough. Light from your headlights hits the road in front of you and reflects in all directions. Some of it reflects back, and

you can see the road clearly (just as you can read this page because the paper is microscopically rough). However, when the road surface is wet, water fills the crevices, turning the road into a relatively smooth reflecting surface (Fig. 1a). Light from the headlights then reflects ahead. The normally diffuse reflection is gone and is replaced by specular reflection. Reflected images of lighted buildings and road lights form, blurring the view of the surface, and the specular reflection of oncoming cars' headlights may make it difficult for you to see the road (Fig. 1b).

Besides wet, slippery surfaces, specular reflection is a major cause of accidents on rainy nights. Thus, extra caution is advised under such conditions.

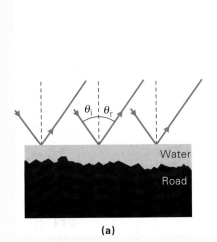

(a)   (b)

FIGURE 1 Diffuse to specular (a) The diffuse reflection from a dry road is turned into specular reflection by water on the road's surface. **(b)** Instead of seeing the road, a driver sees the reflected images of lights, buildings, and so on.

---

speed of light is, in fact, generally less in denser media. For example, the speed of light in water is about 75% of that in air or a vacuum. ▼Fig. 22.9a shows the refraction of light at an air–water boundary.

The change in the direction of wave propagation is described by the **angle of refraction**. In Fig. 22.9b, θ_1 is the angle of incidence and θ_2 is the angle of refraction. We use notations of θ_1 and θ_2 for the angles of incidence and refraction to avoid confusion with θ_i and θ_r for the angles of incidence and reflection. Willebrord Snell (1580–1626), a Dutch physicist, discovered a relationship between the angles (θ) and the speeds (v) of light in two media (Fig. 22.9b):

$$\frac{\sin \theta_1}{\sin \theta_2} = \frac{v_1}{v_2} \quad \textit{Snell's law} \tag{22.2}$$

This expression is known as **Snell's law**. Note that θ_1 and θ_2 are always taken with respect to the normal.

Thus, light is refracted when passing from one medium into another because the speed of light is different in the two media. The speed of light is greatest in a vacuum, and it is therefore convenient to compare the speed of light in other

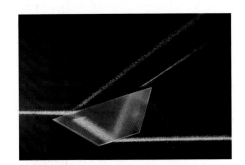

▲ FIGURE 22.8 Reflection and refraction A beam of light is incident on a trapezoidal prism from the left. Part of the beam is reflected, and part is refracted. The refracted beam is partially reflected and partially refracted at the bottom glass–air surface.

▶ **FIGURE 22.9** Refraction **(a)** Light changes direction on entering a different medium. **(b)** The refracted ray is described by the angle of refraction, θ_2, measured from the normal.

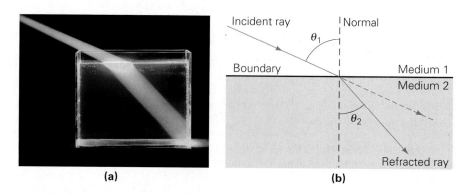

(a)   **(b)**

Illustration 34.1 Huygen's Principle and Refraction

Note: When light is refracted,

- its speed and wavelength are changed;
- its frequency remains unchanged.

media with this constant value (c). This is done by defining a ratio called the **index of refraction (n)**:

$$n = \frac{c}{v}\left(\frac{\text{speed of light in a vacuum}}{\text{speed of light in a medium}}\right) \qquad (22.3)$$

As a ratio of speeds, the index of refraction is a unitless quantity. The indices of refraction of several substances are given in Table 22.1. Note that these values are for a specific wavelength of light. The wavelength is specified because v, and consequently n, are slightly different for different wavelengths. (This is the cause of dispersion, to be discussed later in the chapter.) The values of n given in the table will be used in examples and exercises in this chapter for all wavelengths of light in the visible region, unless otherwise noted. Observe that n is always greater than 1, because the speed of light in a vacuum is greater than the speed of light in any material ($c > v$).

The frequency (f) of light does not change when the light enters another medium, but the wavelength of light in a material (λ_m) differs from the wavelength of that light in a vacuum (λ), as can be easily shown:

$$n = \frac{c}{v} = \frac{\lambda f}{\lambda_m f}$$

or

$$n = \frac{\lambda}{\lambda_m} \qquad (22.4)$$

The wavelength of light in the medium is then $\lambda_m = \lambda/n$. Since $n > 1$, it follows that $\lambda_m < \lambda$.

TABLE 22.1

Indices of Refraction (at $\lambda = 590$ nm)*

Substance	n
Air	1.000 29
Water	1.33
Ice	1.31
Ethyl alcohol	1.36
Fused quartz	1.46
Human eye	1.336–1.406
Polystyrene	1.49
Oil (typical value)	1.50
Glass (by type)†	1.45–1.70
crown	1.52
flint	1.66
Zircon	1.92
Diamond	2.42

*One nanometer (nm) is 10^{-9} m.

†Crown glass is a soda–lime silicate glass; flint glass is a lead–alkali silicate glass. Flint glass is more dispersive than crown glass (Section 22.5).

Example 22.2 ■ The Speed of Light in Water: Index of Refraction

Light from a laser with a wavelength of 632.8 nm travels from air into water. What are the speed and wavelength of the laser light in water?

Thinking It Through. If we know the index of refraction (n) of a medium, the speed and wavelength of light in the medium can be obtained from Eq. 22.3 and Eq. 22.4.

Solution.

Given: $n = 1.33$ (from Table 22.1)   *Find:* v and λ_m (speed and wavelength
$\lambda = 632.8$ nm   of light in water)
$c = 3.00 \times 10^8$ m/s (speed of light in air)

Since $n = c/v$,

$$v = \frac{c}{n} = \frac{3.00 \times 10^8 \text{ m/s}}{1.33} = 2.26 \times 10^8 \text{ m/s}$$

Note that $1/n = v/c = 1/1.33 = 0.75$; therefore, v is 75% of the speed of light in a vacuum. Also, $n = \lambda/\lambda_m$, so

$$\lambda_m = \frac{\lambda}{n} = \frac{632.8 \text{ nm}}{1.33} = 475.8 \text{ nm}$$

Follow-Up Exercise. The speed of light of wavelength 500 nm (in air) in a particular liquid is 2.40×10^8 m/s. What is the index of refraction of the liquid and the wavelength of light in the liquid?

The index of refraction, n, is a measure of the speed of light in a transparent material, or technically, a measure of the *optical density* of the material. For example, the speed of light in water is less than that in air, so water is said to be optically denser than air. (Optical density in general correlates with mass density. However, in some instances, a material with a greater optical density than another can have a lower mass density.) Thus, the greater the index of refraction of a material, the greater is the material's optical density and the smaller is the speed of light in the material.

For practical purposes, the index of refraction is measured in air rather than in a vacuum, since the speed of light in air is very close to c, and

$$n_{air} = \frac{c}{v_{air}} \approx \frac{c}{c} = 1$$

(From Table 22.1, $n_{air} = 1.00029$, and we will usually assume $n_{air} = 1$.)

A more practical form of Snell's law can be rewritten as

$$\frac{\sin \theta_1}{\sin \theta_2} = \frac{v_1}{v_2} = \frac{c/n_1}{c/n_2} = \frac{n_2}{n_1}$$

or

$$n_1 \sin \theta_1 = n_2 \sin \theta_2 \qquad \begin{array}{l} \textit{Snell's law} \\ \textit{(another form)} \end{array} \qquad (22.5)$$

where n_1 and n_2 are the indices of refraction for the first and second media, respectively.

Note that Eq. 22.5 can be used to measure the index of refraction. If the first medium is air, then $n_1 \approx 1$ and $n_2 \approx \sin \theta_1 / \sin \theta_2$. Thus, only the angles of incidence and refraction need to be measured to determine the index of refraction of a material experimentally. On the other hand, if the index of refraction of a material is known, it can be used in Snell's law to find the angle of refraction for any angle of incidence.

Note also that the sine of the refraction angle is inversely proportional to the index of refraction: $\sin \theta_2 \approx \sin \theta_1 / n_2$. Hence, for a given angle of incidence, the greater the index of refraction, the smaller is $\sin \theta_2$ and the smaller is the angle of refraction, θ_2.

More generally, the following relationships hold:

- If the second medium is more optically dense than the first medium ($n_2 > n_1$), the ray is refracted *toward* the normal ($\theta_2 < \theta_1$), as illustrated in ▼Fig. 22.10a.
- If the second medium is less optically dense than the first medium ($n_2 < n_1$), the ray is refracted *away from* the normal ($\theta_2 > \theta_1$), as illustrated in Fig. 22.10b.

▼ **FIGURE 22.10** Index of refraction and ray deviation **(a)** When the second medium is more optically dense than the first ($n_2 > n_1$), the ray is refracted toward the normal, as in the case of light entering water from air. **(b)** When the second medium is less optically dense than the first ($n_2 < n_1$), the ray is refracted away from the normal. [This is the case if the ray in part (a) is traced in reverse, going from medium 2 to medium 1.]

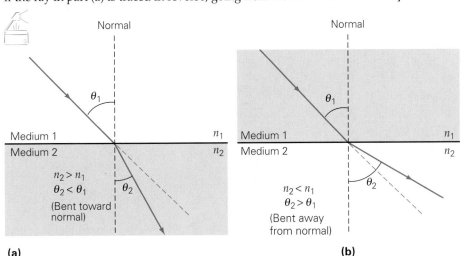

(a)   **(b)**

Demonstration/activity: Use the following methods of demonstrating refraction:
- Shine a laser into an aquarium that contains water and a small amount of powdered milk.
- Shine a laser through a piece of smoked glass.
- Put a ruler into a glass container of water and observe the bending of the image.

PHYSLET®

Exploration 34.4 Fermat's Principle and Snell's Law

Note: During refraction the product of $n \sin \theta$ remains a constant from medium to medium.

Demonstration/activity: Fill a beaker with a solution that has the same index of refraction as a drinking glass. (Try microscope immersion oil or mineral oil.) Then place another, smaller glass in the solution and illuminate it with a light. If the solution has the same index of refraction as the glass, the glass should be invisible.

Integrated Example 22.3 ■ Angle of Refraction: Snell's Law

Light in water is incident on a piece of crown glass at an angle of 37° (relative to the normal). (a) Will the transmitted ray be (1) bent toward the normal, (2) bent away from the normal, or (3) not bent at all? Use a diagram to illustrate. (b) What is the angle of refraction?

(a) Conceptual Reasoning. We can use Table 22.1 to look up the indices of refraction of water and crown glass. According to the alternative form of Snell's law (Eq. 22.5), $n_1 \sin \theta_1 = n_2 \sin \theta_2$, (1) is the correct answer. Since $n_2 > n_1$, the angle of refraction must be smaller than the angle of incidence ($\theta_2 < \theta_1$). Because both θ_1 and θ_2 are measured from the normal, the refracted ray will bend toward the normal. The ray diagram in this case is identical to Fig. 22.10a.

(b) Quantitative Reasoning and Solution. Again, the alternative form of Snell's law (Eq. 22.5) is most practical in this case. (Why?) Listing the given quantities,

Given: $\theta_1 = 37°$ *Find:*   (b) θ_2 (angle of refraction)
            $n_1 = 1.33$ (water, from Table 22.1)
            $n_2 = 1.52$ (crown glass, from Table 22.1)

The angle of refraction is found by using Eq. 22.5,

$$\sin \theta_2 = \frac{n_1 \sin \theta_1}{n_2} = \frac{(1.33)(\sin 37°)}{1.52} = 0.53$$

and

$$\theta_2 = \sin^{-1}(0.53) = 32°$$

Follow-Up Exercise. It is found experimentally that a beam of light entering a liquid from air at an angle of incidence of 37° exhibits an angle of refraction of 29° in the liquid. What is the speed of light in the liquid?

Example 22.4 ■ A Glass Tabletop: More about Refraction

A beam of light traveling in air strikes the glass top of a coffee table at an angle of incidence of 45° (▼ Fig. 22.11). The glass has an index of refraction of 1.5. (a) What is the angle of refraction for the light transmitted into the glass? (b) Prove that the emergent beam is parallel to the incident beam—that is, that $\theta_4 = \theta_1$. (c) If the glass is 2.0 cm thick, what is the lateral displacement between the ray entering and the ray emerging from the glass (the perpendicular distance between the two rays—d in the figure)?

Thinking It Through. Since two refractions are involved in this example, we use Snell's law in parts (a) and (b), and then some geometry and trigonometry in part (c).

Solution. Listing the data:

Given: $\theta_1 = 45°$ *Find:*   (a) θ_2 (angle of refraction)
            $n_1 = 1.0$ (air)   (b) Show that $\theta_4 = \theta_1$
            $n_2 = 1.5$ (c) d (lateral displacement)
            $y = 2.0$ cm

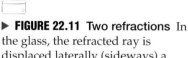

▶ **FIGURE 22.11 Two refractions** In the glass, the refracted ray is displaced laterally (sideways) a distance d from the incident ray, and the emergent ray is parallel to the original ray. (See Example 22.4.)

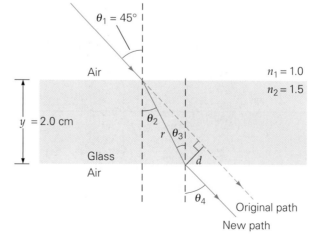

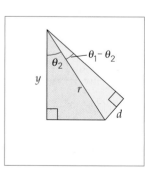

(a) Using the practical form of Snell's law, Eq. 22.5, with $n_1 = 1.0$ for air gives

$$\sin \theta_2 = \frac{n_1 \sin \theta_1}{n_2} = \frac{(1.0) \sin 45°}{1.5} = \frac{0.707}{1.5} = 0.47$$

Thus,

$$\theta_2 = \sin^{-1}(0.47) = 28°$$

Note that the beam is refracted toward the normal.

(b) If $\theta_1 = \theta_4$, then the emergent ray is parallel to the incident ray. Then applying Snell's law to the beam at both surfaces,

$$n_1 \sin \theta_1 = n_2 \sin \theta_2$$

and

$$n_2 \sin \theta_3 = n_1 \sin \theta_4$$

From the figure, $\theta_2 = \theta_3$. Therefore,

$$n_1 \sin \theta_1 = n_1 \sin \theta_4$$

or

$$\theta_1 = \theta_4$$

Thus, the emergent beam is parallel to the incident beam but displaced laterally or perpendicularly to the incident direction at a distance d.

(c) It can be seen from the inset in Fig. 22.11 that, to find d, we need to first find r from the known information in the pink right triangle. Then,

$$\frac{y}{r} = \cos \theta_2 \qquad \text{or} \qquad r = \frac{y}{\cos \theta_2}$$

In the yellow right triangle, $d = r \sin(\theta_1 - \theta_2)$. Substituting r from the previous step yields

$$d = \frac{y \sin(\theta_1 - \theta_2)}{\cos \theta_2} = \frac{(2.0 \text{ cm}) \sin(45° - 28°)}{\cos 28°} = 0.66 \text{ cm}$$

Follow-Up Exercise. If the glass in this Example had $n = 1.6$, would the lateral displacement be the same, larger, or smaller? Explain your answer conceptually, and then calculate the actual value to verify your reasoning.

Conceptual Example 22.5 ■ The Human Eye: Refraction and Wavelength

A simplified representation of the crystalline lens in a human eye shows it to have a cortex (an outer layer) of $n_{\text{cortex}} = 1.386$ and a nucleus (core) of $n_{\text{nucleus}} = 1.406$. (See Fig. 25.1b.) Note that both refraction indices are within the range listed for the human eye in Table 22.1. If a beam of monochromatic (single-frequency or -wavelength) light of wavelength 590 nm is directed from air through the front of the eye and into the crystalline lens, qualitatively compare and list the frequency, speed, and wavelength of light in air, the cortex, and the nucleus. First do the comparison without numbers, and then calculate the actual values to verify your reasoning.

Reasoning and Answer. First, the relative magnitudes of the indices of refraction are needed, where $n_{\text{air}} < n_{\text{cortex}} < n_{\text{nucleus}}$.

As learned earlier in this section, the frequency (f) of light is the same in all three media: air, the cortex, and the nucleus. Thus, the frequency can be calculated by using the speed and the wavelength of light in any of these materials, but it is easiest in air. (Why?) From the wave relationship $c = \lambda f$ (Eq. 13.17),

$$f = f_{\text{air}} = f_{\text{cortex}} = f_{\text{nucleus}} = \frac{c}{\lambda} = \frac{3.00 \times 10^8 \text{ m/s}}{590 \times 10^{-9} \text{ m}} = 5.08 \times 10^{14} \text{ Hz}$$

The speed of light in a medium depends on its index of refraction, since $v = c/n$. The smaller the index of refraction, the higher the speed. Therefore, the speed of light is the highest in air ($n = 1.00$) and lowest in the nucleus ($n = 1.406$).

The speed of light in the cortex is

$$v_{\text{cortex}} = \frac{c}{n_{\text{cortex}}} = \frac{3.00 \times 10^8 \text{ m/s}}{1.386} = 2.16 \times 10^8 \text{ m/s}$$

(continues on next page)

and the speed of light in the nucleus is

$$v_{nucleus} = \frac{3.00 \times 10^8 \text{ m/s}}{1.406} = 2.13 \times 10^8 \text{ m/s}$$

We also know that the wavelength of light in a medium depends on the index of refraction of the medium ($\lambda_m = \lambda/n$). The smaller the index of refraction, the longer the wavelength. Therefore, the wavelength of light is the longest in air ($n = 1.00$ and $\lambda = 590$ nm) and shortest in the nucleus ($n = 1.406$).

The wavelength in the cortex can be calculated from Eq. 22.4:

$$\lambda_{cortex} = \frac{\lambda}{n_{cortex}} = \frac{590 \text{ nm}}{1.386} = 426 \text{ nm}$$

and the wavelength in the nucleus is

$$\lambda_{nucleus} = \frac{590 \text{ nm}}{1.406} = 420 \text{ nm}$$

Finally, a table can be constructed to compare more easily the index of refraction, frequency, speed, and wavelength of light in the three media:

	Index of refraction	Frequency (Hz)	Speed (m/s)	Wavelength (nm)
Air	1.00	5.08×10^{14}	3.00×10^8	590
Cortex	1.386	5.08×10^{14}	2.16×10^8	426
Nucleus	1.406	5.08×10^{14}	2.13×10^8	420

Follow-Up Exercise. A light source of a single frequency is submerged in water in a special fish tank. The beam travels in the water, through double glass panes at the side of the tank (each glass pane has a different n), and into air. In general, what happens to (a) the frequency and (b) the wavelength of the light when it emerges into the outside air?

Refraction is common in everyday life and explains many things we observe. Let's look at refraction in action.

Mirage: A common example of this phenomenon sometimes occurs on a highway on a hot summer day. The refraction of light is caused by layers of air that are at different temperatures (the layer closer to the road is at a higher temperature, lower density, and lower index of refraction). This variation in indices of refraction gives rise to the observed "wet" spot and an inverted image of an object such as a car (▾Fig. 22.12a). The

▼ **FIGURE 22.12** Refraction in action **(a)** An inverted car on a "wet" road, a mirage. **(b)** The mirage is formed when light from the object is refracted by layers of air at different temperatures near the surface of the road.

(a)

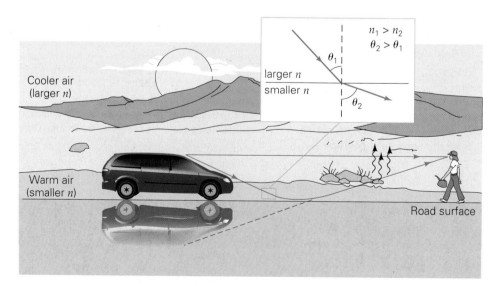

(b)

INSIGHT 22.2 NEGATIVE INDEX OF REFRACTION AND THE "PERFECT" LENS

In 1968 physicists predicted the existence of a material with a negative index of refraction. They expected that, in the presence of such a negative-index material, nearly all known wave propagation and optical phenomena would be substantially altered. Negative-index materials were not known to exist at the time, however.

At the beginning of the twenty-first century, a new class of artificially structured materials was created that were found to have negative indices of refraction. Moreover, a natural ferroelastic material called a *"twinned" alloy* (containing yttrium, vanadium, and oxygen) has also displayed a negative index of refraction (Fig. 1).

Figure 2 illustrates the difference between materials with positive and negative indices of refraction. In Fig. 2a, light incident on a positive-index material is refracted to the other side of the normal to the interface. However, if the material has a negative index of refraction, the same incident light is refracted to the same side of the normal to the interface (Fig. 2b). Due to

this "abnormal" refraction, negative-index slab materials with flat surfaces can even focus light as shown in Fig. 2c, resulting in a new class of lenses (to be discussed in Chapter 23). If a light source is placed on one side of a slab with a refractive index $n = -1$, the light rays are refracted in such a way as to produce a focal point inside the material and then another just outside the material. The "focal length" of such a lens would depend on both the object distance and the thickness of the slab.

The undesirable characteristics of lenses made of materials with a positive index of refraction are energy loss due to reflection, aberrations, and low resolution due to diffraction limit (more on this in Chapter 24). The latest experiments provide strong evidence that negative-index materials have an important future in imaging. This is because negative-index lenses offer a new degree of flexibility that could lead to more compact lenses with reduced lens aberration. The diffraction limit—which is the most fundamental limitation to image resolution—may be circumvented by negative-index materials. Furthermore, total negative refraction—that is, absence of a reflection—has been observed in materials with a negative index of refraction. Such a lens would truly be a "perfect lens."

FIGURE 1 Material with a negative index of refraction This artificial material made from grids of rings and wires has a negative index of refraction.

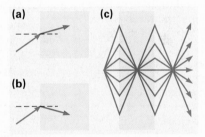

FIGURE 2 Reflection in positive-index versus negative-index materials (a) Light incident on the interface between air and a positive-index material is bent toward the other side of the normal, (b) whereas in a negative-index material light is bent toward the same side of the normal. (c) If a light source is placed on one side of a slab with a refractive index of $n = -1$, the waves are refracted in such a way as to produce a focus inside the material and then another just outside the material.

term *mirage* generally brings to mind a thirsty person in the desert "seeing" a pool of water that really isn't there. This optical illusion plays tricks on the mind, with the image usually seen as in a pool of water and our eye's past experience unconsciously leading us to conclude that there is water on the road.

In Fig. 22.12b, there are two ways for light to get to our eyes from the car. First, the horizontal rays come directly from the car to our eyes, so we see the car above the ground. Also, the rays from the car that travel toward the road surface will be gradually refracted by the layered air. After hitting the surface, these rays will be refracted again and travel toward our eyes. (See the inset in the figure.) Cooler air has a higher density and so a higher index of refraction. A ray traveling toward the road surface will be gradually refracted with increasing angle of refraction until it hits the surface. It will then be refracted again with decreasing angle of refraction, going toward our eyes. As a consequence, we also see an inverted image of the car, below the road surface. In other words, the surface of the road acts almost as a mirror.

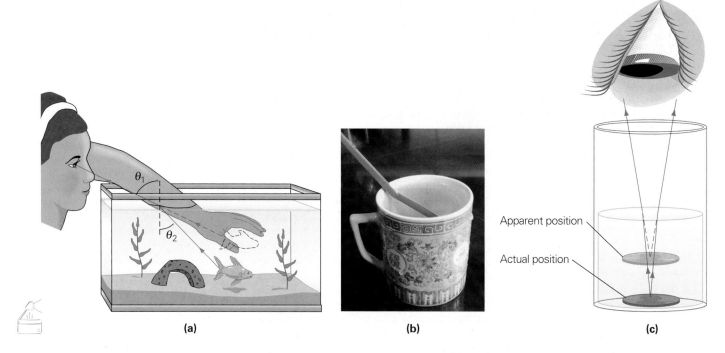

(a) (b) (c)

▲ **FIGURE 22.13 Refractive effects**
(a) The light is refracted, and because we tend to think of light as traveling in straight lines, the fish is below where we think it is.
(b) The chopstick appears bent at the air–water boundary. If the cup is transparent, we see a different refraction. (See Exercise 21.)
(c) Because of refraction, the coin appears to be closer than it actually is.

The "pool of water" is actually sky light being refracted—an image of the sky. This layering of air of different temperatures, creating different indices of refraction, causes us to "see" the rising hot air as a result of continually changing refraction.

The opposite of this is the mirage at sea (looming effect). At the sea, the air above is warmer than below. This causes the light to be refracted opposite as in Fig. 22.12b, causing objects to be seen in the air above sea.

Not where it should be: You may have experienced a refractive effect while trying to reach for something underwater, such as a fish (▲Fig. 22.13a). We are used to light traveling in straight lines from objects to our eyes, but the light reaching our eyes from a submerged object has a directional change at the air–water interface. (Note in the figure that the ray is refracted away from the normal.) As a result, the object appears closer to the surface than it actually is, and therefore we tend to miss the object when reaching for it. For the same reason, a chopstick in a cup appears bent (Fig. 22.13b), a coin in a glass of water will appear closer than it really is (Fig. 22.13c), and the legs of a person standing in water seem shorter than their actual length. The relationship between the true depth and the apparent depth can be calculated. (See Exercise 37.)

Atmospheric effects: The Sun on the horizon sometimes appears flattened, with its horizontal dimension greater than its vertical dimension (▼Fig. 22.14a). This effect is the result of temperature and density variations in the denser air along the horizon. These variations occur predominantly vertically, so light from the top and bottom portions of the Sun are refracted differently as the two sets of beams pass through different atmospheric densities with different indices of refraction.

Atmospheric refraction lengthens the day, so to speak, by allowing us to see the Sun (or the Moon, for that matter) just before it actually rises above the horizon and

▶ **FIGURE 22.14 Atmospheric effects** (a) The Sun on the horizon commonly appears flattened as a result of atmospheric refraction. (b) Before rising and after setting, the Sun can be seen briefly also because of atmospheric refraction. (Exaggerated for illustration).

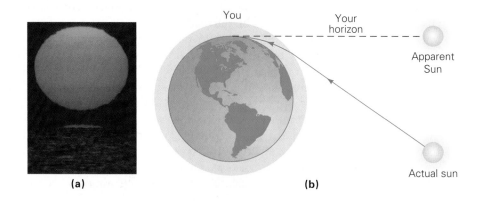

(a) (b)

just after it actually sets below the horizon (as much as 20 min on both ends). The denser air near the Earth refracts the light over the horizon toward us (Fig. 22.14b).

The twinkling of stars is due to atmospheric turbulence, which distorts the light from the stars. The turbulences refract light in random directions and cause the stars to appear to "twinkle." Stars on the horizon will appear to twinkle more than stars directly overhead, because the light has to pass through more of the Earth's atmosphere. However, planets do not "twinkle" as much. This is because stars are much farther away than planets, so they appear as point sources. Outside the Earth's atmosphere, the stars don't twinkle.

22.4 Total Internal Reflection and Fiber Optics

OBJECTIVES: To (a) describe total internal reflection, and (b) understand fiber-optic applications.

An interesting phenomenon occurs when light travels from a more optically dense medium into a less optically dense one, such as when light goes *from* water *into* air. As you know, in such a case a ray will be refracted away from the normal. (The angle of refraction is larger than the angle of incidence.) Furthermore, Snell's law states that the greater the angle of incidence, the greater the angle of refraction. That is, as the angle of incidence increases, the farther the refracted ray diverges from the normal.

However, there is a limit. For a certain angle of incidence called the **critical angle (θ_c)**, the angle of refraction is 90°, and the refracted ray is directed along the boundary between the media. But what happens if the angle of incidence is even larger? If the angle of incidence is greater than the critical angle ($\theta_1 > \theta_c$), the light isn't refracted at all, but is internally reflected (▼Fig. 22.15). This condition is called **total internal reflection**. The reflection process is about 100% efficient. (There is always some absorption of light *in* the materials.) Because of total internal reflection, glass prisms can be used as mirrors (▶Fig. 22.16). In summary, where $n_1 > n_2$, reflection and refraction occur at all angles for $\theta_1 \leq \theta_c$, but the refracted or transmitted ray disappears at $\theta_1 > \theta_c$.

An expression for the critical angle can be obtained from Snell's law. If $\theta_1 = \theta_c$ in the optically denser medium, $\theta_2 = 90°$, and it follows that

$$n_1 \sin \theta_1 = n_2 \sin \theta_2 \quad \text{or} \quad n_1 \sin \theta_c = n_2 \sin 90°$$

Since $\sin 90° = 1$,

$$\sin \theta_c = \frac{n_2}{n_1} \quad \text{where } n_1 > n_2 \tag{22.6}$$

▼ **FIGURE 22.15 Internal reflection** (a) When light enters a less optically dense medium, it is refracted away from the normal. At a critical angle (θ_c), the light is refracted along the interface (common boundary) of the media. At an angle greater than the critical angle ($\theta_1 > \theta_c$), there is total internal reflection. (b) Can you estimate the critical angle in the photograph?

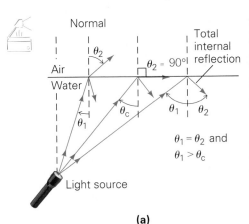
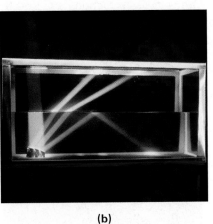

(a)   (b)

Demonstration/activity: Punch a smooth hole in an empty plastic soda bottle, and fill it with water and a pinch of dry milk. Then shine a laser through the bottle into the stream of water as it flows out of the bottle. The light beam will bend with the stream of water.

Exploration 34.2 Snell's Law and Total Internal Reflection

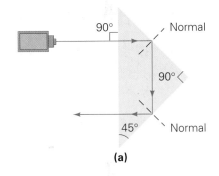

(a)

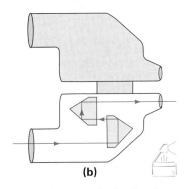

(b)

▲ **FIGURE 22.16 Internal reflection in a prism** (a) Because the critical angle of glass is less than 45°, prisms with 45° and 90° angles can be used to reflect light through 180°. (b) Internal reflection of light by prisms in binoculars makes this instrument much shorter than a telescope because the rays are "folded" by the prisms.

If the second medium is air, $n_2 \approx 1$, and the critical angle at the boundary from a medium into air is given by $\sin \theta_c = 1/n$, where n is the index of refraction of the medium. This is another method that can be used to measure the index of refraction in laboratories.

Example 22.6 ■ A View from the Pool: Critical Angle

(a) What is the critical angle for light traveling in water and incident on a water–air boundary? (b) If a diver submerged in a pool looked up at the surface of the water at an angle of $\theta < \theta_c$, what would she see? (Neglect any thermal or motional effects.)

Thinking It Through. (a) The critical angle is given by Eq. 22.6. (b) As shown in Fig. 22.15a, θ_c forms a cone of vision for viewing from below the water.

Solution.

Given: $n_1 = 1.33$ (for water, from Table 22.1)   *Find:* (a) θ_c (critical angle)
  $n_2 \approx 1$ (why?)   (b) view for $\theta < \theta_c$

(a) The critical angle is

$$\theta_c = \sin^{-1}\left(\frac{n_2}{n_1}\right) = \sin^{-1}\left(\frac{1}{1.33}\right) = 48.8°$$

(b) Using Fig. 22.15a, trace the rays in reverse for light coming from all angles outside the pool. Light coming from the above-water 180° panorama could be viewed only in a cone with a half-angle of 48.8°. As a result, objects above the surface would also appear distorted. An underwater panoramic view is seen in ◄Fig. 22.17. Now can you explain why wading birds like herons usually keep their bodies low while trying to catch a fish?

Follow-Up Exercise. What would the diver see when looking up at the water surface at an angle of $\theta > \theta_c$?

▲ **FIGURE 22.17 Panoramic and distorted** An underwater view of the surface of a swimming pool in Hawaii. (See Example 22.6.)

Internal reflections enhance the brilliance of cut diamonds. (Brilliance is a measure of the amount of light returning straight back to the viewer. Brilliance is reduced if light leaks out the back of a diamond—that is, if the reflection is *not* total.) The critical angle for a diamond–air surface is

$$\theta_c = \sin^{-1}\left(\frac{1}{n}\right) = \sin^{-1}\left(\frac{1}{2.42}\right) = 24.4°$$

A so-called brilliant-cut diamond has many facets, or faces (58 in all—33 on the upper face and 25 on the lower). Light from above hitting the lower facets at angles greater than the critical angle is internally reflected in the diamond. The light then emerges from the upper facets, giving rise to the diamond's brilliance (▼Fig. 22.18).

Fiber Optics

When a fountain is illuminated from below, the light is transmitted along the curved streams of water. This phenomenon was first demonstrated in 1870 by the British scientist John Tyndall (1820–1893), who showed that light was "conducted"

Demonstration/activity: Shine laser light through a piece of fiber-optic cable to show total internal reflection.

▶ **FIGURE 22.18 Diamond brilliance**
(a) Internal reflection gives rise to a diamond's brilliance. **(b)** The "cut," or the depth proportions, of the facets is critical. If a stone is too shallow or too deep, light will be lost (refracted out) through the lower facets.

(a)

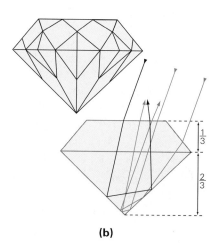

(b)

along the curved path of a stream of water flowing from a hole in the side of a container. The phenomenon is observed because light undergoes total internal reflection along the stream.

Total internal reflection forms the basis of **fiber optics**, a fascinating modern technology field centered on the use of transparent fibers to transmit light. Multiple total internal reflections make it possible to "pipe" light along a transparent rod (as in streams of water), even if the rod is curved (▶Fig. 22.19). Note from the figure that the smaller the diameter of the light pipe, the more total internal reflections it has. A small fiber can produce as many as several hundred total internal reflections per centimeter.

Total internal reflection is an exceptionally efficient process. Optical fibers can be used to transmit light over very long distances with losses of only about 25% per kilometer. These losses are due primarily to impurities in the fiber, which scatter the light. Transparent materials have different degrees of transmission. Fibers are made of special plastics and glasses for maximum transmission efficiency. The greatest efficiency is achieved with infrared radiation, because there is less scattering, as will be learned in Section 24.5.

The greater efficiency of multiple total internal reflections compared with multiple mirror reflections can be illustrated by a good reflecting plane mirror, which has at best a reflectivity of about 95%. After each reflection, the beam intensity is 95% of that of the incident beam from the preceding reflection ($I_1 = 0.95\ I_0$; $I_2 = 0.95\ I_1 = 0.95^2\ I_0; \ldots$). Therefore, the intensity I of the reflected beam after n reflections is given by

$$I = 0.95^n\ I_0$$

where I_0 is the initial intensity of the beam before the first reflection. Thus, after 14 reflections,

$$I = 0.95^{14}\ I_0 = 0.49\ I_0$$

In other words, after 14 reflections, the intensity is reduced to less than half (49%). For 100 reflections, $I = 0.006\ I_0$, and the intensity is only 0.6% of the initial intensity! Compare this to about 75% of the initial intensity in optical fibers over a kilometer in length with *thousands* of reflections, so you can see the advantage of total internal reflection.

Fibers whose diameters are about 10 μm (10^{-5} m) are grouped together in flexible bundles that are 4 to 10 mm in diameter and up to several meters in length, depending on the application (▼Fig. 22.20). A fiber bundle with a

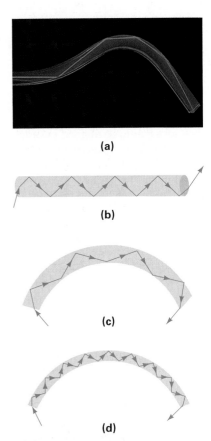

(a)

(b)

(c)

(d)

▲ **FIGURE 22.19** Light pipes
(a) Total internal reflection in an optical fiber. **(b)** When light is incident on the end of a cylindrical form of transparent material such that the internal angle of incidence is greater than the critical angle of the material, the light undergoes total internal reflection down the length of the light pipe. **(c)** Light is also transmitted along curved light pipes by total internal reflection. **(d)** As the diameter of the rod or fiber becomes smaller, the number of reflections per unit length increases.

Illustration 34.2 Fiber Optics

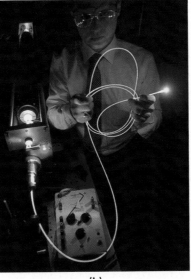

(a) **(b)**

◀ **FIGURE 22.20** Fiber-optic bundle **(a)** Hundreds or even thousands of extremely thin fibers are grouped together **(b)** to make an optical fiber, here colored blue by a laser.

INSIGHT 22.3 FIBER OPTICS: MEDICAL APPLICATIONS

Before fiber optics, *endoscopes*—instruments used to view internal portions of the human body—consisted of lens systems in long, narrow tubes. Some endoscopes contained a dozen or more lenses and produced relatively poor images. Because the lenses had to be aligned in certain ways, the tubes had to have rigid sections, which limited the endoscope's maneuverability. Such an endoscope could be inserted down the patient's throat into the stomach to observe the stomach lining. However, there were blind spots due to the curvature of the stomach and the inflexibility of the instrument.

Fiber-optic bundles have eliminated these problems. Lenses placed at the end of the fiber bundles focus the light, and a prism is used to change the direction for its return. The incident light is usually transmitted by an outer layer of fiber bundles, and the image is returned through a central core of fibers. Me-

chanical linkages allow maneuverability. The end of a fiber endoscope can be equipped with devices to obtain specimens of the viewed tissues for biopsy (diagnostic examination) or even to perform surgical procedures. For example, arthroscopic surgery is performed on injured joints (Fig. 1). The *arthroscope* that is now routinely used for inspecting *and* repairing damaged joints is simply a fiber endoscope fitted with appropriate surgical implements.

A fiber-optic *cardioscope* (for direct observation of heart valves) typically is a fiber bundle about 4 mm in diameter and 30 cm long. Such a cardioscope passes easily to the heart through the jugular vein, which is about 15 mm in diameter, in the neck. To displace the blood and provide a clear field of view for observing and photographing, a transparent balloon at the tip of the cardioscope is inflated with saline (saltwater) solution.

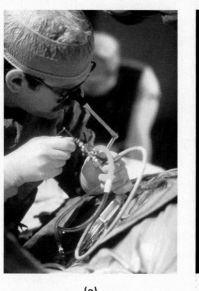

(a)

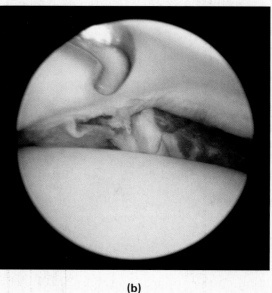

(b)

FIGURE 1 Arthroscopy (a) A fiber-optic arthroscope used to perform surgery. **(b)** An arthroscopic view of a torn knee meniscus.

cross-sectional area of 1 cm² can contain as many as 50 000 individual fibers. (A coating on each fiber is needed to keep the fibers from touching each other.)

There are many important and interesting applications of fiber optics, including communications, computer networking, and medical applications. (See Insight 22.3 on Fiber Optics: Medical Applications.) Light signals, converted from electrical signals, are transmitted through optical telephone lines and computer networks. At the other end, they are converted back to electrical signals. Optical fibers have lower energy losses than electric current–carrying wires, particularly at higher frequencies, and can carry far more data. Also, optical fibers are lighter than metal wires, have greater flexibility, and are not affected by electromagnetic disturbances (electric and magnetic fields), because they are made of materials that are electrical insulators.

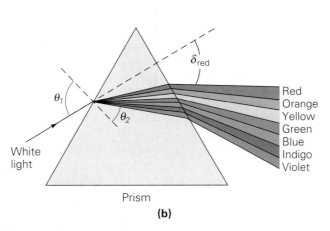

FIGURE 22.21 Dispersion (a) White light is dispersed into a spectrum of colors by glass prisms. **(b)** In a dispersive medium, the index of refraction varies slightly with wavelength. Red light, longest in wavelength, has the smallest index of refraction and is refracted least. The angle between the incident beam and an emergent ray is the angle of deviation (δ) for that ray. (The angles are exaggerated for clarity.) **(c)** Variation in the index of refraction with wavelength for some common transparent media.

22.5 Dispersion

OBJECTIVE: To explain dispersion and some of its effects.

Light of a single frequency, and consequently a single wavelength, is called *monochromatic light* (from the Greek *mono*, meaning "one," and *chroma*, meaning "color"). Visible light that contains all the component frequencies, or colors, at about the same intensities (such as sunlight) is termed *white light*. When a beam of white light passes through a glass prism, as shown in ▲ Fig. 22.21a, it is spread out, or dispersed, into a spectrum of colors. This phenomenon led Newton to believe that sunlight is a mixture of colors. When the beam enters the prism, the component colors, corresponding to different wavelengths of light, are refracted at slightly different angles, so they spread out into a spectrum (Fig. 22.21b).

The emergence of a spectrum indicates that the index of refraction of glass is slightly different for different wavelengths, which is true of many transparent media (Fig. 22.21c). The reason has to do with the fact that in a dispersive medium the speed of light is slightly different for different wavelengths. Since the index of refraction n of a medium is a function of the speed of light in that medium ($n = c/v$), the index of refraction is different for different wavelengths. It follows from Snell's law that light of different wavelengths will be refracted at different angles.

We can summarize the preceding discussion by saying that in a transparent material with different indices of refraction for different wavelengths of light, refraction causes a separation of light according to wavelength, and the material is said to be *dispersive* and exhibit **dispersion**. Dispersion varies with different media (Fig. 22.21c). Also, because the differences in the indices of refraction for different wavelengths are small, a representative value at some specified wavelength can be used for general purposes. (See Table 22.1.)

A good example of a dispersive material is diamond, which is about five times as dispersive as glass. In addition to revealing the brilliance resulting from internal reflections off many facets, a cut diamond shows a display of colors, or "fire," resulting from the dispersion of the refracted light.

Dispersion is a cause of chromatic aberration in lenses, which is described more fully in Chapter 23. Optical systems in cameras often consist of several lenses to minimize this problem (see Section 23.4).

Another dramatic example of dispersion is the production of a rainbow, as discussed in Insight 22.4 on The Rainbow on page 722.

PHYSLET®

Illustration 34.3 Prisms and Dispersion

Note: You can remember the sequence of the colors of the visible spectrum (from the long-wavelength end to the short-wavelength end) by using the name ROY G. BIV, which is an acronym for *r*ed, *o*range, *y*ellow, *g*reen, *b*lue, *i*ndigo, and *v*iolet.

INSIGHT 22.4 THE RAINBOW

Everyone has been fascinated by the beautiful array of colors of a rainbow. With the optical principles learned in this chapter, we are now in a position to understand the formation of this spectacular display.

A rainbow is formed by refraction, dispersion, and internal reflection of light within water droplets. When sunlight shines on millions of water droplets in the air during and after a rain, a multicolored arc is seen whose colors run from violet along the lower part of the bow (in order of wavelength) to red along the upper part. Occasionally, more than one rainbow is seen: The main, or primary, rainbow is sometimes accompanied by a fainter and higher secondary rainbow (Fig. 1) or even a third rainbow. These higher-order rainbows are caused by more than one total internal reflection within the water droplets.

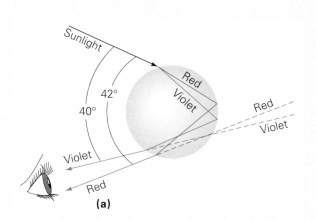

FIGURE 1 Rainbow The colors of the primary rainbow run vertically from red (top) to violet (bottom).

The light that forms the primary rainbow is first refracted and dispersed in each water droplet, then totally reflected once at the back surface of each droplet. Finally, it is refracted and dispersed again upon exiting each droplet, resulting in the light being spread out in different directions into a spectrum of colors (Fig. 2a). However, because of the conditions for refraction and total internal reflection in water, the angles between incoming and outgoing rays for violet to red light lie within a narrow range of 40° to 42°. This means that you can see a rainbow only when the Sun is behind you, so that the dispersed light travels to you only at these angles.

Red appears on the top of the rainbow because light of shorter wavelengths from those water droplets will pass over our eyes (Fig. 2b). Similarly, violet is at the bottom of the rainbow because light of longer wavelengths passes under our eyes.

The secondary rainbow has reversed color orders because of the extra reflection.

Rainbows are generally seen only as arcs, because their formation is cut off at the ground. When on a cliff or on an airplane, you might see a complete circular rainbow (Fig. 2b). Also, the higher the Sun is in the sky, the less of a rainbow will be seen from the ground. In fact, a primary rainbow would not be seen if the Sun's angle above the horizon is greater than 42°. The primary rainbow can still be seen from a height, however. As an observer's elevation increases, more of the arc becomes visible. You may also have seen a circular rainbow in the spray from a garden hose.

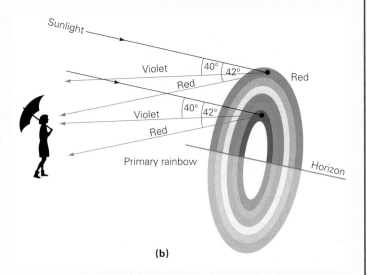

FIGURE 2 The rainbow Rainbows are created by the refraction, dispersion, and internal reflection of sunlight within water droplets. **(a)** Light of different colors emerges from the droplet in different directions. **(b)** An observer sees red light at the top of the rainbow and violet at the bottom.

Integrated Example 22.7 ■ Forming a Spectrum: Dispersion

The index of refraction of a particular transparent material is 1.4503 for the red end ($\lambda_r = 700$ nm) of the visible spectrum and 1.4698 for the blue end ($\lambda_b = 400$ nm). White light is incident on a prism of this material, as in Fig. 22.21b, at an angle of incidence θ_i of 45°. (a) Inside the prism, the angle of refraction of the red light is (a) larger than, (2) smaller than, or (3) the same as the angle of refraction of the blue light. Explain. (b) What is the angular separation of the visible spectrum inside the prism?

(a) Conceptual Reasoning. The angle of refraction is given by Snell's law, $n_1 \sin \theta_1 = n_2 \sin \theta_2$. Since red light has a smaller index of refraction than blue light, the angle of refraction of red light is larger than that of blue light, for the same angle of incidence. Sometimes, we also say that the red light is "refracted less" than blue light because the larger angle of refraction of the red light means it is closer to the direction of the original incident ray. So the answer is (a).

Teaching tip: Make sure students know which color of light is dispersed the most and why.

(b) Quantitative Reasoning and Solution. Again, we use Snell's law to compute the angle of refraction for the red and blue ends of the visible spectrum. The angular separation of the two colors inside the prism is the difference between these angles of refraction.

Given: (red) $n_r = 1.4503$ for $\lambda_r = 700$ nm
(blue) $n_b = 1.4698$ for $\lambda_b = 400$ nm
$\theta_1 = 45°$

Find: $\Delta\theta_2$ (angular separation)

Using Eq. 22.5 with $n_1 = 1.00$ (air),

$$\sin \theta_{2_r} = \frac{\sin \theta_1}{n_{2_r}} = \frac{\sin 45°}{1.4503} = 0.48756 \quad \text{and} \quad \theta_{2_r} = 29.180°$$

Similarly,

$$\sin \theta_{2_b} = \frac{\sin \theta_1}{n_{2_b}} = \frac{\sin 45°}{1.4698} = 0.48109 \quad \text{and} \quad \theta_{2_b} = 28.757°$$

So

$$\Delta\theta_2 = \theta_{2_r} - \theta_{2_b} = 29.180° - 28.757° = 0.423°$$

This is not much of a deviation, but as the light travels to the other side of the prism, it is refracted and dispersed again by the second boundary. Thus the colors spread out even farther. When the light emerges from the prism, the dispersion becomes evident (Fig. 22.21a).

Follow-Up Exercise. In the prism in this Example, if the green light exhibits an angular separation of 0.156° from the red light, what is the index of refraction for green light in the material? Will the green light refract more or less than the red light? Explain.

Chapter Review

• **Law of reflection:** The angle of incidence equals the angle of reflection (as measured from the normal to the reflecting surface):

$$\theta_i = \theta_r \tag{22.1}$$

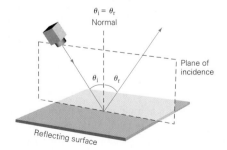

• The **index of refraction (n)** of any medium is the ratio of the speed of light in a vacuum to its speed in that medium:

$$n = \frac{c}{v} = \frac{\lambda}{\lambda_m} \tag{22.3, 22.4}$$

• The refraction of light as it enters one medium from another is given by **Snell's law**. If the second medium is more optically dense, the ray is refracted toward the normal; if the medium is less dense, the ray is refracted away from the normal. Snell's law is

$$\frac{\sin \theta_1}{\sin \theta_2} = \frac{v_1}{v_2} \tag{22.2}$$

$$n_1 \sin \theta_1 = n_2 \sin \theta_2 \tag{22.5}$$

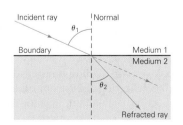

• **Total internal reflection** occurs if the second medium is less dense than the first and the angle of incidence exceeds the critical angle, which is given by

$$\sin \theta_c = \frac{n_2}{n_1} \qquad (n_1 > n_2) \qquad (22.6)$$

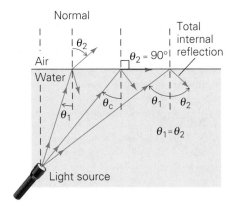

• **Dispersion** of light occurs in some media because different wavelengths have slightly different indices of refraction and hence different speeds. This results in slightly different refraction angles for different wavelengths.

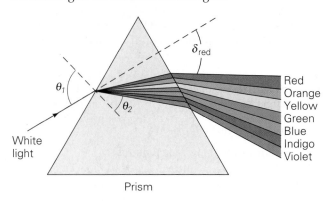

Exercises*

MC = *Multiple Choice Question,* **CQ** = *Conceptual Question, and* **IE** = *Integrated Exercise. Throughout the text, many exercise sections will include "paired" exercises. These exercise pairs, identified with* **red numbers**, *are intended to assist you in problem solving and learning. In a pair, the first exercise (even numbered) is worked out in the Study Guide so that you can consult it should you need assistance in solving it. The second exercise (odd numbered) is similar in nature, and its answer is given at the back of the book.*

22.1 Wave Fronts and Rays *and* 22.2 Reflection

1. **MC** A ray (a) is perpendicular to the direction of energy flow, (b) is always parallel to other rays, (c) is perpendicular to a series of wave fronts, (d) illustrates the wave nature of light. (c)

2. **MC** The angle of incidence is the angle between (a) the incident ray and the reflecting surface, (b) the incident ray and the normal to the surface, (c) the incident ray and the reflected ray, (d) the reflected ray and the normal to the surface. (b)

3. **MC** For both specular (regular) and diffuse (irregular) reflections, (a) the angle of incidence equals the angle of reflection, (b) the incident and reflected rays are on opposite sides of the normal, (c) the incident ray, the reflected ray, and the local normal lie in the same plane, (d) all of the preceding. (d)

4. **CQ** Under what circumstances will the angle of reflection be smaller than the angle of incidence?
   never, they are always equal

5. **CQ** The book you are reading is not, itself, a light source, so it must be reflecting light from other sources. What type of reflection is this? irregular or diffuse, because the paper is microscopically rough

*Assume angles to be exact.

6. **CQ** When you see the Sun over a lake or the ocean, you often observe a long swath of light (▼Fig. 22.22). What causes this effect, sometimes called a "glitter path"? see ISM

◄ **FIGURE 22.22**
A glitter path
See Exercise 6.

7. ● The angle of incidence of a light ray on a mirrored surface is 35°. What is the angle between the incident and reflected rays? 70°

8. ● A beam of light is incident on a plane mirror at an angle of 32° relative to the normal. What is the angle between the reflected rays and the surface of the mirror? 58°

9. **IE** ● A beam of light is incident on a plane mirror at an angle α relative to the surface of the mirror. (a) Will the angle between the reflected ray and the normal be (1) α, (2) 90° − α, or (3) 2α? (b) If α = 43°, what is the angle between the reflected ray and the normal? (a) (2) 90° − α (b) 47°

10. **IE ●●** Two upright plane mirrors touch along one edge, where their planes make an angle of α. A beam of light is directed onto one of the mirrors at an angle of incidence $\beta < \alpha$ and is reflected onto the other mirror. (a) Will the angle of reflection of the beam from the second mirror be (1) α, (2) β, (3) $\alpha + \beta$, or (4) $\alpha - \beta$? (b) If $\alpha = 60°$ and $\beta = 40°$, what will be the angle of reflection of the beam from the second mirror? (a) (4) $\alpha - \beta$ (b) 20°

11. **IE ●●** Two identical plane mirrors of width w are placed a distance d apart with their mirrored surfaces parallel and facing each other. (a) A beam of light is incident at one end of one mirror so that the light just strikes the far end of the other mirror after reflection. Will the angle of incidence be (1) $\sin^{-1}(w/d)$, (2) $\cos^{-1}(w/d)$, or (3) $\tan^{-1}(w/d)$? (b) If $d = 50$ cm and $w = 25$ cm, what is the angle of incidence? (a) (3) $\tan^{-1}(w/d)$ (b) 27°

12. **●●** Two people stand 3.0 m away from a large plane mirror and spaced 5.0 m apart in a dark room. At what angle of incidence should one of them shine a flashlight on the mirror so that the reflected beam directly strikes the other person? 40°

13. **●●** A beam of light is incident on a plane mirror at an angle of incidence of 35°. If the mirror rotates through a small angle of θ, through what angle will the reflected ray rotate? 2θ

14. **●●●** Two plane mirrors, M_1 and M_2, are placed together as illustrated in ▼Fig. 22.23. (a) If the angle α between the mirrors is 70° and the angle of incidence, θ_{i_1}, of a light ray incident on M_1 is 35°, what is the angle of reflection, θ_{r_2}, from M_2? (b) If $\alpha = 115°$ and $\theta_{i_1} = 60°$, what is θ_{r_2}? (a) 35° (b) 55°

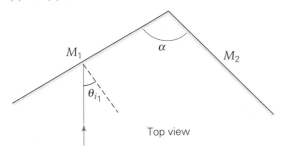

▲ **FIGURE 22.23 Plane mirrors together** See Exercises 14 and 15.

15. **●●●** For the plane mirrors in Fig. 22.23, what angles α and θ_{i_1} would allow a ray to be reflected back in the direction from which it came (parallel to the incident ray)? 90°, any θ_{i_1}

22.3 Refraction *and* 22.4 Total Internal Reflection and Fiber Optics

16. **MC** Light refracted at the boundary of two different media (a) is bent toward the normal when $n_1 > n_2$, (b) is bent away from the normal when $n_1 > n_2$, (c) is bent away from the normal when $n_1 < n_2$, (d) has the same angle of refraction as the angle of incidence. (b)

17. **MC** The index of refraction (a) is always greater than or equal to 1, (b) is inversely proportional to the speed of light in a medium, (c) is inversely proportional to the wavelength of light in the medium, (d) all of the preceding. (d)

18. **MC** Which of the following must be satisfied for total internal reflection to occur: (a) $n_1 > n_2$, (b) $n_2 > n_1$, (c) $\theta_1 > \theta_c$, or (d) $\theta_1 < \theta_c$. (a) and (c)

19. **CQ** Explain the fundamental physical reason for refraction. light speed depends on medium

20. **CQ** As light travels from one medium to another, does its wavelength change? Its frequency? Its speed? yes, no, yes

21. **CQ** Explain why the pencil in ▼Fig. 22.24 appears almost severed. Also, compare this figure with Fig. 22.13b and explain the difference. see ISM

◀ **FIGURE 22.24 Refraction effect** See Exercise 21.

22. **CQ** The photos in ▼Fig. 22.25 were taken with a camera on a tripod at a fixed angle. There is a penny in the container, but only its tip is seen initially. However, when water is added, more of the coin is seen. Why? Use a diagram to explain. see ISM

▲ **FIGURE 22.25 You barely see it, but then you do** See Exercises 22 and 52.

23. **CQ** Two hunters, one with bow and arrow and the other with a laser gun, see a fish under water. They both aim directly where they see it. Which one, the arrow or the laser beam, has a better chance of hitting the fish? Explain. the laser beam

24. **●** The speed of light in the core of the crystalline lens in a human eye is 2.13×10^8 m/s. What is the index of refraction of the core? 1.41

25. **IE ●** The indices of refraction for diamond and zircon can be found in Table 22.1. (a) The speed of light in zircon is (1) greater than, (2) less than, (3) the same as the speed of light in diamond. Explain. (b) Compute the ratio of the speed of light in zircon to that in diamond. (a) (1) greater than, because the index of refraction is smaller (b) 1.26

26. IE ● A beam of light enters water from air. (a) Will the angle of refraction be (1) greater than, (2) equal to, or (3) less than the angle of incidence? Explain. (b) If the beam enters the water at an angle of 60° relative to the normal of the surface, find the angle of refraction. (a) (3) less than (b) 41°

27. IE ● Light passes from a crown glass container into water. (a) Will the angle of refraction be (1) greater than, (2) equal to, or (3) less than the angle of incidence? Explain. (b) If the angle of refraction is 20°, what is the angle of incidence? (a) (1) greater than (b) 17°

28. ● A beam of light traveling in air is incident on a transparent plastic material at an angle of incidence of 50°. The angle of refraction is 35°. What is the index of refraction of the plastic? 1.34

29. IE ● (a) For total internal reflection to occur, should the light be directed from (1) air to a diamond, (2) a diamond to air? Explain. (b) What is the critical angle of the diamond in air? (a) (2) a diamond to air (b) 24.4°

30. ● The critical angle for a certain type of glass in air is 41.8°. What is the index of refraction of the glass? 1.50

31. ●● A beam of light in air is incident on the surface of a slab of fused quartz. Part of the beam is transmitted into the quartz at an angle of refraction of 30° relative to a normal to the surface, and part is reflected. What is the angle of reflection? 47°

32. ●● A beam of light is incident from air onto a flat piece of polystyrene at an angle of 55° relative to a normal to the surface. What angle does the refracted ray make with the plane of the surface? 57°

33. ●● Monochromatic blue light that has a frequency of 6.5×10^{14} Hz enters a piece of flint glass. What are the frequency and wavelength of the light in the glass? 6.5×10^{14} Hz; 2.8×10^{-7} m

34. IE ●● Light passes from material A, which has an index of refraction of $\frac{4}{3}$, into material B, which has an index of refraction of $\frac{5}{4}$. (a) The speed of light in material A is (1) greater than, (2) the same as, (3) less than the speed of light in material B. Explain. (b) Find the ratio of the speed of light in material A to the speed of light in material B. (a) (3) less than (b) $\frac{15}{16}$

35. IE ●● In Exercise 34, (a) the wavelength of light in material A is (1) greater than, (2) the same as, (3) less than the wavelength of light in material B. Explain. (b) What is the ratio of the light's wavelength in material A to that in material B? (a) (3) less than (b) $\frac{15}{16}$

36. ●● The laser used in cornea surgery to treat corneal disease is the *excimer laser*, which emits ultraviolet light at a wavelength of 193 nm in air. The index of refraction of the cornea is 1.376. What are the wavelength and frequency of the light in the cornea? 140 nm, 1.55×10^{15} Hz

37. ●● (a) An object immersed in water appears closer to the surface than it actually is. What is the cause of this illusion? (b) Using ▼ Fig. 22.26, show that the apparent depth for small angles of refraction is d/n, where n is the index of refraction of the water. [*Hint*: Recall that for small angles, $\tan \theta \approx \sin \theta$.] (a) refraction (b) see ISM

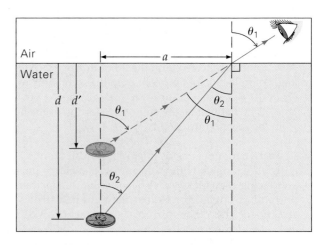

▲ **FIGURE 22.26 Apparent depth?** See Exercise 37. (For small angles only; angles enlarged for clarity.)

38. ●● A person lying at poolside looks over the edge of the pool and sees a bottle cap on the bottom directly below, where the depth is 3.2 m. How far below the water surface does the bottle cap appear to be? (See Exercise 37b.) 2.4 m

39. ●● What percentage of the actual depth is the apparent depth of an object submerged in water if the observer is looking almost straight downward? (See Exercise 37b.) 75.2%

40. ●● A light ray in air is incident on a glass plate 10.0 cm thick at an angle of incidence of 40°. The glass has an index of refraction of 1.65. The emerging ray on the other side of the plate is parallel to the incident ray, but is laterally displaced. What is the perpendicular distance between the original direction of the ray and the direction of the emerging ray? [*Hint*: See Example 22.4.] 3.2 cm

41. IE ●● To a submerged diver looking upward through the water, the altitude of the Sun (the angle between the Sun and the horizon) appears to be 45°. (a) The actual altitude of the Sun is (1) greater than, (2) the same as, (3) less than 45°. Explain. (b) What is the Sun's actual altitude? (a) (3) less than (b) 20°

42. ●● At what angle to the surface must a diver submerged in a lake look toward the surface to see the setting Sun just along the horizon? 41.2°

43. ●● A submerged diver shines a light toward the surface of a body of water at angles of incidence of 40° and 50°. Can a person on the shore see a beam of light emerging from the surface in either case? Justify your answer mathematically. seen for 40° but not for 50°; see ISM

44. IE ●● A beam of light is to undergo total internal reflection through a 45°–90°–45° prism (▶Fig. 22.27). (a) Will this arrangement depend on (1) the index of refraction of the prism, (2) the index of refraction of the surrounding medium, or (3) the indices of refraction of both? Explain. (b) Calculate the minimum index of refraction of the prism if the surrounding medium is air. Repeat if it is water. (a) (3) the indices of refraction of both (b) 1.41; 1.88

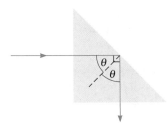

◀ **FIGURE 22.27** Total internal reflection in a prism See Exercises 44 and 45.

45. ●● A 45°–90°–45° prism (Fig. 22.27) is made of a material with an index of refraction of 1.85. Can the prism be used to deflect a beam of light by 90° (a) in air? (b) What about in water? (a) yes (b) no; see ISM

46. ●● A coin lies on the bottom of a pool under 1.5 m of water and 0.90 m from the sidewall (▼Fig. 22.28). If a light beam is incident on the water surface at the wall, at what angle θ relative to the wall must the beam be directed so that it will illuminate the coin? 43°

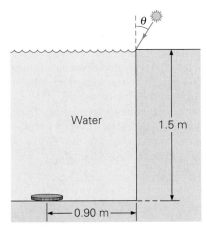

▲ **FIGURE 22.28 Find the coin** See Exercise 46 (not drawn to scale).

47. ●● Can you determine the index of refraction of the fluid in air as shown in Fig. 22.9a? If yes, what is its value? see ISM, about 1.3

48. ●● A crown-glass plate 2.5 cm thick is placed over a newspaper. How far beneath the top surface of the plate would the print appear to be if you were looking almost vertically downward through the plate? (See Exercise 37b.) 1.6 cm

49. ●● A beam of light traveling in water strikes a surface of a transparent material at an angle of incidence of 45°. If the angle of refraction in the material is 35°, what is the index of refraction of the transparent material? 1.64

50. ●● Yellow-green light of wavelength 550 nm is incident on the surface of a flat piece of crown glass at an angle of 40°. What is (a) the angle of refraction of the light? (b) the speed of the light in the glass? (c) the wavelength of the light in the glass? (a) 25° (b) 1.97×10^8 m/s (c) 362 nm

51. IE ●●● A light beam traveling upward in a plastic material with an index of refraction of 1.60 is incident on an upper horizontal air interface. (a) At certain angles of incidence, the light is not transmitted into air. The cause of this is (1) reflection, (2) refraction, (3) total internal reflection. Explain. (b) If the angle of incidence is 45°, is some of the beam transmitted into air? (c) Suppose the upper surface of the plastic material is covered with a layer of liquid with an index of refraction of 1.20. What happens in this case? (a) (3) total internal reflection (b) and (c) see ISM

52. ●●● A 15-cm-deep opaque container is empty except for a single coin resting on its bottom surface. When looking into the container at a viewing angle of 50° relative to the vertical side of the container, you see nothing on the bottom. When the container is filled with water, you see the coin (from the same viewing angle) on the bottom of, and just beyond, the side of the container. (See Fig. 22.25.) How far is the coin from the side of the container? 11 cm

53. ●●● An outdoor circular fish pond has a diameter of 4.00 m and a uniform full depth of 1.50 m. A fish halfway down in the pond and 0.50 m from the near side can just see the full height of a 1.80-m-tall person. How far away from the edge of the pond is the person? 2.0 m

54. ●●● A cube of flint glass sits on a newspaper on a table. The bottom half of the vertical sides of the cube is painted so that portion is opaque but the top half is transparent. By looking into one of the *vertical* sides of the cube, is it possible to see the portion of the newspaper covered by the center of the glass? Prove your answer. [*Hint*: Sketch the light leaving the location of interest.] no, see ISM

55. ●●● Two glass prisms are placed together (▼Fig. 22.29). (a) If a beam of light strikes the face of one of the prisms at normal incidence as shown, at what angle θ does the beam emerge from the other prism? (b) At what angle of incidence would the beam be refracted along the interface of the prisms? (a) 12.5° (b) 26.2°

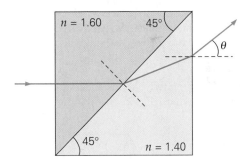

▲ **FIGURE 22.29 Joined prisms** See Exercise 55.

22.5 Dispersion

56. **MC** Dispersion can occur only if the light is (a) monochromatic, (b) polychromatic, (c) white light, (d) both b and c. (d)

57. **MC** Dispersion can occur only during (a) reflection, (b) refraction, (c) total internal reflection, (d) all of the preceding. (b)

58. **MC** Dispersion is caused by (a) the difference in the speed of light in different media, (b) the difference in the speed of light for different wavelengths of light in a given medium, (c) the difference in the angle of incidence for different wavelengths of light in a given medium, (d) the difference in the indices of refraction of light in different media. (b)

59. **CQ** Why is dispersion more prominent when using a triangular prism rather than a square block?
two refractions and two dispersions in a prism

60. **CQ** A glass prism disperses white light into a spectrum. Can a second glass prism be used to recombine the spectral components? Explain. no; see ISM

61. **CQ** You can never walk under a rainbow. Explain why. see ISM

62. **CQ** A light beam consisting of two colors, A and B, is sent through a prism. Color A is refracted more than color B. Which color has a longer wavelength? Explain.
color B; see ISM

63. **CQ** (a) If glass is dispersive, why don't we normally see a spectrum of colors when sunlight passes through a windowpane? (b) Does dispersion occur for polychromatic light incident on a dispersive medium at an angle of 0°? Explain. (Are the speeds of each color of light the same in the medium?) (a) usually $\theta \approx 0°$ (b) no (no); see ISM

64. **IE ●●** The index of refraction of crown glass is 1.515 for red light and 1.523 for blue light. (a) If light is incident on crown glass from air, which color, red or blue, will be refracted more? Explain? (b) Find the angle separating rays of the two colors in a piece of crown glass if their angle of incidence is 37°. (a) blue (b) 0.131°

65. **●●** A beam of light with red and blue components of wavelengths 670 nm and 425 nm, respectively, strikes a slab of fused quartz at an incident angle of 30°. On refraction, the different components are separated by an angle of 0.001 31 rad. If the index of refraction of the red light is 1.4925, what is the index of refraction of the blue light? 1.498

66. **●●** White light passes through a piece of crown glass and strikes an interface with air at an angle of 41.15°. Assume the indices of refraction of crown glass are the same as given in Exercise 64. Which color(s) of light will be refracted out into the air? only red

67. **●●●** A beam of red light is incident on an equilateral prism as shown in ▼Fig. 22.30. (a) If the index of refraction

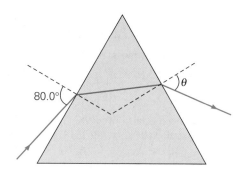

▲ **FIGURE 22.30 Prism revisited** See Exercise 67.

of red light of the prism is 1.400, at what angle θ does the beam emerge from the other face of the prism? (b) Suppose the incident beam were white light. What would be the angular separation of the red and blue components in the emergent beam if the index of refraction of blue light were 1.403? (c) If the index of refraction of blue light were 1.405? (a) 21.7° (b) 0.22° (c) 0.37°

Comprehensive Exercises

68. In Fig. 22.21b, if the glass prism has an index of refraction of 1.5 and the experiment is done in water rather than in air, what happens to the spectrum emerging from the prism? How about in a liquid that also has an index of refraction of 1.5? Explain. less separated; disappear

69. Light passes from medium A into medium B at an angle of incidence of 30°. The index of refraction of A is 1.5 times that of B. (a) What is the angle of refraction? (b) What is the ratio of the speed of light in B to the speed of light in A? (c) What is the ratio of the frequency of the light in B to the frequency of the light in A? (d) At what angle of incidence would the light be internally reflected?
(a) 49° (b) 1.5 (c) 1 (d) 42°

70. For total internal reflection to occur inside an optic fiber as shown in ▼Fig. 22.31, the angle θ must be greater than the critical angle for the fiber–air interface. At the end of the fiber, the incident light undergoes a refraction to enter the fiber. If total internal reflection is to occur for *any* angle of incidence, θ_i, outside the end of the fiber, what is the minimum index of refraction of the fiber? 1.41

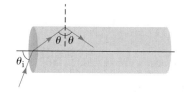

◀ **FIGURE 22.31 Optic fiber** See Exercise 70.

71. **IE** The critical angle for a glass–air interface is 41.11° for red light and 41.04° for blue light. (a) During the time the blue light travels 1.000 m, the red light will travel (1) more than (2) less than, (3) exactly 1.000 m. Explain. (b) Calculate the difference in distance traveled by the two colors. (a) (1) more than (b) 1.3 mm

72. In Exercise 67, if the angle of incidence is too small, light will not emerge from the other side of the prism. How could this happen? Calculate the minimum angle of incidence for the red light so that it does not emerge from the other side of the prism. 20.40°

73. Light in air is incident on a transparent material. It is found that the angle of reflection is twice the angle of refraction. What is the *range* of the index of refraction of the material? 1.41 to 2.00

The following Physlet Physics Problems can be used with this chapter.
34.1, 34.2, 34.3, 34.5, 34.6, 34.7, 34.8, 34.10

PHYSICS FACTS

- The largest refracting optical lens in the world measures 1.827 m (5.99 ft) in diameter. It was constructed by a team at the Optics Shop of the Optical Sciences Center of the University of Arizona in Tucson, Arizona, and completed in January 2000.

- The largest mirror under development for the European Space Agency's Herschel Space Observatory is 3.5 m (11.5 ft) in diameter. It is made from silicon carbide, which reduces its mass by a factor of 5 compared with traditional materials.

- A typical camera lens actually has more than one element (lens) inside. Many camera lenses have seven or more compensating elements to reduce or eliminate various types of lens aberrations. A single lens would produce distorted images.

What would life be like if there were no mirrors in bathrooms or cars, and if eyeglasses did not exist? Imagine a world without optical images of any kind—no photographs, no movies, no TV. Think about how little we'd know about the universe if there were no telescopes to observe distant planets and stars—or how little we'd know about biology and medicine if there were no microscopes to see bacteria and cells. It is often forgotten how dependent we are on mirrors and lenses.

The first mirror was probably the reflecting surface of a pool of water. Later, people discovered that polished metals and glass also have reflective properties. They must also have noticed that when they looked at things through glass, the objects looked different than when viewed directly, depending on the shape of the glass. In some cases, the objects appeared to be reduced or inverted, as the flower in the photo. In time, people learned to shape glass purposefully into lenses, paving the way for the eventual development of the many optical devices we now take for granted.

The optical properties of mirrors and lenses are based on the principles of reflection and refraction of light, as introduced in Chapter 22. In this chapter, you'll learn the principles of mirrors and lenses. Among other things, you'll discover why the images in the photo are upside down and reduced, whereas your image in an ordinary flat mirror is right side up—but the image doesn't seem to comb your hair with the same hand you use!

Plane mirror

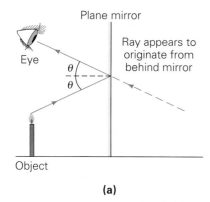

Eye

θ

θ

Ray appears to originate from behind mirror

Object

(a)

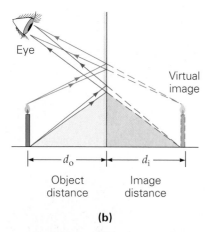

Eye

Virtual image

d_o

d_i

Object distance

Image distance

(b)

▲ **FIGURE 23.1** Image formed by a plane mirror **(a)** A ray from a point on the object is reflected in the mirror according to the law of reflection. **(b)** Rays from various points on the object produce an image. Because the two shaded triangles are identical, the image distance d_i (the distance of the image from the mirror) is equal to the object distance d_o. That is, the image appears to be the same distance behind the mirror as the object is in front of the mirror. The rays appear to diverge from the image position. In this case, the image is said to be virtual.

Teaching tip: It may be necessary to review the properties of similar and identical triangles.

Illustration 33.2 Flat Mirrors

Teaching tip: Point out that the locations of the real object and the virtual image are on opposite sides of a mirror.

23.1 Plane Mirrors

OBJECTIVES: To (a) understand how images are formed, and (b) describe the characteristics of images formed by plane mirrors.

Mirrors are smooth reflecting surfaces, usually made of polished metal or glass that has been coated with some metallic substance. As you know, even an uncoated piece of glass, such as a window pane, can act as a mirror. However, when one side of a piece of glass is coated with a compound of tin, mercury, aluminum or silver, the reflectivity of the glass is increased, as light is not transmitted through the coating. A mirror may be front coated or back coated. Most mirrors are backcoated.

When you look directly into a mirror, you see the images of yourself and objects around you (apparently on the other side of the surface of the mirror). The geometry of a mirror's surface affects the size, orientation, and type of image. In general, an *image* is the visual counterpart of an object, produced by reflection (mirrors) or refraction (lenses).

A mirror with a flat surface is called a **plane mirror**. How images are formed by a plane mirror is illustrated by the ray diagram in ◄Fig. 23.1. An image appears to be behind or "inside" the mirror. This is because when the mirror reflects a ray of light from the object to the eye (Fig. 23.1a), the ray appears to originate from behind the mirror. Reflected rays from the top and bottom of an object are shown in Fig. 23.1b. In actuality, light rays coming from all points on the side of the object facing the mirror are reflected, and an image of the complete object is observed.

The image formed in this way *appears* to be behind the mirror. Such an image is called a **virtual image**. Light rays appear to diverge from virtual images, but do not actually do so. No light energy actually comes from or passes through the image. However, spherical mirrors (discussed in Section 23.2) can project images in front of the mirror where light actually passes through the image. This type of image is called a **real image**. An example of a real image is the image produced by an overhead projector in a classroom.

Notice in Fig. 23.1b the distances of the object and image from the mirror. Quite logically, the distance of an object from a mirror is called the *object distance* (d_o), and the distance its image appears to be behind the mirror is called the *image distance* (d_i). By geometry of identical triangles and the law of reflection, $\theta_i = \theta_r$, it can be shown that $d_o = d_i$, which means that *the image formed by a plane mirror appears to be at a distance behind the mirror that is equal to the distance between the object and the front of the mirror.* (See Exercise 17.)

We are interested in various characteristics of images. Two of these features are the height and orientation of an image compared with those of its object. Both are expressed in terms of the **lateral magnification factor (M)**, which is defined as a ratio of heights of the image (h_i) and object (h_o):

$$M = \frac{\text{image height}}{\text{object height}} = \frac{h_i}{h_o} \qquad (23.1)$$

A lighted candle used as an object allows us to address an important image characteristic: orientation—that is, whether the image is upright or inverted with respect to the orientation of the object. (In sketching ray diagrams, an arrow is a convenient object for this purpose.) For a plane mirror, the image is always upright (or erect). This means that the image is oriented in the same direction as the object. We say that h_i and h_o have *the same sign* (both positive or both negative), so M is positive. Note that M is a dimensionless quantity, as it is a ratio of heights.

In ▶Fig. 23.2, you should also be able to see that the image and object have the same sizes (heights), so $h_i = h_o$. Therefore, $M = +1$ for a plane mirror, the image is upright, and there is no magnification. That is, you and your image in a plane mirror are the same size.

With other types of mirrors, such as spherical mirrors (which we will consider shortly), it is possible to have inverted images where M is negative. In summary, the sign of M tells us the orientation of the image relative to the object, and the absolute value of M gives the magnification.

TABLE 23.1	Characteristics of Images Formed by Plane Mirrors
$d_i = d_o$	The image distance is equal to the object distance; that is, the image appears to be as far behind the mirror as the object is in front.
$M = +1$	The image is virtual, upright, and unmagnified.

Another characteristic of reflected images of plane mirrors is the so-called *right–left reversal*. When you look at yourself in a mirror and raise your right hand, it appears that your image raises its left hand. However, this right–left reversal is actually caused by the front–back reversal. For example, if your front faces south, then your back "faces" north. Your image, on the other hand, has its front to the north and back to the south—a front–back reversal. You can demonstrate this reversal by asking one of your friends to stand facing you (without a mirror). If your friend raises his right hand, you can see that that hand is actually on your left side.

The main characteristics of an image formed by a plane mirror are summarized in Table 23.1. See also Insight 23.1, It's All Done with Mirrors, on page 733.

Example 23.1 ■ All of Me: Minimum Mirror Length

What is the minimum vertical length of a plane mirror needed for a person to be able to see a complete (head-to-toe) image of himself or herself (▼Fig. 23.3)?

Thinking It Through. Applying the law of reflection, we see in the figure that two triangles are formed by the rays needed for the image to be complete. These triangles relate the person's height to the minimum mirror length.

Solution. To determine this length, consider the situation shown in Fig. 23.3. With a mirror of minimum length, a ray from the top of the person's head would be reflected at the top of the mirror, and a ray from the person's feet would be reflected at the bottom of the mirror, to the eyes. The length L of the mirror is then the distance between the dashed horizontal lines perpendicular to the mirror at its top and bottom.

However, these lines are also the normals for the ray reflections. By the law of reflection, they bisect the angles between incident and reflected rays; that is, $\theta_i = \theta_r$. Then, because their respective triangles on each side of the dashed normal are identical, the length of the mirror from its bottom to a point even with the person's eyes is $h_1/2$, where h_1 is the person's height from the feet to the eyes. Similarly, the small upper length of the mirror is $h_2/2$ (the vertical distance between the person's eyes and the top of mirror). Then,

$$ L = \frac{h_1}{2} + \frac{h_2}{2} = \frac{h_1 + h_2}{2} = \frac{h}{2} $$

where h is the person's total height.

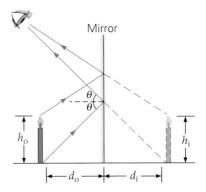

▲ **FIGURE 23.2 Magnification**
The lateral, or height magnification factor is given by $M = h_i/h_o$. For a plane mirror, $M = +1$, which means that $h_i = h_o$, that is, the image is the same height as the object, and that means the image is upright.

PHYSLET®

Exploration 33.1 Image in a Flat Mirror

(continues on next page)

◀ **FIGURE 23.3 Seeing it all** The minimum height, or vertical length, of a plane mirror needed for a person to see his or her complete (head-to-toe) image turns out to be half the person's height. See Example 23.1.

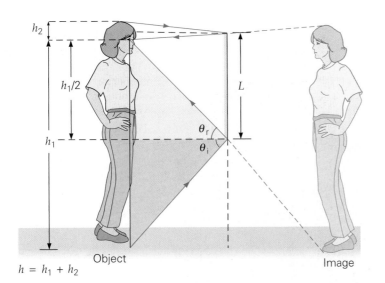

$h = h_1 + h_2$

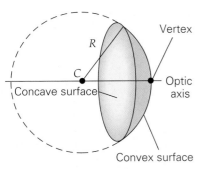

▲ **FIGURE 23.4 Spherical mirrors**
A spherical mirror is a section of a sphere. Either the outside (convex) surface or the inside (concave) surface of the spherical section may be the reflecting surface.

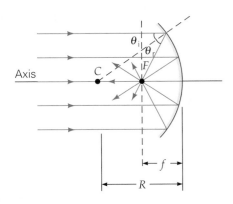

(a) Concave, or converging, mirror

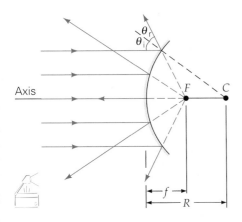

(b) Convex, or diverging, mirror

▲ **FIGURE 23.5 Focal point (a)** Rays parallel and close to the optic axis of a concave spherical mirror converge at the focal point *F*. **(b)** Rays parallel and close to the optic axis of a convex spherical mirror are reflected along paths as though they diverge from a focal point behind the mirror. Note that the law of reflection, $\theta_i = \theta_r$, is satisfied for each ray in each diagram.

Hence, for a person to see his or her complete image in a plane mirror, the minimum height, or vertical length, of the mirror must be half the height of the person.

You can do a simple experiment to prove this conclusion. Get some newspaper and tape, and find a full-length mirror. Gradually cover parts of the mirror with the newspaper until you cannot see your complete image. You will find you need a mirror length that is only half your height to see a complete image.

Follow-Up Exercise. What effect does a person's distance from the mirror have on the minimum mirror length required to produce his or her complete image? (*Answers to all Follow-Up Exercises are at the back of the text.*)

23.2 Spherical Mirrors

OBJECTIVES: To (a) distinguish between converging and diverging spherical mirrors, (b) describe images and their characteristics, and (c) determine image characteristics from ray diagrams and the spherical-mirror equation.

As the name implies, a **spherical mirror** is a reflecting surface with spherical geometry. ◄Figure 23.4 shows that if a portion of a sphere of radius *R* is sliced off along a plane, the severed section has the shape of a spherical mirror. Either the inside or outside of such a section can be the reflecting surface. For reflections on the inside surface, the section behaves as a **concave mirror**. (Think of looking into a *cave* in order to help yourself remember that a con*cave* mirror has a recessed surface.) For reflections from the outside surface, the section behaves as a **convex mirror**.

The radial line through the center of the spherical mirror that intersects the surface of the mirror at the *vertex* of the spherical section (Fig. 23.4) is called the *optic axis*. The point on the optic axis that corresponds to the center of the sphere of which the mirror forms a section is called the **center of curvature (C)**. The distance between the vertex and the center of curvature is equal to the radius of the sphere and is called the **radius of curvature (R)**.

When rays parallel and close to the optic axis are incident on a concave mirror, the reflected rays intersect, or converge, at a common point called the **focal point (F)**. As a result, a concave mirror acts as a **converging mirror** (◄Fig. 23.5a). Note that the law of reflection, $\theta_i = \theta_r$, is satisfied for each ray.

Similarly, rays parallel and close to the optic axis of a convex mirror diverges on reflection, as though the reflected rays came from a focal point behind the mirror's surface (Fig. 23.5b). Thus, a convex mirror acts as a **diverging mirror** (▼Fig. 23.6). When you see diverging rays, your brain interprets the image to mean that there is an object from which the rays *appear* to diverge, even though no such object is actually there.

◄ **FIGURE 23.6 Diverging mirror**
Note by reverse-ray tracing in Fig 23.5b that a diverging (convex) spherical mirror gives an expanded, although distorted, field of view, as can be seen in this store-monitoring mirror.

INSIGHT 23.1 IT'S ALL DONE WITH MIRRORS

FIGURE 1 The Sphinx, an illustration of Tobin's sensational illusion The body was concealed by two plane mirrors.

FIGURE 2 Houdini and Jennie, the disappearing elephant The elephant vanished from view when Houdini fired a pistol.

Most of us are fascinated by a stage magician's sensational tricks that appear to make objects and animals suddenly appear or disappear. Of course they do not really appear or disappear. The magician requires special skills to make the performance quick and smooth so as to "fool" the audience. It's all done with mirrors, as they say.

The very first mirror illusion, "The Sphinx," was invented for stage magicians by Thomas William Tobin in 1876. His invention used mirrors to conceal a person or an object, as shown in Fig. 1, and it was used as the front cover of *Modern Magic* in 1876. Two plane mirrors are placed between the legs of the three-legged table to hide the person's body.

Harry Houdini, the world-famous master of illusion, felt it was too easy to make a pigeon fly out of a hat or a rabbit disappear into thin air. In 1918, Houdini made a 10 000-lb elephant named Jennie "disappear" on the stage of the Hippodrome Theater in New York City (Fig. 2). The act was called "The Vanishing Elephant."

When the time for the elephant's disappearance came, two large plane mirrors at right angles to each other were slid quickly into place. When properly aligned, the mirrors reflected light from the side walls of the stage to form virtual images that

matched the pattern of the stage backdrop. Thus the audience apparently saw the stage with no elephant visible (Fig. 3). A strobe light was used to conceal the brief motion of the mirrors. Unseen by the audience, the elephant was quickly led off stage.

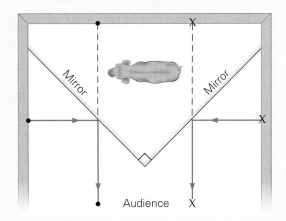

FIGURE 3 The disappearing elephant Two large mirrors at right angles to each other were used to conceal the elephant.

The distance from the vertex to the focal point (F) of parallel rays near the axis of a spherical mirror is called the **focal length** f. (See Fig. 23.5.) The focal length is related to the radius of curvature by the following simple equation:

$$f = \frac{R}{2} \quad \textit{focal length of spherical mirror} \quad (23.2)$$

The preceding result is valid only when the rays are close to the optic axis—that is, for small-angle approximation. Rays far away from the optic axis will focus at different focal points, resulting in some image distortion. In optics, this distortion is an example of an *aberration*. Some telescope mirrors are parabolic in shape, rather than spherical, so *all* rays parallel to the optic axis are focused at the focal point, thus eliminating *spherical aberration*.

Note:

Concave mirror = converging mirror

Convex mirror = diverging mirror

Converging (concave) mirror

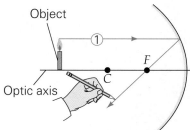

1 **Parallel ray**

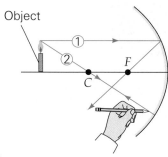

2 **Chief (radial) ray**

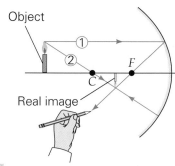

3 **Locating image**

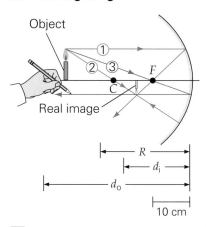

4 **Can also use focal ray to confirm image**

Ray Diagrams

The characteristics of images formed by spherical mirrors can be determined from geometrical optics (Chapter 22). The method involves drawing rays emanating from one or more points on an object. The law of reflection ($\theta_i = \theta_r$) applies, and three key rays are defined with respect to the mirror's geometry as follows:

1. A **parallel ray** is a ray that is incident along a path parallel to the optic axis and is reflected through (or appears to go through) the focal point F (as do all rays near and parallel to the axis).

2. A **chief ray**, or **radial ray**, is a ray that is incident through the center of curvature (C) of the spherical mirror. Since the chief ray is incident normal to the mirror's surface, this ray is reflected back along its incident path, through point C.

3. A **focal ray** is a ray that passes through (or appears to go through) the focal point and is reflected parallel to the optic axis. (It is a reversed parallel ray, so to speak.)

Using any two of these three key rays, we can locate the image (image distance) and determine its size (magnified or reduced), orientation (upright or inverted), and type (real or virtual). It is customary to use the tip of an asymmetrical object (for example, the head of an arrow or the flame of a candle) as the origin point of the rays. The corresponding point of the image is at the point of intersection of the rays. This makes it easy to see whether the image is upright or inverted.

Keep in mind, however, that *properly traced rays from any point on the object can be used to find the image*. Every point on a visible object acts as an emitter of light. For example, for a candle, the flame emits its own light, and many other points on the candle surface reflect light.

Example 23.2 ■ Learn by Drawing: A Mirror Ray Diagram

An object is placed 39.0 cm in front of a concave spherical mirror of radius 24.0 cm. (a) Use a ray diagram to locate the image formed by this mirror. (b) Discuss the characteristics of the image.

Thinking It Through. A ray diagram, drawn accurately, can by itself provide "quantitative" information about image location and image characteristics that might otherwise be determined mathematically.

Solution.

Given: $R = 24.0$ cm $\qquad$ *Find:* (a) image location
$\qquad\quad d_o = 39.0$ cm $\qquad\qquad\qquad$ (b) image characteristics

(a) Since we have been asked to use a ray diagram (drawing) to locate the image, the first thing we need to decide on is a scale for the drawing. If a scale of 1 cm (on the drawing) represents 10 cm, the object would be drawn 3.90 cm in front of the mirror.

First we draw the optic axis, the mirror, the object (a lighted candle), and the center of curvature (C). From Eq. 23.2, $f = 24.0$ cm$/2 = 12.0$ cm, and the focal point (F) is halfway from the vertex to the center of curvature.

To locate the image, follow steps 1–4 in the accompanying Learn by Drawing:

1. The first ray drawn is the parallel ray (① in the drawing). From the tip of the flame, draw a ray parallel to the optic axis. After reflecting, this ray goes through the focal point, F.

2. Then draw the chief ray (② in the drawing). From the tip of the flame, draw a ray going through the center of curvature, C. This ray will be reflected back along the original direction. (Why?)

3. It can be seen that these two rays intersect. The point of intersection is the tip of the *image* of the candle. From this point, draw the image by extending the tip of the flame to the optic axis. The image distance $d_i = 17$ cm, as measured from the diagram.

4. Only two rays are needed to locate the image. However, if the third ray is drawn as a double check, the focal ray in this case (③ in the drawing), it must go through the same point on the image at which the other two rays intersected (if drawn carefully). The focal ray from the tip of the flame going through the focal point, F, after reflection, will travel out parallel to the optic axis.

(b) From the ray diagram drawn in part (a), it can clearly be seen that the image is real (because the reflected rays intersect *in front* of the mirror). They converge and pass through the image. As a result, the real image could be seen on a screen (for example, a piece of white paper) that is positioned at a distance d_i from the concave mirror. The image is also inverted (the image of the candle points downward) and is smaller than the object.

Follow-Up Exercise. In this example, what would the characteristics of the image be if the object were 15.0 cm in front of the mirror? Locate the image and discuss its characteristics.

Exploration 33.3 Ray Diagrams

An example of a ray diagram using the same three rays for a convex (diverging) mirror will be shown in Integrated Example 23.4.

A converging mirror does *not* always form a real image. For a converging spherical mirror, the characteristics of the image change with the distance of the object from the mirror. Dramatic changes take place at two points: C (the center of curvature) and F (the focal point). These points divide the optic axis into three regions (▼Fig. 23.7a): $d_o > R$, $R > d_o > f$, and $d_o < f$.

Teaching tip: Remind students that real images are formed when reflected rays converge, and virtual images are formed when reflected rays diverge.

Teaching tip: Encourage students to be precise in constructing ray diagrams. Use arrows to show direction and so on.

▼ **FIGURE 23.7 Concave mirrors (a)** For a concave, or converging, mirror, the object is located within one of three regions defined by the center of curvature (C) and the focal point (F), or at one of these two points. For $d_o > R$, the image is real, inverted, and smaller than the object, as shown by the ray diagrams in Example 23.2. **(b)** For $R > d_o > f$, the image will also be real and inverted but enlarged, or magnified. **(c)** For an object at the focal point F, or $d_o = f$, the image is said to be formed at infinity. **(d)** For $d_o < f$, the image will be virtual, upright, and enlarged.

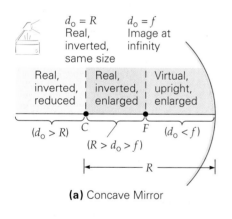

(a) Concave Mirror

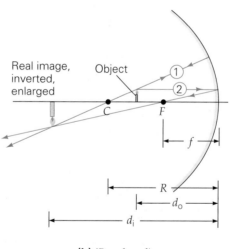

(b) ($R > d_o > f$)

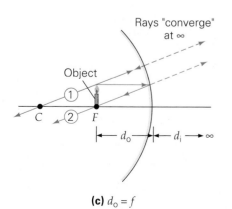

(c) $d_o = f$

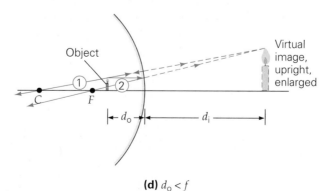

(d) $d_o < f$

Let's start with an object in the region farthest from the mirror ($d_o > R$) and move toward the mirror:

- The case of $d_o > R$ was shown in Example 23.2.
- When $d_o = R = 2f$, the image is real, inverted, and the same size as the object.
- When $R > d_o > f$, an enlarged, inverted, real image is formed (Fig. 23.7b). The image is magnified when the object is inside the center of curvature, C.
- When $d_o = f$, the object is at the focal point (Fig. 23.7c). The reflected rays are parallel, and the image is said to be "formed at infinity." The focal point F is a special "crossover" point, as it divides the space in front of the mirror into two regions.
- When $d_o < f$, the object is inside the focal point (between the focal point and the mirror's surface). A virtual, enlarged, and upright image is formed (Fig. 23.7d).

When $d_o > f$, the image is real; when $d_o < f$, the image is virtual. For $d_o = f$, we say that the image is formed at infinity (Fig. 23.7c). That is, when an object is at "infinity"—it is so far away that rays emanating from it and falling on the mirror are essentially parallel—its image is formed at the focal plane. This fact provides an easy method for determining the focal length of a concave mirror.

As we have seen, the position, orientation, and size of the image can be approximately determined graphically from ray diagrams drawn to scale. However, these characteristics can be determined more accurately by analytical methods. It can be shown by means of geometry that the object distance (d_o), the image distance (d_i), and the focal length (f) are related. That relationship is known as the **spherical-mirror equation**:

$$\frac{1}{d_o} + \frac{1}{d_i} = \frac{1}{f} = \frac{2}{R} \quad \text{spherical-mirror equation} \quad (23.3)$$

Note that this equation can be written in terms of either the radius of curvature, R, or the focal distance, f, since by Eq. 23.2, $f = R/2$. Both R and f can be either positive or negative, as we will discuss shortly.

If d_i is the quantity to be found for a spherical mirror, it may be convenient to use an alternative form of the spherical-mirror equation:

$$d_i = \frac{d_o f}{d_o - f} \quad (23.3a)$$

But, you can always use the reciprocal form of Eq. 23.3.

The signs of the various quantities are very important in the application of Eqs. 23.3. We will use the sign conventions summarized in Table 23.2. For example,

TABLE 23.2	Sign Conventions for Spherical Mirrors
Focal length (f)	
Concave (converging) mirror	f (or R) is positive
Convex (diverging) mirror	f (or R) is negative
Object distance (d_o)	
Object is in front of the mirror (real object)	d_o is positive
Object is behind the mirror (virtual object)*	d_o is negative
Image distance (d_i) and *image type*	
Image is formed in front of the mirror (real image)	d_i is positive
Image is formed behind the mirror (virtual image)	d_i is negative
Image orientation (M)	
Image is upright with respect to the object	M is positive
Image is inverted with respect to the object	M is negative

*In a combination of two (or more) mirrors, the image formed by the first mirror is the object of the second mirror (and so on). If this image–object falls behind the second mirror, it is referred to as a *virtual* object, and the object distance is taken to be negative. This concept is more important for lens combinations, as we will see in Section 23.3, and is mentioned here only for completeness.

for a real object, a positive d_i indicates a real image and a negative d_i corresponds to a virtual image.

The **lateral magnification factor** M defined in Eq. 23.1 can also be found analytically for a spherical mirror. Again, by using geometry, it can be expressed in terms of the image and object distances:

$$M = -\frac{d_i}{d_o} \quad \text{magnification equation} \qquad (23.4)$$

The minus sign is added by convention to indicate the orientation of the image: A positive value for M indicates an upright image, whereas a negative M implies an inverted image. Also, if $|M| > 1$, the image is magnified, or larger than the object. If $|M| < 1$, the image is reduced, or smaller than the object. Note that for mirrors, the lateral magnification M, also called the *magnification factor*, or simply *magnification*, is conveniently expressed in terms of the image distance d_i and the object distance d_o rather than in terms of the image and object heights used in Eq. 23.1. (A description of the origin of Eqs. 23.3 and 23.4 follows as optional content.)

Example 23.3 and Integrated Example 23.4 show how these equations and sign conventions are used for spherical mirrors. In general, this approach usually involves finding the image of an object; you will be asked where the image is formed (d_i) and what the image characteristics are (M). These characteristics tell whether the image is real or virtual, upright or inverted, and larger or smaller than the object (magnified or reduced).

*(Optional) Derivation of the Spherical-Mirror Equation You might wonder from where Equations 23.3 and 23.4 originate. The spherical-mirror equation can be derived with the aid of a little geometry. Consider the ray diagram in ▼Fig. 23.8. The object and image distances (d_o and d_i) and the heights of the object and image (h_o and h_i) are shown. Note that these lengths make up the bases and heights of triangles formed by the ray reflected at the vertex (V). These triangles ($O'VO$ and $I'VI$) are similar, since, by the law of reflection, their angles at V are equal. Hence, we can write

$$\frac{h_i}{h_o} = -\frac{d_i}{d_o} \qquad (1)$$

This equation is Eq. 23.4, from the definition of Eq. 23.1. The negative sign inserted here signifies the fact that the image is inverted, so h_i is negative.

The (focal) ray through F also forms similar triangles, $O'FO$ and VFA, in the approximation that the mirror is small compared with its radius. (Why are the triangles similar?) The bases of these triangles are $VF = f$ and $OF = d_o - f$. Then, if VA is taken to be h_i,

$$\frac{h_i}{h_o} = -\frac{VF}{OF} = -\frac{f}{d_o - f} \qquad (2)$$

Again, the negative sign inserted here signifies the fact that the image is inverted, so h_i is negative.

Equating Eqs. 1 and 2,

$$\frac{d_i}{d_o} = \frac{f}{d_o - f} \qquad (3)$$

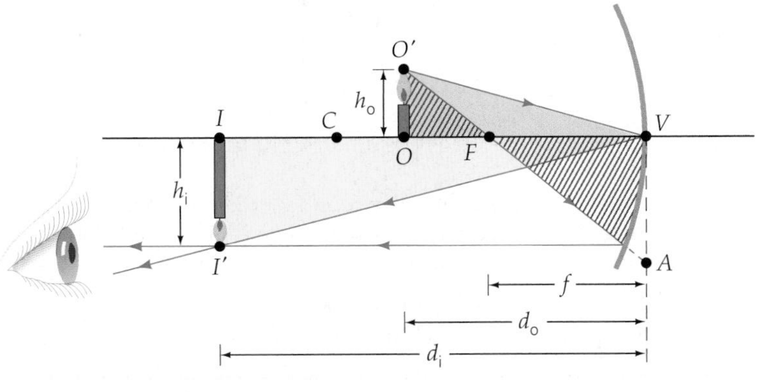

Note: A helpful hint for remembering that the magnification is d_i over d_o is that the ratio is in alphabetical order ("i" over "o").

Note: $|M|$ is the absolute value of M: its magnitude without regard to sign. For example, $|+2| = |-2| = 2$.

Teaching tip: Stress the general rules for determining whether the signs are positive or negative. Urge students to refer to the sign conventions in Table 23.2.

◀ **FIGURE 23.8 Spherical-mirror equation** The rays provide the geometry, through similar triangles, for the derivation of the spherical-mirror equation.

Algebraic manipulation yields

$$\frac{1}{d_o} + \frac{1}{d_i} = \frac{1}{f}$$

which is the spherical-mirror equation (Eq. 23.3).

Example 23.3 ■ What Kind of Image? Characteristics of a Concave Mirror

A concave mirror has a radius of curvature of 30 cm. If an object is placed (a) 45 cm, (b) 20 cm, and (c) 10 cm from the mirror, where is the image formed, and what are its characteristics? (Specify whether each image is real or virtual, upright or inverted, and magnified or reduced.)

Thinking It Through. Here we are given R, from which we can compute the focal length $f = R/2$. Also given are three different object distances, which can be applied in Eqs. 23.3 and 23.4 to determine the image location and characteristics.

Solution.

Given: $R = 30$ cm, so   *Find:* d_i, M, and image characteristics for
               $f = R/2 = 15$ cm   given object distances
               (a) $d_o = 45$ cm
               (b) $d_o = 20$ cm
               (c) $d_o = 10$ cm

Note that the given object distances correspond to the regions shown in Fig. 23.7a. There is no need to convert the distances to meters as long as all distances are expressed in the same unit (centimeters in this case). You could draw representative ray diagrams for each of these cases in order to find the characteristics of each image.

(a) In this case, the object distance is greater than the radius of curvature ($d_o > R$), and

$$\frac{1}{d_o} + \frac{1}{d_i} = \frac{1}{f} \quad \text{or} \quad \frac{1}{d_i} = \frac{1}{f} - \frac{1}{d_o} = \frac{1}{15 \text{ cm}} - \frac{1}{45 \text{ cm}} = \frac{2}{45 \text{ cm}}$$

Then

$$d_i = \frac{45 \text{ cm}}{2} = +22.5 \text{ cm} \quad \text{and} \quad M = -\frac{d_i}{d_o} = -\frac{22.5 \text{ cm}}{45 \text{ cm}} = -\frac{1}{2}$$

Thus, the image is real (positive d_i), inverted (negative M), and half as large as the object $\left(|M| = \frac{1}{2}\right)$.

(b) Here, $R > d_o > f$, and the object is between the focal point and the center of curvature:

$$\frac{1}{d_i} = \frac{1}{15 \text{ cm}} - \frac{1}{20 \text{ cm}} = \frac{1}{60 \text{ cm}}$$

Thus,

$$d_i = +60 \text{ cm} \quad \text{and} \quad M = -\frac{60 \text{ cm}}{20 \text{ cm}} = -3.0$$

In this case, the image is real (positive d_i), inverted (negative M), and three times the size of the object ($|M| = 3$).

(c) For this case, $d_o < f$, and the object is inside the focal point.
Using the alternate form of Eq. 23.3 for illustration:

$$d_i = \frac{d_o f}{d_o - f} = \frac{(10 \text{ cm})(15 \text{ cm})}{10 \text{ cm} - 15 \text{ cm}} = -30 \text{ cm}$$

Then

$$M = -\frac{d_i}{d_o} = -\frac{(-30 \text{ cm})}{10 \text{ cm}} = +3.0$$

In this case, the image is virtual (negative d_i), upright (positive M), and three times the size of the object ($|M| = 3$).
From the denominator of the expression for d_i, you can see that d_i will always be negative when d_o is less than f. Therefore, a virtual image is always formed for an object inside the focal point of a converging mirror.

Follow-Up Exercise. For the converging mirror in this Example, where is the image formed and what are its characteristics if the object is at 30 cm, or $d_o = R$?

Problem-Solving Hint

When using the spherical-mirror equations to find image characteristics, it is helpful to first make a quick sketch (approximate, not necessarily to scale) of the ray diagram for the situation. This sketch shows you the image characteristics and helps you avoid making mistakes when applying the sign conventions. *The ray diagram and the mathematical solution must agree.*

Teaching tip: Stress the importance of constructing ray diagrams with a straightedge, as accurate ray diagrams showing directions can be of great help in determining image characteristics.

Integrated Example 23.4 ■ Similarities and Differences: Characteristics of a Convex Mirror

An object (in this case, a candle) is 20 cm in front of a diverging mirror that has a focal length of −15 cm (see the sign conventions in Table 23.2). (a) Use a ray diagram to determine whether the image formed is (1) real, upright, magnified, (2) virtual, upright, magnified, (3) real, upright, reduced, (4) virtual, upright, reduced, (5) real, inverted, magnified, or (6) virtual, inverted, reduced. (b) Find the location and characteristics of the image by using the mirror equations.

(a) Conceptual Reasoning. Since we know the object distance and the focal length of the convex mirror, a ray diagram can be drawn and the image characteristics can be determined. The first thing to decide on is a scale for the ray diagram. In this example, a scale of 1 cm (on the drawing) could be used to represent 10 cm. That way, the object would be 2.0 cm in front of the mirror in our drawing. Draw the optic axis, the mirror, the object (a lighted candle), and the focal point (F). Since this mirror is convex, the focal point (F) and the center of curvature (C) are behind the mirror. From Eq. 23.2, $R = 2f = 2(-15 \text{ cm}) = -30 \text{ cm}$. So C is drawn at twice the distance of F from the vertex.

Only two out of the three key rays are necessary to locate the image (▼Fig. 23.9). The parallel ray ① starts from the tip of the flame, travels parallel to the optic axis, and then diverges from the mirror after reflection, appearing to come from F. The chief ray ② originates from the tip of the flame, appears to go through C, and then reflects straight back, but appears to come from C. It is clearly seen that these two rays, after reflection, diverge from each other, and there is no chance for them to intersect. However, they appear to start from a common point behind the mirror: the image point of the tip of the flame. We can also draw the focal ray ③ to verify that all three rays appear to emanate from the same image point.

The image is virtual (the reflecting rays don't actually come from a point behind the mirror), upright, and smaller than the object. Therefore, the answer is (4) virtual, upright, reduced. Measuring from the diagram (keep in mind the drawing scale we are using), we find that $d_i \approx -9.0 \text{ cm}$, and the magnification $M = \dfrac{h_i}{h_o} \approx \dfrac{0.5 \text{ cm}}{1.2 \text{ cm}} = +0.4$.

Exploration 35.5 *Convex Mirrors, Focal Point, and Radius of Curvature*

(continues on next page)

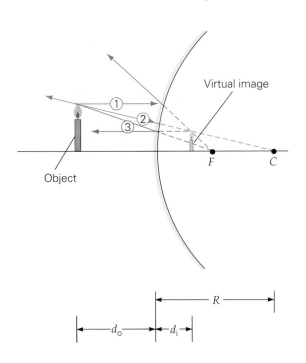

◀ **FIGURE 23.9 Diverging mirror**
Ray diagram of a diverging mirror. See Integrated Example 23.4.

Illustration 33.1 Mirrors and the Small-Angle Approximation

(b) Quantitative Reasoning and Solution. The object distance and focal length are given. The image position and characteristics can be calculated by using the mirror equations.

Given: $d_o = 20$ cm   *Find:* d_i, M, and image characteristics
$f = -15$ cm

Note that the focal length is negative for a convex mirror. (See Table 23.2.) Using Eq. 23.3, we have

$$\frac{1}{20 \text{ cm}} + \frac{1}{d_i} = \frac{1}{-15 \text{ cm}}$$

so

$$d_i = -\frac{60 \text{ cm}}{7} = -8.6 \text{ cm}$$

Then

$$M = -\frac{d_i}{d_o} = -\frac{(-8.6 \text{ cm})}{20 \text{ cm}} = +0.43$$

Thus, the image is virtual (d_i is negative), upright (M is positive), and 0.43 times the size (height) of the object. These results agree well with those from the ray diagram. The image of an object is always virtual for a diverging (convex) mirror. (Can you prove this using either a ray diagram or the mirror equation?)

Follow-Up Exercise. As has been pointed out, a diverging mirror always forms a virtual image of a real object. What about the other characteristics of the image—its orientation and magnification? Can any general statements be made about them?

Spherical-Mirror Aberrations

Technically, our descriptions of image characteristics for spherical mirrors are true only for objects near the optic axis—that is, only for small angles of incidence and reflection. If these conditions do not hold, the images will be blurred (out of focus) or distorted, because not all of the rays will converge in the same plane. As illustrated in ◄Fig. 23.10, incident parallel rays far from the optic axis do not converge at the focal point. The farther the incident ray is from the axis, the farther is the reflected ray from the focal point. This effect is called **spherical aberration**.

   Spherical aberration does not occur with a parabolic mirror. (As the name *parabolic mirror* implies, a parabolic mirror has the form of a paraboloid.) *All* of the incident rays parallel to the optic axis of such a mirror have a common focal point. For this reason, parabolic mirrors are used in most astronomical telescopes (Chapter 24). However, these mirrors are more difficult to make than spherical mirrors and are therefore more expensive.

23.3 Lenses

OBJECTIVES: To (a) distinguish between converging and diverging lenses, (b) describe their images and their characteristics, and (c) find image locations and characteristics by using ray diagrams and the thin-lens equation.

The word *lens* is from the Latin *lentil*, which is a round, flattened, edible seed of a pea-like plant. Its shape is similar to that of a lens. An optical **lens** is made from transparent material (most commonly glass, but sometimes plastic or crystal). One or both surfaces usually have a spherical contour. *Biconvex* spherical lenses (with both surfaces convex) and *biconcave* spherical lenses (with both surfaces concave) are illustrated in ▼Fig. 23.11. Lenses can form images by refracting the light that passes through them.

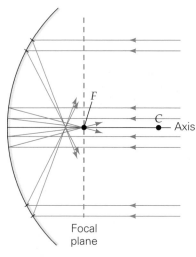

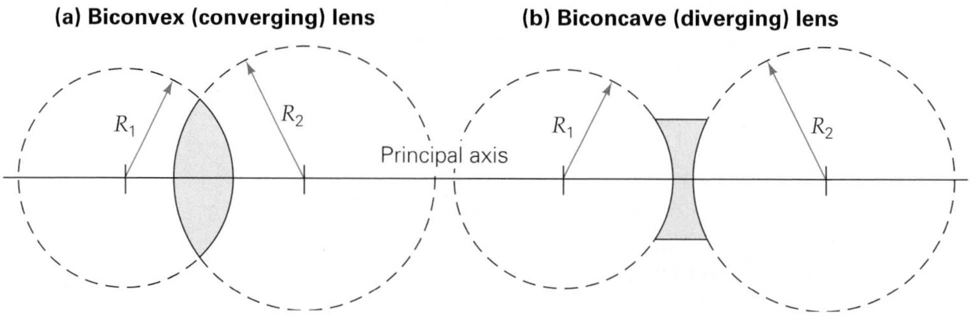

▲ **FIGURE 23.10 Spherical aberration for a mirror** According to the small-angle approximation, rays parallel to and near the mirror's axis converge at the focal point. However, when parallel rays not near the axis are reflected, they converge in front of the focal point. This effect, called *spherical aberration*, gives rise to blurred images.

▶ **FIGURE 23.11 Spherical lenses** Spherical lenses have surfaces defined by two spheres, and the surfaces are either convex or concave. **(a)** Biconvex and **(b)** biconcave lenses are shown here. If $R_1 = R_2$, a lens is spherically symmetric.

(a) Biconvex (converging) lens   **(b) Biconcave (diverging) lens**

R_1 R_2 R_1 R_2

Principal axis

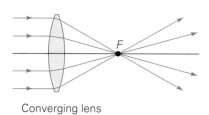

Converging lens

(a) Biconvex (converging) lens

(b)

◀ **FIGURE 23.12** Converging lens **(a)** For a thin biconvex lens, rays parallel to the axis converge at the focal point *F*. **(b)** A magnifying glass (converging lens) can be used to focus the Sun's rays to a spot—with incendiary results. Do not try this at home!

Illustration 35.1 Lenses and the Thin-Lens Approximation

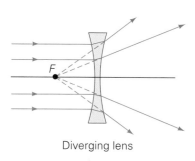

Diverging lens

Biconcave (diverging) lens

▲ **FIGURE 23.13** Diverging lens Rays parallel to the axis of a biconcave, or diverging, lens appear to diverge from a focal point on the incident side of the lens.

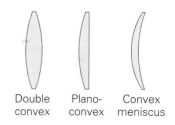

Double convex   Plano-convex   Convex meniscus

Converging lenses

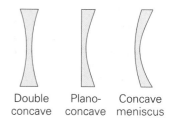

Double concave   Plano-concave   Concave meniscus

Diverging lenses

▲ **FIGURE 23.14** Lens shapes Lens shapes vary widely and are normally categorized as converging or diverging. In general, a converging lens is thicker at its center than at the periphery, and a diverging lens is thinner at its center than at the periphery.

A biconvex lens is an example of a **converging lens**. Incident light rays parallel to the axis of the lens converge at a focal point (*F*) on the opposite side of the lens (▲Fig. 23.12a). This fact provides a way to experimentally determine the focal length of a converging lens. You may have focused the Sun's rays with a magnifying glass (a biconvex, or converging, lens) and thereby witnessed the concentration of radiant energy that results (Fig. 23.12b).

A biconcave lens is an example of a **diverging lens**. Incident parallel rays emerge from the lens as though they emanated from a focal point on the incident side of the lens (▶Fig. 23.13).

There are several types of converging and diverging lenses (▶Fig. 23.14). Convex and concave meniscus lenses are the type most commonly used for corrective eyeglasses. In general, a converging lens is thicker at its center than at its periphery, and a diverging lens is thinner at its center than at its periphery. This discussion will be limited to spherically symmetric biconvex and biconcave lenses, for which both surfaces have the same radius of curvature.

When light passes through a lens, it is refracted and displaced laterally (Example 22.4 and Fig. 22.11). If a lens is thick, this displacement may be fairly large and can complicate analysis of the lens's characteristics. This problem does not arise with thin lenses, for which the refractive displacement of transmitted light is negligible. Our discussion will be limited to thin lenses. A thin lens is a lens for which the thickness of the lens is assumed to be negligible compared with the lens's focal length.

A lens with spherical geometry has, *for each lens surface*, a center of curvature (*C*), a radius of curvature (*R*), a focal point (*F*), and a focal length (*f*). The focal points are at equal distances on either side of a thin lens. However, for a spherical lens, the focal length is *not* simply related to *R* by $f = R/2$ as it is for spherical mirrors. Because the focal length also depends on the lens's index of refraction, the focal length of a lens is usually specified, rather than its radius of curvature. This will be discussed in Section 23.4.

The general rules for drawing ray diagrams for lenses are similar to those for spherical mirrors. But some modifications are necessary, since light passes through a lens. Opposite sides of a lens are generally distinguished as the *object side* and the *image side*. The object side is the side on which an object is positioned, and the image side is the *opposite* side of the lens (where a real image would be formed). The three rays from a point on an object are drawn as follows (see Learn by Drawing for Example 23.5 on page 743):

1. A **parallel ray** is a ray that is parallel to the lens's optic axis on incidence and, after refraction, either (a) passes through the focal point on the image side of a converging lens *or* (b) appears to diverge from the focal point on the object side of a diverging lens.

Exploration 35.2 Ray Diagrams

2. A **chief ray**, or **central ray**, is a ray that passes through the center of the lens and is undeviated because the lens is "thin."

3. A **focal ray** is a ray that (a) passes through the focal point on the object side of a converging lens *or* (b) appears to pass through the focal point on the image side of a diverging lens and, after refraction, is parallel to the lens's optic axis.

As with spherical mirrors, only two rays are needed to determine the image; we will normally use the parallel and chief rays. (As in the case of mirrors, however, it is generally a good idea to include a third ray, the focal ray, in your diagrams as a check.)

Example 23.5 ■ Learn by Drawing: A Lens Diagram

An object is placed 30 cm in front of a thin biconvex lens of focal length 20 cm. (a) Use a ray diagram to locate the image. (b) Discuss the characteristics of the image.

Thinking It Through. Follow the steps for lens ray diagrams, as given previously.

Solution.

Given: $d_o = 30$ cm   *Find:* (a) location of the image (using a ray diagram)
$\quad\quad\quad f = 20$ cm   (b) the image's characteristics

(a) Since we have been asked to use a ray diagram to locate the image (see the accompanying Learn by Drawing), the first thing to decide is a scale for the drawing. In this example, a scale of 1 cm to represent 10 cm is used. That way, the object would be 3.00 cm in front of the mirror in our drawing.

First the optic axis, the lens, the object (a lighted candle), and the focal points (F) are drawn. A vertical dashed line through the center of the lens is drawn because, for simplicity, the refraction is depicted as if it occurs at the center of each lens. In reality, it would occur at the air–glass and glass–air surfaces of each lens.

Follow steps 1–4 in the accompanying Learn by Drawing:

1. The first ray drawn is the parallel ray (① in the drawing). From the tip of the flame, draw a horizontal ray (parallel to the optic axis). After passing through the lens, this ray goes through the focal point F on the image side.

2. Then draw the chief ray (② in the drawing). From the tip of the flame, draw a ray passing through the center of the lens. This ray will go undeviated through the thin lens to the image side.

3. It can be clearly seen that these two rays intersect on the image side. The point of intersection is the image point of the tip of the candle. From this point, draw the image by extending the tip of the flame to the optic axis.

4. Only two rays are needed to locate the image. However, if you draw the third ray, in this case the focal ray (③ in the drawing), it must go through the same point on the image at which the other two rays intersect (if you are drawing the diagram carefully). The ray from the tip of the flame passing through the focal point F on the object side will travel parallel to the optic axis on the image side.

(b) From the ray diagram in part (a), the image is real (because the rays intersect or converge on the image side). As a result, this real image could be seen on a screen (for example, a piece of white paper) that is positioned at a distance d_i from the converging lens. The image is also inverted (the image of the candle points downward) and is larger than the object.

In this case, $d_o = 30$ cm and $f = 20$ cm, so $2f > d_o > f$. Using similar ray diagrams, you can prove that for any d_o in this range, the image is always real, enlarged, and inverted. Actually, the overhead projector in your classroom uses this particular arrangement.

Follow-Up Exercise. In this Example, what does the image look like if the object is 10 cm in front of the lens? Locate the image graphically and discuss the characteristics of the image.

LEARN BY DRAWING   A LENS RAY DIAGRAM (SEE EXAMPLE 23.5)

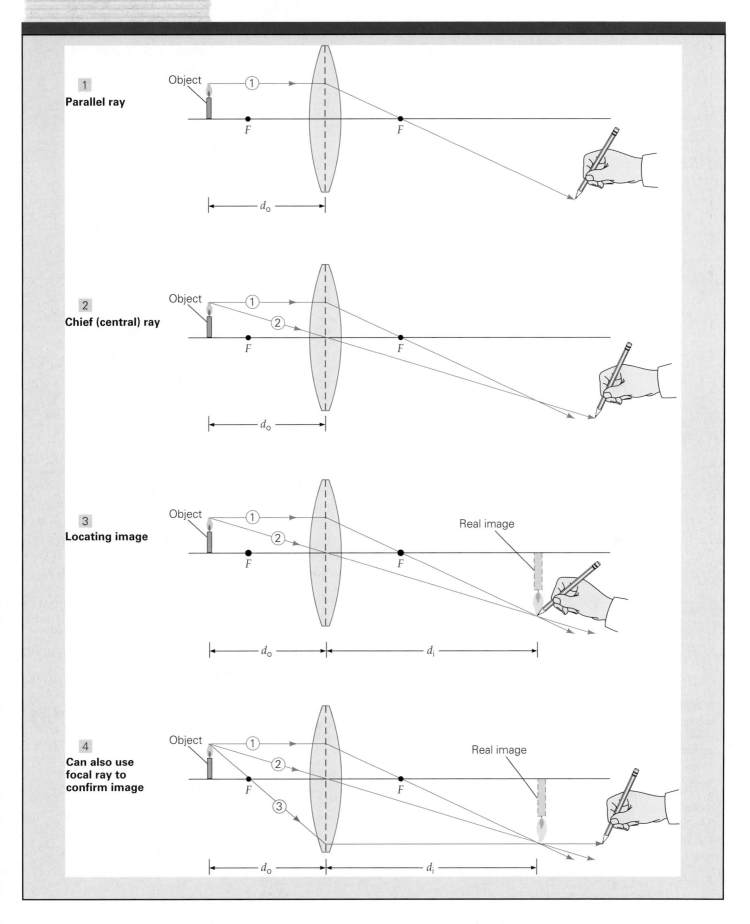

1
Parallel ray

2
Chief (central) ray

3
Locating image

4
Can also use focal ray to confirm image

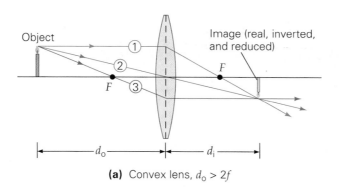

(a) Convex lens, $d_o > 2f$

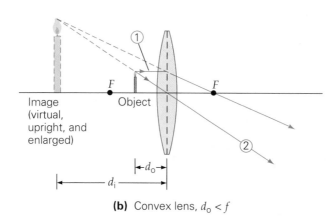

(b) Convex lens, $d_o < f$

▲ **FIGURE 23.15 Ray diagrams for lenses (a)** A converging biconvex lens forms a real object when $d_o > 2f$. The image is real, inverted, and reduced. **(b)** Ray diagram for a converging lens with $d_o < f$. The image is virtual, upright, and magnified. Practical examples are shown for both cases.

To illustrate these procedures, ▲Fig. 23.15 shows other ray diagrams with different object distances for a converging lens, along with real-life applications. The image of an object is real when it is formed on the side of the lens *opposite* the object's side (see Fig. 23.15a). A virtual image is said to be formed on the same side of the lens as the object (see Fig. 23.15b).

Regions could be similarly defined for the object distance for a converging lens as was done for a converging mirror in Fig. 23.7a. Here, an object distance of $d_o = 2f$ for a converging lens has significance similar to that of $d_o = R = 2f$ for a converging mirror (▼Fig. 23.16).

The ray diagram for a diverging lens will be discussed shortly. Like diverging mirrors, diverging lenses can form only virtual images of real objects.

▶ **FIGURE 23.16 Convex lens** For a convex, or converging, lens, the object is located within one of three regions defined by the focal distance (f) and twice the focal distance ($2f$) or at one of these two points. For $d_o > 2f$, the image is real, inverted, and reduced (Fig. 23.15a). For $2f > d_o > f$, the image will also be real and inverted, but enlarged, or magnified, as shown by the ray diagrams in Example 23.5. For $d_o < f$, the image will be virtual, upright, and enlarged (Fig. 23.15b).

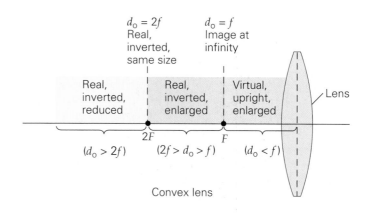

TABLE 23.3	Sign Conventions for Thin Lenses
Focal length (f)	
Converging lens (sometimes called a *positive* lens)	f is positive
Diverging lens (sometimes called a *negative* lens)	f is negative
Object distance (d_o)	
Object is in front of the lens (real object)	d_o is positive
Object is behind the lens (virtual object)*	d_o is negative
Image distance (d_i) and Image type	
Image is formed on the image side of the lens—opposite to the object (real image)	d_i is positive
Image is formed on the object side of the lens—same side as the object (virtual image)	d_i is negative
Image orientation (M)	
Image is upright with respect to the object	M is positive
Image is inverted with respect to the object	M is negative

*In a combination of two (or more) lenses, the image formed by the first lens is taken as the object of the second lens (and so on). If this image–object falls behind the second lens, it is referred to as a virtual object, and the object distance is taken to be negative $(-)$.

PHYSLET®

Exploration 35.3 Moving a Lens

Demonstration/activity: Pass laser light through diverging and converging smoked lenses to show the path the light takes in a lens. Also use a lighted candle or other light source to show how the image changes in size, location, and orientation with the position of the object.

Try to show what happens when the lens is placed under water. Use an aquarium filled with water. Does this make the focal length longer or shorter? Why?

Teaching tip: A summary of sign conventions for the lens equations appears in Table 23.3.

The image distances and characteristics for a spherical lens can also be found analytically. The equations for thin lenses are identical to those for spherical mirrors. The **thin-lens equation** is

$$\frac{1}{d_o} + \frac{1}{d_i} = \frac{1}{f} \quad \textit{thin-lens equation} \tag{23.5}$$

As in the case for spherical mirrors, an alternative form of the thin-lens equation,

$$d_i = \frac{d_o f}{d_o - f} \tag{23.5a}$$

gives a quick and easy way to find d_i.

The **magnification factor**, like that for spherical mirrors, is given by

$$M = -\frac{d_i}{d_o} \tag{23.6}$$

The sign conventions for these thin-lens equations are given in Table 23.3.

Just as when you are working with mirrors, it is helpful to sketch a ray diagram before working a lens problem analytically.

Example 23.6 ■ Three Images: Behavior of a Converging Lens

A biconvex lens has a focal length of 12 cm. For an object (a) 60 cm, (b) 15 cm, and (c) 8.0 cm from the lens, where is the image formed, and what are its characteristics?

Thinking It Through. With the focal length (f) and the object distances (d_o), we can apply Eq. 23.5 to find the image distances (d_i) and Eq. 23.6 to determine the image characteristics. Sketch ray diagrams first to get an idea of the image characteristics. The diagrams should be in good agreement with the calculations.

Solution.

Given: $f = 12$ cm
(a) $d_o = 60$ cm
(b) $d_o = 15$ cm
(c) $d_o = 8.0$ cm

Find: d_i and the image characteristics for all three cases

(continues on next page)

(a) The object distance is greater than twice the focal length ($d_o > 2f$). Using Eq. 23.5,

$$\frac{1}{d_o} + \frac{1}{d_i} = \frac{1}{f}$$

or

$$\frac{1}{d_i} = \frac{1}{f} - \frac{1}{d_o} = \frac{1}{12\text{ cm}} - \frac{1}{60\text{ cm}} = \frac{5}{60\text{ cm}} - \frac{1}{60\text{ cm}} = \frac{4}{60\text{ cm}} = \frac{1}{15\text{ cm}}$$

Then

$$d_i = 15\text{ cm} \qquad \text{and} \qquad M = -\frac{d_i}{d_o} = -\frac{15\text{ cm}}{60\text{ cm}} = -0.25$$

The image is real (positive d_i), inverted (negative M), and one-fourth the object's size ($|M| = 0.25$). A camera uses this arrangement when the object distance is greater than $2f (d_o > 2f)$.

(b) Here, $2f > d_o > f$. Using Eq. 23.5,

$$\frac{1}{d_i} = \frac{1}{12\text{ cm}} - \frac{1}{15\text{ cm}} = \frac{5}{60\text{ cm}} - \frac{4}{60\text{ cm}} = \frac{1}{60\text{ cm}}$$

Then

$$d_i = 60\text{ cm} \qquad \text{and} \qquad M = -\frac{d_i}{d_o} = -\frac{60\text{ cm}}{15\text{ cm}} = -4.0$$

The image is real (positive d_i), inverted (negative M), and four times the object's size ($|M| = 4.0$). This situation applies to the overhead projector and slide projector ($2f > d_o > f$).

(c) For this case, $d_o < f$. Using the alternative form (Eq. 23.5a),

$$d_i = \frac{d_o f}{d_o - f} = \frac{(8.0\text{ cm})(12\text{ cm})}{8.0\text{ cm} - 12\text{ cm}} = -24\text{ cm}$$

Then

$$M = -\frac{d_i}{d_o} = -\frac{(-24\text{ cm})}{8.0\text{ cm}} = +3.0$$

The image is virtual (negative d_i), upright (positive M), and three times the object's size ($|M| = 3.0$). This situation is an example of a simple microscope or magnifying glass ($d_o < f$).

As you can see, a converging lens is versatile. Depending on the object distance (relative to the focal length), the lens can be used as a camera, projector, or magnifying glass.

Follow-Up Exercise. If the object distance of a convex lens is allowed to vary, at what object distance does the real image change from being reduced to being magnified?

PHYSLET®

Exploration 35.1 Image Formation

Conceptual Example 23.7 ■ Half an Image?

A converging lens forms an image on a screen, as shown in ▶Fig. 23.17a. Then the lower half of the lens is blocked, as shown in Fig. 23.17b. As a result, (a) only the top half of the original image will be visible on the screen; (b) only the bottom half of the original image will be visible on the screen; or (c) the entire image will be visible.

Reasoning and Answer. At first thought, you might imagine that blocking off half of the lens would eliminate half of the image. However, rays from *every* point on the object pass through *all parts* of the lens. Thus, the upper half of the lens can form a total image (as could the lower half), so the answer is (c).

You might confirm this conclusion by drawing a chief ray in Fig. 23.17b. Or you might use the scientific method and experiment—particularly if you wear eyeglasses. Block off the bottom part of your glasses, and you will find that you can still read through the top part (unless you wear bifocals).

Follow-Up Exercise. Can you think of any property of the image that *would* be affected by blocking off half of the lens? Explain.

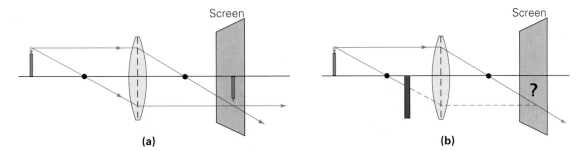

▲ **FIGURE 23.17** Half a lens, half an image? **(a)** A converging lens forms an image on a screen. **(b)** The lower half of the lens is blocked. What happens to the image? See Conceptual Example 23.7.

Integrated Example 23.8 ■ Time for a Change: Behavior of a Diverging Lens

An object is 24 cm in front of a diverging lens that has a focal length of −15 cm. (a) Use a ray diagram to determine whether the image is (1) real and magnified, (2) virtual and reduced, (3) real and upright, or (4) upright and magnified. (b) Find the location and characteristics of the image with the thin-lens equations.

(a) Conceptual Reasoning. (See the sign conventions in Table 23.3.) Use a scale of 1 cm (in our drawing of ▶Fig. 23.18) to represent 10 cm. The object will be 2.4 cm in front of the lens in our drawing. Draw the optic axis, the lens, the object (in this case, a lighted candle), the focal point (F), and a vertical dashed line through the center of the lens.

The parallel ray ① starts from the tip of the flame, travels parallel to the optic axis, diverges from the lens after refraction, and appears to diverge from the F on the object side. The chief ray ② originates from the tip of the flame and goes through the center of the lens, with no direction change. We see that these two rays, after refraction, diverge and do not intersect. However, they appear to come from in front of the lens (object side), and that apparent intersection is the image point of the tip of the flame. We can also draw the focal ray ③ to verify that these rays appear to come from the same image point. The focal ray appears to go through the focal point on the image side and travels parallel to the optic axis after refraction from the lens.

This image is virtual (why?), upright, and smaller than the object, so the answer is (2): virtual and reduced. Measuring from the diagram (keeping in mind the drawing scale we are using), we find that $d_i \approx -9$ cm (virtual image) and $M = \dfrac{h_i}{h_o} \approx \dfrac{0.5 \text{ cm}}{1.4 \text{ cm}} = +0.4$.

(b) Quantitative Reasoning and Solution.

Given: $d_o = 24$ cm
$\qquad\quad f = -15$ cm (diverging lens)

Find: d_i, M, and image characteristics

Note that the focal length is negative for a diverging lens. (See Table 23.3.) From Eq. 23.5,

$$\frac{1}{24 \text{ cm}} + \frac{1}{d_i} = \frac{1}{-15 \text{ cm}} \quad \text{or} \quad \frac{1}{d_i} = \frac{1}{-15 \text{ cm}} - \frac{1}{24 \text{ cm}} = -\frac{13}{120 \text{ cm}}$$

so

$$d_i = -\frac{120 \text{ cm}}{13} = -9.2 \text{ cm}$$

Then

$$M = -\frac{d_i}{d_o} = -\frac{(-9.2 \text{ cm})}{24 \text{ cm}} = +0.38$$

Thus, the image is virtual (d_i is negative) and upright (M is positive), and it is 0.38 times the height of the object. Due to the fact that f is negative for a diverging lens, d_i is always negative for any positive value of d_o, so the image of an object is always virtual.

Follow-Up Exercise. A diverging lens always forms a virtual image of a real object. What general statements can be made about the image's orientation and magnification?

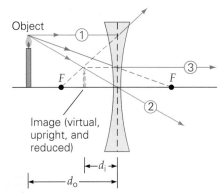

▲ **FIGURE 23.18** Diverging lens Ray diagram of a diverging lens. Here, the image is virtual and in front of the lens, upright, and smaller than the object. See Integrated Example 23.8.

Illustration 35.2 Image from a Diverging Lens

A special type of lens that you may have encountered is discussed in Insight 23.2 on Fresnel Lenses, on page 748.

INSIGHT 23.2 FRESNEL LENSES

To focus parallel light or to produce a large beam of parallel light rays, a sizable converging lens is sometimes necessary. The large mass of glass necessary to form such a lens is bulky and heavy. Moreover, a thick lens absorbs some of the light and is likely to show distortions. A French physicist named Augustin Fresnel (Fray-nel', 1788–1827) developed a solution to this problem for the lenses used in lighthouses. Fresnel recognized that the refraction of light takes place at the surfaces of a lens. Hence, a lens could be made thinner—and almost flat—by removing glass from the interior, as long as the refracting properties of the surfaces were not changed.

This can be accomplished by cutting a series of concentric grooves in the surface of the lens (Fig. 1a). Note that the surface of each remaining curved segment is nearly parallel to the corresponding surface of the original lens. Together, the concentric segments refract light as does the original biconvex converging lens (Fig. 1b). In effect, the lens has simply been slimmed down by the removal of unnecessary glass between the refracting surfaces.

A lens with a series of concentric curved surfaces is called a *Fresnel lens*. Such lenses are widely used in overhead projectors and in beacons (Fig. 1c). A Fresnel lens is very thin and therefore much lighter in weight than a conventional biconvex lens with the same optical properties. Also, Fresnel lenses are easily molded from plastic—often with one flat side (plano-convex) so that the lens can be attached to a glass surface.

One disadvantage of Fresnel lenses is that concentric circles are visible when an observer is looking through such a lens and when an image produced by the lens is projected on a screen, as when we use an overhead projector.

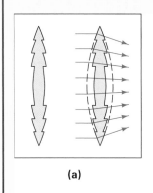

(a)

(b)

(c)

FIGURE 1 Fresnel lens **(a)** The focusing action of a lens comes from refraction at its surfaces. It is therefore possible to reduce the thickness of a lens by cutting away glass in concentric grooves, leaving a set of curved surfaces with the same refractive properties as the lens from which they were derived. **(b)** A flat Fresnel lens with concentric curved surfaces magnifies like a biconvex converging lens. **(c)** An array of Fresnel lenses produces focused beams in this Boston Harbor light. (Fresnel lenses were, in fact, developed for use in lighthouses.)

PHYSLET®

Exploration 35.4 What is Behind the Curtain?

Combinations of Lenses

Many optical instruments, such as microscopes and telescopes (Chapter 25), use a combination of lenses, or a compound-lens system. When two or more lenses are used in combination, we can determine the overall image produced by considering the lenses individually in sequence. That is, the image formed by the first lens becomes the object for the second lens, and so on. For this reason, we present the principles of lens combinations before considering the specifics of their real-life applications.

If the first lens produces an image in front of the second lens, that image is treated as a real object (d_o is positive) for the second lens (▸Fig. 23.19a). If, however, the lenses are close enough so that the image from the first lens is *not* formed before the rays pass through the second lens (Fig. 23.19b), then a modification must be made in the sign conventions. In this case, the image from the first lens is treated as a *virtual* object for the second lens. The virtual object distance is taken to be *negative* in the lens equation (Table 23.3).

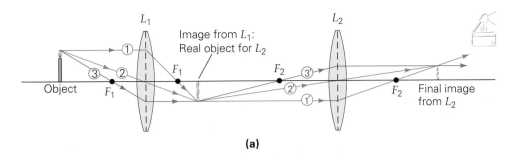

(a)

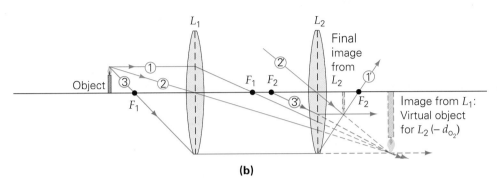

(b)

◀ **FIGURE 23.19** Lens combinations
The final image produced by a compound-lens system can be found by treating the image of one lens as the object for the adjacent lens. **(a)** If the image of the first lens (L_1) is formed in front of the second lens (L_2), the object for the second lens is said to be real. (Note that rays 1', 2', and 3' are the parallel, chief, and focal rays, respectively, for L_2. They are *not* continuations of rays 1, 2, and 3— the parallel, chief, and focal rays, respectively, for L_1.) **(b)** If the rays pass through the second lens before the image is formed, the object for the second lens is said to be virtual, and the object distance for the second lens is taken to be negative.

It can be shown that the total magnification (M_{total}) of a compound-lens system is the product of the individual magnification factors of the component lenses. For example, for a two-lens system, as in Fig. 23.19,

$$M_{total} = M_1 M_2 \qquad (23.7)$$

The conventional signs for M_1 and M_2 carry through the product to indicate, from the sign of M_{total}, whether the final image is upright or inverted. (See Exercise 83.)

Example 23.9 ■ A Special Offer: A Lens Combo and a Virtual Object

Consider two lenses similar to those illustrated in Fig. 23.19b. Suppose the object is 20 cm in front of lens L_1, which has a focal length of 15 cm. Lens L_2, with a focal length of 12 cm, is 26 cm from L_1. What is the location of the final image, and what are its characteristics?

Thinking It Through. This is a double application of the thin-lens equation. The lenses are treated successively. The image of lens L_1 becomes the object of lens L_2. We must keep the quantities distinctly labeled and the distances appropriately referenced (with signs!).

Solution. We have

Given: $d_{o_1} = +20$ cm   *Find:* d_{i_2} and image characteristics
$f_1 = +15$ cm
$f_2 = +12$ cm
$D = 26$ cm (distance between lenses)

The first step is to apply the thin-lens equation (Eq. 23.5) and the magnification factor for thin lenses (Eq. 23.6) to L_1:

$$\frac{1}{d_{i_1}} = \frac{1}{f_1} - \frac{1}{d_{o_1}} = \frac{1}{15 \text{ cm}} - \frac{1}{20 \text{ cm}} = \frac{4}{60 \text{ cm}} - \frac{3}{60 \text{ cm}} = \frac{1}{60 \text{ cm}}$$

or

$$d_{i_1} = 60 \text{ cm (real image from } L_1)$$

and

$$M_1 = -\frac{d_{i_1}}{d_{o_1}} = -\frac{60 \text{ cm}}{20 \text{ cm}} = -3.0 \text{ (inverted and magnified)}$$

The image from lens L_1 becomes the object for lens L_2. This image is then $d_{i_1} - D = 60 \text{ cm} - 26 \text{ cm} = 34 \text{ cm}$ on the right, or image, side of L_2. Therefore, it is a *virtual* object (see Table 23.3), and $d_{o_2} = -34$ cm. (Remember that d_o for virtual objects is taken to be negative.)

(continues on next page)

Then applying the equations to the second lens, L_2:

$$\frac{1}{d_{i_1}} = \frac{1}{f_2} - \frac{1}{d_{o_1}} = \frac{1}{12 \text{ cm}} - \frac{1}{(-34 \text{ cm})} = \frac{23}{204 \text{ cm}}$$

or

$$d_{i_2} = 8.9 \text{ cm (real image)}$$

and

$$M_2 = -\frac{d_{i_1}}{d_{o_2}} = -\frac{8.9 \text{ cm}}{(-34 \text{ cm})} = 0.26 \text{ (upright and reduced)}$$

(*Note*: The virtual object for L_2 was inverted, and thus the term *upright* means that the *final* image is also inverted.) The total magnification M_{total} is then

$$M_{\text{total}} = M_1 M_2 = (-3.0)(0.26) = -0.78$$

The sign is carried through with the magnifications. We determine that the final real image is located at 8.9 cm on the right (image) side of L_2 and that it is inverted (negative sign) relative to the initial object and reduced.

Follow-Up Exercise. Suppose the object in Fig. 23.19b were located 30 cm in front of L_1. Where would the final image be formed in this case, and what would be its characteristics?

23.4 The Lens Maker's Equation

OBJECTIVES: To (a) describe the lens maker's equation, (b) explain how it differs from the thin-lens equation, and (c) understand lens "power" in diopters.

The biconvex and biconcave thin lenses considered so far in this chapter have been relatively easy to analyze. However, there are a variety of other shapes of lenses, as illustrated in Fig. 23.14. For them, the analysis becomes more involved, but it is important to know the focal lengths of such lenses for optical considerations, because lenses are ground for specific purposes or applications.

Lens refraction depends on the shapes of the lens's surfaces and on the index of refraction of the lens. These properties together determine the focal length of a thin lens. The thin-lens focal length is given by the **lens maker's equation**, which enables us to calculate the focal length of a thin lens *in air* ($n_{\text{air}} = 1$) as

$$\frac{1}{f} = (n - 1)\left(\frac{1}{R_1} + \frac{1}{R_2}\right) \qquad \text{(for thin lens in air)} \qquad (23.8)$$

where n is the index of refraction of the lens material and R_1 and R_2 are the radii of curvature of the first (front side) and second (back side) lens surfaces, respectively. (The first surface is the one on which light from an object is first incident.)

A sign convention is required for the lens maker's equation, and a common one is summarized in Table 23.4. The signs depend only on the shape of the surface, that is, convex or concave (▸Fig. 23.20). For the biconvex lens in Fig. 23.20a, both R_1 and R_2 are positive (both surfaces are convex) and for the biconcave lens in Fig. 23.20b, both R_1 and R_2 are negative (both surfaces are concave).

If the lens is surrounded by a medium other than air, then the first term in parentheses in Eq. 23.8 becomes $(n/n_m) - 1$, where n and n_m are the indices of refraction of the lens material and the surrounding medium, respectively. Now we can see why some converging lenses in air become diverging when submerged in water: If $n_m > n$, then f is negative, and the lens is diverging.

Exploration 35.5 Lens Maker's Equation

Exploration 34.1 Lens and a Changing Index of Refraction

TABLE 23.4	Sign Conventions for Lens Maker's Equation
Convex surface	R is positive
Concave surface	R is negative
Plane (flat) surface	$R = \infty$
Converging (positive) lens	f is positive
Diverging (negative) lens	f is negative

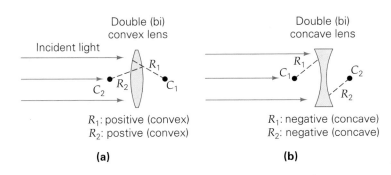

◀ **FIGURE 23.20** Centers of curvature Lenses, such as (a) a biconvex lens and (b) a biconcave lens, have two centers of curvature, which define the signs of the radii of curvature. See Table 23.4.

Lens Power: Diopters

Notice that the lens maker's equation (Eq. 23.8) gives the inverse focal length $1/f$. Optometrists use this inverse relationship to express the *lens power* (P) of a lens in units called **diopters** (abbreviated as D). The lens power is the reciprocal of the focal length of the lens expressed in *meters*:

$$P(\text{expressed in diopters}) = \frac{1}{f(\text{expressed in meters})} \qquad (23.9)$$

Teaching tip: Emphasize that the focal length must be expressed in meters.

So, $1\,D = 1\,m^{-1}$. The lens maker's equation gives a lens's power ($1/f$) in diopters if the radii of curvature are expressed in meters.

If you wear glasses, you may have noticed that the prescription the optometrist gave you for your eyeglass lenses was written in terms of diopters. Converging and diverging lenses are referred to as positive ($+$) and negative ($-$) lenses, respectively. Thus, if an optometrist prescribes a corrective lens with a power of $+2$ diopters, it is a converging lens with a focal length of

Teaching tip: You might want to mention that when examining eyes for glasses, optometrists sometimes talk about adding or subtracting so many "clicks." Each click changes the power by $\frac{1}{4}$ diopter or the focal length by 25 cm.

$$f = \frac{1}{P} = \frac{1}{+2\,D} = \frac{1}{2\,m^{-1}} = 0.50\,m = +50\,cm$$

The greater the power of the lens in diopters, the shorter its focal length is and the more strongly converging or diverging it is. Thus, a "stronger" prescription lens (greater lens power) has a smaller f than does a "weaker" prescription lens (lesser lens power).

Integrated Example 23.10 ■ A Convex Meniscus Lens: Converging or Diverging

The convex meniscus lens shown in Fig. 23.14 has a 15-cm radius for the convex surface and 20 cm for the concave surface. The lens is made of crown glass and is surrounded by air. (a) Is this lens a (1) converging or (2) diverging lens? Explain. (b) What are the focal length and the power of the lens?

(a) Conceptual Reasoning. The index of refraction of crown glass can be obtained from Table 22.1: $n = 1.52$. For a convex meniscus, the first surface is convex, so R_1 is positive; the second surface is concave, so R_2 is negative. Since $R_1 = 15\,cm < |R_2| = 20\,cm$, $1/R_1 + 1/R_2$ will be positive. Therefore, the lens is a converging (positive) lens, according to Eq. 23.8. Thus the answer is (1) converging.

(b) Quantitative Reasoning and Solution.

Given: $R_1 = 15\,cm = 0.15\,m$ *Find:* f and P
        $R_2 = -20\,cm = -0.20\,m$
        $n = 1.52$ (from Table 22.1 for crown glass)

From Eq. 23.8, we have

$$\frac{1}{f} = (n-1)\left(\frac{1}{R_1} + \frac{1}{R_2}\right) = (1.52 - 1)\left(\frac{1}{0.15\,m} + \frac{1}{-0.20\,m}\right) = 0.867\,m^{-1}$$

So $f = \dfrac{1}{0.867\,m^{-1}} = +1.15\,m$.

The power of the lens is $P = \dfrac{1}{f} = +0.867\,D$.

Follow-Up Exercise. In this Example, if this lens is immersed in water, what would your answers be?

*23.5 Lens Aberrations

<u>OBJECTIVES:</u> To (a) describe some common lens aberrations, and (b) explain how they can be reduced or corrected.

Lenses, like mirrors, can also have aberrations. We now discuss some common aberrations.

Spherical Aberration

The discussion of lenses thus far has concentrated on rays that are near the optic axis. Like spherical mirrors, however, converging lenses may show **spherical aberration**, an effect that occurs when parallel rays passing through different regions of a lens do not come together on a common focal plane. In general, rays close to the axis of a converging lens are refracted less and come together at a point farther from the lens than do rays passing through the periphery of the lens (▼Fig. 23.21a).

Spherical aberration can be minimized by using an aperture to reduce the effective area of the lens, so that only light rays near the axis are transmitted. Also, combinations of converging and diverging lenses can be used, as the aberration of one lens can be compensated for by the optical properties of another lens.

Chromatic Aberration

Chromatic aberration is an effect that occurs because the index of refraction of the lens material is *not* the same for all wavelengths of light (that is, the material is dispersive). When white light is incident on a lens, the transmitted rays of different wavelengths (colors) do not have a common focal point, and images of different colors are produced at different locations (Fig. 23.21b).

This dispersive aberration can be minimized, but not eliminated, by using a compound-lens system consisting of lenses of different materials, such as crown glass and flint glass. The lenses are chosen so that the dispersion produced by one is approximately compensated by the opposite dispersion produced by the other. With a properly constructed two-component lens system, called an *achromatic doublet* (*achromatic* means "without color"), the images of any two selected colors can be made to coincide.

Exploration 34.5 *Index of Refraction and Wavelength*

▼ **FIGURE 23.21 Lens aberrations (a)** Spherical aberration. In general, rays closer to the axis of a lens are refracted less and come together at a point farther from the lens than do rays passing through the periphery of the lens. **(b)** Chromatic aberration. Because of dispersion, different wavelengths (colors) of light are focused in different planes, which results in distortion of the overall image.

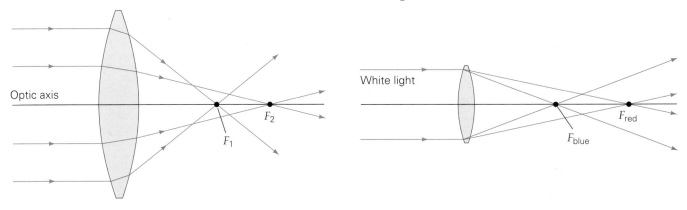

(a) Spherical aberration

(b) Chromatic aberration

Astigmatism

A circular beam of light along the lens axis forms a circular illuminated area on the lens. When incident on a converging lens, the parallel beam converges at the focal point. However, when a circular beam of light from an off-axis source falls on the convex spherical surface of a lens some distance away, the light forms an *elliptical* illuminated area on the lens. The rays entering along the major and minor axes of the ellipse then focus at different points after passing through the lens. This condition is called **astigmatism**.

With different focal points in different planes, the images in both planes are blurred. For example, the image of a point is no longer a point, but rather two separated short-line images (blurred points). Astigmatism can be reduced by decreasing the effective area of the lens with an aperture or by adding a cylindrical lens to compensate.

Chapter Review

- **Plane mirrors** form virtual, upright, and unmagnified images. The object distance is equal to the image distance ($d_o = d_i$).

- The **lateral magnification factor** for all mirrors and lenses is

$$M = -\frac{d_i}{d_o} \qquad (23.4, 23.6)$$

- **Spherical mirrors** are either concave (converging) or convex (diverging). Diverging spherical mirrors always form upright, reduced, virtual images.

Focal length of a spherical mirror:

$$f = \frac{R}{2} \qquad (23.2)$$

Spherical-mirror equation:

$$\frac{1}{d_o} + \frac{1}{d_i} = \frac{1}{f} = \frac{2}{R} \qquad (23.3)$$

Alternative form:

$$d_i = \frac{d_o f}{d_o - f} \qquad (23.3a)$$

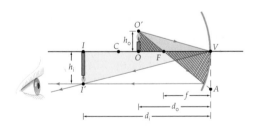

- Bispherical lenses are either convex (converging) or concave (diverging). Diverging spherical lenses always form upright, reduced, virtual images.

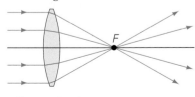

Converging lens

- The **thin-lens equation** relates focal length, object distance, and image distance:

$$\frac{1}{d_o} + \frac{1}{d_i} = \frac{1}{f} \qquad (23.5)$$

Alternative form:

$$d_i = \frac{d_o f}{d_o - f} \qquad (23.5a)$$

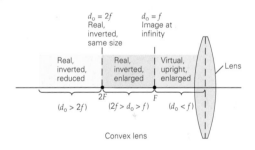

Convex lens

- The **lens maker's equation** is used to compute the grinding radii for the desired focal length of a lens:

$$\frac{1}{f} = (n - 1)\left(\frac{1}{R_1} + \frac{1}{R_2}\right) \quad \text{(thin lens in air only)} \quad (23.8)$$

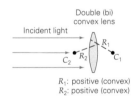

R_1: positive (convex)
R_2: positive (convex)

- **Lens power in diopters** (where f is in meters) is given by

$$P = \frac{1}{f} \qquad (23.9)$$

Exercises

MC = *Multiple Choice Question,* **CQ** = *Conceptual Question, and* **IE** = *Integrated Exercise. Throughout the text, many exercise sections will include "paired" exercises. These exercise pairs, identified with* **red numbers***, are intended to assist you in problem solving and learning. In a pair, the first exercise (even numbered) is worked out in the Study Guide so that you can consult it should you need assistance in solving it. The second exercise (odd numbered) is similar in nature, and its answer is given at the back of the book.*

23.1 Plane Mirrors

1. **MC** A plane mirror (a) has a greater image distance than object distance; (b) produces a virtual, upright, unmagnified image; (c) changes the vertical orientation of an object; (d) reverses an object's top and bottom. (b)

2. **MC** A plane mirror (a) produces both real and virtual images, (b) always produces a virtual image, (c) always produces a real image, (d) forms images by diffuse reflection. (b)

3. **MC** The lateral magnification of a plane mirror is (a) greater than 1, (b) less than 1, (c) equal to +1, (d) equal to −1. (c)

4. **CQ** What is the focal length of a plane mirror? Why? infinite, because it cannot focus light to a point

5. **CQ** Day–night rearview mirrors are common in cars. At night, you tilt the mirror backward, and the intensity and glare of headlights behind you are reduced (▼Fig. 23.22). The mirror is wedge shaped and is silvered on the back. The effect has to do with front-surface and back-surface reflections. The unsilvered front surface reflects about 5% of incident light; the silvered back surface reflects about 90% of the incident light. Explain how the day–night mirror works. see ISM

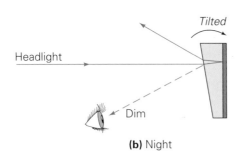

(a) Day

(b) Night

▲ **FIGURE 23.22 Automobile day–night mirror** See Exercise 5.

6. **CQ** When you stand in front of a plane mirror, there is a right–left reversal. (a) Why is there not a top–bottom reversal of your body? (b) Could you affect an apparent top–bottom reversal by positioning your body differently? see ISM

7. **CQ** Why do some emergency vehicles have AMBULANCE (▼Fig. 23.23) printed on the front? see ISM

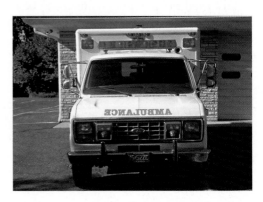

▲ **FIGURE 23.23 Backward and reversed** See Exercise 7.

8. **CQ** Can a virtual image be projected onto a screen? Why or why not? no, because no rays physically intersect at the image

9. ● A person stands 2.0 m away from the reflecting surface of a plane mirror. (a) What is the distance between the person and his or her image? (b) What are the image characteristics? (a) 4.0 m (b) upright, virtual, and same size

10. ● An object 5.0 cm tall is placed 40 cm from a plane mirror. Find (a) the distance from the object to the image, (b) the height of the image, and (c) the image's magnification. (a) 0.80 m (b) 5.0 cm (c) +1.0

11. ● Standing 2.5 m in front of a plane mirror with your camera, you decide to take a picture of yourself. To what distance should the camera be focused to get a sharp image? 5.0 m

12. ●● If you hold a 900-cm² square plane mirror 45 cm from your eyes and can just see the full length of an 8.5-m flagpole behind you, how far are you from the pole? [*Hint:* A diagram is helpful.] 12 m

13. ●● A small dog sits 1.5 m in front of a plane mirror. (a) Where is the dog's image in relation to the mirror? (b) If the dog jumps at the mirror at a speed of 0.50 m/s, how fast does the dog approach its image? (a) 1.5 m behind the mirror (b) 1.0 m/s

14. **IE** ●● A woman fixing the hair on the back of her head holds a plane mirror 30 cm in front of her face so as to look into a plane mirror on the bathroom wall behind her. She is 90 cm from the wall mirror. (a) The image of the back of her head will be from (1) only the front mirror, (2) only the wall mirror, or (3) both mirrors. (b) Approximately how far does the image of the back of her head appear in front of her? (a) both mirrors (b) 2.4 m

15. **IE ●●** (a) When you stand between two plane mirrors on opposite walls in a dance studio, you observe (1) one, (2) two, or (3) multiple images. Explain. (b) If you stand 3.0 m from the mirror on the north wall and 5.0 m from the mirror on the south wall, what are the image distances for the first two images in both mirrors?
(a) multiple images (b) see ISM

16. **●●** A woman 1.7 m tall stands 3.0 m in front of a plane mirror. (a) What is the minimum height the mirror must be to allow the woman to view her complete image from head to foot? Assume that her eyes are 10 cm below the top of her head. (b) What would be the required minimum height of the mirror if she were to stand 5.0 m away?
(a) 0.85 m (b) 0.85 m

17. **●●** Prove that $d_o = d_i$ (equal magnitude) for a plane mirror. [*Hint*: Refer to Fig. 23.2 and use similar and identical triangles.] see ISM

18. **●●●** Draw ray diagrams that show how three images of an object are formed in two plane mirrors at right angles, as shown in ▼Fig. 23.24a. [*Hint*: Consider rays from both ends of the object in the drawing for each image.] Figure 23.24b shows a similar situation from a different point of view that gives four images. Explain the extra image in this case. see ISM

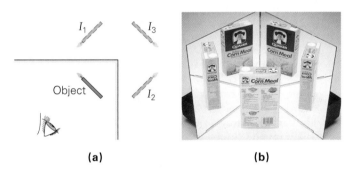

(a)   (b)

▲ **FIGURE 23.24** Two mirrors—multiple images
See Exercise 18.

23.2 Spherical Mirrors

19. **MC** Which of the following statements concerning spherical mirrors is correct? (a) A converging mirror alone can produce an inverted virtual image. (b) A diverging mirror alone can produce an inverted virtual image. (c) A diverging mirror can produce an inverted real image. (d) A converging mirror can produce an inverted real image. (d)

20. **MC** The image produced by a convex mirror is always (a) virtual and upright, (b) real and upright, (c) virtual and inverted, (d) real and inverted. (a)

21. **MC** A shaving/makeup mirror is used to form an image that is larger than the object, so it is a (a) concave, (b) convex, (c) plane mirror. (a) concave

22. **CQ** (a) What is the purpose of using a dual mirror on a car or truck, such as the one shown in ▶Fig. 23.25? (b) Some rearview mirrors on the passenger side of automobiles have the warning "OBJECTS IN MIRROR ARE CLOSER THAN THEY APPEAR." Explain why. (c) Could a TV satellite dish be considered a converging mirror? Explain. see ISM

▲ **FIGURE 23.25** Mirror applications See Exercise 22.

23. **CQ** (a) If you look into a shiny spoon, you see an inverted image on one side and an upright image on the other (▼Fig. 23.26). (Try it.) Why? (b) Could you see upright images on both sides? Explain. see ISM

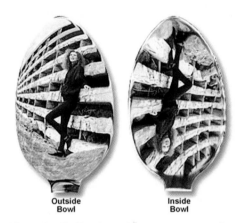

Outside Bowl   Inside Bowl

▲ **FIGURE 23.26** Reflections from concave and convex surfaces See Exercise 23.

24. **CQ** (a) A 10-cm-tall mirror bears the following advertisement: "Full-view mini mirror. See your full body in 10 cm." How can this be? (b) A popular novelty item consists of a concave mirror with a ball suspended at or slightly inside the center of curvature (▼Fig. 23.27). When the ball swings toward the mirror, its image grows larger and suddenly fills the whole mirror. The image appears to be jumping out of the mirror. Explain what is happening. see ISM

▲ **FIGURE 23.27** Spherical-mirror toy See Exercise 24.

25. **CQ** How can the focal length be quickly determined experimentally for a concave mirror? Can you do the same thing for a convex mirror? see ISM

26. **CQ** Can a convex mirror produce an image that is taller than the object? Why or why not?
no, convex mirror produces only reduced images

27. **IE** ● An object is 30 cm in front of a convex mirror that has a focal length of 60 cm. (a) Use a ray diagram to determine whether the image is (1) real or virtual, (2) upright or inverted, and (3) magnified or smaller than the object. (b) Calculate the image distance and image height. (a) virtual, upright, and reduced (b) $d_i = -20$ cm; $h_i = +0.67\,h_o$

28. ● An object 3.0 cm tall is placed 20 cm from the front of a concave mirror with a radius of curvature of 30 cm. Where is the image formed, and how tall is it?
$d_i = 60$ cm; $h_i = 9.0$ cm

29. ● If the object in Exercise 28 is moved to a position 10 cm from the front of the mirror, what will be the characteristics of the image? $d_i = -30$ cm; $h_i = 9.0$ cm; image is virtual, upright, and magnified

30. ● A candle with a flame 1.5 cm tall is placed 5.0 cm from the front of a concave mirror. A virtual image is produced that is 10 cm from the vertex of the mirror. (a) Find the focal length and radius of curvature of the mirror. (b) How tall is the image of the flame?
(a) $f = 10$ cm, $R = 20$ cm (b) 3.0 cm

31. ●● Use the mirror equation and the magnification factor to show that when $d_o = R = 2f$ for a concave mirror, the image is real, inverted, and the same size as the object. see ISM

32. ●● An object 3.0 cm tall is placed at different locations in front of a concave mirror whose radius of curvature is 30 cm. Determine the location of the image and its characteristics when the object distance is 40 cm, 30 cm, 15 cm, and 5.0 cm, using (a) a ray diagram and (b) the mirror equation. see ISM

33. **IE** ●● A virtual image of magnification +0.50 is produced when an object is placed in front of a spherical mirror. (a) The mirror is (1) convex, (2) concave, (3) flat. Explain. (b) Find the radius of curvature of the mirror if the object is 7.0 cm in front of it. (a) convex (b) 14 cm

34. ●● A bottle 6.0 cm tall is located 75 cm from the concave surface of a mirror with a radius of curvature of 50 cm. Where is the image located, and what are its characteristics?
$d_i = 38$ cm, $h_i = 3.0$ cm; real and inverted

35. **IE** ●● A shaving mirror has a magnification of +4.00. (a) The mirror is (1) convex, (2) concave, (3) flat. Explain. (b) What is the focal length of the mirror if your face is 10.0 cm in front of the mirror? (a) concave (b) 13.3 cm

36. ●● Using the spherical-mirror equation and the magnification factor, show that for a concave mirror with $d_o < f$, the image of an object is always virtual, upright, and magnified. see ISM

37. ●● Using the spherical-mirror equation and the magnification factor, show that for a convex mirror, the image of an object is always virtual, upright, and reduced. see ISM

38. ●● A concave makeup mirror produces a virtual image 1.5 times the size of a person whose face is 20 cm from the mirror. (a) Draw a ray diagram of this situation. (b) What is the focal length of the mirror? (a) see ISM (b) 60 cm

39. **IE** ●● The image of an object located 30 cm from a mirror is formed on a screen located 20 cm from the mirror. (a) The mirror is (1) convex, (2) concave, (3) flat. Explain. (b) What is the mirror's radius of curvature?
(a) concave (b) 24 cm

40. **IE** ●● The erect image of an object 18 cm in front of a mirror is half the size of the object. (a) The mirror is (1) convex, (2) concave, (3) flat. Explain. (b) What is the focal length of the mirror? (a) convex, as the erect image is smaller than the object (b) −18 cm

41. **IE** ●● A concave mirror has a magnification of +3.0 for an object placed 50 cm in front of it. (a) The type of image produced is (1) virtual and upright, (2) real and upright, (3) virtual and inverted, (4) real and inverted. Explain. (b) Find the radius of curvature of the mirror.
(a) virtual and upright (b) 1.5 m

42. ●● A concave shaving mirror is constructed so that a man at a distance of 20 cm from the mirror sees his image magnified 1.5 times. What is the radius of curvature of the mirror?
1.2 m

43. ●● A child looks at a reflective Christmas tree ball ornament that has a diameter of 9.0 cm and sees an image of her face that is half the real size. How far is the child's face from the ball? 2.3 cm

44. **IE** ●● A dentist uses a spherical mirror that produces an upright image of a tooth that is magnified four times. (a) The mirror is (1) converging, (2) diverging, (3) flat. Explain. (b) What is the mirror's focal length in terms of the object distance?
(a) converging, as the image is magnified (b) $f = \left(\frac{4}{3}\right)d_o$

45. ●● A 15-cm-long pencil is placed with its eraser on the optic axis of a concave mirror and its point directed upward at a distance of 20 cm in front of the mirror. The radius of curvature of the mirror is 30 cm. Use (a) a ray diagram and (b) the mirror equation to locate the image and determine the image characteristics.
(a) see ISM (b) $d_i = 60$ cm, $M = -3.0$, real and inverted

46. **IE** ●● A pill bottle 3.0 cm tall is placed 12 cm in front of a mirror. A 9.0-cm-tall upright image is formed. (a) The mirror is (1) convex, (2) concave, (3) flat. Explain. (b) What is its radius of curvature?
(a) concave, as the image is magnified (b) 36 cm

47. ●● A spherical mirror at an amusement park shows anyone who stands 2.5 m in front of it an upright image two times the person's height. What is the mirror's radius of curvature? 10 m

48. ●●● For values of d_o from 0 to ∞, (a) sketch graphs of (1) d_i versus d_o and (2) M versus d_o for a converging mirror, and (b) sketch similar graphs for a diverging mirror. see ISM

49. ●●● The front surface of a glass cube 5.00 cm on each side is placed a distance of 30.0 cm in front of a converging mirror that has a focal length of 20.0 cm. (a) Where is the image of the front and back surface of the cube located, and what are the image characteristics? (b) Is the image of the cube still a cube? see ISM

50. ●●● A section of a sphere is mirrored on both sides. If the magnification of an object is +1.8 when the section is used as a concave mirror, what is the magnification of an object at the same distance in front of the convex side? 0.69

51. **IE ●●●** A concave mirror of radius of curvature of 20 cm forms an image of an object that is twice the height of the object. (a) There could be (1) one, (2) two, (3) three object distance(s) that satisfy the image characteristics. Explain. (b) What are the object distances? (a) two (b) 5.0 cm, 15 cm

52. **●●●** A convex mirror is on the exterior of the passenger side of many trucks (Exercise 22a). If the focal length of such a mirror is −40.0 cm, what will be the location and height of the image of a car that is 2.0 m high and (a) 100 m and (b) 10.0 m behind the truck mirror? see ISM

53. **●●●** Two students in a physics laboratory each have a concave mirror with the same radius of curvature, 40 cm. Each student places an object in front of her mirror. The image in both mirrors is three times the size of the object. However, when the students compare notes, they find that the object distances are not the same. Is this possible? If so, what are the object distances? yes: 13 cm; 27 cm

23.3 Lenses

54. **MC** The image produced by a diverging lens is always (a) virtual and magnified, (b) real and magnified, (c) virtual and reduced, (d) real and reduced. (c)

55. **MC** A converging lens (a) must have at least one convex surface, (b) cannot produce a virtual and reduced image, (c) is thicker at its center than at the periphery, (d) all of the preceding. (d)

56. **MC** If an object is placed at the focal point of a converging lens, the image is (a) at zero, (b) also at the focal point, (c) at a distance equal to twice the focal length, (d) at infinity. (d)

57. **CQ** Explain why a fish in a spherical fish bowl, viewed from the side, appears larger than it really is. see ISM

58. **CQ** Can a converging lens ever form a virtual image of a real object? If yes, under what conditions? yes, $d_o < f$

59. **CQ** How can you quickly determine the focal length of a converging lens? Will the same method work for a diverging lens? locate the image of a distant object; no

60. **CQ** If you want to use a converging lens to design a simple overhead projector so as to project the magnified image of some small writing onto a screen on a wall, how far should you place the object in front of the lens? $2f > d_o > f$

61. **●** An object is placed 50.0 cm in front of a converging lens of focal length 10.0 cm. What are the image distance and the lateral magnification? $d_i = 12.5$ cm; $M = -0.250$

62. **●** An object placed 30 cm in front of a converging lens forms an image 15 cm behind the lens. What is the focal length of the lens? 10 cm

63. **●** A converging lens with a focal length of 20 cm is used to produce an image on a screen that is 2.0 m from the lens. What is the object distance? 22 cm

64. **IE ●●** An object 4.0 cm tall is in front of a converging lens of focal length 22 cm. The object is 15 cm away from the lens. (a) Use a ray diagram to determine whether the image is (1) real or virtual, (2) upright or inverted, and (3) magnified or smaller than the object. (b) Calculate the image distance and lateral magnification. (a) virtual, upright, and magnified (b) $d_i = -47$ cm; $M = +3.1$

65. **●●** (a) Design the lens in a single-lens slide projector that will form a sharp image on a screen 4.0 m away with the transparent slides 6.0 cm from the lens. (b) If the object on a slide is 1.0 cm tall, how tall will the image on the screen be, and how should the slide be placed in the projector? (a) $f = 5.9$ cm (b) 67 cm, inverted

66. **●●** Using the thin-lens equation and the magnification factor, show that for a spherical diverging lens, the image of a real object is always virtual, upright, and reduced. see ISM

67. **●●** A biconvex lens has a focal length of 0.12 m. Where on the lens axis should an object be placed in order to get (a) a real, enlarged image with a magnification of 2.0 and (b) a virtual, enlarged image with a magnification of 2.0? (a) 18 cm (b) 6.0 cm

68. **●●** An object is placed in front of a biconcave lens whose focal length is −18 cm. Where is the image located and what are its characteristics, if the object distance is (a) 10 cm and (b) 25 cm? Sketch ray diagrams for each case. (a) $d_i = -6.4$ cm; $M = +0.64$ (b) $d_i = -10$ cm; $M = +0.42$

69. **●●** A biconvex lens produces a real, inverted image of an object that is magnified 2.5 times when the object is 20 cm from the lens. What is the focal length of the lens? 14 cm

70. **●●** A simple single-lens camera (biconvex lens) is used to photograph a man 1.7 m tall who is standing 4.0 m from the camera. If the man's image fills the height of a frame of film (35 mm), what is the focal length of the lens? 8.1 cm

71. **●●** To photograph a full Moon, a photographer uses a single-lens camera having a focal length of 60 mm. What will be the diameter of the Moon's image on the film? (*Note:* Data about the Moon are inside the back cover of this text.) 0.55 mm

72. **●●** (a) For values of d_o from 0 to ∞, sketch graphs of (1) d_i versus d_o and (2) M versus d_o for a converging lens. (b) Sketch similar graphs for a diverging lens. (Compare to Exercise 48.) see ISM

73. **●●** An object is placed 40 cm from a screen. (a) At what point between the object and the screen should a converging lens with a focal length of 10 cm be placed so that it will produce a sharp image on the screen? (b) What is the lens's magnification? (a) 20 cm (b) $M = -1.0$

74. **●●** An object 5.0 cm tall is 10 cm from a concave lens. The resulting image is one fifth as large as the object. What is the focal length of the lens? −2.5 cm

75. **●●** (a) For a biconvex lens, what is the *minimum* distance between an object and its image if the image is real? (b) What is the distance if the image is virtual? (a) $d = 4f$ (b) approaches 0

76. **●●** Using ▼Fig. 23.28, derive (a) the thin-lens equation and (b) the magnification equation for a thin lens. [*Hint:* Use similar triangles.] see ISM

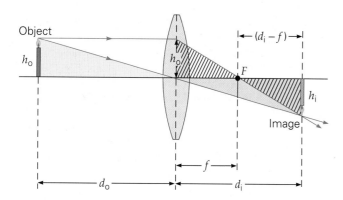

▲ **FIGURE 23.28 The thin-lens equation** The geometry for deriving the thin-lens equation (and magnification factor). Note the two sets of similar triangles. See Exercise 76.

77. ●● (a) If a book is held 30 cm from an eyeglass lens with a focal length of −45 cm, where is the image of the print formed? (b) If an eyeglass lens with a focal length of +57 cm is used, where is the image formed?
(a) −18 cm (b) −63 cm

78. IE ●● A biology student wants to examine a bug at a magnification of +5.00 (a) The lens should be (1) convex, (2) concave, (3) flat. Explain. (b) If the bug is 5.00 cm from the lens, what is the focal length of the lens?
(a) convex, as the image is magnified (b) 6.25 cm

79. ●● With a magnifying glass, a biology student on a field trip views a small insect. If she sees the insect magnified by a factor of 3.5 when the glass is held 3.0 cm from it, what is the focal length of the lens? 4.2 cm

80. ●● The human eye is a complex multiple-lens system. However, it can be approximated to an equivalent single converging lens with an average focal length about 1.7 cm when the eye is relaxed. If an eye is viewing a 2.0-m-tall tree located 15 m in front of the eye, what are the height and orientation of the image of the tree on the retina?
2.3 mm, inverted

81. ●●● The geometry of a compound microscope, which consists of two converging lenses, is shown in ▼Fig. 23.29. (More detail on microscopes is given in Chapter 25.) The objective lens and the eyepiece lens have focal lengths of 2.8 mm and 3.3 cm, respectively. If an object is located 3.0 mm from the objective lens, where is the final image located, and what type of image is it?
18 cm to left of eyepiece; virtual image

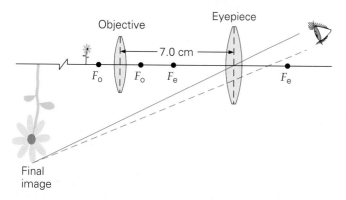

▲ **FIGURE 23.29 Compound microscope** See Exercise 81.

82. ●●● Two converging lenses L_1 and L_2 have focal lengths of 30 cm and 20 cm, respectively. The lenses are placed 60 cm apart along the same axis, and an object is placed 50 cm from L_1 on the side opposite L_2. Where is the image formed relative to L_2, and what are its characteristics?
8.6 cm; real, inverted, $M_{total} = -0.86$

83. ●●● For a lens combination, show that the total magnification $M_{total} = M_1 M_2$. [*Hint*: Think about the definition of magnification.] see ISM

84. ●●● Show that for thin lenses that have focal lengths f_1 and f_2 and are in contact, the effective focal length (f) is given by

$$\frac{1}{f} = \frac{1}{f_1} + \frac{1}{f_2}$$

see ISM

23.4 The Lens Maker's Equation *and* *23.5 Lens Aberrations

85. **MC** The power of a lens is expressed in units of (a) watts, (b) diopters, (c) meters, (d) both b and c. (b)

86. **MC** A lens aberration that is caused by dispersion is called (a) spherical aberration, (b) chromatic aberration, (c) refractive aberration, (d) none of the preceding. (b)

87. **MC** The focal length of a rectangular glass block is (a) zero, (b) infinity, (c) not defined. (b)

88. **CQ** Determine the signs of R_1 and R_2 for each lens shown in Fig. 23.14. +, +; +, ∞; +, −; −, −; ∞, −; +, −.

89. **CQ** When you open your eyes underwater, everything is blurry. However, when you wear goggles, you can see clearly. Explain. see ISM

90. **CQ** A lens that is converging in air is submerged in a fluid whose index of refraction is greater than that of the lens. Is the lens still converging? no, it is diverging

91. **CQ** (a) When a lens with $n = 1.60$ is immersed in water, is there a change in the focal length of the lens? If so, which way? (b) What would be the case for a submerged lens whose index of refraction is less than that of the fluid? (a) yes, increases (b) converging lens becomes diverging, and vice versa

92. ● An optometrist prescribed glasses with a power of −2.0 D for a nearsighted student. What is the focal length of the glass lenses? −0.50 m

93. ● A farsighted senior citizen needs glasses with a focal length of 25 cm. What is the power of the lens? +4.0 D

94. ●● An optometrist prescribes a corrective lens with a power of +1.5 D. The lens maker will start with a glass blank that has an index of refraction of 1.6 and a convex front surface whose radius of curvature is 20 cm. To what radius of curvature should the other surface be ground? 40 cm concave

95. ●● A plastic plano-concave lens has a radius of curvature of 50 cm for its concave surface. If the index of refraction of the plastic is 1.35, what is the power of the lens?
−0.70 D

96. **IE ●●** A plastic convex-meniscus (Fig. 23.14) contact lens is made of plastic of index of refraction 1.55. The lens has a front radius of 2.50 cm and a back radius of 3.00 cm. (a) The signs of R_1 and R_2 are (1) +, +, (2) +, −, (3) −, +, (4) −, −. Explain. (b) What is the focal length of the lens? (a) (2) +, − (b) 27.2 cm

97. **●●** A converging glass lens with an index of refraction of 1.62 has a focal length of 30 cm in air. What is the focal length when the lens is submerged in water? 85 cm

98. **IE ●●●** A biconvex lens is made of glass whose index of refraction is 1.6. The lens has a radius of curvature of 30 cm for one surface and 40 cm for the other. (a) Will the focal length of this lens (1) increase, (2) remain the same, or (3) decrease if the lens is moved from air to under water? Why? (b) Calculate the focal length of this lens as used in air and under water. (a) increase (b) air: 29 cm; water: 84 cm

Comprehensive Exercises

99. A method of determining the focal length of a diverging lens is called *autocollimation*. As ▼ Fig. 23.30 shows, first a sharp image of a light source is projected on a screen by a converging lens. Second, the screen is replaced with a plane mirror. Third, a diverging lens is placed between the converging lens and the mirror. Light will then be reflected by the mirror back through the compound-lens system, and an image will be formed on a screen near the light source. This image is made sharp by adjusting the distance between the diverging lens and the mirror. The distance at which the image is clearest is equal to the focal length of the lens. Explain why this method works. see ISM

100. For the arrangement shown in ▼ Fig. 23.31, an object is placed 0.40 m in front of the converging lens, which has a focal length of 0.15 m. If the concave mirror has a focal length of 0.13 m, where is the final image formed, and what are its characteristics? 0.26 m in front of the concave mirror; $M_{total} = 0.60$, real and upright

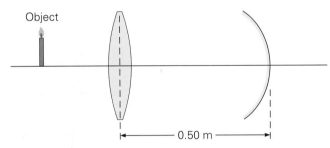

▲ **FIGURE 23.31 Lens–mirror combination** See Exercise 100.

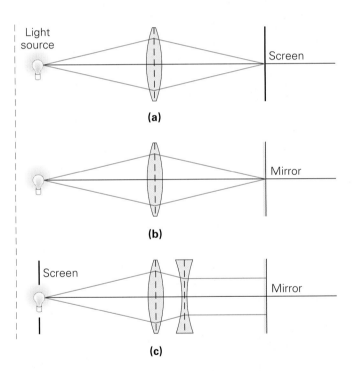

▲ **FIGURE 23.30 Autocollimation** See Exercise 99.

101. Two lenses, each having a power of +10 D, are placed 20 cm apart along the same axis. If an object is 60 cm from the first lens on the side opposite the second lens, where is the final image relative to the first lens, and what are its characteristics? 20 cm on object side of first lens; virtual, inverted, $M_{total} = -1.0$

102. Show that the magnification for objects near the optic axis of a convex mirror is given by $|M| = d_i/d_o$. [*Hint*: Use a ray diagram with rays reflected at the mirror's vertex.] see ISM

103. An object is 15 cm from a converging lens whose focal length is 10 cm. On the opposite side of that lens, at a distance of 60 cm, is a converging lens with a focal length of 20 cm. Where is the final image formed, and what are its characteristics? 60 cm to right of second lens; real, upright, $M_{total} = 4.0$

104. (a) Use ray diagrams to show that a ray parallel to the optic axis of a biconvex lens is refracted toward the axis at the incident surface and again at the exit surface. (b) Show this also holds for a biconcave lens, but with both refractions away from the axis. see ISM

The following Physlet Physics Problems can be used with this chapter.
33.1, 33.2, 33.3, 33.4, 33.5, 33.6, 33.7, 33.8, 34.4, 35.1, 35.2, 35.3, 35.4, 35.5, 35.6, 35.7, 35.8, 35.9, 35.10
Optics Appendix: What's Behind the Curtain? Problems 1–15

PHYSICS FACTS

- Some sources say Thomas Young, who first demonstrated the wave nature of light, could read by age 2 and read the Bible twice in his early years.

- The track-to-track distance is 0.74 μm on a DVD-ROM and 1.6 μm on a CD-ROM. In comparison, the diameter of human hair is on the order of 50–150 μm. DVD-ROM and CD-ROM tracks really split hairs.

- AM radio can be heard better in some areas than FM radio. This is because the longer AM waves are more easily diffracted around buildings and other obstacles.

- Skylight is partially polarized. It is believed that some insects, such as bees, use polarized skylight to determine navigational directions relative to the Sun.

- To an observer on Earth, the "red planet" Mars appears reddish because the surface material contains iron oxide. The rusting of iron on Earth produces iron oxide.

I t's always intriguing to see brilliant colors produced by objects that we know don't have any color of their own. The glass of a prism, for example, which is clear and transparent by itself, nevertheless gives rise to a whole array of colors when white light passes through it. Prisms, like the water droplets that produce rainbows, don't create color. They merely separate the different colors that make up white light.

The phenomena of reflection and refraction are conveniently analyzed by using geometrical optics (Chapter 22). Ray diagrams (Chapter 23) show what happens when light is reflected from a mirror or passed through a lens. However, other phenomena involving light, such as the interference patterns of the soap bubble in the photo, cannot be adequately explained or described using the ray concept, since this technique ignores the wave nature of light. Other wave phenomena include diffraction and polarization.

Physical optics, or **wave optics**, takes into account wave properties that geometrical optics ignores. The wave theory of light leads to satisfactory explanations of those phenomena that cannot be analyzed with rays. Thus, in this chapter, the wave nature of light must be used to analyze phenomena such as interference and diffraction.

Wave optics must be used to explain how light propagates around small objects or through small openings. We see this in our everyday life with the narrow grooves in CDs, DVDs, and other items. An object or opening is considered small if it is in the order of magnitude of the wavelength of light.

24.1 Young's Double-Slit Experiment

OBJECTIVES: To (a) explain how Young's experiment demonstrated the wave nature of light, and (b) compute the wavelength of light from experimental results.

It has been stated that light behaves like a wave, but no proof of this assertion has been discussed. How would you go about demonstrating the wave nature of light? One method that involves the use of interference was first devised in 1801 by the English scientist Thomas Young (1773–1829). **Young's double-slit experiment** not only demonstrates the wave nature of light, but also allows the measurement of its wavelengths. Essentially, light can be shown to be a wave if it exhibits wave properties such as interference and diffraction.

Recall from the discussion of wave interference in Sections 13.4 and 14.4 that superimposed waves may interfere constructively or destructively. Constructive interference occurs when two crests are superimposed. If a crest and a trough are superimposed, then destructive interference occurs. Interference can be easily observed with water waves, for which constructive and destructive interference produce obvious interference patterns (▸Fig. 24.1).

The interference of (visible) light waves is not as easily observed, because of their relatively short wavelengths ($\approx 10^{-7}$ m) and the fact that they usually are not really monochromatic (single frequency). Also, stationary interference patterns are produced only with *coherent sources*—sources that produce light waves having a constant phase relationship to one another. For example, for constructive interference to occur at some point, the waves meeting at that point must be in phase. As the waves meet, a crest must *always* overlap a crest, and a trough must *always* overlap a trough. If a phase difference develops between the waves over time, the interference pattern changes, and a stable or stationary pattern will not be established.

In an ordinary light source, the atoms are excited randomly, and the emitted light waves fluctuate in amplitude and frequency. Thus, light from two such sources is *incoherent* and cannot produce a stationary interference pattern. Interference does occur, but the phase difference between the interfering waves changes so fast that the interference effects are not discernible. To obtain the equivalent of two coherent sources, a barrier with one narrow slit is placed in front of a single light source, and a barrier with two very narrow slits is positioned symmetrically in front of the first barrier (▾Fig. 24.2a).

Waves propagating out from the single slit are in phase, and the double slits then act as two coherent sources by separating each wave into two parts. Any random changes in the light from the original source will thus occur for the light passing through both slits, and the phase difference will be constant. The modern laser beam, a coherent light source, makes the observation of a stable interference pattern much

▲ **FIGURE 24.1** Water-wave interference The constructive and destructive interference of water waves from two coherent sources in a ripple tank produces interference patterns.

Note: Compare Fig. 24.1 with Fig. 14.8a.

Demonstration/activity: Use a ripple tank to demonstrate interference of waves.

PHYSLET®

Illustration 37.1 Ripple Tank

Demonstration/activity: Construct two identical transparencies of concentric circles, representing waves radiating from a point source. Overlaying the transparencies on an overhead projector shows how waves from two point sources interfere with each other.

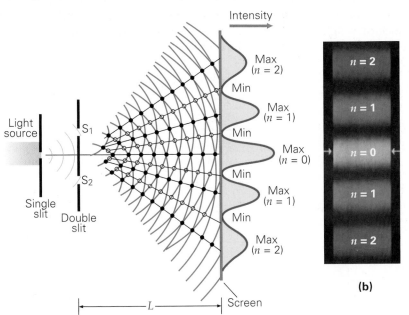

(b)

(a)

◀ **FIGURE 24.2** Double-slit interference **(a)** The coherent waves from two slits are shown in blue (top slit) and red (bottom slit). The waves spread out as a result of diffraction from narrow slits. The waves interfere, producing alternating maxima and minima, on the screen. **(b)** An interference pattern. Note the symmetry of the pattern about the central maximum ($n = 0$).

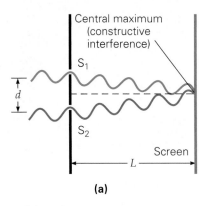

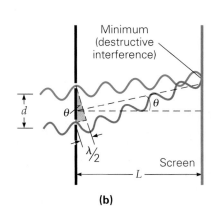

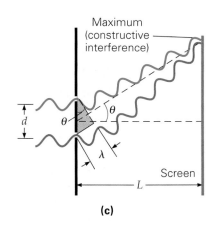

(a)   **(b)**   **(c)**

▲ **FIGURE 24.3 Interference** The interference that produces a maximum or minimum depends on the difference in the path lengths of the light from the two slits. **(a)** The path-length difference at the position of the central maximum is zero, so the waves arrive in phase and interfere constructively. **(b)** At the position of the first minimum, the path-length difference is $\lambda/2$, and the waves interfere destructively. **(c)** At the position of the first maximum, the path-length difference is λ, and the interference is constructive.

▶ **FIGURE 24.4 Geometry of Young's double-slit experiment** The difference in the path lengths for light traveling from the two slits to a point P is $r_2 - r_1 = \Delta L$, which forms a side of the small shaded triangle. Because the barrier with the slits is parallel to the screen, the angle between r_2 and the barrier (at S_2, in the small shaded triangle) is equal to the angle between r_2 and the screen. When L is much greater than y, that angle is almost identical to the angle between the screen and the dashed line, which is an angle in the large shaded triangle. The two shaded triangles are then almost exactly similar, and the angle at S_1 in the small triangle is almost exactly equal to θ. Thus, $\Delta L = d \sin \theta$. (Not drawn to scale. Assume that $d \ll L$.)

easier. A series of maxima or bright positions can be observed on a screen placed relatively far from the slits (Fig. 24.2b).

To help analyze Young's experiment, let's imagine that light with a single wavelength (monochromatic light) is used. Because of diffraction (see Sections 13.4 and 14.4 and, in this chapter, Section 24.3), or the spreading of light as it passes through a slit, the waves spread out and interfere as illustrated in Fig. 24.2a. Coming from two coherent "sources," the interfering waves produce a stable interference pattern on the screen. The pattern consists of a bright central maximum (▲Fig. 24.3a) and a series of symmetrical side minima (Fig. 24.3b) and maxima (Fig. 24.3c), which mark the positions at which destructive and constructive interference occur. The existence of this interference pattern clearly demonstrates the wave nature of light. The intensities of each side maxima decrease with distance from the central maximum.

Measuring the wavelength of light requires us to look at the geometry of Young's experiment, as shown in ▼Fig. 24.4. Let the screen be a distance L from the slits and P be an arbitrary point on the screen. P is located a distance y from the center of the central maximum and at an angle θ relative to a normal line between the slits. The slits S_1 and S_2 are separated by a distance d. Note that the light path from slit S_2 to P is longer than the path from slit S_1 to P. As the figure shows, the path-length difference (ΔL) is approximately

$$\Delta L = d \sin \theta$$

The fact that the angle in the small shaded triangle is almost equal to θ can be shown by a simple geometrical argument involving similar triangles when $d \ll L$, as described in the caption of Fig. 24.4.

The relationship of the phase difference of two waves to their path-length difference was discussed in Chapter 14 for sound waves. These conditions hold for any wave, including light. Constructive interference occurs at any point where the path-length difference between the two waves is an integral number of wavelengths:

$$\Delta L = n\lambda \quad \text{for } n = 0, 1, 2, 3, \ldots \qquad \textit{condition for constructive interference} \qquad (24.1)$$

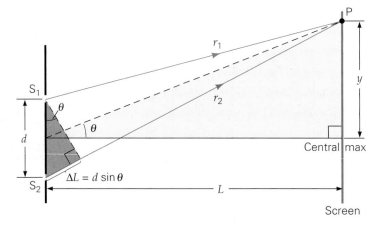

Similarly, for destructive interference, the path-length difference is an odd number of half-wavelengths:

$$\Delta L = \frac{m\lambda}{2} \quad \text{for } m = 1, 3, 5, \ldots \quad \begin{array}{l}\textit{condition for}\\\textit{destructive interference}\end{array} \quad (24.2)$$

Thus, in Fig. 24.4, the maxima (constructive interference) satisfy

$$d \sin \theta = n\lambda \quad \text{for } n = 0, 1, 2, 3, \ldots \quad \begin{array}{l}\textit{condition for}\\\textit{interference maxima}\end{array} \quad (24.3)$$

where n is called the *order number*. The zeroth order ($n = 0$) corresponds to the central maximum; the first order ($n = 1$) is the first maximum on either side of the central maximum; and so on. As the path-length difference varies from point to point, so does the phase difference and the resulting type of interference (constructive or destructive).

The wavelength can therefore be determined by measuring d and θ for a maximum of a particular order (other than the central maximum), because Eq. 24.3 can be solved as $\lambda = (d \sin \theta)/n$.

The angle θ locates a maximum relative to the central maximum. This can be measured from a photograph of the interference pattern, such as shown in Fig. 24.2b. If θ is small ($y \ll L$), $\sin \theta \approx \tan \theta = y/L$. Substituting this y/L for $\sin \theta$ into Eq. 24.3 and solving for y gives a good approximation of the distance of the nth maximum (y_n) from the central maximum on either side:

$$y_n \approx \frac{nL\lambda}{d} \quad \text{for } n = 0, 1, 2, 3, \ldots \quad \begin{array}{l}\textit{lateral distance to}\\\textit{maxima for small } \theta \textit{ only}\end{array} \quad (24.4)$$

A similar analysis gives the locations of the minima. (See Exercise 12a.)

From Eq. 24.3, we see that, except for the zeroth order, $n = 0$ (the central maximum), the positions of the maxima depend on wavelength—different wavelengths (λ) give different values of $\sin \theta$ and therefore of θ and y. Hence, when we use white light, the central maximum is white because all wavelengths are at the same location, but the other orders become a "spread out" spectrum of colors. Because y is proportional to λ ($y \propto \lambda$), we observed, in a given order, that red is farther out than blue or that red has a longer wavelength than blue.

By measuring the positions of the color maxima within a particular order, Young was able to determine the wavelengths of the colors of visible light. Note also that the size or "spread" of the interference pattern, y_n, depends inversely on the slit separation d. The smaller d, the more spread out the pattern. For large d, the interference pattern is so compressed that it appears to us as a single white spot (all maxima together at center).

In this analysis, the word *destructive* does *not* imply that energy is destroyed. Destructive interference is simply a description of a physical fact—that if light energy is not present at a particular location, by energy conservation, it *must* be somewhere else. The mathematical description of Young's double-slit experiment tells you that there is no light energy at the minima. The light energy is redistributed and located at the maxima. This is also observed with sound waves as well.

PHYSLET®

Exploration 37.1 Varying Numbers and Orientations of Sources

PHYSLET®

Exploration 37.2 Changing the Separation Between Sources

Demonstration/activity: Remind students that the sine and tangent functions of small angles are about the same. Suggest that they use their calculators to confirm this situation and review Learn by Drawing on p. 219 to see why.

Teaching tip: You may wish to discuss the effect of the intensity of light and the order number.

Integrated Example 24.1 ■ Measuring the Wavelength of Light: Young's Double-Slit Experiment

In a lab experiment similar to the one shown in Fig. 24.4, monochromatic light (only one wavelength or frequency) passes through two narrow slits that are 0.050 mm apart. The interference pattern is observed on a white wall 1.0 m from the slits, and the second-order maximum is at an angle of $\theta_2 = 1.5°$. (a) If the slit separation decreases, the second-order maximum will be seen at an angle of (1) greater than 1.5°, (2) 1.5°, (3) less than 1.5°. Explain. (b) What is the wavelength of the light and what is the distance between the second-order and third-order maxima? (c) if $d = 0.040$ mm, what is θ_2?

(a) Conceptual Reasoning. According to the condition for constructive interference, $d \sin \theta = n\lambda$, the product of d and $\sin \theta$ is a constant, for a given wavelength λ and order number n. Therefore, if d decreases, $\sin \theta$ will increase and so will θ. Thus the answer is (1).

(continues on next page)

(b) and (c) Quantitative Reasoning and Solution. Eq. 24.3 can be used to find the wavelength. Since $L \gg d$, that is, 1.0 m $\gg$ 0.050 mm, then θ is small. We could compute y_2 and y_3 from Eq. 24.4 and determine the distance between the second-order and third-order maxima $(y_3 - y_2)$. However, the maxima for a given wavelength of light are evenly spaced (for a small θ). That is, the distance between adjacent maxima is a constant.

Given: $\qquad L = 1.0$ m $\qquad\qquad\qquad$ **Find:** (b) λ (wavelength) and $y_3 - y_2$
$\qquad\qquad\qquad n = 2$ $\qquad\qquad\qquad\qquad\qquad\qquad$ (distance between $n = 2$ and $n = 3$)
$\qquad\qquad$ (b) $\theta_2 = 1.5°$ $\qquad\qquad\qquad\qquad\qquad\qquad$ (c) θ_2 if $d = 0.040$ mm
$\qquad\qquad\qquad d = 0.050$ mm $= 5.0 \times 10^{-5}$ m
$\qquad\qquad$ (c) $d = 4.0 \times 10^{-5}$ m

(b) Using Eq. 24.3 gives

$$\lambda = \frac{d \sin \theta}{n} = \frac{(5.0 \times 10^{-5} \text{ m}) \sin 1.5°}{2} = 6.5 \times 10^{-7} \text{ m} = 650 \text{ nm}$$

This value is 650 nm, which is the wavelength of orange-red light (see Fig. 20.23). Using a general approach for n and $n + 1$, we get

$$y_{n+1} - y_n = \frac{(n + 1)L\lambda}{d} - \frac{nL\lambda}{d} = \frac{L\lambda}{d}$$

In this case, the distance between successive fringes is

$$y_3 - y_2 = \frac{L\lambda}{d} = \frac{(1.0 \text{ m})(6.5 \times 10^{-7} \text{ m})}{5.0 \times 10^{-5} \text{ m}} = 1.3 \times 10^{-2} \text{ m} = 1.3 \text{ cm}$$

(c) $\sin \theta_2 = \dfrac{n\lambda}{d} = \dfrac{(2)(650 \times 10^{-9} \text{ m})}{(4.0 \times 10^{-5} \text{ m})} = 0.0325$ so $\theta_2 = \sin^{-1} (0.0325) = 1.9° > 1.5°$.

Follow-Up Exercise. Suppose white light were used instead of monochromatic light in this Example. What would be the separation distance of the red ($\lambda = 700$ nm) and blue ($\lambda = 400$ nm) components in the second-order maximum? *(Answers to all Follow-Up Exercises are at the back of the text.)*

24.2 Thin-Film Interference

<u>OBJECTIVES:</u> To (a) describe how thin films can produce colorful displays, and (b) give some examples of practical applications of thin-film interference.

Have you ever wondered what causes the rainbowlike colors when white light is reflected from a thin film of oil or a soap bubble? This effect—known as *thin-film interference*—is a result of the interference of light reflected from opposite surfaces of the film and may be understood in terms of wave interference.

First, however, you need to know how the phase of a light wave is affected by reflection. Recall from Chapter 13 that a wave pulse on a rope undergoes a 180° phase change [or a *half wave shift* $(\lambda/2)$], when reflected from a rigid support and no phase shift when reflected from a free support (▼ Fig. 24.5). Similarly, as the figure shows, the phase change for the reflection of light waves at a boundary depends on the indices of refraction (n) of the two materials:

- A light wave undergoes a 180° phase change on reflection if $n_1 < n_2$.
- There is no phase change on reflection if $n_1 > n_2$.

▼ FIGURE 24.5 Reflection and phase shifts The phase changes that light waves undergo on reflection are analogous to those for pulses in strings. **(a)** The phase of a pulse in a string is shifted by 180° on reflection from a fixed end, and so is the phase of a light wave when it is reflected from a more optically dense medium. **(b)** A pulse in a string has a phase shift of zero (it is not shifted) when reflected from a free end. Analogously, a light wave is not phase shifted when reflected from a less optically dense medium.

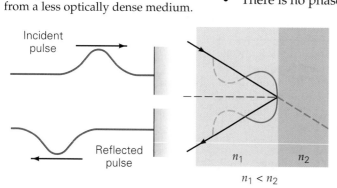

(a) Fixed end: 180° phase shift

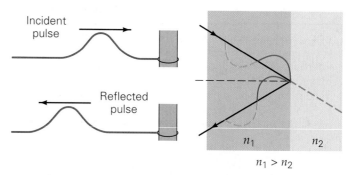

(b) Free end: zero phase shift

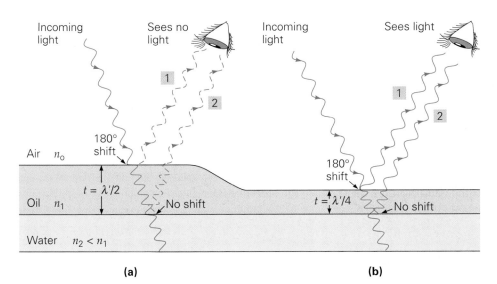

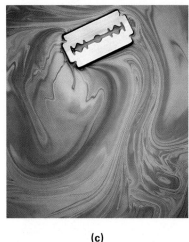

▲ **FIGURE 24.6 Thin-film interference** For an oil film on water, there is a 180° phase shift for light reflected from the air–oil interface and a zero phase shift at the oil–water interface. λ' is the wavelength in the oil. **(a)** Destructive interference occurs if the oil film has a minimum thickness of $\lambda'/2$ for normal incidence. (Waves are displaced and angled for clarity.) **(b)** Constructive interference occurs with a minimum film thickness of $\lambda'/4$. **(c)** Thin-film interference in an oil slick. Different film thicknesses give rise to the reflections of different colors.

To understand why you see colors from a soap bubble or an oil film (for example, floating on water or on a wet road), consider the reflection of monochromatic light from a thin film in ▲Fig. 24.6. The path length of the wave in the film depends on the angle of incidence (why?), but for simplicity, we will assume normal (perpendicular) incidence for the light, even though the rays are drawn at an angle in the figure, for clarity.

The oil film has a greater index of refraction than that of air, and the light reflected from the air–oil interface (wave 1 in the figure) undergoes a 180° phase shift. The transmitted waves pass through the oil film and are reflected at the oil–water interface. In general, the index of refraction of oil is greater than that of water (see Table 22.1)—that is, $n_1 > n_2$—so a reflected wave in this instance (wave 2) does *not* have the phase shift.

You might think that if the path length of the wave in the oil film (2*t*, twice the thickness—down and back) were an integral number of wavelengths—for example, if $2t = 2(\lambda'/2)$ in Fig. 24.6a, where $\lambda' = \lambda/n$ is the wavelength in the oil—then the waves reflected from the two surfaces would interfere constructively. But keep in mind that the wave reflected from the top surface (wave 1) undergoes a 180° phase shift. The reflected waves from the two surfaces are therefore actually *out of phase* and would interfere destructively for this condition. This means that no reflected light for this wavelength would be observed. (The light would be transmitted.)

Similarly, if the path length of the waves in the film were an odd number of half-wavelengths [$2t = 2(\lambda'/4) = \lambda'/2$] in Fig. 24.6b, again where λ' is the wavelength in the oil, then the reflected waves would actually be *in phase* (as a result of a 180° phase shift of wave 1) and would interfere constructively. Reflected light for this wavelength would be observed from above the oil film.

Because oil and soap films generally have different thicknesses in different regions, particular wavelengths (colors) of white light interfere constructively in different regions after reflection. As a result, a vivid display of various colors appears (Fig. 24.6c). This may also change if the film thickness changes with time. Thin-film interference may be seen when two glass slides are stuck together with an air film between them (▼Fig. 24.7a). The bright colors of a peacock, an example of colorful interference in nature, are a result of layers of fibers in its feathers. Light reflected from successive layers interferes constructively, giving bright colors, even though the feather has no pigment of its own. Since the condition for constructive interference depends on the angle of incidence, the color pattern changes somewhat with the viewing angle and motion of the bird (Fig. 24.7b).

Demonstration/activity: View thin-film interference caused by oil on water. Also view the interference in both monochromatic and white light.

Demonstration/activity: Blow soap bubbles and display them in light from an incandescent source. Set up a lens and light source to project reflected light from a soap film onto a large screen. A 1-in.-diameter ring will hold a soap film long enough to observe the change in thickness as the liquid is pulled downward. Just before the film breaks, the reflected light disappears. Why?

A practical application of thin-film interference is nonreflective coatings for lenses. (See Insight 24.1, on page 768, on Nonreflecting Lenses.) In this situation, a film coating is used to create destructive interference between the reflected waves so as to *increase the light transmission* into the glass (▼Fig. 24.8). The index of refraction of the film has a value between that of air and glass ($n_o < n_1 < n_2$). Consequently, phase shifts of incident light take place at the surfaces of both the film and the glass.

In such a case, the condition for constructive interference of the reflected light is

$$\Delta L = 2t = m\lambda' \text{ or } t = \frac{m\lambda'}{2} = \frac{m\lambda}{2n_1} \quad m = 1, 2, \ldots \quad \begin{array}{l}\textit{condition for con-}\\ \textit{structive interference} \\ \textit{when } n_o < n_1 < n_2\end{array} \quad (24.5)$$

and the condition for destructive interference is

$$\Delta L = 2t = \frac{m\lambda'}{2} \text{ or } t = \frac{m\lambda'}{4} = \frac{m\lambda}{4n_1} \quad m = 1, 3, 5, \ldots \quad \begin{array}{l}\textit{condition for de-}\\ \textit{structive interference} \\ \textit{when } n_o < n_1 < n_2\end{array} \quad (24.6)$$

The *minimum* film thickness for destructive interference occurs when $m = 1$, so

$$t_{min} = \frac{\lambda}{4n_1} \quad \begin{array}{l}\textit{minimum film thickness} \\ \textit{(for } n_o < n_1 < n_2\textit{)}\end{array} \quad (24.7)$$

If the index of refraction of the film is greater than that of air and glass, then only the reflection at the air–film interface has the 180° phase shift. Therefore, $2t = m\lambda'$ will actually create destructive interference and $2t = m\lambda'/2$ constructive interference. (Why?)

Example 24.2 ■ Nonreflective Coatings: Thin-Film Interference

A glass lens ($n = 1.60$) is coated with a thin, transparent film of magnesium fluoride ($n = 1.38$) to make the lens nonreflecting. (a) What is the minimum film thickness for the lens to be nonreflecting for normally incident light of wavelength 550 nm? (b) Will a film thickness of 996 nm make the lens nonreflecting?

Thinking It Through. (a) Equation 24.7 can be used directly to get an idea of the minimum film thickness for a nonreflective coating. (b) We need to determine whether 996 nm satisfies the condition in Eq. 24.6.

Solution.

Given: $n_o = 1.00$ (air)
$n_1 = 1.38$ (for film)
$n_2 = 1.60$ (for lens)
$\lambda = 550$ nm

Find: (a) t_{min} (minimum film thickness)
(b) determine whether $t = 996$ nm gives a nonreflecting lens

(a) Because $n_2 > n_1 > n_o$,

$$t_{min} = \frac{\lambda}{4n_1} = \frac{550 \text{ nm}}{4(1.38)} = 99.6 \text{ nm}$$

which is quite thin ($\approx 10^{-5}$ cm). In terms of atoms, which have diameters on the order of 10^{-10} m, or 10^{-1} nm, the film is 10^3 atoms thick.

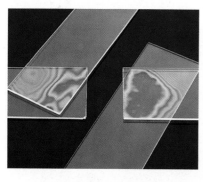

(a)

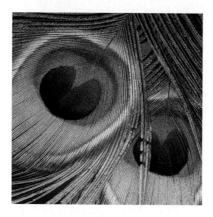

(b)

▲ **FIGURE 24.7 Thin-film interference (a)** A thin air film between microscope slides gives colorful patterns. **(b)** Multilayer interference in a peacock's feathers gives rise to bright colors. The brilliant throat colors of hummingbirds are produced in the same way.

Demonstration/activity: Open a new box of microscope slides. Take out two slides at a time and pass the pairs out to students. Have the students press the two slides together and observe the reflection from the light in the room. If the slides are very clean, the air film will be thin enough to show the phase reversal.

Illustration 37.2 Dielectric Mirrors

▶ **FIGURE 24.8 Thin-film interference** For a thin film on a glass lens, there is a 180° phase shift at each interface when the index of refraction of the film is less than that of the glass. The waves reflected off the top and bottom surfaces of the film interfere. For clarity, the angle of incidence is drawn to be large, but, in reality, it is almost zero.

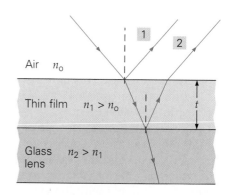

(b)

$$t = 996 \text{ nm} = 10(99.6 \text{ nm}) = 10t_{\min} = 10\left(\frac{\lambda}{4n_1}\right) = 5\left(\frac{\lambda}{2n_1}\right)$$

This means that this film thickness does *not* satisfy the nonreflective condition (destructive interference). Actually, it satisfies the requirement for constructive interference (Eq. 24.5) with $m = 5$. Such a coating specific for infrared radiation on car and house windows could be useful in hot climates, because it maximizes reflection and minimizes transmission.

Follow-Up Exercise. For the glass lens in this Example to reflect, rather than transmit, the incident light through the lens, what would be the minimum film thickness?

Optical Flats and Newton's Rings

The phenomenon of thin-film interference can be used to check the smoothness and uniformity of optical components such as mirrors and lenses. *Optical flats* are made by grinding and polishing glass plates until they are as flat and smooth as possible. (The surface roughness is usually in the order of $\lambda/20$.) The degree of flatness can be checked by putting two such plates together at a slight angle so that a very thin air wedge is between them (►Fig. 24.9a). The reflected waves off the bottom of the top plate (wave 1) and top of the bottom plate (wave 2) interfere. Note that wave 2 has a 180° phase shift as it is reflected from an air–plate interface, whereas wave 1 does not. Therefore, at certain points from where the plates touch (point O), the condition for constructive interference is $2t = m\lambda/2$ ($m = 1, 3, 5, \dots$), and the condition for destructive interference is $2t = m\lambda$ ($m = 0, 1, 2, \dots$). The thickness t determines the type of interference (constructive or destructive). If the plates are smooth and flat, a regular interference pattern of bright and dark bands appears (Fig. 24.9b). This pattern is a result of the uniformly varying differences in path lengths between the plates. Any irregularity in the pattern indicates an irregularity in at least one plate. Once a good optical flat is verified, it can be used to check the flatness of a reflecting surface, such as that of a precision mirror.

Direct evidence of the 180° phase shift can be clearly seen in Fig. 24.9. At the point where the two plates touch ($t = 0$), we see a *dark* band. If there were no phase shift, $t = 0$ would correspond to $\Delta L = 0$, and we would expect a bright band to appear. The fact that it is a dark band proves that there is a phase shift in reflection from a more optically dense material.

A similar technique is used to check the smoothness and symmetry of lenses. When a curved lens is placed on an optical flat, a radially symmetric air wedge is formed between the lens and the optical flat (▼Fig. 24.10a). Since the thickness of the air wedge again determines the condition for constructive and destructive interference, the regular interference pattern in this case is a set of concentric circular bright and dark rings (Fig. 24.10b). They are called *Newton's rings*, after Isaac Newton, who first

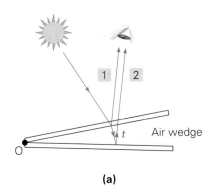

(a)

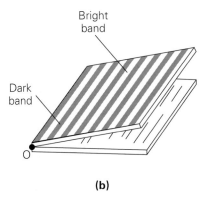

(b)

▲ **FIGURE 24.9 Optical flatness**
(a) An optical flat is used to check the smoothness of a reflecting surface. The flat is placed so that there is an air wedge between it and the surface. The waves reflected from the two plates interfere, and the thickness of the air wedge at certain points determines whether bright or dark bands are seen. **(b)** If the surface is smooth, a regular or symmetrical interference pattern is seen. Note that a dark band is at point O where $t = 0$.

◄ **FIGURE 24.10 Newton's rings**
(a) A lens placed on an optical flat forms a ring-shaped air wedge, which gives rise to interference of the waves reflected from the top (wave 1) and the bottom (wave 2) of the air wedge. **(b)** The resulting interference pattern is a set of concentric rings called *Newton's rings*. Note that at the center of the pattern is a dark spot. Lens irregularities produce a distorted pattern.

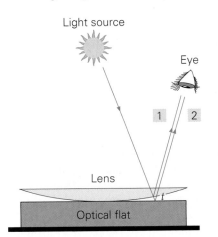

(a)

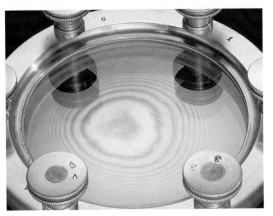

(b)

INSIGHT 24.1 NONREFLECTING LENSES

You may have noticed the blue-purple tint of the coated optical lenses used in cameras and binoculars. The coating makes the lenses almost "nonreflecting." If a lens is a nonreflecting type, then the incident light is mostly transmitted through the lens. Maximum transmission of light is desirable for the exposure of photographic film and for viewing objects with binoculars.

For a typical (reflecting) air–glass interface, about 4% of the light is reflected and 96% is transmitted. A modern camera lens is actually made up of a group of lenses (elements) in order to improve image quality. For instance, a 35-mm–70-mm zoom lens might consist of up to 13 elements, thus having 26 reflective surfaces.

After one reflection, 0.96 = 96% of the light is transmitted. After two reflections, or one element, the transmitted light is only $0.96 \times 0.96 = 0.96^2 = 0.92$, or 92%, of the incident light. Thus after 26 reflections, the transmitted light is only $0.96^{26} = 0.35$, or 35%, of the incident light, if lenses are not coated. Therefore, almost all modern lenses are coated with nonreflecting film.

A lens is made nonreflecting by coating it with a thin-film material with an index of refraction between the indices of refraction of air and glass (Fig. 24.8). If the coating is a quarter-wavelength ($\lambda'/4$) thick, the difference in path length between the reflected rays is $\lambda'/2$, where λ' is the wavelength of light in the coating. In this case, both reflected waves undergo a phase shift, and therefore they are out of phase for a path-length difference of $\lambda'/2$ and interfere destructively. That is, the incident light is transmitted, and the coated lens is nonreflecting.

Note that the actual thickness of a quarter-wavelength thickness of film is specific to the particular wavelength of light. The thickness is usually chosen to be a quarter-wavelength of yellow-green light ($\lambda \approx 550$ nm), to which the human eye is most sensitive. The wavelengths at the red and blue ends of the visible region are still partially reflected, giving the coated lens its bluish-purple tint (Fig. 1). Sometimes other quarter-wavelength thicknesses are chosen, giving rise to other hues, such as amber or reddish purple, depending on the application of the lens.

Nonreflective coatings are also applied to the surfaces of solar cells, which convert light into electrical energy (Chapter 27). Because the thickness of such a coating is wavelength dependent, the losses due to reflection can be decreased from around 30% to only 10%. Even so, the process improves the cell's efficiency.

FIGURE 1 Coated lenses The nonreflective coating on binocular and camera lenses generally produces a characteristic bluish-purple hue. (Why?)

described this interference effect. Note that at the point where the lens and the optical flat touch ($t = 0$), there is, once again, a dark spot. (Why?) Lens irregularities give rise to a distorted fringe pattern, and the radii of these rings can be used to calculate the radius of curvature of the lens.

24.3 Diffraction

OBJECTIVES: To (a) define diffraction, and (b) give examples of diffractive effects.

▼ FIGURE 24.11 Water-wave diffraction This photograph of a beach dramatically shows single-slit diffraction of ocean waves through the barrier openings. Note that the beach has been shaped by the circular wave fronts.

In geometrical optics, light is represented by rays and pictured as traveling in straight lines. If this model were to represent the real nature of light, however, there would be no interference effects in Young's double-slit experiment. Instead, there would be only two bright images of slits on the screen, with a well-defined shadow area where no light enters. But we *do* see interference patterns, which means that the light must deviate from a straight-line path and enter the regions that would otherwise be in shadow. The waves actually "spread out" as they pass through the slits. This spreading is called **diffraction**. Diffraction generally occurs when waves pass through small openings or around sharp edges or corners. The diffraction of water waves is shown in ◄Fig. 24.11. (See also Fig. 13.18.)

As Fig. 13.18 shows, the amount of diffraction depends on the wavelength in relation to the size of the opening or object. In general, *the longer the wavelength compared to the width of the opening or object, the greater the diffraction*. This effect is also shown in ▶Fig. 24.12. For example, in Fig. 24.12a, the width of the opening w is much greater than the wavelength ($w \gg \lambda$), and there is little diffraction—the wave keeps traveling without much spreading. (There is also *some* degree of diffraction due to the edges of the opening.) In Fig. 24.12b, with the wavelength and opening

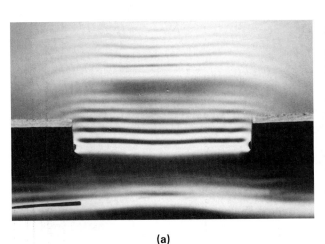

(a)   **(b)**

▲ **FIGURE 24.12 Wavelength and opening dimensions** In general, the narrower the opening compared to the wavelength, the greater the diffraction. **(a)** Without much diffraction ($w \gg \lambda$), the wave would keep traveling in its original direction. **(b)** With noticeable diffraction ($w \approx \lambda$), the wave bends around the opening and spreads out.

width the same order of magnitude ($w \approx \lambda$), there is noticeable diffraction—the wave spreads out and deviates from its original direction. Part of the wave keeps traveling in its original direction but the rest bends *around* the opening and clearly spreads out.

The diffraction of sound is quite evident (Chapter 14). Someone can talk to you from another room or around the corner of a building, and even in the absence of reflections, you can easily hear the person. Recall that audible sound wavelengths are on the order of centimeters to meters. Thus, the widths of ordinary objects and openings are about the same as or narrower than the wavelengths of sound, and diffraction will readily occur under these conditions.

Visible light waves, however, have wavelengths on the order of 10^{-7} m. Therefore diffraction phenomena for these waves often go unnoticed, especially through large openings such as doors where sound readily diffracts. However, close inspection of the area around a sharp razor blade will show a pattern of bright and dark bands (▶Fig. 24.13). Diffraction can lead to interference, and thus these interference patterns are evidence of the diffraction of the light around the edge of the blade.

As an illustration of "single-slit" diffraction, consider a slit in a barrier (▼Fig. 24.14). Suppose that the slit (width w) is illuminated with monochromatic light. A diffraction pattern consisting of a bright central maximum and a symmetrical array of maxima (regions of constructive interference) on both sides is observed on a screen at a distance L from the slit (we will assume $L \gg w$).

Thus a diffraction pattern results from the fact that various points on the wave front passing through the slit can be considered to be small point sources of light. The interference of those waves gives rise to the *diffraction* maxima and minima.

The fairly complex analysis is not done here; however, from geometry, it can be proven that the minima (regions of destructive interference) satisfy the relationship

$$w \sin \theta = m\lambda \qquad \text{for } m = 1, 2, 3, \ldots \quad \text{condition for minima} \qquad (24.8)$$

where θ is the angle of a particular minimum, designated by $m = 1, 2, 3, \ldots$, on either side of the central maximum and m is called the order number. (There is no $m = 0$. Why?)

Although this result is similar in form to that for Young's double-slit experiment (Eq. 24.3), it is extremely important to realize that for the single-slit experiment, minima, rather than maxima, are analyzed. Also, note that the width of the slit (w) is used in diffraction. Physically, this is diffraction from a single slit, *not* interference from two slits.

The small-angle approximation, $\sin \theta \approx y/L$, can be made when $y \ll L$. In this case, the distances of the minima relative to the center of the central maximum are given by

$$y_m = m\left(\frac{L\lambda}{w}\right) \qquad \text{for } m = 1, 2, 3, \ldots \quad \text{location for minima} \qquad (24.9)$$

PHYSLET®

Illustration 38.1 Single Slit
Diffraction

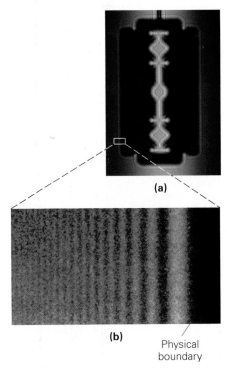

(a)

(b)   Physical
boundary

▲ **FIGURE 24.13 Diffraction in action**
(a) Diffraction patterns produced by a razor blade. **(b)** A close-up view of the diffraction pattern formed at the edge of the blade.

Teaching tip: Compare the expression for a single-slit diffraction pattern with that for a double-slit interference pattern. Students often confuse these two patterns.

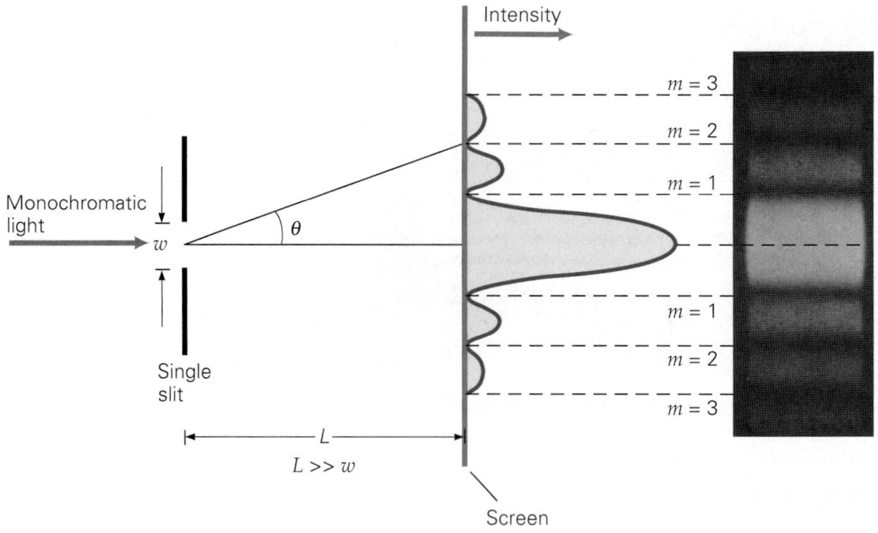

▶ **FIGURE 24.14** **Single-slit diffraction** The diffraction of light by a single slit gives rise to a diffraction pattern consisting of a large and bright central maximum and a symmetric array of side maxima. The order number m corresponds to the minima or dark positions. (See text for description.)

The qualitative predictions from Eq. 24.9 are interesting and instructive:

- For a given slit width (w), the longer the wavelength (λ), the wider (more "spread out") the diffraction pattern.
- For a given wavelength (λ) the narrower the slit width (w), the wider the diffraction pattern.
- The width of the central maximum is twice the width of any side maximum.

Let's look in detail at these results. As the slit is made narrower, the central maximum and the side maxima spread out and become larger. Equation 24.9 is not applicable to very small slit widths (because of the small-angle approximation). If the slit width is decreased until it is of the same order of magnitude as the wavelength of the light, then the central maximum spreads out over the whole screen. That is, diffraction becomes dramatically evident when the width of the slit is about the same as the wavelength of the light used. Diffraction effects are most easily observed when $\lambda/w \approx 1$, or $w \approx \lambda$.

Conversely, if the slit is made wider for a given wavelength, then the diffraction pattern becomes less spread out. The maxima move closer together and eventually become difficult to distinguish when w is much wider than λ ($w \gg \lambda$). The pattern then appears as a fuzzy shadow around the central maximum, which is the illuminated image of the slit. This type of pattern is observed for the image produced by sunlight entering a dark room through a hole in a curtain. Such an observation led early experimenters to investigate the wave nature of light. The acceptance of this concept was, in large part, due to the explanation of diffraction offered by physical optics.

The central maximum is twice as wide as any of the side maxima. Taking the width of the central maximum to be the distance between the bounding minima on each side ($m = 1$), or a value of $2y_1$, we obtain, from Eq. 24.9 with $y_1 = L\lambda/w$,

$$2y_1 = \frac{2L\lambda}{w} \quad \text{width of central maximum} \tag{24.10}$$

Similarly, the width of the side maximum on the sides is given by

$$y_{m+1} - y_m = (m + 1)\left(\frac{L\lambda}{w}\right) - m\left(\frac{L\lambda}{w}\right) = \frac{L\lambda}{w} = y_1 \tag{24.11}$$

Thus, the width of the central maximum is twice that of the side maxima.

Conceptual Example 24.3 ■ Diffraction and Radio Reception

While you drive through a city or mountainous areas, the quality of your radio reception varies sharply from place to place, with stations seeming to fade out and reappear. Could diffraction be a cause of this? Which of the following bands would you expect to be least affected by it: (a) Weather (162 MHz); (b) FM (88–108 MHz); or (c) AM (525–1610 kHz)?

Demonstration/activity: Place a pin or some other small object in the path of laser light to produce a diffraction pattern. Compare this pattern with the diffraction pattern produced by shining laser light through a small slit.

Reasoning and Answer. Radio waves, like visible light, are electromagnetic waves and so tend to travel in straight lines when they are long distances from their sources. They can be blocked by objects in their path—especially if the objects are massive (such as hills and buildings).

However, because of diffraction, radio waves can also "wrap around" obstacles or "fan out" as they pass through obstacles and openings, *provided* their wavelength is at least roughly the size of the obstacle or opening. The longer the wavelength, the greater the amount of diffraction, and so the *less likely* the radio waves are to be obstructed.

To determine which band benefits most by such diffraction, we need the wavelengths that correspond to the given frequencies, as given by $c = \lambda f$. AM radio waves, with $\lambda = 186$–571 m, are the longest of the three bands (by a factor of about 100). Thus AM broadcasts are more likely to be diffracted around such objects as buildings or mountains or through the openings between them, and the answer is (c).

Follow-Up Exercise. Woodwind instruments, such as the clarinet and the flute, usually have smaller openings than brass instruments, such as the trumpet and trombone. During halftime at a football game, when a marching band faces you, you can easily hear both the woodwind instruments and the brass instruments. Yet when the band marches away from you, the brass instruments sound muted, but you can hear the woodwinds quite well. Why?

Integrated Example 24.4 ■ Width of the Central Maximum: Single-Slit Diffraction

Monochromatic light passes through a slit whose width is 0.050 mm. (a) The general spreadout of the diffraction pattern, in general, is (1) larger for longer wavelengths, (2) larger for shorter wavelengths, (3) the same for all wavelengths. Explain. (b) At what angle will the third minimum be seen and what is the width of the central maximum on a screen located 1.0 m from the slit, for $\lambda = 400$ nm and 550 nm, respectively?

(a) Conceptual Reasoning. The general size of the diffraction pattern can be characterized by the position of a particular maximum or minimum. From Eq. 24.8, it can be seen that for a given width w and order number m, the position of a minimum $\sin \theta$ is directly proportional to the wavelength λ. Therefore, a longer wavelength will correspond to a greater $\sin \theta$ or a greater θ, and the answer is (1).

(b) Quantitative Reasoning and Solution. This part is a direct application of Eq. 24.8 and Eq. 24.10.

Given: $\lambda_1 = 400$ nm $= 4.00 \times 10^{-7}$ m **Find:** θ_3 and $2y_1$ (width of central maximum)
$\lambda_2 = 550$ nm $= 5.50 \times 10^{-7}$ m
$w = 0.050$ mm $= 5.0 \times 10^{-5}$ m
$m = 3$
$L = 1.0$ m

For $\lambda = 400$ nm:
From Eq. 24.8, we have

$$\sin \theta_3 = \frac{m\lambda}{w} = \frac{3(4.00 \times 10^{-7}\ \text{m})}{5.0 \times 10^{-5}\ \text{m}} = 0.024 \quad \text{so} \quad \theta_3 = \sin^{-1} 0.024 = 1.4°$$

Equation 24.10 gives

$$2y_1 = \frac{2L\lambda}{w} = \frac{2(1.0\ \text{m})(4.00 \times 10^{-7}\ \text{m})}{5.0 \times 10^{-5}\ \text{m}} = 1.6 \times 10^{-2}\ \text{m} = 1.6\ \text{cm}$$

For $\lambda = 700$ nm:

$$\sin \theta_3 = \frac{m\lambda}{w} = \frac{3(5.50 \times 10^{-7}\ \text{m})}{5.0 \times 10^{-5}\ \text{m}} = 0.033 \quad \text{so} \quad \theta_3 = \sin^{-1} 0.033 = 1.9°$$

$$2y_1 = \frac{2L\lambda}{w} = \frac{2(1.0\ \text{m})(5.50 \times 10^{-7}\ \text{m})}{5.0 \times 10^{-5}\ \text{m}} = 2.2 \times 10^{-2}\ \text{m} = 2.2\ \text{cm}$$

Follow-Up Exercise. By what factor would the width of the central maximum change if red light ($\lambda = 700$ nm) were used instead of 550 nm?

▶ **FIGURE 24.15 Diffraction grating**
A diffraction grating produces a sharply defined interference/diffraction pattern. Two parameters define a grating: the slit separation d and the slit width w. The combination of multiple-slit interference and single-slit diffraction determine the intensity distribution of the various orders of maxima.

Illustration 38.2 Application of Diffraction Gratings

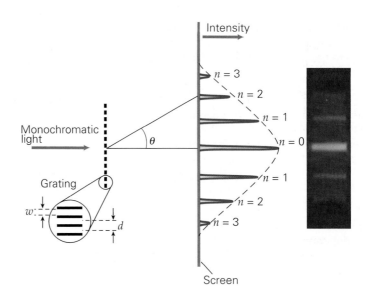

Demonstration/activity: Demonstrate the diffraction pattern that results from a CD-ROM and a DVD. Shine laser light onto the disc and observe the reflection interference/diffraction pattern. Compare the maxima separation between the patterns of the CD and DVD.

Diffraction Gratings

We have seen that maxima and minima result from diffraction followed by interference when monochromatic light passes through a set of double slits. As the number of slits is increased, the maxima become sharper (narrower) and the minima wider. The sharp maxima are very useful in optical analysis of light sources and other applications. ▲Fig. 24.15 shows a typical experiment with monochromatic light incident on a **diffraction grating**, which consists of large numbers of parallel, closely spaced slits. Two parameters define a diffraction grating: the slit separation between successive slits, d, and the individual slit width, w. The resulting pattern of interference and diffraction is shown in ▼Fig. 24.16.

Diffraction gratings were first made of fine strands of wire. They produce effects similar to what can be seen by viewing a candle flame through a feather held close to the eye. Better gratings have a large number of fine lines or grooves on glass or metal surfaces. If light is transmitted through a grating, it is called a *transmission grating*. However, *reflection gratings* are also common. The closely spaced grooves of a compact disc or a DVD act as a reflection grating, giving rise to their familiar iridescent sheen (▶Fig. 24.17). Commercial master gratings are made by depositing a thin film of aluminum on an optically flat surface and then removing some of the reflecting metal by cutting regularly spaced, parallel lines. Precision diffraction gratings

▶ **FIGURE 24.16 Intensity distribution of interference and diffraction** (a) Interference determines the positions of the interference maxima:
$d \sin \theta = n\lambda, n = 0, 1, 2, 3, \ldots$.
(b) Diffraction locates the positions of the diffraction minima:
$w \sin \theta = m\lambda, m = 1, 2, 3, \ldots$, and the relative intensity of the maxima.
(c) The combination (product) of interference and diffraction determine the overall intensity distribution.

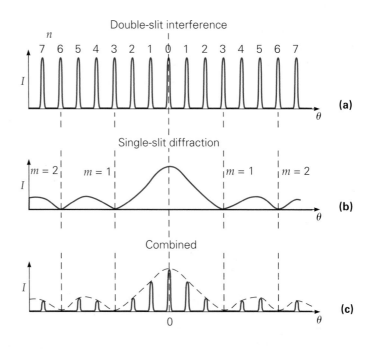

are made using two coherent laser beams that intersect at an angle. The beams expose a layer of photosensitive material, which is then etched. The spacing of the grating lines is determined by the intersection angle of the beams. Precision gratings may have 30 000 lines per centimeter or more and are therefore expensive and difficult to fabricate. Most gratings used in laboratory instruments are *replica gratings*, which are plastic castings of high-precision master gratings.

It can be shown that the condition for interference maxima for a grating illuminated with monochromatic light is identical to that for a double slit. The expression is

$$d \sin \theta = n\lambda \quad \text{for } n = 0, 1, 2, 3, \ldots \quad interference\ maxima \quad (24.12)$$

where n is called the *order-of-interference maximum* and θ is the angle at which that maximum occurs for a particular wavelength. The zeroth-order maximum is coincident with the central maximum. The spacing between adjacent slits (d) is obtained from the number of lines or slits per unit length of the grating: $d = 1/N$. For example, if $N = 5000$ lines/cm, then

$$d = \frac{1}{N} = \frac{1}{5000/\text{cm}} = 2.0 \times 10^{-4} \text{ cm}$$

If the light incident on a grating is white light (polychromatic), then the maxima are multicolored (▼Fig. 24.18a). There is no deviation of the components of the light for the zeroth order ($\sin \theta = 0$ for all wavelengths), so the central maximum is white. However, the colors separate for higher orders, since the position of the maximum depends on wavelength (Eq. 24.12). With the longer wavelength having a larger θ, this produces a spectrum. Note that it is possible for higher orders produced by a diffraction grating to overlap. That is, the angles for different orders may be the same for two different wavelengths.

Only a limited number of spectral orders can be obtained using a diffraction grating. The number depends on the wavelength of the light and on the grating's spacing (d). From Eq. 24.12, because θ cannot exceed 90° (that is, $\sin \theta \leq 1$), we have

$$\sin \theta = \frac{n\lambda}{d} \leq 1 \quad \text{or} \quad n_{max} \leq \frac{d}{\lambda}$$

Diffraction gratings have almost completely replaced prisms in spectroscopy. The creation of a spectrum and the measurement of wavelengths by a grating depend only on geometrical measurements such as lengths and/or angles. Wavelength determination using a prism, in contrast, depends on the dispersive characteristics of the material of which the prism is made. Thus, it is crucial to know precisely how the index of refraction depends on the wavelength of light. In contrast to a prism, which deviates red light least and violet light most, a diffraction grating produces the smallest angle for violet light

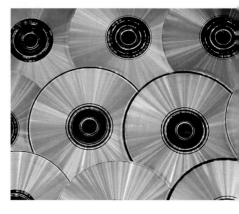

▲ **FIGURE 24.17 Diffraction effects** The narrow grooves of compact discs (CDs) act as reflection diffraction gratings, producing colorful displays.

Note: d is the distance between adjacent slits.

Exploration 38.2 Diffraction Grating

Demonstration/activity: Shine a laser beam through a piece of thin cloth and observe the diffraction pattern on an opposite wall. Does the pattern give any clue as to the weave of the cloth?

▼ **FIGURE 24.18 Spectroscopy** **(a)** In each side maximum, components of different wavelengths (R = red and V = violet) are separated, because the deviation depends on wavelength: $\theta = \sin^{-1}(n\lambda/d)$. **(b)** As a result, gratings are used in spectrometers to determine the wavelengths present in a beam of light by measuring their angles of diffraction and to separate the various wavelengths for further analysis.

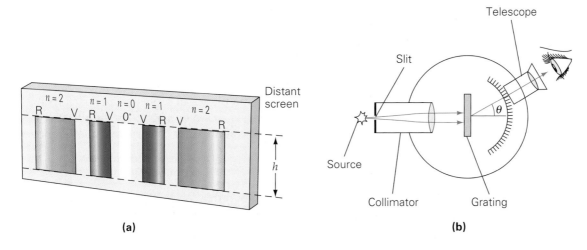

(a)

(b)

(short λ) and the greatest for red light (long λ). Notice that a prism disperses white light into a single spectrum. A diffraction grating, however, produces a number of spectra, one for each order other than $n = 0$, and the higher the order, the more spread out.

The sharp spectra produced by gratings are used in instruments called *spectrometers* (Fig. 24.18b). Using a spectrometer, materials can be illuminated with light of various wavelengths to find which wavelengths are strongly transmitted or reflected. Their absorption can then be measured and material characteristics determined.

Example 24.5 ■ A Diffraction Grating: Line Spacing and Spectral Orders

A particular diffraction grating produces an $n = 2$ spectral order at an angle of 30° for light with a wavelength of 500 nm. (a) How many lines per centimeter does the grating have? (b) At what angle can the $n = 3$ spectral order be seen?

Thinking It Through. (a) To find the number of lines per centimeter (N) the grating has, we need to know the grating spacing (d), since $N = 1/d$. With the given data, we can find d from Eq. 24.12. (b) Using Eq. 24.12 again, we can find θ for $n = 3$.

Solution.

Given: $\lambda = 500\,\text{nm} = 5.00 \times 10^{-7}\,\text{m}$ *Find:* (a) N (lines/cm)
        $n = 2$ (b) θ for $n = 3$
        $\theta = 30°$ for $n = 2$

(a) Using Eq. 24.12, we get the grating spacing:

$$d = \frac{n\lambda}{\sin\theta} = \frac{2(5.00 \times 10^{-7}\,\text{m})}{\sin 30°} = 2.00 \times 10^{-6}\,\text{m} = 2.00 \times 10^{-4}\,\text{cm}$$

Then

$$N = \frac{1}{d} = \frac{1}{2.00 \times 10^{-4}\,\text{cm}} = 5000\,\text{lines/cm}$$

(b)

$$\sin\theta = \frac{n\lambda}{d} = \frac{3(5.00 \times 10^{-7}\,\text{m})}{2.00 \times 10^{-6}\,\text{m}} = 0.75$$

so

$$\theta = \sin^{-1} 0.75 = 48.6°$$

Follow-Up Exercise. If white light of wavelength ranging from 400 to 700 nm were used, what would be the angular width of the spectrum for the second order?

X-Ray Diffraction

In principle, the wavelength of any electromagnetic wave can be determined by using a diffraction grating with the appropriate spacing. Diffraction was used to determine the wavelengths of X-rays early in the twentieth century. Experimental evidence indicated that the wavelengths of X-rays were probably around 10^{-10} m or 0.1 nm, but it is impossible to construct a diffraction grating with this spacing. Around 1913, Max von Laue (1879–1960), a German physicist, suggested that the regular spacing of the atoms in a crystalline solid might make the crystal act as a diffraction grating for X-rays, since the atomic spacing is on the order of 0.1 nm (▶Fig. 24.19). When X-rays were directed at crystals, diffraction patterns were indeed observed. (See Fig. 24.19b.)

Figure 24.19a illustrates diffraction by the planes of atoms in a crystal such as sodium chloride. The path-length difference is $2d\sin\theta$, where d is the distance between the crystal's internal planes. Thus, the condition for constructive interference is

$$2d\sin\theta = n\lambda \qquad \text{for } n = 1, 2, 3, \ldots \qquad \begin{array}{l}\textit{constructive interference}\\ \textit{X-ray diffraction}\end{array} \qquad (24.13)$$

This relationship is known as **Bragg's law**, after W. L. Bragg (1890–1971), the British physicist who first derived it. Note that θ is *not* measured from the normal, as is the convention in optics.

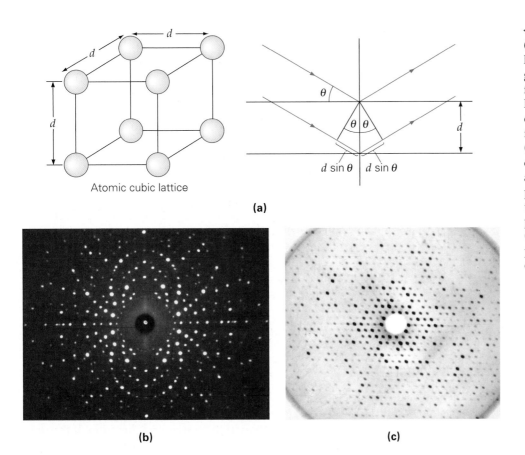

◀ **FIGURE 24.19** Crystal diffraction
(a) The array of atoms in a crystal-lattice structure acts as a diffraction grating, and X-rays are diffracted from the planes of atoms. With a lattice spacing of d, the path-length difference for the X-rays diffracted from adjacent planes is $2d \sin \theta$.
(b) X-ray diffraction pattern of a crystal of potassium sulfate. By analyzing the geometry of such patterns, investigators can deduce the structure of the crystal and the position of its various atoms.
(c) X-ray diffraction pattern of the protein hemoglobin, which carries oxygen in blood.

X-ray diffraction is now routinely used to investigate the internal structure not only of simple crystals, but also of large, complex biological molecules such as proteins and DNA (Fig. 24.19c). Because of their short wavelengths, which are comparable with interatomic distances *within* a molecule, X-rays provide a method for investigating atomic structures within molecules.

24.4 Polarization

OBJECTIVES: To (a) explain light polarization, and (b) give examples of polarization, both in the environment and in commercial applications.

When you think of polarized light, you may visualize polarizing (or Polaroid) sunglasses, since this is one of the more common applications of polarization. When something is polarized, it has a preferential direction, or orientation. In terms of light waves, **polarization** refers to the orientation of the transverse wave (electric field) oscillations.

Recall from Chapter 20 that light is an electromagnetic wave with oscillating electric and magnetic field vectors ($\vec{E}$ and $\vec{B}$, respectively) perpendicular (transverse) to the direction of propagation. Light from most sources consists of a large number of electromagnetic waves emitted by the atoms of the source. Each atom produces a wave with a particular orientation, corresponding to the direction of the atomic vibration. However, since electromagnetic waves from a typical source are produced by many atoms, many random orientations of the $\vec{E}$ and $\vec{B}$ fields are in the emitted composite light. When the field vectors are randomly oriented, the light is said to be *unpolarized*. This situation is commonly represented schematically in terms of the electric field vector as shown in ▼Fig. 24.20a. As viewed along the direction of propagation, the electric field is equally distributed in all directions. However, as viewed parallel to the direction of propagation, this random or equal distribution can be represented by two directions (such as the x- and y-directions in a two-dimensional coordinate system). Here, the vertical arrows denote the electric

Note: In many figures, dots represent a direction perpendicular to the paper of the electric field and arrows denote a direction along the arrow of the electric field.

Illustration 39.2 Polarized Electromagnetic Waves

▶ **FIGURE 24.20** Polarization
Polarization is represented by the orientation of the plane of vibration of the electric field vectors.
(a) When the vectors are randomly oriented, the light is unpolarized. The dots represent a direction perpendicular to the paper of the electric field, and the vertical arrows denote an up-and-down direction of the electric field. Equal numbers of dots and arrows are used to represent unpolarized light.
(b) With preferential orientation of the field vectors, the light is partially polarized. Here, there are fewer dots than arrows. **(c)** When the vectors are in one plane, the light is plane polarized, or linearly polarized. No dots are seen here.

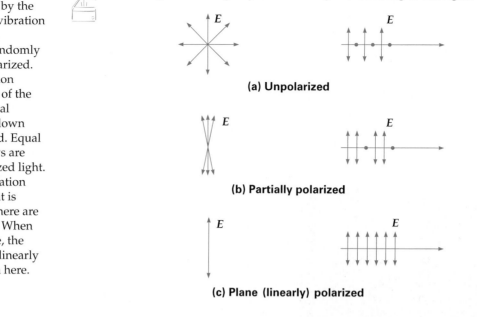

Light is coming at you!   Light is traveling to the right!

(a) Unpolarized

(b) Partially polarized

(c) Plane (linearly) polarized

field components in that direction, and the dots represent the component going in and out of the paper. This notation will be used throughout this section.

If there is some preferential orientation of the field vectors, the light is said to be *partially polarized*. Both representations in Fig. 24.20b show that there are more electric field vectors in the vertical direction than in the horizontal direction. If the field vectors oscillate in only *one* plane the light is *plane polarized* or *linearly polarized* (Fig. 24.20c). Note that polarization is evidence that light is a transverse wave. True longitudinal waves, such as sound waves, cannot be polarized, because the molecules of the media do not vibrate perpendicular to the direction of propagation.

Light can be polarized in many ways. However, polarization by selective absorption, reflection, and double refraction will be discussed here. Polarization by scattering will be considered in Section 24.5.

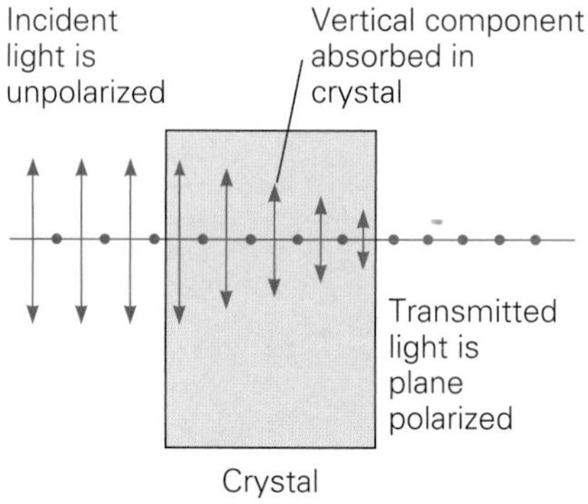

Polarization by Selective Absorption (Dichroism)

Some crystals, such as those of the mineral tourmaline, exhibit the interesting property of absorbing one of the electric field components more than the other. This property is called **dichroism**. If a dichroic crystal is sufficiently thick, the more strongly absorbed component may be completely absorbed. In that case, the emerging beam is plane polarized (◀Fig. 24.21).

Another dichroic crystal is quinine sulfide periodide (commonly called *herapathite*, after W. Herapath, an English physician who discovered its polarizing properties in 1852). This crystal was of great practical importance in the development of modern polarizers. Around 1930, Edwin H. Land (1909–1991), an American scientist, found a way to align tiny, needle-shaped dichroic crystals in sheets of transparent celluloid. The result was a thin sheet of polarizing material that was given the commercial name *Polaroid*.

Better polarizing films have been developed that use synthetic polymer materials instead of celluloid. During the manufacturing process, this kind of film is stretched to align the long molecular chains of the polymer. With proper treatment, the outer (valence) electrons of the molecules can move along the oriented chains. As a result, light with $\vec{E}$ vectors parallel to the oriented chains is readily absorbed, but light with $\vec{E}$ vectors perpendicular to the chains is transmitted. The direction *perpendicular* to the orientation of the molecular chains is called the **transmission axis**, or the **polarization direction**. Thus, when unpolarized light falls on a polarizing sheet, the sheet acts as a polarizer and transmits polarized light (▶Fig. 24.22). Since one of the two electric field components is absorbed, the

Incident light is unpolarized

Vertical component absorbed in crystal

Transmitted light is plane polarized

Crystal

▲ **FIGURE 24.21** Selective absorption (dichroism) Dichroic crystals selectively absorb one polarized component (the vertical component) more than the other. If the crystal is thick enough, the emerging beam is linearly polarized.

Teaching tip: A useful analogy compares a polarizer with a picket fence. A randomly oriented wave in a rope will pass through the fence only if the rope is parallel to the pickets and will be polarized in that direction. However, point out to students that in a polarizing sheet, the polarization direction is *perpendicular* to the oriented molecules.

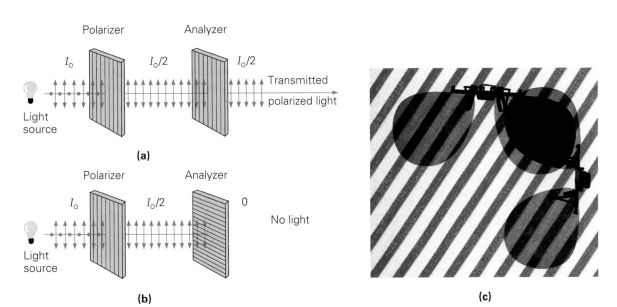

▲ **FIGURE 24.22 Polarizing sheets** **(a)** When polarizing sheets are oriented so that their transmission axes are the same, the emerging light is polarized. The first sheet acts as a polarizer, and the second acts as an analyzer. **(b)** When one of the sheets is rotated 90° and the transmission axes are perpendicular (crossed polarizers), little light (ideally, none) is transmitted. **(c)** Crossed polarizers made using polarizing sunglasses.

light intensity after the polarizer is half of the intensity incident on it ($I_o/2$). The human eye cannot distinguish between polarized and unpolarized light. To tell whether light is polarized, we must use an *analyzer*, which can simply be another sheet of polarizing film. As shown in Fig. 24.22a, if the transmission axis of an analyzer is parallel to the plane of polarization of polarized light, there is maximum transmission. If the transmission axis of the analyzer is perpendicular to the plane of polarization, little light (ideally, none) will be transmitted.

In general, the intensity of the transmitted light is given by

$$I = I_o \cos^2 \theta \quad \textit{Malus's law} \qquad (24.14)$$

where θ is the angle between the transmission axes of the polarizer and analyzer. This expression is known as *Malus's law*, after its discoverer, French physicist E. L. Malus (1775–1812).

Polarizing glasses whose lenses have different transmission axes are used to view some 3D movies. The pictures are projected on the screen by two projectors that transmit slightly different images, photographed by two cameras a short distance apart. The projected light from each projector is linearly polarized, but in mutually perpendicular directions. The lenses of the 3D glasses also have transmission axes that are perpendicular. Thus, one eye sees the image from one projector, and the other eye sees the image from the other projector. The brain receives a slight difference in perspective (or "viewing angle") from the two images, and interprets the image as having depth, or a third dimension, just as in normal vision.

Demonstration/activity: Place two Polaroid sheets on which the polarizing planes are designated on an overhead projector to demonstrate polarization. Then replace one of the sheets with a pair of Polaroid sunglasses.

Demonstration/activity: Most students have a calculator that displays numbers with a liquid crystal. Have students rotate the plane of polarization of a Polaroid filter over the LCD to check for the polarized light.

Integrated Example 24.6 ■ Make Something Out of Nothing: Three Polarizers

In Figs. 24.22b and c, no light is transmitted through the analyzer, because the transmission axes of the polarizer and analyzer are perpendicular. Assume that the unpolarized light incident on the first polarizer has an intensity of I_o. A second polarizer is now inserted between the first polarizer and analyzer, and the transmission axis of the second polarizer makes an angle of θ with the first polarizer. (a) Is it possible for some light to go through this arrangement? If yes, does it occur at (1) $\theta = 0°$, (2) $\theta = 30°$, (3) $\theta = 45°$, or (4) $\theta = 90°$? Explain. What happens if the second polarizer is rotated? (b) When $\theta = 30°$, what is the intensity of the transmitted light in terms of the incident intensity?

Exploration 39.2 Polarizers

(continues on next page)

(a) Conceptual Reasoning. Yes, it is possible for some light to go through this arrangement at any angle other than 0° or 90°. The accompanying Learn by Drawing can help us understand this situation.

With just the first polarizer and the analyzer, no light is transmitted, according to Malus's law (Eq. 24.14), because the angle between the transmission axes is 90°. However, when a second polarizer is inserted in between the first polarizer and the analyzer, some light can actually pass through the system. For example, if the transmission axis of the second polarizer makes an angle of θ with that of the first polarizer, then the angle between the transmission axes of the second polarizer and the analyzer will be $90° - \theta$. (Why?)

When unpolarized light of intensity I_0 is incident on the first polarizer, the transmitted light after the first polarizer is $I_0/2$, because only one of the two electric field components is transmitted. After the second polarizer, the intensity is decreased by a factor of $\cos^2 \theta$. After the analyzer, the intensity is decreased further by a factor of $\cos^2(90° - \theta) = \sin^2 \theta$. So the transmitted intensity is $I = (I_0/2)(\cos^2 \theta)(\sin^2 \theta)$. Therefore, as long as θ is not 0° or 90°, some light will be transmitted through the system.

Since the transmitted light depends on the angle θ, rotating the second polarizer will change the intensity transmitted.

(b) Quantitative Reasoning and Solution. Once this situation is understood, part (b) is a straightforward calculation.

Given: $\theta = 30°$ *Find:* (b) I after three polarizers in terms of I_0

When $\theta = 30°$, $I = \dfrac{I_0}{2}(\cos^2 30°)(\sin^2 30°) = \dfrac{I_0}{2} \cdot \left(\dfrac{\sqrt{3}}{2}\right)^2 \cdot \left(\dfrac{1}{2}\right)^2 = \dfrac{3I_0}{32}$

Follow-Up Exercise. For what value of θ will the transmitted intensity be a maximum in this example?

Polarization by Reflection

When a beam of unpolarized light strikes a smooth, transparent medium such as glass, the beam is partially reflected and partially transmitted. The reflected light may be completely polarized, partially polarized, or unpolarized, depending on the angle of incidence. The unpolarized case occurs for 0°, or normal incidence. As the angle of incidence is changed from 0°, both the reflected and refracted light become partially polarized. For example, the electric field components normal to the surface are reflected more strongly, producing partial polarization (▸Fig. 24.23a).

LEARN BY DRAWING   **THREE POLARIZERS (SEE INTEGRATED EXAMPLE 24.6.)**

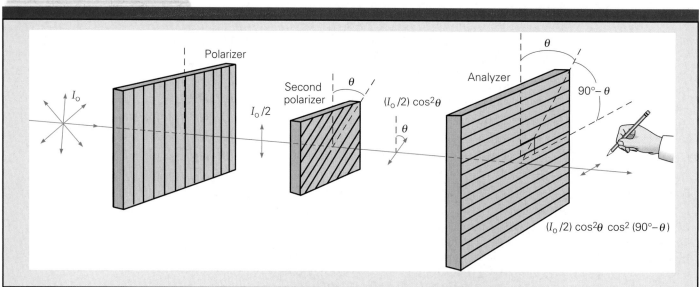

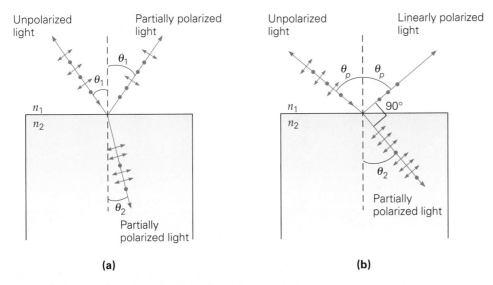

(a)   (b)

 ▲ **FIGURE 24.23 Polarization by reflection** (a) When a beam of light is incident on a boundary, the reflected and refracted beams are normally partially polarized. (b) When the reflected and refracted beams are 90° apart, the reflected beam is linearly polarized, and the refracted beam is partially polarized. This situation occurs when $\theta_1 = \theta_p = \tan^{-1}\left(\dfrac{n_2}{n_1}\right)$.

However, at one particular angle of incidence, the reflected beam is completely polarized on reflection (Fig. 24.23b). (At this angle, though, the refracted beam is still only partially polarized.)

David Brewster (1781–1868), a Scottish physicist, found that the complete polarization of the reflected beam occurs when the reflected and refracted beams are perpendicular. The incident angle at which complete polarization occurs is the **polarizing angle (θ_p)**, or the **Brewster angle**, and it depends on the indices of refraction of the two media. In Fig. 24.23b, the reflected and refracted beams are at 90° and the incident angle θ_1 is thus the polarizing angle θ_p: $\theta_1 = \theta_p$, so

$$\theta_1 + 90° + \theta_2 = 180° \quad \text{or} \quad \theta_2 = 90° - \theta_1$$

By Snell's law (Chapter 22),

$$n_1 \sin \theta_1 = n_2 \sin \theta_2$$

Hence, we have $\sin \theta_2 = \sin(90° - \theta_1) = \cos \theta_1$. Therefore,

$$\frac{\sin \theta_1}{\sin \theta_2} = \frac{\sin \theta_1}{\cos \theta_1} = \tan \theta_1 = \frac{n_2}{n_1}$$

With $\theta_1 = \theta_p$, we get

$$\tan \theta_p = \frac{n_2}{n_1} \quad \text{or} \quad \theta_p = \tan^{-1}\left(\frac{n_2}{n_1}\right) \tag{24.15}$$

If the first medium is air ($n_1 = 1$), then $\tan \theta_p = \dfrac{n_2}{1} = n_2 = n$, where n is the index of refraction of the second medium.

Now you can understand the principle behind polarizing sunglasses. Light reflected from a smooth surface is partially polarized. The direction of polarization is mostly parallel to the surface. (See Fig. 24.23b.) Light reflected from the surface of a road or water can be so intense that it gives rise to visual glare (▼ Fig. 24.24a). To reduce this effect, the polarizing lenses of glasses are oriented with their transmission axes vertical so that some of the partially polarized light from reflective surfaces is absorbed. Polarizing filters also enable cameras to take "clean" pictures without interference from glare (Fig. 24.24b).

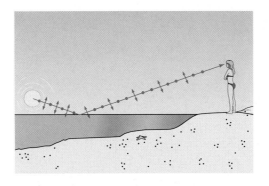

(a)

(b)

▲ **FIGURE 24.24 Glare reduction (a)** Light reflected from a horizontal surface is partially polarized in the horizontal plane. When sunglasses are oriented so that their transmission axis is vertical, the horizontally polarized component of such light is not transmitted, so glare is reduced. **(b)** Polarizing filters for cameras use the same principle. The photo at right was taken with such a filter. Note the reduction in reflections from the store window.

Example 24.7 ■ Sunlight on a Pond: Polarization by Reflection

Sunlight is reflected from the smooth surface of a pond. What is the Sun's altitude (the angle between the Sun and the horizon) when the polarization of the reflected light is greatest?

Thinking It Through. Since the angle of incidence is measured from the normal and the altitude angle is measured from the horizon, the angle of incidence is the angle complementary to the altitude angle (draw a sketch to help yourself visualize this situation). Incident light at the Brewster angle has the greatest polarization upon reflection, so the Sun needs to be at $90° - \theta_p$ from the horizon.

Solution. The index of refraction of water is listed in Table 22.1.

Given: $n_1 = 1$ *Find:* θ (altitude angle for greatest polarization)
$n_2 = 1.33$ (Table 22.1)

The Sun needs to be at an angle of $\theta = 90° - \theta_p$, where θ_p is the Brewster angle. Using Eq. 24.15, we find that

$$\theta_p = \tan^{-1}\left(\frac{n_2}{n_1}\right) = \tan^{-1}\left(\frac{1.33}{1}\right) = 53.1°$$

So

$$\theta = 90° - \theta_p = 90° - 53.1° = 36.9°$$

Follow-Up Exercise. Light is incident on a flat, transparent material with an index of refraction of 1.52. At what angle of refraction would the transmitted light have the greatest polarization if the transparent material is in water?

Polarization by Double Refraction (Birefringence)

When monochromatic light travels through glass, its speed is the same in all directions and is characterized by a single index of refraction. Any material that has such a property is said to be *isotropic*, meaning that it has the same optical charac-

Optic axis

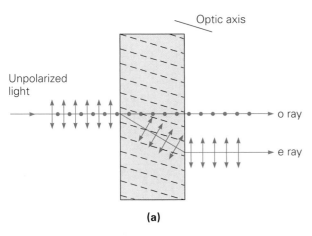

Unpolarized light

o ray

e ray

(a)

(b)

▲ **FIGURE 24.25 Double refraction or birefringence** **(a)** Unpolarized light incident normal to the surface of a birefringent crystal and at an angle to a particular direction in the crystal (dashed lines) is separated into two components. The ordinary (o) ray and the extraordinary (e) ray are plane polarized in mutually perpendicular directions. **(b)** Double refraction seen through a calcite crystal.

teristics in all directions. Some crystalline materials, such as quartz, calcite, and ice, are *anisotropic*; that is, the speed of light, and therefore the index of refraction, is different for different directions within the material. Anisotropy gives rise to some interesting optical properties. These anisotropic materials are said to be doubly refracting, or to exhibit **birefringence**, and polarization is involved.

For example, a beam of unpolarized light incident on a birefringent crystal of calcite ($CaCO_3$, calcium carbonate) is illustrated in ▲Fig. 24.25. When the beam propagates at an angle to a particular crystal axis, the beam is doubly refracted and separated into two components, or rays, upon refraction. These two rays are linearly polarized in mutually perpendicular directions. One ray, called the *ordinary* (o) *ray*, passes straight through the crystal and is characterized by an index of refraction n_o. The second ray, called the *extraordinary* (e) *ray*, is refracted and is characterized by an index of refraction n_e. The particular axis direction indicated by dashed lines in Fig. 24.25a is called the *optic axis*. Along this direction, $n_o = n_e$, and nothing extraordinary is noted about the transmitted light.

Some transparent materials have the ability to *rotate* the plane of polarization of linearly polarized light. This property, called **optical activity**, is due to the molecular structure of the material (▼Fig. 24.26a). The rotation may be clockwise or

Demonstration/activity: Place a calcite crystal over a transparency that has very small writing. View the transmitted light through a Polaroid sheet to demonstrate birefringence.

▼ **FIGURE 24.26 Optical activity and stress detection** **(a)** Some substances have the property of rotating the polarization direction of linearly polarized light. This ability, which depends on the molecular structure of the substance, is called *optical activity*. **(b)** Glasses and plastics become optically active under stress, and the points of greatest stress are apparent when the material is viewed through crossed polarizers. Engineers can thus test plastic models of structural elements to see where the greatest stresses will occur when the models are "loaded." Here, a model of a suspension-bridge strut is being analyzed.

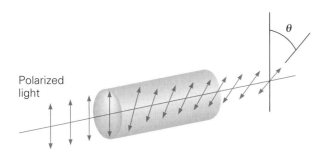

Polarized light

θ

(a)

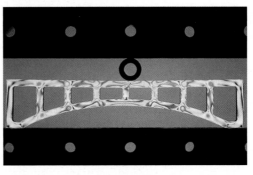

(b)

counterclockwise, depending on the molecular orientation. Optically active molecules include those of certain proteins, amino acids, and sugars.

Glasses and plastics become optically active under stress. The greatest rotation of the direction of polarization of the transmitted light occurs in the regions where the stress is the greatest. Viewing the stressed piece of material through crossed polarizers allows the points of greatest stress to be identified. This determination is called *optical stress analysis* (Fig. 24.26b). Another use of polarizing films, the liquid crystal display (LCD), is described in accompanying Insight 24.2, LCDs and Polarized Light.

*24.5 Atmospheric Scattering of Light

OBJECTIVES: To (a) discuss scattering, and (b) explain why the sky is blue and sunsets are red.

When light is incident on a suspension of particles, such as the molecules in air, some of the light may be absorbed and reradiated in all directions. This process is called *scattering*. The scattering of sunlight in the atmosphere produces some interesting effects, including the polarization of skylight (that is, sunlight that has been scattered by the atmosphere), the blueness of the sky, and the redness of sunsets and sunrises.

Atmospheric scattering causes the skylight to be polarized. When unpolarized sunlight is incident on air molecules, the electric field of the light wave sets electrons of the molecules into vibration. The vibrations are complex, but these accelerated charges emit radiation, like the vibrating electrons in the antenna of a radio broadcast station (see Section 20.4). The intensity of this emitted radiation is strongest along a line perpendicular to the oscillation, and, as illustrated in ◄Fig. 24.27, an observer viewing from an angle of 90° with respect to the direction of the sunlight will receive linearly polarized light, because of the charge oscillations normal to the surface. At other viewing angles, both components are present, and skylight seen through a polarizing filter appears partially polarized, because of the stronger component.

Since the scattering of light with the greatest degree of polarization occurs at a right angle to the direction of the Sun, at sunrise and sunset the scattered light from directly overhead has the greatest degree of polarization. The polarization of skylight can be observed by viewing the sky through a polarizing filter (or a polarizing sunglass lens) and rotating the filter. Light from different regions of the sky will be transmitted in different degrees, depending on its degree of polarization. It is believed that some insects, such as bees, use polarized skylight to determine navigational directions relative to the Sun.

Why the Sky Is Blue

The scattering of sunlight by air molecules is the reason why the sky looks blue. This effect is not due to polarization, but is caused by the selective absorption of light. As oscillators, air molecules have resonant frequencies (at which they scatter most efficiently) in the blue-violet region. Consequently, when sunlight is scattered, the blue end of the visible spectrum is scattered more than in the red end.

For particles such as air molecules, which are much smaller than the wavelength of light, the intensity of the scattered light is inversely proportional to the wavelength to the fourth power $(1/\lambda^4)$. This relationship between wavelength and scattering intensity is called **Rayleigh scattering**, after Lord Rayleigh (1842–1919), a British physicist who derived it. This inverse relationship predicts that light of the shorter-wavelength, or blue, end of the spectrum will be scattered much more than light of the longer-wavelength, or red, end. The scattered blue light is rescattered in the atmosphere and eventually is directed toward the ground. This is why the sky appears blue.

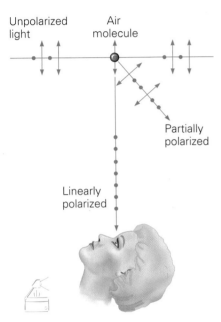

▲ **FIGURE 24.27 Polarization by scattering** When incident unpolarized sunlight is scattered by a gas molecule in the air, the light perpendicular to the direction of the incident ray is linearly polarized. Light scattered at some arbitrary angle is partially polarized. An observer at a right angle (90°) to the direction of the incident sunlight receives linearly polarized light.

Unpolarized light

Air molecule

Partially polarized

Linearly polarized

INSIGHT   24.2 LCDS AND POLARIZED LIGHT

Today, *liquid crystal displays* (LCDs) are commonplace in items such as watches, calculators, televisions, and computer screens. The name "liquid crystal" may seem self-contradictory. Normally, when a crystalline solid melts, the resulting liquid no longer has an orderly atomic or molecular arrangement. Some organic compounds, however, pass through an intermediate state in which the molecules may rearrange somewhat but still maintain the overall order that is characteristic of a crystal.

A common type of LCD, called a *twisted-nematic display*, makes use of the effect of a liquid crystal on polarized light (Fig. 1). These special liquid crystals are optically active and will rotate the direction of polarization of light by 90° if no voltage is applied across them. However, if voltage is applied, the crystals will lose this optical activity.

The liquid crystals are then placed between crossed polarizing sheets and backed with a mirrored surface. With voltage off, light entering and passing through the LCD is polarized, rotated 90°, reflected, and again rotated 90°. After the return trip through the liquid crystal, the direction of polarization of the light is the same as that of the initial polarizer. Thus, the light is transmitted and leaves the display unit. Because of the

reflection and transmission, the display appears to be of a light color (usually light gray) when illuminated with white, unpolarized light.

With voltage on, the polarized light passing through the liquid crystal is absorbed by the second polarizer. Thus, the liquid crystal is opaque and appears dark. Transparent, electrically conductive film coatings arranged in a seven-block pattern are applied to the liquid crystal. Each block, or display segment, has a separate electrical connection. The dark numbers or letters on an LCD are formed by applying an electric voltage to certain blocks of the liquid crystal. Note that all the numerals 0 through 9 can be formed out of pieces of the segmented display.

By using an analyzer, we can readily show that the light from the LCD is polarized (Fig. 2). You can either see or not see the display by rotating the analyzer over the watch. You may have noticed this effect if, while wearing polarizing sunglasses, you have ever tried to see the time on the LCD of a wristwatch.

One of the major advantages of LCDs is their low power consumption. Similar displays, such as those using light-emitting diodes (LEDs), produce light themselves, using relatively large amounts of power. LCDs produce no light but instead use reflected light.

Color flat-panel computer and TV screens, which rely on LCD technology, are popular today. They are about one quarter the size, consume less than half the energy, and are easier on the eyes than CRT monitors and traditional TV screens of the same size. Computer displays and TVs are usually specified in *pixels*, much like the smallest square on graph paper. To produce color, three LCD segments (red, green, and blue) are grouped on each pixel. By controlling the intensities of the three colors, each pixel can generate every color in the visible spectrum.

FIGURE 1 Liquid crystal display (LCD) A twisted-nematic display is an application involving the optical activity of a liquid crystal and crossed polarizing sheets. When the crystalline order is disoriented by an electric field from an applied voltage, the liquid crystal loses its optical activity in that region, and light is not transmitted or reflected. Numerals and letters are formed by applying voltages to segments of a block display.

FIGURE 2 Polarized light The light from an LCD is polarized, as can be shown by using polarizing sunglasses as an analyzer.

Example 24.8 ■ The Red and the Blue: Rayleigh Scattering

How much more is light at the blue end of the visible spectrum scattered by air molecules than is light at the red end?

Thinking It Through. We know that Rayleigh scattering is proportional to $1/\lambda^4$ and that light from the blue end of the spectrum (shorter wavelength) is scattered more than light from the red end. The wording "how much more" implies a factor or ratio.

Solution. The Rayleigh scattering relationship is $I \propto 1/\lambda^4$, where I is the amount, or intensity, of scattering for a particular wavelength. Thus, you can form the ratio

$$\frac{I_{blue}}{I_{red}} = \left(\frac{\lambda_{red}}{\lambda_{blue}}\right)^4$$

The blue end of the spectrum (violet light) has a wavelength of about $\lambda_{blue} = 400$ nm, and red light has a wavelength of about $\lambda_{red} = 700$ nm. Inserting these values gives

$$\frac{I_{blue}}{I_{red}} = \left(\frac{\lambda_{red}}{\lambda_{blue}}\right)^4 = \left(\frac{700 \text{ nm}}{400 \text{ nm}}\right)^4 = 9.4 \qquad \text{or} \qquad I_{blue} = 9.4 I_{red}$$

Thus, blue light is scattered almost 10 times as much as red light.

Follow-Up Exercise. What wavelength of light is scattered twice as much as red light? What color light is this?

Why Sunsets and Sunrises Are Red

Beautiful red sunsets and sunrises are sometimes observed. When the Sun is near the horizon, sunlight travels a greater distance through the denser air near the Earth's surface. Since the light therefore undergoes a great deal of scattering, you might think that only the least scattered light, the red light, would reach observers on the Earth's surface. This would explain red sunsets. However, it has been shown that the dominant color of white light after only molecular scattering is orange. Thus, other types of scattering must shift the light from the setting (or rising) Sun toward the red end of the spectrum (◄Fig. 24.28).

Red sunsets have been found to result from the scattering of sunlight by atmospheric gases *and* by small dust particles. These particles are not necessary for the blueness of the sky, but are compulsory for deep-red sunsets and sunrises. (This is why spectacular red sunsets are observed in the months after a large volcanic eruption that can put tons of particulate matter into the atmosphere.) Red sunsets occur most often when there is a high-pressure air mass to the west, since the concentration of dust particles is generally greater in high-pressure air masses than in low-pressure air masses. Similarly, red sunrises occur most often when there is a high-pressure air mass to the east.

Now you can understand the old saying "Red sky at night, sailors' delight; red sky in the morning, sailors take warning." Fair weather generally accompanies high-pressure air masses, because they are associated with reduced cloud formation. Most of the United States lies in the westerlies wind zone, in which air masses generally move from west to east. A red sky at night is thus likely to indicate a fair-weather, high-pressure air mass to the west that will be coming your way. A red sky in the morning means that the high-pressure air mass has passed and poor weather may set in.

As a final note, how would you like a sky that is normally *red*? Then try Mars, the "red planet." The thin Martian atmosphere is about 95% carbon dioxide (CO_2). The CO_2 molecule is more massive than an oxygen (O_2) or a nitrogen (N_2) molecule. As a result, CO_2 molecules have a lower resonant frequency (longer wavelength) and preferentially scatter the red end of the visible spectrum. Hence, the Martian sky is red during the day. And what of the color of sunrises and sunsets on Mars? Think about it. . . .

And finally, the use of light in biomedical application in Insight 24.3, Optical Biopsy.

▲ **FIGURE 24.28 Red sky at night** A spectacular red sunset over a mountaintop observatory in Chile. The red sky results from the scattering of sunlight by atmospheric gases and small solid particles. A directly observed reddening Sun is due to the scattering of the wavelengths toward the blue end of the spectrum in the direct line of sight.

INSIGHT 24.3 OPTICAL BIOPSY

One of the most reliable ways to detect disease is to perform a biopsy—the removal of tissue samples—and then look for abnormal *changes* in the samples. "Optical biopsy," or biomedical scattering, appears to be a promising tool for diagnosing and monitoring diseases such as cancer *without* such surgery.

Optical biopsies are based on the following physical principle. The particles in the tissues absorb and re-emit light; thus, the scattered light contains information about the makeup of the tissue. Scattering from a tissue depends on the internal structures such as the presence of collagen fibers, and the status of hydration in the tissue. The measurement of the scattered light as a function of wavelength, polarization, or angle can be an important diagnostic tool.

An example of an optical biopsy is the diagnosis and measurement of collagen fibers. A major component of skin and bone, collagen is a fibrous protein found in animal cells. The fibers in the dormant form of collagen (about $2-3$ μm in diameter) are composed of bundles of smaller collagen fibrils, about 0.3 μm in diameter, as shown in Fig. 1. The fibrils are made up of entwined tropocollagen molecules and present a banded pattern of striations with 70-nm periodicity due to the staggered alignment of the tropocollagen molecules. Each of these molecules has an electron-dense "head group" that appears dark in the electron micrograph. This periodic variation in refractive index

at this level scatters light strongly in the visible and ultraviolet regions. The information contained in the scattered light can reveal abnormal conditions in the collagen fibers.

FIGURE 1 An electron micrograph of collagen fibers The details of collagen fibers show the presence of collagen fibrils and tropocollagen molecules.

Chapter Review

- **Young's double-slit experiment** provides evidence of the wave nature of light and a way to measure the wavelength of light ($\approx 10^{-7}$ m).

  The angular position (θ) of the **maxima** satisfies the condition

  $$d \sin \theta = n\lambda \qquad \text{for } n = 0, 1, 2, 3, \ldots \qquad (24.3)$$

  where d is the slit separation.

  For small θ, the distance between the nth maximum and the central maximum is

  $$y_n \approx \frac{nL\lambda}{d} \quad \text{for } n = 0, 1, 2, 3, \ldots \qquad (24.4)$$

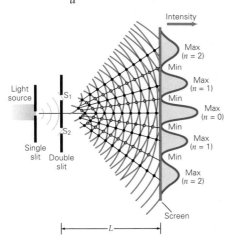

- Light reflected at a media boundary for which $n_2 > n_1$ undergoes a 180° **phase change**. If $n_2 < n_1$, there is no phase change on reflection. The phase changes affect thin-film interference, which also depends on film thickness and index of refraction.

  The **minimum thickness for a nonreflecting film** is

  $$t_{\min} = \frac{\lambda}{4n_1} \text{ (for } n_2 > n_1 > n_o) \qquad (24.7)$$

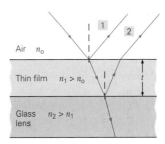

• In a **single-slit diffraction** experiment, the **minima** at location θ satisfy

$$w \sin \theta = m\lambda \quad \text{for } m = 1, 2, 3, \ldots \quad (24.8)$$

where w is the slit width. In general, the longer the wavelength as compared with the width of an opening or object, the greater the diffraction.

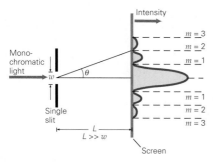

• For a **diffraction grating**, the maxima (bright fringes) satisfy

$$d \sin \theta = n\lambda \quad \text{for } n = 0, 1, 2, \ldots \quad (24.12)$$

where $d = 1/N$ and N is the number of lines per unit length.

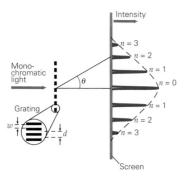

• **Polarization** is the preferential orientation of the electric field vectors that make up a light wave and is evidence that light is a transverse wave. Light can be polarized by selective absorption, reflection, double refraction (**birefringence**), and scattering.

When the transmission axes of a polarizer and an analyzer make an angle of θ, the intensity of the transmitted light is given by **Malus's law**:

$$I = I_o \cos^2 \theta \quad (24.14)$$

In reflection, if the angle of incidence is equal to the **Brewster (polarizing) angle** θ_p, then the reflected light is linearly polarized:

$$\tan \theta_p = \frac{n_2}{n_1} \quad \text{or} \quad \theta_p = \tan^{-1}\left(\frac{n_2}{n_1}\right) \quad (24.15)$$

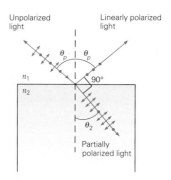

• The intensity of **Rayleigh scattering** is inversely proportional to the fourth power of the wavelength of the light. The blueness of the Earth's sky results from the preferential scattering of sunlight by air molecules.

Exercises

MC = *Multiple Choice Question,* **CQ** = *Conceptual Question, and* **IE** = *Integrated Exercise. Throughout the text, many exercise sections will include "paired" exercises. These exercise pairs, identified with **red numbers**, are intended to assist you in problem solving and learning. In a pair, the first exercise (even numbered) is worked out in the Study Guide so that you can consult it should you need assistance in solving it. The second exercise (odd numbered) is similar in nature, and its answer is given at the back of the book.*

24.1 Young's Double-Slit Experiment

1. **MC** In a Young's double-slit experiment using monochromatic light, if the slit spacing d decreases, the interference maxima spacing will (a) decrease, (b) increase, (c) remain unchanged, (d) disappear. (b)

2. **MC** If the path-length difference between two identical and coherent beams is 2.5λ when they arrive at a point on a screen, the point will be (a) bright, (b) dark, (c) multicolored, (d) gray. (b)

3. **MC** When white light is used in Young's double-slit experiment, many maxima with a spectrum of colors are seen. In a given maximum, the color closest to the central maximum is (a) red, (b) blue, (c) all colors. (b)

4. **CQ** Non–cable television pictures often flutter when an airplane passes by (▶Fig. 24.29). Explain a possible cause of this fluttering, based on interference effects.
path-length difference changes; so does interference

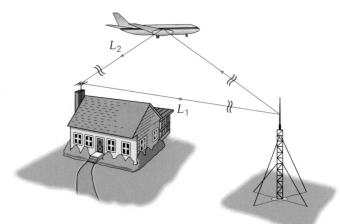

▲ **FIGURE 24.29 Interference** See Exercise 4.

5. **CQ** Describe what would happen to the interference pattern in Young's double-slit experiment if the wavelength of the monochromatic light were to increase.
the spacing between the maxima would increase

6. **CQ** The intensity of the central maximum in the interference pattern of a Young's double-slit experiment is about four times that of either light wave. Is this a violation of the conservation of energy? Explain. no, there are minima

7. ● In the development of Young's double-slit experiment, a small-angle approximation ($\tan \theta \approx \sin \theta$) was used to find the lateral displacements of the bright and dark fringes. How good is this approximation? For example, what is the percentage error for $\theta = 15°$? 3.4%

8. ● To study wave interference, a student uses two speakers driven by the same sound wave of wavelength 0.50 m. If the distances from a point to the speakers differ by 0.75 m, will the waves interfere constructively or destructively at that point? What if the distances differ by 1.0 m? destructively; constructively

9. ● Two parallel slits 0.075 mm apart are illuminated with monochromatic light of wavelength 480 nm. Find the angle between the center of the central maximum and the center of the first side maximum. 0.37°

10. ●● (a) Derive a relationship that gives the locations of the minima in a Young's double-slit experiment. What is the distance between adjacent minima? (b) For a third-order minimum (the third dark-side position from the central maximum), what is the path-length difference between that location and the two slits?
(a) $y_m = m\lambda L/(2d)$, $m = 1, 3, 5, \ldots$; $\Delta y = L\lambda/d$ (b) 2.5λ

11. ●● In a double-slit experiment that uses monochromatic light, the angular separation between the central maximum and the second-order maximum is 0.160°. What is the wavelength of the light if the distance between the slits is 0.350 mm? 489 nm

12. **IE** ●● Monochromatic light passes through two narrow slits and forms an interference pattern on a screen. (a) If the wavelength of light used increases, will the distance between the maxima (1) increase, (2) remain the same, or (3) decrease? Explain. (b) If the slit separation is 0.25 mm, the screen is 1.5 m away from the slits, and light of wavelength 550 nm is used, what is the distance from the center of the central maximum to the center of the third-order maximum? (c) What if the wavelength is 680 nm?
(a) (1) increase (b) 0.99 cm (c) 1.2 cm

13. ●● In a double-slit experiment using monochromatic light, a screen is placed 1.25 m away from the slits, which have a separation distance of 0.0250 mm. The position of the third-order maximum is 6.60 cm from the center of the central maximum. Find (a) the wavelength of the light and (b) the position of the second-order maximum. (a) 440 nm (b) 4.40 cm

14. **IE** ●● (a) If the wavelength used in a double-slit experiment is decreased, the distance between adjacent maxima will (1) increase, (2) also decrease, (3) remain the same. Explain. (b) If the separation between the two slits is 0.20 mm and the adjacent maxima of the interference pattern on a screen 1.5 m away from the slits are 0.45 cm apart, what is the wavelength and color of the light? (c) If the wavelength is 550 nm, what is the distance between adjacent maxima? (a) (2) also decrease (b) 600 nm (yellow-orange color) (c) 0.41 cm

15. **IE** ●● Two parallel slits are illuminated with monochromatic light, and an interference pattern is observed on a screen. (a) If the distance between the slits were decreased, would the distance between the maxima (1) increase, (2) remain the same, or (3) decrease? Explain. (b) If the slit separation is 1.0 mm, the wavelength is 640 nm, and the distance from the slits to the screen is 3.00 m, what is the separation between adjacent interference maxima? (c) What if the slit separation is 0.80 mm? (a) (1) increase (b) 1.9 mm (c) 2.4 mm

16. **IE** ●● (a) In a double-slit experiment, if the distance from the double slits to the screen is increased, the separation between the adjacent maxima will (1) increase, (2) decrease, (3) remain the same. Explain. (b) Yellow-green light ($\lambda = 550$ nm) is illuminated on a double-slit separated by 1.75×10^{-4} m. If the screen is located 2.00 m from the slits, determine the separation between the adjacent maxima. (c) What if the screen is located 3.00 m from the slits?
(a) (1) increase (b) 0.63 cm (c) 0.94 cm

17. ●● In a double-slit experiment with monochromatic light and a screen at a distance of 1.50 m from the slits, the angle between the second-order maximum and the central maximum is 0.0230 rad. If the separation distance of the slits is 0.0350 mm, what are (a) the wavelength and color of the light and (b) the lateral displacement of this maximum? (a) $\lambda = 402$ nm, violet (b) 3.45 cm

18. **IE** ●●● (a) If the apparatus for a Young's double-slit experiment is completely immersed in water, will the spacing of the interference maxima (1) increase, (2) remain the same, or (3) decrease? Explain. (b) What would the lateral displacements in Exercise 12 be if the entire system were immersed in still water? (a) (3) decrease (b) 7.4 mm, 9.0 mm

19. ●●● Light of two different wavelengths is used in a double-slit experiment. The location of the third-order maximum for the first light, yellow-orange light ($\lambda = 600$ nm), coincides with the location of the fourth-order maximum for the other color's light. What is the wavelength of the other light? 450 nm

24.2 Thin-Film Interference

20. **MC** For a thin film with $n_1 > n_o$ and $n_1 > n_2$, where n_1 is the index of refraction of the film, a film thickness for constructive interference of the reflected light is (a) $\lambda'/4$, (b) $\lambda'/2$, (c) λ', (d) both a and b. (a)

21. **MC** For a thin film with $n_o < n_1 < n_2$, where n_1 is the index of refraction of the film, the minimum film thickness for destructive interference of the reflected light is (a) $\lambda'/4$, (b) $\lambda'/2$, (c) λ'. (a)

22. **MC** When a thin film of kerosene spreads out on water, the thinnest part looks bright. The index of refraction of kerosene is (a) greater than, (b) less than, (c) the same as that of water. (b)

23. **CQ** Most lenses used in cameras are coated with thin films and appear bluish-purple when viewed with reflected light. What wavelengths are not visible in the reflected light? all wavelengths except bluish-purple

24. CQ When destructive interference of two waves occurs at a certain location, there is no energy at that location. Is this situation a violation of the conservation of energy? Explain.
no; there are regions where waves interfere constructively

25. CQ At the center of a Newton's rings arrangement (Fig. 24.10a), the air wedge has a thickness of zero. Why is this area always dark?
destructive interference due to 180° phase shift

26. IE ● A film on a lens with an index of refraction of 1.5 is 1.0×10^{-7} m thick and is illuminated with white light. The index of refraction of the film is 1.4. (a) The number of waves that experience the 180° phase shift is (1) zero, (2) one, (3) two. Explain. (b) For what wavelength of visible light will the lens be nonreflecting? (a) (3) two (b) 560 nm

27. ● Light of wavelength 550 nm in air is normally incident on a glass plate ($n = 1.5$) whose thickness is 1.1×10^{-5} m. (a) What is the thickness of the glass in terms of the wavelength of light in glass? (b) Will the reflected light interfere constructively or destructively?
(a) 30λ (b) destructively

28. ● A lens with an index of refraction of 1.60 is to be coated with a material ($n = 1.40$) that will make the lens nonreflecting for red light ($\lambda = 700$ nm) normally incident on the lens. What is the minimum required thickness of the coating? 125 nm

29. ●● Magnesium fluoride ($n = 1.38$) is frequently used as a lens coating to make nonreflecting lenses. What is the difference in the minimum film thickness required for maximum transmission of blue light ($\lambda = 400$ nm) and of red light ($\lambda = 700$ nm)? 54.3 nm

30. ●● A solar cell is designed to have a nonreflective coating of a transparent material. (a) Will the thickness of the coating depend on the index of refraction of the underlying material in the solar cell? Discuss the possible scenarios. (b) If $n_{solar} > n_{film}$ and $n_{film} = 1.22$, what is the minimum thickness of the film for light with a wavelength of 550 nm? see ISM

31. IE ●● A thin layer of oil ($n = 1.50$) floats on water. Destructive interference is observed for light of wavelengths 480 nm and 600 nm, each at a different location. (a) If the order number is the same for both wavelength, which wavelength is at a greater thickness: (1) 480 nm, (2) 600 nm, or (3) both? Explain. (b) Find the two minimum thicknesses of the oil film, assuming normal incidence.
(a) (2) 600 nm (b) 160 nm; 200 nm

32. ●● A camera lens ($n = 1.50$) is coated with a thin layer of a material that has an index of refraction of 1.35. This coating makes the lens nonreflecting for light of wavelength 450 nm (in air) that is normally incident on the lens. What is the thickness of the thinnest film that will make the lens nonreflecting? 83.3 nm

33. ●● Two parallel plates are separated by a small distance as illustrated in ▶Fig. 24.30. If the top plate is illuminated with light from a He–Ne laser ($\lambda = 632.8$ nm), for what minimum separation distances will the light be (a) constructively reflected and (b) destructively reflected? [Note: $t = 0$ is *not* an answer for part (b).] (a) 158.2 nm (b) 316.4 nm

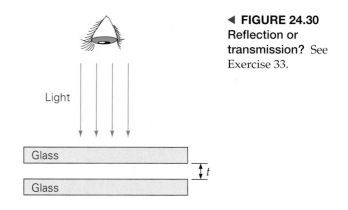

◀ **FIGURE 24.30** Reflection or transmission? See Exercise 33.

34. IE ●●● An air wedge such as that shown in ▼Fig. 24.31 can be used to measure small dimensions, such as the diameter of a thin wire. (a) If the top glass plate is illuminated with monochromatic light, the interference pattern observed will be (1) bright, (2) dark, (3) bright and dark lines based on the air film thickness. Explain. (b) Express the locations of the bright interference maxima in terms of wedge thickness measured from the apex of the wedge.
(a) (3) bright and dark (b) see ISM

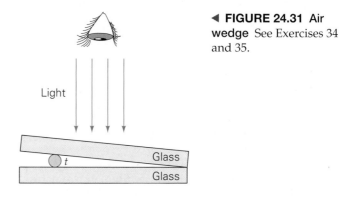

◀ **FIGURE 24.31** Air wedge See Exercises 34 and 35.

35. ●●● The glass plates in Fig. 24.31 are separated by a thin, round filament. When the top plate is illuminated normally with light of wavelength 550 nm, the filament lies directly below the sixth bright maximum. What is the diameter of the filament? 1.51×10^{-6} m

24.3 Diffraction

36. MC In a single-slit diffraction pattern, (a) all maxima have the same width, (b) the central maximum is twice as wide as the side maxima, (c) the side maxima are twice as wide as the central maximum, (d) none of the preceding. (b)

37. MC As the number of lines per unit length of a diffraction grating increases, the spacing between the maxima (a) increases, (b) decreases, (c) remains unchanged. (a)

38. MC In a single-slit diffraction pattern, if the wavelength of light increases, the width of the central maximum will (a) increase, (2) decrease, (c) remain the same. (a)

39. CQ From Eq. 24.8, can the $m = 2$ minimum be seen if $w = \lambda$? How about the $m = 1$ minimum?
no; yes (barely, at $\theta = 90°$)

40. CQ In our discussion of single-slit diffraction, the length of the slit was assumed to be much greater than the width. What changes would be observed in the diffraction pattern if the length were comparable with the width of the slit?
a second diffraction pattern perpendicular to the first

41. CQ In a diffraction grating, the slits are very closely spaced. What is the advantage of this design?
wider diffraction pattern, as d is smaller

42. ● A slit of width 0.20 mm is illuminated with monochromatic light of wavelength 480 nm, and a diffraction pattern is formed on a screen 1.0 m away from the slit. (a) What is the width of the central maximum? (b) What are the widths of the second and third-order maxima?
(a) 4.8 mm (b) 2.4 mm, 2.4 mm

43. ● A slit 0.025 mm wide is illuminated with red light ($\lambda = 680$ nm). How wide are (a) the central maximum and (b) the side maxima of the diffraction pattern formed on a screen 1.0 m from the slit? *(a) 5.4 cm (b) 2.7 cm*

44. ● At what angle will the second-order diffraction maximum be seen from a diffraction grating of spacing 1.25 μm when illuminated by light of wavelength 550 nm? *61.6°*

45. ● A venetian blind is essentially a diffraction grating—not for visible light, but for waves with longer wavelengths. If the spacing between the slats of a blind is 2.5 cm, (a) for what wavelength is there a first-order maximum at an angle of 10°, and (b) what type of radiation is this?
(a) 4.3 mm (b) microwave

46. IE ●● A single slit is illuminated with monochromatic light, and a screen is placed behind the slit to observe the diffraction pattern. (a) If the width of the slit is increased, will the width of the central maximum (1) increase, (2) remain the same, or (3) decrease? Why? (b) If the width of the slit is 0.50 mm, the wavelength is 680 nm, and the screen is 1.80 m from the slit, what is the width of the central maximum? (c) What if the width of the slit is 0.60 mm?
(a) (3) decrease (b) 4.9 mm (c) 4.9 mm (c) 4.1 mm

47. ●● A diffraction grating is designed to have the second-order maxima at 10° from the central maximum for the red end ($\lambda = 700$ nm) of the visible spectrum. How many lines per centimeter does the grating have?
1.24 × 10³ lines/cm

48. ●● A certain crystal gives a deflection angle of 25° for the first-order maximum of monochromatic X-rays with a frequency of 5.0 × 10¹⁷ Hz. What is the lattice spacing of the crystal? *7.1 × 10⁻¹⁰ m*

49. ●● Find the angles of the blue ($\lambda = 420$ nm) and red ($\lambda = 680$ nm) components of the first- and second-order maxima in a pattern produced by a diffraction grating with 7500 lines/cm. *blue: 18.4°, 39.2°; red: 30.7°, not possible*

50. IE ●● (a) Only a limited number of maxima can be observed with a diffraction grating. The factor(s) that limits the number of maxima seen is (a) (1) the wavelength, (2) the grating spacing, (3) both. Explain. (b) How many maxima appear when monochromatic light of wavelength 560 nm illuminates a diffraction grating that has 10 000 lines/cm, and what are their order numbers?
(a) (3) both (b) 3, n = 0, 1 (two of them)

51. ●● In a particular diffraction pattern, the red component (700 nm) in the second-order maximum is deviated at an angle of 20°. (a) How many lines per centimeter does the grating have? (b) If the grating is illuminated with white light, how many maxima of the complete visible spectrum are produced?
(a) 2.44 × 10³ lines/cm (b) 11 (maximum n is 5)

52. ●● White light whose components have wavelengths from 400 to 700 nm illuminates a diffraction grating with 4000 lines/cm. Do the first- and second-order spectra overlap? Justify your answer. *no, see ISM*

53. IE ●● White light ranging from blue (400 nm) to red (700 nm) illuminates a diffraction grating with 8000 lines/cm. (a) For the first maxima, the (1) blue (2) red color is closer to the central maximum. Explain. (b) What are the angles of the first-order maximum for blue and red?
(a) (1) blue (b) blue: 18.7°, red: 34.1°

54. ●● A diffraction grating with 8000 lines/cm is illuminated with a red light from a He–Ne laser ($\lambda = 632.8$ nm). How many side maxima are formed in the diffraction pattern, and at what angles are they observed?
two, at ±30.4°

55. ●●● Show that for a diffraction grating, the violet ($\lambda = 400$ nm) portion of the third-order maximum overlaps the yellow-orange ($\lambda = 600$ nm) portion of the second-order maximum, regardless of the grating's spacing. *see ISM*

56. IE ●●● A teacher standing in a doorway 1.0 m wide blows a whistle with a frequency of 1000 Hz to summon children from the playground (▼Fig. 24.32). Two boys are playing on the swings 100 m away from the school building. One boy is at an angle of 0° and another one at 19.6° from a line normal to the doorway. (a) (1) Only the boy at 0°, (2) only the boy at 19.6°, or (3) both boys may not hear the whistle. Explain. (b) Taking the speed of sound in air to be 335 m/s, does the boy 19.6° hear the whistle?
(a) (1) Only the boy at 0°, (2) only the boy at 19.6° (b) no

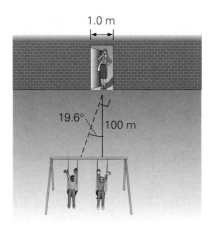

▲ **FIGURE 24.32 Moment of truth** See Exercise 56. (Not drawn to scale.)

24.4 Polarization

57. **MC** Light can be polarized by (a) reflection, (b) refraction, (c) absorption, (d) all of the preceding. (d)

58. **MC** The Brewster angle depends on (a) the indices of refraction of materials, (b) Bragg's law, (c) internal reflection, (d) interference. (a)

59. **MC** A sound wave cannot be polarized. This is because sound is (a) not a light wave, (b) a transverse wave, (c) a longitudinal wave, (d) none of the preceding. (c)

60. **CQ** Given two pairs of sunglasses, could you tell whether one or both were polarizing? see ISM

61. **CQ** Suppose that you held two polarizing sheets in front of you and looked through both of them. How many times would you see the sheets lighten and darken (a) if one were rotated through one complete rotation, (b) if both were rotated through one complete rotation at the same rate in opposite directions, (c) if both were rotated through one complete rotation at the same rate in the same direction, and (d) if one rotates twice as fast as the other and the slower one rotates through one complete rotation?
(a) twice (b) four times (c) none (d) six times

62. **CQ** How does selective absorption produce polarized light? one of the two electric field components is absorbed

63. **CQ** If you place a pair of polarizing sunglasses in front of your calculator's LCD display and rotate them, what do you observe? the numbers appear and disappear as the sunglasses are rotated

64. ● Some types of glass have a range of indices of refraction of about 1.4 to 1.7. What is the range of the polarizing (Brewster) angle for these glasses when light is incident on them from air? 54° to 60°

65. **IE** ● Light is incident on a certain material in air. (a) If the index of refraction of the material increases, the polarizing (Brewster) angle will (1) also increase, (2) decrease, (3) remain the same. Explain. (b) What are the polarizing angles if the index of refraction is 1.6 and 1.8?
(a) (1) also increase (b) 58°, 61°

66. ● A polarizer–analyzer pair can have their transmission axes at either 30° or 45° angles. Which angle will allow more light to be transmitted? 30°

67. **IE** ●● Unpolarized light of intensity I_o is incident on a polarizer–analyzer pair. (a) If the angle between the polarizer and analyzer increases in the range of 0° to 90°, the transmitted light intensity will (1) also increase, (2) decrease, (3) remain the same. Explain. (b) If the angle between the polarizer and analyzer is 30°, what light intensity is transmitted through the polarizer and the analyzer, respectively? (c) What if the angle is 60°?
(a) (2) decrease (b) $0.500I_o$, $0.375I_o$ (c) $0.500I_o$, $0.125I_o$

68. ●● A beam of light is incident on a glass plate ($n = 1.62$) in air and the reflected ray is completely polarized. What is the angle of refraction for the beam? 31.7°

69. ●● The critical angle for internal reflection in a certain medium is 45°. What is the polarizing (Brewster) angle for light externally incident on the medium? 55°

70. ●● The angle of incidence is adjusted so there is maximum linear polarization for the reflection of light from a transparent piece of plastic in air. (a) There is (1) no, (2) maximum, or (3) some light transmitted through the plastic. Explain. (b) If the index of refraction of the plastic is 1.22, what is the angle of refraction in the plastic? (a) (3) some light (b) 39.3°

71. ●● Sunlight is reflected off a vertical plate-glass window ($n = 1.55$). What would the Sun's altitude (angle above the horizon) have to be for the reflected light to be completely polarized? 57.2°

72. **IE** ●● A piece of glass ($n = 1.60$) could be in air or submerged in water. (a) The polarizing (Brewster) angle in water is (1) greater than, (2) less than, (3) the same as that in air. Explain. (b) What is the polarizing angle when it is in air and submerged in water?
(a) (2) less than (b) 58.0°, 50.3°

73. ●●● A plate of crown glass is covered with a layer of water. A beam of light traveling in air is incident on the water and partially transmitted. Is there any angle of incidence for which the light reflected from the water–glass interface will have maximum linear polarization? Justify your answer mathematically. no; see ISM

*24.5 Atmospheric Scattering of Light

74. **MC** Which of the following colors is scattered the most in the atmosphere: (a) blue, (b) yellow, (c) red, or (d) color makes no difference? (a)

75. **MC** Scattering involves (a) the reflection of light off particles, (b) the refraction of light off particles, (c) the absorption and reradiation of light by particles, (d) none of the preceding. (c)

76. **CQ** Explain why the sky is red in the morning and evening and blue during the day.
blue scatters more than red

77. **CQ** (a) On a clear cloudless day why does the sky not have a uniform blueness? (b) What color would an astronaut on the Moon see when looking at the sky or into space? (a) variable air molecule density (b) black

Comprehensive Exercises

78. A thin air wedge between two flat glass plates forms bright and dark interference bands when illuminated with normally incident monochromatic light. (See Fig. 24.9.) (a) Show that the thickness of the air wedge changes by $\lambda/2$ from one bright fringe to the next, where λ is the wavelength of the light. (b) What would be the change in the thickness of the wedge between bright fringes if the space were filled with a liquid with an index of refraction n? see ISM

79. A salesman tries to sell you an optic fiber that claims to give linearly polarized light when light is totally internally reflected off the fiber–air interface. (a) Would you buy it? Explain. (b) If total internal reflection occurs at an angle of 35°, what is the polarizing (Brewster) angle? (a) no, see ISM (b) 29.8°

80. Three parallel slits of width w have a slit separation of d, where $d = 3w$. (a) Will you be able to see all the interference maxima? Explain. (b) If not, which interference maxima will be missing? [*Hint*: See Fig. 24.16.] (a) no (b) 3, 6, 9, etc.

81. If the slit width in a single-slit experiment were doubled, the distance to the screen reduced by one third, and the wavelength of the light changed from 600 nm to 450 nm, how would the width of the maxima be affected? $\Delta y = 0.25 \Delta y_{\mathrm{o}}$

82. Show that when the reflected light is completely polarized, the sum of the angle of incidence and the angle of refraction is equal to 90°. see ISM

83. What is the highest spectral order that can be seen in a diffraction grating with 9000 lines/cm when the grating is illuminated by white light? $n = 1$ for red; $n = 2$ for violet

The following Physlet Physics Problems can be used with this chapter.
37.2, 37.4, 37.7, 37.9, 37.10, 38.1, 38.2, 38.4, 38.5, 38.6, 39.9, 39.10

25

VISION AND OPTICAL INSTRUMENTS

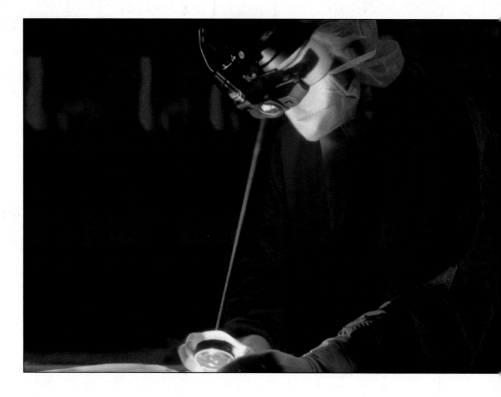

PHYSICS FACTS

- About 80% of the refracting power of a human eye comes from the cornea while the other 20% comes from the crystalline lens. The crystalline lens can change shape to accommodate close or far focusing by means of the ciliary muscles.

- The human eye collects a lot of information. If it were compared to a digital camera, the human eye would be equivalent to 500 megapixels. A common digital camera has 2 to 10 megapixels.

- A red blood cell has a diameter of about $7\,\mu\text{m}$ (7×10^{-6} m). When viewed with a compound microscope at 1000×, it appears to be 7 mm (7×10^{-3} m).

- Some cameras on satellites have excellent resolution. From space they can read the license plates on cars.

Vision is one of our chief means of acquiring information about the world around us. However, the images seen by many eyes are not always clear or in focus, and glasses or some other remedy are needed. Great progress has been made in the last decade in contact lens therapy and surgical correction of vision defects. A popular procedure is laser surgery, as shown in the photo. Laser surgery can be used for such procedures as repairing torn retinas, destroying eye tumors, and stopping abnormal growth of blood vessels that can endanger vision.

Optical instruments, the basic function of which is to improve and extend the power of observation beyond that of the human eye, augment our vision. Mirrors and lenses are used in a variety of optical instruments, including microscopes and telescopes.

The earliest magnifying lenses were drops of water captured in a small hole. By the seventeenth century, artisans were able to grind fair-quality lenses for simple microscopes or magnifying glasses, which were used primarily for botanical studies. (These early lenses also found a use in spectacles.) Soon, the basic compound microscope, which uses two lenses, was developed. Modern compound microscopes, which can magnify an object up to 200 times, extended our vision into the microbe world.

Around 1609, Galileo used lenses to construct an astronomical telescope that allowed him to observe valleys and mountains on the Moon, sunspots, and the four largest moons of Jupiter. Today, huge telescopes that use lenses and mirrors have extended our vision far into the past as we look at farther, and therefore, younger, galaxies.

What would our knowledge be if these instruments had never been invented? Bacteria would still be unknown, and planets, stars, and galaxies would have remained nothing but mysterious points of light.

Mirrors and lenses were discussed in terms of geometrical optics in Chapter 23, and the wave nature of light was investigated in Chapter 24. These principles can be applied to the study of vision and optical instruments. In this chapter, you will learn about our fundamental optical instrument—the human eye, without which all others would be of little use. Also microscopes and telescopes will be discussed, along with the factors that limit their viewing.

25.1 The Human Eye

OBJECTIVES: To (a) describe the optical workings of the eye, and (b) explain some common vision defects and how they are corrected.

The human eye is the most important of all optical instruments, because without it we would know little about our world and the study of optics would not exist. The human eye is analogous to a simple camera in several respects (▼Fig. 25.1). A simple camera consists of a converging lens, which is used to focus images on light-sensitive film (traditional camera) or *charge coupled device*, CCD, (in digital cameras) at the back of the camera's interior chamber. (Recall from Chapter 23 that for relatively distant objects, a converging lens produces a small, inverted, real image.) There is an adjustable diaphragm opening, or aperture, and a shutter to control the amount of light entering the camera.

The eye, too, focuses images onto a light-sensitive lining (the retina) on the rear surface of the eyeball. The eyelid might be thought of as a shutter; however, the shutter of a camera, which controls the exposure time, is generally opened only for a fraction of a second, while the eyelid is normally open for continuous exposure. The human nervous system actually performs a function analogous to a shutter by analyzing image signals from the eye at a rate of 20 to 30 times per second.

Note: Image formation by a converging lens is discussed in Section 23.3; see Figure 23.15a.

Demonstration/activity: Remove the back from a simple camera, and place a piece of Scotch (Magic Plus) tape in the focal plane. Then fix the shutter open and pass the camera around to the students to view a large lighted object in the front of the room. An image of the object will appear on the tape.

◀ **FIGURE 25.1 Camera and eye analogy** In some respects, **(a)** a simple camera is similar to **(b)** the human eye. An image is formed on the film in a camera and on the retina of the eye. (The complex refractive properties of the eye are not shown here, because multiple refractive media are involved.) See text for a comparative description.

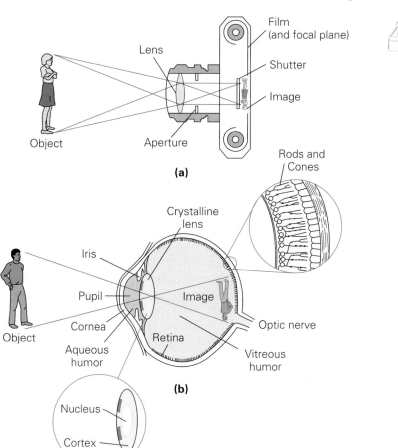

Illustration 36.2 *Camera*

Exploration 36.1 *Camera*

Demonstration/activity: Use a model of the human eye or mount a single lens and screen on an optical bench.

The eye might therefore be better likened to a movie or video camera, which exposes a similar number of frames (images) per second.

Although the optical functions of the eye are relatively simple, its physiological functions are quite complex. As Fig. 25.1b shows, the eyeball is a nearly spherical chamber. It has an internal diameter of about 1.5 cm and is filled with a transparent jellylike substance called the *vitreous humor*. The eyeball has a white outer covering called the *sclera*, part of which is visible as the "white" of the eye. Light enters the eye through a curved, transparent tissue called the *cornea* and passes into a clear fluid known as the *aqueous humor*. Behind the cornea is a circular diaphragm, the *iris*, whose central opening is the *pupil*. The iris contains the pigment that determines eye color. Through muscle action, the area of the pupil can change (from 2 to 8 mm in diameter), thereby controlling the amount of light entering the eye.

Behind the iris is a *crystalline lens*, a converging lens composed of microscopic glassy fibers. (See Conceptual Example 22.5 on page 713 about the internal elements, the nucleous and cortex, inside the crystalline lens.) When tension is exerted on the lens by attached muscles, the glassy fibers slide over each other, causing the shape, and therefore focal length, of the lens to change, to help in focusing the image on the retina properly. Notice that this is an *inverted* image (Fig. 25.1b). We do not see an inverted image, however, because the brain reinterprets this image as being right-side-up.

On the back interior wall of the eyeball is a light-sensitive surface called the **retina**. From the retina, the optic nerve relays retinal signals to the brain. The retina is composed of nerves and two types of light receptors, or photosensitive cells, called **rods** and **cones**, because of their shapes. The rods are more sensitive to light than are the cones and distinguish light from dark in low light intensities (twilight vision). The cones can distinguish frequency ranges but require brighter light. The brain interprets these different frequencies as colors (color vision). Most of the cones are clustered in a central region of the retina called the *macula*. The rods, which are more numerous than the cones, are outside this region and are distributed nonuniformly over the retina.

The focusing adjustment of the eye differs from that of a simple camera. A camera lens has a constant focal length, and the image distance is varied by moving the lens relative to the film to produce sharp images for different object distances. In the eye, the image distance is constant, and the focal length of the lens is varied (as the attached muscles change the lens's shape) to produce sharp images on the retina, regardless of object distance. When the eye is focused on distant objects, the muscles are relaxed, and the crystalline lens is thinnest with a power of about 20 D (diopters). Recall from Chapter 23 that the power (P) of a lens in diopters (D), which is the reciprocal of its focal length *in meters*. So 20 D corresponds to a focal length of $f = 1/(20\,\text{D}) = 0.050\,\text{m} = 5.0\,\text{cm}$. When the eye is focused on closer objects, the lens becomes thicker. Then its radius of curvature and hence its focal length are decreased. For close-up vision, the lens power may increase to 30 D ($f = 0.033\,\text{m}$), or even more in young children. The adjustment of the focal length of the crystalline lens is called *accommodation*. (Look at a nearby object and then at an object in the distance, and notice how fast accommodation takes place. It's practically instantaneous.)

The distance extremes over which sharp focus is possible are known as the *far point* and the *near point*. The *far point* is the greatest distance at which the eye can see objects clearly and is infinity for a normal eye. The *near point* is the position closest to the eye at which objects can be seen clearly. This position depends on the extent to which the lens can be deformed (thickened) by accommodation. The range of accommodation gradually diminishes with age as the crystalline lens loses its elasticity. Generally, in the normal eye the near point gradually recedes with age. The approximate positions of the near point at various ages are listed in Table 25.1.

Children can see sharp images of objects that are within 10 cm of their eyes, and the crystalline lens of a normal young-adult eye can do the same for objects as close as 12 to 15 cm. However, adults at about age 40 normally experience a shift in the near point to beyond 25 cm. You may have noticed middle-aged people holding reading material fairly far from their eyes so as to move it out to be within the range of accommodation. When the print becomes too small (or the arms too short), corrective reading glasses are one solution. The recession of the near point with age is not considered an abnormal defect. Since it proceeds at about the same rate in most normal eyes, it is considered mainly a part of the normal "aging" process.

TABLE 25.1

Approximate Near Points of the Normal Eye at Different Ages

Age (years)	Near Point (centimeters)
10	10
20	12
30	15
40	25
50	40
60	100

Note: The relationship between lens power in diopters and focal length is presented in Eq. 23.9, in Section 23.4.

Note: The eye sees clearly between its far point and near point.

Demonstration/activity: Have students find their near point by moving this book closer to their eyes and then measuring the distance at which the text becomes out of focus.

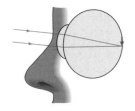
(a) Normal

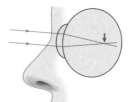
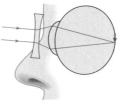
Uncorrected   *Corrected*
(b) Nearsightedness (myopia)

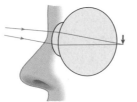
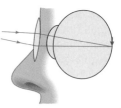
Uncorrected   *Corrected*
(c) Farsightedness (hyperopia)

Vision Defects

The existence of a "normal" eye (▲Fig. 25.2a) implies that some eyes must have defects. This is indeed the case, as is quite apparent from the number of people who wear corrective glasses or contact lenses. Many people have eyes that cannot accommodate within the normal range (25 cm to infinity). These people usually have one of the two most common visual defects: nearsightedness (myopia) or farsightedness (hyperopia). Both of these conditions can usually be corrected with glasses, contact lenses, or surgery.

Nearsightedness (or *myopia*) is the ability to see nearby objects clearly, but not distant objects. That is, the far point is less than infinity. When an object is beyond the far point, the rays focus in *front* of the retina (Fig. 25.2b). As a result, the image on the retina is blurred, or out of focus. As the object is moved closer, its image moves back toward the retina. When the object reaches the far point for that eye, a sharp image is formed on the retina.

Nearsightedness usually arises because the eyeball is too long or the curvature of the cornea is too great. Whatever the reason, the eyeball overconverges the light from distant objects to a spot in front of the retina. Appropriate diverging lenses correct this condition. Such a lens causes the rays to diverge before reaching the cornea. The eye thus focuses the image farther back on the retina.

Farsightedness (or *hyperopia*) is the ability to see distant objects clearly, but not nearby ones. That is, the near point is farther from the eye than normal. The image of an object that is closer than the near point is formed behind the retina (Fig. 25.2c). Farsightedness arises because the eyeball is too short, because of insufficient curvature of the cornea or because of insufficient elasticity of the crystalline lens. If this occurs as part of the aging process as previously discussed, it is called *presbyopia*.

Farsightedness is usually corrected with appropriate converging lenses. Such a lens causes the rays to converge on the retina, and the eye is then able to focus the image on the retina. Converging lenses are also used in middle-aged people to correct presbyopia, a vision condition in which the crystalline lens of the eye loses its flexibility, which makes it difficult to focus on close objects.

▲ **FIGURE 25.2** Nearsightedness and farsightedness
(a) The normal eye produces sharp images on the retina for objects located between its near point and its far point. The image is real, inverted, and always smaller than the object. (Why?) Here, the object is a distant, upward-pointing arrow (not shown) and the light rays come from its tip. **(b)** In a nearsighted eye, the image of a *distant* object is focused *in front of* the retina. This defect is corrected with a diverging lens. **(c)** In a farsighted eye, the image of a *nearby* object is focused *behind* the retina. This defect is corrected with a converging lens. (Not drawn to scale.)

PHYSLET®

Illustration 36.1 The Human Eye

Teaching tip: Explain the effects of aging on the eye.

Integrated Example 25.1 ■ Correcting Nearsightedness: Use of Diverging Lenses

(a) An optometrist has a choice to give a patient either regular glasses or contact lenses to correct nearsightedness (▼Fig. 25.3). Usually, regular glasses sit a few centimeters in front of the eye and contact lenses right on the eye. Should the power of the contact lenses prescribed be (1) the same as, (2) greater than, or (3) less than that of the regular glasses? Explain. (b) A certain nearsighted person cannot see objects clearly when they are more than 78.0 cm from either eye. What power must corrective lenses have, for both regular glasses and contact lenses, if this person is to see distant objects clearly? Assume that the glasses are 3.00 cm in front of the eye.

(a) Conceptual Reasoning. For nearsightedness, the corrective lens is a diverging one (Fig. 25.3). The lens must effectively put the image of a distant object ($d_o = \infty$) at the far point of the eye, that is, d_f from the eye. The image, which acts as an object for the eye, is then within the range of accommodation. Because the image distance is *measured from the lens*, a contact lens will have a *longer* image distance. For a contact lens, $d_i = -(d_f)$. For regular glasses, $d_i = -|d_f - d|$, where d is the distance between the regular glasses and the eye. A minus sign and absolute values are used for the image distance because the image is virtual, being on the object side of the lens. (You may recall from Chapter 23 that diverging lenses can form only virtual images.)

Note: Review Example 23.6 and 23.8.

Note: Image formation by a diverging lens is discussed in Section 23.3; see Figure 23.18.

(continues on next page)

▶ **FIGURE 25.3** Correcting nearsightedness A diverging lens is used. See Integrated Example 25.1. Only regular glasses are shown. For contact lenses, the lens is immediately in front of the eye ($d = 0$).

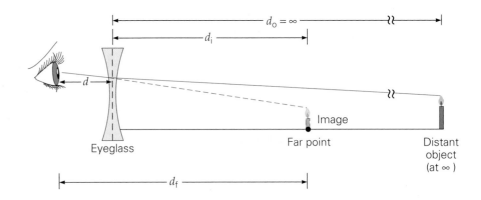

Note that d_i is negative. Recall that the power of a lens is $P = 1/f$ (Eq. 23.9). We can use the thin-lens equation (Eq. 23.5) to find P if we can determine the object and image distances, d_o and d_i:

$$P = \frac{1}{f} = \frac{1}{d_o} + \frac{1}{d_i} = \frac{1}{\infty} + \frac{1}{d_i} = \frac{1}{d_i} = -\frac{1}{|d_i|}$$

That is, a longer $|d_i|$ will yield a smaller P, so the contact lenses should have a lower power than the regular glasses. Thus, the answer is (3).

(b) Quantitative Reasoning and Solution. Once we understand how corrective lenses work, the calculation for part (b) is straightforward.

Given: $d_f = 78$ cm $= 0.780$ m (far point)   *Find:* P (in diopters) for regular glasses
$d = 3.0$ cm $= 0.0300$ m P (in diopters) for contact lenses

For regular glasses,

$$|d_i| = |d_f - d| = 0.780 \text{ m} - 0.0300 \text{ m} = 0.750 \text{ m}$$

(See Fig 25.3, which is not drawn to scale.) So, $d_i = -0.750$ m.

Then, using the thin-lens equation, we get

$$P = \frac{1}{f} = \frac{1}{d_o} + \frac{1}{d_i} = \frac{1}{\infty} + \frac{1}{-0.750 \text{ m}} = -\frac{1}{0.750 \text{ m}} = -1.33 \text{ D}$$

A negative, or diverging, lens with a power of 1.33 D is needed.

For contact lenses,

$$|d_i| = |d_f| = 0.780 \text{ m}$$

($d = 0$.) So, $d_i = -0.78$ m.

Then, using the thin-lens equation, we get

$$P = \frac{1}{\infty} + \frac{1}{-0.780 \text{ m}} = -\frac{1}{0.780 \text{ m}} = -1.28 \text{ D}$$

Follow-Up Exercise. Suppose a mistake was made for regular glasses in this Example such that a "corrective" lens of +1.33 D were used. What happens to the image of objects at infinity? *(Answers to all Follow-Up Exercises are at the back of the text.)*

Demonstration/activity: Sketch ray diagrams illustrating how eye defects cause light to focus for nearsighted and farsighted people. Also, use the eye model or optical bench to demonstrate those conditions.

If the far point for a nearsighted person is changed using diverging lenses (see Integrated Example 25.1), the near point will be affected as well. This causes the close-up vision to worsen, but *bifocal lenses* can be used in this situation to address the problem. Bifocals were invented by Benjamin Franklin, who glued two lenses together. They are now made by grinding or molding lenses with different curvatures in two different regions. Both nearsightedness and farsightedness can be treated at the same time with bifocals. Trifocals are also available, with lenses having three different curvatures. The top lens is for far vision and the bottom lens for near vision. The middle lens is for intermediate vision and is sometimes referred to as a lens for "computer" vision.

More modern techniques involve contact lens therapy or the use of a laser to correct nearsightedness. These are discussed in detail in Insight 25.1, Cornea "Orthodontics" and Surgery, on accompanying page. The purpose of either technique is to change the shape the exposed surface of the cornea, which changes its refractive characteristics. The result, for the nearsighted case, is to make the image of a distant object fall on the retina.

INSIGHT 25.1 CORNEA "ORTHODONTICS" AND SURGERY

The imperfect shape of the cornea of the human eye often causes refractive errors that result in vision defects. For example, a cornea that is curved too much can cause nearsightedness; a flatter-than-normal cornea can cause farsightedness. A cornea that is not spherical can cause astigmatism (Section 23.4).

Recently, a nonsurgical contact lens treatment to improve vision in a matter of hours was developed. This procedure, called *orthokeratology*, or *Ortho-K*, is achieved in a unique way, with the wearing of custom-designed contact lenses. These contact lenses slowly change the shape of the cornea by means of gentle pressure to improve vision safely and quickly. The best analogy to describe Ortho-K is by calling it "orthodontics for the eye."

Laser surgery is also used to reshape the cornea. The surgical procedure corrects the defective shape or irregular surface of the cornea so that it can better focus light on the retina, thereby reducing or even eliminating vision defects (Fig. 1).

In corneal laser surgery, first, a very precise instrument called a *microkeratome* is used to create a thin corneal flap with a hinge on one side of the cornea (Fig. 2a). Once the flap is folded back, a tightly focused ultraviolet pulsed laser is used to reshape the cornea. Each laser pulse accurately removes a microscopic layer of the inner cornea in the targeted area, thus reshaping the cornea to correct vision defects (Fig. 2b). The flap is then placed back in its original position without the need for stitches (Fig. 2c). The procedure is usually painless, and patients typically have only minimal discomfort. Some patients achieve corrected vision within a day after this procedure.

Even more exciting advances in vision correction are on the horizon. For example, researchers have developed techniques for replacing a damaged cornea with freshly bioengineered tissues. If the patient has one healthy eye, stem cells are harvested from it. The cells grow into a sturdy layer of tissue that surgeons can use to replace the bad corneal tissues of the other eye by stitching the new tissue onto that damaged eye. If both of the patient's eyes are damaged, donor tissues may be collected from a close relative.

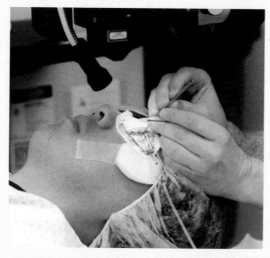

FIGURE 1 Eye surgery Laser surgery is performed to reshape the cornea. Notice that the surgeon is wearing no latex gloves. The fine chalk dust used as a lubricant on the gloves could contaminate the eye.

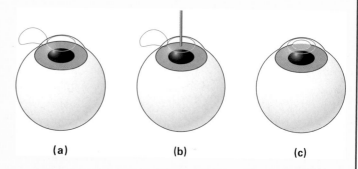

(a)   **(b)**   **(c)**

FIGURE 2 Cornea reshaping **(a)** A flap is made on the corneal surface. **(b)** A laser beam is used to reshape the cornea. **(c)** The flap is placed back.

Integrated Example 25.2 ■ Correcting Farsightedness: Use of a Converging Lens

A farsighted person has a near point of 75 cm for the left eye and a near point of 100 cm for the right one. (a) If the person is prescribed contact lenses, the power of the left lens should be (1) greater than, (2) the same as, (3) less than the power of the right lens. Explain. (b) What powers should contact lenses have to allow the person to see an object clearly at a distance of 25 cm?

(a) Conceptual Reasoning. The normal eye's near point is 25 cm. For farsightedness, the corrective lens must be converging and form the image at its eye's near point of an object at the normal eye's near point. Since the near point of the left eye (75 cm) is closer to the 25-cm normal position than the right eye, the left lens should have less power so the answer is (3).

(b) Quantitative Reasoning and Solution. Let us label the two different eyes as L (left) and R (right). The image distances are negative. (Why?)

Given: $d_{i_L} = -75 \text{ cm} = -0.75 \text{ m}$ *Find:* P_L and P_R (lens power for each eye)
$d_{i_R} = -100 \text{ cm} = -1.0 \text{ m}$
$d_o = 25 \text{ cm} = 0.25 \text{ m}$

(continues on next page)

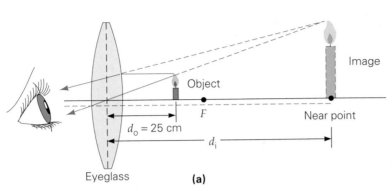

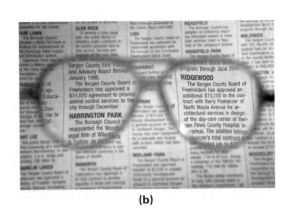

Image

Object

F

Near point

$d_o = 25$ cm

d_i

Eyeglass

(a)

(b)

▲ **FIGURE 25.4** Reading glasses and correcting farsightedness
(a) When an object at the normal near point (25 cm) is viewed through reading glasses with converging lenses, the image is formed farther away, but within the eye's range of accommodation (beyond the receded near point). See Integrated Example 25.2.
(b) Small print as viewed through the lens of reading glasses. The camera used to take this picture is focused past this page onto where the virtual image is.

The optics of these two eyes are usually different (as is typical), as in this problem, and a different lens prescription is usually required for each eye. In this case, each lens is to form an image at its eye's near point of an object that is at a distance (d_o) of 0.25 m. The image will then act as an object within the eye's range of accommodation. This situation is similar to a person wearing reading glasses (▲ Fig. 25.4). (For the sake of clarity, the lens in Fig. 25.4a is not in contact with the eye.)

The image distances are negative, because the images are virtual (that is, the image is on the same side as the object). With contact lenses, the distance from the eye to the object and the distance from the lens to the object are the same. Then

$$P_L = \frac{1}{f_L} = \frac{1}{d_o} + \frac{1}{d_{i_L}} = \frac{1}{0.25 \text{ m}} - \frac{1}{0.75 \text{ m}} = \frac{2}{0.75 \text{ m}} = +2.7 \text{ D}$$

and

$$P_R = \frac{1}{f_R} = \frac{1}{d_o} + \frac{1}{d_{i_R}} = \frac{1}{0.25 \text{ m}} - \frac{1}{1.0 \text{ m}} = \frac{3}{1.0 \text{ m}} = +3.0 \text{ D}$$

Note that the left lens has less power than the right lens, as expected.

Follow-Up Exercise. A mistake is made in grinding or molding the corrective lenses in this Example such that the left lens is made to the prescription intended for the right eye, and vice versa. Discuss what happens to the images of an object at a distance of 25 cm.

Another common defect of vision is **astigmatism**, which is usually due to a refractive surface, normally the cornea or crystalline lens, being out of round (nonspherical). As a result, the eye has different focal lengths in different planes (▼Fig. 25.5a). Points may appear as lines, and the image of a line may be distinct in one direction and blurred in another or blurred in both directions. A test for astigmatism is given in Fig. 25.5b.

▼ **FIGURE 25.5** Astigmatism When one of the eye's refracting components is not spherical, the eye has different focal lengths in different planes. **(a)** The effect occurs because rays in the vertical plane (red) and horizontal plane (blue) are focused at different points: F_v and F_h, respectively. **(b)** To someone with eyes that are astigmatic, some or all of the lines in this diagram will appear blurred. **(c)** Nonspherical lenses, such as plano-convex cylindrical lenses, are used to correct astigmatism.

Astigmatism can be corrected with lenses that have greater curvature in the plane in which the cornea or crystalline lens has deficient curvature (Fig. 25.5c). Astigmatism is lessened in bright light, because the pupil of the eye becomes smaller, so only rays near the axis are entering the eye, thus avoiding the outer edges of the cornea.

You have probably heard of *20/20 vision*. But what is it? *Visual acuity* is a measure of how vision is affected by object distance. This quantity is commonly determined by using a chart of letters placed at a given distance from the eyes. The result is usually expressed as a fraction: The *numerator* is the distance at which the test eye sees a standard symbol, such as the letter E, clearly; the *denominator* is the distance at which

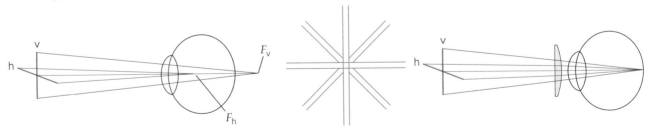

(a) Uncorrected astigmatism   **(b) Test for astigmatism**   **(c) Corrected by lens**

(a) Narrow angle   **(b) Wider angle**

◄ **FIGURE 25.6 Magnification and angle** **(a)** How large an object appears is related to the angle subtended by the object. **(b)** The angle and the size of the virtual image of an object are increased with a converging lens.

the letter is seen clearly by a *normal* eye. A 20/20 (test/normal) rating, which is sometimes called "perfect" vision, means that at a distance of 20 ft, the eye being tested can see standard-sized letters as clearly as can a normal eye.

Teaching tip: Use an eye chart to explain 20/20 vision, 20/30 vision, and so on.

25.2 Microscopes

OBJECTIVES: To (a) distinguish between lateral and angular magnification, and (b) describe simple and compound microscopes and their magnifications.

Microscopes are used to magnify objects so that we can see more detail or see features that are normally indiscernible. Two basic types of microscopes will be considered here.

The Magnifying Glass (A Simple Microscope)

When we look at a distant object, it appears very small. As it moves closer to our eyes, it appears larger. How large an object appears depends on the size of the image on the retina. This size is related to the angle subtended by the object (▲Fig. 25.6): the greater the angle, the larger the image.

When we want to examine detail or look at something closely, we bring it close to our eyes so that it subtends a greater angle. For example, you may examine the detail of a figure in this book by bringing it closer to your eyes. You'll see the greatest amount of detail when the book is at your near point. If your eyes were able to accommodate to shorter distances, an object brought very close to them would appear even larger. However, as you can easily prove by bringing this book very close to your eyes, images are blurred when objects are inside the near point.

A **magnifying glass**, which is simply a single convex lens (sometimes called a *simple microscope*), forms a clear image of an object when it is closer than the near point (Fig. 23.15b). In such a position, the image of an object subtends a greater angle and therefore appears larger, or magnified (▼Fig. 25.7). The lens produces a virtual image beyond the near point on which the eye focuses. If a handheld magnifying glass is used, its position is usually adjusted until this image is seen clearly.

As illustrated in Fig. 25.7, the angle subtended by the virtual image of an object is much greater when a magnifying glass is used. The magnification of an object *viewed through a magnifying glass* is expressed in terms of this angle. This **angular magnification**, or *magnifying power*, is designated by the symbol m. The angular magnification is defined as the ratio of the angular size of the object as viewed

▼ **FIGURE 25.7 Angular magnification** The angular magnification (m) of a lens is defined as the ratio of the angular size of an object viewed through the lens to the angular size of the object viewed without the lens: $m = \theta/\theta_o$.

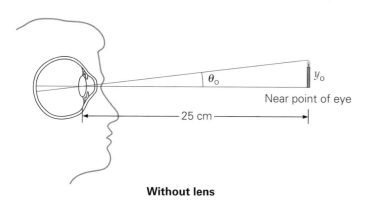

Without lens

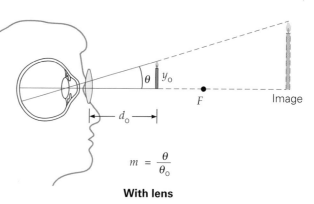

$$m = \frac{\theta}{\theta_o}$$

With lens

Note: Angular magnification is not the same as lateral magnification, which is discussed in Section 23.1. (See Eq. 23.1.)

through the magnifying glass (θ) to the angular size of the object as viewed without the magnifying glass (θ_o):

$$m = \frac{\theta}{\theta_o} \quad \text{angular magnification} \tag{25.1}$$

(This m is not the same as M, the lateral magnification, which is a ratio of heights: $M = h_i/h_o$.)

The maximum angular magnification occurs when the image is at the eye's near point, $d_i = -25$ cm, since this position is as close as it can be seen clearly. (A value of 25 cm will be assumed to be typical for the near point of the normal eye. The minus sign is used because the image is virtual; see Chapter 23.) The corresponding object distance can be calculated from the thin-lens equation, Eq. 23.5, as

$$d_o = \frac{d_i f}{d_i - f} = \frac{(-25 \text{ cm})f}{-25 \text{ cm} - f}$$

or

$$d_o = \frac{(25 \text{ cm})f}{25 \text{ cm} + f} \tag{25.2}$$

where f must be in centimeters.

The angular sizes of the object are related to its height by

$$\tan \theta_o = \frac{y_o}{25} \quad \text{and} \quad \tan \theta = \frac{y_o}{d_o}$$

(See Fig. 25.7.) Assuming that a small-angle approximation ($\tan \theta \approx \theta$) is valid,

$$\theta_o \approx \frac{y_o}{25} \quad \text{and} \quad \theta \approx \frac{y_o}{d_o}$$

Then the maximum angular magnification can be expressed as

$$m = \frac{\theta}{\theta_o} = \frac{y_o/d_o}{y_o/25} = \frac{25}{d_o}$$

Substituting for d_o from Eq. 25.2 gives

$$m = \frac{25}{25f/(25 + f)}$$

which simplifies to

$$m = 1 + \frac{25 \text{ cm}}{f} \quad \begin{array}{l} \text{angular magnification for} \\ \text{image at near point (25 cm)} \end{array} \tag{25.3}$$

where f is in centimeters. Lenses with shorter focal lengths give greater angular magnifications.

In the derivation of Eq. 25.3, the object being viewed by the unaided eye was taken to be at the near point, as was the image viewed through the lens. Actually, the normal eye can focus on an image located anywhere between the near point and infinity. At the extreme at which the image is at infinity, the eye is more relaxed—the muscles attached to the crystalline lens are relaxed, and the lens is thin. For the image to be at infinity, the object must be at the focal point of the lens. In this case,

$$\theta \approx \frac{y_o}{f}$$

and the angular magnification is

$$m = \frac{25 \text{ cm}}{f} \quad \begin{array}{l} \text{angular magnification} \\ \text{for image at infinity} \end{array} \tag{25.4}$$

Mathematically, it seems that the magnifying power can be increased to any desired value by using lenses that have sufficiently short focal lengths. Physically, however, lens aberrations limit the practical range of a single magnifying glass to about 3× or 4×, or a sharp image magnification of three or four times the size of the object when used normally.

Example 25.3 ■ Elementary: Angular Magnification of a Magnifying Glass

Sherlock Holmes uses a converging lens with a focal length of 12 cm to examine the fine detail of some cloth fibers found at the scene of a crime. (a) What is the maximum magnification given by the lens? (b) What is the magnification for relaxed-eye viewing?

Thinking It Through. Equations 25.3 and 25.4 apply here. Part (a) asks for the maximum magnification, which is discussed in the derivation of Eq. 25.3 and occurs when the image formed by the lens is at the near point of the eye. For part (b), note that the eye is most relaxed when viewing distant objects.

Solution.

Given: $f = 12$ cm   *Find:* (a) m (d_i = near point)
                                          (b) m ($d_i = \infty$)

(a) For Equation 25.3, the near point was taken to be 25 cm:

$$m = 1 + \frac{25 \text{ cm}}{f} = 1 + \frac{25 \text{ cm}}{12 \text{ cm}} = 3.1\times$$

(b) Equation 25.4 gives the magnification for the image formed by the lens at infinity:

$$m = \frac{25 \text{ cm}}{f} = \frac{25 \text{ cm}}{12 \text{ cm}} = 2.1\times$$

Follow-Up Exercise. Taking the maximum practical magnification of a magnifying glass to be 4×, which would have the longer focal length, a glass for near-point viewing or one for distant viewing, and how much longer?

The Compound Microscope

A compound microscope provides greater magnification than is attained with a single lens, or a simple microscope. A basic **compound microscope** consists of a pair of converging lenses, each of which contributes to the magnification (▼Fig. 25.8a). The converging lens with a relatively short focal length ($f_o < 1$ cm) is known as the **objective**. It produces a real, inverted, and enlarged image of an object positioned slightly beyond its focal point. The other lens, called the **eyepiece**, or **ocular**, has a longer focal length (f_e is a few centimeters) and is positioned so that the image formed by the objective falls just *inside* its focal point. This lens forms a magnified, inverted, and virtual image that is viewed by the observer. In essence, the objective gives a magnified real image, and the eyepiece is a simple magnifying glass.

The **total magnification (m_{total})** of a lens combination is the *product* of the magnifications produced by the two lenses. The image formed by the objective is larger than its object by a factor M_o equal to the lateral magnification ($M_o = -d_i/d_o$). In Fig. 25.8a, note that the image distance for the objective lens is approximately equal to L, the distance between the lenses—that is, $d_i \approx L$. (The image I_o is formed by the

▼ **FIGURE 25.8 The compound microscope** **(a)** In the optical system of a compound microscope, the real image formed by the objective falls just within the focal point of the eyepiece (F_e) and acts as an object for this lens. An observer looking through the eyepiece sees an enlarged image. **(b)** A compound microscope.

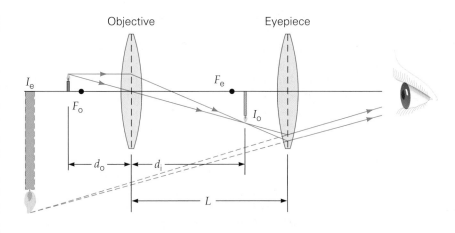

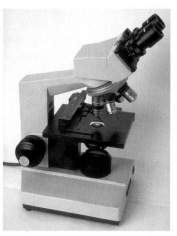

(a)   (b)

objective just inside the focal point of the eyepiece.) Also, because the object is very close to the focal point of the objective, $d_o \approx f_o$. With these approximations,

$$M_o \approx -\frac{L}{f_o}$$

Equation 25.4 gives the angular magnification of the eyepiece for an image at infinity.

$$m_e = \frac{25 \text{ cm}}{f_e}$$

Since the object for the eyepiece (the image formed by the objective) is very near the focal point of the eyepiece, a good approximation is given by

$$m_{\text{total}} = M_o m_e = -\left(\frac{L}{f_o}\right)\left(\frac{25 \text{ cm}}{f_e}\right)$$

or

$$m_{\text{total}} = -\frac{(25 \text{ cm})L}{f_o f_e} \qquad \begin{array}{l}\textit{angular magnification}\\ \textit{of compound microscope}\end{array} \tag{25.5}$$

where f_o, f_e, and L are in centimeters.

The angular magnification of a compound microscope is negative, indicating that the final image is inverted compared to the initial orientation of the object. However, we often state only the magnification ($100\times$, not $-100\times$).

Example 25.4 ■ A Compound Microscope: Finding the Magnification

A microscope has an objective with a focal length of 10 mm and an eyepiece with a focal length of 4.0 cm. The lenses are positioned 20 cm apart in the barrel. Determine the approximate total magnification of the microscope.

Thinking It Through. This is a direct application of Eq. 25.5.

Solution.

Given: $f_o = 10 \text{ mm} = 1.0 \text{ cm}$ *Find:* m_{total} (total magnification)
         $f_e = 4.0 \text{ cm}$
         $L = 20 \text{ cm}$

Using Eq. 25.5, we get

$$m_{\text{total}} = -\frac{(25 \text{ cm})L}{f_o f_e} = -\frac{(25 \text{ cm})(20 \text{ cm})}{(1.0 \text{ cm})(4.0 \text{ cm})} = -125\times$$

Note the relatively short focal length of the objective. The negative sign indicates that the final image is inverted.

Follow-Up Exercise. If the focal length of the eyepiece in this Example were doubled, how would the length of the microscope be affected for the same magnification? (Express the change as a percentage.)

A modern compound microscope is shown in Fig. 25.8b. Interchangeable eyepieces with magnifications from about $5\times$ to more than $100\times$ are available. For standard microscopic work in biology or medical laboratories, $5\times$ and $10\times$ eyepieces are normally used. Microscopes are often equipped with rotating turrets, which usually contain three objectives for different magnifications, such as $10\times$, $43\times$, and $97\times$. These objectives and the $5\times$ and $10\times$ eyepieces can be used in various combinations to provide magnifying powers from $50\times$ to $970\times$. The maximum magnification obtained from a compound microscope is about $2000\times$.

Opaque objects are usually illuminated with a light source placed above them. Specimens that are transparent, such as cells or thin sections of tissues on glass slides, are illuminated with a light source beneath the microscope stage so that light passes through the specimen. A modern microscope is usually equipped with a light condenser (converging lens) and diaphragm below the stage, which are used to concentrate the light and control its intensity. A microscope may have an internal light source. The light is reflected into the condenser from a mirror. Older microscopes have two mirrors with reflecting surfaces; one is a plane mirror for reflecting light from a high-intensity external source, and the other is a concave mirror for converging low-intensity light such as skylight.

25.3 Telescopes

<u>OBJECTIVES:</u> To (a) distinguish between refractive and reflective telescopes, and (b) describe the advantages of each.

Telescopes apply the optical principles of mirrors and lenses to improve our ability to see distant objects. Used for both terrestrial and astronomical observations, telescopes allow some objects to be viewed in greater detail and other fainter or more distant objects simply to be seen. Basically, there are two types of telescopes—refracting and reflecting—characterized by the gathering and converging of light by lenses or mirrors, respectively.

Refracting Telescope

The principle underlying one type of **refracting telescope** is similar to that of a compound microscope. The major components of a refracting telescope are objective and eyepiece lenses, as illustrated in ▼Fig. 25.9. The objective is a large converging lens with a long focal length, and the movable eyepiece has a relatively short focal length. Rays from a distant object are essentially parallel and form an image (I_o) at the focal point (F_o) of the objective. This image acts as an object for the eyepiece, which is moved until the image lies just inside its focal point (F_e). A large, inverted, virtual image (I_e) is seen by an observer.

For relaxed viewing, the eyepiece is adjusted so that its image (I_e) is at infinity, which means that the objective image (I_o) is at the focal point of the eyepiece (f_e). As Fig. 25.9 shows, the distance between the lenses is then the sum of the focal lengths ($f_o + f_e$), which is the length of the telescope tube. The **magnification of a refracting telescope** focused for the final image at infinity can be shown to be

$$m = -\frac{f_o}{f_e} \quad \begin{array}{l} \textit{angular magnification} \\ \textit{of refracting telescope} \end{array} \tag{25.6}$$

where the minus sign is inserted to indicate that the image is inverted, as in our lens sign convention described in Section 23.3. Thus, to achieve the greatest magnification, the focal length of the objective should be made as long as possible and the focal length of the eyepiece as short as possible.

The telescope illustrated in Fig. 25.9 is called an **astronomical telescope**. The final image produced by an astronomical telescope is inverted, but this condition poses little

PHYSLET®

Exploration 36.2 *Telescope*

Note: Astronomical telescopes give an inverted image.

◀ **FIGURE 25.9 The refracting astronomical telescope** In an astronomical telescope, rays from a distant object form an intermediate image (I_o) at the focal point of the objective (F_o). The eyepiece is moved so that the image is at or slightly inside its focal point (F_e). An observer sees an enlarged image at infinity (I_e, shown at a finite distance here for illustration).

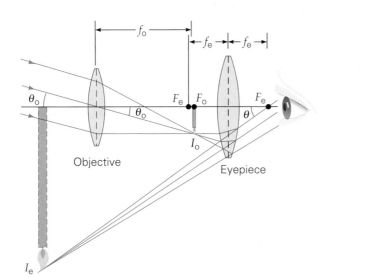

▶ **FIGURE 25.10** Terrestrial telescopes **(a)** A Galilean telescope uses a diverging lens as an eyepiece, producing upright, virtual images. **(b)** Another way to produce upright images is to use a converging "erecting" lens (focal length f_i) between the objective and eyepiece in an astronomical telescope. This addition elongates the telescope, but the length can be shortened by using internally reflecting prisms.

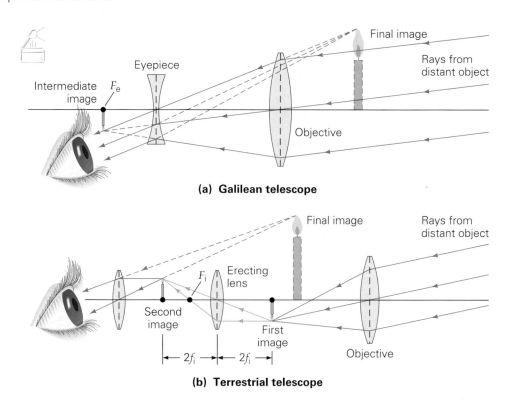

(a) Galilean telescope

(b) Terrestrial telescope

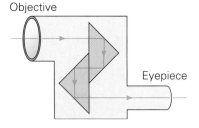

▲ **FIGURE 25.11** Prism binoculars A schematic cutaway view of one ocular (one half of a pair of prism binoculars), showing the internal reflections in the prisms, which reduce the overall physical length.

problem to astronomers. (Why?) However, someone viewing an object on Earth through a telescope finds it more convenient to have an upright image. A telescope in which the final image is upright is called a **terrestrial telescope**. An upright final image can be obtained in several ways; two are illustrated in ▲Fig. 25.10.

In the telescope diagrammed in Fig. 25.10a, a diverging lens is used as an eyepiece. This type of terrestrial telescope is referred to as a *Galilean telescope*, because Galileo built one in 1609. A real image is formed by the objective to the left of the eyepiece, and this image acts as a "virtual" object for the eyepiece. (See Section 23.3.) An observer sees a magnified, upright, virtual image. (Note that with a diverging lens and negative focal length, Eq. 25.6 gives a positive m, indicating an upright image.)

Galilean telescopes have several disadvantages, most notably very narrow fields of view and limited magnification. A better type of terrestrial telescope, illustrated in Fig. 25.10b, uses a third lens, called the *erecting lens*, or *inverting lens*, between converging objective and eyepiece lenses. If the image is formed by the objective at a distance that is twice the focal length of the intermediate erecting lens ($2f_i$), then the lens merely inverts the image without magnification, and the telescope magnification is still given by Eq. 25.6.

However, to achieve the upright image in this way requires a greater telescope length. Using the intermediate erecting lens to invert the image increases the length of the telescope by four times the focal length of the erecting lens ($2f_i$ on each side). The inconvenient length can be avoided by using internally reflecting prisms. This is the principle behind prism binoculars, which are really double telescopes—one for each eye (◀Fig. 25.11).

Example 25.5 ■ An Astronomical Telescope—and a Longer Terrestrial Telescope

An astronomical telescope has an objective lens with a focal length of 30 cm and an eyepiece with a focal length of 9.0 cm. (a) What is the magnification of the telescope? (b) If an erecting lens with a focal length of 7.5 cm is used to convert the telescope to a terrestrial type, what is the overall length of the telescope tube?

Thinking It Through. Equation 25.6 applies directly in part (a). In part (b), the erecting lens elongates the telescope by four times the focal length of the lens ($4f_i$) to the length of the scope (Fig. 25.10b).

Solution. Listing the data, we have

Given: $f_o = 30$ cm   *Find:* (a) *m* (magnification)
$\qquad$ $f_e = 9.0$ cm $\qquad\qquad\qquad\qquad$ (b) *L* (length of telescope tube)
$\qquad$ $f_i = 7.5$ cm (intermediate erecting lens)

(a) The magnification is given by Eq. 25.6 as

$$m = -\frac{f_o}{f_e} = -\frac{30\text{ cm}}{9.0\text{ cm}} = -3.3\times$$

where the minus sign indicates that the final image is inverted.

(b) Taking the length of the astronomical tube to be the distance between the lenses, we find that this length is just the sum of the lenses' focal lengths:

$$L_1 = f_o + f_e = 30\text{ cm} + 9.0\text{ cm} = 39\text{ cm}$$

The overall length is then

$$L = L_1 + L_2 = 39\text{ cm} + 4f_i = 39\text{ cm} + 4(7.5\text{ cm}) = 69\text{ cm}$$

Hence, the telescope length is more than two-thirds of a meter, with an upright image, but the same magnification, 3.3× (why?).

Follow-Up Exercise. A terrestrial telescope 66 cm in length has an intermediate erecting lens with a focal length of 12 cm. What is the focal length of an erecting lens that would reduce the telescope length to a more manageable 50 cm?

Conceptual Example 25.6 ■ Constructing a Telescope

A student is given two converging lenses, one with a focal length of 5.0 cm and the other with a focal length of 20 cm. To construct a telescope to best view distant objects with these lenses, the student should hold the lenses (a) more than 25 cm apart; (b) less than 25 cm, but more than 20 cm, apart; (c) less than 20 cm, but more than 5.0 cm, apart; (d) less than 5.0 cm apart. Specify which lens should be used as the eyepiece.

Reasoning and Answer. First let's see which lens should be used as the eyepiece. The only type of telescope that can be constructed with two converging lenses is an astronomical telescope. In this type of telescope, the lens with the longer focal length is used as an objective lens to produce a real image of a distant object. That image is then viewed with the lens with the shorter focal length, the eyepiece, used as a simple magnifier.

If the object is at a great distance, a real image is formed by the objective lens in the focal plane of the lens (Fig. 25.9). This image acts as the object for the eyepiece, which is positioned so that the image/object lies just inside its focal point so as to produce a large, inverted second image.

The two lenses must be *slightly* less than 25 cm apart, so answer (a) is not correct. Answers (c) and (d) are also not correct, because the eyepiece would be too close to the objective to produce the large secondary image needed for optimal viewing of a distant object. In these cases, the rays would pass through the second lens before the image was formed, and a *reduced* image might be produced. (See Section 23.3.) Thus, answer (b), with the objective image just inside the eyepiece's focal point, is the correct answer.

Follow-Up Exercise. A third converging lens with a focal length of 4.0 cm is used with the aforementioned two lenses to produce a terrestrial telescope in which the third lens does nothing more than invert the image. How should the lenses be positioned and how far apart should they be for the final image to be of maximum size and upright?

Reflecting Telescope

For viewing the Sun, Moon, and nearby planets, large magnifications are important to see details. However, even with the highest feasible magnification, stars appear only as faint points of light. For distant stars and galaxies, it is more important to gather more light than to increase the magnification, so that the object can be seen and its spectrum analyzed more quickly. The intensity of light from a distant source is sometimes very low. In many instances, such a source can be detected only when the light is gathered and focused on a photographic plate over a long period of time.

▶ **FIGURE 25.12** Reflecting
telescopes A concave mirror can
be used in a telescope to converge
light to form an image of a distant
object. **(a)** The image may be at the
prime focus, or **(b)** a small mirror
and lens can be used to focus the
image outside the telescope, a
configuration called a *Newtonian
focus.*

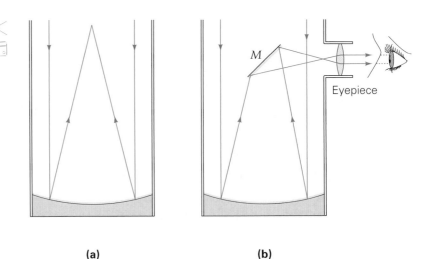

(a)   (b)

Recall from Section 14.3 that intensity is energy per unit time per unit *area*. Thus, more light energy can be gathered if the size of the objective is increased. This increases the distance at which the telescope can detect faint objects, such as distant galaxies. (Recall that the light intensity of a point source is inversely proportional to the *square* of the distance between the source and the observer.) However, producing a large lens involves difficulties associated with glass quality, grinding, and polishing. Compound-lens systems are required to reduce aberrations, and a very large lens may sag under its own weight, producing further aberrations. The largest objective lens in use has a diameter of 40 in. (102 cm) and is part of the refracting telescope of the Yerkes Observatory at Williams Bay, Wisconsin.

These problems can be reduced by using a **reflecting telescope**, which uses a large, concave, parabolic mirror (▲ Fig. 25.12). A parabolic mirror does not exhibit spherical aberration, and a mirror has no inherent chromatic aberration. (Why?) High-quality glass is not needed, because the light is reflected by a mirrored surface. Only one surface has to be ground, polished, and silvered.

The largest single-mirror telescope, with a mirror 8.2 m (323 in.) in diameter, is at the European Southern Observatory in Chile (▼ Fig 25.13a). The largest reflecting telescope in the United States, with a mirror 5.1 m (200 in.) in diameter, is the Hale Observatory's reflecting telescope on Mount Palomar in California.

Even though reflecting telescopes have advantages over refracting telescopes, they do have problems. Like a large lens, a large mirror may sag under its own weight, and the weight necessarily increases with the size of the mirror. The weight factor also increases the costs of construction, because the supporting elements for a heavier mirror must be more massive.

These problems are being addressed by new technologies. One approach is to use an array of small mirrors, coordinated to function as a single large mirror. Examples include

▶ **FIGURE 25.13** European
Southern Observatory, near
Paranal, Chile (a) An 8.2-m-
diameter mirror is undergoing the
final phase of polishing. **(b)** Four
8.2-m telescopes will form a VLT
(Very Large Telescope) with an
equivalent diameter of 16 m.

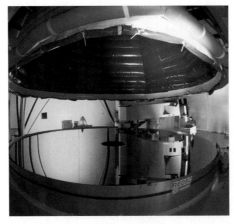

(a)   (b)

the twin Keck telescopes at Mauna Kea in Hawaii. Each has a mirror consisting of 36 hexagonal segments that are computer positioned to be equivalent to a 10-m mirror. The European Southern Observatory plans to have four 8.2-m-diameter mirrors linked to form a VLT (Very Large Telescope,) with an equivalent diameter of 16 m (Fig. 25.13b).

Another way of extending our view into space is to put telescopes into orbit around the Earth. Above the atmosphere, the view is unaffected by the twinkling effect of atmospheric turbulence and refraction, and there is no background problem from city lights. In 1990, the optical Hubble Space Telescope (HST) was launched into orbit (►Fig. 25.14). Even with a mirror diameter of only 2.4 m, its privileged position has allowed the HST to produce images seven times as clear as those formed by Earth-bound telescopes.

Lastly, you may know that not all telescopes are in the visible region. To see some of these, read Insight 25.2 on Telescopes Using Nonvisible Radiation, on page 808.

25.4 Diffraction and Resolution

OBJECTIVES: To (a) describe the relationship of diffraction and resolution, and (b) state and explain Rayleigh's criterion.

The diffraction of light places a limitation on our ability to distinguish objects that are close together when we use microscopes or telescopes. This effect can be understood by considering two point sources located far from a narrow slit of width w (▼Fig. 25.15). The sources could represent distant stars, for example. In the absence of diffraction, two bright spots, or images, would be observed on a screen. As you know from Section 24.3, however, the slit diffracts the light, and each image consists of a central maximum with a pattern of weaker bright and dark fringes on either side. If the sources are close together, the two central maxima may overlap. In this case, the images cannot be distinguished, or are said to be *unresolved*. For the images to be *resolved*, the central maxima must not overlap appreciably.

In general, images of two sources can be resolved if the central maximum of one falls at or beyond the first minimum of the other. This limiting condition for the **resolution** of two images—that is, the ability to distinguish them as separate—was first proposed by Lord Rayleigh (1842–1919), a British physicist. The condition is known as the **Rayleigh criterion**:

> Two images are said to be just resolved when the central maximum of one image falls on the first minimum of the diffraction pattern of the other image.

The Rayleigh criterion can be expressed in terms of the angular separation (θ) of the sources. (See Fig. 25.15.) The first minimum ($m = 1$) for a single-slit diffraction pattern satisfies this relationship:

$$w \sin \theta = m\lambda = \lambda \qquad \text{or} \qquad \sin \theta = \frac{\lambda}{w}$$

According to Fig. 25.15, this is the minimum angular separation for two images to be just resolved according to the Rayleigh criterion. In general, for visible

▲ **FIGURE 25.14** Hubble Space Telescope (HST) Late in 1993, astronauts from the space shuttle *Endeavor* visited the HST in orbit. They installed corrective equipment that compensated for many of the telescope's optical flaws and repaired or replaced other malfunctioning systems. The HST is currently in need of repairs.

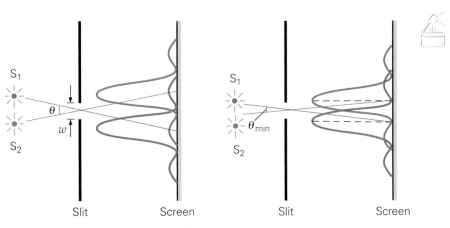

(a) **Resolved**   (b) **Just resolved**

◄ **FIGURE 25.15** Resolution Two light sources in front of a slit produce diffraction patterns. (a) When the angle subtended by the sources at the slit is large enough for the diffraction patterns to be distinguishable, the images are said to be resolved. (b) At smaller angles, the central maxima are closer together. At $\theta_{\min}$, the central maximum of one image's diffraction pattern falls on the first dark fringe of the other image's pattern, and the images are said to be just resolved. For smaller angles, the patterns are unresolved.

INSIGHT 25.2 TELESCOPES USING NONVISIBLE RADIATION

The word *telescope* usually brings to mind visual observations. However, the visible region is only a very small part of the electromagnetic spectrum, and celestial objects emit many other types of radiation, including radio waves. This fact was discovered accidentally in 1931 by an American electrical engineer named Carl Jansky while he was working on static interference with intercontinental radio communications. Jansky found an annoying static hiss that came from a fixed direction in space, apparently from a celestial source. It was soon clear that radio waves could be another valuable source of astronomical information, and radio telescopes were built to investigate this source.

A radio telescope operates similarly to a reflecting visible light telescope. A reflector with a large area collects and focuses the radio waves at a point where a detector picks up the signal (Fig. 1). The parabolic collector, called a *dish*, is covered with metal wire mesh or metal plates. Since the wavelengths of radio waves range from a few millimeters to several meters, wire mesh is "smooth" enough and a good reflecting surface for such waves.

Radio telescopes supplement optical telescopes and provide some definite advantages over them. For instance, radio waves pass freely through the huge clouds of dust that hide a large part of our galaxy from visual observation. Also, radio waves easily penetrate the Earth's atmosphere, which reflects and scatters a large percentage of the incoming visible light.

Infrared light is also affected by the Earth's atmosphere. For example, water vapor is a strong absorber of infrared radiation. Thus, observations with infrared telescopes are sometimes made from high-flying aircraft or from orbiting spacecraft, beyond the influence of atmospheric water vapor. The first *orbiting* infrared observatory was launched in 1983. Not only is atmospheric interference eliminated in space, but the telescope may be cooled to a very low temperature without becoming coated with condensed water vapor. Cooling the telescope helps eliminate the interference of infrared radiation generated by the telescope itself. The orbiting infrared telescope launched

FIGURE 1 Radio telescopes Several of the dish antennae that make up the Very Large Array (VLA) radio telescope near Socorro, New Mexico. There are 27 movable dishes, each 25 m in diameter, forming the array along a Y-shaped railway network. The data from all the antennae are combined to produce a single radio image. In this way, it is possible to attain a resolution equivalent to that of one giant radio dish (a couple hundred feet in diameter).

in 1983 was cooled with liquid helium to about 10 K; it carried out an infrared survey of the entire sky.

The atmosphere is virtually opaque to ultraviolet radiation, X-rays, and gamma rays, so telescopes that detect these types of radiation cannot be Earth based. Orbiting satellites with telescopes sensitive to these types of radiation have mapped out portions of the sky, and other surveys are planned. Observatories by orbiting satellites in the visible region are not affected by air turbulence or refraction. Perhaps in the not-too-distant future, a permanently staffed orbiting observatory carrying a variety of telescopes will replace the uncrewed Hubble Telescope and help expand our knowledge of the universe.

light, the wavelength is much smaller than the slit width ($\lambda < w$), so θ is small and $\sin \theta \approx \theta$. In this case, the limiting, or **minimum angle of resolution (θ_{min})** for a slit of width w is

$$\theta_{min} = \frac{\lambda}{w} \qquad \text{\textit{minimum angle of resolution}} \atop \textit{(for a slit)} \qquad (25.7)$$

(Note that θ_{min} is a pure number and is therefore in radians.) Thus, the images of two sources will be *distinctly* resolved if the angular separation of the sources is greater than λ/w.

The apertures (openings) of cameras, microscopes, and telescopes are generally circular. Thus, there is a *circular* diffraction pattern around the central maximum, in the form of a bright circular disk (▸Fig. 25.16). Detailed analysis shows that the **minimum angle of resolution for a circular aperture** for the images of two objects to be just resolved is similar to, but slightly different from, Eq. 25.7. It is

$$\theta_{min} = \frac{1.22\lambda}{D} \qquad \text{\textit{minimum angle of resolution}} \atop \textit{(for a circular aperture)} \qquad (25.8)$$

where D is the diameter of the aperture and θ_{min} is in radians.

Equation 25.8 applies to the objective lens of a microscope or telescope, or the iris of the eye, all of which may be considered to be circular apertures for light. According to Eqs. 25.7 and 25.8, the smaller θ_{min}, the better the resolution. The minimum angle of resolution, θ_{min}, should be small so that objects close together can be resolved; therefore, the aperture should be as *large* as possible. This is yet another reason for using large lenses (and mirrors) in telescopes.

Example 25.7 ■ The Eye and Telescope: Evaluating Resolution with the Rayleigh Criterion

Determine the minimum angle of resolution by the Rayleigh criterion for (a) the pupil of the eye (daytime diameter of about 4.0 mm) for visible light with a wavelength of 660 nm; (b) the European Southern Observatory refracting telescope (diameter of 8.2 m), for visible light of the same wavelength as in part (a); and (c) a radio telescope 25 m in diameter for radiation with a wavelength of 21 cm.

Thinking It Through. This is a comparison of θ_{min} for apertures with different diameters— a direct application of Eq. 25.8.

Solution.

Given: (a) $D = 4.0 \text{ mm} = 4.0 \times 10^{-3} \text{ m}$ *Find:* (a) θ_{min} (minimum angles
         $\lambda = 660 \text{ nm} = 6.60 \times 10^{-7} \text{ m}$ of resolution)
     (b) $D = 8.2 \text{ m}$ (b) θ_{min}
         $\lambda = 660 \text{ nm} = 6.60 \times 10^{-7} \text{ m}$ (c) θ_{min}
     (c) $D = 25 \text{ m}$
         $\lambda = 21 \text{ cm} = 0.21 \text{ m}$

(a) For the eye,

$$\theta_{min} = \frac{1.22\lambda}{D} = \frac{1.22(6.60 \times 10^{-7} \text{ m})}{4.0 \times 10^{-3} \text{ m}} = 2.0 \times 10^{-4} \text{ rad}$$

(b) For the light telescope,

$$\theta_{min} = \frac{1.22(6.60 \times 10^{-7} \text{ m})}{8.2 \text{ m}} = 9.8 \times 10^{-8} \text{ rad}$$

(*Note*: The resolution of Earth-bound telescopes with large-diameter objectives is usually not limited by diffraction, but rather by other effects, such as atmospheric turbulence. Thus, in actuality, Earth-bound telescopes have a θ_{min} on the order of 10^{-6} rad, or resolution one tenth as good as without the atmosphere.)

(c) For the radio telescope,

$$\theta_{min} = \frac{1.22(0.21 \text{ m})}{25 \text{ m}} = 0.010 \text{ rad}$$

The smaller the angular separation, the better the resolution. What do the results tell you?

Follow-Up Exercise. As noted in Section 25.3, the Hubble Space Telescope has a mirror diameter of 2.4 m. How does its resolution compare with that of the largest Earth-bound telescopes? (See the note in part (b) of this Example.)

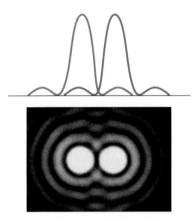
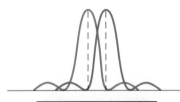

(a)

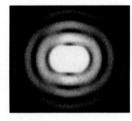

(b)

▲ **FIGURE 25.16** Circular-aperture resolution **(a)** When the angular separation of two objects is large enough, the images are well resolved. (Compare with Fig. 25.15a.) **(b)** Rayleigh criterion: The central maximum of the diffraction pattern of one image falls on the first minimum of the diffraction pattern of the other image. (Compare with Fig. 25.15b.) The images of objects with smaller angular separations cannot be clearly distinguished as individual images.

For a microscope, it is more convenient to specify the actual separation (*s*) between two point sources. Since the objects are usually near the focal point of the objective, to a good approximation,

$$\theta_{min} = \frac{s}{f} \quad \text{or} \quad s = f\theta_{min}$$

where *f* is the focal length of the lens and θ_{min} is expressed in radians. (Here, *s* is taken as the arc length subtended by θ_{min}, and $s = r\theta_{min} = f\theta_{min}$.) Then, using Eq. 25.8, we get

$$s = f\theta_{min} = \frac{1.22\lambda f}{D} \quad \textit{resolving power of a microscope} \quad (25.9)$$

This minimum distance between two points whose images can be just resolved is called the **resolving power** of the microscope. Note that *s* is directly proportional to λ, so shorter wavelength gives better resolution. In practice, the resolving power of a microscope indicates the ability of the objective to distinguish fine detail in specimens' structures. For another real-life example of resolution, see ▼Fig. 25.17.

▶ **FIGURE 25.17** Real-life resolution (a), (b), (c) A sequence of an approaching automobile's headlights. In (a), the headlights are almost unresolved through the circular aperture of the camera (or your eye). As the automobile moves closer, the headlights are resolved.

(a)　　　　　(b)　　　　　(c)

Teaching tip: Have students estimate the diameter of the human iris and determine what features an astronaut in a 100-mi-high orbit could identify in her hometown.

▲ **FIGURE 25.18** The Great Wall The walkway of the Great Wall of China, which was built as a fortification along China's northern border.

Note: The relationship between wavelength and index of refraction is given in Section 22.3; see Eq. 22.4.

Conceptual Example 25.8 ■ Viewing from Space: The Great Wall of China

The Great Wall of China was originally about 2400 km (1500 mi) long, with a base width of about 6.0 m and a top width of about 3.7 m. Several hundred kilometers of the wall remain intact (◀Fig. 25.18). It is sometimes said that the wall is the only human construction that can be seen with the unaided eye by an astronaut orbiting the Earth. Using the result from part (a) of Example 25.7, see if it is visible. (Neglect any atmospheric effects.)

Reasoning and Answer. Despite the length of the wall, it would not be visible from space unless its *width* subtends the minimum angle of resolution for the eye of an observing astronaut ($\theta_{min} = 2.0 \times 10^{-4}$ rad from Example 25.7). Actually, guard towers with roofs as wide as 7.0 m were located every 180 m along the Wall. Let's take the maximum observable width to be 7.0 m. (Actually, it is the circular arc length that subtends the angle, but at such a long radius, the chord length is very nearly equal to the circular arc length. Refer to Example 7.2, and make yourself a sketch.)

Let's assume that the astronaut is just able to distinguish the guard roof. Recall that $s = r\theta$ (Eq. 7.3), where s is the maximum observable width of the wall and r is the radial (height) distance. Then, the astronaut would have to be at, or closer than, a distance of

$$r = \frac{s}{\theta} = \frac{7.0 \text{ m}}{2.0 \times 10^{-4} \text{ (rad)}} = 3.5 \times 10^4 \text{ m} = 35 \text{ km} \ (= 22 \text{ mi})$$

So above 35 km, the wall would *not* be able to be seen with the unaided eye. Orbiting satellites are about 300 km (190 mi) or more above the Earth. The statement about the ability to see the wall from space is false.

Follow-Up Exercise. What would be the minimum diameter of the objective of a telescope that would allow an astronaut orbiting the Earth at an altitude of 300 km to actually see the Great Wall? (Take all conditions to be the same as stated in this Example, and assume that the wavelength of light is 550 nm.)

Note from Eq. 25.8 that higher resolution can be gained by using radiation of a shorter wavelength. Thus, a telescope with an objective of a given size will have greater resolution with violet light than with red light. For microscopes, it is possible to increase resolving power by shortening the wavelengths of the light used to create the image. This can be done with a specialized objective called an *oil immersion lens*. When such a lens is used, a drop of transparent oil fills the space between the objective and the specimen. Recall that the wavelength of light in oil is $\lambda' = \lambda/n$, where n is the index of refraction of the oil and λ is the wavelength of light in air. For values of n about 1.50 or higher, the wavelength is significantly reduced, and the resolving power is increased proportionally.

*25.5 Color

OBJECTIVE: To relate color vision and light.

In general, physical properties are fixed or absolute. For example, a particular type of electromagnetic radiation has a certain frequency or wavelength. However, visual *perception* of this radiation may vary from person to person. How we "see" (or our brain "interprets") radiation gives rise to what we call *color vision*.

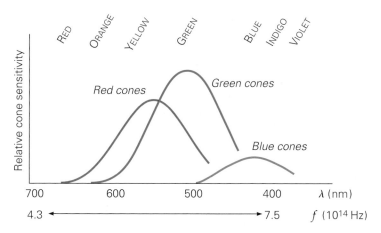

◄ **FIGURE 25.19** Sensitivity of
cones Different types of cones in
the retina of the human eye may
respond to different frequencies of
light to give three general color
responses: red, green, and blue.

Color Vision

Color is perceived because of a physiological response to excitation by light of the cone receptors in the retina of the human eye. (Many animals have no cone cells and thus live in a black-and-white world.) The cones are sensitive to light with frequencies approximately between 7.5×10^{14} Hz and 4.3×10^{14} Hz (wavelengths between 400 and 700 nm). The signals representing different frequencies of light are perceived by the brain as different colors. The association of a color with a particular frequency is subjective and may vary from person to person. The concept of pitch is to sound and hearing as color is to light and vision.

The details of color vision are not well understood. It is known that there are three types of cones responding to different parts of the visible spectrum: the red, green, and blue regions (▲Fig. 25.19). Presumably, each cone absorbs light in a specific range of frequencies and all three functionally overlap to form combinations that are interpreted by the brain as the various colors of the spectrum. For example, when red and green cones are stimulated equally, the brain interprets the two superimposed signals as yellow. But when the red cones are stimulated more strongly than the green cones, the brain senses orange. (That is "yellow" but dominated by red.) *Color blindness* results when one or more type of cone is missing or nonfunctional.

As Fig. 25.19 shows, the human eye is not equally sensitive to all colors. Some colors evoke a greater response than others and therefore appear brighter at the same intensity. The wavelength of maximum visual sensitivity is about 550 nm, in the yellow-green region.

The foregoing theory of color vision (mixing or combining) is based on the experimental fact that beams of varying intensities of red, green, and blue light can be arranged to produce most other colors. The red, blue, and green from which we interpret a full spectrum of colors are called the **additive primary colors**. When light beams of the additive primaries are projected and overlapped on a white screen, other colors are produced, as illustrated in ◄Fig. 25.20. This technique is called the **additive method of color production**. Triad dots consisting of three phosphors that emit the additive primary colors are used in television picture tubes to produce colored images.

Note in Fig. 25.20 that a certain combination of the primary colors appears white to the eye. Also, many *pairs* of colors appear white to the eye when combined. The colors of such pairs are said to be **complementary colors**. The complement of blue is yellow, that of red is cyan, and that of green is magenta. As the figure also shows, the complementary color of a particular primary is the combination, or sum, of the other two primaries. Hence, the primary and its complement together appear white.

Edwin H. Land (the developer of Polaroid film) showed that when the proper mixtures of only two wavelengths (colors) of light are passed through black and white transparencies (no color), the wavelengths produce images of various colors. Land wrote, "In this experiment we are forced to the astonishing conclusion that the rays are not in themselves color-making. Rather they are bearers of information that the eye uses to assign appropriate colors to various objects in an image."*

*From Edwin H. Land, "Experiments in Color Vision," *Scientific American* (May 1959), 84–99.

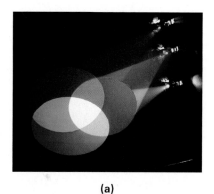

(a)

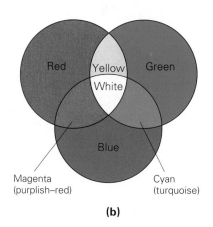

(b)

▲ **FIGURE 25.20** Additive method of color production When light beams of the primary colors (red, blue, and green) are projected onto a white screen, mixtures of them produce other colors. Varying the intensities of the beams allows most colors to be produced.

Objects exhibit a color when they are illuminated with white light because they reflect (scatter) or transmit light predominantly in the frequency range of that color. The other frequencies of the white light are mostly absorbed. For example, when white light strikes a red apple, most of the energy in the red portion of the spectrum is reflected—most of all the others (and thus all other colors) are absorbed. Similarly, when white light passes through a piece of transparent red glass, or a *filter*, mostly the light associated with red is transmitted. This occurs because the color pigments (additives) in the glass are selective absorbers.

Pigments are mixed to form various colors, such as in the production of paints and dyes. You are probably aware that mixing yellow and blue paints produces green. This is because the yellow pigment absorbs most of the wavelengths except those in the yellow and nearby regions (green plus orange) of the visible spectrum, and the blue pigment absorbs most of the wavelengths except those in the blue and nearby regions (violet plus green). The wavelengths in the intermediate (overlap) green region, between the yellow and blue range, are *not* strongly absorbed by either pigment, and therefore the mixture appears green. The same effect can be accomplished by passing white light through stacked yellow and blue filters. The light coming through both filters appears green.

Mixing pigments results in the *subtraction* of colors. The resultant color is created by whatever is *not* absorbed by the pigment—that is, *not* subtracted from the original beam. This is the principle of the **subtractive method of color production**. Three particular pigments—cyan, magenta, and yellow—are the **subtractive primary pigments**. Various combinations of two of the three subtractive primaries produce the three additive primary colors (red, blue, and green), as illustrated in ▾Fig. 25.21. When the subtractive primaries are mixed in the proper proportions, the mixture appears black (because all wavelengths are absorbed). Painters often refer to the subtractive primaries as red, yellow, and blue. They are

▼ **FIGURE 25.21 Subtractive method of color production** **(a)** When the primary pigments (cyan, magenta, and yellow) are mixed, different colors are produced by subtractive absorption; for example, the mixing of yellow and magenta produces red. When all three pigments are mixed and all the wavelengths of visible light are absorbed, the mixture appears black. **(b)** Subtractive color mixing, using filters. The principle is the same as in part (a). Each pigment selectively absorbs certain colors, removing them from the white light. The colors that remain are what we see.

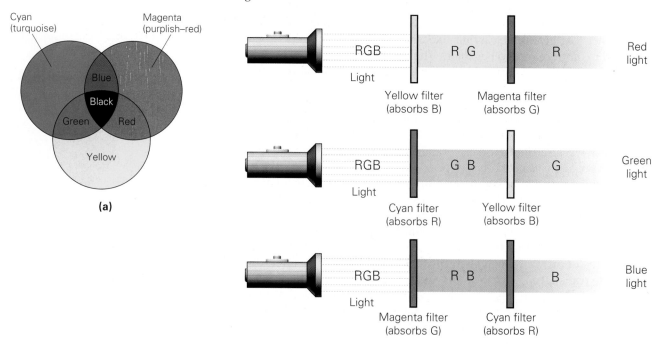

loosely referring to magenta (purplish-red), yellow, and cyan ("true" blue). Mixing these paints in the proper proportions produces a broad spectrum of colors.

Note in Fig. 25.21 that the magenta pigment essentially subtracts the color green where it overlaps with cyan and yellow. As a result, magenta is sometimes referred to as "minus green." If a magenta filter were placed in front of a green light, no light would be transmitted. Similarly, cyan is called "minus red," and yellow is called "minus blue." An example of subtractive color mixing is a photographer's use of a yellow filter to bring out white clouds on black and white film. This filter absorbs blue from the sky, darkening it relative to the clouds, which reflect white light. Hence, the contrast between the two is enhanced. What type of filter would you use to darken green vegetation on black and white film? To lighten it?

Chapter Review

- Nearsighted people cannot see distant objects clearly. Farsighted people cannot see nearby objects clearly. These conditions may be corrected by diverging and converging lenses, respectively.

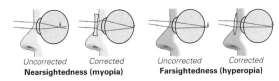

Uncorrected   Corrected   Uncorrected   Corrected
Nearsightedness (myopia)   **Farsightedness (hyperopia)**

- The magnification of a magnifying glass (or simple microscope) is expressed in terms of **angular magnification (m)**, as distinguished from the lateral magnification (M; see Chapter 23):

$$m = \frac{\theta}{\theta_o} \qquad (25.1)$$

Magnification of a magnifying glass with the image at the near point (25 cm) is expressed as

$$m = 1 + \frac{25 \text{ cm}}{f} \qquad (25.3)$$

Magnification of a magnifying glass with the image at infinity is expressed as

$$m = \frac{25 \text{ cm}}{f} \qquad (25.4)$$

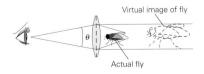

Virtual image of fly

θ

Actual fly

- The objective of a compound microscope has a relatively short focal length, and the eyepiece, or ocular, has a longer focal length. Both contribute to the **total magnification**, m_{total}, given by

$$m_{\text{total}} = M_o m_e = -\frac{(25 \text{ cm})L}{f_o f_e} \qquad (25.5)$$

where L, f_o, and f_e are in centimeters.

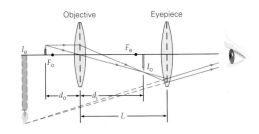

Objective   Eyepiece

- A refracting telescope uses a converging lens to gather light, and a reflecting telescope uses a converging mirror. The image created by either one is magnified by the eyepiece. The **magnification of a refracting telescope** is

$$m = -\frac{f_o}{f_e} \qquad (25.6)$$

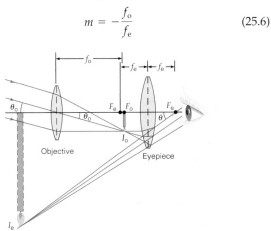

Objective   Eyepiece

- Diffraction places a limit on **resolution**—the ability to resolve, or distinguish, objects that are close together. Two images are said to be just resolved when the central maximum of one image falls on the first minimum of the diffraction pattern of the other image (**Rayleigh criterion**).

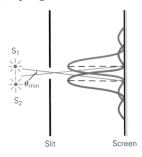

Slit   Screen

- For a rectangular slit, the **minimum angle of resolution** is

$$\theta_{\min} = \frac{\lambda}{w} \qquad (25.7)$$

The **minimum angle of resolution for a circular aperture** of diameter D is

$$\theta_{\min} = \frac{1.22\lambda}{D} \qquad (25.8)$$

The **resolving power** of a microscope is

$$s = f\theta_{\min} = \frac{1.22\lambda f}{D} \qquad (25.9)$$

Exercises

MC = *Multiple Choice Question,* CQ = *Conceptual Question, and* IE = *Integrated Exercise. Throughout the text, many exercise sections will include "paired" exercises. These exercise pairs, identified with* **red numbers***, are intended to assist you in problem solving and learning. In a pair, the first exercise (even numbered) is worked out in the Study Guide so that you can consult it should you need assistance in solving it. The second exercise (odd numbered) is similar in nature, and its answer is given at the back of the book.*

25.1 The Human Eye*

1. **MC** The rods of the retina (a) are responsible for 20/20 vision, (b) are responsible for black-and-white twilight vision, (c) are responsible for color vision, (d) focus light. (b)

2. **MC** An imperfect cornea can cause (a) astigmatism, (b) near-sightedness, (c) farsightedness, (d) all of the preceding. (d)

3. **MC** The image of an object formed on the retina is (a) inverted, (b) upright, (c) the same size as the object, (d) all of the preceding. (a)

4. **MC** The focal length of the crystalline lens of the human eye varies with muscle action. When a distant object is viewed, the radius of the lens is (a) large, (b) small, (c) flat, (d) none of the preceding. (a)

5. **CQ** People and other animals often exhibit "red eye" when photographed with a flash camera. Light reflected from the retina is red because of blood vessels near the surface. Some cameras have an anti-red-eye option, which, when activated, gives a quick flash before the longer picture-taking flash. Explain how this option reduces red eye. see ISM

6. **CQ** Which parts of the camera correspond to the iris, crystalline lens, and retina of the eye? aperture, lens, and film, respectively

7. **CQ** (a) If an eye has a far point of 15 m and a near point of 25 cm, is that eye nearsighted or farsighted? (b) How about an eye with a far point at infinity and a near point at 50 cm? (c) What type of corrective lenses (converging or diverging) would you use to correct the vision defects in parts (a) and (b)? (a) nearsighted (b) farsighted (c) for (a), diverging; for (b), converging

8. **CQ** Will wearing glasses to correct nearsightedness and farsightedness, respectively, affect the size of the image on the retina? Explain. yes, smaller for nearsightedness and larger for farsightedness

9. ● What are the powers of (a) a converging lens of focal length 20 cm and (b) a diverging lens of focal length −50 cm? (a) +5.0 D (b) −2.0 D

10. **IE** ● The far point of a certain nearsighted person is 90 cm. (a) Which type of contact lens, (1) converging, (2) diverging, or (3) bifocal, should be prescribed to enable the person to see more distant objects clearly? Explain. (b) What would the power of the lens be, in diopters? (a) (2) diverging (b) −1.1 D

11. **IE** ● A certain farsighted person has a near point of 50 cm. (a) Which type of contact lens, (1) converging, (2) diverging, or (3) bifocal, should an optometrist prescribe to en-

able the person to see objects clearly as close as 25 cm? Explain. (b) What is the power of the lens, in diopters? (a) (1) converging (b) +2.0 D

12. ●● A woman cannot see objects clearly when they are farther than 12.5 m away. (a) Does she have (1) nearsightedness, (2) farsightedness, or (3) astigmatism? Explain. (b) Which type of lens will allow her to see distant objects clearly, and of what power should the lens be? (a) (1) nearsightedness (b) diverging, −0.0800 D

13. ●● A nearsighted woman has an uncorrected far point of 200 cm. Which type of contact lens would correct this condition, and of what power should it be? diverging, −0.500 D

14. ●● A person can *just* see the print in a book clearly when she holds the book at arm's length (0.80 m from the eyes). (a) Does she have (1) nearsightedness, (2) farsightedness, or (3) astigmatism? Explain. (b) Which type of lens will allow her to read the text at the normal near point and what is the lens's focal length? (a) (2) farsightedness (b) converging, 36 cm

15. ●● To correct a case of hyperopia, an optometrist prescribes positive contact lenses that effectively move the patient's near point from 100 cm to 25 cm. (a) To see distant objects clearly, the patient should wear the contact lenses or take them out. Explain. (b) What is the power of the lenses? (a) take them out (b) +3.0 D

16. ●● A farsighted person with a near point of 0.95 m gets contact lenses and can then read a newspaper held at a distance of 25 cm. What is the power of the lenses? (Assume that the lenses are the same for both eyes.) +2.9 D

17. **IE** ●● A farsighted man is unable to focus on objects nearer than 1.5 m. (a) The type of contact lens that allows him to focus on the print of a book held 25 cm from his eyes should be (1) converging, (2) diverging, (3) flat. Explain. (b) What should be the power of the lens? (a) (1) converging (b) +3.3 D

18. ●● A nearsighted student wears contact lenses to correct for a far point that is 4.00 m from her eyes. When she is not wearing her contact lenses, her near point is 20 cm. What is her near point when she is wearing her contacts? 21 cm

19. ●● A nearsighted woman has a far point located 2.00 m from one eye. (a) If a corrective lens is worn 2.00 cm from the eye, what would be the necessary power of the lens for her to see distant objects? (b) What would be the necessary power if a contact lens were used? (a) −0.505 D (b) −0.500 D

20. ●● A college professor can see objects clearly only if they are between 70 and 500 cm from her eyes. Her optometrist prescribes bifocals (▼Fig. 25.22) that enable her to see distant objects through the top half of the lenses and read students' papers at a distance of 25 cm through the lower half. What are the respective powers of the top and bottom lenses? [Assume that both lenses (right and left) are the same.] top: −0.200 D; bottom: +2.6 D

*Assume that corrective lenses are in contact with the eye (contact lenses) unless otherwise stated.

Nearsightedness
correction

Farsightedness
correction

▲ **FIGURE 25.22 Bifocals** See Exercises 20 and 25.

21. ●● A nearsighted man wears eyeglasses whose lenses have a power of −0.15 D. How far away is his far point? 6.7 m

22. ●● An eyeglass lens with a power of +2.8 D allows a farsighted person to read a book held at a distance of 25 cm from her eyes. At what distance must she hold the book to read it without glasses? 83 cm

23. ●●● A certain myopic man has a far point of 150 cm. (a) What power must a contact lens have to allow him to see distant objects clearly? (b) If he is able to read print at 25 cm while wearing his contacts, is his near point less than 25 cm? If so, what is it? (c) Give an approximation of the man's age, based on the normal rate of recession of the near point. (a) −0.67 D (b) yes, 21 cm (c) 35–40 years old

24. ●●● A middle-aged man starts to wear eyeglasses with lenses of +2.0 D that allow him to read a book held as closely as 25 cm. Several years later, he finds that he must hold a book no closer than 33 cm to read it clearly with the same glasses, so he gets new glasses. What is the power of the new lenses? (Assume that both lenses are the same.) +3.0 D

25. ●●● Bifocal glasses are used to correct both nearsightedness and farsightedness at the same time (Fig. 25.22). If the near points in the right and left eyes are 35.0 cm and 45.0 cm, respectively, and the far point is 220 cm for both eyes, what are the powers of the lenses prescribed for the glasses? (Assume that the glasses are worn 3.00 cm from the eyes.) right: +1.42 D, −0.46 D; left: +2.16 D, −0.46 D

25.2 Microscopes*

26. **MC** A magnifying glass (a) is a concave lens, (b) forms virtual images, (c) magnifies by effectively increasing the angle the object subtends, (d) both b and c. (d)

27. **MC** A compound microscope has (a) unlimited magnification, (b) two lenses of the same focal length, (c) a diverging objective lens, (d) an objective of relatively short focal length. (d)

28. CQ With an object at the focal point of a magnifying glass, the magnification is given by $m = (25 \text{ cm})/f$ (Eq. 25.4). According to this equation, the magnification could be increased indefinitely by using lenses with shorter focal lengths. Why, then, do we need compound microscopes? see ISM

29. CQ When you use a simple convex lens as a magnifying glass, where should you put the object, farther away than the focal length or inside the focal length? Explain. inside; see ISM

*The normal near point should be taken as 25 cm unless otherwise specified.

30. ● Using the small-angle approximation, compare the angular sizes of a car 1.0 m in height when at distances of (a) 500 m and (b) 1025 m. (a) 2.0×10^{-3} rad (b) 9.8×10^{-4} rad

31. ● An object is placed 10 cm in front of a converging lens with a focal length of 18 cm. What are (a) the lateral magnification and (b) the angular magnification? (a) 2.3× (b) 2.5×

32. ● A biology student uses a converging lens to examine the details of a small insect. If the focal length of the lens is 12 cm, what is the maximum angular magnification? 3.1×

33. ● When viewing an object with a magnifying glass whose focal length is 10 cm, a student positions the lens so that there is minimum eyestrain. What is the observed magnification? 2.5×

34. **IE** ● A physics student uses a converging lens with a focal length of 15 cm to read a small measurement scale. (a) Maximum magnification is achieved if the image is at (1) the near point, (2) infinity, (3) either place. Explain. (b) What are the magnifications when the image is at the near point and for viewing with the relaxed eye? (a) (1) the near point (b) 2.7×, 1.7×

35. **IE** ●● A detective wants to achieve maximum magnification when looking at a fingerprint with a magnifying glass. (a) He should use a (1) high-powered, (2) low-powered, or (3) small converging lens. Explain. (b) If he uses lenses of power +3.5 D and +2.5 D, what are the maximum magnifications of the print? (a) (1) high-powered (b) 1.9× and 1.6×

36. ●● What is the maximum magnification of a magnifying glass with a power of +3.0 D for (a) a person with a near point of 25 cm and (b) a person with a near point of 10 cm? (a) 1.8× (b) 1.3×

37. ●● A compound microscope has an objective with a focal length of 4.00 mm and an eyepiece with a magnification of 10.0×. If the objective and eyepiece are 15.0 cm apart, what is the total magnification of the microscope? −375×

38. ●● A compound microscope has a distance of 15 cm between lenses and an ocular with a focal length of 8.0 mm. What power should the objective have to give a total magnification of −360×? +0.77 D

39. **IE** ●● Two lenses of focal length 0.45 cm and 0.35 cm are available for a compound microscope using an eyepiece of focal length 3.0 cm, and the distance between the lenses has to be 15 cm. (a) Which lens, (1) the one with the longer focal length, (2) the one with the shorter focal length, or (3) either should be used as the objective? (b) What are the two possible total magnifications of the microscope. (a) (2) the one with the shorter focal length (b) −280× and −360×

40. ●● The focal length of the objective lens of a compound microscope is 4.5 mm. The eyepiece has a focal length of 3.0 cm. If the distance between the lenses is 18 cm, what is the magnifications of a viewed image? −330×

41. ●● A compound microscope has an objective lens with a focal length of 0.50 cm and an eyepiece with a focal length of 3.25 cm. The separation distance between the lenses is 22 cm. A student with a normal near point uses the microscope. (a) What is the total magnification? (b) Compare the total magnification (as a percentage) with the magnification of the eyepiece alone as a simple magnifying glass. (a) −340× (b) 3900%

42. ●● A −150× microscope has an objective whose focal length is 0.75 cm. If the distance between the lenses is 20 cm, find the focal length of the eyepiece. 4.4 cm

43. ●● A specimen is 5.0 mm from the objective of a compound microscope that has a power of +250 D. What must be the magnifying power of the eyepiece if the total magnification of the specimen is −100×? 25×

44. ●●● A lens with a power of +10 D is used as a simple microscope. (a) For the image of an object to be seen clearly, can the object be placed infinitely close to the lens, or is there a limit on how close it can be? Explain. (b) Calculate how close an object can be brought to the lens. (c) What is the angular magnification at this point?
(a) there is a limit (b) 7.1 cm (c) 3.5×

45. IE ●●● A modern microscope is equipped with a turret that has three objectives with focal lengths of 16 mm, 4.0 mm, and 1.6 mm and interchangeable eyepieces of 5.0× and 10×. A specimen is positioned such that each objective produces an image 150 mm from the objective. (a) Which objective-and-eyepiece combination would you use if you want to have the greatest magnification? How about the least magnification? Explain. (b) What are the greatest and least magnifications possible? (a) greatest: 1.6 mm/10×; least: 16 mm/5× (b) $M_{max} = -930×$; $M_{min} = -42×$

25.3 Telescopes

46. **MC** An astronomical telescope has (a) unlimited magnification, (b) two lenses of the same focal length, (c) an objective of relatively long focal length, (d) an objective of relatively short focal length. (c)

47. **MC** An inverted image is produced by (a) a terrestrial telescope, (b) an astronomical telescope, (c) a Galilean telescope, (d) all of the preceding. (b)

48. **MC** Compared with large refracting telescopes, large reflecting telescopes have the advantage of (a) greater light-gathering capability, (b) freedom from chromatic aberration, (c) lower cost, (d) all of the preceding. (d)

49. **CQ** In Fig. 25.12b, part of the light entering the concave mirror is obstructed by a small plane mirror that is used to redirect the rays to a viewer. Does this mean that only a portion of a star can be seen? How does the size of the obstruction affect the image? no; intensity is affected

50. **CQ** Why is chromatic aberration an important factor in refracting telescopes, but not in reflecting telescopes? reflection is frequency independent

51. **CQ** If you are given two lenses with different focal lengths, which one should you use as the eyepiece for a telescope? Explain.
the one with the shorter focal length; see ISM

52. ● Find the magnification and length of a telescope whose objective has a focal length of 50 cm and whose eyepiece has a focal length of 2.0 cm. −25×, 52 cm

53. ● An astronomical telescope has an objective and an eyepiece whose focal lengths are 60 cm and 15 cm, respectively. What are the telescope's (a) magnifying power and (b) length? (a) −4.0× (b) 75 cm

54. ●● A astronomical telescope has an eyepiece with a focal length of 10.0 mm. If the length of the tube is 1.50 m, what is the angular magnification of the telescope when it is focused for an object at infinity? −149×

55. A telescope has an angular magnification of −50× and a barrel 1.02 m long. What are the focal lengths of the objective and the eyepiece? 1.0 m and 2.0 cm

56. IE ●● A terrestrial telescope has three lenses: an objective, an erecting lens, and an eyepiece. (a) Does the erecting lens (1) increase the magnification, (2) increase the physical length of the telescope, (3) decrease the magnification, or (4) decrease the physical length of the telescope? Explain. (b) The three lenses of this terrestrial telescope have focal lengths of 40 cm, 20 cm, and 15 cm for the objective, erecting lens, and eyepiece, respectively. What is the magnification of the telescope for an object at infinity? (c) What is the length of the telescope barrel? (a) (2) increase the physical length of the telescope (b) 2.7× (c) 135 cm

57. ●● A terrestrial telescope uses an objective and eyepiece with focal lengths of 45 cm and 15 cm, respectively. What should the focal length of the erecting lens be if the overall length of the telescope is to be 0.80 m? 5.0 cm

58. ●● An astronomical telescope uses an objective of power +2.0 D. If the length of the telescope is 52 cm, what is the angular magnification of the telescope? −25×

59. IE ●● You are given two objectives and two eyepieces and are instructed to make a telescope with them. The focal lengths of the objectives are 60.0 cm and 40.0 cm, and the focal lengths of the eyepieces are 0.90 cm and 0.80 cm, respectively. (a) Which lens combination would you pick if you want to have maximum magnification? How about minimum magnification? Explain. (b) Calculate the maximum and minimum magnifications. (a) 60.0 cm and 0.80 cm; 40.0 cm and 0.90 cm (b) −75×; −44×

25.4 Diffraction and Resolution*

60. **MC** The images of two sources are said to be resolved when (a) the central maxima of the diffraction patterns fall on each other, (b) the first maximum of the diffraction patterns fall on each other, (c) the central maximum of one diffraction pattern falls on the first minimum of the other, (d) none of the preceding. (c)

61. **MC** For a telescope with a circular aperture, the minimum angle of resolution is (a) greater for red light than for blue light, (b) independent of the frequency of the light, (c) directly proportional to the radius of the aperture, (d) independent of the area of the aperture. (a)

62. **MC** The purpose of using oil immersion lenses on microscopes is to (a) reduce the size of the microscope, (b) increase the magnification, (c) increase the wavelength of light so as to increase the resolving power, (d) reduce the wavelength of light so as to increase the resolving power. (d)

*Ignore atmospheric blurring unless otherwise stated.

63. **CQ** When an optical instrument is designed, high resolution is often desired so that the instrument may be used to observe fine details. Does higher resolution mean a smaller or larger minimum angle of resolution? Explain.
smaller; see ISM

64. **CQ** A reflecting telescope with a large objective mirror can collect more light from stars than a reflecting telescope with a smaller objective mirror. What other advantage is gained with a large mirror? Explain.
higher resolution, due to large aperture; see ISM

65. **CQ** Modern digital cameras are getting smaller and smaller. Discuss the image resolution of these small cameras.
the smaller camera (lens) has lower resolution

66. **IE** ● (a) For a given wavelength, a wider single slit will give a (1) greater, (2) smaller, (3) the same minimum angle of resolution as a narrower slit, according to the Rayleigh criterion. (b) What are the minimum angles of resolution for two point sources of red light ($\lambda = 680$ nm) in the diffraction pattern produced by single slits with a width of 0.55 mm and 0.45 mm?
(a) (2) smaller (b) 1.2×10^{-3} rad and 1.5×10^{-3} rad

67. ● The minimum angular separation of the images of two identical monochromatic point sources in a single-slit diffraction pattern is 0.0055 rad. If a slit width of 0.10 mm is used, what is the wavelength of the sources? 550 nm

68. ● What is the resolution limit due to diffraction for the European Southern Observatory reflecting telescope (8.20-m, or 323-in., diameter) for light with a wavelength of 550 nm?
8.18×10^{-8} rad

69. ● What is the resolution due to diffraction for the Hale telescope at Mount Palomar, with its 200-in.-diameter mirror, for light with a wavelength of 550 nm? Compare this value with the resolution limit for the European Southern Observatory telescope found in Exercise 68.
1.32×10^{-7} rad; θ_{min} by Hale is 1.6 times as great

70. ●● From a spacecraft in orbit 150 km above the Earth's surface, an astronaut wishes to observe her hometown as she passes over it. What size features will she be able to identify with the unaided eye, neglecting atmospheric effects? [*Hint*: Estimate the diameter of the human iris.]
objects as large as typical houses

71. **IE** ●● A human eye views small objects of different colors, and the eye's resolution is measured. (a) The eye obtains the maximum resolution and sees the finest details for objects of which color: (1) red, (2) yellow, (3) blue, or (4) it does not matter? Explain. (b) The maximum diameter of the eye's pupil at night is about 7.0 mm. What are the minimum angles of separation for sources with wavelengths 550 nm and 650 nm?
(a) (3) blue (b) 9.6×10^{-5} rad and 1.1×10^{-4} rad

72. ●● Some African tribes people claim to be able to see the moons of Jupiter with the unaided eye. If two moons of Jupiter are at a minimum distance of 3.1×10^8 km away from Earth and at a maximum separation distance of 3.0×10^6 km, is this possible in theory? Explain. Assume that the moons reflect sufficient light and that their observation is not restricted by Jupiter. [*Hint*: See Exercise 71b.]
yes, in theory; see ISM

73. ●● Assuming that the headlights of a car are point sources 1.7 m apart, what is the maximum distance from an observer to the car at which the headlights are distinguishable from each other? [*Hint*: See Exercise 71b.]
17 km

74. ●● A refracting telescope with a lens whose diameter is 30.0 cm is used to view a binary star system that emits light in the visible region. (a) What is the minimum angular separation of the two stars for them to be barely resolved? (b) If the binary star is a distance of 6.00×10^{20} km from the Earth, what is the distance between the two stars? (Assume that a line joining the stars is perpendicular to our line of sight.) (a) 1.63×10^{-6} rad (b) 9.76×10^{17} m

75. ●● A radio telescope of diameter 300 m uses a wavelength of 4.0 m to observe a binary star system that is about 2.5×10^{18} km from the Earth. What is the minimum distance of two stars that can be distinguished by the telescope?
4.1×10^{16} km

76. ●● The objective of a microscope is 2.50 cm in diameter and has a focal length of 30.0 mm. (a) If yellow light with a wavelength of 570 nm is used to illuminate a specimen, what is the minimum angular separation of two fine details of the specimen for them to be just resolved? (b) What is the resolving power of the lens?
(a) 2.78×10^{-5} rad (b) 8.34×10^{-4} mm

77. ●●● A microscope with an objective 1.20 cm in diameter is used to view a specimen via light from a mercury source with a wavelength of 546.1 nm. (a) What is the limiting angle of resolution? (b) If details finer than those observable in part (a) are to be observed, what color of light in the visible spectrum would have to be used? (c) If an oil immersion lens were used ($n_{oil} = 1.50$), what would be the change (expressed as a percentage) in the resolving power?
(a) 5.55×10^{-5} rad (b) blue (c) 33.3%

*25.5 Color

78. **MC** An additive primary color is (a) blue, (b) green, (c) red, (d) all of the preceding. (d)

79. **MC** A subtractive primary color is (a) cyan, (b) yellow, (c) magenta, (d) all of the preceding. (d)

80. **MC** White light is incident on two filters as shown in ▼Fig. 25.23. The color of light that emerges from the yellow filter is (a) blue, (b) yellow, (c) red, (d) green. (d)

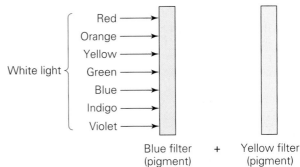

▲ **FIGURE 25.23 Color absorption** See Exercise 80.

81. **CQ** Describe how the American flag would appear if it were illuminated with light of each of the primary colors.
see ISM

82. **CQ** Can white be obtained by the subtractive method of color production? Explain. It is sometimes said that black is the absence of all color or that a black object absorbs all incident light. If so, why do we see black objects? see ISM

83. **CQ** Several beverages, such as root beer, develop a "head" of foam when poured into a glass. Why is the foam generally white or light colored, whereas the liquid is dark? see ISM

Comprehensive Exercises

84. A student uses a magnifying glass to examine the details of a microcircuit in the lab. If the lens has a power of 12.5 D and a virtual image is formed at the student's near point (25 cm), (a) how far from the circuit is the lens held, and (b) what is the angular magnification? (a) 6.1 cm (b) 4.1×

85. Referring to ▼Fig. 25.24, show that the magnifying power of a magnifying glass held at a distance d from the eye is given by

$$m = \left(\frac{25}{f}\right)\left(1 - \frac{d}{D}\right) + \frac{25}{D}$$

when the actual object is located at the near point (25 cm). [Hint: Use a small-angle approximation, and note that $y_i/y_o = -d_i/d_o$, by similar triangles.] see ISM

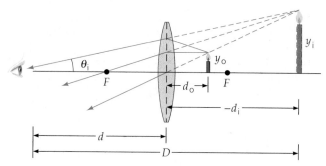

▲ **FIGURE 25.24 Power of a magnifying glass** See Exercise 85.

86. Referring to ▶Fig. 25.25, show that the angular magnification of a refracting telescope focused for the final image at infinity is $m = -f_o/f_e$. (Because telescopes are designed for viewing distant objects, the angular size of an object viewed with the unaided eye is the angular size of the object at its actual location rather than at the near point, as is true for a microscope.) see ISM

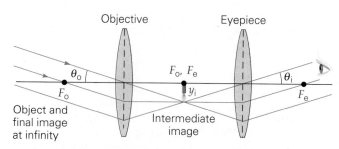

▲ **FIGURE 25.25 Angular modification of a refracting telescope** See Exercise 86.

87. Two astronomical telescopes have the characteristics shown in the following table:

Telescope	Objective Focal Length (cm)	Eyepiece Focal Length (cm)	Objective Diameter (cm)
A	90.0	0.840	75.0
B	85.0	0.770	60.0

(a) Which telescope would you choose (1) for best magnification? (2) for best resolution? Explain. (b) Calculate the maximum magnification and the minimum resolving angle for a wavelength of 550 nm. (a) (1) B, (2) A (b) −110×, 8.95 × 10⁻⁷ rad

88. A refracting telescope has an objective with a focal length of 50 cm and an eyepiece with a focal length of 15 mm. The telescope is used to view an object that is 10 cm high and located 50 m away. What is the apparent angular height of the object as viewed through the telescope? 3.8°

89. The amount of light that reaches the film in a camera depends on the lens aperture (the effective area) as controlled by the diaphragm. The f-number is the ratio of the focal length of the lens to its effective diameter. For example, an f/8 setting means that the diameter of the aperture is one eighth of the focal length of the lens. The lens setting is commonly referred to as the *f-stop*. (a) Determine how much light each of the following lens settings admits to the camera as compared with f/8: (1) f/3.2 and (2) f/16. (b) The exposure time of a camera is controlled by the shutter speed. If a photographer correctly uses a lens setting of f/8 with a film exposure time of 1/60 s, what exposure time should he use to get the same amount of light exposure if he sets the f-stop at f/5.6? see ISM

The following Physlet Physics Problems can be used with this chapter.
PHYSLET® 36.1, 36.2, 36.3, 36.4, 36.5

PHYSICS FACTS

- In one nanosecond (10^{-9} s), light travels about 1 ft.

- The atomic clocks in orbit in the Global Positioning System (GPS) can be used to determine the location of an object on the surface of the Earth to within several meters. To achieve this accuracy, their frequencies must be adjusted for special and general relativistic effects due to their speed and the fact that they are in a weaker gravitational field compared to the clocks at the Earth's surface.

- Near the boundary of a black hole, the gravitational field is so strong that just outside the boundary, light can orbit the black hole similar to satellites orbiting the Earth.

- It is currently believed that at the centers of many galaxies may reside a huge black hole, absorbing nearby stars and growing in size.

- Accurate atomic clocks taken on airplane trips around the world show time differences with respect to ones that "stayed at home," in agreement with predictions of relativity theory.

- The U.S. standard atomic clock is about 1 mile above sea level in Boulder, CO. It gains about 5 microseconds per year compared to identical clocks at sea level because of differences in the Earth's gravitational field between their two locations.

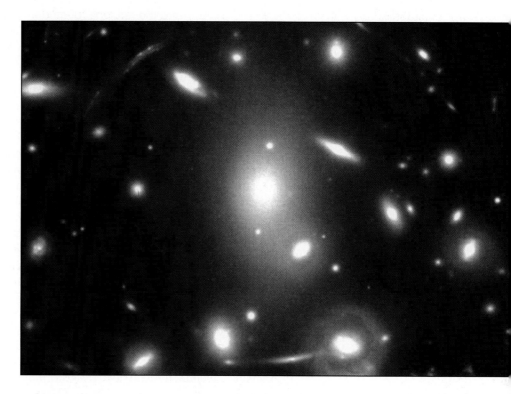

You might not think so, but the photograph tells us something very remarkable about our universe. The bright shapes are galaxies, each consisting of billions or trillions of stars. They are very far from us—billions of light-years away. The faint arcs that make the photograph resemble a spider's web are from galaxies even more distant.

What is remarkable about these wisps of light, however, is not the billions of years they took to reach us, but the paths they have taken. We usually think of light traveling in straight lines. Yet the light from these distant galaxies has had its direction changed by the gravitational fields of the galaxies in the foreground, creating the arcs in the photo.

The fact that light can be affected by gravity was predicted by Albert Einstein as a consequence of his theory of general relativity. Relativity originated from the analysis of physical phenomena involving speeds approaching that of light. Indeed, modern relativity caused us to rethink our understanding of space, time, and gravitation. It successfully challenged Newtonian concepts that had dominated science for 250 years.

The impact of relativity has been especially significant in the branches of science concerned with two extremes of physical reality: the subatomic realm of nuclear and particle physics, in which time intervals and distances are inconceivably small (Chapters 29 and 30); and the cosmic realm, in which time intervals and distances are unimaginably large. All modern theories about the birth, evolution, and ultimate fate of our universe are inextricably linked to our understanding of relativity.

In this chapter, you will learn how Einstein's relativity explains the changes in length and time that are observed for rapidly moving objects, the equivalence of energy and mass, and the bending of light by gravitational fields—phenomena that seem strange from the classical Newtonian view.

26.1 Classical Relativity and the Michelson–Morley Experiment

OBJECTIVES: To (a) summarize the concepts of classical relativity, (b) define inertial and noninertial reference frames, and (c) explain the ether hypothesis and the reasons for its demise.

Physics is concerned with the description of the world around us and depends on observations and measurements (Chapter 1). We expect some aspects of nature to be consistent and unvarying; that is, the ground rules by which nature plays should be consistent, and physical principles should not change from observation to observation. This consistency is emphasized by referring to such principles as *laws*—for example, the laws of motion. Not only have physical laws proved valid over time, but they are the same for all observers.

The last sentence means that a physical principle or law should *not* depend on the observer's frame of reference. When a measurement is made or an experiment performed, reference is usually made to a particular frame or coordinate system—most often the laboratory, which is considered to be "at rest." Now envision the same experiment observed by a passerby (moving relative to the laboratory). Upon comparing notes, the experimenter and observer should find the results of the experiment and the physical principles involved to be the same. Physicists believe that the laws of nature are the same regardless of the observer. Measured quantities may vary and descriptions may be different, but the *laws* that these quantities obey must be the same for all observers.

Suppose that you are at rest and observe two cars traveling in the same direction on a straight road at speeds of 60 km/h and 90 km/h. Even though we rarely say it, it is assumed that these speeds are measured relative to your reference frame—the ground. However, a woman in the car traveling at 60 km/h observes the other car traveling at 30 km/h *relative to her reference frame*—the car in which she is riding. (What does someone in the car traveling at 90 km/h observe?) That is, each person observes a *relative* velocity—the one relative to his or her own reference frame. (See Section 3.4.)

Note: Review the discussion of relative velocities in Section 3.4.

In measuring relative velocities, there seems to be no "true" rest frame. Any reference frame can be considered at rest if the observer moves with it. We can, however, make a distinction between *inertial* and *noninertial* reference frames. An **inertial reference frame** is a reference frame in which Newton's first law of motion holds. That is, in an inertial frame, an object on which there is no net force does not accelerate. Since Newton's first law holds in this frame, the second law of motion ($\vec{F}_{net} = m\vec{a}$) also holds.

Note: The relationship between Newton's first and second laws is discussed in a Section 4.3 footnote.

Conversely, in a **noninertial reference frame** (one that is accelerating as measured from an inertial frame), an object with no net force acting on it would *appear* to accelerate. Note, however, that the frame is accelerating, not the object. If observations are made from a noninertial frame, Newton's second law will not correctly describe motion. One example of a noninertial reference frame is an automobile accelerating forward from rest. A cup on the (frictionless) dashboard may appear, when viewed from the car's (noninertial) reference frame, to accelerate backward without being acted on by any force. In fact, the noninertial observer would have to invoke a *fictitious* backward force to explain the cup's apparent acceleration. From the inertial frame of a sidewalk observer, however, the cup stays put (in accordance with the first law, since, with no appreciable friction, no net force acts on it) as the car accelerates away from it.

Any reference frame moving with a constant velocity relative to an inertial reference frame is itself an inertial frame. Given a constant relative velocity, no acceleration effects are introduced in comparing one frame with another. In such cases, $\vec{F}_{net} = m\vec{a}$ can be used by observers in *either* frame to analyze a situation, and both observers will come to the same conclusions. That is, Newton's second law holds in both frames. Thus, at least with respect to the laws of mechanics, no inertial frame is preferred over another. This is called the **principle of classical, or Newtonian, relativity:**

Teaching tip: Discuss the importance of Newton's first law of motion. Ask students, "Can the classroom in which we are sitting be considered an inertial reference frame? Why?"

Note: Newtonian relativity refers only to laws of mechanics.

The laws of *mechanics* are the same in all inertial reference frames.

The "Absolute" Reference Frame: The Ether

With the development of the theories of electricity and magnetism in the 1800s, some serious questions arose. Maxwell's equations predicted light to be an electromagnetic wave that travels with a speed of $c = 3.00 \times 10^8$ m/s in a vacuum. But relative to what reference frame does light have this speed? Classically, this speed would be expected to be different when measured from different reference frames. For example, you might expect it to be greater than 3.00×10^8 m/s if you were approaching the beam of light, as in the relative case of another approaching you on a highway.

Consider the situation in ▸Fig. 26.1: A person in a reference frame (truck) moving relative to another frame (ground) with a constant velocity $\vec{v}'$ throws a ball with a velocity $\vec{v}_b$ relative to the truck. Then the so-called stationary observer (ground) would say the ball had a velocity of $\vec{v} = \vec{v}' + \vec{v}_b$ relative to the ground. Suppose the truck were moving at 20 m/s east relative to the ground and a ball were thrown at 10 m/s (relative to the truck), also easterly. The ball would have a speed of 20 m/s + 10 m/s = 30 m/s to the east when observed by someone on the ground.

Now suppose that the person on the truck turned on a flashlight, projecting a beam of light to the east. According to Newtonian relativity, $\vec{v} = \vec{v}' + \vec{c}$, and the speed of light measured by the ground observer would be greater than 3.00×10^8 m/s. According to classical relativity then, the speed of light can have *any* value, depending on the observer's reference frame.

Assuming Newtonian relativity to be true, it followed that the particular light speed value of 3.00×10^8 m/s must be referenced to some unique frame, in analogy to other waves. Thus, the assumption of a unique reference frame for light seemed quite natural. Since the Earth receives light from the Sun and from distant stars it was thought that a light-transporting medium must permeate all space. This medium was called the *luminiferous ether*, or simply, **ether**.

The idea of an undetected ether became popular in the latter part of the nineteenth century. Maxwell, whose work laid the foundations for our understanding of electromagnetic waves (Chapter 20), believed in the existence of an etherlike substance, as evidenced by a quote from his writings:

> Whatever difficulties we may have in forming a consistent idea of the constitution of the ether, there can be no doubt that the interplanetary and interstellar spaces are not empty, but are occupied by a material substance or body which is certainly the largest, and probably the most uniform body of which we have any knowledge.

It would seem, then, that Maxwell's equations (the basis of electromagnetic theory that describes the propagation of light; see Section 20.4) did *not* satisfy the Newtonian relativity principle, as did the laws of mechanics. On the basis of the preceding discussion, a preferential reference frame would appear to exist—one that could be considered absolutely at rest—the ether frame.

This was the state of affairs toward the end of the nineteenth century, when scientists set out to investigate whether they had come upon a new dimension of physics or perhaps a flaw in what were considered established principles. One of the first attempts to resolve the situation was to prove that the ether existed. Then, presumably, a truly absolute rest frame could finally be identified. This was the purpose of the famous *Michelson–Morley experiment*.

During the 1880s, two American scientists, A. A. Michelson and E. W. Morley,* carried out a series of experiments designed to measure the Earth's velocity relative to the ether. They sought to do this, in effect, by measuring differences in the speed of light due to the Earth's orbital velocity. According to the ether theory, if you were moving relative to the ether (the absolute reference frame), then you would measure the speed of light to be different from c. Their experimental apparatus, while crude by today's standards, was capable of making such a measurement, and yielded a speed of c, *regardless* of the Earth's velocity—a *null* result.

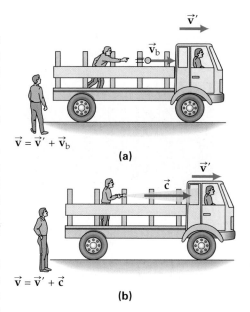

▲ **FIGURE 26.1 Relative velocity**
(a) According to a stationary observer on the ground, $\vec{v} = \vec{v}' + \vec{v}_b$. **(b)** Similarly, the velocity of light would classically be measured to be $\vec{v} = \vec{v}' + \vec{c}$, with a magnitude greater than c. (Velocity vectors are not drawn to scale—can you tell why?)

Note: Maxwell's equations are discussed in Section 20.4.

Note: More precisely, the speed of light has been measured as $2.997\,924\,58 \times 10^8$ m/s. We will assume c to be exact at 3.00×10^8 m/s for simplicity.

*Albert Abraham Michelson (1852–1931) was a German-born physicist who devised the Michelson interferometer in an attempt to detect the motion of the Earth through the ether. Edward Morley (1838–1923) was an American chemist who collaborated with Michelson.

▲ **FIGURE 26.2 Einstein and Michelson** A 1931 photo shows Michelson (left) with Einstein during a meeting in Pasadena, California.

26.2 The Postulates of Special Relativity and the Relativity of Simultaneity

OBJECTIVES: To explain (a) the two postulates of relativity, and (b) how they lead to the relativity of simultaneity.

The failure of the Michelson–Morley experiment to detect the ether left the scientific community in a quandary. The inconsistencies between Newtonian mechanics and electromagnetic theory remained unexplained. Many physicists were convinced that the experiment needed to be more accurate; they could not believe that light did not need a medium in which to propagate. These problems were resolved in a theory advanced by Albert Einstein in 1905 (◄Fig. 26.2). Interestingly, Einstein was apparently *not* motivated by the Michelson–Morley experiment in the development of his theory of relativity. When asked later, Einstein could not recall whether he had even known about the experiment when formulating his theory.

In fact, Einstein's insight was based on his intuition that the laws of mechanics should not be the only ones to obey the relativity principle. He reasoned that nature should be symmetrical. *All* physical laws should be encompassed by the relativity principle. In Einstein's view, the inconsistencies in electromagnetic theory were because of the assumption that an absolute rest frame existed. His theory did away with the need for such a frame (and the "ether") by placing all laws of physics on an equal footing, thus eliminating any way of measuring the absolute speed of an inertial reference frame.

The first of the two postulates on which relativity is based is thus an extension (generalization) of Newtonian relativity. Einstein's **principle of relativity** applies to *all* the laws of physics, including those of electricity and magnetism:

> **Postulate I (principle of relativity):** All the laws of physics are the same in all inertial reference frames.

This postulate means that all inertial reference frames are physically equivalent. That is, all physical laws, not just those of mechanics, are the same in all inertial frames. As a consequence, no experiment performed entirely within an inertial reference frame could enable an observer in that frame to detect its motion. That is, *there is no absolute reference frame.* In hindsight, the first postulate seems reasonable: We have no reason to think that nature would play favorites by picking the laws of mechanics over laws governing other phenomena.

Einstein's second postulate involves the speed of light. Recall that, according to Newtonian relativity, the speed of light can have any value. In fact, if a system were traveling in the same direction as the light and with the same speed, the speed of light in that frame would be zero. This possibility was the source of Einstein's question, "What would I see if I rode a beam of light?" In the same frame as the electromagnetic wave, the electric and magnetic field vectors would not vary with time. However, according to Maxwell's theory of light, the time variation of these two fields is *crucial* to the propagation of light (recall our discussion of time-varying fields and electromagnetic waves in Section 20.4). Hence, *static* fields are *inconsistent* with the propagation or travel of light.

To avoid this inconsistency, Einstein put forth his second postulate, called the **constancy of the speed of light**, as follows:

> **Postulate II (constancy of the speed of light):** The speed of light in a vacuum has the same value in all inertial systems.

These two postulates form the basis of Einstein's **special theory of relativity**. The "special" designation indicates that the theory deals only with the special case of inertial reference frames. The *general* theory of relativity, discussed later in the chapter, deals with the general case of noninertial, or accelerating, frames.

The second postulate is perhaps more difficult to accept than the first. It means that two observers in different inertial reference frames measure the speed of light to be c, *independent of the speed of the source or the observer.* For example, if a person moving toward you at a constant velocity turned on a flashlight, you both would measure the speed of the emitted light to be c, regardless of the relative velocity (►Fig. 26.3).

The second postulate is essential to the validity of the first and consistent with the null result of the Michelson–Morley experiment. By doing away with an absolute reference frame, Einstein could reconcile the apparently fundamental differences between mechanics and electromagnetism. Even so, the second postulate seems to go against "common sense." However, keep in mind that most people have no experience dealing with speeds near that of light. The ultimate test of any theory is provided by the scientific method. What does Einstein's theory predict, and can it be experimentally verified? The answer to the last part of the question is a resounding "yes" for every experiment performed, and we will discuss some of these experiments in the following sections.

With the two postulates in place, let's explore some of their implications. The special relativity postulates can be better understood by imagining simple situations. These situations can often tell us what phenomena the postulates predict. Einstein used this method by employing what he termed *gedanken*, or "thought," experiments—"experiments" done in the mind. Let us begin with a series of famous Einstein *gedanken* experiments related to simultaneity and how we measure length and time. We will then look at some experimental evidence that supports the theory and its predictions.

The Relativity of Simultaneity

In everyday life, we think of two events that are simultaneous to one person as being simultaneous to everyone. That is, we believe the concept of simultaneity is absolute—the same for everyone. What could be more obvious? Simultaneous events occur at the same time, and isn't that the same for all observers? The answer is no—but this result is obvious only for relative velocities near that of light. Thus for all intents and purposes, at everyday speeds this lack of agreement about simultaneity is too small to be observed.

Think of an inertial reference frame (called *O*) in which two events are *designed* to be simultaneous. For example, two firecrackers (located at points A and B on the *x*-axis) could be arranged to explode when a switch controlling a voltage source, placed midway between them, is flipped to the "on" position (▼Fig. 26.4a). Let's equip the observer in this frame with a light receptor at point R (for receptor), exactly midway between the firecrackers. This detector is capable of detecting whether two light flashes from the exploded firecrackers arrive at the same time, that is whether they are simultaneous (detected "in coincidence") or not. (Actually, the receptor could be placed anywhere, but the results would need to be corrected due to unequal travel distances. To avoid these complications, all simultaneity detectors will be placed midway between the two events.)

After detonation, the light receptor records that the two explosions went off simultaneously *in the O frame*. But consider the same two explosions as seen by an observer in a different inertial frame, *O'*. As viewed from *O*, the other frame *O'* is moving to the right at a speed *v*. The observer in *O'* has equipped himself with a series of light receptors on his *x'*-axis, because he is not sure which one will end up midway between the explosion points.

After the explosions, there are burn marks on both the *x*-axis (at A and B) and the *x'*-axis (at *A'* and *B'*), as shown in Fig. 26.4b. These marks can be used to identify the particular *O'* light receptor (call it *R'*) that was, in fact, located midway between *A'* and *B'*. But when the observer in *O'* reviews the data from this receptor, he finds that it did *not* record the explosions simultaneously. This result from *O'* does not cause the observer in *O* to doubt her conclusion, however. She has an explanation of what happened. As she sees the situation, during the time it took for the light to get to *R'*, that receptor had moved toward A and away from B. Consequently, the receptor in *O'* received the flash from A before that from B.

The question then appears to be: Which observer is correct? It's hard to find any objection to the conclusion reached by the observer in *O*—so isn't the observer in *O'* making a mistake? It should be obvious to him that he is moving with respect to the firecrackers. Why doesn't he realize this and take his motion into account? After all, wasn't his light receptor moving toward A and away from B? If so, then it should not surprise him that it recorded the flash from A before the flash

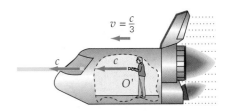

▲ **FIGURE 26.3 Constancy of the speed of light** Two observers in different inertial frames measure the speed of the same beam of light. The observer in frame *O'* measures a speed of *c* inside the ship. According to Newtonian mechanics, the observer in frame *O* would measure a speed of $c + c/3 = 4c/3$ as the beam passes her, but instead, according to Einsteinian mechanics, she measures *c*.

▶ **FIGURE 26.4 The relativity of simultaneity** (a) An observer in reference frame O triggers two explosions (at A and B) simultaneously. A light receptor R, located midway between them, records the two light signals as arriving at the same time. (b) An observer midway between the two explosions, but in frame O', moving with respect to O, sees the burn marks made by the two explosions on the x'-axis, but sees the explosion at A happen before that at B. (c) The situation as viewed from O'. The observer in O' sees the explosion at A before that at B. To him, O is moving to the left.

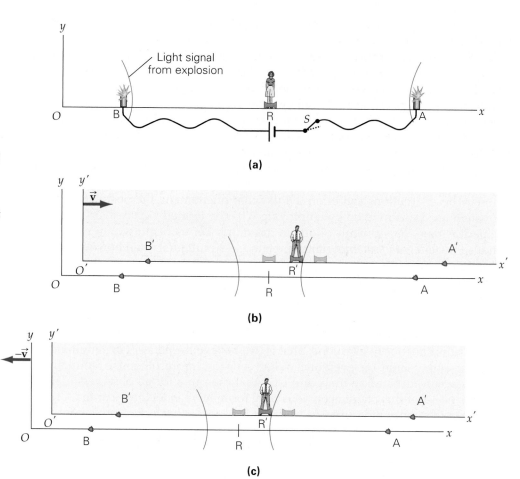

from B. All he has to do is to allow for this motion in his calculations, and he will conclude that the flashes "really" were simultaneous.

But if we reason this way, we are ignoring the postulates of relativity. This line of "logic" assumes that when the situation is viewed from the O frame (as in Figs. 26.4a and b), we are looking at what "really" happened from the vantage point of the frame that is "really" at rest. But according to the first postulate, no inertial reference frame is more valid than any other, and none can be considered absolutely at rest. The observer in O doesn't think of himself as moving. To this observer, *the O frame is moving, and O' is the rest frame.* He observes the firecrackers moving at a speed v to the left (Fig. 26.4c), but this motion would not affect his conclusions. To him, the explosions, equally distant from R', arrived at R' at different times; therefore they were not simultaneous.

You might wonder whether it could be arranged so that the observer in O' would agree that the explosions were simultaneous. However, to accomplish this, the observer in O would have to *delay* the firing of firecracker A relative to B so that R' would receive the two signals at the same time. This does not change the result, because the two observers would still disagree as to whether the events were simultaneous—because now they would no longer be simultaneous in O.

What are we to make of this curious situation? In a nonrelativistic world, one of the observers would have to be wrong. But as we have seen, both observers performed the measurements correctly. Neither one used faulty instruments, or made any errors in logic. So the conclusion must be that *both* are correct.

Notice that there is nothing special about a firecracker explosion. Any "happening" at a particular point in space at a particular time—a karate kick, a soap bubble bursting, a heartbeat—would have done just as well. Such a happening is called an *event* (specified by a location in space *and* a time of occurrence) in the language of relativity. Based on the postulates of relativity, we have the following result:

Events that are simultaneous in one inertial reference frame may not be simultaneous in a different inertial frame.

This kind of *gedanken* experiment convinced Einstein to give up on simultaneity as an absolute concept.

Note that if the relative speed of the reference frames is slow compared to that of light (as is true for everyday speeds), this lack of simultaneity is completely undetectable. This is why we conclude, erroneously, that simultaneity is absolute. *Most relativistic effects have this property.* That is, their departure from familiar experience is *not* apparent when the speeds involved are much less than the speed of light. Since we have no experience with such high speeds, it is hardly surprising that the predictions of special relativity seem strange.

Teaching tip: Explain that the lack of simultaneity here is for events on the x–x'-axes only. What would happen for events on the y–y'-axes?

Conceptual Example 26.1 ■ Agreeing to Disagree: The Relativity of Simultaneity

(a) In Fig. 26.4, estimate the relative speed of the two observers. (b) If the relative speed were only 10 m/s, would there be better agreement on simultaneity? Why?

Reasoning and Answer. (a) To estimate their relative speed, compare the distance between B and B' in Fig. 26.4b with the distance the light has traveled from B. The figure indicates that the O' frame has moved about 25% as far as the light has. Therefore, the relative speed between the two reference frames is approximately 25% of the speed of light, or $v \approx 0.25c$.
(b) At a relative speed of 10 m/s, the two frames would *not* have moved a noticeable distance; thus, both observers would agree on simultaneity.

Follow-Up Exercise. Show that two events that occur simultaneously on the *y*-axis of O are perceived as simultaneous by an observer in O', regardless of the relative speed, as long as the relative motion is along their common $x–x'$-axes. *(Answers to all Follow-Up Exercises are at the back of the text.)*

To grasp the importance of the relativity of simultaneity, think about how important simultaneity is when you are trying to measure the length of a moving object. To measure the length of an object properly, the positions of both ends of it must be marked *simultaneously*. However, two different inertial observers will, in general, *disagree* on simultaneity. Thus, they will also disagree on the object's length.

26.3 The Relativity of Length and Time: Time Dilation and Length Contraction

OBJECTIVES: To (a) understand the origins of time dilation and length contraction, and (b) determine the relationship between time intervals and lengths observed in different inertial frames.

Time Dilation

Another of Einstein's *gedanken* experiments pertained to the measurement of time intervals in different inertial frames. To compare time intervals in different inertial reference frames, he envisioned a *light pulse clock*, illustrated in ▼Fig. 26.5a. A tick (time interval) on the clock corresponds to the time a light pulse would take to make a round trip between the source and the mirror. Let's assume that the observers in the two inertial frames, O and O', have identical light clocks, and the clocks run at the same rate when they are at rest relative to one another. For an observer at rest with respect to one of these clocks, the time interval (Δt_o) for a round trip of a light pulse is the total distance traveled, divided by the speed of light, or

$$\Delta t_o = \frac{2L}{c} \qquad (26.1)$$

Now, suppose O' is moving relative to O with a constant velocity $\vec{v}$ to the right. With his clock (at rest in O') the observer in O' measures the same time interval for

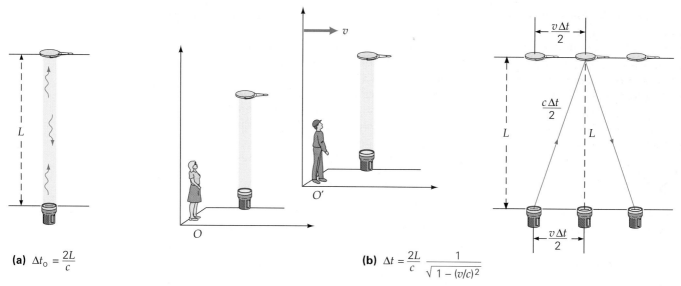

(a) $\Delta t_o = \dfrac{2L}{c}$

(b) $\Delta t = \dfrac{2L}{c}\dfrac{1}{\sqrt{1-(v/c)^2}}$

▲ **FIGURE 26.5 Time dilation** **(a)** A light clock that measures time in units of round-trip reflections of light pulses. The time for light to travel up and back is $\Delta t_o = 2L/c$. **(b)** An observer in O measures a time interval of $\Delta t = (2L/c)\left[1/\sqrt{1-(v/c)^2}\right]$ on the clock in the O' frame. Thus, the moving clock appears to run slowly to the observer in O.

his clock, Δt_o. However, according to the observer in O, the clock in the O' system is moving, and the path of its light pulse forms the sides of two right triangles (Fig. 26.5b). Thus, the observer in O sees the light pulse from the clock in O' take a longer path (and therefore a longer time interval Δt) than the light from her own clock. From the O frame, we can apply the Pythagorean theorem:

$$\left(\frac{c\Delta t}{2}\right)^2 = \left(\frac{v\Delta t}{2}\right)^2 + L^2$$

(Here, Δt is the time interval of the O' clock as measured by the observer in O.) Since the speed of light is the same for all observers, the light in the "moving" clock's reference frame takes a longer time to cover the path, *according to the observer in O.* That is, for the observer in the O frame, the "moving" clock runs slowly, that is, the ticks occur at a slower rate. To determine the relationship between the time intervals, we can solve the preceding equation for Δt:

$$\Delta t = \frac{2L}{c}\left[\frac{1}{\sqrt{1-(v/c)^2}}\right] \tag{26.2}$$

But the time interval measured by an observer at rest with respect to a clock is $\Delta t_o = 2L/c$ (Eq. 26.1). Combining the equations, we obtain

$$\Delta t = \frac{\Delta t_o}{\sqrt{1-(v/c)^2}} \qquad \textit{relativistic time dilation} \tag{26.3}$$

Since $\sqrt{1-(v/c)^2}$ is less than 1 (why?), then $\Delta t > \Delta t_o$. Thus, an observer in O measures a longer time interval (Δt) *on the O' clock* than does the observer in O' on the same clock (Δt_o). This effect is called **time dilation**. With longer time between ticks, the O' clock appears, to an observer in O, to run more slowly than the O clock. The situation is symmetric and relative: The observer in O' would say that the clock in the O frame ran slowly relative to the O' clock:

> Moving clocks are observed to run more slowly than clocks that are at rest in the observer's own frame of reference.

This effect, like all relativistic effects, is significant only if the relative speeds are close to that of light.

To distinguish between the two time intervals, the term **proper time interval** is used. As with most measurements, it is usually "proper" or normal to be at rest with respect to a clock when a time interval is measured. In the preceding development, the proper time interval is Δt_o. Stated another way, the proper time interval between two events is the interval measured by an observer at rest relative to the two events and who sees them occur *at the same location in space*. (What are the two events for the light-pulse clock? Are they at the same location in O' for the O' clock?) In Fig. 26.5, the observer in O sees the events by which the time interval of the O' clock is measured at *different* locations. Because the clock is moving, the starting event (the light pulse leaving) occurs at a different location in O from that of the ending event (the light pulse returning). Thus, Δt, the time interval measured by the observer in O, is *not* the proper time interval.

Many of the equations of relativity can be written more compactly if the expression $1/\sqrt{1 - (v/c)^2}$ is replaced by γ (Greek letter "gamma"), defined as

$$\gamma \equiv \frac{1}{\sqrt{1 - (v/c)^2}} \qquad (26.4)$$

Note that γ is always greater than or equal to 1. (When is it equal to 1?) Also notice that as v approaches c, then γ approaches infinity. Since an infinite time interval is not physically possible, *relative speeds equal to or greater than that of light are not possible*. The values of γ for several values of v (expressed as fractions of c) are listed in Table 26.1. Notice that speeds must be an appreciable fraction of c before relativistic effects can be observed. At $v = 0.10c$, for example, γ differs from 1.00 by only 1%. Using Eq. 26.4, the time dilation relationship (Eq. 26.3) becomes

$$\Delta t = \gamma \Delta t_o \qquad (26.5)$$

Suppose you observed a clock at rest in a system that is moving relative to you at a constant velocity of $v = 0.60c$. For that speed, $\gamma = 1.25$. Thus when 20 min have elapsed on that clock, you would observe an interval of $\Delta t = \gamma \Delta t_o = (1.25)(20 \text{ min}) = 25$ min on *your* clock. The 20 min is the proper time, since the events defining the 20-min interval took place at the same location (that of the "moving" clock—for example, for readings at 8:00 A.M. and then at 8:20 A.M.). Thus, the "moving" clock runs more slowly (20 min elapsed, as opposed to 25 min on your clock) when viewed by an observer (you) moving relative to it.

Finally, note that the time-dilation effect cannot apply just to our artificial light-pulse clock. *It must be true for all clocks and hence all time intervals* (that is, anything that keeps a rhythm or frequency, including the heart). If this were not the case—if a mechanical watch, for example, did not exhibit time dilation—then that watch and a light-pulse clock would run at different rates *in the same inertial frame*. This would mean that observers in that frame would be able to tell whether they were moving by making a comparison *solely within their frame*. Since this violates the first postulate of special relativity, it follows that *all* moving clocks, regardless of their nature, must exhibit time dilation. Let's take a look at an actual situation in nature.

Note: The proper time interval Δt_o is always less than the dilated time interval Δt.

TABLE 26.1

Some Values of

$$\gamma = \frac{1}{\sqrt{1 - (v/c)^2}}$$

v	γ^*
0	1.00
$0.100c$	1.01
$0.200c$	1.02
$0.300c$	1.05
$0.400c$	1.09
$0.500c$	1.15
$0.600c$	1.25
$0.700c$	1.40
$0.800c$	1.67
$0.900c$	2.29
$0.950c$	3.20
$0.990c$	7.09
$0.995c$	10.0
$0.999c$	22.4
c	∞

*To illustrate the dependence of γ as v/c approaches 1.

Example 26.2 ■ Muon Decay Viewed from the Ground: Time Dilation Verified by Experiment

Subatomic particles, called *muons*, can be created in the Earth's atmosphere when cosmic rays (mostly protons) collide with the nuclei of the atoms that compose air molecules. Once created, they approach the Earth's surface with speeds near c (typically, about $0.998c$). However, muons are known to be unstable and decay into other particles. The average lifetime of a muon *at rest* has been measured to be 2.20×10^{-6} s. During this time, the muon would travel a distance of $d = v_o \Delta t = (0.998c)(2.20 \times 10^{-6} \text{ s}) = [0.998(3.00 \times 10^8 \text{ m/s})](2.20 \times 10^{-6} \text{ s}) = 659$ m. This is 0.659 km, or less than half a mile. Since muons are created at altitudes of 5 to 15 km, we should therefore expect very few of them to reach the Earth's surface. However, an appreciable number *actually do* reach the surface. Using time dilation, explain this apparent paradox.

(continues on next page)

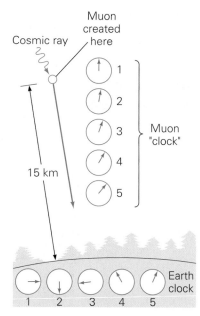

▲ FIGURE 26.6 Experimental evidence of time dilation Muons are observed at the surface of the Earth, as predicted by special relativity. The sequence of integers (1 through 5 on each clock) refers to the order of events. Note that from the Earth's viewpoint, the muon's "clock" runs slower than Earth clocks. (See Example 26.2.)

Thinking It Through. The paradox arises because the preceding calculation does not take time dilation into account. That is, a muon decays by its own "internal clock," as measured in its own reference frame. Its "lifetime" of 2.20×10^{-6} s is a proper time interval, and so is its "proper" lifetime. This is because in the muon's rest frame and "birth" and "death" events take place at the same location. Thus, to an observer on the Earth, any "clock" in the muon's reference frame would appear to run more slowly than a clock on the Earth (◄Fig. 26.6). If the muon's lifetime is actually dilated enough, then perhaps its travel distance will be long enough to explain its presence at the Earth's surface.

Solution. We list the given quantities:

Given: $v = 0.998c$ *Find:*   An explanation, based on time dilation, of
   $\Delta t_o = 2.20 \times 10^{-6}$ s   why muons make it to the Earth's surface
   (muon proper lifetime)

Instead of the proper time interval $\Delta t_o = 2.20 \times 10^{-6}$ s, an observer on the Earth would measure a time interval longer by a factor of γ. From Eq. 26.4,

$$\gamma = \frac{1}{\sqrt{1 - (v/c)^2}} = \frac{1}{\sqrt{1 - (0.998c/c)^2}} = 15.8$$

From Eq. 26.5, the lifetime of the muon, *according to an observer on the Earth*, is

$$\Delta t = \gamma \Delta t_o = (15.8)(2.20 \times 10^{-6}\text{ s}) = 3.48 \times 10^{-5}\text{ s}$$

The distance the muon travels, *according to an observer on the Earth* (using the dilated time interval), is

$$d = v\Delta t = 0.998(3.0 \times 10^8\text{ m/s})(3.48 \times 10^{-5}\text{ s})$$
$$= 1.04 \times 10^4\text{ m} = 10.4\text{ km}$$

This distance is approximately the same altitude at which muons are created. Hence, the detection of more muons than expected is a confirmation of time dilation.

Follow-Up Exercise. In this Example, what speed would enable the muons to travel 20.8 km relative to the Earth (that is, twice as far as the distance in the Example)? Would they have to travel twice as fast? Explain.

Problem-Solving Hint

In working time-dilation problems, the proper time interval Δt_o must be identified. To do this, first identify (1) the events that define the beginning and end of the interval and (2) a clock (real or imagined) that is present at both events. This "clock," and the observer at rest with respect to it, measures the proper time interval Δt_o. The "dilated" time interval can then be determined from $\Delta t = \gamma \Delta t_o$ for any inertial observer moving relative to this "proper" clock.

Length Contraction

To measure the length of a linear object not at rest in our reference frame, we must take care to mark both ends simultaneously. Consider again the two inertial reference frames used in the discussion of simultaneity, and imagine a measuring stick lying on the x-axis at rest in O (▶Fig. 26.7a). If the observer in O marks the ends simultaneously, the observer in O′ will observe end A marked before B. Thus, for the observer in O′ to make a correct length measurement, the observer in O must delay the marking of A relative to that of B (Fig. 26.7b). Imagine the observer in O setting off explosions that create burn marks in both reference frames (on both the x- and x′-axes). When the ends are marked so that the observer in O′ agrees that they were done simultaneously, all that needs to be done is to subtract the two positions to get the length of the stick *as measured in O′*. Notice that it is going to be *less* than the length measured by O.

Again, it might be asked, "Which observer makes the correct measurement?" By now you know the answer: Both are correct. *Both have measured the length correctly*

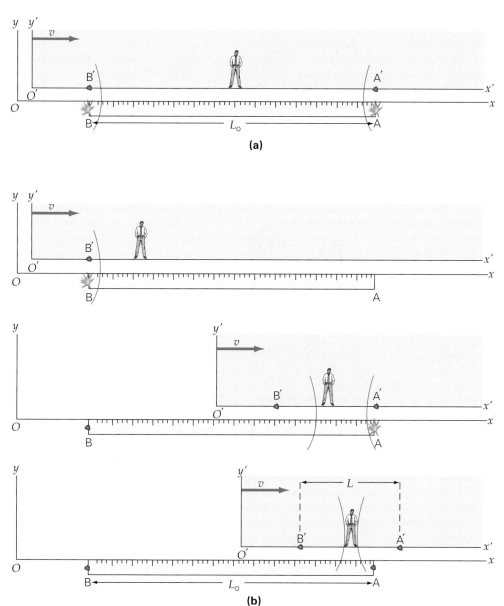

◀ **FIGURE 26.7 Measuring lengths correctly** To measure the length of a moving object correctly, the ends must be marked simultaneously. **(a)** When the observer in frame O marks the ends simultaneously, the observer in frame O' does not agree and observes A marked before B, resulting in too long a length from the viewpoint of O'. **(b)** When O delays the marking of A relative to that of B by just the correct amount (how do we know from the sketch?), the observer in O' measures the correct length from his point of view. The length measured by the observer in O' is less than the rest (or proper) length measured by the observer in O.

in their own frames. Neither thinks that the other has done things correctly, but each is satisfied with his or her own measurement. The observer in O' would have measured the same length as that in O *if he were willing to overlook the fact that the burn marks were not made simultaneously*, as in Fig. 26.7a. However, this is *not* the correct way to measure the length of the moving stick. The lack of agreement on simultaneity leads to the following qualitative statement about length contraction:

> An object's length is largest when measured by an observer at rest with respect to it (the "proper" observer). If the object is moving relative to an inertial observer, that observer measures a smaller length than the proper observer.*

As usual, this effect is entirely negligible at speeds that are slow compared with c.

The distance between two points as measured by the observer at rest with respect to them is designated by L_o and is called the **proper length**. The proper length (also known as the *rest length*) is the largest possible length. The term *proper*

Teaching tip: Impress on students that the phenomena of length contraction and time dilation, among others, are based solely on the two postulates of relativity.

Note: The proper length L_o is always greater than the contracted length L.

---

*Here, "length" refers to the dimension of the object that is in the direction of its relative velocity. That is, if a cylindrical stick is moving parallel to its long axis, then only its long-axis "length"—and not its diameter—would exhibit length contraction.

▶ **FIGURE 26.8 Derivation of length contraction** The observer in O measures the time it takes for the observer in O' to move past the ends of the rod. Similarly, the observer in O' measures the time it takes for the ends of the rod to pass her. The observer in O' is the *proper time measurer*; she measures the shortest possible time between these two events. The measured lengths of the rod are not the same. The observer in O is the *proper length measurer*; he measures the longest possible length.

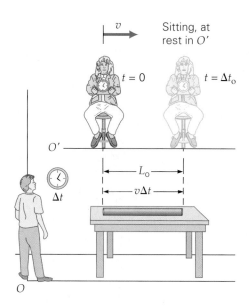

Teaching tip: Ask students how they would measure the length of a moving snake. The word *simultaneous* should pop up naturally, and then it should be clear that length measurements are also relative.

Demonstration/activity: Sketch graphs of L and Δt as functions of relative speed.

has nothing to do with the correctness of the measurement, because each observer is measuring correctly from his or her point of view.

A *gedanken* experiment can help us develop an expression for length contraction. Consider a rod at rest in frame O. This means that the observer in frame O is the proper length measurer for this rod. Thus, the length of the rod that he measures is L_o. An observer in O', traveling at a constant speed v in the direction parallel to the stick, also measures the length of the rod (▲Fig. 26.8). She does this by using the clock she is holding to measure the time interval required for the two ends of the rod to pass her. Since she measures the proper time interval (how do we know this?), the time interval she measures is Δt_o. In her reference frame, the rod is moving to the left with speed v. With that information, she can determine the length L, since $L = v \Delta t_o$ (speed × time).

The observer in O could also measure the length of the rod by the same means. To him, the observer in O' is moving to the right past the rod. If he notes the times on his clock when O' passes the ends of the rod, he measures a time interval Δt. For him, the length of the stick is the proper length; therefore, $L_o = v \Delta t$. Then, dividing one length by the other, we have

$$\frac{L}{L_o} = \frac{v \Delta t_o}{v \Delta t} = \frac{\Delta t_o}{\Delta t} \qquad (26.6)$$

But $\Delta t = \gamma \Delta t_o$ (Eq. 26.5), or $\Delta t_o / \Delta t = 1/\gamma$. Thus Equation 26.6 becomes

$$L = \frac{L_o}{\gamma} = L_o \sqrt{1 - (v/c)^2} \qquad \begin{array}{l} \textit{relativistic} \\ \textit{length contraction} \end{array} \qquad (26.7)$$

Since γ is always greater than 1, then $L < L_o$. This effect is called **relativistic length contraction**.

To see the effects of length contraction and time dilation, consider the following two high-speed Examples.

Example 26.3 ■ Warp Speed? Length Contraction and Time Dilation

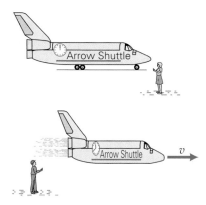

▲ **FIGURE 26.9 Length contraction and time dilation** As a result of length contraction, moving objects are observed to be shorter, or contracted, in the direction of motion, and moving clocks are observed to run more slowly because of time dilation. (See Example 26.3.)

An observer sees a spaceship, measured to be 100 m long when at rest, pass completely by (that is, nose to tail) in uniform motion with a speed of $0.500c$ (◀Fig. 26.9). While this observer is watching the ship, a time of 2.00 s elapses on a clock onboard the ship. (a) What is the length of the ship as measured by the observer? (b) What time interval elapses on the observer's clock during the 2.00-s interval on the ship's clock?

Thinking It Through. One hundred meters is the proper length. (Why?) The time interval of 2.00 s is the proper time interval, because the same clock, in the same location, measures the beginning and end of the interval. In (a), Eq. 26.7 can be used to find the contracted length, and in (b), Eq. 26.5 will enable us to determine the dilated time interval.

Solution. Listing the given quantities:

Given: $L_o = 100$ m (proper length)   **Find:**   (a) L (contracted length)
                $v = 0.500c$ (b) Δt (dilated time interval)
                $\Delta t_o = 2.00$ s (proper time interval)

(a) By calculation or from Table 26.1, $\gamma = 1.15$ for $v = 0.500c$, and the contracted length contraction is given by Eq. 26.7:

$$L = \frac{L_o}{\gamma} = \frac{100 \text{ m}}{1.15} = 87.0 \text{ m}$$

(b) The time interval Δt measured by the observer is longer than the proper time interval Δt_o and is given by Eq. 26.5:

$$\Delta t = \gamma \Delta t_o = 1.15(2.00 \text{ s}) = 2.30 \text{ s}$$

Follow-Up Exercise. In this Example, find the time it takes the spaceship to pass a given point in the observer's reference frame, as seen by (a) a person in the spaceship and (b) the observer watching the ship move by. Explain clearly why these time intervals are *not* the same.

Example 26.4 ■ Muon Decay Revisited: Alternative Explanations

Example 26.2 showed, using relativistic time dilation, how many more muons reach the Earth's surface than can be accounted for without relativistic considerations. Since a hypothetical observer *on the muon* could *not* use this argument (why not?), how would he or she explain the fact that the average muon does make it to the surface? Which explanation is "correct?"

Thinking It Through. A muon traveling at $v = 0.998c$ decays by its own clock (a proper time interval) in $\Delta t_o = 2.20 \times 10^{-6}$ s. In that time, we have seen that it would travel only 659 m, not nearly long enough to reach the Earth's surface. In Example 26.2, this apparent paradox was explained (at least for the Earth observer) by time dilation. According to the Earth observer, the muon's "clock" runs slowly, enabling it to travel farther than expected. But how is the "paradox" explained by a hypothetical observer on the muon? For such an observer, the muon clock is correct—and it registers a time interval that is not sufficient for the muon to reach the Earth's surface! How can this be reconciled with the experimental observation of the Earth observer that finds the muon making it to the surface? After all, two observers *cannot* disagree on this experimental result! To the observer on the muon, the distance to the surface moves by quickly, so the explanation from the muon reference frame must involve length contraction.

Solution.

Given: See Example 26.2   **Find:**   the explanation for muons making it to the
                                                   Earth's surface from the muon reference frame

The apparent paradox disappears when length contraction is taken into account. For the observer on the muon, its "clock" reads correctly ($\Delta t_o = 2.20 \times 10^{-6}$ s), but the travel distance is shorter because of length contraction. With $\gamma = 15.8$ for $v = 0.998c$, a length of 10.0 km in the Earth frame, which is the proper length L_o (why?), is measured by the observer on the muon to be considerably shorter because

$$L = \frac{L_o}{\gamma} = \frac{10.0 \text{ km}}{15.8} = 0.633 \text{ km} = 633 \text{ m}$$

To travel this distance would take a time (according to the muon observer) of

$$\Delta t = \frac{L}{v} = \frac{L}{0.998c} = \frac{633 \text{ m}}{0.998(3.00 \times 10^8 \text{ m/s})} = 2.11 \times 10^{-6} \text{ s}$$

This is approximately equal to the muon lifetime *in the muon's reference frame*. Thus, through relativistic considerations, both observers agree that many muons reach the Earth (the experimental result). The Earth observer explains this result by saying, "The muon clock is running slow" (time dilation). The observer on the muon says, "No, the clock is fine, but the distance we have to travel is considerably less than you claim" (length contraction). Who is correct? Both are. The reasoning is different for different observers, but the experimental result *must be* the same.

Follow-Up Exercise. Muons are actually created with a range of speeds. What is the speed of a muon if it decays 5.00 km from its creation point as measured by an Earth observer?

Teaching tip: Ask students what a trip to the store would be like in terms of distances and time if the speed of light were 100 km/h.

The Twin Paradox

Time dilation gives rise to a popular relativistic topic: the **twin paradox**, or **clock paradox**. According to special relativity, a clock moving relative to an observer runs more slowly than one in that observer's frame. Since heartbeat intervals and ages are proper time intervals, the question arises: Do you age more quickly than does a person moving relative to you?

One way to explore this question is through another *gedanken* experiment. Consider identical twins, one of whom goes on a high-speed journey into space. Will the space traveler come back younger than the Earth-bound twin? Or will the space twin see the Earth twin age more slowly? *Both can't be right*, and therein lies the apparent paradox.

The resolution of this apparent paradox lies in the fact that in leaving and returning to Earth, the space twin *must* experience accelerations and so is not always in an inertial reference frame. The stay-at-home twin does not feel the forces associated with speed and directional changes that the traveling twin experiences. Thus, the two twins are individually "marked," and their experiences are *not* symmetrical. However, if the acceleration periods (speedup at start, velocity reversal at turnaround, and slowdown at return) occupy only a negligible part of the total time of the trip, special relativity gives the correct result. Under these conditions, the result is that the traveling twin does indeed return younger than the Earth-bound twin, and both agree on that fact.

During the (constant-velocity) outward and return trips, the proper length of the trip is measured by the Earth twin with fixed beginning and end points. The traveling twin thus measures a length contraction, or a shorter distance for the trip. Traveling at the same relative speed, the space twin's heart thus beats for a shorter time than that of the Earth twin. The traveler *returns home younger* than the Earth-bound twin, *according to the traveling twin*. Now, according to the Earth twin, the traveling twin's heart beats more slowly (time dilation), so the traveling twin *returns home younger* than the Earth-bound twin, *according to the Earth twin*. There is no disagreement. The traveler returns having aged less than the nontraveler. At high speeds, it is possible for the traveler to return and find many generations of Earthlings long gone.

The twin "paradox" has been experimentally verified with extremely accurate atomic clocks flown around the world. Extreme accuracy is necessary because relativistic effects are very small at speeds much slower than c. The time recorded by the traveling clock was compared with that measured by an identical clock that stayed home. The difference in time intervals agreed with the theoretical predictions (with corrections for accelerations on landings and takeoffs). Thus time dilation in our everyday world, although extremely small, was experimentally verified.*

To see some of the details in the traveling twin situation, consider the next Example.

Example 26.5 ■ To the Stars: Relativity and Space Travel

Consider a high-speed round trip taken by one of a set of twins. Ignore accelerations at the start, end, and turnaround points, and assume that a negligible amount of time is spent at the turnaround. (a) Find the speed at which a space explorer would need to travel to make the round trip to a star 100 light-years away in only 20.0 years of traveler time. (*Note*: A light-year is defined as the distance light travels in a vacuum in 1 year.) (b) How much time elapses on the Earth during this trip?

Thinking It Through. Since the trip is symmetrical, we can calculate a one-way trip and double the result to get the answer for a round trip. The explorer measures a one-way *proper time* of 10.0 years, because he is present at the start and end of the trip. People on the Earth measure a one-way *proper length* of 100 light-years, since the beginning and end markers (the Earth and the star) of the trip are at rest with respect to the Earth. **(a)** To find the speed, either the Earth-bound explanation (time dilation) or that of the traveler (length contraction) can be used. **(b)** To determine the time from the Earth's viewpoint, we use the proper distance and the explorer's speed [from part (a)].

*J. Hafele and R. Keating, "Around-the-World Atomic Clocks: Relativistic Time Gains Observed," *Science*, 117 (July 14, 1972), 166–170.

Solution. Listing the given quantities:

Given: $L_o = 100$ light-years (proper length)
$\Delta t_o = 10.0$ years (one-way proper time)

Find: (a) v (speed of the traveler)
(b) Δt (time elapsed on Earth)

(a) For the traveler, the length is the contracted version of 100 light-years. According to the traveler, a one-way trip takes

$$\Delta t_o = \frac{\text{distance traveled}}{\text{speed}} = \frac{L}{v} = \frac{L_o\sqrt{1 - \left(\frac{v}{c}\right)^2}}{v}$$

This equation can be solved for v:

$$v = \frac{c}{\sqrt{1 + \left(\frac{c\Delta t_o}{L_o}\right)^2}} = \frac{c}{\sqrt{1 + \left(\frac{10.0 \text{ light-years}}{100 \text{ light-years}}\right)^2}} = 0.995c$$

Thus, v is 99.5% the speed of light. Notice that we have used the fact that the distance light travels in 10.0 years $(c\Delta t_o)$ is, by definition, 10.0 light-years. We did *not* have to convert to meters because both distances $(c\Delta t_o$ and $L_o)$ were expressed in light-years.

(b) The people on the Earth observe the traveler covering a total of 200 light-years at $0.995c$, so the round-trip time is 200 light-years$/0.995c$, or about 201 years, compared with 20.0 years for the traveler. The traveler would come back to find that everyone who was alive on Earth when he left had long since died!

Follow-Up Exercise. Verify the same result as that obtained in part (b) of this Example, but from time-dilation considerations. Show, using that approach, that Earth observers find 201 of their years elapsed during the trip. [*Hint*: Carry intermediate results to five decimal places, and round to three significant figures at the end.]

26.4 Relativistic Kinetic Energy, Momentum, Total Energy, and Mass–Energy Equivalence

OBJECTIVES: To (a) understand the relativistically correct expressions for kinetic energy, momentum, and total energy, (b) understand the equivalence of mass and energy, and (c) use the relativistically correct expressions to calculate energy and momentum in elementary particle interactions.

The ramifications of special relativity are particularly important in particle physics, in which speeds routinely approach c. Many of the expressions from classical (low-speed) mechanics are incorrect at these high speeds. Kinetic energy, for example, is different from the familiar $K = \frac{1}{2}mv^2$ we routinely used in classical mechanics. Einstein showed that kinetic energy still increases with speed, but in a different way. He found that the **relativistic kinetic energy** of a particle of mass m moving with speed v is

$$K = \left[\frac{1}{\sqrt{1 - (v/c)^2}} - 1\right]mc^2 = (\gamma - 1)mc^2 \quad \begin{array}{l}\textit{relativistic} \\ \textit{kinetic energy}\end{array} \quad (26.8)$$

It can be shown that this expression becomes the more familiar $K = \frac{1}{2}mv^2$ when $v \ll c$.

According to Eq. 26.8, as v approaches c, the kinetic energy of an object becomes infinite. In other words, to accelerate an object to $v = c$ would require an infinite amount of energy or work, which is not possible. Thus, no object can travel as fast as, or faster than, the speed of light.

Particle accelerators can accelerate charged particles to very high speeds. There is complete agreement between the experimentally measured kinetic energies of these charged particles and Eq. 26.8. A graphical comparison of the relativistically correct expression for kinetic energy and the classical (low-speed) expression is shown in ▶Fig. 26.10. As can be seen, the relativistic and classical expressions agree at low speeds.

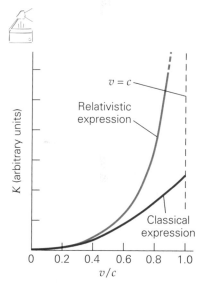

▲ **FIGURE 26.10 Relativistic versus classical kinetic energy** The variation in kinetic energy with particle speed, expressed as a fraction of c, is shown for the relativistically correct expression and for the classical expression. The classical expression becomes negligibly different from the relativistic one for speeds less than about $0.2c$. As objects approach the speed c, their kinetic energy becomes very large. Objects cannot have a speed of exactly c, because their kinetic energy would be infinite.

Relativistic Momentum

Just as for kinetic energy, the momentum of an object is different from the low-speed expression ($\vec{\mathbf{p}} = m\vec{\mathbf{v}}$). The expression for **relativistic momentum** is

$$\vec{\mathbf{p}} = \frac{m\vec{\mathbf{v}}}{\sqrt{1 - \left(\dfrac{v}{c}\right)^2}} = \gamma m \vec{\mathbf{v}} \quad \textit{relativistic momentum} \tag{26.9}$$

Note that momentum is still a vector and total (vector) momentum is a conserved quantity under the proper conditions. (See Section 6.3.)

Relativistic Total Energy and Rest Energy: The Equivalence of Mass and Energy

In classical mechanics, the total mechanical energy of an object is the sum of its kinetic and potential energies ($E = K + U$). When there is no potential energy (a free object), this becomes $E = K$ since the total energy is all kinetic. However, Einstein was able to show that the **relativistic total energy** of such an object (that is, a free one) is instead given by

$$E = \frac{mc^2}{\sqrt{1 - \left(\dfrac{v}{c}\right)^2}} = \gamma mc^2 \quad \textit{relativistic total energy} \tag{26.10}$$

Thus, according to relativity, when an object is at rest and has no kinetic energy ($K = 0, v = 0$) *it still has an energy of mc^2, not zero*. This minimum energy that an object always possesses is called its **rest energy** and is given by

$$E_o = mc^2 \quad \textit{rest energy} \tag{26.11}$$

Demonstration/activity: Sketch graphs of kinetic energy and momentum as functions of speed.

Unlike the classical result, when $v = 0$, the total energy of the particle is *not* zero, but E_o. Since $K = (\gamma - 1)mc^2$, the total energy of a particle can be expressed as the sum of its kinetic and rest energies. To see this, note that the relativistic kinetic energy expression (Eq. 26.8) can be rewritten as

$$K = (\gamma - 1)mc^2 = \gamma mc^2 - mc^2 = E - E_o$$

Thus, an alternative to Eq. 26.10 is

$$E = K + E_o = K + mc^2 \quad \begin{array}{l}\textit{relativistic total energy} \\ \textit{with no potential energy}\end{array} \tag{26.12}$$

(In the more general case, when a particle also has potential energy U, its total energy is given by $E = K + U + mc^2$.)

A relationship between a particle's total energy and its rest energy can be obtained by replacing mc^2 with E_o in Eq. 26.10:

$$E = \gamma E_o \tag{26.13}$$

As a double check, note that when $v = 0$ (that is, $K = 0$), $\gamma = 1$ and $E = E_o$ as expected.

Equation 26.11 expresses Einstein's famous **mass–energy equivalence**. *An object has energy even at rest—its rest energy.* Consequently, mass is a form of energy. In nuclear and particle physics, it is impossible for total energy to be conserved unless mass is treated as a form of energy.

Mass–energy equivalence does not mean that mass can be converted into useful energy at will. If so, our energy problems would be solved, since there is a lot of mass on the Earth. However, significant practical conversion of mass into other forms of energy, such as heat, does take place in nuclear reactors used to generate electric energy (Section 30.2).

The (rest) energy of an object depends on its mass. For example, the mass of an electron is 9.109×10^{-31} kg, and therefore its rest energy is

$$E_o = mc^2 = (9.109 \times 10^{-31} \text{ kg})(2.998 \times 10^8 \text{ m/s})^2 = 8.187 \times 10^{-14} \text{ J}$$

or, converted into electron-volts (eV) and megaelectron-volts (MeV),

$$E_o = (8.187 \times 10^{-14}\,\text{J})\left(\frac{1\,\text{eV}}{1.602 \times 10^{-19}\,\text{J}}\right) = 5.110 \times 10^5\,\text{eV} = 0.5110\,\text{MeV}$$

In high-energy, particle, and nuclear physics, it is common to express rest energies in terms of the electron-volt (eV) or multiples of it, such as the MeV. For example, the rest energy of an electron is said to be 511 keV or 0.511 MeV. This is very convenient because often the particles are accelerated by potential differences measured in volts.

How do you know whether you need to use the relativistic expressions or can "get away" with the classical ones? The rule of thumb is that if an object's speed is 10% of the speed of light or less, then the error in using the nonrelativistic kinetic energy expression is less than 1%. (The classical expression is lower than the relativistic one.) At $v < 0.1c$, the object's kinetic energy is less than 0.5% of its rest energy. Thus, a commonly accepted practice is to use this speed and kinetic-energy region as a dividing line:

> For speeds below 10% of the speed of light or kinetic energies less than 0.5% of an object's rest energy, the error in using the nonrelativistic formulas is less than 1%, and it is then usually acceptable to use the nonrelativistic expressions.

For example, an electron with a kinetic energy of 50 eV would qualify as a "nonrelativistic electron" because 50 eV is (0.01%) of its rest energy. An electron with a kinetic energy of 0.511 MeV would, however, be *highly* relativistic, because its kinetic energy is the same as its rest energy. Thus, what counts is the particle's kinetic energy relative to its rest energy, or its speed relative to that of light—even the dividing line is relative! The following Example shows how particle energies are calculated using relativistically correct expressions.

Example 26.6 ■ A Speedy Electron: Energy Required for Acceleration

(a) How much work is required to accelerate an electron from rest to a speed of 0.900c?
(b) How much error is made by using the nonrelativistic expression?

Thinking It Through. (a) By the work–energy theorem, the work needed is equal to the electron's gain in kinetic energy. The gain in kinetic energy is the same as the final kinetic energy, since the initial kinetic energy is zero. As we have seen, the rest energy of the electron is 0.511 MeV. From its speed, its kinetic energy can be determined with Eq. 26.8. (b) Here we will use the nonrelativistic kinetic energy, $K = \frac{1}{2}mv^2$, and compare it to the answer in part (a).

Solution. The data are as follows:

Given: $v = 0.900c$ *Find:* (a) W (work required)
        $E_o = 0.5110$ MeV (from text)   (b) K (nonrelativistic kinetic energy)

(a) The relativistic kinetic energy is given by Eq. 26.8:

$$K = (\gamma - 1)mc^2 = (\gamma - 1)E_o$$

With $v = 0.900c$, by calculation or from Table 26.1, $\gamma = 2.29$; therefore,

$$K = (\gamma - 1)E_o = (2.29 - 1)(0.511\,\text{MeV}) = 0.659\,\text{MeV}$$

Thus 0.659 MeV of work is required to accelerate an electron to a speed of 0.900c.

(b) Many times, even with nonrelativistic expressions, a shortcut can be applied as follows: Essentially, you can work with mass expressed in energy units. The technique involves first multiplying and then dividing by c^2, which enables you to use E_o. Let's try it here:

$$K_{\text{nonrel}} = \tfrac{1}{2}mv^2 = \tfrac{1}{2}(mc^2)\left(\frac{v}{c}\right)^2 = \tfrac{1}{2}(0.5110\,\text{MeV})(0.900)^2 = 0.207\,\text{MeV}$$

In this case, the nonrelativistic expression for kinetic energy gives an answer that is low by a factor of more than 3.

Follow-Up Exercise. In this Example, (a) what is the relativistic total energy of the electron? (b) What would be the electron's total energy if it were treated nonrelativistically?

To see how relativistic momentum is applied, consider the next Example.

Integrated Example 26.7 ■ When 1 + 1 Doesn't Equal 2: Conservation of Relativistic Momentum and Energy

A particle of mass m, initially moving, collides with an identical particle initially at rest. The two stick together, forming a single particle of mass m'. (a) Do you expect (1) $m' > 2m$, (2) $m' < 2m$, or (3) $m' = 2m$? Explain. (b) If the incoming particle is initially moving at a speed $v = 0.800c$ to the right, what is m' in terms of m?

(a) Conceptual Reasoning. This is an example of an *inelastic* collision (Chapter 6). In such a collision, kinetic energy is not conserved. To conserve linear momentum, some, but not all, of the initial kinetic energy is lost. Since total energy is also conserved, any loss of kinetic energy is converted into mass. If this weren't true, then (3) would be the correct answer. However, the combined mass must include the mass equivalent of the "lost" kinetic energy, or

$$m' = 2m + \text{``some kinetic energy in the form of mass''}$$

Therefore, the correct answer is (1): $m' > 2m$.

(b) Quantitative Reasoning and Solution. Collisions are usually analyzed using conservation of momentum and total energy. After the collision, the combined particle must be moving to the right to conserve the direction of the total momentum. The magnitude of the moving particle's momentum before the collision must equal the magnitude of the single combined particle's momentum afterward. Total relativistic energy must also be conserved. From these considerations, we should be able to determine the combined particle's mass.

Given: $\quad v = 0.800c$ *Find:* m' (mass of combined particle)
$\qquad\quad m = $ mass of one particle

For the incoming particle,

$$\gamma = 1/\sqrt{1 - (v/c)^2} = 1/\sqrt{1 - (0.800)^2} = 1.67$$

The total system momentum P is conserved, that is,

$$\vec{P}_i = \vec{P}_f$$

Using the expression for relativistic momentum (Eq. 26.9) and equating the magnitude of the momentum of the incoming particle to that of the combined particle, we have

$$\gamma m v = \gamma' m' v'$$

where γ' refers to the combined particle after the collision. Putting in the numbers,

$$1.67m(0.800c) = \gamma' m' v'$$

Next, we equate the total relativistic energy before the collision to that after the collision, remembering that the total energy of a particle is related to its rest energy by $E = \gamma E_o$. The initial total energy is the sum of the energy due to the moving particle and the rest energy of the "target," or $E_i = 1.67mc^2 + mc^2$. The final total energy of the combined particle is $E_f = \gamma' m' c^2$. Thus, energy conservation requires that

$$1.67mc^2 + mc^2 = \gamma' m' c^2 \qquad \text{or} \qquad 2.67m = \gamma' m'$$

(In the last step, the speed of light cancels out of both sides of the equation.) Dividing the last result into the momentum result, we obtain

$$\frac{1.67m(0.800c)}{2.67m} = \frac{\gamma' m' v'}{\gamma' m'} = v'$$

or

$$v' = 0.500c$$

Using this result, we find that $\gamma' = 1/\sqrt{1 - (0.500)^2} = 1.15$. Now the energy equation can be used to solve for the mass of the combined particle m':

$$2.67mc^2 = 1.15m'c^2 \qquad \text{or} \qquad m' = 2.32m$$

As expected, the mass of the combined particle is greater than $2m$ because some kinetic energy is converted into mass (energy).

Follow-Up Exercise. (a) In this Example, how much kinetic energy is lost? (b) What would be the mass of the combined particle if the two particles initially approached head-on each with a speed of $0.800c$ and stuck? (Your answers should be in terms of m and c.)

26.5 The General Theory of Relativity

<u>OBJECTIVES:</u> To (a) explain the principle of equivalence, and (b) examine some of the predictions of general relativity.

Special relativity applies to inertial systems, not to accelerating systems. Accelerating systems require a different approach, first described by Einstein about 1915. Called the **general theory of relativity**, it contains many implications about the theory of gravity.

The Principle of Equivalence

An important principle of general relativity was first envisioned by Einstein (after another *gedanken* experiment). He called it the **principle of equivalence**, which can be stated as follows:

> An inertial reference frame in a uniform gravitational field is physically equivalent to a reference frame that is not in a gravitational field, but that is in uniform linear acceleration.

What this means:

> No experiment performed in a closed system can distinguish between the effects of a gravitational field and the effects of an acceleration.

Thus, an observer in an accelerating system would find the effects of a gravitational field and those of their acceleration to be equivalent and indistinguishable. For simplicity, we will restrict consider only systems that are accelerating linearly (that is, those that have no rotational acceleration).

To understand this principle, consider the situations in ▸Fig. 26.11. Imagine yourself as an astronaut in a closed spaceship. Suppose when you drop a pencil, you observe that it accelerates to the floor. What does this mean? According to the principle of equivalence, it could mean (a) that you are in a gravitational field or (b) that you are in an accelerating system (Fig. 26.11a). Any experiment performed entirely inside the ship cannot determine one from the other. Whether the spaceship is in free space and accelerating with an acceleration $a = g$ or whether it is in a gravitational field with $g = -a$, the pencil has the same observed acceleration. In your closed system (remember that you cannot look outside), there is *no experiment* you could perform to distinguish whether the pencil's acceleration is a gravitational or a system's acceleration effect. According to the principle of equivalence, the two are physically indistinguishable.

As another example, suppose the pencil does not accelerate, but instead remains suspended next to you. This could mean that (a) you are in an inertial frame with no gravity or (b) you are in free fall in a gravitational field. Once again, in the closed spaceship (Fig. 26.11b), you have no way of distinguishing between these two possibilities. That is, they are *equivalent*.

From what you know about relativity at this point, you might conclude that it need be considered only at high speeds and in gravitational fields. Would you be surprised to find out that it is crucial to the development of many of our modern technological communications systems? Check out Insight 26.1 on Relativity in Everyday Living on page 838.

Light and Gravitation

The principle of equivalence leads to an important prediction: that a gravitational field bends light. To see how this prediction arises, let's use another *gedanken* experiment. Suppose a beam of light traverses a spaceship that is accelerating

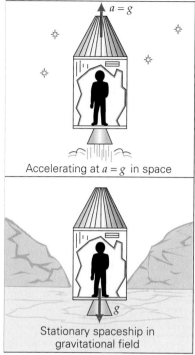

Accelerating at $a = g$ in space

Stationary spaceship in gravitational field

(a)

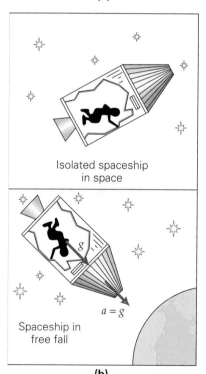

Isolated spaceship in space

Spaceship in free fall

$a = g$

(b)

▲ **FIGURE 26.11 The principle of equivalence (a)** In a closed spaceship, the astronaut can perform no experiment that would determine whether he was in a gravitational field or an accelerating system. **(b)** Similarly, an inertial frame without gravity cannot be distinguished from free fall in a gravitational field.

INSIGHT 26.1 RELATIVITY IN EVERYDAY LIVING

You often hear that relativistic effects are observed only at speeds near the speed of light. This statement implies that these effects are irrelevant in everyday living. However, this is not true in our modern technological world. Relativistic effects, in fact, are crucial in global positioning systems (GPS), used for locating objects on the Earth to an accuracy of several meters. The system is important in many settings, such as modern airplane navigation and military operations. It is now included in many new cars for navigational purposes. The system works properly only because crucial corrections have been made for both special and general relativistic effects.

The GPS consists of an array of satellites, each with a very accurate atomic "clock" onboard. To determine the position of any object on the Earth's surface, that object must have a GPS receiver. The receiver detects light (radio) signals from several satellites. Knowing the speed of light and measuring the travel time, the receiver's computer can determine distances and directions to these satellites. Through triangulation, the GPS

receiver's computer can rapidly calculate its location. As the location changes, the object's velocity can be calculated.

For GPS to work the clocks in orbit must be "in sync" with the corresponding clocks on the Earth. If they are not in sync, then the time of travel will be incorrect and the distances will be wrong. An error of just 100 ns (10^{-7} s) can lead to an error in location of several tens of meters—clearly unacceptable when you are trying to land a plane in bad weather, for example! Due to the satellites' orbital speeds (several kilometers per second) there are special relativistic time-dilation effects to account for. In addition, according to the theory of general relativity, the satellites will run at a faster rate because they are in a weaker gravitational field than the Earth-bound clocks. The net effect of that the orbiting clocks run at a faster rate. To keep them in synchronization with their counterparts on the Earth, these clocks are set to run at a slower rate before they are launched. When they reach the proper orbit, their rate increases, bringing them up to the same rate as that of the surface clocks. Relativity in everyday living—who would have thought it?

rapidly upward. If the spaceship were stationary, light entering the ship at point A would arrive at point B on the far wall; however, because the spaceship is accelerating, the light actually lands at point C (▼Fig. 26.12a).

From the point of view of the person onboard the ship (Fig. 26.12b), the light path is a downward-curving one. To this astronaut, it seems that the gravitational field she is in has bent the light path, much as a baseball's path would be. (Recall that according to the equivalence principle, the acceleration is indistinguishable from a gravitational field.) Although an outside observer "knows" that the rocket's acceleration produces this effect, we conclude by the principle of equivalence that a gravitational field should bend light paths. Notice that if the spaceship were moving up *without* accelerating, the light path in the ship would be a straight line, which is consistent with no gravitational effect. No gravity, no bending. Once again, the principle works.

▶ **FIGURE 26.12 Light bending**
(a) Light traversing an accelerating, closed rocket from point A arrives at point C. **(b)** In the accelerating system, the light path would appear to be bent. Since an acceleration produces this effect, by the principle of equivalence, light should also be bent by a gravitational field.

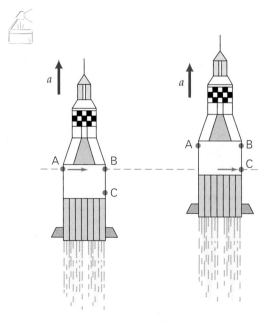
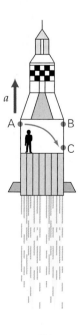

(a)   **(b)**

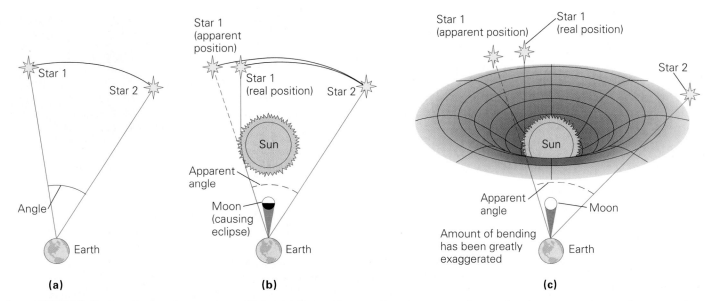

▲ **FIGURE 26.13 Gravitational attraction of light** (a) Normally, two distant stars are observed to have a certain angular separation. (b) During a solar eclipse, the star behind the Sun can still be seen because of the effect of solar gravity on starlight. The star has a larger measured angular separation than it has in the absence of such an eclipse. (c) General relativity views a gravitational field as a warping of space and time. A simplified analogy is the surface of a warped rubber sheet.

Under most conditions, this effect must be very small, since we don't observe everyday evidence of light bending in the Earth's gravitational field. However, this prediction was experimentally verified in 1919 during a solar eclipse. During times of the year when they aren't near the Sun and can be seen at night, distant stars have a constant angular separation between them (▲Fig. 26.13a). During other times of the year, light from some of these stars may pass near the Sun, which has a relatively strong gravitational field. However, any evidence of bending would not usually be observable on Earth because the starlight is almost always masked by the glare of the Sun.

However, stars may be seen during the day under the conditions of a *total solar eclipse*. When the Moon comes between the Earth and the Sun, an observer in the Moon's shadow (the umbra) can see stars not normally visible during the day (Fig. 26.13b). If the light from a star passing near the Sun is bent, then the star will have an *apparent* location that is different from its normal location. As a result, the angular distance between pairs of stars will be measured as slightly larger than normal. Einstein's theory predicted that the angular *difference* between the apparent positions of the stars during the eclipse of 1919 and their positions in the absence of solar gravity effects should be about 1.75 seconds of arc ($\approx 0.00005°$). The experimental difference was 1.61 ± 0.30 seconds of arc, in agreement within the uncertainty!

General relativity conceptually pictures a gravitational field as "warping" space and time, as illustrated in Fig. 26.13c. A light beam follows the curvature of space–time like a ball rolling on a curved surface. The bending of light by gravitational fields has been repeatedly verified during other solar eclipses and also by signals sent back to Earth by space probes passing near the Sun.

Gravitational Lensing

Another effect of gravity on light is called **gravitational lensing**. In the late 1970s, a double quasar was discovered. (A quasar is a powerful astronomical radio source.) The fact that it was a double quasar was not unusual, but everything about the two quasars seemed to be exactly the same, except that one was fainter than the other. It was suggested that perhaps there was only one quasar and that, somewhere between it and the Earth, a massive, but optically faint, object had

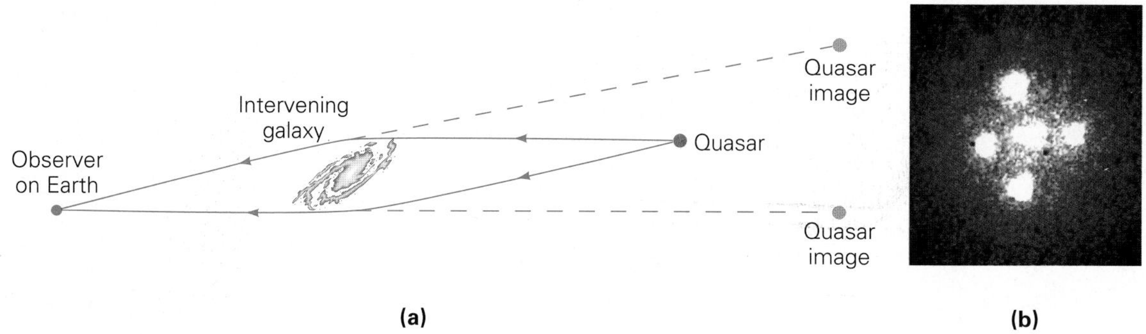

Quasar
image

Intervening
galaxy

Observer
on Earth

Quasar

Quasar
image

(a)

(b)

▲ **FIGURE 26.14 Gravitational lensing (a)** The bending of light by a massive object such as a galaxy or a cluster of galaxies can give rise to multiple images of a more distant object. **(b)** The discovery of what appeared to be four images of the same quasar (the "Einstein Cross") suggested the possibility of gravitational lensing. On investigation, a faint intervening galaxy was found.

Note: A collapsing star becomes a black hole when it collapses to a radius less than its Schwarzschild radius.

bent its light, producing multiple images. The subsequent detection of a faint galaxy between the two quasars confirmed this hypothesis, and other examples have since been discovered (▲ Fig. 26.14).

The discovery of gravitational lenses gave general relativity a new role in modern astronomy. By examining multiple images of a distant galaxy or quasar and their relative brightness, astronomers can gain information about an intervening galaxy or cluster of galaxies whose gravitational field causes the bending of light.

Black Holes

The idea that gravity can affect light finds its most extreme application in the concept of a black hole. A **black hole** is thought to form from the gravitationally collapsed remnant of a massive star, typically many times the mass of our Sun. Such an object has a density so great and a gravitational field so intense that nothing can escape it. In terms of our space–time warp analogy, a black hole is graphically represented as a bottomless pit in the fabric of space–time. Even light can't escape the intense gravitational field of a black hole—hence the blackness.*

An estimate of the size of a black hole can be obtained by using the concept of escape speed. Recall from Section 7.6 that the escape speed from the surface of a spherical body of mass M and radius R is given by

$$v_{esc} = \sqrt{\frac{2GM}{R}} \qquad (26.14)$$

If light does not escape from a black hole, its escape speed must exceed the speed of light. The critical radius of a sphere around a black hole from which light cannot escape is obtained by substituting $v_{esc} = c$ into Eq. 26.14. Solving for R, we obtain

$$R = \frac{2GM}{c^2} \qquad Schwarzschild\ radius \qquad (26.15)$$

The quantity R is called the **Schwarzschild radius**, after Karl Schwarzschild (1873–1916), a German astronomer who developed the concept. The boundary of a sphere of radius R defines the black hole's **event horizon**. Any event occurring within this horizon is invisible to an outside observer, since light cannot escape. The event horizon thus gives the limiting distance within which light cannot escape from a black hole. Information about what goes on inside the event horizon can never reach us, so questions such as, "What does it look like inside the event horizon?" are unanswerable (at least, given the current state of knowledge in physics†).

*It is speculated that black holes also may originate in other ways, such as from the collapse of entire star clusters in the center of a galaxy or at the beginning of the universe during the big bang. Some current theories (combining quantum mechanics and general relativity) hint that black holes might emit material particles and radiation and thus "evaporate," but none have been discovered to be evaporating. In this book, our discussion will be limited to stellar collapse.

†There is a discrepancy here that bothers physicists. According to quantum theory, information cannot be "lost." Theorists have proposed that black holes can actually radiate energy via a quantum phenomenon called "tunneling" (see Section 28.2). If true, this would bring relativity and quantum theory into better agreement, and is currently an active area of theoretical and experimental research.

Example 26.8 ■ If the Sun Were a Black Hole: Schwarzschild Radius

What would be the Schwarzschild radius if our Sun collapsed to a black hole? (The mass of the Sun is $M_S = 2.0 \times 10^{30}$ kg.)

Thinking It Through. This is a straightforward calculation using Eq. 26.15. We will need the gravitational constant and speed of light.

Solution.

Given: $M_S = 2.0 \times 10^{30}$ kg $\qquad\qquad$ *Find:* R (Schwarzschild radius)
$\qquad\quad\ G = 6.67 \times 10^{-11}$ N $\cdot$ m^2/kg^2
$\qquad\quad\ c = 3.00 \times 10^{8}$ m/s

Thus we have

$$R = \frac{2GM_S}{c^2} = \frac{2(6.67 \times 10^{-11}\,\text{N}\cdot\text{m}^2/\text{kg}^2)(2.0 \times 10^{30}\,\text{kg})}{(3.00 \times 10^{8}\,\text{m/s})^2} = 3.0 \times 10^{3}\,\text{m} = 3.0\,\text{km}$$

This is less than 2 miles. The Sun's radius is about 7×10^{5} km. Note however that the Sun will *not* become a black hole. This fate befalls only stars much more massive than the Sun.

Follow-Up Exercise. Once black holes form, they continuously draw in matter, increasing their Schwarzschild radius. How many times more massive would our Sun have to be for its Schwarzschild radius to extend to, and swallow up, Mercury, the innermost planet? (The average distance from the Sun to Mercury is 5.79×10^{10} m.)

If nothing, including radiation, escapes a black hole from inside its event horizon, how, then, might we observe or even merely locate a black hole? This question and some insight into current research into gravitational waves is addressed in Insight 26.2 entitled Black Holes, Gravitational Waves and LIGO, on page 842.

*26.6 Relativistic Velocity Addition

OBJECTIVES: To (a) understand the necessity for a relativistic velocity addition, and (b) investigate relative velocity addition through calculations.

As we have seen, the postulates of special relativity affect our everyday concepts of distance and time. Since velocity involves these quantities, relative velocities should also be affected, and that is indeed the case.

Recall that according to Newtonian relativity, there is a problem with vector addition when light is involved. In Fig. 26.1, the vector addition of velocities predicts a speed greater than c for the beam of light. Such a speed violates the second postulate of relativity.

The same problem occurs with objects moving at appreciable fractions of the speed of light. Consider the rocket separation shown in ▾Fig. 26.15. After separation, the jettisoned stage has a velocity $\vec{v}$ with respect to the Earth, and the rocket has a velocity $\vec{u}'$ with respect to the jettisoned stage. Then, from Newtonian relativity and vector addition, the velocity of the rocket payload with respect to the Earth is $\vec{u} = \vec{v} + \vec{u}'$.

◀ **FIGURE 26.15 Relativistic velocity addition** After jettisoning, the rocket payload has a velocity $\vec{u}'$ with respect to the jettisoned stage, and the jettisoned stage has a velocity $\vec{v}$ with respect to the Earth. For relativistic velocities, the velocity $\vec{u}$ of the rocket with respect to the Earth is obtained from a relativistically correct version.

$\vec{v}$ (velocity of jettisoned stage with respect to Earth)

$\vec{u}'$ (velocity of rocket payload with respect to jettisoned stage)

INSIGHT 26.2 BLACK HOLES, GRAVITATIONAL WAVES, AND LIGO

Try to imagine something so dense that nothing—not even light—can escape from it. According to stellar evolution theory, black holes could result from the collapse of stars having much greater mass than our Sun. But gathering experimental proof of the existence of a black hole is another matter.

The event horizon of a black hole is located at a certain distance (the Schwarzschild radius) from its center. The event horizon marks the location where gravity becomes sufficiently strong to keep even light from escaping. What form or size matter takes inside a black hole is not known and in principle can never be known. Even if a probe could be sent "inside" a black hole (that is, closer than its Schwarzschild radius), the probe could not send data back to us.

How, then, might a black hole be detected? Light passing near it would be bent, but it is unlikely that we would ever observe that, considering the vastness of space. The most likely possibility comes from observing a *binary star system*, which consists of two stars orbiting a common center of mass. Cygnus X-1, the first X-ray source discovered in the constellation Cygnus, provides the best evidence so far for a black hole in our galaxy.

Where Cygnus X-1 is located in the sky, we observe a giant star (visible light *cannot* detect two separate stars) whose spectrum shows periodic Doppler redshifts and blueshifts, indicating a periodic orbital motion away from us and toward us, respectively. This observation indicates a binary star system with an "invisible" companion to the giant star. Measuring the precise orbital data of both stars allow us to compute their masses. In Cygnus X-1, an apparent unseen companion to the giant star *seems* to have enough mass to qualify as a black hole.

Astronomers speculate that, in binary star systems such as Cygnus X-1, one member has evolved to become a black hole. Matter drawn from the other, more normal star would create an *accretion disk* of spiraling matter around the black hole (Fig. 1). The matter falling into the disk would be accelerated and heated. Collisions and deceleration would produce a characteristic spectrum of X-rays, and the black hole would appear to be the source of X-rays we observe. These X-rays are emitted by the hot matter before it gets closer than the Schwarzschild radius.

A similar mechanism may explain the enormous energy output of *active galaxies* and *quasars*. It has been proposed that the centers of these brilliant objects, which produce vast amounts of radiation, might contain black holes with masses millions or even billions of times that of our Sun. In fact, recent observations suggest that even many normal galaxies, including our own Milky Way galaxy, may harbor enormously massive black holes in their cores.

According to general relativity, a gravitationally violent event, such as black holes coalescing, should emit gravitational waves traveling at the speed of light. These waves should cause a displacement of the matter in the regions through which they travel. They are expected to be extremely weak, causing movements on the order of only 10^{-14} m (not much larger than the diameter of an atomic nucleus). As part of current research, U.S. scientists have collaborated in a project called LIGO (*Laser Interferometer Gravitational-Wave Observatory*). Each interferometer consists of two dangling weights with the distance between them monitored by laser. If a gravitational wave passes, the distance, and the laser interference pattern, should change.

When operational, there will be two identical installations, one in Washington and one in Louisiana (see Fig. 2). If a gravitational wave does pass through the Earth, both installations should detect the event simultaneously. Having two simultaneous signals reduces the likelihood that the event resulted from a local disturbance. In the future, a global network of these sensitive interferometers is proposed. If gravitational waves are finally detected, it will be a victory for relativity. Once understood, these waves could then be used as a cosmic probe—that is, to determine the details of the event that gave rise to them.

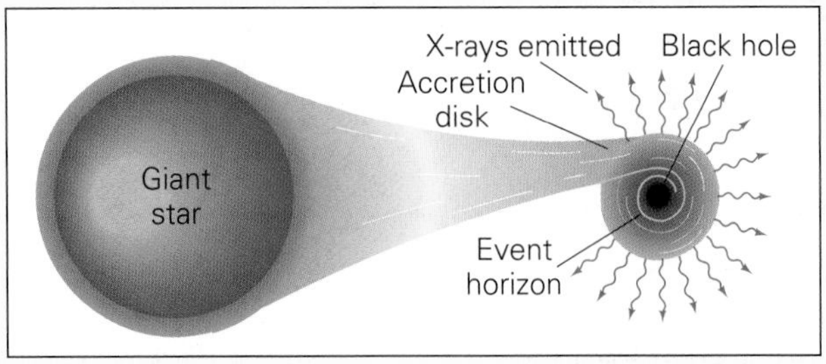

FIGURE 1 X-rays and black holes Matter drawn from the "normal" member of a binary star system forms a spiraling accretion disk around the hole. Matter falling into the disk is accelerated, and collisions give rise to the emission of X-rays.

FIGURE 2 LIGO Sensitive laser interferometers accurately measure distances between weights at various locations in the United States, looking for changes in the distance due to passage of gravitational waves. This is the location in Louisiana.

However, at relativistic speeds, this addition law gives a result contradictory to special relativity. Suppose, for example, that the velocities are all in the same direction with magnitudes of $v = 0.50c$ and $u' = 0.60c$. Then $u = v + u' = 0.50c + 0.60c = 1.10c$. Thus, classical relativity predicts that an observer on the Earth would measure the rocket's speed to be greater than c, which, as we have seen, cannot happen.

Einstein recognized that according to special relativity, lengths and times differ depending on the observer's reference frame. Thus classical velocity vector addition cannot be correct at relativistic speeds. He showed the correct equation (for motion in a straight line) to be

Note: It is usually convenient to express speeds as fractions of c.

$$u = \frac{v + u'}{1 + \dfrac{vu'}{c^2}} \quad \begin{array}{l} \textit{relativistic velocity addition} \\ \textit{(one dimension)} \end{array} \qquad (26.16)$$

where the velocities have the same meanings as in the preceding paragraph and sign notation is used to indicate velocity directions. Notice that the observed velocity u is *reduced* by a factor of $1/[1 + (vu'/c^2)]$ from that of the classical result. At very low speeds (that is, $v/c \ll 1$ and $u'/c \ll 1$), we have $vu'/c^2 = (v/c)(u'/c) \ll 1$. Thus, under low-speed conditions, the denominator in Eq. 26.16 is, for all practical purposes, equal to 1, and the Newtonian result, $u = v + u'$, is obtained.

In working out problems, it is important to identify the velocities clearly:

$\vec{v}$ = velocity of object 1 with respect to an inertial observer

$\vec{u}'$ = velocity of object 2 with respect to object 1

$\vec{u}$ = velocity of object 2 with respect to an inertial observer

Since we will consider only one-dimensional-motion problems, plus/minus signs are used to indicate velocity directions. The next Example shows that Eq. 26.16 gives results that do *not* violate the second postulate.

Example 26.9 ■ Faster Than the Speed of Light? No! Thanks to Relativistic Velocity Addition

For the rocket separation in Fig. 26.15, let the speeds be $v = 0.50c$ and $u' = 0.60c$ (with directions shown in the figure). What is the velocity of the payload, as measured by an observer on Earth?

Thinking It Through. Clearly, Eq. 26.16 needs to be applied here because the speeds are relativistic, and the answer (u) must be less than c.

Solution. The velocities are taken to be in the positive direction.

Given: $v = +0.50c$ (speed of jettisoned stage   *Find:* $\vec{u}$ (velocity of rocket with respect with respect to Earth)   to Earth)
$u' = +0.60c$ (speed of rocket with respect to jettisoned stage)

Since both velocities are to the right, they are designated with plus signs. From Eq. 26.16, we have

$$u = \frac{v + u'}{1 + \dfrac{vu'}{c^2}} = \frac{+0.50c + 0.60c}{1 + \dfrac{(+0.50c)(+0.60c)}{c^2}} = \frac{+1.1c}{1.3} = +0.85c$$

As expected, u is less than c and is directed to the right, as indicated by the plus sign.

Follow-Up Exercise. Repeat this Example with both objects moving to the right at $0.40c$. According to Newtonian relativity, the velocity relative to the Earth is $0.80c$, which does *not* violate the second postulate. What is the correct result? [*Hint:* It should be lower than the Newtonian result of $0.80c$. Why?]

Chapter Review

- **Newtonian** or **classical relativity** supported the belief in the existence of an absolute **inertial reference frame** somewhere in the universe that was "at rest." In this reference frame was the material called the **ether,** the medium through which light could propagate.

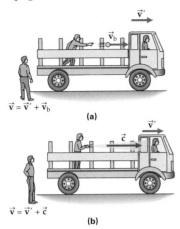

$$\vec{v} = \vec{v}' + \vec{v}_b$$

(a)

$$\vec{v} = \vec{v}' + \vec{c}$$

(b)

- **The Michelson–Morley experiment** attempted to measure the speed of the Earth relative to the ether frame. The results were always zero. Thus, physicists had to give up the idea of a reference frame that was absolutely at rest.

- **The special theory of relativity** involves inertial reference frames moving relative to one another and is based on two postulates:

  Principle of relativity: All the laws of physics are the same in all inertial reference frames.

  Principle of the constancy of the speed of light: The speed of light in a vacuum has the same value in all inertial systems.

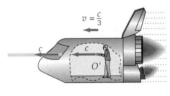

- A time interval measured by a clock that is present at both the starting and stopping events of the interval is a **proper time interval** Δt_o. The time interval Δt of an observer in any other inertial frame is larger than the proper time interval. The two intervals are related by

$$\Delta t = \frac{\Delta t_o}{\sqrt{1 - (v/c)^2}} \qquad (26.3)$$

If γ (Greek letter "gamma") is defined as

$$\gamma \equiv \frac{1}{\sqrt{1 - (v/c)^2}} \qquad (26.4)$$

Equation 26.3 can then be written more compactly as

$$\Delta t = \gamma \Delta t_o \qquad (26.5)$$

This effect on time intervals is called **time dilation.**

- The length of an object, as measured by an observer at rest with respect to it, is called the object's **proper length** L_o. To an observer in any other inertial frame, the length L is smaller than the proper length by a factor of $1/\gamma$, or

$$L = \frac{L_o}{\gamma} = L_o\sqrt{1 - (v/c)^2} \qquad (26.7)$$

This phenomenon is known as **length contraction**.

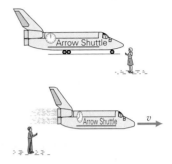

- At speeds near the speed of light, an object's **relativistic kinetic energy** is

$$K = \left[\frac{1}{\sqrt{1 - (v/c)^2}} - 1 \right] mc^2 = (\gamma - 1)mc^2 \qquad (26.8)$$

- At speeds near the speed of light, an object's **relativistic momentum** is

$$\vec{p} = \frac{m\vec{v}}{\sqrt{1 - \left(\frac{v}{c}\right)^2}} = \gamma m\vec{v} \qquad (26.9)$$

- At speeds near the speed of light, the **relativistic total energy** of an object is

$$E = \frac{mc^2}{\sqrt{1 - \left(\frac{v}{c}\right)^2}} = \gamma mc^2 \qquad (26.10)$$

- Even when an object is at rest, it still has an energy of mc^2. This minimum energy called its **rest energy** and is

$$E_o = mc^2 \qquad (26.11)$$

- An object's relativistic total energy can be written in terms of its rest energy as

$$E = K + E_o = K + mc^2 \qquad (26.12)$$

- The **general theory of relativity** expresses how physical quantities are measured by observers in reference frames that are accelerating. The **principle of equivalence** states that

  an inertial reference frame in a uniform gravitational field is physically equivalent to a reference frame that is not in a gravitational field, but that is in uniform linear acceleration.

  this means that no experiment performed in a closed system can distinguish between the effects of a gravitational field and the effects of an acceleration.

- General relativity predicts many interesting phenomena that have been experimentally verified, such as stars called **black holes** and the bending of light by a gravitational field.

Exercises

MC = *Multiple Choice Question,* **CQ** = *Conceptual Question, and* **IE** = *Integrated Exercise. Throughout the text, many exercise sections will include "paired" exercises. These exercise pairs, identified with* **red numbers***, are intended to assist you in problem solving and learning. In a pair, the first exercise (even numbered) is worked out in the Study Guide so that you can consult it should you need assistance in solving it. The second exercise (odd numbered) is similar in nature, and its answer is given at the back of the book.*

Note: Assume c to be exact at 3.00×10^8 m/s. *Consider speeds given in terms of c as having the same number of significant figures as the accompanying numerical coefficient. (For example, 0.85 c has two significant figures.) Gamma (γ) should be calculated explicitly.*

26.1 Classical Relativity and the Michelson–Morley Experiment

1. **MC** An object free of all forces exhibits a changing velocity in a certain reference frame. It follows that (a) the frame is inertial, (b) $\vec{F} = m\vec{a}$ applies in this frame, (c) the laws of mechanics are the same in this reference frame as in all inertial frames, (d) none of the preceding. (d)

2. **MC** Car A is traveling eastward at 85 km/h Car B is traveling westward with a speed of 65 km/h. The velocity of car B as measured by the driver of car A is (a) 150 km/h eastward (b) 20 km/h westward (c) 150 km/h westward (d) 20 km/h eastward (c)

3. **MC** A space probe is moving rapidly away from the Sun at 0.2c. As it passes the probe, the speed of the sunlight measured by the probe has what value? (a) 1.2c (b) exactly c (c) 0.8c (b)

4. **CQ** We live on a rotating Earth and therefore are in an accelerating, noninertial system. How, then, can we apply Newton's laws of motion on the Earth? see ISM

5. **CQ** A car accelerating from rest is a noninertial system. Explain the "forces" that the driver uses to explain why things on the dashboard slide "backward." How does the inertial observer on the roadside explain this? see ISM

6. **CQ** A person rides with her package on the floor of an elevator that is moving upward at constant velocity. How does the free-body diagram of the package as drawn by the elevator person differ from that drawn by the observer remaining on the first floor? Explain. see ISM

7. ● A person 1.20 km away from you fires a gun, and you observe the flash from the muzzle. A wind of 10.0 m/s is blowing. How long will the sound of gunfire take to reach you if the wind is (a) toward you and (b) toward the person who fired the gun? (Take the speed of sound to be 345 m/s.) (a) 3.38 s (b) 3.58 s

8. ● A small airplane has an airspeed (speed with respect to air) of 200 km/h. Find the airplane's ground speed if there is (a) a headwind of 35 km/h and (b) a tailwind of 25 km/h. (a) 165 km/h (b) 225 km/h

9. ● A speedboat can travel with a speed of 50 m/s in still water. If the boat is in a river that has a flow speed of 5.0 m/s, find the maximum and minimum values of the boat's speed relative to an observer on the riverbank. 55 m/s and 45 m/s

10. **IE** ●● A boat can make a round trip between two locations, A and B, on the same side of a river in a time t if there is no current in the river. (a) If there is a constant current in the river, the time the boat takes to make the same round trip will be (1) longer, (2) the same, (3) shorter. Why? (b) If the boat can travel with a speed of 20 m/s in still water, the speed of the river current is 5.0 m/s, and the distance between points A and B is 1.0 km, calculate the times when there is no current and when there is current. (a) (1) longer (b) 1.7 min and 1.8 min

11. ●●● The apparatus used by the French scientist Armand Fizeau in 1849 for measuring the speed of light is illustrated in ▾Fig. 26.16. Teeth on a rotating wheel periodically interrupt a beam of light. The flashes of light travel to a plane mirror and are reflected back to an observer. Show that if the wheel is rotated at just the right frequency f, the light passing through one gap reaches the mirror and is reflected to the observer through the very next gap. When f is adjusted so this happens, show that the speed of light is given by $c = 2fNL$, where N is the number of gaps in the wheel, f is the frequency (in revolutions per second) made by the rotating wheel, and L is the distance between the wheel and the mirror. see ISM

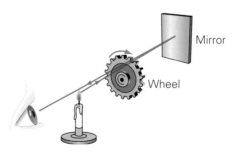

▲ **FIGURE 26.16 Fizeau's apparatus** See Exercise 11.

26.2 The Postulates of Special Relativity and the Relativity of Simultaneity

12. **MC** Events that are simultaneous in one inertial reference frame are (a) always simultaneous in other inertial reference frames, (b) never simultaneous in other inertial reference frames, (c) sometimes simultaneous in other inertial reference frames, (d) none of the preceding. (c)

13. **MC** An object is at rest in an inertial reference frame. What will be the same about the object as measured by an inertial observer moving relative to the object: (a) its free-body diagram, (b) its velocity, (c) its kinetic energy, or (d) none of the preceding? (a)

14. **MC** An object is moving in an inertial reference frame. What will be the same about the object as measured by an inertial observer moving relative to the object? (a) its velocity, (b) its speed, (c) its kinetic energy, (d) its acceleration, or (e) none of the preceding? (d)

15. **CQ** In a space war, a warrior approaching his enemy head on at a speed of 0.85c escapes with a near miss as a high-energy laser beam barely misses him. What speed does the warrior observe for the light as it passes him? c

16. **CQ** Is it possible for either observer on two approaching rocket ships to observe the other ship with a velocity greater than c? Why? no; see ISM

17. **CQ** In the *gedanken* experiment shown in ▼Fig. 26.17, two events in the same inertial reference frame O are related by cause and effect: (1) A gun at the origin fires a bullet along the x-axis with a speed of 300 m/s. (Assume that there are no gravitational or frictional forces.) (2) The bullet hits a target at x = +300 m. Show, using qualitative arguments, that the two events cannot be viewed simultaneously by any inertial observer. (*Note:* This shows that special relativity preserves the time *sequence* of two events if they are related as cause and effect. In this situation, this means that all observers agree that the gun fires before the bullet hits the target.) see ISM

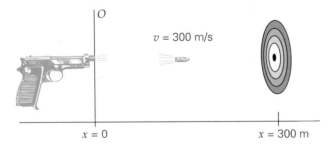

▲ **FIGURE 26.17** A thought experiment See Exercise 17.

18. **CQ** In a *gedanken* experiment (▶Fig. 26.18), two events that cannot be related by cause and effect occur in the same inertial reference frame O: (1) Strobe light A, at the origin of the x-axis, flashes; (2) strobe light B, located at x = +600 m, flashes 1.00 μs later. B's flash cannot be caused by A's flash. (Light travels only 300 m in the time between the two events.) (a) Use qualitative arguments to show that there exists another inertial reference frame, traveling at less than c,

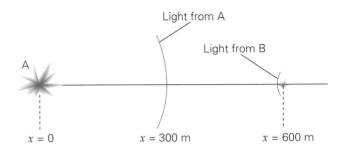

▲ **FIGURE 26.18** Another thought experiment
See Exercise 18.

in which these two events would be observed to occur simultaneously. (b) What is the direction of the velocity of the reference frame in part (a)? (*Note:* This question shows that relativity does *not* have to preserve the time order of events if they are not related by cause and effect. Since one event cannot cause the other, no physical principles are violated if they are seen in reverse order.) see ISM

26.3 The Relativity of Length and Time: Time Dilation and Length Contraction

19. **MC** An observer sees a friend passing her in a rocket ship that has a uniform velocity of 0.90c. The observer knows her friend to be 1.45 m tall, and he is standing such that his length is perpendicular to their relative velocity. To the observer, her friend will appear (a) taller than 1.45 m, (b) shorter than 1.45 m, (c) exactly 1.45 m tall. (c)

20. **MC** An observer sees a friend passing her in a rocket ship that has a uniform velocity of 0.90c. Her friend claims that exactly 10 seconds have elapsed on his clock. How will the observer's identical clock measure the same time interval? The observer's clock will read (a) less than 10 seconds, (b) greater than 10 seconds, (c) exactly 10 seconds. (b)

21. **MC** A speedboat completes one straight leg of a lap that the race organizers have carefully laid out to be exactly 2.55 km long. Which observer(s) measures the leg's proper length: (a) the organizers, (b) the driver of the speedboat, (c) they both do, or (d) neither does? (a)

22. **CQ** You are standing on a road and observe a high-speed car as it passes you. Sketch the shape of the car when it is stationary and when it is moving at 0.50c. see ISM

23. **CQ** A farm boy wants to store a 5-m-long pole in a shed that is only 4 m long (it does have both front and rear doors). He claims that if he runs through the shed sufficiently fast, according to an observer at rest, the pole will fit in the shed (both doors closed at least for an instant) as a result of length contraction. Can this be true? Show that from the boy's reference frame the pole could not possibly fit into the shed. no, see ISM

24. **CQ** You are standing on the Earth and observe a fast-moving spacecraft going by with your professor on board. (a) If both you and your professor are observing your wristwatch, who is measuring the proper time? (b) Who measures the proper length of the spacecraft?
(a) you (b) your professor

25. ● A spacecraft moves past a student with a relative velocity of 0.90c. If the pilot of the spacecraft observes 10 min to elapse on his watch, how much time has elapsed according to the student's watch? 23 min

26. IE ● Your pulse rate is 80 beats/min, and your physics professor in a spacecraft is moving with a speed of 0.85c relative to you. (a) According to your professor, your pulse rate is (1) greater than 80 beats/min, (2) equal to 80 beats/min, (3) less than 80 beats/min. Why? (b) What is your pulse rate, according to your professor? (a) (3) less than 80 beats/min (b) 42 beats/min

27. ● You fly your 15.0-m-long spaceship with a speed of c/3 relative to your friend. Your velocity is parallel to the ship's length. How long is your spaceship, as observed by your friend? 14.1 m

28. IE ● An astronaut in a spacecraft moves past a field 100 m long (according to a person standing on the field) and parallel to the field's length, with a speed of 0.75c. (a) Will the length of the field, according to the astronaut, be (1) longer than 100 m, (2) equal to 100 m, or (3) shorter than 100 m? Why? (b) What is the length as measured by the astronaut? (c) Which length is the proper length?
(a) (3) shorter than 100 m (b) 66 m (c) 100 m

29. ●● The proper lifetime of a muon is 2.20 μs. If the muon has a lifetime of 34.8 μs according to an observer on Earth, what is the muon's speed, as a fraction of c, relative to the observer? 0.998c

30. ●● One of a pair of 25-year-old twins takes a round trip through space while the other twin remains on Earth. The traveling twin moves at a speed of 0.95c for 39 years, according to Earth time. Assuming that special relativity applies, what are the twins' ages when the traveling twin returns to Earth?
Earth twin is 64 years old, traveling twin is 37 years old

31. IE ●● Alpha Centauri, a binary star close to our solar system, is about 4.3 light-years away. Suppose a spaceship traveled this distance with a constant speed of 0.60c relative to Earth. (a) Compared with a clock on the spaceship, an Earth-based clock will measure (1) a longer time, (2) an equal time, (3) a shorter time. Why? (b) How much time would elapse on an Earth-based clock and on a clock on the spaceship?
(a) (1) a longer time (b) 7.2 years and 5.7 years

32. ●● A cylindrical spaceship of length 35.0 m and diameter 8.35 m is traveling in the direction of its cylindrical axis (length). It passes by the Earth with a relative speed of 2.44 × 10⁸ m/s. What are the dimensions of the ship, as measured by an Earth observer?
length 20.4 m, diameter 8.35 m

33. ●● A pole vaulter at the Relativistic Olympics sprints past you to do a vault with a speed of 0.65 c. When he is at rest, his pole is 7.0 m long. What length do you perceive the pole to be as he passes you, assuming his relative velocity is parallel to the length of the pole? 5.3 m

34. ●● A "flying wedge" spaceship is a right triangle in side view. When the ship is at rest, its base and altitude (which form the 90° angle) measure 40.0 m and 15.0 m,

respectively. As the ship moves with a speed of 0.900c past an observer on Earth, what does she (a) measure the area of the side of the ship to be? (b) calculate the angle of the triangular "nose" to be? (a) 131 m² (b) 40.7°

35. ●● How fast must a meterstick be moving, parallel to its length, relative to an observer so that he measures its length to be 50 cm? 0.87c

36. ●● The distance to Planet X is 1.00 light-year. How long does it take a spaceship to reach X, according to the pilot of the spaceship, if the speed of the ship is 0.700c relative to X? 1.02 years

37. ●●● What is the length contraction (ΔL) of an automobile 5.00 m long when it is traveling at 100 km/h? [Hint: For $x \ll 1$, $\sqrt{1 - x^2} \approx 1 - (x^2/2)$.] 2.14 × 10⁻¹⁴ m

38. IE ●●● A student in a certain reference frame cannot understand why there is a problem in converting to the metric system, because she notes that the length of her professor's meterstick in another reference frame is the same length as her yardstick (parallel to the meterstick). (a) Which of the following is true: (1) The student is moving relative to the professor, (2) the professor is moving relative to the student, or (3) the only thing that matters is that they are moving relative to one another? Why? (b) What is their relative speed (assuming them to be moving in a direction parallel to their respective sticks)?
(a) (3) they are moving relative to one another (b) 0.40c

39. ●●● Sirius is about 9.0 light-years from Earth. To reach the star by spaceship in 12 years (ship time), how fast must you travel? 0.60c

26.4 Relativistic Kinetic Energy, Momentum, Total Energy, and Mass–Energy Equivalence

40. MC How does an object's relativistically correct kinetic energy compare to its kinetic energy calculated from the Newtonian expression? (a) The relativistic result is always larger, (b) the Newtonian result is always larger, (c) they are the same, or (d) one can be larger or smaller than the other depending upon the object's speed. (a)

41. MC The total energy E of a free-moving particle of mass m with speed v is given by which of the following: (a) mv^2, (b) γmc^2, (c) $1/2 mc^2$, or (d) $K + \gamma mc^2$. (b)

42. MC How does an object's relativistically correct linear momentum (magnitude) compare to its momentum calculated from the Newtonian expression? (a) The relativistic result is always less, (b) the Newtonian result is always less, (c) they are the same, or (d) one can be larger or smaller than the other, depending on the object's velocity. (b)

43. CQ The special theory of relativity places an upper limit on the speed an object can have. Are there similar limits on energy and momentum? Explain. no, see ISM

44. CQ An object subject to a large constant force approaches the speed of light. Is its acceleration constant? Explain. no, see ISM

45. CQ If an electron has a kinetic energy of 2 keV, could the classical expression for kinetic energy be used to compute its speed accurately? What if its kinetic energy is 2 MeV? Explain. yes; no, see ISM

46. ● An electron travels at a speed of $0.600c$. What is its total energy? 0.639 MeV

47. ● An electron is accelerated from rest through a potential difference of 2.50 MV. Find the electron's (a) speed, (b) kinetic energy, and (c) momentum.
(a) $0.985c$ (b) 2.50 MeV (c) 1.56×10^{-21} kg·m/s

48. ● How fast must an object travel for its total energy to be (a) 1% more than its rest energy and (b) 99% more than its rest energy? (a) $0.14c$ (b) $0.86c$

49. ● An average home uses about 1.5×10^4 kWh of electricity per year. How much matter would have to be converted to energy (assuming 100% efficiency) to supply energy for one year to a city with 250 000 such homes? (Are you surprised by the answer?) 0.15 kg

50. ● The United States uses approximately 3.0 trillion kWh of electricity annually. If this electrical energy were supplied by nuclear generating plants, how much nuclear mass would have to be converted to energy, assuming a production efficiency of 25%? 4.8×10^2 kg

51. ● To travel to a nearby star, a spaceship is accelerated to $0.99c$ in order to take advantage of time dilation. If the ship has a mass of 3.0×10^6 kg, how much work must be done to get it up to speed from rest? Compare this value with the annual electricity usage of the United States. (See Exercise 50.)
1.6×10^{24} J, or 150 000 times more!

52. ●● An electron has a total energy of 2.8 MeV. What is its momentum? 1.5×10^{-21} kg·m/s

53. ●● How much energy, in keV, is required to accelerate an electron from rest to $0.50c$? 79 keV

54. ●● A proton moves with a speed of $0.35c$. What are its (a) total energy, (b) kinetic energy, and (c) momentum?
(a) 1.0×10^3 MeV (b) 63 MeV (c) 1.9×10^{-19} kg·m/s

55. ●● A proton moving with a constant speed has a total energy 2.5 times its rest energy. What are the proton's (a) speed and (b) kinetic energy?
(a) $0.92c$ (b) 1.4×10^3 MeV

56. IE ●● The Sun's mass is 1.989×10^{30} kg and it radiates at a rate of 3.827×10^{23} kW. (a) Over time, must the mass of the Sun (1) increase, (2) remain the same, or (3) decrease? (b) Estimate the lifetime of the Sun from this data, assuming it converts all its mass into energy. (c) The actual lifetime of the Sun is predicted to be much less than the answer to part (b), even though its energy emission rate will remain constant. What does this tell you about the 100% conversion assumption? see ISM

57. ●● A nickel has a mass of 5.00 g. If this mass could be completely converted to electric energy, how long would it keep a 100-W lightbulb lit? 1.43×10^5 years

58. IE ●● Phase changes require energy in the form of latent heat (Chapter 11). (a) If 1 kg of ice at 0°C is converted to water at 0°C, will the water have (1) more, (2) the same, or (3) less mass compared to the ice? Why? (b) What is the difference in mass between the ice and the water? Do you think this difference would be detectable?
(a) (1) more (b) 3.7×10^{-12} kg, no

59. ●● A beam of electrons is accelerated from rest to a speed of $0.950c$ in a particle accelerator. In MeV, what are the (a) kinetic energy and (b) total energy of the electrons?
(a) 1.13 MeV (b) 1.64 MeV

60. IE ●●● A particle of mass m, initially moving with speed v, collides head on elastically with an identical particle initially at rest. (a) Do you expect the total mass of the two particles after the collision to be (1) greater than $2m$, (2) equal to $2m$, or (3) less than $2m$? Why? (b) What are the total energy and momentum of the two particles after the collision, in terms of m, v, and c? (a) (2) equal to $2m$
(b) $[1 + (1/\sqrt{1 - v^2/c^2})]\, mc^2$; $mv/\sqrt{1 - v^2/c^2}$

61. ●●● Using the relativistic expression for total energy E and the magnitude p of the momentum, show that E and p are related by $E^2 = p^2c^2 + (mc^2)^2$. see ISM

62. ●●● In a linear accelerator, protons are accelerated to an energy of 600 MeV. (a) How do we know that this quoted energy value must be kinetic, and not total, energy? (b) What is the speed of the protons? (c) What is their momentum? (a) $E_o = 939$ MeV > 600 MeV (b) $0.792c$
(c) 6.50×10^{-19} kg·m/s

26.5 The General Theory of Relativity

63. MC General relativity (a) provides a theoretical basis for explaining the gravitational force, (b) applies only to rotating systems, (c) applies only to inertial systems. (a)

64. MC One of the predictions of general relativity is (a) the mass–energy equivalence, (b) time dilation, (c) the twin paradox, (d) the bending of light in a gravitational field. (d)

65. MC Black hole A has three times the mass of black hole B. How do their Schwarzschild radii R compare? (a) $R_A = R_B$, (b) $R_B = 3R_A$, (c) $R_A = 3R_B$, (d) $R_A = 9R_B$. (c)

66. CQ Suppose a meterstick is dropped toward a black hole. Describe what happens to its shape lengthwise as it gets close to the event horizon. [Hint: The force of gravity will be a lot different on one end of the stick than the other.] see ISM

67. CQ An apparatus like that in ▸Fig. 26.19 was given to Albert Einstein on his 76th birthday by Eric M. Rogers, a physics professor at Princeton University. The goal is to get the ball into the cup without touching the ball. (Jiggling the pole up and down will not do it.) Einstein solved the puzzle immediately and then confirmed his answer with an experiment. How did Einstein get the ball into the cup? [Hint: He used a fundamental concept of general relativity.*] see ISM

---

*This problem is adapted from R. T. Weidner, *Physics* (Boston: Allyn & Bacon, 1985).

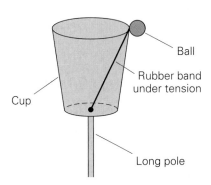

▲ **FIGURE 26.19 How to get the ball into the cup** See Exercise 67.

68. **IE** ●The mass of the planet Jupiter is about 318 times that of the Earth ($M_J = 318M_E$). (a) If these two planets could develop into black holes, would the event horizon for Jupiter be (1) larger than, (2) the same as, or (3) smaller than the Earth's? Why? (b) The mass of the Earth is $M_E = 6.0 \times 10^{24}$ kg; what are the radii of the event horizons for the Earth and Jupiter?
   (a) (1) larger than (b) 8.9 mm and 2.8 m

69. ●● If the Sun became a black hole, what would be its average density? (See Example 26.8.) 1.8×10^{19} kg/m³

70. ●● A black hole has an event horizon of 5.00×10^3 m. (a) What is its mass? (b) Determine the lower limit on the density of the black hole.
   (a) 3.37×10^{30} kg (b) 6.44×10^{18} kg/m³

*26.6 Relativistic Velocity Addition

71. ● After jettisoning a stage, a rocket has a velocity of $+0.20c$ relative to the jettisoned stage. An observer on Earth sees the jettisoned stage moving with a velocity of $+7.5 \times 10^7$ m/s, relative to her, in the same direction as the rocket. What is the velocity of the rocket relative to the Earth observer? $0.43c$

72. ● In moving away from Planet Z, a spacecraft fires a probe with a speed of $0.15c$, relative to the spacecraft, back toward Z. If the speed of the spacecraft is $0.40c$ relative to Z, what is the velocity of the probe as seen by an observer on Z?
   $0.27c$, same direction as spacecraft

73. ●● A rocket launched outward from Earth has a speed of $0.100c$ relative to Earth. The rocket is directed toward an incoming meteor that may hit the planet. If the meteor moves with a speed of $0.250c$ relative to the rocket and directly toward it, what is the velocity of the meteor as observed from Earth?
   $-0.154c$ (toward Earth)

74. **IE** ●● Two spaceships, each with a speed of $0.60c$ relative to Earth, approach each other head on. (a) The speed of one ship relative to the other is (1) greater than c, (2) equal to c, or (3) less than c. Why? (b) What is the speed of one ship relative to the other?
   (a) (3) less than c (b) $0.88c$

75. ●●● In a colliding-beam apparatus, two beams of protons are aimed directly at each other. The first beam contains protons moving with a speed of $0.800c$ to the right, and the second beam's protons have a speed of $0.900c$ to the left. Both speeds are measured relative to the laboratory frame. What are (a) the velocity of the second beam's protons relative to that of the first, and (b) the velocity of the first beam's protons relative to that of the second?
   (a) $0.988c$ to the left (b) $0.988c$ to the right

Comprehensive Exercises

76. **IE** A spaceship containing an astronaut travels at a speed of $0.60c$ relative to a second inertial observer. (a) Which measures proper time intervals in the ship and the proper length of the ship: (1) the astronaut in the ship, (2) the second observer, (3) neither? (b) How much time does a clock onboard the spaceship appear to lose in a day, according to the second observer? (c) If the second observer measures a length of 110 m for the ship, what is its "proper" length"?
   (a) (1) the astronaut in the ship (b) 4.8 h (c) 138 m

77. An electron is accelerated to a speed of 1.5×10^8 m/s. At that speed, compare the relativistic results to the classical results for (a) the electron's kinetic energy, (b) its total energy and (c) the magnitude of its momentum?
   (a) 79 keV (b) 590 keV (c) 1.6×10^{-22} kg·m/s

78. **IE** A relativistic rocket is measured to be 50 m long, 2.5 m high, and 2.0 m wide by its pilot. It is traveling at $0.65c$ (in the direction parallel to its length) relative to an inertial observer. (a) This observer will differ from the pilot in which dimension measurement: (1) length, width, and height, (2) only width and height, or (3) only length? (b) What are the dimensions of the rocket as measured by the inertial observer? (a) (3) only length (b) length, 38 m; height, 2.5 m; width, 2.0 m

79. (a) If the mass of 1.0 kg of coal (or any substance) could be completely converted into energy, how many kilowatt-hours of energy would be produced? (b) Assume that the average U.S. family of four uses 600 kWh of electric energy each month and that modern power plants are 33% efficient in converting this released energy into electric energy. How long would this mass be able to supply the U.S. public with electric energy?
   (a) 2.5×10^{10} kWh (b) 6 days

80. In proton–antiproton annihilation, a proton and an antiproton (which has the same mass as the proton, but carries a negative charge) interact, and both masses are completely converted to electromagnetic radiation. Assuming that both particles are at rest when they annihilate, what is the total energy emitted in the form of radiation? 1.88×10^3 MeV

81. At a typical nuclear power plant, refueling occurs about every 18 months. Assuming that a plant has operated continuously since the last refueling and produces 1.2 GW of electric power at an efficiency of 33%, how much less massive are the fuel rods at the end of the 18 months than at the start? (Assume 30-day months.) 1.9 kg

82. **IE** Many radioactive sources emit neutrons. One way of detecting them is by measuring the light (energy) given off when they are captured by protons (for example, when they enter water, which contains many hydrogen atoms, each one of which has a proton for its nucleus). The released energy is 2.22 MeV and the combined neutron and proton is stable and called a *deuteron*. (a) A very slow neutron (neglect its kinetic energy) is captured by a proton. How should the resulting deuteron's mass compare to the sum of the neutron and proton masses? (1) $m_d = m_p + m_n$ (2) $m_d > m_p + m_n$ (3) $m_d < m_p + m_n$ (b) The proton and neutron masses are $m_p = 1.67262 \times 10^{-27}$ kg and $m_n = 1.67493 \times 10^{-27}$ kg. Determine the mass (in kilograms) of the deuteron (also to six significant figures). (a) (3) $m_d < m_p + m_n$ (b) 3.34360×10^{-27} kg

83. A particle of mass m is initially traveling to the right at $0.900c$. It collides and sticks to a particle of mass $2m$ initially moving to the left at $0.750c$. Determine (a) the total mass (in terms of m) of the composite particle, (b) the amount of kinetic energy lost during the collision, and (c) the velocity (speed and direction) of the composite particle. (a) $5.31m$ (b) $2.31mc^2$ (c) $0.0382c$ left

84. Three inertial observers are all moving in the $+x$-direction. Observer A has a speed of $0.50c$, observer B's speed is $0.90c$ relative to A, and observer C's speed is $0.50c$ relative to B. (a) If C's clock records a time interval of 1.00 hr, how much time has elapsed according to B? (b) In the same time interval, how much time has elapsed according to A? (c) A 10-m pole lies along the x-axis in reference frame A. (d) How long is it as measured by observers in B and C?
(a) 1.15 h (b) 3.84 h (c) 4.36 m (d) 2.60 m

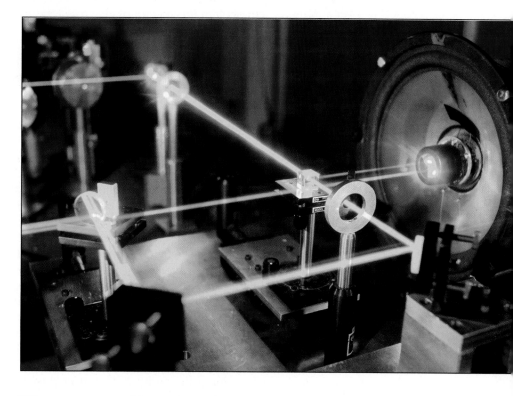

PHYSICS FACTS

- Theoretically, a hot sample of hydrogen gas can give off many different wavelengths. Of those, only four are in the visible range.

- It takes on the order of a million years for the high-energy gamma and X-ray photons released by nuclear fusion at the Sun's center to make their way to the surface of the Sun. This long trip is caused by the many random collisions that result in a reduction of photon energy down into the visible-light range. The trip to Earth from the solar surface takes only about 8 minutes.

- Helium was first discovered in the Sun's atmosphere using its characteristic photon absorption energies. Helium does not exist naturally in the Earth's atmosphere. We have significant amounts of it because it is trapped in deep wells with natural gas.

- Laser beams are so well defined spatially that bouncing them off a mirror located on the Moon (the Apollo astronauts placed the mirrors there in the early 1970s) enables us to determine the distance to the Moon to within a few meters.

Lasers are used in a variety of everyday applications—in bar code scanners at store checkout counters, in computer printers, in various types of surgery, and in laboratory situations, such as the one shown in the chapter-opening photo. You own a laser yourself if you have a CD or DVD player.

The laser is a practical application of principles that revolutionized physics. These principles were first developed in the early twentieth century, one of the most productive eras in the history of physics. For example, special relativity (Chapter 26) helped resolve problems faced by classical (Newtonian) relativity in describing objects moving at speeds near that of light. However, there were other troublesome areas in which classical theories did not agree with experimental results. To address these issues, scientists devised new hypotheses based on nontraditional approaches and ushered in a profound revolution in our understanding of the physical world. Chief among these new theories was the idea that light is *quantized* into discrete amounts of energy. This concept and others like it led to the formulation of a new set of principles and a new branch of physics called *quantum mechanics*.

Quantum theory demonstrated that particles often exhibit wave properties and that waves frequently behave as particles. Thus was born the *wave–particle duality* of matter. As a result of quantum theory, calculations in the realm of the very small—dimensions the sizes of atoms and smaller—must deal with *probabilities* rather than in the precisely determined values associated with classical theory.

A detailed treatment of quantum mechanics requires extremely complex mathematics. However, a general overview of the important results is essential to an understanding of physics as it is known today. Thus, the important developments of "quantum" physics are presented in this chapter, and an introduction to quantum mechanics is provided in Chapter 28.

27.1 Quantization: Planck's Hypothesis

OBJECTIVES: To (a) define blackbody radiation and use Wien's law, and (b) understand how Planck's hypothesis paved the way for quantum ideas.

One of the problems scientists faced at the end of the nineteenth century was how to explain the spectra of electromagnetic radiation emitted by hot objects—solids, liquids, and dense gases. This radiation is sometimes called **thermal radiation**. You learned in Chapter 11 that the total intensity of the emitted radiation from such objects is proportional to the fourth power of the absolute Kelvin temperature (T^4) of the object. Thus, all objects emit thermal radiation to some degree.

However, at everyday temperatures, this radiation is almost all in the infrared (IR) region and not visible to our eyes. However, at temperatures of about 1000 K, a solid object will begin to emit an appreciable amount of radiation in the long-wavelength end of the visible spectrum, observed as a reddish glow. A red-hot electric stove burner is a good example of this. Still-higher temperatures cause the radiation to shift to shorter wavelengths and the color to change to yellow-orange. Above a temperature of about 2000 K, an object glows yellowish-white, like the filament of a light bulb, and gives off appreciable amounts of all the visible colors, but with different percentages.

Although we observe a dominant color with our eyes, in actuality there is a *continuous spectrum*. A spectrum shows how the intensity of emitted energy depends on wavelength, as illustrated in ▼Fig. 27.1a for a hot object. Notice that practically all wavelengths are present, but there is a dominant color (wavelength region), which depends on the object's temperature.

The curves shown in Fig. 27.1a are for an ideal **blackbody**. An ideal blackbody is an ideal object that absorbs all radiation that is incident on it. Although such a blackbody is not attainable, it can be experimentally approximated by a small hole that leads to a cavity inside a block of material (Fig. 27.1b). Radiation falling on the hole enters the cavity and is reflected back and forth off the cavity walls. If the hole is very small, only a small fraction of the incident radiation will make its way back out of the hole. Since nearly all the radiation incident on the hole is absorbed, as viewed from the outside the hole approximates a blackbody.

Two things happen to the spectrum as the temperature increases (Fig. 27.1a). As expected, more radiation is emitted at every wavelength, but also the wavelength

Note: The concept of a blackbody was introduced in Section 11.4.

▶ **FIGURE 27.1 Thermal radiation**
(a) Intensity-versus-wavelength curves for the thermal radiation from a blackbody at different temperatures. The wavelength associated with the maximum intensity (λ_{max}) becomes shorter with increasing temperature. **(b)** A blackbody can be approximated by a small hole leading to an interior cavity in a block of material.

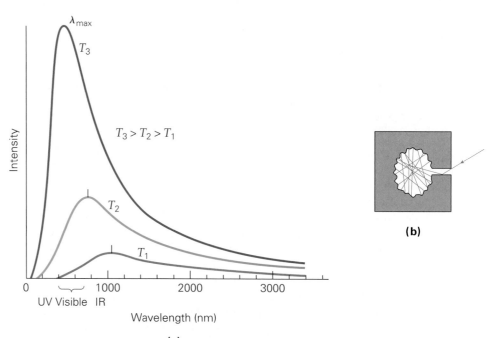

of the maximum-intensity component (λ_{max}) becomes shorter. This wavelength shift is described experimentally by **Wien's displacement law**,

$$\lambda_{max} T = 2.90 \times 10^{-3} \text{ m} \cdot \text{K} \tag{27.1}$$

where λ_{max} is the wavelength of the radiation (in meters) at which maximum intensity occurs and T is the temperature of the body (in kelvins).

Wien's law can be used to determine the wavelength of the maximum spectral component if the temperature of the emitter is known, or the temperature of the emitter if the wavelength of the strongest emission is known. Thus, it can be used to estimate the temperatures of stars (dense gases) from their radiation spectrum, as the following Example shows.

Example 27.1 ■ Solar Colors: Using Wien's Law

The visible surface of our Sun is the gaseous photosphere from which radiation escapes. At the top of the photosphere, the temperature is 4500 K; at a depth of about 260 km, the temperature is 6800 K. Assuming the Sun radiates energy as if it were a blackbody, (a) what are the wavelengths of the radiation of maximum intensity for these temperatures, and (b) to what colors do these wavelengths correspond?

Thinking It Through. Wien's displacement law (Eq. 27.1) enables us to determine the wavelengths.

Solution.

Given: $T_1 = 4500$ K   *Find:* (a) λ_{max} (for the two different temperatures)
$\quad\quad\quad T_2 = 6800$ K   (b) Colors corresponding to these λ_{max} values

(a) At the top of the photosphere,

$$\lambda_{max} = \frac{2.90 \times 10^{-3} \text{ m} \cdot \text{K}}{4500 \text{ K}} = (6.44 \times 10^{-7} \text{ m})(10^9 \text{ nm/m}) = 644 \text{ nm}$$

and at the 260-km depth,

$$\lambda_{max} = \frac{2.90 \times 10^{-3} \text{ m} \cdot \text{K}}{6800 \text{ K}} = (4.26 \times 10^{-7} \text{ m})(10^9 \text{ nm/m}) = 426 \text{ nm}$$

(b) As the temperature increases with depth, the wavelength of the radiation of maximum intensity shifts from red toward the blue end of the spectrum. Thus the Sun's surface is orange-red and shifts towards the blue since temperature increases with depth. (For a discussion of the visible spectrum and color in relation to wavelength, see Section 20.4 and Fig. 20.23.) Combining all the depths and temperatures, we have a spectrum that shows all wavelengths (colors), but is dominated by yellow. Notice that some of the emitted radiation will be in the ultraviolet region, some of which is absorbed by the ozone layer in the Earth's atmosphere.

Follow-Up Exercise. What would be the dominant wavelengths (and corresponding colors) associated with the radiation emitted from the surface of the following stars: (a) Betelgeuse, with an average surface temperature of 3.00×10^3 K; (b) Rigel, with an average surface temperature of 1.00×10^4 K? *(Answers to all Follow-Up Exercises are at the back of the text.)*

The Ultraviolet Catastrophe and Planck's Hypothesis

Classically, thermal radiation results from the oscillations of electric charges associated with the atoms near the surface of an object. Since these charges oscillate at different frequencies, a continuous spectrum of emitted radiation is expected.

Classical calculations describing the radiation spectrum emitted by a blackbody predict an intensity that is inversely related to wavelength (actually $I \propto 1/\lambda^4$). At long wavelengths, the classical theory agrees fairly well with experimental data. However, at short wavelengths, the agreement disappears. Contrary to experimental observations, the classical theory predicts that the radiation intensity should increase without bound as the wavelength gets smaller. This is illustrated in ▸Fig. 27.2. The classical prediction is sometimes called the *ultraviolet catastrophe—ultraviolet*

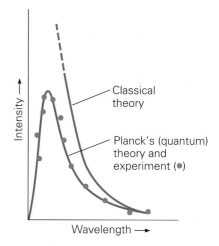

▲ **FIGURE 27.2 The ultraviolet catastrophe** Classical theory predicts that the intensity of thermal radiation emitted by a blackbody should be inversely related to the wavelength of the emitted radiation. If this were true, the intensity would become infinite as the wavelength approaches zero. In contrast, Planck's quantum theory agrees with the observed radiation distribution (solid dots).

because the difficulty occurs for wavelengths shorter than that associated with violet, and *catastrophe* because it predicts that the emitted energy grows without limits at these wavelengths.

The failure of classical electromagnetic theory to explain the characteristics of thermal radiation led Max Planck (1858–1947), a German physicist, to re-examine the phenomenon. In 1900, Planck formulated a theory that correctly predicted the observed distribution of the blackbody radiation spectrum. (Compare the solid blue curve with the data points in Fig. 27.2.) However, his theory depended upon a radical idea. He had to assume that the thermal oscillators (the atoms emitting the radiation) have only *discrete*, or particular, amounts of energy rather than a continuous distribution of energies. Only with this assumption did his theory agree with experiment.

Planck found that these discrete amounts of energy were related to the frequency f of the atomic oscillations by

$$E_n = n(hf) \qquad \text{for } n = 1, 2, 3, \dots \qquad \begin{array}{l}\textit{Planck's quantization}\\ \textit{hypothesis}\end{array} \qquad (27.2)$$

Note: As with other fundamental constants, we will take h to be *exact* at 6.63×10^{-34} J·s, for calculation purposes.

That is, the oscillator energy occurs only in integral multiples of hf. The symbol h is a constant called **Planck's constant**. Its experimental value (to three significant figures) is

$$h = 6.63 \times 10^{-34} \, \text{J·s}$$

The idea expressed in Eq. 27.2 is called **Planck's hypothesis**. Rather than allowing the oscillator energy to have any value, Planck's hypothesis states that the energy is *quantized*; that is, it occurs only in discrete amounts. The smallest possible amount of oscillator energy, according to Eq. 27.2 with $n = 1$, is

$$E_1 = hf \qquad (27.3)$$

All other values of the energy are integral multiples of hf. The quantity hf is called a **quantum** of energy (from the Latin *quantus*, meaning "how much"). As a result, the energy of an atom can change only by the absorption or emission of energy in discrete, or *quantum*, amounts.

Although the theoretical predictions agreed with experiment, Planck himself was not convinced of the validity of his quantum hypothesis. However, the concept of quantization was extended to explain other phenomena that could not be explained classically. Despite Planck's hesitation, the quantum hypothesis earned him a Nobel Prize in 1918.

27.2 Quanta of Light: Photons and the Photoelectric Effect

OBJECTIVES: To (a) describe the photoelectric effect, (b) explain how it can be understood by assuming that light energy is carried by particles, and (c) summarize the properties of photons.

The concept of the quantization of light energy was introduced in 1905 by Albert Einstein in a paper concerning light absorption and emission, at about the same time he published his famous paper on special relativity. Einstein reasoned that energy quantization should be a fundamental property of electromagnetic waves (light). He suggested that if the energy of the thermal oscillators in a hot substance is quantized, then it necessarily followed that, to conserve energy, *the emitted radiation should also be quantized*. For example, suppose an atom initially had an energy of $3hf$ ($n = 3$ in Eq. 27.2) and ended up with a final (lower) energy of $2hf$ ($n = 2$ in Eq. 27.2). Einstein proposed that the atom *must* emit a specific amount (or *quantum*) of light energy—in this case, $3hf - 2hf = hf$. He named this quantum, or package, of light energy the **photon**. Each photon has a definite amount of energy E that depends on the frequency f of the light according to

$$E = hf \qquad \begin{array}{l}\textit{photon energy}\\ \textit{and light frequency}\end{array} \qquad (27.4)$$

This idea suggests that light can behave as discrete quanta (plural of "quantum"), or "particles," of energy rather than as a wave. One way to interpret Eq. 27.4 is as a mathematical "connection" between the *wave nature* of light (a wave of frequency f) and the *particle nature* of light (photons each with an energy E). Given light of a certain frequency or wavelength, Eq. 27.4 enables us to calculate the energy in each photon, or vice versa.

Einstein used the photon concept to explain the **photoelectric effect**, another phenomenon for which the classical description was inadequate. Certain metallic materials are *photosensitive*; that is, when light strikes their surface, electrons may be emitted. The radiant energy supplies the energy necessary to free the electrons from the material. A schematic representation of a photoelectric-effect experiment is shown in ▸Fig. 27.3a. A variable voltage (say by a variable battery or power supply) is maintained between the anode and the cathode. When light strikes the cathode (maintained at ground or zero voltage) which is photosensitive, electrons are emitted. Because they are released by absorption of light energy, these emitted electrons are called *photoelectrons*. They are collected at the anode, which is maintained initially at some positive voltage V to attract the photoelectrons. Thus, in the complete circuit a current is registered on the ammeter.

When a photocell is illuminated with monochromatic (single-wavelength) light of different intensities, characteristic curves are obtained as a function of the applied voltage (Fig. 27.3b). For positive voltages, the anode attracts the electrons. Under these conditions, the *photocurrent* I_p is the flow rate of the photoelectrons, which does *not* vary with voltage. This is because under a positive voltage the electrons are attracted to the anode and *all* the electrons reach it. As expected classically, I_p is proportional to the incident light intensity—the greater the intensity ($I_2 > I_1$ in Fig. 27.3b), the more energy is available to free electrons.

The kinetic energy of the photoelectrons can be measured by *reversing* the voltage across the electrodes, that is, reversing the battery terminals and making $V < 0$, and creating *retarding-voltage* conditions. Now the electrons are repelled from, instead of being attracted to, the anode. The electrons' initial kinetic energies are converted into electric potential energy as they approach the now negatively charged anode. As the retarding voltage is made more and more negative, the photocurrent decreases. This is because (by energy conservation) only electrons with initial kinetic energies greater than $e|V|$ can be collected at the negative anode and thus produce a photocurrent. At some value of retarding voltage, with a magnitude of V_o, called the **stopping potential**, the photocurrent drops to zero. No electrons are collected at that voltage or greater (more negative)—because even the fastest photoelectrons are turned around before reaching the anode. Hence, the maximum kinetic energy (K_{max}) of the photoelectrons is related to the magnitude of the stopping potential (V_o) by

$$K_{max} = eV_o \qquad (27.5)$$

Experimentally, when the frequency of the incident light is varied, this maximum kinetic energy increases linearly with the frequency (▾Fig. 27.4). No emission of electrons is observed for light with a frequency below a certain *cutoff frequency* f_o. Even if the light intensity is very low, the current begins essentially instantaneously with no observable time delay, as long as a material is being illuminated by light with a frequency $f > f_o$.

The important characteristics of the photoelectric effect are summarized in Table 27.1. Notice that only one of the characteristics is predicted correctly by classical wave theory, whereas Einstein's photon concept explains all of the results. When an electron absorbs a quantum of light energy, some of the photon's energy goes to free the electron, with the remainder showing up as kinetic energy. The work required to free the electron is designated by ϕ. So, when a photon of energy E is absorbed, conservation of energy requires that $E = K + \phi$, or, using Eq. 27.4 to replace E, we have

$$hf = K + \phi \qquad (27.6)$$

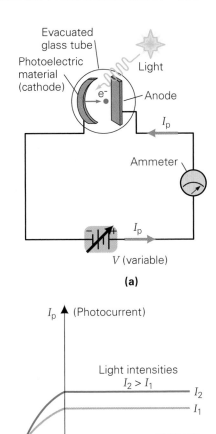

▲ **FIGURE 27.3 The photoelectric effect and characteristic curves** (a) Incident monochromatic light on the photoelectric material in a photocell (or phototube) causes the emission of electrons, which results in a current in the circuit. The applied voltage is variable. (b) As the plots of photocurrent versus voltage for two intensities of light show, the current stays constant as the voltage is increased. However, for negative voltages (using the battery with reversed polarity), the current goes to zero when the stopping potential has a magnitude of $|V_o|$, which, for a fixed frequency, depends only on the type of material, but is independent of intensity.

Note: Recall from Eq. 16.2 that $\Delta U_e = qV$.

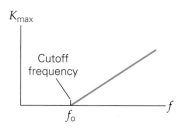

▲ **FIGURE 27.4 Maximum kinetic energy versus light frequency in the photoelectric effect** The maximum kinetic energy (K_{max}) of the photoelectrons is a linear function of the incident-light frequency. Below a certain cutoff frequency f_o, no photoemission occurs, regardless of the intensity of the light.

Teaching tip: Make sure that students have a clear understanding as to why classical theories *cannot* account for the photoelectric effect.

LEARN BY DRAWING

The Photoelectric Effect and Energy Conservation

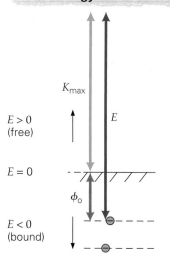

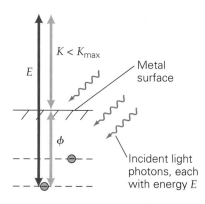

TABLE 27.1	Characteristics of the Photoelectric Effect

Characteristic	Predicted by wave theory?
1. The photocurrent is proportional to the intensity of the light.	Yes
2. The maximum kinetic energy of the emitted electrons depends on the frequency of the light, but not on its intensity.	No
3. No photoemission occurs for light with a frequency below a certain cutoff frequency f_o, regardless of the light intensity.	No
4. A photocurrent is observed immediately when the light frequency is greater than f_o, even if the light intensity is extremely low.	No

Since the energies are very small, the commonly used energy unit is the electron-volt (eV; see Chapter 16). Recall that $1.00 \text{ eV} = 1.60 \times 10^{-19} \text{ J}$.

The least tightly bound electron will have the maximum kinetic energy K_{max}. (Why?) The energy needed to free this electron is called the **work function (ϕ_o)** of the material. For this situation, Eq. 27.6 becomes

$$\underset{\substack{\text{incident photon} \\ \text{energy}}}{hf} = \underset{\substack{\text{maximum kinetic} \\ \text{energy of freed} \\ \text{electron}}}{K_{max}} + \underset{\substack{\text{minimum work} \\ \text{needed to free} \\ \text{the electron}}}{\phi_o} \qquad (27.7)$$

Other electrons require more energy than the minimum to be freed, so their kinetic energy will be less than K_{max}. This concept is explored visually in the Learn by Drawing feature on this page. Some typical numerical values are shown in the next Example.

Example 27.2 ■ The Photoelectric Effect: Electron Speed and Stopping Potential

The work function of a particular metal is known to be 2.00 eV. If the metal is illuminated with light of wavelength 550 nm, what will be (a) the maximum kinetic energy of the emitted electrons and (b) their maximum speed? (c) What is the stopping potential?

Thinking It Through. (a) By energy conservation (Eq. 27.7), the maximum kinetic energy is the difference between the incoming photon energy and the work function. (b) Speed can be determined from kinetic energy, since the mass of an electron is known ($m = 9.11 \times 10^{-31}$ kg). (c) The stopping potential is found by requiring that all of the kinetic energy be converted to electric potential energy (Eq. 27.5).

Solution. First, converting the data into SI units.

Given: $\phi_o = (2.00 \text{ eV})(1.60 \times 10^{-19} \text{ J/eV})$ *Find:* (a) K_{max} (maximum kinetic energy)
$\quad\quad = 3.20 \times 10^{-19} \text{ J}$ (b) v_{max} (maximum speed)
$\quad \lambda = 550 \text{ nm} = 5.50 \times 10^{-7} \text{ m}$ (c) V_o (stopping potential)

(a) Using $\lambda f = c$, we find that the photon energy of light with the given wavelength is

$$E = hf = \frac{hc}{\lambda} = \frac{(6.63 \times 10^{-34} \text{ J·s})(3.00 \times 10^8 \text{ m/s})}{5.50 \times 10^{-7} \text{ m}} = 3.62 \times 10^{-19} \text{ J}$$

Then

$$K_{max} = E - \phi_o = 3.62 \times 10^{-19} \text{ J} - 3.20 \times 10^{-19} \text{ J}$$

$$= (4.20 \times 10^{-20} \text{ J})\left(\frac{1 \text{ eV}}{1.60 \times 10^{-19} \text{ J}}\right) = 0.263 \text{ eV}$$

(b) v_{max} can be found from $K_{max} = \frac{1}{2}mv^2_{max}$:

$$v_{max} = \sqrt{\frac{2K_{max}}{m}} = \sqrt{\frac{2(4.20 \times 10^{-20} \text{ J})}{9.11 \times 10^{-31} \text{ kg}}} = 3.04 \times 10^5 \text{ m/s}$$

(c) The stopping potential is related to K_{max} by $K_{max} = eV_o$; therefore,

$$V_o = \frac{K_{max}}{e} = \frac{0.420 \times 10^{-19}\,J}{1.60 \times 10^{-19}\,C} = 0.23\,V$$

Follow-Up Exercise. In this Example, suppose that a different wavelength of light is used and the new stopping voltage is found to be 0.50 V. What is the wavelength of this new light? Explain why this wavelength requires a larger stopping voltage.

Einstein's photon model of light is, in fact, consistent with *all* the experimental results of the photoelectric effect. In the photon model, an increase in light intensity means an increase in the number of photons and therefore in the number of photoelectrons (that is, the photocurrent). However, an increase in intensity would *not* mean a change in the energy of any one photon, since that energy depends only on the light frequency ($E = hf$). Therefore, K_{max} should be independent of intensity, but linearly dependent on the frequency of the incident light—as is observed experimentally.

Einstein's quantum theory of light also explains the existence of a cutoff frequency. In his interpretation, since photon energy depends on frequency, this means that below a certain (cutoff) frequency (f_o) the photons simply don't have enough energy to dislodge even the most loosely bound electrons. Therefore, no current is observed for those frequencies. Since, at the cutoff frequency, no electrons are emitted, the cutoff frequency can be found by setting $K_{max} = 0$ in Eq. 27.7:

$$hf_o = K_{max} + \phi_o = 0 + \phi_o$$

or

$$f_o = \frac{\phi_o}{h} \quad \begin{matrix} threshold,\ or \\ cutoff,\ frequency \end{matrix} \tag{27.8}$$

Note: A photon of energy hf_o will barely free an electron from a solid, but the electron will have essentially no kinetic energy.

The cutoff frequency f_o is sometimes called the **threshold frequency**. It represents the minimum frequency of light necessary to create photoelectrons. Below the threshold frequency, the binding energy of the least-bound electron exceeds the photon energy. Although the electron may absorb the photon energy, it will not have enough energy to be freed from the material and become a photoelectron. (How would you explain this, using a sketch such as the one in the accompanying Learn by Drawing?)

Example 27.3 ■ The Photoelectric Effect: Threshold Frequency and Wavelength

What are the threshold frequency and corresponding wavelength for the metal described in Example 27.2?

Thinking It Through. Example 27.2 gives the work function, so Eq. 27.8 allows us to determine the threshold frequency. The threshold wavelength can then be computed from $\lambda = c/f$.

Solution. Listing the data, we have

Given: $\phi_o = 2.00\,eV$ *Find:* f_o (threshold frequency)
        $= 3.20 \times 10^{-19}\,J$ λ_o (wavelength at threshold frequency)
    (from Example 27.2)

Solving for the threshold frequency f_o from $\phi_o = hf_o$ (Eq. 27.8), we get

$$f_o = \frac{\phi_o}{h} = \frac{3.20 \times 10^{-19}\,J}{6.63 \times 10^{-34}\,J \cdot s} = 4.83 \times 10^{14}\,Hz$$

The wavelength (the threshold wavelength) corresponding to this frequency is

$$\lambda_o = \frac{c}{f_o} = \frac{3.00 \times 10^8\,m/s}{4.83 \times 10^{14}\,Hz} = 6.21 \times 10^{-7}\,m = 621\,nm$$

Any frequency lower than 4.83×10^{14} Hz, or, alternatively, any wavelength longer than 621 nm, would not yield photoelectrons. Since this wavelength lies in the red-orange end of the electromagnetic spectrum, yellow light, for example, would dislodge electrons, but deep-red light would not.

Follow-Up Exercise. In this Example, what would be the stopping voltage if the frequency of the light were twice the cutoff frequency?

Demonstration/activity: Demonstrate the photoelectric effect by using a UV light source to knock electrons off a negatively charged electroscope. Then, while covering the UV light with glass (which absorbs most of the UV), show that the same source's visible light does not knock electrons off.

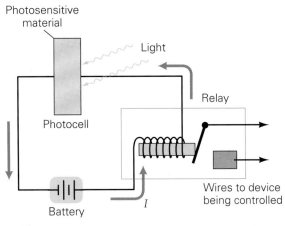

Photosensitive material

Light

Relay

Photocell

Battery

I

Wires to device being controlled

(a)

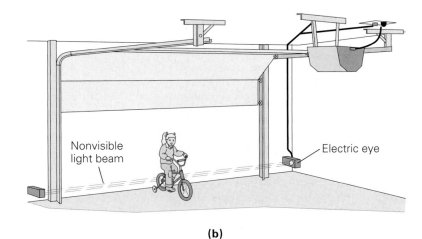

Nonvisible light beam

Electric eye

(b)

▲ **FIGURE 27.5 Photoelectric applications: The electric eye** (a) A diagram of an electric-eye circuit. When light strikes a photocell material (any photosensitive material), it frees electrons from their atoms (but not from the solid as a whole). In effect this lowers the material's resistance by enabling it to conduct current. The result is a current in the circuit. Interruption of the light beam opens the circuit in the relay (a magnetic switch) that controls the particular device. (b) Electric-eye circuits are used in automatic garage-door openers. When the door starts to move downward, any interruption of the electric-eye beam (usually IR light) causes the door to stop, protecting anything that may be under the descending door.

Problem-Solving Hint

In photon calculations, the wavelength of the light is often given rather than the frequency. Typically, what is needed is the photon *energy*. Instead of first calculating the frequency $(f = c/\lambda)$, then the energy in joules $(E = hf)$, and finally converting to electron-volts, this all can be done in one step. To do so, combine these two equations to form $E = hf = hc/\lambda$ and express the product hc in electron-volts times nanometers, or eV · nm. The value of this useful product is

$$hc = (6.63 \times 10^{-34} \text{ J} \cdot \text{s})(3.00 \times 10^8 \text{ m/s}) = 1.99 \times 10^{-25} \text{ J} \cdot \text{m}$$

$$= \frac{(1.99 \times 10^{-25} \text{ J} \cdot \text{m})(10^9 \text{ nm/m})}{1.60 \times 10^{-19} \text{ J/eV}}$$

$$= 1.24 \times 10^3 \text{ eV} \cdot \text{nm}$$

This shortcut can save time and effort in working problems and allows quick estimation of the photon energy associated with light of a given wavelength (or vice versa). Thus, if orange light $(\lambda = 600 \text{ nm})$ is used, you need only divide in your head to realize that each photon carries approximately 2 eV of energy. A more exact result could be calculated if needed, as $E = \dfrac{hc}{\lambda} = \dfrac{1.24 \times 10^3 \text{ eV} \cdot \text{nm}}{600 \text{ nm}} = 2.07 \text{ eV}$.

There are many applications of the photoelectric effect. The fact that the current produced by photocells is proportional to the intensity of the light makes them ideal for use in photographers' light meters. Photocells are also used in solar-energy applications to convert sunlight to electricity.

Another common application of the photocell is the electric eye (▲Fig. 27.5a). As long as light strikes the photocell, there is current in the circuit. Blocking the light opens the circuit in the relay (magnetic switch), which in turn controls some device. A common application of the electric eye is to turn on streetlights automatically at night. A safety application of the electric eye is shown in Fig. 27.5b. Note that in many of these applications (including the garage-door safety mechanism) the IR light photons do not actually free the electrons from the *material*. All is required is that they free them from the *atoms* in the material—thus the electrons stay in the material but are free to move through the material. Once they are freed, the external voltage causes them to flow, resulting in an electrical current that can be detected and put to appropriate use depending on the application.

27.3 Quantum "Particles": The Compton Effect

OBJECTIVES: To (a) understand how the photon model of light explains scattering of light from electrons (the Compton effect), and (b) calculate the wavelength of the scattered light in the Compton effect.

In 1923, the American physicist Arthur H. Compton (1892–1962) explained the scattering of X-rays from a graphite (carbon) block by assuming the radiation to be composed of quanta. His explanation of the observed effect provided additional

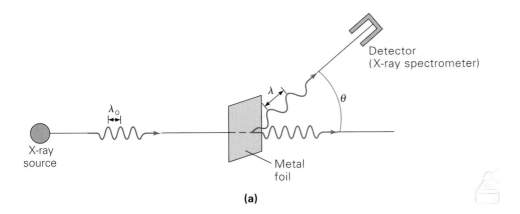

(a)

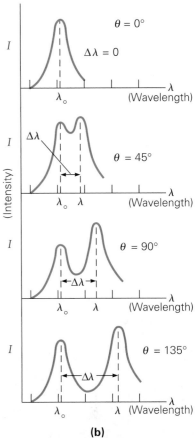

(b)

convincing evidence that, at least in certain types of experiments, light (electromagnetic) energy is carried by *photons*.

Compton had observed that when a beam of monochromatic (single-wavelength) X-rays was scattered by various materials, the wavelength of the scattered X-ray was longer than the wavelength of the incident X-ray. In addition, he noted that the change in the wavelength depended on the angle θ through which the X-rays were scattered, but *not* on the nature of the scattering material (▲Fig. 27.6). This phenomenon came to be known as the **Compton effect**.

According to the wave model, any scattered radiation should have the same frequency (and wavelength) as the incident radiation. In this model, the electrons in the atoms of the scattering material are accelerated by the oscillating electric field of the radiation and therefore oscillate at the same frequency as the incident wave. The scattered (or reradiated) radiation should thus have the same frequency, regardless of direction.

According to Einstein's photon picture, the energy of each photon is proportional to the frequency f of the associated light wave. Therefore a change in frequency or wavelength would indicate a change in photon energy. Because the wavelength increased (and the frequency decreased, because $f = c/\lambda$), the scattered photons had *less* energy than the incident ones. Moreover, the change in photon energy increased with the scattering angle—which reminded Compton of an elastic collision of two particles. Could the same principles apply in the scattering of these quantum "particles" called photons?

Pursuing this idea, Compton assumed that a photon behaves as a particle when it collides with electrons. He reasoned that if an incident photon collides with an electron initially at rest, the photon should transfer some energy and momentum to that electron. Thus we would expect that the energy of the scattered photon, as well as the frequency of the associated scattered light wave, should both decrease (since $E = hf$). Applying conservation of energy and linear momentum, Compton showed, using the photon model, that the shift in the wavelength of the light scattered at an angle θ from an electron is given by

$$\Delta\lambda = \lambda - \lambda_o = \lambda_C (1 - \cos\theta) \quad \textit{Compton scattering} \quad (27.9)$$

where λ_o is the wavelength of the incident light and λ is that of the scattered light. The constant $\boldsymbol{\lambda_C}$ is called the **Compton wavelength** of the electron. It is inversely related to the mass of the electron by $\lambda_C = h/(m_e c)$. The Compton wavelength has a numerical value of $\lambda_C = 2.43 \times 10^{-12}$ m $= 2.43 \times 10^{-3}$ nm.* Equation 27.9 correctly predicts the observed wavelength shift. For his work, Compton was awarded a Nobel Prize in 1927.

Note that the *maximum* wavelength increase occurs when $\theta = 180°$ and has a value of $\Delta\lambda_{max} = 2\lambda_C = 4.86 \times 10^{-3}$ nm. (To see this, use Eq. 27.9 and note that for $\theta = 180°$, $\cos\theta = -1$ and $1 - \cos\theta = 2$.) The scattered wavelength is longest

▲ **FIGURE 27.6 X-ray scattering** **(a)** When X-rays of a single wavelength are scattered by the electrons in metal foil, the scattered wavelength (λ) is longer than the incident wavelength (λ_o). Most of the incident X-rays pass through without interacting (and therefore undergo no change in wavelength). The scattered electrons are not shown, because they remain in the sample. **(b)** The change in wavelength increases with the scattering angle (θ). Note that an unshifted peak at λ_o remains at all angles. This corresponds to photons that scatter off electrons tightly bound to atoms. Thus, since the relatively massive atom (compared to an electron) recoils. In this situation, it takes a negligible energy, leaving the scattered photon with essentially the same energy as the incident one.

Note: Elastic collisions are discussed in Section 6.4.

---

*This numerical value is the Compton wavelength of the *electron*. Compton scattering can occur from any particle; hence, there is a Compton wavelength of the proton, the neutron, and so on. These values are much smaller than the electron's Compton wavelength, because other particles are much more massive than electrons.

when the photon completely reverses direction, and the electron goes forward with the maximum amount of kinetic energy. Since this value is the maximum wavelength *change*, it is difficult to measure for incident wavelengths several thousand times this value or greater, such as for wavelengths larger than several nanometers (strong UV). In other words, the Compton effect is negligible for UV light and any other light with a longer wavelength, such as visible or IR light. It is significant only for X-ray and gamma-ray scattering.

Example 27.4 ■ X-Ray Scattering: The Compton Effect

A monochromatic beam of X-rays of wavelength 1.35×10^{-10} m is scattered by the electrons in a metal foil. By what percentage is the wavelength shifted if the scattered X-rays are observed at an angle of 90°?

Thinking It Through. The change, or shift, in the wavelength, $\Delta\lambda$, is given by Eq. 27.9 with $\theta = 90°$, and the fractional change is $\Delta\lambda/\lambda_o$. The change is positive, because the scattered light has a longer wavelength than that of the incident light.

Solution. Listing the data, we have

Given: $\lambda_o = 1.35 \times 10^{-10}$ m   *Find:* Percentage change in wavelength
$\theta = 90°$

Starting from Eq. 27.9, we can compute the fractional change directly, since $\cos 90° = 0$:

$$\frac{\Delta\lambda}{\lambda_o} = \frac{\lambda_C}{\lambda_o}(1 - \cos\theta) = \frac{2.43 \times 10^{-12}\ \text{m}}{1.35 \times 10^{-10}\ \text{m}}(1 - \cos 90°) = 1.80 \times 10^{-2}$$

So

$$\frac{\Delta\lambda}{\lambda_o} \times 100\% = 1.80\%$$

Follow-Up Exercise. In this Example, (a) what would be the maximum percentage change if gamma rays with a wavelength of 1.50×10^{-14} m were used instead? (b) Is this change larger or smaller than the maximum possible percentage change for the X-rays in the Example? Why?

Einstein's and Compton's success with photons left scientists with two apparently competing theories. Classically, light is a traveling wave, and this satisfactorily explains such phenomena as interference and diffraction. Conversely, quantum theory is necessary to explain the photoelectric and Compton effects. The two theories combined to give rise to a description called the **dual nature** (or **wave–particle duality**) **of light**, as follows:

> To explain all electromagnetic phenomena, light has to be considered sometimes as a wave and other times as a beam of photons. When it interacts with small (quantized) systems, such as atoms, nuclei, and molecules, the photon (quantized-energy) picture must be used. In everyday-sized systems—for example, slit systems causing diffraction and interference—the wave model is applicable.

27.4 The Bohr Theory of the Hydrogen Atom

OBJECTIVES: To (a) understand how the Bohr model of the hydrogen atom explains that atom's emission and absorption spectra, (b) calculate the energies and wavelengths of emitted and absorbed photons for transitions in atomic hydrogen, and (c) understand how the generalized concept of atomic energy levels can explain other atomic phenomena.

In the 1800s, much experimental work was done with gas-discharge tubes—for example, those containing hydrogen, neon, and mercury vapor. Common neon "lights" are actually gas-discharge tubes (◄Fig. 27.7). Recall that light from an incandescent source, such as a lightbulb's hot filament, exhibits a *continuous spectrum* in which all wavelengths are present. However, when light emissions from gas-discharge tubes were

(a)

(b)

▲ **FIGURE 27.7** Gas-discharge tubes **(a)** These luminous glass tubes are gas-discharge tubes, in which atoms of various gases emit light when electrically excited. Each gas radiates its own characteristic wavelengths. **(b)** Only some "neon lights" actually contain neon, which glows with a red hue; other gases produce other colors.

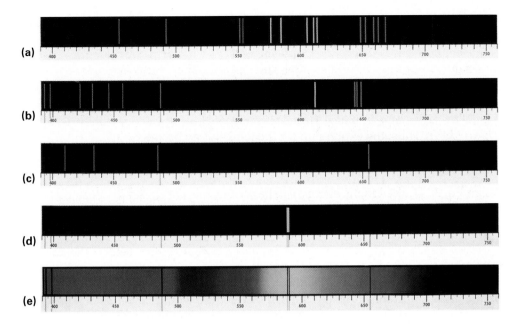

◀ FIGURE 27.8 Emission and absorption spectra of gases When a gas is excited by heat or electricity, the light it emits can be separated into its various wavelengths by a prism or diffraction grating; the result is a bright-line, or emission, spectrum, such as the ones shown here from **(a)** barium, **(b)** calcium, **(c)** hydrogen, and **(d)** sodium, each with its own characteristic pattern. **(e)** When a continuous spectrum from a hot solid or dense gas is viewed after passing through a cool gas, a dark-line, or absorption, spectrum is observed. Each line represents a particular wavelength the gas has absorbed. The absorption spectrum of the Sun provided here shows several prominent absorption lines produced by the gases of the solar atmosphere before the sunlight makes it to the Earth. In fact, the inert gas helium was first discovered to exist on the Sun by this very method.

analyzed, discrete spectra with only certain wavelengths present were observed (▲Fig. 27.8). The spectrum of light coming from such a tube is called a *bright-line spectrum*, or **emission spectrum**. In general, the wavelengths present in an emission spectrum are characteristic of the individual atoms or molecules of the particular gas.

Atoms can absorb light as well as emit it. If white light is passed through a cool gas, the energy at certain frequencies or wavelengths is absorbed. The result is a *dark-line spectrum*, or **absorption spectrum**—a series of dark lines superimposed on a continuous spectrum (Fig. 27.8e). Just as in emission spectra, the missing wavelengths are uniquely related to the type of atom or molecule doing the absorbing. By determining the pattern of emitted and/or absorbed wavelengths, the type of atoms or molecules present in a sample can be identified. This method is called *spectroscopic analysis* and is widely used in physics, astrophysics, biology, and chemistry. For example, the element helium was first discovered on the Sun when scientists found that an absorption-line pattern in the sunlight did not match any known pattern on the Earth. That unknown pattern belonged to the helium atom.

Although the reason for line spectra was not understood in the 1800s, they provided an important clue to the electron structure of atoms. Hydrogen, with its relatively simple visible spectrum, received much of the attention. It is also the simplest atom, consisting of only one electron and one proton. In the late nineteenth century, the Swiss physicist J. J. Balmer found an empirical formula that gives the wavelengths of the four spectral lines of hydrogen in the visible region:

$$\frac{1}{\lambda} = R\left(\frac{1}{2^2} - \frac{1}{n^2}\right) \quad \text{for } n = 3, 4, 5, \text{ and } 6 \quad \begin{array}{l}\textit{visible spectrum}\\\textit{of hydrogen}\end{array} \quad (27.10)$$

R is called the *Rydberg constant*, named after the Swedish physicist Johannes Rydberg (1854–1919), who also studied atomic emission lines. R has an experimental value of $1.097 \times 10^{-2}\,\text{nm}^{-1}$. The four spectral lines of hydrogen in the visible region (note four values for n in Eq. 27.10), are part of the **Balmer series**. There wavelengths fit the formula, but it was not understood why. Similar formulas were found to fit other spectral-line series that were completely in the ultraviolet and infrared regions.

An explanation of the spectral lines was given in a theory of the hydrogen atom put forth in 1913 by the Danish physicist Niels Bohr (1885–1962). Bohr assumed that the electron of the hydrogen atom orbits the proton in a circular orbit analogous to a planet orbiting the Sun. The attractive electrical force between the electron and proton supplies the necessary centripetal force for the circular motion. Recall that the centripetal force is given by $F_c = mv^2/r$, where v is the electron's orbital speed, m is its mass, and r is the radius of its orbit. The force between the proton and electron is given by Coulomb's law as $F_e = kq_1q_2/r^2 = ke^2/r^2$, where

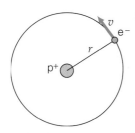

▲ **FIGURE 27.9 The Bohr model of the hydrogen atom** The electron is pictured as revolving around the much more massive proton in a circular orbit. The electric force of attraction provides the centripetal force.

e is the magnitude of the charge of the proton and the electron (◀Fig. 27.9). Equating these two forces, we have

$$\frac{mv^2}{r} = \frac{ke^2}{r^2} \qquad (27.11)$$

The total energy of the atom is the sum of its kinetic and potential energies. Recall from Chapter 16 that the electric potential energy of two point charges is given by $U_e = kq_1q_2/r$. Since the electron and proton are oppositely charged, $U_e = -ke^2/r$. Thus, the expression for the total energy becomes

$$E = K + U_e = \tfrac{1}{2}mv^2 - \frac{ke^2}{r}$$

From Eq. 27.11, the kinetic energy can be written as $\tfrac{1}{2}mv^2 = ke^2/2r$. With this relationship, the total energy becomes

$$E = \frac{ke^2}{2r} - \frac{ke^2}{r} = -\frac{ke^2}{2r} \qquad (27.12)$$

Note that E is negative, indicating that the system is bound. As the radius gets very large, E approaches zero. With $E = 0$, the electron would no longer be bound to the proton, and the atom, having lost its electron, would be *ionized*.

Up to this point, only classical principles had been applied. At this step in the theory, Bohr made a radical assumption—radical in the sense that he introduced a quantum concept to attempt to explain atomic line spectra:

> Bohr assumed that the angular momentum of the electron was quantized and could have only discrete values that were integral multiples of $h/2\pi$, where h is Planck's constant.

Recall that in a circular orbit of radius r, the angular momentum L of an object of mass m is given by mvr (Eq. 8.14). Therefore, Bohr's assumption translates into

$$mvr = n\left(\frac{h}{2\pi}\right) \qquad \text{for } n = 1, 2, 3, 4, \ldots \qquad (27.13)$$

The integer n is an example of a *quantum number*. Specifically, n is the atom's **principal quantum number**.* With this assumption, the orbital speed of the electron (v) can be found. Its (quantized) values are

$$v_n = \frac{nh}{2\pi mr} \qquad \text{for } n = 1, 2, 3, 4, \ldots$$

Putting this expression for v into Eq. 27.11 and solving for r, we obtain

$$r_n = \left(\frac{h^2}{4\pi^2 ke^2 m}\right)n^2 \qquad \text{for } n = 1, 2, 3, 4, \ldots \qquad (27.14)$$

Here, the subscript n on r is used to indicate that only certain radii are possible—that is, the size of the orbit is also quantized. The energy for an orbit can be found by substituting this expression for r into Eq. 27.12, which gives

$$E_n = -\left(\frac{2\pi^2 k^2 e^4 m}{h^2}\right)\frac{1}{n^2} \qquad \text{for } n = 1, 2, 3, 4, \ldots \qquad (27.15)$$

where the energy is also written with a subscript of n to show its dependence on n. The quantities in the parentheses on the right-hand sides of Eqs. 27.14 and 27.15 are constants and can be evaluated numerically. Because the radii are so small, they are typically expressed in nanometers (nm); similarly, energies are expressed in electron-volts (eV):

$$r_n = 0.0529n^2 \text{ nm} \qquad \text{for } n = 1, 2, 3, 4, \ldots \qquad \begin{array}{l}\textit{orbital radii and energies} \\ \textit{for the hydrogen atom}\end{array} \qquad (27.16)$$

$$E_n = \frac{-13.6}{n^2} \text{ eV} \qquad \text{for } n = 1, 2, 3, 4, \ldots \qquad (27.17)$$

The use of these expressions is shown in the next Example.

*The principal quantum number is only one of four quantum numbers necessary to completely describe each electron in an atom. See Chapter 28.

Teaching tip: You may wish to emphasize that the Bohr model of hydrogen can be extended, with reasonable success, to any nucleus with a *single* electron in orbit (ions). E is proportional to Z^2, and r is proportional to $1/Z$, where Z is the number of protons in the nucleus. Thus, for singly ionized helium, $Z = 2$, and we use $13.6(2)^2 = 54.4$ eV in the energy equation and $0.0529/2 = 0.0265$ nm in the radius equation in place of the values for hydrogen.

Example 27.5 ■ A Bohr Orbit: Radius and Energy

Find the orbital radius and energy of an electron in a hydrogen atom characterized by the principal quantum number $n = 2$.

Thinking It Through. Equations 27.16 and 27.17 are used with $n = 2$.

Solution. For $n = 2$,

$$r_2 = 0.0529n^2 \text{ nm} = 0.0529(2)^2 \text{ nm} = 0.212 \text{ nm}$$

and

$$E_2 = \frac{-13.6}{n^2} \text{ eV} = \frac{-13.6}{2^2} \text{ eV} = -3.40 \text{ eV}$$

Follow-Up Exercise. In this Example, what is (a) the speed and (b) the kinetic energy of the orbiting electron?

However, there was still a problem with Bohr's theory. Classically, any accelerating charge should radiate electromagnetic energy (light). For the Bohr circular orbits, the electron is accelerating centripetally. Thus, the orbiting electron should lose energy and spiral into the nucleus. Clearly, this doesn't happen in the hydrogen atom, so Bohr had to make another nonclassical assumption. He postulated that:

> The hydrogen electron does *not* radiate energy when it is in a bound, discrete orbit. It radiates energy only when it makes a *downward transition* to an orbit of lower energy. It makes an *upward transition* to an orbit of higher energy by absorbing energy.

Energy Levels

The "allowed" orbits of the electron in a hydrogen atom are commonly expressed in terms of their energy (▼Fig. 27.10). In this context, the electron is referred to as being in a particular "energy level" or state. The principal quantum number labels the particular energy level. The lowest energy level ($n = 1$) is the **ground state**. The energy levels above the ground state are called **excited states**. For example, $n = 2$ is the *first excited state* (see Example 27.5), and so on.

The electron is normally in the ground state and must be given enough energy to raise it to an excited state. Since the energy levels have specific energies, it follows that the electron can be excited only by absorbing certain discrete amounts of energy. The energy levels can be thought of as the rungs of a ladder.* A person who goes up and down a ladder changes his or her gravitational potential energy by discrete amounts. Similarly, an electron goes up and down its own "energy ladder" in discrete steps. Notice, however, that the energy levels of the hydrogen atom are not evenly spaced, as they are on a ladder (Fig. 27.10).

Note: Review the concept of potential-energy wells, presented in Sections 5.4 and 7.5. (See Figs. 5.14 and 7.18.)

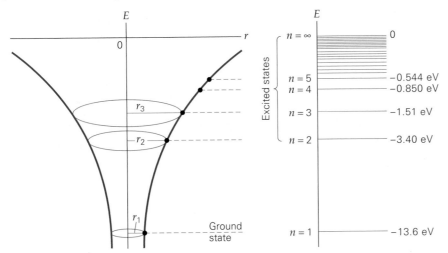

◄ **FIGURE 27.10 Orbits and energy levels of the hydrogen electron** The Bohr theory predicts that the hydrogen electron can occupy only certain orbits having discrete radii. Each allowed orbit has a corresponding total energy, conveniently displayed as an energy-level diagram. The lowest energy level ($n = 1$) is the ground state; the levels above it ($n > 1$) are excited states. The orbits are shown on the left, plotted in the $1/r$ electrical potential of the proton. The electron in the ground state is deepest in the potential-energy well, analogous to the gravitational potential-energy well of Fig. 7.18. (Neither r nor the energy levels are drawn to scale—can you tell why?)

*Care must be taken with this. As we step down or up a ladder, our gravitational potential energy takes on all values in between the end values. However, in an electron's "quantum" jump, it *never* has any intermediate energy value. That is, it has an energy to start and one to end, emitting or absorbing a quantum of energy in making the transition, but it never takes on an energy value in between the initial and final values.

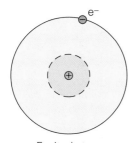

Excited atom

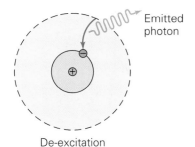

De-excitation

Emitted photon

E_{n_i} —————————

ΔE

E_{n_f} —————————

$\Delta E = hf$

▲ **FIGURE 27.11** Electron transitions and photon emission
When a hydrogen atom emits light, its electron makes a downward transition to a lower orbit (with less energy), and a photon is emitted. The photon's energy is equal to the energy difference between the two levels.

If enough energy is absorbed, it is possible for the electron to no longer be bound to the atom; that is, it is possible for the atom to be *ionized*. For example, to ionize a hydrogen atom initially in its ground state requires a minimum of 13.6 eV of energy. This minimum process makes the final energy of the electron zero (since it is free), and it has a principal quantum number of $n = \infty$. However, if the electron is initially in an excited state, then less energy is needed to ionize the atom. Since the energy of the electron in any state is E_n, the energy needed to free it from the atom is $-E_n$. This energy is called the **binding energy** of the electron. Note that $-E_n$ is positive and represents the energy required to ionize the atom if the electron initially is in a state with a principal quantum number of n.

An electron generally does not remain in an excited state for long; it decays, or makes a downward transition to a lower energy level, in a very short time. The time an electron spends in an excited state is called the **lifetime** of the excited state. For many states, the lifetime is about 10^{-8} s. In making a transition to a lower state, the electron emits a quantum of light energy in the form of a photon (◄Fig. 27.11). The energy ΔE of the photon is equal in magnitude to the energy *difference* of the levels:

$$\Delta E = E_{n_i} - E_{n_f} = \left[\frac{-13.6}{n_i^2} \text{ eV}\right] - \left[\frac{-13.6}{n_f^2} \text{ eV}\right]$$

or

$$\Delta E = 13.6\left(\frac{1}{n_f^2} - \frac{1}{n_i^2}\right) \text{eV} \qquad \begin{array}{l}\textit{photon energy (in eV)}\\ \textit{emitted by an H atom}\end{array} \qquad (27.18)$$

Here, the subscripts i and f refer to the initial and final states, respectively. According to the Bohr theory, this energy difference is emitted as a photon with an energy E. Therefore, $E = \Delta E = hc/\lambda$, or $\lambda = hc/\Delta E$. Thus only particular wavelengths of light are emitted. These particular wavelengths (or alternatively, frequencies) correspond to the various transitions between energy levels and explain the existence of an emission spectrum.

The final principal quantum number n_f refers to the energy level in which the electron ends up during the emission process. The original *Balmer emission series* in the visible region corresponds to $n_f = 2$ and $n_i = 3, 4, 5$ and 6. There is only one emission series entirely in the ultraviolet range, called the *Lyman series*, in which all the transitions end in the $n_f = 1$ state (the ground state). There are many series entirely in the infrared region, most notably the *Paschen series*, which ends with the electron in the second excited state, $n_f = 3$. (These series take their names from their discoverers.)

Usually, the wavelength of the light is what is measured during the emission process. Since photon energy and light wavelength are related by Einstein's equation (Eq. 27.4), the wavelength of the emitted light, λ, can be obtained from

$$\lambda \text{ (in nm)} = \frac{hc}{\Delta E} = \frac{1.24 \times 10^3 \text{ eV} \cdot \text{nm}}{\Delta E \text{ (in eV)}} \qquad (27.19)$$

(See the Problem-Solving Hint on page 858.) Consider Example 27.6.

Example 27.6 ■ Investigating the Balmer Series: Visible Light from Hydrogen

What is the wavelength (and color) of the emitted light when an electron in a hydrogen atom undergoes a transition from the $n = 3$ energy level to the $n = 2$ energy level?

Thinking It Through. The emitted photon has an energy equal to the energy difference between the two energy levels (Eq. 27.18). The wavelength of the light can then be obtained by using Eq. 27.19.

Solution.

Given: $n_i = 3$ *Find:* λ (wavelength of emitted light)
        $n_f = 2$

The energy of the emitted photon is equal to the magnitude of the atom's change in energy. Thus,

$$\Delta E = 13.6\left(\frac{1}{n_f^2} - \frac{1}{n_i^2}\right)\text{eV} = 13.6\left(\frac{1}{4} - \frac{1}{9}\right)\text{eV} = 1.89 \text{ eV}$$

Using Eq. 27.19 (and making sure that ΔE is expressed in electron-volts), we obtain

$$\lambda = \frac{1.24 \times 10^3 \text{ eV} \cdot \text{nm}}{\Delta E} = \frac{1.24 \times 10^3 \text{ eV} \cdot \text{nm}}{1.89 \text{ eV}} = 656 \text{ nm}$$

which is in the red portion of the visible spectrum. Refer to Fig. 27.8c and note the emission line right around 660 nm for hydrogen. The transition in this example is what gives rise to this red line.

Follow-Up Exercise. Light of what wavelength would be just sufficient to ionize a hydrogen atom if it started in its first excited state? Classify this type of light. Is it visible, UV, or IR?

In summary, since the Bohr model of hydrogen requires that the electron make transitions only between discrete energy levels, the atom emits photons of discrete energies (or light of discrete wavelengths), which results in emission spectra. This process is summarized in ▼Fig. 27.12.

Integrated Example 27.7 ■ The Balmer Series: Entirely Visible?

We know that four wavelengths of the Balmer series are in the visible range (Fig. 27.12). (a) There are more than just these four in this series. What type of light are they likely to be: (1) infrared, (2) visible, or (3) ultraviolet? (b) What is the longest wavelength of non-visible light in the Balmer series?

(a) Conceptual Reasoning. The Balmer series of emission lines is given off when the electron ends in the first excited state, that is, $n_f = 2$ (Fig. 27.12). There are four distinct visible wavelengths, corresponding to $n_i = 3, 4, 5,$ and 6. Any other lines in this series must start with $n_i = 7$ or higher and therefore represent a larger energy difference than those for the visible lines. This means that these photons carry more energy than visible-light photons, and the light will have a shorter wavelength than visible light. Wavelengths smaller than visible are UV. Thus the answer is (3); the other Balmer lines must be in the ultraviolet region.

(b) Thinking It Through. The longest nonvisible Balmer series wavelength corresponds to the smallest photon energy above the $n = 6 \rightarrow 2$ transition. (Why?). Hence, we are talking about $n_i = 7$ and $n_f = 2$. The energy difference can be computed from Eq. 27.18. Then λ can be calculated from Eq. 27.19.

Given: The Balmer emission series for hydrogen   *Find:* The longest nonvisible wavelength in the Balmer series

From Eq. 27.18,

$$\Delta E = 13.6\left(\frac{1}{n_f^2} - \frac{1}{n_i^2}\right)\text{eV} = 13.6\left(\frac{1}{4} - \frac{1}{49}\right)\text{eV} = 3.12 \text{ eV}$$

This energy difference corresponds to light of wavelength

$$\lambda = \frac{1.24 \times 10^3 \text{ eV} \cdot \text{nm}}{\Delta E} = \frac{1.24 \times 10^3 \text{ eV} \cdot \text{nm}}{3.12 \text{ eV}} = 397 \text{ nm}$$

This is just below the lower limit of the visible spectrum, which ends at 400 nm.

Follow-Up Exercise. In hydrogen, what is the longest wavelength of light emitted in the Lyman series? In what region of the spectrum is this light?

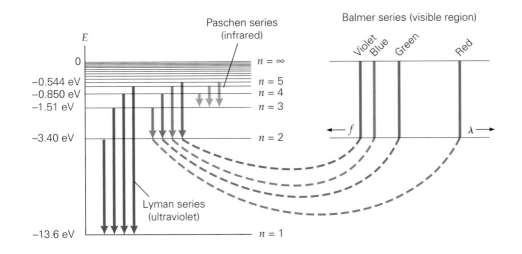

◀ **FIGURE 27.12 Hydrogen spectrum** Transitions may occur between two or more energy levels as the electron returns to the ground state. Transitions to the $n = 2$ state give spectral lines with wavelengths in the visible region (the Balmer series). Transitions to other levels give rise to other series (not in the visible region), as shown.

Conceptual Example 27.8 ■ Up and Down in the Hydrogen Atom: Absorbed and Emitted Photons

Assume that a hydrogen atom, initially in its ground state, absorbs a photon. In general, how many emitted photons would you expect to be associated with the de-excitation process back to the ground state? Can there be (a) more than one photon, or must there be (b) only one photon?

Reasoning and Answer. Since the difference between light emission and absorption is the direction of the transition (down for emission, up for absorption), you might be tempted to answer (b), because only one photon was required for the excitation process. However, if you look at the details of the two processes, you will realize that they are not necessarily symmetrical. In absorbing a photon's energy, a hydrogen atom will jump from its ground state to an excited state—let's say from $n = 1$ to $n = 4$. This process requires a single photon of a unique energy (that is, light of a unique wavelength).

However, in dropping back to the ground state, the atom may take any one of *several* possible routes. For example, the atom could go from the $n = 4$ state to the $n = 3$ state, followed by a transition to the ground state ($n = 1$). This process would involve the emission of two photons. Therefore, the answer is (a). In general, following the absorption of a single photon, several photons can be emitted.

Follow-Up Exercise. (a) How many different-energy photons may a hydrogen atom emit in de-exciting from the third excited state to the ground state? (b) Starting from the ground state, which excitation transition *must* result in only one emitted photon when the hydrogen atom de-excites? Explain your reasoning.

Bohr's theory gave excellent agreement with experiment for hydrogen gas as well as for other ions with just one electron, such as singly ionized helium. However, it could not successfully describe multielectron atoms. Bohr's theory was incomplete in the sense that it patched new quantum ideas into a basically classical framework. The theory contains some correct concepts, but a complete description of the atom did not come until the development of quantum mechanics (Chapter 28). Nevertheless, the idea of discrete energy levels in atoms enables us to qualitatively understand phenomena such as fluorescence.

In **fluorescence**, an electron in an excited state returns to the ground state in two or more steps, like a ball bouncing down a flight of stairs. At each step, a photon is emitted. Each such step represents a smaller energy transition than the original energy required for the upward transition. Therefore, each emitted photon must have a lower energy and a longer wavelength than the original exciting photon. For example, the atoms of many minerals can be excited by absorbing ultraviolet (UV) light and *fluoresce*, or glow, in the visible region when they de-excite (◄Fig. 27.13). A variety of living organisms, from corals to butterflies, manufacture fluorescent pigments that emit visible light.

27.5 A Quantum Success: The Laser

OBJECTIVE: To understand some of the practical applications of the quantum hypothesis—in particular, the laser.

The development of the laser was a major technological success. Unlike the numerous inventions that have come about by trial and error or accident, including Roentgen's discovery of X-rays and Edison's electric lamp, the laser was developed on theoretical grounds. Through the use of quantum physics ideas, the **laser** was first predicted and then designed, built, and, finally, applied. (The word *laser* is an acronym that stands for *light amplification by stimulated emission of radiation*. Stimulated emission will be discussed shortly.) The laser has found widespread applications, some of which are discussed at the end of this section.

The existence of atomic energy levels is of prime importance in understanding the laser's operation. Usually, an electron makes a transition to a lower energy level almost immediately, remaining in an excited state for only about 10^{-8} s. However, the lifetimes of some excited states are appreciably longer than this. An

(a) Visible illumination

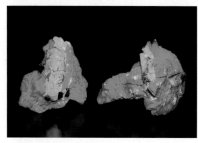

(b) UV illumination

▲ **FIGURE 27.13** Fluorescence Many minerals emit light of visible wavelengths when illuminated by invisible ultraviolet light (so-called black light). The visible light is produced when atoms excited by the UV light de-excite to lower energy levels in several smaller steps, yielding photons of less energy and longer (visible) wavelengths.

atomic state with a relatively long lifetime is called a **metastable state**. For example, **phosphorescent** materials are composed of atoms that have such metastable states. These materials are used on luminous watch dials, toys, and other items that "glow in the dark." When a phosphorescent material is exposed to light, the atoms are excited to higher energy levels. Many of the atoms return to their normal state very quickly. However, there are also metastable states in which the atoms may remain for seconds, minutes, or even more than an hour. Consequently, the material can emit light and glow for some time (▶Fig. 27.14).

A major consideration in laser operation is the emission process. As ▼Fig. 27.15 shows, absorption and spontaneous emission of radiation can occur between two energy levels. That is, a photon is absorbed and a photon is emitted almost immediately. However, when the higher energy state is metastable, there is another possible emission process, called **stimulated emission**. Einstein first proposed this process in 1919. If a photon with an energy equal to an allowed transition strikes an atom already in a metastable state, it may stimulate that atom to make a transition to a lower energy level. This transition yields a second photon that is identical to the first one. Thus, *two* photons with the same frequency and phase will go off in the same direction. Notice that stimulated emission is an amplification process—one photon in, two out. But this process is not a case of getting something for nothing, since the atom must be initially excited, and energy is required to do this.

Ordinarily, when light passes through a material, photons are more likely to be absorbed than to give rise to stimulated emission. This is because there normally are many more atoms in their ground state than in excited states. However, it is possible to prepare a material so that more of its atoms are in an excited metastable state than in the ground state. This condition is known as a **population inversion**. In this case, there may be more stimulated emission than absorption, and the net result is amplification. With the proper instrumentation, the result is a laser.

Today, there are many types of lasers capable of producing light of different wavelengths. The helium–neon (He–Ne) gas laser is probably the most familiar, since it is used for classroom demonstrations and laboratory experiments. The characteristic reddish-pink light produced by the He–Ne laser ($\lambda = 632.8$ nm) is also used in the optical scanning systems at supermarket checkouts. The gas mixture is about 85% helium and 15% neon. Essentially, the helium is used for energizing and the neon for amplification. The gas mixture is subjected to a high-voltage discharge of electrons produced by a radio-frequency power supply converted to direct current (dc). The helium atoms are excited by collision with these electrons (▼Fig. 27.16a). This process is referred to as *pumping*. Energy is pumped into the system, and the helium atoms are pumped into an excited state that is 20.61 eV above its ground state.

This excited state in helium has a relatively long lifetime of about 10^{-4} s and has almost the same energy as an excited state in the Ne atom at 20.66 eV. Because this lifetime is so long, there is a good chance that before an excited He atom can spontaneously emit a photon, it will collide with a Ne atom in its ground state. When such a collision occurs, energy can be transferred to the Ne atom. The lifetime of the 20.66-eV neon state is also relatively long. The delay in this metastable state of neon causes a population inversion in the neon atoms—a higher percentage of atoms in the 20.66-eV state than in ones below it. When the neon drops to the 18.70 eV state, it emits a photon with an energy equal to the difference in energy of the two levels, or about 2.0 eV, which corresponds to the red light with a wavelength of 633 nm that we observe.

▲ **FIGURE 27.14 Phosphorescence and metastable states** When atoms in a phosphorescent material are excited, some of them do not immediately return to the ground state, but remain in metastable states for longer than normal periods of time. In this exhibit at the San Francisco Exploratorium, the phosphorescent walls and floor continue to glow for about 30 seconds after being illuminated, retaining the shadows of children who were present when the phosphors were initially exposed to light.

Teaching tip: Ask students how many of them own a laser. (Refer them to Insight 27.1 on page 869.)

▼ **FIGURE 27.15 Photon absorption and emission (a)** In the absorption of light, once a photon is absorbed, the atom is excited to a higher energy level. **(b)** After a short time, the atom spontaneously decays to a lower energy level with the emission of a photon. **(c)** If another photon with an energy equal to that of the *downward* transition strikes an excited atom, stimulated emission can occur. The result is *two* photons (the original plus the new one) with the same frequency. They travel in the same direction as that of the incident photon and are in phase with one another.

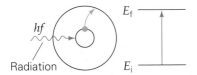

(a) Absorption

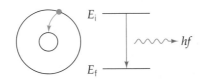

(b) Spontaneous emission

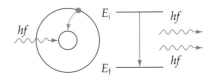

(c) Stimulated emission

▶ **FIGURE 27.16 The helium–neon laser**
(a) Helium atoms are first excited (or "pumped") by collision with electrons. This energy is then transferred from the helium atoms to the neon atoms in a metastable state. All that is needed is one photon from a downward transition to stimulate emission from the other excited neon atoms. When this process occurs, the result is the emission of an intense beam of red laser light. **(b)** End mirrors on the laser tube are used to enhance the light beam (from stimulated emission) in a direction along the tube's axis. One of the mirrors is only partially silvered, allowing for some of the light to come out of the tube and resulting in the beam of red light we observe.

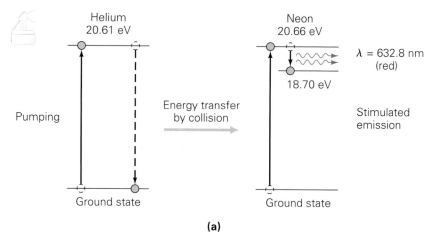

(a)

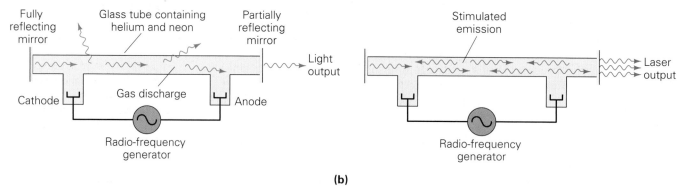

(b)

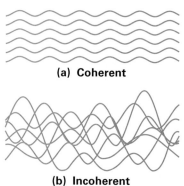

(a) Coherent

(b) Incoherent

▲ **FIGURE 27.17 Coherent light**
(a) Laser light is monochromatic (single-frequency and single-wavelength or color) and coherent, meaning that all the light waves are locked in phase. **(b)** Light waves from sources such as a lightbulb filament are emitted randomly. They consist of many different wavelengths (colors) and are incoherent or, on average, out of phase.

The stimulated emission of the light emitted by these neon atoms is enhanced by reflections from mirrors placed at each end of the laser tube (Fig. 27.16b). Some excited Ne atoms spontaneously emit photons in all directions, and these photons, in turn, induce stimulated emissions. In stimulated emission, the two photons leave the atom in the same direction as that of the incident photon. Photons traveling in the direction of the tube axis are reflected back through the tube by the end mirrors. The photons, in reflecting back and forth, cause even more stimulated emissions. The result is an intense, highly directional, coherent (in phase), monochromatic (single-wavelength) beam of light traveling back and forth along the tube axis. Part of the beam emerges through one of the end mirrors, because it is only partially silvered.

The monochromatic, coherent, and directional properties of laser light are responsible for its unique properties (◀Fig. 27.17). Light from sources such as incandescent lamps is emitted from the atoms randomly and at different frequencies (a result of many different transitions). As a result, the light is out of phase, or incoherent. Such beams spread out and become less intense. The properties of laser light allow the formation of a very narrow beam, which with amplification can be very intense.

Some industrial laser applications are shown in ▼Fig. 27.18, and another is discussed in Insight 27.1 on CD and DVD Systems on page 869. For safety, remember that working with a laser beam can be quite hazardous. If the beam is focused on the retina in a very small area and if its intensity and viewing time are sufficient, the retina can be burned and either damaged or destroyed by the concentrated energy.

▶ **FIGURE 27.18 Some industrial laser applications** **(a)** Lasers are used for accurate car repairs. **(b)** An industrial laser cuts through a steel plate.

(a)

(b)

INSIGHT 27.1 CD AND DVD SYSTEMS

CD's (compact *discs*) and DVD (*digital video discs*) are methods of storing tremendous amounts of data, such as musical recordings and movies. The CD system was introduced around 1980, and the first DVD systems were sold in 1997. Since then, musical CDs and DVD movies have almost completely supplanted audiotapes and videotapes, not to mention vinyl records.

A CD is 12 cm in diameter and can store more than 6 billion bits of information. This capacity is the equivalent of more than 1000 floppy disks, or more than 275 000 pages of text! For audio use, a CD can store 74 min of music, and the sound reproduction is virtually unaffected by dust, scratches, or fingerprints on the disc.

Information on both types of disc is in the form of raised areas called *pits*, which are separated by flat areas called *land*. The surface of the disc is coated with a thin layer of aluminum to reflect the laser beam that "reads" the information (Fig. 1). The pits are arranged in a spiral track like the grooves you may or may not be used to in a phonograph record; however, CD tracks are about $\frac{1}{60}$ as far apart as are the grooves on a record. DVD tracks are even narrower and closer together than that, and can store up to *seven times* the information that is on a CD of the same size. (Software *video compression*, which eliminates the redundant and irrelevant information found on CD systems, also contributes to the increased storage available on a DVD.)

In the readout system which extracts the information, a laser beam from a small semiconductor (solid-state) laser is applied from below the disc and focused on the aluminum coating of the track. The disc rotates at about 3.5 to 8 revolutions per second as the laser beam follows the spiral track. The beam is reflected when it strikes a land area between two pits. When the beam spot overlaps a land area and a pit, the light reflected from the different areas interferes, causing fluctuations in the reflected beam. To make the fluctuations more distinct, the raised-pit thickness (t) is made to be one-fourth of the wavelength of the laser light. Then, reflected light from land areas travels an additional path length of half a wavelength. Destructive interference occurs when the two parts of the reflected beam combine. (See Section 24.1.) As a result, there is less reflected intensity when the beam passes over the edge of a pit than when it passes over a land area alone.

Then the reflected beam of varying intensity strikes a photodiode (a solid-state photocell), which reads the information. The fluctuations of the reflected light convey the coded information as a series of binary numbers (zeros and ones). A pit represents a binary 1, and a land area is read as a binary 0. The signals are then electronically converted into sound or video.

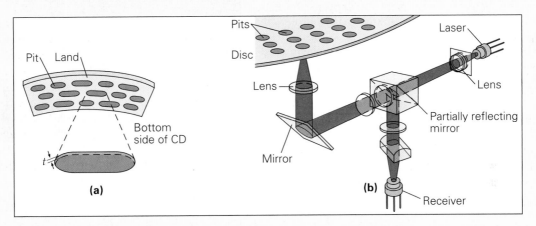

FIGURE 1 The CD and DVD (a) The information on the disc is recorded in the form of raised areas called *pits*, which are separated by flat areas called *land*. The pits are on the bottom of the disc. **(b)** The surface is coated with a thin layer of aluminum to reflect the laser beam, which "reads" the information. DVDs (for movies, for example) are physically similar to the audio CD, but the tracks are narrower and more closely spaced, allowing much more information to be stored.

However, the laser can also be used to promote health, as the applications in Insight 27.2 on Lasers in Modern Medicine on accompanying page.

Another interesting application of laser light is the production of three-dimensional images in a process called **holography**. The process does not use lenses, as ordinary image-forming processes do, yet it re-creates the original scene in three dimensions. The key to holography is the coherent property of laser light, which gives the light waves a definite spatial relationship to each other.

In the photographic process of making a *hologram*, an arrangement such as that illustrated in ▾Fig. 27.19 is used. Part of the light from the laser (the object beam) passes through a partially reflecting mirror to the object. The other part, or reference beam, is reflected to the film. The light incident on the object is also

INSIGHT 27.2 LASERS IN MODERN MEDICINE

The use of lasers has had a large impact on modern medicine, in areas ranging from elective cosmetic surgery to life-saving cancer surgery. You have already read about one of the most well-known medical laser applications in the area of vision correction. (See Chapter 25.) Another application for medical lasers is for *tattoo removal* without harming the surrounding cells. More and more lasers are being chosen over other methods, such as surgical excision, dermabrasion (sanding of the skin), chemical peels, and cryosurgery (freezing). Laser treatment is noninvasive and targets only the inks that make up the tattoo. This can be done by adjusting the wavelength (color) of the laser to match the color of the ink particle, enhancing absorption of the light energy. When the ink particles absorb the laser light, they are heated and fragment. These fragments are then absorbed through the bloodstream and eliminated from the body. This process generally takes a few weeks and may require multiple treatments if the ink particles are large (Figs. 1a and 1b).

Another common use of lasers is to treat painful *varicose veins*. A normal leg has many "one-way" valves in its veins that act to prevent blood from flowing backward (down), thus maintaining an uphill return flow back to the heart. With age (and sometimes pregnancy), these valves can become damaged to the point where they cannot perform. When this happens, the blood will pool in the lower leg, causing the large gnarly purple veins common in many patients (see Fig. 2a). Doppler ultrasound (see Chapter 14) is used to initially diagnose a valve problem by accurately determining the blood flow direction. If left untreated, this can become painful and debilitating. In the past, the only relief was surgical "stripping" of the veins (removal of the surface veins does not appreciably affect the return blood flow, as most of this flow is handled by the deep vein return system). Recently, lasers have been used to *ablate* (that is, thermally destroy) the afflicted vein using a procedure called EVLT (*endovenous laser therapy*). EVLT is considerably less invasive, with little or no scarring, and has a lower complication rate than traditional sur-

gical procedures. The vein usually responsible for the lack of upward blood flow is the *greater saphenous vein*, a large vein that extends from the knee to the hip area. In this procedure, a fiber-optic bundle is attached to a laser (typically an IR laser operating at 810-nm wavelength). After introduction of a local anesthetic, ultrasound images are used to guide the fiber as it is inserted through a small incision near the knee and pushed upward the full length of the saphenous vein. As the fiber is slowly pulled out, the laser is repeatedly fired, heating up the venous tissues and causing irreversible damage. Along with this procedure, it is common to remove the bulged smaller lower veins. Once the "downhill" blood flow is eliminated by destroying the saphenous vein, they should not reappear, and new ones are far less likely to form. After several weeks of continuous compression, the change in the leg can be dramatic, as shown in Figures 2b and 2c.

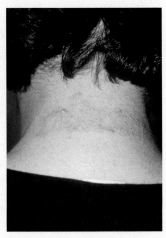

FIGURE 1 Laser and tattoo removal **(a)** After one laser treatment. **(b)** After three treatments.

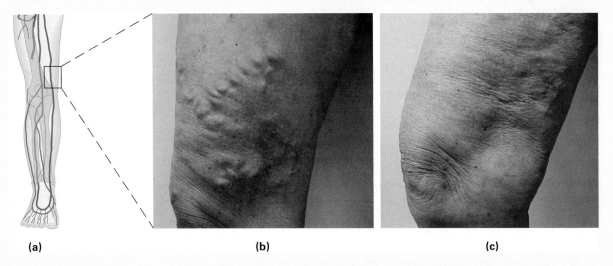

(a)　　　　　　　　　　(b)　　　　　　　　　　(c)

FIGURE 2 Laser and varicose vein treatment **(a)** A sketch of the venous system in the leg. **(b)** Serious and painful varicose veins before destruction by laser. **(c)** Marked improvement after the saphenous vein is destroyed by laser light.

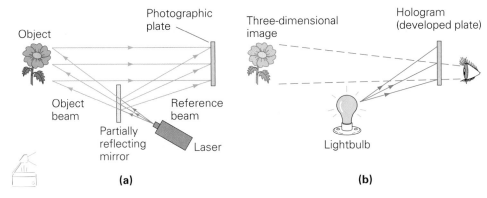

▲ **FIGURE 27.19** Holography **(a)** The coherent light from a laser is split into reference and object beams. The interference pattern between these beams is recorded on a photographic plate. **(b)** When the developed plate, or hologram, is illuminated by normal light, the viewer sees a reconstructed three-dimensional image of the object.

reflected to the film, and it interferes with the reference beam. The film records the interference pattern of the two light beams, which essentially imprints on the film the information carried from the object by the light's wave fronts.

When the film is developed, the interference pattern bears no resemblance to the object and appears as a meaningless pattern of light and dark areas. However, when the wave-front information is reconstructed by passing light through the film, a three-dimensional image is seen (▶Fig. 27.20). If part of the three-dimensional image is hidden from view, you can see it by moving your head to one side, just as you would to see a hidden part of a real object.

▲ **FIGURE 27.20** Holographic images A three-dimensional hologram can be viewed from any angle, just as if it were a real object.

Chapter Review

- All hot solids, liquids, and dense gases emit a **continuous spectrum** of electromagnetic (thermal) radiation. The maximum amount of energy is emitted at a wavelength (λ_{max}) determined by the material's absolute temperature T. **Wien's displacement law** gives the (inverse) relationship of wavelength and absolute temperature:

$$\lambda_{max} T = 2.90 \times 10^{-3} \text{ m} \cdot \text{K} \qquad (27.1)$$

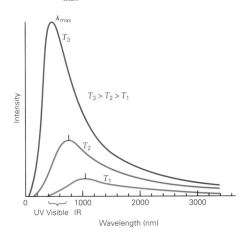

- The classical theory of thermal radiation predicted that an infinite amount of energy is emitted at the short wavelengths, a phenomenon called the **ultraviolet catastrophe**, which did not agree

with experiment. To explain the observed spectrum from hot objects, **Planck's hypothesis** stated that the energy of the atoms in the material was quantized in multiples of their vibrational frequency, or

$$E_n = n(hf) \qquad \text{for } n = 1, 2, 3, \dots \qquad (27.2)$$

where h, **Planck's constant,** has a numerical value of 6.63×10^{-34} J $\cdot$ s.

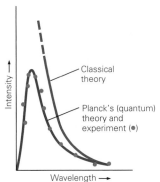

- In the **photoelectric effect**, light incident on a surface causes *photoelectrons* to come off that surface. To explain the experimental results, Einstein had to assume that light consisted of discrete quanta of energy, called **photons,** or particles of light. The energy of a photon associated with light of frequency f is

$$E = hf \qquad (27.4)$$

- Light can also interact with electrons via a scattering process called **Compton scattering**. To explain the details of such scattering, light (frequency f) must be treated as a photon carrying a quantum of energy, hf. When photons scatter off electrons, they impart kinetic energy to the electrons, and the scattered photons have less energy than the incident photons. In terms of waves, the scattered light has a longer wavelength λ than the wavelength of the incident light, λ_o. The relationship between the two wavelengths is given by the Compton scattering equation

$$\Delta\lambda = \lambda - \lambda_o = \lambda_C (1 - \cos\theta) \qquad (27.9)$$

where $\lambda_C = h/m_e c = 2.43 \times 10^{-12}\,\text{m} = 2.43 \times 10^{-3}\,\text{nm}$ is the **Compton wavelength** of the electron.

- The **wave–particle duality of light** means that light must be thought of as having both particle and wave natures.

- The **Bohr theory** of hydrogen treats the electron as a classical particle held in circular orbit around the proton by the electrical force of attraction. In his theory, Bohr made two non-classical, quantum postulates:

> The angular momentum of the electron is quantized and can have only discrete values that are integral multiples of $h/2\pi$, where h is Planck's constant;

and

> The electron does *not* radiate energy when it is in a bound, discrete orbit. It radiates energy only when it makes a downward transition to an orbit of lower energy. It makes an upward transition to an orbit of higher energy by absorbing energy.

Excited atom

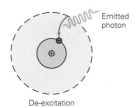

De-excitation

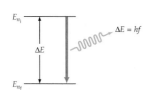

- The Bohr model of hydrogen led directly to the quantization of the radius and energy of the atom. Their values are

$$r_n = 0.0529 n^2 \text{ nm} \qquad \text{for } n = 1, 2, 3, 4, \ldots \qquad (27.16)$$

and

$$E_n = \frac{-13.6}{n^2}\,\text{eV} \qquad \text{for } n = 1, 2, 3, 4, \ldots \qquad (27.17)$$

where n is the **principal quantum number**. When the hydrogen atom is in the $n = 1$ state, it has its smallest size and lowest energy and is in its **ground state**. States of larger size and higher energy are **excited states**.

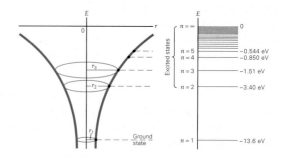

- Atoms emit light in an **emission spectrum,** which shows light emitted only at certain wavelengths characteristic of the atom. Since atomic energies are quantized, the emitted light must come off in quanta (photons) with fixed amounts of energy. Emitted photon energy (equal to the energy difference between the two atomic levels ΔE) is related to the wavelength of the emitted light wavelength by

$$\lambda \text{ (in nm)} = \frac{hc}{\Delta E} = \frac{1.24 \times 10^3\,\text{eV}\cdot\text{nm}}{\Delta E \text{ (in eV)}} \qquad (27.20)$$

- Atoms absorb light in an **absorption spectrum,** which shows light absorbed only at certain wavelengths characteristic of the atom. Since atomic energies are quantized, the absorbed light must be in the form of quanta (photons) with fixed amounts of energy. These amounts of energy correspond to absorbed light of only those certain wavelengths.

- A **metastable state** is a relatively long-lived excited atomic state.

- **Stimulated emission** can occur when a photon prematurely causes a downward atomic transition, yielding a photon identical to itself.

Stimulated emission

- A **laser** uses **population inversion**. This phenomenon is caused by metastable excited atomic states and results in light amplification through stimulated emission.

Exercises

MC = *Multiple Choice Question,* **CQ** = *Conceptual Question, and* **IE** = *Integrated Exercise. Throughout the text, many exercise sections will include "paired" exercises. These exercise pairs, identified with* **red numbers,** *are intended to assist you in problem solving and learning. In a pair, the first exercise (even numbered) is worked out in the Study Guide so that you can consult it should you need assistance in solving it. The second exercise (odd numbered) is similar in nature, and its answer is given at the back of the book.*

Note: Take h to have an exact value of 6.63×10^{-34} J·s for significant-figure purposes, and use $hc = 1.24 \times 10^3$ eV·nm (three significant figures).

27.1 Quantization: Planck's Hypothesis

1. **MC** Blackbody A is at a temperature of 2000 K and blackbody B is at 4000 K. What can you say about the total energy E they radiate: (a) $E_A = \frac{1}{2}E_B$, (b) $E_A = \frac{1}{4}E_B$, (c) $E_A = \frac{1}{8}E_B$, (d) $E_A = \frac{1}{16}E_B$, or (e) you can't tell from the information given? (e)

2. **MC** Blackbody A is at a temperature of 3000 K and blackbody B is at 6000 K. What can you say about the wavelength at which they radiate the maximum energy: (a) $\lambda_{max, A} = \frac{1}{2}\lambda_{max, B}$, (b) $\lambda_{max, A} = 2\lambda_{max, B}$, (c) $\lambda_{max, A} = \lambda_{max, B}$, or (d) you can't tell from the information given? (b)

3. **MC** The absolute temperature of a blackbody radiator is doubled. What can you say about the ratio of total energy E emitted by this object: (a) $E_f/E_i = 2$, (b) $E_f/E_i = 4$, (c) $E_f/E_i = 8$, or (d) $E_f/E_i = 16$. (d)

4. **CQ** Some stars appear reddish, and some others are blue. Which of these stars have the lower surface temperature? red, see ISM

5. **CQ** As a hot piece of iron is heated it begins to glow first red, then orange, then yellow, but then, instead of appearing green or blue as its temperature continues to rise, it becomes white to the eye. Explain. see ISM

6. **CQ** Make a graph showing how the wavelength of the most intense radiation component of blackbody radiation varies with the body's absolute temperature. By what ratio does λ_{max} change (final/initial) if the body's absolute temperature is tripled? see ISM

7. ● The walls of a blackbody cavity are at a temperature of 27°C. What is the frequency of the radiation of maximum intensity? 3.10×10^{13} Hz

8. ● Find the approximate temperature of a red star that emits light with a wavelength of maximum emission of 700 nm (deep red). 4.14×10^3 K

9. ● What are the wavelength and frequency of the most intense radiation component from a blackbody with a temperature of 0°C? 1.06×10^{-5} m, 2.83×10^{13} Hz

10. **IE** ● (a) If you have a fever, will the wavelength of the radiation component of maximum intensity emitted by your body (1) increase, (2) remain the same, or (3) decrease as compared with its value when your temperature is normal? Why? (b) Assume that human skin has a temperature of 32°C. What is the wavelength of the radiation component of maximum intensity emitted by our bodies? In what region of the EM spectrum is this wavelength? (a) (3) decrease (b) 9.51×10^{-6} m; infrared

11. ●● What is the minimum energy of a thermal oscillator in a blackbody producing radiation at λ_{max} at a temperature of 212°F? 2.56×10^{-20} J

12. **IE** ●● The temperature of a blackbody increases from 200°C to 400°C. (a) Will the frequency of the most intense spectral component emitted by this blackbody (1) increase, but not double; (2) double; (3) be reduced in half; or (4) decrease, but not in half? Why? (b) What is the change in the frequency of the most intense spectral component of this blackbody? (a) (1) increase, but not double (b) $\Delta f = 2.07 \times 10^{13}$ Hz

13. ●● The temperature of a blackbody is 1000 K. If the intensity of the emitted radiation, 2.0 W/m², were due entirely to the most intense frequency component, how many quanta of radiation would be emitted per second per square meter?* 2.9×10^{19} quanta/(s·m²)

14. ●●● The wavelength at which the Sun emits its maximum energy light is about 550 nm. Assuming the Sun radiates as a blackbody, estimate (a) its surface temperature and (b) its total emitted power. (a) 5.27×10^3 K (b) 2.7×10^{26} W

27.2 Quanta of Light: Photons and the Photoelectric Effect

15. **MC** In the photoelectric effect, classical theory predicts that (a) no photoemission occurs below a certain frequency; (b) the photocurrent is proportional to the light intensity; (c) the maximum kinetic energy of the emitted electrons depends on the light frequency, or (d) no matter how low the light intensity, photocurrent is observed immediately. (b)

16. **MC** In the photoelectric effect, what happens to the stopping voltage when the light intensity is increased: (a) It increases, (b) it stays the same, or (c) it decreases. (b)

17. **MC** In the photoelectric effect, what happens to the stopping voltage when the light frequency is increased: (a) It increases, (b) it stays the same, or (c) it decreases? (a)

18. **MC** In the photoelectric effect, assuming photoemission continues, what happens to the stopping voltage when the emitting material is changed for one with a larger work function: (a) It increases, (b) it stays the same, or (c) it decreases? (c)

19. **MC** In the SI system, the work function has what units? (a) joules (b) volts (c) coulombs (d) amperes (a)

*Assume that each quantum has the minimum allowed energy.

20. CQ With respect to ionization (which can cause biological damage), is it more dangerous to stand in front of a beam of X-ray radiation with very low total energy or a beam of red light with much more total energy? How does the photon model of light explain this apparent paradox? X-ray, see ISM

21. CQ Is it possible for a beam of IR radiation to contain more total energy than a beam of UV radiation? Explain. yes, see ISM

22. CQ In the photoelectric effect, when the incident light is below the threshold frequency, a significant amount of its energy is still absorbed by the target material, but electrons are not emitted from the surface. Explain where this energy goes. heat (temperature increase)/thermal energy

23. ● Each photon in a beam of light has an energy of 3.3×10^{-15} J. What is the light's wavelength? Use this to classify its type. $\lambda = 6.0 \times 10^{-11}$ m; X-ray

24. IE ● (a) Compared with a quantum of red light ($\lambda = 700$ nm), a quantum of violet light ($\lambda = 400$ nm) has (1) more, (2) the same amount of, (3) less energy. Why? (b) Determine the ratio of the photon energy associated with violet light to that related to red light. (a) (1) more (b) 1.75

25. ● A source of UV light has a wavelength of 150 nm. How much energy does one of its photons have expressed in (a) joules and (b) electron-volts? (a) 1.32×10^{-18} J (b) 8.27 eV

26. IE● The work function of metal A is greater than that of metal B. (a) The threshold wavelength for metal A is (1) shorter than, (2) the same as, (3) longer than that of metal B. Why? (b) If the threshold wavelength for metal B is 620 nm and the work function of metal A is twice that of metal B, what is the threshold wavelength for metal A? (a) (1) shorter than (b) 310 nm

27. ● The photoelectrons ejected from a surface require a stopping voltage of 3.0 V. If the intensity of the light is tripled, what is the stopping voltage now? 3.0 V

28. ●● Assume that a 100-W lightbulb gives off 2.50% of its energy as visible light. How many photons of visible light are given off in 1.00 min? (Use an average visible wavelength of 550 nm.) 4.15×10^{20} photons

29. ●● ▼Figure 27.21 shows a graph of stopping potential versus frequency for a photoelectric material. Determine (a) Planck's constant and (b) the work function of the material from the data contained in the graph. (a) 6.7×10^{-34} J·s (b) 2.9×10^{-19} J

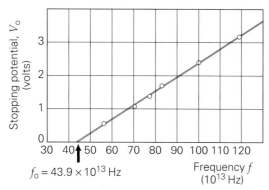

▲ **FIGURE 27.21 Stopping potential versus frequency**
See Exercise 29.

30. ●● A metal with a work function of 2.40 eV is illuminated by a beam of monochromatic light. If the stopping potential is 2.50 V, what is the wavelength of the light? 254 nm

31. ●● What is the lowest frequency of light that can cause the release of electrons from a metal that has a work function of 2.8 eV? 6.8×10^{14} Hz

32. ●● The photoelectric effect threshold wavelength for a certain metal is 500 nm. Calculate the maximum speed of photoelectrons if we use light having a wavelength of (a) 400 nm, (b) 500 nm, and (c) 600 nm. (a) 4.7×10^5 m/s (b) and (c) zero because there is no emission

33. ●● In Exercise 32, what is the work function of the metal in eV? 2.48 eV

34. ●● The work function of a material is 3.5 eV. If the material is illuminated with monochromatic light ($\lambda = 300$ nm), what are (a) the stopping potential and (b) the cutoff frequency? (a) 0.63 V (b) 8.4×10^{14} Hz

35. ●● Blue light with a wavelength of 420 nm is incident on a certain material and causes the emission of photoelectrons with a maximum kinetic energy of 1.00×10^{-19} J. (a) What is the stopping voltage? (b) What is the material's work function? (c) What is the stopping voltage if red light ($\lambda = 700$ nm) is used instead? Explain. (a) 0.625 V (b) 2.33 eV (c) zero

36. ●●● When the surface of a particular material is illuminated with monochromatic light of various frequencies, the stopping potentials for the photoelectrons are determined to be the following:

Frequency (in Hz)

9.9×10^{14}	7.6×10^{14}	6.2×10^{14}	5.0×10^{14}

Stopping potential (in V)

2.6	1.6	1.0	0.60

Plot these data, and from the graph determine Planck's constant and the metal's work function. 6.6×10^{-34} J·s, 1.5 eV, see ISM

37. ●●● When a certain photoelectric material is illuminated with red light ($\lambda = 700$ nm) and then blue light ($\lambda = 400$ nm), it is found that the maximum kinetic energy of the photoelectrons resulting from the blue light is twice that of those from red light. What is the work function of the material? 0.44 eV

27.3 Quantum "Particles": The Compton Effect

38. MC The percentage change in wavelength in the Compton effect would be most easily observed using (a) visible light, (b) infrared radiation, (c) ultraviolet light, (d) X-rays. (d)

39. MC The wavelength shift for Compton scattering is maximum when (a) the photon scattering angle is 90°, (b) the electron receives only a small fraction of its maximum possible recoil energy, (c) the electron is scattered directly forward. (c)

40. MC At what photon-scattering angle will the electron receive the least recoil energy: (a) 20°, (b) 45°, (c) 60°, or (d) 80°? (a)

41. CQ A photon can undergo Compton scattering from a neutron. How does the maximum wavelength shift for Compton scattering from a neutron compare with that from an electron? Explain. it is smaller by a factor of 1836; see ISM

42. **CQ** The Sun's energy production (near its center) is initially X-rays and gamma rays. By the time it reaches the surface, it is mostly in the visible range. Use Compton scattering to explain how this happens. see ISM

43. **CQ** In Compton scattering, the recoiling electron must always be on the other side of the incident beam direction from the scattered photon. Why? to conserve total linear momentum, see ISM

44. **CQ** In Compton scattering, how does the maximum wavelength shift for 0.100-nm X-ray photons compare to that of visible-light photons (500-nm wavelength)? same, see ISM

45. ● What is half the maximum wavelength shift for Compton scattering from a free electron? 2.43×10^{-3} nm

46. ● What is the change in wavelength when monochromatic X-rays are scattered by electrons through an angle of 30°? 3.26×10^{-4} nm

47. ● A monochromatic beam of X-rays with a wavelength of 0.280 nm is scattered by a metal foil. What is the wavelength of the scattered X-rays observed at an angle of 45° from the direction of the incident beam? 0.281 nm

48. ● X-rays with a wavelength of 0.0045 nm are used in a Compton-scattering experiment. If the X-rays are scattered through an angle of 53°, what is the wavelength of the scattered radiation? 0.0055 nm

49. **IE** ●● A photon with an energy of 5.0 keV is scattered by a free electron. (a) The recoiling electron could have an energy of (1) zero, (2) less than 5.0 keV, but not zero, (3) equal to 5.0 keV. Why? (b) If the wavelength of the scattered photon is 0.25 nm, what is the recoiling electron's kinetic energy? (a) (2) less than 5.0 keV, but not zero (b) 20 eV

50. ●● X-rays scattered from a carbon atom show a wavelength shift of 0.000326 nm. What is their scattering angle? 30°

51. ●● If the Compton shift for a photon scattered by an electron is 1.25×10^{-4} nm, what is the scattering angle? 18.5°

52. **IE** ●●● The Compton effect can occur for scattering from any particle—for example, from a proton. (a) Compared with the Compton wavelength for an electron, the Compton wavelength for a proton is (1) longer, (2) the same size, (3) shorter. Why? (b) What is the value of the Compton wavelength for a proton? (c) Determine the ratio of the maximum Compton wavelength shift for scattering by an electron to that for scattering by a proton. (a) (3) shorter (b) 1.32×10^{-15} m (c) 1.84×10^{3}

27.4 The Bohr Theory of the Hydrogen Atom

53. **MC** In his theory of the hydrogen atom, Bohr postulated the quantization of (a) energy, (b) centripetal acceleration, (c) light, (d) angular momentum. (d)

54. **MC** An excited hydrogen atom emits light when its electron (a) makes a transition to a lower energy level, (b) is excited to a higher energy level, (c) is in the ground state. (a)

55. **MC** A hydrogen atom in its first excited state absorbs a photon and makes a transition to a higher excited state. The longest-wavelength photon possible is absorbed. The quantum number of the final state is (a) 1, (b) 2, (c) 3, (d) 4. (c)

56. **CQ** The Bohr theory is applicable only to the hydrogen atom and hydrogen-like atoms, such as singly ionized helium, and other one-electron systems. Why? the theory does not include electron–electron interactions

57. **CQ** Will it take more or less energy to ionize (remove the electron completely from) a hydrogen atom if the electron is in an excited state than if it is in the ground state? Explain. less, see ISM

58. **CQ** Very accurate measurements of the wavelengths emitted by a hydrogen atom indicate that they are all slightly longer than expected from the Bohr theory. Explain how conservation of linear momentum explains this. [*Hint*: Photons carry momentum and energy.] The atom must recoil, see ISM

59. ● Find the energy required to excite a hydrogen electron from (a) the ground state to the first excited state and (b) from the first excited state to the second excited state. (c) Classify the type of light needed to create each of the transitions. (a) 10.2 eV (b) 1.89 eV (c) the first is UV, the second visible (red)

60. ● Find the energy needed to ionize a hydrogen atom whose electron is in the (a) $n = 2$ state and (b) $n = 3$ state. (a) 3.40 eV (b) 1.51 eV

61. ● What is the frequency of light that would excite the electron of a hydrogen atom from (a) a state with a principal quantum number of $n = 2$ to that with a principal quantum number of $n = 5$? (b) What about from $n = 2$ to $n = \infty$? (a) 6.89×10^{14} Hz (b) 8.21×10^{14} Hz

62. ● Find the radius of the electron orbit in a hydrogen atom for states with the following principal quantum numbers: (a) $n = 2$, (b) $n = 4$, (c) $n = 5$. (a) 0.212 nm (b) 0.846 nm (c) 1.32 nm

63. ● Scientists are now beginning to study "large" atoms, that is, atoms with orbits that are almost large enough to be measured in our everyday units of measurement. For what excited state (give an approximate principal quantum number) of a hydrogen atom would the diameter of the orbit be a micron (10^{-6} m)—that is, close to the size of a dust particle? $n \approx 100$

64. ●● Find the binding energy of the hydrogen electron for states with the following principal quantum numbers: (a) $n = 3$, (b) $n = 5$, (c) $n = 10$. (a) 1.51 eV (b) 0.544 eV (c) 0.136 eV

65. **IE** ●● A hydrogen atom has an ionization energy of 13.6 eV. When it absorbs a photon with an energy greater than this energy, the electron will be emitted with some kinetic energy. (a) If the energy of such a photon is doubled, the kinetic energy of the emitted electron will (1) increase, but not necessarily double, (2) remain the same, (3) exactly double, (4) decrease. Why? (b) A photon associated with light of a frequency of 7.00×10^{15} Hz is absorbed by a hydrogen atom. What is the kinetic energy of the emitted electron? (a) (1) increase, but not necessarily double (b) 15.4 eV

66. ●● A hydrogen atom in its ground state is excited to the $n = 5$ level. It then makes a transition directly to the $n = 2$ level before returning to the ground state. (a) What are the wavelengths of the emitted photons? (b) Would any of the emitted light be in the visible region? see ISM

67. **IE** ●● (a) For which of the following transitions in a hydrogen atom is the photon of greatest energy emitted: (1) $n = 5$ to $n = 3$, (2) $n = 6$ to $n = 2$, or (3) $n = 2$ to $n = 1$? (b) Justify your answer mathematically. (a) $n = 2$ to $n = 1$ (b) see ISM

68. •• The hydrogen spectrum has a series of lines called the *Lyman series*, which results from transitions to the ground state. What is the longest wavelength in this series, and in what region of the EM spectrum does it lie?
122 nm; ultraviolet

69. •• What is the binding energy for an electron in the ground state in the following hydrogen-like ions: (a) He^+, (b) Li^{2+}?
(a) 54.4 eV (b) 122 eV

70. •• A hydrogen atom absorbs light of wavelength 486 nm. (a) How much energy did the atom absorb? (b) What are the values of the principal quantum numbers of the initial and final states of this transition? (a) 2.55 eV (b) 2 and 4

71. ••• Show that the speeds of an electron in the Bohr orbits are given (to two significant figures) by $v_n = (2.2 \times 10^6 \text{ m/s})/n$. see ISM

72. ••• By computing and adding the kinetic and electric potential energies, show that for an electron in the ground state of a hydrogen atom the total energy is −13.6 eV. [*Hint*: You will need to use the orbital radius.] see ISM

73. ••• In Exercise 72, (a) how much of the energy is potential, and how much is kinetic? (b) How do the magnitudes of the two forms of energy compare? (a) potential is −27.2 eV and kinetic is +13.6 eV (b) |U| = 2K, see ISM

27.5 A Quantum Success: The Laser

74. **MC** Which of the following is not essential for laser action: (a) population inversion, (b) phosphorescence, (c) pumping, or (d) stimulated emission? (b)

75. **MC** Suppose a hypothetical atom had two metastable excited states, one 2.0 eV above the ground state and one 5.0 eV above it. If used in a laser with transitions only to the ground state, which laser would be in the visible range: (a) 5.0 eV, (b) 2.0 eV, (c) neither, or (d) both? (b)

76. **MC** For the atom in Exercise 75, in order to create only the visible laser, the pumping light should not contain which type of light: (a) UV, (b) visible, or (c) IR? (a)

77. **CQ** You have a choice of several lasers spanning the visible spectrum with which to remove a tattoo ink particle that is blue. What color laser would theoretically work the best? Explain. [*Hint*: See the Insight on Lasers in Modern Medicine, p. 870.] see ISM

78. **CQ** In what sense is a laser an "amplifier" of energy? Explain why this concept does not violate conservation of energy. see ISM

79. **CQ** Explain the difference between spontaneous emission and stimulated emission. see ISM

Comprehensive Exercises

80. Light of wavelength 340 nm is incident on a metal surface and ejects electrons that have a maximum speed of 3.5×10^5 m/s. (a) What is the work function of the metal? (b) What is its stopping voltage? (c) What is its threshold wavelength?
(a) 3.3 eV (b) 0.35 V (c) 3.8×10^2 nm

81. A 10.0-keV X-ray photon is successively scattered by two free electrons initially at rest. In the first case, it is scattered through an angle of 41°; the second scattering is through an angle of 72°. (a) What is the final photon energy? (b) How much kinetic energy does each electron receive?
(a) 9.82 keV (b) first: 48 eV, second: 134 eV

82. **IE** (a) How many transitions in a hydrogen atom result in the absorption of red light: (1) one, (2) two, (3) three, or (4) four? (b) What are the principal quantum numbers of the initial and final states for this process? (c) What are the energy of the required photon and the wavelength of the light associated with it?
(a) (1) one (b) 2 and 3 (c) 1.89 eV and 656 nm

83. Under the right circumstances, if a photon contains above a minimum energy, that energy can be completely converted into creating an electron–positron pair (a positron is identical to an electron except that it has a positive charge). Recall from relativity that the energy equivalent of the electron mass is 0.511 MeV. Determine (a) the minimum-energy photon required to create such a pair, and (b) the wavelength of light associated with these photons. (c) If a photon of twice the minimum energy were used, what would be the total kinetic energy (electron + positron) of the two particles after their creation?
(a) 1.02 MeV (b) 1.21×10^{-3} nm (c) 1.02 MeV

84. Consider an electron in its first excited state in a hydrogen atom. Determine its (a) speed, (b) angular speed, (c) linear momentum, and (d) angular momentum. see ISM

85. A gamma-ray photon scatters off a free proton (initially at rest) at an angle of 45°. The wavelength of the scattered light is measured to be 6.20×10^{-13} m. (a) What was the incoming photon energy? (b) How much kinetic energy did the proton receive? (You may need to carry an extra figure or two in your intermediate answers.)
(a) 2.00 MeV (b) about 1 keV

86. In the nuclear version of the photoelectric effect (called the *photonuclear* effect), a high-energy photon is absorbed by an atomic nucleus and a proton is freed from that nucleus. If the minimum energy needed to free a proton from a particular nucleus is 5.00 MeV, (a) determine the maximum (threshold) wavelength of light that can cause this. (b) If a photon of half this wavelength is used instead, determine the kinetic energy of the ejected proton. (a) 2.48×10^{-4} nm (b) 5.00 MeV

QUANTUM MECHANICS AND ATOMIC PHYSICS

PHYSICS FACTS

- Louis de Broglie's full name was Prince Louis-Victor Pierre Raymond de Broglie. Besides winning the Nobel Prize in physics in 1929 for his discovery of "the wave nature of the electron," he also excelled as a teacher/lecturer. In 1952 the first Kalinga Prize was awarded to him by UNESCO for his efforts to explain aspects of modern physics to laypeople.

- Hitler rose to power in Germany in 1933, the same year that Werner Heisenberg was awarded the Nobel Prize for his contributions to the creation of quantum mechanics. Heisenberg is associated with one of the most famous principles in modern physics: the Heisbenberg uncertainty principle. Protected by the Nobel Prize, Heisenberg became a spokesman for modern physics in Germany and remained there throughout World War II. He was not a Nazi, but he was a German citizen, and felt it his duty to preserve some of German science.

- All elementary particles have associated *antiparticles*. For the familiar electron, the antiparticle is the positive electron, or simply the positron. If electron and positron meet, they annihilate into gamma radiation, the result of converting their mass (energy) into electromagnetic (light) energy. Many of the properties of a particle and its antiparticle are opposite, such as electric charge. However, antiparticles have "regular" mass; that is, they are attracted downward to the Earth, not repelled.

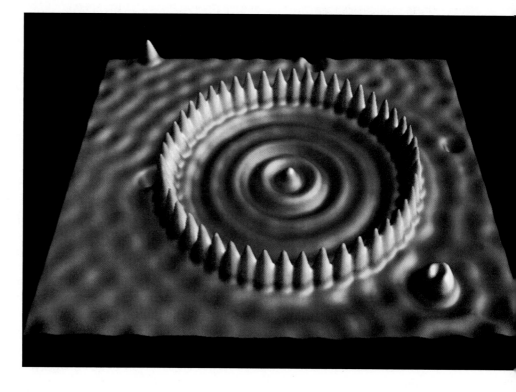

J ust a few decades ago, if someone had claimed to have a photograph of an atom, people would have laughed. Today, a device called the scanning tunneling microscope (STM) routinely produces images such as shown in the photograph. The blue shapes represent iron atoms, neatly arranged on a copper surface.

The STM operates on a quantum-mechanical phenomenon called *tunneling*. Tunneling reflects some of the fundamental features of the subatomic realm: the probabilistic character of quantum processes and the wave nature of particles. These features explain how particles may turn up in places where, according to classical notions, they should not be.

In the 1920s, a new kind of physics, based on the synthesis of wave and quantum ideas, was introduced. This new theory, called **quantum mechanics**, combined the wave–particle duality of matter into a single consistent description. It revolutionized scientific thought and provides the basis of our understanding of phenomena that occur on the scale of molecular sizes and smaller.

In this chapter, some of the basic ideas of quantum mechanics are presented in order to show how they describe matter. Practical applications made possible by the quantum-mechanical view of nature, such as the electron microscope and magnetic resonance imaging (MRI), are also discussed.

28.1 Matter Waves: The de Broglie Hypothesis

OBJECTIVES: To (a) explain de Broglie's hypothesis, (b) calculate the wavelength of a matter wave, and (c) specify under what circumstances the wave nature of matter will be observable.

Note: These relationships are discussed in Section 26.4.

Since a photon travels at the speed of light, it must be treated relativistically as a particle with no mass. If this were not the case (that is, if it were to have mass), it would have infinite energy. This is because, at $v = c$, the relativistic factor γ becomes infinite, and its total energy $E = \gamma mc^2$ would be infinite unless $m = 0$. A photon's energy E and its momentum p are related by $p = E/c$. Recall from the Einstein relationship (Eq. 27.3) that the energy of a photon can be written in terms of associated wave frequency and wavelength as $E = hf = hc/\lambda$. Combining these two, we find that the magnitude of the momentum (p) of a photon is inversely related to the wavelength of the light by

$$p = \frac{E}{c} = \frac{hf}{c} = \frac{h}{\lambda} \quad \text{(photon momentum)} \tag{28.1}$$

In the early part of the twentieth century, French physicist Louis de Broglie (1892–1987) suggested that there might be a symmetry between waves and particles. He conjectured that if light sometimes behaves like particles, then perhaps material particles such as electrons also might have wave properties. In 1924, de Broglie hypothesized that a moving particle has a wave associated with it. He proposed that the particle's wavelength is related to the magnitude of its momentum p (nonrelativistic) by an equation similar to that for a photon (Eq. 28.1), except that momentum is given by the expression for a particle with mass, $p = mv$ (see Chapter 6). The **de Broglie hypothesis** states the following:

> A (nonrelativistic) particle with momentum of magnitude p has a wave associated with it. This wave's wavelength is given by
>
> $$\lambda = \frac{h}{p} = \frac{h}{mv} \quad \text{(material particles only)} \tag{28.2}$$

The waves associated with moving particles were called **matter waves**, or, more commonly, **de Broglie waves**, and were thought to somehow influence or guide the particle's motion. The electromagnetic wave is the associated wave for a photon. However, de Broglie waves associated with particles such as electrons and protons are *not* electromagnetic waves. The de Broglie equation for the wavelength gives no clue as to the nature of the wave associated with a particle that has mass.

Needless to say, de Broglie's hypothesis met with great skepticism. The idea that the motions of photons were somehow governed by the wave properties of light seemed reasonable. But the extension of this idea to the motion of a particle *with mass* was difficult to accept. Moreover, there was no evidence at the time that particles exhibited any wave properties, such as interference and diffraction. (See Section 13.4 and Sections 24.1, 24.3, and 25.4.)

Note: The Bohr model is outlined in Section 27.4.

Note: Compare this standing wave with that in the discussion of standing waves in Section 13.5.

In support of his hypothesis, de Broglie showed how it could give an interpretation of the quantization of the angular momentum postulated by Bohr in his theory of the hydrogen atom. Recall that Bohr had to hypothesize that the angular momentum of the orbiting electron was quantized in integer multiples of $h/2\pi$ (Chapter 27). De Broglie argued that for a *free* particle, the associated wave would be a *traveling* wave. However, the *bound* electron of a hydrogen atom travels repeatedly in discrete circular orbits. The associated matter wave might therefore be expected to be a *standing* wave. For a standing wave to be produced, an integral number of wavelengths has to fit into the orbital circumference. The wave must reinforce itself constructively, much as a standing wave in a circular string (▸Fig. 28.1).

The circumference of a Bohr orbit of radius r_n is $2\pi r_n$, where n is the quantum number. De Broglie equated this circumference to an integer number of electron "wavelengths" in the following manner:

$$2\pi r_n = n\lambda \quad \text{for } n = 1, 2, 3, \ldots$$

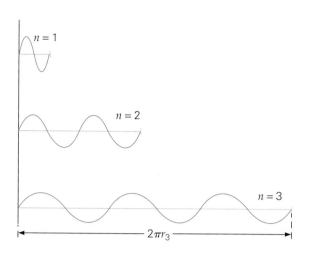

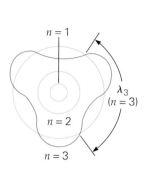

◀ **FIGURE 28.1** de Broglie waves
and Bohr orbits Similar to
standing waves in a stretched
string, de Broglie waves form
circular standing waves on the
circumferences of the Bohr orbits.
The number of wavelengths in a
particular orbit of radius r, shown
here for $n = 3$, is equal to the
principal quantum number of that
orbit. (Not drawn to scale.)

Substituting for the wavelength λ from Eq. 28.2 yields

$$2\pi r_n = \frac{nh}{mv}$$

Now recall that, for a circular orbit, the angular momentum $L = mvr$; therefore we have

$$L_n = mvr_n = n\left(\frac{h}{2\pi}\right) \quad \text{for} \quad n = 1, 2, 3, \ldots$$

Note: See Eq. 8.14 for a reminder about angular momentum.

Thus, Bohr's angular-momentum quantization is equivalent to the de Broglie assumption that, in some way, the electron behaves like a wave.

For orbits other than those allowed by the Bohr theory, the de Broglie wave for the orbiting electron would not close on itself. This observation is consistent with the Bohr postulate of the electron being only in certain "allowed" orbits and implies that *the amplitude of the de Broglie wave might be related to the location of the electron.* This idea is actually a fundamental cornerstone of modern quantum mechanics, as will be seen.

If particles really have wavelike properties, why isn't their wave nature observed in everyday phenomena? It is because effects such as diffraction are significant only when the wavelength λ is on the order of the size of the object or the opening it meets. (See Section 24.3.) If λ is much smaller than these dimensions, then diffraction is negligible. The numbers in Examples 28.1 and 28.2 should convince you of the difference between the atomic world and our everyday world.

Example 28.1 ■ Should Ballplayers Worry about Diffraction? De Broglie Wavelength

A pitcher throws a fastball to the catcher at 40 m/s through a square strike-zone practice target. The square opening is cut in a sheet of canvas and is 50 cm on each side. If the ball's mass is 0.15 kg, (a) what is the wavelength of the de Broglie wave associated with the ball? (b) Should the catcher expect diffraction to occur as the ball passes through the opening to his mitt?

Thinking It Through. (a) The de Broglie relationship (Eq. 28.2) can be used to calculate the wavelength of the matter wave associated with the ball. (b) All that matters is whether the wavelength is much larger than, much smaller than, or approximately the same size as the opening.

Solution.

Given: $v = 40$ m/s   *Find:* (a) λ (de Broglie wavelength)
$d = 50$ cm $= 0.50$ m   (b) whether diffraction is likely
$m = 0.15$ kg

(a) Equation 28.2 gives the wavelength of the baseball:

$$\lambda = \frac{h}{mv} = \frac{6.63 \times 10^{-34} \text{ J} \cdot \text{s}}{(0.15 \text{ kg})(40 \text{ m/s})} = 1.1 \times 10^{-34} \text{ m} \quad \text{(extremely small)}$$

(b) For significant diffraction to occur when a wave passes through an opening, the wave must have a wavelength similar in size to that of the opening. Since 1.1×10^{-34} m $\ll 0.50$ m, the baseball travels straight into the catcher's mitt, with no noticeable wave diffraction.

(continues on next page)

Follow-Up Exercise. In this Example, (a) how fast would the ball have to be thrown for diffractive effects to become important? (b) At that speed, how long would the ball take to travel the 20 m to the plate? (Compare your answer with the age of the universe—about 15 billion years. Would someone watching this ball think that it is moving?) *(Answers to all Follow-Up Exercises are at the back of the text.)*

From Example 28.1, it is little wonder that the wave nature of matter isn't observed in our everyday lives. Notice, however, that a particle's de Broglie wavelength varies inversely with its mass and speed. So particles with very small masses traveling at low speeds might be another story, as the following Example shows.

Example 28.2 ■ A Whole Different Ball Game: De Broglie Wavelength of an Electron

Note: Single-slit diffraction is discussed in Section 24.3.

(a) What is the de Broglie wavelength of the wave associated with an electron that has been accelerated from rest through a potential of 50.0 V? (b) Compare your answer with the typical distance between atoms in a solid crystal, about 10^{-10} m. Would you expect diffraction to occur as these electrons pass between such atoms?

Thinking It Through. (a) We can use Eq. 28.2, but the electron's speed must first be calculated, using the given accelerating voltage. This computation involves consideration of energy conservation—the conversion of electric potential energy into kinetic energy. (b) For diffractive effects to be important, λ must be on the order of 10^{-10} m.

Solution. We list the data, as well as the mass of an electron, which can be found in Table 15.1.

Given: $V = 50.0$ V
$m = 9.11 \times 10^{-31}$ kg (from Table 15.1)

Find: (a) λ (de Broglie wavelength)
(b) whether diffraction is likely

(a) The magnitude of the potential energy lost by the electron is $|\Delta U_e| = eV$ and is equal to its gain in kinetic energy ($\Delta K = \frac{1}{2}mv^2$, because $K_o = 0$). Equating these two quantities enables us to calculate the speed:

$$\frac{1}{2}mv^2 = eV \quad \text{or} \quad v = \sqrt{\frac{2eV}{m}}$$

So,

$$v = \sqrt{\frac{2(1.60 \times 10^{-19}\text{ C})(50.0\text{ V})}{9.11 \times 10^{-31}\text{ kg}}} = 4.19 \times 10^6\text{ m/s}$$

Thus, the electron's de Broglie wavelength is

$$\lambda = \frac{h}{mv} = \frac{6.63 \times 10^{-34}\text{ J}\cdot\text{s}}{(9.11 \times 10^{-31}\text{ kg})(4.19 \times 10^6\text{ m/s})} = 1.74 \times 10^{-10}\text{ m}$$

(b) Since this result is the same order of magnitude as the opening between atoms, diffraction should be observed. Thus, passing electrons through a crystal lattice could prove de Broglie's hypothesis.

Follow-Up Exercise. In this Example, what would change if the particle were a proton? In other words, would the de Broglie wave of a proton be more or less likely to exhibit diffraction effects than would an electron under the same conditions? Explain, and give a numerical answer.

Problem-Solving Hint

In Example 28.2, if the accelerating voltage were changed, the calculation of v and λ would have to be repeated. This process would again involve the use of constants such as the electron's mass, its charge, and Planck's constant. In problems involving accelerating voltages, therefore, it is convenient to use the numerical values of m, e, and h to derive a nonrelativistic expression for the de Broglie wavelength of an electron when it is accelerated through a difference in potential, V. To begin, the kinetic energy is expressed in terms of momentum. Since $p = mv$, the kinetic energy, $\frac{1}{2}mv^2$, can also be written as $p^2/2m$. Then, by energy conservation,

$$\frac{p^2}{2m} = eV \quad \text{or} \quad p = \sqrt{2meV}$$

Thus, the de Broglie wavelength is

$$\lambda = \frac{h}{p} = \frac{h}{\sqrt{2meV}} = \sqrt{\frac{h^2}{2meV}}$$

Inserting the values of h, e, and m and rounding the result to three significant figures,

$$\lambda = \sqrt{\frac{1.50}{V \text{ (in volts)}}} \times 10^{-9} \text{ m} = \sqrt{\frac{1.50}{V \text{ (in volts)}}} \text{ nm} \quad \begin{array}{l}\textit{(nonrelativistic electron} \\ \textit{initially at rest; V in volts)}\end{array} \quad (28.3)$$

Note: This alternative de Broglie wavelength expression for electrons accelerated through a potential difference, in which V must be in volts, applies only to nonrelativistic electrons initially at rest.

If $V = 50.0$ V, then

$$\lambda = \sqrt{\frac{1.50}{50.0}} \times 10^{-9} \text{ m} = 0.173 \text{ nm}$$

Note that this answer differs in the last digit from that found in Example 28.2, because there, the values for h, e, and m were rounded to three significant figures *before* the calculations were performed.

You may wish to derive a comparable formula for a proton, to use in solving similar problems. What quantities would have to be changed?

In 1927, two physicists in the United States, C. J. Davisson and L. H. Germer, used a crystal to diffract a beam of electrons, thereby demonstrating a wavelike property of particles. A single crystal of nickel was cut to expose a spacing of $d = 0.215$ nm between the planes of atoms. When a beam of electrons was directed normally onto the crystal face, a maximum in the intensity of scattered electrons was observed at an angle of 50° relative to the surface normal (►Fig. 28.2). The scattering was most intense for an accelerating potential of 54.0 V.

According to wave theory, constructive interference due to waves reflected from two lattice planes (a distance d apart) should occur at certain scattering angles θ. The theory predicts that the first-order maximum should be observed at an angle given by

$$\sin \theta = \frac{\lambda}{d}$$

This condition for the experimental setup in Fig. 28.2 requires a wavelength of

$$\lambda = d \sin \theta = (0.215 \text{ nm}) \sin 50° = 0.165 \text{ nm}$$

Using Eq. 28.3 to determine the de Broglie wavelength of the electrons after being accelerated through 54.0 V, we obtain

$$\lambda = \sqrt{\frac{1.50}{V}} \text{ nm} = \sqrt{\frac{1.50}{54.0}} \text{ nm} = 0.167 \text{ nm}$$

The agreement was well within the experimental uncertainty. The Davisson–Germer experiment gave convincing proof of de Broglie's matter wave hypothesis. A practical application of the wavelike properties of electrons is discussed in Insight 28.1 on The Electron Microscope on page 883.

Note: Diffraction from crystal lattices is discussed in Section 24.3.

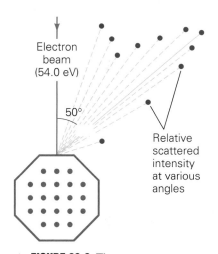

▲ **FIGURE 28.2 The Davisson–Germer experiment** When a beam of electrons of kinetic energy 54.0 eV is incident on the face of a nickel crystal, a maximum in the scattering intensity is observed at an angle of 50°.

28.2 The Schrödinger Wave Equation

OBJECTIVES: To understand qualitatively (a) the reasoning that underlies the Schrödinger wave equation, and (b) the equation's use in finding particle wave functions.

De Broglie's hypothesis predicts that moving particles have associated waves that somehow govern their behavior. However, it does not tell us the *form* of these waves, only their wavelengths. To have a useful theory, an equation that will give the mathematical form of these matter waves is needed. We also need to know how these waves govern particle motion. In 1926, Erwin Schrödinger, an Austrian physicist, presented a general equation that describes the de Broglie matter waves and their interpretation.

A traveling de Broglie wave varies with location and time in a similar way to everyday wave motion (Section 13.2). The de Broglie wave is denoted by ψ (the Greek letter psi, pronounced "sigh") and is called the **wave function ψ**, and is associated with the particle's kinetic, potential, and total energy. Recall that for a conservative mechanical system (Section 5.5), the total mechanical energy E, which is the sum of the kinetic and potential energies, is a constant; that is $K + U = E$. Schrödinger proposed a similar equation for the de Broglie matter waves, involving the wave function ψ. **Schrödinger's wave equation*** has the general form

$$(K + U)\psi = E\psi \tag{28.4}$$

Equation 28.4 can be solved for ψ, but doing so involves complex mathematical operations well beyond the scope of this text. For us, the more important question is related to the physical significance of ψ. During the early development of quantum mechanics, it was not at all clear how ψ should be interpreted. After much thought and investigation, Schrödinger and his colleagues hypothesized the following:

> The square of a particle's wave function is proportional to the probability of finding that particle at a given location.

The interpretation of ψ^2 as a *probability* altered the idea that the electron in a hydrogen atom could be found only in orbits at discrete distances from the nucleus, as described in the Bohr theory. When the Schrödinger equation was solved for the hydrogen atom, there was a nonzero probability of finding the electron at almost any distance from the nucleus. The relative probability of finding an electron [in the ground state ($n = 1$)] at a given distance from the proton is shown in ▾Fig. 28.3a.

The *maximum* probability coincides with the Bohr radius of 0.0529 nm, but it is possible that, for instance, the electron could even be inside the nucleus. Notice that, although the ground-state wave function exists for distances well beyond 0.20 nm from the proton, there is little chance of finding an electron beyond this distance. The probability density distribution gives rise to the idea of an *electron cloud* around the nucleus (Fig. 28.3b). This cloud is actually a probability density cloud, meaning that the electron can be found in many different locations with varying probabilities.

Note: More precisely, the square of the absolute value of the wave-function solution to Schrödinger's equation, $|\psi|^2$, gives the probability of finding a particle at a location.

Teaching tip: Emphasize that this interpretation does not mean that the electron is itself spread out into various locations—it means that the *probability* of finding it is spread out. Thus, if you picture repeated experiments, starting always with the same system, you will find the places of high probability easily, but in any one experiment, the electron can be found in various places.

▶ **FIGURE 28.3** Electron probability for hydrogen-atom orbits **(a)** The square of the wave function is the probability of finding the hydrogen electron at a particular location. Here, it is assumed to be in the ground state ($n = 1$), and its probability is plotted as a function of radial distance from the proton. The electron has the greatest probability of being at a distance of 0.0529 nm, which matches the radius of the first Bohr orbit. **(b)** The probability distribution gives rise to the idea of an electron probability cloud around the nucleus. The cloud's density reflects the probability density.

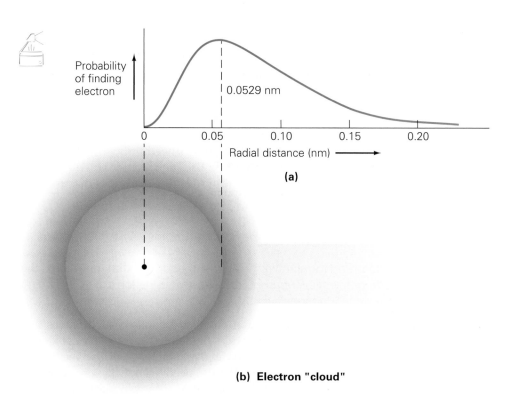

(a)

(b) Electron "cloud"

*Although Eq. 28.4 looks like a multiplication of $E = K + U$ by ψ, it is much more complex. For example, K is no longer $\frac{1}{2}mv^2$, but is instead replaced by a quantity (called an *operator*) that enables us to *extract* the kinetic energy from ψ.

INSIGHT | 28.1 THE ELECTRON MICROSCOPE

The de Broglie hypothesis led to the development of an important practical application—the *electron microscope*. As the Davisson–Germer experiment demonstrated, under the right conditions, electrons experience diffraction, as do light waves. Does this mean that electrons might also be focused like light waves? The answer is yes. In fact, electron "waves" can be focused to form images, though the focusing mechanism is different than those used with light in a visible light microscope. As Example 28.2 shows, moving electrons have very short de Broglie wavelengths. With such short wavelengths, greater magnification and finer resolution can be obtained than with any light microscope. Recall the inverse relationship between resolving power, or the ability to see details, and wavelength—a smaller wavelength means greater resolving power (See Section 25.4.) In fact, the resolving power of standard electron microscopes is on the order of a few nanometers, which is only about 10 times larger than atomic sizes.

Knowledge of how to focus electron beams by using magnetic coils permitted the construction of the first electron microscope in Germany in 1931. In a *transmission electron microscope* (TEM), an electron beam is directed onto a very thin specimen. Different numbers of electrons pass through different parts of the specimen, depending on its structure. The transmitted beam is then brought into focus by a magnetic objective coil. The general components of electron and light microscopes are analogous (Fig. 1), but an electron microscope must be housed in a high-vacuum chamber to prevent the electrons from colliding with air molecules. As a result, an electron microscope looks nothing like a light microscope (Fig. 2). A normal light-microscope image (Fig. 3a) is limited to a magnification of about 2000×, while magnifications up to 100 000× can be achieved with an electron microscope (Fig. 3b).

Another difference between light microscopes and electron microscopes is that the final lens in an electron microscope, called the *projector coil*, has to project a real image onto a fluorescent screen or photographic film, since the eye cannot perceive an electron image directly. Because specimens for transmission electron microscopy must be very thin, special techniques for specimen preparation are required. These methods allow for specimen sections to be sliced as thin as 10 nm (only about 100 atoms thick).

The surfaces of thicker objects can be examined by the *reflection* of the electron beam from the surface. This process is accomplished with the *scanning electron microscope* (SEM). A beam spot is scanned across the specimen by means of deflecting coils, much as is done in a television tube. Surface irregular-

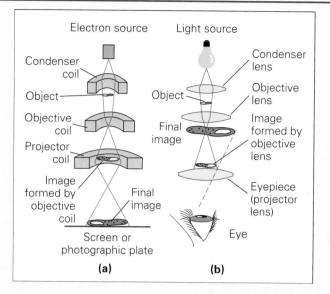

FIGURE 1 Electron and light microscopes A comparison of the elements of **(a)** an electron microscope and **(b)** a light microscope. The light microscope is drawn upside down for a better comparison.

ities cause directional variations in the intensity of the reflected electrons, which gives contrast to the image. Through such techniques, a scanning electron microscope gives pictures with a remarkable three-dimensional quality, such as those in Fig. 3c.

FIGURE 2 An electron microscope The microscope is housed in a cylindrical vacuum chamber, to the left.

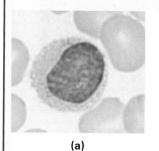

(a)

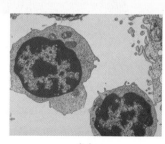

(b)

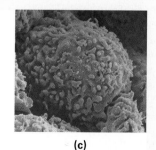

(c)

FIGURE 3 Lymphocytes (white blood cells) Images produced by **(a)** a light microscope, **(b)** a transmission electron microscope (TEM), and **(c)** a scanning electron microscope (SEM). Notice the enormous increase in magnification produced by the electron microscopes.

An interesting quantum-mechanical result that runs counter to our everyday experiences is *tunneling*. In classical physics, there are regions forbidden to particles by energy considerations. These regions are areas where a particle's potential energy would be greater than its total energy. Classically, the particle is not allowed in such regions because it would have a negative kinetic energy there ($E - U = K < 0$), which is impossible. In such situations, we say that the particle's location is limited by a *potential-energy barrier*.

In certain instances, however, quantum mechanics predicts a small, but finite, probability of the particle's wave function penetrating the barrier and thus of the particle being found on the other side of the barrier. Thus, there is a certain probability (which is practically zero for everyday objects) of the particle "tunneling" through the barrier, especially on the atomic level, where the wave nature of particles is exhibited. Such tunneling forms the basis of the scanning tunneling microscope (STM; see the accompanying Insight 28.2 on The Scanning Tunneling Microscope), which creates images with a resolution on the order of the size of a single atom. Barrier penetration also explains certain nuclear decay processes (Chapter 29).

INSIGHT 28.2 THE SCANNING TUNNELING MICROSCOPE (STM)

The *scanning tunneling microscope* (STM) was invented in the late 1970s and promptly revolutionized the field of surface physics. STMs use quantum-mechanical tunneling to produce stunning images of atoms, such as this chapter's opening photograph.

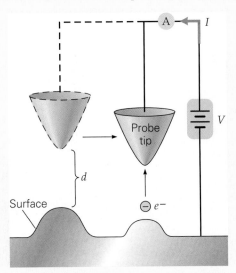

FIGURE 1 Schematic representation of the STM The tip of a probe is moved across the contours of a specimen's surface. A small voltage applied between the probe and the surface causes electrons to tunnel across the vacuum gap. The tunneling current is extremely sensitive to the separation distance between the probe tip and the surface. A feedback circuit (not shown) moves the probe up and down so as to keep constant the tunneling current as measured by the ammeter shown as Ⓐ. Thus, when the probe is scanned across the sample, surface features smaller than atoms (up to 0.001 nm vertically) can be detected. When the probe is passed over the surface in many successive and nearby parallel paths, the resulting data can be processed to produce three-dimensional images.

The STM produces atomic-sized images by positioning its sharp tip very close (about 1 nm) to a surface. A voltage is applied between the tip and the surface, causing electron tunneling through the vacuum gap (Fig. 1). This tunneling current is extremely sensitive to the separation distance. A feedback circuit monitors this current and moves the probe vertically to keep the current constant. The separation distance is digitized, recorded, and processed by computers for display. When the probe is passed over the surface of the specimen in successive nearby parallel movements, a three-dimensional image of the surface can be displayed. Typical vertical resolution for STMs is 0.001 nm, and lateral resolutions are about 0.1 nm. Because the diameters of individual atoms are on the order of several tenths of a nanometer, STMs allow detailed images of atoms to be created. Figure 2 shows the result of a surface scan on gallium arsenide, a semiconductor material.

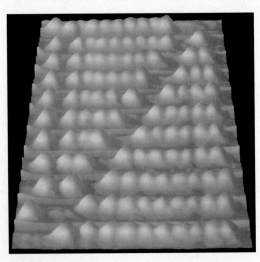

FIGURE 2 Scanning tunneling microscopy results Atoms of the semiconductor gallium arsenide.

28.3 Atomic Quantum Numbers and the Periodic Table

OBJECTIVE: To understand the structure of the periodic table in terms of quantum-mechanical electron orbits and the Pauli exclusion principle.

The Hydrogen Atom

When the Schrödinger equation was solved for the hydrogen atom, the results predicted the energy levels to be the same as those from the Bohr theory (Section 27.4). Recall that the Bohr-model energy values depended only on the **principal quantum number** n. However, in addition, the solution to the Schrödinger equation gave two other quantum numbers, designated as ℓ and m_ℓ. Three quantum numbers are needed because the electron can move in three dimensions.

The quantum number ℓ is called the **orbital quantum number**. It is associated with the orbital angular momentum of the electron. For each value of n, the ℓ quantum number has integer values from zero up to a maximum value of $n - 1$. For example, if $n = 3$, the three possible values of ℓ are 0, 1, and 2. ▶Figure 28.4 shows three orbits with different angular momenta, but the same energy. The number of different ℓ values for a given n value is equal to n. In the hydrogen atom, the energy of the electron depends only on n. Thus, orbits with the same n value, but different ℓ values, have the same energy and are said to be *degenerate*.

The quantum number m_ℓ is called the **magnetic quantum number**. The name originated from experiments in which an external *magnetic* field was applied to a sample. They showed that a particular energy level (one with given values of n and ℓ) of a hydrogen atom actually consists of several orbits that differ slightly in energy *only when in a magnetic field*. Thus, in the absence of a field, there was additional energy degeneracy. Clearly, there must be more to the description of the orbit than just n and ℓ. The quantum number m_ℓ was introduced to enumerate the number of levels that existed for a given orbital quantum number ℓ. Under zero-magnetic-field conditions, the energy of the atom does not depend on either of these quantum numbers.

The magnetic quantum number m_ℓ is associated with the orientation of the orbital angular-momentum vector $\vec{L}$ in space (▶Fig. 28.5). If there is no external magnetic field, then all orientations of $\vec{L}$ have the same energy. For each value of ℓ, m_ℓ is an integer that can range from zero to $\pm\ell$. That is, $m_\ell = 0, \pm 1, \pm 2, \ldots, \pm\ell$. For example, an orbit described by $n = 3$ and $\ell = 2$ can have m_ℓ values of $-2, -1, 0, +1$, and $+2$. In this case, the orbital angular-momentum vector $\vec{L}$ has five possible orientations, all with the same energy if no magnetic field is present. In general, for a given value of ℓ, there are $2\ell + 1$ possible values of m_ℓ. For example, with $\ell = 2$, there are five values of m_ℓ, since $2\ell + 1 = (2 \times 2) + 1 = 5$.

However, this finding was not the end of the story. The use of high-resolution optical spectrometers showed that each emission line of hydrogen is, in fact, two very closely spaced lines. Thus, each emitted wavelength is actually two. This splitting is called *spectral fine structure*. Hence, a fourth quantum number was necessary in order to describe each atomic state completely. This number is called the **spin quantum number** m_s of the electron. It is associated with the *intrinsic* angular momentum of the electron. This property, called *electron spin*, is sometimes described by analogy to the angular momentum associated with a spinning object (▼Fig. 28.6). Because each energy level is split into only two levels, the electron's intrinsic angular momentum (or, more simply, its "spin") can possess only two orientations, called "spin up" and "spin down."

Thus, the fine structure of an atom's energy levels results from the electron spin's having two orientations with respect to the atom's *internal* magnetic field, produced by the electron's orbital motion. Analogous to a magnetic moment (such as a compass) in a magnetic field, the atom possesses slightly less energy when its electron's spin is "lined up" with the field than when the spin is aligned opposite to the field. However, keep in mind that spin is fundamentally a purely quantum-mechanical concept; it is not really analogous to a spinning top. This is because, as far as we know now, the electron possesses no size—that is, it is truly a point particle.

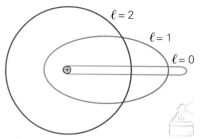

▲ **FIGURE 28.4 The orbital quantum number (ℓ)** The orbits of an electron are shown for the second excited state in hydrogen. For the principal quantum number $n = 3$, there are three different values of angular momentum (corresponding to the three differently shaped orbits and three different values of ℓ), but they all have the same total energy. The circular orbit has the maximum angular momentum; the narrowest orbit would actually pass through the proton classically and thus has zero angular momentum.

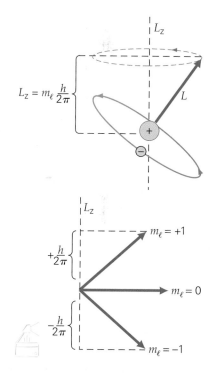

▲ **FIGURE 28.5 The magnetic quantum number m_ℓ** Here, $\vec{L}$ is the vector angular momentum associated with the orbit of the electron. Because the energy of an orbit is independent of the orientation of its plane, orbits with the same ℓ that differ only in m_ℓ have the same energy. There are $2\ell + 1$ possible orientations for a given ℓ. The value of m_ℓ tells the *component* of the angular-momentum vector in a given direction, as shown (below) for $\ell = 1$.

TABLE 28.1	Quantum Numbers for the Hydrogen Atom		
Quantum Number	Symbol	Allowed Values	Number of Allowed Values
Principal	n	$1, 2, 3, \ldots$	no limit
Orbital angular momentum	ℓ	$0, 1, 2, 3, \ldots, (n - 1)$	n (for each n)
Orbital magnetic	m_ℓ	$0, \pm 1, \pm 2, \pm 3, \ldots, \pm \ell$	$2\ell + 1$ (for each ℓ)
Spin	m_s	$\pm \frac{1}{2}$	2 (for each m_ℓ)

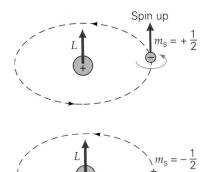

▲ **FIGURE 28.6 The electron-spin quantum number (m_s)** The electron spin can be either up or down. Electron spin is strictly a quantum mechanical property and should not be identified with the physical spin of a macroscopic body, except with respect to conceptual reasoning.

Note: The n quantum number designates orbital shells; the ℓ quantum number designates orbital subshells.

TABLE 28.2	
Subshell Designations	
Value of ℓ	Designation
$\ell = 0$	s
$\ell = 1$	p
$\ell = 2$	d
$\ell = 3$	f
$\ell = 4$	g
$\ell = 5$	h
$\ldots$	$\ldots$

Thus, for the excited state of hydrogen with $n = 3$ and $\ell = 2$, each value of m_ℓ also has two possible spin orientations. For example, when $m_\ell = +1$, there are two possible sets of the four quantum numbers: $n = 3$, $\ell = 2$, and $m_\ell = +1$, with $m_s = +\frac{1}{2}$; and $n = 3$, $\ell = 2$, and $m_\ell = +1$, with $m_s = -\frac{1}{2}$. Both sets would have nearly the same energy. The orbit's energy is therefore almost independent of the electron's spin direction. This condition results in yet another (approximate) energy degeneracy. In summary, the energy of the various states of the hydrogen atom are, to a very high degree, determined solely by the primary quantum number n.

The four quantum numbers for the hydrogen atom are summarized in Table 28.1. Other particles, such a protons, neutrons, and composites of them called *atomic nuclei* also possess spin. A particularly useful property of the spin of a proton, and of atomic nuclei in general, is discussed in Insight 28.3 on Magnetic Resonance Imaging (MRI) on page 888.

Multielectron Atoms

The Schrödinger equation cannot be solved exactly for atoms with more than one electron (multielectron atoms). However, a solution can be found, to a workable approximation, in which each electron occupies a state characterized by a set of quantum numbers similar to those for hydrogen. Because of the repulsive forces between the electrons, the description of a multielectron atom is much more complicated. For one, the energy depends not only on the principal quantum number n, but also on the orbital quantum number ℓ. This condition gives rise to a subdivision (or "splitting") of the degeneracy seen in hydrogen atoms. In multielectron atoms, the energies of the atomic levels generally depend on all four quantum numbers.

It is common to refer to all the electron levels that share the same n value as making up a **shell** and to all the electron levels that share the same ℓ values within that shell as **subshells**. That is, electron levels with the same n value are said to be "in the same shell." Similarly, electrons with the same n and ℓ values are said to be "in the same subshell."

The ℓ subshells can be designated by integers; however, it is common to use letters instead. The letters $s, p, d, f, g, \ldots$ correspond to the values of $\ell = 0, 1, 2, 3, 4, \ldots$ respectively. After f, the letters go alphabetically (Table 28.2).

Because in multielectron atoms an electron's energy depends on both n and ℓ, both quantum numbers are used to label atomic energy levels (a shell and subshell, respectively). The labeling convention is as follows: n is written as a number, followed by the letter that stands for the value of ℓ. For example, $1s$ denotes an energy level with $n = 1$ and $\ell = 0$; $2p$ is for $n = 2$ and $\ell = 1$; $3d$ is for $n = 3$ and $\ell = 2$; and so on. Also, it is common to refer to the m_ℓ values as representing *orbitals*. For example, a $2p$ energy level has three orbitals, corresponding to the m_ℓ values of -1, 0, and $+1$ (because $\ell = 1$).

The hydrogen atom energy levels are not evenly spaced, but do increase sequentially. In multielectron atoms, not only are the energy levels unevenly spaced, but their numerical sequence is also, in general, out of order. The shell–subshell (n–ℓ notation) energy-level sequence for a multielectron atom is shown in ▶Fig. 28.7a. Notice, for example, that the $4s$ level is, energywise, below the $3d$ level. Such variations result in part from electrical forces between the electrons. Furthermore, note

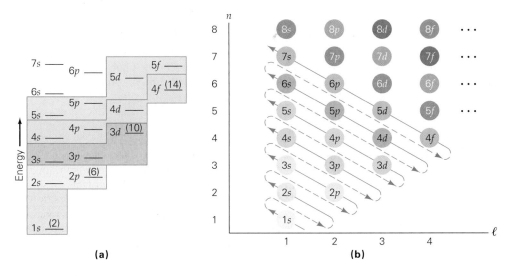

(a)

(b)

▲ **FIGURE 28.7 Energy levels of a multielectron atom (a)** The shell–subshell (n–ℓ) sequence shows that the energy levels are not evenly spaced and that the sequence of energy levels has numbers out of order. For example, the 4s level lies below the 3d level. The maximum number of electrons for a subshell, $2(2\ell + 1)$, is shown in parentheses on representative levels. (The vertical energy differences may not be drawn to scale.) **(b)** A convenient way to remember the energy-level order of a multielectron atom is to list the n-versus-ℓ values as shown here. The diagonal lines then give the energy levels in ascending order.

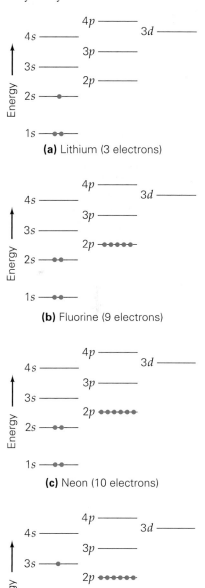

▼ **FIGURE 28.8 Filling subshells**
The electron subshell distributions for several unexcited atoms (in their ground state) according to the Pauli exclusion principle. Because of spin, any s subshell can hold a maximum of two electrons, and any p subshell can hold a maximum of six electrons. What can you say about the d subshells?

(a) Lithium (3 electrons)

(b) Fluorine (9 electrons)

(c) Neon (10 electrons)

(d) Sodium (11 electrons)

that the electrons in the outer orbits are "shielded" from the attractive force of the nucleus by the electrons that are closer to the nucleus. For example, consider the highly elliptical orbit (Fig. 28.4) of an electron in the 4s ($\ell = 0$) orbit. The electron clearly spends more time near the nucleus, and hence is more tightly bound, than if it were in the more circular 3d ($\ell = 2$) orbit. A convenient way to remember the order of the levels is given in Fig. 28.7b.

The ground state of a multielectron atom has some similarities to that of the hydrogen atom, with its one electron in the 1s, or lowest energy, level. In a multielectron atom, the ground state still has the lowest total energy, but is a combination of energy levels in this case. That is, the electrons are in the lowest possible energy levels. But to identify how the electrons fill these levels, we must know how many electrons can occupy a particular energy level. For example, the lithium (Li) atom has three electrons. Can they all be in the 1s level? As we shall see, the answer, given by the Pauli exclusion principle, is no.

The Pauli Exclusion Principle

Exactly how the electrons of a multielectron atom distribute themselves in the ground-state energy levels is governed by a principle set forth in 1928 by the Austrian physicist Wolfgang Pauli. The **Pauli exclusion principle** states the following:

| No two electrons in an atom can have the same set of quantum numbers (n, ℓ, m_ℓ, m_s). That is, no two electrons in an atom can be in the same quantum state.

This principle limits the number of electrons that can occupy a given energy level. For example, the 1s ($n = 1$ and $\ell = 0$) level can have only one m_ℓ value, $m_\ell = 0$, along with only two m_s values, $m_s = \pm\frac{1}{2}$. Thus, there are only two unique sets of quantum numbers (n, ℓ, m_ℓ, m_s) for the 1s level—$\left(1, 0, 0, +\frac{1}{2}\right)$ and $\left(1, 0, 0, -\frac{1}{2}\right)$—so only two electrons can occupy the 1s level. If this situation is indeed the case, then we say that such a shell is *full*; all other electrons are excluded from it by Pauli's principle. Thus, for a Li atom, with three electrons, the third electron must occupy the next higher level (2s) when the atom is in the ground state. This case is illustrated in ▶Fig. 28.8, along with the ground-state energy levels for some other atoms.

INSIGHT 28.3 MAGNETIC RESONANCE IMAGING (MRI)

Magnetic resonance imaging, or MRI, has become a common and important medical technique for the noninvasive examination of the human body (Fig. 1). Originally known as NMR (for *nuclear magnetic resonance*), MRI is based on the quantum-mechanical concept of spin.

A current-carrying loop in a magnetic field experiences a torque that tends to orient the loop's magnetic moment parallel to the field, much like a compass (Chapter 20). Electrons possess an intrinsic angular momentum called *spin*. This condition gives rise to a *spin magnetic moment*, which aligns in a similar manner to the compass. When atoms are placed in a magnetic field, each of their energy levels is split into two levels (for the two possible spin orientations—parallel and antiparallel to the magnetic field), each with a slightly different energy, which results in the fine structure in the atoms' emission spectra.

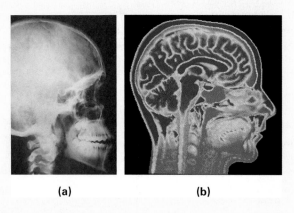

FIGURE 1 Diagnostic images (a) An X-ray of a human head. **(b)** A magnetic resonance image (MRI) of a human head. The amount of detail captured, especially in the soft tissues of the brain, makes such images very useful for medical diagnosis.

Nuclei exhibit similar spin effects when placed in magnetic fields. Because neutrons and protons possess spin, nuclei, which are composed of neutrons and protons, also possess magnetic moments. To understand the basics of MRI, let us concentrate on the simplest atom, hydrogen, with a nucleus composed of a single proton. The magnetic resonance of hydrogen is commonly measured in an MRI apparatus, since hydrogen is the most abundant element in the human body. The spin angular momentum of the proton can take on only two values, similar to that of electron spin, called "spin up" and "spin down." These values describe the orientation of the proton's magnetic moment relative to the direction of an external magnetic field (Fig. 2a). With spin up, the magnetic moment is parallel to the field, with spin down, it points opposite to the field. The spin-down orientation has a slightly higher energy, and the energy difference ΔE between the two levels is proportional to the magnitude of the magnetic field. The transition to the higher energy level can be made by allowing the proton to absorb a photon with energy equal to ΔE.

In a typical MRI apparatus, the sample (usually a region of the human body) is placed in a magnetic field $\vec{B}$. The magnitude of the field determines the energy E (or light frequency f) of the photon needed to cause a transition. This is because the photon energy absorption requires that $E = hf = \Delta E$, which is proportional to the strength of the magnetic field, B. The photons that trigger this transition are supplied by a pulsed beam of radiofrequency (RF) radiation applied to the sample. If the frequency of the radiation is adjusted so that the photon energy equals the energy level difference, many nuclei will absorb the photon energy and be excited into the higher energy level. The frequency that creates such excitation is known as the *resonance frequency* ($f = \Delta E/h$), hence the name magnetic *resonance* imaging.

Figure 3 shows a typical MRI device. Large coils produce the magnetic field. (Notice the solenoid arrangement in Fig. 3a.) Other coils produce the RF signals ("RF photons") that cause the nuclei to

Integrated Example 28.3 ■ The Quantum Shell Game: How Many States?

(a) How does the number of possible electron states compare between the 2*p* and the 4*p* subshells: (1) The number of states in the 2*p* subshell is greater than that in the 4*p* subshell; (2) the number of states in the 2*p* subshell is less than that in the 4*p* subshell; or (3) the number of states in the 2*p* subshell is the same as that in the 4*p* subshell? (b) Compare the number of possible electron states in the 3*p* subshell to the number in the 4*d* subshell.

(a) Conceptual Reasoning. The term *subshell* refers to the states that share the same ℓ quantum number within a shell. That is, they all have the same principal quantum number n. All that matters is that their ℓ values are the same. Therefore, the number of states in the two subshells is the same, so the correct answer is (3).

(b) Thinking It Through. This part is a matter of following the quantum-mechanical counting rules. Also remember that each letter stands for a specific value of ℓ. The *p* level means that $\ell = 1$, and the *d* level means that $\ell = 2$. In each subshell, we replace the letter designation for ℓ by its number. Thus, the data given are as follows:

Given: 3*p* level means $n = 3$ and $\ell = 1$ *Find:* the number of quantum states in
4*d* level means $n = 4$ and $\ell = 2$ the 3*p* subshell as compared with the number in the 4*d* subshell

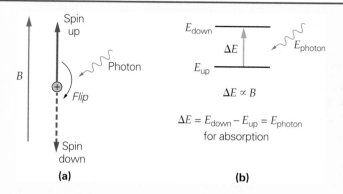

(a) **(b)**

FIGURE 2 Nuclear spin **(a)** In a uniform magnetic field, the spin angular momentum, or spin magnetic moment, of a hydrogen nucleus (a proton) can have only two values—called "spin up" and "spin down," in reference to the direction of the external magnetic field. **(b)** This condition gives rise to two energy levels for the nucleus. Energy must be absorbed in order to "flip" the proton spin.

"flip their spin"—that is, to be excited from the lower to the upper energy level. The resulting absorption of energy is detected, as is emitted radiation coming from a return transition to the lower state. Regions that produce the greatest absorption (or re-emission) are those with the greatest concentration of the particular nucleus to which the apparatus is "tuned" by the choice of B and f. (Other nuclei besides hydrogen have their own characteristic intrinsic spins and can be imaged by tuning the frequency to match their resonance frequency.) Images are produced by means of computerized tomography, similar to that used in X-ray CT scans (see Section 20.4). The result is a two- or three-dimensional image (Fig. 1b) that can provide a great deal of diagnostic medical information.

Although a variety of atomic nuclei exhibit nuclear magnetic resonance, most MRI work is done with hydrogen, because body tissues vary in water content. For example, muscle tissue has more water than does fatty tissue, so there is a distinct contrast in the radiation intensity of the two materials. Similarly, fatty deposits in blood vessels are perceived distinctly from the tissue of the vessel walls. A tumor with a water content different than that of the surrounding tissue would also show up in an MRI image.

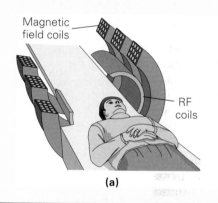

(a) **(b)**

FIGURE 3 MRI **(a)** A diagram and **(b)** a photograph of the apparatus used for magnetic resonance imaging.

For a particular subshell, the ℓ value determines the number of states. Recall that there are $(2\ell + 1)$ possible m_ℓ values for a given ℓ. Thus, for $\ell = 1$, there are $[(2 \times 1) + 1] = 3$ values for m_ℓ. (They are +1, 0, and −1.) Each of these values can have two m_s values $\left(\pm\frac{1}{2}\right)$, making six different combinations of (n, ℓ, m_ℓ, m_s), or six states.

In general, the number of possible states for a given value of ℓ is $2(2\ell + 1)$, taking into account the two possible "spin states" for each orbital state:

Number of electron states for $\ell = 1$ is $2(2\ell + 1) = 2[(2 \times 1) + 1] = 6$

Number of electron states for $\ell = 2$ is $2(2\ell + 1) = 2[(2 \times 2) + 1] = 10$

Comparison shows that the d subshell has more possible electron states than does the p subshell, regardless of which shell they are in (that is, their n value).

These results are summarized in Table 28.3. Notice that the total number of states in a given *shell* (designated by n) is $2n^2$. For example, for the $n = 2$ shell, the total number of states for its combined s and p subshells ($\ell = 0, 1$) is $2n^2 = 2(2)^2 = 8$. This means that up to eight electrons can be accommodated in the $n = 2$ shell: two in the $2s$ subshell and six in the $2p$ subshell.

Follow-Up Exercise. How many electrons could be accommodated in the $3d$ subshell if there were no spin quantum number?

TABLE 28.3	Possible Sets of Quantum Numbers and States					
Electron Shell n	Subshell ℓ	Subshell Notation	Orbitals (m_ℓ)	Number of Orbitals (m_ℓ) in Subshell = $(2\ell + 1)$	Number of States m_s in Subshell = $2(2\ell + 1)$	Total Electron States for Shell = $2n^2$
1	0	$1s$	0	1	2	2
2	0	$2s$	0	1	2	8
	1	$2p$	1, 0, −1	3	6	
3	0	$3s$	0	1	2	18
	1	$3p$	1, 0, −1	3	6	
	2	$3d$	2, 1, 0, −1, −2	5	10	
4	0	$4s$	0	1	2	32
	1	$4p$	1, 0, −1	3	6	
	2	$4d$	2, 1, 0, −1, −2	5	10	
	3	$4f$	3, 2, 1, 0, −1, −2, −3	7	14	

Electron Configurations   The electron structure of the ground state of atoms can be determined, so to speak, by putting an increasing number of electrons in the lower energy subshells [hydrogen (H), 1 electron; helium (He), 2 electrons; lithium (Li), 3 electrons; and so on], as was done for four elements in Fig. 28.8. However, rather than drawing diagrams, a shorthand notation called the **electron configuration** is widely used.

In this notation, the subshells are written in order of increasing energy, and the number of electrons in each subshell designated with a superscript. For example, $3p^5$ means that a $3p$ subshell is occupied by five electrons. The electron configurations for the atoms shown in Fig. 28.8 can thus be written as follows:

Li	(3 electrons)	$1s^2 2s^1$
F	(9 electrons)	$1s^2 2s^2 2p^5$
Ne	(10 electrons)	$1s^2 2s^2 2p^6$
Na	(11 electrons)	$1s^2 2s^2 2p^6 3s^1$

In writing an electron configuration, when one subshell is filled, you go on to the next higher one. The total of all the superscripts in any configuration must add up to the number of electrons in the atom.

The energy spacing between adjacent subshells is not uniform, as Figs. 28.7a and 28.8 show. In general, there are relatively large energy gaps between the s subshells and the subshells immediately below them. (Compare the $4s$ subshell with the $3p$ one in Fig. 28.7a.) The subshells just below the s subshells are usually p subshells, with the exception of the lowest subshell—the $1s$ subshell is below the $2s$ subshell. The gaps between other subshells—for example, between the $3s$ subshell and the $3p$ subshell above it, or between the $4d$ and $5p$ subshells—are considerably smaller.

This unevenness in energy differences gives rise to periodic large energy gaps, represented by vertical lines between certain subshells in the electron configuration:

$$1s^2 \mid 2s^2 2p^6 \mid 3s^2 3p^6 \mid 4s^2 3d^{10} 4p^6 \mid 5s^2 4d^{10} 5p^6 \mid 6s^2 4f^{14} 5d^{10} 6p^6 \mid \dots$$

(*number of states*)   (2)   (8)   (8)   (18)   (18)   (32)

The subshells *between* the lines have only slightly different energies. The grouping of subshells (for example, $2s^2 2p^6$) that have about the same energy is referred to as an **electron period**.

Electron periods are the basis of the periodic table of elements. With your present knowledge of electron configurations, you are now in a position to understand the periodic table better than the person who originally developed it.

The Periodic Table of Elements

By 1860, more than sixty chemical elements had been discovered. Several attempts had been made to classify the elements into some orderly arrangement, but none were satisfactory. It had been noted in the early 1800s that the elements could be listed in such a way that similar chemical properties recurred periodically throughout the list. With this idea, in 1869, a Russian chemist, Dmitri Mendeleev (pronounced men-duh-*lay*-eff), created an arrangement of the elements, based on this periodic property. The modern version of his **periodic table of elements** is used today and can be seen on the walls of just about every science building (▼Fig. 28.9).

Mendeleev arranged the known elements in rows, called **periods**, in order of increasing atomic mass. When he came to an element that had chemical properties similar to those of one of the previous elements, he put this element below the previous similar one. In this manner, he formed both horizontal rows of elements and vertical columns called **groups**, or families of elements with similar chemical properties. The table was later rearranged in order of increasing atomic, or proton, number (the number of protons in the nucleus of an atom is the number at the top left of each of the element boxes in Fig. 28.9) in order to resolve some inconsistencies. Notice that if atomic masses were used, cobalt and nickel, atomic numbers 27 and 28, respectively, would fall in reversed columns.

With only 65 elements known at the time, there were vacant spaces in Mendeleev's table. The elements for these spaces were yet to be discovered. Because the missing elements were part of a sequence and had properties similar to those of other elements in a group, Mendeleev could predict their masses and chemical properties. Less than 20 years after Mendeleev devised his table, which showed chemists what to look for in order to find the undiscovered elements, three of the missing elements were, in fact, discovered.

The periodic table puts the elements into seven horizontal rows, or periods. The first period has only two elements. Periods 2 and 3 each have 8 elements, and periods 4 and 5 each have 18 elements. Recall that the s, p, d, and f subshells can contain a maximum of 2, 6, 10, and 14 electrons $[2(2\ell + 1)]$, respectively. You should begin to see a correlation between these numbers and the arrangements of elements in the periodic table.

The periodicity of the periodic table can be understood in terms of the electron configurations of the atoms. For $n = 1$, the electrons are in one of two s states ($1s$); for $n = 2$, electrons can fill the $2s$ and $2p$ states, which gives a total of 10 electrons; and so on. Thus, the period number for a given element is equal to the highest n shell containing electrons in the atom. Notice the electron configurations for the elements in Fig. 28.9. Also, compare the electron periods given earlier, as defined by energy gaps, and the periods in the periodic table (▼Fig. 28.10, p. 893). There is a one-to-one correlation, so the periodicity comes from energy-level considerations in atoms.

Chemists refer to *main group elements* as elements in which the last (least bound) electron enters an s or p subshell. In *transition elements*, the last electron enters a d subshell; and in *inner transition elements*, the last electron enters an f subshell. So that the periodic table is not unmanageably wide, the f subshell elements are usually placed in two rows at the bottom of the table. Each of the two rows is given a name—the *lanthanide series* and the *actinide series*—based on its position within the period.

Finally, we can also understand why elements in vertical columns, or groups, have similar chemical properties. The chemical properties of an atom, such as its ability to react and form compounds, depend almost entirely on the atom's outermost electrons—that is, those electrons in the outermost *unfilled* shell. It is these electrons, called *valence electrons*, that form chemical bonds with other atoms. Because of the way in which the elements are arranged in the table, the outermost electron configurations of all the atoms in any one group are similar. The atoms in such a group would thus be expected to have similar chemical properties, and they do. For example, notice the first two groups at the left of the table. They have one and two outermost electrons in an s subshell, respectively. These elements are all highly reactive metals that form compounds that have many similarities. The group at the far right,

▲ **FIGURE 28.9 The periodic table of elements** The elements are arranged in order of increasing atomic, or proton, number. Horizontal rows are called *periods*, and vertical columns are called *groups*. The elements in a group have similar chemical properties. Each atomic mass represents an average of that element's isotopes, weighted to reflect their relative abundance in our immediate environment. The masses have been rounded to two decimal places; more precise values are given in Appendices IV and V. (A value in parentheses represents the mass number of the best-known or longest-lived isotope of an unstable element. See Appendix IV for an alphabetical listing of elements.

Shell (last to be filled)	Subshells	Number of electrons in subshell, $2(2\ell + 1)$	Corresponding period in periodic table
$n = 7$	$7p$	6	
	$6d$	10	Period 7
	$5f$	14	(32 elements)
	$7s$	2	
$n = 6$	$6p$	6	
	$5d$	10	Period 6
	$4f$	14	(32 elements)
	$6s$	2	
$n = 5$	$5p$	6	
	$4d$	10	Period 5
	$5s$	2	(18 elements)
$n = 4$	$4p$	6	
	$3d$	10	Period 4
	$4s$	2	(18 elements)
$n = 3$	$3p$	6	Period 3
	$3s$	2	(8 elements)
$n = 2$	$2p$	6	Period 2
	$2s$	2	(8 elements)
$n = 1$	$1s$	2	Period 1 (2 elements)

Energy →

◀ **FIGURE 28.10** Electron periods
The periods of the periodic table are related to electron configurations. The last n shell to be filled is equal to the period number. The electron periods and the corresponding periods of the table are defined by relatively large energy gaps between successive subshells (such as between $4s$ and $3p$) of the atoms.

the noble gases, includes elements with completely filled subshells (and thus a full shell). These elements are at the ends of electron periods, or just before a large energy gap. These gases are nonreactive and can form compounds (by chemical bonding) only under very special conditions.

Conceptual Example 28.4 ■ Combining Atoms: Performing Chemistry on the Periodic Table

Combinations of atoms, called *molecules,* can form if atoms come together and share outer electrons. This sharing process is called *covalent bonding.* In this "shared-custody" scheme, both atoms find it energetically beneficial (that is, they lower their combined total energies) to have the equivalent of a filled outer shell of electrons, if only on a part-time basis. Using your knowledge of electron shells and the periodic chart, determine which of the following atoms would most likely form a covalent arrangement with oxygen: (a) neon (Ne); (b) calcium (Ca); or (c) hydrogen (H).

Reasoning and Answer. Choice (a), neon, with a total of 10 electrons, can be eliminated immediately, because it has a full outer shell of 8 electrons and, as such, has nothing to be gained by losing or adding electrons. Looking at the periodic table, we see that oxygen, with 6 outer-shell electrons, is 2 electrons shy of having a full complement of 8 electrons. It *could* occasionally have those 2 electrons by sharing electrons with another atom or atoms. Choice (b), calcium, with its 20 electrons, is 2 electrons beyond the previous full shell of 10. You might think, therefore, that calcium is a possible covalent partner. However, you must remember that the covalent arrangement is a two-way street. In other words, the arrangement would also require calcium sometimes to have two *more* electrons than normal.

This situation would put the calcium atom in the awkward position of being 4 electrons beyond the full shell of 10 and 14 electrons away from the next complete shell. Hence, even though this attempt at covalent bonding might seem to work for oxygen, it certainly won't work well for calcium. The two species do, however, combine to form calcium oxide (CaO). The bonding that keeps calcium oxide together is based on the electrical attraction between the positive calcium ion, Ca^{+2}, and the negative oxygen ion, O^{-2}. In this type of bonding, the two atoms permanently exchange two electrons, making each a doubly charged ion of opposite signs. This bond is called an *ionic bond* and is *not* covalent. Thus, answer (b), calcium, is not correct.

The remaining candidate, hydrogen, has one electron fewer than a full shell of two. Thus, if a hydrogen atom could add one electron, it would attain an electron configuration like that of the lightest inert gas, helium. Since each hydrogen atom needs to share only one electron, two of them can accomplish this by sharing with a single oxygen atom. Part

(continues on next page)

of the time, the hydrogen atoms must share their electrons with the oxygen atom in order to create the latter's full outer shell of eight electrons. So, the correct answer is (c)—two hydrogen atoms covalently bound to a single oxygen atom. This combination has the molecular formula H_2O—water.

Follow-Up Exercise. In this Example, (a) what would be the electron configuration of the oxygen in the water molecule at some instant when it has "custody" of the electrons from both hydrogen atoms? (b) What would be the net charge on the oxygen in this case?

28.4 The Heisenberg Uncertainty Principle

OBJECTIVE: To understand the inherent quantum-mechanical limits on the accuracy of physical observations.

An important aspect of quantum mechanics has to do with measurement and accuracy. In classical mechanics, there is no limit to the accuracy of a measurement. Theoretically, by continual refinement of a measurement instrument and procedure, the accuracy could be improved to any degree so as to give *exact* values. This theoretical approach results in a *deterministic* view of nature. For example, if both the position and the velocity of an object are known *exactly* at a particular time, you can determine exactly where the object will be in the future and where it was in the past (assuming that you know all the forces that act on it).

However, quantum theory predicts otherwise and sets limits on the accuracy of measurements. This idea was introduced in 1927 by the German physicist Werner Heisenberg, who had developed another approach to quantum mechanics that complemented Schrödinger's wave theory. The **Heisenberg uncertainty principle** as applied to position and momentum (or velocity) can be stated as follows:

> It is impossible to know simultaneously an object's exact position and momentum.

This concept is often illustrated with a simple thought experiment. Suppose that you want to measure the position and momentum (actually, the velocity) of an electron. In order for you to "see," or locate, the electron, at least one photon must bounce off the electron and come to your eye (or detector), as illustrated in ◄Fig. 28.11. However, in the collision process, some of the photon's energy and momentum are transferred to the electron.

After the collision, the electron recoils. Thus, in the very process of locating the electron's position accurately, uncertainty is introduced into the electron's velocity (or momentum, because $\Delta \vec{p} = m\Delta \vec{v}$). This effect isn't noticed in our everyday macroscopic world, because the recoil produced by viewing an object with light is negligible. This is because the pressure exerted by the light cannot appreciably alter the motion or position of an object of everyday mass.

According to wave optics, the position of an electron can be measured *at best* to an uncertainty Δx of about the wavelength λ of the light used—that is, $\Delta x \approx \lambda$. The photon "particle" used for this location has a momentum of $p = h/\lambda$. Because the amount of momentum transferred during collision isn't determined, the final momentum of the electron would have an uncertainty on the order of the momentum of the photon, or $\Delta p \approx h/\lambda$.

Notice that the product of these two uncertainties is *at least* as large as h, because

$$(\Delta p)(\Delta x) \approx \left(\frac{h}{\lambda}\right)(\lambda) = h$$

This equation relates the *minimum* uncertainties, or maximum accuracies, of *simultaneous* measurements of the momentum and position. In actuality, the uncertainties could be worse, depending on the amount of light (number of photons) used, the apparatus, and the technique. Using more detailed considerations, Heisenberg found that the product of the two uncertainties would equal, *at a minimum*, $h/2\pi$. However, it could be higher. Hence, we can write the following:

$$(\Delta p)(\Delta x) \geq \frac{h}{2\pi} \tag{28.5}$$

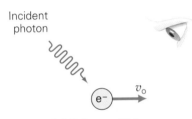

Incident photon

v_0

e⁻

(a) Before collision

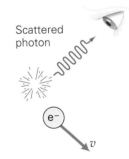

Scattered photon

e⁻

v

(b) After collision

▲ **FIGURE 28.11 Measurement-induced uncertainty** (a) To measure the position and momentum (or velocity) of an electron, at least one photon must collide with the electron and be scattered toward the eye or detector. (b) In the collision process, energy and momentum are transferred to the electron, which induces uncertainty in its velocity.

That is, the product of the *minimum* uncertainties of simultaneous momentum and position measurements is on the order of Planck's constant divided by 2π (that is, about 10^{-34} J·s).

In order to locate the position of the particle accurately (that is, to make Δx as small as possible), a photon with a very short wavelength must be used. However, this type of photon carries a lot of momentum, which results in an increased uncertainty in the momentum. To take the extreme case, if the location of a particle could be measured exactly (that is, $\Delta x \to 0$), we would have no idea about its momentum ($\Delta p \to \infty$). Thus, it is the measurement process itself that limits the accuracy to which position and momentum can be measured simultaneously. In Heisenberg's words, "Since the measuring device has been constructed by the observer . . . we have to remember that what we observe is not nature in itself but nature exposed to our method of questioning."

To see how the Heisenberg uncertainty principle affects the microscopic and macroscopic worlds, consider the following Integrated Example.

Integrated Example 28.5 ■ An Electron versus a Bullet: The Uncertainty Principle

An electron and a bullet both have the same speed, measured to the same accuracy. (a) How would their minimum uncertainties in position compare: (1) The electron's location would be more uncertain than that of the bullet; (2) the bullet's location would be more uncertain than that of the electron; or (3) their location uncertainties would be the same? (b) If the bullet's mass is 20.0 g and both the bullet and the electron have a speed of 300 m/s, with an uncertainty of $\pm 0.010\%$, determine the minimum uncertainty in the position of each.

(a) Conceptual Reasoning. The minimum uncertainty in location is related to the minimum uncertainty in momentum. Since both the bullet and the electron have the same uncertainty in velocity, the bullet's momentum is much more uncertain (momentum is proportional to mass, as is $\Delta \vec{p}$, since $\Delta \vec{p} = m\Delta \vec{v}$). Therefore, the bullet's location will be much less uncertain than that of the electron, so the answer is (1).

(b) Thinking It Through. The uncertainty principle (Eq. 28.5) can be solved for Δx in each case, because $\Delta \vec{p}$ can be determined from the uncertainty in velocity, $\Delta \vec{v}$. The electron is affected much more, because of its very small mass, as seen in part (a). Listing the quantities, including the known electron mass,

Given: $m_e = 9.11 \times 10^{-31}$ kg   *Find:* Δx_e and Δx_b (minimum uncertainties in position)
$\quad\quad\quad m_b = 20$ g $= 0.020$ kg
$\quad\quad\quad v_b = v_e = 300$ m/s $\pm 0.01\%$

The uncertainty in speed is 0.01% (or 0.00010) for both the electron and the bullet. This uncertainty is $(300 \text{ m/s})(0.00010) = 0.030$ m/s. Thus we have the same uncertainty for both:

$$v = 300 \text{ m/s} \pm 0.030 \text{ m/s}$$

The *total* uncertainty in speed is *twice* this amount, because the measurements can be off both above and below the measured values; hence, $\Delta v = 0.060$ m/s.

For the electron, the minimum uncertainty in position is

$$\Delta x_e = \frac{h}{2\pi \Delta p} = \frac{h}{2\pi m_e \Delta v} = \frac{6.63 \times 10^{-34} \text{ J·s}}{2\pi (9.11 \times 10^{-31} \text{ kg})(0.060 \text{ m/s})} = 0.0019 \text{ m} = 1.9 \text{ mm}$$

Similarly, the bullet's minimum uncertainty in position is

$$\Delta x_b = \frac{h}{2\pi m_b \Delta v} = \frac{6.63 \times 10^{-34} \text{ J·s}}{2\pi (0.020 \text{ kg})(0.060 \text{ m/s})} = 8.8 \times 10^{-32} \text{ m}$$

Notice that the uncertainty in the bullet's position is much smaller than the diameter of a nucleus. Its position uncertainty is also many orders of magnitude less than that of the electron. The lesson is that uncertainty in location for everyday objects traveling at ordinary speeds is negligible. However, for electrons, 1.9 mm is significant and measurable.

Follow-Up Exercise. In this Example, what would the minimum uncertainty in the electron's speed have to be for the minimum uncertainty in its position to be on the order of atomic dimensions, or 0.10 nm?

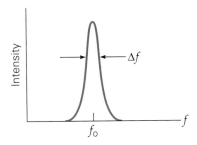

▲ FIGURE 28.12 Natural line broadening Because a measurement must be carried out in an amount of time comparable with the lifetime (Δt) of an excited atomic state, the energy of that state is uncertain by an amount $\Delta E = h\Delta f$. The observed emission line has a width of Δf, rather than being a line of single frequency f_0 with zero width.

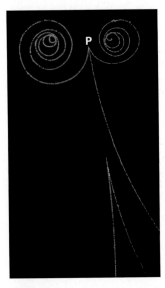

▲ FIGURE 28.13 Cloud-chamber photograph of pair production In this false-color cloud-chamber photograph, a gamma-ray photon (not visible, but it enters the region in a downward direction) interacts with a nearby atomic nucleus (point P) to produce an electron and a positron (green and red spiral tracks, respectively, at the top). In the process, the photon also dislodges an orbital electron (the nearly straight and vertical green track). An external magnetic field causes the electron and positron to be deflected in paths of opposite curvature. A similar event is recorded in the bottom half of the photo. (Why might the paths of the particles created in this case show less deflection?)

An equivalent form of the uncertainty principle relates uncertainties in energy and time. As with the position and momentum, a detailed analysis shows that, *at best*, the product of the uncertainties in energy and time is $h/2\pi$. Of course, it could be larger; hence, this form of the uncertainty principle is written as

$$(\Delta E)(\Delta t) \geq \frac{h}{2\pi} \tag{28.6}$$

This form of the Heisenberg uncertainty principle shows that the energy of an object may be uncertain by an amount ΔE, depending on the time taken to measure it, Δt. For longer times, the energy measurement becomes increasingly more accurate. Once again, these uncertainties are important only for very light objects. Such uncertainties are of particular importance in nuclear physics and elementary particle interactions (Chapter 30).

Notice that the energy of a particle cannot be measured exactly unless an infinite amount of time is taken to do so. If a measurement of energy is carried out in a time Δt, then the energy is uncertain by an amount ΔE. For example, the measurement of the frequency of light emitted by an atom is really the measurement of the energy of the photon associated with the transition from an excited state to the ground state. The measurement must be carried out in an amount of time comparable with the lifetime of the excited state. As a result, the observed emission line (Section 27.4) has a nonzero energy width, since $\Delta E = h\Delta f$ is *not* zero (◄Fig. 28.12). This *natural broadening* is generally small for atomic emissions and was therefore ignored in Chapter 27, in which spectral lines were considered to have exact frequencies.

28.5 Particles and Antiparticles

OBJECTIVES: To understand (a) the relationship between particles and antiparticles, and (b) the energy requirements for pair production.

When the British physicist Paul A. M. Dirac, in 1928, extended quantum mechanics to include relativistic considerations, something new and very different was predicted—a particle called the **positron**. The positron was predicted to have the same mass as the electron, but to carry a *positive* charge. The oppositely charged positron is the **antiparticle** of the electron. (Antiparticles will be discussed in more depth in Chapters 29 and 30.)

The positron was first observed experimentally in 1932 by the American physicist C. D. Anderson in cloud-chamber experiments with cosmic rays. The curvature of the particle tracks in a magnetic field showed two types of particles, with opposite charge and the same mass (◄Fig. 28.13). Anderson had, in fact, discovered the positron.

Because electric charge is conserved, a positron can be created only with the simultaneous creation of an electron (so that the net charge created is zero). This process is called **pair production**. In Anderson's experiment, positrons were observed to be emitted from a thin lead plate exposed to cosmic rays from outer space, which contain highly energetic X-rays. Pair production occurs when an X-ray photon comes near a nucleus—the nuclei of the lead atoms of the plate in the Anderson experiment. In this process, the photon goes out of existence, and an *electron–positron pair* (an electron and a positron) is created, as illustrated in ►Fig. 28.14. This result represents a direct conversion of electromagnetic (photon) energy into mass. By the conservation of energy (neglecting the small recoil kinetic energy of the massive nearby nucleus),

$$hf = 2m_e c^2 + K_{e^-} + K_{e^+}$$

where hf is energy of the photon, $2m_e c^2$ is the total mass–energy equivalent of the electron–positron pair, and the Ks represent the kinetic energies of the produced particles.

The minimum energy to produce such a pair occurs when they are produced at rest—when K_{e^-} and K_{e^+} are zero. Thus, the *minimum* photon energy to produce an electron–positron pair is

$$E_{\min} = hf = 2m_e c^2 = 1.022 \text{ MeV} \tag{28.7}$$

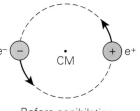

Magnetic *B* field
(into page)

Photon

Nucleus

◀ **FIGURE 28.14 Pair production** An electron (e⁻)
and a positron (e⁺) can be created when an
energetic photon passes near a heavy nucleus.

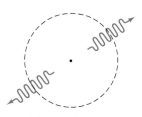

Before annihilation
(positronium atom)

After annihilation
(photons)

▲ **FIGURE 28.15 Pair annihilation**
A slow positron and electron can
form a system called a *positronium
atom*. The disappearance of a
positronium atom is signaled by the
appearance of two photons, each
with an energy of 0.511 MeV. Why
would we not expect one photon
with an energy of 1.022 MeV? (*CM*
stands for *center of mass*.)

(Here, we have used the result from Section 26.5 that says that the mass of an electron is equivalent to an energy of $m_e c^2 = 0.511$ MeV.) This minimum photon energy is called the **threshold energy** for pair production.

But if they are created by cosmic rays, why aren't positrons commonly found in nature? This is because, almost immediately after their creation, positrons go out of existence by a process called **pair annihilation**. When an energetic positron appears, it loses kinetic energy by collision as it passes through matter. Finally, almost at rest, it combines with an electron and forms a hydrogen-like atom, called a *positronium atom*, in which a positron substitutes for a proton. The positronium atom is unstable and quickly decays ($\approx 10^{-10}$ s) into two photons, each with an energy of 0.511 MeV (▶Fig. 28.15). Pair annihilation is then a direct conversion of mass into electromagnetic energy—the inverse of pair production, so to speak. Pair annihilation is the basis for a medical diagnostic tool called a *positron emission tomography* (or PET) *scan*. This application will be discussed in more detail in Chapter 29 after you learn more about nuclear decay processes.

More generally, all particles have antiparticles. For example, there is an antiproton with the same mass as a proton, but with a negative charge. Even a neutral particle such as the neutron has an antiparticle—the antineutron. It is even conceivable that antiparticles predominate in some parts of the universe. If so, atoms made of the **antimatter** in these regions would consist of negatively charged nuclei composed of antiprotons and antineutrons, surrounded by orbiting positively charged positrons (antielectrons). It would be difficult to distinguish a region of antimatter visibly, since the physical behavior of antimatter atoms would presumably be the same as that of ordinary matter.

Chapter Review

- The magnitude of the momentum (*p*) of a photon carrying energy *E* is inversely related to the wavelength of the associated light wave by

$$p = \frac{E}{c} = \frac{hf}{c} = \frac{h}{\lambda} \tag{28.1}$$

- The **de Broglie hypothesis** assigns a wavelength to material particles, by analogy with the assignment of momentum to photons. The **de Broglie wavelength** of a particle is

$$\lambda = \frac{h}{p} = \frac{h}{mv} \tag{28.2}$$

- A quantum-mechanical **wave function** ψ is a "probability wave" associated with a particle. The probability density, the square of the wave function, gives the relative probability of finding a particle at a particular location.

- Electron-orbital energies are determined primarily by the **principal quantum number** *n*.

- The quantum number ℓ is called the **orbital quantum number** and is associated with the orbital angular momentum of the electron's orbit. For each value of *n*, the ℓ quantum number has one of *n* possible integer values from zero up to a maximum value of $n - 1$.

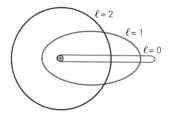

• The quantum number m_ℓ is called the **magnetic quantum number** and is associated with the z-component of the electron's orbital angular momentum. For a given ℓ, there are $2\ell + 1$ possible m_ℓ values (integers), given by $m_\ell = 0, \pm1, \pm2, \ldots, \pm\ell$.

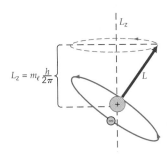

• The quantum number that describes an electron's intrinsic angular momentum is the **spin quantum number** m_s, which, for electrons, protons, and neutrons, can have only two values: $m_s = \pm\frac{1}{2}$. These values correspond to the two angular momentum directions of spin: up and down.

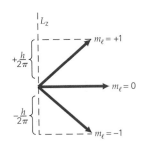

• Orbits with different quantum numbers that have the same energy are **degenerate**.

• Orbits that share a common principal quantum number n are in the same **shell**.

• Orbits that share a common orbital quantum number ℓ are in the same **subshell**.

• The **Pauli exclusion principle** states that in a given atom, no two electrons can have exactly the same set of quantum numbers.

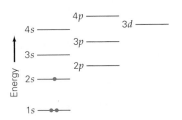

• The **Heisenberg uncertainty principle** states that you cannot simultaneously measure both the position and the momentum (or velocity) of a particle exactly. The same condition holds true for the particle's energy and the period of time during which that energy is measured. The uncertainties satisfy

$$(\Delta p)(\Delta x) \geq \frac{h}{2\pi} \qquad (28.5)$$

and

$$(\Delta E)(\Delta t) \geq \frac{h}{2\pi} \qquad (28.6)$$

• **Pair production** refers to the creation of a particle and its **antiparticle**. The reverse process is **pair annihilation**, in which a particle and its antiparticle annihilate and their energy is converted into two photons.

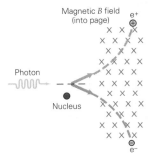

Exercises*

MC = *Multiple Choice Question,* **CQ** = *Conceptual Question, and* **IE** = *Integrated Exercise. Throughout the text, many exercise sections will include "paired" exercises. These exercise pairs, identified with* **red numbers***, are intended to assist you in problem solving and learning. In a pair, the first exercise (even numbered) is worked out in the Study Guide so that you can consult it should you need assistance in solving it. The second exercise (odd numbered) is similar in nature, and its answer is given at the back of the book.*

28.1 Matter Waves: The de Broglie Hypothesis

1. **MC** Photons associated with which color light would have the largest momentum: (a) red, (b) green, or (c) violet? (c)

2. **MC** If the following were traveling at the same speed which would have the shortest de Broglie wavelength: (a) an electron, (b) a proton, (c) a carbon atom, or (d) a hockey puck? (d)

3. **MC** You have three neutrons traveling at different speeds. Which speed would be associated with the shortest neutron de Broglie wavelength: (a) 10^3 m/s, (b) 10^4 m/s, (c) 10^2 m/s, or (d) the wavelengths are the same at all speeds? (b)

*Note: Use Eq. 28.3 for λ where appropriate.

4. **CQ** The de Broglie hypothesis predicts that a wave is associated with any object that has momentum. Why don't we observe the wave associated with a moving car? its wavelength is too short compared with everyday dimensions

5. **CQ** If a baseball and a bowling ball were traveling at the same speed, which one would have a longer de Broglie wavelength? Why? baseball, as its momentum is smaller

6. **CQ** An electron is accelerated from rest through an electric-potential difference. Will increasing the difference in potential result in a longer or shorter de Broglie wavelength? Why? shorter, as a higher difference in potential causes a higher momentum

7. ● What is the de Broglie wavelength associated with a 1000-kg car moving at 25 m/s? 2.7×10^{-38} m

8. **IE** ● An electron and a proton are moving with the same speed. (a) Compared with the proton, will the electron have (1) a shorter, (2) an equal, or (3) a longer de Broglie wavelength? Why? (b) If the speed of the electron and proton is 100 m/s, what are their de Broglie wavelengths? (a) (3) a longer (b) electron: 7.28×10^{-6} m; proton: 3.97×10^{-9} m

9. ●● A proton and an electron are accelerated from rest through the same difference in potential, V. What is the ratio of the de Broglie wavelength of an electron to that of a proton (to two significant figures)? $\lambda_e/\lambda_p = 43$

10. **IE** ●● Electrons are accelerated from rest through an electric-potential difference. (a) If this potential difference increases to nine times the original value, the new de Broglie wavelength will be (1) nine times, (2) three times, (3) one ninth, (4) one third that of the original. Why? (b) If the original potential is 250 kV and the new potential is 600 kV, what is the ratio of the new de Broglie wavelength to the original? (a) (4) one third (b) 0.645

11. ●● An electron is accelerated from rest through a potential difference so that its de Broglie wavelength is 0.010 nm. What is the potential difference? 1.5×10^4 V

12. ●● A charged particle is accelerated through a potential difference V. If the voltage were doubled, what would be the ratio of the new de Broglie wavelength to the original value? $\lambda_2/\lambda_1 = 1/\sqrt{2} \approx 0.71$

13. **IE** ●● A proton traveling at a speed of 4.5×10^4 m/s is accelerated through a potential difference of 37 V. (a) Will its de Broglie wavelength (1) increase, (2) remain the same, or (3) decrease, due to the potential difference? Why? (b) By what percentage does the de Broglie wavelength of the proton change? (a) (3) decrease (b) -53% (a decrease)

14. ●● What is the energy of a beam of electrons that exhibits a first-order maximum at an angle of 25° when diffracted by a crystal grating with a lattice plane spacing of 0.15 nm? 3.7×10^2 eV

15. ●● A scientist wants to use an electron microscope to observe details on the order of 0.25 nm. Through what potential difference must the electrons be accelerated from rest so that they have a de Broglie wavelength of this magnitude? 24 V

16. ●●● According to the Bohr theory of the hydrogen atom, the speed of the electron in the first Bohr orbit is 2.19×10^6 m/s. (a) What is the wavelength of the matter wave associated with the electron? (b) How does this wavelength compare with the circumference of the first Bohr orbit? (a) 3.32×10^{-10} m (b) same, $\lambda = 2\pi r_1$, because it is a standing wave

17. ●●● (a) What is the de Broglie wavelength of the Earth in its orbit about the Sun? (b) Treating the Earth as a de Broglie wave in a large "gravitational" atom, what would be the principal quantum number, n, of its orbit? (c) If the principal quantum number increased by 1, how would the radius of the orbit change? (Assume a circular orbit.) see ISM

28.2 The Schrödinger Wave Equation

18. **MC** The wave-function solution to the Schrödinger equation's description of a single particle (a) is the particle's de Broglie wavelength, (b) tells us the probability of finding a particle at a given location, (c) functionally describes the de Broglie wave of a particle, (d) none of the preceding. (c)

19. **MC** The square of a particle's wave function (a) is the energy of the particle, (b) is the probability of locating the particle, (c) tells us the quantum number of its state, (d) provides the basis of the Pauli exclusion principle. (b)

20. **MC** In the scanning tunneling microscope (STM), how does the tunneling current change if the tip gets nearer to the surface? (a) It increases, (b) it decreases, (c) it doesn't change, or (d) its change depends on the type of surface being explored. (a)

21. **CQ** Explain how you would program the tip of an scanning tunnel microscope (STM) to move if it encounters a dip in the surface being explored and you want to keep the tunneling current constant. Move toward the surface to make up for decreasing current

22. **CQ** How would the maximum of the graph in Fig. 28.3 a change if the charge on the proton in the hydrogen atom were suddenly increased? Explain your reasoning see ISM

23. **CQ** According to modern quantum theory and the Schrödinger equation, there is a probability that if you ran into a wall you could end up on the other side. (Don't try this!) Explain the idea behind this and discuss why this has never been observed to happen (except in comic books and cartoons). see ISM

24. ●● A particle in box is constrained to move in one dimension, like a bead on a wire, as illustrated in ▼Fig. 28.16. Assume that no forces act on the particle in the interval $0 < x < L$ and that it hits a perfectly rigid wall. The particle will exist only in states of a certain kinetic energies that can be determined in analogy to a standing wave on a string (Chapter 13, Section 13.5). This means an integral number n of half-wavelengths "fit" into the box's length L, or $n\left(\dfrac{\lambda_n}{2}\right) = L$ where $n = 1, 2, 3, \ldots$. Using this relationship, show that the "allowed" kinetic energies K_n of the particle are given by $K_n = n^2\left[h^2/\left(8mL^2\right)\right]$ where $n = 1, 2, 3, \ldots$ and m is the particle's mass. [*Hint*: Recall that kinetic energy is related to momentum by $K = p^2/(2m)$ and that the de Broglie wavelength of the particle is also related to its momentum.] see ISM

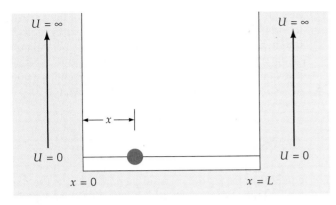

▲ **FIGURE 28.16 Particle in a box** See Exercises 24 and 25.

25. ●● Let's model a nucleus as a particle trapped in the one-dimensional box. Assume the particle is a proton and it is in a one dimensional nucleus of length of 7.11 fm (the approximate diameter of a Pb-208 nucleus). (a) Using the results of Exercise 24, determine the energies of the proton in the ground state and first two excited states. (b) The nucleus is to absorb a photon of just the right energy to enable the proton to make an upward transition from the ground state to the second excited state. How much energy would this be and what type of photon would it be classified as? (Neglect recoil of the absorbing nucleus after the photon energy is absorbed.) see ISM

28.3 Atomic Quantum Numbers and the Periodic Table

26. **MC** The quantum number ℓ (a) determines the total energy of a state, (b) is associated with the angular momentum of the electron in orbit, (c) is associated with the orientation of the angular-momentum vector, (d) is associated with electron spin. (b)

27. **MC** The quantum number m_ℓ (a) determines the energy of the electron, (b) tells whether the electron is spinning up or down, (c) tells the orientation of the angular-momentum vector of the electron in orbit, (d) all of the preceding. (c)

28. **MC** The quantum number m_s (a) can take on only two different values for an electron, (b) arises from the orbital motion of an electron, (c) is needed because the electron actually spins like a ball, (d) all of the preceding. (a)

29. **CQ** What information does the quantum number n give for a hydrogen atom? total energy and orbital radius

30. **CQ** Niels Bohr set forth a *correspondence principle*, which states that the results of quantum mechanics and classical physics approach agreement when quantum numbers become very large. Discuss this principle in terms of the hydrogen atom. see ISM

31. **CQ** What is the basis of the periodic table of elements in terms of quantum theory, and what do the elements in a particular group have in common? see ISM

32. ● (a) How many possible sets of quantum numbers are there for $n = 2$ and $n = 3$ shells? (b) Write the explicit values of all the quantum numbers (n, ℓ, m_ℓ, m_s) for these levels. (a) 8, 18 (b) see ISM

33. ● How many possible sets of quantum numbers are there for the subshells with (a) $\ell = 0$ and (b) $\ell = 3$? (a) 2 (b) 14

34. **IE** ● (a) Which has more possible sets of quantum numbers associated with it, $n = 2$ or $\ell = 2$? (b) Prove your answer to part (a). (a) $\ell = 2$ (b) 8 for $n = 2$; 10 for $\ell = 2$

35. ● An electron in an atom is in an orbit that has a magnetic quantum number of $m_\ell = 2$. What are the minimum values that (a) ℓ and (b) n could be for that orbit? (a) $\ell = 2$ (b) $n = 3$

36. ●● Draw the ground-state energy-level diagrams like those in Fig. 28.8 for (a) nitrogen (N) and (b) potassium (K). see ISM

37. ●● Draw schematic diagrams for the electrons in the subshells of (a) sodium (Na) and (b) argon (Ar) atoms in the ground state. see ISM

38. ●● Identify the atoms of each of the following ground-state electron configurations: (a) $1s^2 2s^2$; (b) $1s^2 2s^2 2p^3$; (c) $1s^2 2s^2 2p^6$; (d) $1s^2 2s^2 2p^6 3s^2 3p^4$. (a) Be (b) N (c) Ne (d) S

39. ●● Write the ground-state electron configurations for each of the following atoms: (a) boron (B), (b) calcium (Ca), (c) zinc (Zn), and (d) tin (Sn). see ISM

40. **IE** ●●● (a) If there were no electron spin, the $1s$ state would contain a maximum of (1) zero, (2) one, (3) two electrons. Why? (b) What would be the first two inert or noble gases if there were no electron spin? (a) (2) one (b) hydrogen and boron

41. ●●● How would the electronic structure of lithium differ if electron spin were to have three possible orientations instead of just two? it would be $1s^3$, see ISM

28.4 The Heisenberg Uncertainty Principle

42. **MC** If the uncertainty in the position of a moving particle increases, (a) the particle may be located more exactly, (b) the uncertainty in its momentum decreases, (c) the uncertainty in its velocity increases, (d) none of the preceding. (b)

43. **MC** According to the uncertainty principle, measurement of the exact energy of a particle requires (a) special equipment, (b) an infinite amount of time, (c) uncertainty in the momentum, (d) none of the preceding. (b)

44. **CQ** Why is it impossible to simultaneously and accurately measure the position and velocity of a particle? see ISM

45. **CQ** A bowling ball has well-defined position and speed, whereas an electron does not. Why? see ISM

46. ● A 0.50-kg ball has a position of 5.0 m $\pm$ 0.01 m. To what minimum uncertainty can its momentum be measured? 5.3×10^{-33} kg·m/s

47. **IE** ● An electron and a proton each have a momentum of 3.28470×10^{-30} kg·m/s $\pm$ 0.00025 $\times 10^{-30}$ kg·m/s. (a) The minimum uncertainty in the position of the electron compared with that of the proton will be (1) larger, (2) the same, (3) smaller. Why? (b) Calculate the minimum uncertainty in the position for each. (a) (2) the same (b) both 0.21 m

48. ● What is the minimum uncertainty in the velocity of an electron that is known to be somewhere between 0.050 nm and 0.10 nm from a proton? 2.3×10^6 m/s

49. ● What is the minimum uncertainty in the velocity of a 0.50-kg ball that is known to be at 1.0000 cm $\pm$ 0.0005 cm from the edge of a table? 2.1×10^{-29} m/s

50. ●● The energy of a 2.00-keV electron is known to within $\pm 3.00\%$. How accurately can its position be measured? within 1.5×10^{-10} m

51. ●● If an excited state of an atom has a lifetime of 1.0×10^{-7} s, what is the minimum error associated with the measurement of the energy of this state? 1.1×10^{-27} J

52. ●● The energy of the first excited state of a hydrogen atom is -0.34 eV $\pm$ 0.0003 eV. What is the average lifetime for this state? 1.1×10^{-12} s

53. IE ●● (a) If the lifetime of excited state A is longer than that of state B, then the width of a spectral line due to natural broadening for a transition from state A to state B will be (1) smaller than, (2) the same as, (3) greater than that for a transition from state B. Why? (b) Calculate the ratio of the width of a spectral line due to natural broadening for a transition from an excited state with a lifetime of 10^{-12} s to that for a state with a lifetime of 10^{-8} s. (a) (1) smaller than (b) 10^4

28.5 Particles and Antiparticles

54. **MC** Pair production involves (a) the production of two electrons, (b) the production of two positrons, (c) a positronium atom, (d) the production of a particle and its antiparticle. (d)

55. **MC** Due to momentum considerations, pair annihilation cannot result in the emission of how many photons? (a) one (b) two (c) three (a)

56. **MC** A new particle/antiparticle is found to require a minimum photon energy of 25 MeV to occur. How do their masses compare to that of an electron: (a) They are less massive than an electron, (b) they are more massive than an electron, (c) they would have the same mass as an electron, or (d) you can't tell from the data given? (b)

57. **CQ** Why is the energy threshold for electron–positron pair production actually higher than the sum of their two masses (1.022 MeV in energy terms)? [*Hint*: Linear momentum must be conserved.] see ISM

58. **CQ** Can the production of two electrons and two positrons be accomplished using a high-energy photon? Explain. Why can't two electrons and one positron result? see ISM

59. IE ● A photon with an energy of 1.94 MeV comes near a heavy nucleus. (a) What will happen? (1) Pair production will not happen, (2) pair production can happen with the resulting particles at rest, or (3) pair production can happen but the resulting particles must be moving. (b) How much kinetic energy will each particle have after creation? (a) see ISM (b) 0.46 MeV

60. ● What is the energy of the photons produced in electron–positron pair annihilation, assuming that both particles are essentially at rest initially? 0.511 MeV

61. ● What is the threshold energy for the production of a proton–antiproton pair? 1.9 GeV

62. IE ●● A muon, or μ meson, has the same charge as an electron, but is 207 times as massive. (a) Compared with electron–positron pair production, the pair production of a muon and an antimuon requires a photon of (1) more, (2) the same amount of, (3) less energy. Why? (b) What would be the minimum energy for such a photon? (a) (1) more (b) 212 MeV

Comprehensive Exercises

63. An electron traveling at 3.00×10^4 m/s is further accelerated by a potential difference so as to reduce its de Broglie wavelength to one third of its original value. How much voltage is required to accomplish this? 27.6 V

64. There are negatively charged particles that have a spin quantum number, $m_s = 3/2$. (a) How many different spin orientations would these have: (1) two, (2) three, or (3) four? (b) If you were building an atom with these particles orbiting the nucleus, how many would the "atom" have if it were to be the first "inert" atom in the periodic chart? (a) (3) four (b) four

65. Suppose a starship had a mass of 1.25×10^9 kg and was initially at rest. If its "matter–antimatter engines" produced photons from electron–positron annihilation and focused them to travel backward out from the ship, how many photons would they have to emit to reach a speed of 3.00×10^5 m/s (0.100% the speed of light)? [*Hint*: Use conservation of linear momentum and remember that relativity is not needed here. (Why?)] 1.37×10^{36} photons

66. Using a typical nuclear diameter of 4.25×10^{-15} m as its location uncertainty, compute the uncertainty in momentum and kinetic energy associated with an electron if it were part of the nucleus. For energies greater than a few MeV, particles such as electrons would escape the nucleus. What does this tell you about the likelihood that an electron resides in the nucleus of an atom? $\Delta p = 2.48 \times 10^{-20}$ kg·m/s; $\Delta K = 2.11 \times 10^3$ MeV so would not stay

PHYSICS FACTS

- Spent nuclear fuel rods from nuclear reactors are laden with radioactive nuclei. Many of these are chemically separated and used in medical and industrial applications.

- Many fission fragments are potentially harmful to living things if ingested at high levels. For example, I-131, used as a diagnostic tool for thyroid cancer, can actually cause thyroid cancer at high levels of exposure.

- The radioactive nuclide americium-241, used in most smoke detectors, is actually created artificially. None exists naturally as its half-life is only about 400 years.

- A lengthy plane flight at high altitude can expose passengers to the amount of radiation energy dosage (from cosmic rays) comparable to that of a chest X-ray.

- Of the yearly "dose" of radiation, more than half is due to natural background radiation, the rest from sources such as medical X-rays.

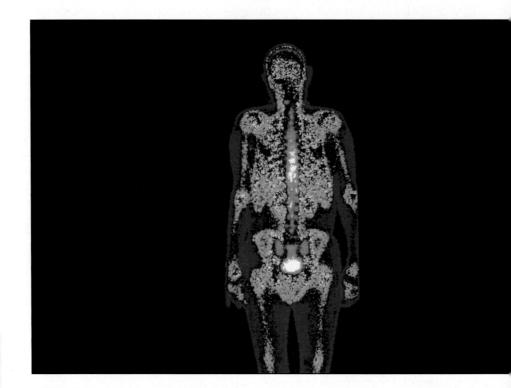

The skeletal image (a bone scan) in the chapter-opening photograph was created by radiation from a radioactive source. *Radiation* and *radioactivity* are words that sometimes produce anxiety, but the beneficial uses of radiation are often overlooked. For instance, exposure to high-energy radiation can cause cancer—yet precisely the same sort of radiation, in relatively small doses, can be useful in the diagnosis and treatment of cancer.

The bone scan in the chapter-opening photo was created by radiation released when unstable nuclei spontaneously broke apart after being administered to and taken up by the body—a process we call *radioactive decay*. But what makes some nuclei stable, while others decay? What determines the rate at which they break down and the particles that they emit? These are some of the questions that will be explored in this chapter. We'll also learn how radiation is detected and measured, as well as more about its dangers and uses.

In addition, the study of radioactivity and nuclear stability helps us understand the nature of the nucleus, its structure, its energy, and how this energy can be released. Nuclear energy is one of our major energy sources, and will be considered in Chapter 30. In this chapter we concentrate on understanding the nucleus itself.

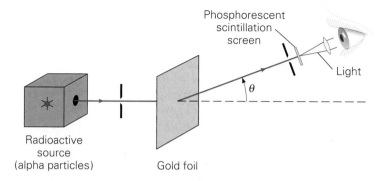

◀ **FIGURE 29.1** Rutherford's
scattering experiment A beam of
alpha particles from a radioactive
source was scattered by gold nuclei
in a thin foil, and the scattering was
observed as a function of the
scattering angle θ. The observer
detects the light (viewed through a
lens) given off by a phosphorescent
scintillation screen.

29.1 Nuclear Structure and the Nuclear Force

OBJECTIVES: To (a) distinguish between the Thomson and Rutherford–Bohr models
of the atom, (b) specify some of the basic properties of the strong
nuclear force, and (c) understand nuclear notation.

It is evident from the emission of electrons from heated filaments (called *thermionic
emission*) and the photoelectric effect that atoms contain electrons. Since an atom is
normally electrically neutral, it must contain a positive charge equal in magnitude
to the total charge of its electrons. Since the electron's mass is small compared with
the mass of even the lightest of atoms, most of an atom's mass appears to be associated with that positive charge.

Based on these observations, J. J. Thomson (1856–1940), a British physicist who
had experimentally proven the existence of the electron in 1897, proposed a model
of the atom. In his model, the electrons are uniformly distributed within a continuous
sphere of positive charge. It was called a "plum pudding" model, because the electrons in the positive charge are analogous to raisins in a plum pudding. The region of
positive charge was assumed to have a radius on the order of 10^{-10} m, or 0.1 nm, roughly the diameter of an atom.

As you probably know, our modern model of the atom is quite different. This
model concentrates all the positive charge, and practically all of the mass, in a central *nucleus*, surrounded by orbiting electrons. The existence of such a nucleus was
first proposed by British physicist Ernest Rutherford (1871–1937). Combining this
idea with the Bohr theory of electron orbits (Section 27.4) led to the simplistic
"solar system" model, or **Rutherford–Bohr model**, of the atom.

Rutherford's insight came from the results of alpha-particle scattering experiments
performed in his laboratory about 1911. An alpha (α) particle is a doubly positively
charged particle ($q_\alpha = +2e$) that is naturally emitted from some radioactive materials.
(See Section 29.2.) A beam of these particles was directed at a thin gold-foil "target," and
the deflection angles and percentage of scattered particles were observed (▲Fig. 29.1).

An alpha particle is more than 7000 times as massive as an electron. Thus, the
Thomson model predicts only tiny deflections—the result of collisions with the light
electrons as an alpha particle passes through such a model of a gold atom (▶Fig. 29.2).
Surprisingly, however, Rutherford observed alpha particles scattered at appreciable
angles. In about 1 in every 8000 scatterings, the alpha particles were actually
backscattered; that is, they were scattered through angles greater than 90° (▼Fig. 29.3).

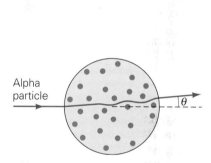

▲ **FIGURE 29.2 The plum pudding
model** In Thomson's plum
pudding model of the atom,
massive alpha particles were
expected to be only slightly
deflected by collisions with the
electrons (blue dots) in the atom.
The experimental results were quite
different.

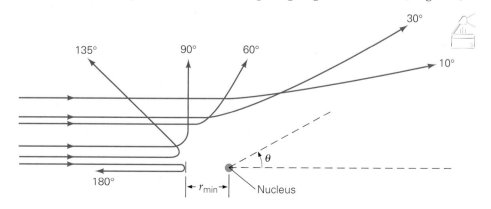

◀ **FIGURE 29.3** Rutherford
scattering A compact, dense
atomic nucleus with a positive
charge accounts for the observed
scattering. An alpha particle in a
head-on collision with the nucleus
would be scattered directly
backward ($\theta = 180°$) after coming
within a distance r_{min} of the nucleus.
At this scale, the electron orbits
(about the nucleus) are too far away
to be seen.

Calculations showed that the probability of backscattering in the Thomson model was minuscule—certainly much, much less than 1 in 8000. As Rutherford described the backscattering, "It was almost as incredible as if you had fired a 15-inch shell at a piece of tissue paper and it came back and hit you."

The experimental results led Rutherford to the concept of a nucleus: "On consideration, I realized that this scattering backward must be the result of a single collision, and when I made calculations I saw that it was impossible to get anything of that order of magnitude unless you took a system in which the greater part of the mass of the atom was concentrated in a minute nucleus. It was then that I had the idea of an atom with a minute massive center carrying a charge."

Teaching tip: A review of electrostatics may be necessary.

If all of the positive charge of a target atom were concentrated in a small region, then an alpha particle coming close to this region would experience a large deflecting (electrical repulsion) force. The mass of this positive "nucleus" would be larger than that of the alpha particle, and in this model backscattering is much more likely to occur than in the plum pudding model.

A simple estimate can give an idea of the approximate size of a nucleus. It is during a head-on collision that an alpha particle comes closest to the nucleus (a distance labeled as r_{min} in Fig. 29.3). That is, the alpha particle approaching the nucleus stops at r_{min} and is accelerated back along its original path. Assuming a spherical charge distribution, the electric potential energy of the alpha particle (α) and nucleus (n) when separated by a center-to-center distance r is $U = kq_\alpha q_n/r = k(2e)(Ze)/r$ (Eq. 16.5). Here, Z is the **atomic number**, or the number of protons in the nucleus. Therefore the charge of the nucleus is $q_n = +Ze$. By conservation of energy, the kinetic energy of the incoming alpha particle is completely converted into electric potential energy at the turnaround point, r_{min}. Using $q_\alpha = +2e$, we equate the two energies, which results in

$$\tfrac{1}{2}mv^2 = \frac{k(2e)Ze}{r_{min}}$$

or, solving for r_{min}, we obtain

$$r_{min} = \frac{4kZe^2}{mv^2} \tag{29.1}$$

In Rutherford's experiment, the kinetic energy of the alpha particles from the particular source had been measured, and Z was known to be 79 for gold. Using these values, along with the constants in Eq. 29.1, Rutherford found r_{min} to be on the order of 10^{-14} m.

Although the nuclear model of the atom is useful, the nucleus is much more than a volume of positive charge. The nucleus is actually composed of two types of particles—protons and neutrons—collectively referred to as **nucleons**. The nucleus of the hydrogen atom is a single proton. Rutherford suggested that the hydrogen nucleus be named *proton* (from the Greek meaning "first") after he became convinced that no nucleus could be less massive than the hydrogen nucleus. A **neutron** is an electrically neutral particle with a mass slightly greater than that of a proton. The existence of the neutron was not experimentally verified until 1932.

The Nuclear Force

Of the forces in the nucleus, there is certainly the attractive gravitational force between nucleons. But in Chapter 15, this gravitational force was shown to be negligible compared with the repulsive electrical force between the positive protons. Taking only these repulsive forces into account, it would be predicted that the nucleus should fly apart. Yet the nuclei of many atoms are stable. Therefore, there must be an *attractive* force between nucleons that overcomes the electrical repulsion and thereby holds the nucleus together. This strong attractive force is called the **strong nuclear force**, or simply the *nuclear force*.

The exact expression for the nuclear force is extremely complex. However, some general features of it are as follows:

Teaching tip: You may wish to mention that there is also a weak nuclear force, which is weaker than the strong nuclear force, but much stronger than gravity. It is involved in certain types of radioactive decay and is discussed in Chapter 30.

- The nuclear force is strongly attractive and much larger in magnitude than both the electrostatic force and the gravitational force between nucleons.

- The nuclear force is very short-ranged; that is, a nucleon interacts only with its nearest neighbors, over distances on the order of 10^{-15} m.
- The nuclear force is independent of electric charge; that is, it acts between *any* two nucleons—two protons, a proton and a neutron, or two neutrons.

Thus, nearby protons repel each other electrically, but attract each other (and nearby neutrons) by the strong force, with the latter winning the battle. Having no electric charge, neutrons only attract nearby protons and neutrons.

Nuclear Notation

To describe the nuclei of different atoms, it is convenient to use the notation illustrated in ▶Fig. 29.4a. The chemical symbol of the element is used with subscripts and a superscript. The subscript on the left is called the *atomic number* (Z), which indicates the number of protons in the nucleus. A more descriptive name for the symbol Z is **proton number**, which will be used in this book. For electrically neutral atoms, Z is equal to the number of orbital electrons. (Why?)

The number of protons in the nucleus of an atom determines the species of the atom—that is, the element to which the atom belongs. In Fig. 29.4b, the proton number $Z = 6$ indicates that the nucleus belongs to a carbon atom. The proton number thus defines which chemical symbol is used. Electrons can be removed from (or added to) an atom to form an ion, *but this does not change the atom's species*. For example, a nitrogen atom with an electron removed, N^+, is still nitrogen—a nitrogen *ion*. It is the proton number, rather than the electron number, that determines the species of atom.

The superscript to the left of the chemical symbol is called the **mass number (A)**—the total number of protons and neutrons in the nucleus. Since protons and neutrons have roughly equal masses, the mass numbers of nuclei give a relative comparison of nuclear masses. For the carbon nucleus in Fig. 29.4b, the mass number is $A = 12$, because there are six protons and six neutrons. The number of neutrons, called the **neutron number (N)**, is sometimes indicated by a subscript on the right side of the chemical symbol. However, this subscript is usually omitted, because it can be calculated from A and Z; that is, $N = A - Z$. Similarly, the proton number is routinely omitted, because the chemical symbol uniquely specifies the value of Z.

Even though the atoms of an element all have the same number of protons in their nuclei, they may have different numbers of neutrons. For example, nuclei of different carbon atoms ($Z = 6$) may contain six, seven, or eight neutrons. In nuclear notation, these atoms would be written as $^{12}_{6}C_6$, $^{13}_{6}C_7$, and $^{14}_{6}C_8$, respectively. Atoms whose nuclei have the same number of protons, but different numbers of neutrons, are called **isotopes**. The three atoms we just listed are three isotopes of carbon.

Isotopes are like members of a family. They all have the same Z number and the same surname (element name), but they are distinguishable by the number of neutrons in their nuclei and, therefore by their mass. Isotopes are referred to by their mass numbers; for example, these isotopes of carbon are called *carbon-12*, *carbon-13*, and *carbon-14*, respectively. There are other isotopes of carbon that are unstable: ^{11}C, ^{15}C, and ^{16}C. A particular nuclear species or isotope of any element is also called a **nuclide**. So far, we have mentioned six nuclides of carbon. Generally, only a few isotopes of a given species are stable. But this number can vary from none to several or more. In fact, in our carbon example, ^{14}C is unstable, although it is long lived. Only ^{12}C and ^{13}C, in fact, are truly stable isotopes of carbon.

Another important family of isotopes is that of hydrogen, which has three isotopes: ^{1}H, ^{2}H, and ^{3}H. These isotopes are given special names. ^{1}H is called *ordinary hydrogen*, or simply *hydrogen*; ^{2}H is called *deuterium*. Deuterium, which is stable, is sometimes known as *heavy hydrogen*. It can combine with oxygen to form heavy water (written D_2O). The third isotope of hydrogen, ^{3}H, called *tritium*, is unstable.

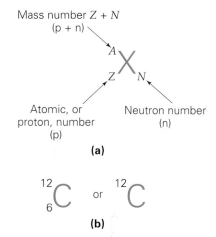

▲ **FIGURE 29.4** Nuclear notation **(a)** The composition of a nucleus is shown by the chemical symbol of the element with the mass number A (sum of protons and neutrons) as a left superscript and the proton (atomic) number Z as a left subscript. The neutron number N may be shown as a right subscript, but both Z and N are routinely omitted, because the letter symbol tells you Z, and $N = A - Z$. **(b)** The two most common nuclear notations for a nucleus of one of the stable isotopes of carbon—carbon-12.

Note: All the isotopes in one family have almost the same orbital-electron structure and thus very similar chemical properties.

29.2 Radioactivity

To (a) define the term *radioactivity*; (b) distinguish among alpha, beta, and gamma decay; and (c) write nuclear-decay equations.

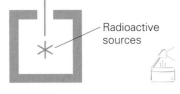

Photographic plate

α   γ   β

Magnetic field (into page)

Radioactive sources

▲ **FIGURE 29.5 Nuclear radiation**
Different types of radiation from radioactive sources can be distinguished by passing them through a magnetic field. Alpha and beta particles are deflected. From the right-hand magnetic-force rule, alpha particles are positively charged and beta particles are negatively charged. The radii of curvature (not drawn to scale) allow the particles to be distinguished by mass. Gamma rays are not deflected and thus are uncharged; they are quanta of electromagnetic energy.

Teaching tip: Make sure that students can distinguish clearly among the three types of radioactive decay.

Most elements have at least one stable isotope. It is atoms with stable nuclides with which we are most familiar in the environment. However, some nuclei are unstable and disintegrate spontaneously (or decay), emitting energetic particles and photons. Unstable isotopes are said to be *radioactive* or to exhibit **radioactivity**. For example, tritium ($^3_1$H) has a radioactive nucleus. Of all the unstable nuclides, only a small number occur naturally. Others can be produced artificially (Chapter 30).

Radioactivity is unaffected by normal physical or chemical processes, such as heat, pressure, and chemical reactions. Processes such as these simply do *not* affect the source of the radioactivity—the nucleus. Nor can nuclear instability be explained by a simple imbalance of attractive and repulsive forces within the nucleus. This is because, experimentally, nuclear disintegrations (of a given isotope) occur at a fixed rate. That is, the nuclei in a given sample do not all decay at the same time. According to classical theories, identical nuclei *should* decay at the same time. Therefore, radioactive decay suggests that the probability effects of quantum mechanics might be in play.

The discovery of radioactivity is credited to the French scientist Henri Becquerel. In 1896, while studying the fluorescence of a uranium compound, he discovered that a photographic plate near a sample had been darkened, even though the compound had not been activated by exposure to light and was not fluorescing. Apparently, this darkening was caused by some new type of radiation emitted from the compound itself. In 1898, Pierre and Marie Curie announced the discovery of two radioactive elements, radium and polonium, which they had isolated from uranium pitchblende ore.

Experiment shows that the radiation emitted by radioactive isotopes is of three different kinds. When a radioactive isotope is placed in a chamber so that the emitted radiation passes through a magnetic field to a photographic plate (◄Fig. 29.5), the various types of radiation expose the plate, producing characteristic spots by which the types of radiation may be identified. The positions of the spots show that some isotopes emit radiation that is deflected to the left; some emit radiation that is deflected to the right; and some emit radiation that is undeflected. These spots are characteristic of what came to be known as *alpha, beta,* and *gamma* radiations.

From the opposite deflections of two of the types of radiation in the magnetic field, it is evident that positively charged particles are associated with alpha decay and that negatively charged particles are emitted during beta decay. Because of their much smaller deflection, alpha particles must be considerably more massive than beta particles. The undeflected gamma radiation must be electrically neutral. (Why?)

Detailed investigations of the three different radiation types revealed the following:

- **Alpha particles** are actually doubly charged ($+2e$) particles that contain two protons and two neutrons. They are identical to the nucleus of the helium atom ($^4_2$He).
- **Beta particles** are electrons (positive electrons or *positrons* were discovered later).
- **Gamma rays** are high-energy quanta of electromagnetic energy (photons).

For a few radioactive elements, two spots are found on the film, indicating that the elements decay by two different modes. Let's now look at some details of each of these three modes of decay.

Alpha Decay

When an alpha particle is ejected from a radioactive nucleus, the nucleus loses two protons and two neutrons, so the mass number (A) is decreased by four ($\Delta A = -4$). The proton number (Z) is also decreased by two ($\Delta Z = -2$). Because the *parent nucleus* (the original nucleus) loses two protons, the *daughter nucleus* (the resulting nucleus) is the nucleus of a different element, defined by the new proton number. Thus, the **alpha-decay** process is one of nuclear *transmutation*, in which the nuclei of one element change into the nuclei of a lighter element.

An example of an isotope, or nuclide, that undergoes alpha decay is polonium-214. The decay process is represented as a nuclear equation (usually written without neutron numbers):

$$^{214}_{84}\text{Po} \longrightarrow {}^{210}_{82}\text{Pb} + {}^{4}_{2}\text{He}$$
$$\text{\textit{polonium}} \qquad \text{\textit{lead}} \qquad \text{\textit{alpha particle}}$$
$$\text{\textit{(helium nucleus)}}$$

Notice that both the mass-number and proton-number totals are equal on each side of the equation: $(214 = 210 + 4)$ and $(84 = 82 + 2)$, respectively. This condition reflects the experimental facts that *two conservation laws apply to all nuclear processes*. The first is the **conservation of nucleons**:

| The total number of nucleons (A) remains constant in any nuclear process. |

The second is the familiar **conservation of charge**:

| The total charge remains constant in any nuclear process. |

These conservation laws allow us to predict the composition of the daughter nucleus, as the following Example illustrates.

Example 29.1 ■ Uranium's Daughter: Alpha Decay

A $^{238}_{92}\text{U}$ nucleus undergoes alpha decay. What is the resulting daughter nucleus?

Thinking It Through. Nucleon conservation allows the prediction of the daughter's proton number. From that, the element's name can be determined from the periodic table.

Solution. Since $\Delta Z = -2$ for alpha decay, the parent uranium-238 (^{238}U) nucleus loses two protons, and the daughter nucleus has a proton number $Z = 92 - 2 = 90$, which is thorium's proton number (see the periodic table, Fig. 28.9). The equation for this decay can therefore be written as

$$^{238}_{92}\text{U} \rightarrow {}^{234}_{90}\text{Th} + {}^{4}_{2}\text{He} \qquad \text{or} \qquad {}^{238}_{92}\text{U} \rightarrow {}^{234}_{90}\text{Th} + {}^{4}_{2}\alpha$$

where the helium nucleus is finally written as $^{4}_{2}\alpha$ (sometimes just α).

Follow-Up Exercise. Using high-energy accelerators, it is possible to *add* an alpha particle to a nucleus—essentially the reverse of the reaction in this Example. Write the equation for this nuclear reaction, and predict the identity of the resulting nucleus if an alpha particle is added to a ^{12}C nucleus. *(Answers to all Follow-Up Exercises are at the back of the text.)*

From experiments, it is found that the kinetic energies of alpha particles from radioactive sources are typically a few MeV. (See Section 16.2.) For example, the energy of the alpha particle from the decay of ^{214}Po is about 7.7 MeV, and that from ^{238}U decay is about 4.14 MeV. Alpha particles from such sources were used in the scattering experiments that led to the Rutherford nuclear model.

Outside the nucleus, the repulsive electric force increases as an alpha particle approaches the nucleus. Inside the nucleus, however, the strongly attractive nuclear force dominates. These conditions are depicted in ▸Fig. 29.6, which shows a graph of the potential energy U as a function of r, the distance from the center of the nucleus. Consider alpha particles (with kinetic energy of 7.7 MeV) from a ^{214}Po source incident on ^{238}U (Fig. 29.6). The alpha particles don't have enough kinetic energy to overcome the *electric potential-energy "barrier,"* whose height exceeds 7.7 MeV. Thus, Rutherford scattering occurs. On the other hand, we do know that the ^{238}U nucleus does undergo alpha decay, emitting an alpha particle with an energy of 4.4 MeV, which is below the height of the barrier. How can these lower-energy alpha particles cross a barrier from the inside to the outside, when higher-energy alpha particles cannot cross from outside to the inside? According to classical theory, this is impossible, since it violates the conservation of energy. However, quantum mechanics offers an explanation.

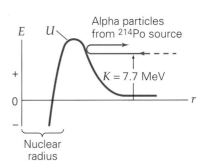

▲ **FIGURE 29.6 Potential-energy barrier for alpha particles** Alpha particles from radioactive polonium with energies of 7.7 MeV do not have enough energy to overcome the electrostatic potential-energy barrier of the ^{238}U nucleus and are scattered.

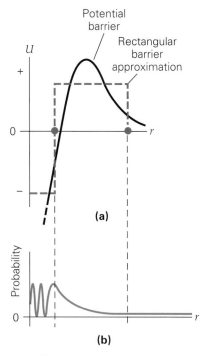

U

+

0 ———————————————— r

−

(a)

Potential barrier

Rectangular barrier approximation

Probability

0 ———————————————— r

(b)

▲ **FIGURE 29.7** Tunneling or barrier penetration **(a)** The potential-energy barrier presented by a nucleus to an alpha particle can be approximated by a rectangular barrier. **(b)** The probability of finding the alpha particle at a given location, according to quantum-mechanical calculations, is shown. If the particle is initially inside the nucleus, it has a likelihood of "tunneling" through the barrier and appearing outside the nucleus. Typically, this event has a very small, but nonzero, probability of occurring for elements above lead on the periodic table.

Teaching tip: Make sure the students realize that if an electron were actually to exist inside a neutron prior to beta decay, the neutron's spin would have to be either zero or one. Since the neutron is a spin-$\frac{1}{2}$ particle, the electron does not exist prior to decay.

Note: Remember that the positron or electron emitted during $\beta^{\pm}$ decay is *not* initially present in the neutron or proton that decays. Among other things, its presence before decay would violate conservation of energy, the uncertainty principle, and conservation of angular momentum. The electron or positron is *created at the time of the decay* and does not exist before that. See Chapter 30 for further details.

Quantum mechanics predicts a nonzero probability of an alpha particle, initially inside the nucleus, to be found *outside* the nucleus (◄Fig 29.7). This phenomenon is called **tunneling**, or **barrier penetration**, since the alpha particle has a certain probability of tunneling through the barrier. (As an example, recall that electrons do this in the scanning tunneling microscope [STM], in Chapter 28.)

Beta Decay

The emission of an electron (a beta particle) in a nuclear-decay process might seem contradictory to the proton–neutron model of the nucleus. Note, however, that the electron emitted in **beta decay** is *not* part of the original nucleus. *The electron is created during the decay.* There are several types of beta decay. When a negative electron is emitted, the process is called β^{-} **decay**. An example of this type of beta decay is that of ^{14}C:

$$\underset{carbon}{^{14}_{6}\text{C}} \quad \rightarrow \quad \underset{nitrogen}{^{14}_{7}\text{N}} \quad + \quad \underset{\substack{beta\ particle\\(electron)}}{^{0}_{-1}\text{e}}$$

The parent nucleus (carbon-14) has six protons and eight neutrons, whereas the daughter nucleus (nitrogen) has seven protons and seven neutrons. Notice that the electron symbol has a nucleon number of zero (because the electron is not a nucleon) and a charge number of −1. Thus, both nucleon number (14) and electric charge (+6) are conserved.

In this type of beta decay, the neutron number of the parent nucleus decreases by one, and the proton number of the daughter nucleus increases by one. Thus, the nucleon number remains unchanged. In essence, it would appear that *a neutron within such an unstable nucleus decays into a proton and an electron (which is then emitted)*:

$$\underset{neutron}{^{1}_{0}\text{n}} \quad \rightarrow \quad \underset{proton}{^{1}_{1}\text{p}} \quad + \quad \underset{electron}{^{0}_{-1}\text{e}} \quad (basic\ \beta^{-}\ decay)$$

Beta decay generally happens when a nucleus is unstable because of having too many neutrons compared to protons. (See Section 29.5, which discusses nuclear stability.) The most massive stable isotope of carbon is ^{13}C, with only seven neutrons. But ^{14}C has too many neutrons for a nucleus with six protons and is unstable. Since beta decay simultaneously decreases the neutron number *and* increases the proton number, the product is more stable. In this case, the product nucleus is ^{14}N, which is stable. For completeness, we note that another elementary particle, called a *neutrino*, is emitted in beta decay. For simplicity, it will not be shown in the nuclear-decay equations here. Its important role in beta decay will be discussed more fully in Chapter 30.

There are actually two modes of beta decay, β^{-} and β^{+}, as well as a third process called *electron capture*. Whereas β^{-} decay involves the emission of an electron, $\boldsymbol{\beta^{+}}$ **decay**, or *positron decay*, involves the emission of a positron. The positron is a positive electron—the antiparticle of the electron (see Section 28.5). A positron is symbolized as $^{0}_{+1}\text{e}$. Nuclei that undergo β^{+} decay have too many protons relative to the number of neutrons. The net effect of β^{+} decay is to convert a proton into a neutron. As in β^{-} decay, this process serves to create a more stable daughter nucleus. An example of β^{+} decay is the following:

$$\underset{oxygen}{^{15}_{8}\text{O}_{7}} \quad \rightarrow \quad \underset{nitrogen}{^{15}_{7}\text{N}_{8}} \quad + \quad \underset{positron}{^{0}_{+1}\text{e}}$$

Positron emission is also accompanied by a neutrino (but a different type from that associated with β^{-} decay), which we will also discuss in Chapter 30.

A process that competes with β^{+} decay is called **electron capture** (abbreviated as **EC**). This process involves the absorption of *orbital* electrons by a nucleus. The net result is the same daughter nucleus that would have been produced by positron decay—hence describes as *competing*. That is, there is usually a certain probability that *both* can happen. A specific example of electron capture is as follows:

$$_{-1}^{0}e + {}_{4}^{7}Be \rightarrow {}_{3}^{7}Li$$

orbital electron   *beryllium*   *lithium*

As in β^+ decay, a proton changes into a neutron, but no beta particle is emitted in electron capture.

Gamma Decay

In **gamma decay**, the nucleus emits a gamma (γ) ray, a high-energy photon of electromagnetic energy. The emission of a gamma ray by a nucleus in an excited state is analogous to the emission of a photon by an excited atom. Most commonly, the nucleus emitting the gamma ray is a daughter nucleus left in an excited state after alpha decay, beta decay, or electron capture.

Nuclei possess energy levels analogous to those of atoms. However, nuclear energy levels are much farther apart and more complicated than those of an atom. The nuclear energy levels are typically separated by *kilo*electron-volts (keV) and *mega*electron-volts (MeV), rather than the few electron-volts (eV) that separate energy levels in atoms. As a result, gamma rays are very energetic, having energies larger than those of visible light and, thus, gamma rays have extremely short wavelengths. It is common to indicate a nucleus in an excited state with a superscript asterisk. For example, the decay of ^{61}Ni from an excited nuclear state (indicated by the asterisk) to one of lesser energy would be written as follows:

$$_{28}^{61}Ni^* \rightarrow {}_{28}^{61}Ni + \gamma$$

nickel   *nickel*   *gamma ray*
(excited)

Note that *in gamma decay, the mass and proton numbers do not change*. The daughter nucleus is simply the parent nucleus with less energy. As an example of gamma emission following beta decay, consider the following Integrated Example.

Integrated Example 29.2 ■ Two for One: Beta Decay and Gamma Decay

Naturally occurring cesium has only one stable isotope, $_{55}^{133}$Cs. However, the unstable isotope $_{55}^{137}$Cs is a common nucleus found in used nuclear fuel rods at power plants after their original uranium fuel has become depleted. (See Chapter 30.) When $_{55}^{137}$Cs decays, its daughter nucleus is sometimes left in an excited state. After the initial decay, the daughter emits a gamma ray to produce a final stable nucleus. (a) Does $_{55}^{137}$Cs first decay by (1) β^+ decay, (2) β^- decay, or (3) electron capture? Explain. (b) Find the final daughter product by writing the chain of decay equations. Show all the steps leading to the final stable nucleus.

(a) Conceptual Reasoning. The $_{55}^{137}$Cs isotope has too many neutrons to be stable, as $_{55}^{133}$Cs, with four fewer neutrons, is stable. Choices (1) and (3) both increase the number of neutrons relative to the number of protons. Lowering the number of neutrons calls for β^- decay, so the correct choice is (2), β^- decay.

(b) Thinking It Through. Since we know that $_{55}^{137}$Cs must decay by emitting a β^- particle, its daughter (in an excited state) can be determined from charge and nucleon conservation. The final state of the daughter will result after a gamma-ray photon is emitted.

The data is as follows:

Given: Initial nucleus of $_{55}^{137}$Cs   *Find:* The decay schemes that lead to the stable nucleus

During β^- decay, the proton number increases by one; thus, the daughter will be barium ($Z = 56$). (See the periodic table, Fig. 28.9.) The decay equation should indicate that barium is left in an excited state that is ready to decay via gamma emission. (As usual in this chapter, the neutrino is omitted.) Thus, the decay equation is

$$_{55}^{137}Cs \rightarrow {}_{56}^{137}Ba^* + {}_{-1}^{0}e$$

cesium   *barium*   *electron*
(excited)

(continues on next page)

Teaching tip: Warn students that these beta-decay reactions are incomplete. For clarity, they do not show the neutrinos, as these particles have no significant interactions with matter. You may wish to explain that alpha decay, which produces a two-body final state, predicts a unique monoenergetic spectrum for emitted alpha particles. The beta-decay process, which yields three final-state products that can share energy in many ways, produces an energy spectrum for the beta particles that is continuous from zero to some maximum value; thus, the beta energy for any given decay process is not unique. The beta-decay equations will be written in their complete form, with emphasis on the neutrinos, in Chapter 30, where elementary particles are discussed.

This process is then quickly followed by the emission of a gamma ray from the excited barium nucleus:

$$\underset{\substack{barium \\ (excited)}}{^{137}_{55}\text{Ba}^*} \quad \rightarrow \quad \underset{barium}{^{137}_{55}\text{Ba}} \quad + \quad \underset{gamma\ ray}{\gamma}$$

Sometimes this process is written as a combined equation to show the sequential behavior:

$$\underset{cesium}{^{137}_{55}\text{Cs}} \quad \rightarrow \quad \underset{barium\ (excited)}{^{137}_{56}\text{Ba}^*} \quad + \quad \underset{electron}{^{0}_{-1}\text{e}}$$

$$\downarrow$$

$$\underset{barium}{^{137}_{56}\text{Ba}} \quad + \quad \underset{gamma\ ray}{\gamma}$$

Follow-Up Exercise. An unstable isotope of sodium, ^{22}Na, can be produced in nuclear reactors. The only stable isotope of sodium is ^{23}Na. ^{22}Na is known to decay by one type of beta decay. (a) Which type of beta decay is it? Explain. (b) Write down the beta-decay scheme, and predict the daughter nucleus.

Radiation Penetration

The absorption, or degree of penetration, of nuclear radiation is an important consideration in many modern applications. A familiar use of radiation is the radioisotope treatment of cancer. Radiation penetration is also important, for example, in determining the amount of nuclear shielding needed around a nuclear reactor. In our food industry, gamma radiation is now used to penetrate some foods in order to kill bacteria and thus sterilize the food. In industry, the absorption of radiation is used to monitor and control the thickness of metal and plastic sheets in fabrication processes.

The three types of radiation (alpha, beta, and gamma) are absorbed quite differently. As they move along their penetration paths, the electrically charged alpha and beta particles interact with the electrons of the atoms of a material and may ionize some of them. The charge and speed of the particle determine the rate at which it loses energy along its path (remember that ionizing an atom takes energy) and, thus, the degree of penetration. The degree of penetration also depends on properties of the material, such as its density. In general, what happens when the various particles enter a material is as follows:

- Alpha particles are doubly charged, have a relatively large mass, and move relatively slowly. Thus a few centimeters of air or a sheet of paper will usually completely stop them.

- Beta particles are much less massive and are singly charged. They can travel a few meters in air or a few millimeters in aluminum before being stopped.

- Gamma rays are uncharged and are therefore more penetrating than alpha and beta particles. A significant portion of a beam of high-energy gamma rays can penetrate a centimeter or more of a dense material, such as lead. Lead is commonly used as shielding against harmful X-rays and gamma rays. Photons can lose energy or be removed from a beam of gamma rays by a combination of Compton scattering, the photoelectric effect, and pair production (the latter occurs only for photon energies above about 1 MeV). (See Ch. 27.)

Radiation passing through matter can do considerable damage. Structural materials can become brittle and lose their strength when exposed to strong radiation, such as can happen in nuclear reactors (Chapter 30) and to space vehicles exposed to cosmic radiation. In biological tissue, the radiation damage is chiefly due to ionizations in living cells (Section 29.5). We are continually exposed to normal background radiation from radioisotopes in the environment and cosmic radiation from outer space. The energy we absorb and the damage inflicted to cells from exposure to everyday levels of such radiation is usually too low to be harmful. However, concern has been expressed about the radiation exposure of

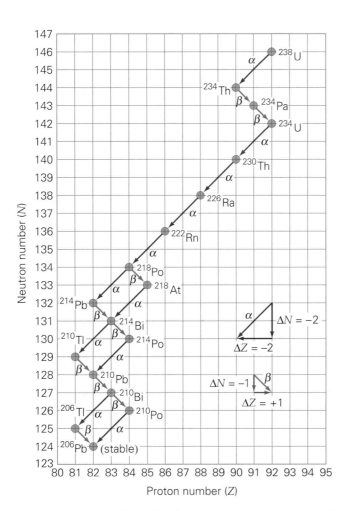

◀ **FIGURE 29.8 Decay series of uranium-238** On this plot of N versus Z, a diagonal transition from right to left is an alpha-decay process, and a diagonal transition from left to right is a β^--decay process. (How can you tell?) The decay series continues until the stable nucleus ^{206}Pb is reached.

people employed in jobs in which radiation levels may be considerably higher. For example, workers at nuclear power plants are constantly monitored for absorbed radiation and subject to strict rules that govern the amount of time for which they can work in a given period. Also, airplane flight crews who spend many hours aboard high-flying jet aircraft may receive significant exposure to radiation from cosmic rays. (Cosmic rays are discussed in more detail on page 915.)

Of the many unstable nuclides, only a small number occur naturally. Most of the radioactive nuclides found in nature are products of the decay series of heavy nuclei. There is continual radioactive decay progressing in a series into successively lighter elements. For example, the ^{238}U decay series (or "chain") is shown in ▲Fig. 29.8. It stops when the stable isotope of lead, ^{206}Pb, is reached. Note that some nuclides in the series decay by two modes and that radon (^{222}Rn) is part of this decay series. This radioactive gas has received a great deal of attention in the last few decades because it can accumulate in significant amounts in poorly ventilated buildings.

Note: See also Figs. 29.20 (^{237}Np) and 29.21 (^{239}Pu).

29.3 Decay Rate and Half-Life

OBJECTIVES: To (a) explain the concepts of activity, decay constant, and half-life of a radioactive sample; and (b) use radioactive decay to find the age of objects.

The nuclei in a sample of radioactive material do *not* decay all at once, but rather do so randomly at a rate characteristic of the particular nucleus and unaffected by external influences. It is impossible to tell exactly when a *particular* unstable nucleus will decay. What can be determined, however, is how many nuclei in a sample will decay during a given period of time.

The **activity (R)** of a sample of radioactive nuclide is defined as the number of nuclear disintegrations, or decays, per second. For a given amount of material, activity decreases with time, as fewer and fewer radioactive nuclei remain. Each nuclide has its own characteristic rate of decrease. The rate at which the number of parent nuclei (N) decreases is proportional to the number present, or $\Delta N/\Delta t \propto N$. This can be rewritten in equation form (using a constant of proportionality called λ) as follows:

$$\frac{\Delta N}{\Delta t} = -\lambda N$$

Teaching tip: Point out that the SI unit of λ (not wavelength!) is s^{-1}, but in many of the radioactive-decay problems in this text, it will appear in more convenient forms, such as day^{-1} or $year^{-1}$.

where the constant λ is called the **decay constant**. This quantity has SI units of s^{-1} (why?) and depends on the particular nuclide. The larger the decay constant λ, the greater is the rate of decay. The minus sign in the previous equation indicates that N is decreasing. The activity (R) of a radioactive sample is the magnitude of $\Delta N/\Delta t$, or the *decay rate*, expressed in decays per second, but without the minus sign (see the usage in Example 29.3):

$$R = \text{activity} = \left| \frac{\Delta N}{\Delta t} \right| = \lambda N \qquad (29.2)$$

Note: Activity = number of decays per second = $|\Delta N/\Delta t|$ and is a positive number.

Using calculus, Eq. 29.2 (with the minus sign put back in) can be solved for the number of the remaining (or undecayed) parent nuclei (N) at any time t compared with the number N_o present at $t = 0$. The result is:

$$N = N_o e^{-\lambda t} \qquad (29.3)$$

Thus the number of undecayed (parent) nuclei expressed as a fraction of the number initially present (N/N_o) decreases *exponentially* with time, as illustrated in ▼Fig. 29.9. This graph follows an exponentially decaying function $e^{-\lambda t}$. (Remember that $e \approx 2.718$ is the base of natural logarithms and should be available on your calculator.)

Teaching tip: Explain how to determine the half-life from Fig. 29.9.

The decay rate of a nuclide is commonly expressed in terms of its *half-life* rather than the decay constant. The **half-life ($t_{1/2}$)** is defined as the time it takes for half of the radioactive nuclei in a sample to decay. This is the time corresponding to $N/N_o = \frac{1}{2}$ in Fig. 29.9. In the same amount of time, activity (decays per second) also is cut in half, since the activity is proportional to the number of undecayed nuclei present. Because of this proportionality, the decay rate is usually measured to determine half-life. In other words, what is usually measured is the rate at which the decay particles are emitted, and the time for that rate to drop in half.

For example, by plotting measured decay rates, ▶Fig. 29.10 illustrates that the half-life of strontium-90 (^{90}Sr) can be determined to be 28 years. An alternative way to view the concept of half-life is to consider the mass of parent material. Thus, if there were initially 100 micrograms (μg) of ^{90}Sr, only 50 μg of ^{90}Sr would remain after 28 years. The other 50 μg would have decayed by the following beta-decay process:

$$^{90}_{38}\text{Sr} \rightarrow ^{90}_{39}\text{Y} + ^{0}_{-1}\text{e}$$
$$\quad\text{strontium} \qquad\quad \text{yttrium} \qquad\quad \text{electron}$$

▶ **FIGURE 29.9 Radioactive decay** The fraction of the remaining parent nuclei (N/N_o) in a radioactive sample plotted as a function of time follows an exponential-decay curve. The curve's shape and steepness depend on the decay constant λ or the half-life $t_{1/2}$.

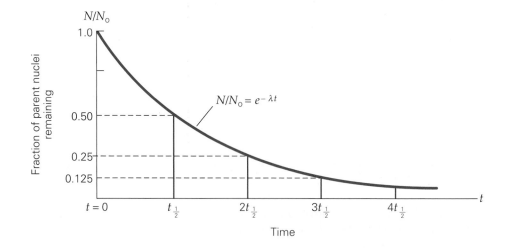

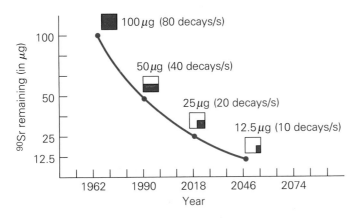

◀ **FIGURE 29.10 Radioactive decay and half-life** As shown here for strontium-90, after each half-life ($t_{1/2}$ = 28 y), only half of the amount of ^{90}Sr present at the start of that period of time remains, with the other half having decayed to ^{90}Y via beta decay. Similarly, the activity (decays per second) has also decreased by half after 28 years.

Thus, the sample would contain a mixture of both strontium and yttrium (and any decay products of yttrium). After another 28 years, half of these strontium nuclei would decay, leaving only 25 μg of ^{90}Sr, and so on.

The half-lives of radioactive nuclides vary greatly, as Table 29.1 shows. Nuclides with very short half-lives are generally created in nuclear reactions (Chapter 30). If these nuclides had existed when the Earth was formed (about 5 billion years ago), they would have long since decayed. In fact, this is the case for technetium (Tc) and promethium (Pm, not shown in Table 29.1). These elements do *not* exist naturally, as they have no stable configurations and their half-lives are short. However, they can be produced in laboratories. Conversely, the half-life of the naturally occurring ^{238}U isotope is about 4.5 billion years. This means that about half of the original ^{238}U present when the Earth was formed exists today. The longer the half-life of a nuclide, the more slowly it decays and the smaller is the decay constant λ. Thus the half-life and the decay constant have an inverse relationship, or $t_{1/2} \propto 1/\lambda$. To show the numerical relationship, consider Eq. 29.3. When $t = t_{1/2}$, then $N/N_0 = \frac{1}{2}$. Therefore,

$$\frac{N}{N_o} = \frac{1}{2} = e^{-\lambda t_{1/2}}$$

TABLE 29.1	The Half-Lives of Some Radioactive Nuclides (in Order of Increasing Half-Life)	
Nuclide	*Primary Decay Mode*	*Half-Life of Decay Mode*
Beryllium-8 ($^{8}_{4}$Be)	α	1×10^{-16} s
Polonium-213 ($^{213}_{84}$Po)	α	4×10^{-16} s
Oxygen-19 ($^{19}_{8}$O)	β^-	27 s
Fluorine-17 ($^{17}_{9}$F)	β^+, EC	66 s
Polonium-218 ($^{218}_{84}$Po)	α, β^-	3.05 min
Technetium-104 ($^{104}_{43}$Tc)	β^-	18 min
Iodine-123 ($^{123}_{53}$I)	EC	13.3 h
Krypton-76 ($^{76}_{36}$Kr)	EC	14.8 h
Magnesium-28 ($^{28}_{12}$Mg)	β^-	21 h
Radon-222 ($^{222}_{86}$Rn)	α	3.82 days
Iodine-131 ($^{131}_{53}$I)	β^-	8.0 days
Cobalt-60 ($^{60}_{27}$Co)	β^-	5.3 y
Strontium-90 ($^{90}_{38}$Sr)	β^-	28 y
Radium-226 ($^{226}_{88}$Ra)	α	1600 y
Carbon-14 ($^{14}_{6}$C)	β^-	5730 y
Plutonium-239 ($^{239}_{94}$Pu)	α	2.4×10^4 y
Uranium-238 ($^{238}_{92}$U)	α	4.5×10^9 y
Rubidium-87 ($^{87}_{37}$Rb)	β^-	4.7×10^{10} y

But because $e^{-0.693} \approx \frac{1}{2}$ (check this on your calculator), we can compare the exponents and determine (to three significant figures) the following result:

$$t_{1/2} = \frac{0.693}{\lambda} \qquad (29.4)$$

The concept of half-life is important in medical applications, as is shown in Example 29.3.

Example 29.3 ■ An "Active" Thyroid: Half-Life and Activity

The half-life of iodine-131 (^{131}I), used in thyroid treatments, is 8.0 days. At a certain time, about 4.0×10^{14} iodine-131 nuclei are in a hospital patient's thyroid gland. (a) What is the ^{131}I activity in the thyroid at that time? (b) How many ^{131}I nuclei remain after 1.0 day?

Thinking It Through. (a) Equation 29.4 enables us to determine the decay constant λ from the half-life, and then use Eq. 29.2 to find the initial activity. (b) To get N, Eq. 29.3 can be used in connection with the e^x button on a calculator.

Solution. Listing the data and converting the half-life into seconds,

Given: $\quad t_{1/2} = 8.0 \text{ days} = 6.9 \times 10^5 \text{ s}$ *Find:*   (a) R_o (activity at $t = 0$)
$\qquad\qquad N_o = 4.0 \times 10^{14}$ nuclei (initially)   (b) N (number of undecayed nuclei
$\qquad\qquad t = 1.0$ day   after 1.0 day)

(a) The decay constant is determined from its relationship to the half-life (Eq. 29.4) as follows:

$$\lambda = \frac{0.693}{t_{1/2}} = \frac{0.693}{6.9 \times 10^5 \text{ s}} = 1.0 \times 10^{-6} \text{ s}^{-1}$$

Using the initial number of undecayed nuclei, N_o, we find that the initial activity R_o is

$$R_o = \left| \frac{\Delta N}{\Delta t} \right| = \lambda N_o = (1.0 \times 10^{-6} \text{ s}^{-1})(4.0 \times 10^{14}) = 4.0 \times 10^8 \text{ decays/s}$$

(b) With $t = 1.0$ day and $\lambda = 0.693/t_{1/2} = 0.693/8.0 \text{ days} = 0.087 \text{ day}^{-1}$,

$$N = N_o e^{-\lambda t} = (4.0 \times 10^{14} \text{ nuclei})e^{-(0.087 \text{ day}^{-1})(1.0 \text{ day})}$$
$$= (4.0 \times 10^{14} \text{ nuclei})e^{-0.087} = (4.0 \times 10^{14} \text{ nuclei})(0.917) = 3.7 \times 10^{14} \text{ nuclei}$$

The e^x-function calculator button is sometimes labeled as the inverse of the ln x function. Become familiar with it. Here we have $e^{-0.087} \approx 0.917$ to three significant figures.

Follow-Up Exercise. In this Example, suppose that the attending physician will not allow the patient to go home until the activity is $\frac{1}{64}$ of its original level. (a) How long would the patient have to remain in observation? (b) In practice, the amount of time is much shorter than your answer to part (a). Can you think of a possible biological reason(s) for this?

By the "strength" of a radioactive sample we really mean its activity R. A common unit of radioactivity is named in honor of Pierre and Marie Curie.* One **curie (Ci)** is defined as

$$1 \text{ Ci} \equiv 3.70 \times 10^{10} \text{ decays/s}$$

This definition is historical and is based on the known activity of 1.00 g of pure radium. However, the modern SI unit is the **becquerel (Bq)**, which is defined as

$$1 \text{ Bq} \equiv 1 \text{ decay/s}$$

Therefore,

$$1 \text{ Ci} = 3.70 \times 10^{10} \text{ Bq}$$

Even with the present-day emphasis on SI units, the "strengths" of radioactive sources are still commonly specified in curies. The curie is a relatively large unit, however, so the *millicurie* (mCi), the *microcurie* (μCi), and even smaller multiples such as the *nanocurie* (nCi) and *picocurie* (pCi) are used. Teaching laboratories, for

---

*Marie Sklodowska Curie (1867–1934) was born in Poland and studied in France, where she met and married physicist Pierre Curie (1859–1906). In 1903, Madame Curie (as she is commonly known) and Pierre shared the Nobel Prize in physics with Henri Becquerel (1852–1908) for their work on radioactivity. She was also awarded the Nobel Prize in chemistry in 1911 for the discovery of radium.

example, typically use samples with activities of one microcurie or less. The strength of a source is calculated in the following Example.

Example 29.4 ■ Declining Source Strength: Get a Half-Life!

A ^{90}Sr beta source has an initial activity of 10.0 mCi. How many decays per second will be taking place after 84.0 years?

Thinking It Through. Table 29.1 lists the half-life for the source. In this Example, we can use the fact that in each successive half-life, the activity decreases by half from what it was at the start of that interval. Thus, Eq. 29.3 need not be used, because the elapsed time is exactly three half-lives. (This approach is advisable only when the elapsed time is an integral multiple of the half-life, as it is here.)

Solution.

Given: Initial activity = 10.0 mCi   *Find:*   R (activity after 84.0 years)
$t = 84.0$ y
$t_{1/2} = 28.0$ y (from Table 29.1)

Since 84 years is exactly three half-lives, the activity after that amount of time has elapsed will be one-eighth as great $\left(\frac{1}{2} \times \frac{1}{2} \times \frac{1}{2} = \frac{1}{8}\right)$, and the strength of the source will then be

$$R = \left|\frac{\Delta N}{\Delta t}\right| = 10.0 \text{ mCi} \times \frac{1}{8} = 1.25 \text{ mCi} = 1.25 \times 10^{-3} \text{ Ci}$$

In terms of decays per second, or becquerels, we have

$$R = \left|\frac{\Delta N}{\Delta t}\right| = (1.25 \times 10^{-3} \text{ Ci})\left(3.70 \times 10^{10} \frac{\text{decays/s}}{\text{Ci}}\right)$$
$$= 4.63 \times 10^7 \text{ decays/s} = 4.63 \times 10^7 \text{ Bq}$$

Follow-Up Exercise. For the material in this Example, suppose a radiation safety officer tells you that this sample can go into the low-level waste disposal only when its activity drops to one millionth of its initial activity. Estimate, to two significant figures, how long the sample must be kept before it can be disposed. [*Hint:* Your calculator may save some time: 2 raised to what power produces about a million?]

Radioactive Dating

Because their decay rates are constant, radioactive nuclides can be used as nuclear clocks. In the previous Example, the half-life of a radioactive nuclide was used to determine how much of the sample will exist in the future. Similarly, by using the half-life to calculate backward in time, scientists can determine the age of objects that contain known radioactive nuclides. As you might surmise, some idea of initial amount of the nuclide present must be known.

To illustrate the principle of radioactive dating, let's look at how it is done with ^{14}C, a very common method used in archeology. **Carbon-14 dating** is used on materials that were once part of living things and on the remnants of objects made from or containing such materials (such as wood, bone, leather, or parchment). The process depends on the fact that living things (including yourself) contain a known amount of radioactive ^{14}C. The concentration is very small—about one ^{14}C atom for every 7.2×10^{11} atoms of ordinary ^{12}C. Even so, the ^{14}C present in our bodies *cannot* be due to ^{14}C that was present when the Earth was formed. This is because the half-life of ^{14}C is $t_{1/2} = 5730$ yr, which is very short in comparison with the age of the Earth.

The ^{14}C nuclei that exist in living things are there because that isotope is continuously being produced in the atmosphere by cosmic rays. *Cosmic rays* are high-speed charged particles that reach us from various sources such as the Sun and nearby exploding stars called supernovae. These "rays" are actually primarily protons. When they enter our upper atmosphere they can cause reactions that produce neutrons (▸Fig. 29.11). These neutrons are then absorbed by the nuclei of the nitrogen atoms of the air, which, in turn, decay by emitting a proton (written as p or $^{1}_{1}$H) to produce ^{14}C by the reaction

$$^{14}_{7}\text{N} + ^{1}_{0}\text{n} \rightarrow ^{14}_{6}\text{C} + ^{1}_{1}\text{H}$$

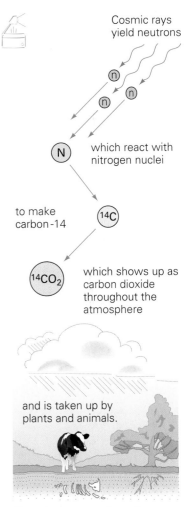

Cosmic rays yield neutrons

which react with nitrogen nuclei

to make carbon-14

which shows up as carbon dioxide throughout the atmosphere

and is taken up by plants and animals.

But when an organism dies, no fresh carbon-14 replaces the carbon-14 decaying in its tissues, and the carbon-14 radioactivity decreases by half every 5730 years.

▲ **FIGURE 29.11** Carbon-14 radioactive dating The formation of carbon-14 in the atmosphere and its entry into the biosphere.

Note: The cosmic-ray production of ^{14}C is an example of a nuclear reaction that induces a nuclear transmutation. Such reactions will be studied in more detail in Chapter 30.

^{14}C eventually decays by β^- decay ($^{14}_{6}C \rightarrow {}^{14}_{7}N + {}^{0}_{-1}e$), because it is neutron rich. Although the intensity of incident cosmic rays may not be exactly constant over time, the concentration of ^{14}C in the atmosphere is relatively constant, because of atmospheric mixing and the fixed decay rate.

The ^{14}C atoms are oxidized into carbon dioxide (CO_2), so a small fraction of the CO_2 molecules in the air is radioactive. Plants take in this radioactive CO_2 by photosynthesis, and animals ingest the plant material. As a result, the concentration of ^{14}C in living organic matter is the same as the concentration in the atmosphere, one part in 7.2×10^{11}. However, once an organism dies, the ^{14}C in that organism is *no longer* replenished, and thus the ^{14}C concentration decreases. *Thus, the concentration of ^{14}C in dead matter relative to that in living things can be used to establish when the organism died.* Since radioactivity is generally measured in terms of activity, the ^{14}C activity in organisms now alive must somehow be found. The following Example shows how this is done.

Example 29.5 ■ Living Organisms: Natural Carbon-14 Activity

For ^{14}C, determine the average activity R in decays per minute per gram of natural carbon, found in living organisms, if the concentration of ^{14}C in the organisms is the same as that in the atmosphere.

Thinking It Through. From the previous discussion, we know the concentration of ^{14}C relative to that of ^{12}C. To calculate the ^{14}C activity, we need the decay constant (λ), which can be computed from the half-life of ^{14}C ($t_{1/2} = 5730$ years; see Table 29.1) and the number of ^{14}C atoms (N) per gram. Carbon has an atomic mass of 12.0, so N can be found from Avogadro's number (recall that $N_A = 6.02 \times 10^{23}$ atoms/mole) and the number of moles, $n = N/N_A$ (see Section 10.3).

Solution. Listing the known ratio and the half-life from Table 29.1 (and converting the half-life into minutes), we have

Given:
$\dfrac{^{14}C}{^{12}C} = \dfrac{1}{7.2 \times 10^{11}} = 1.4 \times 10^{-12}$
$\qquad$ *Find:* Average activity R per gram

$\quad t_{1/2} = (5730 \text{ years})(5.26 \times 10^5 \text{ min/year})$
$\qquad\; = 3.01 \times 10^9 \text{ min}$

Carbon has 12.0 g per mole.

From the half-life, the decay constant is

$$\lambda = \frac{0.693}{t_{1/2}} = \frac{0.693}{3.01 \times 10^9 \text{ min}} = 2.30 \times 10^{-10} \text{ min}^{-1}$$

For 1.0 g of carbon, the number of moles is $n = 1.0 \text{ g}/(12 \text{ g/mol}) = \frac{1}{12}$ mol, so the number of atoms (N) is

$$N = nN_A = \left(\frac{1}{12}\text{mol}\right)(6.02 \times 10^{23} \text{ C nuclei/mol}) = 5.0 \times 10^{22} \text{ C nuclei (per gram)}$$

The number of ^{14}C nuclei per gram is given by the concentration factor

$$N\left(\frac{^{14}C}{^{12}C}\right) = (5.0 \times 10^{22} \text{ C nuclei/g})\left(1.4 \times 10^{-12} \frac{^{14}C \text{ nuclei}}{C \text{ nuclei}}\right) = 7.0 \times 10^{10} \; (^{14}C \text{ nuclei/g})$$

The activity in decays per gram of carbon per minute (to two significant figures) is

$$\left|\frac{\Delta N}{\Delta t}\right| = \lambda N = (2.30 \times 10^{-10} \text{ min}^{-1})(7.0 \times 10^{10} \; {}^{14}C/\text{g}) = 16\frac{^{14}C \text{ decays}}{\text{g} \cdot \text{min}}$$

Thus, if an artifact such as a bone or a piece of cloth has a current activity of 8.0 decays per gram of carbon per minute, then the original living organism would have died about one half-life, or about 5700 years, ago. This would put the date of the artifact near 3700 B.C.

Follow-Up Exercise. Suppose that your instruments could measure ^{14}C beta emissions only down to 1.0 decays/min. How far back (to two significant figures) could you estimate the ages of dead organisms?

Now consider how the activity calculated in Example 29.5 can be used to date ancient organic finds.

Example 29.6 ■ Old Bones: Carbon-14 Dating

A bone is unearthed in an archeological dig. Laboratory analysis determines that there are 20 beta emissions per minute from 10 g of carbon in the bone. What is the approximate age of the bone?

Thinking It Through. Since the initial activity of a living sample is known (Example 29.5), we can work backward to determine the amount of time elapsed.

Solution. For comparison purposes, the activity *per gram* is the relevant number.

Given: Activity = 20 decays/min in 10 g of carbon   *Find:*   Age of the bone
 = 2.0 decays/g·min

Assuming that the bone had the normal concentration of ^{14}C when the organism died, the ^{14}C activity at the time of death would have been 16 decays/g·min (Example 29.5). Afterward, the decay rate would decrease by half for each half-life:

$$16 \xrightarrow{t_{1/2}} 8 \xrightarrow{t_{1/2}} 4 \xrightarrow{t_{1/2}} 2 \text{ decay/g·min}$$

So, with an observed activity of 2.0 decays/g·min, the ^{14}C in the bone has gone through approximately three half-lives. Thus, the bone is three half-lives old, or, to two significant figures,

$$\text{Age} \approx 3.0 t_{1/2} = (3.0)(5730 \text{ y}) = 1.7 \times 10^4 \text{ y}$$
$$\approx 17\,000 \text{ y}$$

Follow-Up Exercise. Studies indicate that on Earth the stable isotope ^{39}K represents about 93.2% of all the potassium. A long-lived (but unstable) nuclide, ^{40}K, represents only about 0.010%. ^{40}K has a half-life of 1.28×10^9 y. (a) The remainder of the existing potassium (6.8%) is all one other isotope of potassium. What isotope is this most likely to be? (b) What would the percentage ^{40}K abundance have been when the Earth was first formed, assumed to be 4.7×10^9 years ago?

The limit of radioactive carbon dating depends on the ability to measure the very low activity in old samples. Current techniques give an age-dating limit of about 40 000–50 000 years, depending on the sample size. After about ten half-lives, the radioactivity is barely measurable (less than two decays per gram per *hour*).

Another radioactive dating process uses lead-206 (^{206}Pb) and uranium-238 (^{238}U). This dating method is used extensively in geology, because of the long half-life of ^{238}U. Lead-206 is the stable end isotope of the ^{238}U decay series. (See Fig. 29.8.) If a rock sample contains both of these isotopes, the lead is assumed to be a decay product of the uranium that was there when the rock first formed. Thus, the ratio of $^{206}Pb/^{238}U$ can be used to determine the age of the rock.

29.4 Nuclear Stability and Binding Energy

OBJECTIVES: To (a) state which proton- and neutron-number combinations result in stable nuclei, (b) explain the pairing effect and magic numbers in relation to nuclear stability, and (c) calculate nuclear binding energies.

Now that we have considered properties of unstable isotopes, let's turn to the stable ones. Stable isotopes exist naturally for all elements having proton numbers from 1 to 83, *except* those with $Z = 43$ (technetium) and $Z = 61$ (promethium). The nuclear interactions (forces) that determine nuclear stability are extremely complicated. However, by looking at some of the properties of stable nuclei, it is possible to obtain general criteria for nuclear stability.

Nucleon Populations

One of the first considerations is the relative number of protons and neutrons in stable nuclei. Nuclear stability depends on the dominance of the attractive nuclear

▶ **FIGURE 29.12 A plot of *N* versus *Z* for stable nuclei** For nuclei with mass numbers $A < 40$ ($Z < 20$ and $N < 20$), the number of protons and the number of neutrons are equal or nearly equal. For nuclei with $A > 40$, the number of neutrons exceeds the number of protons, so these nuclei lie above the $N = Z$ line.

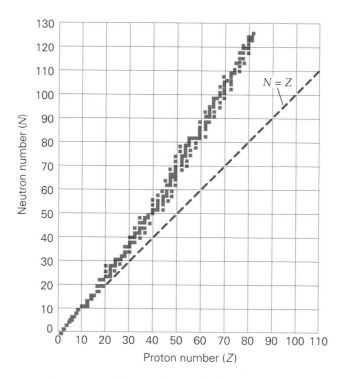

force between nucleons over the repulsive Coulomb force between protons. This force dominance depends on the ratio of protons to neutrons.

For stable nuclei of low mass numbers (about $A < 40$), the ratio of neutrons to protons (N/Z) is approximately 1. That is, the number of protons and the number of neutrons are equal or nearly equal. As examples of this, we have ^4_2He, $^{12}_6\text{C}$, $^{23}_{11}\text{Na}$, and $^{27}_{13}\text{Al}$. For stable nuclei of higher mass numbers ($A > 40$), the number of neutrons exceeds the number of protons ($N/Z > 1$). The heavier the nuclei, the higher is this ratio, that is, the more the neutrons outnumber the protons and the N/Z ratio increases with A.

This trend is illustrated in ▲Fig. 29.12, a plot of neutron number (N) versus proton number (Z) for stable nuclei. The heavier stable nuclei lie above the $N = Z$ line ($N > Z$). Examples of heavy stable nuclei include $^{62}_{28}\text{Ni}$, $^{114}_{50}\text{Sn}$, $^{208}_{82}\text{Pb}$, and $^{209}_{83}\text{Bi}$. In fact, bismuth-209 is the heaviest element that has a stable isotope.*

Radioactive decay "adjusts" the proton and neutron numbers of an unstable nuclide until a stable nuclide is produced—that is, until the product nucleus lands on the stability curve in Fig. 29.12. Since alpha decay decreases the numbers of protons and neutrons by equal amounts, alpha decay alone would give nuclei with neutron populations that are *larger* than those of the stable nuclides on the curve. However, β^- decay *following* alpha decay can lead to a stable combination, since the effect of β^- decay is to convert a neutron into a proton. Thus, very heavy unstable nuclei undergo a chain, or sequence, of alpha and beta decays until a stable nucleus is reached (recall Fig. 29.8 for ^{238}U).

Pairing

Many stable nuclei have even numbers of both protons and neutrons, and very few have odd numbers of *both* protons and neutrons. A survey of the stable isotopes (Table 29.2) shows that 168 stable nuclei have even–even combination, while 107 have even–odd or odd–even arrangements, and only 4 contain odd numbers of both protons and neutrons. These four are isotopes of the elements with the four lowest odd proton numbers: ^2_1H, ^6_3Li, $^{10}_5\text{B}$, and $^{14}_7\text{N}$.

The dominance of even–even combinations indicates that the protons and neutrons in nuclei tend to "pair up." That is, two protons pair up and, separately, two neutrons pair up. Aside from the four nuclei mentioned above, all odd–odd nuclei are unstable. Also, odd–even and even–odd nuclei tend to be less stable than the even–even variety.

TABLE 29.2

Pairing Effect of Stable Nuclei

Proton Number	Neutron Number	Number of Stable Nuclei
Even	Even	168
Even	Odd	} 107
Odd	Even	
Odd	Odd	4

*Bismuth-209 alpha decays, but with a half-life of 2×10^{18} years; for practical purposes, it is considered to be stable.

This **pairing effect** provides a qualitative criterion for stability. For example, you might expect the aluminum isotope $^{27}_{13}Al$ to be stable (even–odd), but not $^{26}_{13}Al$ (odd–odd). This is the case.

The *general criteria for nuclear stability* can be summarized as follows:

1. All isotopes with a proton number greater than 83 ($Z > 83$) are unstable.

2. (a) Most even–even nuclei are stable.
   (b) Many odd–even and even–odd nuclei are stable.
   (c) Only four odd–odd nuclei are stable ($^{2}_{1}H$, $^{6}_{3}Li$, $^{10}_{5}B$, and $^{14}_{7}N$).

3. (a) Stable nuclei with mass numbers less than 40 ($A < 40$) have approximately the same number of protons and neutrons.
   (b) Stable nuclei with mass numbers greater than 40 ($A > 40$) have more neutrons than protons.

Conceptual Example 29.7 ▪ Running Down the Checklist: Nuclear Stability

Is the sulfur isotope $^{38}_{16}S$ likely to be stable?

Reasoning and Answer. We use the general criteria for nuclear stability to analyze this case:

1. *Satisfied*. Isotopes with $Z > 83$ are unstable. With $Z = 16$, this criterion is satisfied.
2. *Satisfied*. The isotope $^{38}_{16}S_{22}$ has an even–even nucleus, so it could be stable.
3. *Not satisfied*. Here, $A < 40$, but $Z = 16$ and $N = 22$ are not approximately equal.

Therefore, the ^{38}S isotope is likely to be unstable. (The nucleus is unstable and decays by β^- emission, since it is neutron rich.)

Follow-Up Exercise. (a) List likely isotopes of copper ($Z = 29$). (b) Apply the criteria for nuclear stability to see which of those isotopes should be stable. Use Appendix V to check your conclusions.

Binding Energy

An important quantitative aspect of nuclear stability is the *binding energy* of the nucleons. Binding energy can be calculated by considering the mass–energy equivalence along with known nuclear masses. Since nuclear masses are so small in relation to the kilogram, another standard, the **atomic mass unit (u)**, is used to measure them. The conversion factor (to six significant figures) between the atomic mass unit and the kilogram is

$$1\,u = 1.66054 \times 10^{-27}\,kg$$

The masses of the various particles are typically expressed in atomic mass units. (See Table 29.3.) The listed energy equivalents reflect Einstein's $E = mc^2$ mass–energy equivalence relationship from Eq. 26.11. Thus, a mass of 1 u has an energy equivalent to

$$mc^2 = (1.66054 \times 10^{-27}\,kg)(2.9977 \times 10^8\,m/s)^2 = 1.4922 \times 10^{-10}\,J$$

$$= \frac{1.4922 \times 10^{-10}\,J}{1.602 \times 10^{-13}\,J/MeV} = 931.5\,MeV$$

Note: The concept of binding energy was introduced in Section 27.4.

TABLE 29.3	The Atomic Mass Unit (u), Particle Masses, and Their Energy Equivalents		
Particle	*u*	*Mass (kg)*	*Equivalent Energy (MeV)*
—	1 (exact)	1.66054×10^{-27}	931.5
Electron	5.48578×10^{-4}	9.10935×10^{-31}	0.511
Proton	1.007276	1.67262×10^{-27}	938.27
Hydrogen atom	1.007825	1.67356×10^{-27}	938.79
Neutron	1.008665	1.67500×10^{-27}	939.57

as given in the first entry of Table 29.3. We will use 931.5 MeV/u (to four significant figures) as a handy conversion factor (mass into its energy equivalent) to avoid having to multiply by c^2.

Note in Table 29.3, the proton and hydrogen atom ($^1_1$H) are listed separately, as they have slightly different masses. This is due to the mass of the atomic electron. We experimentally measure the masses of neutral *atoms* (nucleons plus Z electrons) rather than of their *nuclei*. Keep this factor in mind. Since nuclear-energy calculations usually involve very small differences in mass, the mass of the electron can be significant.

Nuclear stability can be looked at in terms of energy. For example, if the mass of a helium-4 nucleus is compared with the total mass of nucleons that compose it, a significant inequity emerges: A neutral helium atom (including its *two* electrons) has a mass of 4.002 603 u. (Atomic masses of various atoms are given in Appendix V.) The total mass of two hydrogen atoms (^{1}H) (including *two* electrons) and two neutrons is

$$2m(^1\text{H}) = 2.015\,650 \text{ u}$$
$$2m_n = \underline{2.017\,330 \text{ u}}$$
$$\text{Total} = 4.032\,980 \text{ u}$$

This total is greater than the mass of the helium atom (4.002 603 u). The helium nucleus is less massive than the sum of its parts by an amount

$$\Delta m = [2m(^1\text{H}) + 2m_n] - m(^4\text{He})$$
$$= 4.032\,980 \text{ u} - 4.002\,603 \text{ u} = 0.030\,377 \text{ u}$$

(Note that the two electron masses of helium subtract out, since the mass of two hydrogen atoms also included two electrons.) This difference in mass, called the **mass defect**, has an energy equivalent of

$$(0.030\,377 \text{ u})(931.5 \text{ MeV/u}) = 28.30 \text{ MeV}$$

This energy is the *total binding energy* (E_b) of the ^{4}He nucleus.

In general, for any nucleus, the **total binding energy (E_b)** is related to the mass defect by

$$E_b = (\Delta m)c^2 \quad \textit{total binding energy} \qquad (29.5)$$

where Δm is the mass defect. An alternative interpretation of binding energy is that it represents the energy required to separate the constituent nucleons completely into free particles. This concept is illustrated in ◀Fig. 29.13 for the helium nucleus, for which 28.30 MeV of energy is necessary to separate it into four nucleons.

An insight into the nature of the nuclear force can be gained by considering the *average binding energy per nucleon* for stable nuclei. This quantity is the total binding energy of a nucleus, divided by the total number of nucleons, or E_b/A, where A is the mass number. For the helium nucleus (^{4}He) in Fig. 29.13, the average binding energy per nucleon is

$$\frac{E_b}{A} = \frac{28.30 \text{ MeV}}{4} = 7.075 \text{ MeV/nucleon}$$

Compared with the binding energy of atomic electrons (13.6 eV for a hydrogen electron in the ground state), nuclear binding energies are millions of times larger, indicative of a very strong binding force.

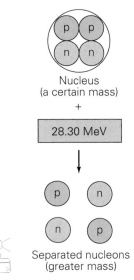

Nucleus
(a certain mass)

+

28.30 MeV

Separated nucleons
(greater mass)

▲ **FIGURE 29.13 Binding energy** 28.30 MeV is required to separate a helium nucleus into its constituent protons and neutrons. Conversely, if two protons and two neutrons combine to form a helium nucleus, 28.30 MeV of energy would be released.

Example 29.8 ■ The Stablest of the Stable: Binding Energy per Nucleon

Compute the average binding energy per nucleon of the iron-56 nucleus ($^{56}_{26}$Fe).

Thinking It Through. The atomic mass of iron-56 is found in Appendix V, and the other needed masses are in Table 29.3. We can then compute the mass defect Δm, the total binding energy E_b, and, lastly, the average binding energy per nucleon, E_b/A.

Solution.

Given: $^{56}_{26}$Fe mass = 55.934 939 u   *Find:* E_b/A (average binding energy
$^1_1$H mass = $m(^1$H) = 1.007 825 u   per nucleon)
$^1_0$n mass = m_n = 1.008 665 u

(Notice the use of the masses of the iron *atom* and the hydrogen *atom* rather than the nuclear masses.)

The mass defect is the difference between the mass of the iron atom and the mass of its separated constituents. The total mass of the constituents (here, 26 hydrogen atoms and 30 neutrons) is found as follows:

$$26m(^1\text{H}) = 26(1.007\,825\text{ u}) = 26.203\,450\text{ u}$$
$$30m_n = 30(1.008\,665\text{ u}) = \underline{30.259\,950\text{ u}}$$
$$\text{Total} = 56.463\,400\text{ u}$$

Thus, the mass defect is

$$\Delta m = [26m(^1\text{H}) + 30m_n] - m(^{56}\text{Fe}) = 56.463\,400\text{ u} - 55.934\,939\text{ u} = 0.528\,461\text{ u}$$

The total binding energy is easily calculated using the energy equivalence of 1 u:

$$E_b = (\Delta m)c^2 = (0.528\,461\text{ u})(931.5\text{ MeV/u}) = 492.3\text{ MeV}$$

This iron nuclide has 56 nucleons, so the average binding energy per nucleon is

$$\frac{E_b}{A} = \frac{492.3\text{ MeV}}{56} = 8.791\text{ MeV/nucleon}$$

Follow-Up Exercise. (a) To illustrate the pairing effect, compare the average binding energy per nucleon for ^4He (calculated previously to be 7.075 MeV) with that of ^3He. (Find the atomic masses in Appendix V.) (b) Which one is more tightly bound, on average, and how does your answer reflect pairing?

If E_b/A is calculated for various nuclei and plotted versus mass number, the values lie along the curve shown in ▼Fig. 29.14. The value of E_b/A rises rapidly with increasing A for light nuclei and starts to level off (around $A = 15$) at about 8.0 MeV/nucleon, with a maximum value of about 8.8 MeV/nucleon in the vicinity of iron, which has the most stable nucleus. (See Example 29.8.) For $A > 60$, the E_b/A values decrease slowly, indicating that the nucleons are, on average, less tightly bound.

The importance of the maximum in the curve cannot be understated. Consider what would happen on either side of it. If a massive nucleus split, or *fissioned*, into two lighter nuclei, the nucleons would be more tightly bound, and energy would be released. On the low-mass side of the maximum, if two nuclei could be fused, in a process called **fusion**, a more tightly bound nucleus would be created, and energy would be released. The details and application of these processes will be discussed in detail in Chapter 30.

The E_b/A curve shows that, except for very light nuclei, the binding energy per nucleon does not change a great deal and has a value of $E_b/A \approx 8$ MeV/nucleon. This means that, to a good approximation, we can write $E_b \propto A$. In other words, the *total* binding energy is (approximately) proportional to the total number of nucleons.

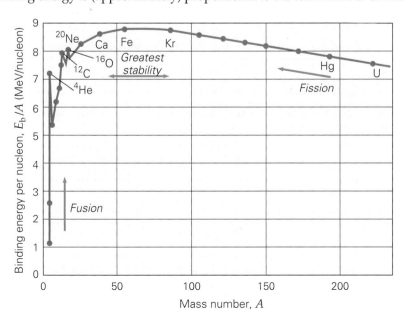

◄ **FIGURE 29.14 A plot of binding energy per nucleon versus mass number** If the binding energy per nucleon (E_b/A) is plotted versus mass number (A), the curve has a maximum near iron (Fe). This indicates that the nuclei in this region are, on average, the most tightly bound and have the greatest stability. Extremely heavy nuclei can release energy by splitting (fissioning). Extremely light nuclei can combine by fusion to release energy.

This proportionality indicates a characteristic of the nuclear force that is quite different from the electrical force. Suppose that the attractive nuclear force *did* act between all the pairs of nucleons in a nucleus. Each pair of nucleons would then contribute to the total binding energy. Considering all combinations, statistics tell us that in a nucleus containing A nucleons, there are $A(A - 1)/2$ pairs. Thus there would be $A(A - 1)/2$ contributions to the total binding energy. For nuclei with $A \gg 1$ (heavy nuclei), $A(A - 1) \approx A^2$, and we would expect the binding energy to be proportional to the *square* of A, or $E_b \propto A^2$ *if the nucleon–nucleon force were to act over a long range.* But as we have seen, in actuality, $E_b \propto A$. This relationship indicates that a given nucleon is *not* bound to all the other nucleons. This phenomenon, called *saturation*, implies that the nuclear forces act over a short range and that any particular nucleon interacts only with its nearest neighbors.

Magic Numbers

We are familiar with the concept of filled shells in atoms. There is an analogous effect in the nucleus. Although the concept of individual nucleon "orbits" inside the nucleus is hard to visualize, experimental evidence does indicate the existence of "closed nuclear shells" when the number of protons *or* neutrons is 2, 8, 20, 28, 50, 82, or 126. Important work on the nuclear-shell model was done by Nobel Prize winner Maria Goeppert-Mayer (1906–1972), a German-born physicist.

The number of stable isotopes of various elements provides solid evidence of the existence of such **magic numbers**. If an element has a magic number of protons, it has an unusually high number of stable isotopes. Elements whose proton number is far away numerically from a magic number may have only 1 or 2 (or even no) stable isotopes. Aluminum, for example, with 13 protons, has just 1 stable isotope, ^{27}Al. But tin, with $Z = 50$ (a magic number), has 10 stable isotopes, ranging from $N = 62$ to $N = 74$. Neighboring indium, with $Z = 49$, has only 2 stable isotopes, and antimony, with $Z = 51$, also has only 2.

Another piece of experimental evidence for magic numbers is related to binding energies. High-energy gamma-ray photons can be used to knock out single nucleons from a nucleus (a phenomenon called the *photonuclear effect*) in a manner analogous to the photoelectric effect in metals. Experimentally, a nuclide with a proton magic number, such as tin, requires about 2 MeV *more* photon energy to eject a proton from its nucleus than does a nuclide that does not have a magic number of protons. Thus, magic numbers are associated with extra large binding energies, another sign of higher-than-average stability.

29.5 Radiation Detection, Dosage, and Applications

OBJECTIVES: To (a) gain insight into the operating principles of various nuclear-radiation detectors, (b) investigate the medical and biological effects of radiation exposure, and (c) study some of the practical uses and applications of radiation.

Detecting Radiation

Since, in general, our senses cannot detect radioactive decay directly, detection must be accomplished through indirect means. For example, people who work with radioactive materials or in nuclear reactors usually wear film badges that indicate cumulative exposure to radiation by the degree of darkening of the film when developed. If more immediate ("real time") and quantitative methods are needed to detect radiation, a variety of instruments is available.

These instruments, as a group, are known as **radiation detectors**. Fundamentally, they are all based on the ionization or excitation of atoms, a phenomenon caused by the passage of energetic particles through matter. The electrically charged alpha and beta particles transfer energy to atoms by electrical interactions, removing electrons and creating ions. Gamma-ray photons can produce ionization by the photoelectric effect and Compton scattering (Section 27.3). They may also produce electrons and positrons by pair production (Section 28.5), if their energy is large enough. Regardless of the source, the particles produced by these interactions, not the actual radiated particles, are the objects "detected" by a radiation detector.

One of the most common radiation detectors is the *Geiger counter*, developed by Hans Geiger (1882–1945), a student and then colleague of Ernest Rutherford. The principle of the Geiger counter is illustrated in ▶Fig. 29.15. A voltage of about 1000 V is applied across the wire electrode and outer electrode (a metallic tube) of the Geiger tube that contains a gas (such as argon) at low pressure. When an ionizing particle enters the tube through a thin window, the particle ionizes some gas atoms. The freed electrons are accelerated toward the positive anode. On their way, they strike and ionize other atoms. This process snowballs, and the resulting "avalanche" produces a current pulse. The pulse is amplified and sent to an electronic counter that counts the pulses, or the number of particles detected. The pulses are sometimes used to drive a loudspeaker so that particle detection is heard as a click.

Another method of detection is the *scintillation counter* (▼Fig. 29.16). Here, the atoms of a phosphor material [such as sodium iodide (NaI)] are excited by an incident particle. A visible-light pulse is emitted when the atoms return to their ground state. The light pulse is converted to an electrical pulse by a photoelectric material. The pulse is then amplified in a *photomultiplier tube*, which consists of a series of electrodes of successively higher potential. The photoelectrons are accelerated toward the first electrode and acquire sufficient energy to cause several secondary electrons from ionization to be emitted when they strike the electrode. This process continues, and relatively weak scintillations are converted into sizable electrical pulses, which are then counted electronically.

In a *solid-state*, or *semiconductor, detector*, charged particles passing through a semiconductor material produce electrons, because of ionization. When a voltage is applied across the material, the electrons are collected as an electric current, which can be amplified and counted.

The three previous detectors determine the number of particles that interact in their material. Other different methods allow the actual trajectory, or "tracks," of charged particles to be seen and/or recorded. Among this type of detector are the cloud chamber, the bubble chamber, and the spark chamber. In the first two, vapors and liquids are supercooled and superheated, respectively, by suddenly varying the volume and pressure.

The *cloud chamber* was developed early in the 1900s by C. T. R. Wilson, a British atmospheric physicist. In the chamber, supercooled vapor condenses into droplets on the sites of ionized molecules created along the path of an energetic particle. When the chamber is illuminated, the droplets scatter the light, making the path visible (▶Fig. 29.17).

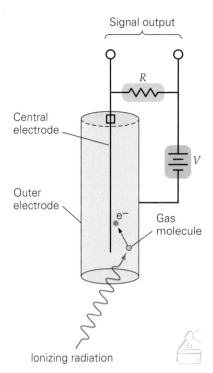

▲ FIGURE 29.15 The Geiger counter Incident radiation ionizes a gas atom, freeing an electron that is, in turn, accelerated toward the central (positive) electrode. On the way, this electron produces additional electrons through ionization, resulting in a current pulse that is detected as a voltage across the external resistor R.

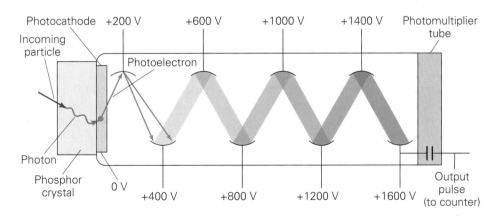

▲ FIGURE 29.16 The scintillation counter A photon emitted by a phosphor atom excited by an incoming particle causes the emission of a photoelectron from the photocathode. Accelerated through a difference in potential in a photomultiplier tube, the photoelectrons free secondary electrons when they collide with successive electrodes at higher potentials. After several steps, a relatively weak scintillation is converted into a measurable electric pulse.

▲ FIGURE 29.17 Cloud-chamber tracks The circular track in this photograph of a cloud chamber was made by a positron in a strong magnetic field. (Can you explain the approximately circular path of the particle in terms of the orientation of the magnetic field relative to the positron's velocity?)

The *bubble chamber*, which was invented by the American physicist D. A. Glazer in 1952, uses a similar principle. A reduction in pressure causes a liquid to be superheated and able to boil. Ions produced along the path of an energetic particle become sites for bubble formation, and a trail of bubbles is created. Since the bubble chamber uses a liquid, commonly liquid hydrogen, the density of atoms in it is much greater than in the vapor of a cloud chamber. Thus, tracks are more readily observable in bubble chambers, and hence bubble chambers have largely replaced cloud chambers.

In a *spark chamber*, the path of a charged particle is registered by a series of sparks. The charged particle passes between a pair of electrodes that have a high difference in potential and are immersed in an inert (noble) gas. The charged particle causes the ionization of gas molecules, giving rise to a visible spark (flash of light) between the electrodes as the released electrons travel to the positive electrode. A spark chamber is merely an array of such electrodes in the form of parallel plates or wires. A series of sparks, which can be photographed, then marks the particle's path.

Once the particle's trajectory is displayed, the particle's energy can be determined. Typically, a magnetic field is applied across the chamber, any charged particles are deflected, and the energy of a particle can be calculated from the radius of curvature of its path. Gamma rays, of course, will not leave visible tracks in any of these detectors. However, their presence can be detected indirectly because they are able to produce electrons by such processes as the photoelectric effect and pair production. The gamma-ray energy can be determined from the measured energy of these electrons.

Biological Effects and Medical Applications of Radiation

In medicine, nuclear radiation can be used beneficially in the diagnosis and treatment of some diseases, but it also is potentially harmful if not handled and administered properly. Nuclear radiation and X-rays can penetrate human tissue without pain or any other sensation. However, early investigators quickly learned that large doses or repeated small doses can lead to reddened skin, lesions, and other conditions. It is now known that certain types of cancers can be caused by excessive exposure to radiation.

The chief hazard of radiation is damage to living cells, due primarily to ionization. Ions, particularly complex ions or radicals produced by radiation, may be highly reactive (for example, a hydroxyl ion [OH^-] produced from water). Such reactive ions interfere with the normal chemical operations of the cell. If enough cells are damaged or killed, cell reproduction might not be fast enough, and the irradiated tissue could eventually die. In other instances, genetic damage, or mutation, may occur in a chromosome in the cell nucleus. If the affected cells are sperm or egg cells (or their precursors), any children that they produce may have various birth defects. If the damaged cells are ordinary body cells, they may become cancerous, reproducing in a rapid and uncontrolled manner and eventually becoming a malignant tumor. The human cells most susceptible to radiation damage are those of the reproductive organs, bone marrow, and lymph nodes. To begin our discussion of radiation damage and applications, let us investigate how the radiation "dose" is quantified.

Radiation Dosage An important consideration in radiation therapy and radiation safety is the amount, or *dose,* of radiation energy absorbed. Several quantities are used to describe this amount in terms of *exposure, absorbed dose,* or *equivalent dose.* The earliest unit of dosage, the **roentgen (R)**, based on exposure and defined in terms of ionization produced in air. One roentgen is the quantity of X-rays or gamma rays required to produce an ionization charge of 2.58×10^{-4} C/kg in air.

The **rad** (**ra**diation **a**bsorbed **d**ose) is an *absorbed dose* unit. One rad is an absorbed dose of radiation energy of 10^{-2} J/kg of absorbing *material.* Note that the rad is based on energy absorbed from the radiation rather than simply ionization caused by the radiation in air (as the roentgen is). As such, it is more directly related to the biological damage caused by the radiation. Because of this characteristic, the rad has largely replaced the roentgen.

The rad is not an SI unit. The SI unit for absorbed dose is the **gray (Gy)**, defined as

$$1 \text{ Gy} = 1 \text{ J/kg} = 100 \text{ rad}$$

Note: The gray was named in honor of Louis Harold Gray, a British radiobiologist whose studies laid the foundation for measuring absorbed dose.

TABLE 29.4	Typical Relative Biological Effectiveness (RBE) Values of Various Types of Radiations	
Type		RBE (or QF)
X-rays and gamma rays		1
Beta particles		1.2
Slow neutrons		4
Fast neutrons and protons		10
Alpha particles		20

However, the most meaningful assessment of the effects of radiation must involve measuring the *biological damage* produced, because it is well known that equal doses (in rads) of different types of radiation produce *different* effects. For example, a relatively massive alpha particle with a charge of $+2e$ moves through the tissue rather slowly, with a great deal of electrical interaction. The ionizing collisions thus occur close together along a short penetration path and are more localized. Therefore, potentially more dangerous damage is done by alpha particles than by electrons or gamma rays.

This *effective dose* is measured in terms of the **rem** (**r**ad **e**quivalent **m**an). The various degrees of effectiveness of different particles are characterized by a factor called **relative biological effectiveness (RBE)**, or *quality factor* (QF), which has been tabulated for various particles in Table 29.4. (Note in Table 29.4 that X-rays and gamma rays have, by definition, an RBE of 1.)

The effective dose is given by the product of the dose in rads and the appropriate RBE:

$$\text{effective dose (in rems)} = \text{dose (in rads)} \times \text{RBE} \qquad (29.6)$$

Thus, 1 rem of *any* type of radiation does approximately the same amount of biological damage. For example, a 20-rem effective dose of alpha particles does the same amount of damage as a 20-rem effective dose of X-rays. However, note that to administer these doses, 20 rad of X-rays is needed, compared with only 1 rad of alpha particles.

Remember that the SI unit of *absorbed dose* is the gray. The SI unit of *effective dose* is the **sievert (Sv)**:

$$\text{effective dose (in sieverts)} = \text{dose (in grays)} \times \text{RBE} \qquad (29.7)$$

Since 1 Gy = 100 rad, it follows that 1 Sv = 100 rem.

It is difficult to set a maximum permissible radiation dosage, but the general standard for humans is an average dose of 5 rem/yr after age 18, with no more than 3 rem in any three-month period. In the United States, the normal average annual dose per capita is about 200 mrem (millirem). About 125 mrem comes from the natural background of cosmic rays and naturally occurring radioactive isotopes in soil, building materials, and so on. The remainder is chiefly from diagnostic medical applications, mostly X-rays.

Medical Treatment Using Radiation   Some radioactive isotopes can be used for medical treatment, typically for cancerous conditions. Since a radioactive isotope, or a *radioisotope* as it is sometimes called, behaves chemically like a stable isotope of the element, it can participate in chemical reactions associated with normal bodily functions. One such radioisotope, used to treat thyroid cancer, is ^{131}I. Under usual conditions, the thyroid gland absorbs normal iodine. However, if ^{131}I is absorbed in a large enough dose, it can kill cancer cells. To see how a dose of radiation to the thyroid from ^{131}I can be estimated, consider Example 29.9. For a discussion of further uses of radioisotopes, see Insight 29.1 on Biological and Medical Applications of Radiation (page 927).

Example 29.9 ■ Radiation Dosage: Iodine-131 and Thyroid Cancer

One method of treating a cancerous thyroid is to administer a hefty amount of the radioactive isotope ^{131}I. The thyroid absorbs this iodine, and the iodine's gamma rays kill cells in the thyroid. (For data on ^{131}I, see Example 29.3.) (a) Write down the decay scheme of ^{131}I, and predict the identity of the daughter nucleus, which, in this case, is stable after emitting a gamma ray. (b) The charged particle (part (a) tells the type) has an average kinetic energy of 200 keV. Assume that the patient was given 0.0500 mCi of ^{131}I and that the thyroid absorbs only 25% of this. Further assume that only 40% of that 25% actually decays in the thyroid. If all of the energy carried by the charged particles is deposited in the thyroid, estimate the dose received by the thyroid (50.0 g) due to the ionization created by the charged particle radiation only. (Do not include the effect of the gamma rays.)

Thinking It Through. (a) The decay will be β^- decay, because the initial nucleus contains too many neutrons (78 neutrons compared with the 74 for stable iodine). The daughter nucleus is determined by its proton number. The daughter is left in an excited state and emits a gamma ray in order to become stable. (b) The dose depends on the energy deposited per kilogram of thyroid. Hence, we need to know how many β^- particles are emitted (and therefore absorbed). This number is determined by the initial number of ^{131}I nuclei that are actually present in the thyroid. The effective dose will depend on the RBE for β^- particles, found in Table 29.4.

Solution.

Given: 0.0500 mCi of ^{131}I ingested   *Find:* (a) the decay scheme for ^{131}I
25% of the ^{131}I makes it to the thyroid   (b) the dose (in rem) from
40% of the ^{131}I that makes it to   the emitted particles
    the thyroid decays there
    $\overline{K}_\beta = 200$ keV
    $m = 50.0$ g $= 0.0500$ kg
    RBE (see Table 29.4)

(a) Looking up the element with $Z = 54$, we find that it is xenon (Xe). The decay scheme is therefore

$$^{131}_{53}\text{I} \quad \rightarrow \quad ^{131}_{54}\text{Xe}^* \quad + \quad \beta^-$$
$$\text{iodine} \qquad\qquad \text{xenon} \qquad\qquad \text{beta}$$
$$\text{(excited)}$$
$$\downarrow$$
$$^{131}_{54}\text{Xe} \quad + \quad \gamma$$
$$\text{xenon} \qquad \text{gamma ray}$$

(b) Of the 0.0500 mCi, only 0.0125 mCi makes it to the thyroid. Of that, only 40%, or 0.00500 mCi (5.00×10^{-6} Ci), actually decays in the thyroid. From this information and Eq. 29.2 the number of ^{131}I nuclei that decay in the thyroid can be found. From Example 29.3, the decay constant for ^{131}I is $\lambda = 1.0 \times 10^{-6}$ s^{-1}. Thus, the number of ^{131}I nuclei, N, that decays in the thyroid is

$$N = \frac{R}{\lambda} = \frac{(5.00 \times 10^{-6}\ \text{Ci})\left(3.7 \times 10^{10}\ \dfrac{\text{nuclei/s}}{\text{Ci}}\right)}{1.0 \times 10^{-6}\ \text{s}^{-1}} = 1.85 \times 10^{11}\ ^{131}\text{I nuclei}$$

Each ^{131}I nucleus releases one β^- particle, with an average kinetic energy of 200 keV. Remembering that there is 1.60×10^{-19} J/eV, or 1.60×10^{-16} J/keV, we find that the energy, E, deposited in the thyroid is

$$E = (1.85 \times 10^{11}\ ^{131}\text{I nuclei})\left(200\ \frac{\text{keV}}{^{131}\text{I nuclei}}\right)\left(1.60 \times 10^{-16}\ \frac{\text{J}}{\text{keV}}\right)$$
$$= 5.92 \times 10^{-3}\ \text{J}$$

The absorbed dose is

$$\text{absorbed dose} = \frac{5.92 \times 10^{-3}\ \text{J}}{0.0500\ \text{kg}} = 0.118\ \frac{\text{J}}{\text{kg}}$$
$$= 0.118\ \text{Gy or } 11.8\ \text{rad}$$

The effective dose (in sieverts and rems) is this multiplied by the RBE for beta particles (1.2):

$$\text{effective dose} = (0.118\ \text{Gy})(1.2) = 0.142\ \text{Sv} = 14.2\ \text{rem}$$

Follow-Up Exercise. In this Example, determine the absorbed and effective dose from the gamma rays, assuming that 10% of the gamma rays are absorbed in the thyroid tissue and that their energy is 364 keV. Compare these results with the dose from the β^- radiation.

INSIGHT 29.1 BIOLOGICAL AND MEDICAL APPLICATIONS OF RADIATION

Radiation has always been a double-edged sword. We know of its potentially harmful side, yet sources of radiation can also provide solutions to problems. Exposure to high levels of gamma radiation is now an approved method of food sterilization in the United States. Chicken and beef are commonly sterilized this way, thus reducing the threat of *Salmonella* and *E. coli* contamination. In the aftermath of the terrorist attacks on the United States in the fall of 2001, some of this gamma-radiation technology is being retooled so that it can kill anthrax spores and other weapons based on organisms.

External radiation sources such as ^{60}Co are also used to treat cancer. ^{60}Co emits energetic gamma rays with energies of 1.17 and 1.33 MeV. Thus a sample of ^{60}Co, with its relatively long half-life, can provide an inexpensive and convenient source of penetrating radiation. Essentially, all you need is the ^{60}Co sample in a lead box with a hole to allow gamma rays to exit in one direction. One problem, however, with this "single-beam" method is that the gamma rays deposit energy in the healthy flesh both in front of and behind the targeted tumor.

An improved version of ^{60}Co treatment, called the *gamma knife*, is in use at several research hospitals (Fig. 1). Once the tumor is located, beams of gamma rays from ^{60}Co sources arranged in a ring are accurately aimed at it. Any one source is relatively weak and thus does not do too much damage outside the tumor itself. However, where the beams meet at the tumor site, a lot of energy is

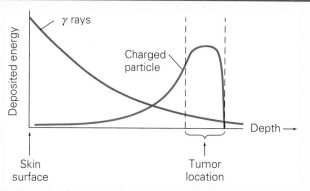

FIGURE 2 A comparison of the energy deposited for a gamma-ray beam with that of a charged particle, such as a negative pion, passing through tissue.

deposited. Thus, a large dose can be deposited at the tumor, with minimal damage to surrounding tissue. This instrument requires careful computer calculations and is still under development.

Even more exotic techniques are being tested in an effort to kill inoperable tumors. One such method is *pion therapy*. A pion is an unstable elementary particle (Chapter 30) that can be produced in accelerators by bombarding a target, such as carbon, with high-energy protons. Of medical interest are the negatively charged pions π^- (positive and neutral ones exist also). Pions of a specific kinetic energy can be selected and focused by magnetic fields onto a region of the body where a tumor exists. Unlike photons (gamma rays and X-rays), charged particles create most of their ionization "damage" at the end of their path, when they are moving slowly. By adjusting their kinetic energy, researchers can end the pion's path right at the tumor site (Fig. 2), thus causing maximum damage to the cancer cells. As a bonus, since the pions are unstable, they give off gamma rays that will do even more damage from inside the tumor (Fig. 3). Research on this technology-intensive technique is ongoing, with some success.

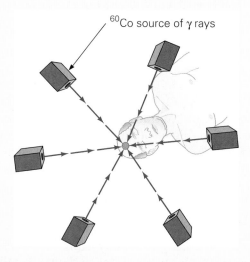

FIGURE 1 The gamma knife consists of many relatively weak beams of gamma rays (usually from ^{60}Co) that are calculated to converge on the location of an inoperable tumor. Thus, unlike in the traditional single-beam treatment, much more of the energy ends up at the tumor site, and less damage is done in front of or behind the tumor.

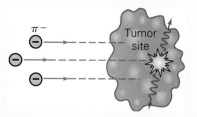

FIGURE 3 Pionic cancer therapy consists of focusing a beam of negative pions onto a tumor. Unlike energy from beams of gamma rays, most of the pion energy is deposited at the tumor site. In addition, when pions decay, gamma rays are released at the tumor, causing further destruction of cancer cells.

Medical Diagnosis Applications That Use Radiation Besides theraputic use, such as that previously described for iodine-131, radioactive isotopes can be used for diagnostic procedures. Since the radioisotope behaves chemically like a stable isotope, attaching radioisotopes to molecules enables the molecules to be used as tracers as they travel to different organs and regions of the body.

Many bodily functions can be studied by monitoring the location and activity of tracer molecules as they are absorbed during body processes. For example, the

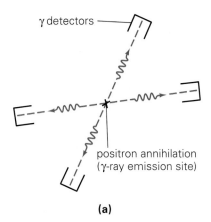

γ detectors

positron annihilation
(γ-ray emission site)

(a)

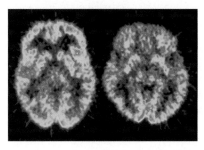

(b)

(c)

▶ **FIGURE 29.18 PET scan (a)** and **(b)** A PET scanner can, for example, monitor brain activity after the administration of glucose-containing radioactive isotopes. Note the use of an array of detectors for the gamma radiation produced when a positron is annihilated. Any one detector pair pinpoints a line on which the source of the gamma emission was located. With many such pairs working together, the site can be determined to an accuracy of about a centimeter. **(c)** PET scans of a normal brain (left) and the brain of a schizophrenic patient (right).

activity of the thyroid gland can be determined by monitoring its iodine uptake with small amounts of radioactive iodine-123. This isotope emits gamma rays and has a half-life of 13.3 h. The uptake of radioactive iodine by a person's thyroid can be monitored by a gamma detector and compared with the function of a normal thyroid to check for abnormalities. Similarly, radioactive solutions of iodine and gold are quickly absorbed by the liver.

One of the most commonly used diagnostic tracers is technetium-99 (^{99}Tc). It has a convenient half-life of 6 h, emits gamma rays, and combines with a large variety of compounds. When injected into the bloodstream, ^{99}Tc will not be absorbed by the brain, because of the blood–brain barrier. However, tumors do not have this barrier, and brain tumors readily absorb the ^{99}Tc. These tumors then show up as gamma-ray emitting sites using detectors external to the body. Similarly, other areas of the body can be scanned and unusual activities noted and measured.

It is possible to image gamma-ray activity in a single plane, or "slice," through the body. A gamma detector is moved around the patient to measure the emission intensity from many angles. A complete image can then be constructed by using computer-assisted tomography, as in X-ray CT. This process is referred to as *single-photon emission tomography* (SPET). Another technique, *positron emission tomography* (PET), uses tracers that are positron emitters, such as ^{11}C and ^{15}O. When a positron is emitted, it is quickly annihilated, and two gamma rays are produced that travel in opposite directions. The gamma rays are recorded simultaneously by a ring of detectors surrounding the patient (◀Fig. 29.18).

In a common application, PET technology is used to detect fast-growing cancer cells. The positron emitter ^{18}F is chemically attached to glucose molecules and administered to the patient. Actively growing cells absorb glucose, but *very* active cancer cells absorb considerably more. By comparing the emissions coming from a given region on a potentially sick patient to that from a normal, healthy person, these "overactive" cancer cells can be detected. Such PET scans are now routinely done, for example, as a follow-up to chemotherapy treatment of lymphoma (cancer of the lymph system). A PET scan can detect even tiny leftover active tumors that can then be targeted for further treatment.

Domestic and Industrial Applications of Radiation

A common application of radioactivity in the home is the smoke detector. In this detector, a weak radioactive source ionizes air molecules. The freed electrons and the positive ions are collected using the voltage of a battery, thus setting up a small current in the detector circuit. If smoke enters the detector, the ions there become attached to the smoke particles, causing a reduction in the current. The drop in current is sensed electronically, which triggers an alarm (▶Fig. 29.19).

Industry also makes good use of radioactive isotopes. Radioactive tracers are used to determine flow rates in pipes, to detect leaks, and to study corrosion and wear. Also, it is possible to radioactivate certain compounds at a particular stage in a process by irradiating them with particles, generally neutrons. This technique is called **neutron activation analysis** and is an important method of identifying elements in a sample. Before the development of this procedure, the chief methods of identification were chemical and spectral analyses. In both of these methods, a fairly large amount of a sample has to be destroyed during the procedure. As a result, a sample may not be large enough for analysis, or small traces of elements in a sample may go undetected. Neutron activation analysis has the advantage over these methods on both scores. Only minute samples are needed, and the method can detect very minute trace amounts of an element.

A typical neutron activation process might start with californium-252, an unstable neutron emitter that can be produced artificially:

$$^{252}_{98}\text{Cf} \rightarrow ^{251}_{98}\text{Cf} + ^{1}_{0}\text{n}$$

(source)

These neutrons are used to bombard a sample and create characteristic gamma rays. A common target is nitrogen, consisting mostly of ^{14}N. When ^{14}N absorbs a neutron, the ^{15}N nucleus is usually created in an excited state. The excited nitrogen-15

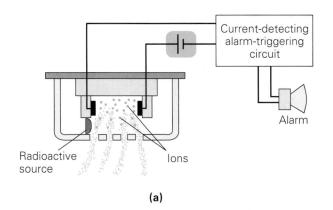

(a)　　　　　　　　　　　　　　**(b)**

(a) A weak radioactive source ionizes the air and sets up a small current. Smoke particles that enter the detector attach to some of the ionized electrons, thereby reducing the current, causing an alarm to sound. **(b)** Inside a real smoke detector, the ionization chamber is the aluminum "can" containing the americium-241. The slots allow airflow, and the can acts as one of the charged plates. Inside is a ceramic holder that contains the oppositely charged plate. Under that plate is the americium-241 source. Even though the activity of the source is small, caution should be taken never to touch the source if the detector is opened like this.

nucleus decays with the emission of a gamma ray with a distinctive energy. This reaction is shown as follows:

$$\ce{^{1}_{0}n} \ + \ \ce{^{14}_{7}N} \ \rightarrow \ \underset{(excited\ nucleus)}{\ce{^{15}_{7}N^*}} \ \rightarrow \ \ce{^{15}_{7}N} \ + \ \underset{(gamma\ ray)}{\gamma}$$

When an energy-sensitive gamma-ray detector is placed to the side of the sample, the presence of nitrogen can be determined.

Nitrogen activation is commonly used as an important antiterrorist tool at airports. Virtually all explosives contain nitrogen. Thus, by using neutron activation and analyzing the energy of any gamma-ray emission coming from a suitcase, we can check for an explosive device in the suitcase. Other materials in the suitcase may contain nitrogen too, so manual checks are made to confirm any suspicious findings.

Recently, the U.S. government has given permission for the use of gamma radiation in the processing of poultry. The radiation kills bacteria, helps preserve the food, and in no way makes the food radioactive. There are, however, continuing concerns with this process from food health professionals. Even though the gamma-ray emission cannot make the meat radioactive, it *can* change some of the *chemical* bonding through ionization effects. This possibility has prompted enough concerns about whether this process can affect the chemical structure of the meat—making it unsafe to eat—to warrant further study.

Chapter Review

- The **nuclear force** is the short-range attractive force between nucleons that is responsible for holding the nucleus together.

- The nucleus of an atom contains protons and neutrons, collectively called **nucleons**.

- The nucleus is characterized by its **proton number Z** and its **neutron number N**. Its **mass number**, A, is the total number of nucleons, so $A = N + Z$.

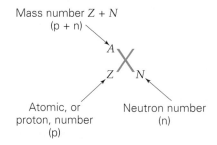

- **Isotopes** of a given element differ only in the number of neutrons in their nucleus.

- Nuclei may undergo **radioactive decay** by the emission of an **alpha particle** (a helium nucleus) **(α)**; a **beta particle**, which can be either an electron **(β^-)** or a positron **(β^+)**; or a **gamma ray (γ)**, a high-energy photon of electromagnetic radiation.

Some nuclei become more stable by capturing an orbital electron (electron capture, or **EC**).

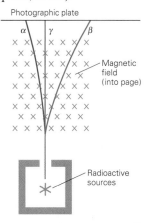

- In any nuclear process, two conservation rules pertain: **Conservation of nuclear number** and **conservation of charge**.

- The **half-life** of a nuclide is the time required for the number of undecayed nuclei in a sample to fall to half of its initial value. The number of undecayed nuclei remaining after a time t is given by the exponential-decay relationship

$$N = N_0 e^{-\lambda t} \qquad (29.3)$$

where the **decay constant (λ)** is inversely related to the half-life by

$$t_{1/2} = \frac{0.693}{\lambda} \tag{29.4}$$

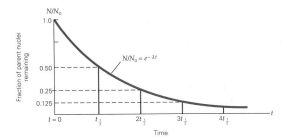

The **activity (R)** of a radioactive sample is the rate at which the nuclei in the sample decay. It is proportional to the number of undecayed nuclei and is given by

$$R = \text{activity} = \left| \frac{\Delta N}{\Delta t} \right| = \lambda N \tag{29.2}$$

Activity is measured in units of the **curie (Ci)** or the **becquerel (Bq)**. $1\ \text{Ci} \equiv 3.70 \times 10^{10}$ decays/s, and $1\ \text{Bq} \equiv 1$ decay/s; thus, $1\ \text{Ci} = 3.70 \times 10^{10}$ Bq.

- Nuclear masses are usually measured in terms of the **atomic mass unit (u)**. The atomic mass unit is related to the kilogram by $1\ \text{u} = 1.66054 \times 10^{-27}$ kg. In light of Einstein's mass–energy equivalence, the complete annihilation of 1 u of mass releases 931.5 MeV of energy.

- The **total binding energy (E_b)** of a nucleus is the minimum amount of energy needed to separate a nucleus into its constituent nucleons:

$$E_b = (\Delta m)c^2 \tag{29.5}$$

where Δm is the **mass defect**. The mass defect and is the difference between the sum of the masses of the constituent nucleons and the mass of the nucleus.

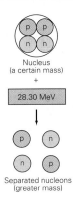

- The **effective dose** of radiation [in **rems** or **sieverts (Sv)**] is determined by the energy deposited per kilogram of material [in **rads** or **grays (Gy)**] and the type of particle depositing that energy (as expressed by the **relative biological effectiveness**, or **RBE**):

$$\text{effective dose (in rem)} = \text{dose (in rad)} \times \text{RBE} \tag{29.6}$$

$$\text{effective dose (in Sv)} = \text{dose (in Gy)} \times \text{RBE} \tag{29.7}$$

Exercises

MC = *Multiple Choice Question,* **CQ** = *Conceptual Question, and* **IE** = *Integrated Exercise. Throughout the text, many exercise sections will include "paired" exercises. These exercise pairs, identified with* **red numbers**, *are intended to assist you in problem solving and learning. In a pair, the first exercise (even numbered) is worked out in the Study Guide so that you can consult it should you need assistance in solving it. The second exercise (odd numbered) is similar in nature, and its answer is given at the back of the book.*

29.1 Nuclear Structure and the Nuclear Force

1. **MC** In the Rutherford scattering experiment, for which target nucleus would alpha particles of a given kinetic energy approach more closely: (a) carbon, (b) iron, or (c) uranium? (a)

2. **MC** The nuclei of carbon-12 and carbon-13 (a) have the same number of nucleons, (b) have the same number of neutrons, (c) have the same number of protons, or (d) none of the preceding. (c)

3. **MC** At the same close distance, between which pair of particles is the nuclear force the largest: (a) neutron–proton, (b) neutron–neutron, (c) proton–proton, or (d) the force is the same for all pairs? (d)

4. **CQ** In the Rutherford scattering experiment, the minimum distance of approach for the alpha particle is given by Eq. 29.1. Explain why this does not necessarily represent the nuclear radius. Is it larger or smaller than the nuclear radius? see ISM

5. **CQ** Nuclei with the same number of neutrons are called *isotones*. What nitrogen nucleus is an isotone of carbon-13? nitrogen-14

6. **CQ** Nuclei with the same number of nucleons are called *isobars*. What nuclide of nitrogen is an isobar of carbon-13? nitrogen-13

7. ● Determine the number of protons, neutrons, and electrons in a neutral atom with the following nuclei: (a) ^{43}Ca and (b) ^{206}Pb. (a) 20p, 23n, 20e (b) 82p, 124n, 82e

8. ● Oxygen has three stable isotopes, with 8, 9, and 10 neutrons, respectively. Write these isotopes in nuclear notation. $^{16}_{8}$O, $^{17}_{8}$O, $^{18}_{8}$O

9. ● An isotope of potassium has the same number of neutrons as the nuclide argon-40. Write the nuclear notation for this potassium isotope. $^{41}_{19}$K

10. ● ^{24}Mg and ^{25}Mg are two isotopes of magnesium. What are the numbers of protons, neutrons, and electrons in each if (a) the atom is electrically neutral, (b) the ion has a -2 charge, and (c) the ion has a $+1$ charge? see ISM

11. ● One isotope of uranium has a mass number of 235. What are the numbers of protons, neutrons, and electrons in a neutral atom of this isotope? 92p, 143n, 92e

12. **IE** ● (a) Isotopes of an element have the same (1) atomic number, (2) neutron number, (3) mass number. (b) Write two possible isotopes for ^{197}Au.
(a) (1) atomic number (b) ^{196}Au, ^{198}Au

13. ●● An approximate expression for the nuclear radius (R) is $R = R_o A^{1/3}$, where $R_o = 1.2 \times 10^{-15}$ m and A is the mass number of the nucleus. (a) Find the nuclear radii of atoms of the noble gases: He, Ne, Ar, Kr, Xe, and Rn. (b) Determine the mass density of each of these species. Does your answer surprise you? see ISM

14. **IE** ●●● Assume Rutherford used alpha particles with a kinetic energy of 5.25 MeV. (a) To which of the following nuclei would the alpha particle come closest in a head-on collision: (1) aluminum, (2) iron, or (3) lead? (b) Determine the distance of closest approach for the three nuclei in part (a) and compare them to the nuclear radii given in Exercise 13. Are any of these distance comparable to the radius of the target nucleus? [*Hint*: In Equation 29.1, $mv^2 = 2K_\alpha$; why?]
(a) (1) aluminum (b) see ISM, all are larger than radii

29.2 Radioactivity

15. **MC** The conservation of nucleons and the conservation of charge apply to (a) only alpha decay, (b) only beta decay, (c) only gamma decay, or (d) all nuclear decay processes. (d)

16. **MC** β^- decay can occur only in nuclei with what Z values: (a) $Z > 82$, (b) $Z \le 82$, or (c) it can occur regardless of the Z value? (c)

17. **MC** Aluminum has only one stable isotope, ^{27}Al. ^{26}Al would be expected to decay by which beta decay mode: (a) β^+, (b) β^-, or (c) beta decay would not be an option for ^{26}Al? (a)

18. **CQ** The neutron number is not conserved in a beta decay. Is this a violation of the conservation of nucleons? Explain.
no, see ISM

19. **CQ** ^{19}F is the only stable isotope of fluorine. What two possible decay modes would you expect ^{18}F to decay by? What would be the resulting nucleus in both cases?
β^+ and electron capture, ^{18}O

20. **CQ** When an excited nucleus decays to a lower energy level by gamma-ray emission, the actual energy of the gamma-ray photon is a bit less than the difference in energy between the two levels involved in the transition. Explain why this is true. to conserve momentum, the nucleus recoils, taking some of the energy

21. **IE** ● Tritium is radioactive. (a) Would you expect it to (1) β^+, (2) β^-, or (3) alpha decay? Why? (b) What is the identity of the daughter nucleus? Is it stable? [*Hint*: Write the nuclear equation for each.]
(a) (2) β^-; see ISM (b) helium-3; yes, it is stable

22. ● Write the nuclear equations expressing (a) the beta decay of $^{60}_{27}$Co and (b) the alpha decay of $^{226}_{88}$Ra.
(a) $^{60}_{27}$Co → $^{60}_{28}$Ni + $^{0}_{-1}$e (b) $^{226}_{88}$Ra → $^{222}_{86}$Rn + $^{4}_{2}$He

23. ● Write the nuclear equations for (a) the alpha decay of neptunium-237, (b) the β^- decay of phosphorus-32, (c) the β^+ decay of cobalt-56, (d) electron capture in cobalt-56, and (e) the γ decay of potassium-42. see ISM

24. **IE** ● Polonium-214 can decay by alpha decay. (a) The product of its decay has how many fewer protons than polonium-214: (1) zero, (2) one, (3) two, or (4) four? Why? (b) Write the nuclear equation for this decay.
(a) (3) two (b) $^{214}_{84}$Po → $^{210}_{82}$Pb + $^{4}_{2}$He

25. ● A lead-209 nucleus results from both alpha–beta sequential decays and beta–alpha sequential decays. What was the grandparent nucleus? (Show this result for both decay routes by writing the nuclear equations for both decay processes.)
$^{213}_{83}$Bi, see ISM

26. ●● Complete the following nuclear-decay equations:
(a) $^{8}_{4}$Be → $^{4}_{2}$He + $^{4}_{2}$He
(b) $^{240}_{98}$Pu → $^{97}_{38}$Sr → $^{139}_{56}$Pu + ($^{1}_{0}$n)
(c) $^{47}_{21}$Sc* → $^{47}_{21}$Sc + γ
(d) $^{29}_{11}$Na → $^{0}_{-1}$e + $^{29}_{12}$Mg

27. ●● Complete the following nuclear-decay equations:
(a) $^{238}_{92}$U → $^{234}_{90}$Th + $^{4}_{2}$He
(b) $^{40}_{19}$K → $^{40}_{20}$Ca + $^{0}_{-1}$e
(c) $^{236}_{92}$U → $^{131}_{53}$I + 3 ($^{1}_{0}$n) + $^{102}_{39}$Y
(d) $^{23}_{11}$Na + γ → $^{23}_{11}$Na*
(e) $^{22}_{11}$Ne + $^{0}_{-1}$e → $^{22}_{10}$Ne

28. ●● Actinium-227 ($^{227}_{89}$Ac) decays by alpha decay or by beta decay and is part of a decay sequence like the one shown in Fig. 29.8. Each daughter nucleus then decays to radium-223 ($^{223}_{88}$Ra), which subsequently decays to polonium-215 ($^{215}_{84}$Po). Write the nuclear equations for the decay process in the decay series from ^{227}Ac to ^{215}Po. see ISM

29. ●●● The decay series for neptunium-237 is shown in ▼ Fig. 29.20. Identify the decay modes and each of the nuclei in the sequence. see ISM

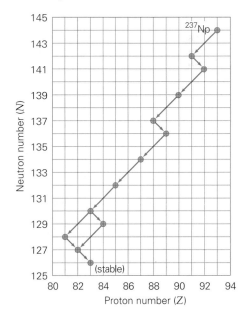

▲ **FIGURE 29.20 Neptunium-237 decay series** See Exercise 29.

29.3 Decay Rate and Half-Life

30. **MC** After one half-life, a sample of a particular radioactive material (a) is half as massive, (b) has its half-life reduced by half, (c) is no longer radioactive, (d) has its activity reduced by half. (d)

31. **MC** In two half-lives, the activity of a radioactive sample will have decreased by what percent: (a) 25%, (b) 50%, (c) 75%, or (d) 87.5%? (c)

32. **MC** An alpha emitter gives off alpha particles with an energy of 4.4 MeV. After three half-lives, what is the energy of the alpha particles being given off: (a) 2.2 MeV, (b) 1.1 MeV, (c) 8.8 MeV, or (d) 4.4 MeV? (d)

33. **CQ** What physical or chemical properties affect the decay rate, or half-life, of a radioactive isotope? none

34. **CQ** Nuclide A has a decay constant that is half that of nuclide B. If the two nuclides start with the same number of undecayed nuclei, will twice as many of nuclide A's nuclei as the number of nuclide B's nuclei decay in a given time? Explain. no, decay is exponential

35. **CQ** What are the (a) half-life and (b) decay constant for a stable isotope? (a) infinity (b) zero

36. ● A particular radioactive sample undergoes 2.50×10^6 decays/s. What is the activity of the sample in (a) curies and (b) becquerels? (a) 67.6 μCi (b) 2.50×10^6 Bq

37. ● At present, a laboratory radioactive beta source has an activity of 20 mCi. (a) What is the present decay rate in decays per second? (b) Assuming that one beta particle is emitted per decay, how many are currently emitted per minute? (a) 7.4×10^8 decays/s (b) 4.4×10^{10} betas/min

38. **IE** ● The half-life of a radioactive isotope is 1 h. (a) What fraction of a sample would be left after 3 h: (1) one third, (2) one eighth, or (3) one ninth? Why? (b) What fraction of a sample would be left after 1 day? (a) (2) one eighth (b) $\left(\frac{1}{2}\right)^{24}$, or approximately 6×10^{-6} % of the original

39. ● A 1.25-μCi alpha source gives off alpha particles with a kinetic energy of 2.78 MeV. At what rate (in watts) is kinetic energy being produced? 2.06×10^{-8} W

40. ●● A sample of technetium-104 has an activity of 10 mCi. Estimate the activity of the sample after one hour has elapsed ($t_{1/2} = 18$ min). 0.99 mCi

41. ●● What period of time is required for a sample of radioactive tritium (^{3}H) to lose 80.0% of its activity? Tritium has a half-life of 12.3 years. 28.6 years

42. ●● A certain amount of ^{131}I is introduced into a patient in a medical diagnostic procedure for her thyroid. What percentage of the sample remains after 1.00 day, assuming that all of the ^{131}I is retained in the patient's thyroid gland? (Answer to three significant figures.) 91.7%

43. **IE** ●● Carbon-14 dating is used to determine the age of some buried bones. (a) If the activity of bone A is higher than that of bone B, then bone A is (1) older than, (2) younger than, (3) the same age as bone B. Why? (b) A sample of old bone is found to have 4.0 beta decays/min for each gram of carbon. Approximately how old is the bone? (a) (2) younger than (b) 1.1×10^4 years

44. ●● Show that the number N of radioactive nuclei remaining in a sample after n half-lives is given by

$$N = \frac{N_o}{2^n} = \left(\frac{1}{2}\right)^n N_o$$

where N_o is the initial number of nuclei. see ISM

45. ●● Some ancient writings on parchment are found sealed in a jar in a cave. If carbon-14 dating shows the parchment to be 28 650 years old, what percentage of the carbon-14 atoms still remains in the sample compared with the number that was present when the parchment was made? 3.1%

46. ●● How long would it take (to the nearest whole year) for the activity of a sample of cobalt-60 to reduce to 20% of the original activity? 12 years

47. ●● A soil sample contains 40 μg of ^{90}Sr. Approximately how much ^{90}Sr will be in the sample 150 years from now? 0.98 μg

48. ●● (a) What is the decay constant of fluorine-17 ($t_{1/2} = 66.0$ s)? (b) How long will it take for the activity of a sample of ^{17}F to decrease to 10% of its initial value? (a) 1.05×10^{-2} s^{-1} (b) 219 s

49. ●● Francium-223 ($^{223}_{87}$Fr) has a half-life of 21.8 min. (a) How many nuclei are initially present in a 25.0-mg sample of this isotope? (b) How many nuclei will be present 1 h and 49 min later? (a) 6.75×10^{19} nuclei (b) 2.11×10^{18} nuclei

50. ●● A basement room containing radon gas ($t_{1/2} = 3.82$ days) is sealed to be airtight. (a) If 7.50×10^{10} radon atoms are trapped in the room, estimate how many radon atoms remain in the room after one week. (b) Radon undergoes alpha decay. After 30 days, is the number of its daughter nuclei equal to the number of radon parents that have decayed? Explain. (a) 2.11×10^{10} (b) no, the daughters themselves are radioactive

51. ●● In 1898, Pierre and Marie Curie isolated about 10 mg of radium-226 from eight tons of uranium ore. If this sample had been placed in a museum, how much of the radium would remain in the year 2100? 9.2 mg

52. ●● An ancient artifact is found to contain 250 g of carbon and has an activity of 475 decays per minute. What is the approximate age of the artifact, to the nearest thousand years? 17 000 years

53. ●●● The recoverable U.S. reserves of high-grade uranium-238 ore (high-grade ore contains about 10 kg of ^{238}U$_3$O$_8$ per ton) are estimated to be about 500 000 tons. Neglecting any geological changes, what mass of ^{238}U existed in this high-grade ore when the Earth was formed, about 4.6 billion years ago? [*Hint*: See Appendix V.] 7.6×10^6 kg

54. **IE** ●●● Nitrogen-13, with a half-life of 10 min, decays by positron emission. (a) The end product is (1) ^{13}N, (2) ^{13}C, or (3) ^{13}O. (b) If a sample of pure ^{13}N has a mass of 1.5 g at a certain time, what is the activity 35 min later? (c) What percentage of the sample is ^{13}N at this time? (a) (2) ^{13}C (b) 4.2×10^{20} decays/min (c) 8.8%

29.4 Nuclear Stability and Binding Energy

55. **MC** For nuclei with a mass number greater than 40, which of the following statements is correct? (a) The number of protons is approximately equal to the number of neutrons; (b) the number of protons exceeds the number of neutrons; (c) all such nuclei are stable up to $Z = 92$; (d) none of the preceding. (d)

56. **MC** The average binding energy per nucleon of the daughter nucleus in a decay process is (a) greater than, (b) less than, or (c) equal to that of the parent nucleus. (a)

57. **MC** From which nucleus is it easier to remove a neutron: (a) ^{25}Mg, (b) ^{24}Mg, or (c) both require the same amount of energy? (a)

58. **CQ** Why does aluminum have only one stable isotope (^{27}Al)? see ISM

59. **CQ** Explain why, of the two main uranium isotopes, ^{238}U is more abundant than ^{235}U. [*Hint*: Although they both are unstable, ^{238}U is closer to stability; why?] see ISM

60. **CQ** Compared to ^{3}He, the probability of absorbing a neutron is much less likely for ^{4}He. Explain this using what you know about odd/even proton and neutron numbers. see ISM

61. ● From which of the following pairs of nuclei would you expect it to be easier to remove a neutron: (a) $^{16}_{8}$O or $^{17}_{8}$O; (b) $^{40}_{20}$Ca or $^{42}_{20}$Ca; (c) $^{10}_{5}$B or $^{11}_{5}$B; (d) $^{208}_{82}$Pb or $^{209}_{83}$Bi? State your reasoning for your choice in each case. (a) $^{17}_{8}$O (b) $^{42}_{20}$Ca (c) $^{10}_{5}$B (d) approximately the same, see ISM

62. ● Only two isotopes of Sb (antimony, $Z = 51$) are stable. Pick the two stable isotopes from the following list: (a) ^{120}Sb; (b) ^{121}Sb; (c) ^{122}Sb; (d) ^{123}Sb; (e) ^{124}Sb. (b) and (d); others are odd–odd

63. ● The total binding energy of $^2_1$H is 2.224 MeV. Use this information to compute the mass of a ^{2}H nucleus from the known mass of a proton and a neutron. 2.013 553 u

64. ●● Use Avogadro's number (see Section 10.3), $N_A = 6.02 \times 10^{23}$ atoms/mole, to show that 1 u $= 1.66 \times 10^{-27}$ kg. (Recall that a ^{12}C atom has a mass of exactly 12 u.) see ISM

65. ●● The mass of $^{12}_{6}$C is exactly 12 u. (a) What is the total binding energy of this nucleus? (b) What is the average binding energy per nucleon? (a) 92.2 MeV (b) 7.68 MeV/nucleon

66. ●● The mass of $^{16}_{8}$O is 15.994 915 u. What is the average binding energy per nucleon (E_b/A) for this nucleus? 7.98 MeV/nucleon

67. ●● Which isotope of hydrogen has the lower average binding energy per nucleon, deuterium or tritium? Justify your answer mathematically. deuterium; see ISM

68. ●● Near high-neutron areas, such as a nuclear reactor, neutrons will be absorbed by protons (the hydrogen nucleus in water molecules) and will give off a gamma ray of a characteristic energy in the process. What is the energy of the gamma ray (to three significant figures)? 2.22 MeV

69. ●● How much energy (to four significant figures) would be required to completely separate all the nucleons of a nitrogen-14 nucleus, the atom of which has a mass of 14.003 074 u? 104.7 MeV

70. ●● Calculate the binding energy of the last neutron in the $^{40}_{19}$K nucleus. [*Hint*: Compare the mass of $^{40}_{19}$K with the mass of $^{39}_{19}$K plus the mass of a neutron.] 7.80 MeV

71. ●● If an alpha particle could be removed intact from an aluminum-27 nucleus ($m = 26.981 541$ u), a sodium-23 nucleus ($m = 22.989 770$ u) would remain. How much energy would be required to do this operation? 10.1 MeV

72. ●● On average, are nucleons more tightly bound in an ^{27}Al nucleus or in a ^{23}Na nucleus?
^{27}Al (Al: 8.33 MeV/nucleon; Na: 8.11 MeV/nucleon)

73. ●● The atomic mass of $^{235}_{92}$U is 235.043 925 u. Find the average binding energy per nucleon for this isotope. 7.59 MeV/nucleon

74. ●●● The mass of $^8_4$Be is 8.005 305 u. (a) Which is less, the total mass of two alpha particles or the mass of the ^{8}Be nucleus? (b) Which is greater, the total binding energy of the ^{8}Be nucleus or the total binding energy of two alpha particles? (c) On the basis of your answers to parts (a) and (b) alone, do you expect the ^{8}Be nucleus to decay spontaneously into two alpha particles?
(a) two alphas (b) two alphas (c) yes

29.5 Radiation Detection, Dosage, and Applications

75. **MC** Which type of detector records the trajectory of charged particles: (a) Geiger counter, (b) scintillation counter, (c) solid-state detector, or (d) spark chamber? (d)

76. **MC** A bubble chamber (in a magnetic field) event shows two tracks of equal curvature but in opposite directions, emanating from a point in space with no apparent incoming particle. More than likely this event is (a) alpha decay, (b) beta decay, or (c) pair production. (c)

77. **MC** The same effective radiation dose (in rems) is given by a source of slow neutrons and an X-ray machine. What is the ratio of the dose (in rads) from the neutrons to that from the X-rays (also in rads): (a) 1:2 (b) 1:4, (c) 2:1, or (d) 4:1? (b)

78. **CQ** A basic assumption of radiocarbon dating is that the cosmic-ray intensity has been generally constant for the last 40 000 years or so. Suppose it were found that the intensity was much less 100 000 years ago than it is today. How would this finding affect the results of carbon-14 dating? see ISM

79. **CQ** If X-ray and alpha particles give the same dose (in grays), how will their effective doses (in sieverts) compare? alpha is twenty times as effective, see ISM

80. **CQ** PET scans require extremely fast computers coupled with gamma-ray detectors capable of accurate energy measurements. Explain why both energy accuracy and comparison of arrival times are crucial to the success of a PET scan. [*Hint*: In a PET scan, two opposing detectors pick up pair-annihilation gamma-ray photons. See page 928.] see ISM

81. ● In a diagnostic procedure, a patient in a hospital ingests 80 mCi of gold-198 ($t_{1/2} = 2.7$ days). What is the activity at the end of one month if none of the gold is eliminated from the body by biological functions? 36 μCi

82. ● A technician working at a nuclear reactor facility is exposed to a slow neutron radiation and receives a dose of 1.25 rad. (a) How much energy is absorbed by 200 g of the worker's tissue? (b) Was the maximum permissible radiation dosage exceeded? (a) 0.250 J (b) yes; see ISM

83. ● A person working with nuclear isotopes for a two-month period receives a 0.5-rad dose from a gamma source, a 0.3-rad dose from a slow-neutron source, and a 0.1-rad dose from an alpha source. Was the maximum permissible radiation dosage exceeded? yes; see ISM

84. ●● Neutron activation analysis was performed on small pieces of hair that had been taken from the exiled Napoleon after he died on the island of St. Helena in 1821. The samples were found to contain abnormally high levels of arsenic, which supported a theory that his death was not due to natural causes. If this evidence was derived by studying beta emissions coming from arsenic-76 nuclei, what was the arsenic isotope present in the hair, and what was the final nucleus after the beta decay? $^{75}_{33}$As, $^{76}_{34}$Se

85. ●●● A cancer treatment called the gamma knife (see Insight 29.1 on page 927) uses a large number of ^{60}Co sources to treat tumors. ^{60}Co emits two gamma rays of energy 1.33 MeV and 1.17 MeV in quick succession. Assume that 50.0% of the total gamma-ray energy is absorbed by a tumor. The total activity of the ^{60}Co sources is 1.00 mCi, the tumor's mass is 0.100 kg, and the patient is exposed for an hour. Calculate the effective radiation dose received by the tumor. (Since the ^{60}Co half-life is 5.3 years, changes in its activity during treatment can be neglected.)
0.266 Sv or 26.6 rem

Comprehensive Exercises

86. The radioactive source in most smoke detectors is ^{241}Am, which has a half-life of 432 years. In a typical detector, only about 10^{-4} g of this material is needed. (a) Write down its alpha decay equation and predict the product nucleus. (b) What is the source activity? (c) What would be the source's activity after 20 years in operation? (a) $^{241}_{95}$Ar $\rightarrow$ $^{237}_{93}$Np $+$ $^{4}_{2}$He
(b) 1.27×10^7 decays/s (c) 1.23×10^7 decays/s

87. A sample of ^{215}Bi, which beta-decays ($t_{1/2} = 2.4$ min), contains Avogadro's number of nuclei. (a) Write down the decay equation and predict the product nucleus. (b) How many bismuth nuclei are present after 10 min? (c) After 1.0 h? (d) What are the activities, in curies and becquerels, at these times? see ISM

88. **IE** High-energy photons (gamma rays) can remove nucleons from nuclei in a process called the *photonuclear effect*. (a) If you wanted to remove a single neutron from a nucleus, which nucleus would most likely require the higher energy photon: (1) ^{12}C; (2) ^{13}C; or (3) the energies would be the same? (b) Calculate the minimum energy of each photon required to eject a neutron from the two isotopes in part (a). (c) Determine the wavelengths of the light associated with the photons in part (b). see ISM

89. A very slow neutron (neglect its kinetic energy) is captured by an ^{10}B nucleus, which ends up in its ground state. Calculate the (a) energy and (b) wavelength of the photon given off. (Neglect the recoil kinetic energy of the daughter nucleus.) (a) 11.5 MeV (b) 1.08×10^{-13} nm

90. The approximate radius of a nucleus is given by $R = R_o A^{1/3}$, where $R_o = 1.2 \times 10^{-15}$ m and A is the mass number of the nucleus. Assuming that nuclei are spherical (they are approximately so in many cases), (a) show

that the average nucleon density in a nucleus is 1.4×10^{44} nucleons/m^3 and (b) estimate the nuclear density in kilograms per cubic meter.
(a) 1.4×10^{44} nucleons/m^3 (b) 2.3×10^{17} kg/m^3

91. ▼Figure 29.21 shows the decay series for plutonium-239. Use the information in the figure to (a) determine the two different decay schemes by which the isotope of radon associated with this chain can be formed, and (b) determine the subsequent decay scheme for that isotope of radon.
see ISM

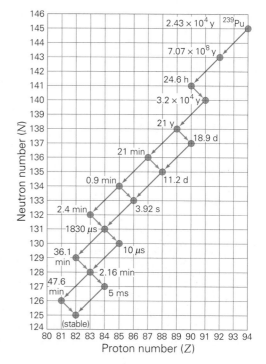

▲ **FIGURE 29.21 Plutonium-239 decay series** See Exercise 91.

92. Determine which of the following isotopes are likely to be stable and explain clearly and fully how you came to your conclusion, based on the "rules of stability." (a) ^{15}O; (b) ^{8}Li; (c) ^{222}Rn; (d) ^{27}Mg; (e) ^{41}Ca.
(a) not likely (b) no (c) no (d) no (e) stable

93. $^{3}_{1}$H (tritium) can be produced in water surrounding a strong source of neutrons, such as nuclear reactors. One of the ways is by neutron capture onto deuterium. (a) Write down the equation for this capture reaction. (b) Tritium has a half-life of 12.33 years. What percentage of a sample containing $^{3}_{1}$H will remain after 6.00 years? (c) Determine the gamma ray energy emitted during the capture (assuming the tritium ends up in its ground state and the incoming neutron kinetic energy is negligible). (d) Write down the reaction for the subsequent beta decay of the tritium and determine the stable daughter identity. (e) If all the energy released in the beta decay went into the beta particle, determine its energy.
see ISM

NUCLEAR REACTIONS AND ELEMENTARY PARTICLES

PHYSICS FACTS

- Although nuclear power plants produce radioactive waste that presents long-term storage problems, they have advantages over fossil fuel plants. For example, they emit no greenhouse gases (which cause global warming) or oxides of sulfur and nitrogen (which cause acid rain). Nor do they require obtaining fossil fuels in environmentally sensitive areas. In fact, coal-fired plants emit much more radioactive material than nuclear plants due to the uranium in the coal.

- Photons are now known to be the only massless elementary particle. In the 1980s, the family of neutrinos, originally thought to all be massless, were determined to have mass. Recent experiments indicate that their mass is on the order of one millionth of the electron's mass. This has vast implications about the mass of the universe, and thus its (Big Bang) evolution.

- The smallest amount (quantum) of electric charge is not that of the electron (which has charge $-e$). Quarks have fractional charges of $\pm\frac{1}{3}e$ and $\pm\frac{2}{3}e$.

- Protons and neutrons are not elementary particles, but are each composed of three quarks. Because of the type of binding between quarks, current theories predict that free quarks may never be seen experimentally.

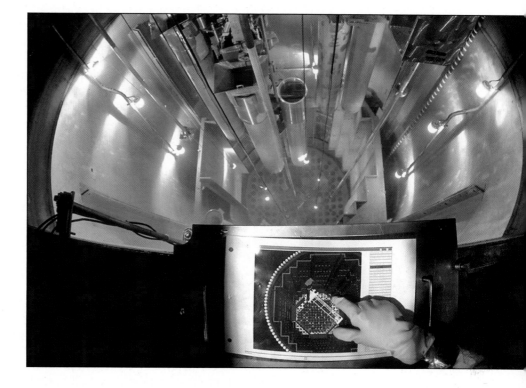

Today, more than half a century after the first nuclear reactor was built, these facilities remain a center of controversy. Some people think they are essential to solving the world's energy problems. To others, they embody all that is wrong and dangerous about modern technology. The reactor shown in the photo is a symbol of this ongoing debate. It is part of the Three Mile Island (TMI) plant in Pennsylvania, the site of one of the largest and most publicized nuclear accidents in the world, second perhaps only to Chernobyl (more on that later). Here, you are looking into the heart of one of the two remaining TMI reactors, both of which are still running safely.

In this chapter the nuclear fission reactions that power such reactors will be studied. Also we will study how scientists are trying to harness fusion reactions, which occur within the Sun and other stars and might one day provide a safe and virtually inexhaustible supply of clean energy for our planet.

Also considered are the elementary particles that make up our universe and whose interactions govern the structure and evolution of that universe. Investigations have shown that the proton and neutron are not elementary particles, but are composites of elementary particles. The discovery of families of elementary particles (and their properties and interactions) has given us broad insight into the structure of the nucleus and provided a clearer understanding of the nature, origin, and evolution of our universe.

30.1 Nuclear Reactions

OBJECTIVES: To (a) use charge and nucleon conservation to write nuclear reaction equations, and (b) understand and use the concepts of Q value and threshold energy to analyze nuclear reactions.

Ordinary chemical reactions between atoms and molecules involve only orbital electrons. The nuclei of these atoms do not participate in the process, so the atoms retain their identity. Conversely, in **nuclear reactions**, the original nuclei are converted into the nuclei of other elements. Scientists first became aware of this type of reaction during experimental studies that involved bombarding nuclei with energetic particles.

The first artificially induced nuclear reaction was produced by Ernest Rutherford in 1919. Nitrogen was bombarded with alpha particles from a natural source (^{214}Bi). The particles produced by the reactions were identified as protons. Rutherford reasoned that an alpha particle colliding with a nitrogen *nucleus* must sometimes be able to induce a reaction that produces a proton. We say that the nitrogen nucleus is *artificially transmuted* into an oxygen nucleus by the following reaction:

$$^{14}_{7}\text{N} \quad + \quad ^{4}_{2}\text{He} \quad \rightarrow \quad ^{17}_{8}\text{O} \quad + \quad ^{1}_{1}\text{H}$$

nitrogen	*alpha particle*	*oxygen*	*proton*
(14.003 074 u)	(4.002 603 u)	(16.991 33 u)	(1.007 825 u)

(The atomic masses are given for later use.)

This reaction and many others like it actually form a short-lived (intermediate or temporary) *compound nucleus* in an excited state. For example, the preceding reaction can be written more correctly as

$$^{14}_{7}\text{N} + ^{4}_{2}\text{He} \rightarrow (^{18}_{9}\text{F}^*) \rightarrow ^{17}_{8}\text{O} + ^{1}_{1}\text{H}$$

The intermediate nucleus is a fluorine nucleus, $^{18}_{9}\text{F}^*$, formed in an excited state indicated by the asterisk. A compound nucleus typically loses excess energy by ejecting a particle (or particles)— in this case, a proton. Since a compound nucleus lasts only a very short time, it is commonly omitted from the nuclear reaction equation.

What Rutherford discovered was a way to change one element into another. This was the age-old dream of alchemists, although their main goal was to change common metals, such as mercury and lead, into gold. This seemingly profitable metamorphosis and many other transmutations can be initiated today with *particle accelerators*, machines that accelerate charged particles to very high speeds. When these particles strike target nuclei, they can initiate nuclear reactions. One reaction that can occur when a proton strikes a nucleus of mercury is

$$^{200}_{80}\text{Hg} \quad + \quad ^{1}_{1}\text{H} \quad \rightarrow \quad ^{197}_{79}\text{Au} \quad + \quad ^{4}_{2}\text{He}$$

mercury	*proton*	*gold*	*alpha particle*
(199.968 321 u)	(1.007 825 u)	(196.966 56 u)	(4.002 603 u)

In this reaction, mercury is converted into gold, so it would seem that modern physics has fulfilled the alchemists' dream. However, making such tiny amounts of gold in an accelerator costs far more than the gold is worth.

Reactions such as the foregoing ones have the general form

$$A + a \rightarrow B + b$$

where the uppercase letters represent the nuclei and the lowercase letters represent the particles. Such reactions are often written in a shorthand notation:

$$A(a, b)B$$

For example, in this form, the two previous reactions can be rewritten more compactly as

$$^{14}\text{N}(\alpha, p)^{17}\text{O} \quad \text{and} \quad ^{200}\text{Hg}(p, \alpha)^{197}\text{Au}$$

The periodic table (see Fig. 28.9 and back cover of this text) lists more than 100 elements, but only 90 stable elements occur naturally on Earth. There are elements with unstable nuclei that exist from $Z = 83$ (bismuth) to $Z = 92$ (uranium) due to the decay chains that continually create them from heavier elements. Those with proton numbers greater than uranium ($Z = 92$), such as plutonium-239, as well as technetium ($Z = 43$) and promethium ($Z = 61$), are created artificially by nuclear reactions. (If technetium and promethium had been present when the Earth was formed, they would have long since decayed away.) The name *technetium* comes from the Greek word *technetos*, meaning "artificial"; technetium was the first unknown element to be created by artificial means. Elements with Z values up to about $Z = 114$ have been created artificially.*

Conservation of Mass–Energy and the Q Value

In every nuclear reaction, total (relativistic) energy ($E = K + mc^2$) must be conserved. (See Chapter 26.) Consider the reaction in which nitrogen is converted into oxygen: $^{14}N(\alpha, p)^{17}O$. By the conservation of total relativistic energy,

$$(K_N + m_N c^2) + (K_\alpha + m_\alpha c^2) = (K_O + m_O c^2) + (K_p + m_p c^2)$$

where the subscripts refer to the particular particle or nucleus. Rearranging the equation, we have

$$K_O + K_p - (K_N + K_\alpha) = (m_N + m_\alpha - m_O - m_p)c^2$$

The **Q value** of the reaction is defined as the change in kinetic energy, and

$$Q = \Delta K = (K_O + K_p) - (K_N + K_\alpha) \tag{30.1}$$

Q can be positive or negative, depending on whether the total kinetic energy of the system increases or decreases. Thus, the Q value is a measure of the kinetic energy released or absorbed in a reaction. Equation 30.1 can alternatively be expressed in terms of the masses:

$$Q = (m_N + m_\alpha - m_O - m_p)c^2 \tag{30.2}$$

In terms of a general reaction of the form A + a → B + b,

$$Q = \Delta K = (m_A + m_a - m_B - m_b)c^2 = (\Delta m)c^2 \tag{30.3}$$

Note: Here $\Delta m = m_i - m_f$.

An alternative interpretation of Q is that it is the difference in the mass-equivalent energies of the reactants (initial) and the products (final) of a reaction. This reflects the fact that mass can be converted into kinetic energy and vice versa. Note that the mass difference Δm can be positive or negative. Notice also that $\Delta m = m_i - m_f$, the opposite of our usual convention. This is to guarantee that the value of Q is correctly related to ΔK. Thus, if the total mass of the system increases during the reaction and the kinetic energy therefore decreases, Q must be negative. Similarly, if the total mass decreases and the kinetic energy thus increases, then Q is positive.

If Q is negative, the reaction requires a minimum amount of kinetic energy before the reaction can happen. To see this, let us look at Rutherford's original reaction in some detail. Using the masses given under the $^{14}N(\alpha, p)^{17}O$ reaction equation on page 936, we have

$$Q = (m_N + m_\alpha - m_O - m_p)c^2$$
$$= [(14.003\,074\text{ u} + 4.002\,603\text{ u}) - (16.999\,133\text{ u} + 1.007\,825\text{ u})]c^2$$
$$= (-0.001\,281\text{ u})c^2$$

or, using the mass–energy equivalence factor from Section 29.4, we obtain

$$Q = (-0.001\,281\text{ u})(931.5\text{ MeV/u}) = -1.193\text{ MeV}$$

A reaction with a negative Q value is said to be **endoergic** (or *endothermic*). In endoergic reactions, the kinetic energy of the reacting particles is partially converted into mass.

When the Q value of a reaction is positive, energy is released, and the reaction is said to be **exoergic** (or *exothermic*). That is, energy is produced (*exo*) by the reaction. In this case, some mass is converted into energy in the form of increased kinetic energy of the reaction products.

Teaching tip: Point out other examples of endoergic and exoergic reactions.

*Beyond $Z = 112$, there are gaps. Extremely short-lived nuclei with $Z = 114$ (and possibly 116) have tentatively been discovered, but are yet unnamed. Can you explain why the even values of atomic numbers have been discovered and not the odd ones in between?

Example 30.1 ■ A Possible Energy Source: Q Value of a Reaction

Determine whether the following reaction is endoergic or exoergic, and calculate its Q value.

$$\underset{\substack{deuteron \\ (2.014\,102\ u)}}{^2_1\mathrm{H}} \quad + \quad \underset{\substack{deuteron \\ (2.014\,102\ u)}}{^2_1\mathrm{H}} \quad \rightarrow \quad \underset{\substack{helium \\ (3.016\,029\ u)}}{^3_2\mathrm{He}} \quad + \quad \underset{\substack{neutron \\ (1.008\,665\ u)}}{^1_0\mathrm{n}}$$

Thinking It Through. The reaction is endoergic if $Q < 0$, and exoergic if $Q > 0$. We need the mass difference (Δm) to determine Q from Eq. 30.3.

Solution. Δm is calculated by subtracting the final masses from the initial masses. Therefore,

$$\begin{aligned} \Delta m &= 2m_\mathrm{D} - m_\mathrm{He} - m_\mathrm{n} \\ &= 2(2.014\,102\ u) - 3.016\,029\ u - 1.008\,665\ u = +0.003\,51\ u \end{aligned}$$

Thus, mass has been lost, the total kinetic energy has increased, and the reaction is exoergic. The Q value is

$$Q = (0.003\,51\ u)(931.5\ \mathrm{MeV/u}) = +3.27\ \mathrm{MeV}$$

Follow-Up Exercise. Determine whether the following reaction is endoergic or exoergic, and calculate its Q value:

$$\underset{\substack{carbon \\ (12.000\,000\ u)}}{^{12}_6\mathrm{C}} \quad + \quad \underset{\substack{helium \\ (4.002\,603\ u)}}{^4_2\mathrm{He}} \quad \rightarrow \quad \underset{\substack{carbon \\ (13.003\,355\ u)}}{^{13}_6\mathrm{C}} \quad + \quad \underset{\substack{helium \\ (3.016\,029\ u)}}{^3_2\mathrm{He}}$$

(Answers to all Follow-Up Exercises are at the back of the text.)

Problem-Solving Hint

Note in Example 30.1 that Q values are computed from the mass difference, expressed in atomic mass units, by using the mass–energy conversion factor derived in Chapter 29. (See Table 29.3.) This method eliminates the need to use c^2 and gives Q directly in MeV.

TABLE 30.1

Interpretation of Q Values

Q Value	Effect
Positive $(Q > 0)$	Exoergic, some mass converted into energy (mass of reactants greater than mass of products)
Negative $(Q < 0)$	Endoergic, some kinetic energy converted into mass (mass of products greater than mass of reactants)

Radioactive decay (Chapter 29) is a special type of nuclear reaction with one reactant nucleus and two (or more) products. The Q value of radioactive decay is always positive, because there is a gain in kinetic energy. For decay reactions, Q is called the *disintegration energy*. The interpretation of the sign of Q is summarized in Table 30.1.

When a reaction's Q value is negative, you might think that the reaction could occur if the incident particle had a kinetic energy at least equal to Q—that is, if it were to have $K_{\min} = |Q|$.* However, if all the kinetic energy were converted to mass, the particles would be at rest after the reaction, which violates the conservation of linear momentum (Chapter 6).

Hence, in an endoergic reaction, to conserve linear momentum, the kinetic energy of the incident particle must be *greater* than $|Q|$. The minimum kinetic energy that a particle needs to initiate an endoergic reaction is called the **threshold energy** ($K_{\min}$). For nonrelativistic energies, the threshold energy is

$$K_{\min} = \left(1 + \frac{m_\mathrm{a}}{M_\mathrm{A}}\right)|Q| \qquad \substack{\text{(stationary} \\ \text{target only)}} \qquad (30.4)$$

where m_a and M_A are the masses of the incident particle and the stationary target nucleus, respectively. In Eq. 30.4, the factor by which $|Q|$ is multiplied is greater than 1 (why?), so, as expected, $K_{\min} > |Q|$. The calculation of a threshold energy is shown in the following Example.

---

*Kinetic energy is written in terms of $|Q|$, the absolute value of Q, because kinetic energy cannot be negative. The sign of Q arises from the mass difference and indicates the gain or loss of mass during the reaction. If Q is negative, we want $K_{\min}$ to be positive—hence, the use of absolute value.

Example 30.2 ■ Nitrogen into Oxygen: Threshold Energy

What is the threshold energy for the reaction $^{14}\text{N}(\alpha, \text{p})^{17}\text{O}$?

Thinking It Through. The Q value for this reaction was calculated previously in the text. To get the threshold energy, we use Eq. 30.4.

Solution. The following data are taken from the text:

Given: $m_a = m_\alpha = 4.002\,603$ u   *Find:* K_{min} (threshold energy)
          $M_A = m_N = 14.003\,074$ u
          $Q = -1.193$ MeV

From Eq. 30.4,

$$K_{min} = \left(1 + \frac{m_\alpha}{M_N}\right)|Q| = \left(1 + \frac{4.002\,603\ u}{14.003\,074\ u}\right)|-1.193\ \text{MeV}| = 1.534\ \text{MeV}$$

Follow-Up Exercise. In this Example, how much of the threshold kinetic energy goes into increasing the mass of the system, and how much shows up as kinetic energy in the final state? Explain your reasoning.

Reaction Cross-Sections

In an endoergic reaction, when the incident particle has more than the threshold energies of several reactions, any of the reactions may occur, usually with differing probabilities according to the rules of quantum mechanics. A measure of the probability that a particular reaction will occur is called the **cross-section** for that reaction. The probability for a particular reaction depends on many factors. Usually, it depends on the kinetic energy of the initiating particle, sometimes very dramatically. For positively charged incident particles, the presence of the (repulsive) Coulomb barrier means that the probability of a given reaction occurring generally increases with the kinetic energy of the incident particle.

Being electrically neutral, neutrons are unaffected by the Coulomb barrier. As a result, the cross-section for a given reaction involving neutrons can be quite large, even for low-energy neutrons. Reactions involving neutrons such as $^{27}\text{Al}(\text{n}, \gamma)^{28}\text{Al}$ are called *neutron-capture reactions*. As the energy of the neutron increases, the cross-section can vary a great deal, as ▶Fig. 30.1 shows. The peaks in the curve, called *resonances*, are associated with nuclear energy levels in the nucleus being formed. If the neutron's energy is "just right" to create the final nucleus in one of its energy levels, there is a relatively high probability that neutron absorption will occur.

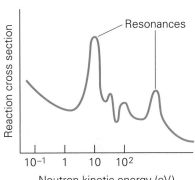

▲ **FIGURE 30.1 Reaction cross-section** A typical graph of a neutron reaction's cross-section versus energy. The peaks where the probabilities of reactions are greatest are called *resonances*. They correspond to energy levels in the compound nucleus formed when the neutron is temporarily captured.

30.2 Nuclear Fission

OBJECTIVES: To understand (a) the process of nuclear fission, (b) the nature and cause of a nuclear chain reaction, and (c) the basic principles involved in the operation of nuclear reactors.

In early attempts to make heavier elements artificially, uranium, the heaviest element known at the time, was bombarded with neutrons. An unexpected result was that the uranium nuclei sometimes split into fragments. These fragments were identified as the nuclei of lighter elements. The process was dubbed *nuclear fission*, after the biological fission process of cell division.

In a **fission reaction**, a heavy nucleus divides into two lighter nuclei with the emission of neutrons. Some of the initial mass is converted into kinetic energy of the neutron and fragments. Some heavy nuclei undergo *spontaneous fission*, but at very slow rates. However, fission can be *induced*, and this is the important process in practical energy production. For example, when a ^{235}U nucleus absorbs a neutron, it can fission into xenon and strontium by the reaction

$$^{235}_{92}\text{U} + {}^{1}_{0}\text{n} \rightarrow ({}^{236}_{92}\text{U}^*) \rightarrow {}^{140}_{54}\text{Xe} + {}^{94}_{38}\text{Sr} + 2({}^{1}_{0}\text{n})$$

According to the *liquid-drop model*, due to the absorbed energy, this intermediate nucleus (^{236}U) undergoes oscillations and becomes distorted like a liquid drop (▼Fig. 30.2).

Teaching tip: Use of the word *resonance* in this context is not a coincidence. In mechanical and electrical systems, a system parameter becomes very large when exactly the correct, or resonant, frequency of the system is reinforced. For example, the amplitude of a driven oscillator becomes very large when the oscillator is driven at its resonance frequency. The effect is similar in nuclear reactions: If the energy of the incoming particle is exactly right, we say that we have hit a "nuclear resonance," or a natural state of the system.

▶ **FIGURE 30.2 Liquid-drop model of fission** When an incident neutron is absorbed by a fissionable nucleus, such as ^{235}U, the unstable compound nucleus (^{236}U) undergoes violent oscillations and breaks apart like a liquid drop, typically emitting two or more neutrons and yielding two radioactive fragments.

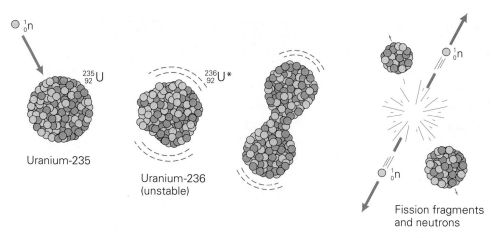

Uranium-235

Uranium-236
(unstable)

Fission fragments
and neutrons

The separation of the nucleons into different parts of the "drop" weakens the nuclear force, and the repulsive electrical force between the two parts of the "nuclear drop" causes it to split, or fission.

Note that the preceding reaction involving ^{235}U is *not* unique. There are many other possible outcomes, including the following (the compound nuclei are omitted):

$$^{1}_{0}n + ^{235}_{92}U \rightarrow ^{141}_{56}Ba + ^{92}_{36}Kr + 3(^{1}_{0}n)$$

and

$$^{1}_{0}n + ^{235}_{92}U \rightarrow ^{150}_{60}Nd + ^{81}_{32}Ge + 5(^{1}_{0}n)$$

Only certain nuclei undergo fission. For them, the probability of fissioning depends on the energy of the incident neutrons. For example, the largest probabilities for fission of ^{235}U and ^{239}Pu occur for "slow" neutrons, that is, neutrons with kinetic energies less than about 1 eV. However, for ^{232}Th, "fast" neutrons with energies of 1 MeV or greater are more likely to trigger a fission reaction.

An estimate of the energy released in a fission reaction can be obtained by considering the E_b/A curve for stable nuclei (Fig. 29.14). When a nucleus with a high mass number (A), such as uranium, splits into two nuclei, it is, in effect, moving inward and upward along the sloping tail of this curve toward more stable nuclei. As a result, the average binding energy per nucleon increases from about 7.8 MeV to approximately 8.8 MeV. Thus energy liberated is on the order of 1 MeV per nucleon in the fission products. In the reaction at the top of this page, $140 + 94 = 234$ nucleons are bound in the products. Thus, the energy release is approximately (1 MeV/nucleon) $\times$ 234 nucleons $\approx$ 234 MeV.

At first glance, this amount might not seem like much energy. 234 MeV is only about 3.7×10^{-11} J, which pales in comparison with everyday energies. In fact, 234 MeV is only about 0.1 percent of the energy equivalent of the mass of the ^{235}U nucleus, which is approximately (235 nucleons)(939 MeV/nucleon) $= 2.2 \times 10^{5}$ MeV. Nevertheless, on a percentage basis, it is many times larger than the amount of energy released in ordinary chemical reactions, such as in the burning of oil or coal.

Practical amounts of energy from fission can, however, be obtained when huge numbers of these fissions occur per second. One way of accomplishing this is by a **chain reaction**. For example, suppose a ^{235}U nucleus fissions (on its own or triggered by an external neutron) with the release of two neutrons (▶Fig. 30.3). Ideally, the released neutrons can then initiate two more fission reactions, a process that, in turn, releases four neutrons. These neutrons may initiate more reactions, and so on. Thus, the process can multiply, with the number of neutrons doubling with each generation. When this occurs, the neutron production rate (and, hence, the energy released from the sample) grows exponentially.

To maintain a sustained chain reaction, there must be an adequate quantity of fissionable material. The minimum mass required to produce a sustained chain reaction is called the **critical mass**. When the critical mass is attained, there is enough fissionable material such that at least one neutron from each fission event, on average, goes on to fission another nucleus.

Several factors determine critical mass. Most evident is the amount of fissionable material. If the quantity of material is small, many neutrons will escape from

Note: A chain reaction requires a critical mass.

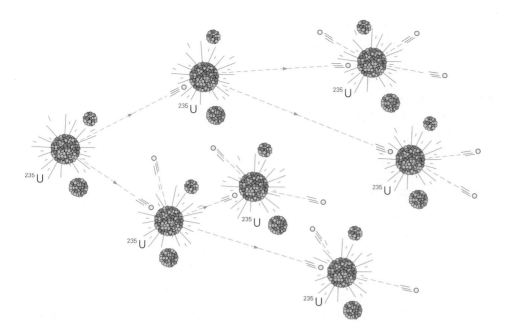

◀ **FIGURE 30.3 Fission chain reaction** The neutrons that result from one fission event can initiate other fission reactions, which, in turn, initiate further fission reactions, and so on. When enough fissionable material is present, the sequence of reactions can be adjusted to be self-sustaining (a chain reaction).

the sample (through the surface) before inducing a fission, and the chain reaction will die out. Also, nuclides other than ^{235}U in the sample may absorb neutrons, thereby limiting the chain reaction. As a result, the purity of the fissionable isotope affects the critical mass.

Natural uranium consists of the isotopes ^{238}U and ^{235}U. The natural concentration of ^{235}U is only about 0.7%. The remaining 99.3% is ^{238}U, which can absorb neutrons without fissioning, thereby inhibiting the chain reaction fission of ^{235}U. To have more fissionable ^{235}U nuclei in a sample and thus reduce the critical mass, the ^{235}U can be concentrated. This enrichment varies from 3% to 5% ^{235}U for nuclear reactor–grade material to more than 99% for weapons-grade material. This difference is important, because it is highly desirable that a nuclear reactor *not* explode like an atomic bomb nor use fuel capable of being made into a bomb.

Chain reactions take place almost instantly. If such a reaction proceeds uncontrolled, the quick and enormous release of energy can cause an explosion. This is the principle of the so-called atomic bomb. (A more descriptive name is the *fission or nuclear bomb.*) In such a bomb, several subcritical pieces of fuel are suddenly imploded to form a critical mass. The resulting chain reaction is then out of control, releasing an enormous amount of energy in a short period of time. For the steady production of energy from the fission process, the chain-reaction process must be controlled. We shall now see how this task is accomplished in nuclear reactors.

Nuclear Reactors

The Power Reactor Currently, the only practical design for generating electrical power from nuclear energy is based on the fission chain reaction. A typical design for a nuclear reactor is shown in ▼Fig. 30.4. There are five key elements to a reactor: fuel rods, core, coolant, control rods, and moderator.

Tubes packed with pellets of uranium oxide form the **fuel rods**, located in the central portion of the reactor called the **core**. A typical commercial reactor contains fuel rods bundled into assemblies of approximately 200 rods each. Coolant flows around the rods to remove the heat energy generated during the chain reaction. Reactors used in the United States are light-water reactors, which means that ordinary water is used as a **coolant** to remove heat. However, the hydrogen nuclei of ordinary water can capture neutrons to form deuterium, thus removing neutrons from the chain reaction. Hence, enriched uranium with 3–5% ^{235}U must be used.*

---

*Some Canadian reactors use heavy water, D_2O, as a coolant in place of light water. The advantage is that the deuterium (D) does not readily absorb neutrons, so the fuel can be of lower uranium enrichment. However, D_2O must first be separated from normal water, H_2O, an operation that takes energy.

▶ **FIGURE 30.4** Nuclear reactor **(a)** A schematic diagram of a reactor vessel. **(b)** A fuel rod and its assembly.

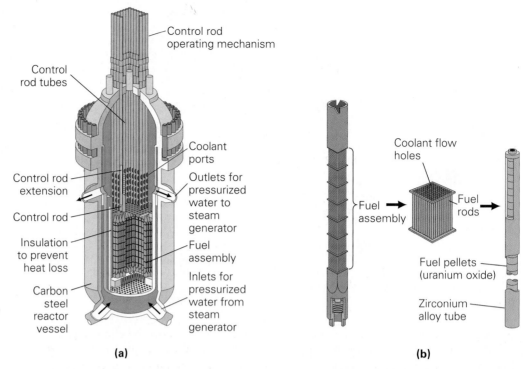

(a)

(b)

The chain-reaction rate and, therefore, the energy output of a reactor are controlled by boron or cadmium **control rods**, which can be inserted into or withdrawn from the reactor core. Cadmium and boron have a very high probability (cross-section) for absorbing neutrons. When these rods are inserted between the fuel-rod assemblies, neutrons are removed from the chain reaction. The control rods are adjusted so the chain reaction proceeds at a steady rate. The idea is to create a sustainable fission chain reaction in which the average fission produces only one more fission. For refueling, or in an emergency, the control rods can be fully inserted, and enough neutrons are removed to curtail the chain reaction and shut down the reactor. However, even with the chain reaction shut down, water must continue to circulate to prevent heat buildup due to the continuing decay of radioactive fission products in the fuel rods. If not, damage to the fuel rods can result. Melting and cracking of fuel rods due to inadequate cooling was the cause of the accident at Three Mile Island in 1979. The core of that reactor is still highly radioactive, because fission fragments were released from the broken rods.

The water flowing through the fuel-rod assemblies acts not only as a coolant, but also as a **moderator**. The fission cross-section for ^{235}U is largest for slow neutrons (sometimes called *thermal* neutrons with kinetic energies less than 1 eV). However, neutrons emitted from a fission are actually fast neutrons (with kinetic energies of about 2 MeV). Their speed is reduced, or moderated, by collisions with the water molecules. It takes about 20 collisions to moderate fast neutrons down to energies of about 1 eV.

The Breeder Reactor In a commercial power reactor, the ^{238}U goes along for the ride, so to speak. However, while it is unlikely to fission by absorbing a slow neutron, ^{238}U can be involved in reactions caused by *fast* neutrons. If a neutron's energy has not been completely moderated, reactions with ^{238}U can occur. For example, a conversion of ^{238}U to ^{239}Pu via successive beta decays after a fast neutron absorption can happen as follows:

$$\underset{(fast)}{{}^{1}_{0}n} + {}^{238}_{92}U \rightarrow ({}^{239}_{92}U^*) \rightarrow {}^{239}_{93}Np + \beta^-$$
$$\downarrow$$
$$^{239}_{94}Pu + \beta^-$$

Note: A breeder reactor doesn't produce something from nothing. It uses the kinetic energy of the fast neutrons to create fissionable ^{239}Pu from the nonfissionable ^{238}U component of uranium.

^{239}Pu, with a half-life of 24 000 years, *is* fissionable. Since it is possible to actively promote the conversion of ^{238}U to ^{239}Pu in a reactor by reducing the degree of moderation, the same amount of fissionable fuel (or more) can be produced (^{239}Pu) as is consumed (^{235}U). This is the principle behind the **breeder reactor**.

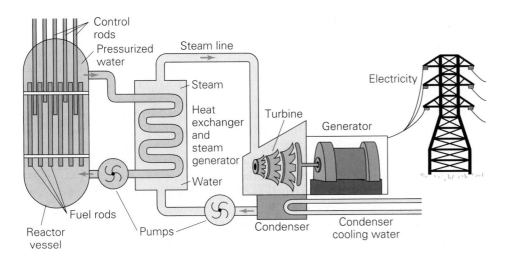

◀ **FIGURE 30.5 Pressurized water reactor** The components of a pressurized water reactor. The heat energy from the reactor core is carried away by the circulating water. The water in the reactor is pressurized so that it can be heated to high temperatures for more efficient heat removal. The energy is used to generate steam, which drives the turbine that turns the generator to produce electrical energy.

Notice that this isn't a case of getting something for nothing. Rather, the reactor is converting the unfissionable ^{238}U part of the fuel to the fissionable ^{239}Pu, while continuing to produce energy by ^{235}U fission.

Developmental work on the breeder reactor in the United States was essentially stopped in the 1970s. However, France went on to develop operational breeder reactors that provide nuclear fuel (^{239}Pu) for power reactors. Countries such as France that do not have huge natural resources of fossil fuels are highly dependent on nuclear energy and breeder reactors for their electrical energy needs.

Electricity Generation The components of a typical pressurized water reactor used in the United States are shown in ▲ Fig. 30.5. The heat generated by the controlled chain reaction is carried away by the water passing through the rods in the fuel assembly. The water is pressurized to several hundred atmospheres so that it can reach temperatures over 300°C for more efficient heat removal. The hot water is then pumped to a heat exchanger, where the heat energy is transferred to the water of a steam generator. Notice that the reactor coolant and the exchanger water are in two separate and distinct closed systems. (Why?)

Next, high-pressure steam turns a turbine that operates an electrical-energy generator, as is the case for any nonnuclear power plant. The steam is then cooled and condensed after turning the turbine. This final loop of coolant water typically carries the heat to a nearby ocean, lake, or river.

Note: Recall from Section 11.3 that the boiling point of water increases with increasing pressure.

Nuclear Reactor Safety

Nuclear energy is used to generate a substantial amount of the electricity in the world. More than 25 countries now produce electricity by this method. Several hundred nuclear reactor units are in operation throughout the world, with more than 100 units in the United States. With the increasing number of nuclear facilities comes the fear of nuclear accidents and the subsequent release of radioactive materials into the environment.

If the coolant of a light-water reactor is lost, the chain reaction stops, because the coolant is also the moderator. However, the decay of the fission fragments, some with half-lives of hundreds of years, continues. In such a **LOCA** (an acronym for *loss-of-coolant accident*), the fuel rods might become hot enough (several thousand degrees Celsius) for the cladding (outside covering) to melt and fracture. Once this occurs, the hot, fissioning mass could fall into the water on the floor of the containment vessel and cause a steam explosion, a hydrogen explosion, or both. This explosion could rupture the walls of the containment vessel and allow radioactive fragments into the environment. Even if the walls were not breached, the hot "melt" of the fuel rods could burn through the floor of the building, eventually reaching ground level and the atmosphere (a situation called *China syndrome* because of the "melt" heading downward "towards China"). A partial **meltdown** did occur at the TMI generating plant in 1979. This meltdown was a LOCA, and a small amount of radioactive steam

▲ **FIGURE 30.6** Chernobyl, April **1986** An aerial photo showing damage to the reactor at Chernobyl. This accident released large amounts of radioactive materials into the environment, with dire consequences.

was vented to the atmosphere. Inside one reactor vessel, now sealed, electronic robots have discovered heavy damage to the fuel rods.

The April 1986 nuclear accident at Chernobyl was a meltdown following a LOCA caused by human error and magnified by an inherent instability resulting from the use of carbon as a moderator instead of water. When the flow of cooling water was inadvertently removed, the chain reaction went out of control—something that could not happen in a light-water reactor—producing a huge rise in temperature. The resulting explosions blew the top off the building (◄Fig. 30.6). When the fuel rods melted, the graphite blocks burned like a massive charcoal barbecue, spewing radioactive smoke into the air. Winds carried this radioactive smoke over much of Europe and over the North Pole into Canada and the United States, where significant amounts of core fission fragments (such as ^{131}I, ^{90}Sr, and ^{137}Cs) were detected.

Even if nuclear reactors operate safely (and their safety record, particularly in the United States, is a very good one), and although they emit virtually no pollutants such as greenhouse gases into the environment, there remains the problem of radioactive waste. Fission fragments are radioactive and have long half-lives. As a result, the safe handling of nuclear waste will be a problem for centuries to come. The United States is just beginning to come to grips with this. The consensus is that the only viable storage plan is one that buries the waste deep underground in geologically stable formations that keep it isolated from the atmosphere and ground water.

With its first shipment of low-level radioactive waste from the defense industry (*not* commercial power plants) in 1999, the United States opened its nuclear-waste site, in the New Mexico desert. Named the Water Isolation Pilot Plant, this plant stores such radioactive debris as plutonium-contaminated clothing, tools, and sludge. This material is housed several thousand feet below ground level in a hollowed-out salt formation. In 1987, Congress designated Yucca Mountain, about 90 miles northwest of Las Vegas, Nevada, to be the primary potential nuclear waste disposal site for the United States. A vertical shaft and excavated waste repository have been constructed there and the Department of Energy is proceeding with its application for a storage license. At the earliest, however, nuclear waste from commercial plants will not be received there until 2010. In fact there are proposals to locate waste storage in Utah rather than Nevada. Clearly, where and how to seal, safely transport, bury, and guard nuclear waste will be important decisions for generations to come.

30.3 Nuclear Fusion

OBJECTIVES: To (a) explain the fundamental difference between fusion and fission, (b) calculate energy releases in fusion reactions, and (c) understand how fusion might eventually provide a source of electric energy.

Another type of nuclear reaction that can release energy is *fusion*. In a **fusion reaction**, light nuclei fuse to form a more massive nucleus, releasing energy in the process ($Q > 0$). A simple fusion reaction—the fusion of two deuterium nuclei ($^{2}_{1}$H), sometimes called a D–D reaction—was examined in Example 30.1. There, it was shown that this reaction releases 3.27 MeV of energy per fusion.

Another example is the fusion of deuterium and tritium (a D–T reaction):

$$^{2}_{1}\text{H} + ^{3}_{1}\text{H} \rightarrow ^{4}_{2}\text{He} + ^{1}_{0}\text{n}$$
$$\text{(2.014 102 u)} \quad \text{(3.016 049 u)} \quad \text{(4.002 603 u)} \quad \text{(1.008 665 u)}$$

Using the given masses, you should be able to show that this reaction involves a release of 17.6 MeV per fusion.

A fusion reaction releases much less energy in comparison with the more than 200 MeV released from a typical single fission. However, equal-mass samples of hydrogen and uranium have many, many more hydrogen nuclei than uranium nuclei. As a result, *per kilogram*, the fusion of hydrogen gives almost three times the energy released from uranium fission.

In a sense, our lives depend crucially on nuclear fusion, because it is the source of energy for most stars, including our Sun. One sequence of fusion reactions that is believed to be responsible for the Sun's energy output is as follows. First, there is proton–proton fusion,

$$^{1}_{1}\text{H} + ^{1}_{1}\text{H} \rightarrow ^{2}_{1}\text{H} + \beta^{+} + \nu$$

(where ν represents a neutrino, a particle to be discussed in the next section). Then another proton fuses with the deuteron:

$$_1^1\text{H} + \,_1^2\text{H} \rightarrow \,_2^3\text{He} + \gamma$$

Finally, two of the ^{3}He nuclei fuse:

$$_2^3\text{He} + \,_2^3\text{He} \rightarrow \,_2^4\text{He} + \,_1^1\text{H} + \,_1^1\text{H}$$

The net effect of this sequence, called the *proton–proton cycle*, is that four protons ($_1^1\text{H}$) combine to form one helium nucleus ($_2^4\text{He}$) plus two positrons (β^+), two gamma rays (γ), and two neutrinos (ν) with a release of about 25 MeV of energy:

$$4(_1^1\text{H}) \rightarrow \,_2^4\text{He} + 2\beta^+ + 2\gamma + 2\nu + Q \quad (Q = +24.7\ \text{MeV})$$

In a star such as our Sun, gamma-ray photons scatter off nuclei on their way to the surface. Each scattering results in a reduction in energy, until each photon has only a few electron-volts of energy. On reaching the surface, the photons are mostly visible-light photons. In our Sun, fusion involves only the central 10% of the Sun's mass. It has been going on for about 5 billion years and should continue, approximately as is, for another 5 billion years. As the next Example shows, an enormous number of fusion reactions per second is required to power the Sun.

Example 30.3 ■ Still Going: The Fusion Power of the Sun

Incoming sunlight energy falls on the Earth at the rate of 1.40×10^3 W/m^2. Assuming that the Sun's energy is produced by the proton–proton cycle, calculate the mass lost by the Sun per second.

Thinking It Through. To find the mass-loss rate, we need to know the Sun's total power output. Imagine the power flow through a sphere centered on the Sun, with a radius equal to the distance between the Earth and Sun ($R_{\text{E-S}}$), and then calculate that sphere's area. The total power can then be determined from the power per square meter and the total area in square meters. The Sun's mass-loss rate can be calculated from the total power, based on the mass–energy equivalence (Table 29.3).

Solution. The data are as follows (using some solar system data given in Appendix III):

Given: $\quad R_{\text{E-S}} = 1.50 \times 10^8$ km $= 1.50 \times 10^{11}$ m $\quad$ *Find:* $\quad \Delta m/\Delta t$ (overall mass-loss rate)
$\qquad\qquad M_\text{S} = 2.00 \times 10^{30}$ kg
$\qquad\qquad P_\text{S}/A = 1.40 \times 10^3$ W/m^2

The surface area of the imaginary sphere that intercepts all the Sun's energy is

$$A = 4\pi R_{\text{E-S}}^2 = 4\pi(1.50 \times 10^{11}\ \text{m})^2 = 2.83 \times 10^{23}\ \text{m}^2$$

Thus, the total power output of the Sun is

$$P_\text{S} = (1.40 \times 10^3\ \text{W/m}^2)(2.83 \times 10^{23}\ \text{m}^2) = 3.96 \times 10^{26}\ \text{W} = 3.96 \times 10^{26}\ \text{J/s}$$

To find the equivalent mass-loss rate, we convert this power to MeV per second:

$$\frac{3.96 \times 10^{26}\ \text{J/s}}{1.60 \times 10^{-13}\ \text{J/MeV}} = 2.48 \times 10^{39}\ \text{MeV/s}$$

Then, using the mass–energy equivalence (931.5 MeV/u), we find that the mass-loss rate is

$$\frac{\Delta m}{\Delta t} = \frac{(2.48 \times 10^{39}\ \text{MeV/s})(1.66 \times 10^{-27}\ \text{kg/u})}{931.5\ \text{MeV/u}} = 4.42 \times 10^9\ \text{kg/s}$$

Follow-Up Exercise. In this Example, how many proton–proton cycles happen per second? [*Hint*: Each cycle releases 24.7 MeV.]

Fusion as a Source of Energy Produced on the Earth

In several ways, fusion appears to be an ideal energy source for the future. Enough deuterium exists in the oceans, in the form of heavy water (D$_2$O), to supply our needs for centuries. In addition, fusion does not depend on a chain reaction, so there is less danger of the release of radioactive material. Also, fusion products tend to have relatively short half-lives. For example, tritium has a half-life of only 12.3 years, compared with hundreds or thousands of years for some fission products.

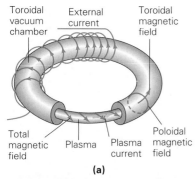

▲ **FIGURE 30.7** Magnetic confinement Magnetic confinement is one method by which controlled nuclear fusion might be achieved. **(a)** Tokamak configuration, showing the *B* field generated by external currents. The magnetic field confines the plasma in the ring. **(b)** The Princeton Tokamak Fusion Test Reactor (TFTR).

However, many unresolved technical problems must be overcome before controlled fusion can be used commercially to produce electric energy. A primary problem is that very high temperatures are needed to *initiate* fusion reactions, because of the electrical repulsion between the nuclei. Temperatures on the order of millions of degrees are needed to initiate these *thermonuclear fusion reactions*. The problem is in confining sufficient energy in a reaction region to maintain these high temperatures. Because of this difficulty, practical fusion reactors have not yet been achieved. However, uncontrolled fusion has been demonstrated in the form of the *hydrogen (H) bomb*. In this case, the fusion reaction is initiated by an implosion created by a small atomic (fission) bomb. This implosion provides the necessary density and temperatures to begin the fusion process.

At the high temperatures required for fusion, electrons are stripped from their nuclei, and a gas is created that consists of positively charged ions and free, negatively charged electrons. Such a "gas" of charged particles is called a **plasma**.* Plasmas have a number of special physical properties and are sometimes referred to as the *fourth phase of matter*.

The technological problems involved in confining a plasma are being approached in at least two different ways: magnetic confinement and inertial confinement. Since a plasma is a gas of charged particles, it can be controlled and manipulated by using electric and magnetic fields. In **magnetic confinement**, magnetic fields are used to hold the plasma in a confined space, a so-called magnetic bottle. (See Fig. 19.33b.)

Once a plasma is confined, electric fields can be used to produce electric currents in it. The currents, in turn, raise the plasma's temperature. Temperatures of 100 million kelvins have been achieved in a design called a *tokomak* (Fig. 30.7). This design uses a magnetic field arranged in a donut shape to trap the charged particles.

In addition to high temperatures, the initiation of fusion has minimum requirements on plasma density and confinement time. The trick is to meet all these requirements at the same time. The generation of several megawatts of power for less than a second in a magnetically confined plasma is typical of the best results so far. Clearly, for commercial applications, much higher power levels must be attained at a continuous level.

Inertial confinement depends on implosion techniques. Hydrogen fuel pellets are either dropped or positioned in a reactor chamber. Pulses of laser, electron, or ion beams are then used to implode the pellet, producing compression and high densities and temperatures. Fusion can occur if the pellet stays together for a sufficient time, which depends on its inertia (hence the name *inertial confinement*). At present, laser and particle beams are not powerful enough to produce sustainable fusion by inertial confinement. In summary, practical energy production from fusion is not expected to be accomplished until well into this century, if at all.

30.4 Beta Decay and the Neutrino

OBJECTIVES: To (a) explain why the neutrino is necessary in order to account for observed beta-decay data, (b) specify some of the physical properties of neutrinos, and (c) write complete beta-decay equations.

At first glance, beta decay *appears* to be a two-body decay process in which unstable nuclei emit an electron ($_{-1}^{0}e$) or a positron ($_{+1}^{0}e$). Examples of both types of decay are

$$\begin{array}{ccccccc} {}^{14}_{6}\text{C} & \rightarrow & {}^{14}_{7}\text{N} & + & {}^{0}_{-1}e & & \beta^{-} \; decay \\ (14.003\,242\,u) & & (14.003\,074\,u) & & & & \end{array}$$

$$\begin{array}{ccccccc} {}^{13}_{7}\text{N} & \rightarrow & {}^{13}_{6}\text{C} & + & {}^{0}_{+1}e & & \beta^{+} \; decay \\ (13.005\,739\,u) & & (13.003\,355\,u) & & & & \end{array}$$

However, when analyzed in detail, these equations appear to violate the conservation of energy and linear momentum, as well as other conservation laws.

*Plasmas exist in our everyday world, such as in fluorescent lamps and lightning strokes.

The energy released (or disintegration energy) in the foregoing β^- process, as calculated from the mass defect, using the given masses*, is $Q = 0.156$ MeV. (You should show this.) Therefore, if the decay involves only two particles (electron and daughter), the electron, being much lighter than the daughter, should always have a kinetic energy of just slightly less than 0.156 MeV. However, this is *not* what happens. When the electron's kinetic energies are measured, a *continuous spectrum* of energies is observed up to $K_{max} \approx Q$, ▸Fig. 30.8. (See Conceptual Example 30.4.) That is, not all of the released energy is accounted for by the electron's kinetic energy. Nor is this the only difficulty. The emitted β^- and the daughter nucleus hardly ever leave the disintegration site in opposite directions. Thus, linear-momentum conservation appears to be violated as well.[†]

What, then, is the problem? It would be hoped that the conservation laws are not invalid. An alternative explanation is that these apparent violations are telling us something about nature that we do not yet recognize. All of the apparent difficulties can be resolved if it is assumed that an unobserved particle is also emitted during the decay. This explanation and the existence of such a particle were first proposed in 1930 by Wolfgang Pauli. Fermi christened this particle the **neutrino** (meaning "little neutral one" in Italian). For charge to be conserved, the neutrino has to be electrically neutral. Because the neutrino had been virtually impossible to observe, it must interact very weakly with matter. In fact, scientists eventually discovered that the neutrino interacts with matter through a second nuclear force, much weaker than the strong force, called the *weak interaction*, or the *weak nuclear force*. (See Section 30.5.)

Details of the initial experimental observations of beta decay suggested that the neutrino had zero mass and therefore traveled with the speed of light. It also had linear momentum p related to its total energy E by $E = pc$. Furthermore, it had a spin quantum number of $\frac{1}{2}$. In 1956, a particle with these properties was finally detected, and the neutrino's existence was firmly established.

The previous beta-decay equations can now be written correctly and completely as

and

$$^{14}_{6}\text{C} \rightarrow {}^{14}_{7}\text{N} + \beta^- + \bar{\nu}_e$$

$$^{13}_{7}\text{N} \rightarrow {}^{13}_{6}\text{C} + \beta^+ + \nu_e$$

where ν (the Greek letter nu) symbolizes the neutrino. The symbol with a bar over it represents an *antineutrino*. The overbar notation is a common way of indicating an antiparticle. Two *different* neutrinos are associated with beta decay. A neutrino is emitted in β^+ decay (ν_e), and an antineutrino is emitted in β^- decay $(\bar{\nu}_e)$. The subscript e identifies the neutrinos as associated with electron–positron beta decay. As we will see in Section 30.5, additional types of neutrinos are associated with other decays triggered by the weak interaction.

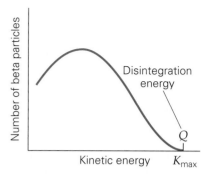

▲ **FIGURE 30.8** Beta ray spectrum For a typical beta-decay process, all beta particles are emitted with $K < Q$, leaving unaccounted-for energy.

Conceptual Example 30.4 ■ Having It All? Maximum Kinetic Energy in Beta Decay

Consider the decay of ^{14}C initially at rest:

$$^{14}_{6}\text{C} \rightarrow {}^{14}_{7}\text{N} + \beta^- + \bar{\nu}_e$$

What can be said about the maximum possible kinetic energy (K_{max}) of the beta particle: (a) $K_{max} > Q$, (b) $K_{max} = Q$, or (c) $K_{max} < Q$? Explain your reasoning clearly.

Reasoning and Answer. Answer (a) certainly cannot be correct, as it violates energy conservation. No one particle can have more than the total amount of energy available. So, can answer (b) be correct? Suppose that the beta particle did get all the released energy.

(continues on next page)

---

*Use of the atomic mass of the daughter ^{14}N (with seven electrons) is necessary in order to take into account the emitted electron, since the ^{14}N resulting from beta decay would have only the six electrons that orbit the parent ^{14}C nucleus.

[†]This process also violates the conservation of angular momentum. Careful analysis of the nuclear spin before the decay shows that it does *not* match the total spin (angular momentum) of the daughter plus the emitted electron.

That would mean that neither the neutrino nor the daughter nucleus had any energy, and therefore, neither would have any momentum. But to conserve momentum, at least one of these two particles must move off in the direction opposite to that of the beta particle. Hence, at least the neutrino or the daughter must have some energy. Therefore, the beta particle can't have it all, and answer (c) must be the correct one: $K_{max} < Q$.

Follow-Up Exercise. A typical energy spectrum of emitted beta particles is shown in Fig. 30.8. This spectrum indicates that there is a small probability of the beta particle having almost no kinetic energy. What would the decay products' trajectories look like after the decay if this were indeed the case?

30.5 Fundamental Forces and Exchange Particles

OBJECTIVES: To (a) understand the quantum-mechanical description of forces, and (b) classify the various forces according to their strengths, properties, ranges, and virtual particles.

The forces involved in everyday activities are complicated, because of the large numbers of atoms that make up ordinary objects. Contact forces between two hard objects, for example, are due to the repulsive electromagnetic forces between the atomic electrons. Looking at the *fundamental* interactions between particles makes things simpler. On this level, there are only four known **fundamental forces**: the *gravitational force*, the *electromagnetic force*, the *strong nuclear force*, and the *weak nuclear force*.

The most familiar of these forces are the gravitational and electromagnetic forces. Gravity acts between all particles, while the electromagnetic force is restricted to charged particles.* Both forces decrease with increasing particle separation distance and have a very large (essentially infinite) range.

To describe these forces classically, the concept of a field was employed. Modern physics, however, provides an alternative, more fundamental, description of how forces are transmitted. The force transmittal process is viewed as an exchange of particles. For example, a repulsive force would be analogous to you and another person interacting by tossing a ball back and forth. As you throw the ball and the other person catches it, for example, each of you feels a backward force. An observer who couldn't see the ball might conclude that there was a repulsive force between you and the other person.

The creation of such force-carrying particles would seem to violate energy conservation. However, because of the uncertainty principle (Section 28.4), a particle can be created for a *short time* with no outside energy input without violating energy conservation. Thus, over long time intervals, energy is conserved. For *extremely* short time intervals, the uncertainty principle *permits* a large *uncertainty* in energy ($\Delta E \propto 1/\Delta t$); creation of a particle is allowed; and energy conservation can be briefly violated. However, the created particle is absorbed before it is ever detected, so we never *observe* energy nonconservation. A particle created and absorbed in such a manner is called a **virtual particle**. In this sense, *virtual* means "undetected."

Thus, in the modern view, the fundamental forces are carried, or transmitted, by virtual **exchange particles**. The exchange particles for the four forces must differ in mass. Recall that the greater the mass of the particle, the greater is the energy ΔE required to create it, and thus the shorter is the time Δt for which it exists. Since a massive particle can exist for only a short time, the distance it can travel, and hence the range of the associated force, must be small. That is, *the range of a force associated with an exchange particle is inversely proportional to the mass of that exchange particle.*

The Electromagnetic Force and the Photon

The exchange particle of the electromagnetic force is a **virtual photon**. Here "virtual" refers to a photon that is exchanged but never detected directly. As a "massless" particle, it has an infinite range, as is required for the electromagnetic force.

*Both the electric and magnetic forces are actually components of a single force—the electromagnetic force.

A particle exchange can be visualized graphically by using a *Feynman diagram*, as in ▸Fig. 30.9. This graph shows the specific example of how the exchange idea can explain the electrical repulsion between two electrons. Such *space–time diagrams* are named after American scientist and Nobel Prize winner Richard Feynman (1918–1988), who used them to analyze electromagnetic interactions in his *quantum electrodynamics* theory. The important points are the vertices (intersections) of the diagram. One electron creates a virtual photon at point A, and the other electron absorbs it at point B. Each of the electrons undergoes a change in energy and in momentum (including direction) by virtue of the photon exchange and resulting force.

The Strong Nuclear Force and Mesons

Japanese physicist Hideki Yukawa (1907–1981) proposed in 1935 that the short-range strong nuclear force between two nucleons is associated with an exchange particle called the **meson**. An estimate of the meson's mass can be made from the uncertainty principle. If a nucleon were to create a meson, conservation of energy would be (undetectably) violated by an amount of energy at least equal to the energy equivalent of the meson's mass, or

$$\Delta E = (\Delta m)c^2 = m_m c^2$$

where m_m is the mass of the meson.

By the uncertainty principle, the meson would have to be absorbed in the exchange process in an amount of time on the order of

$$\Delta t \approx \frac{h}{2\pi \Delta E} = \frac{h}{2\pi m_m c^2}$$

In this amount of time, the meson could travel a distance R (which stands for *range*) that must be less than that traveled by light in the same time; thus,

$$R < c\Delta t = \frac{h}{2\pi m_m c} \tag{30.5}$$

Taking this distance to be the experimentally known range of the nuclear force ($R \approx 1.4 \times 10^{-15}$ m) and solving for m_m gives $m_m \approx 270 m_e$, where m_e is the electron's mass. If Yukawa's meson exists, it should have a mass on the order of 270 times that of an electron.

Virtual mesons in an exchange process cannot be directly observed, because they are emitted and reabsorbed during the nucleon–nucleon interaction. Physicists reasoned, however, that if sufficient energy were involved in the *collision* of nucleons, real mesons might be created from the energy available in the collision. These mesons could then be detectable. At the time of Yukawa's prediction, there were no known particles with masses between that of the electron (m_e) and the proton ($m_p = 1836 m_e$).

In 1936, Yukawa's prediction seemed to come true when a new particle with a mass of about $200 m_e$ was discovered in cosmic rays. Originally called the μ (Greek letter mu) meson, and now just the **muon**, it was shown to have two charge varieties, $\pm e$, with a mass of $m_{\mu^\pm} = 207 m_e$. However, further investigations showed that the muon did *not* behave like the strongly interacting particle of Yukawa's theory.

This situation was a source of controversy and confusion for years. But, in 1947, more particles in this mass range were discovered in cosmic radiation. These particles (one positive, one negative, and one with no charge) were called π (pi) mesons (for *primary mesons*) and now are more commonly called **pions**. Measurement showed the masses of the pions to be $m_{\pi^\pm} = 273 m_e$ and $m_{\pi^0} = 264 m_e$. Moreover, pions were found to interact strongly with matter. The pion fulfilled the requirements of Yukawa's theory. This meson was generally accepted as the particle primarily responsible for the transmission of the strong nuclear force. The Feynman diagram for a neutron and proton interacting via the exchange of a negative pion (the strong nuclear force) is shown in ▸Fig. 30.10.

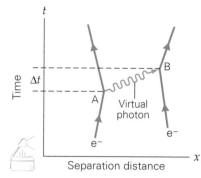

▲ **FIGURE 30.9** Feynman diagram of an electron–electron interaction The interacting electrons undergo a change in energy and momentum, due to the exchange of a virtual photon, which is created at A and absorbed at B in an amount of time (Δt) that is consistent with the uncertainty principle.

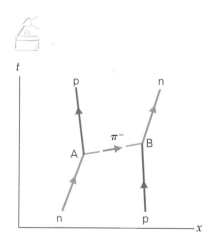

▲ **FIGURE 30.10** A Feynman diagram for nucleon–nucleon interactions via pion exchange Nucleon–nucleon interactions via the strong nuclear force occur through the exchange of virtual pions, here a negative one. This diagram shows one of many possible n–p interactions. It is called a *charge exchange* reaction; can you tell why? What other possible diagrams for neutron–proton interactions are there?

Note: Muon decay was treated in Examples 26.2 and 26.4 as evidence of time dilation.

Free pions and muons are unstable. For example, the π^+ particle decays in about 10^{-8} s into a muon and another type of neutrino:

$$\pi^+ \rightarrow \mu^+ + \nu_\mu$$

The ν_μ is called a *muon neutrino,* and it differs from the electron neutrino (ν_e) produced in beta decay. The muons can also decay into positrons and electrons with the emission of both types of neutrinos. For example, the positive-muon decay scheme is

$$\mu^+ \rightarrow \beta^+ + \nu_e + \bar{\nu}_\mu$$

The Weak Nuclear Force and the W Particle

The discrepancies in beta decay discussed in the preceding section led to another discovery. Electrons and neutrinos are emitted from unstable nuclei, but they do *not* exist inside the nucleus before the decay takes place. Enrico Fermi proposed that these particles are actually created at the time the nucleus decays. For β^- decay, this means that a neutron is in some way changed, or *transmuted,* into three particles:

$$n \rightarrow p^+ + \beta^- + \bar{\nu}_e$$

Experiments confirmed that free neutrons disintegrate by this decay scheme, with a half-life of about 10.4 min. But which force could cause a neutron to disintegrate in this manner? Since the neutron is outside the nucleus (free) when it decays, none of the known forces, including the strong nuclear force, seemed applicable. Thus, some other fundamental force must be acting in beta decay. Decay-rate measurements indicated that the force was extraordinarily weak—weaker than the electromagnetic force, but still much stronger than the gravitational force. This force was dubbed the **weak nuclear force**.

Originally, it was thought that this *weak interaction* was extremely localized, without any measurable range. We now know that the weak force has a range of about 10^{-17} m. While this range is much smaller than that of the strong force, it isn't zero. This means that the exchange particles associated with the weak force (the virtual force carriers) must be much more massive than the pions of the strong force. The virtual exchange particles associated with the weak force were named **W particles**.* W (*weak*) particles have masses about 100 times that of a proton, a fact that correlates with the extremely short range of the weak force. The existence of the W particle was confirmed in the 1980s when accelerators were built with enough energy to create the first real (nonvirtual) W particles.

The weak force is the only force that acts on neutrinos, which explains why these particles are so difficult to detect. Research has shown that the weak force is involved in the transmutation of other subatomic particles as well. In general, the weak force is limited to transmuting the identities of particles within the nucleus. The only way it manifests its existence in the outside world is through the emitted neutrinos. For example, the Sun's fusion reactions create neutrinos that continuously pass through the Earth.

Teaching tip: You may wish to discuss with students the problem of the universe's "missing mass" and the recent tentative experimental results that neutrinos may possess a very small amount of mass. If these results are correct, then the ocean of neutrinos that pervades the universe might account for some of the (apparently) missing universe mass.

One highly noticeable, but infrequent, announcement of the weak force at work occurs during the explosion of a stellar *supernova*. In a supernova, the collapse of the core of an aging star gives rise to a huge energy release, accompanied by a great number of neutrinos. In a relatively "nearby" supernova (a mere 1.5×10^{18} km away) observed in 1987, a burst of neutrinos was detected.

Whereas the original thought was that neutrinos were massless, recent experiments have determined that they do, in fact, have some very small amount of mass, on the order of a millionth of the electron's mass. This fact could have vast implications to our understanding of the Big Bang theory and the unexplained predominance of regular matter over antimatter in our universe. At this point, neutrino research is a growing field that will play an important role in physics for years to come.

---

*The weak force is actually carried by *three* exchange particles: W^+, W^-, and Z^0 (neutral).

TABLE 30.2   Fundamental Forces

Force	Relative Strength	Action Distance	Exchange Particle	Particles That Experience the Interaction
Strong nuclear	1	Short range ($\approx 10^{-15}$ m)	Pion (π meson)	Hadrons*
Electromagnetic	10^{-3}	Inverse square (infinite)	Photon	Electrically charged
Weak nuclear	10^{-8}	Extremely short range ($\approx 10^{-17}$ m)	W particle†	All
Gravitational	10^{-45}	Inverse square (infinite)	Graviton	All

*Hadrons are discussed in Section 30.6.
†Three particles are involved, as described in Section 30.8.

The Gravitational Force and Gravitons

The exchange particles associated with the gravitational force are called **gravitons**. There is still no firm evidence of the existence of this massless particle. (Why must gravitons be massless?) Ongoing experiments to detect the graviton have as yet proven unsuccessful, due to the relative weakness of its interaction.

A comparison of the relative strengths of the four fundamental forces is given in Table 30.2.

30.6 Elementary Particles

OBJECTIVES: To (a) classify the elementary particles into families and, (b) understand the different properties of the various families of elementary particles.

The fundamental building blocks of matter are referred to as **elementary particles**. Simplicity reigned when it was thought that an atom was an indivisible particle and therefore was *the* elementary particle. Early in the twentieth century, the proton, neutron, and electron were discovered to be constituents of atoms. It was hoped that these three particles were nature's elementary particles. However, scientists now know of a huge variety of subatomic particles and are working to simplify and reduce this list to a smaller set of truly elementary particles—building blocks for all the other "composite" particles—if this task is indeed possible.

Several systems classify elementary particles on the basis of their various properties. One classification uses the distinction of nuclear-force interactions. Particles that interact via the weak nuclear force, but not the strong force, are called leptons ("light ones"). The lepton family includes the electron, the muon, and their neutrinos. Other particles, called **hadrons**, are the only particles to interact by the strong nuclear force. These particles include the proton, neutron, and pion. Let's look briefly at the lepton and hadron families.

Leptons

The most familiar **lepton** is the electron. It is the only lepton that exists naturally in atoms. There is no evidence that it has any internal structure, at least down to 10^{-17} m. Thus, at present, the electron is considered to be a point particle.

Muons were first observed in cosmic rays. They appear not to have any internal structure and are therefore sometimes referred to as *heavy electrons*. Muons are unstable and decay in about 2×10^{-6} s, according to the following scheme:

$$\mu^- \rightarrow \beta^- + \bar{\nu}_e + \nu_\mu$$

A third charged lepton is known as a tau (τ^-) particle, or **tauon**. It has a mass about twice that of a proton. The electron, muon, and tauon are all negatively charged, have no apparent internal structure, and have positively charged antiparticles.

The remaining leptons are neutrinos, which are present in cosmic rays, emitted by the Sun, and appear in some radioactive decays. Neutrinos are now known to have a very small mass and travel close to, but slightly less than, the speed of light. They feel neither the electromagnetic force nor the strong force and pass through matter with very little interaction.

There are three types of neutrinos, each associated with a different charged lepton ($e^{\pm}, \mu^{\pm}, \tau^{\pm}$). They are named, not surprisingly, the *electron neutrino* (ν_e), the *muon neutrino* (ν_μ), and the *tau neutrino* (ν_τ). There is also an antineutrino for each of them, for a total of six different neutrinos. This completes the list of leptons. With a total of 6 leptons plus their antiparticles, there are 12 different leptons in all. Current theories predict that there should be no others.

Hadrons

Another family of elementary particles is the **hadrons**. All hadrons interact by the strong force, the weak force, and gravity. The electrically charged members can also interact by the electromagnetic force.

The hadrons are subdivided into *baryons* and *mesons*. **Baryons** include the familiar nucleons—the proton and neutron. They are distinguished from mesons in that they possess half-integer intrinsic spin values ($\frac{1}{2}, \frac{3}{2}, \dots$). Except for the stable proton, baryons decay into products that eventually include a proton. For example, recall from Section 29.2 the beta decay of a neutron into a proton.

Mesons, which include pions, have integer spin values ($0, 1, 2, \dots$) and eventually decay into leptons and photons. For example, the neutral pion decays into two gamma rays. (Why must there be at least two?)

The large number of hadrons suggests that they may be composites of other truly elementary particles. Some help in sorting out the hadron "zoo" came in 1963 when Murray Gell-Mann and George Zweig of the California Institute of Technology put forth the *quark theory*, which is discussed in the next section.

Leptons and hadrons and their properties are summarized in Table 30.3.

TABLE 30.3 Some Elementary Particles and Their Properties

Family Name	Particle Type	Particle Symbol	Antiparticle Symbol	Rest Energy (MeV)	Lifetime* (s)
Lepton					
	Electron	e^-	e^+	0.511	stable
	Muon	μ^-	μ^+	105.7	2.2×10^{-6}
	Tauon	τ^-	τ^+	1784	$\approx 3 \times 10^{-13}$
	Electron neutrino	ν_e	$\bar{\nu}_e$	$0^\dagger$	stable
	Muon neutrino	ν_μ	$\bar{\nu}_\mu$	$0^\dagger$	stable
	Tauon neutrino	ν_τ	$\bar{\nu}_\tau$	$0^\dagger$	stable
Hadron					
Mesons	Pion	π^+	π^-	139.6	2.6×10^{-8}
		π^0	same	135.0	8.4×10^{-17}
	Kaon	K^+	K^-	493.7	1.2×10^{-8}
		K^0	$\bar{K}^0$	497.7	8.9×10^{-11}
Baryons	Proton	p	$\bar{p}$	938.3	stable (?)§
	Neutron	n	$\bar{n}$	939.6	0.9×10^2
	Lambda	Λ^0	$\overline{\Lambda}^0$	1116	2.6×10^{-10}
	Sigma	Σ^+	$\overline{\Sigma}^-$	1189	8.0×10^{-10}
		Σ^0	$\overline{\Sigma}^0$	1192	0.6×10^{-20}
		Σ^-	$\overline{\Sigma}^+$	1197	1.5×10^{-10}
	Xi	Ξ^0	$\overline{\Xi}^0$	1315	2.9×10^{-10}
		Ξ^-	$\overline{\Xi}^+$	1321	1.6×10^{-10}
	Omega	Ω^-	Ω^+	1672	8.2×10^{-11}

*Lifetimes are expressed to two significant figures or fewer.
†Neutrinos are now known to possess a very small amount. Experiments yield upper limits and the mass of the electron neutrino is known to be less than 7×10^{-6} MeV.
§Electroweak theory predicts that the proton is unstable, with a half-life of 1000 trillion times the age of the universe.

30.7 The Quark Model

OBJECTIVES: To (a) become familiar with the quark model and properties of quarks, and (b) understand how the quark model accounts for the properties of baryons and mesons.

Gell-Mann and Zweig proposed that, in fact, hadrons are *not* elementary particles and therefore are not fundamental building blocks. Hadrons, they theorized, are composite particles composed of truly elementary (fundamental) particles. They named these particles **quarks** (taken from James Joyce's novel *Finnegan's Wake**). However, Gell-Mann and Zweig noted that because leptons and photons are essentially point particles, they *are* likely to be truly elementary particles with no internal structure.

Noting that some hadrons are electrically charged, Gell-Mann and Zweig reasoned that quarks must also possess charge. Their initial **quark model** consisted of three different quarks (with fractional charges) and their antiparticles (called *antiquarks*). Table 30.4 shows that, to account for hadrons that were undiscovered at the time, the list of quarks eventually had to be expanded to include six types. The original quark idea required only three quarks (plus three antiquarks) to construct the hadrons known at that time. These quarks were named the *up* quark (*u*), the *down* quark (*d*), and the *strange* quark (*s*). By using various combinations of these three types of quarks, the relatively heavy hadrons, the baryons (whose name means "heavy ones") could be built. In addition, quark–antiquark pairs could account for the lighter hadrons, the mesons. Thus, Gell-Mann and Zweig had proposed a radical new idea:

| Quarks are the fundamental particles of the hadron family. |

Since several quarks have to combine to give the charge on the hadron, the quarks must have fractions of an electron charge *e*. The theory proposed that *u*, *d*, and *s* quarks have charges of $+\frac{2}{3}e$, $\pm\frac{1}{3}e$, and $-\frac{1}{3}e$, respectively. The antiquarks, designated by overbars, such as $\bar{u}$, have opposite charges. Thus, three quark combinations could produce any baryon. For instance, the quark composition of the proton and neutron would be *uud* and *udd*, respectively. Mesons could be constructed from various pairs of a quark and an antiquark, such as $u\bar{d}$ for the positive pion, π^+ (▶Fig. 30.11).

Conceptual Example 30.5 ■ Building Mesons: Quark Engineering, Inc.

Using the data in Tables 30.3 and 30.4, explain why it is not possible to build a meson from two quarks (in other words, without using any antiquarks).

Reasoning and Answer. It can be seen from the list of mesons in Table 30.3 that mesons are either neutral or if they are charged, it is an integral multiple of *e*. In Table 30.4, note that all the positively charged quarks have a charge of $+\frac{2}{3}e$. Thus, any two of them would add up to a meson charge of $+\frac{4}{3}e$, which is contrary to observation. Similar reasoning holds if we use two negatively charged quarks: Since they all have the same charge $\left(-\frac{1}{3}e\right)$, any two of them add up to a total charge of $-\frac{2}{3}e$, again in disagreement with experiment. Finally, consider combining one positively charged quark and one negatively charged quark. The net charge of this combination is $+\frac{1}{3}e$, again not in agreement with observation. Thus, no combination of two quarks (or two antiquarks) can be used to produce a meson. The combinations must include at least one antiquark and one quark.

Follow-Up Exercise. The antiparticle of the positive pion (π^+) is the negative pion (π^-). What is the quark structure of the negative pion? Show that it is composed of the antiquarks of the quarks that make up the positive pion.

The discovery of new subatomic particles in the 1970s led to the addition of the last three quark types: *charm* (*c*); *top*, or truth (*t*); and *bottom*, or beauty (*b*). Today, there is firm experimental evidence of the existence of all six quarks and their six antiquarks. How many more such particles will be needed to keep up

*A line in the novel exclaims, "Three quarks for Muster Mark!" The "three quarks" denote the three children of a character in the novel, Mister (Muster) Mark, who is also known as Mr. Finn.

TABLE 30.4

Types of Quarks*

Name	Symbol	Charge
Up	*u*	$+\frac{2}{3}e$
Down	*d*	$-\frac{1}{3}e$
Strange	*s*	$-\frac{1}{3}e$
Charm	*c*	$+\frac{2}{3}e$
Top (Truth)	*t*	$+\frac{2}{3}e$
Bottom (Beauty)	*b*	$-\frac{1}{3}e$

*Antiquarks are designated by an overbar and have opposite charges compared with those of the corresponding quarks.

Note: Quark names (such as *up*, *down*, and *strange*) are arbitrary and therefore should not be taken literally. They serve simply to identify the quarks.

Note: Three quarks in proper combination account for all known baryons. Quark–antiquark pairs in proper combination account for all known mesons.

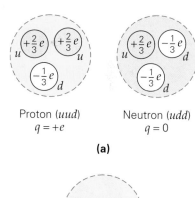

Proton (*uud*)
$q = +e$

Neutron (*udd*)
$q = 0$

(a)

Positive pion, π^+ ($u\bar{d}$)
$q = +e$

(b)

▲ **FIGURE 30.11 Hadronic quark structure (a)** Three combinations of quarks can be used to construct all the baryons, such as the proton and neutron. **(b)** Quark–antiquark combinations can be used to construct all the mesons, such as the positive pion.

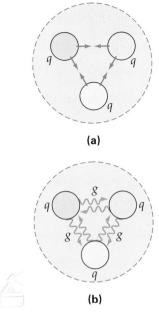

▲ **FIGURE 30.12 Quarks, color charge, confinement, and gluons** (a) Quarks of different color charge attract each other via the color force, which keeps them confined (here, inside a baryon—how do you know this is a baryon?). (b) Gluons (wiggly arrows) are exchanged between quarks of different color charge, creating the color force—analogous to virtual photons and the electric force.

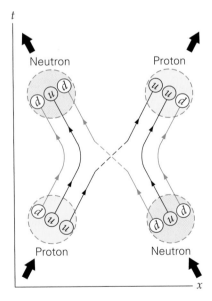

▲ **FIGURE 30.13 Quark depiction of the strong nuclear force** Instead of envisioning the n–p interaction as an exchange of a virtual π^- particle, we can use a model of quark exchange. In this exchange, a pair of quarks (equivalent to a π^- particle) is transferred. A proton becomes a neutron, and vice versa.

with our growing "zoo" of elementary particles? The hope is, of course, that this list will not need to be expanded. In summary, the present picture includes the following truly elementary particles: leptons (and antileptons), quarks (and antiquarks), and exchange particles such as the photon.

Quark Confinement, Color Charge, and Gluons

Thus far in our discussion on quarks, one thing has been missing—direct experimental observation of a quark. Unfortunately, even in the most energetic particle collisions, *a free quark has never been observed*. Physicists now believe that quarks are permanently confined within their particles by a springlike force. That is, a force exists between quarks that grows as they separate from one another. This force grows very rapidly with distance and prevents the ejection of a quark from its particle. This phenomenon is called **quark confinement**.

In order to explain the force between quarks and to clear up some problems with apparent violation of the Pauli exclusion principle (see the discussion of atomic structure in Section 28.3), quarks were endowed with another characteristic called **color charge**, or simply *color*. There are three types of color charge: red, green, and blue. (These names have nothing to do with visual color.) In analogy to electric charge, the quark confinement force exists because different color charges attract each other. Recall from Section 30.5 that the electromagnetic force is due to virtual photon exchanges between charged particles. Similarly, the force between quarks of different color is due to exchanges of virtual particles called **gluons** (◄Fig. 30.12). This force is sometimes called the **color force**. This theory is named *quantum chromodynamics (QCD)*; *chromo* for "color", in analogy to Feynman's quantum *electrodynamics (QED)*, which successfully explains the electromagnetic force.

The concept of the force between quarks can be extended to explain the force between hadrons—the strong nuclear force. Consider our previous explanation of the strong force between a neutron and proton as shown in Fig. 30.10a: the exchange of a virtual negative pion. More fundamentally, we can qualitatively depict this force in terms of an exchange of quarks between the hadrons (◄Fig. 30.13).

30.8 Force Unification Theories, the Standard Model, and the Early Universe

OBJECTIVES: To (a) become familiar with current attempts to unify the four fundamental forces, and (b) understand why elementary-particle interactions might hold the key to understanding the very early evolution of the universe.

Unification Theories

Early in the twentieth century, Einstein was one of the first theorists to conjecture that it might be possible to unify the fundamental forces of nature—that these four, apparently very different forces, might really be just different manifestations of one force. Each manifestation would appear under a different set of physical conditions. For example, an electrically neutral particle would not exhibit the electromagnetic portion. Since then, it has been the dream of physicists to understand the "fundamental interactions" in the universe, using just one force. Attempts to unify the various forces are called **unification theories**.

Actually, the first step toward unification was taken in the nineteenth century by Maxwell when he combined the electric and magnetic forces into a single electromagnetic force. Einstein later showed that the two are connected by relativity. The next major step occurred in the 1960s when Sheldon Glashow, Abdus Salam, and Steven Weinberg successfully combined the electromagnetic force with the weak force into a single **electroweak force**. For their efforts, they were awarded a Nobel Prize in 1979.

How can such apparently very different forces be unified? After all, their exchange particles are so different—recall the massless photon (γ) for the electromagnetic force and the massive W particle for the weak force. However, it was reasoned that, at extremely high energies, the mass–energy of the W particle is negligible compared with its total energy, in effect making it massless, like the photon. To understand this reasoning, recall from Section 26.5 that the total energy of a particle is the sum of its kinetic and rest energies ($E = K + mc^2$). When $K \gg mc^2$ (that is, at high energies), the particle's rest energy is negligible compared with its kinetic energy—perhaps, Glashow, Salam, and Weinberg reasoned, making the W particle more "photonlike" than was previously thought.

In the electroweak theory, weakly interacting particles such as electrons and neutrinos carry a *weak charge*. This weak charge is responsible for the exchange of particles, thus creating the combined electroweak force. The electroweak unification theory predicted the existence of three electroweak exchange particles: W^+ and W^- when there is charge exchange, and the neutral Z^0 when there is no charge transfer (▸Fig. 30.14). The eventual discovery of these particles led to a Nobel Prize for Carlo Rubbia and Simon van der Meer in 1984.

Scientists are now attempting to unify the electroweak force with the strong nuclear force. If successful, this **grand unified theory (GUT)** would reduce the number of fundamental forces to two. Most GUT attempts require several dozen exchange particles. In addition, leptons and quarks are combined into one family and can change into each other through the exchange of these particles. Scientists are mildly optimistic that experimental verification of one of the GUT candidates might occur in this century. For example, most of the current GUT theories predict that the proton should be unstable and decay with a half-life of about 10^{32} years, which is 10^{22} times the age of the universe. Experiments looking for this decay are currently underway, so far with no success. The experiments require huge amounts of water (protons) in order to make up for the expected very slow decay rate.

The ultimate unification would be to fold the gravitational force into the GUT, creating a single **superforce**. How this would be done, or even *if* it can ever be done, is not clear at this point. One problem is that, although the three components of the GUT can be represented as force fields in space and time, in our current view, gravity *is* space and time!

The Standard Model

Taken together, the electroweak theory and the QCD model for the strong interaction are referred to as the **standard model**. In this model, the gluons carry the strong force. This force keeps quarks together to form composite particles such as protons and pions. Leptons do not experience the strong force and participate only in the gravitational and electroweak interactions. Presumably, the gravitational interaction is carried by the graviton. The electromagnetic part of the electroweak interaction is carried by the photon, and the W and Z bosons carry the weak portion of the electroweak force.

Evolution of the Universe and the Superforce

An interesting connection now exists between elementary particle physicists and astrophysicists interested in the evolution of the universe—in particular, its very early evolution. This connection occurs because, according to present ideas, the universe began 10 billion to 15 billion years ago with the *Big Bang*. It is theorized that temperatures during the first 10^{-45} second of the universe were on the order of 10^{32} K. This corresponds to particle kinetic energies of about 10^{22} MeV, high enough that the rest energy of even the most massive elementary particles would be negligible. In effect, the particles would be massless, perhaps placing the four forces on an equal footing.

As the universe expanded and cooled, these early elementary particles condensed into what we see today. First, protons, neutrons, and electrons formed; in turn, they combined into atoms and, eventually, molecules. Over the billions of years, the average temperature of the universe has cooled off to its present value of about 3 K. In the process, scientists think that the superforce symmetry (the equal footing of all four forces) has been lost, leaving us with four very different-looking forces that are really components of a superforce.

It is hoped that future experiments might tell us more about the early moments of the universe and thus about the superforce. The ultimate goal of physics might even be within reach—to understand the basic interactions that govern the universe. While great strides are being made, it is likely to take well into this century, if not further, to achieve this goal.

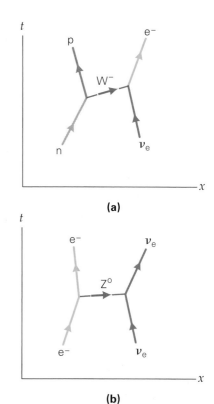

▲ **FIGURE 30.14** Weak-force interactions The Feynman diagrams for **(a)** the neutrino-induced conversion of a neutron into a proton and an electron through the exchange of a W^- particle and **(b)** the scattering of a neutrino by an electron through the exchange of a neutral Z^0 particle.

Chapter Review

- A **nuclear reaction** (including decay) involves the interactions of nuclei and particles, usually resulting in different product nuclei or particles.

- The **Q value** of a reaction (or decay) is the energy released or absorbed in the process. For a two-body reaction of the form $A + a \rightarrow B + b$,

$$Q = (m_A + m_a - m_B - m_b)c^2 = (\Delta m)c^2 \qquad (30.3)$$

  If $Q > 0$, the reaction is **exoergic** and energy is released.

  If $Q < 0$, the reaction is **endoergic** and energy is absorbed.

- In **fission**, an unstable heavy nucleus decays by splitting into two fragments and several neutrons, which together have less total mass than that of the original nucleus; kinetic energy is thus released.

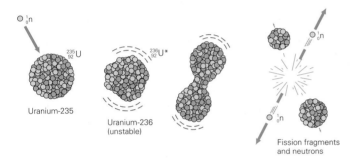

- A **chain reaction** occurs when the neutrons released from one fission trigger other fissions, which trigger further fissions, and so on.

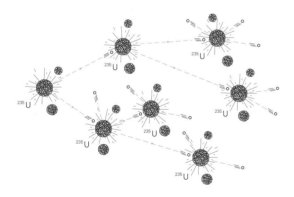

- In a **nuclear power reactor**, the fission chain reaction is kept from going out of control by a set of control rods. The heat energy released is usually used to create steam, which eventually turns generator turbine blades to create electricity.

- In **fusion**, two light nuclei fuse, producing a nucleus with less total mass than that of the original nuclei; energy is thus released. Controlled fusion reactions are, at present, not commercially feasible, because no one has achieved confinement of the gas of ionized atoms and free electrons, called a **plasma**, at the proper density and temperature.

- The **beta decay** of an unstable nucleus produces a daughter nucleus, an electron or positron, and an antineutrino or neutrino.

- **Exchange particles** are virtual particles associated with various forces. The **pion** (a meson) is the exchange particle primarily responsible for the strong nuclear force.

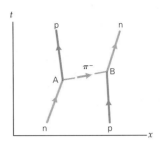

- The **weak nuclear force**, transmitted by the **W particle**, is primarily responsible for beta decay and the instability of the neutron.

- **Leptons** are the family of elementary particles that interact through the weak force, electromagnetism, and gravity, but not the strong force. Electrons, muons, tauons, and neutrinos make up the lepton family.

- **Hadrons** are the family of elementary particles that interact by the strong force, the weak force, and gravity. If electrically charged, hadrons also interact by the electromagnetic force. The hadrons are subdivided into baryons and mesons. **Baryons** include the familiar nucleons—the proton and neutron. Except for the stable proton, baryons decay into products that eventually include a proton. **Mesons**, which include pions, eventually decay into leptons and photons.

- **Quarks** are elementary, fractionally charged particles that make up hadrons.

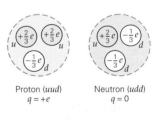

- **Charge color** describes the color force between quarks. Color force explains why a free quark is never likely to be seen, a phenomenon called **quark confinement**. Quark exchange is the fundamental explanation for the strong nuclear force.

- The **electroweak force** is the name given to the unified electromagnetic and weak forces.

- The **grand unified theory (GUT)** is an attempt to unify the electroweak force with the strong nuclear force.

- The **superforce** is the single force that will result if the long-pursued unification of the "fundamental" forces ever materializes.

- In the **standard model**, gluons carry the strong force, which keeps quarks together in composite particles such as protons. Leptons participate only in the gravitational and electroweak interactions. The former is carried by the graviton. The electroweak interaction is carried by the photon and the W and Z bosons.

Exercises

MC = *Multiple Choice Question,* **CQ** = *Conceptual Question, and* **IE** = *Integrated Exercise. Throughout the text, many exercise sections will include "paired" exercises. These exercise pairs, identified with* **red numbers***, are intended to assist you in problem solving and learning. In a pair, the first exercise (even numbered) is worked out in the Study Guide so that you can consult it should you need assistance in solving it. The second exercise (odd numbered) is similar in nature, and its answer is given at the back of the book.*

30.1 Nuclear Reactions

1. **MC** In a particular nuclear reaction, the masses of the products add up to be less than the masses of all the initial nuclei involved. The total kinetic energy has (a) increased, (b) decreased, (c) stayed the same. (a)

2. **MC** To initiate an endoergic reaction, a particle incident on a stationary nucleus must have (a) a kinetic energy equal to the Q value, (b) a kinetic energy less than a certain threshold energy, (c) a kinetic energy larger than the Q value. (c)

3. **MC** The absorption of a slow neutron by a ^{238}U temporarily results in a compound nucleus which would be (a) ^{238}U, (b) ^{238}U*, (c) ^{239}U*, (d) ^{239}Np*. (c)

4. **CQ** Using conservation of momentum, explain how a colliding beam reaction (in which both particles move toward a head-on collision) requires less incident kinetic energy than if one of them is at rest. [*Hint:* Linear momentum must be conserved.] see ISM

5. **CQ** For a given Q-value and incident particle (called a), how does the threshold energy required to initiate a specific nuclear reaction vary as the mass of the target (called A) changes? Sketch a graph of threshold energy versus target mass ranging from target masses on the order of the incident mass to target masses much more than the incident mass. What is the threshold energy if $M_A \gg m_a$? Explain what this means physically. see ISM

6. **CQ** In the capture of a slow neutron, which nucleus would probably have a larger capture cross-section: (a) ^{24}Mg, (b) ^{25}Mg, or (c) they would be about the same? ^{25}Mg, see ISM

7. **CQ** What is the threshold energy for an alpha particle elastically scattering off a Ca-40 nucleus? Explain. zero, see ISM

8. ● Complete the following nuclear reactions:
   (a) $^1_0 n + ^{40}_{18}Ar \rightarrow ^{37}_{16}S + \alpha$
   (b) $^1_0 n + ^{235}_{92}U \rightarrow ^{98}_{40}Zr + ^{135}_{52}Te + 3(^1_0 n)$
   (c) $^1_0 n + ^{235}_{92}U \rightarrow ^{133}_{51}Sb + ^{99}_{41}Nb + 4(^1_0 n) (^1_0 n)$
   (d) $^{14}_7 N(\alpha, \underline{p})^{17}_8 O$
   (e) $^{137}_{56}Ba (n, p)^{137}_{55}Cs$
   see ISM

9. ● Complete the following nuclear reactions:
   (a) $^{13}_6 C + ^1_1 H \rightarrow \underline{\gamma} + ^{14}_7 N$
   (b) $^{10}_5 B + ^4_2 He \rightarrow ^{10}_6 C + ^2_1 H$
   (c) $^{27}_{13}Al(\alpha, ^1_0 n)^{30}_{15}P$
   (d) $^{14}_7 N(\alpha, p)^{17}_8 O$
   (e) $^{13}_6 C(\underline{p}, \alpha)^{10}_5 B$
   see ISM

10. **IE** ● (a) Consider the reaction $^{13}_6 C + ^1_1 H \rightarrow ^4_2 He + ^{10}_5 B$. Use the masses of the nuclides involved to determine whether it is endoergic or exoergic. (b) If it is exoergic, find the amount of energy released; if it is endoergic, find the threshold energy. (a) endoergic (b) 4.32 MeV

11. ● What would the daughter nuclei in the following decay equations be? Can these decays actually occur spontaneously? Explain your reasoning in each case.
    (a) $^{22}_{10}Ne \rightarrow \underline{^{22}_{11}Na} + ^0_{-1}e$
    (b) $^{226}_{88}Ra \rightarrow \underline{^{222}_{86}Rn} + ^4_2 He$
    (c) $^{16}_8 O \rightarrow \underline{^{12}_6 C} + ^4_2 He$
    (a) $^{22}_{11}Na$, no (b) $^{222}_{86}Rn$, yes (c) $^{12}_6 C$, no; see ISM

12. ● Show that the Q value for the reaction $^1_1 H + ^2_1 H \rightarrow ^3_2 He + \gamma$ is 5.49 MeV. see ISM

13. **IE** ● Uranium-238 undergoes alpha decay as follows:

    $$^{238}_{92}U \quad \rightarrow \quad ^{234}_{90}Th \quad + \quad ^4_2 He$$
    $$(238.050\,786\ u) \qquad (234.043\,583\ u) \qquad (4.002\,603\ u)$$

    (a) Would you expect the Q value to be (1) positive, (2) negative, or (3) zero? Why? (b) Find the Q value. (a) (1) positive (b) +4.28 MeV

14. ●● Find the threshold energy for the following reaction:

    $$^{16}_8 O \quad + \quad ^1_0 n \quad \rightarrow \quad ^{13}_6 C \quad + \quad ^4_2 He$$
    $$(15.994\,915\ u) \quad (1.008\,665\ u) \quad (13.003\,355\ u) \quad (4.002\,603\ u)$$
    2.35 MeV

15. ●● Find the threshold energy for the following reaction:

    $$^3_2 He \quad + \quad ^1_0 n \quad \rightarrow \quad ^2_1 H \quad + \quad ^2_1 H$$
    $$(3.016\,029\ u) \quad (1.008\,665\ u) \quad (2.014\,102\ u) \quad (2.014\,102\ u)$$
    4.36 MeV

16. ●● Is the given reaction endoergic or exoergic? Prove your answer.

    $$^7_3 Li \quad + \quad ^1_1 H \quad \rightarrow \quad ^4_2 He \quad + \quad ^4_2 He$$
    $$(7.016\,005\ u) \quad (1.007\,825\ u) \quad (4.002\,603\ u) \quad (4.002\,603\ u)$$
    exoergic; $Q = +17.3$ MeV

17. ●● Is the reaction ^{200}Hg(p, α)^{197}Au endoergic or exoergic? Prove your answer. (The reaction on page 936 gives the mass values.) exoergic; $Q = +6.50$ MeV

18. ●● Determine the Q value of the following reaction:

    $$^9_4 Be \quad + \quad ^4_2 He \quad \rightarrow \quad ^{12}_6 C \quad + \quad ^1_0 n$$
    $$(9.012\,183\ u) \quad (4.002\,603\ u) \quad (12.000\,000\ u) \quad (1.008\,665\ u)$$
    +5.70 MeV

19. ●● What is the minimum kinetic energy a proton must have in order to initiate the reaction $^3_1 H(p, d)^2_1 H$? ("d" stands for a deuterium nucleus and is called the *deuteron*.) 5.38 MeV

20. ●● ^{226}Ra decays and emits a 4.706-MeV alpha particle. Find the kinetic energy of the recoiling daughter nucleus from the decay of a stationary radium-226 nucleus. 0.165 MeV

21. IE ●●● The same type of incident particle is used for two endoergic reactions. In one reaction, the mass of the target nucleus is 15 times that of the incident particle, and in the other reaction, it is 20 times the mass. The Q value of the first reaction is known to be three times that of the second. (a) Compared with the second reaction, the first reaction has (1) greater, (2) the same amount of, or (3) less minimum threshold energy. (b) Prove your answer to part (a) by calculating the ratio of the minimum threshold energy for the first reaction to that for the second. (a) (1) greater (b) 3.05

22. ●●● Consider n (where n is an integer ≥ 1) ceramic pie plates of radius R randomly fixed (but none overlapping) on a rectangular wall with dimensions L and W. If you were to throw a very small (point) object at the wall, what would be the percentage probability of hitting a pie plate, in terms of the given parameters? Note that this is a crude analogy to the concept of a nuclear cross-section in determining, for example, the likelihood that a neutron might be absorbed by a given target. However, the nuclear cross-section is not necessarily related to the area of the target nucleus as it is for the pie plate. [*Hint*: Think in terms of the total area of plates compared to the total wall area.] $n\pi R^2/(LW)$

30.2 Nuclear Fission *and*
30.3 Nuclear Fusion

23. MC Nuclear fission (a) is endoergic, (b) occurs only for uranium-235, (c) releases about 500 MeV per fission, (d) requires a critical mass for a sustained reaction. (d)

24. MC A nuclear reactor (a) can operate on natural (unenriched) uranium, (b) has its chain reaction controlled by neutron-absorbing materials, (c) can be partially controlled by the amount of moderator, (d) all of the preceding. (d)

25. MC A nuclear fusion reaction (a) has a positive Q value, (b) may occur spontaneously, (c) is an example of "splitting" the atom. (a)

26. CQ Explain clearly why the cooling water in a U.S. nuclear reactor helps keep the chain reaction from multiplying in addition to serving as a neutron moderator. see ISM

27. CQ Spent (used) fuel rods from most U.S. reactors are currently stored on site in pools of circulating cold water that contains a salt of boron (it is borated water). Why is the boron needed? see ISM

28. ● Find the approximate energy released in the following fission reactions: (a) $^{235}_{92}\text{U} + ^{1}_{0}\text{n} \rightarrow$ fission products plus the release of five neutrons, (b) $^{235}_{94}\text{Pu} + ^{1}_{0}\text{n} \rightarrow$ fission products plus the release of three neutrons. (a) 231 MeV (b) 233 MeV

29. ● Calculate the amounts of energy released in the following fusion reactions: (a) $^{1}_{1}\text{H} + ^{1}_{0}\text{n} \rightarrow ^{2}_{1}\text{H} + \gamma$, (b) $^{3}_{2}\text{He} + ^{3}_{2}\text{He} \rightarrow ^{4}_{2}\text{He} + 2^{1}_{1}\text{H}$. (a) 2.22 MeV (b) 12.9 MeV

30. ● Calculate the amounts of energy released in the following fusion reactions: (a) $^{2}_{1}\text{H} + ^{2}_{1}\text{H} \rightarrow ^{3}_{2}\text{He} + ^{1}_{0}\text{n}$, (b) $^{2}_{1}\text{H} + ^{3}_{1}\text{H} \rightarrow ^{4}_{2}\text{He} + ^{1}_{0}\text{n}$. (a) 3.27 MeV (b) 17.6 MeV

31. ●● In power reactors, using water as a moderator works well, because the proton and neutron have nearly the same mass. For a head-on elastic collision, we might expect a neutron to lose all of its kinetic energy in one collision, whereas for an "almost miss," we might expect it to lose essentially none. Assume that, on average, the neutron loses 60% of its kinetic energy during each collision. Estimate how many collisions are needed to reduce a 2.0-MeV neutron to a neutron with a kinetic energy of only 0.02 eV (approximately "thermal"). 36 collisions

30.4 Beta Decay and the Neutrino

32. MC In the absence of a neutrino, which of the following quantities would not be conserved in beta decay: (a) total energy, (b) total linear momentum, (c) total angular momentum, or (d) all of the preceding? (d)

33. MC A neutrino interacts with matter by (a) the electromagnetic interaction, (b) the strong interaction, (c) the weak interaction, (d) both b and c. (c)

34. MC In β^- decay, if the daughter nucleus is left stationary, how do the momenta (magnitude) of the β^- particle and the antineutrino $\bar{\nu}_e$ compare: (a) the β^- has more momentum, (b) the $\bar{\nu}_e$ has more momentum, or (c) they have the same momentum? (c)

35. CQ A $\bar{\nu}_e$ can interact with a proton in a target of water (although this is very rare) and the reaction is essentially the inverse of β^- decay; that is, the proton is converted into a neutron. Neutrino experimenters can detect that this reaction happened by using gamma-ray detectors. The neutrino "signature" occurs when the detector simultaneously records a gamma ray of 2.22 MeV and two others each with an energy of 0.511 MeV. Write the reaction and explain where the three gamma rays come from. [*Hint*: 2.22 MeV is the binding energy of the deuteron and 0.511 MeV is the rest energy of an electron.] $p + \bar{\nu}_e \rightarrow n + \beta^+$, see ISM

36. CQ Recently it has been determined that neutrinos actually possess a small amount of mass (which depends on the kind of neutrino). If they were massless (as previously assumed), neutrinos of all energies would travel at the same speed, c. Since they have mass, as we have seen from relativity, their speed is a bit less than c. Is it still true that all neutrinos of the same type would travel at the same speed regardless of their energy? Explain. no, see ISM

37. ● A neutrino created in a beta-decay process has an energy of 2.65 MeV. Assume that it is massless, and thus has the same relationship between its energy and momentum as a photon (that is, $E = pc$). Determine the de Broglie wavelength of the neutrino. 4.69×10^{-13} m

38. IE ●● In Exercise 37 with the same neutrino energy, what would be the maximum possible energy of the beta particle if the disintegration energy were 5.35 MeV? (a) (1) zero, (2) less than 5.35 MeV but not zero, (3) 5.35 MeV, or (4) greater than 5.35 MeV. (b) Under these conditions, determine the total and kinetic energy of the beta particle. (c) What is the direction, relative to the neutrino direction, of the momentum of the beta particle after the decay? Explain your reasoning. see ISM

39. ●● Show that the disintegration energy for β^- decay is $Q = (m_P - m_D - m_e)c^2 = (M_P - M_D)c^2$, where the m's represent the masses of the parent and daughter *nuclei* and the M's represent the masses of the neutral *atoms*. see ISM

40. ●● What is the maximum kinetic energy of the electron emitted when a ^{12}B nucleus beta decays into a ^{12}C nucleus? (See Exercise 39.) 13.37 MeV

41. ●● The kinetic energy of an electron emitted from a ^{32}P nucleus that beta decays into a ^{32}S nucleus is observed to be 1.00 MeV. What is the energy of the accompanying neutrino of the decay process? Neglect the recoil energy of the daughter nucleus. (See Exercise 39.) 0.71 MeV

42. ●●● Show that the disintegration energy for β^+ decay is $Q = (m_P - m_D - m_e)c^2 = (M_P - M_D - 2m_e)c^2$, where the m's represent the masses of the parent and daughter *nuclei* and the M's represent the masses of the neutral *atoms*. see ISM

43. ●●● The kinetic energy of a positron emitted from the β^+ decay of a ^{13}N nucleus into a ^{13}C nucleus is measured to be 1.190 MeV. What is the energy of the accompanying neutrino of the decay process? Neglect the recoil energy of the daughter nucleus. (See Exercise 42.) 9 keV

44. IE ●●● The expressions for the Q values associated with both β^- and β^+ decay are given in Exercises 39 and 42. Assume that the daughter's atomic mass M_D is the same in both processes. (a) If both expressions are to be energetically possible, the mass of the parent atom, M_P, for β^- decay must be (1) greater than, (2) equal to, (3) less than that for β^+ decay. Why? (b) Write the mass requirements of the parent atoms for β^- and β^+ processes, respectively, in terms of the atomic masses of the daughter, parent, and electron. (a) (2) less than (b) β^-: $M_P > M_D$; β^+: $M_P > M_D + 2m_e$

30.5 Fundamental Forces and Exchange Particles

45. MC Virtual particles (a) form virtual images, (b) exist only in an amount of time as permitted by the uncertainty principle, (c) make up positrons, (d) can be observed in exchange processes. (b)

46. MC The exchange particle for the strong nuclear force is the (a) pion, (b) W particle, (c) muon, (d) positron. (a)

47. CQ If virtual exchange particles are unobservable by themselves, how is their existence verified? see ISM

48. CQ When a proton interacts with another proton, which of the four fundamental forces would be involved? How about when an electron interacts with another electron? see ISM

49. ● Assuming the range of the nuclear force to be on the order of 10^{-15} m, predict the mass (in kilograms) of the exchange particle that is related to this force. 3.5×10^{-28} kg

50. ●● In a certain type of reaction, a high-speed proton collides with a nucleus and travels, on average, a distance of 5.0×10^{-16} m within the nucleus before the reaction takes place. What type of fundamental interaction is this most likely to be, and how much time elapses before the interaction takes place? strong nuclear; 1.7×10^{-24} s

51. IE ●● (a) In an interaction using the virtual-particle model, the range of the interaction (1) increases, (2) remains the same, (3) decreases as energy of the exchange particle increases. Why? (b) A W particle in a weak interaction is found to have rest energy of 1.00 GeV. What is the approximate range for the weak interaction with this exchange particle? (a) (3) decreases (b) 1.98×10^{-16} m

52. ●● By what minimum amount of energy is the conservation of energy "violated" during a neutral pi-meson exchange process? 135 MeV

53. ●● How long is the conservation of energy "violated" in a neutral pi-meson exchange process? 4.71×10^{-24} s

30.6 Elementary Particles *and*
30.7 The Quark Model

54. MC Particles that interact by the strong nuclear force are called (a) muons, (b) hadrons, (c) W particles, (d) leptons. (b)

55. MC Quarks are thought to make up which of the following particles: (a) hadrons, (b) muons, (c) Z particles, or (d) all of the preceding? (a)

56. CQ What is meant by quark flavor and color? Can these attributes be changed? Explain. see ISM

57. CQ With so many types of hadrons, why aren't fractional electronic charges observed? see ISM

58. What are some of the important differences and distinctions between baryons and mesons? see ISM

59. ● Out of each of the following pairs, which particle has more mass: (a) π^+ and π^0; (b) K^+ and K^0; (c) Σ^+ and Σ^0; (d) Ξ^0 and Ξ^-? [*Hint*: See Table 30.3.] (a) π^+ (b) K^0 (c) Σ^0 (d) Ξ^-

60. IE ● (a) The quark combination for a proton is (1) *uuu*, (2) *uud*, (3) *udd*, (4) *ddd*. (b) Prove your answer to part (a) given the correct electric charge. (a) (2) *uud* (b) see ISM

61. IE ● (a) The quark combination for a neutron is (1) *uuu*, (2) *uud*, (3) *udd*, (4) *ddd*. (b) Prove your answer to part (a) given the correct electric charge. (a) (3) *udd* (b) see ISM

30.8 Force Unification Theories, the Standard Model, and the Early Universe

62. MC The *grand unified theory* would reduce the number of fundamental forces to (a) one, (b) two, (c) three, (d) four. (b)

63. MC The magnetic force is part of the (a) electroweak force, (b) weak force, (c) strong force, (d) superforce. (a)

Comprehensive Exercises

64. Complete the following nuclear reactions:
    (a) $^6_3$Li + $\underline{^1_1\text{H}}$ → $^3_2$He + $^4_2$He
    (b) $^{58}_{28}$Ni + $^2_1$H → $\underline{^{59}_{28}\text{Ni}}$ + $^1_1$H
    (c) $^{235}_{92}$U + $^1_0$n → $^{138}_{54}$Xe + 5^1_0n + $\underline{^{23}_{38}\text{Sr}}$
    (d) $^9_4$Be(α, $\underline{\text{n}}$)$^{12}_6$C
    (e) $\underline{^{16}_8\text{O}}$ (n, p)$^{16}_7$N
    see ISM

65. Compute the Q value (to three significant figures) of the following fusion reaction: $^2_1H + ^2_1H \rightarrow ^3_1H + ^1_1H$.
    4.03 MeV

66. (a) Write down the decay equation for the decay of ^{14}C. (b) What kind of neutrino is released in this decay? (c) Find the maximum kinetic energy of that neutrino.
    (a) see ISM (b) $\bar{\nu}_e$ (c) 0.156 MeV

67. Show that the Q value for electron capture is given by $Q = (m_P + m_e - m_D)c^2 = (M_P - M_D)c^2$, where the m's represent the masses of the parent and daughter *nuclei* and the M's represent the masses of the neutral *atoms*.
    see ISM

68. IE Theoretically a ^7Be nucleus can be converted into a ^7Li nucleus by capturing an electron (see Exercise 67) or through β^+ decay (see Exercise 42). (a) From an energy point of view, which is more likely to occur:

(1) electron capture, (2) β^+ decay, or (3) both are equally likely to occur? Why? (b) What is the energy of the emitted neutrino (neglect daughter-nucleus recoil) in electron capture?
(a) (1) electron capture (β^+ decay is not possible) (b) 0.86 MeV

69. Determine the threshold energy of the following reaction:

$$^{16}_8O \quad + \quad ^1_0n \quad \rightarrow \quad ^{13}_6C \quad + \quad ^4_2He$$

(15.994 915 u) (1.008 665 u) (13.003 355 u) (4.002 603 u)
2.35 MeV

70. Assume that the average kinetic energy of ions in a plasma is the same as the average kinetic energy of the atoms in an ideal gas $\left(\frac{1}{2}mv^2 = \frac{3}{2}k_BT\right)$. If fusion can occur when the ions approach to within a distance of about 10^{-14} m, estimate the temperature required for fusion of two deuterium ions. 5.6×10^8 K

APPENDICES

APPENDIX I* Mathematical Review (with Examples) for College Physics

A Symbols, Arithmetic Operations, Exponents, and Scientific Notation

Commonly Used Symbols in Relationships

$=$ means two quantities are equal, such as $2x = y$

$\equiv$ means "defined as," such as the definition of pi:

$$\pi \equiv \frac{\text{circumference of a circle}}{\text{the diameter of that circle}}.$$

$\approx$ means approximately equal, as in $30\,\frac{\text{m}}{\text{s}} \approx 60\,\frac{\text{mi}}{\text{h}}$.

$\neq$ means inequality, such as $\pi \neq \frac{22}{7}$.

$\geq$ means that one quantity is greater than or equal to another. For example, the age of the universe ≥ 5 billion years.

$\leq$ means that a quantity is less than or equal to another. For example, if a lecture room holds ≤ 45 students, the maximum is 45 students.

$>$ means that one quantity is greater than another, such as 14 eggs > 1 dozen eggs.

$\gg$ means that one quantity is *much* greater than another. For example, number of people on Earth $\gg 1$ million.

$<$ means that one quantity is less than another, such as $3 \times 10^{22} <$ Avogadro's number.

$\ll$ one quantity is *much* less than another, such as $10 \ll$ Avogadro's number.

$\propto$ means linearly proportional to. That is, if $y = 2x$ then $y \propto x$. This means that if x is increased by a certain multiplicative factor, y is also increased the same way. For example, if $y = 3x$, then if x is changed by a factor of n (that is, if x becomes nx), then so is y, because $y' = 3x' = 3(nx) = n(3x) = ny$.

*This appendix does not include a discussion of significant figures, since a thorough discussion is presented in Chapter 1, Section 1.6.

ΔQ means "change in the quantity Q." This means "final minus initial." Hence if the value of an investor's stock portfolio in the morning is $V_i = \$10\,100$ and at the close of trading it is $V_f = \$10\,050$, then $\Delta V = \$10\,050 - \$10\,100 = -\$50$

The Greek letter capital sigma (Σ) indicates the sum of a series of values for the quantity Q_i where $i = 1, 2, 3, \ldots, N$, that is,

$$\sum_{i=1}^{N} Q_i = Q_1 + Q_2 + Q_3 + \cdots Q_N.$$

$|Q|$ denotes the absolute value of a quantity Q without a sign. If Q is positive then $|Q| = Q$; if Q is negative then $|Q| = -Q$. Thus $|-3| = 3$.

■ Appendix I-A Exercises

1. What values of x satisfy $3 \leq |x| \leq 8$?
2. What integer y comes closest to $y \approx |\sqrt{10}|$?
3. If at the end of the weekend you count your widgets and find $\Delta w = -10$ and the number of widgets on Friday was 500, how many do you have on Monday morning?
4. Give a reasonable number z that satisfies $1 < z \ll 100$.
5. If $y \propto x^2$ and the value of x doubles, what happens to the value of y?
6. What is $\dfrac{\sum_{i=1}^{3} 3^i}{10}$?

Arithmetic Operations and Their Order of Usage

Basic arithmetic operations are addition ($+$), subtraction ($-$), multiplication ($\times$ or $\cdot$), and division ($/$ or $\div$). Another common operation, exponentiation (x^n), involves raising a quantity (x) to a given power (n). If several of these operations are included in one equation, they are performed in this order: (a) parentheses, (b) exponentiation, (c) division, (d) multiplication, (e) addition and subtraction.

A handy mnemonic used to remember this order is: "**P**lease **E**xcuse **M**y **D**ear **A**unt **S**ally," where the capital letters stand for the various operations: **P**arentheses, **E**xponents,

A-1

Multiplication/Division, Addition/Subtraction. Note that operations within parentheses are always first, so to be on the safe side, appropriate use of parentheses is encouraged. For example, $24^2/8 \cdot 4 + 12$ could be evaluated several ways. However, according to the agreed-on order, it has a unique value: $24^2/8 \cdot 4 + 12 = 576/8 \cdot 4 + 12 = 576/32 + 12 = 18 + 12 = 30$. To avoid possible confusion, the quantity could be written using two sets of parentheses as follows: $(24^2/(8 \cdot 4)) + 12 = (576/(32)) + 12 = 18 + 12 = 30$.

■ Appendix I-B Exercises

1. Insert parentheses so $3^2 + 4^2 \cdot 1^3 - \sqrt{4} + 7$ yields 30 without any questions.
2. Evaluate $2 \cdot 3/4 + 5/2 \times 4 - 1$.
3. Evaluate $2 \times 4 + 7 - 6/3 \times 2$.
4. How would you use parentheses to write $3^2 + 4^2 \cdot 1^3 - \sqrt{4} + 7$ to guarantee that anyone evaluating the expression would obtain zero even if he or she didn't know the ordering rules?

Exponents and Exponential Notation

Exponents and exponential notation are very important when employing scientific notation (see the next section). You should be familiar with power and exponential notation (both positive and negative, fractional and integral) such as the following:

$$x^0 = 1$$

$$x^1 = x \qquad\qquad x^{-1} = \frac{1}{x}$$

$$x^2 = x \cdot x \qquad\quad x^{-2} = \frac{1}{x^2} \qquad x^{\frac{1}{2}} = \sqrt{x}$$

$$x^3 = x \cdot x \cdot x \qquad x^{-3} = \frac{1}{x^3} \qquad x^{\frac{1}{3}} = \sqrt[3]{x} \qquad \text{etc.}$$

Exponents combine according to the following rules:

$$x^a \cdot x^b = x^{(a+b)} \qquad x^a/x^b = x^{(a-b)} \qquad (x^a)^b = x^{ab}$$

■ Appendix I-C Exercises

1. What is the value of $\dfrac{2^3}{2^4}$?
2. Evaluate $3^3 \times |9^{-1/2}|$.
3. Find the value(s) of $3^4 \times \sqrt{4^6}$.
4. What is $\left(\sqrt{10}\right)^4$?

Scientific Notation (Also Known as Powers-of-10 Notation)

In physics, many quantities have values that are very large or very small. To express them, **scientific notation** is frequently used. This notation is sometimes referred to as powers-of-10 notation for obvious reasons. (See the previous section for a discussion of exponents.) When the number 10 is squared or cubed, we have $10^2 = 10 \times 10 = 100$ or $10^3 = 10 \times 10 \times 10 = 1000$. You can see that the number of zeros is equal to the power of 10. Thus 10^{23} is a compact way of expressing the number 1 followed by 23 zeros.

A number can be represented in many different ways—all of which are correct. For example, the distance from the Earth to the Sun is 93 million miles. This value can be written as 93 000 000 miles. Expressed in a more compact scientific nota-

tion, there are many correct forms, such as 93×10^6 miles, 9.3×10^7 miles, or 0.93×10^8 miles. Any of these is correct, although 9.3×10^7 is preferred, because in expressing powers-of-10 notation, it is customary to leave only one digit to the left of the decimal point, in this case 9. (This is called customary or standard form.) And so the exponent, or power of 10, changes when the decimal point of the prefix number is shifted.

Negative powers of 10 also can be used. For example, $10^{-2} = \dfrac{1}{10^2} = \dfrac{1}{100} = 0.01$. So, if a power of 10 has a negative exponent, the decimal point may be shifted to the left once for each power of 10. For example, 5.0×10^{-2} is equal to 0.050 (two shifts to the left).

The decimal point of a quantity expressed in powers-of-10 notation may be shifted to the right or left irrespective of whether the power of 10 is positive or negative. General rules for shifting the decimal point are as follows:

1. The exponent, or power of 10, is *increased* by 1 for every place the decimal point is shifted to the *left*.
2. The exponent, or power of 10, is *decreased* by 1 for every place the decimal point is shifted to the *right*.

This is simply a way of saying that as the coefficient (prefix number) gets smaller, the exponent gets correspondingly larger, and vice versa. Overall, the number is the same.

■ Appendix I-D Exercises

1. Express your weight (in pounds) in scientific notation.
2. The circumference of the Earth is about 40 000 km. Express this in scientific notation.
3. Evaluate and express the answer in scientific notation: $\dfrac{12.1}{1.10 \times 10^{-1}}$.
4. Find the value of $(1.44 \times 10^2)^{1/2}$ in scientific notation.
5. What is $(3.0 \times 10^8)^2$ in scientific notation?

B Algebra and Common Algebraic Relationships

General

The basic rule of algebra, used for solving equations, is that if you perform any legitimate operation on both sides of an equation, it remains an equation, or equality. (An example of an illegal operation is dividing by zero; why?) Thus adding a number to both sides, taking the square root of both sides, cubing both sides, and dividing both sides by the same number all maintain the equality.

For example, suppose you want to solve $\dfrac{x^2 + 6}{2} = 11$ for x. To do this, first multiply both sides by 2, giving $\left(\dfrac{x^2 + 6}{\cancel{2}}\right) \times \cancel{2} = 11 \times 2 = 22$ or $x^2 + 6 = 22$. Then subtract 6 from both sides to obtain $x^2 + 6 - 6 = 22 - 6 = 16$ or $x^2 = 16$. Finally taking the square root of both sides, the solution(s) are $x = \pm 4$ (two roots were expected; why?)

Some Useful Results

Many times *the square of the sum and/or difference of two numbers* is required. For any numbers a and b:

$$(a \pm b)^2 = a^2 \pm 2ab + b^2$$

Similarly *the difference of two squares* can be factored:

$$(a^2 - b^2) = (a + b)(a - b)$$

A quadratic equation is one that can be expressed in the form $ax^2 + bx + c = 0$. In this form it can always be solved (usually for two different roots) using the *quadratic formula*: $x = \dfrac{-b \pm \sqrt{b^2 - 4ac}}{2a}$. In kinematics this result can be especially useful as it is common to have equations of this form to solve: $4.9t^2 - 10t - 20 = 0$. Just insert the coefficients (making sure to include the sign) and solve for t (here t represents the time for a ball to reach the ground when thrown upward from a cliff; see Chapter 2). The result is

$$x = \frac{10 \pm \sqrt{10^2 - 4(4.9)(-20)}}{2(4.9)} = \frac{10 \pm 22.2}{9.8}$$

$$= +3.3\,\text{s} \quad \text{or} \quad -1.2\,\text{s}$$

In all such problems, time is "stopwatch" time and starts at zero; hence the negative answer can be ignored as physically unreasonable although it is a solution to the equation.

Solving Simultaneous Equations

Occasionally solving a problem might require solving two or more equations simultaneously. In general if you have N unknowns in a problem, you will need exactly N independent equations. If you have less than N equations, there are not enough for a complete solutions. If you have more than N equations, then some are redundant, and a solution is usually still possible, although more complicated. In general in this textbook, we will be concerned with two simultaneous equations, and both will be linear. Linear equations are of the form $y = mx + b$. Recall that when plotted on an x–y Cartesian coordinate system, the result is a straight line with a slope of m ($\Delta y / \Delta x$) and a y-intercept of b, as shown for the red line here.

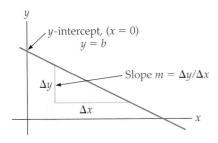

To solve two linear equations simultaneously *graphically*, simply plot them on the axes and evaluate the coordinates at their intersection point. While this can always be done in principle, it is only an approximate answer and usually takes quite a bit of time.

The most common (and exact) method of solving simultaneous equations involves the use of algebra. Essentially you solve one equation for an unknown and substitute the result into the other equation, ending up with one equation and one unknown. Suppose you have two equations and two unknown quantities (x and y), but in general, any two unknown quantities):

$$3y + 4x = 4 \quad \text{and} \quad 2x - y = 2$$

Solving the second equation for y, we have $y = 2x - 2$. Substituting this value for y into the first equation, we have $3(2x - 2) + 4x = 4$. Thus, $10x = 10$ and $x = 1$. Putting this value into the second of the original two equations, we have $2(1) - y = 2$ and therefore $y = 0$. (Of course, at this point a good double check is to substitute the answers and see if they solve both equations.)

Appendix I-E Exercises

1. Expand $(y - 2x)^2$.
2. Express $x^2 - 4x + 4$ as a product of two factors.
3. Solve the following equation for t: $4.9t^2 - 30t + 10 = 0$. How many physically reasonable roots are there?
4. Show that a quadratic equation has real roots only if $b^2 \geq 4ac$. Under what conditions (for a, b, and c) are the two roots identical?
5. Solve these equations simultaneously using algebra: $2x - 3y = 2$ and $3y + 5x = 7$.
6. Solve the two equations in Exercise 5 approximately using graphing methods.

C Geometric Relationships

In physics and many other areas of science, it is important to know how to find circumferences, areas, and volumes of some common shapes. Here are some equations for such shapes.

Circumference (c), Area (A), and Volume (V)

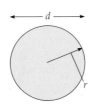

Circle: $c = 2\pi r = \pi d$

$A = \pi r^2 = \dfrac{\pi d^2}{4}$

Rectangle: $c = 2l + 2w$

$A = l \times w$

Triangle: $A = \dfrac{1}{2}ab$

Sphere: $A = 4\pi r^2$

$V = \dfrac{4}{3}\pi r^3$

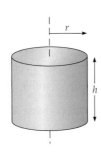

Cylinder: $A = \pi r^2$ (end)

$A = 2\pi rh$ (body)

$V = \pi r^2 h$

For practice, try the following exercises.

Appendix I-F Exercises

1. Estimate the volume of a bowling ball in cubic centimeters and cubic inches.

2. A square hole has a side measuring 5.0 cm. What is the area of the end of a cylindrical rod that will barely fit into this hole?

3. A glass of water has an interior diameter of 4.5 cm and contains a column of water 4.0 in. high. What volume of water does it contain in liters?

4. What is the total surface area of a pancake that is 16 cm in diameter and 8.0 mm thick?

5. Compute the volume of the pancake in Exercise 4 in cubic centimeters.

D Trigonometric Relationships

Understanding elementary trigonometry is crucial in physics, especially as many of the quantities are vectors. Here is a brief summary of common definitions, the first few of which you should commit to memory.

Definitions of Trigonometric Functions

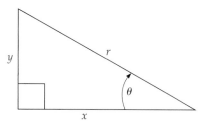

$$\sin \theta = \frac{y}{r} \qquad \cos \theta = \frac{x}{r} \qquad \tan \theta = \frac{\sin \theta}{\cos \theta} = \frac{y}{x}$$

$\theta°$ (rad)	$\sin \theta$	$\cos \theta$	$\tan \theta$
0°(0)	0	1	0
30° ($\pi/6$)	0.500	0.866	0.577
45° ($\pi/4$)	0.707	0.707	1.00
60° ($\pi/3$)	0.866	0.500	1.73
90° ($\pi/2$)	1	0	$\to \infty$

For very small angles,

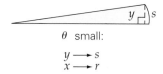

θ small:

$$y \longrightarrow s$$
$$x \longrightarrow r$$

$$\theta \text{ (in rad)} = \frac{s}{r} \approx \frac{y}{r} \approx \frac{y}{x}$$

$$\theta \text{ (in rad)} \approx \sin \theta \approx \tan \theta$$

$$\cos \theta \approx 1 \qquad \sin \theta \approx \theta \text{ (radians)}$$

$$\tan \theta = \frac{\sin \theta}{\cos \theta} \approx \theta \text{ (radians)}$$

The sign of a trigonometric function depends on the quadrant, or the signs of x and y. For example, in the second quadrant x is negative and y is positive, therefore $\cos \theta = x/r$ is negative and $\sin \theta = y/r$ is positive. (Note that r is always taken as positive.) In the figure, the red lines are positive and the blue lines negative.

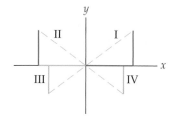

Some Useful Trigonometric Identities

$$\sin^2 \theta + \cos^2 \theta = 1$$
$$\sin 2\theta = 2 \sin \theta \cos \theta$$
$$\cos 2\theta = \cos^2 \theta - \sin^2 \theta = 2 \cos^2 \theta - 1 = 1 - 2 \sin^2 \theta$$
$$\sin^2 \theta = \frac{1}{2}(1 - \cos 2\theta)$$
$$\cos^2 \theta = \frac{1}{2}(1 + \cos 2\theta)$$

For half-angle ($\theta/2$) identities, simply replace θ with $\theta/2$; for example,

$$\sin^2 \theta/2 = \frac{1}{2}(1 - \cos \theta)$$
$$\cos^2 \theta/2 = \frac{1}{2}(1 + \cos \theta)$$

Trigonometric values of sums and differences of angles are sometimes of interest. Here are several basic relationships.

$$\sin(\alpha \pm \beta) = \sin \alpha \cos \beta \pm \cos \alpha \sin \beta$$
$$\cos(\alpha \pm \beta) = \cos \alpha \cos \beta \mp \sin \alpha \sin \beta$$
$$\tan(\alpha \pm \beta) = \frac{\tan \alpha \pm \tan \beta}{1 \mp \tan \alpha \tan \beta}$$

Law of Cosines

For a triangle with angles A, B, and C with opposite sides a, b, and c, respectively:

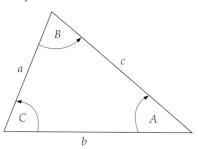

$$a^2 = b^2 + c^2 - 2bc \cos A \qquad \begin{array}{l} \textit{(with similar results for} \\ b^2 = \cdots \textit{ and } c^2 = \cdots). \end{array}$$

If $A = 90°$, this equation reduces to the Pythagorean theorem as it should:

$$a^2 = b^2 + c^2 \quad \textit{(of the form } r^2 = x^2 + y^2)$$

Law of Sines

For a triangle with angles A, B, and C with opposite sides a, b, and c, respectively:

$$\frac{a}{\sin A} = \frac{b}{\sin B} = \frac{c}{\sin C}$$

■ Appendix I-G Exercises

1. From ground level you find you must look at an upward angle of 60 degrees to see the very top of a building that is 50 m away from you. How high is the building? How far is the top of the building from you?

2. On an x–y Cartesian set of axes, a point is at $x = -2.5$ and $y = -4.2$. What quadrant is it in? What is the angle of the line drawn between it and the origin? (Express the answer in degrees and radians.)

3. Use the sine equation for the sum of two angles, one angle being 30°, the other 60°, to show that the sine of a 90° angle comes out to be 1.00.

4. Assume that the radius of the Earth's orbit about the Sun is a circle with a radius of 150 million km. Calculate the arc-length distance traveled by the Earth about the Sun in four months. Using the laws of sines and cosines, determine the straight-line distance between the beginning and end points of this arc.

5. A right triangle has a hypotenuse of length 11 cm and one angle of 25°. Determine the two sides of the triangle, its area, and its perimeter.

E   Logarithms

The following are fundamental definitions and relationships for logarithms. The logarithm is common in physics; you should know what they are and how to use them. Logarithms are very useful because they allow you to multiply and divide very large and very small numbers by adding and subtracting exponents (which we call the logarithms of the numbers).

General Definition of Logarithms

If a number x is written as another number a to some power n, as $x = a^n$, then n is defined to be the *logarithm of the number x to the base a*. This is written compactly as

$$n \equiv \log_a x.$$

Common Logarithms

If the base a is 10, the logarithms are called *common logarithms*. When the abbreviation *log* is used, without a base specified, base 10 is assumed. If another base is being used, it will be specifically shown. For example, $1000 = 10^3$; therefore $3 = \log_{10} 1000$, or simply $3 = \log 1000$. This is read "3 is the log of 1000."

Identities for Common Logarithms

For any two numbers x and y:

$$\log(10^x) = x$$
$$\log(xy) = \log x + \log y$$
$$\log\left(\frac{x}{y}\right) = \log x - \log y$$
$$\log(x^y) = y \log x$$

Natural Logarithms

The natural logarithm uses as its base the irrational number e. To six significant figures, its value is $e \approx 2.71828\ldots$. Fortunately most calculators have this number (along with other irrational numbers, such as pi) in their memories. (You should be able to find both e and π on yours.) The natural logarithm received its name because it occurs naturally when describing a quantity that grows or decays at a constant percentage (rate). The natural logarithm is abbreviated *ln* to distinguish it from the common logarithm, *log*. That is, $\log_e x \equiv \ln x$, and if $n = \ln x$, then $x = e^n$. Similarly to the common logarithm, we have the following relationships for any two numbers x and y:

$$\ln(e^x) = x$$
$$\ln(xy) = \ln x + \ln y$$
$$\ln\left(\frac{x}{y}\right) = \ln x - \ln y$$
$$\ln(x^y) = y \ln x$$

Occasionally you must convert between the two types of logarithms. For that, the following relationships can be handy:

$$\log x = 0.43429 \ln x$$
$$\ln x = 2.3026 \log x$$

For practice with logarithms of both types, try the following exercises:

■ Appendix I-H Exercises:

1. Use your calculator to find the following: log 20, log 50, log 2500, and log 3.

2. Explain why numbers less than 10 have a negative logarithm. Does it make sense to talk about log(−100)? Explain.

3. Use your calculator to find the following: ln 20, log 2, ln 100, and log 3.

4. Double check your answers for ln 2 and log 2 from Exercises 1 and 3 using the relationships $\log x = 0.43429 \ln x$ and $\ln x = 2.3026 \log x$.

5. Show that the rules for combining logarithms work for the following by evaluating each side and showing an equivalence: $\log 1500 = \log(15 \times 100)$, $\log 6400 = \log\left(\frac{64}{0.01}\right)$, and $\log 8 = \log(2^3)$.

6. Show that the rules for combining logarithms work for the following by evaluating each side and showing an equivalence: $\ln 4 = \ln(2 \times 2)$, $\ln 20 = 2.3026 \log(2 \times 10)$, and $\log 49 = 0.43429 \ln(7^2)$.

7. In describing the growth of a bacteria colony, the number of bacteria N at any given time t (from the start of observation) can be written in terms of the number at the start, N_0, as follows: $N = N_0 e^{0.020t}$, where t is in minutes. How many minutes does it take the colony to double in size?

8. In describing the decay of a radioactive sample of atomic nuclei, the number of undecayed nuclei N at any given time t (from the start of observation) can be written in terms of the number at the start, N_0, as follows: $N = N_0 e^{-0.050t}$, where t is in years. How many years does it take until one-tenth the original number of nuclei remain?

APPENDIX II   Kinetic Theory of Gases

The basic assumptions are as follows:

1. All the molecules of a pure gas have the same mass (m) and are in continuous and completely random motion. (The mass of each molecule is so small that the effect of gravity on it is negligible.)

2. The gas molecules are separated by large distances and occupy a volume that is negligible compared with these distances.

3. The molecules exert no forces on each other except when they collide.

4. Collisions of the molecules with one another and with the walls of the container are perfectly elastic.

The magnitude of the force exerted on the wall of the container by a gas molecule colliding with it is $F = \Delta p / \Delta t$. Assuming that the direction of the velocity (v_x) is normal to the wall, the magnitude of the average force is

$$F = \frac{\Delta(mv)}{\Delta t} = \frac{mv_x - (-mv_x)}{\Delta t} = \frac{2mv_x}{\Delta t} \tag{1}$$

After striking one wall of the container, which, for convenience, is assumed to be a cube with sides of dimensions L, the molecule recoils in a straight line. Suppose that the molecule reaches the opposite wall without colliding with any other molecules along the way. The molecule then travels the distance L in a time equal to L/v_x. After the collision with that wall, again assuming no collisions on the return trip, the round trip will take $\Delta t = 2L/v_x$. Thus, the number of collisions per unit time a molecule makes with a particular wall is $v_x/2L$, and the average force of the wall from successive collisions is

$$F = \frac{2mv_x}{\Delta t} = \frac{2mv_x}{2L/v_x} = \frac{mv_x^2}{L} \tag{2}$$

The random motions of the many molecules produce a relatively constant force on the walls, and the pressure (p) is the total force on a wall divided by the wall's area:

$$p = \frac{\Sigma F_i}{L^2} = \frac{m(v_{x_1}^2 + v_{x_2}^2 + v_{x_3}^2 + \cdots)}{L^3} \tag{3}$$

The subscripts refer to individual molecules.

The average of the squares of the speeds is given by

$$\overline{v_x^2} = \frac{v_{x_1}^2 + v_{x_2}^2 + v_{x_3}^2 + \cdots}{N}$$

where N is the number of molecules in the container. In terms of this average, Eq. 3 can be written as

$$p = \frac{Nm\overline{v_x^2}}{L^3} \tag{4}$$

However, the molecules' motions occur with equal frequency along any one of the three axes, so $\overline{v_x^2} = \overline{v_y^2} = \overline{v_z^2}$ and $\overline{v^2} = \overline{v_x^2} + \overline{v_y^2} + \overline{v_z^2} = 3\overline{v_x^2}$. Then

$$\sqrt{\overline{v^2}} = v_{\text{rms}}$$

where v_{rms} is called the root-mean-square (rms) speed. Substituting this result into Eq. 4 and replacing L^3 with V (since L^3 is the volume of the cubical container) gives

$$pV = \tfrac{1}{3}Nmv_{\text{rms}}^2 \tag{5}$$

This result is correct even though collisions between molecules were ignored. Statistically, these collisions average out, so the number of collisions with each wall is as described. This result is also independent of the shape of the container. A cube merely simplifies the derivation.

We now combine this result with the empirical perfect gas law:

$$pV = Nk_{\text{B}}T = \tfrac{1}{3}Nmv_{\text{rms}}^2$$

The average kinetic energy per gas molecule is thus proportional to the absolute temperature of the gas:

$$\overline{K} = \tfrac{1}{2}mv_{\text{rms}}^2 = \tfrac{3}{2}k_{\text{B}}T \tag{6}$$

The collision time is negligible compared with the time between collisions. Some kinetic energy will be momentarily converted to potential energy during a collision; however, this potential energy can be ignored, because each molecule spends a negligible amount of time in collisions. Therefore, by this approximation, the total kinetic energy is the internal energy of the gas, and the internal energy of a perfect gas is directly proportional to its absolute temperature.

APPENDIX III   Planetary Data

Name	Equatorial Radius (km)	Mass (Compared with Earth's)*	Mean Density ($\times 10^3$ kg/m³)	Surface Gravity (Compared with Earth's)	Semimajor Axis		Orbital Period		Eccentricity	Inclination to Ecliptic
					$\times 10^6$ km	AU†	Years	Days		
Mercury	2439	0.0553	5.43	0.378	57.9	0.3871	0.24084	87.96	0.2056	7°00'26"
Venus	6052	0.8150	5.24	0.894	108.2	0.7233	0.61515	224.68	0.0068	3°23'40"
Earth	6378.140	1	5.515	1	149.6	1	1.00004	365.25	0.0167	0°00'14"
Mars	3397.2	0.1074	3.93	0.379	227.9	1.5237	1.8808	686.95	0.0934	1°51'09"
Jupiter	71398	317.89	1.36	2.54	778.3	5.2028	11.862	4337	0.0483	1°18'29"
Saturn	60000	95.17	0.71	1.07	1427.0	9.5388	29.456	10760	0.0560	2°29'17"
Uranus	26145	14.56	1.30	0.8	2871.0	19.1914	84.07	30700	0.0461	0°48'26"
Neptune	24300	17.24	1.8	1.2	4497.1	30.0611	164.81	60200	0.0100	1°46'27"
Pluto	1500–1800	0.02	0.5–0.8	~0.03	5913.5	39.5294	248.53	90780	0.2484	17°09'03"

*Planet's mass/Earth's mass, where $M_{\text{E}} = 6.0 \times 10^{24}$ kg.
†Astronomical unit: 1 AU = 1.5×10^8 km, the average distance between the Earth and the Sun.

APPENDIX IV Alphabetical Listing of the Chemical Elements (The periodic table is provided inside the back cover.)

Element	Symbol	Atomic Number (Proton Number)	Atomic Mass	Element	Symbol	Atomic Number (Proton Number)	Atomic Mass	Element	Symbol	Atomic Number (Proton Number)	Atomic Mass
Actinium	Ac	89	227.0278	Hafnium	Hf	72	178.49	Praseodymium	Pr	159	140.9077
Aluminum	Al	13	26.98154	Hahnium	Ha	105	(262)	Promethium	Pm	61	(145)
Americium	Am	95	(243)	Hassium	Hs	108	(265)	Protactinium	Pa	91	231.0359
Antimony	Sb	51	121.757	Helium	He	2	4.00260	Radium	Ra	88	226.0254
Argon	Ar	18	39.948	Holmium	Ho	67	164.9304	Radon	Rn	86	(222)
Arsenic	As	33	74.9216	Hydrogen	H	1	1.00794	Rhenium	Re	75	186.207
Astatine	At	85	(210)	Indium	In	49	114.82	Rhodium	Rh	45	102.9055
Barium	Ba	56	137.33	Iodine	I	53	126.9045	Rubidium	Rb	37	85.4678
Berkelium	Bk	97	(247)	Iridium	Ir	77	192.22	Ruthenium	Ru	44	101.07
Beryllium	Be	4	9.01218	Iron	Fe	26	55.847	Rutherfordium	Rf	104	(261)
Bismuth	Bi	83	208.9804	Krypton	Kr	36	83.80	Samarium	Sm	62	150.36
Bohrium	Bh	107	(264)	Lanthanum	La	57	138.9055	Scandium	Sc	21	44.9559
Boron	B	5	10.81	Lawrencium	Lr	103	(260)	Seaborgium	Sg	106	(263)
Bromine	Br	35	79.904	Lead	Pb	82	207.2	Selenium	Se	34	78.96
Cadmium	Cd	48	112.41	Lithium	Li	3	6.941	Silicon	Si	14	28.0855
Calcium	Ca	20	40.078	Lutetium	Lu	71	174.967	Silver	Ag	47	107.8682
Californium	Cf	98	(251)	Magnesium	Mg	12	24.305	Sodium	Na	11	22.98977
Carbon	C	6	12.011	Manganese	Mn	25	54.9380	Strontium	Sr	38	87.62
Cerium	Ce	58	140.12	Meitnerium	Mt	109	(268)	Sulfur	S	16	32.066
Cesium	Cs	55	132.9054	Mendelevium	Md	101	(258)	Tantalum	Ta	73	180.9479
Chlorine	Cl	17	35.453	Mercury	Hg	80	200.59	Technetium	Tc	43	(98)
Chromium	Cr	24	51.996	Molybdenum	Mo	42	95.94	Tellurium	Te	52	127.60
Cobalt	Co	27	58.9332	Neodymium	Nd	60	144.24	Terbium	Tb	65	158.9254
Copper	Cu	29	63.546	Neon	Ne	10	20.1797	Thallium	Tl	81	204.383
Curium	Cm	96	(247)	Neptunium	Np	93	237.048	Thorium	Th	90	232.0381
Dubnium	Db	105	(262)	Nickel	Ni	28	58.69	Thulium	Tm	69	168.9342
Dysprosium	Dy	66	162.50	Niobium	Nb	41	92.9064	Tin	Sn	50	118.710
Einsteinium	Es	99	(252)	Nitrogen	N	7	14.0067	Titanium	Ti	22	47.88
Erbium	Er	68	167.26	Nobelium	No	102	(259)	Tungsten	W	74	183.85
Europium	Eu	63	151.96	Osmium	Os	76	190.2	Uranium	U	92	238.0289
Fermium	Fm	100	(257)	Oxygen	O	8	15.9994	Vanadium	V	23	50.9415
Fluorine	F	9	18.998403	Palladium	Pd	46	106.42	Xenon	Xe	54	131.29
Francium	Fr	87	(223)	Phosphorus	P	15	30.97376	Ytterbium	Yb	70	173.04
Gadolinium	Gd	64	157.25	Platinum	Pt	78	195.08	Yttrium	Y	39	88.9059
Gallium	Ga	31	69.72	Plutonium	Pu	94	(244)	Zinc	Zn	30	65.39
Germanium	Ge	32	72.561	Polonium	Po	84	(209)	Zirconium	Zr	40	91.22
Gold	Au	79	196.9665	Potassium	K	19	39.0983				

APPENDIX V Properties of Selected Isotopes

Atomic Number (Z)	Element	Symbol	Mass Number (A)	Atomic Mass*	Abundance (%) or Decay Mode† (if Radioactive)	Half-Life (if Radioactive)
0	(Neutron)	n	1	1.008665	β^-	10.6 min
1	Hydrogen	H	1	1.007825	99.985	
	Deuterium	D	2	2.014102	0.015	
	Tritium	T	3	3.016049	β^-	12.33 y
2	Helium	He	3	3.016029	0.00014	
			4	4.002603	≈ 100	
3	Lithium	Li	6	6.015123	7.5	
			7	7.016005	92.5	

Atomic Number (Z)	Element	Symbol	Mass Number (A)	Atomic Mass*	Abundance (%) or Decay Mode† (if Radioactive)	Half-Life (if Radioactive)
4	Beryllium	Be	7	7.016930	EC, γ	53.3 d
			8	8.005305	2α	6.7×10^{-17} s
			9	9.012183	100	
5	Boron	B	10	10.012938	19.8	
			11	11.009305	80.2	
			12	12.014353	β^-	20.4 ms
6	Carbon	C	11	11.011433	β^+, EC	20.4 ms
			12	12.000000	98.89	
			13	13.003355	1.11	
			14	14.003242	β^-	5730 y
7	Nitrogen	N	13	13.005739	β^-	9.96 min
			14	14.003074	99.63	
			15	15.000109	0.37	
8	Oxygen	O	15	15.003065	β^+, EC	122 s
			16	15.994915	99.76	
			18	17.999159	0.204	
9	Fluorine	F	19	18.998403	100	
10	Neon	Ne	20	19.992439	90.51	
			22	21.991384	9.22	
11	Sodium	Na	22	21.994435	β^+, EC, γ	2.602 y
			23	22.989770	100	
			24	23.990964	β^-, γ	15.0 h
12	Magnesium	Mg	24	23.985045	78.99	
13	Aluminum	Al	27	26.981541	100	
14	Silicon	Si	28	27.976928	92.23	
			31	30.975364	β^-, γ	2.62 h
15	Phosphorus	P	31	30.973763	100	
			32	31.973908	β^-	14.28 d
16	Sulfur	S	32	31.972072	95.0	
			35	34.969033	β^-	87.4 d
17	Chlorine	Cl	35	34.968853	75.77	
			37	36.965903	24.23	
18	Argon	Ar	40	39.962383	99.60	
19	Potassium	K	39	38.963708	93.26	
			40	39.964000	β^-, EC, γ, β^+	1.28×10^9 y
20	Calcium	Ca	30	39.962591	96.94	
24	Chromium	Cr	52	51.940510	83.79	
25	Manganese	Mn	55	54.938046	100	
26	Iron	Fe	56	55.934939	91.8	
27	Cobalt	Co	59	58.933198	100	
			60	59.933820	β^-, γ	5.271 y
28	Nickel	Ni	58	57.935347	68.3	
			60	59.930789	26.1	
			64	63.927968	0.91	
29	Copper	Cu	63	62.929599	69.2	
			64	63.929766	β^-, β^+	12.7 h
			65	64.927792	30.8	
30	Zinc	Zn	64	63.929145	48.6	
			66	65.926035	27.9	
33	Arsenic	As	75	74.921596	100	
35	Bromine	Br	79	78.918336	50.69	
36	Krypton	Kr	84	83.911506	57.0	
			89	88.917563	β^-	3.2 min
38	Strontium	Sr	86	85.909273	9.8	
			88	87.905625	82.6	
			90	89.907746	β^-	28.8 y
39	Yttrium	Y	89	89.905856	100	
43	Technetium	Tc	98	97.907210	β^-, γ	4.2×10^6 y

Atomic Number (Z)	Element	Symbol	Mass Number (A)	Atomic Mass*	Abundance (%) or Decay Mode† (if Radioactive)	Half-Life (if Radioactive)
47	Silver	Ag	107	106.905 095	51.83	
			109	108.904 754	48.17	
48	Cadmium	Cd	114	113.903 361	28.7	
49	Indium	In	115	114.903 88	95.7; β^-	5.1×10^{14} y
50	Tin	Sn	120	119.902 199	32.4	
53	Iodine	I	127	126.904 477	100	
			131	130.906 118	β^-, γ	8.04 d
54	Xenon	Xe	132	131.904 15	26.9	
			136	135.907 22	8.9	
55	Cesium	Cs	133	132.905 43	100	
56	Barium	Ba	137	136.905 82	11.2	
			138	137.905 24	71.7	
			144	143.922 73	β^-	11.9 s
61	Promethium	Pm	145	144.912 75	EC, α, γ	17.7 y
74	Tungsten	W	184	183.950 95	30.7	
76	Osmium	Os	191	190.960 94	β^-, γ	15.4 d
			192	191.961 49	41.0	
78	Platinum	Pt	195	194.964 79	33.8	
79	Gold	Au	197	196.966 56	100	
80	Mercury	Hg	202	201.970 63	29.8	
81	Thallium	Tl	205	204.974 41	70.5	
			210	209.990 069	β^-	1.3 min
82	Lead	Pb	204	203.973 044	β^-, 1.48	1.4×10^{17} y
			206	205.974 46	24.1	
			207	206.975 89	22.1	
			208	207.976 64	52.3	
			210	209.984 18	α, β^-, γ	22.3 y
			211	210.988 74	β^-, γ	36.1 min
			212	211.991 88	β^-, γ	10.64 h
			214	213.999 80	β^-, γ	26.8 min
83	Bismuth	Bi	209	208.980 39	100	
			211	210.987 26	α, β^-, γ	2.15 min
84	Polonium	Po	210	209.982 86	α, γ	138.38 d
			214	213.995 19	α, γ	164 μs
86	Radon	Rn	222	222.017 574	α, β	3.8235 d
87	Francium	Fr	223	223.019 734	α, β^-, γ	21.8 min
88	Radium	Ra	226	226.025 406	α, γ	1.60×10^3 y
			228	228.031 069	β^-	5.76 y
89	Actinium	Ac	227	227.027 751	α, β^-, γ	21.773 y
90	Thorium	Th	228	228.028 73	α, γ	1.9131 y
			232	232.038 054	100; α, γ	1.41×10^{10} y
92	Uranium	U	232	232.037 14	α, γ	72 y
			233	233.039 629	α, γ	1.592×10^5 y
			235	235.043 925	0.72; α, γ	7.038×10^8 y
			236	236.045 563	α, γ	2.342×10^7 y
			238	238.050 786	99.275; α, γ	4.468×10^9 y
			239	239.054 291	β^-, γ	23.5 min
93	Neptunium	Np	239	239.052 932	β^-, γ	2.35 d
94	Plutonium	Pu	239	239.052 158	α, γ	2.41×10^4 y
95	Americium	Am	243	243.061 374	α, γ	7.37×10^3 y
96	Curium	Cm	245	245.065 487	α, γ	8.5×10^3 y
97	Berkelium	Bk	247	247.070 03	α, γ	1.4×10^3 y
98	Californium	Cf	249	249.074 849	α, γ	351 y
99	Einsteinium	Es	254	254.088 02	α, γ, β^-	276 d
100	Fermium	Fm	253	253.085 18	EC, α, γ	3.0 d

*The masses given throughout this table are those for the neutral atom, including the Z electrons.

†"EC" stands for electron capture.

ANSWERS TO FOLLOW-UP EXERCISES

Chapter 1

1.1 $L = 10$ m.

1.2 Yes, $[L] = [L]$, or m = m.

1.3 (a) 50 mi/h $[(0.447$ m/s$)/($mi/h$)] = 22$ m/s.
(b) $(1$ mi/h$)(1609$ km/mi$)(1$ h/3600 s$) = 0.477$ m/s.

1.4 13.3 times.

1.5 1 m$^3 = 10^6$ cm^3.

1.6 European. 10 mi/gal ≈ 16 km/4 L $= 4$ km/L, as compared with 10 km/L.

1.7 (a) 7.0×10^5 kg^2 (b) 3.02×10^2 (no units).

1.8 (a) 23.70. (b) 22.09.

1.9 $V = \pi r^2 h = \pi(0.490$ m$)^2(1.28$ m$) = 0.965$ m^3

1.10 11.6 m.

1.11 A little steeper, $\theta = 31.3°$.

1.12 750 cm$^3 = 7.50 \times 10^{-4}$ m$^3 \approx 10^{-3}$ m^3,
$m = \rho V \approx (10^3$ kg/m$^3)(10^{-3}$ m$^3) = 1$ kg.
(By direct calculation, $m = 0.79$ kg.)

1.13 $V \approx 10^{-2}$ m^3, cells/vol $\approx 10^4$ cells/mm^3 $(10^9$ mm^3/m$^3) = 10^{13}$ cells/m^3, and (cells/vol)(vol) $\approx 10^{11}$ white cells.

Chapter 2

2.1 $\Delta t = (8 \times 5.0$ s$) + (7 \times 10$ s$) = 110$ s.

2.2 $s_1 = 2.00$ m/s; $s_2 = 1.52$ m/s; $s_3 = 1.72$ m/s $\neq 0$, although the velocity is zero.

2.3 No. If the velocity is also in the negative direction, the object will speed up.

2.4 9.0 m/s in the direction of the original motion.

2.5 Yes, 96 m. (A lot quicker, isn't it?)

2.6 No, always more than one unknown variable.

2.7 No, changes x_o positions, but separation distance is the same.

2.8 $x = v^2/2a$, $x_B = 48.6$ m, and $x_C = 39.6$ m; the Blazer should not tailgate within at least 9.0 m.

2.9 1.16 s longer.

2.10 Time for bill to fall its length = 0.179 s. This time is less than the average reaction time (0.192 s) computed in the Example, so most people cannot catch the bill.

2.11 $y_u = y_d = 5.12$ m, as measured from reference $y = 0$ at the release point.

2.12 Eq. 2.8′, $t = 4.6$ s; Eq. 2.10′, $t = 4.6$ s.

Chapter 3

3.1 $v_x = -0.40$ m/s, $v_y = +0.30$ m/s; the distance is unchanged.

3.2 $x = 9.00$ m, $y = 12.6$ m (same).

3.3 $\vec{v} = (0)\,\hat{x} + (3.7$ m/s$)\,\hat{y}$.

3.4 $\vec{C} = (-7.7$ m$)\,\hat{x} + (-4.3$ m$)\,\hat{y}$.

3.5 (a) $y_o = +25$ m and $y = 0$; the equation is the same.
(b) $\vec{v} = (8.25$ m/s$)\,\hat{x} + (-22.1$ m/s$)\,\hat{y}$.

3.6 Both increase six fold.

3.7 (a) If not, the stone would hit to the side of the block.
(b) Eq. 3.11 does not apply; the initial and final heights are not the same. $R = 15$ m, which is way off the 27-m answer.

3.8 The ball thrown at 45°. It would have a greater initial velocity.

3.9 At the top of the parabolic arc, the player's vertical motion is zero and is very small on either side of this maximum height. Here, the player's horizontal velocity component dominates, and he moves horizontally, with little motion in the vertical direction. This gives the illusion of "hanging" in the air.

3.10 4.15 m from the net.

3.11 $v_{bs}t = (2.33$ m/s$)(225$ m$) = 524$ m

3.12 14.5° W of N.

Chapter 4

4.1 6.0 m/s in the direction of the net force.

4.2 (a) 11 lb. (b) Weight in pounds ≈ 2.2 lb/kg.

4.3 8.3 N

4.4 (a) 50° above the $+x$-axis. (b) x- and y-components reversed: $\vec{v} = (9.8$ m/s$)\,\hat{x} + (4.5$ m/s$)\,\hat{y}$.

4.5 Yes, mutual gravitational attractions between the briefcase and the Earth.

4.6 (a) $m_2 > 1.7$ kg. (b) $\theta < 17.5°$.

4.7 (a) 7.35 N. (b) Neglecting air resistance, 7.35 N, downward.

4.8 Increase. $\tan\theta = \dfrac{T}{mg} = \dfrac{55 \text{ N}}{(5.0 \text{ kg})(9.8 \text{ m/s}^2)} = 1.1$, $\theta = 48°$

4.9 (a) $F_1 = 3.5w$. Even greater than F_2. (b) $\Sigma F_y = ma$, and F_1 and F_2 would both increase.

4.10 $\mu_s = 1.41\mu_k$ (for three cases in Table 4.1).

4.11 No. F varies with angle, with the angle for minimum applied force being around 33° in this case. (Greater forces are required for 20° and 50°.) In general, the optimum angle depends on the coefficient of friction.

4.12 Friction is kinetic, and f_k is in the $+x$ direction. Acceleration in the $-x$ direction.

4.13 Air resistance depends not only on speed, but also on size and shape. If the heavier ball were larger, it would have more exposed area to collide with air molecules, and the retarding force would increase faster. Depending on the size difference, the heavier ball might reach terminal velocity first, and the lighter ball would strike the ground first. Alternatively, the balls might reach terminal velocity together.

Chapter 5

5.1 -2.0 J

5.2 $d = \dfrac{W}{F\cos\theta} = \dfrac{3.80 \times 10^4 \text{ J}}{(189 \text{ N})(0.866)} = 232$ m

5.3 No, speed would decrease and it would stop moving.

5.4 $W_{x_1} = 0.034$ J, $W_x = 0.64$ J (measured from x_o)

5.5 No, $W_2/W_1 = 4$, or 4 times as much

5.6 Here we have $m_s = m_g/2$ as before. However, $v_s/\vec{v}_g = (6.0$ m/s$)/(4.0$ m/s$) = \frac{3}{2}$. Using a ratio, $K_s/K_g = \frac{9}{8}$, and the safety still has more kinetic energy than the guard. (Answer could also be obtained from direct calculations of kinetic energies, but for a relative comparison, a ratio is usually quicker.)

5.7 $W_3/W_2 = 1.4$, or 40% larger. More work, but a smaller percentage increase.

5.8 $\Delta U = mgh = (60$ kg$)(9.8$ m/s$^2)(1000$ m$)\sin 10° = 10.2 \times 10^4$ J, yes doubled.

5.9 $\Delta K_{total} = 0$, $\Delta U_{total} = 0$

5.10 Without friction, the liquid would oscillate back and forth between the containers.

5.11 9.9 m/s

5.12 No. $E_o = E$ or $\frac{1}{2}mv_o^2 + mgh = \frac{1}{2}mv^2$. The mass cancels and the speed is independent of mass. (Recall that in free fall, all objects or projectiles fall with the same vertical acceleration g—see Section 2.5.)

5.13 0.025 m

5.14 (a) 59% (b) $E_{\text{loss}}/t = mg(y/t) = mgv = (60\ mg)$ J/s

5.15 Block would stop in rough area.

5.16 52%

5.17 (a) Same work in twice the time. (b) Same work in half the time.

5.18 (a) No. (b) Creation of energy.

Chapter 6

6.1 5.0 m/s. Yes, this is 18 km/h or 11 mi/h, a speed at which humans can run.

6.2 (1) Ship the greatest KE. (2) Bullet the least KE.

6.3 $(-3.0\ \text{kg} \cdot \text{m/s})\ \hat{\mathbf{x}} + (4.0\ \text{kg} \cdot \text{m/s})\ \hat{\mathbf{y}}$

6.4 It would increase to 60 m/s: greater speed, longer drive, ideally. (There is also a directional consideration.)

6.5 $F_{\text{avg}} = \dfrac{\Delta p}{\Delta t} = \dfrac{-310\ \text{kg} \cdot \text{m/s}}{0.600\ \text{s}} = -517\ \text{N}$

6.6 (a) No, for the m_1/m_2 system, external force on block. Yes, for the m_1/m_2 Earth system. But with m_2 attached to the Earth, the mass of this part of the system would be vastly greater than that of m_2, so its change in velocity would be negligible. (b) Assuming the ball is tossed in the +direction: for the tosser, $v_t = -0.50$ m/s; for the catcher, $v_c = 0.48$ m/s. For the ball: $p = 0, +25\ \text{kg} \cdot \text{m/s}, +1.2\ \text{kg} \cdot \text{m/s}$.

6.7 No. Energy went into work of breaking the brick, and some lost as heat and sound.

6.8 No.

6.9 No; all of the kinetic energy cannot be lost to make the dent. The momentum after the collision cannot be zero, since it was not zero initially. Thus, the balls must be moving and have kinetic energy. This can also be seen from Eq. 6.11: $K_f/K_i = m_1/(m_1 + m_2)$, and K_f cannot be zero (unless m_1 is zero, which is not possible).

6.10 $x_1 = v_1 t = (-0.80\ \text{m/s})(2.5\ \text{s}) = -2.0\ \text{m}, x_2 = v_2 t = (1.2\ \text{m/s})(2.5\ \text{s}) = 3.0\ \text{m}$

$\Delta x = x_2 - x_1 = 3.0\ \text{m} - (-2.0\ \text{m}) = 5.0\ \text{m}$. Objects 5.0 m apart.

6.11 (a) $\Delta p_1 = p_{1_f} - p_{1_0} = 32\ \text{kg} \cdot \text{m/s} - 40\ \text{kg} \cdot \text{m/s} = -8.0\ \text{kg} \cdot \text{m/s}$

$\Delta p_2 = p_{2_f} - p_{2_0} = 13\ \text{kg} \cdot \text{m/s} - 5.0\ \text{kg} \cdot \text{m/s} = +8.0\ \text{kg} \cdot \text{m/s}$

(b) $\Delta p_1 = p_{1_f} - p_{1_0} = (-20\ \text{kg} \cdot \text{m/s}) - (12\ \text{kg} \cdot \text{m/s}) = -32\ \text{kg} \cdot \text{m/s}$

$\Delta p_2 = p_{2_f} - p_{2_0} = (8.0\ \text{kg} \cdot \text{m/s}) - (-24\ \text{kg} \cdot \text{m/s}) = +32\ \text{kg} \cdot \text{m/s}$

6.12 $p_{1_0} = mv_{1_0}, p_{2_0} = -mv_{2_0}$ and $p_1 = mv_1 = -mv_{2_0}$,

$p_2 = mv_2 = mv_{1_0}$, so conserved $K_i = \dfrac{m}{2}(v_{1_0}^2 + v_{2_0}^2)$ and

$K_f = \dfrac{m}{2}(v_1^2 + v_2^2) = \dfrac{m}{2}[(-v_{2_0})^2 + (v_{1_0})^2]$, so conserved

6.13 All of the balls swing out, but to different degrees. With $m_1 > m_2$, the stationary ball (m_2) moves off with a greater speed after collision than the incoming, heavier ball (m_1), and the heavier ball's speed is reduced after collision, in accordance with Eq. 6.16 (see Fig. 6.14b). Hence, a "shot" of momentum is passed along the row of balls with equal mass (see Fig. 6.14a), and the end ball swings out with the same speed as was imparted to m_2. Then, the process is repeated: m_1, *now moving more slowly*, collides again with the initial ball in the row (m_2), and another, but smaller, shot of momentum is passed down the row. The new end ball in the row receives less kinetic energy than the one that swung out just a moment previously, and so doesn't swing as high. This process repeats itself instantaneously for each ball, with the observed result that all of the balls swing out to different degrees.

6.14 $X_{\text{CM}} = \dfrac{(\text{same as in example}) + (8.0\ \text{kg})x_4}{(\text{same as in example}) + (8.0\ \text{kg})} =$

$= \dfrac{0 + (8.0\ \text{kg})x_4}{19\ \text{kg}} = +1.0\ \text{m}$

$x_4 = \left(\dfrac{19}{8}\right)\ \text{m} = 2.4\ \text{m}$

6.15 $(X_{\text{CM}}, Y_{\text{CM}}) = (0.47\ \text{m}, 0.10\ \text{m})$; same location as in Example, two-thirds of the length of the bar from m_1. Note: The location of the CM does not depend on the frame of reference.

6.16 Yes, the CM does not move.

Chapter 7

7.1 $1.61 \times 10^3\ \text{m} = 1.61\ \text{km}$ (about a mile)

7.2 (a) 0.35% for 10° (b) 1.2% for 20°

7.3 (a) 4.7 rad/s, 0.38 m/s; 4.7 rad/s, 0.24 m/s (b) To equalize the running distances, because the curved sections of the track have different radii and thus different lengths.

7.4 120 rpm

7.5 (a) 106 rpm (b) $a = \sqrt{2}g = 13.9\ \text{m/s}^2$, at 45° below plane of centrifuge.

7.6 The string cannot be exactly horizontal; it must make some small angle to the horizontal so that there will be an upward component of the tension force to balance the ball's weight.

7.7 No; it depends on mass: $F_c = \mu_s mg$.

7.8 No. Both masses have the same angular frequency or speed ω, and $a_c = r\omega^2$, so actually $a_c \propto r$. Remember, $v = 2\pi r/T$, and note that $v_2 > v_1$, with $a_c = v^2/r$.

7.9 $T = 5.2\ \text{N}$

7.10 (a) The directions of ω and α would be downward, perpendicular to the plane of the CD. (b) Negative α, which means it is the opposite direction of ω.

7.11 $-0.031\ \text{rad/s}^2$

7.12 $2.8 \times 10^{-3}\ \text{m/s}^2$ (a large force, but a small acceleration)

7.13 $T^2 = \left(\dfrac{4\pi^2}{GM_E}\right)r^3 = \left(\dfrac{4\pi^2}{GM_E}\right)(R_E + h)^3 \approx \left(\dfrac{4\pi^2}{g}\right)R_E \approx 4R_E$

$T = 2\sqrt{R_E} = 2(6.4 \times 10^6\ \text{m})^{\frac{1}{2}} = 5.1 \times 10^3\ \text{s}$ (Why are the units not consistent?)

7.14 No, they do not vary linearly; $\Delta U = 2.4 \times 10^9\ \text{J}$, only a 9.1% increase.

7.15 This is the amount of *negative* work done by an external force or agent when the masses are brought together. To separate the masses by infinite distances, an equal amount of positive work (against gravity) would have to be done.

7.16 $T^2 = \left(\dfrac{4\pi^2}{GM_S}\right)r^3$ and $M_S = \dfrac{4\pi^2 r^3}{GT^2} =$

$\dfrac{4\pi^2(1.50 \times 10^{11}\ \text{m})^3}{(6.67 \times 10^{-11}\ \text{N} \cdot \text{m}^2/\text{kg}^2)(3.16 \times 10^7\ \text{s})^2} = 2.00 \times 10^{30}\ \text{kg}$

Chapter 8

8.1 $s = r\omega = 5(0.12\ \text{m})(1.7) = 0.20\ \text{m}$;

$s = v_{\text{CM}}t = (0.10\ \text{m/s})(2.00\ \text{s}) = 0.20\ \text{m}$

8.2 The weights of the balls and the forearm produce torques that tend to cause rotation in the direction opposite that of the applied torque.

8.3 More strain.

8.4 $T \propto 1/\sin\theta$, and as θ gets smaller, so does $\sin\theta$ and T increases. In the limit, $\sin\theta \to 0$ and $T \to$ infinity (unrealistic).

8.5 $\Sigma\tau$: $Nx - m_1gx_1 - m_2gx_2 - m_3gx_3 = (200\text{ g})g(50\text{ cm}) - (25\text{ g})g(0\text{ cm}) - (75\text{ g})g(20\text{ cm}) - (100\text{ g})g(85\text{ cm}) = 0$, where $N = Mg$.

8.6 No. With f_{s_1}, the reaction force N would not generally be the same (f_{s_2} and N are perpendicular components of the force exerted on the ladder by the wall). In this case, we still have $N = f_{s_1}$, but $Ny - (m_1g)x_1 - (m_mg)x_m - f_{s_2}$, and $x_3 = 0$.

8.7 Hanging vertically.

8.8 Male: lighter upper torso. Female: heavier lower torso.

8.9 5 bricks

8.10 (d) no (equal masses) (e) Yes; with larger mass farther from axis of rotation, $I = 360\text{ kg}\cdot\text{m}^2$.

8.11 The long pole (or your extended arms) increases the moment of inertia by placing more mass farther from the axis of rotation (the tightrope or rail). When the walker leans to the side, a gravitational torque tends to produce a rotation about the axis of rotation, causing a fall. However, with a greater rotational inertia (greater I), the walker has time to shift his or her body so that the center of gravity is again over the rope or rail and thus again in (unstable) equilibrium. With very flexible poles, the CG may be below the wire, thus ensuring stability.

8.12 $t = 0.63$ s

8.13 $\alpha = \dfrac{2\,mg - (2\tau_f R)}{(2m + M)R}$; $\dfrac{N}{\text{kg}\cdot\text{m}}$; and $\dfrac{N}{\text{kg}\cdot\text{m}} = \dfrac{\text{kg}\cdot\text{m/s}^2}{\text{kg}\cdot\text{m}} = \dfrac{1}{\text{s}^2}$

8.14 The yo-yo would roll back and forth, oscillating about the critical angle.

8.15 (a) 0.24 m (b) The force of *static* friction, f_s, acts at the point of contact, which is always instantaneously at rest and so does no work. Some frictional work may be done due to rolling friction, but this is considered negligible for hard objects and surfaces.

8.16 $v_{CM} = 2.2$ m/s; using a ratio, 1.4 times greater; no rotational energy.

8.17 You already know the answer: 5.6 m/s. (It doesn't depend on the mass of the ball.)

8.18 $M_a = 75\text{ kg }(0.75) = 56$ kg. Then $L_1 = 13\text{ kg}\cdot\text{m}^2/\text{s}$ and $L_2 = (1.3\text{ kg}\cdot\text{m}^2)\omega$ [math not shown]. $L_2 = L_1$ or $(1.3\text{ kg}\cdot\text{m}^2)\omega = 13\text{ kg}\cdot\text{m}^2/\text{s}$ and $\omega = 10$ rad/s

Chapter 9

9.1 (a) +0.10% (b) 39 kg

9.2 2.3×10^{-4} L, or 2.3×10^{-7} m³

9.3 (1) Having enough nails, and (2) having them all of equal height and not so sharp a point. This could be achieved by filing off the tips of the nails so as to have a "uniform" surface. Also, this would increase the effective area.

9.4 3.03×10^4 N (or 6.82×10^3 lb—about 3.4 tons!) This is roughly the force on your back right now. Our bodies don't collapse under atmospheric pressure because cells are filled with incompressible fluids (mostly water), bone, and muscle, which react with an equal outward pressure (equal and opposite forces). As with forces, it is a pressure *difference* that gives rise to dynamic effects.

9.5 $d_o = \sqrt{\dfrac{F_o}{F_i}}d_i = \sqrt{\dfrac{1}{10}}(8.0\text{ cm}) = 2.5$ cm

9.6 Pressure in veins is lower than that in arteries (120/80).

9.7 As the balloon rises, the buoyant force decreases as a result of the temperature decrease (less helium pressure, less volume) and the less dense air ($F_b = m_fg = \rho_f g V_f$). When the net force is zero, the velocity is constant. The cooling effect continues with altitude and the balloon will start to sink when the net force is negative.

9.8 $r \approx 1.0$ m. $F_b = \rho g V = \rho g\left(\dfrac{4}{3}\pi r^3\right) =$
$(0.18\text{ kg/m}^3)\left(\dfrac{4g\pi}{3}\right)(1.0\text{ m})^3 = 7.4$ N, much more.

9.9 (a) The object would sink, so the buoyant force is less than the object's weight. Hence, the scale would have a reading greater than 40 N. Note that with a greater density, the object would not be as large and less water would be displaced. (b) 41.8 N.

9.10 11%

9.11 -18%

9.12 $r = 9.00 \times 10^{-3}$ m, $v = \dfrac{\text{constant}}{A} = \dfrac{8.33 \times 10^{-5}\text{ m}^3/\text{s}}{\pi(9.00 \times 10^{-3}\text{ m})^2} =$ 0.327 m/s; 23%

9.13 69%

9.14 As the water falls, speed (v) increases and area (A) must decrease to have $Av = $ a constant.

9.15 0.38 m

Chapter 10

10.1 (a) 40° C (b) You should immediately know the answer—this is the temperature at which the Fahrenheit and Celsius temperatures are numerically equal.

10.2 (a) $T_R = T_F + 460$ (b) $T_R = \frac{9}{5}T_C + 492$ (c) $T_R = \frac{9}{5}T_K$

10.3 96°C

10.4 273° C; no, not on Earth

10.5 50 C°

10.6 It depends on the metal of the bar. If the thermal expansion coefficient (α) of the bar is less than that of iron, it will not expand as much and not be as long as the diameter of the circular ring after heating. However, if the bar's α is greater than that of iron, the bar will expand more than the ring and the ring will be distorted.

10.7 Basically, the situations would be reversed. Faster cooling would be achieved by submerging the ice in Example 10.7—the cooler water would be less dense and would rise, promoting mixing. For a lake with cooling at the surface, cooler, less-dense water would remain at the surface until minimum density was achieved. With further cooling, the denser water would sink and freezing would occur from the bottom up.

10.8 v_{rms}, 1.69%; K, 3.41%

10.9 The rotational kinetic energy for oxygen is the difference between the total energies, 2.44×10^3 J. The oxygen is less massive and has the higher v_{rms}.

Chapter 11

11.1 2.84×10^3 m

11.2 12.5 kg

11.3 (a) The ratio will be smaller because the specific heat of aluminum is greater than that of copper. (b) $Q_w/Q_{pot} = 15.2$

11.4 The final temperature T_f is expected to be higher because the water was at a higher initial temperature. $T_f = 34.4°$C

11.5 -1.09×10^5 J (negative because heat is lost)

11.6 (a) 2.64×10^{-2} kg or 26.4 g of ice melts. (b) The final temperature is still 0°C because the liver cannot lose enough heat to melt all the ice, even if the ice started at 0°C. The final result is an ice/water/liver system at 0°C, but with more water than in the Example.

11.7 1.1×10^5 J/s (difference due to rounding)

11.8 No, since the air spaces provide good insulation because air is a poor thermal conductor. The many small pockets of air between the body and the outer garment form an insulating layer that minimizes conduction and so retards the loss of body heat. (There is little convection in the small spaces.)

11.9 (a) -1.5×10^2 J/s or -1.5×10^2 W (b) The huge ear flaps have large surface area so more heat can be radiated out.
11.10 Drapes reduce heat loss by limiting radiation through the window and by keeping convection current away from the glass.

Chapter 12

12.1 0.20 kg
12.2 In both cases, heat flow is into the gas. During the isothermal expansion, $Q = W = +3.14 \times 10^3$ J. During the isobaric expansion, $W = +4.53 \times 10^3$ J and $\Delta U = +6.80 \times 10^3$ J, therefore $Q = \Delta U + W = +1.13 \times 10^4$ J.
12.3 753° C
12.4 As the air comes to lower elevations and higher pressures, it quickly compresses. This is approximately an adiabatic process, resulting in a temperature rise of the air.
12.5 (a) 142 K or $-131°$ C. (b) For monatomic gas, $\Delta U = (3/2)nR\Delta T = -3.76 \times 10^3$ J. This should be the same as $-W$ since, for an adiabatic process, $Q = 0 = \Delta U + W$; therefore, $\Delta U = -W$. The slight difference is due to rounding.
12.6 -1.22×10^3 J/K
12.7 Overall zero entropy change requires $|\Delta S_w| = |\Delta S_m|$ or $|Q_w/T_w| = |Q_m/T_m|$. Because the system is isolated, the magnitudes of the two heat flows *must* be the same $|Q_w| = |Q_m|$. Thus no overall entropy change requires the water and the metal to have the same average temperature $\overline{T}_w = \overline{T}_m$. This is not possible, unless they are initially at the *same* temperature. Thus, this can only happen if there is no net heat flow.
12.8 If the basics of the cycles are kept, that is, the triangular shape and the volume doubling, then a way to increase the net work (the area inside the cycle) is to drop the pressure even further at the end of the isometric segment. If you allow the volume to more than double during the isobaric expansion, that would achieve the same end. Anything that increases the net area (work) will do.
12.9 (a) 150 J/cycle (b) 850 J/cycle
12.10 $Q_{34} = 610$ J and $Q_{23} = 730$ J, therefore $Q_c = Q_{23} + Q_{34} = 1.34 \times 10^3$ J. This agrees with $Q_c = Q_h - W_{net} = 59 \times 10^3$ J $- 245$ J $= 1.35 \times 10^3$ J (to within rounding errors).
12.11 (a) The new values are: $COP_{ref} = 3.3$ and $COP_{hp} = 4.3$. (b) The COP of the air conditioner has the largest percentage increase.
12.12 It would show an increase of 7.5%.

Chapter 13

13.1 No, its maximum speed is $\left(\sqrt{k/m}\right)A = 4.0$ m/s. Thus it is traveling at 75% of the maximum speed.
13.2 0.49 J
13.3 (1) $y = -0.0881$ m, up. $n = 0.90$. (2) $y = 0$, going up. $n = 1.5$.
13.4 9.76 m/s^2; no. Since this is less than the accepted value at sea level, the park is probably at an altitude above sea level.
13.5 (a) 0.50 m (b) 0.10 Hz
13.6 440 Hz
13.7 increase the tension (by 44% as can be calculated).

Chapter 14

14.1 (a) 2.3 (b) 10.2
14.2 $v = (331 + 0.6 T_C)$ m/s $= [331 + 0.6(38°)] = 354$ m/s Increase.
14.3 It would be greatest in He, because it has the smallest molecular mass. (It would be lowest in oxygen, which has the largest molecular mass.)

14.4 (a) The dB scale is logarithmic, not linear.
(b) 3.16×10^{-6} W/m^2
14.5 No, $I_2 = (316)I_1$
14.6 65 dB
14.7 Destructive interference: $\Delta L = 2.5\lambda = 5(\lambda/2)$, and $m = 5$. No sound would be heard if the waves from the speakers had equal amplitudes. Of course, during a concert the sound would not be single-frequency tones but would have a variety of frequencies and amplitudes. Listeners at certain locations might not hear certain parts of the audible spectrum, but this probably wouldn't be noticed.
14.8 Toward, 431 Hz; past, 369 Hz
14.9 With the source and the observer traveling in the same direction at the same speed, their relative velocity would be zero. That is, the observer would consider the source to be stationary. Since the speed of the source and observer is subsonic, the sound from the source would overtake the observer without a shift in frequency. Generally, for motions involved in a Doppler shift, the word *toward* is associated with an *increase* in frequency and *away* with a *decrease* in frequency. Here, the source and observer remain a constant distance apart. (What would be the case if the speeds were supersonic?)
14.10 768 Hz; yes
14.11 $f_1 = \dfrac{v}{4L} = \dfrac{353 \text{ m/s}}{4(0.0130 \text{ m})} = 6790$ Hz

Chapter 15

15.1 1.52×10^{-20} %
15.2 No, if the comb were positive, it would polarize the paper in the reverse way and still attract it.
15.3 $\vec{F}_1$ has a magnitude of 3.8×10^{-7} N at an angle of 57° above the positive x axis. In unit vector notation: $\vec{F}_1 = (-0.22 \,\mu N) \,\hat{x} + (0.32 \,\mu N) \,\hat{y}$.
15.4 0.12 m or 12 cm.
15.5 $\dfrac{F_e}{F_g} = \dfrac{ke^2}{Gm_e^2} = 4.2 \times 10^{42}$ or $F_e = 4.2 \times 10^{42} F_g$. The magnitude of the electrical force is the same as that between a proton and electron (in the Example) because they have the same (magnitude) charge on them. However the gravitational force is reduced because the attracting masses are two electrons rather than an electron and a much more massive proton.
15.6 The field is zero to the left of q_1 at $x = -0.60$ m.
15.7 $\vec{E} = (-797 \text{ N/C}) \,\hat{x} + (359 \text{ N/C}) \,\hat{y}$ or $E = 874$ N/C at an angle of 24.2° above the negative x-axis.
15.8 In all three locations there are two fields to consider and add vectorially, one from the positive end and one from the negative end of the dipole. (a) Here the larger of the two fields is from the closer positive end and points upward. The smaller field due to the negative end points downward, thus the field direction is upward away from the positive end. (b) Here the larger of the two fields is from the closer negative end and points upward. The smaller field due to the positive end points downward, thus the field direction is upward towards the negative end. (c) Here both fields point downward, thus the net field is downward, away from the positive end and towards the negative end.
15.9 (a) The electric field is upward from ground to cloud. (b) 2.3×10^3 C.
15.10 Positive charge would reside completely on the outside surface, thus only the electroscope attached to the outside surface would show deflection.
15.11 Their sign is negative since the electric field lines end at negative charges, they are all inward relative to the Gaussian surface.

Chapter 16

16.1 (a) ΔU_e would double to $+7.20 \times 10^{-18}$ J because the particle's charge is doubled. (b) ΔV is unchanged because it is not related to the particle. (c) $v = 4.65 \times 10^4$ m/s.

16.2 6.63×10^7 m/s

16.3 (a) It has moved further from a positive charge (the proton) and thus has moved to a region of lower electric potential. (b) $\Delta U_e = +3.27 \times 10^{-18}$ J.

16.4 $U_{CO} = -3.27 \times 10^{-19}$ J. It is less stable, because it would take less work to break it apart than a water molecule.

16.5 (a) 2.22 m. (b) The one closest to the Earth's surface is at a higher potential. (c) No, you can only tell the separation distance between the two surfaces, not their absolute location.

16.6 (a) Surface 1 is at a higher electric potential than surface 2 because it is closer to the positively charged surface. (b) At large distances, the charged object would "look like" a point charge, thus the equipotential surfaces gradually becomes spherical as the distance from the object gets larger.

16.7 $d = 8.9 \times 10^{-16}$ m which is much smaller than the size of an atom (or a nucleus for that matter) and thus this design is entirely unfeasible.

16.8 7.90×10^3 V

16.9 The capacitance decreases as the spacing d increases. Since the voltage across the capacitor remains constant, this means the charge on the capacitor would have to decrease, thus charge would flow off of the capacitor. $\Delta Q = -3.30 \times 10^{-12}$ C.

16.10 $U_{parallel} = 1.20 \times 10^{-4}$ J and $U_{series} = 5.40 \times 10^{-4}$ J so the parallel arrangement stores more energy.

16.11 (a) $Q_1 = 8.0 \times 10^{-7}$ C; $Q_2 = 1.6 \times 10^{-6}$ C; $Q_3 = 2.4 \times 10^{-6}$ C. (b) $U_1 = 3.2 \times 10^{-6}$ J; $U_2 = 6.4 \times 10^{-6}$ J; $U_3 = 4.8 \times 10^{-6}$ J

Chapter 17

17.1 The result is the same, $V_{AB} = V$.

17.2 About 32 years.

17.3 100 V.

17.4 Our assumption is that $R = \dfrac{\rho L}{A}$. Thus, if resistivity is doubled and length halved, the numerator stays the same. If the diameter is halved, the area decreases by a factor of 4. The net result of these changes is that the resistance increases by a factor of 4, up to $3.0 \times 10^3 \ \Omega$. Thus $I = \dfrac{V}{R} = \dfrac{400 \text{ V}}{3.0 \times 10^3 \ \Omega} = 0.133$ A.

17.5 $R = 0.67 \ \Omega$. The material with the largest temperature coefficient of resistivity makes a more sensitive thermometer because it produces a larger (and therefore more accurate to measure) change in resistance for a given temperature change.

17.6 The heat needed is $Q = mc\Delta T = 1.67 \times 10^5$ J. Thus the power output of the heater needs to be

$P = \dfrac{Q}{t} = \dfrac{1.67 \times 10^5 \text{ J}}{180 \text{ s}} = 930$ W. Since this is supplied by the joule heating, we have $R = \dfrac{V^2}{P} = \dfrac{(120 \text{ V})^2}{1.67 \times 10^5 \text{ J}} = 15.5 \ \Omega$.

17.7 (a) $R_1 = \dfrac{V^2}{P_1} = \dfrac{(115 \text{ V})^2}{1200 \text{ W}} = 11.0 \ \Omega$ and

$R_2 = 0.900 R_1 = 9.92 \ \Omega$. (b) $I_1 = \dfrac{V}{R_1} = \dfrac{115 \text{ V}}{11.0 \ \Omega} = 10.5$ A

and $I_2 = 1.11 I_1 = 11.6$ A.

17.8 8.3 hours.

17.9 At best, power plants produce electric energy with efficiencies of 35% (ignoring transmission losses). Thus in terms of primary fuels, the maximum efficiency of any electrical appliance is 35%. However, natural gas is delivered at essentially no energy cost. At the point of delivery, it is burned and can deliver, at least theoretically, up to 100% of its heat content to the task at hand. For example, a well-insulated water heater will be able to absorb about 95% of the energy heat delivered to it. Thus the overall electrical efficiency would be 0.95 (35%) or about 34%. For the gas version, it would be 95% efficient.

Chapter 18

18.1 (a) Series: $P_1 = 4.0$ W, $P_2 = 8.0$ W, $P_3 = 12$ W. Parallel: $P_1 = 1.4 \times 10^2$ W, $P_2 = 72$ W, $P_3 = 48$ W. (b) In series, the most power is dissipated in the largest resistance. In parallel, the most power is dissipated in the least resistance. (c) Series: total resistor power is 24 W, and $P_b = I_b V_b = (2.0 \text{ A})(12 \text{ V}) = 24$ W, so yes, as required by energy conservation. Parallel: total resistor power is $P_{tot} = 2.6 \times 10^2$ W, and $P_b = I_b V_b = (22 \text{ A})(12 \text{ V}) = 2.6 \times 10^2$ W (to two significant figures), so yes, as required by energy conservation

18.2 (a) The voltage across the open socket will be 120 V. (b) The voltage across the remaining bulbs will be zero.

18.3 $P_1 = I_1^2 R_1 = 54.0$ W, $P_2 = I_2^2 R_2 = 9.0$ W, $P_3 = I_3^2 R_3 = 0.87$ W, $P_4 = I_4^2 R_4 = 2.55$ W, and $P_5 = I_5^2 R_5 = 5.63$ W. Their sum is 72.1 W to three significant figures. We have agreement with the power output of the battery (difference due to rounding), that is, $P_b = I_b V_b = (3.00 \text{ A})(24.0 \text{ V}) = 72.0$ W.

18.4 (a) If R_2 is increased then the equivalent parallel resistance of R_2 and R_1 will increase. Thus the total circuit resistance should increase, resulting in a reduction in the total circuit current. Since the current in R_3 is the same as the total current, I_3 will decrease. From this, V_3 should decrease. Therefore V_1 and V_2 should increase since they are equal $V = V_2 + V_3 = $ a constant. Since R_1 has not changed, due to the voltage increase, I_1 should increase. Since I_3 decreases and I_1 increases, it must be (since $I_3 = I_1 + I_2$) that I_2 must decrease. (b) Recalculation confirms these predictions: $I_1 = 0.51$ A (increase), $I_2 = 0.38$ A (decrease), and $I_3 = 0.89$ A (decrease).

18.5 At the junction, we still have $I_1 = I_2 + I_3$ (Eq. 1). Using the loop theorem around loop 3 in the clockwise direction (all numbers are volts, deleted for convenience): $6 - 6I_1 - 9I_2 = 0$ (Eq. 2). For loop 1, the result is $6 - 6I_1 - 12 - 2I_3 = 0$ (Eq. 3). Solve Eq. 1 for I_2 and substitute into Eq. 2. Then solve Eq. 2 and Eq. 3 simultaneously for I_1 and I_3. All answers are the same as in the Example as they should be.

18.6 (a) The maximum energy storage at 9.00 V is 4.05 J. At 7.20 V, the capacitor stores only 2.59 J or 64% of the maximum. This is because the energy storage varies as the *square* of the voltage across the capacitor and $0.8^2 = 0.64$. (b) 8.64 V, because the voltage does not rise linearly, but levels off in an exponential fashion.

18.7 10 A.

18.8 0.20 mA.

Chapter 19

19.1 East, since reversing both the velocity direction and the sign of the charge leaves the direction the same.

19.2 (a) Using the force right-hand rule, the proton would initially deflect in the negative x direction. (b) 0.10 T

19.3 0.500 V

19.4 (a) At the poles the magnetic field is perpendicular to the ground. Since the current is parallel to the ground, according to the force right-hand rule the force on the wire would be in a plane parallel to the ground. Thus it would not be able to cancel the downward force of gravity. (b) The wire's mass is 0.041 g, which is unrealistically low.

19.5 (a) At 45°, the torque is 0.269 m · N or 70.7% of the maximum torque. (b) 30°.

19.6 (a) South (b) 38 A.

19.7 1500 turns

19.8 (a) The force becomes repulsive. You should be able to show this by using the right-hand source and force rules. (b) 0.027 m or 27 mm.

19.9 The permeability would only have to be 40% of the value in the Example, or $\mu \geq 480\mu_o = 6.0 \times 10^{-4}$ T · m/A.

Chapter 20

20.1 (a) Clockwise. (b) 0.335 mA.

20.2 Any way that will increase the flux, such as increasing the loop area or the number of loops. Changing to a lower resistance would also help.

20.3 7.36×10^{-4} T

20.4 1.5 m/s

20.5 0.28 m

20.6 (a) 6.1×10^3 J (b) 5.0×10^3 J so about 12 times more energy is used during startup.

20.7 (a) She would use a step-up transformer because European appliances are designed to work at 240 V, which is twice the U.S. voltage of 120 V. (b) The output current would be 1500 W/240 V or 6.25 A. Thus the input current would be 12.5 A. (Voltage would be stepped up by a factor of two, and thus the input current is twice as large as the output current.)

20.8 (a) Higher voltages allow for lower current usage. This in turn reduces joule heat losses in the delivery wires and in the motor windings, making more energy available for doing mechanical work and therefore a higher efficiency. (b) Since the voltage is doubled, the current is halved. The heat loss in the wire is proportional to the *square* of the current. Thus, losses will be cut by a factor of 4 or be reduced to 25% of their value at 120 V.

20.9 0.38 cm/s.

20.10 (a) With increasing distance, the Sun's light intensity (energy per second per unit area) drops. Therefore so would the force due to the light pressure on the sail. In turn, the ship's acceleration would be reduced. (b) You would need to somehow enlarge the sail area to catch more light.

Chapter 21

21.1 (a) 0.25A (b) 0.35 A (c) $9.6 \times 10^2 \, \Omega$, larger than the 240 Ω required for a bulb of the same power in the United States. The voltage in Great Britain is larger than that in the United States. Thus to keep the power constant, the current must be reduced by using a larger resistance.

21.2 If the resistance of the appliance is constant, the power will quadruple since $P \propto V^2$. Even if the resistance increased, the power would probably be much more than the appliance was designed for and it would likely burn out, or at least blow a fuse.

21.3 (a) $\sqrt{2}$(120 V) = 170 V (b) 120 Hz

21.4 (a) $\sqrt{2}$(2.55 A) = 3.61 A (b) 180 Hz

21.5 (a) The current would increase to 0.896 A. (b) The capacitor is responsible; with a frequency increase, X_C decreases.

Since resistance is independent of frequency, it remains constant and overall Z decreases.

21.6 (a) In an RLC circuit, the phase angle ϕ depends on the difference $X_L - X_C$. If you increase the frequency, X_L would increase and X_C would decrease, thus their difference would increase and so would ϕ. (b) $\phi = 84.0°$, an increase as expected.

21.7 6.98 W

21.8 (a) If you have a receiver tuned to a frequency *between* the two station frequencies, you would not receive the maximum strength signal from either station but there might be enough power from each to hear them simultaneously. (b) 651 kHz

Chapter 22

22.1 Light travels in straight lines and is reversible. If you can see someone in a mirror, that person can see you. Conversely, if you can't see the trucker's mirror, then he or she can't see your image in that mirror and won't know that your car is behind the truck.

22.2 $n = 1.25$ and $\lambda_m = 400$ nm

22.3 By Snell's law, $n_2 = 1.24$ so $v = c/n_2 = 2.42 \times 10^8$ m/s.

22.4 With a greater n, θ_2 is smaller so the refracted light inside the glass is toward the lower-left. Therefore the lateral displacement is larger. 0.72 cm.

22.5 (a) The frequency of the light is unchanged in the different media, so the emerging light has the same frequency as that of the source. (b) The wavelength in air is independent of the water and glass media, as can be shown by adding another step (medium) to the Example solution. By reverse analysis, $\lambda_{air} = n_{water}\lambda_{water} = (c/v_{water})\lambda_{water} = c/f$. Thus, the wavelength in air is c/f.

22.6 Because of total internal reflections, the diver could not see anything above water. Instead, he would see the reflection of something on the sides and/or bottom of the pool. (Use reverse ray tracing.)

22.7 $n = 1.4574$. Green light will be refracted more than red light as green has a shorter wavelength, thus greater n than red light. By Snell's law, green will have a smaller angle of refraction so it is refracted more.

Chapter 23

23.1 No effect. Note that the solution to the Example does not include the distance. The geometry of the situation is the same regardless of the distance from the mirror.

23.2 $d_i \approx 60$ cm; real, inverted, and magnified.

23.3 $d_i = d_o$ and $M = -1$; real, inverted, and same size

23.4 The image is also always upright and smaller than the object.

23.5 $d_i = -20$ cm (in front of the lens); virtual, upright, and magnified.

23.6 $d_o = 2f = 24$ cm

23.7 Blocking off half of the lens would result in half the *amount* of light focused at the image plane, so the resulting image would be less bright but still full size.

23.8 The image is also always upright and smaller than the object

23.9 3 cm behind L_2; real, inverted, and smaller than the object ($M_{total} = -0.75$)

23.10 If the lens is immersed in water, Eq. 23.8 should be modified to $\dfrac{1}{f} = (n/n_m - 1)\left(\dfrac{1}{R_1} + \dfrac{1}{R_2}\right)$, where $n_m = 1.33$ (water). Since $n = 1.52 > n_m = 1.33$, the lens is still converging.

$$P = \frac{1}{f} = (1.52/1.33 - 1)\left(\frac{1}{0.15 \text{ m}} + \frac{1}{-0.20 \text{ m}}\right)$$

$$= 0.238 \; 1/\text{m} = 0.238 \text{ D}. \; f = \frac{1}{0.238 \; 1/\text{m}} = 4.20 \text{ m}.$$

Chapter 24

24.1 $\Delta y = y_r - y_b = 1.2 \times 10^{-2} = 1.2$ cm
24.2 twice as thick, $t = 199$ nm
24.3 In brass instruments, the sound comes from a relatively large, flared opening. Thus there is little diffraction, so most of the energy is radiated in the forward direction. In wood-wind instruments, much of the sound comes from tone holes along the column of the instrument. These holes are small compared to the wavelength of the sound, so there is appreciable diffraction. As a result, the sound is radiated in nearly all directions, even backward.
24.4 The width would increase by a factor of $700/550 = 1.27$
24.5 $\Delta\theta_2 = \theta_2(700 \text{ nm}) - \theta_2(400 \text{ nm}) = 44.4° - 23.6° = 20.8°$.
24.6 45°
24.7 $\theta_2 = 41.2°$
24.8 589 nm; yellow

Chapter 25

25.1 It wouldn't work; a real image would form on person's side of lens. ($d_i = +0.75$ m).
25.2 For an object at $d_o = 25$ cm, the image for eye 1 would be formed at 1.0 m; this is beyond the near point for that eye, so the object could be seen clearly. The image for eye 2 would be formed at 0.77 m; this is inside the near point for that eye, so the object would not be seen clearly.
25.3 glass for near-point viewing, 2.0 cm longer
25.4 length doubles
25.5 $f_i = 8.0$ cm
25.6 The erecting lens (of focal length f_e) should go between the objective and the eyepiece, positioned a distance of $2f_e$ from the image formed by the objective, which acts as an object. The erecting lens then produces an inverted image of the same size at $2f_e$ on the opposite side of the lens, which acts as an object for the eyepiece. The use of the erecting lens lengthens the telescope by $4f_e$.
25.7 3.4×10^{-7} rad, an order of magnitude better than the typical 10^{-6} rad
25.8 2.9 cm

Chapter 26

26.1 Light waves from two simultaneous events on the y-axis meet at some midpoint receptor on the y-axis. Since there is no relative motion along that axis, a simultaneous recording of the two events will also be recorded along the y'-axis. Hence the two observers agree on simultaneity for this situation.
26.2 $v = 0.9995c$; no, not twice as fast. Travel is limited to less than c, so this is only about a 0.15% increase.
26.3 (a) 0.667 μs (b) 0.580 μs The observer watching the ship measures the proper time interval. To the person on the ship, that time interval is dilated.
26.4 $v = 0.991c$
26.5 The traveler measures the proper time interval of 20.0 y, but Earth inhabitants measure the dilated version of this (why?). The gamma factor is based on a recalculated value for traveler speed $v = 0.90504c$. Keeping five places after the decimal and then rounding to three significant figures we get

$\gamma = 1/\sqrt{1 - (0.99504c/c)^2} = 10.04988$ and $\Delta t = \gamma\Delta t_o = (10.04988)(20.0 \text{ y}) = 200.99 \text{ y} \approx 201$ y.

26.6 (a) 1.17 MeV (b) 0.207 MeV
26.7 (a) $0.319mc^2$ (b) $3.33mc^2$
26.8 1.95×10^7 more massive
26.9 $u = +0.69c$ which, as expected, is lower in magnitude than the nonrelativistic (and wrong) result of $0.80c$.

Chapter 27

27.1 (a) The wavelength is 967 nm, which is infrared. With some emissions in the red region, Betelgeuse would appear reddish. (b) The wavelength is 290 nm, which is ultraviolet. With significant visible emissions, Rigel would appear bluish-white.
27.2 496 nm, which is shorter than the Example, indicating a higher photon energy. Thus the maximum kinetic energy of the photoelectrons is higher, requiring an increased stopping voltage.
27.3 2.00 V
27.4 (a) The ratio is $\Delta\lambda/\lambda_o = 324$ meaning a wavelength increase of 3.24×10^4%. (b) Percentagewise this is a much larger increase than in the Example because the wavelength of the incoming light (λ_o) is much smaller for gamma rays.
27.5 (a) 1.09×10^6 m/s. (b) 5.41×10^{-19} J = 3.40 eV.
27.6 365 nm (UV).
27.7 The least energetic photon in the Lyman series results in a transition from $n = 2$ (first excited state) to the ground state $n = 1$. This results in a photon of energy 10.2 eV. The wavelength of this light is 122 nm, which is UV.
27.8 (a) There are six possible transitions from the $n = 4$ state to the ground state, and thus the emitted light has six different possible wavelengths. (b) If the atom is excited from the ground state to the first excited state ($n = 1$ to $n = 2$), then it has no choice but to emit a single photon during de-excitation state ($n = 2$ to $n = 1$) because there are no intermediate states.

Chapter 28

28.1 (a) 8.8×10^{-33} m/s (b) 2.3×10^{33} s, or 7.2×10^{25} y. This is about 4.8×10^{15} times longer than the age of the universe. This movement would definitely not be noticeable.
28.2 The proton's de Broglie wavelength is 4.1×10^{-12} m, or about twenty times smaller than atomic spacing distances. With a wavelength much smaller than the atomic spacing, these protons would *not* be expected to exhibit significant diffraction effects.
28.3 Only five electrons could be accommodated in the 3d subshell if there were no spin (with spin there can be ten).
28.4 (a) $1s^22s^22p^6$ (b) $-2e$ or -3.2×10^{-19} C
28.5 1.2×10^6 m/s

Chapter 29

29.1 $^{12}_{6}\text{C} + ^{4}_{2}\text{He} \longrightarrow ^{16}_{8}\text{O}$, thus the resulting nucleus is oxygen-16.
29.2 (a) Since $^{23}_{11}\text{Na}$ is the stable isotope with 11 protons and 12 neutrons, $^{22}_{11}\text{Na}$ is one neutron shy of being stable. In other words, it is proton-rich or neutron-poor. Thus, the expected decay mode is β^+ or positron decay. (b) Neglecting the emitted neutrino, the decay is $^{22}_{11}\text{Na} \longrightarrow ^{22}_{10}\text{Ne} + ^{0}_{+1}\text{e}$. The daughter nucleus is neon-22.
29.3 (a) 48 d because reducing the activity by a factor of 64 requires six half lives; $1/2^6 = 1/64$. (b) The process of excretion from the body can also remove ^{131}I.
29.4 The closest integer is 20, since $2^{20} \approx 1.05 \times 10^6$. Thus it takes 20 half-lives, or about 560 y.
29.5 The measurement can be made to four ^{14}C half-lives, or 2.3×10^4 y.
29.6 (a) ^{40}K is an odd-odd nucleus (19 protons, 21 neutrons) and thus is unstable. ^{41}K, an odd-even potassium isotope (19 protons, 22 neutrons) is the likely candidate for the remainder of the stable potassium. ^{43}K would have too many neutrons

(24) compared to protons (19) for this region of the periodic chart. (b) Using $N = N_o e^{-\lambda t}$, it follows that $\dfrac{N_o}{N} = e^{\lambda t} = 12.8$. Thus there would have been about 13 times more ^{40}K (than exists now) at the time of the formation of the Earth.

29.7 (a) Starting with 29 protons and 29 neutrons (why?), we have the following candidates: $^{58}_{29}Cu$, $^{59}_{29}Cu$, $^{60}_{29}Cu$, $^{61}_{29}Cu$, $^{62}_{29}Cu$, $^{63}_{29}Cu$, $^{64}_{29}Cu$, etc. Now delete the odd–odd isotopes (why?) to get the most likely (stable) isotopes: $^{59}_{29}Cu$, $^{61}_{29}Cu$, $^{63}_{29}Cu$, $^{65}_{29}Cu$, etc. (b) Further trimming of the list can be done by deleting those with $N \approx Z$ (why?) and those with N significantly larger than Z (why?). Since Z should be just a bit smaller than N in this mass region, we expect neutron numbers in the mid-30s. Hence a good guess would be just $^{63}_{29}Cu$ and $^{65}_{29}Cu$. According to Appendix V, these are, in fact, the only two stable isotopes of copper.

29.8 (a) The result for ^3He is 2.573 MeV/nucleon, which is considerably smaller than the 7.075 MeV/nucleon for ^4He. (b) Thus ^4He is the more tightly bound of the two. Unlike ^3He, all protons and neutrons in ^4He *are* paired, resulting in a more tightly bound nucleus.

29.9 The absorbed dose is 0.0215 Gy or 2.15 rad. Since the RBE for gamma rays is 1, the effective dose is 0.0215 Sv or 2.15 rem.

Both of these are about one-seventh of the dose from the beta radiation.

Chapter 30

30.1 $Q = -15.63$ MeV, so it is endoergic (takes energy to happen).

30.2 The increase in mass has an energy equivalent of 1.193 MeV. The rest of the incident kinetic energy (1.534 MeV − 1.193 MeV, or 0.341 MeV) must be distributed between the kinetic energies of the proton and of the oxygen-17.

30.3 There are 1.00×10^{38} proton cycles per second.

30.4 Because the beta particle has little energy and therefore little momentum, the neutrino and the daughter nucleus would have to recoil in opposite directions in order to conserve linear momentum. This assumes the original nucleus had zero linear momentum.

30.5 Since the negative pion (π^-) has a charge of $-e$, its quark structure must be $\bar{u}d$. Using similar reasoning, the quark structure of the positive pion (π^+) is $u\bar{d}$. Thus the antiparticle of the π^+ would have the composition of $\bar{u}\bar{\bar{d}}$. However, the antiquark of an antiquark is just the original quark—for example $(\bar{\bar{u}}) = u$. Hence the antiparticle of the π^+ has the quark structure $\bar{u}\bar{\bar{d}} = \bar{u}d$, which is the π^- quark structure, as expected.

ANSWERS TO ODD-NUMBERED EXERCISES

Chapter 1

1. (c)

3. (b)

5. Because there are no more fundamental quantities and all other quantities can be derived from the fundamental ones.

7. Mean solar day replaced the original second definition. No, atomic clocks are now used.

9. (b)

11. (a)

13. no, yes

15. Metric ton is defined as the mass of 1 m^3 of water. 1 m^3 = 1000 L and 1 L of water has a mass of 1 kg. So one metric ton is equivalent to 1000 kg.

17. (a) Different ounces are used for volume and weight measurements. 16 oz = 1 pt is a volume measure and 16 oz = 1 lb is a weight measure. (b) Two different pound units are used. Avoirdupois lb = 16 oz, troy lb = 12 oz.

19. (d)

21. (a)

23. No, unit analysis can only tell if it is dimensionally correct.

25. (Length)
= (Length) + (Length)/(Time)× (Time)
= (Length) + (Length)

27. m^2 = (m)^2 = m^2

29. no, $V = \pi d^3/6$

31. a: 1/m; b: dimensionless; c: m

33. kg/m^3

35. The first student, because
m/s = $\sqrt{(m/s^2)(m)}$ = $\sqrt{m^2/s^2}$ = m/s.

37. (a) kg·m^2/s (b) The unit of $L^2/(2mr^2)$ is
(kg·m^2/s)^2/(kg·m^2) = kg·m^2/s^2, which is the unit of kinetic energy, K. (c) kg·m^2

39. (c)

41. (c)

43. Yes, whether you multiply or divide should be consistent with units.

45. 39.6 m

47. 37 000 000 times

49. (a) 91.5 m by 48.8 m (b) 27.9 cm to 28.6 cm

51. 0.78 mi

53. (a) (1) 1 m/s (b) 33.6 mi/h

55. (a) 77.3 kg (b) 0.0773 m^3 or about 77.3 L

57. 6.5 × 10^3 L/day

59. (a) 59.1 mL (b) 3.53 oz

61. 6.1 cm

63. (a) 1.5 × 10^5 m^3 (b) 1.5 × 10^8 kg
(c) 3.3 × 10^8 lb

65. (a)

67. (b)

69. No, there is always one doubtful digit, the last digit.

71. 5.05 cm; 5.05 × 10^{-1} dm; 5.05 × 10^{-2} m.

73. (a) 4 (b) 3 (c) 5 (d) 2

75. (b) and (d). (a) has four and (c) has six.

77. 32 ft^3

79. (a) (2) three, since the height has only three significant figures. (b) 469 cm^2

81. (a) (1) zero, since 38 m has zero decimal place. (b) 15 m.

83. (a)

85. (d)

87. all six steps as listed in the chapter

89. The accuracy of the answer is expected to be within an order of 10.

91. 100 kg

93. (a) 8.72 × 10^{-3} cm (b) about 10^{-2} cm

95. 0.87 m

97. same area for both, 1.3 cm^2

99. 3.5 × 10^{10} white cells, 1.3 × 10^{12} platelets

101. 4.7 × 10^2 lb

103. about 10^{12} m^3

105. 17 m

107. (a) 283 mi (b) 45° north of east

109. 3.0 × 10^8 km

111. (a) (3) less than the 190 mi/h because more time spent at lower speeds, so affect average speed to be below average of all speeds. (b) 187 mi/h

Chapter 2

1. (a)

3. (c)

5. Yes, for a round-trip. No; distance is always greater than or equal to the magnitude of displacement.

7. The distance traveled is greater than or equal to 300 m. The object could travel a variety of ways as long as it ends up at 300 m north. If the object travels straight north, then the minimum distance is 300 m.

9. Yes, this is possible. The jogger can jog in the opposite direction during the jog (negative instantaneous velocity) as long as the overall jog is in the forward direction (positive average velocity).

11. 1.65 m down

13. (a) 0.50 m/s (b) 8.3 min

15. 0.17 m/s

17. (a) (2) greater than R but less than $2R$
(b) 71 m

19. (a) (3) between 40 m and 60 m (b) 45 m at 27° west of north

21. (a) 2.7 cm/s (b) 1.9 cm/s

23. (a) 90.6 ft at 6.3° above horizontal
(b) 36.2 ft/s at 6.3° (c) Average speed depends on the total path length, which is not given. The ball might take a curved path.

25. (a) $\bar{s}_{0-2.0\,s}$ = 1.0 m/s; $\bar{s}_{2.0\,s-3.0\,s}$ = 0;
$\bar{s}_{3.0\,s-4.5\,s}$ = 1.3 m/s; $\bar{s}_{4.5\,s-6.5\,s}$ = 2.8 m/s;
$\bar{s}_{6.5\,s-7.5\,s}$ = 0; $\bar{s}_{7.5\,s-9.0\,s}$ = 1.0 m/s
(b) $\bar{v}_{0-2.0\,s}$ = 1.0 m/s; $\bar{v}_{2.0\,s-3.0\,s}$ = 0;
$\bar{v}_{3.0\,s-4.5\,s}$ = 1.3 m/s; $\bar{v}_{4.5\,s-6.5\,s}$ = −2.8 m/s;
$\bar{v}_{6.5\,s-7.5\,s}$ = 0; $\bar{v}_{7.5\,s-9.0\,s}$ = 1.0 m/s
(c) $v_{1.0\,s}$ = $\bar{s}_{0-2.0\,s}$ = 1.0 m/s;
$v_{2.5\,s}$ = $\bar{s}_{2.0\,s-3.0\,s}$ = 0; $v_{4.5\,s}$ = 0;
$v_{6.0\,s}$ = $\bar{s}_{4.5\,s-6.5\,s}$ = −2.8 m/s
(d) $v_{4.5\,s-9.0\,s}$ = −0.89 m/s

27. 1 month

29. (a) 500 km at 37° east of north
(b) 400 km/h at 37° east of north (c) 560 km/h
(d) Since speed involves total distance, which is greater than the magnitude of the displacement, the average speed does not equal the magnitude of the average velocity.

31. (d)

33. (c)

35. Yes, although the speed of the car is constant, its velocity is not, because of the change in direction. A change in velocity signifies an acceleration.

37. Not necessarily. A negative acceleration can also speed up objects if the velocity is also negative (that is, in the same direction as the acceleration).

39. v_o. Since an equal amount of time is spent on acceleration and deceleration of the same magnitude.

41. 6.9 m/s^2

43. (a) (2) opposite to velocity as the object slows down (b) −2.2 m/s each second, opposite direction of velocity

45. −2.0 m/s^2

47. −70.0 km/h or −19.4 m/s, +2.78 m/s^2
(deceleration because the velocity is negative)

49. 4.8 × 10^2 m/s^2, this is a very large acceleration due to the change in direction of the velocity and the short contact time.

51. (a) $\bar{a}_{0-1.0\,s}$ = 0; $\bar{a}_{1.0\,s-3.0\,s}$ = 4.0 m/s^2;
$\bar{a}_{3.0\,s-8.0\,s}$ = −4.0 m/s^2; $\bar{a}_{8.0\,s-9.0\,s}$ = 8.0 m/s^2;
$\bar{a}_{9.0\,s-13.0\,s}$ = 0 (b) Constant velocity of −4.0 m/s

53. 150 s

55. (d)

57. It is zero because the velocity is a constant.

59. Consider the displacement ($x - x_o$) as one quantity; there are four quantities involved in each kinematic equation (Eqs. 2.8, 2.10, 2.11, and 2.12). So three must be known before one can solve for any unknown. Or equivalently all but one have to be known.

61. no, acceleration must be 9.9 m/s^2

63. (a) 1.8 m/s^2 (b) 6.3 s

65. (a) 81.4 km/h (b) 0.794 s

67. 3.09 s and 13.7 s. The 13.7 s answer is physically possible but not likely in reality. After 3.09 s, it is 175 m from where the reverse thrust was applied, but the rocket keeps traveling forward while slowing down. Finally it stops. However, if the reverse thrust is continuously applied (which is possible, but not likely), it will reverse its direction and be back to 175 m from the point where the initial reverse thrust was applied; a process that would take 13.7 s.

69. no, a = 3.33 m/s^2 < 4.90 m/s^2

71. 2.2 × 10^5 m/s^2

73. no, 13.3 m > 13 m

75. (b) 96 m

77. (a) (3) $v_1 > \frac{1}{2}v_2$ (b) 9.22 m/s, 13.0 m/s

79. (a) −12 m/s; −4.0 m/s (b) −18 m (c) 50 m

81. (a) 12.2 m/s, 16.4 m/s (b) 24.8 m (c) 4.07 s

83. (d)

85. (c)

87. (c)

89. The ball moves with a constant velocity because there is no gravitational acceleration in deep space. If the gravitational acceleration is zero, g = 0, then v = constant.

91. First of all, the gravitational acceleration on the Moon is only 1/6 of that on the Earth.

Or $g_M = g_E/6$. Secondly, there is no air resistance on the Moon.

93. (a) (3) four times, height is proportional to the time squared. **(b)** 15.9 m, 3.97 m

95. no, not a good deal (0.18 s < 0.20 s)

97. 67 m

99. $\Delta t = 0.096$ s

101. (a) (1) less than 95%, as the height depends on the initial velocity squared **(b)** 3.61 m

103. (a) 1.64 m/s^2 **(b)** 2.07 m/s

105. (a) 5.00 s **(b)** 36.5 m/s

107. 1.49 m above the top of the window

109. (a) 155 m/s **(b)** 2.22×10^3 m **(c)** 28.7 s

111. (a) 8.45 s **(b)** $x_M = 157$ m; $x_C = 132$ m **(c)** 13 m

113. (a) 38.7 m/s **(b)** 15.5 s **(c)** 19.2 m/s

115. (a) 119 m **(b)** 4.92 s **(c)** Lois: 48.2 m/s; Superman: 73.8 m/s

117. (a) -297 m/s **(b)** 3.66 m/s^2 **(c)** 108 s

Chapter 3

1. (a)

3. (c)

5. Yes, this is possible. For example, if an object is in circular motion, the velocity (along a tangent) is perpendicular to the acceleration (toward the center of the circle).

7. (a) (1) greater then because for $\theta < 45°$, $\cos\theta > \sin\theta$ and $v_x = v\cos\theta$ and $v_y = v\sin\theta$. **(b)** 28 m/s, 21 m/s

9. ± 6.3 m/s, there are two possible answers because the vector could be either in the first or the fourth quadrant.

11. (a) (2) north of east **(b)** 1.1×10^2 m, 27° north of east

13. $x = 1.75$ m, $y = -1.75$ m

15. (a) 75.2 m **(b)** 99.8 m

17. (a) $\theta = 56.3°$ below horizontal **(b)** 18.0 m/s

19. (a) 1.2 m/s **(b)** 49 m

21. (c)

23. (d)

25. Yes, when the vector is in the y-direction, it has a zero x-component.

27. Yes, if equal and opposite

29.

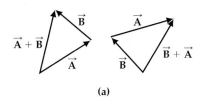

(a)

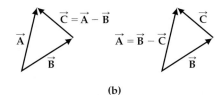

(b)

31. 4.9 m, 59° above $-x$-axis

33. 113 mi/h

35. (a) $(-3.4 \text{ cm})\,\hat{x} + (-2.9 \text{ cm})\,\hat{y}$ **(b)** 4.5 cm, 63° above $-x$-axis **(c)** $(4.0 \text{ cm})\,\hat{x} + (-6.9 \text{ cm})\,\hat{y}$

37. (a) $(14.4 \text{ N})\,\hat{y}$ **(b)** 12.7 N at 85.0° above $+x$-axis

39. (a) $\vec{v}_2 = (-4.0 \text{ m/s})\,\hat{x} + (8.0 \text{ m/s})\,\hat{y}$ **(b)** 8.9 m/s

41. 21 m/s at 51° below the $+x$-axis

43. (a) $(-9.0 \text{ cm})\,\hat{x} + (6.0 \text{ cm})\,\hat{y}$ **(b)** 33.7°, relative to $-x$ axis

45. 8.5 N at 21° below $-x$-axis

47. parallel 30 N; perpendicular 40 N

49. The forces act on different objects (one on horse, one on cart) and therefore cannot cancel.

51. (a) (2) north of west **(b)** 102 mi/h at 61.1° north of west

53. (a) 42.8° south of west **(b)** 0.91 m **(c)** The reason is due to the fact that the ball might travel in a curve.

55. (b)

57. (b)

59. The horizontal motion does not affect the vertical motion. The vertical motion of the ball projected horizontally is identical to that of the ball dropped.

61. (a) 0.64 s **(b)** 0.64 m

63. 6.4 m

65. 40 m

67. (a) (2) Ball B collides with ball A because they have the same horizontal velocity. **(b)** 0.11 m, 0.11 m

69. (a) 0.77 m **(b)** ball would not fall back in.

71. 35° or 55°

73. 3.65 m/s^2

75. (a) 26 m **(b)** 23 m/s at 68° below horizontal

77. 1.4°

79. yes, at $x = 15$ m, $y = 0.87$ m < 1.2 m

81. 8.7 m/s

83. The pass is short.

85. (a)

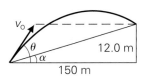

(b) 66.0 m/s **(c)** The shot is too long for the hole.

87. (d)

89. No, the Earth undergoes several movements such as orbiting the Sun and rotating about itself.

91. Since the rain is coming down at an angle relative to you, you should hold the umbrella so it is tilted forward.

93. You throw the ball straight up. This way, both you and the ball have the same horizontal velocity relative to the ground or you have zero horizontal velocity relative to each other, so the object returns to your hand.

95. 6.7 s

97. (a) +85 km/h **(b)** -5 km/h

99. 146 s = 2.43 min

101. (a) Same both ways, 4.25° upstream **(b)** 44.6 s

103.

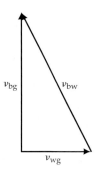

Use the following subscripts: b = boat, w = water, and g = ground. For the boat to make the trip straight across, v_{bw} must be the hypotenuse of the right-angle triangle. So it must be greater in magnitude than v_{wg}. So if the reverse is true, that is, if $v_{wg} > v_{bw}$, the boat cannot make the trip directly across the river.

105. 1.21 m/s

107. (a) 24° east of south **(b)** 1.5 h

111. (a) 47.8 s **(b)** 5.32×10^3 m **(c)** 310 m/s

Chapter 4

1. (d)

3. (c)

5. (d)

7. According to Newton's first law, your tendency is to remain at rest or move with constant velocity. However, the plane is accelerating to a velocity faster than yours so you are "behind" and feel "pushed" into the seat. The seat actually supplies a forward force to accelerate you to the same velocity as the plane.

9. (a) The bubble moves forward in the direction of velocity or acceleration, because the inertia of the liquid will resist the forward acceleration. So the bubble of negligible mass or inertia moves forward relative to the liquid. Then it moves backward opposite the velocity (or in the direction of acceleration) for the same reason. **(b)** The principle is based on the liquid inertia.

11. According to Newton's first law, or the law of inertia, the dishes at rest tend to remain at rest. The quick pull of the tablecloth required a force that exceeds the maximum static friction (discussed in Section 4.6) so the cloth can move relative to the dishes.

13. 0.40 kg

15. 0.64 m/s^2

17. (a) (3) the upward force is the same in both situations. In both situations, there is no acceleration in the vertical direction so the net force in the vertical direction is zero or the normal force is equal to the weight. **(b)** 0.50 lb

19. (a) (3) either (1) or (2) is possible, because "at rest" or "constant velocity" both have zero acceleration. **(b)** yes, 2.5 N at 36° above the $+x$ axis

21. (a) no **(b)** $F_5 = 4.1$ N at 13° above the $-x$ axis

23. (b)

25. (d)

27. There will be extra acceleration. A pickup truck in snow (mass increases) will have less acceleration due to the extra mass and launched rocket (mass decreases) will have greater acceleration.

29. "Soft hands" here result in longer contact time between the ball and the hands. The increase in contact time decreases the magnitude of acceleration. From Newton's second law, this, in turn, decreases the force required to stop the ball and its reaction force, the force on the hands.

31. 1.7 kg

33. (a) (3) 6.0 kg, because mass is a measure of inertia, and it does not change. (b) 9.8 N

35. (a) (1) on the Earth. 1 lb is equivalent to 454 g, or 454 g has a weight of 1 lb. (b) 5.4 kg (2.0 lb)

37. (a) (4) one-fourth as great. (b) 4.0 m/s^2

39. 2.40 m/s^2

41. (a) 30 N (b) -4.60 m/s^2

43. 8.9×10^4 N

45. Lois is saved.

47. (c)

49. The forces act on different objects (one on horse, one on cart) and therefore cannot cancel.

51. (a) (2) two forces acting on the book, the gravitational force (weight, w) and the normal force by the surface, N. (b) The reaction of w is an upward force on the Earth by the book, and the reaction force of N is a downward force on the horizontal surface by the book.

53. (a) (3) the force the blocks exerted forward on him. (b) 3.08 m/s^2

55. (a) (4) the pull of the rope on the girl. The pull of the rope on the girl is the reaction of the pull of the girl on the rope. (b) 264 N

57. (d)

59.

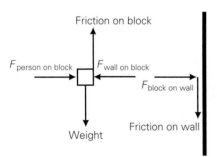

The force on the block by the wall $F_{\text{wall on block}}$ and the force on the wall by the block $F_{\text{block on wall}}$ are an action-reaction pair. The friction on the block and on the wall are also an action-reaction pair.

61. (a) (1) less than the weight of the object, $N = w \cos \theta < w$ (for any $\theta \neq 0$). (b) 98 N and 85 N

63. 585 N

65. (a) 1.7 m/s^2 at 19° north of east (b) 1.2 m/s^2 at 30° south of east

67. (a) 0.96 m/s^2 (b) 2.6×10^2 N

69. 123 N up the incline

71. 64 m

73. (a) (3) both the tree separation and sag. (b) 6.1×10^2 N

75. 2.63 m/s^2

77. 6.25×10^5 N

79. 2.0 m/s^2

81. 1.1 m/s^2 up

83. (a) (1) $T > w_2$ and $T < F$ (b) 1.70×10^3 N (c) 1.13×10^3 N

85. (a) 1.2 m/s^2, m_1 up and m_2 down (b) 21 N

87. (c)

89. (b)

91. This is because kinetic friction (sliding) is less than static friction (rolling). A greater friction force can decrease the stopping distance.

93. (a) No, there is no inconsistency. Here the friction force OPPOSES slipping. (b) Wind can increase or decrease air friction depending on wind directions. If wind is in the direction of motion, friction decreases, and vice versa.

95. (a) (3) increase, but more than double. There is constant friction involved in this exercise so the net force more than doubles. So the acceleration is more than double. (b) 7.0 m/s^2

97. 2.7×10^2 N

99. (a) $\theta_{\min} = \tan^{-1} 0.65 = 33° > 20°$, so it will not move.

101. 0.064

103. (a) (2) pulling at the same angle. (b) 296 N; 748 N

105. (a) 30° (b) 22°

107. (a) 6.0 kg (b) 1.2 m/s^2

109. (a)

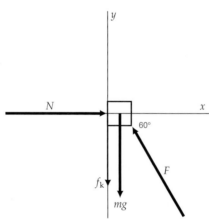

(b) 30 N (c) 23 N

111. (a) 0.179 kg (b) 0.862 m/s^2

113. (a) (4) the force of static friction on A due to the top surface of B. (b) 5.00 N, 17.5 N

115. (a) 5.5 m/s^2 (b) 173 N

117. (a) 10.3 N (b) 0.954 kg

Chapter 5

1. (d)

3. (b)

5. (a) No, the weight is not moving, so there is no displacement and therefore, no work. (b) Yes, positive work is done by the force exerted by the weightlifter. (c) No, as in (a), no work is done. (d) Yes, but the positive work is done by gravity, not the weightlifter.

7. Positive on the way down and negative on the way up. No, it is not constant.

9. -98 J

11. 8.31 m

13. 3.7 J

15. 2.3×10^3 J

17. (a) (2) one, the only force that does non-zero work is the kinetic friction force. (b) -62.5 J

19. (a) 1.48×10^5 J (b) -1.23×10^5 J (c) 2.50×10^4 J.

21. (a) Pulling requires the student to do (1) less work. Compared with pushing, pulling decreases the normal force on the crate. This, in turn, decreases the kinetic friction force. (b) pulling: 1.3×10^3 J; pushing: 1.7×10^3 J

23. No, it takes more work. This is because the force increases as the spring stretches, according to Hooke's law. Also the displacement is greater.

25. 80 N/m

27. 1.25×10^5 N/m

29. (a) (1) $\sqrt{2}$, because when W doubles, x becomes $\sqrt{2}$ as much. Therefore, it will stretch by a factor of $\sqrt{2}$. (b) 900 J

31. (a) (1) more than, because the force required is greater (while the displacement is the same) to stretch from 10 cm to 20 cm, according to Hooke's law. (b) 0–10 cm: 0.25 J; 10–20 cm: 0.75 J

33. (a) 4.5 J (b) 3.5 J

35. 6.0 J

37. (d)

39. (c)

41. Reducing speed by half will reduce kinetic energy by $\frac{3}{4}$, whereas reducing the mass by half will only reduce kinetic energy by half.

43. $\sqrt{2}\, v$

45. -1.3×10^3 N

47. (a) 45 J (b) 21 m/s

49. 200 m

51. 2.0×10^3 m

53. (d)

55. (d)

57. They will have the same potential energy at the top because they have the same height.

59. (a) (4) only the difference between the two heights, because the change in potential energy depends only on the height difference, not positions. (b) position is lowered 0.51 m

61. (a) (4) all the same, because the change in potential energy is independent of the reference level. (b) $U_b = -44$ J, $U_a = 66$ J (c) -1.1×10^2 J

63. 1/25

65. (a) 0.154 J (b) -0.309 J

67. (d)

69. The initial potential energy is equal to the final potential energy so the final height is equal to the initial height.

71. Yes. When the thrown-up ball is at its maximum height, its velocity is zero so it has the same energy as the dropped ball. Since the total energy of each ball is conserved, both balls will have the same mechanical energy at half the height of the window. As a matter of fact, both balls will have the same kinetic and potential energy at the half height position.

73. 5.10 m

75. (a) (3) at the bottom of the swing, because the lower the potential energy (at the bottom is the lowest), the higher the kinetic energy, therefore speed. **(b)** 5.42 m/s

77. 0.176 m

79. (a) 1.03 m **(b)** 0.841 m **(c)** 2.32 m/s

81. (a) 11 m/s **(b)** no **(c)** 7.7 m/s

83. (a) 2.7 m/s **(b)** 0.38 m **(c)** 29°

85. -7.4×10^2 J

87. 12 m/s

89. (b)

91. No, paying for energy because kWh is the unit of power $\times$ time = energy. 9.0×10^6 J

93. They are doing the same amount of work (same mass, same height). So the one that arrives first will have expended more power due to shorter time interval.

95. 97 W

97. 5.7×10^{-5} W

99. 6.0×10^2 J

101. (a) 5.5×10^2 W **(b)** 0.74 hp

103. 48.7%

105. 5.0×10^2 W

107. 0.536

109. 10 m

Chapter 6

1. (b)

3. (d)

5. Not necessarily. Even if the momentum is the same, different mass can still have different kinetic energies.

7. (a) 1.5×10^3 kg·m/s **(b)** zero

9. (a) 85 kg·m/s **(b)** 3.0×10^4 kg·m/s

11. 31 m/s

13. 4.05 kg·m/s in the direction opposite v_o

15. (a) 13 kg·m/s **(b)** 43 kg·m/s

17. $\Delta \vec{p} = (-3.0 \text{ kg·m/s}) \, \hat{\mathbf{y}}$

19. 16 N

21. 68 N

23. (a) 10.6 kg·m/s opposite v_o **(b)** 2.26×10^3 N

25. (a) 2.09 kg·m/s **(b)** 24.0 N upward

27. (c)

29. By stopping, the contact time is short. From the impulse momentum theorem ($F_{avg}\Delta t = \Delta p = mv - mv_o$), a shorter contact time will result in a greater force if all other factors (m, v_o, v) remain the same.

31. In (a), (b), and (c), it is to have greater contact time—less average force. This is because ($F_{avg}\Delta t = \Delta p = mv - mv_o$ so the greater the Δt, the less the $\overline{F}$. It can also decrease the pressure on the body because the force is spread over a larger area.

33. 6.0×10^3 N

35. (a) 1.2×10^3 N **(b)** 1.2×10^4 N

37. (a) (2) the driver putting on the brakes, because it is to reduce the speed. **(b)** 28.7 m/s

39. (a) Hitting it back requires a greater force. Force is proportional to the change in momentum. When a ball changes its direction, the change in momentum is greater. **(b)** 1.2×10^2 N in direction opposite v_o

41. 1.1×10^3 N, 4.7×10^2 N

43. 15 N upward

45. (a) 36 km/h, 56° north of east **(b)** 49% lost

47. 8.06 kg·m/s, 29.7° above the $-x$ axis

49. (d)

51. Air moves backward and the boat moves forward according to momentum conservation. If a sail were installed behind the fan on the boat, the boat would not go forward because the forces between the fan and the sail are internal forces of the system.

53. No, it is impossible. Before the hit, there is some initial momentum of the two-object system due to the one moving. According to momentum conservation, the system should also have momentum after the hit. Therefore it is not possible for both to be at rest (zero total system momentum).

55. moves 0.083 m/s in opposite direction

57. 1.08×10^3 s = 18.1 min

59. 7.6 m/s, 12° above $+x$-axis

61. (a) 45 km/h **(b)** 15 km/h **(c)** 105 km/h

63. 0.33 m/s

65. 0.78 m/s

67. (a) 4.0 m/s **(b)** 4.0 m/s

69. 82.8 m

71. (a) 9.7° **(b)** 0.10 m/s

73. (c)

75. (a)

77. This is due to the fact that momentum is a vector and kinetic energy is a scalar. For example, two objects of equal mass traveling with the same speed in opposite directions have positive total kinetic energy but zero total momentum. After they collide inelastically, both stop, resulting in zero total kinetic energy and zero total momentum. Therefore, kinetic energy is lost and momentum is conserved.

79. Yes, it can. For example, when two objects of equal masses are approaching each other with equal speeds, the total initial momentum is zero. After they collide, if they both are stationary, the total momentum is still zero. However, after the collision, the two-object system has no kinetic energy left.

81. $v_p = -1.8 \times 10^6$ m/s; $v_a = 1.2 \times 10^6$ m/s

83. $v_1 = -0.48$ m/s; $v_2 = +0.020$ m/s

85. $v_p = 38.2$ m/s, $v_c = 40.2$ m/s

87. 0.94 m

89. 1.1×10^2 J

91. (a) (1) south of east, according to momentum conservation. The initial momentum of the minivan is to the south, and the initial momentum of the car is to the east, so the two-vehicle system has a total momentum to the southeast after the collision. **(b)** 13.9 m/s, 53.1° south of east

93. no, 1.1 kg·m/s, 249°

95. (a) 28% **(b)** 1.3×10^7 m/s

97. $\theta_1 = 9.94°$, $\theta_2 = 19.8°$

99. (d)

101. The flamingo's center of mass is directly above the foot on the ground for it to be in equilibrium.

103. (a) (0, -0.45 m) **(b)** no, only that they are equidistant from CM due to the equal masses of the particles.

105. (a) 4.6×10^6 m from the center of the Earth **(b)** 1.8×10^6 m below the surface of the Earth

107. 82.8 m

109. The CM of both the square sheet and the circle are at the center of the square. So from symmetry, the CM of the remaining portion is still at center of sheet.

111. 0.175 m

113. (a) the 65 kg travels 3.3 m and the 45 kg travels 4.7 m **(b)** same distances as in part (a)

115. massive end: 5.59 m; less massive end: 6.19 m

117. (a) 0.222° **(b)** 199 m/s

119. (a) 4.65 m/s, $\theta = 2.13°$ **(b)** $K < K_o$, inelastic

Chapter 7

1. (c)

3. 2π rad = 360°, so 1 rad = 57.3°

5. (2.5 m, 53°)

7. 1.4×10^9 m

9. (a) 30° **(b)** 75° **(c)** 135° **(d)** 180°

11. (a) 4.00 rad **(b)** 229°

13. 10.7 rad

15. (a) 129° **(b)** about 1.4×10^4 km

17. 28.3 m

19. (a) 3.0×10^2 rad **(b)** 91 m

21. (b)

23. (d)

25. Viewing from opposite sides would give different circular senses, i.e., make clockwise counterclockwise and vice versa.

27. 21 rad/s to 47 rad/s

29. 1.8 s

31. particle B

33. (a) 0.84 rad/s **(b)** 3.4 m/s, 4.2 m/s

35. (a) The rotating angular speed is greater because the time is less (angular displacement is the same). **(b)** 7.27×10^{-5} rad/s for rotation, 1.99×10^{-7} rad/s for revolution

37. (a) 60 rad **(b)** 3.6×10^3 m

39. (d)

41. (d)

43. The floats of the little mass will move in direction of acceleration, inward. It works the same way as the accelerometer in Fig. 4.24. No, it does not make a difference since the centripetal acceleration is always inward.

45. Centripetal force is required for a car to maintain its circular path. When a car is on a banked turn, the horizontal component of the normal force on the car is pointing toward the center of the circular path. This component will enable the car to negotiate the turn even when there is no friction.

47. 1.3 m/s

49. 2.69×10^{-3} m/s^2

51. 11.3°

53. (a) weight is supplying the centripetal force **(b)** 3.1 m/s

55. 29.5 N, string will work

57. (a) $v = \sqrt{rg}$ **(b)** $h = (5/2)r$

61. (d)

63. Yes, a car in circular motion always has centripetal acceleration. Yes, it also has angular acceleration as its speed is increasing.

65. The answer is no. When the tangential acceleration increases, the tangential speed increases, resulting in an increase in centripetal acceleration.

67. $1.1 \times 10^{-3} \text{ rad/s}^2$

69. (a) (3) both angular and centripetal accelerations. There is always centripetal acceleration for any car in circular motion. When the car increases its speed on a circular track, there is also angular acceleration. (b) 53 s (c) $\vec{a} = -(8.5 \text{ m/s}^2)\,\hat{r} + (1.4 \text{ m/s}^2)\,\hat{t}$

71. 6.69 rad/s^2

73. (a) 2.45 rad/s^2 (b) 17.0 m/s^2 (c) 38.2 N

75. (d)

77. No, these terms are not correct. Gravity acts on the astronauts and the spacecraft, providing the necessary centripetal force for the orbit, so g is not zero and there is weight by definition ($w = mg$). The "floating" occurs because the spacecraft and astronauts are "falling" ("accelerating") toward Earth at the same rate).

79. Yes, if you also know the radius of the Earth. The acceleration due to gravity near the surface of the Earth can be written as $a_g = GM_E/R_E^2$. By simply measuring a_g, you can determine $M_E = a_g R_E^2/G$.

81. $2.0 \times 10^{20} \text{ N}$

83. $8.0 \times 10^{-10} \text{ N}$, toward opposite corner

85. $3.4 \times 10^5 \text{ m}$

87. 1.5 m/s^2

89. (a) $-2.5 \times 10^{-10} \text{ J}$ (b) 0

91. (c)

93. (d)

95. (a) Rockets are launched eastward to get more velocity relative to space because the Earth rotates toward the east. (b) The tangential speed of the Earth is higher in Florida because Florida is closer to the equator than California hence a greater distance from the axis of rotation. Also, the launch is over the ocean for safety.

97. (a) $3.7 \times 10^3 \text{ m/s}$ (b) 34%

99. $4.4 \times 10^{11} \text{ m}$

101. $1.53 \times 10^9 \text{ m}$

103. (a) 8.91 m/s^2, toward center (b) $1.40 \times 10^4 \text{ N}$, toward car (c) 2.27 m/s^2, opposite velocity (d) 9.19 m/s^2, $\theta = 75.7°$

105. 31 s (b) 19 rev

107. $2.97 \times 10^{30} \text{ kg}$

Chapter 8

1. (a)

3. (b)

5. (b)

7. If v is less than $R\omega$, the object is slipping. Yes, it is possible for v to be greater than $R\omega$ when the object is sliding.

9. At the nine-o'clock position, the velocity is straight upward. So it is a "free-fall" with an initial upward velocity. It will rise, reach a maximum height, and then fall back down.

11. 0.10 m

13. 1.7 rad/s

15. (a) 0.0331 rad/s^2 (b) $1.99 \times 10^{-3} \text{ m/s}^2$

17. (b)

19. (a)

21. This is to lower the center of gravity so the weight will have a shorter lever arm therefore smaller torque.

23. Frictional torque causes the motorcycle to rotate upward until balanced by the torque of weight.

25. $5.6 \times 10^2 \text{ N}$

27. $3.3 \times 10^2 \text{ N}$

29. (a) Yes, the seesaw can be balanced if the lever arms are appropriate for the weights of the children because torque is equal to force times the lever arm. (b) 2.3 m

33. $1.6 \times 10^2 \text{ N}$

35. (a) clockwise. (b) $4.66 \text{ m} \cdot \text{N}$

37. 0, $9.80 \text{ m} \cdot \text{N}$, $17.0 \text{ m} \cdot \text{N}$, $19.6 \text{ m} \cdot \text{N}$

39. (a) (2) toward the scale at the person's head, because the mass distribution of the human body is more toward the upper body than the lower body. (b) 0.87 m from the feet

41. Yes. The center of gravity of every stick is at or to the left of the edge of the table.

43. 1.2 m from left end of board

45. $T_1 = 21 \text{ N}$, $T_2 = 15 \text{ N}$

47. (a) (2) the board with the clown on it. At first glance, there seem to be two factors resulting in a greater tension in the rope. One is the extra weight of the crown. The other is the shortened lever arm of the tension in the rope when it is at an angle. However, the weight of the board and crown will also have shortened the lever arms by the same factor, so the effects cancel. (b) 172 N, 539 N

49. (d)

51. (a)

53. (a) Yes. Moment of inertia has a minimum value calculated about the center of mass. (b) No, mass would have to be negative.

55. The hard-boiled egg is a rigid body, while the raw egg is not.

57. This is to increase the moment of inertia. If walker starts to rotate (fall), the angular acceleration will be smaller, thus giving more time to recover.

59. $0.64 \text{ m} \cdot \text{N}$

61. (a) $2.4 \text{ kg} \cdot \text{m}^2$ (b) $0.27 \text{ kg} \cdot \text{m}^2$ (c) $2.4 \text{ kg} \cdot \text{m}^2$ (same)

63. 1.1 rad

65. 1.2 m/s^2

67. (a) $2.93 \text{ m} \cdot \text{N}$ (b) 58.7 N

69. (a) $1.5g$ (b) 67-cm position

71. 6.5 m/s^2

73. (a) (3) $7\mu_s/2$ (b) 63.8°

75. (c)

77. Yes. The rotational kinetic energy depends on the moment of inertia, which depends on both the mass and the mass distribution. Translational kinetic energy depends only on the mass.

79. According to the work-energy theorem, rotational work is required to produce a change in rotational kinetic energy. Rotational work (W) is done by a torque (τ) acting through an angular displacement (θ).

81. (a) 28 J (b) 14 W

83. $0.47 \text{ m} \cdot \text{N}$

85. 0.16 m

87. 78.5 N

89. cylinder goes higher by 7.1%

91. (a) $1.31 \times 10^8 \text{ J}$ (b) $1.46 \times 10^6 \text{ W}$

93. (a) 29% (b) 40% (c) 50%

95. (a) $v = \sqrt{gR}$ (b) $h = 2.7R$ (c) weightlessness

97. (d)

99. Walking toward the center decreases the moment of inertia and so increases the rotational speed.

101. In each case, the change in the wheel's angular momentum vector is compensated by the rotation of the person to conserve the total angular momentum, so the vertical angular momentum remains constant.

103. (a) zero (b) Linear kinetic energy is converted into rotational kinetic energy

105. 1.4 rad/s

107. $L_{rot} = 2.4 \times 10^{29} \text{ kg} \cdot \text{m}^2/\text{s}$; $L_{rev} = 2.8 \times 10^{34} \text{ kg} \cdot \text{m}^2/\text{s}$

109. 1.18 rad/s

111. (a) 4.3 rad/s (b) $K = 1.1K_o$ (c) work done by skater

113. $d = b(v_o/v)$

115. (a) (2) rotate in the direction opposite that in which the cat is walking, because from angular momentum conservation, the lazy Susan will rotate in the opposite direction. (b) 0.56 rad/s (c) no, 2.1 rad

117. (a) 33.1 rad/s (b) 5.29 m/s (c) 28.6 rad/s, 4.58 m/s

119. 0.104 m/s

Chapter 9

1. (c)

3. (d)

5. Steel wire has a greater Young's modulus. Young's modulus is a measure of the ratio of stress over strain. For a given stress, a greater Young's modulus will have a smaller strain. Steel will have smaller strain here.

7. Through capillary action, the wooden peg absorbs water and it swells and splits the rock.

9. $3.1 \times 10^4 \text{ N/m}^2$

11. (a) $9.4 \times 10^4 \text{ N/m}^2$ (b) $1.2 \times 10^5 \text{ N/m}^2$

13. 47 N

15. (a) (1) a cold day, because tracks expand when temperature increases. (b) $1.9 \times 10^5 \text{ N}$

17. (a) bends toward brass, because the stresses are the same for both, and brass has a smaller Young's modulus. Brass will have a greater strain $\Delta L/L_o$, so it will be compressed more. Therefore, the brass will be shorter than the copper. (b) brass: $\Delta L/L_o = 2.8 \times 10^{-3}$; copper: $\Delta L/L_o = 2.3 \times 10^{-3}$

19. $4.2 \times 10^{-7} \text{ m}$

21. (a) Ethyl alcohol has the greatest compressibility, because it has the smallest bulk modulus B. The smaller the B, the greater the compressibility. (b) $\Delta p_w/\Delta p_{ea} = 2.2$

23. $7.28 \times 10^5 \text{ N/m}^2$

25. (d)

27. (a)

29. Bicycle tires have a much smaller contact area with the ground so they need a higher pressure to balance the weight of the bicycle and the rider.

31. (a) The pressure is determined by only depth so the effect is none—same depth, same pressure. (b) The dams are usually thicker at the bottom because of the greater pressure at greater depth.

33. (a) The pressure inside the can is equal to the atmospheric pressure outside. When the liquid is poured from an unvented can, a partial vacuum develops inside, and the pressure difference causes the pouring to be difficult. By opening the vent, you are allowing air to go into the can and the pressures are equalized so the liquid can be easily poured. (b) When you squeeze a medicine dropper before inserting it into a liquid, you are forcing the air out and reducing the pressure inside the dropper. When you release the top with the dropper in a liquid, the liquid rises in the dropper due to atmospheric pressure. (c) To inhale, the lungs physically expand, the internal pressure decreases, and air flows into the lungs. To exhale, the lungs contract, the internal pressure increases, and air is forced out.

35. (a) (1) a higher height than the mercury barometer due to its low density. (b) 10 m

37. (a) (1) a higher column height because it has a lower density. (b) 22 cm

39. $6.39 \times 10^{-4}\,\text{m}^2$

41. (a) (3) A lower pressure inside the can as steam condenses. Since there is a partial vacuum inside the can, the net force exerted by the atmospheric pressure on the outside crushes the can. (b) $1.6 \times 10^4\,\text{N} = 3600\,\text{lb}$

43. $p = -1.1 \times 10^4\,\text{Pa}$. Obviously, pressure cannot be negative, so this calculation shows that the air density is not a constant, but air density decreases rapidly with altitude.

45. 0.51 N

47. $2.2 \times 10^5\,\text{N}$ (about 50000 lb)

49. $1.9 \times 10^2\,\text{m/s}$

51. (a) $1.1 \times 10^8\,\text{Pa}$ (b) $1.9 \times 10^6\,\text{N}$

53. 549 N, $1.37 \times 10^6\,\text{Pa}$

55. 0.173 N

57. (c)

59. (a)

61. Ice has a larger volume than water of equal mass. The level does not change. As the ice melts, the volume of the newly converted water decreases; however, the ice, which was initially above the water surface, is now under the water. This compensates for the decrease in volume. It does not matter whether the ice is hollow or not.

63. Same buoyant force, because buoyant force depends only on the volume of the fluids displaced and is independent of the mass of the object.

65. (a) (3) stay at any height in the fluid, because the weight is exactly balanced by the buoyant force. Wherever the object is placed, it stays at that place. (b) sink, $W > F_b$

67. $2.6 \times 10^3\,\text{kg}$

69. no, $14.5 \times 10^3\,\text{kg/m}^3 < \rho_g = 19.3 \times 10^3\,\text{kg/m}^3$

71. (a) 0.09 m (b) 8.1 kg

73. $1.00 \times 10^3\,\text{kg/m}^3$ (probably H_2O)

75. 17.7 m

77. $8.1 \times 10^2\,\text{N}$

79. (a)

81. (d)

83. There are many capillaries and only a few arteries. The total area of the capillaries is greater than that of the arteries. So if A increases, v decreases.

85. (a) The concave bottom makes the air travel faster under the car. This increase in speed will reduce the pressure under the car. The pressure difference forces the car on to the ground more to provide a greater normal force and friction for traction. (b) The spoiler is to produce a downward force for better traction or more friction.

87. 0.98 m/s

89. (a) $3.5\,\text{cm}^3/\text{s}$ (b) 0.031% (c) It is a physiological need. The slow speed is needed to give time for the exchange of substances such as oxygen between the blood and the tissues.

91. 53.6 Pa

93. (a) $0.13\,\text{m}^3/\text{s}$ (b) 1.8 m/s

95. 2.2 Pa

99. (c)

101. The 10 and 40 measure viscosity and the "W" means winter.

103. $3.5 \times 10^2\,\text{Pa}$

105. 13.5 s

107. $8.0 \times 10^2\,\text{kg/m}^3$

109. (a) 3.83 m/s (b) 399

Chapter 10

1. (b)

3. (a)

5. incandescent lamp filament, up to 3000°C

7. Celsius

9. (a) 302°F (b) 90°F (c) −13°F (d) −459°F

11. (a) 245°F (b) 375°F

13. 56.7°C and −62°C

15. (a) (3) $T_F = T_C$, because we want to find the one temperature at which the Celsius and Fahrenheit scales have the same reading. (b) −40°C = −40°F

17. (a) −101 F° (b) 558 F°

19. (a) (2) $T_C = 0$, because $T_F = 9/5\,T_C + 32$, compared to $y = ax + b$. To find the y-intercept, we set $x\,(T_C) = 0$. (b) 32°F (c) 5/9; −18°C

21. (a)

23. The volume of the gas is held constant. So if the temperature increases, so does the pressure and vice versa, according to the ideal gas law. Therefore, temperature can be determined by measuring pressure.

25. Absolute zero implies zero pressure or volume. Negative absolute temperature implies negative pressure or volume.

27. The same, because mole is defined in terms of the number of molecules.

29. (a) −273°C (b) −23°C (c) 0°C (d) 52°C

31. (a) 53541°F; 29727°C (b) 0.910%

33. (a) (2) decrease, because $p_1V_1/T_1 = p_2V_2/T_2$. With $V_1 = V_2$, $T_2/T_1 = p_2/p_1$ or temperature is proportional to pressure. (b) 167°C

35. 1.7×10^{23}

37. $0.0370\,\text{m}^3$

39. 0.16 L

41. $33.4\,\text{lb/in.}^2$

43. 2.31 atm

45. (a) (1) increase, because with $p = p_o$, $p_oV_o/T_o = pV/T$ becomes $V/V_o = Tp_o/(T_op) = T/T_o$ or volume is proportional to temperature. (b) 10.6%

47. $5.1\,\text{cm}^3$

49. (c)

51. (a) ice moves upward (b) ice moves downward (c) copper

53. When the ball alone is heated, it expands and cannot go through the ring. When the ring is heated, it expands and the hole gets larger so the ball can go through again.

55. Metal has a higher coefficient of thermal expansion than glass. The lid expands more than glass so it is easier to loosen the lid.

57. (a) (1) high, because the tape shrinks. One division on the tape (it is now less than one division due to shrinkage) still reads one division. (b) 0.060%

59. 0.0027 cm

61. (a) 60.1 cm (b) $3.91 \times 10^{-3}\,\text{cm}^2$, yes

63. (a) (1) the ring, so it expands, then the ball can go through. (b) 353°C

65. $5.52 \times 10^{-4}/\text{C}°$

67. (a) larger, because it expands. (b) $5.5 \times 10^{-6}\,\text{m}^3$

69. (a) 116°C (b) no

71. yes, 79°C

73. (a)

75. The gases diffuse through the porous membrane, but the helium gas diffuses faster because its atoms have a smaller mass. Eventually, there will be equal concentrations of gases on both sides of the container.

77. (a) $6.07 \times 10^{-21}\,\text{J}$ (b) $7.72 \times 10^{-21}\,\text{J}$

79. (a) $6.21 \times 10^{-21}\,\text{J}$ (b) $1.37 \times 10^3\,\text{m/s}$

81. increases by a factor of $\sqrt{2}$

83. (a) $1.82 \times 10^7\,\text{J}$ (b) $1.55 \times 10^3\,\text{m}$

85. 273°C

87. 899°C

89. (a) (1) $^{235}\text{UF}_6$, because at the same temperature, the smaller the mass, the greater the average rms speed. (b) 1.00429

91. (b)

93. nRT

95. $6.1 \times 10^3\,\text{J}$

97. (a) $1.21 \times 10^7\,\text{J}$ (b) $3.03 \times 10^7\,\text{J}$

99. 0.272%

101. (a) (3) helium, because at the same temperature, helium has the smallest mass so the highest rms speed. (b) 425 m/s < 1100 m/s

Chapter 11

1. (d)

3. 1 Cal = 1000 cal

5. $6.279 \times 10^6\,\text{J}$

7. 4

9. (a)

11. (b)

13. Since $Q = cm\Delta T = cm(T_f - T_i)$, specific heat and mass can cause the final temperature of the two objects to be different, if Q and T_i are the same.

15. (a) (1) more heat, because copper has a higher specific heat. (b) copper requires $2.1 \times 10^4\,\text{J}$ more

17. $1.7 \times 10^6\,\text{J}$

19. (a) (3) less than, because aluminum has a higher specific heat than copper. From $Q = mc\Delta T$, if Q and ΔT are the same, a higher specific heat results in a lower mass. (b) 1.27 kg

21. 84°C

23. 0.13 kg

25. **(a)** (1) more heat than the iron, because aluminum has a higher specific heat, when all other factors (m and ΔT) are the same. **(b)** Al by 1.8×10^4 J more

27. **(a)** (1) higher, because if some water splashed out, there will be less water to absorb the heat. The final temperature will be higher and the measured specific heat value will be in error and appear to be higher than the value calculated for the case in which the water does not splash out. **(b)** 3.1×10^2 J/(kg·C°)

29. **(a)** 7.0×10^2 W **(b)** 9.4×10^2 W

31. 20.0°C

33. **(d)**

35. **(c)**

37. This is due to the high value of the latent heat of vaporization. When steam condenses, it releases 2.26×10^6 J/kg of heat. When 100°C water drops its temperature by 1 C°, it releases only 4186 J/kg.

39. 1.13×10^6 J

41. 2.5×10^6 J and 2.5×10^5 J; yes

43. 1.2×10^6 J

45. **(a)** (2) only latent heat, because the boiling point of mercury is 357°C = 630 K so it is already at the boiling temperature. **(b)** 4.1×10^3 J

47. 11°C

49. 1.8×10^{-2} kg

51. 2.1×10^5 J

53. **(a)** (2) some of the ice will melt. Answer (3) is eliminated because water at 10°C will give off enough heat to raise the temperature of the ice, and melt some of the ice because ice has a smaller specific heat than water. Answer (1) is also eliminated because of the high value of the latent heat of fusion for ice (3.3×10^5 J/kg). The 10 C° drop in temperature of the water does not release enough heat to melt the ice completely. **(b)** 0.5032 kg

55. 0.17 L

57. **(c)**

59. Metal has a higher heat conductivity, so metal conducts heat away from your hand more quickly.

61. Air is a poor heat conductor so the hollow hair minimizes heat loss.

63. 4.54×10^6 J

65. 13 J

67. **(a)** (1) longer, because copper has a higher thermal conductivity **(b)** 1.63

69. **(a)** 5.5×10^5 J/s **(b)** 73 kg; yes, see ISM

71. 411°C

73. **(a)** (3) thinner than, because glass wool has a lower thermal conductivity. **(b)** 4.9 in.

75. 1.0×10^8 J

77. **(a)** 1.5×10^3 J/s **(b)** 46 J/s

79. 2.3 cm

81. 23°C

83. 7.8 h

85. 4.0×10^2 m/s

87. 0.49 kg

Chapter 12

1. **(a)**

3. **(d)**

5. No. All it means is that the intermediate steps are non-equilibrium states so you cannot retrace the process exactly.

7. **(c)**

9. 1: isothermal expansion; 2: isobaric compression; 3: isometric pressure increase

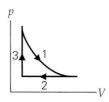

11. When you play a game of basketball, you lost heat, did work, and decreased your internal energy.

13. Work: 1, 2, 3. Work is equal to the area under the curve in p-V diagram. The area under 1 is the greatest and the area under 3 is the smallest. Final temperature: 1, 2, 3. According to the ideal gas law, the temperature of a gas is proportional to the product of pressure and volume, $pV = nRT$. Since the final volume is the same for all three processes, the higher the pressure, the higher the final temperature.

15. **(a)** (2) the same as, because $\Delta U = 0$ for a cyclic process. **(b)** added, 400 J

17. **(a)** (3) decreases, as $Q = 0$ and W is positive **(b)** −500 J

19. **(a)** 3.3×10^3 J **(b)** yes, 5.1×10^3 J

21. **(a)** (2) zero, because $\Delta T = 0$. **(b)** $-p_1 V_1$ (on the gas) **(c)** $-p_1 V_1$ (out of the gas)

23. 3.6×10^4 J

25. **(a)** (2) isobaric, because the pressure is maintained at 1.00 atm. **(b)** 146 J

27. **(a)** (2) process 2, because more heat is added and the change in internal energy is the same (the initial and final temperatures are the same). **(b)** $\Delta U = 2.5 \times 10^3$ J for both; $W_1 = 0$, $W_2 = 5.0 \times 10^2$ J

29. **(a)** Path AB, -1.66×10^3 J; path BC, 0; path CD, 3.31×10^3 J; path DA, 0 **(b)** $\Delta U = 0$, $Q = W = 1.65 \times 10^3$ J **(c)** 800 K

31. **(c)**

33. **(a)** increases since heat is added. **(b)** decreases since heat is removed. **(c)** increases since heat is added. **(d)** decreases since heat is removed.

35. No, this is not a valid challenge, because ice or water itself is not an isolated system. When water freezes into ice, it gives off heat and that causes the entropy of the surroundings to increase. This increase actually is more than the decrease that occurred in the water-ice phase change. So the net change in entropy of the system (ice plus water) still increases.

37. **(a)** (1) positive, because heat is added in the process (positive heat). **(b)** $+1.2 \times 10^3$ J/K

39. -2.1×10^2 J/K

41. 126°C

43. **(a)** (1) increase, because $Q > 0$. **(b)** +11 J/K

45. **(a)** (1) positive, according to the second law of thermodynamics. **(b)** +1.33 J/K

47. **(a)** (2) zero, $\Delta S = 0$ **(b)** 2.73×10^4 J

49. **(a)** 61.0 J/K **(b)** −57.8 J/K **(c)** 3.2 J/K

51. **(b)**

53. It is unchanged, because it returns to its original value for a cyclic process. This is true for many quantities such as temperature, pressure, and volume.

55. This is important because heat can be completely converted to work for a single process (not a cycle), such as an isothermal expansion process of an ideal gas.

57. No, as the warm air rises to the higher altitude, both gravity and buoyancy forces do work. Since it is a natural process with work input, the entropy increases and the second law is not violated.

59. 25%

61. 1.47×10^5 J

63. **(a)** 6.6×10^8 J **(b)** 27%

65. **(a)** (1) increases, because $\varepsilon = 1 - Q_c/Q_h$. When ε increases, the ratio of Q_c/Q_h decreases or the ratio Q_h/Q_c increases. **(b)** +0.024

67. **(a)** 6.1×10^5 J **(b)** 1.9×10^6 J

69. 3.0 kW

71. 6.0 h

73. **(a)** 1800 **(b)** 3.4×10^7 J **(c)** 2.7×10^7 J

75. **(a)**

77. The more efficient one is water cooled, because efficiency of cooling depends on ΔT, and water can maintain a large ΔT. Also water has high specific heat so it can absorb more heat.

79. Diesel engines run hotter because diesel fuel has a higher spontaneous combustion temperature. According to Carnot efficiency, the higher the hot reservoir temperature, the higher the efficiency, for a fixed low temperature reservoir.

81. 0°C

83. **(a)** 6.7% **(b)** Probably not at the moment, due to low efficiency and still relatively cheap fossil fuels.

85. 9.1×10^3 J

87. **(a)** (3) higher than 327°C. From $\varepsilon_C = 1 - T_c/T_h$, we can see that T_h must be higher for ε_C to increase while T_c is kept constant. **(b)** 427°C

89. **(a)** no, $\varepsilon_C = 57\%$ while $\varepsilon = 67\%$ **(b)** 17.5 kW

91. **(a)** 42% **(b)** 39 kW

93. 53%

95. **(a)** (2) zero, because many quantities such as temperature, pressure, volume, internal energy, and entropy return to original value after each cycle. **(b)** 3750 J

97. **(a)** 64% **(b)** ε_C is the upper limit of efficiency. In reality, a lot more energy is lost than in the ideal situation.

99. **(a)** 13 **(b)** no, $COP_C = 11$

101. 20 mi/gal

103. 2.7 kg

105. 0.157

Chapter 13

1. **(b)**

3. **(b)**

5. **(a)** four times as large **(b)** twice as large

7. $T/4, T/2$

9. $4A$

11. 0.025 s

13. 41 N/m

15. (a) 10^{-12} s (b) 63 m/s

17. (a) (1) $x = 0$, because at $x = 0$ there is no elastic potential energy, so all the energy of the system is kinetic, thus maximum speed. (b) 2.0 m/s

19. (a) 0.77 m/s (b) 1.2 N

21. 1.08 m/s

23. (a) 2.5 m/s (b) 2.5 m/s (c) 2.7 m/s, equilibrium position

25. (a) 17.6 N/m (b) 1.04 m/s

27. (d)

29. (b)

31. This could be done by tracing out the path of the object on a scrolling horizontal paper.

33. In an upward accelerating elevator, the effective gravitational acceleration increases. According to $T = 2\pi\sqrt{L/g}$, the period would be decreased.

35. 10 kg

37. (a) 1.7 s (b) 0.57 Hz

39. (a) $x = A \sin \omega t$ (b) $x = A \cos \omega t$

41. (a) 5.0 cm (b) 10 Hz (c) 0.10 s

43. (a) (3) less, because $E \propto 1/T^2$ and System A has a longer period. (b) 1.8×10^2

45. (a) 0.188 m (b) 3.00 m/s^2

47. (a) (3) $1/\sqrt{3}$, because $T = 2\pi\sqrt{m/k}$ so $T_2/T_1 = \sqrt{k_1/k_2} = \sqrt{1/3}$ (b) 2.8 s

49. (a) using the period of vibration (b) 76 kg

51. 2.70 J

53. 0.279 m/s, 0.897 m/s^2 = $(0.0915)g$

55. (a) (1) increase, because $T = 2\pi\sqrt{L/g}$, a smaller g will have a longer T. (b) 4.9 s

57. (a) $y = (-0.10\,\text{m})\sin(10\pi/3)t$ (b) $k = 27$ N/m

59. (a) 1.21 m (b) 0.301 m/s (c) 0.248 rad/s

61. (d)

63. (c)

65. The one on the top is transverse and the one on the bottom is longitudinal.

67. 0.340 m

69. 0.47 m/s

71. 1.7 cm to 17 m

73. No, it is not in vacuum.
$v = \lambda f = (500 \times 10^{-9}\,\text{m})(4.00 \times 10^{14}\,\text{Hz})$
$= 2.00 \times 10^8$ m/s $< 3.00 \times 10^8$ m/s.

75. 6.00 km

77. (a) 3.8×10^2 s (b) yes,
1.9×10^3 km > 30 km (c) 1.6×10^3 s

79. (a) 0.20 s (b) 0.40 s

81. (d)

83. (d)

85. Reflection (this is called echolocation), because the sound is reflected by the prey.

87. (d)

89. (c)

91. (a) This is caused by the glass vibrating in a resonance mode. (b) The frequency will increase because the wavelength will decrease due to the shortened air column. This results in an increase in frequency as $v = \lambda f$ and v is a constant.

93. A thinner string will sound higher frequency. Since $v = \sqrt{F_\text{T}/\mu}$, and a thinner string has a smaller μ, the speed is higher so is frequency $(v = \lambda f)$.

95. 150 Hz

97. (a) (1) increases by $\sqrt{2}$, because $v = \sqrt{F_\text{T}/\mu}$, so $v_2/v_1 = \sqrt{F_{\text{T}2}/F_{\text{T}1}} = \sqrt{2/1} = \sqrt{2}$.
(b) 8.49 m/s (c) $f_n = (0.425)n$ Hz; $n = 1, 2, 3, \ldots$

99. $n = 5$

101. 16.5 N

103. 1/4

105. 0.016 kg

107. (a) released form rest, downward (b) 1.05 s (c) $y = (0.100\,\text{m})\cos(6t)$ (d) 3.6 m/s^2

109. (a) 5.00 N (b) 12.5 Hz (c) 0.40 m from one end

111. 3.0 s

Chapter 14

1. (b)

3. (a)

5. Some insects produce sounds with frequencies that are not all in our audible range.

7. They arrive at the same time because sound is not dispersive, i.e., speed does not depend on frequency.

9. (a) 1.0 km (b) 0.60 mi

11. 32°C

13. The unit of v in a liquid is
$$\sqrt{\frac{\text{N/m}^2}{\text{kg/m}^3}} = \sqrt{\frac{\text{N}\cdot\text{m}}{\text{kg}}} = \sqrt{\frac{\text{kg}\cdot\text{m}^2/\text{s}^2}{\text{kg}}}$$
$$= \sqrt{\frac{\text{m}^2}{\text{s}^2}} = \text{m/s}.\ Y \text{ has the same unit as } B,$$
so the unit of v in a solid is also m/s.

15. (a) (1) increases), because the speed of sound increases with temperature and $v = \lambda f$. So if v increases and f remains the same, λ increases. (b) +0.047 m

17. (a) 7.5×10^{-5} m (b) 1.5×10^{-2} m

19. (a) 1.08 s (b) 1.04 s

21. 90 m

23. (a) 0.107 m (b) 1.43×10^{-4} s (c) 4.29×10^{-4} m

25. (a) (1) less than double, because the total time is the sum of the time it takes for the stone to hit the ground (free fall motion) and the time it takes sound to travel back that distance. While the time for sound is directly proportional to the distance, the time for free fall is not. Since $d = \frac{1}{2}gt^2$ or $t = \sqrt{2d/g}$ (see Chapter 2), doubling the distance d will only increase the time of fall by a factor of $\sqrt{2}$. (b) 1.0×10^2 m (c) 8.7 s

27. 4.5%

29. (b)

31. (1) by a factor of 2

33. Yes. Since $\beta = 10 \log I/I_\text{o}$ and $\log x < 0$ for $x < 1$, if $I < I_\text{o}$. So for an intensity below the intensity of threshold of hearing, β is negative.

35. (a) (4) 1/9, because I is inversely proportional to the square of R. Tripling R will reduce I to $1/3^2 = 1/9$. (b) 1.4 times

37. (a) 3.0 Hz (b) not enough information given

39. (a) 100 dB (b) 60 dB (c) -30 dB

41. (a) 1.1×10^{-5} W/m^2

43. (a) 3.72×10^{-4} W/m^2; 1.00×10^{-1} W/m^2 (b) 9.55×10^{-3} W/m^2; 6.03×10^{-2} W/m^2

45. (a) 2.82 m (b) 2.82×10^3 m, unreasonable

47. five

49. 10 bands

51. $I_\text{B} = 0.563I_\text{A}$, $I_\text{C} = 0.250I_\text{A}$, $I_\text{D} = 0.173I_\text{A}$

53. (a) 2.5×10^2 m (b) 2.5×10^5 m away from the charge

55. 10^5 bees

57. (a)

59. No. The beat of music has to do with tempo. Beats are physical phenomena related to the frequency difference between two tones.

61. The varying sound intensity is caused by the interference effect. At certain locations there is constructive interference and at other locations, there is destructive interference.

63. 0.172 m

65. (a) (4) both (1) and (3), because the beat frequency measures only the frequency difference between the two, and it does not specify which frequency is higher. So the frequency of the violin can be either higher or lower than that of the instrument. (b) 267 Hz or 261 Hz

67. (a) (1) moving toward the siren, because the frequency heard is higher than the siren's. (b) 14 m/s

69. 3.3 Hz

71. (a) (2) less than 300 Hz, because the observer and the source are moving toward each other and the frequency heard is greater than the source frequency. (b) 251 Hz

73. 30°, yes

75. (a) 2.0 (b) 638 m/s

77. (a) 103 Hz approaching, 97.0 Hz receding (b) 6 Hz

79. (a) 36.3 kHz (b) 37.6 kHz (c) yes

81. (b)

83. (d)

85. (a) The snow absorbs sound so there is little reflection. (b) In an empty room, there is less absorption. So the reflections die out more slowly; and therefore, the sound seems hollow and echoing. (c) Sound is reflected by the shower walls, and standing waves are set up, giving rise to more harmonics and therefore richer sound quality.

87. For an open pipe, $f_n = nv/(2L)$ for $n = 1, 2, 3, \ldots$. For a closed pipe, $f_m = mv/(4L)$ for $m = 1, 3, 5, \ldots$. So $f_n/f_m = (n/2)/(m/4) = 2n/m$. For $f_n = f_m$, $2n = m$. No, this is not possible because $2n$ is an even integer and m is an odd integer.

89. For a pipe closed at one end, the closed end must be a node. So only odd harmonics are possible.

91. 510 Hz

93. (a) f_2 does not exist, only odd harmonics (b) 0.30 m

95. (a) 0.635 m (b) 265 Hz

97. 0°C

99. $f_\text{air} = 552$ Hz, $f_\text{He} = 1610$ Hz

101. 0.249 m and 0.251 m

103. 5.55×10^4 Hz

105. (a) (1) yes, because the observer will hear beats between the source and the reflection. (b) 1.03×10^3 Hz (c) 12 Hz

Chapter 15

1. (c)

3. (c)

5. No. Charges are simply moved from the object to another object.

7. Bring the charged object near the electroscope and observe how the leaves move. If the repulsion between the leaves increases, the charge on the object has the same sign as the one on the electroscope; if the repulsion between the leaves decreases, then the charge on the object has opposite sign as the one on the electroscope.

9. -1.6×10^{-13} C

11. $+6.40 \times 10^{-19}$ C

13. (a) (1) positive, because of the conservation of charge. When one object becomes negatively charged, it gains electrons. These same electrons must be lost by another object, and therefore it is positively charged. **(b)** $+4.8 \times 10^{-9}$ C, 2.7×10^{-20} kg **(c)** 2.7×10^{-20} kg

15. (d)

17. This is to remove excess charge due to friction of rubber on road. If the excess charge is not removed, this could result in a spark, causing a gasoline explosion.

19. If you bring a negatively charged object near the electroscope, the induction process will charge the electroscope with positive charges. You can prove the charges are positive by bringing the negatively charged object near the leaves and see if the leaves are attracted by the negatively charged object.

21. (a)

23. Although the electric force is fundamentally much stronger than the gravitational force, both the Earth, our bodies, and other objects are electrically neutral so there are no noticeable electric forces.

25. 9

27. (a) 1 **(b)** 1/4 **(c)** 1/2

29. (a) 5.8×10^{-11} N **(b)** zero

31. 2.24 m

33. (a) 50 cm **(b)** 50 cm

35. (a) $x = 0.25$ m **(b)** nowhere **(c)** $x = -0.94$ m for $\pm q_3$

37. (a) 8.2×10^{-8} N **(b)** 2.2×10^{6} m/s **(c)** 9.2×10^{21} g

39. (a) 96 N, 39° below positive x-axis **(b)** 61 N, 84° above negative x-axis

41. (c)

43. (a)

45. It is determined by the relative density or spacing of the field lines. The closer the lines, the greater the magnitude.

47. If a positive charge is at center of the spherical shell, the electric field is *not* zero inside. The field lines run radially outward to the inside surface of the shell where they stop at the induced negative charges on this surface. The field lines reappear on the outside shell surface (positively charged) and continue radially outward as if emanating from the point charge at the center. If the charge were negative, the field lines would reverse their directions.

49. (a) Yes, this is possible. For example, when the electric fields created by the two charges are equal in magnitude and opposite in direction at some locations. At the midway point in between and along a line joining two charges of the same type and magnitude, the electric field is zero. **(b)** No, this is not possible.

51. 2.0×10^5 N/C

53. 1.2×10^{-7} m away from the charge

55. 1.0×10^{-7} N/C upward, 5.6×10^{-11} N/C downward

57. $\vec{E} = (2.2 \times 10^5\,\text{N/C})\hat{x} + (-4.1 \times 10^5\,\text{N/C})\hat{y}$

59. 5.4×10^6 N/C toward the charge of $-4.0\ \mu$C

61. 3.8×10^7 N/C in the $+y$-direction

63. $15\ \mu$C/m^2

65. $\vec{E} = (-4.4 \times 10^6\,\text{N/C})\hat{x} + (7.3 \times 10^7\,\text{N/C})\hat{y}$

67. (b)

69. (b)

71. The surface must be spherical.

73. (a) (1) negative due to induction **(b)** zero **(c)** $+Q$ **(d)** $-Q$ **(e)** $+Q$

75. (a) zero **(b)** kQ/r^2 **(c)** zero **(d)** kQ/r^2

77.

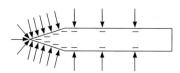

79. (c)

81. Since the number of lines is proportional to charge, the net charges are equal and opposite in sign.

83. -6 lines, or net of 6 lines entering it

85. 10 field lines entering (negative)

87. (a) bottom **(b)** 4.90×10^{-4} kg

89. (a) positive on right plate and negative on left plate **(b)** from right to left **(c)** 1.13×10^{-13} C

91. 5.5×10^3 N/C at 66° below positive x-axis

93. (a) 2.02×10^4 N/C **(b)** 1.79×10^{-7} C/m^2

Chapter 16

1. (d)

3. (b)

5. (a) The kinetic energy of the approaching proton decreases as its electric potential energy increases since its total energy is constant. **(b)** The electric potential energy of the system increases because the distance between the charges decreases. **(c)** The total energy of the system remains the same because of the conservation of energy.

7. It will move to the right or toward the higher potential region because the electron has negative charge. The higher the potential region, for the electron, the lower the potential energy.

9. It takes zero work. Since $W = q_o \Delta V$, if $\Delta V = 0$, $W = 0$.

11. (a) 2.7 μC **(b)** negative to positive

13. 1.6×10^{-15} J

15. (a) 5.9×10^5 m/s, down **(b)** lose potential energy

17. (a) (2) 3, because electric potential is inversely proportional to the distance. **(b)** 0.90 m **(c)** -6.7 kV

19. (a) gains 6.2×10^{-19} J **(b)** loses 6.2×10^{-19} J **(c)** gains 4.8×10^{-19} J

21. 1.1 J

23. (a) $+0.27$ J **(b)** no

25. -0.72 J

27. (a) 3.1×10^5 V **(b)** 2.1×10^5 V

29. (a) (3) a lower, because electrons have a negative charge. They move toward higher potential regions where they have lower potential energy. **(b)** 4.2×10^7 m/s **(c)** 6.0×10^{-9} s

31. (b)

33. The ball would accelerate in the direction from the beach to the ocean (from higher potential energy to lower potential energy).

Higher gravitational potential energy

Beach ————————

———————— | Ball accelerating

Ocean ————————

Lower gravitational potential energy

35. It takes zero work, because there is no change in kinetic or potential energy. The net work is zero.

37. (a) cylindrical **(b)** near the outer surface **(c)** near the inner surface

39. (a) 1.60×10^{-13} J **(b)** it would double

41. 12.6 m

43. 70 cm

45. 1.7 mm away from the positive plate, toward the negative plate

47. (a) (1) concentric spheres, because the electric potential depends only on the distance from the charge. **(b)** $+298$ eV

49. (a) 2.0×10^7 eV **(b)** 2.0×10^4 keV **(c)** 20 MeV **(d)** 2.0×10^{-2} GeV **(e)** 3.2×10^{-12} J

51. 6.2×10^7 m/s (proton), 4.4×10^7 m/s (alpha)

53. (a) 3.5 V, 1.1×10^6 m/s **(b)** 4.1 kV, 3.8×10^7 m/s **(c)** 5.0 kV, 4.2×10^7 m/s

55. (a) $+0.40$ V **(b)** -0.40 V **(c)** zero

57. (c)

59. (a)

61. (b)

63. (a) Since $Q = CV$, it doubles. **(b)** Since $U_C = \frac{1}{2}CV^2$, it quadruples.

65. 2.4×10^{-5} C

67. 0.71 mm

69. (a) 4.2×10^{-9} C **(b)** 2.5×10^{-8} J

71. 2.2 V

73. (a) 2.2×10^4 V/m **(b)** 1.1×10^{-5} C **(c)** 5.7×10^{-4} J **(d)** $E = 6.7 \times 10^4$ V/m, $\Delta Q = 0$, $\Delta U_C = -1.7 \times 10^{-3}$ J

75. (d)

77. We cannot maintain a nonzero voltage on a conductor; charges will move from the positive to the negative immediately so they cannot be stored.

79. When the power supply is not connected, the charge remains the same but the capacitance increases once the dielectric material is inserted. Therefore the potential difference decreases ($V = Q/C$) and so does the electric field ($E = V/d$). When the power supply remains connected, the potential difference remains constant, so does the electric field.

81. 3.1×10^{-9} C; 3.7×10^{-8} J
83. **(a)** $\kappa = 2.4$ **(b)** decreased **(c)** -6.3×10^{-5} J
85. **(b)**
87. **(b)**
89. They have the same charge when they have equal capacitance.
91. **(a)** C/N **(b)** NC **(c)** $4C/N$
93. **(a)** (1) more, because the equivalent capacitance is higher and the energy stored (drawn) is proportional to capacitance. **(b)** 6.0 μF
95. **(a)** (3) $Q/3$, because $Q_{total} = Q_1 + Q_2 + Q_3$. Also $Q_1 = Q_2 = Q_3$ because the capacitors have the same capacitance. Therefore each capacitor has only 1/3 of the total charge. **(b)** 3.0 μC **(c)** 9.0 μC
97. max. 6.5 μF; min. 0.67 μF
99. C_1: 2.4 μC, 6.0 V; C_2: 2.4 μC, 6.0 V; C_3: 1.2 μC, 6.0 V; C_4: 3.6 μC, 6.0 V
101. **(a)** $K_o = 29.2$ eV, total $\Delta U = 75$ eV, so it can't **(b)** 30.6 cm from bottom
103. **(a)** -1.7×10^{-17} J **(b)** 6.9×10^{23} m/s^2 **(c)** -8.5×10^{-18} J **(d)** 8.5×10^{-18} J
105. **(a)** (1) a higher potential, because the electron has negative charge. It will experience an upward force if the potential is higher at the top. **(b)** 8.37×10^{-13} V **(c)** any location
107. **(a)** 2.9 pF **(b)** 0.20 pC

Chapter 17

1. **(b)**
3. **(b)**
5. **(c)**
7. No. Any battery has internal resistance, and there will be a voltage across the internal resistance when the battery is in use. The terminal voltage is lower than the emf of the battery when it is in use.
9. **(a)** 4.5 V **(b)** 1.5 V
11. **(a)** 24 V **(b)** two 6.0-V in series, together in parallel with the 12-V
13. **(a)** (2) the same, because the total voltage of identical batteries in parallel is the same as the voltage of each individual battery, and the total voltage of the batteries in series is the sum of the voltages of each individual battery. Each arrangement has one parallel and one series so they have the same total voltage. **(b)** 3.0 V, 3.0 V
15. **(a)**
17. **(a)** upward **(b)** downward **(c)** upward
19. 0.25 A
21. **(a)** 0.30 C **(b)** 0.90 J
23. 56 s
25. **(a)** (2) to the left, because the current due to the protons will be to the left, and the current due to the electrons will also be to the left because electrons have negative charge. **(b)** 3.3 A
27. **(a)**
29. **(a)**
31. From $V = (R)I$ ($y = mx$ is the equation for a straight line where m is the slope) we conclude that the one with the shallower slope is less resistive.

33. **(a)** same **(b)** one-quarter the current
35. **(a)** 11.4 V **(b)** 0.32 Ω
37. **(a)** (1) a greater diameter, because aluminum has a higher resistivity. Its area (diameter) must be greater, if the length of the wire is the same, to have the same resistance as copper according to $R = \rho L/A$. **(b)** 1.29
39. 1.0 V
41. 1.3×10^{-2} Ω
43. **(a)** 4 **(b)** 4
45. **(a)** 0.13 Ω **(b)** 0.038 Ω
47. **(a)** 4.6 mΩ **(b)** 8.5 mA
49. 5.4×10^{-2} m
51. **(a)** (1) greater than, because after the stretch, the length L increases and the cross-sectional area A decreases, so R increases according to $R = \rho L/A$. **(b)** 1.6
53. **(a)** 7.8 Ω **(b)** 0.77 A **(c)** 16.4°C
55. **(d)**
57. **(d)**
59. Since $P = V^2/R$, the bulb of higher power has smaller resistance or thicker wire. So the wire in the 60-W bulb would be thicker.
61. 144 Ω
63. 2.0×10^3 W
65. 1.2 Ω
67. **(a)** (4) 1/4, because if the voltage is halved, the current is also halved. Power is equal to voltage times current, so power becomes 1/4 of its original value. **(b)** 10 W
69. **(a)** 4.3×10^3 W **(b)** 13 Ω
71. **(a)** 58 Ω **(b)** 86 Ω
73. **(a)** 0.60 kWh **(b)** $0.09
75. **(a)** 0.15 A **(b)** 1.4×10^{-4} $\Omega \cdot$ m **(c)** 2.3 W
77. **(a)** 1.1×10^2 J **(b)** 6.8 J
79. 21 Ω
81. $R_{120}/R_{60} = 4/3$
83. $152
85. 117°C for copper or −72.6°C for aluminum
87. yes, R is a constant ($R + r = 6.00$ Ω)
89. about one-half (1 power plant delivers about 1000 MW)
91. 6.6×10^{-6} m/s
93. 1.6×10^3 Ω
95. **(a)** 400 A **(b)** 4.5×10^{-3} Ω **(c)** 1.8 V **(d)** 250 kV

Chapter 18

1. **(b)**
3. **(b)**
5. **(a)**
7. No, not generally. However if all resistors are equal, the voltages across them are the same.
9. If they are in series, the effective resistance will be closer in value to that of the large resistance because $R_s = R_1 + R_2$. If $R_1 \gg R_2$, then $R_s \approx R_1$. If they are in parallel, the effective resistance will be closer in value to that of the small resistance because $R_p = R_1 R_2/(R_1 + R_2)$. If $R_1 \gg R_2$, then $R_p \approx R_1 R_2/R_1 = R_2$.
11. **(a)** The third resistor has the largest current, because the total current through the two other resistors is equal to the current through the third resistor. **(b)** The third resistor also has the largest voltage, because the current through it is the largest and all the resistors have the same resistance value ($V = IR$).

(c) The third resistor also has the largest power output, because it has the largest voltage and current, and power is equal to the product of current and voltage.
13. **(a)** in series, 60 Ω **(b)** in parallel, 5.5 Ω
15. 30 Ω
17. **(a)** 30 Ω **(b)** 0.30 A **(c)** 1.4 W
19. **(a)** 0.57 Ω **(b)** 6.0 V **(c)** 9.0 W
21. **(a)** (1) $R/4$. Each shortened segment has a resistance of $R/2$ because resistance is proportional to length (Chapter 17). Then two $R/2$ resistors in parallel gives $R/4$. **(b)** 3.0 $\mu\Omega$
23. **(a)** 1.0 A **(b)** 1.0 A **(c)** 2.0 W, 4.0 W, 6.0 W **(d)** $P_{sum} = P_{total} = 12$ W
25. 1.0 A (for all); $V_{8.0} = 8.0$ V; $V_{4.0} = 4.0$ V
27. 2.7 Ω
29. **(a)**

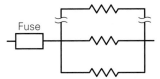

(b) 31
31. **(a)** 1.0 A; 0.50 A; 0.50 A **(b)** 20 V; 10 V; 10 V **(c)** 30 W
33. no, since $I = 14.6$ A < 15 A
35. 100 s = 1.7 min
37. **(a)** 0.085 A **(b)** 7.0 W, 2.6 W, 0.24 W, 0.41 W
39. **(a)** 0.67 A, 0.67 A, 1.0 A, 0.40 A, 0.40 A **(b)** 6.7 V, 3.3 V, 10 V, 2.0 V, 8.0 V
41. 8.1 Ω
43. **(a)**
45. **(d)**
47. No, it does not have to be. An example is charging a battery. When a battery is connected to a charger (with a higher electromotive force), the current is forced through the battery.
49. The 60 W bulb has a higher resistance than the 100 W bulb. When these are in series, they have the same current. Therefore, the 60 W bulb will have a higher voltage. Thus, the 60 W bulb has more power because $P = IV$.
51. Around loop 1 (reverse), $-V_1 + I_3 R_3 + V_2 + I_1 R_1 = 0$. If we multiply by −1 on both sides, it is the same as the equation for loop 1 (forward). Around loop 2 (reverse), $I_2 R_2 - V_2 - I_3 R_3 = 0$. Again, if we multiply by −1 on both sides, it is the same as the equation for loop 2 (forward).
53. $I_1 = 1.0$ A; $I_2 = I_3 = 0.50$ A
55. $I_1 = 0.33$ A (left); $I_2 = 0.33$ A (right)
57. $I_1 = 3.75$ A (up); $I_2 = 1.25$ A (left); $I_3 = 1.25$ A (right)
59. $I_1 = 0.664$ A (left); $I_2 = 0.786$ A (right); $I_3 = 1.450$ A (up); $I_4 = 0.770$ A (down); $I_5 = 0.016$ A (down); $I_6 = 0.664$ A (right)
61. **(c)**
63. **(b)**
65. It takes shorter than one time constant because the time constant is defined as the time it takes to charge the capacitor charged to 63% of its maximum charge.
67. **(a)** $V_C = 0$; $V_R = V_o$ **(b)** $V_C = 0.86V_o$; $V_R = 0.14V_o$ **(c)** $V_C = V_o$; $V_R = 0$

69. (a) (1) increase the capacitance, because $\tau = RC$ **(b)** 2.0 MΩ

71. (a) 1.50 MΩ **(b)** 11.4 V

73. (a) 9.4×10^{-4} C **(b)** $V_C = 24$ V; $V_R = 0$

75. (a) 2.0×10^{-3} A at $t = 0$ **(b)** 0.080% **(c)** 1.7×10^{-6} C at a very long time after connection **(d)** 99.9%

77. (b)

79. (a) An ammeter has very low resistance, so if it were connected in parallel in a circuit, the circuit current would be very high and the galvanometer could burn out.
(b) A voltmeter has very high resistance, so if it were connected in series in a circuit, it would read the voltage of the source because it has the highest resistance (most probably) and therefore the most voltage drop among the circuit elements.

81. An ammeter is used to measure current when it is connected in series to a circuit element. If it has very small resistance, there will be very little voltage across it, so it will not affect the voltage across the circuit element, nor its current.

83. (a) (3) a multiplier resistor, because a galvanometer cannot have a large voltage across it, the large voltage has to be across a series resistor (multiplier). **(b)** 7.4 kΩ

85. 50 kΩ

87. 0.20 mA

89. (a) (1) zero, because an ammeter is connected in series with a circuit element. If its resistance is zero, it will not affect the current through the circuit element.
(b) The current reading I is the current through R, and the voltage reading is the total voltage across R and R_a so V/I gives the resistance of the series combination.
(c) The voltage reading is $V = I(R + R_a)$, so $R = V/I - R_a$.
(d) An ideal ammeter has R_a approaching 0, then $R = V/I$, i.e., the measurement is "perfect."

91. (c)

93. No, a high voltage can produce high harmful current, even if resistance is high because current is caused by voltage (potential difference).

95. It is safer to jump. If you step off the car one foot at a time, there will be a high voltage between your feet. If you jump, the voltage between your feet is zero because your feet will be at the same potential all the time.

97. $I_1 = 2.6$ A (right); $I_2 = 1.7$ A (left); $I_3 = 0.86$ A (down)

99. (a) $I_1 = 1.0$ A; $I_2 = 0.40$ A; $I_3 = 0.20$ A; $I_4 = 0.40$ A **(b)** $P_1 = 100$ W; $P_2 = 4.0$ W; $P_3 = 2.0$ W; $P_4 = 4.0$ W

101. 6.0 Ω

103. 10 mΩ, 2.0 mΩ, and 1.0 mΩ

105. two in parallel with each other and in series with the other resistor

107. (a) 12.1 ms **(b)** 1.21 kΩ **(c)** 13.0 ms

Chapter 19

1. (a)

3. (c)

5. Near the north pole of a permanent bar magnet, the north pole of a compass will point away from the bar magnet so the field lines leave the north pole. Near the south pole of a permanent bar magnet, the south pole of a compass will point toward the bar magnet so the field lines enter the south pole.

7. (a)

9. (d)

11. Not necessarily, because there still could be a magnetic field. If the magnetic field and the velocity of the charged particle make an angle of either 0° or 180°, there is no magnetic force because $F = qvB \sin \theta$.

13. (a) The bottom half would have a magnetic field directed into the page and the top half would have a magnetic field directed out of the page. **(b)** They are the same since centripetal force does not change the speed of the particle.

15. 3.5×10^3 m/s

17. 2.0×10^{-14} T, left, looking in the direction of the velocity

19. (a) 3.8×10^{-18} N **(b)** 2.7×10^{-18} N **(c)** zero **(d)** zero

21. (a) 8.6×10^{12} m/s^2, horizontal and south **(b)** 8.6×10^{12} m/s^2, horizontal and north **(c)** same magnitude but the direction is horizontal and north

23. (b)

25. The magnetic force on the electron beam, which "prints" pictures, causes the deflection of the electrons.

27. The electric force is $F_e = qE$ and the magnetic force is $F_B = qvB$. The purpose of the velocity selector is for the electric force to equal the magnetic force. Since $qE = qvB$, $v = E/B$, independent of the charge.

29. (a) 1.8×10^3 V **(b)** same voltage, independent of charge

31. 5.3×10^{-4} T

33. (a) 4.8×10^{-26} kg **(b)** 2.4×10^{-18} J **(c)** no, work equals zero

35. (d)

37. (b)

39. It shortens because the coils of the spring attract each other due to the magnetic fields created in the coils. (Parallel wires with current in same direction will attract each other.)

41. Pushing the button in both cases completes the circuit. The current in the wires activates the electromagnet, causing the clapper to be attracted and ring the bell. However, this breaks the armature contact and opens the circuit. Holding the button causes this to repeat, and the bell rings continuously. For the chimes, when the circuit is completed, the electromagnet attracts the core and compresses the spring. Inertia causes it to hit one tone bar, and the spring force then sends the core in the opposite direction to strike the other bar.

43. 1.2 N perpendicular to the plane of $\vec{B}$ and I.

45. 5.0×10^{-3} T north to south

47. (a) zero **(b)** 4.0 N/m in $+z$ **(c)** 4.0 N/m in $-y$ **(d)** 4.0 N/m in $-z$ **(e)** 4.0 N/m in $+y$

49. 0.40 N/m; $+z$

51. (a) (1) attractive **(b)** 6.7×10^{-6} N/m

53. 0.53 N northward at an angle of 45° above the horizontal

55. 2.7×10^{-5} N/m toward wire 1

57. 7.5 N upward in the plane of the paper

59. zero, yes

61. (a)

63. (b)

65. Because $B = \mu_o I/(2\pi d)$, you need to double the current and reverse the direction.

67. There are two wires carrying the current into and out of the appliances. These two currents are in opposite directions. When the two wires are very close to each other, the magnetic fields created by the two opposite currents essentially cancel.

69. 3.8 A

71. 0.25 m

73. (a) 2.0×10^{-5} T **(b)** 9.6 cm from wire 1

75. both 2.9×10^{-6} T

77. 3.3×10^{-5} T

79. 1.0×10^{-4} T, away from the observer

81. 4.0 A

83. (a) 8.8×10^{-2} T **(b)** to the right

85. $\sqrt{2}\,\mu_o I/(\pi a)$ at 45° toward the lower left wire

87. (b)

89. The direction of the magnetic field is away from you, according to the right-hand source rule (electron has negative charge).

91. You can destroy or reduce the magnetic field of a permanent magnet by hitting or heating it.

93. 12 T

95. (b)

97. (a)

99. It will be the north magnetic pole. Right now, the pole near the Earth's geographical North pole is the south magnetic pole.

101. 0.44 T

103. (a) 5.9×10^{-21} kg·m/s **(b)** 1.0×10^{-14} J

105. 0.682 V

107. (a) (2) into the page **(b)** 0.030 m from left wire

109. (a) 3.74×10^{-3} T·m/A **(b)** 3.0×10^3

Chapter 20

1. (d)

3. (d)

5. (d)

7. The direction is counterclockwise (in head-on view).

9. No, it does not depend on the magnetic flux. It depends on the rate of the flux change with time.

11. Sound waves cause the resistance of the button to change as described. This results in a change in the current, so the sound waves produce electrical pulses. These pulses travel through the phone lines and to a receiver. The receiver has a coil wrapped around a magnet, and the pulses create a varying magnetic field as they pass through the coil causing the diaphragm to vibrate and thus produce sound waves as the diaphragm vibrates in the air.

13. 42° or 138°

15. 3.3×10^{-2} T·m²

17. 1.3×10^{-6} T·m²

19. 1.6 V

21. 0.30 s

23. (a) (1) counterclockwise (b) 0.35 V

25. (a) (1) at the equator, because the velocity of the metal rod is parallel to the magnetic field at the equator. (b) 0.50 mV at the pole, zero at the equator

27. (a) 0.60 V (b) 0 A

29. 4.0 V

31. (a) 0.037 T·m² (lower incline); 0.034 T·m² (upper incline) (b) −0.071 T·m² (c) zero (d) zero; this means that the net flux is equal to zero or there are as much flux leaving the box as entering.

33. (c)

35. The magnet moving through the coil produces a current. As the magnet moves up and down in the coil it will induce a current in the coil that will light the bulb. However, the magnet produces the current (Faraday's law of induction) at the expense of its kinetic energy and potential energy. The magnet's motion will therefore damp out.

37. $\mathscr{E} = \mathscr{E}_o = \sin \omega t$, where $\mathscr{E}_o = NBA\omega$. (N is the number of turns, B is the magnetic field strength, and ω is the angular speed) You could increase N, B, or ω.

39. (a) 0.057 V (b) 0.57 V

41. (a) (2) two, because the initial voltage direction was not specified, thus, there are two possible directions. (b) ±104 V

43. (a) 100 V (b) 0 V

45. 16 Hz

47. (a) (3) lower than 44 A (110 V/2.50 Ω = 44 A), because of the back emf induced when the motor turns. The back emf lowers the effective voltage of the motor, thus, the current is lower than 44 A. (b) 4.00 A

49. (a) 216 V (b) 160 A (c) 8.1 Ω

51. (b)

53. Yes, a step-up transformer can be used as a step-down transformer. You just need to reverse the roles primary and secondary coils, so there are more turns on the high voltage side.

55. (a) 16 (b) 5.0×10^2 A

57. 24 : 1

59. (a) 17.5 A (b) 15.7 V

61. (a) (2) non-ideal, because $P_s < P_p$ (the power in the secondary is lower than that in the primary). (b) 45%

63. (a) N_s/N_p is 1 : 20 (b) 2.5×10^{-2} A

65. (a) 128 kWh (b) $1840

67. (a) 1 : 2, 1 : 14, 1 : 30 (b) 2.0, 14, 30 (c) 833

69. (a) 53 W (b) $N_p/N_s = 200$

71. (d)

73. (d)

75. UV radiation causes sunburn and it can still go through the clouds. You feel cool

because infrared (heat) radiation is absorbed by clouds (water molecules).

77. According to $c = \lambda f$, wavelength and frequency are inversely proportional to each other. Thus, radar frequencies are much higher, because wavelengths are much shorter, speeds are the same.

79. 326 m and 234 m

81. 2.6 s

83. AM: 67 m; FM: 0.77 m

85. (a) (1) up, according to Lenz's law. (b) 25 mA

87. (a) no, input power is greater than output power (b) 90.9%

89. (a) (1) clockwise (b) 5.00×10^{-3} V (c) 0.0879 s

91. 3.79 m, no

93. 0.159 Ω

Chapter 21

1. (a)

3. (a)

5. That means the voltage and current both reach maximum or minimum at the same time.

7. No, the circuit element cannot be a resistor. Voltage and current should be in phase for a resistor. Yes, the frequency is 60 Hz, because $\omega = 2\pi f = 120\pi$.

9. 7.1 A

11. 1.2 A

13. (a) 10.0 A (b) 14.1 A (c) 12.0 Ω

15. (a) 4.47 A, 6.32 A (b) 112 V, 158 V

17. $V = (170$ V$) \sin(119\pi t)$

19. 0.33 A; 0.47 A

21. (a) 20 Hz, 0.050 s (b) 2.4×10^2 W

23. (a) 60 Hz (b) 1.4 A (c) 1.2×10^2 W (d) $V = (120$ V$) \sin 380t$ (e) $P = (240$ W$) \sin^2 380t$ (f) $P = (240$ W$)[1 - \cos 2(380t)]/2 = 120$ W $- (120$ W$) \cos 2(380t)$. The average of a sine or cosine function is zero. So $\bar{P} = 120$ W, the same as in (c).

25. (b)

27. For a capacitor, the *lower the frequency*, the longer the charging time in each cycle. If the frequency is very low (dc), then the charging time is very long, so it acts as an ac open circuit. For an inductor, the *lower the frequency*, the more slowly the current changes in the inductor The more slowly the current changes, the less back emf is induced in the inductor, resulting in less impedance to current.

29. At $t = 0$, $I = 120$ A, or at maximum. The voltage is then zero, because current leads voltage by 90° in a capacitor. When current is maximum, voltage is 1/4 period behind, or at zero. They are out of phase.

31. 1.3×10^3 Ω

33. (a) 19 Ω (b) 6.4 A (c) Voltage leads current by 90°

35. an increase of 60%

37. 255 Hz

39. (a) 90 V (b) voltage leads current by 90°

41. 4.4 μF

43. (d)

45. (d)

47. No, there is no power delivered to capacitors or inductors in an ac circuit. For either a pure capacitive or inductive circuit, the phase angle $\phi = 90°$, and so the power factor is $\cos \phi = 0$.

49. (a) 1.7×10^2 Ω (b) 2.0×10^2 Ω

51. (a) 38 Ω; 1.1×10^2 Ω (b) 1.1 A

53. (a) (3) negative, because this is a capacitive circuit. (b) −27°

55. (a) (3) in resonance, because $X_L = X_C$, so $Z = R$. (b) 72 Ω

57. 50 W

59. 5.3×10^{-11} F

61. ab: 1.3 A; ac: 1.2 A; bc: 4.0 A; cd: 1.8 A; bd: 1.6 A; ad: 2.9 A

63. 13 A

65. $(V_{rms})_R = 12$ V; $(V_{rms})_L = 2.7 \times 10^2$ V; $(V_{rms})_C = 2.7 \times 10^2$ V

67. (a) (2) equal to 25 Ω. At resonance, $X_L = X_C$, so $Z = R$. (b) 362 Ω

69. 30%

71. (a) 38 Ω (b) 63 Ω (c) 1.8 A (d) zero (e) 37°

73. (a) (2) zero, as $X_L = X_C$, according to $\phi = \tan^{-1}[(X_L - X_C)/R]$. (b) 9.4 μF

Chapter 22

1. (c)

3. (d)

5. This is irregular or diffuse reflection, because the paper is microscopically rough.

7. 70°

9. (a) (2) 90° − α, because $\theta_i = \theta_r$ and $\alpha + \theta_i = 90°$, $\theta_r = \theta_i = 90° − \alpha$ (b) 47°

11. (a) (3) $\tan^{-1}(w/d)$ (b) 27°

13. When the mirror rotates through a small angle of θ, the normal will rotate through an angle of θ and the angle of incidence is 35° + θ. The angle of reflection is also 35° + θ. Since the original angle of reflection is 35°, the reflected ray will rotate through an angle of 2θ. If the mirror rotates in the opposite direction, the angle of reflection will be 35° − θ. However, the normal will again rotate the through an angle of θ but also in the opposite direction. Thus, the reflected ray still rotates through an angle of 2θ.

15. 90°, any θ_{i_1}

17. (d)

19. It is because light speed depends on the medium. For example, light speed is different in air than in water. Because of the speed difference, light changes direction when entering a different medium at an angle of incidence that is not zero.

21. This severed look is because the angle of refraction is different for air-glass interface than for water-glass interface. The top portion refracts from air to glass, and the bottom portion refracts from water to glass. This is different from what's on Fig. 22.13b. In that figure, we see the top portion in air directly and the bottom portion in water through refraction from water to air.

The angle of refraction made the pencil appear to be bent.

23. The laser beam has a better chance to hit the fish. The fish appears to the hunter at a location different from its true location due to refraction. The laser beam obeys the same law of refraction and retraces the light the hunter sees from the fish. The arrow goes into the water in a near-straight line path and thus passes above the fish.

25. (a) (1) greater than, because its index of refraction is smaller. **(b)** 1.26

27. (a) (1) greater than, because water has a lower index of refraction. **(b)** 17°

29. (a) (2) a diamond to air, because diamond has a higher index of refraction. **(b)** 24.4°

31. 47°

33. 6.5×10^{14} Hz, 2.8×10^{-7} m

35. (a) (3) less than, because its index of refraction is higher. **(b)** 15/16

37. (a) This is caused by refraction of light in the water-air interface. The angle of refraction in air is greater than the angle of incidence in water so the object immersed in water appears closer to the surface.

39. 75.2%

41. (a) (3) less than, because it is equal to $90° - \theta_1$. $\theta_1 > 45° = \theta_2$ and $n_1 < n_2$. **(b)** 20°

43. seen for 40° but not for 50°, $\theta_c = 49°$

45. (a) yes, $\theta_c = 32° < 45°$ **(b)** no, $\theta_c = 46° > 45°$

47. We can measure the angles of incidence and refraction from the photography and calculate the index of refraction of the fluid from the law of refraction. It is about 1.3.

49. 1.64

51. (a) (3) total internal reflection **(b)** $\theta_c = 39° < 45°$, no **(c)** $\theta_c = 56° < 71°$, still not transmitted

53. 2.0 m

55. (a) 12.5° **(b)** 26.2°

57. (b)

59. In a prism, there are two refractions and two dispersions because both refractions cause the refracted light to bend downward, therefore doubling the effect or dispersion.

61. To see a rainbow, the light has to be behind you. Actually, you won't see a primary rainbow if the Sun's angle above the horizon is greater than 42°. Therefore you cannot look up to find a rainbow, thus you cannot walk under a rainbow.

63. (a) Usually $\theta \approx 0°$, so there is no dispersion because the angle of refraction for all colors is also zero. **(b)** as explained in (a). With $\theta \approx 0$, light of any wavelength will experience no refraction. (No, the speeds are actually different.)

65. 1.498

67. (a) 21.7° **(b)** 0.22° **(c)** 0.37°

69. (a) 49° **(b)** 1.5 **(c)** 1 **(d)** 42°

71. (a) (1) more than, because red light will have a smaller index of refraction and thus a higher speed of light than blue light. **(b)** 1.3 mm

73. 1.41 to 2.00

Chapter 23

1. (b)

3. (c)

5. During the day, the reflection is mainly from the silvered back surface. During the night, when the switch is flipped, the reflection comes from the front side. And so there is a reduction of intensity and glare because the front side reflects only about 5% of the light, which is more than enough to see due to the dark background.

7. When viewed by a driver through a rear view mirror, the right-left reversal property of the image formed by a plane mirror will make it read "AMBULANCE."

9. (a) 4.0 m **(b)** upright, virtual, and same size

11. 5.0 m

13. (a) 1.5 m behind the mirror **(b)** 1.0 m/s

15. (a) You see multiple images caused by reflections off two mirrors. **(b)** 3.0 m behind the *north* mirror, 11 m behind the *south* mirror, 5.0 m behind the *south* mirror, 13 m behind the *north* mirror

17. The two triangles (with d_o and d_i as base, respectively) are similar to each other because all three angles of one triangle are the same as those of the other triangle due to the law of reflection. Furthermore, the two triangles share the same height, the common vertical side. Therefore the two triangles are identical. Hence $d_o = d_i$.

19. (d)

21. (a)

23. (a) A spoon can behave as either a concave or a convex mirror depending on which side you use for reflection. If you use the concave side, you normally see an inverted image. If you use the convex side, you always see an upright image. **(b)** In theory, the answer is yes. If you are very close (inside the focal point) to the spoon on the concave side, an upright image exists. However, it might be difficult for you to see the image in practice, because your eyes might be too close to the image. Eyes cannot see things that are closer than the near point (Chapter 25).

25. The image of a distant object (at infinity) is formed on a screen in the focal plane. The distance from the vertex of the mirror to the plane is the focal length. The same cannot be done for a convex mirror because the image is a virtual one and it cannot be formed on a screen.

27. (a) It is seen from the ray diagram that the image is virtual, upright, and reduced.

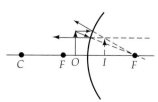

(b) $d_i = -20$ cm; $h_i = +0.67h_o$

29. $d_i = -30$ cm; $h_i = 9.0$ cm; image is virtual, upright, and magnified

31. From the mirror equation, $1/(2f) + 1/d_i = 1/f$. So $1/d_i = 1/f - 1/(2f) = 1/(2f)$, or $d_i = 2f$. $M = -d_i/d_o = -(2f)/(2f) = -1$. Therefore, the image is inverted ($M < 0$), and the same size as the object ($|M| = 1$).

33. (a) convex, because a concave mirror can only form magnified virtual images. **(b)** 14 cm

35. (a) concave, because only concave mirror can form magnified images. **(b)** 13.3 cm

37. f is negative for a convex mirror. So $d_i = d_o f/(d_o - f) = d_o(-|f|)/(d_o + |f|) < 0$. Also $M = -d_i/d_o = -d_o(-|f|)/[d_o(d_o + |f|)] = |f|/(d_o + |f|) < +1$. Therefore the image is virtual (negative d_i), upright (positive M, and reduced ($|M| < 1$).

39. (a) concave, because only concave mirror can form real images (formed on a screen). **(b)** 24 cm

41. (a) virtual and upright **(b)** 1.5 m

43. 2.3 cm

45. (a)

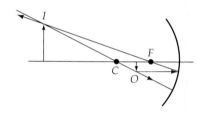

(b) $d_i = 60$ cm, $M = -3.0$, real and inverted

47. 10 m

49. (a) Front surface: 60 cm, real, inverted, and magnified; back surface: 46.7 cm, real, inverted, and magnified **(b)** no, the image of the cube is no longer a cube because different parts of the cube have different magnifications.

51. (a) two, one for a real image and another for a virtual image. **(b)** 5.0 cm, 15 cm

53. Yes, it is possible. One is a real image and the other is a virtual image. 13 cm; 27 cm

55. (d)

57. When the fish is inside the focal point, the image is upright, virtual, and magnified.

59. You can locate the image of a distant object. The distance from the converging lens to the image is the focal length. No, the same method won't work for a diverging lens because a diverging lens does not form real images of real objects.

61. $d_i = 12.5$ cm; $M = -0.250$

63. 22 cm

65. (a) $f = 5.9$ cm **(b)** 67 cm, inverted

67. (a) 18 cm **(b)** 6.0 cm

69. 14 cm

71. 0.55 mm

73. (a) 20 cm **(b)** $M = -1.0$

75. (a) $d = 4f$ **(b)** approaches 0

77. (a) −18 cm **(b)** −63 cm

79. 4.2 cm

81. 18 cm to the left of the eyepiece; virtual image

83. $M_1 = -h_{i1}/h_{o1}$, $M_2 = -h_{i2}/h_{o2}$, and $M = h_{i2}/h_{o1}$. Since $h_{o2} = h_{i1}$ (the image formed by the first lens is the object for the second lens), $M_1 M_2 = (h_{i1}/h_{o1})(h_{i2}/h_{o2}) = h_{i2}/h_{o1} = M_{total}$.

85. (b)

87. (b)

89. Our eyes are "designed" or used to seeing things clearly when our surroundings are air. When you are underwater, the index of refraction of the surroundings (water now) changes. From Exercise 23.80(a), the focal length of the eyes changes so everything is blurry. When you wear goggles, the surroundings of the eyes are again air, so you can see things clearly.

91. (a) yes, increases **(b)** diverging lens becomes converging lens and vice versa

93. +4.0 D

95. −0.70 D

97. 85 cm

99. The image formed by the converging lens is at the mirror. This image is the object for the diverging lens. If the mirror is at the focal point of the diverging lens, the rays refracted after the diverging lens will be parallel to the axis. These rays will be reflected back parallel to the axis by the mirror and will form another image at the mirror. This second image is now the object for the converging lens. By reversing the rays, a sharp image is formed on the screen located where the original object is. Therefore the distance from the diverging lens to the mirror is the focal length of the diverging lens.

101. 20 cm on object side of first lens, inverted, $M_{total} = -1.0$

103. 60 cm to right of second lens; real and upright, $M_{total} = 4.0$

Chapter 24

1. (b)

3. (b)

5. Since $\Delta y = L\lambda/d \propto \lambda$, the spacing between the maxima would increase if wavelength increases.

7. 3.4%

9. 0.37°

11. 489 nm

13. (a) 440 nm **(b)** 4.40 cm

15. (a) (1) increase, because $\Delta y = L\lambda/d \propto 1/d$, the distance between the maxima would decrease if the distance between the slits were increased. **(b)** 1.9 mm **(c)** 2.4 mm

17. (a) $\lambda = 402$ nm, violet **(b)** 3.45 cm

19. 450 nm

21. (a)

23. The wavelengths that are not visible in the reflected light are all wavelengths except bluish purple.

25. It is always dark because of destructive interference due to the 180° phase shift. If there had not been the 180° phase shift, zero thickness would have corresponded to constructive interference.

27. (a) 30λ **(b)** destructively

29. 54.3 nm

31. (a) (2) 600 nm, because $t_{min} = \lambda/(4n_1)$ or $t \propto \lambda$. **(b)** 160 nm; 200 nm

33. (a) 158.2 nm **(b)** 316.4 nm

35. 1.51×10^{-6} m

37. (a)

39. no; yes (barely, at $\theta = 90°$)

41. According to $d \sin \theta = n\lambda$, the advantage is a wider diffraction pattern, as d is smaller.

43. (a) 5.4 cm **(b)** 2.7 cm

45. (a) 4.3 mm **(b)** microwave

47. 1.24×10^3 lines/cm

49. blue, 18.4°, 39.2°; red: 30.7°, not possible

51. (a) 2.44×10^3 lines/cm **(b)** 11 (maximum n is 5)

53. (a) (1) blue, because it has a shorter wavelength. From $d \sin \theta = n\lambda$, we can see that the shorter the wavelength, the smaller the $\sin \theta$ or θ. **(b)** blue: 18.7°; red: 34.1°

55. From $d \sin \theta = n\lambda$, $\theta = \sin^{-1} n\lambda/d$. For violet, $\theta_{3v} = \sin^{-1}(3)(400 \text{ nm})/d = \sin^{-1}(1200 \text{ nm})/d$. For yellow-orange, $\theta_{2y} = \sin^{-1}(2)(600 \text{ nm})/d = \sin^{-1}(1200 \text{ nm})/d$. So $\theta_{3v} = \theta_{2y}$, that is, they overlap.

57. (d)

59. (c)

61. (a) twice **(b)** four times **(c)** none **(d)** six times

63. The numbers appear and disappear as the sunglasses are rotated because the light from the numbers on a calculator is polarized.

65. (a) (1) also increase, because $\tan \theta_p = n_2/n_1 = n_2 (n_1 = 1)$. If n_2 increases, so does θ_p. **(b)** 58°, 61°

67. (a) (2) decrease, because the transmitted light intensity depends on $\cos^2 \theta$. As θ increases from 0° to 90°, $\cos \theta$ decreases. **(b)** $0.500I_o$, $0.375I_o$ **(c)** $0.500I_o$, $0.125I_o$

69. 55°

71. 57.2°

73. In water, $\theta_p = \tan^{-1}(n_2/n_1)$. The angle of incidence at the water–glass interface must be $\theta_p = \tan^{-1}(1.52/1.33) = 48.8°$. For the air–water interface, $n_1 \sin \theta_1 = n_2 \sin \theta_2$, so $\sin \theta_1 = n_2 \sin \theta_2/n_1 = (1.33) \sin 48.8° > 1$. Since the maximum of $\sin \theta_1$ is 1, the answer is no.

75. (c)

77. (a) This is caused by the variable air molecule density. **(b)** There is no air on the surface of the Moon, and so an astronaut would see a black sky.

79. no, because $n_2 = 1$ (air), $\tan \theta_p = n_2/n_1 = 1/n_1$. For total internal reflection, $\sin \theta_c = n_2/n_1 = 1/n_1$. That means $\tan \theta = \sin \theta$. This is not possible for any angle that is not equal to zero. **(b)** 29.8°

81. $\Delta y = 0.25 \Delta y_o$

83. $n = 1$ for red; $n = 2$ for violet

Chapter 25

1. (b)

3. (a)

5. The pre-flash occurs before the aperture is open and the film exposed. The bright light causes the iris to reduce down (giving a small pupil) so that when the second flash comes momentarily, you don't have a wide opening through which you get the red-eye reflection from the retina.

7. (a) The eye is nearsighted because the far point is not at infinity. **(b)** The eye is farsighted because the near point is not 25 cm. **(c)** (a), diverging; for (b), converging

9. (a) +5.0 D **(b)** −2.0 D

11. (a) (1) converging, because the person is farsighted. **(b)** +2.0 D

13. diverging, −0.500 D

15. (a) take them out **(b)** +3.0 D

17. (a) (1) converging, because he is farsighted. **(b)** +3.3 D

19. (a) −0.505 D **(b)** −0.500 D

21. 6.7 m

23. (a) −0.67 D **(b)** yes, 21 cm **(c)** 30–40 years old

25. right: +1.42 D, −0.46 D; left: +2.16 D, −0.46 D

27. (d)

29. The object should be inside the focal length. When the object is inside the focal length, the image is virtual, upright, and magnified.

31. (a) 2.3× **(b)** 2.5×

33. 2.5×

35. (a) (1) high powered, because a high powered lens has short focal length and the magnification is $1 + (25 \text{ cm})/f$. **(b)** 1.9× and 1.6×

37. −375×

39. (a) (2) The one with the shorter focal length, because the total magnification is inversely proportional to the focal length of the objective. **(b)** −280× and −360×

41. (a) −340× **(b)** 3900%

43. 25×

45. (a) greatest: 1.6 mm/10×; least: 16 mm/5× **(b)** $M_{max} = -930×$; $M_{min} = -42×$;

47. (b)

49. No, the whole star can still be seen. The obstruction will reduce the intensity or brightness of the image.

51. The one with the shorter focal length should be used as the eyepiece for a telescope. The magnification of the telescope is inversely proportional to the focal length of the eyepiece $(m = -f_o/f_e)$.

53. (a) −4.0× **(b)** 75 cm

55. 1.00 m and 2.0 cm

57. 5.0 cm

59. (a) 60.0 cm and 80.0 cm; 40.0 cm and 90.0 cm **(b)** −75×; −44×

61. (a)

63. Smaller minimum angle of resolution corresponds to higher resolution because smaller angle of resolution means more details can be resolved.

65. From a resolution point of view, the smaller camera (lens) has lower resolution. The smaller the lens, the greater the minimum angle of resolution, the lower the resolving power.

67. 550 nm

69. 1.32×10^{-7} rad; θ_{min} by Hale is 1.6 times as great

71. (a) (3) blue, because the minimum angle of resolution is proportional to wavelength and blue has the shorter wavelength. **(b)** 9.6×10^{-5} rad and 1.1×10^{-4} rad

73. 17 km

75. 4.1×10^{16} km

77. (a) 5.55×10^{-5} rad (b) blue (c) 33.3%

79. (d)

81. With red light, red and white appear red; blue appears black. With green light, only white appears green; both red and blue appear black. With blue light, red appears black; white and blue appear blue.

83. The liquid is dark or colored because it absorbs all light except that color. The amount of light absorbed by an object always depends on how much material is absorbing the light. Foam has very low material density and it can only absorb very little light or almost all light is reflected; therefore, the foam is generally white.

85. From thin lens equation:
$d_o = d_i f/(d_i - f)$, we have
$d_i/d_o = (d_i - f)/f = [-(D - d) - f]/f$. By small angle approximation:
$m = \theta_i/\theta_o = (y_i/D)/[y_o/(25\text{ cm})]$
$= (y_i/y_o) \times [(25\text{ cm})/D]$. By similar triangles: $y_i/y_o = -d_i/d_o$, the negative sign is introduced because d_i is negative (virtual image). So
$m = \{[(D - d) + f]/f\} \times [(25\text{ cm})/D]$
$= (25/f) \times (1 - d/D) + 25/D$.

87. (a) (1) B, (2) A (b) $-110\times$, 8.95×10^{-7} rad

89. (a) 6.3 and 0.25 (b) 1/120 s

Chapter 26

1. (d)

3. (b)

5. The driver invokes a "fictitious" force in the backward direction so the objects appeared to accelerate backward and Newton's law is still valid. Observer on street simply observes the car accelerates out from under the objects which stay at rest since the net force on them is zero.

7. (a) 3.38 s (b) 3.58 s

9. 55 m/s and 45 m/s

13. (a)

15. The speed is c, since the speed of light is the same regardless of motion of the source or observer.

17. In frame O, the bullet takes 1 s to hit target. Light takes 10^{-6} s to get to the target (the time for light of speed 3.00×10^8 m/s to travel 300 m). Frame O' would have to travel to the right. The light flashes from the gun reaches the target in 10^{-6} s. The observer in frame O' would have to cover the 300 m in less than 10^{-6} s to intercept the signals at the same time, which means $v > c$. Since $v > c$ is not possible, all observers agree that the gun fires before the bullet hits the target.

19. (c)

21. (a)

23. No, this is not possible. From the boy's view, the barn is moving at the same speed, so it would appear to contract and be even shorter than 4.0 m.

25. 23 min

27. 14.1 m

29. 0.998c

31. (a) (1) a longer time due to time dilation (b) 7.2 years and 5.7 years

33. 5.3 m

35. 0.87c

37. 2.14×10^{-14} m

39. 0.60c

41. (b)

43. No, there are no such limits on momentum and energy as $p = \gamma m v$ and $E = \gamma m c^2$. γ and m can be anything from 0 to ∞.

45. Yes, because the kinetic energy is much less than the rest-energy of the electron (2 keV $\ll$ 511 keV). No, because the kinetic energy is considerably greater than the rest-energy of the electron (2 MeV $\gg$ 0.511 MeV).

47. (a) 0.985c (b) 2.50 MeV (c) 1.56×10^{-21} kg·m/s

49. 0.15 kg

51. 1.6×10^{24} J, or 150000 times more!

53. 79 keV

55. (a) 0.92c (b) 1.4×10^3 MeV

57. 1.43×10^5 years

59. (a) 1.13 MeV (b) 1.64 MeV

63. (a)

65. (c)

67. Drop the cup with the pole vertical. By the principle of equivalence, the weight of the ball appears to be zero in the falling reference frame. The ball is then subject only to the tension force of the stretched rubber band and is pulled inside the cup.

69. 1.8×10^{19} kg/m³

71. 0.43c

73. $-0.154c$ (toward Earth)

75. (a) 0.988c to the left (b) 0.988c to the right (c) 1.6×10^{-22} kg·m/s

77. (a) 79 keV (b) 590 keV

79. (a) 2.5×10^{10} kWh (b) 6 days

81. 1.9 kg

83. (a) $5.31m$ (b) $2.31mc^2$ (c) $0.0382c$ left

Chapter 27

1. (e)

3. (d)

5. When it is really hot, it actually radiates in all wavelengths (although not in equal amounts). The brain interprets the combination of all wavelengths as "white."

7. 3.10×10^{13} Hz

9. 1.06×10^{-5} m, 2.83×10^{13} Hz

11. 2.56×10^{-20} J

13. 2.9×10^{19} quanta/(s·m²)

15. (b)

17. (a)

19. (a)

21. Yes, it is possible. Energy depends on two things: the number of photons and frequency. A beam of lower frequency can have more energy if it has a lot more photons.

23. 6.0×10^{-11} m; X-ray

25. (a) 1.32×10^{-18} J (b) 8.27 eV

27. 3.0 V

29. (a) 6.7×10^{-34} J·s (b) 2.9×10^{-19} J

31. 6.8×10^{14} Hz

33. 2.48 eV

35. (a) 0.625 V (b) 2.33 eV (c) zero

37. 0.44 eV

39. (c)

41. The maximum wavelength shift for Compton scattering from a neutron is smaller compared with that from a free electron because the Compton wavelength is inversely proportional to the mass of the scattering par-

ticle and $\Delta\lambda_{max} = 2\lambda_C = 2h/(mc) \propto 1/m$. Since the mass of a neutron is about 1836 times that of electron, it is smaller by a factor of 1836.

43. to conserve total linear momentum

45. 2.43×10^{-3} nm

47. 0.281 nm

49. (a) (2) less than 5.0 keV but not zero, according to the conservation of momentum and energy, the electron must recoil so it has some kinetic energy at the expense of the photon energy. The photon has some energy but less than its initial amount. (b) 20 eV

51. 18.5°

53. (d)

55. (c)

57. It takes less energy to ionize the electron that is in an excited state than in the ground state. The excited state already has more energy.

59. (a) 10.2 eV (b) 1.89 eV (c) The first is UV; the second visible (red)

61. (a) 6.89×10^{14} Hz (b) 8.21×10^{14} Hz

63. $n \approx 100$

65. (a) (1) increase, but not necessarily double, because part of the energy is used to overcome the ionization energy which does not double. Actually the energy of the emitted electron will more than double because the portion of energy left for the electron (after ionization energy) will be more than double. (b) 15.4 eV

67. (a) $n = 2$ to $n = 1$ (b) $\Delta E_{53} = 0.967$ eV; $\Delta E_{62} = 2.97$ eV; $\Delta E_{21} = 10.2$ eV

69. (a) 54.4 eV (b) 122 eV

73. (a) potential is -27.2 eV and kinetic is $+13.6$ eV (b) $|U| = 2K$, potential energy is twice as large in magnitude.

75. (b)

77. Since tattoo is blue it reflects blue, red end of spectrum would work the best to maximize absorption.

79. In a spontaneous emission, electrons jump from a higher-energy state to a lower-energy state without any external stimulation, and a photon is released in the process.

Stimulated emission is an induced emission. The electron in the higher-energy orbit can jump to a lower-energy orbit when a photon of energy equals the difference of the energy between the two orbits is introduced. Once atoms are prepared with enough electrons in the higher-energy state, stimulating photons trigger them to jump down to the lower-energy state. The emitted photons trigger the rest of the electrons and eventually all the electrons will be in the lower energy state.

81. (a) 9.82 keV (b) first: 48 eV, second: 134 eV

83. (a) 1.02 MeV (b) 1.21×10^{-3} nm (c) 1.02 MeV

85. (a) 2.00 MeV (b) about 1 keV

Chapter 28

1. (c)

3. (b)

5. The baseball, as its momentum is smaller due to its smaller mass. The de Broglie wavelength is inversely proportional to the mass, $\lambda = h/mv$.

7. 2.7×10^{-38} m

9. $\lambda_e / \lambda_p = 43$

11. 1.5×10^4 V

13. **(a)** (3) decrease due to the potential difference. The proton gains speed from the potential difference and the de Broglie wavelength is inversely proportional to speed, $\lambda = h/mv$. **(b)** −53% (a decrease)

15. 24 V

17. **(a)** 3.71×10^{-63} m **(b)** 1.18×10^{72} **(c)** There would be an increase but undetectable because 1 compared to 1.18×10^{72} is negligible.

19. **(b)**

21. When the distance increases due to the dip, the current decreases. To keep the current constant, the tip should move toward the surface to make up for decreasing current.

23. According to modern quantum theory and the Schrödinger equation, there is a probability that if you ran into a wall you could end up on the other side. (Don't try this!)

25. **(a)** 4.07 MeV, 16.3 MeV, 36.6 MeV **(b)** 32.5 MeV, gamma ray

27. **(c)**

29. The principle quantum number, n, gives information about the total energy and orbit radius of an orbit in a hydrogen atom.

31. The periodic table groups elements according to the values of quantum numbers n and l. Within a group, the elements have the same or very similar electronic configurations for the outmost electrons.

33. **(a)** 2 **(b)** 14

35. **(a)** $\ell = 2$ **(b)** $n = 3$

37.

Na has 11 electrons

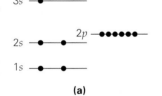

(a)

Ar has 18 electrons

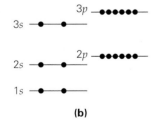

(b)

39. **(a)** $1s^2 2s^2 2p^1$ **(b)** $1s^2 2s^2 2p^6 3s^2 3p^6 4s^2$ **(c)** $1s^2 2s^2 2p^6 3s^2 3p^6 3d^{10} 4s^2$ **(d)** $1s^2 2s^2 2p^6 3s^2 3p^6 3d^{10} 4s^2 4p^6 4d^{10} 5s^2 5p^2$

41. It would be a $1s^3$ and would be the first closed shell inert gas. With three spin orientations, s states can contain three electrons without violating the Pauli exclusion principle.

43. **(b)**

45. According to the uncertainty principle, the product of uncertainty in position and the uncertainty in momentum is in the order of Plank's constant. A bowling ball's large diameter and momentum (mostly due to its large mass) make the uncertainty in them (determined by the extremely small value of Planck's constant) negligible. However for an electron, with its very small size and momentum (mostly due to its very small mass), the uncertainties cannot be ignored.

47. **(a)** (2) the same, because both have the same uncertainty in momentum. **(b)** both 0.21 m

49. 2.1×10^{-29} m/s

51. 1.1×10^{-27} J

53. **(a)** (1) smaller than, according to the uncertainty principle, $\Delta E \Delta t \geq h/(2\pi)$. The greater the lifetime, Δt, the smaller the energy uncertainty, therefore the smaller the width of spectral line. **(b)** 10^4

55. **(a)**

57. Linear momentum needs to be conserved so the particles will be moving afterwards because the initial photon has momentum. Therefore the electron-positron pair carry kinetic energy. The supplied energy needs to create the masses plus the kinetic energy.

59. **(a)** (3) Pair production can happen but the resulting particles must be moving to conserve momentum. **(b)** 0.46 MeV

61. 1.9 GeV

63. 27.6 V

65. 1.37×10^{36} photons

Chapter 29

1. **(a)**

3. **(d)**

5. Carbon-13 has 7 neutrons. Since nitrogen has 7 protons, the nitrogen isotone of carbon-13 should have 7 neutrons so it is nitrogen-14.

7. **(a)** 20p, 23n, 20e **(b)** 82p, 124n, 82e

9. $^{41}_{19}$K

11. 92p, 143n, 92e

13. **(a)** $R_{He} = 1.9 \times 10^{-15}$ m; $R_{Ne} = 3.3 \times 10^{-15}$ m; $R_{Ar} = 4.1 \times 10^{-15}$ m; $R_{Kr} = 5.3 \times 10^{-15}$ m; $R_{Xe} = 6.1 \times 10^{-15}$ m; $R_{Rn} = 7.3 \times 10^{-15}$ m **(b)** $M_{He} = 6.68 \times 10^{-27}$ kg; $M_{Ne} = 3.34 \times 10^{-26}$ kg; $M_{Ar} = 6.68 \times 10^{-26}$ kg; $M_{Kr} = 1.40 \times 10^{-25}$ kg; $M_{Xe} = 2.20 \times 10^{-25}$ kg; 3.71×10^{-25} kg **(c)** 2.3×10^{17} kg/m³; yes, the answer surprises because the density is huge.

15. **(d)**

17. **(a)**

19. β^+ decay and/or electron capture; in β^+ decay, $^{18}_9$F $\rightarrow$ $^{18}_8$O + $^0_1$e, so the resulting nucleus is ^{18}O; in an electron capture, $^{18}_9$F + $^0_{-1}$e $\rightarrow$ $^{18}_8$O, so the resulting nucleus is ^{18}O.

21. **(a)** Tritium, $^3_1$H, has one extra neutron compared to the stable deuterium ($^2_1$H). We expect (2) β^- decay. **(b)** helium-3; yes, it's stable

23. **(a)** $^{237}_{93}$Np $\rightarrow$ $^{233}_{91}$Pa + $^4_2$He **(b)** $^{32}_{15}$P $\rightarrow$ $^{32}_{16}$S + $^0_{-1}$e **(c)** $^{56}_{27}$Co $\rightarrow$ $^{56}_{26}$Fe + $^0_{+1}$e **(d)** $^{56}_{27}$Co + $^0_{-1}$e $\rightarrow$ $^{56}_{26}$Fe **(e)** $^{42}_{19}$K* $\rightarrow$ $^{42}_{19}$K + γ

25. **(a)** $\alpha - \beta$: $^{209}_{82}$Pb + $^0_1$e $\rightarrow$ $^{209}_{81}$Tl; $^{209}_{81}$Tl + $^4_2$He $\rightarrow$ $^{213}_{83}$Bi. **(b)** $\beta - \alpha$: $^{209}_{82}$Pb + $^4_2$He $\rightarrow$ $^{213}_{84}$Po; $^{213}_{84}$Po + $^0_{-1}$e $\rightarrow$ $^{213}_{83}$Bi.

27. **(a)** $^{234}_{90}$Th **(b)** $^0_{-1}$e **(c)** 3 **(d)** γ **(e)** $^{22}_{11}$Na

29. α to ^{233}Pa; β^- to ^{233}U; α to ^{229}Th; α to ^{225}Ra; β^- to ^{225}Ac; α to ^{221}Fr; α to ^{217}At; α to ^{213}Bi; α to ^{209}Tl; β^- to ^{209}Pb; β^- to ^{209}Bi. Or α to ^{233}Pa; β^- to ^{233}U; α to ^{229}Th; α to ^{225}Ra; β^- to ^{225}Ac; α to ^{221}Fr; α to ^{217}At; α to ^{213}Bi; β^- to ^{213}Po; α to ^{209}Pb; β^- to ^{209}Bi.

31. **(c)**

33. None, it is totally independent of temperature, environment, and chemistry.

35. **(a)** infinite **(b)** zero

37. **(a)** 7.4×10^8 decays/s **(b)** 4.4×10^{10} betas/min

39. 2.06×10^{-8} W

41. 28.6 years

43. **(a)** (2) younger than, because the activity decreases as time passes so a higher activity indicates a short time, that is, younger. **(b)** 1.1×10^4 years

45. 3.1%

47. 0.98 μg

49. **(a)** 6.75×10^{19} nuclei **(b)** 2.11×10^{18} nuclei

51. 9.2 mg

53. 7.6×10^6 kg

55. **(d)**

57. **(a)**

59. $^{238}_{92}$U is even-even so tends to be more stable than $^{235}_{92}$U which is even-odd. Even so, both are unstable because $Z > 82$, but $^{238}_{92}$U has a longer half-life, that is, it is closer to stability.

61. **(a)** $^{17}_8$O **(b)** $^{42}_{20}$Ca **(c)** $^{10}_5$B **(d)** approximately the same

63. 2.013553 u

65. **(a)** 92.2 MeV **(b)** 7.68 MeV/nucleon

67. deuterium; H-2: 1.11 MeV/nucleon; H-3: 2.83 MeV/nucleon

69. 104.7 MeV

71. 10.1 MeV

73. 7.59 MeV/nucleon

75. **(d)**

77. **(b)**

79. RBE for X-rays is 1 and 20 for alphas from Table 29.4. Equal absorbed doses mean 20 times more effective dose for alphas, thus the effective dose of an alpha particle would be 20 times that of an X-ray.

81. 36 μCi

83. yes

85. 0.266 Sv or 26.6 rem

87. **(a)** $^{215}_{83}$Bi $\rightarrow$ $^{215}_{84}$Po + $^0_{-1}$e + $\bar{\nu}_e$ **(b)** 3.4×10^{22} nuclei **(c)** 1.8×10^{16} nuclei **(d)** 10 min: 1.6×10^{20} Bq = 4.4×10^9 Ci; 1 h: 8.6×10^{13} Bq = 2.3×10^3 Ci

89. **(a)** 11.5 MeV **(b)** 1.08×10^{-13} m

91. **(a)** $^{223}_{88}$Ra $\rightarrow$ $^{219}_{86}$Rn + $^4_2$He and $^{219}_{85}$At $\rightarrow$ $^{219}_{86}$Rn + $^0_{-1}$e + $\bar{\nu}_e$ **(b)** $^{219}_{86}$Rn $\rightarrow$ $^{215}_{84}$Po + $^4_2$He.

93. **(a)** $^2_1$H + $^1_0$n $\rightarrow$ $^3_1$H + γ **(b)** 71.4% **(c)** 6.26 MeV **(d)** $^3_1$H $\rightarrow$ $^3_2$He + $^0_{-1}$e + $\bar{\nu}_e$ **(e)** 18.6 keV

Chapter 30

1. (a)
3. (c)
5.

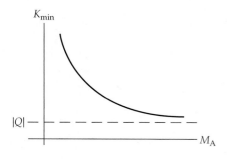

As the target mass M_A becomes very large $(M_A \gg m_a)$, $K_{min} \rightarrow |Q|$. Physically, the target would remain at rest so no energy is "lost" to recoil.

7. Elastic scattering occurs at all energies so $K_{th} = 0$. Another way to look at this is that there is no mass change in the scattering.

9. (a) γ (b) ^2_1H (c) n or ^1_0n (d) $^{17}_8O$ (e) p or ^1_1H
11. For a reaction to occur spontaneously, the Q value must be greater than zero.
(a) $^{22}_{11}Na$, no (b) $^{222}_{86}Rn$, yes (c) $^{12}_6C$, no
13. (a) (1) positive, because this is a decay (spontaneous) process. (b) +4.28 MeV
15. 4.36 MeV
17. exoergic, $Q = +6.50$ MeV
19. 5.38 MeV
21. (a) (1) greater, because $K_{min} = (1 + m_a/M_A)|Q|$. The difference in Q value (3 times) is greater than the difference caused by the target mass (15 and 20). (b) 3.05
23. (d)
25. (a)
27. This is because boron has a high probability of absorbing neutrons and it absorbs many neutrons that might create fission in other fuel rods.
29. (a) 2.22 MeV (b) 12.9 MeV
31. 36 collisions
33. (c)
35. $p + \bar{\nu}_e \rightarrow n + \beta^+$ and $p + n + \rightarrow ^2_1H + \gamma$. The 2.22 MeV gamma ray is from the forma-

tion of deuterium. The two 0.511 MeV gamma rays are the result of the pair annihilation when the positron meets an electron.
37. 4.69×10^{-13} m
41. 0.71 MeV
43. 9 keV
45. (b)
47. The forces created by the virtual exchange particles can be predicted and these predictions agree with experiment (scientific method).
49. 3.5×10^{-28} kg
51. (a) (3) decreases, because if energy increases, the mass increases, and the range is inversely proportional to the mass, $R = h/(2\pi m_m c)$. (b) 1.98×10^{-16} m
53. 4.71×10^{-24} s
55. (a)
57. All hadrons contain quarks and/or anti-quarks. Quarks are not believed to exist freely outside the nucleus.
59. (a) π^+ (b) K° (c) Σ° (d) Ξ^-
61. (a) (3) udd (b) the neutron has no charge, $\frac{2}{3}e - \frac{1}{3}e - \frac{1}{3}e = 0$
63. (a)
65. 4.03 MeV
69. 2.35 MeV

PHOTO CREDITS

FM—**AU1** Jerry Wilson **AU2** Anthony Buffa **AU3** Bo Lou

Chapter 1—CO.1 Donald Miralle/Donald Miralle/Allsport Concepts/Getty Images **Fig. 1.2b** International Bureau of Weights and Measures, Sevres, France **Fig. 1.3b** National Institute of Standards and Technology (NIST) **Fig. 1.4** IBM Research, Almaden Research Center. **Fig. 1.6** Frank Labua/Pearson Education/PH College **Fig. 1.8a** Jerry Wilson/Lander University **Fig. 1.8b** Jerry Wilson/Jerry Wilson **Fig. 1.9** T. Kuwabara/Don W. Fawcett/Visuals Unlimited **Fig. 1.9.1** JPL/NASA **Fig. 1.10** John Smith/Jerry Wilson **Fig. 1.15** Peter Brock/Getty Images, Inc./Liaison **Fig. 1.17** Frank Labua/Pearson Education/PH College **Fig. 1.18** Dennis Kunkel/©Dennis Kunkel Microscopy Inc.

Chapter 2—CO.2 David A. Northcott/Corbis/Bettmann **Fig. 2.2** John Smith/Jerry Wilson **Fig. 2.3** JPL/NASA Headquarters **Fig. 2.5** Royalty-Free/© Royalty-Free/CORBIS **Fig. 2.14b** James Sugar/Stockphoto.com/Black Star **Fig. 2.14.1** North Wind Picture Archives **Fig. 2.14.2** AP Wide World Photos **Fig. 2.15a** Frank Labua/Pearson Education/PH College **Fig. 2.15b** Frank Labua/Pearson Education/PH College **Fig. 2.27** AP Wide World Photos

Chapter 3—CO.3 C. E. Nagele/Edmund Nagele F. R. P. S. **Fig. 3.10b** Richard Megna/Educational Development Center/© Richard Megna Fundamental Photographs, NYC **Fig. 3.12** Paul Thompson; Ecoscene/© Paul Thompson; Ecoscene/CORBIS **Fig. 3.16b** John Garrett/John Garrett © Dorling Kindersley **Fig. 3.17a** JAVIER SORIANO/AFP/Javier Soriano/Agence France Presse/Getty Images **Fig. 3.17b** Andrew D. Bernstein/Photo by Andrew D. Bernstein/NBAE/Allsport Concepts/Getty Images. **Fig. 3.19c** Photri-Microstock, Inc.

Chapter 4—CO.4 AP Wide World Photos **Fig. 4.1** The Granger Collection **Fig. 4.4** John Smith/Jerry Wilson **Fig. 4.6.1** Science Photo Library/Photo Researchers, Inc. **Fig. 4.12.2a** AP Wide World Photos **Fig. 4.12.2b** Francois Moussis/Sygma/Corbis/Sygma **Fig. 4.14a** Ronald Brown/Arnold & Brown Photography **Fig. 4.14b** Ronald Brown/Arnold & Brown Photography **Fig. 4.18a** Vandystadt/Photo Researchers, Inc. **Fig. 4.18b** Charles Krebs/Corbis/Stock Market **Fig. 4.23** M. Ferguson/PhotoEdit **Fig. 4.25** Jump Run Productions/Getty Images Inc./Image Bank **Fig. 4.27a** Prentice Hall School Division **Fig. 4.27b** Prentice Hall School Division **Fig. 4.29** Tom Hauck/Allsport Concepts/Getty Images **Fig. 4.30** Tony Buffa **Fig. 4.38** Michael Dunn/Corbis/Stock Market **Fig. 4.39L** The Goodyear Tire & Rubber Company **Fig. 4.39R** The Goodyear Tire & Rubber Company **Fig. 4.40** Agence Zoom/Agence Zoom/Allsport Concepts/Getty Images

Chapter 5—CO.5 Harold E. Edgerton/© Harold & Esther Edgerton Foundation, 2002, courtesy of Palm Press, Inc. **Fig. 5.10** Rae Cooper/Construction Photography.com **Fig. 5.12a** Reuters/Corbis/Reuters America LLC **Fig. 5.12b** Vince Streano/The Image Works **Fig. 5.20** The Image Works **Fig. 5.23** SuperStock, Inc. **Fig. 5.24a** Bob Daemmrich/Stock Boston **Fig. 5.24b** Bob Daemmrich/Stock Boston

Chapter 6—CO.6 Jed Jacobsohn/Allsport Concepts/Getty Images **Fig. 6.1a** Gary S. Settles/Photo Researchers, Inc. **Fig. 6.1b** David G. Curran/Rainbow **Fig. 6.1c** Tom & Susan Bean, Inc. **Fig. 6.5a** Omikron/Science Scource/Photo Researchers, Inc. **Fig. 6.8a** Vandystadt/Photo Researchers, Inc. **Fig. 6.8b** Globus Brothers/Corbis/Stock Market **Fig. 6.8.1** Llewellyn/Pictor/ImageState/International Stock Photography Ltd. **Fig. 6.8.2a** NASA Headquarters **Fig. 6.8.2b** Dan Maas/Animation by Dan Maas, Maas Digital LLC ©2002 Cornell University. All rights reserved. This work was performed for the Jet Propulsion Laboratory, California Institute of Technology, sponsored by the United States Government under Prime Contract # NAS7-1 **Fig. 6.8.2c** Dan Maas/Animation by Dan Maas, Maas Digital LLC ©2002 Cornell University. All rights reserved. This work was performed for the Jet Propulsion Laboratory, California Institute of Technology, sponsored by the United States Government under Prime Contract # NAS7-1 **Fig. 6.11a** Carl Purcell/Photo Researchers, Inc. **Fig. 6.11b** H.P. Merten/Corbis/Stock Market **Fig. 6.15a** Richard Megna/Fundamental Photographs, NYC **Fig. 6.15b** Richard Megna/Fundamental Photographs, NYC **Fig. 6.16** Richard Megna/Fundamental Photographs, NYC **Fig. 6.20L** Paul Silverman/Fundamental Photographs, NYC **Fig. 6.20C** Paul Silverman/Fundamental Photographs, NYC **Fig. 6.20R** Paul Silverman/Fundamental Photographs, NYC **Fig. 6.22** John McDermott Photography **Fig. 6.23** Jeff Rotman/Nature Picture Library **Fig. 6.25b** NASA Headquarters **Fig. 6.25c** NASA Headquarters **Fig. 6.28** Jonathan Watts/Science Photo Library/Photo Researchers, Inc. **Fig. 6.30** Runk/Schoenberger/Grant Heilman Photography, Inc. **Fig. 6.34** Fritz Polking/© Fritz Polking/Peter Arnold, Inc. **Fig. 6.37** Charles Krebs/Getty Images Inc./Stone Allstock

Chapter 7—CO.7 Tony Savino/The Image Works **Fig. 7.6b** Tom Tracy/Corbis/Stock Market **Fig. 7.10** Chris Priest/Science Photo Library/Photo Researchers, Inc. **Fig. 7.21** NASA Headquarters **Fig. 7.23a** NASA/Science Source/Photo Researchers, Inc. **Fig. 7.24T** Photo Researchers, Inc. **Fig. 7.33** NASA Headquarters **Fig. 7.34** Frank LaBua/Pearson Education/PH College

Chapter 8—CO.8 AP Wide World Photos **Fig. 8.5a** STEVE SMITH/Getty Images, Inc./Taxi **Fig. 8.10a** David Madison/Getty Images Inc./Stone Allstock **Fig. 8.13a** Jean-Marc Loubat/Agence Vandystadt/Photo Researchers, Inc. **Fig. 8.13b** Richard Hutchings/Photo Researchers Inc. **Fig. 8.14a** Holly Drummond/Holly Drummond **Fig. 8.14b** Holly Drummond/Holly Drummond **Fig. 8.14.1** Albert Einstein(TM) Represented by The Roger Richman Agency, Inc. Beverly Hills, CA 90212. www.hollywoodlegends.com. Photo courtesy of the Archives, California Institute of Technology. **Fig. 8.15a** Gavin Hellier/Nature Picture Library **Fig. 8.15b** Owen Franken/CORBIS/NY **Fig. 8.16b** Beaura Kathy Ringrose/Michael Tobin **Fig. 8.28.a1** Frank LaBua/ Pearson Education/PH College **Fig. 8.28.a2** Frank LaBua/Pearson Education/PH College **Fig. 8.28b1** Brian Bahr/Brian Bahr/Allsport Concepts/Getty Images **Fig. 8.28b2** Brian Bahr/Brian Bahr/Allsport Concepts/Getty Images **Fig. 8.28c** AFP Photo/NOAA/Getty Images, Inc./Agence France Presse **Fig. 8.32.a** Vince Streano/Corbis/Stock Market **Fig. 8.32.b** NASA Headquarters **8.33L** Jerry Wilson **8.33R** Courtesy of Arbor Scientific **Fig. 8.49b** Tony Savino Photography/The Image Works **Fig. 8.51** Gerard Lacz/NHPA Limited

Chapter 9—CO.9 Masterfile Corporation **Fig. 9.6.1** ESRF-CREATIS/Photo Researchers, Inc. **Fig. 9.6.3** Yoav Levy/Phototake NYC **Fig. 9.11** Science Photo Library/Photo Researchers, Inc. **Fig. 9.11.2** Blair Seitz/Photo Researchers, Inc. **Fig. 9.12** REUTERS/Alexander Demianchuk/Getty Images, Inc./Liaison **Fig. 9.14** Compliments of Clearly Canadian Beverage Corporation **Fig. 9.15** Ralph A. Clevenger/© Ralph A. Clevenger/CORBIS **Fig. 9.16b** Tom Pantages **Fig. 9.22c** Hermann Eisenbeiss/Photo Researchers, Inc. **Fig. 9.23a** David Spears/Science Photo Library/Photo Researchers, Inc. **Fig. 9.23b** Richard Steedman/Corbis/Stock Market **Fig. 9.25** Patrick Watson/Stockphoto.com/Medichrome/The Stock Shop, Inc. **Fig. 9.27** PORNCHAI KITTIWONGSAKUL/AFP/PORNCHAI KITTIWONGSAKUL/Agence France Presse/Getty Images **Fig. 9.28** Raymond Gehman/CORBIS-NY **Fig. 9.29** Frank LaBua/Pearson Education/PH College **Fig. 9.30L** Charles D. Winters/Photo Researchers, Inc. **Fig. 9.30R** Charles D. Winters/Photo Researchers, Inc. **Fig. 9.37a** Stephen T. Thornton **Fig. 9.37R** Stephen T. Thornton **Fig. 9.38** Michael J. Howell/Stock Boston **Fig. 9.39a** Stephen T. Thornton **Fig. 9.39b** John Smith/Jerry Wilson **Fig. 9.41** Stephen T. Thornton

Chapter 10—CO.10 Jim Corwin/Photo Researchers, Inc. **Fig. 10.2b** Frank LaBua/Pearson Education/PH College **Fig. 10.3a** Leonard Lessin/© Leonard Lessin/Peter Arnold Inc. **Fig. 10.3b** Richard Megna/Fundamental Photographs, NYC **Fig. 10.5.1** Maximilian Stock Ltd./Science Photo Library/Photo Researchers, Inc. **Fig. 10.5.2** Spitzer Science Center, California Institute of Technology (NASA Infrared Processing and Analysis Center). **Fig. 10.6a** Sinclair Stammers/Science Photo Library/Photo Researchers, Inc. **Fig. 10.6b** Sinclair Stammers/Science Photo Library/Photo Researchers, Inc. **Fig. 10.11a** Richard Choy/© Richard Choy/Peter Arnold Inc. **Fig. 10.11b** Joe Sohm/The Image Works **Fig. 10.15a** Paul Silverman/Fundamental Photographs, NYC **Fig. 10.15b** Paul Silverman/Fundamental Photographs, NYC **Fig. 10.15c** Paul Silverman/Fundamental Photographs, NYC **Fig. 10.20a,b,c** Stephen T. Thornton

Chapter 11—CO.11 AP Wide World Photos **Fig. 11.2** Jerry Wilson **Fig. 11.4** John Smith/Jerry Wilson **Fig. 11.8** Frank LaBua/Pearson Education/PH College **Fig. 11.9c** Richard Lowenberg/Science Source/Photo Researchers, Inc. **Fig. 11.9.1** McClain Finlon Advertising Inc. **Fig. 11.13** Jerry Wilson **Fig. 11.14** AFP/STR/Agence France Presse/Getty Images **Fig. 11.15T** Collection CNRI/Phototake NYC **Fig. 11.15B** Collection CNRI/Phototake NYC **Fig. 11.16** John Smith/Jerry Wilson **Fig. 11.18** Carl Glassman/The Image Works **Fig. 11.19b** Bo Lou/Bo Lou **Fig. 11.21** John Smith/Jerry Wilson

Chapter 12—CO.12 Helen Marcus/Photo Researchers, Inc. **Fig. 12.5** Michael Marzelli/Index Stock Imagery, Inc. Royaly Free

Franzon, and Jean Claude Thierr, Atlas of Optical Phenomena. New York: Springer-Verlag, 1962. © 1962 by Springer-Verlag GmbH & Co.
Fig. 25.16b Reproduced by permission from Michel Cagnet, Maurice Franzon, and Jean Claude Thierr, Atlas of Optical Phenomena. New York: Springer-Verlag, 1962. © 1962 by Springer-Verlag GmbH & Co.
Fig. 25.17a The Image Finders 2003 **Fig. 25.17b** The Image Finders 2003 **Fig. 25.17c** The Image Finders 2003 **Fig. 25.18** Bill Bachmann/Photo Researchers, Inc. **Fig. 25.20a** Fritz Goro/Fritz Goro, Life Magazine ©TimePix

Chapter 26—CO.26 W. Couch/W. Couch, University of New South Wales, Australia/Space Telescope Science Institute OPO/NASA
Fig. 26.2 Science Photo Library/Photo Researchers, Inc.
Fig. 26.14b Space Telescope Science Institute/Photo Researchers, Inc.
Fig. 26.15.2 NASA/Ames Research Center/Ligo Project

Chapter 27—CO.27 Hank Morgan/Photo Researchers, Inc.
Fig. 27.7a Larry Albright/Larry Albright Etc **Fig. 27.7b** Ann Purcell/ Photo Researchers, Inc. **Fig. 27.8a** Wabash Instrument Corp./ Fundamental Photographs, NYC **Fig. 27.8b** Wabash Instrument Corp./ Fundamental Photographs, NYC **Fig. 27.8c** Wabash Instrument Corp./ Fundamental Photographs, NYC **Fig. 27.8d** Wabash Instrument Corp./ Fundamental Photographs, NYC **Fig. 27.8e** Wabash Instrument Corp./ Fundamental Photographs, NYC **Fig. 27.13a** Mark A. Schneider/©Mark A. Schneider/Photo Researchers, Inc. **Fig. 27.13b** Mark A. Schneider/ ©Mark A. Schneider/Photo Researchers, Inc. **Fig. 27.14** Dan McCoy/ Rainbow **Fig. 27.18a** AP Wide World Photos **Fig. 27.18b** Rosenfeld Images Ltd/Science Photo Library/Photo Researchers, Inc.

Fig. 27.18.2L JPD/Custom Medical Stock Photo/Custom Medical Stock Photo, Inc. **Fig. 27.18.2R** JPD/Custom Medical Stock Photo/Custom Medical Stock Photo, Inc. **Fig. 27.18.3b** Steven E. Zimmet/Photos courtesy of Steven E. Zimmet, MD FACPh. **Fig. 27.18.3c** Steven E. Zimmet/Photos courtesy of Steven E. Zimmet, MD FACPh. **Fig. 27.20** Philippe Plailly/ SPL/Photo Researchers, Inc.

Chapter 28—CO.28 Almaden Research Center/Research Division/NASA/Media Services **Fig. 28.3.3** Bob Thomason/Stone/Getty Images Inc./Stone Allstock **Fig. 28.3.4a** Manfred Kage/© Manfred Kage/Peter Arnold Inc. **Fig. 28.3.4b** Dr. R. Kessel/© Dr. R. Kessel/Peter Arnold Inc. **Fig. 28.3.4c** David Scharf/© David Scharf/Peter Arnold Inc. **Fig. 28.3.5** Courtesy of International Business Machines Corporation. Unauthorized use not permitted. **Fig. 28.8.1a** Omikron/Science Source/ Photo Researchers, Inc. **Fig. 28.8.1b** Mehau Kulyk/Science Photo Library/Photo Researchers, Inc. **Fig. 28.8.3c** Will & Deni McIntyre/Photo Researchers, Inc. **Fig. 28.13** Lawrence Berkeley National Laboratory/ Science Photo Library/Photo Researchers, Inc.

Chapter 29—CO.29 Roger Tully/Getty Images Inc./Stone Allstock **Fig. 29.17** Lawrence Berkeley National Laboratory/Photo Researchers, Inc. **Fig. 29.18b** Dan McCoy/Rainbow **Fig. 29.18c** Monte S. Buchsbaum, M.D., Mt. Sinai School of Medicine, New York, NY. **Fig. 29.19b** Jeff J. Daly/ Fundamental Photographs, NYC

Chapter 30—CO.30 Alexander Tsiaras/Stock Boston **Fig. 30.6** Shone/ Getty Images, Inc./Liaison **Fig. 30.7b** Plasma Physics Laboratory, Princeton University

INDEX

Wolfgang Christian and Mario Belloni
Physlet Physics: Interactive Illustrations, Explorations, and Problems for Introductory Physics
CD-ROM
0-13-140334-6
© 2004 Pearson Education, Inc.
Pearson Prentice Hall
Pearson Education, Inc.
Upper Saddle River, NJ 07458
All rights Reserved.
Pearson Prentice Hall™ is a trademark of Pearson Education, Inc.

YOU SHOULD CAREFULLY READ THE TERMS AND CONDITIONS BEFORE USING THE CD-ROM PACKAGE. USING THIS CD-ROM PACKAGE INDICATES YOUR ACCEPTANCE OF THESE TERMS AND CONDITIONS.

Pearson Education, Inc. provides this program and licenses its use. You assume responsibility for the selection of the program to achieve your intended results, and for the installation, use, and results obtained from the program. This license extends only to use of the program in the United States or countries in which the program is marketed by authorized distributors.

LICENSE GRANT

You hereby accept a nonexclusive, nontransferable, permanent license to install and use the program ON A SINGLE COMPUTER at any given time. You may copy the program solely for backup or archival purposes in support of your use of the program on the single computer. You may not modify, translate, disassemble, decompile, or reverse engineer the program, in whole or in part.

TERM

The License is effective until terminated. Pearson Education, Inc. reserves the right to terminate this License automatically if any provision of the License is violated. You may terminate the License at any time. To terminate this License, you must return the program, including documentation, along with a written warranty stating that all copies in your possession have been returned or destroyed.

LIMITED WARRANTY

THE PROGRAM IS PROVIDED "AS IS" WITHOUT WARRANTY OF ANY KIND, EITHER EXPRESSED OR IMPLIED, INCLUDING, BUT NOT LIMITED TO, THE IMPLIED WARRANTIES OR MERCHANTABILITY AND FITNESS FOR A PARTICULAR PURPOSE. THE ENTIRE RISK AS TO THE QUALITY AND PERFORMANCE OF THE PROGRAM IS WITH YOU. SHOULD THE PROGRAM PROVE DEFECTIVE, YOU (AND NOT PEARSON EDUCATION, INC. OR ANY AUTHORIZED DEALER) ASSUME THE ENTIRE COST OF ALL NECESSARY SERVICING, REPAIR, OR CORRECTION. NO ORAL OR WRITTEN INFORMATION OR ADVICE GIVEN BY PEARSON EDUCATION, INC., ITS DEALERS, DISTRIBUTORS, OR AGENTS SHALL CREATE A WARRANTY OR INCREASE THE SCOPE OF THIS WARRANTY.

SOME STATES DO NOT ALLOW THE EXCLUSION OF IMPLIED WARRANTIES, SO THE ABOVE EXCLUSION MAY NOT APPLY TO YOU. THIS WARRANTY GIVES YOU SPECIFIC LEGAL RIGHTS AND YOU MAY ALSO HAVE OTHER LEGAL RIGHTS THAT VARY FROM STATE TO STATE.

Pearson Education, Inc. does not warrant that the functions contained in the program will meet your requirements or that the operation of the program will be uninterrupted or error-free. However, Pearson Education, Inc. warrants the CD-ROM(s) on which the program is furnished to be free from defects in material and workmanship under normal use for a period of ninety (90) days from the date of delivery to you as evidenced by a copy of your receipt. The program should not be relied on as the sole basis to solve a problem whose incorrect solution could result in injury to person or property. If the program is employed in such a manner, it is at the user's own risk and Pearson Education, Inc. explicitly disclaims all liability for such misuse.

LIMITATION OF REMEDIES

Pearson Education, Inc.'s entire liability and your exclusive remedy shall be:
1. the replacement of any CD-ROM not meeting Pearson Education, Inc.'s "LIMITED WARRANTY" and that is returned to Pearson Education, or
2. if Pearson Education is unable to deliver a replacement CD-ROM that is free of defects in materials or workmanship, you may terminate this agreement by returning the program.

IN NO EVENT WILL PEARSON EDUCATION, INC. BE LIABLE TO YOU FOR ANY DAMAGES, INCLUDING ANY LOST PROFITS, LOST SAVINGS, OR OTHER INCIDENTAL OR CONSEQUENTIAL DAMAGES ARISING OUT OF THE USE OR INABILITY TO USE SUCH PROGRAM EVEN IF PEARSON EDUCATION, INC. OR AN AUTHORIZED DISTRIBUTOR HAS BEEN ADVISED OF THE POSSIBILITY OF SUCH DAMAGES, OR FOR ANY CLAIM BY ANY OTHER PARTY.

SOME STATES DO NOT ALLOW FOR THE LIMITATION OR EXCLUSION OF LIABILITY FOR INCIDENTAL OR CONSEQUENTIAL DAMAGES, SO THE ABOVE LIMITATION OR EXCLUSION MAY NOT APPLY TO YOU.

GENERAL

You may not sublicense, assign, or transfer the license of the program.

Any attempt to sublicense, assign or transfer any of the rights, duties, or obligations hereunder is void.

This Agreement will be governed by the laws of the State of New York.

Should you have any questions concerning this Agreement, you may contact Pearson Education, Inc. by writing to:
ESM Media Development
Higher Education Division
Pearson Education, Inc.
1 Lake Street
Upper Saddle River, NJ 07458

Should you have any questions concerning technical support, you may write to:
New Media Production
Higher Education Division
Pearson Education, Inc.
1 Lake Street
Upper Saddle River, NJ 07458

YOU ACKNOWLEDGE THAT YOU HAVE READ THIS AGREEMENT, UNDERSTAND IT, AND AGREE TO BE BOUND BY ITS TERMS AND CONDITIONS. YOU FURTHER AGREE THAT IT IS THE COMPLETE AND EXCLUSIVE STATEMENT OF THE AGREEMENT BETWEEN US THAT SUPERSEDES ANY PROPOSAL OR PRIOR AGREEMENT, ORAL OR WRITTEN, AND ANY OTHER COMMUNICATIONS BETWEEN US RELATING TO THE SUBJECT MATTER OF THIS AGREEMENT.

Program Instructions
–Windows:
To access the Physlet Physics curricular material content, follow the following steps:
    –Insert the "Physlet Physics" CD into your CD-ROM drive.
    –Browse to the "Physlet_Physics" CD-ROM using your Windows explorer window.
    –Double-click the "start.html" file.
To access the Physlet Physics Exploration Worksheets, double-click the "exploration_worksheets" folder, choose a chapter folder to view the selection of associated Exploration PDF worksheets, and double-click a PDF file to view.

Minimum System Requirements
–Windows:
400MHz Intel Pentium processor
Windows 2000/XP
32 MB or more of available RAM
800×600 monitor resolution set to 16 bit color
Mouse or other pointing device (?)
4× CD-ROM Drive
Active Internet connection (optional)
Requires a browser supporting the Java 1.4 Virtual Machine and JavaScript to Java communication. Internet Explorer 5.5 or higher with the Sun Java plugin 1.4 or higher, or Mozilla 1.3 or higher with the Sun Java plugin 1.4 or higher are recommended.
–Macintosh
PowerPC G3, G4, or G5 processor
Mac OS X.3/X.4
Safari 1.2
Macintosh Java plugin 1.4.2 or higher
Adobe Acrobat Reader 5.0 © 2002 or above is required to view and print the pdfs of the Exploration Worksheets.

TECHNICAL SUPPORT

If you are having problems with this software, call (800) 677-6337 between 8:00 a.m. and 8:00 p.m. Monday through Friday, and 5:00 p.m. to 12:00 a.m. on Sunday (all times listed are Eastern). You can also get support by filling out the web form located at: http://247.prenhall.com/mediaform

Our technical staff will need to know certain things about your system in order to help us solve your problems more quickly and efficiently. If possible, please be at your computer when you call for support. You should have the following information ready:

- Textbook ISBN
- CD-ROM ISBN
- corresponding product and title
- computer make and model
- Operating System (Windows or Macintosh) and Version
- RAM available
- hard disk space available
- Sound card? Yes or No
- printer make and model
- network connection
- detailed description of the problem, including the exact wording of any error messages

NOTE: Pearson does not support and/or assist with the following:

- third-party software (i.e. Microsoft including Microsoft Office suite, Apple, Borland, etc.)
- homework assistance
- Textbooks and CD-ROMs purchased used are not supported and are non-replaceable. To purchase a new CD-ROM, contact Pearson Individual Order Copies at 1-800-282-0693

The Periodic Table of Elements

(See Chapter 28.3)

Main-group elements

Legend:
- Metals
- Nonmetals
- Noble gases

Transition elements

Period	1 / 1A	2 / 2A	3 / 3B	4 / 4B	5 / 5B	6 / 6B	7 / 7B	8 / 8B	9 / 8B	10 / 8B	11 / 1B	12 / 2B	13 / 3A	14 / 4A	15 / 5A	16 / 6A	17 / 7A	18 / 8A
1	1 **H** 1.00794																	2 **He** 4.00260
2	3 **Li** 6.941	4 **Be** 9.01218											5 **B** 10.811	6 **C** 12.011	7 **N** 14.0067	8 **O** 15.9994	9 **F** 18.9984	10 **Ne** 20.1797
3	11 **Na** 22.9898	12 **Mg** 24.3050											13 **Al** 26.9815	14 **Si** 28.0855	15 **P** 30.9738	16 **S** 32.066	17 **Cl** 35.4527	18 **Ar** 39.948
4	19 **K** 39.0983	20 **Ca** 40.078	21 **Sc** 44.9559	22 **Ti** 47.88	23 **V** 50.9415	24 **Cr** 51.9961	25 **Mn** 54.9381	26 **Fe** 55.847	27 **Co** 58.9332	28 **Ni** 58.693	29 **Cu** 63.546	30 **Zn** 65.39	31 **Ga** 69.723	32 **Ge** 72.61	33 **As** 74.9216	34 **Se** 78.96	35 **Br** 79.904	36 **Kr** 83.80
5	37 **Rb** 85.4678	38 **Sr** 87.62	39 **Y** 88.9059	40 **Zr** 91.224	41 **Nb** 92.9064	42 **Mo** 95.94	43 **Tc** (98)	44 **Ru** 101.07	45 **Rh** 102.906	46 **Pd** 106.42	47 **Ag** 107.868	48 **Cd** 112.411	49 **In** 114.818	50 **Sn** 118.710	51 **Sb** 121.76	52 **Te** 127.60	53 **I** 126.904	54 **Xe** 131.29
6	55 **Cs** 132.905	56 **Ba** 137.327	57 *__La__ 138.906	72 **Hf** 178.49	73 **Ta** 180.948	74 **W** 183.84	75 **Re** 186.207	76 **Os** 190.23	77 **Ir** 192.22	78 **Pt** 195.08	79 **Au** 196.967	80 **Hg** 200.59	81 **Tl** 204.383	82 **Pb** 207.2	83 **Bi** 208.980	84 **Po** (209)	85 **At** (210)	86 **Rn** (222)
7	87 **Fr** (223)	88 **Ra** 226.025	89 †**Ac** 227.028	104 **Rf** (261)	105 **Db** (262)	106 **Sg** (263)	107 **Bh** (262)	108 **Hs** (265)	109 **Mt** (266)	110 **Ds** (281)	111 ** (272)	112 ** (285)		114 ** (289)		116 ** (292)		

*Lanthanide series

58 **Ce** 140.115	59 **Pr** 140.908	60 **Nd** 144.24	61 **Pm** (145)	62 **Sm** 150.36	63 **Eu** 151.965	64 **Gd** 157.25	65 **Tb** 158.925	66 **Dy** 162.50	67 **Ho** 164.930	68 **Er** 167.26	69 **Tm** 168.934	70 **Yb** 173.04	71 **Lu** 174.967

†Actinide series

90 **Th** 232.038	91 **Pa** 231.036	92 **U** 238.029	93 **Np** 237.048	94 **Pu** (244)	95 **Am** (243)	96 **Cm** (247)	97 **Bk** (247)	98 **Cf** (251)	99 **Es** (252)	100 **Fm** (257)	101 **Md** (258)	102 **No** (259)	103 **Lr** (260)

** Not yet named

Notes: (1) Values in parentheses are the mass numbers of the most common or most stable isotopes of radioactive elements. (2) Some elements adjacent to the stair-step line between the metals and nonmetals have a metallic appearance but some nonmetallic properties. These elements are often called metalloids or semimetals. There is no general agreement on just which elements are so designated. Almost every list includes Si, Ge, As, Sb, and Te. Some also include B, At, and/or Po.